STUDY GUIDE AND SOLUTIONS MANUAL

Organic Chemistry: Principles and Mechanisms

Joel M. Karty

Joel M. Karty
ELON UNIVERSITY

Marie M. Melzer
OLD DOMINION UNIVERSITY

W · W · NORTON & COMPANY · NEW YORK · LONDON

W. W. Norton & Company has been independent since its founding in 1923, when William Warder Norton and Mary D. Herter Norton first published lectures delivered at the People's Institute, the adult education division of New York City's Cooper Union. The firm soon expanded its program beyond the Institute, publishing books by celebrated academics from America and abroad. By mid-century, the two major pillars of Norton's publishing program—trade books and college texts—were firmly established. In the 1950s, the Norton family transferred control of the company to its employees and today—with a staff of 400 and a comparable number of trade, college, and professional titles published each year—W. W. Norton & Company stands as the largest and oldest publishing house owned wholly by its employees.

Associate Media Editor: Jennifer Barnhardt
Assistant Editor: Renee Cotton
Assistant Media Editor: Paula Iborra
Project Editor: Christine D'Antonio
Managing Editor, College: Marian Johnson
Production Manager: Eric Pier-Hocking

ISBN: 978-0-393-92293-6

W. W. Norton & Company, Inc., 500 Fifth Avenue, New York, N.Y. 10110
www.wwnorton.com

W. W. Norton & Company Ltd., Castle House, 75/76 Wells Street, London W1T 3QT

2 3 4 5 6 7 8 9 0

CONTENTS

PREFACE

We have written this *Study Guide and Solutions Manual* to be a valuable resource to help you excel in organic chemistry. More than just providing the answers to the Your Turn exercises and the Problems in each chapter (both the in-chapter and end-of-chapter problems), it can help you become a better learner and a better problem solver. One major way this can be done is with the Think/Solve approach that every solution in this *Manual* has, which is the same approach exhibited in the Solved Problems throughout the textbook. In the Think part of every solution, we provide questions that relate to the particular problem at hand; when answered, those Think questions help guide you to the problem's solution. In the Solve part, we answer the questions that were posed and show how those answers are used to solve the problem.

Before you read the Solve part of the solution, make sure that you read the Think part and make an honest effort to answer the questions that we have posed. Doing so will provide two benefits. One is that it will help enhance your approach to solving problems in general. Organic chemistry problems are not solved algorithmically; there is no specific set of steps to memorize that can be used to solve every problem you encounter. To the contrary, solving organic chemistry problems requires you to bring together various concepts you've learned, often in widely different ways. The Think questions help you determine which concepts are relevant to the particular problem at hand and how you should consider bringing them together to solve it. The more that you practice this vital step in problem solving, the better you will become at it.

The second benefit of spending time on the Think questions comes from realizing which ones you can answer easily and which ones you can't. If you have trouble answering a particular Think question, this could be an indicator that you have a gap in your understanding of a specific concept or idea. Take the time to review that concept or idea where it is discussed in the textbook. That way, you will be better poised to solve other problems that require you to apply the same concept (but perhaps in a different way).

Ideally, you should attempt all Your Turn exercises, in-chapter problems, and end-of-chapter problems, but not all at once. Carry out the Your Turn exercises as you are reading the material in the textbook for the first time (even before the material is discussed in class). Those exercises are designed to be relatively straightforward and take little time, and they provide you quick feedback as to whether you "get" the concept at hand. Work through each in-chapter problem after you have had sufficient time to digest the relevant material that is being discussed and have gone through any related Solved Problems. Finally, challenge yourself with as many end-of-chapter problems as possible immediately after you have completed a chapter.

Don't put the end-of-chapter problems off to the days leading up to an exam. Instead, work on them regularly. Do your best to work through about 10 each day. That way, you will be able to internalize and learn from each problem much better. Importantly, if you get stuck on a particular problem, this will give you plenty of time to seek help with any questions or confusion you have, perhaps by rereading portions of the textbook, asking a study partner or teaching assistant, or visiting your professor. If you procrastinate on the end-of-chapter problems, you will find yourself rushing through them, you won't learn as much from them, and you will not have sufficient time to seek the proper help.

There are two ways to view the usefulness of the end-of-chapter problems. One way is simply as practice. There are certain types of problems you will be expected to solve on exams, and by working through more end-of-chapter problems, you will become more familiar with the ways that certain types of problems might look. You will become more proficient in how to begin and complete them. You will become more accurate and more efficient, and these are certainly things that will result in better performances on exams.

A much better way to use the end-of-chapter problems, however, is as *self assessment*. After you have completed a chapter (along with the Your Turn exercises and the in-chapter problems), try to tackle each end-of-chapter problem with no assistance—that is, without consulting the relevant material in the textbook, your notes, or your study partners. If you successfully solve a problem in little time, take that as a message that you are relatively strong in that particular area. If you get the right answer but find that it took you a long time to solve the problem, take what you have learned from that problem and apply it to a similar problem. This time, challenge yourself to solve it faster. If your answer to a problem is incorrect or if you needed help working through the problem (perhaps by reading the Think questions or by asking a study partner), then this is an indicator that you have a weakness in a particular area that needs to be addressed before the next exam. Take the time to strengthen that area of weakness. Reread the relevant portions of the textbook, discuss things with study partners, or visit your instructor. Then try another, similar problem unassisted and repeat the process until you are no longer struggling.

We wish you all the best in your studying.

Joel M. Karty
Marie M. Melzer

CHAPTER 1 | Atomic and Molecular Structure

Your Turn Exercises

Your Turn 1.1

Think

When designating orbitals as 1*s*, 2*s*, 2*p*, etc., what corresponds to the shell number? Which orbitals share that designation? Which electrons are in this shell?

Solve

The 2*s* and 2*p* orbitals are in the second shell, noted by the dashed box. Electrons (3)–(10) are in the second shell.

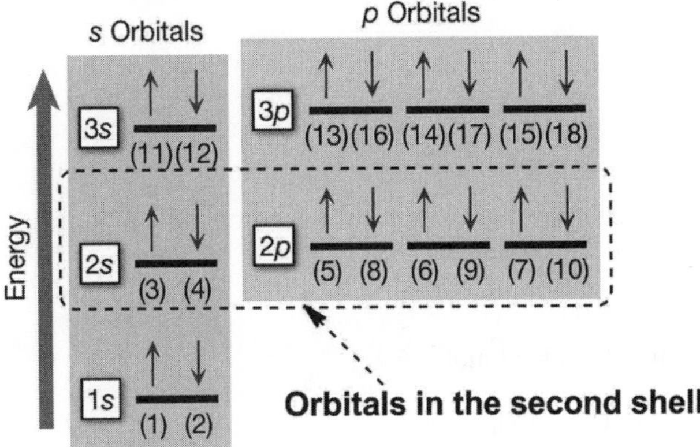

Your Turn 1.2

Think

How many total electrons does carbon have? What is the electron configuration for carbon? What distinguishes valence electrons from core electrons?

Solve

The valence electrons are in the highest energy shell—that is, the $n = 2$ shell. The core electrons are in all the other (lower energy) shells. Carbon has six total electrons: $1s^2\ 2s^2\ 2p^2$. Therefore, the valence electrons are second shell, $2s^2$ and $2p^2$ and the core electrons are in the first shell, $1s^2$ (noted in the figure by the circle and rectangle, respectively).

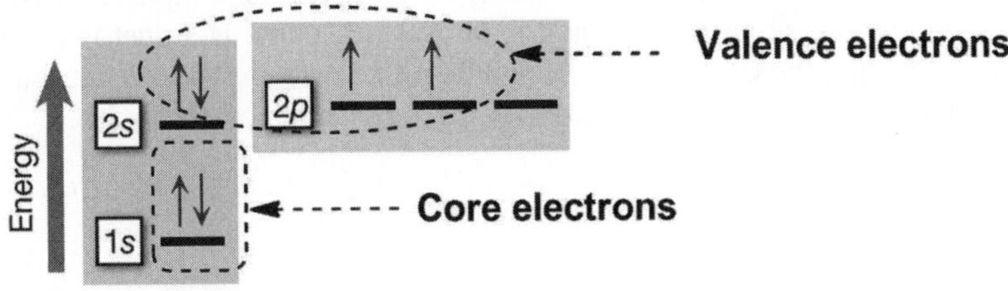

Your Turn 1.3

Think

What is the definition of bond energy? Looking at the plot of energy versus internuclear distance, what is the energy of the separate H atoms? What is the energy of the bonded H–H molecule? How does the bond energy correlate to these two values?

Solve

The bond energy represents the change in energy (on the vertical scale) needed to increase the bond length from its equilibrium value (on the horizontal scale) to infinity. This can be calculated from the potential energy diagram (Fig. 1-10a). Using the energy difference of the separated H atoms (E = 450 kJ/mol) and bonded H–H molecule (E = 0 kJ/mol), a bond energy of 450 kJ/mol is calculated.

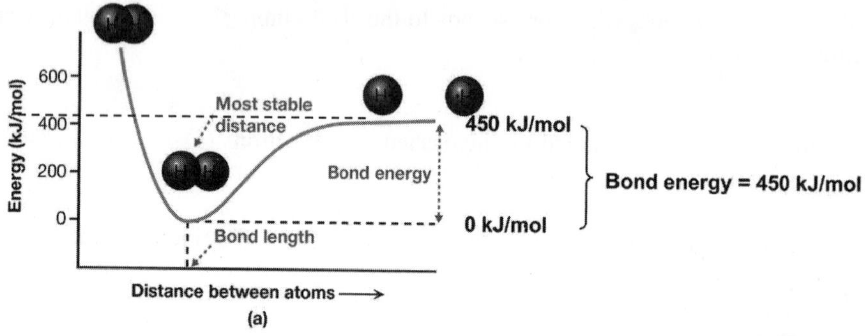

Your Turn 1.4

Think

Look at Tables 1-2 and 1-3. Does the strongest bond have the largest or smallest kJ/mol value? In general, are single bonds stronger or weaker compared to double and triple bonds?

Solve

(a) The strongest *single* bond has a value of 586 kJ/mol (140 kcal/mol).
(b) The strongest *single* bond is the Si–F bond.
(c) The weakest *single* bond has a value of 138 kJ/mol (33 kcal/mol).
(d) The weakest *single* bond is the O–O bond.
(e) The strongest bond of any type has a value of 1072 kJ/mol (256 kcal/mol).
(f) The strongest bond of any type is the C≡O triple bond.

Your Turn 1.5

Think

Lewis structures take into account only which type of electrons? How are bonding and nonbonding electrons represented? How do you determine whether a pair of electrons counts as part of an atom's octet?

Solve.

For the electrons to be counted around any atom, the lines (bonding) or dots (nonbonding) have to touch the atom. The C atom's octet is composed of the four sets of bonding electrons (lines) around the C atom. Each bond is composed of two electrons shared between the atoms it touches. O's octet is composed of two sets of bonding electrons, one to the C atom and one to an H atom, and then two lone pairs. H's duet is composed of the two shared electrons in the bond to either the C or O atom. You will notice that each bond contributes to the octet or duet simultaneously for each atom that touches the bond—hence, the overlap of the circles.

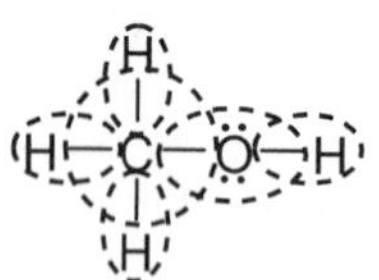

Your Turn 1.6

Think

Consult Figure 1-16 to determine the electronegativity value for boron (B) and hydrogen (H). How is a dipole arrow drawn? To which atom, more or less electronegative, does the arrow point? Which end of the dipole arrow receives the the positive delta (δ^+) and negative delta (δ^-) symbols?

Solve

The electronegativity of B is 2.04 and H is 2.20. This leads to the conclusion that H is *slightly* more electronegative than B. The difference in electronegativity is 0.16. The dipole arrow (⟼) points to the more electronegative atom (each H atom in this example). This leads to a partial negative charge on each H atom (δ^-) and a partial positive on the B atom (δ^+).

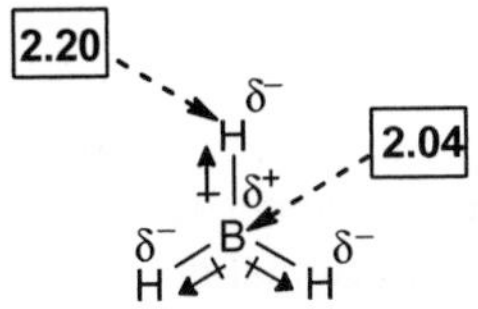

Your Turn 1.7

Think

Consult Figure 1-16 to determine the electronegativity value for sodium (Na), chlorine (Cl), carbon (C), and hydrogen (H). To what does the *difference* in electronegativity correspond in terms of bond type (i.e., ionic vs. covalent)? Are any metals present? What type of bonding is likely to result?

Solve

For NaCl, the difference is $3.16 - 0.93 = 2.23$. The electronegativity difference of 2.23 correlates to an ionic bond. This means that Cl acquires the one extra electron needed to fill its octet from the less electronegative Na atom. This results in the formation of the chloride anion, Cl^-, and the sodium cation, Na^+. For CH_4, the difference is $2.55 - 2.20 = 0.35$. This corresponds to a small difference in electronegativity. Therefore the C–H bond is a covalent bond that is not very polar.

Your Turn 1.8

Think

What gives the atom its identity (protons or electrons)? What is the charge of a proton? What is the charge of an electron? How does the overall number of protons and electrons contribute to the charge on the atom? How many core electrons does carbon have?

Solve

For a carbon nucleus, the number of protons is always six. Add the core electrons (always two for a carbon atom) to the number of valence electrons given to get the total number of electrons. The charge is determined by comparing the number of protons and electrons. If the number of protons and electrons is equal, the charge is 0. It is +1 if there is one more proton than electron, and −1 if there is one more electron than proton.

Number of Valence Electrons	Total Number of Electrons	Number of Protons	Charge
3	5	6	+1
4	6	6	0
5	7	6	−1

Your Turn 1.9

Think

How does the sum of formal charges compare to the sum of the oxidation states? How do those values compare to the total charge?

Solve

For HCO_2^-

Sum of formal charges: $0 + 0 + 0 + (-1) = -1$.

Sum of oxidation states: $(+2) + (+1) + (-2) + (-2) = -1$

The two must always be the same and add up to the charge of the molecule. In this case, both values must sum to −1, the total charge of HCO_2^-.

Your Turn 1.10

Think

Remember that the curved arrow illustrates electron movement necessary to change one resonance structure into another. In the structure on the left, which electrons move to form the double bond in the structure on the right? Which electrons move in the structure on the left to form the lone pair on the C atom in the structure on the right?

Solve

The first curved arrow is used to convert the lone pair into a covalent double bond, C=C. The formal charge becomes more positive by one. Without any other changes, this would lead to the center C atom having 10 valence electrons. The second arrow is used to convert a pair of electrons from the original double bond, C=C, into a lone pair on the other end C atom, :CH_2. The formal charge on this C atom becomes more negative by one.

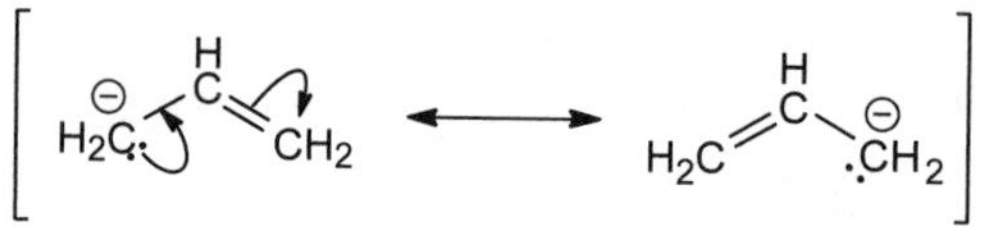

Your Turn 1.11

Think

Are there any lone pairs adjacent to a double or triple bond? Do any of the double bonds convert into lone pairs? How can you show via a curved arrow the electrons in the lone pair moving to form a double bond? How can you show via a curved arrow the electrons in a double bond moving to form a lone pair?

Solve

Recall that atoms do not move in resonance structures; only nonbonding electrons or electrons from double or triple bonds move. The first curved arrow is used to convert the lone pair on the fifth C atom into a covalent double bond, C=C. Without any other changes, this would lead to 10 valence electrons on the sixth C atom. The second arrow is used to convert a pair of electrons from the double bond, C=C, into a lone pair on the seventh C atom, :CH.

Your Turn 1.12

Think

Do any of the atoms have an unfilled valence shell? If so, is the atom with the unfilled valence shell adjacent to a double or triple bond? How can you show via a curved arrow the electrons in the triple bond moving to form a double bond?

Solve

The carbocation has an unfilled valence shell and is adjacent to the triple bond. The curved arrow is used to convert the triple bond into a double bond. This leaves an unfilled valence on the left carbon.

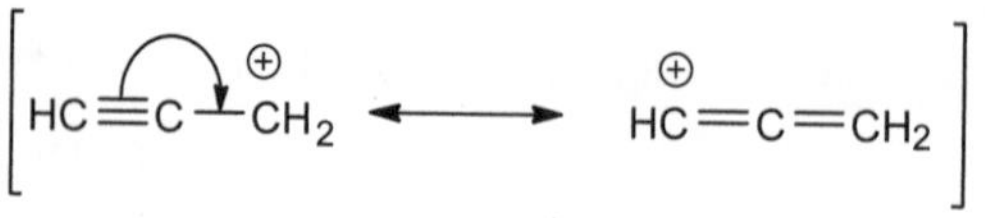

Your Turn 1.13

Think

Do any of the atoms have an unfilled valence shell? If so, is the atom with the unfilled valence shell adjacent to a double or triple bond? How can you show via a curved arrow the electrons in the double bond moving to fill the unfilled valence?

Solve

The carbocation has an unfilled valence shell and is adjacent to a double bond. A single curved arrow is used to show the movement of the electrons in the double bond. This leaves an unfilled valence on the C atom bearing the positive charge.

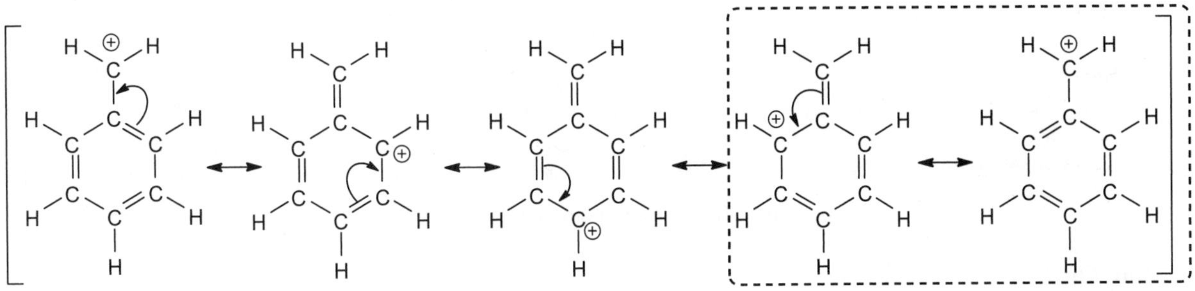

Your Turn 1.14

Think

Where do the double bonds move based on the two curved arrows shown? How do you count electrons in a Lewis structure to evaluate whether an atom satisfies an octet?

Solve

Based on the curved arrow electron movement, one C atom is left with an unfilled valence and another C atom is left with an exceeded octet. This leads to a Lewis structure that is not valid. To draw a correct resonance structure of benzene, six electrons must be shifted at the same time.

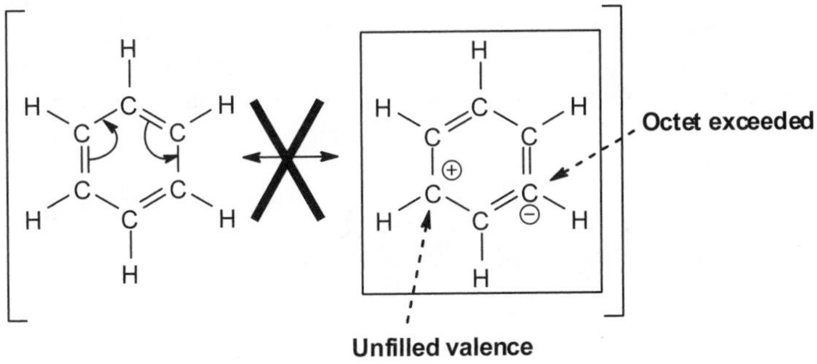

Octet exceeded

Unfilled valence

To draw a correct resonance structure of benzene, six electrons must be shifted at the same time.

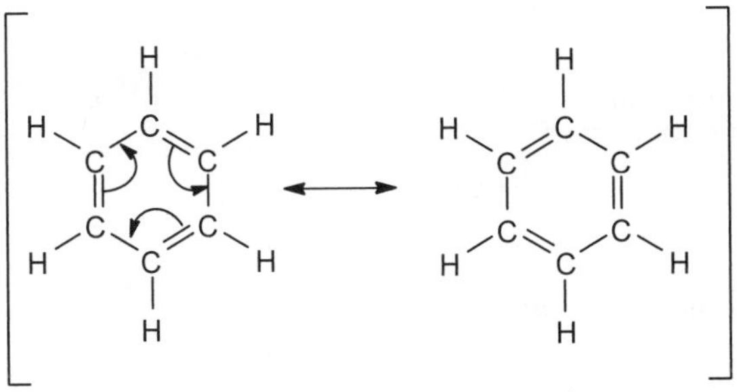

Your Turn 1.15

Think

Do you see any atoms other than carbon? Do you see any triple or double bonds? What atoms are attached by the double or triple bond? On the C=O, what atoms are bonded to either side of the carbon? What functional group does this represent?

Solve

The C=O group has the only non-C/H atoms and the only non-single bond in both compounds. On either side of the C=O is a C atom. This is the ketone functional group.

Ketone

Ketone

General ketone functional group

Cyclohexanone

Hexan-3-one

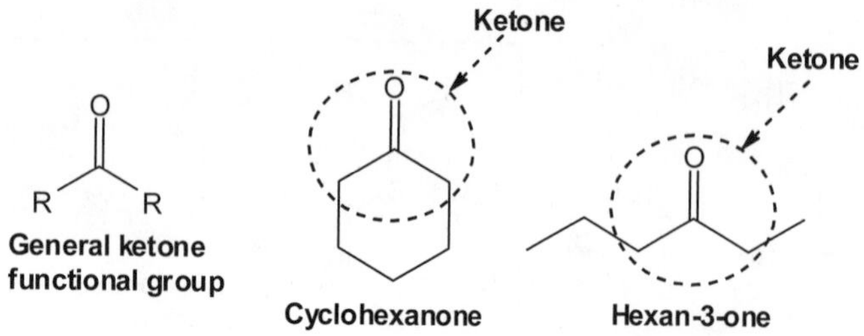

Your Turn 1.16

Think

Consult Table 1-7 and draw each of the four amino acids in the question. What atoms/bonds compose an alcohol, carboxylic acid, amine, and aryl ring? Consult Table 1-6 if you need to look up the functional groups. Be mindful of atoms other than carbon and hydrogen and the presence of multiple bonds.

Solve

(a) The alcohol functional group is composed of a C atom single bonded to an OH group, R–OH.

(b) The carboxylic acid functional group is composed of a central C atom double bonded to an O atom, single bonded to an OH, and a general R group.

(c) The amine functional group is composed of a C atom single bonded to an NR_2, NHR, or NH_2 group.

(d) An aryl ring is a six-membered carbon ring with alternating single and double bonds.

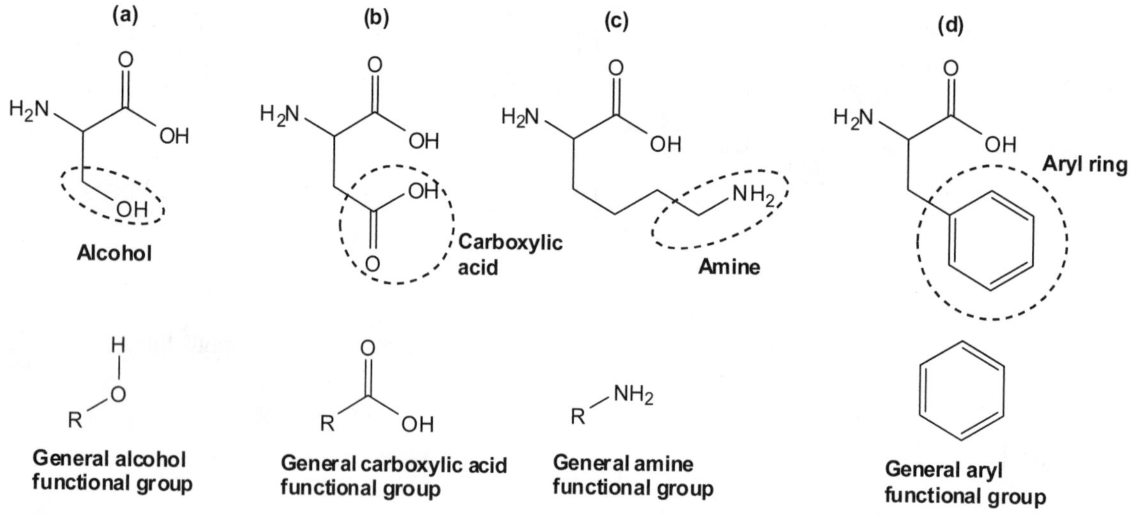

(a) Alcohol

(b) Carboxylic acid

(c) Amine

(d) Aryl ring

General alcohol functional group

General carboxylic acid functional group

General amine functional group

General aryl functional group

Your Turn 1.17

Think

What atoms make up an alcohol functional group? What distinguishes the structure of a ketone from an aldehyde?

Solve

The alcohol functional group is composed of a C atom single bonded to an OH group, R–OH (circled). The C=O carbonyl group in glucose is RC(O)H. The presence of the H atom indicates the functional group is an aldehyde.

Alcohol groups — Aldehyde

Glucose, $C_6H_{12}O_6$

General alcohol functional group

General aldehyde functional group

Your Turn 1.18

Think

Look at Table 1-6 if you need to look up the functional groups. What are the general structures of alkene and amide functional groups?

Solve

The alkene functional group is a carbon–carbon double bond, C=C, and the amide functional group is a C=O group with an N atom attached to a C atom, $RC(O)NH_2$. Ovals are placed around alkenes; circles are placed around amides.

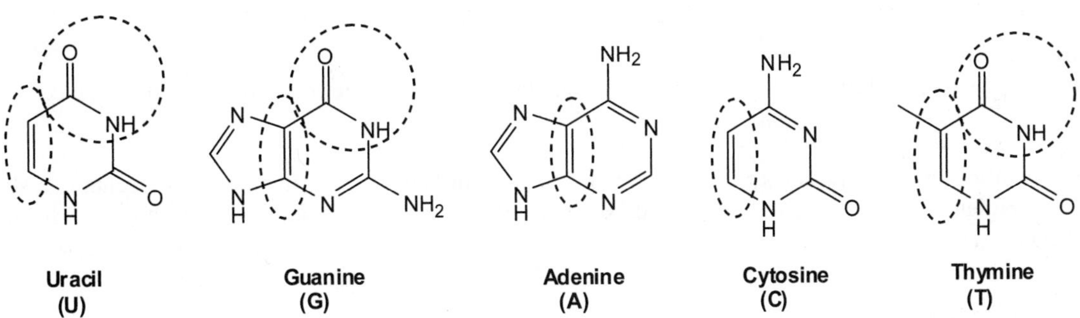

Chapter Problems

Problem 1.2

Think

How many protons are there in the nucleus of a carbon atom? Oxygen atom? Does a cation have more protons than electrons, or vice versa? What does the charge of the species indicate about the number of protons compared to the number of electrons?

Solve

(a) The carbon nucleus always has six protons. A carbon anion with a −1 charge should have one more electron than it has protons; thus, this species has seven electrons.

(b) The oxygen nucleus always has eight protons. An oxygen cation with a +1 charge should have one more proton than it has electrons; thus this species has seven electrons.

(c) An oxygen anion with a −1 charge should have one more electron than it has protons; thus this species has nine electrons.

Problem 1.4

Think

How many total electrons are there in an oxygen atom? What is the order in which the atomic orbitals should be filled (see Fig. 1-7)? What is the valence shell and where do the core electrons reside?

Solve

There are eight total electrons (Z = 8 for O). The first two are placed in the 1*s* orbital and the next two in the 2*s* orbital, leaving four electrons for the three 2*p* orbitals. The electron configuration is $1s^2 2s^2 2p^4$. The valence shell is the second shell, so there are six valence electrons and two core electrons.

Problem 1.6

Think

How is bond length represented on each curve? Which one occurs at a larger internuclear bond distance?

Solve

The red (bottom) curve represents the longer bond. The minimum energy for the red curve occurs at a slightly longer bond length than it does for the blue curve. To see this, draw a line from the minimum on each curve to the *x* axis.

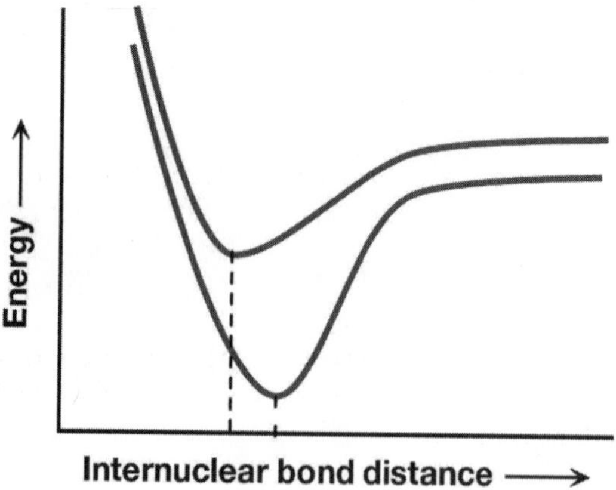

Problem 1.8

Think

Consider the steps for drawing Lewis structures. Which atoms must be bonded together?

Solve

First count the valence electrons for each atom and the total number of valence electrons for the compound C_2H_3N.

3 H atoms:	3 × 1 valence electrons	=	3 valence electrons
2 C atoms:	2 × 4 valence electrons	=	8 valence electrons
1 N atom:	1 × 5 valence electrons	=	5 valence electrons
	Total	=	16 valence electrons

A C atom is bonded to three H atoms and a second C atom. The second C atom is bonded to the N atom. This accounts for only 10 valence electrons. The remaining six electrons are added and the N atom has an incomplete valence. A C≡N triple bond and a lone pair on N are required to give all atoms their octets.

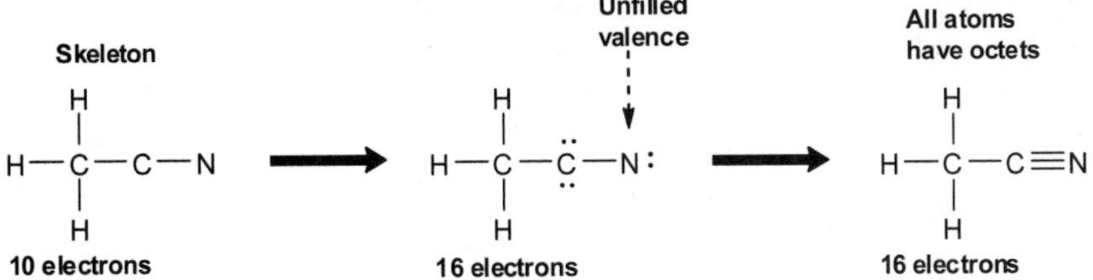

Problem 1.9

Think

Consider the steps for drawing Lewis structures. For which atoms is a deficient or expanded octet permissible? Which row in the periodic table is forbidden to exceed the octet?

Solve

Atoms in the second row are strictly forbidden to exceed the octet.

Acceptable

Not acceptable; C's octet is exceeded

Acceptable; S allowed to have expanded octet

Acceptable; P allowed to have expanded octet

Not acceptable; C's octet is exceeded

Problem 1.11

Think

Which atoms (other than H) have an octet and which atoms don't? How many bonds and lone pairs are typical for each element? Refer to Table 1-4 for the common number of covalent bonds and lone pairs for H, C, N, and O.

Solve

The atoms not shown with an octet are in boldface type on the left. Based on Table 1-4, we need to convert the C–C single bond and C–N single bond in the ring each to double bonds, convert the C–O to a double bond, add a lone pair to the N atom, and two lone pairs to each of the O atoms.

Problem 1.12

Think

Which atoms (other than H) need lone pairs or additional bonds to complete their octet? Which atom in each bond is more electronegative? To which atom does the bond dipole (⟼) point? What does the length of the arrow indicate about the difference in electronegativity?

Solve

The C–F bond dipole is the largest because the C and F have the greatest difference in electronegativity. The C–H bond dipole is the smallest. Bond dipoles point toward the more electronegative atom.

(a) (b) (c)

Problem 1.13

Think

What does an electrostatic potential map depict? Which color (red or blue) corresponds to a buildup of negative charge? Draw the bond dipoles and add the δ^+ and δ^- symbols. Match the electrostatic potential map to one of the four compounds.

Solve

The red region on an electrostatic potential map corresponds to excess negative charge (δ^-), whereas each blue region corresponds to excess positive charge (δ^+). The electrostatic potential map, therefore, is consistent with $H_2C{=}O$.

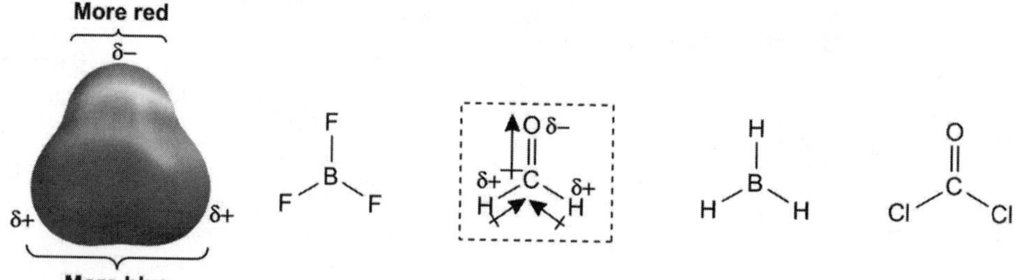

Problem 1.15

Think

Does the compound contain elements from both the left and right sides of the periodic table? Does the compound contain any recognizable polyatomic cations?

Solve

The first structure is ionic; its sodium cation Na^+ is easiest to spot, and its phenoxide anion has a formal charge of −1 on the O atom. The middle structure is covalent; all atoms have their "preferred" number of covalent bonds. The third structure contains an N atom that must have four bonds. This is a type of ammonium cation, $CH_3CH_2NH_3^+Cl^-$. The Cl^- anion is not covalently bonded to any other atom in this structure.

(a) Ionic (b) Covalent (c) Ionic

Problem 1.17

Think

How are lone pairs assigned? In determining how to assign bonding pairs of electrons, when is electronegativity relevant and when is it not?

Solve

The figure below on the left shows how valence electrons are assigned according to the formal charge method: Each pair of bonding electrons is split evenly. The number of electrons on the atom is then compared to the group number. All atoms have a formal charge of 0 in this compound. The oxidation-state method for assigning valence electrons is shown on the right. The oxidation state of each H atom is +1 because, being bonded to a more electronegative atom, each is assigned zero valence electrons. All other oxidation states are shown below. N (−1) is assigned all four electrons in the double bond plus its lone pair. O (−2) is assigned eight valence electrons (two bonds and two lone pairs). The middle C atom (+2) is assigned only two valence electrons (one electron from each of its single bonds). Each CH_3 carbon is assigned seven valence electrons (one electron from its C–C bond and both electrons from each of its C–H bonds). All 12 oxidation states sum to zero because the molecule is uncharged overall.

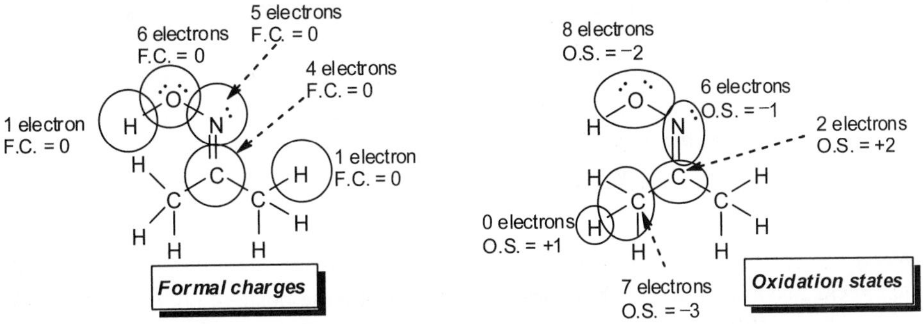

Problem 1.19

Think

Are both resonance structures equivalent? Add in formal charges. Which structure has fewer atoms with formal charges other than zero?

Solve

The resonance structures are inequivalent. The structure on the right has a formal charge of −1 on the O atom and +1 on the Cl atom. Therefore, the structure on the left, with all formal charges equal to zero, makes a much greater contribution to the resonance hybrid.

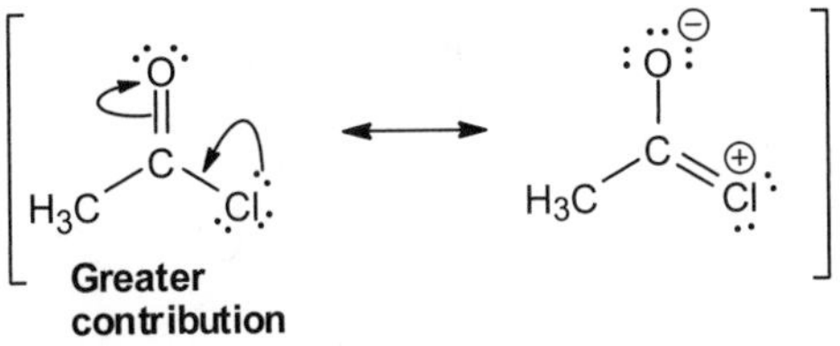

Problem 1.21

Think

Which atom corresponds to W in the figure in Section 1.11a (Page 27)? To X? To Y? What happens to the respective formal charges? Are any lone pairs adjacent to a double bond?

Solve

The N atom with the lone pair corresponds to W; the C=C atoms to the right of the N atom correspond to X and Y, respectively. The curved arrows are applied, yielding the additional resonance structure shown on the right in the figure on the next page.

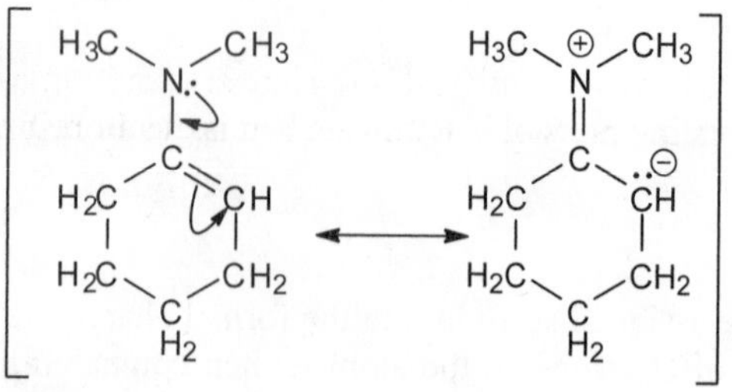

Problem 1.22

Think

Do any of the atoms have an unfilled valence shell? If so, is the atom with the unfilled valence shell adjacent to a double or triple bond? How can you show, via a curved arrow, the electrons in the triple bond moving to form a double bond?

Solve

There is only one additional resonance structure. The original structure has an atom deficient of an octet (B) adjacent to a triple bond, requiring the movement of only two electrons to arrive at the second resonance structure.

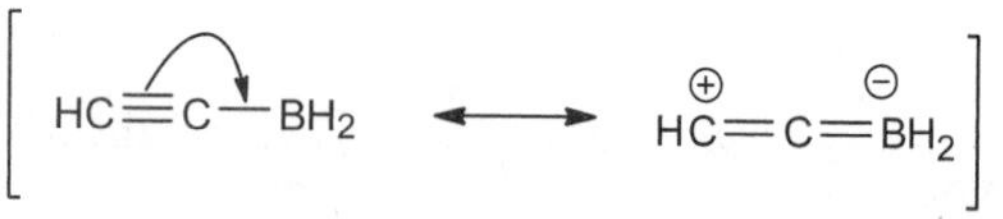

Problem 1.23

Think

Refer to the individual resonance structures in Your Turn 1.13. Where are the positive charges located? Which bonds change between resonance structures? How do you represent partial bonds and partial charges?

Solve

The resonance hybrid is shown below. The bond between each pair of C atoms appears as a single bond in at least one resonance structure and as a double bond in at least one structure. Thus, each C–C bond behaves as a hybrid between a single and a double bond, and is represented as a single bond plus a partial bond (------). Each C–H bond is a single bond in every resonance structure, and is thus a single bond in the resonance hybrid. The four C atoms with a partial positive charge (δ^+) have a formal charge of 0 in at least one resonance structure and a formal charge of +1 in at least one structure.

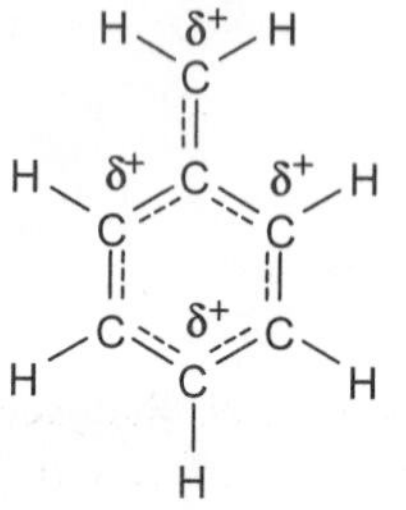

Problem 1.24

Think

Do any of the atoms have an unfilled valence shell? If so, is the atom with the unfilled valence shell adjacent to a double or triple bond? Are there any lone pairs adjacent to a double or triple bond? How can you show, via a curved arrow, the electrons in the triple bond moving to form a double bond? Are any atoms unable to participate in resonance? In evaluating the major contributor, consider formal charges and filled valences.

Solve

The resonance structures are shown below. That with the greatest contribution to the resonance hybrid has all atoms with octets. The C atom on the far right has four single bonds. Single bonds do not move in resonance structures, and thus this atom does not participate in resonance.

Greatest contribution because all atoms have octets

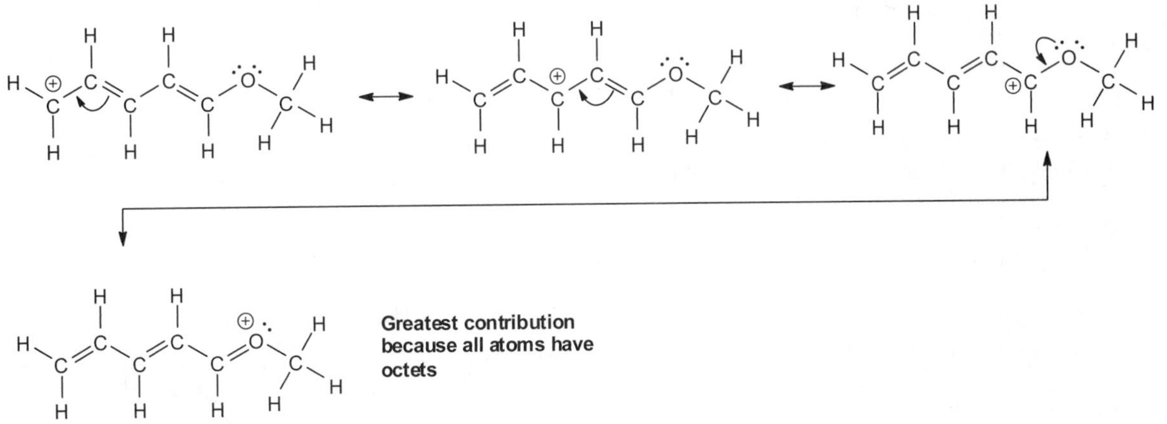

Problem 1.25

Think

Refer to the figure in Section 11.11d that shows a ring of alternating single and multiple bonds. Are the multiple and single bonds in this example alternating around the entire ring? If so, how can you show, using curved arrows, that the pair of electrons from each multiple bond can shift around the entire ring?

Solve

With a ring of alternating single and double bonds, a pair of electrons from each multiple bond can be shifted around the ring. This compound has nine C=C bonds and nine C–C bonds. Each C=C bond converts into a C–C and vice versa. Therefore, nine curved arrows are needed to draw the new resonance structure.

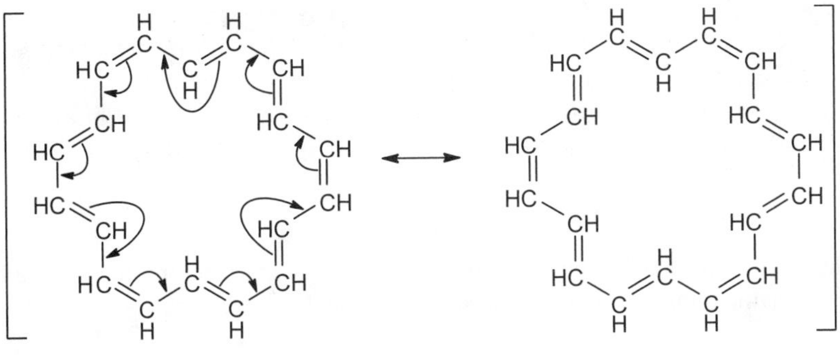

Problem 1.26

Think

Do any of the atoms have an unfilled valence shell? If so, is the atom with the unfilled valence shell adjacent to a double or triple bond? Are there any lone pairs adjacent to a double or triple bond? Are there any rings of alternating single and multiple bonds? How can you show, via a curved arrow, the electrons in the triple bond moving to form a double bond? Are any atoms unable to participate in resonance?

Solve

(a) The electrons in the double bond move around the ring on the left as shown on the next page. The ring on the right has two carbons with four single bonds; therefore, it does not participate in resonance.

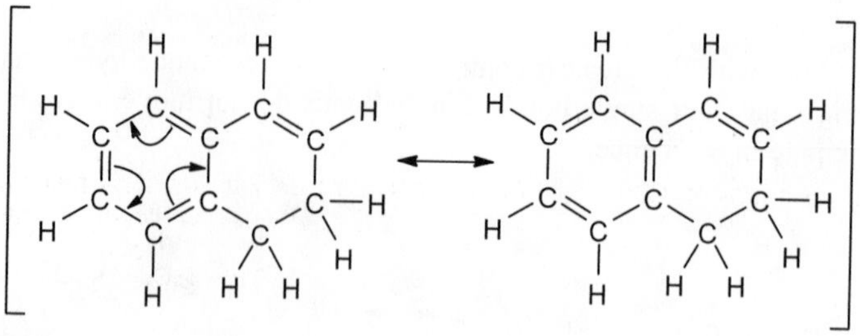

(b) The alternating double bonds are located around both rings. Therefore, the electrons in the double bond can move around the entire ring. Two additional resonance structures are possible.

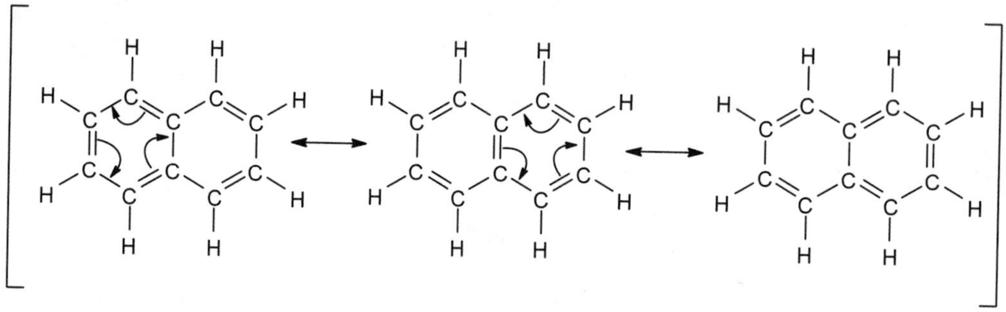

(c) The O atom has a lone pair and is attached to C=C, so a lone pair folds down to make an O=C bond and the C=C electrons fold over to become a lone pair on the C atom; the formal charge on O atom becomes more positive by 1 and that on the C atom becomes more negative by 1. This same lone-pair-to-double-bond and double-bond-to-lone-pair conversion occurs on each additional resonance structure until resonance is completed over the entirety of the ring.

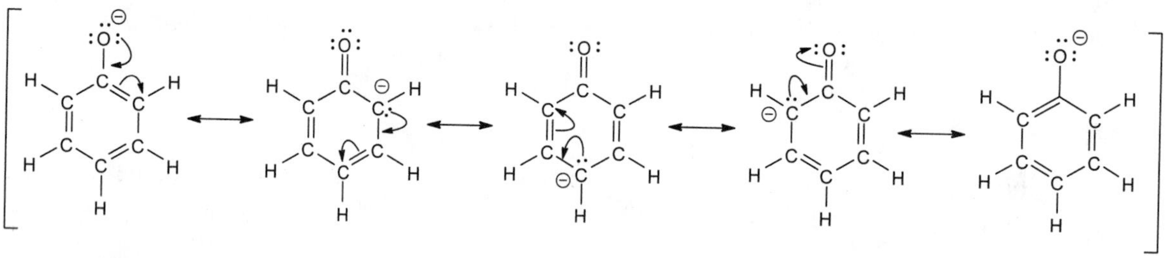

(d) The double bond on the right is adjacent to a C atom with an unfilled valence. Therefore, the double bond moves over one to the right. The same thing occurs from the second to the third structure.

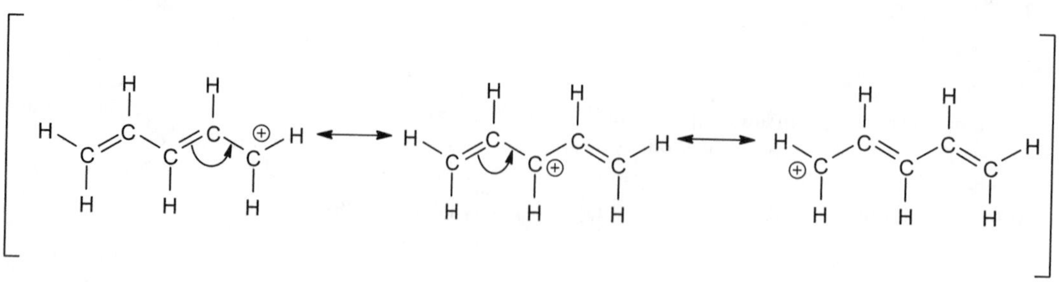

Problem 1.27

Think

Refer to Table 1-5 to review formal charges on atoms in specific bonding scenarios. Try to fill the valences on each atom and consider how the formal charge reflects the number of bonds and lone pairs.

Solve
The Lewis structures are completed below.

(a)	(b)	(c)	(d)	(e)
Fill in two lone pairs on the O atom and one lone pair on the N atom	Fill in two lone pairs on the N atom to make N⁻	Already complete	Already complete	Fill in a lone pair on the C atom at far right to make C⁻

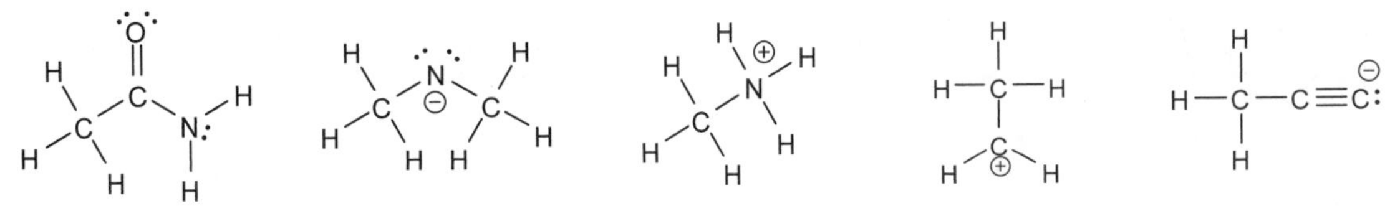

Problem 1.29

Think
Review the rules for drawing line structures. How are C atoms represented? Is it necessary to draw all lone pairs? What are the rules for drawing H atoms on C atoms? H atoms on heteroatoms?

Solve
The line structures are drawn below. C atoms are not drawn explicitly, nor are H atoms bonded to C atoms; all other atoms are drawn, as are their attached H atoms. Bonds between heteroatoms and H can be omitted.

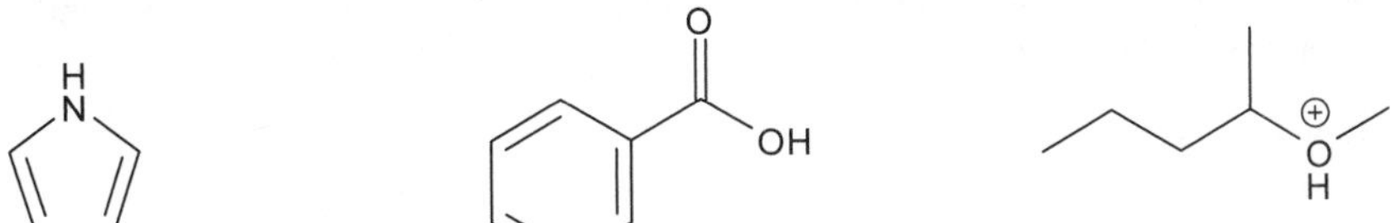

(a) (b) (c)

Problem 1.30

Think
What atom occurs at the intersection of two lines? Draw in the appropriate C–H bonds and lone pair electrons that accurately represent the formal charge shown.

Solve
(a) No formal charges; therefore, each C has four bonds. **(b)** The center C atom has a (+) charge and, therefore, has only three bonds. **(c)** The last C atom on the right has a negative charge and, therefore, has only three bonds and a lone pair.

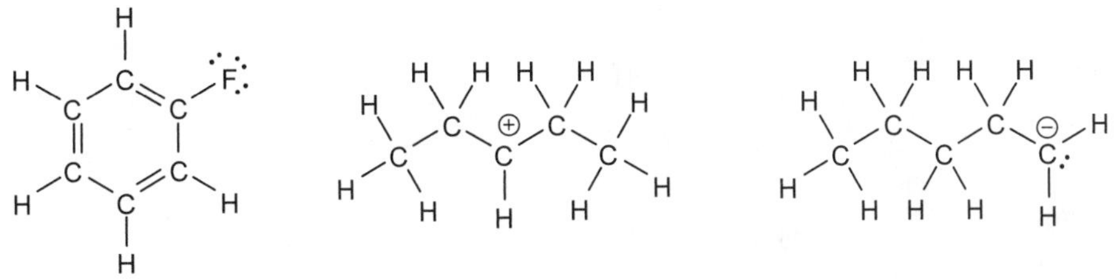

(a) (b) (c)

Problem 1.32

Think
What would the structures look like as complete Lewis structures (i.e., with all hydrogen atoms and lone pairs drawn in)? Which carbon atoms do not participate in resonance?

Solve

(a) The two C atoms that have four single bonds and no formal charge do not participate in resonance. Therefore, resonance is only possible with the other five C atoms.

(b) No resonance structures are possible with this compound (without introducing formal charges) because there are two C atoms between the C=C and the C^+ that have four bonds. These "block" the possibility of resonance.

(c) All of the C atoms in the ring can participate in resonance. There are five resonance structures possible; each resonance structure transforms into the other when a lone pair converts into a double bond and a double bond converts into a lone pair on the adjacent atom.

Problem 1.33

Think

Which groups contain the bonding arrangement of smaller functional groups?

Solve

The aryl ring has what appears to be an alkene. The acetal contains two ether groups, the hemiacetal contains an ether and an alcohol, the ester contains an ether and carbonyl, and an amide contains an amine and carbonyl.

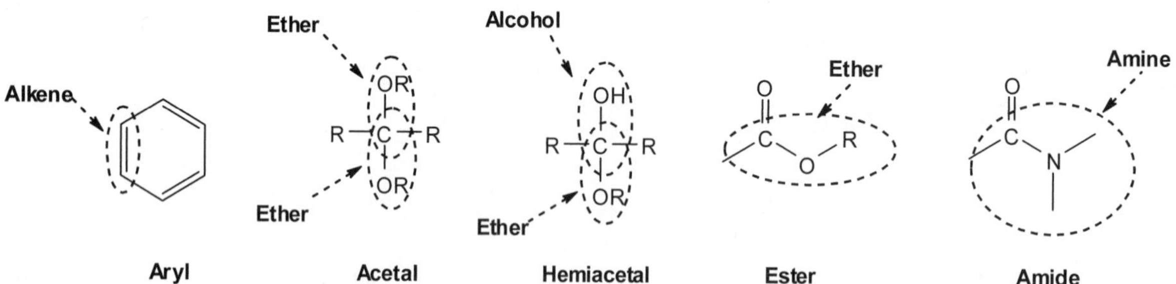

Problem 1.35

Think

Are the functional groups the same or different? What impact will the ring have on the reactivity?

Solve
The compound on the left is a cyclic ester and the compound on the right is a carboxylic acid. Since the functional groups are different the chemical reactivity will also be different.

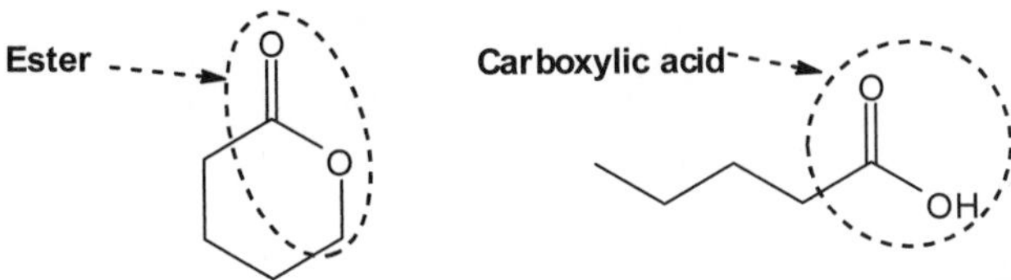

Problem 1.37

Think
Are there bonds present other than C–C and C–H single bonds? Are there *atoms* present other than carbon and hydrogen? Are there any special rings present? What arrangements of bonds and atoms can you find that match with the entries in Table 1-6?

Solve
The functional groups are circled and labeled in the figure below.

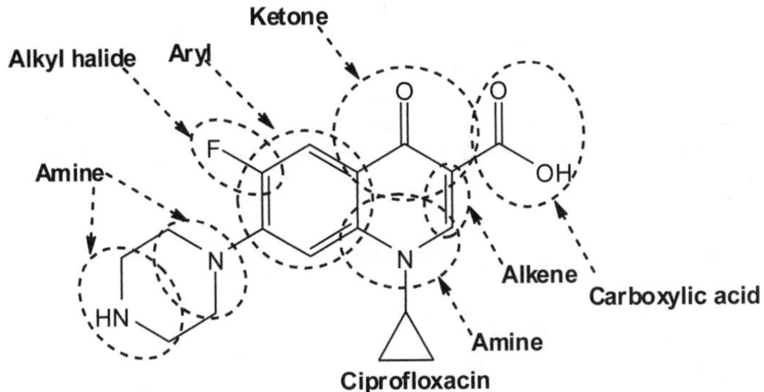

Problem 1.38

Think
Refer to Table 1-6 to review and identify the functional groups present in the amino acids (Table 1-7).

Solve
(a) Alcohol functional group: serine, threonine, tyrosine.

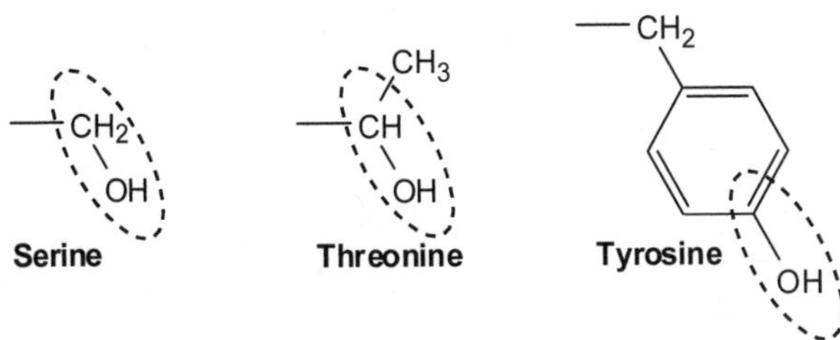

(b) Amide functional group: asparagine, glutamine.

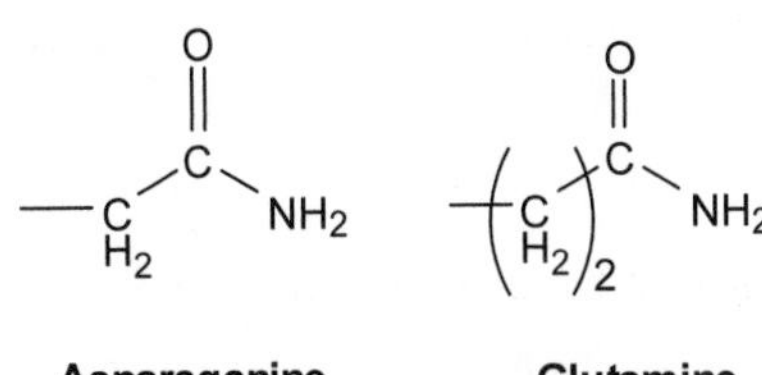

(c) Carboxylic acid functional group: aspartic acid, glutamic acid.

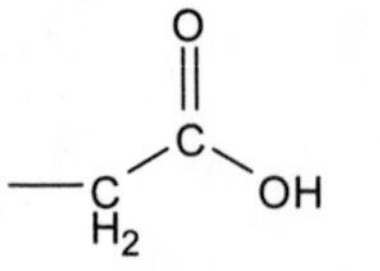

Aspartic acid 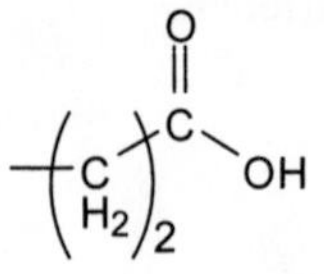 **Glutamic acid**

(d) Amine functional group: lysine, histidine, arginine, tryptophan.

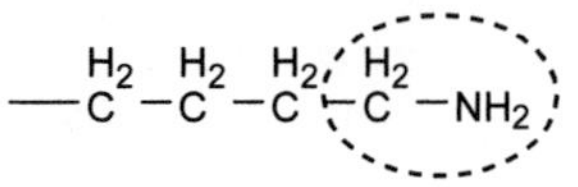

Lysine 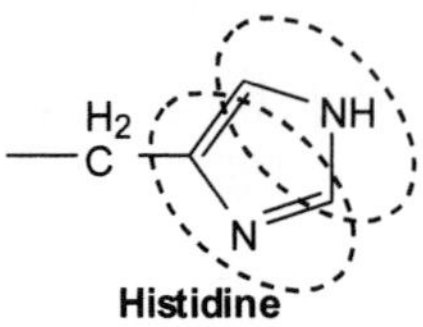 **Histidine** 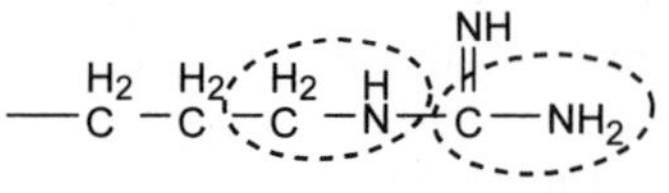 **Arginine** 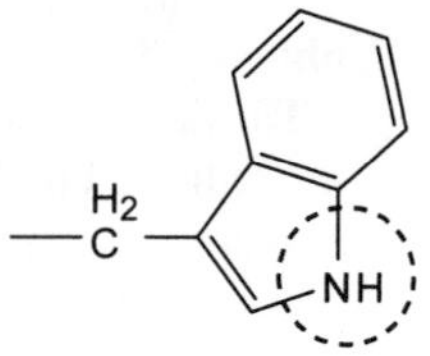 **Tryptophan**

(e) Aryl ring functional group: phenylalanine, tryptophan, tyrosine.

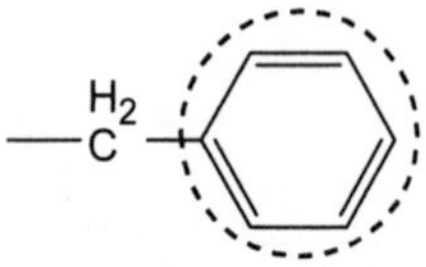

Phenylalanine 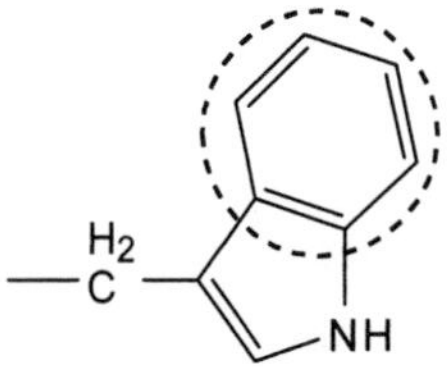 **Tryptophan** 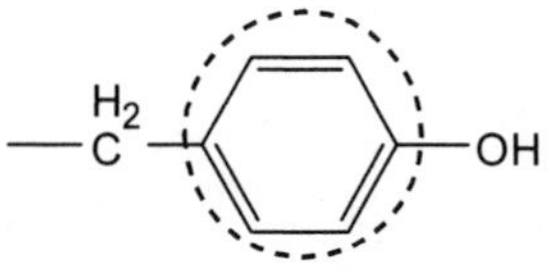 **Tyrosine**

Problem 1.39

Think

What atoms are present in a carbohydrate? What is the general formula for a carbohydrate? What is the general formula for a monosaccharide? Write the formula for each compound and compare it to the definitions.

Solve

In a carbohydrate, carbon, hydrogen, and oxygen are present in a general formula of $C_xH_{2y}O_y$. A monosaccharide is a carbohydrate with one additional restriction to the chemical formula, $C_xH_{2x}O_x$.

(a) Both a carbohydrate and a monosaccharide.

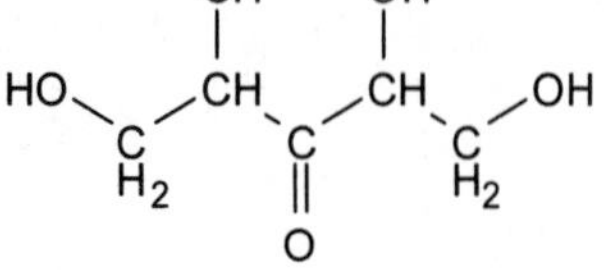

Chemical formula: $C_5H_{10}O_5$

(b) Neither.

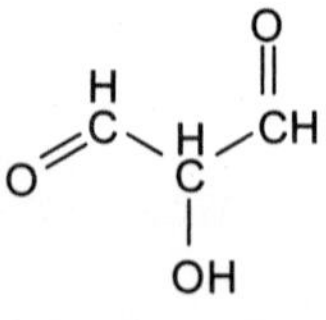

Chemical formula: $C_3H_4O_3$

(c) Both a carbohydrate and a monosaccharide.

Chemical formula: $C_6H_{12}O_6$

(d) Carbohydrate only.

Chemical formula: $C_{12}H_{22}O_{11}$

Problem 1.40

Think

Refer to Figure 1-38. Which atoms compose the phosphate, sugar, and nitrogen base? Refer to Figure 1-39 to identify the nitrogen bases in DNA and RNA. What is the difference between a ribose and deoxyribose sugar?

Solve

(a) The portions are circled and labeled below.

(b) The first one can be part of DNA because the sugar unit is deoxyribose, given that it is missing an oxygen atom at the location indicated. The second one can be part of RNA because it has the O atom at that location.

(c) The nitrogen bases are labeled. Notice that uracil is part of RNA, not DNA.

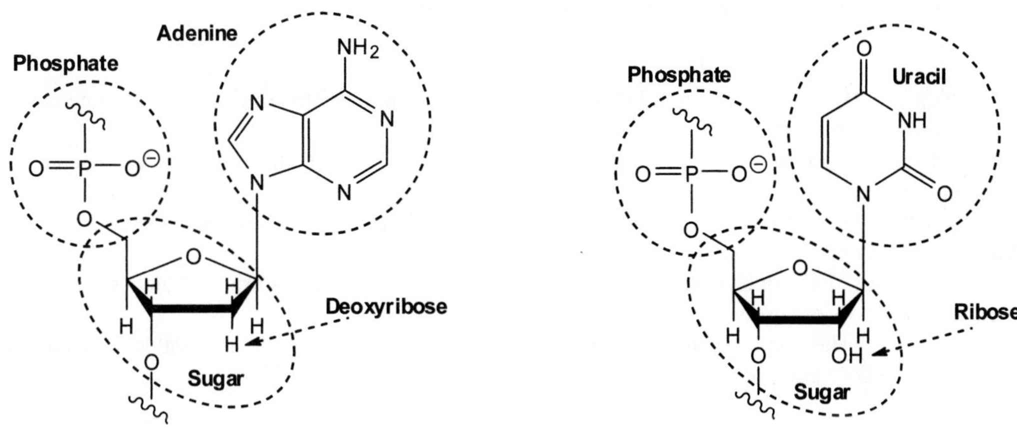

Problem 1.41

Think

Refer to Figure 1-7 to view the orbital filling diagram. Which orbital is filled last? Which orbital is at the highest energy? How is the principal quantum number related to the potential energy?

Solve

The 4*s* orbital possesses the most energy. It has the highest principal quantum number, n=4 (i.e., it is in the fourth shell).

Problem 1.42

Think

Refer to Figure 1-7 to view the orbital filling diagram. Which orbital is filled last? Which orbital is at the highest energy? How is the principal quantum number related to the potential energy?

Solve

The 5*d* orbital possesses the most energy. When the principal quantum number (n) is the same for different orbitals, energy is dictated by the subshell, l. In this case, for the three orbitals given that are in the n=5 shell, energy increases in the order: (lowest) $5s < 5p < 5d$ (highest).

Problem 1.43

Think

What is the charge on a cation? Do like charges repel or attract? What happens to the potential energy of the two cations as they get closer?

Solve

The most energy (least stable) would be at 0.1 nm. Positive charges repel each other, and it requires adding energy to move two cations together.

Problem 1.44

Think

What is the charge on a cation? On an anion? Do opposite charges repel or attract? What happens to the potential energy of the two ions as they get closer?

Solve

The most energy (least stable) would be at 1000 nm. Opposite charges attract each other and it requires energy to separate them.

Problem 1.45

Think

How many electrons does each element listed possess? What is the valence shell and where do the core electrons reside? What distinguishes valence electrons from core electrons?

Solve

(a) Al has 13 electrons. The configuration is $1s^22s^22p^63s^23p^1$; the 3*s* and 3*p* electrons are the valence electrons; the 1*s*, 2*s*, and 2*p* electrons are core electrons.
(b) S has 16 electrons. The configuration is $1s^22s^22p^63s^23p^4$; the 3*s* and 3*p* electrons are the valence electrons; the 1*s*, 2*s*, and 2*p* electrons are core electrons.
(c) O has eight electrons. The configuration is $1s^22s^22p^4$; the 2*s* and 2*p* electrons are the valence electrons; the 1*s* electrons are the core electrons.
(d) N has seven electrons. The configuration is $1s^22s^22p^3$; the 2*s* and 2*p* electrons are the valence electrons; the 1*s* electrons are the core electrons.
(e) F has nine electrons. The configuration is $1s^22s^22p^5$; the 2*s* and 2*p* electrons are the valence electrons; the 1*s* electrons are the core electrons.

Problem 1.46

Think

Does the compound contain elements from both the left and right sides of the periodic table? Does the compound contain any recognizable polyatomic cations? Are any metals present?

Solve

(a), **(e)**, and **(f)** contain only nonmetals and have no recognizable polyatomic ions; they have only covalent bonds.
(b), **(c)**, **(d)**, **(g)**, **(h)**, and **(i)** are ionic compounds.
(b) consists of Na^+ and Cl^-.
(c) consists of Na^+ and HO^-.
(d) consists of Na^+ and CH_3O^-.
(g) consists of Li^+ and CH_3NH^-.
(h) consists of K^+ and $CH_3CH_2CO_2^-$.
(i) consists of $C_6H_5NH_3^+$ and Cl^-. Even though **(i)** has only nonmetals, a positive charge appears due to N having four bonds and no lone pairs.

Problem 1.47

Think

Refer to Figure 1-16 to determine the electronegativity value for each C and the atoms that are in the C–X bond indicated. To what does the *difference* in electronegativity correspond in terms of bond type (i.e., ionic vs. covalent)? Are any metals present? What type of bonding is likely to result?

Solve

The greater the electronegativity of the atom bonded to C, the smaller the concentration of negative charge on C. F has the greatest electronegativity, giving C the smallest concentration of negative charge. Li has the smallest electronegativity, giving C the greatest concentration of negative charge.

$$\mathbf{CH_3–F < CH_3–OH < CH_3–NH_2 < CH_3–CH_3 < CH_3–MgBr < CH_3–Li}$$

Problem 1.48

Think

Review the rules for drawing Lewis structures and draw the Lewis structure for CH_3NO_2. Is there an atom with a lone pair attached to a multiple bond? If so, how many curved arrows must be drawn to indicate electron movement to arrive at another resonance structure? In the resonance hybrid, where are the partial positive and negative charges, δ^+ and δ^-? Can you match the hybrid to one of the electrostatic potential maps? Which color (red or blue) corresponds to a buildup of negative charge? Of positive charge?

Solve

Electrostatic potential map **(C)** most accurately represents nitromethane. Another resonance structure exists due to the lone pair on O^- attached to N=O. Two curved arrows show the movement of electrons to arrive at the second resonance structure. The resonance hybrid (shown below) exhibits a substantial and equal partial negative charge on each O atom (shown in red in the electrostatic potential map) and a full positive charge on N (shown in blue in the electrostatic potential map).

H3C
H3C
H3C
δ−
δ−
Resonance hybrid
C

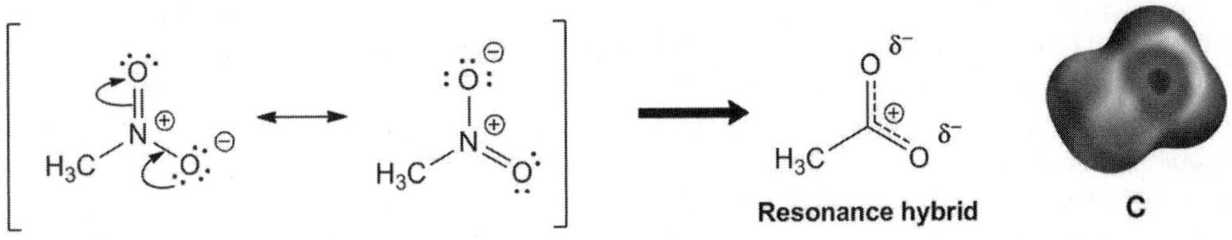

Problem 1.49

Think

Consider the steps for drawing Lewis structures. Which atoms must be bonded together?

Solve

First count the total valence electrons. Then, write the skeleton of the molecule, distribute the remaining electrons as lone pairs in an attempt to fill each atom's valence, and convert lone pairs into double or triple bonds as necessary. The first row below shows the results before converting lone pairs to multiple bonds. The next row shows the completed Lewis structures. See the figure on the next page.

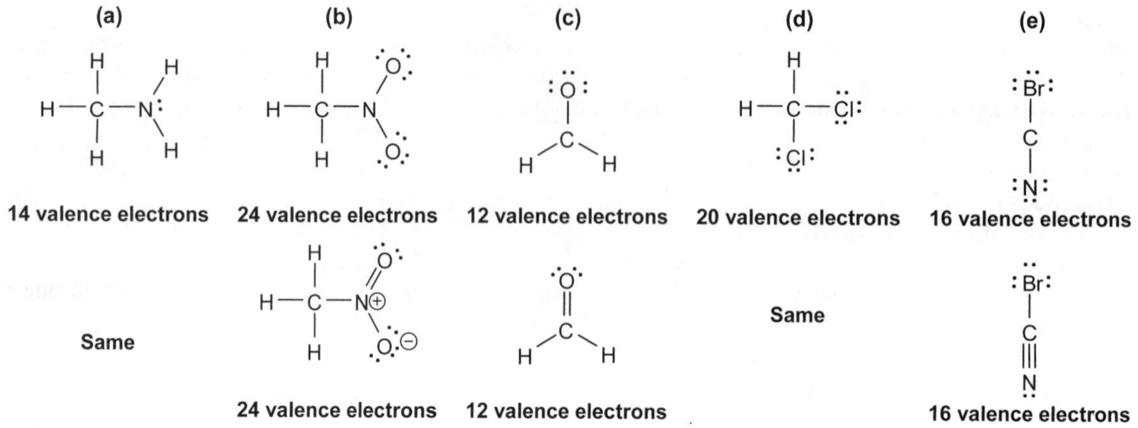

(a)	(b)	(c)	(d)	(e)
14 valence electrons	24 valence electrons	12 valence electrons	20 valence electrons	16 valence electrons
Same	24 valence electrons	12 valence electrons	Same	16 valence electrons

Problem 1.50

Think

Refer to Table 1–4 for common bonding patterns for oxygen and carbon. To have a negative charge on the O, what type of bond must the C–O be? How many lone pairs would O need to have?

Solve

No, it is not possible for a C=O double bond to exist in the methoxide anion. If the O atom has a negative charge, it must have three lone pairs and a single bond to the C atom.

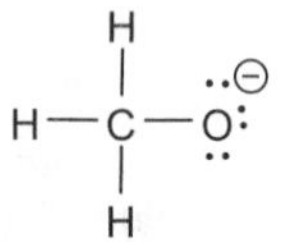

Problem 1.51

Think

Consider the steps for drawing Lewis structures. Which atoms must be bonded together? How is the total number of valence electrons changed when the molecule is a cation or an anion? Refer to Table 1–4 for common bonding patterns for oxygen, nitrogen, and carbon. How are electrons assigned in the formal charge method?

Solve

(a) C_2H_5 anion = 14 valence electrons. The negative charge is located on one of the C atoms. That C^- should have three bonds and one lone pair, and the other C should have four bonds.

(b) CH_3O cation = 12 valence electrons. The positive charge is located on the O, so it should have three bonds and one lone pair. The C should have four bonds: two single bonds and a double bond.

(c) CH_6N cation = 14 electrons. The positive charge is located on the N, so it should have four bonds, and the C should have four bonds.

(d) CH_5O cation = 14 valence electrons. The positive charge is located on the O, so it should have three bonds and one lone pair, and the C should have four bonds.

(e) C_3H_3 anion = 16 valence electrons. The negative charge is located on the last C, so it should have three bonds and one lone pair. The other two C atoms should each have four total bonds.

Note that **(a)** and **(e)** both have a carbon with a negative charge that bears three bonds and one lone pair. It does not matter that **(a)** has three single bonds and **(e)** has a triple bond.

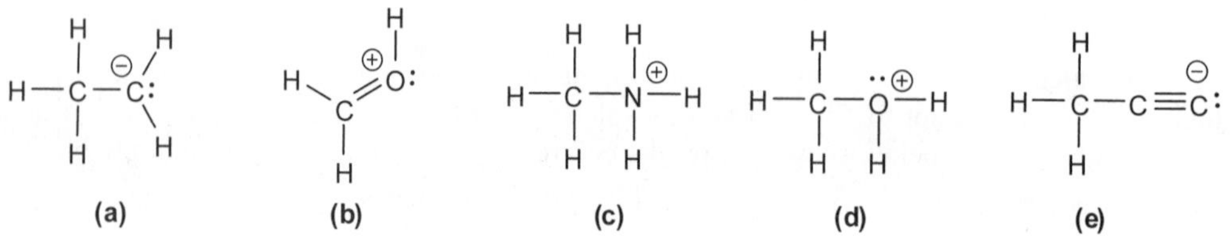

(a) (b) (c) (d) (e)

Problem 1.52

Think

Which atoms (other than hydrogen) have an octet and which atoms don't? How many bonds and lone pairs are typical for each element? Refer to Table 1-4 for the common number of covalent bonds and lone pairs for H, C, N, and O.

Solve

The atoms without an octet are circled on the left. Based on Table 1-4, we need to:

(a) Convert the C–O single bond and one C–N single bond each to double bonds, and convert the terminal C–N to a triple bond.Aadd a lone pair to each N, two lone pairs to each O, and three lone pairs to the F.

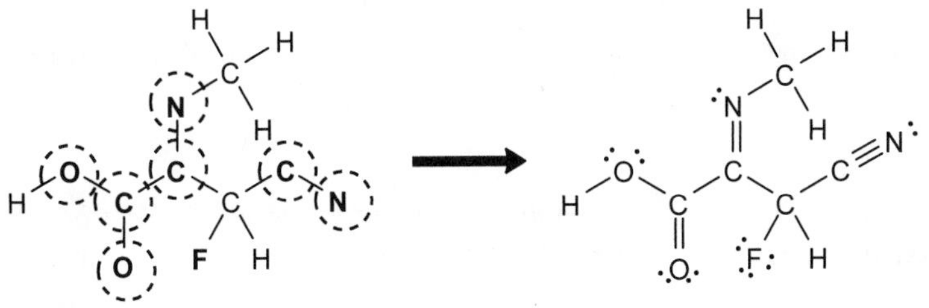

(b) Convert alternating C–C single bonds in the ring to double bonds and the terminal C–C outside of the ring to a triple bond.

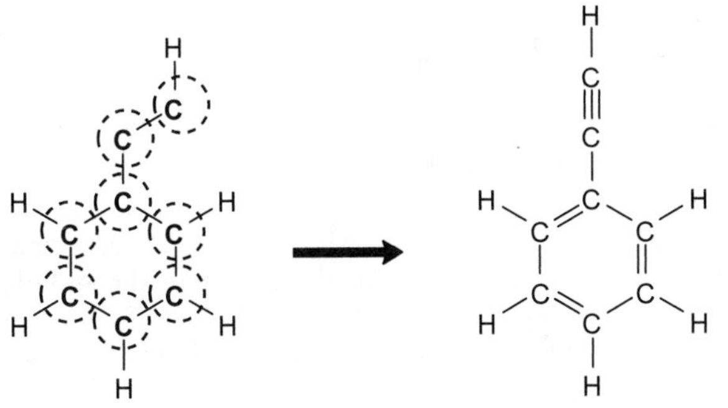

(c) Convert the second C–C single bond to a double bond, and the N–O single bond to a double bond. Add a lone pair to the N and two lone pairs to the O.

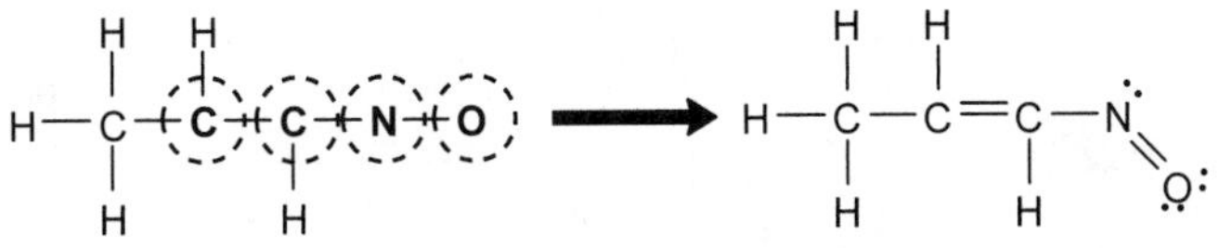

Problem 1.53

Think

How are covalently bonded electrons assigned in the oxidation state method? Consider the electronegativity of each atom in the bond. How are lone pairs of electrons assigned? How many valence electrons should each atom be assigned to have an oxidation state of 0?

Solve

The oxidation state method for assigning valence electrons assigns all electrons in a bond to the more electronegative atom. If the atoms are identical, the electrons are split evenly. All lone pairs on an atom are assigned to that atom. To determine an atom's oxidation state, compare the number of assigned electrons to the atom's group number. The oxidation state on each H atom is +1 because, being bonded to a more electronegative atom, is assigned zero valence electrons, whereas its group number is 1. All other oxidation states are shown below. Carbon's group number is 4, so it has an oxidation state of 0 when it is assigned four electrons. Each additional electron results in a more negative oxidation state and each fewer electron results in a more positive oxidation state. Oxygen's group number is 6, so its oxidation state is −2 when assigned eight valence electrons. The sum of the oxidation states of each atom must equal the overall charge of the compound.

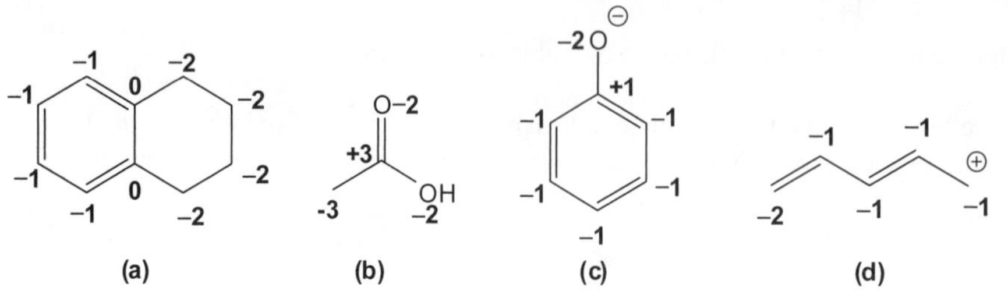

(a) (b) (c) (d)

Problem 1.54

Think

How are lone pairs assigned? In determining how to assign bonding pairs of electrons, is electronegativity relevant in the formal charge method or the oxidation state method? How many valence electrons should each atom be assigned to have a formal charge or oxidation state of 0?

Solve

The left figure below shows the formal charges—each atom is assigned half of its bonding electrons and all of its lone pairs. To have a formal charge of 0, an atom must be assigned the same number of valence electrons as its group number. Negative charges results when there are additional electrons, and positive charges result when there are fewer. There are two atoms with a nonzero formal charge. The oxygen with three lone pairs has a formal charge of −1, and the N that has a triple bond and a single bond has a formal charge of +1. The oxidation states are shown on the right. Each bonding pair of electrons is assigned entirely to the more electronegative atom, or is shared equally if the two atoms are the same. The oxidation state on each H atom is +1 because, being bonded to a more electronegative atom, is assigned zero valence electrons, compared to its group number of 1. For the oxidation states on the other atoms, compare the assigned valence electrons to the group number. Remember, the sum of the formal charges and oxidation states must equal the total charge on the compound—in this case zero.

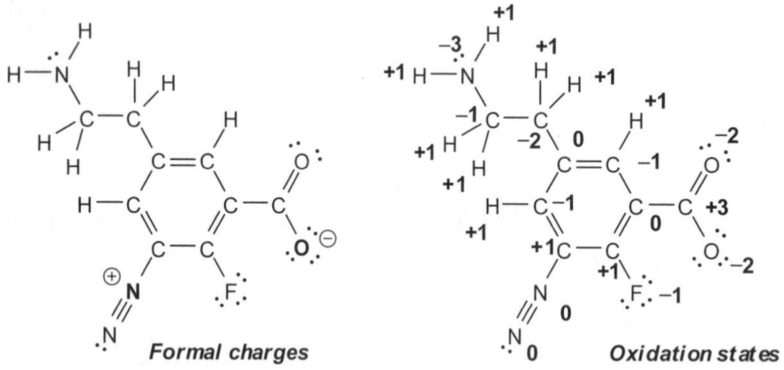

Formal charges *Oxidation states*

Problem 1.55

Think

Refer to Table 1-4 for common bonding patterns for O, N, and C. How many bonds and lone pairs does each atom possess with a −1 formal charge? How many bonds and lone pairs should the remaining atoms have if they are to be uncharged? Be sure to fill in the octet for the other non-hydrogen atoms in the molecule.

Solve

(a) A −1 formal charge on N requires two bonds and two lone pairs.
(b) A −1 formal charge on O requires one bond and three lone pairs.
(c) A −1 formal charge on C requires three bonds and one lone pair.

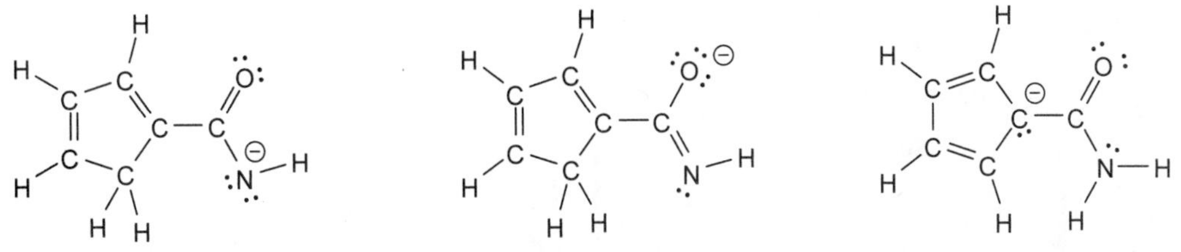

(a) −1 on N

(b) −1 on O

(c) −1 on C

Problem 1.56

Think

Review the rules for drawing line structures. How are C atoms represented? Is it necessary to draw all lone pairs? What are the rules for drawing H atoms on C atoms? Heteroatoms? Are there bonds present other than C–C and C–H single bonds? Are there *atoms* present other than carbon and hydrogen? Are there any special rings present?

Solve

The line structure is shown below. There are four alcohol groups and one hemiacetal. Notice that the hemiacetal is the result of fusing an alcohol and an ether together.

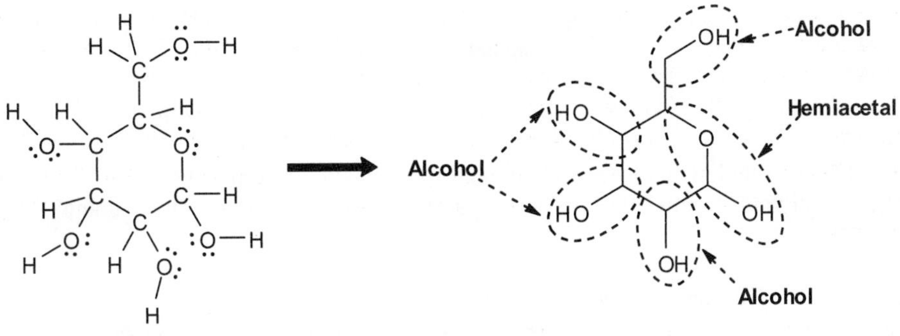

Glucose

Alcohol

Hemiacetal

Alcohol

Alcohol

Problem 1.57

Think

What atom occurs at the intersection of two lines? Draw in the appropriate C–H bonds and lone pair electrons that accurately represent the formal charges shown. Consider Table 1-5 when doing so. To identify functional groups, are there bonds present other than C–C and C–H single bonds? Are there *atoms* present other than carbon and hydrogen? Are there any special rings present?

Solve

(a) To redraw the line structure into the complete Lewis structure, all of the C and H atoms and lone pair electrons must be explicitly shown (left figure next page). All C atoms are uncharged, so each should have a total of four bonds and no lone pairs. If a carbon has fewer than four bonds shown, add bonds to H.
(b) There are eight alcohol functional groups and two acetals. Each acetal consists of a C atom with two separate C–O–C bonds (right figure next page).

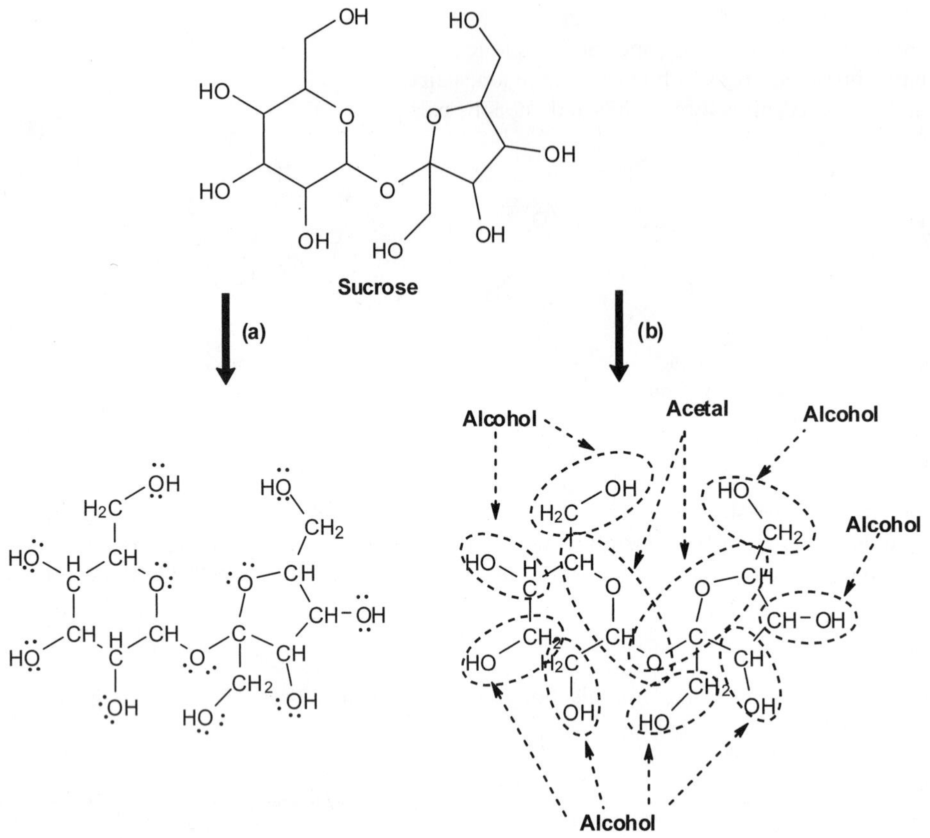

Sucrose

(a)

(b)

Alcohol Acetal Alcohol Alcohol Alcohol

Problem 1.58

Think

Refer to Sections 1.10 and 1.11 to review the rules and strategies for drawing resonance structures. Are atoms allowed to move? Do the line structures show all of the atoms and lone pairs that you need to evaluate?

Solve

A, **B**, **F**, **G**, and **H** are not pairs of resonance structures because they are different by the movement of atoms, not just by the movement of valence electrons. Notice that in **F**, the F atoms are in different positions relative to the double bond. Also, notice that the pair of molecules in **H** are different by the movement of H atoms (in boldface type). **C**, **D**, and **E** are pairs of resonance structures. Lone pair electrons and curved arrows are shown below and on the next page.

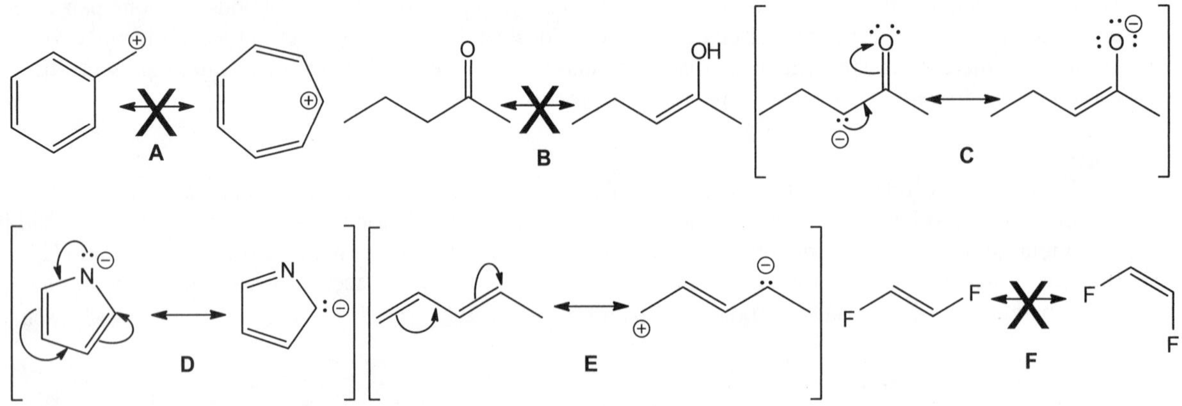

A B C

D E F

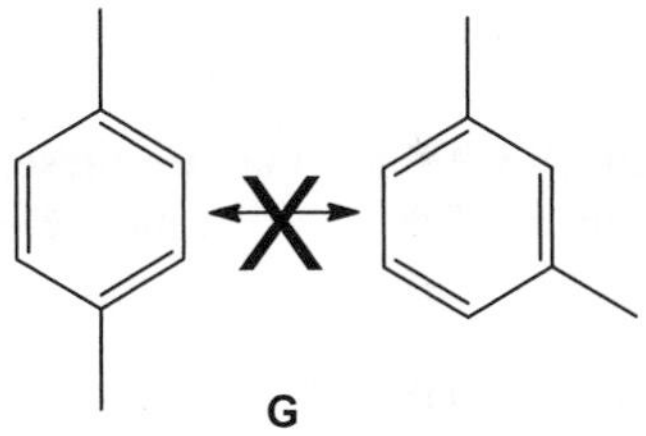

G

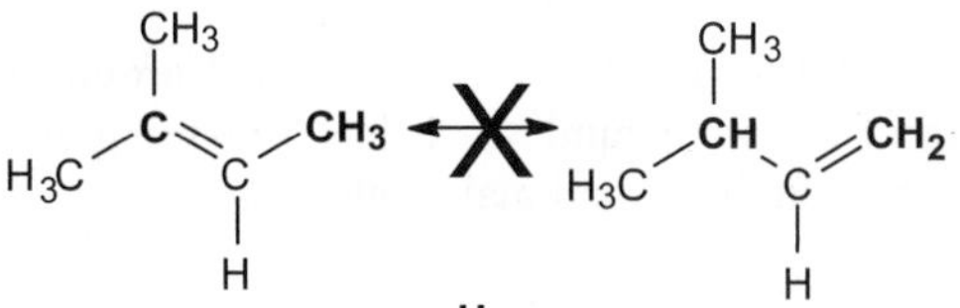

H

Problem 1.59

Think

Review the rules for drawing resonance structures and draw the resonance hybrid for each compound below. In the structure you are given, is there a multiple bond attached to an atom with a lone pair? To an atom lacking an octet? Is there a ring of alternating single and multiple bonds? What does an electrostatic potential map depict? Which color (red or blue) corresponds to a buildup of negative charge? Match up the electrostatic potential map to one of the five compounds.

Solve

The electrostatic potential map indicates a concentration of negative charge shared among three different carbon atoms. This is consistent with **C**, which has three total resonance structures, as shown below. Each resonance structure has a C^- with a lone pair attached to C=C, so the curved arrows are drawn away from the lone pair.

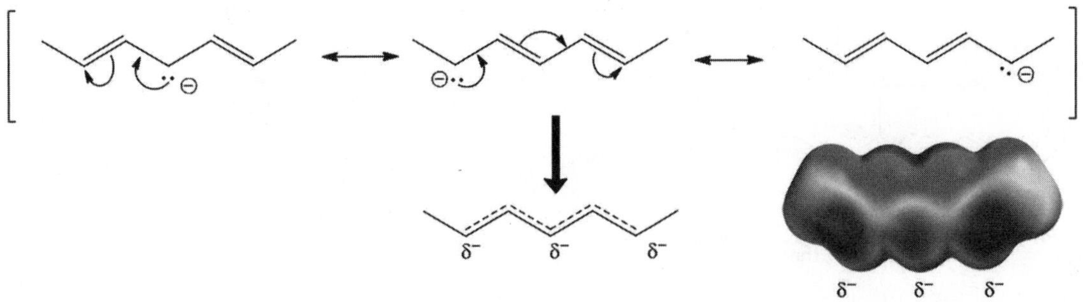

B and **D** have no additional resonance structures, so the negative charge is localized on just one C atom. **A** has three additional resonance structures, allowing the negative charge to be shared among four C atoms. **E** has one additional resonance structure, allowing the negative charge to be shared among two C atoms.

A

Resonance hybrid

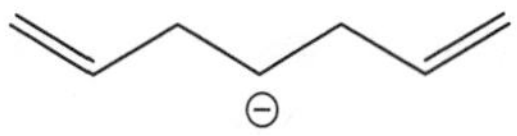

B
No resonance structures possible

E

Resonance hybrid

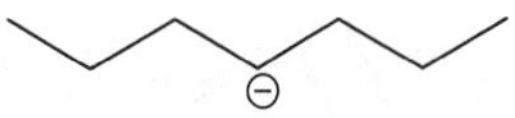

D
No resonance structures possible

Problem 1.60

Think

Refer to Table 1-6. Which group has a heteroatom between two singly bonded carbon groups? Where is sulfur located in the periodic table? How do groups in the same column behave chemically?

Solve

CH_3–S–CH_3 should most resemble the behavior of an ether (R–O–R) because S is located below O in the periodic table. Moreover, S and O each have two bonds and two lone pairs when uncharged.

Problem 1.61

Think

What functional group does ethanol have? Does that functional group match up with any that are present in molecules **A–E**? If not, which of the compounds **A–E** has a bonding arrangement of atoms that most closely resembles the bonding arrangement in that functional group?

Solve

The functional group in ethanol is an alcohol, C–O–H. (b) is an ether, (c) is a carboxylic acid, and (d) is an aldehyde. H_2O does not have a functional group, but its bonding arrangement, H–O–H, is very similar to C–O–H, the only difference being H exchanged for C. Because H and C have very similar electronegativities, ethanol will behave most like H_2O.

Problem 1.62

Think

Are there bonds present other than C–C and C–H single bonds? Are there *atoms* present other than carbon and hydrogen? Are there any special rings present? Refer to Table 1-6 for common functional groups.

Solve

The functional groups are circled and labeled below.

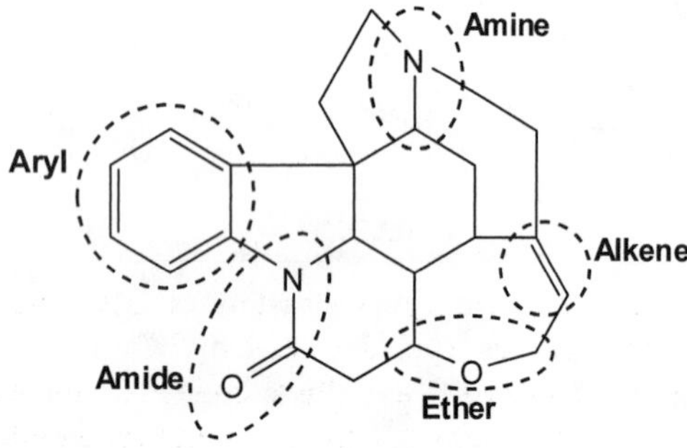

Problem 1.63

Think

Are there bonds present other than C–C and C–H single bonds? Are there *atoms* present other than carbon and hydrogen? Are there any special rings present? Refer to Table 1-6 for common functional groups.

Solve

The functional groups are circled and labeled below.

Problem 1.64

Think

Refer to Table 1-6 to review the functional groups. Draw in the necessary functional groups first and then fill in the rest of the formula. Be sure that all atoms (other than H) in your structures have octets and that there are nine H atoms.

Solve

(a) An amine is R–NH_2, R–NHR, or R–NR_2 and an alkyne is a C≡C.
(b) An alkene is a C=C.
(c) A nitrile is a C≡N. Two examples for each problem are shown below.
Note: There are additional correct answers.

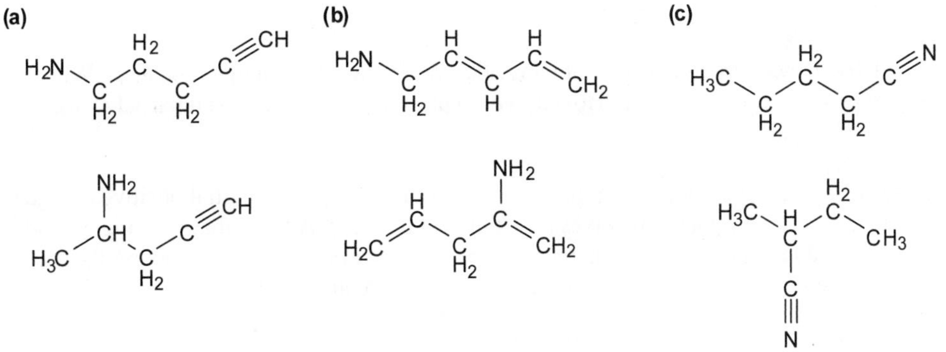

(a) H_2N–CH_2–CH_2–CH_2–C≡CH; H_3C–CH(NH_2)–CH_2–C≡CH

(b) H_2N–CH_2–CH=CH–CH=CH_2; H_2C=CH–CH_2–C(NH_2)=CH_2

(c) H_3C–CH_2–CH_2–CH_2–C≡N; H_3C–CH(C≡N)–CH_2–CH_3

Problem 1.65

Think

Before you can draw resonance structures, you must begin with a complete Lewis structure or line structure. Which atoms are bonded together? How many total valence electrons must each species have? Do any of the completed molecules have atoms that have an unfilled valence shell attached to a multiple bond? Are there any lone pairs on an atom adjacent to a double or triple bond? Are there any rings of alternating single and multiple bonds? Are any atoms unable to participate in resonance?

Solve

The resonance contributors are shown below and on the next page. For **(e)**, resonance contributors are minor because they have additional formal charges.

(a) H_3C–$N^{\oplus}$(=O)–$O^{\ominus}$ ⟷ H_3C–$N^{\oplus}$($O^{\ominus}$)=O

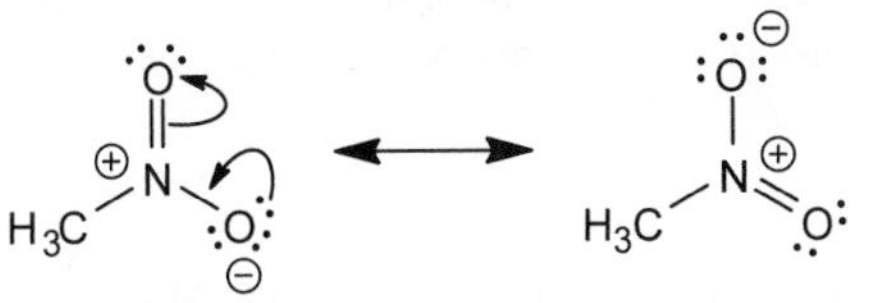

Lone pair adjacent to a double bond
CH_3 unable to participate in resonance (has four single bonds)

(b) H_3C–C(=O)–$O^{\ominus}$ ⟷ H_3C–C($O^{\ominus}$)=O

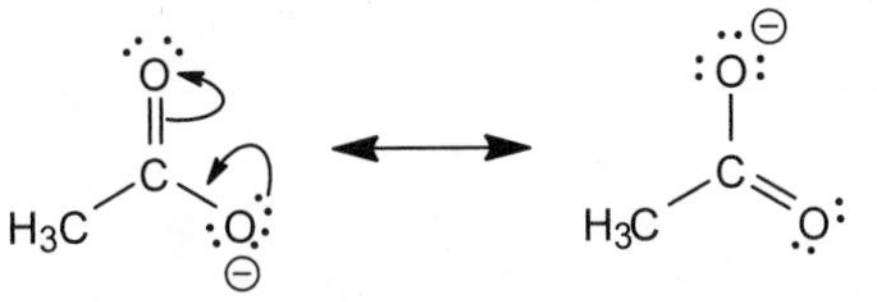

Lone pair adjacent to a double bond
CH_3 unable to participate in resonance (has four single bonds)

(c) H_3C–CH=CH–$CH_2^{\oplus}$ ⟷ H_3C–$CH^{\oplus}$–CH=CH_2

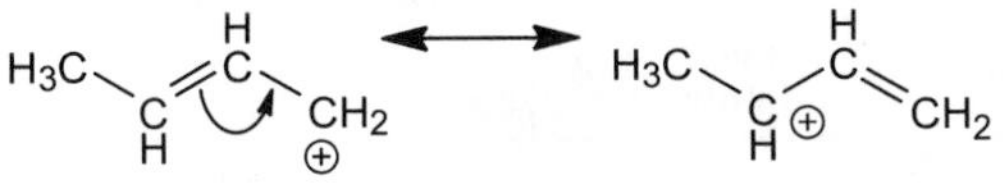

Incomplete octet adjacent to a double bond
CH_3 unable to participate in resonance (has four single bonds)

(d) 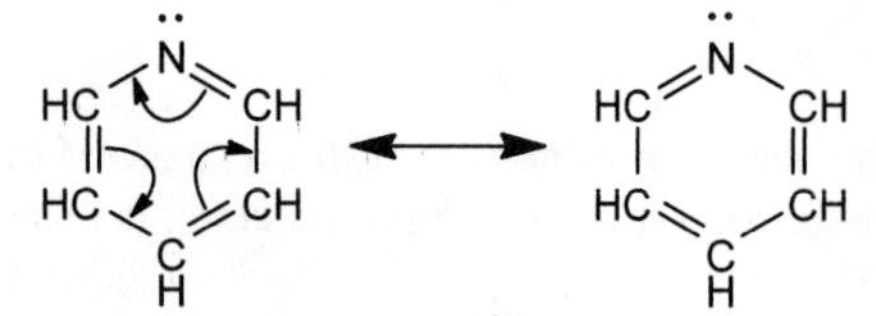 Ring of alternating single and multiple bonds

(e) 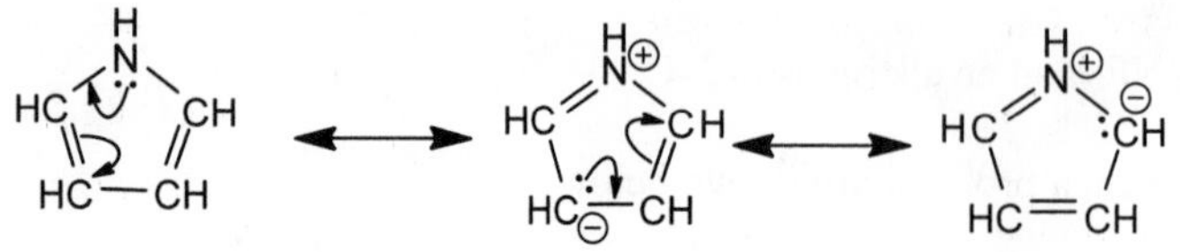 Lone pair adjacent to a double bond

Minor resonance contributors

Problem 1.66

Think

Refer to the individual resonance structures in Problem 1.65. Where are the charges located? Which bonds change from one resonance structure to the next? How do you represent partial bonds and partial charges?

Solve

Each atom that shares a +1 charge in the individual resonance structures receives a partial positive charge (δ^+) in the hybrid; each atom that shares a −1 charge receives a partial negative (δ^-). A bond that is a single bond in one resonance structure but a double bond in another is represented as ----- in the hybrid. Showing the location of lone pairs involved in resonance is impractical with resonance hybrid structures.

(a) 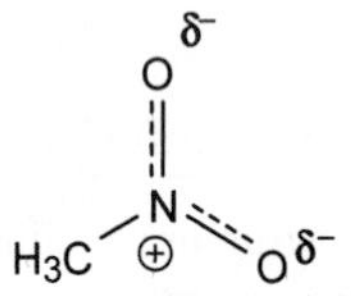 (b) 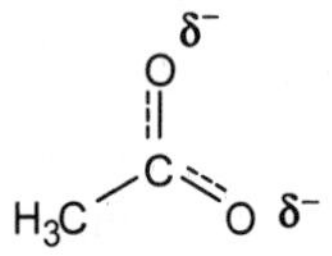 (c) 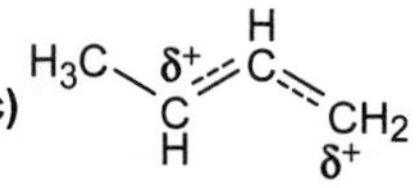 (d) 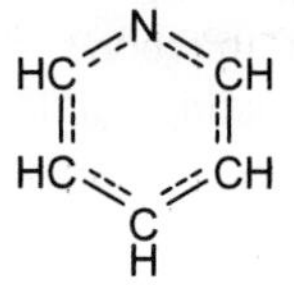 (e)

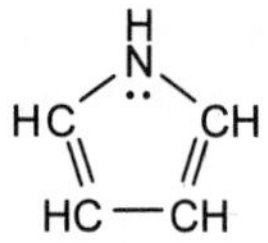

Problem 1.67

Think

How many bonds and lone pairs are typical of O and H? Are there any atoms allowed to have an expanded octet? Are there any lone pairs adjacent to a double or triple bond? If so, how many curved arrows are needed to show electron movement from one resonance structure to the next?

Solve

The resonance structures differ in the number of bonds and the number of formal charges. With fewer bonds and more formal charges, the contribution toward the resonance hybrid is diminished.

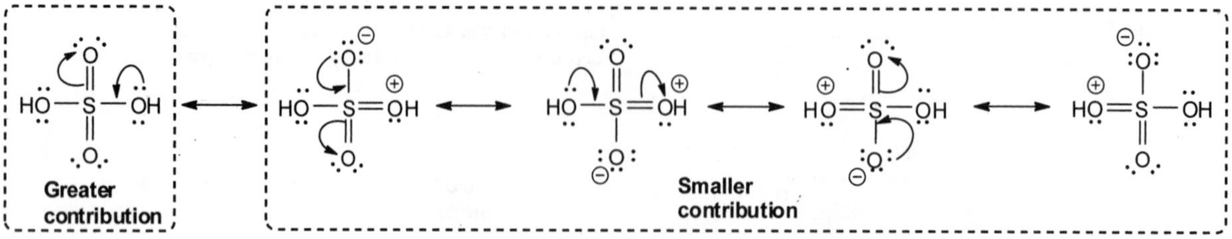

Greater contribution

Smaller contribution

Problem 1.68

Think

Are there any atoms with an unfilled valence shell adjacent to a double or triple bond, or adjacent to an atom with a lone pair? Are there any lone pairs adjacent to a double or triple bond? Are there any rings of alternating single and multiple bonds? Once you identify one of these features, how many curved arrows should be used to show electron movement? Are any atoms unable to participate in resonance? To draw each resonance hybrid, where are the charges located in each species' resonance structures? Which bonds change between resonance structures? How do you represent partial bonds and partial charges?

Solve

Resonance structures and resonance hybrids for **(a)–(d)** are shown below.

(a) Incomplete octet adjacent to double bond

H_3C :OH H_3C :OH H_3C :OH H_3C OH

δ+ OH δ+ H_3C δ+ δ+

Resonance hybrid

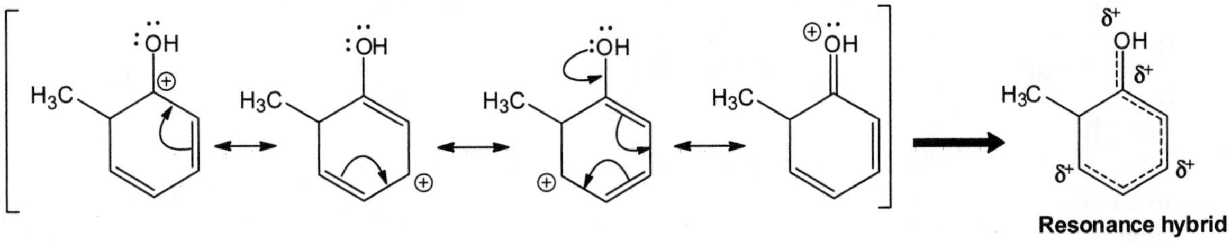

(b) Incomplete octet adjacent to double bond

δ+ δ+

Resonance hybrid

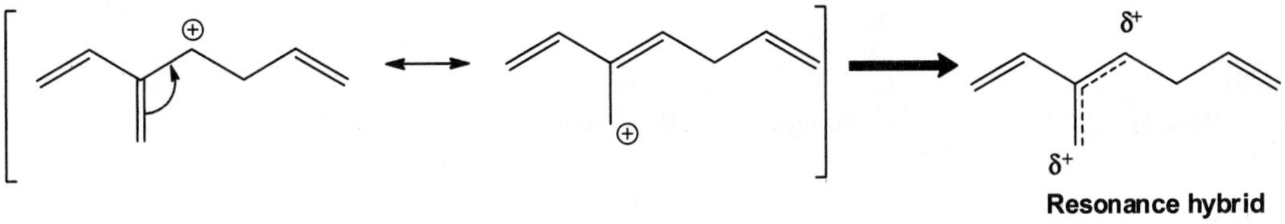

(c) Lone pair adjacent to double bond

δ− δ− δ−

Resonance Hybrid

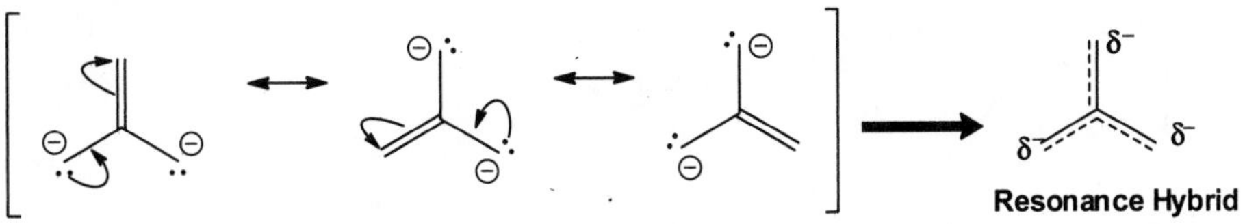

(d) Lone pair adjacent to double bond

O O O O O O O O

δ− O δ− δ− O δ−

Resonance hybrid

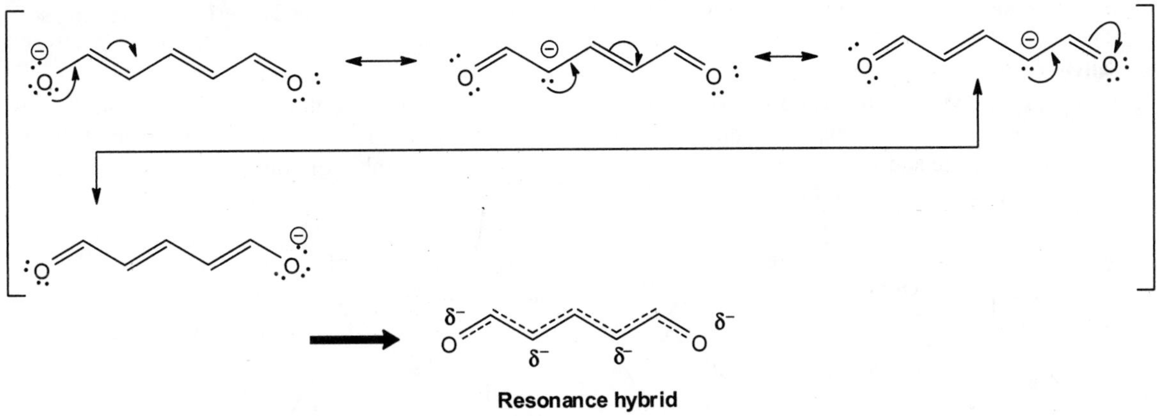

Problem 1.69

Think

Review the rules for drawing condensed formulas. How do you represent multiple H atoms bonded to the same atom? How are rings drawn? How are branches handled?

Solve

Condensed formulas are shown below. Note that rings must be shown explicitly.

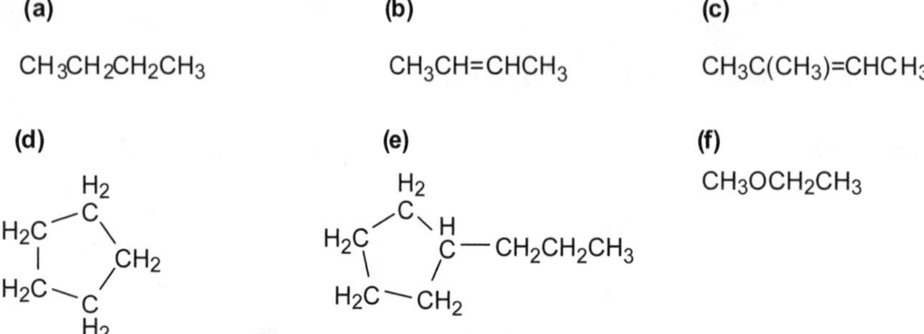

Problem 1.70

Think

Review the rules for drawing line structures. How are C atoms represented? Is it necessary to draw all lone pairs? What are the rules for drawing H atoms on C atoms? Heteroatoms? H atoms bonded to heteroatoms?

Solve

The line structures are drawn below. C atoms are not drawn explicitly and H atoms on C are not drawn. All other atoms are drawn, including H atoms bonded to heteroatoms.

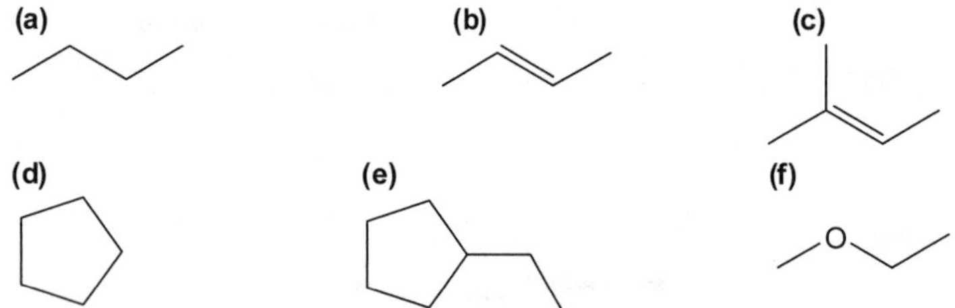

Problem 1.71

Think

What atom occurs at the intersection of two lines? At the end of a line? Draw in the appropriate C–H bonds and lone pair electrons that accurately represent the formal charges shown. Consider Table 1-5 when doing so.

Solve

Completed Lewis structures are shown below and on the next page. All atoms, bonds, formal charges, and nonbonding electrons are explicitly drawn. C atoms are located at the intersections of lines and at the ends of lines. H atoms are added to C atoms to give each C the appropriate number of bonds: Four bonds for uncharged C, and three bonds for C^+ and C^-.

(a) (b) (c)

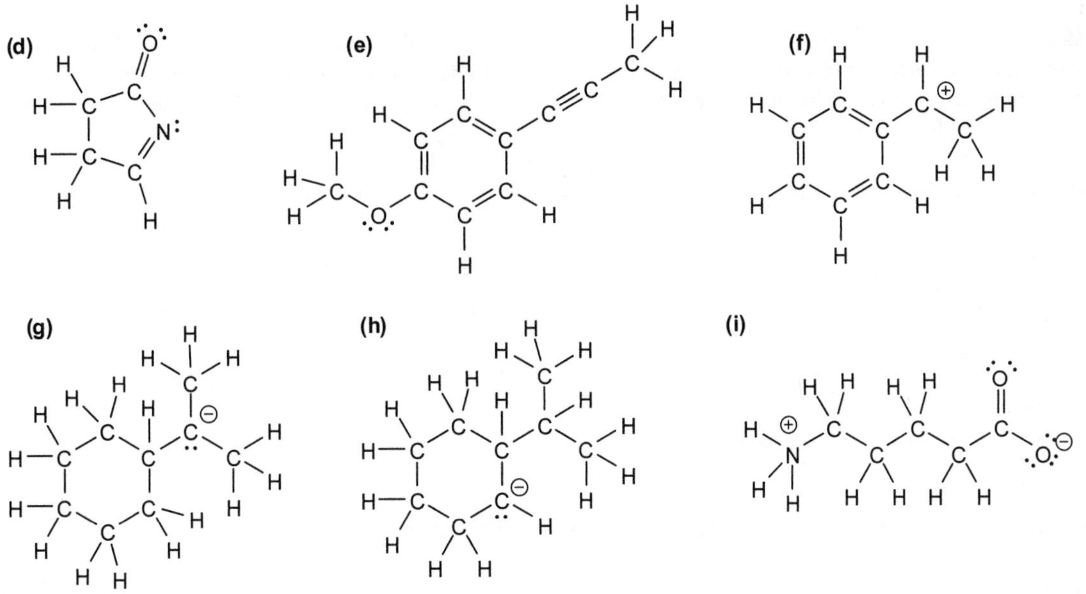

Problem 1.72

Think

Review the rules for drawing condensed formulas. How do you represent H atoms attached to a non-H atom? How are rings drawn? How are branches handled?

Solve

Condensed formulas are shown below. H atoms typically follow the atom to which they are bonded. Note that rings must be shown explicitly.

(a) $CH_3CH{=}CHCH(OH)CH_3$ **(b)** $CH_3CH_2CH_2CCl_2CH_3$ **(c)** $(CH_3)_2CHCH_2CO_2CH_2CH_3$

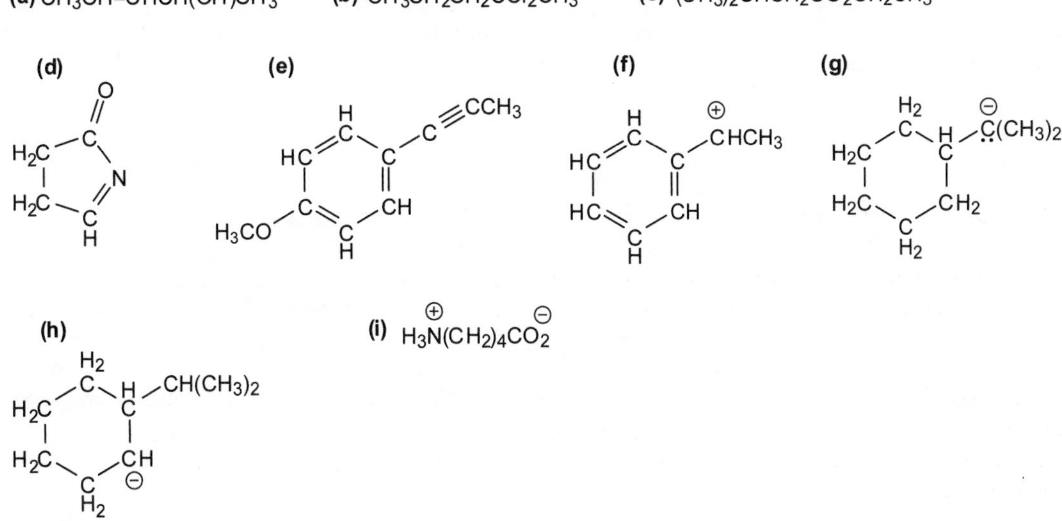

Problem 1.73

Think

Draw out the Lewis structure (hint: there are no structures that avoid charged atoms). Are there any lone pairs adjacent to a double or triple bond? Any other features that indicate the existence of another resonance structure? How should curved arrows be added to show electron movement to convert one resonance structure to another? In the resonance structures, what do you notice about filled valences and formal charges? What does Table 1-3 (bond strengths) tell you about the stability of the structures?

Solve

In Structure **A**, there is a lone pair on C adjacent to N≡N, so two curved arrows are drawn to show how electrons move to arrive at the second resonance structure, **B**. Both structures have the same number of bonds, and all atoms have octets. This is why you must use Table 1-3 to evaluate the stability of these two structures. To dissociate each molecule into atoms, you would have to break:

Resonance structure A: 2 C–H, 1 C–N, and 1 N≡N bond = 2(436) + 389 + 946 = **2207 kJ** (larger bond energy).

Resonance structure B: 2 C–H, 1 C=N, and 1 N=N bond = 2(436) + 619 + 418 = **1909 kJ**.

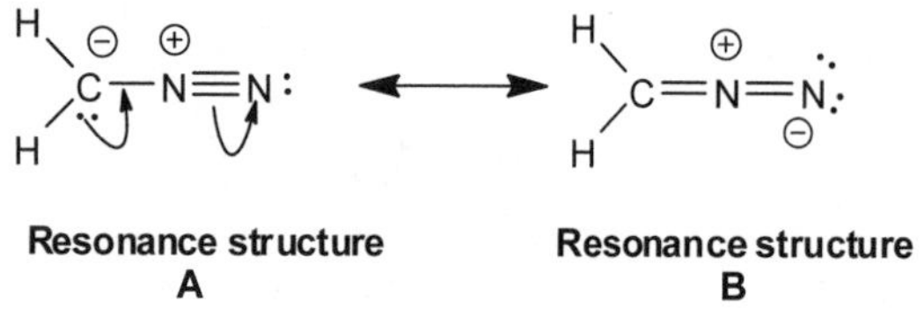

Resonance structure A

Resonance structure B

Thus, **Structure A** is more stable and contributes more to the resonance hybrid.

Problem 1.74

Think

Are there any lone pairs adjacent to a double or triple bond? Any atoms lacking an octet adjacent to a double or triple bond? Any rings of alternating single and multiple bonds? After identifying one of these features, how many curved arrows should you add to show electron movement? Are any atoms unable to participate in resonance? To draw the resonance hybrid, where are the charges located? Which bonds change between resonance structures? How do you represent partial bonds and partial charges?

Solve

A lone pair on C^- is adjacent to C=C, so two curved arrows are added to move electrons, as shown below. This can be repeated two more times to arrive at four total resonance structures. The CH_3 is not able to participate in resonance because it possesses four single bonds. The C–C bond at the far left is the longest. In all resonance contributors, this bond is a single bond; every other C–C bond has some double bond character and will be shorter, as shown in the resonance hybrid (right) below.

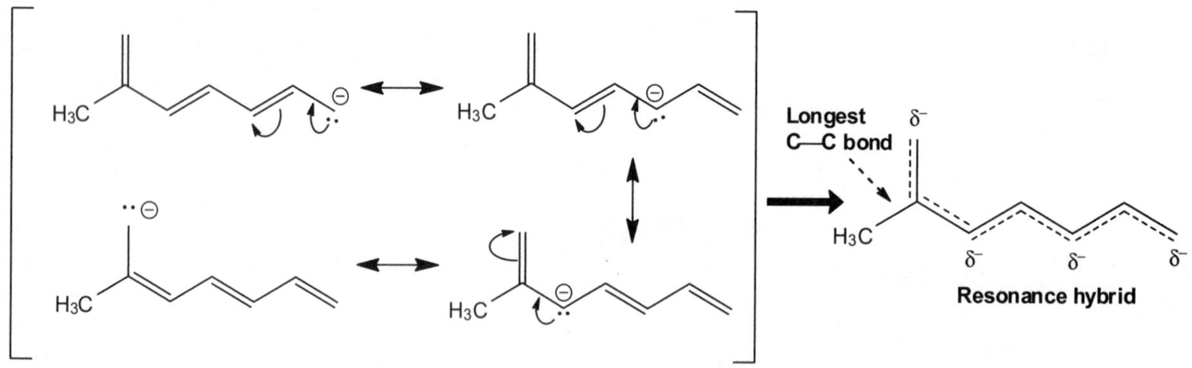

Longest C—C bond

Resonance hybrid

Problem 1.75

Think

Are there any lone pairs adjacent to a double or triple bond? Any atoms lacking an octet adjacent to a double or triple bond? Any rings of alternating single and multiple bonds? After identifying one of these features, how many curved arrows should you add to show electron movement? Are any atoms unable to participate in resonance? To draw the resonance hybrid, where are the charges located? Which bonds change between resonance structures? How do you represent partial bonds and partial charges?

Solve

This is an example of C^+ that has an incomplete octet and is adjacent to a double bond. One curved arrow is used to move electrons to arrive at the second resonance structure. This is repeated five more times, producing a total of seven resonance structures. Resonance is possible for all of the C atoms except the two $-CH_3$, groups which already have four single bonds. In the resonance hybrid, all atoms that share the +1 charge receive a partial positive charge (δ^+). All bonds that appear as a single bond in one resonance structure and a double bond in another are represented as a single bond plus a partial bond in the hybrid (=====).

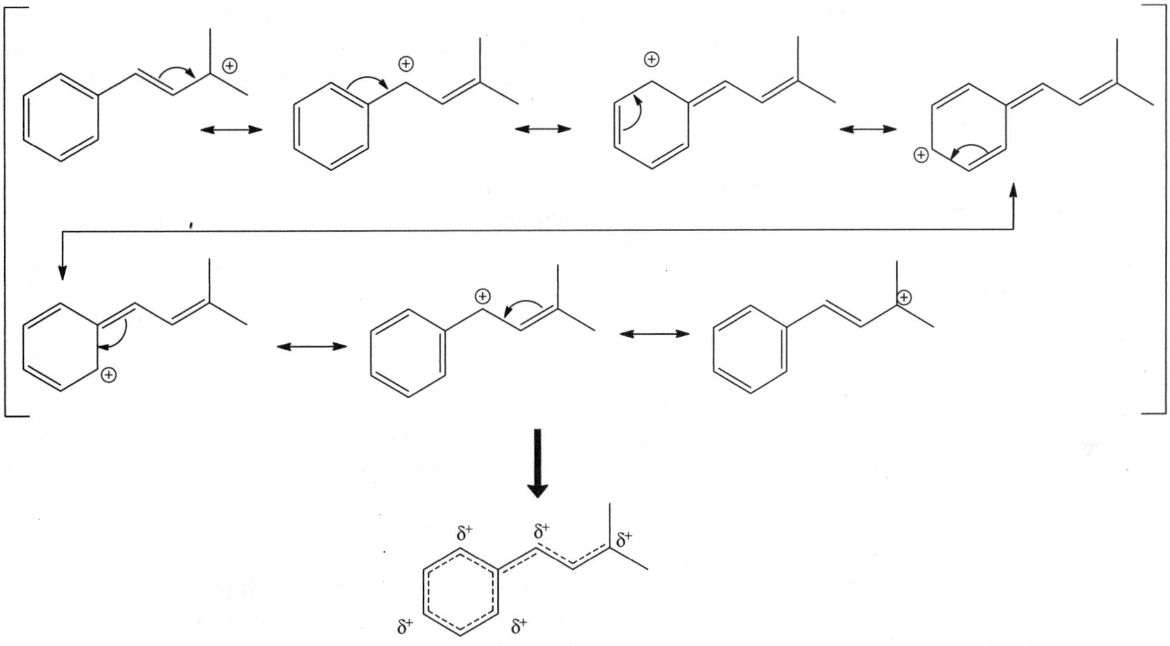

Resonance hybrid

Problem 1.76

Think

Review the rules for drawing line structures and condensed formulas. For line structures, what atom occurs at the intersection of two lines? At the end of a line? Draw in the appropriate C–H bonds and lone pair electrons that accurately represent the formal charges shown. Consider Table 1-5 when doing so. In condensed formulas, how do you represent H atoms attached to a non-H atom? How are rings drawn? How are branches handled?

Solve

For the line structure given, C atoms are located at the intersections of lines and at the ends of lines. H atoms are added to C atoms to give each C the appropriate number of bonds—in this case, four bonds for each uncharged C. In a condensed formula, H atoms attached to a particular C are written after that C. There is no practical way to type a condensed formula for cholesterol. All rings must be shown explicitly. Writing a condensed formula with all Cs and Hs explicitly drawn results in a very congested structure. The line structure is less congested, easier to view all parts, and simpler to draw.

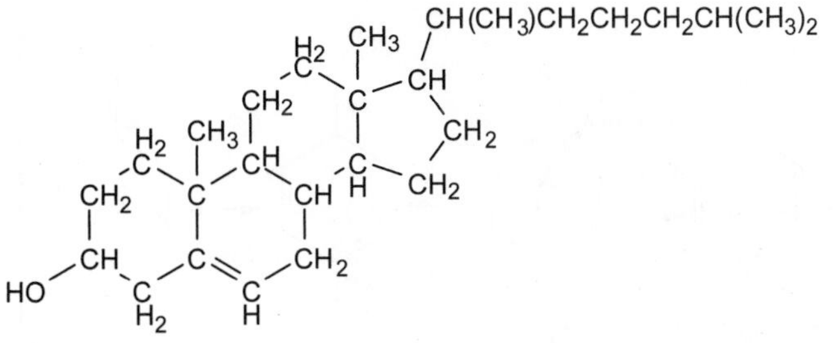

Problem 1.77

Think

If there were another resonance structure, how would you add curved arrows to show electron movement? What would the resonance structure look like? Given the different bond lengths of the single versus double bonds, could all the atoms stay in the same locations they began?

Solve

The molecule does not have a resonance structure, even though it has a ring of alternating single and double bonds. If we attempt to draw the resonance structure by shifting the electrons around the ring in the same way that we do for benzene, we do indeed arrive at a new structure (below). However, because the double bonds are shorter than the single bonds in this molecule, doing so requires that the atoms are also moved. The reason that cyclobutadiene does not possess resonance structures will be discussed in Chapter 14.

HC=CH, HC=CH ⟷ HC—CH, HC—CH

Note the location of the longer and shorter C—C bonds.

Problem 1.78

Think

Do any of the atoms have an unfilled valence shell? If so, is the atom with the unfilled valence shell adjacent to a double or triple bond? How many curved arrows do you need to add to show electron movement to produce a new resonance structure? Are any atoms unable to participate in resonance? How many resonance structures are possible for each structure?

Solve

Structure **A** is more stable because it has more resonance structures. To draw each resonance structure, notice that C^+ lacks an octet and is attached to C=C, so one curved arrow is added to move electrons. This electron movement can be repeated three more times to produce a total of five resonance structures. A sixth resonance structure can be produced because the lone pair on oxygen can be involved in resonance without giving rise to additional formal charges (far right). This is not true for structure **B** (resonance structures are shown on the next page), leaving it with only five total resonance structures.

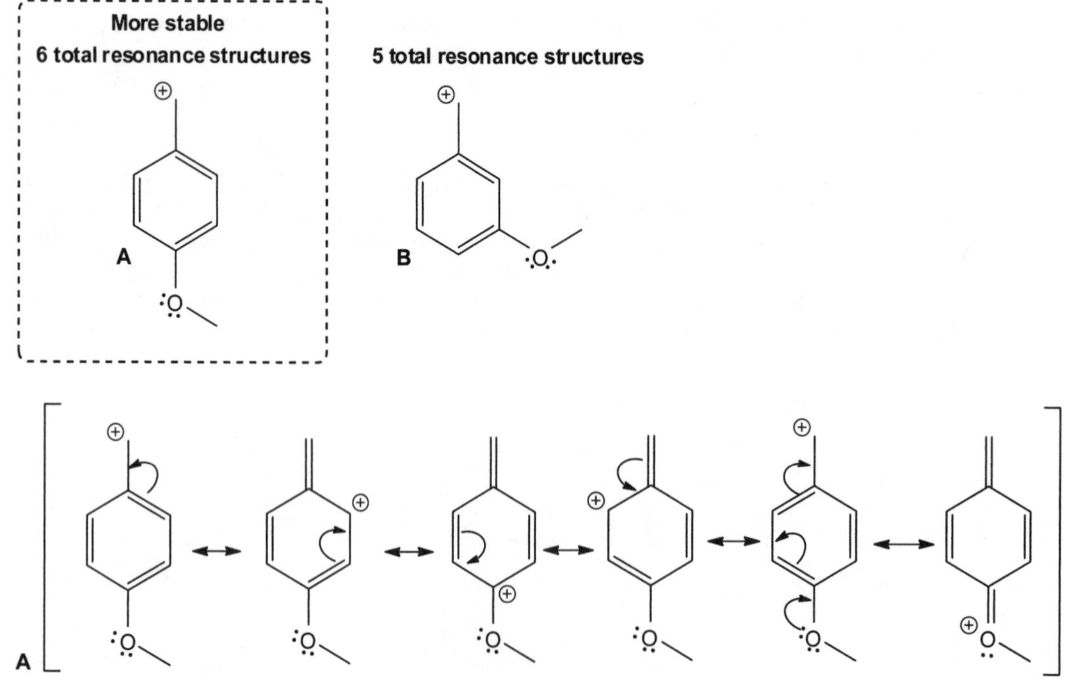

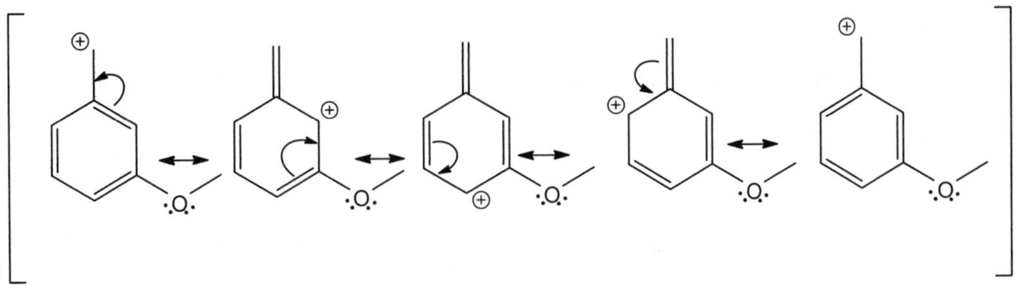

B

Problem 1.79

Think

Do any of the atoms have lone pairs? If so, is the atom with the lone pair adjacent to a double or triple bond? How many curved arrows should be added to show electron movment? Are any atoms unable to participate in resonance? How many resonance structures are possible for each structure?

Solve

Structure **D** is more stable because it has more resonance structures. To draw each resonance structure, notice that the O^- has a lone pair and is adjacent to C=C. Two curved arrows are therefore added to move electrons to produce the second resonance structure. In **D**, this electron movement can take place three more times to produce five resonance structures. Then, the C=O can participate in resonance to produce a sixth (far right). In structure **C**, the C=O double bond cannot participate in resonance, so only five resonance structures are possible.

5 resonance structures

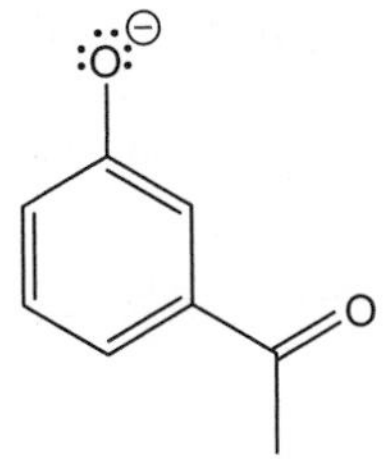

More stable
6 resonance structures

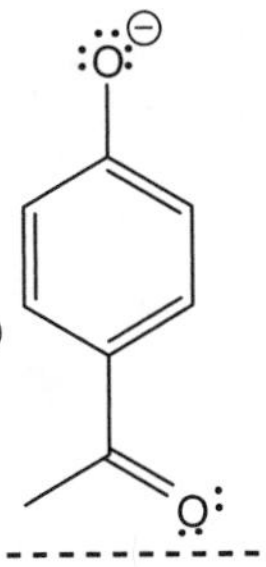

C

D

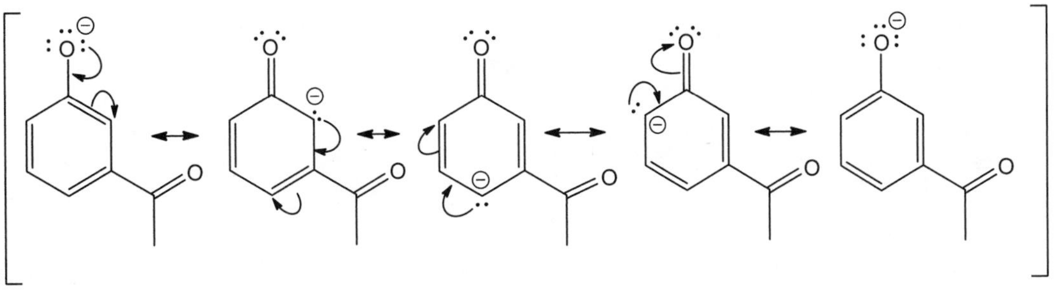

C

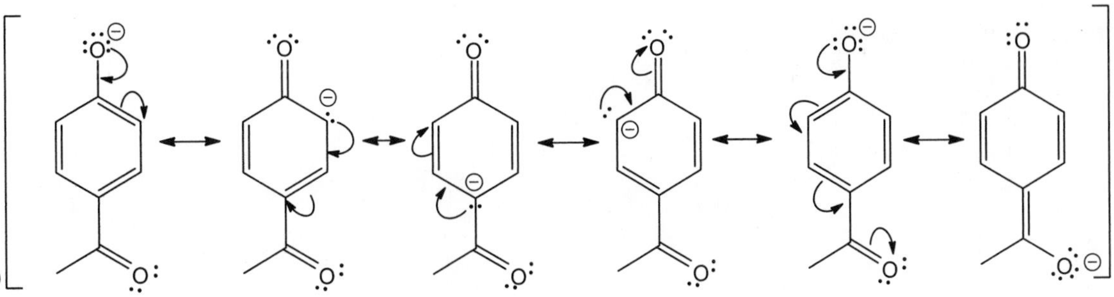

D

NOMENCLATURE 1 | Introduction: The Basic System for Naming Simple Organic Compounds: Alkanes, Cycloalkanes, Haloalkanes, Nitroalkanes, and Ethers

Problem N1.1

Think

How many carbon atoms are in a structure with the prefix "non"? What does the prefix "cyclo" indicate about the structure?

Solve

Cyclononane is a cyclic alkane with nine C atoms.

Cyclononane

Problem N1.2

Think

How many C atoms are in the ring? What prefix do you need to indicate that the structure is in a ring? Are there any functional groups present?

Solve

This is a cyclic alkane, so the name requires the prefix "cyclo." It contains 10 C atoms, so it is requires the root "decane." The name is ***cyclodecane***.

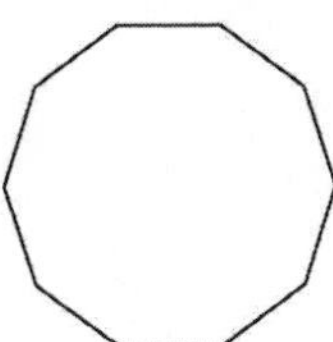

Cyclodecane

Problem N1.3

Think

Is there a substituent present? What would the name of the alkane or cycloalkane be without the generic substituent? How many C atoms are present in the main chain?

Solve

(a) A five-carbon cycloalkane with a substituent = substituted cyclopentane.
(b) A six-carbon alkane with a substituent = substituted hexane.

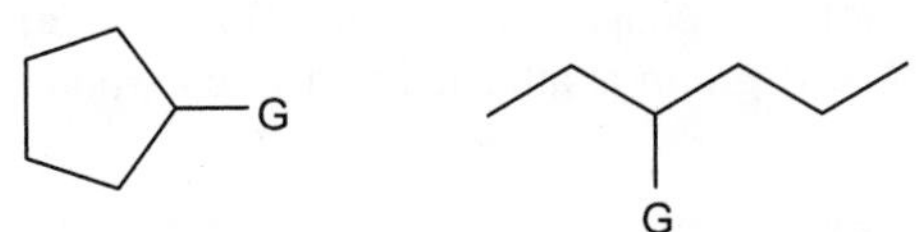

(a) Substituted cyclopentane **(b) Substituted hexane**

Problem N1.4

Think

How many carbon atoms are in a structure with the prefix "but"? With the prefix "hept"? What does the prefix "cyclo" indicate about the structure? Do these molecules have substituents?

Solve

(a) The substituted cyclobutane is a cyclic alkane with four carbon atoms and a generic substituent.

(b) The substituted heptane is a linear chain alkane with seven carbon atoms and a generic substituent. Note: Other substituted heptane structures are possible.

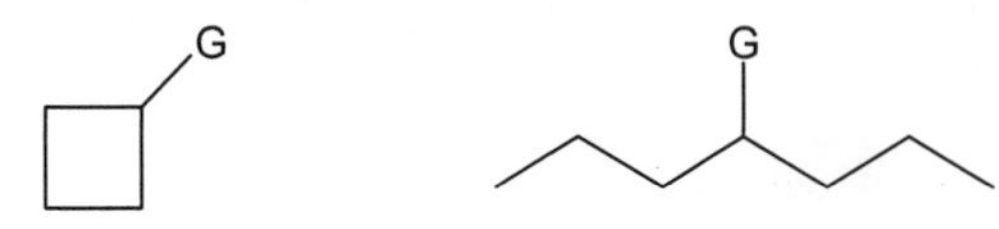

(a) Substituted cyclobutane **(b) Substituted heptane**

Problem N1.5

Think

How do you name this alkyl substituent? How many C atoms are in the ring? Name the group from which it was constructed, by removing an H atom at the C atom where the branch occurs. After you name the main group, how do you denote that it is a substituent? What suffix do you use?

Solve

Alkyl substituents have the suffix "yl." Name the alkane group from which it was constructed, drop the "ane," and add "yl."

(a) This is a cyclic alkyl substituent with eight C atoms, so it is ***cyclooctyl***.

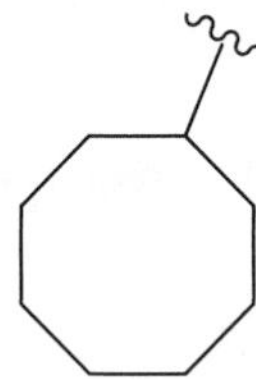

(b) A cycloheptyl substituent is a cyclic alkyl substituent with seven C atoms.

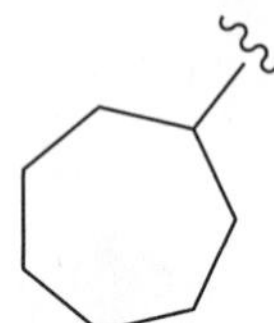

Problem N1.6

Think

What does R represent? How many C atoms are present in the main chain? Is there a substituent? What would the name of the alkane or cycloalkane be without the R group?

Solve

R is a generic alkyl group that represents the rest of the molecule consisting of C and H atoms. (a) The R group is attached to a two-carbon chain, which is ethane. (b) The R group is attached to a five-carbon ring, which is cyclopentane.

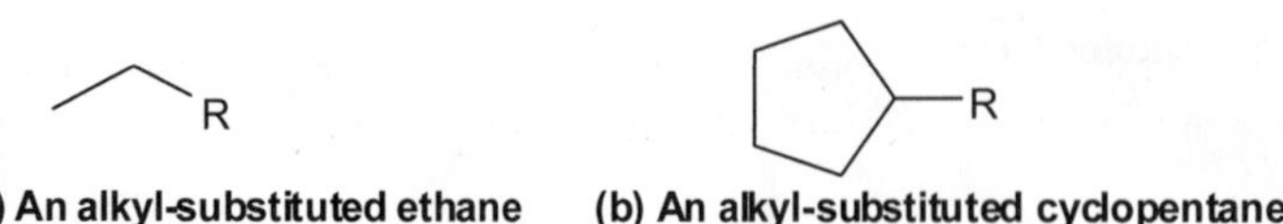

(a) An alkyl-substituted ethane **(b) An alkyl-substituted cyclopentane**

Problem N1.7

Think

How many carbon atoms are in a structure with the prefix "pent"? With the prefix "prop"? What does the prefix "cyclo" indicate about the structure? What letter is used to represent an alkyl substituent?

Solve

R is used to represent alkyl substitution.

(a) A generic alkyl-substituted pentane is an alkane with five carbon atoms and an alkyl substituent R. Note: other alkyl-substituted pentane structures are possible.

(b) A generic alkyl-substituted cyclopropane is a cyclic alkane with three carbon atoms and an alkyl substituent R.

(a) An alkyl-substituted pentane **(b) An alkyl-substituted cyclopropane**

Problem N1.8

Think

Count the C atoms in the longest continuous chain. Does the chain need to proceed across, or can it wrap around? Refer to Figure N1-3 for some examples.

Solve

The longest chain for each example is circled below.

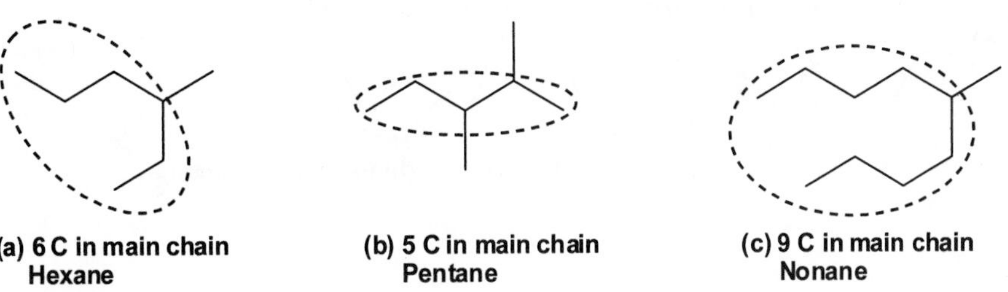

(a) 6 C in main chain Hexane **(b) 5 C in main chain Pentane** **(c) 9 C in main chain Nonane**

Problem N1.9

Think

How many C atoms are in the longest continuous chain? Name the longest continuous chain. How many C atoms are in the alkyl substituent? How do you name the alkyl substituent using the "yl" suffix?

Solve

(a) Three C atoms in the main chain = propane, one C atom in the alkyl substituent = methyl.

(b) Seven C atoms in the main chain = heptane, three C atoms in the alkyl substitutent = propyl.

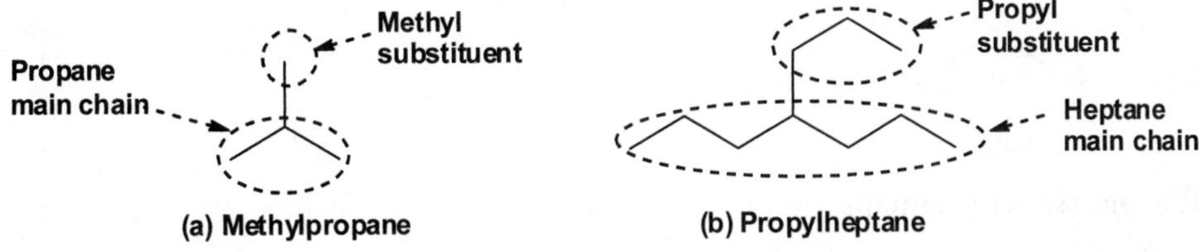

(a) Methylpropane **(b) Propylheptane**

Problem N1.10

Think

What is the group with the largest number of continuous C atoms? How many carbon atoms are in the substituent? Is the substituent a chain or ring? How do you name the substituent using the "yl" suffix?

Solve

(a) Six C atoms in the cyclic main group = cyclohexane, three C atoms in the cyclic alkyl substituent = cyclopropyl.

(b) Four C atoms in the cyclic main group = cyclobutane, two C atoms in the linear alkyl substituent = ethyl.

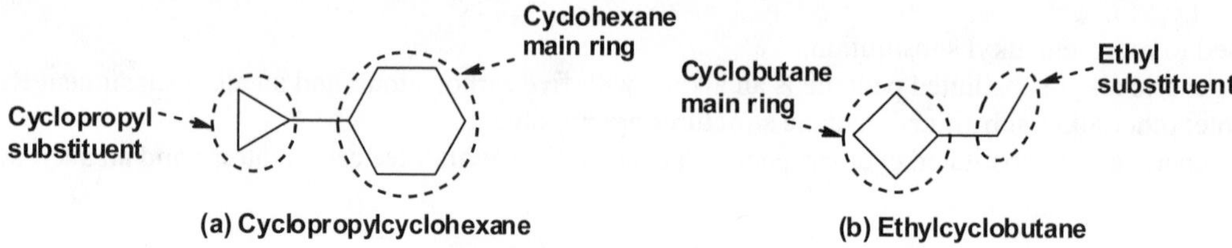

Problem N1.11

Think

What is the root? How many carbon atoms does it indicate? Is the main group cyclic? How many carbon atoms are in the alkyl substituent? Is the substituent linear or cyclic?

Solve

(a) Cyclobutylcyclohexane = a six-carbon cyclic alkane with a four-carbon cyclic alkyl substituent.

(b) Pentylcycloheptane = a seven-carbon cyclic alkane with a five-carbon linear alkyl substituent.

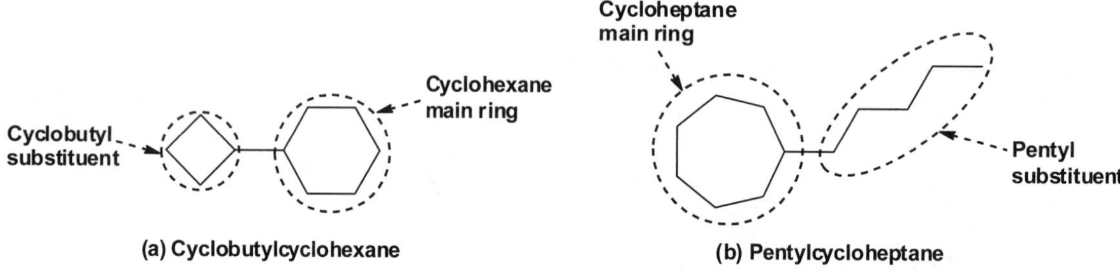

Problem N1.12

Think

How many C atoms are in the longest continuous chain? What is the name of the main chain? What are the names of each of the substituents? How many substituents are there of each type? Use the prefixes given in Table N1-3. Do these prefixes factor into alphabetical order when naming the molecule?

Solve

The longest chain is five C atoms = pentane. There are five CH_3 methyl substituents = pentamethyl, and one CH_3CH_2 ethyl substituent. Alphabetically, "ethyl" comes before "methyl," so in the IUPAC name of the molecule, "ethyl" appears before "pentamethyl."

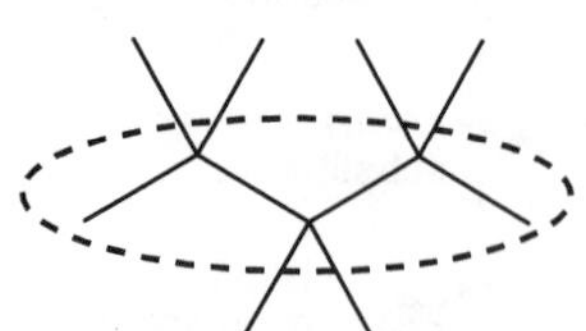

Ethylpentamethylpentane

Problem N1.13

Think

Based on the root, what is the structure of the main chain or ring? What are the substituents that are attached? How many of each? Is there more than one way the substituents can be arranged?

Solve

(a) Cyclopropane = a three-carbon cyclic alkane, hexamethyl = six CH_3 substituents.

(b) Methane = a one-carbon alkane, cyclobutyl = a four-carbon cyclic alkyl substituent, dicyclopentyl = two five-carbon cyclic alkyl substituents.

In both examples, there is only one way to arrange the substituents around the main chain.

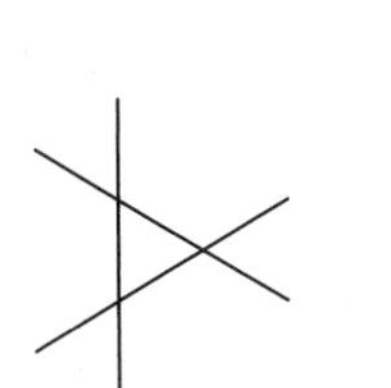

(a) Hexamethylcyclopropane (b) Cyclobutyldicyclopentylmethane

Problem N1.14

Think

Which name minimizes the number locating the first methyl substituent? Which name minimizes the number locating the second one? The third one?

Solve

1,1,3-trimethylcyclopentane is the correct name because the number prefixes are minimized. The "1,1,3" set of locators, in particular, is better than "1,1,4" because the "3" is a smaller number than "4."

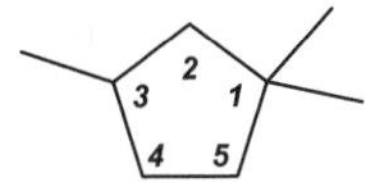

1,1,3-Trimethylcyclopentane

Problem N1.15

Think

How many carbon atoms are in the main group? Is it cyclic? What is the name of the main group and of each substituent? How can you minimize the numbers that locate the substituents? Is there more than one of any particular substituent? How do you denote this in the name? When do you use dashes and commas?

Solve

(a) The root is cyclopropane because the longest continuous set of carbon atoms is a three-membered ring.

(b) The root is cyclobutane because the longest continuous set of carbon atoms is a four-membered ring.

In both cases, commas separate numbers from each other and dashes separate numbers from letters. Each cycloalkane below contains two CH_3 groups. Both CH_3 groups are located at C1 = 1,1-dimethyl.

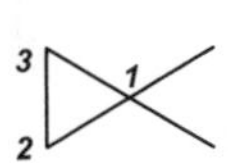

(a) 1,1-Dimethylcyclopropane

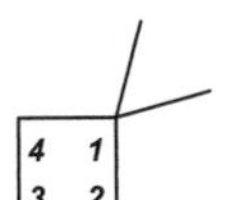

(b) 1,1-Dimethylcyclobutane

Problem N1.16

Think

How many carbon atoms are in the main group? Is it cyclic? What is the name of the main group and of each substituent? How can you minimize the numbers that locate the substituents? Is there more than one of any particular substituent? How do you denote this in the name?

Solve

Names are listed below. The number locators are minimized for the alkyl substituents.

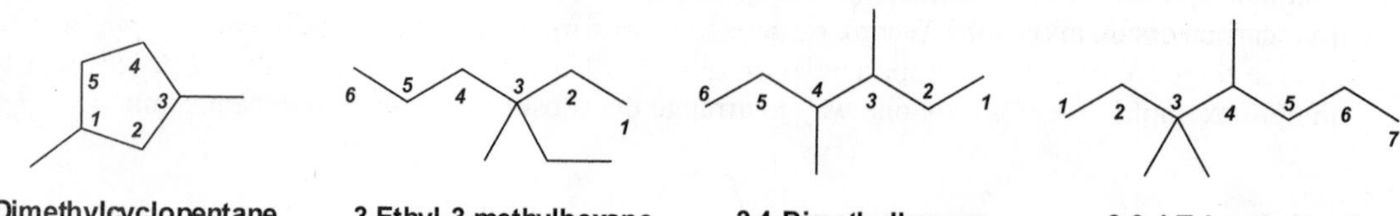

1,3-Dimethylcyclopentane **3-Ethyl-3-methylhexane** **3,4-Dimethylhexane** **3,3,4-Trimethylheptane**

Problem N1.17

Think

Based on the root, how many carbon atoms are in the longest continuous chain or ring? How do you denote the generic alkyl substituent? On which carbon atoms are these substituents located?

Solve

The generic alkyl substituent is denoted by an R. The structures are shown below.

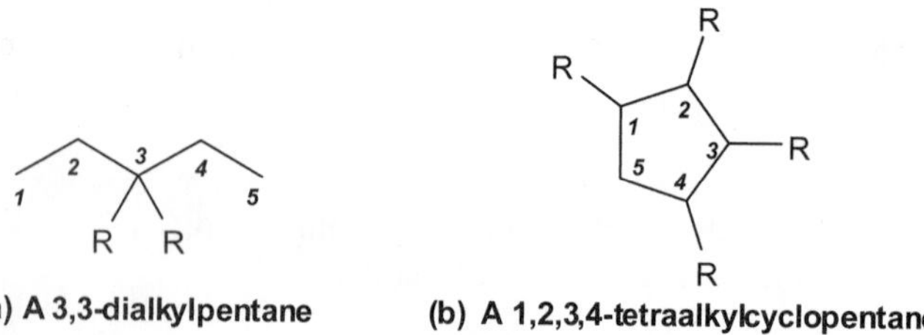

(a) A 3,3-dialkylpentane **(b) A 1,2,3,4-tetraalkylcyclopentane**

Problem N1.18

Think

How many carbon atoms are in the longest continuous chain or ring? What substituents are attached? How many of each kind? Where does numbering begin for a chain? For a ring? Add in the substituents at the designated locations.

Solve

See structures and names below.

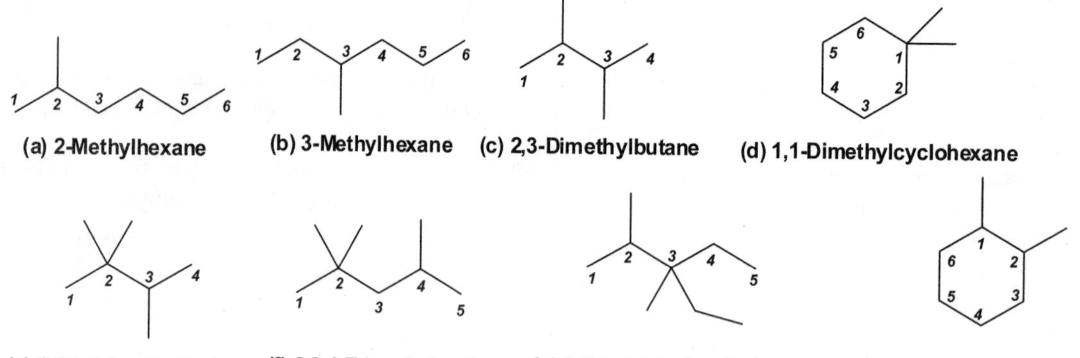

(a) 2-Methylhexane **(b) 3-Methylhexane** **(c) 2,3-Dimethylbutane** **(d) 1,1-Dimethylcyclohexane**

(e) 2,2,3-Trimethylbutane **(f) 2,2,4-Trimethylpentane** **(g) 3-Ethyl-2,3-dimethylpentane** **(h) 1,2-Dimethylcyclohexane**

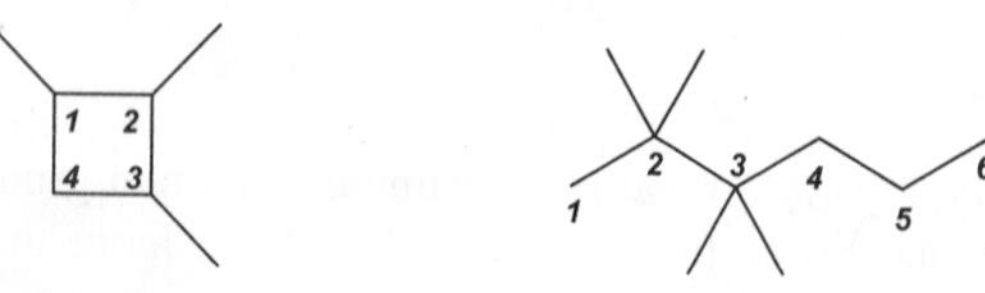

(i) 1,2,3-Trimethylcyclobutane **(j) 2,2,3,3-Tetramethylhexane**

Problem N1.19

Think

Review the rules for naming alkanes and cycloalkanes that you have learned thus far. What is the longest continuous carbon chain or ring? What root does that correspond to? What substituents are present? How many of each are there? Where does numbering begin on a chain? On a ring? How you do put the pieces of the name together?

Solve

Names are given below for each of the compounds.

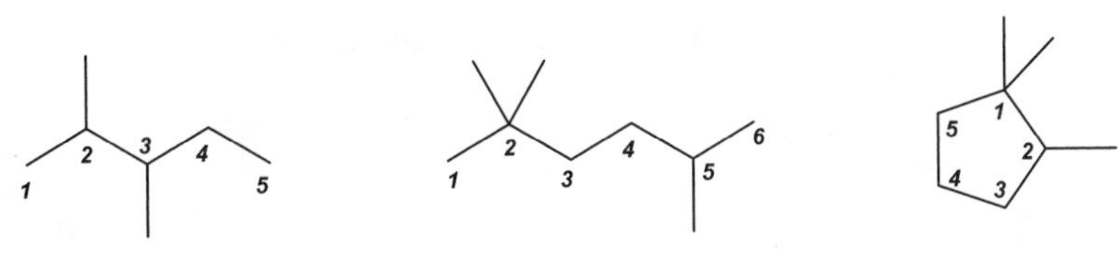

(a) 2,3-Dimethylpentane (b) 2,2,5-Trimethylhexane c) 1,1,2-Trimethylcyclopentane

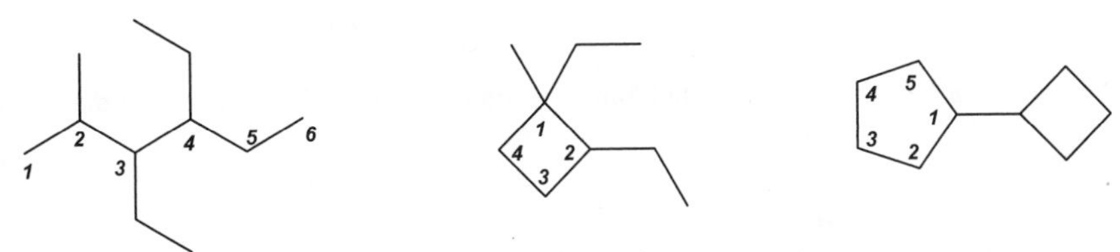

(d) 3,4-Diethyl-2-methylhexane (e) 1,2-Diethyl-1-methylcyclobutane (f) Cyclobutylcyclopentane

(g) 3-Cyclobutyl-2-cyclopropylpentane (h) 2,2-Dicyclopropyl-3,3-dimethylbutane

Problem N1.20

Think

For each C atom in question, how many other C atoms are bonded directly to it?

Solve

Primary (1°) = one C atom, secondary (2°) = two C atoms, tertiary (3°) = three C atoms, quaternary (4°) = four C atoms. Terminal (end) C atoms are always primary.

2° C 1° C 1° C

1° C 1° C 1° C

(*sec*-Butyl) (*s*-Butyl) (*tert*-Butyl) (*t*-Butyl)

Problem N1.21

Think

Review the rules for naming alkanes and cycloalkanes that you have learned thus far. What is the longest continuous carbon chain or ring? What common substituents are attached to it? How is the longest chain or ring numbered? How you do put the pieces of the name together? Which parts of the substituents' names are considered when arranging them in the alphabetical order and which are not?

Solve

Names for the compounds are given below. Notice how the numbering of the longest carbon chain or ring minimizes the numbers in each case. *Iso* is included in alphabetical order, whereas *t* or *tert* is not. That's because "iso" is part of the substituent name, whereas "tert" is a type of classification.

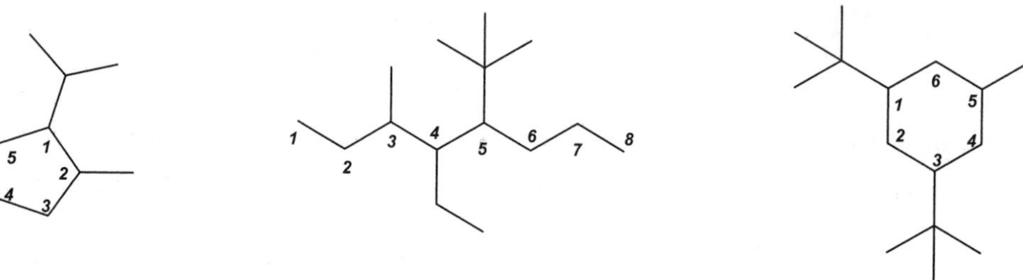

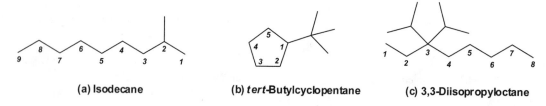

(a) 1-Isopropyl-2-methylcyclopentane **(b) 5-*t*-Butyl-4-ethyl-3-methyloctane** **(c) 1,3-Di-*t*-butyl-5-methylcyclohexane**

Problem N1.22

Think

How many carbon atoms are in the longest continuous carbon chain or ring? What is the name of the main chain or ring, and how should the carbons be numbered? Add in the substituents. What common alkyl substituents are present? What structural feature does "iso" represent?

Solve

The structures are given below. Notice how "iso" represents $CH(CH_3)_2$ in **(c)**.

(a) Isodecane **(b) *tert*-Butylcyclopentane** **(c) 3,3-Diisopropyloctane**

Problem N1.23

Think

Draw the compound given the name listed. Using the name given, what carbon atom must have been numbered C1? Looking at the molecule, is there another way to number the main chain to give the carbon atoms with substituents lower numbers?

Solve

If the longest carbon chain (still six C atoms) is numbered as shown below, the lowest substituent now is on C2. The correct name is 3-ethyl-2,2-dimethylhexane.

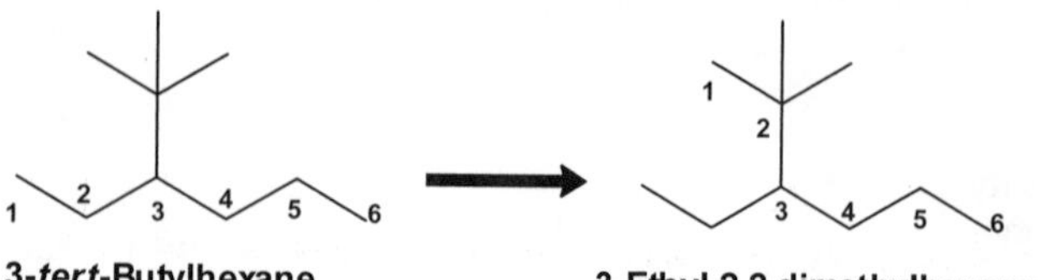

3-*tert*-Butylhexane **3-Ethyl-2,2-dimethylhexane**

Problem N1.24

Think

Draw the structure of 3-*tert*-butylhexane. How many carbons are bonded to each carbon atom?

Solve

There are five 1° C atoms, three 2° C atoms, one 3° C atom, and one 4° C atom (labeled below).

3-*tert*-Butylhexane

Problem N1.25

Think

How many carbon atoms are in the longest continuous chain or ring? Based on the root name, what is the structure of the main chain or ring? On what carbon atom does numbering begin? What substituents are present? Do any of them have trivial names? Add in the substituents.

Solve

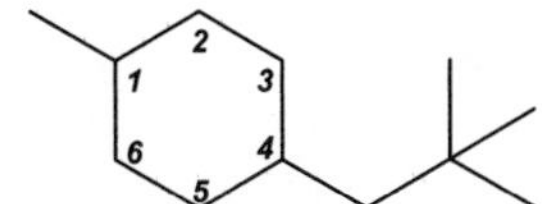

(a) 1-Methyl-4-neopentylcyclohexane

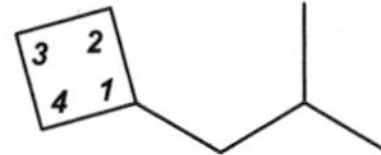

(b) Isobutylcyclobutane

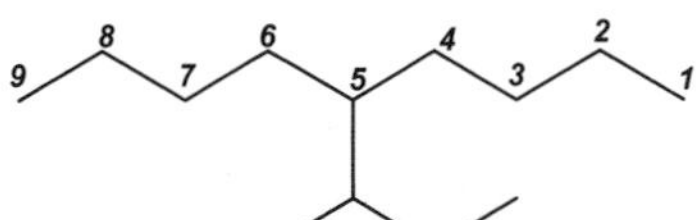

(c) 5-*sec*-Butylnonane

Problem N1.26

Think

Review the rules for naming alkanes and cycloalkanes that you have learned thus far. What is the longest carbon chain or ring? What substituents are present? How is the longest chain or ring numbered? How you do put the pieces of the name together?

Solve

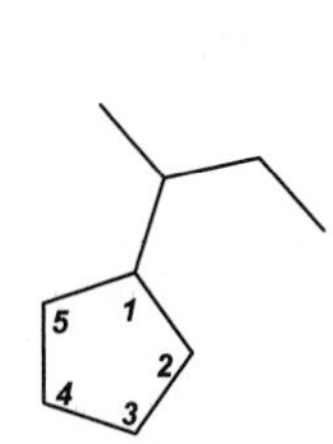

(a) *sec*-Butylcyclopentane

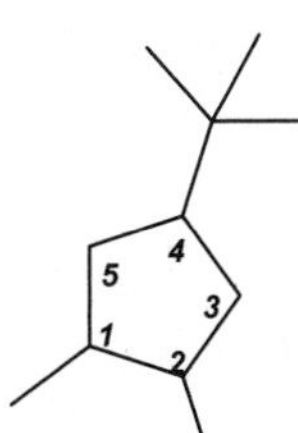

(b) 4-(*tert*-Butyl)-1,2-dimethylcyclopentane

(c) 3-Methyl-4-propylnonane

(d) 3-Cyclopropyl-4-isopropylheptane

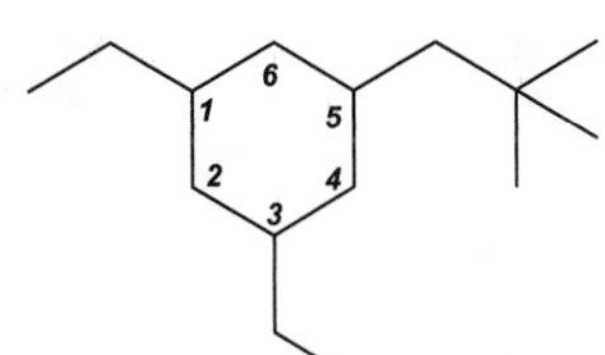

(e) 1,3-Diethyl-5-neopentylcyclohexane

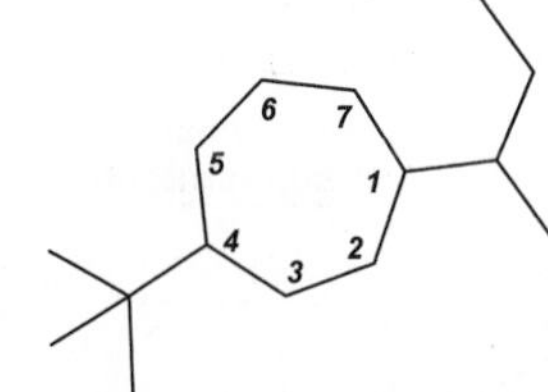

(f) 1-*sec*-Butyl-4-*tert*-butylcycloheptane

Problem N1.27

Think

What new rules are added for haloalkanes and nitroalkanes? Review the previous rules for naming alkanes. What is the longest continuous carbon chain or ring in each case? What is the name of the main chain or ring? How do you add number locators to minimize the numbers? What are the names of the substituents present, and how many of each are there?

Solve

The only new rules are the prefixes for the haloalkanes (fluoro, chloro, bromo, iodo) and nitroalkanes (nitro). All of the other naming rules are the same.

(a) 2-Chlorobutane **(b) 1,5-Dinitrohexane** **(c) 1,3-Dibromo-1,4-diiodocyclopentane**

Problem N1.28

Think

Based on the root, what is the longest carbon chain or ring? What substituents are present? How many are there of each? How is the longest chain or ring numbered? How you do put the pieces of the name together?

Solve

(a) 1,1-Dichloro-2-nitrocyclohexane **(b) 1,1,1,2-Tetrabromopropane**

Problem N1.29

Think

What is the longest continuous carbon chain or ring? What functional group is R–O–R? How do you name an OR substituent as an alkoxy group? How do you add number locators to the main chain or ring to minimize the numbers?

Solve

For **(b)**, the ethoxy gets the lower number due to alphabetical order considerations. Notice that in **(c)**, no number locator is necessary in the name because there is only one way to attach the substituent to the carbon ring.

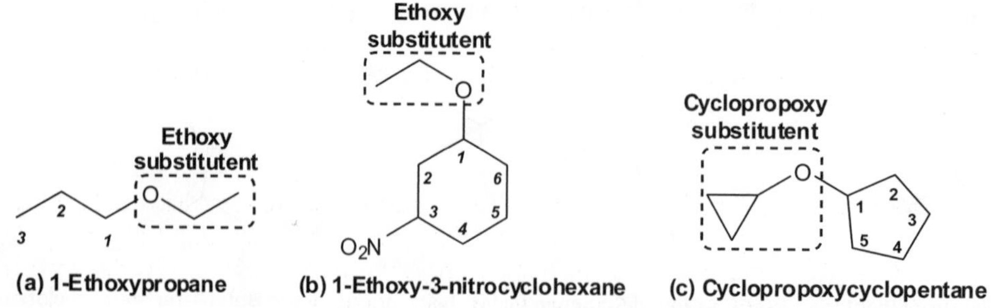

(a) 1-Ethoxypropane **(b) 1-Ethoxy-3-nitrocyclohexane** **(c) Cyclopropoxycyclopentane**

Problem N1.30

Think

Based on the root, what is the longest continuous chain or ring of carbon atoms? Where does the numbering of the chain or ring begin? What substituents are present and how many are there of each? Where are they located along the main chain or ring?

Solve

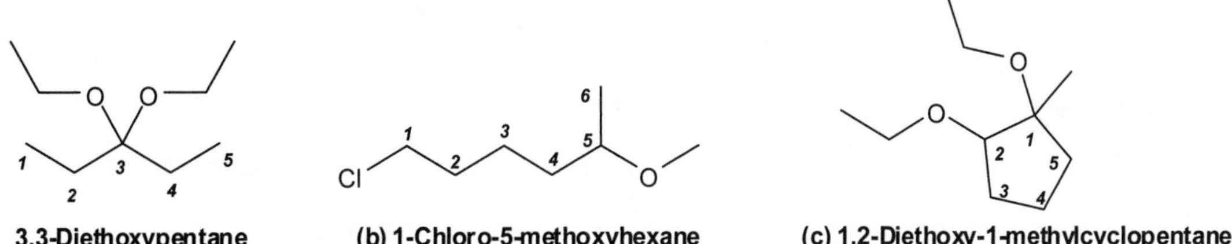

3,3-Diethoxypentane | (b) 1-Chloro-5-methoxyhexane | (c) 1,2-Diethoxy-1-methylcyclopentane

Problem N1.31

Think

Use the trivial name to draw the structure for each compound. For the IUPAC name, how is the main chain or ring named? Are the halogen groups treated as the main chain or a substituent? Where should numbering begin along the main chain or ring?

Solve

In the IUPAC rules, the halogens are treated as substituents. Note that **(a)** doesn't require a number locator because there is only one way to number the ring.

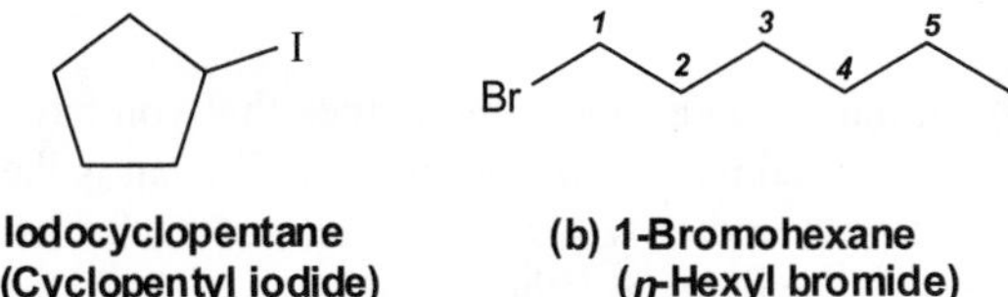

(a) Iodocyclopentane (Cyclopentyl iodide)

(b) 1-Bromohexane (*n*-Hexyl bromide)

Problem N1.32

Think

Use the trivial name to draw the structure for each compound. What alkyl groups are bonded to each side of the ether oxygen atom? For the IUPAC name, how is the main chain or ring named? How are alkoxy groups identified and named as substituents? Where should numbering begin along the main chain or ring?

Solve

Notice that a number locator is not required in **(b)** because there is only one way to number the ring.

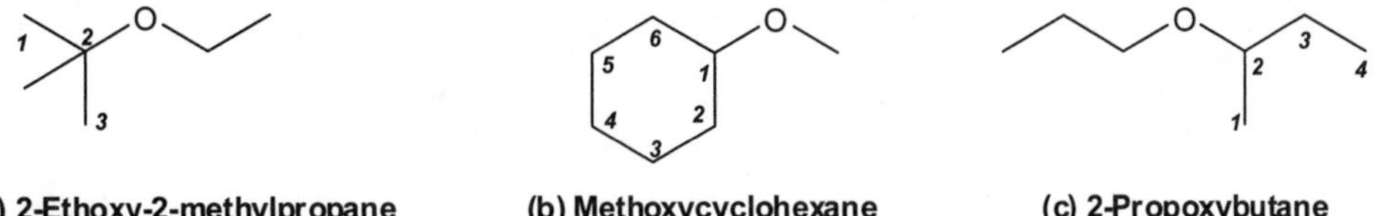

(a) 2-Ethoxy-2-methylpropane (*tert*-Butyl ethyl ether)

(b) Methoxycyclohexane (Cyclohexyl methyl ether)

(c) 2-Propoxybutane (*sec*-Butyl propyl ether)

Problem N1.33

Think

Review the rules for naming alkanes, cycloalkanes, ethers, haloalkanes, and nitroalkanes that you have learned thus far. Are any common alkyl groups present? Do you use trivial names for these groups? What is the longest continuous carbon chain or ring? What substituents are present, and how many are there of each? How is the longest chain or ring numbered? How you do put the pieces of the name together?

Solve

(a) 1,2,3-Tribromohexane (b) 2,2,3,3,4-Pentachlorohexane (c) 1,2-Dichloro-4-nitrohexane

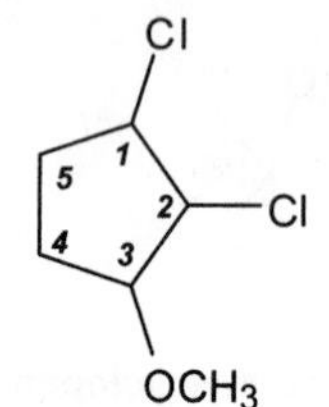

(d) 1,2-Dichloro-3-methoxycyclopentane (e) 2,2-Dichloro-3-cyclopropylbutane (f) 1-Bromo-1-chloro-1-iodobutane

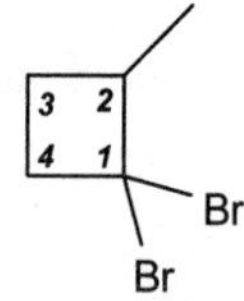

(g) 1,1-Dibromo-2-methylcyclobutane (h) 1,1,2,2-Tetrabromo-3-propoxypropane

Problem N1.34

Think

Review the rules for naming alkanes, cycloalkanes, ethers, haloalkanes, and nitroalkanes that you have learned thus far. Are any common alkyl groups present? Do you use trivial names for these groups? What is the longest continuous carbon chain or ring? What substituents are present, and how many are there of each? How is the longest chain or ring numbered? How you do put the pieces of the name together?

Solve

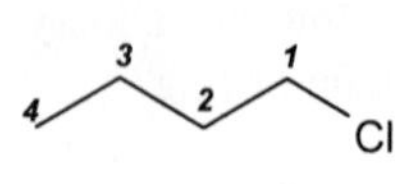

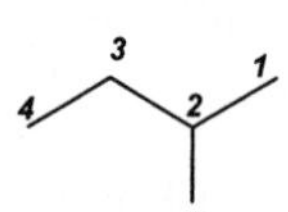

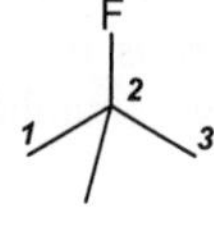

(a) 1-Chlorobutane (*n*-Butyl chloride) (b) 2-Iodobutane (*sec*-Butyl iodide) (c) 2-Fluoro-2-methylpropane (*tert*-Butyl fluoride)

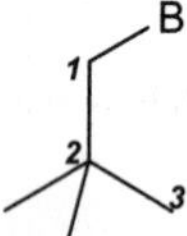

(d) 1-Bromo-2,2-dimethylpropane (Neopentyl bromide) (e) 2-Isopropoxypropane or 2-(1-Methylethoxy)-propane (Diisopropyl ether)

Problem N1.35

Think

Draw the structure using the IUPAC name. Based on the root, how many carbon atoms are in the main chain or ring? What substituents are present, and how many are there of each? How should the main chain or ring be numbered? What functional group is present? What is the name of the alkyl group on each side of the ether oxygen atom? How do you put the names together?

Solve

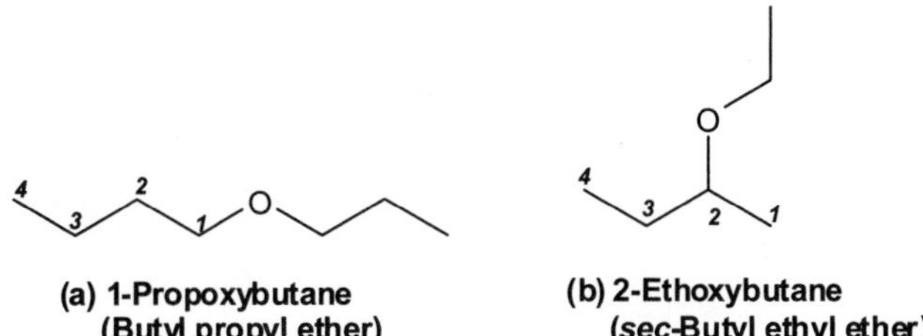

(a) 1-Propoxybutane (Butyl propyl ether)

(b) 2-Ethoxybutane (*sec*-Butyl ethyl ether)

Problem N1.36

Think

How many carbon atoms are in the longest continuous chain or ring of carbon atoms? How does that establish the root? Where should the numbering of the longest chain or ring begin? What substituents are present, and how many are there of each? In what order should the substituents appear in the molecule's name?

Solve

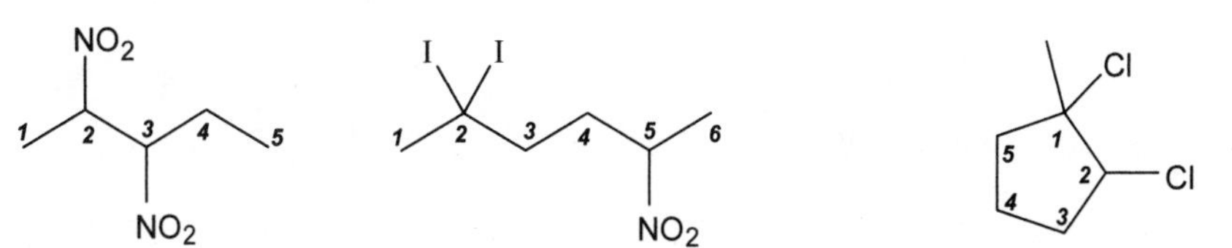

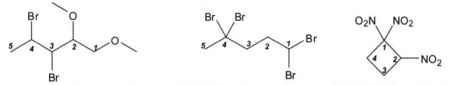

(a) 2,3-Dinitropentane

(b) 2,2-Diiodo-5-nitrohexane

(c) 1,2-Dichloro-1-methylcyclopentane

(d) 3,4-Dibromo-1,2-dimethoxypentane

(e) 1,1,4,4-Tetrabromopentane

(f) 1,1,2-Trinitrocyclobutane

(g) 1-Cyclobutyl-2-fluorocyclopentane

(h) 2,2,3,3,4,4-Hexafluoropentane

(i) (1,2,2,2-Tetrabromoethyl)cyclopropane

(j) 1,2,3-Tricyclopropoxycyclopropane

CHAPTER 2 | Three-Dimensional Geometry, Intermolecular Interactions, and Physical Properties

Your Turn Exercises

Your Turn 2.1

Think

In a Lewis structure, what constitutes a group of electrons? Consult Table 2-1.

Solve

In a Lewis structure, a group of electrons is a lone pair of electrons, a single bond, a double bond, or a triple bond. CH_3^+ has three single bonds and no lone pairs, thus three groups of electrons. CH_3^- has three single bonds and one lone pair, thus four groups of electrons.

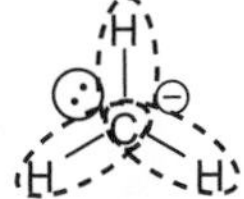

Your Turn 2.2

Think

Which N–H bonds are in the plane of the paper? Which N–H bonds come out of the plane of paper? Which N–H bonds point away from you? How are each of these kinds of bonds represented in dash–wedge notation?

Solve

N–H bonds that are in the plane of paper are represented by a flat line (—), N–H bonds that come out of the plane of paper are represented by a wedge (▬), and N–H bonds that point away from you are represented by a dash (''''''''). The structure on the left has two bonds in the plane of the paper, one out of the plane, and one pointed away. The structure on the right has the two horizontal bonds that are pointed away and two vertical bonds that are out of the plane of paper.

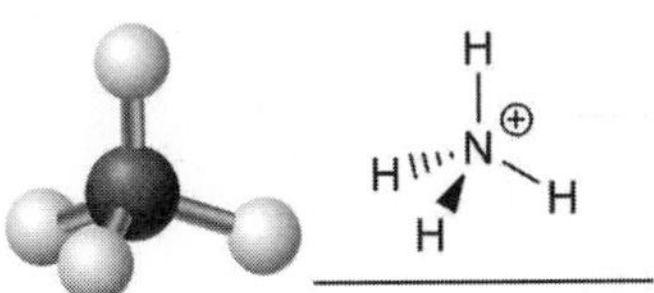
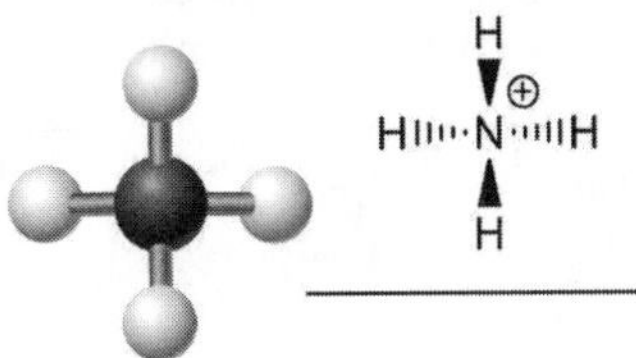

Your Turn 2.3

Think

Are the wedge/dash bonds or the flat lines in the plane of the paper? Which bonds are perpendicular to the plane of the paper? In which direction does the V open?

Solve

The flat lines are in the plane of the paper and the V opens pointing to the top right corner (in this figure). The wedge/dash bonds are perpendicular to the plane of the paper and the V opens to the bottom left (in this figure). See below.

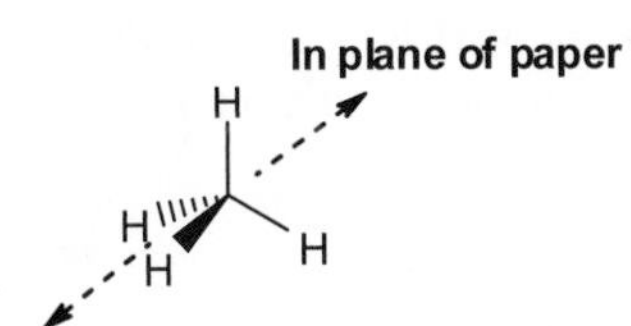

Your Turn 2.4

Think

In this structure, which bonds (C–C, C–H, or C–O) are in the plane of the paper? Which bonds come out of the plane of paper? Which bonds point away? How are each of these kinds of bonds represented in dash–wedge notation?

Solve

By convention, the line structure drawings represent the carbon–carbon chain as a zigzag line that is planar. The four bonds in a tetrahedron can be thought of as two perpendicular Vs (Fig. 2-6). The carbon chain in more complex molecules is represented by alternating flat Vs (i.e., ΛVΛV). At the point of each V is the perpendicular V. In line structures, the C–H bond is not drawn. In butan-2-ol, the C–O bond points away from you and this is the only bond where it is necessary to draw the dash notation.

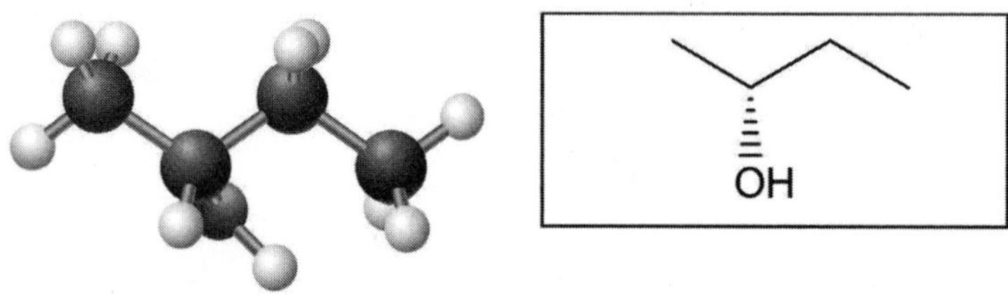

Your Turn 2.5

Think

What effect does a 180° turn have on a bond in the plane of the paper? What effect does a 180° turn have on a bond out of the plane of paper? What effect does a 180° turn have on a bond pointed away from you? In the cyclopentane ring shown, bonds on the right point where after a 180° turn?

Solve

A 180° turn does not have any effect on bonds in the plane of the paper (flat lines in both) and inverts wedges to dashes and dashes to wedges. A 180° turn about the vertical axis flips bonds on the right to bonds on the left, while a 180° turn about the horizontal axis flips bonds from top to bottom.

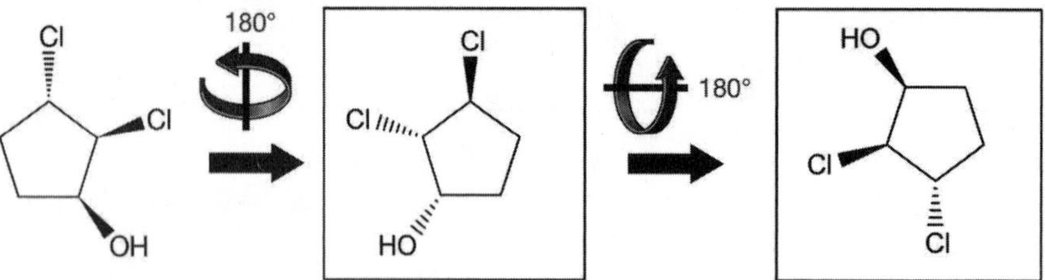

Your Turn 2.6

Think

Consult Figure 1-16 to determine the electronegativity value for beryllium and hydrogen. How is a dipole arrow drawn? To which atom, more or less electronegative, does the arrow point?

Solve

The electronegativity of Be is 1.57 and the electronegativity of H is 2.20 (see Fig. 1-16). H is more electronegative than Be and the dipole arrow (+⟶) points to the more electronegative atom (each H atom in this example). BeH_2 is a linear nonpolar molecule.

⟵+ +⟶

H—Be—H

Your Turn 2.7

Think

Consult Figure 1-16 to determine the electronegativity value for carbon and chlorine. How is a dipole arrow drawn? To which atom, more or less electronegative, does the arrow point? Recall that dipole moments are vectors. In which direction does the net dipole point for each V (two C–Cl bonds)?

Solve

The electronegativity of C is 2.55 and the electronegativity of Cl is 3.16 (see Fig. 1-16). Cl is more electronegative than C and the dipole arrow (+→) points to the more electronegative atom (each Cl in this example). The individual C–Cl bond dipoles can be added to yield a net dipole for each V, which points from the C to toward the Cl atoms. The nonboldface arrows indicate the bond dipoles. The boldface arrows indicate the vector sum of each pair of bond dipoles. Notice that the boldface arrows are equal in magnitude and point in opposite directions. The molecule has no net dipole moment and is, therefore, nonpolar.

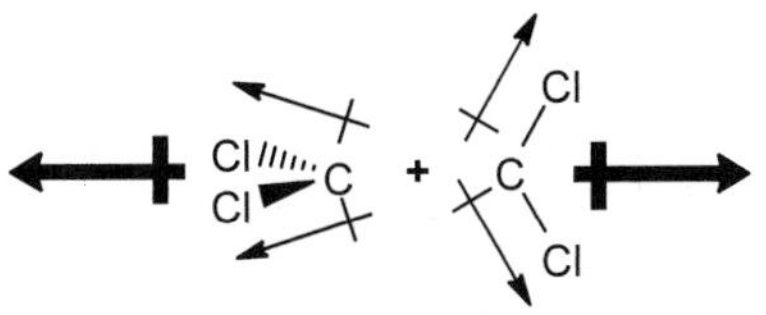

Your Turn 2.8

Think

Consult Table 1-6 for a list of functional group name and structures. Look for multiple bonds and bonds other than C–C and C–H.

Solve

Functional groups are provided below.

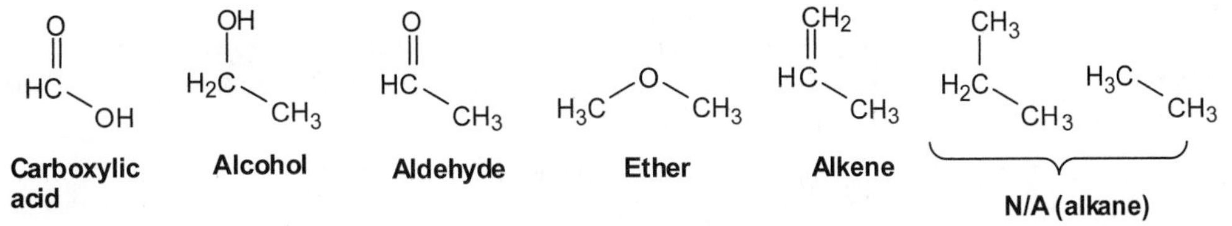

Carboxylic acid | Alcohol | Aldehyde | Ether | Alkene | N/A (alkane)

Your Turn 2.9

Think

Consult Table 2-4 to determine the boiling point of $CH_3CH{=}O$. Which compound has the higher boiling point? What intermolecular forces are present? How is the strength of the intermolecular forces reflected in the physical property of boiling point?

Solve

Both CH_3CH_2F and $CH_3CH{=}O$ have permanent dipoles and thus have dipole–dipole interactions as the dominant intermolecular interaction. $CH_3CH{=}O$ has a higher boiling point than CH_3CH_2F (−19 °C vs. −37.1 °C). The stronger the attraction one molecule has to itself, the higher the boiling point. $CH_3CH{=}O$, therefore, must have stronger dipole–dipole interactions and a greater dipole moment.

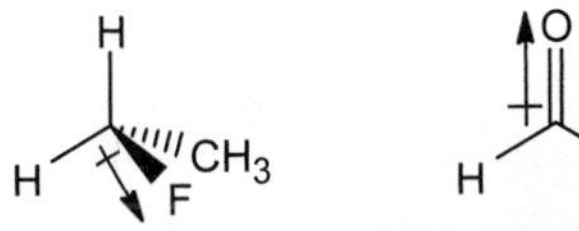

Higher boiling point, larger dipole moment

Your Turn 2.10

Think

Which atoms must be covalently bonded to a hydrogen atom for it to be considered a H-bond donor? Which atoms are considered H-bond acceptors? In a given H bond, how many donors and acceptors are there?

Solve

Any H bond consists of one H-bond donor and one H-bond acceptor. In the H bonds below, the donor is from the O–H covalent bond and the acceptor is the O atom from the other molecule.

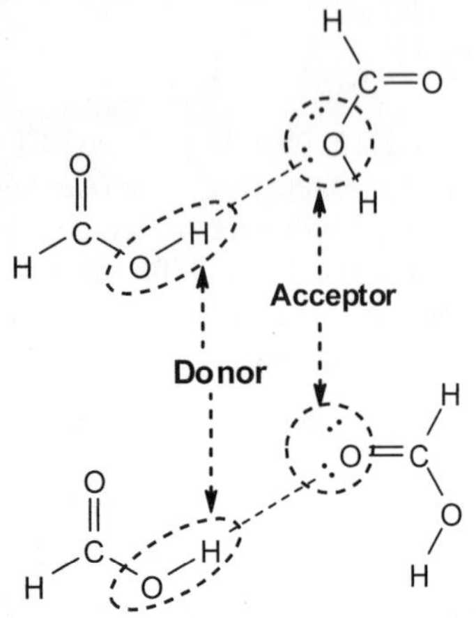

Acceptor

Donor

Your Turn 2.11

Think Which atoms must be covalently bonded to a hydrogen atom for it to be considered a H-bond donor? Which atoms are considered H-bond acceptors?

Solve

All O atoms in both compounds are potential H-bond acceptors. The potential H-bond donors are the O–H hydrogen atoms.

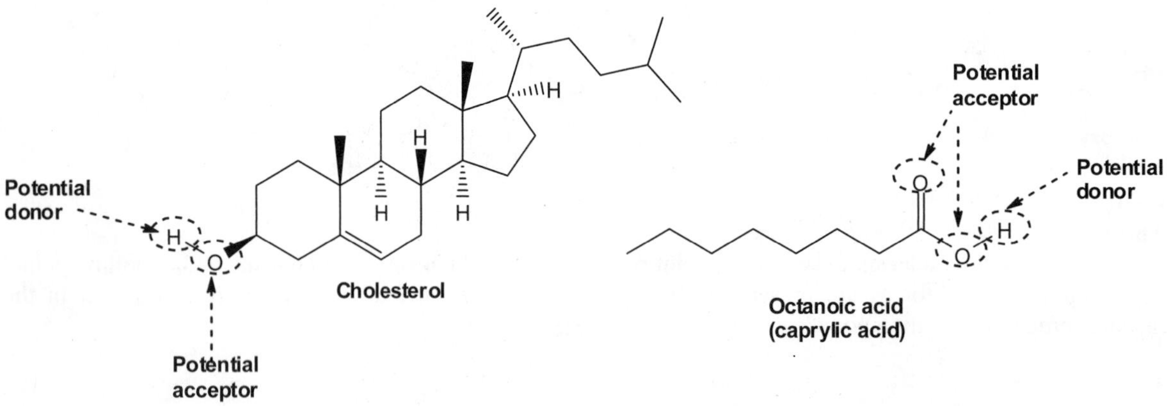

Potential donor

Potential acceptor

Cholesterol

Potential acceptor

Potential donor

Octanoic acid (caprylic acid)

Your Turn 2.12

Think

Consult Table 2-5 for a list of the total number of electrons and the boiling points for the various straight-chain alkanes.

Solve

The graph on the next page is a plot of boiling points against the total number of electrons. Positions of points in the graph are approximate, but a clear trend is indicated (greater number of total electrons leads to an increase in boiling point for these nonbranching hydrocarbons). As the total number of electrons increases, so does the strength of the London dispersion forces.

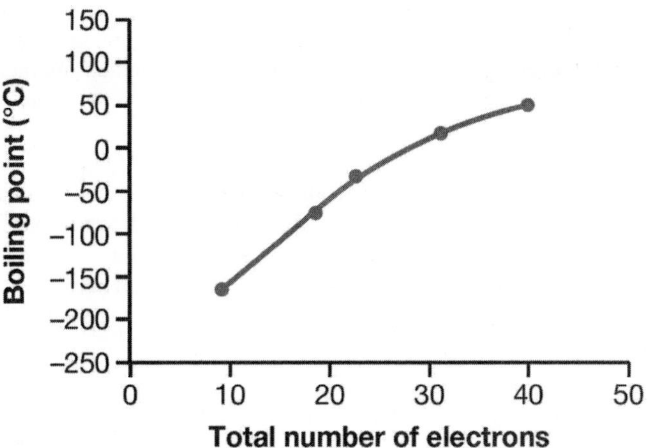

Your Turn 2.13

Think

Where do the molecules contact each other? How does the shape of the molecule affect the amount of contact surface area?

Solve

The molecules contact each other on the outside of the molecule. Pentane (left) has a greater contact surface area due to the long linear shape (no branching). This leads to an increased boiling point.

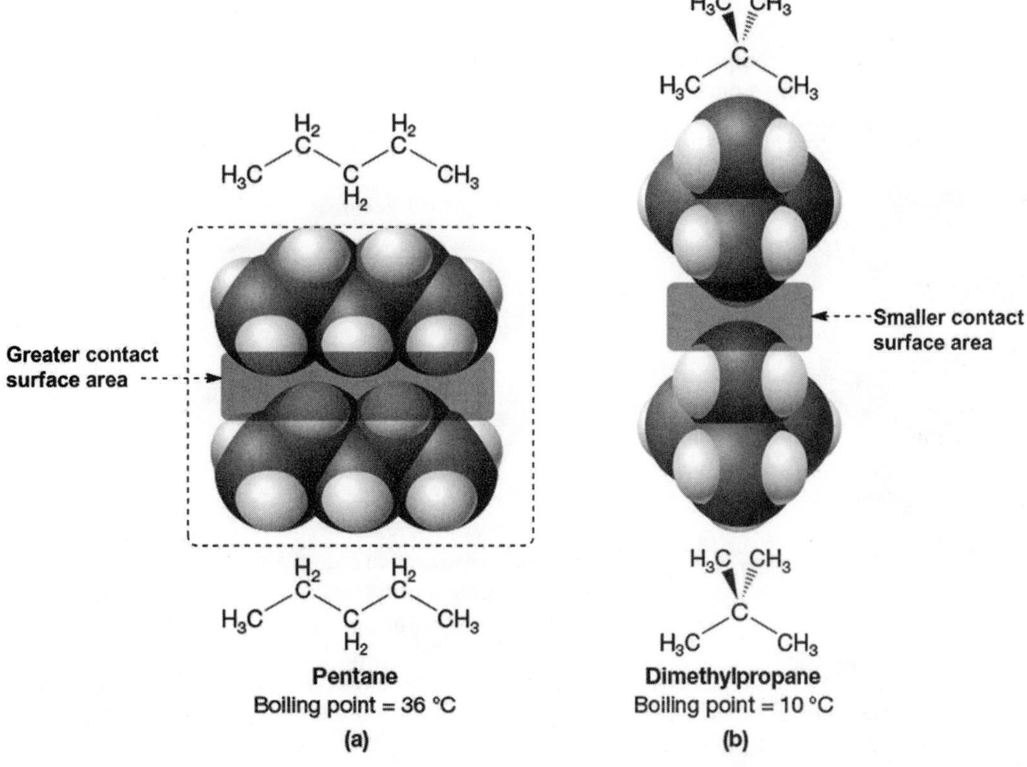

Your Turn 2.14

Think

What intermolecular interactions do each of the compounds exhibit? What intermolecular interactions does water exhibit? Upon mixing, what solute–solvent interactions are possible?

Solve

Compound **A** is nonpolar and the only interactions that can exist would come from a weak induced dipole. Compound **B** has a nonpolar region (hydrocarbon chain) and a region that exhibits hydrogen bonding (NH_2). Water exhibits hydrogen bonding. In general, two compounds are more soluble in each other if the intermolecular interactions between the two substances are roughly the same strength. Therefore, **B** is more soluble in water than **A**.

NH_2

A

B
More soluble in H_2O

Your Turn 2.15

Think

What are the requirements for an ion–dipole interaction? Locate the partial charge on each polar molecule that will be attracted to the ion.

Solve

An ion–dipole interaction occurs between a molecule with a permanent dipole and an ion. In **(a)**, HCO_2^- is the anion and interacts with the δ^+ near H in water, and in **(b)** Na^+ is the cation and interacts with the δ^- near O in water.

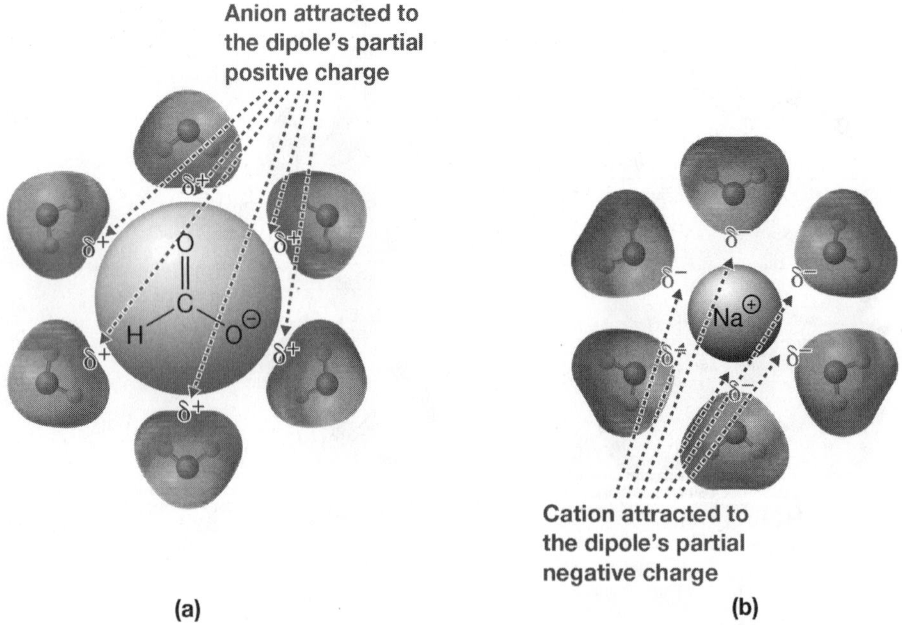

Your Turn 2.16

Think

What are the requirements for hydrogen bonding? Which end of the H–Y covalent bond is the H-bond donor: the partial positive or negative?

Solve

A molecule must possess an H–O, H–N, or H–F bond to be capable of H-bonding interactions. The partial positive end—namely, the H–is the H-bond donor region.

Polar Protic Solvents		Polar Aprotic Solvents	
Structure	**Name**	**Structure**	**Name**
H–O–H (δ^+ on H)	Water	H_3C–S(=O)–CH_3 (δ^+ on S)	Dimethyl sulfoxide (DMSO)
H_3C–CH_2–O–H (δ^+ on H)	Ethanol	H_3C–C(=O)–CH_3 (δ^+ on C)	Propanone (acetone)
H_3C–C(=O)–O–H (δ^+ on H)	Ethanoic acid (acetic acid)	H–C(=O)–N(CH_3)–CH_3 (δ^+ on C)	*N,N*–Dimethyl formamide (DMF)

H-bond donors (the H atoms of the protic solvents); No H-bond donors (the aprotic solvents)

Your Turn 2.17

Think

What group in a fatty acid carboxylate is hydrophilic? What group in a fatty acid carboxylate is hydrophobic? From the electrostatic potential map, is the red region electron rich or poor?

Solve

The hydrophilic group in a fatty acid carboxylate is the CO_2^- (ionic head group) and is very electron rich; it is represented by the red area on an electrostatic potential map. The hydrophobic group in a fatty acid carboxylate is the hydrocarbon tail (all C and H atoms), represented by the blue area on the electrostatic potential map.

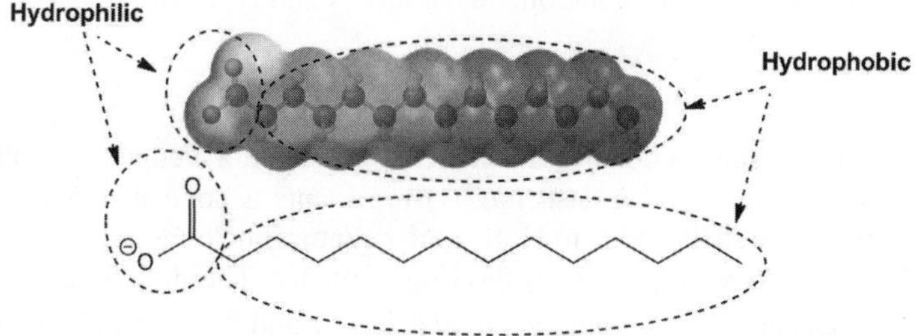

Your Turn 2.18

Think

What constitutes an ester functional group? Notice what bonds/atoms are labeled already in the fat molecule.

Solve

An ester consists of a carbonyl C=O adjacent to an ether linkage R–O. There are a total of three ester groups in triglyceride fat molecules.

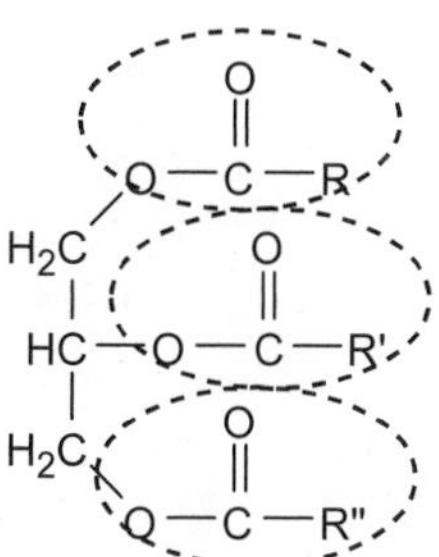

Chapter Problems

Problem 2.2

Think

How many electron groups are surrounded by each nonhydrogen atom? Are any of these electron groups lone pairs? When a lone pair is present, can an atom's electron and molecular geometries be the same? Refer to Table 2-3 for electron and molecular geometries.

Solve

When no lone pairs are present, the molecular geometry is the same as the electron geometry. The first two C atoms are linear in both electron geometry and molecular geometry (two electron groups: one single bond and one triple bond). The third C atom is tetrahedral in both electron and molecular geometry (four electron groups: four single bonds). The O atom is tetrahedral in electron geometry and bent in molecular geometry (four electron groups: two single bonds and two lone pairs).

Electron geometry = linear
Molecular geometry = linear

Electron geometry = tetrahedral
Molecular geometry = bent

Electron geometry = tetrahedral
Molecular geometry = tetrahedral

Prop-2-yn-1-ol
(Propargyl alcohol)

Problem 2.3

Think

For each nonhydrogen atom, what is the molecular geometry? Review the rules for dash–wedge notation. How are linear carbon chains drawn in line structure notation? What atoms and types of bonds should be omitted in line structures?

Solve

In a line structure, a flat line (—) represents a bond in the plane of the paper, a wedge (◀) represents a bond that comes out of the plane of the paper, and a dash (''''''''') represents a bond that points away from you. Remember to omit C and H atoms, as well as bonds to H. See the structures below.

(a) The six C atoms of the ring are trigonal planar, so the ring is planar. The three C atoms of the chain are tetrahedral. The first Cl is attached to the carbon chain by a wedge and the second Cl is attached also by a wedge.

(b) The four doubly bonded C atoms are trigonal planar and the remaining one is tetrahedral. The ring, therefore, is essentially planar. The C atom of C≡N is linear. The substituent comes off the ring pointed toward you and, therefore, is attached by a wedge.

(c) The C atoms are all tetrahedral and the C–C bonds in the linear carbon chain are drawn as zigzag. The NH_2 group points away from you and is, therefore, attached to the carbon chain by a dash.

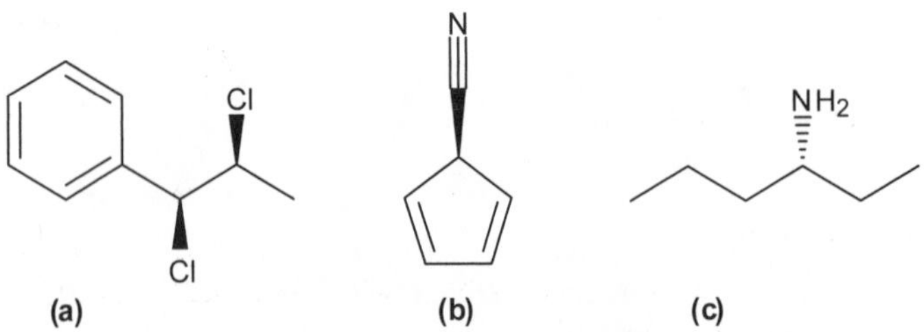

Problem 2.4

Think

How should the four bonds of a tetrahedral atom be represented? Refer to Figure 2-6. Should the two Vs be in the same plane or different planes? Should they open in the same direction or opposite directions?

Solve

The four bonds that make up the tetrahedral geometry must consist of two perpendicular Vs that open in opposite directions. On the left, the V in the plane of the paper opens upward, but the V constructed from the two C–Br bonds does not open downward. The problem is fixed in the structure on the right, where the V made from the C–Br bonds is perpendicular to the first V and opens downward.

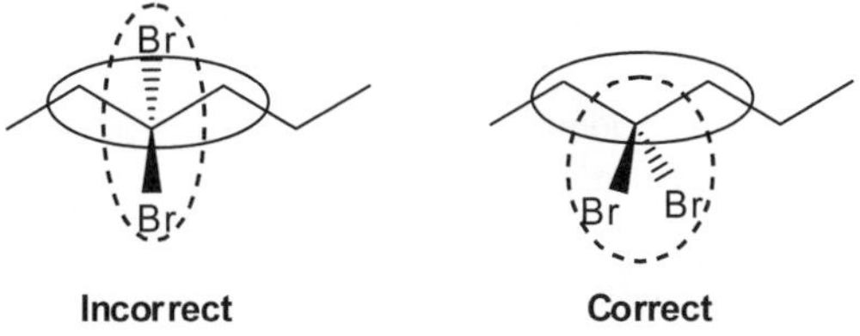

Problem 2.5

Think

What is the molecular geometry about each nonhydrogen atom? Consult Figure 1-16 to determine the electronegativity value for each atom. How is a bond dipole arrow drawn? To which atom, more or less electronegative, does each bond dipole arrow point? Do the bond dipoles perfectly cancel? If not, in which direction does the net dipole point?

Solve

Molecular geometry is very important in drawing the net dipole moment. The dipole arrow (+⟶) points to the more electronegative atom in each covalent bond. The net dipole moment is the sum of the individual bond dipoles. The arrows in regular type indicate the bond dipoles. The arrows in boldface indicate the vector sum of each set of bond dipoles. Structure **(c)** is the only one that is not polar, because it is symmetric—the two C=O bond dipoles point in opposite directions and, therefore, cancel out. From its Lewis structure, **(a)** appears not to be polar, but it is because of its tetrahedral geometry. You can decompose CBr_2Cl_2 into two "Vs" in the same way as CH_2Cl_2. The resultant dipoles point in opposite directions but with different magnitudes, so they **do not** cancel. Similarly, the bond dipoles in NH_3 **(b)** appear to cancel in the Lewis structure, but they do not, because NH_3 is pyramidal. Structure **(d)** is polar because the C=O bond dipoles do not point in opposite directions and, therefore, do not cancel.

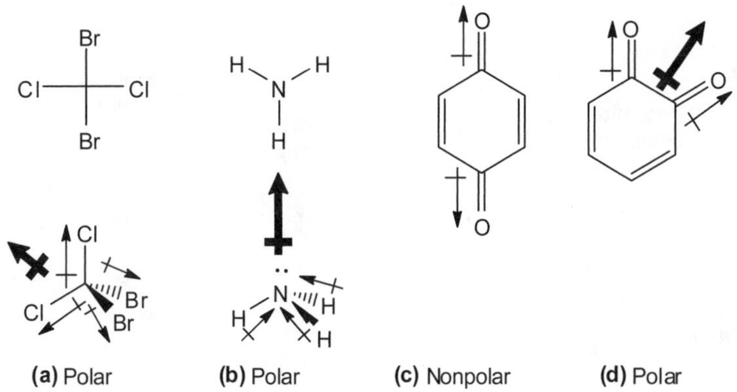

Problem 2.7

Think

What is the strongest intermolecular interaction in a sample of $NaOCH_2CH_3$? In a sample of $CH_3CH{=}O$? Which type of interaction is stronger? How does this affect the boiling point?

Solve

$NaOCH_2CH_3$ is an ionic compound made of Na^+ and $^-OCH_2CH_3$ ions, whereas $CH_3CH{=}O$ is polar covalent. Therefore, CH_3CH_2ONa is held together by ion–ion interactions, the strongest of the intermolecular forces, whereas $CH_3CH{=}O$ is not. The strongest intermolecular interaction between molecules of $CH_3CH{=}O$ is dipole–dipole interactions. With stronger intermolecular interactions, it takes more heat energy for CH_3CH_2ONa to boil. Thus, CH_3CH_2ONa has a higher boiling point.

Problem 2.8

Think

What is the strongest intermolecular interaction among molecules of CH_4? Among molecules of CH_3F? Which type of interaction is stronger? How does this affect the boiling point?

Solve

CH_3F has a higher boiling point than CH_4 because CH_3F is a polar compound, whereas CH_4 is nonpolar. Thus, CH_3F molecules are held together by dipole–dipole interactions, whereas molecules of CH_4 are not.

Problem 2.9

Think

What are the types of intermolecular interactions that hold each species together? Are any ions present? Do any molecules have net dipoles? Can hydrogen bonds be formed?

Solve

In **A**, $(CH_3)_2CHNH^-$ and Li^+ are oppositely charged ions and are held together by ion–ion interactions. The pair of molecules in **B** are polar, so they are held together by dipole–dipole interactions. No hydrogen bonds can be formed in **B** because there are no F–H, O–H, or N–H covalent bonds. The molecules in **C** are polar, so they, too, are held together by dipole–dipole interactions. Furthermore, they undergo hydrogen bonding because of the N–H donor that exists in one molecule and the N acceptor that exists in the other. Of these interactions, the ion–ion interactions in **A** are the strongest and the dipole–dipole interactions in **B** are the weakest.

Problem 2.10

Think

Which atoms must be covalently bonded to a hydrogen atom for it to be considered a hydrogen-bond (H-bond) donor? Which atoms are considered H-bond acceptors?

Solve

Each potential H-bond donor (which includes H bonded to either F, O, or N) and each potential H-bond acceptor (either F, O, or N) are labeled below.

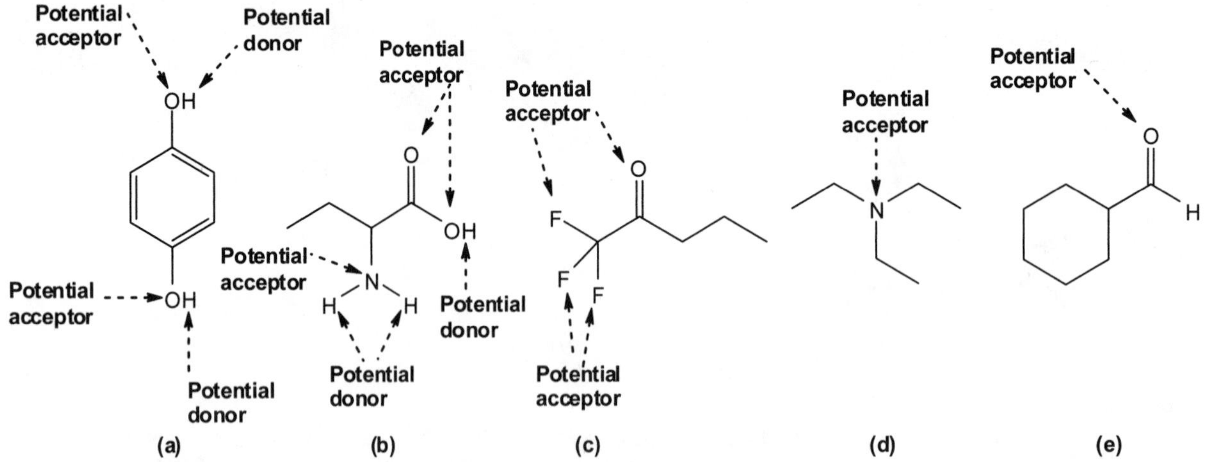

Problem 2.11

Think

Which atoms must be covalently bonded to a hydrogen atom for it to be considered a H-bond donor? Which atoms are considered H-bond acceptors?

Solve

H-bond donors exist in covalent F–H, O–H, and N–H bonds. H-bond acceptors include F, O, and N atoms. Functional groups that possess at least one H-bond acceptor but no H-bond donors include alkyl fluorides, ethers, acetals, epoxides, nitriles, ketones, aldehydes, and esters. Also included are amines and amides if only C atoms are bonded to the N atom. Functional groups that contain at least one H-bond donor and one H-bond acceptor include alcohols, hemiacetals, and carboxylic acids. Also included are amines and amides if at least one H atom is bonded to N. Functional groups that contain no H-bond donors and no H-bond acceptors include alkyl chlorides and bromides, alkenes, alkynes, aryl groups, and thiols.

Problem 2.13

Think

What is the most important intermolecular interaction that will occur between two molecules of **C**? Between two molecules of **D**? Can the extent of hydrogen bonding be distinguished by the number of H-bond donors and acceptors? How would that affect the boiling points of **C** and **D**?

Solve

C has a higher boiling point than **D**. For both compounds, the most important intermolecular interaction is hydrogen bonding. For compound **C**, a pair of molecules has a total of four H-bond acceptors (N atoms) and eight H-bond donors (NH bonds). For compound **D**, a pair of molecules has a total of four H-bond acceptors (N atoms) and four H-bond donors (NH bonds). Hydrogen bonding, therefore, is stronger in **C** than in **D**.

H_2N NH_2

C

Higher boiling point

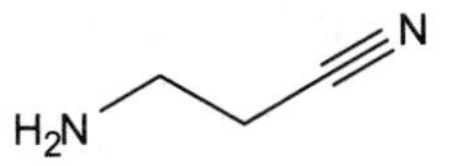

H_2N N

D

Problem 2.14

Think

Which atom, F or N, is more electronegative? How does that affect the concentration of negative charge on that atom? What is the effect on the strength of the hydrogen bond?

Solve

We should expect the F–H···F hydrogen bond to be stronger than the N–H···N hydrogen bond. Because F is more electronegative than N, the concentration of negative charge on F is greater than that on N, and the concentration of positive charge on H in an F–H bond is greater than that in an N–H bond. So the attraction between the opposite charges is stronger in the F–H···F hydrogen bond.

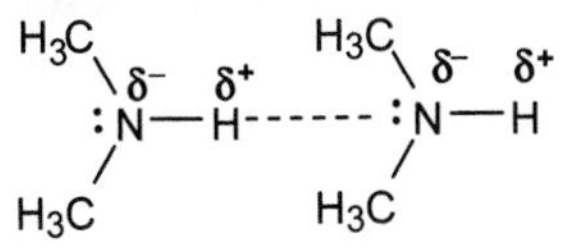

H_3C δ^- δ^+ H_3C δ^- δ^+
:N—H------:N—H
H_3C H_3C

E

δ^- δ^+ δ^- δ^+
:F—H------:F—H

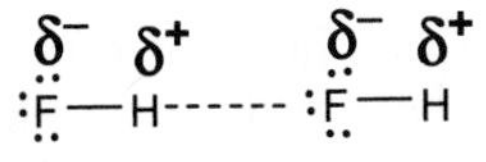

F

Stronger hydrogen bond

Problem 2.15

Think

What is the most important intermolecular interaction in each of the molecules in **A**? In each of the molecules in **B**? Do the molecules in **A** have the same number of electrons? Do they have the same contact surface area? How about the molecules in **B**?

Solve

For the two molecules in **A**, 2-methylpentane should have a higher boiling point than 2,2-dimethylbutane. Both molecules are nonpolar, so the most important intermolecular interaction in each sample is induced dipole–induced dipole interactions. Both molecules have precisely the same number of electrons, but 2-methylpentane is less compact; therefore, it has more contact surface area, making the induced dipole–induced dipole interactions stronger. For the two molecules in **B**, 1,2-dimethylcyclopropane should have a higher boiling point than 1,1-dimethylcyclopropane for the same reason.

Higher boiling point

2,2-Dimethylbutane 2-Methylpentane

A

Higher boiling point

1,1-Dimethylcyclopropane 1,2-Dimethylcyclopropane

B

Problem 2.17

Think

What are the relative strengths of the intermolecular interactions in the pure substances that would be disrupted upon mixing? How do these compare to the strengths of the intermolecular interactions gained?

Solve

We should expect **C** to be more soluble than **D** in toluene. Because toluene is nonpolar, the interactions between each compound and toluene involves induced dipoles, and so are quite weak. To dissolve, H bonding is destroyed in the first compound, whereas ion–ion interactions are destroyed in the second compound. Because ion–ion interactions are stronger than H bonding, it is more favorable for the second compound to remain intact (i.e., not dissolve).

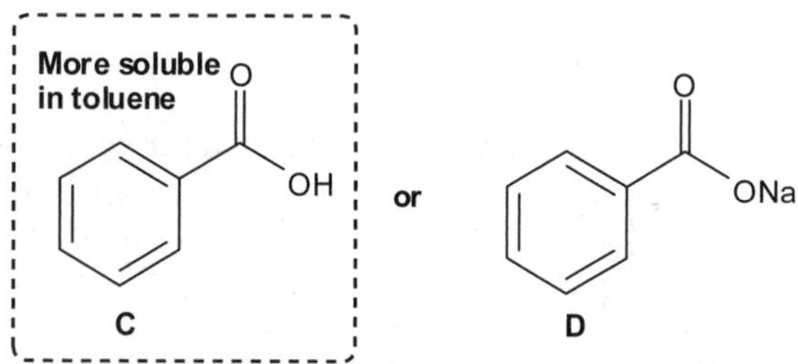

Problem 2.19

Think

How is the solubility of an ionic compound related to how strongly solvated the constituent ions are in solution? What is the structural difference between the two solvent molecules that makes one better than the other at solvating ions?

Solve

The more strongly solvated the ions are by the solvent, the more soluble the ionic compound is in that solvent. Therefore, CH_3OH is better at solvating ions. CH_3OH has only one carbon in the hydrophobic region and is more polar than $CH_3CH_2CH_2OH$, which has three carbons in the hydrophobic region. This allows CH_3OH to have stronger ion–dipole interactions with the Na^+ and Cl^- ions.

Problem 2.20

Think

What are the intermolecular interactions in pure water that would be disrupted upon mixing with a solute? How do these compare to the intermolecular interactions gained between water and each functional group in Table 1-6? Which intermolecular interactions gained upon dissolving would be comparable to the ones lost from pure water?

Solve

Hydrogen bonding interactions among water molecules would be disrupted upon dissolving a solute. For a functional group to be considered hydrophilic, therefore, it should be able to undergo comparable intermolecular interactions with water, such as significant H bonding, strong dipole–dipole interactions, or ion–dipole interactions. Those include some alkyl halides, alcohols, hemiacetals, nitriles, ketones, aldehydes, and carboxylic acids. They also include amines and amides if the N atom is bonded to at least one H atom. Functional groups are hydrophobic if such interactions with water are not possible. Those include alkenes, alkynes, aryl groups, some alkyl halides, thiols, ethers, acetals, epoxides, and esters. They also include amines and amides if N is bonded only to C.

Problem 2.22

Think

Do aldehydes possess H-bond donors, acceptors, or both? What about alcohols? Would that make an aldehyde group more or less hydrophilic than an alcohol group?

Solve

We should expect that the maximum number of carbon atoms a water-soluble aldehyde contains should be less than that for water-soluble alcohols. Whereas the OH group in alcohols has a H-bond donor and acceptor that can participate in H bonding with water, an aldehyde has only a single H-bond acceptor and no donor. Thus the H bonding that takes place between an aldehyde and water is less favorable than that between an alcohol and water, making an aldehyde less hydrophilic than a comparable alcohol.

Problem 2.24

Think

What intermolecular interactions are present in each compound? Are they of similar strength in each compound? What is the total number of electrons in each? How does the total number of electrons affect the boiling point?

Solve

Without concern for differences in polarizability, we would normally expect **A** to have a higher boiling point (b.p.) than **B**. Whereas **A** has a highly polar C=O bond that can give rise to relatively strong dipole–dipole interactions, **B** is nonpolar. However, there are many more electrons in a molecule of **B** (146) compared to **A** (56), meaning that **B** is much more polarizable. Therefore, the induced dipole–induced dipole interactions are much stronger in **B**, giving it a significantly higher boiling point (285 °C) than **A** (178 °C).

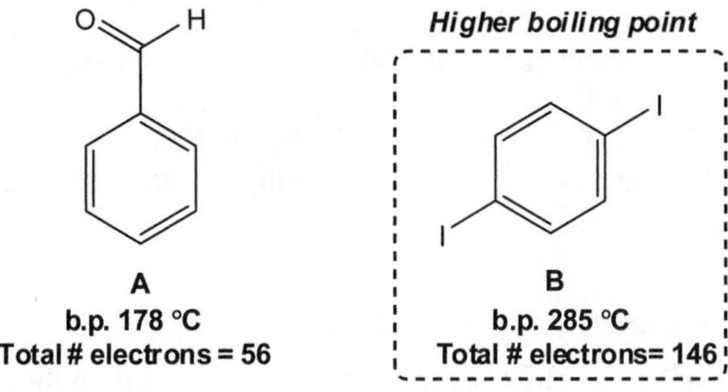

Problem 2.26

Think

What are the most important intermolecular interactions that exist in the isolated substance and what are their relative strengths? What are the most important intermolecular interactions that exist between each solute molecule and water? How does the solubility of the compounds in water compare with the solubility in hexane?

Solve

Hydrogen bonding is disrupted among water molecules when each solute is mixed in. Therefore, solutes that can undergo relatively strong intermolecular interactions with water will be soluble. The weaker that interaction is, the less soluble the compound will be. Water solubility increases in the order **D** < **A** < **E** < **B** < **C**.

Water solubility increases in the order:

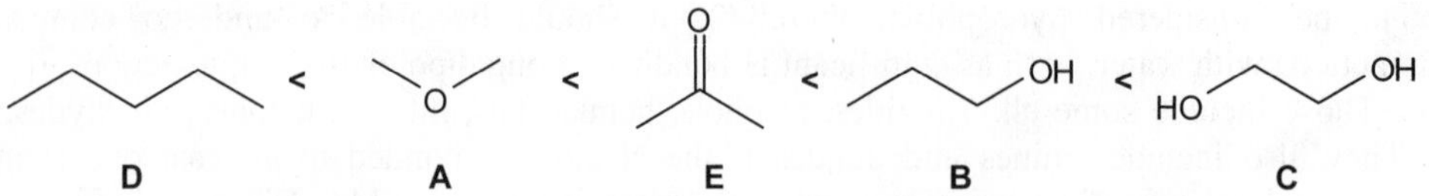

D < A < E < B < C

A, **B**, **C** and **E** can all undergo H bonding with water. To distinguish the strengths of these interactions, consider the numbers of H-bond donors and acceptors. **C** has two donors and two acceptors; **B** has one donor and one acceptor; and both **A** and **E** have one acceptor only. This makes **C** the most soluble, followed by **B**. **E** is more polar than **A**, so it will undergo stronger dipole–dipole interactions with water, making **E** more soluble than **A**. **D** is the least soluble because, being nonpolar, it has the weakest interactions with water (induced dipoles). Notice that the order of solubility of these compounds in water is the reverse of the order in Solved Problem 2.25.

Problem 2.27

Think

What is the strongest intermolecular interaction in a sample of $KSCH_3$? Is ethanol a protic or aprotic solvent? What about acetone? How well can ethanol and acetone solvate K^+? How well can the two solvents solvate CH_3S^-?

Solve

$KSCH_3$ is an ionic compound made of K^+ and CH_3S^- ions, so they are held together by ion–ion interactions, the strongest of the intermolecular interactions. $KSCH_3$, therefore, will be soluble only in solvents that can solvate ions well. Ethanol (CH_3CH_2OH) is a polar protic solvent, so it can solvate both K^+ and CH_3S^- ions very well. Acetone is a polar aprotic solvent, so it can solvate K^+ ions well, but does not solvate CH_3S^- ions very well. Consequently, $KSCH_3$ will be more soluble in ethanol.

Problem 2.28

Think

Based on the solubility of ethyl acetate, is an ester functional group strongly or weakly hydrophilic? Is the long hydrocarbon chain of the compound shown hydrophilic or hydrophobic? Do the two ends of the molecule have dramatically different affinities for water?

Solve

We would not expect the given compound to form micelles in water. Both the large compound given and ethyl acetate have a single ester functional group. The fact that even a relatively small ester like ethyl acetate is not very soluble in water suggests that an ester functional group is not very hydrophilic. The long hydrocarbon chain is hydrophobic, so the two ends of the molecule do *not* have dramatically different affinities for water—something that would be necessary to form micelles.

Problem 2.29

Think

Does the compound have a very hydrophilic end? A very hydrophobic end? What is required for a compound to act as a soap?

Solve

We should not expect this compound to act as a soap. The two ends are ionic and, therefore, are very hydrophilic. The molecule, however, has no hydrophobic end. Although the hydrocarbon rings between the two ionic groups are hydrophilic, a particle of dirt or oil cannot dissolve that portion without at least partially dissolving an ionic group.

Problem 2.30

Think

How many electron groups are surrounded by each nonhydrogen atom? Are any of these electron groups lone pairs? How does the number of electron groups relate to the electron geometries in Table 2-2? Refer to Table 2-3 for molecular geometries.

Solve

Bond angles and electron and molecular geometries are listed below. When no lone pairs are present, the molecular geometry is the same as the electron geometry. Lone pair electrons and bonds to H are added in to explicitly show each electron and molecular geometry.

Electron = tetrahedral
Molecular = tetrahedral
Bond angle ~109.5°

Electron = tetrahedral
Molecular = bent
Bond angle ~109.5°

Electron = tetrahedral
Molecular = bent
Bond angle ~109.5°

Electron = trigonal planar
Molecular = N/A
Bond angle = N/A

Electron = trigonal planar
Molecular = trigonal planar
Bond angle ~120°

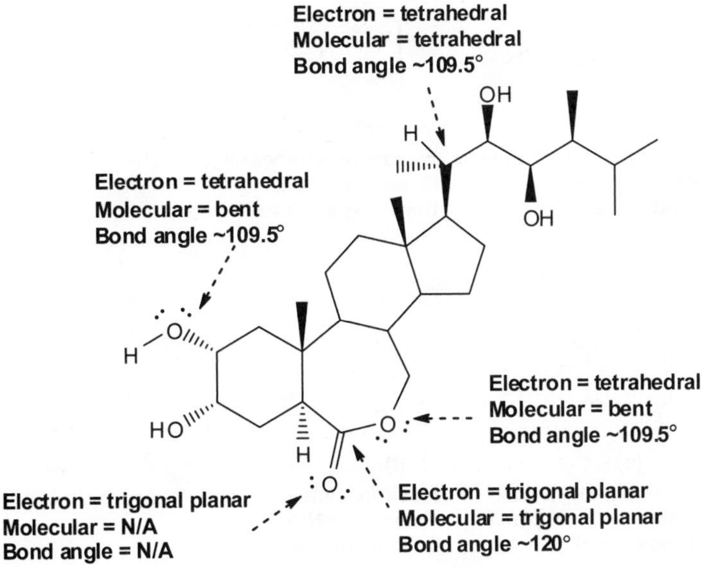

Problem 2.31

Think

How many electron groups are surrounded by each nonhydrogen atom? Are any of these electron groups lone pairs? How does the number of electron groups relate to the electron geometries in Table 2-2? Refer to Table 2-3 for molecular geometries.

Solve

Bond angles, and electron and molecular geometries are listed below. When no lone pairs are present, the molecular geometry is the same as the electron geometry. Lone pair electrons and bonds to H are added in to explicitly show each electron and molecular geometry.

Electron = tetrahedral
Molecular = bent
Bond angle ~109.5°

Electron = linear
Molecular = linear
Bond angle = 180°

Electron = trigonal planar
Molecular = trigonal planar
Bond angle ~ 120°

Electron = trigonal planar
Molecular = trigonal planar
Bond angle ~120°

Electron = tetrahedral
Molecular = tetrahedral
Bond angle ~ 109.5°

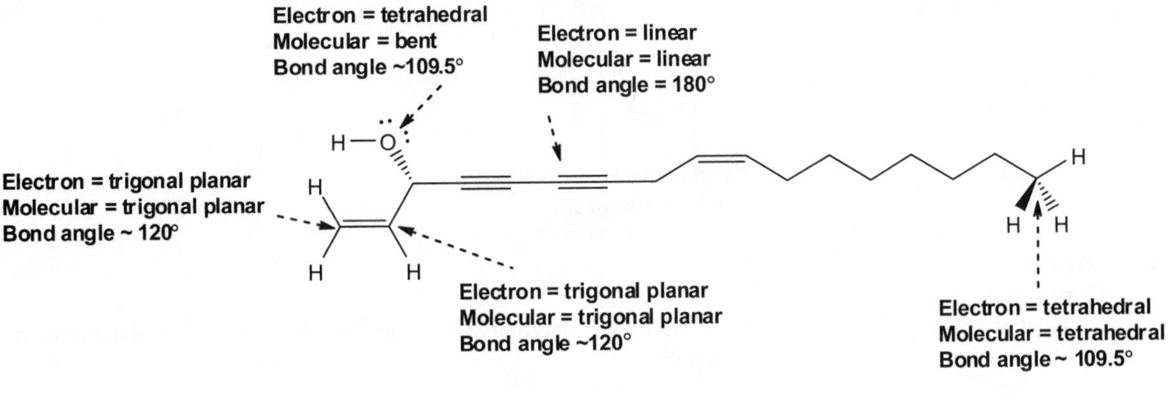

Problem 2.32

Think

How many electron groups are surrounded by each nonhydrogen atom? Are any of these electron groups lone pairs? How does the number of electron groups relate to the electron geometries in Table 2-2? Refer to Table 2-3 for molecular geometries.

Solve

Bond angles and electron and molecular geometries are listed below. When no lone pairs are present, the molecular geometry is the same as the electron geometry. Lone pair electrons and bonds to H are added in to explicitly show each electron and molecular geometry.

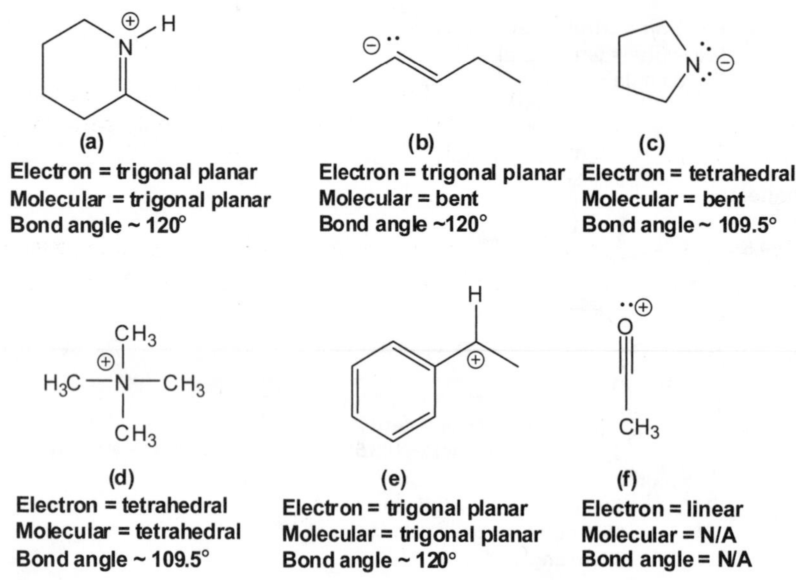

Problem 2.33

Think

What is the ideal molecular geometry and angle for each carbon? What is the angle of each C–C bond in a three-membered ring? How does the difference between ideal angle and forced angle affect the angle strain?

Solve

In a three-carbon ring, the bond angles are forced to be 60° on average. For **A**, cyclopropane, each C atom's ideal geometry is tetrahedral and should have a bond angle of 109.5°. For **B**, cyclopropene, the C atom with four single bonds should have tetrahedral geometry and a bond angle of 109.5°; the ideal geometry of the two C atoms in the C=C bond is trigonal planar and should have a bond angle of 120°. The C atoms of the C=C bond, therefore, are forced to be farther away from their ideal geometry than are the tetrahedral C atoms, giving **B** more angle strain overall.

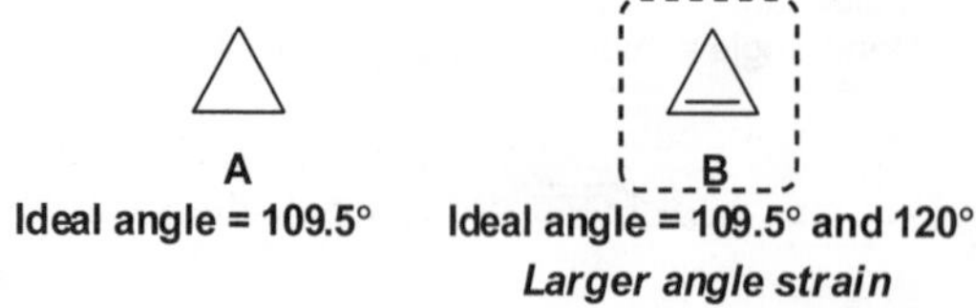

Problem 2.34

Think

For each nonhydrogen atom, what is the molecular geometry? Review the rules for dash–wedge notation. How are linear carbon chains drawn in line notation?

Solve
A flat line (—) represents a bond in the plane of the paper, a wedge (◀) represents a bond that comes out of the plane of the paper, and a dash (''''''''') represents a bond that points away from you. See the structures below.

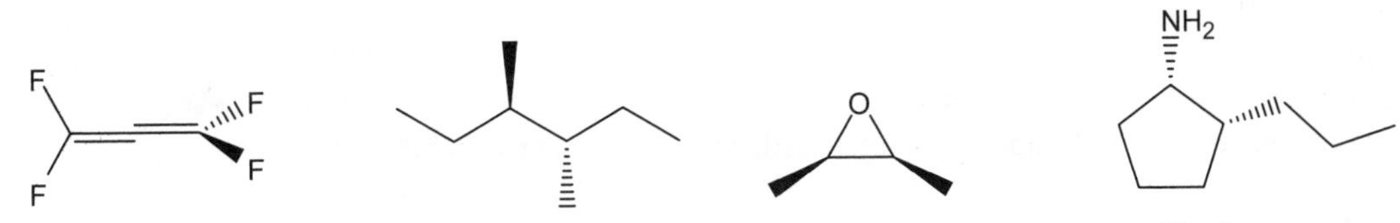

(a) (b) (c) (d)

Problem 2.35

Think
How does this rotation affect the wedge bonds? Dash bonds? Planar bonds? How does the rotation affect whether bonds point toward the top, bottom, left, or right of the figure? Be mindful of only performing the rotation indicated. Building a molecular model might be helpful.

Solve

(a) The 180° rotation indicated inverts all wedges to dash bonds and vice versa. All bonds pointing toward the top of the figure end up pointing toward the bottom and vice versa. The zigzag inverts all V to Λ and all Λ to V.

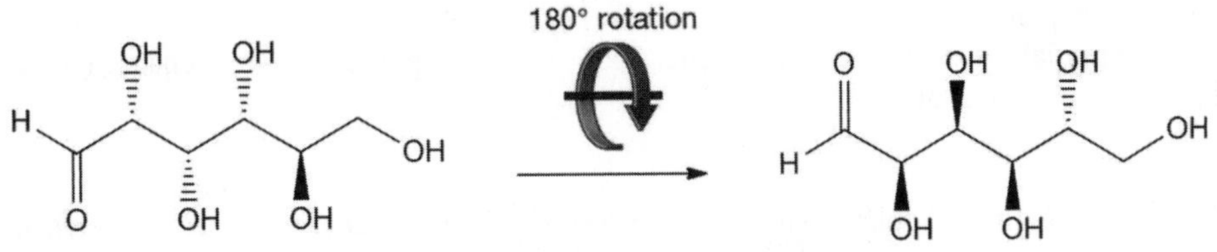

180° rotation

(a) D-Glucose

(b) The 180° rotation indicated inverts all wedges to dash bonds and vice versa. All groups on the left before the transformation end up on the right after the transformation.

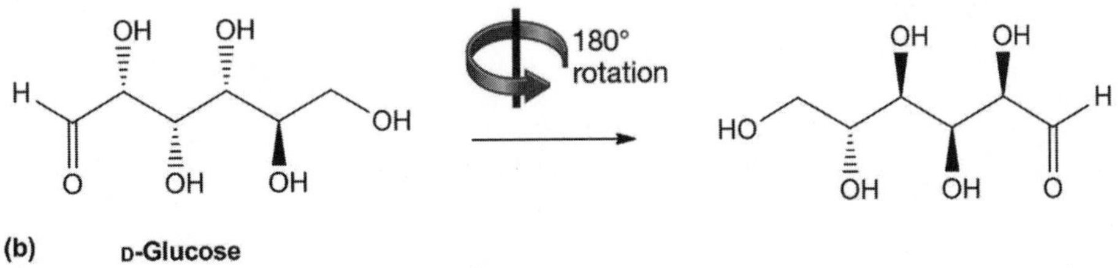

180° rotation

(b) D-Glucose

Problem 2.36

Think
In CH_3^+, what type of charge is concentrated on the carbon? Is it a partial or full charge? What type of charge would be attracted to the carbon? In the species that are given, are there partial or full charges?

Solve
CH_3^+ is a cation having a full positive charge concentrated on the C atom, and would be attracted to full or partial negative charges. Both Cl^- and F^- anions will be attracted to CH_3^+ via ion–ion interactions. Both H_2O and $H_2C{=}O$ are polar, so they have a partial negative charge that will be attracted via ion–dipole interactions. Na^+ has a positive charge and will thus be repelled by CH_3^+.

Problem 2.37

Think

Which ions have the greatest concentration of charge? How do charge differences affect the attraction between two ions?

Solve

B will have the strongest attraction (Al^{3+} and O^{2-}). These two ions have the greatest concentration of charge. Larger concentration of opposite charges leads to stronger attraction between two ions.

Problem 2.38

Think

Are there any differences in charge from one pair of species to the next? Are there differences in the sizes of the ions? How does size affect the concentration of charge and, thus, the attraction between two ions?

Solve

The charges do not change from one pair of ions to the next, but the sizes of the ions do change. Smaller ions have larger concentration of charge, and larger concentration of opposite charges leads to stronger attraction between two ions. Mg^{2+} and O^{2-} will attract each other most strongly, therefore, because Mg^{2+} is the smallest of the metal cations given and thus has the greatest concentration of positive charge.

Problem 2.39

Think

Which equation in this chapter incorporates dipole moment (μ), charge (q), and internuclear distance (r)? Which atom in HBr bears the partial positive charge and which atom bears the partial negative charge (see Fig. 1-16)? Will the units that are given cancel appropriately?

Solve

Equation 2-2 on page 84 gives the relationship between dipole moment, charges, and internuclear distance: $\mu = qr$. Because q is in units of m, the value of r (141 pm) must be converted to m, giving a distance of 141×10^{-12} m.

$$\mu = 0.82 \text{ D} = q \times (141 \times 10^{-12} \text{ m}) \times \left(\frac{1 \text{ D}}{3.33564 \times 10^{-30} \text{ C} \cdot \text{m}}\right) = q \times (4.23 \times 10^{19} \frac{\text{D}}{\text{C}})$$

Solve for q

$$q = \frac{0.82 \text{ D}}{4.23 \times 10^{19} \frac{D}{C}} = 1.93 \times 10^{-20} \text{ C}$$

$$\text{H} = +1.93 \times 10^{-20} \text{ C}$$
$$\text{Br} = -1.93 \times 10^{-20} \text{ C}$$

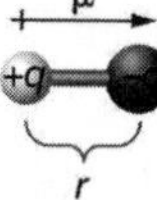

A full charge is 1.602×10^{-19} C, so the amount of charge centered on each atom is $(1.93 \times 10^{-20} \text{ C}) / (1.602 \times 10^{-19} \text{ C}) = 0.120 = 12.0\%$ of a full charge.

Problem 2.40

Think

What is the molecular geometry for each of the molecules? Which directions do the bond dipoles point along each bond? Draw in the dipole arrows for each polar bond. Consider the difference in electronegativity (Fig. 1-16) for each atom to determine the magnitude of each bond dipole. Do the bond dipoles cancel? How does the magnitude of the net dipole relate to the polarity of the molecule?

Solve

Polarity is: (least) $BF_3 < BFH_2 < BF_2H$ (most).

All these compounds are trigonal planar. Fluorine and hydrogen are both more electronegative than boron, so each B–F bond dipole points toward F and each B–H bond dipole points toward H. Fluorine is more electronegative than hydrogen, however, so the B–F bond dipoles are larger. Bond dipoles in BF_3 cancel entirely, as shown below, and, therefore, BF_3 is nonpolar. In BFH_2, the B–F bond dipole partially cancels with that from the two B–H bonds, resulting in a small net dipole (in boldface type) in the direction of the B–F bond. BF_2H has an even larger dipole moment because of the additional F.

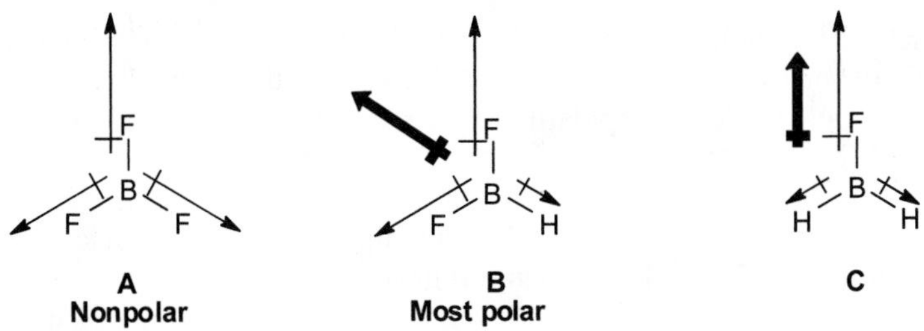

Problem 2.41

Think

In each compound shown below, where do the bond dipoles point? Do the additional bond dipoles in the ester point in the same direction as the C=O bond dipole?

Solve

Oxygen is more electronegative than carbon, so each C=O and C–O bond dipole points toward O. As we can see below, the C–O bond dipoles (not in boldface) in an ester do not point in the same direction as the C=O bond dipole, so they partially cancel with the C=O bond dipole to give the net dipole moment (boldface arrow). The only significant bond dipole in a ketone, on the other hand, is from the C=O bond. Therefore, cancellation of the bond dipole does not occur in a ketone.

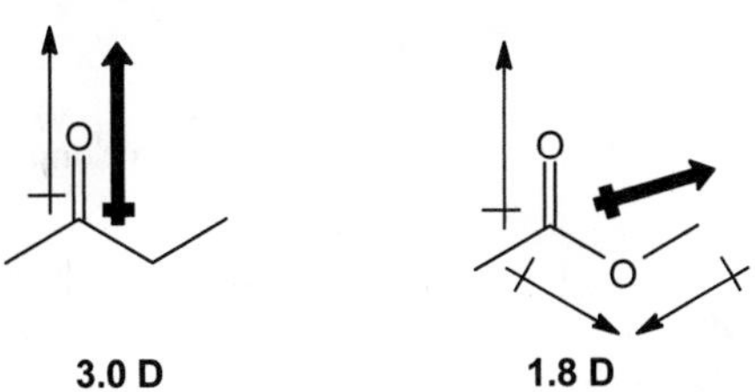

Problem 2.42

Think

What colors represent concentrations of negative and positive charge? By looking at an electrostatic potential map, what does the presence of both red and blue indicate? If the charge distribution is symmetric, what does that indicate about the polarity of the molecule?

Solve

(a), **(b)**, and **(f)** are nonpolar, because the concentrations of negative charge (red) and concentrations of positive charge (blue) are symmetrically distributed about the center of the molecule; thus, any bond dipoles must cancel completely. For **(c)**, **(d)**, and **(e)**, the net dipole points from the blue region (excess positive charge) to the red region (excess negative charge), as shown below.

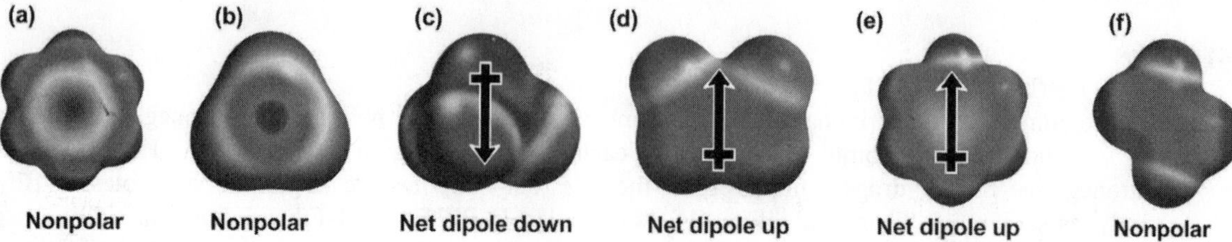

Problem 2.43

Think

What is the molecular geometry for each of the molecules? Draw in the dipole arrows for each polar bond. Consider the difference in electronegativity for each atom (Fig. 1-16) to determine the magnitude of the net dipole. How does the magnitude of the net dipole relate to the polarity of the molecule?

Solve

Molecules **(b)**, **(d)**, **(e)**, **(g)**, **(j)**, **(l)**, and **(n)** are polar. The direction of the net dipole moment is indicated by the boldface arrow. The nonpolar molecules, **(a)**, **(c)**, **(h)**, **(i)**, **(k)**, and **(m)** all have polar bonds, but because of the symmetry of those bonds the dipoles cancel and no net dipole exists. In general, hydrocarbons such as molecule **(f)** are considered nonpolar.

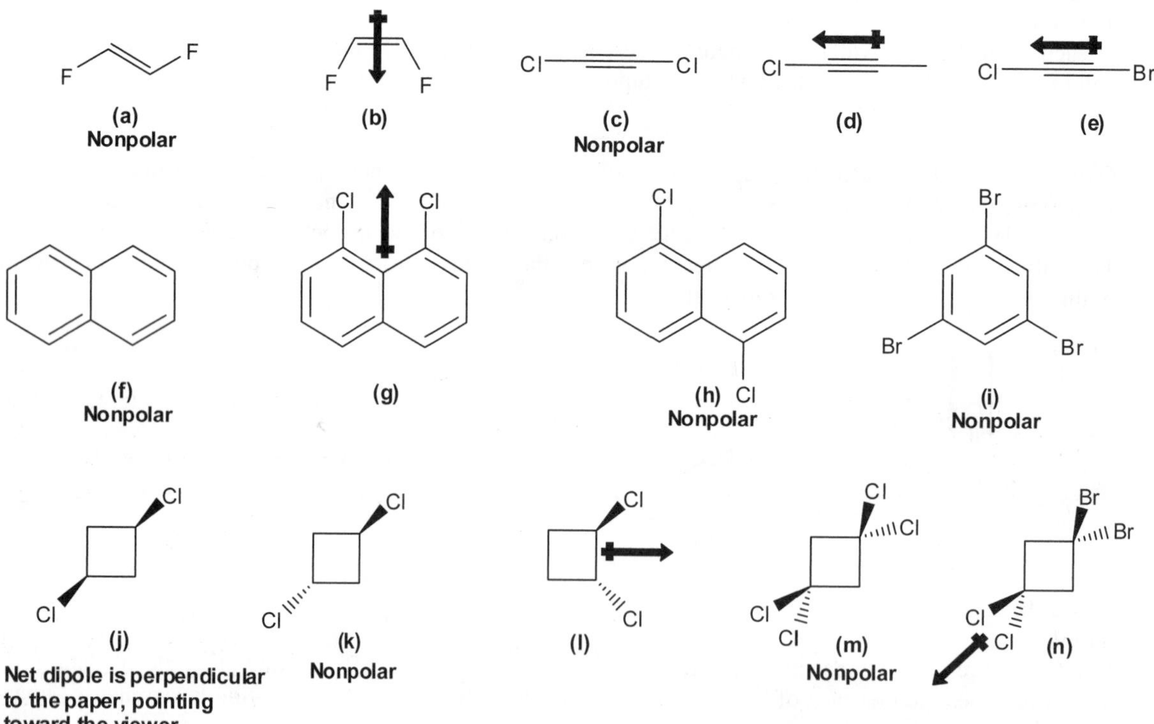

Problem 2.44

Think

What are possible ways you could draw CX_2Y_2 in a square planar geometry? Is there more than one way? Does the polarity differ in each drawing? How does this compare to a tetrahedral geometry?

Solve

If CX_2CY_2 were planar, there could be two different compounds, as shown below. The one on the left would be polar, whereas the one on the right would be nonpolar. However, a tetrahedral geometry of such a compound, like CH_2Cl_2, is consistent with the fact that the compound is always found to be polar. This is because the net dipole from the V of the CCl_2 is not the same as that from the V of the CH_2.

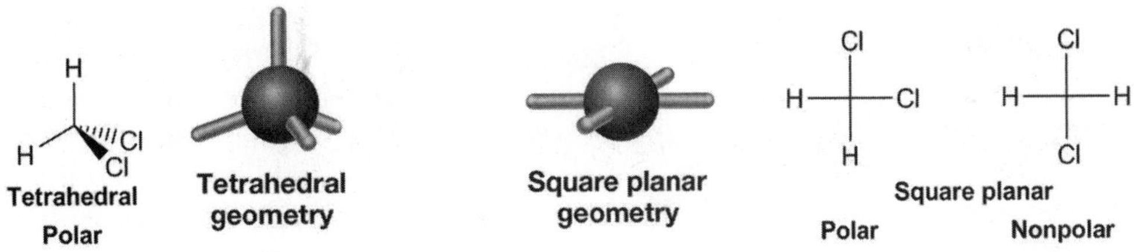

Tetrahedral Polar | Tetrahedral geometry | Square planar geometry | Square planar: Polar, Nonpolar

Problem 2.45

Think

Which atoms must be covalently bonded to the H atom for it to be considered a H-bond donor? Which atoms are considered H-bond acceptors?

Solve

Each potential H-bond donor (which includes H covalently bonded to either F, O, or N) and each potential H-bond acceptor (either F, O, or N) are labeled below.

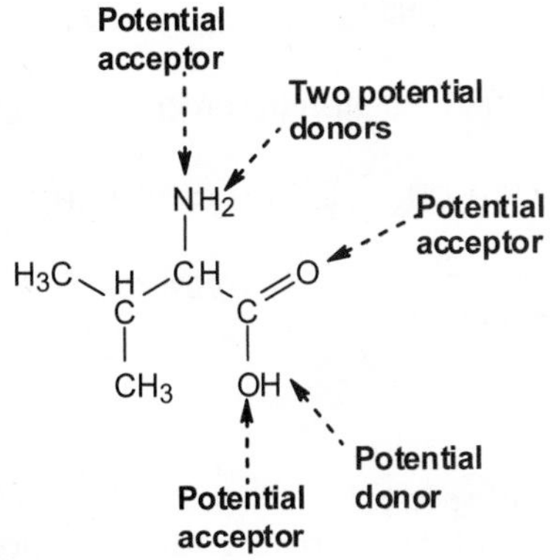

Potential acceptor; Two potential donors; Potential acceptor; Potential donor; Potential acceptor

Problem 2.46

Think

What are the types of intermolecular interactions in a pure sample of each compound? Of those, what are the most important? Are any ions present? Any polar bonds? Any H-bond donors and acceptors? Are the net dipoles of the same magnitude? Do the compounds differ in the number of H-bond donors and acceptors? Do they differ in contact surface area? How does the strength of the intermolecular interactions affect the boiling point?

Solve

Order of increasing boiling point: **A < H < F < G < B < C < E < D**

The dominant interactions for each compound are shown on the next page. Induced dipole–induced dipole interactions are the weakest, giving rise to the lowest boiling points. Ion–ion interactions are the strongest, giving rise to the highest boiling point. The two nonpolar compounds **A** and **H** have the same number of electrons, so they are differentiated by their contact surface area: the greater the contact surface area, the higher the boiling point. The two polar compounds **F** and **G** are differentiated by their magnitude of the dipole moment; the ketone **G** has a larger dipole moment than the ether **F**. The H-bonding compounds **B**, **C**, and **E** are differentiated by the number of H-bond donors and acceptors in a pair of molecules: As the number of these features increases, so does the strength of H-bonding, and therefore the boing point increases.

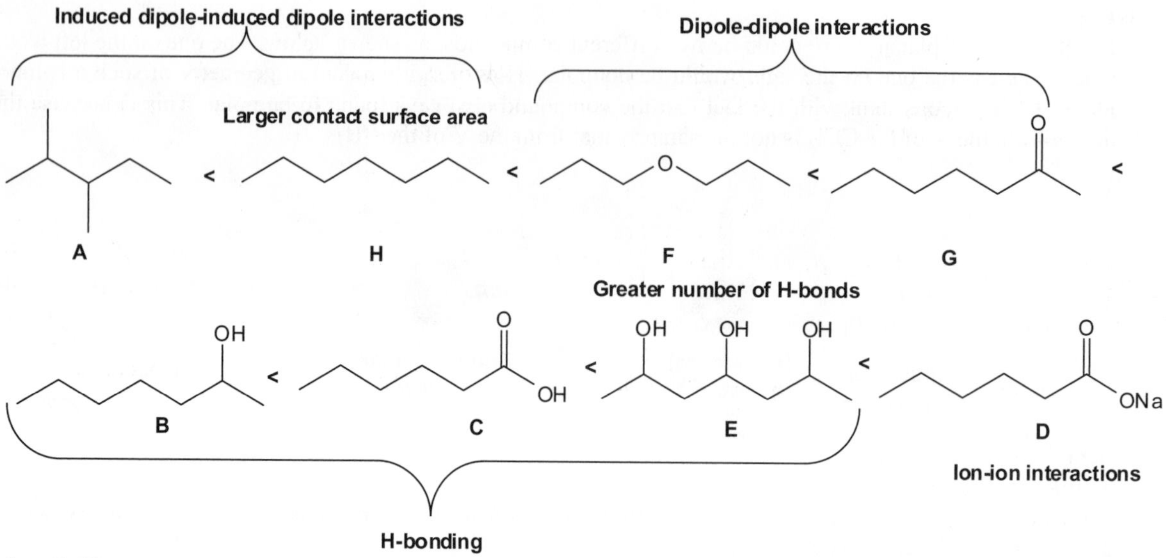

Induced dipole-induced dipole interactions

Dipole-dipole interactions

Larger contact surface area

A < H < F < G <

Greater number of H-bonds

OH OH OH OH O ONa

B < C < E < D

H-bonding

Ion-ion interactions

Problem 2.47

Think

What are the types of intermolecular interactions between each pair of species? Are any ions present? H-bond donors and acceptors? Significant net dipoles? How does the strength of the intermolecular interactions affect the boiling point?

Solve

The dominant intermolecular interactions (indicated below each molecule) establish the order of boiling points.

$CH_3CH_2CH_2CH_2CH_3$	<	$CH_3CH_2OCH_2CH_3$	<	$FCH_2CH_2OCH_2CH_3$	<	$CH_3CH(OH)CH_2CH_3$
B		**A**		**D**		**C**
Disperson forces only		**One dipole**		**Two dipoles** **CF greater than CO**		**Hydrogen bonding**

Problem 2.48

Think

What are the types of intermolecular interactions between each pair of species? Are any ions present? H-bond donors and acceptors? Significant net dipoles? How does the strength of the intermolecular interactions affect the boiling point?

Solve

The dominant intermolecular interactions (indicated below each molecule) establish the order of boiling points.

$FCH_2CH_2CH_2CH_3$	<	$CH_3CH_2CH_2CH_2OH$	<	$CH_3CH_2CH(OH)CH_2OH$	<	$CH_3CH(OH)CH(OH)CH_2OH$
D		**B**		**A**		**C**
Dipole-dipole **(No H bonding)**		**Hydrogen bonding** **1 H-bond donor** **1 H-bond acceptor**		**Hydrogen bonding** **2 H-bond donors** **2 H-bond acceptors**		**Hydrogen bonding** **3 H-bond donors** **3 H-bond acceptors**

Problem 2.49

Think

Are any ions present? Any H-bond donors or acceptors? Any significant bond dipoles? Do the species have the same number of electrons? Same contact surface area?

Solve

Being uncharged and nonpolar, the only intermolecular interaction that exists between two H_2 molecules or two He atoms is induced dipole–induced dipole interactions. Therefore, H_2 must be more polarizable than He. This should make sense because, even though both species have two electrons, the electrons in H_2 are distributed between two nuclei, so they occupy a greater region of space than electrons around a single He nucleus. Thus, H_2 has a greater contact surface area than He, giving H_2 the higher boiling point.

Problem 2.50

Think

What are the dominant intermolecular interactions in samples of the two pure compounds? Are they the same or different? If they are the same, how can the shapes of the molecules affect the strengths of those interactions? It may help to draw each pair of molecules undergoing that type of interaction.

Solve

The two molecules are made from exactly the same atoms (i.e., have the same formula) and each possesses both a H-bond donor and acceptor. Thus, the dominant intermolecular interaction in each compound is H bonding. H bonding is stronger in propan-1-ol than in propan-2-ol, giving rise to the higher boiling point for propan-1-ol. The strength of H bonding is made weaker in propan-2-ol because of the bulkiness of each alkyl group (i.e., CH_3 group) bonded to the C atom to which the OH group is attached. That bulkiness causes each H bond to be less stable. In propan-1-ol, on the other hand, the C atom bonded to the OH group is connected to only one alkyl group (i.e., a CH_2CH_3 group) and a H atom, which is very small.

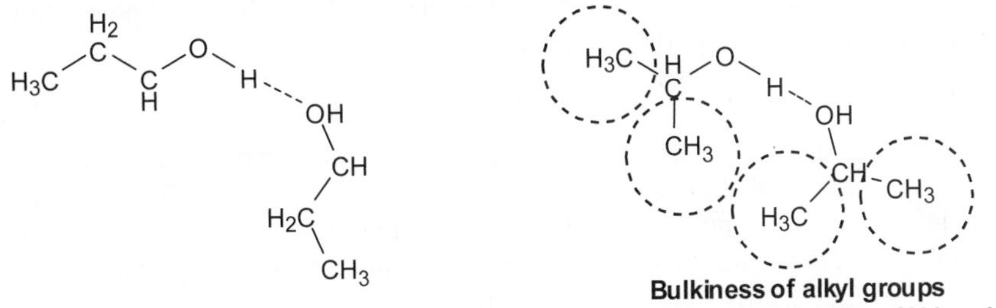

Bulkiness of alkyl groups *decreases* stability of H bond

Problem 2.51

Think

What is the dominant type of intermolecular interaction disrupted among water molecules when each organic solvent is mixed in? What intermolecular interactions exist between water and each organic solvent molecule? Which organic solvent has the strongest intermolecular interaction with water?

Solve

Strong hydrogen bonding is disrupted among water molecules when the organic solvents are mixed in. Only **B** and **C** will undergo hydrogen bonding with water, because they each have at least one O atom to act as a H-bond acceptor, so they will be more soluble than **A** or **D**. Molecule **C** ($CH_3OCH_2CH_2OCH_3$) has two acceptors, whereas **B** has only one, so **C** would be more soluble.

Problem 2.52

Think

What is the charge on the carboxylate group? What intermolecular interactions are occurring? What difference does the +1 charge versus +2 charge have on the ability of the ions to participate in that kind of interaction with the carboxylate group in soap?

Solve

The attraction is due to ion–ion interactions involving a metal cation and a carboxylate anion (RCO_2^-). Ca^{2+} and Mg^{2+} both have a +2 charge and so will attract more strongly to the −1 charge on the carboxylate anion than will Na^+ or K^+.

Problem 2.53

Think

Draw out each structure. Consider the concentration of charge for each. Are resonance structures possible? If so, how does the number of resonance structures affect charge localization?

Solve

In a carboxylate anion (RCO_2^-), the negative charge is resonance delocalized over two O atoms, whereas in a sulfate anion ($ROSO_3^-$), the negative charge is resonance delocalized over three O atoms. Thus, in a sulfate anion, the concentration of negative charge is smaller than in a carboxylate anion, so its attraction to a positively charged ion is weaker.

Carboxylate hybrid

Alkyl sulfate hybrid

Problem 2.54

Think

What are the dominant intermolecular interactions in a pure sample of each compound? What are the important bond dipoles? In which directions do they point? Determine the magnitude of the net dipole. How does the magnitude of the net dipole relate to the boiling point of the molecule?

Solve

For both compounds, the dominant intermolecular interaction is dipole–dipole interactions, so we would expect that the one with the larger dipole moment should have a higher boiling point. In 1,2-difluorobenzene, the two bond dipoles (arrows not in boldface) are pointing nearly in the same direction, so the net dipole moment (boldface arrows) is substantial. In 1,3-difluorobenzene, on the other hand, the two bond dipoles are pointing in nearly the opposite direction, so they nearly cancel each other out. Thus, 1,2-difluorobenzene has the larger dipole moment and, hence, should have the higher boiling point.

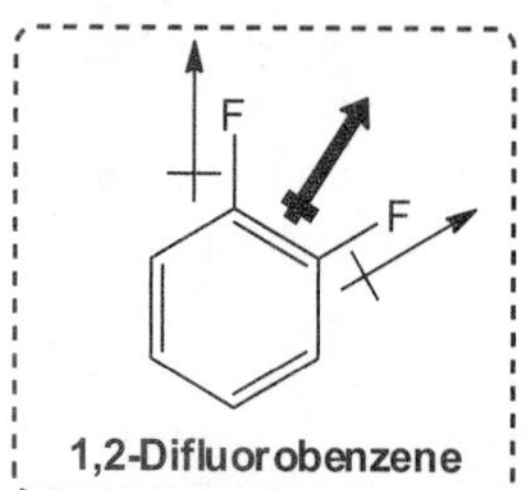

1,2-Difluorobenzene

1,3-Difluorobenzene

Problem 2.55

Think

What are the dominant intermolecular interactions in a pure sample of each compound? What are the important bond dipoles? In which directions do they point? Determine the magnitude of the net dipole. How does the magnitude of the net dipole relate to the boiling point of the molecule?

Solve

As shown below, for the small cyclic ethers, the two C–O bond dipoles (arrows not in boldface) point in nearly the same directions and so add together to give a substantial net dipole (boldface arrows). For these compounds, therefore, there are relatively strong dipole–dipole interactions. Small cyclic alkanes, on the other hand, are nonpolar. Therefore, the significant dipole moments in the small cyclic ethers give rise to higher boiling points. For larger rings, the vectors of the two C–O bond dipoles diverge, allowing them to more efficiently cancel each other, making the larger rings less polar. Thus, the polarity of the larger cyclic ethers more closely resembles the polarity of the larger cyclic alkanes, and the boiling points of the larger cyclic ethers more closely resemble the boiling points of the larger cyclic alkanes.

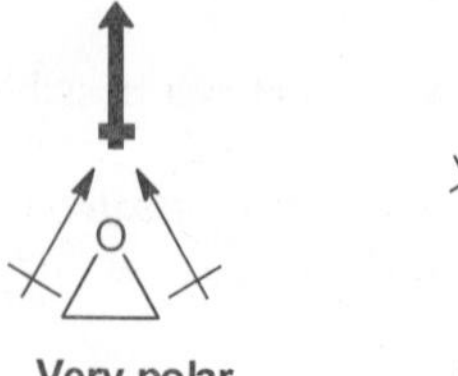

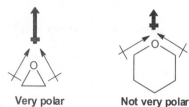

Very polar

Not very polar

Problem 2.56

Think

What are the intermolecular interactions in a pair of **A** molecules? A pair of **B** molecules? How many H-bond donors and acceptors are there? Are the number of electrons significantly different? Consider the surface area of contact between each molecule.

Solve

Hydrogen bonding exists between a pair of each molecule, and the strength of hydrogen bonding is expected to be very similar because molecules **A** and **B** have the same number of H-bond donors and acceptors and are also structurally very similar. The difference must be in the hydrocarbon tails, which are highly nonpolar. Being nonpolar, those tails participate mainly in induced dipole–induced dipole interactions. Such interactions must be stronger with the saturated fatty acid because it has a higher melting point. Since the two molecules have nearly the same number of total electrons, the two molecules have about the same polarizability. Thus, the stronger interactions with the saturated fatty acid can be explained by a greater contact surface area.

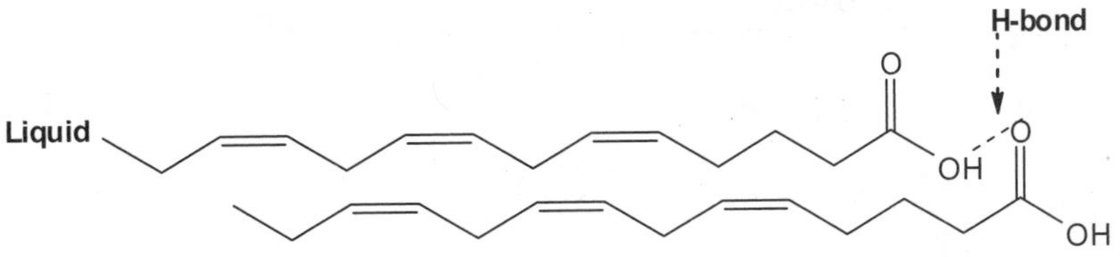

H-bond

Liquid

A

H-bond

Solid

Stronger inducted dipole-induced dipole interactions

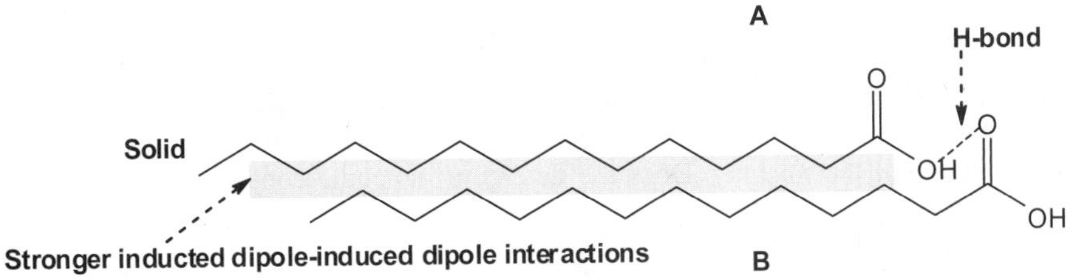

B

Problem 2.57

Think

Given that CH_3O^- carries a full negative charge, what type of charge would Br_2 need to undergo attraction? Does it initially have that type of charge? If not, how could it develop that kind of charge? Would I_2 or Br_2 develop that type of charge more easily?

Solve

(a) To be attracted to CH_3O^-, Br_2 would need to develop a partial positive charge. It does not have one initially, but would develop one via an induced dipole when in the presence of CH_3O^-. The name of the attraction should be ion–induced dipole interactions.

(b) The attraction between CH_3O^- and I_2 would be stronger because I_2, containing more electrons than Br_2, is more polarizable.

Problem 2.58

Think

What are the relative strengths of the intermolecular interactions in the pure substances that would be disrupted upon mixing? How do these compare to the strengths of the intermolecular interactions gained when the solute is mixed in? Would a larger hydrocarbon portion in the solvent molecule lead to stronger or weaker intermolecular interactions with the solute?

Solve

H bonding between solvent molecules and ion–ion interactions in NaCl are given up when the solute mixes in. Ion–dipole interactions are gained between the solute ions and solvent molecules. These ion–dipole interactions are compromised as the size of the nonpolar hydrocarbon portion increases, making the intermolecular interactions weaker. Therefore, **A** (CH_3OH), having the smallest nonpolar hydrocarbon portion, would best dissolve NaCl.

Problem 2.59

Think

What types of intermolecular interactions contribute to a functional group being hydrophilic? What types of intermolecular interactions contribute to a functional group being hydrophobic? Refer to Figure 2-27 as a guide for how the micelle is drawn. With water being the solvent, should the hydrophilic or hydrophobic portions of the molecules be on the outside of the micelle? What is the strongest intermolecular interaction between $CH_3(CH_2)_{15}N(CH_3)_3^+Cl^-$ and a molecule of water?

Solve

Functional groups are hydrophilic if they can interact strongly with water through H bonding, dipole–dipole interactions, or ion–dipole interactions. Functional groups are hydrophobic if such interactions with water are not possible (i.e., nonpolar). **(a)** See below. **(b)** Same as Figure 2-27, except the exterior of the micelle is the $N^+(CH_3)_3$ group. **(c)** Ion–dipole interactions.

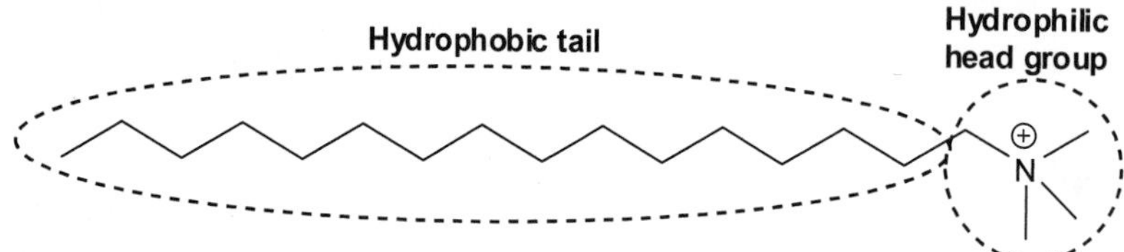

Problem 2.60

Think

What types of intermolecular interactions contribute to a functional group being hydrophilic? What types of intermolecular interactions contribute to a functional group being hydrophobic? Refer to Figure 2-27 as a guide for how the micelle is drawn. With water being the solvent, should the hydrophilic or hydrophobic portions of the molecules be on the outside of the micelle? Does the detergent molecule consist of ions? Does it have H-bond donors and acceptors?

Solve

Functional groups are hydrophilic if they can interact strongly with water through H bonding, dipole–dipole interactions, or ion–dipole interactions. Functional groups are hydrophobic if such interactions with water are not possible (i.e., nonpolar).

(a) See below.

(b) Same as Figure 2-27, with the OH groups on the outside.

(c) H-bonding interactions dominate between the OH groups of the head group and the OH groups of water.

(d) Because they do not consist of ions, these detergents have less tendency to attract metal cations, which would otherwise cause the detergent to precipitate out of solution.

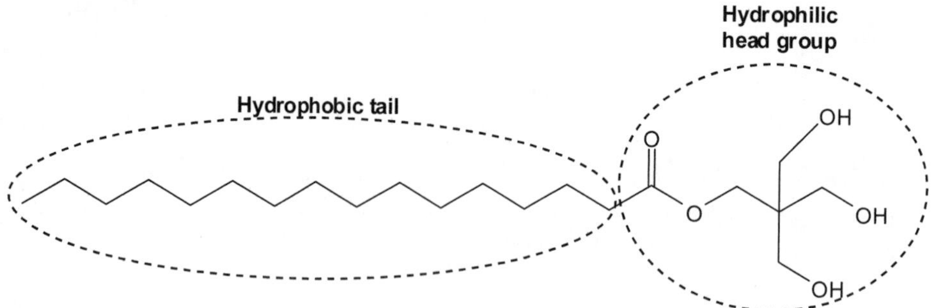

Problem 2.61

Think

What are the dominant intermolecular interactions between a pair of each molecules? Are ions present? Are strong bond dipoles present? Are there H-bond donors and acceptors present? How can you predict the extent of H bonding based on the numbers of donors and acceptors? How does the extent of H bonding affect boiling point (b.p.)?

Solve

The first two compounds can undergo H bonding because in a pair of each of them, there is at least one H-bond donor (NH) and one H-bond acceptor (N). For the third compound, a pair of molecules cannot undergo H bonding, because there are no H-bond donors. Therefore, the third compound has the lowest b.p. The first two compounds differ in the number of H-bond donor groups: two for the first, and one for the second. Thus, H bonding is stronger for the first compound, giving it a higher b.p.

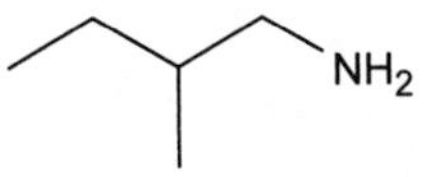

NH$_2$ H N N

2-Methylbutan-1-amine
b.p. = 97 °C
2 H-bond donors

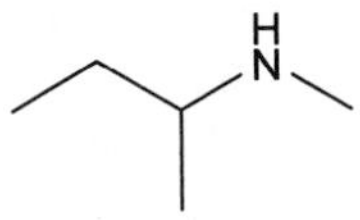

N-Methylbutan-2-amine
b.p. = 84 °C
1 H-bond donor

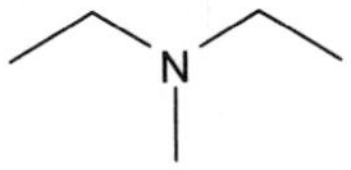

N-Ethyl-N-methylethan-1-amine
b.p. = 65 °C
Zero h-bond donors

Problem 2.62

Think

What are the major intermolecular interactions in each? Would you expect hexafluorobenzene to be more polarizable than benzene? To be more polarizable, what must the electrons be able to do easily? How do you think that the strong electronegativity of F would affect the electrons' ability to do this?

Solve

Because both compounds are nonpolar, induced dipole–induced dipole interactions govern their boiling points (b.p.). The fact that they have about the same boiling points means that they have about the same polarizability. We would normally expect hexafluorobenzene to have a greater polarizability (and hence a higher boiling point) because it has so many more electrons. However, F is very highly electronegative, which prevents its electrons from being moved around easily (i.e., F's high electronegativity decreases its polarizability).

F F F F F F

Benzene
b.p. = 80 °C

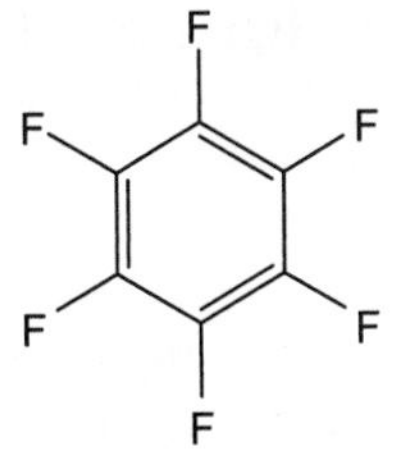

Hexafluorobenzene
b.p. = 81 °C

Problem 2.63

Think

How many electron groups are present about the O atom in H_2O? What is the electron geometry about the O in H_2O? What is the bond angle? How do you think the bond angle compares to the angle of the two H bonds?

Solve

The O atom has four electron groups and, therefore, a tetrahedral electron geometry, so its lone pairs are about 109.5° apart. Thus, the H bonds are about 109.5° apart.

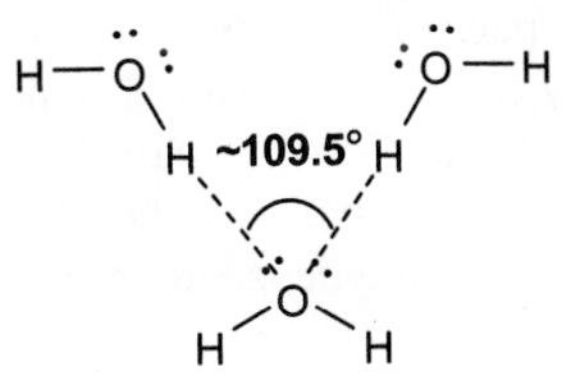

H—O: :O—H
H ~109.5° H
O
H H

Problem 2.64

Think

In the depiction given in the problem, does the compound have a hydrophilic end? Hydrophobic end? What is required for a compound to act as a soap? If the chain is flexible, as described in the problem, can the molecule attain a shape in which one end is very hydrophilic and the other is very hydrophobic?

Solve

If the carbon chain were rigid, the compound could not be an effective soap, for the same reason as in Problem 2.29. However, because the chain is flexible, it can attain the geometry below, where one end is highly hydrophilic (the ionic end) and the other end is highly hydrophobic (the hydrocarbon portion). Thus, it can act as a soap.

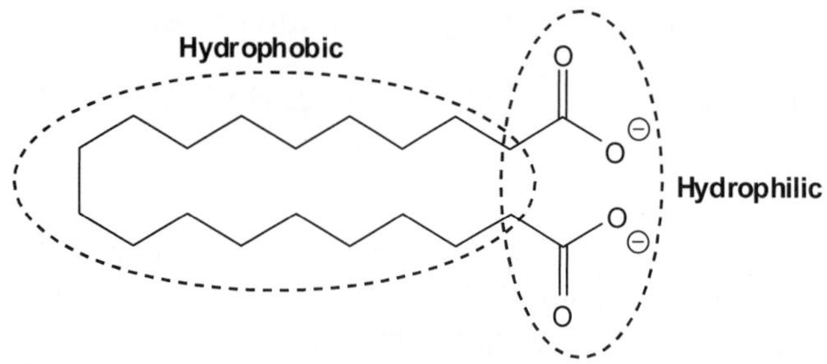

Problem 2.65

Think

What are the intermolecular interactions in each alcohol below? Are ions present? Strong bond dipoles? Are there H-bond donors and acceptors? How do you think the bulkiness of the alkyl groups affects the strength of the dominant intermolecular interaction?

Solve

For each compound, the dominant intermolecular interaction between a pair of molecules is H bonding, due to the presence of an H-bond donor (OH) and an H-bond acceptor (O). The molecules have the same number of donors and acceptors, but with greater bulkiness from the alkyl groups surrounding the OH group, H bonding is more difficult, and with weaker H bonding comes lower boiling points.

OH	OH	HO
Pentan-1-ol	**Pentan-3-ol**	**2-Methylbutan-2-ol**
b.p. = 136–138 °C	**b.p. = 114–115 °C**	**b.p. = 102 °C**

Problem 2.66

Think

What are the intermolecular interactions in each alcohol below? How would the OH groups on the same molecule interact when they are in close proximity? What impact would that have on the molecules' ability to interact with other molecules?

Solve

For each compound, the dominant intermolecular interaction between a pair of molecules is H bonding. Because the OH groups in 1,2-dihydroxybenzene are so close together, they form an *internal* H bond (see the next page). In 1,3-dihydroxybenzene this cannot happen because the OH groups are too far apart. With the OH groups in 1,2-dihydroxybenzene tied up in internal H bonding, they are less available to H bond with other molecules. Therefore, the intermolecular forces between molecules in 1,2-dihydroxybenzene are less than in 1,3-dihydroxybenzene, giving the latter a lower boiling point.

Internal H-bond

b.p. = 281 °C — 1,3-Dihydroxybenzene

b.p. = 245 °C — 1,2-Dihydroxybenzene

Problem 2.67

Think

What do you think the word *protic* indicates? What types of intermolecular interactions should protic solvents be able to undergo with solute species?

Solve

For a solvent to be considered protic, it must possess a H-bond donor. This allows it to form strong ion–dipole interactions with dissolved ions, especially anions. Therefore, it must contain an O–H, N–H, or F–H bond. It cannot solely possess a H-bond acceptor. The word *protic* indicates that there is some kind of proton (H^+) character to the solvent. The H-bond donors are circled below.

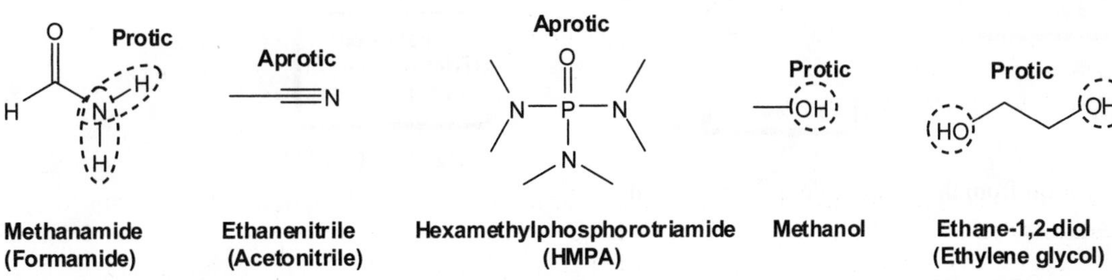

Problem 2.68

Think

What is the structure of an isoprene unit? Can you assign each C to an isoprene unit without having a C part of two isoprene units? How many isoprene units are present?

Solve

An isoprene unit has five carbon atoms, consisting of a four-carbon chain and the fifth carbon attached to C2.

That is, it appears as [isoprene skeletal structure].

The figure below shows how the C atoms can be assigned to different isoprene units. There are four isoprene units in all, making this compound a derivative of a diterpene.

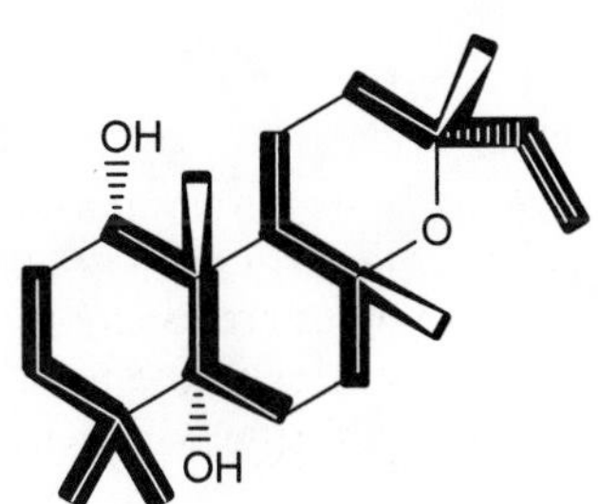

Problem 2.69

Think

What is the general structure of a fat or oil (Section 2.12a)? How does the structure change when you have a specific fatty acid group? Consult Table 2-8 for the structures of various fatty acids.

Solve

The generic structure for a fat or oil is shown below. A fat or oil contains three adjacent ester groups produced from fatty acids and the alcohol group of glycerol. Fats and oils are distinguished by the identities of the R groups, which are governed by the fatty acids from which they come.

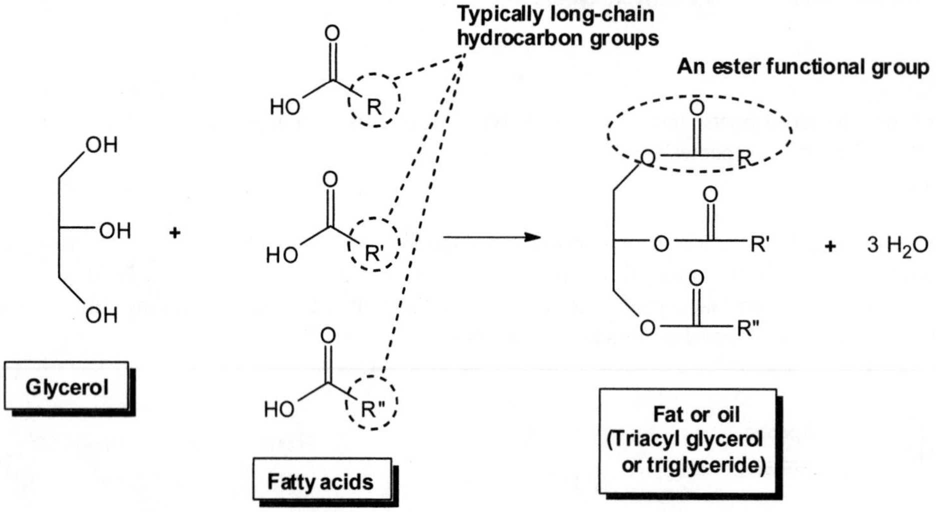

Typically long-chain hydrocarbon groups

An ester functional group

Glycerol

Fatty acids

Fat or oil (Triacyl glycerol or triglyceride)

+ 3 H_2O

(a) Fat or oil from three molecules of stearic acid:

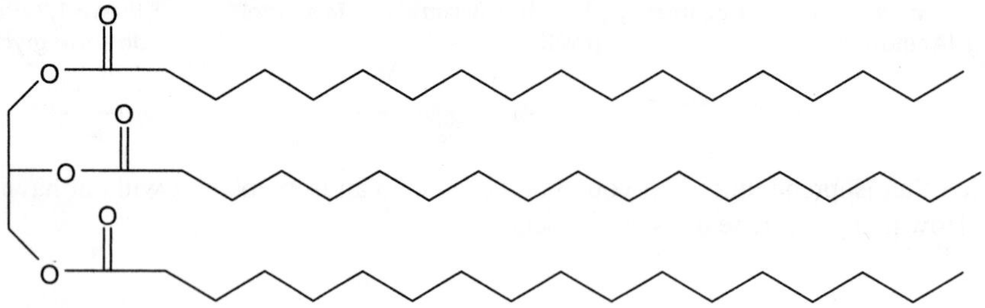

(b) Fat or oil from two molecules of oleic acid and one molecule of linolenic acid:

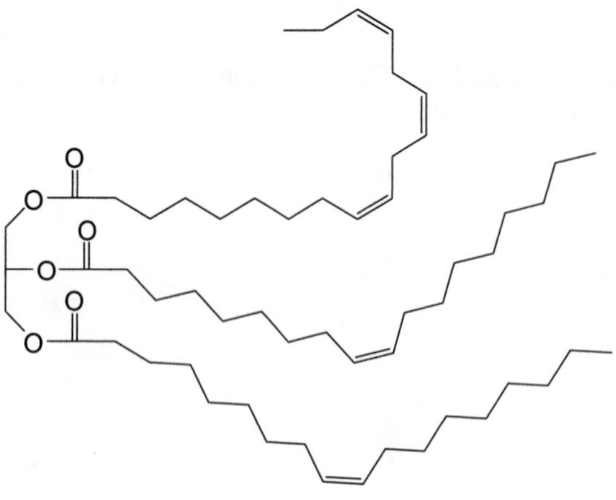

Problem 2.70

Think

Review sections 2.12a, 2.12c, and 2.12d to study the basic structures of fatty acids, steroids, and waxes, respectively.

Solve

(a) Wax. Waxes are typically mixtures of compounds consisting mainly of long-chain esters and alkanes.

(b) Steroid. Steroids consist of three six-membered rings and one five-membered ring fused together.

(c) Fatty acid. A fatty acid is a carboxylic acid with a long hydrophobic hydrocarbon tail (alkene or alkanes).

(a)

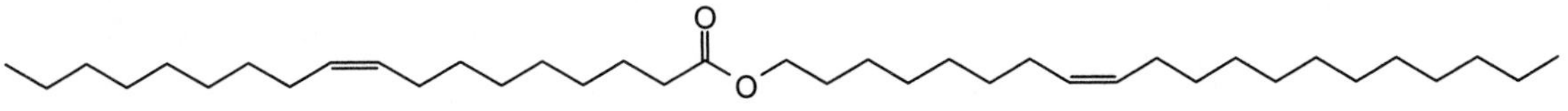

Wax

Isolated from the seeds of the jojoba plant

(b)

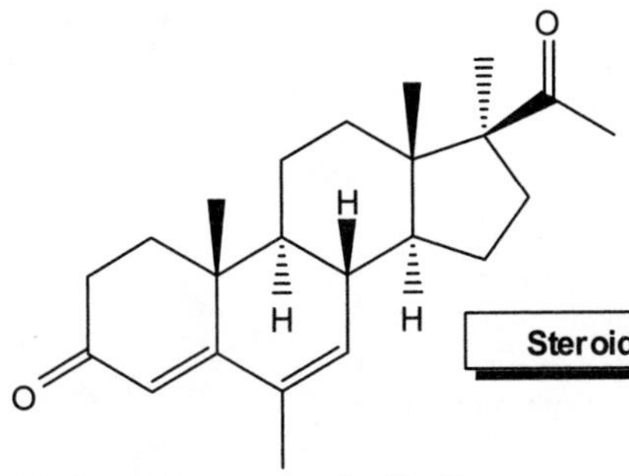

Steroid

Medrogestone, a synthetic drug

(c)

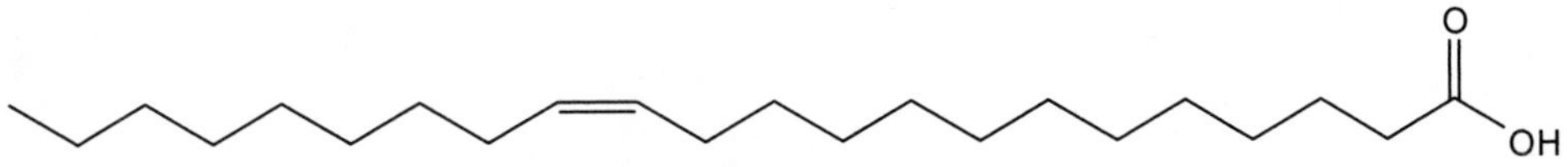

Fatty acid

Erucic acid, isolated from mustard seed

CHAPTER 3 | Orbital Interactions 1: Hybridization and Two-Center Molecular Orbitals

Your Turn Exercises

Your Turn 3.1

Think

What is significant about encompassing 90% of the dots? What do the dots represent?

Solve

The electron density is the probability of finding an electron and is governed by the orbital to which the electron belongs. An electron in the 1*s* orbital has a 90% probability of being found inside a sphere, represented by the circle.

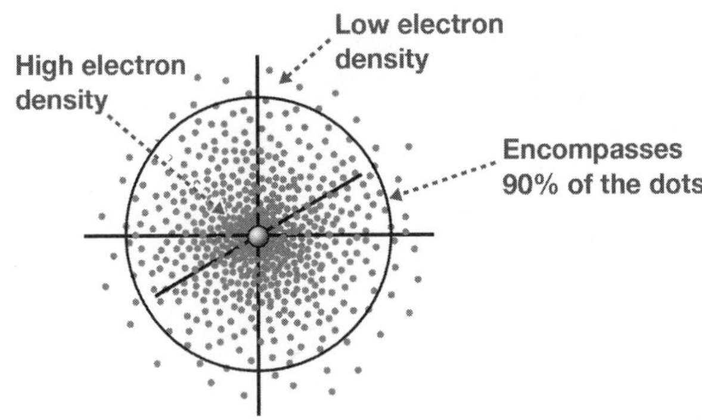

Your Turn 3.2

Think

Why are two lobes drawn? In the figure, how is one lobe distinguished from the other? Where is the dividing line between the lobes? What is the division between the two lobes called?

Solve

Two lobes are drawn for the 2*p* orbital because at the nucleus of the orbital lies a nodal plane. The lobes are distinguished from each other by a different shading for the dots representing electron density. The two lobes are of equal size because the electron density around the nodal plane (vertical plane in this example) is symmetrical.

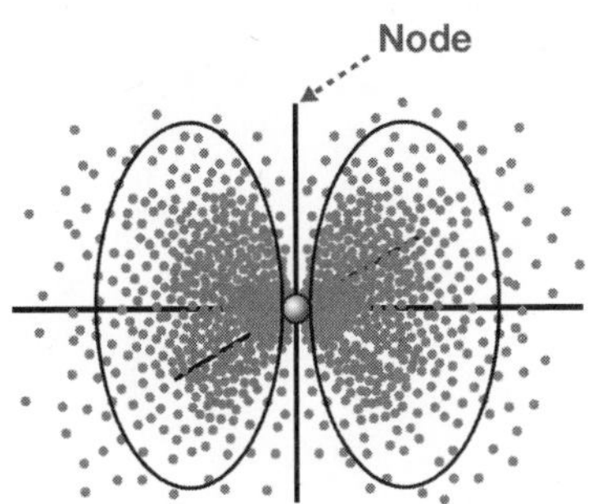

Your Turn 3.3

Think

If both waves are generated on the same side of the rope, what type of interference occurs when the waves meet? With that kind of interference, what happens to the displacement of the rope when the waves meet?

Solve

Both initial waves are generated on the left side. This is analogous to two orbitals having the same phase. The waves propagate towards each other and when the centers of the two waves meet, the result is a new wave that displaces the rope twice as much as each of the individual waves. This is **constructive interference.** Constructive interference results from addition of the two waves.

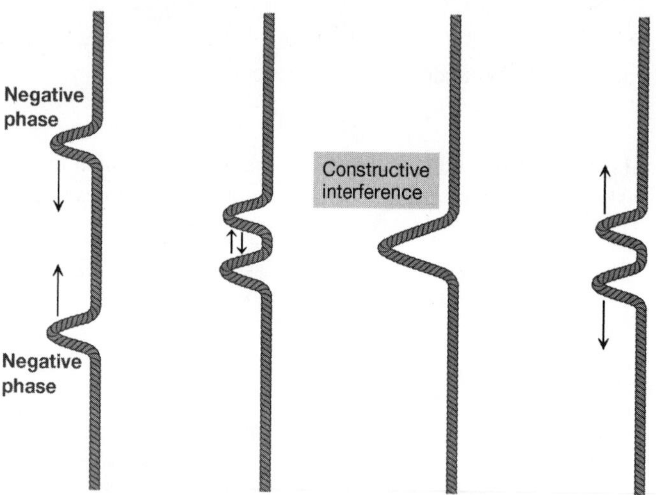

Your Turn 3.4

Think

If the waves are generated on opposite sides of the rope, what type of interference occurs when the waves meet? With that kind of interference, what happens to the displacement of the rope when the waves meet?

Solve

The initial waves are generated on opposite sides of the rope with equal amplitudes. When the two waves meet in the middle, they cancel and yield zero displacement along the entire rope. This is **destructive interference.** Destructive interference results from subtraction of the two waves.

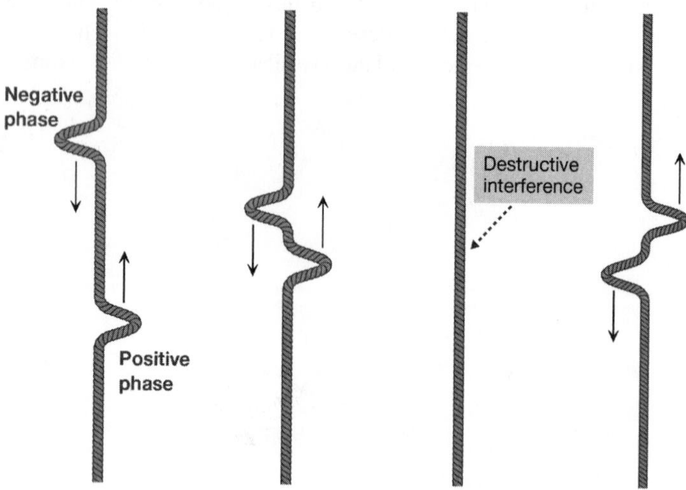

Your Turn 3.5

Think

Solved Problem 3.3 shows the molecular orbital (MO) energy diagram for H_2 that results from promoting one electron from the σ MO to the σ* MO. In that MO energy diagram, which MOs are occupied? Of those, which is the highest in energy?

Solve

When one electron from the σ MO is promoted to the σ* MO, the σ* MO now has the highest energy electron. Therefore, the highest occupied MO (HOMO) is the σ* MO.

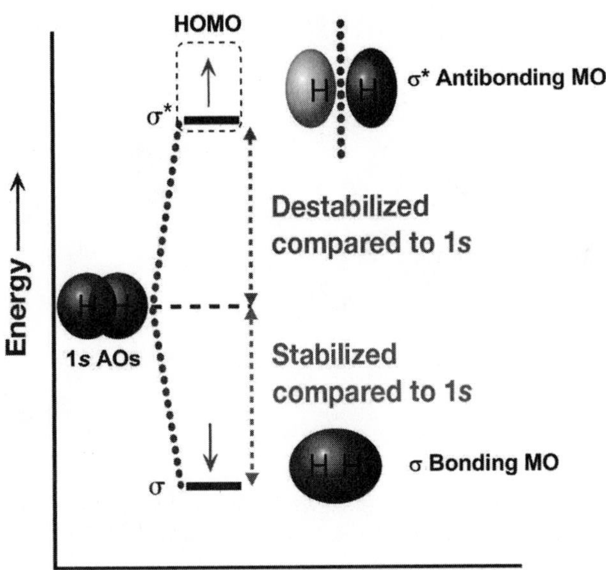

Your Turn 3.6

Think

What does the shading represent in orbital pictures? What type of interference results from the overlap of two shaded regions or two unshaded regions? What type of interference results from the overlap of an unshaded region and a shaded region?

Solve

Regions of constructive interference are those in which the lobes from the overlapping orbitals are the same phase—in this case, both lobes are shaded. Regions of destructive interference are those in which the lobes from overlapping orbitals are opposite in phase—that is, one is shaded and the other is unshaded.

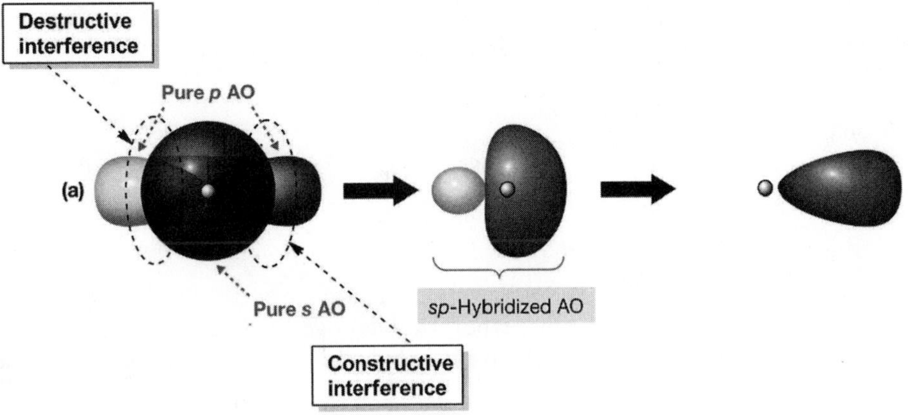

Your Turn 3.7

Think

What does the shading represent in orbital pictures? What type of interference results from the overlap of two shaded regions or two unshaded regions? What type of interference results from the overlap of an unshaded region and a shaded region?

Solve

Regions of constructive interference are those in which the lobes from the overlapping orbitals are the same phase—in this case, both lobes are shaded. Regions of destructive interference are those in which the lobes from overlapping orbitals are opposite in phase—that is, one is shaded and the other is unshaded.

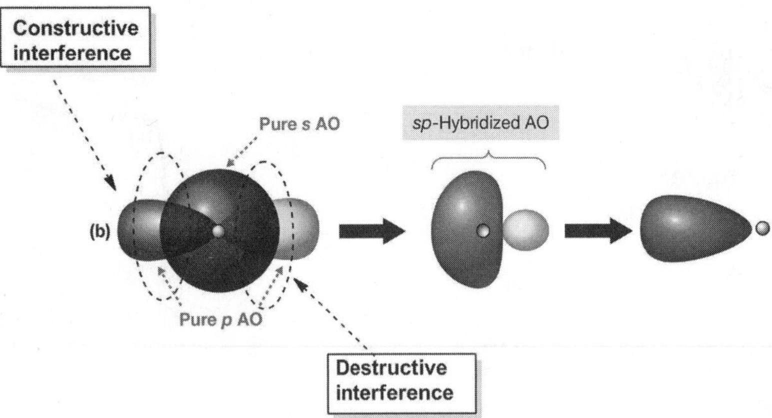

Your Turn 3.8

Think

Use Figure 3-11a and 3-11b as a guide. How many hybridized orbitals are present? How many unhybridized orbitals remain? In which plane do the hybrid orbitals lie?

Solve

There are four orbitals shown: three sp^2 hybridized orbitals (from the s, p_y, and p_z orbitals), and one unhybridized p_x orbital. The three sp^2 orbitals point 120° apart and are all in the yz plane. The unhybridized p_x orbital is perpendicular to the plane that contains the hybridized orbitals.

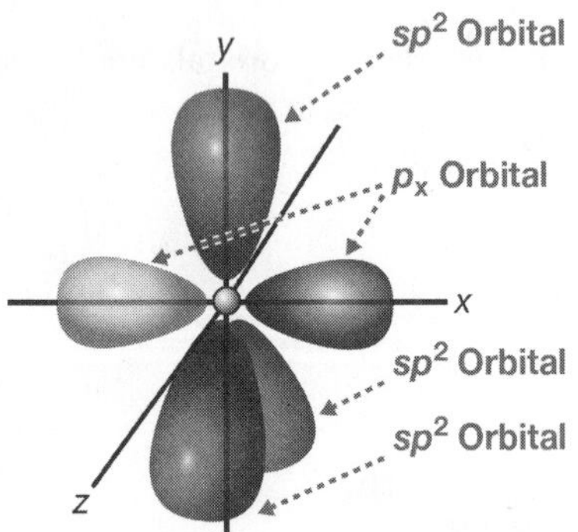

Your Turn 3.9

Think

What type of orbital overlap results in a π MO? What type of orbital overlap results in a π^* MO? If the molecular plane of ethene is perpendicular to the molecular plane in Figure 3-19, what is the effect on the nodal plane perpendicular to the bonding axis?

Solve

The first orbital is the π^* MO. Two *p* orbitals pointing in and out of the plane of the page overlap side by side with opposite phases, leading to destructive interference. The new orbital consists of four separate lobes—two in front of the molecular plane and two behind. The nodal plane that is perpendicular to the bonding axis points out of the plane of the paper and splits the C–C bond. The second orbital is the π MO. The lobes of the two *p* orbitals overlap with the same phase in front and behind the plane of the molecule. The new orbital consists of two lobes—one in front of and one behind the molecular plane.

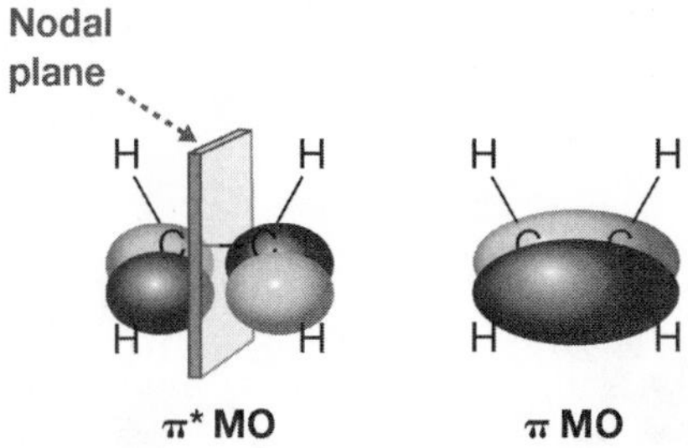

Your Turn 3.10

Think

Which orbitals contain electrons? Of those orbitals, which is the highest in energy (HOMO)? Which unfilled orbital is lowest in energy (lowest unoccupied MO [LUMO])?

Solve

The electrons in Figure 3-19 reside in the σ bonding orbitals and the π bonding orbital. Of those, the π bonding orbital is the highest in energy, so it is the HOMO. The LUMO is the π^* antibonding orbital.

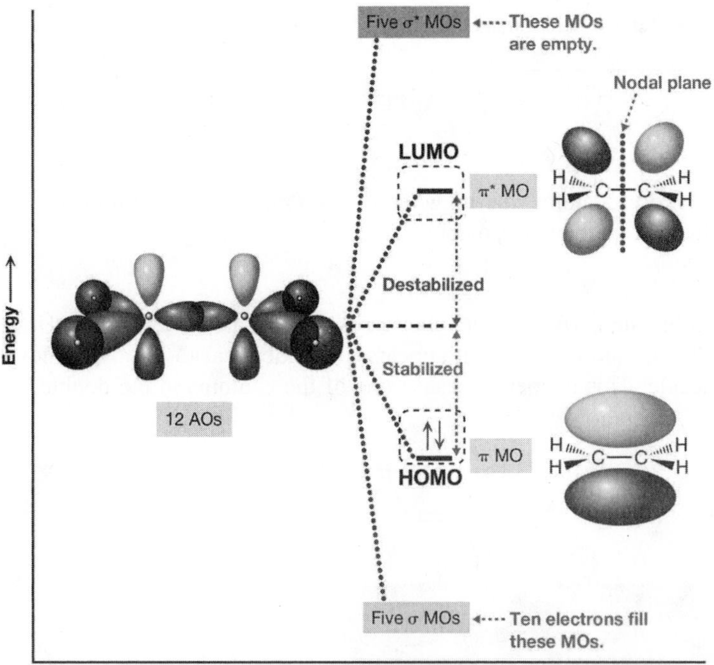

Your Turn 3.11

Think

Which orbitals contain electrons? Of those orbitals, which is the highest in energy (HOMO)? Which unfilled orbital is lowest in energy (LUMO)?

Solve

The electrons in Figure 3-25 reside in the σ bonding orbitals and the two π bonding orbitals. Of those, one of the π bonding orbitals is the highest in energy, making it the HOMO. The LUMO is one of the π* antibonding orbitals.

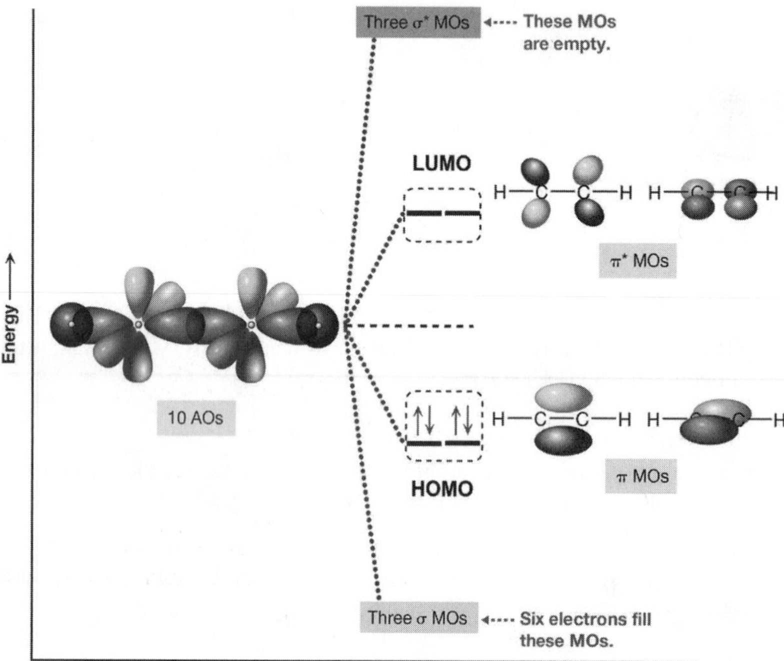

Your Turn 3.12

Think

In the orientations given for the molecules, do all atoms line up when the two molecules are brought together? Can you reorient one molecule so that all atoms in the two molecules line up?

Solve

In the orientations given, all atoms do not line up perfectly when the two molecules are held together. However, if the molecule on the left is flipped 180°, as shown below, it appears to be identical to the molecule on the right, so they are, indeed, the same molecule. This happens because one of the C atoms in the double bond is bonded to two H atoms.

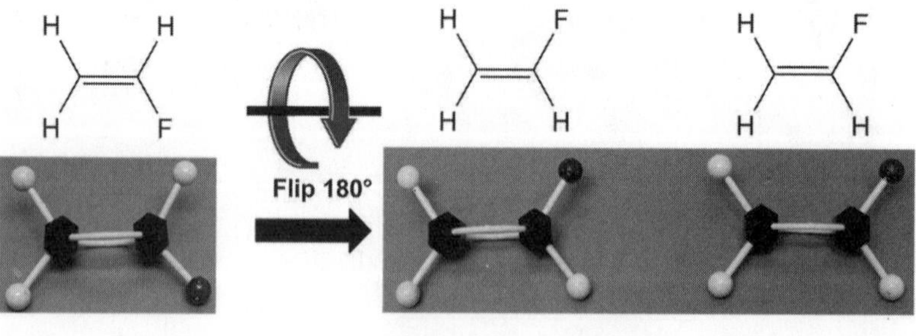

Your Turn 3.13

Think

A model kit is really helpful in understanding this problem. What is the hybridization of the second and third carbons in the first structure? What is the hybridization of the second, third, and fourth carbons in the second structure? Draw the orbitals that form the π bonds. What do you notice about the orientation of alternating π bonds? Look at Figure 3-32 for assistance in setting up your structure.

Solve

The internal carbons are *sp* hybridized. This means that there are two *p* orbitals that are unhybridized and are perpendicular. In a system like this, every π bond is perpendicular to the one next to it. The first molecule is planar. The second molecule is not; one CH_2 plane is perpendicular to the other.

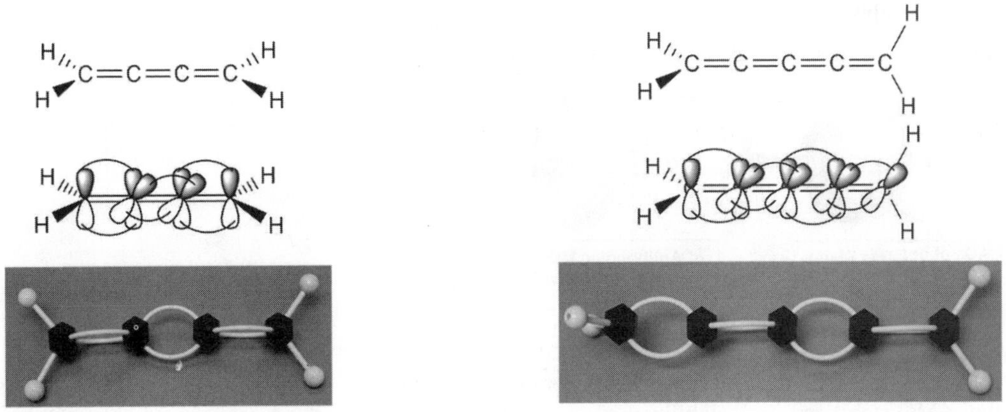

Your Turn 3.14

Think

How many orbitals were hybridized to form the sp^3 orbital? How many of those orbitals were *s* orbitals? What fraction does that correspond to? How many orbitals were hybridized to form the sp^2 orbital? How many of those were *s* orbitals? How many orbitals were hybridized to form the *sp* orbital? How many of those were *s* orbitals?

Solve

Four orbitals (*p*, *p*, *p*, and *s*) were hybridized to form the sp^3 orbital. Only one of those was an *s* orbital, giving a fraction of one-fourth, or 25%, *s* character. Three orbitals (*p*, *p*, and *s*) were hybridized to form the sp^2 orbital, only one of which was an *s* orbital. Thus the *s* character is one-third, or 33.33%. Two orbitals (*p* and *s*) were hybridized to form the *sp* orbital. Thus the *s* character is one-half, or 50%.

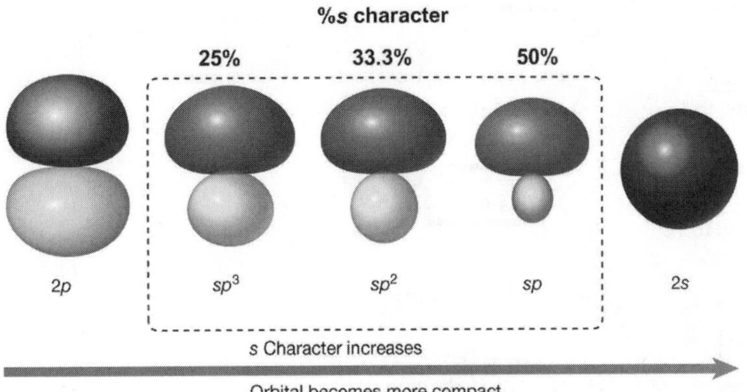

Chapter Problems

Problem 3.2

Think

In the region in which the *s* orbitals overlap, are the phases the same or opposite? Will that lead to constructive interference or destructive interference? How does the resulting orbital shape compare to the one at the right of Figure 3-7b?

Solve

One orbital is shaded and the other orbital is unshaded, indicating that they have opposite phases. Therefore, destructive interference will take place, and the result is a MO that has been diminished in the internuclear region and possesses a nodal plane between the two nuclei. This MO is designated by a sigma star, σ*, and indicates an antibonding MO. This MO differs from the one at the right of Figure 3-7b only by the phases of the orbitals, so it is not unique.

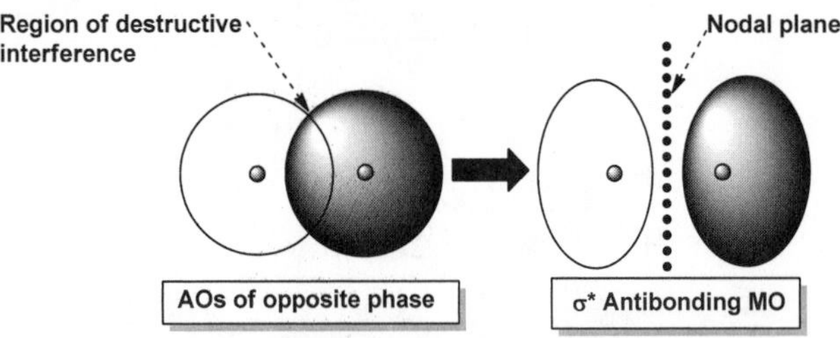

Problem 3.4

Think

How many electrons remain after one electron is removed? In which MO will that electron reside? Is an electron in the σ MO stabilized or destabilized compared to an electron in the 1*s* atomic orbital (AO)? Is the HOMO filled or unfilled?

Solve

The electron will be removed from the σ MO, leaving one electron in the lower-energy σ MO. Compared to the electron in the 1*s* H-atom AO, the one electron is stabilized in the σ MO for H_2^+. Because only stabilized orbitals have electrons, it should be more stable than the isolated H atom and H^+ ion. However, H_2^+ is not as stable as the H_2 atom, because H_2 has one more electron stabilized in the σ MO.

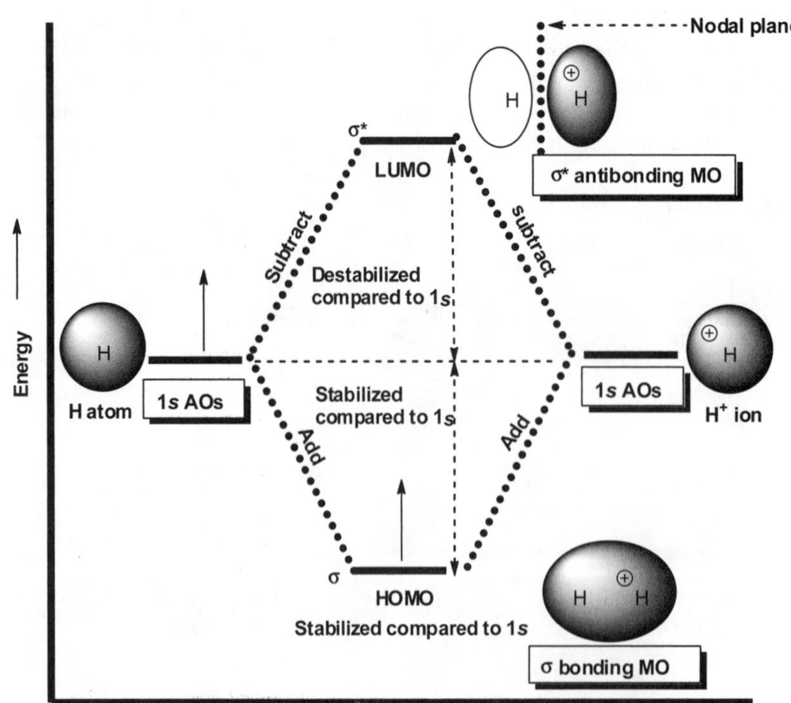

Problem 3.5

Think

Which regions of the orbitals overlap in the same phase? What type of interference results? Which regions of the orbitals overlap in opposite phases? What type of interference results? How does the shape of the resulting orbital compare to the one in Figure 3-10? What is different about the two orbitals?

Solve

The unshaded *s* orbital overlaps in the same phase as the unshaded *p* orbital lobe (right) and results in constructive interference. The unshaded *s* orbital overlaps in the opposite phase of the shaded *p* orbital lobe (left) and results in destructive interference. This is the same as Figure 3-10a but the phases are inverted.

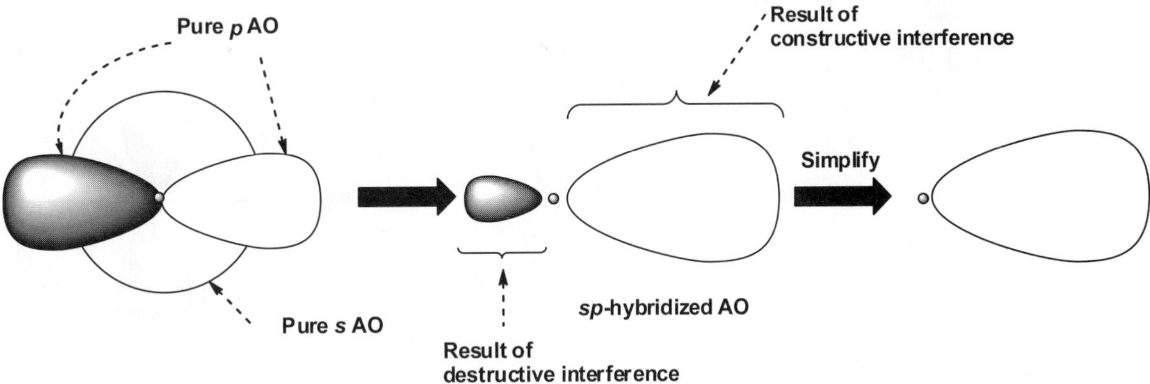

Problem 3.6

Think

Which regions of the orbitals overlap in the same phase? What type of interference results? Which regions of the orbitals overlap in opposite phases? What type of interference results? How does the shape of the resulting orbital compare to the one in Figure 3-10b? What is different about the two orbitals?

Solve

The unshaded *s* orbital overlaps in the same phase as the unshaded *p* orbital lobe (left) and results in constructive interference. The unshaded *s* orbital overlaps in the opposite phase of the shaded *p* orbital lobe (right) and results in destructive interference. This is the same as Figure 3-10b but the phases are inverted.

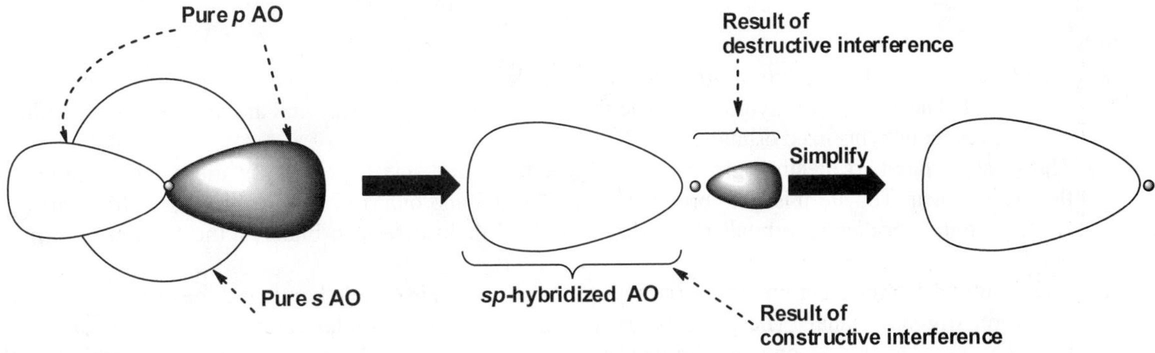

Problem 3.8

Think

Along which axis are these *sp*-hybridized orbitals aligned? What orbitals, therefore, must interact? What is the phase of the large lobe of each hybridized orbital?

Solve

The *sp*-hybridized orbitals are aligned along the *y* axis. Therefore, it must be the p_y orbital that has been used for hybridization, leaving the p_x and the p_z orbitals unhybridized. The large lobe is shaded in both hybrid orbitals, so the *s* and p_y orbitals must both be shaded where the overlap occurs. The respective interactions are as follows:

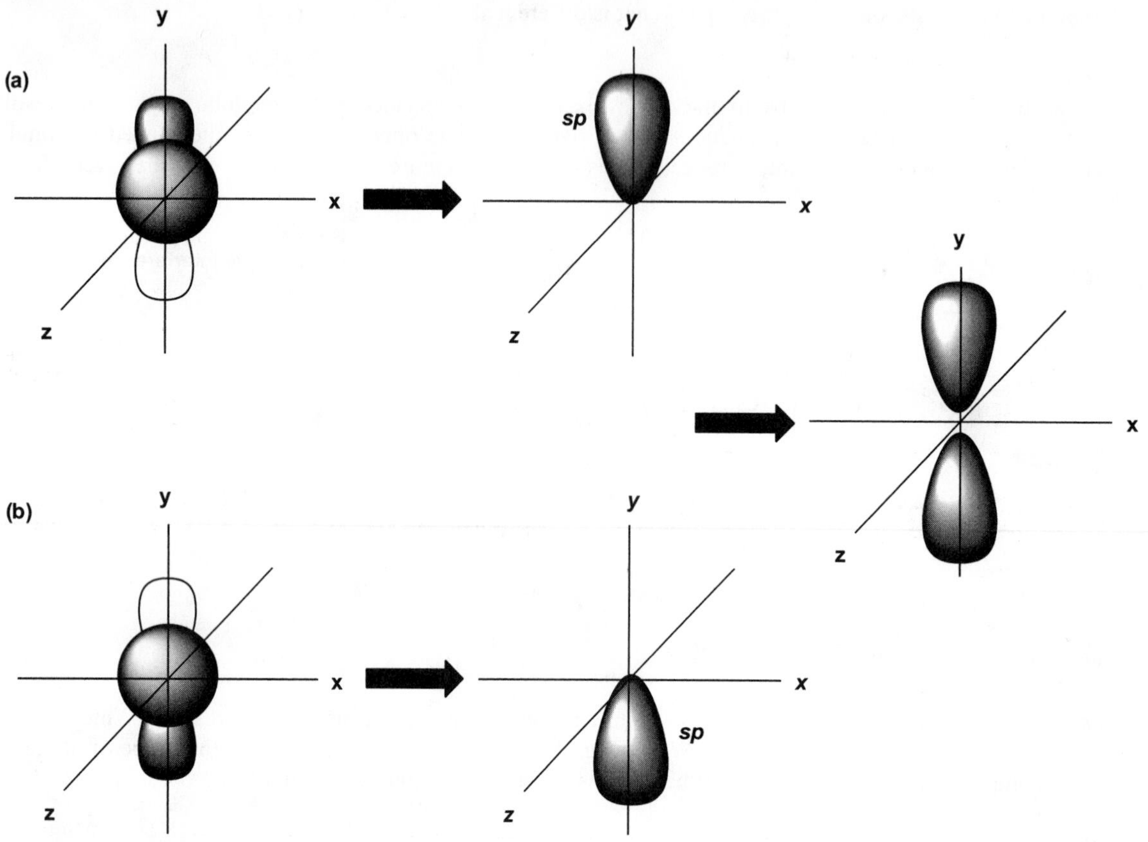

Problem 3.10

Think

What is the electron geometry about each atom indicated? What is the relationship between electron geometry and hybridization?

Solve

(a) The C atom has four electron groups (four single bonds) and the electron geometry is, therefore, tetrahedral. The atom is sp^3 hybridized. The three *p* orbitals and one *s* orbital are all used for hybridization. This leaves no unhybridized orbitals.

(b) The C atom has two electron groups (one single bond and one triple bond); the electron geometry is, therefore, linear. The atom is *sp* hybridized. One *p* orbital and one *s* orbital are used for hybridization. This leaves two unhybridized *p* orbitals to overlap with the unhybridized *p* orbitals on the N atom to form two π bonds.

(c) The C atom has three electron groups (two single bonds and one double bond) and the electron geometry is, therefore, trigonal planar. The atom is sp^2 hybridized. Two *p* orbitals and one *s* orbital are used for hybridization. This leaves one unhybridized *p* orbital to overlap with with unhybridized *p* orbital on the O atom to form one π bond.

(d) The O atom has three electron groups (one double bond and two lone pairs). The atom is sp^2 hybridized. Two *p* orbitals and one *s* orbital are used for hybridization. This leaves one unhybridized *p* orbital to overlap with the unhybridized *p* orbital on the C atom to form one π bond.

(e) The C atom has four electron groups (four single bonds) and the electron geometry is, therefore, tetrahedral. The atom is sp^3 hybridized. The three *p* orbitals and one *s* orbital are all used for hybridization. This leaves no unhybridized orbitals.

(a) (b) (c) (d) (e)

Problem 3.11

Think

How many electron groups are around an *sp*-hybridized atom? An sp^2-hybridized atom? With the formula C_5H_8 that is given, can any of those electron groups be lone pairs? What types of bonds could those be for a carbon atom to have *sp* hybridization? sp^2 Hybridization?

Solve

(a) Two electron groups must surround an *sp*-hybridized atom. For uncharged carbon atoms, those groups can be two double bonds or one single bond and one triple bond. Two examples are shown below. Other molecules are possible.

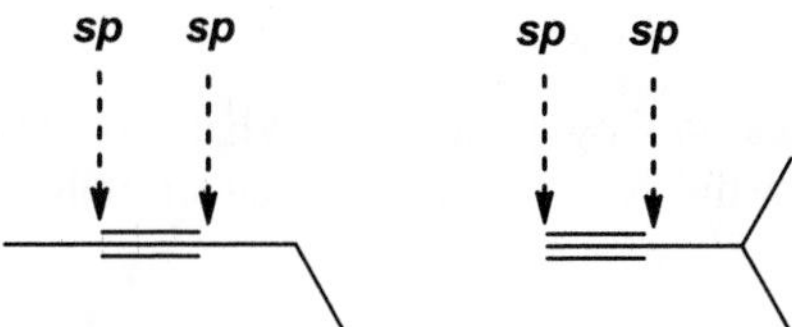

(b) Three electron groups must surround an sp^2-hybridized carbon. For an uncharged carbon atom, those groups can be a double bond and two single bonds. Two examples are shown below. Other molecules are possible.

Problem 3.13

Think

What is the hybridization of N? What valence shell orbitals does it contribute? What orbitals do the H atoms contribute? How many total orbitals should be produced from AO mixing?

Solve

(a) The N atom is sp^3 hybridized, so it contributes four sp^3-hybridized orbitals in the valence shell. Each H atom contributes a 1*s* orbital. The overlap of AOs appears on the left in the diagram on the next page.

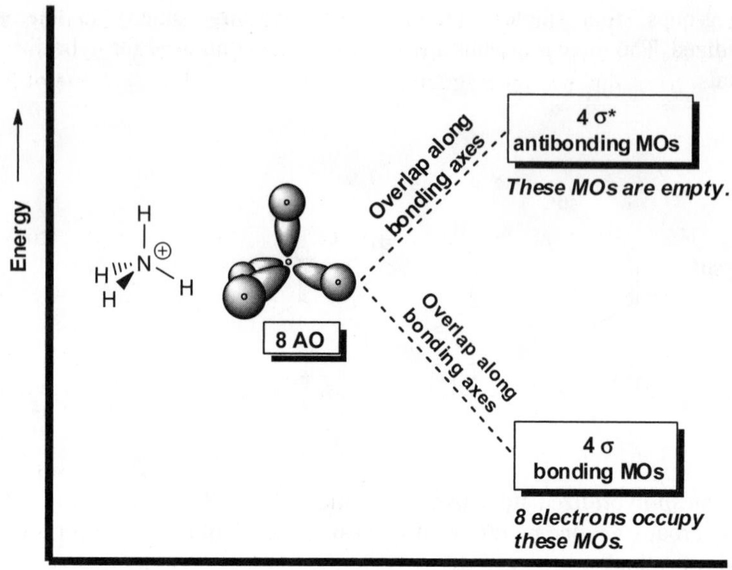

(b) The energies of the MOs appear on the right. The eight AOs that overlap do so along the bonding axes, generating four σ and four σ* MOs. Eight valence electrons completely fill the σ MOs, leaving the σ* empty. This is the same orbital picture shown in Solved Problem 3.12, except that the C atom is replaced with an N^+.

Problem 3.14

Think

What is the hybridization of N and C? What valence shell orbitals do they contribute? What orbitals do the H atoms contribute? What type of bond exists between the C and the N? How many total orbitals should be produced from AO mixing?

Solve

(a) The N atom and the C atom are both sp^3 hybridized and each contributes four sp^3-hybridized orbitals in the valence shell. Each H atom contributes a 1*s* orbital. The overlap of AOs appears on the left in the following diagram:

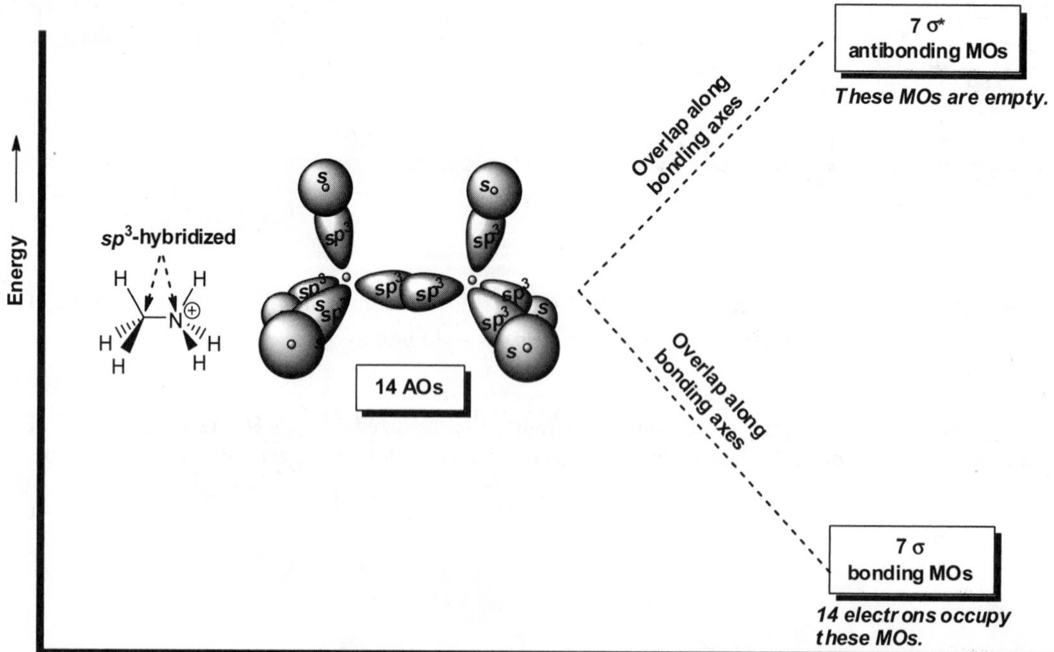

(b) The energies of the MOs appear on the right. The 14 AOs that overlap do so along the bonding axes, generating seven σ and seven σ* MOs. Fourteen valence electrons completely fill the σ MOs, leaving the σ* empty. This is nearly identical to the orbital energy diagram shown in Figure 3-15, except one of the C atoms is replaced by an N^+.

Problem 3.16

Think

How many double bonds are present? What is each double bond composed of? How many single bonds are present? What is each single bond composed of?

Solve

(a) There is one C=C double bond; it is composed of one σ bond and one π bond. Additionally, there are three C–C single bonds and 10 C–H single bonds. Each single bond is a σ bond. This gives a total of one π bond and 14 σ bonds.

(b) There are three C=C double bonds and each is composed of one σ bond, and one π bond for a total of three σ bonds and three π bonds. Additionally, there are two C–C single bonds and eight C–H single bonds. Each single bond is a σ bond. This gives a total of three π bonds and 13 σ bonds.

(c) There are three C=C double bonds; each is composed of one σ bond and one π bond, for a total of three σ bonds and three π bonds. Additionally, there are eight C–C single bonds and 12 C–H single bonds. Each single bond is a σ bond. This gives a total of three π bonds and 23 σ bonds.

(d) There is one N=C double bond; it is composed of one σ bond and one π bond. Additionally, there are four C–C single bonds, one N–C single bond, one N–H single bond, and 13 C–H single bonds. Each single bond is a σ bond. This gives a total of one π bond and 20 σ bonds.

(a) 1 π, 14 σ **(b)** 3 π, 13 σ **(c)** 3 π, 23 σ **(d)** 1 π, 20 σ

Problem 3.17

Think

What is the hybridization of N and C? What valence shell orbitals does each contribute? What orbitals do the H atoms contribute? What type of bond exists between the C and the N? What types of orbitals are associated with lone pairs of electrons?

Solve

The C and N atoms are sp^2 hybridized, just as the C and O atoms are in Figure 3-22. The picture, therefore, is nearly identical to Figure 3-22. The only differences arise from (1) a N atom in place of an O atom, and (2) an N–H bond in place of a lone pair on O. Consequently, we should expect one nonbonding MO instead of the two that appear in Figure 3-22.

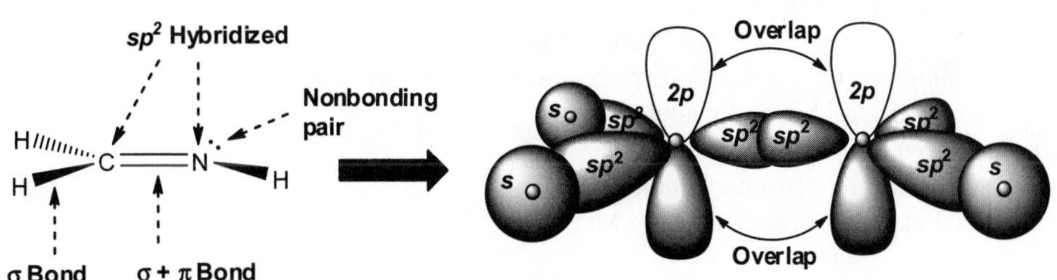

Problem 3.18

Think

How many double bonds are present? What is each double bond composed of? How many single bonds are present? What is each single bond composed of? How many lone pairs are present? What types of orbitals are associated with lone pairs of electrons?

Solve

Each double bond represents one σ bond and one π bond. Each single bond (including those to H) represents one σ bond. Each lone pair is associated with a nonbonding MO.

(a) 2 π bonds, 16 σ bonds, 4 electrons (2 pairs) in nonbonding MOs.
(b) 2 π bonds, 16 σ bonds, 8 electrons (4 pairs) in nonbonding MOs.
(c) 2 π bonds, 12 σ bonds, 4 electrons (2 pairs) in nonbonding MOs.
(d) 1 π bond, 19 σ bonds, 2 electrons (1 pair) in nonbonding MOs.
(e) 2 π bonds, 11 σ bonds, 8 electrons (4 pairs) in nonbonding MOs.

Problem 3.19

Think

What is the hybridization of N and C? What valence shell orbitals does each contribute? What orbitals do the H atoms contribute? What type of bond exists between the C and the N?

Solve

The hybridization of N and C are both *sp*, the same as the two C atoms in Figure 3-24. The orbital picture, therefore, is identical to that in Figure 3-24, except for (1) an N atom in place of the C atom and (2) a lone pair on N in place of a C–H bond. Without overlap between the N and H atoms, there is one fewer σ bond and one fewer σ* MO.

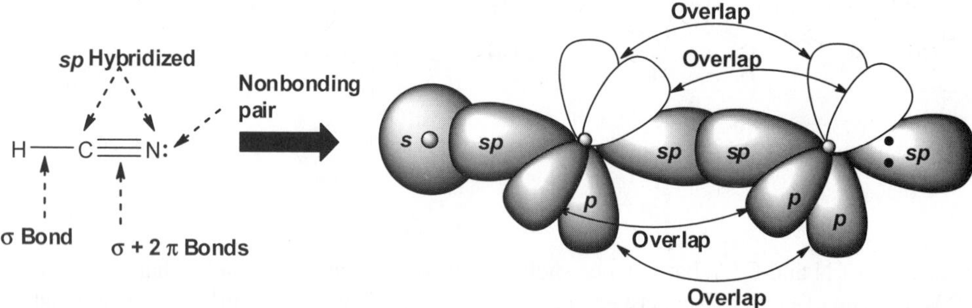

Problem 3.20

Think

How many triple bonds are present? What is each triple bond composed of? How many double bonds are present? What is each double bond composed of? How many single bonds are present? What is each single bond composed of? How many lone pairs are present? What types of orbitals are associated with a lone pair of electrons?

Solve

Each triple bond represents two π bonds and one σ bond. Each double bond represents one π bond and one σ bond. Each single bond represents one σ bond. Lone pairs of electrons are associated with nonbonding MOs.

(a) 2 π bonds, 20 σ bonds, 0 electrons in nonbonding MOs.
(b) 2 π bonds, 15 σ bonds, 6 electrons in nonbonding MOs.
(c) 2 π bonds, 11 σ bonds, 2 electrons in a nonbonding MO.
(d) 4 π bonds, 7 σ bonds, 2 electrons in a nonbonding MO.

Problem 3.21

Think

What is the hybridization about each nonhydrogen atom? What is the molecular geometry about each nonhydrogen atom? Which geometries are planar? Are the planes of any adjacent atoms required to be parallel?

Solve

The atoms in the same plane (circled) are all connected to the sp^2-hybridized C atom or the sp^2-hybridized N atom. Atoms that are sp^2-hybridized have trigonal planar or bent molecular geometries and are always planar. When two sp^2-hybridized atoms are connected by a double bond, the π bond locks the electron groups on those atoms in the same plane. Therefore, atoms that are doubly bonded together and any atoms to which they are directly bonded must lie in the same plane.

(a) (b) (c)

Problem 3.22

Think

Are any ions present? Are polar bonds present? Are there any H-bond donors or acceptors? Do the bond dipoles completely cancel when they are summed?

Solve

No ions are present, so no ion–ion interactions are possible. There are no H-bond donors or acceptors, so no H bonding is present. Polar C–Cl bonds are present in both compounds. In *cis*-1,2-dichloroethane, those bond dipoles sum to give a net molecular dipole moment, making the molecule polar. A pair of molecules of *cis*-1,2-dichloroethane, therefore, has dipole–dipole interactions in addition to induced dipole–induced dipole interactions. In *trans*-1,2-dichloroethane, on the other hand, the bond dipoles cancel perfectly, making the molecule nonpolar. A pair of molecules of *trans*-1,2-dichloroethane, therefore, only has induced dipole–induced dipole interactions. The presence of dipole–dipole interactions gives the cis form a higher boiling point (b.p.) than the trans form.

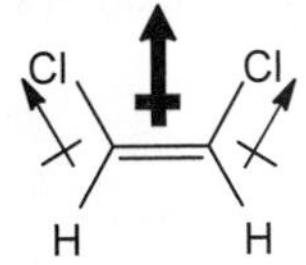

cis-1,2-Dichloroethene
b.p. = 60.3 °C

trans-1,2-Dichloroethene
b.p. = 47.5 °C

Problem 3.24

Think

Does a hypothetical rotation of 180° about the double bond give rise to a different molecule? How can you tell?

Solve

Yes. The easiest way to see this is to build a model with two C atoms arranged trans around the C=C bond:

trans-Cyclooctene

Then, using a bridge of four more C atoms, *make a connection behind* the double bond. The three-dimensional molecule forms a figure "8." This is different from the cis configuration, as shown on the next page.

Problem 3.25

Think

What is the structure of α-linolenic acid? What makes a double bond cis or trans? How can you arrange each of these double bonds to be in the trans configuration? How many double bonds are there and how many configurations are possible for each double bond?

Solve

The all-trans form is shown below. In each trans double bond, the non-H groups are attached to opposite sides of the C=C double bond. In all, because there are two possible configurations for each double bond (cis and trans), there are 2^3 (or 8) total possibilities.

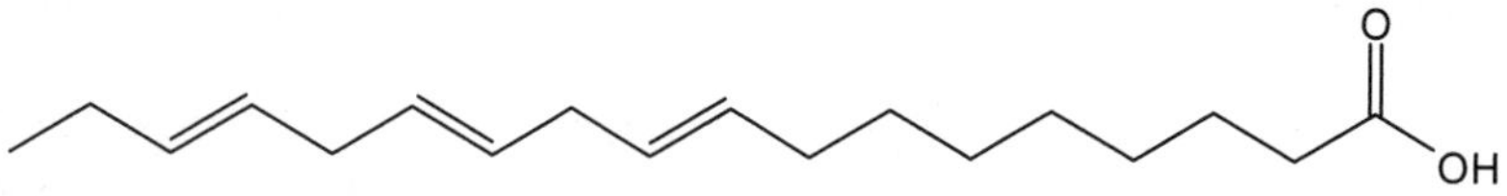

Problem 3.27

Think

What is the hybridization of each atom involved in the C–C bond? How does that affect the length and strength of bonds involving the atom?

Solve

The molecule on the left has the shorter C–C single bond. In the molecule on the left, the C–C bond is between an sp^3-hybridized C atom and an *sp*-hybridized C atom, whereas in the second molecule it is between an sp^3-hybridized C atom and an sp^2-hybridized C atom. With greater *s* character in the *sp*-hybridized C than in sp^2, the bond becomes shorter and stronger.

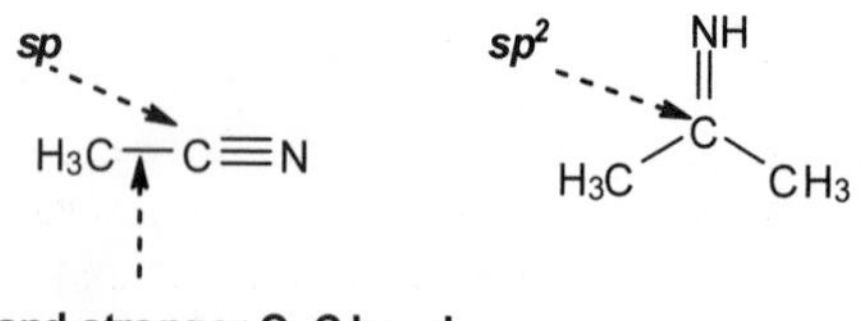

Problem 3.28

Think

How many electron groups are on each atom? What is the electron geometry about each atom indicated? What is the relationship between electron geometry and hybridization?

Solve

The number of electron groups is directly related to the hybridization: tetrahedral (4) = sp^3, trigonal planar (3) = sp^2, linear (2) = sp. The electron geometry (not the molecular geometry) and the hybridization of all nonhydrogen atoms are listed below.

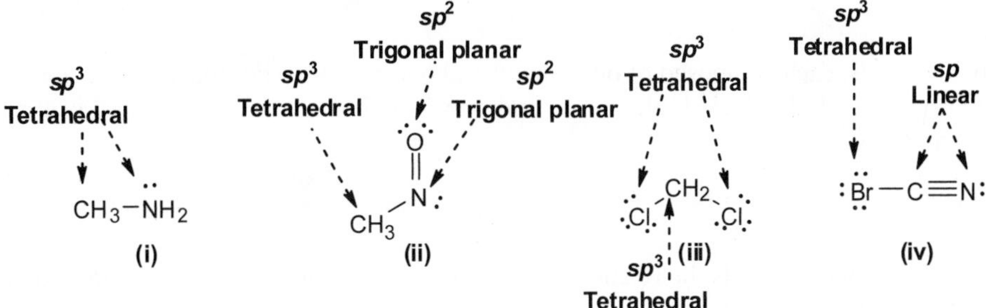

Problem 3.29

Think

How many electron groups are on each atom? What is the electron geometry about each atom indicated? What is the relationship between electron geometry and hybridization?

Solve

The number of electron groups is directly related to the hybridization: tetrahedral (4) = sp^3, trigonal planar (3) = sp^2, linear (2) = sp. The electron geometry (not the molecular geometry) and the hybridization of all nonhydrogen atoms are listed below.

Problem 3.30

Think

How many electron groups are around each nonhydrogen atom? How does the number of groups correspond to electron geometry? What is the relationship between electron geometry and hybridization? Depending on the number of lone pairs around each atom, how does electron geometry relate to molecular geometry?

Solve

The number of electron groups is directly related to the hybridization: tetrahedral (4) = sp^3, trigonal planar (3) = sp^2, linear (2) = sp. The electron geometry, the molecular geometry, and the hybridization of the sp^2-hybridized C atoms and the sp^3-hybridized O atom are listed below. All other C atoms (circled) are sp^3 hybridized and have tetrahedral molecular and electron geometries. The electron geometry and the molecular geometry are identical if all electron groups are bonds. If at least one electron group is a lone pair, the molecular and electron geometries are different.

Problem 3.31

Think

How many double bonds are present? What is each double bond composed of? How many single bonds are present? What is each single bond composed of?

Solve

There are two C=C double bonds, each composed of one σ bond and one π bond. Additionally there are 13 C–C single bonds, one C–O single bond, 25 C–H single bonds, and one O–H bond. Each single bond is a σ bond. This gives a total of two π bonds and 42 σ bonds.

Problem 3.32

Think

Where are the double bonds located? What is the relationship between the H atoms on each C atom of the C=C: Are they on the same side or different sides?

Solve

If the H atoms are on the same side of the double bond, the configuration is cis; if the H atoms are on opposite sides of the double bond, the configuration is trans. See below.

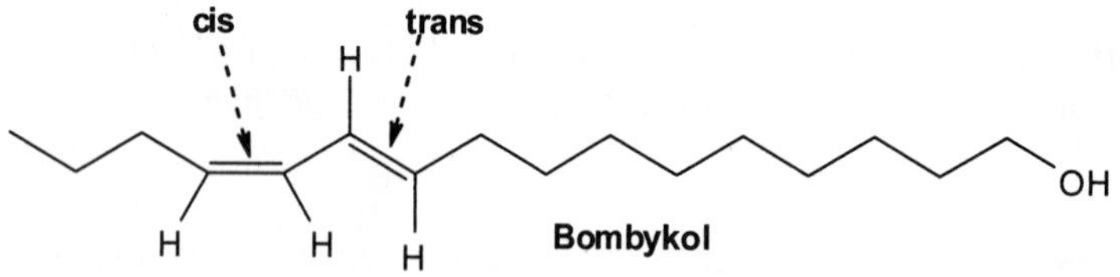

Bombykol

Problem 3.33

Think

How many electron groups are on each nonhydrogen atom? What is the electron geometry about each of those atoms? What is the relationship between electron geometry and hybridization? How many single/double/triple bonds are present? What is each bond composed of?

Solve

(a) All hybridizations are labeled on the figure below. The sp^3-hybridized C atoms each have four electron groups and a tetrahedral electron geometry, sp^2-hybridized C atoms each have three electron groups and a trigonal planar electron geometry, and *sp*-hybridized C atoms each have two electron groups and a linear electron geometry. The ROH oxygen is sp^3 hybridized as it has four electron groups and a tetrahedral electron geometry. The ketone O is sp^2 hybridized, as it has three electrons groups and a trigonal planar electron geometry.

Norethynodrel

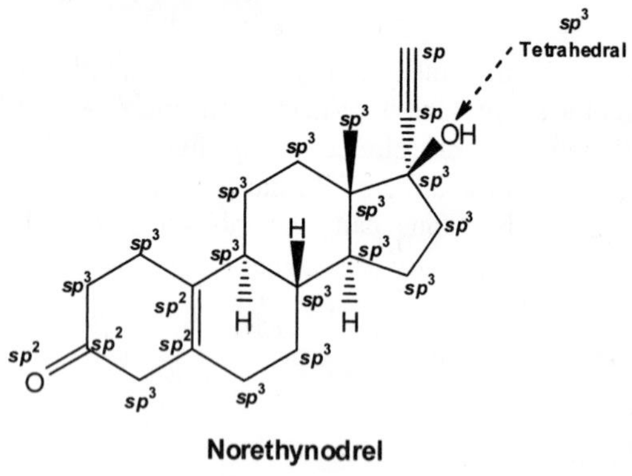

(b) There are 50 σ bonds and 4 π bonds.

Problem 3.34

Think

How many valence orbitals does each atom in the second row of the periodic table contribute? How many MOs can be produced by mixing together that number of AOs?

Solve

Each atom from the second row of the periodic table contributes four AOs: 2*s*, $2p_x$, $2p_y$, and $2p_z$. If there are three of these atoms, then 12 AOs would be contributed in all. Because the number of orbitals must be conserved, those 12 AOs would produce 12 MOs.

Problem 3.35

Think

What is the relationship between the number of AOs and the number of MOs? How many MOs would be produced in all? When two orbitals from different atoms interact, how many bonding MOs are produced? Antibonding MOs? Nonbonding MOs?

Solve

As explained in the solution to Problem 3.34, there are 12 MOs in all. Every time one AO mixes with another AO on a different atom, one bonding and one antibonding MO result. The problem describes four pairs of AOs interacting, which will result in four bonding MOs and four antibonding MOs. The remaining four MOs must be nonbonding.

Problem 3.36

Think

Draw the Lewis structure for each molecule. Does a hypothetical rotation of 180° about the double bond give rise to a different molecule? How can you tell?

Solve

Two unique configurations exist about the double bond in **(c)** ClHC=CHBr and **(d)** HC≡CCH=CHCl. In both cases, each doubly bonded C is itself bonded to two different atoms or groups. Molecule **(a)** has two identical CH_3 groups on one end of its double bond; molecule **(b)** has two identical H atoms on one end.

(c)

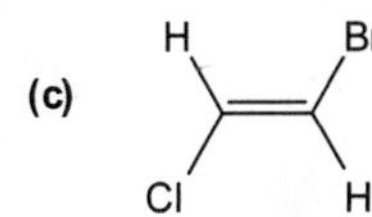

***trans*-1-Bromo-2-chloroethene**

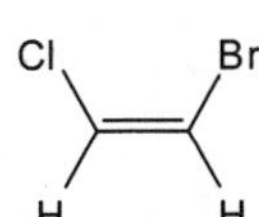

***cis*-1-Bromo-2-chloroethene**

(d)

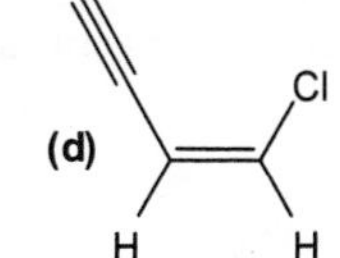

***cis*-1-Chlorobut-1-en-3-yne**

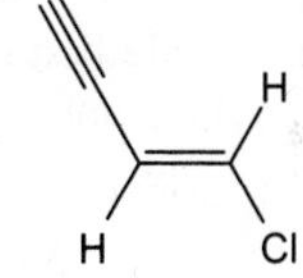

***trans*-1-Chlorobut-1-en-3-yne**

Problem 3.37

Think

How many double bonds are present? What kinds of bonds make up each double bond? How many π bonds does that correspond to? How many electrons are in each π bond?

Solve

There are two double bonds; each double bond consists of one σ bond and one π bond, for a total of two π bonds. There are two electrons in each π bonds for a total of four π electrons.

Problem 3.38

Think

How many double bonds are present? How many π bonds are in each double bond?

Solve

There are 11 double bonds present; each double bond consists of one σ bond and one π bond. All 11 π bonds are labeled on the next page.

β-Carotene

Problem 3.39

Think

What is the hybridization of each carbon atom? How can you tell what its hybridization is, based on the number of electron groups? What types of bonds are represented by single bonds? Double bonds? Triple bonds?

Solve

The hybridization of each carbon atom is listed below. Molecule **(b)** contains a single bond that forms from the overlap of an *sp*- and *sp*^{2}-hybridized orbital (boxed), which is also a σ bond.

(a) (b) (c) (d)

Problem 3.40

Think

What is the hybridization of each nonhydrogen atom? What is the molecular geometry about each nonhydrogen atom? What does a π bond require of the orientations of atoms connected by a double bond?

Solve

All atoms that are *sp*^{2} hybridized (i.e., have three electron groups) have a trigonal planar geometry. The planes of *sp*^{2}-hybridized atoms that are connected by a double bond must be the same. Because of the alternating single and double bonds in these molecules, several atoms must lie in the same plane. The atoms that are required to be in the plane are in highlighted.

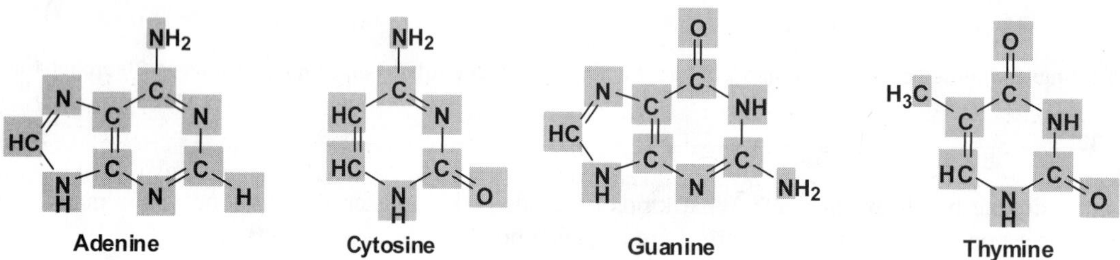

Problem 3.41

Think

What is the hybridization of C and O on the carbon monoxide molecule? How can you tell what its hybridization is, based on the number of electron groups? What AOs does each atom contribute to the molecule? Which orbitals overlap along bonding axes to form the σ bonds and which orbitals overlap on opposite sides of bonding axes to form the two π bonds? Are there any nonbonding electrons?

Solve

The C and O atoms each have two electron groups in the Lewis structure, and, therefore, are *sp* hybridized. Each atom contributes two *sp*-hybridized AOs and two unhybridized *p* AOs to the molecule. The orbital overlap picture is shown on the next page. Two *sp* AOs overlap between the C atoms, leading to one σ MO and one σ*

MO. There is overlap between adjacent p_y orbitals and between adjacent p_z orbitals. Each overlap involves two orbitals, giving rise to one π MO and one π* MO, or two MOs of π symmetry. In all, there are four MOs of π symmetry. The two *sp*-hybridized AOs on the outside of the molecule don't mix with other AOs, so they remain nonbonding. The 10 valence electrons are filled in the MOs from lowest to highest energy. The four nonbonding electrons represent the two lone pairs.

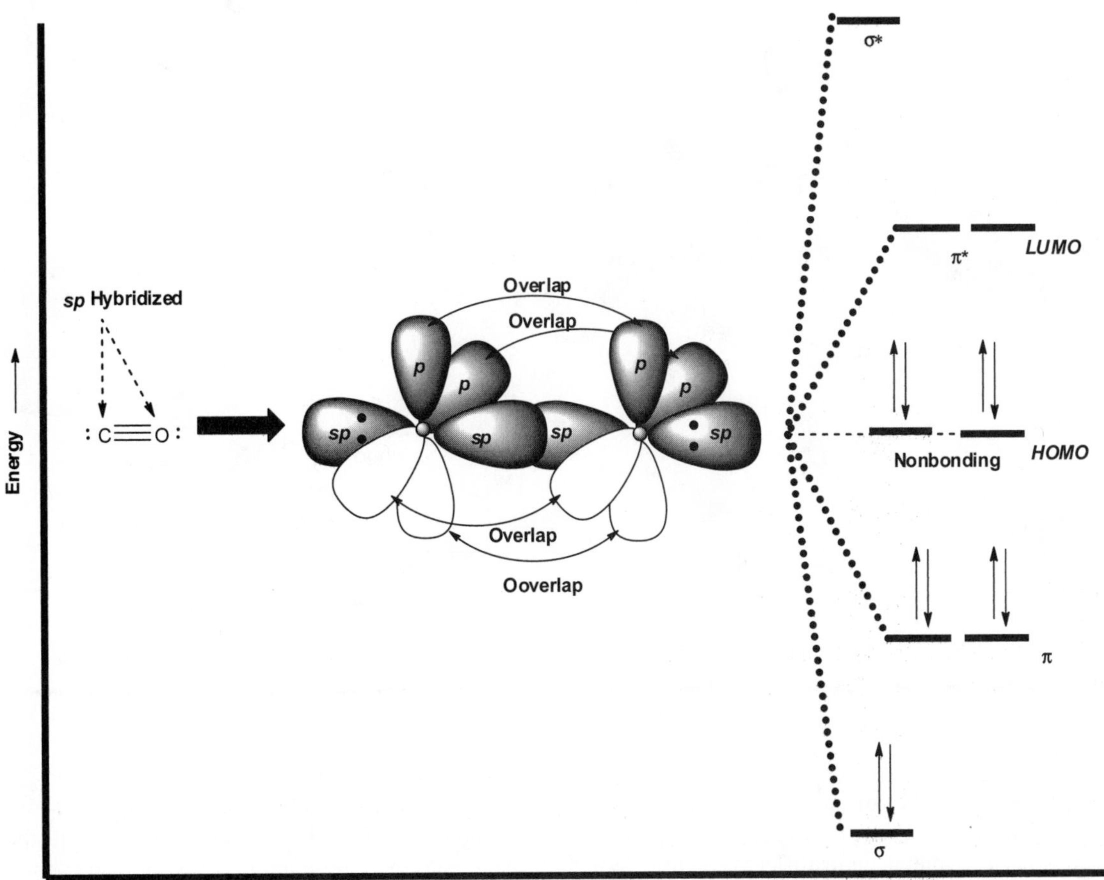

Problem 3.42

Think

Based on the number of electron groups in the Lewis structure, what is the hybridization of each C atom in propa-1,2-diene? What AOs does each atom contribute to the molecule? Based on the molecular model in Figure 3-32, do the CH_2 groups occupy the same plane or different planes? Which orbitals overlap along bonding axes to form σ bonds and which orbitals overlap on opposite sides of bonding axes to form the π bonds?

Solve

The two end C atoms each have three electron groups in the Lewis structure, and, therefore, are sp^2 hybridized. Each of these atoms contributes three sp^2-hybridized AOs and one unhybridized *p* AO to the molecule. The central C atom has two electron groups in the Lewis structure, making it *sp* hybridized. This atom contributes two *sp*-hybridized AOs and two unhybridized *p* AOs to the molecule. Figure 3-32 shows that the two CH_2 groups are in perpendicular planes. The orbital picture and the energy diagram are shown on the next page. Each terminal sp^2-hybridized AO overlaps with an *s* AO from an H atom to form one σ MO and one σ* MO. Each *sp*-hybridized AO overlaps with an sp^2-hybridized AO to form one σ MO and one σ* MO. Two pairs of *p* AOs overlap in perpendicular planes, each pair forming one π MO and one π* MO. The 16 valence electrons are filled in the MOs from lowest energy to highest. From the energy diagram on the next page, we can see that one of the π MOs is the HOMO and one of the π* MOs is the LUMO.

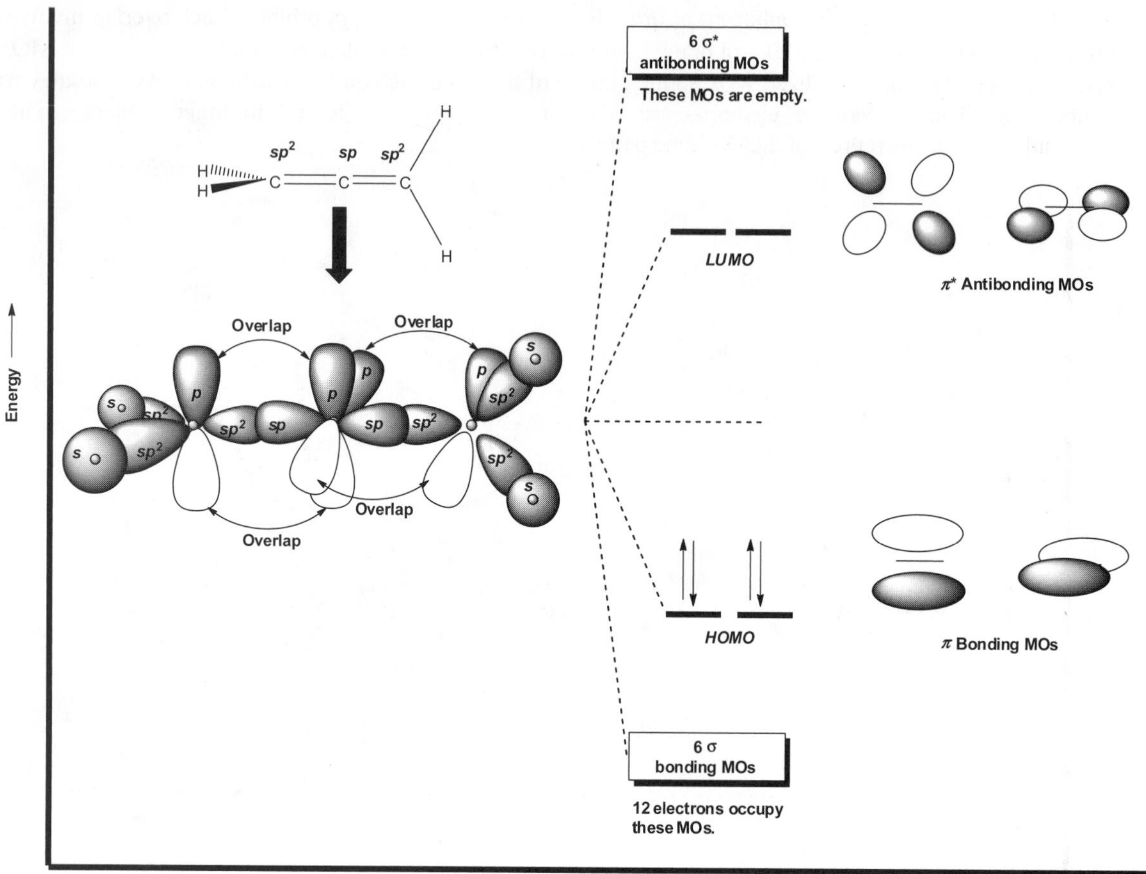

Problem 3.43

Think

Based on the number of electron groups in the Lewis structure, what is the hybridization of each C in buta-1,2,3-triene? What AOs does each atom contribute to the molecule? Based on Your Turn 3.13, are the CH_2 groups in the same plane or different planes? Which orbitals overlap along bonding axes to form σ bonds and which orbitals overlap on opposite sides of bonding axes to form π bonds?

Solve

The two end C atoms each have three electron groups in the Lewis structure and are sp^2 hybridized, so they each contribute three sp^2-hybridized AOs and one unhybridized *p* AO. The two central C atoms each have two electron groups and are *sp* hybridized, so they each contribute two *sp*-hybridized AOs and two unhybridized *p* AOs. According to Your Turn 3.13, the two CH_2 groups occupy the same plane. The orbital picture and the energy diagram are shown on the next page. Each overlap of an *s* AO with an sp^2 AO results in one σ MO and one σ* MO. The same is true for each overlap of two hybridized AOs. Each overlap of adjacent *p* AOs results in one π MO and one π* MO. After filling in the 20 valence electrons, we can see that one of the π MOs is the HOMO and one of the π* MOs is the LUMO.

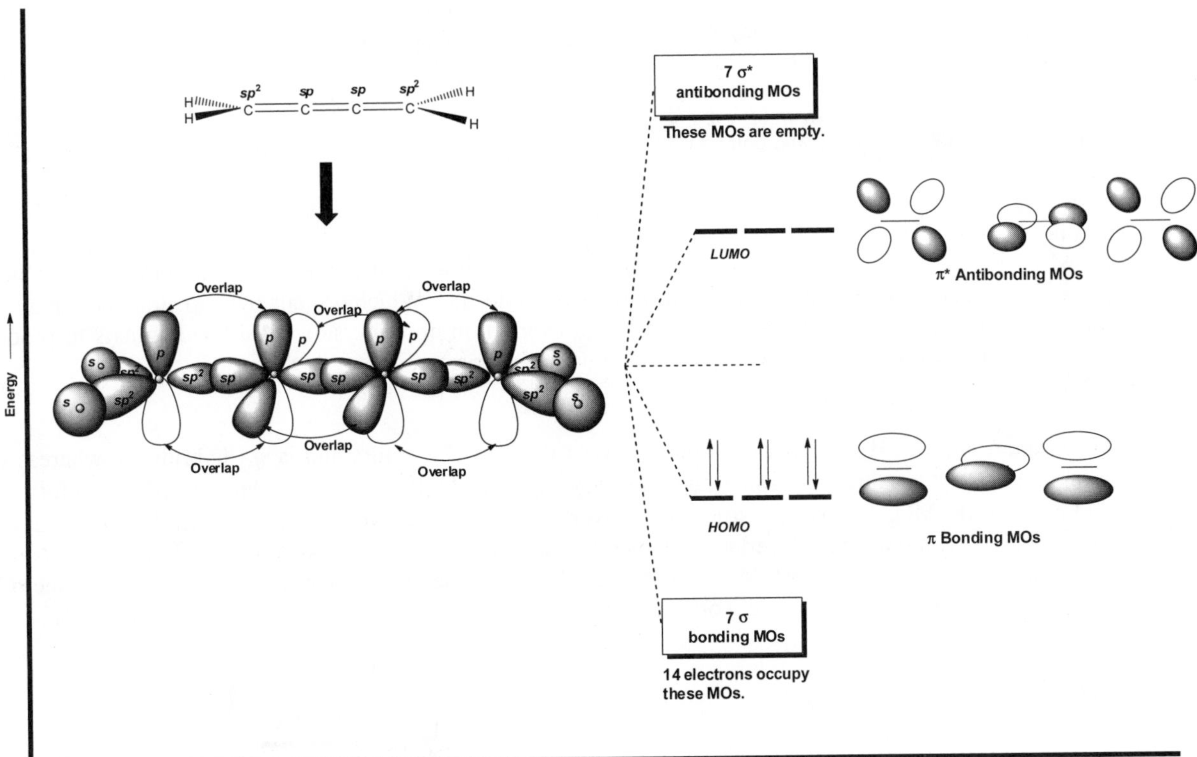

Problem 3.44

Think

What type of bond exists between carbons 2 and 3? Is free rotation permitted about this type of bond? What does this mean in regard to the atoms residing in the same plane?

Solve

A single bond exists between carbons 2 and 3, so the groups attached by that single bond are not locked into place. Single bonds can rotate; this means that the atoms around the single bond are not limited to being in the same plane.

Rotation about single bond

Problem 3.45

Think

If the sp^2 C atom has a greater effective electronegativity than an sp^3 C atom, what does this mean in regard to the bond dipoles along the C–C bonds? Do those bond dipoles cancel or do they result in an overall polar molecule? What is the strongest intermolecular interaction in a pair of each molecule?

Solve

The cis alkene is slightly polar because an sp^2 C atom has a greater effective electronegativity than an sp^3 C atom. This gives rise to bond dipoles (arrows not in boldface), as shown on the next page, and to the net dipole moment, indicated by the boldface arrow. For the trans alkene, these bond dipoles cancel out, making the molecule nonpolar. Thus, the cis alkene can participate in dipole–dipole interactions, whereas the trans alkene cannot.

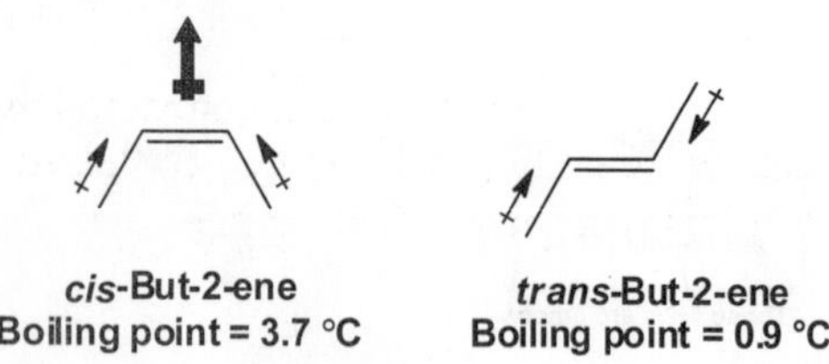

Problem 3.46

Think

Based on the number of electron groups in the Lewis structure, what is the hybridization of each C atom in $CH_3CH_2^+$? What AOs does each atom contribute to the molecule? Which orbitals overlap along bonding axes to form σ bonds? Are there adjacent orbitals that can overlap on opposite sides of a bonding axis to produce π MOs? Based on the Lewis structure, should any π bonds be present?

Solve

The C atom on the CH_3 has four electron groups in the Lewis structure and is sp^3 hybridized, whereas the C atom on the CH_2^+ has three electron groups and is sp^2 hybridized. Each *s* AO overlaps an sp^3 AO to form one σ MO and one σ* MO. The same is true for the overlap between the two hybridized AOs. The *p* orbital on the sp^2-hybridized C atom cannot overlap another *p* orbital and, therefore, remains unhybridized. The 12 valence electrons fill the six MOs that are lowest in energy. As we can see from the energy diagram, one of the σ MOs is the HOMO, and the unhybridized *p* orbital is the LUMO.

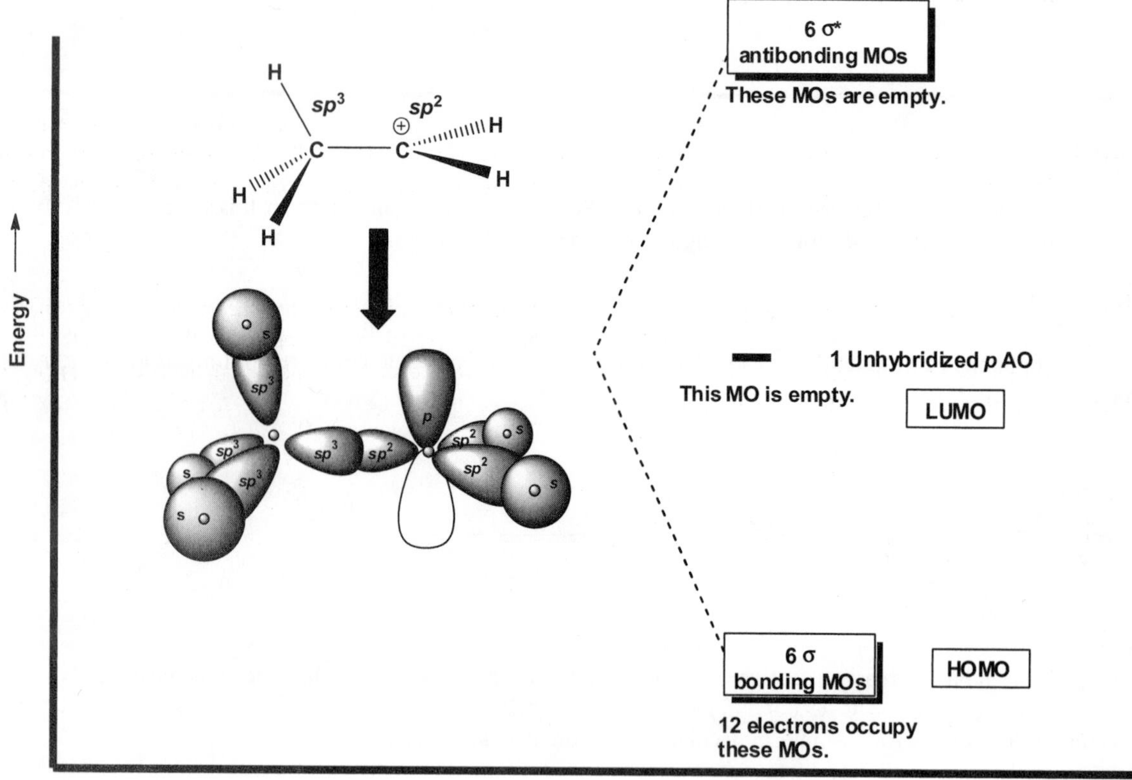

Problem 3.47

Think

Which MOs shown in Problem 3.46 have room for an additional electron? Of those, which MO is lowest in energy? After the electron is added, which occupied MO is highest in energy? Which empty MO is lowest in energy?

Solve

MO energy diagrams are filled from the low energy orbital to the high energy orbital (i.e., bottom to top). The additional electron is added to the lowest energy orbital available, which is the nonbonding *p* orbital. After the electron is added, that nonbonding *p* orbital becomes the HOMO and a σ* becomes the LUMO.

Problem 3.48

Think

Is there a region of constructive interference? Destructive interference? How does each type of interference affect the energy of the resulting orbital? Is one type of interference more prominent than the other?

Solve

Although the *s* orbital is in phase with one lobe of the *p* orbital (constructive interference), it is out of phase with the other lobe (destructive interference). The two types of interactions are equally prominent, so any stabilizing interaction with the top lobe is canceled by the destabilizing interaction with the bottom lobe.

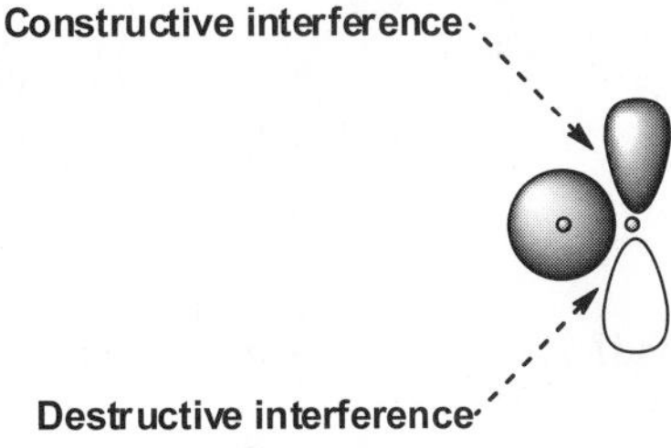

Problem 3.49

Think

How many MOs should result from the interaction of two AOs? How should those AOs mix together differently to give rise to different MOs? When the AOs are in phase, what type of MO results? When the AOs are out of phase, what type of MO results? Does the overlap occur along the bonding axis or on opposite sides of the bonding axis?

Solve

The two AOs should mix to form two MOs. They do so by mixing with different phase combinations of the AOs, as shown below. When the AOs are in phase in the region of overlap, constructive interference takes place and a bonding MO results; when they are out of phase in the overlap region, destructive interference takes place and an antibonding MO results. In both cases, the symmetry is σ, because overlap occurs *along* the bonding axis. Notice that in the antibonding MO, there is an additional node perpendicular to the bonding axis.

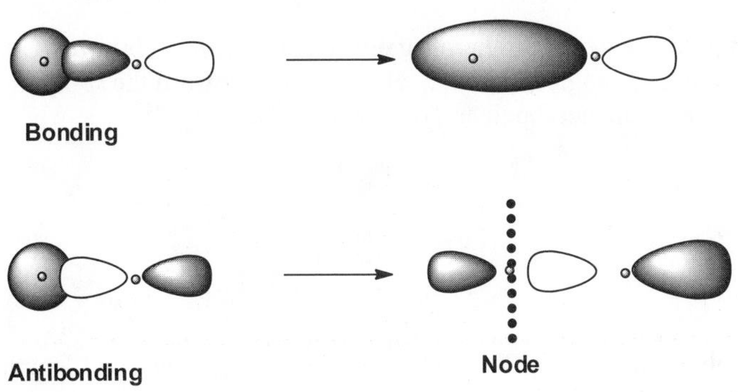

Problem 3.50

Think

How many MOs should result from the interaction of two AOs? How should those AOs mix together differently to give rise to different MOs? When the AOs are in phase, what type of MO results? When the AOs are out of phase, what type of MO results? Does the overlap occur along the bonding axis or on opposite sides of the bonding axis?

Solve

The two AOs should mix to form two MOs. They do so by mixing with different phase combinations of the AOs, as shown below. When the AOs are in phase in the region of overlap, constructive interference takes place and a bonding MO results; when they are out of phase in the overlap region, destructive interference takes place and an antibonding MO results. In both cases, the symmetry is σ, because overlap occurs *along* the bonding axis. Notice that in the antibonding MO, there is an additional node perpendicular to the bonding axis.

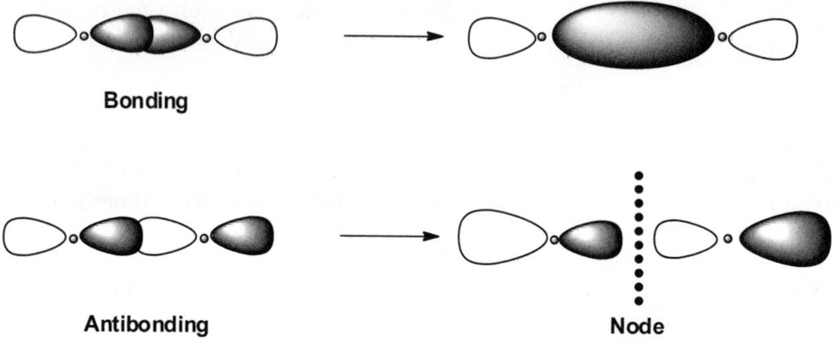

Problem 3.51

Think

What is the hybridization of each atom involved in each C–C bond? How does that affect bond length and strength? How does hybridization relate to effective electronegativity?

Solve

(a) and **(b)** The single bond on the left is between an sp^3-hybridized C atom and an *sp*-hybridized C atom, whereas that on the right is between two *sp*-hybridized C atoms. Because the sp^3 C atom has less *s* character in its hybridized orbitals, it tends to form longer and weaker bonds than an *sp*-hybridized C atom. Thus, we should expect the single bond on the right to be shorter and stronger.

(c) Because an *sp*-hybridized C atom has a greater effective electronegativity than an sp^3-hybridized C atom, there will be a significant bond dipole in the C–C single bond on the left, pointing toward the *sp*-hybridized C atom (shown below). Thus, we should expect the net dipole to point to the right.

Problem 3.52

Think

What is the hybridization of each atom involved in each C–Cl bond? How does that affect bond length and strength? How does hybridization affect electronegativity?

Solve

(a) and **(b)** In molecules **A** and **B**, the Cl atom is sp^3 hybridized, but the hybridization of the C atom to which it is bonded is different in the two molecules. The C atom is sp^3 hybridized in molecule **A**, but is *sp* hybridized in molecule **B**. Thus, the C atom in molecule **B** has more *s* character in its hybridized orbitals, giving rise to a shorter and stronger bond. **(c)** With greater *s* character, the C atom in molecule **B** has a greater effective electronegativity and, therefore, draws electron density toward itself more than does the C atom in molecule **A**. Thus, the Cl atom in molecule **B** bears a lower concentration of negative charge and the one in **A** bears a higher concentration.

sp^3 Cl **Longer, weaker**

sp Cl **Shorter, stronger**

Problem 3.53

Think

How many electron groups are present on each nonhydrogen atom? How does that relate to each atom's electron geometry and hybridization? Which bonds are not permitted to undergo free rotation? Does a hypothetical rotation of 180° about the double bond give rise to a different molecule? How can you tell? How does hybridization affect bond length and strength?

Solve

(a) As shown below, all atoms with two electron groups are *sp* hybridized. All atoms with three electron groups are sp^2 hybridized. All atoms with four electron groups are sp^3-hybridized. (As we will learn in Chapter 14, the singly bonded O atom is an exception in this case, due to the resonance involving its lone pairs and the attached C=O bond.)

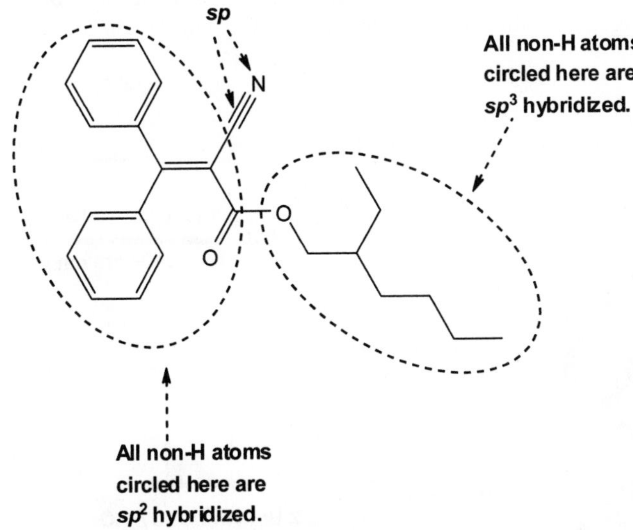

(b) The atoms included in the dashed lines must be in the same plane. Even though the N atom is not directly bonded to the C atoms of the double bond, it must be in the same plane because the C atom to which it is bonded is linear and is also bonded to a C atom in a double bond.

(c) cis and trans configurations are not possible, because two benzene rings are attached to one end of the double bond. A hypothetical 180° rotation about that C=C bond results in exactly the same structure.

(d) We should expect the single bond to the CN group to be shorter. The C atom of the CN group is *sp* hybridized and so has a greater *s* character than the C atom of the CO_2R group, which is sp^2 hybridized. With greater *s* character comes a shorter and stronger bond.

Problem 3.54

Think

Based on the number of electron groups in the Lewis structure, what is the hybridization of each carbon? What AOs does each atom contribute to the molecule? What orbitals overlap along bonding axes to form σ MOs? In twisted ethene, in which plane is each unhybridized *p* orbital? In this orientation, do you expect overlap between the two unhybridized *p* orbitals? Will that result in a π bond?

Solve

(a) The orbital overlap picture looks nearly the same as that for planar ethene (Fig. 3-17), except the one CH_2 group is perpendicular to the other. Thus, the unhybridized *p* orbitals are perpendicular to each other and do not interact.

(b) The energy diagram is shown below. The formation of σ MOs and σ* MOs is identical to that of planar ethene in Figure 3-19. Because the *p* orbitals do not interact, they remain as unhybridized *p* orbitals, or nonbonding MOs. There are a total of 12 valence electrons. The first 10 are filled in σ MOs. The last two are placed in the nonbonding *p* orbitals—one is filled in each, according to Hund's rule. Therefore, one of the nonbonding *p* orbitals is the HOMO, and one of the σ* MOs is the LUMO. In Figure 3-19 the two electrons that are highest in energy are in a π MO, which is substantially lower than the energy of the two highest electrons here. Therefore, twisting about the double bond raises the energy of the electrons significantly, which precludes rotation.

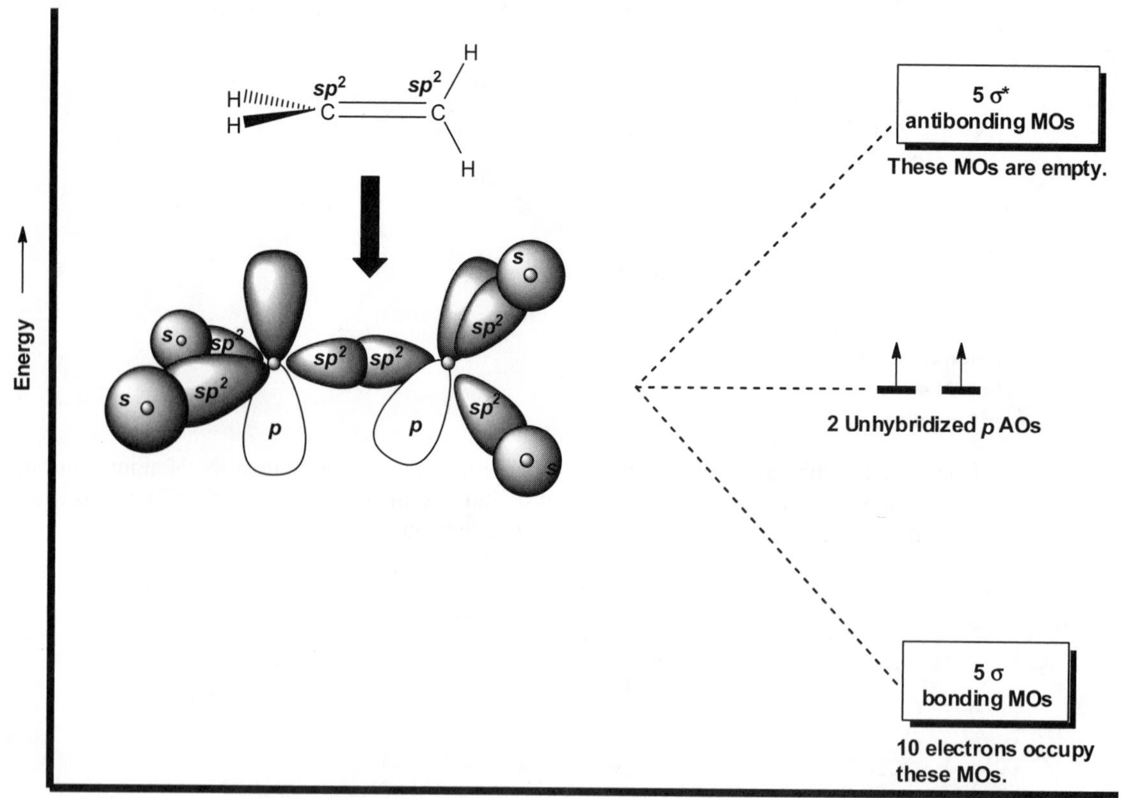

Problem 3.55

Think

Are there lone pairs adjacent to the double bond? Is resonance possible? If so, draw the resonance structure and consider the types of bonds and geometries suggested by it.

Solve

When the N atom with lone pairs is adjacent to a C=O group, a lone pair on N is able to participate in resonance. In the resonance structure, the N atom has three electron groups, suggesting a trigonal planar geometry and an sp^2 hybridization. The resonance structure also shows that the N and C atoms form a C=N double bond, which suggests that free rotation about this bond is not permitted.

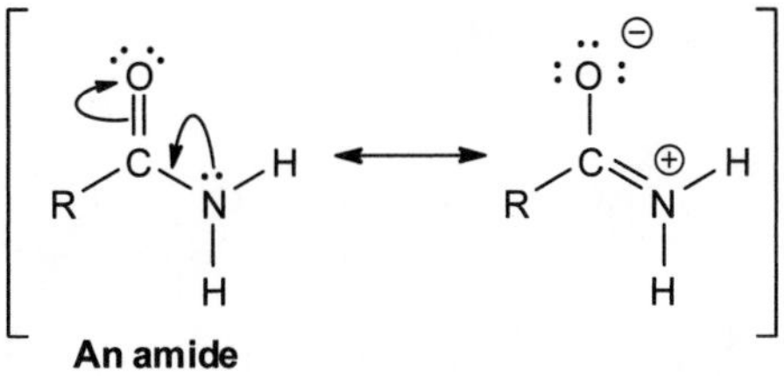

An amide

NOMENCLATURE 2 | Naming Alkenes, Alkynes, and Benzene Derivatives

Problem N2.1

Think

What would the root name of each molecule be without the double bond? Are any rings present? How should the suffix be changed to indicate the presence of a double bond?

Solve

No numbers are necessary for these molecules because the location of the multiple bond is implied between C1 and C2. Notice how each name is derived by changing the corresponding alkane or cycloalkane name to an alkene or cycloalkene.

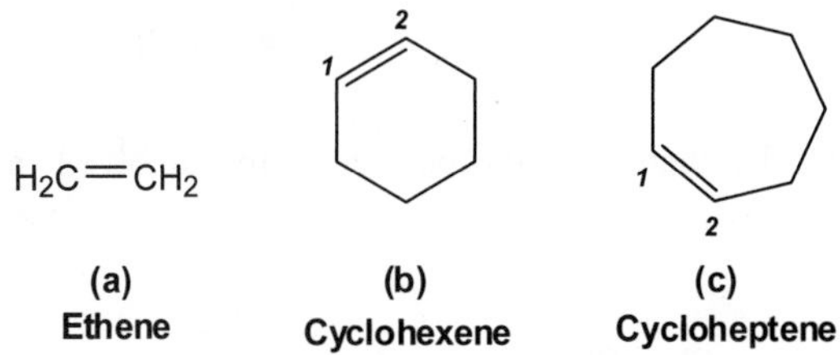

Problem N2.2

Think

What would the root name of each molecule be without the double bond? Are any rings present? Which number C gets assigned to the double bond? What is the suffix for a C=C double bond? When do you need to include a number to indicate the location of the double bond?

Solve

Numbers for the multiple bond in a ring are not necessary, as it is implied that the C=C is located at C1 and C2. Numbers are needed in the name when a different location of the multiple bond results in a different molecule or the ring contains substituents.

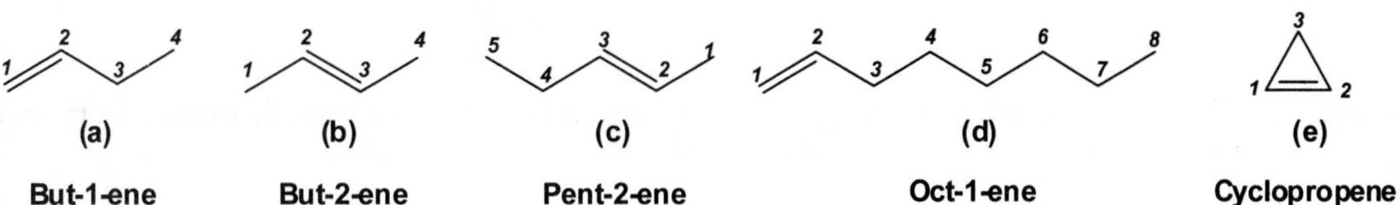

Problem N2.3

Think

Based on the root, how many carbon atoms are in the longest continuous carbon chain or ring? Is there a double or triple bond present? If so, where is it located?

Solve

Numbers for the multiple bond in a ring are not necessary when there are no substituents, as it is understood that the multiple bond is located at C1 and C2. Numbers are needed when a different location of the multiple bond results in a different molecule or the ring contains substituents. Numbers are required in the acyclic compounds to identify where along the chain the multiple bond is located.

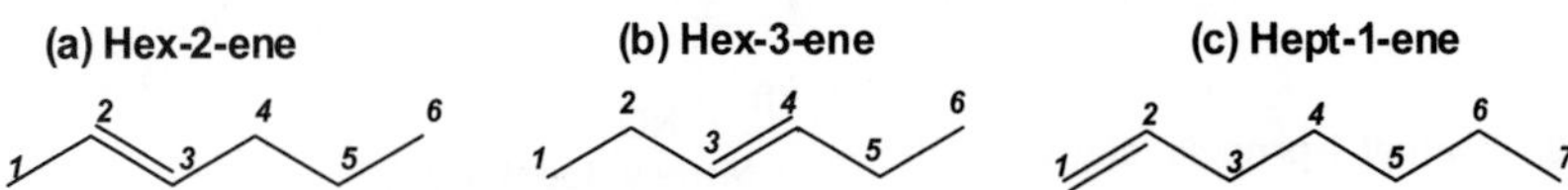

(d) Oct-2-yne

(e) Cycloheptene

Problem N2.4

Think

Based on the number of carbon atoms in the longest continuous carbon chain or ring, what is the root name of each molecule? Are any rings present? Which number C gets assigned to the double or triple bond? What suffix indicates the presence of a double or triple bond? What substituents are present? How should the carbon chain or ring be numbered to give the higher priority (lower carbon number) to the carbon atoms of the multiple bond?

Solve

The double or triple bond has a higher priority than the substituents, so the carbon atoms involved in the multiple bond receive the lowest numbers. See numbering below.

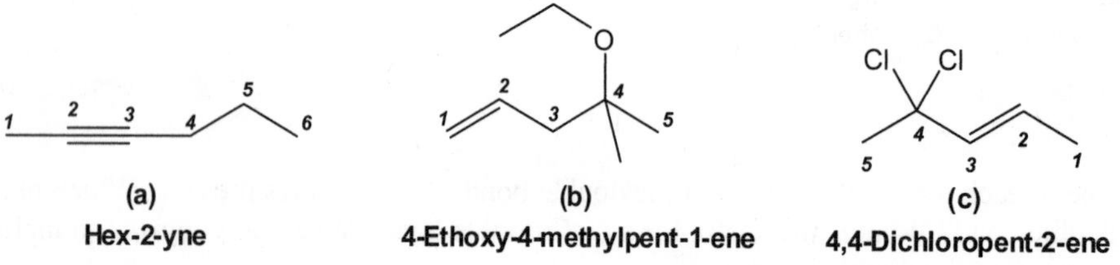

(a) Hex-2-yne

(b) 4-Ethoxy-4-methylpent-1-ene

(c) 4,4-Dichloropent-2-ene

Problem N2.5

Think

Based on the number of carbon atoms in the longest continuous carbon chain or ring, what is the root name of each molecule? Are any rings present? Which number C gets assigned to the double or triple bond? What is the suffix for a double or triple bond? What substituents are present? Which group has the higher priority (lower carbon number), the multiple bond or the substituent?

Solve

The double or triple bond is assigned the higher priority (or lower numbers). In a ring, this typically means that the carbon atoms in the multiple bond are assigned C1 and C2. See numbering below.

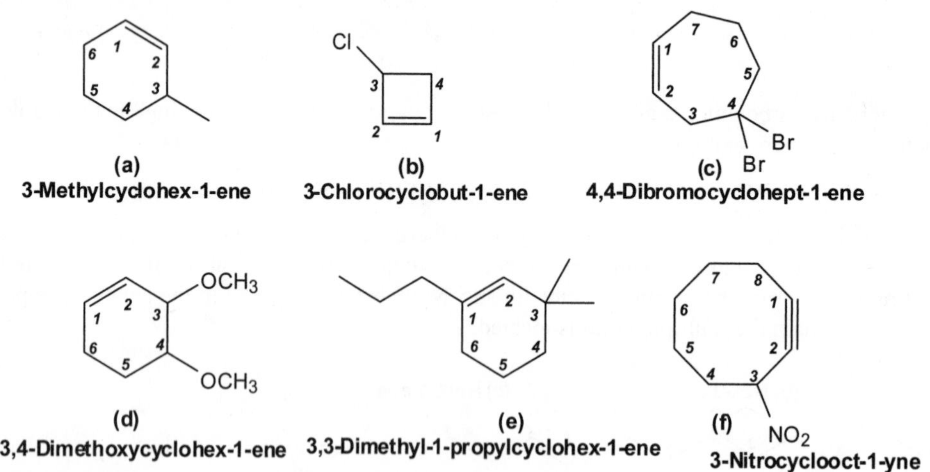

(a) 3-Methylcyclohex-1-ene

(b) 3-Chlorocyclobut-1-ene

(c) 4,4-Dibromocyclohept-1-ene

(d) 3,4-Dimethoxycyclohex-1-ene

(e) 3,3-Dimethyl-1-propylcyclohex-1-ene

(f) 3-Nitrocyclooct-1-yne

Problem N2.6

Think

What is the longest continuous carbon chain that contains the multiple bond? Is the multiple bond part of the root or is it a substituent? What substituents are attached to the main chain, and how many are there of each? How should the main chain be numbered to give the multiple bonded carbon atoms the lowest numbers?

Solve

In each of the following examples, the multiple bond is part of the root and also part of the longest continuous carbon chain. Notice in **(b)** and **(c)** that the longest continuous carbon chain is not drawn entirely horizontal.

(a) 2,3-Dimethylpent-1-ene

(b) 3-Ethylhex-2-ene

(c) 3-Methylhept-1-yne

Problem N2.7

Think

What does the structure of the compound look like given the name provided? Does the longest continuous carbon chain contain the multiple bond? If not, how many carbon atoms are in the longest chain that contains the multiple bond? Is the multiple bond part of the root or is it a substituent?

Solve

The structure of the molecule is shown below. The longest continuous carbon chain has six C atoms, but that chain does not entirely contain the double bond. The chain that contains the double bond has five carbon atoms. Thus 2-propylbut-1-ene should be 2-ethylpent-1-ene.

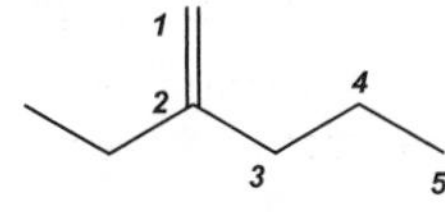

~~2-Propylbut-1-ene~~
2-Ethylpent-1-ene

Problem N2.8

Think

Based on the root, how many carbon atoms are in the longest continuous carbon chain? Where is the multiple bond located? Is it part of the main carbon chain or is it a substituent? What other substituents are present?

Solve

The methylene group is the $=CH_2$ substituent.

(a) 2-Chloro-1-methylenecyclopentane (b) 1-Methylene-2,4-dinitrocycloheptane

Problem N2.9

Think

Looking at the numbers in the names in Figure N2-5, on which numbered carbons should the double bonds be located to give them the lowest possible numbers? Start with C1 and work your way to the other carbons to make sure that your C=C numbering correlates with the name.

Solve

In cyclohepta-1,3,5-triene, the sp^3-hybridized C atom in the ring is C7. In 4,4-dimethylpenta-1,2-diene, the C atoms with the double bonds receive the lower numbers (higher priority).

Cyclohepta-1,3,5-triene

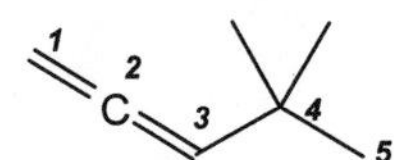

4,4-Dimethylpenta-1,2-diene

Problem N2.10

Think

How many carbons are in the ring? How many C=C double bonds are there and where are the C=C bonds located? In the name given, the numbers are located at the beginning of the name. Where else can the C=C bond numbers be located later in the name?

Solve

The C=C bond numbers (1,4) can be located in the name after the six-carbon cyclo root, cyclohexa. Note, the "a" is added to hex (hexa) due to the presence of "di" in "diene."

Cyclohexa-1,4-diene

Problem N2.11

Think

How many carbons are in the longest continuous carbon chain or ring? Are there double bonds present? If so, how many, and where are they located? Are they part of the main carbon chain or are they substituents? What other substituents are present?

Solve

The two double bonds are assigned the higher priority (or lower numbers). When two double bonds are present, the name includes the number of the first C atom of each double bond and the suffix diene. See below.

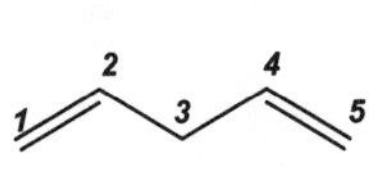

Penta-1,4-diene

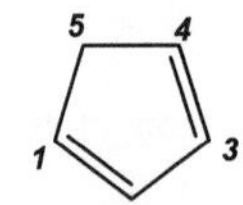

Cyclopenta-1,3-diene

Problem N2.12

Think

How many C atoms are in the main chain? How many C=C bonds are present? On which C atoms are the C=C bonds located? How do you denote the location and number of C=C bonds? Do you need to add an "a" that appears before the suffix?

Solve

There are six C atoms in the main chain, which are part of three double bonds located at C1, C3, and C5. The three double bonds are denoted by 1,3,5-triene. Therefore, the name is hexa-1,3,5-triene. Notice the extra "a" added after "hex" and before the suffix, due to the presence of the three double bonds.

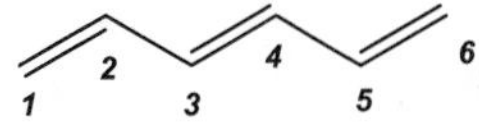

Hexa-1,3,5-triene

Problem N2.13

Think

How many carbon atoms are in the main chain? Are there C=C double bonds present? C≡C triple bonds? On which carbon atoms are the multiple bonds located? Are there any substituents present? Which gets a lower number (higher priority), a double bond or a triple bond?

Solve

Double bonds get lower numbers, but come first in the name when a double and triple bond are both present. Note that the "e" is dropped on diene when there is more to follow for the name.

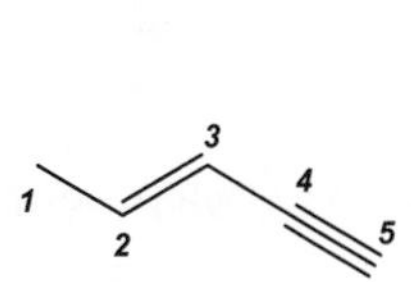

Pent-2-en-4-yne

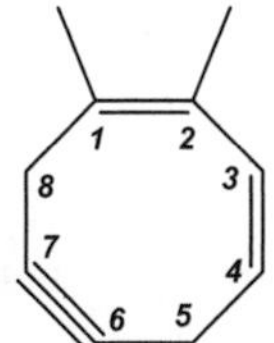

1,2-Dimethylcycloocta-1,3-dien-6-yne

Problem N2.14

Think

How many C atoms are in the longest continuous carbon chain? Which functional group, C=C or C≡C, has the higher priority in the numbering system? How do you denote two C=C bonds? Which suffix comes last? What other substituents are present, and where are they located?

Solve

There are seven C atoms in the longest continuous carbon chain. The C=C takes priority and, therefore, gets a lower number, C1. The "e" in the "ene" suffix is dropped if another suffix, in this case "yne," follows.

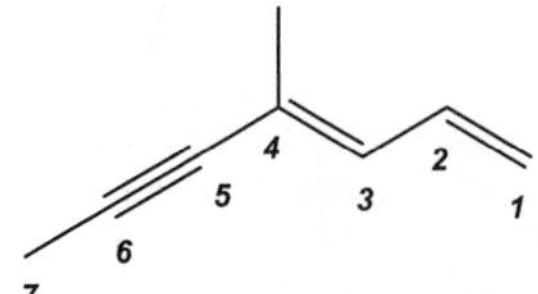

4-Methylhepta-1,3-dien-5-yne

Problem N2.15

Think

Refer to Figure N2-7 to familiarize yourself with the common names for alkenes and alkynes. Based on the root, how many carbon atoms make up the longest continuous carbon chain or ring? What substituents are present, how many are there of each, and where are they located?

Solve

Vinyl is the –CH=CH$_2$ substituent, allyl is the –CH$_2$-CH=CH$_2$ substituent, and propargyl is the –CH$_2$-C≡CH substituent.

(a) 1,4-Divinylcyclohexane

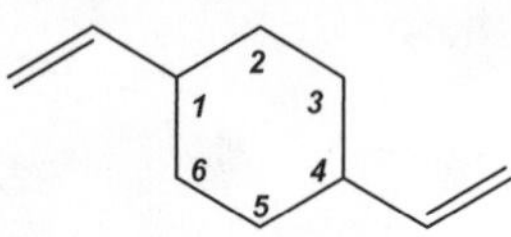

(b) Allyl vinyl ether

O

(c) 1-Chloro-3-propargylcyclopentane

Cl 1 2 3 4 5

Problem N2.16

Think

Based on the root, how many carbon atoms make up the longest continuous carbon chain or ring? Are there any C=C or C≡C bonds present? If so, how many are there of each and where are they located? What substituents are present? How many are there of each and where are they located?

Solve

(a) 2-Chloropropene

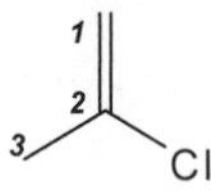

(b) 3-Methylbut-1-ene

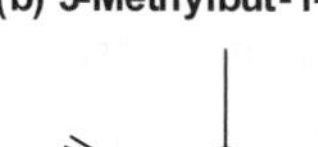

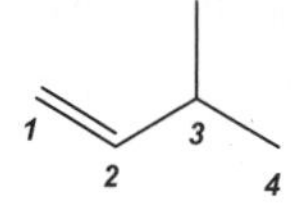

(c) 2,3-Dimethyl-2-butene

1 2 3 4

(d) 2-Ethoxy-3,3-dimethylcyclohexene

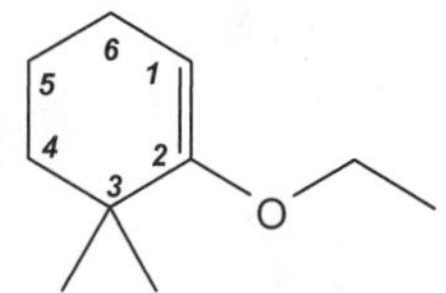

(e) 3,4,5-Trimethoxycycloheptene

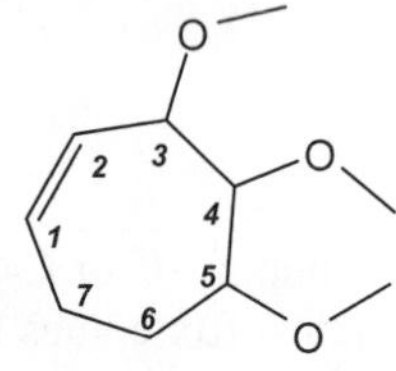

(f) 3-Bromo-2-methyl-4-nitrocyclopentene

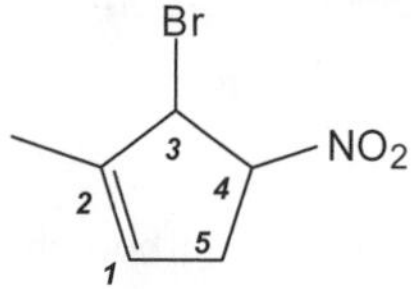

(g) 3,3-Dibromo-4-methylcyclopentene

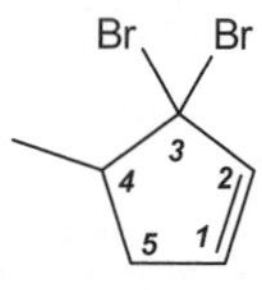

(h) 1,6-Dimethyoxyhexa-1,5-diene

O 1 2 3 4 5 6 O

(i) 2-Methyl-1,3,5-hexatriene

1 2 3 4 5 6

(j) 4-Methyl-2-pentyne

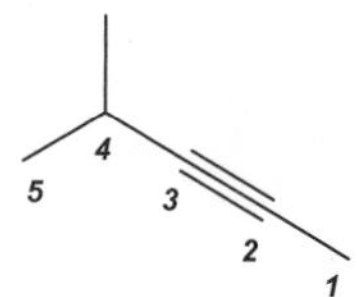

Problem N2.17

Think

Review the rules for naming alkenes and alkanes. What is the longest continuous carbon chain or ring? What functional groups and substituents are present and how many are there of each? How do you number the carbon atoms to give the highest priority carbon atoms the lowest numbers?

Solve

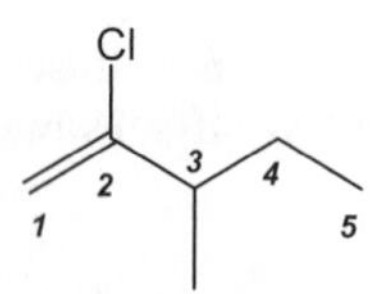

(a)
2-Chloro-3-methylpent-1-ene

1 2 3 4 5 6

(b)
5-Methylhex-1-ene

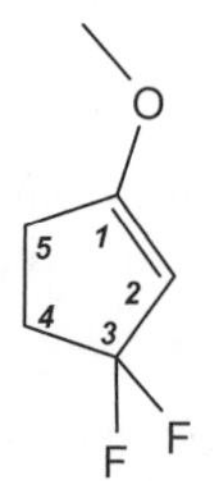

(c)
3,3-Difluoro-1-methoxycyclopent-1-ene

(d) 2,3,4,5-Tetrabromopent-2-ene

(e) 2,2,5-Trimethylhex-3-ene

(f) 1,2-Dimethylcyclobut-1-ene

(g) 1-Methylene-2-methylcyclopentane

(h) 3-Ethoxypenta-1,4-diyne

(i) 2-Cyclopropyl-3-fluoropropene

Problem N2.18

Think

What is the structure of benzene? What does the substituent look like in each case? Are any numbers necessary?

Solve

Benzene is a six-membered carbon ring with alternating single and double bonds. No numbers are necessary, as there is only one substituent and it is implied that the substituent is located at C1.

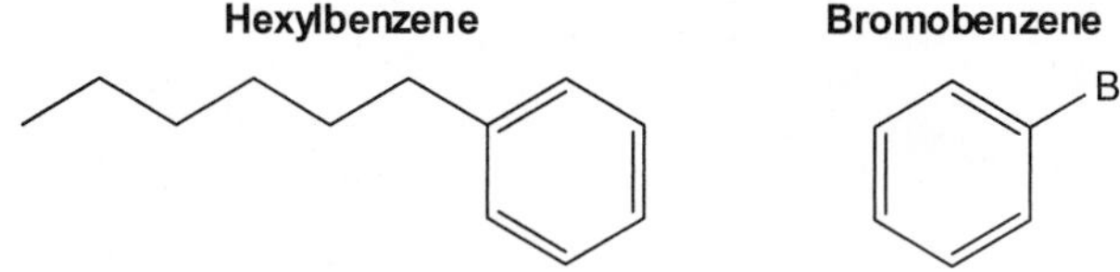

Problem N2.19

Think

What is the name of the six-membered carbon ring with alternating single and double bonds? How many C atoms are on the substituent?

Solve

Benzene is the name of the six-membered carbon ring with alternating single and double bonds and there are five C atoms in the substituent alkyl chain. Thus, the name of the molecule is pentylbenzene.

Pentylbenzene

Problem N2.20

Think

What is the structure of benzene? How many substituents are present? What is the identity and location of each substituent?

Solve

(a) C1, C2, and C3 each bear one methyl group.

(b) C1 has the nitro NO_2 group, Cl is on C2, and Br is on C4.

(a) 1,2,3-Trimethylbenzene **(b) 4-Bromo-2-chloro-1-nitrobenzene**

Problem N2.21

Think

What is the root name of each molecule? What is the name of each substituent? How many of each substituent are present? How do you decide which carbon is C1? Which group takes priority?

Solve

In **(a)**, the carbon atoms are numbered to give the lowest set of numbers. In **(b)**, the Cl group gets assigned C1 due to alphabetical naming rules. Halogens and alkyl groups have equal priorities in IUPAC naming rules and, thus, alphabetical order determines the numbering.

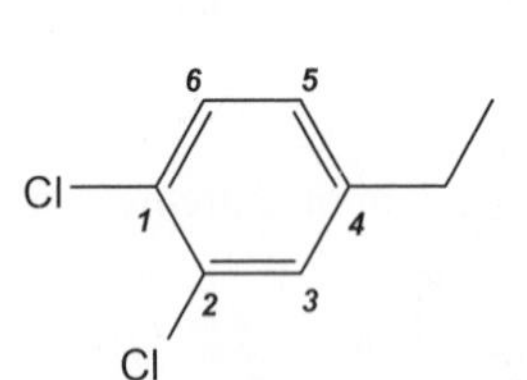

(a)
1,2-Dichloro-4-ethylbenzene

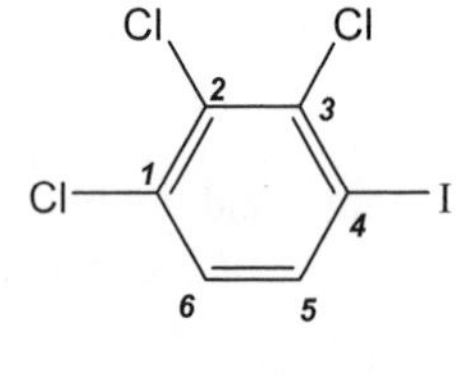

(b)
1,2,3-Trichloro-4-iodobenzene

Problem N2.22

Think

What structure corresponds to the root? What substituents are present and how many are there of each? Which positioning, 1,2-, 1,3-, or 1,4-, correlates with *p*-, *m*-, and *o*- nomenclature?

Solve

The benzene root corresponds to the six-membered rings below. Para, *p*-, is 1,4-; ortho, *o*-, is 1,2-; and meta, *m*-, is 1,3- positioning.

p-Dichlorobenzene

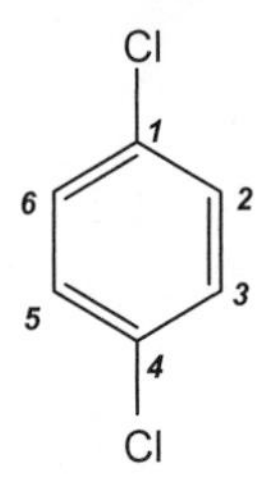

m-Bromoethoxybenzene

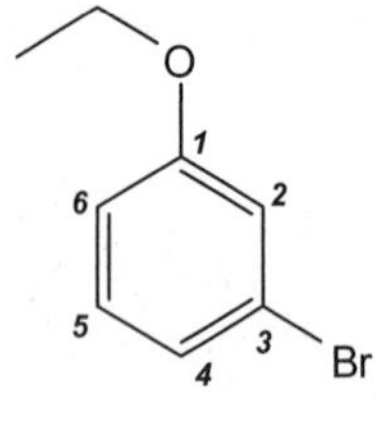

Problem N2.23

Think

Review the trivial names involving the benzene ring and draw the structures starting from the trivial name. What substituent must be present in toluene, and which carbon atom is assigned C1? What about anisole?

Solve

In **(a)** and **(b)**, the root name is toluene, which is $C_6H_5{-}CH_3$, and the methyl group gets assigned to C1. In **(c)**, the root name is anisole, which is $C_6H_5{-}OCH_3$, and the methoxy group gets assigned to C1.

(a) *m*-Bromotoluene

(b) 2,5-Dinitrotoluene

(c) *p*-Chloroanisole

Problem N2.24

Think

What does the phenyl (Ph) substituent look like? How many carbon atoms are in the main parent chain? Are any functional groups present? Which groups have the highest priority in establishing the numbering system of the carbon chain?

Solve

When benzene is not the root, the C_6H_5 becomes the phenyl substituent.

(a) 2-Phenyl-1-hexene

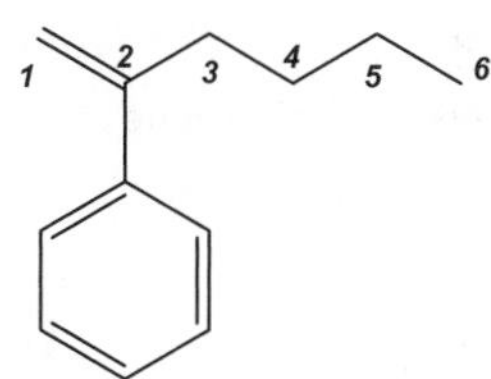

(b) 1,5-Diphenylpentane

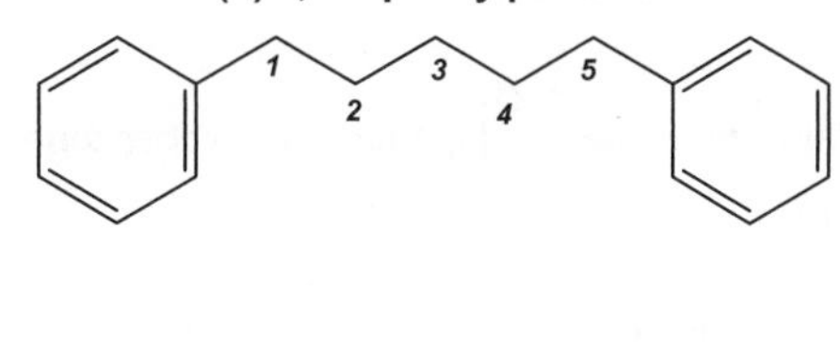

Problem N2.25

Think

Based on the number of carbon atoms, what is the name of the main carbon chain? What is the name of the Ph group, how many Ph groups are present, and on which C atoms are the Ph groups located?

Solve

The main carbon chain consists of two C atoms, so it is ethane. The Ph group is phenyl, and there are six Ph groups: three on C1 and three on C2.

1,1,1,2,2,2-Hexaphenylethane

Problem N2.26

Think

Is the C_6H_5 group the main parent chain or a substituent? Is the $C_6H_5CH_2$ group the main parent chain or a substituent? What are the names of the C_6H_5 and $C_6H_5CH_2$ groups? Based on the number of C atoms, what is the name of the main parent chain? Which C atom is assigned C1?

Solve

The main parent chain is the seven-membered carbon chain, so it is heptane. The numbering goes left to right to assign the substituents the lowest possible carbon numbers.

Cl 1 2 3 4 5 6 7

4-Benzyl-2-chloro-5-phenylheptane

Problem N2.27

Think

Review the nomenclature rules covered in this chapter and the first nomenclature chapter. How many carbon atoms are in the longest continuous carbon chain or ring? Is there an alkene or alkyne functional group present? If so, how many are there of each? How was the numbering system established? Is a benzene ring present? If so, is it assigned as the root, or is it treated as a substituent?

Solve

(a) Fluorobenzene F

(b) 1-Chloro-2-fluorobenzene Cl F

(c) 1-Iodo-4-nitrobenzene

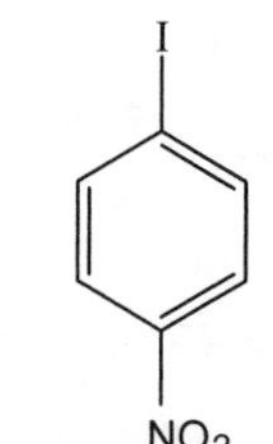

(d) 1,3-Dibromobenzene Br Br

(e) 2-Fluorotoluene F

(f) 4-Ethoxytoluene O

(g) 2-Ethoxyanisole O O

(h) 2,3-Dimethyl-1-cyclopentylbenzene

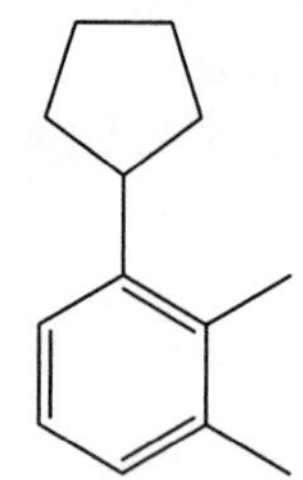

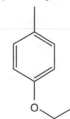

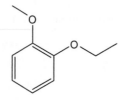

(i) 1,3-Diphenylheptane

(j) 4,4-Diphenyl-1-octene

Problem N2.28

Think

Review the nomenclature rules covered in this chapter and the first nomenclature chapter. How many carbon atoms are in the longest continuous carbon chain or ring? Is there an alkene or alkyne functional group present? If so, how many are there of each? How do you number the main chain or ring to give the alkene or alkyne carbon atoms the lowest numbers? Is a benzene ring present? If so, should it be assigned as the root, or should it be treated as a substituent? What types of molecules can be described as *ortho*, *meta*, or *para* disubstituted?

Solve

The names of the molecules are shown below. Only **(c)** and **(d)** can be named using the *ortho*, *meta*, or *para* system because they are disubstituted benzenes.

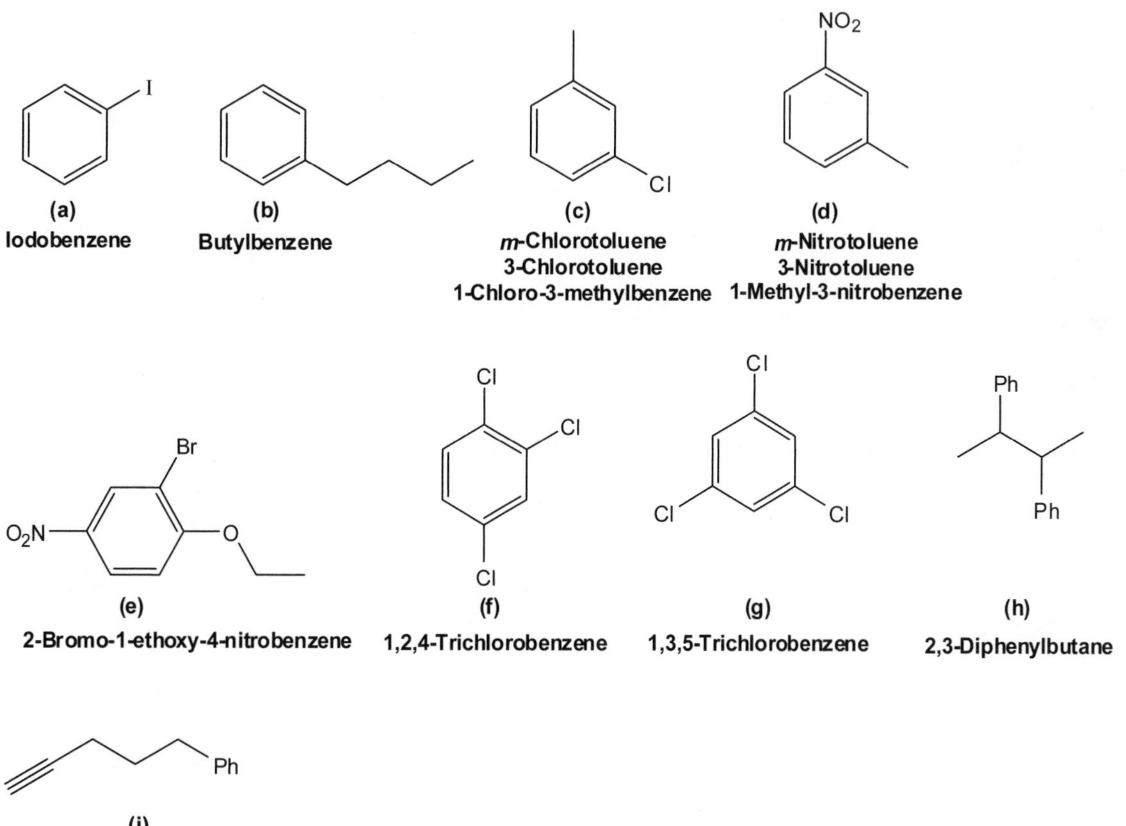

CHAPTER 4 | Isomerism 1: Conformational and Constitutional Isomers

Your Turn Exercises

Your Turn 4.1

Think

If you perform a 90° counterclockwise rotation of the molecule CX_3–CY_3 (as viewed from the top), which group is in the front? Which group is in the back? In which plane is the C–C bond? Do bonds intersecting at the dot represent bonds to the front C atom or back C atom? What about bonds drawn to the circle?

Solve

If the molecule is rotated 90° counterclockwise, the CX_3 group is in the front and the CY_3 group is in the back. The C–C bond is not visible in the Newman projection because it is perpendicular to the plane of the paper. The front C atom is depicted by a point and the three other bonds to the front C atom converge at this point. The back C atom is depicted by a circle and the three other bonds to the back C atom connect to the circle. Make sure to draw your three other bonds approximately equally spaced around the circle or point.

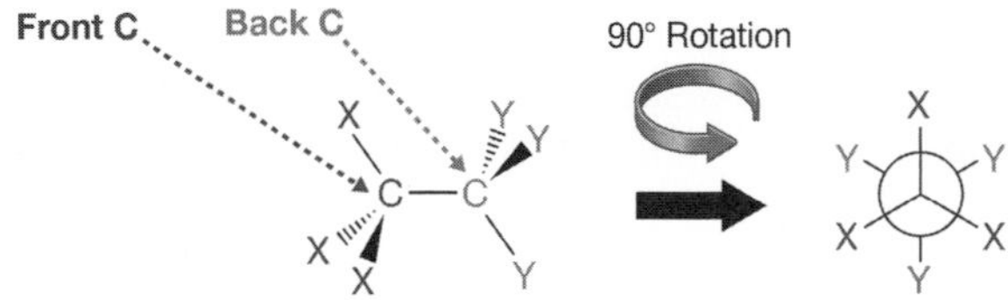

Your Turn 4.2

Think

Each carbon is sp^3 hybridized and the CH_3 and CBr_3 groups all rotate freely. Once you have your model constructed, draw the C–C bond first and fill in with the other three bonds—wedge, dash, and planar. If you perform a 90° clockwise rotation of the molecule CH_3–CBr_3, which group is in the front? Which group is in the back? In which plane is the C–C bond? How do you represent the three other bonds on the front C atom? The back C atom?

Solve

If the molecule in **(a)** drawn below is rotated 90° clockwise (as viewed from the top), the CBr_3 group is in the front and the CH_3 group is in the back; the Newman projection is drawn following the steps in Your Turn 4.1. You may have instead elected to draw the C–Br bonds on the left and the C–H bonds on the right, as in the molecule shown in **(b)** on the next page. The Newman projection that results is 180° opposite of the first Newman projection. Both Newman projections are identical.

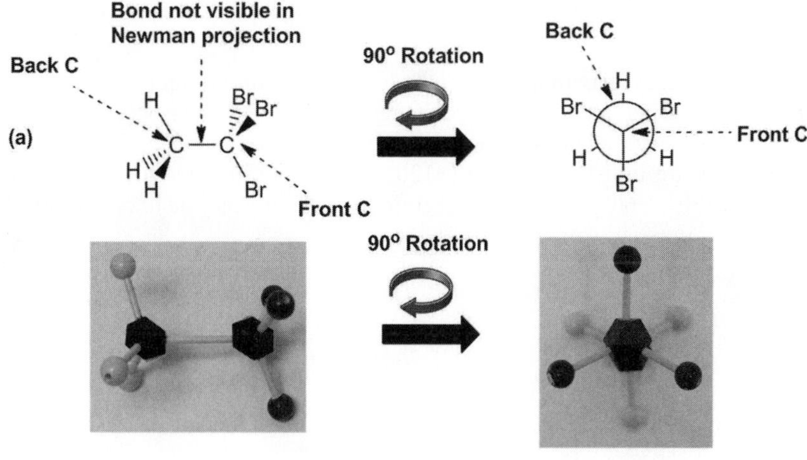

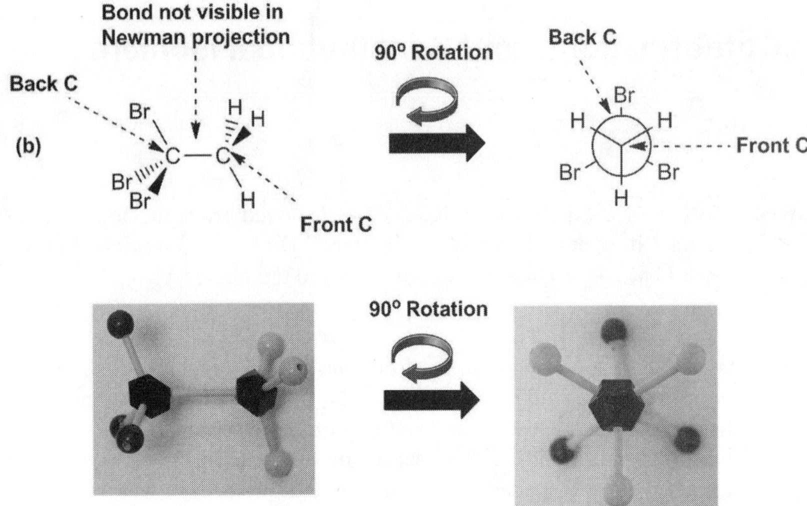

Your Turn 4.3

Think

Use Figure 4-3 as a guide. Keep the front CH_2Cl group frozen and rotate the back CH_2Cl group. Where should the Cl on the back C atom appear after a +60° rotation? After a +120° rotation? Do you notice any patterns forming in each successive 60° rotation? Are the back bonds directly behind the front bonds or in between?

Solve

The CH_2Cl group remains frozen in place on the front C atom as the back CH_2Cl is rotated. At 0°, the back bonds are directly behind the front bonds. Upon a +60° rotation clockwise, the back bonds are now in between the front bonds. Upon another +60° rotation clockwise (+120° from the start), the back bonds are again behind the front bonds.

0° +60° +120°

Your Turn 4.4

Think

Do eclipsed conformations pass through an energy maximum (higher energy) or an energy minimum (lower energy)? In eclipsed conformations, do the front C–H bonds "cover" or "bisect" the back C–H bonds? What about staggered conformations?

Solve

The eclipsed conformations occur at an energy maximum due to the torsional strain brought about by electron repulsion between the bonds directly covering each other. The staggered conformations occur at an energy minimum because the bonds in the front bisect the bonds in the back and the electron repulsion is minimized. In this example, all eclipsed conformations are indistinguishable (−120°, 0°, and +120°) and all three staggered conformations are indistinguishable (±180°, −60°, and +60°). Therefore, there are two distinct conformations for ethane. See the figure on the next page.

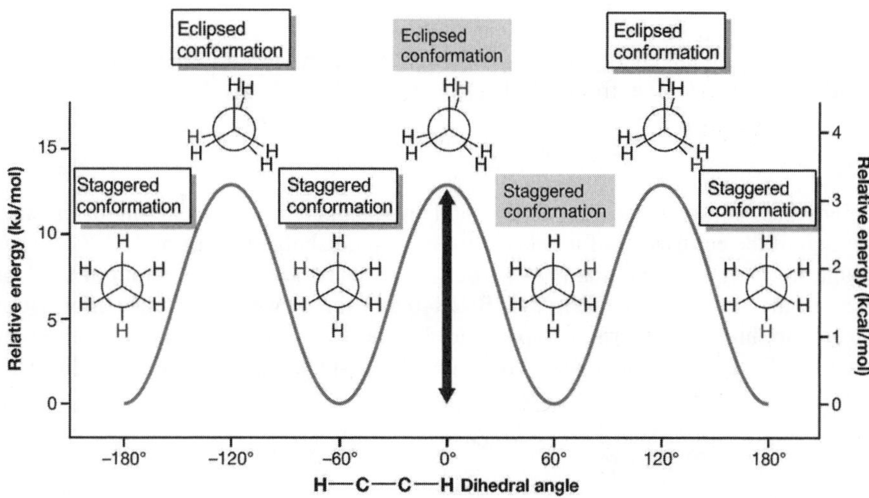

Your Turn 4.5

Think

After you construct your model of $H_3C–CH_3$, try to hold the CH_3 groups in place as you examine the molecule from different angles. When $\theta = 0°$, are the front and back C–H bonds at a maximum or minimum distance? To which conformation (staggered or eclipsed) does this correlate?

Solve

Your model should resemble the structures in Figure 4-5. The front and back C–H bonds are closer to each other in an eclipsed conformation, and thus are *higher* in energy.

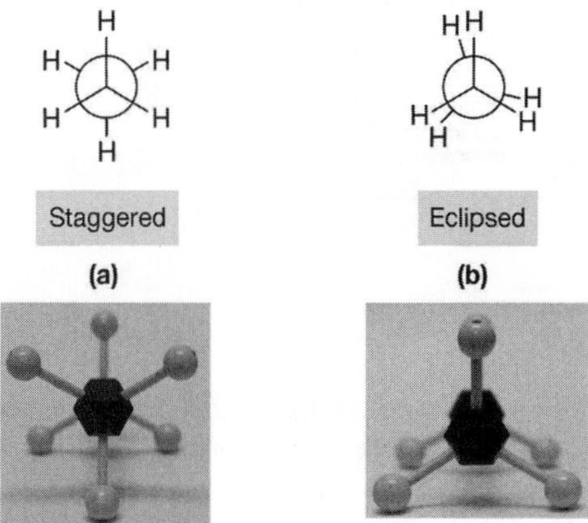

Your Turn 4.6

Think

Consult Figure 4-4. What does the double-headed arrow represent? What is the energy value for the eclipsed conformations? What is the energy value for the staggered conformations? What is the difference in energy between these two conformations?

Solve

The double-headed arrow represents the rotational energy barrier, which is the energy difference between the eclipsed and staggered conformations. The energy for the eclipsed conformations is about 13 kJ/mol and the energy for the staggered conformation is 0 kJ/mol (according to Fig. 4-4). Thus, the energy difference is about 13 kJ/mol. This energy difference is the torsional strain of the ethane molecule in the eclipsed conformation—namely, the energy the molecule must acquire to pass through the less stable, higher energy conformation.

Your Turn 4.7

Think

Do eclipsed conformations occur at an energy maximum (higher energy) or an energy minimum (lower energy)? In eclipsed conformations, do the front bonds "cover" or "bisect" the back bonds? What about staggered conformations?

Solve

The eclipsed conformations pass through energy maxima and the staggered conformations occur at energy minima. In this example, two of the eclipsed conformations are indistinguishable (−120° and +120°). These two conformations pass through lower energy maxima (~24 kJ/mol), due to the lower steric strain of the H and Br atoms eclipsed compared to the steric strain of Br and Br (~40 kJ/mol). Two staggered conformations are gauche (−60° and +60°) and are higher in energy compared to the anti (±180°) conformation. Therefore, there are four distinct conformations for 1,2-dibromoethane: two staggered (anti and gauche) and two eclipsed.

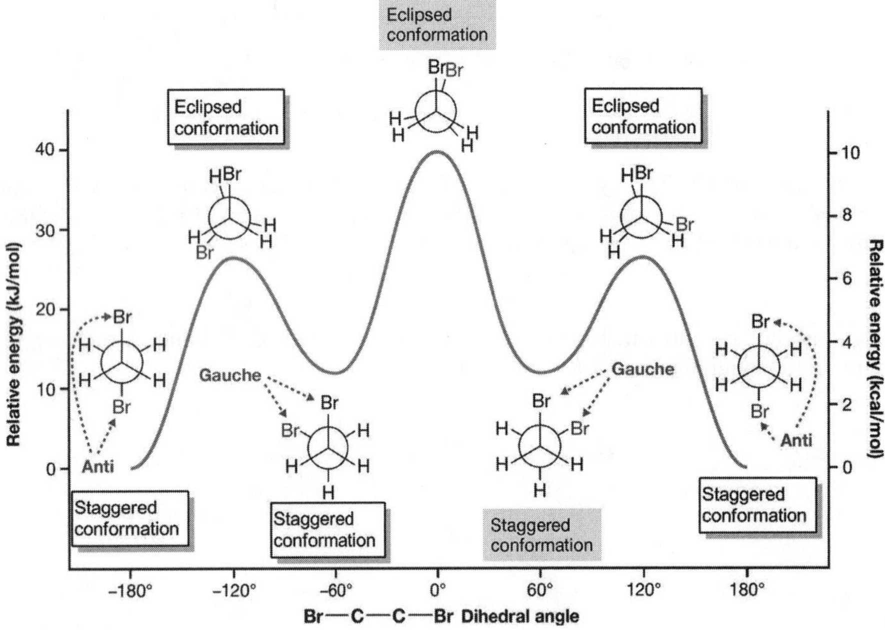

Your Turn 4.8

Think

What is the energy value for the two distinguishable eclipsed conformations? What is the energy difference between the two? What is the energy value for two distinguishable staggered conformations? What is the difference in energy between these two conformations?

Solve

The energy for the eclipsed conformation at 0° is about 40 kJ/mol and the energy for the eclipsed conformation at ±120° is about 24 kJ/mol. Therefore, the eclipsed conformation at 0° is about 16 kJ/mol higher in energy than the other two. The energy for the gauche staggered conformation is about 12 kJ/mol and the energy for the anti staggered conformation is about 0 kJ/mol. Therefore, the anti staggered conformation at ±180° is about 12 kJ/mol lower than the other two (gauche).

Your Turn 4.9

Think

Which group is considered the bulky group (H or Cl)? In which conformation are the bulky groups 180° apart (anti)? In which conformation are the bulky groups 60° apart (gauche)?

Solve

Cl is a larger atom compared to H and, therefore, is considered the bulky group in the molecule. The conformation on the left has the two Cl groups 60° apart, so it is *gauche*. The conformation on the right has the two Cl groups 180° apart, so it is *anti*.

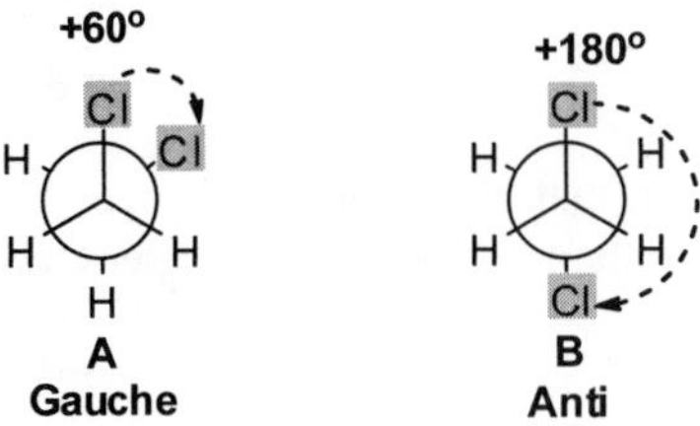

Your Turn 4.10

Think

At what energy is the anti staggered conformation (consult the *y* axis)? At what energy is the gauche staggered conformation? What conformation occurs at an energy maximum between the anti and gauche conformations? Draw the arrow from the anti staggered energy to the energy of that conformation at the energy maximum.

Solve

The anti conformation is located at 0 kJ/mol, and both gauche conformations are ~12 kJ/mol higher. The energy of the eclipsed conformation between the two is ~24 kJ/mol higher. The rotational energy barrier to go from anti to gauche is, therefore, the difference in energy between the eclipsed conformation and the anti conformation, indicated by the double-headed arrow below, and is about 24 kJ/mol (5.7 kcal/mol).

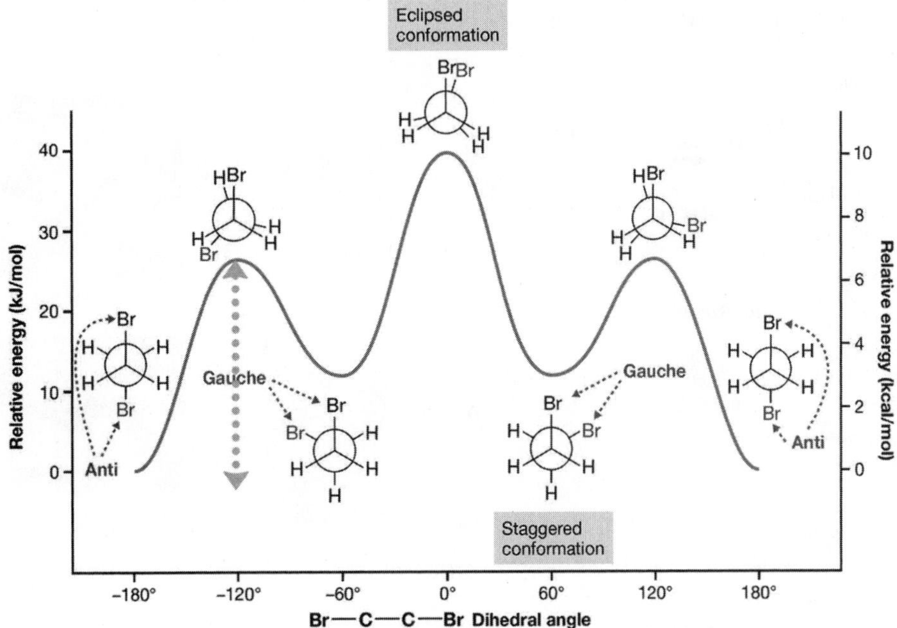

Your Turn 4.11

Think

Draw out each CH_2 on each point on the ring. The heat of combustion for cyclooctane is 4962.2 kJ/mol. How do you calculate the heat of combustion per CH_2? Once you calculate the heat of combustion (ΔH°) per CH_2 for cyclooctane, subtract the heat of combustion per CH_2 for cyclohexane (consult Table 4-1) to obtain the ring strain per CH_2 group. Multiply the ring strain per CH_2 group by eight to obtain the total ring strain for the molecule.

Solve

Eight CH_2 groups compose cyclooctane. The $\Delta H°$ per carbon in cyclooctane is $\frac{4962.2\frac{\text{kJ}}{\text{mol}}}{8\ \text{CH}_2\ \text{groups}} = 620.3\ \text{kJ/mol}$.
Ring strain is the difference between $\Delta H°$ per CH_2 group in cyclooctane versus cyclohexane, or:
$620.3\frac{\text{kJ}}{\text{mol}} - 615.1\frac{\text{kJ}}{\text{mol}} = 5.2\frac{\text{kJ}}{\text{mol}}$.
Total ring strain is the ring strain per CH_2 group times the number of CH_2 group, or:
$5.2\frac{\text{kJ}}{\text{mol}} \times 8\ \text{CH}_2\ \text{groups} = 41.6\frac{\text{kJ}}{\text{mol}}$.
This value is less than that of cyclopropane and cyclobutane, but more than that of cyclopentane, cyclohexane, and cycloheptane, so per CH_2, cyclobutane and cyclopropane are less stable than cyclooctane, but (per CH_2) cyclopentane and cyclohexane are more stable than cyclooctane.

Your Turn 4.12

Think

First, form your six-membered ring of tetrahedral carbons and then add in two C–H bonds per C atom. How should the ring look in the chair conformation? In the planar conformation, the hexagon ring should be flat. Can you see how the staggered conformation in the chair conformation changes into an eclipsed conformation once the ring is planar?

Solve

In the chair conformation, one CH_2 group points up (head of the chair), four are in the middle (body of chair), and one points down (foot of chair). Rotate the model so that the two C–C bonds are perpendicular to the plane of the paper—one involving a head CH_2 carbon, and one involving the foot CH_2 carbon (bonds are in boldface for emphasis). Can you observe the staggered conformation about each C–C bond? It may help to draw a Newman projection for each side of the ring. You can force all of the C atoms to be planar by making the six-membered ring as flat as possible. Can you observe the eclipsed conformation about each C–C bond? It may help to draw a Newman projection for each side of the ring.

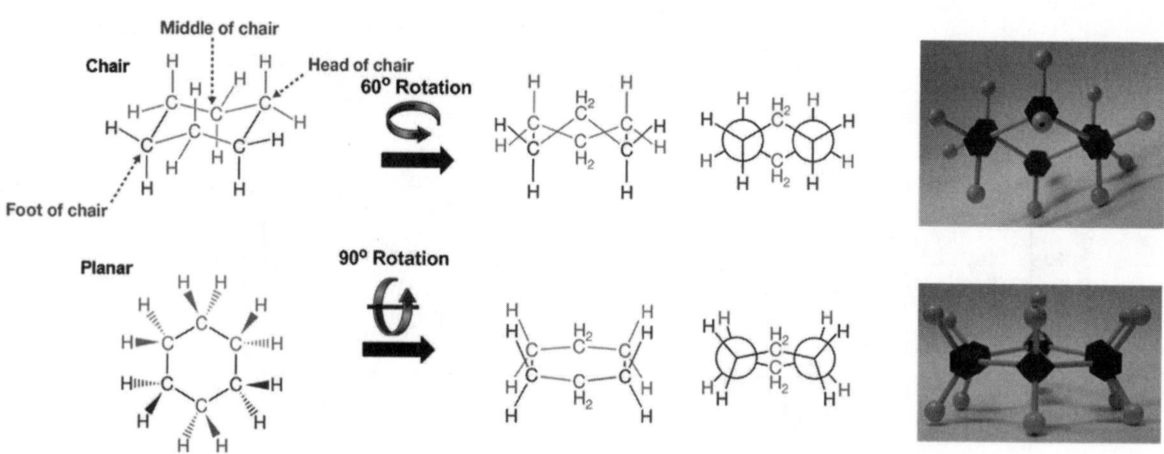

Your Turn 4.13

Think

One C–C has eclipsed H atoms, and the other four have H atoms that are intermediate between staggered and eclipsed. Can you identify the eclipsed C–C bond as the one where, when you look down the bond, the front bonds appear to cover the back bonds?

Solve

In Figure 4-14, the C atom on the left perpendicular to the plane of the page has the front C–H bonds covering the back C–H bonds. This is the eclipsed C–C bond. The structure on the far right on the next page has the eclipsed C–C bolded for emphasis.

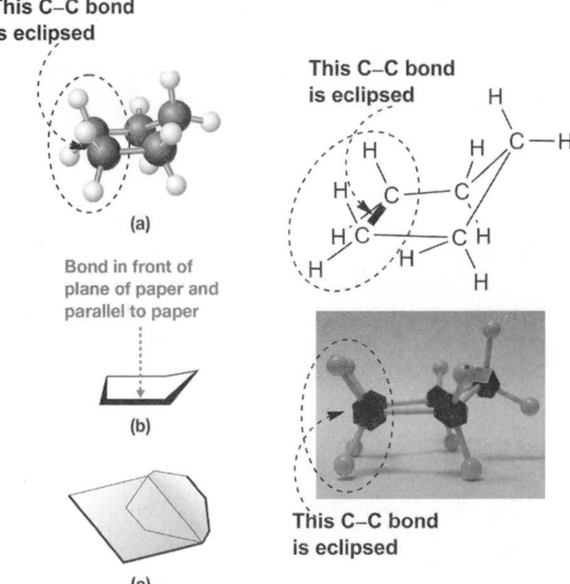
This C–C bond is eclipsed
(a)
Bond in front of plane of paper and parallel to paper
(b)
(c)
This C–C bond is eclipsed
This C–C bond is eclipsed

Your Turn 4.14

Think

What type of rotation is allowed in a ring structure? How does moving the orange C atom from its out-of-plane position into the plane and the green C atom from in the plane into an out-of-plane position affect the other C–C bonds? Which bonds are eclipsed before and after the rotation?

Solve

The bonds indicated below are the ones that undergo the most significant rotation. In this example the two CH_2 groups on the left side of the molecule are the eclipsed groups in both conformations.

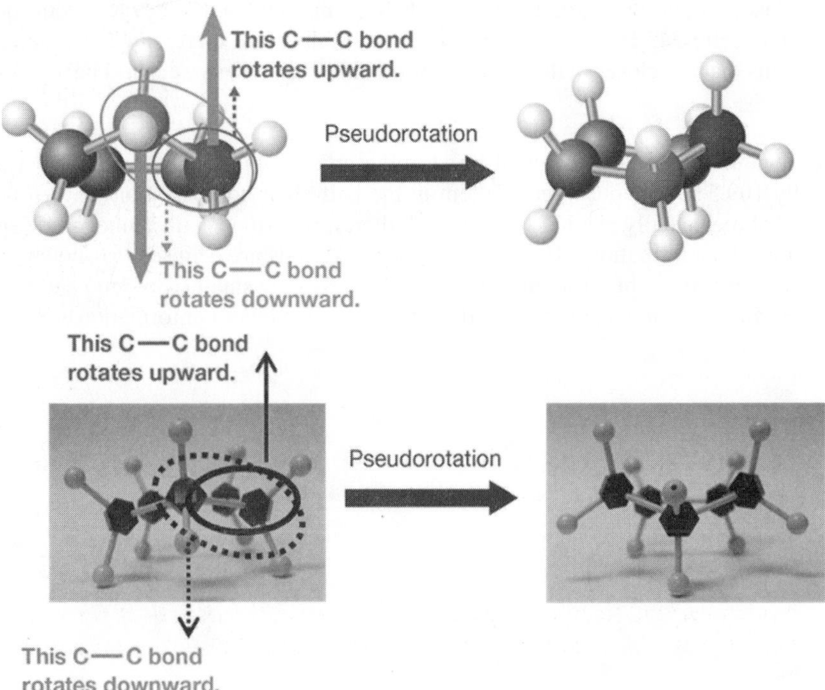
This C—C bond rotates upward.
Pseudorotation
This C—C bond rotates downward.
This C—C bond rotates upward.
Pseudorotation
This C—C bond rotates downward.

Your Turn 4.15

Think

When you build your cyclohexane model, make sure you include all C–C and C–H bonds. It might also be helpful to mark the H atoms in a blue or red marker to differentiate axial (all red initially) from equatorial (all blue initially) C–H bonds.

Solve

All of the C–C bonds undergo rotation as axial and equatorial groups change their positions. Every pair of adjacent C–C bonds undergo rotation in the opposite directions. The "head" of the chair (C′ below) rotates to become the foot of the chair and the foot of the chair (C″ highlighted below) rotates to become the head of the chair. All axial C–H bonds convert into equatorial and vice versa. The axial H atoms are circled in the structure on the left and the same H atoms that change into equatorial H atoms are circled on the right.

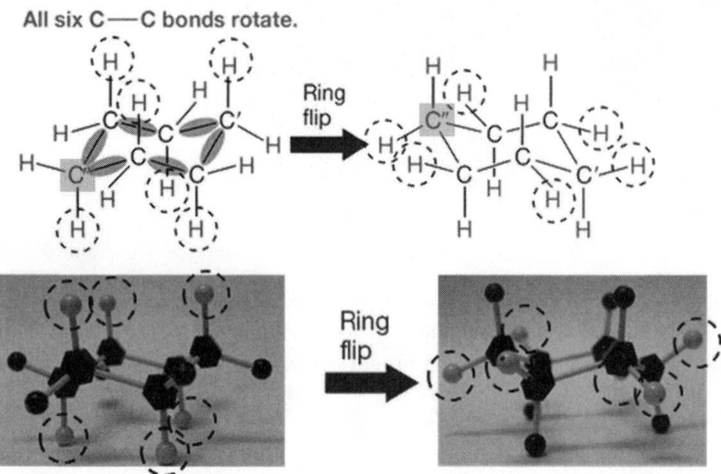

Your Turn 4.16

Think

What constitutes angle strain? Based on the drawing, attempt to identify any angles that deviate from the sp^3 109.5°. Are any of the C–H bonds eclipsed? If so, what type of strain results? Are there any substituents not directly bonded to adjacent C atoms that are closer in this conformation than they otherwise could be?

Solve

Angle strain comes from the deviation between the actual bond angle and the ideal bond angle. All carbon atoms of cyclohexane are ideally 109.5° apart, but three of them in the half-chair conformation (shown on the next page) have C–C–C angles that are actually at ~120° angles and, therefore, provide substantial angle strain. Torsional strain comes from eclipsed conformations. H atoms on two C–C bonds are eclipsed, as shown on the next page. There is no substantial steric strain because the substituents are all very small (H atoms) and none of the ones bonded to nonadjacent C atoms are much closer than they would be in a chair conformation.

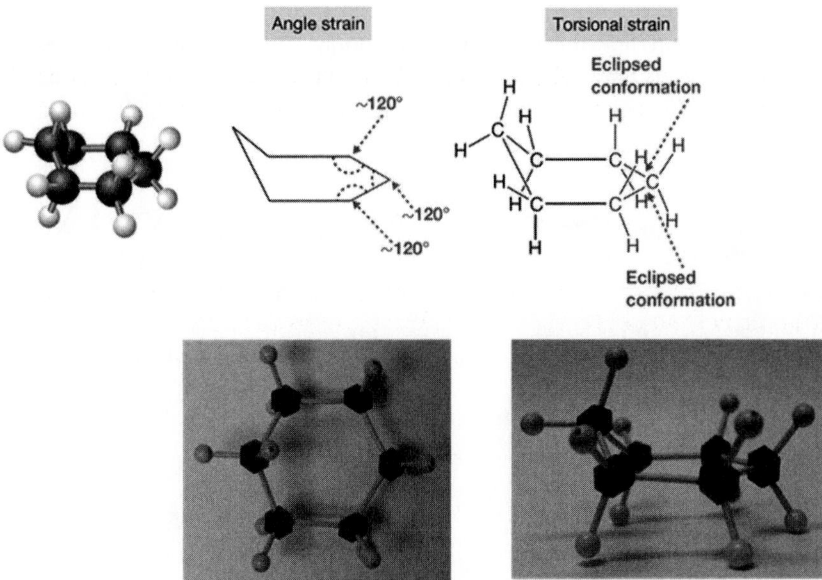
Angle strain
Torsional strain
~120°
~120°
~120°
Eclipsed conformation
Eclipsed conformation

Your Turn 4.17

Think

Which two bonds are drawn first? Adding the V and Λ adds how many more bonds? Now the cyclohexane chair ring is drawn. Axial bonds point either up or down. How do you know in which direction you should draw the bond? Equatorial bonds point slightly up or slightly down. How do you know in which direction you should draw the bond?

Solve

Work with the models, referring to Figure 4-28. The progression of steps to draw this new chair conformation appears below, which is similar to Figure 4-28. Added bonds are in boldface for emphasis. Whether an axial bond points up or down depends on whether the V points up or down. Where there is an axial bond pointing up, there must be an equatorial bond pointing slightly down, and vice versa. These structures take time to draw accurately, so practice drawing the chair conformation of cyclohexane until you can draw it accurately.

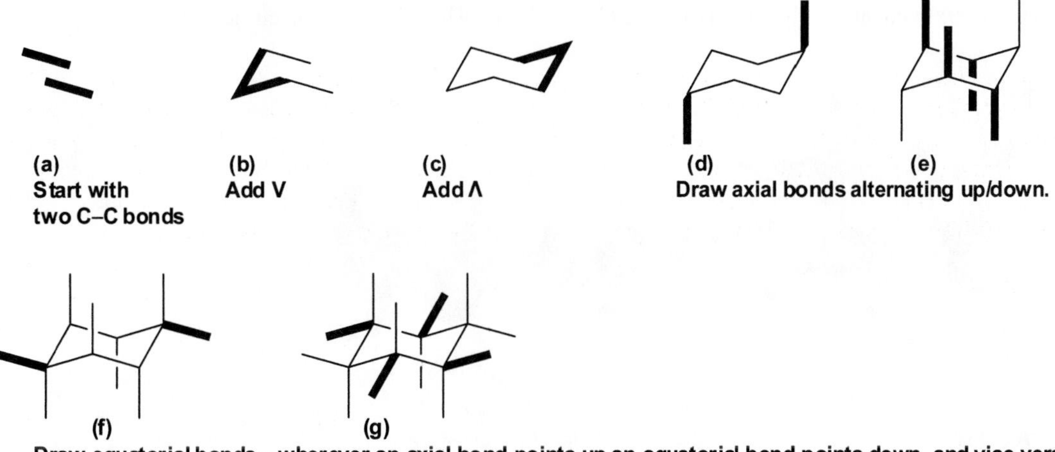
(a) Start with two C–C bonds
(b) Add V
(c) Add Λ
(d)
(e)
Draw axial bonds alternating up/down.
(f)
(g)
Draw equatorial bonds—wherever an axial bond points up an equatorial bond points down, and vice versa.

Your Turn 4.18

Think

The two bonds that are drawn first are drawn in which direction? Follow the steps outlined in Your Turn 4.17 to add the V, Λ, axial, and equatorial C–H bonds.

Solve

The two bonds that are drawn first are drawn from top right to bottom left. These bonds serve as the framework upon which the rest of the structure is drawn. Adding in the V, Λ, axial, and equatorial C–H bonds follows the same steps as previously described.

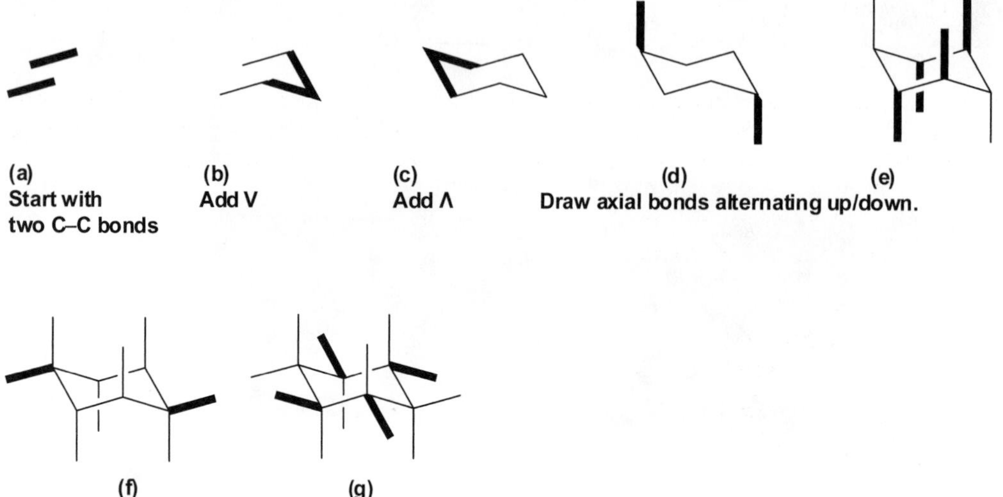
(a)
Start with
two C–C bonds
(b)
Add V
(c)
Add Λ
(d)
(e)
Draw axial bonds alternating up/down.
(f)
(g)
Draw equatorial bonds—wherever an axial bond points up an equatorial bond points down, and vice versa

Your Turn 4.19

Think

Do axial C–H bonds lie almost in the plane defined by the ring or perpendicular to the plane? What about equatorial C–H bonds? How many total axial bonds should there be in each chair? Equatorial bonds? What position is occupied by the CH_3 group in each of the conformations given?

Solve

Axial bonds are perpendicular to the plane defined by the ring, whereas equatorial bonds are closer to parallel. In each structure there should be a total of six axial bonds and six equatorial bonds. Axial H atoms are labeled below. All unlabeled H atoms are equatorial. The conformation shown in **(a)** has five axial C–H bonds and six equatorial, and the conformation shown in **(b)** has six axial C–H bonds and five equatorial. The sixth axial bond in **(a)** and the sixth equatorial bond in **(b)** is the C–CH_3 bond. The CH_3 group is circled for clarity.

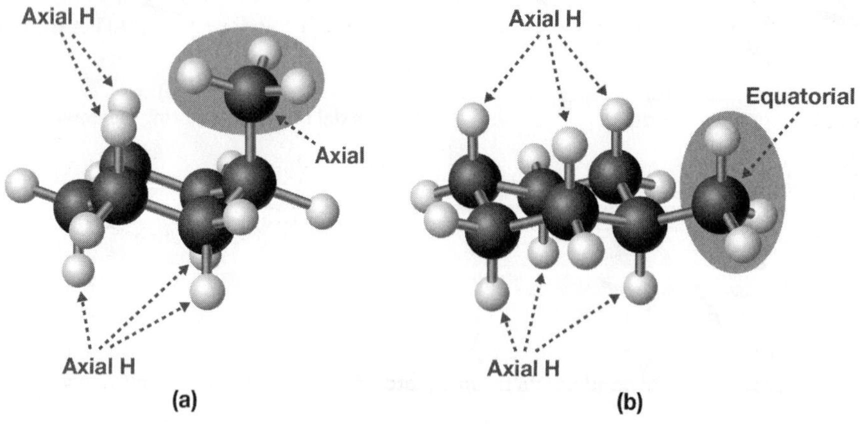
Axial H
Axial
Axial H
(a)
Axial H
Equatorial
Axial H
(b)

Axial H atoms are labeled. All other H atoms are equatorial.

Your Turn 4.20

Think

The H atoms on the ring involving 1,3-diaxial interactions in Figure 4-30a are on which C atoms? Count the C–CH_3 as C1 and then locate the 1,3-diaxial interactions. The CH_2 that is gauche to the CH_3 in Figure 4-30a is C3. This carbon is two C atoms away from the C–CH_3. Following this pattern, which other carbon number would have the CH_2 gauche to the CH_3?

Solve

The H atoms involved in 1,3-diaxial interactions are the C–H bonds on C3 and C5, as shown below. The other CH_2 that is gauche to the CH_3 is the one that contains C5. This is difficult to see if viewing only the Newman projection in Figure 4-30a, so you are encouraged to build the model and look down the C1–C6 bond.

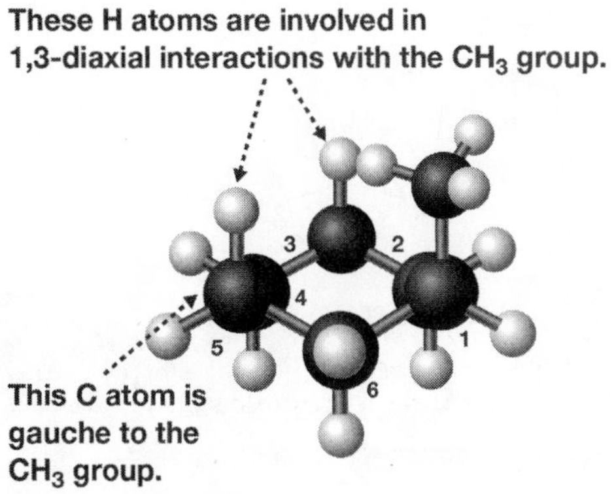

Your Turn 4.21

Think

How is the cyclohexane ring represented in a Haworth projection? How do you transform a planar cyclohexane ring with wedges/dashes into a Haworth projection and vice versa? In a Haworth projection, in which directions should the bonds to the substituents point?

Solve

In both representations, the cyclohexane ring is depicted as planar and not a chair conformation. In **(a)** and **(b)** the ring in the plane of the page is rotated 90° to depict the Haworth projection. The wedge bonds in **(a)** and **(b)** are drawn up and the dash bonds are drawn down. To transform a Haworth projection into a wedge/dash drawing, rotate the plane of the ring to be in the plane of the page, so that you view the Haworth projection from the top. Bonds that are pointed up in the Haworth projection become wedges in the wedge/dash drawing and bonds that point down become dashes.

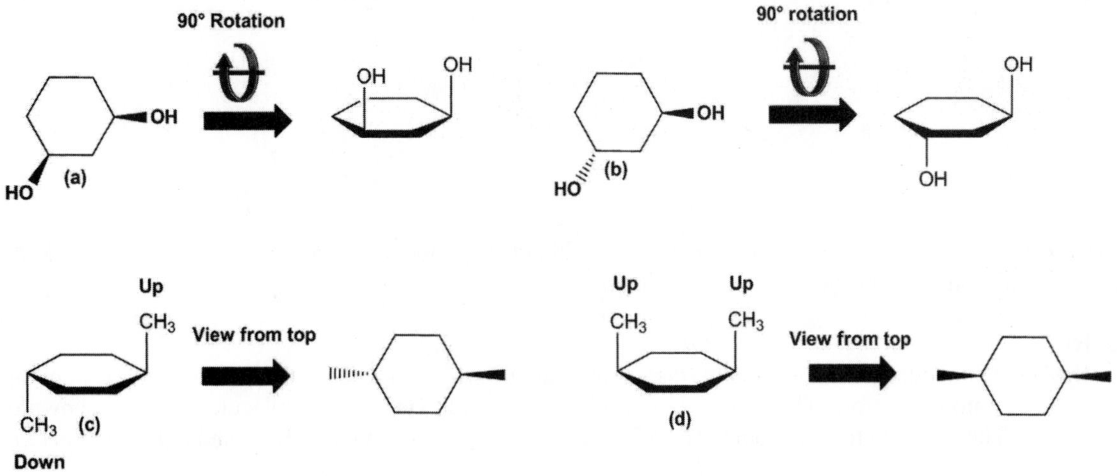

Your Turn 4.22

Think

Use your model kit to construct *trans*-1,3-dimethylcyclohexane exactly as shown in the structure on the left. In both structures given, there is one axial CH_3 and one equatorial CH_3. Without carrying out a chair flip, can you reorient the molecule to match up the axial and equatorial CH_3 groups with how they appear in the structure on the right?

Solve

To translate the conformation of *trans*-1,3-dimethylcyclohexane on the left into the "ring-flipped" conformation on the right, without flipping the chair, carry out the following rotations of the molecule.

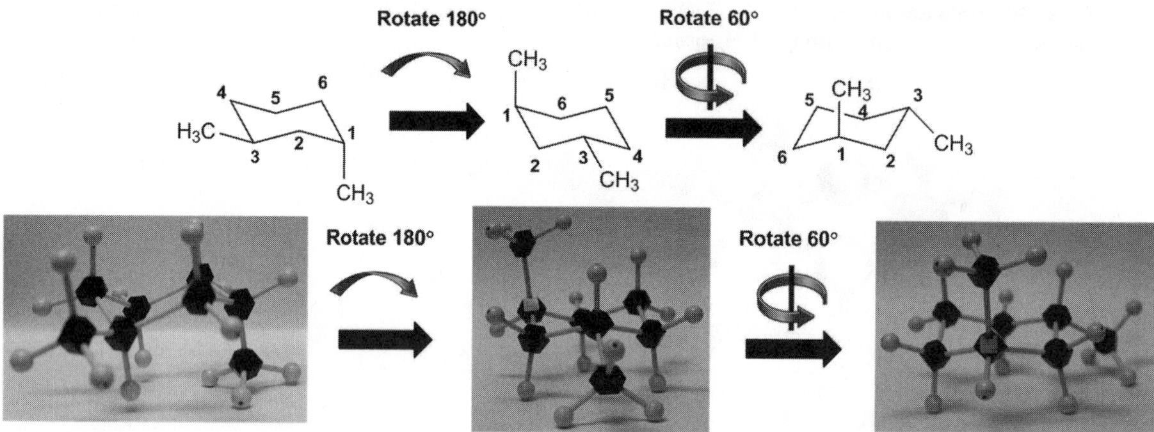

Your Turn 4.23

Think

Find the longest continuous chain of carbon atoms and number the chain so that the first substituent receives the lowest number. Is that carbon chain the same length as the ones in the previous two molecules? On what carbon atoms of that chain are the substituents located? Are these carbon atoms the same as in the previous two molecules?

Solve

The longest continuous chain of carbon atoms has six carbons. The first substituent is a CH_3 group located on C2. The second substituent is a CH_3 located on C4. This is the same connectivity as the previous two molecules. Recall that there is free rotation about single C–C bonds. These three molecules are *conformational isomers*.

CH_3 groups on C2 and C4

Your Turn 4.24

Think

What cannot be present in a saturated molecule? Without those features present, can you add more hydrogen atoms to the nonhydrogen atoms given in each molecule?

Solve

Each saturated molecule must have only single bonds and no rings, allowing it to contain the maximum number of hydrogen atoms possible. The saturated molecules that correspond to each molecule given are shown on the next page. The formula changes from C_2H_4 to C_2H_6, C_2H_2 to C_2H_6, C_2H_4O to C_2H_6O, and C_2H_4O to C_2H_6O.

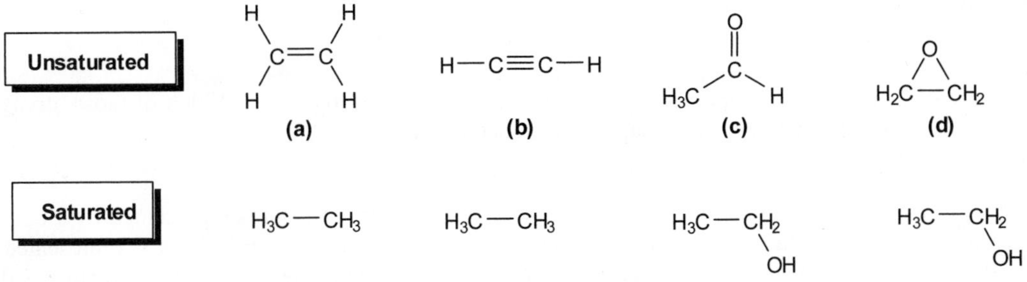

Unsaturated	$H_2C{=}CH_2$ (a)	$H{-}C{\equiv}C{-}H$ (b)	$H_3C{-}C(=O){-}H$ (c)	$H_2C{-}CH_2$ with O bridge (d)
Saturated	$H_3C{-}CH_3$	$H_3C{-}CH_3$	$H_3C{-}CH_2{-}OH$	$H_3C{-}CH_2{-}OH$

Your Turn 4.25

Think

Look at Table 1-6 if you need to look up the functional groups. Be mindful of atoms other than carbon and hydrogen and the presence of multiple bonds.

Solve

The ROH functional group is an alcohol, the ROR functional group is an ether, the C=C functional group is an alkene, the RC(O)H functional group is an aldehyde, and the RC(O)R functional group is a ketone.

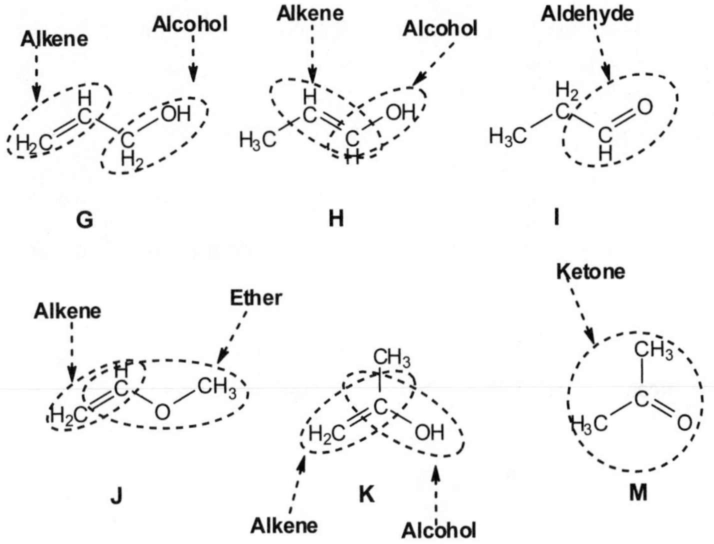

Alkene
Alcohol
G
Alkene
Alcohol
H
Aldehyde
I
Alkene
Ether
J
Alkene
Alcohol
K
Ketone
M

Chapter Problems

Problem 4.2

Think

What atoms are connected by the bond not observable in the Newman projection? Which of those atoms is in front and which is behind? What atoms or groups are attached to those atoms?

Solve

The bond not shown in each example is a C–C single bond. The C atom in front is represented by a point and the three groups connected to the front C atom all attach at the point. The C atom in back is represented by a circle and the three groups connected to the back C atom all attach to the circle. Each Lewis structure is drawn from a 90° counterclockwise rotation. The prime (′) labels below are used to help visualize the rotation. Each rotation below is counterclockwise as viewed from above, but the rotation could instead be clockwise.

(a) The C atom in front is connected to two CH_3 groups and one H atom. The C atom in back is connected to two H atoms and a CH_3 group.

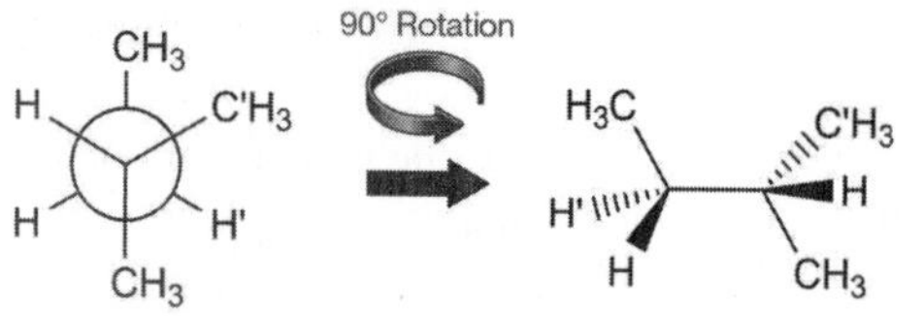

(b) The C atom in front is connected to three H atoms. The C atom in back is connected to three Cl atoms.

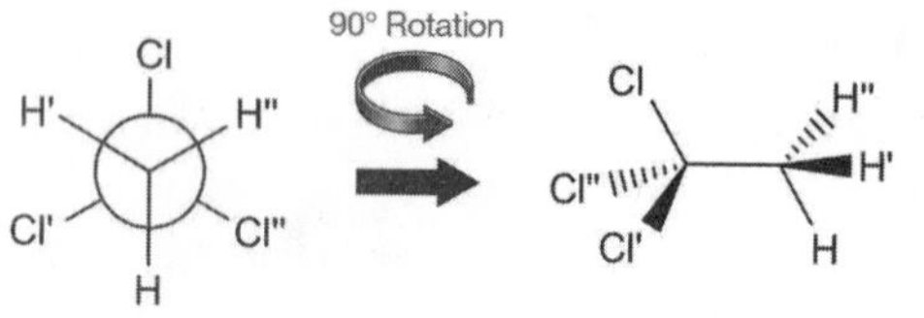

(c) The C atom in front is connected to two F atoms and one CF_3 group. The C atom in back is connected to three F atoms.

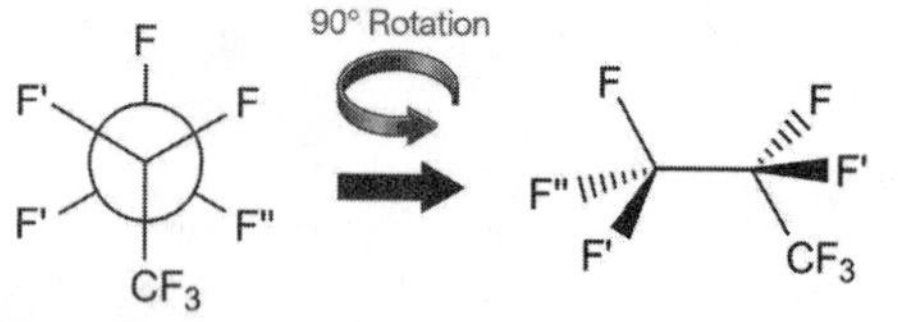

Problem 4.3

Think

What groups are connected by the bond not observable in the Newman projection? Which one is in back and which is in front? Which group are you asked to rotate, and which should remain frozen in place? Using Figure 4-3 and Your Turn 4.3 as guides, what is the angle between each of the three groups on a C atom in a Newman projection? How many 60° rotations does it take for one group to get to the location of the next group? How many 60° rotations does it take to get all the way around the circle?

Solve

Each bonding group is 120° apart. It takes two 60° rotations to go from group to group and it takes six 60° rotations to go all the way around the circle. See the rotations below. The back group is locked in place and thus the **X** label will be pointed down for the entire rotation sequence. The front group is rotating around the circle clockwise and thus the **Y** label will move 60° clockwise with each rotation. Both are in boldface for emphasis. Note: All rotations are performed clockwise.

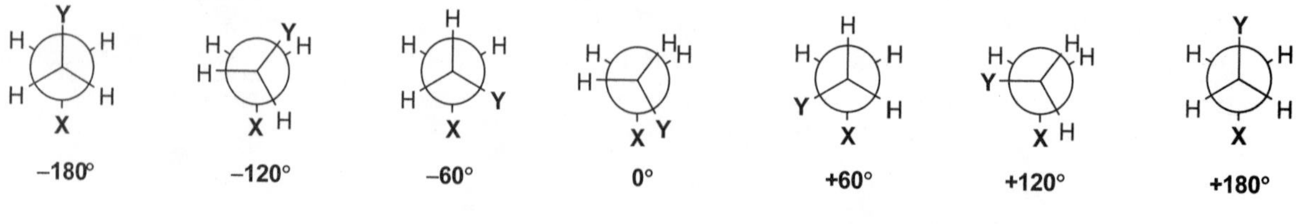

–180° –120° –60° 0° +60° +120° +180°

Problem 4.4

Think

How is 1,2-dichloroethane related to 1,2-dibromoethane, whose conformational analysis is shown in Figure 4-7? How is steric strain related to the relative energy of the conformations? Which interactions (staggered or eclipsed) are higher in energy? Which staggered interactions, anti or gauche, are higher in energy?

Solve

1,2-Dichloroethane differs from 1,2-dibromoethane only by the exchange of the two Br atoms for two Cl atoms. Therefore, the conformational analysis should look qualitatively the same as Figure 4-7. Staggered conformations minimize the torsional strain because the groups have a dihedral angle of 60° and, therefore, are lower in energy. Staggered anti conformations are lowest in energy due to less steric strain, followed by staggered gauche conformations. Eclipsed conformations are higher in energy than staggered conformations because of torsional strain. In the conformational analysis shown below, conformations **A**, **C**, **E**, and **G** are staggered and, therefore, are lower in energy. **A** and **G**, which are the same, are staggered anti and, therefore, have the *lowest* energy. **B**, **D**, and **F** are eclipsed and have the higher energy. **D** has the highest energy because the two larger Cl atoms are eclipsed and, therefore, have the largest steric strain.

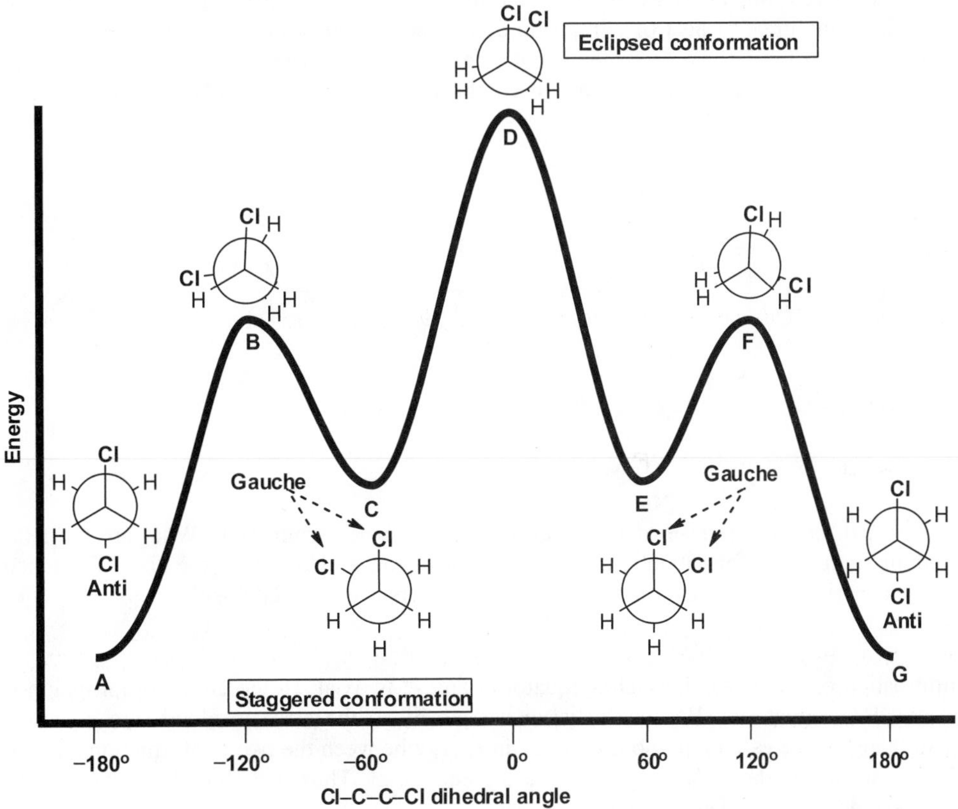

Problem 4.5

Think

Draw out each structure. Which compound has larger steric strain when rotated about the C–C bond? How does steric strain relate to the rotational energy barrier?

Solve

Because Br is much larger than F, steric strain in the eclipsed conformations in 1,2-dibromoethane is much greater than the steric strain in the eclipsed conformations of 1,2-difluoroethane. Thus, the rotational energy barrier is larger in 1,2-dibromoethane. See the structures on the next page.

Eclipsed conformation

Br Br H H H H Br Br H H H H

1,2-Dibromoethane

More steric strain = larger rotational energy barrier

F F H H H H F F H H H H

1,2-Difluoroethane

Problem 4.7

Think

What are the molecular formulas of these molecules? What are the structural differences among them? How do those differences translate into heats of combustion? Which compound has the largest ring strain?

Solve

All four of these compounds have the molecular formula C_7H_{14} and contain only C–C and C–H single bonds. Difference in heats of combustion, therefore, will largely reflect differences in ring strain. The structure that contains two three-membered rings, **D**, is the most strained, so it will give off the most heat during combustion—it will have the greatest heat of combustion. Next comes the structure with the three- and four-membered rings, **C**, followed by the two four-membered ring structure, **A**. The four- and five-membered ring structure, **B**, is the least strained. The order of heats of combustion is **B** < **A** < **C** < **D**. The ring numbers are listed below each structure.

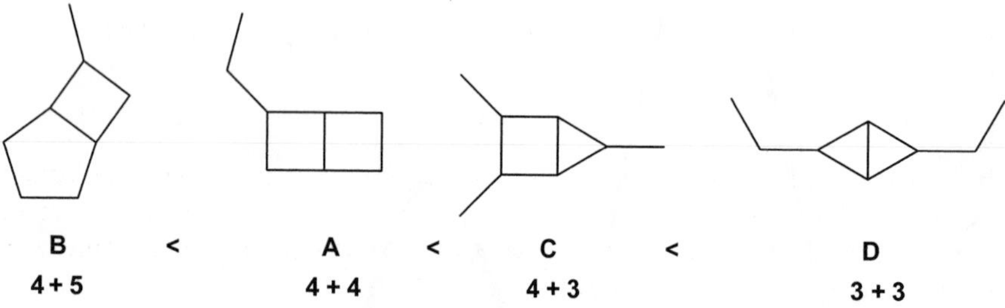

Problem 4.9

Think

What is the difference, if any, in the size of hydrogen compared to deuterium (D)? What are the axial and equatorial interactions in each conformation? Are there any differences? How does the size of a substituent attached to a cyclohexane ring govern the preference for the axial versus equatorial position?

Solve

In the chair conformation on the left below, D is equatorial and H is axial. In the chair conformation on the right, D is axial and H is equatorial. Because D is an isotope of H, the electron clouds are essentially no different in size, so there will be essentially no difference in energy between the two conformations. Remember that the bulkier substituent has the stronger preference for equatorial. Thus, we should expect that the two conformations will be present in equal abundance.

Equatorial D H Axial

Axial D H Equatorial

Problem 4.10

Think

What is the difference in size of I compared to F? Which group is bulkier, CI_3 or CF_3? Does the larger substituent have a greater preference for an axial or equatorial position?

Solve

The I atom is much larger than the F atom, making the CI_3 group bulkier than CF_3. The CI_3 group, therefore, has a stronger preference for the equatorial position than does the CF_3 group. Trifluromethylcyclohexane, consequently, will have a greater percentage of molecules in the axial position compared to triiodomethylcylcohexane.

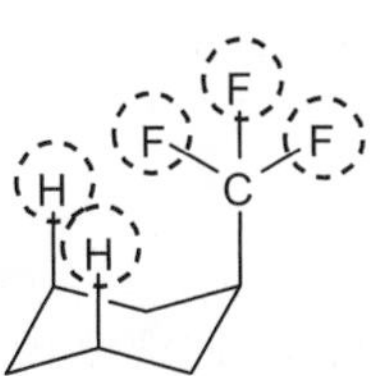

CF_3 smaller
More stable in axial position compared to CI_3.

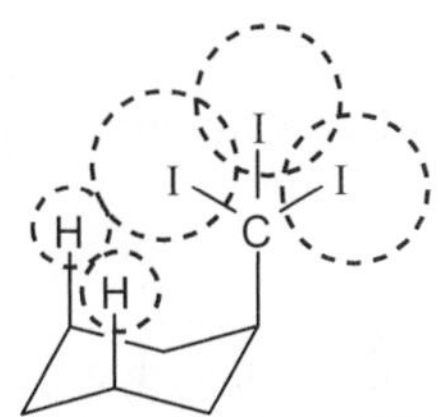

CI_3 larger
Less stable in axial position compared to CF_3.

Problem 4.12

Think

Does one substituent on the ring require more room than the other? Which position, axial or equatorial, offers more room? Can both substituents achieve that position?

Solve

The two methyl (CH_3) groups are the largest substituents on the ring and each is more stable in the equatorial position. Draw out *trans*-1,2-dimethylcyclohexane (in a Haworth projection or a line structure with dash–wedge) and then transfer to a chair. Perform a ring flip. As shown below, *trans*-1,2-dimethylcyclohexane can achieve equatorial positions for both CH_3 groups simultaneously.

CH_3 / CH_3 or H_3C / CH_3 → CH_3 / CH_3 ⇌ Ring flip H_3C / CH_3

More stable, CH_3 groups both equatorial

***trans*-1,2-Dimethylcyclohexane**

Problem 4.13

Think

Which substituent requires more room, methyl or *t*-butyl? Which position, axial or equatorial, offers more room? Can both substituents achieve that position?

Solve

Because the *t*-butyl group is so bulky, the ring is essentially "locked" in the equatorial position. Two carbons away on the ring, the CH_3 group must be on the opposite side of the ring—if the *t*-butyl group is up, the methyl group must be down. The position occupied by CH_3, therefore, must be axial. See structures on the next page.

More stable, $C(CH_3)_3$ equatorial

Problem 4.14

Think

Which substituent requires more room, methyl or trichloromethyl? Which position, axial or equatorial, offers more room? Can both substituents achieve that position?

Solve

Both CH_3 and CCl_3 must be on the same side of the ring—for example, both pointed up (as shown below). Only one of those positions can be equatorial, whereas the other is axial. Because CCl_3 is larger than CH_3, the compound is more stable with CCl_3 in the equatorial position, forcing CH_3 into the axial position, as shown below.

More stable, CCl_3 equatorial

Problem 4.16

Think

Do both compounds have the same molecular formula? Is the largest continuous chain or ring in each molecule the same? Are the C atoms involved in the double bonds assigned the same numbers in each molecule? In each molecule, are the substituents attached to the numbered C atoms in the same order?

Solve

Pairs **(a)**, **(b)**, and **(c)** have different chemical formulas and, therefore, are not isomers.

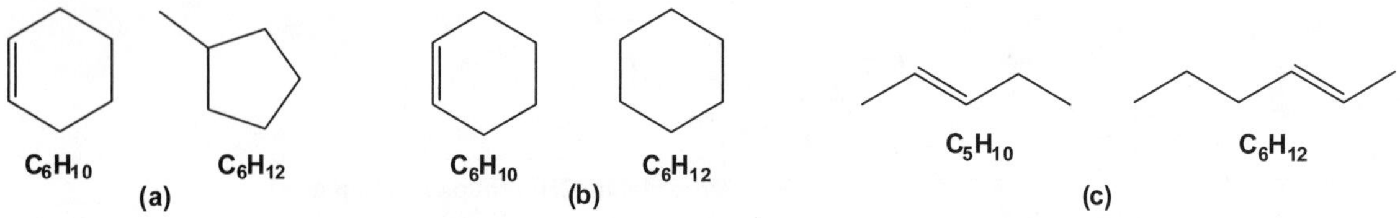

Pairs **(d)** and **(i)** have the same molecular formula but different connectivity, making them constitutional isomers. For pair **(d)**, both have the formula C_7H_{14}, but the methyl groups are on different C atoms. In the first molecule, the methyl groups are on C1 and C2; in the second molecule, the methyl groups are on C1 and C3. For pair **(i)**, both have the formula C_9H_{16}, but the propyl group is on a different C atom in each molecule. In the first molecule, the propyl group is on C3; in the second molecule, the propyl group is on C4.

Pairs **(e)**, **(f)**, **(g)**, and **(h)** have the same formula and same connectivity and, therefore are not constitutional isomers. See numbering below.

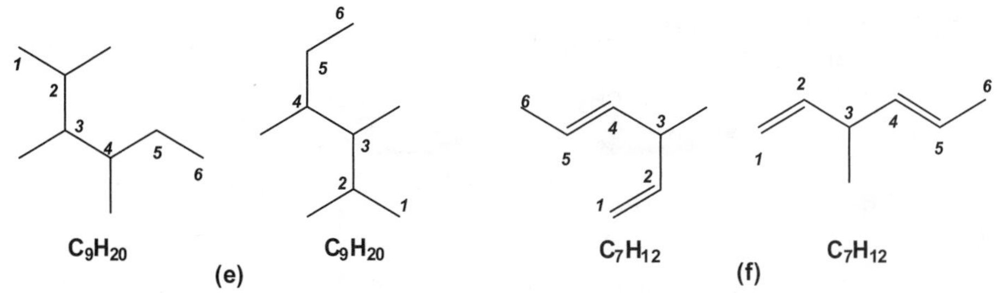

C_9H_{20} **(e)** C_9H_{20} C_7H_{12} **(f)** C_7H_{12}

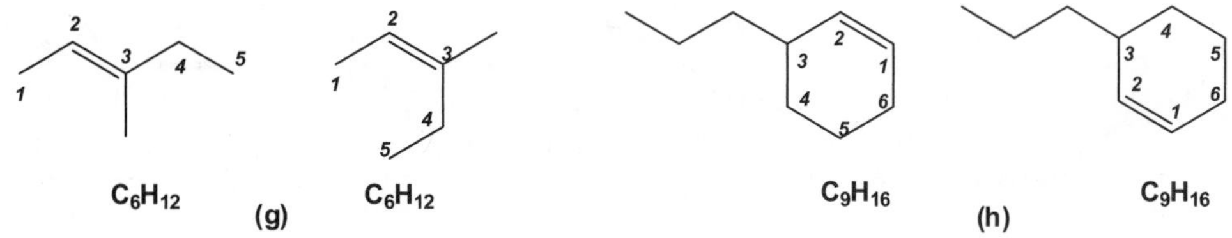

C_6H_{12} **(g)** C_6H_{12} C_9H_{16} **(h)** C_9H_{16}

Problem 4.17

Think

Do both compounds have the same molecular formula? Is the largest continuous chain or ring in each molecule the same? Are the C atoms involved in the double bonds assigned the same numbers in each molecule? In each molecule, are the substituents attached to the numbered C atoms in the same order?

Solve

Pairs **(b)**, **(e)**, and **(g)** have the same molecular formula and different connectivities, and, therefore, are constitutional isomers. For pair **(b)**, both have the formula C_8H_{16}, but the methyl groups are on different C atoms. In the first molecule, the methyl groups are on C1 and C3 but in the second molecule the methyl groups are on C1 and C2. For pair **(e)**, the methyl groups are on different carbon atoms—C3 or C4, respectively. For pair **(g)** (line structures drawn on the right), both have the same formula, C_5H_{12}, but the length of the carbon chains differ, as do the methyl and substituent number and location.

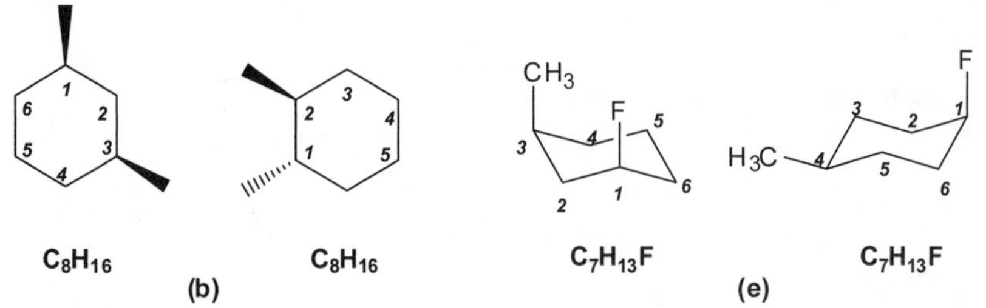

C_8H_{16} **(b)** C_8H_{16} $C_7H_{13}F$ **(e)** $C_7H_{13}F$

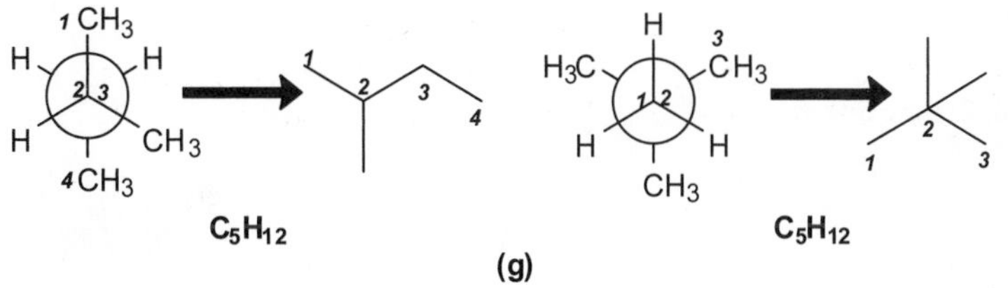

C_5H_{12} **(g)** C_5H_{12}

Pairs **(a)**, **(c)**, **(d)**, and **(f)** have the same formula and same connectivity and, therefore, are not constitutional isomers. See numbering below.

Problem 4.19

Think

How many rings, double bonds, and triple bonds are present? How much does each contribute to the overall IHD?

Solve

A double bond and a ring each contribute one to the index of hydrogen deficiency (IHD) and a triple bond contributes two to the IHD.

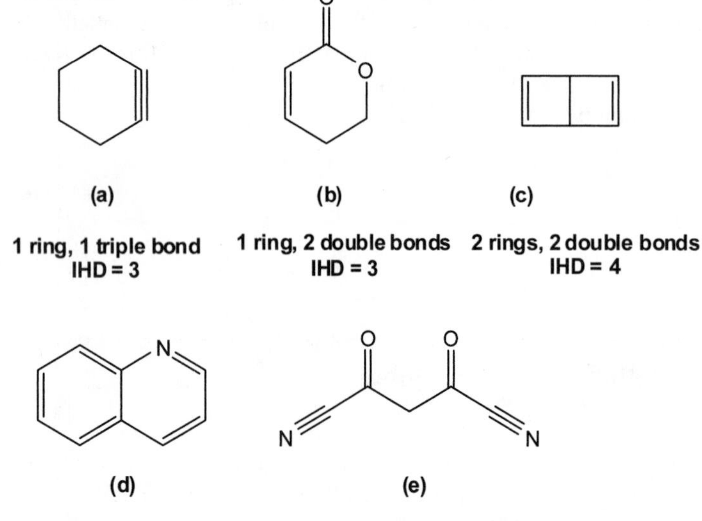

Problem 4.20

Think

In each functional group in Table 1-6, how many rings, double bonds, and triple bonds are present? How much does each of these features contribute to the overall IHD?

Solve
Contributing to an IHD of 1 are alkene (one double bond), epoxides (one ring), ketones (one double bond), aldehydes (one double bond), carboxylic acids (one double bond), esters (one double bond), and amides (one double bond). Contributing to an IHD of 2 are alkynes and nitriles (one triple bond each). No functional groups from the table contribute to an IHD of 3. Aryl groups contribute to an IHD of 4.

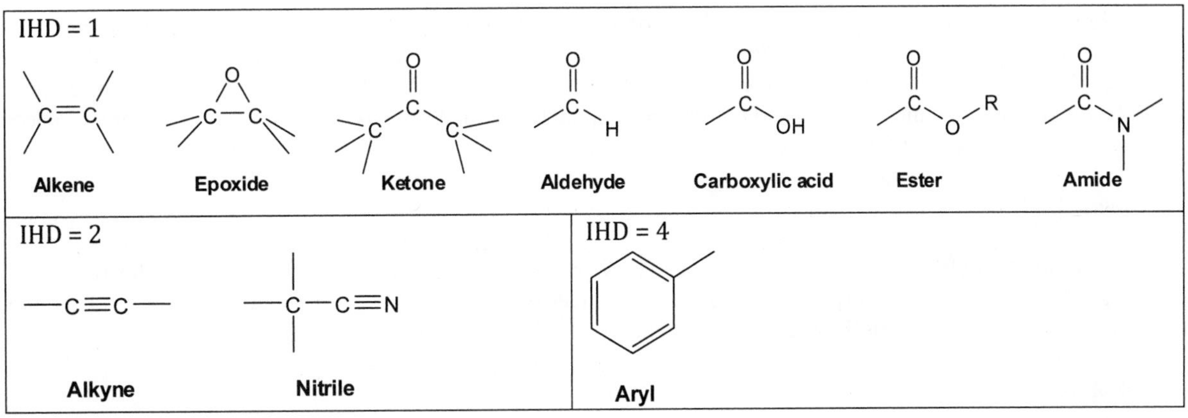

IHD = 1

Alkene Epoxide Ketone Aldehyde Carboxylic acid Ester Amide

IHD = 2

Alkyne Nitrile

IHD = 4

Aryl

Problem 4.22

Think
What is the formula for an analogous saturated compound? How many H atoms are missing from the formula given?

Solve
We can construct a (hypothetical) saturated molecule as shown below, having only single bonds and no rings. This molecule has 11 H atoms, so a compound with the formula $C_4H_7N_2OF$ is deficient by four H atoms, or two H_2 molecules. Thus, the IHD = 2.

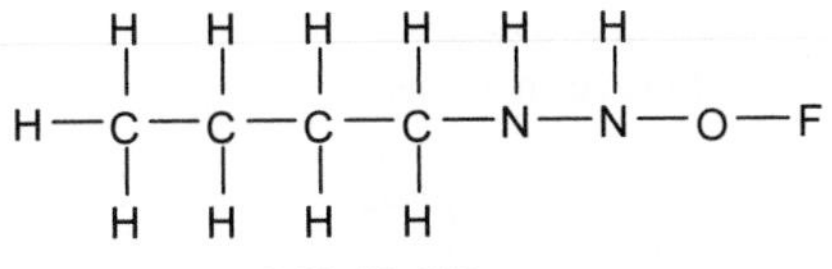

$C_4H_{11}N_2OF$

Problem 4.23

Think
What is the formula for each saturated compound described? What is n and $2n + 2$ equal to in each case? How does $2n + 2$ compare to the number of H atoms?

Solve
The formulas for the three saturated compounds are: C_4H_{10}, $C_4H_{10}O$, and $C_4H_{10}O_2$. Each has $n = 4$ and $2n + 2 = 10$. Therefore, each additional O atom does not change the number of H atoms required to achieve a saturated compound.

Problem 4.24

Think
What is the formula for each saturated compound described? What is n and $2n + 2$ equal to in each case? How does $2n + 2$ compare to the number of H atoms in each molecule? What is the effect of adding a N atom? How many bonds does an uncharged N have? What is the effect of adding an F atom? How many bonds does an uncharged F have?

Solve

Nitrogen. The formulas for the saturated compounds are: C_4H_{10}, $C_4H_{11}N$, and $C_4H_{12}N_2$. In each case, $n = 4$ and $2n + 2 = 10$. With no N atoms, the number of H atoms is $2n + 2$; with one N atom, the number of H atoms is $(2n + 2) + 1$. With two N atoms, the number of H atoms is $(2n + 2) + 2$. Each additional N atom requires one additional H atom to achieve a saturated compound. Uncharged N atoms have three bonds. Two bonds are required to continue the chain, and the additional bond is made up by adding in a H atom.

Fluorine. The formulas for the saturated compounds are: C_4H_{10}, C_4H_9F, and $C_4H_8F_2$. In each case, $n = 4$ and $2n + 2 = 10$. With no F atoms, the number of H atoms is $2n + 2$; with one F atom, the number of H atoms is $(2n + 2) - 1$. With two F atoms, the number of H atoms is $(2n + 2) - 2$. Each additional F atom requires one fewer H atom to achieve a saturated compound. An uncharged F atom has one bond and, therefore, replaces a H atom. This is the case for all halogen atoms.

Problem 4.25

Think

What do the prefixes "aldo" and "keto" specify? What tells you the number of C atoms in each structure? If the molecule is a monosaccharide, how should the number of hydrogen and oxygen atoms compare to the number of carbon atoms? Refer to Figure 4-48 as a general reference to each acyclic sugar.

Solve

The "aldo" and "keto" prefixes specify whether the C=O group is part of an aldehyde, requiring it to be at the end of the chain, or part of a ketone, requiring it to be internal. The number of carbons in the chain is specified in the name, too: "tri" = 3; "tetr" = 4; "pent" = 5; "hex" = 6. For example, **(a)** aldotetrose has a terminal carbonyl with the aldehyde functional group and has four C atoms total. See structures for **(a)**–**(e)** below.

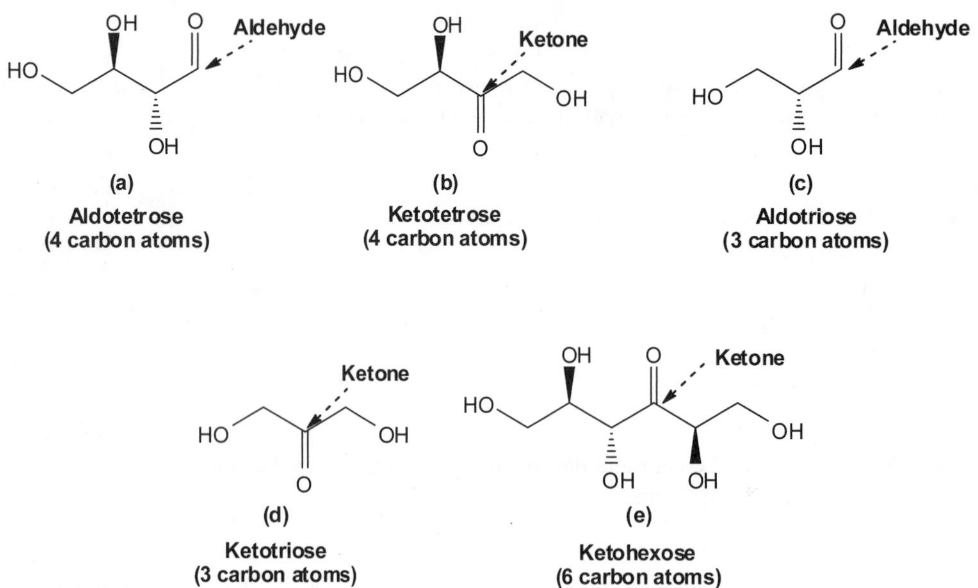

Problem 4.26

Think

How many rings, double bonds, and triple bonds are present? How much does each contribute to the overall IHD?

Solve

A double bond and a ring each contribute one to the IHD and a triple bond contributes two to the IHD.

(a)
No rings, no multiple bonds
IHD = 0

(b)
1 double bond
IHD = 1

(c)
1 ring, 1 triple bond
IHD = 3

(d)
1 ring, 3 double bonds
IHD = 4

(e)
1 ring
IHD = 1

(f)
2 rings, 2 double bonds
IHD = 4

(g)
2 rings, 4 double bonds
IHD = 5

(h)
1 double bond, 2 triple bonds
IHD = 5

Problem 4.27

Think

What is the formula for each saturated compound listed? For each IHD given, how many hydrogen atoms should be removed from the saturated compound?

Solve

For the nonhydrogen atoms given, the formula for a corresponding saturated molecule is provided. For each IHD, one H_2 molecule should be missing, so two H atoms need to be removed.

	Condition	Saturated Molecule	Formula of Saturated Molecule	IHD	Number of H Atoms to Remove	Formula of Described Molecule
(a)	4 C	$CH_3CH_2CH_2CH_3$	C_4H_{10}	0	0	C_4H_{10}
(b)	4 C	$CH_3CH_2CH_2CH_3$	C_4H_{10}	2	4	C_4H_6
(c)	3 C, 2 O	$CH_3CH_2CH_2OOH$	$C_3H_8O_2$	1	2	C_3H_6O
(d)	5 C, 2 Cl, 1 N	$CH_3(CH_2)_4NCl_2$	$C_5H_{11}NCl_2$	3	6	$C_5H_5NCl_2$
(e)	1 C, 1 N	CH_3NH_2	CH_5N	2	4	CHN
(f)	6 C, 2 N, 1 O, 3 F	$CF_3(CH_2)_5NHNHOH$	$C_6H_{13}N_2OF_3$	4	8	$C_6H_5N_2OF_3$

Problem 4.28

Think

What is the formula for each saturated compound listed? Which atom does Si behave like? Which atom does S behave like? How does a negative charge affect the number of bonds and lone pairs on an atom? What is the difference in the saturated formula compared to the formula given? How many H atoms does each IHD represent?

Solve

See the filled in table below. Si behaves like C and should be added to the value of n in calculating $2n + 2$. S behaves like O and, therefore, there is no change to the number of H atoms. A negative charge decreases the number of H atoms and a positive charge increases the number of H atoms.

	Given Formula	Saturated Molecule	Formula of Saturated Molecule	Number of H Atoms Missing	IHD
(a)	C_6H_6	$CH_3(CH_2)_4CH_3$	C_6H_{14}	8	4
(b)	$C_6H_5NO_2$	$CH_3(CH_2)_5NHOOH$	$C_6H_{15}NO_2$	10	5
(c)	$C_8H_{13}NOF_2$	$CH_3(CH_2)_7N(F)O(F)$	$C_8H_{17}NOF_2$	4	2
(d)	$C_4H_{12}Si$	$CH_3(CH_2)_3SiH_3$	$C_4H_{12}Si$	0	0
(e)	$C_6H_5O^-$	$CH_3(CH_2)_5O^-$	$C_6H_{13}O^-$	8	4
(f)	$C_4H_6O_3S$	$CH_3(CH_2)_3OOOSH$	$C_4H_{10}O_3S$	4	2

Problem 4.29

Think

Which substituent attached to the cyclohexane ring requires more room? Which position, axial or equatorial, offers more room? Can both substituents achieve that position?

Solve

A *t*-butyl group $[C(CH_3)_3]$ requires more room than a methyl (CH_3). The equatorial position offers more room.

(a) One CH_3 group will be axial and the other CH_3 group will be equatorial. There is no way to achieve both CH_3 groups in the equatorial position.

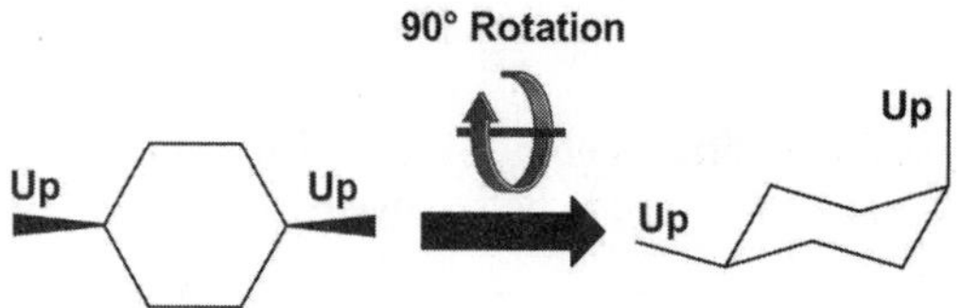

(b) Both CH_3 groups are able to be in the equatorial position.

(c) Both the CH_3 and the $C(CH_3)_3$ groups are able to be in the equatorial position.

(d) Only one group is able to be in the equatorial position. The $C(CH_3)_3$ is larger and, therefore, will be more stable in the equatorial position over the CH_3 group.

90° Rotation

Down Down Down Down

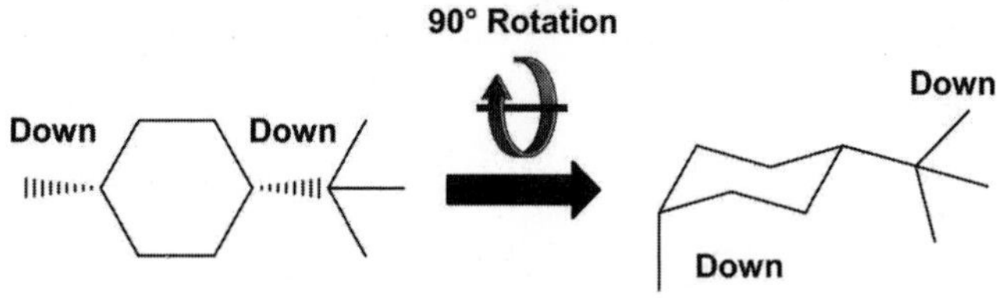

(e) Both the CH_3 and the $C(CH_3)_3$ groups are able to be in the equatorial position.

90° Rotation

Down Down Down Down

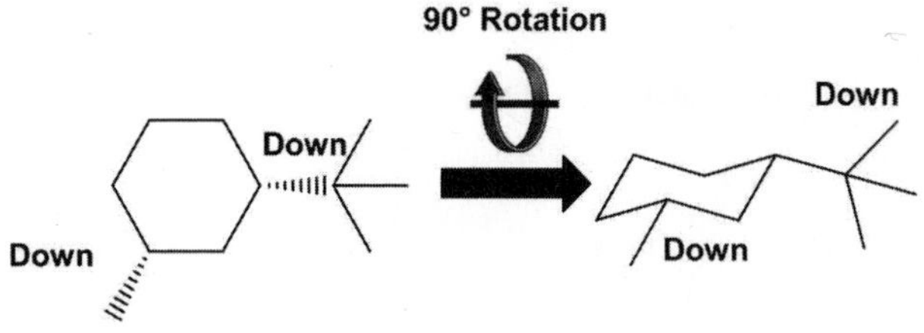

(f) Only one group is able to be in the equatorial position. The $C(CH_3)_3$ is larger and, therefore, will be more stable in the equatorial position over the CH_3 group.

90° Rotation

Up Up Up Up

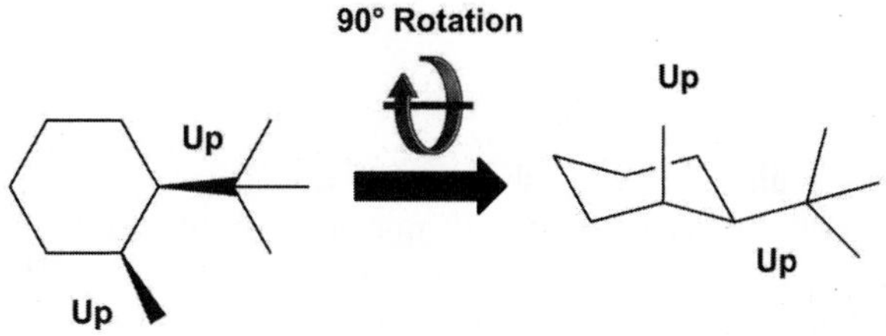

(g) Both the CH_3 and the $C(CH_3)_3$ are able to be in the equatorial position.

90° Rotation

Up Down Up Down

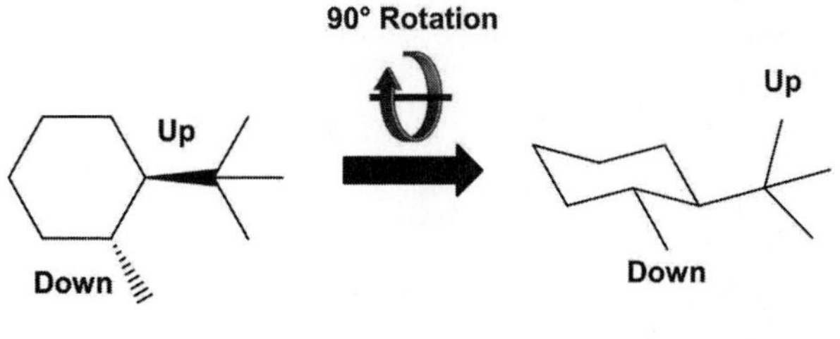

Problem 4.30

Think

Construct a molecular model of each molecule, exactly as shown, and orient it so you are looking down the bond indicated. How do you represent the front C atom in a Newman projection? How do you represent the back C atom? Should the C–C bond indicated be visible? How many atoms should be shown bonded to each C atom? As drawn, is the staggered or eclipsed conformation shown?

Solve

The front C atom is represented by a dot and the back C atom is represented by a circle. The C–C bond is not shown, as it is perpendicular to the plane of the paper and is not visible. The remaining three groups on the front C atom converge at the point and the remaining three groups on the back C atom connect to the circle.

These Newman projections result from looking down the structures in the problem from the left side as shown by **(a)** on the next page. As you look from the left, the front C atom has H, H′, and Br, and the back C atom has CH_3, H, and H′. The flat bonds in the dash–wedge structure become up in the Newman projection, the wedges become bonds to the right, and dashes become bonds to the left. See the figure on the next page.

Eclipsed

(b) Eclipsed

(c) Staggered

(d) Staggered (gauche)

(e) Staggered (anti)

(f) Staggered

(g) Staggered

(h) Eclipsed

Problem 4.31

Think

Construct a molecular model of each molecule, exactly as shown, and orient it so you are looking down the bond indicated. Looking down the C–C bond indicated, which C atom is in front and which C atom is in the back? On the front C atom, are there groups pointing up, down, left, or right? On the back C atom, are there groups pointing up, down, left, or right? On which side is the ring connected? Which molecule minimizes steric strain?

Solve

As we can see in the Newman projections below, there is more steric strain in the cis compound than in the trans due to the eclipsing of the CH_3 groups.

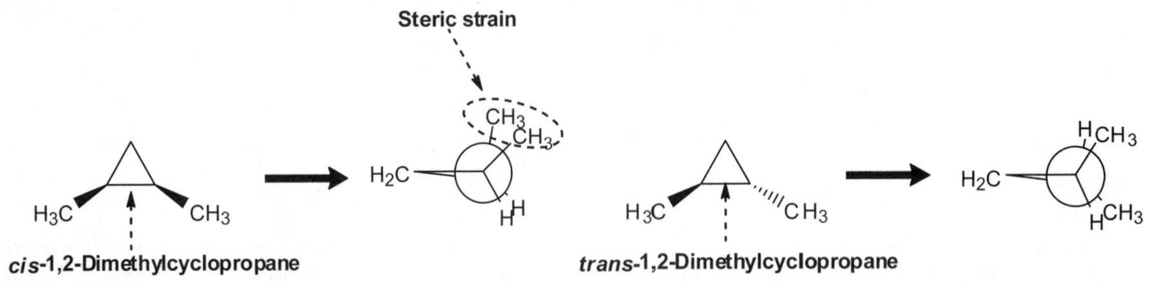

Problem 4.32

Think

Construct a molecular model of each molecule, exactly as shown, and orient it so you are looking down the bond not visible in the Newman projection. If the Newman projection undergoes a 90° rotation as indicated by the arrow, , which bonds are wedge, which bonds are dash, and which bonds are in the plane of the paper?

Solve

These dash–wedge structures are the result of a 90° rotation as indicated by the arrow, ↻, as shown by **(a)** below.

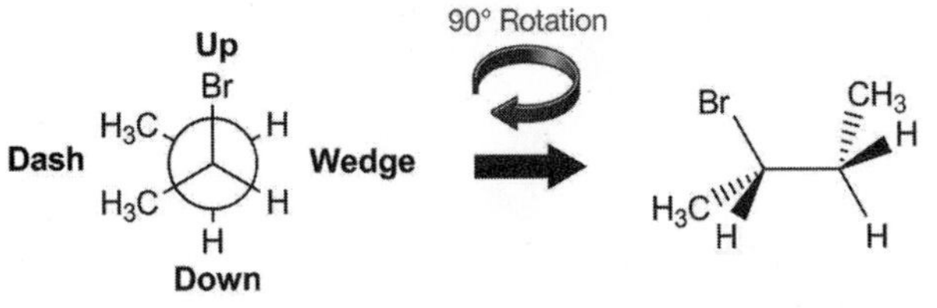

Problem 4.33

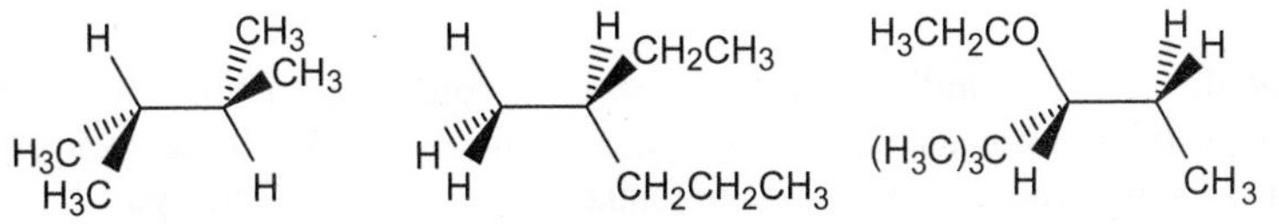

Think

Which interactions (staggered or eclipsed) are higher in energy? Which staggered interactions, anti or gauche, are higher in energy? Which group, CH_3, Cl, CBr_3, or H, takes up the most space?

Solve

Staggered anti conformations are lowest in energy (most stable), followed by staggered gauche conformations. Eclipsed conformations are higher in energy compared to staggered conformations because they have substantial torsional strain. Compounds **D**, **B**, and **C** are all eclipsed and, therefore, are the least stable. **D** is the least stable of these three because the largest groups, CH_3 and CBr_3, have a dihedral angle of 0° and are closest in space. The staggered conformations are more stable. **F** is the most stable because the largest groups, CH_3 and CBr_3, are anti.

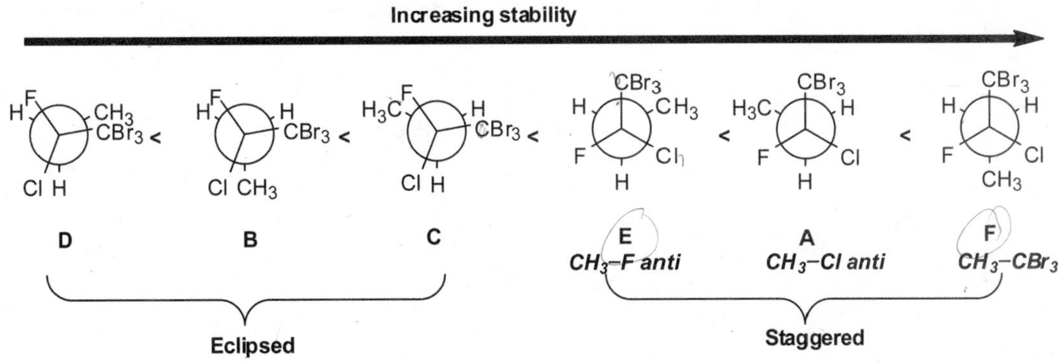

Problem 4.34

Think

Build a molecular model of cyclohexane and orient it so you are looking down the bonds indicated by the Newman projections. Can you identify the plane that is roughly defined by the cyclohexane ring? Which C–H bonds are straight up and down, perpendicular to the plane? Which atoms point outward from the center of the ring?

Solve

From the Newman projection, we can identify four axial H atoms (up and down) and four equatorial H atoms (point outward from the center of the ring). The other four H atoms that are labeled CH_2 are not distinguishable as either axial or equatorial. See the figure on the next page.

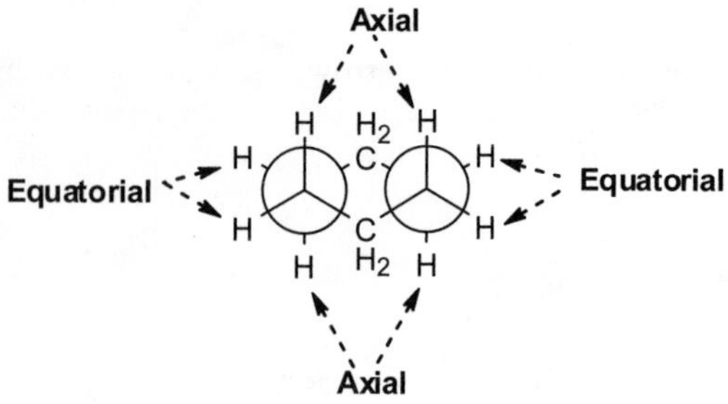

Axial

Equatorial

Equatorial

Axial

Problem 4.35

Think

Build a molecular model of each molecule exactly as indicated in the Newman projections given. Which groups take up the most space? Are these groups axial or equatorial in the conformation given? Are those groups axial or equatorial after a chair flip? Is there more room in an axial or equatorial position? Once you have the molecular model in the more stable chair conformation, reorient it in space so you can use it as a guide to draw the dash–wedge and Haworth projections.

Solve

(i) is not in the most stable conformation because both Cl atoms are in axial positions. The more stable conformation, in which both Cl atoms are in equatorial positions, is shown below. **(ii)** is not in the most stable conformation because the very bulky *t*-butyl group is in an axial position. After a chair flip, the *t*-butyl group is in an equatorial position. Only **(iii)** is in the most stable chair conformation, because CI_3, which is bulkier than CBr_3, is already in the equatorial position.

	(i)	(ii)	(iii)
Given Structure	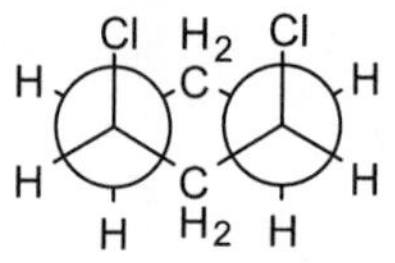	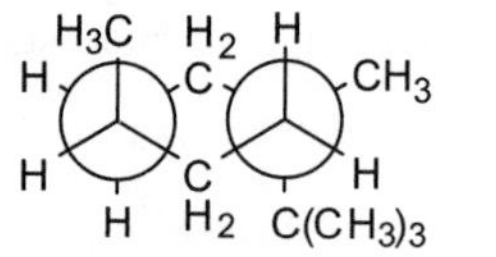	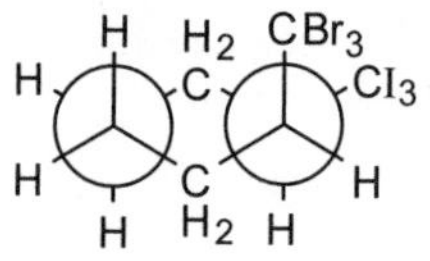
Dash–Wedge Structure	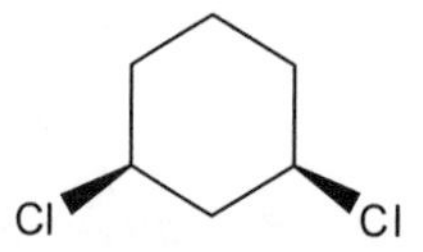	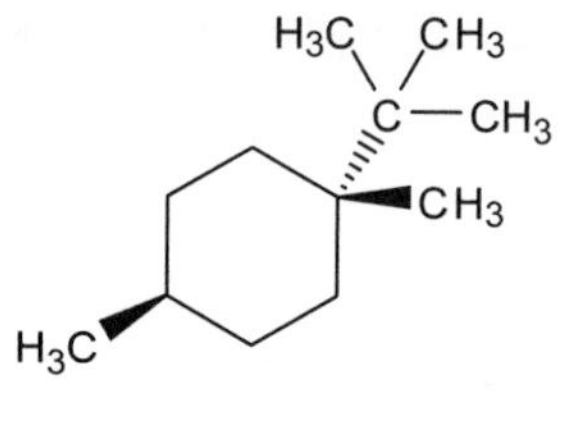	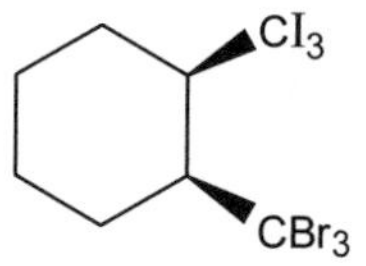
Haworth Projection	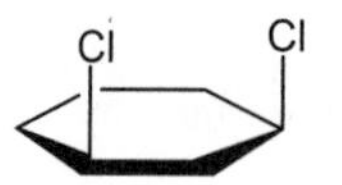	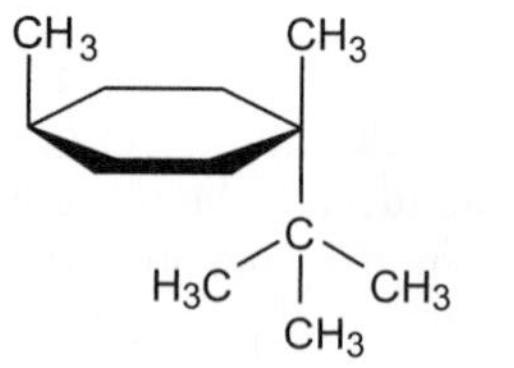	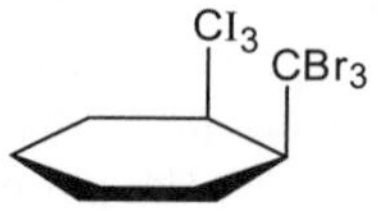
Most Stable Conformation	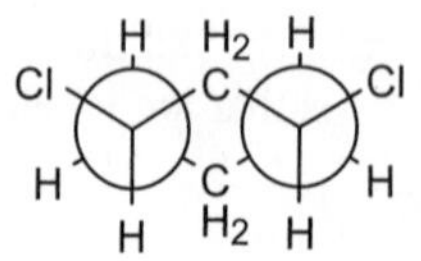	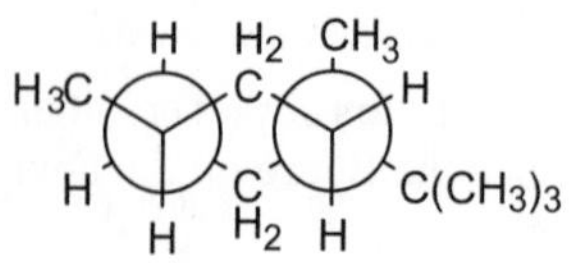	***Already most stable conformation***

Problem 4.36

Think

Build a model of the molecule, look down the bond about which the rotation is to be performed, and carry out the 360° rotation. Can you identify each staggered and eclipsed conformation? In each conformation, is there any significant steric strain? If so, how is steric strain related to the relative energy of the conformations? Which conformations (staggered or eclipsed) are higher in energy? Which staggered conformations, anti or gauche, are higher in energy?

Solve

The conformational analysis is performed from the direction indicated below.

H_3C H H H H H

Front **Back**

The back group is locked in place and the front group rotates. Eclipsed conformations are higher than the staggered ones. All three staggered conformations are identical in energy because the steric strain between the front and back substituents is identical in each, and all three eclipsed conformations are identical in energy for the same reason.

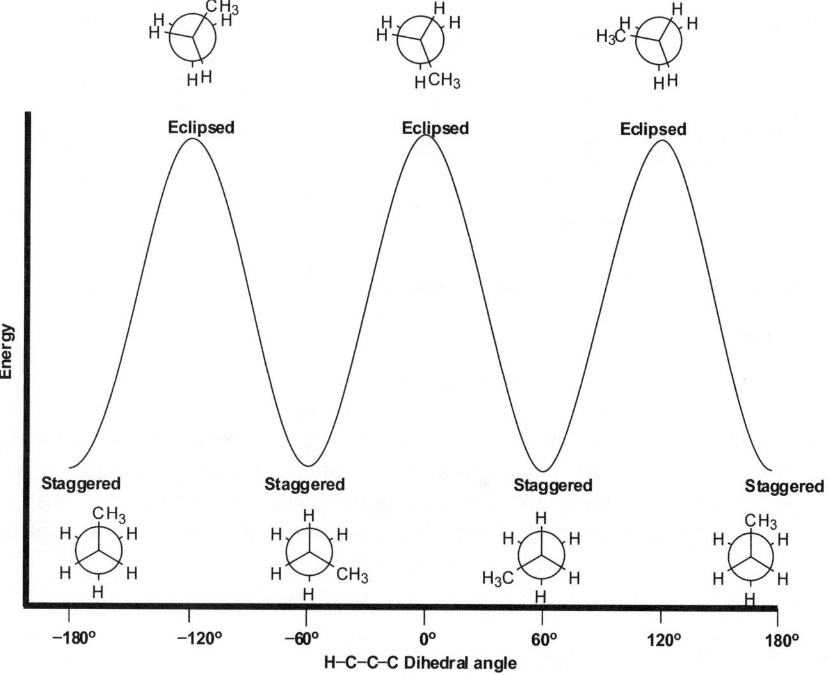

Problem 4.37

Think

Build a model of the molecule, look down the bond about which the rotation is to be performed, and carry out the 360° rotation. Can you identify each staggered and eclipsed conformation? In each conformation, is there any significant steric strain? If so, how is steric strain related to the relative energy of the conformations? Which conformations (staggered or eclipsed) are higher in energy? Which staggered conformations, anti or gauche, are higher in energy?

Solve

The conformational analysis is performed from the direction indicated below.

H_3C H H H H CH_3

Front Back

The back group is locked in place and the front group rotates. The plot is qualitatively identical to that for Br–CH_2–CH_2–Br (Fig. 4-7). The two gauche conformations at +60° and −60° are higher in energy than the anti conformation at ±180° due to steric strain between the two CH_3 groups. The eclipsed conformation at 0° is higher in energy than those at +120° and −120°, due to steric strain between the two CH_3 groups.

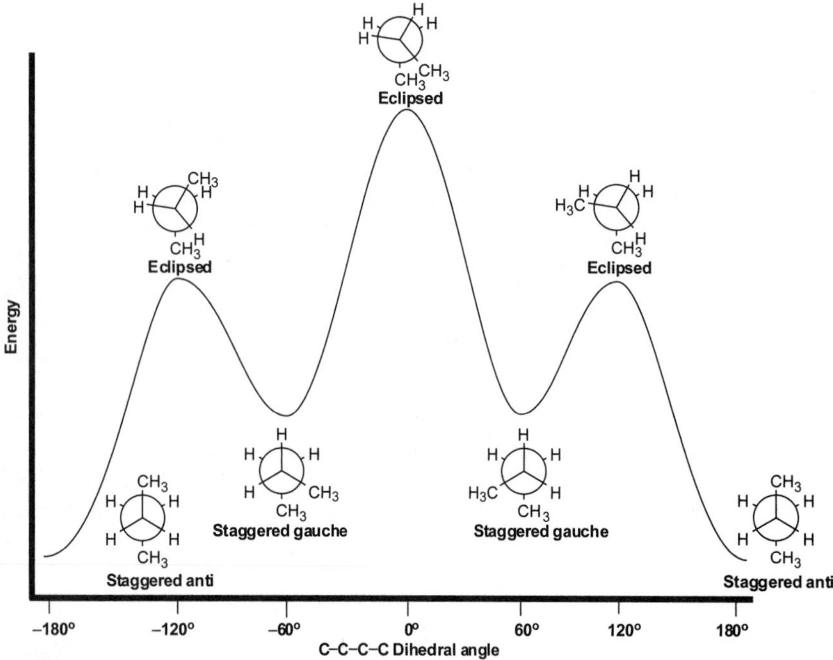

Problem 4.38

Think

Build a model of the molecule, look down the bond about which the rotation is to be performed, and carry out the 360° rotation. Can you identify each staggered and eclipsed conformation? In each conformation, is there any significant steric strain? If so, how is steric strain related to the relative energy of the conformations? Which conformations (staggered or eclipsed) are higher in energy? Which staggered conformations, anti or gauche, are higher in energy?

Solve

The conformational analysis is performed from the direction indicated below.

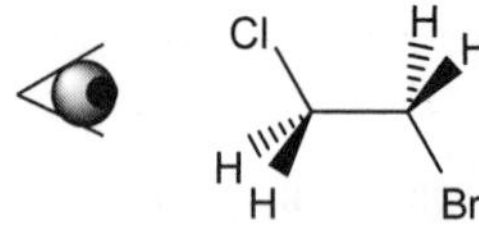

Front Back

The back group is locked in place and the front group rotates. The plot is qualitatively identical to that for Br–CH_2–CH_2–Br (Fig. 4-7). The two gauche conformations at +60° and −60° are higher in energy than the anti conformation at ±180° due to steric strain between the Br and Cl substituents. The eclipsed conformation at 0° is higher in energy than those at +120° and −120° due to steric strain between the Br and Cl groups. See the figure on the next page.

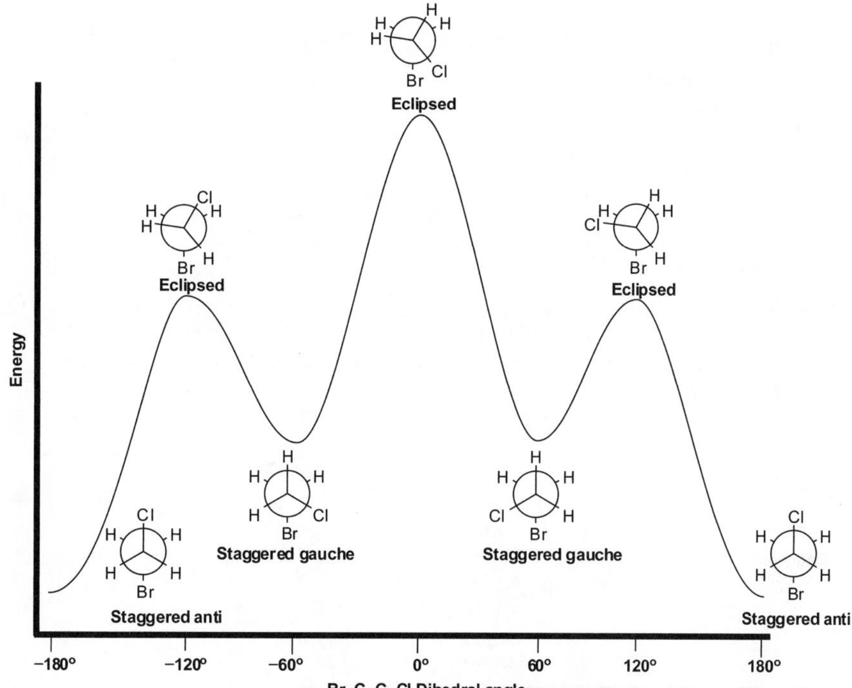

Problem 4.39

Think

Build a model of the molecule, look down the bond about which the rotation is to be performed, and carry out the 360° rotation. Can you identify each staggered and eclipsed conformation? In each conformation, is there any significant steric strain? If so, how is steric strain related to the relative energy of the conformations? Which conformations (staggered or eclipsed) are higher in energy? Which staggered conformations, anti or gauche, are higher in energy?

Solve

The conformational analysis is performed from the direction indicated below.

H_3C H H H H_3C CH_3

Front **Back**

The back group is locked in place and the front group rotates. The eclipsed conformations are higher in energy than the staggered ones, due to torsional strain. The staggered conformation at −60° is higher in energy than the other two because both CH_3 groups on the front carbon atom are gauche to the CH_3 group on the back carbon atom. Each gauche interaction represents steric strain. The other two staggered conformations each have only one such gauche interaction. The eclipsed conformations at −120° and 0° are higher in energy than that at +120° because the two bulky CH_3 groups are eclipsed. In the eclipsed conformation at +120°, CH_3 groups are not eclipsed. See the figure on the next page.

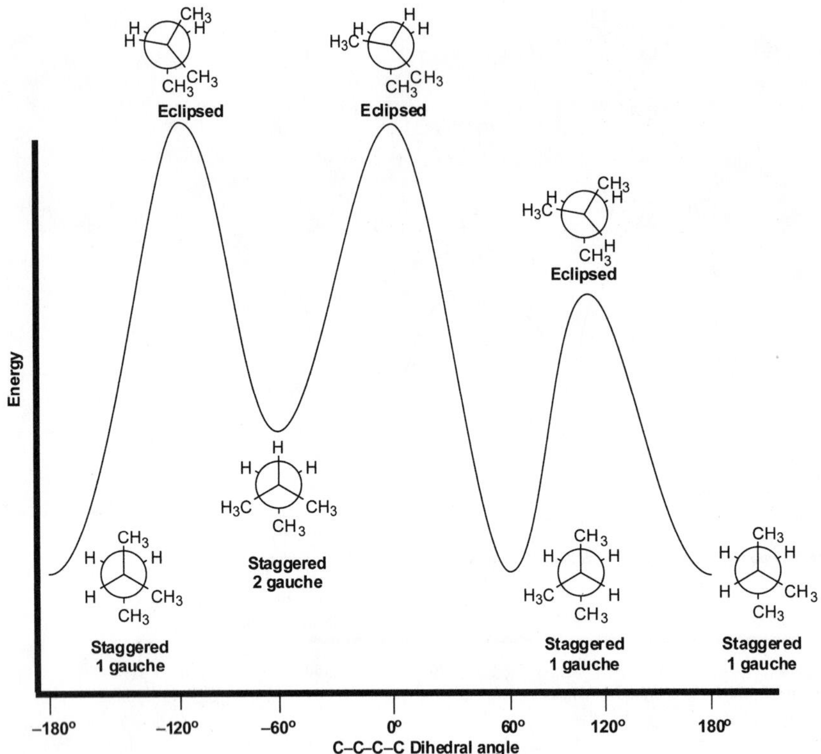

Problem 4.40

Think

Build a model of the molecule, look down the bond about which the rotation is to be performed, and carry out the 360° rotation. Can you identify each staggered and eclipsed conformation? In each conformation, is there any significant steric strain? If so, how is steric strain related to the relative energy of the conformations? Which conformations (staggered or eclipsed) are higher in energy? Which staggered conformations, anti or gauche, are higher in energy?

Solve

The conformational analysis is performed from the direction indicated below.

Br F H H H Br

Front **Back**

The three staggered conformations are all at different energies, and the three eclipsed conformations are all at different energies. The staggered conformation at −60° is higher in energy than that at ±180° because the gauche interaction between two Br atoms is greater than that between a Br atom and a F atom. The staggered conformation at +60° is higher in energy than either of those because the Br atom on the back C atom is gauche to both the F and Br atoms on the front C atom. The lowest-energy eclipsed conformation is at −120° because all of the halogen atoms are eclipsed only with a small H atom. That at +120° is higher in energy due to the eclipsing interaction between F and Br atoms. That at 0° is highest in energy due to eclipsing between the two Br atoms, and the Br atom is the largest of the substituents. See the figure on the next page.

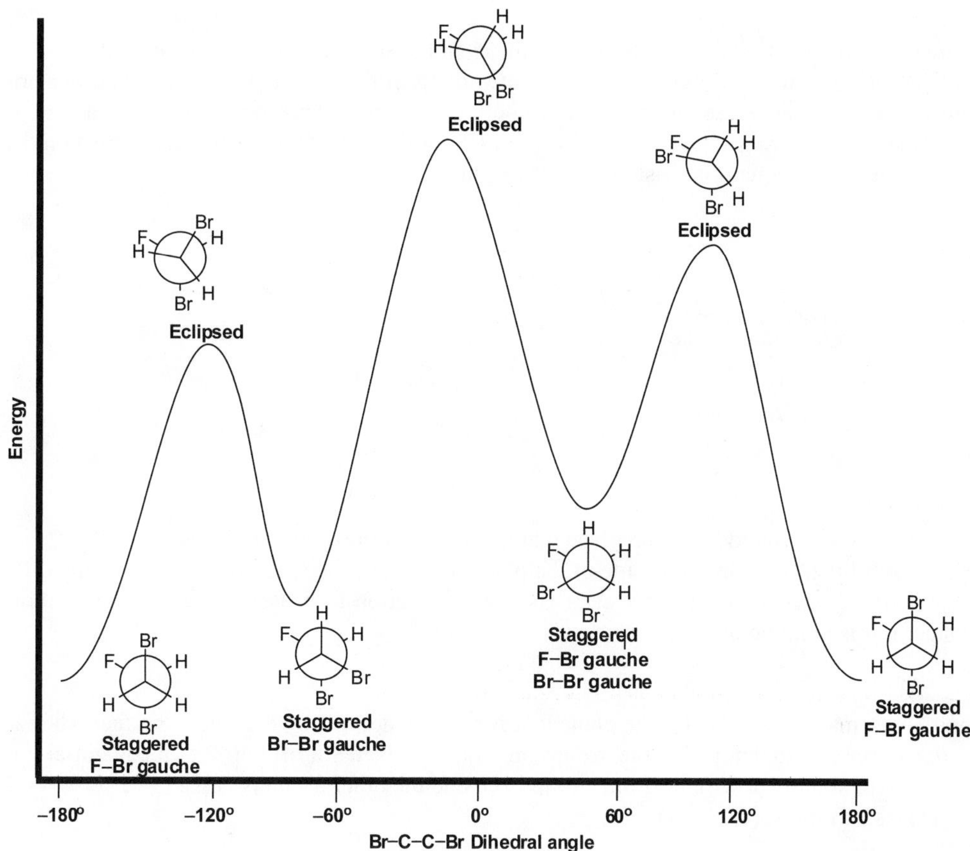

Problem 4.41

Think

How many CH_2 groups are present? The heat of combustion for cyclononane is 5586 kJ/mol. How do you calculate the heat of combustion per CH_2? How does that compare to the heat of combustion per CH_2 for cyclohexane (consult Table 4.1)? With that difference representing the ring strain per CH_2 group, how do you calculate the total ring strain? Compare this value to cycloheptane.

Solve

Nine CH_2 groups compose cyclooctane. The $\Delta H°$ per carbon in cyclooctane is $\frac{5586 \frac{\text{kJ}}{\text{mol}}}{9\ \text{CH}_2\ \text{groups}} = 620.67 \frac{\text{kJ}}{\text{mol}}$.

Ring strain per CH_2 is the difference between $\Delta H°$ per CH_2 group in cyclononane versus cyclohexane, or: $620.67 \frac{\text{kJ}}{\text{mol}} - 615.1 \frac{\text{kJ}}{\text{mol}} = 5.57 \frac{\text{kJ}}{\text{mol}}$.

Total ring strain is the ring strain per CH_2 group times the number of CH_2 groups, or: $5.57 \frac{\text{kJ}}{\text{mol}} \times 9\ \text{CH}_2\ \text{groups} = 50.13 \frac{\text{kJ}}{\text{mol}}$.

This value is more than that of cycloheptane, whose total ring strain is 26.6 kJ/mol.

Problem 4.42

Think

How are equatorial groups oriented relative to the plane defined by a chair conformation? How are axial groups oriented? Are axial positions on the same side of the ring? Are all equatorial groups on the same side of the ring?

Solve

Axial positions are perpendicular to the plane defined by the cyclohexane chair, and, going around the ring, they alternate sides of the plane. Equatorial positions point outward from the center of the ring, and are oriented slightly up or down. They, too, alternate up and down. In both cases, each adjacent pair of adjacent CH_3 groups must be on opposite sides of the ring. The molecule with all CH_3 groups in axial positions is shown on the left. That in which the groups are all equatorial is shown on the right.

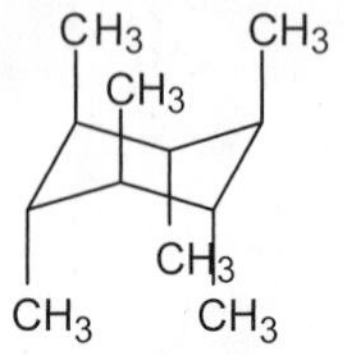

All axial

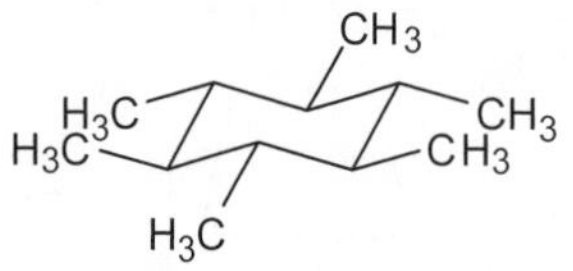

All equatorial

Problem 4.43

Think

In the chair conformations, can you identify the plane that is roughly defined by the cyclohexane chair? Relative to that plane, are the substituents on the same side of the plane or on opposite sides? If you are having difficulty working with the chair representation, transfer each chair conformation to a dash–wedge representation. For each group, consider if it is pointing up or down.

Solve

The way the chair conformations are drawn, the plane that is roughly defined by the ring is perfectly horizontal. The key is to notice if, relative to that plane, the groups are both up (**even slightly** up for an equatorial group is sufficient), both down (or slightly down), or if one group is up and the other group is down.

(a) cis; **(b)** trans; **(c)** cis; **(d)** trans; **(e)** cis; **(f)** trans.

(a) cis

(b) trans

(c) cis

(d) trans

(e) cis

(f) trans

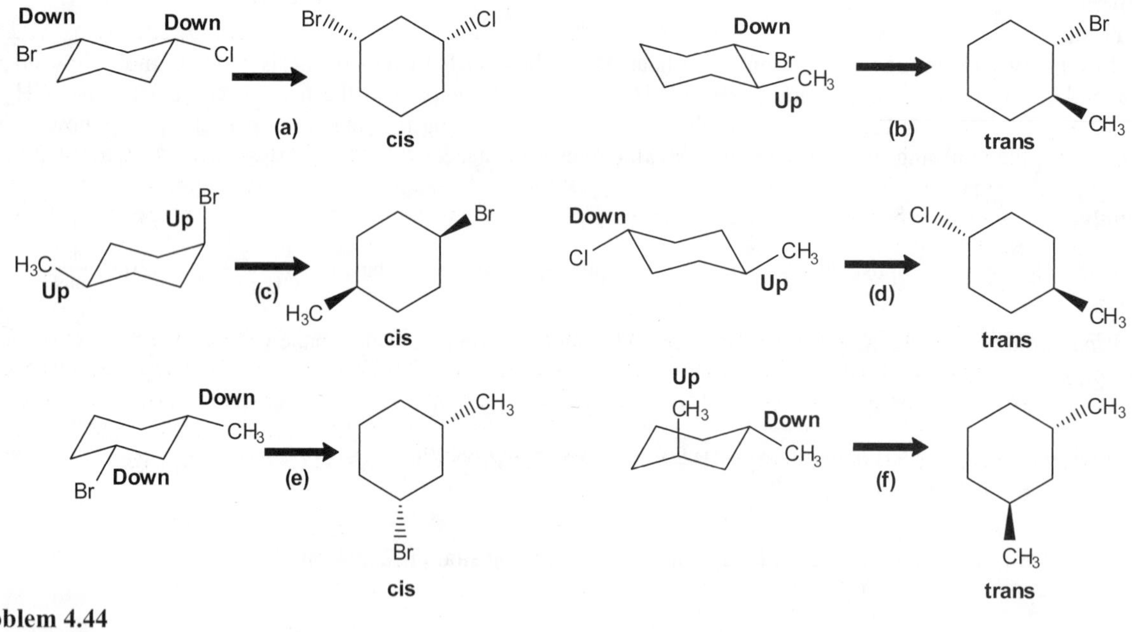

Problem 4.44

Think

Write the chair structure for the cis and trans of each molecule. It can help to draw each molecule in a dash–wedge line structure first and then, using a molecular model to help, transfer each structure to a chair. Which substituents require the most room? Which position, axial or equatorial, offers more room? Can both substituents achieve that position?

Solve

For each compound, both the cis and trans isomer forms are shown in chair structures, and a box is drawn around the more stable form. In each case, in the more stable compound, both CH_3 groups can attain equatorial positions; in the less stable form, one CH_3 group is forced into an axial position.

(a) cis | trans

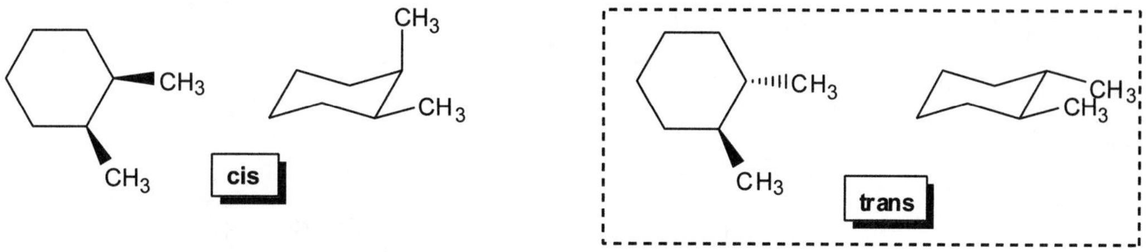

(b) cis | trans

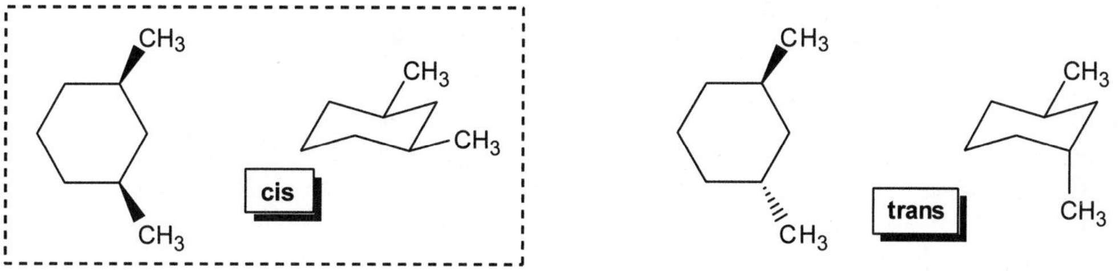

(c) cis | trans

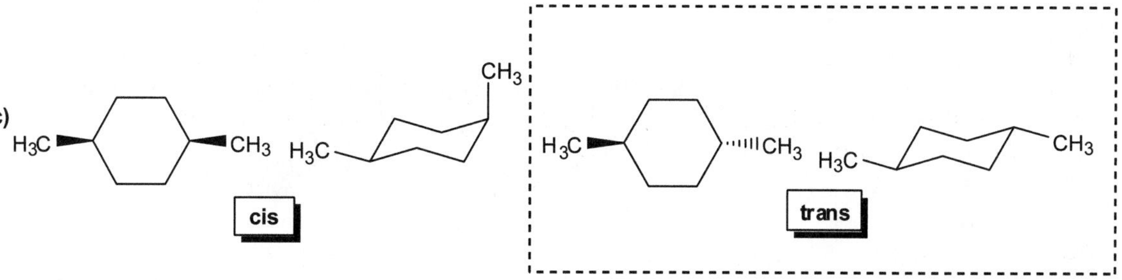

Problem 4.45

Think

The molecular formula is the same for each structure. What would contribute to the differences in the heat of combustion? Draw each structure in a chair conformation. Perform any necessary ring flip to get each structure in its most stable conformation. What is the order of stability for these four structures? How does stability relate to heat of combustion value?

Solve

The least stable isomer has the greatest heat of combustion. The *t*-butyl group is much larger than the CH_3 groups so it must go equatorial in the most stable conformation of each structure. The remaining three CH_3 groups must follow the dash–wedge notation that is given. **B** has all four substituents equatorial and, therefore, is the most stable and has the smallest heat of combustion. Following the same logic, the order of heat of combustion from smallest to greatest is **B** < **A** < **C** < **D**. See structures below and on the next page.

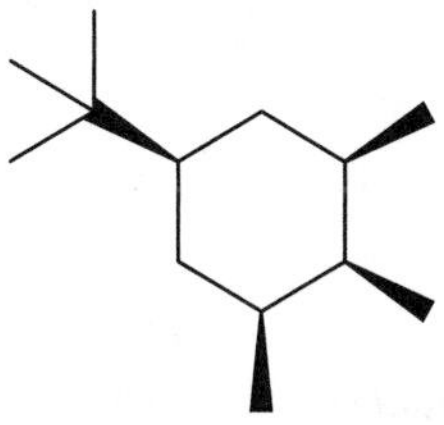

A

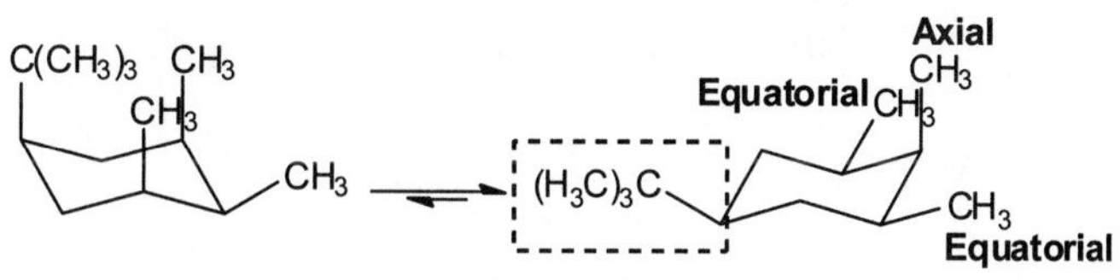

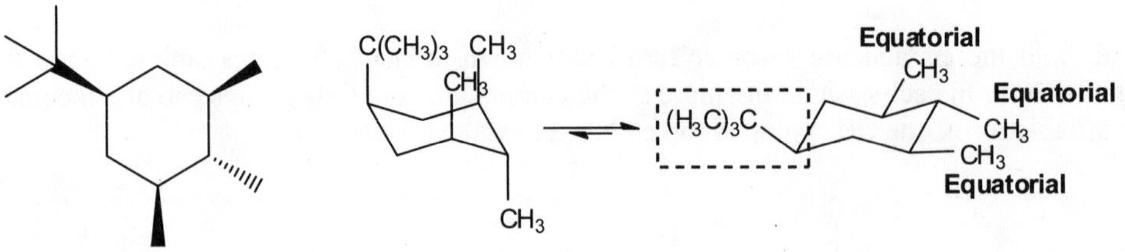

Problem 4.46

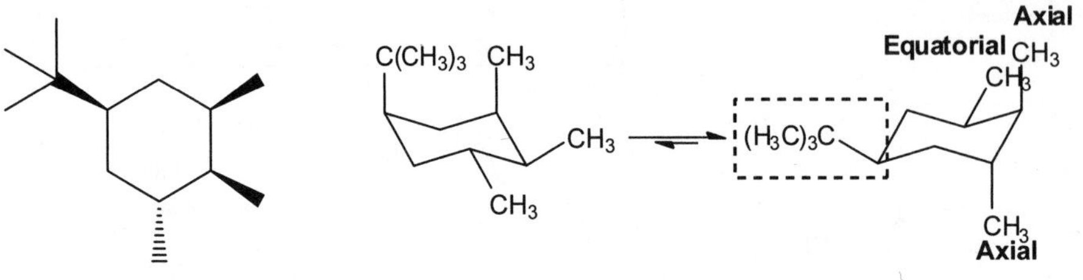

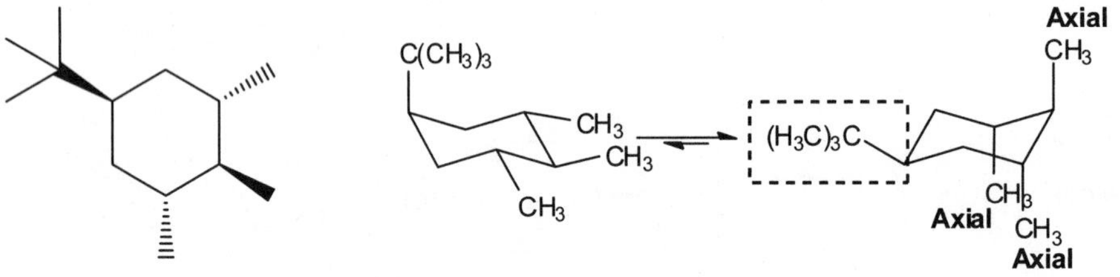

Think

Draw a chair for each ring. Can one ring occupy the axial position of the second ring? Equatorial? Can the second ring occupy the axial position of the first ring? Equatorial? How many combinations are possible?

Solve

The three conformations are shown below. Each ring can occupy the axial or equatorial position of the other ring. The most stable conformation is the one in which the bond connecting the two rings is equatorial with regard to each ring. The bond connecting the two rings is bolded for emphasis.

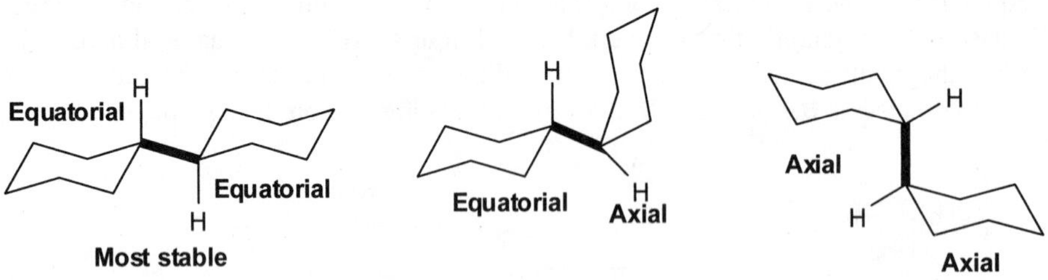

Problem 4.47

Think

Refer to Section 4.13 to follow the steps to draw all constitutional isomers for each formula.

Solve

The IHD of $C_5H_{11}Br$ is 0, so there are no rings, double bonds, or triple bonds. There are eight constitutional isomers in all, which are shown below.

Problem 4.48

Think

Refer to Section 4.13 to follow the steps to draw all constitutional isomers for each formula.

Solve

An analogous saturated molecule is $CH_3CH_2CH_2NH_2$, which has a formula of C_3H_9N. Therefore, the formula C_3H_5N has an IHD of 2, so there are either two double bonds, two rings, one of each, or one triple bond.

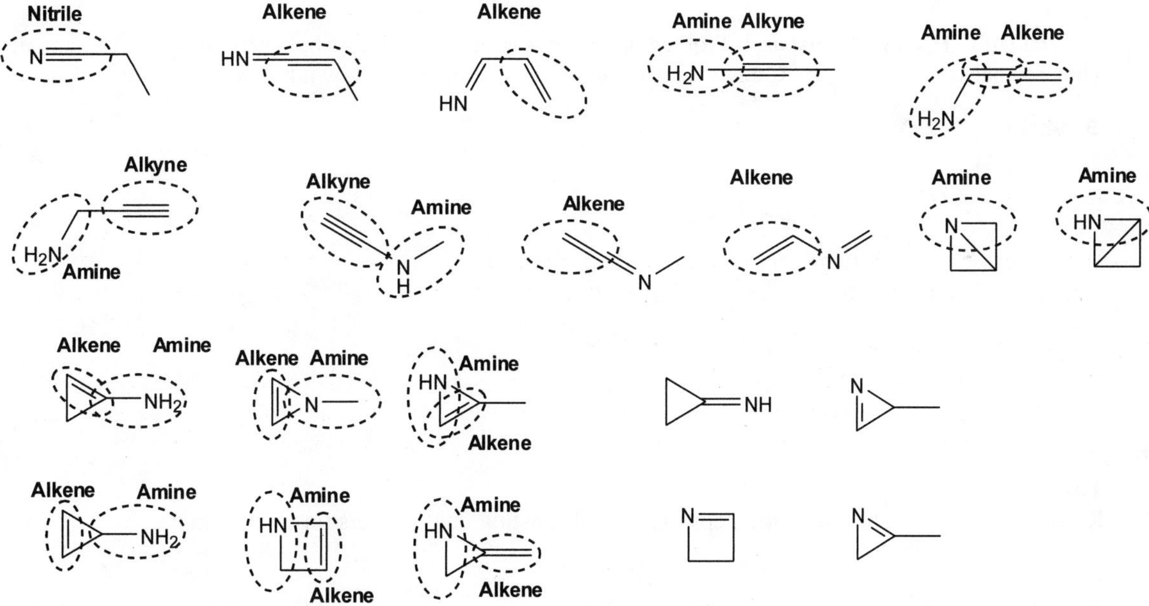

Problem 4.49

Think

Refer to Section 4.13 to follow the steps to draw all constitutional isomers for each formula.

Solve

An analogous saturated molecule is $CHF_2CH_2CH_2OH$, which has the formula $C_3H_6OF_2$, so the IHD for $C_3H_6OF_2$ = 0. Therefore, there should be no rings, double bonds, or triple bonds. There are nine possibilities with the backbone C–C–C–O, and five possibilities with the backbone C–C(O)–C. The constitutional isomers are shown on the next page.

Backbone C–C–C–O:

Backbone C–C(O)–C:

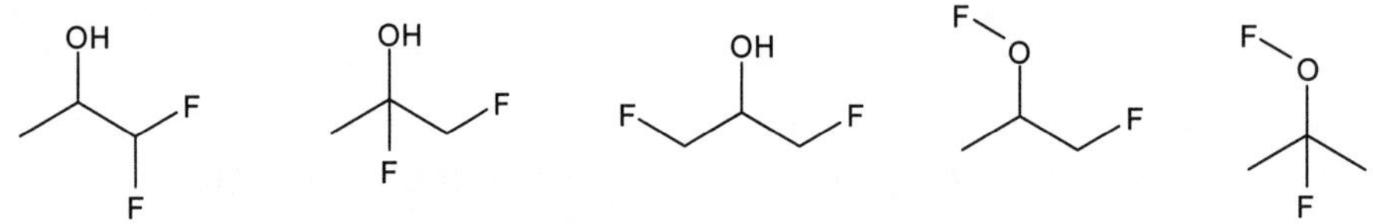

Problem 4.50

Think

Refer to Section 4.13 to follow the steps to draw all constitutional isomers for each formula.

Solve

The IHD of $C_3H_6OF_2$ = 0, just as in Problem 4.49. There can be no double bonds, no triple bonds, and no rings. The only backbone is C–C–O–C and the six constitutional isomers are shown below.

Backbone C–C–O–C:

Problem 4.51

Think

Refer to Section 4.13 to follow the steps to draw all constitutional isomers for each formula.

Solve

An analogous saturated molecule is $CH_3CH_2CH_2CH_3$, whose formula is C_4H_{10}.The IHD of C_4H_6 = 2. There are two rings, two double bonds, one of each, or one triple bond. Four isomers have no ring and four other isomers have one ring. There are two isomers with a triple bond and one bicyclic isomer.

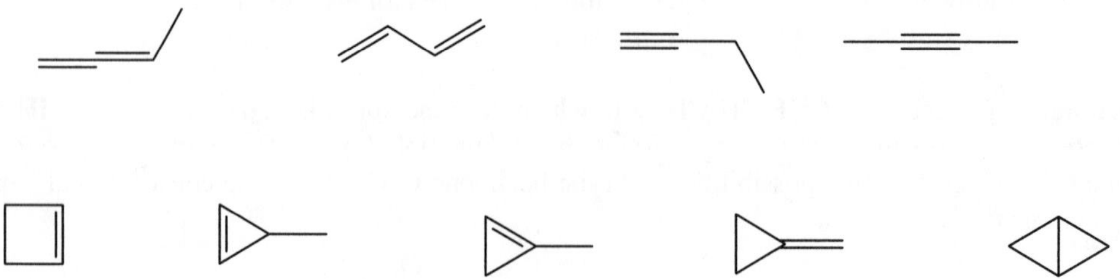

Problem 4.52

Think

Refer to Section 4.13 to follow the steps to draw all constitutional isomers for each formula. What is a benzene ring? How much of the IHD does a benzene ring account for?

Solve

An analogous saturated compound is $CH_3\,(CH_2)_7CH_3$. The IHD for C_9H_{12} is 4. A benzene ring has three formal double bonds and a ring, which accounts for the entire IHD of 4. So aside from the benzene ring, there should be no double bonds, triple bonds, or rings.

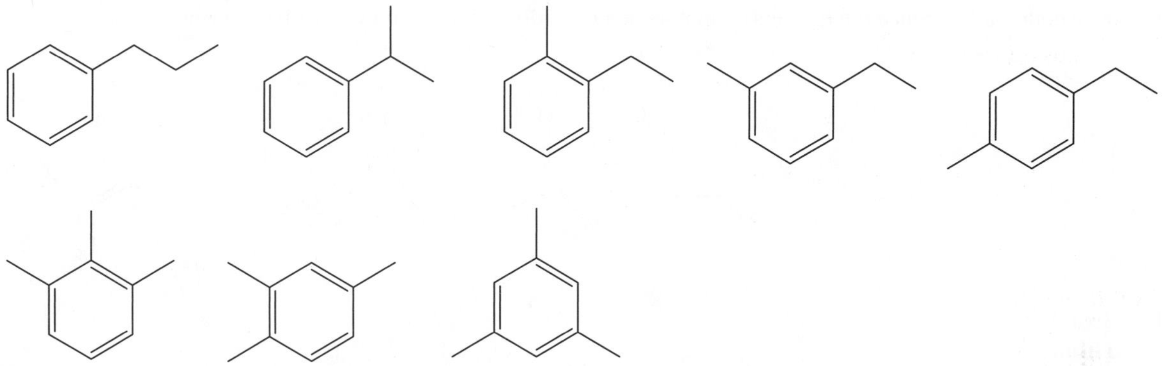

Problem 4.53

Think

Refer to Section 4.13 to follow the steps to draw all constitutional isomers for each formula. What is a carboxylic acid functional group? How much of an IHD does a carboxylic acid account for? Can a carboxylic acid functional group be at the end of a chain, internal to the chain, or either?

Solve

An analogous saturated molecule is $CH_3(CH_2)_4OOH$, which has 12 H atoms. The formula $C_5H_8O_2$ has an IHD of 2. A carboxylic acid functional group is CO_2H and contains a C=O double bond. So the carboxylic acid functional group accounts for an IHD of 1, leaving another IHD of 1 yet to be accounted for. This can be either a second double bond or a ring. A carboxylic acid carbon atom has only one available bond, so it must be located at the end of a chain.

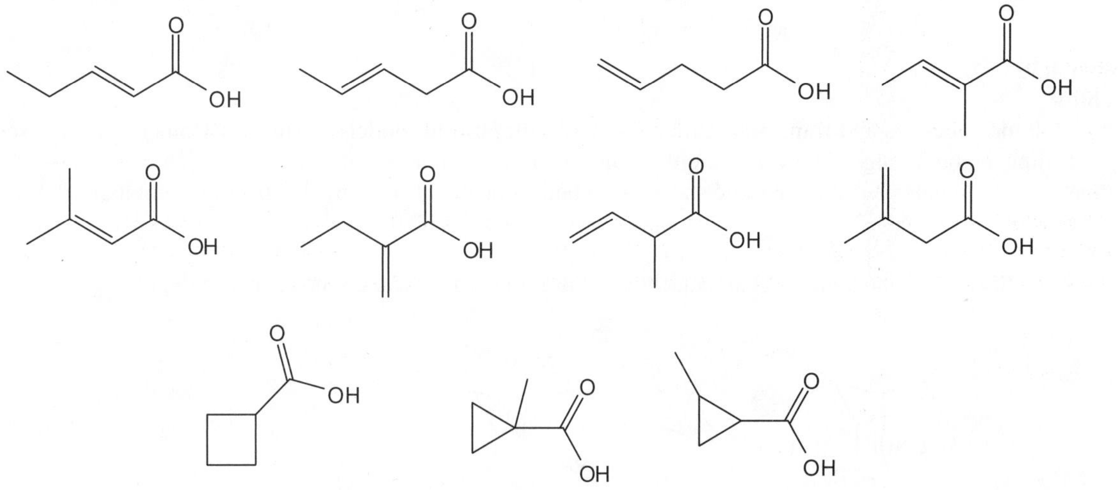

Problem 4.54

Think

Refer to Section 4.13 to follow the steps to draw all constitutional isomers for each formula. What is a ketone functional group? An alkene? How much of an IHD does a ketone account for? Can a ketone functional group be at the end of a chain, internal to the chain, or either?

Solve

An analogous saturated compound is $CH_3(CH_2)_4OH$, which has 12 H atoms. Compounds of the formula C_5H_8O have an IHD of 2. A ketone functional group is of the form C–C(O) –C, in which the C and O atoms are joined by a double bond (C=O). So the ketone functional group accounts for an IHD of 1, leaving an IHD of 1 yet to be accounted for. Since the isomers cannot have a C=C double bond, they must have a ring.

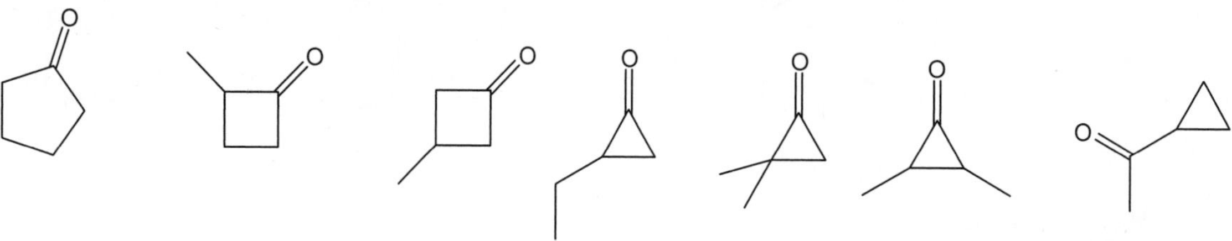

Problem 4.55

Think

Can you draw both chair conformations of each compound? If a bonding pair of electrons on N occupies an axial position, what kind of position does the lone pair occupy? If a bonding pair of electrons on N occupies an equatorial position, what position does the lone pair occupy? Which position offers more room, an axial or an equatorial position?

Solve

In the molecule on the left, a lone pair occupies the equatorial position, forcing the H atom into the axial position. This suggests that a lone pair occupies more room than a N–H bond. In the molecule on the right, the CH_3 group occupies the equatorial position, forcing the lone pair into the axial position. So a CH_3 group takes up more space than a lone pair.

Problem 4.56

Think

Which bonds does cis and trans refer to for this molecule? Build models of the two isomers, and try several chair flips in both rings. Draw the more stable chair conformation of each. In which position—axial or equatorial—is a bulky substituent more stable? Can both substituents occupy that position simultaneously?

Solve

Cis and trans refer to the bonds that are attached to the atoms that fuse the second ring to the first.

Down
CH_2
Down
CH_2
Axial
cis

Down
CH_2
CH_2
Up
trans

In the cis isomer (left), a carbon in the right-hand ring cannot avoid being axial. In the trans isomer, the two CH_2 groups in the ring on the right occupy equatorial positions. Thus, the trans isomer is more stable.

Problem 4.57

Think

Which substituents require the most room? Which position, axial or equatorial, offers more room? Can both substituents achieve that position simultaneously?

Solve

The more stable conformation of each molecule is drawn below. Bulky substituents prefer to occupy the equatorial position. If both groups are not able to occupy an equatorial position, the larger group will take priority over the smaller group.

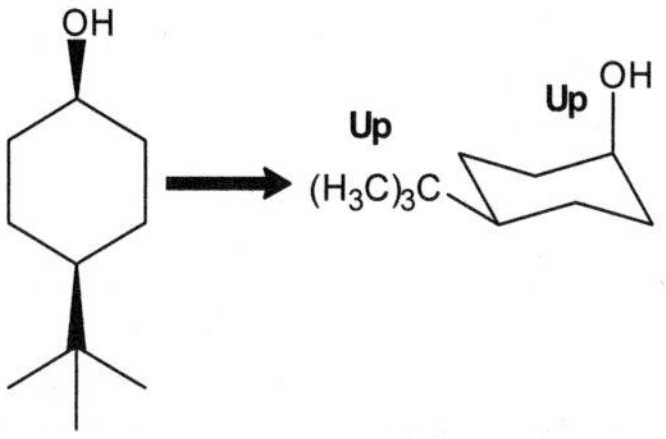

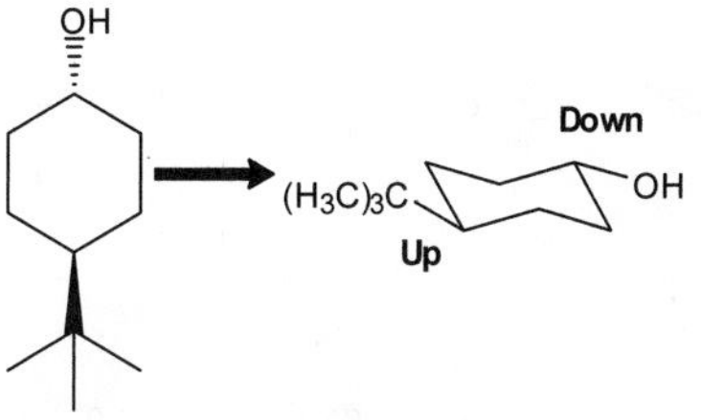

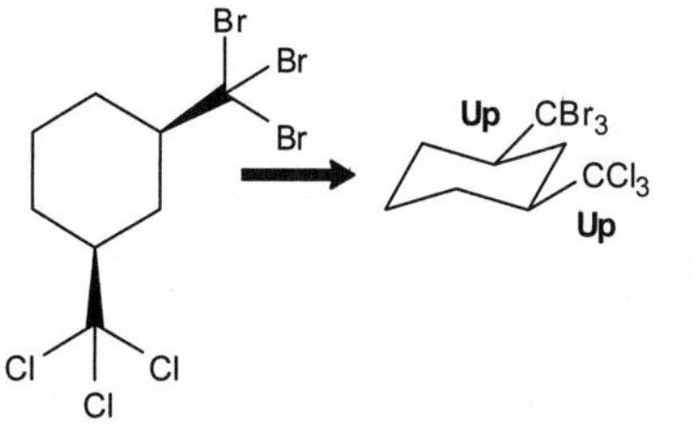

Problem 4.58

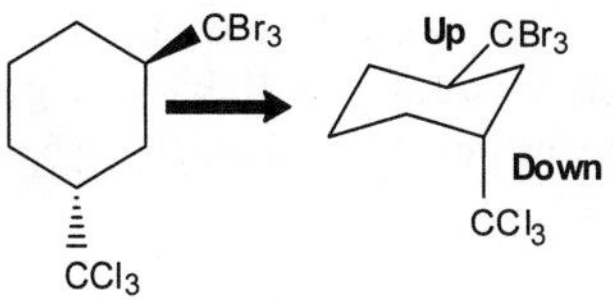

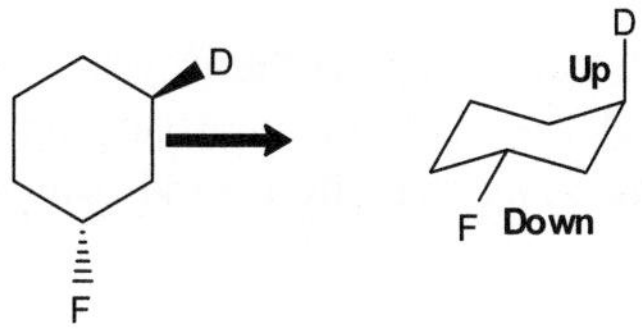

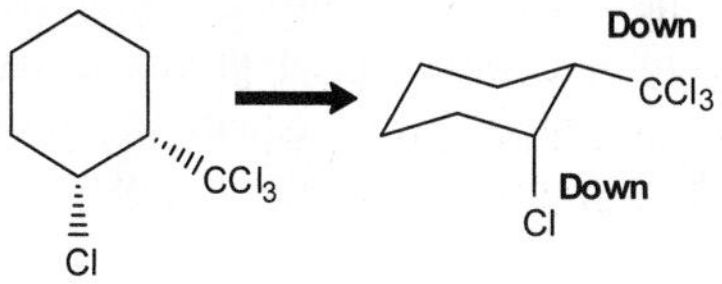

Think

Draw a dash–wedge line structure of each molecule and then draw both chair conformations of each line structure. It will help to build a molecular model of each and carry out chair flips on the model. What positions—axial or equatorial—will each of the three substituents occupy in each constitutional isomer? Which position does each bulky substituent prefer? Can all bulky substituents occupy that position?

Solve

The various constitutional isomers are shown below and on the next page. Because equatorial positions alternate up and down around a cyclohexane ring, only the isomer boxed can have all three of its substituents occupy equatorial positions. In each of the other isomers, at least one of the alkyl substituents will be forced to occupy an axial position.

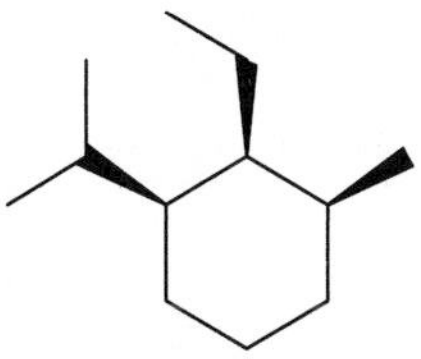

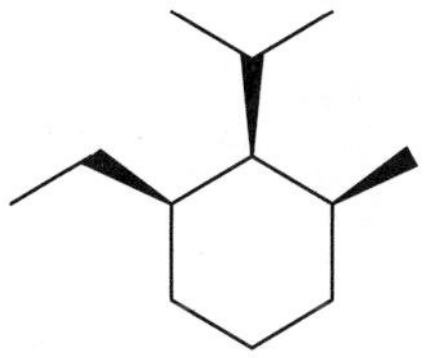

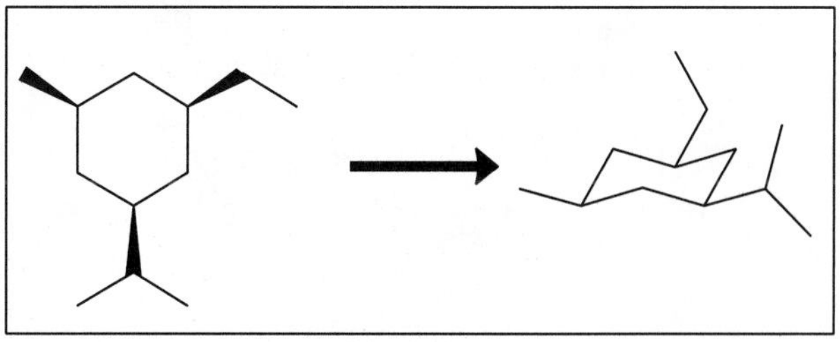

Problem 4.59

Think

Which group is bulkier, isopropyl or propyl? How does bulkiness affect a substituent's preference for equatorial versus axial?

Solve

The isopropyl group is bulkier. The C atom attached to the ring is attached to one other alkyl group in propyl and two other alkyl groups in isopropyl (see below). The isopropyl group, therefore, will have a greater preference for the equatorial position, meaning that the propyl group will be more likely than the isopropyl group to occupy the axial position.

Problem 4.60

Think

Which substituents require the most room? Which position, axial or equatorial, offers more room? Can all substituents achieve that position simultaneously? It will help to build a molecular model and carry out multiple chair flips on it.

Solve

The OH and CH_2OH groups attached to the ring all prefer the equatorial position. In the most stable conformation, all of the substituents occupy the equatorial position.

β-D-Glucopyranose

Problem 4.61

Think

What is the size of the ring in each compound? Which ring size has more strain associated with it?

Solve

β-D-Glucofuranose has a five-membered ring, whereas β-D-glucopyranose has a six-membered ring. Because there is greater ring strain with a five-membered ring, β-D-glucopyranose is more stable.

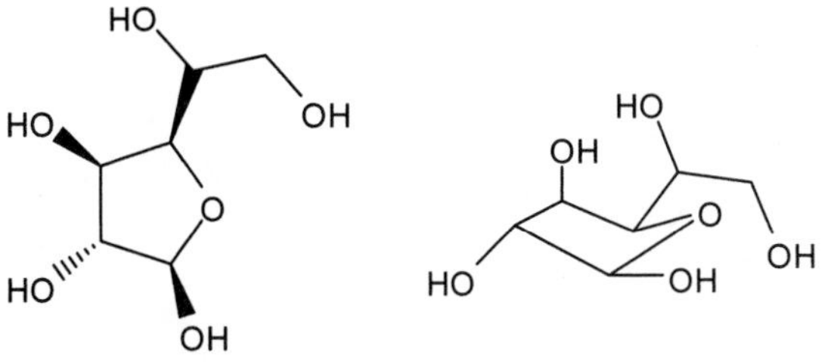

β-D-Glucofuranose

Problem 4.62

What is the electron geometry for each sp^3-hybridized heteroatom? Is the chair conformation still possible? Which substituents require the most room? Which position, axial or equatorial, offers more room? Can all substituents achieve that position?

Solve

The electron geometry of each heteroatom is tetrahedral and yes, the chair conformation is still possible. In the most stable conformation for **(a)**, the larger substituent $–C(CH_3)_3$ occupies the equatorial position preferentially over the smaller substituent $–CH_3$. In **(b)** and **(c)**, both substituents occupy the equatorial position.

(a) **(b)** **(c)**

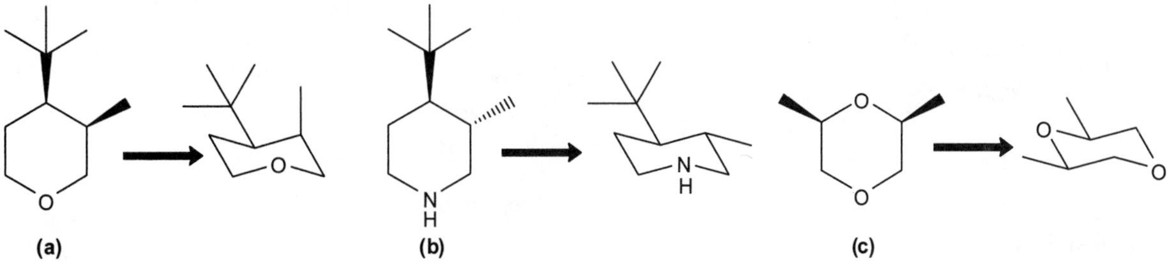

Problem 4.63

What is the electron geometry for the sp^2-hybridized carbon? For the O atom that has two single bonds? Is the chair conformation still possible? Which substituents require the most room? Which position, axial or equatorial, offers more room? Can all substituents achieve that position?

Solve

The electron geometry for the sp^2-hybridized C is trigonal planar. Each ring, therefore, will still be able to attain the chair conformation, though with a geometry that is slightly altered from one composed entirely of sp^3-hybridized C atoms. In **(e)**, the sp^3-hybridized O can be treated just like an sp^3-hybridized C. The most stable conformations are shown below and on the next page.

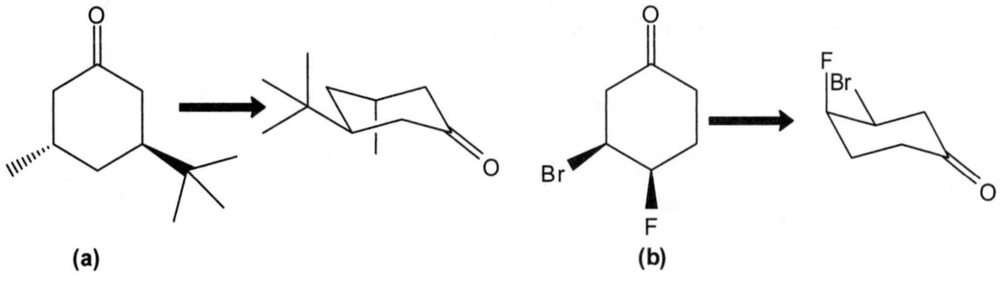

(a) **(b)**

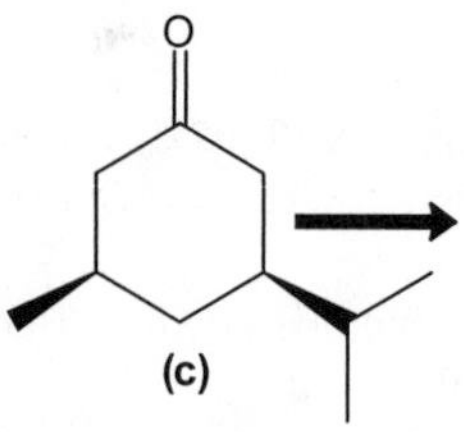

(c)

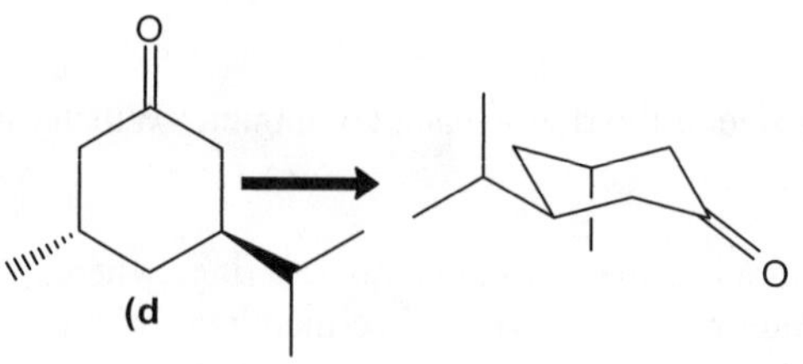

(d

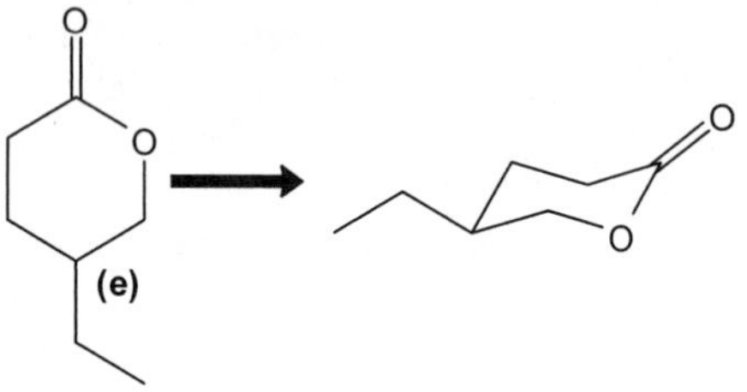

(e)

Problem 4.64

Think

The molecular formula is the same for each structure. What would then contribute to the difference in the heat of combustion? Do they have the same strain? Perform any necessary ring flips to get each structure in its most stable conformation. How does stability relate to heat of combustion value?

Solve

The least stable isomer has the greatest heat of combustion. β-D-Allopyranose has the greater heat of combustion due to the presence of an axial OH (circled). This makes the structure less stable and, therefore, it has a greater heat of combustion.

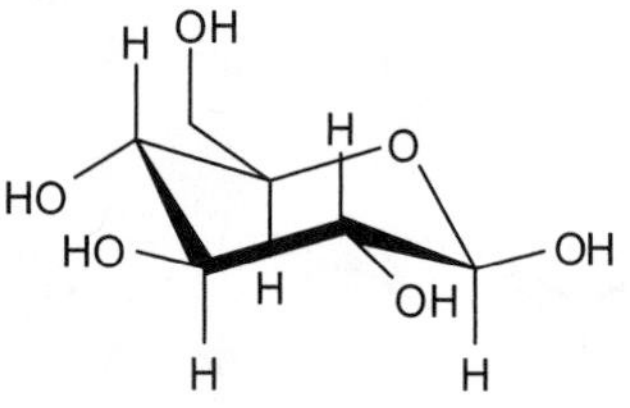

β-D-Glucopyranose 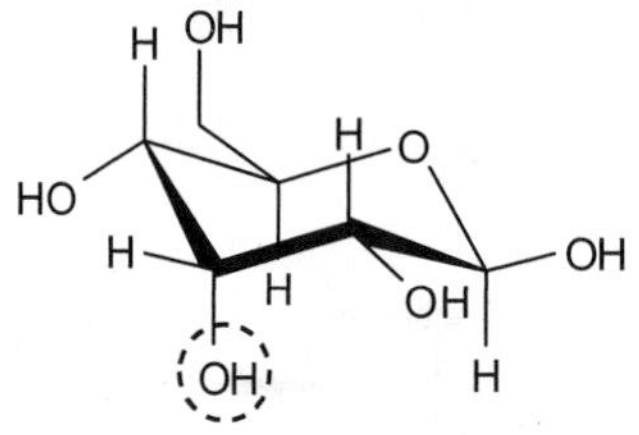 **β-D-Allopyranose**

Problem 4.65

Think

The molecular formula is the same for each structure. What would contribute to the difference in the heat of combustion? Which ring size is the most stable? Least stable? How does stability relate to heat of combustion value?

Solve

All five of these compounds have the molecular formula C_9H_{16} and contain only C–C and C–H single bonds. Differences in heats of combustion, therefore, will largely reflect differences in ring strain. The structure that contains two three-membered rings, **B**, is the most strained, so it will give off the most heat during combustion—it will have the greatest heat of combustion. Next comes the structure with one three- and one four-membered ring, **D**, followed by the structure with two four-membered rings, **A**. The structure with one four- and one five-membered ring, **E**, is next, and the least strained structure is the compound with the two five-membered rings, **C**. The order of heats of combustion are **C** < **E** < **A** < **D** < **B**.

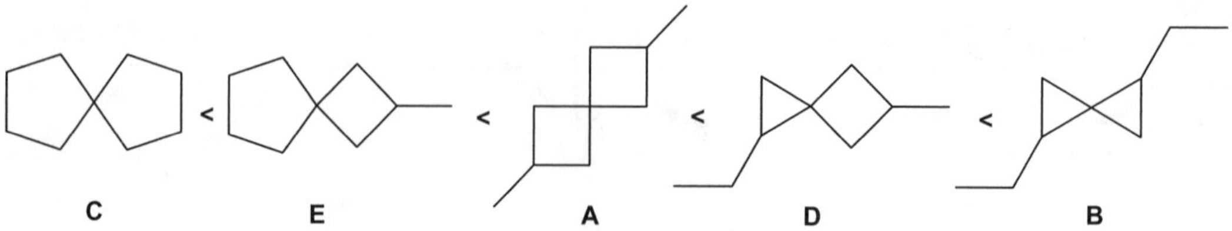

C < E < A < D < B

Problem 4.66

Think

What intermolecular forces are available to each fatty acid? Which fatty acid has a fully saturated hydrophobic region. Which fatty acid can pack closer together in the crystalline lattice? How does tight packing affect melting point?

Solve

Both fatty acids have a long hydrophobic chain and a carboxylic acid end, so induced dipole–induced dipole and hydrogen bonding interactions are available to both. The difference between behenic acid and erucic acid is that behenic acid has all single bonds making up the carbon chain (it is saturated), whereas erucic acid has a cis double bond (making it unsaturated). This allows for tighter/closer packing in the crystalline lattice of behenic acid, giving it the higher melting point. The cis double bond in erucic acid puts a kink in the carbon chain, which disrupts the dispersion forces. Saturated fatty acids are often solids at room temperature, whereas unsaturated fatty acids are often oils.

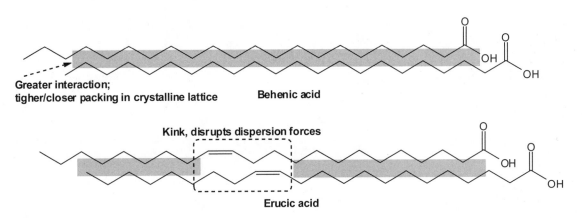

Greater interaction;
tigher/closer packing in crystalline lattice

Behenic acid

Kink, disrupts dispersion forces

Erucic acid

Problem 4.67

Think

How does the size of the atom relate to the length of the bond? Which bond is longer? How does the length of the bond affect the 1,3-diaxial interactions?

Solve

I is much larger than Br. The C–I bond, therefore, will be longer than the C–Br bond. Even though the I atom is larger and you might expect the 1,3-diaxial repulsions to be stronger, the longer C–I bond allows the I atom to be farther from the axial H atoms. The two factors (larger I atom and longer C–I bond distance) essentially cancel each other out in this case.

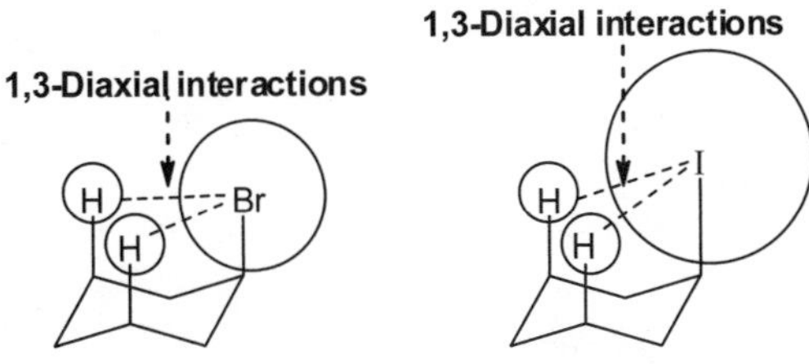

1,3-Diaxial interactions

Problem 4.68

Think

What stabilizing interactions are available between the O and OH groups in the axial position but not available in the equatorial position? Draw out the structure in the chair conformation to visualize this interaction. How is this a stabilizing interaction?

Solve

Hydrogen bonding is available to OH (H-bond donor) and O (H-bond accepter). In the equatorial position, this interaction is not possible because the donor and acceptors are too far apart. In the axial position, H-bonding is available because the donor and acceptors are close enough.

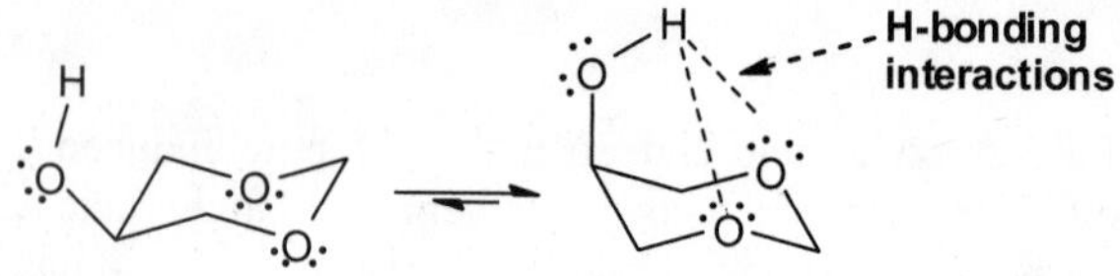

H-bonding interactions

CHAPTER 5 | Isomerism 2: Chirality, Enantiomers, and Diastereomers

Your Turn Exercises

Your Turn 5.1

Think

First, construct your model given the orientation of the atoms in the structure on the right of Figure 5-2a. How can you orient this molecule to have the F atom on the left and pointing toward you and the Br atom on the left and pointing away from you? Draw the Cl and H atoms in the boxes and compare to the structure on the left. If two atoms are swapped, can it be the same molecule? If not, how are these molecules related?

Solve

Construct your model of the molecule on the right in Figure 5-2a so that the H atom is up, the Cl atom is to the left and planar, the F atom is to the right and pointed towards you, and the Br atom is to the back and pointed away from you. Rotate the molecule 109.5° and make sure you keep the C–H and C–Cl bonds flat, the F atom pointed toward you, and the Br atom pointed away from you. You should see that the Cl atom is now up and the Br atom is to the left and pointed away from you. If you compare this molecule to the molecule on the left in Figure 5-2a (drawn below), you will note that the H and Cl atoms are swapped. This confirms that the two molecules are nonsuperimposable.

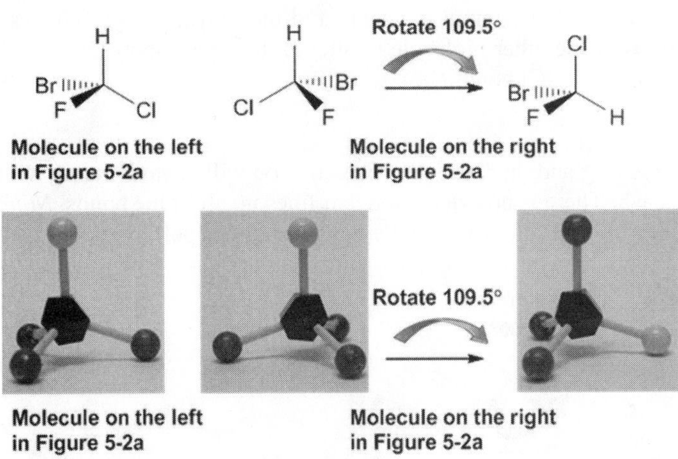

Your Turn 5.2

Think

When a carbon atom has four different groups, such as CHBrClF in Your Turn 5.1, is the mirror image superimposable? What is the effect of having two of the same groups (CH_2BrCl)? Do you anticipate that the mirror images are or are not superimposable?

Solve

Construct your models given the orientation of the atoms in Figure 5-3a. Rotate the mirror image molecule 109.5° as shown on the next page, and make sure you keep both C–H bonds flat, the F atom pointed toward you, and the Cl atom pointed away from you. You should see that one H atom in the resulting orientation (on the next page on the right) is up, the other H atom is pointed to the left and down, the F atom is to the right and pointed toward you, and the Cl atom is to the right and pointed away from you. If you compare this molecule to the molecule on the right in Figure 5-3a (i.e., the original molecule) drawn on the next page, you will notice that the two are identical. This illustrates that CH_2FCl has a mirror image that is the same as itself. *Note*: H_A and H_B are labels assigned to the mirror image to help you visualize the transformations performed.

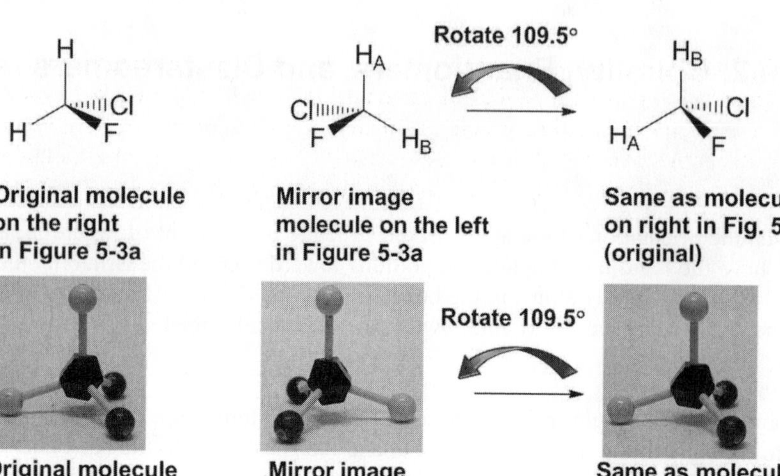

Original molecule on the right in Figure 5-3a

Mirror image molecule on the left in Figure 5-3a

Same as molecule on right in Fig. 5-3a (original)

Your Turn 5.3

Think

Consult Solved Problem 5.6 to build your model of the original molecule and the mirror image. Always keep one molecule in place as you perform rotations of the other molecule to attempt to superimpose the two. What do you notice as you attempt to line up the C–C and C–Cl bonds?

Solve

As you attempt to line up the C–C bonds in the ring and the C–Cl substituents, you will always end up with one C–Cl bond wedge and the other C–Cl bond dash. There is no orientation that lines up all of the bonds. *Note*: Cl_A and Cl_B are labels assigned to the mirror image to help you visualize the transformations performed.

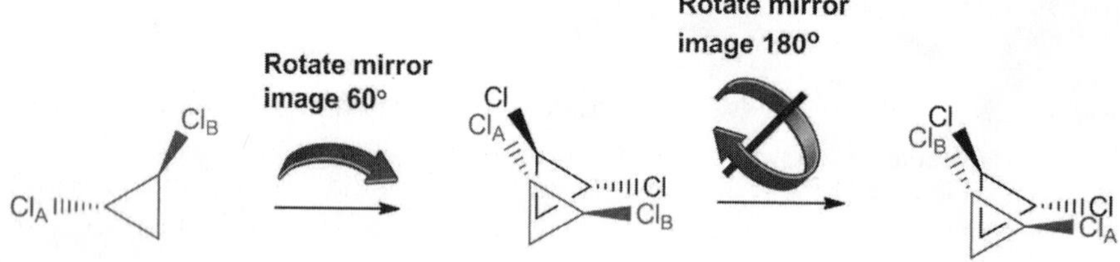

Your Turn 5.4

Think

In Solved Problem 5.6, is the mirror image of the red molecule shown? Are the two structures superimposable? Does that make the structures enantiomers?

Solve

The molecule is chiral. The red and black molecules in Solved Problem 5.6 are mirror images of each other. From Solved Problem 5.6, we have already verified that the two molecules are not the same and are, therefore, enantiomers.

Your Turn 5.5

Think

In a Newman projection, which bond is not visible? Build your model to visualize the gauche interaction. Do you think the presence of two identical CH_2 groups in this example will lead to mirror images that are or are not superimposable?

Solve

Once you have your models built, rotate them 90° so that the C–C bond that was once pointing towards you is now flat. Then, keep the original molecule as your reference point and attempt to superimpose the mirror image. The CH_2Br group on the right is in the same orientation in both molecules. Then, rotate the CH_2Br group on the left 120° about the C–C bond to move the Br atom to point away. Now the two conformations are superimposable. *Note*: H_A and H_B are labels assigned to the mirror image to help you visualize the transformations performed.

Rotate 90°

Original molecule

Rotate 90°

Rotate about bond

Mmirror image

Mirror image rotated to return to original

Your Turn 5.6

Think

You have read that *cis*-1,2-difluorocyclohexane is *achiral*. Therefore, what must be true regarding the mirror image? Upon initial examination of Figure 5-7b, it appears that the two molecules are *nonsuperimposable*. Attempt to superimpose the two molecules. Are you able to do so? If you ring flip the mirror image, are you now able to superimpose the two molecules?

Solve

The two models for *cis*-1,2-difluorocyclohexane are in two different conformations. If you attempt to superimpose the two conformations, you will not be able to do so. Ring flipping the mirror will now lead to a conformation that is superimposable on the original structure. Construct molecular models and carry out the rotations/chair flips shown in Figure 5-7. *Note*: F_A and F_B are labels assigned to the mirror image to help you visualize the transformations performed.

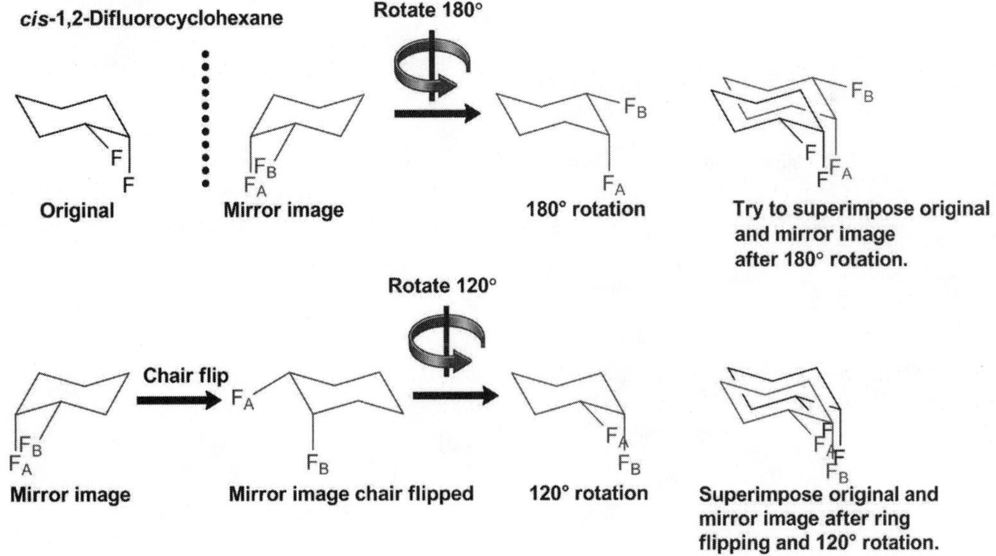

Your Turn 5.7

Think

Are the C–F bonds shown in the figure axial or equatorial? Would a ring flip aid your attempt to superimpose the two molecules?

Solve

Consult the steps in Figure 5–7 for assistance in your attempt to superimpose the two molecules. Every time you attempt to line up one of the C–F bonds, you will notice that the other C–F bond will not also line up. This illustrates that the two mirror images of *trans*-1,2-difluorocyclohexane are *nonsuperimposable*. Ring flipping the two molecules turns the 1,2-diequatorial into the 1,2-diaxial conformation. The two molecules are still *nonsuperimposable*. *Note*: F_A and F_B are labels assigned to the mirror image to help you visualize the transformations performed.

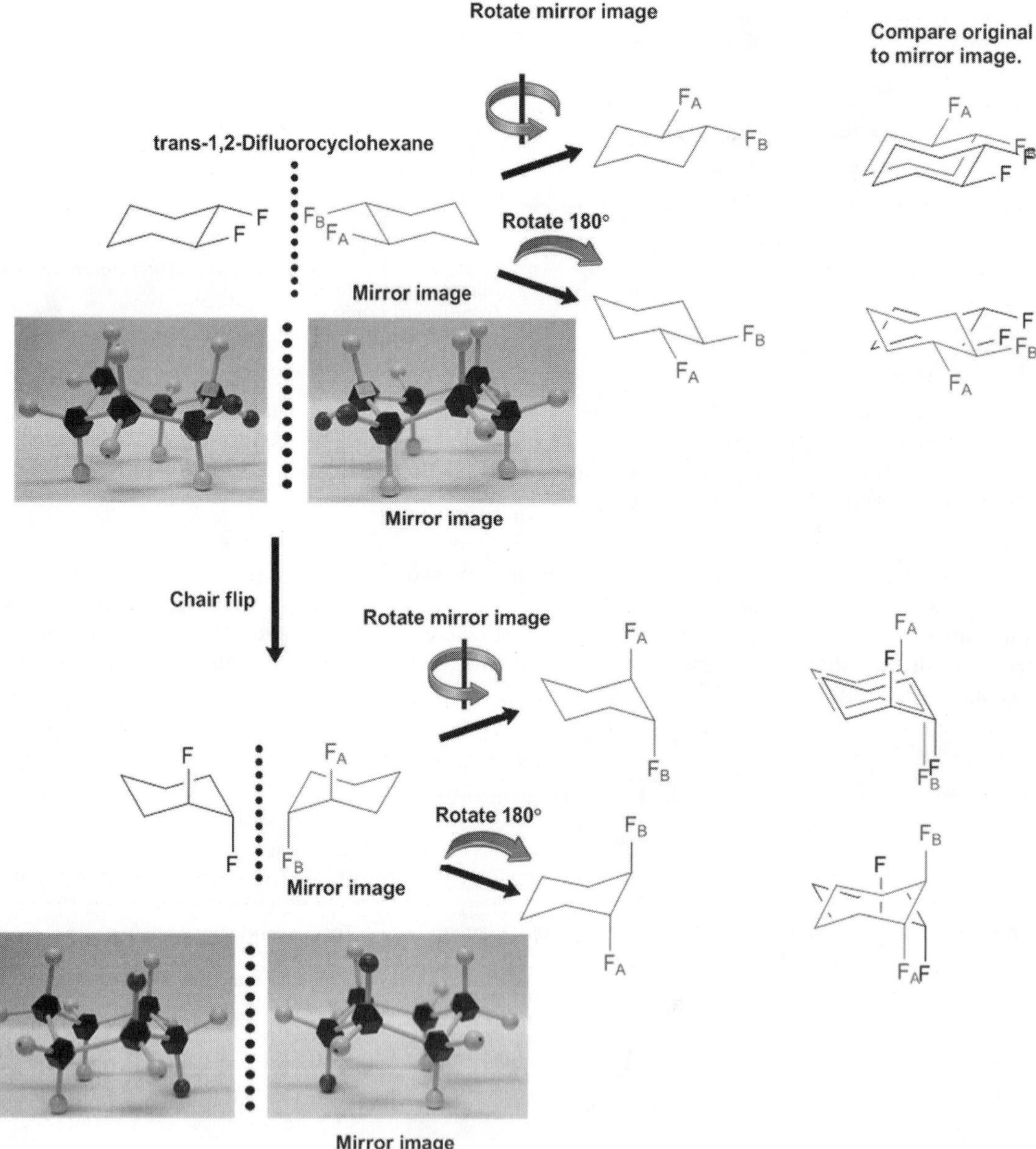

Your Turn 5.8

Think

Can you divide the molecule into two halves so that one half is the mirror image of the other? Where should the plane of symmetry be with respect to the two C–F bonds both pointed down?

Solve

The plane of symmetry is shown below. It reflects C1 and C2, C6 and C3, C5 and C4, and the two C–F bonds that are both pointed down.

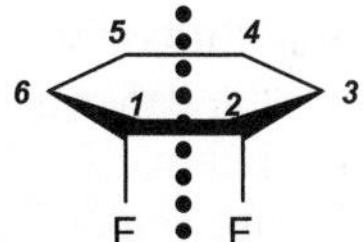

cis-1,2-Difluorocyclohexane

Your Turn 5.9

Think

What are the requirements for a plane of symmetry? Where should the reflection of an atom appear with respect to a mirror?

Solve

A plane of symmetry requires one half of the molecule to be the mirror image of the other half. Below, an "O" is drawn where each F atom's reflection should be if the plane that is drawn were actually a plane of symmetry. At each "O" that is drawn, there is no F atom already there. So neither of the planes shown is a plane of symmetry.

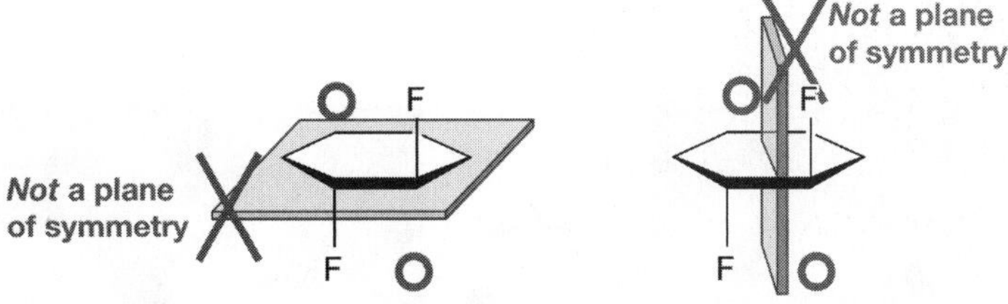

***trans*-1,2-Difluorocyclohexane**

Your Turn 5.10

Think

Which atoms are tetrahedral (*sp*^{3} hybridized)? For each *sp*^{3} atom, list the groups that are bonded to it. If you have four different groups, what does that mean? If you end up with two or more of the same group, what does that mean? For the ring structure, how do you look at the groups attached to an atom of the ring to determine if they are the same or different?

Solve

The tetrahedral stereocenters are marked below with an asterisk (*). Each stereocenter has four different groups on it. In butan-2-ol, the four groups are H, CH_3, CH_2CH_3, and OH; in the halogenated molecule, the groups are H, Cl, $-CH_2CH_2CH_2-$(ring), and $-CH_2C(CH_3)_2CH_2-$(ring). When looking at groups on a ring, go around the ring until to you come to a different substituent. If all substituents on the ring are the same in either direction, these are considered the same group.

OH

Butan-2-ol

Cl

3-Chloro-1,1-dimethylcyclohexane

Your Turn 5.11

Think

Draw the molecule of butan-2-ol that is given and its mirror image, build both molecules, and orient the molecules so their carbon chains line up. When they do, do all atoms from the two molecules line up? Does rotating about any of the C–C bonds help? Are four different groups attached to any of the carbon atoms? How many stereocenters are present? Can you draw any planes of symmetry?

Solve

The molecule of butan-2-ol that is given and its mirror image are the two structures on the left. After rotation, the zigzag orientations of the carbon chains line up and the two molecules are compared (at bottom). The OH group is in front of the plane of the paper in one molecule, but it is behind the plane of the paper in its mirror image. Therefore, the two mirror images are not superimposable. If you rotate about the internal C–C bond for one molecule, its carbon chain will no longer be zigzag, so this will not help with superimposing the two molecules. As with any molecule containing exactly one stereocenter, the molecule has no planes of symmetry and must be chiral.

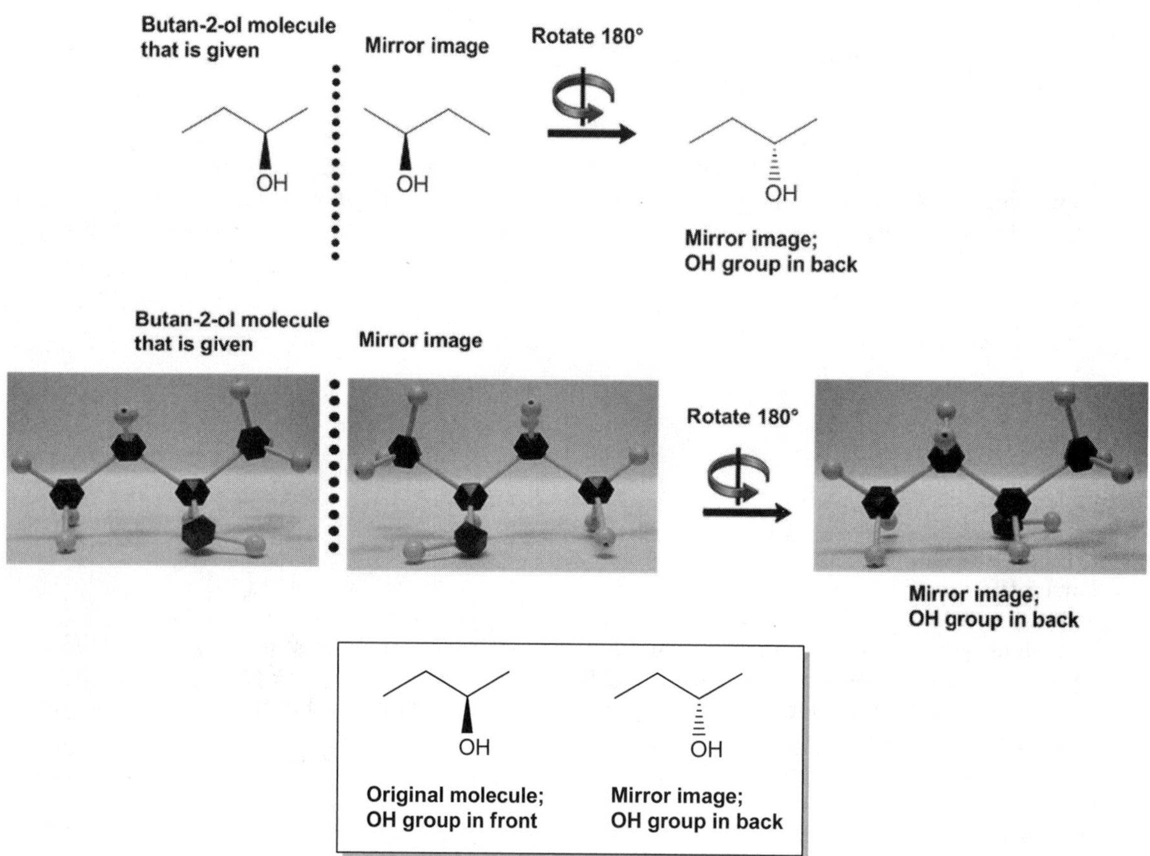

Your Turn 5.12

Think

As drawn, do the structures appear to be superimposable? Is there another orientation you can consider to convert a wedge bond into a dash bond? What would the resulting structure look like if you rotated about the C–C bond where the OH and Cl are bonded?

Solve

If the two structures are compared as given (left), they do not appear to be superimposable. Whereas Cl is in back in the molecule on the left, **A**, it is in front in the molecule on the right, **B**. If the left molecule is rotated 180° about the C–C bond, we arrive at conformation **A′**. In that new conformation, the OH is in the back and the Cl atom is in the front (like in molecule **B**). However, you will notice that the carbon chain has been rotated and is not superimposable on **B**. If you flip the molecule over horizontally, you will obtain orientation **A″**. Comparison to molecule **B** once again shows that the two molecules are not superimposable: Whereas OH is in front in **A″**, it is in back in the molecule **A**.

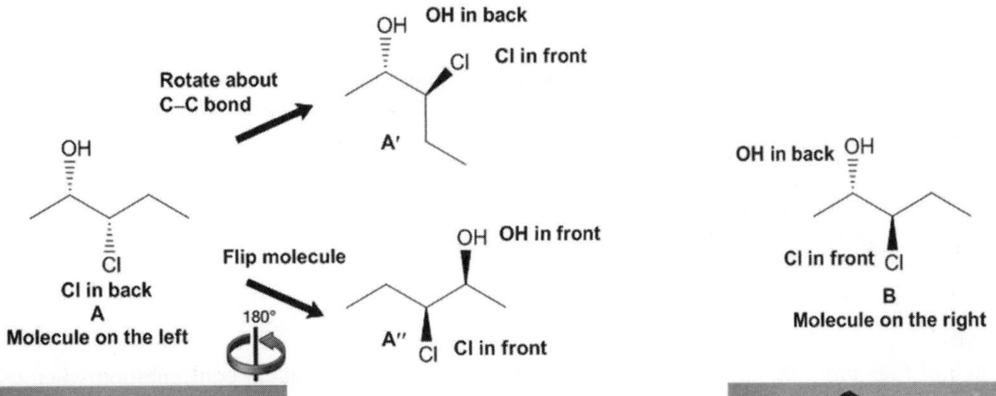

OH in back
Cl in front
Rotate about C–C bond
A′
Cl in back
A
Molecule on the left
Flip molecule
180°
OH in front
A″
Cl in front
OH in back
Cl in front
B
Molecule on the right

Your Turn 5.13

Think

What constitutes a stereocenter for a carbon atom? For each sp^3 carbon atom, list the four groups that are attached. If you have four different groups, what does that mean? If you end up with two or more of the same group, what does that mean?

Solve

The three stereocenters are marked below with an asterisk (*). Each stereocenter has four different groups bonded to it. In 2,3-dibromo-4-methylhexane, C2 is bonded to CH_3, H, Br, and CHBrR; C3 is bonded to Br, $CHBrCH_3$, H, and $CH(CH_3)R$; C4 is bonded to CH_3, H, CH_2CH_3, and CHBrR. The remaining C atoms are each bonded to at least two H atoms and are therefore not stereocenters.

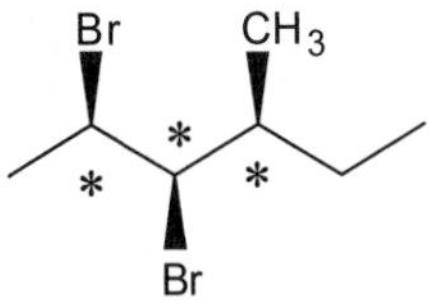

Your Turn 5.14

Think

In a Fischer projection, are the horizontal bonds pointed toward you or away from you, wedge or dash, respectively? What about the vertical bonds? When you perform the 90° rotation, does group X point toward you or away from you? Group Z?

Solve

Horizontal bonds in a Fischer projection point toward you (wedge) and vertical bonds point away from you (dash). The W and Y groups both go from dash to planar bonds. After rotation 90°, the X group points to the left and toward you (wedge), whereas the Z group points to the left and away from you (dash).

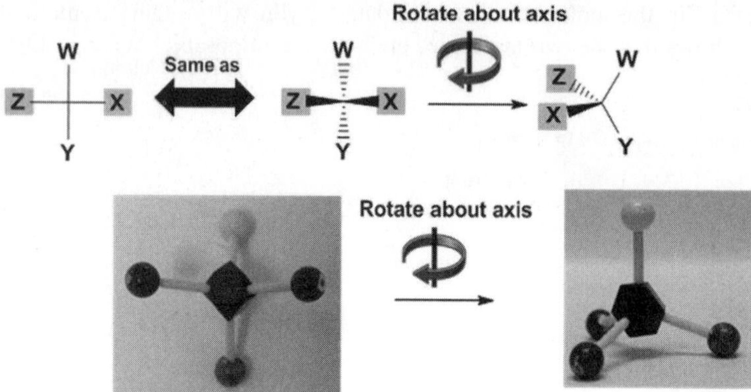

Your Turn 5.15

Think

Which of the atoms are stereocenters? What happens to a stereochemical configuration when two substituents are exchanged?

Solve

There are four carbon stereocenters, represented by each of the intersecting horizontal and vertical lines in the Fischer projection. The stereochemical configurations are different at the C atoms circled below, because they are related by the exchange of the two substituents—in this case, H and OH.

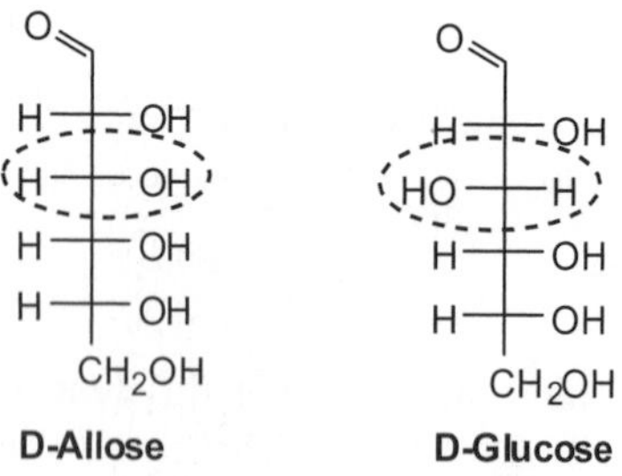

Your Turn 5.16

Think

In a Fischer projection, are the horizontal bonds pointed toward you or away from you? Which ball-and-stick model in Figure 5-25b should you use as a guide when completing the Fischer projection for C5? For C6?

Solve

To fill in the positions at C5, we must examine the model on the right because in that model, the substituents on the horizontal point toward us. In that model, OH at C5 is on the left, so the same should be true in the Fischer projection. To fill in the positions on C6, we must use the model on the left, in which the substituents on the horizontal point toward us. In that model, OH is on the right, so the same should be true in the Fischer projection. See the figure on the next page.

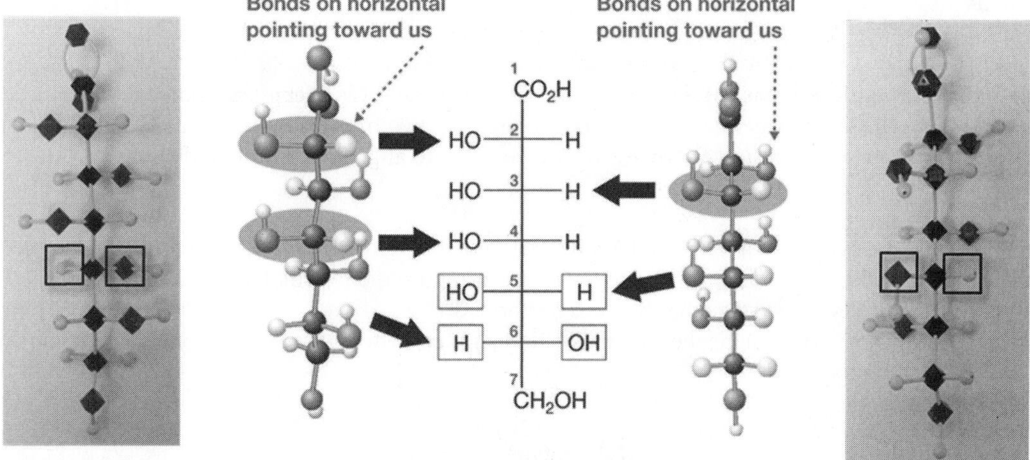

Your Turn 5.17

Think

In the model on the left, at which C atom—C4 or C5—are the bonds pointing toward you? In the model on the right, at which C atom—C4 or C5—are the bonds pointing toward you? Recall that Fischer projections depict the horizontal bonds as pointed toward you.

Solve

The substituents on C4 in the Fischer projection should be written as they appear in the model on the left, where the horizontal bonds point toward us. Therefore, at C4, the H atom should be on the left and OH should be on the right. The substituents on C5 in the Fischer projection should be written as they appear in the model on the right, where the horizontal bonds point toward us. So at C5, the H atom should be on the left and OH should be on the right.

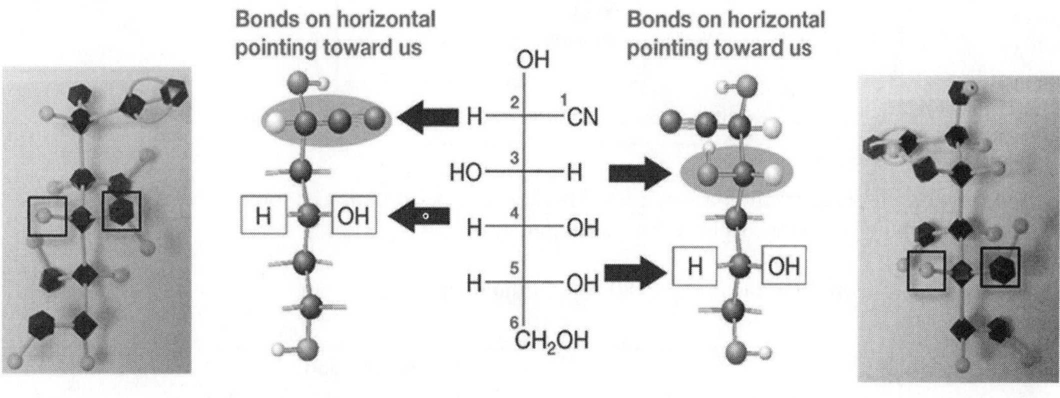

Chapter Problems

Problem 5.1

Think

It might be helpful to build a model of each molecule using a model kit. Keep one molecule stationary and orient the other molecule every way imaginable to try to align the two molecules. It is usually helpful to pick a point of reference. When you perform your rotations, are you able to align ALL bonds and atoms?

Solve

In all of the solutions below, the molecule on the left is the one that remains stationary and the one on the right is reoriented in space (when necessary) in an attempt to superimpose the two molecules. Prime symbols (′) are used to label atoms to help you visualize the rotations performed.

(a) Superimposable. The C–H″ bond is the axis of rotation and the bottom three bonds rotate counterclockwise 120° as viewed from the top.

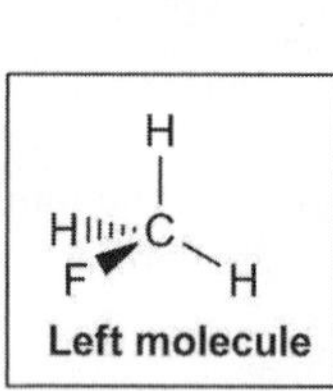

120° rotation

H"
F
C
H'
H
H"
H
C
F
H'
Right molecule

(b) Superimposable. The molecule is rotated 180° out of the plane of the paper.

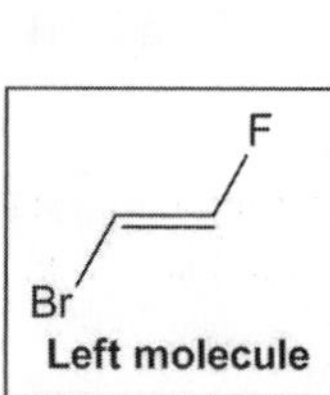

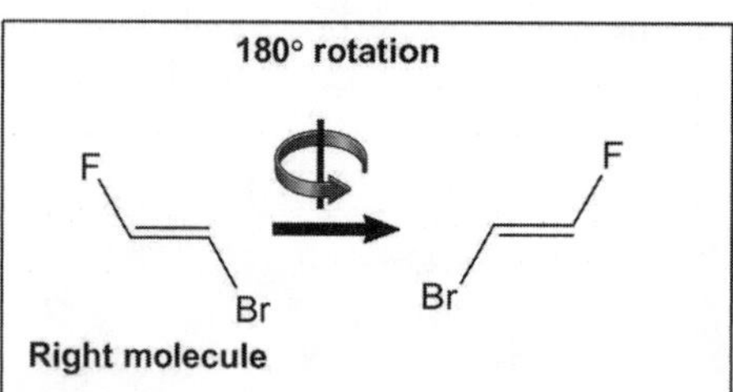

(c) Nonsuperimposable. Free rotation about a C=C bond is not permitted. The molecule on the left is the trans isomer and the molecule on the right is the cis isomer.

(d) Superimposable. The molecule is rotated 180° out of the plane of the paper.

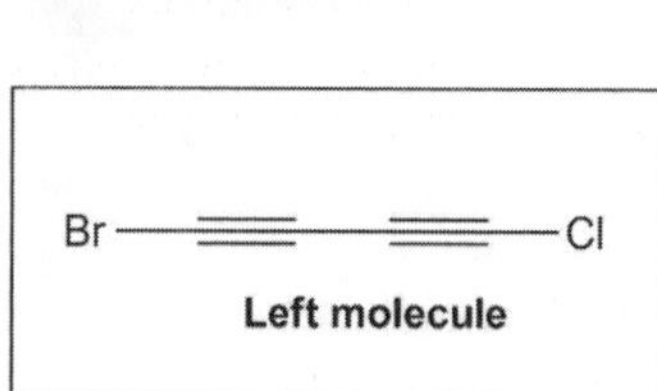

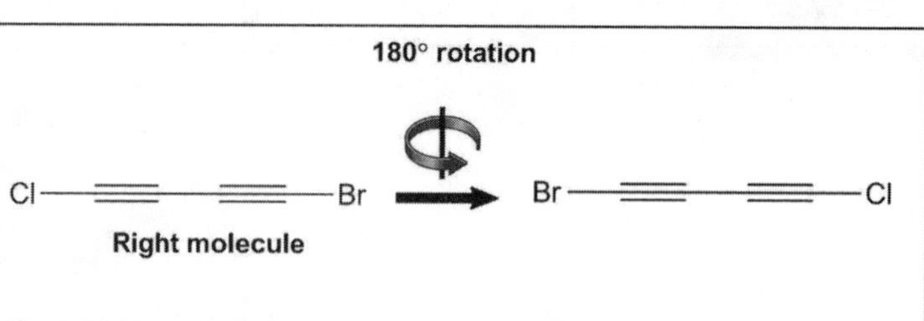

(e) Superimposable. A 180° rotation is performed about the diagonal on the cyclobutane ring as shown. The dash C–Cl bonds convert to wedge C–Cl bonds, and thus the molecules are shown to be superimposable.

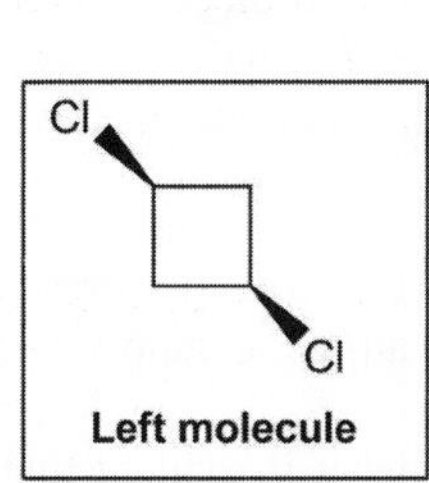

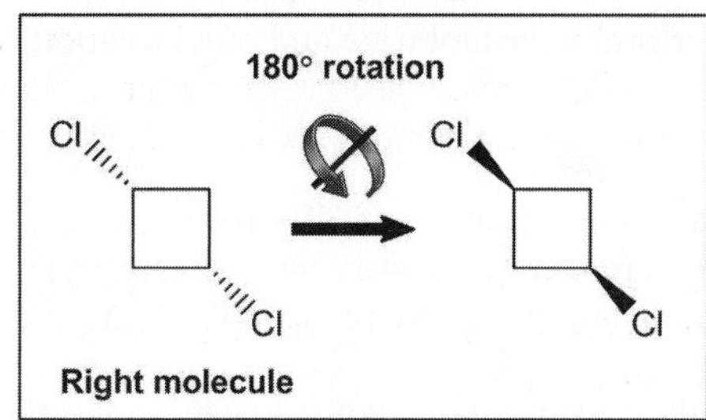

(f) Nonsuperimposable. Full rotation about single bonds in a ring is not permitted. The molecule on the left has the C–Cl bonds both wedge (cis) and the molecule on the right has one wedge and one dash (trans).

(g) Nonsuperimposable. Both molecules have one wedge and one dash C–Cl bond. However, if you attempt to superimpose the two molecules, you will see it is not possible; where there is a dash bond in the first molecule there is a wedge bond in the second, and vice versa. A 180° rotation of the second molecule does not change the situation, as shown below.

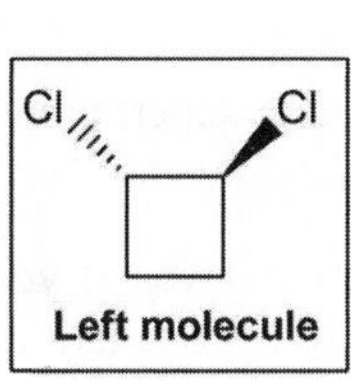

(h) Nonsuperimposable. Full rotation about single bonds in a ring is not permitted. The molecule on the left has the C–Cl bonds both wedge (cis) and the molecule on the right has one wedge and one dash (trans).

Problem 5.3

Think

For each molecule, draw a mirror next to the molecule. For each atom or group in the original molecule, where should its mirror image be with respect to the mirror? How should dash–wedge notation be treated in the mirror image?

Solve

Mirrors are represented by the dotted lines below, and the mirror images are drawn to the right of each mirror. If a bond is a wedge in the original, it will be a wedge in the mirror and, if a bond is a dash in the original, it will be a dash in the mirror. Make sure that each atom has a mirror image that is the same distance from the mirror, at the same position along the mirror, and has the same position relative to the plane of the paper.

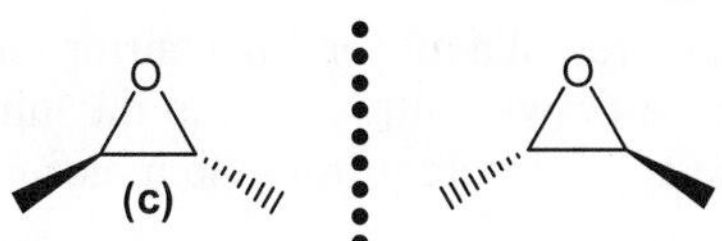

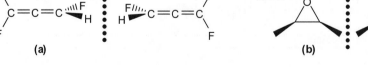

Problem 5.4

It might be helpful to build a model of each molecule and its mirror image using a model kit. Keep one molecule stationary and try to align the other molecule to make it superimposable. It is usually helpful to pick a point of reference. When you perform your molecule and bond rotations, are you able to align ALL bonds and atoms?

Solve

Each molecule and mirror image from Problem 5.3 is repeated below. In all of the solutions below, the molecule on the left is the one that remains stationary and the one on the right has rotations performed on it in an attempt to superimpose the two molecules. Prime symbols (′) are used to label atoms to help you visualize the rotations performed.

(a) Nonsuperimposable. From the 180° rotation shown below on the right molecule, it is evident that the two molecules are nonsuperimposable. The flat C–H and C–F bonds align, but the wedge and dash C–H and C–F bonds do not align.

Left molecule

180° rotation

Right molecule

(b) Superimposable. It is evident from the mirror image that the two molecules are superimposable. No rotations are necessary.

(c) Nonsuperimposable. Both molecules have one wedge and one dash C–CH_3 bond. However, if you attempt to superimpose the two molecules you will see it is not possible because the dash–wedge notation does not agree. The rotation of the molecule shown below still does not allow the dash–wedge notation to agree.

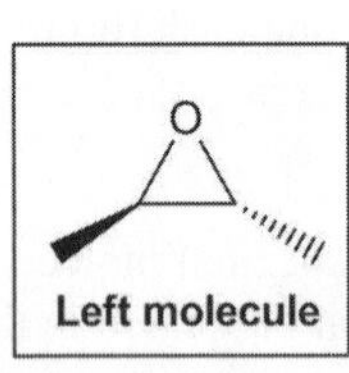

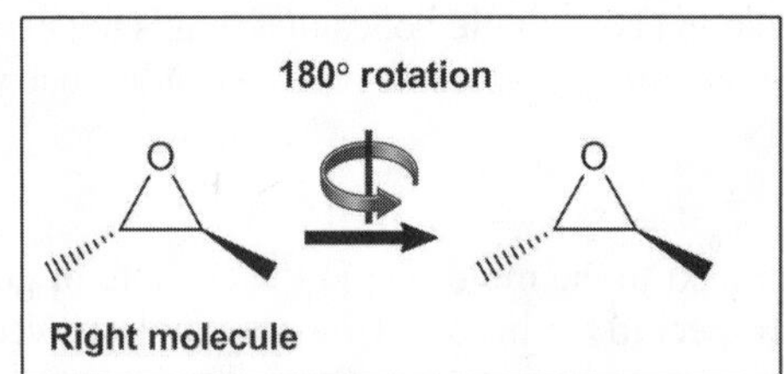

(d) Superimposable. Keep the front C of the Newman projection and its bonds frozen in place because they are already superimposable. Rotate the back CHF_2 group 120° clockwise, as shown below, to make the two molecules entirely superimposable.

Left molecule

120° rotation about C–C bond

Right molecule

Problem 5.5

Think

Can you draw the mirror image of the molecule on the right? How do you represent a mirror? In a mirror image, how is dash–wedge notation treated? How is each atom in the mirror image drawn with respect to the mirror? Does the mirror image of the molecule on the right appear to be the same as or different from the molecule on the left?

Solve

A mirror can be represented by a dotted line adjacent to the molecule. Dash–wedge notation is conserved in the mirror image. Make sure that each atom has a mirror image that is the same distance from the mirror, at the same position along the mirror, and has the same position relative to the plane of the paper.

(a) The two molecules are not mirror images. The molecule on the left has two wedge C–Cl bonds and the molecule on the right has one wedge and one dash C–Cl bond. The mirror image drawn is not the same.

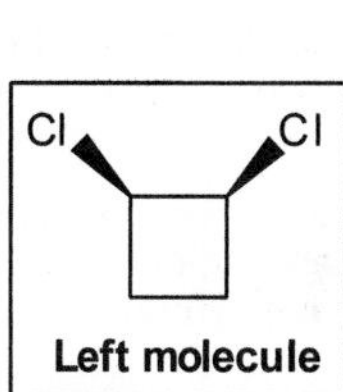

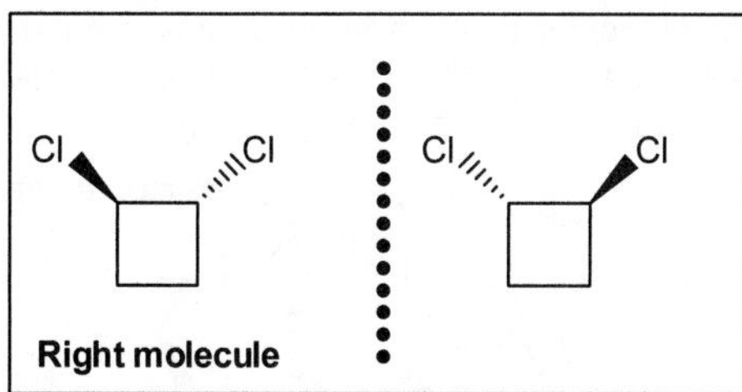

(b) The two molecules are mirror images.

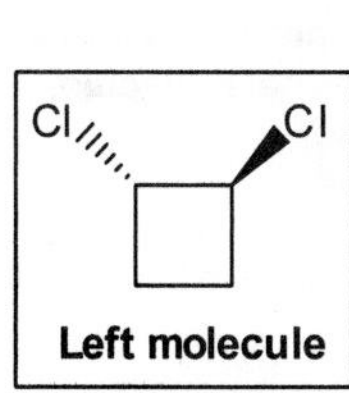

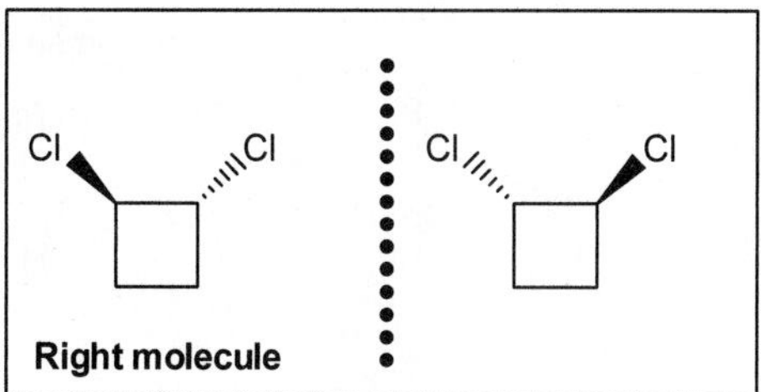

(c) The two molecules are not mirror images. The two molecules originally drawn are actually the same molecule. When the mirror image for the right molecule is drawn, it is evident that the front group with the CH_3, Cl, and H groups will not align with the left molecule.

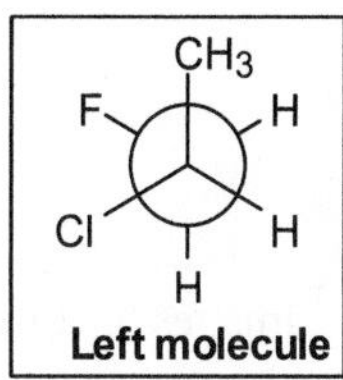

CH3 H F Cl H H CH3 F H H Cl H

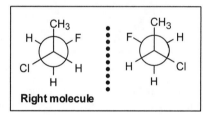

Right molecule

Problem 5.7

Think

Can you draw the mirror image of the molecule? Is the mirror image different from the original molecule, or are the two superimposable? It may help to construct models of the molecule and its mirror image and orient them in every way imaginable.

Solve

Trans-1,2-dichloroethene is not chiral, because **all the atoms lie in the same plane.** Below we show the mirror image, and after it is rotated 180°, we can see that it is the same as the original molecule.

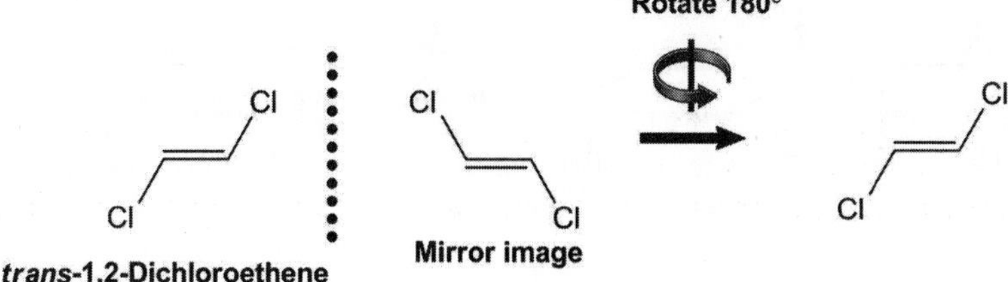

Problem 5.8

Think

Can you draw the mirror image of the molecule? Is the mirror image different from the original molecule, or are the two superimposable? It may help to construct models of the molecule and its mirror image and orient them in every way imaginable. Make sure to try rotations about single bonds.

Solve

(a) and **(c)** are achiral. For each molecule, rotation about a single bond yields a conformation whose mirror image is the same as itself (see below). For **(b)**, there is not a conformation for which this is true, so it is chiral.

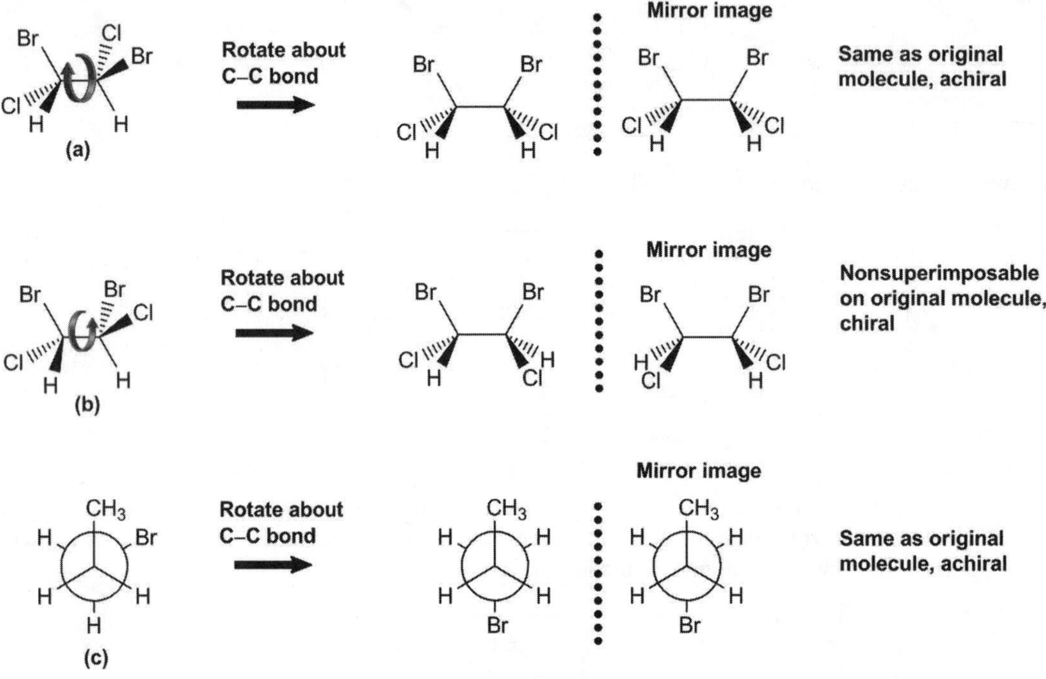

Problem 5.9

Think

Transform each chair to a Haworth projection. Sometimes it is easier to see mirror images in this type of structure. Does the molecule have an enantiomer or is its mirror image superimposable?

Solve

We can more clearly identify chirality in cyclohexane rings by working with Haworth projections instead of chair representations. As we can see below and on the next page, the mirror image of **(b)** is exactly the same as the original molecule. The mirror images of **(a)**, **(d)**, and **(e)** are the same after rotation of the resulting molecules. The mirror image of **(c)** is not superimposable on the original molecule, so it is chiral.

(a) Achiral.

Rotate 180°

Mirror image

Down

CH_3 CH_3 CH_3

(b) Achiral.

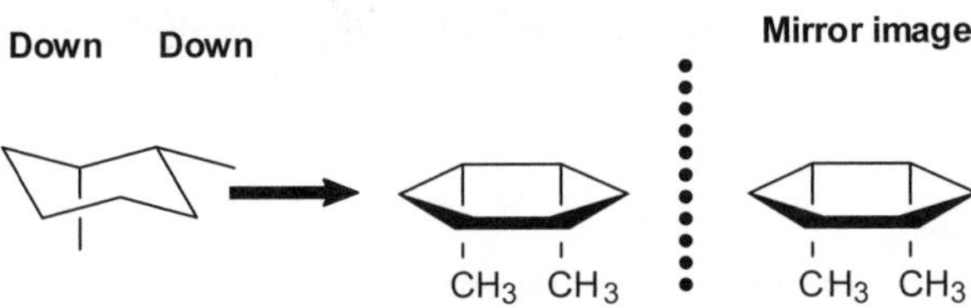

(c) Chiral.

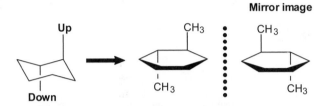

(d) Achiral.

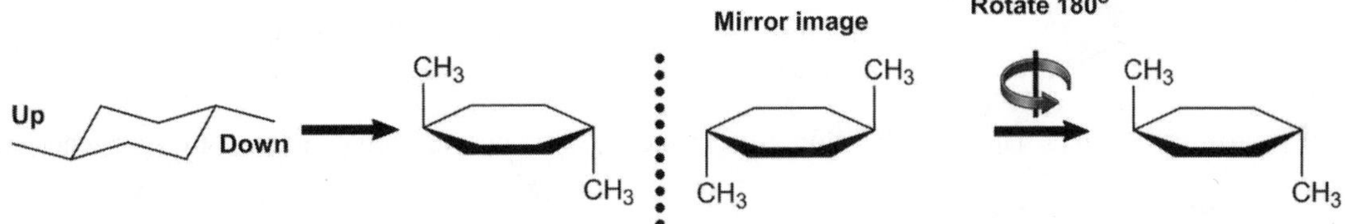

(e) Achiral.

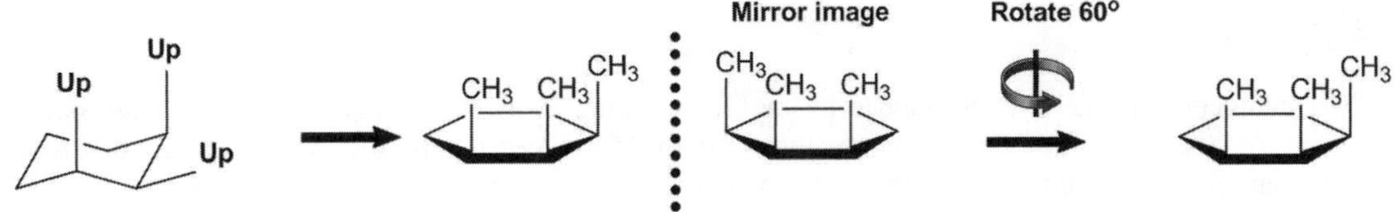

Problem 5.10

Think

Is there a way to bisect the molecule such that one half of the molecule is the mirror image of the other half? If so, draw in the dotted line that represents this internal plane of symmetry.

Solve

Molecules **(a)**, **(c)**, and **(f)** each have a plane of symmetry, as shown below. The others do not.

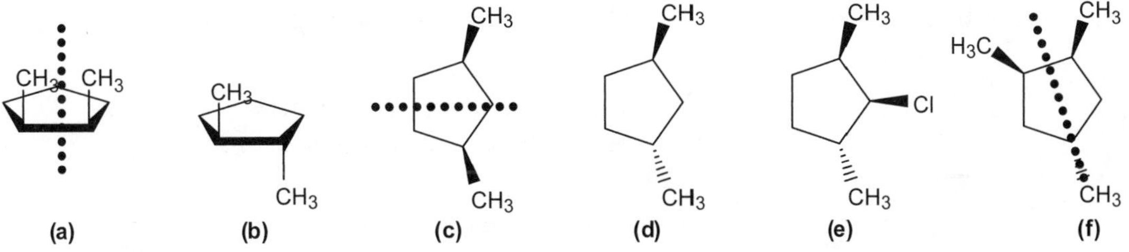

Problem 5.12

Think

Which atoms are bonded to four different groups? How can you determine whether groups that are part of a ring are different from each other?

Solve

For **(a)–(e)** stereocenters are indicated by an asterisk.

(a) One stereocenter. The second C atom from the left in is not a stereocenter because it is bonded to two CH_3 groups (one is explicitly labeled and the other is not).

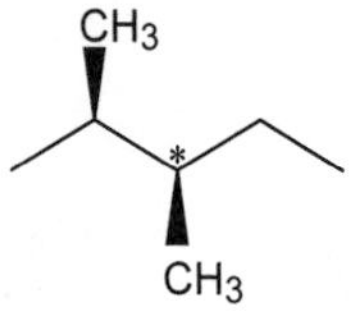

(b) Zero stereocenters. C2 and C4 are not stereocenters because each is bonded to two CH_3 groups.

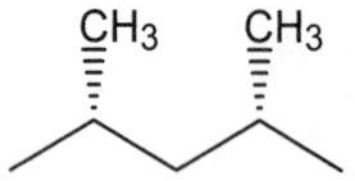

(c) Zero stereocenters. The central C atom is not a stereocenter because it is bonded to two benzene rings.

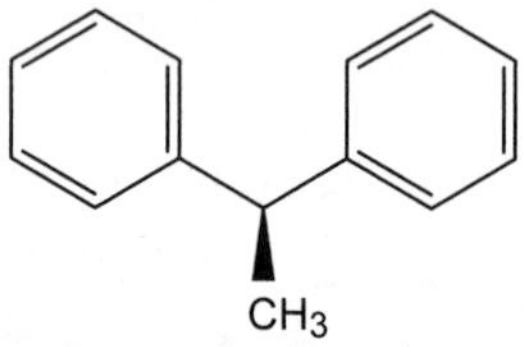

(d) Two stereocenters. The C atom bonded to the CH_3 pointing away from us is not a stereocenter because it lies on a plane of symmetry that contains that C atom. More specifically, the two groups bonded to that C atom that are part of the ring are identical to each other.

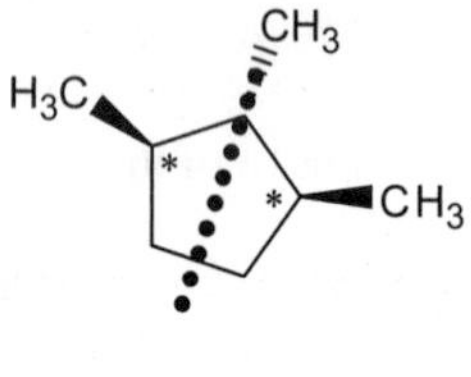

(e) No plane of symmetry exists, so the C atom bonded to the CH_3 group pointing away from us *is* a stereocenter. Alternatively, notice that that C atom is bonded to four different groups: H, CH_3, $CH(CH_3)R$, and $CH(CH_2Cl)R$.

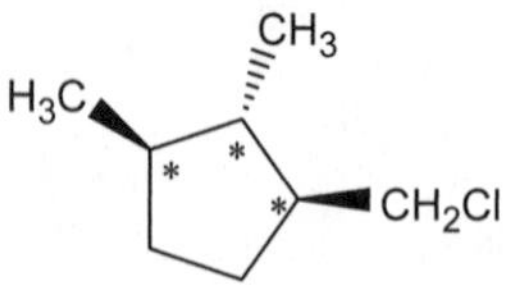

Problem 5.13

Think

How many stereocenters are in each molecule? Do any of the molecules possess an internal mirror plane (i.e., a plane of symmetry)?

Solve

To be meso, a compound must contain at least two stereocenters and have a plane of symmetry. **(a)** is chiral, so it is not meso. **(b)** is achiral and contains two stereocenters, so it is meso. **(c)** does not contain any stereocenters, so it is not meso, even though it has a plane of symmetry. **(d)** does not have a plane of symmetry, so it is chiral and, therefore, cannot be meso. **(e)** is achiral (which we can see after rotation about the C–C bond), and contains two stereocenters, so it is meso.

Br Cl Br Br Cl Cl

(a) (b) *meso* (c)

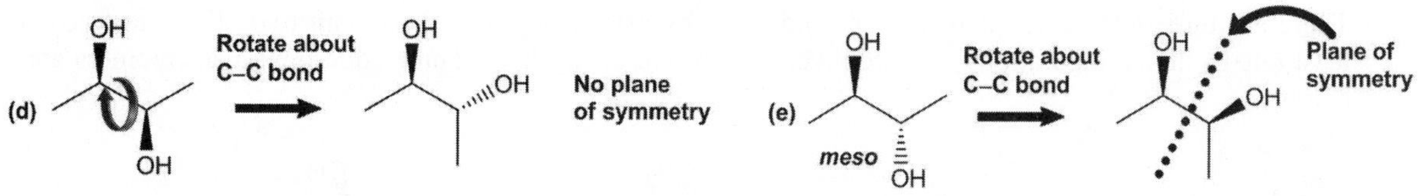

(d) OH OH Rotate about C–C bond OH OH No plane of symmetry (e) OH *meso* OH Rotate about C–C bond OH OH Plane of symmetry

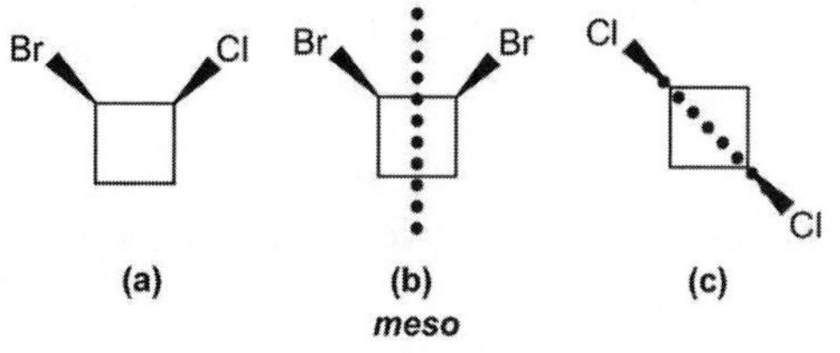

Problem 5.14

Think

Use Figure 5-15 as a guide to exchange W and Y. Are you able to align all four bonds in the resulting structure with the appropriate ones in the mirror image of the original molecule?

Solve

Groups W and Y are exchanged and the resulting molecule (gray) is superimposable on the blue structure from Figure 15-5. To view how the two are superimposable, two 109.5° rotations are necessary (see below).

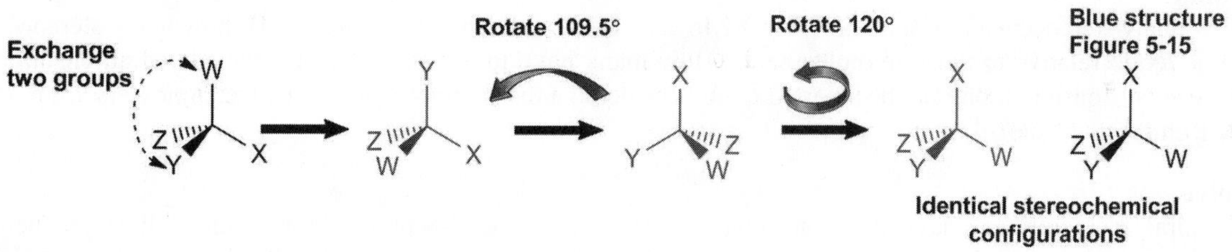

Exchange two groups W Z Y X → Y Z W X Rotate 109.5° → X Y Z W Rotate 120° → X Z Y W Blue structure Figure 5-15 X Z Y W

Identical stereochemical configurations

Problem 5.15

Think

Which C and N atoms are bonded to four different groups? How can you determine whether groups that are part of a ring are different from each other?

Solve

Molecules **(a)**, **(c)**, and **(d)** have no tetrahedral stereocenters. In **(a)**, each planar C atom is bonded to only three groups total. The tetrahedral C atoms are each bonded to at least two H atoms. In **(c)**, all planar C atoms are bonded to only three groups total, and the same is true of the N atom. The four tetrahedral C atoms are each bonded to at two H atoms. In **(d)**, each tetrahedral C atom is bonded to at least two H atoms. The N atom is bonded to two of the same ethyl groups. There are two stereocenters in molecule **(b)**. A nitrogen atom is a stereocenter if four different groups are attached to it: H, CH_3, $–CH_2–$(ring), and $–CH(CH_3)–$(ring). The four groups on the carbon stereocenter are H, CH_3, N, and the ring. See the structures on the next page.

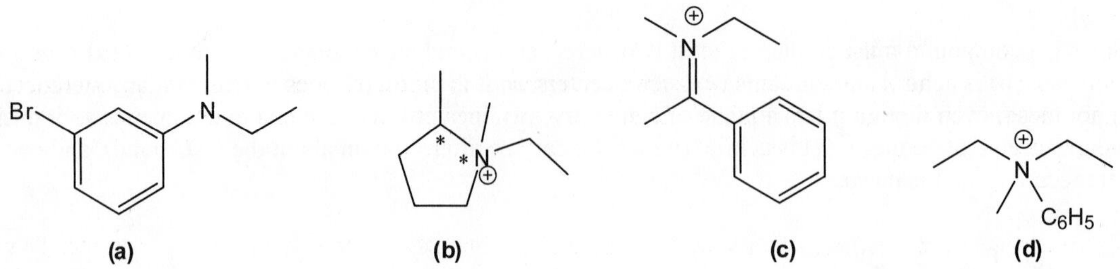

(a) (b) (c) (d)

Problem 5.17

Think

Which atoms are the tetrahedral stereocenters? To obtain a diastereomer of the given molecule, how many of those configurations can be reversed?

Solve

There are three tetrahedral stereocenters, indicated by asterisks below. To obtain two other diastereomers, at least one of the configurations must be reversed, but not all of them. Three additional diastereomers are given below.

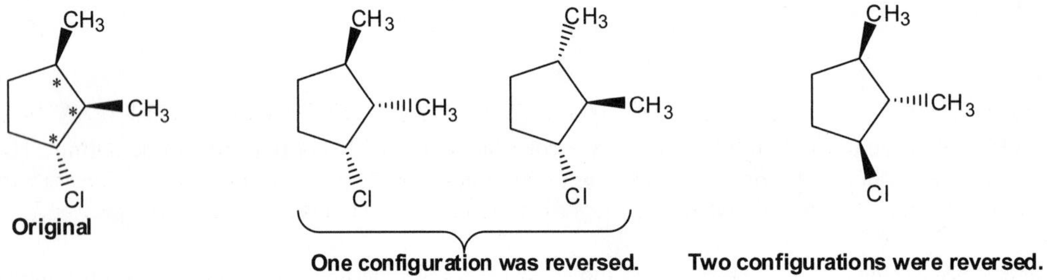

Original

One configuration was reversed. Two configurations were reversed.

Problem 5.18

Think

How many stereocenters exist in molecule **A**? In each of the following structures **B–H**, how many stereocenters are inverted relative to those in molecule **A**? How many have to be inverted to be considered an enantiomer? How many configurations can be inverted to be considered a diastereomer? How many enantiomers are possible for a chiral molecule?

Solve

A chiral molecule can have only one enantiomer. Here, it is the molecule in which all stereochemical configurations are the reverse of those in molecule **A**, which is molecule **H**. In all of the other molecules, at least one stereochemical configuration is reversed, but not all of them are. So, all of the other molecules—**B, C, D, E, F,** and **G**—are diastereomers of **A**.

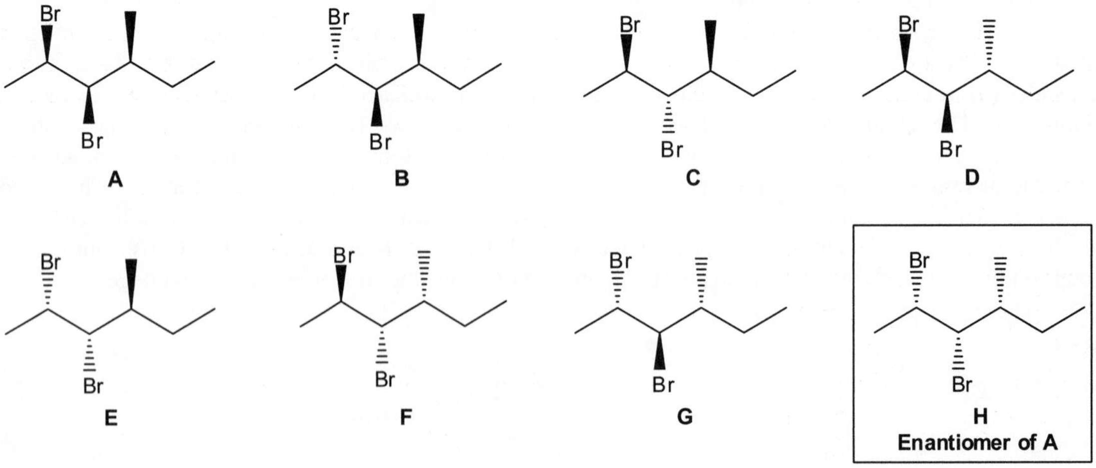

A B C D

E F G H

Enantiomer of A

Problem 5.20

Think

How many stereocenters are present? How can we change each stereocenter to convert from one configurational isomer to another? How does the formula 2^n assist in determining the total number of stereoisomers? Are any of the isomers meso?

Solve

(a) There are two stereocenters, thus the maximum number of configurational isomers is $2^n = 2^2 = 4$. To obtain all possible configurations, systematically reverse the configurations at the different stereocenters. Notice that **A** and **B** both possess a plane of symmetry and are exactly the same (rotate **B** 180° to visualize this); they are meso. Therefore, there are **three** configurational isomers possible.

(b) There are two stereocenters: one at C2 and one at C4. C3 is not bonded to four different groups. All of the permutations of reversing the configurations at just those stereocenters are shown below. There are four of them, which does equal $2^n = 2^2 = 4$. Notice that **A** and **B** are not the same in this example because the C3 C–OH bond is denoted as a wedge and a 180° rotation would not interconvert these two molecules.

Problem 5.21

Think

How many stereocenters do D-allose and L-glucose have? How many stereocenters are reversed between the two? How many must be reversed for the molecules to be enantiomers? How many can be reversed for the molecules to be diastereomers?

Solve

They cannot be enantiomers of each other, because any molecule can have only one mirror image, and D- and L-glucose are enantiomers. Because D-allose and L-glucose have the same connectivity but are not the same, they must be diastereomers. This is consistent with the fact that three of the four stereocenters are reversed, as seen below.

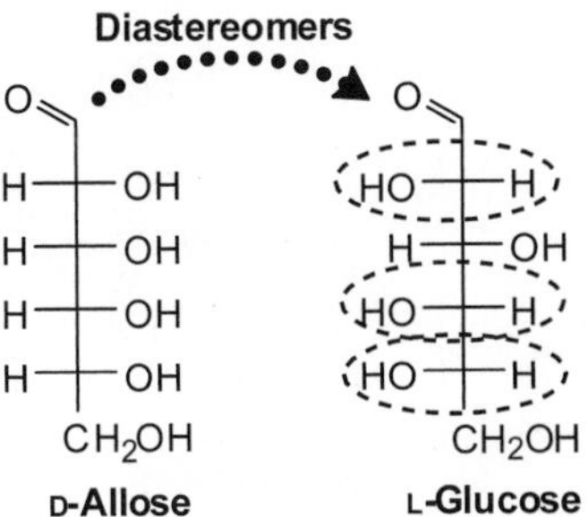

Problem 5.22

Think

How many stereocenters are present? How many stereocenters must be reversed to draw the enantiomer? How is the mirror image of a chiral molecule related to the original molecule?

Solve

All four stereocenters must be reversed and/or you can draw the mirror image (as shown below) to obtain the enantiomer.

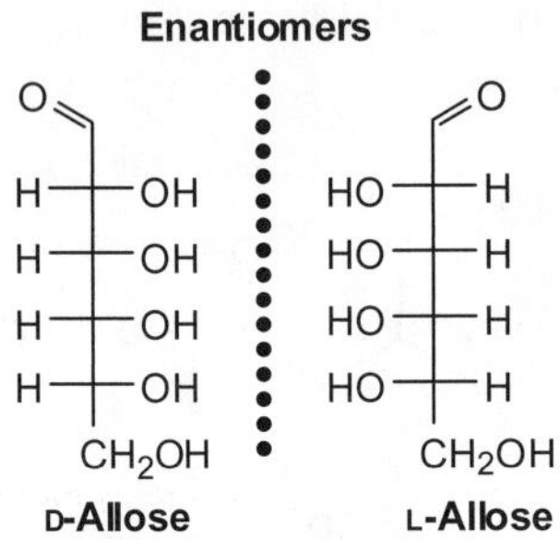

Problem 5.23

Think

How many stereoisomers are present? Review the steps for converting a zigzag structure to a Fischer projection.

Solve

(a) There are three stereocenters: C3, C4, and C5. Draw in the Fischer framework to denote the presence of stereocenters at these three C atoms, but temporarily leave the substituents on the horizontal bonds blank. C1 should be on top.

Use a model kit to build a molecular model of the molecule you are given, paying special attention to the dash–wedge notation. Orient the molecule vertically so that C1 is on top and C6 is on the bottom, such as is shown on the left below. In that orientation, the horizontal bonds of C3 and C5 point toward you, so add the H and OH substituents to C3 and C5 of the Fischer projection as they appear in the model. Then, flip the model over so it appears as on the right below. The horizontal bonds of C4 point toward you, so add the H and OH substituents to C4 of the Fischer projection as they appear in the model.

(b) Same methodology is followed.

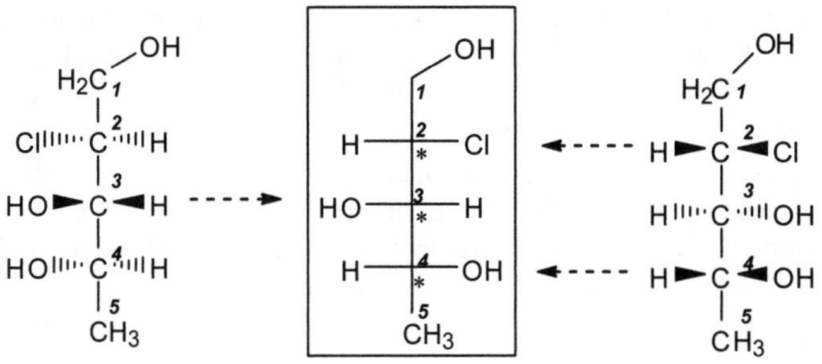

Problem 5.24

Think

How many stereoisomers are present? Review the steps for converting a Fischer projection to a zigzag structure.

Solve

(a) There are four stereocenters: C2, C3, C4, and C5. Construct a molecular model with a six-carbon chain so that the chain is zigzag. Hold the molecule so that zigzag is vertical. With the horizontal bonds pointed toward you on C2 and C4, add the H and Cl atoms as they appear in the Fischer projection. This is shown on the left below. Then, flip the molecule over so that the horizontal bonds on C3 and C5 are pointed toward you and add the H and Cl atoms as they appear in the Fischer projection.

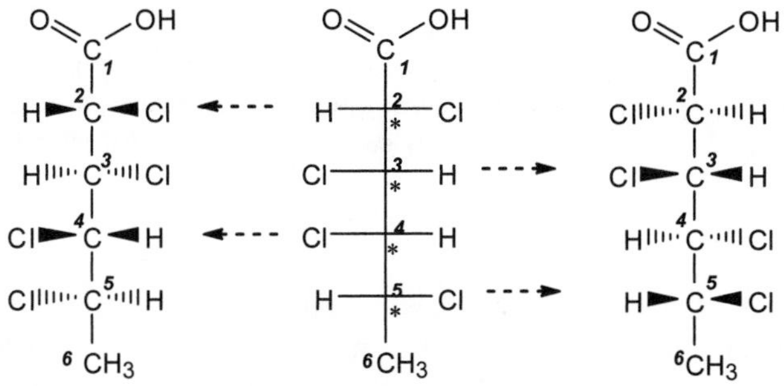

Once all the substituents have been added, you can reorient the completed model so the zigzag is in the plane of the page, as shown below.

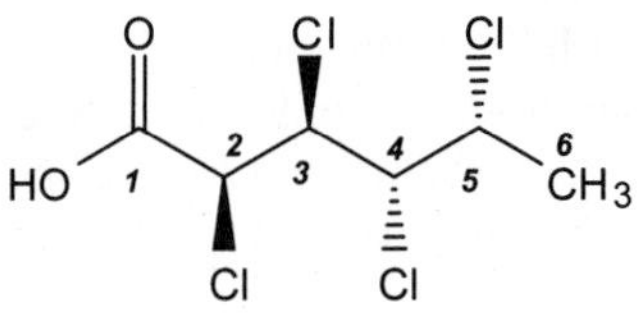

View from the side

(b) Same methodology is followed.

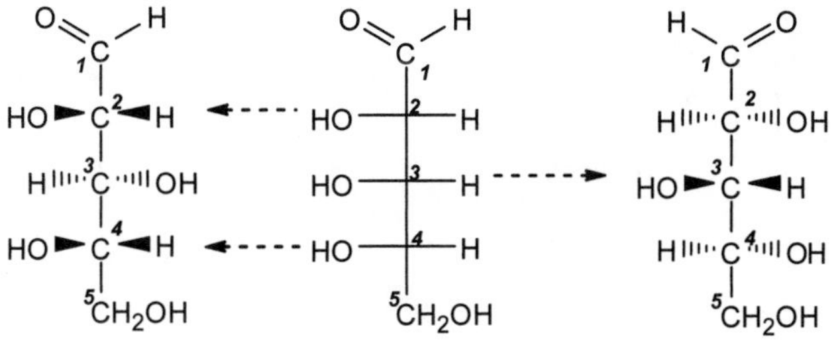

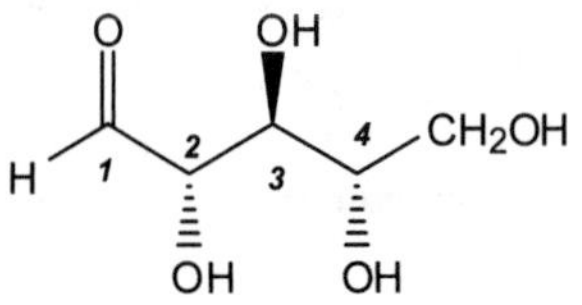

View from the side

Problem 5.26

Think

How is each molecule—**A** through **E**—related to **Y**? Are they constitutional isomers, enantiomers, diastereomers, unrelated, or the same molecule? How do these relationships translate into relative behavior? Explain.

Solve

The acidities will be different unless the molecules are enantiomers or are identical molecules. Since molecule **E** is its enantiomer, it will have exactly the same acidity. All of the remaining molecules will have acidities different from the molecule given. The relationships are given below.

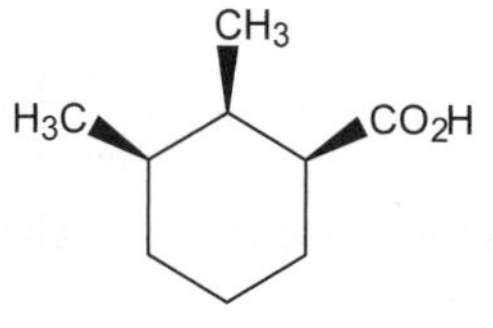

Y
Chemical Formula: $C_9H_{16}O_2$

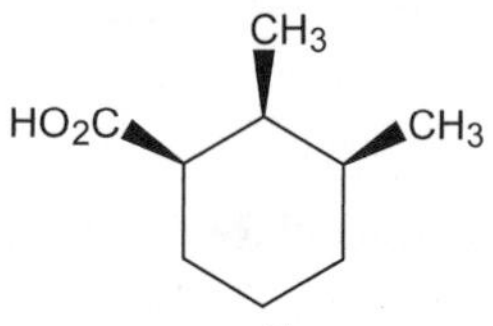

E
Chemical Formula: $C_9H_{16}O_2$
Enantiomer of Y

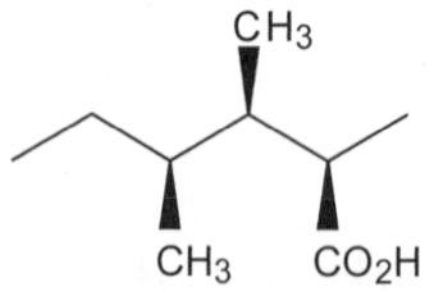

A
Chemical Formula: $C_9H_{18}O_2$
Not an isomer

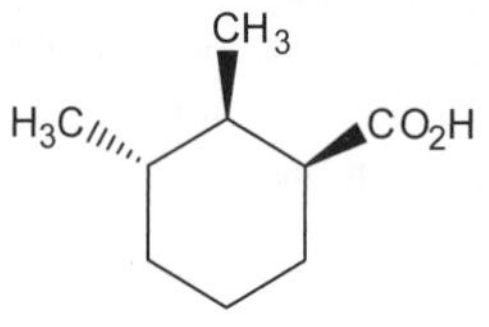

B
Chemical Formula: $C_9H_{16}O_2$
Diastereomer

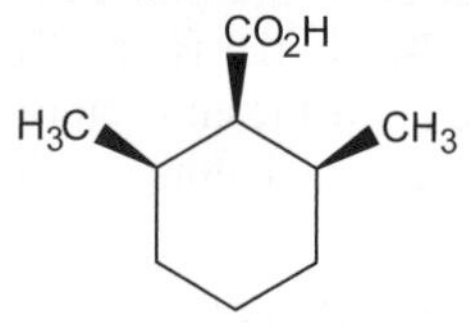

C
Chemical Formula: $C_9H_{16}O_2$
Constitutional isomer

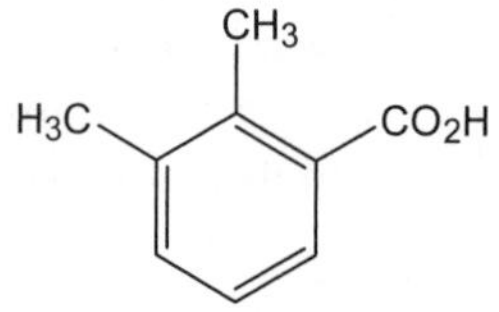

D
Chemical Formula: $C_9H_{10}O_2$
Not an isomer

Problem 5.28

Think

What is the degree of alkyl substitution on each of the C=C bonds? How does that translate into stability? How many of each kind of C=C bond are there? What is the relationship between relative stability and relative heats of combustion?

Solve

All three have the same molecular formula (C_8H_{12}), two double bonds, and one six-membered ring. Therefore, the difference in heats of combustion will be due solely to the stability of the C=C bonds. **D** will have the greatest heat of combustion because it is the least stable, having the least highly substituted alkene functional groups. Its alkene groups are both disubstituted. **E** will be in the middle because it has one disubstituted and one trisubstituted. **C** will have the smallest heat of combustion, as it is the most stable, having the most highly substituted alkene groups. Its alkene groups are both trisubstituted. The increasing order of heats of combustion is as follows: **C** < **E** < **D**.

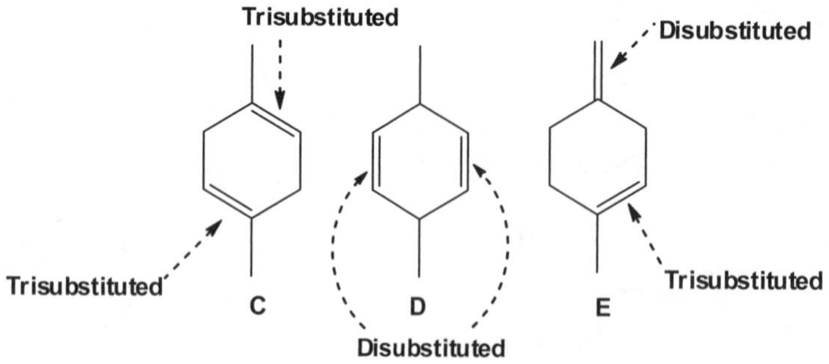

Trisubstituted
Disubstituted
Trisubstituted
Trisubstituted
C
D
E
Disubstituted

Problem 5.30

Think

For which variable in Equation 5-2 are we solving? Are the units for concentration correct? Are the units for the cell length correct?

Solve

We are asked to solve for the measured angle of rotation α. The units for concentration are correct (g/mL), but the units for the cell length need to be converted from cm to dm. 10.0 cm $\times \frac{1\text{ dm}}{10\text{ cm}}$ = 1.0 dm.

$$\alpha = ([\alpha]_D^{20})(l)(c) = (+223°)(1.0\text{ dm})\left(0.00300\frac{\text{g}}{\text{mL}}\right) = +0.669°$$

Problem 5.32

Think

What is the sign of the specific rotation of the mixture of the two enantiomers? Which enantiomer is in excess? How can you tell from the specific rotation of the mixture the extent by which that enantiomer in excess? What percentage of the solution is optically inactive?

Solve

The specific rotation of the mixture of the two enantiomers is +12°. This means that the (+) enantiomer is in excess.

$$\text{The \% enantiomeric excess} = \frac{(\text{specific rotation of mixture})}{(\text{specific rotation of pure enantiomer})} \times 100 = \frac{+12°}{+49°} \times 100 = 24\%.$$

Thus 76% is optically inactive, meaning that half of this percentage is the (+) enantiomer and the other half is the (−) enantiomer. Thus, 38% is the (−) enantiomer and 38% + 24% = 62% is the (+) enantiomer.

Problem 5.33

Think

How many stereochemical configurations have to be different for two sugars to be considered epimers? How do you know which configuration(s) to change? Consult Figure 5-34 for names and structures of the D family of aldoses.

Solve

Epimers differ only by the stereochemical configuration at one carbon atom. Reverse the configuration at the carbon atom indicated.

(a) The C2 epimer of D-glucose

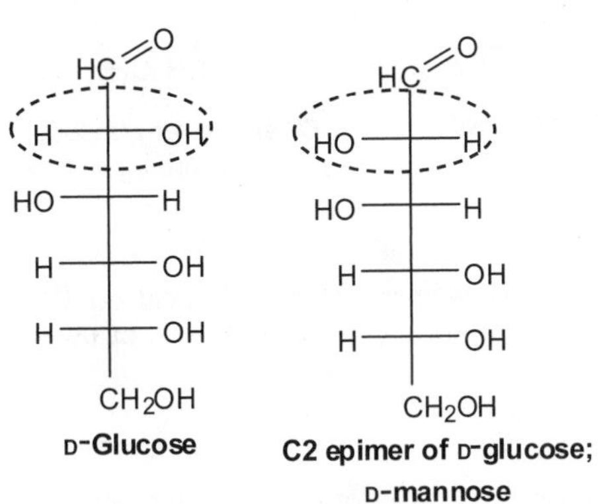

(b) The C3 epimer of D-glucose

D-Glucose

C3 epimer of D-glucose; D-allose

(c) The C4 epimer of D-talose

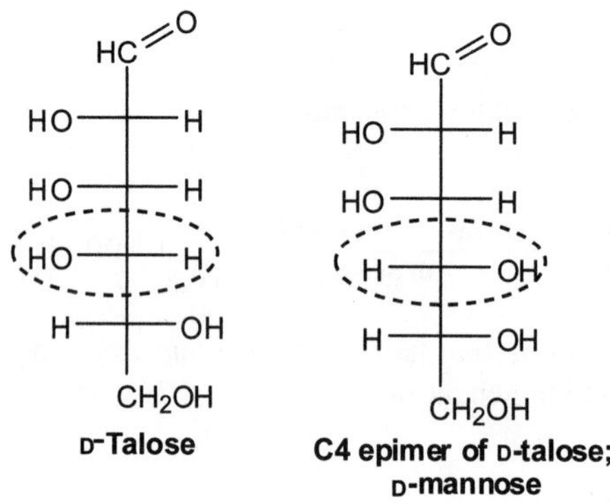

D-Talose

C4 epimer of D-talose; D-mannose

(d) The C3 epimer of D-xylose

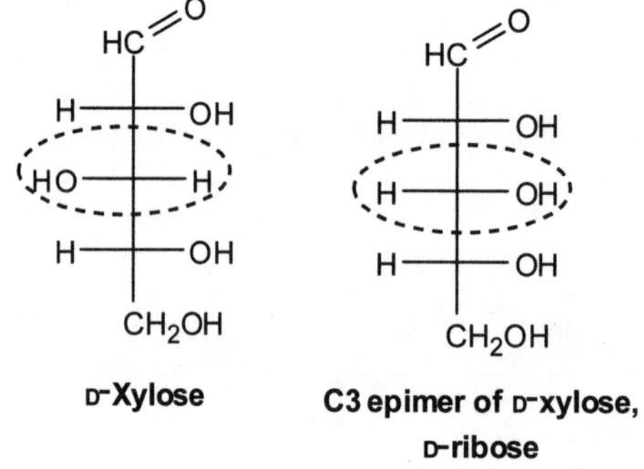

D-Xylose

C3 epimer of D-xylose, D-ribose

Problem 5.34

Think

How are the D and L sugars related? How many stereochemical configurations have to be different for two sugars to be considered enantiomers? Epimers? Consult Figure 5-34 for names and structures of the D family of aldoses.

Solve

D and L structures are enantiomers. Epimers differ only by the stereochemical configuration at one carbon atom.

(a) L-Mannose is the enantiomer of D-mannose, so reverse the configuration at all four stereocenters.
(b) L-Arabinose is the enantiomer of D-arabinose, so reverse the configuration at all three stereocenters.
(c) L-Threose is the enantiomer of D-threose, so reverse the configuration at both of the stereocenters.
(d) The C2 epimer of L-arabinose should differ from L-arabinose only by reversing the configuration at C2 of L-arabinose.

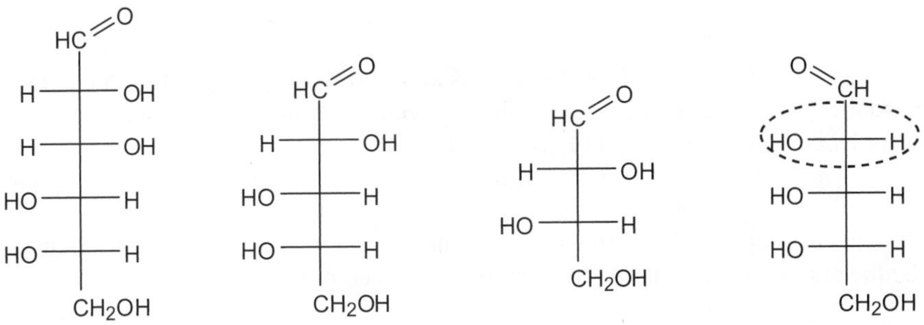

(a) L-Mannose

(b) L-Arabinose

(c) L-Threose

(d) C2 epimer of L-arabinose

Problem 5.35

Think

Consult Figure 5-1 to review the requirements of each type of isomer.

Solve

If **A** and **B** are isomers, it is implied that they have the same molecular formula.

(a) They could be enantiomers, diastereomers, or constitutional isomers since they have the same index of hydrogen deficiency (IHD). No other information is given.

(b) Constitutional isomers only because they will have different connectivity.

(c) Constitutional isomers only because they will have different connectivity.

(d) Enantiomers, diastereomers, or constitutional isomers.

(e) Diastereomers or constitutional isomers, because they cannot be enantiomers (mirror images) if one has a plane of symmetry and the other does not.

Problem 5.36

Think

Which of these objects possess a plane of symmetry? How is the presence of a plane of symmetry related to an object being chiral or achiral?

Solve

Chiral objects (no plane of symmetry): **(b)**, **(d)**, **(e)**, **(h)**, and **(i)**.

Achiral objects (plane of symmetry): **(a)**, **(c)**, **(f)**, **(g)**, and **(j)**.

Problem 5.37

Think

How do you represent the C–C bond in a Newman projection? What is the dihedral angle for the Cl groups for the *anti* and *gauche* conformations? For a plane of symmetry to exist, what must be true of atoms on either side of the plane?

Solve

Newman projections for 1,2-dichloroethane are shown below. Only the *anti* conformation has a plane of symmetry, as shown. The others do not. That plane is not a plane of symmetry in either of the gauche conformations because the Cl atom that does not lie in the plane does not have a reflection on the other side of the plane. Where another Cl would need to appear, an H atom appears instead.

Plane of symmetry

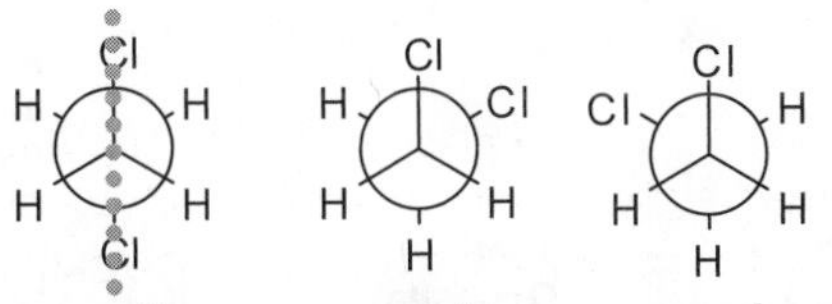

anti *gauche* *gauche*

Problem 5.38

Think

Can you identify the stereocenter in each molecule? Do the groups attached to the stereocenter in one molecule superimpose on the same groups of the stereocenter in the other molecule? Or are the stereocenters related by the exchange of two groups or by being mirror images of each other?

Solve

(a) Opposite configuration. This molecule is the mirror image of the original molecule, so the configuration of its stereocenter must be the mirror image of the stereocenter in the given molecule.

OH OH

Original

Mirror image;
opposite configuration

(b) Same configuration. A 180° rotation shows that the two molecules are the same.

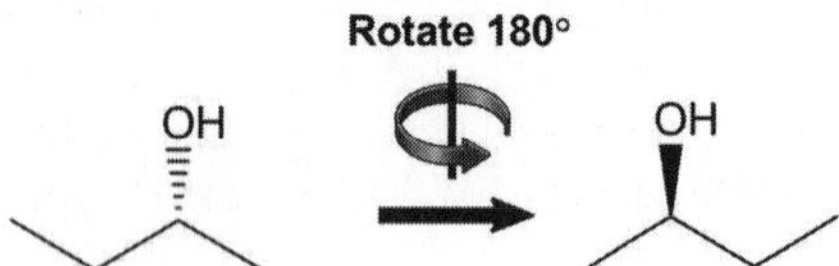

(c) Opposite configuration. The OH group is wedge in the original and now the CH_3 group is wedge; in both cases, the H atom is dash. Therefore, two groups attached to the stereocenter must have been exchanged.

Rotate about bond

H CH_3 HO H

HO H_3C

(d) and **(e):** The original structure is translated to a Newman projection, with the back C atom being the stereocenter. The CH_3 groups are anti to each other and **(d)** and **(e)** both need to have a 120° rotation performed in order to obtain the same CH_3 anti conformation as the original. In **(d)**, rotating the back carbon yields the original molecule, so the configuration of the stereocenter is the same in both. In **(e)**, rotating the back carbon 120° shows that the OH and CH_3 groups attached to the stereocenter were switched, so **(e)** has the opposite configuration.

H OH
Back
Front
Original

CH_3
HO H
H H
CH_3
CH_3 groups are anti.

(d)
Same

CH_3
H CH_3
H H
OH

Rotate 120°
to obtain anti
conformation for
CH_3 groups.

CH_3
HO H
H H
CH_3

(e)
Opposite

CH_3
HO CH_3
H H
H

Rotate 120°
to obtain anti
conformation for
CH_3 groups.

CH_3
H OH
H H
CH_3

(f) and **(g):** The original is structure is converted to a Fischer projection to make direct comparisons.

H OH
Original
CH3
HO H
CH2CH3

(f) The CH_2CH_3 is held in place because it is in the same location as the CH_2CH_3 on the original structure; a counterclockwise 120° rotation of the molecule is performed about C2–C3 to get the CH_3 in the vertical position. It is now evident that **(f)** has the same configuration at the original.

Rotate 120° about C2-C3
OH
H CH3
CH2CH3
(f)
CH3
HO H
CH2CH3
Same configuration

(g) None of the four groups is in the same location compared to the original structure. Thus, first perform a 120° clockwise rotation to get the CH_2CH_3 in the same location as the original ethyl group. Then perform a counterclockwise 120° rotation about C2–C3 to move the OH to the left. It is now evident that the two structures have opposite configurations because they differ by the exchange of the CH_3 and H groups.

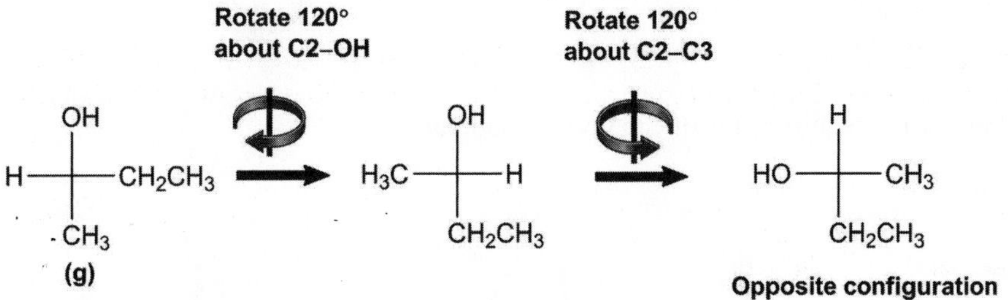

(h) In both the original and molecule **(h)**, the CH_2CH_3 group is flat and pointed down to the right. Rotate **(h)** about the C2–C3 bond to get the flat bond on the bottom left. It is evident that the configurations are the same.

OH
H
H3C
CH2CH3
Rotate about C–C bond
H OH
H3C CH2CH3
Same configuration

Problem 5.39

Think

Consult Figure 5-1 to review the requirements of each type of isomer.

Solve

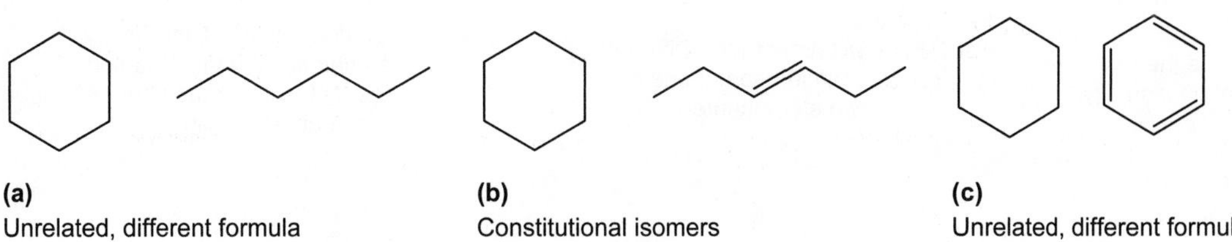

(a)
Unrelated, different formula

(b)
Constitutional isomers

(c)
Unrelated, different formula

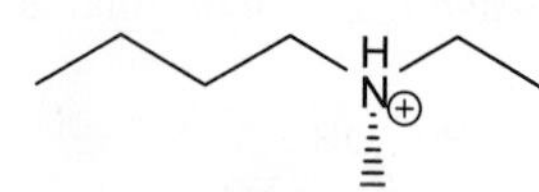

(d)
Constitutional isomers
Different locations of the CH_2CH_3 substituent along the carbon chain

(e)
Enantiomers

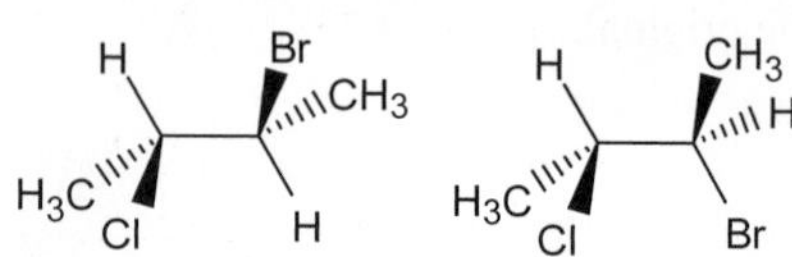

(f)
Conformational Isomers
Different by the rotation of the C atom in the plane on the right

(g)
Enantiomers

(h)
Diastereomers

Problem 5.40

Think

Consult Figure 5-1 to review the requirements of each type of isomer.

Solve

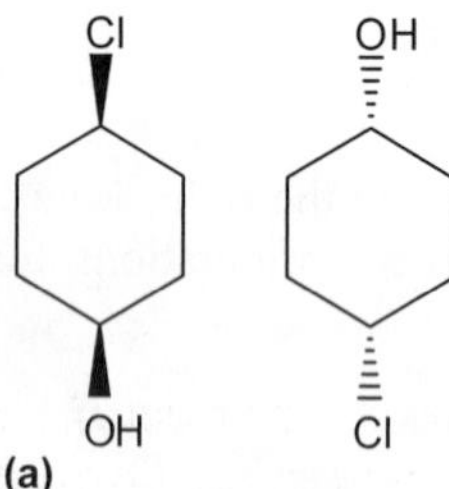

(a)
Same. Flip the molecule on the right vertically.

(b)
Constitutional Isomers.
Different locations of the two Substituents along the ring

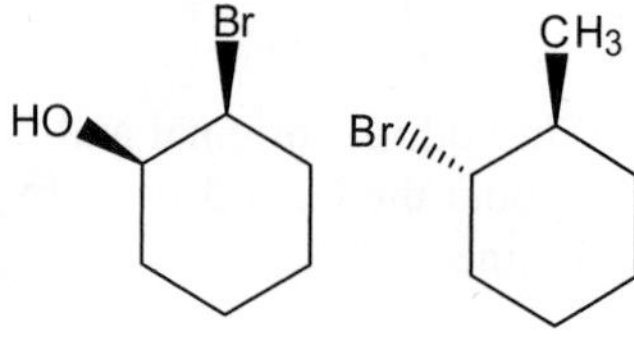

(c)
Unrelated
Different formulas

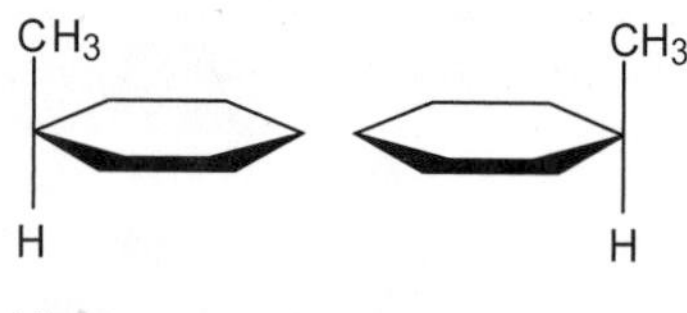

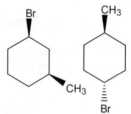

(d)
Same. Rotate the molecule on the right horizontally.

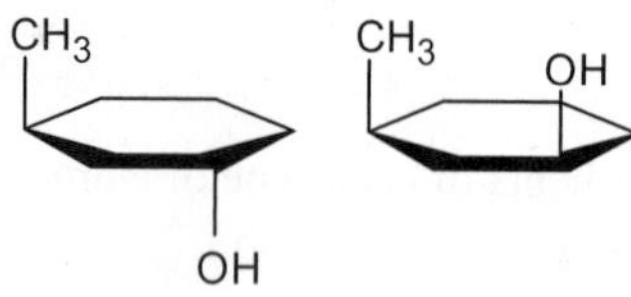

(e)
Diastereomers. Different by the configuration of one out of two stereocenters

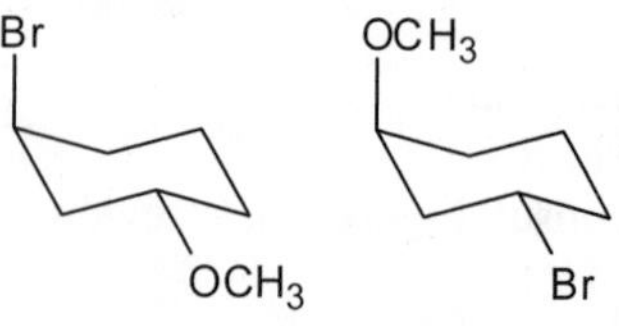

(f)
Conformational. Different by a chair flip and then rotation of the molecule on the right 180° about the horitzontal axis

Problem 5.41

Think

Consult Figure 5-1 to review the requirements of each type of isomer.

Solve

(a)

Same. Rotate the entire molecule on the right about the C–C bond.

(b)

Same. Rotate the Fischer projection on the right by 180°, which leaves the configuration unchanged.

(c)

Enantiomers. Different by a 90° rotation of the Fischer projection, which reverses the configuration.

(d)

Enantiomers. Mirror images of each other, but not superimposable

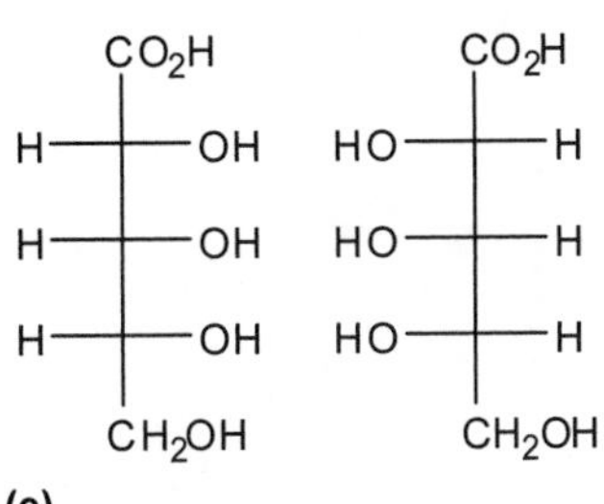

(e)

Enantiomers. Mirror images of each other, but not superimposable

(f)

Diastereomers. Different by the inversion of one tetrahedral stereocenter out of three.

(g)

Same. Rotate the Fischer projection on the right by 180°, which leaves the configurations the same.

(h)

Same. Rotate the Fischer projection on the right by 180°, which leaves the configurations the same.

Problem 5.42/5.43

Think

Which C and N atoms are bonded to four different groups? Which molecules possess a plane of symmetry and which do not? What is the minimum number of stereocenters a meso compound must have? Can a meso compound be chiral?

Solve

Molecules **(b)**, **(e)**, **(f)**, and **(h)** are chiral. The plane of symmetry in **(a)** is the plane of the page. **(c)** is achiral due to nitrogen inversion. The plane of symmetry in **(d)** is the plane of the page. **(g)** has a plane of symmetry after rotation of the C–C bond, as shown on the next page. Stereocenters are identified with * below. Only **(g)** is meso because it has two tetrahedral stereocenters but has a plane of symmetry, making it achiral.

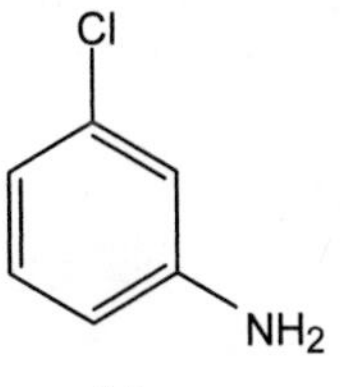

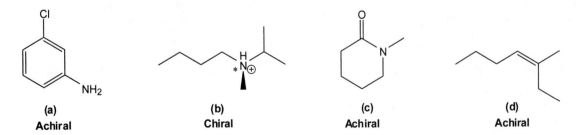

(a)
Achiral

(b)
Chiral

(c)
Achiral

(d)
Achiral

(e) **Chiral**

(f) **Chiral**

(g) **Achiral/meso**

(h) **Chiral**

Alternative view of **(g)** to view the plane of symmetry:

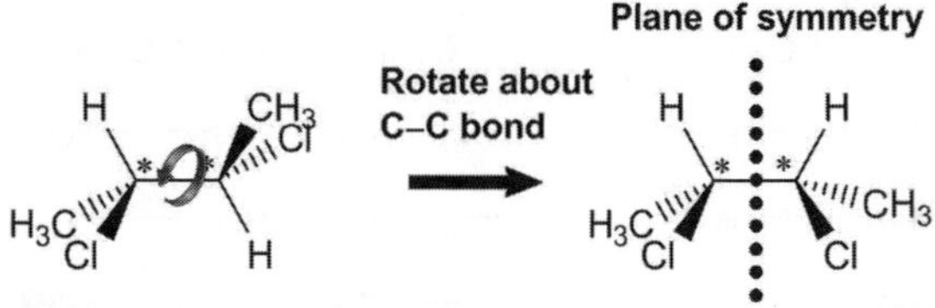

Problem 5.44/5.45

Think

Are stereocenters present? Do any of the molecules have a plane of symmetry? What is the minimum number of stereocenters a meso compound must have? Can a meso compound be chiral?

Solve

Molecules **(a)**, **(b)**, and **(d)** are chiral. **(a)** is chiral because it has exactly one tetrahedral stereocenter. **(b)** does not have a plane of symmetry. **(c)** is achiral due to the plane of symmetry shown. **(d)** does not have a plane of symmetry. Stereocenters are indicated below using *. Only **(c)** is meso, as it has two tetrahedral stereocenters but has a plane of symmetry, making it achiral.

(a) **(b)** **(c)** **(d)**

Problem 5.46

Think

Which C atoms are bonded to four different groups? What is the formula for determining the maximum number of stereocenters? What can you do to a stereocenter to produce a new stereoisomer? How can you determine if one of the stereoisomers is meso?

Solve

There are six stereocenters, indicated by * below. Since no plane of symmetry can exist in the molecule, we are guaranteed that each permutation in which a configuration is reversed at one of those stereocenters will yield a new stereoisomer. Therefore, there are $2^6 = 64$ different configurational isomers for this molecule (stereoisomers are not shown).

Problem 5.47

Think

Do the following Lewis structures actually depict the three-dimensional geometry? Is there free rotation about C=C bonds? Are these molecules all planar? Which molecules have mirror images that are nonsuperimposable on the original molecule?

Solve

In each of these molecules, the plane of one trigonal planar C is perpendicular to the plane of the other (shown below). Molecules **(b)**, **(c)**, and **(d)** are chiral because there is no plane of symmetry. Molecules **(a)** and **(e)** are achiral because there is a plane of symmetry that lies in the plane of the paper (not shown).

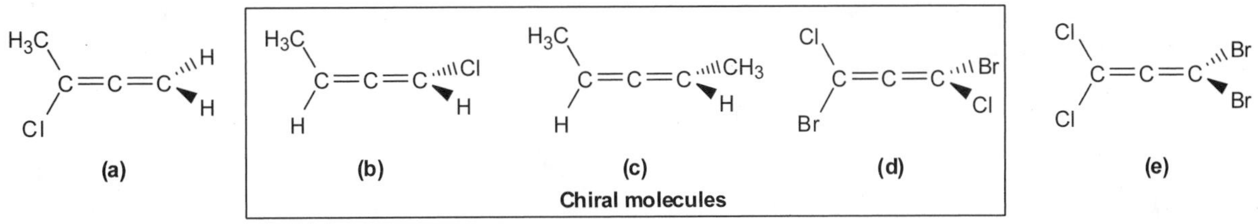

(a) (b) (c) (d) (e)

Chiral molecules

Problem 5.48

Think

With an odd number of stereocenters, is it possible for a molecule to possess a plane of symmetry? Can one tetrahedral stereocenter be the mirror image of a second? Can a tetrahedral stereocenter be the mirror image of itself?

Solve

No, it is not possible for a meso compound to contain exactly three stereocenters. To be meso, the compound must be achiral and possess a plane of symmetry. To have a plane of symmetry, each stereocenter on one side of the plane of symmetry must have a corresponding stereocenter on the other side of that plane of symmetry. With an odd number of stereocenters, this is impossible unless the plane of symmetry contains one stereocenter. However, by its very nature, a stereocenter does not have a plane of symmetry, so it cannot lie in a molecule's plane of symmetry. Thus, a meso compound must contain an even number of stereocenters.

Problem 5.49

Think

How many stereoisomers are present? Review the steps for converting a Fischer projection to a zigzag structure.

Solve

Zigzag structures are shown below. Each Fischer projection is first converted to dash–wedge notation by adding substituents to horizontal bonds as they appear in the Fischer projection, but only when the horizontal bonds are pointing toward you.

(a)

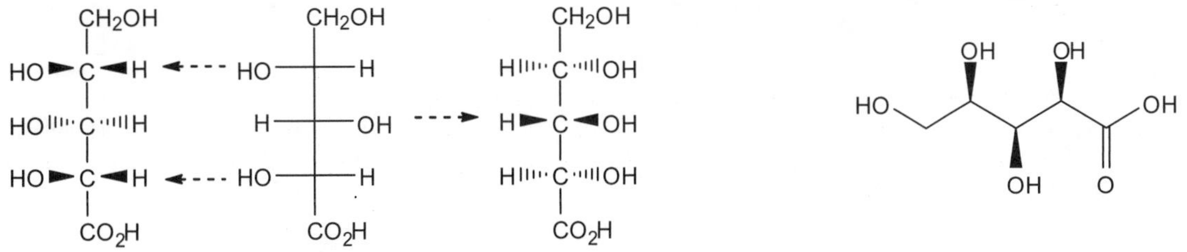

(b)

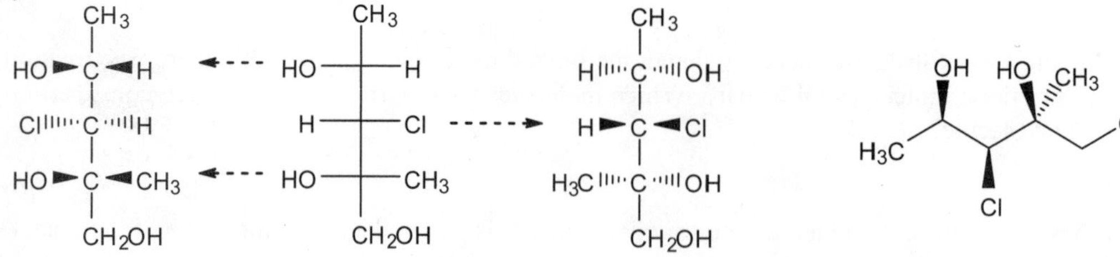

(c)

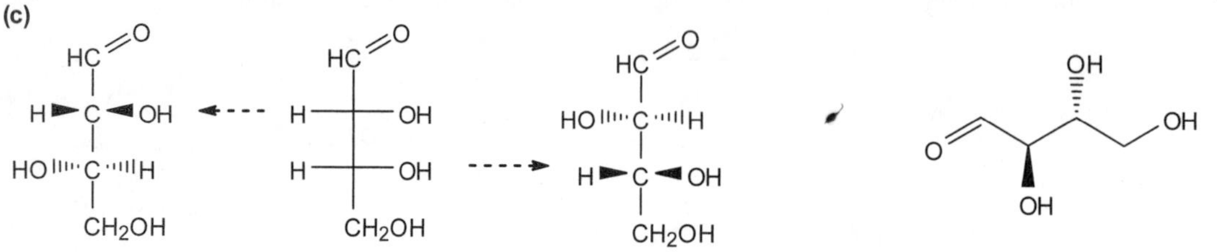

(d)

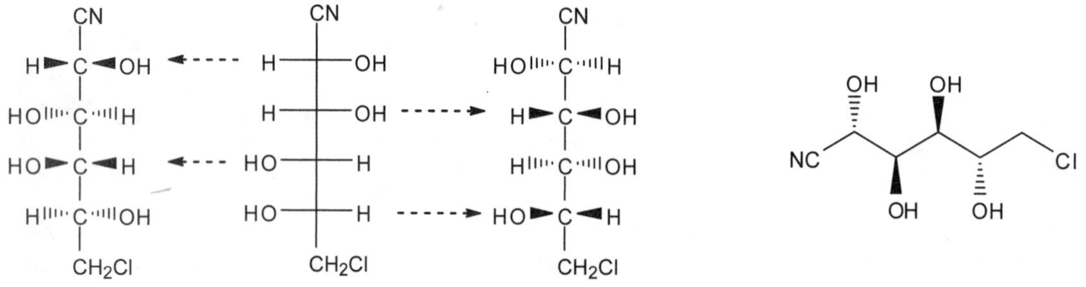

Problem 5.50

Think

How many stereoisomers are present? Review the steps for converting a zigzag structure to a Fischer projection. When the carbon chain is oriented vertically, should the substituents on the horizontal bonds in a molecule match the Fischer projection when the horizontal bonds are pointed toward you or away from you?

Solve

The Fischer projections are shown below. With the carbon chain vertical, the substituents on the horizontal bonds should match how the Fischer projection appears only when the horizontal bonds are pointed toward you.

(a)

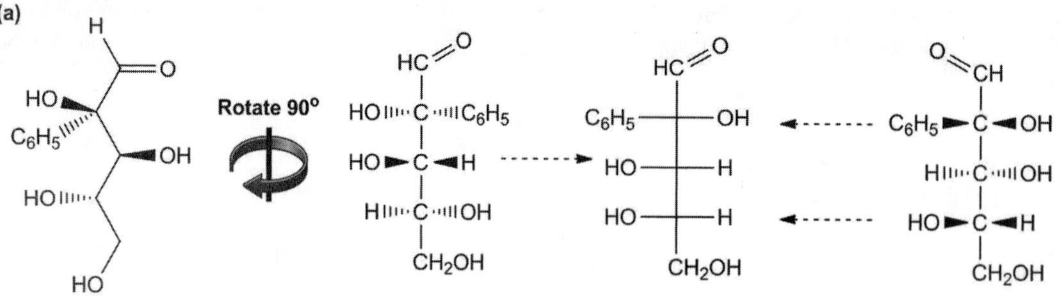

Rotate 90°

(b)

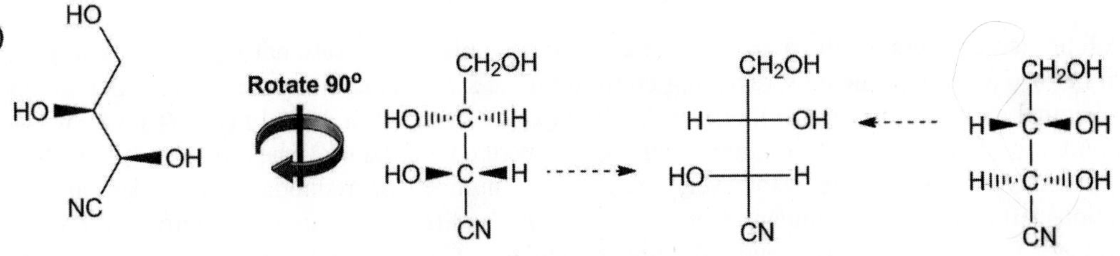

(c)

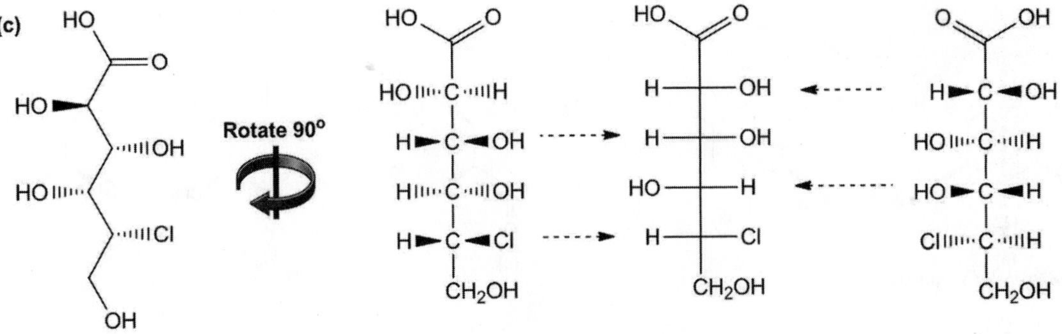

Problem 5.51

Think

Which atoms are bonded to four different groups? To identify a stereocenter, make sure to consider the entire rest of the molecule that is attached, not just the immediate atoms attached.

Solve

The tetrahedral stereocenters are indicated by * below:

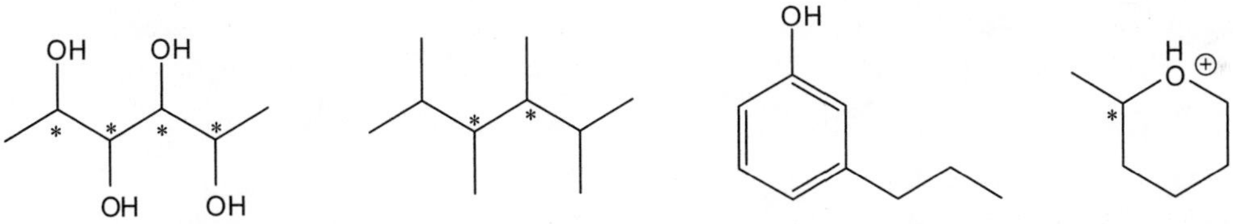

(a) 4 stereocenters

(b) 2 stereocenters

(c) 0 stereocenters

(d) 1 stereocenter

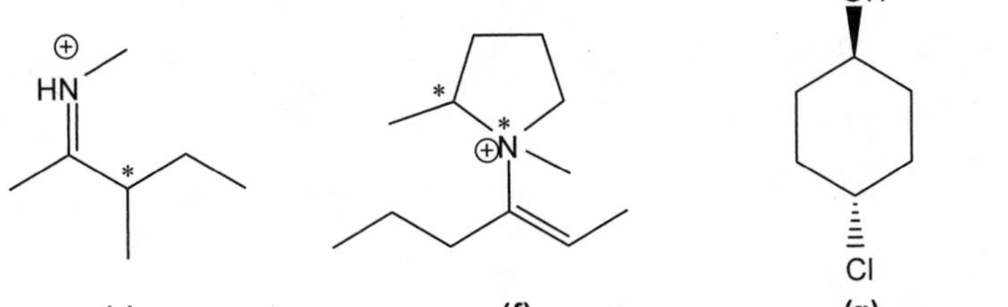

(e) 1 stereocenter

(f) 2 stereocenters

(g) 0 stereocenters

Problem 5.52

Think

How many C atoms are bonded to four different groups? What is the formula to determine the maximum number of stereoisomers? Are any planes of symmetry present? Systematically draw all isomers by reversing the configuration at each stereocenter.

Solve

There are four stereocenters, so there are at most $2^4 = 16$ possible configurational isomers. Those possibilities are shown below. We obtain them by reversing configurations systematically at each of the four stereocenters. **(a)** is the original molecule. In **(b)** – **(e),** one wedge bond was converted to dash. In **(f)** – **(k)**, two wedge bonds were converted to dashes. In **(l)** – **(o)**, three wedge bonds were converted to dashes. And in **(p)**, all four wedge bonds were converted to dashes. However, there are a number of redundancies, which are X'ed out. Configurations **(p)** and **(a)** are redundant. Configurations **(m)** and **(b)** are redundant. Configurations **(l)** and **(c)** are redundant. Configurations **(o)** and **(d)** are redundant. Configurations **(n)** and **(e)** are redundant. Configurations **(h)** and **(f)** are redundant. So in all, there are 10 different configurational isomers. The meso compounds are **(a)** and **(f)**, as they each have a plane of symmetry (shown below). None of the remaining compounds do.

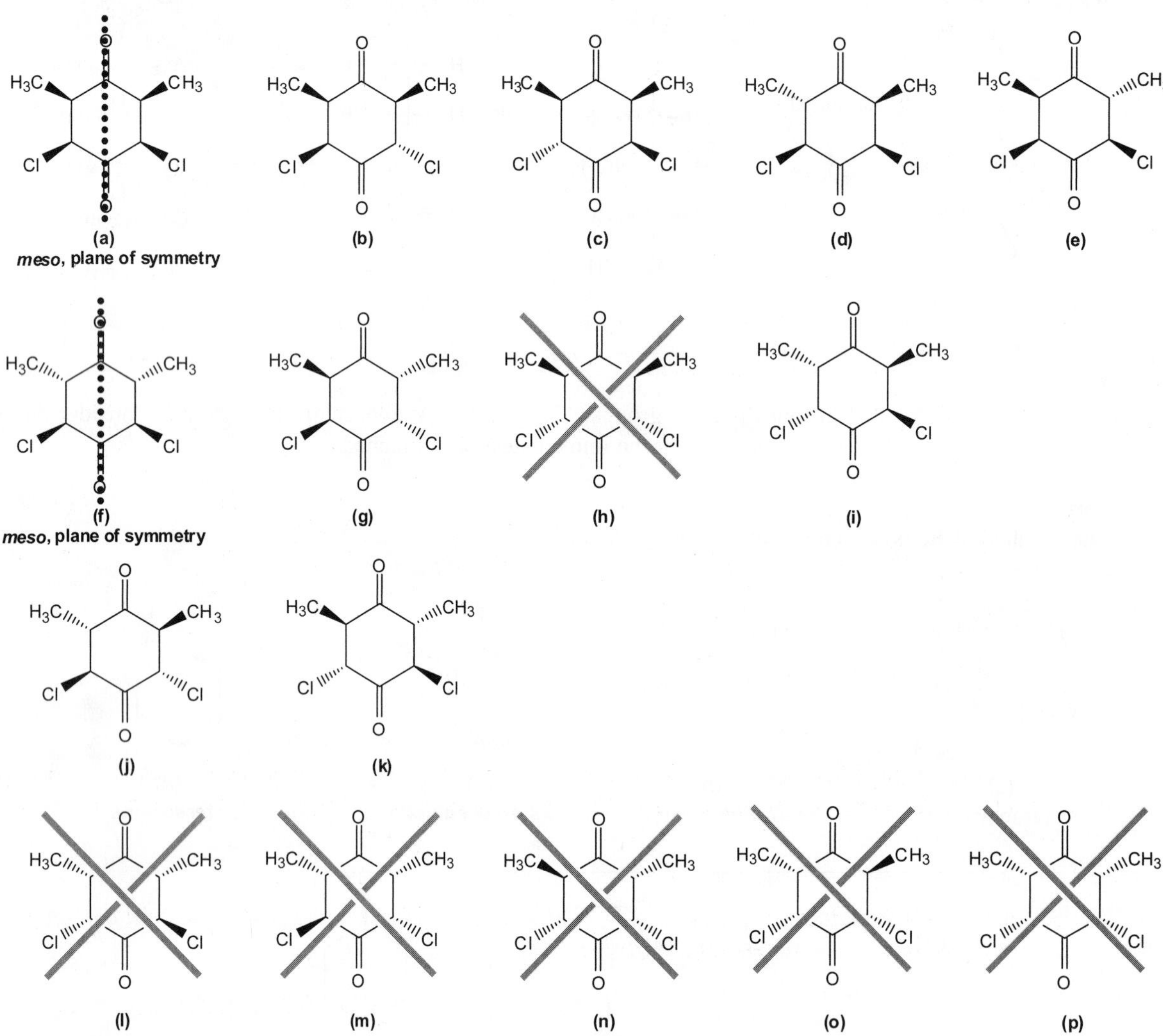

Problem 5.53

Think

How many C atoms are bonded to four different groups? What is the formula to determine the maximum number of stereoisomers? Are any planes of symmetry present? What does a plane of symmetry mean in terms of optical activity? Systematically draw all isomers by reversing the configuration at each stereocenter.

Solve

There are four stereocenters, so there are up to $2^4 = 16$ possible configurational isomers. Those possibilities are shown below. We obtain them by reversing configurations systematically at each of the four stereocenters. There are no planes of symmetry, thus no meso compounds; therefore, all 16 configurational isomers are optically active. See all 16 below. **(a)** is the original molecule. In **(b)** – **(e),** one wedge bond was converted to dash. In **(f)** – **(k)**, two wedge bonds were converted to dashes. In **(l)** – **(o)**, three wedge bonds were converted to dashes. And in **(p)**, all four wedge bonds were converted to dashes.

Cl Br Cl Br **(a)** Cl Br Cl Br **(b)** Cl Br Cl Br **(c)** Cl Br Cl Br **(d)**

Cl Br Cl Br **(e)** Cl Br Cl Br **(f)** Cl Br Cl Br **(g)** Cl Br Cl Br **(h)**

Cl Br Cl Br **(i)** Cl Br Cl Br **(j)** Cl Br Cl Br **(k)** Cl Br Cl Br **(l)**

Cl Br Cl Br **(m)** Cl Br Cl Br **(n)** Cl Br Cl Br **(o)** Cl Br Cl Br **(p)**

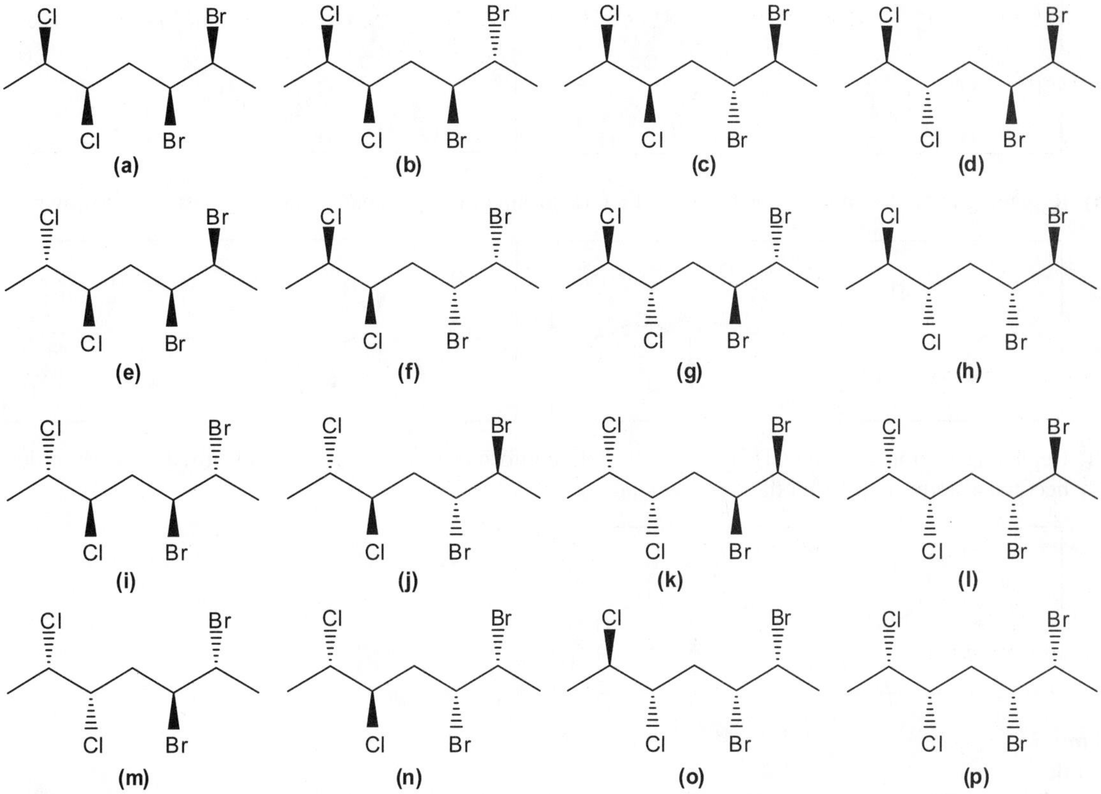

Problem 5.54

Think

What is the IHD? How many constitutional isomers are possible? To be optically active, should a molecule be chiral or achiral? How can you determine chirality by the plane-of-symmetry test? By the number of tetrahedral stereocenters?

Solve

The IHD is 0, thus there are no rings and no multiple bonds. All isomers are shown below. The two optically active isomers are shown with dash–wedge notation.

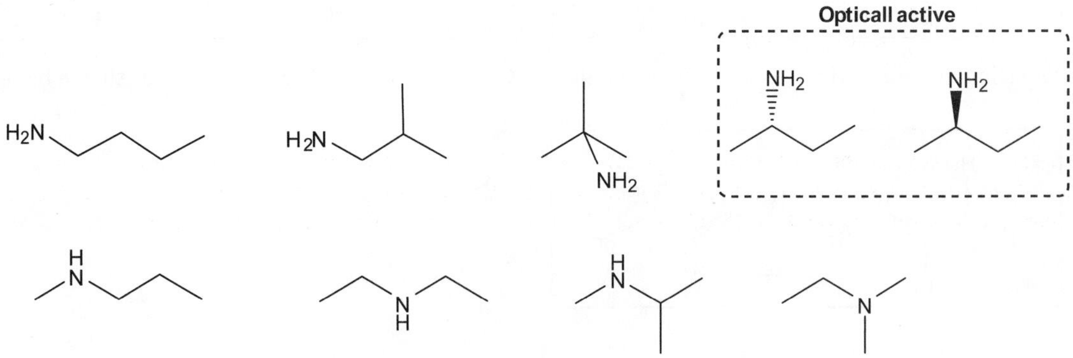

H_2N H_2N NH_2

Opticall active

NH_2 NH_2

H N N H H N N

Problem 5.55

Think

Consult Figure 5-1 to review the requirements of each type of isomer.

Solve

(a) Constitutional isomers have different connectivity. Replacing any two H atoms on nonadjacent C atoms will yield constitutional isomers. Two examples are shown below.

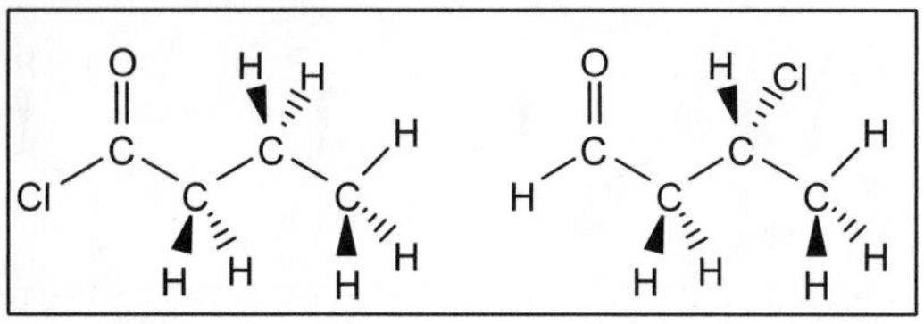

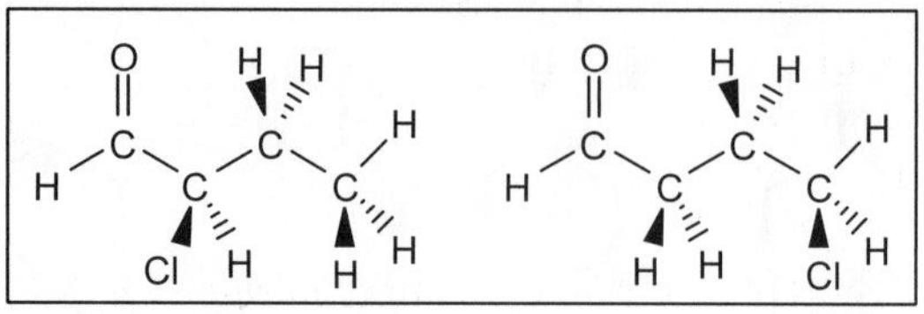

(b) Replacing H atoms on the same C atom of a CH_2 group will yield enantiomers. Examples are shown below.

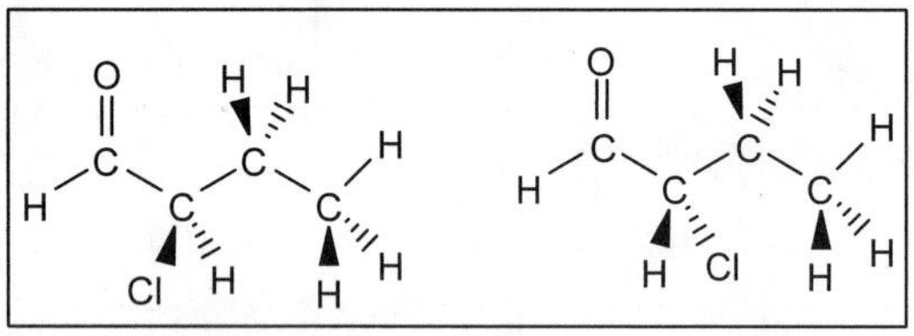

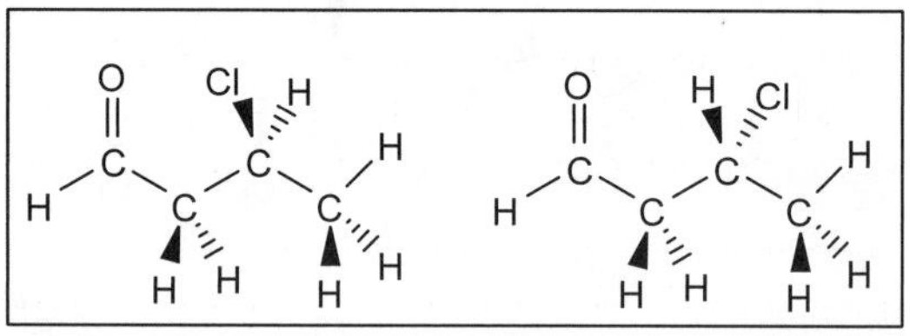

(c) Replacing H atoms on the CH_3 group will yield conformational isomers because the CH_3 carbon does not become a stereocenter after the replacement.

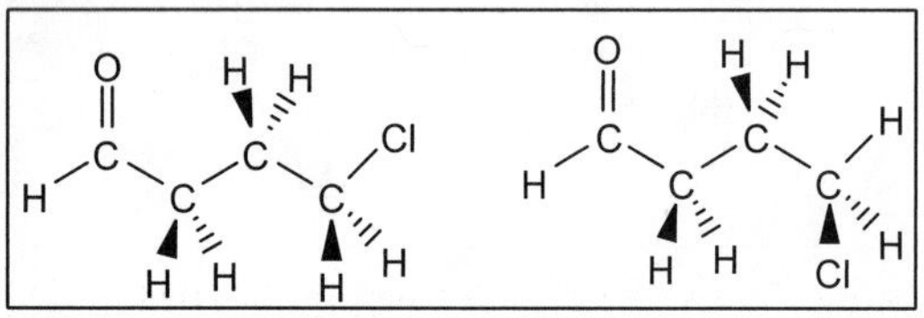

Problem 5.56

Think

Consult Figure 5-1 to review the requirements of each type of isomer.

Solve

(a) Replacing any pair of H atoms that are not on the same C atom will yield constitutional isomers. Examples are shown below.

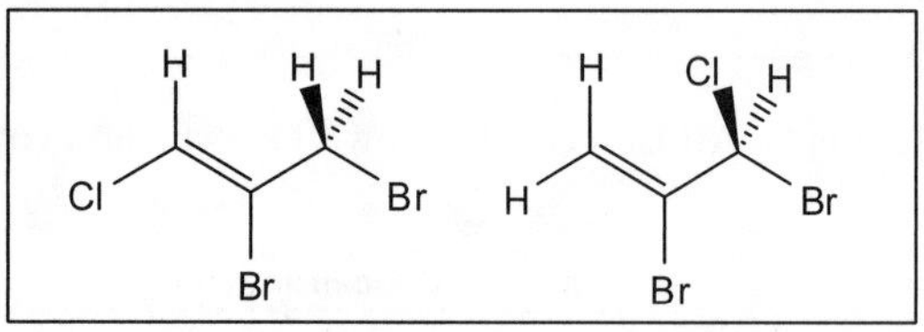

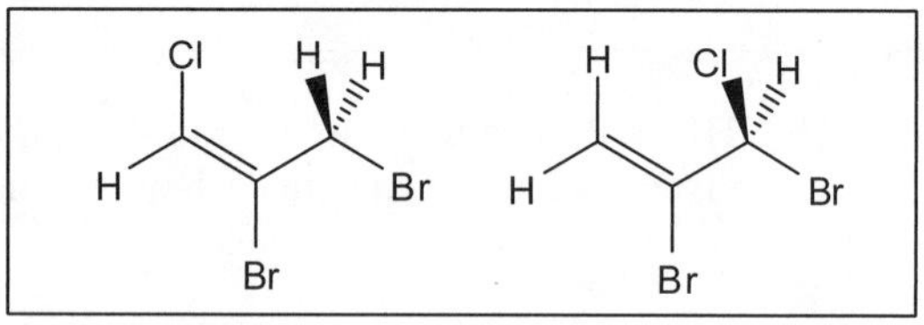

(b) Replacing the pair of H atoms only on the tetrahedral CH_2 group will yield enantiomers, as shown below.

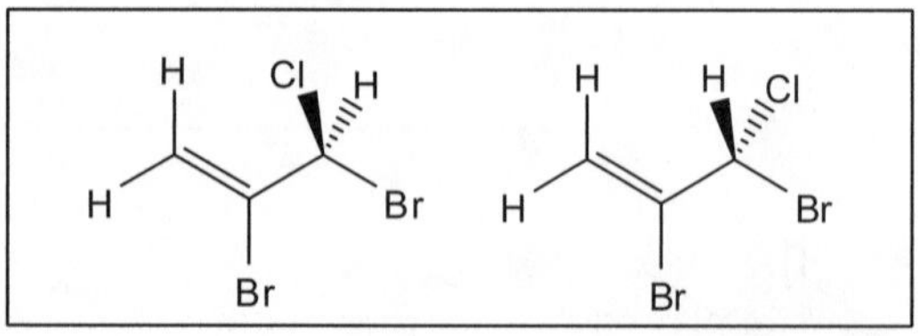

(c) Replacing the H atoms attached to the same C atom of the C=C bond will produce diastereomers. More precisely, this will produce Z/E isomers.

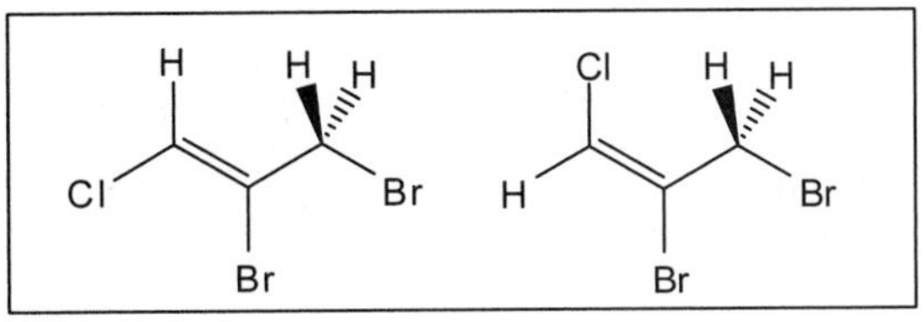

Problem 5.57

Think

What type of isomer relationship exists between each pair? What is true about the physical properties of enantiomers, diastereomers, and constitutional isomers?

Solve

Enantiomers must have the same physical properties in an achiral environment. Diastereomers and constitutional isomers must have different physical properties. To determine the relationships between the molecules, apply the flow chart in Figure 5-1.

	Structures	Relationship	Boiling Point
(a)	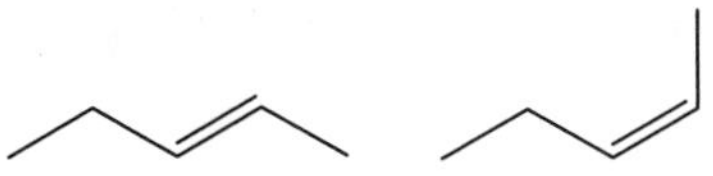	Diastereomers	Different
(b)	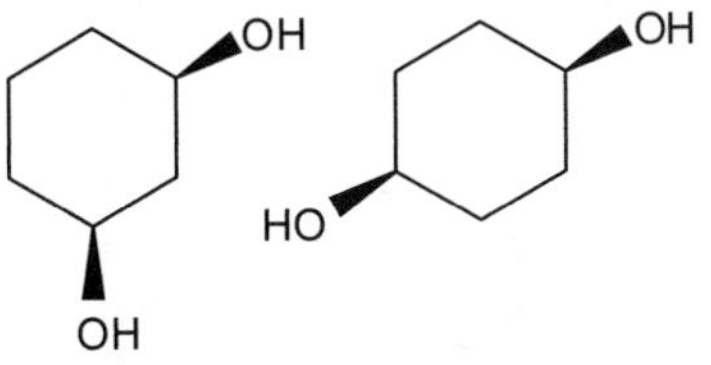	Constitutional Isomers	Different
(c)	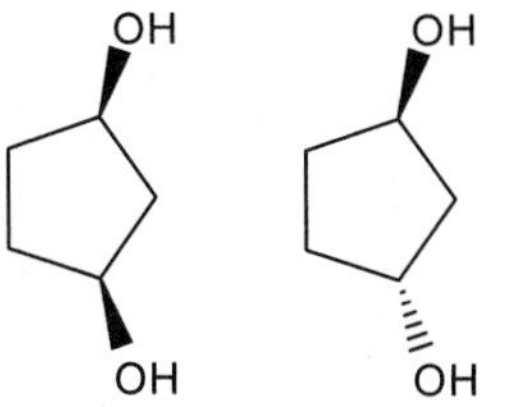	Diastereomers	Different
(d)	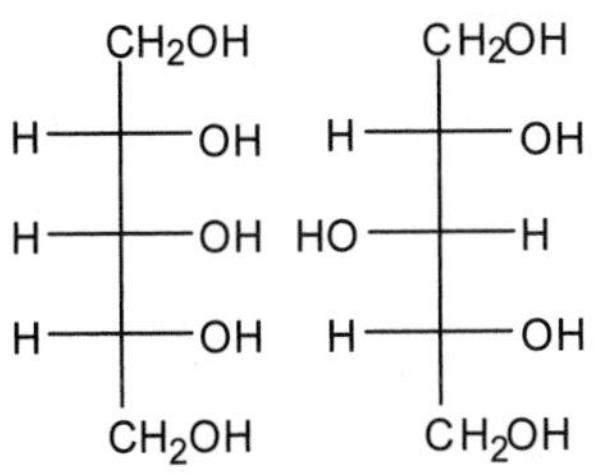	Diastereomers	Different
(e)	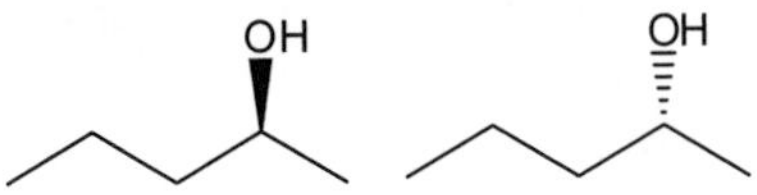	Enantiomers	Same

Problem 5.58

Think

What is the IHD for C_6H_{12}? For a compound to be optically active, is it chiral or achiral? Should the molecule have a plane of symmetry?

Solve

The IHD for C_6H_{12} is 1. Therefore, there is either one ring or one double bond. The optically active isomers are drawn below. There are quite a few more isomers of C_6H_{12}, but those are not optically active. Each molecule below lacks a plane of symmetry.

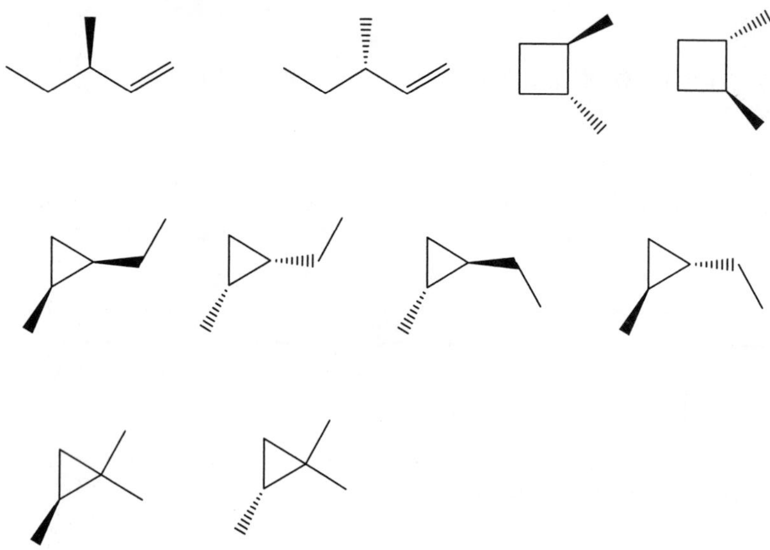

Problem 5.59

Think

Which atoms are bonded to four different groups? Remember that, when identifying groups on a potential stereocenter, look at the entire rest of the molecule that is attached, not just the immediate atoms that are attached.

Solve

Aldosterone has seven stereocenters, shown below using *.

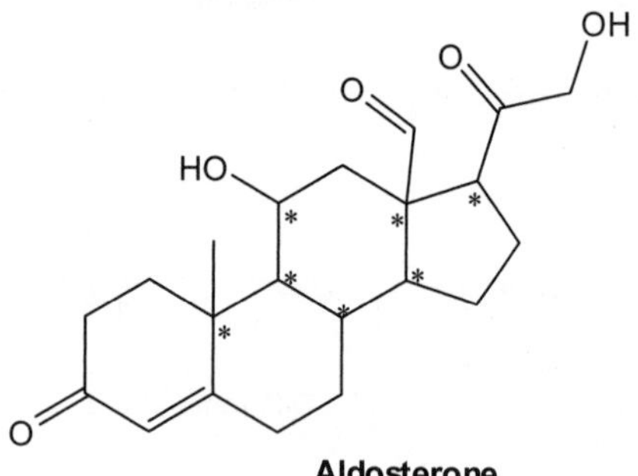

Problem 5.60

Think

Which atoms are bonded to four different groups? Remember that, when identifying groups on a potential stereocenter, look at the entire rest of the molecule that is attached, not just the immediate atoms that are attached.

Solve
Taxol has 11 stereocenters, shown below using *.

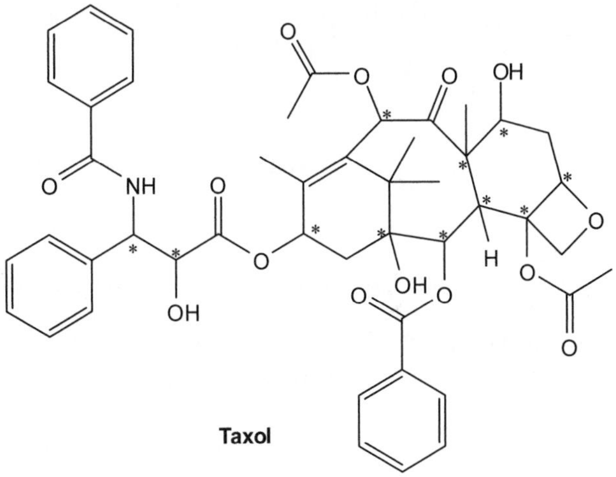

Taxol

Problem 5.61

Think
What is true about the magnitude and sign of the specific rotation of enantiomers?

Solve
Enantiomers have optical rotations of the same magnitude but opposite signs. Thus, if taxol has a specific rotation of −49°, its enantiomer will have a specific rotation of +49°.

Problem 5.62

Think
Which enantiomer (+) or (−) is in excess? Which way does that enantiomer rotate plane-polarized light? What percentage of the mixture is racemic? What is the remaining percentage of the mixture?

Solve
In a mixture that is 60% (+) enantiomer and 40% (−) enantiomer, the (+) enantiomer is in excess, so optical rotation will be in the (+) direction, or clockwise. The amount of the mixture that is racemic is 80%, or 40% (+) and 40% (−), which leaves a 20% excess of the (+) enantiomer.

Problem 5.63

Think
What is the concentration of the solution? For which variable in Equation 5-2 are we solving? Are the units for concentration correct? Are the units for the cell length correct?

Solve
The concentration (c) needs to be in units of g/mL. Thus, $c = \frac{20.0\ \text{g}}{1.00\ \text{L}} \times \frac{1.0\ \text{L}}{1000\ \text{mL}} = 2.00\ \times 10^{-2}\frac{\text{g}}{\text{mL}}$. The path length needs to be in dm. Thus, $l = 10.0\ \text{cm}\ \times \frac{1\ \text{dm}}{10\ \text{cm}} = 1.0\ \text{dm}$. The specific rotation $[\alpha]_D^{20} = +\ 23.1°$.
The measured rotation, $\alpha = ([\alpha]_D^{20})(l)(c) = (+23.1°)(1.00\ \text{dm})\left(2.00\ \times 10^{-2}\frac{\text{g}}{\text{mL}}\right) = +0.462°$

Problem 5.64

Think
If the enantiomeric excess is 84%, what percentage is left as racemic? What makes up a racemic mixture? What is the percentage of *R* and *S* enantiomers? What percentage is optically active?

Solve

If the enantiomeric excess is 84%, 16% is left. This leftover percentage is racemic and is made up of equal amounts of *R* and *S* enantiomers (8% each). Therefore, the *S* enantiomer is 84% + 8% = 92% and the *R* enantiomer is 8%. The enantiomeric excess is the only optically active part of the solution. Thus, 84% is optically active. The specific rotation, $[\alpha]_D$, therefore, is equal to 84% of +25.0°. Thus, $[\alpha]_D = +21°$.

Problem 5.65

Think

Which of the C=C bonds is most highly alkyl substituted? Most stable? What is the relationship between relative stability and relative heats of combustion?

Solve

All three have the same molecular formula (C_8H_{14}), one double bond, and one six-membered ring. Therefore, the difference in heats of combustion will be due solely to the stability of the C=C bonds. The tetrasubstituted alkene is the most stable, so it has the smallest heat of combustion. The disubstituted alkene is the least stable, so it has the greatest heat of combustion. The order of increasing heat of combustion is **B** < **A** < **C**.

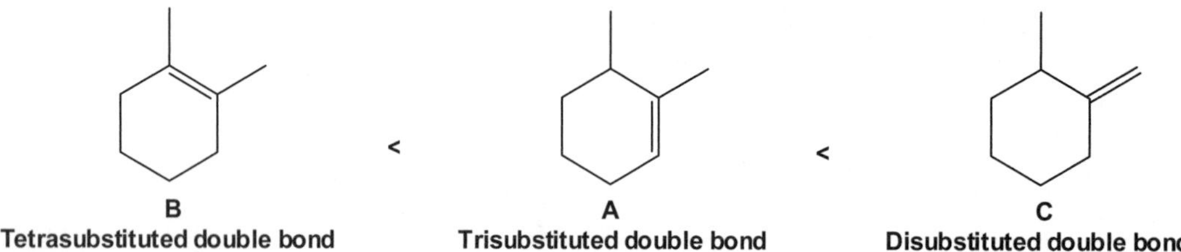

Problem 5.66

Think

What is the difference in stability of the two alkenes. Is one more highly substituted? What about cis/trans relationships? Does one have more steric strain?

Solve

Both molecules are disubstituted alkenes, and both are cis. **A** has the greater heat of combustion, however, due to steric crowding between the isopropyl groups. With greater crowding, it becomes less stable (i.e., contains more energy).

H_3C CH_3
CH_3 H_3C
A

Problem 5.67

Think

Is the monosubstituted or disubstituted C≡C alkyne more stable? What is the relationship between alkyne stability and heat of combustion?

Solve

Just as the greater alkyl substitution provides more stability to alkene functional groups, a greater degree of alkyl substitution stabilizes the alkyne functional group. The alkyne group in but-1-yne is monosubstituted, whereas that in but-2-yne is disubstituted, as shown below. Therefore, but-2-yne is more stable (i.e., lower in energy) and has the smaller (less negative) heat of combustion.

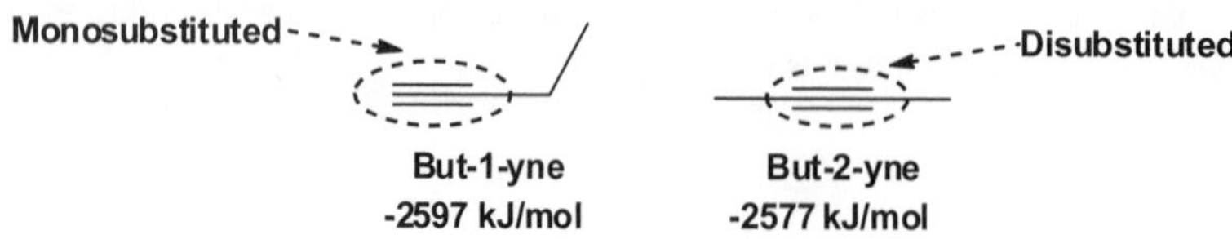

Problem 5.68

Think

Are any of the atoms bonded to four different groups? Remember that, when identifying groups on a potential stereocenter, look at the entire rest of the molecule that is attached, not just the immediate atoms that are attached. What is the hybridization of each C atom? If 1,1′-bi-2-naphthol is chiral, what must be true about its mirror image? What is true about the magnitude and sign of the specific rotation of enantiomers?

Solve

(a) The molecule does not contain any tetrahedral stereocenters. All C atoms are sp^2-hybridized and are bonded to only three groups.

(b) The enantiomer pairs are shown below; they are mirror images. Enantiomers have optical rotations of the same magnitude but opposite signs. The specific rotation of the enantiomer is +32.70°.

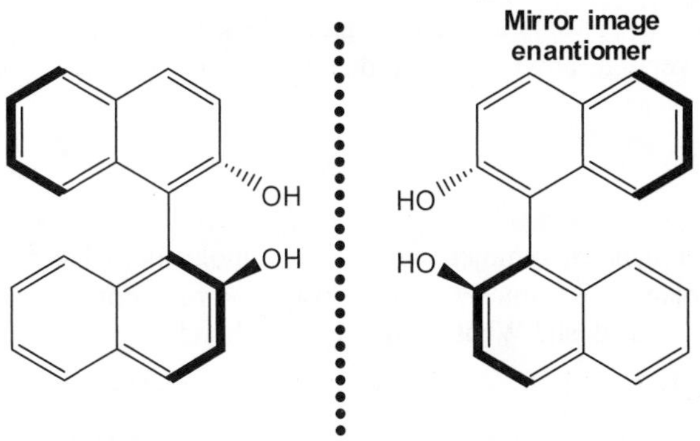

(c) The two molecules are also related by the rotation about the C–C single bond that connects each pair of fused rings, as shown below. So they are conformational isomers, too. They do not interconvert readily, however, because during that rotation, there is excessive steric hindrance resulting from the H atoms and the OH groups crashing into each other, as indicated.

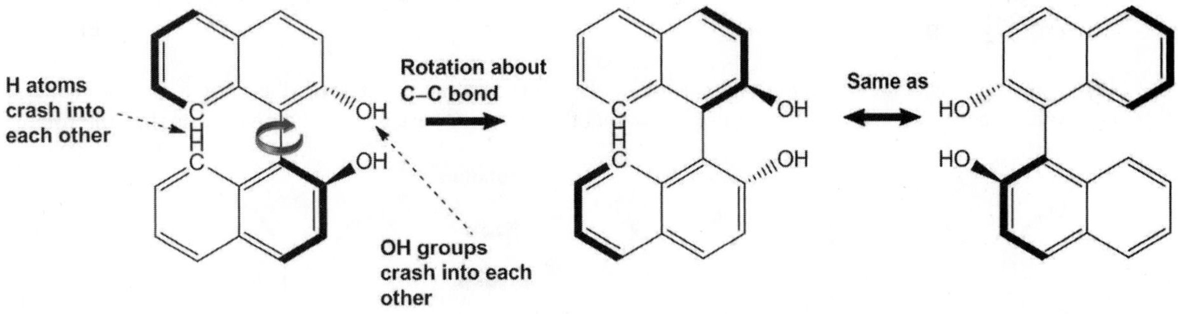

Problem 5.69

Think

What is the relationship between any two enantiomers? How does the limited rotation in a ring affect the ability to interconvert the two mirror images? For nitrogen inversion to take place, the N atom is instantaneously rehybridized to sp^2. What do you notice about the strain in Tröger's base if N were to be sp^2 hybridized?

Solve

(a) The enantiomer is the mirror image, as shown below.

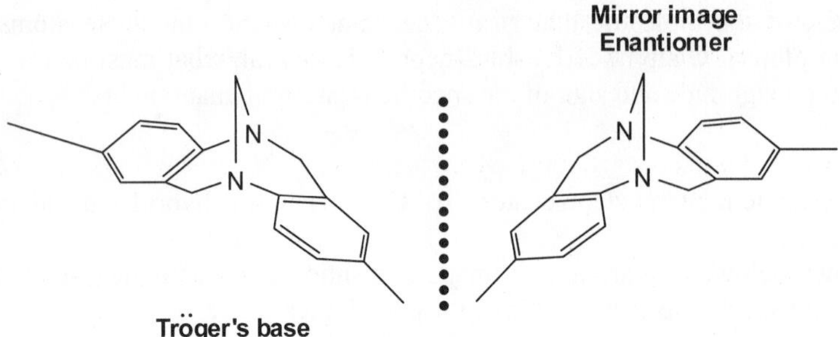

(b) Nitrogen inversion doesn't take place because the N–C bonds to the CH_2 group lock the N atoms in their respective configurations. To undergo nitrogen inversion, N would need to attain a planar geometry, which would severely strain the ring system.

Problem 5.70

Think

It might help to build a model of this compound. For a plane of symmetry to exist in a molecule, what must be true of the molecule on either side of that plane? If a plane of symmetry were to bisect the ring, could the C–Cl bonds reflect into each other if one is wedge and the other is dash? What about the C–Br bonds?

Solve

(a) Attempts at drawing a plane of symmetry are shown below. All of the attempts below **DO NOT** show a plane of symmetry. In situations where the dash–wedge notation reflects across the plane appropriately, the atoms do not, and vice versa.

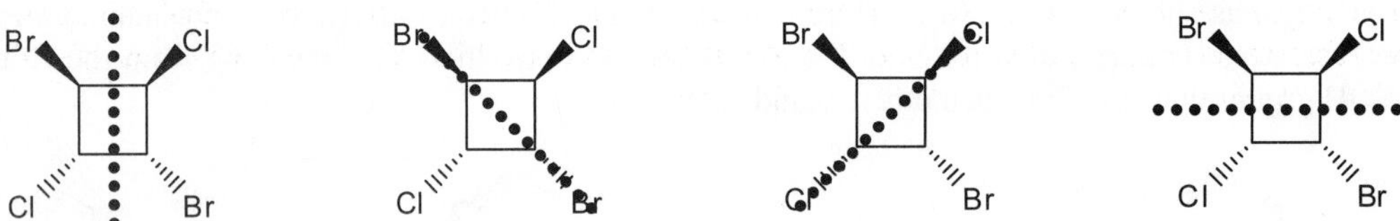

(b) Shown below is the molecule's mirror image. If a180° rotation is performed as shown, the molecule is seen to be exactly the same as the original. Note that the primes (') are used to help you visualize the rotation.

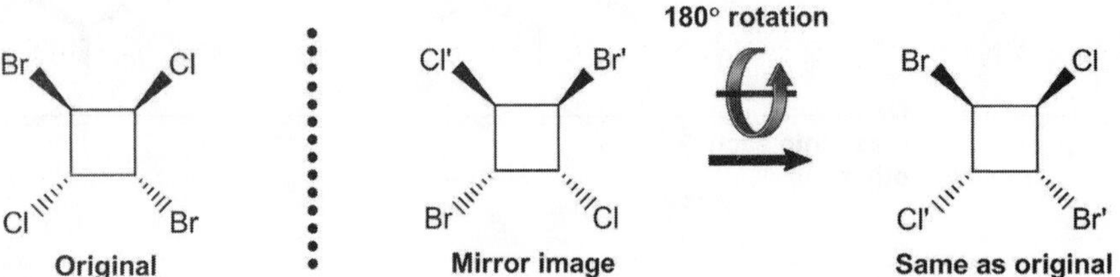

(c) The point of symmetry is in the center of the molecule, as shown below. To reflect an atom through the point of symmetry, draw a line from the atom to the point. The reflection of the atom is located along the same line on the other side of the point. Both atoms are found the same distance from the point. The Cl atoms, for example, are reflections through the point of symmetry, and so, too, are the Br atoms. Notice that a reflection through a point of symmetry reverses the dash–wedge notation.

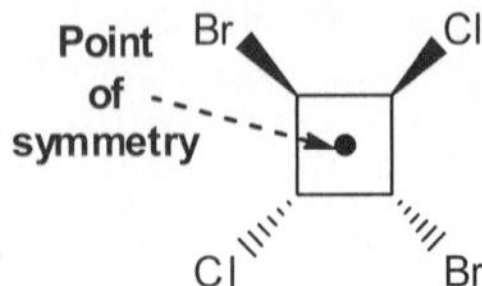

Problem 5.71

Think

Which C atom in **A** is the stereocenter? Draw the two enantiomers. What does salt **C** look like with one enantiomer of **A**? The other enantiomer of **A**? What is the isomeric relationship between these two salts? What is true of physical and chemical properties of enantiomer pairs? Of diastereomer pairs?

Solve

(a) The dash–wedge notation is shown below for the carbon stereocenters for the racemic mixture and the pure base enantiomer. The salt products are also shown below.

Racemic mixture　　Pure enantiomer　　Two salts are produced

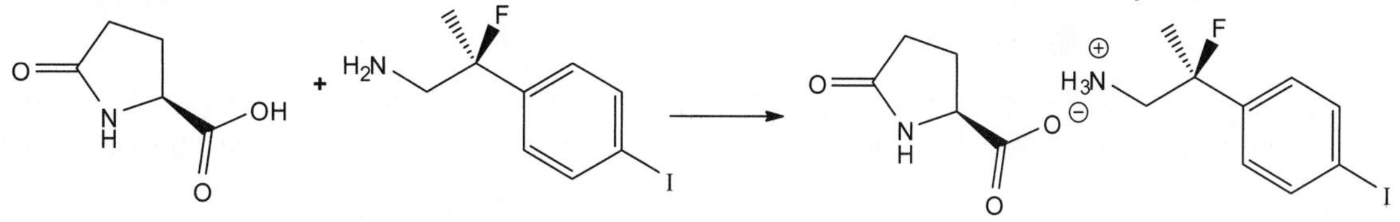

A　　B　　C

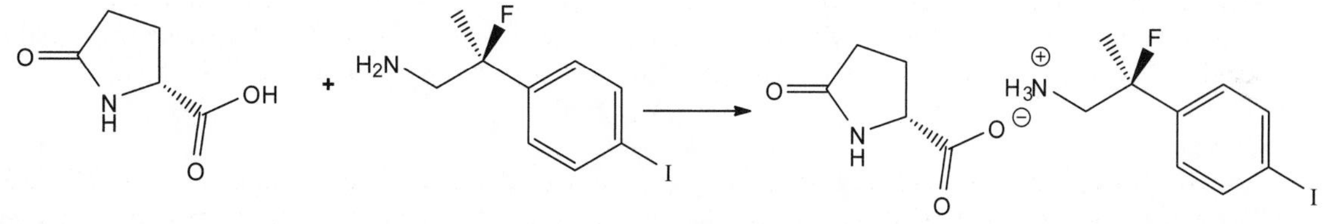

(b) The two salts have exactly the same connectivity. The cations of each salt have the same configuration but the anions have opposite configuration and are enantiomers. This means that the two salts are diastereomers. Diastereomers have different physical and chemical properties, which is why they can be separated readily.

Problem 5.72

Think

Consult Table 1-7 to review the structures of the amino acids. Which α carbon does not have four different groups?

Solve

The α carbon of glycine (shown below) has two H atoms and, therefore, is not a stereocenter, and the molecule is achiral.

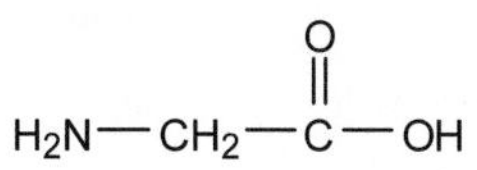

$$H_2N—CH_2—\overset{\overset{\large O}{\|}}{C}—OH$$

NOMENCLATURE 3 | Considerations of Stereochemistry: *R* and *S* Configurations about Tetrahedral Stereocenters and *Z* and *E* Configurations about Double Bonds

Problem N3.1

Think

Which atoms have the highest atomic number? Does the highest atomic number get the highest or lowest priority? To which numbers does this correspond?

Solve

The atoms with the highest atomic number are assigned the highest priority, which corresponds to the lowest number. The atoms are arranged below from *highest* to *lowest* priority.

(a) Br, Cl, F, CH_3 **(b)** I, Br, CH_3, H **(c)** F, O, N, CH_3

Problem N3.2

Think

What atoms attach each group to the stereocenter and what are their atomic numbers? What are the priority assignments for each of the four groups attached to the stereocenter? Is the lowest priority group pointed toward or away from you? Are the substituents arranged clockwise (*R*) or counterclockwise (*S*)? Review the rules for IUPAC nomenclature from Nomenclature 1 and Nomenclature 2.

Solve

The numbers listed below are the priority 1–4 numbers, not the nomenclature carbon chain numbers.

(a) There are two CH_3 groups and therefore this is not a stereocenter.

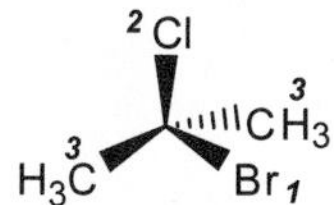

Achiral
2-Bromo-2-chloropropane

(b) H is the lowest priority group and is pointed away, while the 1, 2, and 3 priority groups are arranged clockwise, so the configuration is *R*. The group priorities are distinguished by the atomic numbers of the atoms at the points of attachment: H < C < Br < I. The name is (*R*)-1-bromo-1-iodoethane.

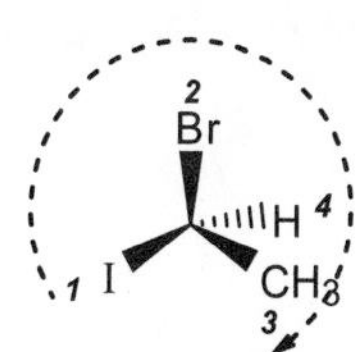

(*R*)-1-Bromo-1-iodoethane

(c) The top three priority groups are F, O, N, which are arranged counterclockwise with the lowest priority group pointing away, so the configuration is *S*. The group priorities are distinguished by the atomic numbers of the atoms at the points of attachment: C < N < O < F. The name is (*S*)-1-ethoxy-1-fluoro-1-nitroethane.

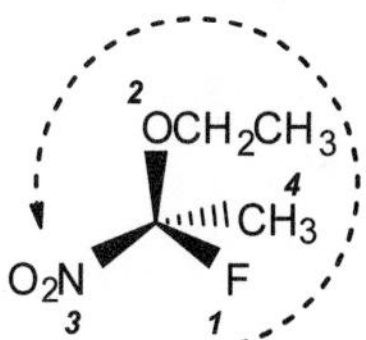

(*S*)-1-Ethoxy-1-fluoro-1-nitroethane

(d) The lowest priority group is H, which is pointing away. The 1, 2, and 3 groups are arranged clockwise, so it is an *R* configuration. The group priorities are distinguished by the atomic numbers of the atoms at the points of attachment: H < C < O < F. The name is (*R*)-1-fluoro-1,4-dimethoxybutane.

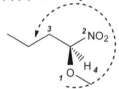

(*R*)-1-Fluoro-1,4-dimethoxybutane

Problem N3.3

Think

Draw the molecule that corresponds to each name without drawing the configuration. Where is the C stereocenter? What are the 1–4 priority group assignments? How are the group priorities determined according to the atoms at the points of attachment? Draw the molecule in such a way that the 1–4 arrangement matches the configuration given.

Solve

In these examples, the substituents are distinguished by the atoms at their points of attachment. The order of priority from highest to lowest is labeled: F > O > N > C > H.

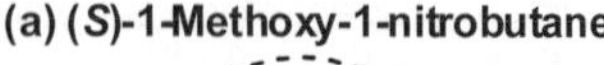

(a) (*S*)-1-Methoxy-1-nitrobutane

(b) (*R*)-1-Methoxy-1-nitrobutane

(c) (*R*)-1-Fluoro-1-methoxy-1-nitropropane

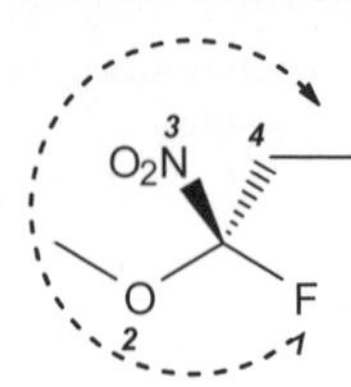

(d) (*S*)-3,3-Dichloro-1-ethoxy-1-fluorohexane

Problem N3.4

Think

What are the priority assignments for each of the four groups? Which weighs more: ^{12}C or ^{13}C or ^{14}C, ^{16}O or ^{18}O? What effect does mass have on priority labeling? Is the lowest priority group pointed toward or away from you? Are the substituents arranged clockwise (*R*) or counterclockwise (*S*)?

Solve

(a) Because of their different atomic numbers, F has priority over ^{13}C and ^{12}C, which have priority over H. ^{13}C has priority of ^{12}C due to a greater atomic mass. The configuration is *R*.

(b) NH_2 has priority over $^{14}CH_3$ because N has the higher atomic number. D and H have the same atomic number, but D has priority over H because it has a higher atomic mass. #4 is neither in the front nor in the back; therefore, you need to rotate the molecule counterclockwise to get the #4 H in the back. The configuration is *R*.

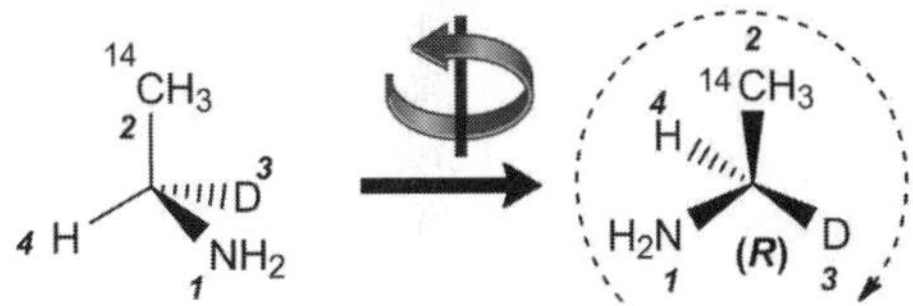

(c) ^{18}O has priority over $^{16}OCH_2CH_3$. Both of those have priority over C, which has priority over H. The lowest-priority group (H) is in the back and the 1–3 groups are arranged clockwise. The configuration is *R*.

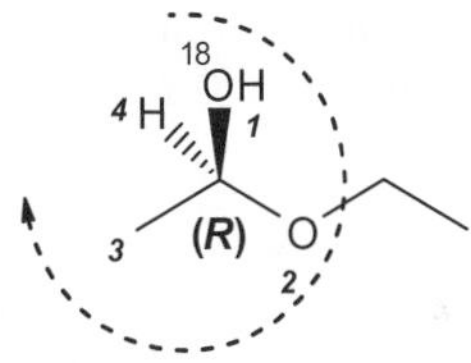

Problem N3.5

Think

What are the priority assignments for each of the four groups? Are the priorities distinguished by the atoms at the points of attachment? If there is a tie between two groups, do you need to compare atoms an additional bond farther away from the points of attachment? Is the lowest priority group pointed toward or away from you? Are the 1–3 substituents arranged clockwise (*R*) or counterclockwise (*S*)? Review the rules for IUPAC nomenclature from Nomenclature 1 and Nomenclature 2.

Solve

The numbers listed are the priority 1–4 numbers not the nomenclature carbon chain numbers.

(a) H has the lowest priority because the atomic number of H is lower than that of C. The other three substituents are attached by C, and to break the tie we look at the sets of atoms one bond away from the points of attachment. CH_3 has lower priority than either CH_2CH_3 or $CH_2CH_2CH_3$ because {C,H,H} beats {H,H,H}. $CH_3CH_2CH_2$ and CH_3CH_2 are still tied, however, but looking at the sets of atoms two bonds away, $CH_3CH_2CH_2$ has higher priority than CH_3CH_2 because {C,H,H} beats {H,H,H}. The configuration is *S*.

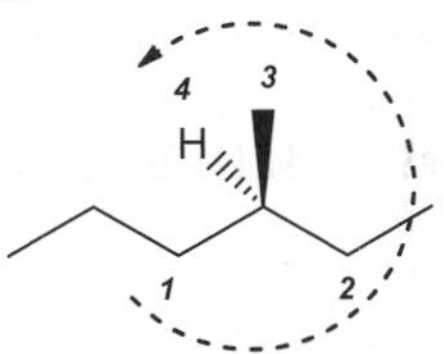

(*S*)-3-Methylhexane

(b) Cl has the #1 priority and H has the #4 priority because of atomic number of those atoms. The other two substituents are both attached by C, and to break the tie, we look at the sets of atoms one bond away from the points of attachment. CH_3OCH_2 has higher priority because {O,H,H} beats {C,H,H}. So the configuration is *S*.

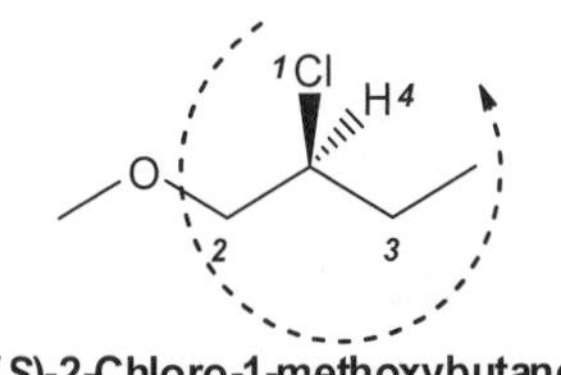

(*S*)-2-Chloro-1-methoxybutane

(c) F has the #1 priority and H has the #4 priority because of atomic number of those atoms. The other two substituents are attached by C, and to break the tie, we look at the sets of atoms one bond away from the points of attachment. NO_2CH_2 has higher priority because {N,H,H} beats {C,H,H}. #4 is in the front and it appears that the configuration is *S*, but you need to reverse the arrangement, so the configuration is *R*.

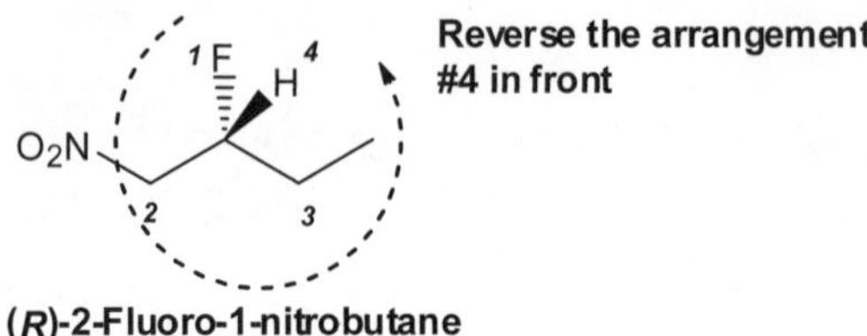

(*R*)-2-Fluoro-1-nitrobutane

(d) F has the #1 priority and H has the #4 priority because of atomic number. The other two substituents are attached by C, and to break the tie, we look at the sets of atoms one bond away from the points of attachment. $BrCH_2$ has higher priority because {Br,H,H} beats {C,H,H}. #4 is in the front and it appears that the configuration is *S*, but you need to reverse the arrangement, so the configuration is *R*.

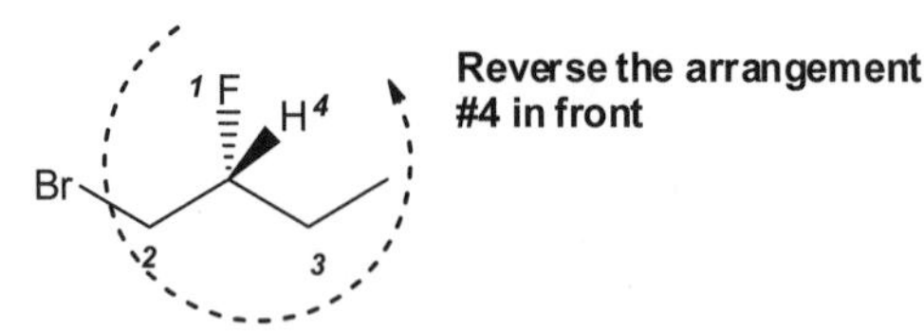

(*R*)-1-Bromo-2-fluorobutane

Problem N3.6

Think

Draw the molecule that corresponds to each name without drawing the configuration. Where is the C stereocenter? What are the 1–4 priority group assignments? Are substituents attached by the same type of atom or different atoms? If there is a tie at the points of attachment, how do you break the tie looking at atoms one bond farther away from the points of attachment? Draw the molecule in such a way that the 1–4 arrangement matches the configuration given.

Solve

In these examples, the order of priority from highest to lowest is labeled.

(a) Group #2 beats group #3 because, one bond away from the points of attachment, {C,H,H} beats {H,H,H}.
(b) Group #2 beats group #3 because, one bond away from the points of attachment, {F,F,F} beats {C,H,H}.
(c) Group #2 beats group #3 because, one bond away from the points of attachment, {C,H,H} beats {H,H,H}.
(d) Group #1 beats group #2 because, two bonds away from the points of attachment, {C,H,H} beats {H,H,H}. Group #3 beats group #4 because, one bond away from the points of attachment, {C,H,H} beats {H,H,H}.

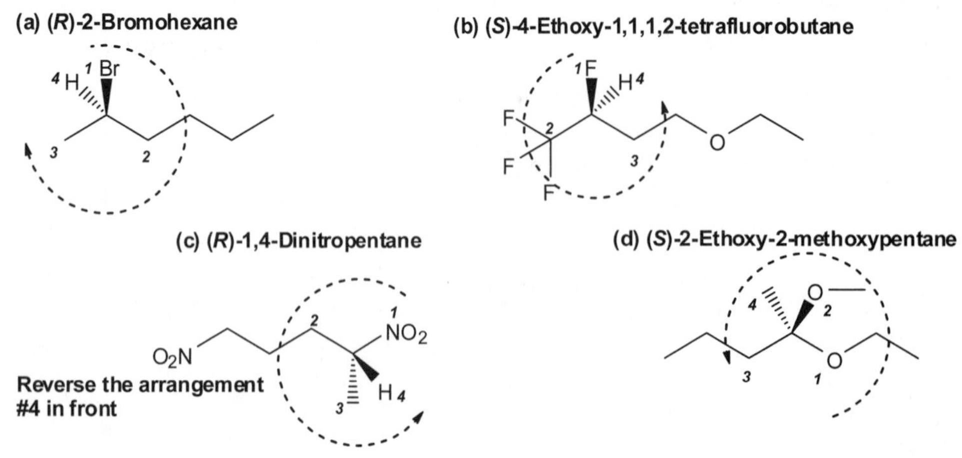

Problem N3.7

Think

How do you treat the atoms in a double or triple bond? It might be helpful to draw out single bond representations for each multiple bond. Do the groups differ at the points of attachment? By the sets of atoms one bond farther away from the points of attachment?

Solve

Each of the multiple bonds is replaced by single bonds and the configuration for each stereocenter is labeled.

(a) *S* configuration. Group #2 beats group #3 because, one bond away from the points of attachment, {C,C,H} beats {C,H,H}.

(b) *R* configuration. Groups #1, #2 and #3 have that order because, one bond away from the points of attachment, {C,C,C} beats {C,C,H}, which beats {H,H,H}.

(c) *R* configuration. Group #2 beats group #3 because, one bond away from the points of attachment, {C,C,H} beats {C,H,H}.

(d) *S* configuration. Groups #1, #2 and #3 have that order because, one bond away from the points of attachment, {N,N,N} beats {C,C,C}, which beats {H,H,H}

(e) *R* configuration. Group #2 beats group #3 because, one bond away from the points of attachment, {C,C,H} beats {C,H,H}

(a) Replace the multiple bond with single bonds → (S)

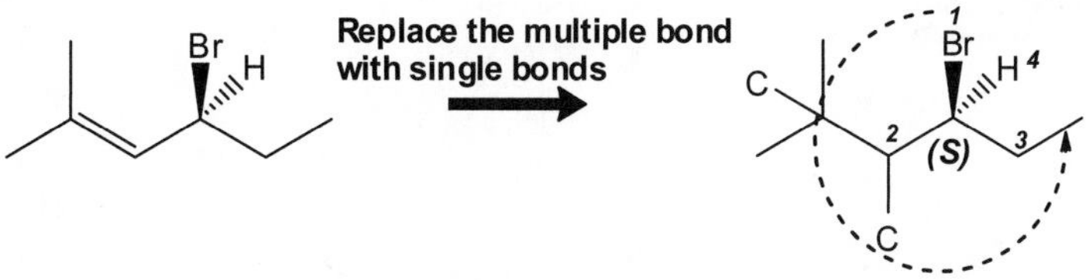

(b) Replace the multiple bond with single bonds → (R)
Reverse the arrangement #4 in front

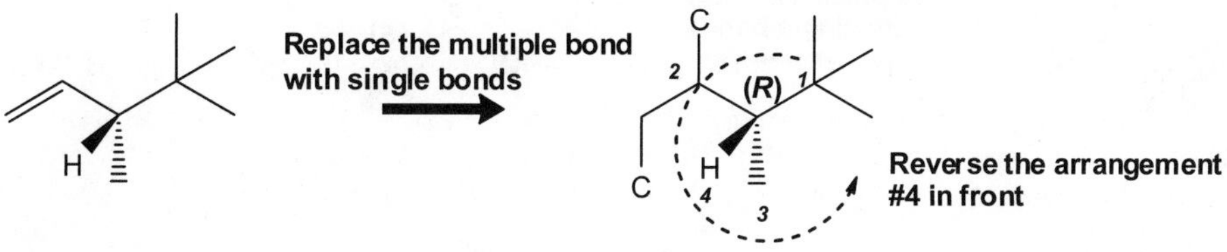

(c) Replace the multiple bond with single bonds → (R)

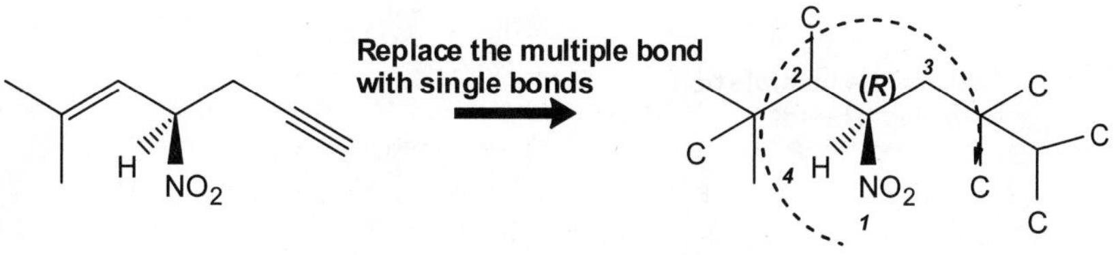

(d) Replace the multiple bond with single bonds → (S)
Reverse the arrangement #4 in front

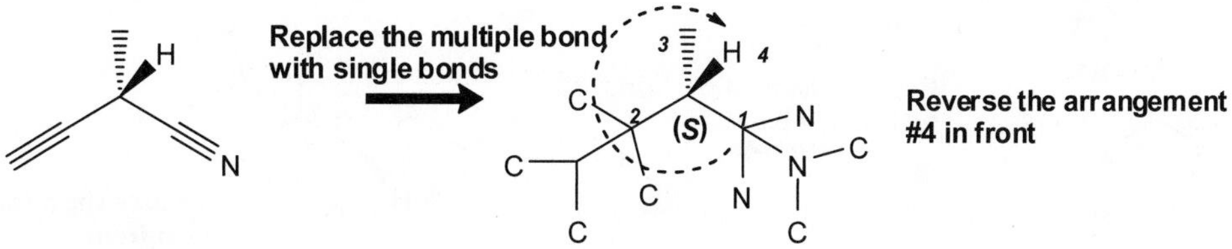

(e) Replace the mutiple bond with single bonds → (R)

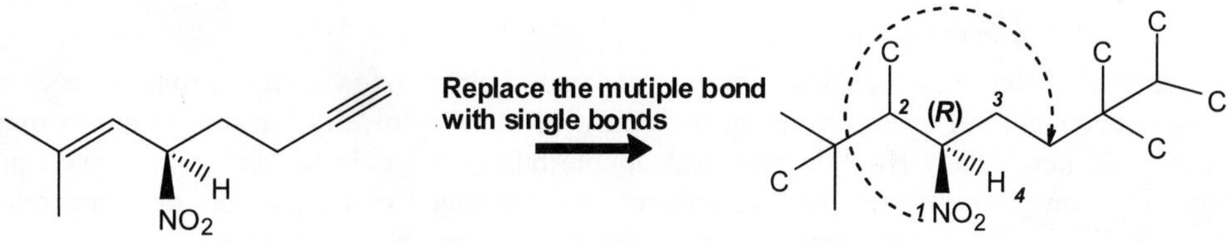

Problem N3.8

Think

How do you treat the atoms in a double or triple bond? It might be helpful to draw out single bond representations for each multiple bond. Do the groups differ at the points of attachment? By the sets of atoms one bond farther away from the points of attachment?

Solve

Each of the multiple bonds is replaced by a single bond and the configuration for each stereocenter is labeled.

(a) *S* configuration. Group #2 beats group #3 because, one bond away from the points of attachment, {C,C,C} beats {H,H,H}.

(b) *S* configuration. Group #2 beats group #3 because, one bond away from the points of attachment, {O,O,C} beats {O,H,H}.

(c) *R* configuration. Group #2 beats group #3 because, one bond away from the points of attachment, {N,N,C} beats {C,H,H}.

(d) *R* configuration. Group #2 beats group #3 because, one bond away from the points of attachment, {C,C,C} beats {C,H,H}.

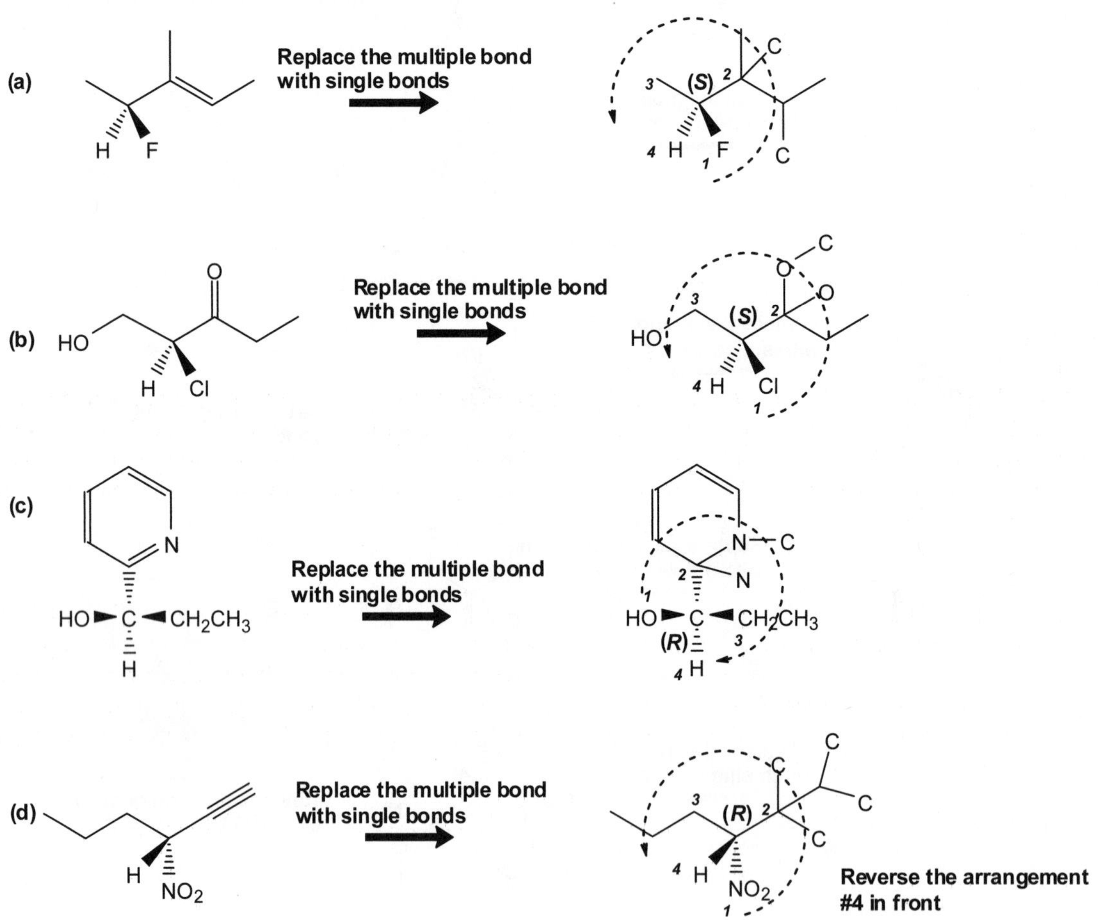

Problem N3.9

Think

Identify all tetrahedral stereocenters. What are the priority assignments for the four groups in each stereocenter? Are the groups distinguished by the atoms at the points of attachment? By the sets of atoms one bond away from the points of attachment? How do you treat double bonds as single bonds? Is the lowest priority group pointed toward or away from you? Are substituents 1–3 arranged clockwise (*R*) or counterclockwise (*S*)? Review the rules for IUPAC nomenclature from Nomenclature 1 and Nomenclature 2.

Solve

R and *S* configurations are shown below for each stereocenter carbon. Prime labels are used to distinguish the 1–4 priority numbering for each stereocenter. In **(a)**, **(b)**, and **(d)** the ties at the points of attachment are broken one bond away from the points of attachment. In the stereocenter on the left in **(c)**, group #1 beats group #2 because the set of atoms one bond away for the alkene and alkane are both {C,C,H}. The set of atoms two bonds away for both the alkene and alkyl group are {C,H,H}. There are no atoms another bond away for the alkene, but the next atoms out for the alkyl group are {H,H,H}. Therefore the alkyl group wins.

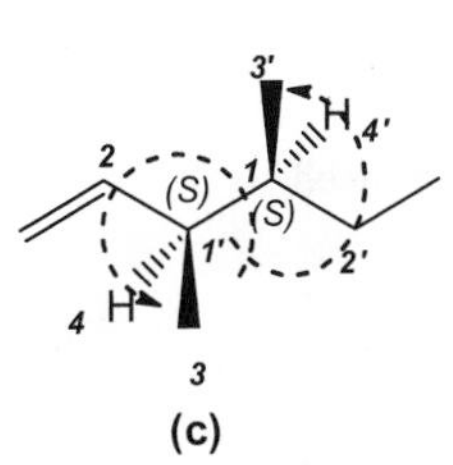

(a)

(2*R*,4*S*)-2-Chloro-4-nitropentane

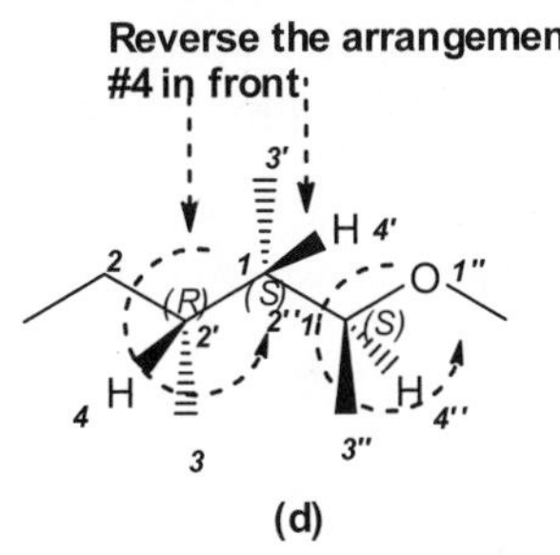

(b)

(2*R*,4*R*)-2-Chloro-4-nitropentane

Reverse the arrangement
#4 in front

(c)

(3S,4*S*)-3,4-Dimethylhex-1-ene

(d)

(2S,3*S*,4R)-2-Methoxy-3,4-dimethylhexane

Problem N3.10

Think

Identify all stereocenters. What are the priority assignments for each of the four groups? Are the groups distinguished by the atoms at the points of attachment? By the sets of atoms one bond away from the points of attachment? How do you treat double bonds as single bonds? Is the lowest priority group pointed toward or away from you? Are substituents 1–3 arranged clockwise (*R*) or counterclockwise (*S*)? Review the rules for IUPAC nomenclature from Nomenclature 1 and Nomenclature 2.

Solve

R and *S* configurations are shown below and on the next page for each stereocenter carbon. Prime labels are used to distinguish the 1–4 priority numbering for each stereocenter. All ties at the points of attachment are broken with the sets of atoms one bond away from the points of attachment, with the exception of the middle stereocenter in **(c)**. For that stereocenter, groups 2′ and 3′ have {Cl,C,H} one bond away from the points of attachment, resulting in a tie. Two bonds away from the points of attachment, the group on the left has {C,C,C}, whereas the one on the right has {H,H,H}.

(a) (*S*)-2-Chloro-(*S*)-3-ethoxypentane

Reverse the arrangement
#4 in front

(b) (4*R*,5*S*)-2,4-Dimethyl-5-nitrohex-2-ene

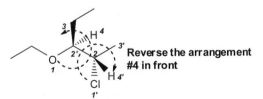

(c) (4*R*,5*R*,6*S*)-4,5,6-Trichloro-2-methyl-3-phenylhept-2-ene

Problem N3.11

Think

Identify all stereocenters. What are the priority assignments for each of the four groups? Are the groups distinguished by the atoms at the points of attachment? By the sets of atoms one bond away from the points of attachment? How do you treat double bonds as single bonds? Is the lowest priority group pointed toward or away from you? Are the substituents 1–3 arranged clockwise (*R*) or counterclockwise (*S*)? Review the rules for IUPAC nomenclature from Nomenclature 1 and Nomenclature 2.

Solve

R and *S* configurations are shown below for each stereocenter carbon. Prime labels are used to distinguish the 1–4 priority numbering for each stereocenter.

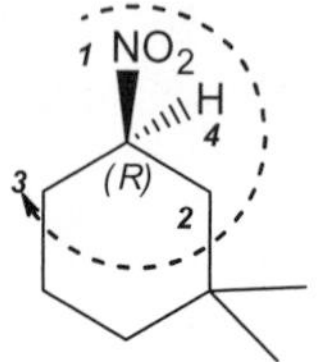

(a) (*R*)-1,1-Dimethyl-3-nitrocyclohexane

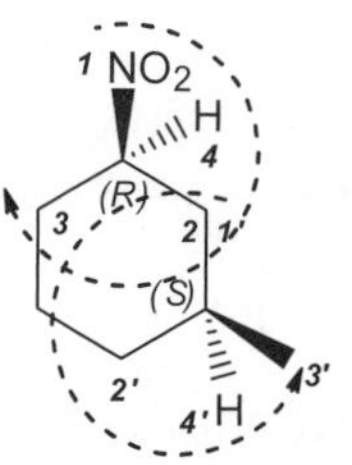

(b) (1*S*,3*R*)-1-methyl-3-nitrocyclohexane

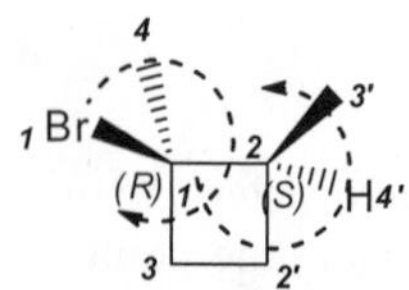

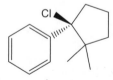

(c) (1*R*,2*S*)-1-bromo-1,2-dimethylcyclobutane

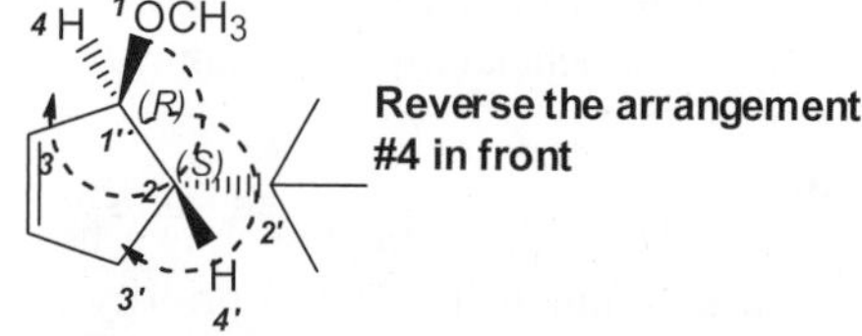

(d) (3*R*,4*S*)-4-(*tert*-butyl)-3-methoxycyclopent-1-ene

Problem N3.12

Think

Draw the molecule that corresponds to each name without drawing the configuration. Where is the C stereocenter? What are the 1–4 priority group assignments? Are the groups distinguished by the atoms at the points of attachment? By the sets of atoms one bond away from the points of attachment? Draw the molecule in such a way that the 1–4 arrangement matches the configuration given.

Solve

(a) (*S*)-1-Chloro-2,2-dimethyl-1-phenylcyclopentane

(b) (1*R*,2*S*)-1-Methyl-1,2-dinitrocyclopropane

(c) (*R*)-4-Ethoxycyclohexene

H O

Reverse the arrangement
#4 in front

(d) (3*S*,4*S*)-3-Chloro-4-fluoro-2-methylhepta-1,6-diene

Reverse the arrangement
#4 in front

H F H Cl

Reverse the arrangement
#4 in front

Problem N3.13

Think

Which molecules have a plane of symmetry? Identify the chiral and achiral molecules. Which molecules can be unambiguously described by cis and trans?

Solve

(a) The molecule has a plane of symmetry and therefore is achiral. Cis is an unambiguous way to describe the molecule.

***cis*-1,2-Difluorocyclohexane**

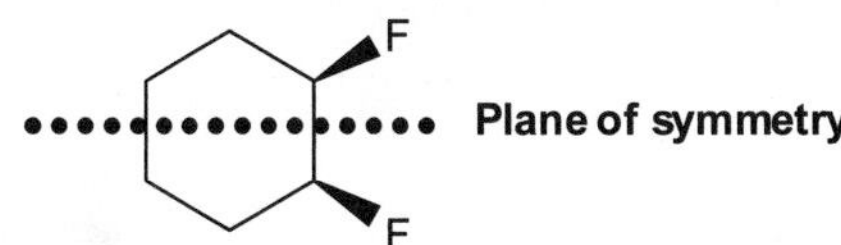

(b) The molecule does not have a plane of symmetry, is chiral, and *trans*-1,2-difluorocyclohexane is ambiguous. The two possible stereoisomers are shown below.

***trans*-1,2-Difluorocyclohexane**

(1*R*,2*R*)-1,2-Difluorocyclohexane **(1*S*,2*S*)-1,2-Difluorocyclohexane**

(c) The molecule has a plane of symmetry and therefore is achiral. Trans is an unambiguous way to describe the molecule.

***trans*-1,4-Difluorocyclohexane**

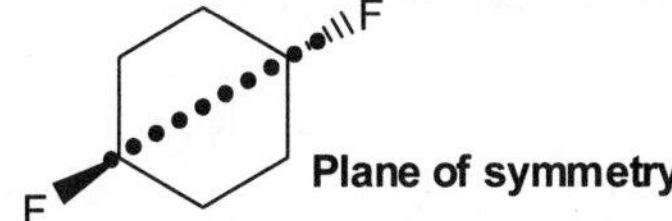

(d) The molecule does not have a plane of symmetry, is chiral, and *cis*-1-chloro-2-fluorocyclohexane is ambiguous. The two possible stereoisomers are shown below.

***cis*-1-Chloro-2-fluorocyclohexane**

Cl F Cl F

(1*R*,2*S*)-1-Chloro-2-fluorocyclohexane **(1*S*,2*R*)-1-Chloro-2-fluorocyclohexane**

(e) The molecule does not have a plane of symmetry, is chiral, and *trans*-1,4-dimethylcycloheptane is ambiguous. The two possible stereoisomers are shown below.

(1*R*,4*R*)-1,4-Dimethylcycloheptane (1*S*,4*S*)-1,4-Dimethylcycloheptane

Problem N3.14

Think

What are the priority assignments for each of the four groups? Is the lowest priority group pointed toward you (horizontal bond) or away from you (vertical bond)? Are the substituents 1–3 arranged clockwise (*R*) or counterclockwise (*S*) when the lowest priority substituent points away? Review the rules for IUPAC nomenclature from Nomenclature 1 and Nomenclature 2.

Solve

If priority #4 is on a horizontal bond, it points toward you instead of away. The configuration, therefore, is the reverse of what it appears to be.

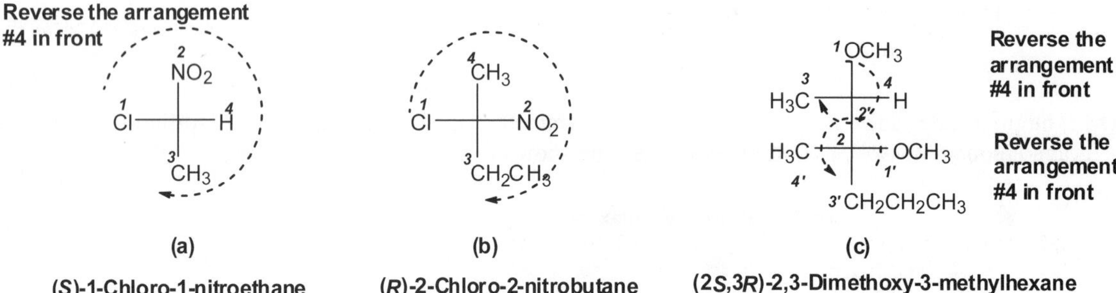

(*S*)-1-Chloro-1-nitroethane (*R*)-2-Chloro-2-nitrobutane (2*S*,3*R*)-2,3-Dimethoxy-3-methylhexane

Problem N3.15

Think

How many of the configurations are reversed in an enantiomer pair? How many are reversed in a diasteriomer relationship?

Solve

Enantiomers have all stereocenter configurations reversed and diastereomers have some, but not all, configurations reversed. For a molecule with (2*R*, 3*S*, 5*R*) configuration, the enantiomer would have (2*S*, 3*R*, 5*S*). The diastereomer example would have (2*R*, 3*S*, 5*S*) or (2*R*, 3*R*, 5*S*), and so on.

Problem N3.16

Think

Are the two molecules stereoisomers or constitutional isomers? If the molecules are stereoisomers, compare the configuration of each stereocenter. How many of the configurations are reversed in an enantiomer pair? How many are reversed in a diastereomer relationship? Can cis/trans isomers be mirror images?

Solve

Two different molecules have the same connectivity and are stereoisomers if they have the same name with the exception of the stereochemical configuration designations. Enantiomers have all stereocenter configurations reversed and diastereomers have some, but not all, configurations reversed.

(a) (1*R*,3*S*)-1-methyl-3-nitrocyclohexane and (1*S*,3*R*)-1-methyl-3-nitrocyclohexane are enantiomers.

(b) (1*R*,3*S*)-1-methyl-3-nitrocyclohexane and (1*S*,3*S*)-1-methyl-3-nitrocyclohexane are diastereomers.

(c) (1*R*,3*S*)-1-methyl-3-nitrocyclohexane and (1*S*,2*R*)-1-methyl-2-nitrocyclohexane are constitutional isomers (neither).

(d) *cis*-1,2-dimethylcyclobutane and *trans*-1,2-dimethylcyclobutane are diastereomers.

Problem N3.17

Think

Draw the molecule that corresponds to each name without drawing the configuration. Where is the C stereocenter? What are the 1–4 priority group assignments? Draw the molecule in such a way that the 1–4 arrangement matches the configuration given. Review the rules for IUPAC nomenclature from Nomenclature 1 and Nomenclature 2.

Solve

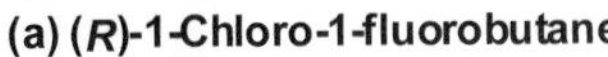

(a) (*R*)-1-Chloro-1-fluorobutane

(b) (*S*)-2-Chloropentane

(c) (*R*)-2-Chloro-2-methoxypentane

(d) (*R*)-2,2,3-Trichlorobutane

(e) (*S*)-3-Methylhexane

(f) (*S*)-2-Bromo-1-nitropentane

(g) (*R*)-3-Chloropent-1-ene

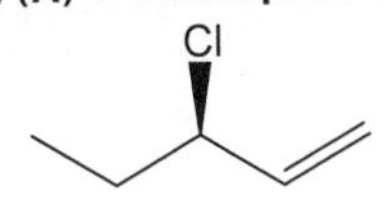

(h) (2*S*,3*S*)-2-Bromo-3-chloropentane

(i) (*R*)-1-Bromo-(*R*)-2-iodocyclopentane

(j) (*S*)-3-Chlorocyclohexene

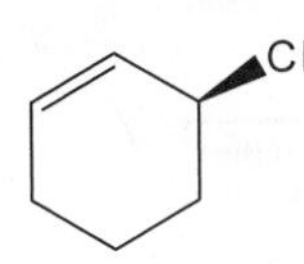

(k) (1*R*,2*S*)-1,2-Dibromocyclopentane

Problem N3.18

Think

What are the priority assignments for each of the four groups? Is the lowest priority group pointed toward you or away from you? Which bonds in a Fischer projection are pointed toward you and which are pointed away? Are the substituents arranged clockwise (R) or counterclockwise (S)? Review the rules for IUPAC nomenclature from Nomenclature 1 and Nomenclature 2.

Solve

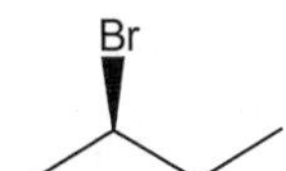

(a) (*R*)-2-Bromobutane

(b) (*S*)-1-Chloro-1-methoxypropane

(c) (*S*)-1-Chloro-1-phenylpropane

(d) (*R*)-3-Chlorohexane

(e) (*R*)-2-methyl-4-nitrohexane

(f) (*S*)-1,2-Diethoxy-3-phenylpropane

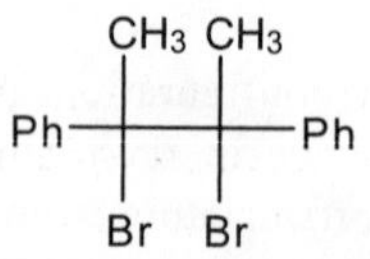

(g) (2*R*,3*S*)-2,3-Dibromo-2,3-diphenylbutane

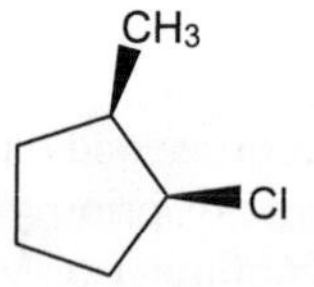

(h) (1*S*,2*R*)-1-Chloro-2-methylcyclopentane

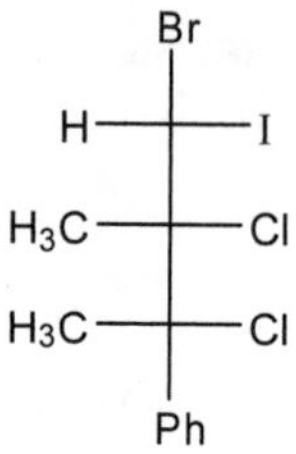

(i) (1*R*,2*R*,3*R*)-1-Bromo-2,3-dichloro-1-iodo-2-methyl-3-phenylbutane

Problem N3.19

Think

Of the two groups attached to one C atom of the C=C, which has higher priority? Are the groups distinguished by the atoms at the points of attachment or by sets of atoms farther away from the points of attachment? Which group has higher priority on the other C atom of the C=C? Are the higher priority groups on the same side or opposite sides?

Solve

Higher priorities on the same side = (*Z*), higher priorities on opposite sides = (*E*).

(a) (b) (c) (d)

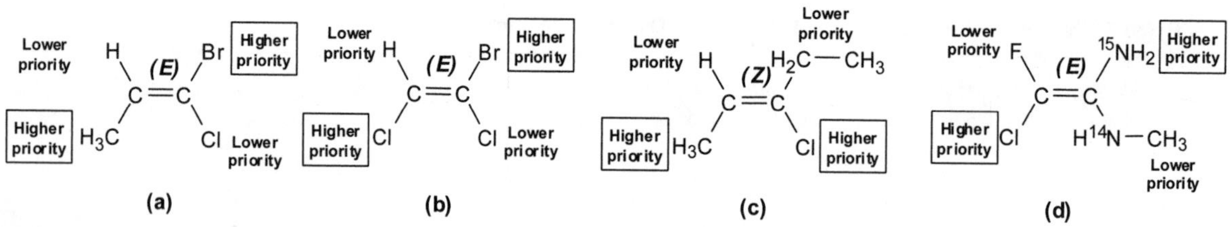

Problem N3.20

Think

Of the two groups attached to one C atom of the C=C, which has higher priority? Are the groups distinguished by the atoms at the points of attachment or by sets of atoms farther away from the points of attachment? Which group has higher priority on the other C atom of the C=C? Are the higher priority groups on the same side or opposite sides?

Solve

Higher priorities on the same side = (*Z*), higher priorities on opposite sides = (*E*).

(a) (b) (c) (d)

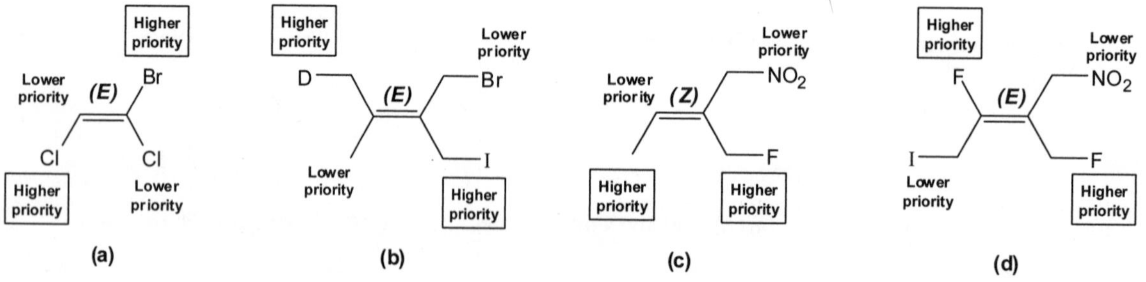

Problem N3.21

Think

Of the two groups attached to one C atom of the C=C, which has higher priority? Are the groups distinguished by the atoms at the points of attachment or by sets of atoms farther away from the points of attachment? Which group has higher priority on the other C atom of the C=C? Are the higher priority groups on the same side or opposite sides?

Solve

Higher priorities on the same size = (*Z*), higher priorities on opposite side = (*E*).

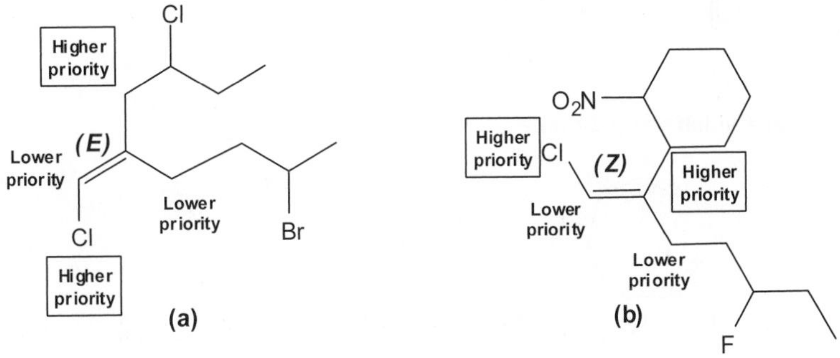

(a)

(b)

Problem N3.22

Think

Assign higher and lower priority for each group on both sides of the double bond. Are the higher priority groups on the same side or opposite sides? How many alkene bonds require specifying the *E*/*Z* configuration? Review the rules for IUPAC nomenclature from Nomenclature 1 and Nomenclature 2.

Solve

Higher priorities on the same size = (*Z*), higher priorities on opposite side = (*E*).

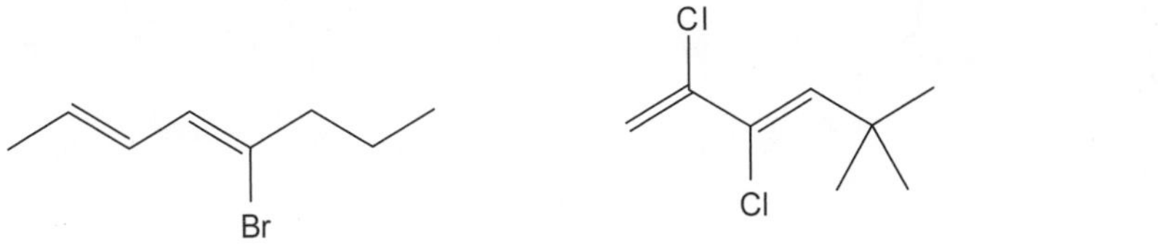

(a)
(2*E*,4*Z*)-5-Bromoocta-2,4-diene

(b)
(*Z*)-2,3-Dichloro-5,5-dimethylhexa-1,3-diene

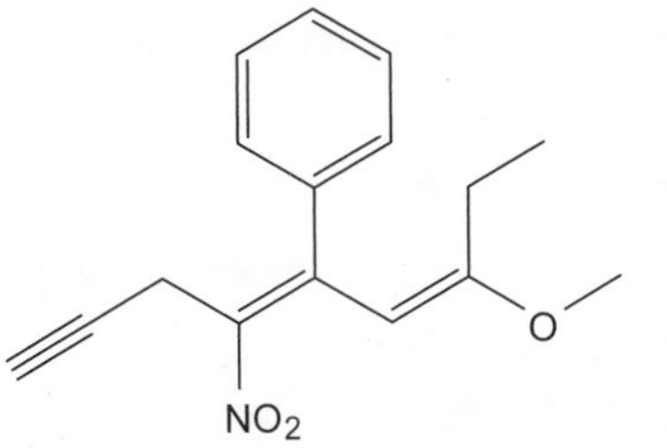

(c)
(4*E*,6*E*)-7-Methoxy-4-nitro-5-phenylnona-4,6-dien-1-yne

Problem N3.23

Think

Draw the molecule that corresponds to each name without drawing the configuration. Consider the rules for *R*/*S* and *E*/*Z* configurations. When is it necessary to designate the configuration at the C=C in a ring?

Solve

(a) 3,3-Dichlorocyclohexene (b) 1-Chlorocyclohexene (c) (*E*)-4,4-Dinitrocyclodecene

(d) (*R*)-3-Fluorocycloheptene

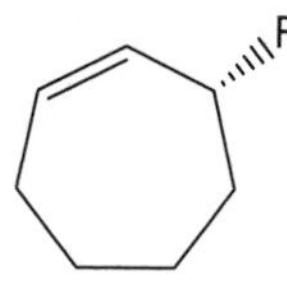

(e) Cyclohepta-1,3-diene

Problem N3.24

Think

What are the priority assignments for each of the four groups attached to the different tetrahedral stereocenters? Is the lowest priority group pointed toward you or away from you? Are substituents 1–3 arranged clockwise (*R*) or counterclockwise (*S*)? Assign higher and lower priority for each group on both sides of the double bond. Are the higher priority groups on the same side or opposite sides? How many alkene bonds require specifying the *E*/*Z* configuration? Review the rules for IUPAC nomenclature from Nomenclature 1 and Nomenclature 2.

Solve

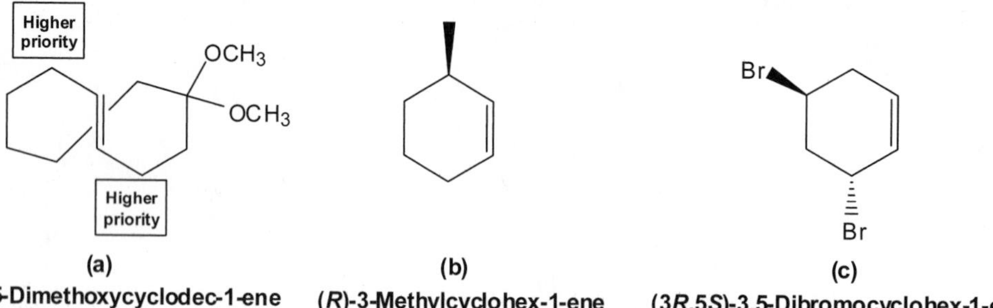

(a) (*E*)-5,5-Dimethoxycyclodec-1-ene (b) (*R*)-3-Methylcyclohex-1-ene (c) (3*R*,5*S*)-3,5-Dibromocyclohex-1-ene

Problem N3.25

Think

Draw the molecule that corresponds to each name without drawing the configuration. Consider the rules for *R*/*S* and *E*/*Z* configurations.

Solve

(a) (*Z*)-2-Methoxypent-2-ene

(b) (*E*)-3-Methylpent-2-ene (c) (*Z*)-1-Chloro-2-methylpent-1-ene

(d) (*E*)-2-Chloro-3-methoxybut-2-ene

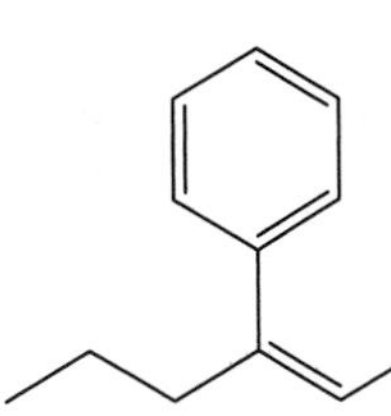

(e) (*Z*)-1-Bromo-1-chloropent-1-ene

(f) (*Z*)-3-Methylpent-2-ene

(g) (*Z*)-3-Phenylhex-2-ene

(h) 1,2-Dichlorocyclopentene

(i) (2*E*,4*E*)-2-Ethoxyhexa-2,4-diene

(j) (1*E*,3*E*,5*E*)-1,3,4,6-Tetrachlorohexa-1,3,5-triene

Problem N3.26

Think

What are the priority assignments for each of the four groups attached to the different tetrahedral stereocenters? Is the lowest priority group pointed toward you or away from you? Are substituents 1–3 arranged clockwise (*R*) or counterclockwise (*S*)? Assign higher and lower priority for each group on both sides of the double bond. Are the higher priority groups on the same side or opposite sides? How many alkene bonds require the *E*/*Z* configuration? Review the rules for IUPAC nomenclature from Nomenclature 1 and Nomenclature 2.

Solve

(a) (*Z*)-2-Bromo-3-methylpent-2-ene

(b) (*E*)-2-Chloro-3-ethoxypent-2-ene

(c) (*E*)-2-Fluoro-3-methoxypent-2-ene

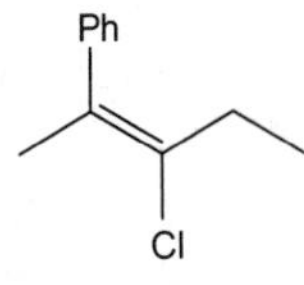

(d) (*E*)-3-Chloro-2-phenylpent-2-ene

(e) (*Z*)-1-Cyclopropyl-2-methyl-3-phenylbut-2-ene

(f) 1,1-Dichloro-2-phenylbut-1-ene

(g) (*E*)-5,5-Dimethyl-2,3-diphenylhex-2-ene

(h) (*Z*)-4-Methyl-3-pentylhex-3-en-1-yne

(i) (*E*)-1,2-Dichlorocyclooct-1-ene

Note: Make a model; the diagonal bond is behind the double bond.

$H-\ddot{N}(-H)-H + H-\ddot{O}-H$

$NH_4^+ + OH^-$

CHAPTER 6 | The Proton Transfer Reaction: An Introduction to Mechanisms, Thermodynamics, and Charge Stability

Your Turn Exercises

Your Turn 6.1

Think

What does the double-barbed curved arrow represent? Is bond breaking shown at the tail or head of the arrow? Is bond making shown at the tail or head of the arrow? In comparing the reactants and products, what bonds were broken? What bonds were formed?

Solve

Each double-barbed curved arrow represents the movement of two valence electrons. The head of the first arrow shows bond making (HO–H) and the tail of the second arrow shows bond breaking (H–Cl). The electrons in the newly formed HO–H bond were initially the lone pair on the hydroxide oxygen. The electrons in the broken H–Cl end up as an additional lone pair on Cl. Each pair of electrons that are involved in the bond making/breaking steps is circled with a dashed oval.

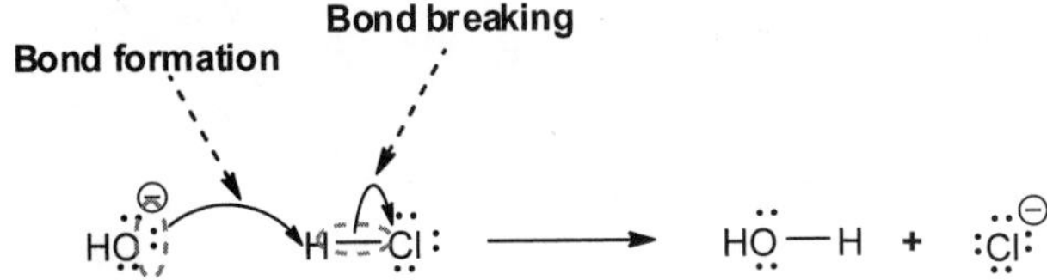

Your Turn 6.2

Think

In comparing the reactants and products, what bonds were broken? What bonds were formed? To indicate the conversion of a lone pair into a bond, where should the curved arrow originate? Where should it point? To indicate the conversion of a bond into a lone pair, where should the curved arrow originate? Where should it point? How does the magnitude of the K_{eq} inform you of the extent of product formation at equilibrium?

Solve

The lone pair on the O atom of the alcohol is used to form a bond to the H atom of the carboxylic acid, and the H–O bond in the carboxylic acid is broken, with the electrons from that bond ending up as an additional lone pair on the carboxylic acid's O. Therefore, one curved arrow is drawn from a lone pair on the alcohol O atom to the H atom on the carboxylic acid, and a second curved arrow originates from the center of the carboxylic acid's O–H bond and points to the carboxylic acid's O atom. The larger equilibrium constant favors more product formation, as K_{eq} = [products]/[reactants]. In this case, it is the second reaction that forms more products because the $K_{eq} = 4.0 \times 10^{-3}$ is five orders of magnitude larger than that of the first reaction with $K_{eq} = 7.1 \times 10^{-8}$.

Acid Base $K_{eq} = 7.1 \times 10^{-8}$

Acid Base $K_{eq} = 4.0 \times 10^{-3}$

Your Turn 6.3

Think

Consult Table 6-1. Is the stronger acid the one with the more negative or more positive pK_a?

Solve

The acid with the more negative pK_a is the stronger acid. Therefore, HCl (pK_a = −7) is a stronger acid than H_3O^+ (pK_a = −1.7). The difference in pK_a values is −1.7 − (−7) = 5.3, which corresponds to a difference in acid strength of >10^5. Thus, HCl is >100,000 times stronger an acid than H_3O^+.

Acid is stronger than H_3O^+

This side is heavily favored

$H_2\ddot{O}$: + H—Cl: ⇌ H_2O^+—H + :Cl:⁻

pK_a = −7   pK_a = −1.7

Your Turn 6.4

Think

Consult Table 6-1. Is the stronger acid the one with the more negative or more positive pK_a?

Solve

The acid with the more negative pK_a is the stronger acid. Therefore, H_2O (pK_a = 15.7) is a stronger acid than $(CH_3)_2NH$ (pK_a = 38). The difference in pK_a values is 38 − 15.7 = 22.3, which corresponds to a difference in acid strength of $10^{22.3}$. Thus H_2O is 2.0 × 10^{22} times stronger as an acid than $(CH_3)_2NH$. Moreover, HO^- is a weaker base than $(CH_3)_2N:^-$ because the stronger the acid, the weaker the conjugate base.

Base is stronger than HO⁻

$(H_3C)_2N:^-$ + H—OH ⇌ $(H_3C)_2N$—H + :OH⁻

pK_a = 15.7   pK_a = 38

This side is heavily favored.

Your Turn 6.5

Think

Consult Table 6-1. Is the stronger acid the one with the more negative or more positive pK_a? For diethyl ether to be a suitable solvent, should it react with the solute? Should the proton transfer given favor the reactant side or the product side?

Solve

The pK_a values are written below. The reactant side of the reaction is favored because the stronger acid is on the product side. This is not the same side that is favored in Equation 6-13. Therefore, diethyl ether is a suitable solvent for $(CH_3)_2N^-$ because the equilibrium lies to left, indicating that diethyl ether is relatively inert in the presence of $(CH_3)_2N^-$.

$(H_3C)_2N:^-$ + H_3C–CH(H)–O–CH_2–CH_3 ⇌ $(H_3C)_2N$—H + H_3C–CH^-–O–CH_2–CH_3

This side is favored.

pK_a = ~45   pK_a = 38

Your Turn 6.6

Think

Consult Figure 6-1. What is the pH at the pK_a? At what pH is the acid 100% dissociated? How many pH units above the pK_a is this value? At what pH is the acid 0% dissociated (100% associated)? How many pH units below the pK_a is this value?

Solve

The acid is nearly 100% dissociated at around two pH units above the pK_a, or pH = 7. It is nearly 100% associated around two pH units below the pK_a, or pH = 3.

Your Turn 6.7

Think

Consult Figure 6-1 as a guide. What is the pH when an acid whose pK_a = 9 is dissociated 50%? How many pH units above the acid's pK_a of 9 must the solution be to cause the acid to dissociate nearly 100%? How many pH units below the acid's pK_a of 9 must the solution be to cause the acid to dissociate roughly 0%? How does the increase in pK_a by four units affect the appearance of the graph?

Solve

The pH curve shifts four units higher so that 50% dissociation takes place at pH = pK_a = 9. The acid is nearly 100% dissociated at around two pH units above the pK_a, or pH = 11. It is nearly 100% associated around two pH units below the pK_a, or pH = 7.

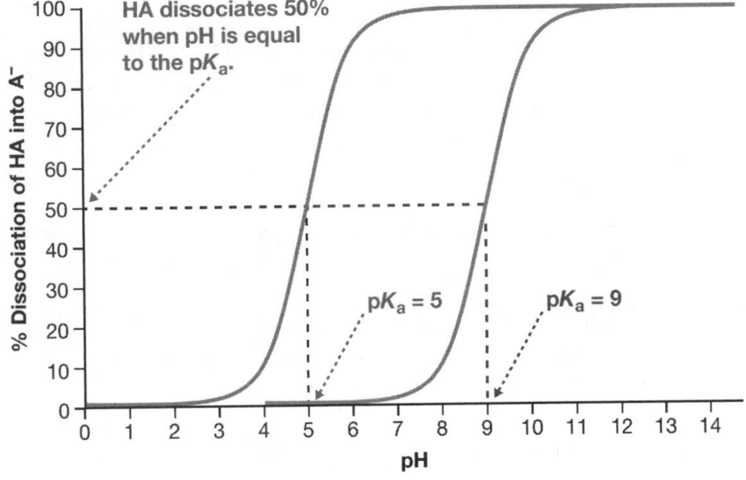

Your Turn 6.8

Think

As the reaction coordinate increases (going from left to right in the figure), which bonds are broken? Which bonds are formed? How does the distance between the atoms correlate to bond breaking and bond forming?

Solve

As the reaction coordinate increases, the distance between Cl and H decreases, because as the reaction proceeds, the H–Cl bond is forming; the distance between the O and H increases because the O–H bond is breaking.

Your Turn 6.9

Think

Draw an arrow to represent the difference between the free energy of products and reactants (ΔG°_{rxn}). Draw an arrow to represent the difference between the free energy of the reactants and the transition state ($\Delta G^{\circ\ddagger}$). Compare the length of the arrow you drew to the length of the arrow in Figure 6-2a. Which one is longer? What does that mean for the value of $\Delta G^{\circ\ddagger}$?

Solve

The free energy quantities are indicated in the diagram below. The $\Delta G^{\circ\ddagger}$ is larger in Figure 6-2b compared to the reaction in Figure 6-2a.

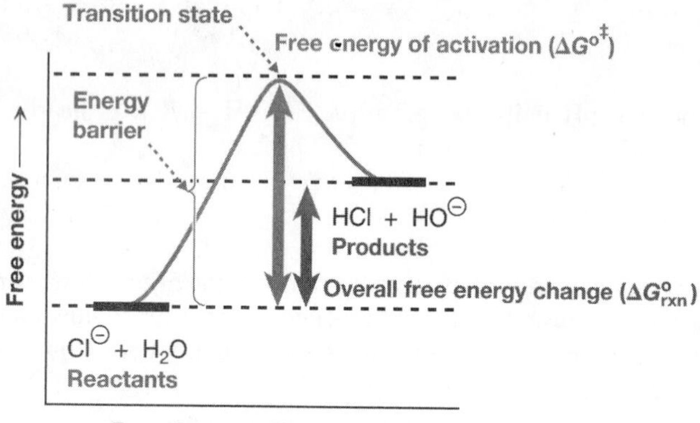

Your Turn 6.10

Think

Consult Table 6-1. Is the stronger acid the one with the more negative or more positive pK_a?

Solve

The pK_a of HCl is −7, and that of H_2S is approximately the same as that of CH_3CH_2SH, which is 10.6. HCl, having the more negative pK_a, is a stronger acid. Cl is more electronegative than S, which allows Cl to better accommodate a negative charge. Thus, Cl^- is a more stable anion and a weaker conjugate base.

Your Turn 6.11

Think

Consult Table 6-1. Is the stronger acid the one with the more negative or more positive pK_a?

Solve

The pK_a of H_3O^+ is −1.7, and that of NH_4^+ is 9.4. H_3O^+ is a stronger acid because it has the more negative pK_a. O is more electronegative than N, which allows O to better accommodate a negative charge, but N to better accommodate a positive charge. Thus, NH_4^+ will not give up its proton as easily.

Your Turn 6.12

Think

Consult Table 6-1. Is the stronger acid the one with the more negative or more positive pK_a?

Solve

The pK_a of $H_3C{-}CH_3$ is about 50, that of $H_2C{=}CH_2$ is about 44, and that of $HC{\equiv}CH$ is about 25. The acid strength goes in the following increasing order: C_2H_6 (sp^3) < C_2H_4 (sp^2) < C_2H_2 (sp). The difference in acid strength is due to the stability of the negative charge that develops on the conjugate base. Charge stability depends, in turn, on the effective electronegativity of the C atom, which depends on its hybridization—$sp^3 < sp^2 < sp$.

Your Turn 6.13

Think

Consult Table 6-1. Is the stronger acid the one with the more negative or more positive pK_a?

Solve

The pK_a of ethanoic acid (acetic acid) is 4.75. That of ethanol is 16. Ethanoic acid, having the less positive pK_a, is a stronger acid due to resonance stabilization of the anion conjugate base.

Your Turn 6.14

Think

Remember that the curved arrow illustrates electron movement necessary to change one resonance structure into another. In the structure on the left, which electrons move to form the double bond in the structure on the right? Which electrons move in the structure on the left to form the lone pair on the O atom in the structure on the right? How do you draw the average of the two structures (i.e., the hybrid)? How do you represent partial charges?

Solve

In the structure on the left, a lone pair from O^- is used to make the C=O double bond in the structure on the right. This requires a curved arrow from the lone pair to the center of the initial C–O bond. Without any other changes, this would lead to 10 valence electrons on the C atom. Therefore, in addition, a bonding pair of electrons from the C=O double bond on the left is converted to an additional lone pair on the O atom that is initially uncharged. This requires a second curved arrow to be drawn from the center of the C=O double bond to the O atom at the top. Because each O has a −1 charge in one structure and a 0 charge in the other, the hybrid charge is represented by a partial charge symbol, δ−. Because each C–O bond is a single bond in one structure and a double bond in the other, they are both represented as a bond intermediate between a single and double bond.

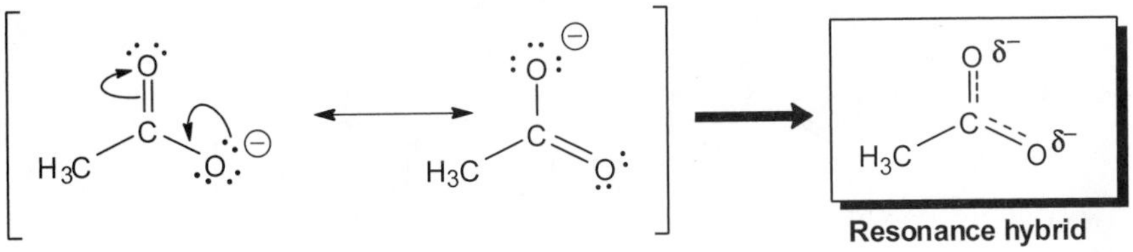

Resonance hybrid

Your Turn 6.15

Think

What structural feature in HSO_4^- indicates that there should be another resonance structure—a lone pair attached to a double/triple bond, an atom lacking an octet attached to a double/triple bond, or a ring of alternating single and double bonds? With that structural feature, how many curved arrows do you need to convert one resonance structure into another? When a lone pair is converted into a bonding pair (and vice versa), how does that change the formal charge?

Solve

In HSO_4^-, the negatively charged O has three lone pairs and is attached to an S=O bond. Therefore, two curved arrows are required to arrive at the next resonance structure: one from a lone pair on O^- that points to the S–O bond, and one from an S=O bond that points to the corresponding O. This is done twice to obtain the two other resonance structures, which are shown below. You may have chosen to draw the third resonance structure second. This does not matter as long as your arrow movement is consistent with the structure that results.

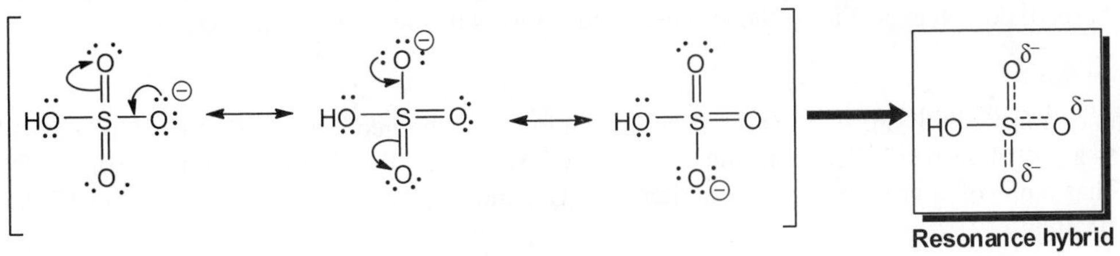

Resonance hybrid

Your Turn 6.16

Think

Are there any lone pairs adjacent to a double or triple bond? Do any of the double bonds convert to lone pairs? How can you show via a curved arrow the electrons in the lone pair moving to form a double bond? How can you show via a curved arrow the electrons in a double bond moving to form a lone pair?

Solve

Recall that atoms do not move in resonance structures. Only nonbonding electrons or electrons from multiple bonds move. *Top structure*: The first curved arrow is used to convert the lone pair on the fifth C atom into a covalent double bond, C=C. Without any other changes, this would lead to 10 valence electrons on the fourth C atom. The second arrow is used to convert a pair of electrons from the double bond, C=O, into a lone pair on the O. *Bottom structure*: The first curved arrow is used to convert the lone pair on the third C atom into a covalent double bond, C=C. Without any other changes, this would lead to 10 valence electrons on the fourth C atom. The second arrow is used to convert a pair of electrons from the double bond, C=O, into a lone pair on the O atom. This yields the resonance structure in the middle. A third resonance structure can be produced from structure on the left by repeating the electron movement on the left side of the ion instead of the right. To arrive at that resonance structure from the one in the middle, the three curved arrows shown are required.

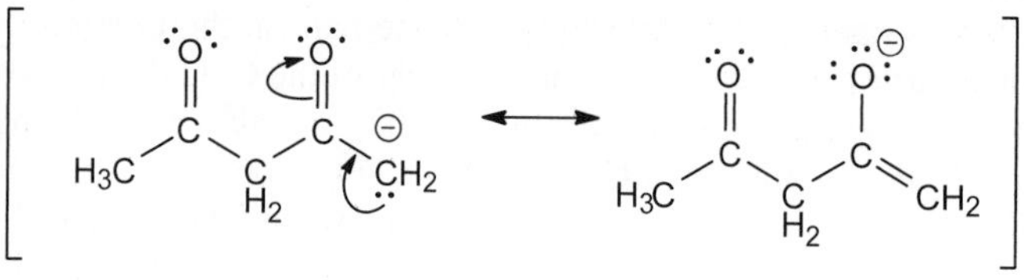

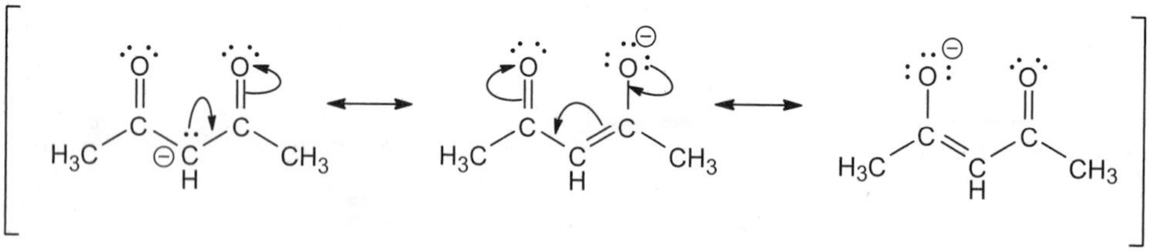

Your Turn 6.17

Think

Is the CH_3 group electron donating or electron withdrawing relative to the H atom? In which direction does the arrow point?

Solve

The CH_3 group is electron donating compared to the H atom, so an arrow is drawn from the CH_3 group toward the N^+.

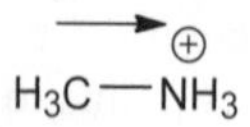

$H_3C—NH_3^{\oplus}$

Your Turn 6.18

Think

Consult Table 6-1. Is the stronger acid the one with the more negative or more positive pK_a? Which species (charge or uncharged) do you expect to be the stronger acid? How is this reflected in the pK_a values?

Solve

Protonated amines and alcohols are stronger acids compared to the uncharged species. R–NH_3^+ has a pK_a of 9.4 and R–NH_2 has a pK_a of 38 (estimated from the 2° amine R_2NH), so R–NH_3^+ is more acidic by almost 30 pK_a units. R–OH_2^+ has a pK_a of approximately −1 (similar to H_3O^+) and R–OH has a pK_a of 16, so R–OH_2^+ is more acidic by ~17 pK_a units.

Your Turn 6.19

Think

Consult Table 6-1. Is the stronger acid the one with the more negative or more positive pK_a? How does the type of atom that the acidic proton is bound to affect the strength of the acid? How does hybridization of the C atom bound to the acidic proton affect the strength of the acid?

Solve

R–OH (pK_a = 16), R–NH_2 (pK_a = 38, estimated from the 2° amine R_2NH), R–CH_3 (pK_a = 50), R–C≡CH (pK_a =25). Therefore, R–OH is >20 pK_a units more acidic than R–NH_2, and R–C≡CH is about 25 pK_a units more acidic than R–CH_3. O, N, and C are in the same row and are about the same size. O is more electronegative than N, which is more electronegative than C. This allows O to accommodate negative charge more easily and the conjugate base is stabilized. For comparison of the two CH species, the effective electronegativity depends on hybridization of the C atom ($sp^3 < sp^2 < sp$). Therefore, the *sp*-hybridized conjugate base is more stable and the acid is stronger.

Your Turn 6.20

Think

Are there any lone pairs adjacent to a double or triple bond? Can any of the double bonds convert into lone pairs? How can you show via a curved arrow the electrons in the lone pair moving to form a double bond? How can you show via a curved arrow the electrons in a double bond moving to form a lone pair? How do you draw the average of the four structures (hybrid)? How do you represent partial charges?

Solve

In each resonance structure, there is an atom with a lone pair attached to a double bond, so two curved arrows are necessary to convert from one to the other. In each case, the first arrow moves a lone pair to form a double bond and the second arrow shows the double bond converting into a lone pair. When bonding electrons convert to nonbonding electrons, the formal charge decreases by 1. When nonbonding electrons convert to bonding electrons, the formal charge becomes more positive by 1. The hybrid structure shows the partial negative charge on each of the four atoms that bears a full negative charge on one of the resonance structures. You will note that the negative charge skips every other atom and goes around the entire ring. Also, each C–C and C–N bond in the hybrid is intermediate between a single and double bond.

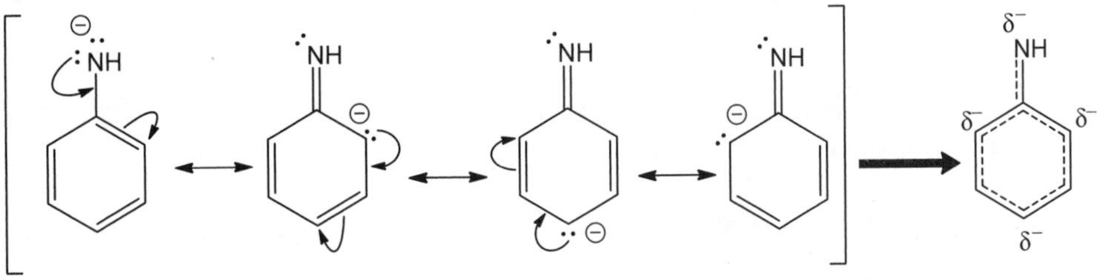

Chapter Problems

Problem 6.2

Think

What does it mean to be an acid? A base? What are the important electrons to keep track of during the course of the reaction? What bonds are broken? What bonds are formed? What does a curved arrow represent? How many curved arrows are needed?

Solve

In the reverse reaction, hydroxide acts as the base (proton acceptor) and ammonium acts as the acid (proton donor). The H–OH bond forms and the H–$^{+}NH_3$ bond breaks. A curved arrow represents electron movement to show the breaking of the H–$^{+}NH_3$ bond and formation of the H–OH bond. The reaction mechanism is as follows:

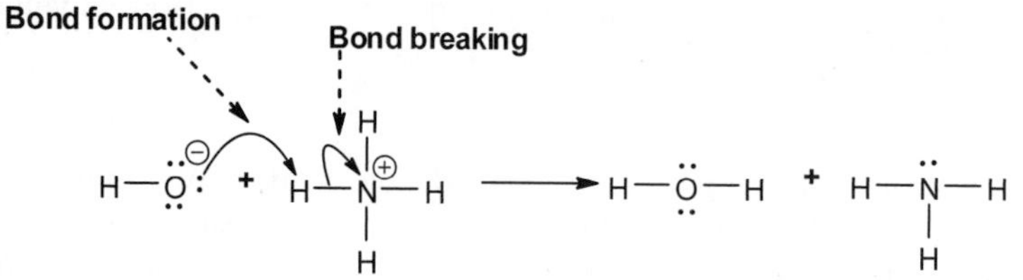

Bond formation

Bond breaking

Problem 6.3

Think

What does it mean to be an acid? A base? What are the important electrons to keep track of during the course of the reaction? What bonds are broken? What bonds are formed? What does a curved arrow represent? How many curved arrows are needed? How can you derive the products by moving the electrons according to the curved arrows?

Solve

When NH_3 acts as an acid, it is a proton donor, and when H_2O acts as a base, it is a proton acceptor. The products of the reaction are NH_2^- and H_3O^+. The H_2N–H bond breaks and the H–O^+H_2 bond forms.

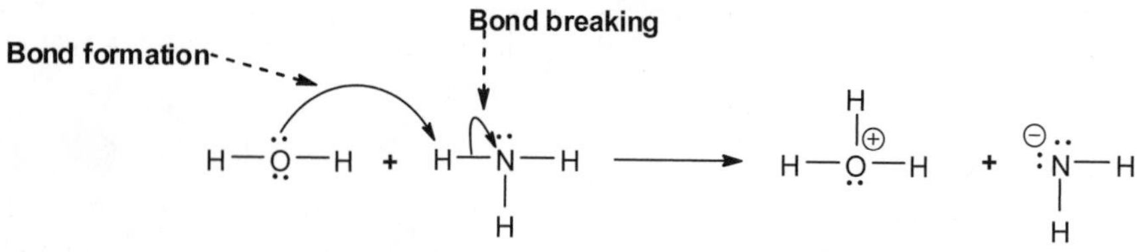

Bond formation

Bond breaking

Problem 6.5

Think

Is the stronger acid the one with the more positive pK_a or the less positive pK_a? What is the difference between pK_a values, and how can that difference be used to calculate the difference in their acid strengths?

Solve

The compound with the less positive pK_a value, phenol (pK_a = 10.00), is the stronger acid. The difference in pK_a values is 10.26 − 10.00 = 0.26, which corresponds to a difference in acid strength of $10^{0.26}$ = 1.8. Thus, phenol is 1.8 times stronger as an acid than 4-methylphenol.

Phenol
pK_a = 10.00

4-Methylphenol
pK_a = 10.26

Problem 6.6

Think

What is the pK_a of water? What is the K_a of water? What is the expression for K_a? How does the expression for K_a relate to the expression for K_w? What is the concentration of water? What is the concentration of $[H_3O^+]$ and $[HO^-]$?

Solve

The expression for K_a is: $K_a = [H_3O^+(aq)][HO^-(aq)]/[H_2O]$.
The expression for K_w is: $K_w = [H_3O^+(aq)]_{eq}[HO^-(aq)]_{eq}$.
Therefore, $K_w = K_a\,[H_2O] = (2.00 \times 10^{-16})(55.5) = 1 \times 10^{-14}$

Problem 6.7

Think

What is the general expression for an equilibrium equation? Do you include water in this expression? Use Equation 6-7 for the relationship between K_a and K_{eq}. Use Equation 6-8 for the relationship between pK_a and K_a.

Solve

(a) For the following proton transfer reaction, $K_{eq} = 1$. The products and the reactants are the same.

$$H_3O^+ + H_2O \rightleftharpoons H_2O + H_3O^+ \qquad K_{eq} = \frac{[H_2O][H_3O^+]}{[H_2O][H_3O^+]} = 1$$

(b) Use the value of $K_{eq} = 1$ and plug this value into Equation 6-7 to obtain the value for K_a.

$$K_a = K_{eq}[H_2O]_{eq} = K_{eq}\left(55.5\frac{\text{mol}}{\text{L}}\right) = 1 \times 55.5\frac{\text{mol}}{\text{L}} = 55.5\frac{\text{mol}}{\text{L}}$$

$$pK_a = -\log(K_a) = -\log(55.5) = -1.74$$

This matches the value of −1.7 in Table 6-1.

Problem 6.8

Think

What is the relationship between the strength of an acid and the strength of its conjugate base? What is the conjugate acid of Cl^-? Of $C_6H_5O^-$? Consulting Table 6-1, what is the pK_a of each of those acids? Which acid is stronger?

Solve

The strength of a base decreases as the strength of its conjugate acid increases. Thus, the stronger the acid, the weaker the conjugate base, and vice versa. Phenoxide, $C_6H_5O^-$, is a stronger base than chloride, Cl^-, because its conjugate acid, phenol (C_6H_5OH; pK_a = 10.0), is weaker than the conjugate acid of Cl^-, which is HCl (pK_a = −7). Because the difference in pK_a is 17, phenoxide is 10^{17} times stronger as a base compared to chloride.

Problem 6.10

Think

What are the products of the reaction? What acid is present on each side of the reaction? Which one is stronger? What is the difference in their pK_a values?

Solve

The stronger acid (lower pK_a) is on the reactant side, as shown below, so the product side is favored. The extent by which the product side is favored is $10^{(38-20)} = 10^{18}$. The equilibrium arrow ⇀ is drawn to show that the reaction favors the products and the equilibrium lies to the right.

$$(CH_3)_2N^- + CH_3C(O)CH_3\ (pK_a = 20) \rightleftharpoons (CH_3)_2NH\ (pK_a = 38) + {}^-CH_2C(O)CH_3$$

Problem 6.11

Think

You will need to use Equation 6-8 to substitute pK_a. You will also need to use the following two logarithmic properties: $10^{\log x} = x$ and $\log x + \log y = \log xy$.

Solve

For an acid–base equilibrium involving two acids, HA_1 and HA_2, we have the general equilibrium and equilibrium constant expression:

$$HA_1 + A_2^- \rightleftharpoons A_1^- + HA_2 \text{ and } K_{eq} = ([A_1^-][HA_2])/([HA_1][A_2^-])$$

We also have expressions for K_a of each acid:

$$K_a(HA_1) = ([H_3O^+][A_1^-])/[HA_1] \text{ and } K_a(HA_2) = ([H_3O^+][A_2^-])/[HA_2]$$

We divide the first expression by the second:

$$K_a(HA_1)/K_a(HA_2) = \{([H_3O^+][A_1^-])([HA_2])\}/\{([H_3O^+][A_2^-])([HA_1])\}$$

$$= ([A_1^-][HA_2])/([HA_1][A_2^-]) \qquad = K_{eq}$$

We then take the −log of both sides:

$$-\log\{K_a(HA_1)/K_a(HA_2))\} = -\log K_{eq}$$

Applying a property of log functions:

$$\{-\log K_a(HA_1)\} - \{-\log K_a(HA_2)\} = -\log K_{eq}$$

$$pK_a(HA_1) - pK_a(HA_2) = -\log K_{eq}$$

$$pK_a(HA_2) - pK_a(HA_1) = \log K_{eq}$$

We next apply exponentiation to both sides of the equation, and realize that $10^{\log(x)} = x$

$$\mathbf{10^{\{pKa(HA2) - pKa(HA1)\}} = K_{eq}}$$ This is the expression in Equation 6-11.

Problem 6.12

Think

For the solvent not to react with the solute, should the solvent be a weaker or stronger acid than the carbanion's conjugate acid, HC≡CH? What is the pK_a of HC≡CH? What is the pK_a of each solvent given?

Solve

The solvent must be a weaker acid than the conjugate acid of HC≡C:$^-$ so the solvent will not be deprotonated. The pK_a value, therefore, must be more positive (weaker acid) than that of HC≡CH, which is ~25. The only solvents that qualify are **(d)** and **(e)**, whose pK_a values are 35 and 45, respectively. The pK_a values of **(a)**, **(b)**, and **(c)** are 15.7, 16, and 17, respectively.

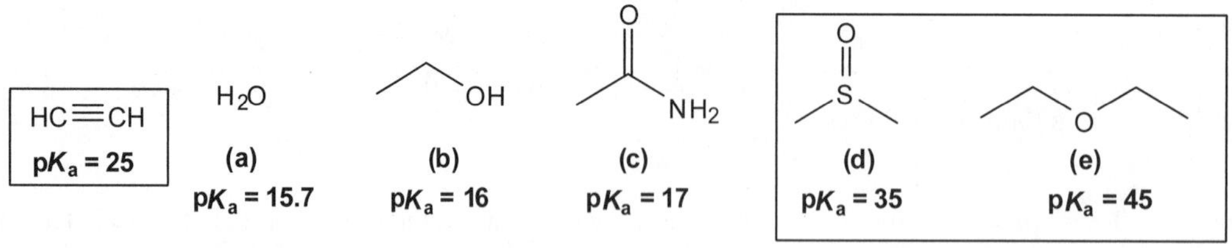

Problem 6.14

Think

On what functional group does each acidic H atom appear? What molecule(s) in Table 6-1 have the same functional group? Are there any nearby electronegative atoms or adjacent double bonds?

Solve

(a) C–H adjacent to C=O that is part of a ketone; pK_a ≈ 20.
(b) C–H adjacent to C=O that is part of an aldehyde; pK_a ≈20.
(c) C–H adjacent to O on ether; pK_a ≈45.
(d) H on sp^3-hybridized O^+; pK_a ≈ −1.7.
(e) H on O that is part of alcohol; pK_a ≈16.

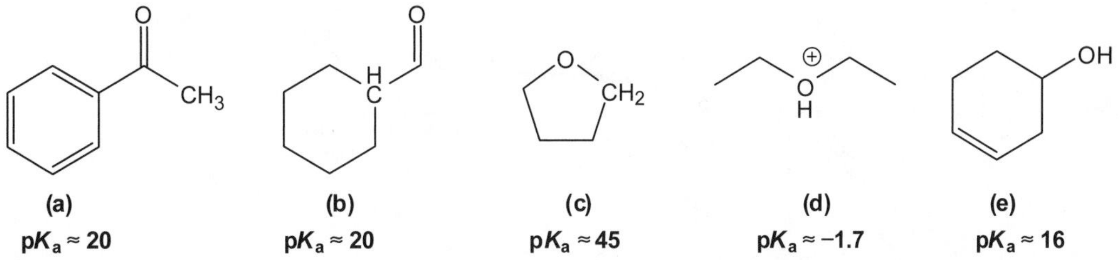

(a) pK_a ≈ 20 (b) pK_a ≈ 20 (c) pK_a ≈ 45 (d) pK_a ≈ −1.7 (e) pK_a ≈ 16

Problem 6.15

Think

What are the products of each reaction? Is there better charge stabilization on the reactant side or the product side? What is the pK_a of each acid? What is the relationship between acid strength and free energy?

Solve

The reactions and corresponding pK_a values are as follows:

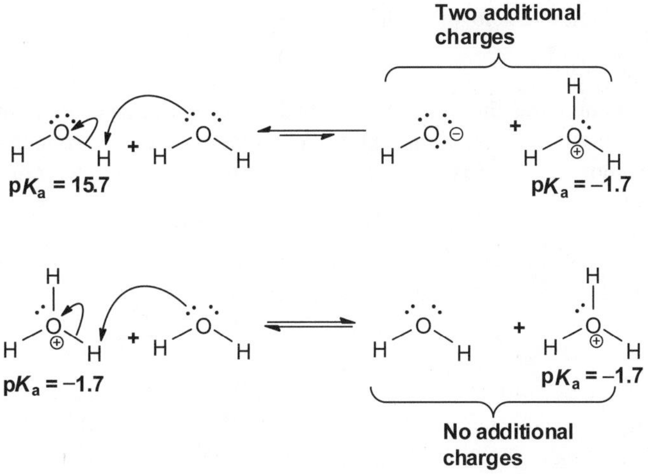

Two additional charges

pK_a = 15.7 pK_a = −1.7

pK_a = −1.7 pK_a = −1.7

No additional charges

Creating a figure similar to Figure 6-6, we obtain the following. Notice that with water as an acid, two additional charges are created, so the products are higher in energy than the reactants. With H_3O^+ as the acid, no additional charges are created, so the reactants and products have the same energy and are the same species. The latter is more energetically favorable, therefore, making H_3O^+ a stronger acid, consistent with the much lower pK_a of H_3O^+ than of H_2O. See the energy diagram on the next page.

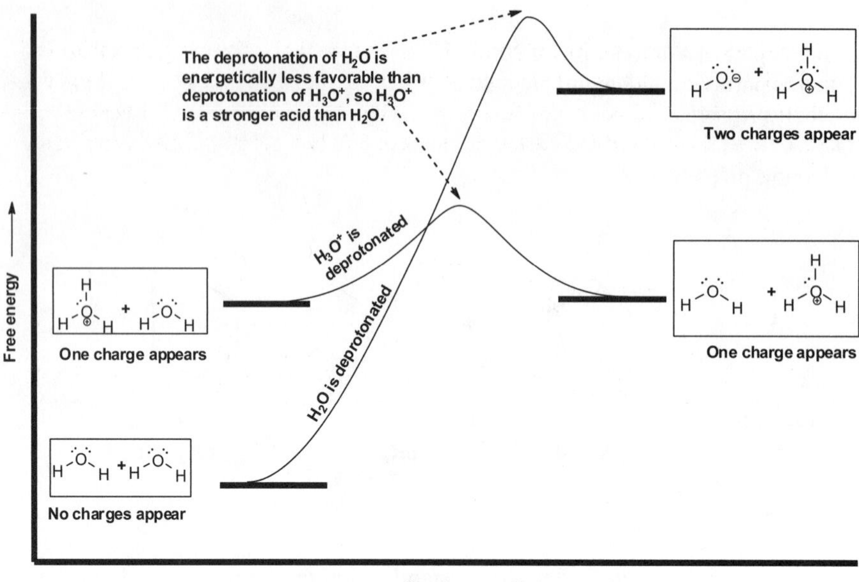

Problem 6.17

Think

What are the products of the reaction of each acid with water? What are the relative stabilities of the reactants? Which is more stable, Br^- or I^-? Based on their relative stabilities, which anion's conjugate acid is deprotonated more favorably? How does that correspond to the relative acid strength?

Solve

HI is the stronger acid. The energy diagram that compares the reactions in which they behave as acids is shown below. I^- is lower in energy than Br^- because I is the larger atom. The larger ion has the less concentrated charge and is more stable. As we can see, deprotonation of HI is the more energetically favorable reaction.

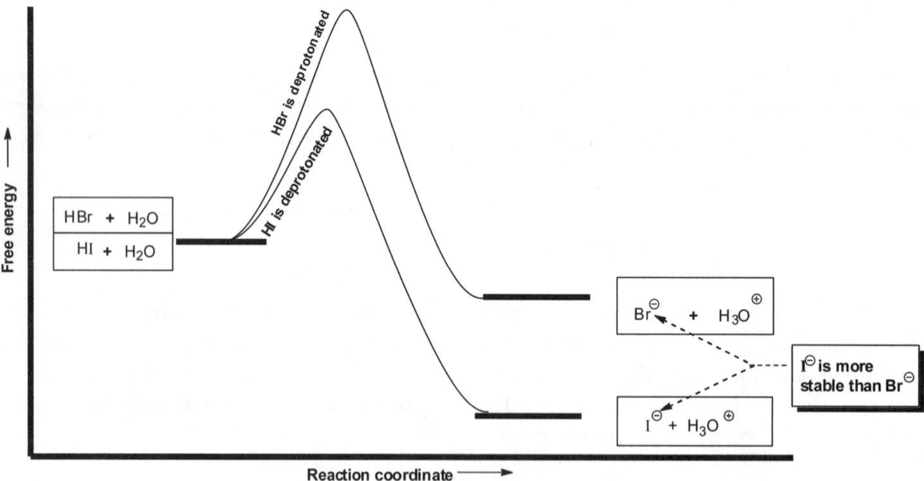

Problem 6.18

Think

What are the products of the reaction of each acid with water? What are the relative stabilities of the reactants? Which is more stable, H_2P^- or CH_3^-? Based on their relative stabilities, which anion's conjugate acid is deprotonated more favorably? How does that correspond to the relative acid strength?

Solve
The energy diagram comparing the deprotonation of CH_4 and PH_3 is below. H_2P^- is more stable than H_3C^-. We know that H_2P^- is more stable than H_2N^- because P, being below N in the periodic table, is larger and can accommodate the negative charge better. And we know that H_2N^- is more stable than H_3C^- because N is more electronegative than C. Because deprotonation of PH_3 is more energetically favorable, PH_3 is the stronger acid.

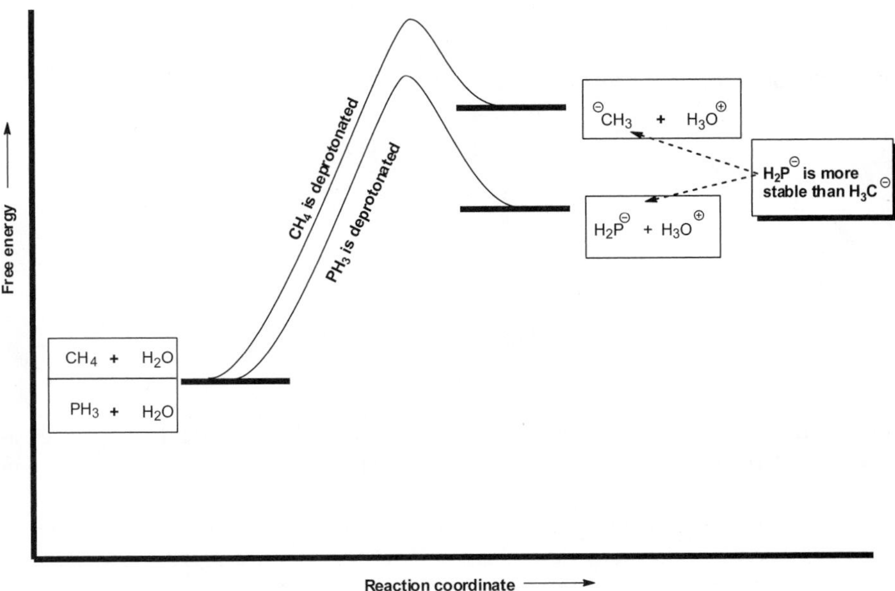

Problem 6.20

Think
What are the complete reactions? What are the relative stabilities of the two sets of reactants? Products? Do the species have the same charges? Are the charges on the same atoms? How does the hybridization of each N atom affect the stability of the charge? What is the effect of hybridization on acid strength?

Solve
$HC{\equiv}NH^+$ is the stronger acid. In the figure below, $HC{\equiv}NH^+$ is less stable than $H_3C–NH_3^+$ because the positive charge cannot be accommodated as easily on an *sp*-hybridized N atom. The *sp*-hybridized N atom has a greater effective electronegativity than the sp^3-hybridized N atom. Therefore, deprotonating $HC{\equiv}NH^+$ is more energetically favorable than deprotonating $H_3C–NH_3^+$.

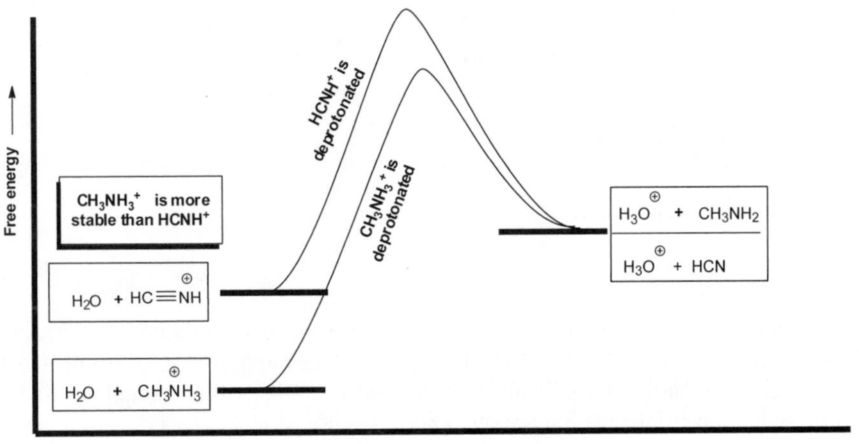

Problem 6.22

Think

Write reactions that depict each acid being deprotonated by a base. Do you expect a significant difference in energy between the reactants of one and the reactants of the other? Between the products of one and the products of the other? Are the charge-bearing atoms different in these two acids? In the conjugate bases, do they have different effective electronegativities? Do the ions differ in resonance delocalization of the charge?

Solve

Being uncharged acids, the reactions have the same forms as Equations 6-21a and 6-21b (p. 314).

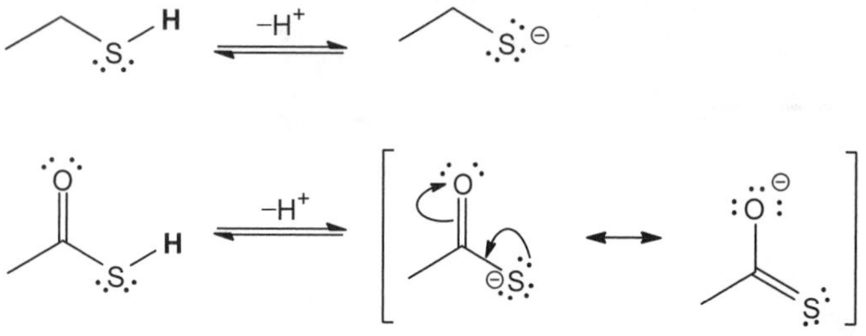

Therefore, according to Figure 6-7 (p. 315), the stronger acid is the one with the more stable conjugate base. In this case, the stronger acid is $CH_3C(O)SH$, because the negative charge on sulfur in the product anion is resonance stabilized.

Problem 6.24

Think

Being uncharged acids, the reactions have the same forms as Equations 6-21a and 6-21b (p. 314). Therefore, according to Figure 6-7 (p. 315), the stronger acid is the one with the more stable conjugate base. Draw the conjugate base for each acid. Which conjugate base experiences more resonance stabilization of the charge that develops?

Solve

HNO_3 is a stronger acid than CH_3CO_2H. Both product anions are resonance stabilized, with the negative charge being shared over different O atoms. But NO_3^- has one additional resonance structure than $CH_3CO_2^-$, making NO_3^- more stable. So deprotonating HNO_3 is more energetically favorable than deprotonating CH_3CO_2H.

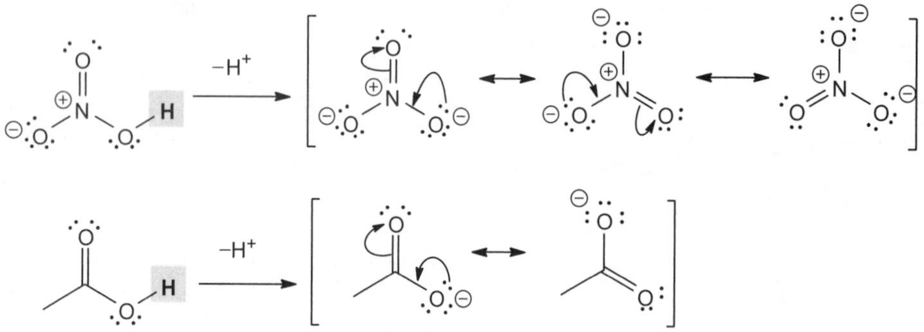

Problem 6.25

Think

Draw the complete reactions. Should the acidities of these acids be governed by the relative stabilities of the acids themselves or of their respective conjugate bases? Do the same charges appear? Are they on the same types of atoms? Is there a difference in charge delocalization by resonance? What is the inductive effect? Are there electronegative atoms present in one molecule but not the other?

Solve

Being charged acids, the complete reactions are analogous to those in Equations 6-22a and 6-22b (p. 316), so their relative acidities are governed by the relative stabilities of the acids themselves, as shown in Figure 6-8 (p. 316). In this case, the stronger acid is **B** because inductive effects make **B** less stable than **A**. The Cl atom is electron withdrawing compared to H and removes negative charge from the positively charged O atom. This intensifies the positive charge on the O atom and destabilizes the species.

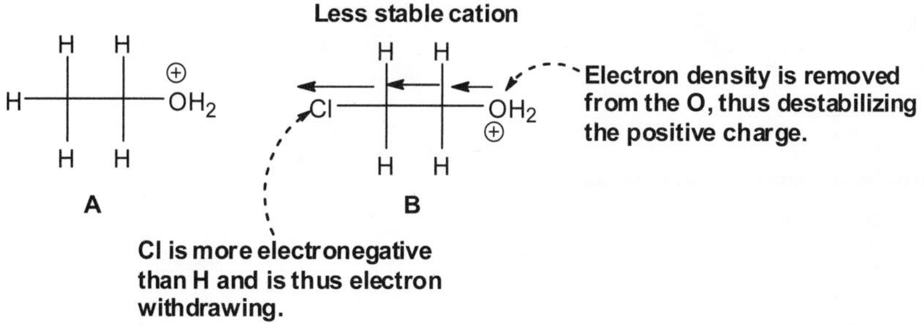

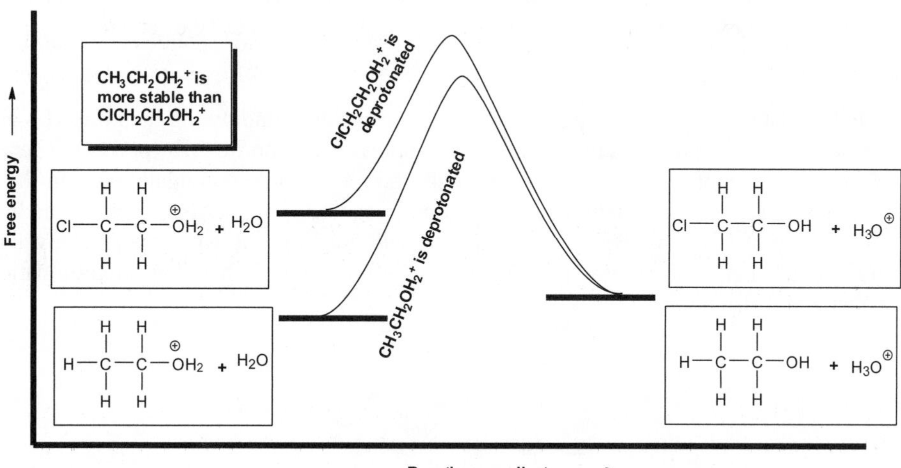

Problem 6.26

Think

Write reactions that depict each acid being deprotonated by a base. Do you expect a significant difference in energy between the reactants of one and the reactants of the other? Between the products of one and the products of the other? Are the charge bearing atoms different in these two acids? Do the ions differ in resonance delocalization of the charge? Do they have different effective electronegativities?

Solve

The stronger acid is H_2S due to inductive effects. Being uncharged acids, the reactions have the same forms as Equations 6-21a and 6-21b (p. 314). Therefore, according to Figure 6-7 (p. 315), the stronger acid is the one with the more stable conjugate base. The CH_3CH_2 group in CH_3CH_2SH is electron donating compared to H and, therefore, destabilizes the nearby negative charge in the product anion. Therefore, deprotonating H_2S is more energetically favorable than deprotonating CH_3CH_2SH, as shown on the next page.

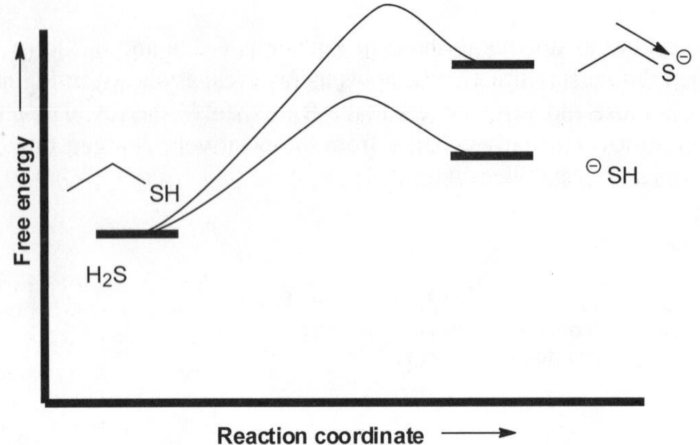

Problem 6.28

Think

Draw out the Lewis structure for each acid. Does the stability of the acid or the conjugate base dictate the pK_a? Do electron-donating or electron-withdrawing effects stabilize those species? Are NO_2 and NH_2 substituents electron donating or electron withdrawing? Which substituent invokes stronger inductive effects?

Solve

The stronger acid is $O_2NCH_2CH_2OH$ because the NO_2 group is more electron withdrawing than the NH_2 group. Being uncharged acids, the reactions have the same forms as Equations 6-21a and 6-21b (p. 314). Therefore, according to Figure 6-7 (p. 315), the stronger acid is the one with the more stable conjugate base. In the NO_2 group, the N atom is very highly electron deficient, bearing a formal positive charge. Also, there are additional electronegative O atoms in place of H atoms. Being more electron withdrawing, the NO_2 group removes more negative charge from O^-, which better stabilizes the species. Therefore, as shown below, deprotonation of $O_2NCH_2CH_2OH$ is more energetically favorable.

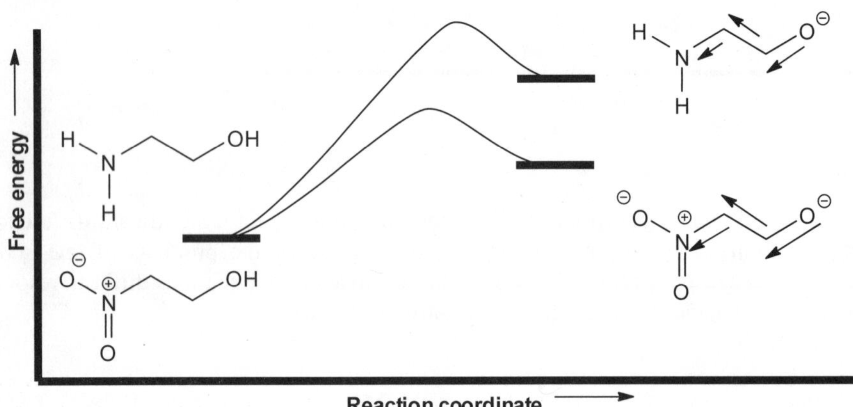

Problem 6.29

Think

Draw the conjugate base for each carboxylic acid. Does the stability of the acid or the conjugate base dictate the pK_a? Do electron-donating or electron-withdrawing effects stabilize those species? Is the C=O substituent electron donating or electron withdrawing? How does the location of the C=O affect the stability of the conjugate base?

Solve

In both molecules, the most acidic functional group is the carboxylic acid. Being uncharged acids, the reactions have the same forms as Equations 6-21a and 6-21b (p. 314). Therefore, according to Figure 6-7 (p. 315), the stronger acid is the one with the more stable conjugate base. The nearby O atom that is part of the C=O bond is electron withdrawing, helps to stabilize the resulting negative charge in the product anion, and increases the acid strength. The stronger acid is **B**, because the C=O group at the top of the ring is closer to the carboxyl group.

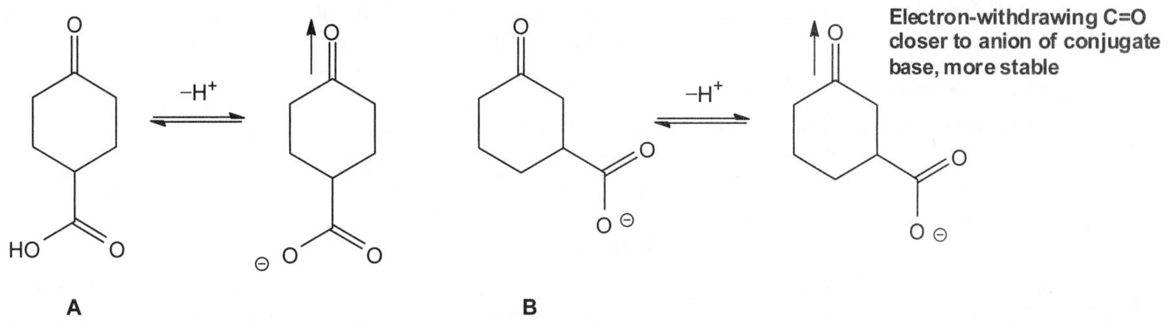

Electron-withdrawing C=O closer to anion of conjugate base, more stable

A B

Problem 6.30

Think

Which electrons/bonds move between the two resonance structures? How can you show this movement using curved arrows? Do all nonhydrogen atoms have an octet? Do additional atoms with nonzero formal charges lead to a greater or lesser contribution of a resonance structure?

Solve

One lone pair on the Cl atom becomes a π bond and the π bond between the C≡C becomes a lone pair on the C atom. The contributor on the right is less stable because it has two formal charges, whereas that on the left has no formal charges. Therefore the structure on the left has the greater contribution.

$$\left[H_3C{-}C{\equiv}C{-}\ddot{C}l{:} \longleftrightarrow H_3C{-}\overset{\ominus}{C}{=}C{=}\overset{\oplus}{C}l \right]$$

Problem 6.31

Think

Do all nonhydrogen atoms have an octet? Does one structure have fewer atoms with nonzero formal charges? Are there more covalent bonds in one structure? What is the stability of the positive charge on the O atom versus the N atom?

Solve

The more stable contributor is **B**. The positive charge is on the N atom in one resonance structure and on the O atom in the other. The structure is more stable with the positive charge on N because N is the less electronegative atom. All other stability factors are the same (octet, formal charges, and number of covalent bonds).

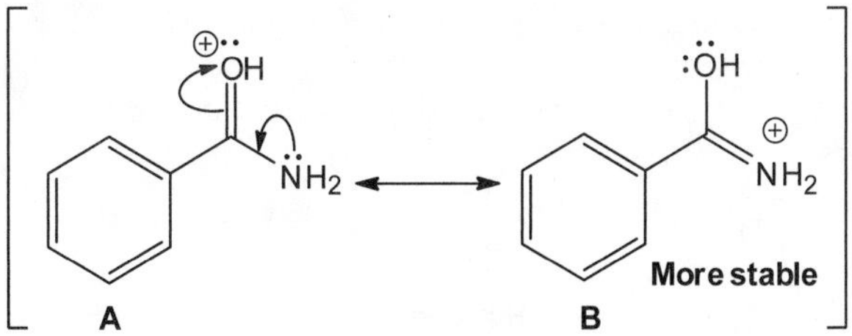

More stable

A B

Problem 6.32

Think

Do all atoms have an octet? Does one structure have fewer atoms with nonzero formal charges? Are there more covalent bonds in one structure? What is the stability of the positive charge on the O atom versus the C atom?

Solve

(a) It is counterintuitive because the resonance structure is more stable with the positive charge on the *less* electronegative atom. Thus, you would normally think that C^+ is more stable than an O^+.

(b) The contributor on the right is more stable because every atom in the structure has a complete octet. In the contributor on the left, the C atom with the positive charge is deficient of an octet.

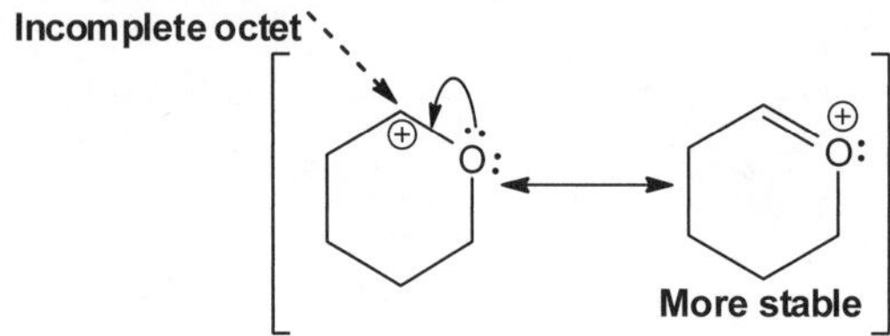

Problem 6.33

Think

Do all atoms have an octet? Does one structure have fewer atoms with nonzero formal charges? Are there more covalent bonds in one structure? Is CF_3 an electron-withdrawing or electron-donating group? What is the stability of the positive charge on the C atom next to the CF_3 in molecule **A**, or farther away from the CF_3, as in molecule **B**?

Solve

CF_3 is an electron-withdrawing group. Electron-withdrawing groups destabilize a positive charge. The carbocation C^+ is more destabilized when the CF_3 is in closer proximity. Therefore, **B** is the more important resonance contributor.

CF_3 CF_3

A **B**
More stable

Problem 6.34

Think

What is the structure of alanine in its fully protonated form (i.e., at the most acidic pH)? What are the pK_a values of alanine? What is the relationship between the pH and the pK_a values? Which proton will be removed first, and at what pH? Second?

Solve

Alanine's structures at various pH values are shown below. The pK_a values from Table 6-2 are 2.35 and 9.87, so when the solution pH is significantly above those values, the most acidic proton present will almost entirely be dissociated.

pK_{a2} = 9.87 pK_{a1} = 2.35

pH = 1 **pH = 4 and 8** **pH = 11**

Problem 6.35

Think

What are the structures of glutamic acid and cysteine in their fully protonated forms (i.e., at the most acidic pH)? What are the pK_a values of each? What is the relationship between the pH and the pK_a values? Which proton will be removed first, and at what pH? Second? Third?

Solve

The fully protonated form of glutamic acid is shown below on the left. Deprotonations for glutamic acid occur at the pH = pK_a values 2.10, 4.07, and 9.47. So the species below are the most abundant at the given pH values:

pK_{a3} = 9.47 pK_{a1} = 2.10 pK_{a2} = 4.07

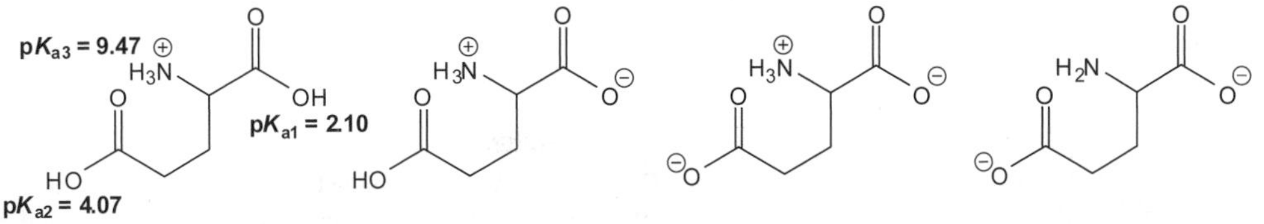

pH = 1 pH = 3 pH = 5 pH = 11

The fully protonated form of cysteine is shown below on the left. Deprotonations for cysteine occur at the pH = pK_a values 2.05, 8.00, and 10.25. So the species below are the most abundant at the given pH values:

pK_{a3} = 10.25 pK_{a1} = 2.05 pK_{a2} = 8.00

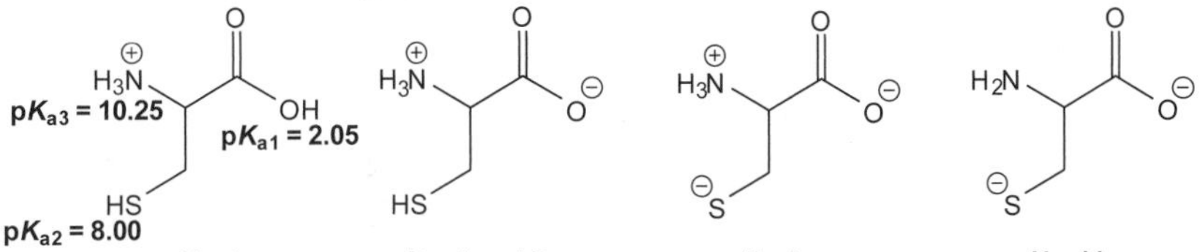

pH = 1 pH = 4 and 6 pH = 9 pH = 11

Problem 6.36

Think

What are the structures of arginine and histidine in their fully protonated forms (i.e., most acidic pH)? What are the pK_a values of each? What is the relationship between the pH and the pK_a values? Which proton will be removed first, and at what pH? Second? Third?

Solve

The fully protonated form of arginine is shown below on the left. Deprotonations for arginine occur at the pH = pK_a values 2.01, 9.04, and 12.48. So the species below are the most abundant at the given pH values:

pK_{a2} = 9.04 pK_{a3} = 12.48 pK_{a1} = 2.01

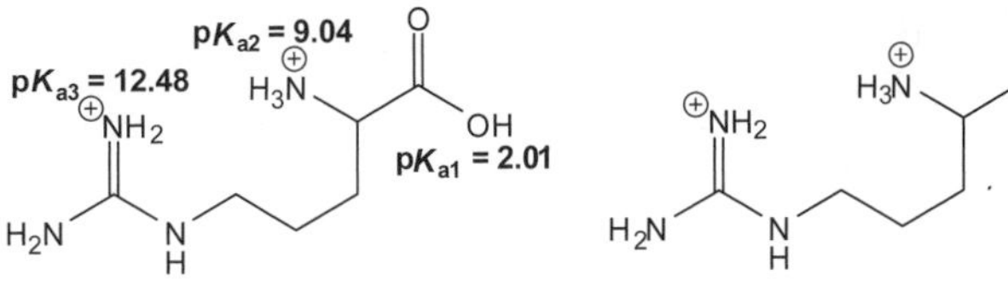

pH = 1 pH = 4 and 8

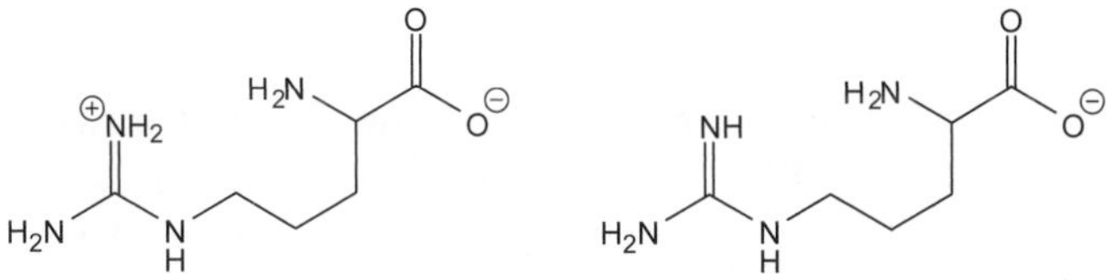

pH = 10 pH = 14

The fully protonated form of histidine is shown below on the left. Deprotonations for histidine occur at the pH = pK_a values 1.77, 6.10, and 9.18. Therefore, the species below are the most abundant at the given pH values:

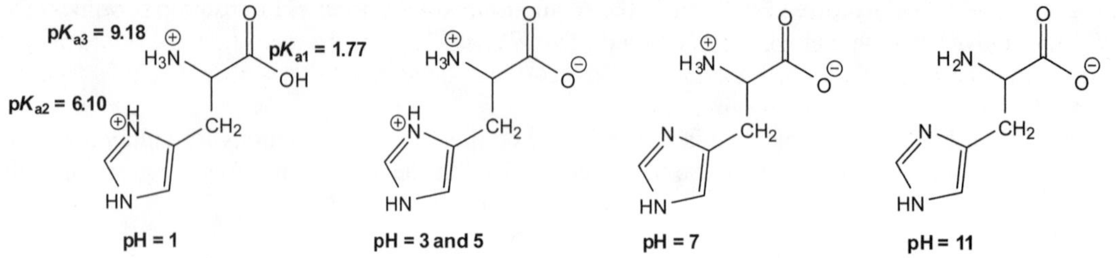

pK_{a3} = 9.18 pK_{a1} = 1.77 pK_{a2} = 6.10

pH = 1 pH = 3 and 5 pH = 7 pH = 11

Problem 6.37

Think

What are the pK_a values for glycine? What is the isoelectric point (pI) of glycine? Is the pH greater than, less than, or equal to the pI? Is this species positive, negative, or uncharged on average? Is the cathode positive or negative? What is the charge at the anode?

Solve

The pI of glycine is 6.07, so at a pH of 7, glycine will be deprotonated, on average, and the average charge on glycine will be negative, so glycine will migrate toward the positively charged anode.

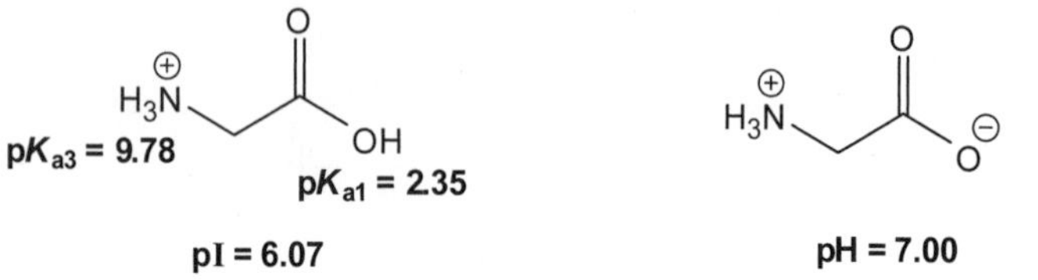

pK_{a3} = 9.78 pK_{a1} = 2.35

pI = 6.07 pH = 7.00

Problem 6.38

Think

What are the pK_a values for alanine? How do you calculate the pI of alanine from those values? Is the pH greater than, less than, or equal to the pI? Is this species positive, negative, or uncharged on average? Is the cathode positive or negative? What is the charge at the anode?

Solve

The pI of alanine is the average of its two pK_a values: (2.35 + 9.87)/2 = 6.11. At a pH of 4, alanine will be protonated, on average, and the average charge on alanine will be positive, so alanine will migrate toward the negatively charged cathode.

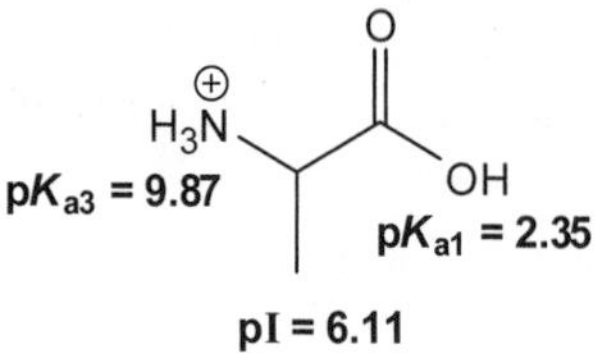

pK_{a3} = 9.87 pK_{a1} = 2.35

pI = 6.11

Problem 6.39

Think

What are the pK_a values for glutamic acid and tyrosine? For each, which two pK_a values involve the zwitterion? How do you calculate pI from those pK_a values? Is the pH greater than, less than, or equal to the pI? Is this species positive, negative, or uncharged on average? Is the cathode positive or negative? What is the charge at the anode?

Solve

The species that exist in solution for glutamic acid are shown below (Problem 6.35). The zwitterion is the second species from the left. The pK_a of the proton transfer equilibrium that leads to the formation of the zwitterion is 2.10, and the pK_a of the equilibrium that involves the deprotonation of the zwitterion is 4.07. So, the pI = (2.10 + 4.07)/2 = 3.09. This acidic pI is consistent with the fact that the side chain has an acidic group. At a pH of 7, glutamic acid is deprotonated, on average, and the average charge on glutamic acid is negative, so glutamic acid will migrate toward the positive anode.

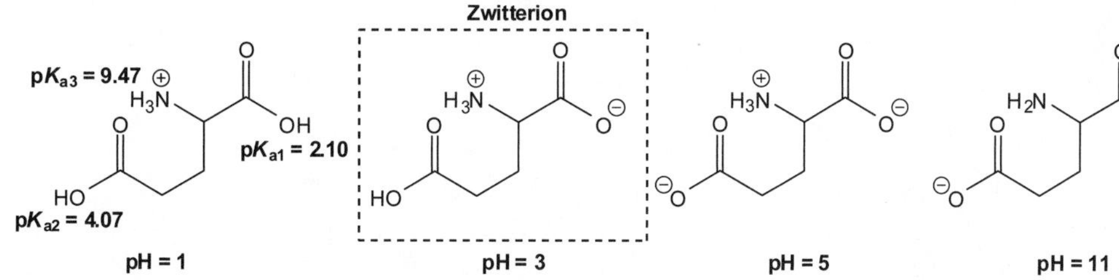

Zwitterion

pK_{a3} = 9.47, pK_{a1} = 2.10, pK_{a2} = 4.07

pH = 1 pH = 3 pH = 5 pH = 11

The species that exist in solution and the pK_a values for tyrosine are shown below:

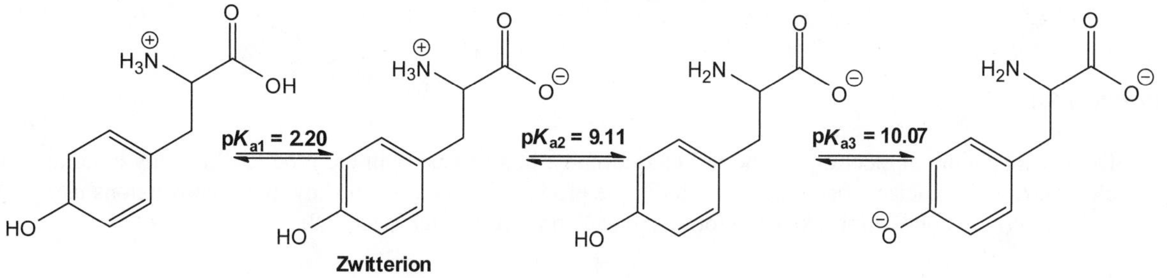

pK_{a1} = 2.20 pK_{a2} = 9.11 pK_{a3} = 10.07

Zwitterion

The zwitterion is the second molecule from the left. The pK_a values of the equilibria involving the zwitterion are 2.20 and 9.11, so the pI = (2.20 + 9.11)/2 = 5.66. This value is consistent with the fact that the side chain is neutral. At a pH of 7, tyrosine is deprotonated, on average, and the average charge on tyrosine is negative, so tyrosine will migrate toward the positive anode.

Problem 6.40

Think

Is the pH greater than, less than, or equal to the pI? Is this species positive, negative, or uncharged on average? Is the cathode positive or negative? What is the charge at the anode? At what pH is lysine uncharged?

Solve

At a pH of 12, the pH is above the pI, so lysine will be deprotonated, on average, and the average charge on the species will be negative, so the species will migrate toward the positively charged anode. At a pH values of 1 or 7, the opposite is true. The species will not migrate at a pH of 9.74, the value of its pI.

Problem 6.41

Think

What do the curved arrows show? Which bonds are broken and formed in each proton transfer reaction?

Solve

Curved arrows show movement of two electrons. In each reaction, the base accepts a proton to form the conjugate acid and the acid donates a proton to form the conjugate base. Products for reactions **(a)–(d)** are shown on the next page.

(a)

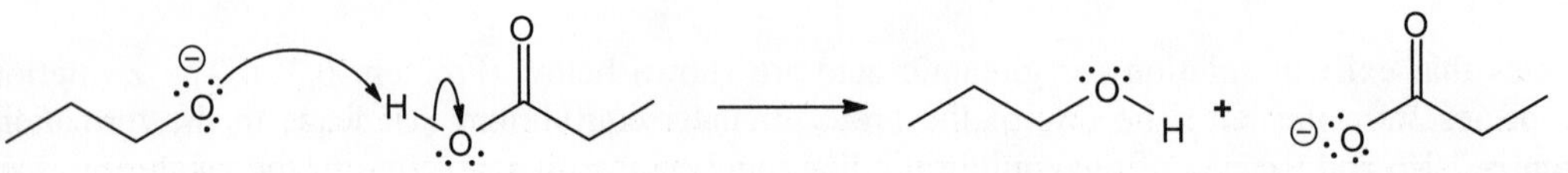

(b)

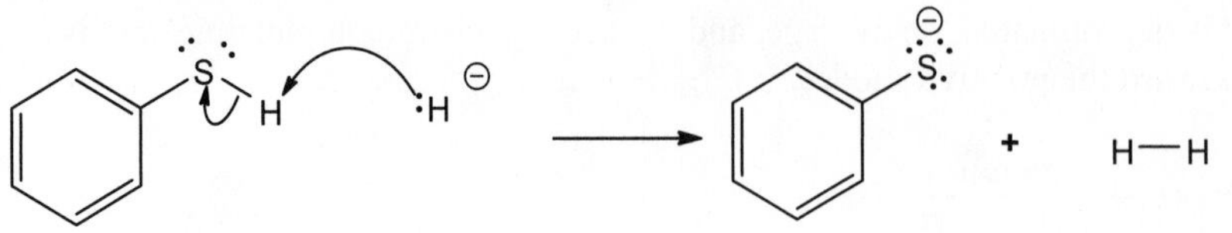

(c)

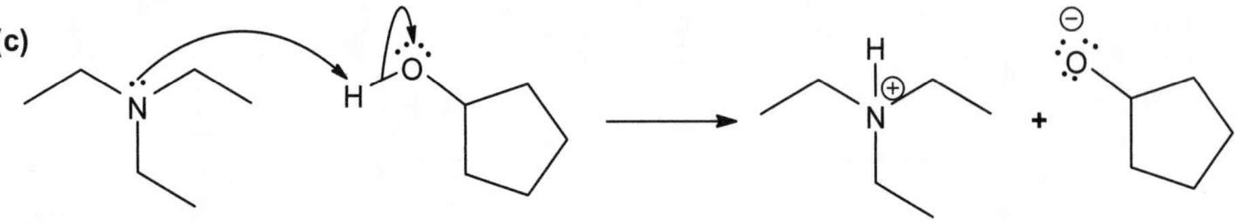

(d)

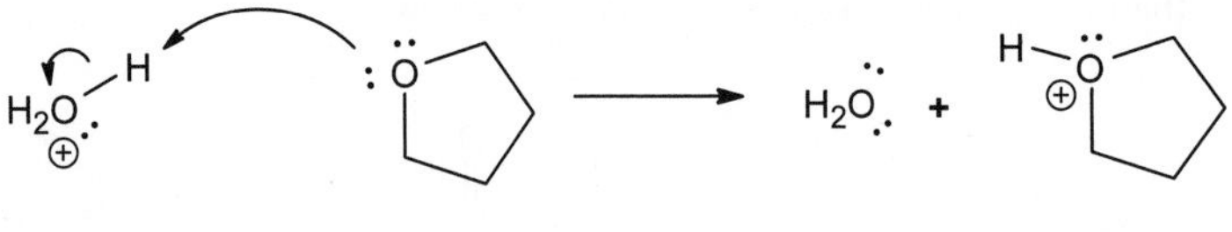

Problem 6.42

Think

Do any lone pairs of electrons in the reactants become bonding pairs in the products? Do any bonding pairs of electrons in the reactants become lone pairs in the products? How do you show these conversions using curved arrows? Which bonds are broken and formed in each proton transfer reaction?

Solve

Curved arrows and missing nonbonding electrons are shown below. Each proton transfer reaction requires two arrows. One arrow shows the bond forming from the lone pair of the base to the H atom of the acid and the other arrow shows the bond breaking between the H atom and the acid, with the bonding pair of electrons ending up as a lone pair.

(a)

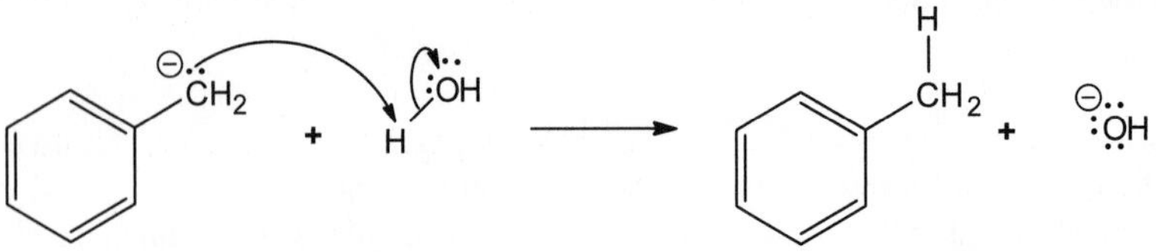

(b)

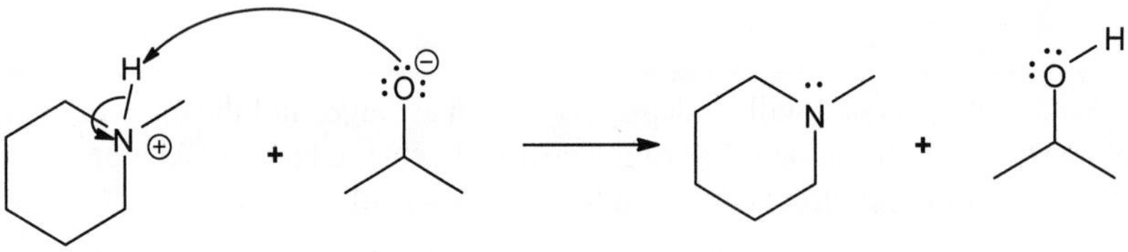

(c)

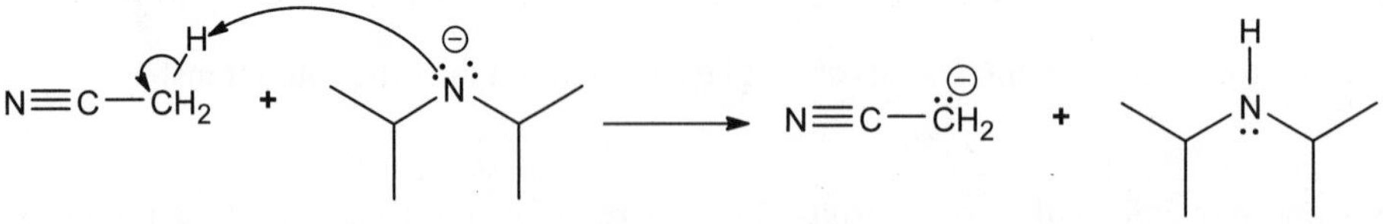

(d)

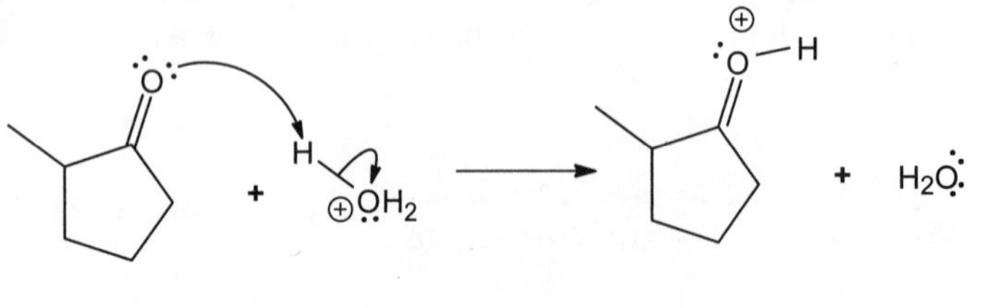

Problem 6.43

Think

To what functional group does the acidic proton belong? Use Table 6-1 or Appenix A to look up or estimate the pK_a based on similar functional groups. If there is more than one acidic proton, which one is the most acidic?

Solve

The most acidic protons are circled below, the functional groups to which they belong are provided, and their pK_a values are estimated.

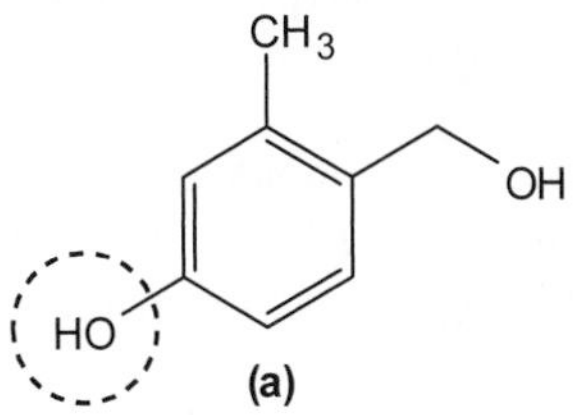

(a) Phenol pK_a ~10.0

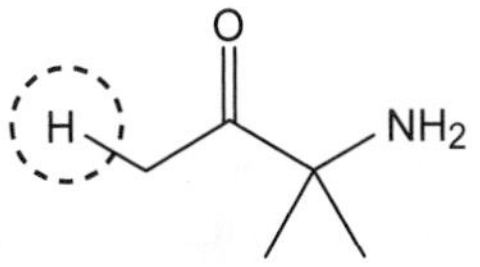

(b) Ketone pK_a ~20

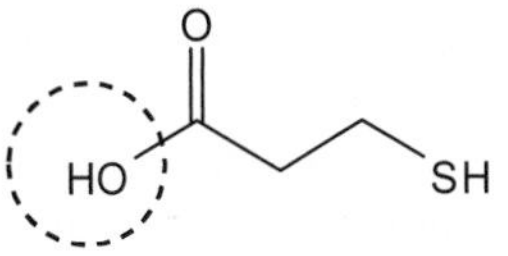

(c) Carboxylic Acid pK_a ~4.75

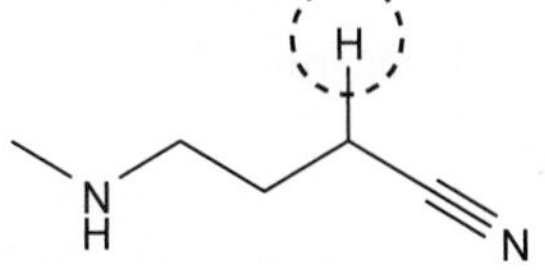

(d) Nitrile pK_a ~25

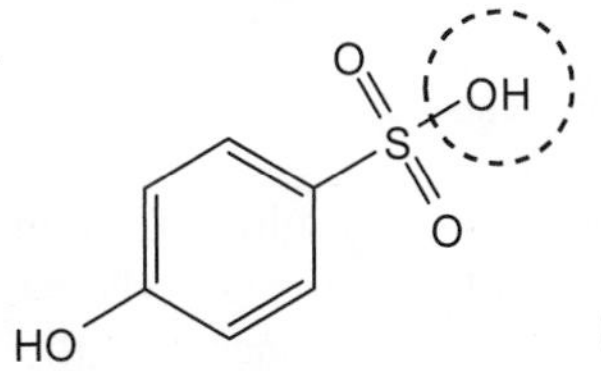

(e) Sulfonic acid pK_a ~ −2

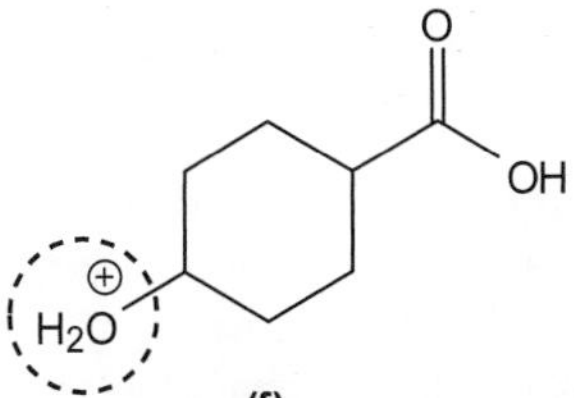

(f) Protonated alcohol pK_a ~ −1.7

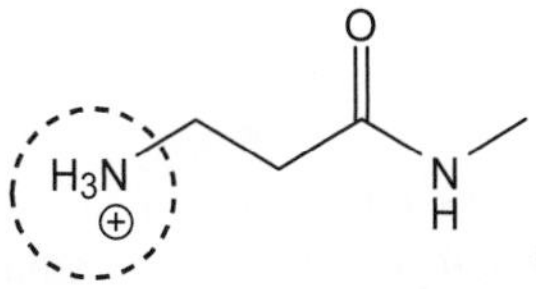

(g) Protonated amine pK_a ~9.4

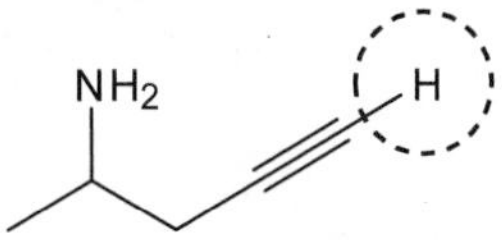

(h) Terminal alkyne pK_a ~25

Problem 6.44

Think

Which sites can accept a proton? What would the conjugate acid be? What is the relationship between the pK_a of the conjugate acid and base strength? How can you estimate the pK_a value of each conjugate acid?

Solve

To determine relative base strength, we can use pK_a values of their conjugate acids. The pK_a relationship is such that the stronger base is associated with the weaker conjugate acid.

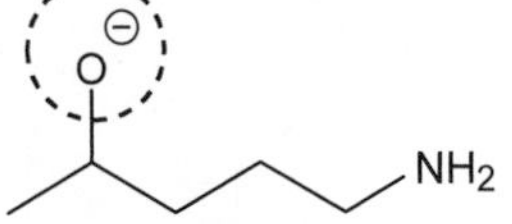

(a) ROH (pK_a ~16) is a weaker acid than RNH_3^+ (pK_a ~10.6)

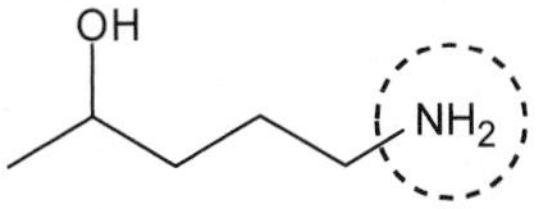

(b) RNH_3^+ (pK_a ~10.6) is a weaker acid than ROH_2^+ (pK_a ~ −1.7)

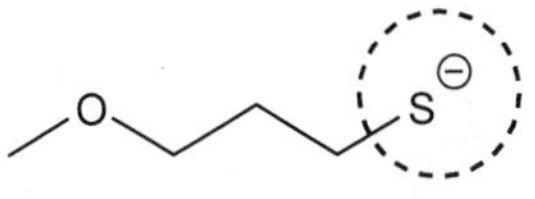

(c) RSH (pK_a ~7.2) is a weaker acid than ROH_2^+ (pK_a ~ −1.7)

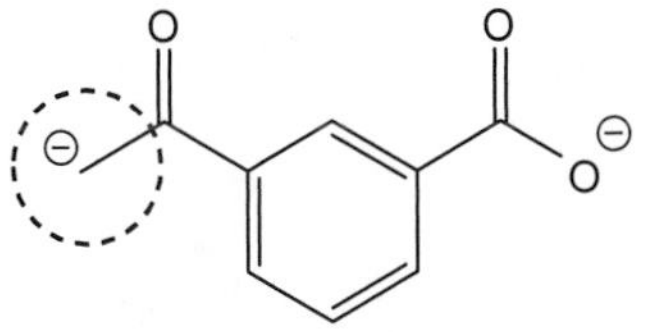

(d) A ketone (pK_a ~20) is a weaker acid than a carboxylic acid (pK_a ~4.75)

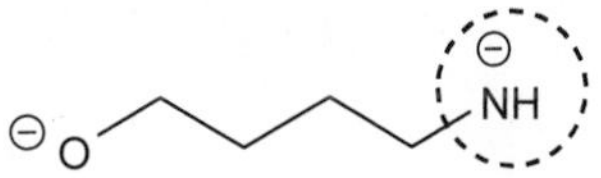

(e) RNH_2 (pK_a ~ 38) is a weaker acid than R-OH (pK_a ~16).

Problem 6.45

Think

Identify the most acidic proton in each species and draw the complete deprotonation reactions. For uncharged acids, is acidity governed primarily by the stability of the acid itself or of its conjugate base? Does one structure have fewer atoms with nonzero formal charges? What is the stability of the ion (inductive effect, resonance, electronegativity)?

Solve

Being uncharged acids, the reactions have the same forms as Equations 6-21a and 6-21b (p. 314). Therefore, according to Figure 6-7 (p. 315), the stronger acid is the one with the more stable conjugate base.

(a) The molecule on the right is more acidic. In both cases, the acidic H is on a CH_3 group adjacent to a C=O bond. In both cases, there is another resonance structure that allows the negative charge to be shared on the O of the C=O bond. But the highly electronegative F atoms nearby are inductively electron withdrawing and stabilize anions.

O, C, H_3C, CH_3, $-H^+$, O, C, H_3C, CH_2, O, C, F_3C, CH_3, $-H^+$, O, C, F_3C, CH_2

More stable conjugate base

(b) The molecule on the left is more acidic. In both cases, the acidic H is on an sp^3-hybridized C atom. In the molecule on the left, that C atom is adjacent to two C=C double bonds (part of the ring), whereas in the molecule on the right, it is adjacent to only one. As we learned in the chapter, each adjacent multiple bond increases the strength of the acid as a result of the resonance delocalization of the charge that develops upon deprotonation.

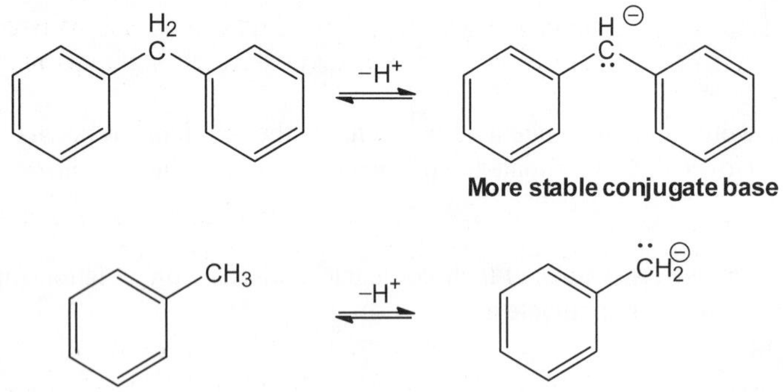

H_2, C, $-H^+$, H, C

More stable conjugate base

CH_3, $-H^+$, CH_2

Problem 6.46

Think

Draw the complete protonation reactions. Which base has more charge stability? Which acid has more charge stability? Consider the presence of a charge, the type of atom on which the charge appears, resonance, and inductive effects. What is the strength of the resulting conjugate acid and how does that relate to base strength?

Solve

	Stronger Base (boxed)	Reasoning
(a)	OH or O⊖	The charged base on the right is less stable than the one on the left due to the presence of the charge. The opposite is true of their conjugate acids. Therefore, it is more energetically favorable for the species on the right to accept a proton, making it a stronger base.
(b)	OH or ⊕OH2	After accepting a proton the base on the left will have an atom with a +1 charge, whereas the one on the right will have an atom with a +2 charge.
(c)	or	Both bases are negatively charged and become uncharged upon protonation, so the difference in base strength is determined primarily by the stability of the bases themselves. The first base has a negative charge on an sp^2-hybridized C atom and the second base has a negative charge on an *sp*-hybridized C atom. Decreased *s* character leads to increased energy, less stability, and thus a stronger base.
(d)	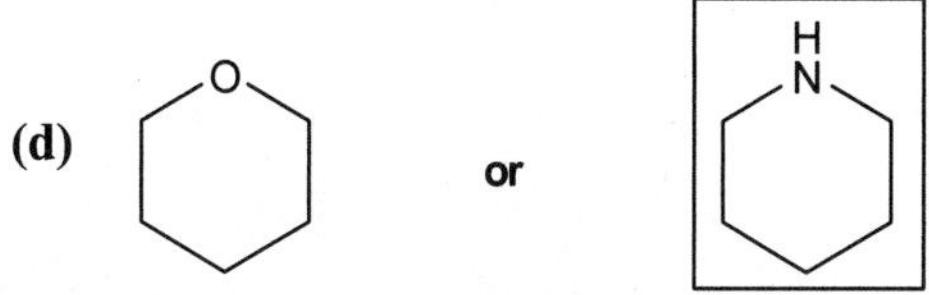	Of the two conjugate acids, R_2OH^+ is a stronger acid than R_2NH^+. The R_2OH^+ conjugate acid is less stable because the positive charge is on a more electronegative atom.
(e)	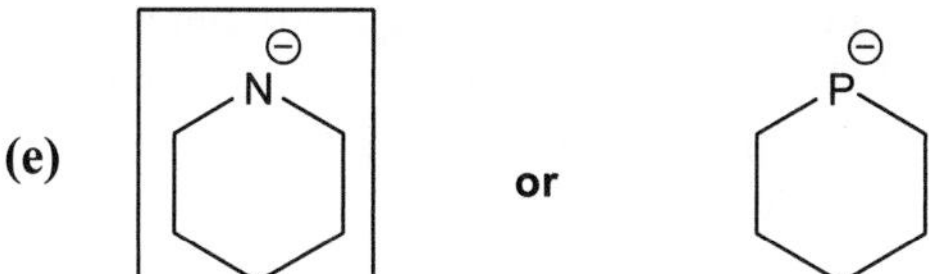	A negative charge is more stable on P than on N because P is lower in the periodic table, and thus a larger atom.
(f)	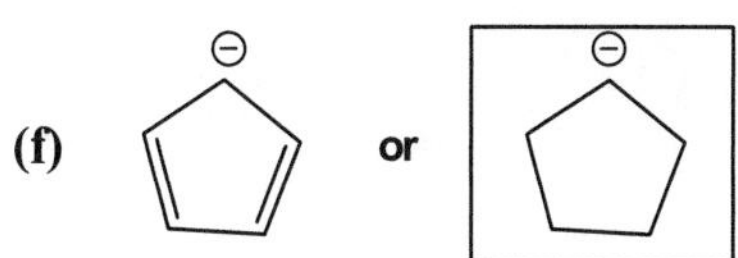	The negative charge on the first anion is stabilized by resonance, but is not on the second anion.
(g)	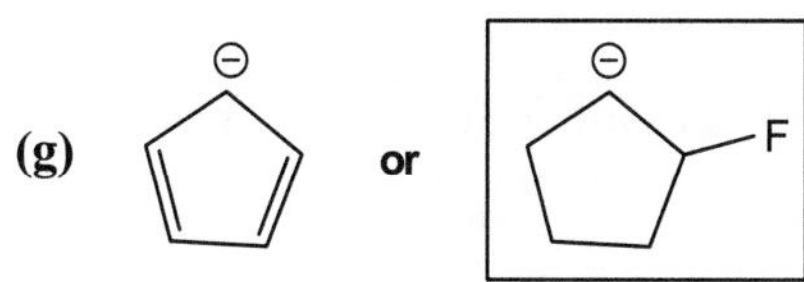	The first anion is stabilized by resonance and the second anion is stabilized by the inductive effect. F is electron withdrawing. The inductive effect is not as important as resonance stabilization.
(h)	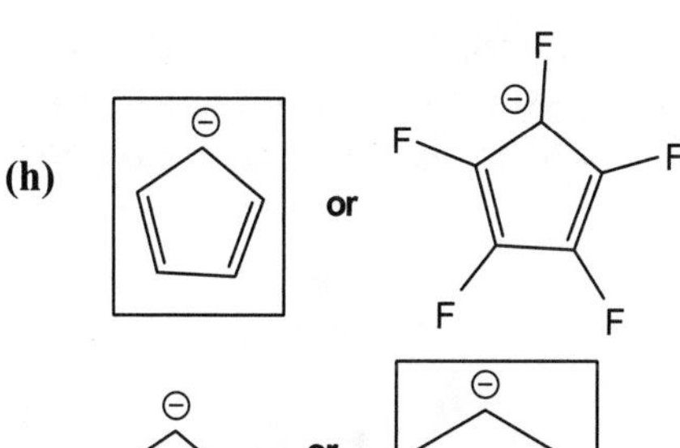	The F atoms are electron withdrawing, which stabilizes the anion.
(i)	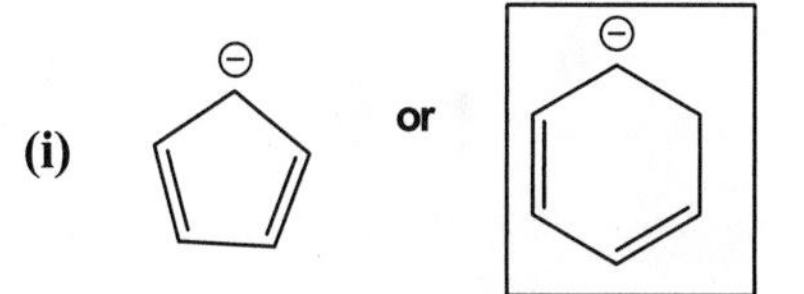	There are more resonance structures in the first anion to stabilize the negative charge.

Problem 6.47

Think

What is the conjugate base that results after each successive deprotonation? Do you think it is easier to remove a proton from an uncharged species or a negatively charged species?

Solve

The two deprotonations are shown below. In each reaction, a new negative charge is generated. But in the second deprotonation, generation of that additional charge introduces charge repulsion among the two -1 formal charges that doesn't exist in HSO_4^-. This additional destabilization in SO_4^{2-} makes the second deprotonation less energetically favorable than the first.

$HO-SO_2-OH$ $\xrightarrow{-H^+}$ $HO-SO_2-O^-$ $\xrightarrow{-H^+}$ $^-O-SO_2-O^-$

pK_a = −9

pK_a = 2
One negative charge

Two negative charges

Problem 6.48

Think

Which acid is stronger? How does the strength of the electron-withdrawing group affect the strength of the acid? Consider the stability of the conjugate base.

Solve

Because the pK_a is lower (stronger acid) with the NO_2 group, the NO_2 group must better stabilize the negative charge that develops in the conjugate base. This can happen only if the NO_2 group is a stronger electron-withdrawing group, more effectively reducing the concentration of negative charge that is produced. As we can see in the structures below, the N atom has a formal positive charge, making it very electron deficient. The C=O carbon has just a partial positive charge, so it is not as electron deficient. Stronger electron-withdrawing groups, in general, increase the stability of the anion conjugate base, and thus the acid is stronger.

pK_a = 1.68

δ+

pK_a = 2.83

Problem 6.49

Think

Which color (red or blue) indicates more electron density? Which O^-, according to the electrostatic potential maps, bears more electron density? What does the increase in electron density indicate about the strength of the electron withdrawing group?

Solve

The red region (representing a buildup of negative charge or electron density) is slightly larger in $CF_3CH_2O^-$ than in $NCCH_2O^-$. This suggests that the CN group more effectively delocalizes the negative charge over the rest of the molecule, so the CN group is more electron withdrawing.

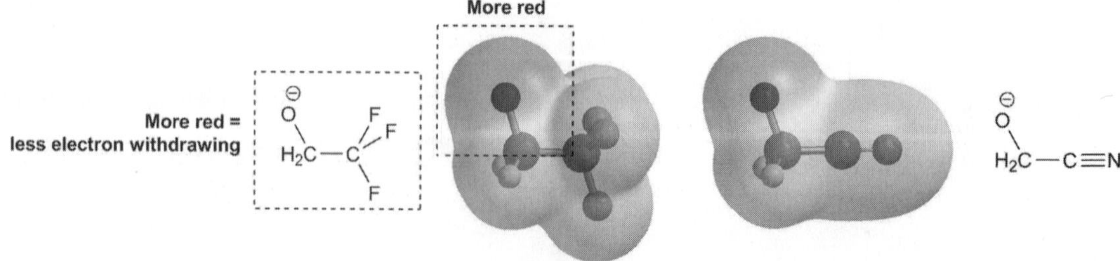

Problem 6.50

Think

Which base is stronger, HO^- or $(CH_3)_3CO^-$? Is an R group electron donating or withdrawing? Is the conjugate base stabilized or destabilized by an electron-donating group?

Solve

$NaOC(CH_3)_3$ is a stronger base than NaOH. A stronger base is required to deprotonate the alkyl-substituted derivative because it is less acidic than the unsubstituted molecule. The alkyl group is electron donating and destabilizes the negative charge in the resulting conjugate base, as shown below.

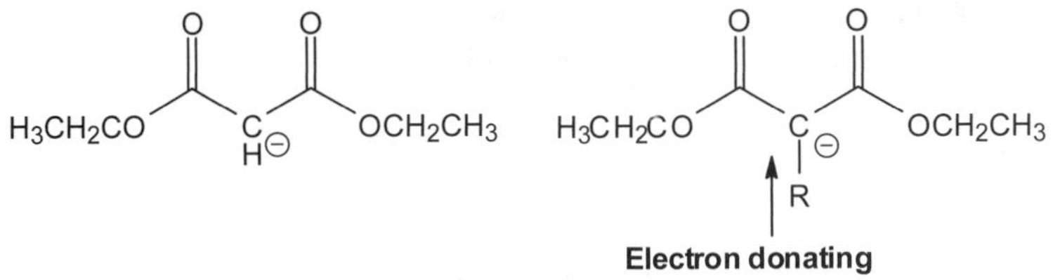

H3CH2CO OCH2CH3 H3CH2CO OCH2CH3 R

Electron donating

Problem 6.51

Think

Is the NO_2 an electron-donating group or electron-withdrawing group (EWG)? What is the effect of the NO_2 group on the stability of the conjugate base anion? How does the position of the NO_2 group affect the stability of the conjugate base? Consider the number of possible resonance structures.

Solve

(a) All of the nitrophenols are more acidic than phenol because the NO_2 group stabilizes each of the conjugate bases through inductive effects, as shown below.

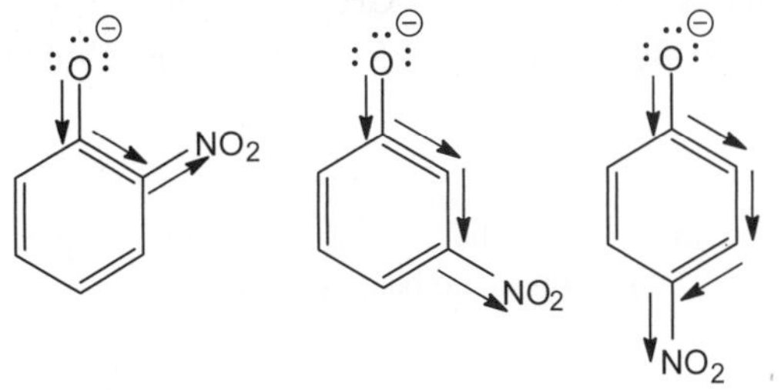

NO2 NO2 NO2

(b) The ortho and para nitrophenols are more acidic than the meta because in the conjugate bases of the ortho and para compounds, the NO_2 group can participate in resonance with the negative charge that is generated in the conjugate base. This is exemplified by the ortho conjugate base below.

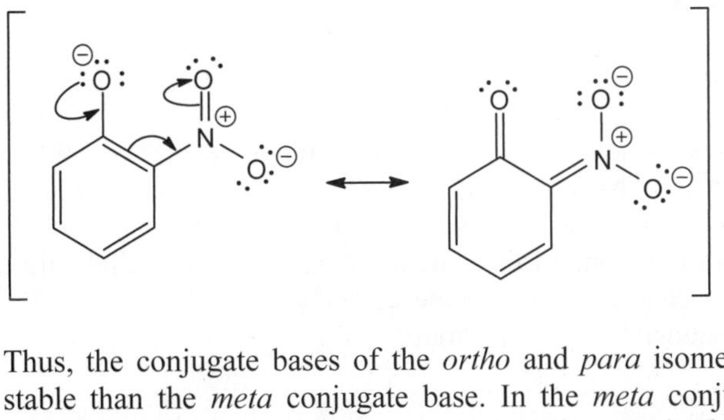

Thus, the conjugate bases of the *ortho* and *para* isomers have more resonance structures and are more stable than the *meta* conjugate base. In the *meta* conjugate base, the NO_2 group cannot participate in resonance with the negative charge that is developed. The total number of resonance structures is specified below each conjugate base, and the resonance hybrids are provided, too. See the figures on the next page.

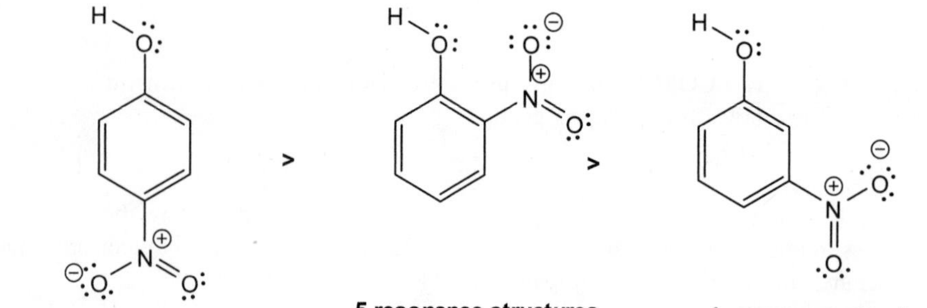

5 resonance structures, also EWG (inductive) $pK_a = 7.14$ > 5 resonance structures, also EWG (inductive) $pK_a = 7.22$ > 4 resonance structures, also EWG (inductive) $pK_a = 8.36$

Hybrid structures

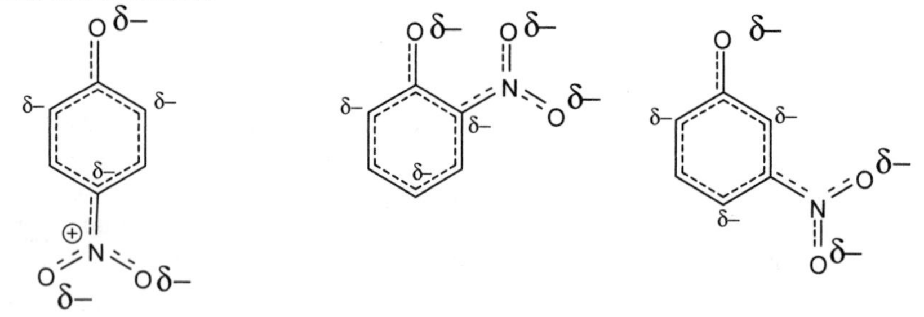

Problem 6.52

Think

Is the CH_3 group electron donating or withdrawing? What is the effect of the CH_3 group on the stability of the conjugate base anion? How does the position of the CH_3 group affect the stability of the conjugate base? Consider the number of possible resonance structures.

Solve

(a) All three methyl-substituted phenols are less acidic than phenol because the methyl group is electron donating and destabilizes the negative charge generated in the conjugate base, shown below.

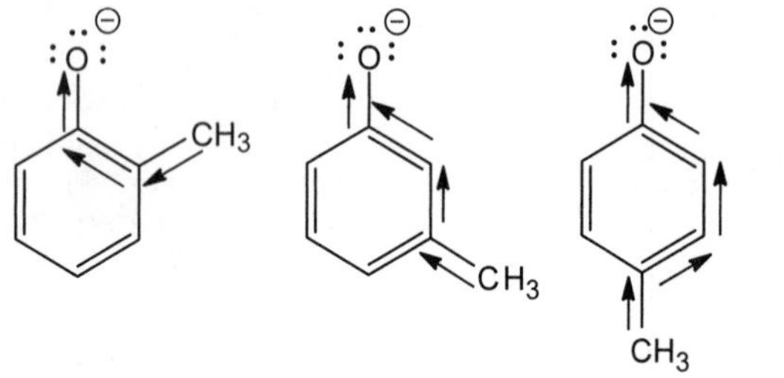

(b) The ortho and para compounds are less acidic than the meta compound because this inductive destabilization is more pronounced in ortho and para than it is in ortho. We can see why by looking at the resonance hybrid of phenoxide itself (shown on the next page), which delocalizes the negative charge onto carbons 2, 4, and 6. When a methyl group is on one of these carbons, as in the ortho and para conjugate bases, the negative charge becomes more concentrated, and the anion becomes less stable. In the meta conjugate base, none of these C atoms is bonded to a methyl group.

Problem 6.53

Think

Should the relative acidities be governed by the stabilities of the acids themselves or of their respective conjugate bases? Are there differences in the charges that appear? The types of atoms on which the charges appear? Resonance effects? What is the difference in the inductive effect of Cl, Br, and I? Which one is the strongest electron-withdrawing group? What is the effect of an electron-withdrawing group on the stability of the conjugate base anion?

Solve

Being uncharged acids, the reactions have the same forms as Equations 6-21a and 6-21b (p. 314). Therefore, according to Figure 6-7 (p. 315), the stronger acid is the one with the more stable conjugate base. In each case, OH becomes O^-, so the same charge is produced. The negative charge is delocalized by the same number of resonance structures involving the benzene ring. The main difference is inductive effects. Each halogen is electron withdrawing, and thus stabilizes the negative charge generated in the conjugate base and increases the strength of the acid. Because electronegativity increases in the order I < Br < Cl, the acid on the left should have the most stable conjugate base and, thus, should be the most acidic. Its pK_a is 9.0. The molecule on the right should have the least stable conjugate base and should be the weakest acid. Its pK_a is 9.2. The bromo compound has a pK_a of 9.1.

$pK_a = 9.0$ $pK_a = 9.1$ $pK_a = 9.2$

Problem 6.54

Think

Do all atoms have an octet? Does one structure have fewer atoms with nonzero formal charges? Are there more covalent bonds in one structure? What is the stability of the positive charge on the C atom versus the N atom?

Solve

D has the greatest contribution to the resonance hybrid because it is the only resonance structure in which all atoms have a complete octet. Even though a C atom is less electronegative than an N atom, suggesting that **D** might have a lesser contribution, having complete octets is generally more important than charge stability.

D **E** **F** **G**

Greatest contributor
All atoms have octets.

Problem 6.55

Think

Are there any π bonds adjacent to an atom with an incomplete octet? Do all resonance structures have the same number of atoms lacking an octet? Is the charge located on the same type of atom in each resonance structure? How do inductive effects impact the stability of the different C^+ species?

Solve

The species has three resonance structures, shown below. All resonance structures have a single C^+ atom lacking an octet. The rightmost structure is the most stable due to electron-donating effects. The rightmost structures has two electron-donating alkyl groups directly attached to C^+, whereas the other structures do not.

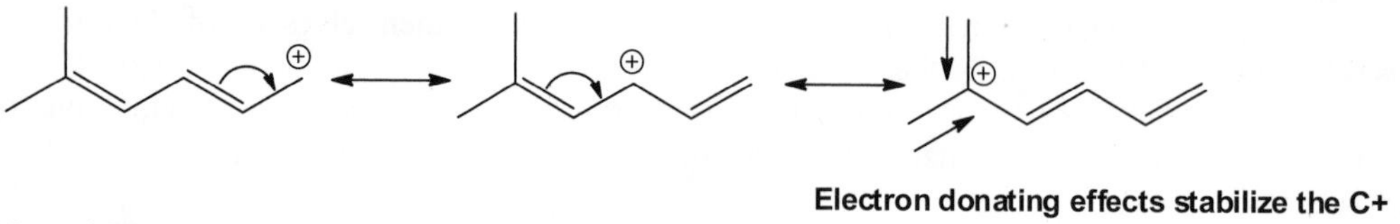

Electron donating effects stabilize the C+

Problem 6.56

Think

Which acid is stronger? Does the stronger acid have the F atom or Cl atom? How does the acid strength change when the F is changed to a Cl? How does the acid strength change when the F and Cl atoms are moved farther away?

Solve

Distance from the reaction center is more important. Going from F to Cl, the pK_a changes by fewer than 0.3 units.

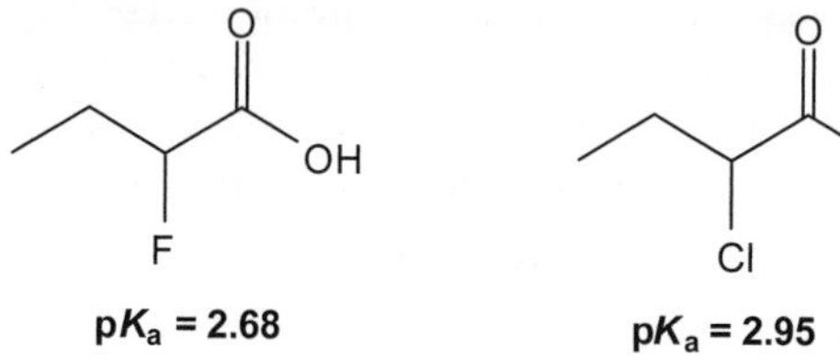

pK_a = 2.68 **pK_a = 2.95**

Going from two atoms away from the reaction center to three atoms away, the pK_a changes by more than one unit.

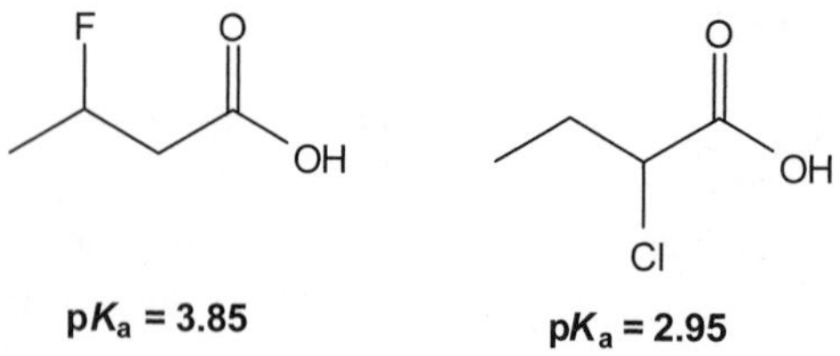

pK_a = 3.85 **pK_a = 2.95**

Problem 6.57

Think

Draw the complete deprotonation reactions. Should the differences in acidity be due primarily to the stabilities of the acids themselves or of the conjugate bases? Which anion, Cl^-, Br^-, or I^- is most stable? Where is each located on the periodic table relative to the others? Which anion is the largest?

Solve

All reactions are of the form $HX \rightarrow H^+ + X^-$, where X changes. Being uncharged acids, the reactions have the same forms as Equations 6-21a and 6-21b (p. 314). Therefore, according to Figure 6-7 (p. 315), the stronger acid is the one with the more stable conjugate base. The major difference in stability is with X^- on the product side. The more stable X^-, the stronger the acid HX. I^- is the largest anion (farthest down in the periodic table) and, therefore, is the most stable. Thus, acidity increases going down the periodic table column, HCl < HBr < HI.

Problem 6.58

Think

Draw each complete protonation reaction. Does charge stability affect the reactant side? The product side? What factors affect the stability of a charged species? What is the nature of the atom with the negative charge? Which anions can participate in resonance? Are inductive effects present? Which protonation reaction is most energetically favorable? Least?

Solve

As shown below, bases **A**–**F** are negatively charged and become uncharged upon protonation, whereas **G** is uncharged and becomes positively charged.

A – F: $\text{Base}^{\ominus}\!:\ \longrightarrow\ \text{Base—H}$

G: $\text{Base}\!:\ \longrightarrow\ \text{Base}^{\oplus}\text{—H}$

In the free energy diagram below, therefore, base **G** appears lower in energy than bases **A**–**F**. Bases **B** and **D** are lower in energy than **C**, **E**, **A**, and **F** because of the atom on which the negative charge appears. A negative charge on O is more stable than a negative charge on C. Base **D** is more stable than base **B** because the F atoms inductively stabilize the nearby negative charge. Of bases **C**, **E**, **A**, and **F**, **F** is the least stable. Although the negative charge on bases **C**, **E**, **A**, and **F** are all resonance delocalized, the negative charge on **F** is delocalized onto N, whereas that on **C**, **E**, and **A** are delocalized onto the O atom. N is less electronegative than O, so it cannot handle a negative charge as well. Base **C** is more stable than **A** or **E** due to resonance. In base **C**, the negative charge is resonance delocalized onto two O atoms, whereas in **A** and **E**, the negative charge is delocalized onto only one O atom. Base **A** is less stable than base **E** because the CH_3 groups inductively destabilize the negative charge. On the product side, conjugate acid **G** is positively charged, making it less stable than conjugate acids **A**–**F**. With the relative order of the bases and the conjugate acids established, we can rank the base strengths according to how energetically favorable their protonation reactions are: **G** < **D** < **B** < **C** < **E** < **A** < **F**.

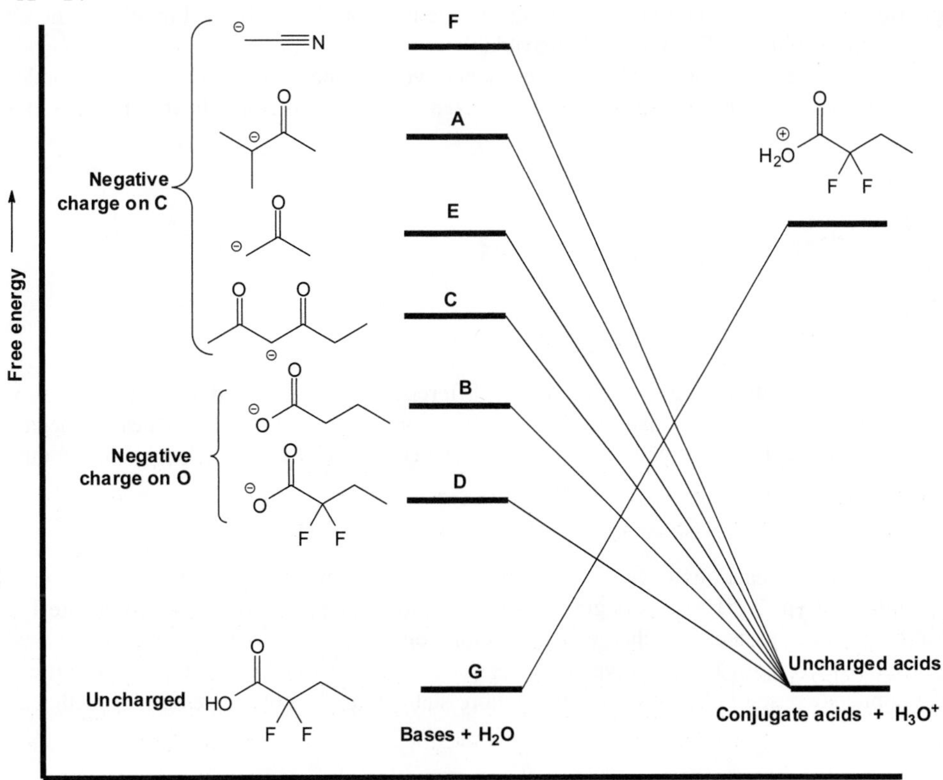

Problem 6.59

Think

Draw the complete deprotonation reactions. Should the differences in acidity be due primarily to the stabilities of the acids themselves or of the conjugate bases? Draw the conjugate base for each after one OH is deprotonated. What do you notice about the proximity of the O^- and the adjacent OH in cis compared to trans?

Solve

Being uncharged acids, the reactions have the same forms as Equations 6-21a and 6-21b (p. 314). Therefore, according to Figure 6-7 (p. 315), the stronger acid is the one with the more stable conjugate base. When the *cis* diol is deprotonated, a strong internal hydrogen bond is formed between OH and O^-. This stabilizes the conjugate base and increases the strength of the acid. That H bond does not exist in the conjugate base of the *trans* diol.

Internal hydrogen bond

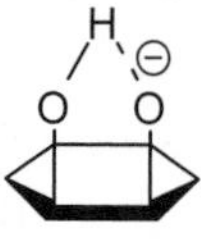

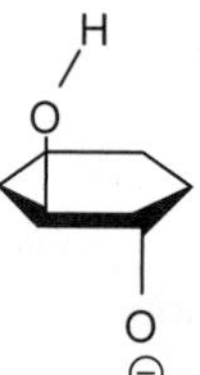

Problem 6.60

Think

Which proton is acidic? Should the differences in acidity be due primarily to the stabilities of the acids themselves or of the conjugate bases? In the conjugate base, does resonance serve to delocalize the negative charge of the same atoms or different atoms? Does the OR group affect the ability of resonance to delocalize that negative charge? How does this impact the stability of the conjugate base and the strength of the acid?

Solve

Being uncharged acids, the reactions have the same forms as Equations 6-21a and 6-21b (p. 314). Therefore, according to Figure 6-7 (p. 315), the stronger acid is the one with the more stable conjugate base. In both a ketone and an ester, the negative charge in the conjugate base is resonance stabilized involving the C=O group. In an ester, the electrons in the C=O bond are less available to participate in resonance with the negative charge on the C atom because they are also involved in resonance with a lone pair from the singly bonded O atom. Therefore, the negative charge that develops is more localized on C and less stable in an ester than in a ketone.

Problem 6.61

Think

Draw the complete deprotonation reactions. Should the differences in acidity be due primarily to the stabilities of the acids themselves or of the conjugate bases? Draw the conjugate base of each. Does the negative charge appear on the same atom? Does resonance delocalize the negative charge equally? What about inductive effects?

Solve

Being uncharged acids, the reactions have the same forms as Equations 6-21a and 6-21b (p. 314). Therefore, according to Figure 6-7 (p. 315), the stronger acid is the one with the more stable conjugate base. In the conjugate base of phenol, the negative charge that develops on the O atom is resonance delocalized onto the benzene ring (see the next page). The negative charge is shared between one O and the C atoms around the ring. This makes the conjugate base of phenol substantially more stable than the conjugate base of methanol, CH_3O^-.

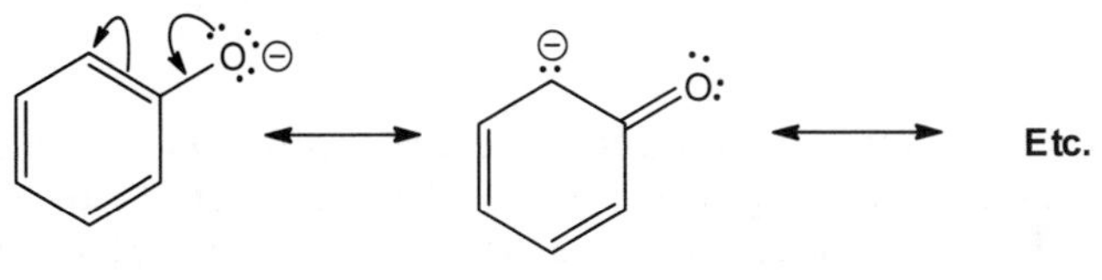

Etc.

In benzoic acid, the benzene ring does not participate in resonance with the negative charge that develops. The conjugate bases of benzoic acid and acetic acid therefore have similar charge stability, so the acids have similar strengths.

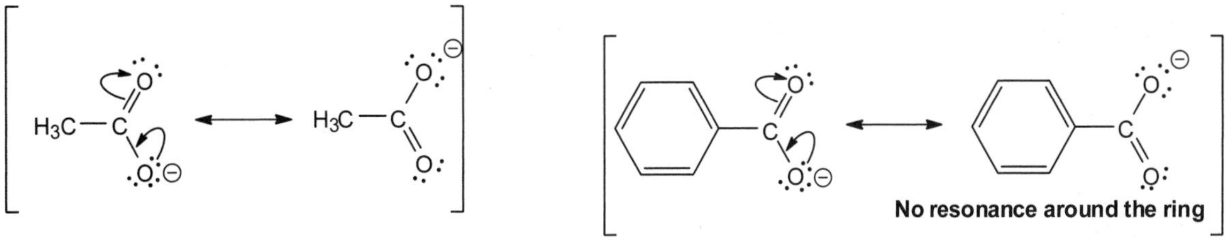

No resonance around the ring

Benzoic acid and acetic acid are stronger acids compared to phenol because the negative charge is shared among two O atoms that are much more electronegative compared to C and, therefore, can handle the negative charge better.

Problem 6.62

Think

Draw the complete deprotonation reactions. Should the differences in acidity be due primarily to the stabilities of the acids themselves or of the conjugate bases? Draw the conjugate base of each. Does the negative charge appear on the same atom? Does resonance delocalize the negative charge equally? What can be said about inductive effects?

Solve

Being uncharged acids, the reactions have the same forms as Equations 6-21a and 6-21b (p. 314). Therefore, according to Figure 6-7 (p. 315), the stronger acid is the one with the more stable conjugate base. The conjugate base of the para compound is more stable because the C=O bond can participate in resonance to provide additional resonance delocalization to the negative charge that develops (see below). This cannot happen in the meta compound. Therefore, the para compound is the stronger acid.

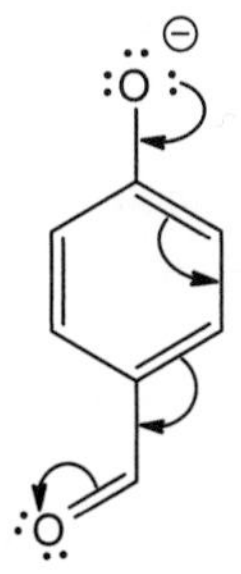

Problem 6.63

Think

What is the definition of a base? Draw the conjugate acid for each. Being uncharged bases, should base strength be governed by the stabilities of the bases themselves or of the respective conjugate acids? Do the same kinds of charges appear? Do the charges appear on the same kinds of atoms? How do resonance and inductive effects impact charge stability?

Solve

Because the bases are uncharged and the conjugate acids pick up a positive charge, the stabilities of the conjugate acids govern the base strengths, as shown below; the stronger base is the one with the more stable conjugate acid. In this case, the conjugate acid of the left reaction is less stable than the conjugate acid of the right reaction. In the left reaction, a positive charge develops on an sp^2-hybridized N atom, whereas in the right reaction, it is on an sp^3-hybridized N atom. So, the N atom in the first reaction has a greater effective electronegativity and cannot accommodate the positive charge as well. Therefore, the second reaction is more energetically favorable, making the base in the second reaction stronger.

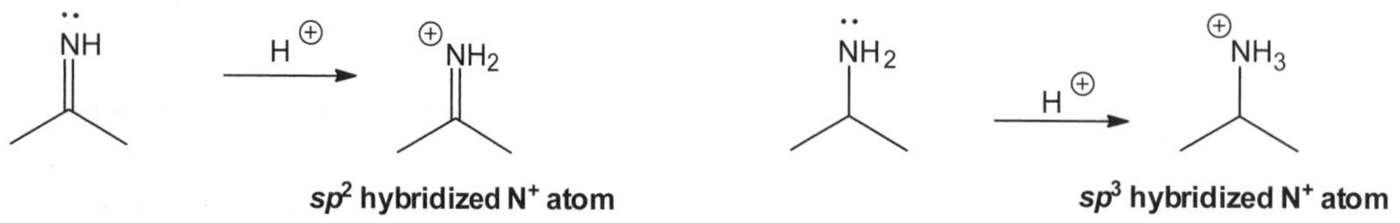

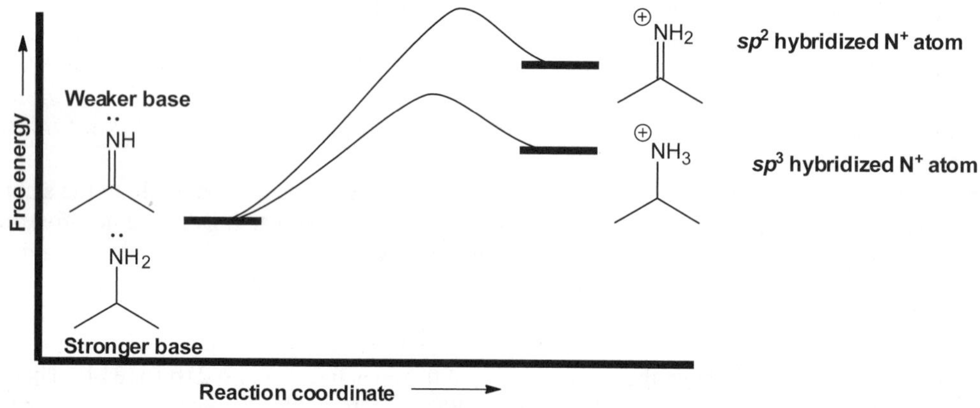

*sp*² hybridized N⁺ atom

*sp*³ hybridized N⁺ atom

Weaker base

Stronger base

Free energy

Reaction coordinate

*sp*² hybridized N⁺ atom

*sp*³ hybridized N⁺ atom

Problem 6.64

Think

Draw out the Lewis structure for each. Draw out the reaction where each base picks up a proton. Does the basic atom begin with a formal charge? Does it pick up a formal charge?

Solve

HNC is the stronger base because it has a formal negative charge on the C atom (see below). When it picks up a proton, that negative charge becomes 0. In HCN, a 0 charge on the N atom becomes positive when it picks up a proton.

Stronger base

Problem 6.65

Think

Draw the complete deprotonation reactions. Should the differences in acidity be due primarily to the stabilities of the acids themselves or of the conjugate bases? What is the conjugate base for each? Are the charges the same? Do the charges appear on the same types of atoms? How do resonance and inductive effects impact charge stability?

Solve

CH_3NC is the stronger acid. Being uncharged acids, the reactions have the same forms as Equations 6-21a and 6-21b (p. 314). Therefore, according to Figure 6-7 (p. 315), the stronger acid is the one with the more stable conjugate base. In both cases, a negative charge is formed on the C atom upon deprotonation. In CH_3NC, a formal positive charge exists on the N atom, which very strongly inductively stabilizes the negative charge that is formed. In CH_3CN, the neighboring C atom has a 0 formal charge.

$H_3C—C≡N: \xrightarrow{-H^+} H_2\overset{\ominus}{C}—C≡N:$ $\qquad$ $H_3C—\overset{\oplus}{N}≡\overset{\ominus}{C}: \xrightarrow{-H^+} H_2\overset{\ominus}{C}—\overset{\oplus}{N}≡\overset{\ominus}{C}:$

Stronger acid

Problem 6.66

Think

Identify the acidic proton in formaldehyde and the acidic proton in each of the structures listed for comparison. Draw the complete deprotonation reactions. Should the differences in acidity be due primarily to the stabilities of the acids themselves or of the conjugate bases? What is the conjugate base for each? Are the charges the same? Do the charges appear on the same types of atoms, including hybridization? How do resonance and inductive effects impact charge stability?

Solve

(a) The pK_a of formaldehyde should be closest to $H_2C{=}CH_2$, **D**, whose pK_a is 44. Both molecules are uncharged, so their relative acidities should be governed primarily by the stabilities of their conjugate bases. In both conjugate bases, the negative charge develops on the same type of atom (an sp^2-hybridized C atom) with no resonance.

(b) The only difference is inductive effects from the nearby O atom, which stabilizes the negative charge in the conjugate base and, hence, increases the acid strength. But recall that inductive effects typically have a fairly small effect on pKa. We should, therefore, estimate the pK_a to be ~40.

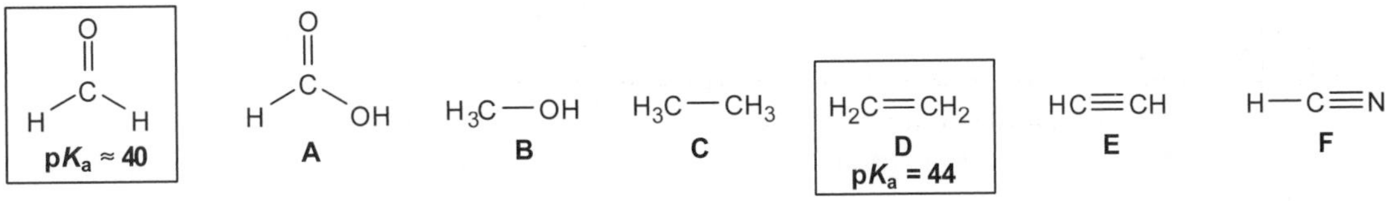

H(C=O)H	HC(=O)OH	$H_3C—OH$	$H_3C—CH_3$	$H_2C{=}CH_2$	HC≡CH	H—C≡N
pK_a ≈ 40	**A**	**B**	**C**	**D** pK_a = 44	**E**	**F**

Problem 6.67

Think

Draw the complete protonation reactions. Should the differences in base strength be due primarily to the stabilities of the bases themselves or of the conjugate acids? What is the conjugate acid for each? Are the charges the same? Do the charges appear on the same types of atoms? How do resonance and inductive effects impact charge stability?

Solve

Because the basic sites are uncharged and the conjugate acids pick up a positive charge, the stabilities of the conjugate acids govern the base strengths. When the O atom of the C=O group is protonated, a lone pair from the other O atom can participate in resonance to delocalize the positive charge that is generated (see below). When the OR group is protonated, the positive charge that develops cannot be resonance delocalized.

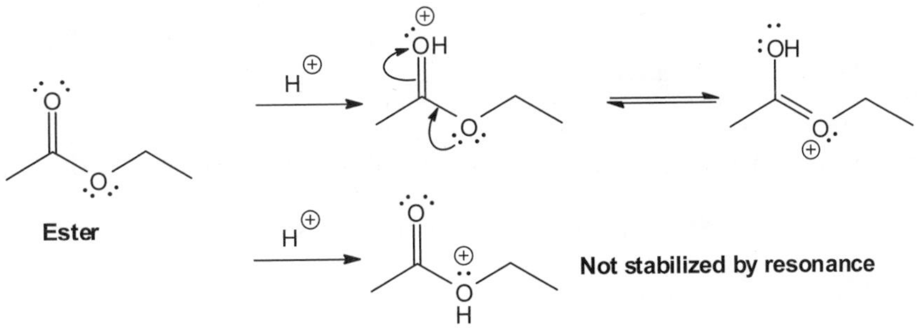

Ester $\xrightarrow{H^+}$ (C=O protonated) ⇌ (resonance structure)

Ester $\xrightarrow{H^+}$ (OR protonated) **Not stabilized by resonance**

Problem 6.68

Think

Draw the complete deprotonation reactions. Should the differences in acidity be due primarily to the stabilities of the acids themselves or of the conjugate bases? Which alcohol shares similar resonance and inductive properties?

Solve

Being uncharged acids, the reactions have the same forms as Equations 6-21a and 6-21b (p. 314). Therefore, according to Figure 6-7 (p. 315), the stronger acid is the one with the more stable conjugate base. The acidity of molecule **A** should be the most similar to cyclohexanol. In both cases, the negative charge that develops in the conjugate base on the O atom is part of an alcohol attached to an sp^3-hybridized C atom, and that C atom is adjacent to two alkyl groups. In **B,** the analogous C atom is bonded to only one alkyl group, and there is a nearby electronegative atom (part of the C=O) that will increase the acidity. In **C**, the nearby Cl will increase the acidity as well. The molecule whose acidity is most different from cyclohexanol is **D**, because the OH group is part of a different functional group altogether—a carboxylic acid.

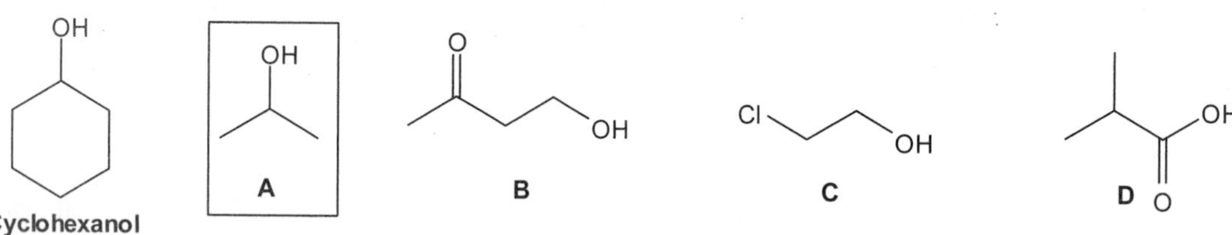

Problem 6.69

Think

What is the concentration of the acid initially? What is the K_a of the acid? Does the K_a expression involve initial concentrations or equilibrium concentrations? How can you relate the two concentrations using a variable? How do the initial and equilibrium concentrations relate to the percent ionization?

Solve

The K_a expression involves equilibrium concentrations, but the initial and equilibrium concentrations are related by a difference x. The percent ionization is the amount of acid that has dissociated, x, divided by the initial amount of the acid.

	$[C_6H_5OH]$		$[C_6H_5O^-]$	$[H_3O^+]$
Initial	0.100 M	$\rightleftharpoons$	0 M	0 M
Change	$-x$		$+x$	$+x$
Equilibrium	$0.100 - x$		x	x

So at equilibrium, $x^2/(0.100 - x) = K_a = 10^{-pKa} = 10^{-10.0}$
Solving, $x = 3.16 \times 10^{-6} = [C_6H_5O^-]_{eq}$
% dissociation $= (3.16 \times 10^{-6})/(0.100) \times 100\% = 0.00316\%$

Problem 6.70

Think

What bonds are broken and formed in an acid/base reaction? What is the relationship between K_{eq} and pK_a? When K_{eq} is >1, does that mean the products or reactants are favored?

Solve

(a) The products of each reaction are shown on the next page, as are the pK_a for each acid involved in the respective equilibria.

(b) Products are favored for the reactions in **(iii)**, **(iv)**, **(v)**, **(vi)**, and **(vii)**, because the acid on the reactant side is stronger (lower pK_a) than the one on the product side.

(c) The K_{eq} values are computed from Equation 6-11 as $K_{eq} = 10^{(pKa,prod - pKa,react)}$.

(i) + $^{\ominus}$OH ⇌ + H_2O $K_{eq} = 10^{-1.3} = 5.0 \times 10^{-2}$

pK_a~17 $pK_a = 15.7$

(ii) + $Cl^{\ominus}$ ⇌ + H—Cl $K_{eq} = 10^{-23.5} = 3.2 \times 10^{-24}$

pK_a ~ 16.5 $pK_a = -7$

(iii) $K_{eq} = 10^{22}$

pK_a ~16 $pK_a = 38$

(iv) $H^{\ominus}$ + ⇌ H_2 + $K_{eq} = 10^{10}$

pK_a ~ 25 $pK_a = 35$

(v) + $H_3O^{\oplus}$ ⇌ + H_2O $K_{eq} = 10^{6.7} = 5 \times 10^6$

$pK_a = -1.7$ pK_a ~5

(vi) + HO ⇌ + $K_{eq} = 10^{27}$

pK_a ~ 16 $pK_a = 43$

(vii) + $H^{\ominus}$ ⇌ + H_2 $K_{eq} = 10^{30.25} = 1.8 \times 10^{30}$

$pK_a = 4.75$ $pK_a = 35$

Problem 6.71

Think

What species is the acid and what species is the base? How do you calculate K_{eq} from pK_a values? When K_{eq} is >1, does that mean the products or reactants are favored?

Solve

The pK_a values in Table 6-1 are 35 for H_2 and 16 for CH_3CH_2OH, so the latter is the stronger acid (lower pK_a). The equilibrium constant for the reaction is computed from Equation 6-11 as: $K_{eq} = 10^{(35-16)} = 10^{19}$, which is very large. Thus, the reaction heavily favors products.

$$H^- + CH_3CH_2OH \longrightarrow H{-}H + CH_3CH_2O^-$$

Hydride anion — Ethanol — Ethoxide

Problem 6.72

Think

How does the leveling effect apply when ethanol is the solvent? What is the pK_a of ethanol? What is the pK_a of the conjugate acid of each species listed for comparison? Is it desirable for the solvent to react with the solute or to remain unreacted? Are you looking for a stronger or weaker acid compared to ethanol?

Solve

With ethanol as the solvent, the strongest base that can exist in solution to any appreciable extent is the conjugate base of ethanol, $CH_3CH_2O^-$. Any base that is stronger than $CH_3CH_2O^-$ will deprotonate ethanol readily. Our desire, however, is for the bases to remain unreacted when dissolved in ethanol, so only those bases that will not deprotonate ethanol (pK_a = 16) are acceptable. Their conjugate acids, therefore, must be stronger than ethanol, or <16. Only **(b)**, **(c)**, **(d)**, and **(e)** are acceptable.

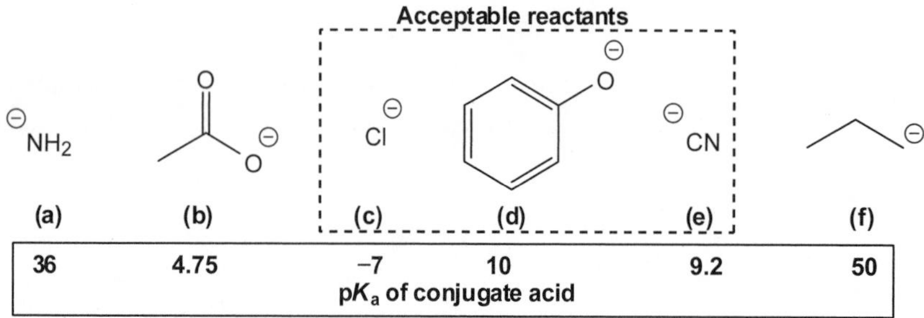

Problem 6.73

Think

How does the leveling effect apply when ethanamine is the solvent? What is the pK_a of ethanamine? What is the pK_a of the conjugate acid of ethanamine, $CH_3CH_2NH_3^+$? For comparison what is the pK_a of each acid listed and the pK_a of each base's conjugate acid? Is it desirable for the solvent to react with the solute or to remain unreacted? Are you looking for a stronger or weaker acid compared to ethananime? A stronger or weaker acid compared to the conjugate acid of ethanamine?

Solve

With ethanamine as the solvent, the strongest base that can exist is the conjugate base of ethanamine, $CH_3CH_2NH^-$, and the strongest acid that can exist is the conjugate acid of ethanamine, $CH_3CH_2NH_3^+$. Only those bases that will not deprotonate ethanamine (pK_a ≈38) are acceptable, and only those acids that will not protonate ethanamine to produce $CH_3CH_2NH_3^+$ (pK_a = 10.6) are acceptable. The conjugate acids of the bases, therefore, must be stronger than ethanamine, or <38. **(a)**, **(d)**, and **(f)** are acceptable bases (dash boxes). **(b)** and **(e)** are acids, and to remain unreacted they must be less acidic than $CH_3CH_2NH_3^+$, or >10.6. Both are more acidic, however, so they are unacceptable.

	(a)	(b)	(c)	(d)	(e)	(f)
pK_a of acid		−7			9.4	
Species	Cl^-	HCl	CH_3^-	NH_2^-	NH_4^+	OH^-
pK_a of conjugate acid	−7		48	36		15.7

Problem 6.74

Think

When the pH equals the pK_a, what is the percent ionization? What is the ratio of $[A^-]/[HA]$ when the percent ionization is 90% and 10%?

Solve

An acid dissociates 50% (i.e., $[A^-] = [HA]$) when the pH equals the pK_a. That is, pH = 0.77. The acid dissociates 90% (i.e., $[A^-]/[HA] = 9$) when pH is more basic than pK_a by one unit (i.e., pH = 1.77). The acid dissociates 10% (i.e., $[A^-]/[HA] = 0.111$) when the pH is more acidic than pK_a by one unit (i.e., pH = −0.23).

Problem 6.75

Think

What does the pK_a indicate about the pH at which the proton is lost? How many pH units above the pK_a does the solution need to be for the acid to be 99% deprotonated? How many pH units below the pK_a does the solution need to be for the acid to be 99% protonated? Use Figure 6-1 as a guide.

Solve

We expect the dominant form to be the protonated form (>99%) when pH is more acidic than pK_a by at least two units, or pH <8.7. We expect the deprotonated form to be dominant (>99%) when pH is more basic than pK_a by more than two units, or pH >12.7. We expect equal amounts of the two forms when pH = pK_a = 10.7.

Problem 6.76

Think

Which species is the acid and which species is the base? What are the products of the reaction? Based on pK_a values, which acid is stronger, the one on the reactant side or the conjugate acid on the product side? What does the acid strength indicate about the stability of the species?

Solve

(a) The products and reaction mechanism are given below.

H + $:\ddot{N}H_2^{\ominus}$ ⟶ + $\ddot{N}H_3$

pK_a = 25 pK_a = 38

(b) The reaction heavily favors products because the acid on the reactant side (RC≡CH) is much stronger than that on the product side (NH_3). Therefore, the products are lower in energy; the reaction is exothermic.

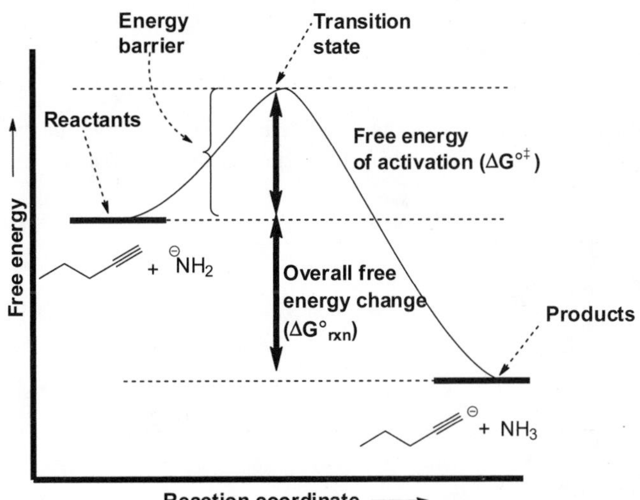

Problem 6.77

Think

Which species is the acid and which species is the base? What are the products of the reaction? Based on pK_a values, which acid is stronger, the one on the reactant side or the conjugate acid on the product side? What does the acid strength indicate about the stability of the species?

Solve

(a) The products and reaction mechanism are given below.

$pK_a = 16$ $pK_a = 20$

(b) The reaction is significantly product-favored because the acid on the reactant side (CH_3CH_2OH) is much stronger than that on the product side (ketone). Therefore, the products are lower in energy; the reaction is exothermic.

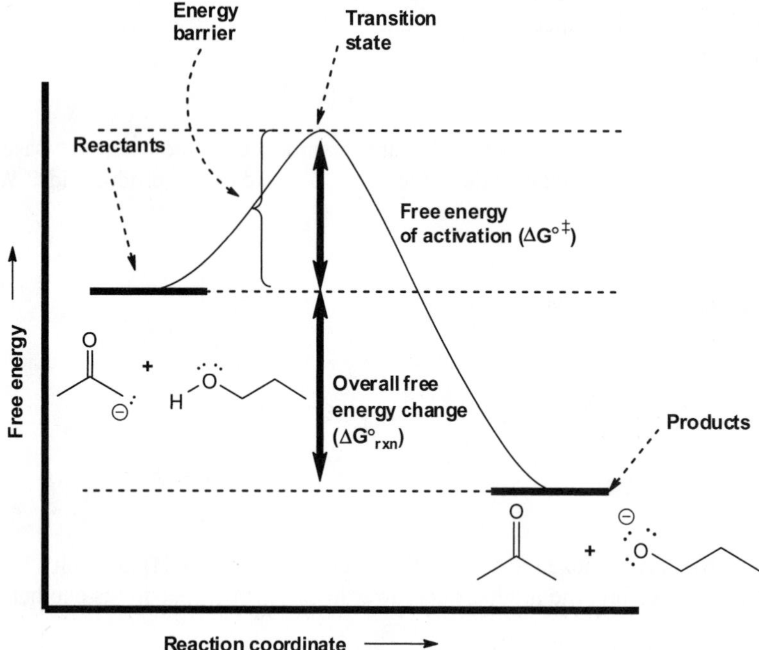

Problem 6.78

Think

In each species, what are the possible acidic protons? What are the possible basic sites? Draw out the products for each reaction. On which atom is a negative charge more stable? On which atom is a positive charge more stable?

Solve

The products for the reaction are listed below. Oxygen is more electronegative compared to nitrogen. Therefore, the O atom stabilizes the negative charge better and the N atom accommodates the positive charge better. Thus, the more favorable reaction is the first one, in which the alcohol acts as the acid and the amine acts as the base.

Problem 6.79

Think

Which product is lower in energy? Which product is more stable? What does the stability and energy content of the product indicate about the extent of formation of the product at equilibrium?

Solve

Y is in greater abundance at equilibrium because it is more stable (lower in energy).

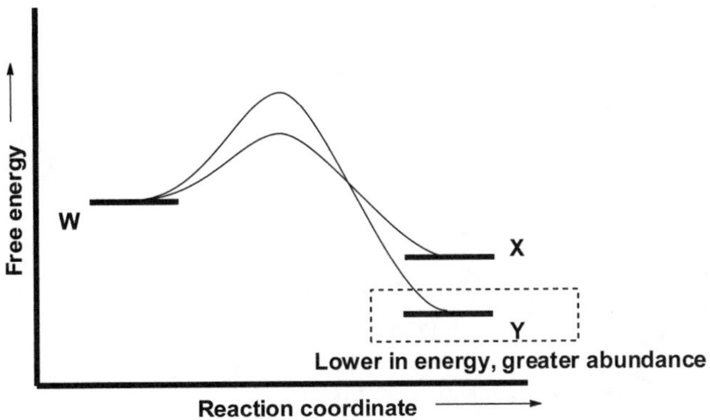

Problem 6.80

Think

Which product is lower in energy? Which product is more stable? What does the stability and energy content of the product indicate about the extent of formation of the product at equilibrium?

Solve

Y is in greater abundance at equilibrium because it is more stable (lower in energy).

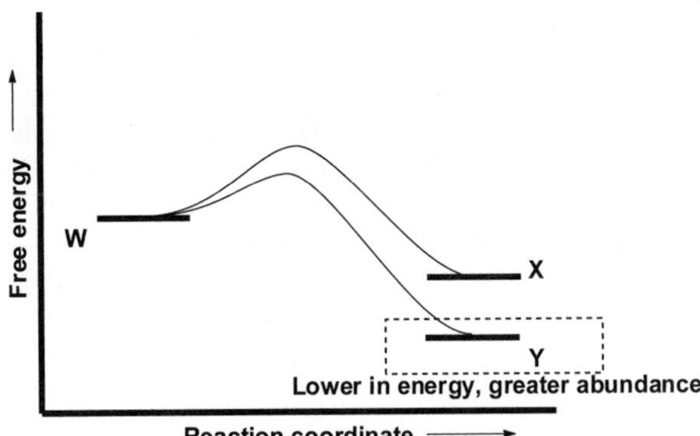

Problem 6.81

Think

What was the conjugate base that must have picked up a D^+ from D_2O? From the conjugate base, what was the structure of the original acid? What is the pK_a of D_2O and each original acid? What base can deprotonate the original acids?

Solve

In all cases, a D^+ must replace an H^+. This can be done in two successive proton transfer reactions. The first reaction removes the appropriate H^+, and the second reaction deposits a D^+. For each of these successive reactions to be efficient, K_{eq} should favor products. The proper choice of base in the first step, therefore, is that in which the pK_a of its conjugate base is significantly greater than the pK_a of the reactant acid. H^- works in all three cases because the pK_a of H_2 is 35. In the second proton transfer, the acid is D_2O, whose pK_a is similar to H_2O, 15.7. So, in all three cases, the second proton transfer does, indeed, favor products.

(a) H; pK_a = 20; $-H^{\oplus}$; D_2O, pK_a = 15.7; D

(b) H; pK_a = 25; $-H^{\oplus}$; D_2O, pK_a = 15.7; D

(c) OH; pK_a = 19; $-H^{\oplus}$; O$^{\ominus}$; D_2O, pK_a = 15.7; OD

Problem 6.82

Think

Draw the complete deprotonation reactions. Should the differences in acidity be due primarily to the stabilities of the acids themselves or of the conjugate bases? Review Section 2.9 and consider the ability of each solvent to solvate cations and anions. Draw the structure of water and dimethyl sulfoxide (DMSO). Which solvent stabilizes the conjugate base of acetic acid better?

Solve

Being an uncharged acid, the reaction has the same form as Equations 6-21a and 6-21b (p. 314). Therefore, according to Figure 6-7 (p. 315), the acid becomes stronger as the conjugate base becomes more stable. The acetate anion conjugate base that forms is solvated—and therefore stabilized—better by water than by DMSO. Water solvates anions better than DMSO because water is a polar protic solvent, whereas DMSO is an aprotic solvent. With the conjugate base stabilized better in water, the acid becomes stronger. This explains why the pK_a is 4.75 in water and 12.6 in DMSO.

Problem 6.83

Think

Draw the complete deprotonation reactions. Should the acidity be due primarily to the stability of the acid itself or of the conjugate base? Review Section 2.9 and consider the ability of each solvent to solvate cations and anions. Draw the structure of water and DMSO. Why is this effect not felt in NH_4^+?

Solve

Being a charged acid, the complete reaction is analogous to those in Equations 6-22a and 6-22b (p. 316), so the relative acidity is governed by the relative stability of the acid itself, as shown in Figure 6-8 (p. 316). Water is a protic solvent, whereas DMSO is an aprotic solvent. Even though protic solvents solvate anions much better than polar aprotic solvents, they both solvate cations well. Because cations are present in this case, changing the solvent doesn't have a dramatic effect on charge stability, and therefore doesn't have a dramatic effect on acidity.

NH_4^+ **pK_a = 9.4 (in water)**

NH_4^+ **pK_a = 10.5 (in DMSO)**

Dimethyl sulfoxide DMSO

Problem 6.84

Think

Is the strength of CH_3NH_2 as a base governed primarily by the stability of the base itself or of the conjugate acid? Which solvent has the greatest effect on the stability of that species?

Solve

Being an uncharged base, its strength is governed primarily by the stability of its conjugate acid, $CH_3NH_3^+$, due to the presence of the positive charge. The conjugate base is better stabilized as the intermolecular interactions with the solvent become stronger. All the solvents except **E** (CCl_4) can undergo hydrogen bonding and/or ion–dipole interactions with the conjugate base. Therefore, $CH_3NH_3^+$ will be the least stable in **E**, making CH_3NH_2 the weakest base in that solvent.

Problem 6.85

Think

Draw the complete reactions. Should the acidities of these acids be governed by the relative stabilities of the acids themselves or of their respective conjugate bases? Are alkyl groups electron withdrawing or donating? What is the effect of the alkyl group on the stability of the N^+? How would the additional alkyl groups affect the solvation by water?

Solve

(a) Being charged acids, the complete reactions are analogous to those in Equations 6-22a and 6-22b (p. 316), so their relative acidities are governed by the relative stabilities of the acids themselves, as shown in Figure 6-8 (p. 316). Alkyl groups are electron donating and should stabilize the N cation, making the acid weaker with each additional alkyl group. This is what we observe with the addition of the first two alkyl groups, where the pK_a increases slightly.

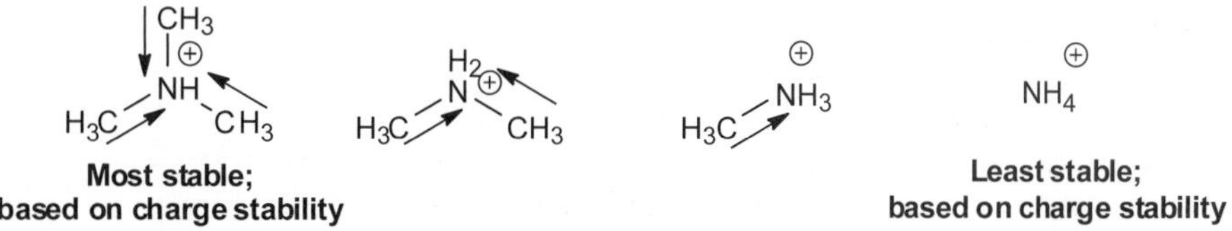

(b) The addition of the third alkyl group should increase the stability of the cation, weakening the base. However, the pK_a decreases, signifying a stronger acid. This has to do with differences in solvation by water. The substantial bulkiness surrounding the N^+ prevents solvation by water, which decreases stability and increases acidity.

CHAPTER 7 | An Overview of the Most Common Elementary Steps

Your Turn Exercises

Your Turn 7.1

Think

What kinds of charges characterize an electron-rich atom? Are any formal charges present? Any strong partial charges? In which direction do electrons flow—from electron rich to electron poor, or the opposite? How is a curved arrow used to denote the flow of electrons? Are Lewis bases electron-pair donors or acceptors? Lewis acids?

Solve

When the S atom bears a full negative charge, it is quite electron rich and not particularly stable due to the mutual repulsion of the extra electron density. The HS^- is, therefore, a good electron-pair donor and is the Lewis base. The H in H–Br bears a partial positive charge, δ^+, because Br is more electronegative than H. The electrons in HS^- are attracted to the proton in H–Br. The H–Br accepts the two electrons from HS^- and is, therefore, the Lewis acid. The bond between H–Br simultaneously breaks as the HS–H bond forms.

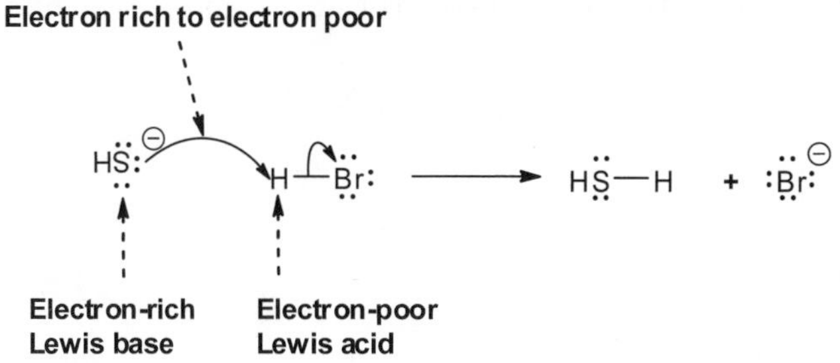

Electron rich to electron poor

HS:⁻ + H–Br: ⟶ HS–H + :Br:⁻

Electron-rich Lewis base

Electron-poor Lewis acid

Your Turn 7.2

Think

According to the curved arrow notation, what bond is formed? What atom receives a share of an additional pair of electrons? Do any bonds break from that atom?

Solve

The HO^- acts as a Lewis base and donates an electron pair to the $H^{\delta+}$ in H–Cl. A new HO–HCl bond is formed. No bonds are shown to break, however. This is unacceptable because H cannot form two covalent bonds. To have an acceptable product, the H–Cl bond would simultaneously need to break and the two electrons would become an additional lone pair on the Cl to yield the chloride anion, Cl^-.

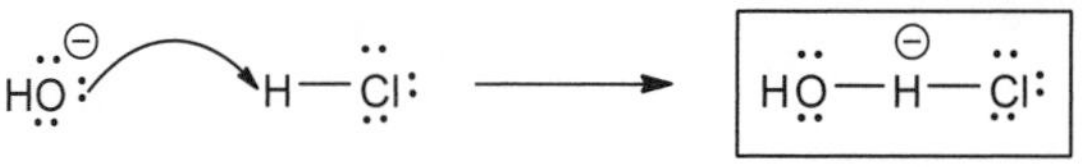

HO:⁻ + H–Cl: ⟶ [HO–H⁻–Cl:]

Your Turn 7.3

Think

Draw out the bonds connecting each atom and fill in lone-pair electrons to fulfill each atom's octet. How do you determine formal charge? (Consult Section 1.9, if necessary.) For the uncharged nucleophiles, consider how electronegativity differences in atoms lead to a polar bond.

Solve

Negatively charged nucleophiles:

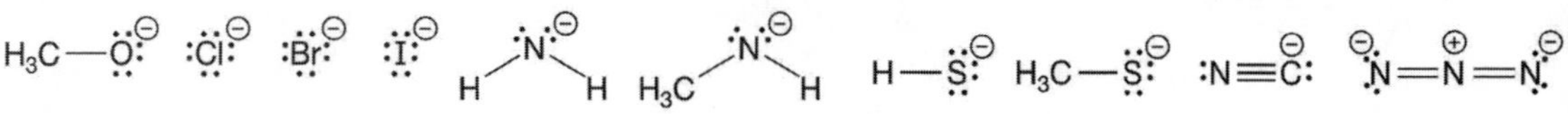

H_3C–O:⁻   :Cl:⁻   :Br:⁻   :I:⁻   H–N⁻–H H_3C–N⁻–H   H–S:⁻ H_3C–S:⁻   :N≡C:⁻   ⁻N=N⁺=N⁻

Uncharged nucleophiles:

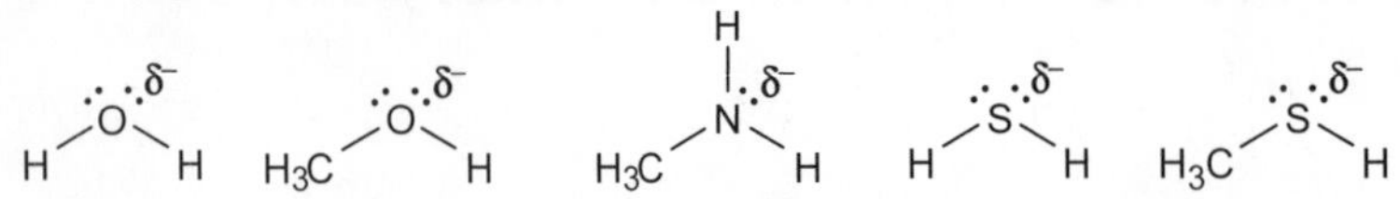

Your Turn 7.4

Think

What kinds of charges characterize an electron-rich atom? An electron-poor atom? Are any formal charges present? Any strong partial charges? In which direction do electrons flow—from electron rich to electron poor, or the opposite? How is a curved arrow used to denote the flow of electrons? Are Lewis bases electron-pair donors or acceptors? Lewis acids? Are any new bonds formed from lone-pair electrons?

Solve

When the Cl atom bears a full negative charge, it is quite electron rich and not particularly stable due to the mutual repulsion of the extra electron density. Therefore, the Cl^- is a good electron-pair donor and is the Lewis base. The C atom in H_3C–Br bears a partial positive charge, δ^+, because Br is more electronegative than C. The electrons in Cl^- are attracted to the C atom in H_3C–Br and a new Cl–CH_3 bond forms. The C atom accepts the two electrons from Cl^- and is, therefore, the Lewis acid. The bond between H_3C–Br simultaneously breaks as the new bond forms.

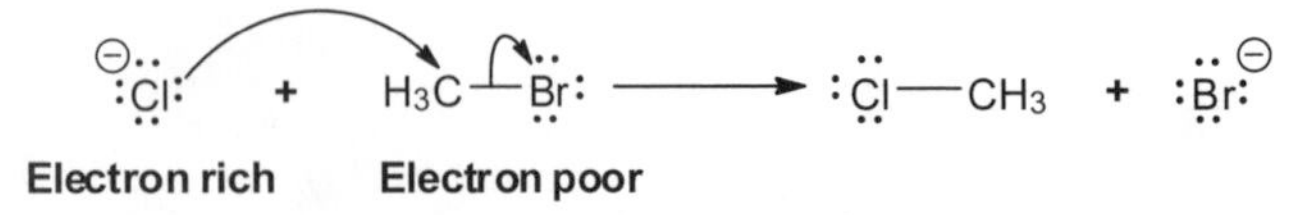

Your Turn 7.5

Think

In looking at an atom, how can you tell if the atom lacks an octet? How do you count bonding electrons? Nonbonding electrons? When a (+) formal charge is drawn on a carbon atom, how many bonds and lone pairs does the carbon atom have?

Solve

Each bond contributes two electrons to an atom's octet and each lone pair contributes two electrons. A positively charged C atom has three bonds and no lone pairs, for a total of six electrons, so each C^+ lacks an octet. The same is true for an uncharged Al. The atoms lacking octets are shown below.

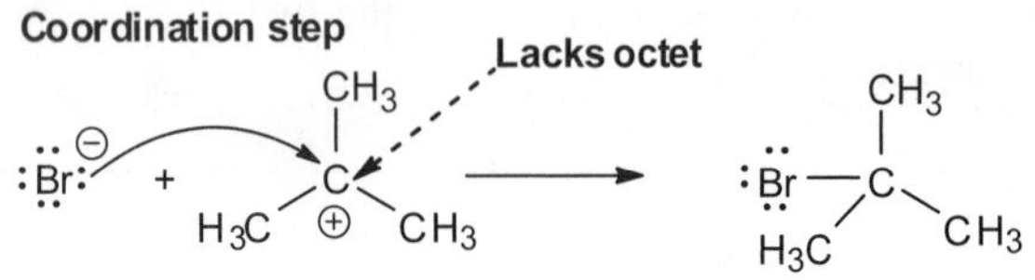

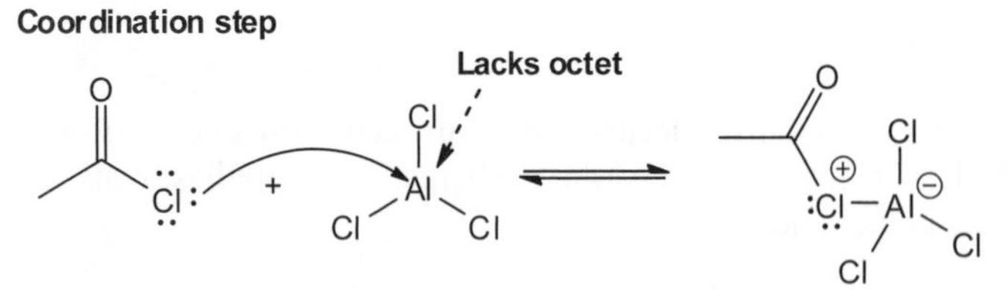

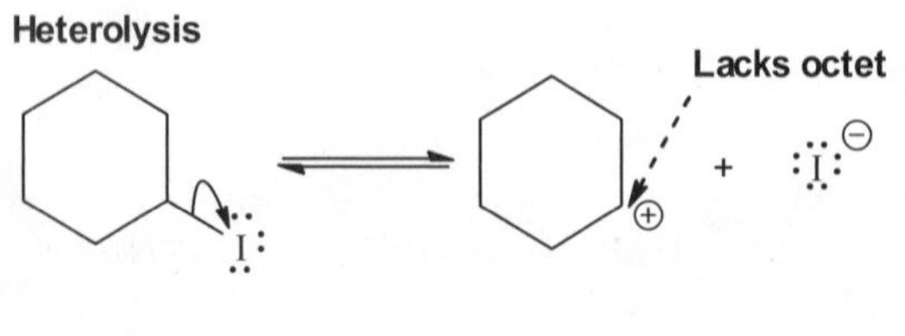

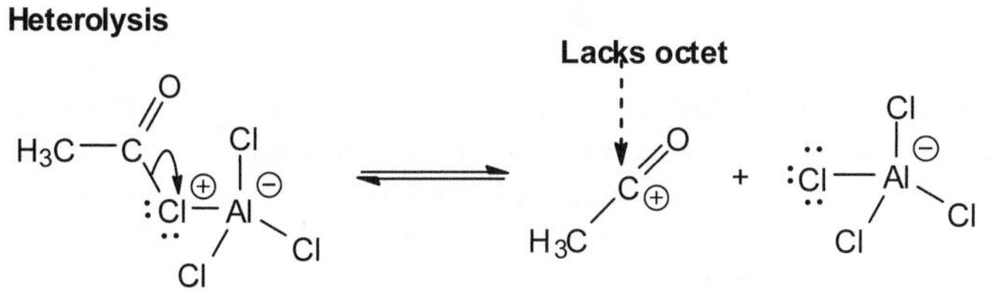

Heterolysis

Lacks octet

(7-8)

Your Turn 7.6

Think

What are the electron-rich and electron-poor sites in each compound? How does lacking an octet affect the electron richness of an atom? Do you think it can more easily accept or donate electrons? Are Lewis bases electron-pair donors or acceptors? Lewis acids? Are any new bonds formed from lone-pair electrons?

Solve

The Cl in the $CH_3C(O)Cl$ compound is electron rich because Cl has lone pairs of electrons and, being more electronegative compared to the C atom, bears a partial negative charge. The Al in $AlCl_3$ lacks an octet and is, therefore, electron poor and accepts an electron pair from the Cl atom. The electron movement is shown below. Because electrons flow from the electron-rich Cl, $CH_3C(O)Cl$ is the Lewis base, and because electrons flow to the Al, $AlCl_3$ is the Lewis acid.

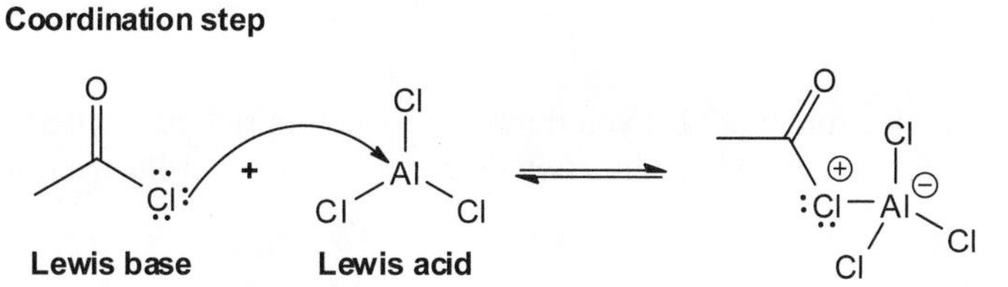

Coordination step

Lewis base

Lewis acid

Your Turn 7.7

Think

What kinds of charges characterize an electron-rich atom? Are any formal charges present? Any strong partial charges? Which bonds are broken and which bonds are formed? How is a curved arrow used to denote the flow of electrons? In which direction do electrons flow—from electron rich to electron poor, or the opposite?

Solve

A C atom bearing a full negative charge is quite electron rich and not particularly stable due to the mutual repulsion of the extra electron density. The C atom in a C=O bears a large, partial positive δ^+ due to the electronegativity difference of C and O. The electrons flow from the $H_3C:^-$ to the C atom of the C=O bond. A second curved arrow (to illustrate the breaking of the π bond between the C and O atoms) is necessary to avoid exceeding an octet on the less electronegative C atom.

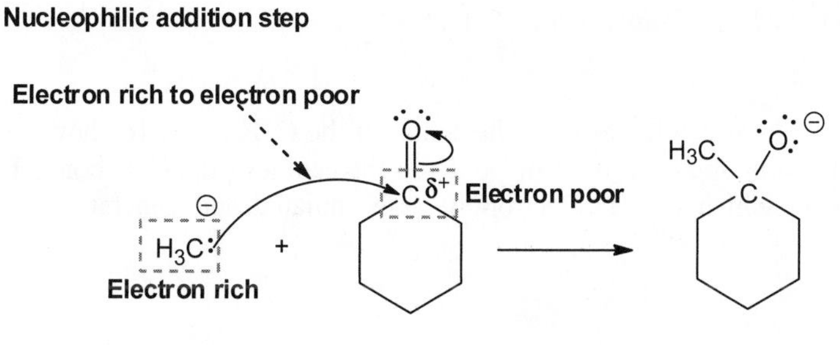

Nucleophilic addition step

Electron rich to electron poor

Electron poor

Electron rich

Your Turn 7.8

Think

What kinds of charges characterize an electron-rich atom? Are any formal charges present? Any strong partial charges? Which bonds are broken and which bonds are formed? How is a curved arrow used to denote the flow of electrons? In which direction do electrons flow—from electron rich to electron poor, or the opposite?

Solve

The O^- is the electron-rich atom and the C atom bonded to two O and two C atoms is electron poor because it bears a partial positive charge, δ^+. A lone pair of electrons on O^- folds down to make a C=O. A second curved arrow is necessary to avoid exceeding an octet on C and the C–OCH_3 bond is broken, leaving CH_3O^- as the charged product.

Nucleophile elimination step

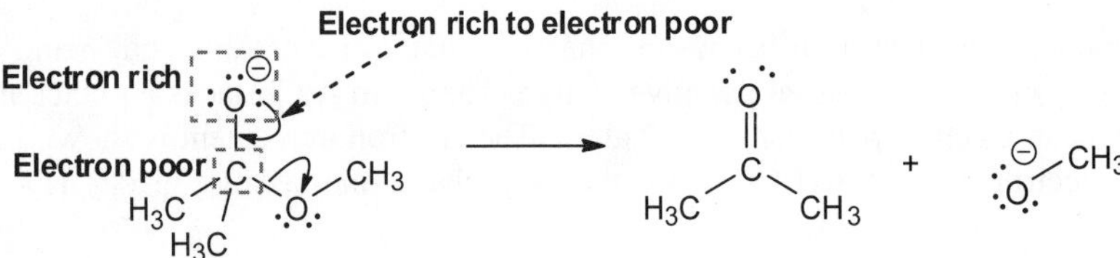

Your Turn 7.9

Think

Are π bond electrons in a C=C electron rich or electron poor? Do you think H^+ is electron rich or electron poor? Which bonds are broken and which bonds are formed? In which direction do electrons flow—from electron rich to electron poor, or the opposite?

Solve

The curved arrow originates from the electron-rich double bond and points to the electron-poor H^+. This is an example of electrophilic addition: The nonpolar C=C π bond gains a strongly electron-deficient species, H^+. The π bond in C=C breaks and is used to form a C–H bond, leaving the other C atom lacking an octet, C^+.

Electrophilic addition step

Your Turn 7.10

Think

Which bond has to break to reform the electrophile? Where is a new bond formed? How do you used curved arrow notation to show the breaking of a bond? The formation of a bond?

Solve

To show the C–H bond breaking, a curved arrow originates from the center of the C–H bond. To show the pair of electrons ending up in the C=C double bond, the curved arrow points to the center of the C–C bond. This is an example of electrophile elimination, in which the H^+ electrophile is eliminated and generates a stable, uncharged organic species.

Electrophile elimination step

Your Turn 7.11

Think

What kinds of charges characterize an electron-rich atom? An electron-poor atom? Are any formal charges present? Are there any nonpolar double or triple bonds? Is the electrophile electron rich or electron poor? Do the curved arrow directions show movement from electron rich to electron poor, or vice versa?

Solve

In electrophilic addition, the H^+ electrophile is the electron-poor site, characterized by the positive charge. The π electrons in the C=C are the electron-rich site, because there are four electrons confined between the two atoms.

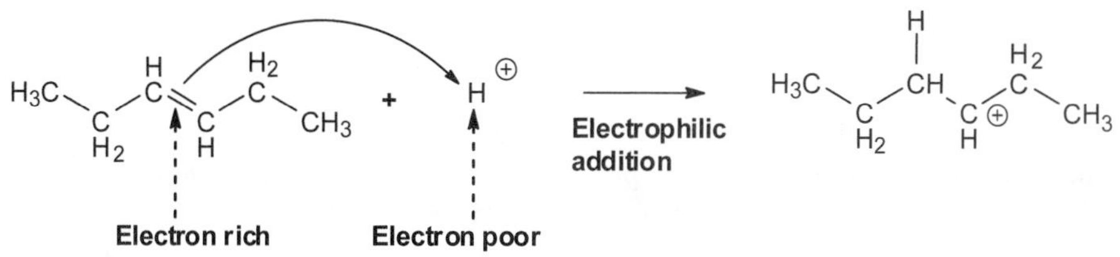

Electron rich   Electron poor   Electrophilic addition

In electrophile elimination, the carbocation C^+ is the electron-poor site and the C–H is the electron-rich site.

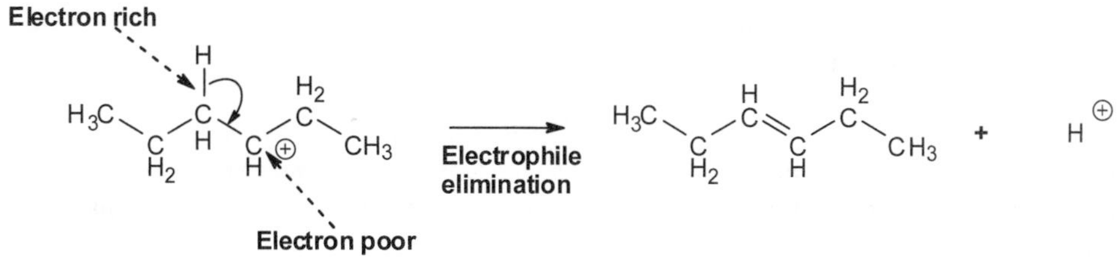

Electron rich   Electron poor   Electrophile elimination

Your Turn 7.12

Think

On what type of C atom—1°, 2°, or 3°—is the carbocation located in each structure? It might be helpful to draw out implied H atoms. Which H atom shifted? How do you represent the breaking of one C–H with the formation of another C–H using curved arrows? In which direction do electrons flow—from electron rich to electron poor, or the opposite?

Solve

The single curved arrow shows a C–H bond breaking from the 3° C atom and simultaneously forming a bond to the 2° C atom. The curved arrow points from the electron-rich single bond to the electron-poor C^+.

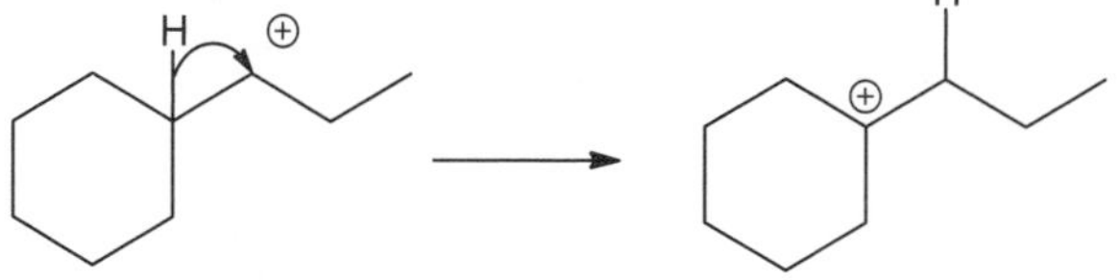

Your Turn 7.13

Think

Consult Table 6-1 for the pK_a values of HCl and HF. Does a more negative pK_a indicate a stronger or weaker acid? Label the stronger and weaker acids. From the proton transfer reaction in Equation 7-29, does the reaction favor the reactants or products? Do you expect K_{eq} to be >1 or <1? Use $K_{eq}(\text{proton transfer}) = 10^{\Delta pK_a}$ to solve for the equilibrium constant.

Solve

HCl's pK_a = −7, and HF's pK_a = 3.2. The difference in pK_a values is: 3.2 − (−7) = 10.2. $K_{eq} = 10^{10.2} = 1.6 \times 10^{10}$. HCl is a stronger acid than HF (it has a more negative pK_a) and the reaction favors the product side (away from the stronger acid). You should expect a large, positive K_{eq} with the ΔpK_a >10.

Your Turn 7.14

Think

Look up the pK_a values for HCl and NH_3 in Table 6-1. Does a more negative pK_a indicate a stronger or weaker acid? Label the stronger and weaker acid. From the proton transfer reaction in Equation 7-29, does the reaction favor the reactants or products? Do you expect K_{eq} to be >1 or <1? Use $K_{eq}(\text{proton transfer}) = 10^{\Delta pK_a}$ to solve for the equilibrium constant.

Solve

HCl's pK_a = −7, and NH_3's pK_a = 36. The difference in pK_a values is 36 − (−7) = 43. $K_{eq} = 10^{43}$.
HCl is a stronger acid than NH_3 (it has a more negative pK_a) and the reaction favors the product side (away from the stronger acid). You should expect a large, positive K_{eq} with the ΔpK_a = 43.

Your Turn 7.15

Think

Consult Figure 7-4 to determine the bond strength for each of the three bonds highlighted in the enol structure (Fig. 7-4a) and the keto structure (Fig. 7-4b). Which bond energy difference is greatest? How does this lead do the conclusion that the keto form is more stable?

Solve

The completed table is below. From this it appears that, because the difference is greatest between the C=O and C=C bond energies, the C=O bond (being the stronger of the two) is the one that has the most influence on the outcome of the reaction.

Bond in Keto Form	Bond in Enol Form	Difference in Bond Energy
C=O	C=C	720 – 619 = 101 kJ/mol
C–C	C–O	339 – 351 = −12 kJ/mol
C–H	O–H	418 – 460 = −42 kJ/mol

Chapter Problems

Problem 7.2

Think

What kinds of charges characterize an electron-poor atom? Are any formal charges present? Any strong partial charges? When the $(CH_3)_3N$ and H_2O are combined, what curved arrow can we draw to depict the flow of electrons from an electron-rich site to an electron-poor site?

Solve

Neither molecule bears any full charges. O is much more electronegative than H and, thus, the H atom is an electron-poor site and the O atom is an electron-rich site. N is more electronegative than C, so the C atom is an electron-poor site and the N atom is an electron-rich site. Only one proton transfer reaction makes sense in terms of "electron rich to electron poor," as shown below.

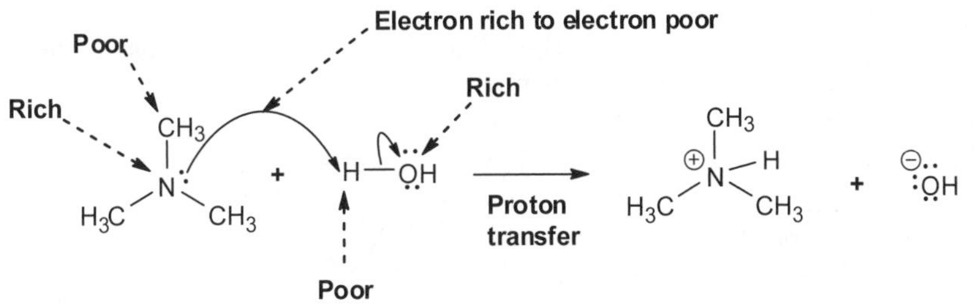

Electron rich to electron poor

Poor

Rich

Rich

Proton transfer

Poor

Problem 7.4

Think

What are the electron-rich and electron-poor sites in each compound? Are there any simplifying assumptions we can make? Are any spectator ions present?

Solve

NaSH is an ionic compound that dissolves in solution as Na^+ and HS^-. Na^+ is treated as a spectator ion and HS^- is treated as an electron-rich Lewis base. CH_3CO_2H has an electron-poor H atom due to the high electronegativity of the O atoms. Thus, CH_3CO_2H is a Lewis acid. A curved arrow is drawn from the electron-rich S atom to the electron-poor H atom to indicate bond formation in a proton transfer. A second curved arrow is drawn to break the O–H bond to avoid two bonds to the H atom.

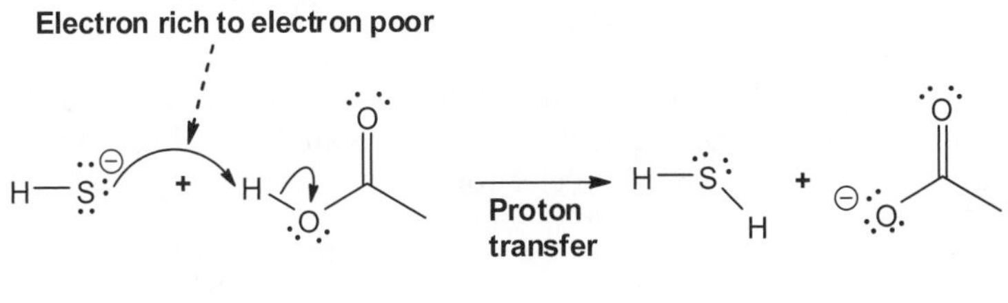

Electron rich to electron poor

Proton transfer

Problem 7.6

Think

What are the electron-rich and electron-poor sites in each compound? Are there any simplifying assumptions we can make? Are any spectator ions present?

Solve

Although the C–Li bond is polar covalent, Li^+ can be treated as a spectator ion and is omitted from the scheme, leaving H_3C^-. Therefore, H_3C^- is highly electron rich and the H of $H–OCH_3$ is electron poor because of the highly electronegative O atom. A CH bond forms and an OH bond breaks.

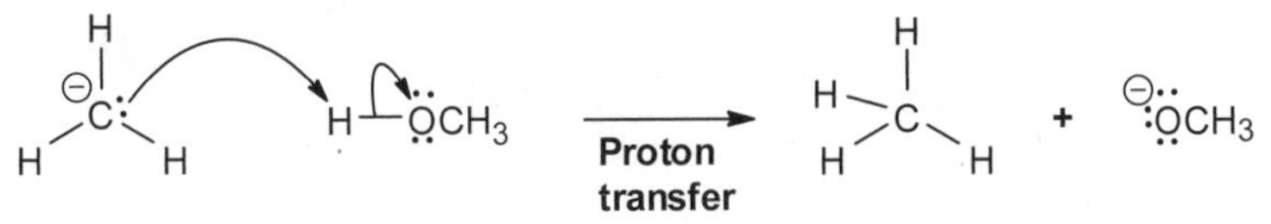

Proton transfer

Problem 7.7

Think

What are the electron-rich and electron-poor sites in each compound? Are there any simplifying assumptions we can make? Are any spectator ions present?

Solve

(a) $LiAlH_4$ may be treated as H^-, which is electron rich. H in H_2O is electron poor because of the highly electronegative O atom. An H–H bond forms and an H–O bond breaks.

Proton transfer

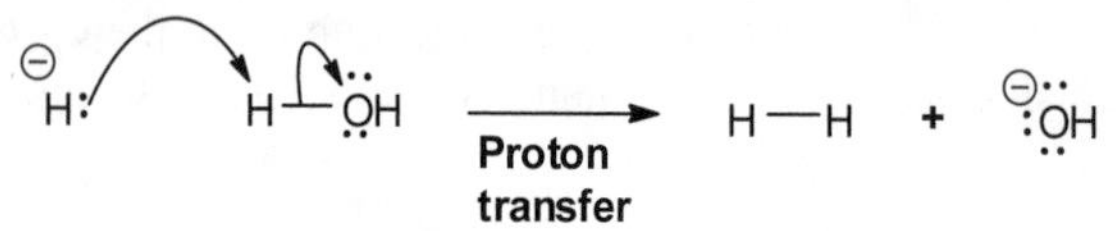

(b) $NaBH_4$ may be treated as H^-, which is electron rich. H in H–O–C_6H_5 is electron poor because of the highly electronegative O atom. An H–H bond forms and an H–O bond breaks.

Proton transfer

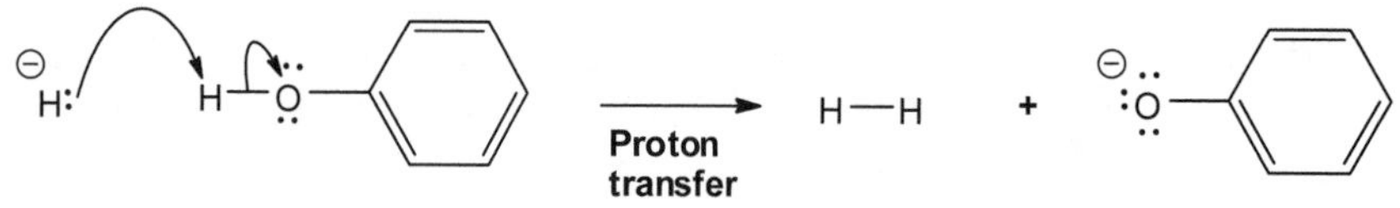

Problem 7.9

Think

Which atoms carry a partial or full negative charge? Does that atom have a pair of electrons that can be used to form a bond to another atom?

Solve

The Lewis structures, omitting the metal atoms, appear as follows.

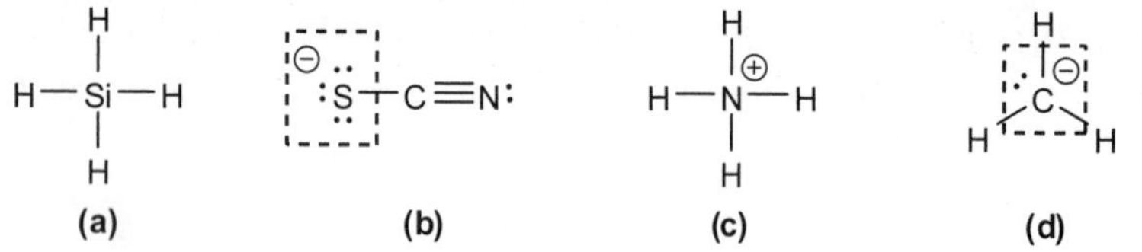

(a) (b) (c) (d)

Only **(b)** and **(d)** can behave as nucleophiles because they have an atom with a lone pair of electrons and some excess negative charge (in these cases, a formal −1 charge). The nucleophilic atoms are boxed. In **(a)** and **(c)**, no atom is especially electron rich and no atom possesses a lone pair of electrons.

Problem 7.11

Think

Which species is the nucleophile? Which is the substrate? What do we do with the metal atom? Which species is electron rich? Electron poor?

Solve

The Cl atom is electronegative and pulls electron density away from the C atom. Thus, the C atom is electron poor and bears a δ^+ charge, and the Cl atom bears a δ^- and is a good leaving group. The curved arrow notation is shown below. The NC^- nucleophile forms a bond to the C atom at the same time the C–Cl bond breaks and the Cl^- leaves.

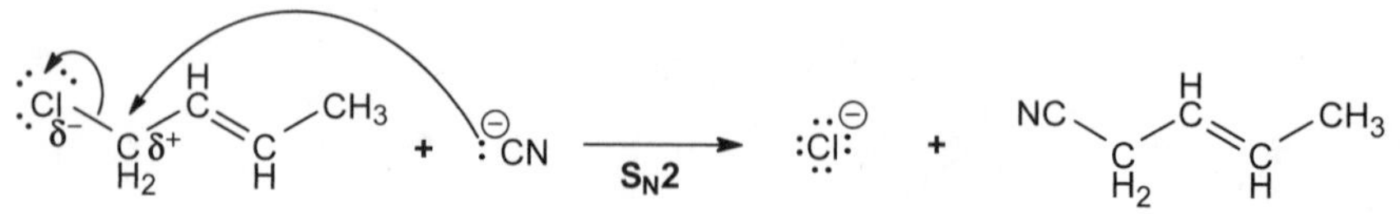

S_N2

Problem 7.12

Think

What sites are electron rich? Electron poor? How many curved arrows are needed to show the coordination step? Are curved arrows drawn from electron rich to electron poor, or vice versa? Which species is donating a pair of electrons and which species is accepting a pair? How are the heterolysis step and coordination step related? How many arrows are needed to show the heterolysis step? Which bond is involved?

Solve

(a) The reaction is shown below. A single curved arrow is used to show the bond formation, and is drawn from the electron-rich Cl^- toward the electron-poor Fe. Cl^- is the Lewis base because it is donating a pair of electrons to form the bond. $FeCl_3$ is the Lewis acid because it accepts that pair of electrons.

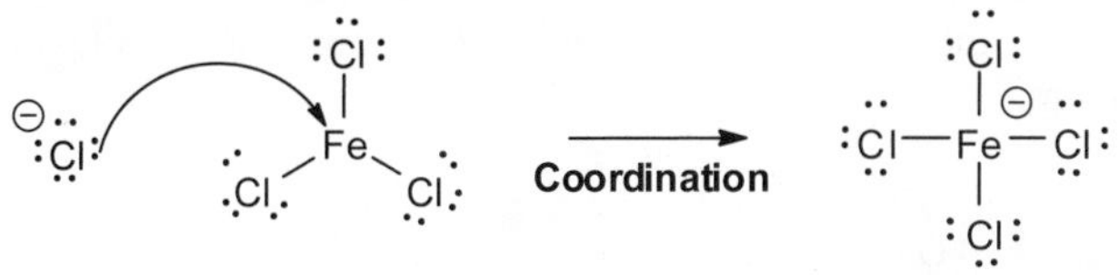

Coordination

(b) The heterolysis step is shown below and is the opposite of the coordination step for this example. The Cl–Fe bond is the bond forming in **(a)** and breaking in **(b)**. To show the bond breaking, a single curved arrow is drawn from the center of the Fe–Cl bond and points toward Cl.

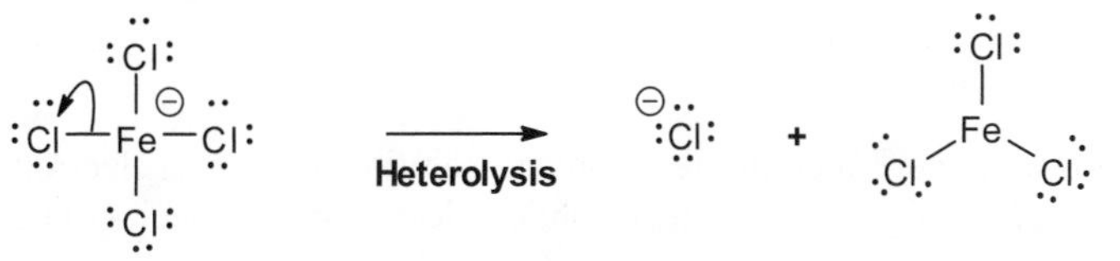

Heterolysis

Problem 7.13

Think

Identify the electron-rich and the electron-poor sites. Which species is the Lewis acid? Lewis base? How many curved arrows are needed to show the bimolecular elimination (E2) step? What can act as the leaving group? On which two C atoms will the double bond form? Refer to Equation 7-12 as a guide.

Solve

The E2 reaction requires three arrows:

1. Base–H bond formation
2. H–C bond breaking and C=C bond forming
3. C–L breaking

(a) Br^- is the leaving group; HO^- is the base and attacks the H atom on the carbon next to the carbon bearing the Br atom.

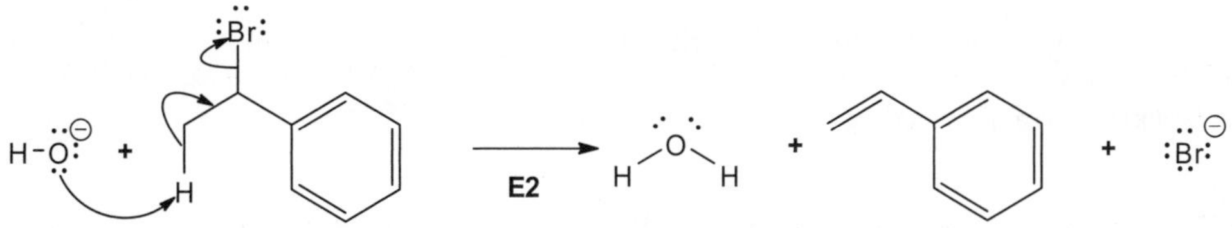

E2

(b) Cl^- is the leaving group; H_3C^- is the base and attacks the H atom on the carbon next to the carbon bearing the Cl atom. A triple bond is formed.

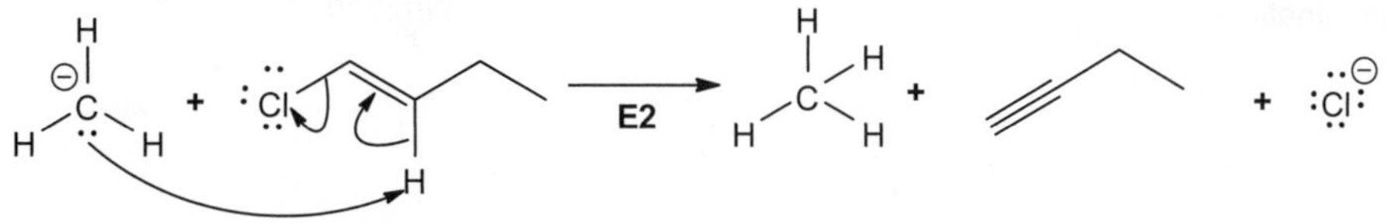

E2

Problem 7.14

Think

Which sites are electron rich? Electron poor? Are curved arrows drawn from electron rich to electron poor, or vice versa? Which species donates a pair of electrons? Which species accepts a pair? How many curved arrows are needed to show the nucleophilic addition step?

Solve

Two arrows are necessary to show the nucleophilic addition step:
1. Nu–C bond formation (addition of the nucleophile to the electron-poor site)
2. C=X π bond breaking (avoids exceeding an octet at the C atom)

(a) PhMgBr can be thought of as a source of C:$^-$ and, thus, the $MgBr^+$ combination is left out. The phenyl carbanion is the nucleophile and the carbon of the C=O is the electrophile (electron-poor species) because it is bonded to a highly electronegative O atom.

Nucleophilic addition

(b) The Na^+ in $NaOCH_3$ is a spectator ion and, therefore, omitted from the solution. CH_3O^- is electron rich and, thus, is the nucleophile. The carbon of the C≡N is the electrophile (electron-poor species) because it is bonded to an electronegative N atom.

:OCH3 Nucleophilic addition OCH3

Problem 7.15

Think

Which species are electron rich? Electron poor? Are curved arrows drawn from electron rich to electron poor, or vice versa? Which species donates a pair of electrons? Which species accepts a pair? How many curved arrows are needed to show the nucleophile elimination step?

Solve

Two arrows are necessary to show the nucleophile elimination step:
1. C=X π bond formation
2. C–L bond breaking (avoids exceeding an octet at the C atom)

(a) OH Nucleophile elimination + :OH

(b) H3CO Nucleophile elimination + :OCH3

Problem 7.16

Think

Which sites are electron rich? Electron-poor? Are curved arrows drawn from electron rich to electron poor, or vice versa? Which species accepts a pair of electrons? Which species donates a pair? How many curved arrows are needed to show the electrophilic addition step?

Solve

One or two arrows are necessary to show the electrophilic addition step:

1. Nonpolar π electrons and electrophile (E^+) bond formation
2. E–Y bond breaking (only necessary to avoid exceeding an octet or duet at E)

(a) The electron-rich, nonpolar, C=C π bond electrons attack the electron-poor carbon of the carbocation. Only one arrow is necessary, as the electrophile lacks an octet. The carbocation is now on the phenyl ring.

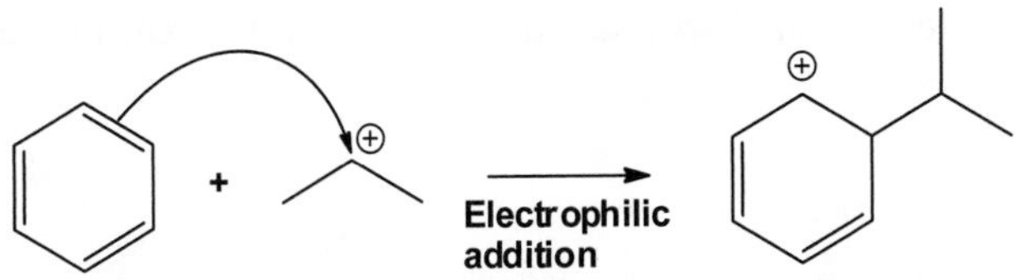

Electrophilic addition

(b) The electron-rich, nonpolar, C≡C π bond electrons attack the electron-poor H atom of H–Br. The H–Br bond must break to avoid exceeding the duet for H, so two arrows are necessary.

Electrophilic addition

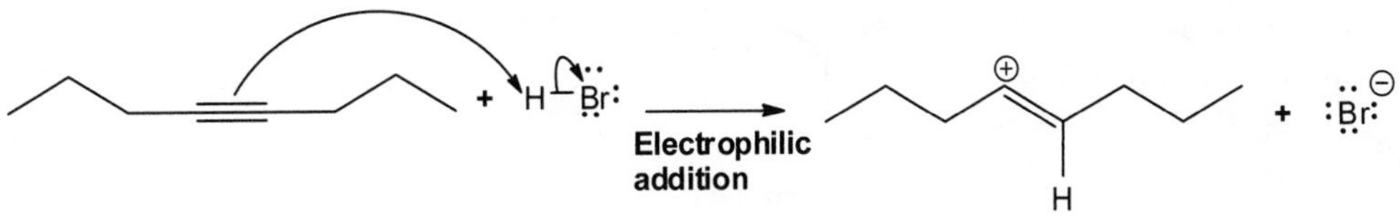

Problem 7.17

Think

Which sites are electron rich? Electron poor? Are curved arrows drawn from electron rich to electron poor, or vice versa? Which species accepts a pair of electrons? Which species donates a pair? How many curved arrows are needed to show the electrophile elimination step?

Solve

Two arrows are necessary to show the electrophile elimination step when the electrophile leaves as H^+.

1. Breaking the C–E bond and forming the C=C π bond.
2. Formation of a H–base bond.

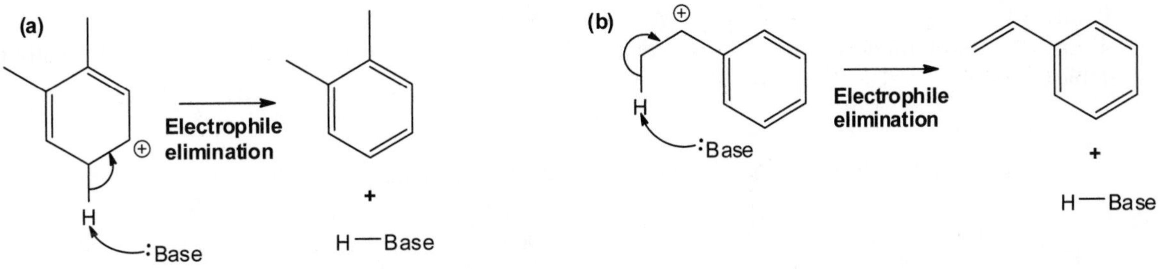

(a) Electrophile elimination + H—Base

(b) Electrophile elimination + H—Base

Problem 7.18

Think

Which bond is broken and formed in a 1,2-hydride shift? A 1,2-methyl shift? How do you show the electron movement via curved arrows? Are curved arrows drawn from electron rich to electron poor, or vice versa?

Solve

In a 1,2-hydride shift, the bond that breaks is the C–H bond adjacent to C^+, and the new bond that forms is the C^+ and hydride C–H bond. In a 1,2-methyl shift, the bond that breaks is the C–CH_3 bond adjacent to C^+ and the new bond that forms is the C^+ and CH_3 C–CH_3 bond.

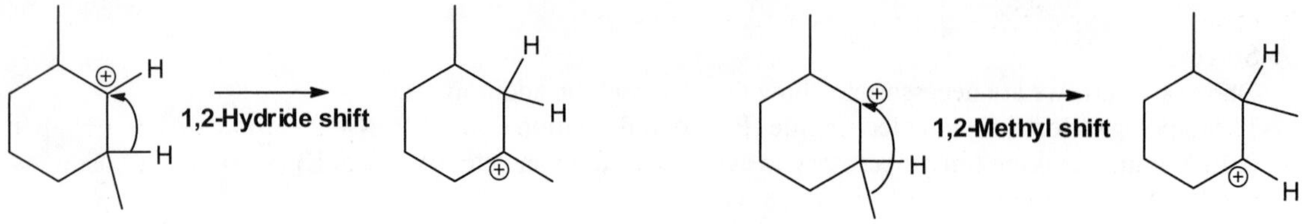

1,2-Hydride shift

1,2-Methyl shift

Problem 7.20

Think

On the two sides of the reaction, is there a difference in charge stability? Is there a difference in total bond energy?

Solve

The favored side of each reaction is boxed.

(a) The + charge on the left is secondary and the + on the right is secondary benzylic. Benzylic carbocations are more stable due to resonance delocalization of the charge around the ring.

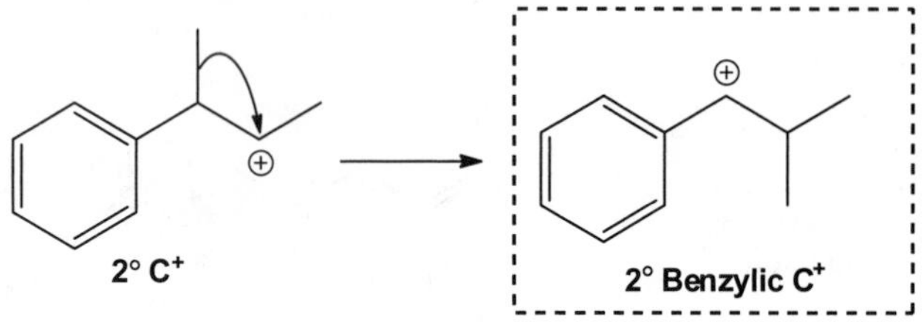

2° C^+ 2° Benzylic C^+

(b) The + charge on the left is secondary and the + on the right is tertiary. Tertiary carbocations are more stable due to electron-donation by the alkyl groups.

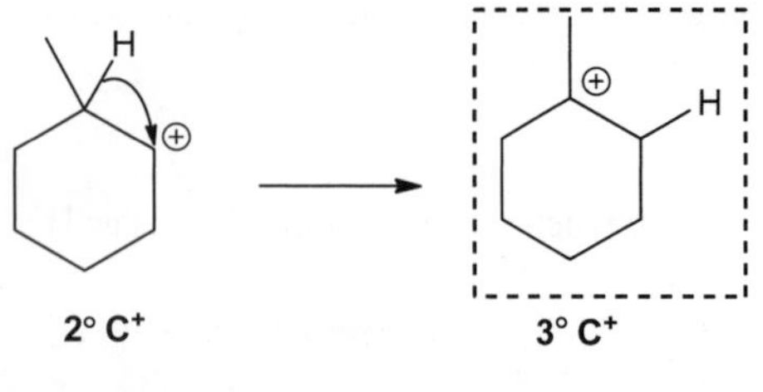

2° C^+ 3° C^+

Problem 7.21

Think

What is the keto form of the enol that immediately forms after decarboxylation? Which form is typically more stable?

Solve

The enol form and keto form are shown below. Most keto forms are more stable than their enol forms.

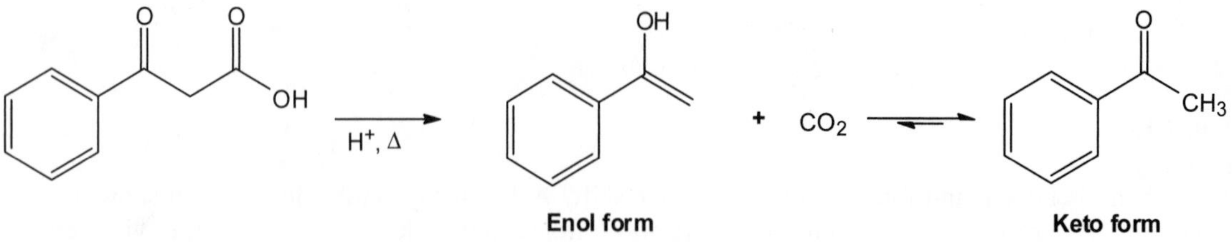

H^+, Δ + CO_2

Enol form Keto form

Problem 7.22

Think

Which sites are electron rich? Electron poor? Are curved arrows drawn from electron rich to electron poor, or vice versa? Which species donates an electron pair? Which species accepts an electron pair? How many curved arrows are needed to show the coordination step? Which bond forms in the coordination step? Which bond breaks in the heterolysis (reverse of coordination) step?

Solve

The Lewis acid is the electron-poor species and accepts the electron pair from the Lewis base, which is the electron-rich species (labeled below). The bond that forms in the Lewis acid/base coordination step is the same bond that breaks in the heterolysis step.

(a) Coordination Step **(b) Heterolysis Step** **Product**

(i)

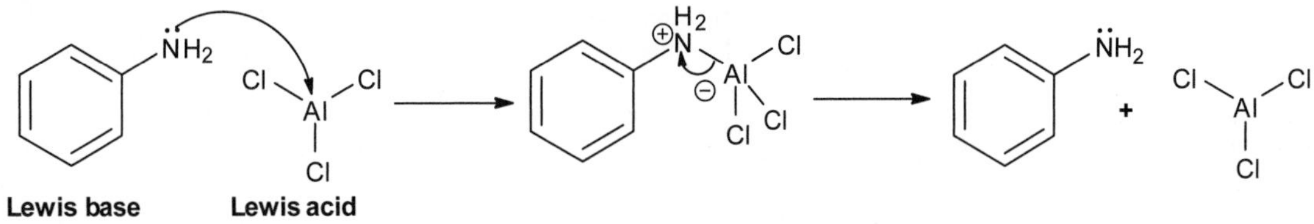

Lewis base Lewis acid

(ii)

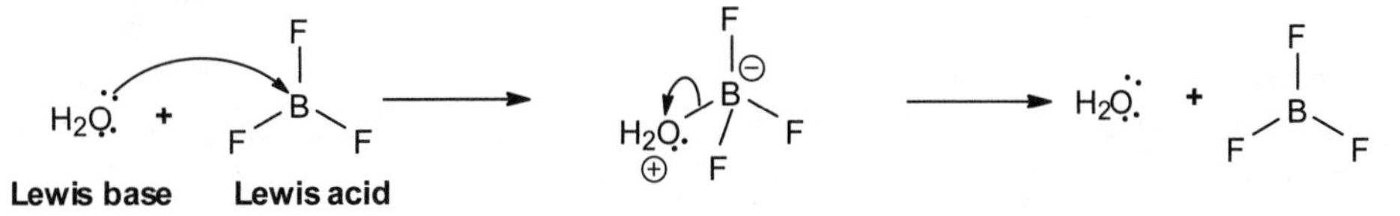

Lewis base Lewis acid

(iii)

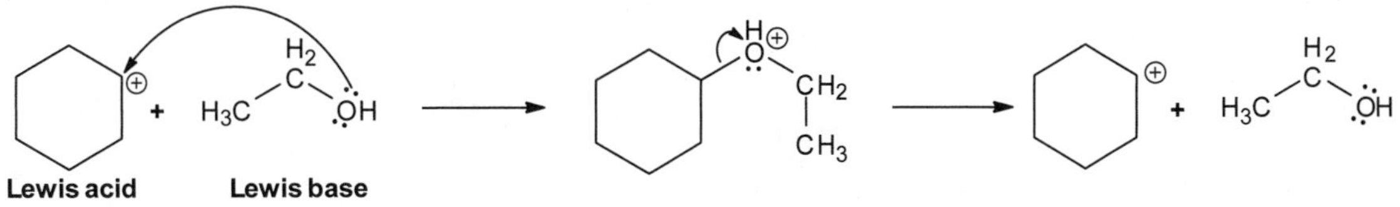

Lewis acid Lewis base

Problem 7.23

Think

Which sites have partial or full negative charges? Partial or full positive charges? Which atoms have a lone pair of electrons? Which atoms lack an octet? Which species is the Lewis acid? Lewis base? What type of bond forms/breaks? How many curved arrows are needed to show the elementary step? Name the elementary step.

Solve

The pertinent electron-rich and electron-poor sites are labeled below and on the next page as "rich" or "poor." The type of elementary step is identified under each reaction arrow.

(i) Rich Poor Proton transfer

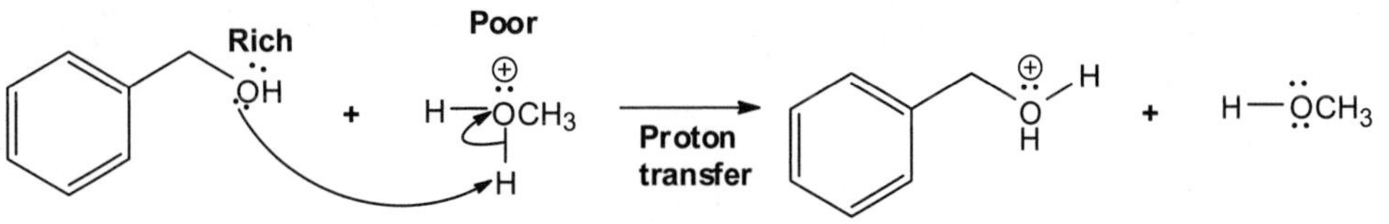

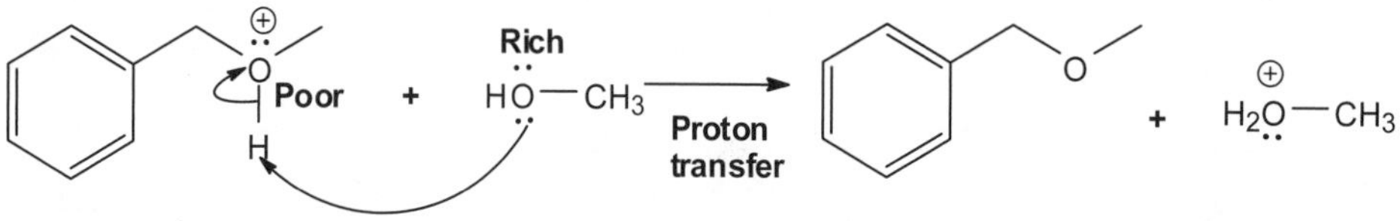

Problem 7.24

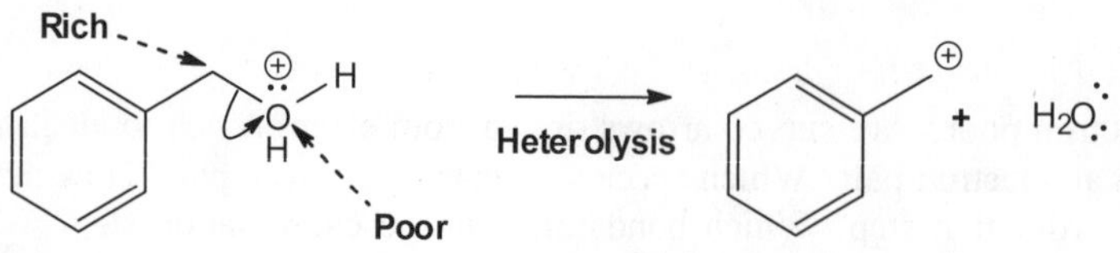

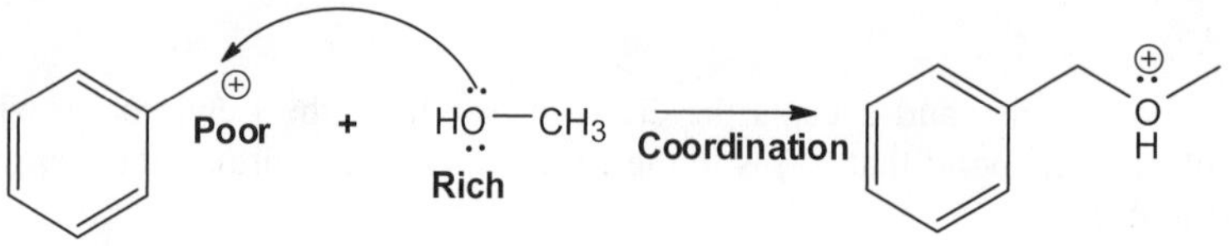

Think

Which sites have partial or full negative charges? Partial or full positive charges? Which atoms have a lone pair of electrons? Which atoms lack an octet? Which species is the Lewis acid? Lewis base? What type of bond forms/breaks? How many curved arrows are needed to show the elementary step? Name the elementary step.

Solve

The pertinent electron-rich and electron-poor sites are labeled below as "rich" or "poor." The type of elementary step is identified under each reaction arrow.

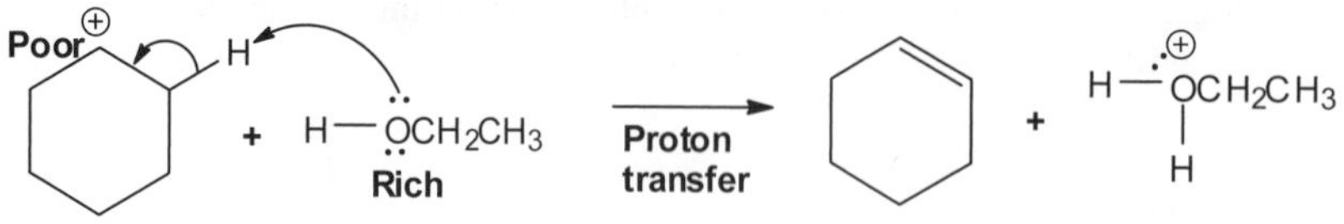

Problem 7.25

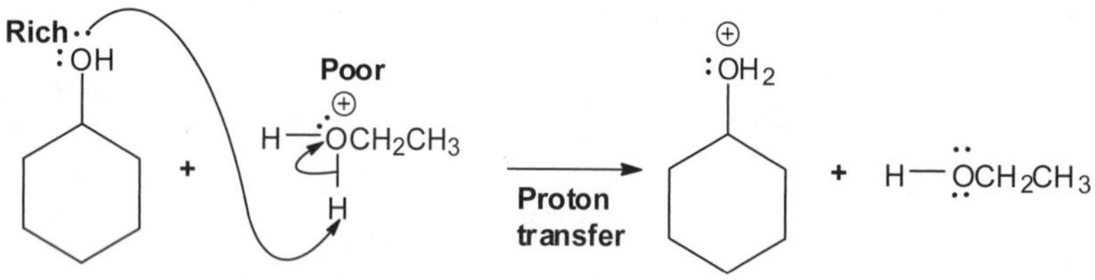

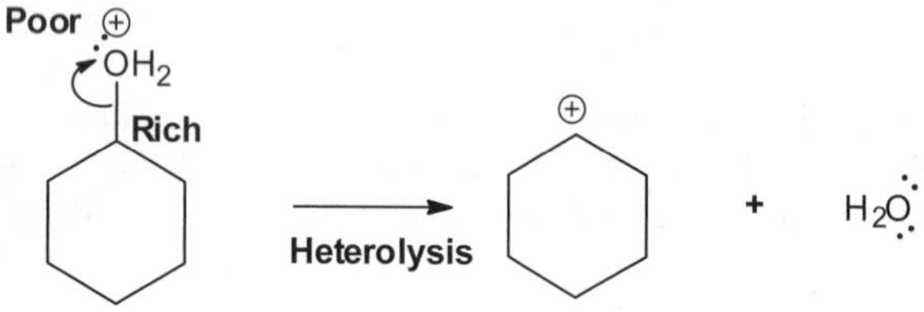

Think

Which sites have partial or full negative charges? Partial or full positive charges? Which atoms have a lone pair of electrons? Which atoms lack an octet? Which species is the Lewis acid? Lewis base? What type of bond forms/breaks? How many curved arrows are needed to show the elementary step? Name the elementary step.

Solve

The pertinent electron-rich and electron-poor sites are labeled below as "rich" or "poor." The type of elementary step is identified under each reaction arrow.

(a) Rich Poor Proton transfer

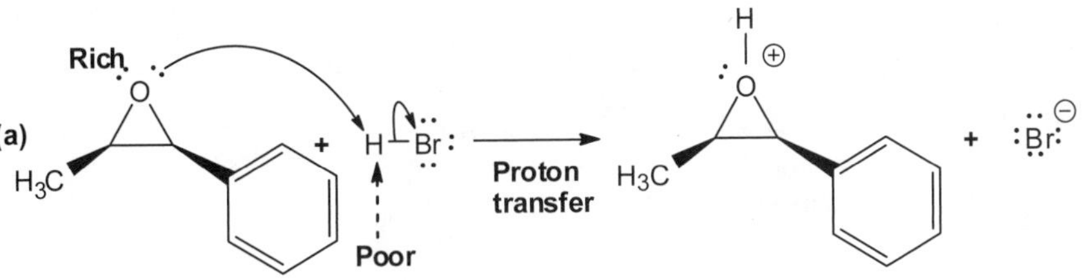

(b) Poor Rich S_N2

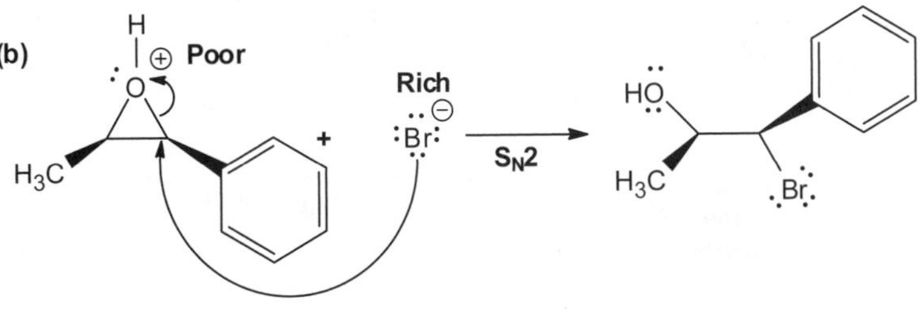

Problem 7.26

Think

Which sites have partial or full negative charges? Partial or full positive charges? Which atoms have a lone pair of electrons? Which atoms lack an octet? Which species is the Lewis acid? Lewis base? What type of bond forms/breaks? How many curved arrows are needed to show the elementary step? Name the elementary step.

Solve

The pertinent electron-rich and electron-poor sites are labeled below as "rich" or "poor." The type of elementary step is identified under each reaction arrow.

(i) Poor Rich S_N2

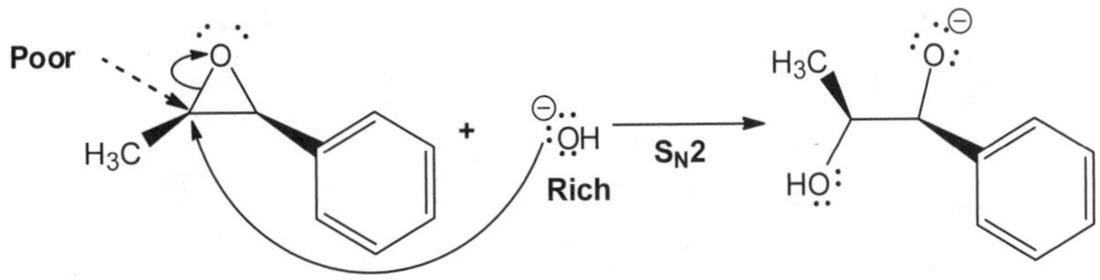

(ii) Rich Poor Proton transfer

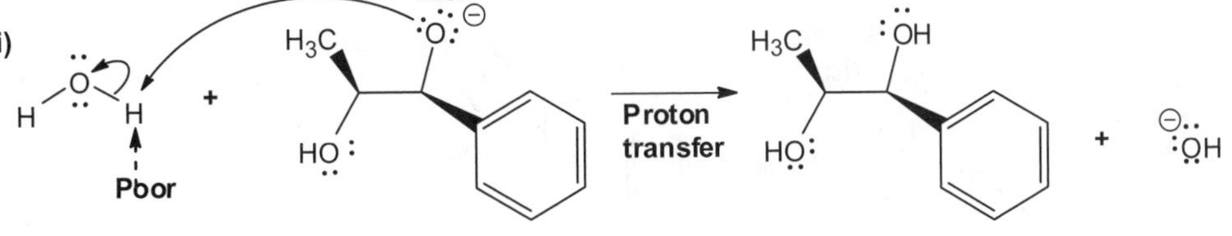

Problem 7.27

Think

Which sites have partial or full negative charges? Partial or full positive charges? Which atoms have a lone pair of electrons? Which atoms lack an octet? Which species is the Lewis acid? Lewis base? What type of bond forms/breaks? How many curved arrows are needed to show the elementary step? Name the elementary step.

Solve

The pertinent electron-rich and electron-poor sites are labeled below "rich" or "poor." The type of elementary step is identified under each reaction arrow.

(i) Rich, Poor — Proton transfer

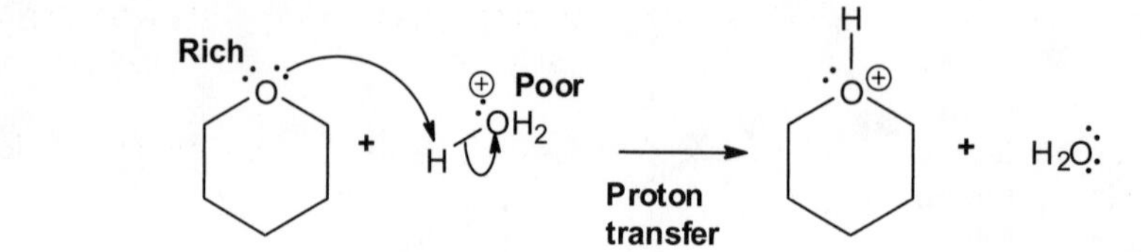

(ii) Poor, Rich — S_N2

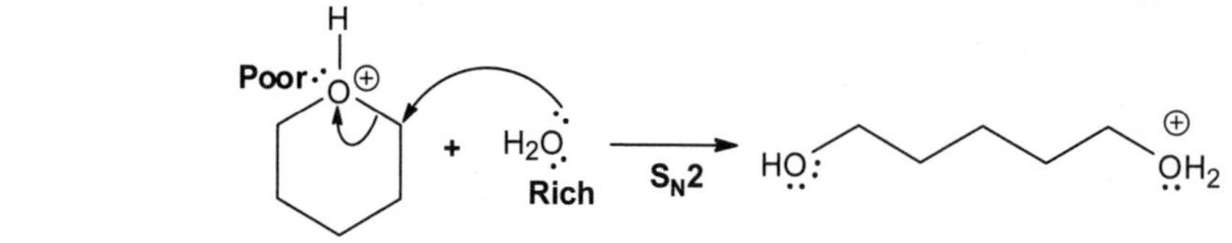

(iii) Poor, Rich — Proton transfer

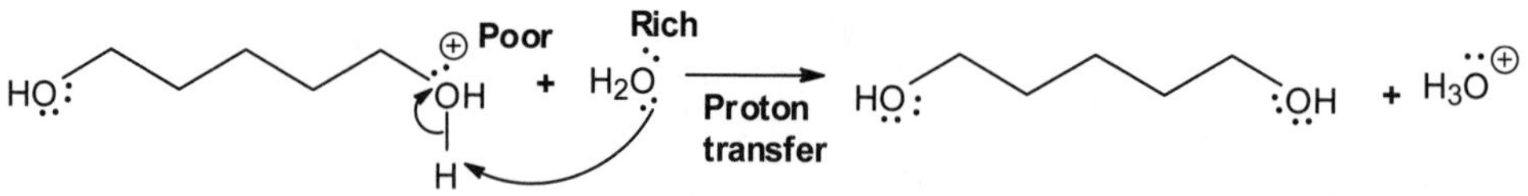

Problem 7.28

Think

Which sites have partial or full negative charges? Partial or full positive charges? Which atoms have a lone pair of electrons? Which atoms lack an octet? Which species is the Lewis acid? Lewis base? What type of bond forms/breaks? How many curved arrows are needed to show the elementary step? Name the elementary step.

Solve

The pertinent electron-rich and electron-poor sites are labeled below as "rich" or "poor." The type of elementary step is identified under each reaction arrow.

(i) Rich, Poor — S_N2

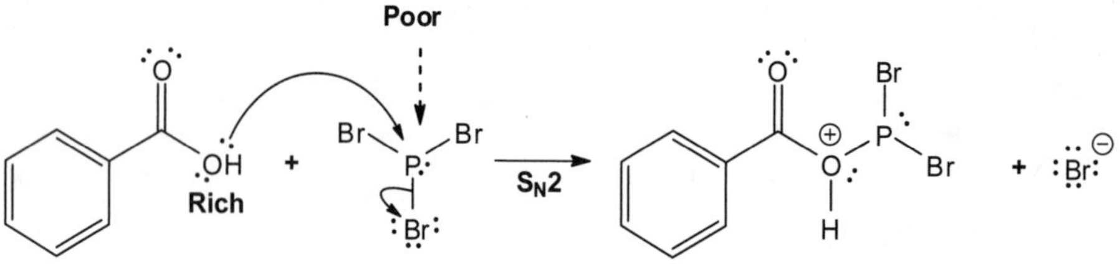

(ii) Poor, Rich — Nucleophilic addition

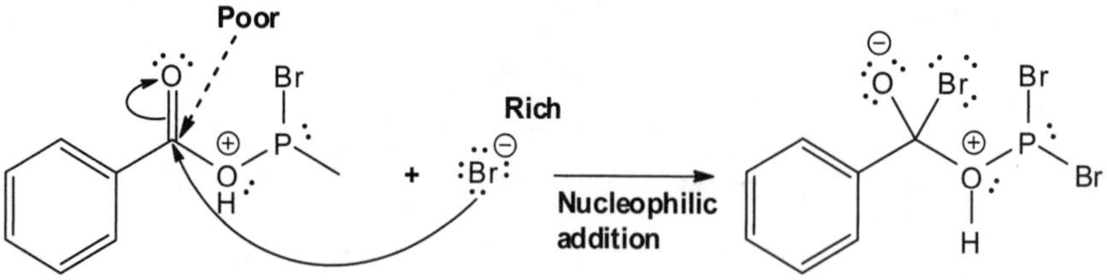

(iii) Rich, Poor — Nucleophile elimination

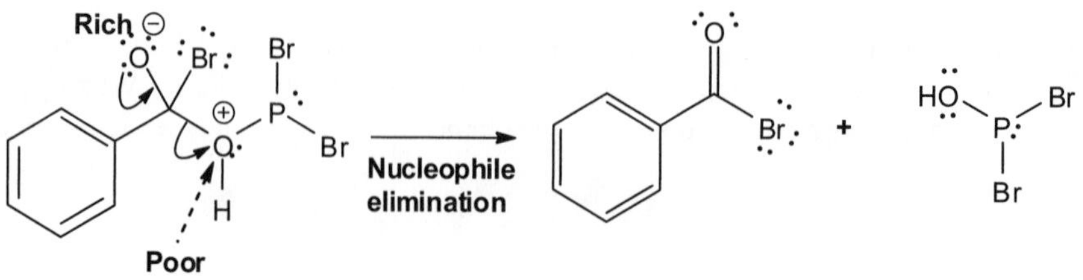

Problem 7.29

Think

Which sites have partial or full negative charges? Partial or full positive charges? Which atoms have a lone pair of electrons? Which atoms lack an octet? Which species is the Lewis acid? Lewis base? What type of bond forms/breaks? How many curved arrows are needed to show the elementary step? Name the elementary step.

Solve

The pertinent electron-rich and electron-poor sites are labeled below as "rich" or "poor." The type of elementary step is identified under each reaction arrow.

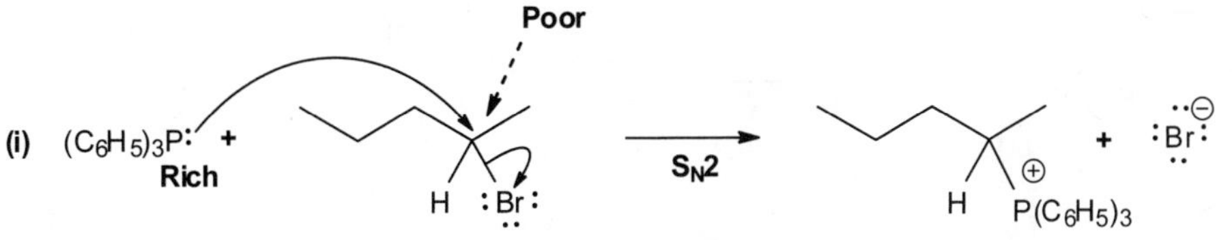
(i)
Poor
Rich
S_N2

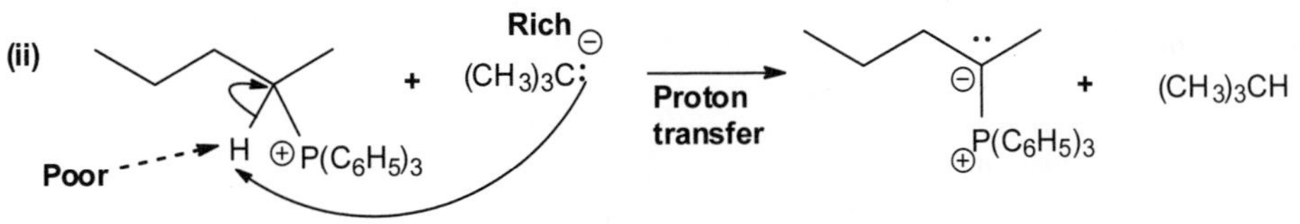
(ii)
Rich
Poor
Proton transfer
$(CH_3)_3CH$

Problem 7.30

Think

Which sites have partial or full negative charges? Partial or full positive charges? Which atoms have a lone pair of electrons? Which atoms lack an octet? Which species is the Lewis acid? Lewis base? What type of bond forms/breaks? How many curved arrows are needed to show the elementary step? Name the elementary step.

Solve

The pertinent electron-rich and electron-poor sites are labeled below and on the next page as "rich" or "poor." The type of elementary step is identified under each reaction arrow.

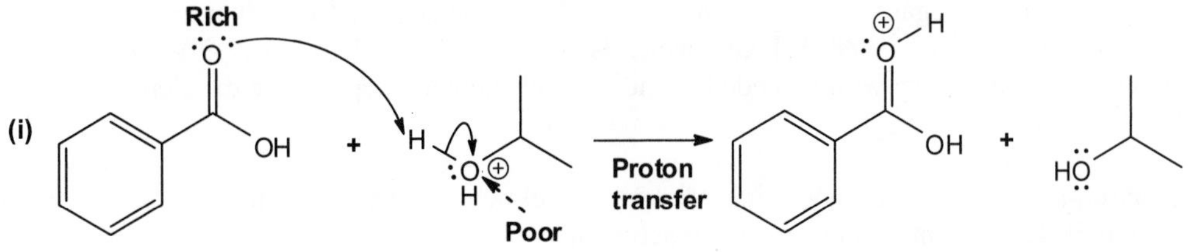
(i)
Rich
Poor
Proton transfer

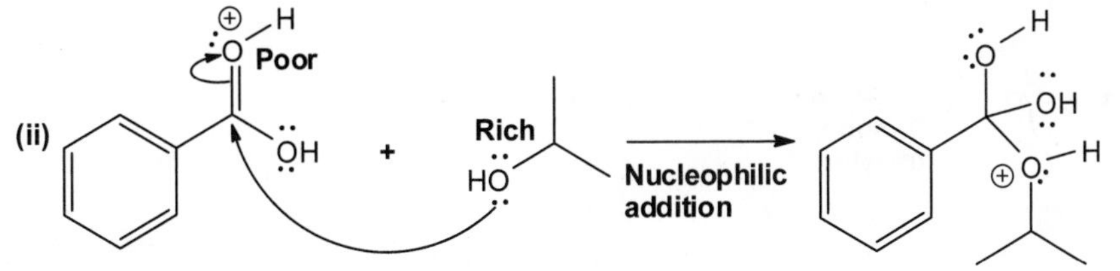
(ii)
Poor
Rich
Nucleophilic addition

(iii) Poor; Rich; Proton transfer

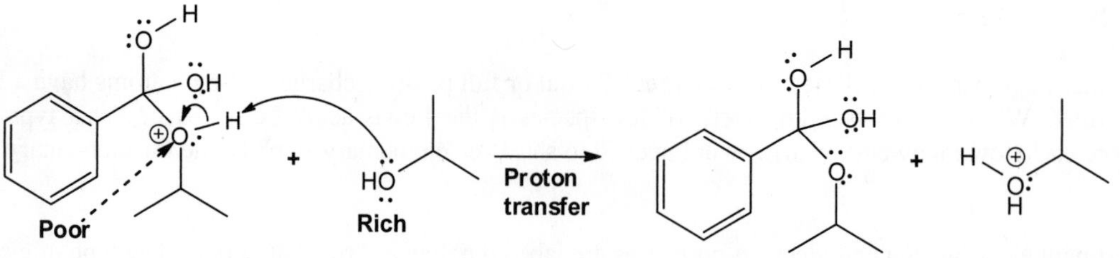

(iv) Rich; Poor; Proton transfer

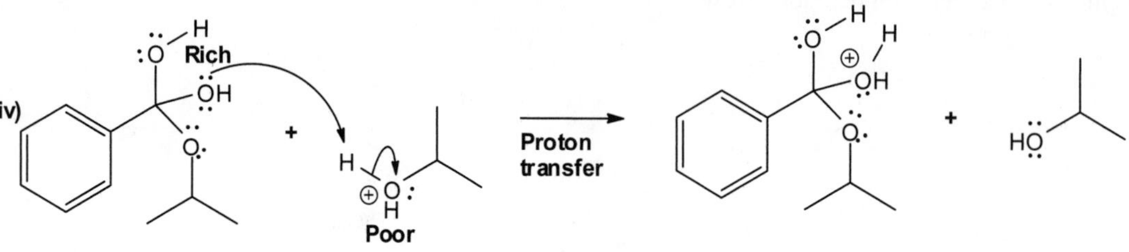

(v) Rich; Poor; Nucleophile elimination

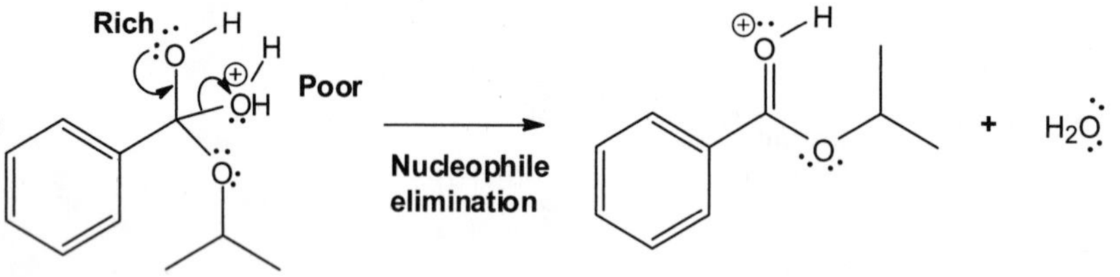

(vi) Poor; Rich; Proton transfer

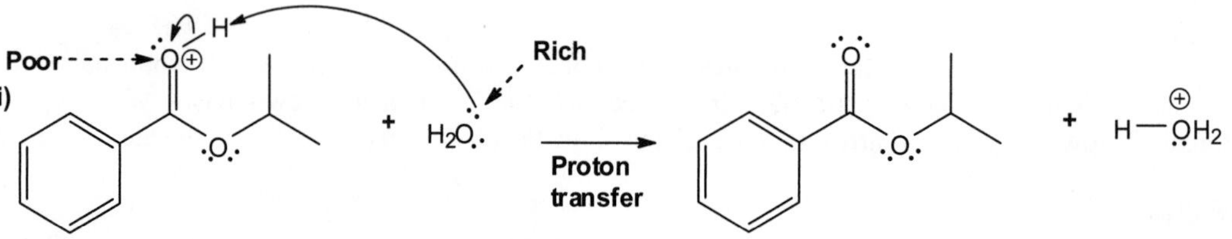

Problem 7.31

Think

Which sites have partial or full negative charges? Partial or full positive charges? Which atoms have a lone pair of electrons? Which atoms lack an octet? Which species is the Lewis acid? Lewis base? What type of bond forms/breaks? How many curved arrows are needed to show the elementary step? Name the elementary step.

Solve

The pertinent electron-rich and electron-poor sites are labeled below and on the next page as "rich" or "poor." The type of elementary step is identified under each reaction arrow.

(i) Rich; Poor; Proton transfer

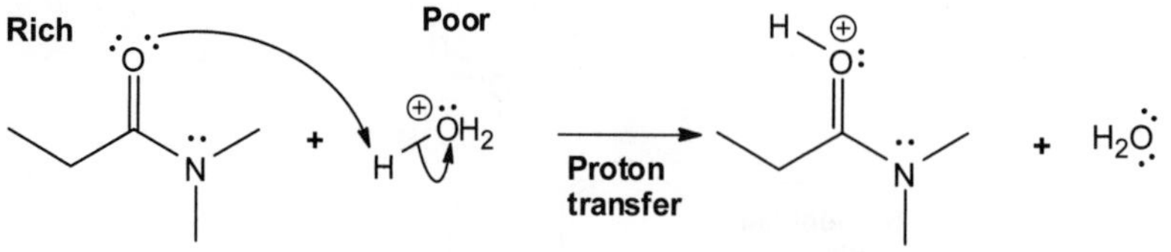

(ii) Poor; Rich; Nucleophilic addition

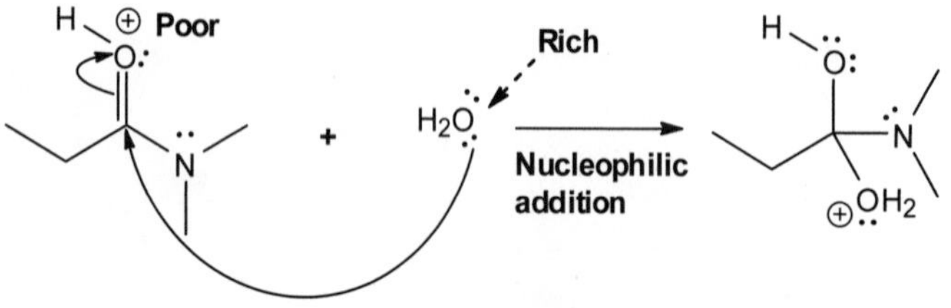

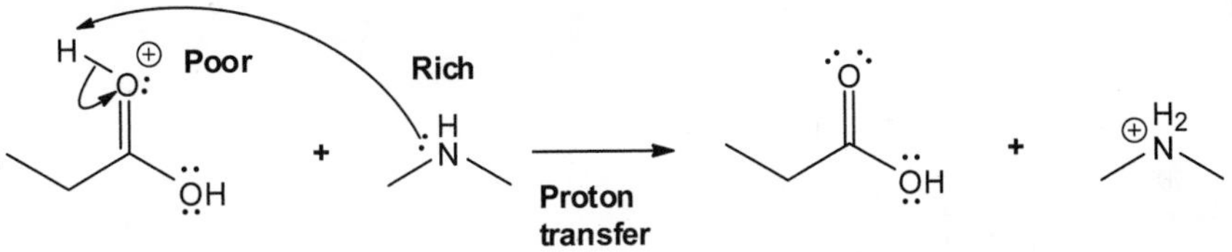

Problem 7.32

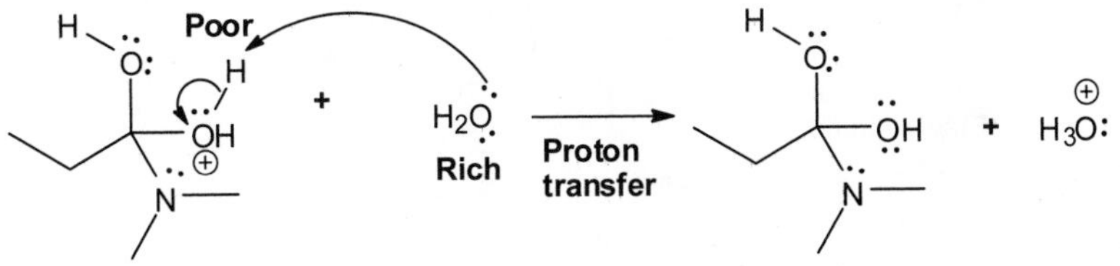

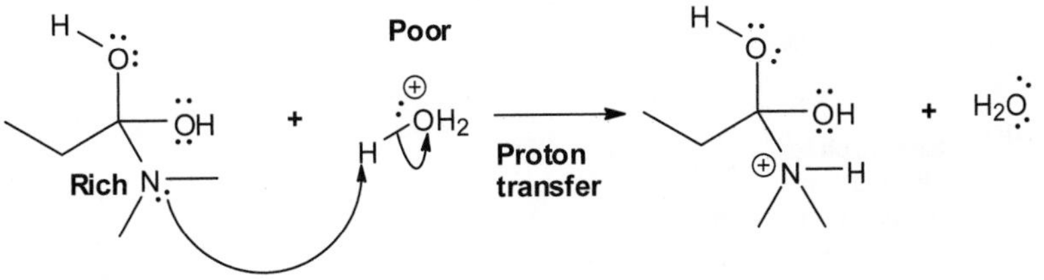

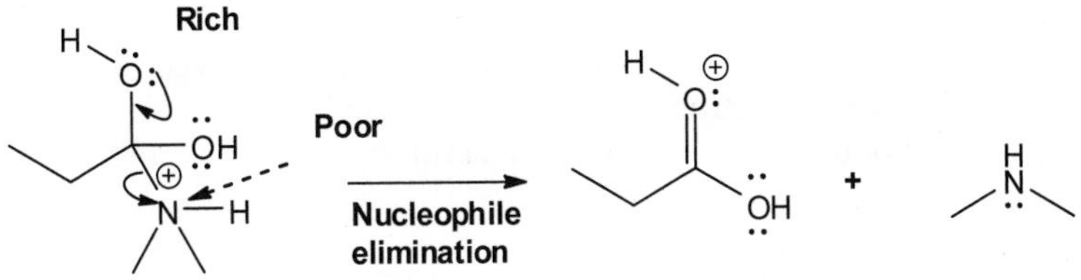

Think

Which sites have partial or full negative charges? Partial or full positive charges? Which atoms have a lone pair of electrons? Which atoms lack an octet? Which species is the Lewis acid? Lewis base? What type of bond forms/breaks? How many curved arrows are needed to show the elementary step? Name the elementary step.

Solve

The pertinent electron-rich and electron-poor sites are labeled below and on the next page as "rich" or "poor." The type of elementary step is identified under each reaction arrow.

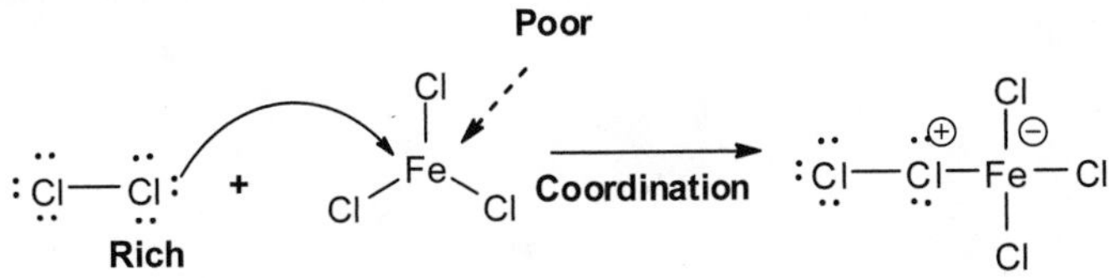

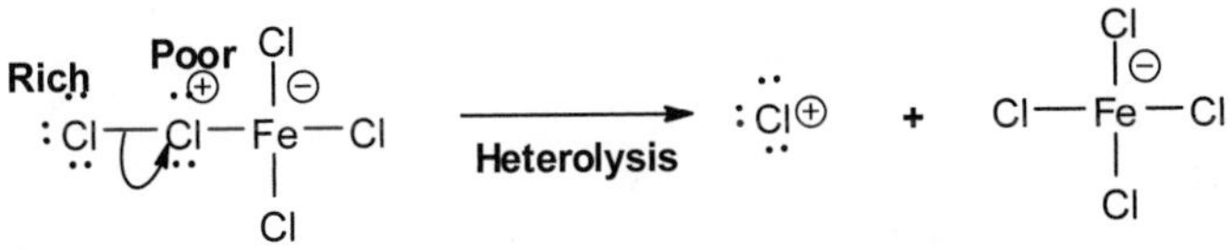

(iii) Rich + $:Cl^{\oplus}$ Poor → Electrophilic addition

(iv) Poor + H_2O Rich → Electrophile elimination (or proton transfer) + $H_3O^{\oplus}$

Problem 7.33

Think

Use the curved arrow notation to follow the bond making and breaking. Do the curved arrows show the movement of electrons? Do the arrows flow from electron rich to electron poor? Which of the 10 elementary steps (if any) is represented? Do any of the products violate the duet/octet rule?

Solve

Products are drawn for all reactions. The unacceptable reactions are marked with an X.

(a) No. The curved arrow is drawn from electron poor to electron rich. The curved arrow shows the movement of atoms, not electrons.

Rich ... $H—OH_2^{\oplus}$ Poor → (X) ... $H—OH_2^{\oplus}$

(b) No. In the products, there would be four electrons on the H atom—a bond and a lone pair.

Poor H ... Rich $:OH^{\ominus}$ → (X) ... + $:H^{-2}—OH$

(c) No. A bond is formed to a C atom with four bonds, so a bond must be broken simultaneously. Otherwise, the C atom would end up with five bonds.

$HO^{\ominus}$ Rich ... Poor → (X) HO ...

(d) Yes. This is an S_N2 reaction.

$H_3N:$ Rich ... Poor ... Br → S_N2 → $NH_3^{\oplus}$ ... + $:Br:^{\ominus}$

(e) Yes. This is an electrophilic addition step.

Electrophilic addition

Problem 7.34

Think

If a charged species results as a product, compare the charge stability of the ions in each reaction. What types of bonds broke? Formed? Compare the bond energy in the products and reactants for each reaction. Which driving force is in play in each example?

Solve

(a) There is greater charge stability in the products of the second reaction. The primary carbocation in the first reaction is too unstable because only one alkyl group inductively stabilizes the positive charge. In the second reaction, a tertiary carbocation is produced, in which three alkyl groups stabilize the positive charge.

CH_3CH_2OH 1° C+ CH_3CH_2OH 3° C+

(b) Charge stability and bond energies both favor the second reaction substantially more than the first. In the first reaction, the C atom that has the leaving group is sp^2-hybridized, so the C–Cl bond is too strong and the positive charge that is formed is too unstable (the C atom has a fairly high effective electronegativity). In the second reaction, the bond that breaks is weaker because it involves an sp^3-hybridized C atom, and the product is a benzylic carbocation in which the positive charge is delocalized by resonance.

CH_3CH_2OH *sp*² C+ CH_3CH_2OH *sp*³ C+ **Benzylic**

(c) DMSO is an aprotic solvent and does not stabilize the resulting ions very well via solvation. CH_3CH_2OH (or any alcohol) is a protic solvent and stabilizes carbocations via solvation.

DMSO **Aprotic solvent** 3° C+ CH_3CH_2OH **Protic solvent** 3° C+

(d) The second reaction is favored much more by charge stability. H is a very small atom and not very electronegative. The negative charge is more stable on I^- because the I atom is much larger. I^- is the conjugate base of the strong acid H–I.

CH_3CH_2OH CH_3CH_2OH

Problem 7.35

Think

If the indicated H atom leaves as H^+, where do the electrons go? What type of bond results? What is the definition of a diastereomer?

Solve

Because free rotation can occur about the central C–C bond, both cis and trans isomers are formed, as shown below. The curved arrow notation, by itself, does not specify the formation of one stereoisomer or the other.

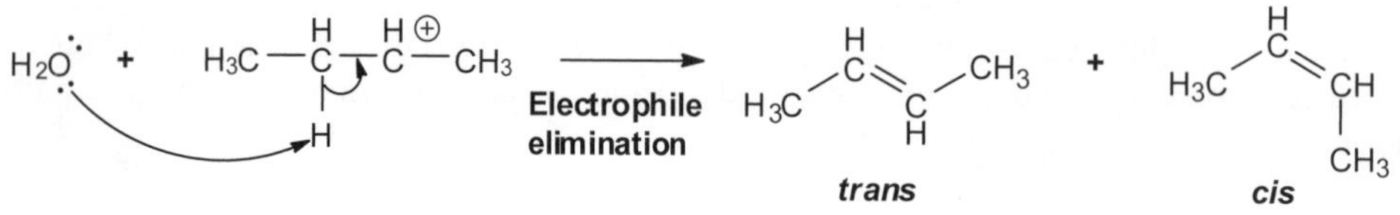

Electrophile elimination

trans *cis*

Problem 7.36

Think

On which carbons can the new C–H bond form? If the proton adds to one C atom, what is generated on the other C atom? On which C atom is a positive charge better stabilized? Why?

Solve

The H^+ can be deposited onto either of the alkene C atoms. A positive charge results on the other C atom. The two carbocations that can be formed are shown below. The second product is more stable because it can participate in resonance with the lone pairs on the O atom (shown).

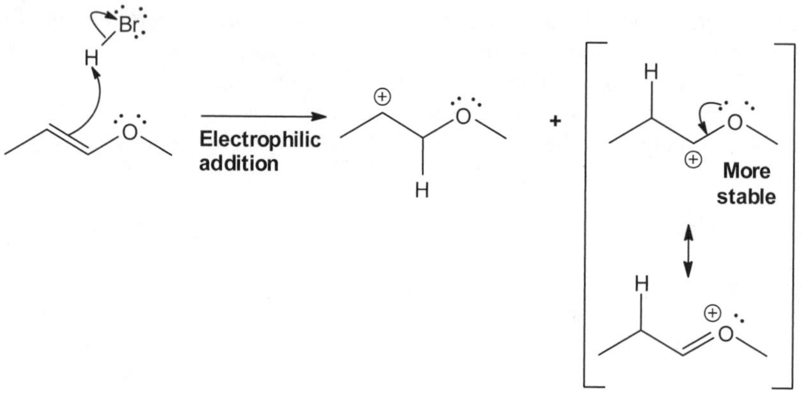

Electrophilic addition

More stable

Problem 7.37

Think

Where are the three H atoms that can be eliminated located? What product results from an electrophile elimination step? How many curved arrows are necessary when a proton is eliminated?

Solve

The three reactions and their products are shown below and on the next page. Each reaction is the result of eliminating a different proton. Notice that two curved arrows are necessary to show the elimination of a proton: One curved arrow is used to show the breaking of the H–C bond and the simultaneous formation of the C–C bond. The second curved arrow shows a base forming a bond to the proton. The tetrasubstituted alkene is the most stable and, therefore, will be the major product.

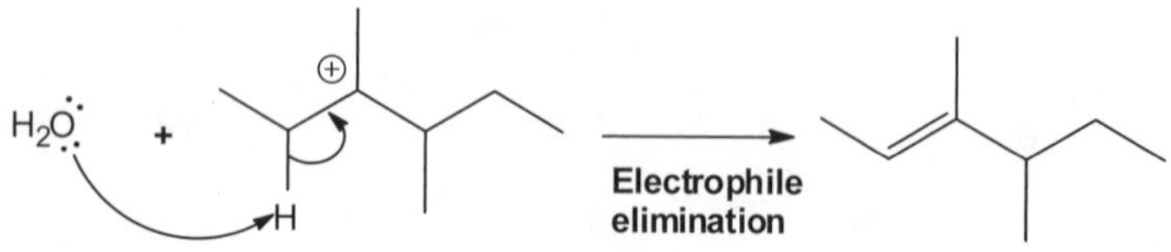

Electrophile elimination

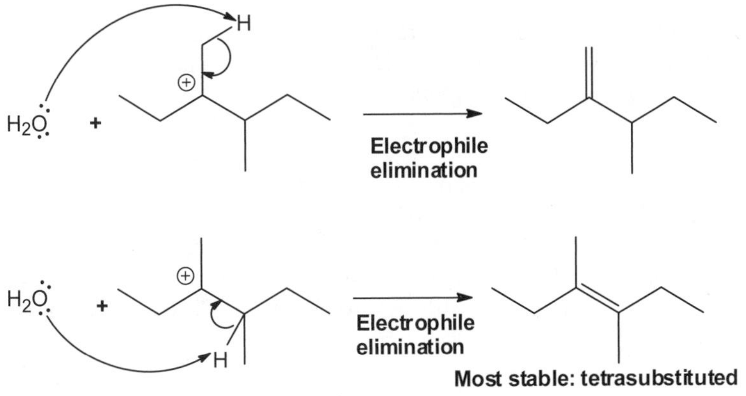

Electrophile elimination

Electrophile elimination

Most stable: tetrasubstituted

Problem 7.38

Think

Which sites are electron rich? Electron poor? Are curved arrows drawn from electron rich to electron poor, or vice versa? How many curved arrows are used to show electrophilic addition? If the electrophile adds to one C atom of a C=C, then what appears on the other C? What is the difference in the stabilities of the carbocations from electrophilic addition at the para, ortho, and meta sites?

Solve

The reactions and their products are shown below. The C=C is electron rich and NO_2^+ is electron poor. A single curved arrow is drawn from the center of C=C and points to NO_2^+. When NO_2^+ adds to one C atom of the C=C, a + charge is generated on the other C atom. Electrophilic addition at the ortho and para sites results in additional resonance structures. These two sites are, therefore, preferred over the meta site.

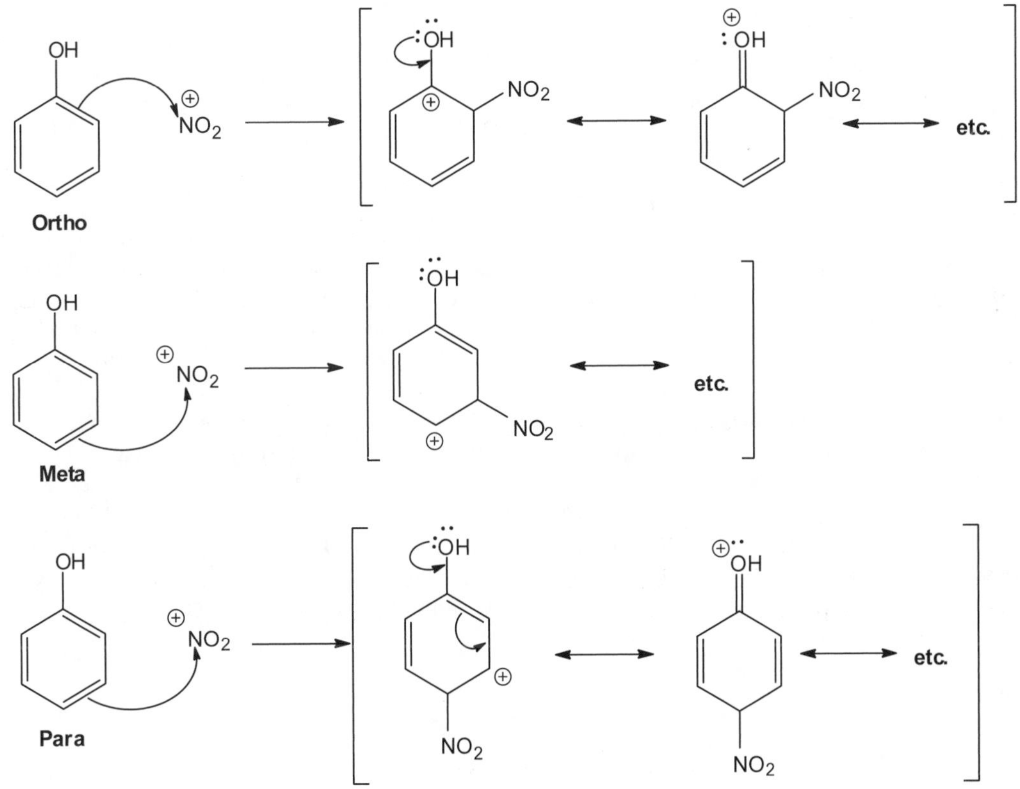

Ortho

Meta

Para

Problem 7.39

Think

Which sites are electron rich? Electron poor? Are curved arrows drawn from electron rich to electron poor, or vice versa? How many curved arrows are used to show nucleophile elimination? Which anion stabilizes a charge better? How does that charge stability affect the major product outcome?

Solve

The possible products are shown below. In each case, a curved arrow originates from O^- and points to the center of the C–O bond because O^- is electron rich and the C atom is electron poor. The second curved arrow is necessary to avoid exceeding the octet on carbon. The products of the second reaction are favored because Cl is substantially larger than C or O, so the negative charge is more stable on Cl than it is on either C or O. Moreover, Cl^- is a weak base because it is the conjugate base of the strong acid H–Cl. C^- and O^- are strong bases.

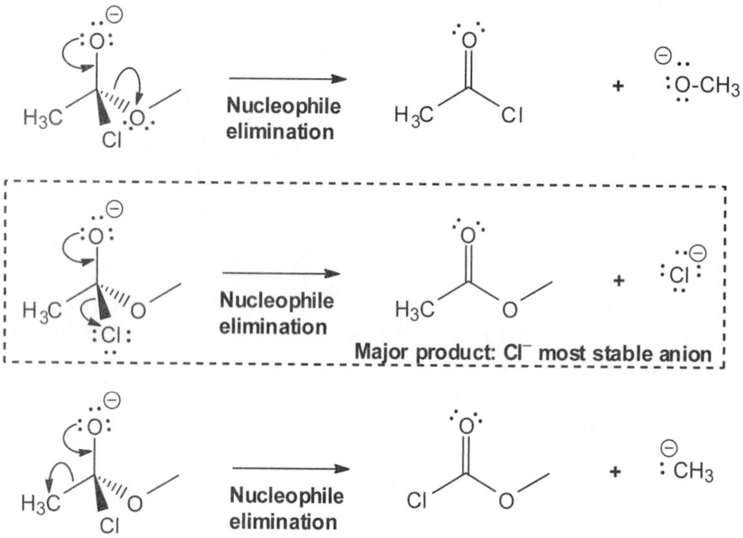

Nucleophile elimination

Nucleophile elimination

Major product: Cl^- most stable anion

Nucleophile elimination

Problem 7.40

Think

Relative to the position of the C^+, where must a H or CH_3 group be located to undergo a 1,2-hydride or 1,2-methyl shift? How many curved arrows are used to depict a 1,2-hydride or 1,2-methyl shift? Are curved arrows drawn from electron rich to electron poor, or vice versa? What type of carbocation is initially present? What are the effects of resonance on charge stability? What are the effects of electron donation by the alkyl groups on charge stability? Is a more stable carbocation produced from a 1,2-hydride or 1,2-alkyl shift?

Solve

Only **(b)**, **(e)**, and **(f)** would undergo rearrangements, as shown below and on the next page. Feasible carbocation rearrangements, in general, are 1,2-hydride shifts or 1,2-methyl shifts that take a less stable carbocation to a more stable carbocation. To depict a 1,2-hydride or 1,2-methyl shift, a single curved arrow originates from the center of a C–H or C–CH_3 bond that is adjacent to a C^+, and points to the C^+. In **(b)**, a 1,2-hydride shift converts a secondary carbocation into a tertiary one. In both **(e)** and **(f)**, the resulting carbocations are resonance stabilized, whereas the initial carbocations have the positive charge localized on a single C atom. In none of the other species does a 1,2-hydride or 1,2-methyl shift produce a more stable carbocation.

(b) 2° C^+ → 3° C^+

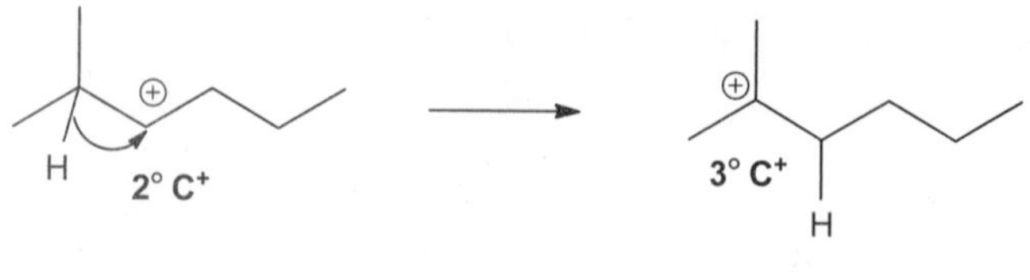

(e)

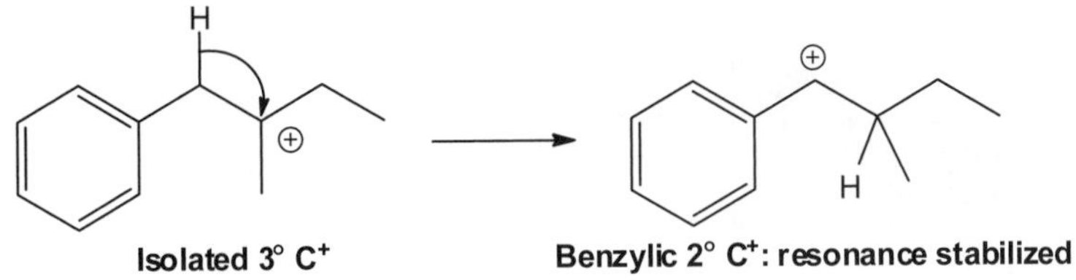

Isolated 3° C+

Benzylic 2° C+: resonance stabilized

(f)

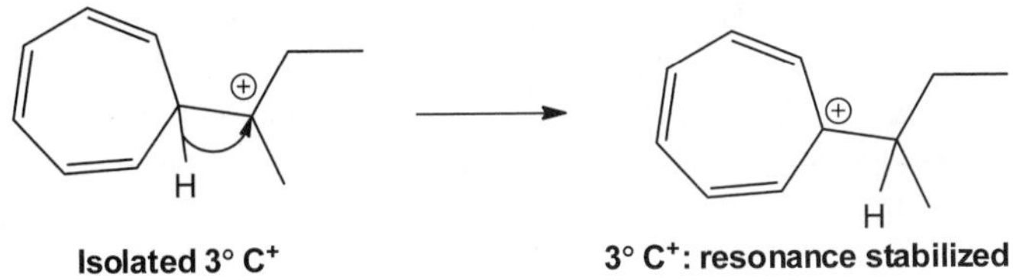

Isolated 3° C+

3° C+: resonance stabilized

Problem 7.41

Think

Can you make any simplifying assumptions to metal-containing species? Which species is the nucleophile? Which is the substrate? Which species is electron rich? Electron poor? How many curved arrows are necessary to show the S_N2 step? Are curved arrows drawn from electron rich to electron poor, or vice versa?

Solve

The reactions are shown below. The Na^+ and K^+ are spectator ions and, therefore, are omitted from the reaction mechanism. Two curved arrows are necessary in an S_N2 reaction mechanism—one to show the Nu–C bond formation and another to show the C–L bond breaking.

(a) + SH $\xrightarrow{S_N2}$ SH + I

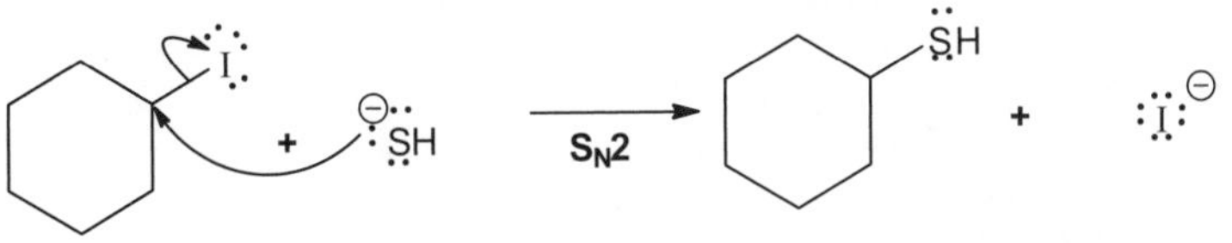

(b) Br + CN $\xrightarrow{S_N2}$ CN + Br

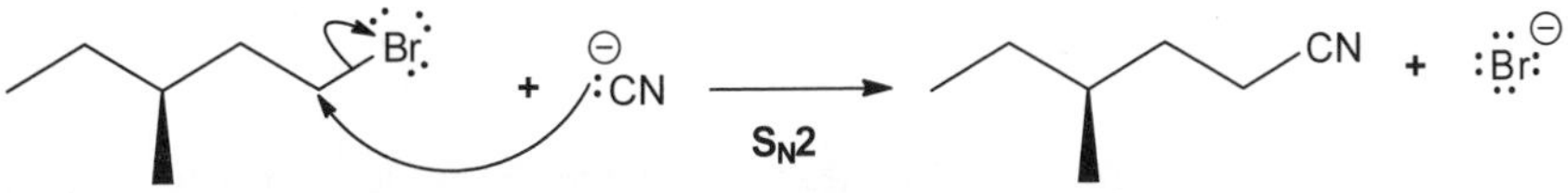

Problem 7.42

Think

Can you make any simplifying assumptions to metal-containing species? Which species is the nucleophile? Which is the substrate? Which species is electron rich? Electron poor? How many curved arrows are necessary to show the nucleophilic addition step? Are curved arrows drawn from electron rich to electron poor, or vice versa?

Solve

The products of each nucleophilic addition are shown below and on the next page. In **(a)**, CH_3COK is treated as CH_3CO^-. In **(b)**, CH_3Li is treated as CH_3^-. In (c), C_6H_5MgBr is treated as $C_6H_5^-$. In **(d)**, $NaBH_4$ is treated as H^-. In **(e)**, NaOH is treated as HO^-. In each case, two curved arrows are necessary. One curved arrow originates from the electron-rich nucleophile and points to the C atom of C=O or C≡N. The second curved arrow originates from the middle of the C=O or C≡N bond and points to the heteroatom.

(a) O + CH_3O^- $\xrightarrow{\text{Nucleophilic addition}}$ O, OCH_3

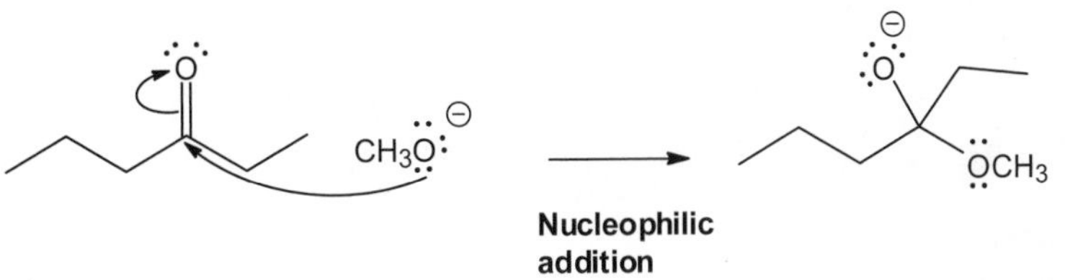

(b) Nucleophilic addition

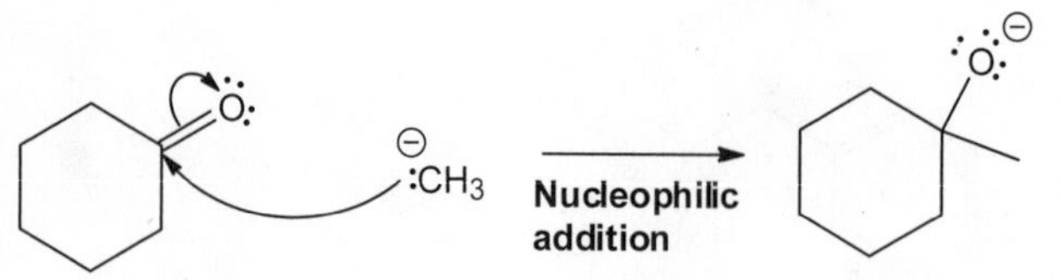

(c) Nucleophilic addition

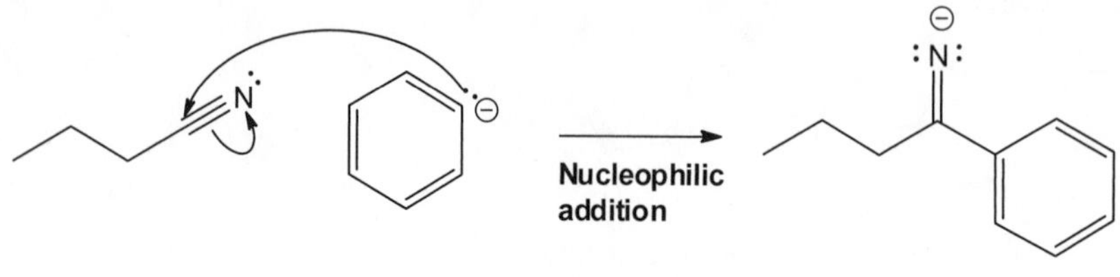

(d) Nucleophilic addition

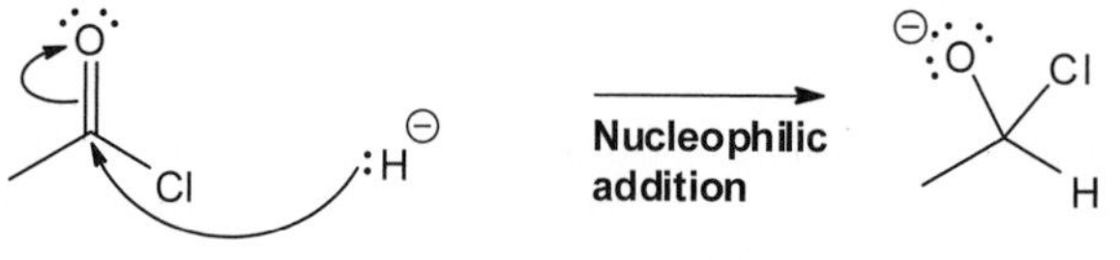

(e) Nucleophilic addition

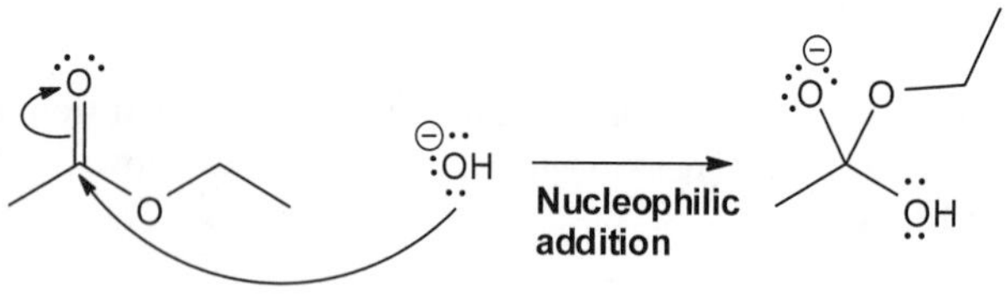

(f) Nucleophilic addition

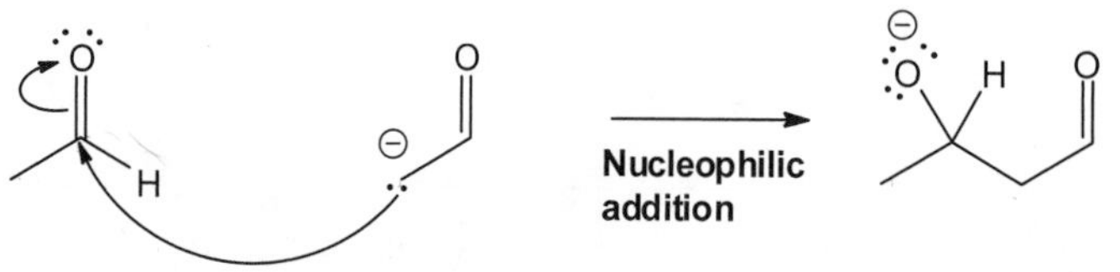

Problem 7.43

Think

Which sites are electron rich? Electron poor? Are curved arrows drawn from electron rich to electron poor, or vice versa? How many curved arrows are required to depict a nucleophile elimination? What are the possible leaving groups for each case? Is the leaving group different from the nucleophile in the previous problem?

Solve

The nucleophile elimination steps shown below and on the next page produce a compound that is different from the reactants. These steps involve expelling a leaving group that is different from the nucleophile that attacks in the previous problem. *Note*: As we will discuss further in Chapter 21, a nucleophile elimination is unfeasible when H^- or R^- is eliminated. For these reactions, an X is drawn in front of the reaction arrow.

(a) Nucleophile elimination

Nucleophile elimination

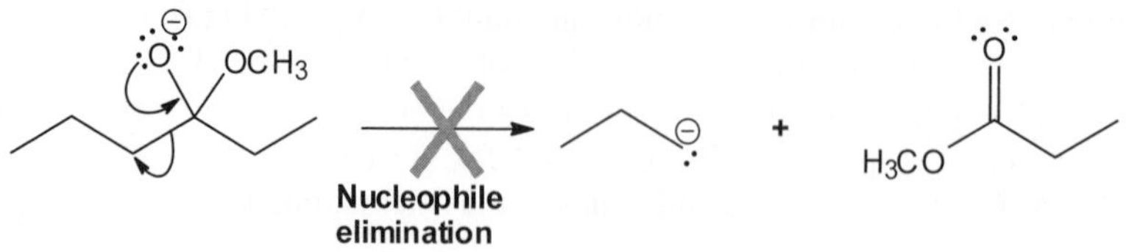

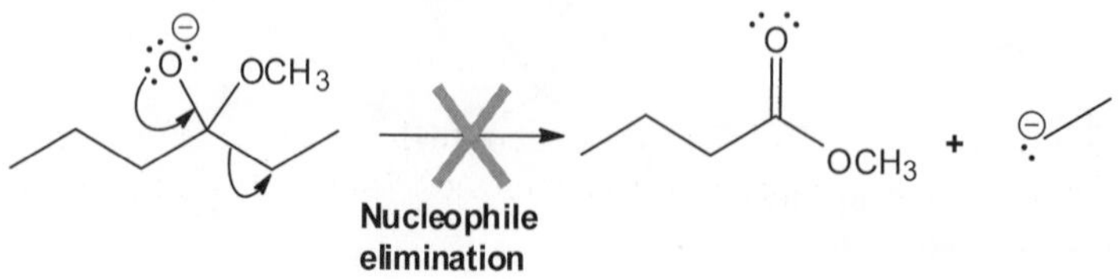

(b)

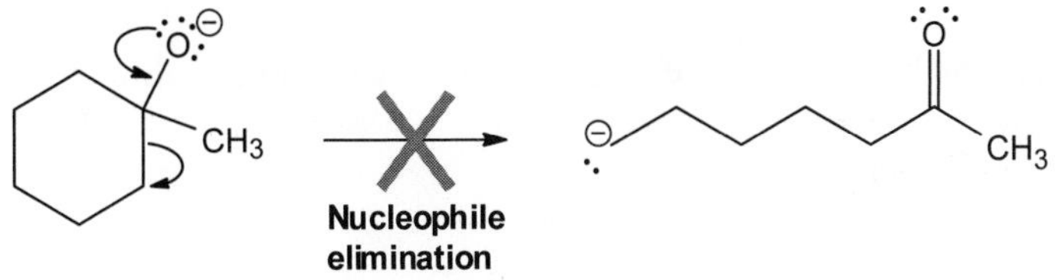

Nucleophile elimination

(c)

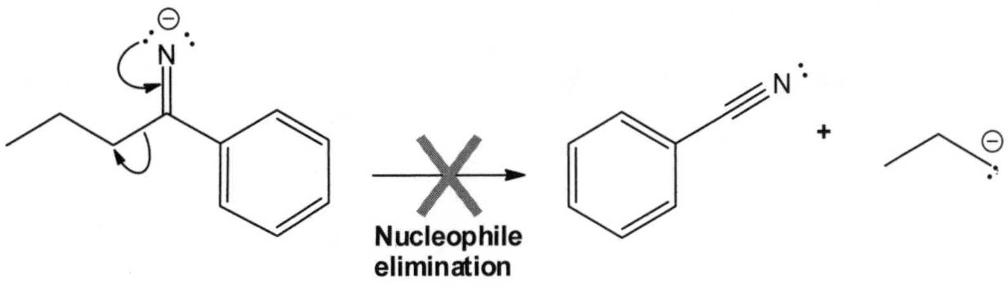

Nucleophile elimination

(d)

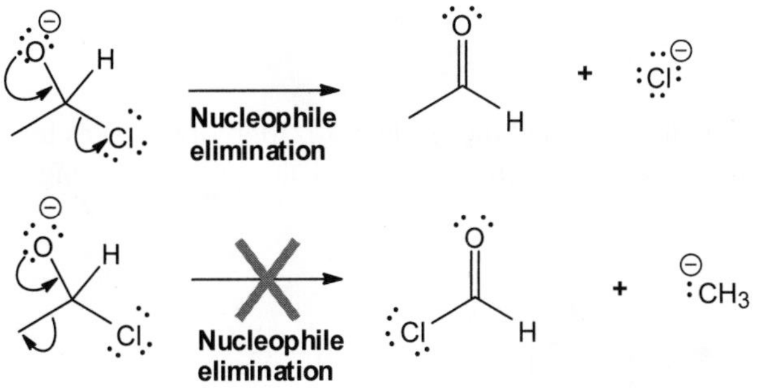

Nucleophile elimination

Nucleophile elimination

(e)

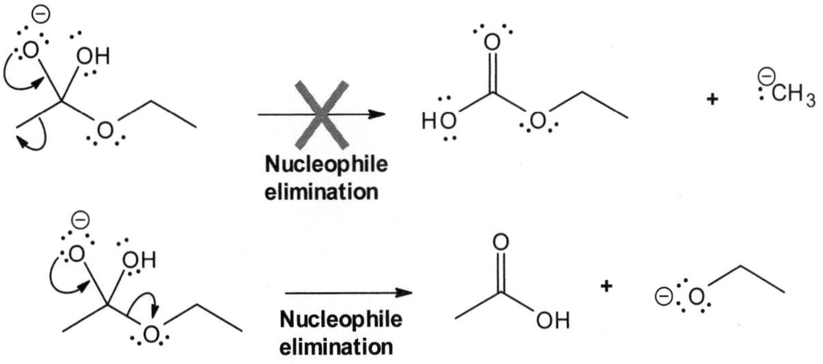

Nucleophile elimination

Nucleophile elimination

(f)

Nucleophile elimination

Nucleophile elimination

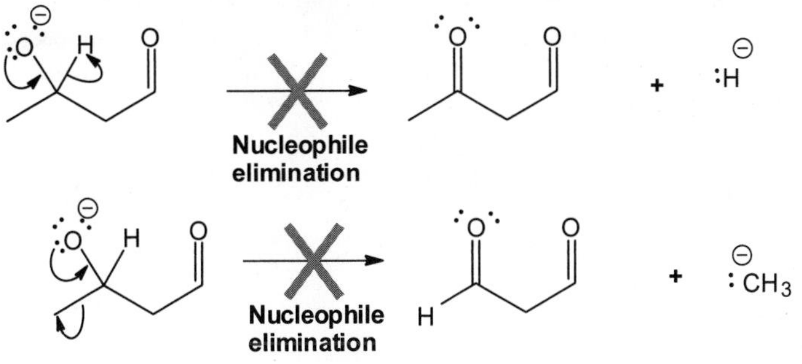

Problem 7.44

Think

Are the species listed electron rich or electron poor? Toward which kinds of sites are electron-rich species attracted? Toward which kinds of sites are electron-poor species attracted? Which site on phenol is most likely to react with each of the follow species?

Solve

The compounds given in the problem can be treated as the electron-rich species shown below.

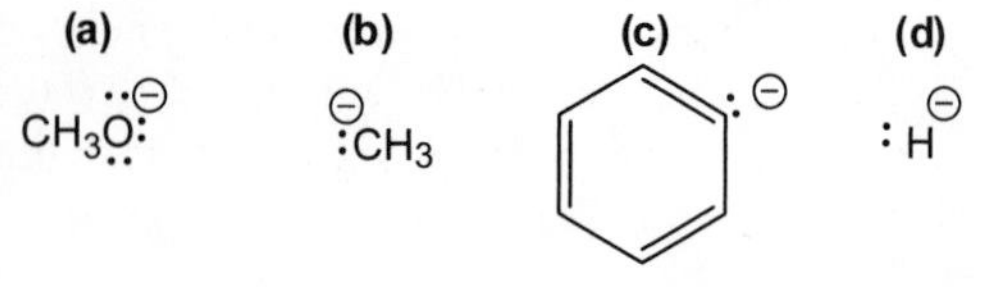

(a) CH_3O^- (b) $^-CH_3$ (c) phenyl anion (d) H^-

They can react with an electron-poor site on C_6H_5OH. As we can see below, there are two such sites.

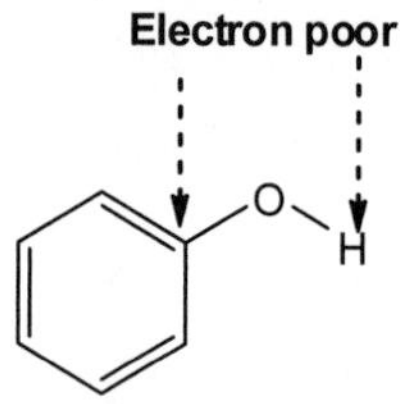

Electron poor

However, the proton is more easily accessed than the C is, and the sp^2 hybridization of C makes the C–O bond stronger than usual. Therefore, the predominant reaction is from attack on the proton, leading to a proton transfer reaction.

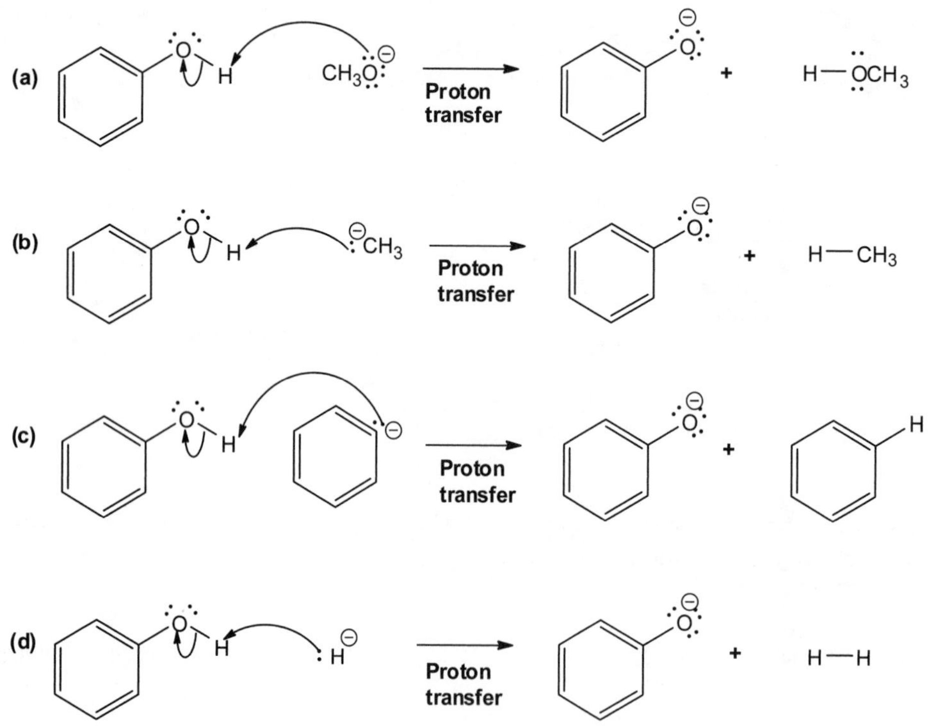

(a) C_6H_5OH + CH_3O^- → (Proton transfer) $C_6H_5O^-$ + H—OCH_3

(b) C_6H_5OH + $^-CH_3$ → (Proton transfer) $C_6H_5O^-$ + H—CH_3

(c) C_6H_5OH + $C_6H_5^-$ → (Proton transfer) $C_6H_5O^-$ + C_6H_5—H

(d) C_6H_5OH + H^- → (Proton transfer) $C_6H_5O^-$ + H—H

Problem 7.45

Think

Is the alkyne electron rich or electron poor? From where do the electrons originate to form the new bond with the H^+ from $CH_3OH_2^+$? One C atom of the alkyne forms the new C–H bond; what is generated on the other C atom? Which product exhibits greater charge stability?

Solve

The curved arrow notation and products are shown below. This electrophilic addition step requires two curved arrows. One originates from the center of the C≡C bond and points to the electron-poor H. The second curved arrow indicates the breaking of the H–O bond, originating from the center of the H–O bond and pointing to O. When one C atom of C≡C picks up a proton, the other C atom gains a positive charge. The first carbocation is more stable because the positively charged carbon is attached to one additional alkyl group, which, as indicated, is electron donating and thus provides stability to the carbocation.

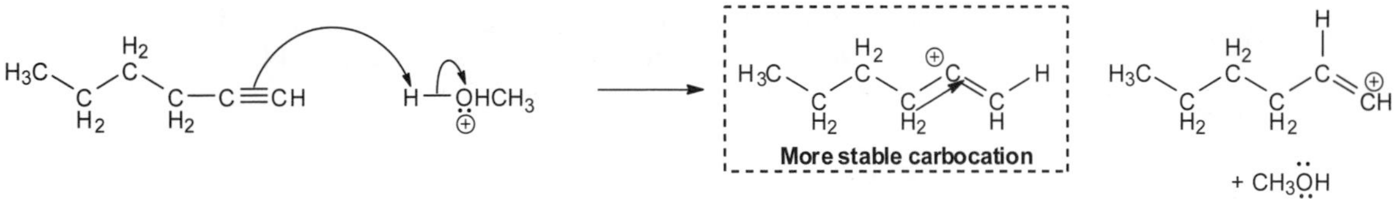

More stable carbocation

$+ CH_3OH$

Problem 7.46

Think

What is the elementary step to eliminate H^+ or SO_3? What type of bond forms on the ring? Which sites are electron rich? Electron poor? Are curved arrows drawn from electron rich to electron poor, or vice versa?

Solve

The electrophile elimination mechanisms are shown below. In **(a)** a proton is eliminated in the presence of H_2O, a weak base, so two curved arrows are necessary. One curved arrow indicates the elimination of H^+ and the second curved arrow indicates water picking up the proton simultaneously. In **(b)** SO_3 is eliminated to break the S–C bond and form a C=C double bond. Two curved arrows are necessary. One curved arrow indicates the formation of the S=O double bond and the second curved arrow indicates the breaking of the S–C bond.

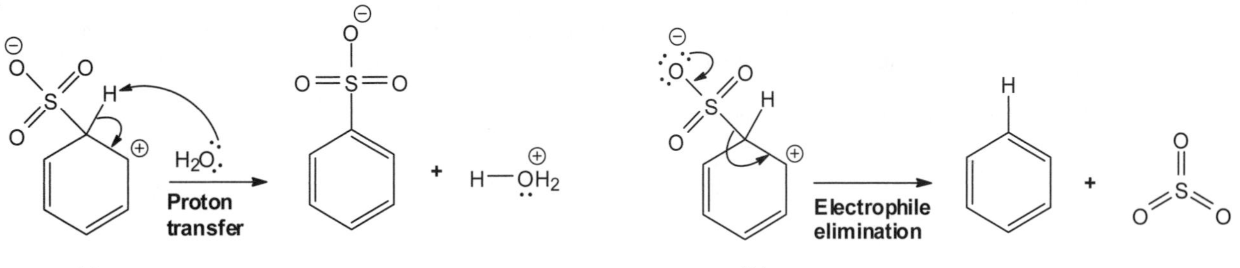

Proton transfer

(a)

Electrophile elimination

(b)

Problem 7.47

Think

The keto form of each species was given. What are their corresponding enol forms? Do the alkene portions of the enols have the same degree of alkyl substitution?

Solve

Greater alkyl substitution of the alkene group (C=C) in the enol form provides more stability to the molecule. Because the enol of the first molecule is better stabilized, more of that enol will be present at equilibrium than the enol of the second molecule, so less of the keto form will be present.

More highly substitued alkene group

99.986% 0.014% 99.99994% 0.00006%

Problem 7.48

Think

The keto form of each species was given. What are their corresponding enol forms? What impact does ring size have on the stability of the molecule? Will ring size affect the keto and enol forms equally?

Solve

The keto and enol forms of each molecule are shown below. The interior angle of a five-membered ring is smaller than it is for a six-membered ring, so the five-membered ring is more strained. That ring strain is more pronounced for the enol forms because of the additional sp^2-hybridized C atom; the ideal angle of an sp^2-hybridized C is larger than for an sp^3-hybridized C. With greater ring strain, the enol form of cyclopentanone is less abundant at equilibrium than the enol form of cyclohexanone.

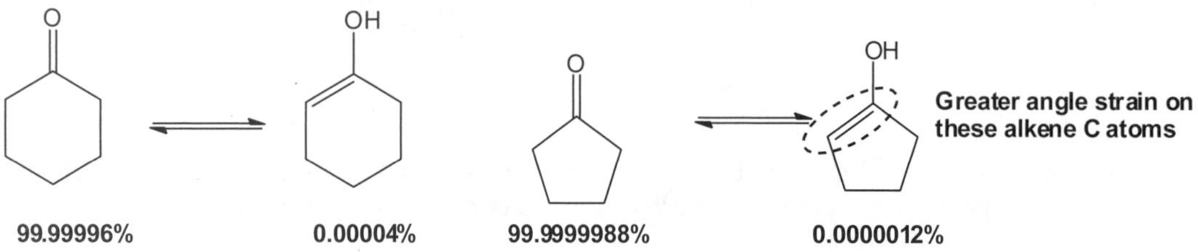

O OH O OH Greater angle strain on these alkene C atoms

99.99996% 0.00004% 99.9999988% 0.0000012%

Problem 7.49

Think

The keto form of the species was given. What is its corresponding enol form? Draw the tautomeric product. What type of intermolecular interaction is possible with an OH? What does the presence of an OH and an O permit? Is that interaction possible in the keto form?

Solve

In the enol form (below), an internal hydrogen bond can form, making the enol more stable than usual. That internal hydrogen bond is not possible in the keto form, which has two H-bond acceptors only.

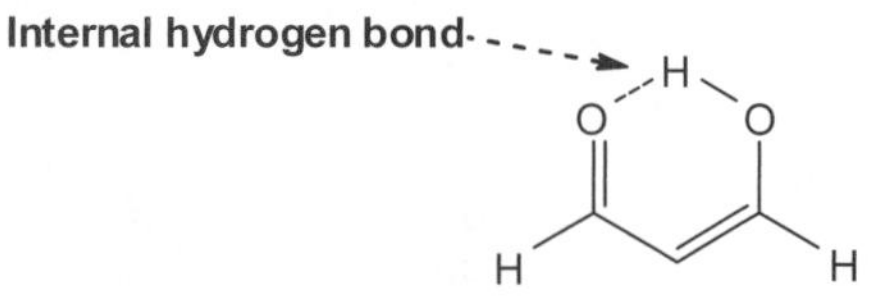

Internal hydrogen bond H O O H H

Problem 7.50

Think

Which bonds appear in the imine that are not in the enamine? Which bonds appear in the enamine that are not in the imine? What is the difference in bond energy between the imine and enamine form? Which form is favored?

Solve

The bonds that are different between the reactants and products are shown below. The enthalpy of the reaction is estimated by taking the sum of the energies of the bonds that are lost from the reactants minus the sum of the energies of the bonds that are gained on the product side.

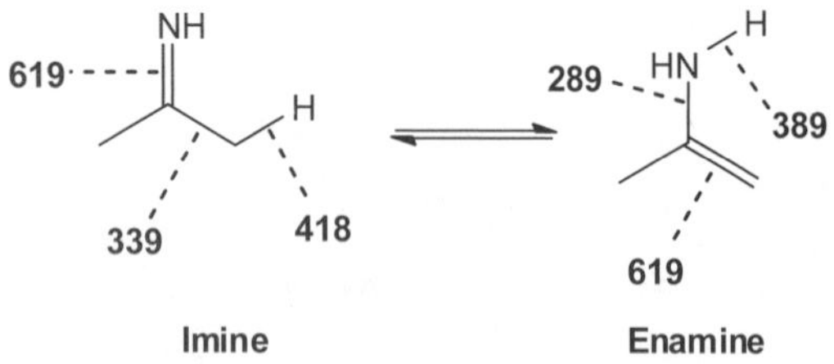

NH 619 H 339 418 Imine

H HN 289 389 619 Enamine

So ΔH°_{rxn} is approximately (619 + 339 + 418) – (389 + 289 + 619) = +79 kJ/mol. Thus, the product (the enamine) is significantly less stable than the reactant (the imine).

Problem 7.51

Think

What impact does an O atom attached to C^+ have on charge stability? Resonance? Inductive effects? What impact does a CH_2 group attached to C^+ have on charge stability?

Solve

The O atom is electron-withdrawing, which destabilizes a carbocation C^+. However, O also has lone pairs that can participate in resonance with C^+. In the resonance structure, moreover, all atoms have octets, which provides significant stabilization. Resonance and inductive effects work in opposite directions in this case but, as we learned in Chapter 6, resonance is generally more important than inductive effects, so the O atom provides substantial stability. This kind of resonance stabilization is not available in the product of the second reaction.

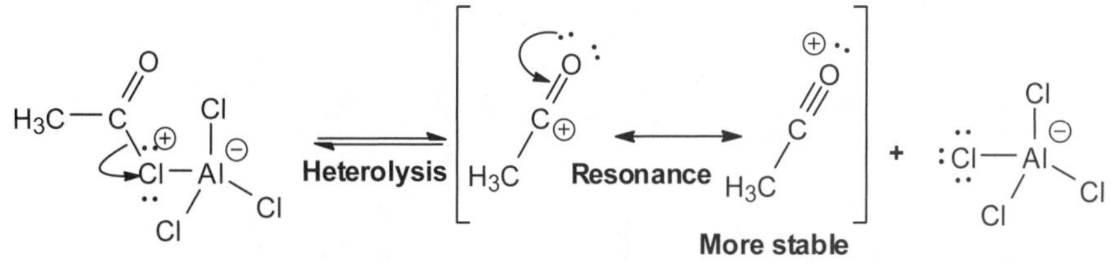

Heterolysis

Resonance

More stable

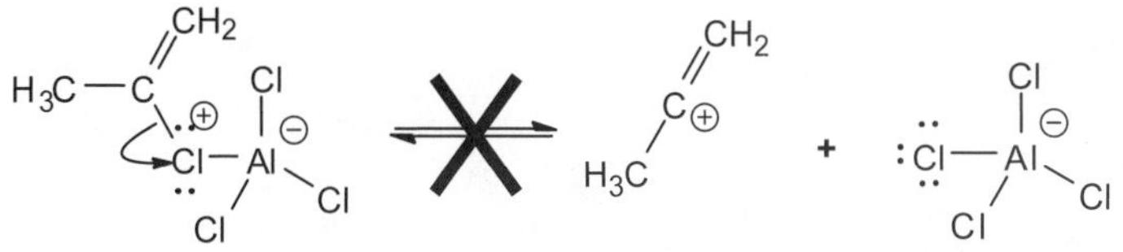

INTERCHAPTER 1| Molecular Orbital Theory and Chemical Reactions

Your Turn Exercises

Your Turn IC1.1

Think

What is the hybridization of the H, O, and Cl atoms and what orbitals are contributed from each atom? What type of overlap occurs between H and O? Between H and Cl? What type of bonding correlates to this type of overlap? How many total orbitals should be produced from the atomic orbital (AO) mixing?

Solve

H is not hybridized and contributes a 1*s* orbital, whereas O and Cl are both sp^3 hybridized. End-on overlap occurs in both H–O$^-$ and H–Cl and, therefore, results in a sigma (σ) bond. When the two AOs overlap, two molecular orbitals (MOs) are produced: σ bonding and σ antibonding (σ*). The three sets of lone pairs on Cl and O are in sp^3-hybridized nonbonding orbitals. The highest occupied MO (HOMO) from HO$^-$ is, therefore, a nonbonding orbital, and the lowest unoccupied MO (LUMO) from HCl is a σ* MO, in agreement with Figure IC1-3.

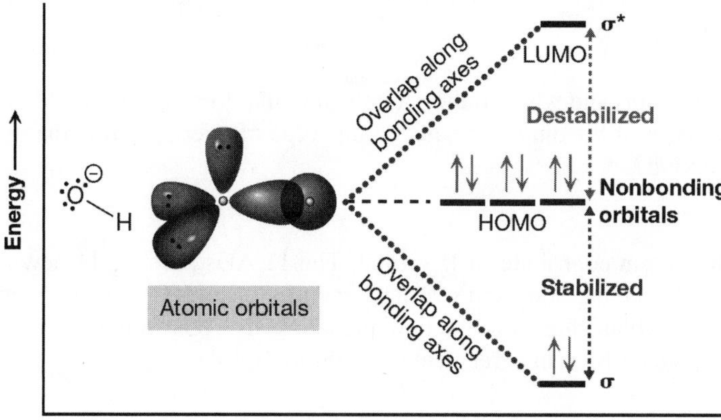

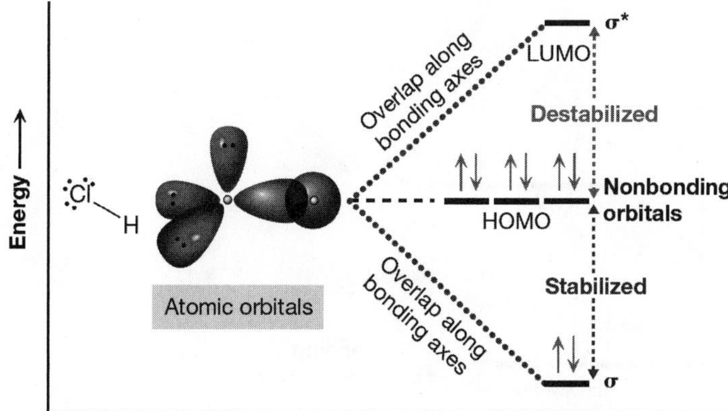

The nonbonding orbital of HO$^-$ comes from a hybridized AO. The σ* MO of HCl is the result of destructive interference, leaving a node in the internuclear region.

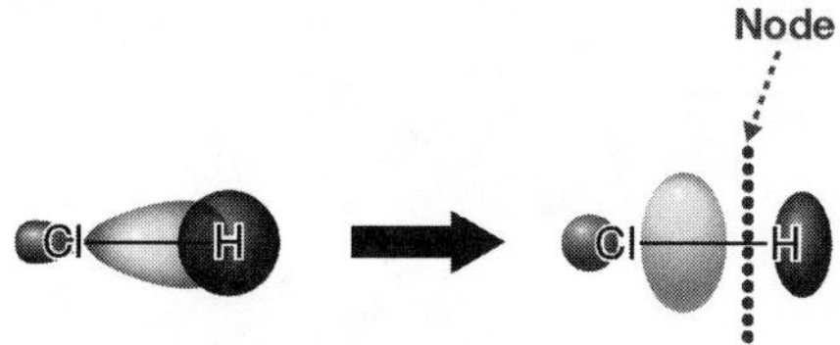

Your Turn IC1.2

Think

When looking at AO overlap, does *constructive interference* have overlap of the same phase or opposite phase of the AO? What about *destructive interference*?

Solve

Constructive interference occurs when two orbitals overlap with the same phases. Destructive interference occurs when two orbitals overlap with opposite phases.

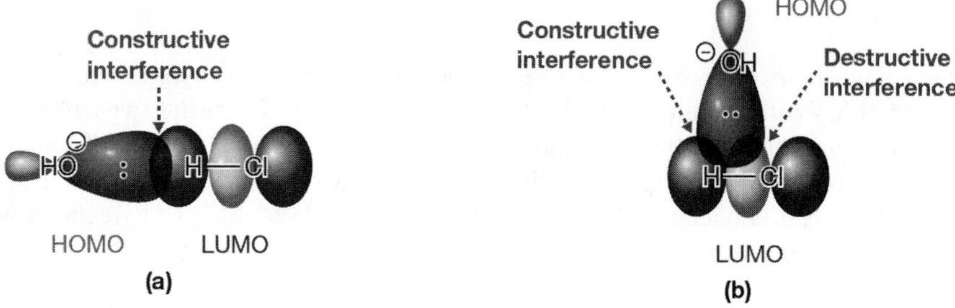

Your Turn IC1.3

Think

What is the hybridization of the H, C, and Cl atoms and what do the orbitals look like for each atom? What type of overlap occurs between C and Cl? What type of bonding correlates to this type of overlap? How many total orbitals should be produced from the AO mixing?

Solve

C and Cl are both sp^3 hybridized and each H atom contributes a 1*s* orbital. The 11 AOs produce 11 new MOs. End-on overlap occurs in the C–Cl and C–H bonding and each overlap produces two MOs: σ and σ*, which accounts for eight MOs. The remaining three orbitals are from the noninteracting sp^3 AOs from Cl, producing nonbonding orbitals. The LUMO is, therefore, a σ* MO, in agreement with Figure IC1-4.

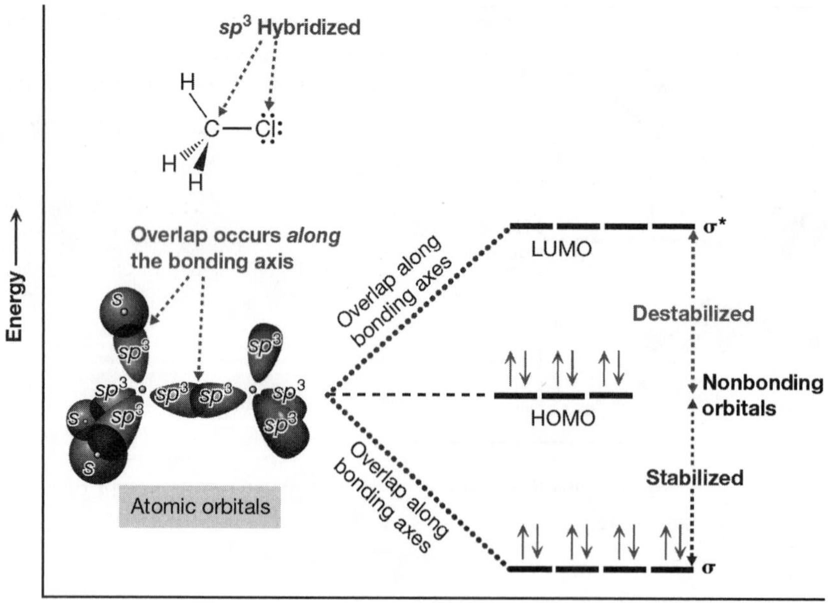

The C–Cl σ* can be thought of as resulting from the destructive interference between two sp^3 AOs, producing a node in the internuclear region:

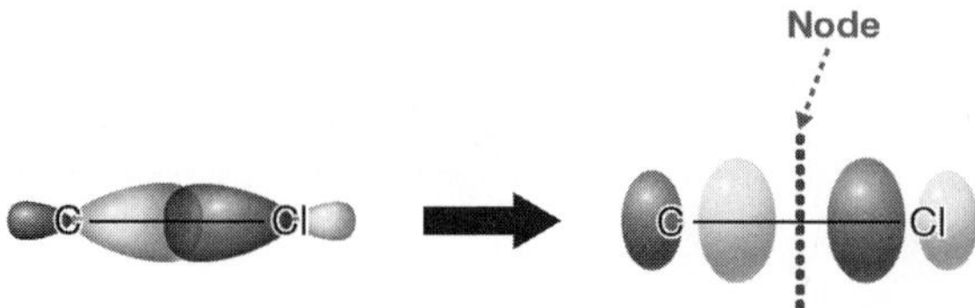

Your Turn IC1.4

Think

When looking at AO overlap, does *constructive interference* have overlap of the same phase or opposite phase of the AO? What about *destructive interference*?

Solve

Constructive interference occurs when two orbitals overlap with the same phases. Destructive interference occurs when two orbitals overlap with opposite phases.

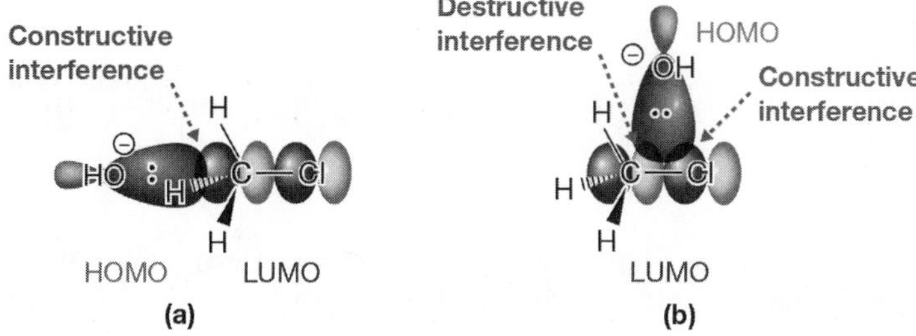

Your Turn IC1.5

Think

What is the hybridization of each of the C atoms? What do the orbitals look like for each atom? What type of overlap occurs between the C–C and C–H? What type of bonding correlates to this type of overlap? How many total orbitals should be produced from the AO mixing? Are there any unhybridized orbitals on the C atom?

Solve

The central C is sp^2 hybridized and has an unhybridized pure *p* orbital. The CH_3 carbon atoms are all sp^3 hybridized and each H atom contributes a 1*s* orbital. End-on overlap occurs in the C–H and C–C bonding and, therefore, results in 12 σ bonds. When the 24 AOs overlap, 24 MOs are produced: 12 σ and 12 σ*. Yes, these figures are in agreement with Figure IC1-5, which shows the LUMO of $(H_3C)_3C^+$ as a nonbonding *p* orbital as shown here.

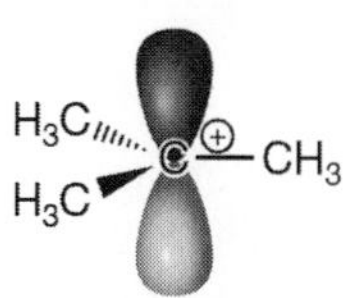

See the energy diagram on the next page.

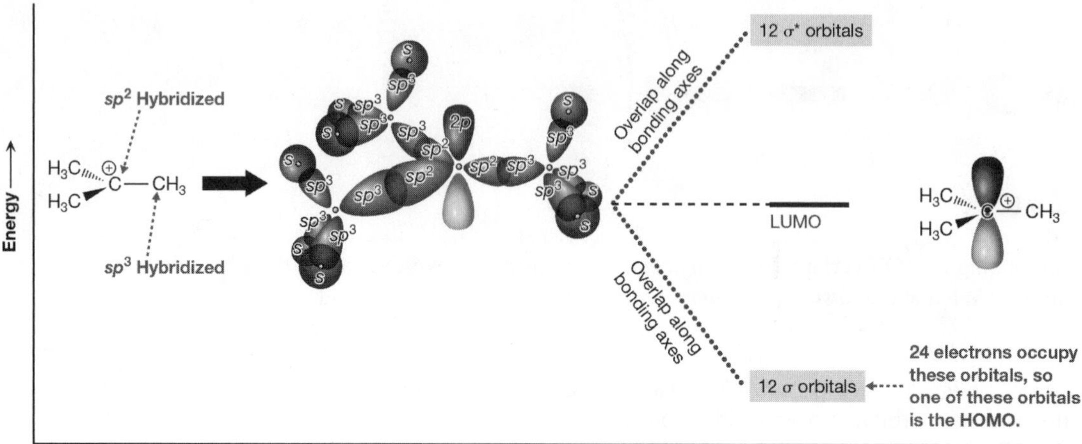

Your Turn IC1.6

Think

When looking at AO overlap, does *constructive interference* have overlap of the same phase or opposite phase of the AO? What about *destructive interference*?

Solve

Constructive interference occurs when two orbitals overlap with the same phases. Destructive interference occurs when two orbitals overlap with opposite phases. Both IC1-5a and 5b show constructive interference.

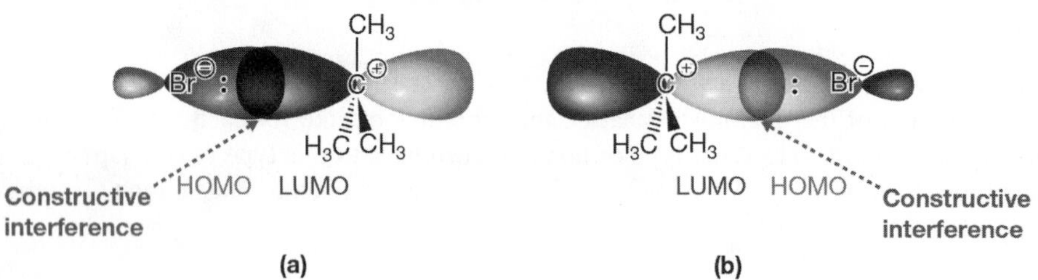

Your Turn IC1.7

Think

What is the hybridization of each of the C atoms, the H atoms, and the Br atom? What do the contributed orbitals look like for each atom? What type of overlap occurs between the C–CH_3, H–C, and C–Br? What type of bonding correlates to this type of overlap? How many total orbitals should be produced from the AO mixing?

Solve

The two C atoms and the Br atom are all sp^3 hybridized, and each H atom contributes a 1*s* orbital. End-on overlap occurs in the C–H, C–C, and C–Br bonding and, therefore, results in seven σ bonds. When the 14 AOs overlap, 14 MOs are produced: seven σ and seven σ*. This leaves three nonbonding orbitals. See the energy diagram on the next page.

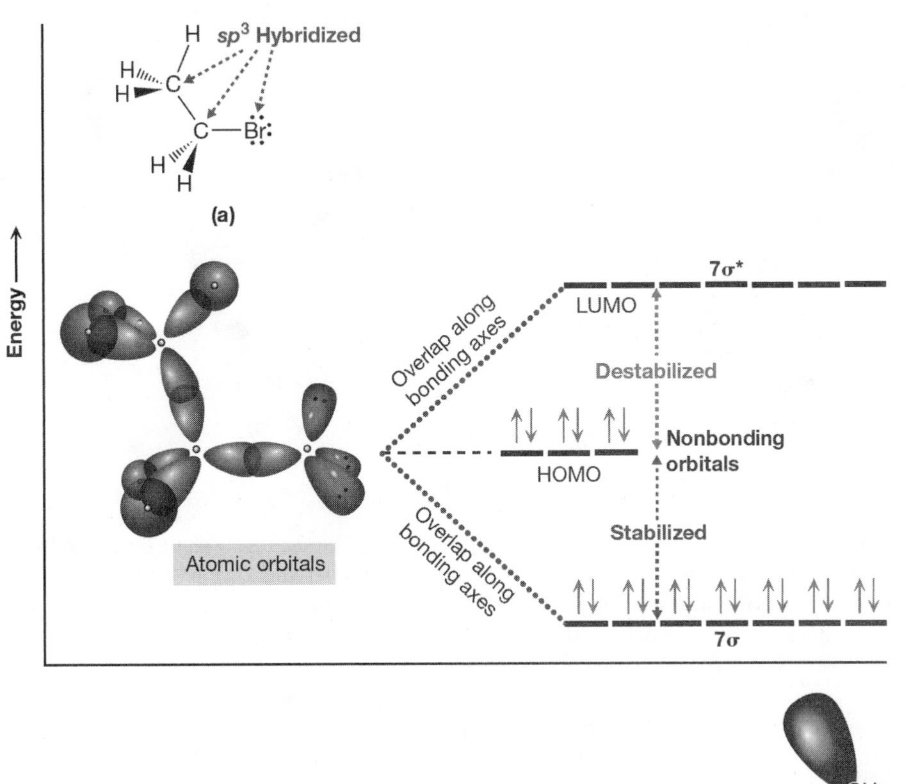

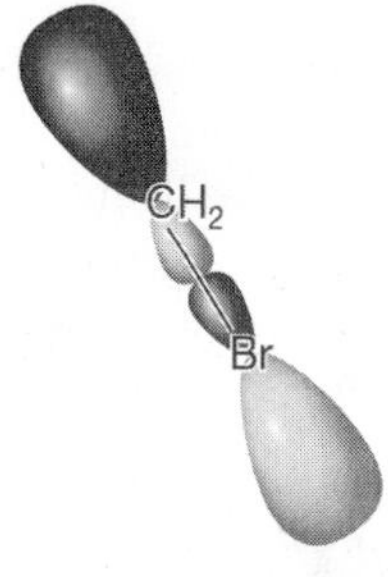

The LUMO of the molecule is a σ* MO, in agreement with Figure IC1-7a:

Although the HOMO of the entire molecule is a nonbonding orbital, the HOMO involving the CH_3 group is a σ

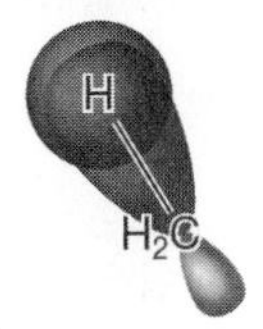

MO, in agreement with Figure IC1-7a:

Your Turn IC1.8

Think

When looking at AO overlap, does *constructive interference* have overlap of the same phase or opposite phase of the AO? What about *destructive interference*?

Solve

Constructive interference occurs when two orbitals overlap with the same phases. Destructive interference occurs when two orbitals overlap with opposite phases. Predominantly constructive interference takes place in both IC1-7a and 7b. See the figure on the next page.

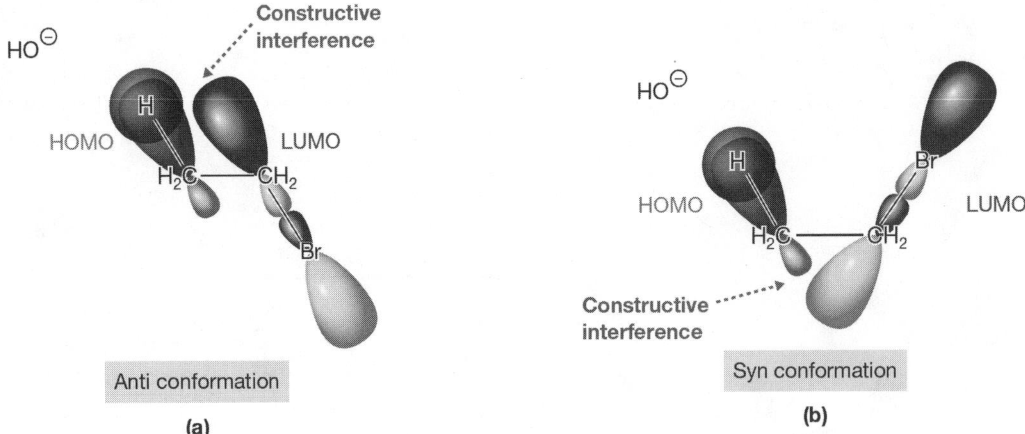

Your Turn IC1.9

Think

What is the hybridization of each of the H atoms, the C atoms, the Cl atom, and the O atom? What do the contributed orbitals look like for each atom? What type of overlap occurs between the H–C, C–CH_3, C–Cl, and the C–O? What type of bonding correlates to this type of overlap? How many total orbitals should be produced from the AO mixing? Are there any unhybridized orbitals on the C and O atoms?

Solve

The central C and O atoms are both sp^2 hybridized and each has an unhybridized pure p orbital. The C atom of CH_3 and the Cl atom are both sp^3 hybridized and each H atom contributes a 1s orbital. End-on overlap occurs in the C–H and C–C, C–Cl, and C–O bonding and, therefore, results in six σ bonds. Side-by-side overlap occurs between the two pure p orbitals of C and O. This results in a π bond. When the 14 AOs overlap, 14 MOs are produced: six σ, six σ*, one π bonding, and one π* antibonding. The LUMO is the π* MO, in agreement with the LUMO shown in Figure IC1-8a:

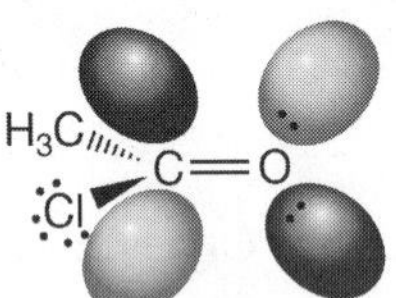

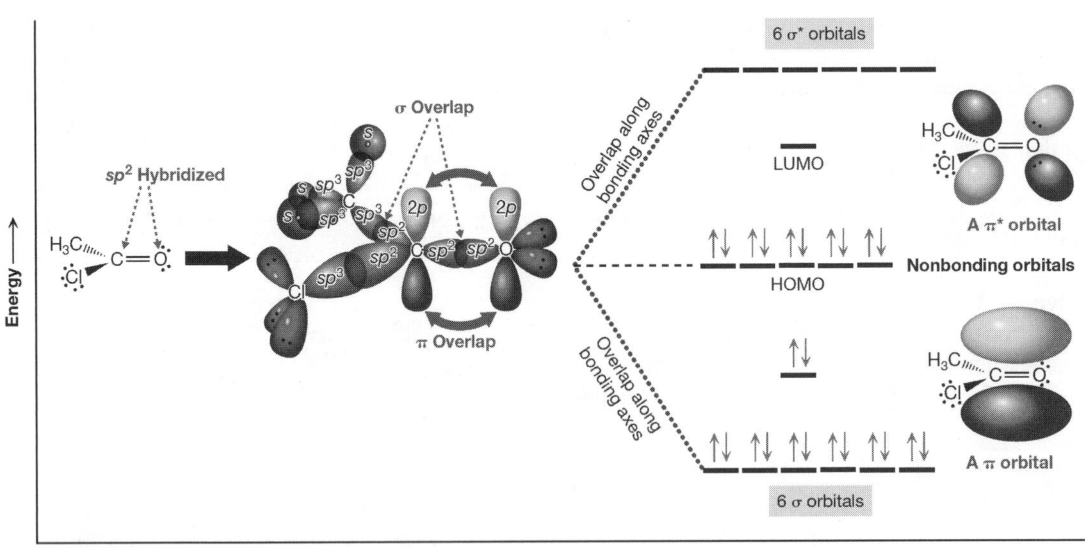

Your Turn IC1.10

Think

When looking at AO overlap, does *constructive interference* have overlap of the same phase or opposite phase of the AO? What about *destructive interference*?

Solve

Constructive interference occurs when two orbitals overlap with the same phases. Destructive interference occurs when two orbitals overlap with opposite phases. Constructive interference predominantly takes place in both IC1-8a and 8b.

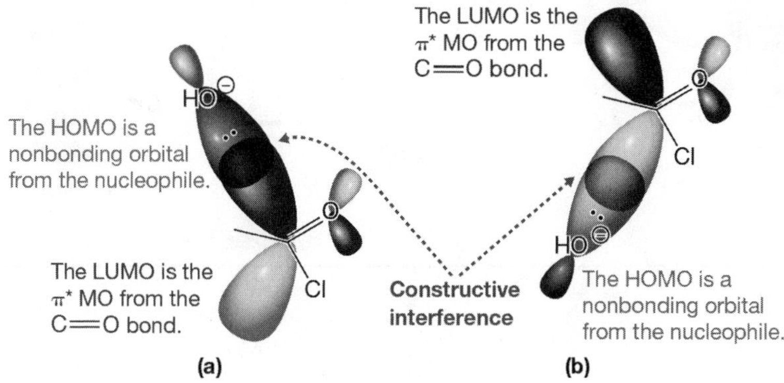

Your Turn IC1.11

Think

What is the hybridization of each of the H atoms and the C atoms? What do the contributed orbitals look like for each atom? What type of overlap occurs between the H–C and C–C? What type of bonding correlates to this type of overlap? How many total orbitals should be produced from the AO mixing? Are there any unhybridized orbitals on the C atoms?

Solve

The two central C atoms are both sp^2 hybridized and each has an unhybridized pure *p* orbital. The CH_3 carbon atoms are sp^3 hybridized and each H atom contributes a 1*s* orbital. End-on overlap occurs in the C–H and C–C bonding and, therefore, results in 11 σ bonds. Side-on overlap occurs between the two pure *p* orbitals of the two sp^2-hybridized C atoms. This results in a π bond. When the 24 AOs overlap, 24 MOs are produced: 11 σ, 11 σ*, one π, and one π*. When the 24 total valence electrons fill the MOs, the π MO is the HOMO and the π* MO is the LUMO. Yes, these figures are in agreement with Figure IC1-9, which shows the HOMO of $(CH_3)_2C{=}C(CH_3)_2$.

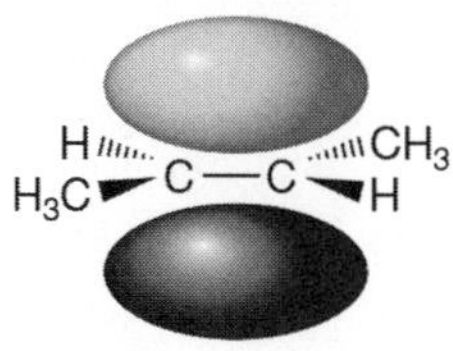

See the energy diagram on the next page.

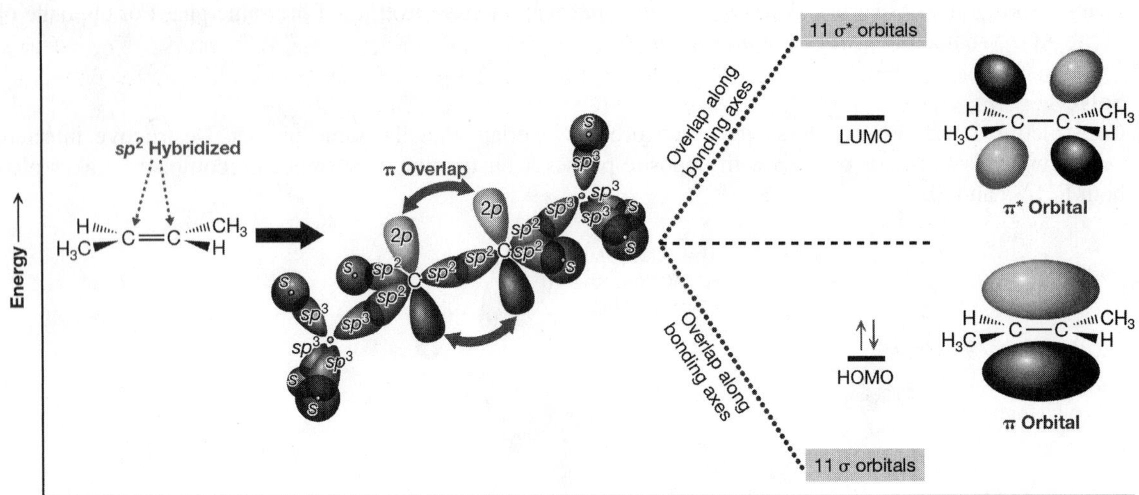

Your Turn IC1.12

Think

When looking at AO overlap, does *constructive interference* have overlap of the same phase or opposite phase of the AO? What about *destructive interference*?

Solve

Constructive interference occurs when two orbitals overlap with the same phases. Destructive interference occurs when two orbitals overlap with opposite phases. Constructive interference predominantly takes place in Figure IC1-9.

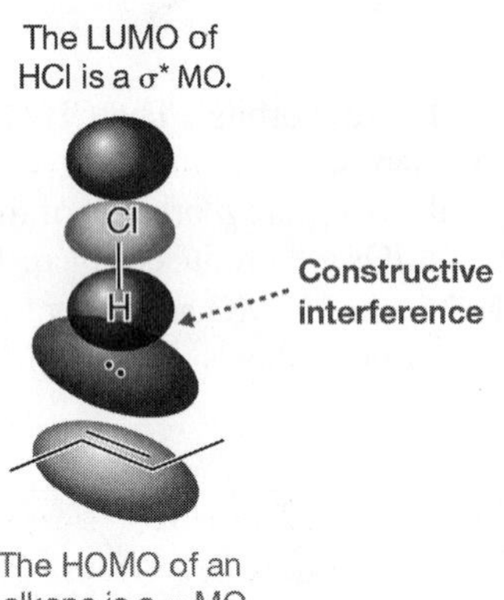

Your Turn IC1.13

Think

What is the hybridization of the each of the C atoms? What do the contributed orbitals look like for each atom? What type of overlap occurs between the C–C and C–H? What type of bonding correlates to this type of overlap? How many total orbitals should be produced from the AO mixing? Are there any unhybridized orbitals on the C atom?

Solve

The carbocation C atom is sp^2 hybridized and has an unhybridized pure p orbital. All the other C atoms are sp^3 hybridized and each H atom contributes a 1s orbital. End-on overlap occurs in the C–H and C–C bonding and, therefore, results in 15 σ bonds. The unhybridized p AO remains as a nonbonding orbital. The 30 valence electrons fill the 15 σ MOs, so the HOMO is a σ MO and the p AO is the LUMO. Yes, these figures are in agreement with Figure IC1-10, which shows the HOMO and LUMO of $(H_3C)_3C^+$ as

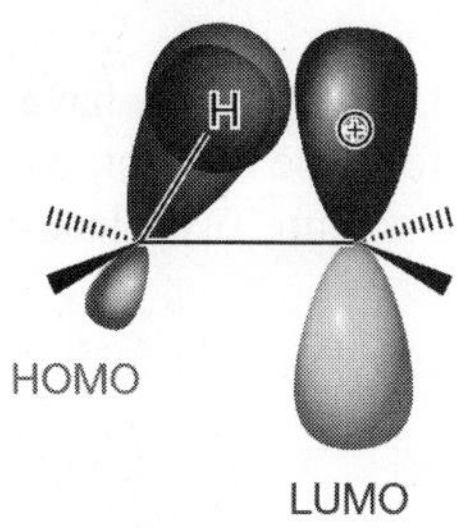

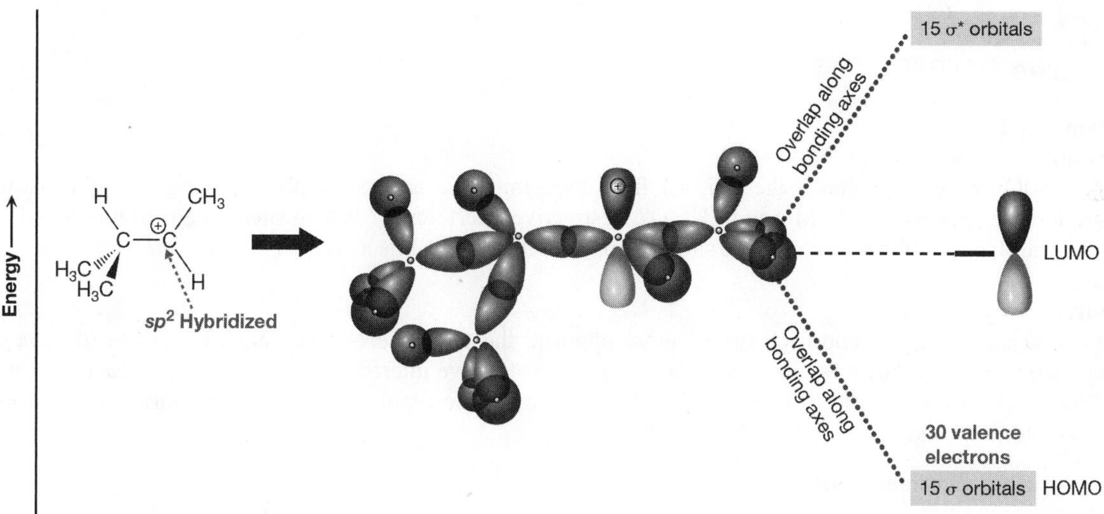

Your Turn IC1.14

Think

When looking at AO overlap, does *constructive interference* have overlap of the same phase or opposite phase of the AO? What about *destructive interference*?

Solve

Constructive interference occurs when two orbitals overlap with the same phases. Destructive interference occurs when two orbitals overlap with opposite phases. Predominantly constructive interference takes place in Figure IC1-10.

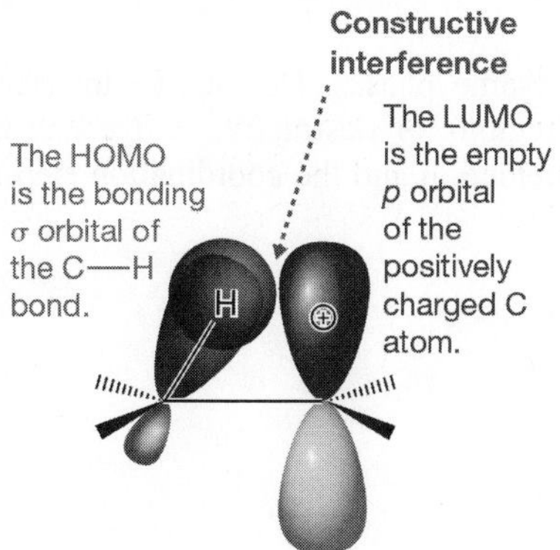

Chapter Problems

Problem IC1.1

Think

If the HO^- were to approach the H–Cl from the same side as the Cl, are there regions of constructive interference between the HOMO and LUMO? Destructive interference? Which sites are electron rich and which are electron poor? If two electron-rich sites approach each other, is this a favored interaction?

Solve

If the O atom were to approach the Cl atom opposite the H, the HOMO and LUMO would have a net interaction. This is shown below, where there is just constructive interference. Therefore, the reaction would be allowed. However, it will not occur readily due to electrostatic repulsion—it would require two electron-rich sites to form a bond.

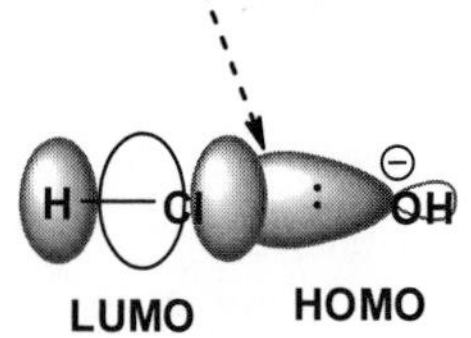

Problem IC1.2

Think

If the HO^- were to approach the CH_3–Cl from the same side as the Cl, are there regions of constructive interference between the HOMO and LUMO? Destructive interference? Which sites are electron rich and which are electron poor? If two electron-rich sites approach each other, is this a favored interaction?

Solve

If the O atom were to approach the Cl atom opposite the CH_3, the HOMO and LUMO would have a net interaction. This is shown below, where there is just constructive interference. Therefore, the reaction would be allowed. However, it will not occur readily due to electrostatic repulsion—it would require two electron-rich sites to form a bond.

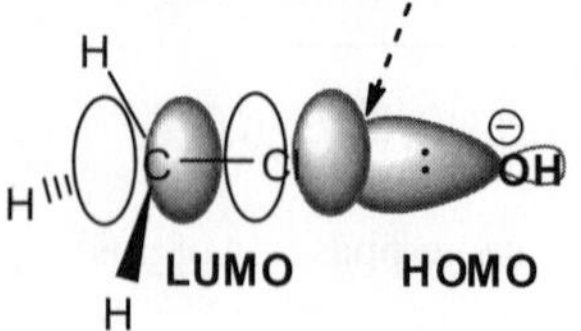

Problem IC1.3

Think

If the Br^- were to approach the $(CH_3)_3C^+$ from within the plane of the central C atom, are there regions of constructive interference between the HOMO and the LUMO? Destructive interference? Is there a net interaction?

Solve

Constructive interference occurs when two orbitals overlap with the same phases. Destructive interference occurs when two orbitals overlap with opposite phases. There are regions of destructive and constructive interference, which cancel each other out. Therefore, there is no net interaction and the coordination step from this type of approach does not occur.

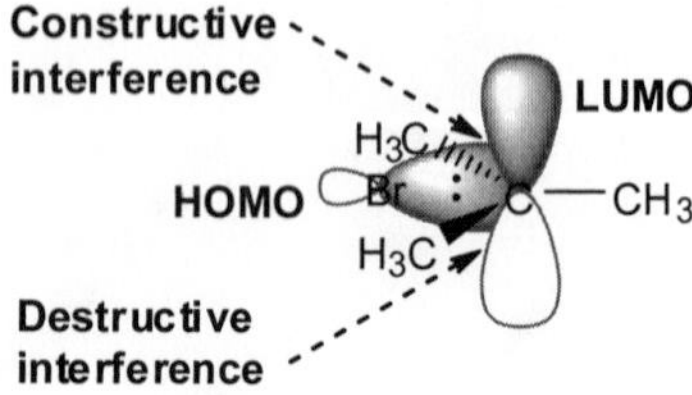

Problem IC1.4

Think

If the HO^- were to approach the carbonyl C atom from directly along the C=O bond, are there regions of constructive interference between the HOMO and the LUMO? Destructive interference? Is there a net interaction?

Solve

Constructive interference occurs when two orbitals overlap with the same phases. Destructive interference occurs when two orbitals overlap with opposite phases. There are regions of destructive and constructive interference, as shown below, which cancel each other out. Therefore, there is no net interaction and the nucleophilic addition step from this type of approach does not occur.

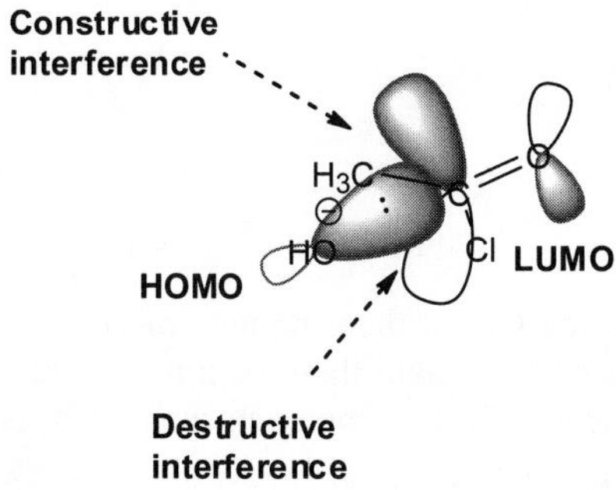

Problem IC1.5

Think

If $(CH_3)_3C^+$ were to approach the $CH_3CH{=}CHCH_3$ from above the C=C bond, are there regions of constructive interference between the HOMO and the LUMO? Destructive interference? Is there a net interaction?

Solve

Constructive interference occurs when two orbitals overlap with the same phases. Destructive interference occurs when two orbitals overlap with opposite phases. When $(CH_3)_3C^+$ approaches from above the C=C bond, there is a region of constructive interference and essentially no destructive interference. Therefore, there is a net interaction and the reaction is allowed.

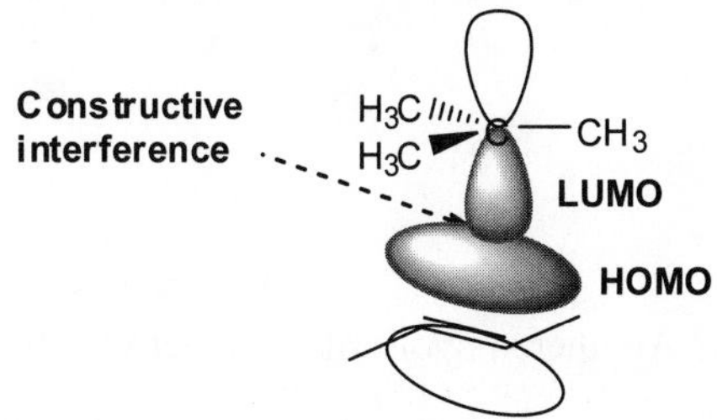

Problem IC1.6

Think

If CH_3 were to shift to the C^+, are there regions of constructive interference between the HOMO and the LUMO? Destructive interference? Will there be a net interaction?

Solve

The HOMO is the sigma orbital of the H_3C–C bond and the LUMO is the empty *p* orbital of the carbocation. The HOMO and the LUMO are in phase; they will have a net interaction and the shift is allowed. Moreover, the shift will occur readily because a more stable tertiary carbocation is formed.

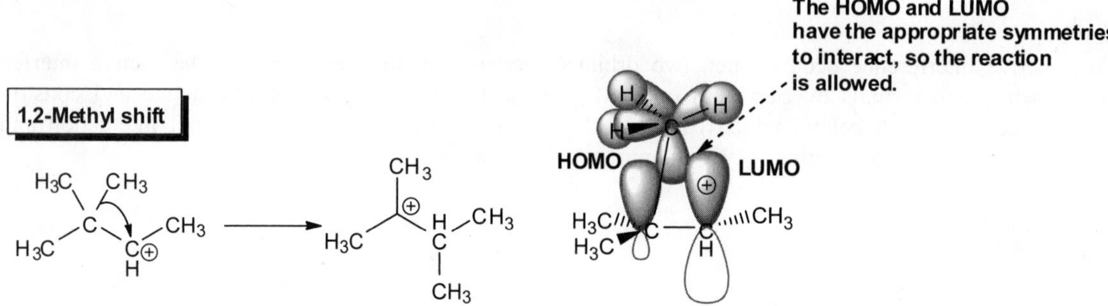
1,2-Methyl shift
The HOMO and LUMO have the appropriate symmetries to interact, so the reaction is allowed.
HOMO
LUMO

Problem IC1.7

Think

If the nucleophile were to approach the C=O from the same side as the O, are there regions of constructive interference between the HOMO and the LUMO? Destructive interference? Would there be a net interaction? Which sites are electron rich and which are electron poor? If two electron-rich sites approach each other, is this a favored interaction?

Solve

If the nucleophile were to approach the C=O from the side of the O, the HOMO and LUMO would have a net interaction. This is shown below, where there is just constructive interference. Therefore, the reaction would be allowed. However, it will not occur due to electrostatic repulsion—it would require two electron-rich sites to form a bond.

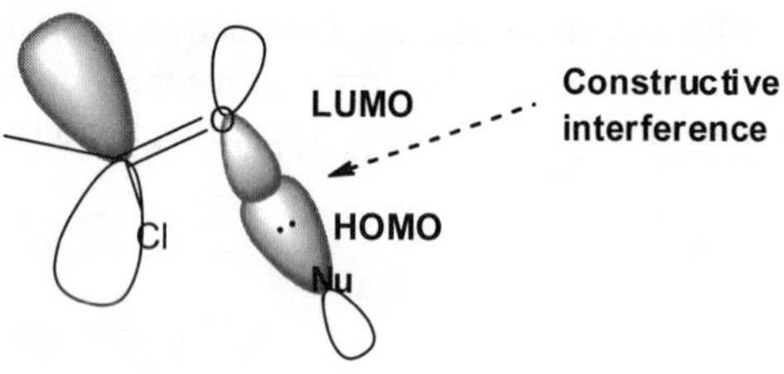
LUMO
Constructive interference
HOMO
Cl
Nu

Problem IC1.8

Think

What is the HOMO and the LUMO for this type of interaction? Are there regions of constructive interference? Destructive interference? Will there be a net interaction?

Solve

Constructive interference occurs when two orbitals overlap with the same phases. Destructive interference occurs when two orbitals overlap with opposite phases. There are regions of destructive and constructive interference, which cancel each other out. Therefore, there is no net interaction and this type of rearrangement does not occur.

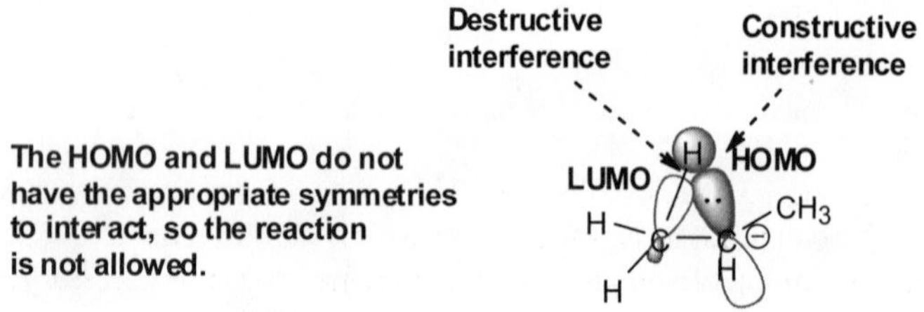
Destructive interference
Constructive interference
The HOMO and LUMO do not have the appropriate symmetries to interact, so the reaction is not allowed.
LUMO
HOMO

NOMENCLATURE 4 | Naming Compounds with Common Functional Groups: Alcohols, Amines, Ketones, Aldehydes, Carboxylic Acids, Acid Halides, Acid Anhydrides, Nitriles, and Esters

Problem N4.1

Think

What is the root name? What functional group is present? What is the suffix for the functional group present? Does the "e" need to be dropped from "ane?" Refer to Table N4-1.

Solve

The longest carbon chain (root) is numbered below. The functional group is indicated by the dashed box and each name is given below the compound.

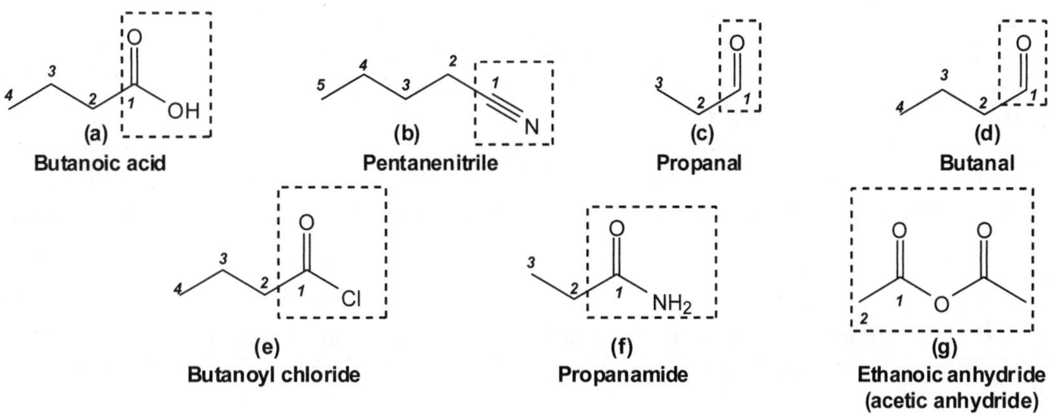

Problem N4.2

Think

How many C atoms are indicated by the root? Based on the suffix, what functional group is present? Refer to Table N4-1.

Solve

The structures are drawn below each given name.

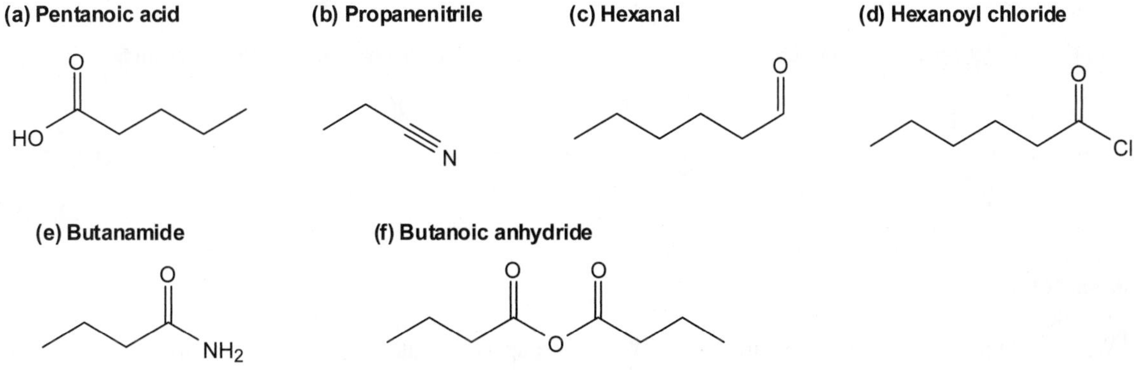

Problem N4.3

Think

What is the root name? Identify the functional group present. What is the suffix for the functional group present? Should "e" be dropped from "ane?" Refer to Table N4-1. How many of each functional group are present? What suffix is used to indicate the number of functional groups? When are numbers not necessary?

Solve

The longest carbon chain (root) is numbered on each compound below. The functional group is indicated by the dashed box and each name is given below the compound. Numbers are not necessary for terminal functional groups—carboxylic acid, nitrile, acid chloride, acid anhydride, aldehyde, and amide. Because there are multiple functional groups specified by the suffix, "e" is not dropped.

(a) Ethanedioic acid

(b) Hexanedial

(c) Pentanedinitrile

(d) Cyclohexane-1,2,3-triamine

(e) Cyclohexane-1,2,4,5-tetraone

(f) Propane-1,2-diol

(g) Hexanedioyl chloride

(h) Ethanediamide

Problem N4.4

Think

How many C atoms are indicated by the root? Identify the functional group present from the suffix. Refer to Table N4-1. How many of each functional group are present? At which C atoms are those functional groups located?

Solve

The structures are drawn below each given name.

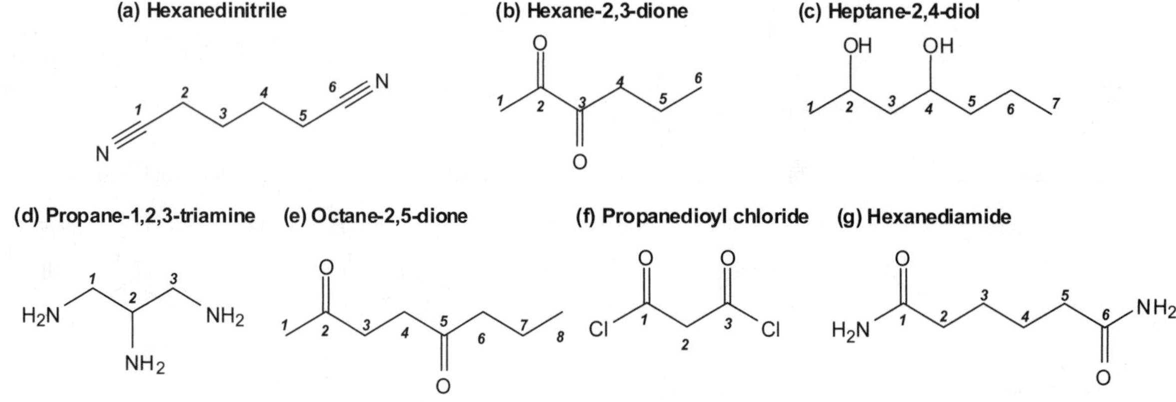

Problem N4.5

Think

Which group is the RCO_2 alkanoate group? Which group is the alkyl group? How many C atoms are in the longest carbon chain of each group? How do you put the two groups together to name the ester?

Solve

The alkanoate group contains the O=C–O and the alkyl group is singly bonded to the O atom. The ester's name is put together with the alkyl group first and the alkanoate group second—alkyl alkanoate. The alkanoate is numbered and the alkyl group is boxed.

(a) Methyl propanoate

(b) Butyl methanoate

(c) Ethyl hexanoate

Problem N4.6

Think

Which group is the alkanoate group? Which group is the alkyl group? How many C atoms are in the longest carbon chain of each group?

Solve

The alkanoate group contains the O=C–O and the alkyl group is singly bonded to the O atom. The alkanoate is numbered and the alkyl group is boxed.

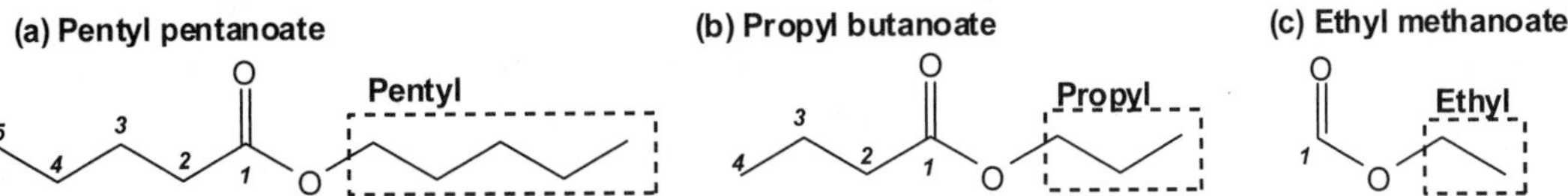

Problem N4.7

Think

Identify the functional group present. What is the longest continuous carbon chain that contains the functional group? What is the root? What is the suffix? How many functional groups corresponding to the suffix are present? On what C atoms are they located? Should the "e" be dropped from "ane?"

Solve

The longest continuous carbon chain containing the functional group corresponding to the suffix is numbered in each compound below and the name is written below the compound. Notice that "e" is not dropped from "ane" when there are multiple functional groups corresponding to the suffix. Also, notice in (g) that no number locators are necessary because there is only one C atom along the main chain to which substituents can be bonded.

(a) 3-Ethylhexan-2-one

(b) 1-Phenylethan-1-ol

(c) 2-Phenylethan-1-ol

(d) 1-Cyclopentyl-2-methoxyethan-1-amine

(e) 2-Butyl-2-propylbutane-1,4-diamine

(f) 2-Butyl-3-ethoxybutane-1,4-diol

(g) Pentylphenylpropanedioic acid

Problem N4.8

Think

How many C atoms compose the longest continuous carbon chain that contains the functional group corresponding to the suffix? How should they be numbered? What functional group is present, as indicated by the suffix, and on which C atom is this functional group located? What substituents are present and on which C atoms are the substituents located? In the molecule you construct, how many C atoms compose the longest continuous carbon chain, irrespective of whether it contains the main functional group?

Solve

The longest continuous carbon chain containing the main functional group is numbered and the structures are drawn below each given name. The longest continuous carbon chain, regardless of whether it contains the main functional group, is highlighted.

(a) 3,3-Dibutylpentane-2,4-dione (b) 2-Nitro-2-pentylpropanedial (c) 2,2-Dichloro-3-hexylbutanedioic acid

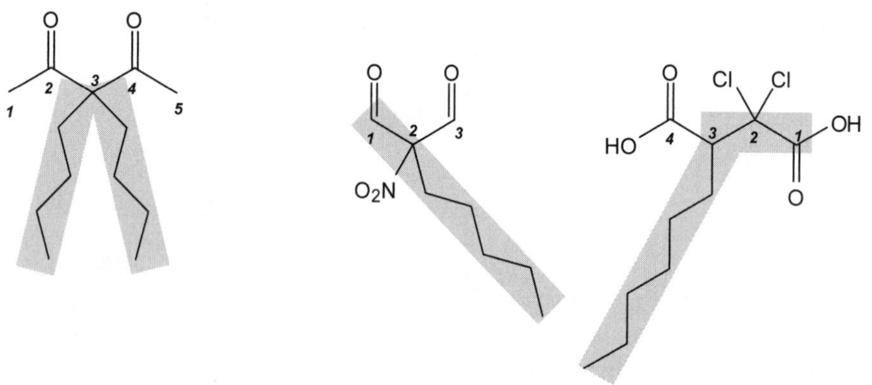

(d) 3-Butyl-2-heptanamine (e) 3,4-Dimethylhexanamide (f) Propyl 5-ethoxypentanoate

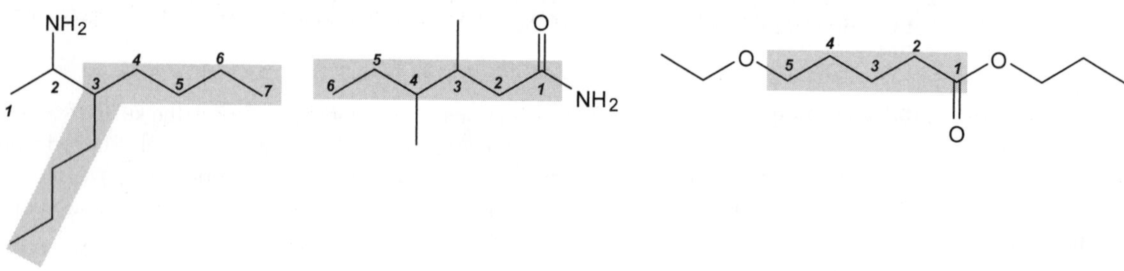

Problem N4.9

Think

How many C atoms compose the longest continuous carbon chain that contains the main functional group? How should they be numbered? What substituents are present on the main chain and on what C atoms are they located? What substituents are present on the N atom? How do you indicate in the name the presence of substituents on a N atom?

Solve

N precedes the name of each alkyl group attached to the N atom of an amine or amide. The longest carbon chain that contains the main functional group is numbered and the *N* alkyl groups are boxed.

(a) ***N,N*-Dimethylbutan-2-amine**

(b) ***N*,2-Dimethylpentan-1-amine**

(c) ***N*,2-Diethylbutanamide**

(d) ***N*-Propylmethanamide (or *N*-Propylformamide)**

Problem N4.10

Think

How many C atoms compose the longest continuous carbon chain? How should those C atoms be numbered? What substituents are present on the main chain and on what C atoms are they located? What substituents are present on the N atom? How do you indicate in the molecule's name the presence of substituents on a N atom?

Solve

N precedes the name of each alkyl group attached to the N atom of an amine or amide. The longest carbon chain is numbered and the *N* alkyl groups are boxed.

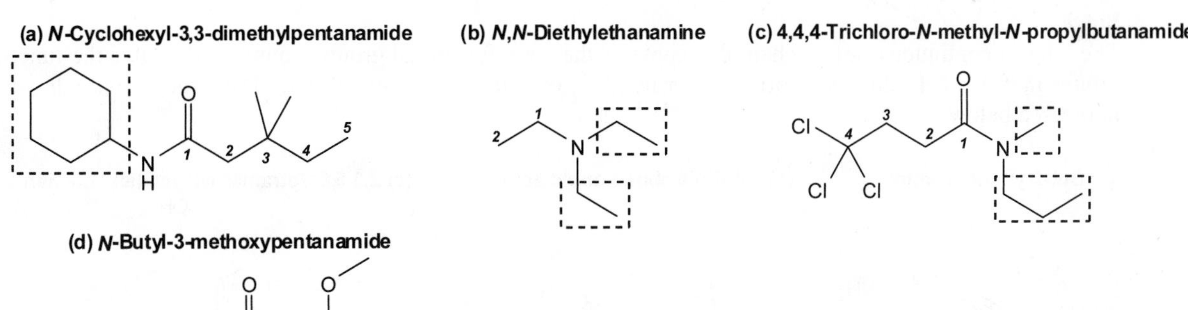

Problem N4.11

Think

How many C atoms compose the longest continuous carbon chain that contains the main functional group? How should those C atoms be numbered? What functional group is present, as indicated by the suffix, and how many of them are present? On which C atoms are those functional groups located? Is there an alkene or alkyne present? How do you determine the numbering of the carbon chain? What substituents are present and on which C atoms are the substituents located? Should the "e" be dropped from "ene" or "yne?"

Solve

The longest continuous carbon chain is numbered to give the functional groups in Table N4-1 higher priority (lower number) over the alkene and alkyne. When there are multiple functional groups corresponding to the suffix, the "e" should not be dropped from "ene" or "yne." The names and structures are below.

(a) Cyclohex-2-ene-1,4-diol

(b) Hex-5-yn-1-amine

(c) 3-Ethylpent-2-enoic acid

(d) Hepta-4,6-diyn-3-one

(e) 3-Methylhept-6-enenitrile

(f) 4,4-Dimethoxybut-3-enal

Problem N4.12

Think

How many C atoms compose the longest continuous carbon chain that contains the main functional group? How should those C atoms be numbered? What functional group is present, as indicated by the suffix, and how many of them are there? On which C atoms are those functional groups located? Is there an alkene or alkyne present? How do you determine the numbering of the carbon chain? What substituents are present and on which C atoms are the substituents located?

Solve

The longest continuous carbon chain that contains the main functional group is numbered to give the functional groups in Table N4-1 higher priority (lower number) over the alkene and alkyne. The structures are drawn and numbered below.

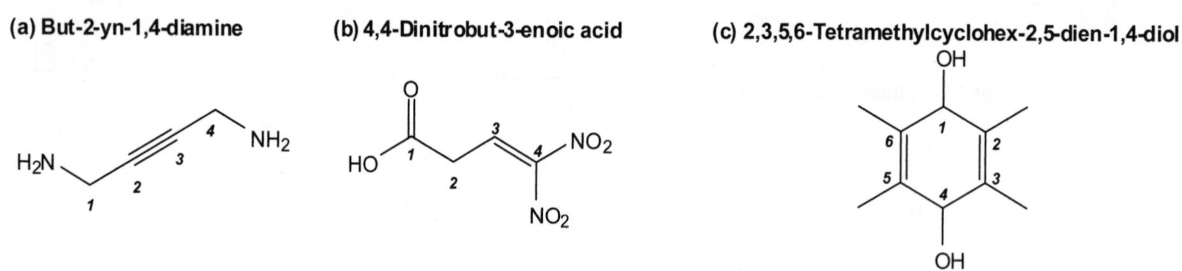

(a) But-2-yn-1,4-diamine

(b) 4,4-Dinitrobut-3-enoic acid

(c) 2,3,5,6-Tetramethylcyclohex-2,5-dien-1,4-diol

Problem N4.13

Think

Draw the structure given the incorrect name. What is incorrect about the name? Look for correct priority numbering, the longest continuous carbon chain to contain the functional group, and correct ordering and placement of commas, dashes, and alphabetical order.

Solve

(a) The alkyne should be located at C3, not C4.

Incorrect numbering

Correct numbering

Hept-3-ynedioic acid

(b) The alkene takes priority over the two nitro groups.

Incorrect numbering

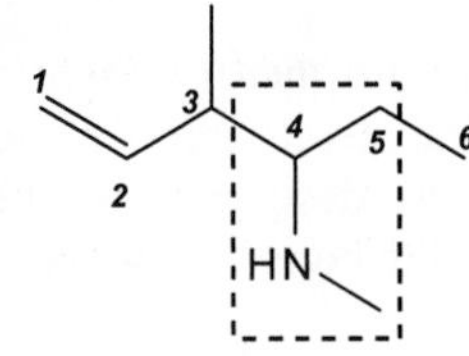

Correct numbering

5,5-Dinitrocyclopent-2-enone

(c) The amine takes priority over the alkene.

Incorrect numbering

Correct numbering

***N*,4-dimethylhex-5-en-3-amine**

(d) The alcohol takes priority over the alkene.

Incorrect numbering

Correct numbering

2-Methoxy-3,4-dimethylpent-3-en-1-ol

Problem N4.14

Think

Is a tetrahedral stereocenter present? Is there a double bond that can have *E*/*Z* configurations? What are the priorities of the groups around the tetrahedral stereocenter? On each atom of the double bond? On which C atom is the tetrahedral stereocenter or double bond located? Is more than one present? When are numbers necessary?

Solve

The names are given below each compound. The longest continuous carbon chain that contains the main functional group is numbered (these are not priority group assignments). A number before the *R*/*S* or *E*/*Z* is necessary if there is more than one stereocenter of each type. Review the rules for assigning *R*/*S* and *E*/*Z* in Nomenclature 3.

(a) **(*E*)-Hex-3-en-2-ol**

(b) **(*E*)-3-Chloro-2-ethylpent-2-enoic acid**

(c) **(3*R*,4*S*)-3,4-Dinitrohexan-1-ol**

(d) **(1*R*,4*S*)-2,3-Dimethylcyclohex-2-ene-1,4-diamine**

(e) **(*S*)-2-Methylhept-5-ynenitrile**

Problem N4.15

Think

What functional groups are present, as indicated by the suffix, and how many are there? Where are they located? What substituents are present and where are they located? Are tetrahedral stereocenters present? Are there double bonds that have possible *E*/*Z* configurations? What configurations are designated by the name? What are the priorities of the groups attached to the tetrahedral stereocenter or double bond? Draw the structure in a way to achieve the configuration indicated.

Solve

The structures of each compound are drawn below. The longest continuous carbon chain that contains the main functional group is numbered. Review the rules for assigning *R*/*S* and *E*/*Z* in Nomenclature 3.

(a) (1*R*,3*S*)-Cyclopent-4-ene-1,3-diol **(b) (2*E*,4*Z*)-Hepta-2,4,6-trienoic acid**

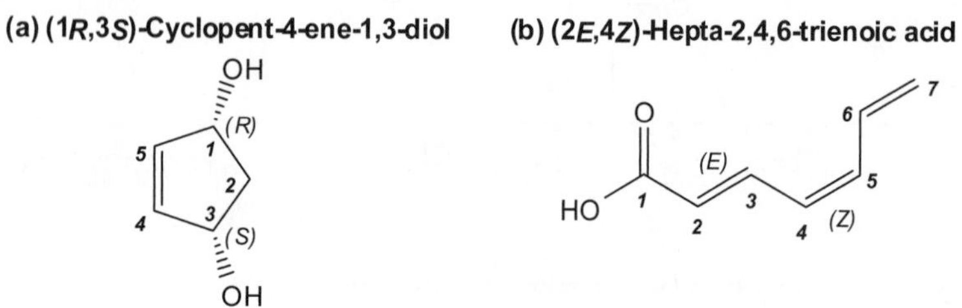

(c) (2*S*,5*S*)-2,5-diethoxyhexane-3,4-dione **(d) (2*R*, 3*Z*)-3-Chloro-2-phenylhex-3-enal**

Problem N4.16

Think

Which functional group has the highest priority? What is the longest continuous carbon chain that contains that functional group? What is the name of the other functional groups when considered a substituent?

Solve

Names of the compounds are given below the structure and the longest continuous carbon chain that contains the main functional group is numbered. The substituents are boxed.

(a) The carboxylic acid has higher priority over the nitrile. Thus the suffix is "oic acid" for the carboxylic acid and the nitrile prefix is "cyano." Note that the carbon of the nitrile is not considered part of the root.

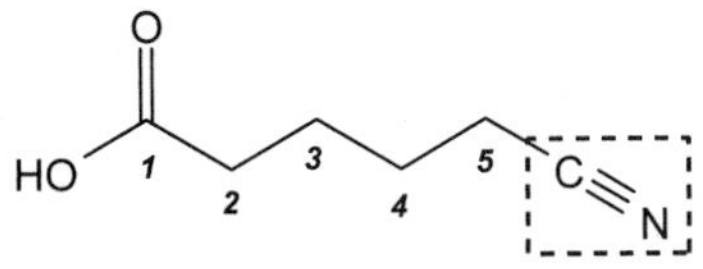

5-Cyanopentanoic acid

(b) The ketone has higher priority over the alcohol. Thus the suffix "one" is used and the alcohol has the prefix "hydroxy."

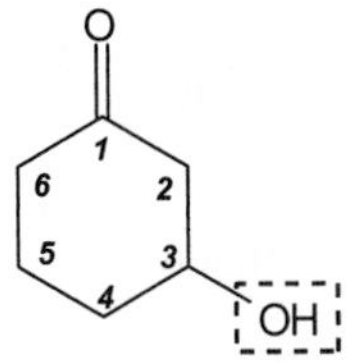

3-Hydroxycyclohexanone

(c) The functional group with the highest priority is the aldehyde functional group. Thus the suffix is "al." The ketone has the prefix "oxo," the amine has the prefix "amino," and the chlorine groups are "chloro." There is a stereocenter at C4, which is *S*. The name is arranged alphabetically.

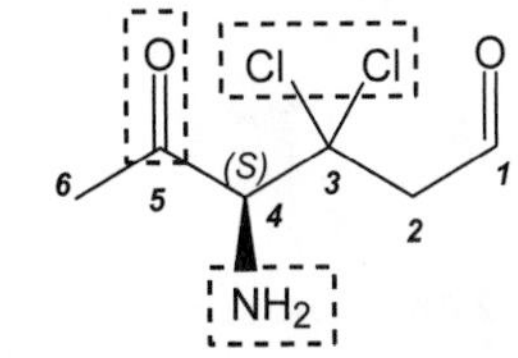

(*S*)-4-Amino-3,3-dichloro-5-oxohexanal

(d) The ester has the highest priority. The alkyl group singly bonded to the O atom is an ethyl and the alcohol prefix is "hydroxy."

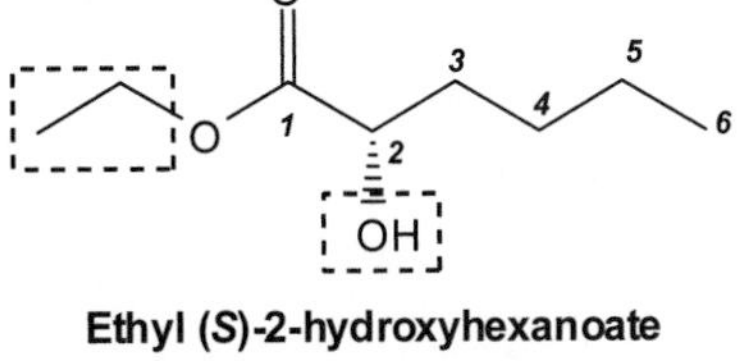

Ethyl (*S*)-2-hydroxyhexanoate

Problem N4.17

Think

What is the functional group indicated by the suffix? On which C atom is this group located? How many C atoms are in the main chain, as indicated by the root? What substituents are present and on which C atoms are they located?

Solve

The structures are given below with the longest continuous carbon chain numbered that contains the main functional group and the substituents boxed.

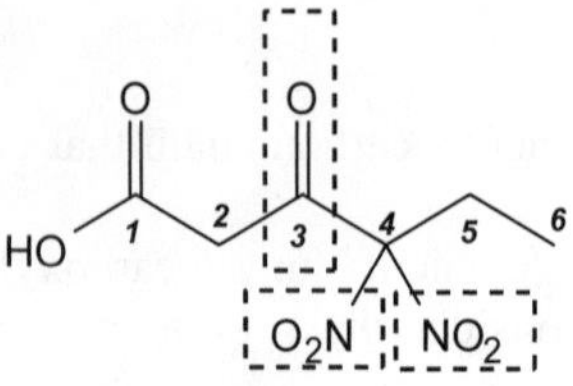

(a)

4,4-Dinitro-3-oxohexanoic acid

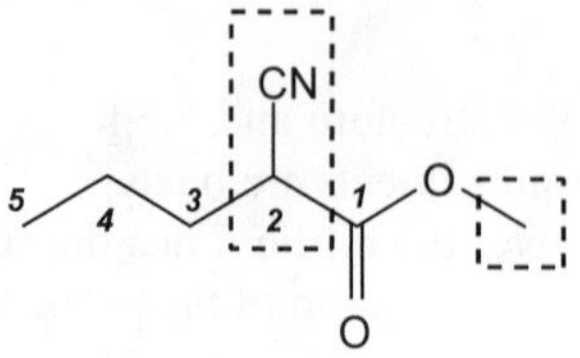

(b)

Methyl 2-cyanopentanoate

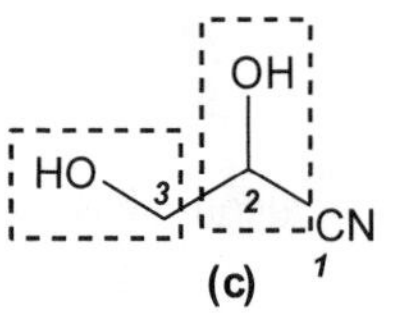

(c)

2,3-Dihydroxypropanenitrile

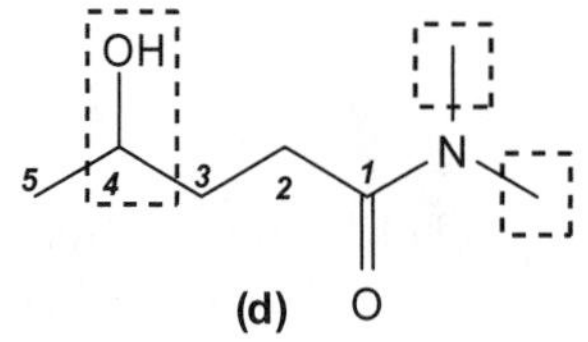

(d)

4-Hydroxy-*N*,*N*-dimethylpentanamide

Problem N4.18

Think

For each trivial name, how many C atoms are present in the main chain? Are other functional groups present? How many carboxylic acid groups are present?

Solve

The structure and IUPAC names are drawn below each trivial name.

(a) Trichloroacetic acid

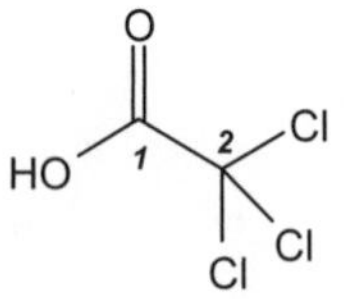

Trichloroethanoic acid

(b) 2,2-Dimethylbutyric acid

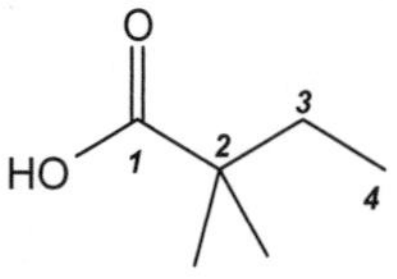

2,2-Dimethylbutanoic acid

(c) 2-Aminopropionic acid

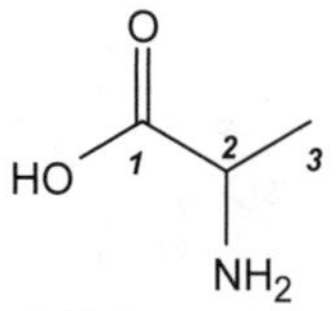

2-Aminopropanoic acid

(d) Dimethylmaleic acid

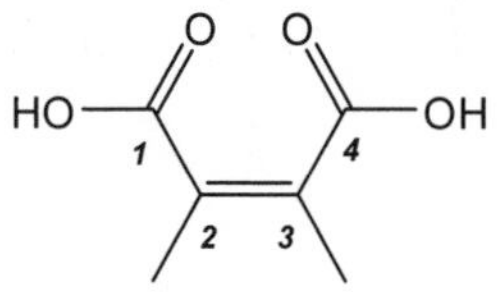

(*Z*)-Dimethylbutenedioic acid

(e) Diethylmalonic acid

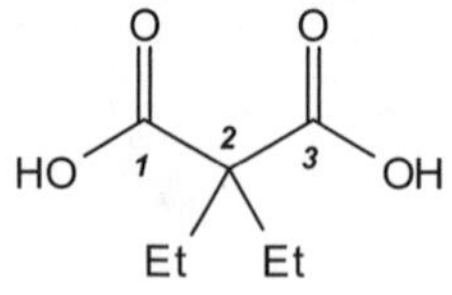

Diethylpropanedioic acid

Problem N4.19

Think

For each trivial name, how many C atoms are present in the main chain? Are other functional groups present?

Solve

The structure and IUPAC names are drawn below each trivial name.

(a) ***N,N*-Dimethylacetamide**

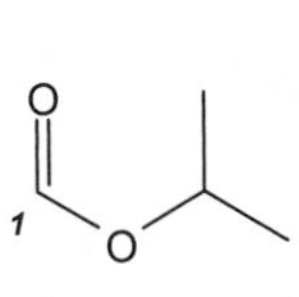

***N,N*-Dimethylethanamide**

(b) **Methoxyacetonitrile** (c) **Ethyl trichloroacetate**

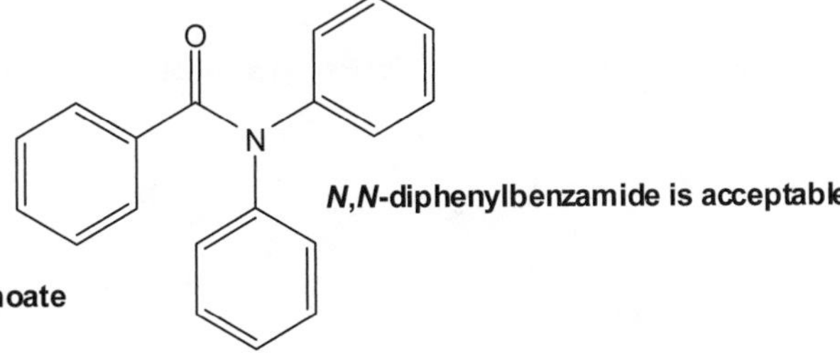

Methoxyethanenitrile **Ethyl trichloroethanoate**

(d) **Isopropyl formate**

Isopropyl methanoate
or (1-methylethyl)-methanoate

(e) ***N,N*-Diphenylbenzamide**

***N,N*-diphenylbenzamide is acceptable**

Problem N4.20

Think

For each trivial name, how many C atoms are present in the main chain? Are other functional groups present?

Solve

The structure and IUPAC names are drawn below each trivial name.

(a) **Phenylacetaldehyde**

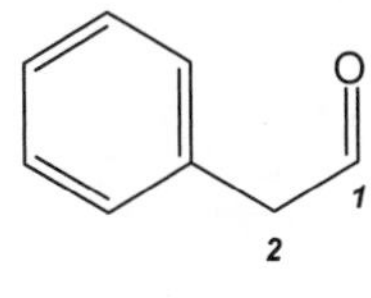

Phenylethanal

(b) **2,3-Dichloropropionaldehyde**

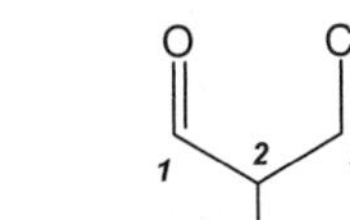

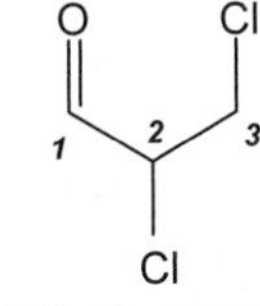

2,3-Dichloropropanal

(c) **4-Nitrobutyraldehyde**

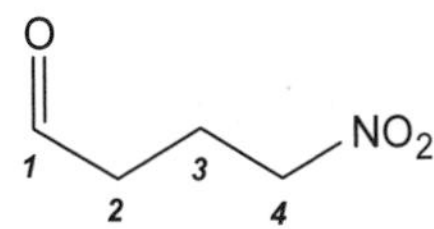

4-Nitrobutanal

Problem N4.21

Think

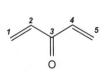

For each trivial name, how many C atoms are present in the main chain? Are other functional groups present?

Solve

The structure and IUPAC names are drawn below each trivial name.

(a) **Divinyl ketone** (b) **Benzyl isopropyl ketone**

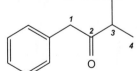

Penta-1,4-dien-3-one **3-Methyl-1-phenylbutan-2-one**

(c) **Cyclohexyl methyl ketone**

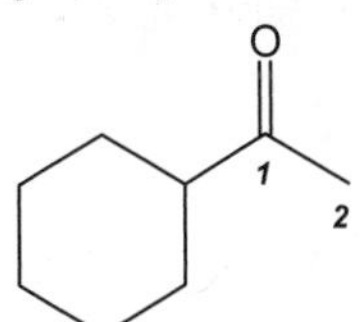

Cyclohexylethanone

(d) Diisopropyl ketone

2,4-Dimethylpentan-3-one

(e) Isobutyl phenyl ketone

3-Methyl-1-phenylbutan-1-one

Problem N4.22

Think

For each trivial name, how many C atoms are present in the main chain? Are other functional groups present?

Solve

The structure and IUPAC names are drawn below each trivial name.

(a) Trichloromethyl alcohol

Trichloromethanol

(b) Isobutyl alcohol

2-Methylpropan-1-ol

(c) Pentyl alcohol

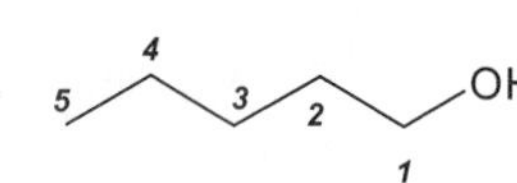

Pentan-1-ol

(d) *sec*-Butyl alcohol

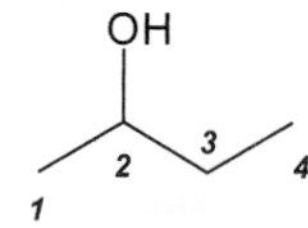

Butan-2-ol

Problem N4.23

Think

Identify the carbon attached to the alcohol. How many C atoms are attached to that carbon? Is the carbon methyl, primary, secondary, or tertiary? What does that mean for the type of alcohol?

Solve

The type of carbon (methyl, primary, secondary, or tertiary) bound to the OH of the alcohol dictates the type of alcohol (methyl, primary, secondary, or tertiary). The C atom attached to the alcohol is in boldface and the neighboring C atoms are also indicated.

(a) Methyl

(b) Primary

(c) Primary

(d) Secondary

Problem N4.24

Think

For each trivial name, how many C atoms are present in the main chain? Are other functional groups present?

Solve

The structure and IUPAC names are drawn below each trivial name.

(a) Diisopropylamine

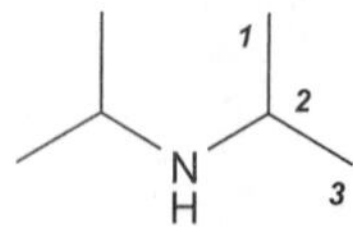

N-isopropylpropan-2-amine
or *N*-(1-methylethyl)-propan-2-amine

(b) *sec*-Butylisopropylamine

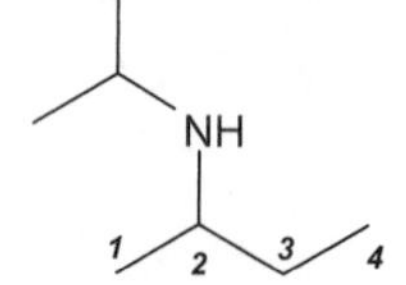

N-Isopropylbutan-2-amine

(c) *tert*-Butyldimethylamine

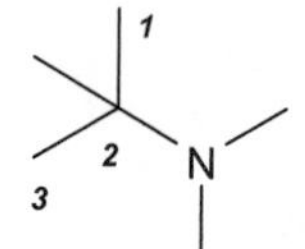

N,N,2-trimethylpropan-2-amine

(d) Triethylamine

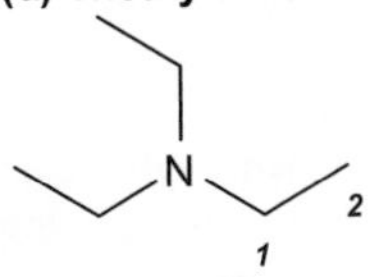

N,N-diethylethanamine

(e) Diphenylamine

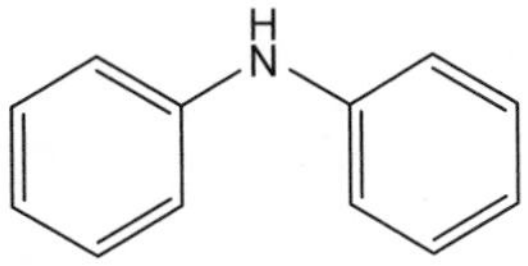

N-phenylaniline

Problem N4.25

Think

How many alkyl groups are directly attached to the N atom? How does the number of alkyl groups directly attached to the N atom classify the type of amine?

Solve

In a primary amine there is one alkyl group attached to the N atom, in a secondary amine there are two alkyl groups attached to the amine, and in a tertiary amine there are three alkyl groups attached to the amine.

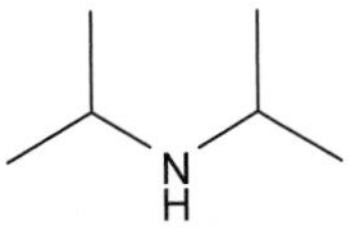

(a) Secondary 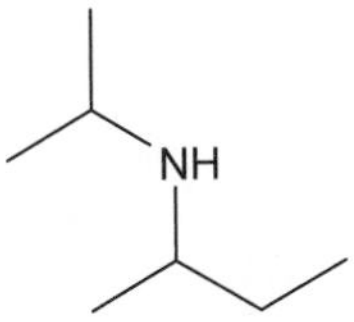 (b) Secondary 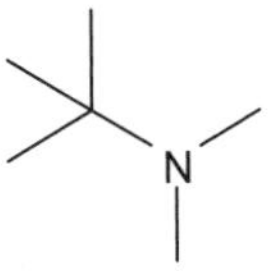 (c) Tertiary 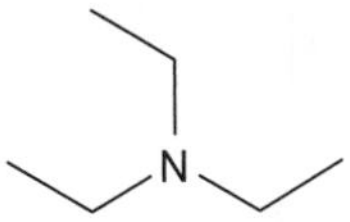 (d) Tertiary 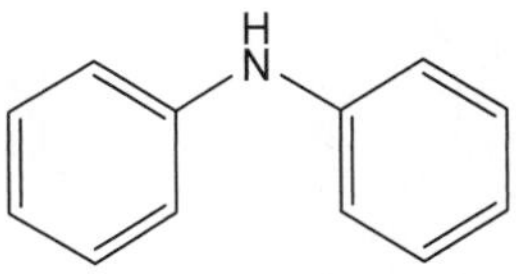 (e) Secondary

Problem N4.26

Think

What is the highest priority functional group attached to the benzene ring? Can that group be combined with benzene to establish a root that is recgonzied by IUPAC? Are other functional groups present?

Solve

The structure and IUPAC names are drawn below each structure.

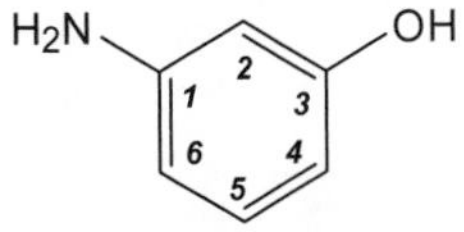

(a) 3-Aminophenol 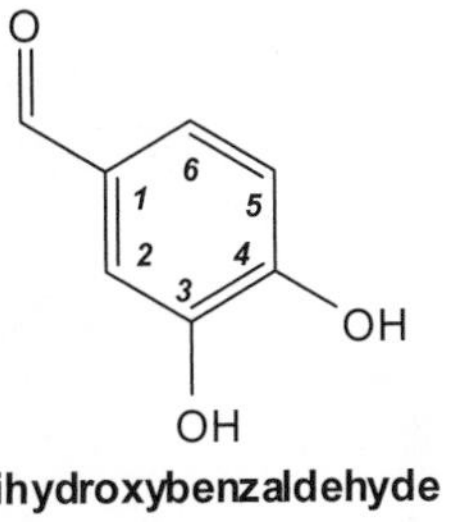 (b) 3,4-Dihydroxybenzaldehyde 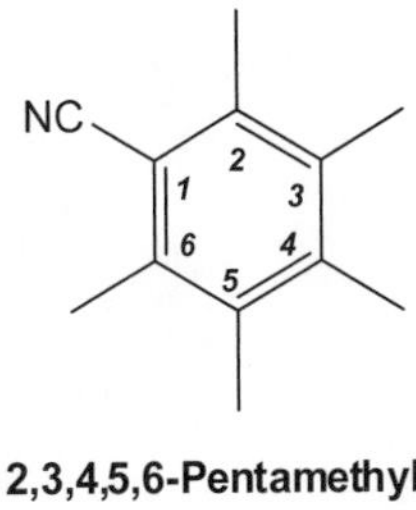 (c) 2,3,4,5,6-Pentamethylbenzonitrile 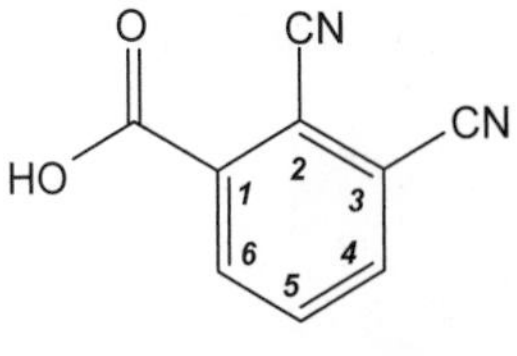 (d) 2,3-Dicyanobenzoic acid

Problem N4.27

Think

According to the root, what is the highest-priority functional group attached to the benzene ring? Which carbon is C1? Are other functional groups present?

Solve

The structure is drawn below each name.

(a) 2,3-Diaminoaniline

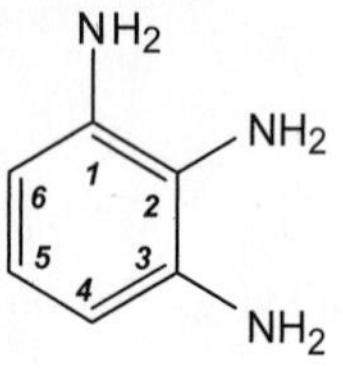

(b) 2,4-Dinitrophenol

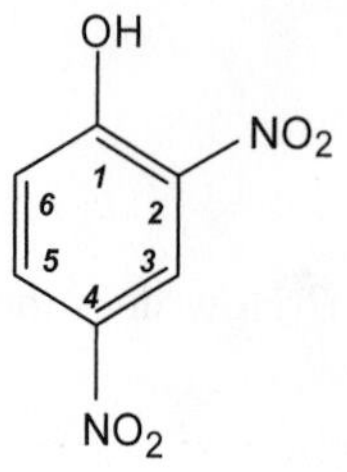

(c) 2,4,5-Trihydroxybenzoic acid

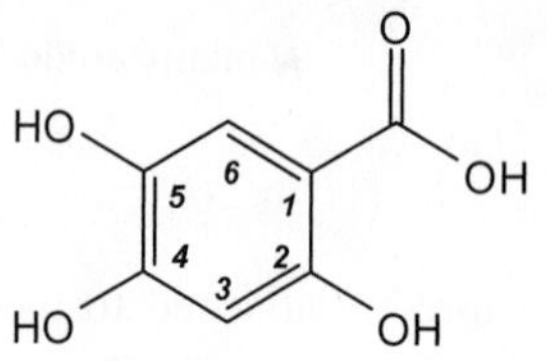

(d) 3-Hydroxybenzonitrile

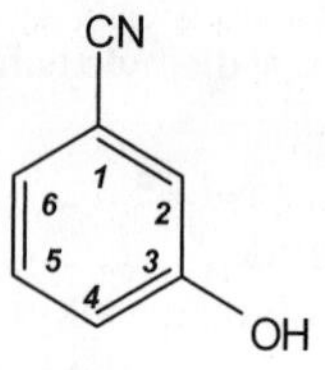

(e) 4-Cyanobenzonitrile

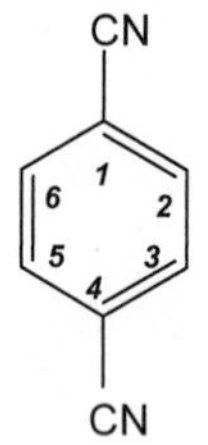

(f) Ethyl 4-methylbenzoate

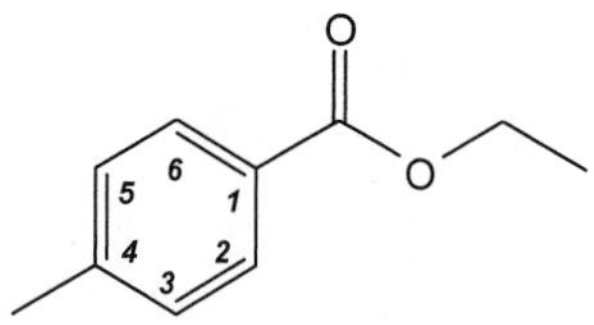

CHAPTER 8 | An Introduction to Multistep Mechanisms: S_N1 and E1 Reactions

Your Turn Exercises

Your Turn 8.1

Think

Consult Chapters 6 and 7 to review the 10 elementary steps. What kinds of bonds are breaking or forming? Are any protons or π bonds involved?

Solve

In Equation 8-2a, the C–L bond breaks. This is a heterolysis step that forms a carbocation and the L^- anion. In Equation 8-2b, the Nu–C bond forms. This is a coordination step in which electrons flow directly from the electron-rich $Nu:^-$ to the electron-poor C^+.

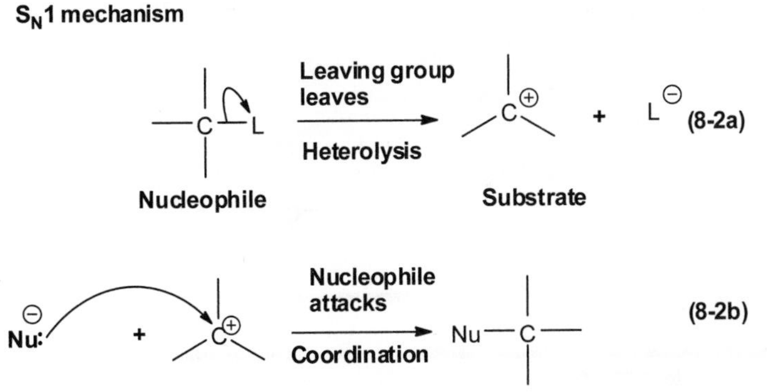

S_N1 mechanism

Leaving group leaves → Heterolysis (8-2a)

Nucleophile Substrate

Nucleophile attacks → Coordination (8-2b)

Your Turn 8.2

Think

Sum up the steps in the reaction first. Which species appear as a product in one step and a reactant in another, and, therefore, get canceled? Which species appear in the overall reaction? Which species appear in the mechanism but not in the overall reaction?

Solve

In the first step (heterolysis), the C–Br bond is broken, producing an allylic carbocation and Br^-. In the second step (coordination), the electron-rich I^- donates two electrons to the electron-poor allylic C^+, producing the C–I bond. The allylic C^+ is a product in the first step and a reactant in the second, so it is canceled when you sum the steps. Because the allylic C^+ appears only in the reaction mechanism and not in the overall reaction, it is an intermediate.

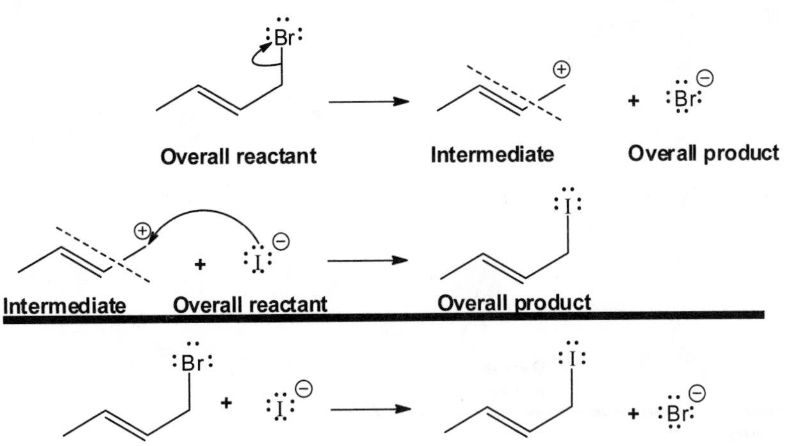

Overall reactant Intermediate Overall product

Intermediate Overall reactant Overall product

Your Turn 8.3

Think

Do transition states appear at energy maxima or minima? How many transition states appear for the two-step reaction shown? Do intermediates appear at energy minima or maxima? How many intermediate stages appear for the two-step reaction shown?

Solve

Transition states appear at energy maxima and show the bond breaking and bond making in progress. Intermediates appear at energy minima. There are two transition states and one intermediate stage for a two-step mechanism.

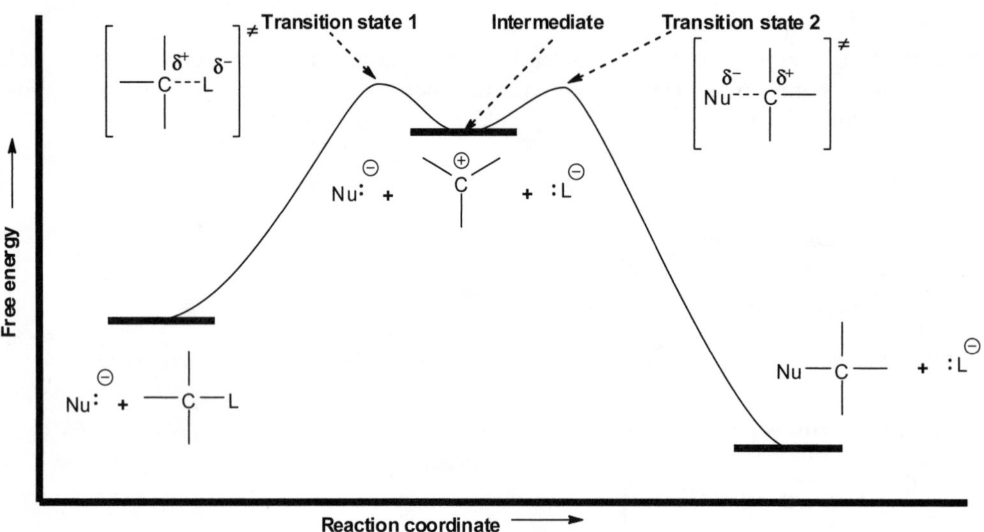

Your Turn 8.4

Think

Do transition states appear at energy maxima or minima? Do intermediates appear at energy minima or maxima? Should overall reactants and products appear in a free energy diagram at the end points of the curve or in between? How is the number of transition states and/or number of intermediates related to the number of steps in a mechanism?

Solve

For a mechanism that contains *n* number of steps, there must be *n* number of transition states and $n-1$ number of intermediates. Transition states occur at energy maxima and there are three transition states for this mechanism. Intermediates occur at energy minima and there are two intermediates for this mechanism. Therefore, there are three elementary steps in this mechanism.

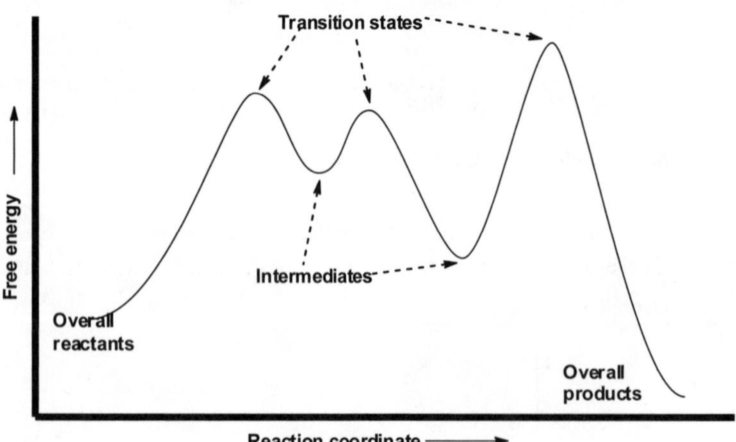

Your Turn 8.5

Think

Consult Chapters 6 and 7 to review the 10 elementary steps. What kinds of bonds are breaking or forming? Are any protons or π bonds involved?

Solve

In Equation 8-5a, the C–L bond is breaking. This is a heterolysis step that forms a carbocation and the L^- anion. In Equation 8-5b, the H–C bond is breaking and the C=C π bond forms. This is an electrophile elimination step in which electrons flow directly from the electron-rich generic base $B:^-$ to the electron-poor H atom and form the B–H bond. Simultaneously, the H–C bond breaks and the electrons fold over to form a C=C π bond.

E1 mechanism

H—C—C—L —(Leaving group leaves / Heterolysis)→ H—C—C⊕ + :L⊖ (8-5a)

Overall reactant → Intermediate, Overall product

B:⊖ + H—C—C⊕ —(Base attacks / Electrophile elimination (proton transfer))→ B—H + C=C (8-5b)

Overall reactant, Intermediate → Overall product, Overall product

H—C—C—L + B⊖ ⟶ B—H + C=C + :L⊖

Your Turn 8.6

Think

Sum up the steps in the reaction first. Which species appear as a product in one step and a reactant in another, and, therefore, get canceled? Which species appear in the overall reaction? Which species appear in the mechanism but not in the overall reaction?

Solve

The carbocation C^+ appears as a product in the first step and a reactant in the second, so it is canceled when the two steps are summed to yield the overall reaction. The carbocation appears only in the reaction mechanism and not in the overall reaction; therefore, it is an intermediate. See the solution of Your Turn 8.5 for the labeled figure.

Your Turn 8.7

Think

Do transition states appear at energy maxima or minima? How many transition states appear for the two-step reaction shown? Do intermediates appear at energy minima or maxima? How many intermediate stages appear for the two-step reaction shown?

Solve

Transition states appear at energy maxima and show the bond breaking and bond making in progress. Intermediates appear at energy minima. There are two transition states (drawn on the next page) and one intermediate for a two-step mechanism.

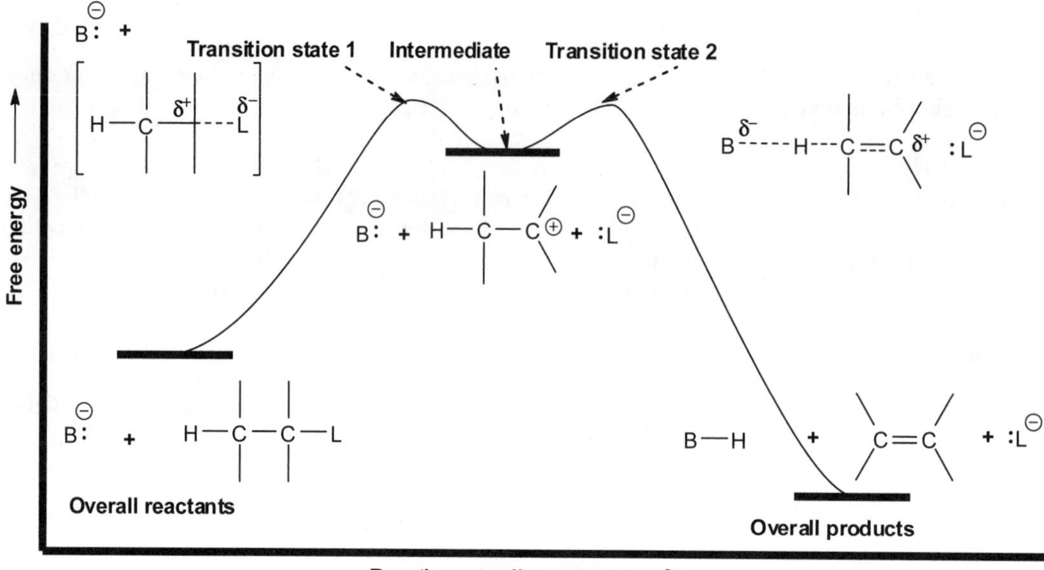

Your Turn 8.8

Think

Consult Figure 8-5. Which line corresponds to 100 °C? At a 15 kJ/mol energy barrier (on the *x* axis), what is the corresponding value on the *y* axis (molecules able to surmount the energy barrier, %)? What is the *y*-axis value when the *x*-axis value is 25 kJ/mol? Do you think that the larger energy barrier will lead to a larger or smaller percentage of molecules that can surmount the energy barrier?

Solve

The higher temperature corresponds to the top (red) curve. At a 15 kJ/mol energy barrier, the percentage of molecules that can surmount the energy barrier is 2%. At a 25 kJ/mol energy barrier, the percentage of molecules that can surmount the energy barrier is ~0.1%. When the energy barrier increases by 10 kJ/mol, the percentage of molecules that are able to surmount the energy barrier decreases by a factor of ~20.

Your Turn 8.9

Think

Consult Figure 8-5. Which line corresponds to 100 °C? To 0 °C? At a 20 kJ/mol energy barrier (on the *x* axis), what is the corresponding value on the *y* axis for each curve? Do you think that increasing the temperature from 0 °C to 100 °C will lead to a larger or smaller percentage of molecules that can surmount the energy barrier?

Solve

The bottom (blue) curve corresponds to 0 °C. At a 20 kJ/mol energy barrier (*x*-axis), <0.1% of the molecules are able to surmount the energy barrier. The top (red) curve corresponds to 100 °C. At a 20 kJ/mol energy barrier, ~0.4% of the molecules are able to surmount the energy barrier. Increasing the temperature leads to a larger percentage of the molecules with enough kinetic energy to surmount the energy barrier (~0.4% vs. <0.1%).

Your Turn 8.10

Think

Refer back to Chapter 5 to recall what makes a tetrahedral carbon stereocenter. In the first step of a unimolecular substitution (S_N1) reaction, what happens to the number of groups bonded to the carbon with the leaving group? What happens to the geometry of that carbon? When the nucleophile attacks the C^+ from in front and behind, what happens to the number of groups bonded to that carbon? What happens to its geometry?

Solve

Each tetrahedral stereocenter is bonded to four different groups. The overall reactant has one stereocenter and, therefore, is chiral. In the first step of an S_N1 reaction, the leaving group departs and leaves behind a carbocation C^+. The carbocation is no longer a stereocenter and is trigonal planar, so it becomes achiral. The nucleophile I^-, therefore, can attack from either side of the C^+ plane. This regenerates the carbon stereocenter and leads to formation of both the *R* and *S* enantiomers in this example. Both products are chiral.

S_N1

I^- can attack from in front of or from behind the plane of C^+

Chiral (S) — Achiral, Planar C — Attack behind → (R) Chiral (8-18a); Attack in front → (S) Chiral (8-18b)

Your Turn 8.11

Think

The reaction takes place under basic conditions. What kinds of species are not compatible with basic conditions: weak acids, strong acids, weak bases, or strong bases? How do you determine whether an acid is strong or weak? How do you determine whether a base is strong or weak?

Solve

If this reaction occurred as shown, the first step yields R_2OH^+, which is an acid that is about as strong as H_3O^+, so it is a strong acid. This strong acid is incompatible in the basic HO^- conditions. Strong acids are incompatible with basic conditions because proton transfer reactions are fast and the acid–base reaction will take place before the organic reaction. This leads to an unreasonable mechanism.

1. S_N2 — 2. Proton transfer

This product is incompatible in basic solution because the product R_2OH^+ is a strong acid.

Your Turn 8.12

Think

The reaction takes place under acidic conditions. What kinds of species are not compatible with acidic conditions: weak acids, strong acids, weak bases, or strong bases? How do you determine whether an acid is strong or weak? How do you determine whether a base is strong or weak?

Solve

If this reaction occurs as shown, the first step yields RO^-, which is a base that is slightly stronger than HO^-, so it is a strong base. This strong base is incompatible with the acidic H_3O^+ conditions. Strong bases are incompatible with acidic conditions because proton transfer reactions are fast and the acid–base reaction will take place before the organic reaction. This leads to an unreasonable mechanism.

1. Heterolysis

This is strongly basic.

2. Proton transfer

3. Coordination

4. Proton transfer

Your Turn 8.13

Think

Are the reagents or reaction conditions acidic, basic, or neutral? Are there any incompatible species as a result? Does the proton transfer reaction shown take place intramolecularly or is it a solvent-mediated step?

Solve

The reagents and reaction conditions are neutral, so neither strong acids nor strong bases would be incompatible. The second step is an intramolecular proton transfer. This is an unreasonable step because there are typically solvent molecules that reside between the acidic and basic sites at any given time. This makes *direct* transfer of protons within a molecule difficult and not likely to occur.

This is an intramolecular proton transfer and is unreasonable.

S_N2

Proton transfer

Chapter Problems

Problem 8.1

Think

How many steps take place in a bimolecular nucleophilic substitution (S_N2) reaction? An S_N1? Which bonds break and form in the overall reaction for each? How many curved arrows are required for each step? What are the products of the reaction?

Solve

The S_N2 mechanism takes place in a single step, requiring two curved arrows. The S_N1 mechanism has two steps—heterolysis followed by coordination—and each requires one curved arrow. The overall products for each reaction are the same: $CH_3CH_2CH(Br)CH_2CH_3$ and Cl^-.

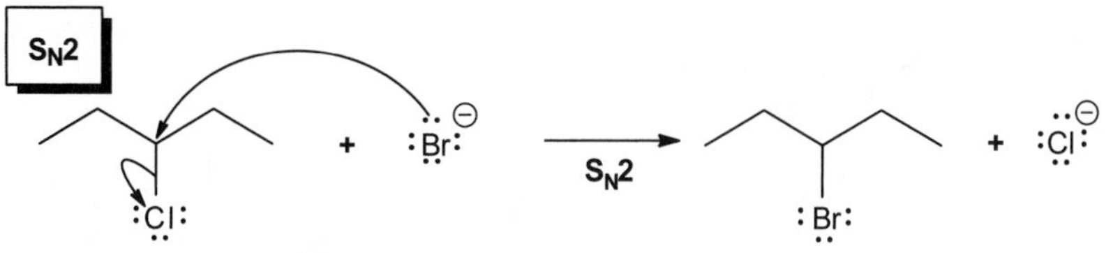

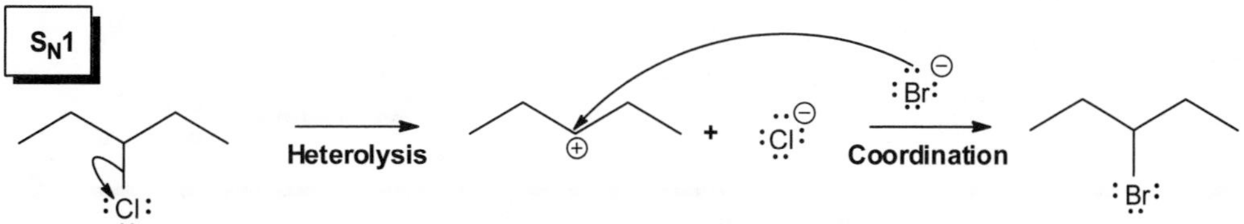

SN2

+ :Br:⁻ → (SN2) + :Cl:⁻

SN1

Heterolysis + :Cl:⁻ ; :Br:⁻ Coordination

Problem 8.2

Think

What is the overall reaction? Which species appears in one of the reaction steps, but does not appear in the overall reaction? How can you tell if a species gets canceled out and does not appear in the overall reaction?

Solve

Overall reactants and products appear in the net (overall) reaction, whereas intermediates do not. The carbocation is the only intermediate in these two reactions. The intermediate gets canceled out because it is a product of one elementary step and a reactant of another elementary step.

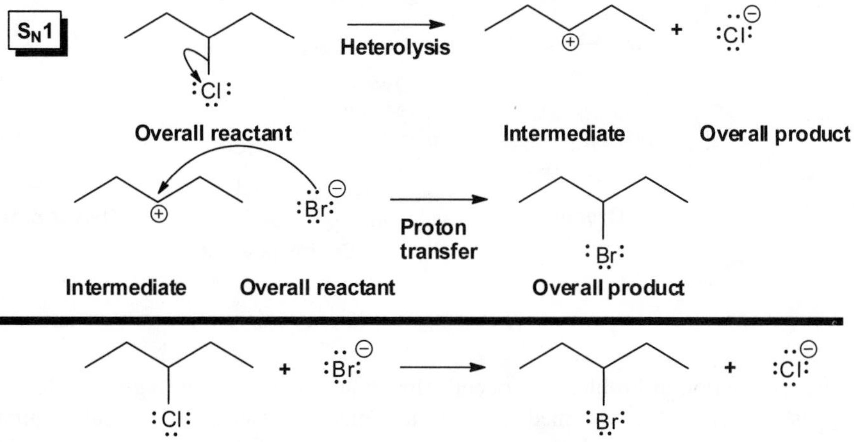

SN1

Heterolysis

Overall reactant | Intermediate | Overall product

Proton transfer

Intermediate | Overall reactant | Overall product

Problem 8.3

Think

In how many steps does the S_N1 reaction in Your Turn 8.2 occur? How many intermediates are there? How many transition states? What is the intermediate? Do intermediates occur at a local minimum or maximum? Is the energy of the intermediate higher or lower than the energy of the reactants/products? Where does a transition state lie on a free energy diagram?

Solve

The S_N1 reaction occurs in two steps, has two transition states (local maxima), one intermediate (local minimum), and the intermediate is higher in energy than the reactants or products. The energy diagram is shown below.

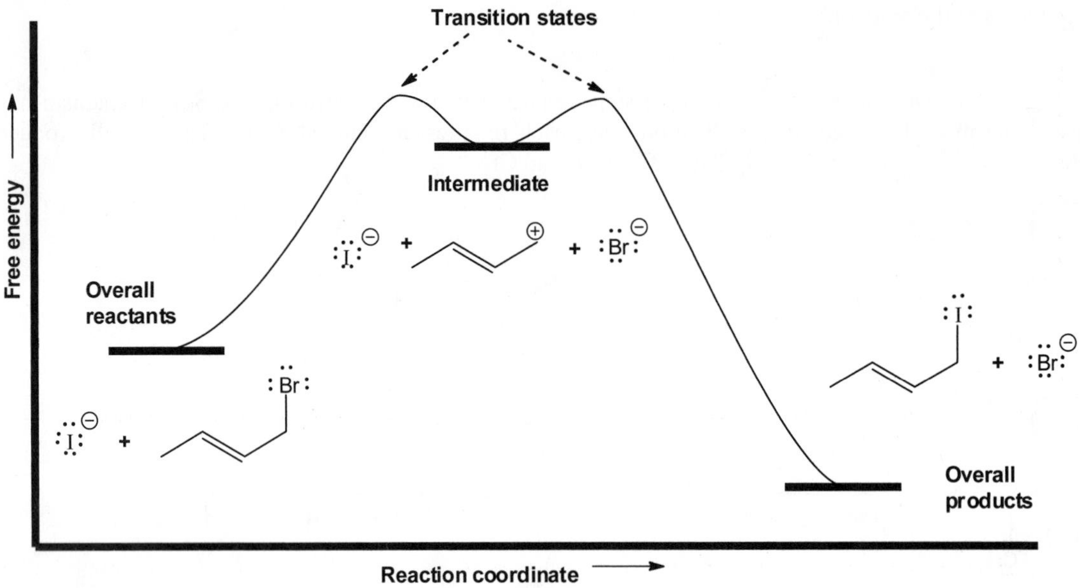

Problem 8.4

Think

In how many steps does the E1 reaction occur? What steps make up an E1 mechanism? What is the leaving group? What is the intermediate? What acts as the base?

Solve

The E1 reaction occurs in two steps:

1. The Br leaving group leaves and forms a tertiary carbocation intermediate on the ring.
2. The $HOCH_3$ acts as a base and deprotonates the adjacent hydrogen, forming a C=C on the ring.

See below:

$C(CH_3)_3$
Heterolysis
$C(CH_3)_3$
Overall reactant
HOCH$_3$
$C(CH_3)_3$
+ H_2OCH_3
Electrophile elimination
Overall reactant
Intermediate
Overall product
Overall product
Overall product

Problem 8.5

Think

In how many steps does the E1 reaction in Problem 8.4 occur? How many intermediate stages should there be? How many transition states? What is the intermediate? Do intermediates occur at a local minimum or maximum? Is the energy of the intermediate higher or lower than the energy of the reactants/products? Where does the transition state lie on the free energy diagram? What functional group results in the product?

Solve

The E1 reaction occurs in two steps, has two transition states (local maxima), one intermediate (local minimum), and the intermediate is higher in energy than the reactants or products. An alkene is the product that results. See the energy diagram on the next page.

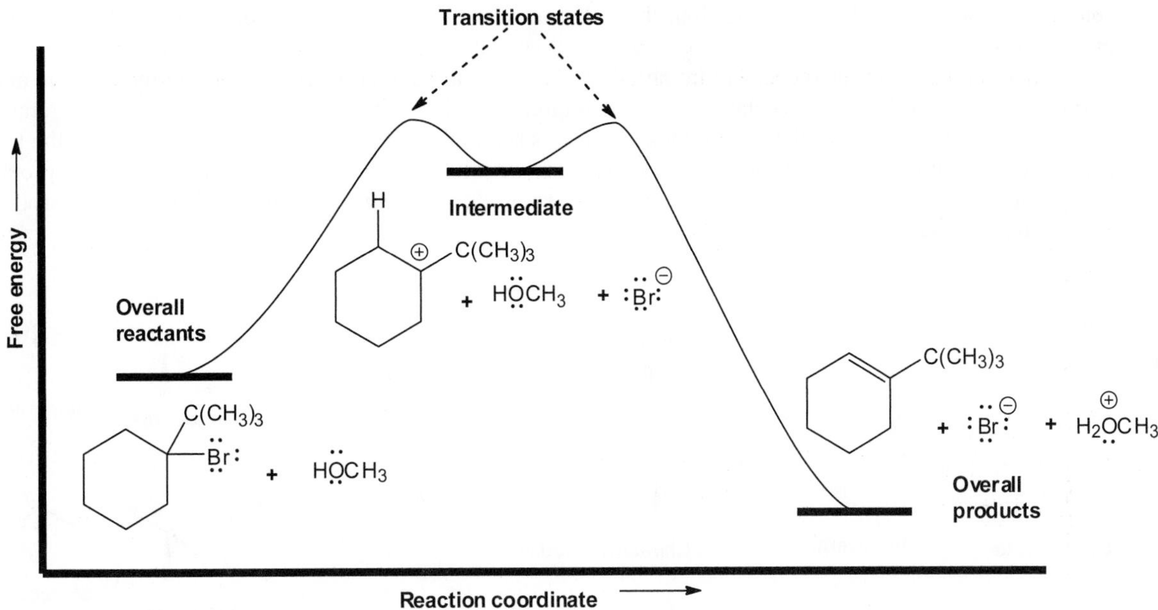

Problem 8.7

Think

How is the concentration of HO^- changed upon going from Trial 1 to Trial 2? What happens to the reaction rate? How is the concentration of R–Br changed upon going from Trial 2 to Trial 3? What happens to the reaction rate? What is the corresponding empirical rate law? Is it consistent with the rate law for an S_N1 or S_N2 reaction?

Solve

Going from Trial 1 to Trial 2, the concentration of the HO^- halves and the concentration of R–Br remains the same. This causes no change in the reaction rate, suggesting that the rate is *not* dependent on HO^-. Going from Trial 1 to Trial 3, the concentration of HO^- remains constant and the concentration of R–Br doubles. This causes the rate to double—from 5.5×10^{-7} M/s to 1.0×10^{-6} M/s. Thus, the rate is directly proportional to the concentration of R–Br. These results indicate an empirical rate law of the form, rate = k[R–Br], which is consistent with an S_N1 reaction.

Problem 8.9

Think

What would the rate-determining step be under this assumption? What is the theoretical rate law for this step?

Solve

If the second step of an E1 reaction were much slower than the first step, then the second step would be the rate-determining step. In that step, the base and the carbocation are reactants, so the rate of that step would be rate = $k[B^-][R^+]$, as shown in Equation 8-15b. Because this is the rate law for the rate-determining step, the rate law would describe the overall reaction as well, and would suggest that the rate of the overall reaction should be directly proportional to the concentration of the base. This disagrees with the actual rate dependence.

Problem 8.11

Think

What can act as the leaving group? What can act as the nucleophile? What are the steps that compose an S_N1 mechanism? Does the reaction take place at a stereocenter?

Solve

The Br substituent can act as a leaving group, coming off as Br^-, which is a relatively stable anion. The HS^- ion can act as a nucleophile. The S_N1 mechanism takes place in two steps. First, the leaving group leaves via heterolysis, yielding a planar carbocation, then HS^- attacks C^+ via a coordination step, yielding the overall products.

Notice that there are two stereocenters in the alkyl halide reactant—one is affected during the reaction and the other one is left alone. The stereochemical configuration at C5 is retained. The stereochemical configuration at C2, however, which is where the reaction takes place, is scrambled. In the carbocation intermediate, the C^+ that is attacked by the nucleophile is planar, so both R and S configurations of the new stereocenter are produced. The product is a mixture of diastereomers, and they are produced in *unequal* amounts because the carbocation intermediate is chiral.

Stereocenter Stereocenter Planar C^+ Chiral intermediate Can attack in front or behind Stereocenter Stereocenter Stereocenter Stereocenter Diastereomers of each other

Problem 8.13

Think

What can act as the leaving group? What can act as the nucleophile? How many steps make up an S_N2 mechanism? How does the nucleophile approach the substrate during attack?

Solve

In each S_N2 reaction, the nucleophile attacks from the side opposite the leaving group. Notice in **(c)** that the Br attached to the ring is *not* the leaving group, because it is bonded to an sp^2-hybridized C atom.

(a) (S) + $:I:^-$ $\xrightarrow{S_N2}$ (R) + $:Br:^-$

(b) (R) (S) $N(CH_3)_3^+$ + $:Br:^-$ $\xrightarrow{S_N2}$ (S) (S) + $:N(CH_3)_3$

(c) Br (S) + $(C_6H_5)_3P:$ $\longrightarrow$ Br $P(C_6H_5)_3^+$ (R) + $:Br:^-$

Problem 8.15

Think

What can act as the leaving group? What can act as the base? What proton can be removed in the second step of an E1 mechanism? Do *E* and *Z* configurations exist for this product? If so, can both be formed?

Solve

In both reactions, Cl is the leaving group and H_2O acts as the base to remove the proton adjacent to the C^+ that forms in Step 1. The products are the alkene, H_3O^+, and Cl^-.

(a) Notice that both *E* and *Z* isomers exist for the product of this first reaction. Because this is an E1 mechanism and the single bonds to C^+ can rotate, both stereoisomers are produced.

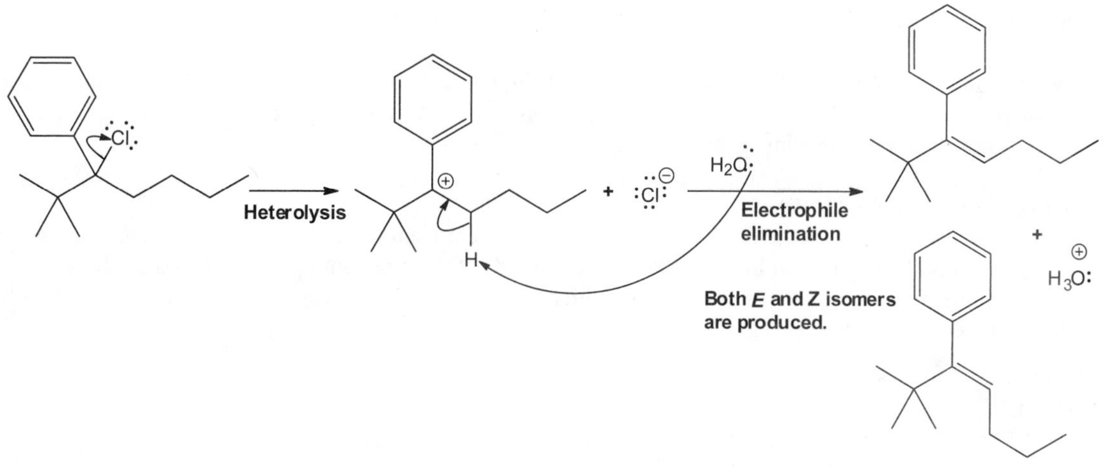

Heterolysis

Electrophile elimination

Both *E* and Z isomers are produced.

(b) In the second reaction, no *E*/*Z* isomers exist about the double bond, so stereochemistry is not a concern for this reaction.

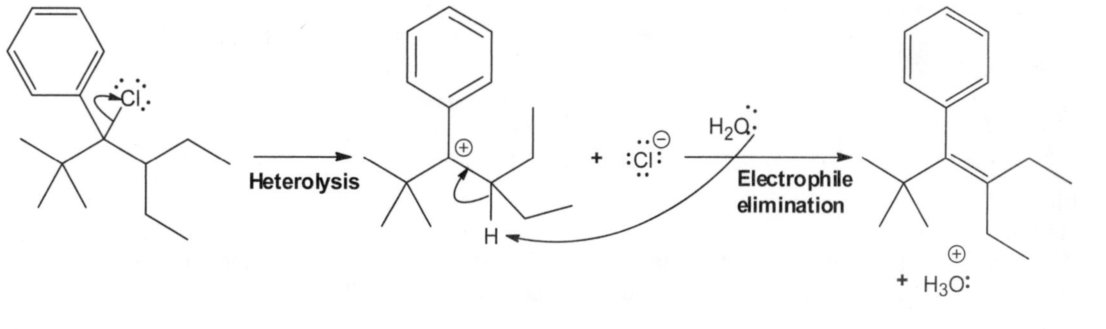

Heterolysis

Electrophile elimination

Problem 8.16

Think

Which group on each side of the C=C is larger? Which isomer has less steric strain? How does steric strain relate to stability? How does the stability of the product relate to the major product?

Solve

In Problem 8.15, only **(a)** has a mixture of diastereomers possible. On the left side of the C=C, the *t*-butyl group is bulkier than the Ph group; on the right side of the C=C, the propyl group is bulkier than the H atom. The more stable isomer has the larger groups on opposite sides of the C=C. Thus in this example, the *Z* isomer is more stable. The more stable isomer is the major product. See the structures on the next page.

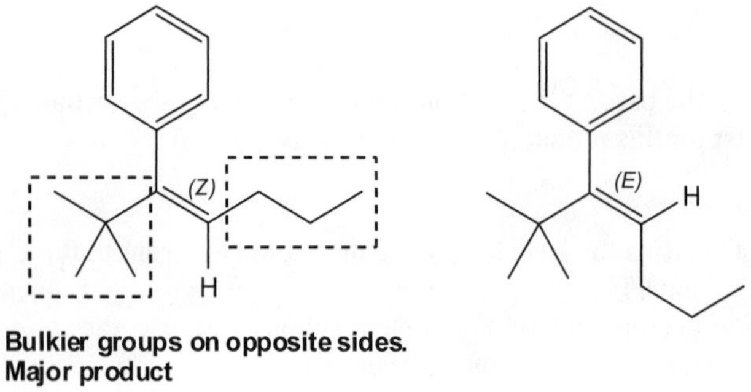

Bulkier groups on opposite sides.
Major product

Problem 8.17

Think

What is the diastereomer of the alkene product in Equation 8-24? What is the preferred conformation of the substrate for E2? Which groups have to be in this conformation? Which group is the leaving group and which H atom is adjacent to the leaving group?

Solve

E2 occurs in one step, and the Br leaving group and adjacent H atom have to be anti-coplanar (shown below). In comparison to the reaction in Equation 8-24, the other diastereomer (both phenyl groups on the same side of the C=C) can be produced from the substrate in which the CH_3 and C_6H_5 groups on the left-hand carbon have been interchanged. Thus, these two groups are interchanged in the product as well.

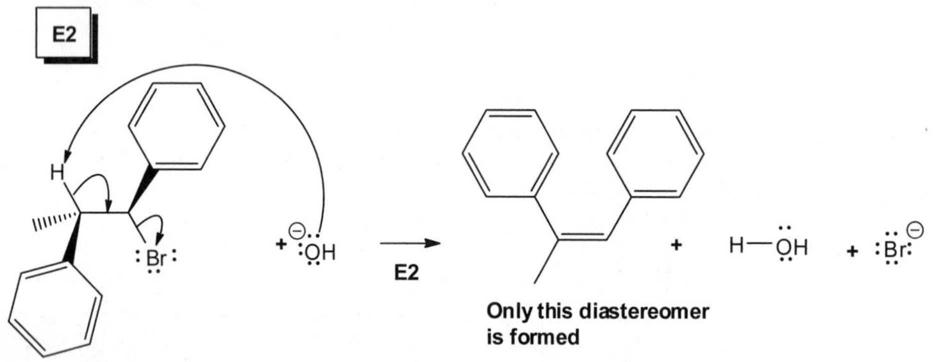

Only this diastereomer is formed

Problem 8.19

Think

What conformations of the substrate facilitate an E2 reaction? In those conformations, what dictates which atoms/groups are on the same side of the double bond in the products?

Solve

An E2 reaction is facilitated by an anti-coplanar conformation involving H (or D), Br (the leaving group), and the two C atoms to which they are bonded. There are two such conformations, as shown on the left in the reaction below and on the next page: One involves H and the other involves D. In **(a)**, the H and Br atoms are anti-coplanar, so they can be eliminated in an E2 step. In that conformation, the CH_3 groups point in opposite directions, so they end up on opposite sides of the double bond in the product. In **(b)**, the D and Br atoms are anti-coplanar, so they will be eliminated in an E2 step. In that conformation, the CH_3 groups both point toward you, so they end up on the same side of the double bond in the product.

(a)

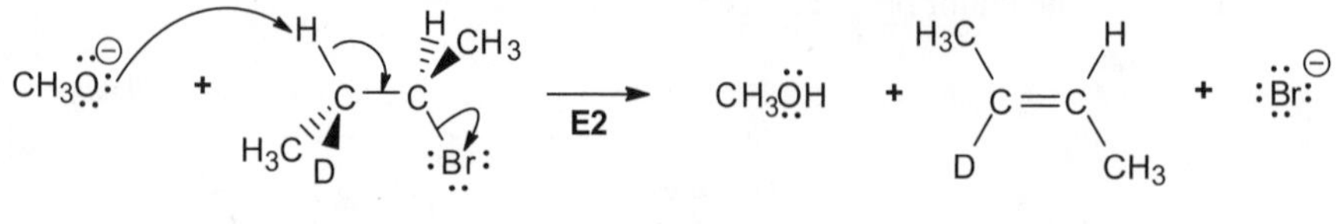

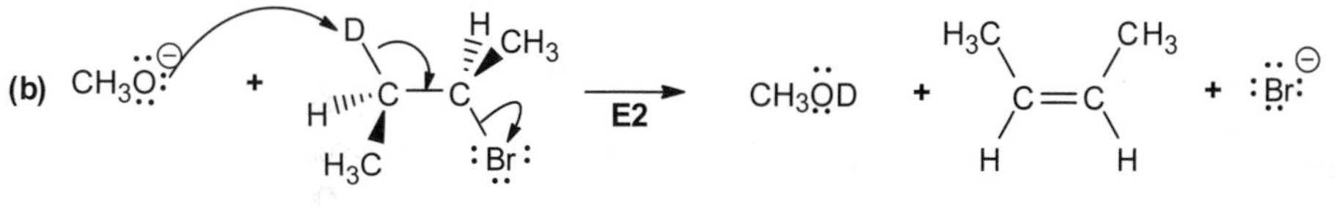

(b) CH_3O^- + [substrate] $\xrightarrow{E2}$ CH_3OD + [alkene] + Br^-

Problem 8.20

Think

On which C atom is the leaving group located? On the two neighboring C atoms, are there H atoms present in the correct orientation to be eliminated? What is the correct orientation of the leaving group and the eliminated H for E2?

Solve

Molecule **A** should undergo E2 elimination more readily because the Cl substituent in molecule **B** is not anti to any H atoms on an adjacent carbon. The only H atom on an adjacent atom is on the *same* side of the ring. By contrast, in molecule **A**, the H and Cl atoms indicated are on opposite sides of the ring and, therefore, can attain the anti-coplanar arrangement.

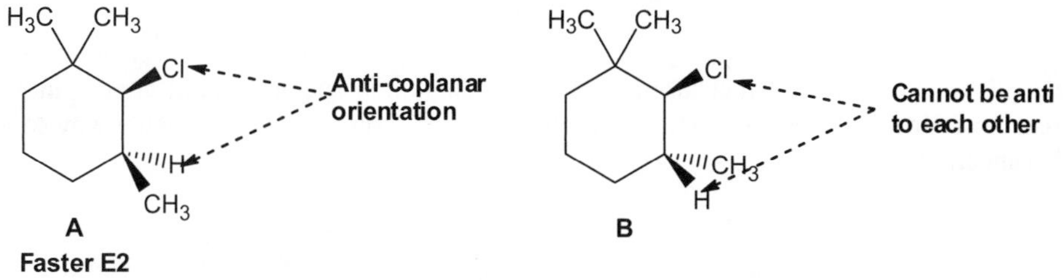

Anti-coplanar orientation

A

Faster E2

Cannot be anti to each other

B

Problem 8.21

Think

Is the reaction carried out in acidic or basic conditions? What species appears that makes the reaction mechanism unreasonable? How can you change the order of the steps in the mechanism to make the reaction reasonable? What type of species are you trying to avoid forming?

Solve

The mechanism is carried out in basic media and, therefore, the presence of a strong acid (such as R_2OH^+) should be avoided. We can accomplish the same net reaction, but avoid the generation of a strong acid, by first deprotonating cyclohexanol. Then, in the second step, the S_N2 reaction can take place.

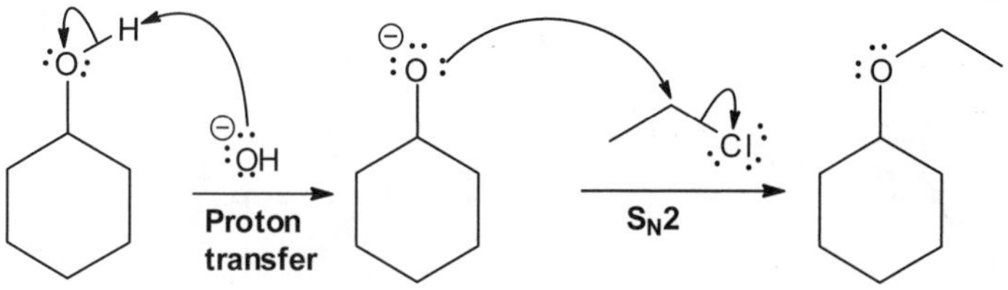

Proton transfer

S_N2

Problem 8.22

Think

Is the reaction carried out in acidic or basic conditions? What species appears that makes the reaction mechanism unreasonable? How can you change the order of the steps in the mechanism to make the reaction reasonable? What type of species are you trying to avoid forming?

Solve

The mechanism is carried out in acidic media and, therefore, the presence of a strong base (such as RO^-) should be avoided. We can avoid the generation of a strong base by reversing the first two steps. Proton transfer occurs first; this allows the C–O to be broken without forming a strong base. See the mechanism on the next page.

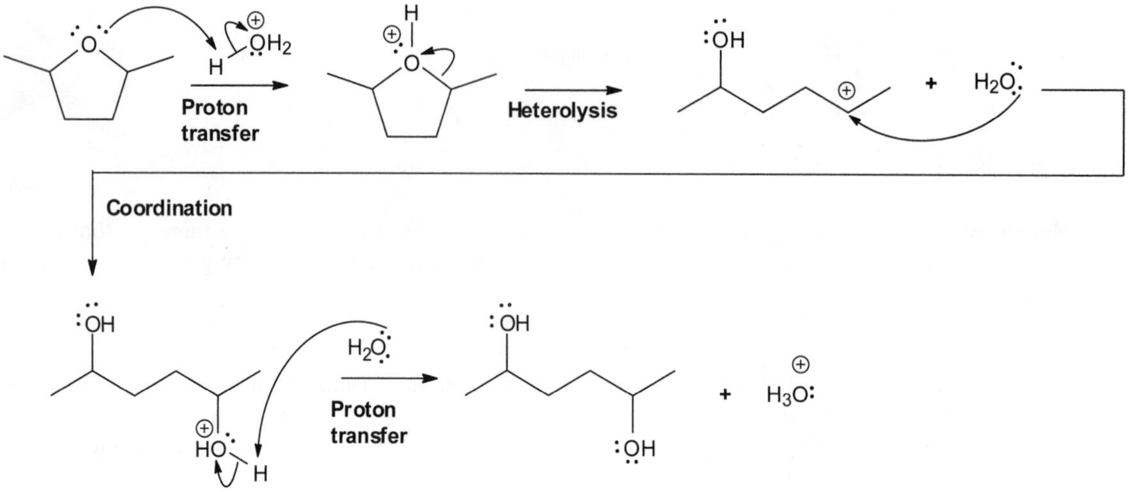

Proton transfer

Heterolysis

Coordination

Proton transfer

Problem 8.24

Think

How many steps generally make up an E2 mechanism? What incompatible species would appear in such a mechanism for this reaction? How can you incorporate a proton transfer reaction in a reasonable way to avoid the formation of such a species?

Solve

An E2 mechanism generally takes place in a single step. In this case, if the E2 step were to take place first, CH_3O^- would be produced, which is strongly basic.

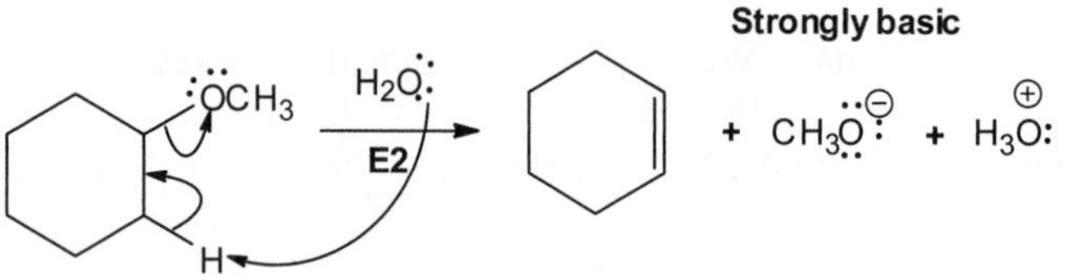

Strongly basic

The reaction, however, takes place under acidic conditions, making CH_3O^- incompatible. A proton transfer step needs to occur first to avoid the generation of a strong base when the leaving group leaves. By protonating the oxygen in the ether first, you allow it to leave as an uncharged alcohol and avoid the formation of a basic alkoxide.

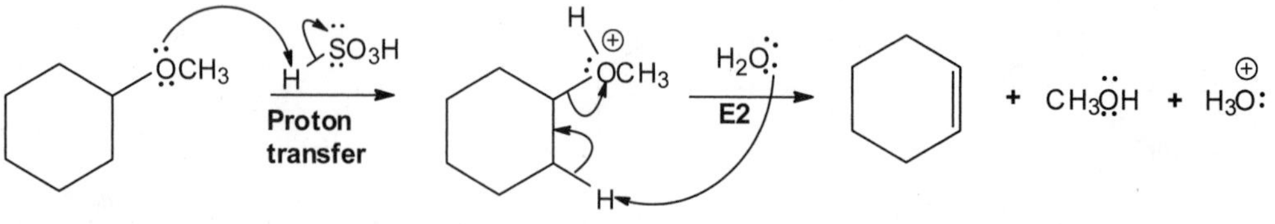

Proton transfer

Problem 8.25

Think

Is proton transfer occurring via intramolecular proton transfer or solvent-mediated proton transfer? Which type of proton transfer is reasonable? Why?

Solve

An intramolecular proton transfer is proposed and is unreasonable because the solvent molecules are abundantly present and exist in between the two sites involved in the proton transfer. Therefore, a solvent-mediated proton transfer is more reasonable, as shown on the next page.

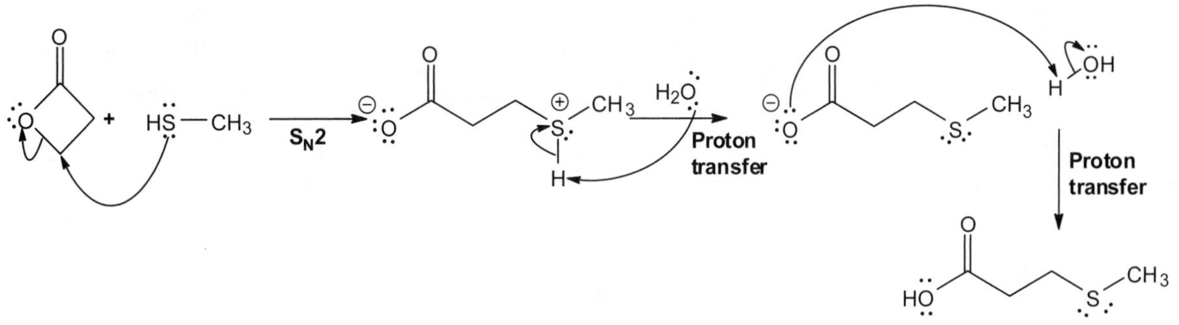

Problem 8.27

Think

What is the molecularity of the mechanism proposed? Is this likely to occur? How can you propose a more reasonable mechanism?

Solve

The step is unreasonable because it is a termolecular step—that is, there are three reactants in this step. It is more reasonable to split it up into two steps, while avoiding the generation of a strongly basic species, given that the reaction takes place under acidic conditions. This requires the OH to be protonated in the first step.

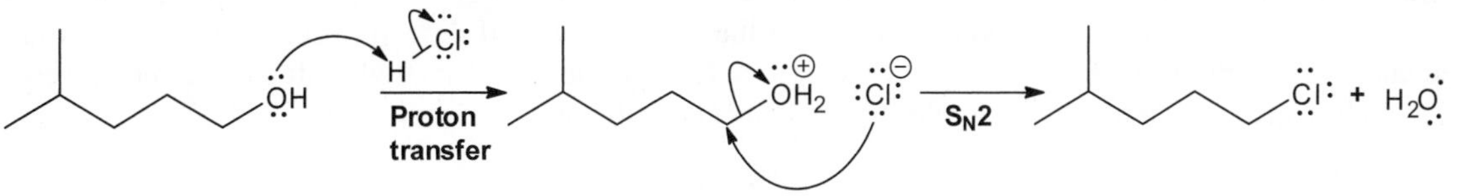

Problem 8.29

Think

What are the normal steps of an S_N1 mechanism? Is a carbocation generated as an intermediate? If so, can it undergo a 1,2-hydride or 1,2-methyl shift to become more stable?

Solve

The normal S_N1 reaction with an uncharged nucleophile takes place in three steps. The leaving group leaves in the first step, the nucleophile coordinates to the carbocation in the second step, and proton transfer takes place in the third step to form an uncharged organic product.

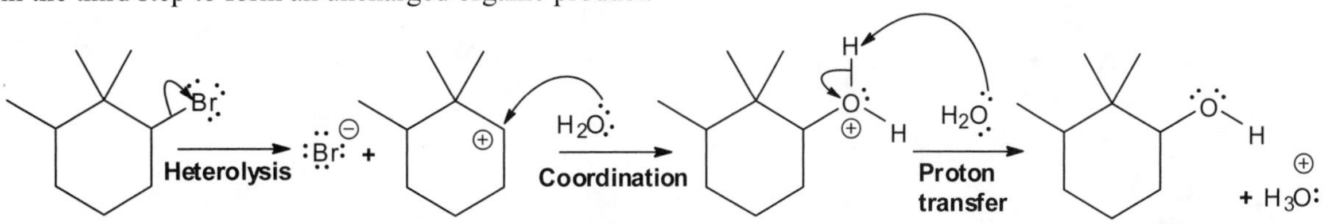

Notice, however, that there is a secondary carbocation intermediate that can rearrange to become more stable, via 1,2-methyl shift, as shown on the next page. After this rearrangement, coordination and proton transfer follow to form the tertiary alcohol product. See the mechanism on the next page.

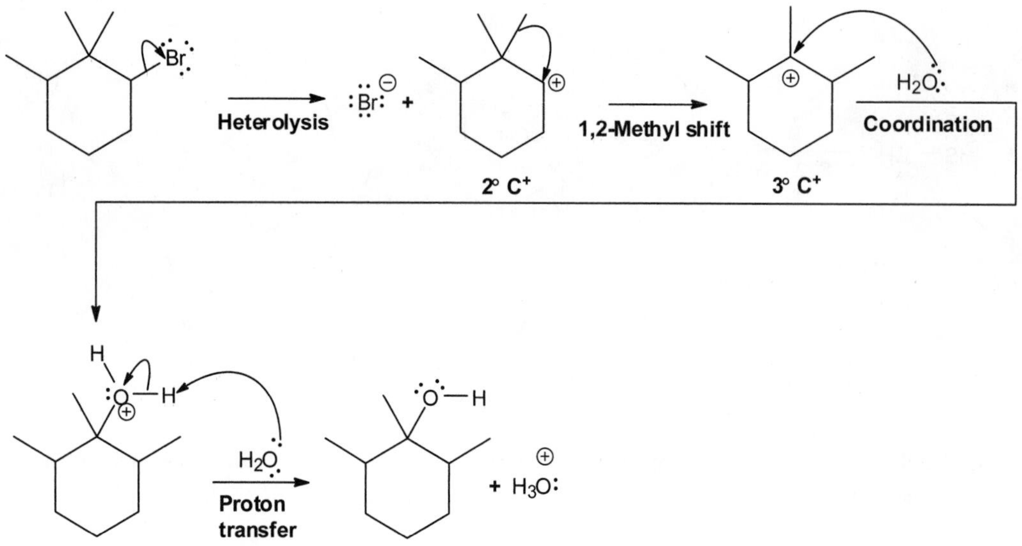

Heterolysis

1,2-Methyl shift

Coordination

2° C⁺

3° C⁺

Proton transfer

Problem 8.30

Think

How are the metal-containing reactants simplified? What are the normal steps of an S_N2 mechanism? Does the reaction take place at a stereocenter? If so, how does the approach of the nucleophile govern which configuration is formed in the product? What are the normal steps of an S_N1 mechanism? Is a carbocation generated as an intermediate? If so, can it undergo a 1,2-hydride or 1,2-methyl shift to become more stable? Does an S_N1 mechanism produce a single stereochemical configuration or a mixture of both?

Solve

Na^+ and K^+ are treated as spectator ions. Each S_N2 mechanism consists of a single step, whereas each S_N1 mechanism consists of two steps. In an S_N2 reaction, the nucleophile attacks the substrate from the side opposite the leaving group to produce only a single stereoisomer; in an S_N1 mechanism, a mixture of stereoisomers can form. Notice that stereochemistry is important in **(ii)**, **(iii)**, **(iv)**, and **(v)**. Notice also in **(iv)** that a carbocation rearrangement will take place in an S_N1 mechanism but not in an S_N2 mechanism. The mechanisms for **(i)**–**(v)** are shown below and on the next page.

(i)

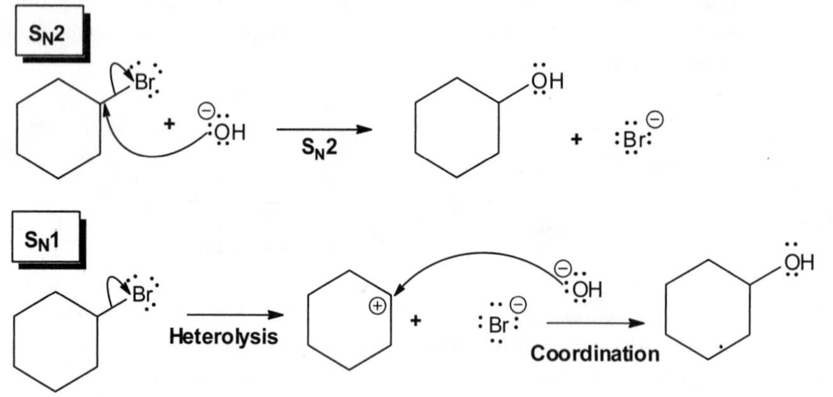

S_N2

S_N2

S_N1

Heterolysis

Coordination

(ii) S_N2

S_N2

S_N1

Heterolysis

Coordination

(iii) S_N2

S_N2

S_N1

Heterolysis

Coordination

(iv) S_N2

S_N2

S_N1

Heterolysis

Coordination

(v) S_N2

Br ... :Cl: + :OCH3 → S_N2 → Br ... OCH3 + :Cl:

S_N1

Br ... :Cl: → Heterolysis → Br ... (+) + :Cl: + :OCH3 → Coordination → Br ... :OCH3 / Br ... :OCH3

Problem 8.31

Think

How are the metal-containing species simplified? What are the normal steps of an E2 mechanism? What are the normal steps of an E1 mechanism? Is a carbocation generated as an intermediate? If so, can it undergo a 1,2-hydride or 1,2-methyl shift to become more stable? If *E* and *Z* configurations exist for the alkene product, is one produced exclusively, or is a mixture of the two produced?

Solve

E2 reactions consist of a single step, whereas E1 reactions consist of two steps. E2 reactions are favored when the substrate has a H atom and leaving group in an anti-coplanar conformation. When *E*/*Z* isomers are possible, an E1 reaction produces a mixture of both. Frequently, an E2 reaction produces either the *E* or *Z* isomer exclusively. But when there are two conformations in which the anticoplanar arrangement can be attained, as in **(i)**—because there are two H atoms on the adjacent C atom—then both *E* and *Z* isomers are produced. Also, whereas carbocation rearrangement is possible for an E1 mechanism, as in **(iv)**, **(v)**, and **(vi)**, it is not possible for an E2 mechanism. In **(vi)**, notice that only the proton shown being removed can attain an anti-coplanar arrangement with the leaving group. The mechanisms for **(i)**–**(vi)** are shown below and on the next page.

(i) E2

H H ... :Cl: + :OH → E2 → + :Cl:

Rotation about C–C bond

H H ... :Cl: + :OH → E2 → + :Cl:

E1

Heterolysis

Electrophile elimination

(ii) E2

E2

E1

Heterolysis

Electrophile elimination

(iii) E2

E2

E2

E1

Heterolysis

Electrophile elimination

Plus enantiomer

(iv) **E2**

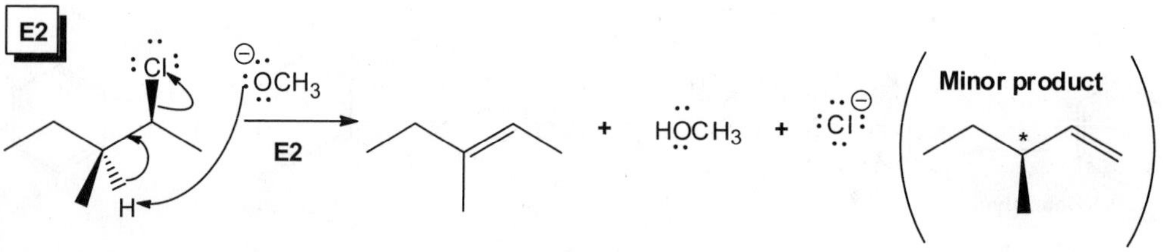

:Cl: ⊖:OCH3 → E2 → + HOCH3 + :Cl:⊖ (Minor product *)

E1

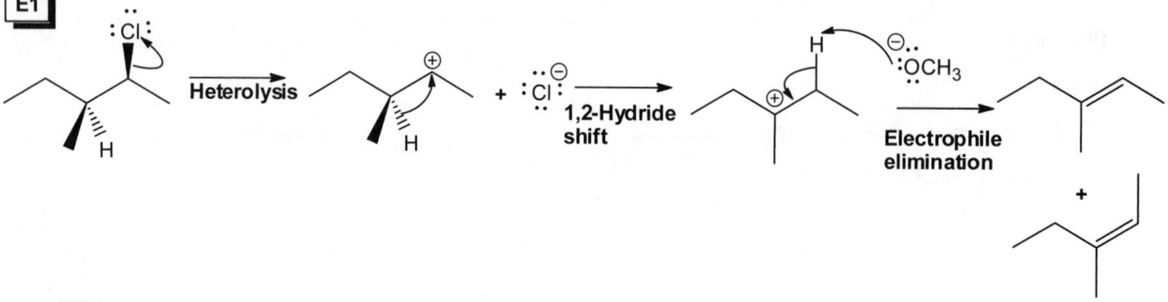

:Cl: → Heterolysis → ⊕ + :Cl:⊖ → 1,2-Hydride shift → ⊕ H ⊖:OCH3 → Electrophile elimination → +

(v) **E2**

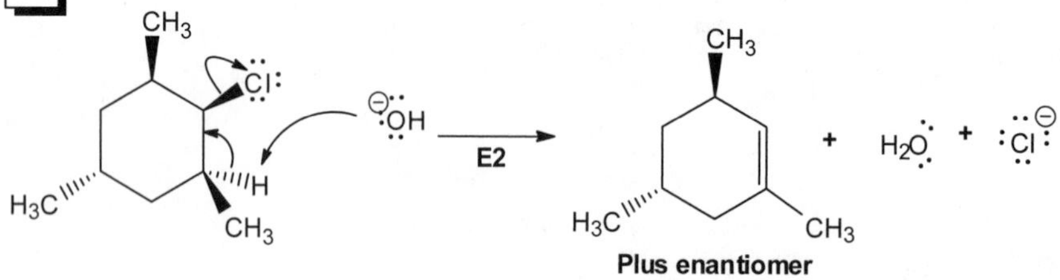

CH3 :Cl: ⊖:OH → E2 → CH3 + H_2O + :Cl:⊖
H3C H CH3 H3C CH3
Plus enantiomer

E1

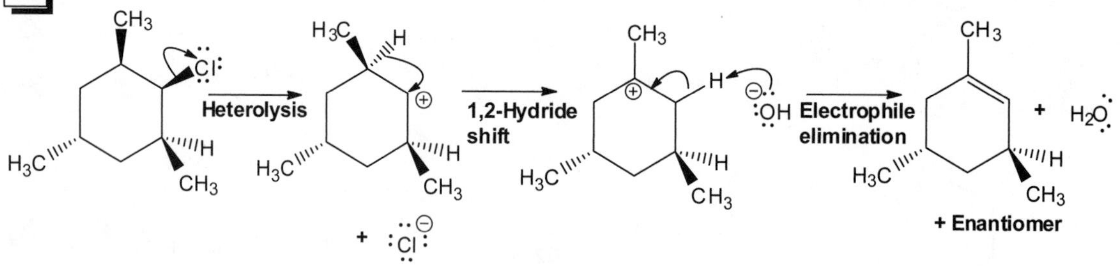

CH3 :Cl: → Heterolysis → H3C H ⊕ → 1,2-Hydride shift → CH3 ⊕ H ⊖:OH → Electrophile elimination → CH3 + H_2O
H3C H CH3 H3C H CH3 H3C H CH3 H3C H CH3
+ :Cl:⊖
+ Enantiomer

(vi) **E2**

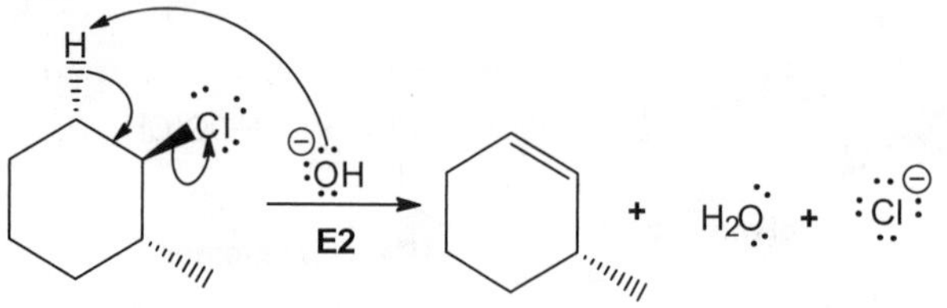

H :Cl: ⊖:OH → E2 → + H_2O + :Cl:⊖

E1

:Cl: → Heterolysis → ⊕ H → 1,2-Hydride shift → H ⊕ ⊖:OH → Electrophile elimination → + H_2O
+ :Cl:⊖

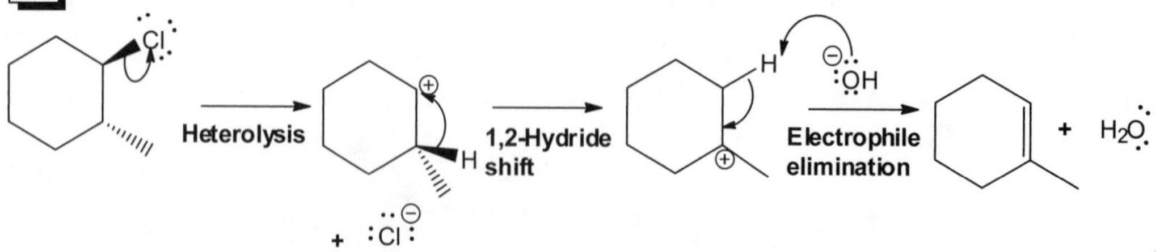

Problem 8.32

Think

What is the most stable chair conformation of each isomer? What is the favored orientation for the Br atom and the adjacent H atom in an E2 reaction? Which chair conformation allows the Br and H atoms to be in the correct orientation and still maintain the stable chair conformation?

Solve

The most stable chair conformations of the two compounds are shown below. In both cases, the *t*-butyl group is in the equatorial position, because it is so bulky. In the first molecule, the Br atom is essentially locked in the axial position, whereas in the second molecule, the Br atom is essentially locked in the equatorial position. Notice that in the first molecule, Br is anti to an adjacent H atom, which favors E2 reactions. By contrast, in the second molecule, the H atoms on the opposite side of the ring from Br are equatorial, so the H and Br atoms are gauche to each other. That is, there is not an H atom that is anti to the Br atom on an adjacent carbon. This slows the reaction.

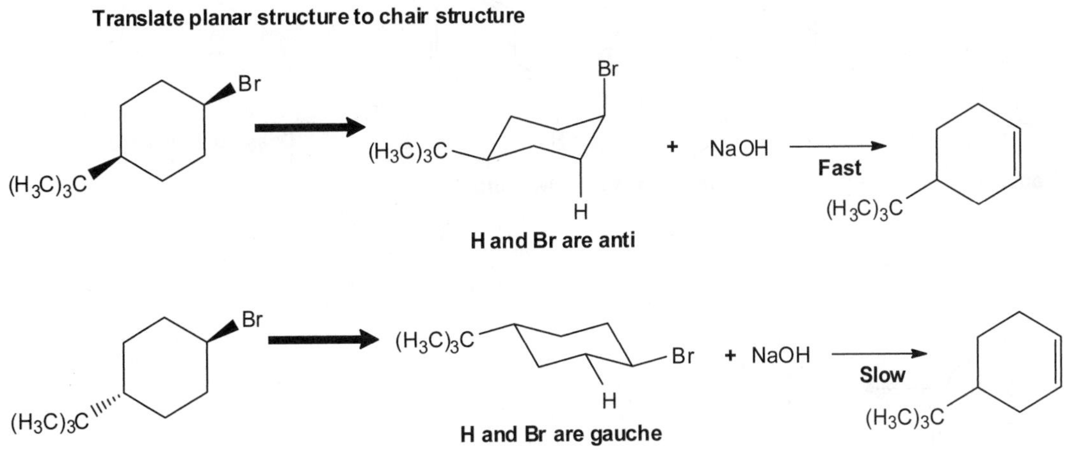

Translate planar structure to chair structure

Br
$(H_3C)_3C$
Br
$(H_3C)_3C$
H
+ NaOH
Fast
$(H_3C)_3C$
H and Br are anti

Br
$(H_3C)_3C$
$(H_3C)_3C$
Br
H
+ NaOH
Slow
$(H_3C)_3C$
H and Br are gauche

Problem 8.33

Think

How is a carbocation produced in an S_N1 or E1 mechanism? What type of carbocation is formed in each case? Is the formation of a more stable carbocation possible via a 1,2-hydride or a 1,2-alkyl shift?

Solve

This problem is easiest if we draw the carbocation intermediate that would be produced after the leaving group has departed. These are shown below and on the next page. Carbocation rearrangements are shown for each one that can attain greater stability via a 1,2-hydride shift or a 1,2-methyl shift. In **(a)**, a secondary carbocation rearranges to a tertiary one. The same is true in **(d)** and **(i)**. In **(e)**, a localized carbocation rearranges to a resonance-delocalized one. In **(f)**, a primary carbocation rearranges to a secondary one. In **(g)**, a carbocation that is resonance stabilized by one phenyl ring can rearrange to one that is stabilized by two phenyl rings. Rearrangement does not occur in **(h)**, as ethyl shifts are not common.

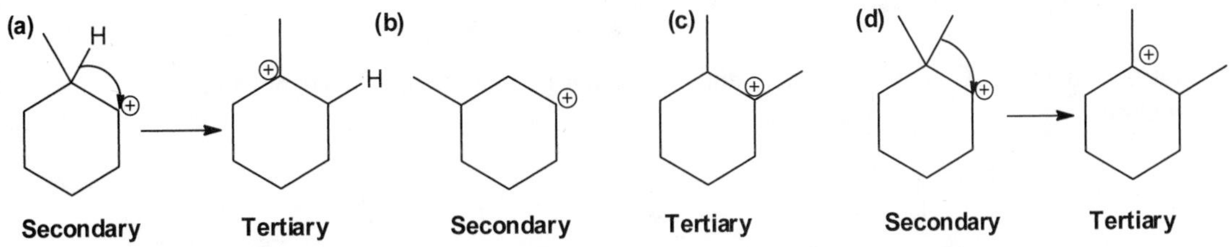

(a) H — **Secondary** → **Tertiary**

(b) H — **Secondary**

(c) **Tertiary**

(d) **Secondary** → **Tertiary**

(e) Localized → Resonance-delocalized

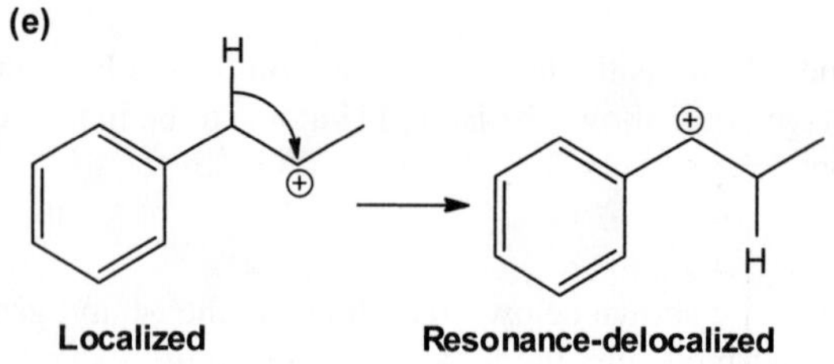

(f) Primary → Secondary

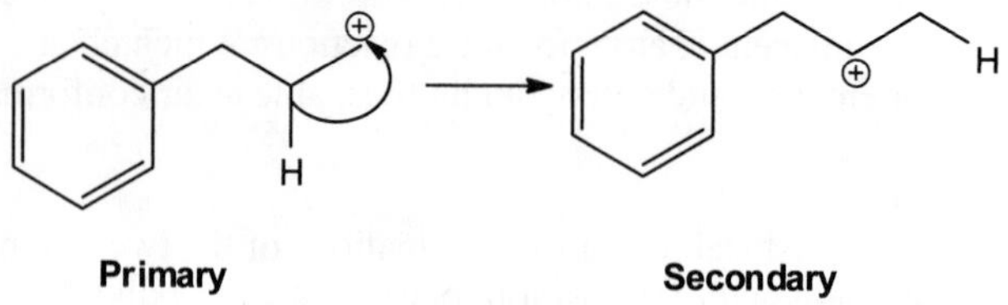

(g) Resonance from one ring → Resonance from two rings

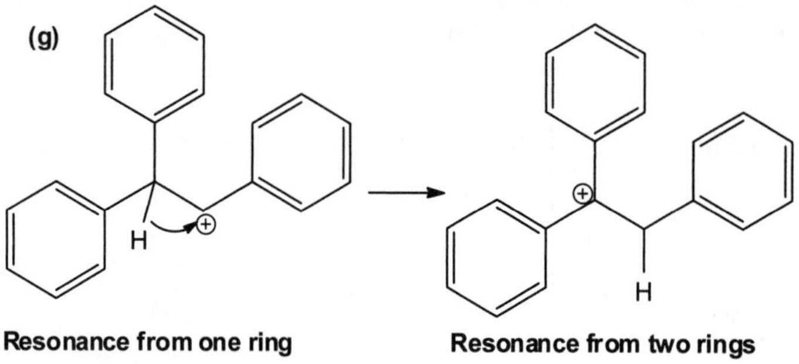

(h) H_3CH_2C CH_2CH_3 Does not rearrange

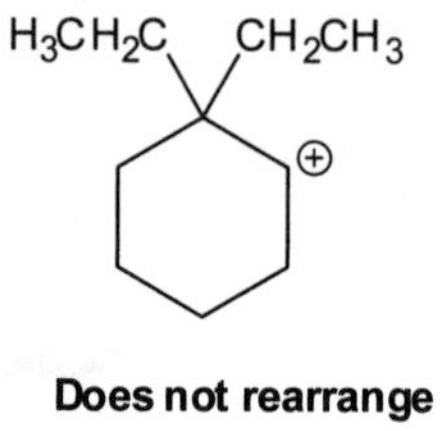

(i) Secondary → Tertiary

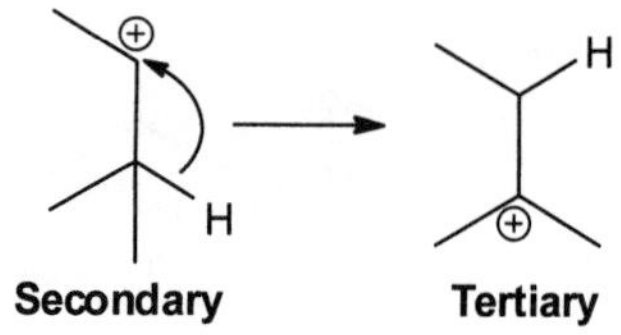

Problem 8.34

Think

Consider all possible elementary steps from Chapter 7. Review the rules in Section 8.6 for reasonable or unreasonable mechanisms. Are there any strong acids or bases in media that will not allow it? Are carbocation rearrangements possible? Are there any termolecular steps? What is reasonable for proton transfer steps? Are the curved arrows correct?

Solve

(i) Reasonable. This is a nucleophilic addition.

$H_3C—C≡N:$:OH → Nucleophilic addition → H_3C–C(OH)=N:⁻

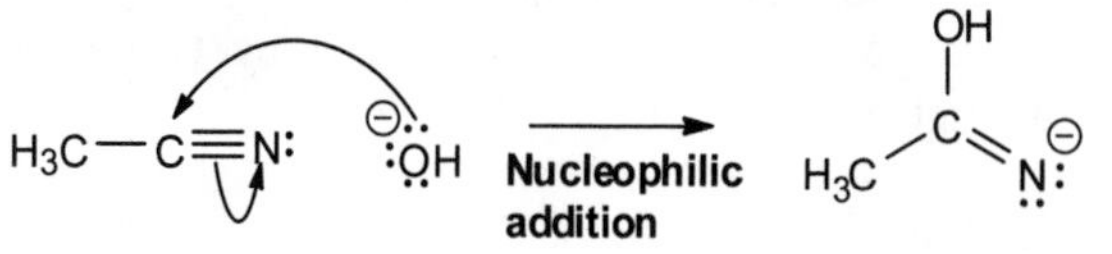

(ii) Not reasonable. H_3O^+ is a strong acid and should not participate in a mechanism that takes place in basic solution.

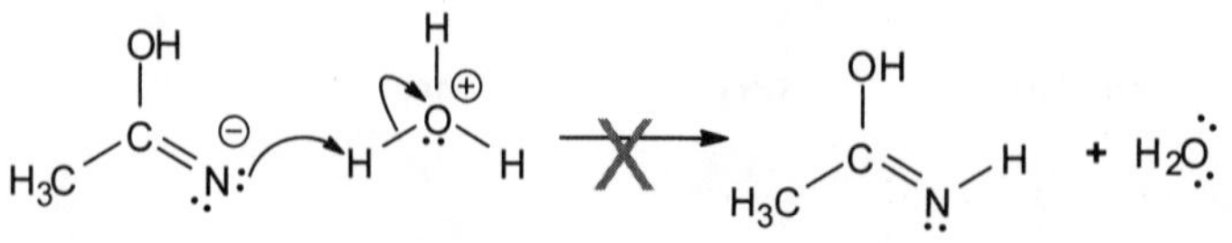

H_3C–C(OH)=N:⁻ + H_3O^+ ✗→ H_3C–C(OH)=N–H + H_2O

(iii) Not reasonable. This is a termolecular step.

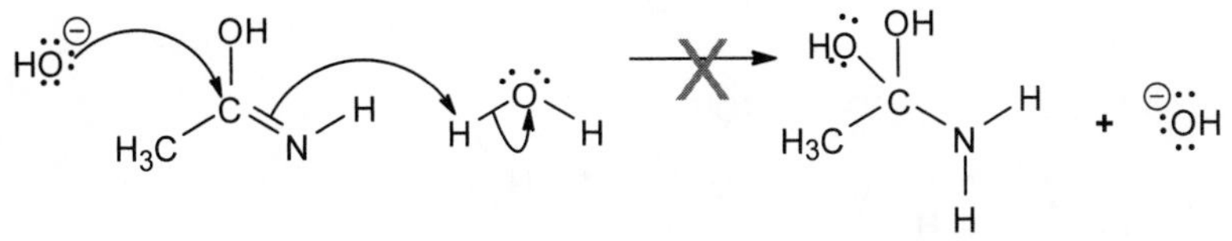

(iv) Not reasonable. The curved arrow shows the movement of an atom instead of electrons.

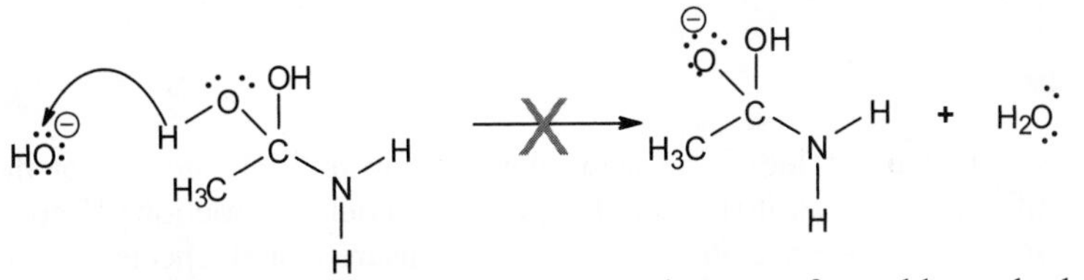

(v) Reasonable, although the leaving of H_2N^- is not as favorable as the leaving of HO^-. This is a nucleophile elimination step.

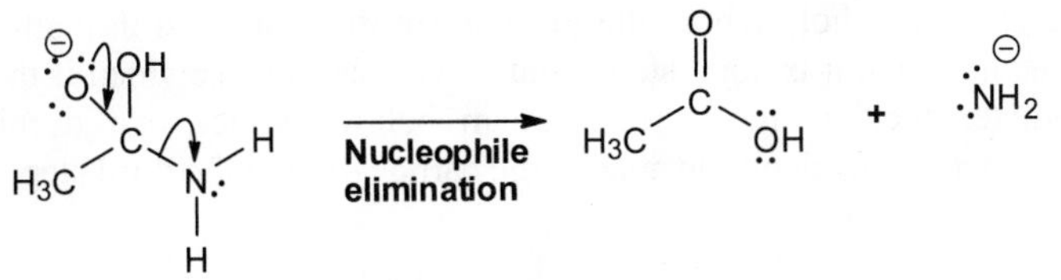

Nucleophile elimination

(vi) Reasonable. A very strong base removes a proton from an acid.

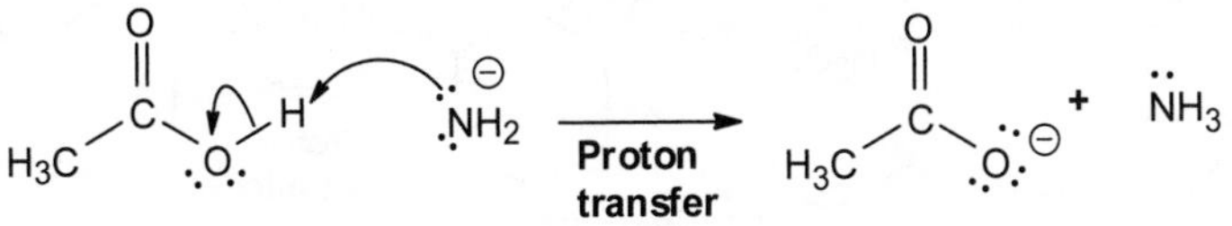

Proton transfer

Problem 8.35

Think

Consider all possible elementary steps from Chapter 7. Review the rules in Section 8.6 for reasonable or unreasonable mechanisms. Are there any strong acids or bases in media that will not allow it? Are carbocation rearrangements possible? Are there any termolecular steps? What is reasonable for proton transfer steps? Are the curved arrows correct?

Solve

(i) Not reasonable. This is an internal proton transfer; it is more reasonable for this proton transfer to be solvent mediated.

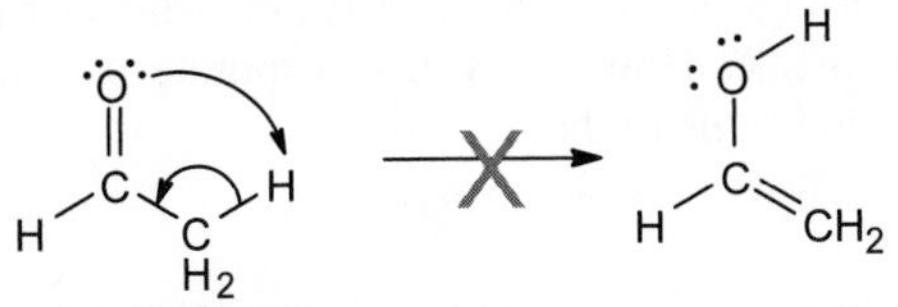

(ii) Not reasonable. The product contains a strongly basic O atom, which is not acceptable in acidic solution.

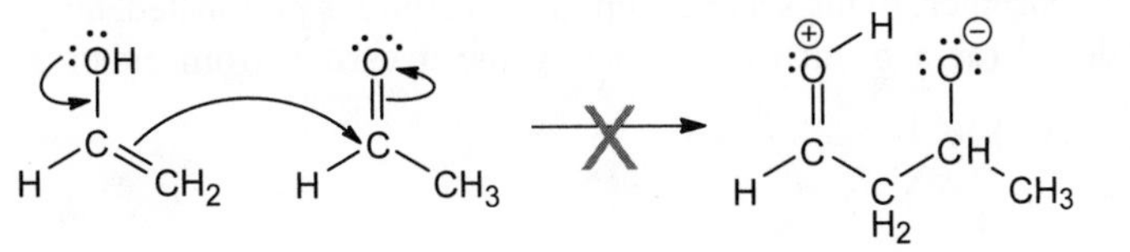

(iii) Not reasonable. This is a termolecular step.

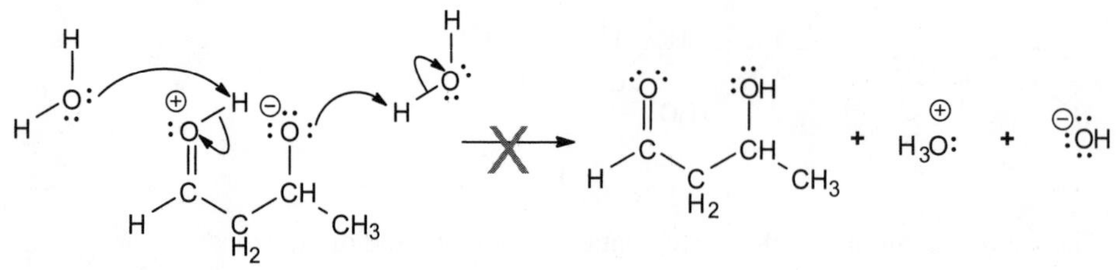

Problem 8.36

Think

Can the reaction take place in a single step, or does it need to take place in multiple steps? For the alcohol carbon to become a mixture of configurations, can it stay tetrahedral throughout the reaction? If not, which of its bonds should break? How can you incorporate proton transfer steps to ensure that all species are compatible with the acidic conditions of the reaction?

Solve

To produce a mixture of configurations at the alcohol carbon, the mechanism must proceed through a step in which that carbon atom is planar—that is, in which it is not a stereocenter. This can be explained if that carbon atom is a carbocation in an intermediate, such as in the mechanism below. Notice that a mixture of configurations at the C atom is produced when the nucleophile attacks the carbocation in the third step.

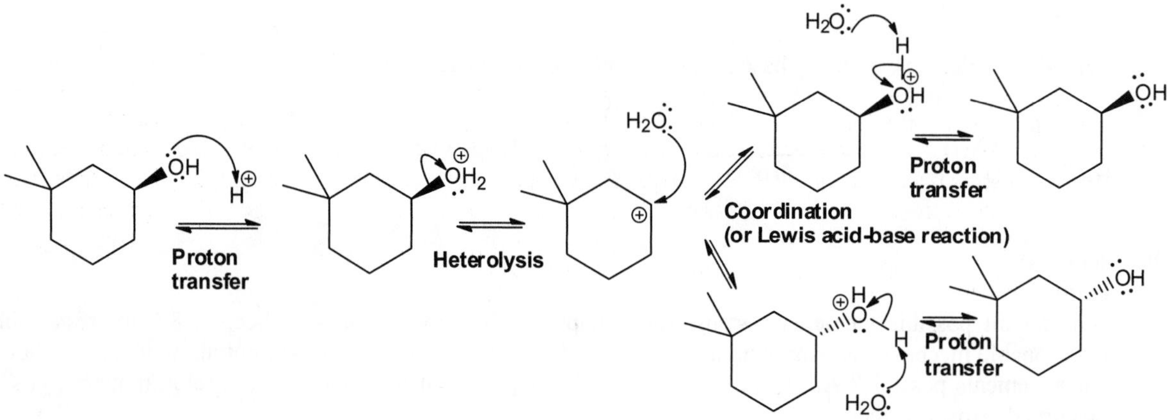

Proton transfer

Heterolysis

Coordination (or Lewis acid-base reaction)

Proton transfer

Proton transfer

Problem 8.37

Think

Can the reaction take place in a single step, or does it need to take place in multiple steps? For the α carbon to become a mixture of configurations, can it stay tetrahedral throughout the reaction? If not, which of its bonds should break? What species present can cause that bond to break? How can you incorporate proton transfer steps to ensure that all species are compatible with the basic conditions of the reaction?

Solve

Because there are equal amounts of enantiomers produced (a racemic mixture), the stereocenter that we see must be formed from an achiral precursor. That achiral intermediate can be formed by deprotonating the α carbon, as shown on the next page. In doing so, the α carbon becomes planar due to the contribution by its resonance structure, and thus is not a stereocenter. In the second step, the α carbon is protonated, and when it is, both R and S configurations are produced (as a result of protonating the α carbon from either side of the carbon's plane).

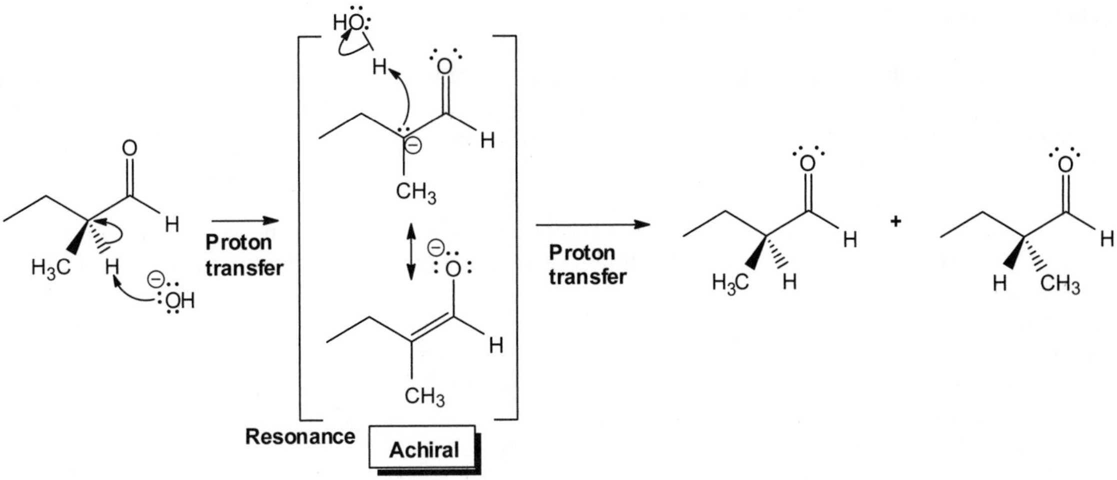

Proton transfer

Resonance

Achiral

Proton transfer

Problem 8.38

Think

Can the reaction take place in a single step, or does it need to take place in multiple steps? For the α carbon to become a mixture of configurations, can it stay tetrahedral throughout the reaction? If not, which of its bonds should break? What species present can cause that bond to break? How can you incorporate proton transfer steps to ensure that all species are compatible with the acidic conditions of the reaction?

Solve

The process is very similar to the previous problem. To form a racemic mixture of enantiomers, the stereocenter must be formed from an achiral precursor. That precursor is formed from two successive proton transfer reactions, shown below. Deprotonation cannot take place first, because that would generate a negative charge on the C atom (or elsewhere on the molecule), which corresponds to a strong base. This cannot happen because the reaction takes place in acidic conditions. So the O atom is protonated first, followed by deprotonation at the α carbon. Once the achiral intermediate is formed, the first two steps are reversed—first protonation of the C atom, followed by deprotonation at the O atom. When protonation at the C atom takes place, the stereocenter is regenerated, and a mixture of configurations is produced.

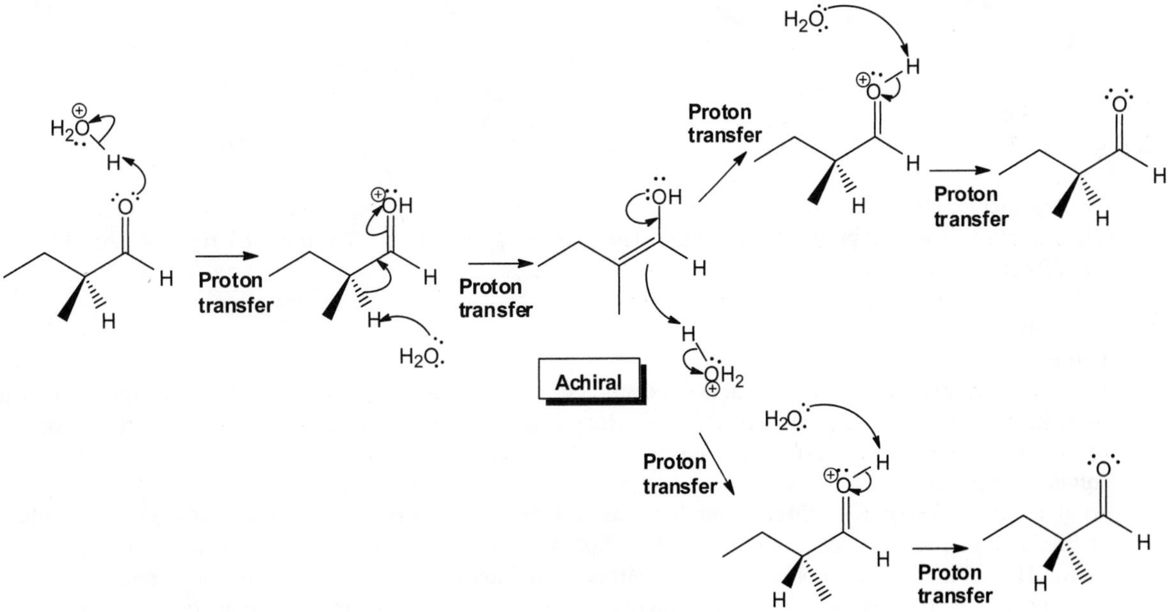

Proton transfer

Proton transfer

Achiral

Proton transfer

Proton transfer

Proton transfer

Proton transfer

Problem 8.39

Think

On which carbon is the leaving group located? Is rearrangement likely to occur in an E1 mechanism? In an E2? Which neighboring C atoms have H or D atoms available for elimination? Is an anti-coplanar arrangement of the appropriate atoms possible for an E2 reaction? How many products are possible for E2? For E1?

Solve

(a) The E2 mechanism and product are shown below. On the C atom adjacent to the one with the leaving group, the D can be anti to Cl, but the H atom cannot. So only D can be eliminated along with Cl in an E2 reaction.

Only D is anti to Cl

D H Cl: :Base E2 H C_8H_{14}

(b) The E1 mechanism and products are shown below. Both the D and the H atoms can be eliminated from the carbocation intermediate, so two products are formed. The carbocation intermediate is already tertiary and rearrangement is unlikely.

D H Cl: Heterolysis D H + :Cl: :Base Electrophile elimination D $C_8H_{13}D$

D H Cl: Heterolysis D H + :Cl: :Base Electrophile elimination H C_8H_{14}

(c) The mass of the E2 product is 110 amu. The masses of the products from the E1 reaction are 111 amu and 110 amu, respectively.

Problem 8.40

Think

In how many steps does an E1 reaction occur? Do proton transfer steps need to be incorporated to avoid incompatible species? Is rearrangement likely? How many transition states and intermediates are present?

Solve

(a) The leaving group (i.e., OH) cannot leave as HO^- because this is a strong base that would be produced in acidic conditions. Instead, the OH group is first protonated to create OH_2^+, which would leave in the form of H_2O. Finally, deprotonation of the carbocation intermediate forms the alkene product. The Roman numerals on the next page are used to identify the intermediate structures on the energy diagram for **(b)**.

Proton transfer — I — Heterolysis — II — Electrophile elimination — + H_3O^+

(b) Since there are three steps, there are three transition states and two intermediates. The overall reaction is endothermic, because the C=C π bond and O–H σ bonds that form are collectively weaker than the H–C and C–O σ bonds that break.

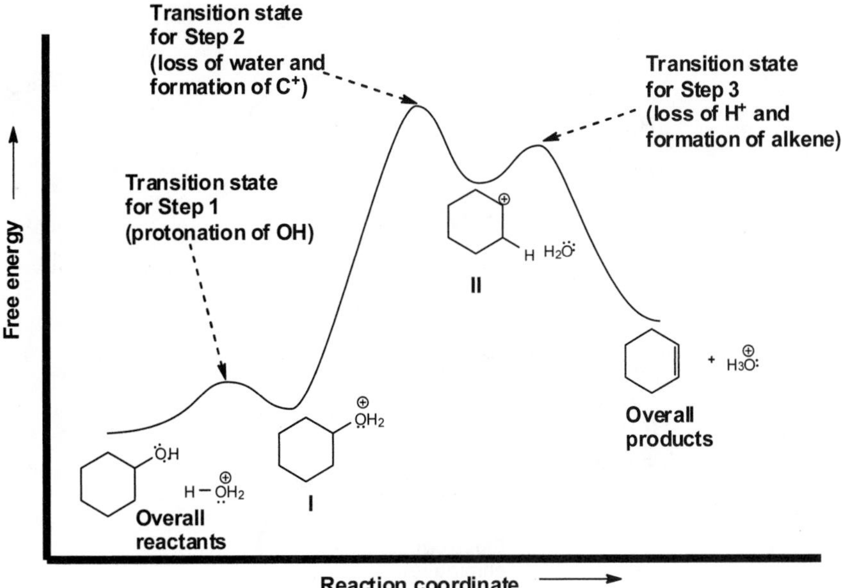

Problem 8.41

Think

How many steps make up an E1 reaction? Which proton is shifted in a 1,2-hydride shift involving the carbocation intermediate? Which proton is removed from that intermediate to complete the mechanism? If you were to attempt to distinguish the mechanisms using ^{13}C labeling, do you think a carbon involved in the reaction should be labeled, or one not involved in the reaction? For deuterium labeling, do you think a hydrogen involved in the reaction should be labeled, or one not involved in the reaction?

Solve

(a) The first step of an E1 is the leaving of the leaving group, as shown below. The secondary carbocation can then undergo a fast 1,2-hydride shift to yield a more stable tertiary carbocation. In the final step, elimination of H^+ yields the alkene product.

Heterolysis — 1,2-Hydride shift — Electrophile elimination

(b) The first step of an E1 is the leaving of the leaving group, as shown below. Without a carbocation rearrangement, the second step would then be elimination of H^+ to yield the alkene product.

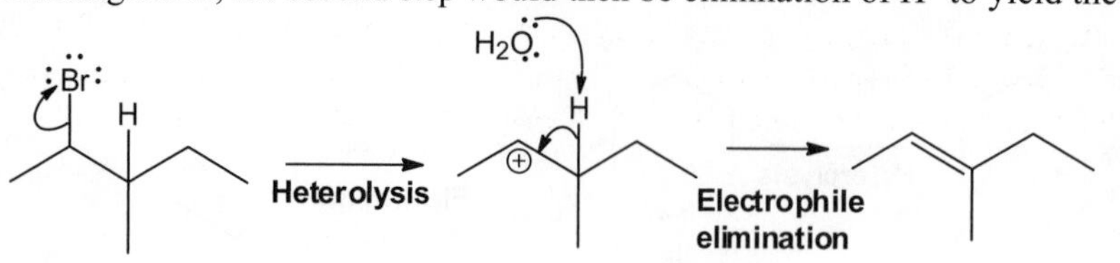

The same alkene product is produced in **(a)** and **(b)** because the C atom on which the H atom is eliminated and the C^+ occur on the same two C atoms that make up the C=C in both mechanisms.

(c) If the C atom bonded to the Br atom is the ^{13}C isotope, then a carbocation rearrangement would yield two different isomeric alkene products, as shown below.

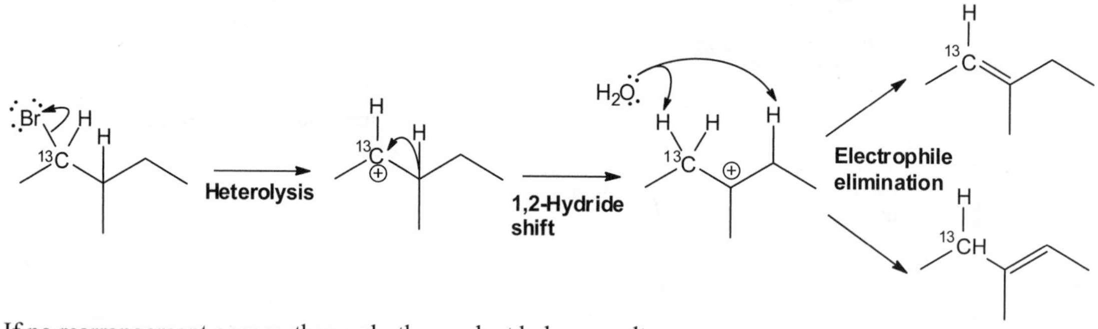

If no rearrangement occurs, then only the product below results.

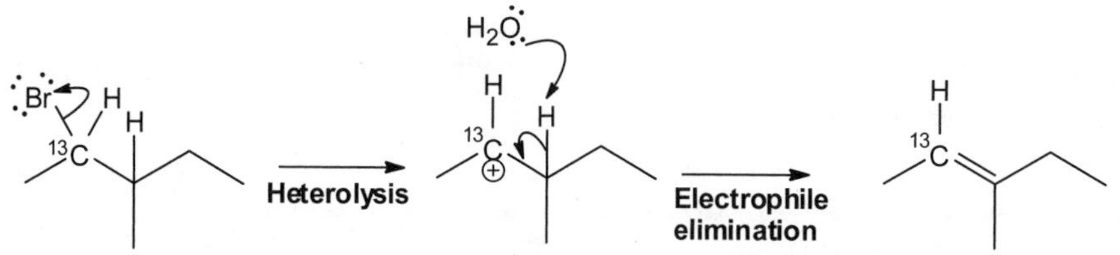

(d) If the H atom that is proposed to undergo migration is replaced by D, we can tell if the migration occurs by whether we observe any D in the alkene product. If we observe D in a portion of the alkene product, then migration must have occurred.

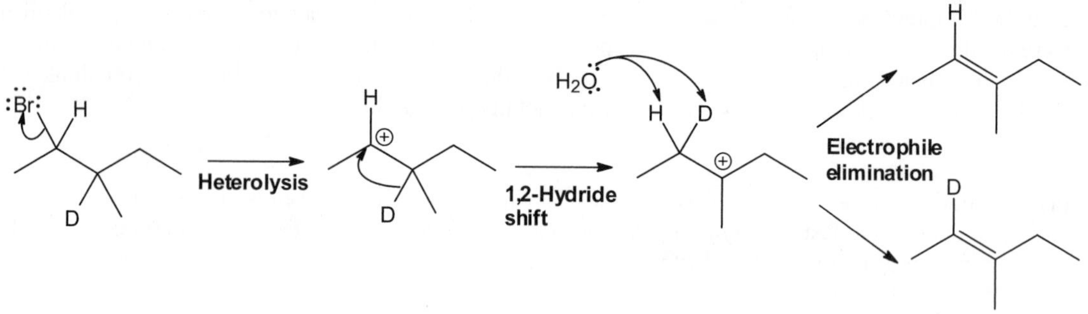

If no D remains, then migration must not have occurred.

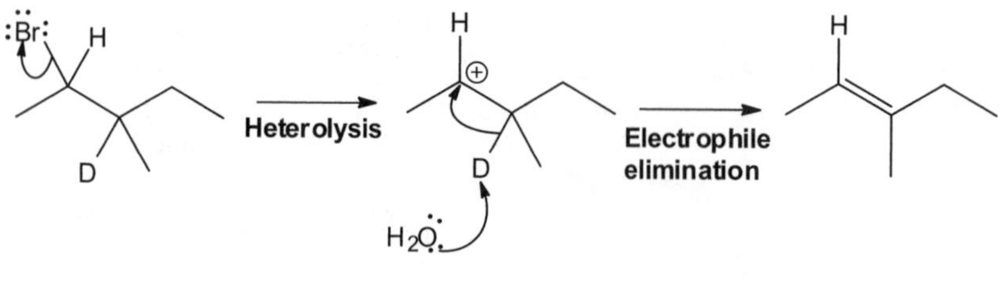

Problem 8.42

Think

On which C atom is the leaving group located? Is that C atom a stereocenter in the product? If so, is one stereochemical configuration produced, or is a mixture produced? What mechanism does this suggest? Do proton transfer steps need to be incorporated to avoid species that are incompatible with the conditions of the reaction?

Solve

The C atom that was initially bonded to the Br leaving group is a stereocenter in the product, and only the configuration shown is produced. This suggests S_N2. Furthermore, a careful examination of the product shows that the C atom on the left has been stereochemically inverted. This is normal for an S_N2 mechanism. Once you are aware of this, you know that backside attack must occur, and the molecule must adopt the conformation shown in the third molecule below for this to happen. Notice also that it would be unreasonable for this S_N2 step to take place first, because it would have produced an R_2OH^+ species, which is a strong acid and is incompatible with the basic conditions of the reaction. Instead, the OH is first deprotonated.

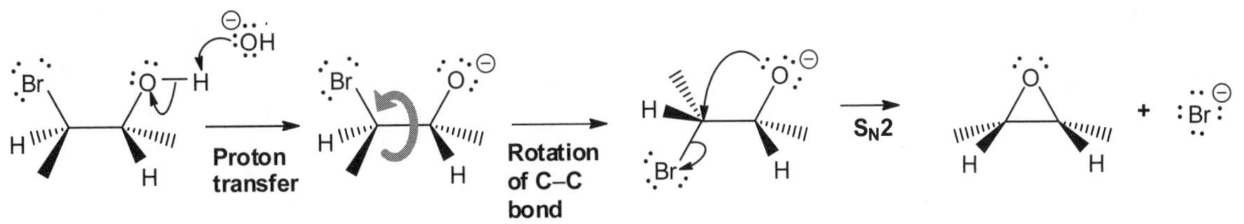

Proton transfer

Rotation of C–C bond

S_N2

Problem 8.43

Think

How many steps make up an S_N2 mechanism? Where must substitution take place in an S_N2? Can both products be produced by substitution at that site? How many steps make up an S_N1 mechanism? In the carbocation intermediate, is there an atom lacking an octet adjacent to a double or triple bond? As a result, how many electron-poor sites does the carbocation intermediate have?

Solve

(a) If an S_N2 reaction occurred, then only the first of the two products would have been formed, as substitution takes place directly (and only) at the C atom bonded to the leaving group. However, because there is a second substitution product, the mechanism must be an S_N1.

(b) An S_N1 mechanism proceeds through a carbocation intermediate, shown below. As indicated, that carbocation has the positive charge resonance delocalized over two different C atoms. The different products are formed by attack of the nucleophile at each of those two C atoms.

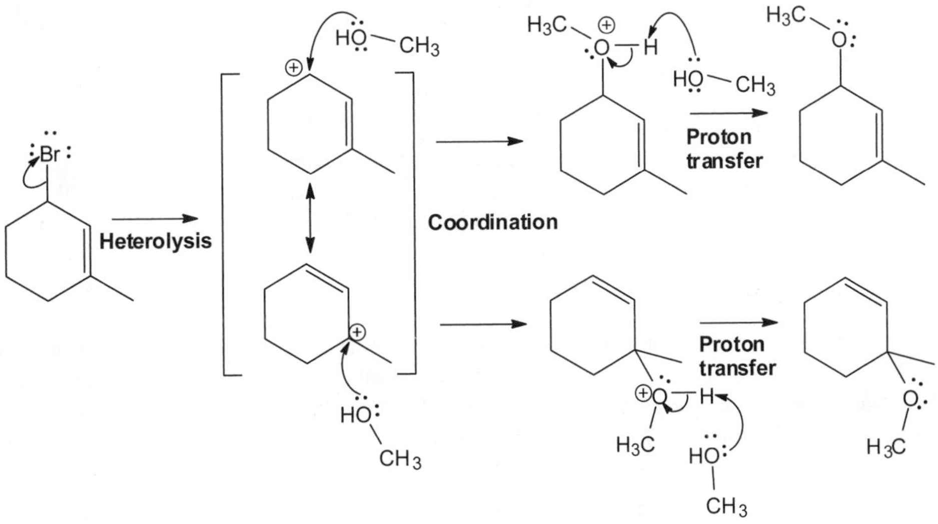

Heterolysis

Coordination

Proton transfer

Proton transfer

Problem 8.44

Think

How many H atoms are on the carbon adjacent to the carbon with the leaving group? How many of those H atoms are able to adopt the anti-coplanar conformation? How many products result from an E2? An E1?

Solve

No, it is not. Either H atom on the C atom on the left can become anti-coplanar with the Br atom, so either diastereomer of the alkene can be formed in an E2 reaction. A similar result is expected from an E1 reaction, which generally produces both *E* and *Z* isomers.

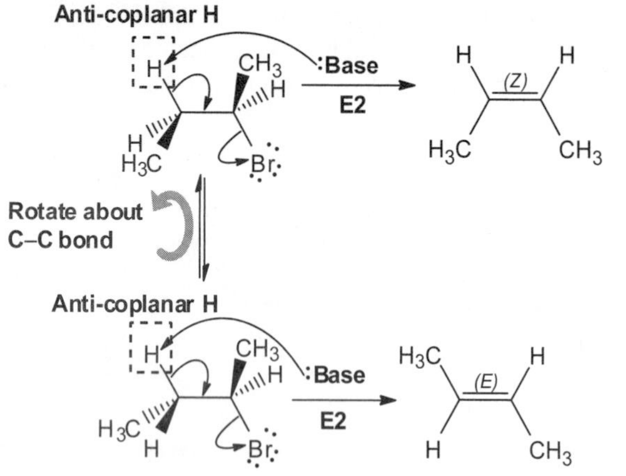

Anti-coplanar H

Rotate about C–C bond

Anti-coplanar H

Problem 8.45

Think

On which C atom is the leaving group located? Did the nucleophile attach to the same C atom? Has rearrangement occurred? What mechanism does this suggest?

Solve

A rearrangement of the carbon skeleton has occurred, indicating that a carbocation rearrangement has taken place. Formation of a carbocation can occur only with an S_N1, not an S_N2.

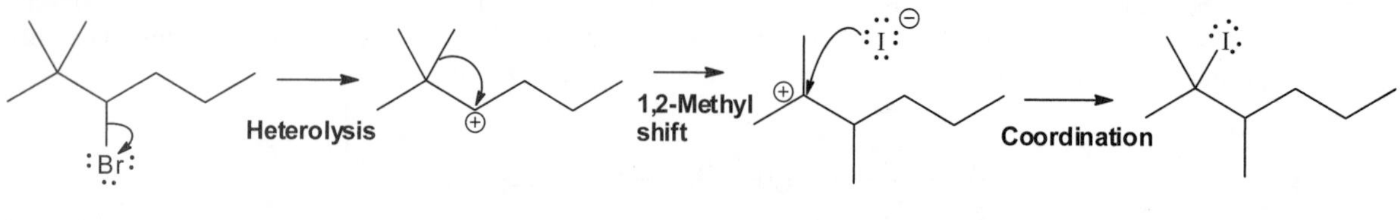

Heterolysis

1,2-Methyl shift

Coordination

Problem 8.46

Think

How many steps make up an S_N2 mechanism? Where must substitution take place in an S_N2? Can both products be produced by substitution at that site? How many steps make up an S_N1 mechanism? In the carbocation intermediate, is there an atom lacking an octet adjacent to a double or triple bond? As a result, how many electron-poor sites does the carbocation intermediate have?

Solve

An S_N2 mechanism consists of just a single step, so substitution must take place at the C atom that is initially bonded to the Br leaving group. Therefore, an S_N1 mechanism involving a resonance-stabilized carbocation must have taken place. As we can see on the next page, the positive charge of the carbocation intermediate is delocalized by resonance over three different C atoms, so nucleophilic attack can occur at all three of those C atoms.

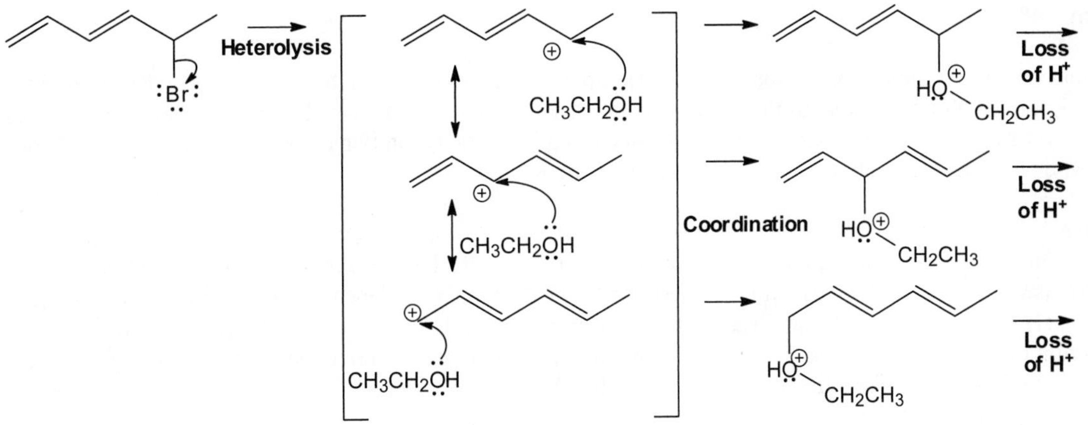

After loss of H^+ in the final step, the three products are as follows:

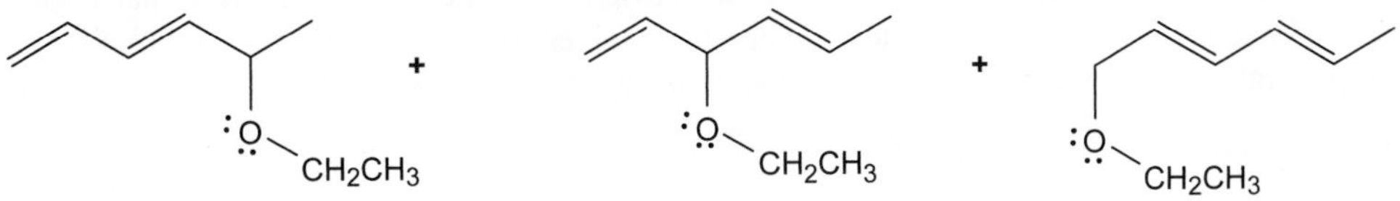

Problem 8.47

Think

Which mechanism, S_N1 or S_N2, does not depend on the nucleophile? If the mechanism does not depend on the nucleophile, does changing the concentration change the rate?

Solve

(a) The mechanism is S_N1, because the reaction kinetics do not appear to depend on the nature of the nucleophile.

(b) The mechanisms are shown below. Notice that a proton transfer step is required after attack of methanol, so that the final product is uncharged.

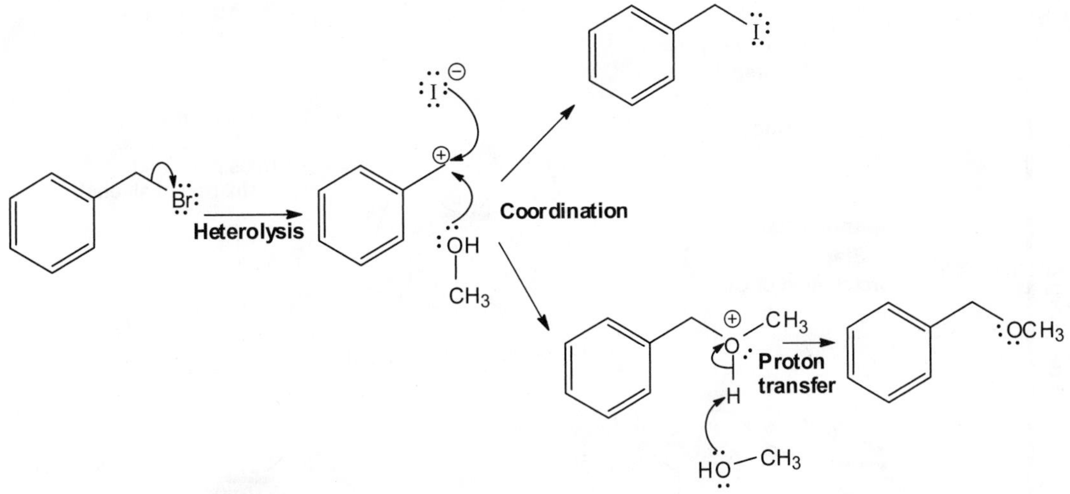

(c) Because the reaction proceeds by an S_N1 mechanism, the reaction rate is independent of the concentration of the nucleophile. So if the concentration of KI were doubled, the rate of the substitution would stay the same.

Problem 8.48

Think

What happens to the concentration of R–OCH_3 upon going from Trial 1 to Trial 2, and what is the effect on reaction rate? What happens to the concentration of H_2O upon going from Trial 2 to Trial 3, and what is the effect on reaction rate? On which species' concentration, R–OCH_3 and/or H_2O, does the reaction rate depend? What does this suggest about the reaction mechanism?

Solve

Between Trial 1 and Trial 2, the concentration of R–OCH_3 doubles (the concentration of H_2O is constant) and the reaction rate also doubles. This suggests that the reaction rate depends directly on the concentration of R–OCH_3. Between Trial 2 and Trial 3, the concentration of H_2O is halved (R–OCH_3 is constant) and the reaction rate remains constant. This suggests that the reaction rate does not depend on the concentration of H_2O. Thus rate= k[R–OCH_3] and the mechanism is likely E1.

Problem 8.49

Think

In how many steps does an E1 reaction mechanism occur? How many intermediates? How many transition states? Any rearrangements? Do transition states appear at local energy maxima or minima? What about intermediates? In general, what is the energy of a carbocation intermediate relative to the overall reactants or products?

Solve

The full E1 mechanism is given below. The reaction occurs in three steps: The OCH_3 is protonated first to avoid the leaving group departing as a strong base, which is incompatible with the acidic conditions of the reaction. This is followed by heterolysis, and finally electrophile elimination. There are two intermediates and three transition states. The carbocation intermediate is higher in energy than the overall reactants or products. Each intermediate appears at a local energy minimum, and each transition state appears at a local energy maximum.

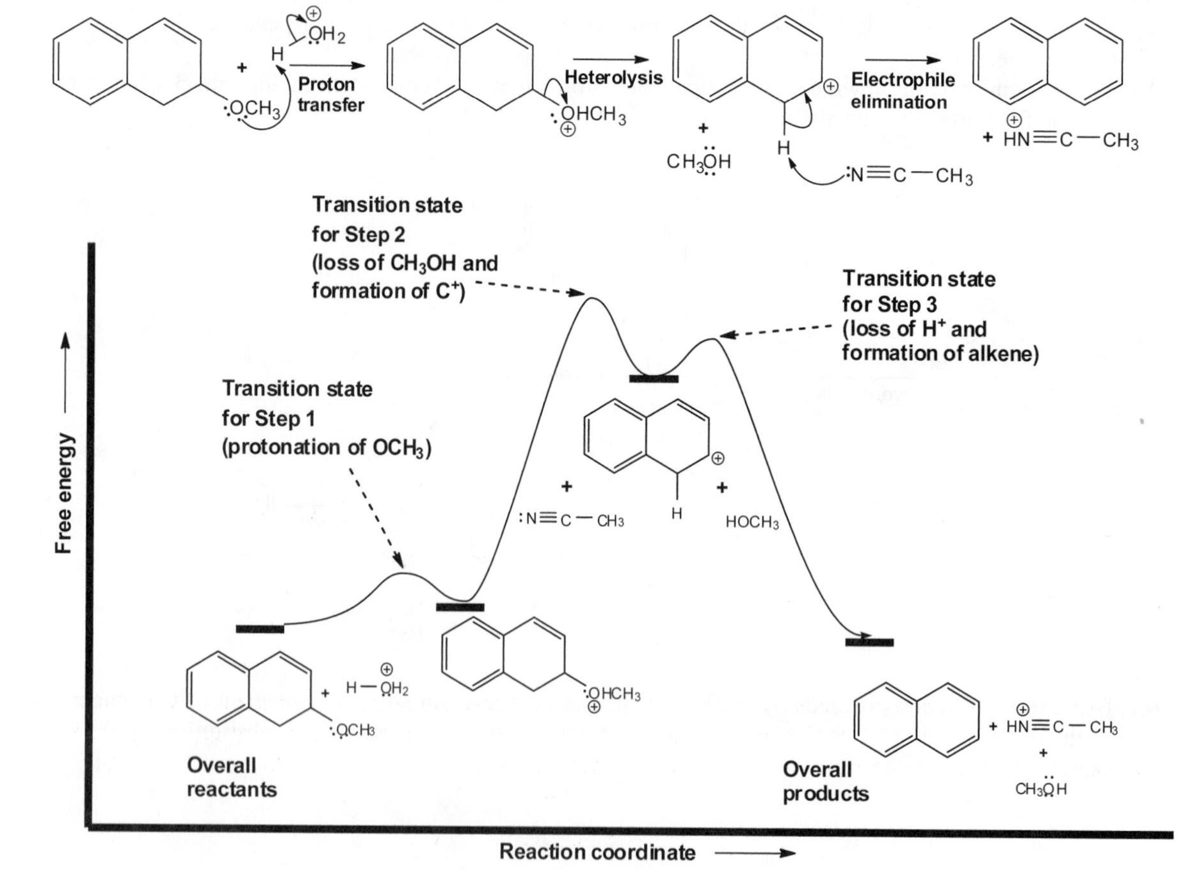

Proton transfer

Heterolysis

Electrophile elimination

Transition state for Step 2 (loss of CH_3OH and formation of C^+)

Transition state for Step 3 (loss of H^+ and formation of alkene)

Transition state for Step 1 (protonation of OCH_3)

Free energy

Overall reactants

Overall products

Reaction coordinate

Problem 8.50

Think

What happens to the concentration of R–Br upon going from Trial 1 to Trial 3, and what is the effect on reaction rate? What happens to the concentration of $KOCH_2CH_3$ upon going from Trial 2 to Trial 3, and what is the effect on reaction rate? On which species' concentration, R–Br and/or $KOCH_2CH_3$, does the reaction rate depend? What does this suggest about the reaction mechanism?

Solve

Between Trial 1 and Trial 3, the concentration of R–Br is halved (the concentration of $KOCH_2CH_3$ is constant) and the reaction rate also is halved. This suggests that the reaction mechanism depends directly on the concentration of R–Br. Between Trial 2 and Trial 3, the concentration of $KOCH_2CH_3$ is doubled (the concentration of R–Br is constant) and the reaction rate doubles. This suggests that the reaction mechanism also directly depends on the concentration of $KOCH_2CH_3$. Thus, rate= $k[\text{R–Br}][KOCH_2CH_3]$ and the mechanism is likely E2.

Problem 8.51

Think

In how many steps does an E2 reaction mechanism occur? How many intermediates? How many transition states? Any rearrangements? Do transition states appear at local energy maxima or minima? What about intermediates?

Solve

The full E2 mechanism is drawn below. The reaction occurs in one step. There are no intermediates and one transition state. Transition states appear at local energy maxima.

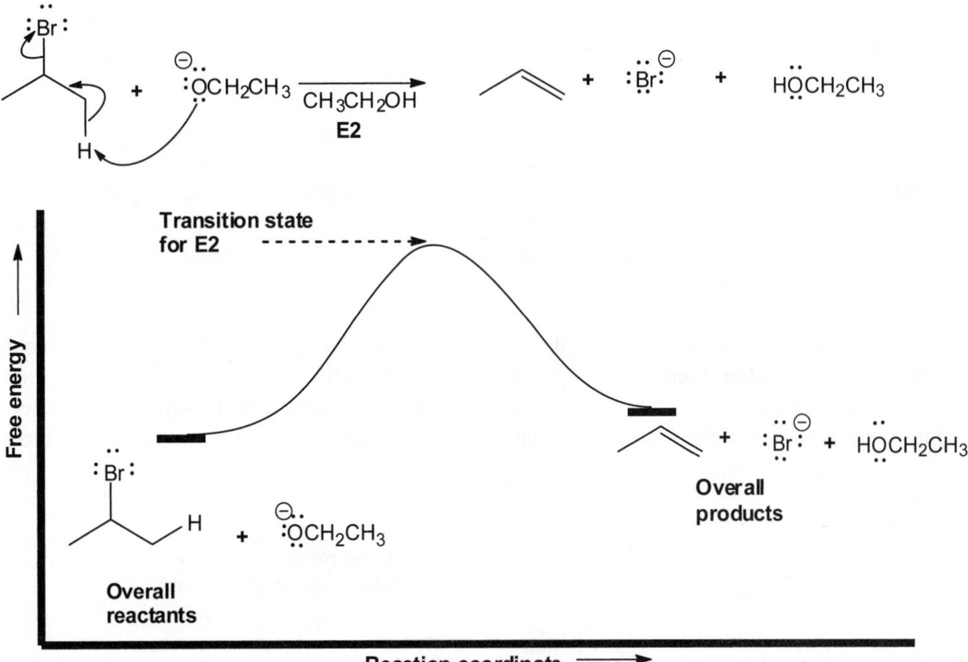

Problem 8.52

Think

If the reaction has a carbocation rearrangement, does that suggest a unimolecular or bimolecular reaction? What intermediate forms? What contributes to charge stability (resonance, inductive effects, etc.) in the carbocation intermediate that is initially formed? What contributes to charge stability in the rearranged intermediate?

Solve

This is a substitution reaction where, overall, Br has been replaced by OCH_3. It is, furthermore, unimolecular, making it S_N1, because a carbocation must have formed prior to the rearrangement. A carbocation rearrangement occurs because after an H atom has migrated, the resulting cation can be stabilized by resonance, involving a lone pair of electrons from the adjacent O atom. That resonance doesn't exist in the carbocation intermediate that initially formed.

Heterolysis

1,2-Hydride shift

Coordination

Proton transfer

Problem 8.53

Think

Would the mechanism be a substitution or elimination? What is the nucleophile? If the leaving group departs in the first step, will the resulting species be compatible with the conditions under which the reaction is carried out? If not, how can you incorporate proton transfer steps to ensure that the species remain compatible?

Solve

This must be a substitution reaction, where, overall, OH has been replaced by OCH_2CH_3. The mechanism involves proton transfer, an S_N2 step, then another proton transfer. Protonation of the OH group is necessary first because the reaction takes place in acidic conditions. If nucleophilic substitution were to take place first, instead, the leaving group would come off as HO^-, which is a strong base and is not compatible with the acidic conditions.

Very low boiling point
Leaves reaction mixture
Shifts equilibrium to right

Proton transfer

S_N2

Proton transfer

Problem 8.54

Think

Is this a substitution or elimination? On which carbon is the leaving group located? Is the leaving group attached to one C atom or two? How many different C atoms can be attacked by the nucleophile? What happens to the stereochemistry at that carbon? What mechanism does this suggest?

Solve

The reaction proceeds by an S_N2 mechanism, followed by a proton transfer. The leaving group can be viewed as Ring–O⁻. As we can see below, the nucleophile in the S_N2 step must attack the C atom from the side opposite the leaving group, forcing the other substituents on that C atom to flip over to the other side. After protonation, we can see that the product is exactly the stereoisomer that is shown in the problem.

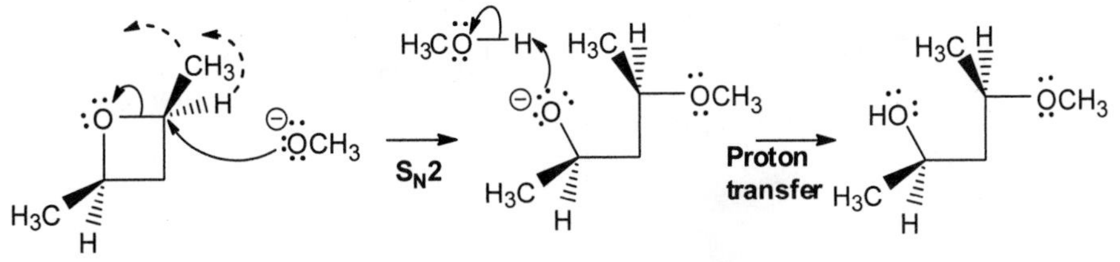

S_N2 Proton transfer

The enantiomer is produced by attack of the nucleophile at the other C atom on the ring, as shown below.

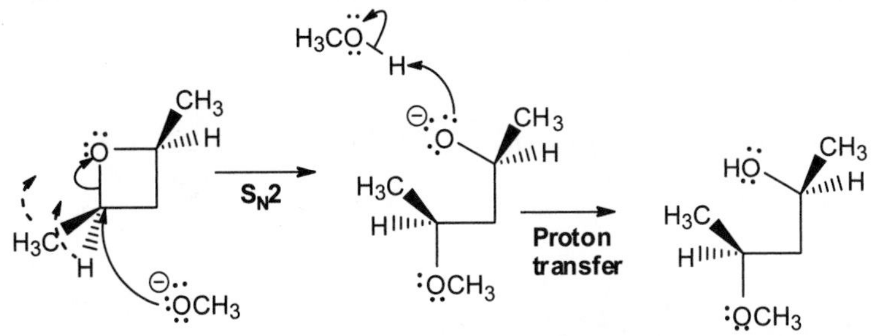

S_N2 Proton transfer

An S_N1 step did not take place, because other diastereomers were not produced. As shown below, a C atom from which the leaving group leaves would become planar, and when the stereocenter is regenerated upon attack of the nucleophile, both stereochemical configurations would be produced. An example is shown below. Notice that the product is not one of those given in the problem.

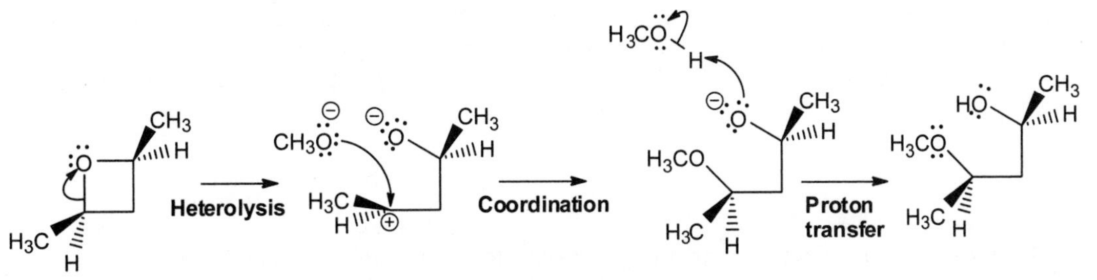

Heterolysis Coordination Proton transfer

Problem 8.55

Think

Is this a substitution or elimination reaction? On which carbon does the leaving group leave? Which C atom must have been attacked by a nucleophile to produce the product shown? Is that the same C atom from which the leaving group departed, or is it a different one? Does this suggest that the reaction is unimolecular or bimolecular? Are any proton transfer reactions involved?

Solve

This is an internal S_N1 reaction. It is a substitution reaction because, overall, Br has been replaced by O–Ring. Br is the leaving group on the terminal C atom, but the new O–Ring bond formed at a different C atom, so this must be unimolecular, making it S_N1. The intermediate carbocation produced in the mechanism is resonance stabilized, and the positive charge is shared over two different C atoms. Thus, the OH group can attack either C atom. It preferentially attacks the one shown because it produces a relatively stable, five-membered ring. Attack of the other C atom would yield a seven-membered ring. A final proton transfer is necessary to produce the uncharged product.

HO Br: Heterolysis HO HO Coordination CH_3CH_2OH H O Proton transfer O

Problem 8.56

Think

Which bond formed during the reaction? Which bond broke? How do you show this bond formation and bond breaking using curved arrows?

Solve

(a) This is an S_N2 reaction in which the $(PO_3)_3O^{5-}$ leaving group leaves the 1° carbon.

H_3N S CO_2 Methionine + O P O P O P O H_2N N N N N O OH OH Adenosine triphosphate (ATP) S_N2 H_3N CO_2 S H_2N N N N N O OH OH (*S*)-Adenosyl methionine (SAM)

O P O P O P O Leaving group

(b) The O atoms on the methionine are part of a carboxylate group and they share a negative charge through resonance. Notice that the triphosphate leaving group has several negative charges. Therefore, charge repulsion prevents the CO_2^- portion from acting as the nucleophile. The S atom has no formal charge, so there is less charge repulsion during nucleophilic attack, allowing it to bind to the electrophilic carbon in ATP.

Problem 8.57

Think

For guanidoacetate to gain a N–C bond, is the reaction a substitution or elimination? To make that bond, which species needs to be the electron-rich nucleophile, guanidoacetate or SAM? Which species needs to be the electron-poor substrate? What must be the leaving group? Are any proton transfer reactions involved?

Solve

The reaction must be a substitution because a new single bond is formed and no new double or triple bonds are formed. The CH_3 bound to the S^+ on SAM is the electrophilic (i.e., electron-poor) carbon and the S^+ on SAM is part of the leaving group. The N atom of the NH group is the electron-rich, nucleophilic site. The S_N2 reaction is shown below.

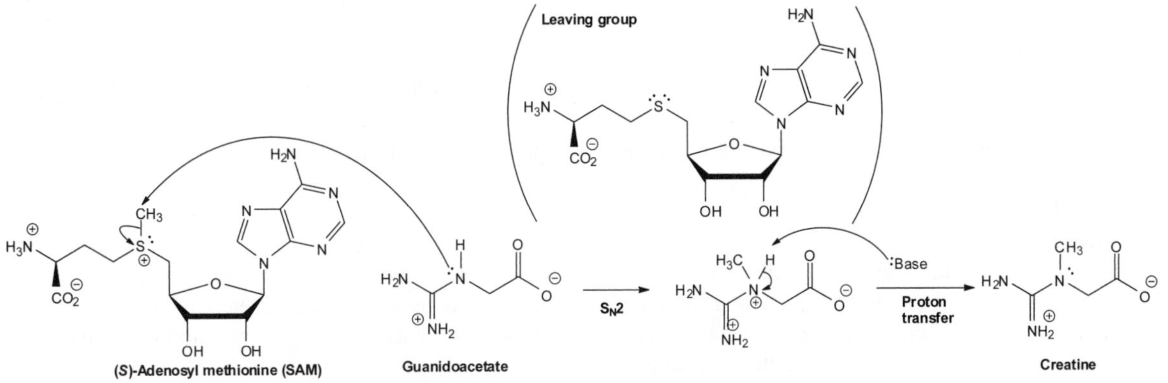

Leaving group

:Base

Proton transfer

S_N2

(*S*)-Adenosyl methionine (SAM)

Guanidoacetate

Creatine

Problem 8.58

Think

What happens to the stereochemistry at the electrophilic carbon in an S_N2 reaction? Is the CH_3 (R) group a dash or wedge in the product? What must it be, therefore, in the reactant?

Solve

The CH_3 group on L-alanine is dash and S_N2 reactions invert the stereochemistry at the electrophilic carbon. Therefore, in the starting α-bromo acid, the CH_3 needs to be wedge.

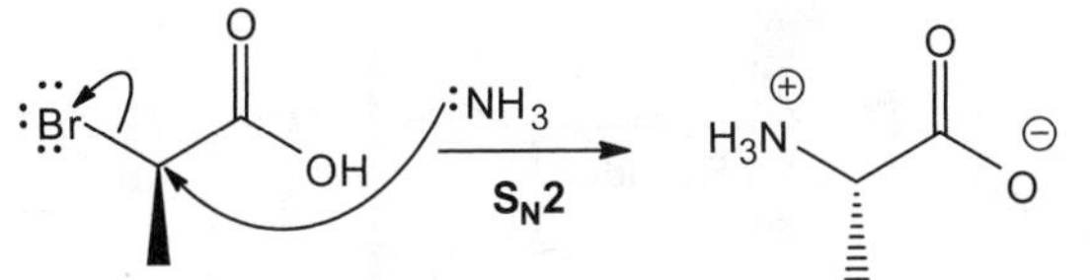

S_N2

Problem 8.59

Think

When there are *E* and *Z* isomers possible in the product, does an E1 reaction typically produce one isomer exclusively or a mixture? How many H atoms are available for elimination on the adjacent C atom? Can each of them undergo an E2 elimination? What does this mean for the number of stereoisomers possible as products of E2?

Solve

(a) E2 will produce a mixture of *E* and *Z* stereoisomers because there are two H atoms on the C atom adjacent to the C–Br, and each can attain the anti-coplanar arrangement that favors E2. This allows for both *E* and *Z* stereoisomers to be produced.

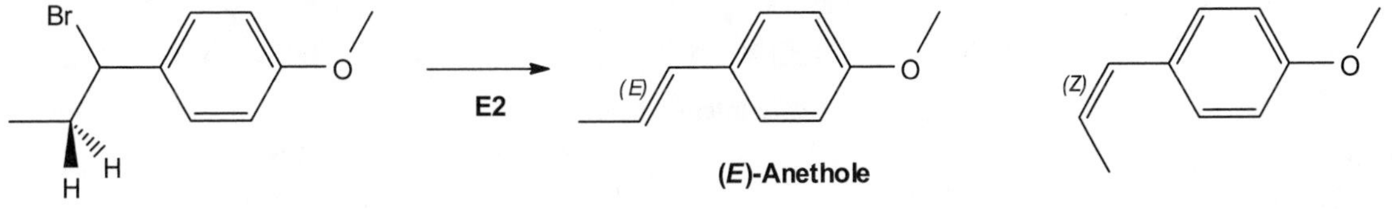

E2

(*E*)-Anethole

(b) E1 will produce a mixture of *E* and *Z* stereoisomers. In Step 1 of the E1 mechanism, the leaving group leaves via heterolysis and forms a carbocation. The electrophile elimination in Step 2 forms the more stable stereoisomer, but also some of the less stable alkene isomer.

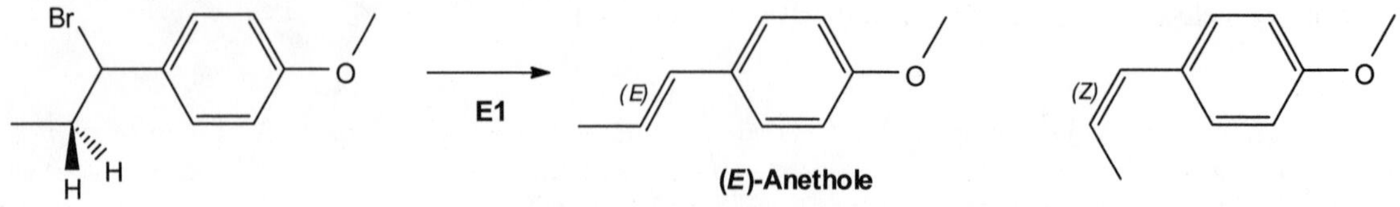

Br H H E1 (E) (E)-Anethole (Z)

(c) In both E2 and E1, the more stable isomer is the major product. In this example, it is the *E* isomer.

Problem 8.60

Think

Identify the H atoms and CH_3 atoms involved in the 1,2-hydride and 1,2-methyl shifts. Where is the positive charge on the carbon located in each shift? How many curved arrows are used for each of these steps? Are the curved arrows drawn from electron rich to electron poor, or vice versa? For the elimination of H^+ involving the base, identify the new C=C bond that is formed.

Solve

The 1,2-hydride shifts, 1,2-methyl shifts, and the electrophile elimination step are shown below. Each 1,2-hydride and 1,2-methyl shift requires one curved arrow, originating from the electron-rich bond and pointing to the electron-poor C^+. The final elimination of H^+ requires two curved arrows.

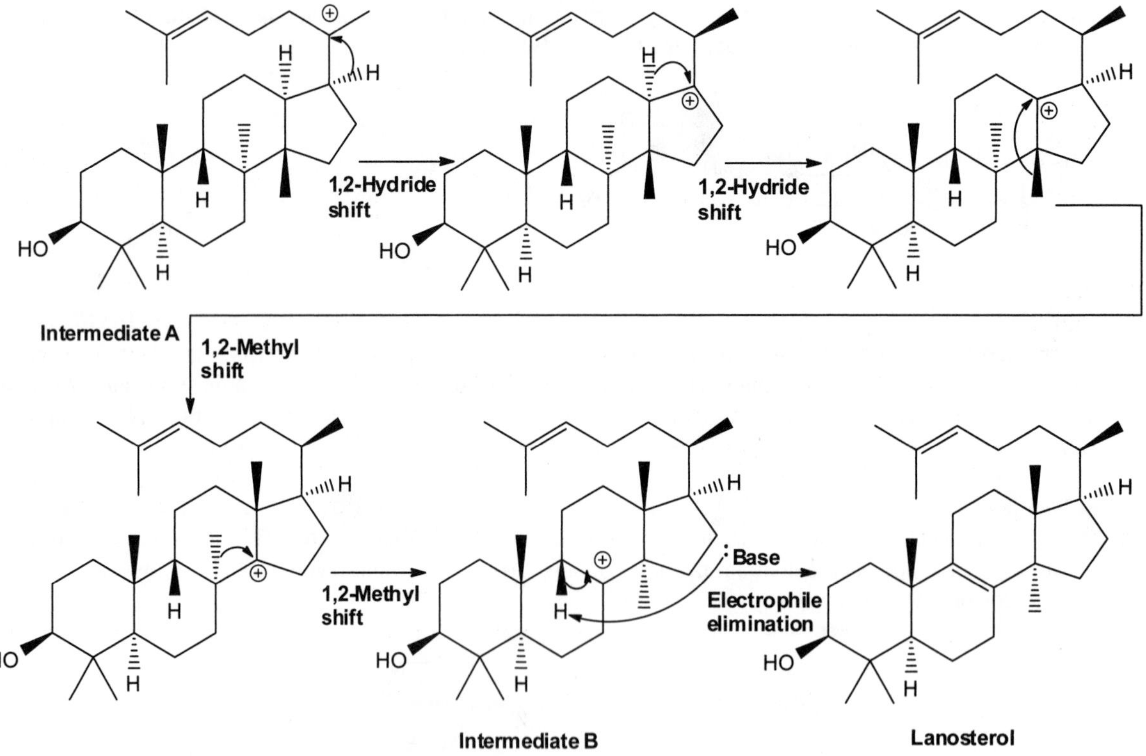

HO H H H H 1,2-Hydride shift 1,2-Hydride shift Intermediate A 1,2-Methyl shift 1,2-Methyl shift :Base Electrophile elimination Intermediate B Lanosterol

Problem 8.61

Think

Which atom is capable of being protonated? Is this a substitution or elimination reaction? On which C atom does the leaving group leave? For a rearrangement to occur, is the mechanism unimolecular or bimolecular? Are any proton transfer reactions involved?

Solve

The mechanism involves an internal E1 reaction. It is an elimination because an OH group is missing from the products and a new double bond between C and O has been formed. It must be unimolecular because, for a carbocation rearrangement to take place, a carbocation must have formed. The 3° carbocation rearranges to a resonance stabilized 2° carbocation, and the resonance provides substantial charge stability. Notice that a proton transfer is necessary in the first step because the leaving group does not come off as HO^- under acidic conditions.

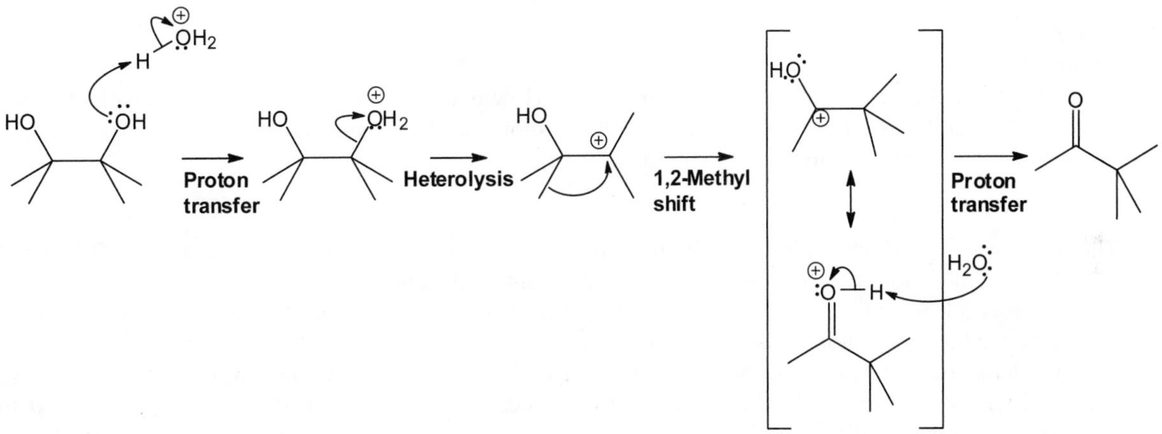

Proton transfer → Heterolysis → 1,2-Methyl shift → Proton transfer

Problem 8.62

Think

Is this a substitution or an elimination reaction? From which C atom does the leaving group leave? Are any proton transfer reactions involved to ensure that the species are compatible with the conditions of the reaction? Does rearrangement occur?

Solve

This reaction involves four substitutions, each of which involves Br being replaced by an O-containing group. Proton transfer steps are necessary in Steps 2 and 5 to ensure that the species remain compatible with the basic solution. Without those proton transfer steps, the product would be a strongly acidic R_2OH^+.

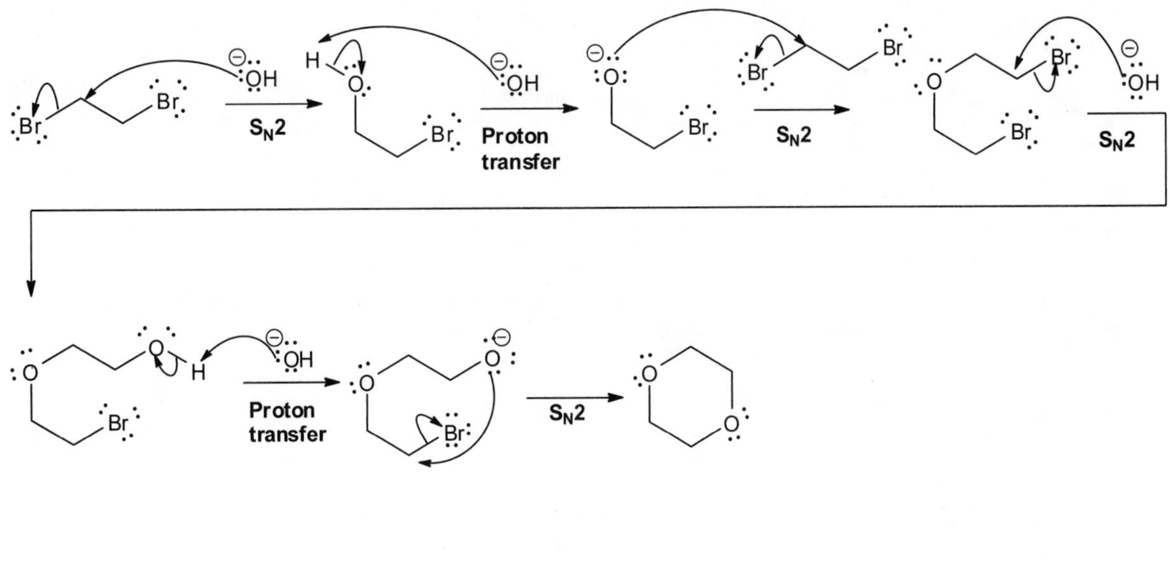

S_N2 → Proton transfer → S_N2 → S_N2 → Proton transfer → S_N2

Problem 8.63

Think

Which mechanism, S_N2 or S_N1, is dependent upon the concentration of the attacking species? How do you account for the optically inactive sample?

Solve

The fact that the rate of the reaction is directly proportional to the concentration of the attacking species, Br^-, suggests S_N2. The reason that the optical rotation goes to zero has to do with the fact that the S_N2 reaction produces the enantiomer of the reactant, resulting in a racemic mixture of enantiomers. This is shown below.

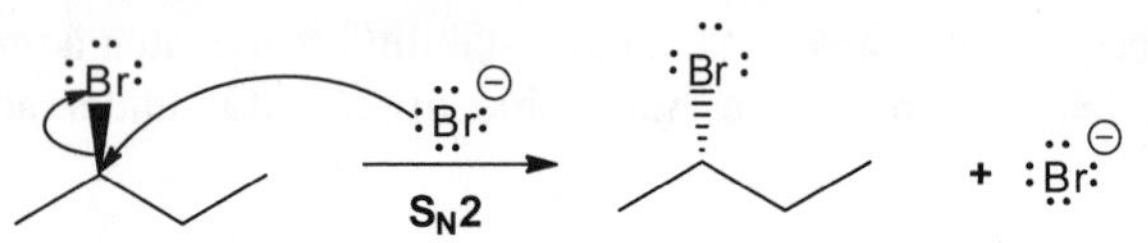

S_N2

Problem 8.64

Think

Does this reaction involve substitution or elimination? How many such substitutions or eliminations must take place? How many leaving groups are present? Are proton transfers involved? What is the attacking species? Are any rearrangements involved?

Solve

The reaction involves substitution, not elimination, because two C–Br bonds are replaced by C–O bonds. Note also that a carbocation rearrangement is necessary because substitution takes place at a benzylic carbon that does *not* have a leaving group. This suggests S_N1. An important detail to notice is that in the product, an O atom will be bonded to two C atoms adjacent to the aromatic ring. These are benzylic carbons, which can stabilize a positive charge quite well via resonance. And the S_N1 conditions favor the formation of such carbocations. The choice of which Br atom leaves first is ***not*** arbitrary, because loss of the Br atom that is already next to the aromatic ring leads directly to a resonance-stabilized carbocation. Heterolysis of the other C–Br bond does not have this kind of stabilization initially.

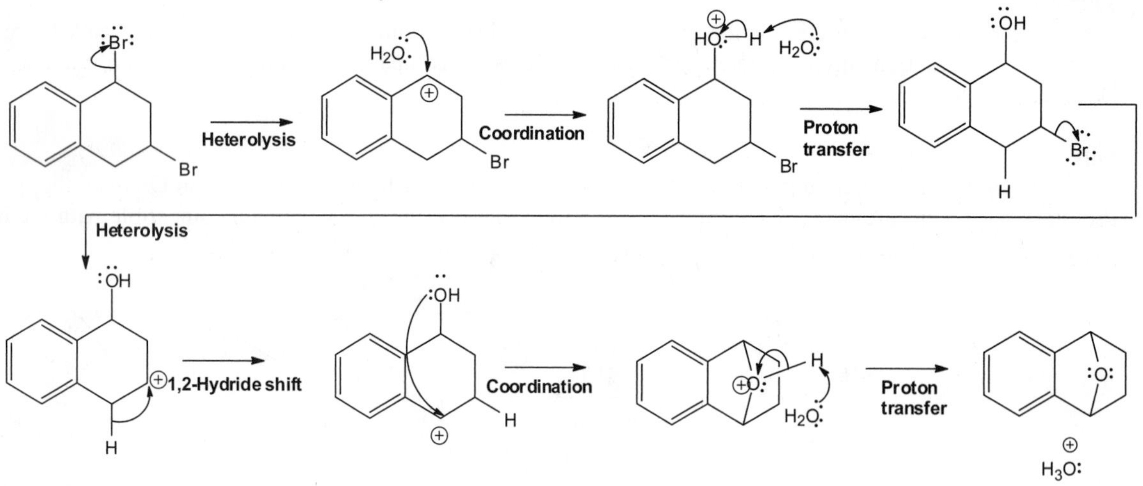

Heterolysis → Coordination → Proton transfer → Heterolysis → 1,2-Hydride shift → Coordination → Proton transfer

Problem 8.65

Think

What steps make up an S_N1 mechanism? On which C atom is the leaving group located? Does rearrangement occur? Are proton transfer reactions involved?

Solve

A carbocation rearrangement must take place because the carbon backbone in the product is different from that in the reactants. Therefore, a carbocation must be formed. The product will form as a mixture of stereoisomers, because in Steps 2 and 3, new stereocenters are formed from the addition of a nucleophile to a planar C^+. More specifically, either side of the cyclopropyl ring can open during rearrangement in Step 2, and subsequent attack of water in Step 3 can occur on either side of the carbocation.

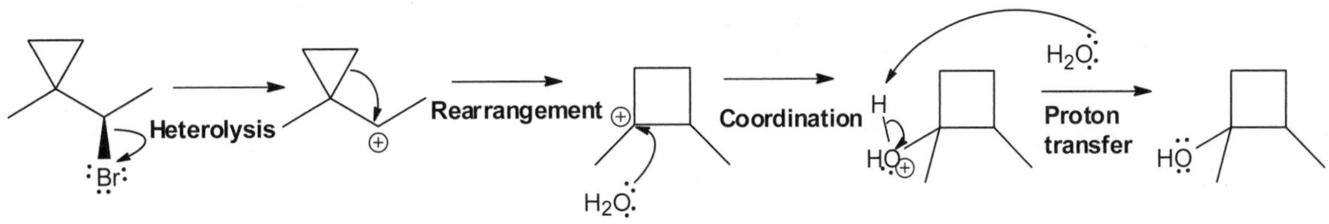

Heterolysis → Rearrangement → Coordination → Proton transfer

Problem 8.66

Think

What does a measured angle of rotation suggest about the optical activity of the product? Is the product of the reaction chiral or achiral? If the product is chiral, how can the product mixture be optically inactive? Which mechanism forms a racemic mixture? Why?

Solve

Optical rotation starts out as nonzero because the reactant is chiral, consisting of just one enantiomer. Each product molecule is chiral, too, but the fact that optical rotation goes to zero suggests that the product is a racemic mixture of enantiomers. This suggests that the reaction is S_N1 instead of S_N2.

NaBr, S_N1 → AND

Problem 8.67

Think

Draw out the mechanism for the S_N2 reaction in which $H^{18}O^-$ is the nucleophile. On which species does the ^{18}O label appear if the reaction proceeds via S_N2? Does this agree with the products actually observed?

Solve

These results suggest that the mechanism is *not* S_N2. The S_N2 mechanism that would take place is shown below and, as we can see, the labeled O atom would end up in CH_3OH. The actual mechanism that takes place to account for the experimental results will be discussed in Chapter 17.

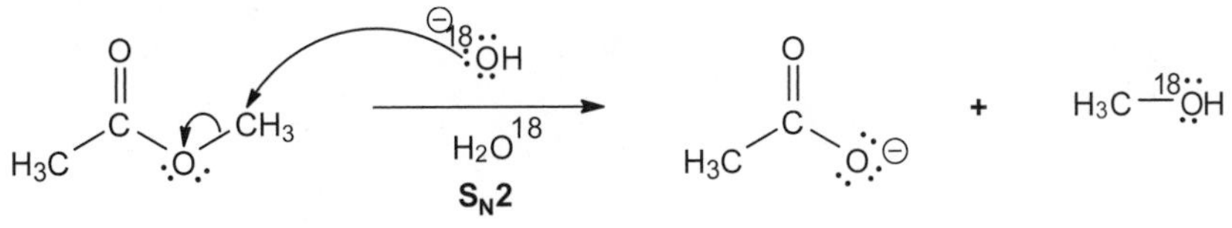

H_2O^{18}, S_N2 → + $H_3C-^{18}OH$

Problem 8.68

Think

Draw out the mechanism for the S_N2 reaction in which $H^{18}O^-$ is the nucleophile. On which species does the ^{18}O label appear if the reaction proceeds via S_N2? Does this agree with the products actually observed?

Solve

These results suggest that the mechanism is S_N2, as shown below. The explanation is provided in the previous problem.

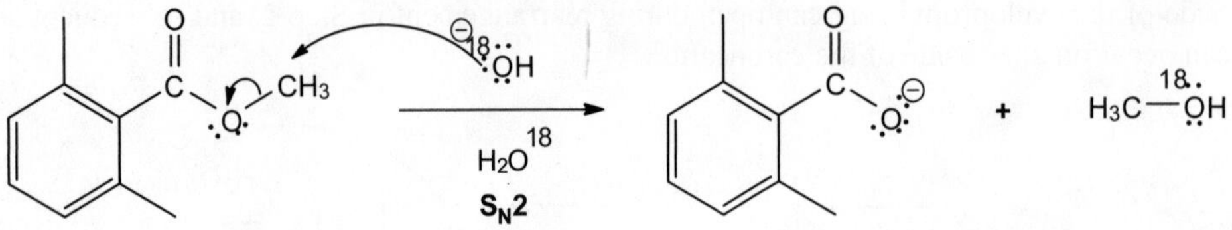

CHAPTER 9 | Nucleophilic Substitution and Elimination Reactions 1: Competition among S_N2, S_N1, E2, and E1 Reactions

Your Turn Exercises

Your Turn 9.1

Think

When an attacking species acts as a base, to what type of electron-poor atom does it bond? When an attacking species acts as a nucleophile, to what type of electron-poor atom does it bond?

Solve

A base uses its lone pair of electrons to form a bond to a hydrogen atom, which is what takes place in the elementary step on the left. A nucleophile uses its lone pair of electrons to form a bond to an electron-poor nonhydrogen atom, which is what takes place in the elementary step on the right. In this case, C is attacked by NH_3.

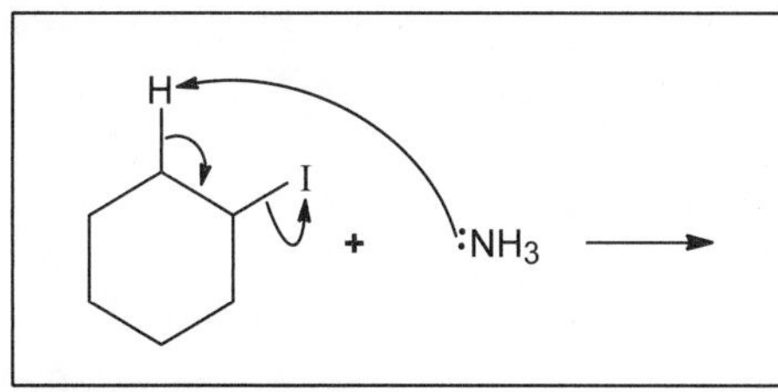

NH_3 acting as a base

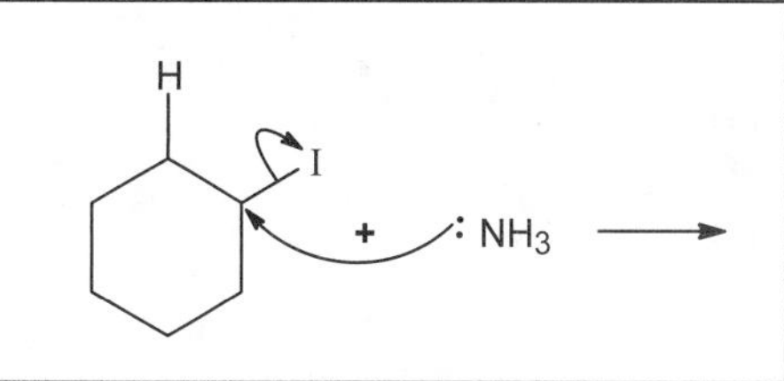

NH_3 acting as a nucleophile

Your Turn 9.2

Think

Looking at the curved arrows, identify which bonds are broken and formed. Is a nucleophilic atom electron rich or electron poor? What kinds of charges are associated with a nucleophile?

Solve

In the reaction below, HO^- acts as a base and attacks the H. The C–H bond breaks and the electrons fold onto the C atom, forming a carbanion. The C:$^-$ is electron rich due to its lone pair and −1 formal charge and, therefore, is a nucleophile.

Nucleophilic atom

:N≡C—H + $^-$:OH → :N≡C:$^-$ + H_2O

Hydrocyanic acid

Proton transfer

Your Turn 9.3

Think

Which groups are bulky? Sometimes it helps to draw out the C–H bonds to get a better understanding of the size of a CH_3 group versus a $C(CH_3)_3$ group. How is bulkiness related to steric hindrance?

Solve

Tert-butyl and isopropyl groups are bulky and contribute to the steric hindrance of the bases in this problem.

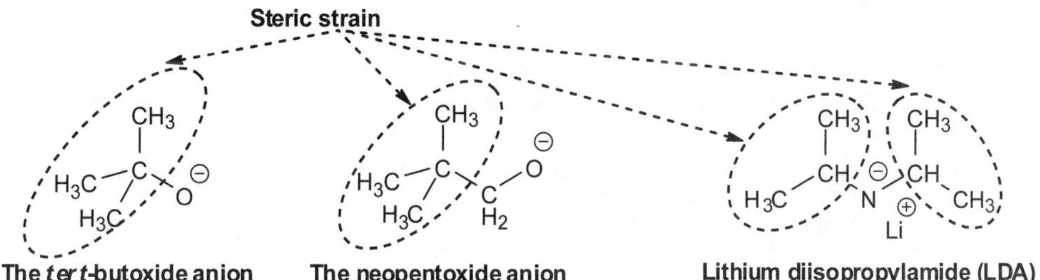

The *tert*-butoxide anion The neopentoxide anion Lithium diisopropylamide (LDA)

Your Turn 9.4

Think

Are there any lone pairs adjacent to a double or triple bond? Can any of the double bonds convert into lone pairs? How can you show via a curved arrow the electrons in the lone pair moving to form a double bond? How can you show via a curved arrow the electrons in a double bond moving to form a lone pair?

Solve

Recall that atoms do not move in resonance structures. Only nonbonding electrons or electrons from multiple bonds move. In the first resonance structure below, the first curved arrow is used to convert the lone pair on the O^- into a covalent double bond, S=O. The second arrow is used to convert a pair of electrons from the double bond, S=O, into a lone pair on the O, resulting in a formal charge of −1. This same pattern is used to convert the second resonance structure into the third.

H_3C—S—O: ⟷ H_3C—S=O ⟷ H_3C—S=O

Methyl sulfonate (Mesylate, MsO^-)

Your Turn 9.5

Think

What criteria do you use to judge the ability of a leaving group to depart? What is the importance of charge stability? What types of formal charges are most favored (positive, negative, or uncharged) in good leaving groups?

Solve

The more stable a leaving group is in the form in which it has left, the better its leaving group ability. Good leaving groups are ones that depart as uncharged molecules or as anions in which the −1 charge is stabilized on a large atom or by resonance. Leaving groups tend to be better when their departed forms are weaker bases. In the example below, HO^- or H_2O are the possible leaving groups, respectively. HO^- is a strong base and, therefore, is a very poor leaving group. H_2O is an excellent leaving group because it is uncharged and a weak base.

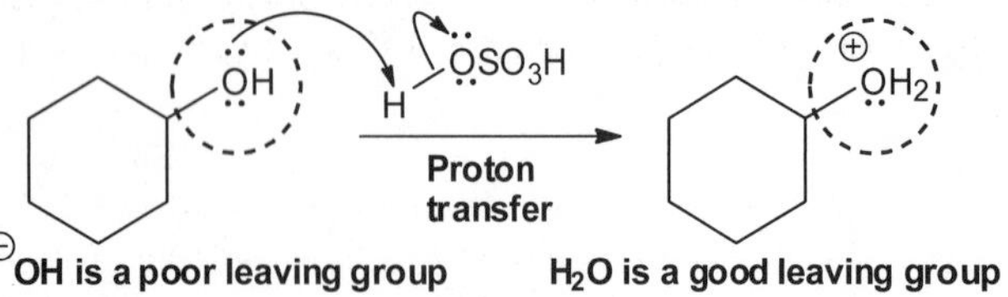

Your Turn 9.6

Think

Consult Figure 3-33 on p. 155 and write down the bond energy for each C–H bond. What is the percent *s* character in an sp^3, sp^2, and *sp* C–H bond? Why do you think that increasing percentage of *s* character increases the bond strength?

Solve

Increasing percentage of *s* character increases the bond strength because pure 2*s* orbitals hold electrons closer to the nucleus than pure 2*p* orbitals. This leads to more compact orbitals as the percentage of *s* character increases.

sp^3	sp^2	sp
C—H	C—H	C—H
410 kJ/mol	431 kJ/mol	523 kJ/mol

Your Turn 9.7

Think

Do any of the atoms have an unfilled valence shell? If so, is the atom with the unfilled valence shell adjacent to a double or triple bond? With this kind of structural feature, how many curved arrows are necessary to arrive at the next resonance structure?

Solve

The carbocation has an unfilled valence shell and is adjacent to a double bond. One curved arrow is used to shift a pair of π electrons from the double bond to form a new π bond to the initial C^+. This leaves an unfilled valence shell on a C atom two atoms away from the original carbocation. The resonance can continue around the entire ring because all the C atoms are sp^2 hybridized.

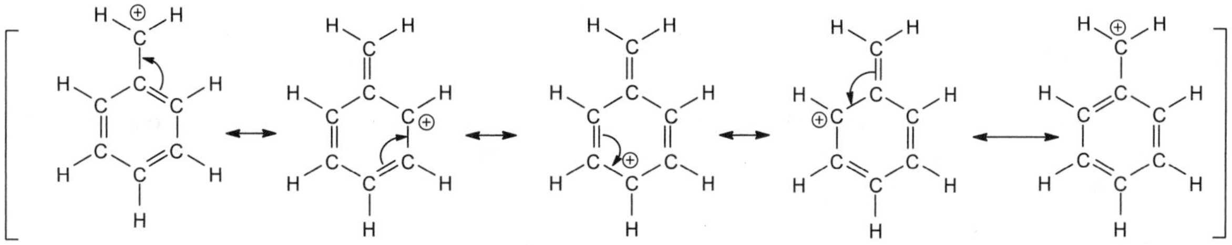

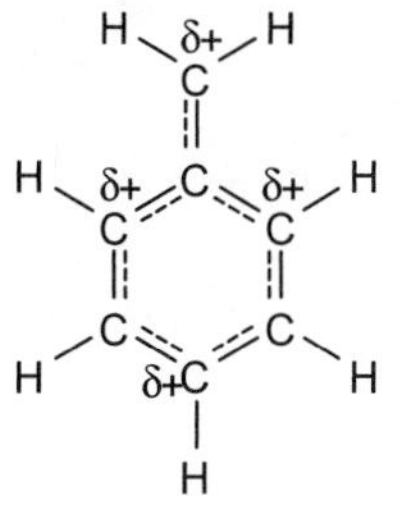

Hybrid

Your Turn 9.8

Think

Compare *relative* reaction rates and select a pair of nucleophiles that are reversed in the protic ethanol solvent compared to the aprotic DMF solvent. How are reaction rates related to the strength of the nucleophile? Why do you think a protic solvent results in a reversal of the reaction rates for certain pairs of nucleophiles?

Solve

Br^- and N_3^- are reversed in ethanol compared to DMF. In DMF, N_3^- is a stronger nucleophile, but in ethanol, Br^- is the stronger nucleophile. This suggests that N_3^- is more strongly solvated by ethanol than Br^- is. Similarly, $CH_3CO_2^-$ is a stronger nucleophile in DMF than either Br^- or Cl^-, but in ethanol, the opposite is true, suggesting that the ethanol solvates $CH_3CO_2^-$ stronger than it solvates Br^- or Cl^-.

Your Turn 9.9

Think

What is entropy? Count the number of product species. How does the number of product species relate to the entropy?

Solve

Entropy is a measure of energy dispersal in a system and is generally thought of as a measure of "disorder." Greater entropy exists in the elimination products because of the greater number of species—three species in elimination and two species in substitution. See the figure on the next page.

OTs + HO$^{\ominus}$ → OH + TsO$^{\ominus}$

Substitution products (2 species)

→ + TsO$^{\ominus}$ + H_2O

Elimination products (3 species)
Greater entropy

Your Turn 9.10

Think

Is there a suitable leaving group? Does the type of carbon bonded to the leaving group rule out any of the four reactions? What is the attacking species (strong/weak nucleophile/base)? What is the solvent (protic/aprotic)? Fill in the chart based on these factors.

Solve

OTs is a good leaving group suitable for both substitution and elimination. The leaving group is bonded to a secondary carbon, so none of the four reactions can be ruled out. The attacking species, $(CH_3)_3CONa$, is negatively charged and is a strong base, which would normally favor both S_N2 and E2, but because it is bulky, too, it favors E2 over S_N2. It is not indicated that the concentration is low and, therefore, bimolecular mechanisms are favored. However, in this case, the attacking species has a large amount of steric hindrance and E2 is favored. The leaving group ability of OTs is excellent, which promotes all four reactions, but unimolecular mechanisms (S_N1 and E1) are more sensitive to the leaving group ability. The solvent DMF is an aprotic solvent that favors S_N2 and E2.

Factor	S_N1	S_N2	E1	E2
Strength				✓
Concentration				✓
Leaving group	✓		✓	
Solvent		✓		✓
Total	1	1	1	**3**

CH_3, H_3C, CH_3, OTs $\xrightarrow[\text{DMF}]{(CH_3)_3CONa}$ CH_3, H_3C, CH_3

Chapter Problems

Problem 9.1

Think

Review the elementary steps for each mechanism, S_N1, S_N2, E1, and E2. In how many steps does each occur? What is the attacking species? What is the substrate? What is different about the site of attack in substitution versus elimination? If an uncharged nucleophile is involved, are proton transfer steps necessary?

Solve

In the substitution mechanisms, the attacking species (NH_3) acts as a nucleophile and attacks the electrophilic C atom. After a subsequent deprotonation, an amine results as the product. In the elimination mechanisms, the attacking species (NH_3) acts as a base and attacks the H atom adjacent to the electrophilic C atom. An alkene product results.

(a) S_N2: one step followed by a proton transfer, amine product.

S_N2

Proton transfer

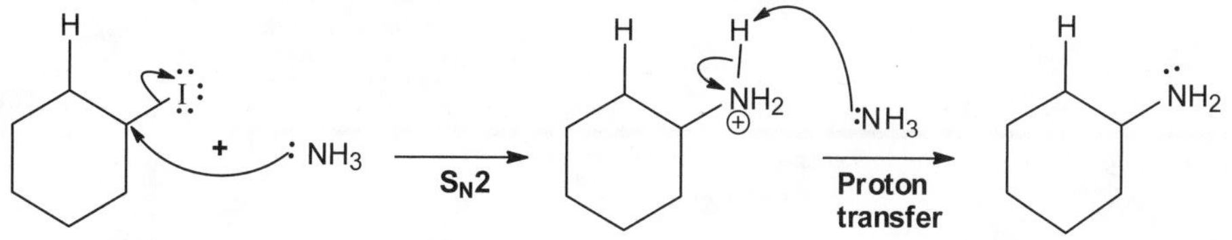

(b) S_N1: two steps (heterolysis, coordination) followed by proton transfer, amine product.

Heterolysis

Coordination

Proton transfer

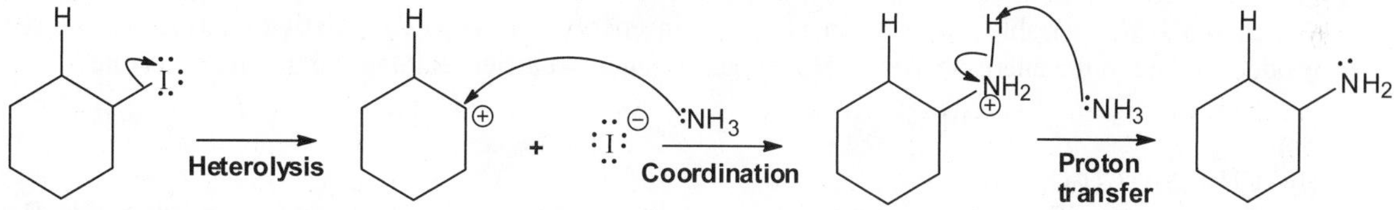

(c) E2: one step, alkene product.

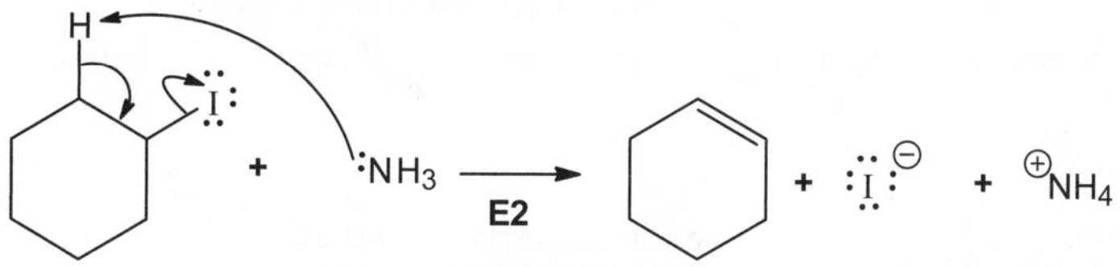

E2

(d) E1: two steps (heterolysis, electrophile elimination), alkene product.

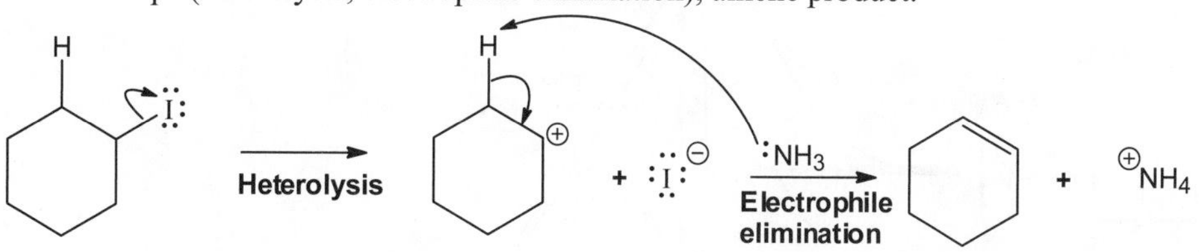

Heterolysis

Electrophile elimination

Problem 9.2

Think

Which nucleophile has a faster S_N2 reaction rate? What does that mean about the energy barrier? Which transition state will have a lower activation energy? Should that lower activation energy correspond to a difference in energy of the reactants or of the products?

Solve

Because the reaction involving NC^- is much faster (250 times faster), the free energy of activation must be much smaller. Because the same charged species appears in both sets of products, that difference in activation energy should correspond to a difference in reactant energies. According to the Hammond postulate, for the attack by NC^- to have the lower energy barrier, NC^- is higher in energy than Br^-. See the energy diagram on the next page.

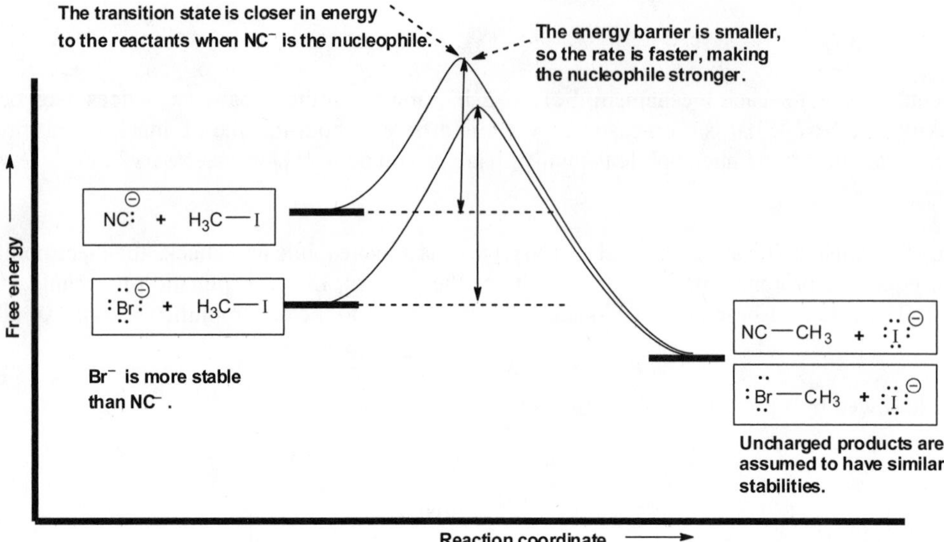

Problem 9.4

Think

Are there differences in charge stability between the two reactions on the reactant side? On the product side? Which ion is more stable? Which ion is higher in energy? How do the relative energies of the reactants or products relate to the energy barriers? How does the energy barrier relate to the S_N2 reaction rate?

Solve

(i) CH_3O^- or $CH_3O_2^-$

(a) There is a significant difference in charge stability on the reactant side but not the product side. Because the negative charge in $CH_3CO_2^-$ is resonance delocalized, it is more stable than CH_3O^-. Therefore, the pair of free energy diagrams appears as follows. As we can see, the energy barrier involving CH_3O^- is smaller, consistent with what would be suggested by the Hammond postulate, so the reaction is faster. CH_3O^- is a stronger nucleophile.

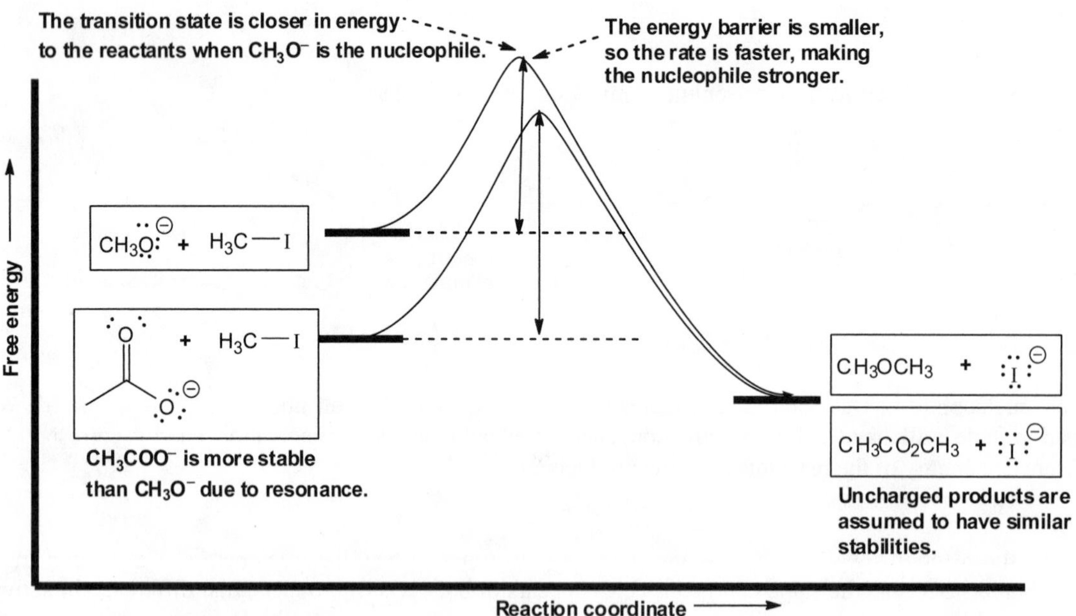

(b) CH_3O^- is a stronger nucleophile compared to $CH_3O_2^-$.

(ii) H_3N or H_3P

(a) There is a significant difference in charge stability on the product side but not the reactant side. Because a positive charge is more stable on P than on N (P is the bigger atom), the product of nucleophilic substitution involving PH_3 is more stable, and the Hammond postulate suggests that its reaction should have a smaller energy barrier. Thus it proceeds faster.

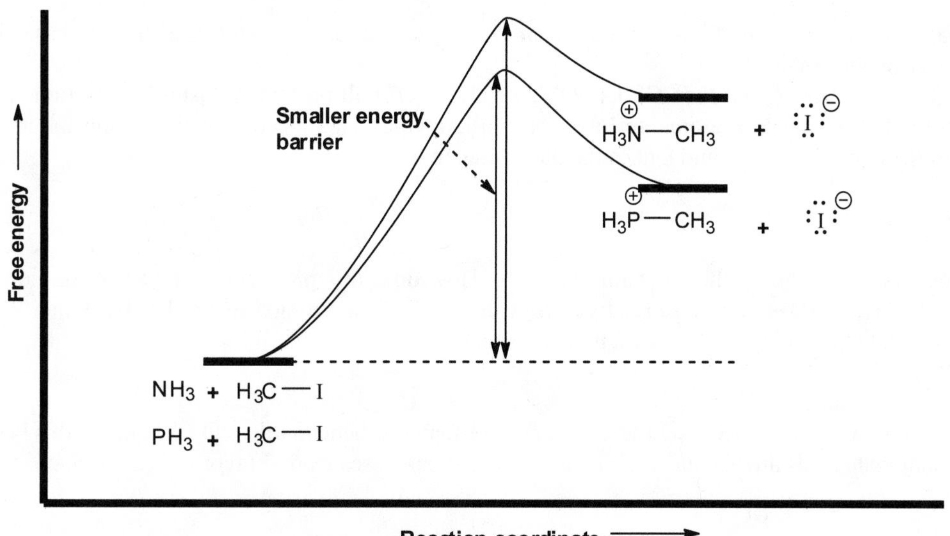

(b) H_3N is a stronger nucleophile compared to H_3P.

Problem 9.5

Think

What kind of charge does this nucleophile have? Is the charge resonance delocalized? What kind of charge is generally associated with a strong nucleophile? Are S_N2 reactions favored by strong or weak nucleophiles? What about S_N1 reactions?

Solve

H_2P^- is a strong nucleophile because it has a negative charge localized on a single atom (resonance is not possible). Therefore, it will favor the S_N2 mechanism. Strong nucleophiles promote fast S_N2 reactions by forcing the leaving group out. Being a strong nucleophile, H_2P^- will not wait until the leaving group leaves (slow step in S_N1).

H–P⁻–H

Strong nucleophile

Problem 9.6

Think

Does the alkyne have a significantly nucleophilic atom? How does the H:⁻ react with the terminal alkyne? What is the identity of your nucleophile after H:⁻ reacts? Which species acts as the electrophile in the S_N2 reaction? How many steps are in the S_N2 reaction?

Solve

The purpose of the NaH step is to deprotonate the terminal alkyne and produce the alkynide ion. In a second step, the S_N2 reaction (one step) takes place, in which the nucleophilic C^- attacks the electrophilic C atom bonded to Br. The six-carbon terminal alkyne is transformed into oct-3-yne.

H :H⁻ → **Proton transfer** → Br: → **S_N2** →

Oct-3-yne

Problem 9.8

Think

How does the strength of the base relate to the rate of the E2 reaction? Which base is stronger? How can you predict base strength (see Chapter 6)?

Solve

(a) HO^- is a stronger base than F^-, so HO^- will promote a faster E2 reaction. HO^- is a stronger base because O is less electronegative than F^-.

(b) $CH_3CH_2O^-$ is a stronger base than $CF_3CH_2O^-$, so $CH_3CH_2O^-$ will promote a faster E2 reaction. $CF_3CH_2O^-$ is a weaker base because F atoms are electron withdrawing. This electron-withdrawing inductive effect stabilizes the negative charge and makes the base weaker.

Problem 9.10

Think

Consult Appendix A to look up the pK_a value of HCN. How does this pK_a value compare to that of H_2O, the conjugate acid of HO^-? Based on these relative pK_a values, is NC^- a stronger or weaker base than HO^-? How does base strength help predict E1 versus E2 mechanisms?

Solve

NC^- is considered a weak base because it is a weaker base than our benchmark, HO^-. We know this because the pK_a of their conjugate acids are 9.2 and 15.7, respectively. Weak bases tend to favor the E1 mechanism.

Problem 9.11

Think

Can charge stability distinguish among these three nucleophiles? If not, what other factor is present? What is the effect of nearby methyl groups on the strength of the nucleophile?

Solve

The nucleophilic sites are all negatively charged O atoms. Thus, the charge on the atom is not a factor in this example. The difference is in the proximity of steric hindrance surrounding those sites, represented by the *t*-butyl groups. The nucleophile in **B** has the *t*-butyl group closest to the O atom, so **B** will give rise to the slowest reaction. The nucleophile in **C** has the *t*-butyl group farthest away from O, so it will proceed the fastest. The reaction rate thus increases in the order: **B** < **A** < **C.**

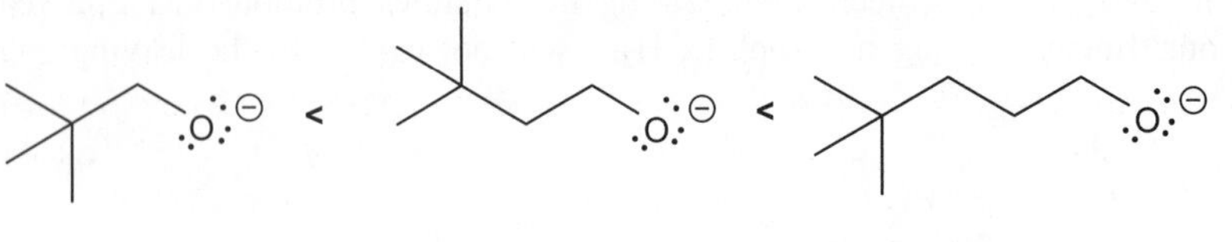

B A C

Slowest, *t*-butyl group closest to nucleophilic site

Fastest, *t*-butyl group farthest away from nucleophilic site

Problem 9.13

Think

Is the anion in question a strong nucleophile or a weak nucleophile? Is the anion in question a strong base or a weak base? What relative concentrations favor each reaction? What are the exceptions?

Solve

(a) S_N2, because HO^- is a strong nucleophile in high concentration; E2, because HO^- is a strong base in high concentration.

(b) S_N1 and E1, even though the attacking species is a strong nucleophile and a strong base, because the low concentrations of the nucleophile/base slows S_N2 and E2 a lot.

(c) S_N2, because Br^- is a strong nucleophile in high concentration; E1, because Br^- is a weak base.

(d) S_N1, even though Br^- is a strong nucleophile, because the low concentration of the nucleophile slows S_N2 a lot. E1, because Br^- is a weak base.
(e) S_N2, because $(CH_3)_3CO^-$ is a strong nucleophile in high concentration; E2, because it is a strong base in high concentration. E2 will beat out S_N2, however, because the base is very bulky.
(f) S_N1 and E1, for the same reasons as in **(b)**.

Problem 9.14

Think

What is the pK_a of each base's conjugate acid? What does that mean for the relative strength of the base? How is the strength of the base related to the leaving group ability? What if you could not look up the pK_a; how else could you evaluate the stability of the anion?

Solve

HCO_2^- is a better leaving group because it is a weaker base. The pK_a of HCO_2H is 3.75, whereas that of C_6H_5OH is 10.0. It is evident that HCO_2^- is a more stable anion because the negative charge can be delocalized over two O atoms. A more stable anion leads to a weaker, less reactive base. $C_6H_5O^-$ does have a delocalized charge, but the charge is delocalized over C atoms, so it is less stable.

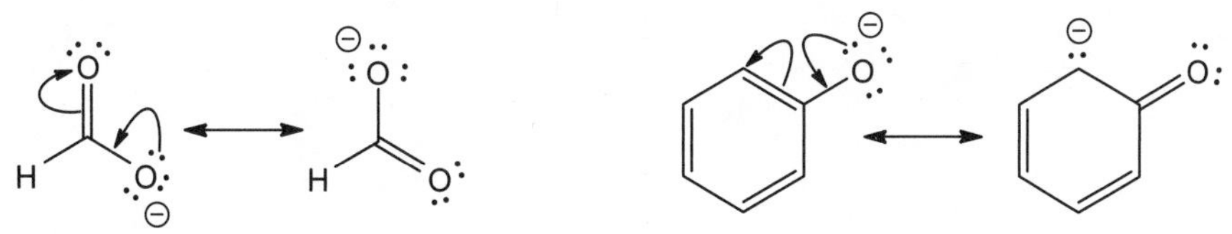

Negative charge delocalization onto oxygen.
More stable anion

Negative charge delocalization onto carbon.
Less stable anion

Problem 9.16

Think

What are the relative strengths of the bases in D–F, and how are the E1 and E2 reactions affected by base strength? What are the relative leaving group abilities of the leaving groups that appear in D–F, and how are the E1 and E2 reactions affected by leaving group ability?

Solve

The rate of an E1 reaction depends only on the rate of the departure of the leaving group. The difference among the substrates is only in the identity of the leaving group. Because Br^- is not as good a leaving group as TsO^-, **F** will proceed the slowest via an E1 mechanism. By contrast, an E2 reaction rate depends both on the strength of the base and the ability of the leaving group to leave. **E** will proceed faster than **D** by E2 because HO^- is a stronger base than NH_3. And **E** will proceed faster than **F** by an E2 because TsO^- is a better leaving group than Br^-.

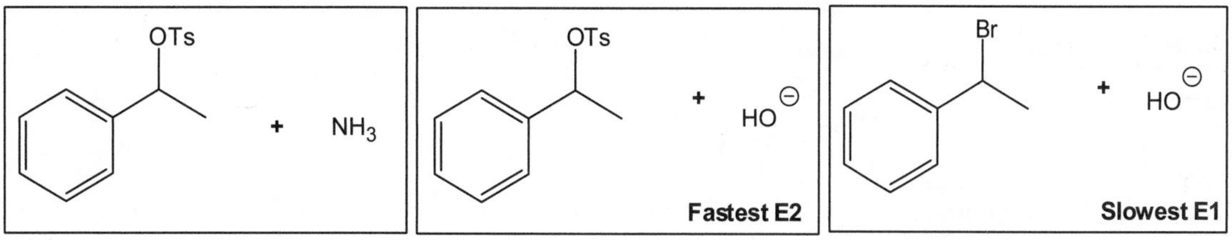

Fastest E2 (E)

Slowest E1 (F)

D **E** **F**

Problem 9.17

Think

How does the basicity of $C_6H_5O^-$ compare to that of F^-? How does basicity relate to leaving group ability? Is the bimolecular or unimolecular mechanism more sensitive to leaving group ability?

Solve

$C_6H_5O^-$ is a stronger base than F^- (because the pK_a of C_6H_5OH is more positive than that of HF). So the leaving group ability of $C_6H_5O^-$ should be worse than F^-, making it a poor leaving group. This should favor S_N2 and E2 over S_N1 and E1.

Problem 9.18

Think

Are the leaving groups the same or different? What is the hybridization of the C atoms attached to each Br atom? How does the hybridization of the reaction affect the ability of the leaving group to leave?

Solve

Nucleophilic substitution will occur most readily at the sp^3-hybridized C shown below. The leaving group on each C atom is Br^-, so the leaving group identity does not come into play. The other Br leaving groups are on sp^2- and sp-hybridized C atoms, respectively, which tend not to undergo nucleophilic substitution or elimination.

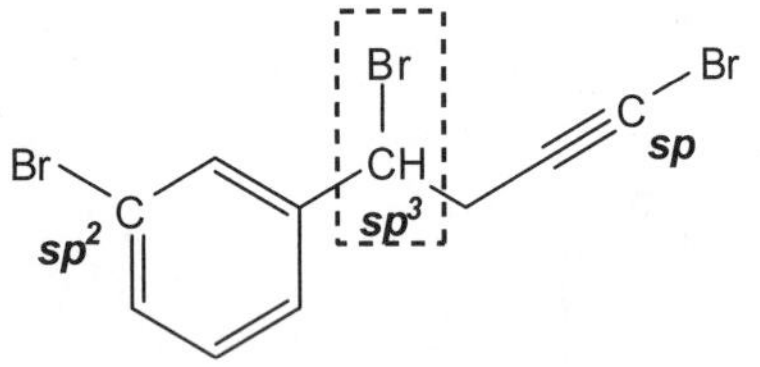

Br C sp^2 Br CH sp^3 C Br sp

Problem 9.20

Think

What is the rate-determining step in an E1 reaction? Are there any differences in the leaving group? What reactive intermediate forms from the slow step? What stabilizes/destabilizes this reactive intermediate? What happens to the rate when the intermediate is stabilized?

Solve

In an E1 reaction, the rate-determining step is the departure of the leaving group, heterolysis, to form the carbocation. The leaving group, Br^-, is the same in both substrates, so the difference is in the nature of the C atom to which the leaving group is bonded. In the first molecule, the C is secondary, whereas in the second molecule, the C is tertiary. As shown below, the molecule with the tertiary carbon produces the more stable carbocation in the rate-determining step of an E1 reaction, so it will be involved in the faster E1 reaction.

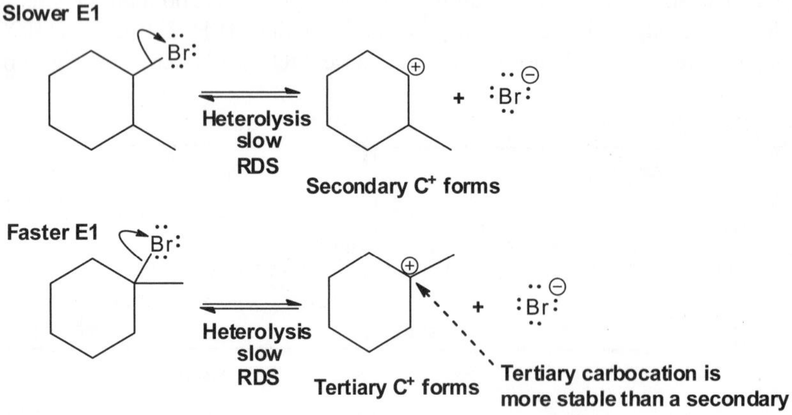

Slower E1

Heterolysis slow RDS

Secondary C^+ forms

Faster E1

Heterolysis slow RDS

Tertiary C^+ forms

Tertiary carbocation is more stable than a secondary

Problem 9.21

Think

Which atom is attacked by the base in an E2? What is the resulting functional group? What is the minimum number of carbon atoms?

Solve

E2 reactions require an H and a leaving group attached to adjacent atoms. Usually they are on adjacent C atoms, producing a new C=C bond. With a substrate of the form CH_3–L, there is no such adjacent C atom, so no double bond can be formed. See the general examples below.

$L{-}CH_2{-}CH_3 \xrightarrow{\text{base}} H_2C{=}CH_2$ **Primary**

$L{-}CH(CH_3)_2 \xrightarrow{\text{base}} (H_3C)HC{=}CH_2$ **Secondary**

$L{-}C(CH_3)_3 \xrightarrow{\text{base}} (H_3C)_2C{=}CH_2$ **Tertiary**

$L{-}CH_3 \xrightarrow{\text{base}}$ **No adjacent CH, cannot form a C=C double bond** **Methyl**

Problem 9.22

Think

Do the substrates differ in their leaving group? To what type of carbons (primary, secondary, tertiary) is the leaving group attached? What is the rate-determining step for each reaction? For S_N2 reactions, how does the sterics of the C–L affect the rate of the reaction? For S_N1 and E1, how does the stability of the C^+ formed affect the rate of the reaction?

Solve

X reacts fastest by an S_N2 mechanism. The leaving group, Cl^-, is the same in all three cases, so it does not factor in, but the primary carbon is the least hindered. **Z** reacts fastest by an S_N1 and E1 mechanism (same rate-determining step), because it proceeds through the more stable tertiary carbocation, as shown below. **X** and **Y** form less stable primary and secondary carbocations and, therefore, are slower via a unimolecular mechanism.

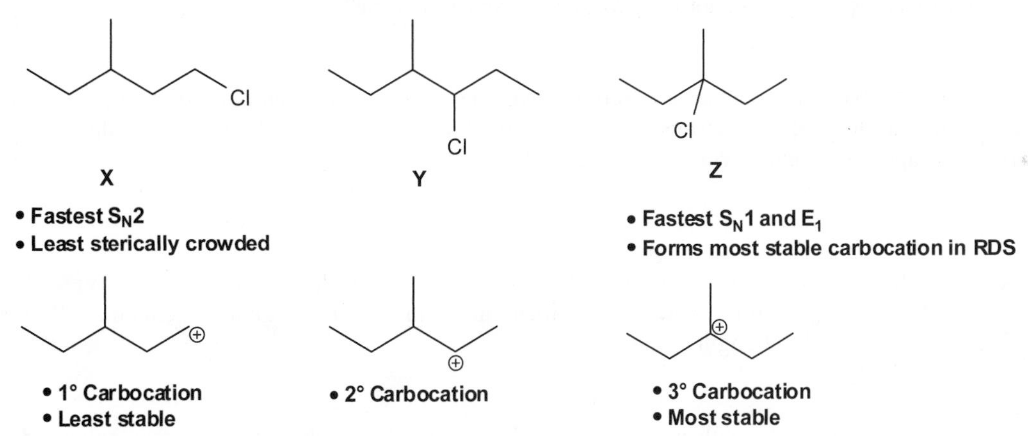

Problem 9.24

Think

Are the leaving groups the same or different in this example? What is the carbocation intermediate involved in each E1 mechanism? Which one is more stable? Why? How does the stability of the carbocation affect the E1 reaction rate?

Solve

In both **C** and **D**, the carbocation is generated by the loss of the same leaving group, Br^-. **C** reacts faster than **D.** The carbocation formed from **C** is resonance stabilized by the adjacent O atom, as shown on the next page. This forms a strong resonance contributor with all atoms having their octet. The carbocation formed from molecule **D** is not resonance stabilized (isolated on C only), and thus is less stable. Formation of the carbocation is the rate-determining step in an E1 reaction and thus the formation of a more stable carbocation leads to a faster reaction rate.

Faster E1 reaction rate

Heterolysis

C

All atoms have their octet

Heterolysis

No resonance structures possible

D

Problem 9.25

Think

Which solvent is protic? Which solvent is aprotic? Which mechanism favors protic solvents? Which mechanism favors aprotic solvents? Which mechanism forms ionic intermediates? Which mechanism's rate depends on the strength of the nucleophile?

Solve

Ethanol is a protic solvent and DMSO is an aprotic solvent. Protic solvents promote the formation of ions better, which is necessary in an S_N1 (or E1) reaction. Protic solvents weaken the strength of nucleophiles and the rate of an S_N2 reaction depends on the strength of the nucleophile. Therefore, the reaction will occur faster via an S_N1 mechanism in ethanol and faster via an S_N2 mechanism in DMSO.

Problem 9.26

Think

Do protic or aprotic solvents favor S_N2 reactions? How is the strength of the nucleophile affected by solvation in a protic solvent? Is acetone protic or aprotic? Is ethanol protic or aprotic?

Solve

Acetone is an aprotic solvent and ethanol is a protic solvent (possesses hydrogen-bond donors). The reaction will proceed faster in acetone because ethanol will solvate (and, therefore, weaken) the nucleophile to a greater degree. Remember, aprotic solvents favor S_N2 and E2 reactions.

Problem 9.27

Think

Which solvent is protic? Which solvent is aprotic? Which mechanisms are favored by protic solvents? Which mechanisms are favored by aprotic solvents? Which mechanism forms ionic intermediates? Which mechanism's rate depends on the strength of the nucleophile?

Solve

Solvent **Y** is protic (it has an N–H covalent bond), and will favor the E1 mechanism more than solvent **Z** will, which is aprotic (it has no N–H or O–H bonds). Protic solvents solvate ions (especially anions) well, which helps promote the rate-determining step of the E1.

Problem 9.29

Think

Which is a stronger nucleophile *intrinsically* (i.e., without considerations of solvation)? Is ethanol a protic or an aprotic solvent? Is the solvation of F^- in ethanol much different from the solvation of I^-?

Solve

Absent of any solvent, we would predict that F^- would be a stronger nucleophile than I^-, because I is a larger atom than F, and thus can stabilize the negative charge better. In a protic solvent like ethanol, however, F^- will be solvated much more strongly than I^- will, because F^- is much smaller and thus has a more highly concentrated negative charge. That solvation of F^- weakens F^- much more than it does I^-. This makes F^- a weaker nucleophile than I^- in ethanol.

Problem 9.30

Think

Which is a stronger nucleophile *intrinsically* (i.e., without considerations of solvation)? Is ethanol a protic or an aprotic solvent? Is the solvation of H_2S in ethanol much different from the solvation of NH_3?

Solve

The nucleophiles are uncharged, so solvation is not a huge factor. Thus, we should expect the relative nucleophilicities to be the same as what they would be absent of any solvent. That is, H_2S is a stronger nucleophile than H_3N because S, being a larger atom, can accommodate a positive charge better than N can. That positive charge develops in the substitution product, as shown in Solved Problem 9.3 on p. 474.

Problem 9.31

Think

Hydroxide is a strong nucleophile (S_N2) as well as a strong base (E2). This leads to a mixture of the two products at room temperature. Which bimolecular reaction mechanism has more entropy? Does an increase in heat favor the reaction with greater entropy or lesser entropy? Consider the equation $\Delta G^\circ_{rxn} = \Delta H^\circ_{rxn} - T\Delta S^\circ_{rxn}$.

Solve

ΔS°_{rxn} is more positive for an elimination reaction (three products vs. two products). Raising the temperature will favor elimination products (due to greater entropy, more positive ΔS°_{rxn}), which leads to the second reaction product. Lowering the temperature will do the reverse, favoring the first product (substitution, S_N2).

S_N2

Lesser entropy, favored at lowered temperatures

E2

Greater entropy, favored at raised temperatures

Problem 9.33

Think

Is there a suitable leaving group? Does the type of C atom bonded to the leaving group rule out any of the four reactions? What is the attacking species? What is the solvent? Which reaction does each of the factors favor?

Solve

The leaving group is on a benzyl carbon, so elimination (E1 or E2) is not feasible due to the fact that the benzyl carbon cannot be involved in a new C=C bond. Even though the leaving group is on a primary carbon, we must consider S_N1 along with S_N2 because a benzyl carbocation is relatively stable. CH_3O^- is a strong nucleophile, which favors S_N2. The concentration of the nucleophile is assumed to be high, which also favors S_N2. The leaving group is very good, which favors S_N1. The solvent is aprotic, which favors S_N2. So three factors favor S_N2, whereas only one factor favors S_N1. Thus, the S_N2 reaction below yields the major product. Note the stereochemistry, which is governed by the attack of the nucleophile from the side opposite the leaving group.

S_N2

Acetone

Factor	S_N1	S_N2	E1	E2
Strength		✓		
Concentration		✓		
Leaving group	✓			
Solvent		✓		
Total	1	3		

Problem 9.35
Think
In an E2 reaction, which CH protons can be attacked by the base? Draw out the C–H bond for all adjacent CH groups. Where does the double bond in the product appear relative to the leaving group and an adjacent proton? What alkenes form? Which one is most stable?

Solve

(a) The possible E2 products, not considering *E*/*Z* isomerism, are shown below. These are the products of eliminating one of the highlighted H atoms along with the leaving group. The major product is the most highly substituted alkene (most stable), which is tetrasubstituted in this case.

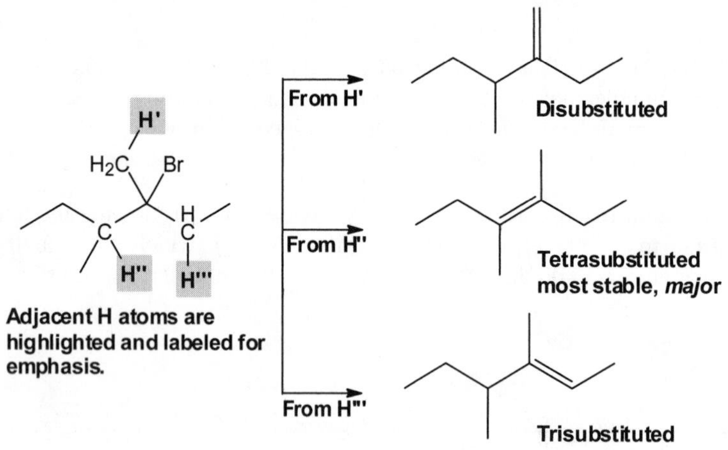

H'
H_2C Br
H C
C
H'' H'''
Adjacent H atoms are highlighted and labeled for emphasis.

From H' → Disubstituted
From H'' → Tetrasubstituted most stable, *major*
From H''' → Trisubstituted

(b) There are two possible E2 products, as shown below. The disubstituted alkene is the major product.

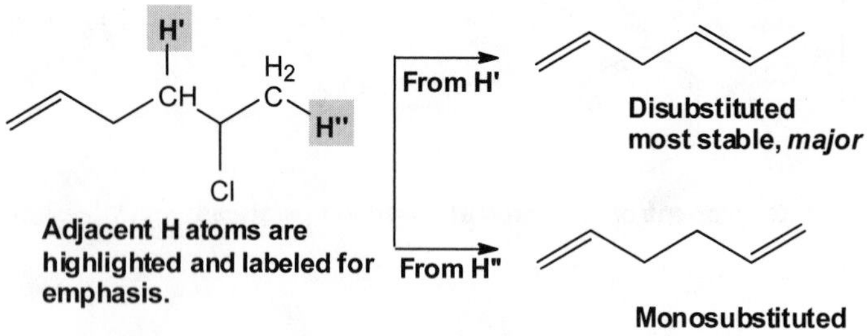

H'
CH
H_2 C
H''
Cl
Adjacent H atoms are highlighted and labeled for emphasis.

From H' → Disubstituted most stable, *major*
From H'' → Monosubstituted

(c) The two possible E2 products are shown below. The alkene on the top is the major product. Both alkenes are dialkyl-substituted, but the one on top has an additional resonance structure that provides stabilization.

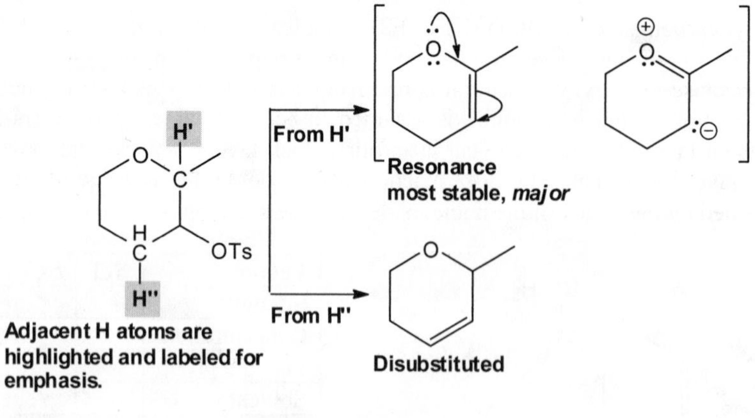

H'
O C
H C
OTs
H''
Adjacent H atoms are highlighted and labeled for emphasis.

From H' → Resonance most stable, *major*
From H'' → Disubstituted

Problem 9.37

Think

Under what conditions does the reaction take place—acidic, basic, or neutral? For an intramolecular nucleophilic substitution reaction under these conditions, what can act as the nucleophile? What can act as the leaving group? For each intramolecular nucleophilic substitution reaction, what is the size of the ring that is formed?

Solve

The reaction takes place under basic conditions, so the alcohol will be deprotonated rapidly. After the alcohol is deprotonated, two different intramolecular S_N2 reactions are possible, as shown below. The one that leads to the formation of a six-membered ring gives the major product.

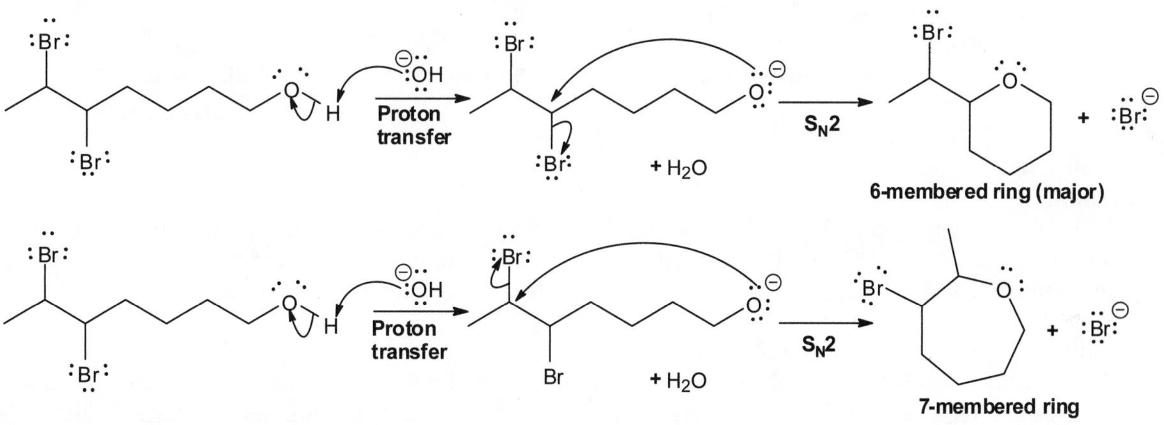

Proton transfer

S_N2

+ H_2O

6-membered ring (major)

Proton transfer

S_N2

+ H_2O

7-membered ring

Problem 9.38

Think

What is the charge stability of Br^- compared to $CH_3CH_2O^-$? How does charge stability of the reactants and products affect the reversibility of the reaction?

Solve

It is irreversible because charge stability heavily favors the product side of the reaction. This is because Br^- is much more stable than $CH_3CH_2O^-$, due to the fact that Br is a significantly larger atom than O, and can stabilize the negative charge better.

Problem 9.39

Think

Are the leaving groups the same or different? What is the rate-determining step in an E1 reaction? Consider the stability of the carbocation intermediate from each compound. How does the stability of the carbocation affect the E1 reaction rate?

Solve

Reaction rate increases with increasing stability of the leaving group and of the carbocation that would be formed upon the departure of the leaving group. In this example, the leaving group (Br^-) is the same in each compound and, therefore, does not affect the reaction rate. Reaction occurs slowest at the aromatic sp^2-hybridized carbon (**D**), because of effective electronegativity of C and bond strength. Reaction occurs fastest at the resonance stabilized benzylic tertiary carbon (**C**). See the figures on the next page.

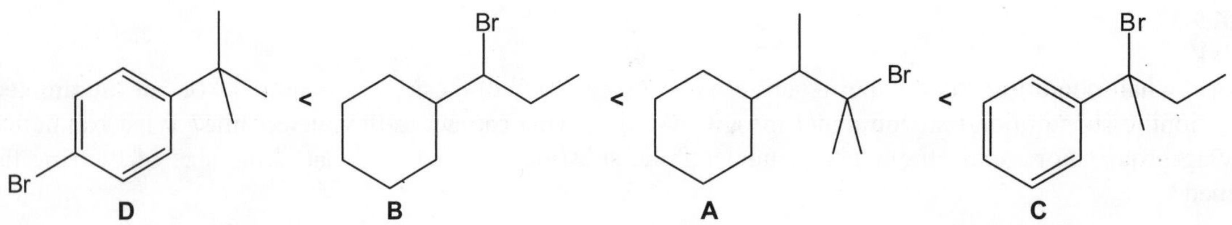

D < B < A < C

D — Slowest; C — Fastest

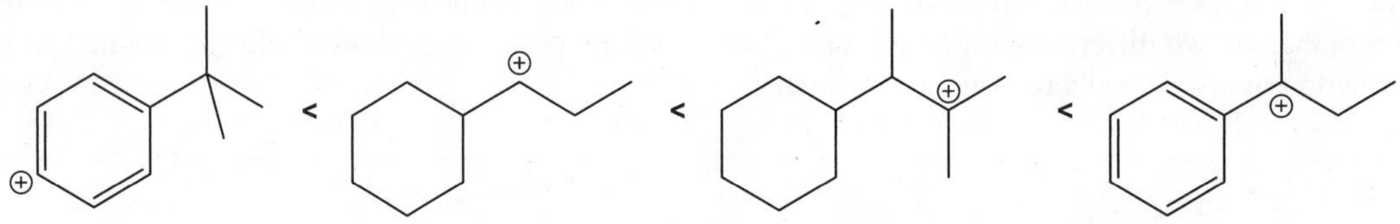

sp^2 C < Secondary C^+ < Tertiary C^+ < Tertiary benzylic C^+

Most stable C^+, fastest

Problem 9.40

Think

Is the leaving group in each substrate the same or different? To what type of carbon in the substrate is each leaving group attached? How does the hybridization and steric crowding affect the reaction rate of an S_N2 reaction? In how many steps does an S_N2 reaction occur? Are any intermediates formed?

Solve

The reaction occurs slowest at the sp^2-hybridized carbon because of bond strength and the additional electrostatic repulsion encountered during attack (**B**). For the others, the S_N2 rate increases with less steric hindrance from surrounding alkyl groups, **C** < **D** < **A**.

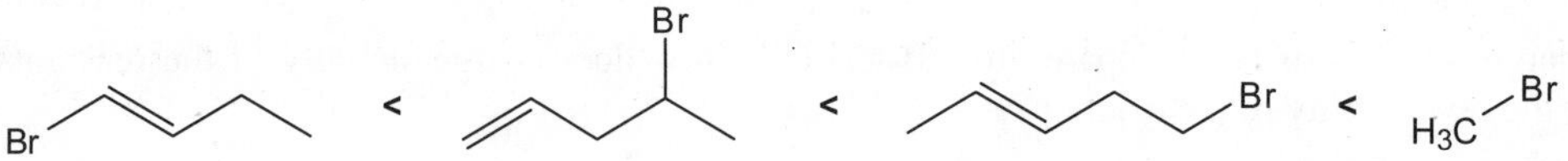

B < C < D < A (H_3C–Br)

B — sp^2 C–Br, Slowest

C — Secondary C–Br

D — Primary C–Br

A — Methyl C–Br, Fastest, least crowded

Problem 9.41

Think

What is the basic site in each compound? How should you treat the metal cation species? How does base strength affect the rate of the E2 reaction? Which base is the strongest? Weakest?

Solve

Each Na^+ and K^+ species should be treated as a spectator ion, leaving a negatively charged base. The E2 reaction rate, in general, increases with increasing base strength. The pK_a values of the conjugate acids are shown below.

Decreasing E2 reaction rate →

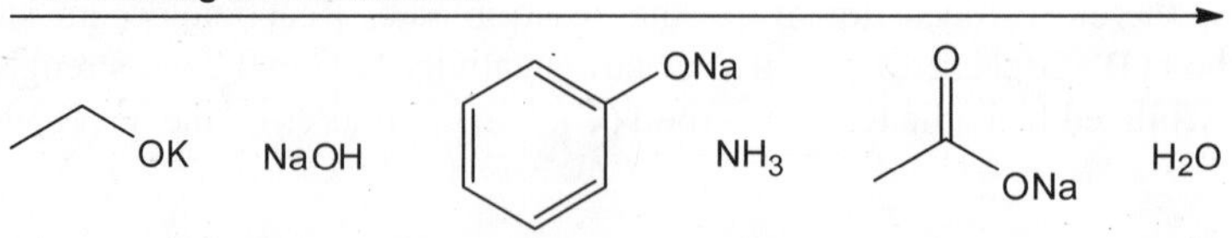

	OK (ethoxide)	NaOH	ONa (phenoxide)	NH_3	ONa (acetate)	H_2O
	F	D	C	E	A	B
pK_a of conjugate acid:	16	15.7	10.0	9.4	4.75	–1.7
	Fastest					Slowest

Problem 9.42

Think

Are the leaving groups the same or different in this example? What is the carbocation intermediate involved in each S_N1 mechanism? Which one is more stable? Why? How does the stability of the carbocation affect the S_N1 reaction rate?

Solve

The S_N1 reaction rate is governed by the rate at which the leaving group leaves. Because all of the leaving groups are the same (Br^-), the main differences are in the stabilities of the carbocations that are formed. Carbocations that are formed from **A** and **D** are resonance stabilized, so the reactions involving those cations are faster than those involving **B** and **C**. Because **D** is also a tertiary carbocation, it is more stable than **A**, which is a secondary carbocation. Likewise, **B** is a tertiary carbocation, so it is more stable than the carbocation from **C**, a secondary carbocation. Overall, then, stability of the carbocations increases in the order: **C** < **B** < **A** < **D**, which is the same as the order of the reaction rates.

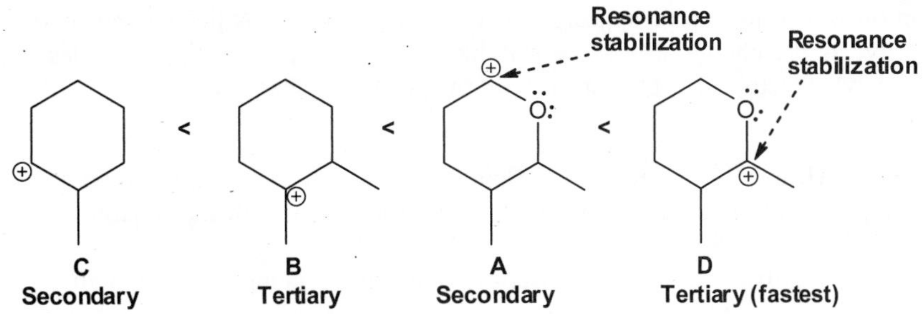

Resonance stabilization
Resonance stabilization
C Secondary < B Tertiary < A Secondary < D Tertiary (fastest)

Problem 9.43

Think

Did the products arise from substitution or elimination? Which mechanism gives rise to a stereospecific reaction route? In that reaction, does the stereochemical configuration of the C atom initially bonded to the leaving group remain the same or does it undergo inversion? On which carbon must the leaving group be located?

Solve

The desired starting compounds are shown below. The reactions should occur by the S_N2 mechanism to ensure that inversion occurs and only one stereoisomer is produced.

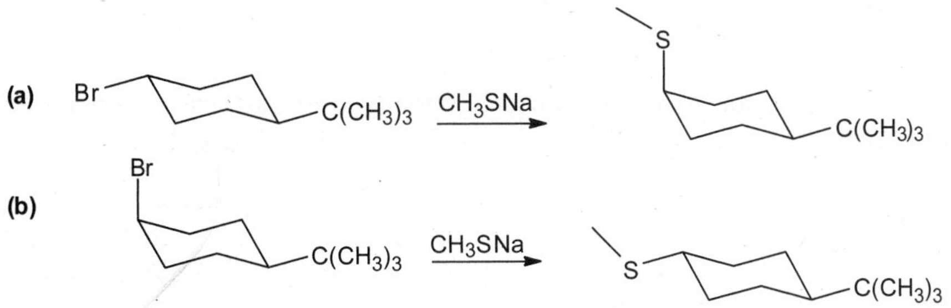

(a) Br–(cyclohexane)–$C(CH_3)_3$ $\xrightarrow{CH_3SNa}$ S–(cyclohexane)–$C(CH_3)_3$

(b) Br–(cyclohexane)–$C(CH_3)_3$ $\xrightarrow{CH_3SNa}$ S–(cyclohexane)–$C(CH_3)_3$

Problem 9.44

Think

To what type of carbon is each leaving group attached? How does the steric crowding of the electrophilic carbon affect the reaction rate of an S_N2 reaction? How does steric crowding of the nucleophile affect the reaction rate? On which species does such steric crowding play a greater role?

Solve

Reaction **(a)** is more efficient due to less steric hindrance surrounding the C atom with the leaving group. The C–L in reaction **(a)** is methyl, and in reaction **(b)** is secondary. With secondary carbons, S_N2 reactions proceed somewhat slowly due to the reduced accessibility of the carbon. There is also steric crowding surrounding the nucleophilic O^- in **(a)**, but those sterics have less of an impact on reaction rate because the bulkiness is farther away from the reaction site. See the mechanisms on the next page.

(a)

H_3C–Cl + alkoxide → ether + Cl^-

Methyl, less crowded

(b)

H_3C–O^- + secondary alkyl Br → ether + Br^-

Secondary, more crowded

Problem 9.45

Think

Is acetone protic or aprotic? Is solvent a factor in these examples? Is a nucleophile stronger when it is positively charged, negatively charged, or uncharged? In the examples where a charge is present on each nucleophile, how does the size of the atom on which the charge appears affect stability? How does the electronegativity of that atom affect charge stability? Is the charge able to be stabilized by resonance? In the examples where both nucleophiles are uncharged, what factors affect the stability of the positively charged product?

Solve

The stronger nucleophile of each pair is circled below. Being in an aprotic solvent, the relative nucleophile strengths should be the same as with no solvent, so we can use arguments of intrinsic charge stability as we learned in Section 9.3a.

In **(b)** the species that has a negative charge is the stronger nucleophile compared to the species with the positive charge.

(b)

H_3C-O^- (circled) or $H_3C-OH_2^+$

In **(d)**, **(f)**, and **(g)**, the atoms with the negative charge are in the same row of the periodic table, so the stronger nucleophile is the less electronegative atom.

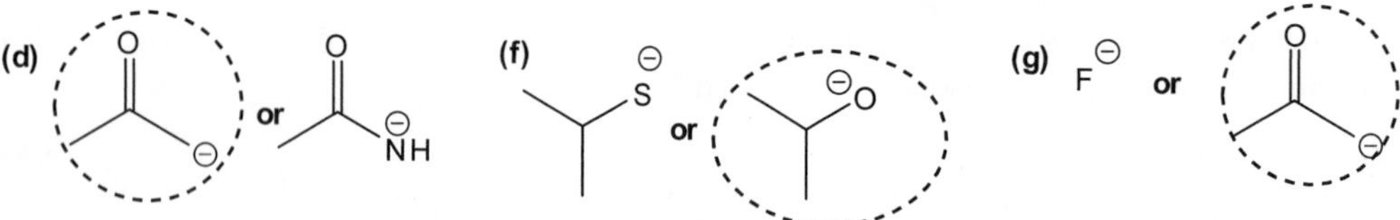

In **(c)**, both nucleophiles are uncharged, so the greater charge stability in the products has the positive charge appear on the less electronegative atom.

(c)

H_3C-OH or H_3C-NH_2 (circled)

In **(a)**, the negatively charged nucleophile is less stable.

(a)

H_3C-OH or H_3C-O^- (circled)

In **(h)**, **(i)**, and **(j)**, the one that has less charge stability is the one in which the negative charge appears on the smaller atom.

(h) acetamide anion (NH^-, circled) or ethyl PH^-

(i) CH_3S^- (circled) or CH_3Se^-

(j) CH_3Se^- (circled) or Br^-

In **(e)**, the less stabile charge is the one that has the charge localized instead of resonance delocalized.

(e)
or

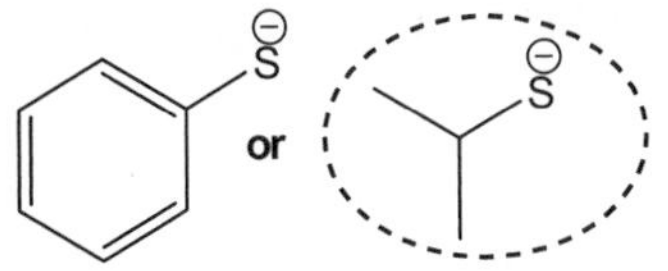

Problem 9.46

Think

Is ethanol protic or aprotic? Is solvent a factor in these examples? Is a nucleophile stronger when it is positively charged, negatively charge, or uncharged? In the examples where a charge is present on each nucleophile, how does the size of the atom on which the charge appears affect stability? How does the electronegativity of that atom affect charge stability? Is the charge able to be stabilized by resonance? In the examples where both nucleophiles are uncharged, what factors affect the stability of the positively charged product?

Solve

The results are determined the same way they were in Problem 9.45, except when the nucleophile has a negative charge on atoms of significantly different size—that is, in different rows of the periodic table. Under these circumstances, the protic solvent (ethanol) solvates the nucleophile much more strongly and, hence, weakens the nucleophile strength much more, with the negative charge on the *smaller* atom. So only in **(d)** and **(e)** are the relative nucleophilicities reversed from what we would observe in an aprotic solvent.

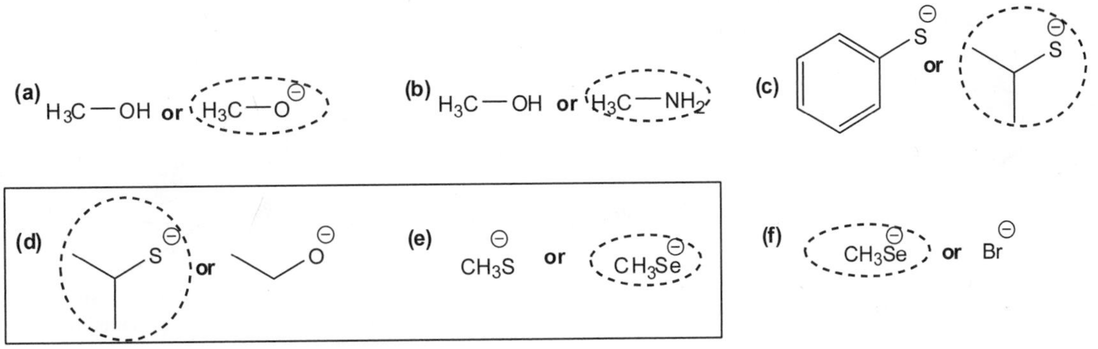

(a) $H_3C—OH$ or $H_3C—O^{\ominus}$ (b) $H_3C—OH$ or $H_3C—NH_2$ (c) or

(d) or (e) $CH_3S^{\ominus}$ or $CH_3Se^{\ominus}$ (f) $CH_3Se^{\ominus}$ or $Br^{\ominus}$

In a protic solvent the smaller negatively charged atom is a weaker nucleophile.

Problem 9.47

Think

Did the product arise from a bimolecular or unimolecular mechanism? Did you get inversion of stereochemistry or a racemic mixture? How does knowing the mechanism (S_N1 vs. S_N2) affect the choice of nucleophile (strong or weak) and choice of solvent (protic or aprotic)?

Solve

(a) This is a stereospecific nucleophilic substitution reaction in which there is inversion of the configuration at the stereocenter—this is S_N2. A polar aprotic solvent such as DMSO or DMF will favor such a reaction. Also, a strong nucleophile such as $NaOCH_3$ will favor the reaction.

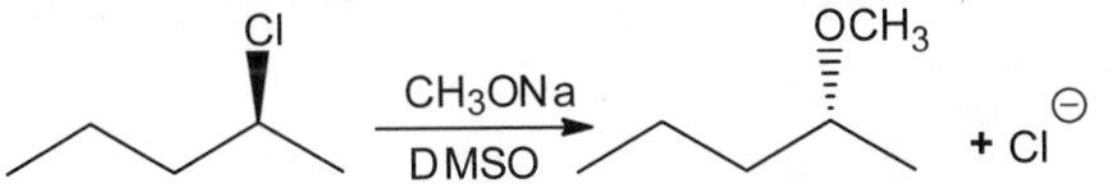

Cl → (CH_3ONa, DMSO) → OCH_3 + $Cl^{\ominus}$

Inversion of stereochemistry
S_N2

(b) This nucleophilic substitution generates a mixture of configurations (equal *R* and *S*, racemic), so it is S_N1. A protic solvent will favor such a reaction, as will a weak nucleophile. This can simply be a solvolysis reaction that takes place in CH_3OH, which can act as both the weak nucleophile and the protic solvent.

CH_3OH Solvolysis; OCH_3 (*S*) and OCH_3 (*R*) + $Cl^{\ominus}$

Racemic
S_N1

Problem 9.48

Think

Do the reaction conditions favor S_N1 or S_N2? How many leaving groups are present on the substrate? How many nucleophilic sites are present on the nucleophile? Will an intramolecular reaction occur?

Solve

Each S atom displaces bromide in an S_N2 mechanism. This is an intramolecular reaction that forms a six-membered ring.

S_N2 → Proton transfer → S_N2 → Proton transfer

Problem 9.49

Think

If an alkynide anion ($RC\equiv C:^-$) is needed to synthesize pent-2-yne, which carbon–carbon bond is formed? Are there two possible bonds that can be formed? How many carbon atoms must the alkyl bromide then possess?

Solve

Pent-2-yne is not symmetrical and there are two alkynide anions that can be used as a nucleophile. The C atoms necessary for the alkynide anion are boxed below. In the first example, the alkynide anion is $CH_3C\equiv C:^-$ and the alkyl bromide, therefore, is CH_3CH_2Br. In the second example, the alkynide anion is $CH_3CH_2C\equiv C:^-$ and the alkyl bromide is, therefore, CH_3Br. The C–C bond formed is in boldface.

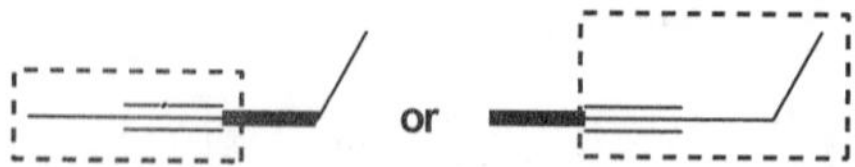

The key to forming the alkynide anion is to convert each terminal alkyne into a strong nucleophile by deprotonating it with a strong base like NaH.

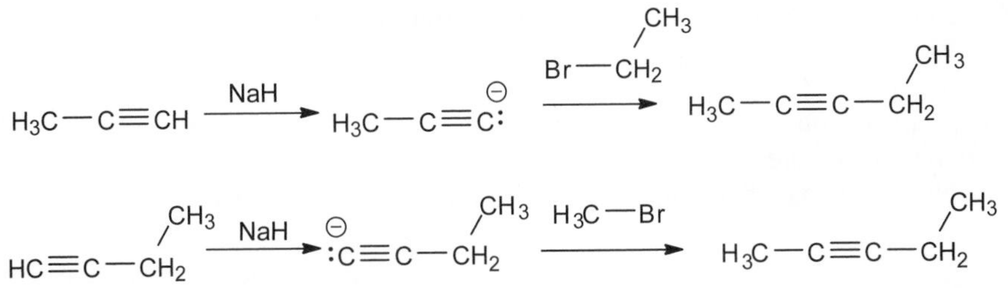

$$H_3C{-}C{\equiv}CH \xrightarrow{NaH} H_3C{-}C{\equiv}C{:}^{\ominus} \xrightarrow{Br{-}CH_2{-}CH_3} H_3C{-}C{\equiv}C{-}CH_2{-}CH_3$$

$$HC{\equiv}C{-}CH_2{-}CH_3 \xrightarrow{NaH} {}^{\ominus}{:}C{\equiv}C{-}CH_2{-}CH_3 \xrightarrow{H_3C{-}Br} H_3C{-}C{\equiv}C{-}CH_2{-}CH_3$$

Problem 9.50

Think

Are the products a result of substitution or elimination? Did the product arise from a bimolecular or unimolecular mechanism? Did you get inversion of stereochemistry or a mixture of configurations? Does the reaction form a reactive carbocation intermediate?

Solve

The S_N1 mechanism was dominant—it is a nucleophilic substitution reaction that yields a mixture of configurations at the electrophilic carbon. The methyl group at C4 remains wedge in both products because no bonds were broken or formed at that carbon. Formation of the trigonal planar carbocation intermediate allows the ethanol nucleophile to attack the electrophilic site from the front or back. This leads to a mixture of configurations at that carbon.

Step 1: A carbocation forms.

Step 2: Ethanol attacks ***either side*** of the carbocation's plane, forming protonated ethers with a mixture of configurations.

Step 3: Each protonated ether loses a proton to a weak base present, such as ethanol, forming the products given.

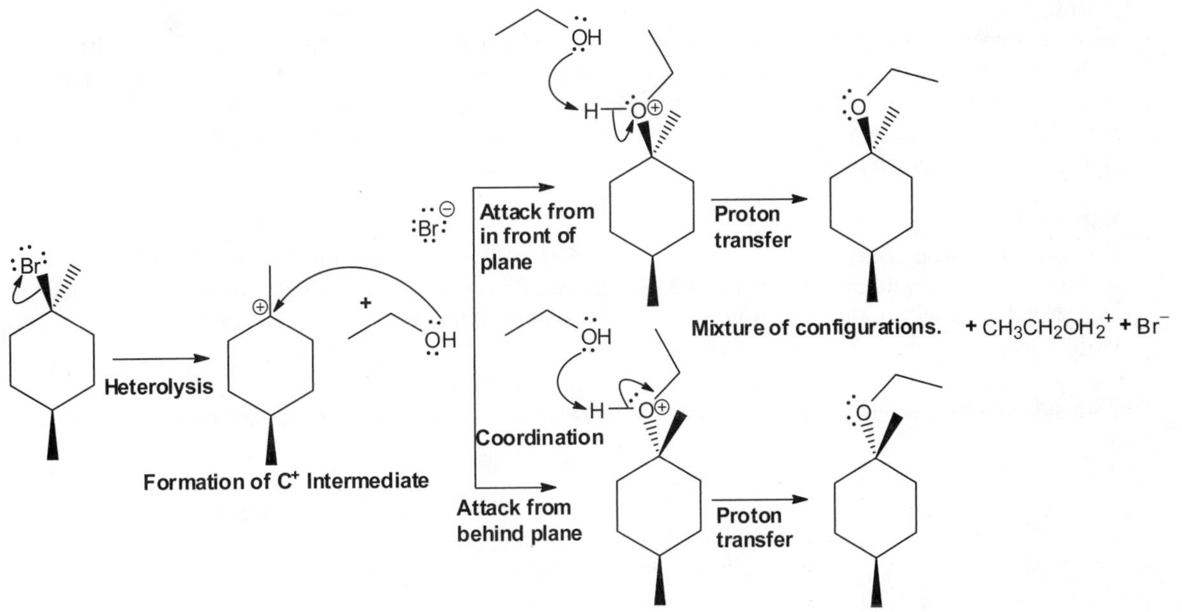

Heterolysis

Formation of C^+ Intermediate

Attack from in front of plane

Coordination

Attack from behind plane

Proton transfer

Proton transfer

Mixture of configurations. $+ CH_3CH_2OH_2^+ + Br^-$

Problem 9.51

Think

Are the products a result of substitution or elimination? Did the products arise from a bimolecular or unimolecular mechanism? Does the reaction form a reactive carbocation intermediate? Does the double bond that has formed involve the C atom that was initially attached to the leaving group?

Solve

An E1 mechanism dominated because a carbocation rearrangement must have occurred. We know this because among the products is an alkene in which the double bond does *not* involve the C atom that is initially bonded to the leaving group. For a carbocation rearrangement to happen, a carbocation must be formed. This describes an E1 reaction, not an E2.

Step 1: The alcohol is protonated.
Step 2: Water is lost to form a carbocation.
Step 3: The carbocation rearranges to a more stable one.
Step 4: A proton is lost from either of two carbons adjacent to C^+ to form the final products.

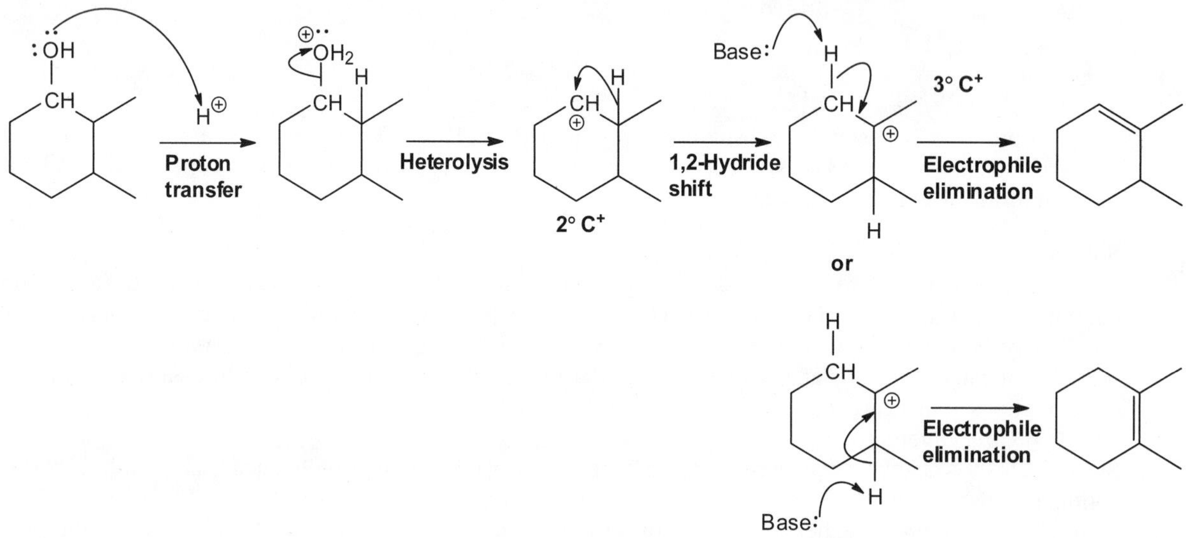

Proton transfer

Heterolysis

2° C⁺

1,2-Hydride shift

3° C⁺

Electrophile elimination

or

Electrophile elimination

Problem 9.52

Think

In which mechanisms are carbocation rearrangements possible? To favor these kinds of reactions, should the attacking species be strong or weak? Should the leaving group be excellent, moderate, or poor? Should the type of carbon attached to the leaving group be primary, secondary, or tertiary? Should the solvent be protic or aprotic? If a carbocation forms, when is it capable of rearrangement? What type of carbocation is formed? Is a more stable carbocation possible?

Solve

Carbocation rearrangement can occur only in the S_N1 and E1 mechanisms because those are the ones that proceed through a carbocation intermediate. S_N1/E1 reactions will take place with **(a)**, **(b)**, **(c)**, **(d)**, **(f)**, and **(g)**. Rearrangements will occur if a carbocation becomes more stable through a 1,2-hydride shift or a 1,2-methyl shift. These reactions are **(a)**, **(d)**, and **(f)**.

The carbocation generated in **(a)** is secondary and can convert to tertiary via a 1,2-hydride shift.

(a)

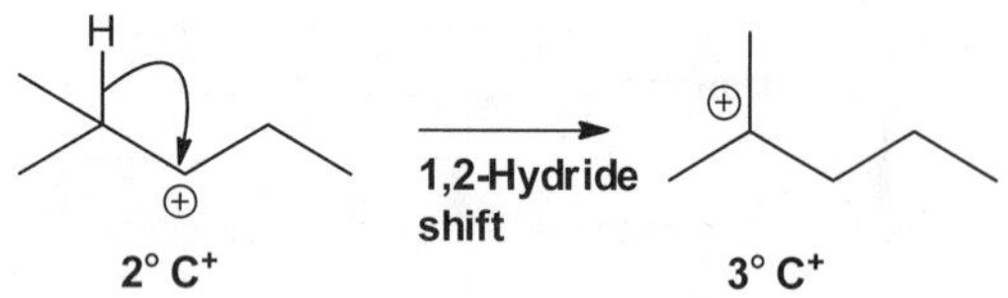

2° C⁺

1,2-Hydride shift

3° C⁺

In **(d)**, a 1,2-hydride shift occurs to produce a resonance-stabilized carbocation.

(d)

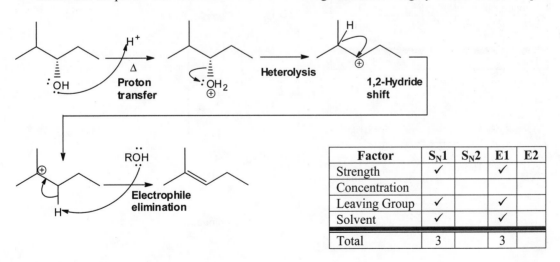

1,2-Hydride shift

3° C^+ Resonanced stabilized 3° C^+

In **(f)** a 1,2-hydride shift occurs to produce a secondary carbocation that is resonance stabilized.

(f)

1,2-Hydride shift

2° C^+ Benzylic C^+

Rearrangements will not occur in **(b)**, **(c)**, **(e)**, and **(g)**. Formation of a carbocation is not expected in reaction **(e)** because the reaction conditions favor S_N2; in the other reactions, rearrangement will not produce a more stable carbocation through a 1,2-hydride shift or 1,2-methyl shift.

Problem 9.53

Think

Is there a suitable leaving group? Does the type of carbon bonded to the leaving group rule out any of the four reactions? What is the attacking species? What is the solvent? Which reaction does each of the factors favor?

Solve

(a) The predominant mechanism is E1. The attacking species is an alcohol, which is a weak base and weak nucleophile, favoring S_N1 and E1. The concentration does not matter because the attacking species is weak. The leaving group is water (under acidic conditions), which is an excellent leaving group, and favors S_N1 and E1. The carbon is secondary, which favors all four mechanisms. The solvent is an alcohol, which favors S_N1 and E1. Heat favors elimination over substitution. The complete mechanism is shown below. Notice that the proton that is eliminated is the one that gives the most highly substituted alkene product.

H^+, Δ, Proton transfer → Heterolysis → 1,2-Hydride shift → ROH, Electrophile elimination

Factor	S_N1	S_N2	E1	E2
Strength	✓		✓	
Concentration				
Leaving Group	✓		✓	
Solvent	✓		✓	
Total	3		3	

(b) E1 is favored. The factors are the same as in **(a)**, but the S_N2 column is omitted because the leaving group is on a tertiary carbon. The H atom that is eliminated in the final step yields the most highly substituted alkene product.

(c) The factors are identical to **(a)**, so E1 is favored, but no carbocation rearrangement is feasible. Notice, however, that both cis and trans isomers are formed.

(d) S_N1 is favored. The attacking species is weak as a nucleophile and base, favoring S_N1 and E1. Concentration doesn't matter, because the attacking species is weak. The leaving group is very good, favoring S_N1 and E1. The carbon is tertiary, shutting down S_N2. The solvent is an alcohol, which is protic and favors S_N1 and E1. The reaction is not heated, so substitution is favored over elimination, though we can expect some elimination product as well. Notice that the carbocation is resonance delocalized, so the nucleophile can attack either carbon bearing the positive charge, producing isomeric products.

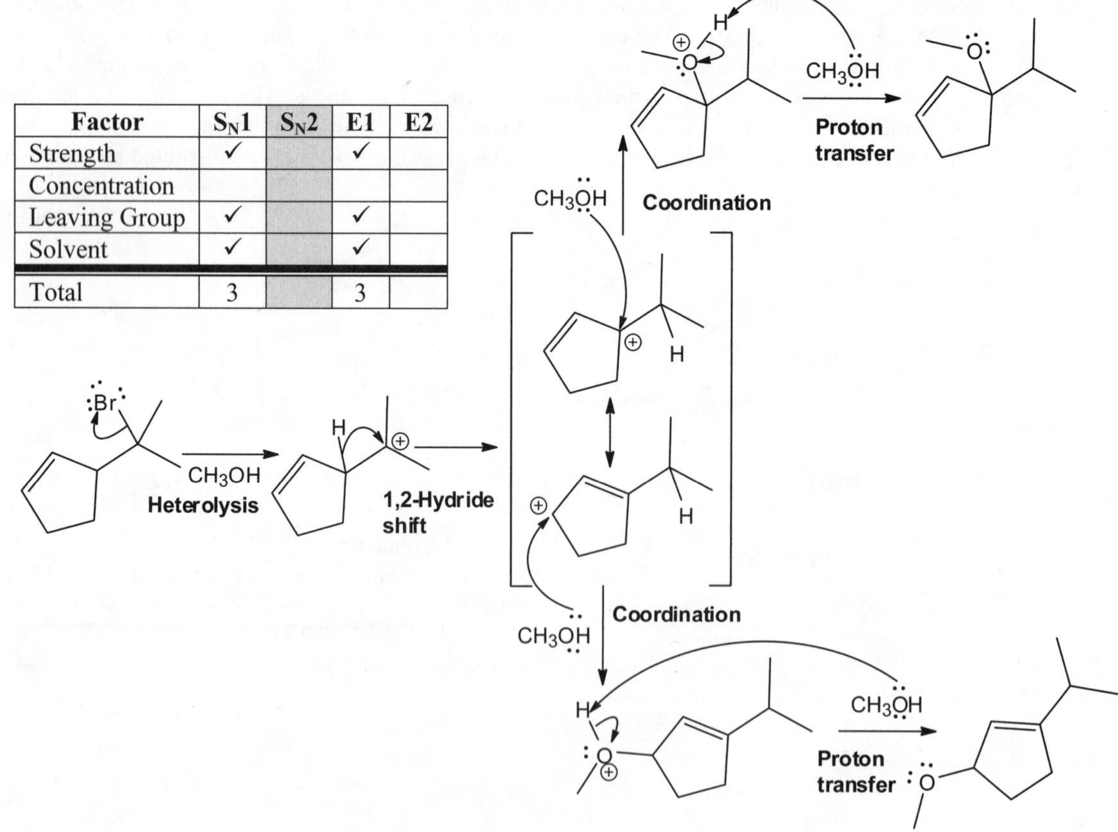

Factor	S_N1	S_N2	E1	E2
Strength	✓		✓	
Concentration				
Leaving Group	✓		✓	
Solvent	✓		✓	
Total	3		3	

(e) S_N2 is favored. Br^- is a strong nucleophile but a weak base, favoring S_N2 and E1. The concentration of Br^- is high, also favoring S_N2. Cl^- is a moderate leaving group, favoring all four reactions. The carbon is secondary, favoring all four reactions. The solvent, DMSO, is aprotic, which favors S_N2 and E2. Notice the inversion of configuration that takes place.

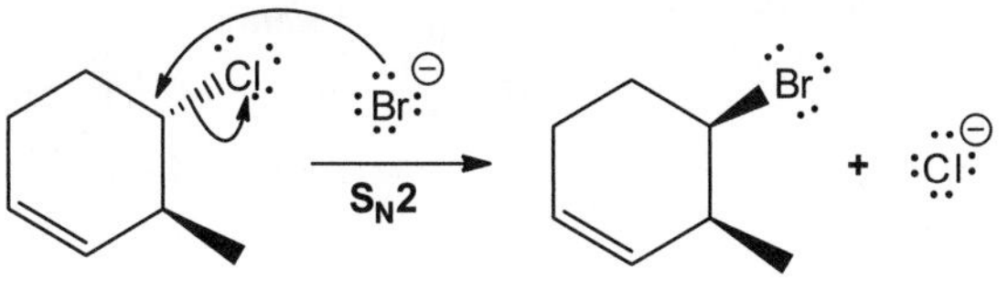

Factor	S_N1	S_N2	E1	E2
Strength		✓	✓	
Concentration		✓		
Leaving Group	✓	✓	✓	✓
Solvent		✓		✓
Total	1	**4**	2	2

(f) S_N1 and E1 are favored, but S_N1 is favored more. The attacking species is a strong bulky base, which favors S_N2 and E2 (though it favors E2 more than S_N2), but the concentration is weak, which favors S_N1 and E1. The leaving group is excellent, which favors S_N1 and E1. The carbon is secondary, which favors all four reactions. The solvent is protic, which favors S_N1 and E1. The reaction is not heated, which tips the balance in the favor of substitution over elimination. Notice that both *R* and *S* configurations are produced for the substitution products, and both cis and trans isomers are produced for the elimination products.

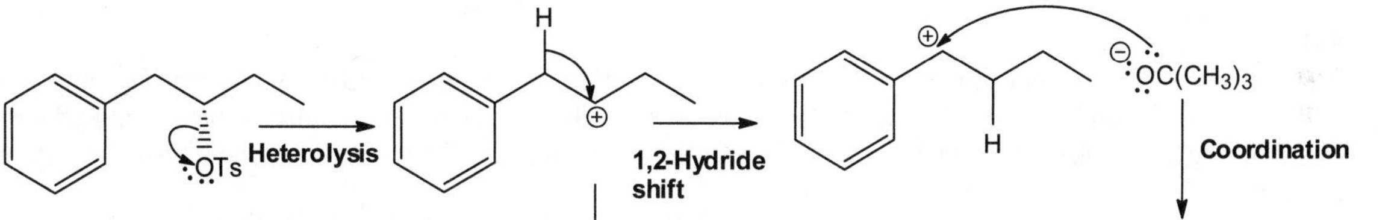

Factor	S_N1	S_N2	E1	E2
Strength		✓		✓
Concentration	✓		✓	
Leaving Group	✓		✓	
Solvent	✓		✓	
Total	3	1	**3**	1

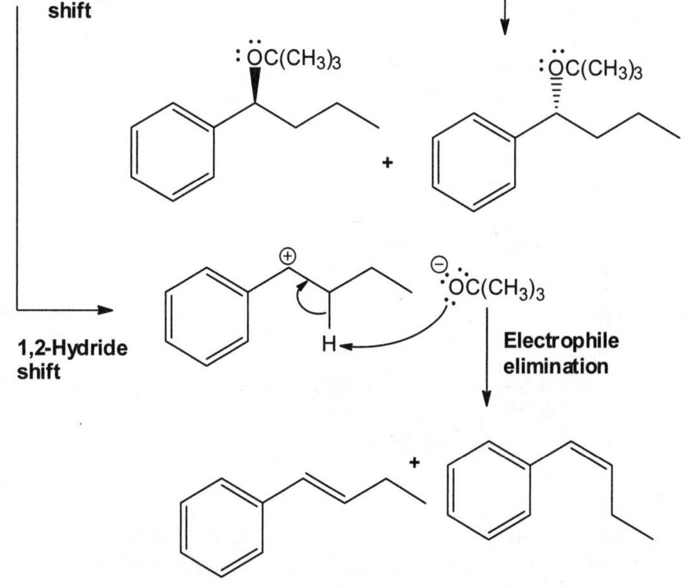

(g) S_N1 is favored. The attacking species is a weak nucleophile and weak base, favoring S_N1 and E1. The concentration does not matter. The leaving group is excellent, which favors S_N1 and E1. The carbon is benzylic, which favors S_N1 and E1. The solvent is protic, which favors S_N1 and E1. The reaction is not heated, so we can expect this to favor S_N1, but E1 products will be mixed in. Notice that the substitution products form a racemic mixture because the ethanol nucleophile can attack the carbocation from either side of the plane of the C^+. See the mechanism and table on the next page.

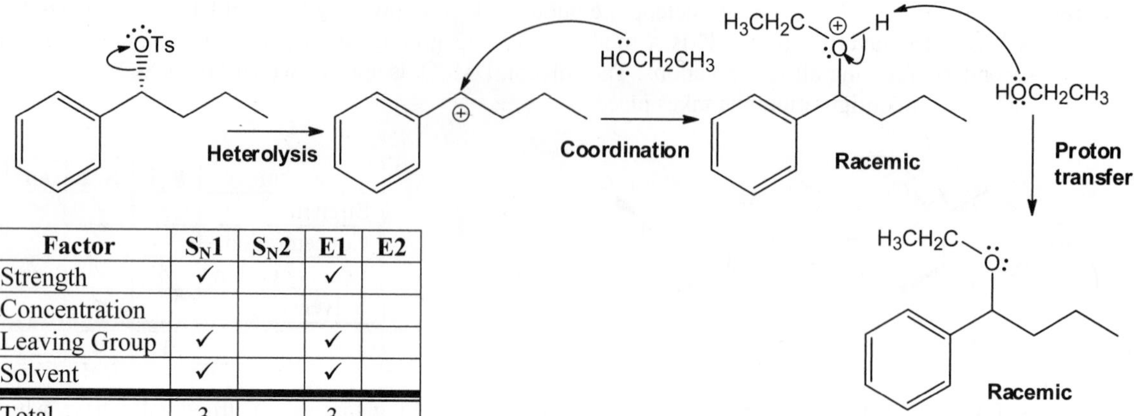

Factor	S_N1	S_N2	E1	E2
Strength	✓		✓	
Concentration				
Leaving Group	✓		✓	
Solvent	✓		✓	
Total	3		3	

Problem 9.54

Think

What must be true about the orientation of the leaving group and the adjacent H atom to be eliminated in an E2? Are there H atoms available in this orientation to the Cl leaving group?

Solve

The only H atoms that could be eliminated in an E2 reaction are highlighted below, as they are on a C atom adjacent to that bonded to the leaving group. However, neither of them can be anti to the leaving group. The dihedral angle that the top C–H bond forms with the C–Cl bond is ~120°, but to be anti, that dihedral angle must be 180°.

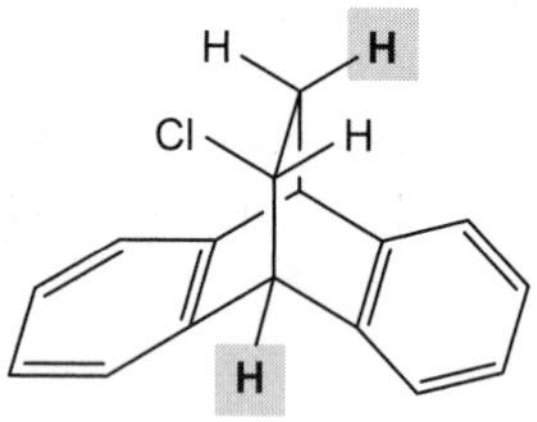

Problem 9.55

Think

Is there a good leaving group present without the acid? In the presence of an acid, does a good leaving group form? Did the product result via substitution or elimination? Was the mechanism unimolecular or bimolecular? How many times did this reaction occur?

Solve

The mechanism consists of two S_N2 steps and some proton transfer steps. The reactions must be substitution, and because the leaving group is on a primary carbon, the S_N1 reaction is precluded, leaving only S_N2.
Step 1: The ether is protonated.
Step 2: Ethanol opens the ring (first S_N2).
Steps 3 and 4: Solvent-mediated proton transfers.
Step 5: Ethanol displaces water (second S_N2).
Step 6: A final proton transfer completes the reaction.
See the mechanism on the next page.

$CH_3CH_2\ddot{O}H$ Proton transfer S_N2 Proton transfer

CH_3CH_2OH Proton transfer $CH_3CH_2\ddot{O}H$ $+ H_2O$ S_N2

$H_2\ddot{O}$ Proton transfer

Problem 9.56

Think

Consider the C–L first. Is there a suitable leaving group? To what type of carbon is the leaving group attached? Which mechanisms does this favor or rule out? What is the attacking species? Strong base, weak base, strong nucleophile, weak nucleophile, sterically bulky? What is the solvent? Which reaction does each of the factors favor?

Solve

(a) Reaction conditions favor S_N2 and E2 (leaving group is on a secondary C, so we must consider all four mechanisms; attacking species is a strong nucleophile and a strong base; high concentration of the attacking species; aprotic solvent; moderate leaving group). There is a single S_N2 product, but four different E2 products we should expect in significant abundance. The different locations of the double bonds are an outcome of the H atoms on C2 and C4 that can be removed, and both alkenes are disubstituted. The cis/trans isomers are possible because on both C2 and C4 there are two different H atoms that can attain an anti-coplanar conformation with the leaving group. The E2 reaction mechanism is shown on the next page.

Factor	S_N1	S_N2	E1	E2
Strength		✓		✓
Concentration		✓		✓
Leaving Group	✓	✓	✓	✓
Solvent		✓		✓
Total	1	4	1	4

S_N2

$CH_3CH_2\ddot{O}:^-$ $:\ddot{Cl}:$ S_N2 $:\ddot{O}CH_2CH_3$ $+ :\ddot{Cl}:^-$

E2

(b) S_N2 reactions are ruled out because the leaving group is on a tertiary C. A strong nucleophile/base favors S_N2 and E2. High concentrations of the attacking species favor S_N2 and E2. The Cl leaving group favors S_N1, S_N2, E1, and E2. Protic solvent favors S_N1 and E1. So overall, the major products should be E2. *E* and *Z* isomers are produced because both H atoms on the C adjacent to the C–Cl can attain the anti-coplanar conformation.

E2

Factor	S_N1	S_N2	E1	E2
Strength		✓		✓
Concentration		✓		✓
Leaving Group	✓	✓	✓	✓
Solvent	✓		✓	
Total	2		2	**3**

(c) The substrate has the leaving group on a secondary carbon, so all four mechanisms must be considered. H_3PO_4 or H_2O would be the nucleophile/base, which is weak, so S_N1 and E1 are favored. Concentration of the attacking species is not considered because it is weak. The leaving group, water, has a very good leaving group ability, favoring S_N1 and E1. The leaving group is on a secondary allylic C, which can stabilize a carbocation by resonance, thus favoring S_N1 and E1. And the solvent is protic, favoring S_N1 and E1. So overall, S_N1 and E1 are favored, but heat tips the balance in favor of E1.

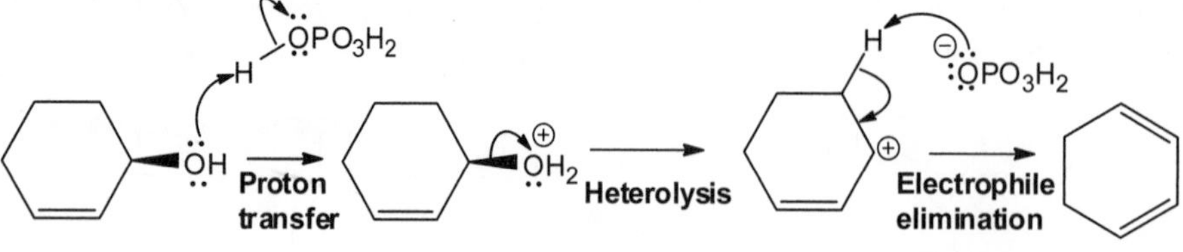

Factor	S_N1	S_N2	E1	E2
Strength	✓		✓	
Concentration				
Leaving Group	✓		✓	
Solvent	✓		✓	
Total	**3**		**3**	

(d) The substrate has the leaving group on a secondary C, so all four mechanisms must be considered. The attacking species is $(CH_3)_3CO^-$, which favors E2. The concentration of the attacking species is high, which favors E2. The leaving group is Cl^-, which favors all four mechanisms. The leaving group is on a secondary C, which favors all four mechanisms. And the solvent is aprotic, which favors S_N2 and E2. Overall, E2 is favored. The proton that is removed is one that can achieve an anti-coplanar conformation with the leaving group. Normally, the more highly substituted alkene product would be favored, but the bulkiness of the base directs deprotonation to the more accessible side.

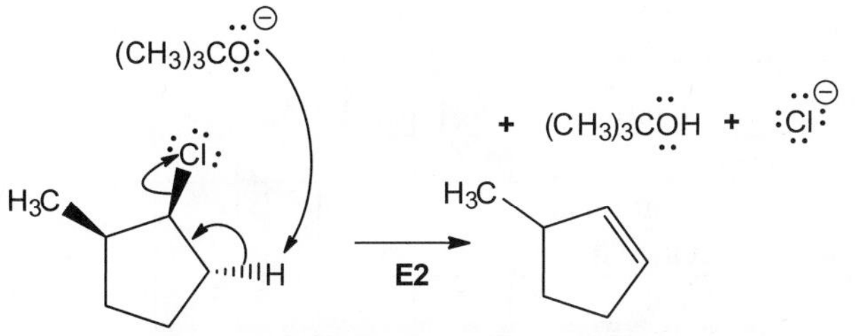

Factor	S_N1	S_N2	E1	E2
Strength				✓
Concentration				✓
Leaving Group	✓	✓	✓	✓
Solvent		✓		✓
Total	1	2	1	**4**

(e) The only feasible leaving group in the substrate is TfO^-, which is on a primary carbon, so S_N1 and E1 reactions are not considered. The attacking species is a strong nucleophile but a weak base, favoring S_N2, but not E2. The high concentration of the attacking species also favors S_N2. The leaving group is excellent, favoring S_N1 and E1, but those reactions aren't considered. The leaving group is on a primary C, which favors both S_N2 and E2. The solvent is protic, which would normally favor S_N1 and E1, but those are not considered. So, overall, S_N2 is the winner. Since there are no stereocenters, there is no stereochemistry to consider.

Factor	S_N1	S_N2	E1	E2
Strength		✓		
Concentration		✓		
Leaving Group	✓		✓	
Solvent	✓		✓	
Total		2		

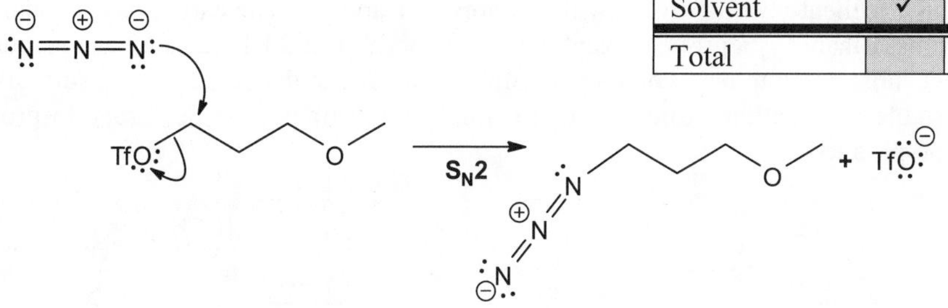

(f) There is no H atom on the C atom adjacent to that with the leaving group, so we can rule out E1 and E2. The leaving group is on a primary carbon, which would normally have us not consider the S_N1 or E1 mechanisms, but that primary C atom is at the benzylic position, which can accommodate a positive charge quite well. So we must consider S_N1 as well. The attacking species is ethanol, which is a weak nucleophile, favoring S_N1 and E1. We ignore the concentration factor because the nucleophile is weak. The leaving group is MsO^-, which is excellent, favoring S_N1 and E1. The leaving group is on a benzylic C atom, which favors S_N1 and E1. And the solvent is ethanol, which is protic and favors S_N1 and E1. So overall, this is an S_N1 reaction. See the mechanism and table on the next page.

Factor	S_N1	S_N2	E1	E2
Strength	✓			
Concentration				
Leaving Group	✓			
Solvent	✓			
Total	3			

Problem 9.57

Think

Consider the C–L first. Is there a suitable leaving group? To what type of carbon is the leaving group attached? Which mechanisms does this favor or rule out? What is the attacking species? Strong base, weak base, strong nucleophile, weak nucleophile, sterically bulky? What is the solvent? Which reaction does each of the factors favor?

Solve

(a) S_N1 is favored. CH_3CH_2OH is a weak nucleophile and weak base, favoring S_N1 and E1. Concentration does not matter. TsO^- is an excellent leaving group, which favors S_N1 and E1. The carbon is secondary, which favors all four reactions. Ethanol is a protic solvent, which favors S_N1 and E1. Because the reaction is not heated, we can expect substitution to be favored over elimination. Notice that a 1,2-methyl shift takes place to produce a more stable carbocation. Note also that a final proton transfer is necessary to produce the uncharged substitution product.

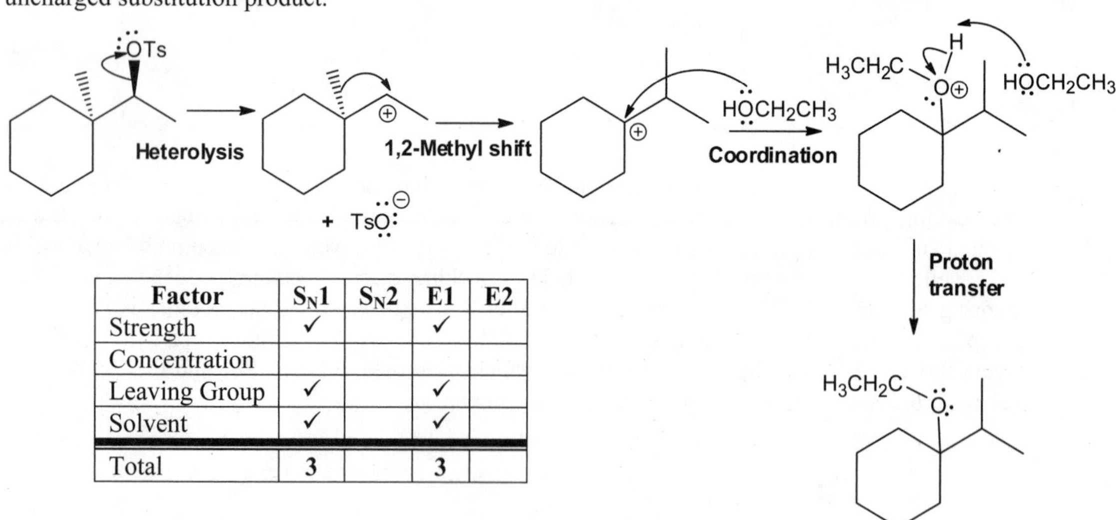

Factor	S_N1	S_N2	E1	E2
Strength	✓		✓	
Concentration				
Leaving Group	✓		✓	
Solvent	✓		✓	
Total	3		3	

(b) S_N2 is favored. NC^- is a strong nucleophile but a weak base, favoring S_N2 and E1. A high concentration of NC^- favors S_N2. TsO^- is an excellent leaving group, favoring S_N1 and E1. The carbon is secondary, favoring all four reactions. DMF is an aprotic solvent, favoring S_N2 and E2. Notice that the S_N2 reaction leads to inversion of stereochemistry.

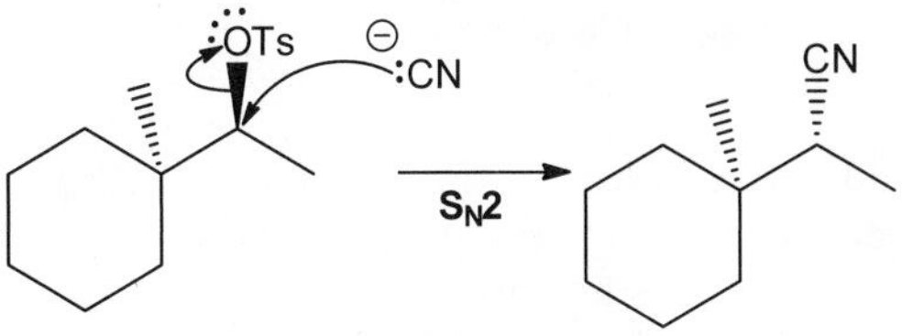

Factor	S_N1	S_N2	E1	E2
Strength		✓	✓	
Concentration		✓		
Leaving Group	✓		✓	
Solvent		✓		✓
Total	1	**3**	2	1

(c) E2 is favored. The strong bulky base favors E2, as does its high concentration. TsO^- is an excellent leaving group, which favors S_N1 and E1. The carbon is secondary, which favors all four reactions. DMSO is an aprotic solvent, favoring S_N2 and E2. Notice that the terminal CH_3 has the only H atom that can be eliminated.

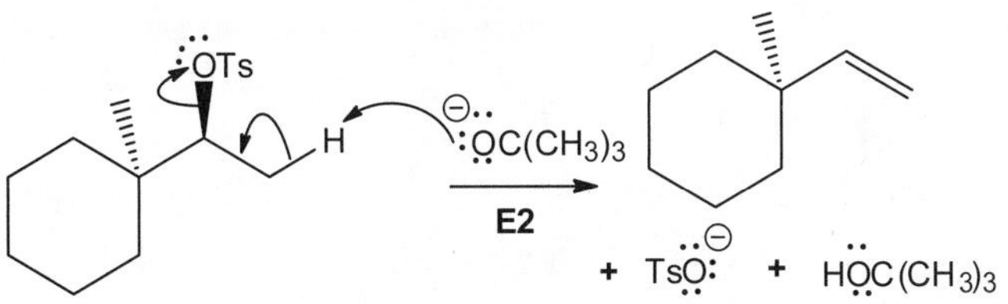

Factor	S_N1	S_N2	E1	E2
Strength				✓
Concentration				✓
Leaving Group	✓		✓	
Solvent		✓		✓
Total	1	1	1	**3**

(d) E1 is favored. CO_3^{2-} is a weak base and weak nucleophile, favoring S_N1 and E1. Concentration is ignored. TsO^- is an excellent leaving group, favoring S_N1 and E1. The carbon is secondary, favoring all four reactions. Ethanol is a protic solvent, favoring S_N1 and E1. The reaction is heated, which favors elimination over substitution. A 1,2-methyl shift takes place, producing a more stable carbocation. Also, the H atom that is eliminated is the one that gives the most highly substituted alkene.

Factor	S_N1	S_N2	E1	E2
Strength	✓		✓	
Concentration				
Leaving Group	✓		✓	
Solvent	✓		✓	
Total	**3**		**3**	

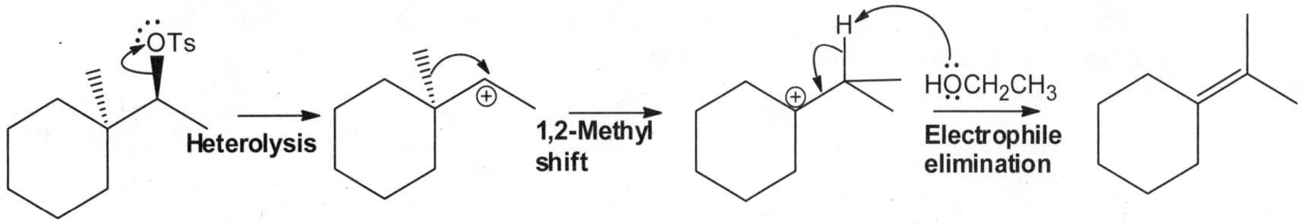

(e) E1 is favored. CH_3CH_2OH is a weak nucleophile and a weak base, favoring S_N1 and E1, and concentration is ignored. Cl^- is a moderate leaving group, favoring all four reactions. The carbon is benzylic, favoring S_N1 and E1. Ethanol is a protic solvent, favoring S_N1 and E1. The reaction is heated to favor elimination over substitution. See the mechanism on the next page.

Factor	S_N1	S_N2	E1	E2
Strength	✓		✓	
Concentration				
Leaving Group	✓	✓	✓	✓
Solvent	✓		✓	
Total	**3**	1	**3**	1

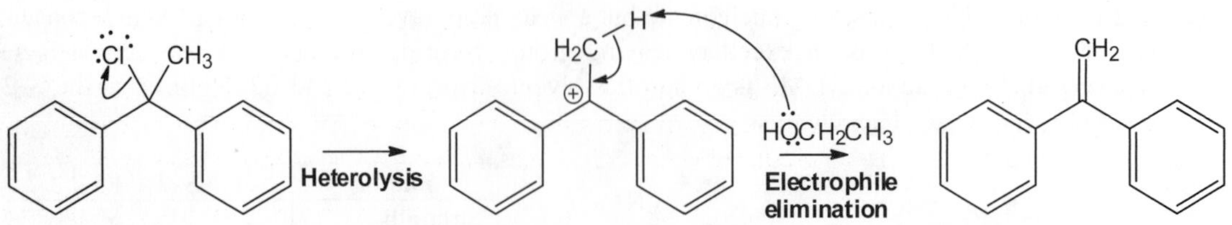

(f) E2 is favored. $CH_3CH_2O^-$ is a strong nucleophile and strong base, favoring S_N2 and E2. Concentration is high, favoring S_N2 and E2. The leaving group is moderate, favoring all four reactions. The carbon attached to the leaving group is benzylic, favoring S_N1 and E1. DMSO is an aprotic solvent, favoring S_N2 and E2. Heat favors elimination over substitution. Notice that there is only one H atom that can be eliminated, and because E2 is the mechanism, it favors the anti-coplanar conformation. This gives rise to one configuration about the C=C in the alkene product, exclusively.

Factor	S_N1	S_N2	E1	E2
Strength		✓		✓
Concentration		✓		✓
Leaving Group	✓	✓	✓	✓
Solvent		✓		✓
Total	1	4	1	4

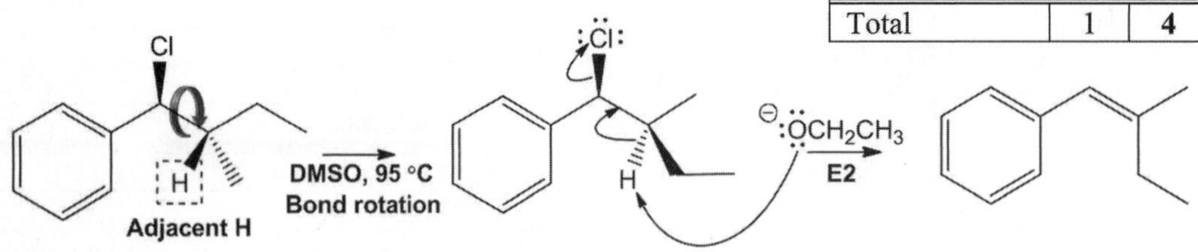

Problem 9.58

Think

Consider the C–L first. Is there a suitable leaving group? To what type of carbon is the leaving group attached? Which mechanisms does this favor or rule out? What is the attacking species? Strong base, weak base, strong nucleophile, weak nucleophile, sterically bulky? What is the solvent? Which reaction does each of the factors favor?

Solve

(a) There are two sites for substitution or elimination because there are two separate Cl leaving groups. The conditions favor S_N2, which governs the site of reaction. Br^- is a strong nucleophile but a weak base, which favors S_N2 and E1. The high concentration also favors S_N2. Also, DMSO is an aprotic solvent, which favors S_N2 and E2. The leaving group is moderate, which favors all four reactions. Because S_N2 is favored, the nucleophile will attack the least sterically hindered site, which is at the secondary carbon. The tertiary C–Cl does not undergo S_N2 and, therefore, is unchanged. Notice that stereochemistry is important here, and the S_N2 reaction leads to inversion at the electrophilic carbon.

Factor	S_N1	S_N2	E1	E2
Strength		✓		
Concentration		✓		✓
Leaving Group	✓	✓	✓	✓
Solvent		✓		✓
Total	1	4	1	3

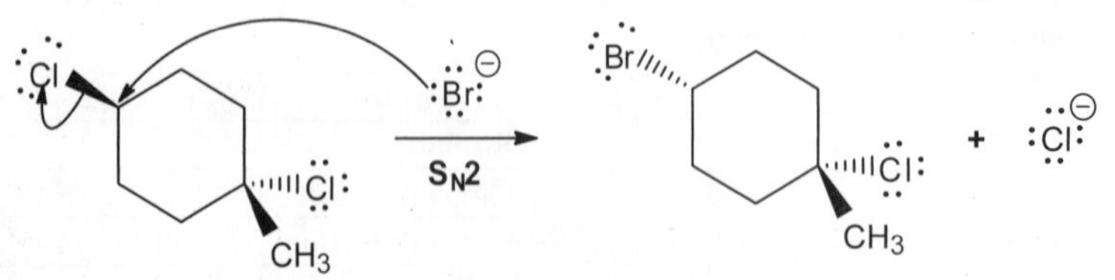

(b) This reaction favors E2. The attacking species is a strong, bulky base in high concentration. Also, the solvent is aprotic. Only the Cl and F are feasible leaving groups, because the Br atom is bonded to an sp^2-hybridized carbon. Of those, the E2 reaction will favor the Cl leaving group because it has better leaving group ability. Notice that there are two H atoms that can be eliminated from the adjacent C atom, so both cis and trans isomers can be produced from an anti-coplanar conformation of the substrate.

Factor	S_N1	S_N2	E1	E2
Strength	✓			✓
Concentration				✓
Leaving Group	✓	✓	✓	✓
Solvent		✓		✓
Total	2	2	1	**4**

Br Cl F H :OC(CH3)3 DMF **E2** Br F + :Cl: + HOC(CH3)3 Br F

Problem 9.59

Think

Are any proton transfer reactions possible? Consider the C–L first. Is there a suitable leaving group? To what type of carbon is the leaving group attached? Which mechanisms does this favor or rule out? What is the attacking species? Are intramolecular reactions possible?

Solve

Both reactions favor the S_N2 mechanism, because the nucleophile is strong and in high concentrations, and the leaving group is on a primary carbon. The first compound is produced by an intramolecular S_N2 reaction after a fast deprotonation of the phenolic OH.

HO: H O Cl **Proton transfer** O Cl $\mathbf{S_N2}$ O + H_2O + :Cl:

Chemical formula: C_8H_8O

The second compound has difficulty undergoing such an intramolecular S_N2 reaction because of the ring strain the product would possess. Instead, an intermolecular S_N2 reaction takes place.

HO Cl HO: $\mathbf{S_N2}$ HO OH + :Cl:

Chemical formula: $C_8H_{10}O_2$

Problem 9.60

Think

Consider the C–L first. Is there a suitable leaving group? In the presence of HBr, is HO able to be made into a suitable leaving group? To what type of carbon is the leaving group attached? What type of carbocation is formed? Is resonance possible?

Solve

To produce products of the given formula, substitution must take place. The reaction conditions favor S_N1 over S_N2. Under the acidic conditions, water is the leaving group, which favors S_N1 over S_N2. The leaving group is on an allylic carbon, which makes a resonance-delocalized carbocation, thus favoring S_N1 over S_N2. And the solvent is an alcohol (the reactant) or water (HBr's solvent), both of which are protic and favor S_N1 over S_N2. After the leaving group leaves, the resonance-delocalized carbocation can be attacked by Br^- at two different locations, leading to the two different products.

H3C OH H Br: Proton transfer H3C OH2 Heterolysis H3C Resonance H3C Br: Coordination H3C Br: H3C Br:

Problem 9.61

Think

What was the major mechanism in Problem 9.60? Which mechanism shares the same rate-determining step? How does heating up the reaction change the major mechanism?

Solve

S_N1 and E1 reactions are favored by the same factors. Heating the reaction mixture favors elimination over substitution, which is why the products given are elimination products. For the same reasons that S_N1 is favored in the previous problem, E1 is favored under these conditions. Once again, a resonance-delocalized carbocation allows for reaction at two different sites. As shown below, the products that are given in the problem are the result of elimination from one of the carbocation's resonance structures. *Note*: H′ and H″ are used to distinguish the mechanisms that form the two isomeric alkenes, as are the dash and solid mechanism arrows.

H3C OH H Br: Proton transfer H3C OH2 Heterolysis H3C Resonance Br: H' Br: H" C H2 Electrophile elimination, from H' Electrophile elimination, from H" + HBr H3C H2C

Problem 9.62

Think

What is the mechanism that produced **A**, **B**, and the product shown? What do you notice about the methyl group attached to the benzylic carbon in the starting material and product? What mechanism involves inversion of stereochemistry? Are proton transfer reactions involved?

Solve

The first reaction to produce **A** is an S_N2 reaction. The secondary benzylic carbon in the presence of a strong nucleophile in an aprotic solvent favors S_N2 with inversion of stereochemistry. The second reaction to produce **B** first involves a proton transfer to deprotonate the alcohol using NaH, which produces a strongly nucleophilic alkoxide anion (RO^-), and then an S_N2 follows. The third reaction to produce the final product first involves a proton transfer to deprotonate the alkyne using *n*-butyl lithium, which produces a strongly nucleophilic alkynide anion ($RC{\equiv}C{:}^-$), and then an S_N2 follows. The reaction scheme is given below.

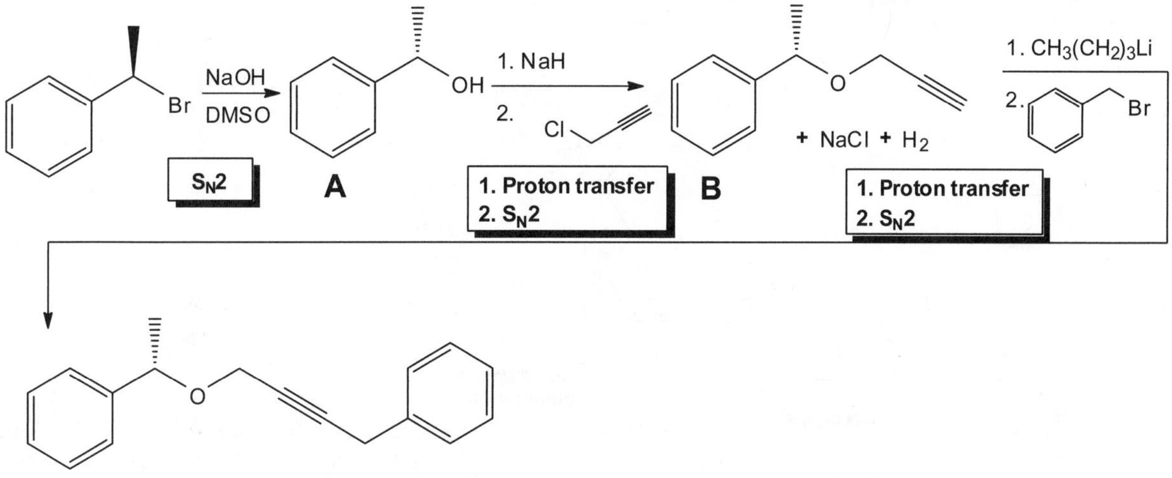

NaOH / DMSO
S_N2
A
1. NaH
2. Cl
1. Proton transfer
2. S_N2
B
+ NaCl + H_2
1. $CH_3(CH_2)_3Li$
2. Br
1. Proton transfer
2. S_N2

Problem 9.63

Think

What reaction occurs when acetic acid is in the presence of a strong base? What then forms? Consider the C–L in the substrate. Is there a suitable leaving group? To what type of carbon is the leaving group attached? Which mechanisms does this favor or rule out? What is the attacking species?

Solve

Acetic acid in the presence of a strong base, HO^- undergoes a fast deprotonation. The carboxylate anion is then formed, which produces a strongly nucleophilic carboxylate anion, RCO_2^-. Elimination reactions are ruled out as there is no other C or H atom to form a C=C. The reaction conditions favor an S_N2 reaction, as shown below.

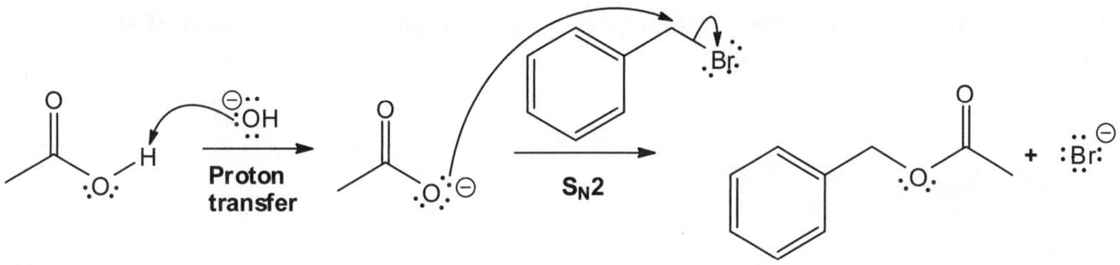

Proton transfer
S_N2
+ :Br:⊖

Problem 9.64

Think

What type of attacking species is $(CH_3)_3CO^-$? Is it bulky? What type of solvent is DMF? Ethanol? Which mechanism does each one favor? Why? How does varying the concentration of the reactant alter the mechanism? Are there any carbocation intermediates with resonance delocalization of the charge?

Solve

The first reaction favors an E2 (strong, bulky base; high concentration; aprotic solvent DMF), whereas the second favors an E1 (strong base; low concentration; excellent leaving group; protic solvent; heat). The carbocation that is formed in the E1 mechanism is allylic, so the positive charge is delocalized by resonance over two carbon atoms. This allows the D to be eliminated in the second step.

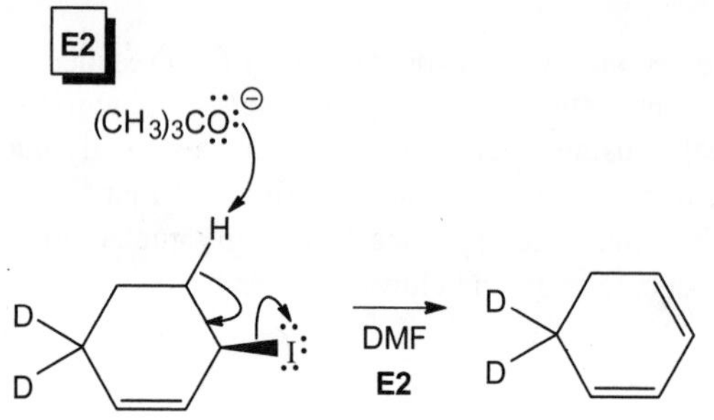

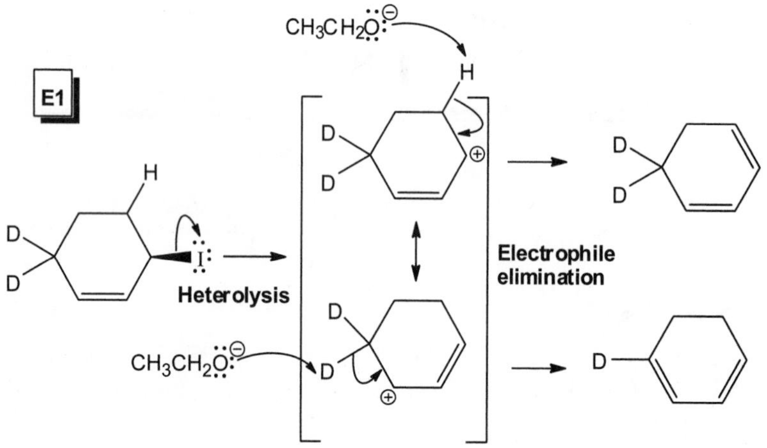

E2

$(CH_3)_3CO^-$

DMF

E2

E1

$CH_3CH_2O^-$

Heterolysis

Electrophile elimination

Problem 9.65

Think

Did substitution or elimination occur? Did a stereospecific product result or a mixture of products? What mechanisms do the attacking species, concentration, and solvent favor?

Solve

(a) Because the product's formula is $C_6H_{11}OCl$, substitution must have taken place. The conditions with a high concentration of a strong nucleophile and an aprotic solvent favor S_N2. So we need to choose the appropriate stereochemical configuration in the substrate to account for the inversion of stereochemistry that takes place in an S_N2 reaction.

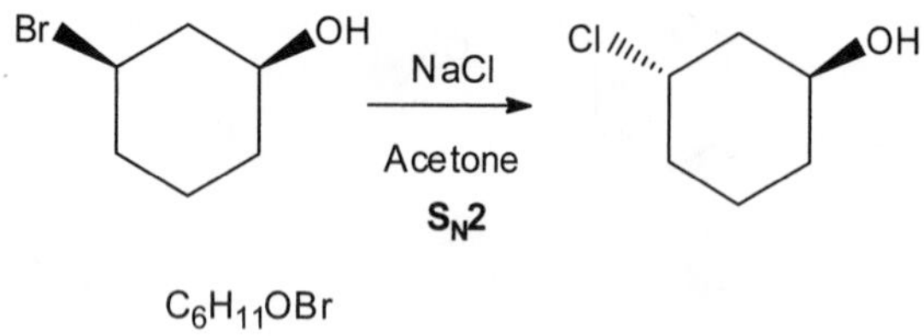

NaCl

Acetone

S_N2

$C_6H_{11}OBr$

(b) The product given is the result of elimination, recognized by the fact that the attacking species does not end up in the product. The reaction conditions favor E2, given that the attacking species is a strong, bulky base in high concentration, and the solvent is aprotic. Because there are E/Z isomers possible for the product, we have to choose the stereochemistry appropriately for the substrate. That stereochemistry is chosen so that the H atom and the leaving group (Cl in this case) that are eliminated are in an anti-coplanar conformation prior to elimination.

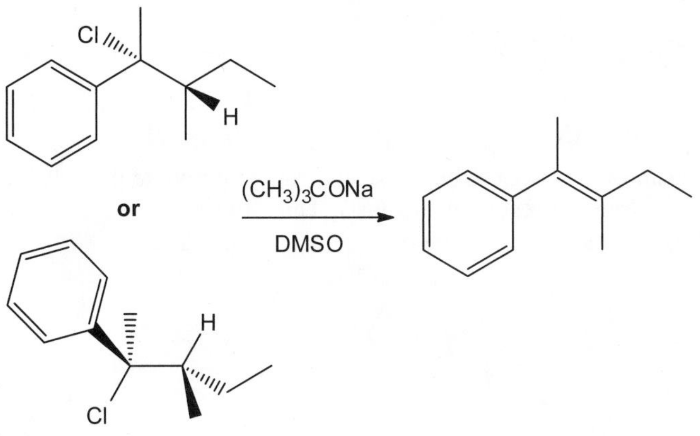

Cl H or H Cl $C_{12}H_{17}Cl$ $(CH_3)_3CONa$ DMSO

Problem 9.66

Think

Did substitution or elimination occur? Did a stereospecific product result or a mixture of products? Does the attacking species match the predicted reaction mechanism? Did proton transfer reactions occur? Is there more than one reactive site?

Solve

(a) The intended leaving group is the OH after protonation, but protonation of the ether oxygen will also produce a CH_3OH leaving group. Thus reaction can take place at the ether site as well, producing an unwanted product.

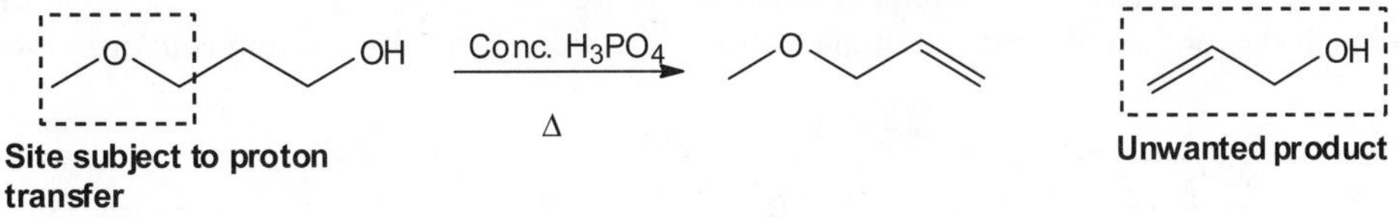

O OH Conc. H_3PO_4 Δ O OH

Site subject to proton transfer

Unwanted product

(b) The reaction conditions favor S_N2 (strong nucleophile, high concentration, aprotic solvent), which causes inversion of stereochemistry. The product that is given shows retention of configuration instead.

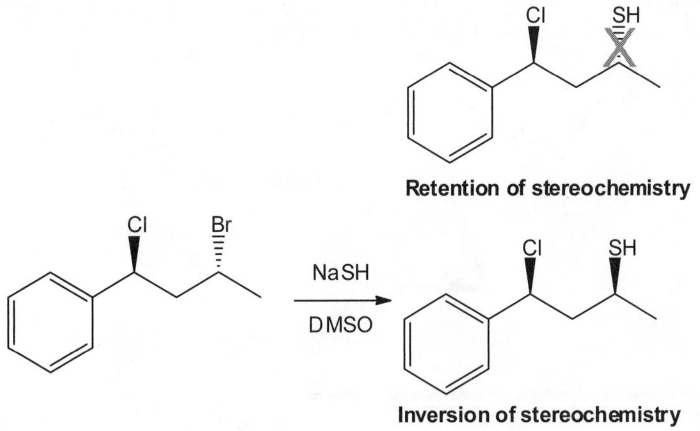

Cl SH

Retention of stereochemistry

Cl Br NaSH DMSO Cl SH

Inversion of stereochemistry

(c) The NaH base deprotonates the terminal carbon to make a negatively charged carbon nucleophile. That nucleophile is intended to react in a substitution involving the chloropropanoic acid as the substrate. However, that negatively charged carbon in the nucleophile is strongly basic and will deprotonate the carboxylic acid instead (remember, proton transfer reactions are fast). The alkyne is then reformed (see below).

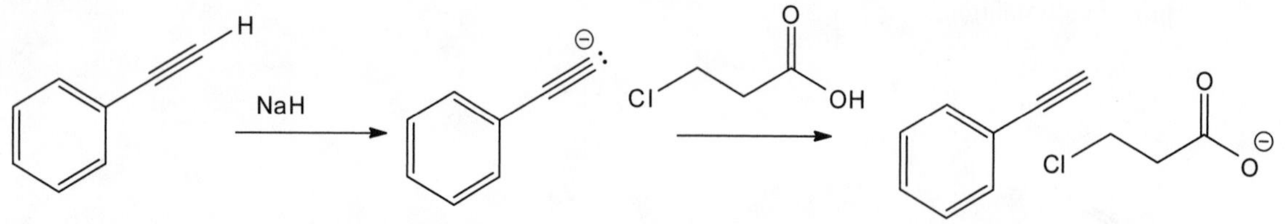

(d) The intended nucleophile, an alkoxide anion, is also a strong base and can deprotonate the solvent, ethanol, to make $CH_3CH_2O^-$ (remember, proton transfer reactions are fast). Thus, there will be a mixture of two different strong nucleophiles present. This leads to a mixture of two different products.

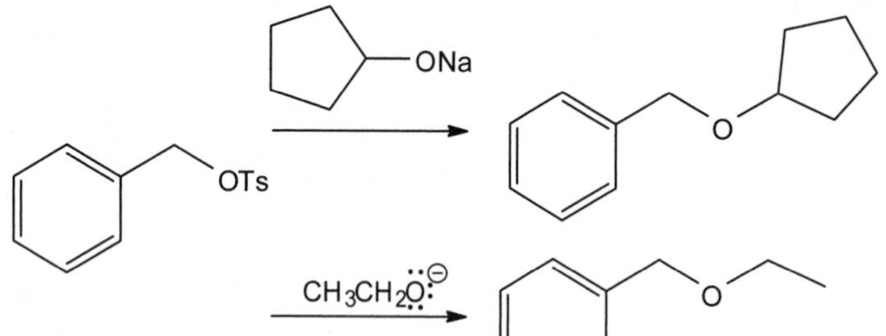

Problem 9.67

Think

What determines whether a reaction is reversible or irreversible? Complete each reaction by drawing the products. Draw a free energy diagram. Does charge stability favor the reactant or product side? A little or a lot?

Solve

(a) Br^- has greater charge stability compared to CH_3S^- because Br is a larger atom and the charge is less concentrated. This makes $\Delta G°_{rxn}$ substantially negative. Therefore, $\Delta G^{\circ\neq}_{forward} << \Delta G^{\circ\neq}_{reverse}$, so the reverse reaction is much slower than the reaction in the forward direction. Thus, the reaction is *irreversible*.

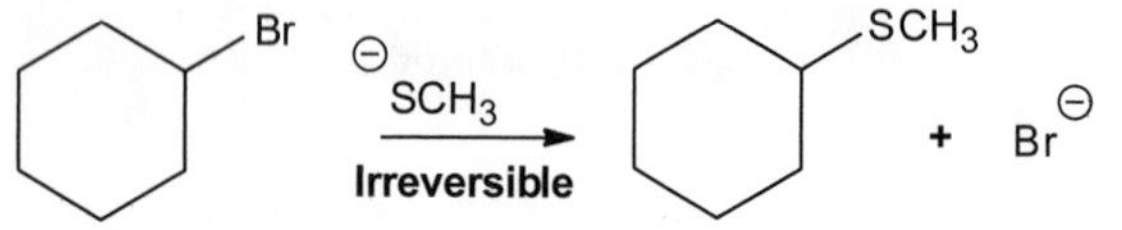

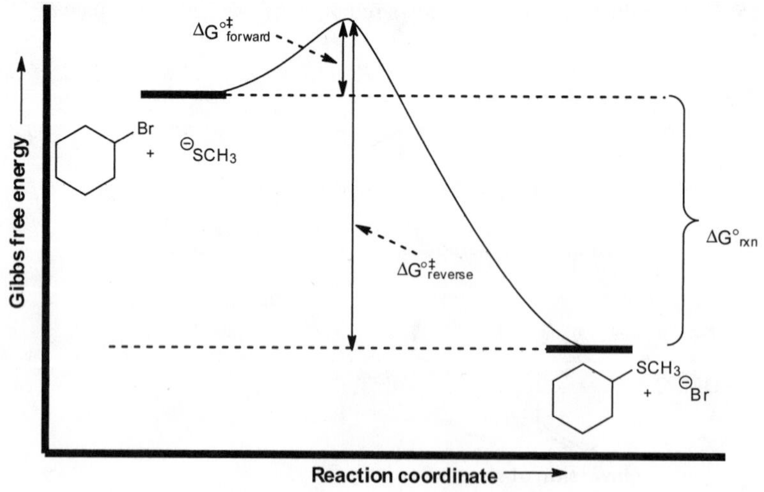

(b) $\Delta G°_{rxn}$ is not substantially negative. To the contrary, Cl^- is slightly less stable than Br^- due to the smaller atomic size of Cl. This makes the reaction faster in the reverse direction than in the forward direction under standard conditions ($\Delta G^{o\neq}_{forward} > \Delta G^{o\neq}_{reverse}$), so the reaction is *reversible*.

Cl $\overset{\ominus}{}$Br ⇌ Br + $Cl^{\ominus}$

Reversible

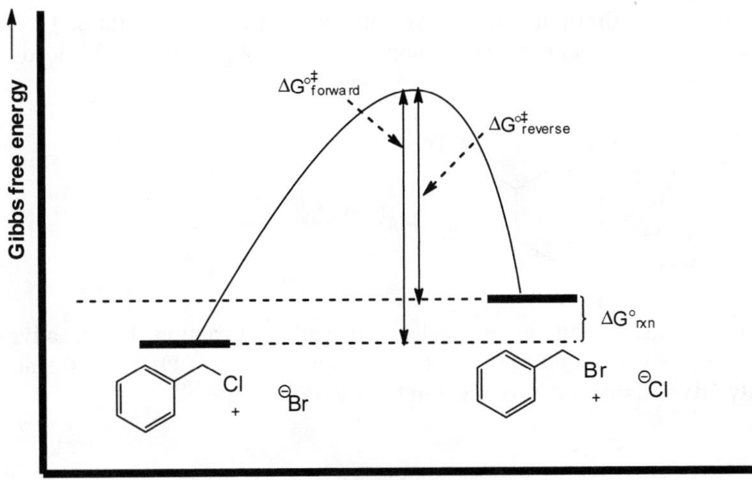

(c) TsO^- has greater charge stability compared to NC^-. This can be evaluated by consideration of the inductive and resonance effects on TsO^-. TsO^- has electron-withdrawing groups and the negative charge is highly delocalized over the structure of the compound. Thus, TsO^- is a much more stable anion, making $\Delta G°_{rxn}$ substantially negative. Therefore, $\Delta G^{o\ddagger}_{forward} << \Delta G^{o\ddagger}_{reverse}$, so the reverse reaction is much slower than the reaction in the forward direction. Thus the reaction is *irreversible*. The reaction free energy diagram is similar to the free energy diagram in **(a)**.

OTs $\xrightarrow{\ominus CN}$ CN + $\overset{\ominus}{}$OTs

Irreversible

Problem 9.68

Think

If NaBr and NaCl are insoluble in acetone, how does that affect the concentration of Br^- and Cl^- in solution? If the ions are not present in the reaction mixture, is the reverse reaction possible?

Solve

Because NaBr and NaCl are insoluble in acetone, they precipitate out of the reaction mixture. This essentially makes their concentration zero or negligible. If the ion is not present in solution, the reverse reaction is not possible. Thus, the reaction is irreversible.

R—Cl **or** R—Br $\xrightarrow[\text{acetone}]{\text{NaI}}$ R—I + [$Cl^{\ominus}$ **or** $Br^{\ominus}$] **Insoluble in acetone and precipitates out of the solution**

Irreversible

Problem 9.69

Think

Write out the complete reactions. How do you evaluate charge stability? What factors stabilize/destabilize an anion? How does base/acid strength factor into charge stability? Consider octet, formal charge location, and resonance.

Solve

(a) In this proton transfer reaction, the reactants and products both have the negative charge on O. No resonance delocalization of the charge is possible, but there are some minor effects on charge stability from the alkyl group. Thus, the reactants and products have similar charge stability, making the reaction reversible. This is reflected by the equilibrium constant we can compute from pK_a values. The reactant is hydroxide (H_2O; pK_a = 15.7) and the product is cyclohexanolate (2° OH; pK_a = ~16.5). This leads to a K_{eq} = ~0.16.

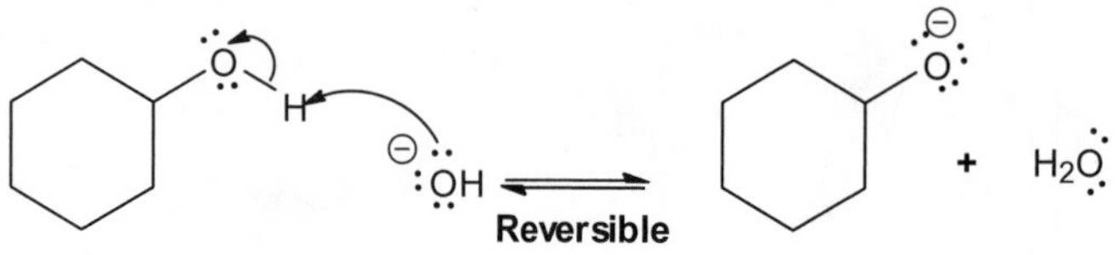

Reversible

(b) An sp^3-hybridized carbanion is the nucleophile in this nucleophilic addition reaction. The negative charge is much more stable on O in the products, due to the greater electronegativity of O. Thus, the nucleophilic addition products are heavily favored, making this reaction *irreversible*.

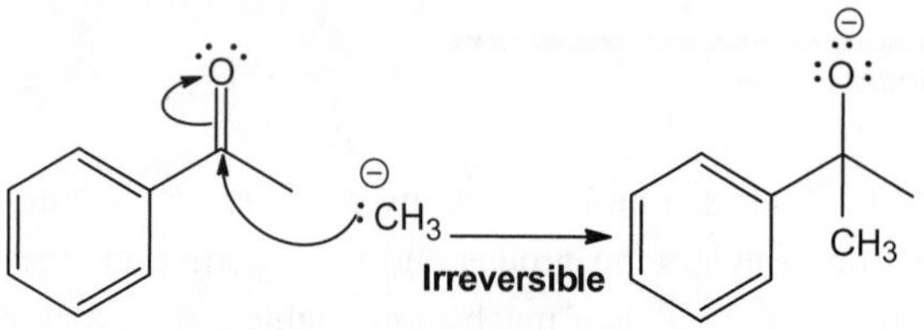

Irreversible

(c) This elementary step is a heterolysis. The structure on the left has all atoms with an octet and the product on the right has the C atom with an incomplete octet. This leads to a positive $\Delta G°_{rxn}$ and the reaction is thus said to be *reversible*.

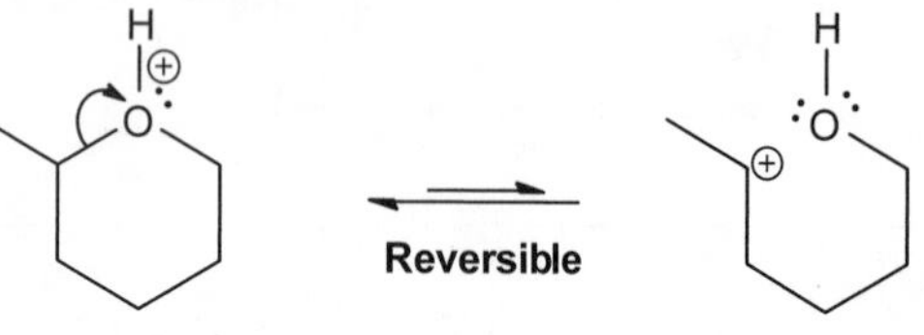

Reversible

Problem 9.70

Think

This is an acid-catalyzed E1 reaction mechanism (Section 8.6). What step must occur before the heterolysis step? Is OH a good leaving group? What species are incompatible in acidic conditions?

Solve

D-Fructose to A:

Step 1: Protonate OH to form water, a good leaving group.
Step 2: Heterolysis; water leaves and forms a carbocation.
Step 3: Adjacent H atom is removed by the HSO_4^- conjugate base to reform the acid catalyst.

D-Fructose

Proton transfer

$-H_2O$ Heterolysis

Electrophile elimination

$+ H_2SO_4$

A

A to B:

Step 1: Protonate OH to form water, a good leaving group.
Step 2: Heterolysis; water leaves and forms a carbocation.
Resonance: Nonbonding electrons from the oxygen fold down to form another resonance structure where all atoms have an octet.
Step 3: Adjacent H atom is removed by the HSO_4^- conjugate base to reform the acid catalyst.

A

Proton transfer

$-H_2O$ Heterolysis

Resonance

Electrophile elimination

$+ H_2SO_4$

B

B to HMF:
Step 1: Protonate OH to form water, a good leaving group.
Step 2: Heterolysis; water leaves and forms a carbocation.
Step 3: Adjacent H atom is removed by the HSO_4^- conjugate base to reform the acid catalyst.

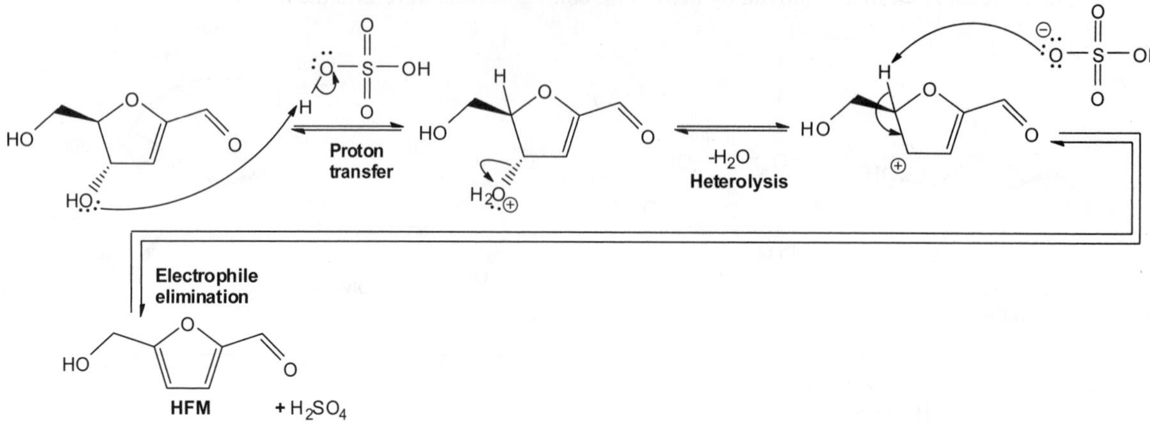

Proton transfer

$-H_2O$
Heterolysis

Electrophile elimination

HFM + H_2SO_4

Problem 9.71

Think

This is an S_N1 mechanism with a weak protic nucleophile in acidic conditions. What step must occur before the heterolysis step? Is OH a good leaving group? What species are incompatible in acidic conditions? In considering stereochemistry of the product, what does an S_N1 mechanism do to the configuration of the carbon stereocenter initially attached to the leaving group?

Solve

The mechanism is drawn and written out below.
Step 1: Protonate OH to form water, a good leaving group.
Step 2: Heterolysis; water leaves and forms a carbocation.
Step 3: Coordination; CH_3CH_2OH nucleophile can attack from above or below the planar C+.
Step 4: Proton transfer. Deprotonation removes the positive charge from oxygen.

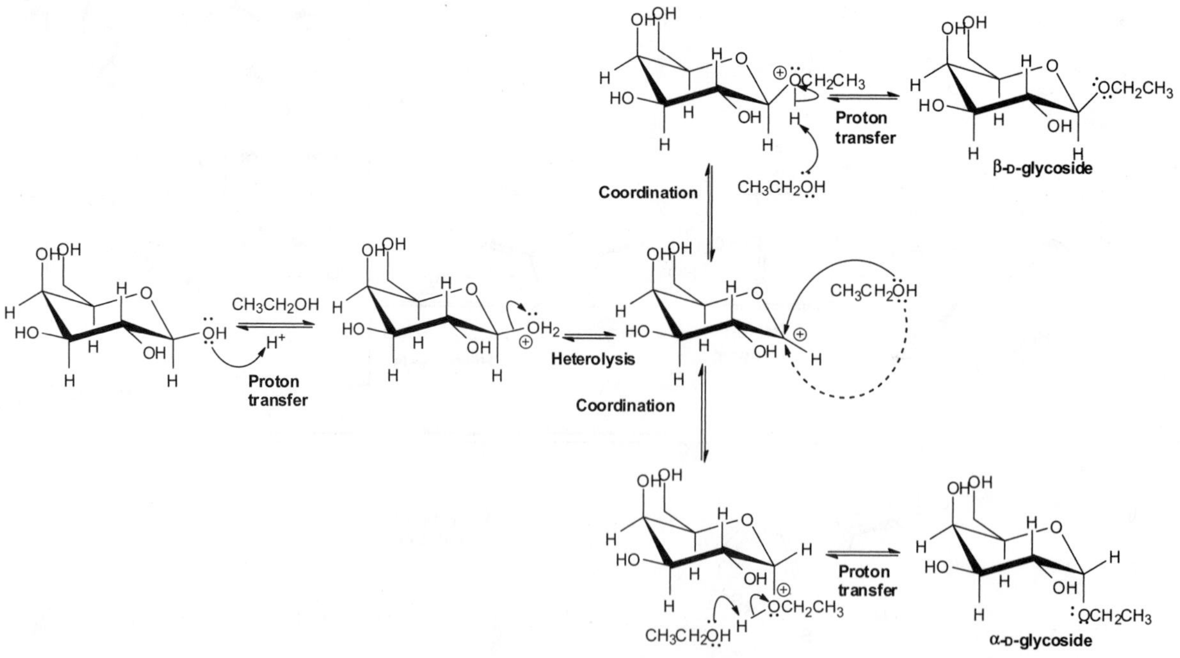

CH_3CH_2OH, H^+
Proton transfer

Heterolysis

Coordination

Proton transfer

β-D-glycoside

α-D-glycoside

Problem 9.72

Think

Can you identify the acetal group, $C(OR)_2$? Does one of those R groups belong to a sugar that is different from the sugar to which C belongs? Are the groups at the glycoside linkage axial (α) or equatorial (β)? The mechanism for the acid-catalyzed hydrolysis of lactose is essentially the reverse of the mechanism shown in Equation 9-54 except the roles of water and alcohol are reversed. What is the nucleophile? What is the leaving group? What monosaccharide products are formed?

Solve

(a) See labels below.

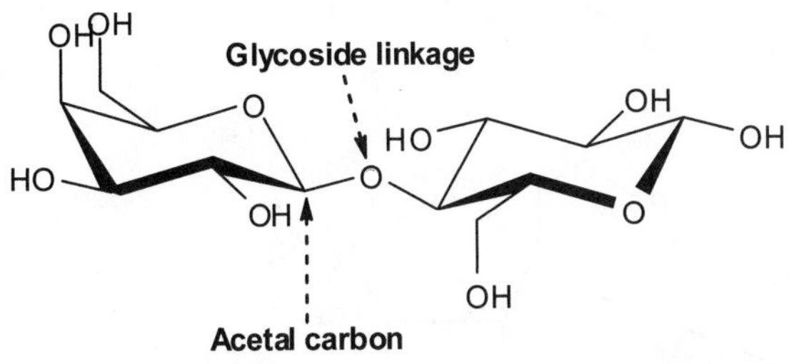

Glycoside linkage

Acetal carbon

(b) β-1,4′-glycosidic linkage. It is β because the OR group on the acetal C is equatorial and is on the same side of the ring as the CH_2OH group. It is 1,4′ because in the ring on the right, the glycosidic linkage is bonded to C4.

(c) Acid-catalyzed mechanism of acetal hydrolysis; hydrolysis of lactose into D-galactose and D-glucose.

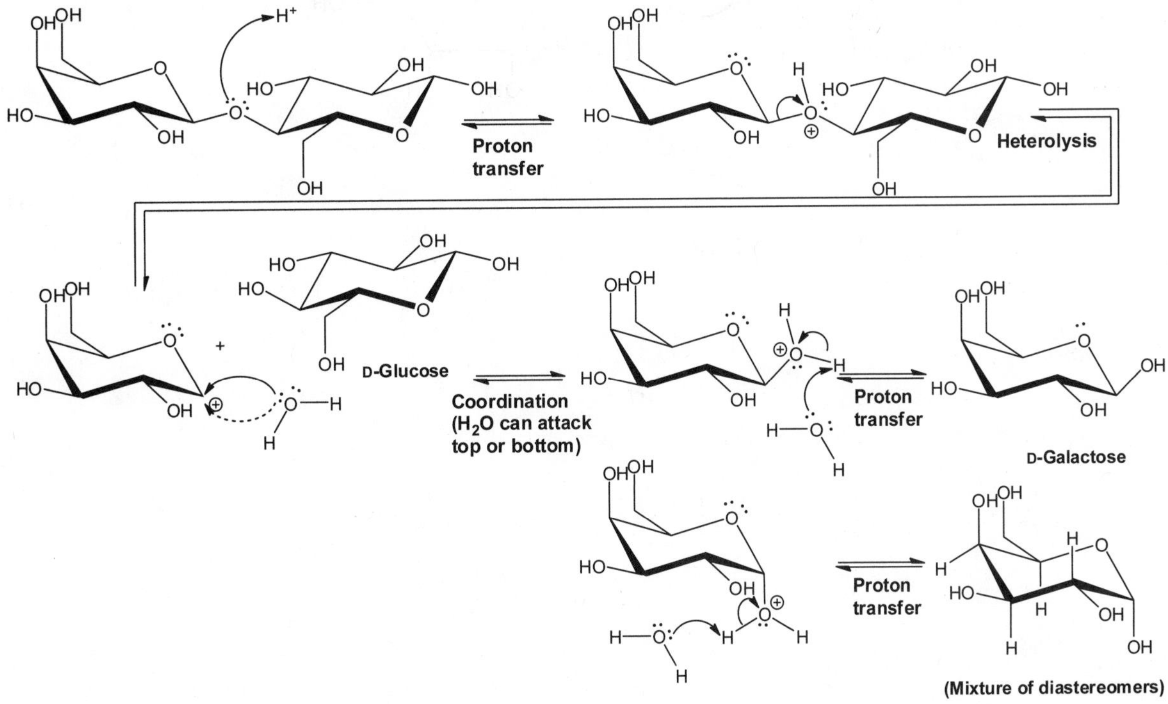

Proton transfer

Heterolysis

D-Glucose

Coordination (H_2O can attack top or bottom)

Proton transfer

D-Galactose

Proton transfer

(Mixture of diastereomers)

Problem 9.73

Think

Which mechanism (S_N1, S_N2, E1, or E2) is occurring in this reaction? What is the nucleophile? What is the leaving group?

Solve

This is an S_N1 mechanism, aided by the fact that the carbocation that is produced has significant resonance stabilization involving the oxygen. Alkylation of the ring nitrogen has converted the nitrogen base into an excellent leaving group, which gets the S_N1 mechanism started. And the solvent, water, is protic.

Step 1: The nitrogen base leaves, forming a resonance-stabilized carbocation (most stable contributor shown).

Step 2: A water molecule attacks the carbocation.

Step 3: A proton transfer completes the reaction.

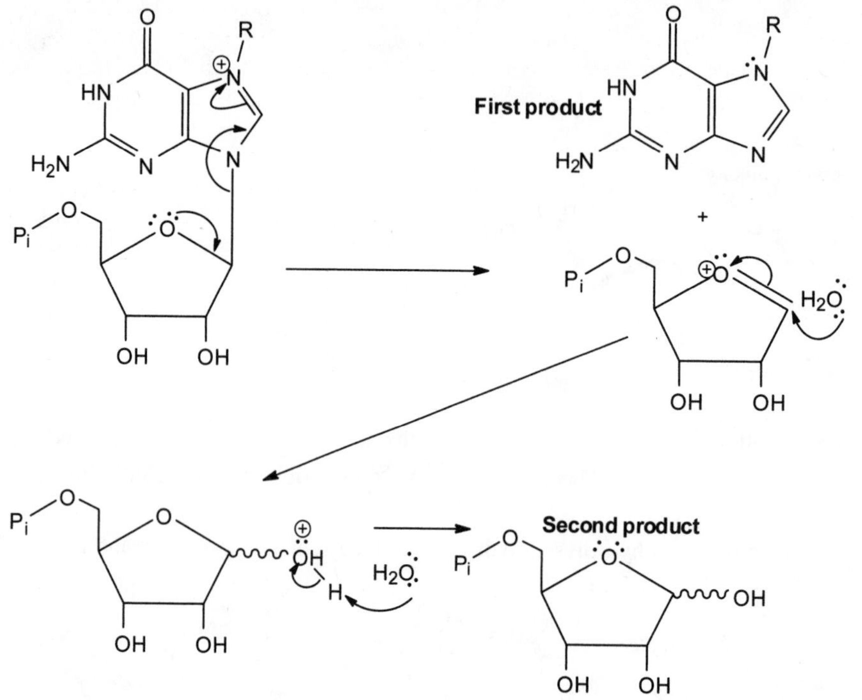

First product

Second product

CHAPTER 10 | Nucleophilic Substitution and Elimination Reactions 2: Reactions That Are Useful for Synthesis

Your Turn Exercises

Your Turn 10.1

Think

What are the electron-rich and electron-poor species in the reactants? What species will act as the nucleophile? The leaving group? Using the mechanism in Equation 10-8 as a guide, draw curved arrows to show the first step in the reaction. Which elementary step is shown in the first step? Draw the products. Follow the same procedure for Step 2. How is the stereochemistry of the C◄O affected in each step of the reaction?

Solve

The O atom in ROH is electron rich and acts as the nucleophile and the P atom in PCl_3 is electron poor and acts as the electrophile. The first step is an S_N2 reaction to form the O–P bond and break the P–Cl bond. The C◄O bond is not affected in the first step. In the second step, Cl^- acts as the nucleophile and attacks the electrophilic C◄O via an S_N2 backside attack. This inverts the configuration. The R◄OH is converted to an R''''''Cl via two back-to-back S_N2 reactions.

Backside attack inverts the configuration

Electron rich

Electron poor

1. S_N2

2. S_N2

Electron poor

Electron rich

+ $HOPCl_2$

Your Turn 10.2

Think

Is Cl^- a weak, moderate, or strong nucleophile? Consult Section 9.7b. How does the relative rate of the reaction relate to relative nucleophilicities?

Solve

Cl^- is an example of a moderate nucleophile. The relative nucleophilicities of NH_3 and Cl^- are 320,000 and 23,000, respectively. So NH_3 is, in fact, a stronger nucleophile in water than Cl^- is, making NH_3 a moderate nucleophile, too.

Your Turn 10.3

Think

Identify the bonds that are broken and formed in each step. How do you draw curved arrows to show bond formation? Bond breaking? Which steps involve a Nuc–C bond formation and which steps involve electron-poor H atoms?

Solve

The first step is an S_N2 in which the secondary amine R_2NH acts as a nucleophile and attacks the electrophilic C, causing Br to leave as Br^-. In the second step, the positively charged tertiary ammonium ion R_3NH^+ is deprotonated by another equivalent of R_2NH and a tertiary amine results: R_3N. The tertiary amine acts as a nucleophile to attack the electrophilic C and Br leaves, forming Br^- and a quaternary ammonium ion, R_4N^+.
See the mechanism on the next page.

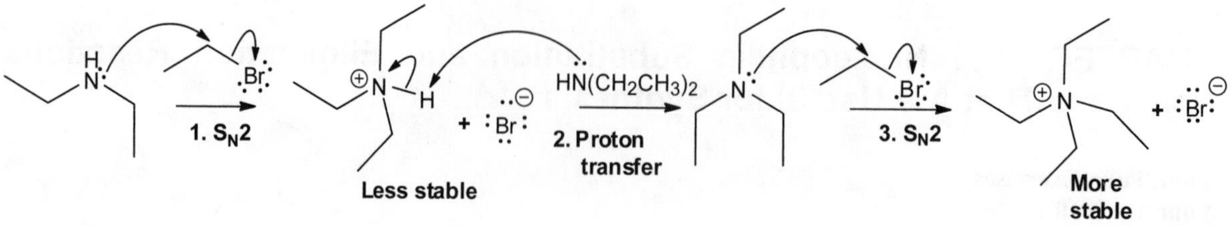

1. S_N2 — Less stable — 2. Proton transfer — 3. S_N2 — More stable

Your Turn 10.4

Think

Are alkyl groups electron donating or withdrawing? Does that effect stabilize or destabilize a nearby positive charge? What is the effect of having one more alkyl group on a quaternary ammonium ion compared to a tertiary ammonium ion?

Solve

Alkyl groups are electron donating and, therefore, stabilize a nearby positive charge. Having four versus three electron-donating groups stabilizes the N^+ more in R_4N^+ compared to R_3NH^+. This is an example of the inductive effect. See labels in the solution to Your Turn 10.3.

Your Turn 10.5

Think

Identify the bonds that are broken and formed in each step. How do you draw curved arrows to show bond formation? Bond breaking? Which steps involve a Nuc–C bond formation and which steps involve electron-poor H atoms?

Solve

Hydride, $H:^-$, acts as a base in the first step to deprotonate the α C–H to generate the enolate anion. This is a proton transfer step. The enolate anion acts as a nucleophile in step two and attacks the electrophilic C, causing Br to leave as Br^-. This is an S_N2 step that forms a new C–C bond.

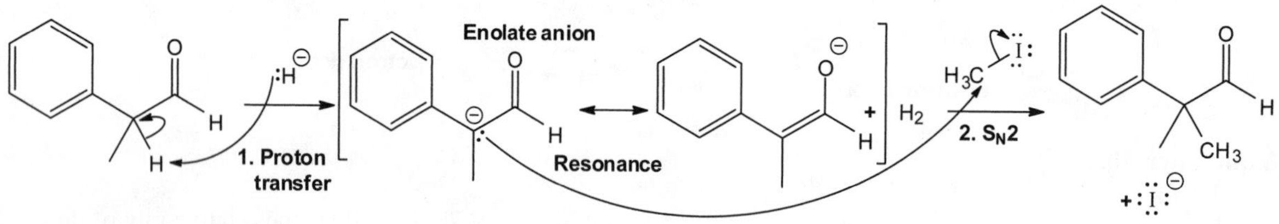

1. Proton transfer — Enolate anion — Resonance — 2. S_N2

Your Turn 10.6

Think

Using Equation 10-27 as a guide, what type of reaction occurs in the first step? Are there any protons on the α carbon? What bonds broke? Formed? How is the enolate anion formed? Is resonance possible? What type of reaction occurs in the second step? What bonds broke? Formed?

Solve

In the first step, HO^- acts as a base and deprotonates the α C to form the nucleophilic enolate anion. In the second step, the nucleophilic enolate anion attacks the molecular I–I in an S_N2 reaction, in which I^- departs as the leaving group. Even though I_2 appears not to be an electron-poor substrate, it is polarizable and an induced dipole is generated in the presence of the nucleophilic enolate anion. See the mechanism on the next page.

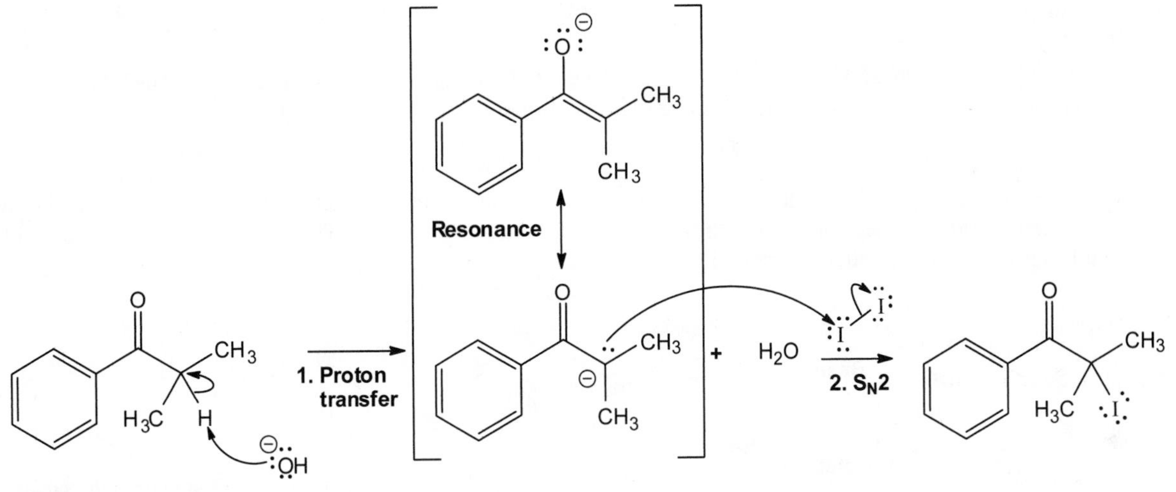

Resonance

1. Proton transfer

2. S_N2

Your Turn 10.7

Think

Identify which bonds broke and formed in each reaction step. When are protons involved (proton transfer)? When does a nucleophile attack an electron-poor nonhydrogen atom (S_N2)?

Solve

The α carbon in this example has two H atoms. In the first step, HO:⁻ acts as a base and deprotonates the α carbon to form the nucleophilic enolate anion. In the second step, the nucleophilic enolate anion attacks the molecular Br–Br in an S_N2 reaction, in which Br^- departs as the leaving group. In the third step, HO:⁻ acts as a base and deprotonates the α carbon to form the nucleophilic enolate anion. In the fourth step, the nucleophilic enolate anion attacks the molecular Br–Br in an S_N2 reaction.

Resonance

Halogen atom increases acidity

1. Proton transfer

2. S_N2

3. Proton transfer

4. S_N2

Br^- stabilizes the negative charge on the enolate anion.

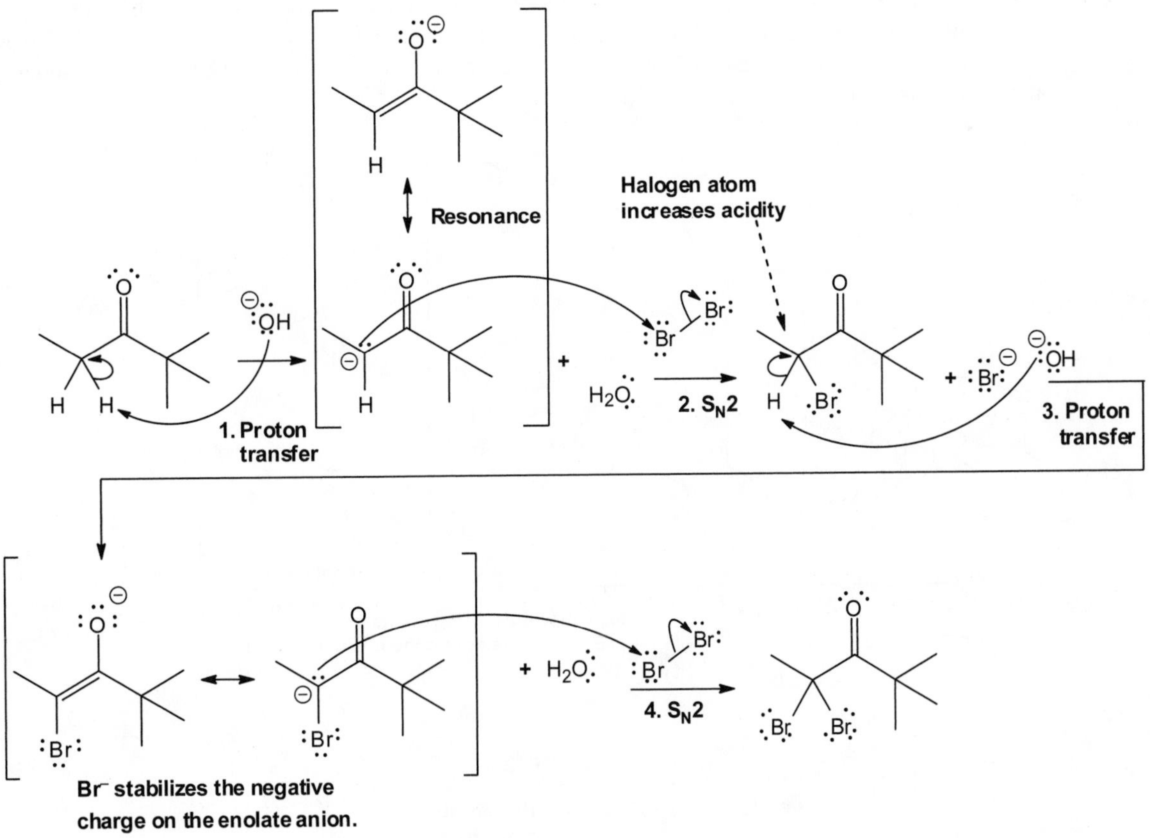

Your Turn 10.8

Think

Under basic conditions, what reaction takes place at an α carbon? What product forms? Is Cl electron withdrawing or donating? Will that inductive effect stabilize or destabilize that product?

Solve

Under basic conditions, the enolate anion forms after the proton transfer reaction. Cl is an electron-withdrawing group and withdraws electron density from the enolate's negative charge. This stabilizes the enolate and leads to a faster chlorination under basic conditions.

Base
Proton transfer
A
B
Faster
Cl stabilizes the negative charge on the enolate anion.

Your Turn 10.9

Think

In each step, identify the bonds that broke and formed. Use curved arrows to show electron movement. Using Equation 10-31 as a guide, in the presence of acid H–B, where is the basic site on the organic molecule? What atom gets protonated? Are there any protons on the α carbon? What reaction takes place at the α carbon? What type of reaction occurs in the third step? Fourth step?

Solve

The ketone O is protonated in the presence of an acid (proton transfer). The conjugate base deprotonates the α carbon (proton transfer). The enol form acts as the nucleophile because the negatively charged enolate anion cannot exist in any substantial concentration in acidic conditions due to its basic nature. The enol is electron rich and can polarize the Br–Br. This sets up the electron-rich to electron-poor driving force of the S_N2 reaction. The O–H is deprotonated in the last step by the conjugate base to re-form the C=O. A second halogenation reaction does not take place.

1. Proton transfer
Electron rich C=C bond
2. Proton transfer
Enol form
3. S_N2
4. Proton transfer
Electron withdrawing effects from the Br atom destabilize the positive charge.
Halogenation stops here.

Your Turn 10.10

Think

Under acidic conditions, what reaction takes place at the C=O? What product forms? Is Br electron withdrawing or donating? Will that inductive effect stabilize or destabilize the product?

Solve

In acidic conditions, the C=O is protonated. **A** is faster due to the electron-withdrawing Br that exists in **B**. Br destabilizes the positive charge that develops on the carbonyl O.

A
Faster

B

Br destabilizes the positive charge on the enol cation.

Your Turn 10.11

Think

In each step, identify the bonds that broke and formed. How are curved arrows used to indicate bond breaking and bond forming? Using Equation 10-33 as a guide, in the presence of the electron-rich C atom of CH_2=N=N, where is the acidic site on the organic molecule? What atom gets protonated? What molecule becomes nucleophilic after the first step?

Solve

The C atom of diazomethane, CH_2=N=N, is basic and protonated by the acidic proton of the carboxylic acid. The carboxylate anion, $CH_3CO_2^-$, is electron rich and acts as a nucleophile in an S_N2 reaction. N_2 is the leaving group (stable and gaseous).

1. Proton transfer

2. S_N2

+ $N_2(g)$

Your Turn 10.12

Think

In each step, identify the bonds that broke and formed. Use curved arrows to show electron movement. Using Equation 10-39 as a guide, what type of reaction occurs between ROH and a HO^-? In the resulting species, which site is nucleophilic? Where is the leaving group?

Solve

In the first step, the ROH is deprotonated by basic HO^- to form an alkoxide nucleophile. The leaving group Br is on the same molecule that contains the nucleophile O^-. An intramolecular S_N2 results in a six-membered ring ether.

1. Proton transfer

2. S_N2

+ H—OH

Your Turn 10.13

Think

Identify the bonds that broke and formed. What inorganic products result that are not shown? Use curved arrows to show electron movement. Using Equation 10-39 as a guide, what type of reaction occurs between ROH and HO^-? In the resulting species, which site is nucleophilic? Where is the leaving group?

Solve

In the first step, the ROH is deprotonated by basic HO^- to form an alkoxide nucleophile and a molecule of H_2O. The leaving group Br is on the same molecule that contains the nucleophile O^-. An intramolecular S_N2 results in a three-membered ring ether, an epoxide.

1. Proton transfer

$+ H_2O$

2. S_N2

$+ Br^-$

Your Turn 10.14

Think

In each step, identify the bonds that broke and formed. How are curved arrows used to show bond breaking and bond forming? What type of reaction results from ROH and HCl? What is the leaving group that results? Why does a C^+ not form?

Solve

The acidic conditions cause the ROH group to be protonated, producing ROH_2^+ and generating a good leaving group, H_2O. The leaving group is on a primary C and thus the nucleophile ROH attacks the electrophilic C in the step where the leaving group departs, S_N2. A final proton transfer step ensues to form the ether product.

1. Proton transfer

$+ Cl^-$

2. S_N2

H_2O

3. Proton transfer

$+ H_3O^+$

Your Turn 10.15

Think

What species is electron rich? What site is electron poor? In what kind of step is the C–Cl bond formed? In what kind of step is the RO–H bond formed? Use curved arrows to show electron movement.

Solve

Cl^- is a good nucleophile and attacks the electron-poor C of the epoxide. The S_N2 reaction causes the highly strained epoxide ring to open. This compensates for the poor leaving group ability of the RO^-. The alkoxide RO^- is a strong base and abstracts a proton from the methanol solvent.

Primary C

Very poor leaving group

1. S_N2

$H-OCH_3$

2. Proton transfer

$+ {}^-OCH_3$

Ring opening relieves ring strain.

Your Turn 10.16

Think

Consult Table 6-1 to find the pK_a values. Does a lower pK_a indicate a stronger or weaker acid? Is a proton transfer reaction favored on the same side as or opposite from the stronger acid? How do you use pK_a values of the two acids to determine the extent to which the proton transfer reaction is favored, and how is that related to reversibility?

Solve

The pK_a of NH_3 is 36, and the pK_a of RC≡CH is about the same as that of HC≡CH, which is 25. A lower pK_a indicates a stronger acid; thus, RC≡CH is a stronger acid compared to NH_3 by a factor of 10^9. Therefore, the side opposite the terminal alkyne—the product side—is favored by a factor of 10^9. This makes the reverse reaction difficult, so the forward reaction is irreversible.

Your Turn 10.17

Think

Consult Table 6-1 to find the pK_a values. Does a lower pK_a indicate a stronger or weaker acid? Is a proton transfer reaction favored on the same side as or opposite from the stronger acid? How do you use pK_a values of the two acids to determine the extent to which the proton transfer reaction is favored, and how is that related to reversibility?

Solve

The pK_a of H_2O is 15.7, and that of RC≡CH is about 25. RC≡CH is a weaker acid than H_2O by a factor of about 10^9. Therefore, the side opposite H_2O—the reactant side—is favored, which makes the reverse reaction easier than the forward reaction. Thus, the forward reaction is reversible.

Your Turn 10.18

Think

On which C is the leaving group? Draw out the C–H groups on C2 and C4. What is the requirement for two groups to be anti-coplanar? What is the bulkiest group that will be involved with the most severe steric strain? How many gauche interactions (60° apart in dihedral angle) are there between that group and alkyl groups in each Newman projection?

Solve

The quaternary R–$N^+(CH_3)_3$ is the leaving group on C3. Therefore, there are adjacent H atoms on C2 or C4 that could be abstracted by a base in an elimination reaction. In an E2, the adjacent H atom must be anti-coplanar. This can be seen in a Newman projection where the leaving group and the H atom are 180° apart. The $N^+(CH_3)_3$ group is the bulkiest group and will be involved in the most severe steric strain. Comparison of the two Newman projections (**A:** C2–C3 bond or **B:** C3–C4 bond) shows two gauche interactions involving that bulky group with alkyl groups in **A**, as well as another gauche interaction between two alkyl groups. In **B**, there is just one gauche interaction between the $N^+(CH_3)_3$ group and an alkyl group, so there is less steric strain in **B**.

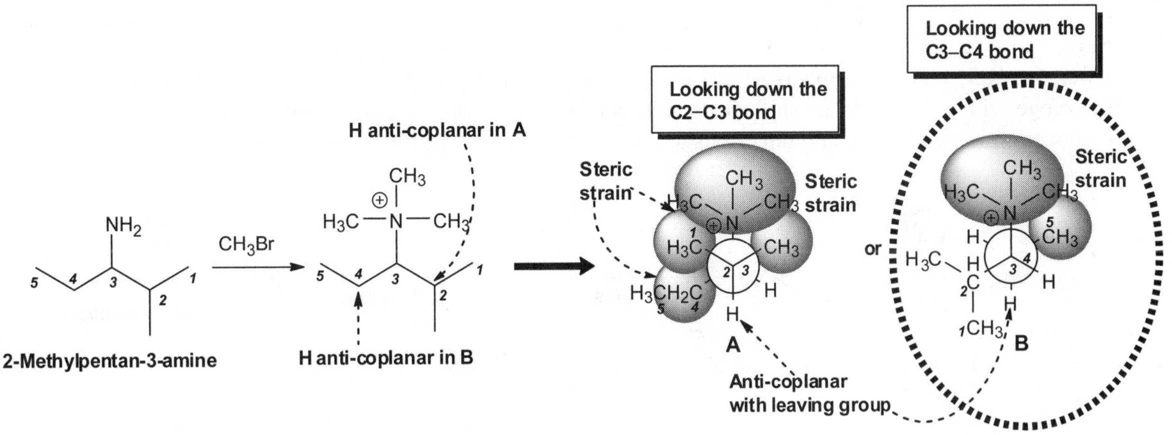

2-Methylpentan-3-amine

CH_3Br

H anti-coplanar in A

H anti-coplanar in B

Looking down the C2–C3 bond

Steric strain

Steric strain

A

Anti-coplanar with leaving group

or

Looking down the C3–C4 bond

Steric strain

B

Chapter Problems

Problem 10.1

Think

What is the first step in an S_N1 mechanism? In the intermediate that forms, what is the geometry of the C that was initially bonded to the leaving group? From how many directions can the nucleophile attack the electrophilic carbocation?

Solve

The mechanism for Equation 10-3 is shown below. When the nucleophile attacks the positively charged C atom, both *R* and *S* configurations can be formed because that C atom is planar.

Proton transfer — Heterolysis — + H_2O — Attack top — Coordination — Attack bottom

A similar story arises with the mechanism for the reaction in Equation 10-4.

Proton transfer — Heterolysis — + H_2O — Attack top — Coordination — Attack bottom

Problem 10.2

Think

What is the first step in an S_N1 mechanism? What type of carbocation forms? Can a more stable intermediate be formed via a 1,2-hydride shift or a 1,2-methyl shift?

Solve

After the HO^- leaving group is turned into a good leaving group, H_2O, the carbocation that is formed can rearrange via a 1,2-hydride shift from a secondary carbocation into a tertiary carbocation. This tertiary carbocation is more stable and thus leads to the major product. The mechanism is given below.

Proton transfer — Heterolysis — + Br^- — + H_2O — 1,2-Hydride shift — Coordination

Problem 10.3

Think

Is HO^- a good leaving group? How does PCl_3 or PBr_3 turn HO^- into a good leaving group? What two elementary steps are involved in the reaction? What happens to the configuration at the electrophilic carbon?

Solve

Recall that D is deuterium. This is an isotope of hydrogen. It has identical chemical behavior to H, but serves as a label in reactions. HO^- is not a good leaving group and PCl_3 and PBr_3 serve to turn HO^- into $HOPX_2$, which is an uncharged leaving group. The reaction occurs in two back-to-back S_N2 steps. This inverts the configuration at the electrophilic carbon. Notice in **(a)** that the C–Br is unaffected because sp^2-hybridized C atoms are resistant to S_N2 reactions.

(a)

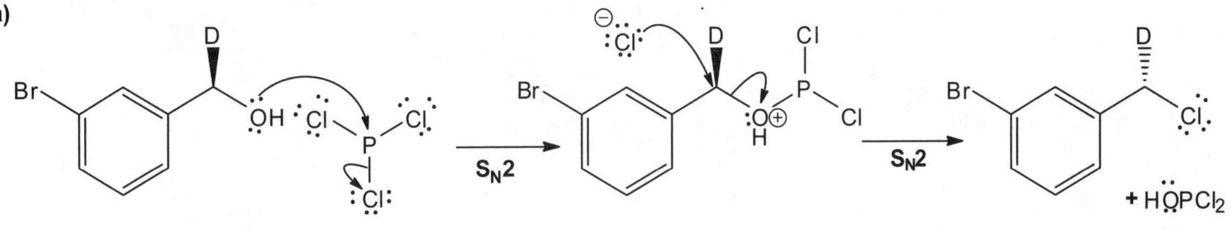

S_N2 S_N2 + $HOPCl_2$

(b)

S_N2 S_N2 + $HOPBr_2$

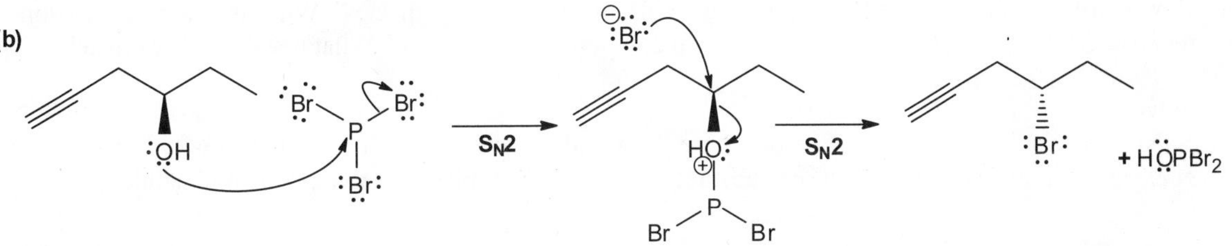

Problem 10.5

Think

On what type of carbon is the OH group attached? What is the mechanism by which the PCl_3 reagent reacts with an alcohol? What types of carbon atoms attached to the leaving group prevent S_N2 reactions?

Solve

Only **(a)** and **(c)** will react with PCl_3. **(b)** will not react because the leaving group (OH) is on an sp^2-hybridized C. **(d)** will not react because the leaving group is on a tertiary carbon. For the reactions that do proceed, keep in mind that the second S_N2 step causes inversion of the configuration.

(a)

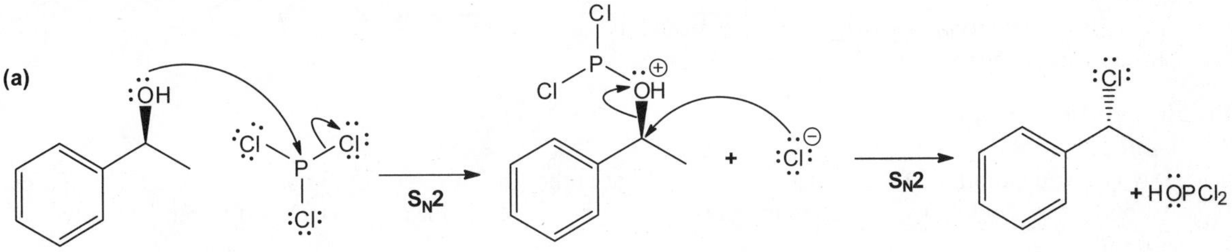

S_N2 S_N2 + $HOPCl_2$

(c)

S_N2 S_N2 + $HOPCl_2$

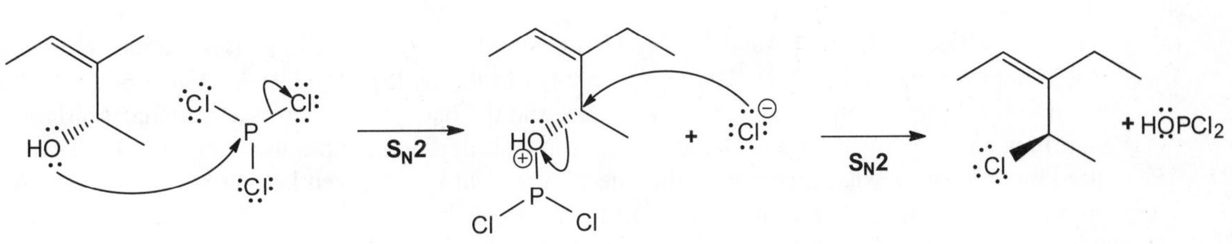

Problem 10.7

Think

Which species is electron rich? Which is electron poor? What kind of reaction will take place between the two? Can the reaction take place a second time?

Solve

The amine N is electron rich (nucleophile) and the C attached to the Br is electron poor (electrophile). The resulting reaction is an S_N2, yielding a protonated tertiary amine. The product can be deprotonated by another molecule of the amine to yield an uncharged tertiary amine. Because there is excess CH_3Br, a further alkylation generates a quaternary ammonium ion as the major product. No further reaction can take place because there are no acidic H atoms on the ammonium ion.

Problem 10.8

Think

How can you simplify $NaNH_2$? What is the role of the resulting species? What is the acidic proton that is removed? What mechanism takes place in Step 2 with the R–Br reagent? What new bond is formed?

Solve

$NaNH_2$ is simplified to $^-NH_2$, a strong base, which deprotonates the α carbon to make a relatively strong nucleophile as an enolate anion. Then substitution via S_N2 takes place involving the alkyl halide.

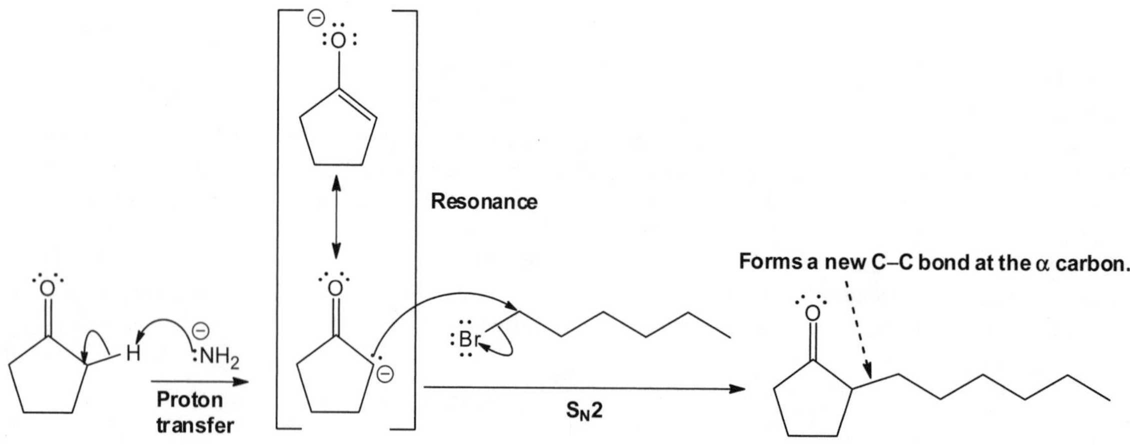

Problem 10.10

Think

How can you simplify LDA and $KOC(CH_3)_3$? What is the role of the resulting species? Are the two α carbons distinct? How does LDA differentiate between α carbons? How does $KOC(CH_3)_3$ differentiate between those carbons?

Solve

(a) LDA is simplified to $(C_3H_7)_2N^-$ and $KOC(CH_3)_3$ is simplified to $^-OC(CH_3)_3$. Both anions are strong bases and will deprotonate an α carbon. The two α carbons in the molecule are, indeed, distinct. The one on the left is a tertiary carbon, part of an isopropyl group, and the one on the right is a secondary carbon, part of a CH_2 group. LDA is a very strong, bulky base, so it will deprotonate the less sterically hindered α carbon under kinetic control. That α carbon is the one on the right in the given ketone because it has fewer alkyl groups attached. The enolate anion that is formed then attacks allyl bromide in an S_N2 reaction, yielding an alkylated ketone. See the mechanism on the next page.

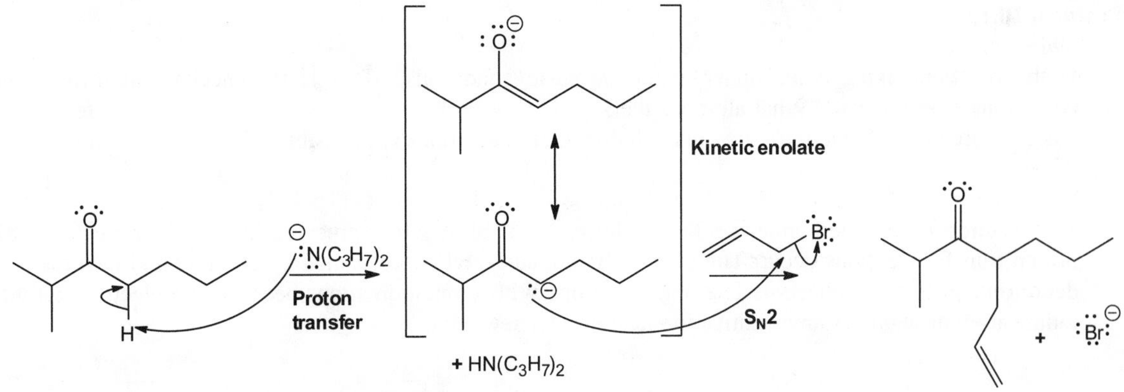

Kinetic enolate

Proton transfer

S_N2

$+ HN(C_3H_7)_2$

(b) If $NaOC(CH_3)_3$ used as the base instead, then the α carbon on the left would be deprotonated reversibly to yield the more stable (thermodynamic) enolate anion. Therefore, alkylation would take place at the α carbon on the left, yielding the product shown below.

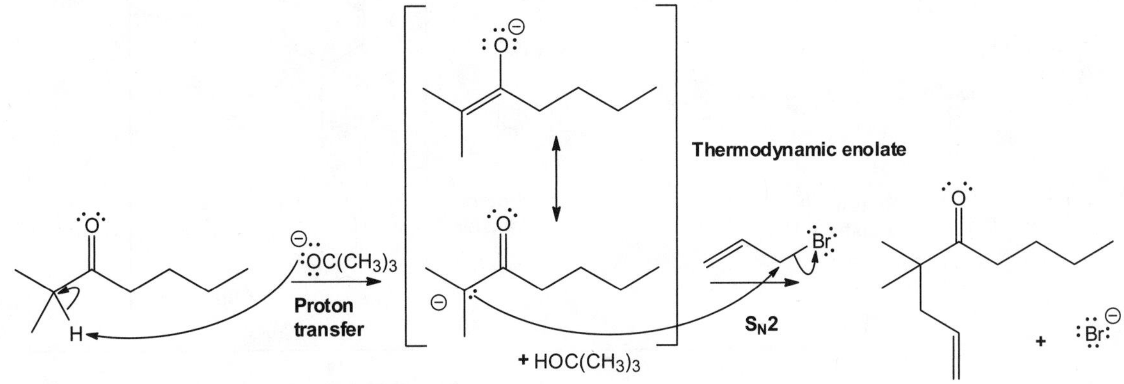

Thermodynamic enolate

Proton transfer

S_N2

$+ HOC(CH_3)_3$

Problem 10.12

Think

In the presence of a strong base, what reaction takes place at an α carbon? How does this affect the chemical properties at that carbon? What species will be attacked as a result? How many times will such a reaction take place at that α carbon?

Solve

In the presence of a strong base, an α carbon can be deprotonated to generate a strongly nucleophilic C atom. Subsequently, I_2 will be attacked in an S_N2 reaction. Because this reaction takes place under basic conditions, all α H atoms will be replaced. In this case, there is one on each of the two α carbons.

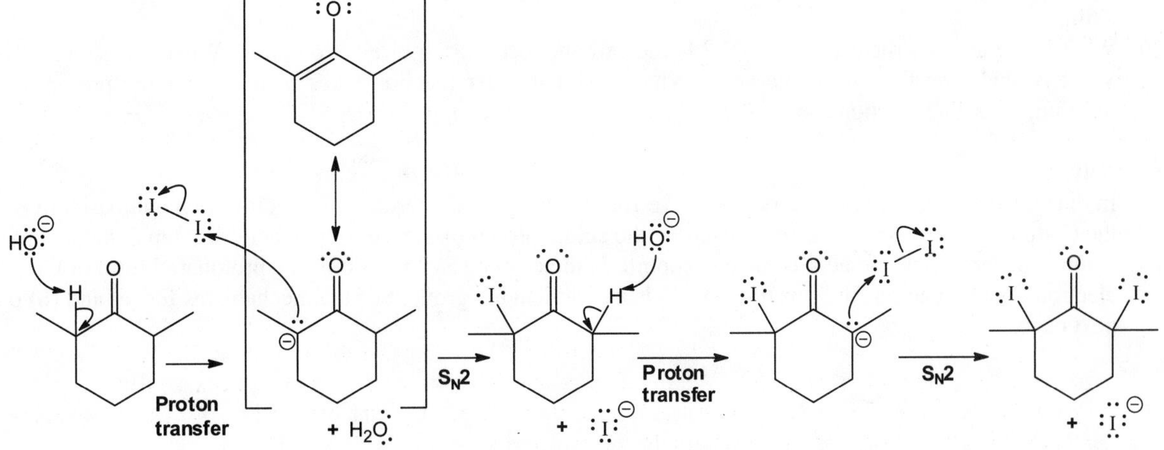

Proton transfer

$+ H_2O$

S_N2

Proton transfer

S_N2

Problem 10.13

Think

Is the reaction taking place under acidic or basic conditions? How is the mechanism different in acidic conditions than in basic? What atom on the ketone acts as the base? How many proton transfer reactions take place before the S_N2 reaction? How many halogenation reactions are possible?

Solve

The reaction takes place under acidic conditions and follows the same mechanism as Equation 10-31. Two proton transfer reactions occur: The O atom on the carbonyl is protonated by the acid and then the α carbon is deprotonated to form an enol. The reaction stops with monoiodination, because the electronegativity of the iodine atom destabilizes any positive ion leading to a new enol.

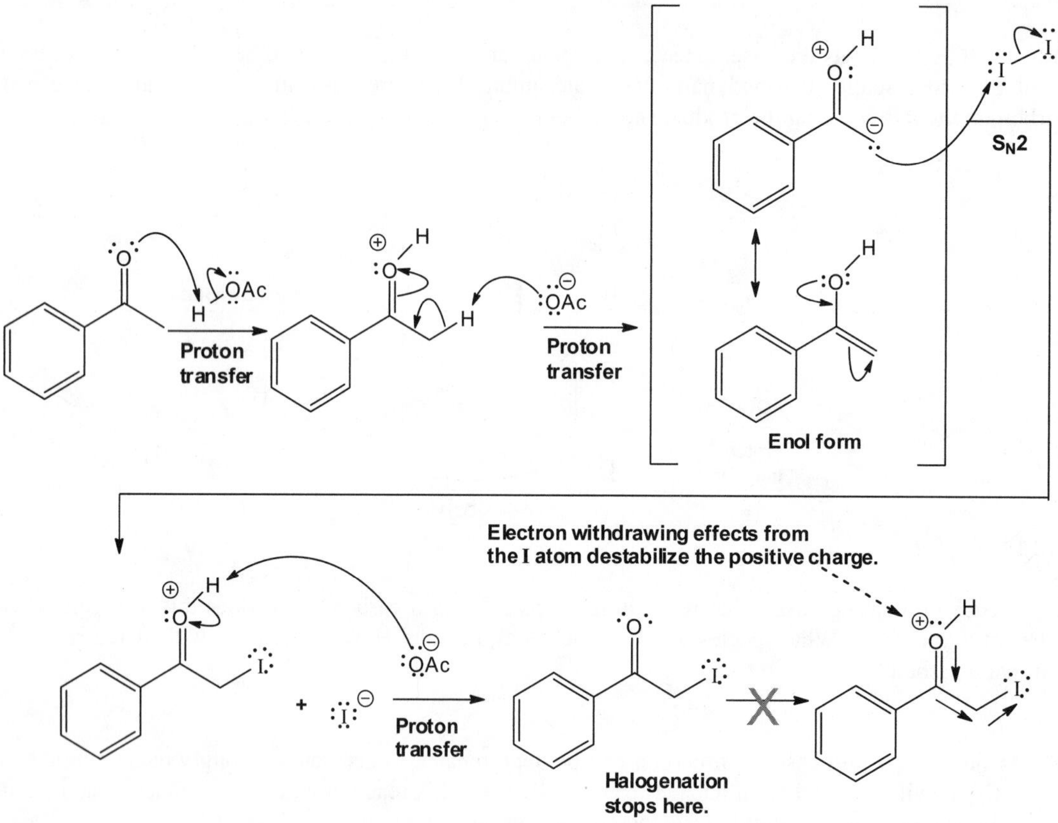

Problem 10.14

Think

What is the Lewis structure for CH_2N_2? Is the carbon electron rich or electron poor? What is basic on CH_2N_2? What is acidic on the carboxylic acid? What proton transfer reaction takes place? After the proton transfer reaction, what S_N2 reaction can occur?

Solve

In diazomethane, the C atom bears a negative formal charge and, therefore, is electron rich and basic. In Step 1, the C atom of diazomethane is protonated by the acidic proton of the carboxylic acid. In Step 2, the carboxylate anion is electron rich and acts as the nucleophile in the ensuing S_N2 reaction. The protonated form of $CH_3N_2^+$ is electron poor and acts as the substrate. $N_2(g)$ is a great leaving group. See the mechanisms for **(a)** and **(b)** on the next page.

(a) After diazomethane deprotonates the carboxylic acid, an S_N2 step causes N_2 to be displaced, producing the methyl ester.

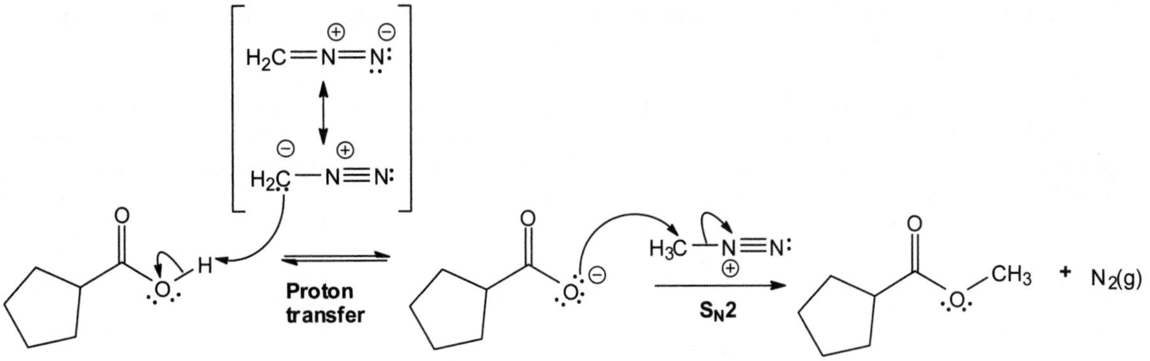

(b) This mechanism takes place twice, once for each carboxylic acid functional group.

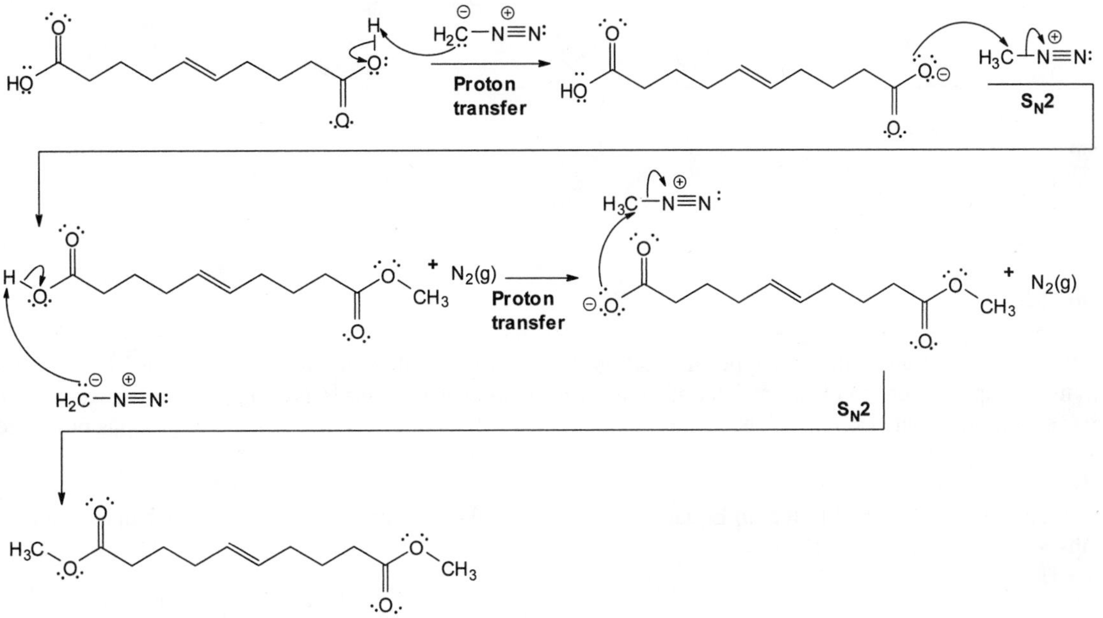

Problem 10.15

Think

In the presence of a strong base, what type of reaction occurs with an alcohol? What then forms? What are the resulting electron-rich and an electron-poor species? Are these species located on the same molecule? Will an intramolecular reaction result?

Solve

In the presence of a strong base, the alcohol is deprotonated to form an alkoxide, which is an electron-rich nucleophile. The carbon of the C–Br is electron poor and serves as the electrophile. An intramolecular S_N2 reaction results and produces a cyclic ether.

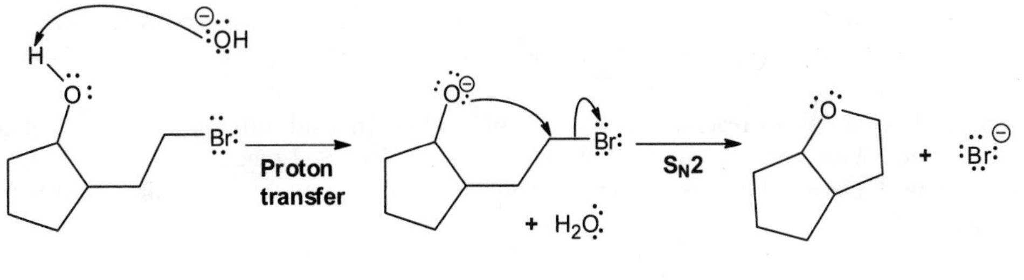

Problem 10.16

Think

In the presence of an acid, what type of reaction occurs with an alcohol? What then forms? Is there a good leaving group? Is there an electron-rich and an electron-poor species? What functional group is produced?

Solve

Formation of an ether is expected. Because the leaving group is on a benzylic carbon and the solvent is protic, an S_N1 reaction takes place. (However, formation of a primary benzylic carbocation makes other reactions possible; namely, electrophilic aromatic substitution and polymerization, which are covered in future chapters.)

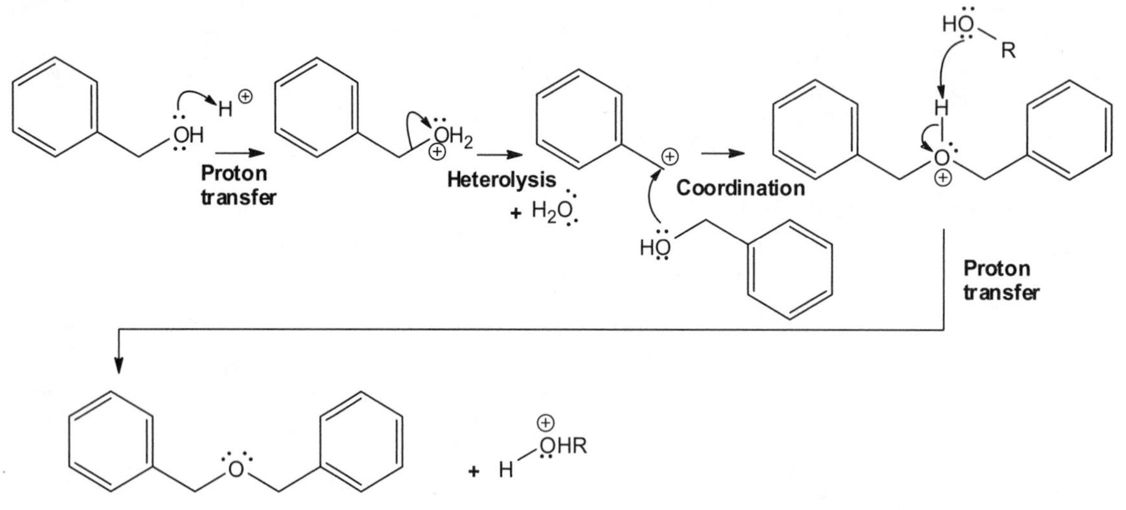

Proton transfer

Heterolysis + H_2O

Coordination

Proton transfer

Problem 10.17

Think

In the presence of an acid, what type of reaction occurs with an alcohol? What then forms? Is there a good leaving group? Is there an electron-rich and an electron-poor species? To produce an unsymmetric ether, should those species originate from the same alcohol or from different alcohols? What functional group is produced?

Solve

The mechanism is identical to that in Equation 10-42 except that propan-1-ol acts as the nucleophile rather than propan-2-ol in the coordination step.

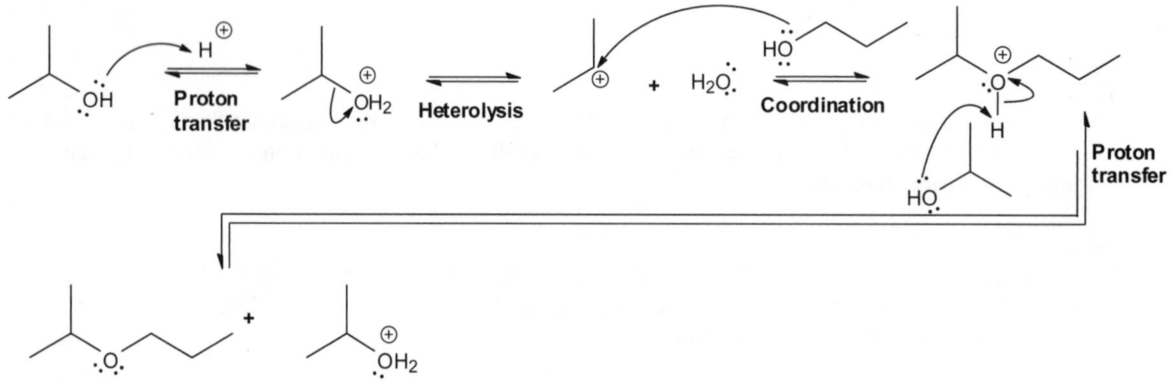

Proton transfer

Heterolysis

Coordination

Proton transfer

Problem 10.19

Think

In the presence of an acid, what type of reaction occurs with an alcohol? In a substitution reaction, what could then act as the nucleophile? What could act as the substrate? Is an S_N1 or S_N2 mechanism favored? Can an intramolecular reaction take place? If so, is it more favorable or less favorable than the corresponding intermolecular reaction?

Solve

The acidic conditions cause an OH group to be protonated, generating a good leaving group (H_2O). The second OH group in the molecule can act as a nucleophile in an intramolecular S_N2 reaction. The intramolecular reaction is favored over the intermolecular one, due to the formation of the five-membered ring.

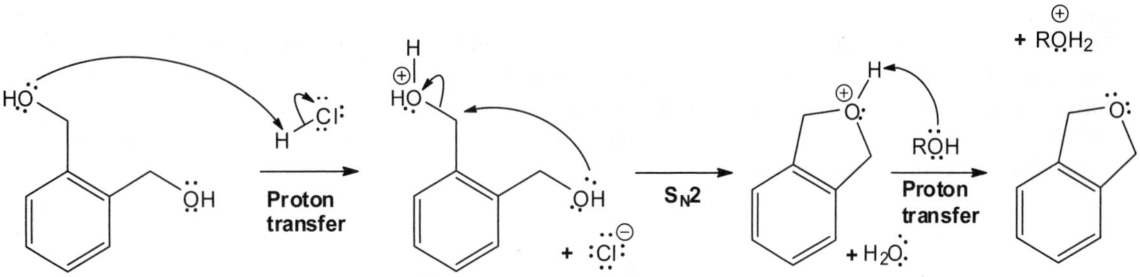

Problem 10.20

Think

For each reaction, consider what is the electron-rich and the electron-poor species in each step of the reaction. Are any proton transfer reactions involved? At what step does the epoxide ring open? What is acting as the nucleophile in this step for each reaction? What is the purpose of the acid workup step?

Solve

The mechanisms for each reaction are shown below.

Equation 10-52: The NC^- acts as the nucleophile to open the epoxide ring in an S_N2 reaction. A new C–C bond is formed. The second step is a proton transfer reaction to form an uncharged alcohol product.

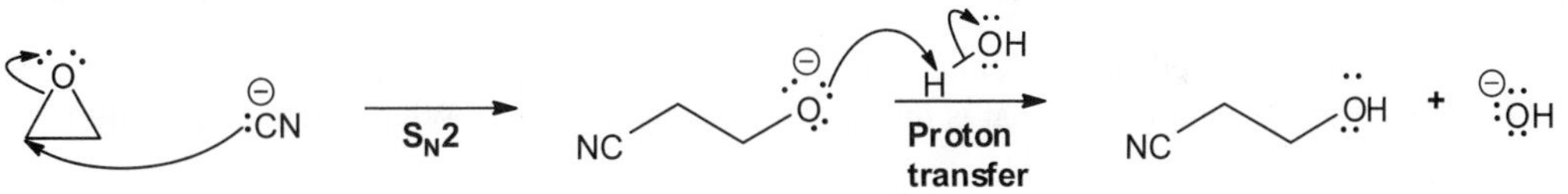

Equation 10-53: The NaH serves as a source of hydride base $H:^-$ to deprotonate the alkyne C–H and form the alkynide anion. The $RC{\equiv}C^-$ acts as the nucleophile to open the epoxide ring in an S_N2 reaction. A new C–C bond is formed. The second step is a proton transfer reaction to form an uncharged alcohol product.

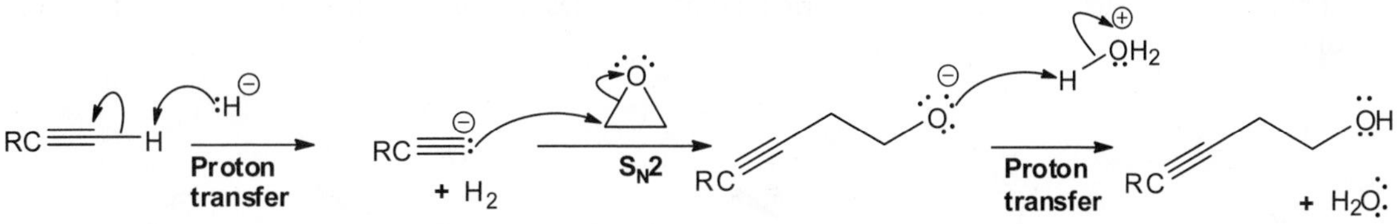

Equation 10-54: $LiAlH_4$ or $NaBH_4$ serves as a source of hydride nucleophile $H:^-$ to open the epoxide ring in an S_N2 reaction. A new C–H bond is formed. The second step is a proton transfer reaction to form an uncharged alcohol product.

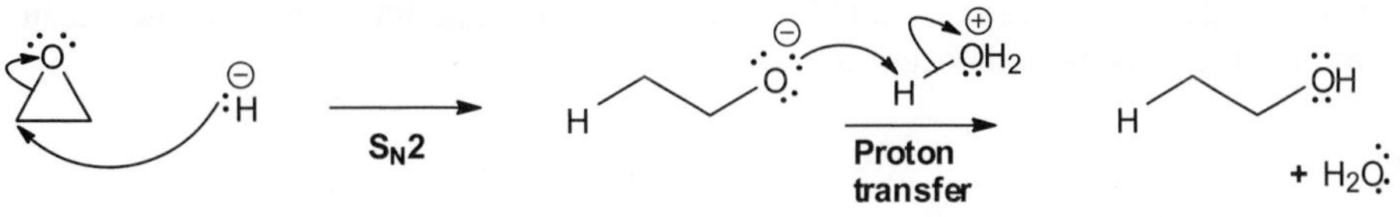

Problem 10.22

Think

What is the nucleophile? Which C atom of the epoxide ring will it attack? What is the stereochemistry of such a reaction?

Solve

The nucleophile is the cyanide anion, NC^-. It will undergo an S_N2 reaction with the epoxide ring, attacking the less sterically hindered C atom, which is the one on the left. The nucleophile attacks from the bottom of the epoxide (i.e., opposite the epoxide O atom), forcing the H and CH_3 substituents upward. The more substituted carbon with the CH_3 and the CH_2CH_3 groups does not undergo bond breaking or formation, so its stereochemical configuration remains the same.

(a)

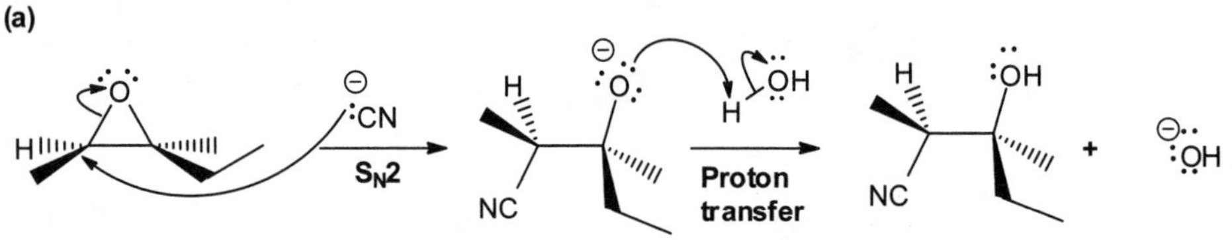

:CN ⊖ S_N2 → Proton transfer → + :OH⊖

(b)

:CN ⊖ S_N2 → Proton transfer → + :OH⊖

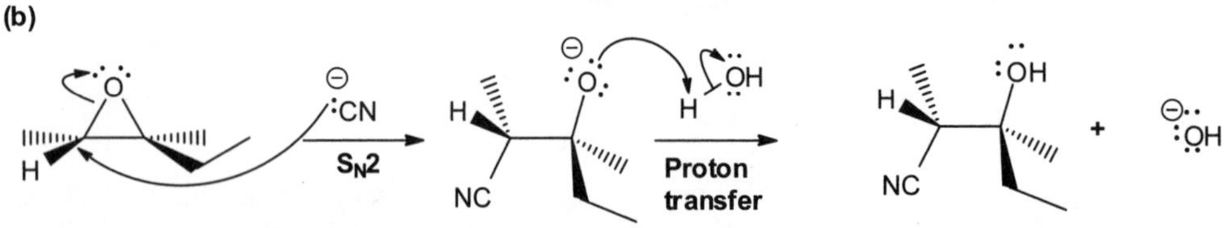

Problem 10.24

Think

What types of species are allowed under acidic conditions and what types are not allowed? Under neutral conditions? Under each set of conditions, what is the substrate being attacked? What is the nucleophile?

Solve

The regioselectivity is different for these reactions.

(a) Under neutral conditions, the nucleophile is Br^- and the substrate is the uncharged epoxide. Because Br^- is a stronger nucleophile than water and is high in concentration, we can predict that bromide will open the strained ring to form a bromo alcohol. The mechanism follows the one in Equation 10-56. Because the epoxide being attacked is uncharged, steric hindrance guides the nucleophile to the less substituted carbon of the ring.

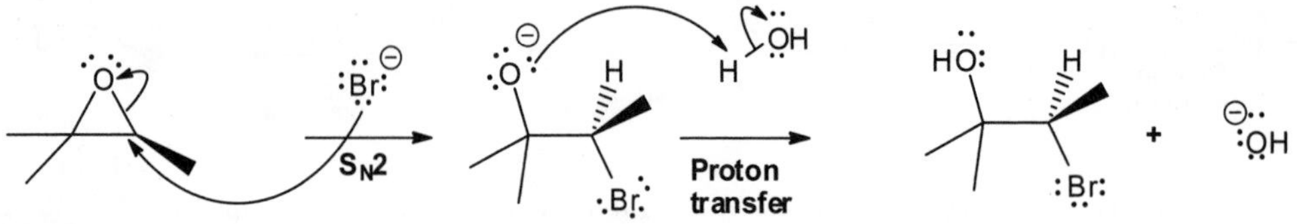

:Br:⊖ S_N2 → Proton transfer → + :OH⊖

(b) Under acidic conditions, however, strong bases are not allowed, so we may *not* include species with a localized negative charge on O. Instead, the mechanism in Equation 10-58 takes place, as shown below. Unlike in **(a)**, the nucleophile attacks the more highly alkyl substituted C atom of the epoxide because the positive charge from the O atom is more delocalized on that C.

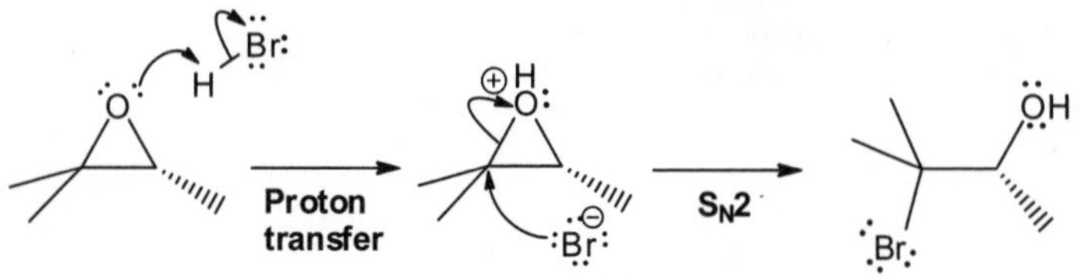

Proton transfer → S_N2 →

Problem 10.25

Think

How many leaving groups are present? In the presence of a strong base, what reaction mechanism will take place? How many times will this reaction take place? What functional group results? Are proton transfer reactions involved?

Solve

Each of these reactions is a back-to-back E2 set of reactions. For **(a)**, it is also acceptable if the other Br is eliminated first.

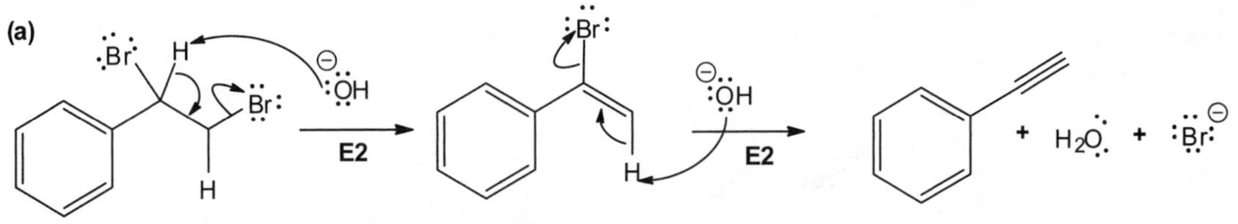

(a)

The reaction in **(b)** requires three equivalents of base because the initial product is a terminal alkyne that is deprotonated by NH_2^-, so H^+ is added in an acid workup to replenish that H^+.

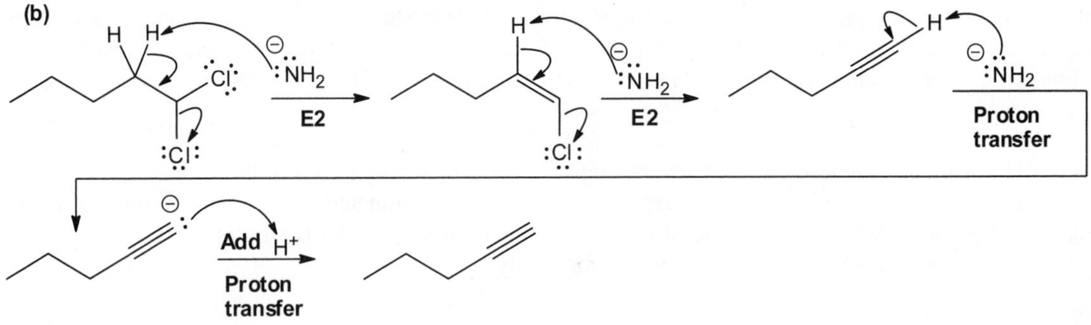

(b)

Problem 10.26

Think

How many leaving groups are present? In the presence of a strong base, what reaction mechanism will take place? How many times will this reaction take place? What functional group results? Are proton transfer reactions involved?

Solve

In each problem, three equivalents of H_2N^- are required because after the first two are used for E2 steps, the terminal alkyne produced is quickly deprotonated by the strong base. Acid workup is necessary to replenish the proton. See the mechanisms below and on the next page.

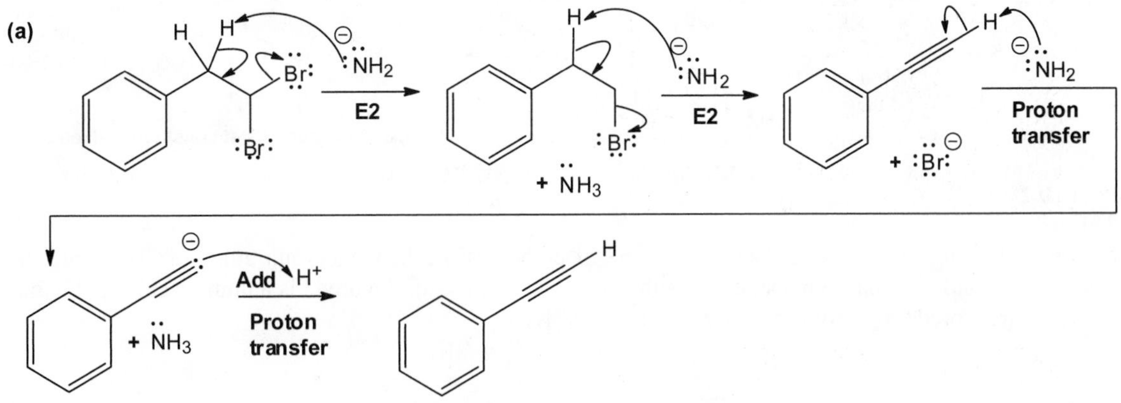

(a)

(b)

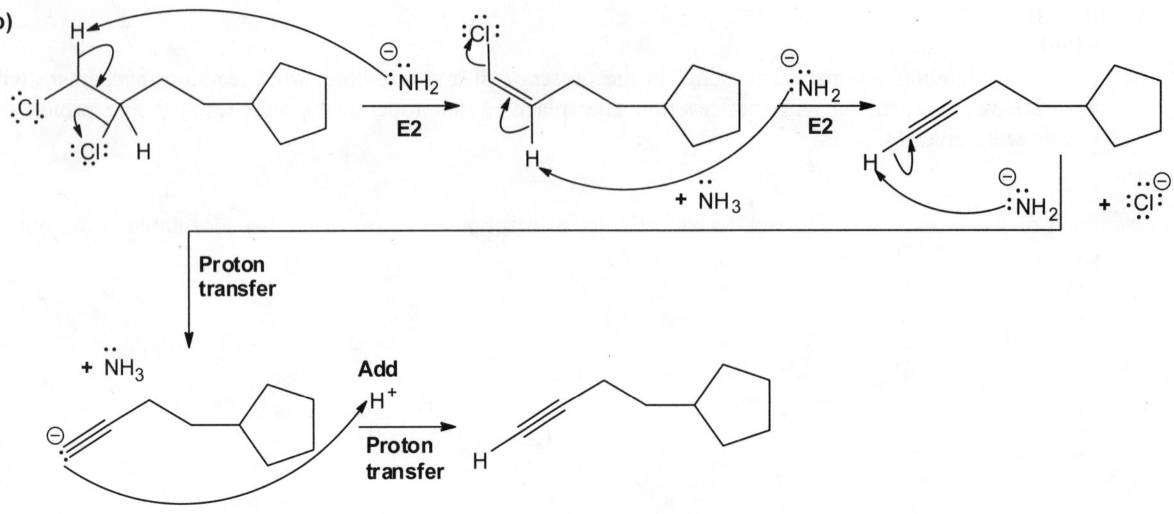

Proton transfer

Add H^+

Proton transfer

Problem 10.27

Think

Is there a good leaving group? In the presence of excess CH_3Br, what results on the N atom? Is there now a good leaving group? A strong base in the presence of heat results in what major reaction? Which H atom is eliminated? Can one H attain the anti-coplanar conformation more easily than others?

Solve

This is a Hofmann elimination. The first several steps are sequential alkylations of the amine N, which produces a quaternary ammonium ion. The final step is E2. H can be eliminated from two different C atoms, but predominantly takes place to give the least substituted alkene product. The reason is explained in Your Turn 10.18. Note: By-products are omitted in steps 2–4 for clarity.

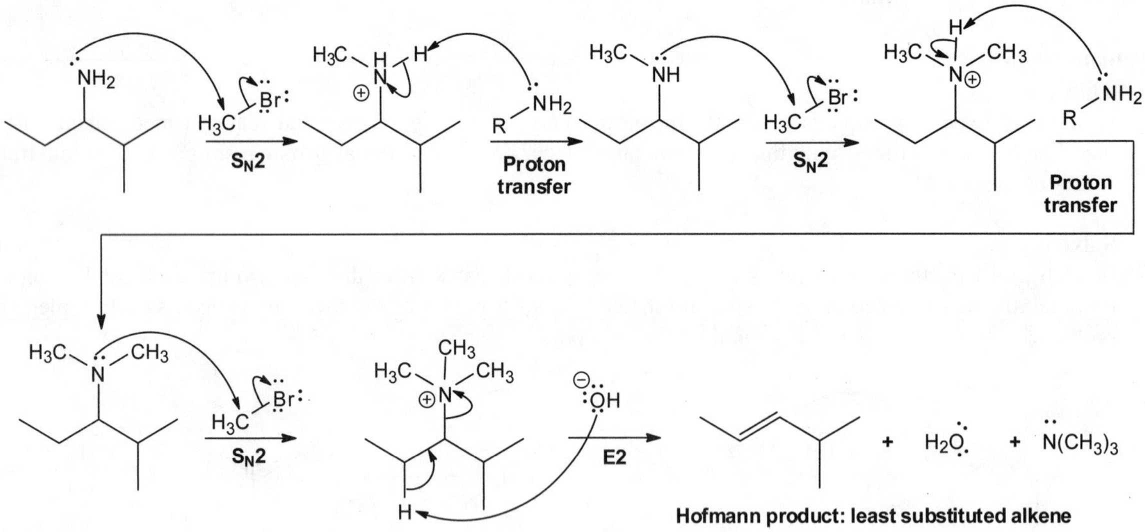

Proton transfer

Proton transfer

Hofmann product: least substituted alkene

Problem 10.28

Think

Is a good leaving group present? On what type of carbon is the leaving group attached? What are the key electron-rich and electron-poor species? Is substitution or elimination favored? Is a unimolecular or bimolecular mechanism favored? Are proton transfer reactions involved?

Solve

(a) HO^- is not a good leaving group, but an S_N2 reaction with PBr_3 will turn HO^- into an $HOPBr_2$ leaving group (a good leaving group). The Br^- then attacks the electrophilic carbon via an S_N2 reaction and inverts the stereochemistry. NC is a good nucleophile and another S_N2 reaction results, which also inverts the stereochemistry at the electrophilic carbon.

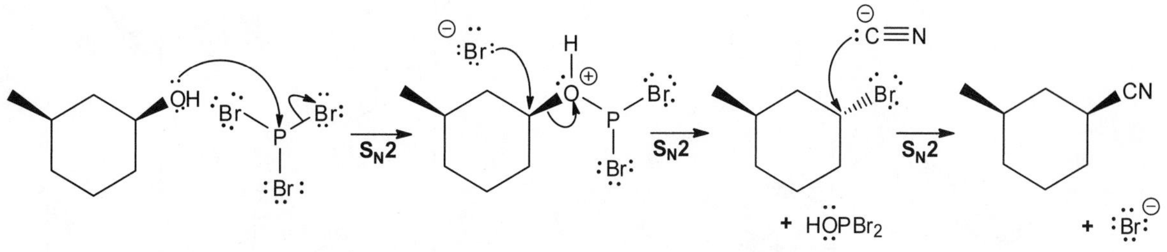

(b) In a basic medium, the α H is deprotonated to form the enolate anion. The enolate anion acts as a nucleophile to attack the Br–Br and form a C–Br bond via an S_N2 mechanism. The last two steps are repeated to form another C–Br bond.

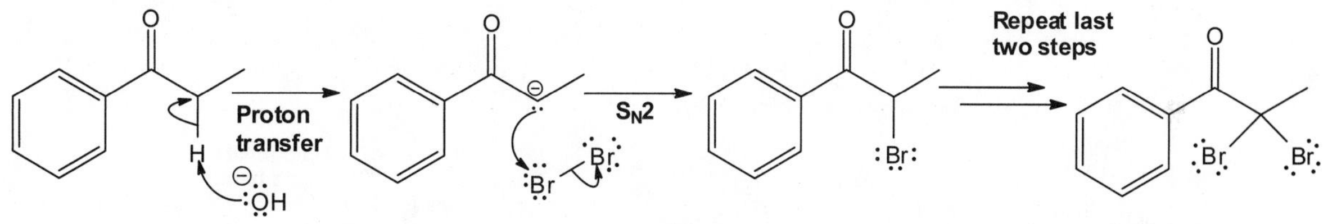

(c) In the presence of a strong base, the HO group is deprotonated to form an alkoxide. The alkoxide acts as a nucleophile to form an epoxide via an S_N2 mechanism with Br^- as the leaving group.

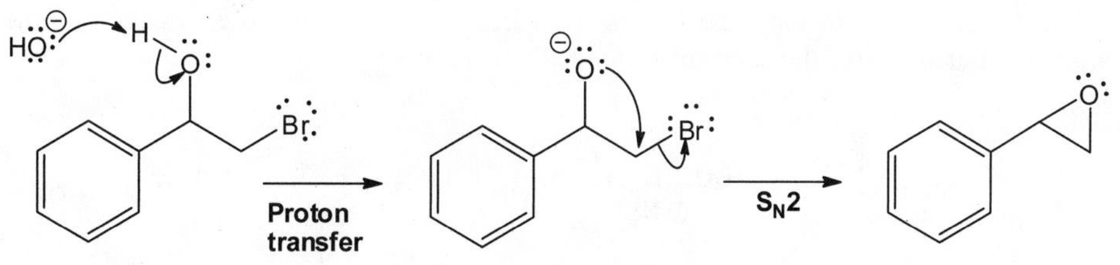

In a neutral reaction medium, the less substituted carbon is attacked by the Cl^- to open the epoxide ring. In the presence of a weakly acidic solvent (water), the alcohol is the final product.

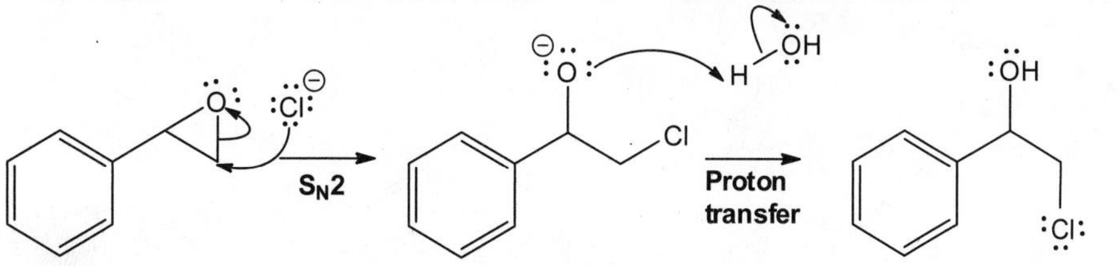

In an acidic reaction medium, the epoxide O is protonated and the more substituted carbon is attacked by the Cl^-.

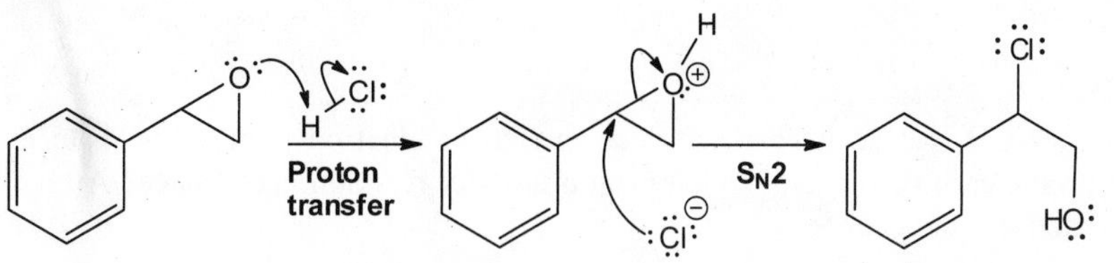

(d) In an acidic medium, the O is protonated. Water then acts as a base to remove the α H and form an enol. The enol acts as a nucleophile to attack the Br–Br and form a C–Br bond via an S_N2 mechanism. The last step is a proton transfer to form an uncharged product.

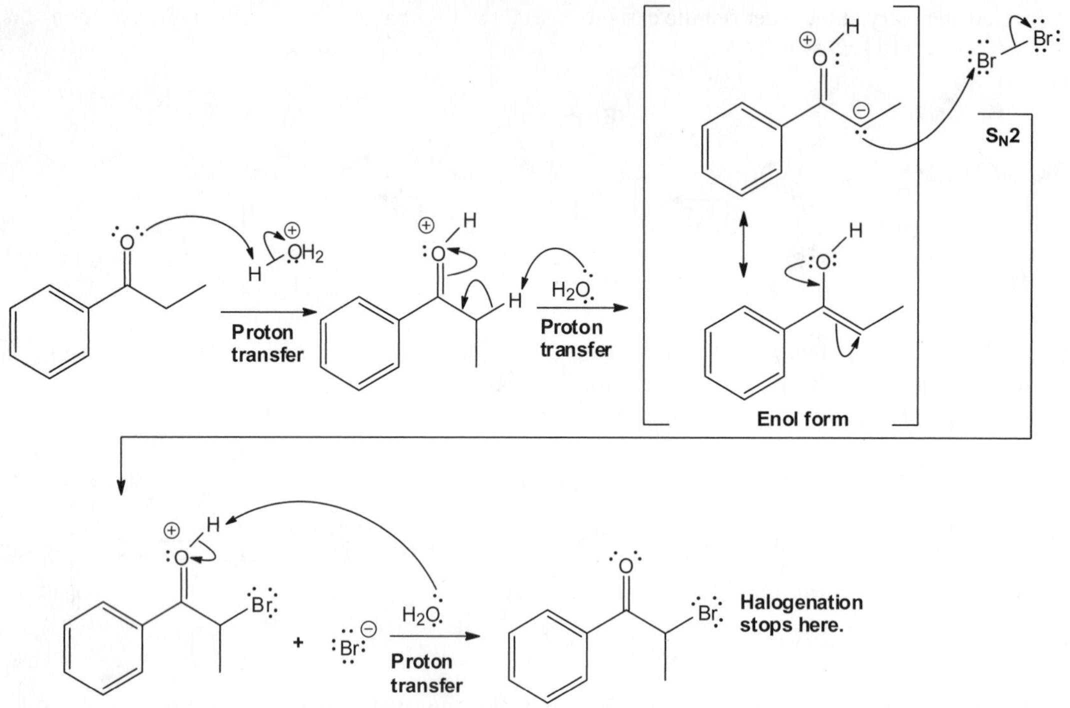

(e) This is a Hofmann elimination. The first several steps are sequential alkylations of the amine N, which produce a quaternary ammonium ion. The final step is E2, which due to the excessive bulkiness of the leaving group, takes place to give the least substituted alkene product.

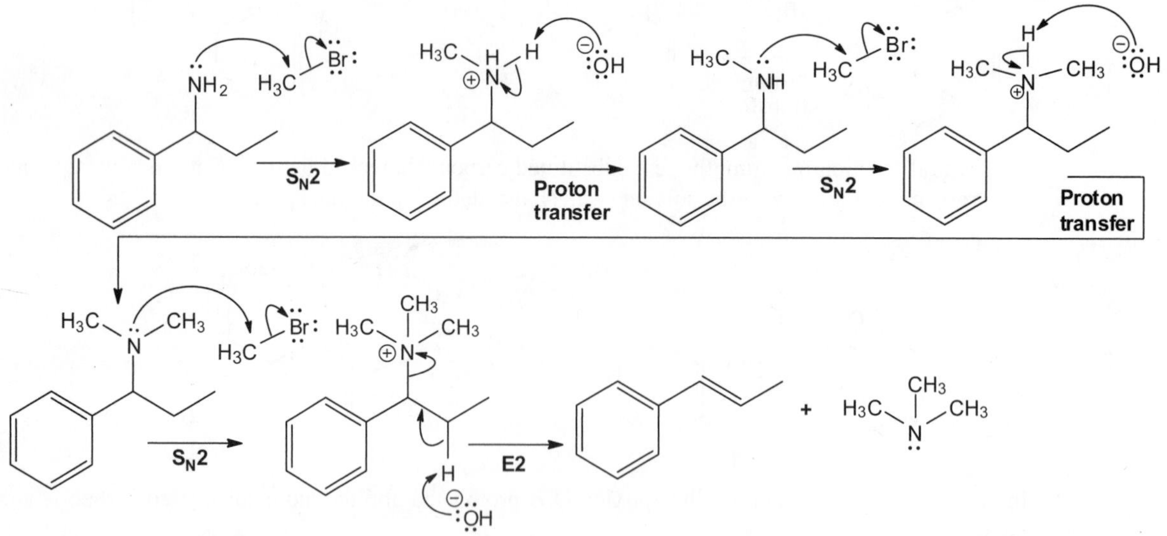

Problem 10.29

Think

What is the Lewis structure for CH_2N_2? Is the carbon electron rich or electron poor? What site is basic on CH_2N_2? What site is acidic on the carboxylic acid? What proton transfer reaction takes place? After the proton transfer reaction, what S_N2 reaction can occur?

Solve

In diazomethane, the C atom bears a negative formal charge and, therefore, is electron rich and basic. In Step 1, the C atom of diazomethane is protonated by the acidic proton of the carboxylic acid. In Step 2 the carboxylate anion is electron rich and acts as the nucleophile in the ensuing S_N2 reaction. The protonated form of $CH_3N_2^+$ is electron poor and acts as the substrate. $N_2(g)$ is a great leaving group. Each of these reactions converts a carboxylic acid into a methyl ester.

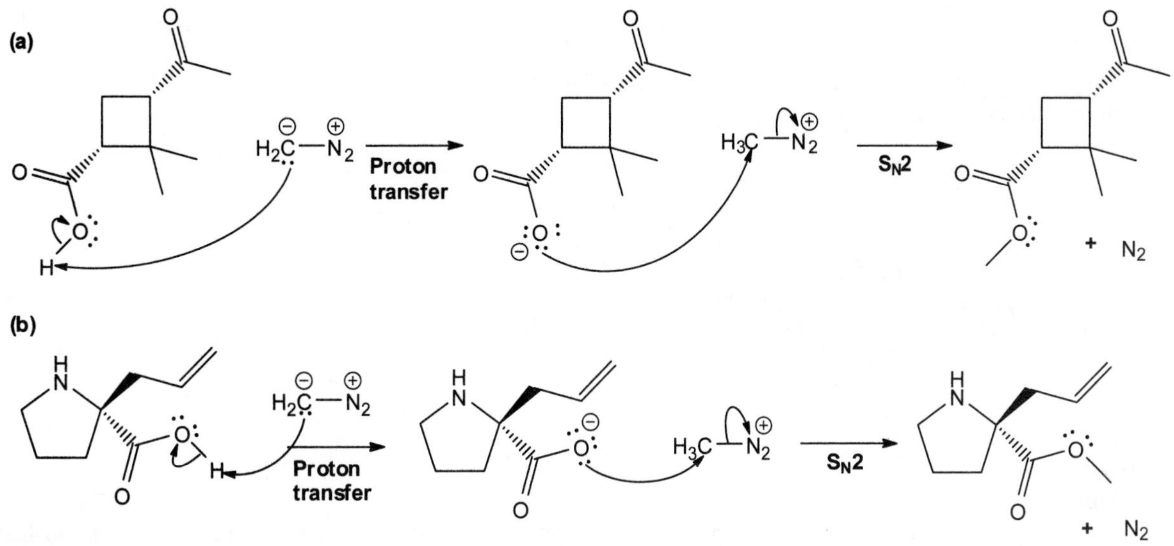

(a) H_2C^{-}—N_2^{+} → Proton transfer → H_3C—N_2^{+} → S_N2 → + N_2

(b) H_2C^{-}—N_2^{+} → Proton transfer → H_3C—N_2^{+} → S_N2 → + N_2

Problem 10.30

Think

What functional group does diazomethane form with a carboxylic acid? Identify that functional group in each product listed. What bond formed? What carboxylic acid formed the product shown?

Solve

Diazomethane converts a carboxylic acid to a methyl ester (boxed). So each methyl ester can be produced starting with a carboxylic acid.

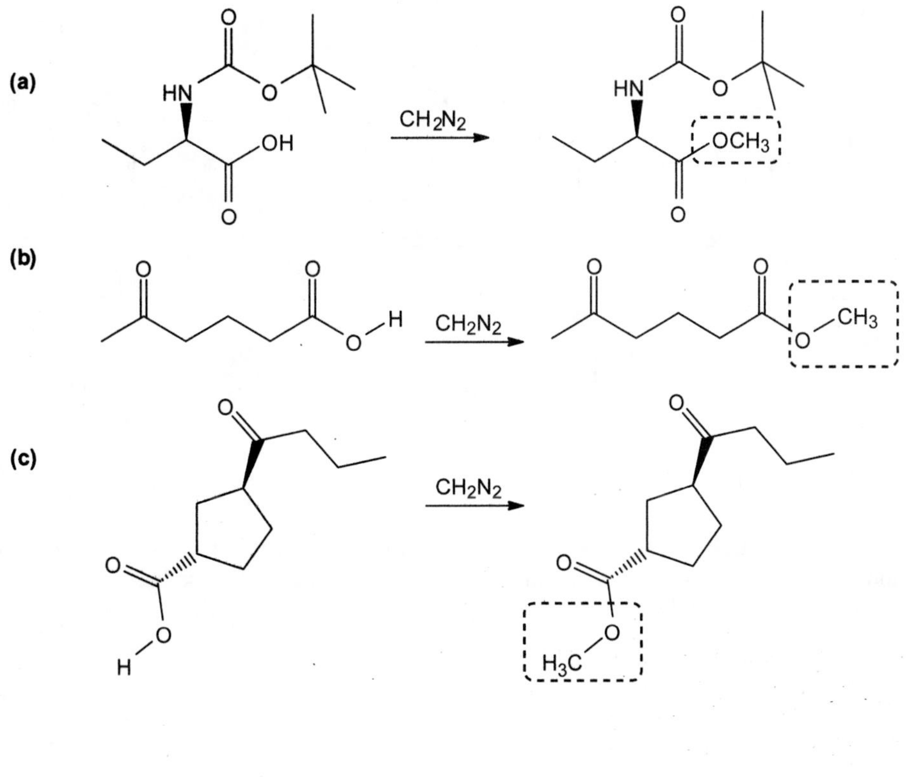

(a) HN, OH → CH_2N_2 → HN, OCH_3

(b) → CH_2N_2 → CH_3

(c) → CH_2N_2 → H_3C

Problem 10.31

Think

Is HO⁻ a good leaving group? How does PCl_3 or PBr_3 turn HO⁻ into a good leaving group? What two elementary steps are involved in the reaction? What happens to the stereochemical configuration at the electrophilic carbon?

Solve

Hydroxide is not a good leaving group and PCl_3 turns the HO⁻ into $HOPCl_2$, which leaves as an uncharged compound. The reaction occurs via two back-to-back S_N2 reactions. The first S_N2 breaks the P–Cl bond and forms the O–P bond. The second S_N2 breaks the C–O bond, forms the C–Cl bond, and inverts the stereochemistry of the carbon.

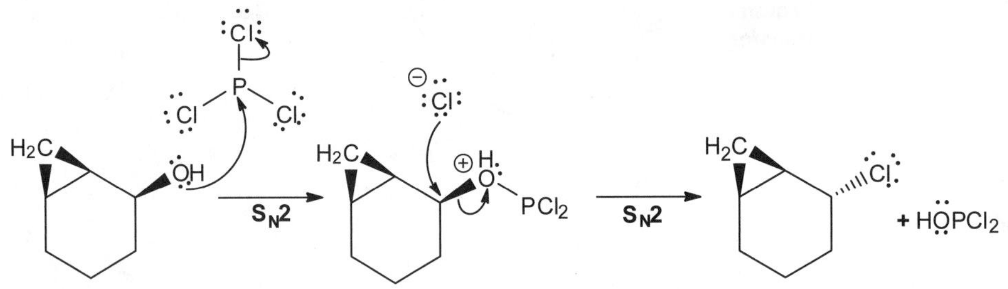

Problem 10.32

Think

Are RO⁻ and HO⁻ good leaving groups? How does an acidic medium turn RO⁻ and HO⁻ into good leaving groups? On what type of carbon are the leaving groups attached? Does this occur via S_N1 or S_N2?

Solve

All steps are equilibria. Protonation of the O atom makes a good leaving group, ROH, which is replaced by a CH_3CH_2OH nucleophile in an S_N2 step. After two more proton transfers, H_2O is produced as a leaving group, which is then displaced by a second CH_3CH_2OH nucleophile in an S_N2 step. A final proton transfer produces the uncharged product.

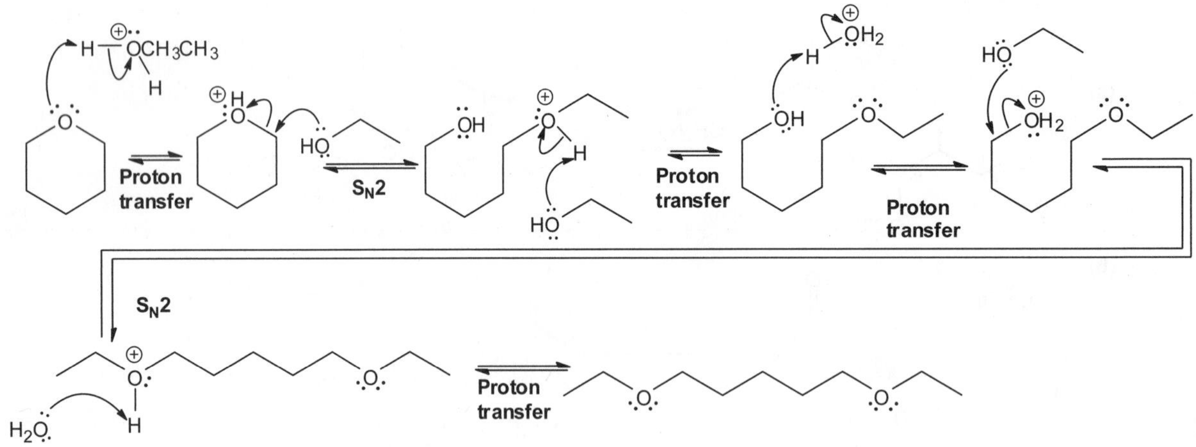

Problem 10.33

Think

Is sodium carbonate a strong or weak base? Is H_2N^- a strong or weak base? What is the pK_a of the C–H of a monoketone and diketone? Which C–H is depprontated in each reaction? Why? Will these bases deprotonate a monoketone or diketone reversibly or irreversibly?

Solve

Sodium carbonate is a weak base (the pK_a of HA is 6.3). The pK_a of a monoketone is about 20, and that of a diketone is 8.9 (Appendix A). Therefore, deprotonation takes place reversibly with sodium carbonate, producing the more highly substituted enoate anion. That proton is on the central carbon, which is why the subsequent alkylation takes place there. By contrast, H_2N^- deprotonates irreversibly to produce the kinetic enolate anion. So deprotonation takes place at the less sterically hindered terminal carbon, as does the subsequent S_N2 reaction.

Problem 10.34

Think

To form a stereospecific product, is S_N2 or S_N1 required? In this example, is the stereochemistry inverted or retained? How many reactions need to take place?

Solve

To replace an atom stereospecifically, we need an S_N2 reaction, but S_N2 occurs with inversion. The solution is to do two inversions. First, treat the bromoalcohol with base to form an epoxide, then let the mixture cool and allow the epoxide to open with aqueous acid.

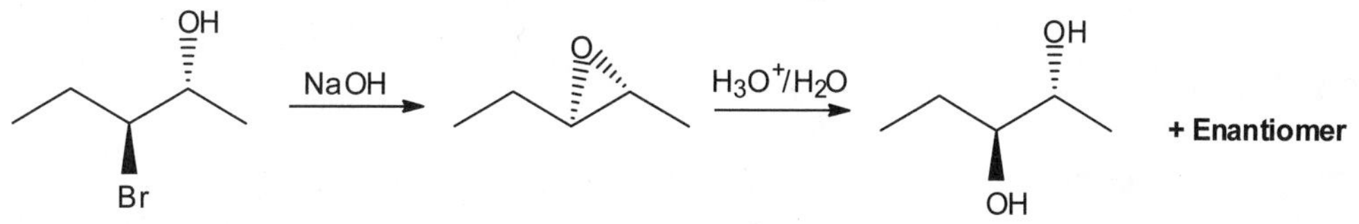

OH Br NaOH O H_3O^+/H_2O OH OH + Enantiomer

Problem 10.35

Think

Under what reaction conditions does alkylation at the α carbon occur? As more alkyl groups are added to the α carbon, what happens to the sterics of that carbon? How does this inhibit multiple alkylation steps?

Solve

Once an alkyl group has been attached to the α carbon, it provides steric hinderance to further alkylation. Alkylations take place by the S_N2 mechanism, which is highly sensitive to steric hindrance, so each subsequent alkylation would take place more slowly.

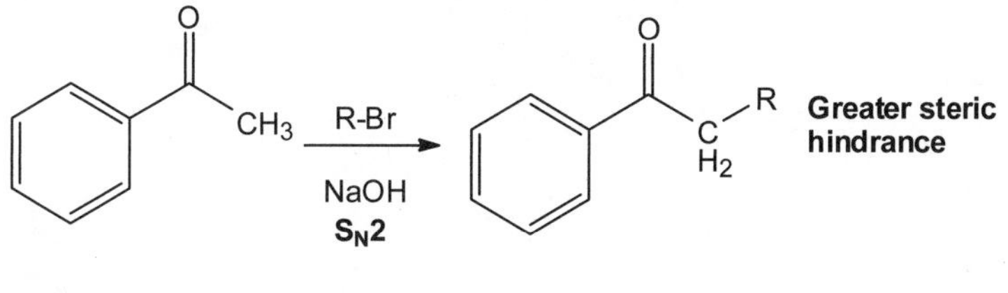

O CH_3 R-Br NaOH S_N2 O C H_2 R **Greater steric hindrance**

Problem 10.36

Think

How many E2 reactions take place for a vicinal dihalide in the presence of a $NaNH_2$? What is at the end of a terminal alkyne that is not present in an internal alkyne? What reaction takes place between $NaNH_2$ and a terminal alkyne?

Solve

One equivalent of $NaNH_2$ is used for each elimination of HBr, as shown below. Because there are two HBr eliminations, two equivalents of $NaNH_2$ are required.

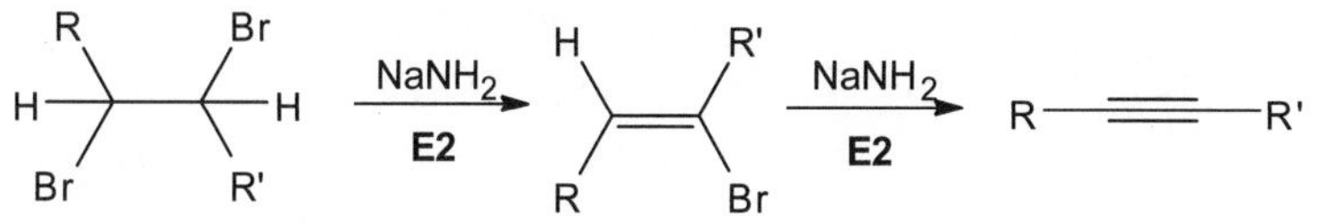

R Br H H Br R' $NaNH_2$ **E2** H R' R Br $NaNH_2$ **E2** R—≡—R'

If the product is a terminal alkyne, an acidic H is present, whose pK_a is about 25. This is easily deprotonated by $NaNH_2$. Thus, a third equivalent of $NaNH_2$ is spent in such a deprotonation.

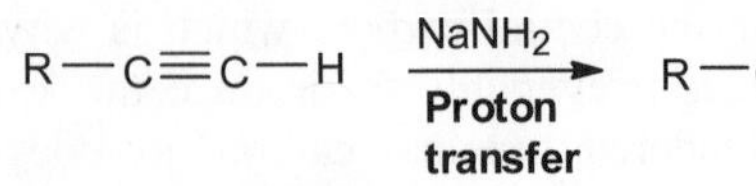

$$R-C\equiv C-H \xrightarrow[\text{Proton transfer}]{NaNH_2} R-C\equiv C:^{\ominus}$$

Problem 10.37

Think

What must be true about the orientation of the leaving group and the adjacent H in an E2 reaction? Can this occur in an epoxide? Is rotation limited in rings?

Solve

An E2 mechanism is favored when the proton on the first carbon and the leaving group on the second carbon are anti-coplanar, giving a 180° dihedral angle. The presence of the small epoxide ring forces a dihedral angle of 120° between these two groups, preventing the anti-coplanar conformation, thus disfavoring elimination.

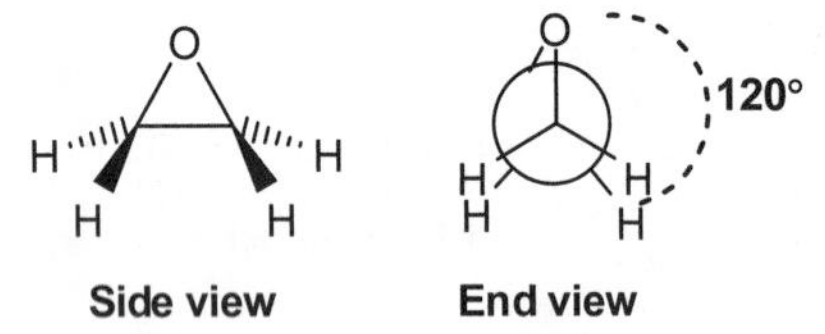

O, H, H, H, H — 120°

Side view End view

Problem 10.38

Think

What functional group results from this reaction? What is the geometry of the carbon atoms in that functional group? How does ring strain factor into the stability of the products? Will this increase or decrease the rate of product formation?

Solve

These reaction conditions normally convert an alkene into an alkyne via E2. The intended products for each reaction are shown below. In the case of the cyclic alkene reactant, a cyclic alkyne would be produced, and with only six atoms making up the ring, the structure is extremely strained. This will significantly decrease the rate of its formation.

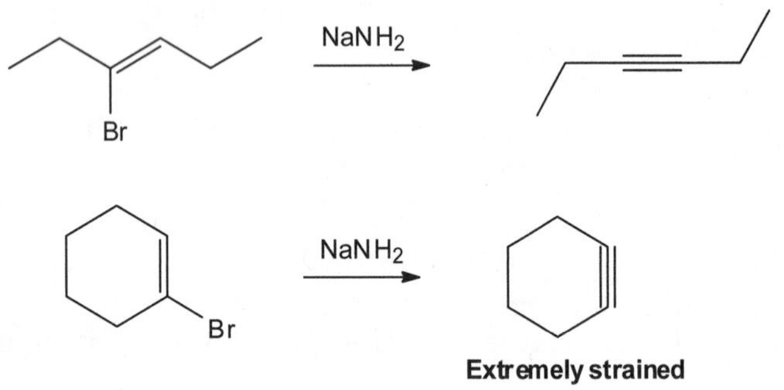

Br $\xrightarrow{NaNH_2}$

Br $\xrightarrow{NaNH_2}$

Extremely strained

Problem 10.39

Think

In the presence of a strong base, what reaction "normally" occurs with a vinyl bromide? In the second product, how is the deuterium removed? Are proton transfer reactions involved? Is a resonance structure possible in any of the intermediates?

Solve

The first product results from a "normal" E2 reaction, with loss of H on C2 and bromine on C3.

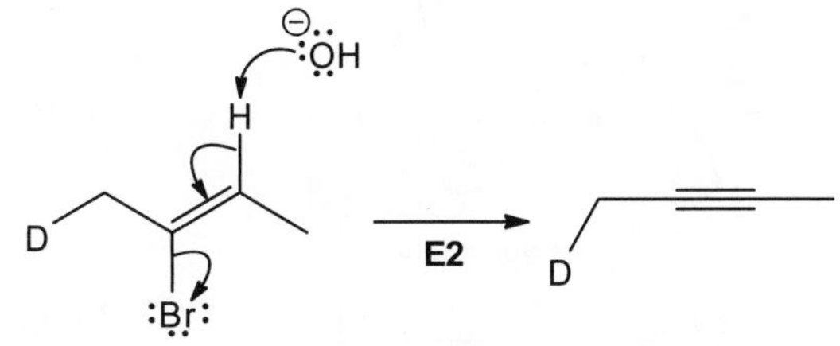

The second product must be the result of an acid–base exchange reaction, where H is traded for D on C1. Then the E2 elimination occurs normally. This acid–base exchange is facilitated by the resonance stabilization in the allylic anion as well as by the heat that is supplied. This is typically a very endothermic proton transfer.

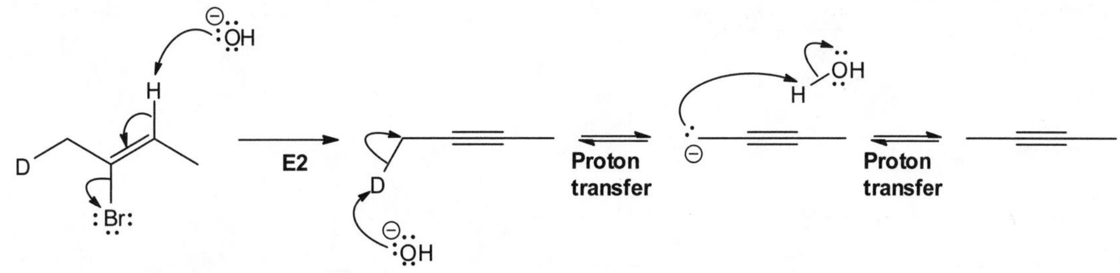

Problem 10.40

Think

Does a ring inhibit the nitrogen methylation step? What forms on the N atom in the presence of excess CH_3Br? Is this sterically bulky? In a cylcohexane ring, what is the most stable conformation of a bulky group? How does this conformation limit/shut down E2?

Solve

Although the methylation of N proceeds normally (several S_N2 steps), in the resulting chair conformation of the ammonium ion, the trimethylammonium substituent is essentially locked in the equatorial position, just as a *t*-butyl group would be. In that conformation, there are no H atoms on adjacent C atoms that can be anti-coplanar to the N leaving group, so E2 is shut down.

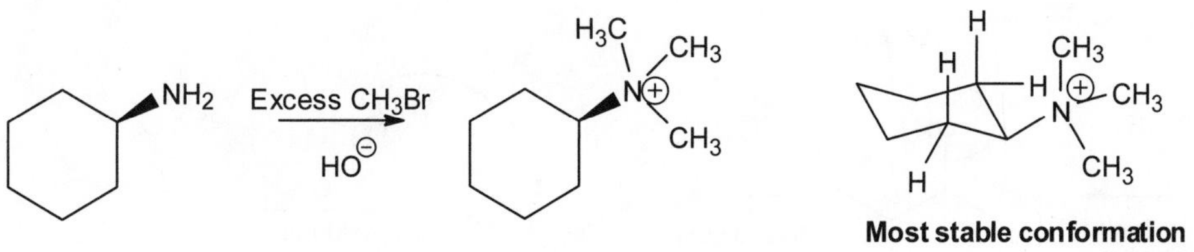

Most stable conformation

Problem 10.41

Think

What is the mechanism by which alcohols form ethers in acidic media? If two alcohols are mixed together in acidic media, is R–O–R′ the only ether possible? What is in solution that can act as a nucleophile? What is in solution that can act as a substrate with a sufficient leaving group?

Solve

In addition to the mixed ether product that is desired, there are two ethers that are produced by self-reaction among the alcohols. The reactions occur via S_N1.

Proton transfer

Heterolysis

Coordination

ROH

Proton transfer

Proton transfer

Heterolysis

Coordination

ROH

Proton transfer

Proton transfer

Heterolysis

Coordination

ROH

Proton transfer

Problem 10.42

Think

What is another possible route to synthesize an ether? What is the mechanism of that reaction? Which R group should be the alkyl halide and which R group should be the alkoxide? Is more than one route possible?

Solve

The Williamson synthesis of ethers is ideal for this reaction. The reaction requires an alkoxide anion (which can be produced from an alcohol) and an alkyl halide (synthesized from the alcohol using PBr_3), as shown below, and can be done in two different ways.

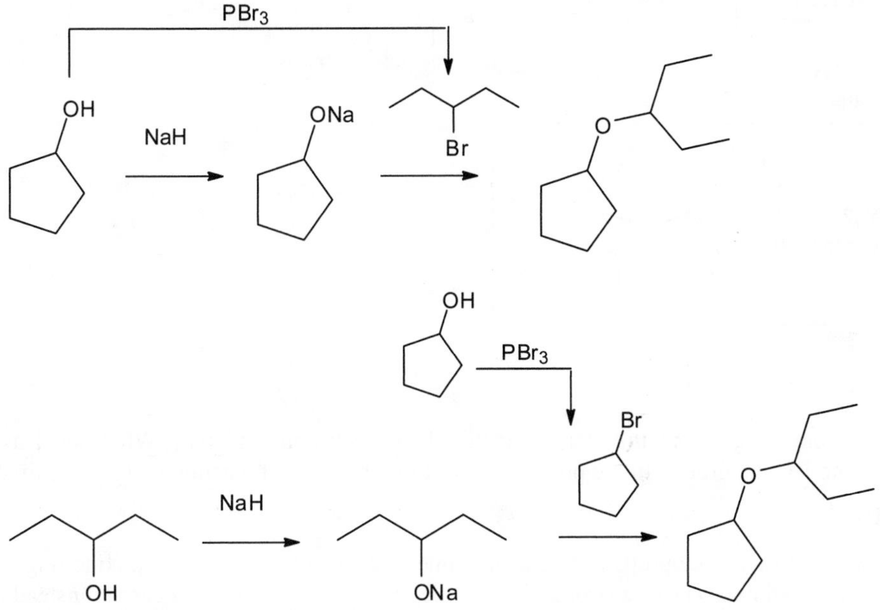

PBr3
OH NaH ONa Br

OH
PBr3
Br
NaH
OH ONa

Problem 10.43

Think

In each product, what bonds broke and formed to make the product from the starting material? In the presence of a strong base, how does an alcohol react? What electron-rich and electron-poor products result? How does the epoxide form with the labeled ^{18}O?

Solve

The first "product" is the original compound, so the mechanism simply needs to explain how the second product forms. A proton transfer creates a nucleophilic labeled O, which then displaces the unlabeled O, trading one strained ring for another. All steps are equilibria, so a mixture of the two compounds results.

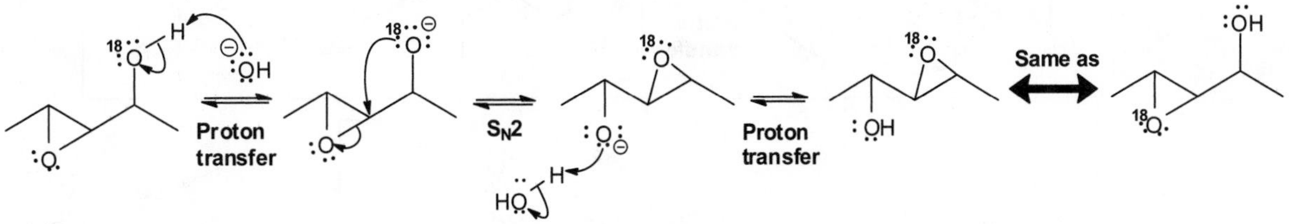

Proton transfer
S_N2
Proton transfer
Same as

Problem 10.44

Think

For the O atom to displace only the bromide attached to the labeled ^{13}C, what must be true about the conformation of the O and the Br? What is the mechanism that takes place to form this product?

Solve

For O to displace only the Br on the unlabeled carbon, and not the other bromide, an S_N2 reaction must take place, not an S_N1. For an S_N2 reaction, the molecule must attain a conformation that allows backside attack on the unlabeled carbon. Thus, the Br atom that is lost must be trans to the O atom. The Br on the labeled C must be cis to the O atom. Either the stereoisomer shown or its enantiomer is correct.

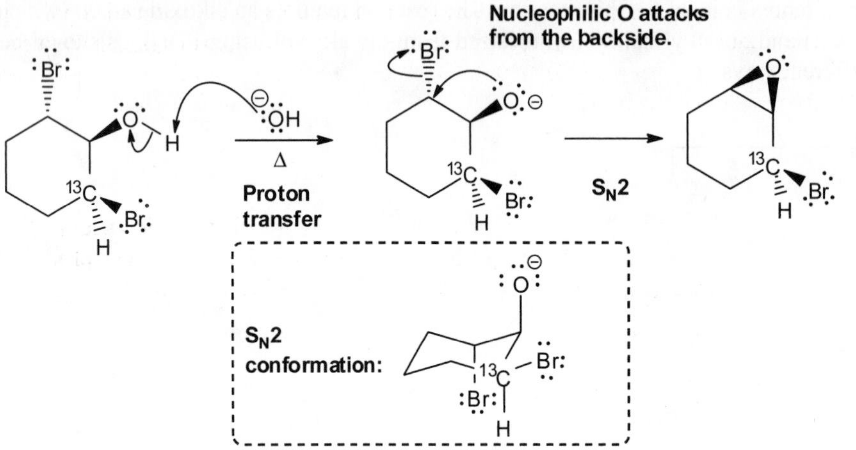

Problem 10.45

Think

What bonds formed and broke to go from the starting material to the product shown? What acted as the nucleophile, electrophile, and leaving group? In the presence of a strong base, what reaction takes place first?

Solve

Fundamentally, the formation of each cyclopropane ring is the same as the one that forms an epoxide ring under basic conditions. First, deprotonation produces a strong nucleophile (but in this case at an α carbon instead of at an OH oxygen), and then the leaving group is displaced in an S_N2 reaction.

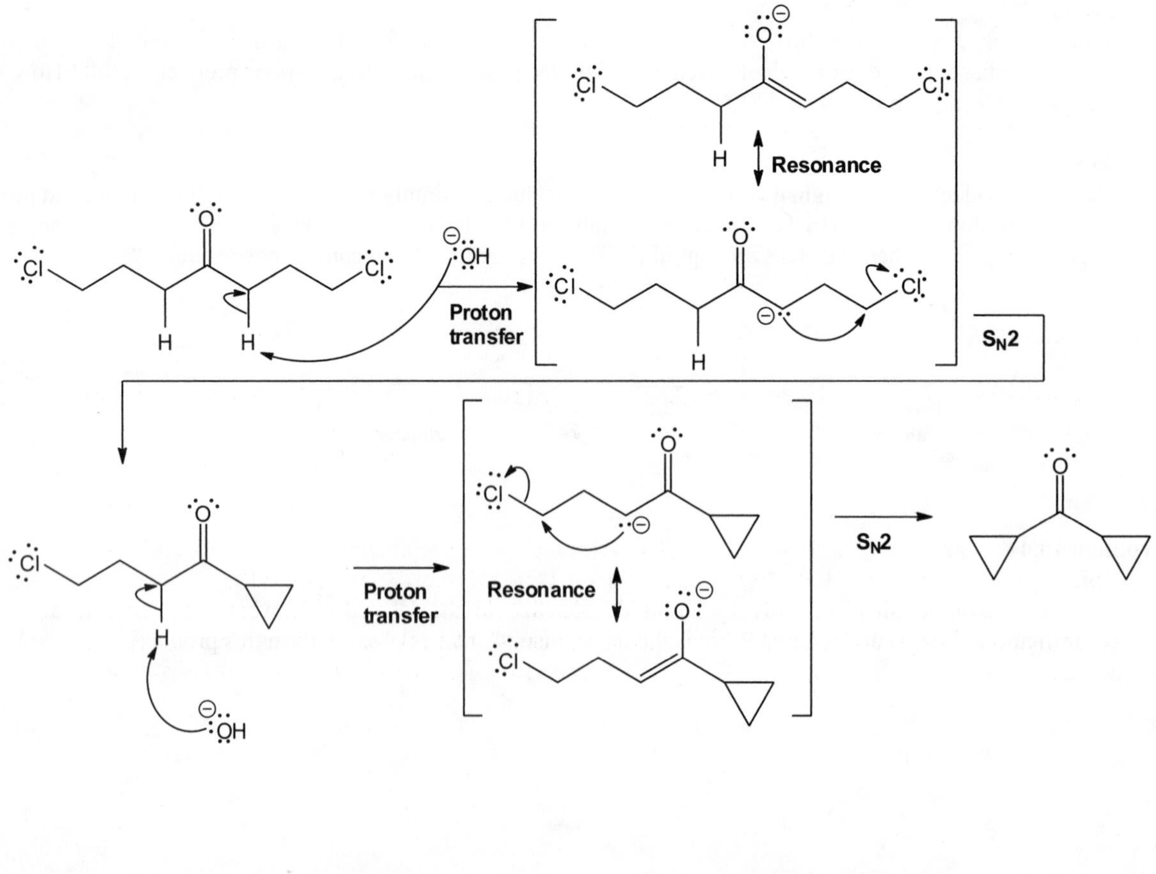

Problem 10.46

Think

Is F_2 polar or nonpolar? Compare the size of F_2 to Cl_2, Br_2, and I_2. How does size affect polarizability? Why is polarizability necessary for halogenation to occur at the α carbon?

Solve

The electron-rich to electron-poor driving force for such reactions originates from the induced dipole on the molecular halogen atom, as depicted in Figure 10-4. Because F_2 has relatively few electrons, and also because F is so highly electronegative, F_2 is not very polarizable. Therefore, the induced dipole on F_2 is relatively small, making the driving force for the reaction rather weak.

Problem 10.47

Think

What is the connectivity of the epoxide formed from heating 3-bromobutan-2-ol in basic media? For the epoxide to have no optical activity, are the CH_3 groups cis or trans? In the S_N2 reaction that produces the epoxide, from what direction must the nucleophilic atom attack the C that has the leaving group?

Solve

The epoxide that will be formed has the following connectivity:

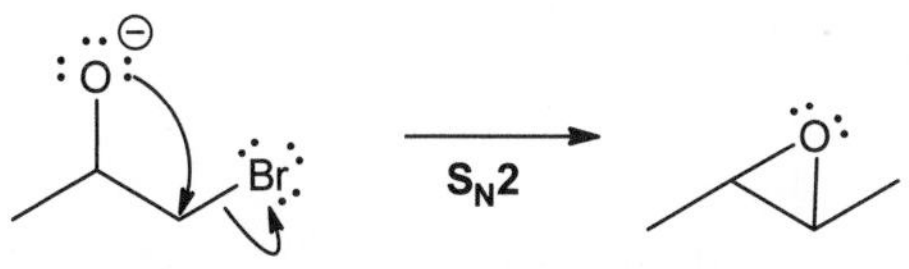

S_N2

To be optically *inactive*, the epoxide must have a plane of symmetry, as shown below. One haloalcohol that can be used to produce the epoxide is shown below. To arrive at this precursor, we have to take into consideration the fact that, in an S_N2 step, the alkoxide nucleophile must attack the C–Br from the side opposite the Br leaving group.

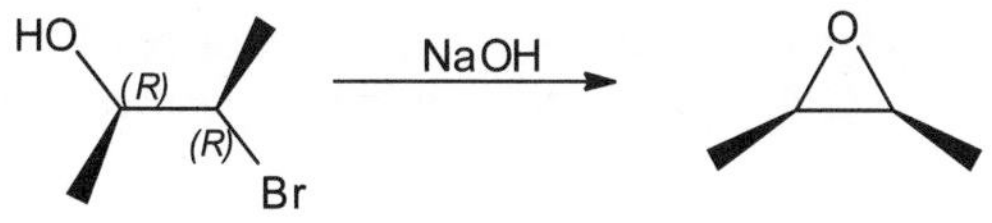

HO (R) (R) Br — NaOH → O

The enantiomer can also be used.

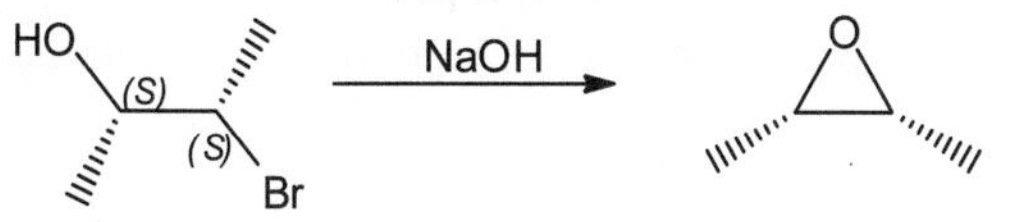

HO (S) (S) Br — NaOH → O

Problem 10.48

Think

Does the reaction occur in a basic, neutral, or acidic medium? Does the nucleophilic attack occur at the more or less substituted carbon? Does the nucleophilic substitution reaction occur first or does a proton transfer reaction occur first? When nucleophilic substitution does occur, from which direction does the nucleophile attack?

Solve

When nucleophilic attack takes place under neutral or basic conditions, the least sterically hindered C is attacked, which is the one on the right. This is the case for **(a)**, **(b)**, **(d)**, **(f)**, **(g)** and **(h)**. Under acidic conditions, the O atom is protonated first, and the carbon that can best handle a positive charge is the one that is attacked subsequently. This is the benzylic carbon, as shown in **(c)** and **(e)**. In all cases, the nucleophile attacks the epoxide C from the side opposite the C–O bond to open the ring. See the reactions on the next page.

(a) NaN_3, H_2O → HO, N_3

(b) KCN, $HOCH_2CH_2OH$ → HO, CN

(c) HCN, $HOCH_2CH_2OH$ → OH, NC

(d) 1. $LiAlH_4$ 2. $H_3O^{\oplus}$ → HO, H

(e) CH_3OH, H_2SO_4 → OH, H_3CO

(f) CH_3ONa, CH_3OH → HO, OCH_3

(g) $C_6H_5CH_2NH_2$, CH_3OH → HO, $HN—CH_2C_6H_5$

(h) 1. CH_3Li 2. $H_3O^{\oplus}$ → HO, CH_3

Problem 10.49

Think

Under acidic conditions, what reaction takes place first? What nucleophile is produced as a result? Does nucleophilic attack occur at the less sterically hindered C or the C that can better stabilize a positive charge? How does the allylic C help stabilize the positive charge?

Solve

The epoxide O is protonated first under these acidic conditions. Then, nucleophilic attack takes place at the C that can better stabilize a positive charge. The one on the left is allylic. Similar to how allylic carbocations are stabilized by resonance delocalization of the positive charge, the partial positive charge, δ^+, which builds up is better handled on the allyic C here.

H–Cl: → Proton transfer → δ^+ δ^+ :Cl:⊖ → S_N2 → :Cl: :OH

On allyl C, a positive charge is better stabilized.

Problem 10.50

Think

Does the cyclohexane ring exist as a planar structure or a chair? Which group on the ring is the bulkiest? On what position, axial or equatorial, should this group be positioned? How does that affect the OH and Br groups? Why is the S_N2 reaction shut down?

Solve

Formation of an epoxide via an S_N2 reaction is favored by the attack of the O^- from the side opposite the leaving group, Br. In this case, the *t*-butyl group locks the Br and OH groups in equatorial positions because of the *t*-butyl group's bulkiness. In this configuration, the OH and Br groups are gauche to each other, making it impossible for attack to occur.

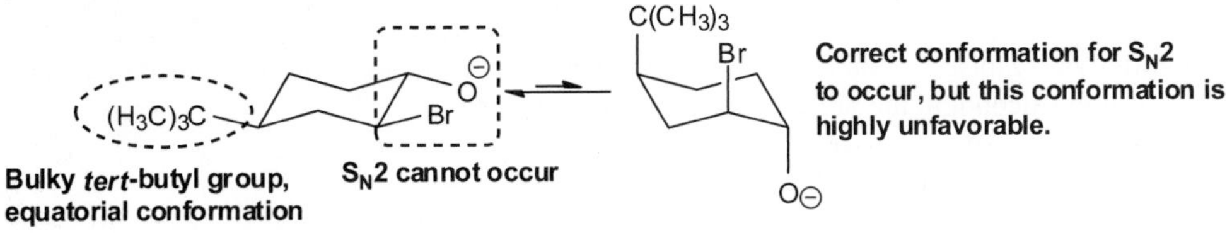

Bulky *tert*-butyl group, equatorial conformation

S_N2 cannot occur

Correct conformation for S_N2 to occur, but this conformation is highly unfavorable.

Problem 10.51

Think

What reagent and conditions will permit alkylation at the N atom to occur? How many times must this alkylation reaction occur? What must be true about the alkyl halide reagent to form a ring at the N?

Solve

Alkylation of the amine N takes place twice. To form a ring, the dibromide below is required.

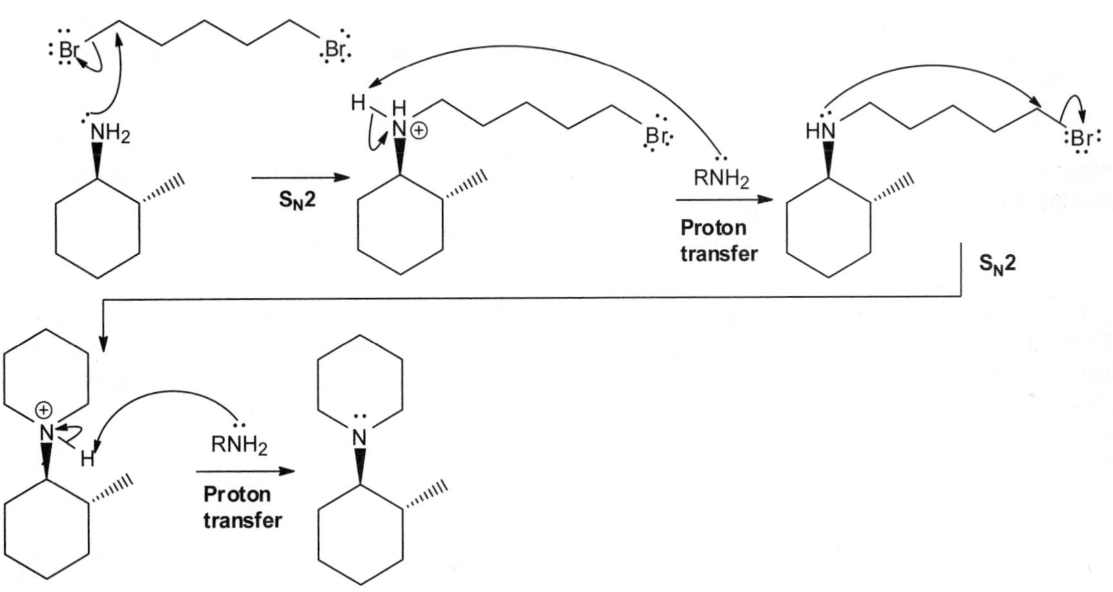

S_N2

RNH_2

Proton transfer

S_N2

RNH_2

Proton transfer

Problem 10.52

Think

Does the reaction occur in a basic, neutral, or acidic medium? Does the nucleophilic attack occur at the more or less substituted carbon? Does the nucleophilic substitution reaction occur first or does a proton transfer reaction occur first?

Solve

Nucleophilic attack does *not* take place under acidic conditions, so we should expect attack to take place at the less sterically hindered carbon of the aziridine ring. Notice that the proton transfer from S to N takes place intermolecularly instead of intramolecularly.

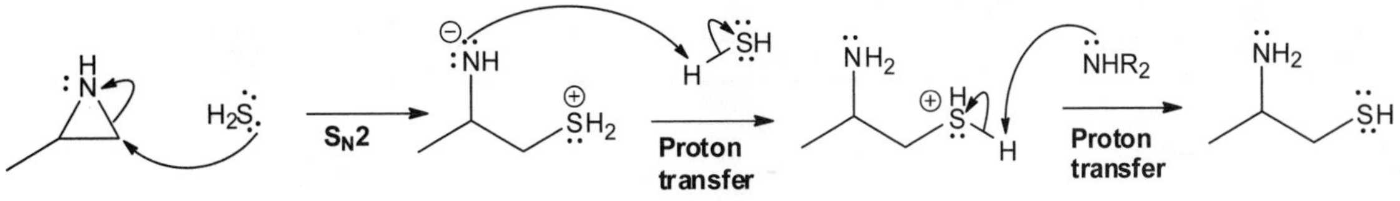

S_N2

Proton transfer

Proton transfer

Problem 10.53

Think

Does substitution or elimination have to occur to form the final alkene product? The starting material is the same. How is it possible to carry out an elimination reaction to put the alkene in two different locations? Which alkene forms from Zaitsev elimination and which alkene forms from Hofmann elimination?

Solve

(a) This is a straightforward Zaitsev elimination, in which the product is the most highly substituted alkene. So we can simply heat the alcohol in the presence of concentrated acid.

OH — conc. H_3PO_4, Δ →

(b) The product is the least highly substituted alkene, which is the product of Hofmann elimination. Hofmann elimination requires an amine precursor. That can be produced from the starting alcohol after using PBr_3 to convert the alcohol to the bromide. The synthesis appears as follows:

OH — PBr_3 → Br — NH_3 (1 equiv) → NH_2 — CH_3Br (excess), NaOH, Δ →

The alkylation of ammonia will not be an ideal step because of the mixture of the polyalkylated products.

Problem 10.54

Think

What substrate is present initially? What nucleophile is present? What product forms from the reaction of a primary amine in the presence of excess CH_3I? What is the leaving group that is produced as a result of that reaction? Strong base and heat results in what reaction? How many times does this reaction occur?

Solve

This reaction consists of back-to-back Hofmann eliminations. An amine is alkylated by CH_3I three times to form a quaternary ammonium ion. Then, $N(CH_3)_3$ acts as a leaving group. In the presence of a strong base and heat, E2 is the dominant reaction mechanism. An alkene is formed. The whole reaction occurs again to form the alkyne. (It would also be acceptable to show both amines converted to quaternary ammonium ions first, followed by back-to-back E2 reactions.)

NH_2, NH_2 — H_3C–I, S_N2 → NH_2 — $\ddot{R}NH_2$, Proton transfer → NH, NH_2 — H_3C–I, S_N2 → NH_2 — RNH_2, Proton transfer →

H_3C, CH_3, N, NH_2 — H_3C–I, S_N2 → CH_3, H_3C, CH_3, H_2N, H — $^{\ominus}OH$, E2 → NH_2 — H_3C–I, S_N2 → H, NH, CH_3 — RNH_2, Proton transfer →

HN, CH_3 — H_3C–I, S_N2 → H_3C, N, CH_3, H — RNH_2, Proton transfer → H_3C, N, CH_3 — S_N2 → H, H_3C, N, CH_3, CH_3 — $^{\ominus}OH$, E2 →

Problem 10.55

Think

What substrate is present initially? What nucleophile is present? What product forms from the reaction of a primary amine in the presence of excess CH_3I? What is the leaving group produced in that reaction? Strong base and heat results in what reaction? How many different adjacent H atoms are present? How many alkene products can result?

Solve

The first part of the Hofmann elimination is alkylation of the amine N to produce a quaternary ammonium ion.

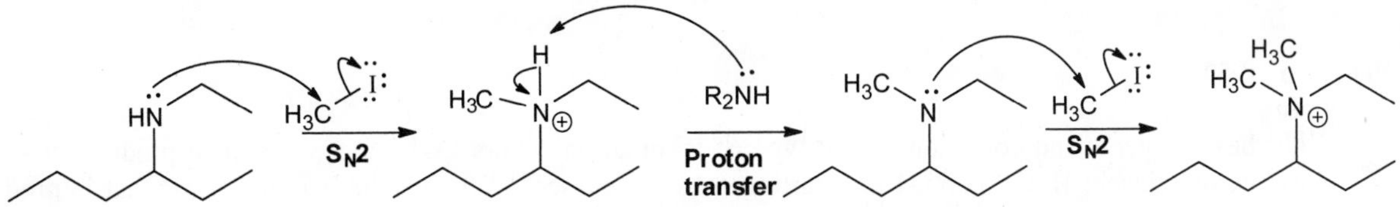

An E2 step can then take place by eliminating the N-containing leaving group along with a H atom on a C atom adjacent to the one attached to N. There are three such C atoms, which, as shown below, can lead to three different alkene products.

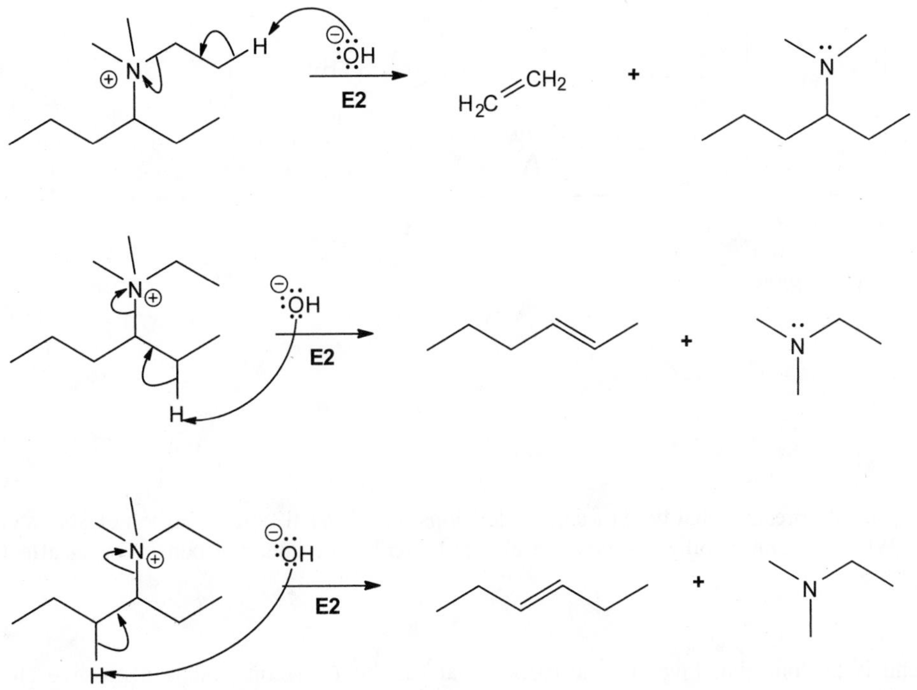

The major alkene product will be the one that is the least substituted, which is that from the first reaction above: ethene.

Problem 10.56

Think

Is the major alkene product the one that is more highly substituted or less substituted? With these reagents and conditions, what type of elimination occurs? Why are two alkene products possible? How many adjacent H atoms must the starting material possess? What is the leaving group?

Solve

These are the conditions for a Hofmann elimination. The two alkene products that are given are the products of E2 elimination, and notice that the major product is the less highly substituted alkene. The precursor is given below with the adjacent H atoms drawn out explicitly.

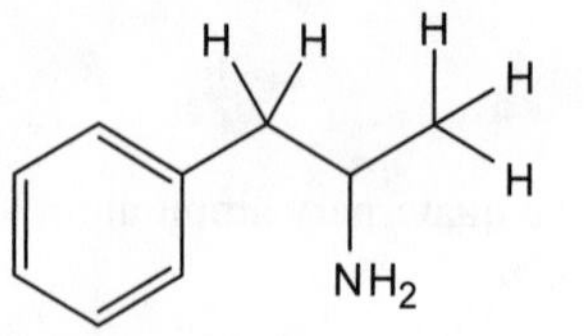

H H H H H NH$_2$

Problem 10.57

Think

With these reagents and conditions, what type of elimination occurs? Why are two alkene products possible? How many adjacent H atoms must the starting material possess? What is the leaving group that is produced from the reaction of excess CH_3I and an amine?

Solve

Each of these provides the conditions necessary for Hofmann elimination. The quaternary ammonium ion intermediate through which each elimination proceeds is given below.

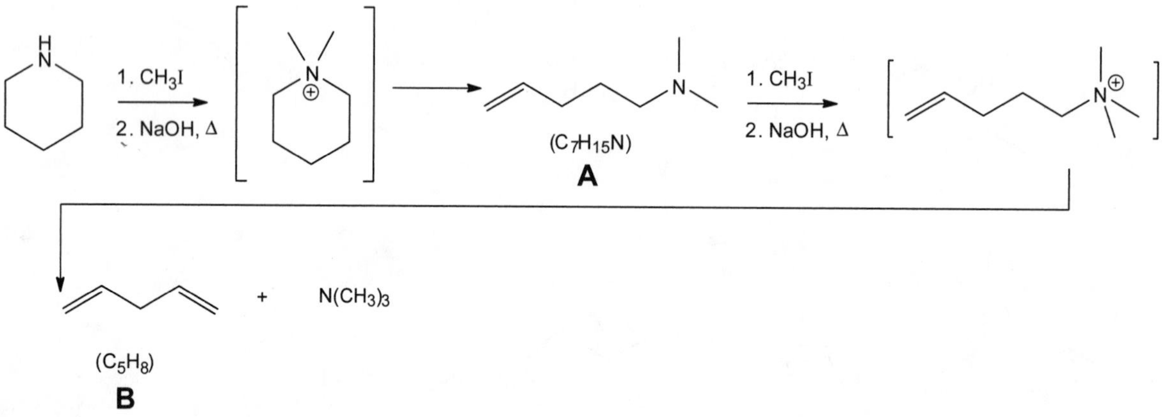

1. CH_3I 2. NaOH, Δ ($C_7H_{15}N$) **A** 1. CH_3I 2. NaOH, Δ

(C_5H_8) **B** + $N(CH_3)_3$

Problem 10.58

Think

When the C–O bond partially breaks, what type of charge develops on C? What type of carbon exists on either side of the epoxide? Which carbon stabilizes a positive charge better? How does the benzene ring affect the charge stability?

Solve

The longer C–O bond is the one that involves the carbon that can better accommodate a positive charge, because a partial breaking of the C–O bond places a partial positive charge on C. Both carbons are secondary, but the one on the left is a benzylic carbon as well. So the C–O bond on the left is longer.

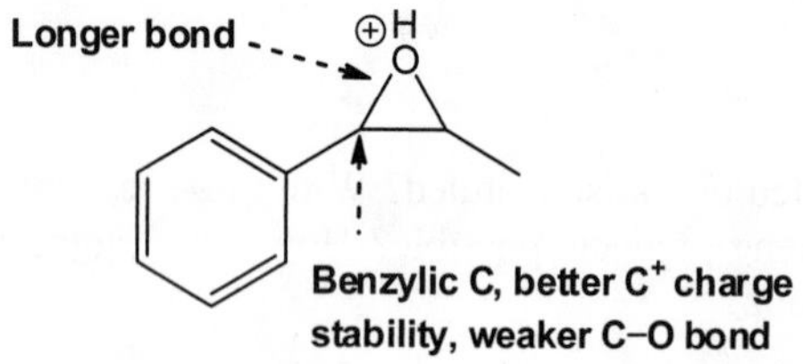

Longer bond ⊕H O

Benzylic C, better C$^+$ charge stability, weaker C–O bond

Problem 10.59

Think

In each reaction, what are the electron-rich and the electron-poor species? Is the reaction set up to do elimination or substitution? Are rearrangements possible? Are any proton transfer reactions involved? In how many elementary steps does each reaction occur?

Solve

In the first reaction, the strong Grignard reagent opens the strained ring, producing the alcohol. The alcohol is converted to Br with PBr_3.

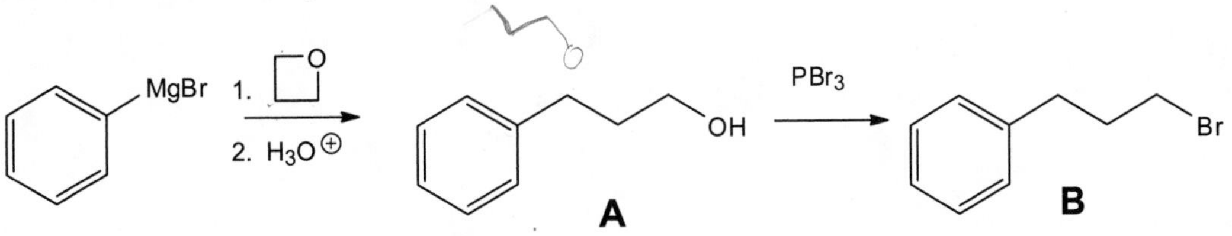

MgBr 1. (oxetane) 2. $H_3O^{\oplus}$ → OH (A) —PBr_3→ Br (B)

$NaOC(CH_3)_3$ in the next reaction deprotonates the ketone reversibly at the more highly substituted α carbon, making an enolate nucleophile. That nucleophile attacks the alkyl halide carbon of compound **B**, which alkylates the α carbon. The next step is bromination of the α carbon, and the alkyllithium reagent in the final step replaces Br in a nucleophilic substitution.

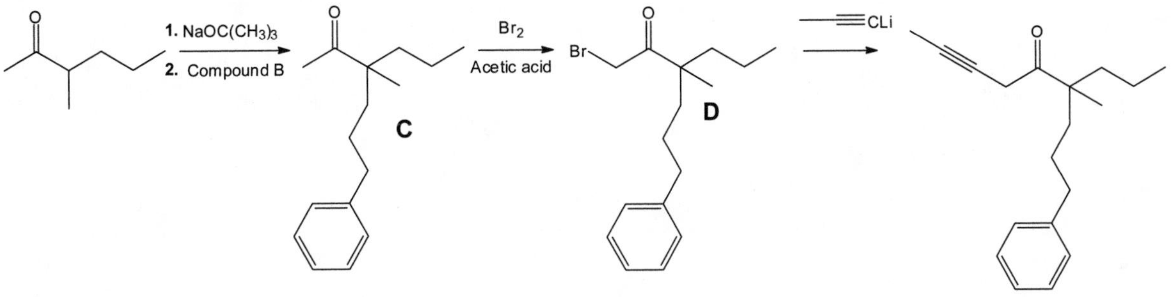

1. $NaOC(CH_3)_3$ 2. Compound B → C —Br_2, Acetic acid→ D —≡CLi→

Problem 10.60

Think

How is this mechanism similar to the alkylation of the α carbon of a ketone? What does the NaH reagent form in the first step? What atom is alkylated?

Solve

This is the same mechanism as alkylation of a ketone. First, the NH proton is removed with the strong base, making the N atom strongly nucleophilic. Then the nucleophile displaces Cl of the alkyl halide.

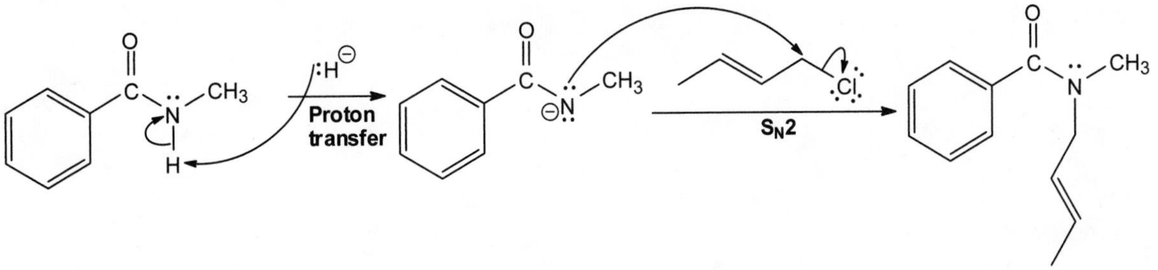

:$H^{\ominus}$ Proton transfer → S_N2 →

CHAPTER 11 | Electrophilic Addition to Nonpolar π Bonds 1: Addition of a Brønsted Acid

Your Turn Exercises

Your Turn 11.1

Think

Consult Equations 11-1 and 11-2. Which bonds are broken and formed? What is the difference in the number and type of atoms between the reactant and product? What happens to the π bond of the C=C?

Solve

In the product, there is an additional H atom and a Br atom that does not appear in the reactant. HBr adds across the C=C double bond of but-2-ene to produce 2-bromobutane. In doing so, a new C–Br and C–H bond form, the π bond of the C=C breaks, and the H–Br bond breaks.

Your Turn 11.2

Think

Which bonds broke and formed to result in the formation of a C^+ and Br^-? What is the electrophile? Nucleophile? How can you use curved arrow notation to show the electron movement to form the new C–Br bond in Step 2? If π bond electrons from the nonpolar C=C are involved in the first step, what is the name of the elementary step? Are π bonds electron rich or poor?

Solve

Step 1 is an electrophilic addition in which a pair of electrons from the electron-rich C=C π bond forms a bond to the acid's electron-poor H atom. This leaves a C^+ (electron poor) on the other C of the C=C and a Br^- anion (electron rich). Step 2 is a coordination step whereby Br^- forms a bond to the C^+.

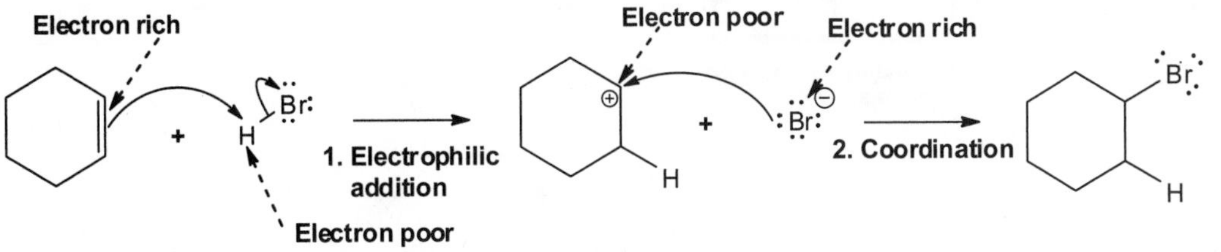

Electron rich | H–Br: | Electron poor | 1. Electrophilic addition | Electron poor | Electron rich | :Br: | H | 2. Coordination | Br | H

Your Turn 11.3

Think

Identify the bonds that broke and the bonds that formed. Consult Table 1-2 and Table 1-3 to fill in the appropriate energy values for each. Is a reaction energetically favorable if the enthalpy is negative (exothermic) or positive (endothermic?)

Solve

The C=C π and H–Cl bonds break. The C–H, C–C σ, and C–Cl bonds form. From the bond energy values found in Table 1-1, the reaction is exothermic by 38 kJ/mol, which favors products.

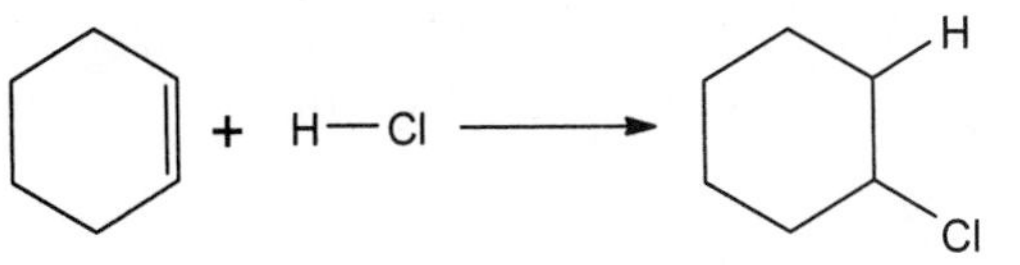

+ H—Cl ⟶ (H, Cl)

$$\Delta H^\circ_{rxn} = (\underset{C=C}{619}\ \text{kJ/mol} + \underset{H-Cl}{431}\ \text{kJ/mol}) - (\underset{C-C}{339}\ \text{kJ/mol} + \underset{H-C}{418}\ \text{kJ/mol} + \underset{C-Cl}{331}\ \text{kJ/mol})$$

Sum of energies of bonds lost | Sum of energies of bonds gained

$$= -38\ \text{kJ/mol}$$

Your Turn 11.4

Think

Draw the carbocation that results from formation of the C–H bond on either end of the C=C. Are alkyl groups electron donating or withdrawing? How does an additional alkyl group affect the stability of a nearby positive charge? Is the more stable carbocation higher or lower energy?

Solve

Each carbocation intermediate is produced by the addition of an H^+ to one of the C=C carbons, resulting in a positive charge on the other C. The primary carbocation is less stable (higher energy) and the secondary carbocation is more stable (lower energy). Alkyl groups are electron donating and stabilize the carbocation. A secondary C^+ has two alkyl groups, whereas a primary C^+ only has one.

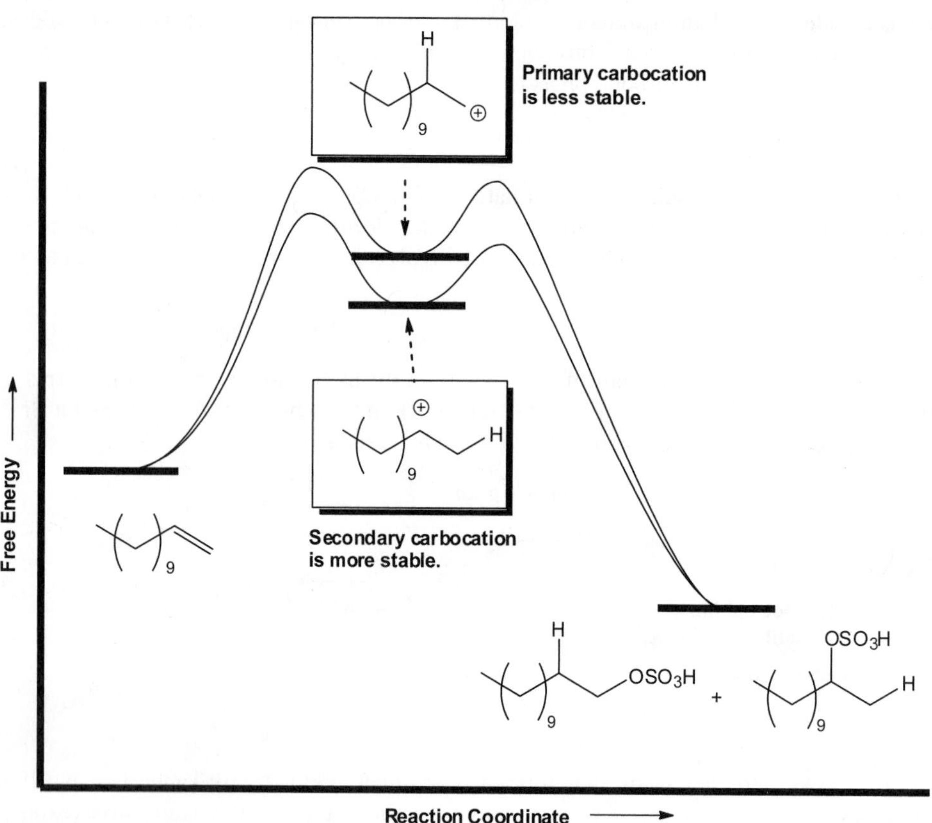

Your Turn 11.5

Think

What are the requirements for a stereocenter? How many unique groups must be bonded to the C?

Solve

Four unique groups must be bonded to the C to make it a tetrahedral stereocenter. The two possible products from the addition of H_2SO_4 across the double bond are shown. The major product (right) results in the formation of a new stereocenter (C*).

New stereocenter

H_2SO_4

OSO_3H + OSO_3H

Your Turn 11.6

Think

What is electron rich (nucleophilic)? Electron poor (electrophilic)? What are the two elementary steps involved in the addition of a Brönsted acid like H–OH across the double bond? What two charged species form in the first step? How do the stabilities of these species compare to the species produced in electrophilic addition reactions encountered previously?

Solve

The C=C bond is electron rich, and the proton on H_2O is electron poor, leading to an electrophilic addition step. The HO^- that is produced then adds to the carbocation in a coordination step, resulting in the overall addition of water across the double bond. The electrophilic addition products in Step 1 are a carbocation C^+ and hydroxide HO^-. Hydroxide is significantly less stable than anions produced in other electrophilic addition reactions, such as Cl^- or Br^-, because O, being a significantly smaller atom, cannot accommodate the negative charge as well.

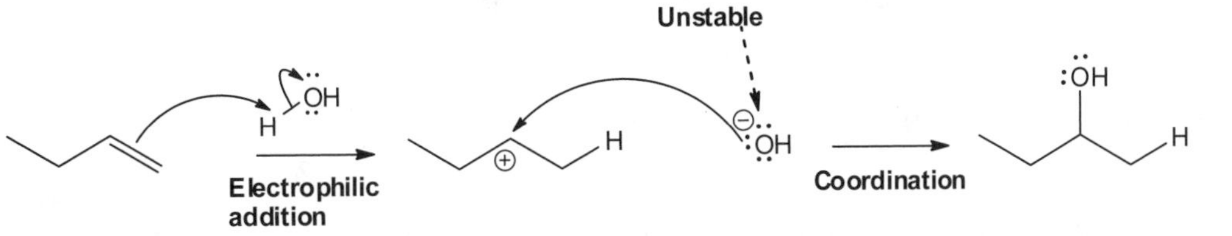

Unstable

Electrophilic addition

Coordination

Your Turn 11.7

Think

Which bonds are broken and formed to result in the formation of a C^+ and Cl^-? What is electron rich (nuclephilic)? Electron poor (electrophilic)? How can you use curved arrows to show the electron movement? If π bond electrons from the nonpolar C≡C are involved in the first step, what is the name of the elementary step?

Solve

Step 1 is an electrophilic addition where a pair of electrons from the electron-rich C≡C π bond forms a bond to the acid's electron-poor H atom. This leaves a vinylic C^+ on the other C of the C≡C and a Cl^- anion. The H^+ adds to the terminal C in Step 1 for the reaction to proceed through the more stable of the two possible carbocation intermediates. In Step 2, Cl^- attacks the positively charged C in a coordination step.

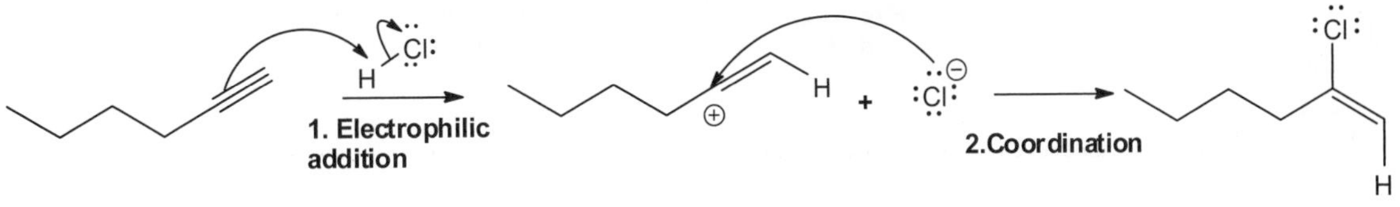

1. Electrophilic addition

2.Coordination

Your Turn 11.8

Think

Which bonds are broken and formed to result in the formation of a C^+ and a Cl^-? What is electron rich (nucleophilic)? Electron poor (electrophilic)? How can you use curved arrows to show the electron movement? If π bond electrons from the nonpolar C≡C are involved in the first step, what is the name of the elementary step? In the presence of a hydrogen halide acid, how does an alkene react?

Solve

The C≡C bond is electron rich and the H atom of HCl is electron poor, which leads to an electrophilic addition. This is followed by coordination between Cl^- and the carbocation to produce a vinyl halide. With a second equivalent of HCl, the C=C double bond is electron rich and can undergo a second addition of HCl (electrophilic addition followed by coordination) to yield a geminal dihalide. In Step 3, H^+ adds to the terminal C of the alkene to yield a carbocation that is resonance stabilized by the halide. See the mechanism on the next page.

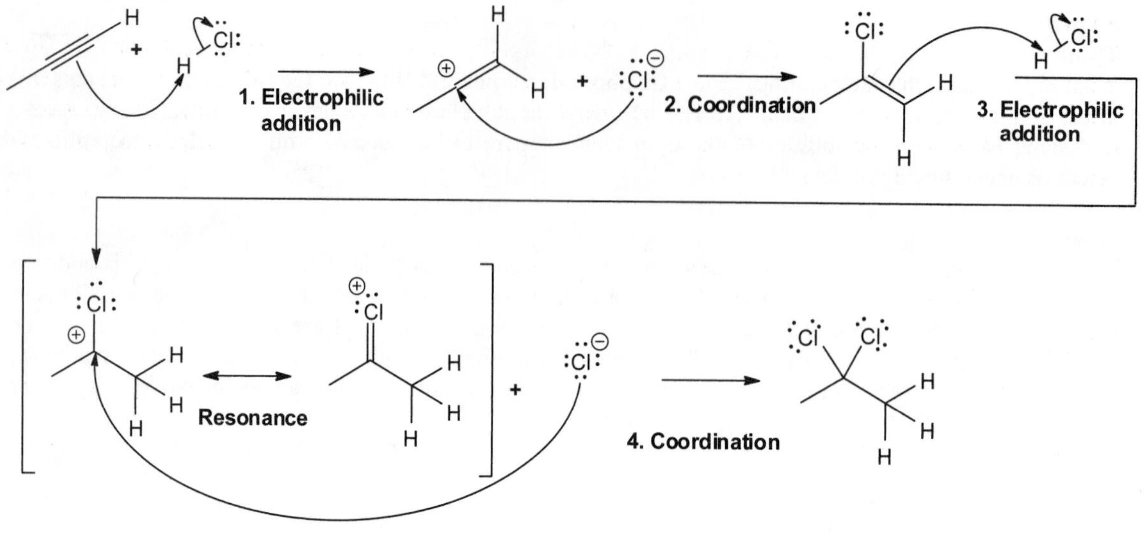

1. Electrophilic addition
2. Coordination
3. Electrophilic addition

Resonance

4. Coordination

Your Turn 11.9

Think

Identify which C atom of a terminal alkyne C≡C has to form the bond to O to produce an aldehyde. For that to happen, which C (terminal or internal) in the first step of electrophilic addition must gain the positive charge? Which C atom must gain the C–H bond (terminal or internal)? Compare the stability of this C^+ to that in Equation 11-23.

Solve

To produce an aldehyde, the O atom from water would have to form a bond to the terminal C, meaning that the terminal C would have to have the positive charge in a previous step, as shown below. This is a less stable carbocation than the one produced in Equation 11-23 because that charge is stabilized by fewer alkyl groups.

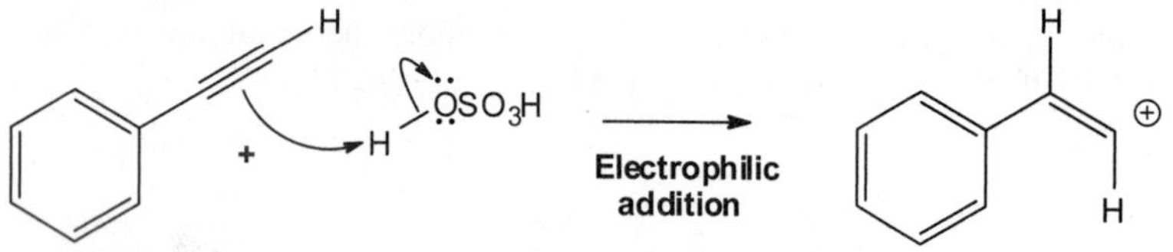

Electrophilic addition

Your Turn 11.10

Think

Consult Table 6-1 to determine the pK_a values for H_2SO_4 and TfOH. Does a lower pK_a indicate a stronger or weaker acid? What is the significance of increasing the acid strength in the mechanism of electrophilic addition of an acid to an alkyne to produce a ketone?

Solve

The pK_a values are: H_2SO_4: −9; TfOH: −13.

The pK_a of TfOH is more negative than that of H_2SO_4 by four units, so TfOH is a stronger acid by four orders of magnitude. Increasing the acid strength makes the first step in Equation 11-23 more favorable by using a higher energy/less stable reactant and forming a more stable conjugate base.

Your Turn 11.11

Think

Which C atom gains the H from HCl? Identify the C atom that is C1 in both 1,2- and 1,4-electrophilic addition. Which C atoms would then be numbered C2, C3, and C4? (Remember these are just relative numbering labels and do not correlate with nomenclature numbering rules.) Which of those C atoms gains the Cl?

Solve

The first is from 1,2-addition and the second is from 1,4-addition. In both cases, the C that gains the H atom is assigned C1. In 1,2-addition, C2 gains the Cl, and the remaining double bond will be in the same location as in the overall reactant. In 1,4-addition, C4 gains the Cl, and the double bond will be in a different location. The double bond moves due to the resonance that exists in the intermediate as a result of a C atom with an incomplete octet next to a double bond.

CH_3 H–Cl: 1. Electrophilic addition → [resonance-stabilized carbocation intermediates] → 2. Coordination → 1,2-addition (Cl on C2); 1,4-addition (Cl on C4)

Your Turn 11.12

Think

Do warm temperatures or cold temperatures facilitate a kinetically controlled reaction? Thermodynamically controlled? Which product, that from 1,2-addition or 1,4-addition, is the major product at cold temperatures? Warm temperatures?

Solve

The reaction in Equation 11-30 takes place under thermodynamic control (warm temperatures), so its major product must be the thermodynamic product—that is, the most stable product. The reaction in Equation 11-31 takes place under kinetic control (cold temperatures), so its major product must be the kinetic product—that is, the product that forms the fastest. This is indicated below:

Buta-1,3-diene —HCl, 20 °C→ 3-chlorobut-1-ene (25% of adduct mixture) + 1-chlorobut-2-ene (75% of adduct mixture)

Major product at warm temperatures: 1-chlorobut-2-ene ← Thermodynamic product

Buta-1,3-diene —HCl, -80 °C→ 3-chlorobut-1-ene (80% of adduct mixture) + 1-chlorobut-2-ene (20% of adduct mixture)

Major product at cold temperatures: 3-chlorobut-1-ene ← Kinetic product

Your Turn 11.13

Think

Label each product 1,2-addition or 1,4-addition. Is 1,2- or 1,4-addition faster? Which alkene is the most stable (mono-, di-, tri-, or tetrasubstituted)?

Solve

The first is from 1,4-addition and the second is from 1,2-addition. In 1,2-addition, the remaining double bond will be in the same location as in the overall reactant. In 1,4-addition, it will be in a different location. 1,2-Addition is faster, so it gives the kinetic product. The 1,4-adduct has the more highly substituted alkene product, so it is more stable and is the thermodynamic product.

HCl

1,4-addition
Thermodynamic product

1,2-addition
Kinetic product

Chapter Problems

Problem 11.1

Think

What is electron rich in the first step? What is electron poor (electrophilic)? What elementary step occurs in each step of this two-step reaction mechanism? What functional group is formed?

Solve

The reaction occurs in two steps. The first elementary step is electrophilic addition, where the π electrons of the C=C define the electron-rich species and the H atom of the H–Y of the acid defines the electron-poor species (electrophile). A carbocation forms in the first step. The second elementary step is coordination, where the conjugate base of the H–Y acid acts as the nucleophile.

(a)

H–Cl: Electrophilic addition → Cl⁻ Coordination → Cl, H

(b)

H–I: Electrophilic addition → I⁻ Coordination → I, H

(c)

H–OSO_3H Electrophilic addition → $^{-}OSO_3H$ Coordination → OSO_3H, H

(d)

H–Br Electrophilic addition → Br⁻ Coordination → Br, H

Problem 11.2

Think

What type of reaction starts with an alkene and ends with the functional group shown? On which two C atoms must the C=C be located in the starting material?

Solve

Each compound can be made from an alkene, as shown below and on the next page. Each alkene is devised by removing the nucleophile that would have added, along with a neighboring H atom, and connecting the two C atoms with an additional π bond.

(a)

HCl → H, Cl

(b) [cyclopropene + HBr → bromocyclopropane; labels: HBr, Br, H]

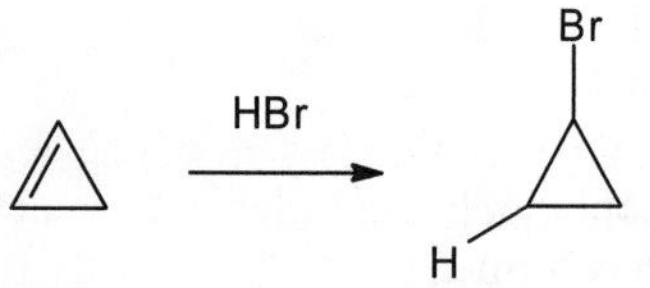

(c) [1-methylcyclohexene + H_2SO_4 → product; labels: H_2SO_4, OSO_3H, H]

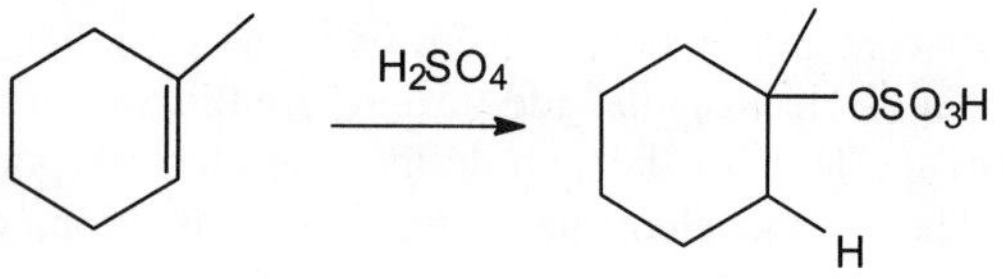

Problem 11.4

Think

Which C=C double bond will undergo electrophilic addition? What are the *possible* products and the corresponding carbocation intermediates from which they derive? Which carbocation intermediate is more stable?

Solve

Mechanisms are shown below. In the first step of each reaction, electrophilic addition of H^+ takes place to produce the more stable carbocation. In **(a)**, the more highly substituted carbocation is more stable. In **(b)** and **(c)**, the carbocation is resonance stabilized. In **(d)**, the NO_2 group would destabilize a neighboring positive charge, so the better choice is having the positive charge farther away. Notice that the phenyl rings in **(b)** and **(d)** do not undergo reaction in the electrophilic addition steps.

(a) H–Cl; Electrophilic addition; More substituted carbocation; Cl⁻; Coordination; Cl

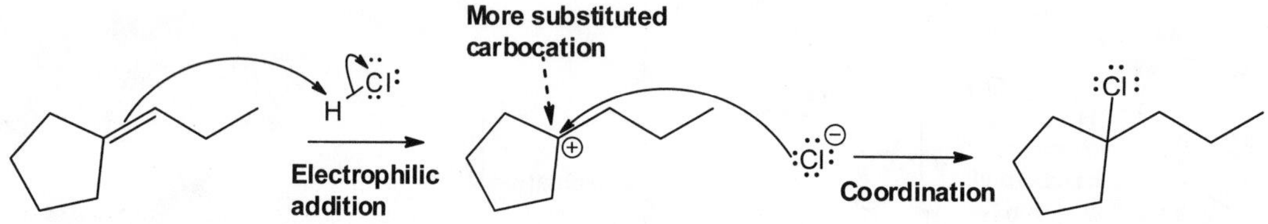

(b) H–Cl; Electrophilic addition; Resonance stabilized carbocation; Cl⁻; Coordination; Cl

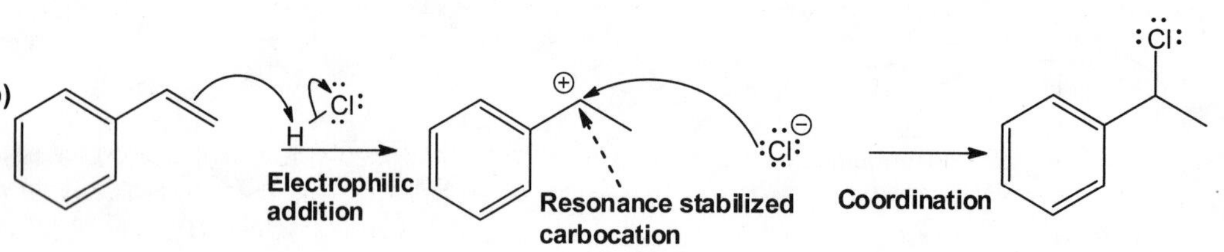

(c) H–Cl; O; Electrophilic addition; Resonance stabilization of carbocation; Cl⁻; Coordination; O; Cl

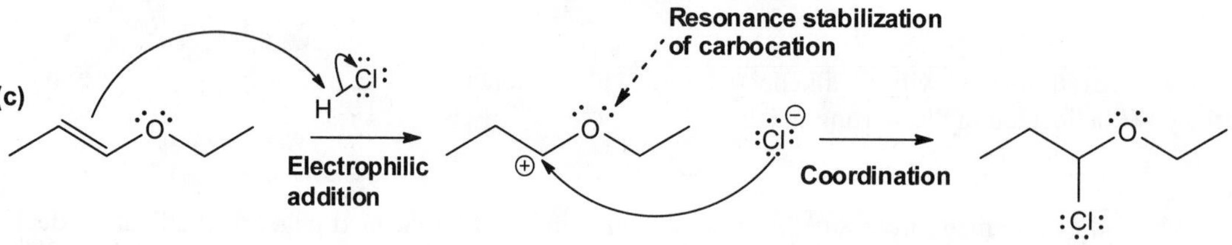

(d) NO_2; H–Cl; Electrophilic addition; Inductive destabilization of positive charge; NO_2; Cl⁻; Coordination; NO_2; Cl

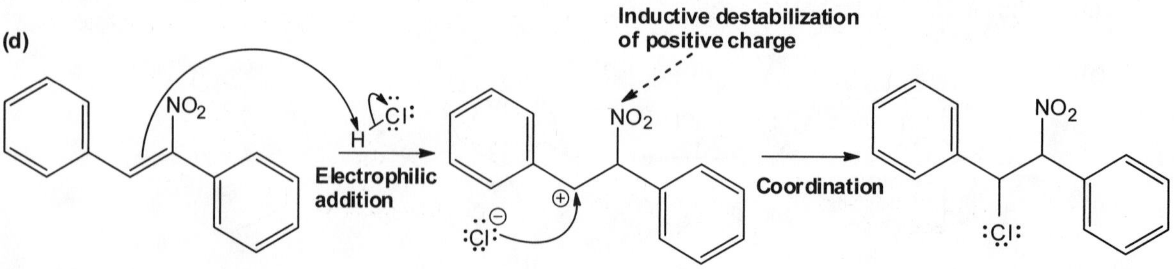

Problem 11.5

Think

What type of reaction starts with an alkene and ends with the functional group shown? On which C atoms could the C=C be located? When the C=C is in that location, to which C will a proton add?

Solve

The alkene precursors have the nucleophile and an adjacent proton removed. In each case, the two different alkenes produce the same product because the most stable carbocation that is produced is the same.

(a) or HCl Cl

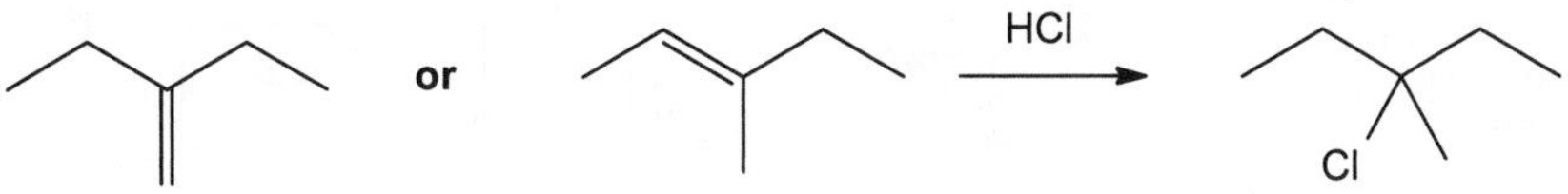

(b) O or O HBr Br O

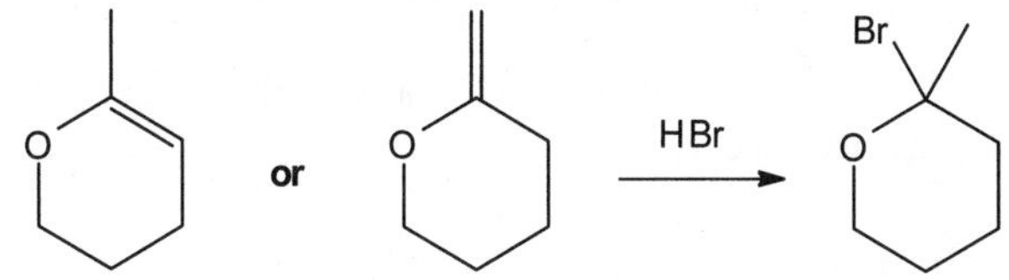

(c) or HCl Cl

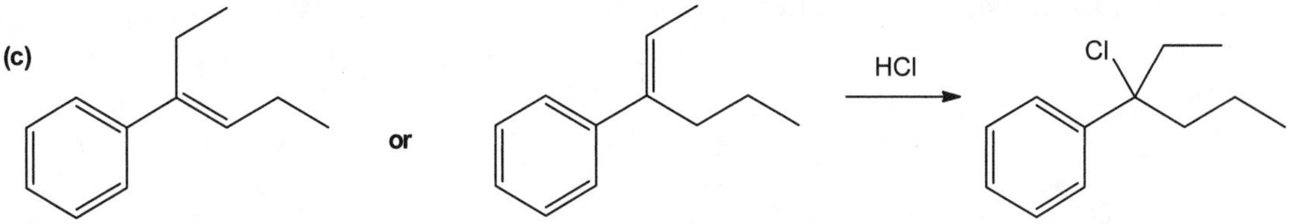

Problem 11.7

Think

What carbocation intermediate is produced upon addition of H^+? Can that carbocation intermediate undergo a 1,2-hydride shift or a 1,2-methyl shift to attain greater stability? What species produced from the first step will react with the carbocation intermediate?

Solve

(a) In Step 1 of this electrophilic addition reaction, H^+ adds to either alkene C to produce a secondary carbocation intermediate, as shown below. The secondary carbocation intermediate converts to a more stable tertiary one via a 1,2-methyl shift in Step 2. In Step 3, the I^- coordinates to the carbocation to complete the reaction.

H I Electrophilic addition 1,2-Methyl shift I Coordination I

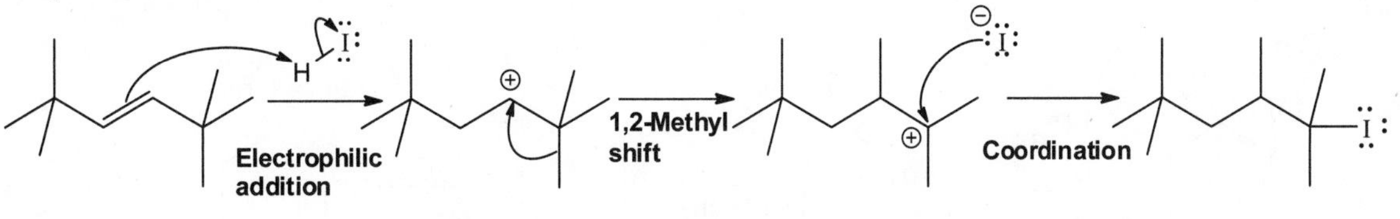

(b) In Step 1 of this electrophilic addition reaction, H^+ adds to either alkene C to produce a secondary carbocation intermediate. The secondary carbocation intermediate converts to a more stable tertiary one via a 1,2-hydride shift in Step 2. In Step 3, the Cl^- coordinates to the carbocation to complete the reaction.

H Cl Electrophilic addition H 1,2-Hydride shift H Cl Coordination Cl

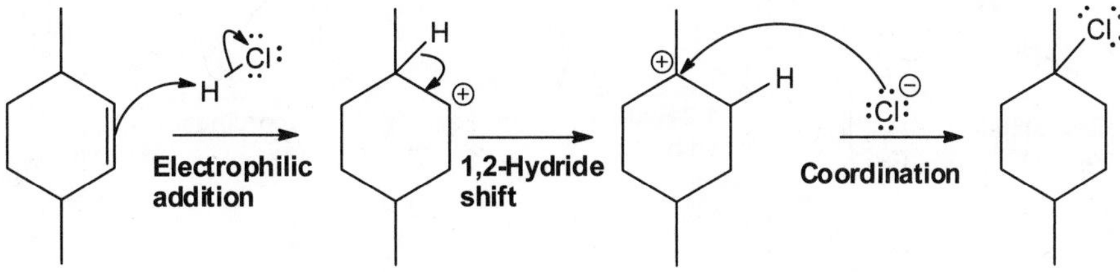

(c) In Step 1 of this electrophilic addition reaction, H^+ adds to the terminal alkene C to produce a secondary carbocation intermediate, as shown below. (Addition of H^+ to the internal alkene C would, instead, produce a less stable primary carbocation intermediate.) The secondary carbocation intermediate converts to a more stable benzylic resonance-stabilized one via a 1,2-hydride shift in Step 2. In Step 3, the HSO_4^- coordinates to the carbocation to complete the reaction.

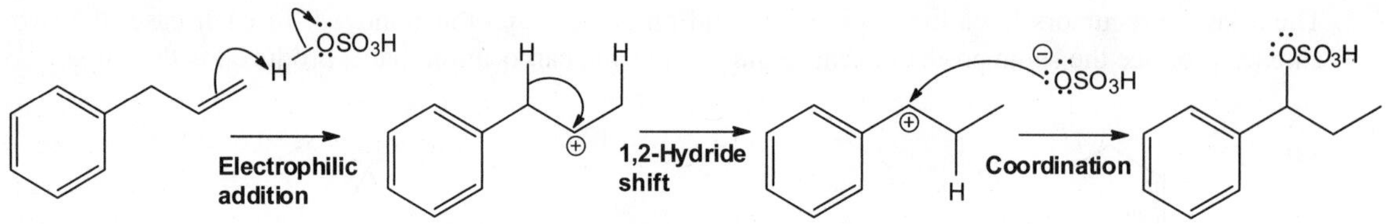

Electrophilic addition → 1,2-Hydride shift → Coordination

Problem 11.8

Think

On which carbon atoms are the Br and D located? Are they on adjacent C atoms in the product? On which carbon atoms could the C=C be located? Did rearrangement occur?

Solve

After addition of D^+, a carbocation rearrangement takes place in each case prior to the coordination of Br^-. In **(a)**, **(b)**, and **(d)**, a 1,2-hydride shift takes place, whereas in **(c)**, a 1,2-methyl shift takes place.

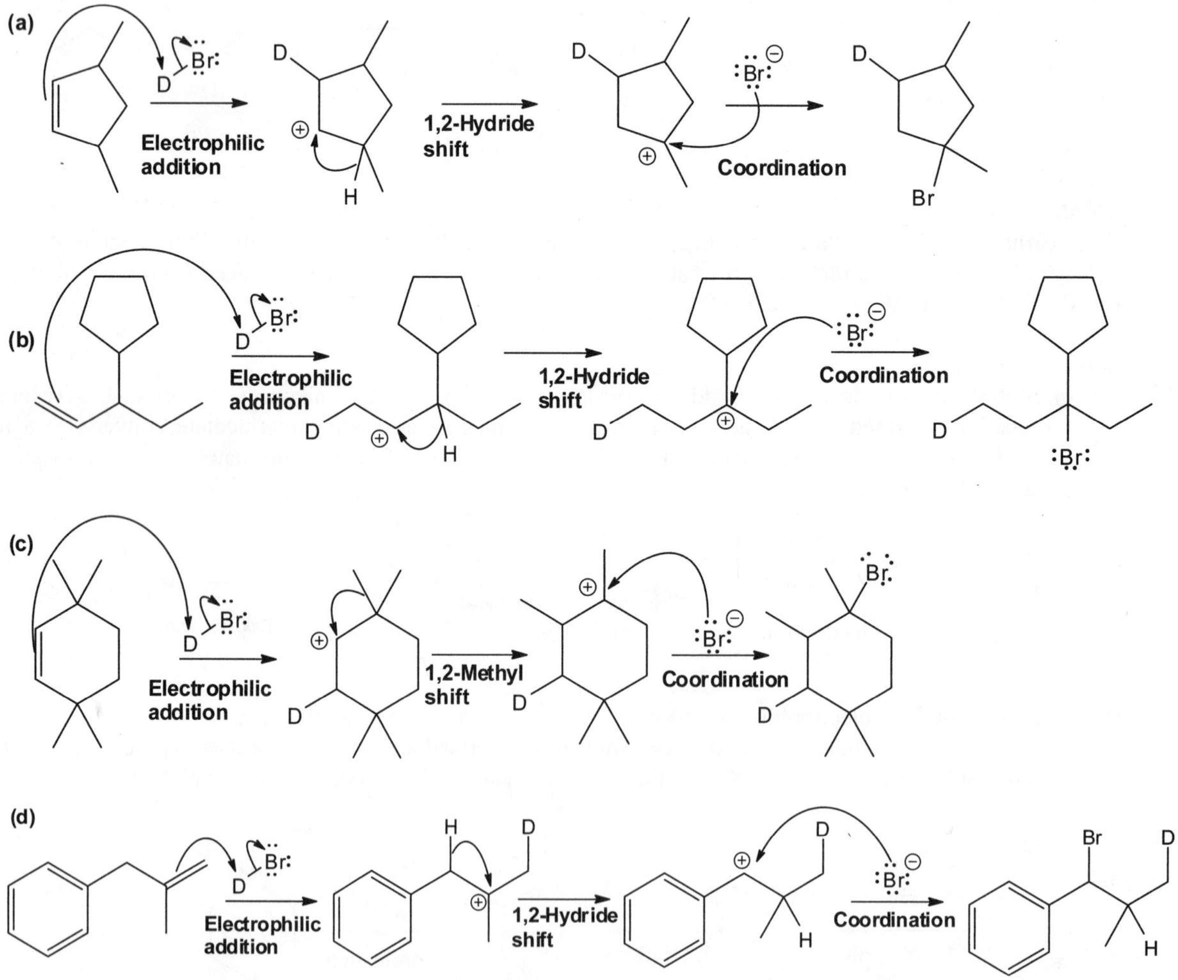

(a) Electrophilic addition → 1,2-Hydride shift → Coordination

(b) Electrophilic addition → 1,2-Hydride shift → Coordination

(c) Electrophilic addition → 1,2-Methyl shift → Coordination

(d) Electrophilic addition → 1,2-Hydride shift → Coordination

Problem 11.9

Think

How many stereocenters are produced? How many stereoisomers are possible? From how many directions can the electrophile approach the C=C? What intermediate is formed first? What is the relationship between the stereoisomers?

Solve

There are two stereocenters formed in the reaction, and each stereocenter is produced as a mixture of *R* and *S* configurations. This is because the electrophile, H^+, can approach the alkene C from either side of the carbon's plane, and the Cl^- can approach the C^+ from either side of the plane. Therefore, there are four total stereoisomers as products—two pairs of enantiomers—as shown below. Each pair of enantiomers must be produced in equal amounts. Notice that the diastereomers that are produced differ by the exchange of H and D. Because H and D have nearly identical behavior, those diastereomers should be produced in nearly equal amounts.

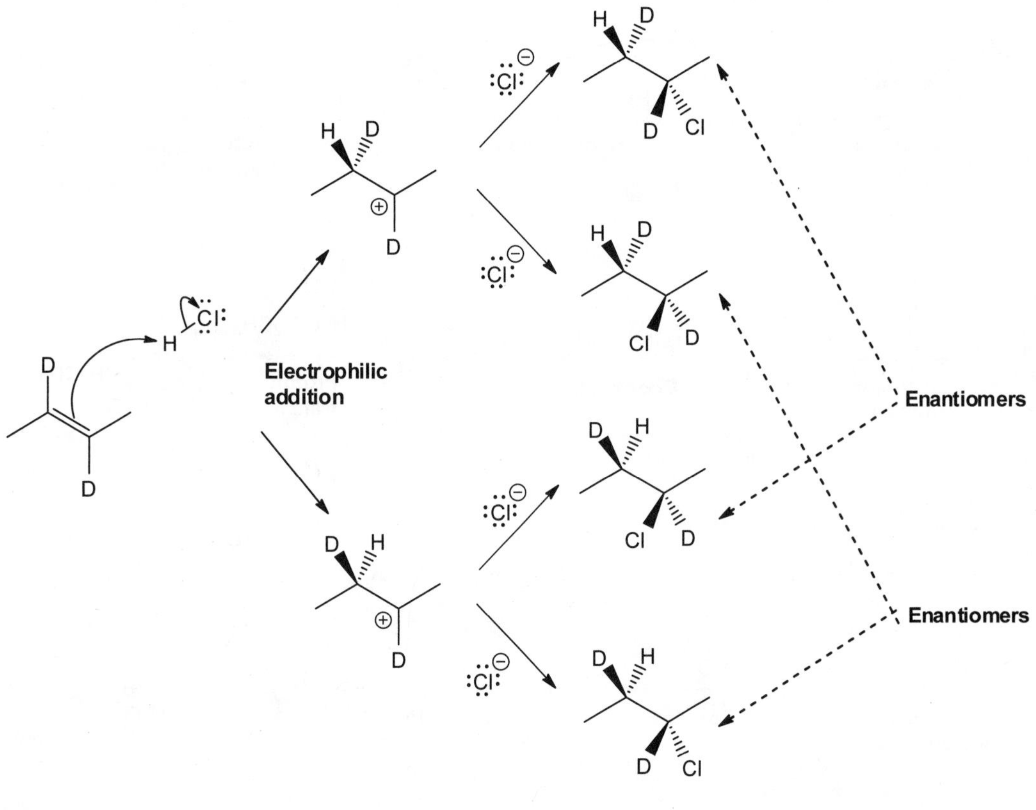

Electrophilic addition

Enantiomers

Enantiomers

Problem 11.11

Think

What is the electrophile? What is the nucleophile? To which alkene carbon does the electrophile add? Where is the C^+ formed? Are rearrangements possible?

Solve

In each case, the strong acid adds H$^+$ to the double bond in the first step—electrophilic addition. The product is the most stable carbocation. In the second step, the weak nucleophile undergoes coordination to the carbocation, and in the final step, a weak base deprotonates the positively charged O atom.

(a)

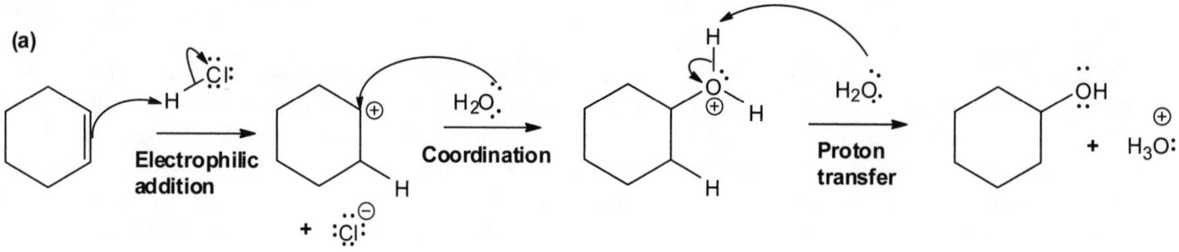

(b)

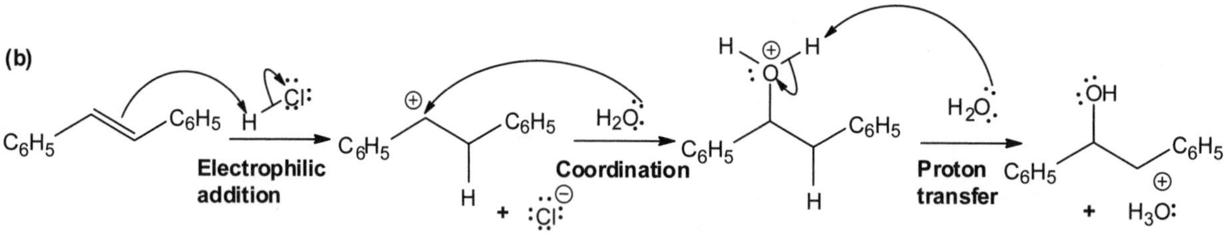

(c)

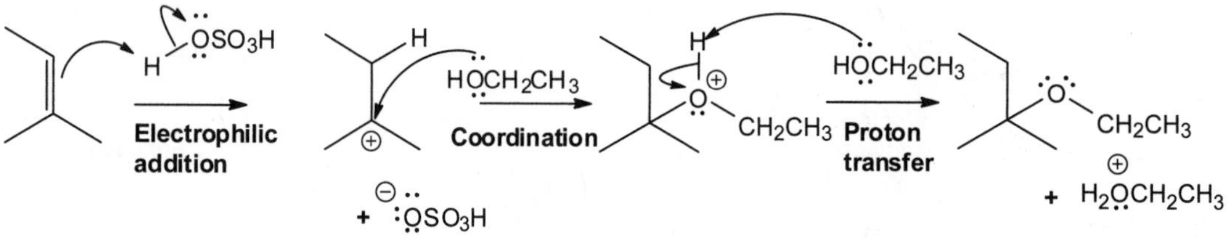

(d)

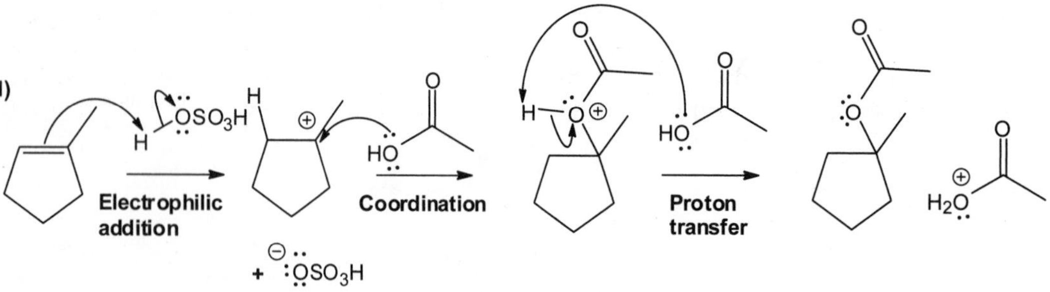

Problem 11.12

Think

What type of reaction starts with an alkene and ends with the functional group shown? On which C atoms could the C=C be located? Is there more than one option? When a proton adds to a C=C, to which C should it add? What nucleophile that is present could coordinate to a C$^+$?

Solve

Addition of the H^+ occurs at the less substituted carbon to form the more stable C^+. The nucleophile is water (to form an alcohol) or an alcohol (to form an ether). Rearrangements did not occur in these examples. In all cases, a final proton transfer is necessary to produce the uncharged product.

(a)

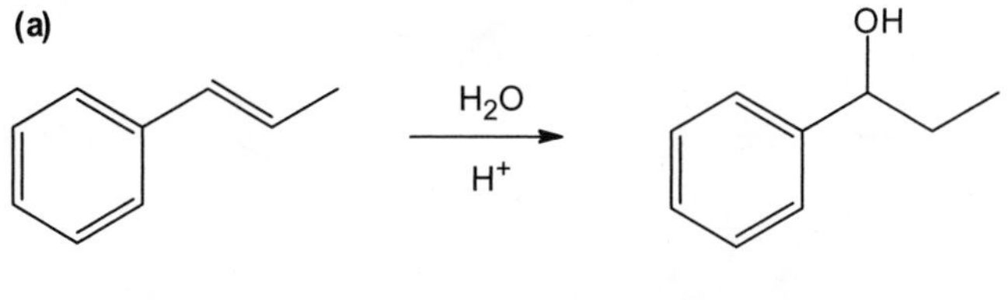

H_2O, H^+

OH

(b)

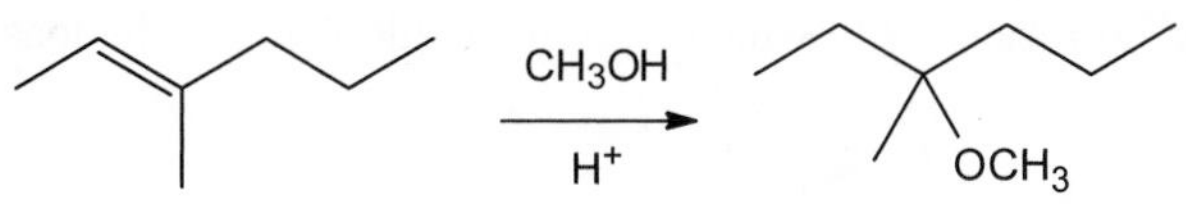

CH_3OH, H^+

OCH_3

(c)

HO, H^+

O

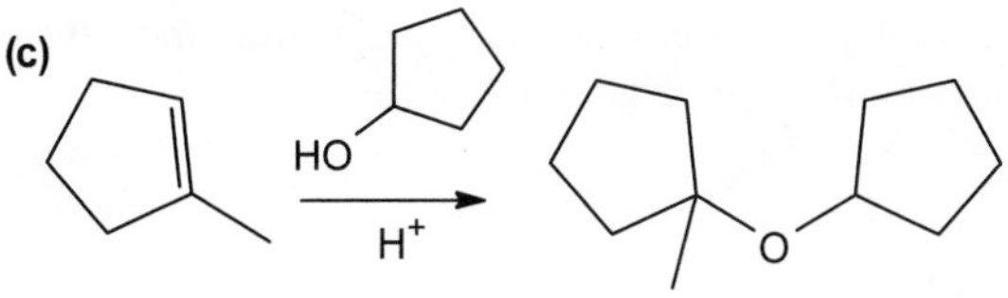

Problem 11.14

Think

Based on the electron-rich and electron-poor sites in the reactants, what type of reaction will take place? How many times will that reaction take place? What aspects of regiochemistry should be considered?

Solve

The alkyne is relatively electron rich and the H atom of HCl or HBr is relatively electron poor. So, the conditions favor an electrophilic addition reaction in which HCl or HBr adds across the multiple bond. In **(a)**, the product of that addition to the alkyne is a vinyl halide, which, in the presence of excess HCl, can undergo a second electrophilic addition. In **(b)**, the addition is to the C=C of the vinyl halide. Notice that in each addition to the vinyl halide, the proton adds to the C that is not bonded to the halogen so that the resulting C^+ is resonance stabilized by the halogen, so the product is generated in high yield.

(a)

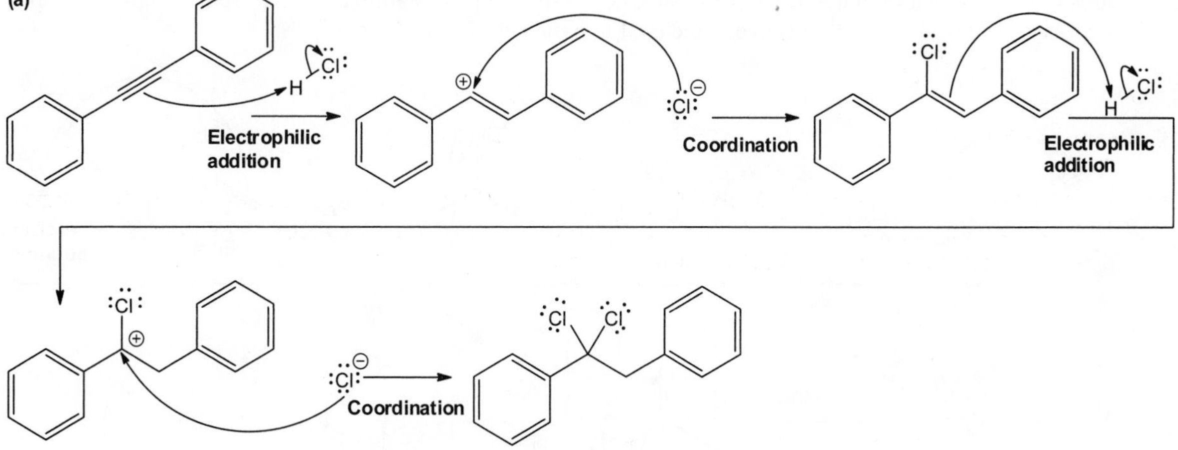

H—Cl

Electrophilic addition

Cl

Coordination

Cl

H—Cl

Electrophilic addition

Cl

Cl

Coordination

Cl Cl

(b)

Electrophilic addition

Coordination

Resonance

Problem 11.15

Think

On which C atoms are the halogen atom and H (or D) located? On which C atoms could the C≡C be located? How many times did the addition reaction occur?

Solve

In **(a)**, the alkyne must be a terminal alkyne to control the reaction via Markovnikov's rule. HCl adds twice, and the resonance stabilization of the carbocation that is produced in the second addition is responsible for generating the product in high yield. In **(b)**, DBr has added just once. The carbocation intermediate is not delocalized via resonance and is relatively unstable, so the product is generated in low yield.

(a) Excess HCl

(b) 1 eq. DBr

Problem 11.16

Think

What is electron rich (nucleophilic)? Electron poor (electrophilic)? Did proton transfer reactions occur? What is the product that forms after the acid-catalyzed addition of water across the triple bond? Is this product stable? If not, what type of rearrangement occurs?

Solve

The first few steps compose the acid-catalyzed addition of water across the triple bond. Notice that the proton adds to the terminal C so that the positive charge that develops is stabilized by the attached alkyl group. Then, acid-catalyzed tautomerization converts the enol into the keto form.

Electrophilic addition

Coordination

Proton transfer

Enol

Electrophilic addition

Resonance

Proton transfer

Ketone

Problem 11.18

Think

How many distinct carbocation intermediates are possible from protonation of a double bond? Which one is the most stable? What are the species that can be produced upon nucleophilic attack of that carbocation intermediate?

Solve

Each of the alkene groups can gain a proton at either of its C atoms, giving rise to two possible carbocation intermediates (**A** and **B**). Intermediate **A** is more stable than intermediate **B** due to resonance delocalization of the positive charge. With the most stable intermediate identified, the 1,2- and 1,4-addition products are obtained by attack of the nucleophile, Br^-, on the C atoms sharing the positive charge.

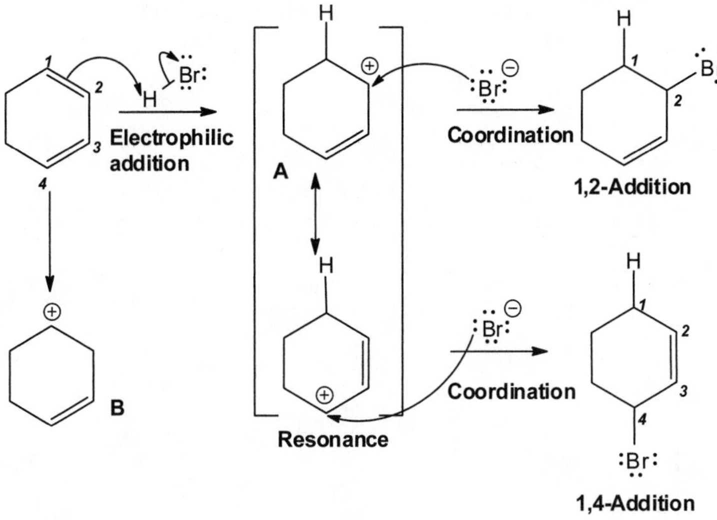

Electrophilic addition

A

B

Resonance

Coordination

1,2-Addition

Coordination

1,4-Addition

Problem 11.19

Think

Which product was formed from 1,2-addition and which product was formed from 1,4-addition? On which C atom was the H^+ added? Which product results from resonance movement of the C=C?

Solve

The diene below could produce the mixture of products yielded by 1,2-addition (first product) and 1,4-addition (second product). However, it would have to proceed through a carbocation intermediate that is not the most stable (secondary vs. tertiary).

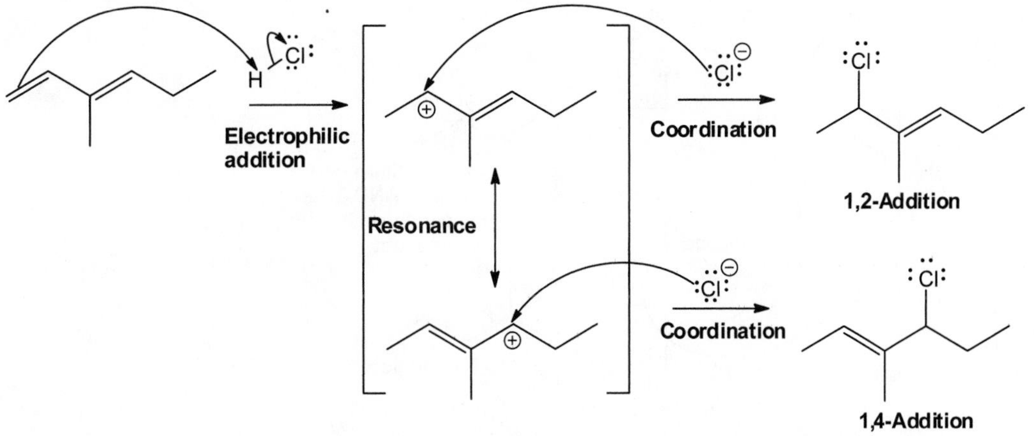

Electrophilic addition

Resonance

Coordination

1,2-Addition

Coordination

1,4-Addition

Problem 11.21

Think

What are the *possible* carbocation intermediates? What is the most stable carbocation intermediate? What are the products that can be produced upon nucleophilic attack of that carbocation intermediate? Which of those products is produced faster? Which is the more stable product?

Solve

Each of the alkene groups can gain a proton at either of its C atoms, giving rise to four possible carbocation intermediates (**A**, **B**, **C**, and **D**). Intermediates **B** and **D** are more stable than **A** and **C** due to resonance delocalization of the positive charge. Intermediate **D** is more stable than intermediate **B** (and is the most stable of all four) because its positive charge is shared on a tertiary C atom. Adding Cl^- to that C^+ yields the 1,2- and 1,4-adducts. As usual, the 1,2-adduct is the kinetic product. It is also the most stable alkene product because it is the most highly alkyl substituted. Therefore, the 1,2-adduct is the thermodynamic product, too.

H
Cl:
A
B
Resonance
H
Cl:
C
D
Resonance
:Cl:
:Cl:
1,2-addition
kinetic AND
thermodynamic
product
:Cl:
:Cl:
1,4-addition

Problem 11.22

Think

What bonds broke and formed in the reaction of neryl phosphate to form α-terpinenol. What C=C underwent electrophilic addition? What acted as the nucleophile to form the C–OH bond? Are proton transfer reactions involved? What bonds broke and formed in the reaction of α-terpinenol to form terpin hydrate?

Solve

The mechanisms are drawn below.

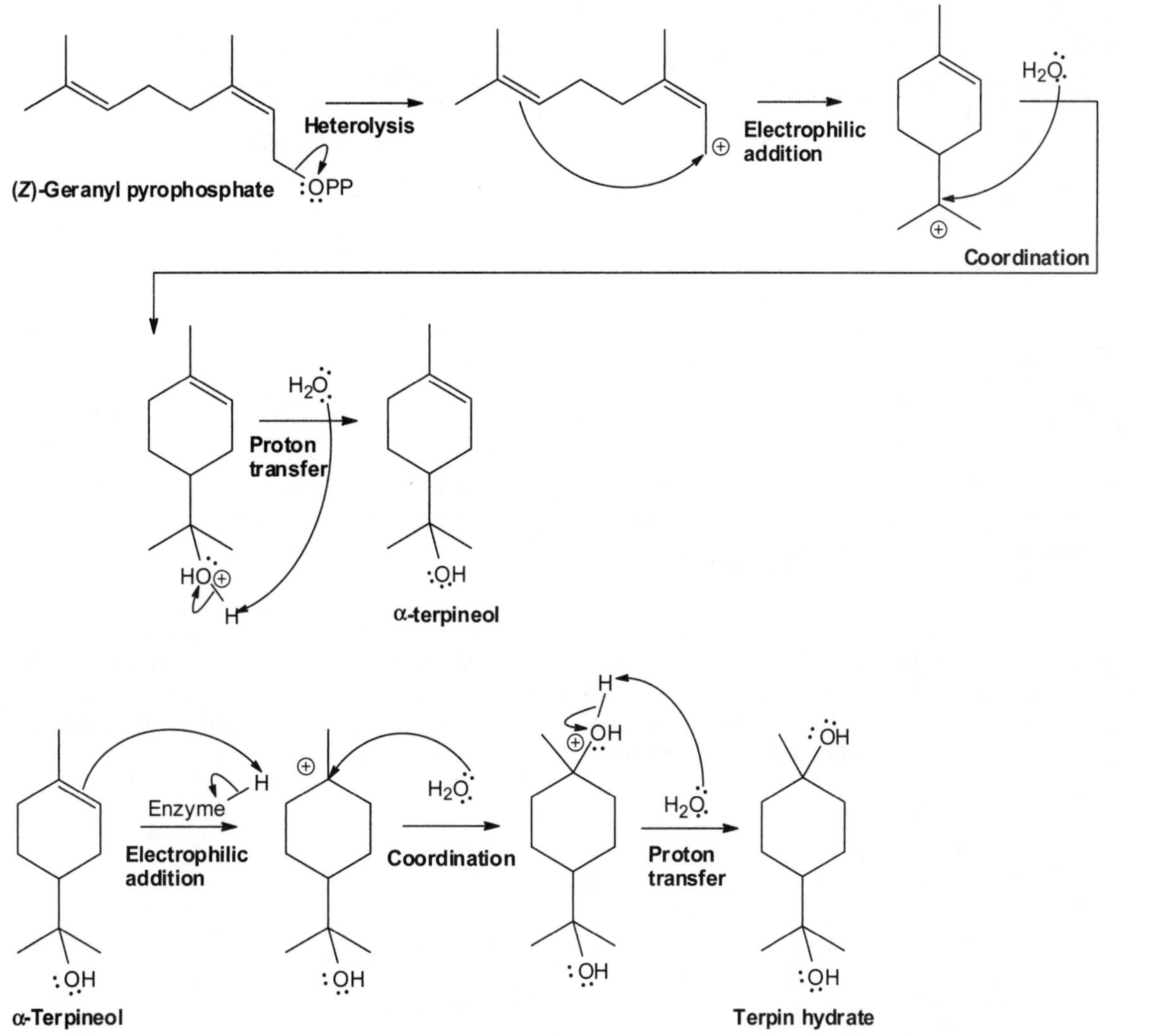

(Z)-Geranyl pyrophosphate

Heterolysis

Electrophilic addition

Coordination

Proton transfer

α-terpineol

α-Terpineol

Enzyme

Electrophilic addition

Coordination

Proton transfer

Terpin hydrate

Problem 11.23

Think

What are the two steps in the addition of an HX acid to an alkene? Of the two steps, which step produces new charges? Which step reduces the number of charges? Are the charges stabilized the same? Which atom is largest?

Solve

The reaction rate is determined by how fast the H–X bond breaks in the first (rate-determining) step, generating X^-. The more stable the anion that is generated, the lower the energy barrier, as suggested by the Hammond postulate. Since the order of charge stability is $F^- < Cl^- < Br^- < I^-$, the order of reactivity is HF < HCl < HBr < HI.

Problem 11.24

Think

What is the first step in addition of an HX acid to an alkene? What intermediate forms? Which intermediate is most stable? How does the stability of the intermediate affect the rate of the reaction?

Solve

The reaction rate is determined by how fast the carbocation forms in the first (rate-determining) step. The more stable the carbocation, ***the more stable the transition state leading to it*** (according to the Hammond postulate) and the faster the reaction. Because of resonance and the (+) charge on a tertiary carbon, the carbocation **C** is by far the most stable. **A** < **B** < **C**.

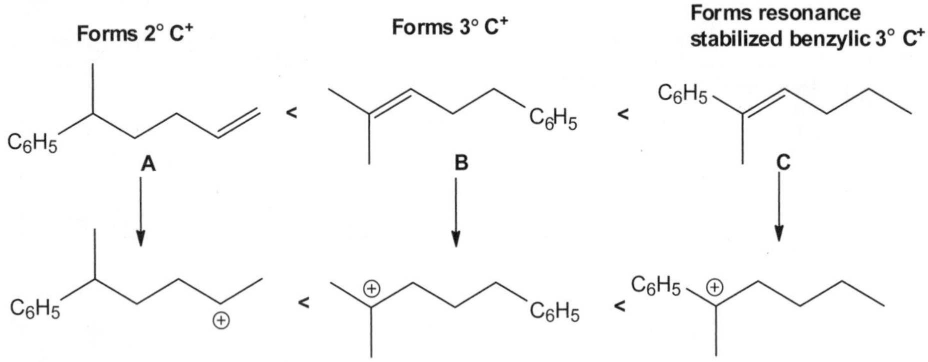

Forms 2° C$^+$ | Forms 3° C$^+$ | Forms resonance stabilized benzylic 3° C$^+$

C_6H_5 A < B C_6H_5 < C_6H_5 C

Problem 11.25

Think

Draw out 2-chloro-3-methyl-2-phenylbutane. What carbocation intermediate forms in the first step when each of the reactants is treated with HCl? If there are two carbocations that could form in the first step, which carbocation would be favored? Are rearrangements possible?

Solve

The best way to answer this question is to draw the mechanism for each reaction. **D**, **E**, and **F** give 2-chloro-3-methyl-2-phenylbutane, because Cl^- attacks the exact same carbocation in each case. In **D** and **F**, the initial carbocation formed undergoes a fast rearrangement to a more stable carbocation prior to attack of Cl^-. **G** and **H** have a longer carbon chain than three carbon atoms and will not form 2-chloro-3-methyl-2-phenylbutane. The mechanisms are shown below and on the next page.

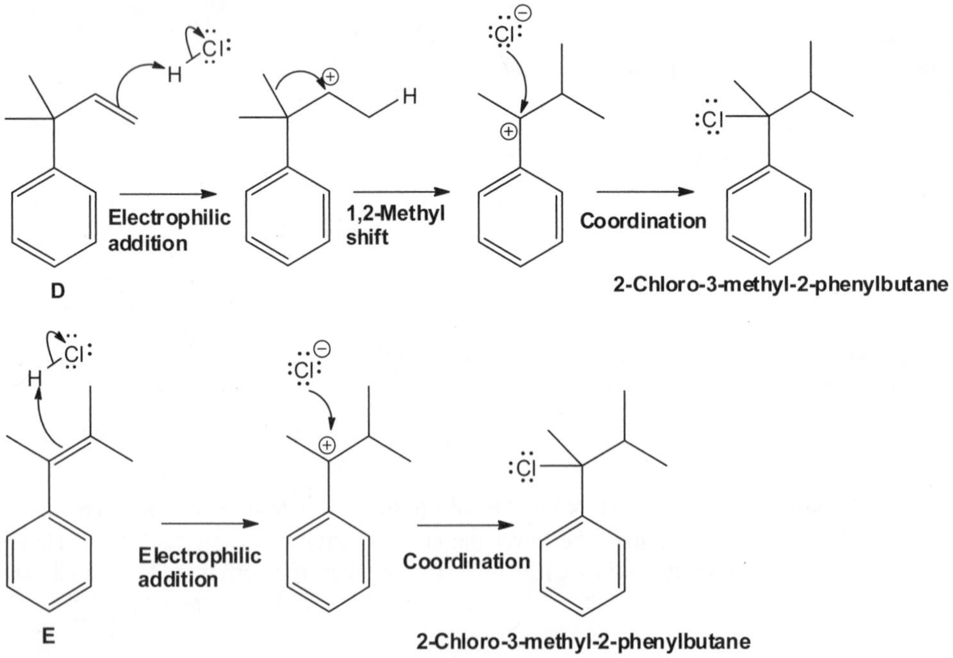

D → Electrophilic addition → 1,2-Methyl shift → Coordination → 2-Chloro-3-methyl-2-phenylbutane

E → Electrophilic addition → Coordination → 2-Chloro-3-methyl-2-phenylbutane

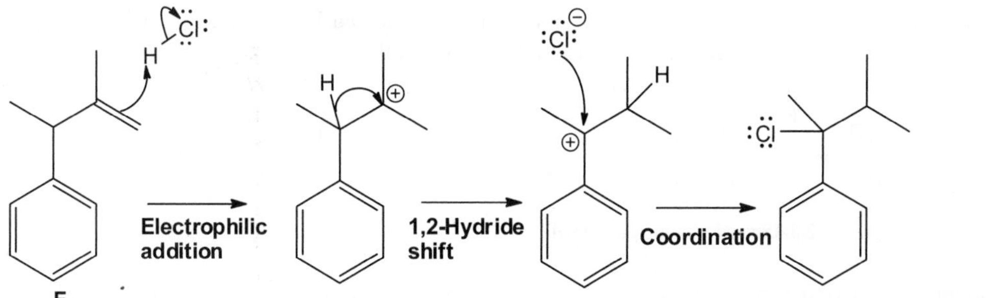

F

Electrophilic addition

1,2-Hydride shift

Coordination

2-Chloro-3-methyl-2-phenylbutane

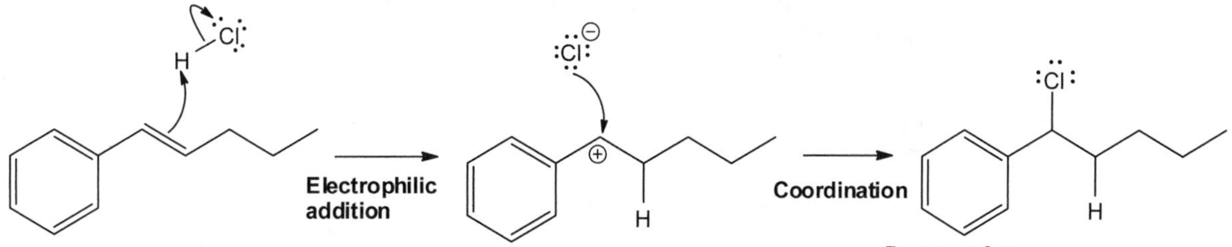

G

Electrophilic addition

Coordination

Does not form 2-chloro-3-methyl-2-phenylbutane

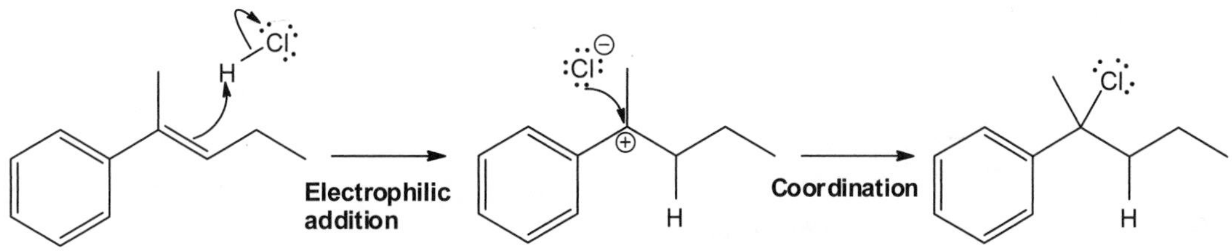

H

Electrophilic addition

Coordination

Does not form 2-chloro-3-methyl-2-phenylbutane

Problem 11.26

Think

What is the structure of the carbocation intermediate that forms in the first step? Do any rearrangements occur? What acts as the nucleophile in the second step? Are proton transfer reactions involved? What functional group forms?

Solve

(a) The secondary carbocation forms because it is more stable, and thus the Br^- attacks the secondary carbon of the alkene.

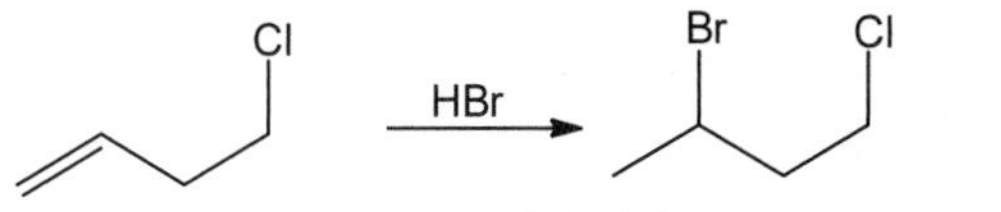

HBr

3-Bromo-1-chlorobutane

(b) The carbocation forms on the first carbon because it forms a resonance-stabilized carbocation, and thus the Br^- attacks the first carbon.

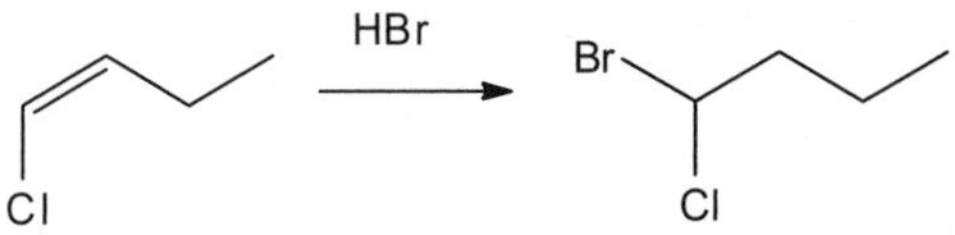

HBr

1-Bromo-1-chlorobutane

(c) The alkene is symmetrical, and the carbocation can form on either carbon to produce the same product. Water attacks that C^+, and the O is deprotonated to yield the alcohol.

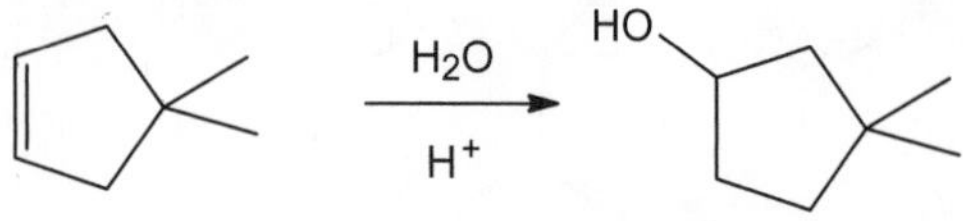

H_2O, H^+

HO

3,3-Dimethylcyclopentanol

(d) One equivalent of HCl first forms the vinyl chloride, and then the geminal dichloride is formed from the second equivalent of HCl. Both protons add to the terminal C of the alkyne to produce the more stable C^+ each time. The first C^+ is stabilized by the adjacent alkyl group, and the second C^+ is stabilized by resonance with the adjacent Cl.

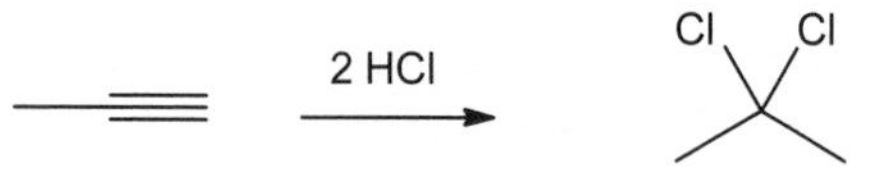

2 HCl

Cl Cl

2,2-Dichloropropane

(e) The carbocation forms on the secondary carbon, Then, a 1,2-hydride shift results in a more stable tertiary carbocation. The tertiary carbocation is then attacked by H_2O. A final deprotonation of the O yields the tertiary alcohol product.

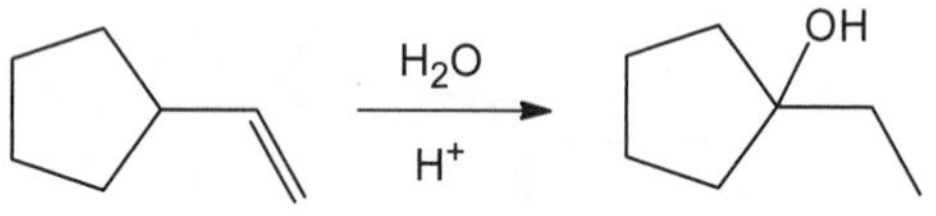

H_2O, H^+

OH

1-Ethylcyclopentanol

Problem 11.27

Think

What is the structure of the compound given? What functional group is present, and what nucleophile formed that functional group? On which carbon atoms are the halogen and H from the addition reaction located? On which carbon atoms could the C=C be located?

Solve

The overall reaction and mechanism are both shown below and on the next page for each of the compounds given. In these examples, the product is an alkyl halide formed from an electrophilic addition reaction of H^+ followed by coordination of the halide anion.

(a)

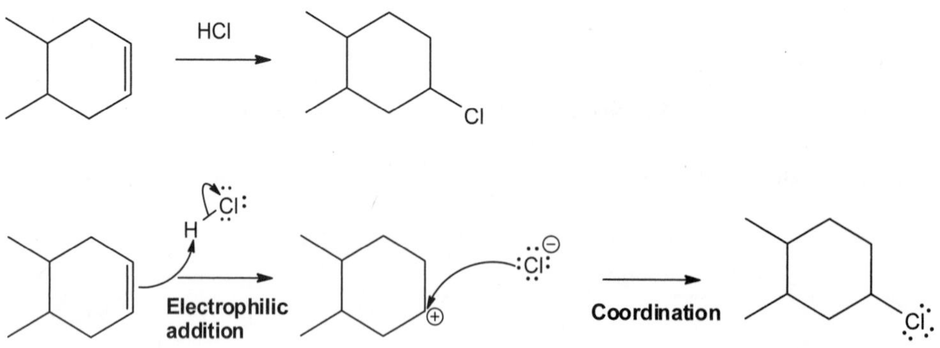

HCl

Cl

H, :Cl:

Electrophilic addition

:Cl:⊖

Coordination

:Cl:

(b)

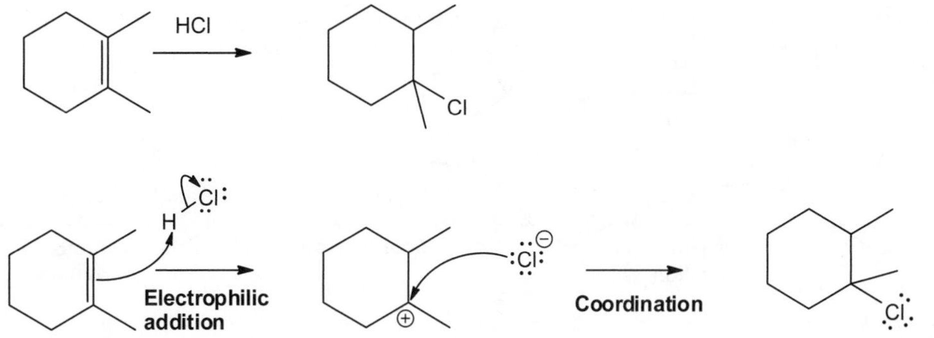

HCl

Cl

H

:Cl:

Electrophilic addition

:Cl:

Coordination

:Cl:

(c)

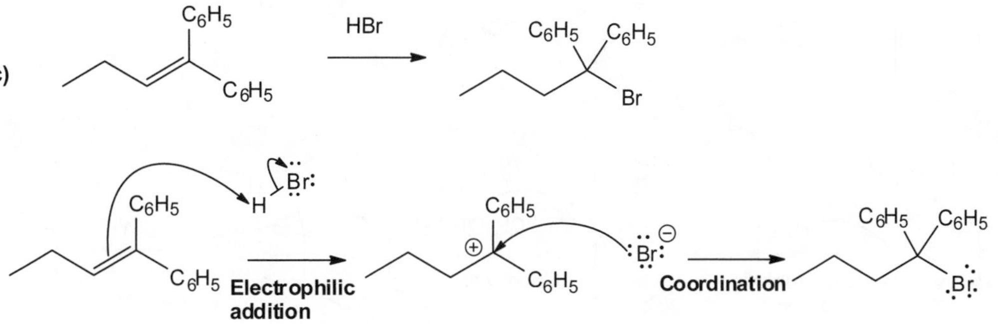

C_6H_5

C_6H_5

HBr

C_6H_5 C_6H_5

Br

H

:Br:

C_6H_5 Electrophilic addition

C_6H_5

:Br:

Coordination

C_6H_5 C_6H_5

:Br:

(d)

HCl

Cl

Cl Cl

:Cl:

H

Electrophilic addition

Cl

:Cl:

:Cl:

Coordination

:Cl: :Cl:

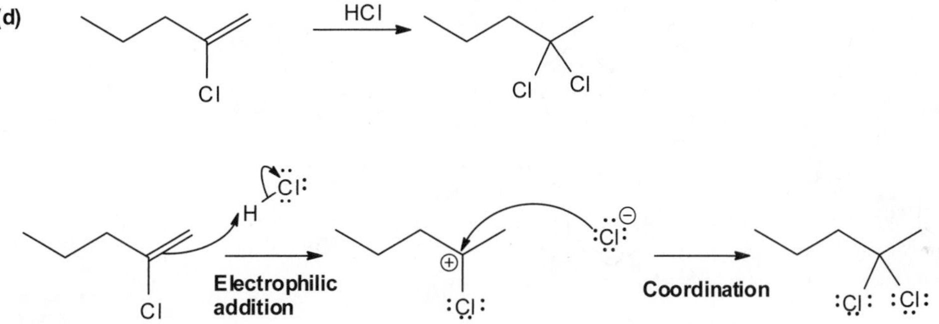

Problem 11.28

Think

What is the structure of the compound given? What functional group is present, and what nucleophile formed that functional group? On which carbon atoms are the OH/OR and H from the addition reaction located? On which carbon atoms could the C=C be located?

Solve

The overall reaction and mechanism are both shown below and on the next page for each of the compounds given. In these examples, the product is an alcohol or ether formed from an electrophilic addition reaction of H^+ followed by coordination of the water or an alcohol. A final proton transfer reaction takes place to form an uncharged product.

(a)

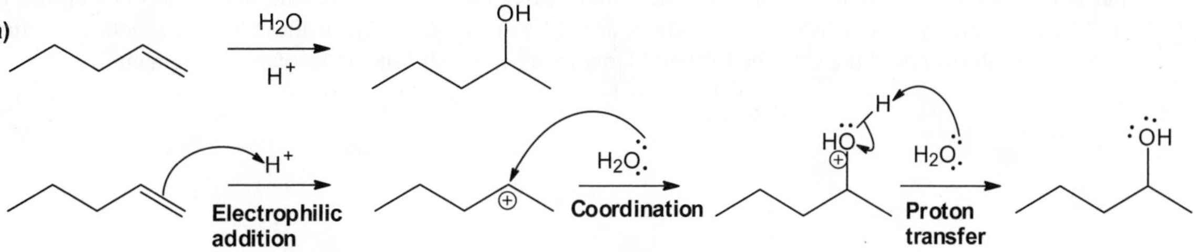

H_2O

H^+

OH

H^+

Electrophilic addition

H_2O:

Coordination

HO

H

H_2O:

Proton transfer

:OH

(b)

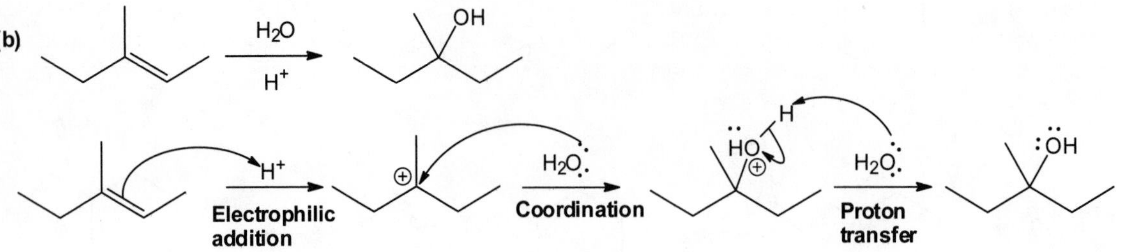

H_2O, H^+ → OH

H^+ — Electrophilic addition → H_2O: — Coordination → HO, H — H_2O: — Proton transfer → :OH

(c)

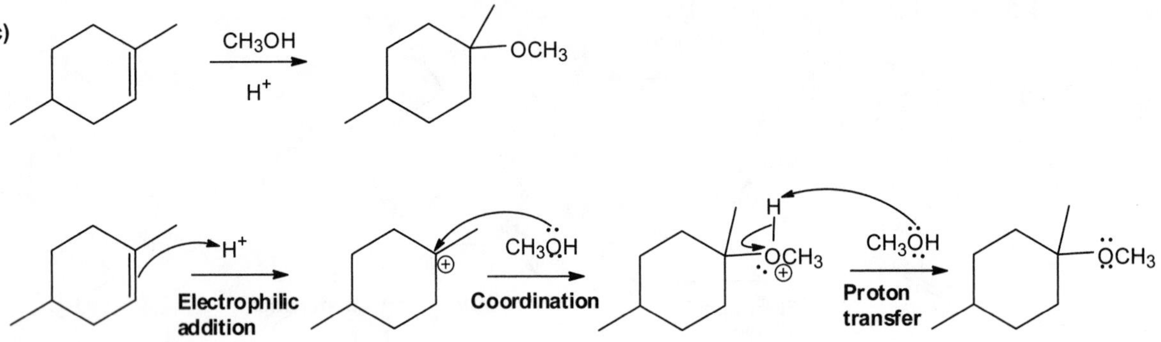

CH_3OH, H^+ → OCH_3

H^+ — Electrophilic addition → CH_3OH — Coordination → H, OCH_3 — CH_3OH — Proton transfer → OCH_3

(d)

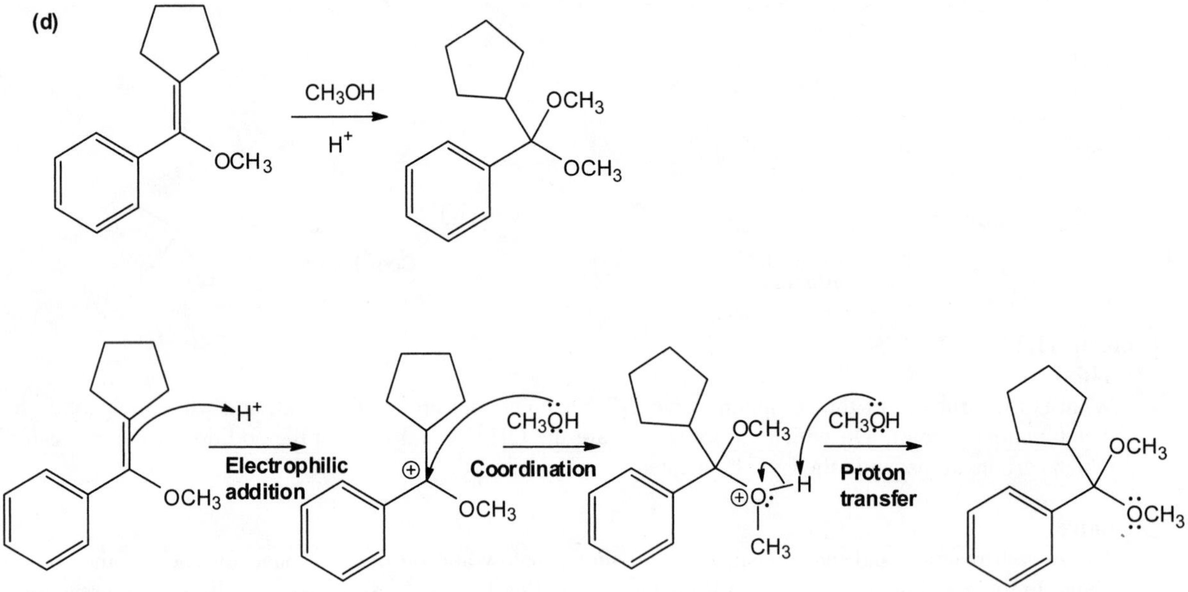

OCH_3 — CH_3OH, H^+ → OCH_3, OCH_3

H^+ — OCH_3 Electrophilic addition → OCH_3 — CH_3OH — Coordination → OCH_3, H, O, CH_3 — CH_3OH — Proton transfer → OCH_3, OCH_3

Problem 11.29

Think

What is the structure of the compound given? What functional group is present, and what nucleophile formed that functional group? On which carbon atoms are the halide and H from the addition reaction located? On which carbon atoms could the C≡C be located? How many times did the addition reaction occur?

Solve

The overall reaction and mechanism are both shown below and on the next page for each of the compounds given. In these examples, the product is an alkyl halide (vinyl or geminal) formed from an electrophilic addition reaction of H^+ followed by coordination of the halide anion. In the geminal dihalide products **(a)**, **(b)**, and **(d)**, the addition reaction occurred twice.

(a)

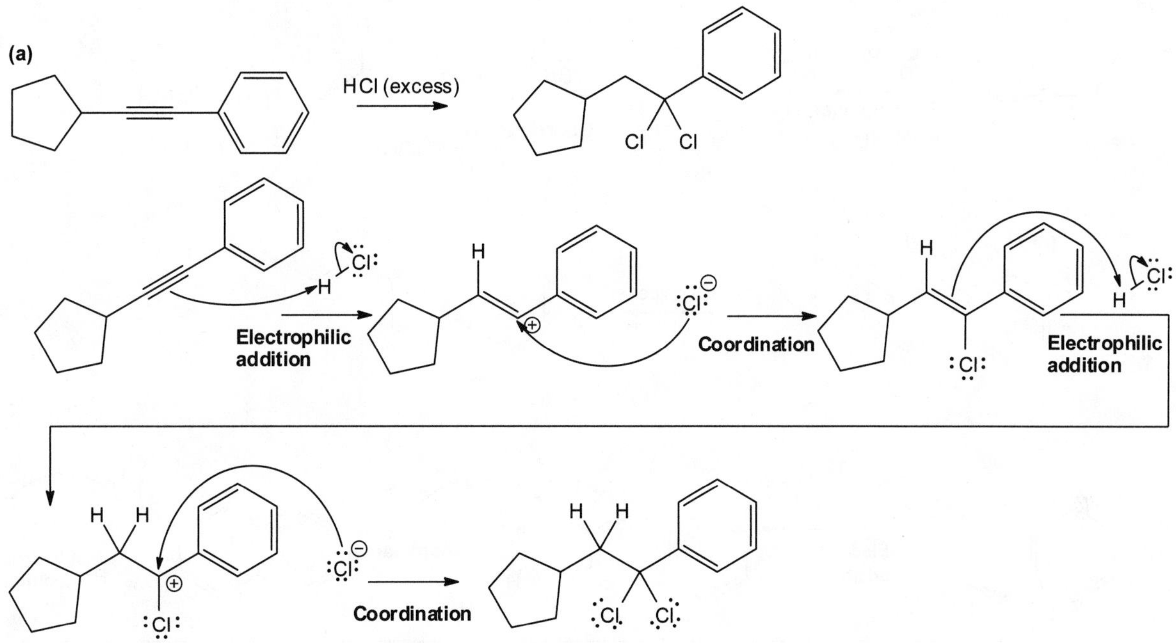

HCl (excess)

Cl Cl

Electrophilic addition

Coordination

Electrophilic addition

Coordination

(b)

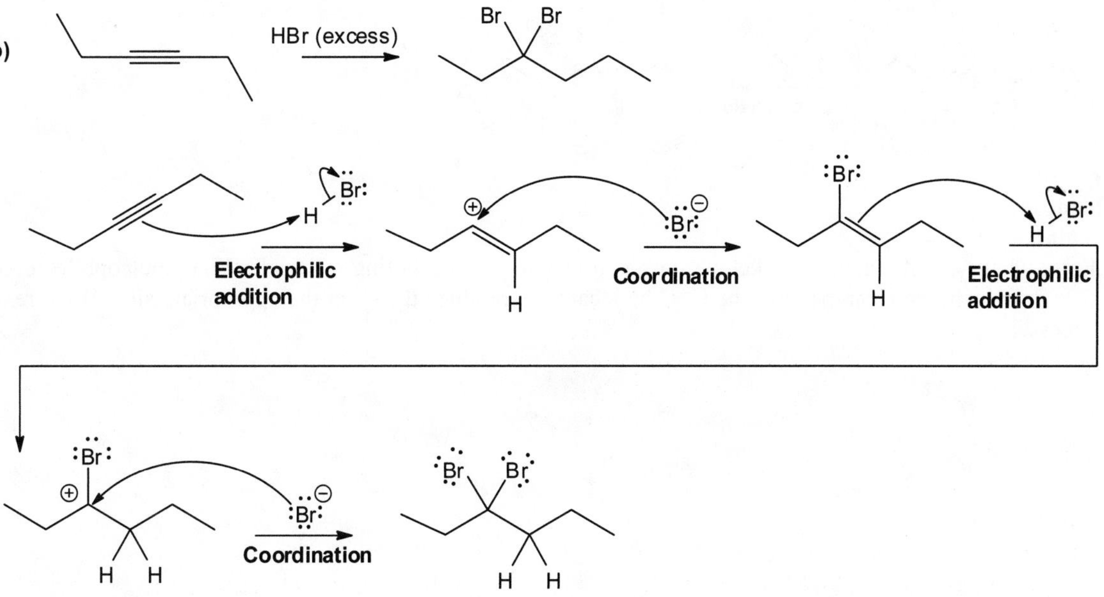

HBr (excess)

Br Br

Electrophilic addition

Coordination

Electrophilic addition

Coordination

(c) DBr (1 eq.)

Electrophilic addition

Coordination

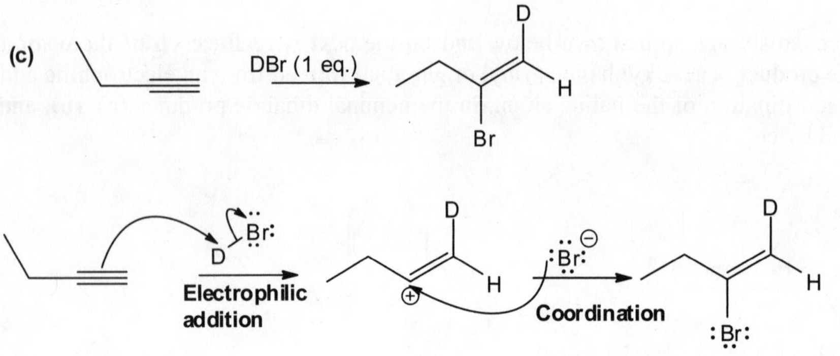

(d) HCl (excess)

Electrophilic addition

Coordination

Electrophilic addition

Coordination

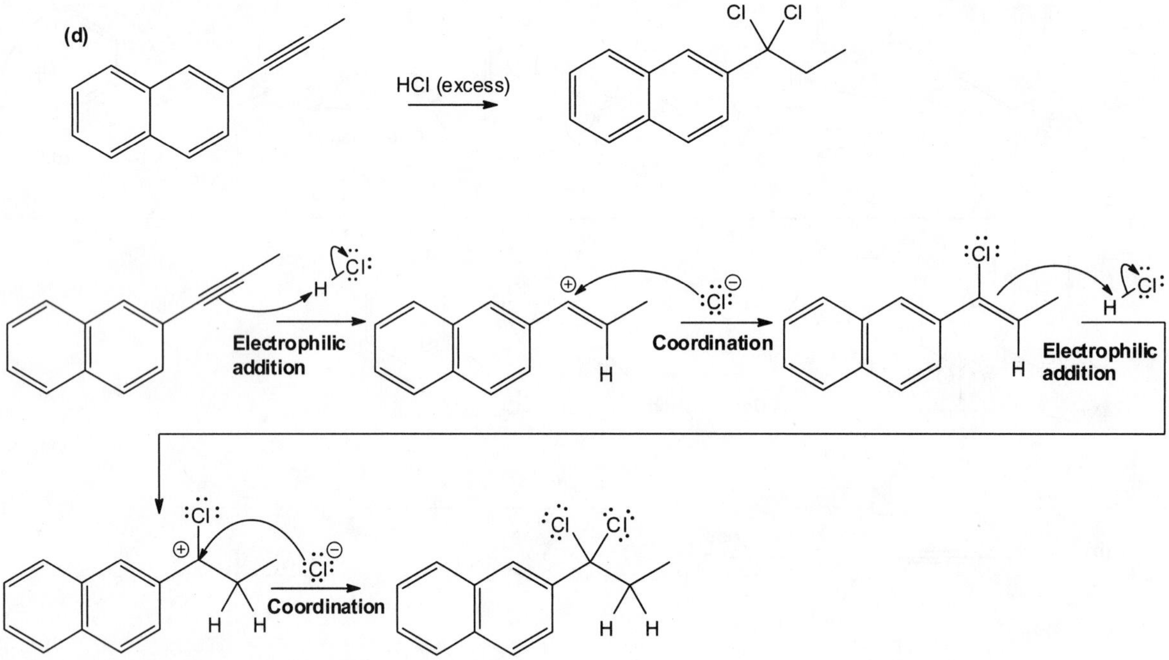

Problem 11.30

Think

From what type of reaction is a ketone formed given an alkyne starting material? What nucleophile adds to the C^+? On which two C atoms must the C≡C be located to produce the C^+ at the appropriate site? What reagent is necessary?

Solve

Each overall reaction and mechanism is shown below. Each is simply an acid-catalyzed hydration of water across the triple bond, followed by an acid-catalyzed keto–enol tautomerization.

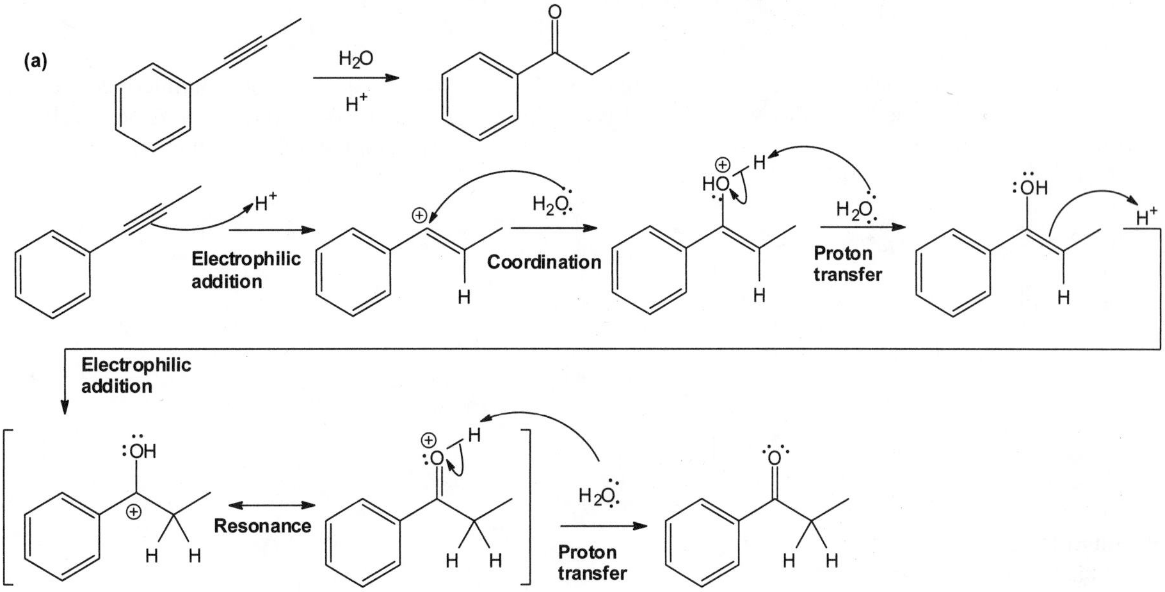

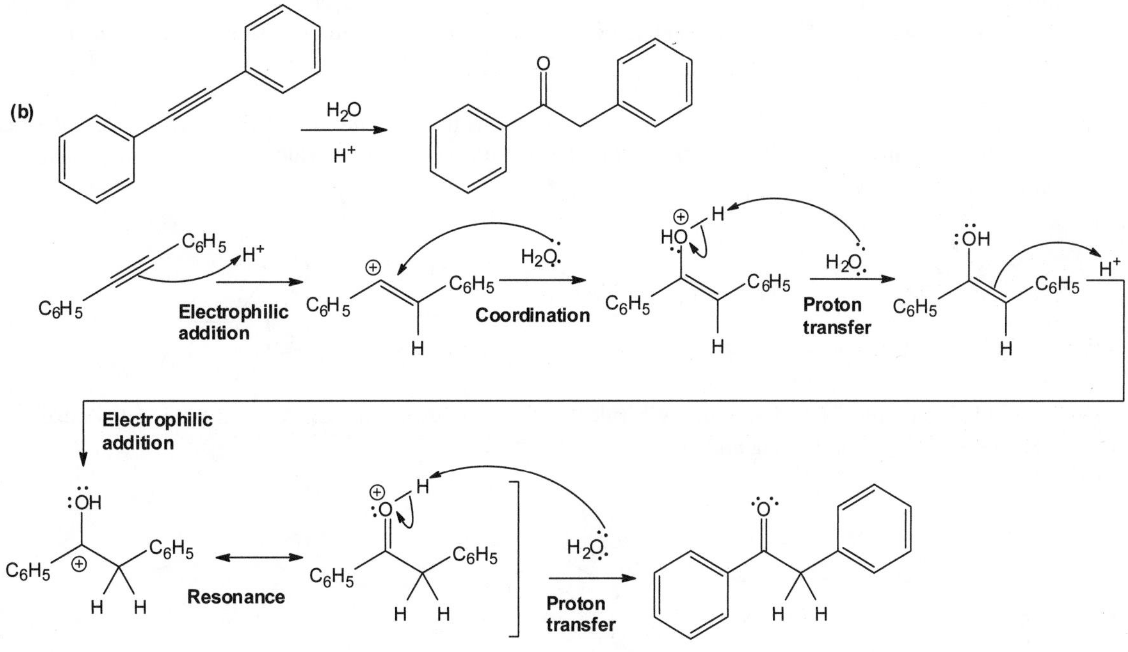

(a) H_2O, H^+

Electrophilic addition; Coordination; Proton transfer; Electrophilic addition; Resonance; Proton transfer

(b) H_2O, H^+

Electrophilic addition; Coordination; Proton transfer; Electrophilic addition; Resonance; Proton transfer

Problem 11.31

Think

What cation forms in the first step of the electrophilic addition reaction? Are two cation options possible in an unsymmetrical alkyne? Which cation is more stable? What nucleophile adds to the C^+? What rearrangement subsequently occurs?

Solve

The most stable carbocation results from addition of a proton to C2 of 1-phenylpropyne, because it places a positive charge on C1, where it can be resonance stabilized. The ketone is $C_6H_5{-}CO{-}CH_2CH_3$. This can be seen in the mechanism in **(a)** of Problem 11.30.

Problem 11.32

Think

What is the structure of 1-methylcyclohexanol? On which C atom is the HO located? Does this C atom have to have been part of the C=C? What other C atom(s) could compose the other half of the C=C?

Solve

Addition of water in acid to either of the two alkenes yields exactly the same carbocation intermediate, and thus the same product. The C atom to which the OH is attached in the product is the one that was positively charged in the carbocation intermediate. That C^+ is produced, in turn, by the addition of H^+ on the other C of the initial C=C.

H_3O^+

OH

H_3O^+

OH

Problem 11.33

Think

If the reaction starts and ends with an alkene, what two types of reactions must have occurred? Did the carbon skeleton get altered? Did a rearrangement occur? What reagents could accomplish the addition reaction? What reagents could accomplish the elimination reaction? Is the least substituted or most substituted alkene formed?

Solve

(a) Addition of water follows Markovnikov's rule (proceeds through the most stable carbocation intermediate), and elimination of water under acidic conditions follows Zaitsev's rule (forms the most stable alkene product).

H_3O^+

OH

H_2SO_4, Δ

(b) HBr adds according to Markovnikov's rule, and elimination with LDA yields the anti-Zaitsev product because LDA is a very strong and bulky base.

HBr

Br

LDA

Problem 11.34

Think

How many stereocenters are produced? How many stereoisomers are possible? Can H^+ add to either C atom of the C=C? What intermediate is formed? Can the nucleophile add to either side of the C^+ plane? What is the relationship between the stereoisomers that are produced? Which isomer(s) has/have an internal plane of symmetry? How does this affect the optical activity of the compound?

Solve

The alkene is not symmetrical, and addition of H^+ can form two different, secondary carbocation intermediates that are of similar stability. The carbocation intermediate is trigonal planar and the Cl^- nucleophile can attack the C^+ from in front (wedge) or from behind (dash). This leads to four products. The original wedge C–Cl is not broken and, therefore, has to remain the same configuration in the product. Only *trans*-1,3-dichlorocyclohexane is optically active. All of the other isomers contain an internal plane of symmetry.

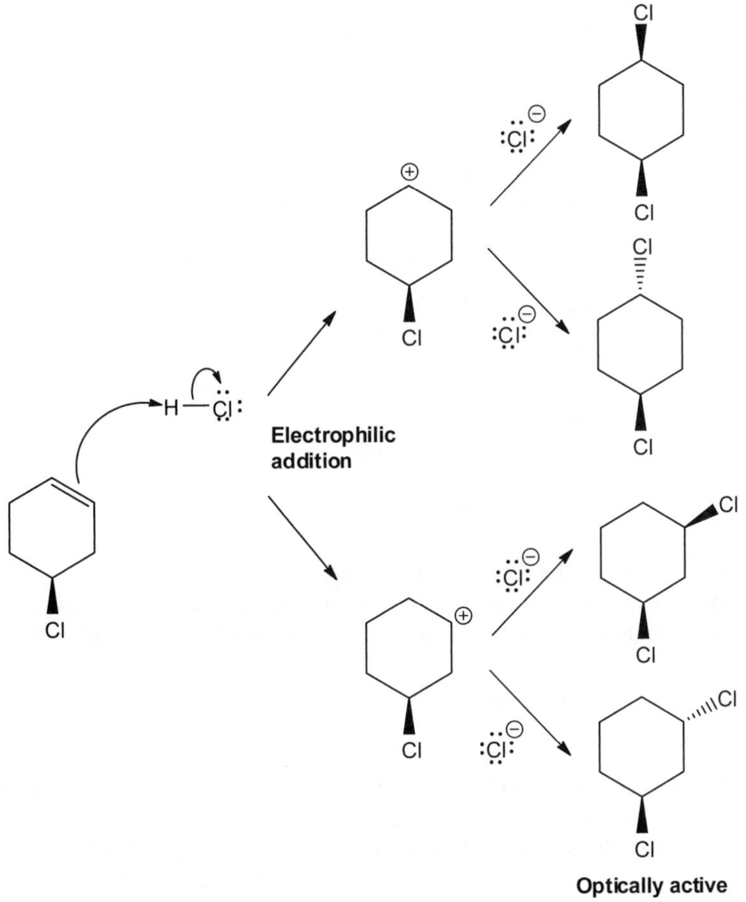

Electrophilic addition

Optically active

Problem 11.35

Think

How many C atoms are part of a C=C? Where does the C^+ intermediate form? Which intermediate is most stable? What factors contribute to stability? Are resonance structures possible? In the most stable carbocation intermediate, is there more than one site for the Br^- to attack? Which alkene product is most stable? How does temperature affect the product distribution?

Solve

(a) H^+ can add to any of the four C atoms that are part of the two C=C bonds, producing carbocations **A** – **D**. **D** is the most stable cation because it is resonance stabilized, with a tertiary carbocation resonance contributor. **A** is also resonance stabilized, but both contributors are secondary carbocations.

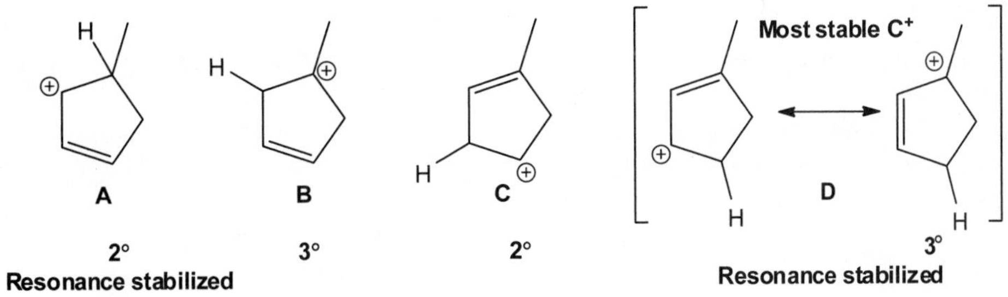

Most stable C^+

A 2° Resonance stabilized

B 3°

C 2°

D 3° Resonance stabilized

(b) Now that the most stable intermediate is identified, the nucleophile attack occurs on that intermediate. Attack of Br^- forms two (racemic) products. One pair of enantiomers results from 1,2-addition and another pair of enantiomers results from 1,4-addition.

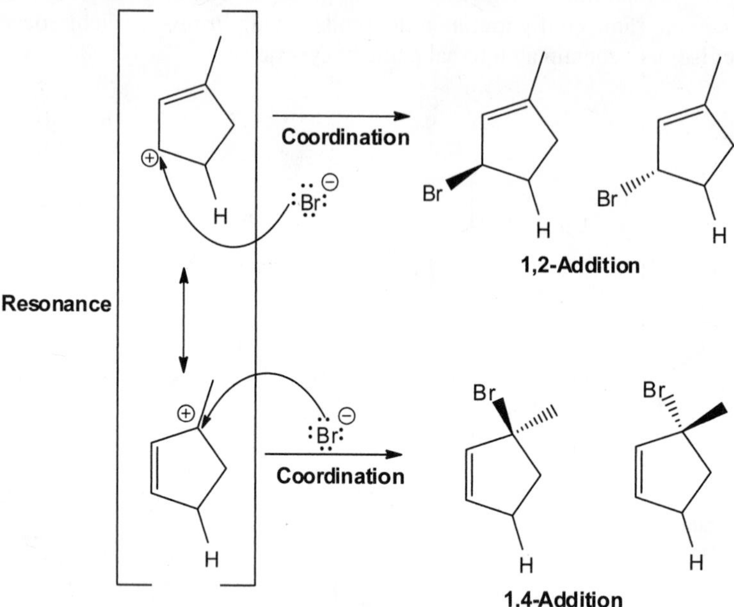

(c) The 1,2-adduct is the kinetic product and will be the major product at low temperatures.

(d) The 1,2-adduct is also the thermodynamic product and will be the major product at high temperatures. This is because the 1,2-adduct has the more highly alkyl-substituted C=C bond than does the 1,4-adduct, so the 1,2-adduct is the more stable one.

Problem 11.36

Think

How many C atoms are part of a C=C? Where does the C^+ intermediate form? Which intermediate is most stable? What factors contribute to stability? Are resonance structures possible? In the most stable carbocation intermediate, is there more than one site for the Br^- to attack? Which alkene product is most stable? How does temperature affect the product distribution?

Solve

(a) A proton can add to three different C atoms, giving rise to three different carbocation intermediates, **A** – **C**. The most stable ion has the greatest resonance stabilization of the positive charge (**B**). You should draw the resonance contributors for practice.

Most stable

H H H

A B C

(b) Br^- can attack any of the three C atoms that share the positive charge via resonance delocalization, giving rise to the three products below. The first and second products below will be racemic because a new stereocenter is formed.

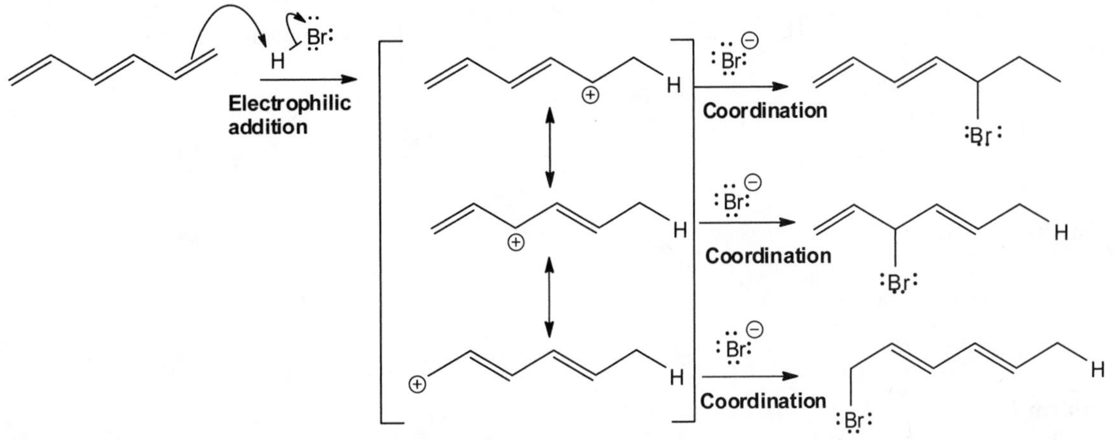

Electrophilic addition

Coordination

Coordination

Coordination

(c) The 1,2-adduct (the first product) will be the major product at low temperatures, because it is the kinetic product.

(d) The 1,6-adduct (the third product) will be the major product at high temperatures because it has the most stable double bonds (they are the most highly substituted and, as we will learn in Chapter 14, they are conjugated) and is the thermodynamic product.

Problem 11.37

Think

To have resonance movement of the C=C, what intermediate must form? What elementary step must, therefore, take place? What mechanism allows for this isomerization reaction? Which of the two alkenes is more stable? Should that alkene be favored at low temperatures or high temperatures?

Solve

It is an S_N1 mechanism, taking place in two steps. Bromide leaves, forming a resonance-stabilized cation. Bromide attacks either positively charged C in the cation, but favors attack at the primary C at high temperatures, because it gives rise to the more stable (thermodynamic) product.

Heterolysis

Resonance

Coordination

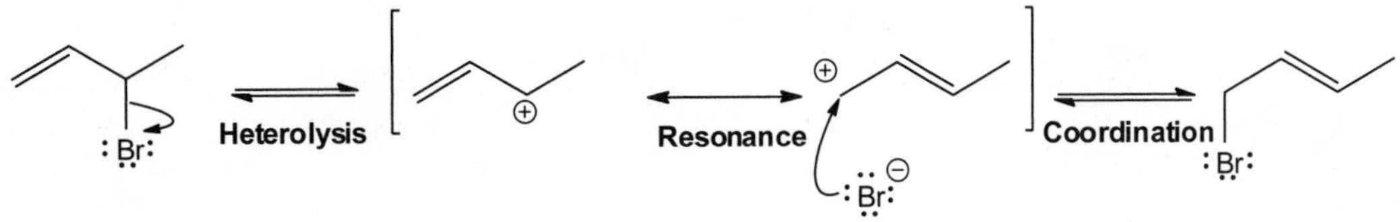

Problem 11.38

Think

Upon addition of H^+, how many carbocations can form? Which carbocation is more stable? Are resonance structures possible? What acts as the nucleophile in the second step?

Solve

There are two possible carbocations that can be formed by addition of H^+ to the C=C double bond (**A** or **B**). **B** is the more stable carbocation intermediate because it is resonance stabilized by the lone pair from O. This leads to the major product. Then chloride attacks the carbocation to give the overall products shown on the next page.

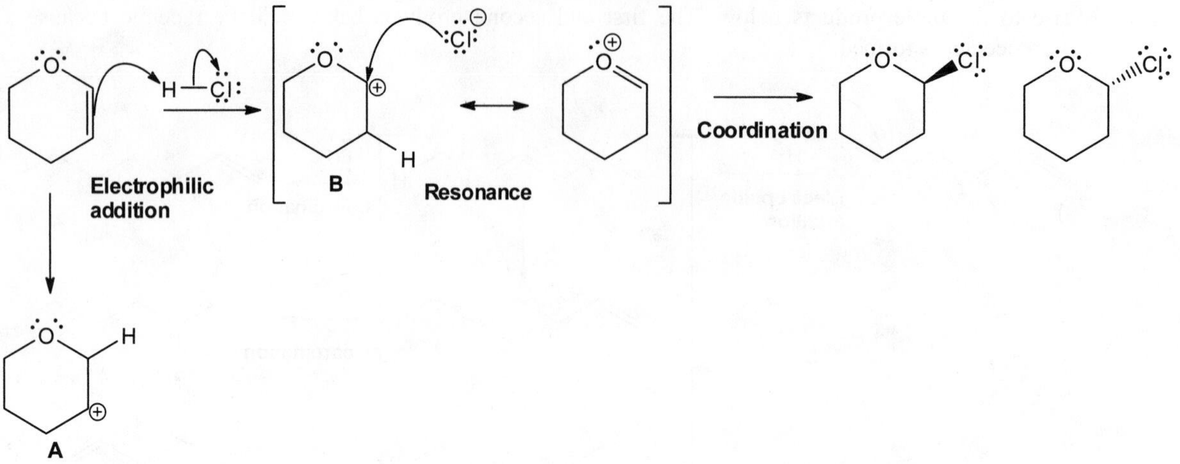

Electrophilic addition

B

Resonance

Coordination

A

Problem 11.39

Think

Upon addition of H^+, how many carbocations can form? Which carbocation is more stable? Are resonance structures possible? What acts as the nucleophile in the second step? Are proton transfers incorporated into the mechanism?

Solve

There are two possible carbocations that can be formed by addition of H^+ to the C=C double bond (**A** or **B**). **B** is the more stable carbocation intermediate because it is resonance stabilized by the lone pair from O. This leads to the major product. Then the alcohol O atom attacks the carbocation and a deprotonation step follows to give the overall uncharged products shown below.

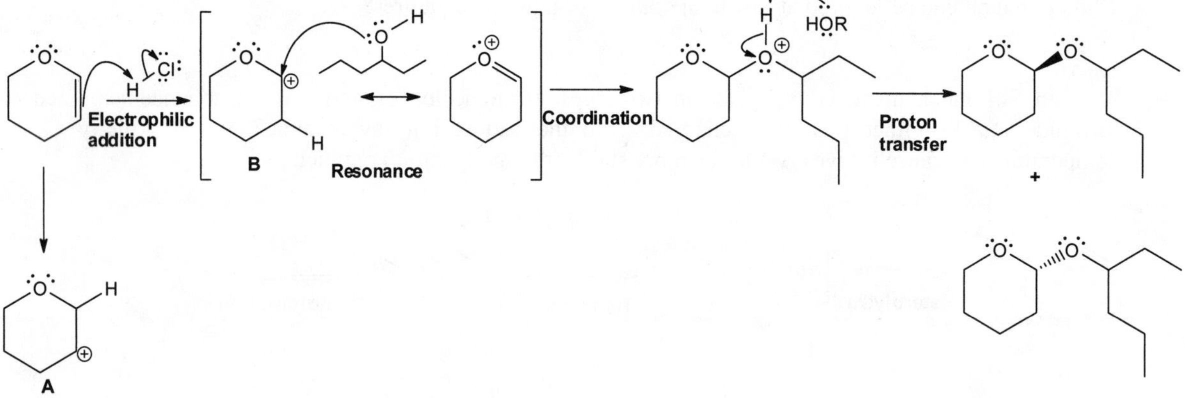

Electrophilic addition

B

Resonance

Coordination

HOR

Proton transfer

+

A

Problem 11.40

Think

What is the electron-rich and electron-poor species in each reaction? Does addition, substitution, or elimination occur? Are rearrangements possible?

Solve

A results from electrophilic addition of water to a C=C bond to form an alcohol on the benzylic C atom (most stable C^+ intermediate). **B** results from deprotonation of the O–H bond by hydride in the first step to form the alkoxide nucleophile. The epoxide ring is opened by attack of the RO^- at the less substituted C atom. **C** results from two S_N2 reactions to turn the ROH into an RBr. **D** results from an E2 reaction to form the most substituted alkene. See the reaction on the next page.

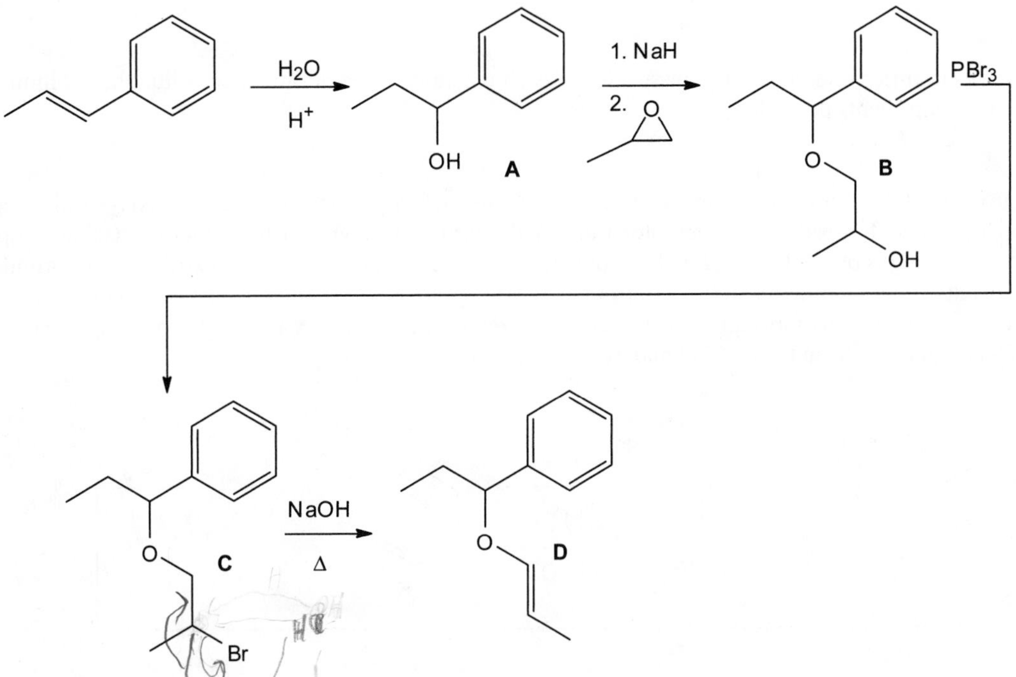

H2O
H+
A
1. NaH
2.
B
PBr3
C
NaOH
Δ
D

Problem 11.41

Think

What is the electron-rich and electron-poor species in each reaction? Does addition, substitution, or elimination occur? Are rearrangements possible?

Solve

E results from electrophilic addition of H_2O to the alkyne in acid to form an enol that tautomerizes to the ketone. **F** results from deprotonation of the α C–H to form the enolate anion that undergoes an S_N2 reaction with the allylic C–Br. **G** results from electrophilic addition of methanol to the C=C to form the ether on the secondary carbon. **H** results from deprotonation of the α C–H to form the enolate anion that undergoes an S_N2 reaction with the Br–Br. **I** results from an E2 reaction in base to form an alkene.

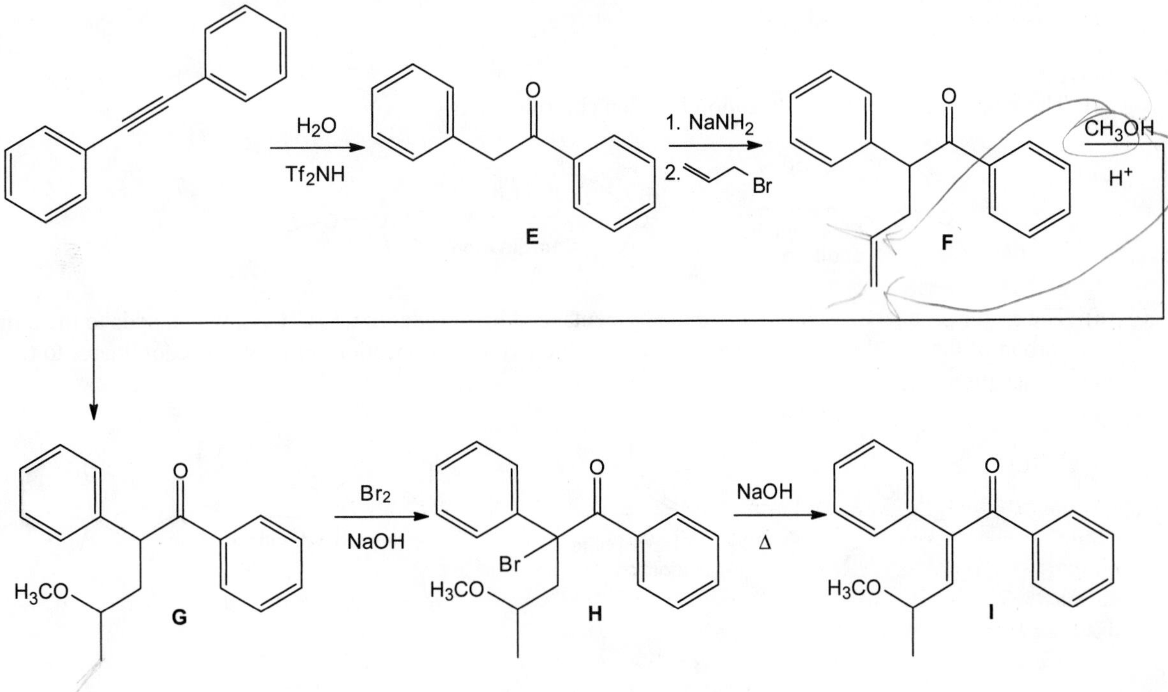

H2O
Tf2NH
E
1. NaNH2
2.
F
CH3OH
H+
G
Br2
NaOH
H
NaOH
Δ
I

Problem 11.42

Think

What is the electron-rich and electron-poor species in each reaction? Does addition, substitution, or elimination occur? Are rearrangements possible?

Solve

J results from an E2 reaction to form the terminal alkyne, which is deprotonated by NH_2^-, followed by protonation by water. **K** results from deprotonation of the terminal alkyne to form the RC≡C:⁻ nucleophile. Then the epoxide ring is opened via S_N2 and protonated to extend the carbon chain by two C atoms, forming the alcohol. **L** results from two S_N2 reactions to turn the ROH into an RBr. **M** results from electrophilic addition of H_2O to the alkyne in acid to form an enol that tautomerizes to the ketone. **N** results from an S_N2 reaction to break the C–Br bond and form the C–CN bond.

Cl; 1. 2 equiv. $NaNH_2$; 2. H_2O; J; 1. $NaNH_2$; 2. O (epoxide); 3. H_2O; K; OH; PBr_3; L; Br; H_2O; Tf_2NH; M; O; Br; NaCN; N; O; CN

Problem 11.43

Think

What is the electrophile in the electrophilic addition step? In Step 2, what acts as the nucleophile to coordinate to the C^+?

Solve

(a) The π bond captures the carbocation R^+, then chloride attacks.

C=C; $R^{\oplus}$; Electrophilic addition; R; C–C⊕; :Cl:⊖; Coordination; R; Cl; C–C

(a) The major product is 2-chloro-2,4-dimethylpentane. $AlCl_3$ produces $CH_3C^+CH_3$, which adds to the terminal carbon of the double bond to yield the more stable tertiary carbocation. Finally, Cl^- coordinates to the new carbocation.

:Cl:; $AlCl_3$; ⊕; Electrophilic addition; ⊕; :Cl:⊖; Coordination; :Cl:

Problem 11.44

Think

In acid, what is the electrophile that adds to the C=C? To which C atom does it add to produce the most stable carbocation? If the carbocation that is produced reacts with an alkene, what reaction can take place?

Solve

The π bond in styrene captures a proton, forming a resonance-stabilized carbocation. Successive molecules of styrene add to the cation, lengthening the polymer chain.

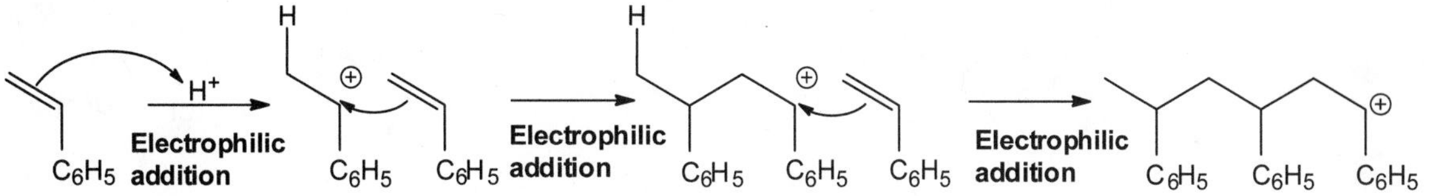

Problem 11.45

Think

Which π bond undergoes addition with H^+? To which C does the H^+ add to form the most stable C^+? Are resonance structures possible?

Solve

This is a two-step addition of HBr to a multiple bond. Addition of H^+ to the terminal C of the triple bond yields a resonance-delocalized carbocation, which then undergoes coordination with Br^-.

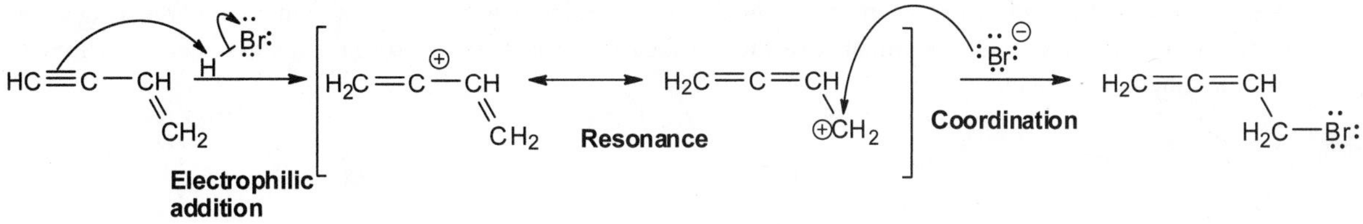

Problem 11.46

Think

What carbocation intermediate forms with reaction of hepta-1,6-diene and H^+? To which C atom of the C=C does H^+ add to produce the most stable C^+? What intramolecular reaction can then take place to form the ring? What is the structure of the carbocation intermediate now? What can then act as the nucleophile?

Solve

After a π bond captures the proton in HCl, the newly formed R^+ electrophile adds to the second C=C double bond, resulting in ring formation, and giving another carbocation intermediate. Attack by Cl^- finishes the mechanism.

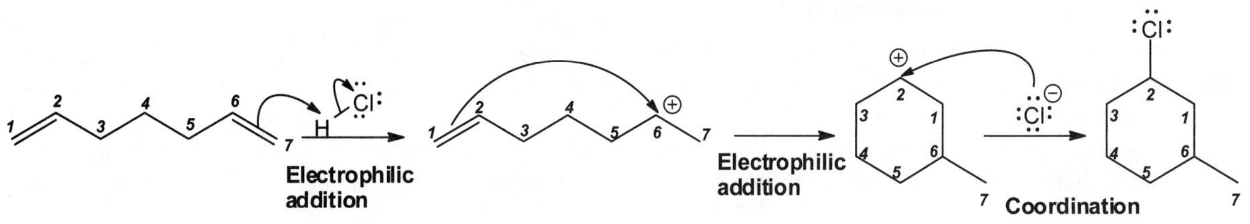

Problem 11.47

Think

What carbocation intermediate forms with reaction of penta-1,4-diene and H^+? Is the carbocation able to rearrange? What factor drives this rearrangement? How is the C–Cl bond formed?

Solve

After a π bond captures the proton in HCl, the secondary C^+ undergoes a 1,2-hydride shift to form a resonance-stabilized secondary C^+. The Cl^- then coordinates to the terminal C^+ to yield the internal alkene, which is more highly substituted, and therefore more stable than the terminal alkene.

Electrophilic addition

1,2-Hydride shift

Resonance

Coordination

Problem 11.48

Think

What carbocation intermediate forms with reaction of the alkene and H^+? Is the carbocation able to rearrange? What factor drives this rearrangement? Are four-carbon rings strained? Are proton transfer reactions involved?

Solve

After a π bond captures the proton in HBr, the tertiary C^+ undergoes a rearrangement twice. Each rearrangement is a 1,2-alkyl shift and is driven by the relief of ring strain in going from a four-membered ring to a five-membered ring. The water coordinates to the carbocation, and a final proton transfer reaction is involved to form an uncharged alcohol product.

Electrophilic addition

1,2-Alkyl shift

1,2-Alkyl shift

Coordination

Proton transfer

Problem 11.49

Think

What two carbocation intermediates are possible when the alkene reacts with HCl? Which carbocation intermediate is more stable? What is the effect of nearby electron-withdrawing groups on a positive charge?

Solve

The alkene is not symmetrical, and two carbocation intermediates are possible, a secondary and primary C^+. Carbocation intermediates are stabilized by electron-donating groups (alkyl groups) and destabilized by electron-withdrawing groups (CF_3 in this example). The primary carbocation is typically less stable than a secondary; however, with a neighboring CF_3, the secondary carbocation is destabilized and not favored. The mechanism and product are shown on the next page.

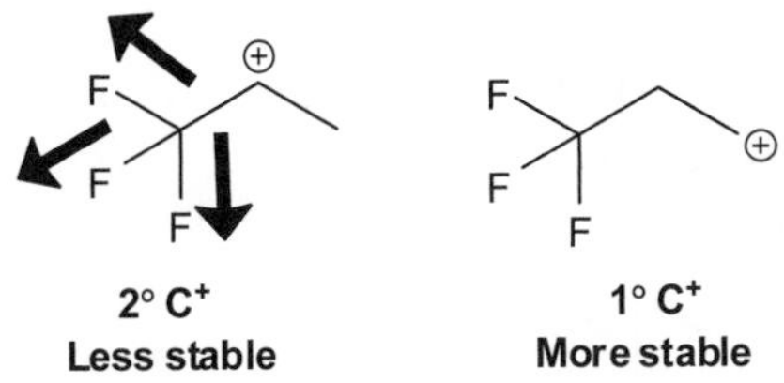

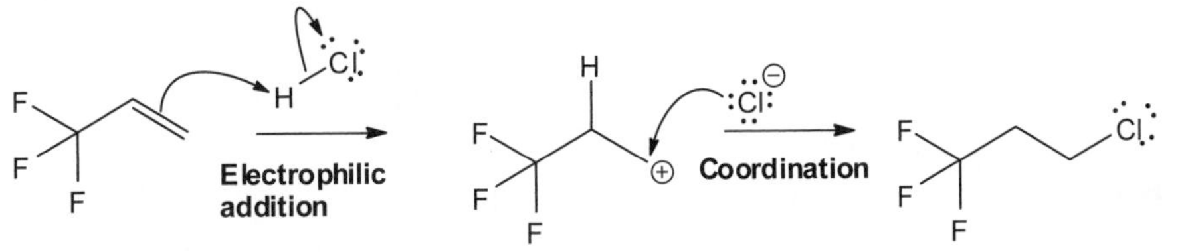

2° C^+
Less stable

1° C^+
More stable

Electrophilic addition

Coordination

Problem 11.50

Think

Looking at the mechanism arrows drawn, what bonds are broken and formed? Can the new C–C bond form to either atom of the C=C? What is left on the other C atom? What is the effect of a nearby O atom with lone pairs?

Solve

The new C–C bond could form to either atom of the C=C. A carbocation is left on the other C atom. The neighboring O atom with lone pairs provides the ability of the C^+ to be resonance stabilized. The mechanism is shown below.

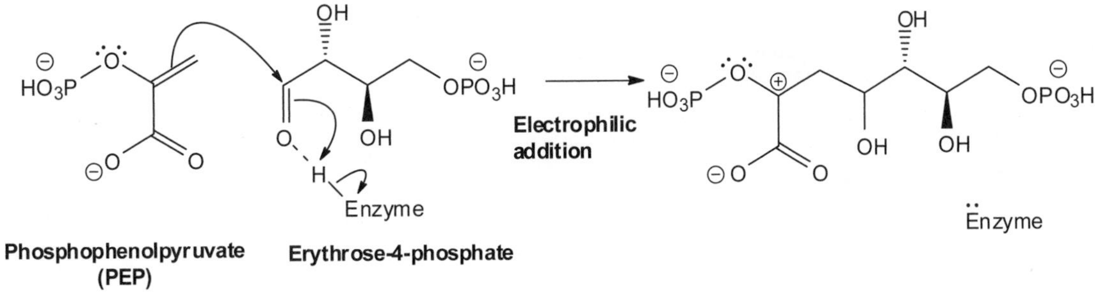

Electrophilic addition

Phosphophenolpyruvate (PEP)

Erythrose-4-phosphate

Problem 11.51

Think

On which carbon of the C=C does the C^+ form when a proton adds? What acts as the nucleophile in the second step? Are proton transfer reactions involved? In the intermediate that forms the 1,4-cineole product, are any rearrangements possible? If so, why?

Solve

The carbocation forms on the tertiary C^+ when the proton adds to the secondary C. The alcohol acts as the nucleophile and attacks the tertiary C^+ in an intramolecular reaction to produce a new ring. A proton transfer reaction follows to form the uncharged 1,8-cineole product. To form the 1,4-cineole product, the intermediate undergoes a 1,2-hydride shift. The alcohol acts as the nucleophile and attacks the tertiary C^+. A proton transfer reaction follows to form the uncharged 1,4-cineole product. See the mechanism on the next page.

OSO3H
H
Electrophilic addition
Coordination
OSO3H
Proton transfer
OH
α-Terpineol
H
1,8-Cineole
H
OSO3H
H
Electrophilic addition
Coordination
1,2-Hydride shift
H
OH
α-Terpineol
H
Coordination
H
OSO3H
Proton transfer
1,4-Cineole

CHAPTER 12 | Electrophilic Addition to Nonpolar π Bonds 2: Reactions Involving Cyclic Transition States

Your Turn Exercises

Your Turn 12.1

Think

Is there free rotation about the C–C bond in the intermediate? What should be the cis/trans relationship of the CH_3 groups in the alkene compared to that of the product: conserved, inverted, or scrambled?

Solve

The cis/trans relationship for the CH_3 groups attached to the C=C should be conserved. Groups that are on the same side of the C=C (cis) should both be wedge (or both dash) in the product. For groups that are on opposite sides, one bond should be wedge and one should be dash. The cyclic product for the trans alkene is chiral—specifically, enantiomers are produced (the other possible product is shown below). Therefore, the reaction product mixture is racemic.

E

E

E

E

E Racemic

E (Enantiomer)

Your Turn 12.2

Think

In the formal charge method, how many electrons in a bond are assigned to each atom? Lone pair? How many electrons does C have in the carbene molecule? How does this compare to the number of valence electrons in an isolated carbon atom?

Solve

Carbon has four valence electrons. In the formal charge method, half of the electrons are assigned to each atom in the bond. The carbene C has two bonds, thus two electrons. Each nonbonding electron (dot) represents one electron. There are two electrons in the lone pair. Thus, the carbene C is assigned four valence electrons, the same as in an isolated C atom, so the formal charge is 0.

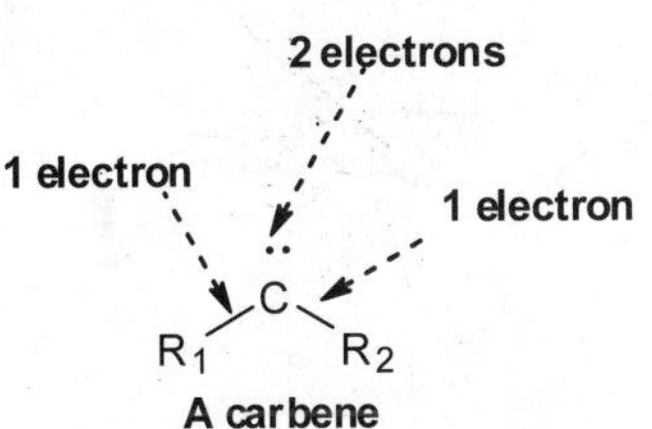

Your Turn 12.3

Think

Using Equation 12-7 as a guide, what bonds broke and formed in each step? How do you use curved arrows to show bonds breaking and forming? Are protons involved in any of the elementary steps? Are π bonds involved?

Solve

Step 1 is a *proton transfer* step where HO^- acts as a base and abstracts a proton from CBr_3H to form a carbanion. In Step 2, Br^- is eliminated via *heterolysis* to generate the carbene, CBr_2. Step 3 is electrophilic addition of CBr_2 to the C=C to produce the cyclopropane ring.

This step is slow because the product carbene is highly unstable.

Dibromocarbene

A new cyclopropane ring is produced.

1. Proton transfer

2. Heterolysis

3. Electrophilic addition

Your Turn 12.4

Think

Using Equation 12-9 as a guide, what bonds broke and formed in each step? How do you use curved arrows to indicate bond breaking and bond formation? Are π bonds involved? What species are electron rich and electron poor in each step?

Solve

In Step 1, the C=C π bond is electron rich and polarizes the Cl–Cl to have an electron-rich and electron-poor site. In a concerted electrophilic addition reaction, the chloronium ion intermediate and Cl^- ion are formed. In Step 2, the Cl^- acts as a nucleophile and the positively charged Cl atom in the ring acts as the leaving group. This is an S_N2 reaction where the nucleophile attacks from the side opposite the leaving group. It is just as likely that Cl^- attacks the C on the right side of C=C (shown first) as the C on the left (shown second). This produces the enantiomer and, therefore, the mixture is racemic.

Chloronium ion intermediate

1. Electrophilic addition

2. S_N2

Same as

Attack left C atom:

Chloronium ion intermediate

1. Electrophilic addition

2. S_N2

Same as

Your Turn 12.5

Think

Using Equation 12-17 as a guide, identify which bonds broke and formed in each step. How do you use curved arrows to show bond formation/breaking? What species are electron rich and electron poor in each step? What nucleophiles are present? Do any steps involve proton transfers?

Solve

Step 1 is an electrophilic addition that produces the chloronium ion intermediate and Step 2 is an S_N2 reaction that opens the ring as H_2O acts as the nucleophile and attacks the electrophilic C of the ring. *Note*: Either the left or right side of the ring could be attacked, but this example shows just the right-side C being attacked. In this example, H_2O is the solvent and is present in a much larger concentration than Cl^- and, therefore, wins out over Cl^- as the nucleophile. Step 3 is a proton transfer step to form the chlorohydrin.

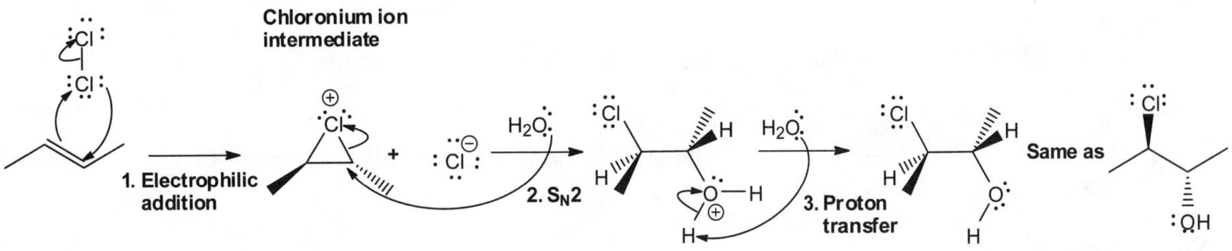

Chloronium ion intermediate

1. Electrophilic addition

2. S_N2

3. Proton transfer

Same as

Your Turn 12.6

Think

It is helpful to draw out the C–H bonds that are not shown in the line structure. Recall from Section 11-1 an alkene reaction with an hydrogen halide acid. To which C does the proton add? What is the product that results? How can you show the carbocation rearrangement? Why does the carbocation rearrangement take place? If water is the solvent, what acts as the nucleophile in the third step?

Solve

Step 1 is electrophilic addition of H^+ to C=C to result in the formation of a secondary carbocation. Step 2 is the rearrangement of the secondary carbocation to a tertiary carbocation via a 1,2-hydride shift, and is driven by the greater stability of the tertiary carbocation. Step 3 is coordination where H_2O acts as the nucleophile to attack the tertiary carbocation. Step 4 is proton transfer to form the ROH as the final product.

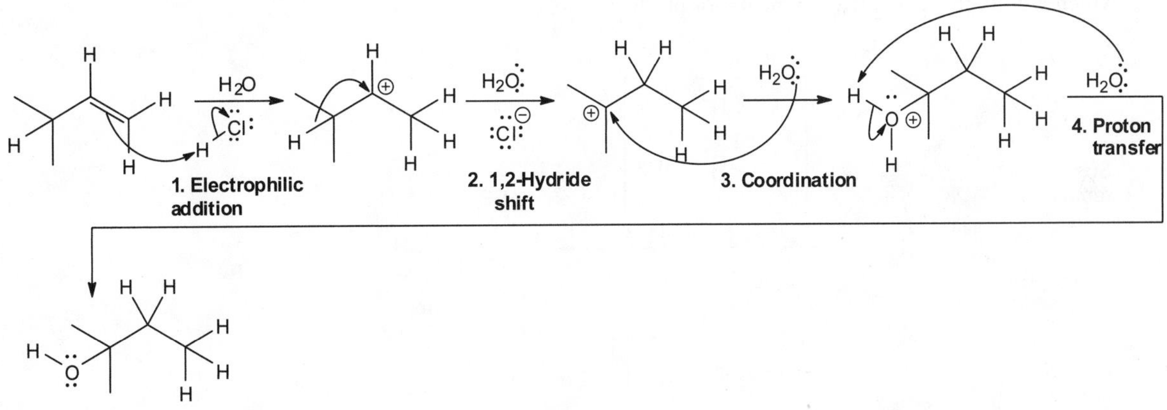

1. Electrophilic addition

2. 1,2-Hydride shift

3. Coordination

4. Proton transfer

Your Turn 12.7

Think

Consult Equation 12-22 to look at the mechanism for oxymercuration–reduction. Which C atom of the C≡C can accommodate the partial positive charge better? What determines the regiochemistry? Look at the ketone that is formed and take the reaction one step back. Where is the OH located? What characterizes an enol?

Solve

A δ+ charge on the first structure would be better accommodated by the benzylic carbon. This is the C atom that the H_2O attacks and is the site of the enol—a functional group in which a C atom is simultaneously part of an alkene and alcohol group. A δ+ charge on the second structure would be better accommodated by C2 because it is more highly substituted. This is the carbon that the H_2O attacks and is the site of the enol. See the figure on the next page.

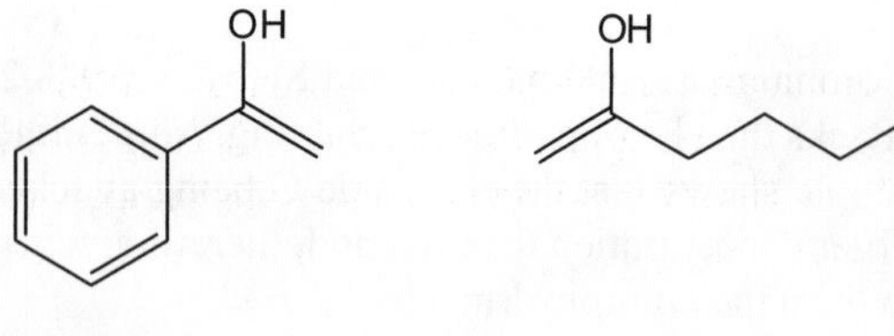

OH OH

Your Turn 12.8

Think

Consult Table 1-2 to look up the value of the O–O and the C–C bond strengths.

Solve

O–O = 138 kJ/mol; C–C = 339 kJ/mol. The O–O bond is less than half the strength of a C–C bond, and this verifies that the O–O bond is weak.

Your Turn 12.9

Think

Consult Equations 12-35 and 12-36 as guides. In how many steps does the reaction occur? Is borane (BH_3) electron rich or electron poor? Is the π bond electron rich or electron poor? Does an intermediate form? What is the influence of the proximity of BH_2–H to the CH_3 group on the double bond? On which C atom does a partial positive charge develop in the transition state?

Solve

The reaction occurs in one *concerted* step. The B atom of borane is electron poor and the C=C π bond is electron rich. Steric repulsion directs the larger group, BH_2, to the C atom with the fewer number of alkyl groups. Also, in the transition state, a partial positive charge develops on the C atom that gains the H. In the hydroboration steps below, the first reaction has significant steric repulsion, and a partial positive develops on the less alkyl-substituted C, which would, therefore, lead to the minor product. The second hydroboration minimizes the steric repulsion and a partial positive charge develops on the more alkyl-substituted carbon, which would, therefore, lead to the major product.

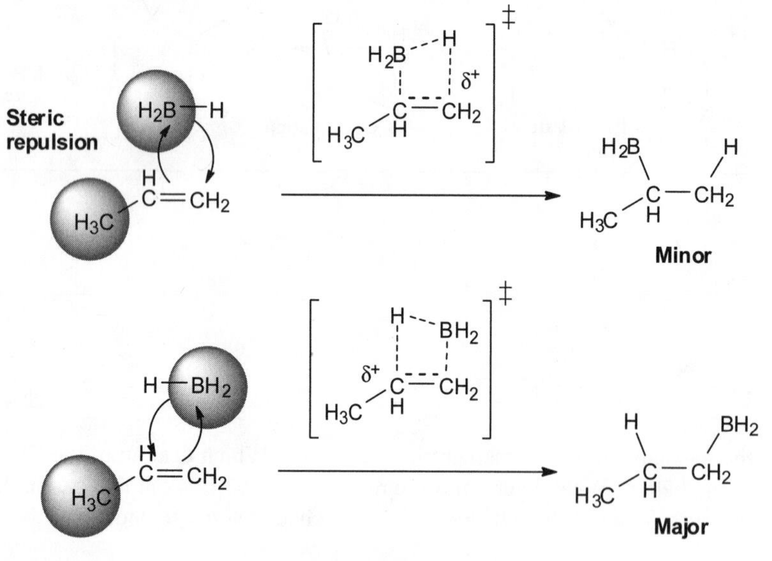

Steric repulsion

Minor

Major

Your Turn 12.10

Think

In the oxidation step, what group replaces B for each B–C bond? How does that replacement take place? Is there retention or inversion of the stereochemical configuration at each of those C atoms?

Solve

In Step 2 of the mechanism, the C–B bond breaks at the same time the C–O bond forms. In this concerted process, the O assumes the same position about the C atom as the original B atom did. The configuration about the C atom does not change. Each C–B bond is replaced by a C–OH bond, and the stereochemistry is retained.

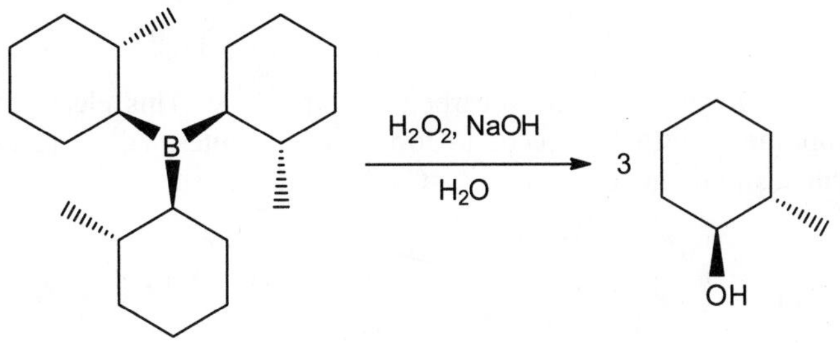

Chapter Problems

Problem 12.1

Think

What is the Lewis structure of CH_2N_2? In the presence of heat, what bond is broken? What reactive intermediate forms? Does the electrophile have a lone pair of electrons? How does this affect the intermediate formed?

Solve

The $H_2C–N_2$ bond breaks in a heterolysis step to form the reactive carbene intermediate. This electrophilic species has a lone pair and, therefore, electrophilic addition proceeds to form a cyclic intermediate to avoid losing an octet. Two new σ C–C bonds are formed simultaneously as the C=C π bond is broken.

H_2C–N≡N —Δ (Heterolysis)→ H_2C: + $N_2(g)$↑ —Electrophilic addition→

Problem 12.2

Think

What reagent reacts with an alkene to form a cyclopropane ring? In how many steps does the ring form? What must be true about the stereochemistry of the substituents in the alkene relative to the product?

Solve

Diazomethane in the presence of heat produces the H_2C: carbene reactive intermediate. This carbene intermediate reacts with an alkene in a concerted one-step reaction to form a cyclopropane ring. The stereochemistry of the substituents in the alkene is conserved in the cyclopropane product.

(a) Electrophilic addition

(c) Electrophilic addition

(b) Electrophilic addition

(d) Electrophilic addition

Problem 12.3

Think

How is the carbene Br_2C: formed? What is the role of the base? Once the carbene forms, how does it react with the C=C bond?

Solve

The $(CH_3)_3CO^-$ acts as a base to deprotonate the $CHBr_3$ reagent and form the conjugate base. A heterolysis step follows to break the C–Br bond, producing the reactive Br_2C: carbene. The electrophilic addition step is next to form the cyclopropane product.

Proton transfer

Slow Heterolysis

Fast Electrophilic addition

These are diastereomers.

Problem 12.4

Think

How many cyclopropane rings are present? What is the starting carbene-producing reagent when just CH_2 is added to form a cyclopropane? If a CH_2 group is disconnected from a cyclopropane ring, then what atoms are left for the alkene? What is the starting carbene-producing reagent when CCl_2 is added to form a cyclopropane? If a CCl_2 group is disconnected from a cyclopropane ring, then what atoms are left for the alkene?

Solve

Because the product has two cyclopropane rings, we could envision each one as the result of a carbene addition. The first reaction is the result of the addition of CCl_2, produced from $CHCl_3$. The second reaction is the result of the addition of CH_2, produced from CH_2N_2.

$CHCl_3$ / $NaOC(CH_3)_3$

CH_2N_2 / light

Problem 12.6

Think

What reaction takes place between Br_2 and an alkene? Do the Br atoms add to the C=C bond in a syn or anti fashion? Are the product molecules chiral? How does chirality relate to optical activity?

Solve

Br_2 undergoes anti addition across the double bond, yielding mixtures of products. The products are shown below and on the next page. Only the product mixture of **(b)** and **(d)** will be optically active because the products are diastereomers, whereas **(a)** and **(c)** products are equal mixtures of enantiomers (i.e., racemic mixtures).

(a) HO_2C–CH=CH–CO_2H → (Br_2, CCl_4)

Mixture of enantiomers, optically inactive

(b) Br2 / CCl4

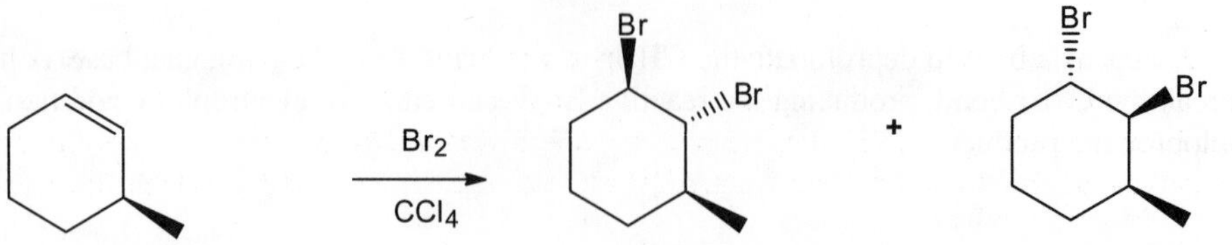

Mixture of diastereomers, optically active

(c) Br2 / CCl4

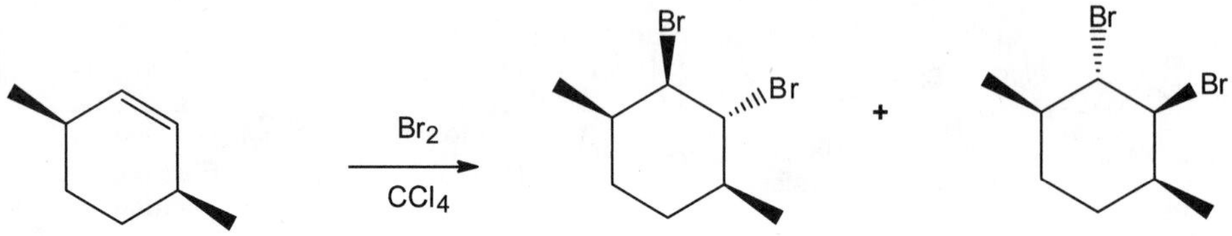

Mixture of enantiomers, optically inactive

(d) Br2 / CCl4

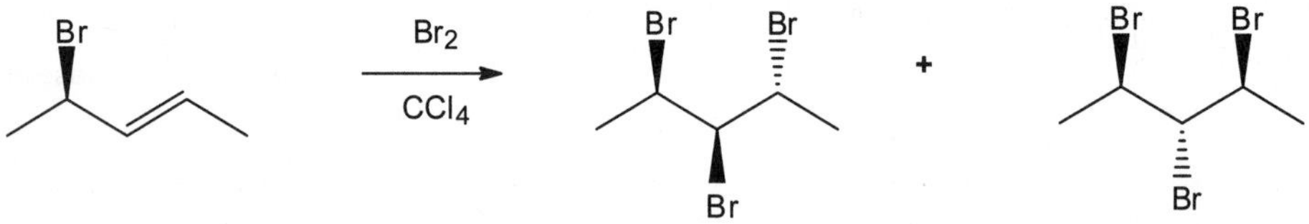

Mixture of diastereomers, optically active

Problem 12.7

Think

What is the structure of (*R*,*R*)-3,4-dichloro-2-methylhexane? Its enantiomer? Are rearrangements permitted with addition of Cl_2 to an alkene? Does the addition reaction occur syn or anti? On what two C atoms must the C=C be located in the alkene precursor? What is the stereochemistry of the substituents on the C=C in the alkene precursor?

Solve

The *R*,*R* and *S*,*S* products are shown below on the right. When the C–C bond is rotated so the Cl atoms are anti to each other (which is what is directly produced in the chlorination reactions), the two alkyl groups appear on the same side of the Cl–C–C–Cl plane, and so do the two H atoms. The starting alkene, therefore, has the two H atoms and the two alkyl groups cis to each other.

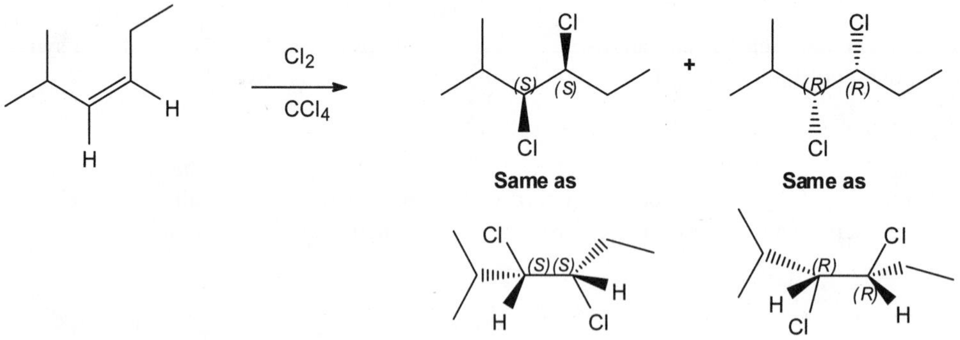

Cl2 / CCl4

Same as

Same as

Problem 12.8

Think

Consult Equation 12-17. What acts as the nucleophile and electrophile in the electrophilic addition step? In the subsequent S_N2 step, what competes with Br^- as the nucleophile? Which C atom was attacked in Equation 12-17? Can another C atom be attacked in that step? With a cyclic bromonium ion intermediate, what is the stereochemistry of the atoms to which the Br^+ and nucleophile attach? Are proton transfer reactions involved?

Solve

In the first step, electrophilic addition to the C=C bond produces the bromonium ion intermediate. In H_2O, which is a nucleophilic solvent, the water nucleophile competes with the Br^- to attack the electrophilic C atom in the intermediate. In Equation 12-17 the C atom on the left was attacked in the S_N2, but the one on the right can also be attacked, as shown below. The Br and OH are anti to each other in the final product.

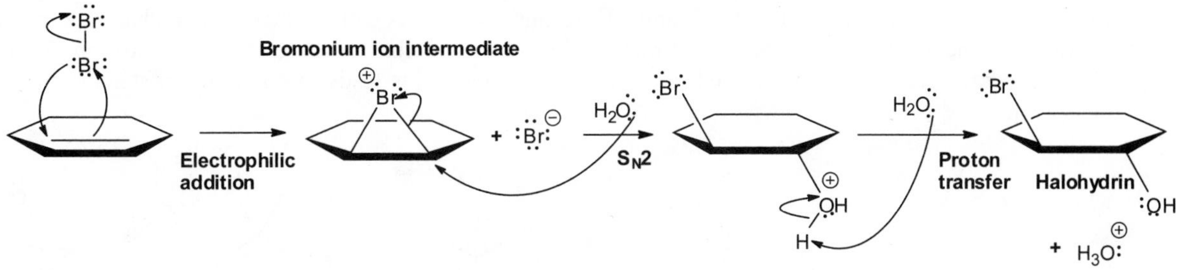

Bromonium ion intermediate

Electrophilic addition

S_N2

Proton transfer

Halohydrin

Problem 12.10

Think

For the reaction in Equation 12-15, what was the role of H_2O? Can ethanol assume the same role? Is the reaction stereoselective?

Solve

This mechanism is identical to the mechanism in Problem 12.8 except that the nucleophilic solvent is ethanol instead of water. A haloether forms rather than a halohydrin. The enantiomer of the product is also generated by attack at the other C atom in the second step, producing a racemic mixture.

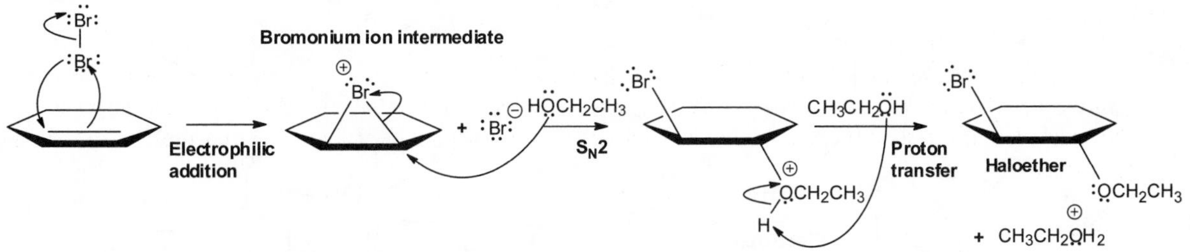

Bromonium ion intermediate

Electrophilic addition

S_N2

Proton transfer

Haloether

Problem 12.11

Think

Are the two C atoms in the C=C the same or distinct? Which C atom can accommodate the partial positive charge better? Why? Which C atom is the nucleophile more likely to attack?

Solve

In the chloronium ion intermediate, there is greater concentration of positive charge on the tertiary carbon than on the secondary carbon. Therefore, H_2O attacks the tertiary carbon.

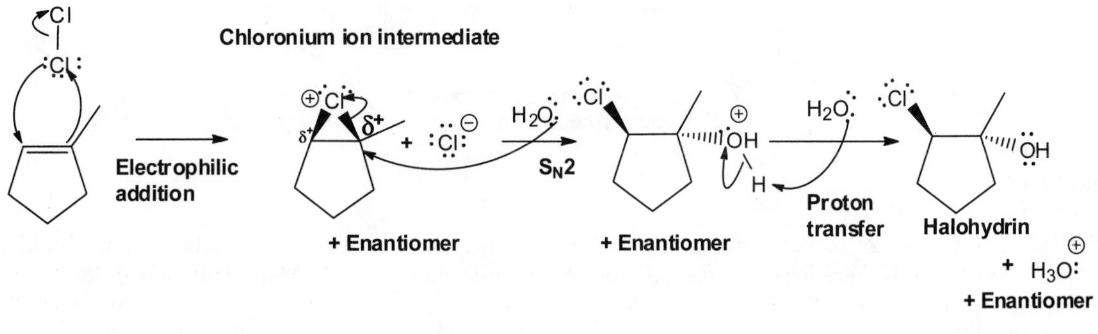

Chloronium ion intermediate

Electrophilic addition

S_N2

+ Enantiomer

+ Enantiomer

Proton transfer

Halohydrin

+ Enantiomer

Problem 12.12

Think

For each reaction, what intermediate forms after the electrophilic addition step? Are rearrangements possible? Is a new stereocenter generated? Is the addition reaction syn, anti, or a mixture of both?

Solve

(a) The addition reaction of the C=C and H^+ results in a carbocation intermediate that is capable of rearrangement. In this example, a 1,2-methyl shift occurs to form a more stable carbocation. Two new tetrahedral stereocenters are formed (indicated by an *) and this leads to a mixture of stereoisomers (not shown).

Electrophilic addition → 1,2-Methyl shfit → Coordination → Proton transfer

(b) The addition reaction of the C=C and $Hg(OAc)_2$ results in a cyclic mercurinium ion intermediate that is not capable of rearrangement. In Step 2 of the mechanism, H_2O acts as a nucleophile to open the three-membered ring, and in Step 3, the positively charged O atom is deprotonated. Subsequent reduction with sodium borohydride ($NaBH_4$) replaces the Hg-containing substituent with H. One new tetrahedral stereocenter is formed (indicated by an *) and this leads to a mixture of stereoisomers (not shown).

Electrophilic addition → S_N2 → Proton transfer → $NaBH_4$

$NaBH_4$ reduces the C atom, replacing the Hg group with an H.

Problem 12.13

Think

What is the structure of 2-methylpentan-2-ol? Where is the OH located? What carbon atoms could have composed the C=C? Is there more than one option? Do both alkenes undergo Markovnikov addition of H_2O to yield the same product?

Solve

The alcohol is located on C2, which is the most substituted carbon. Therefore, the alkene could have occurred at C1 and C2 or at C2 and C3. In the Markovnikov addition of H_2O to each of these alkenes, the OH adds to C2 to yield the same alcohol product.

or

1. $Hg(OAc)_2$, H_2O/THF

2. $NaBH_4$, ethanol

OH

Problem 12.14

Think

On what C atom did the OH add? Does this represent Markovnikov addition? If this formed a carbocation intermediate, is rearrangement likely to occur to produce a more stable carbocation? If rearrangement is not desired, what reagents are needed?

Solve

This is the result of Markovnikov addition of water across the C=C double bond. We cannot carry out this transformation with just an acid-catalyzed hydration because the carbocation that would be formed would rearrange to a more stable benzylic carbocation, as shown below

H H^+ Electrophilic addition H ⊕ 1,2-Hydride shift ⊕ H

Benzylic carbocation

We, therefore, must perform an oxymercuration–reduction addition of water in which the water adds to the most substituted carbon, but rearrangements do not occur.

1. $Hg(OAc)_2$, H_2O/THF

2. $NaBH_4$

OH

Problem 12.16

Think

How does the mechanism for an alkoxymercuration–reduction reaction of an alkyne compare to the one shown for an alkene in Solved Problem 12.15? What considerations should be made about *regiochemistry*? What considerations should be made about rearrangements?

Solve

The mechanism for this alkoxymercuration–reduction is identical to the one shown in Solved Problem 12.15, the only difference being the presence of an alkyne instead of an alkene. In this case, H will add to the terminal C, and OCH_2CH_3 will add to the adjacent C. There will be no keto–enol tautomerization, as is observed with oxymercuration–reduction, because the product of this reaction is not an enol. It is a vinyl ether.

1. $Hg(OAc)_2$, EtOH/THF

2. $NaBH_4$

O

Vinyl ether

Problem 12.17

Think

What is the structure of MCPBA? What functional group is present? How does this functional group react with an alkene? What functional group is produced? In how many steps does the reaction occur? If the reaction is concerted, what happens to the orientation of the substituents on the C=C?

Solve

MCPBA is a peroxyacid and forms an epoxide in a one-step concerted reaction with an alkene. Rearrangements and rotations about the C=C are not permitted. Therefore, the stereochemistry of the substituents on the C=C remain the same in the product. In this case, the trans configuration about the double bond produces a trans configuration about the epoxide ring. A racemic mixture of enantiomers is produced.

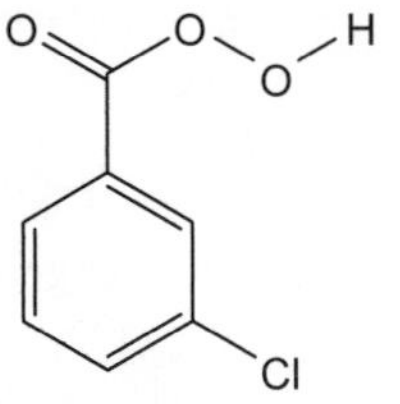

***m*-Chloroperbenzoic acid**
MCPBA

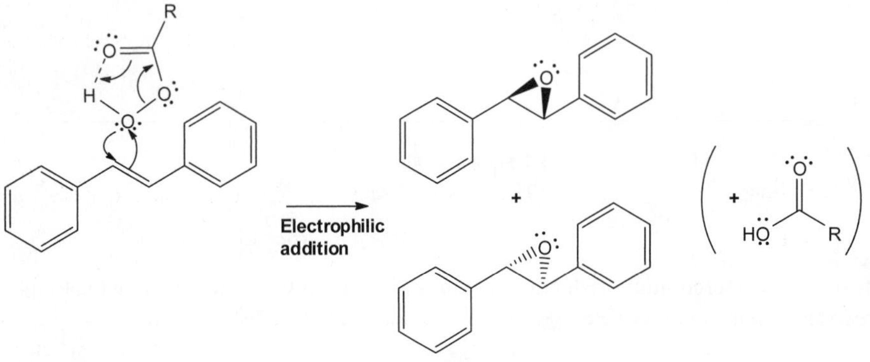

Electrophilic addition

Problem 12.18

Think

What is the orientation of the substituents in the epoxide? Cis or trans? What must be the orientation of the same substituents in the alkene?

Solve

Because the two groups are on the same side of the epoxide ring, the starting alkene must be cis.

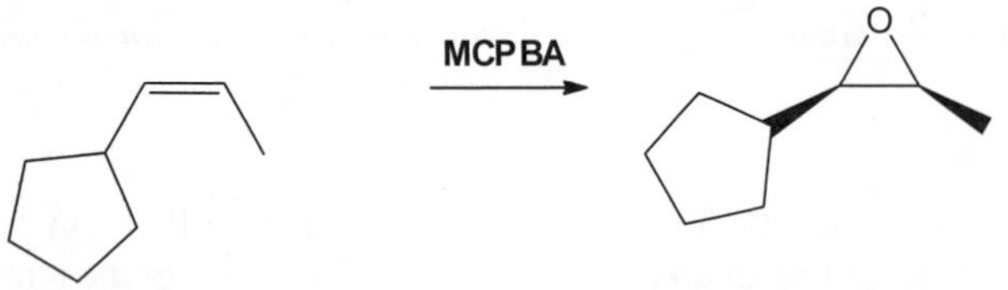

MCPBA

Problem 12.20

Think

In the first reaction, what is the nature of the leaving group? The attacking species? The solvent? What type of reaction does heat promote? For the second reaction, what type of reaction takes place between an alkene and a peroxyacid? In the third reaction, what acts as the nucleophile? The substrate?

Solve

In the first reaction, E2 is the major mechanism due to the high concentration of the strong base and the heat that is added. The E2 reaction forms the most substituted alkene. Because E2 requires the groups to be anti-coplanar, the ethyl groups are cis in product **D**. In the second reaction, MCPBA adds to the C=C in a one-step concerted reaction to form the epoxide **E** where the ethyl groups are still cis. CH_3MgBr is a source of nucleophilic carbon, $H_3C:^-$. In an S_N2 reaction, the $H_3C:^-$ attacks the *less* substituted C atom of the epoxide and the ring is opened where the O and the CH_3 become anti to each other. An acid workup step is required to form the uncharged alcohol product.

Br H H' Rotate to get Br and H' anti coplanar :Br: H H' NaOCH3 Δ D

Br NaOCH3 Δ D MCPBA O H E 1. CH3MgBr 2. H3O+ + Enantiomer HO H (3S,4S)-3,4-dimethylhexan-3-ol F

Problem 12.21

Think

In how many steps does the monoalkyl borane compound form? Does the boron go on the more or less substituted carbon? Why?

Solve

The monoalkyl borane is formed in one step. The BH_2 goes on the less substituted C atom due to sterics of the transition state and the fact that the partial positive charge in the transition state is more stable on the more substituted C atom.

H BH2 H BH2 B

Problem 12.22

Think

What is the reactive species in a basic solution of H_2O_2? How does that react with the electron-poor trialkylborane? How is the C–O bond formed and the B–O bond broken? How many times does this occur? How is the O–H bond formed and the B–O bond broken? Compare this mechanism to the one shown in Equation 12-39.

Solve

This reaction mechanism takes 15 steps overall. The R group represents the naphthyl portion, $C_{10}H_7$. The first two steps are coordination of the HOO^- to the electron-poor boron atom, followed by a 1,2-alkyl shift. This is then repeated two more times to form the trialkylborate ester (Steps 1–6). The HO^- coordinates to the electron-poor boron atom followed by heterolysis of the B–OR bond. A proton transfer reaction occurs on the alkoxide to form the uncharged alcohol product (Steps 7–9). This is then repeated two more times (Steps 10–15) to form three equivalents of the ROH and $B(OH)_4^-$. The alcohol forms on the less substituted carbon because the O–C bond forms at the same C where the C–B bond broke. In this example, the alcohol is not chiral, but it should be noted that stereochemistry is retained for the C–O bond. This leads to ultimate syn addition of the H and OH.

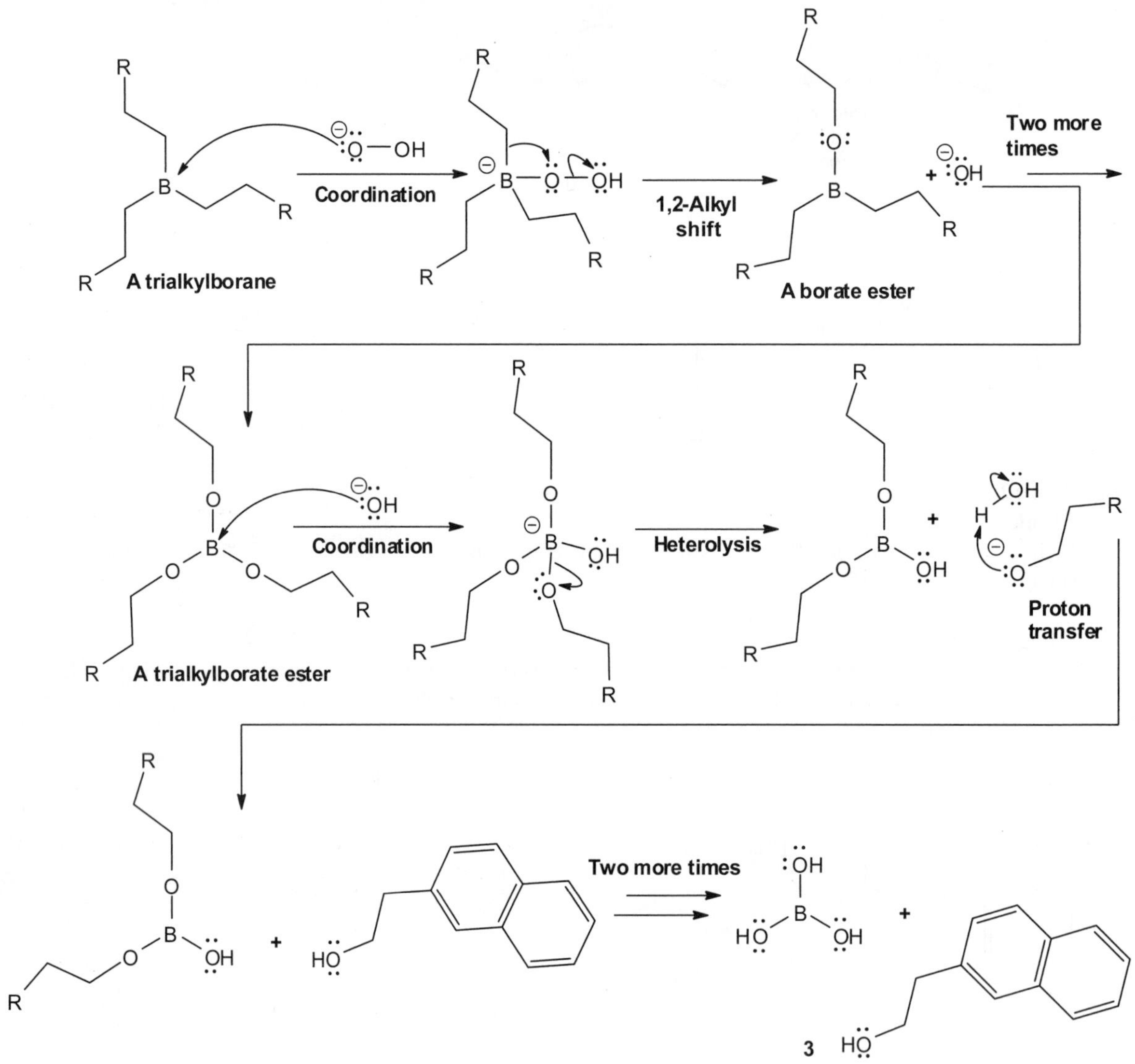

Problem 12.24

Think

In each case, there is a net addition of what molecule across the C=C? What is the regiochemistry required in each reaction? Are carbocation rearrangements a concern?

Solve

(a) The net addition is H_2O across the C=C and the OH adds to the less substituted carbon. This is an example of anti-Markovnikov addition and can be carried out with hydroboration–oxidation.

1. BH_3•THF
2. H_2O_2, NaOH, H_2O

OH

(b) The net addition is H_2O across the C=C. In this example, a rearrangement occurred because the OH appears on the benzylic C atom, which was not previously part of the double bond. The reaction conditions must be hydration in acidic water.

OH

H_3O^+/H_2O

Problem 12.25

Think

From what functional group does an aldehyde tautomerize? Where must the OH be located? In the presence of disiamylborane, does the more- or less-substituted C–B bond form? Is the alkyne terminal or internal?

Solve

Treatment of a terminal alkyne with $(C_5H_{11})_2BH$ results in the C–B bond forming at the less-substituted C atom. Treatment of **B** in basic solution results in a terminal enol that tautomerizes to the aldehyde.

$(C_5H_{11})_2BH$ / THF; $B(C_5H_{11})_2$; $HO^{\ominus}$, H_2O_2; OH; O; H

A **B** **Enol** **Aldehyde**

Problem 12.26

Think

In 1,2-dibromination reactions, is an alkene relatively electron rich or electron poor? Are C=O groups electron-donating or electron-withdrawing groups? What is the effect of these group on the extent to which the C=C is electron rich? Does this increase or decrease the driving force for the electrophilic addition step?

Solve

In the electrophilic addition of Br_2 across a C=C double bond, the C=C double bond is considered electron rich. In maleic anhydride, the two C=O groups are electron withdrawing, making the C=C double bond less electron rich than the C=C double bond in cyclopentene. This diminishes the driving force for the electrophilic addition step.

Br_2

O O O; Br Br

Less electron rich, slower addition of Br_2

Problem 12.27

Think

In 1,2-dibromination reactions, is an alkene relatively electron rich or electron poor? Are CH_3 alkyl groups electron donating or withdrawing? What is the effect of these groups on the extent to which the C=C bond is electron rich? Does this increase or decrease the driving force for the electrophilic addition step?

Solve

In the electrophilic addition of Br_2 across a C=C double bond, the C=C double bond is considered electron rich. Each methyl group is electron donating, making the C=C double bond more electron rich, which helps facilitate the reaction.

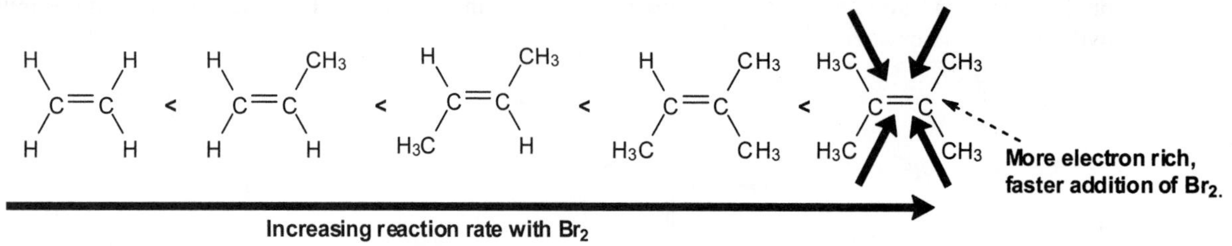

More electron rich, faster addition of Br_2.

Increasing reaction rate with Br_2

Problem 12.28

Think

In each case, there is a net addition of what molecule across the C=C? What is the regiochemistry required in each reaction? The stereochemistry? Are carbocation rearrangements a concern? How many times does the addition reaction occur?

Solve

(a) HBr adds to the C=C. The H^+ adds first to form the secondary carbocation and the Br^- adds second to form the C–Br at the C^+.

HBr

(b) HBr adds to the C=C. The H^+ adds first to form the terminal carbocation. The terminal carbocation is more stable in this example due to the adjacent Cl substituent, which can resonance stabilize the C^+. The Br^- adds second to form the C–Br at the C^+.

HBr

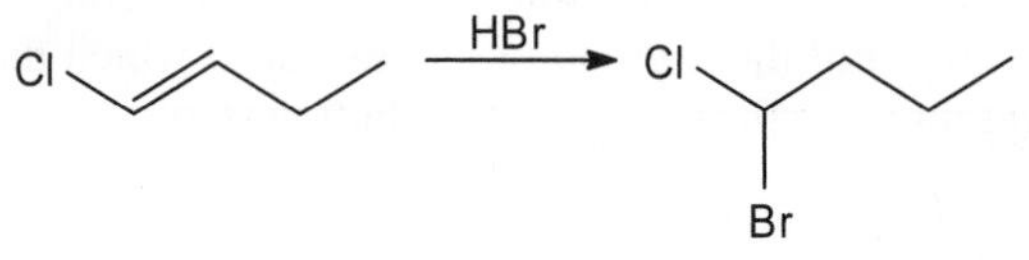

(c) Water adds to the C=C. In this example, the alkene is symmetrical, so regiochemistry is not a concern, and rearrangements are not permitted as the CH_3 substituents are not adjacent to the C=C bond.

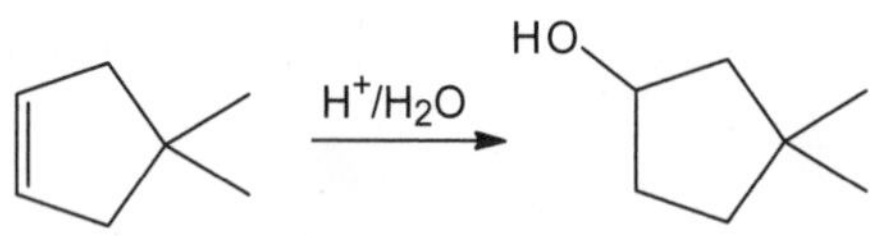

H^+/H_2O

(d) Diazomethane in the presence of light ($h\nu$) is a source of H_2C: carbene. Addition of H_2C: to an alkene results in the formation of a cyclopropane. The C–CH_3 remains in a wedge configuration because no bonds were broken or formed at that carbon. A mixture of diastereomers is formed.

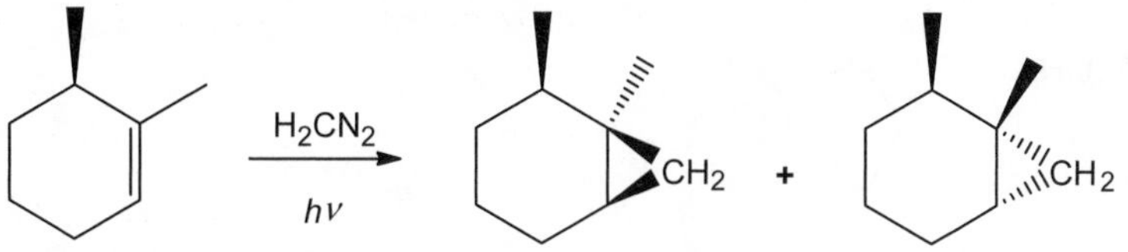

H_2CN_2, $h\nu$ → CH_2 + CH_2

(e) Two equivalents of Cl_2 will each add to the C=C to form C–Cl bonds (four total). A mixture of configurations will be produced at each of the two new stereocenters, resulting in a mixture of three stereoisomers (one is meso).

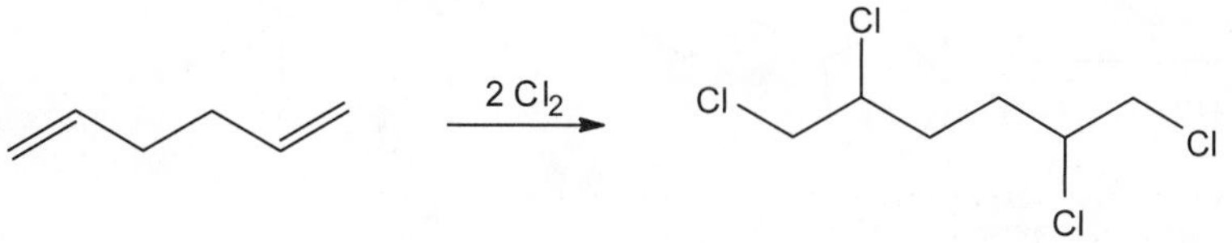

2 Cl_2 → Cl, Cl, Cl, Cl

(f) Br_2 adds to the C=C to form the bromonium ion intermediate. The H_2O acts as the nucleophile and adds to the more substituted C atom.

Br_2/H_2O → OH, Br

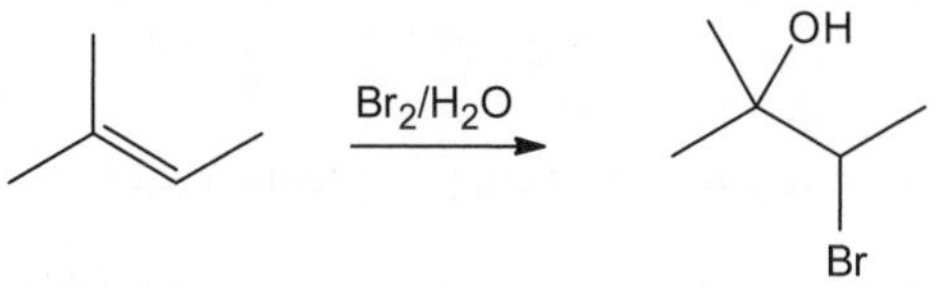

(g) Two equivalents of HCl will add to the C≡C. The more stable C^+ forms in each addition reaction and thus the Cl is added to the secondary carbon in both addition reactions.

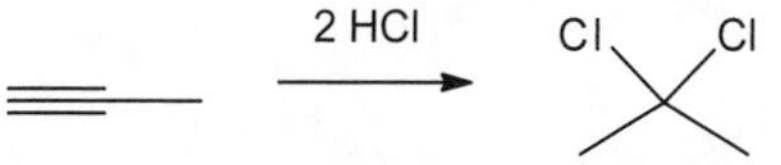

2 HCl → Cl, Cl

(h) Br_2 adds to the C≡C via the formation of the bromonium ion intermediate. The Br^- acts as the nucleophile and adds to the more substituted C atom. Rearrangements are not possible. The reaction occurs two times to form the tetrabrominated alkane product.

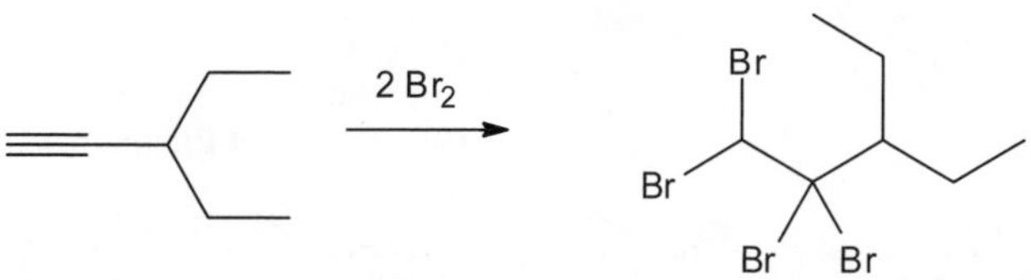

2 Br_2 → Br, Br, Br, Br

(i) Water adds to the C=C. The mechanism goes through a C^+ intermediate and rearrangements are permitted. The C^+ is initially formed on C2, which is a secondary carbon, but a 1,2-hydride shift then produces a tertiary C^+. Thus, the OH appears on the tertiary C.

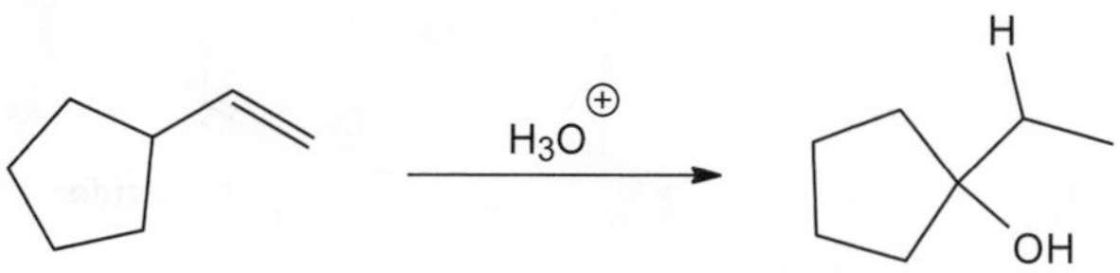

$H_3O^{\oplus}$ → H, OH

(j) Water adds to the C=C. The mechanism goes through a mercurinium ion intermediate and rearrangements are not permitted. The reaction follows Markovnikov addition. Thus, the water adds to the more substituted secondary C.

1. $Hg(OAc)_2/H_2O$
2. $NaBH_4/OH^-$

(k) Water adds to the C=C. The first step of the reaction forms an alkyl borane where the C–B bond forms at the less substituted C. Rearrangements are not permitted. The oxidation occurs to replace the C–B bond with a C–OH bond. Thus, the OH forms on the primary C.

1. BH_3•THF
2. NaOH / H_2O_2, H_2O

Problem 12.29

Think

In each case, there is a net addition of what molecule across the C=C or C≡C? What is electron rich and what is electron poor? What is the regiochemistry required in each reaction? The stereochemistry? Are carbocation rearrangements a concern? How many times does the addition reaction occur?

Solve

(a) The net addition is HCl across the C=C. The H^+ adds first to form the benzylic secondary carbocation and the Cl^- adds second to form the C–Cl.

Electrophilic addition

Coordination

+ Enantiomer

(b) Overall, water adds to the C=C. The mechanism goes through a mercurinium ion intermediate and rearrangements are not permitted. Thus, the water adds to the more substituted secondary C.

Electrophilic addition

S_N2

+ Enantiomer

Proton transfer

+ Enantiomer

$NaBH_4$

$^{\ominus}OH$

+ Enantiomer

(c) The net addition is H_2O across the C=C. The H^+ adds first to form a secondary carbocation. The alkene is not symmetrical and two secondary C^+ intermediates are possible. The bottom reaction scheme shows the rearrangement of the secondary C^+ to a tertiary C^+. Water then acts as the nucleophile to attack the C^+ and a proton transfer reaction follows to form the uncharged alcohol.

Electrophilic addition

Coordination

Proton transfer

+ Enantiomer

1,2-Hydride shift

Coordination

Proton transfer

(d) Br_2 adds to the C=C to form the bromonium ion intermediate. The CH_3OH acts as the nucleophile and adds to the more substituted C atom. Rearrangements are not permitted.

Electrophilic addition

S_N2

Proton transfer

+ Enantiomer

(e) BH_3 adds to the less sterically hindered carbon of the C=C and H adds to the other carbon. The triakylborane forms and then is oxidized to form a syn alcohol.

Electrophilic addition

Coordination

1,2-Alkyl shift

Coordination

Heterolysis

Proton transfer

+ Enantiomer

(f) Diazomethane in the presence of heat produces the H_2C: carbene reactive intermediate. This carbene intermediate reacts with an alkene in a concerted one-step reaction to form a cyclopropane ring. The stereochemistry of the substituents in the alkene is conserved in the cyclopropane product.

Heterolysis

Electrophilic addition

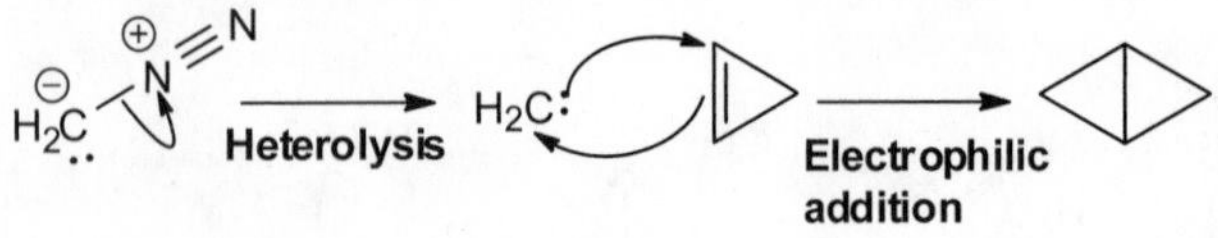

(g) The CH_3O^- acts as a base to deprotonate the $CHBr_3$ reagent and form the conjugate base. A heterolysis step follows to break the C–Br bond, resulting in the reactive Br_2C: carbene. An electrophilic addition step is next to form the cyclopropane product. The trans configuration about the C=C results in a trans configuration about the plane of the cyclopropane ring in the product.

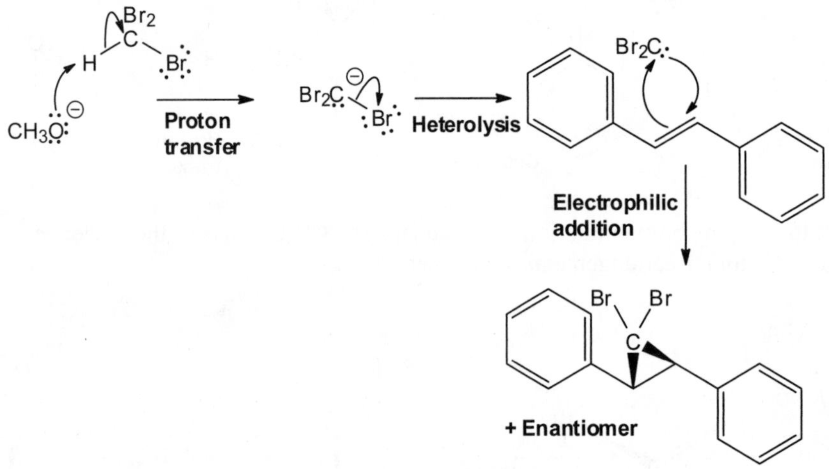

Proton transfer

Heterolysis

Electrophilic addition

+ Enantiomer

(h) MCPBA is a peroxyacid and forms an epoxide in a one-step concerted reaction with an alkene. Rearrangements and rotations about the C=C are not permitted. Therefore, the stereochemistry of the substituents on the C=C remain the same in the product.

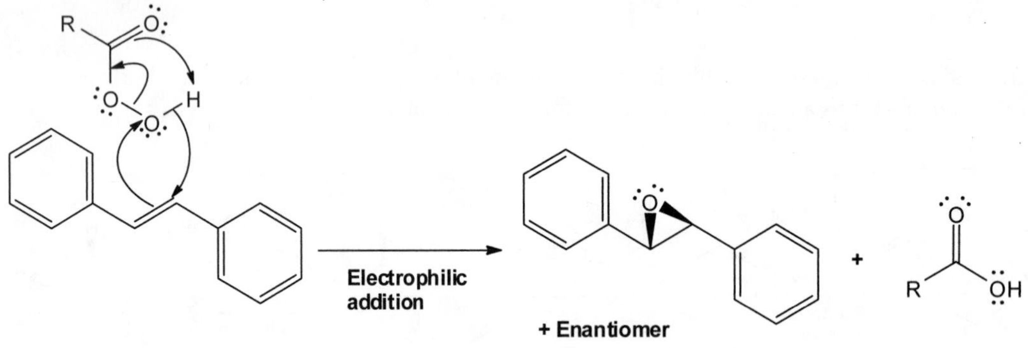

Electrophilic addition

+ Enantiomer

(i) Br_2 in a non-nucleophilic solvent adds two Br atoms across the C=C. The bromonium ion intermediate is formed and the second Br^- attacks the more substituted C atom anti to the Br atom in the ring.

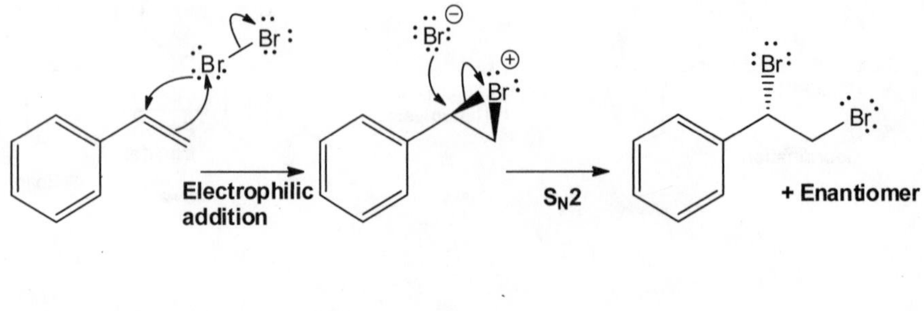

Electrophilic addition

S_N2

+ Enantiomer

(j) 1,2- or 1,4-addition is possible for a diene in the presence of HBr. At higher temperatures, the thermodynamic (most stable) product is the major product. The tertiary carbocation is the most stable intermediate. 1,4-addition leads to most highly substituted double bond.

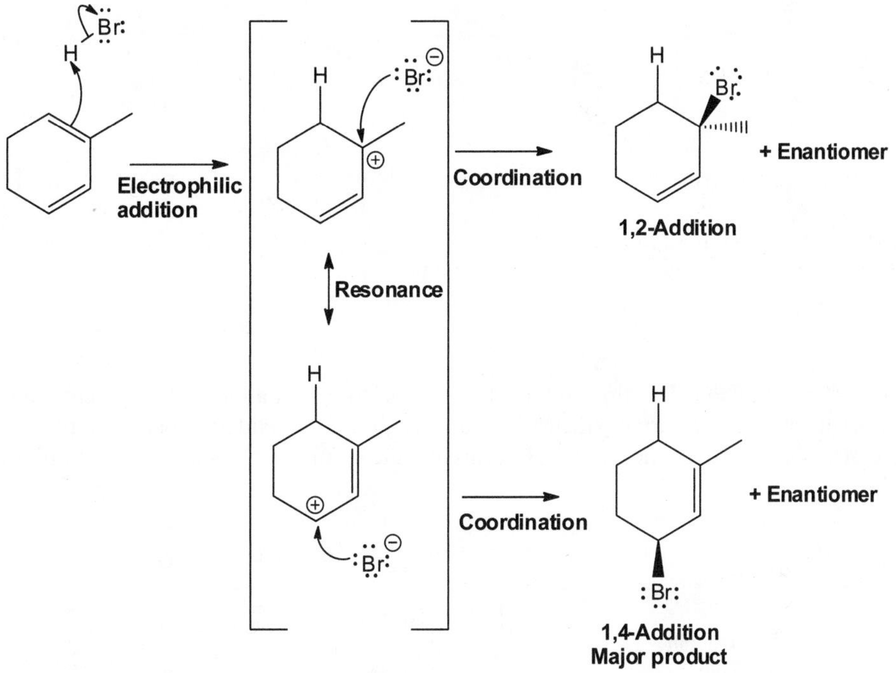

Electrophilic addition

Resonance

Coordination

+ Enantiomer

1,2-Addition

Coordination

+ Enantiomer

1,4-Addition
Major product

(k) 1,2- or 1,4-addition is possible for a diene in the presence of HBr. At cold temperatures, the reaction takes place under kinetic control and the 1,2-addition product is the major product. A tertiary carbocation is the most stable intermediate. 1,2-addition takes place faster than 1,4-addition.

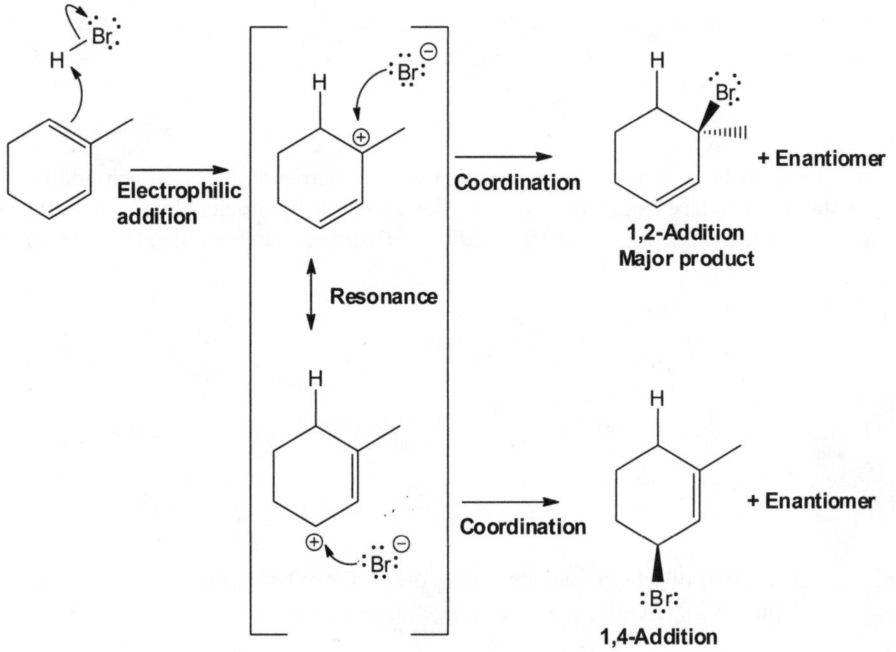

Electrophilic addition

Resonance

Coordination

+ Enantiomer

1,2-Addition
Major product

Coordination

+ Enantiomer

1,4-Addition

(l) The net addition is H_2O across the C≡C. The H^+ adds first to form an internal carbocation. Water acts as a nucleophile and an enol is formed. The enol tautomerizes to form the ketone.

Electrophilic addition → Coordination → Proton transfer → Tautomerization

(m) Overall, water adds to the C≡C. The first step of the reaction forms an alkyl borane where the C–B bond forms at the less substituted C. Rearrangements are not permitted. The oxidation occurs to replace the C–B bond with a C–OH bond. Thus, the enol OH forms on the terminal C. The enol tautomerizes to the aldehyde.

Electrophilic addition → NaOH, H_2O_2, H_2O → Tautomerization

(n) An excess of HBr leads to two addition reactions of H and Br across the C≡C. Each addition proceeds through the more stable secondary carbocation intermediate, so both Br^- nucleophiles add to the secondary carbocation. The carbocation formed in the second addition is, furthermore, stabilized by resonance.

Electrophilic addition → Coordination → Electrophilic addition → Coordination

Problem 12.30

Think

In each case, there is a net addition of what molecule across the C=C? What is the regiochemistry required in each reaction? The stereochemistry? Are carbocation rearrangements a concern?

Solve
Syntheses are shown below.

(a) Br_2, CCl_4

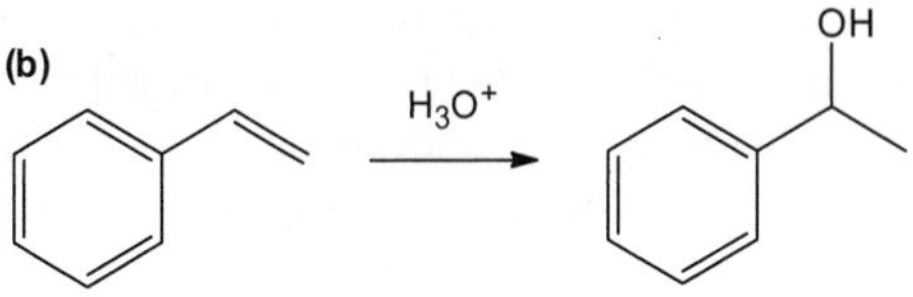

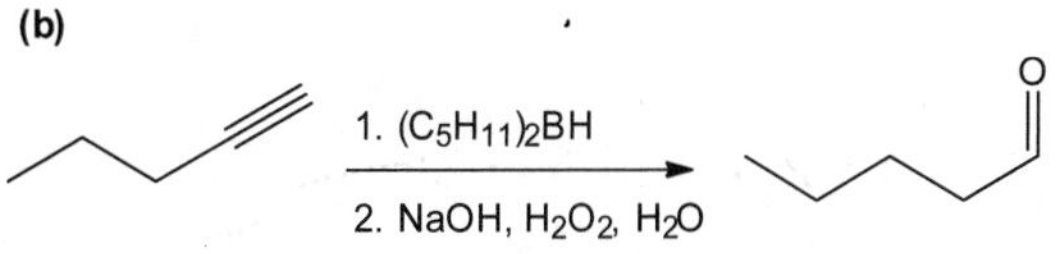

(b) H_3O^+

(c) 1. BH_3•THF 2. NaOH, H_2O_2, H_2O

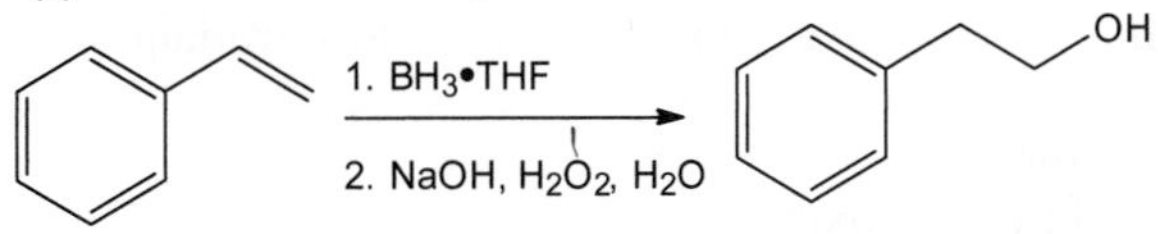

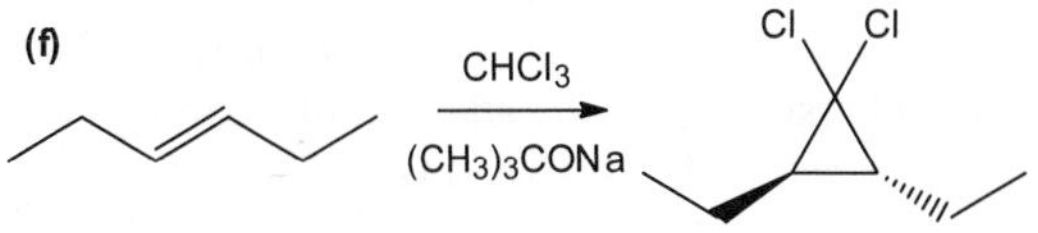

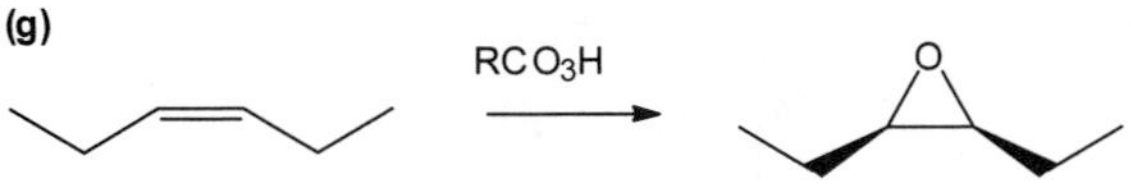

(d) 1. MCPBA 2. NaOH, H_2O 3. H_3O^+, H_2O

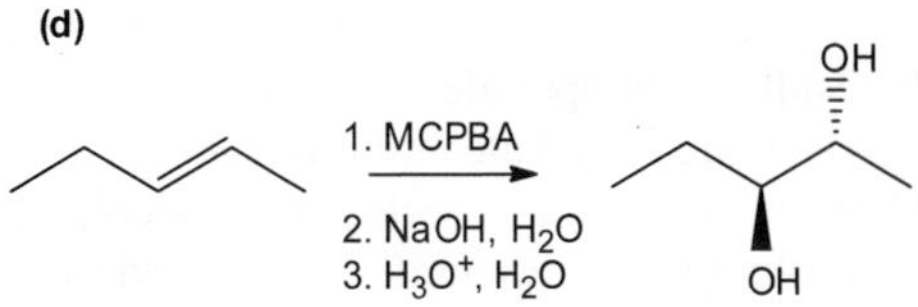

(e) CH_2N_2, Δ/hν

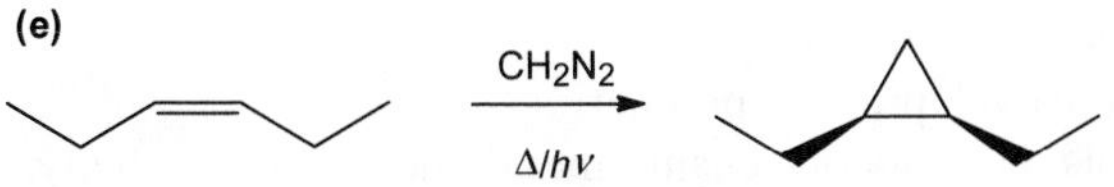

(f) $CHCl_3$, $(CH_3)_3CONa$

(g) RCO_3H

(h) Br_2, H_2O

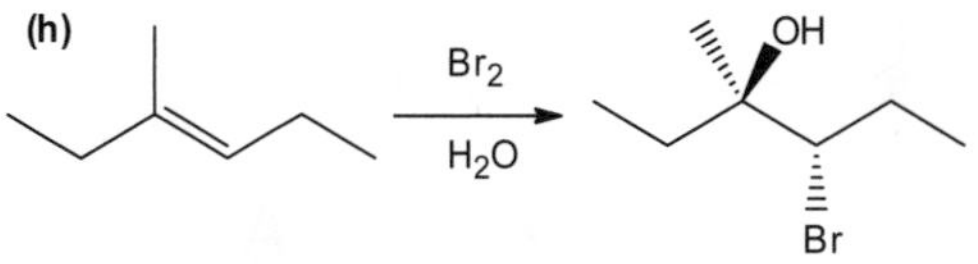

Problem 12.31

Think

In each case, there is a net addition of what molecule across the C≡C? What is the regiochemistry required in each reaction? Are carbocation rearrangements a concern? How many times does the addition reaction occur?

Solve
Syntheses are shown below.

(a) H_2O, $H^{\oplus}$

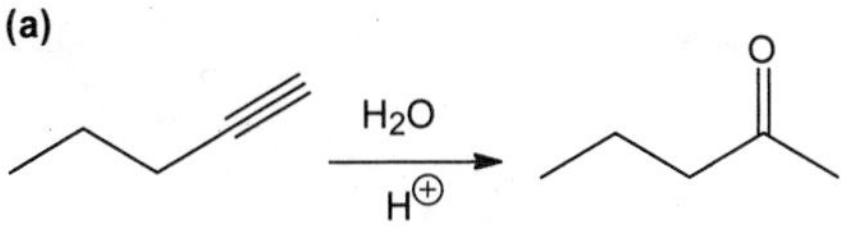

(b) 1. $(C_5H_{11})_2BH$ 2. NaOH, H_2O_2, H_2O

(c) 1. $HB(C_5H_{11})_2$ 2. H_2O_2, $HO^{\ominus}$

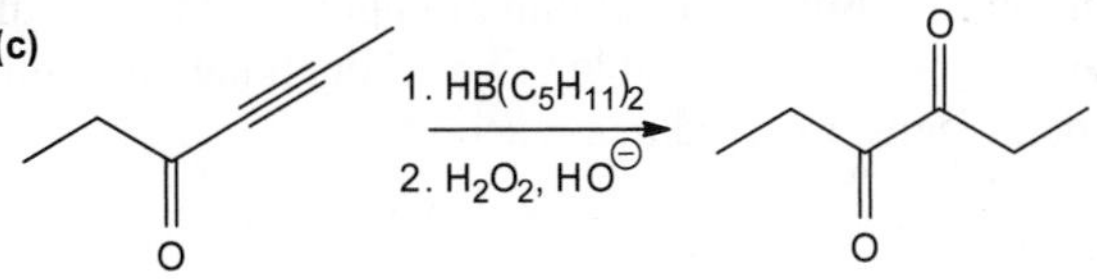

(d) CH_2N_2, Δ/hν

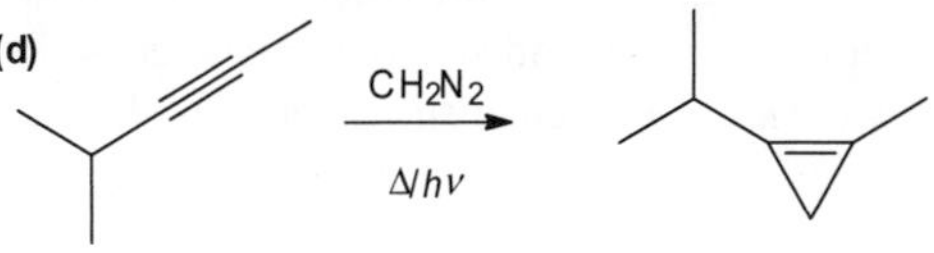

(e) 1 eq. Br_2, CCl_4

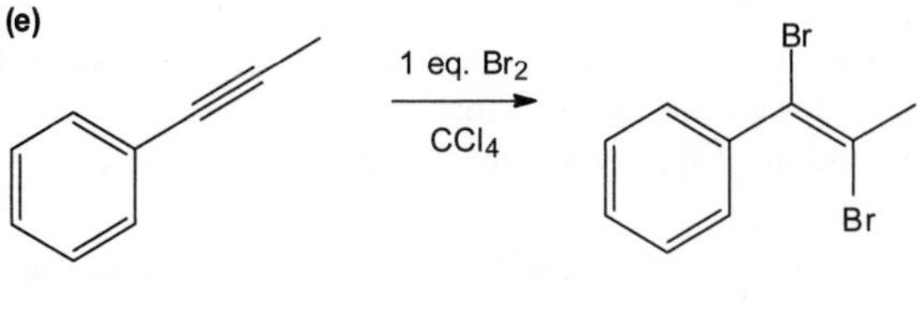

(f) Excess Br_2, CCl_4

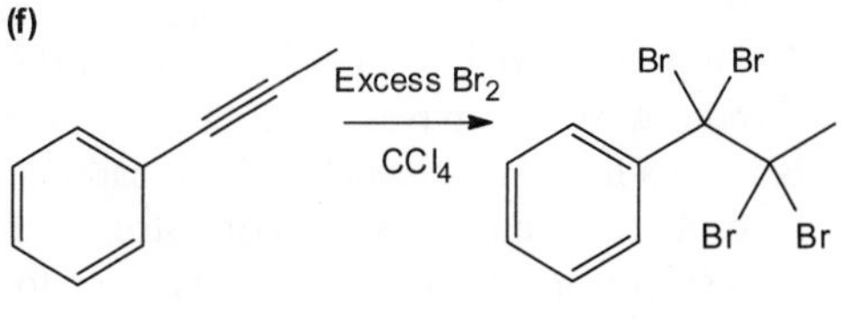

Problem 12.32

Think

In each step, consider what is the electron-rich species (nucleophile) and what is the electron-poor species (electrophile). Will the reaction proceed to give an addition, substitution, or elimination product? Be mindful of regiochemistry and stereochemical considerations. Are rearrangements possible? Even though the question does not ask for a mechanism, always think about the elementary steps involved in a reaction.

Solve

A results from addition of a peroxyacid to an alkene to form an epoxide.
B results from epoxide ring opening in basic conditions (R group anti to OH) followed by workup in acid to form an uncharged alcohol product.
C results from an E1 reaction of an alcohol heated in acidic water.
D results from addition of a peroxyacid to an alkene to form an epoxide.
E results from epoxide ring opening in neutral conditions (less substituted C atom attacked and anti to OH) followed by protonation to form an uncharged alcohol product.
F results from the elimination of H and Br to form the alkene product.
G results from the addition of a carbene H_2C: to form the cyclopropane product.
Note: For **B**, **D**, **F**, and **G**, the enantiomer is also possible, but only one enantiomer is shown for simplicity.

MCPBA; 1. CH_3CH_2Li 2. H_3O^+; H_3O^+, Δ
A B C
MCPBA; NaBr, CH_3CH_2OH; NaOH, Δ; CH_2N_2, $h\nu$
D E F G

Problem 12.33

Think

In each step, consider what is the electron-rich species (nucleophile) and what is the electron-poor species (electrophile). Will the reaction proceed to give an addition, substitution, or elimination product? Be mindful of regiochemistry and stereochemical considerations. Are rearrangements possible? Even though the question does not ask for a mechanism, always think about the elementary steps involved in a reaction.

Solve

H results from addition of Br_2 to an alkene to form a vicinal dibromide.
I results from two E2 reactions followed by a proton transfer to produce the alkynide anion ($RC\equiv C:^-$). An acid workup follows to form an uncharged alkyne product.
J results from oxymercuration to form the mercurinium ion intermediate; water attacks the more substituted carbon. A reduction reaction follows to give an enol that tautomerizes to the ketone.
K results from bromination at the α carbon. Under acidic conditions, only a single bromination occurs.
L results from an E2 reaction of a strong base in a heated solution.
M results from Br_2 addition of an alkene to yield the bromonium ion intermediate where water attacks the more substituted C atom.
Note: For **H**, **K**, and **M**, the enantiomer is also possible, but only one enantiomer is shown for simplicity.
The missing compounds in the synthesis scheme are shown on the next page.

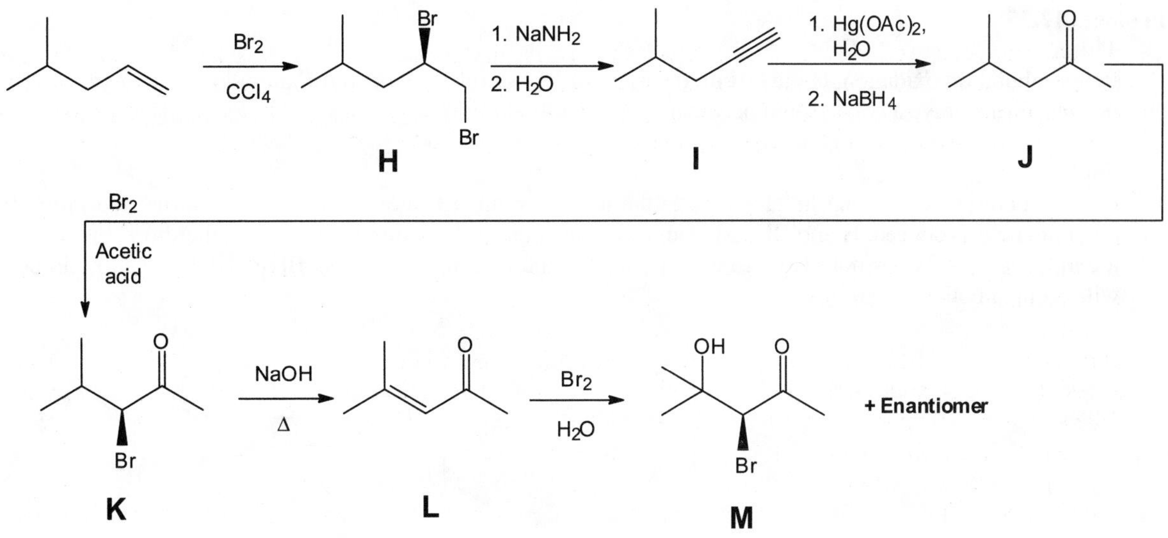

Br2 CCl4 H 1. NaNH2 2. H2O I 1. Hg(OAc)2, H2O 2. NaBH4 J Br2 Acetic acid K NaOH Δ L Br2 H2O M + Enantiomer

Problem 12.34

Think

When X_2 (Br_2 or Cl_2) adds to a C=C, what intermediate is formed? What nucleophile opens up the ring? Which carbon can better accommodate a partial positive charge? Both are secondary, but is there another factor to consider?

Solve

(a) Water attacks the carbon adjacent to the ring because it can stabilize a positive charge better than the other alkene carbon, due to resonance with the ring.

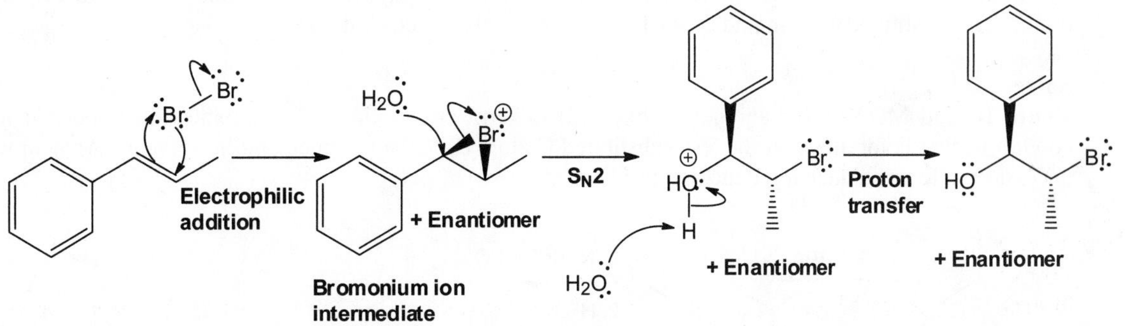

Electrophilic addition → Bromonium ion intermediate + Enantiomer → SN2 → + Enantiomer → Proton transfer → + Enantiomer

(b) Water attacks the alkene carbon adjacent to the O atom because the lone pairs on O can help stabilize the positive charge due to resonance with the lone pair electrons on the O atom.

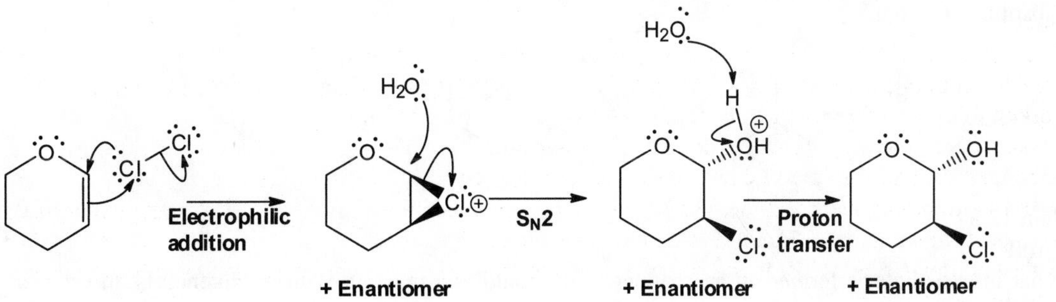

Electrophilic addition → + Enantiomer → SN2 → + Enantiomer → Proton transfer → + Enantiomer

Problem 12.35

Think

In hydroboration oxidation, is the OH on the more or less sterically hindered C atom in the final product? Does this reaction involve a competition between 1,2- and 1,4-addition?

Solve

In hydroboration oxidation, the HO adds to the less substituted C atom syn to the H. In the step in which the alkylborane is produced, H and BH_2 add simultaneously in a 1,2-positioning only and, therefore, the presence of a conjugated C=C does not lead to a 1,4-product. In the presence of excess BH_3•THF reagent, hydroboration will occur at both C=C groups.

1. BH_3•THF

BH_2 H OH H

1. BH_3•THF

OH H H_2B OH H HO OH

Problem 12.36

Think

What reactions will add Br and OH to adjacent C atoms of an alkene? Is the Br in the target on the more or less substituted C atom? Is the OH on the more or less substituted C atom? Which conditions will favor putting a Br on the less substituted C atom and the OH on the more substituted C atom?

Solve

Route 1: Add MCPBA (or another peroxyacid) to form an epoxide. The epoxide ring is opened in neutral conditions that favor attack at the less substituted C atom (Br^- in a nonnucleophilic solvent). An acid workup is necessary to form an uncharged alcohol product.

MCPBA O 1. NaBr/DMSO 2. H_3O^+ HO Br

Route 2: Add Br_2 to form the brominium ion intermediate. In the presence of water, the H_2O attacks the more substituted C atom.

Br_2 Br ⊕ H_2O HO Br

Problem 12.37

Think

What intermediate is formed in an oxymercuration addition reaction with an alkene? Is an ester an electron-donating or electron-withdrawing group? What is the effect of the ester on the ability of the α carbon atom in the mercurinium ion ring to accommodate a partial positive charge?

Solve

The ester group is an electron-withdrawing group and, therefore, destabilizes nearby positive charges. This hinders the ability of the α carbon atom in the mercurinium ion intermediate to accommodate a partial positive charge. Therefore, the β carbon atom has a larger δ^+ and will be the site that water attacks. The partial mechanism is shown below.

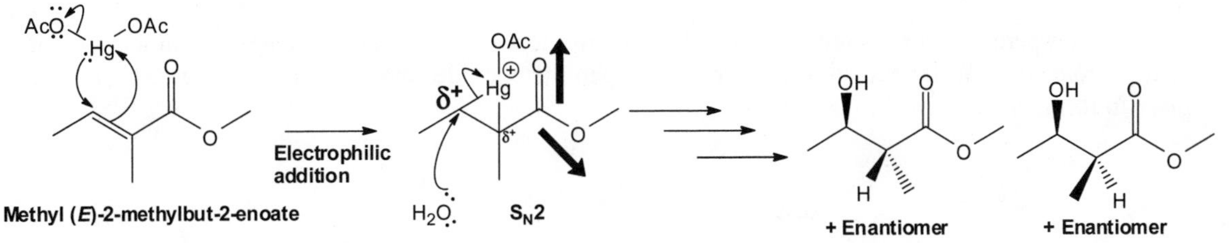

Methyl (*E*)-2-methylbut-2-enoate — Electrophilic addition — S_N2 — + Enantiomer — + Enantiomer

Problem 12.38

Think

Are both C–Br bonds of the intermediate formed on the same side of the six-membered ring or on opposite sides? What is the second elementary step in the addition reaction of Br_2 to an alkene? What is true about the stereochemistry of the two groups in this type of mechanism? Does a mixture of cis and trans support this?

Solve

(a) In the first step, both bonds to Br must form on the same side of the ring's plane. In the second step, which is an S_N2 step, the nucleophile must come in from the opposite side of the C–Br bond and so must yield a trans dibromide. This is shown below.

(b) The observations from experiment, therefore, discredit the proposed mechanism.

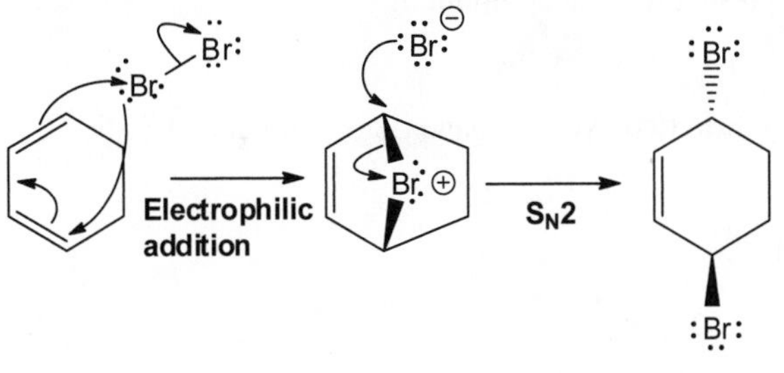

Electrophilic addition — S_N2

Problem 12.39

Think

How are the Br and $OC(CH_3)_3$ groups arranged relative to each other? How are the ethyl groups arranged in the starting material and the product? What does this suggest about the type of intermediate that must have formed? To what mechanism is this similar?

Solve

The anti-addition of the Br and the OR groups suggest a bromonium ion intermediate. Therefore, the mechanism is the same as the reaction with Br_2, except RO^- is the nucleophile that is generated instead of Br^-.

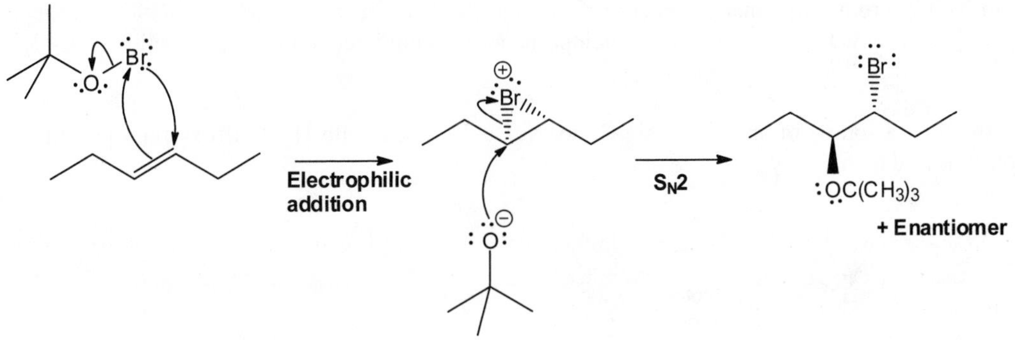

Electrophilic addition — S_N2 — $OC(CH_3)_3$ + Enantiomer

Problem 12.40

Think

What product is formed with the reaction of a peroxyacid in an alkene? For water to attack an epoxide, what must first happen to the epoxide? What reagent is present that can accomplish this reaction? Why are the two OH groups anti?

Solve

MCPBA is a peroxyacid and forms an epoxide when reacted with an alkene. The epoxide in the presence of an acid is protonated. Water acts as a nucleophile to open the epoxide ring. A final proton transfer reaction takes place to form the uncharged trans diol product.

Electrophilic addition → Proton transfer → S_N2 → Proton transfer

Problem 12.41

Think

What reactive intermediate reacts to produce a cyclopropane ring? To what type of functional group does that intermediate add? What is the starting material that reacts to form a carbene? How many C atoms are added per reaction with a carbene? How many C atoms are present in each compound?

Solve

(a) The carbene addition reaction occurs twice to form two cyclopropane rings from buta-1,3-diene.

2 equiv. CH_2N_2, hν

(b) The carbene addition reaction occurs twice to form two cyclopropane rings from propa-1,2-diene.

$H_2C{=}C{=}CH_2$ —2 equiv. CH_2N_2, hν→

Problem 12.42

Think

Is Li positive or negative? What is the charge on the C atom in the C–Li bond? How does this react with the H_2CCl_2 to form the carbene? Compare the acidity of $CHCl_3$ to CH_2Cl_2? What is the base strength of a carbanion compared to HO^-? From how many directions can the :CHCl carbene attack the C=C? What is the stereochemistry of the new bonds that form the cyclopropane ring compared to the wedge methyl on C4?

Solve

(a) Butyl lithium is a source of anionic, basic C atoms. Bu–Li reacts with H_2CCl_2 first via a proton transfer reaction followed by heterolysis.

Proton transfer → Heterolysis

(b) With fewer Cl atoms, CH_2Cl_2 is less acidic than $CHCl_3$, so the base must be stronger. A negative sp^3 C atom is more basic than a negative sp^3 O atom.

(c) The four products are shown below. The carbene attack can occur from the same side as the wedge methyl (top two molecules) or opposite from the wedge methyl (bottom two molecules).

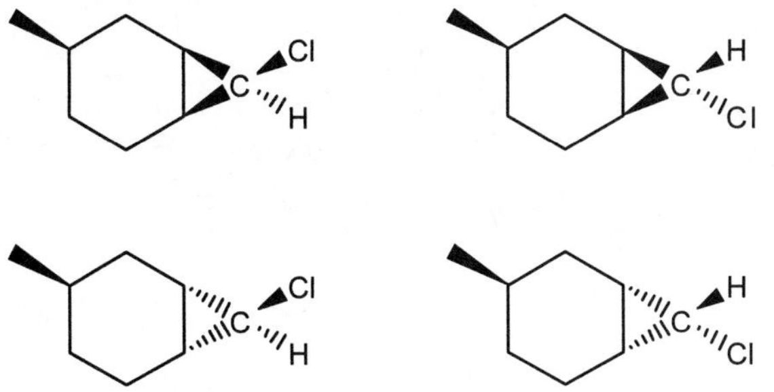

Problem 12.43

Think

Overall, what gets added to the alkyne? If you want the two Br atoms on the same carbon, what mechanism must the reaction go through? What type of intermediate must form? If you want the two Br atoms on different carbon atoms, what intermediate structure must the reaction go through?

Solve

In both cases, two H atoms and two Br atoms are added to the alkyne.

(a) For the Br atoms to add to the same C atom, the reaction is carried out with excess HBr to form the carbocation on the internal carbon and have the Br coordinate to this carbon two times.

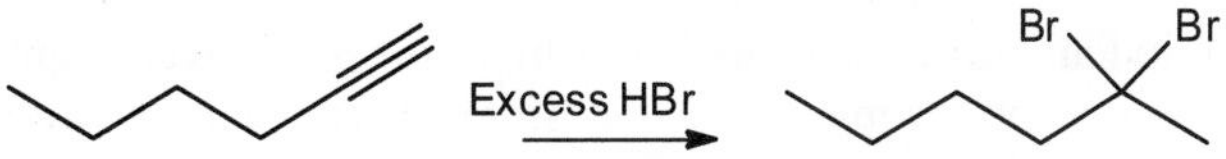

Excess HBr

(b) For the Br atoms to add to two different C atoms, the alkyne is first converted to the alkene. This transformation occurs in three synthetic reactions—hydroboration oxidation of the alkyne to form the aldehyde, reduction to form the alcohol, and dehydration to form the alkene. The addition of Br_2 to the alkene forms the vicinal dibromide.

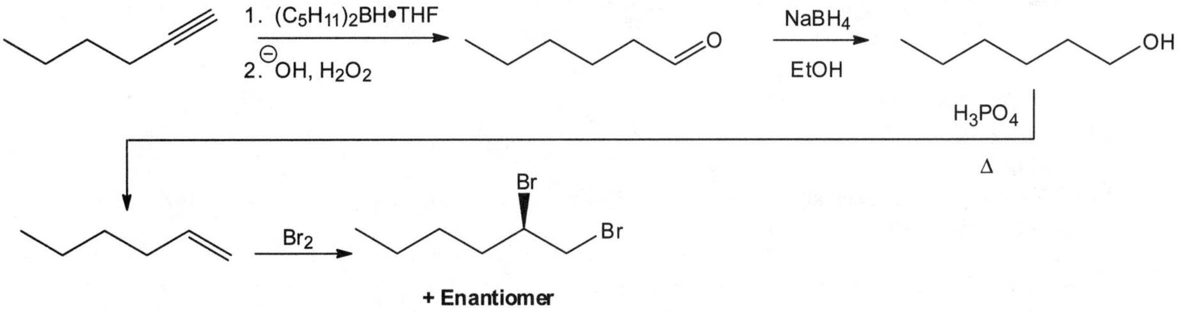

1. $(C_5H_{11})_2BH$•THF
2. $^{\ominus}OH$, H_2O_2

$NaBH_4$
EtOH

H_3PO_4
Δ

Br_2

+ Enantiomer

Problem 12.44

Think

What intermediate forms first in the reaction of Br_2 with an alkene? What do you know about this intermediate in the presence of a good nucleophile? Does the ROH compete with the Br^- for nucleophilic attack? Will an intramolecular or intermolecular nucleophilic attack be favored?

Solve

Attack of the alkene by Br_2 leads to the bromonium ion intermediate. In the presence of a good nucleophile (ROH), the more substituted C atom is attacked to form the five-membered ether ring. This is the same mechanism by which a halohydrin is formed.

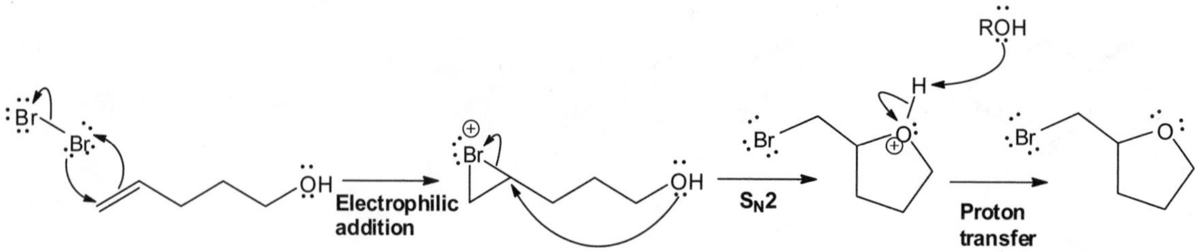

Problem 12.45

Think

Which atom, I or Cl, is more electronegative? In which direction does the dipole point? Which atom will then act as the electrophile to engage in an electrophilic addition reaction with the alkene? What intermediate forms? What acts as the nucleophile in the second step? Is the more- or less-substituted C atom attacked?

Solve

The I atom is more electrophilic because the more highly electronegative Cl atom leaves it with a partial positive charge.

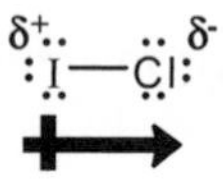

The mechanism proceeds through an iodonium ion intermediate, which forces the overall anti-addition of I–Cl. Cl^- attacks the tertiary carbon because it better accommodates a positive charge than the secondary carbon will. The mechanism is shown below.

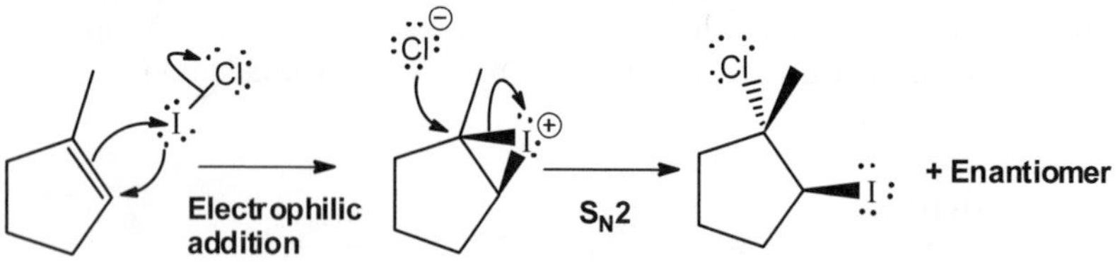

Problem 12.46

Think

What bonds are broken and formed in I–CH_2–Zn–I to form a cyclopropane ring and I–Zn–I? How is this similar to the peroxyacid mechanism?

Solve

The mechanism is shown below.

Broken bonds are: Zn–C, C=C, and C–I. Bonds formed are: two C–C, and Zn–I.

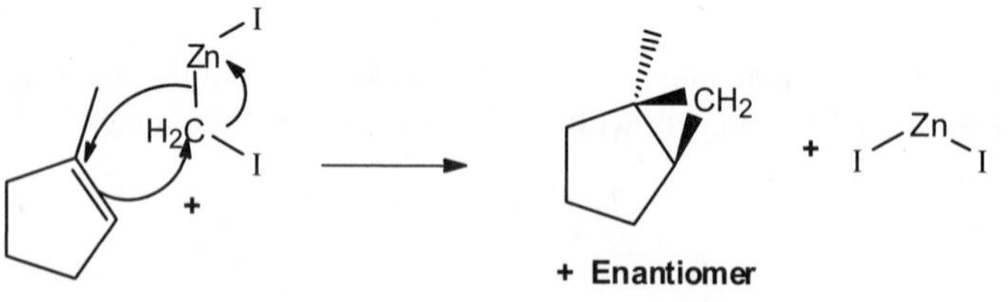

Problem 12.47

Think

How many C atoms are in the starting material compared to the product? What forms from the reaction of diazomethane and light? How would this reagent react with the C=C on benzene? What bonds would need to form and break in that intermediate to form the product?

Solve

The $H_2C–N_2$ bond breaks in a heterolysis step to form the reactive carbene reagent. This electrophilic reagent has a lone pair, and electrophilic addition, therefore, proceeds to form an intermediate that avoids losing an octet. Two new σ C–C bonds are formed simultaneously as the C=C π bond is broken. This forms a cyclopropane ring on the same side of the six-membered ring. This ring undergoes rearrangement to relieve the strain of the three-carbon ring.

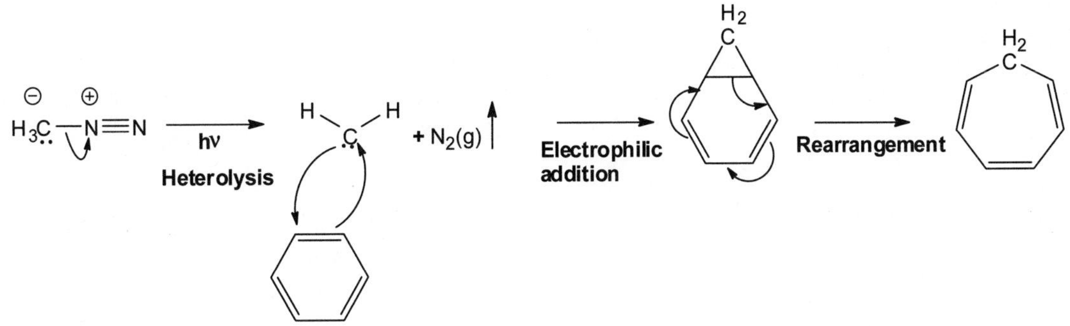

H3C–N≡N —hν→ (Heterolysis) H2C: + N2(g)↑ —→ (Electrophilic addition) —→ (Rearrangement)

CHAPTER 13 | Organic Synthesis 1: Beginning Concepts

Your Turn Exercises

Your Turn 13.1

Think

What is the abbreviation commonly used for tosyl? Draw out the organic reactant and product using nomenclature rules previously discussed. Is it necessary to include the inorganic byproduct? In drawing out the mechanism, think about the type of carbon to which the leaving group is bonded (1°, 2°, 3°). What type of attacking species is Cl^-? Identify the most likely mechanism.

Solve

(a) The common abbreviation for tosyl is OTs. The synthetic step does not need to show the NaOTs inorganic byproduct.

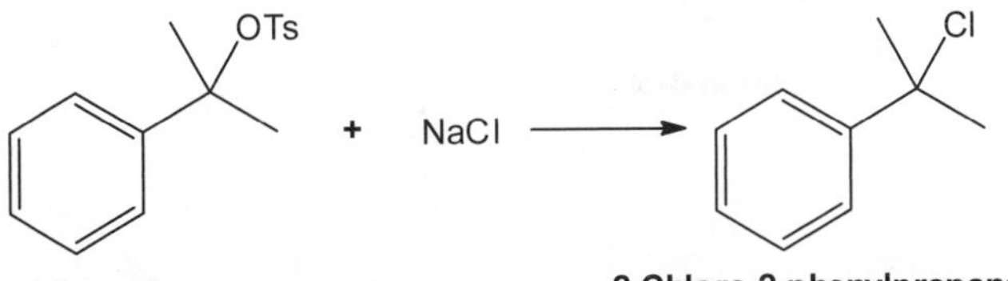

OTs + NaCl → Cl

2-Phenyl-2-tosylpropane   2-Chloro-2-phenylpropane

(b) The OTs is located on a 3° C and Cl^- is a weak base but a good nucleophile. The reaction, therefore, follows an S_N1 mechanism. The first step is heterolysis to form the benzylic 3° C^+ and TsO^-. The Cl^- then attacks the C^+ via a coordination step.

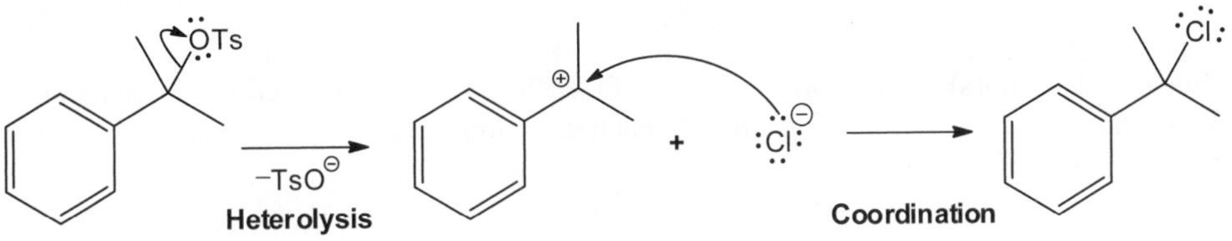

OTs → $-TsO^{\ominus}$ Heterolysis → ⊕ + :Cl:⊖ → Coordination → Cl

Your Turn 13.2

Think

From the description, identify the reagent and reaction conditions. Relative to the arrow (above or below), where should each be written in a synthetic step?

Solve

Water, H_2O, is the solvent and the temperature, 70 °C, is a reaction condition. Both are written below the arrow.

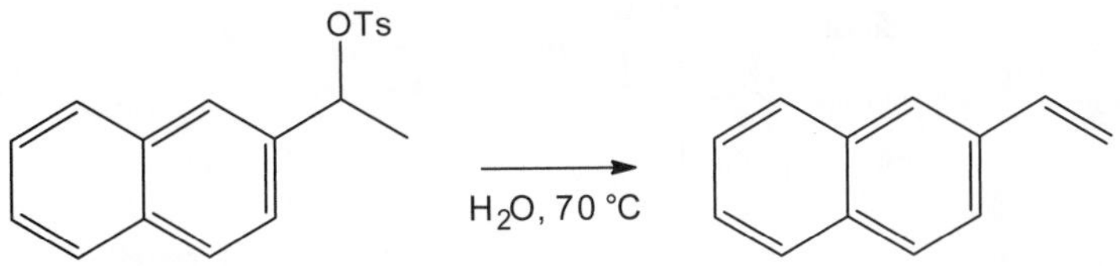

OTs → H_2O, 70 °C

Your Turn 13.3

Think

What do the numbers in front of the reagents indicate? How many separate, overall reactions are occurring to accomplish this synthesis?

Solve

The numbers indicate that one reaction is carried out to completion before the next one is initiated. In this case, two separate, overall reactions take place. First, butanoic acid is treated with sodium hydroxide (to yield sodium butanoate). Once that reaction has come to completion, bromoethane is then added (to yield ethylbutanoate). This is a two-reaction synthetic scheme.

Your Turn 13.4

Think

Draw the organic reactant and product using nomenclature rules previously discussed, and draw the two inorganic reagents. In how many separate, overall reactions is this synthesis occurring? What are the reagents, solvents, and conditions? How do you specify each complete reaction?

Solve

There are two separate, overall reactions occurring, which can be distinguished using numbers above and below the reaction arrow. First, (*Z*)-hex-3-ene (drawn on the left) is treated with $Hg(OAc)_2$, H_2O (mercury(II) acetate in water), and this can be indicated with a "1." above the reaction arrow. Once that reaction has come to completion, $NaBH_4$ (sodium borohydride) is added in a second step, and this can be indicated with a "2." below the reaction arrow. The final product, hexan-3-ol, is written on the right of the arrow.

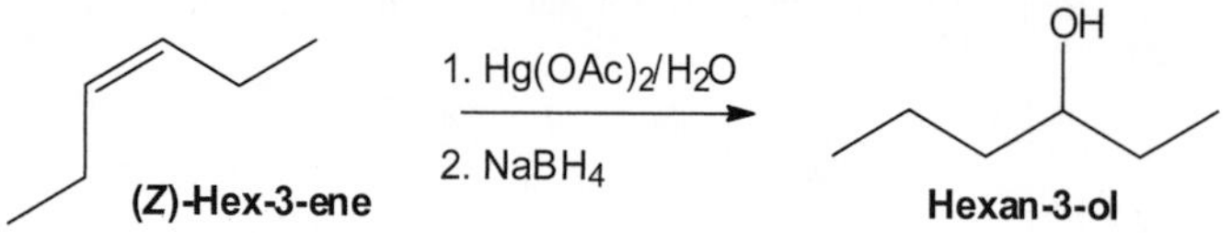

(Z)-Hex-3-ene

1. $Hg(OAc)_2/H_2O$

2. $NaBH_4$

OH

Hexan-3-ol

Your Turn 13.5

Think

Review the reaction tables in Chapters 9–12 and note any difficulties you have.

Solve

Review the reaction tables in Chapters 9–12 and note any difficulties you have.

Your Turn 13.6

Think

Identify the functional group(s) present in each step. In each reaction, consider if the functional group is different in the products than it is in the reactants. In each step, look at the carbon skeleton. Were any C–C σ bonds broken and/or formed?

Solve

In the first reaction, an epoxide is transformed into an alcohol and ether, but no C–C σ bonds are formed or broken. In the second reaction, the alcohol is converted into an alkylbromide, but no C–C σ bonds are formed or broken. In the third reaction, a new C–C bond is formed and the carbon skeleton is increased by two carbons, C≡CH. In the fourth reaction, the alkyne is transformed into an aldehyde, but no C–C σ bonds are formed or broken. In the fifth reaction, the α C is alkylated by CH_2CH_3, which forms a new C–C σ bond.

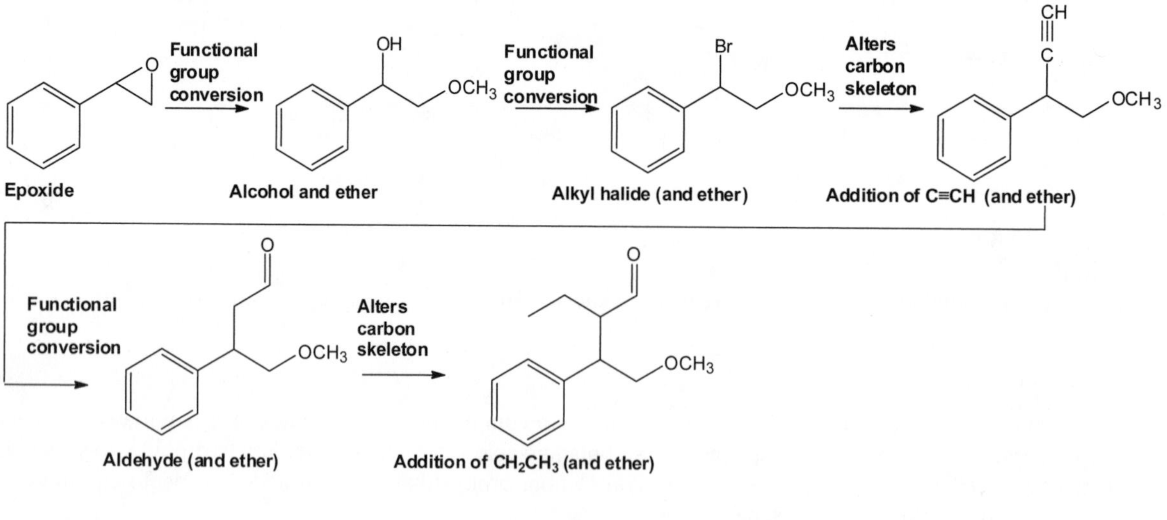

Epoxide — Functional group conversion → Alcohol and ether — Functional group conversion → Alkyl halide (and ether) — Alters carbon skeleton → Addition of C≡CH (and ether)

Functional group conversion → Aldehyde (and ether) — Alters carbon skeleton → Addition of CH_2CH_3 (and ether)

O

OH

OCH_3

Br

CH

C

O

Your Turn 13.7

Think

The squiggly line indicates that the C–C bond between the benzylic C and the alkyne C should be disconnected. Once this occurs, one C atom in the precursors should be electron rich and the other C atom should be electron poor. What are the necessary functional groups in the precursors that accomplish this? (*Hint*: consult Table 13-1 and Equation 13-10).

Solve

One precursor should be an alkyl halide (electron-poor C) and the other precursor should be a terminal alkyne (electron-rich C.) The alkyl halide is benzyl bromide, $PhCH_2Br$.

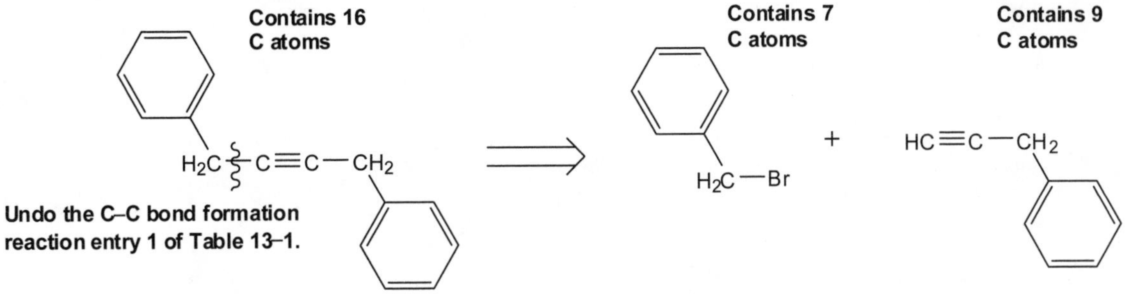

Contains 16 C atoms

Contains 7 C atoms

Contains 9 C atoms

H_2C–$C{\equiv}C$–CH_2 ⟹ H_2C–Br + $HC{\equiv}C$–CH_2

Undo the C–C bond formation reaction entry 1 of Table 13–1.

Your Turn 13.8

Think

Consult Table 13-2. What are the alkoxide and alkyl bromide products if the other C–O bond is disconnected? Does this affect the reaction in Step 2 of the synthesis in Equation 13-12?

Solve

If the other C–O bond is disconnected, CH_3Br and $HC{\equiv}CCH_2ONa$ are the precursors. The synthesis in Step 2 is not altered.

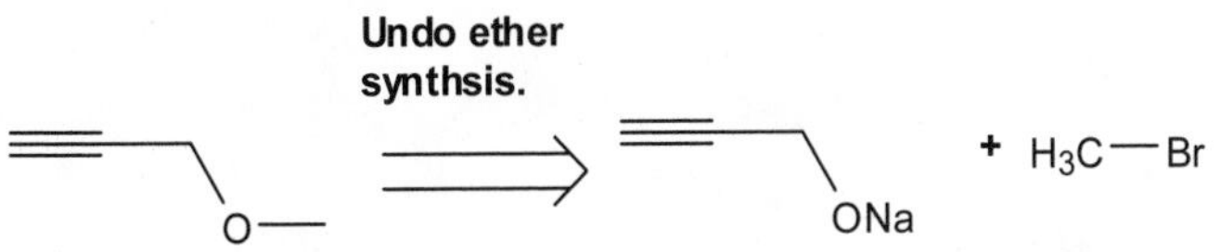

Undo ether synthsis.

O— ⟹ ONa + H_3C—Br

The synthesis would proceed in the forward direction as shown below:

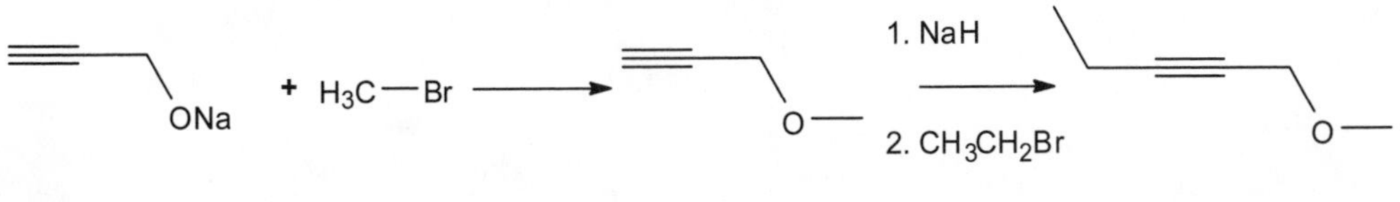

ONa + H_3C—Br ⟶ O— ⟶ (1. NaH; 2. CH_3CH_2Br) O—

Your Turn 13.9

Think

What undesirable reaction would be a potential concern when an alcohol solvent (ROH) is used with $CH_3CH_2O^-$? Is there an alcohol solvent that, when added to $CH_3CH_2O^-$, does not produce a new nucleophile even after that reaction takes place?

Solve

Ethanol, CH_3CH_2OH, would be an appropriate solvent. The undesirable reaction is a proton transfer between the alcohol solvent and $CH_3CH_2O^-$. With ethanol as the solvent, a proton transfer would produce species that are identical to the reactant species, as shown below.

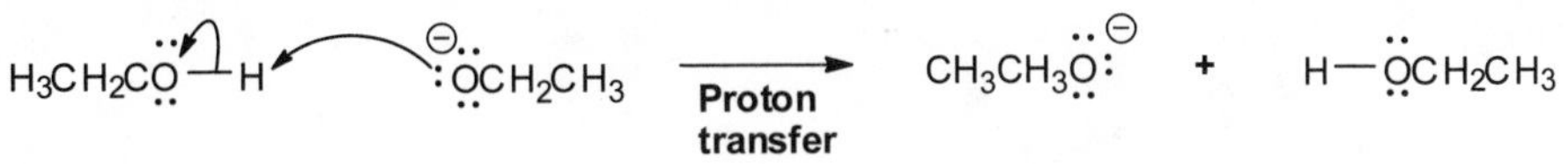

H_3CH_2CO—H $^{\ominus}$:OCH_2CH_3 ⟶ (Proton transfer) CH_3CH_3O:$^{\ominus}$ + H—OCH_2CH_3

Your Turn 13.10

Think

What is the yield for each step? For a linear synthesis, what is the relationship between the overall product yield and the product yield for each step?

Solve

For a linear synthesis, the overall product yield is the product of each step in the sequence. If each step's yield is 80%, the yield of a six-step synthesis would be $(0.80)^6 = 0.26$, or 26%.

Chapter Problems

Problem 13.1

Think

Refer to Your Turn 13.6 to determine if the reaction is a functional group conversion or a reaction that alters the carbon skeleton. In each reaction, determine which bonds formed. Did an electrophile or nucleophile react with the organic reactant in each step? Is regiochemistry a concern? Consult the reaction tables indicated in Your Turn 13.5 if necessary.

Solve

The first reaction has CH_3O^- attack the epoxide ring at the less substituted C atom. Therefore, the reaction occurs in basic conditions. An acid workup step follows to form the uncharged alcohol product. The second reaction is substitution of the HO for Br. Since hydroxide is not a good leaving group, PBr_3 changes it into a good leaving group via back-to-back S_N2 reactions. The third reaction is nucleophilic substitution of the Br for the alkynide anion. The fourth reaction is hydroboration–oxidation to form the terminal enol that tautomerizes to the aldehyde. The fifth and final reaction is alkylation of the α C atom.

1. $NaOCH_3$, CH_3OH 2. H_3O^+ — OH — OCH_3 — PBr_3 — Br — OCH_3 — NaC≡CH — CH ≡ C — OCH_3

1. $(C_5H_{11})_2BH$ 2. NaOH, H_2O_2 — O — OCH_3 — 1. $NaNH_2$ 2. CH_3CH_2Br — O — OCH_3

Problem 13.2

Think

Which C–C bond should be disconnected? Is there more than one choice? From what type of functional group did the OH originate? Was the more- or less-substituted C atom attacked?

Solve

Reaction entry 2 in Table 13-1 involves deprotonation of the alkyne to form the alkynide anion nucleophile. This nucleophile opens the epoxide ring by attack at the less substituted C. An acid workup step is necessary to form the uncharged alcohol product.

Retrosynthsis — OH — **Undo an epoxide ring opening.** — + — O

Synthesis in forward direction — NaH — 1. O — 2. H_3O^+ — OH

Problem 13.4

Think

If you want to make pent-2-yne and have to start with compounds containing two or fewer C atoms, how many reactions that alter the carbon skeleton must you include? How can you form a new C–C using an alkynide anion? To what type of atom must a carbon be attached to be an electrophilic carbon?

Solve

Because the target contains five carbons, we will need to form two new carbon–carbon bonds if the starting material is to contain two or fewer carbons. So we must apply transforms that will disconnect two carbon–carbon bonds. An electrophilic carbon is one that is electron poor and, therefore, can be bonded to an electronegative atom that is also a good leaving group. *Note*: Instead of the order shown below, the CH_3CH_2Br could be added first and the CH_3Br second.

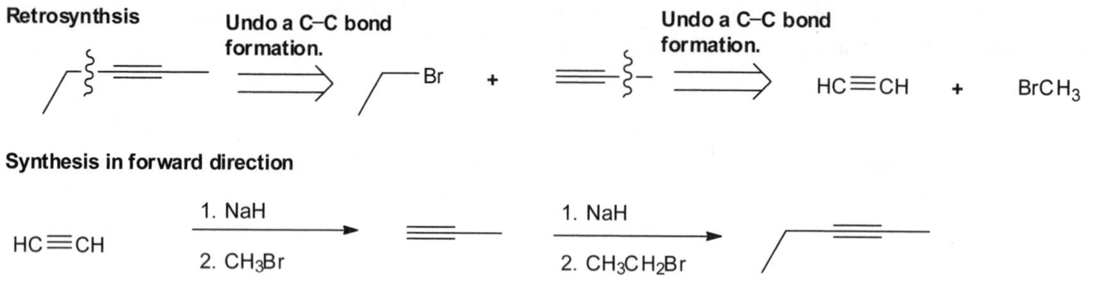

Retrosynthesis

Undo a C–C bond formation.

Br +

Undo a C–C bond formation.

HC≡CH + $BrCH_3$

Synthesis in forward direction

HC≡CH

1. NaH
2. CH_3Br

1. NaH
2. CH_3CH_2Br

Problem 13.5

Think

How do you alkylate the α C atom of a C=O compound? How many times does this alkylation need to take place? What is the identity of your C=O compound after undoing these alkylations?

Solve

Alkylation of the α C atom of a C=O compound first involves a proton transfer to form an enolate anion and then the enolate anion acts as a nucleophile with an R–L via S_N2. The reaction sequence is performed twice.

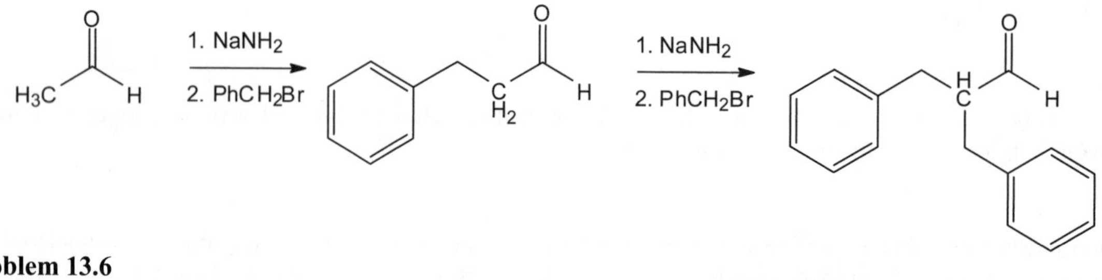

O, H_3C, H

1. $NaNH_2$
2. $PhCH_2Br$

O, C, H_2, H

1. $NaNH_2$
2. $PhCH_2Br$

O, H, C, H

Problem 13.6

Think

What reagent and reaction conditions are necessary to reverse the transform indicated to carry out the reaction in the forward direction? What other functional groups are present? Does that functional group react with the reagent you used? If so, is this desired?

Solve

The first three can proceed as planned.

(a) NaCN can be used as a nucleophile for an S_N2 reaction. A concern might be the alcohol group, but NC^- is not a strong enough base to deprotonate an alcohol, and HO^- is not a sufficient leaving group to be displaced in an S_N2 reaction. Therefore, the synthesis will proceed as planned.

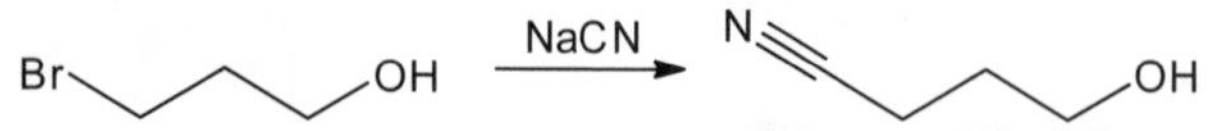

Br, OH

NaCN

N, OH

(b) This presents a similar situation to **(a)**. Therefore, the synthesis will proceed as planned.

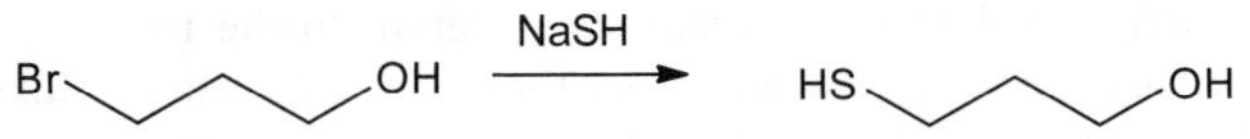

(c) The conversion can take place using PBr_3. A concern might be the phenolic OH, but nucleophilic substitution does not take place on an sp^2-hybridized carbon. Therefore, the synthesis will proceed as planned.

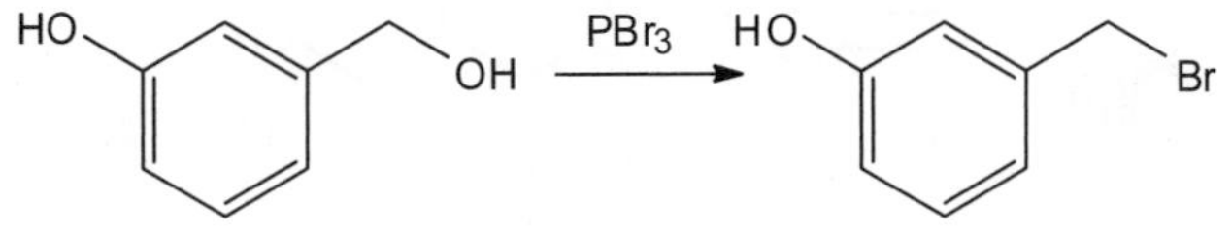

(d) This is a synthetic trap. We could attempt to convert the OH group to Br using PBr_3, but both OH groups could be converted, as both are attached to sp^3-hybridized carbons. It is likely, however, that the intended product would be the major product, because the functional group is on a primary carbon. The other is on a secondary carbon, which impedes S_N2 reactions.

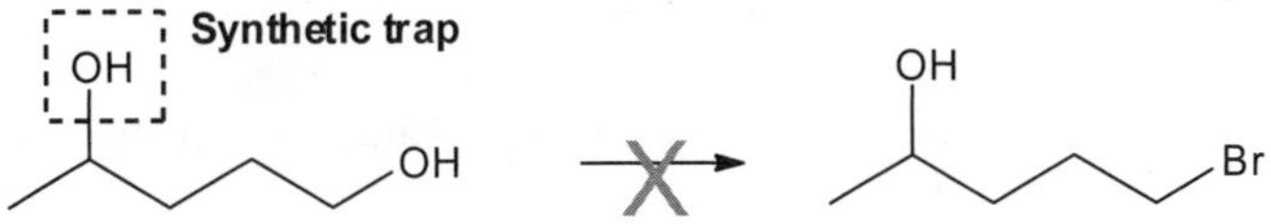

Synthetic trap

(e) This is a synthetic trap. We could attempt to convert the alkene C=C into the alkyl halide using HCl. This could put the Cl at the more substituted C atom. However, the ether O atom is likely protonated in a strong acid and could lead to an S_N2 in which the Cl opens the protonated ether ring.

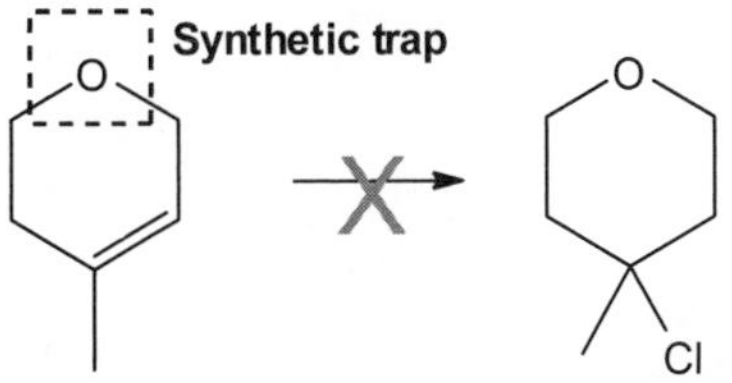

Synthetic trap

(f) This is a synthetic trap. We could attempt to do oxymercuration–reduction of the alkyne to form the internal enol that tautomerizes to the ketone. The problem is that an alkene is also reactive under oxymercuration–reduction reaction conditions and a secondary alcohol would form.

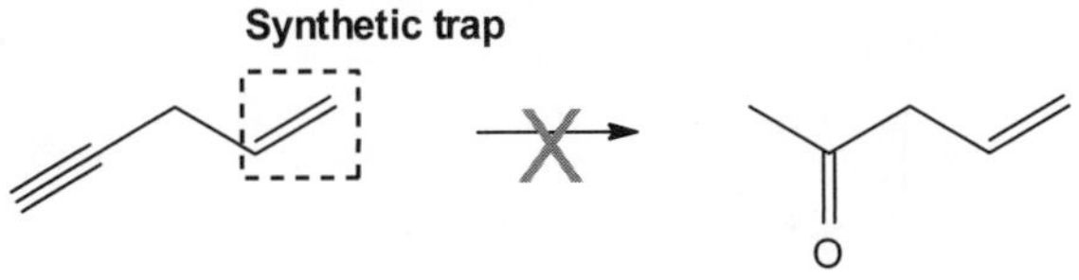

Synthetic trap

Problem 13.8

Think

What precursor can be used to produce the alkyl bromide target? Is regiochemistry a concern? If so, can the alkyl bromide be produced directly with the desired regiochemistry? If not, what other reaction can be carried out to give the desired regiochemistry?

Solve

The bromide is added to the more substituted C atom when the HBr addition reaction is undone. If the elimination reaction is also undone, we can arrive at the intended starting material. In the forward direction, elimination proceeds through an elimination dehydration mechanism to form the disubstituted alkene. The subsequent electrophilic addition of HBr will produce the target with the desired regiochemistry.

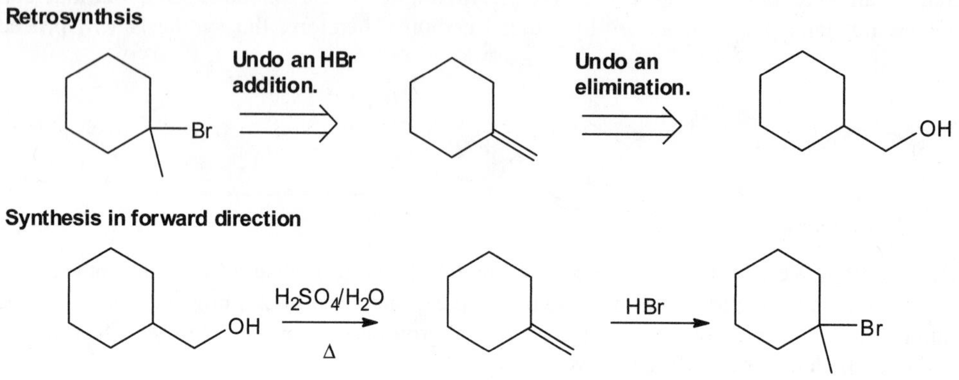

Retrosynthesis

Undo an HBr addition. ⟹ Undo an elimination. ⟹ OH

Br

Synthesis in forward direction

OH —H_2SO_4/H_2O, Δ→ —HBr→ Br

Problem 13.9

Think

Should alkylation occur at the more or less substituted C atom? What is the alkyl group that adds? What reaction conditions are needed to form the regioisomer indicated?

Solve

In both reactions the benzyl group is the alkyl group added.

(a) Alkylation occurs at the less substituted C atom. LDA is used because a very strong bulky base is required to produce the kinetic enolate intermediate.

O —1. LDA; 2. Br→ O

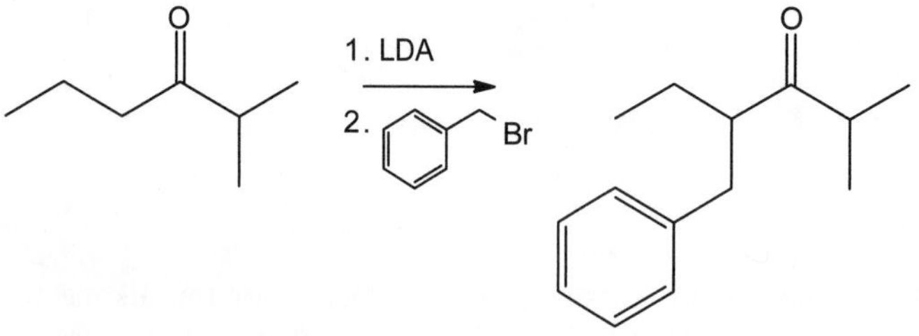

(b) Alkylation occurs at the more substituted C atom. *tert*-Butoxide is used because a moderately strong base is required to produce the thermodynamic enolate intermediate.

O —1. $(CH_3)_3CO^{\ominus}$; 2. Br→ O

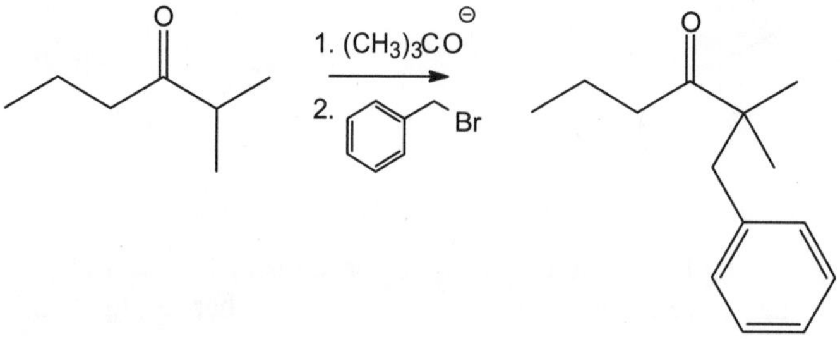

Problem 13.10

Think

Does bromination occur at the more or less substituted C atom? What reaction conditions are needed to form the regioisomer indicated? What reagents are used?

Solve

(a) Bromination occurs at the less substituted C atom, so the reaction must take place under neutral or basic conditions. An acid workup step is necessary to form the uncharged alcohol product.

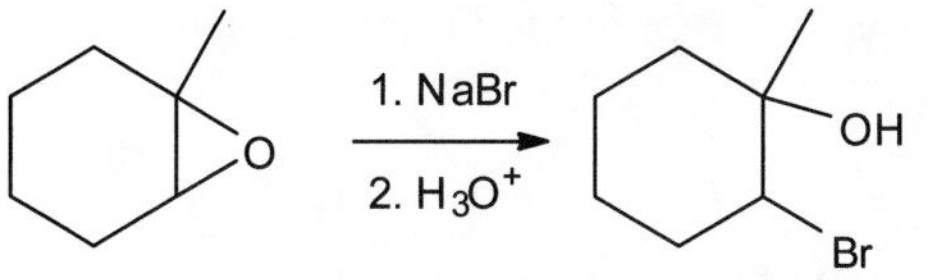

1. NaBr
2. H_3O^+

(b) Bromination occurs at the more substituted C atom, so the reaction must take place under acidic conditions.

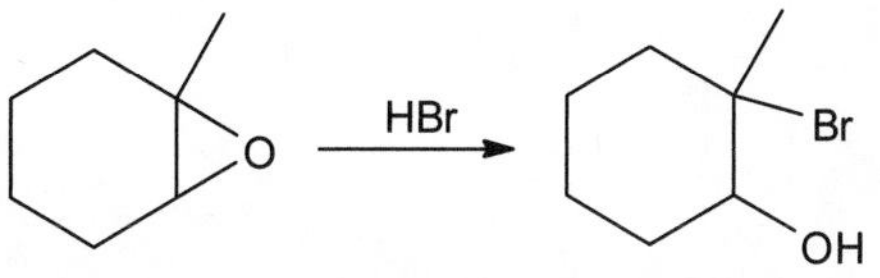

HBr

Problem 13.12

Think

What does the starting material look like when the leaving group is on C2? On C3? Which protons can be eliminated along with the leaving group? What would be the elimination product in each case? Which is the Zaitsev product, which is the Hofmann product, and which corresponds to the target?

Solve

Leaving group on C2: This is a normal E2 reaction, which will produce the Zaitsev (i.e., more highly substituted alkene) product. The reaction conditions are given below.

This H will be deprotonated to produce the more substituted alkene.

CH_3CH_2ONa

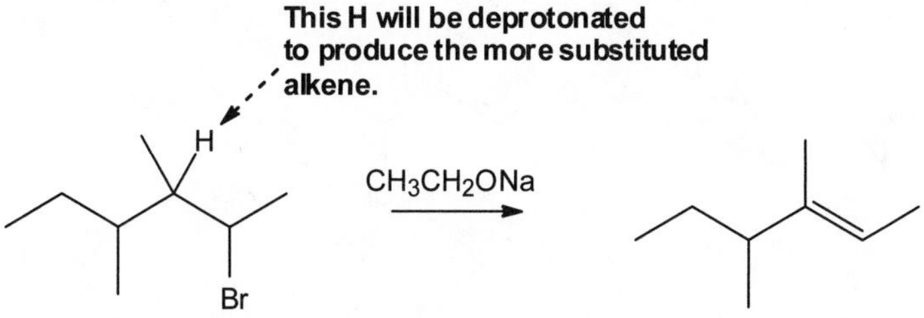

Leaving group on C3: This is a Hofmann elimination, which will produce the Hofmann product, or anti-Zaitsev (i.e., less highly substituted alkene) product. The reaction conditions are given below.

1. excess CH_3I
2. NaOH, Δ

This H will be deprotonated to produce the less substituted alkene.

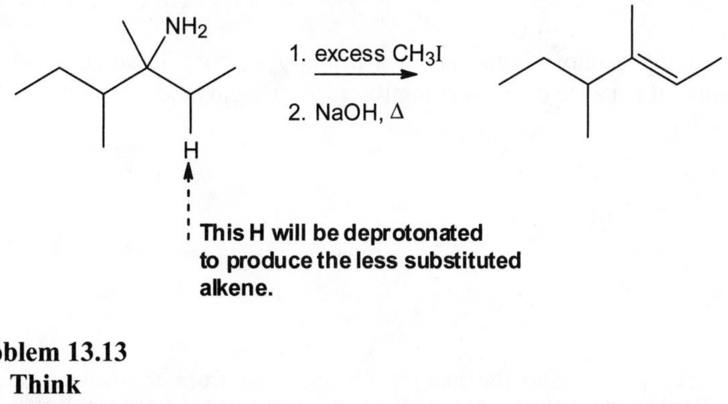

Problem 13.13

Think

What is added to the alkene in each reaction? Did 1,2-addition or 1,4 addition occur? How does temperature affect the regiochemistry of the product? In **(c)** and **(d)**, how many addition reactions occurred? What was the second reagent needed to accomplish that transformation?

Solve

(a) 1,4-Addition occurred. Warm temperatures are required. HBr adds to the alkene. H adds at C1, and Br adds at C4. An S_N2 reaction follows in which Br is substituted by CN.

HBr, High *T*; NaCN

(b) 1,2-Addition occurred. Cold temperatures are required. HBr adds to the alkene. H adds at C1, and Br adds at C2. An S_N2 reaction follows in which Br is substituted by NC.

HBr, Low *T*; NaCN

(c) 1,4-Addition occurred. Warm temperatures are required. HCl adds to the alkene. H adds at C1 and Cl adds at C4. A second addition reaction occurs where a carbene $:CH_2$ adds to the C=C to form the cyclopropane ring.

HCl, High *T*; H_2CN_2, Δ

(d) 1,2-Addition occurred. Cold temperatures are required. HCl adds to the alkene. H adds at C1, and Cl adds at C2. A second addition reaction occurs where Br_2 adds to the C=C to form the 1,2 dibromo product.

HCl, Low *T*; Br_2

Problem 13.15

Think

Should the reaction be S_N1 or S_N2? To ensure that this reaction takes place, should you use a strong nucleophile or a weak nucleophile? What protic solvent could be used that would not react with that nucleophile to form a new nucleophile?

Solve

The reaction should be S_N2 because the leaving group is attached to a primary carbon. The nucleophile is ethoxide $CH_3CH_2O^-$. Ethanol is used as a solvent because deprotonating the solvent yields the same nucleophile that was added.

$NaOCH_2CH_3$, $HOCH_2CH_3$

Problem 13.16

Think

In how many steps does an S_N2 reaction take place? Can the leaving group leave from the same side the nucleophile attacks the electrophilic carbon? If not, from what side must the nucleophile attack? What does this do to the stereochemistry at the electrophilic carbon?

Solve

S_N2 takes place in one step and the leaving group leaves opposite the side from which the nucleophile attacks. This leads to inversion of configuration at the electrophilic carbon. The substrate must have the leaving group pointing toward us if the nucleophile is to form a bond pointing away from us.

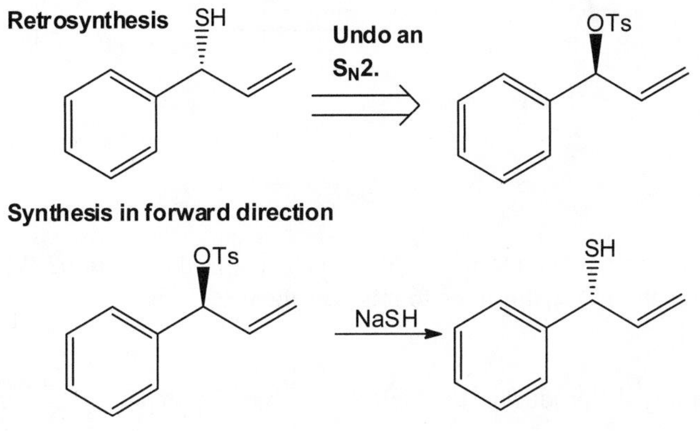

Retrosynthesis

Undo an S_N2.

Synthesis in forward direction

NaSH

Problem 13.17

Think

For E2 to occur, the Cl and H must be in what orientation? Are they in this orientation in the molecule given? If so, which groups are cis to each other in the alkene? Which groups are trans? What does this mean for the orientation of those groups in the wedge/dash structure?

Solve

If the carbon with the Cl atom in the precursor is the same carbon as the left-hand alkene carbon in the target, then it must also be bonded to H, as shown below.

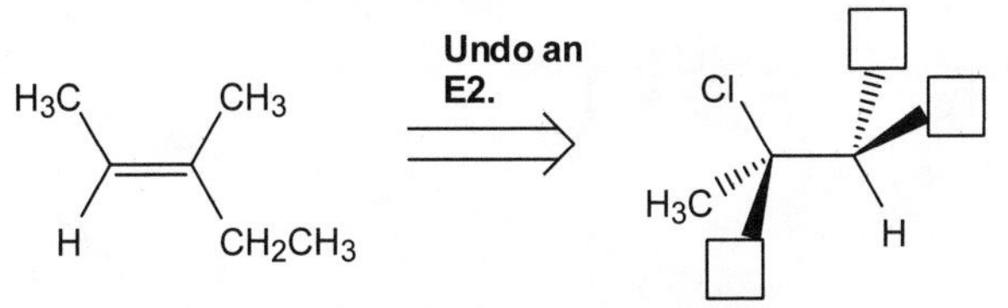

Undo an E2.

We then recognize that in the target, the two CH_3 groups are on the same side of the double bond. Thus, in the precursor, if one CH_3 group is pointing behind the plane of the paper, so must the CH_3 group bonded to the other carbon. This means that the CH_2CH_3 group must point in front. Notice that this precursor is the enantiomer of the one in Equation 13-42.

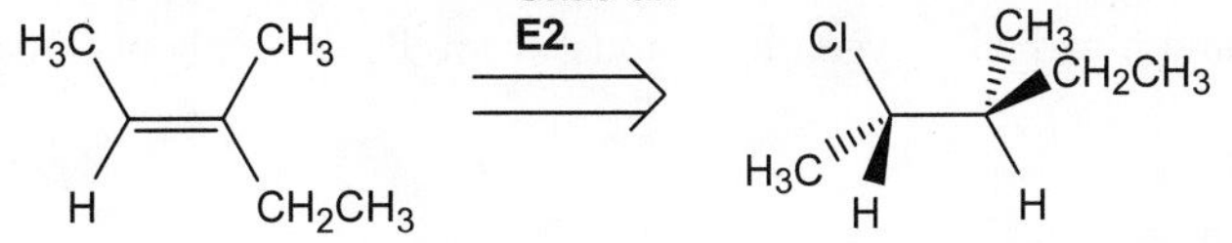

Undo an E2.

Problem 13.18

Think

Does hydroboration–oxidation add the HO to the more- or less-substituted C? Do the H and the HO add syn or anti? What is the orientation of those groups in (2*S*, 3*R*)-3-phenylbutan-2-ol? Is another conformation necessary to view the target as the immediate product of hydroboration–oxidation?

Solve

Hydroboration–oxidation adds the HO to the less substituted C, which indicates that the C=C must be attached to the benzene ring. The H and the HO add syn in hydroboration–oxidation and, therefore, another conformation is necessary to view the H and HO as syn. Rotation about the C–C accomplishes this and the methyl groups are, therefore, cis. See the figure on the next page.

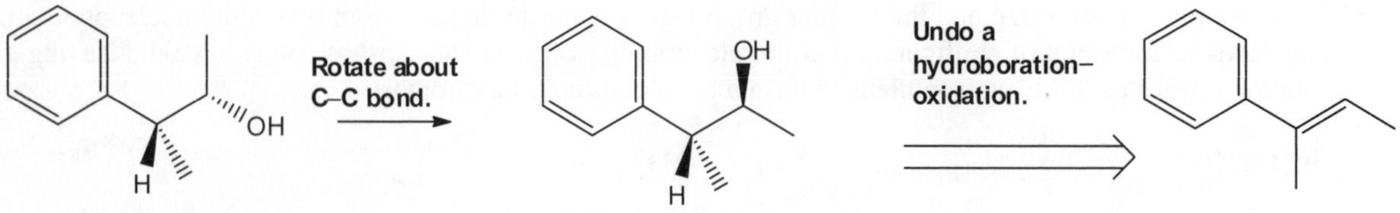

Rotate about C–C bond.

Undo a hydroboration–oxidation.

(2*S*,3*R*)-3-Phenylbutan-2-ol

H and OH same side

Problem 13.19

Think

For a linear synthesis, how is the overall product yield related to the yield of each step in the reaction? What is the percent yield of each step of the eight-step reaction? What is done to this percentage?

Solve

Percent yield in a linear synthesis like this is simply the product of the yields of each of the steps.
$\%\ \text{yield} = (0.75)^8 = 0.100 = 10.0\%$

Problem 13.20

Think

For a linear synthesis, how is the overall product yield related to the yield of each step in the reaction? What is the percent yield of each step of the seven-step reaction? What is done to these percentages?

Solve

Percent yield in a linear synthesis like this is simply the product of the yields of each of the steps.
$\%\ \text{yield} = (0.90)(0.81)(0.85)(0.98)(0.82)(0.72)(0.94) = 0.337 = 33.7\%$

$$\text{S} \xrightarrow{90\%} \text{A} \xrightarrow{81\%} \text{B} \xrightarrow{85\%} \text{C} \xrightarrow{98\%} \text{D} \xrightarrow{82\%} \text{E} \xrightarrow{72\%} \text{F} \xrightarrow{94\%} \text{T}$$

Problem 13.21

Think

How many C atoms are in the product? How many C atoms are in the reactant? Did the number of C atoms change? Are any carbon–carbon σ bonds formed or broken?

Solve

(a) Yes. A carbon–carbon σ bond must break.
(b) No. A carbon–oxygen bond must form, but no carbon–carbon σ bonds are broken or formed.
(c) Yes. A carbon–carbon σ bond must form.
(d) No. A carbon–carbon π bond must break and a carbon–oxygen bond must form, but no carbon–carbon σ bonds must form or break.
(e) Yes. A carbon–carbon σ bond must form.
(f) Yes. Two carbon–carbon σ bonds must form.
(g) No. A carbon–carbon π bond must break and two carbon–oxygen bonds must form, but no carbon–carbon σ bonds must form or break.
(h) Yes. At least two carbon–carbon σ bonds must form.

Problem 13.22

Think

What is the structure of each organic compound named? What is the structure of each inorganic reagent named? In how many separate steps is each synthesis? How do you denote separate synthetic steps?

Solve

The syntheses are listed below.

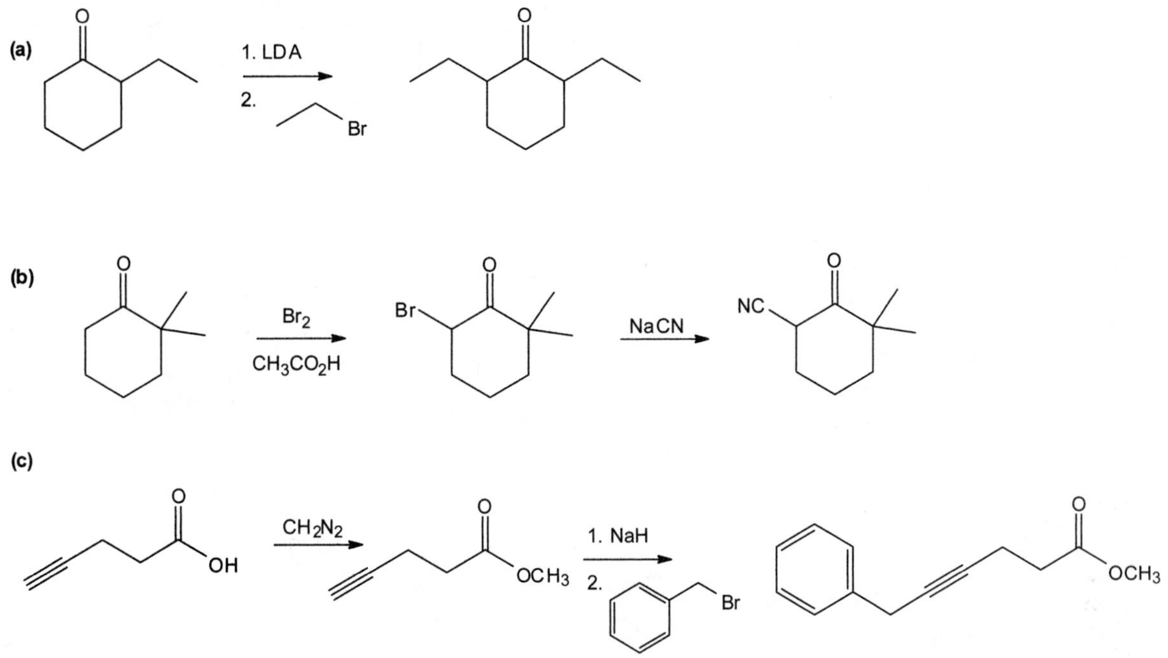

(a) 1. LDA 2. Br

(b) Br_2, CH_3CO_2H; Br; NaCN; NC

(c) OH; CH_2N_2; OCH_3; 1. NaH 2. Br; OCH_3

Problem 13.23

Think

What is the name of each organic compound? What is the name of each inorganic reagent? Do any of the reaction conditions need to be specified (e.g., temperature, solvent)? In how many steps does each reaction occur? What products form during intermediate steps?

Solve

(a) To the starting material epoxide, add phenylmagnesium bromide, followed by aqueous acid to yield 2-phenylcyclopentanol. Treat 2-phenylcyclopentanol product with concentrated phosphoric acid at 100 °C to yield 1-phenylcyclopentene.

(b) Treat phenylethanone with lithium diisopropylamide, followed by bromomethane to yield phenylpropanone. Then add molecular bromine in the presence of acetic acid to yield 2-bromo-1-phenylpropanone. Next add sodium acetate to yield the final product.

(c) Heat 1-(ethoxymethyl)-3-nitrobenzene with aqueous acid. Upon completion of that reaction, add phosphorus tribromide to yield 1-bromomethyl-3-nitrobenzene. Finally, add sodium azide to yield the target.

Problem 13.24

Think

What is the target compound in each reaction? What is the starting material? How do you deconstruct the product in each step? What type of arrow should you use in a retrosynthetic analysis?

Solve

Retrosyntheses from Problem 13.22:

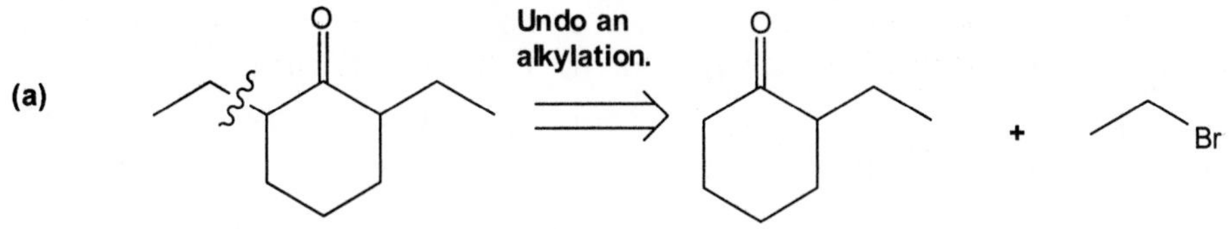

(a) Undo an alkylation. + Br

(b) NC — Undo a substitution. ⟹ NaCN + Br — Undo a bromination. ⟹

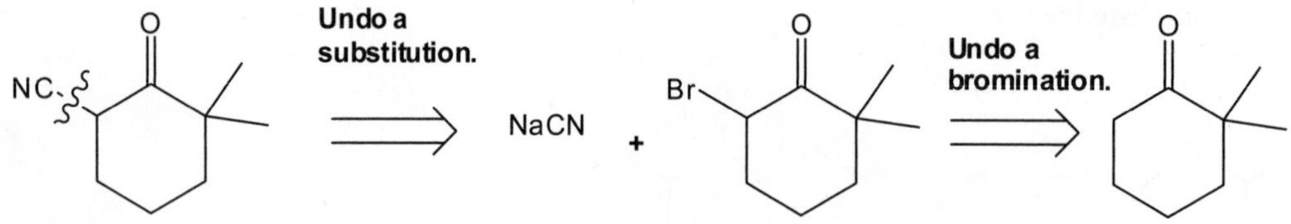

(c) Undo a substitution. ⟹ Br + OCH_3 — Undo an esterification. ⟹ OH + CH_2N_2

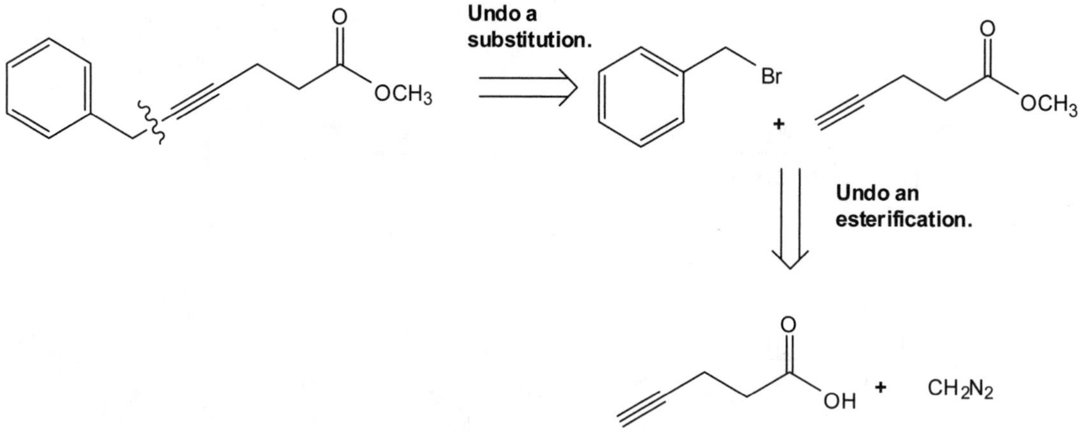

Retrosyntheses from Problem 13.23:

(a) Undo a dehydration. ⟹ OH — Undo a Grignard. ⟹ MgBr + O

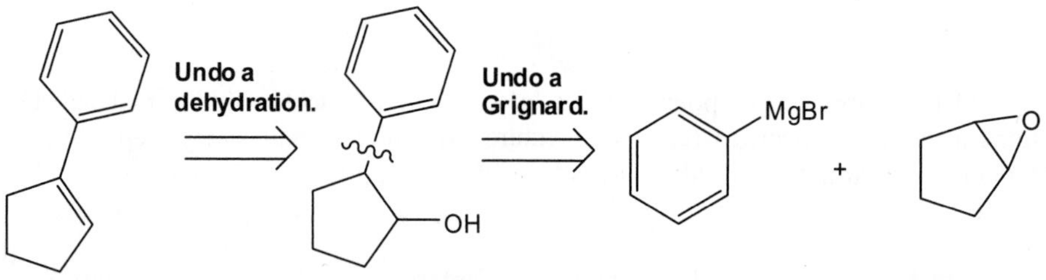

(b) Undo an S_N2. ⟹ Br — Undo a bromination. ⟹ Undo an alkylation. ⟹

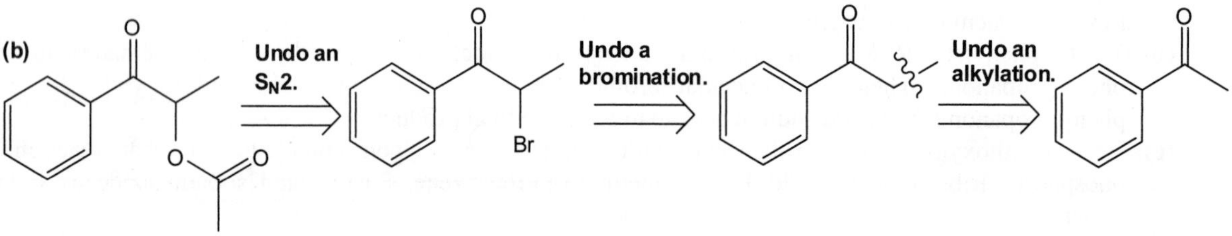

(c) NO_2, N_3 — Undo an S_N2. ⟹ NO_2, Br — Undo a bromination. ⟹ NO_2, HO — Undo a hydrolysis. ⟹ NO_2, O, CH_2CH_3

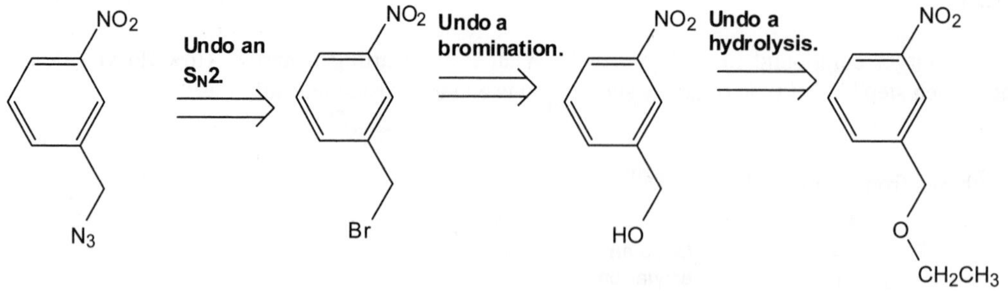

Problem 13.25

Think

Does the solvent react with anything in the reaction mixture? If so, does it form a product that competes with other reagents in the intended reaction? If so, what solvent avoids this problem?

Solve

(a) The solvent is appropriate. This is an E2 reaction, which favors the Zaitsev product. If the solvent is deprotonated, the result is $CH_3CH_2O^-$, the same as the base that is already shown.

(b) The solvent is not appropriate. The alkoxide shown, $(CH_3)_3CO^-$, is strong enough to deprotonate the solvent, introducing $CH_3CH_2O^-$ as a nucleophile. Use *tert*-butanol as a solvent instead.

(c) The solvent is not appropriate, for the same reason as in **(b)**. Use cyclohexanol as a solvent instead.

(d) The solvent is appropriate. The nucleophile shown, phenoxide anion, is not a strong enough base to deprotonate the solvent to generate a different alkoxide as a nucleophile.

Problem 13.26

Think

What reagent and reaction conditions are necessary to reverse the transform indicated and carry out the reaction in the forward direction? What other functional groups are present? Does that functional group react with the reagent you used? If so, is this desired? Is regiochemistry a concern?

Solve

(a) This is a synthetic trap. Instead of attacking the epoxide, the Grignard reagent could attack the alkyl halide as a base in an E2 reaction, or as a nucleophile in an S_N2 reaction.

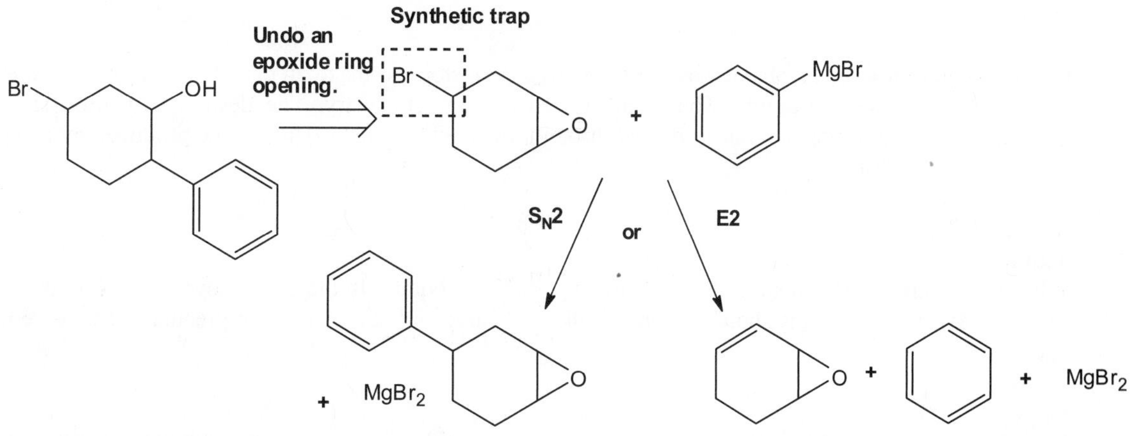

Synthetic trap

Undo an epoxide ring opening.

Br OH

Br O + MgBr

S_N2 or E2

+ $MgBr_2$ O

O + + $MgBr_2$

(b) This is a synthetic trap. If the epoxide is treated with NaCN, which would be under neutral or basic conditions, then NC^- would attack the less sterically hindered C atom. To obtain the target, however, the more hindered C atom must be attacked.

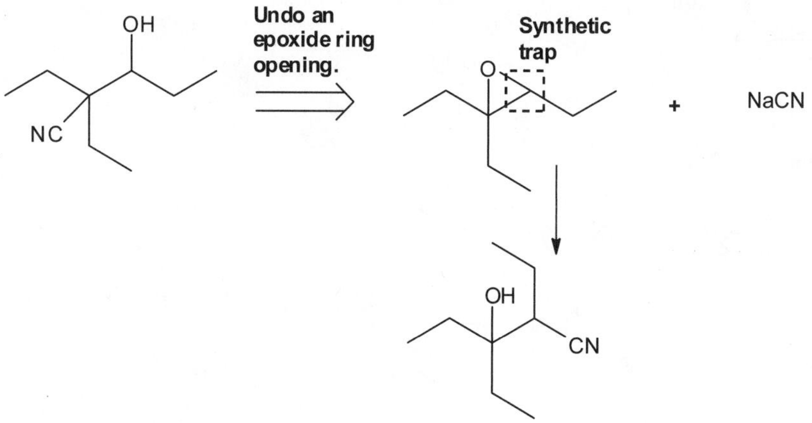

OH NC

Undo an epoxide ring opening.

Synthetic trap

O + NaCN

OH CN

(c) This could proceed as planned. The terminal alkyne would have to be treated with a strong base such as NaH to produce the alkynide anion as a nucleophile. When the ethyoxyalkyl halide is added, the alkynide anion will attack only the alkyl halide group. This is because the ether group is resistant to substitution and elimination under basic conditions.

(d) This is a synthetic trap. When the alkoxide anion is added to the bromoalcohol, the alcohol can be deprotonated. This would set up an intramolecular S_N2 reaction, which would produce a six-membered ring, and would thus be more favorable than the desired intermolecular S_N2 reaction.

Undo a Williamson synthesis.

HO ... O (cyclohexyl) ⟹ HO ... Br + cyclohexyl ONa

Synthetic trap

→ tetrahydropyran (O) + cyclohexyl OH + NaBr

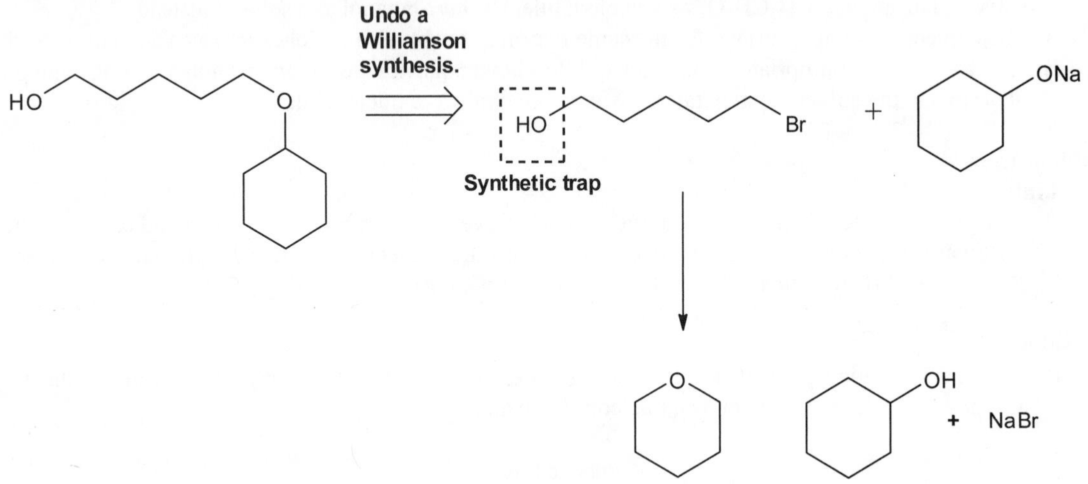

(e) This could proceed as planned by deprotonating the ketone to make an enolate anion nucleophile and then adding the benzyl bromide. The main concern is regiochemistry. The desired alkylation is at the less substituted carbon, which must proceed through the kinetic enolate. This can be produced by using LDA as the base, not NaOH.

Problem 13.27

Think

What is the structure of pent-2-yne? Identify the C–C≡C bonds. Is the alkyne symmetric? If the C–C≡C is formed, how many C atoms should be on the alkyl bromide and alkynide anion precursors? Is there more than one option?

Solve

In a retrosynthetic analysis, the carbon–carbon bond on either side of the triple bond can be disconnected, giving us a terminal alkyne and an alkyl halide as precursors.

Route A

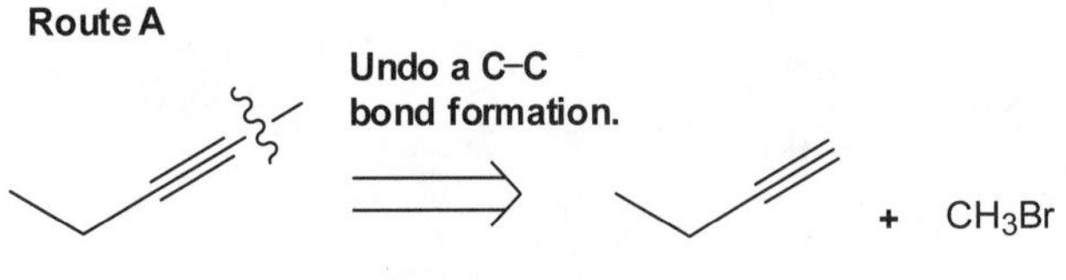

Undo a C–C bond formation.

+ CH_3Br

Route B

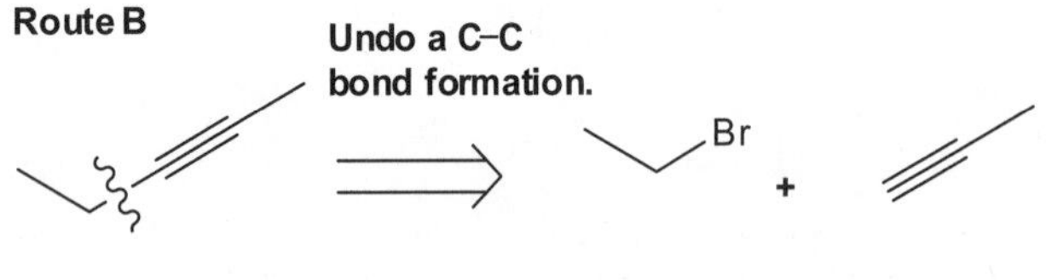

Undo a C–C bond formation.

Br

+

The syntheses in the forward direction would appear as follows:

Route A

1. NaH
2. CH_3Br

Route B

1. NaH
2. CH_3CH_2Br

Problem 13.28

Think

What reaction can be used to form an ether? What are the reagents that would be required? How can an alcohol be converted into an alkylbromide?

Solve

As shown in the retrosynthetic analysis below, the final ether can be produced using a Williamson synthesis, which requires an alkoxide anion and an alkyl halide. The alkoxide would be the phenoxide anion, which can be produced by deprotonating phenol. The alkyl halide can be produced from an alcohol via bromination with PBr_3.

Undo a Williamson synthesis. ONa + Br **Undo a bromination.** HO

In the forward direction, the synthesis would appear as follows.

OH NaH O

HO PBr_3 Br

Problem 13.29

Think

How many C atoms appear in your target compound? Did you have to alter the carbon skeleton? What functional group is in your target compound? What functional groups are in your starting compound? Did you have to do an elimination, addition, or substitution? How many steps are needed to accomplish your transformation? Is stereochemistry a concern? Regiochemistry?

Solve

(a) We can disconnect the carbon–carbon bond as shown.

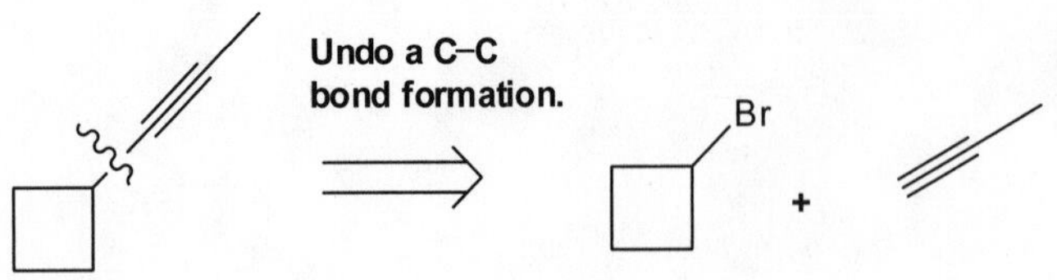

Undo a C–C bond formation.

Br

+

In the forward direction, the synthesis would appear as follows:

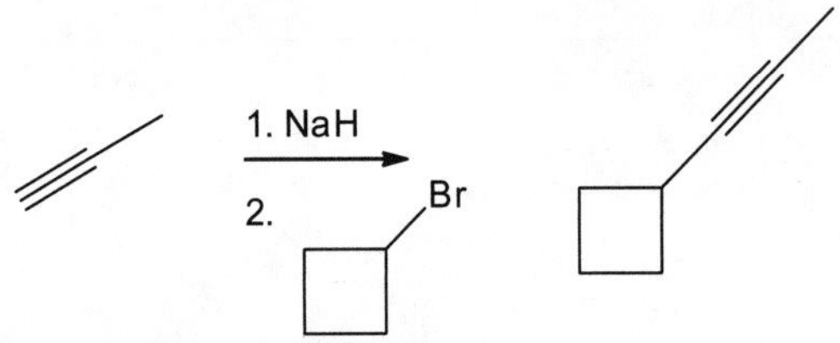

1. NaH

2.

Br

(b) An ether can be synthesized from an alcohol and an alkyl halide via a Williamson synthesis. Therefore, we need to reverse a Williamson synthesis as shown. The alkyl halide precursor can be obtained by brominating an alcohol. In each of these reactions, we must take into account the fact that an S_N2 reaction inverts the stereochemical configuration at the C atom with the leaving group.

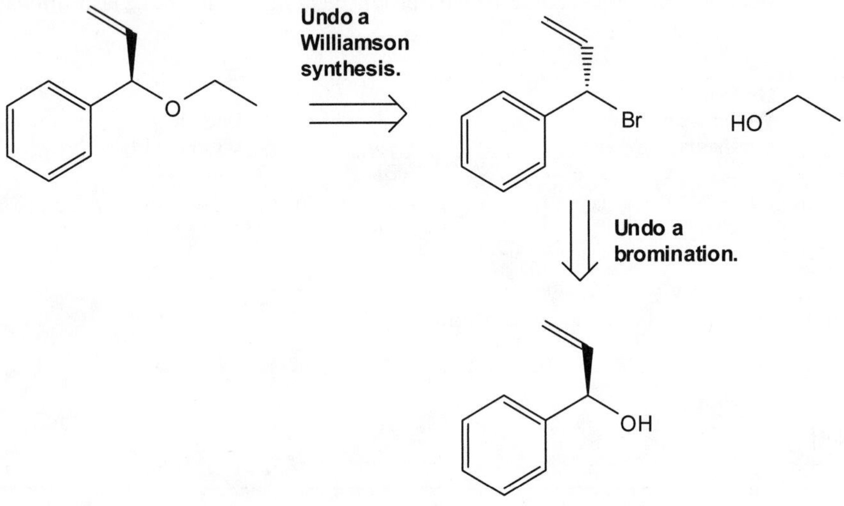

O

Undo a Williamson synthesis.

Br

HO

Undo a bromination.

OH

In the forward direction, we simply need to deprotonate the alcohol first, to convert it into an alkoxide nucleophile.

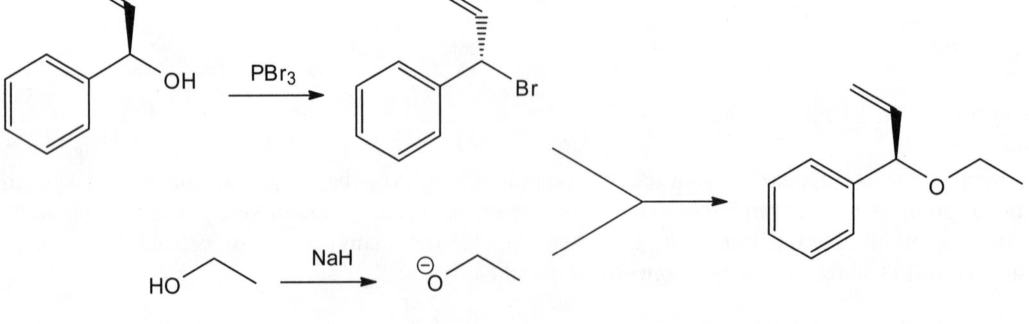

OH

PBr_3

Br

O

HO

NaH

$^{\ominus}$O

(c) We can disconnect two carbon–carbon bonds involving alkyne carbons to arrive at precursors with five or fewer carbons.

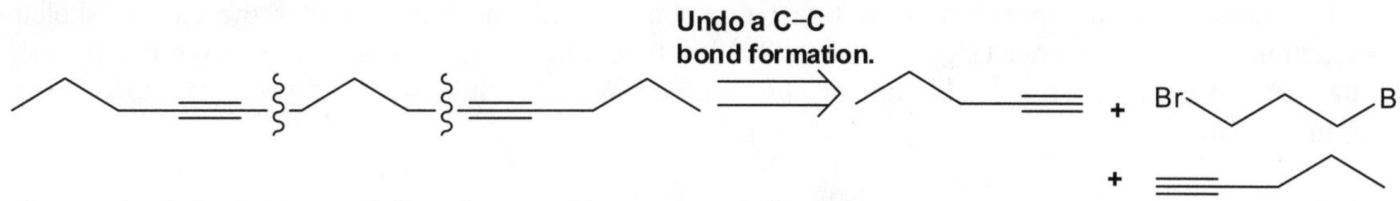

Undo a C–C bond formation.

Br Br

The synthesis in the forward direction would appear as follows:

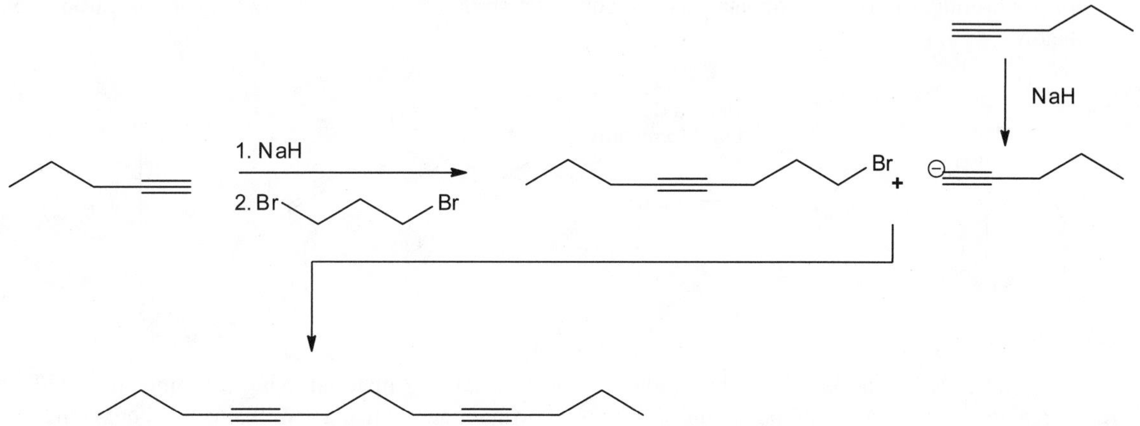

NaH

1. NaH
2. Br Br

Br

Problem 13.30

Think

In how many elementary steps does an E2 reaction occur? Is the Hofmann or Zaitsev product shown? On which C was the leaving group and on which C was the adjacent H? Is a bulky base needed? What is the configuration of the R groups in the alkyl halide starting material?

Solve

E2 occurs in one elementary step. The leaving group and the adjacent H must be anticoplanar. To arrive back at the alkyl halide, we must add H and a halogen on the alkene carbons in an anti fashion.

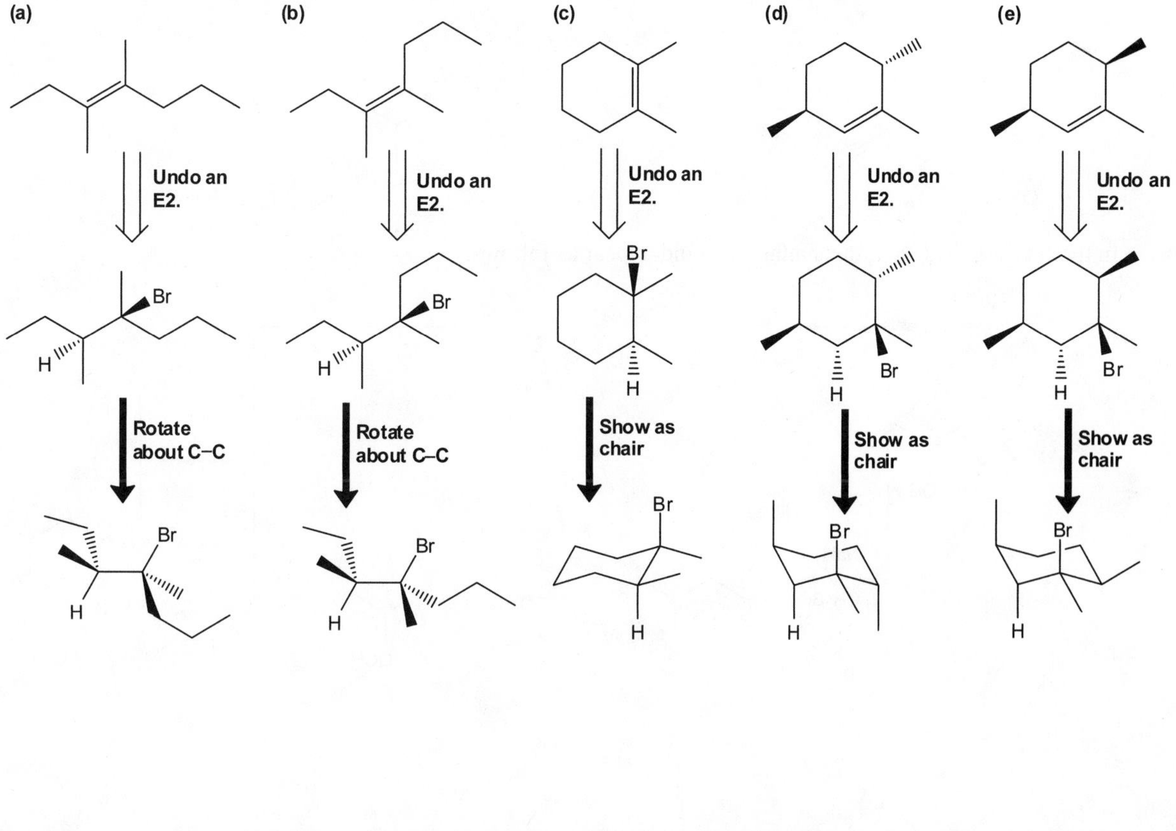

(a) Undo an E2. Br H Rotate about C–C Br H

(b) Undo an E2. Br H Rotate about C–C Br H

(c) Undo an E2. Br H Show as chair Br H

(d) Undo an E2. Br H Show as chair Br H

(e) Undo an E2. Br H Show as chair Br H

Problem 13.31

Think

How many C atoms appear in your target compound? Did you have to alter the carbon skeleton? What functional group is in your target compound? What functional groups are in your starting compound? Did you have to do an elimination, addition, or substitution? Will multiple steps be necessary to accomplish your transformation?

Solve

In performing a retrosynthetic analysis, we could get started by disconnecting the carbon–carbon bond shown below.

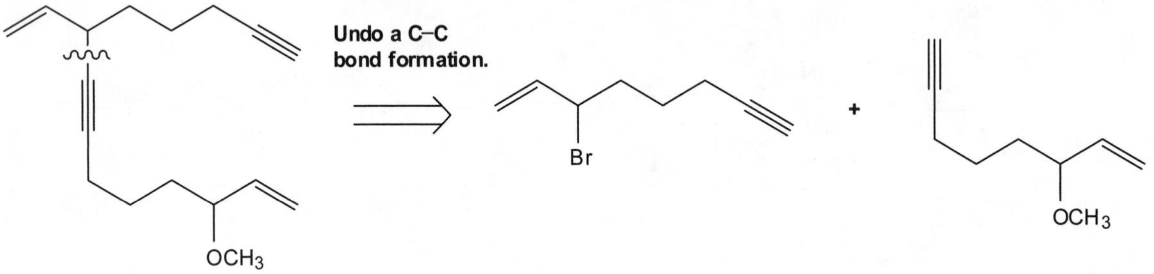

Undo a C–C bond formation.

Br

+

OCH_3

OCH_3

The precursor on the left could be produced from the starting material using a bromination with PBr_3. The precursor on the right could be produced from the starting material using a Williamson ether synthesis.

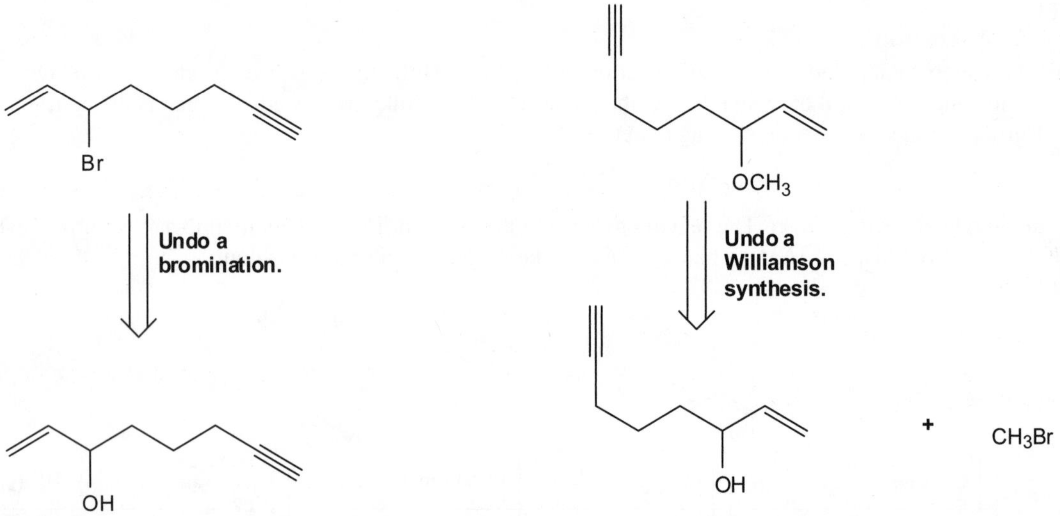

Br

OCH_3

Undo a bromination.

Undo a Williamson synthesis.

+ CH_3Br

OH

OH

In the forward direction, the synthesis would appear as follows:

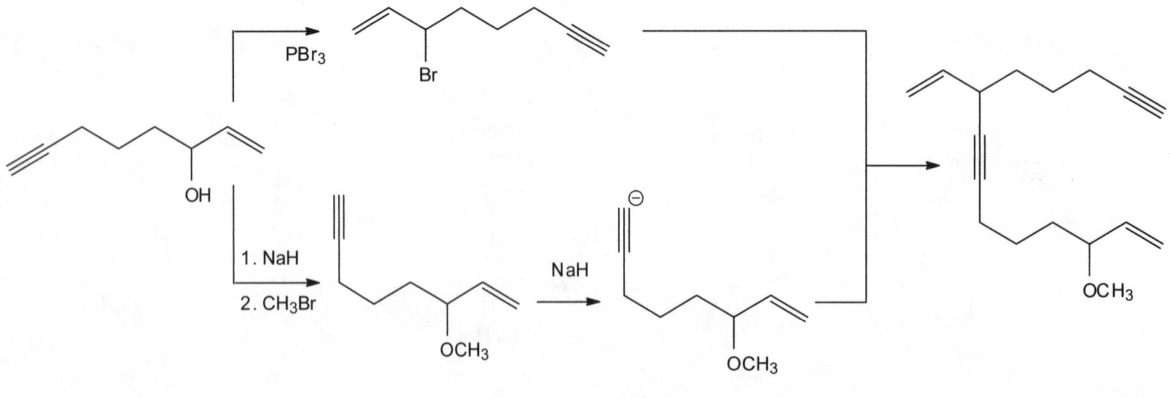

PBr_3

Br

OH

1. NaH

2. CH_3Br

NaH

OCH_3

OCH_3

OCH_3

Problem 13.32

Think

How many C atoms appear in your target compound? Did you have to alter the carbon skeleton? What functional group is in your target compound? What functional groups are in your starting compound? Did you have to do an elimination, addition, or substitution? Will multiple steps be needed to accomplish your transformation?

Solve

The retrosynthesis is shown below.

Undo an elimination.

OH

Undo an epoxide ring opening.

O

In the forward direction, the synthesis would appear as follows:

H—C≡— NaH → NaC≡— + H_2

1. (epoxide) O
2. H_3O^+

OH

H_2SO_4, Δ

Problem 13.33

Think

How many C atoms appear in your target compound? Did you have to alter the carbon skeleton? What functional group is in your target compound? What functional groups are in your starting compound? Did you have to do an elimination, addition, or substitution? Are multiple steps needed to accomplish your transformation? Is regiochemistry a concern?

Solve

(a) This is the product of elimination. That precursor can then be obtained from bromination of the corresponding methylcyclohexanone.

O

Undo an elimination.

O

Br

Undo a halogenation.

O

Undo an alkylation.

O

In the forward direction, we must be concerned with regiochemistry in the halogenation step. To ensure that bromination takes place at the more highly substituted α C and to prevent polyhalogenation, we must carry out the halogenation under acidic conditions.

O

CH_3Br, NaOH

O

Br_2, Acetic acid

O

Br

CH_3ONa, Δ

O

(b) The only difference from part **(a)** is in the regiochemistry of the halogenation step. Halogenation must instead take place at the less substituted α C, so we must choose a base to produce the kinetic enolate anion. LDA would work.

CH_3Br, NaOH → ; 1.LDA, 2. Br_2 → ; $(CH_3)_3CONa$, Δ →

Problem 13.34

Think

For a linear synthesis, how is the overall product yield related to the number of steps in the reaction? What is the percent yield of each step of the seven-step reaction? What is done to this percentage?

Solve

The overall yield of a linear synthesis is the product of the yields of each step.

% yield = (0.92)(0.84)(0.80)(0.80)(0.84)(0.91)(0.77) = 0.291 = 29.1%

Problem 13.35

Think

How can you disconnect the target to form two large pieces? Can those large pieces be produced from the same precursor? What is the percent yield for the synthesis of each of those large pieces? When you add the two large pieces together, what is the percent yield of that reaction? How do you calculate the overall yield?

Solve

To design a convergent synthesis, we can begin by disconnecting the target into two relatively large pieces.

SCH_3 — **Undo a C–C bond formation.** ⇒ Br + SCH_3

The first large piece is already shown in Problem 13.34 to be produced in three synthetic steps in a yield of (0.92)(0.84)(0.80) = 0.618 = 61.8%. We can perform a retrosynthetic analysis of the second precursor as follows.

SCH_3 — **Undo a substitution.** ⇒ Br + $NaSCH_3$ — **Undo a C–C bond formation.** ⇒ OH — **Undo an epoxide ring opening.** ⇒ +

In the forward direction, the total synthesis would appear as follows. If we assume the reactions that are similar to the ones in Problem 13.34 also have the same yields, the second precursor could be produced in a 51.7% yield. Taking this to be the limiting reagent, an 80% yield in the final step would give an overall yield of (0.517)(0.80) = 0.4136 = 41.4%. This is much higher than the 29.1% yield calculated in the previous problem.

84% 1. $NaNH_2$ 2. (epoxide) 3. HCl → OH — **80%** PBr_3 → Br — **77%** CH_3SH → SCH_3 **51.7%**

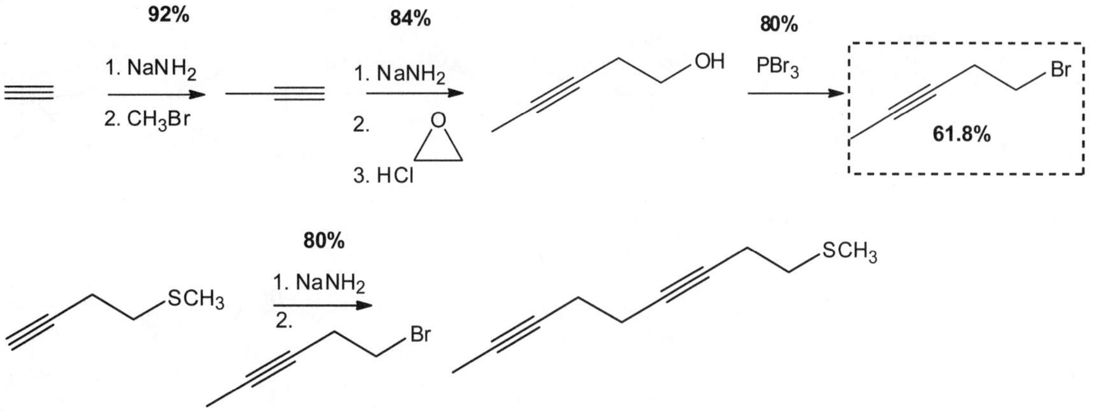

Problem 13.36

Think

How many C atoms appear in your target compound? Did you have to alter the carbon skeleton? What functional group is in your target compound? What functional groups are in your starting compound? Did you have to do an elimination, addition, or substitution? Are multiple steps needed to accomplish your transformation? Is stereochemistry a concern?

Solve

In performing a retrosynthetic analysis, we start by disconnecting one of the CN groups. The fact that the CN groups are anti suggests that the reaction at the end must be stereospecific and, thus, an S_N2 occurred. The full retrosynthesis is given below.

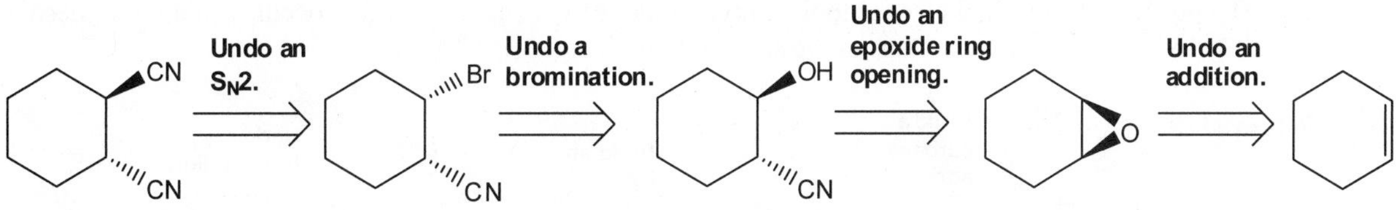

The synthesis in the forward direction is given below.

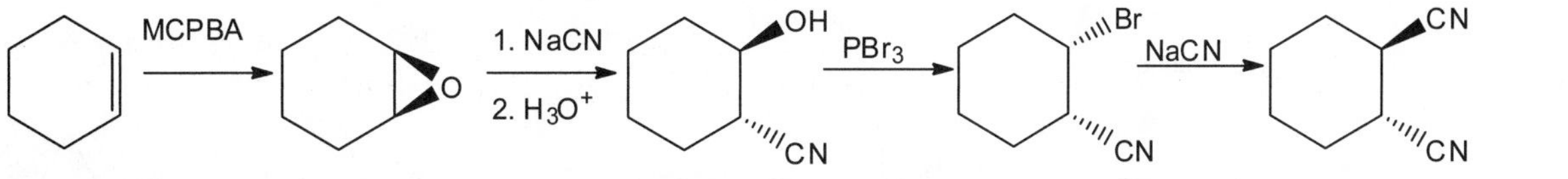

Problem 13.37

Think

How many C atoms appear in your target compound? Did you have to alter the carbon skeleton? What functional group is in your target compound? What functional groups are in your starting compound? Did you have to do an elimination, addition, or substitution? Are multiple steps needed to accomplish your transformation? Is stereochemistry a concern?

Solve

In performing a retrosynthetic analysis, we start by disconnecting one of the CN groups. The fact that the CN groups are syn suggests that the reaction at the end must be stereospecific and, thus, an S_N2 occurred. The full retrosynthesis is given on the next page.

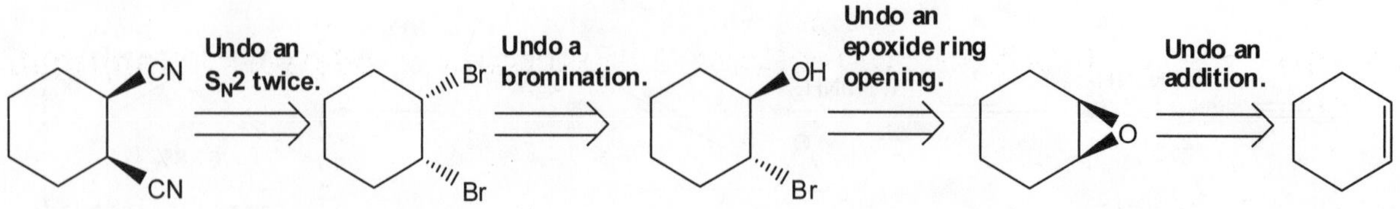

Undo an S_N2 twice. Undo a bromination. Undo an epoxide ring opening. Undo an addition.

The forward reaction is given below.

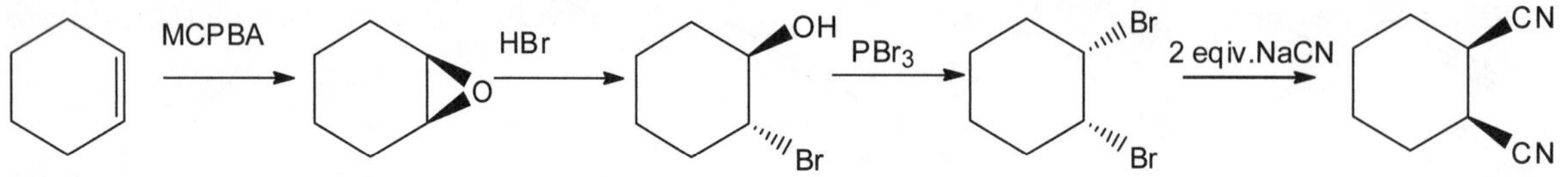

MCPBA → HBr → PBr_3 → 2 eqiv.NaCN

Problem 13.38

Think

How many C atoms appear in your target compound? Did you have to alter the carbon skeleton? What functional group is in your target compound? What functional groups are in your starting compound? Did you have to do an elimination, addition, or substitution? Are multiple steps needed to accomplish your transformation? What reactions can you use to obtain the desired stereochemistry?

Solve

In performing a retrosynthetic analysis where stereochemistry is considered, notice that the target compound has the methyl groups trans to each other. This indicates that, when we carry out a transform that undoes a carbene addition, the methyl groups also were trans to each other in the alkene precursor. E2 is a stereospecific reaction; the H and Br in the alkyl halide have to be anticoplanar, and the methyl groups also still need to be trans. To go from the original alcohol to the alkyl halide, PBr_3 is necessary; this occurs via an S_N2 mechanism, so the methyl groups need to be trans in the starting material.

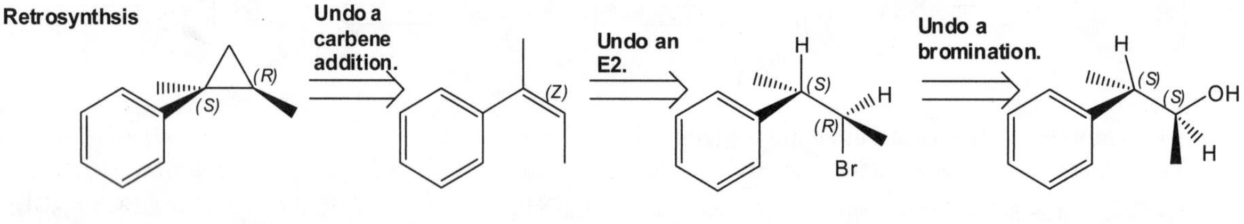

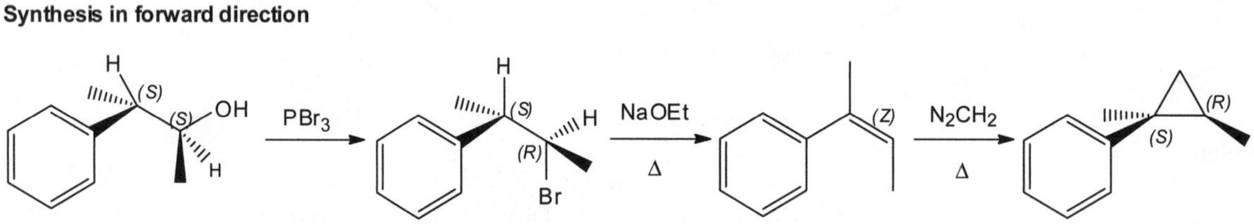

Retrosynthesis

Undo a carbene addition. Undo an E2. Undo a bromination.

Synthesis in forward direction

PBr_3 → NaOEt, Δ → N_2CH_2, Δ

CHAPTER 14 | Orbital Interactions 2: Extended π Systems, Conjugation, and Aromaticity

Your Turn Exercises

Your Turn 14.1

Think

From the valence bond (VB) picture shown below on the right, does the π bond occur between C1 and C2 or between C2 and C3? How do you represent a π bond in a Lewis structure? How is the π bond orbital different from an empty *p* orbital in the picture?

Solve

The resonance structure is shown below. The π bond orbital is shown by the overlap between two *p* orbitals and is spread out over C2 and C3. The π bond on the right suggests a double bond between C2 and C3. The empty *p* orbital belonging to C1 suggests that that is where the positive charge is located.

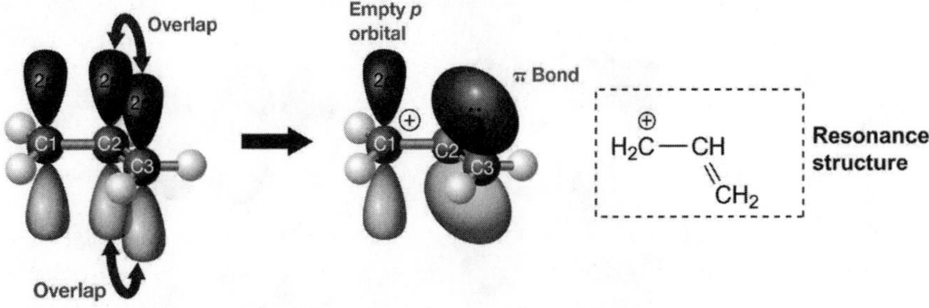

Your Turn 14.2

Think

From the phases of the *p* orbitals, how do you identify a nodal plane? Does constructive or destructive interference result from that type of interaction between those *p* orbitals? In the resulting molecular orbital (MO), what characterizes a nodal plane?

Solve

The nodes are indicated by a vertical dashed line in each MO below. There is one nodal plane in π_2 and two nodal planes in π_3. Notice that on either side of each nodal plane, the contributing orbitals have opposite phase. This results in destructive interference. In the resulting MOs, there is complete cancellation at each nodal plane, indicating that an electron in the orbital has zero probability of being found there.

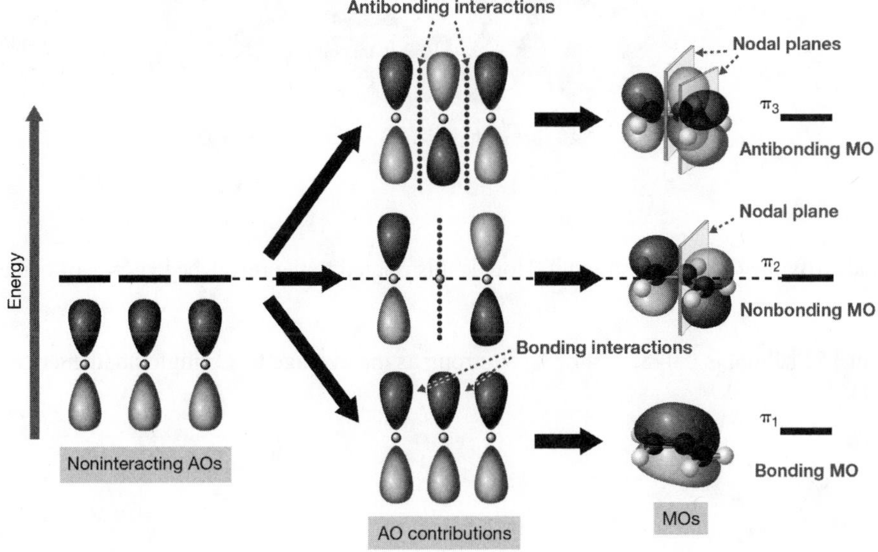

Your Turn 14.3

Think

Do adjacent orbitals with the same phase give rise to a bonding interaction or an antibonding one? Once you identify these interactions, consider what the net interaction is relative to the *p* AO. For example, one antibonding and one bonding interaction have a net interaction that essentially leaves the energy unchanged from the *p* AO (nonbonding), and two antibonding and one bonding have a net of one antibonding (higher in energy relative to the *p* AO).

Solve

Adjacent orbitals with the same phase give rise to a bonding interaction and those with opposite phase give rise to an antibonding interaction. In this case, all the antibonding interactions among AOs are indicated by the presence of a nodal plane (dashed line). The bonding interactions are indicated below by dashed ovals. Notice that π_1 has three bonding interactions and no antibonding interactions (net: three bonding); π_2 has two bonding interactions and one antibonding interaction (net: one bonding); π_3 has one bonding interaction and two antibonding interactions (net: one antibonding); and π_4 has no bonding interactions and three antibonding interactions (net: three antibonding). Each increase in antibonding interactions leads to a higher energy MO.

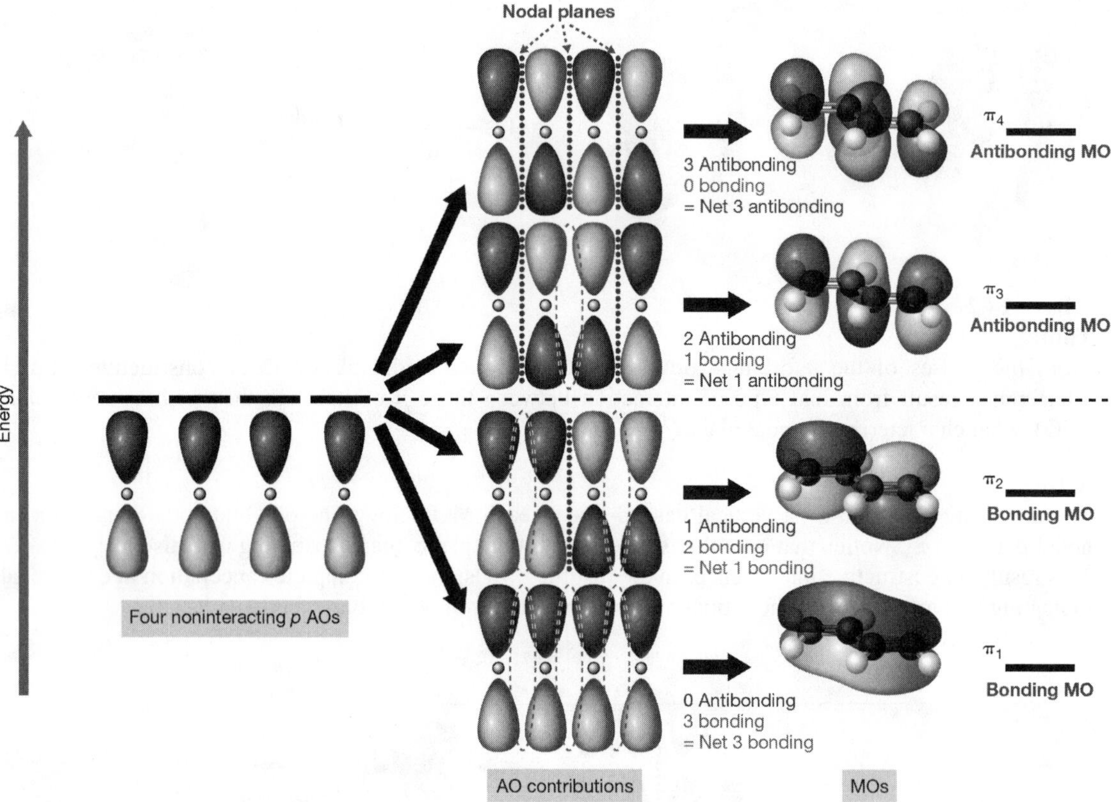

Your Turn 14.4

Think

Consult Table 1-2 and identify the C–C bond strength. How is this related to the resonance energy of benzene?

Solve

The resonance energy of 152 kJ/mol is 0.448, or 44.8%, as strong as the average C–C single bond energy of 339 kJ/mol (from Chapter 1).

Your Turn 14.5

Think

Consult Figure 14-2 and Figure 14-17 for bond lengths. If cyclobutadiene was significantly resonance stabilized, would you expect the single bond lengths in this molecule to be significantly longer or shorter than an isolated C–C?

Solve

If cyclobutadiene were significantly resonance stabilized, we would expect the C–C single bond to have significant double-bond character and be substantially shorter than a typical single bond. The bond length values are 132 pm for a normal C=C bond and 154 pm for a normal C–C bond. The single bonds in cyclobutadiene are 4% longer than average and the double bonds in cyclobutadiene are the same length as in ethylene. Overall, the numbers are very similar, suggesting that the electrons are *not* significantly resonance delocalized.

Your Turn 14.6

Think

Relative to the *p* AO energies, are bonding MOs lower energy, higher energy, or the same energy? Antibonding MOs? Nonbonding MOs? For orbitals to be degenerate, what is required of their energy levels?

Solve

Bonding orbitals are lower in energy than the *p* AOs (dotted line), whereas antibonding are higher in energy. There are two pairs of degenerate orbitals, which have the same energy (circled).

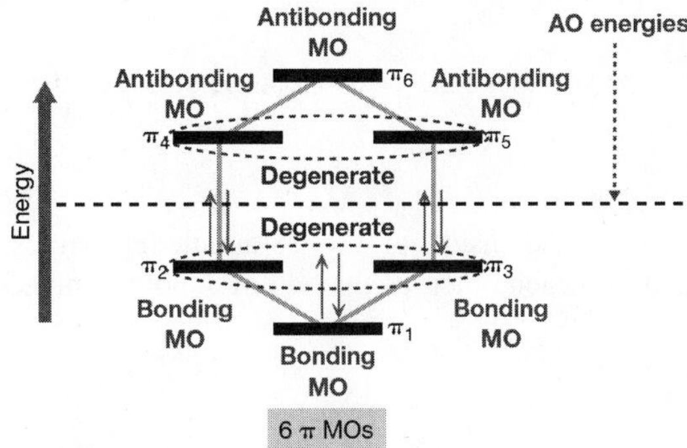

Your Turn 14.7

Think

How many nodal planes are there in each of the first five π MO energy levels? What happens to the number of nodal planes with each increase in MO energy? Is π_6 the highest or lowest energy orbital?

Solve

Number of nodes: $\pi_1 = 0$; $\pi_2 = 1$; $\pi_3 = 1$; $\pi_4 = 2$; $\pi_5 = 2$; $\pi_6 = 3$ (highest energy, greatest number of nodal planes). This is because π_6 is higher in energy than π_5 and should have one additional node. Two of the nodal planes are in the same locations as the ones in Figure 14-20d. The third is the one shown in Figure 14-20e that is parallel to the plane of the paper.

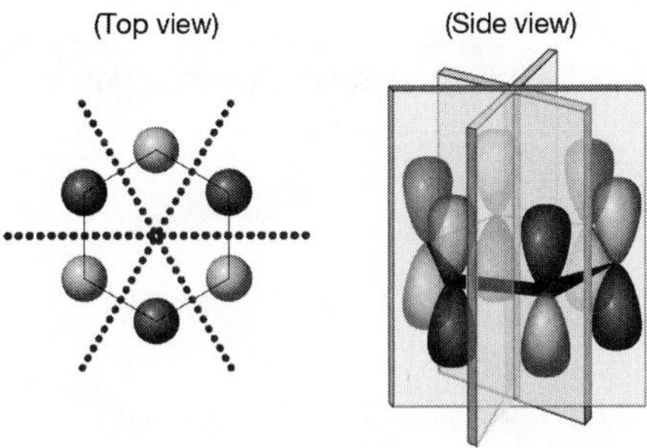

Your Turn 14.8

Think

Relative to the *p* AO energies, are bonding MOs lower energy, higher energy, or the same energy? Antibonding MOs? Nonbonding MOs?

Solve

As shown below, there is one bonding MO (π_1), two nonbonding MOs (π_2 and π_3), and one antibonding MO (π_4). Bonding orbitals are lower in energy than the *p* AOs (dotted line), whereas antibonding MOs are higher in energy and nonbonding MOs appear at the same energy as the contributing AOs.

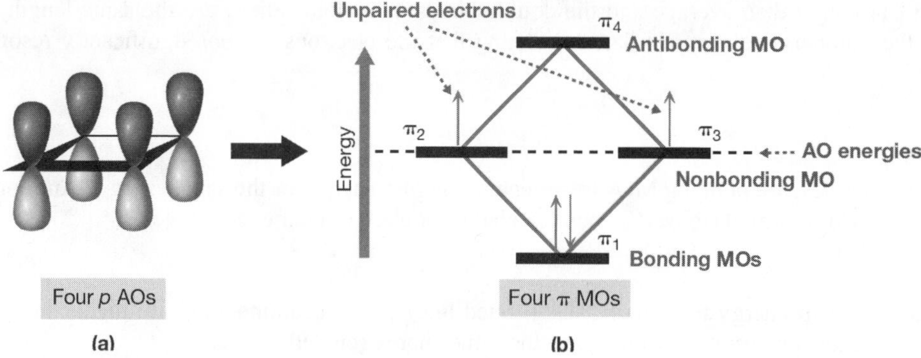

Your Turn 14.9

Think

After you draw in your C–H bonds, how can you represent the possible steric strain of the H atoms? How do you think the ring size affects that steric strain?

Solve

The H atoms are included in the figure below. The inside of the 18[annulene] ring is large enough to accommodate the H atoms so they do not crash into each other, as they do in the other molecules in Figure 14-24.

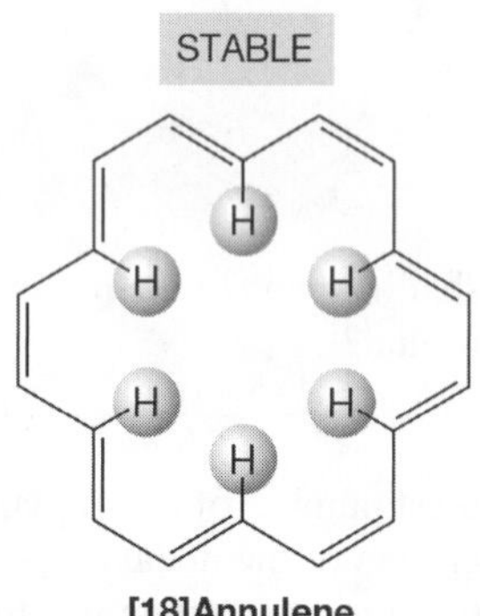

Your Turn 14.10

Think

How can you connect the shaded lobes to draw one closed loop of orbitals? How can you view the fused ring in a single aromatic π system?

Solve

The dotted lines indicate a closed loop of orbitals. Each of these, therefore, is viewed as a single aromatic π system.

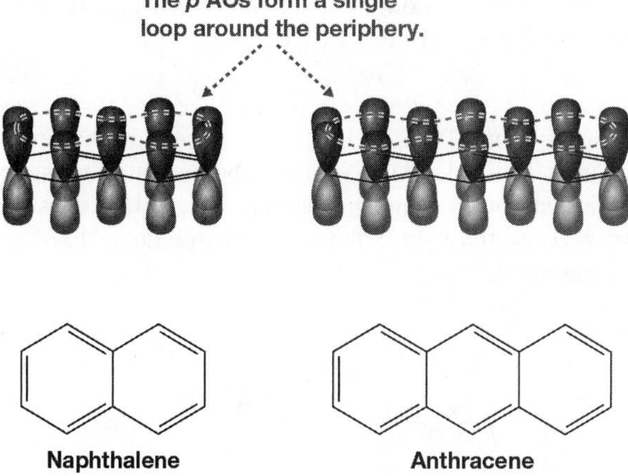
The *p* AOs form a single loop around the periphery.
Naphthalene
Anthracene

Your Turn 14.11

Think

For naphthalene and anthracene, can you identify a ring of alternating double and single bonds that does not include all π electrons of the system? After you identify such a ring, how can the π electrons be shifted to arrive at a new resonance structure?

Solve

In naphthalene, there is a six-carbon ring of alternating single and double bonds and the six π electrons can be shifted around the ring to arrive at the new resonance structure shown below. In anthracene, there is a 10-carbon ring of alternating single and double bonds and the 10 π electrons can be shifted around the ring to arrive at a new resonance structure. In each new resonance structure, the π electrons appear to make two separate rings, but the *p* orbitals remain in the same locations, so there is still one complete loop of *p* orbitals on the periphery of the molecule.

Naphthalene Anthracene

Chapter Problems

Problem 14.1

Think

What does the VB picture for the allyl cation look like in Your Turn 14.1? What is the hybridization of each C atom in the allyl cation? How does the VB picture of the allyl anion compare to the VB picture of the allyl cation? Draw the resonance hybrid. Is the negative charge localized or delocalized?

Solve

The VB picture is identical to that of the allyl cation in Your Turn 14.1 with the exception that in the allyl anion, there is a lone pair of electrons occupying the unhybridized *p* AO. As shown below, this corresponds to a C=C double bond involving two carbons, with the third carbon bearing a −1 formal charge. This does *not* agree with the resonance hybrid shown below at the right, because the hybrid shows the charge delocalized and both C–C bonds as having partial single and double bond character.

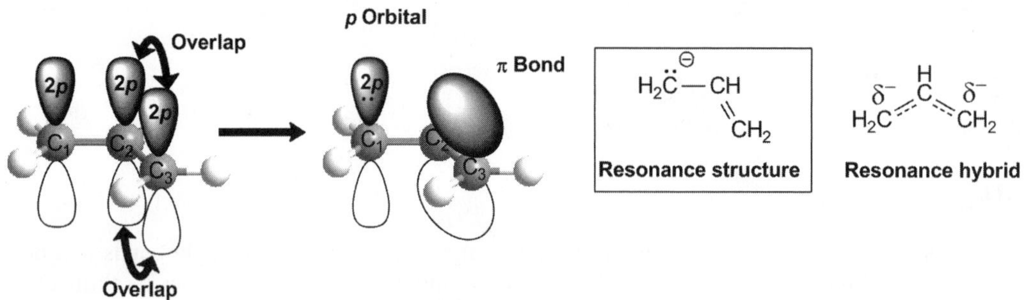

Problem 14.3

Think

How many *p* orbitals are conjugated? How many π MOs should result? How do those π MOs differ from each other? How many total σ and σ* MOs should there be? How many total valence electrons are there and how should they be arranged in the orbitals?

Solve

As shown below, the MO energy diagram is identical to that for the allyl anion shown in Solved Problem 14.2, with the exception that the allyl radical has one fewer electron to make the species uncharged overall. So there is only one electron in π_2 instead of two. That leaves π_2 as the HOMO and π_3 as the LUMO.

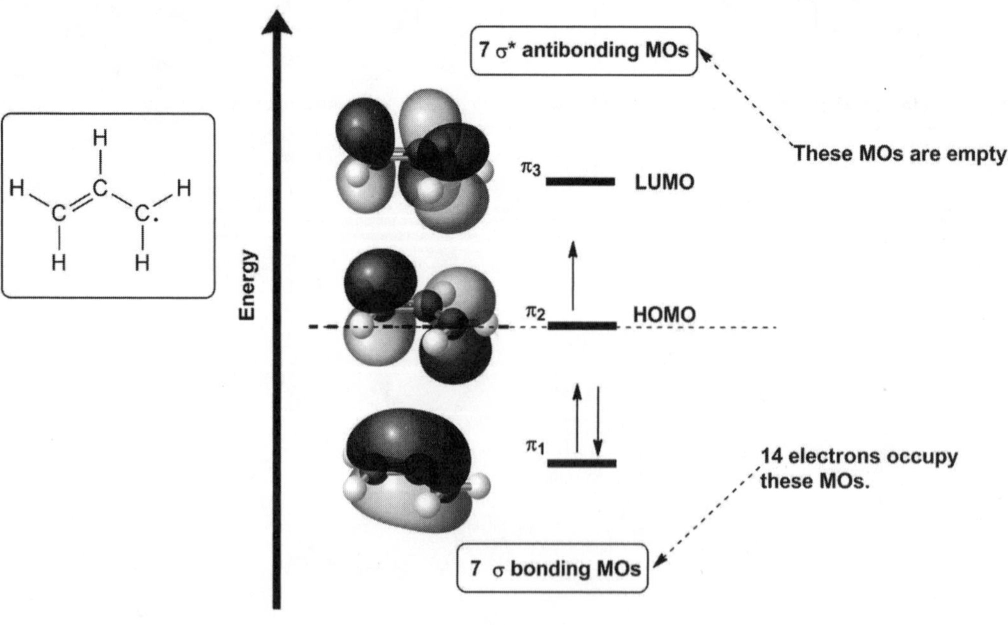

Problem 14.5

Think

What does the MO energy diagram look like if one electron from π_1 is promoted to π_2? Does an electron occupying π_1 contribute to bonding character between the carbon atoms? Does an electron occupying π_2 contribute to bonding character between the carbon atoms? When π_2 gains an electron, on which carbon atoms does that electron largely reside?

Solve

The energy diagram for the ground-state electron configuration is shown at left below, and the result of promoting the electron from π_1 to π_2 is shown at right. Notice that an electron in π_1 is delocalized over both bonding regions, whereas an electron in π_2 is more localized on the terminal carbon atoms. So removing an electron from π_1 and adding it to π_2 will decrease the bonding character and increase the bond length between the carbon atoms. Also, adding an electron to π_2 will increase electron density (i.e., some negative charge) on the terminal carbons and will somewhat decrease the concentration of positive charge on the terminal carbons.

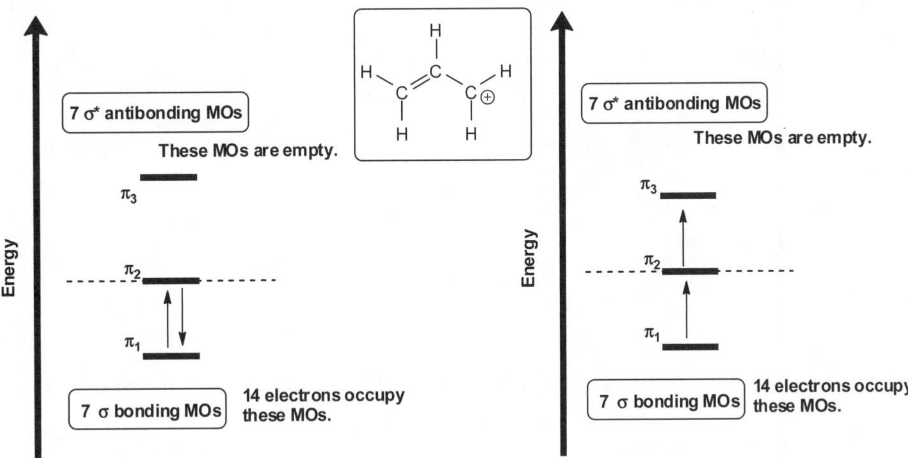

Problem 14.6

Think

Is a π-bonding interaction represented by constructive or destructive interference? Locate each type of interference in the structure drawn. Where are the nodal planes? Do bonding contributions raise or lower the π MO energy relative to the *p* AO? Antibonding? Count up the number of each type of interaction. What is your net?

Solve

The nodes are drawn below between *p* AOs undergoing destructive interference (opposite phases of *p* orbitals). These represent antibonding interactions which raise the MO energy relative to the *p* AOs. Constructive interference has the same phase of *p* orbitals. These represent bonding interactions that lower the MO energy relative to the *p* AOs. There are three bonding interactions and two antibonding interactions and, thus, a net of one bonding interaction (one more bonding than antibonding). The orbital, therefore, is a bonding MO.

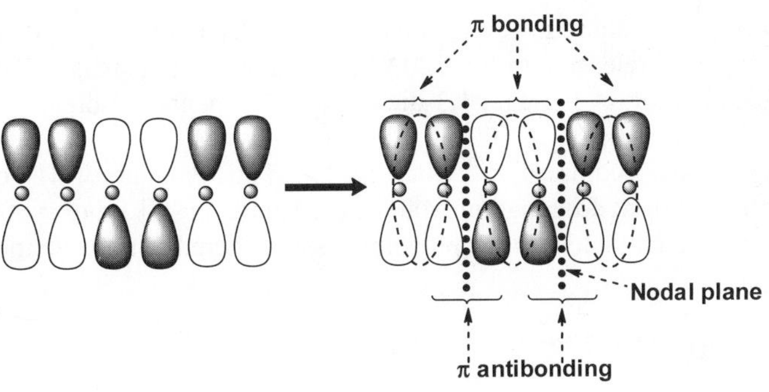

Problem 14.7

Think

If all four carbon atoms are sp^2 hybridized, how many *p* orbitals exist? How many π MOs, therefore, must exist? How many σ and σ* MOs are present? How many total valence electrons does the butadienyl dication have? How many electrons are in the σ MOs? How many are left? How does this compare to the MO energy diagram for butadiene?

Solve

The MO energy diagram of the butadienyl dication is similar to that for 1,3-butadiene in Figure 14-12. The difference is that the butadienyl dication has two fewer electrons (20 valence electrons total). Therefore, the π_1 MO ends up doubly occupied, but π_2 ends up empty. This makes π_1 the HOMO and π_2 the LUMO, as indicated below.

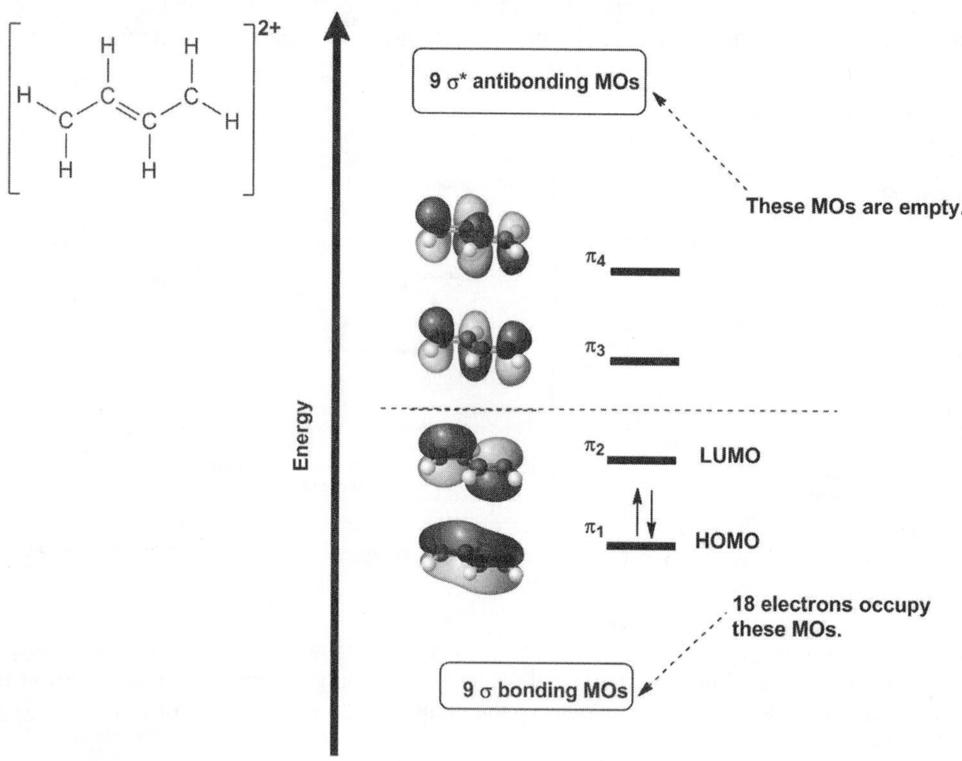

Problem 14.9

Think

How many moles of H_2 are required to completely hydrogenate cyclopenta-1,3-diene and naphthalene? How much heat is released when that number of moles of hydrogen reacts with cyclohexene instead? How does that quantity compare to the heat of hydrogenation for cyclopenta-1,3-diene and naphthalene?

Solve

Hydrogenation of one mole of cyclopenta-1,3-diene would require two moles of H_2. This amount of H_2 can also hydrogenate two moles of cyclohexene, which would release 2 mol × 120 kJ/mol = 240 kJ/mol. This is only 29 kJ/mol different from the heat of hydrogenation of cyclopenta-1,3-diene, so cyclopenta-1,3-diene is not considered to be aromatic.

Hydrogenation of one mole of naphthalene would require five moles of H_2. This amount of H_2 could hydrogenate five moles of cyclohexene, whose heat of hydrogenation would total 5 mol × 120 kJ/mol = 600 kJ. This is nearly 300 kJ more heat than is required by naphthalene, so naphthalene is considered to be aromatic.

Problem 14.11

Think

Is there a ring of conjugated *p* AOs? Are there any sp^3 C atoms that break up the conjugated π system? Is the structure planar? If any of these questions is answered negatively, how is the structure categorized? If all of the questions are answered affirmatively, how many electrons are present in the cyclic π system? Does it contain a Hückel or anti-Hückel number of π electrons?

Solve

(a) Nonaromatic; although there are six total π electrons (a Hückel number), the π system is not completely cyclic due to the presence of four sp^3 C atoms (labeled).

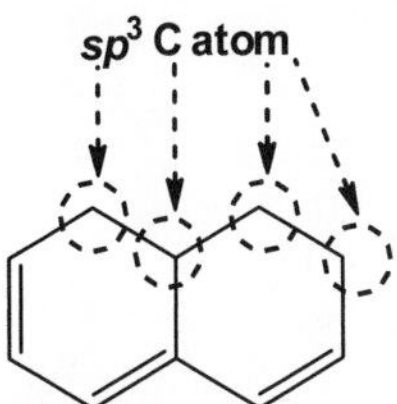

6 π electrons
Nonaromatic

(b) Aromatic; six π electrons (a Hückel number) are completely conjugated in a ring. Even though there are sp^3 C atoms present, these are outside the ring and not part of the conjugated π system.

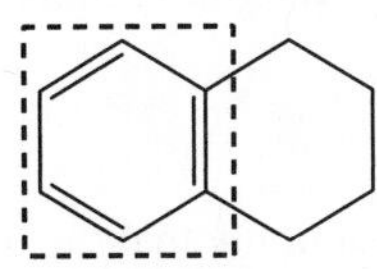

6 π electrons
Aromatic

(c) Nonaromatic; the π system is not conjugated in a complete ring because of the sp^3-hybridized C atom.

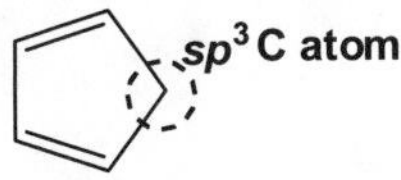

4 π electrons
Nonaromatic

(d) Nonaromatic, for the same reason as in **(c)**.

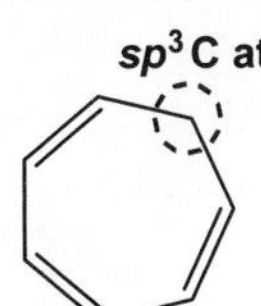

6 π electrons
Nonaromatic

(e) Nonaromatic; there are two sp^3-hybridized C atoms that prevent conjugation in a complete ring.

sp^3 C atom

2 π electrons
Nonaromatic

(f) Nonaromatic; the π system is not cyclic.

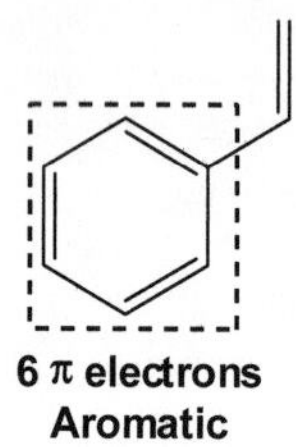

(g) Aromatic; although there are eight total π electrons, they are in two different π systems—one that is part of the benzene ring, which has six π electrons completely conjugated around a ring, and the C=C double bond that has two π electrons.

Problem 14.12

Think

How many valence electrons are in benzene, C_6H_6? How many σ bonds are present in benzene? How many σ and σ* MOs do they represent? Where should the σ and σ* MOs appear in the energy diagram relative to the π MOs for benzene (Fig. 14-19)? How many electrons should fill the σ MOs?

Solve

The diagram is shown below. There are six C–C σ interactions and six C–H σ interactions, making 12 total σ MOs and 12 total σ* MOs. The π MOs from Figure 14-19 should appear between the σ and σ* MOs. The six C atoms contribute four valence electrons each; the six H atoms contribute one electron each. The 30 valence electrons are distributed into 12 σ MOs and three π MOs. The π antibonding MOs are empty. The existence of the all paired π electrons in the bonding MOs makes benzene very stable.

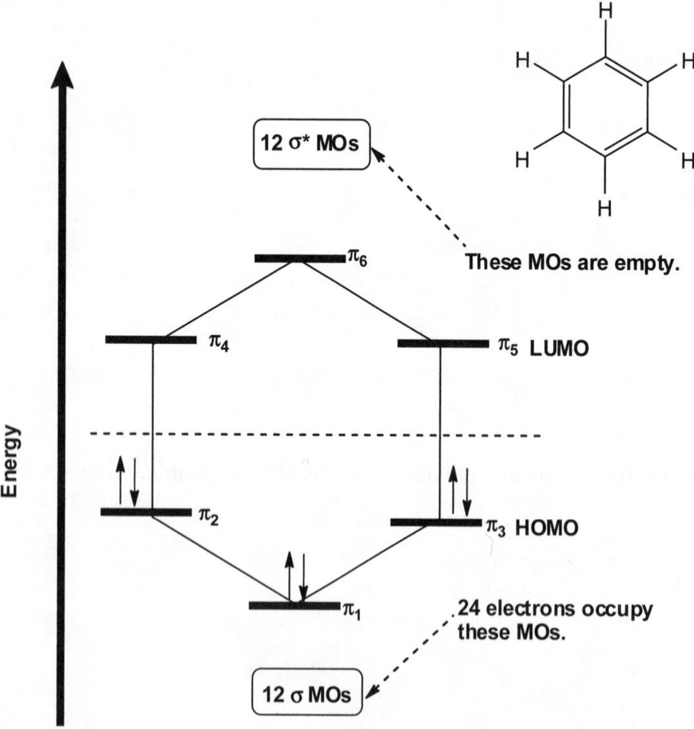

Problem 14.13

Think

How many valence electrons are in cyclobutadiene, C_4H_4? How many σ bonds are present in cyclobutadiene? How many σ and σ* MOs do they represent? Where should the σ and σ* MOs appear in the energy diagram relative to the π MOs for cyclobutadiene (Fig. 14-21)? How many electrons should fill the σ MOs?

Solve

The diagram is shown below. There are four C–C σ interactions and four C–H σ interactions, making eight total σ MOs and eight total σ* MOs. The π MOs from Figure 14-21 should appear between the σ and σ* MOs. The four C atoms contribute four valence electrons each; the four H atoms contribute one electron each. The 20 valence electrons are distributed into eight σ MOs, one π MO, and two π nonbonding MOs. The existence of the unpaired π electrons in the nonbonding MOs makes cyclobutadiene unstable. (Cyclobutadiene distorts from square to rectangular which lowers the energy of one non-bonding orbital and raises the energy of the other in order to become more stable.)

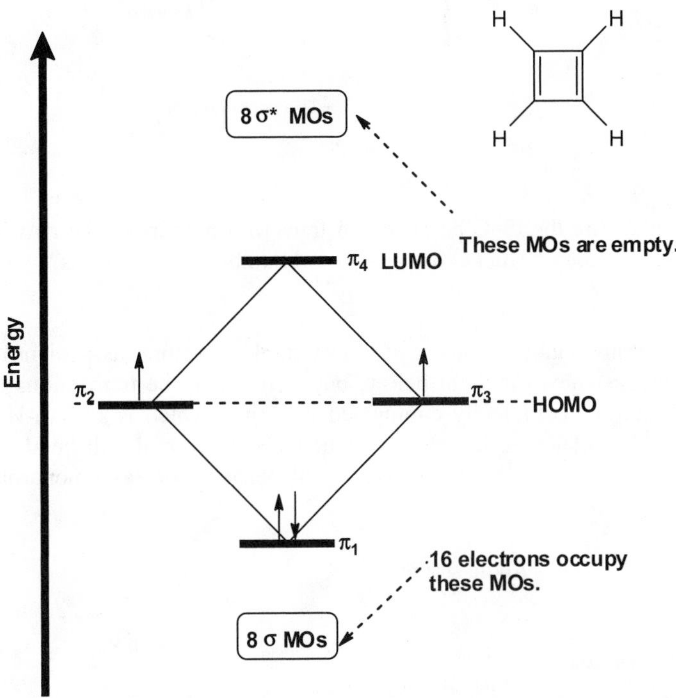

Problem 14.15

Think

Do the benzene dication and benzene dianion have cyclic and fully conjugated π systems? Can the Frost method be applied to these species to derive their π MO energy diagrams? How many π electrons are present? How do you fill in the electrons in the MO diagram? What does the HOMO look like? Is it filled?

Solve

The energy diagrams for the π systems of the two species are shown on the next page. We begin with the energy diagram from Figure 14-19 (Frost method), in which there are six π electrons. The dication, therefore, must have four π electrons. Hund's rule says that one should be in π_2 and the other in π_3. Even though all electrons are in bonding MOs, the instability of the unpaired electrons makes the dication antiaromatic. By the same token, the dianion should be antiaromatic. It has eight π electrons. The three lowest-energy π MOs are completely filled, but π_4 and π_5 each have one unpaired electron.

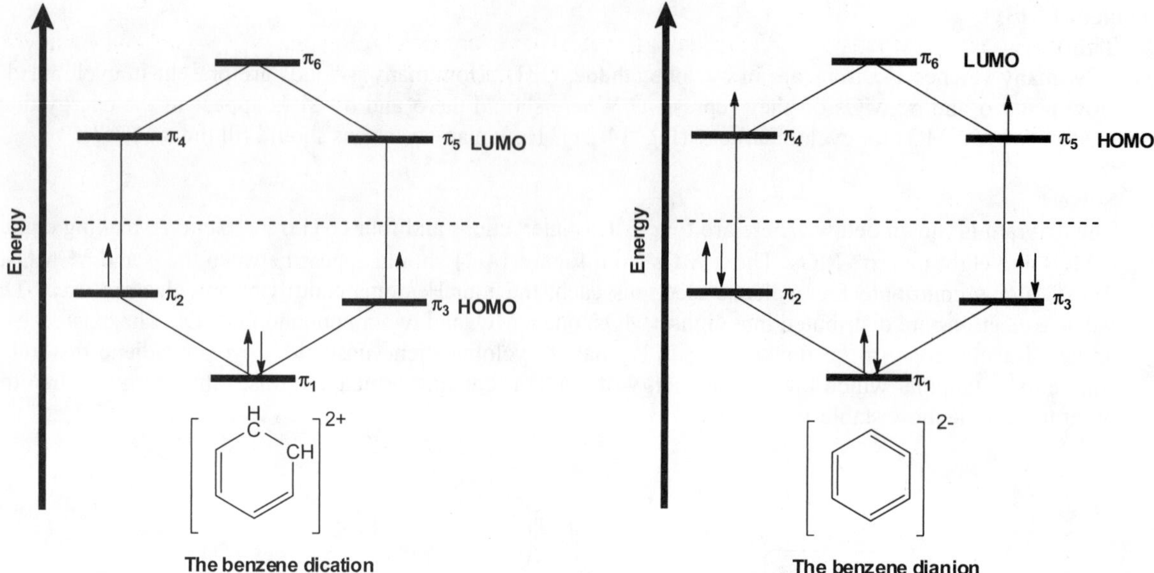

Problem 14.16

Think

How many carbon atoms are in [16] annulene? Are the C=C bonds cis or trans or a mixture of the two? How many π electrons are in [16] annulene? Does it contain a Hückel or anti-Hückel number of π electrons?

Solve

[16]annulene will have some cis and some trans double bonds in its most stable structure, as shown at left below. The structure can be drawn as all cis, as shown at right below, but there would be tremendous angle strain. Either way, however, shows 16 π electrons completely conjugated in a ring, which is an anti-Hückel number, making it antiaromatic, so long as it is planar. Like cylcooctatetraene, however, it will bend out of plane to avoid the complete conjugation of the *p* AOs, in which case it will behave more as a nonaromatic compound.

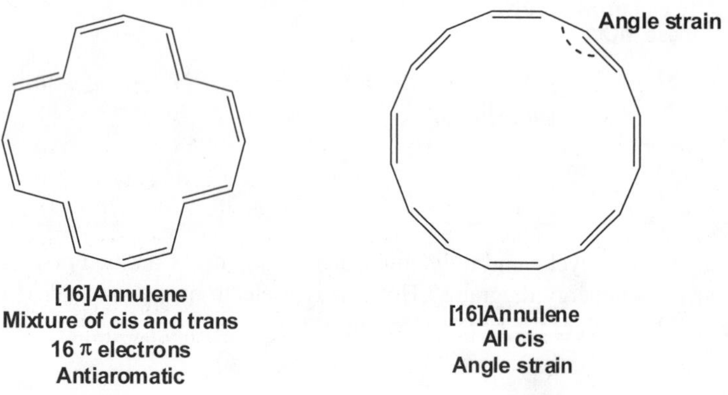

Problem 14.18

Think

Does the molecule contain a π system formed from a ring of fully conjugated *p* orbitals? How many electrons are in that π system?

Solve

(a) Antiaromatic; the outside of the fused ring system has a completely conjugated π system, and contains eight π electrons, which is an anti-Hückel number (an even number of pairs).

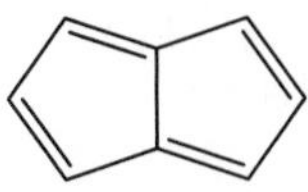

8 π electrons
Antiaromatic

(b) Aromatic; the outside of the fused ring system is a completely conjugated π system, which contains 14 electrons, a Hückel number (an odd number of pairs).

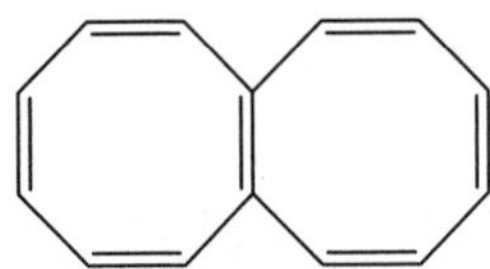

14 π electrons
Aromatic

(c) Nonaromatic; the π system is not completely conjugated around the outside of the fused ring system, due to the presence of two sp^3 C atoms.

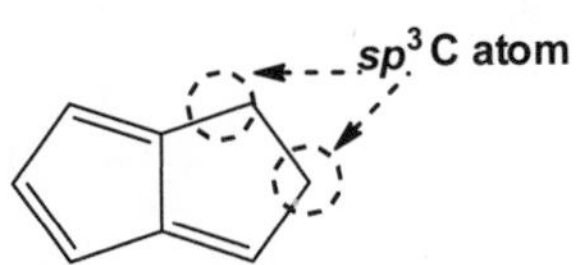

6 π electrons
Nonaromatic

Problem 14.20

Think

How many lone pairs from the heteroatom are part of the π system? How many total π electrons are there?

Solve

The answers are given below and on the next page. For each that is aromatic or antiaromatic, the electron pairs that are part of the π system are indicated by arrows. In **(a)** and **(c)**, the O is sp^2 hybridized, so one lone pair is part of the π system and the other lone pair is not (it is perpendicular). In **(b)**, the lone pair on N is not part of the π system, because it resides in an sp^2 orbital that is perpendicular to the π system. In **(f)**, the lone pair is part of the π system, because the sp^2-hybridized orbitals, used to make the three bonds to N, are in the plane perpendicular to the π system. In **(g)**, there is a set of *p* AOs conjugated around the benzene ring only, composing a system that has just the three π bonds, for six total π electrons (the other double bond and lone pair involving the N are not part of that π system). The remaining ones, **(d)** and **(e)**, are nonaromatic because the π systems do not form a complete ring. **(c)** would be antiaromatic if planar, but, like cyclooctatetraene, it bends out of plane to avoid being antiaromatic. So it is more accurately described as nonaromatic.

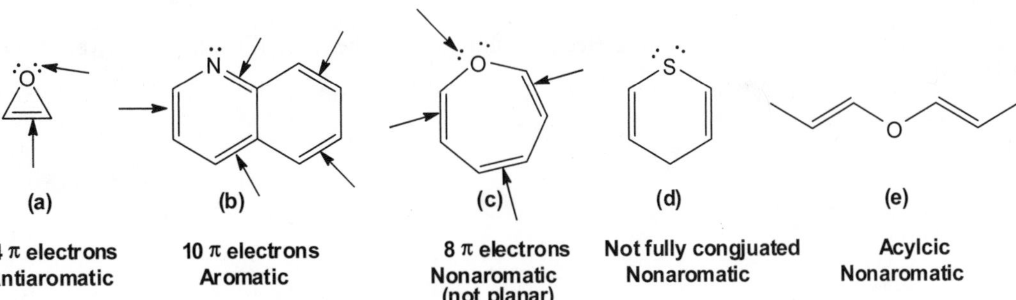

(f)
10 π electrons
Aromatic

(g)
6 π electrons
Aromatic

Problem 14.21

Think

What is the hybridization of the C^+ atom? Draw in any lone pairs not shown. How many lone pairs from the heteroatom are part of the π system? How many total π electrons are there?

Solve

Cation **(a)** is aromatic, with two π electrons and a completely conjugated ring of *p* AOs (all C atoms are sp^2 hybridized). Molecule **(b)** is antiaromatic, with four π electrons (two π electrons from the double bond, and two π electrons from the pair of electrons shown on N).

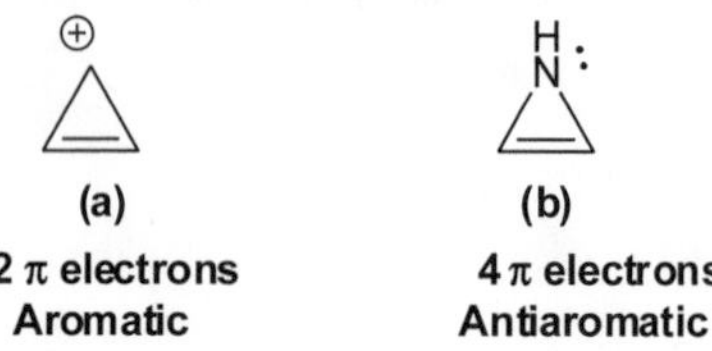

(a)
2 π electrons
Aromatic

(b)
4 π electrons
Antiaromatic

Problem 14.22

Think

How many double bonds are there? How many π electrons are in each C=C? Can all those electrons be shifted simultaneously in resonance?

Solve

The system contains one π system because a resonance structure can be drawn in which all the π electrons are shifted at once, as shown below. That system contains 22 electrons—two for each C=C double bond shown.

β-Carotene
22 π electrons
1 π system

Problem 14.24

Think

Can we draw a resonance structure by shifting electrons fully around the ring? If so, how many electrons must be shifted?

Solve

If the molecule is planar, it would be antiaromatic because the π system that is completely conjugated in a ring contains 12 electrons. This is indicated by the resonance structures below.

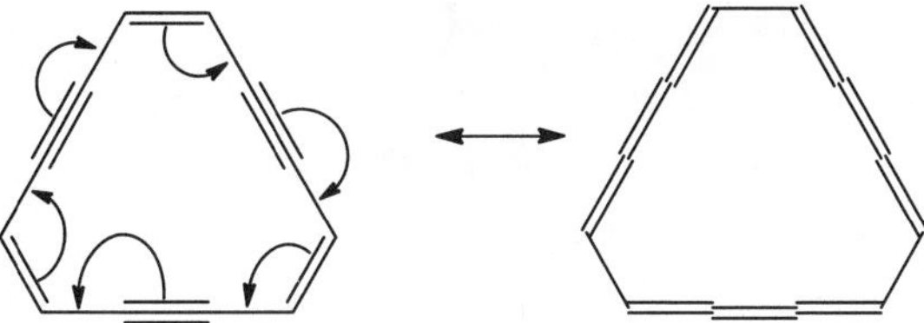

Problem 14.25

Think

How many lone pairs from the heteroatom are part of the π system? How many total π electrons are there? How many electrons can be shifted completely around the ring via resonance?

Solve

(a) and **(b)** are both aromatic because the π systems that are conjugated in a ring each contain six total electrons, which is a Hückel number. In both molecules, the lone pairs on N are not part of the π system, evidenced by the fact that resonance structures can be drawn that shift electrons around the ring, and the lone pairs are unaffected. An example is shown below with **(a)**.

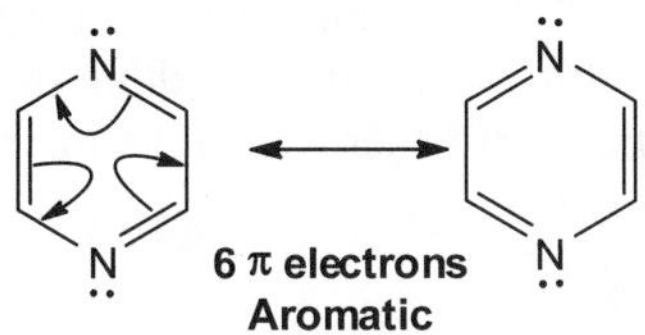

(c) is also aromatic. The lone pair on the N atom with three single bonds shown is part of the π system, indicated by the fact that a resonance structure can be drawn in which that lone pair and the electrons from the double bonds are shifted around the ring. The lone pair on the doubly bonded N atom are untouched, so they are not part of the π system. Thus, there are six total π electrons in the ring, which is a Hückel number.

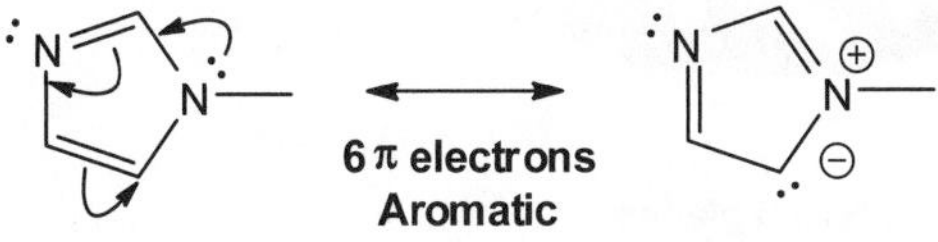

Problem 14.26

Think

From the phases of the *p* orbitals, how do you identify a nodal plane? Does constructive or destructive interference result from that type of interaction between those *p* orbitals? In the resulting MO, what characterizes a nodal plane? Do bonding contributions raise or lower the π MO energy relative to the *p* AO? Antibonding? Count the number of each type of interaction. What is your net?

Solve

(a) The nodes are indicated by a vertical dashed line in each MO below, where there is complete cancellation of the MO. There are three nodal planes in **A**, two nodal planes in **B**, and four nodal planes in **C**. Energy increases with increasing number of nodes; therefore, the order of MOs in increasing energy is **B** < **A** < **C**.

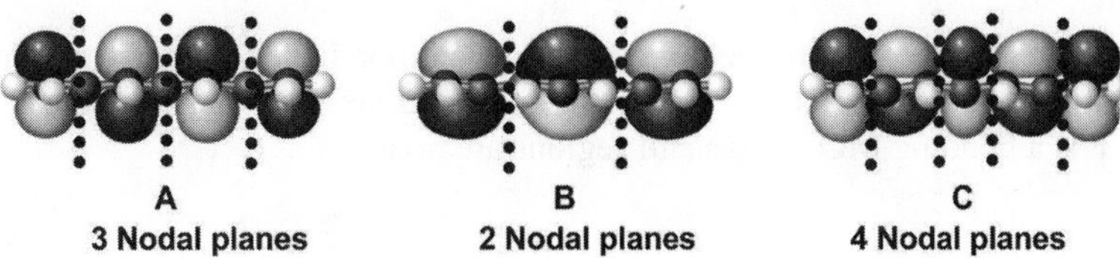

(b) The *p* orbital contributions are shown below. No *p* orbital contribution appears on any atom lying in a nodal plane. Notice that the *p* orbitals on either side of each nodal plane have opposite phases.

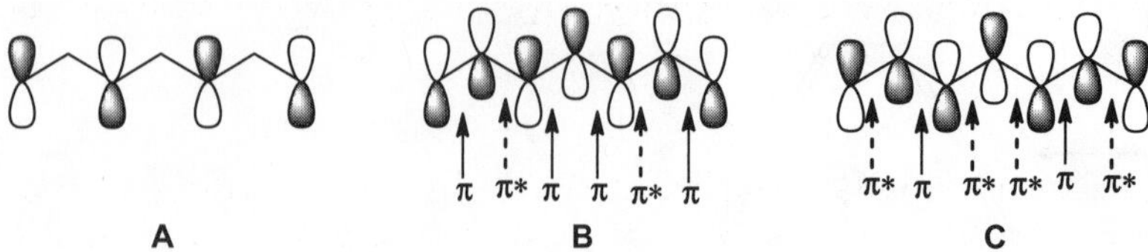

(c) The bonding regions are indicated by the solid arrows above. The antibonding regions are indicated by the dashed arrows above. In MO **A**, no adjacent *p* orbitals interact, so there are no bonding or antibonding interactions. In MO **B** and MO **C**, bonding interactions are identified as regions where the *p*-orbital contribution on either side is of the same phase. In each antibonding interaction, the *p* orbitals contribute opposite phases.

(d) MO **A** is nonbonding because there are no bonding or antibonding interactions at all. MO **B** is a bonding MO because there are four bonding interactions but only two antibonding interactions (net: two bonding interactions). MO **C** is an antibonding MO because there are only two bonding interactions but four antibonding interactions (net: two antibonding interactions).

Problem 14.27

Think

From the phases of the *p* orbitals, how do you identify a nodal plane? Does constructive or destructive interference result from that type of interaction between those *p* orbitals? In the resulting MO, what characterizes a nodal plane? Do bonding contributions raise or lower the π MO energy relative to the *p* AO? Antibonding? Count the number of each type of interaction. What is your net?

Solve

(a) The nodes are indicated by the vertical dashed lines below, where there is complete cancellation of the MO. There is one node in the first MO, six nodes in the second, and no nodes in the third. Energy increases with increasing number of nodes; therefore, the order of MOs in increasing energy is **F** < **D** < **E**.

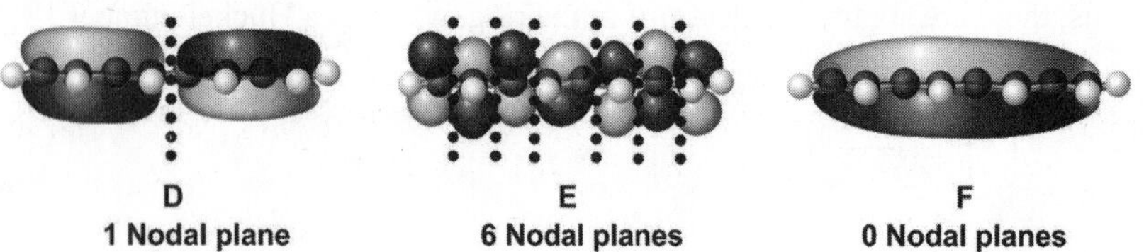

(b) The *p*-orbital contributions are indicated below. Notice that the *p* orbitals on either side of each nodal plane have opposite phases.

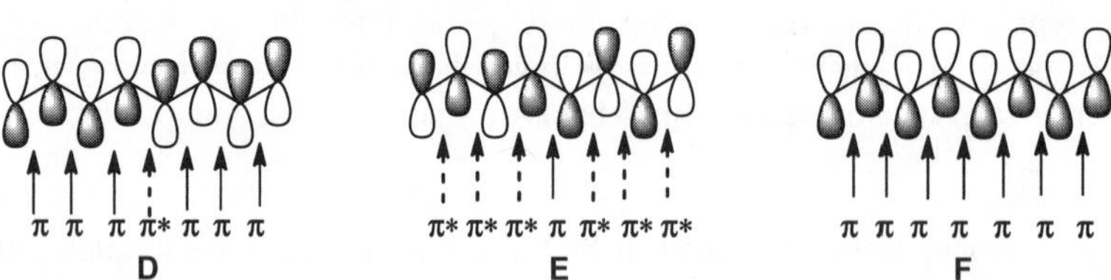

(c) The bonding regions are indicated by solid arrows above. The antibonding regions are indicated by dashed arrows.

(d) MO **D** is bonding because it has six bonding regions and one antibonding region (net: five bonding regions). MO **E** is antibonding because it has one bonding region and six antibonding regions (net: five antibonding regions). MO **F** is a bonding MO because all regions are bonding.

Problem 14.28

Think

What is the hybridization of each C atom? Where are the *p* AOs located? Where do the *p* AOs overlap to form a π bond? Where is there no *p* orbital overlap? In a VB picture, what does a positively charged C indicate? A negatively charged C?

Solve

All the C atoms are sp^2 hybridized and, therefore, have an unhybridized *p* AO. The carbocation and carbanion on the ends do not contribute to any π bonding interaction. The interaction between the *p* orbitals on C2 and C3 results in a π bond between C2 and C3. A positively charged C is associated with an empty *p* AO on C, whereas a negatively charged C is associated with a filled *p* AO on C. See the valence bond picture for each of the two weak resonance contributors below.

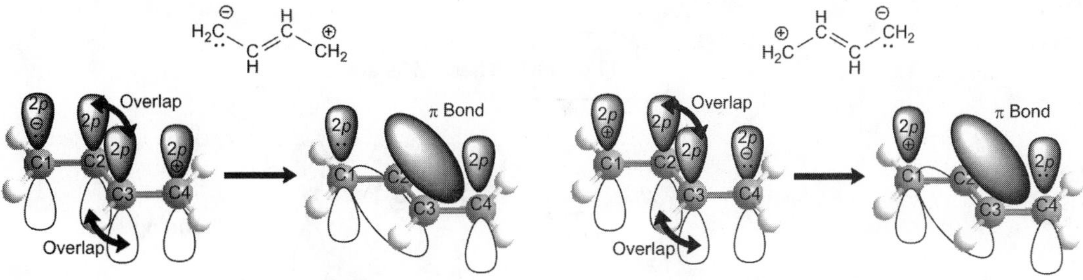

Problem 14.29

Think

What does the MO energy diagram look like if one electron from π_2 is promoted to π_3? Where does an electron occupying π_2 predominantly reside? What impact does such an electron have on the C1–C2 bond? The C2–C3 bond? Where does an electron occupying π_3 predominantly reside? What impact does such an electron have on the C1–C2 bond? The C2–C3 bond?

Solve

If an electron is promoted out of π_2 into π_3, the bond length between C1–C2 increases. This is because an electron in π_2 resides about half the time in the C1–C2 bonding region and thus contributes toward an increased C1–C2 bond character. An electron in π_3, on the other hand, does not reside much in the C1–C2 bonding region because there is a node there. So, overall, the electron density between C1–C2 decreases with this electron transition. Because the opposite is true for the C2–C3 bonding region, the electron density increases between C2–C3, which will increase the bond character and decrease the bond length.

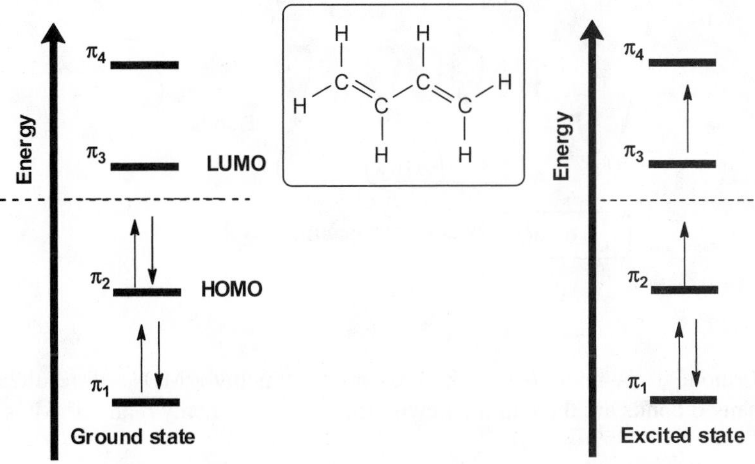

Problem 14.30

Think

What is the hybridization of each C atom? How many *p* AOs are present? How many π MOs will result? How many π electrons are there? How many σ bonds are there in the Lewis structure? How many σ and σ* MOs does this represent?

Solve

There will be five *p* AOs combining to make five π MOs. Two will be bonding, one will be nonbonding, and two will be antibonding. There are 11 σ bonds in the Lewis structure, corresponding to 11 σ MOs and 11 σ * MOs. The 28 total electrons fill the 11 σ MOs and the first three π MOs. The six π electrons are observable from the Lewis structure—two π bonds and one lone pair of electrons conjugated to it. The HOMO is π_3 and the LUMO is π_4.

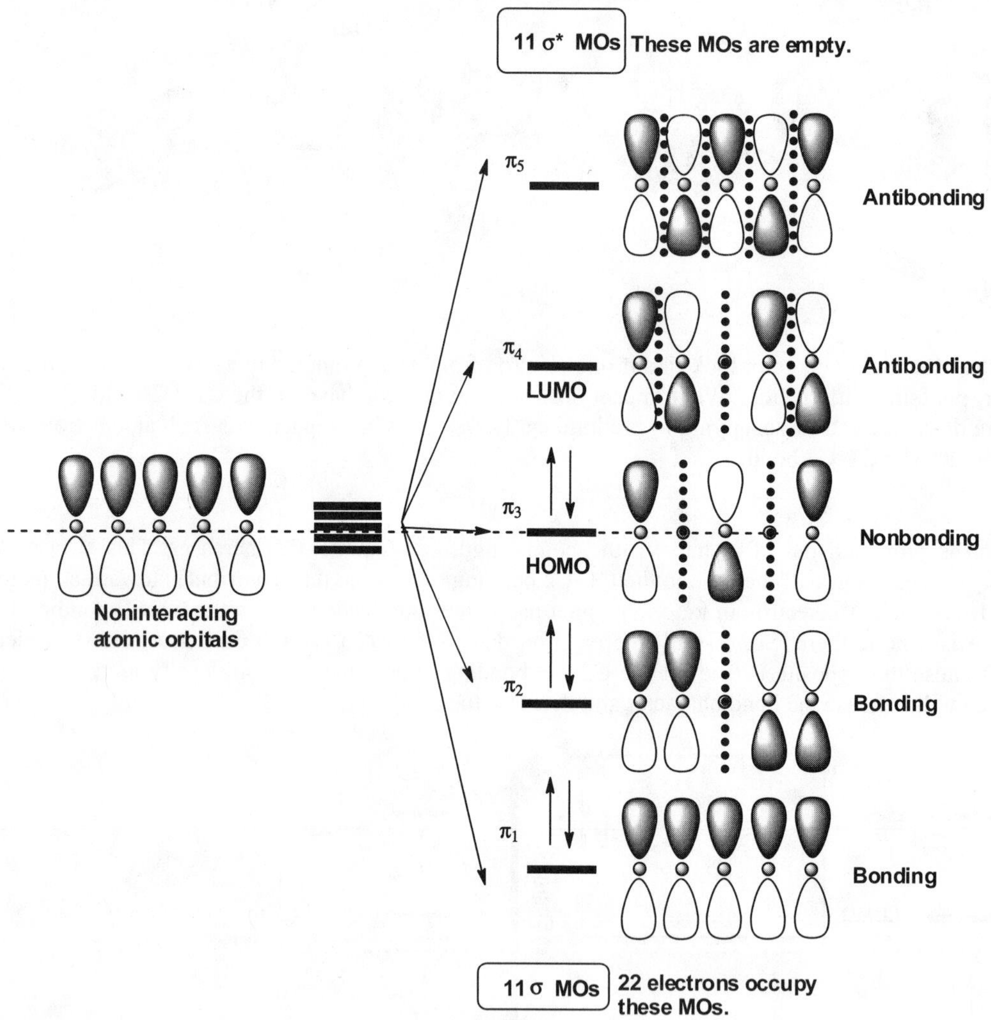

Problem 14.31

Think

What is the hybridization of each C atom? How many *p* AOs are present? How many π MOs will result? How many π electrons are there? How many σ bonds are there in the Lewis structure? How many σ and σ* MOs does this represent?

Solve

There will be five p AOs combining to make five π MOs. Two will be bonding, one will be nonbonding, and two will be antibonding. There are 11 σ bonds in the Lewis structure, corresponding to 11 σ MOs and 11 σ^* MOs. The 26 total electrons fill the 11 σ MOs and the first two π MOs. The four π electrons are observable from the Lewis structure—the two π bonds. The HOMO is π_2 and the LUMO is π_3.

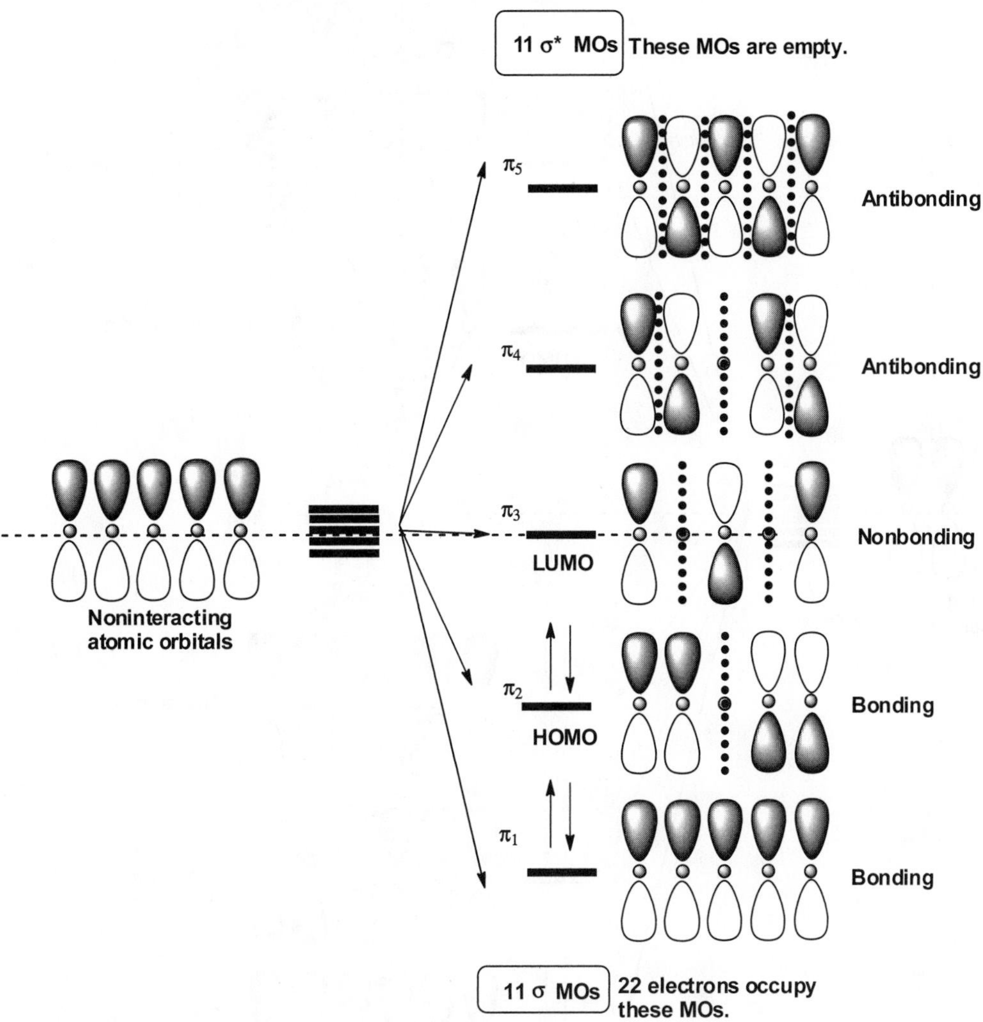

Problem 14.32

Think

What is the hybridization of each C atom? How many p AOs are present? How many π MOs will result? How many π electrons are there? How many σ bonds are there in the Lewis structure? How many σ and σ^* MOs does this represent?

Solve

There will be six p AOs combining to make six π MOs. Three will be bonding and three will be antibonding. There are 13 σ bonds in the Lewis structure, corresponding to 13 σ MOs and 13 σ^* MOs. The 32 total electrons fill the 13 σ MOs and the first three π MOs. The six π electrons are observable from the Lewis structure—three π bonds. The HOMO is π_3 and the LUMO is π_4. See the figure on the next page.

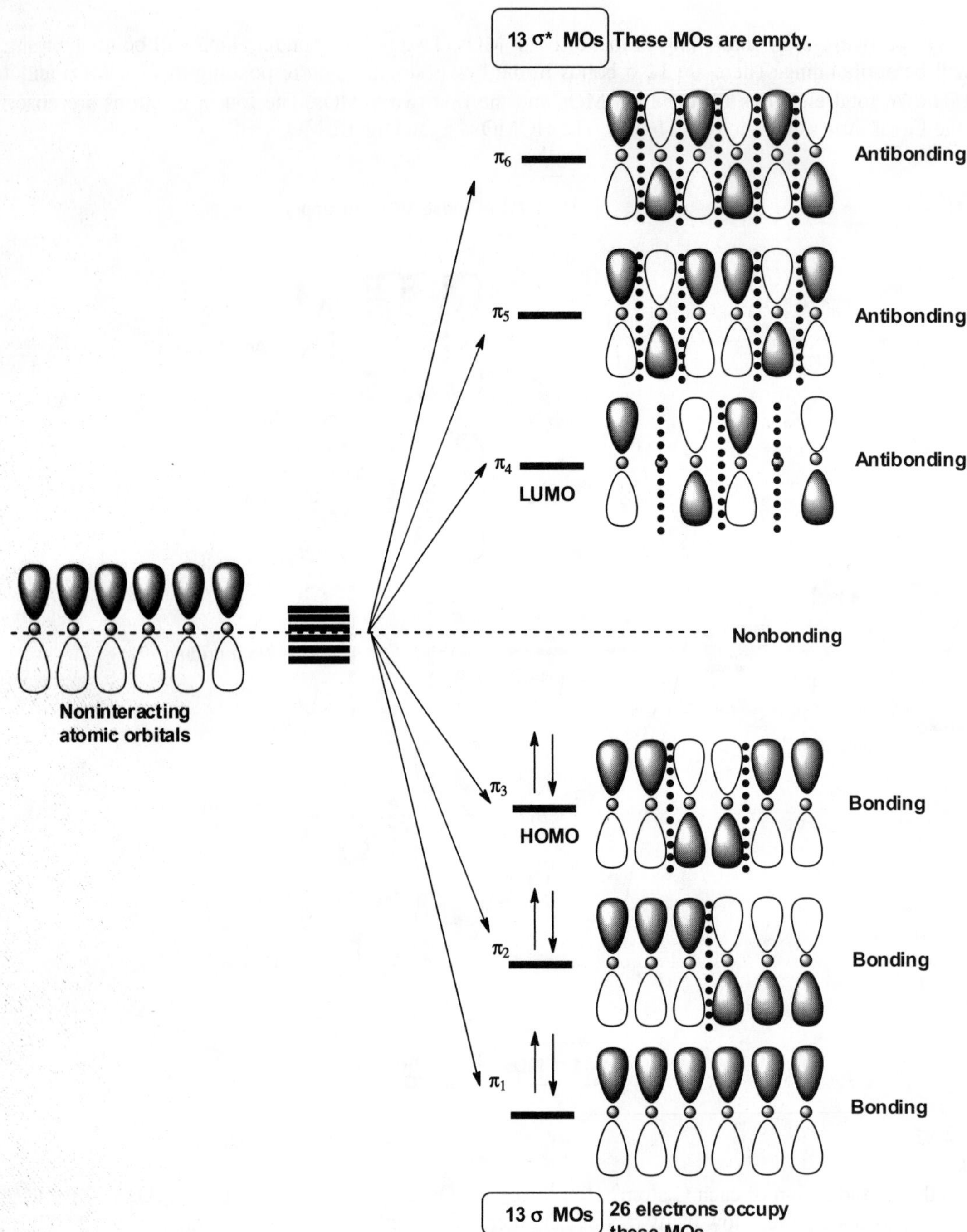

Problem 14.33

Think

See Figure 14-2 for the C–C bond length of ethane. Which π MOs are occupied with electrons? In the MO picture from Problem 14.32, what type of interaction occurs between C2 and C3 in each occupied π MO? How does putting an electron in each of these MOs affect the bond and length?

Solve

(a) The C–C bond shown in hexa-1,3,5-triene is 146 pm, whereas that of a normal C–C single bond is 154 pm. In 1,3,5-hexatriene, that single bond is shortened by about 5%.

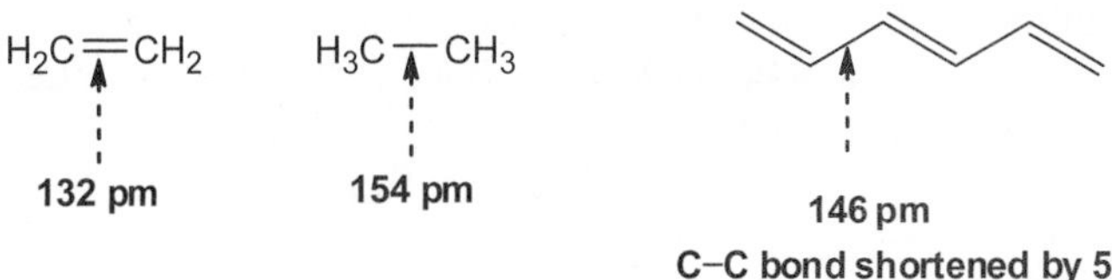

(b) The occupied π MOs that contribute to shortening that single bond are the ones in which there is a bonding interaction between C2 and C3. Those include π_1 and π_2. The three highest-energy MOs don't contribute because they are not occupied.

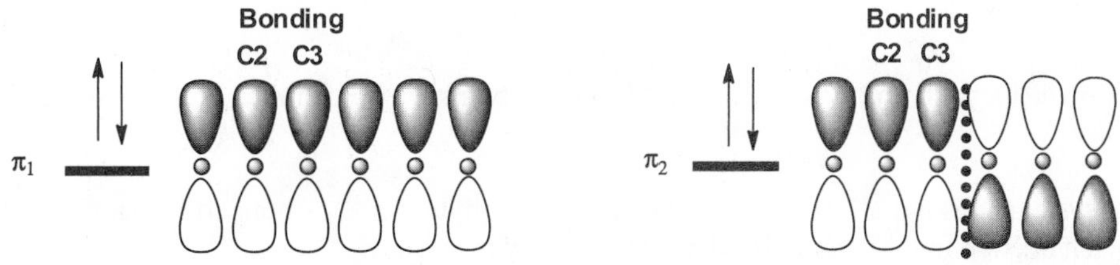

(c) If an electron were promoted from π_3 to π_4, the C2–C3 bond should shorten slightly. The occupancy of π_4 has essentially no effect because in the C2–C3 internuclear region, there is neither a bonding nor an antibonding interaction. However, the C2–C3 internuclear region of π_3 is an antibonding interaction, so removal of an electron from that orbital contributes to a shorter C2–C3 bond.

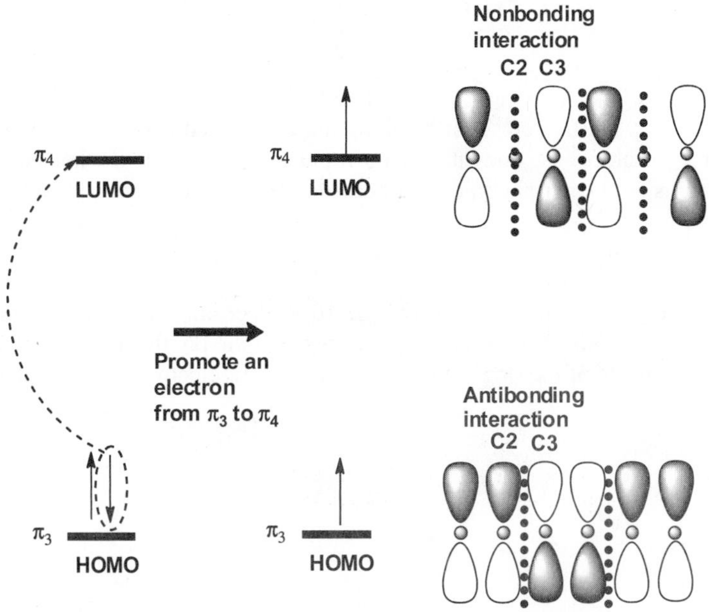

Problem 14.34

Think

How many double bonds are present? In each C=C, how many π electrons are present? Are there any sp^3 C atoms in between double bonds? If so, how does this affect the conjugation of the π system? In a triple bond, what is the orientation of the four *p* orbitals? If the *p* orbitals are not parallel, are they conjugated?

Solve

There are 12 total π electrons—two from each double bond and four from the triple bond. These will occupy four separate π systems. The double bonds between C2–C3 and C4–C5 are conjugated and make up one π system. The π system between C1–C2 is a separate π system because the *p* orbitals are perpendicular to the first π system. The third π system is made up of the double bond between C7–C8 and one of the π bonds between C9–C10, as those are conjugated together. The second π bond of the triple bond is a separate π system because its *p* orbitals are perpendicular to the first π bond of the triple bond.

Problem 14.35

Think

Are the *p* orbitals in the bicyclic ring system parallel? If the *p* orbitals are not parallel, can they be conjugated?

Solve

The bicyclic ring system does not allow the *p* orbitals on the sp^2-hybridized carbons to be parallel. Thus, there can be no significant overlap to form conjugated π MOs, as shown below.

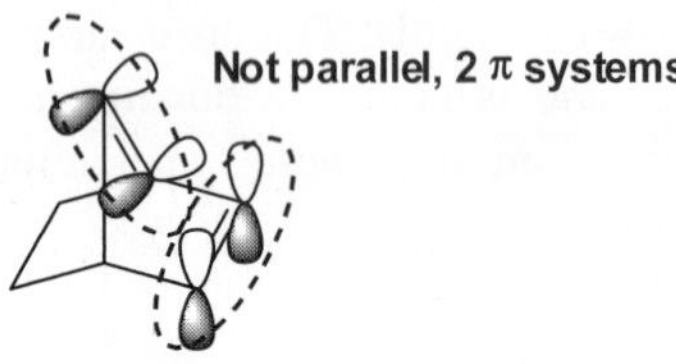

Problem 14.36

Think

Is the ring fully conjugated? Are there any sp^3 C atoms that break up the conjugated π system? Is the structure planar? Counting the total number of π electrons, does it contain a Hückel or an anti-Hückel number of π electrons? Counting only the π electrons on the periphery, does it contain a Hückel or an anti-Hückel number of π electrons?

Solve

It might appear to break Hückel's rule because there are a total of 16 π electrons, which is an anti-Hückel number. However, these 16 electrons can compose separate π systems—one on the outside of the ring, containing 14 electrons, and one on the inside of the ring, containing two. The 14 electrons on the outside of the ring are a Hückel number, making it aromatic.

Separate π system

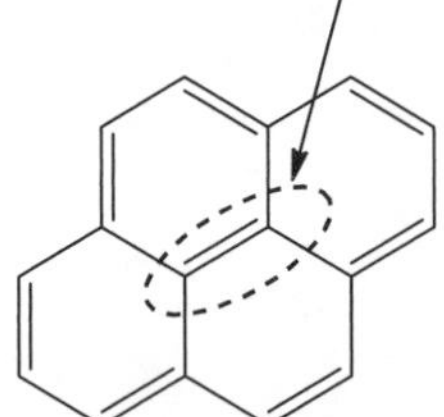

14 π electrons around periphery
Aromatic

Problem 14.37

Think

Considering the Frost method, what is the polygon shape of [10]annulene? How is the shape related to the MO diagram? How many π electrons are in [10]annulene? Is the HOMO filled or half filled? How is this related to aromaticity?

Solve

The molecule, with five pairs of (or 10 total) π electrons, should be aromatic, as long as the molecule is planar. The MO energy diagram agrees. All MOs are completely filled, and all occupied orbitals are bonding. The π MO energies are derived from the Frost method by orienting the decagon so a vertex points straight down, the center of the decagon is taken to be the *p* AO energies, and each π MO energy is located at a vertex.

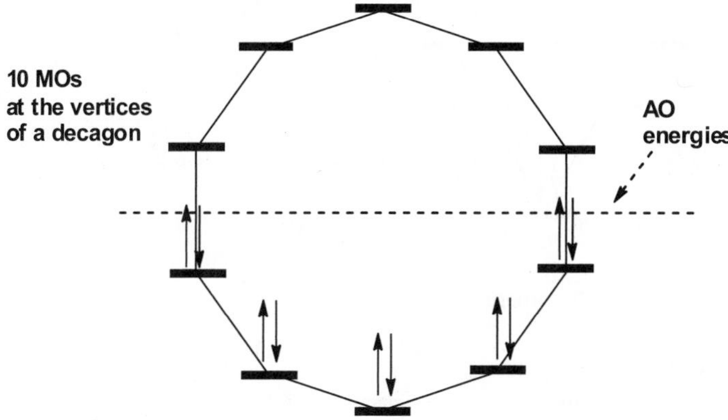

Problem 14.38

Think

Considering the Frost method, what is the polygon shape of [8]annulene? How is the shape related to the MO diagram? How many π electrons are in [8]annulene? Is the HOMO filled or half filled? How is this related to aromaticity?

Solve

With four pairs of (or eight total) π electrons, this molecule should be antiaromatic, as long as the molecule is planar. The MO energy diagram agrees. Not all MOs are filled, leaving very reactive unpaired electrons. And not all occupied orbitals are bonding. The π MO energies are derived from the Frost method by orienting the octagon so a vertex points straight down, the center of the octagon is taken to be the *p* AO energies, and each π MO energy is located at a vertex.

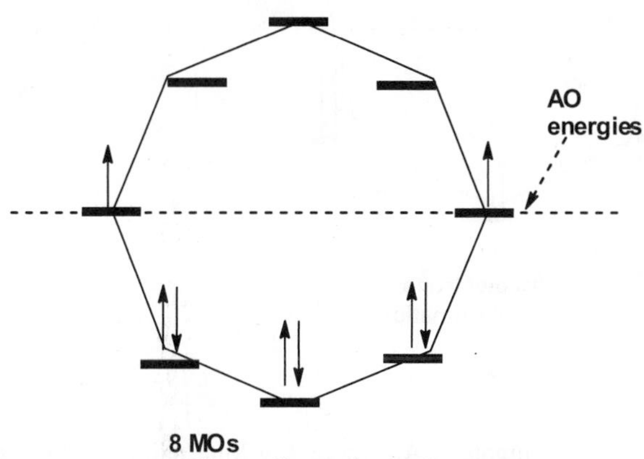

Problem 14.39

Think

Considering the Frost method, what is the polygon shape of each compound? How is the shape related to the MO diagram? How many π electrons are in each compound? Is the HOMO filled or half filled? How is this related to aromaticity?

Solve

The number of π electrons in each system and the aromatic or antiaromatic character are listed below each MO diagram. The π MO energies of each molecule are derived from the Frost method by orienting the polygon so a vertex points straight down, the center of the polygon is taken to be the *p* AO energies, and each π MO energy is located at a vertex.

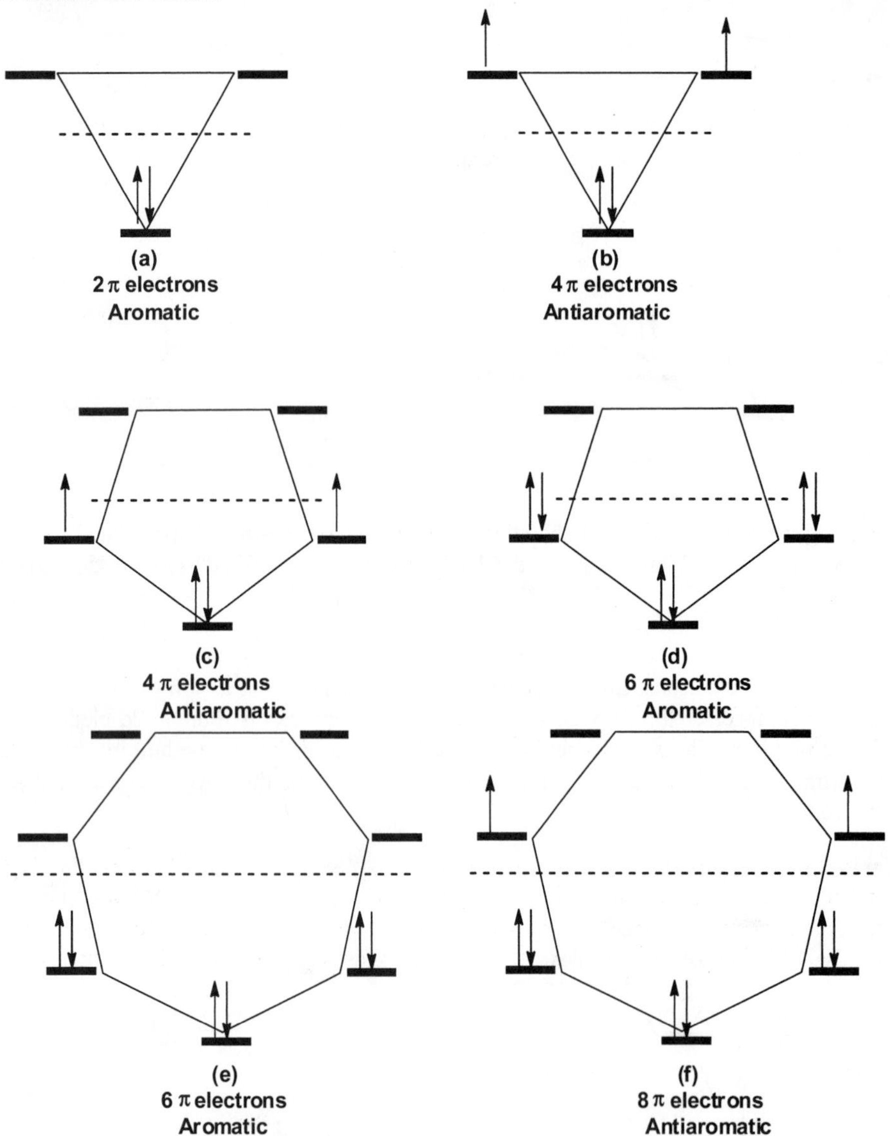

Problem 14.40

Think

What is the hybridization of each atom? Is the ring fully conjugated? Are there any lone pairs on the N or B atoms? Should those lone pairs be part of the same π system as the double bonds or are they perpendicular to it? How many π electrons are in the conjugated system?

Solve

Boron has no lone pairs and is sp^2 hybridized. The lone pair on each N atom is in an sp^2-hybridized orbital, which is perpendicular to the π system of the ring, and so each lone pair cannot be part of the π system. The molecule, with three pairs of (or six total) π electrons, is aromatic. It is cyclic and conjugated just like benzene.

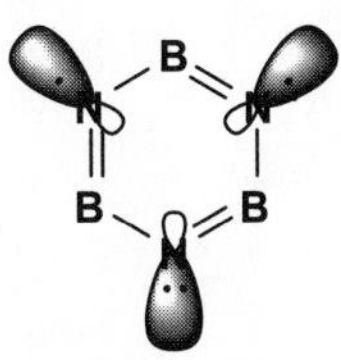

6 π electrons
Aromatic

Problem 14.41

Think

What is the hybridization of the N atom? What valence AOs does the N atom contribute to the molecule? Which of those AOs are used for bonding? What AO is left over to contain the lone pair?

Solve

Each N atom is sp^2 hybridized and contributes three sp^2-hybridized AOs and one unhybridized *p* AO to the molecule. Two of the sp^2-hybridized AOs are used to make the σ bonds involving N, and the *p* AO is used to make the π bond. Therefore, each lone pair on N resides in a leftover sp^2-hybridized AO, as shown below.

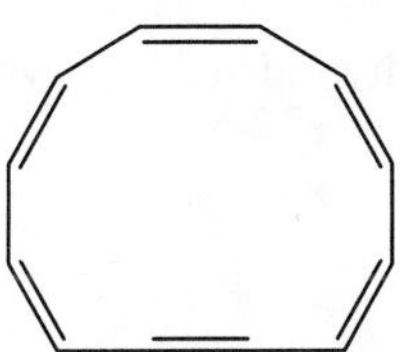

Lone pairs reside in
sp^2-hybridized orbitals

Problem 14.42

Think

Does the molecule contain a π system formed from a ring of fully conjugated *p* orbitals? How many electrons from the heteroatom are part of the π system? How many electrons are in that π system? Will the π system remain planar?

Solve

(a) Nonaromatic: It has 14 total π electrons, but 12 that appear to be completely conjugated around the ring (six pairs: an anti-Hückel number). Two of the π electrons of the triple bond are not in orbitals that can be parallel to the other *p* AOs, so they compose a separate π system of electrons. Therefore, if the molecule were planar, it would be antiaromatic. The ring is large enough, however, to avoid being antiaromatic by bending out of plane.

14 π electrons
Nonaromatic
(not planar)

(b) Nonaromatic: It has 12 π electrons completely conjugated around the periphery of the ring system, which is an anti-Hückel number (six pairs); the center carbon cannot participate in resonance, so there is no way to change the number of π electrons around the outer ring carbons. Therefore, if the molecule were planar, it would be antiaromatic. The ring is large enough, however, to avoid being antiaromatic by bending out of plane.

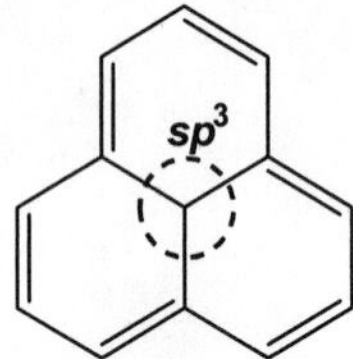

12 π electrons
Nonaromatic
(not planar)

(c) Aromatic: There are 10 total π electrons completely conjugated around the periphery of the molecule, which is a Hückel number (five pairs).

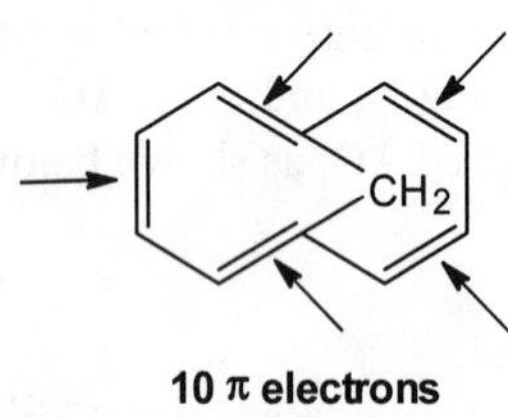

10 π electrons
Aromatic

(d) Nonaromatic: Although conjugated with six total π electrons, it is not cyclic.

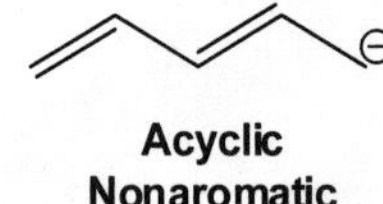

Acyclic
Nonaromatic

(e) Antiaromatic: Two of the lone pair electrons on the O atom can participate in resonance, bringing the total number of π electrons to four, which is an anti-Hückel number. And the π electrons are completely conjugated around the ring.

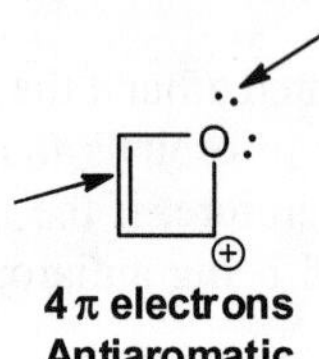

4 π electrons
Antiaromatic

(f) Nonaromatic: There are two total π electrons from the double bond, but the sp^3-hybridized N prevents the π system from being completely conjugated around a ring.

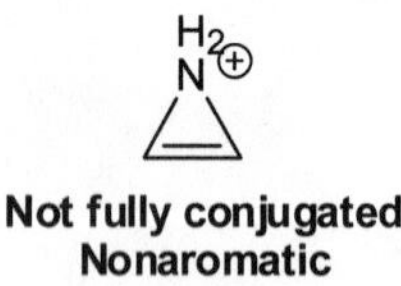

Not fully conjugated
Nonaromatic

(g) Nonaromatic: Two of the lone pair electrons on the O atom can participate in resonance, bringing the total number of π electrons to six; however, the sp^3-hybridized C atom prevents those electrons from being completely conjugated around the ring.

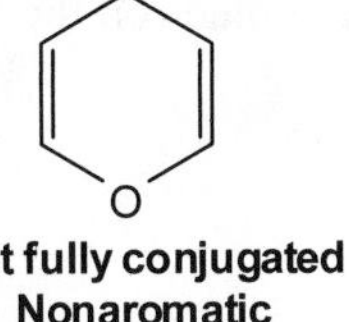

Problem 14.43

Think

What is the formula for each compound? What is the product of the hydrogenation reaction? Which reactant is most stable? Least stable? How does aromaticity factor in?

Solve

In all cases, the product of complete hydrogenation is the following bicyclo compound.

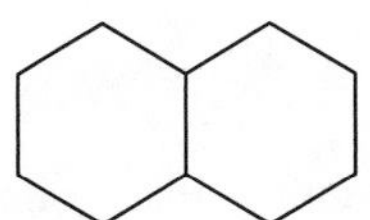

Bicyclo[3.3.0]decane

So the more stable the reactant, the smaller the heat of hydrogenation (the less exothermic the reaction). **A** is aromatic and, therefore, the most stable, and will have the smallest heat of combustion. The other three compounds are nonaromatic. **B** has three conjugated double bonds, **C** has two conjugated double bonds, and **D** has no conjugated double bonds. The order of increasing heat of hydrogenation is **A** < **B** < **C** < **D**.

Problem 14.44

Think

When the double bond is outside the ring, are its electrons considered part of the conjugated π system? Consider another resonance structure for each species.

Solve

E has a resonance contributor that has a cyclopentidenyl cation. This cation has four π electrons completely conjugated about the ring and, therefore, is antiaromatic. **F** has a resonance contributor that has a cycloheptatrienyl cation. This cation has six π electrons completely conjugated about the ring and, therefore, is aromatic. Consequently, **F** has the more stable π system.

E
4 π electrons
Antiaromatic

F
6 π electrons
Aromatic

Problem 14.45

Think

Consider other resonance structures for the compound. Is there a way to move the electrons in the middle double bond to allow each structure to be aromatic? In that resonance structure, are there opposite charges that appear at either end of the molecule?

Solve

Another resonance contributor can be drawn for this molecule. Even though it has significant charge separation (less stable), both rings become aromatic, with six π electrons and two π electrons, respectively (each a Hückel number). This separation of charge is consistent with the dipole shown, as the red (on the left side of the molecule) in the figure in the textbook indicates an excess of negative charge and the blue (on the right side of the molecule) indicates an excess of positive charge.

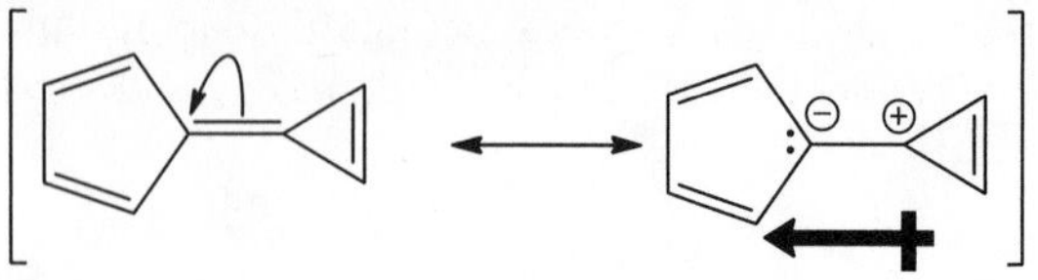

6 π electrons Aromatic　　2 π electrons Aromatic

Problem 14.46

Think

Consider other resonance structures for the compound. Is there a way to move the electrons in the middle double bond to allow each structure to be aromatic? In that resonance structure, are there opposite charges that appear at either end of the molecule?

Solve

As shown below, a resonance structure can be drawn in which both rings are aromatic, giving that resonance structure a relatively high contribution to the overall resonance hybrid. In that resonance structure, the ring on the left has a formal negative charge and the ring on the right has a formal positive charge. So this separation of charge creates a strong dipole moment pointing to the left, as indicated below.

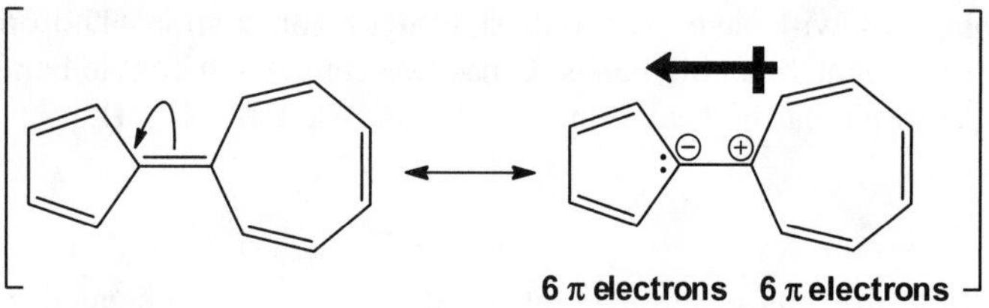

6 π electrons Aromatic　　6 π electrons Aromatic

Problem 14.47

Think

Consider other resonance structures for the compounds. Is there a way to move the electrons to form a resonance structure that is aromatic? In that resonance structure, is there a dipole?

Solve

The dipole moment of the species **X** is primarily due to the highly polar C=O bond. This is true of the compound **W** as well, but in addition, there is a relatively strong resonance contributor that adds to that dipole moment. As shown below, that resonance contributor has a cyclopropenyl cation, which has two electrons completely conjugated about the ring and, therefore, is aromatic. This contributor places an additional positive charge on the ring and an additional negative charge on O. Species **X** also has a similar resonance contributor, but that structure is not aromatic.

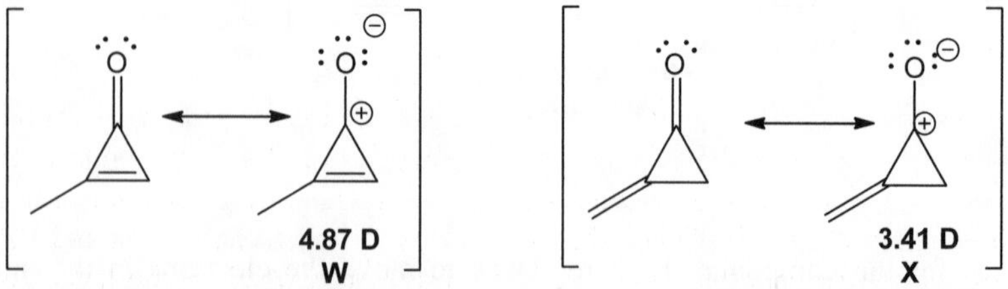

4.87 D
W
2 π electrons
Aromatic

3.41 D
X
Nonaromatic

Problem 14.48

Think

How is the measure of the heat of hydrogenation related to the stability of a compound's π system? Which compound(s) is/are aromatic? Does aromaticity increase or decrease a π system's stability?

Solve

Molecule **Y** is aromatic because it has 10 π electrons (five pairs) completely conjugated around the outside of the ring structure. Molecule **Z** is nonaromatic because the 10 π electrons are not completely conjugated—the sp^3-hybridized C atom prevents complete conjugation. So the molecule on the right should have the less stable π system, which would give it the greater heat of hydrogenation.

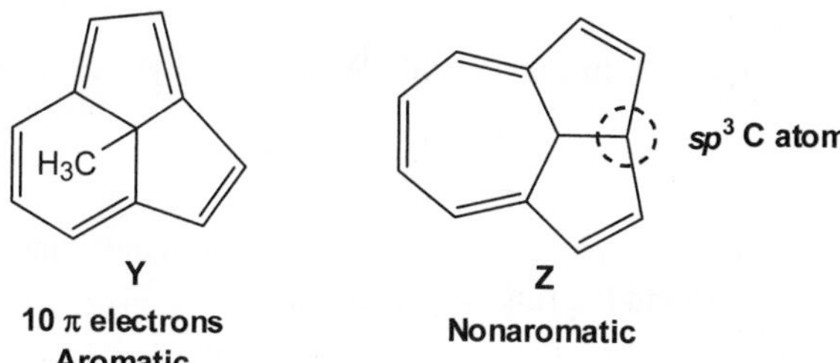

Problem 14.49

Think

How many total π electrons are present in coronene? How many of these π electrons are on the periphery? Is there complete conjugation around the outside of the molecule? Around the inside of the molecule?

Solve

There are 24 total π electrons (12 pairs). According to Hückel's rule, this should be antiaromatic. The observation that it is aromatic, however, is reconciled by the fact that there are two separate aromatic π systems. There is a central benzene-type ring with six π electrons, highlighted by a dashed square below. There is also an aromatic outer ring of 18 electrons (nine pairs).

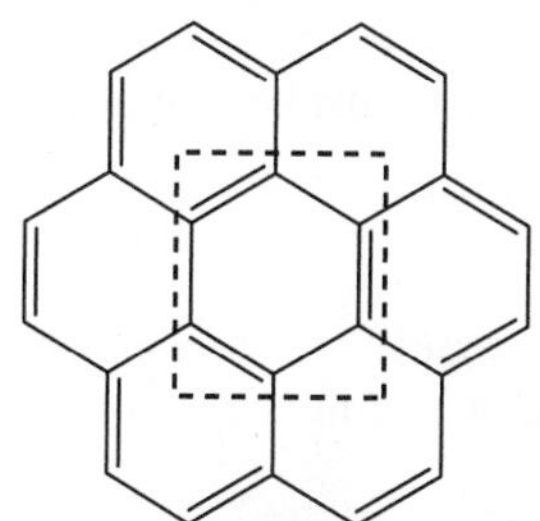

Coronene
24 total π electrons
18 π electrons on periphery (aromatic)
6 π electrons inside (aromatic)

Problem 14.50

Think

For an organic compound to be water soluble, what intermolecular forces could be present between the compound and water? With the structure shown, is this possible? Is Br a good leaving group? What results after a heterolysis step?

Solve

For the compound with several C atoms to be soluble in water, there needs to be extensive hydrogen bonding and/or significant ion–dipole interactions. Neither of these interactions is possible with the compounds as given. However, the C–Br bond in the compound on the right dissociates easily because the product cation is aromatic, and the positive charge is delocalized around the ring. When this takes place in water, that cation and Br^- are formed, which are solvated well by water. By contrast, the C–Br bond in the compound on the left does *not* dissociate easily in water, because the product has a localized positive charge on the C atom. See the figure on the next page.

Isolated positive charge

Heterolysis

Positive charge delocalized around an aromatic ring

Heterolysis

Problem 14.51

Think

How many σ bonds does the N atom possess? Are any lone pair electrons present? What is the hybridization of the N atom? Is molecule **X** fully conjugated?

Solve

The N atom has four bonds to it and is, therefore, sp^3-hybridized. Thus, it has no *p* orbital to contribute, and there is no complete conjugation of the π system completely around the ring.

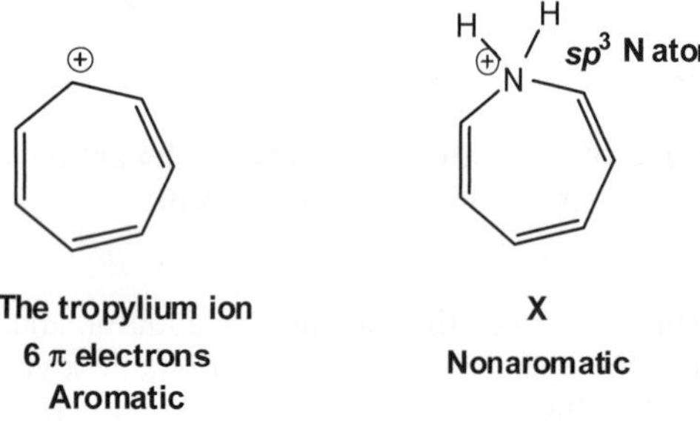

Problem 14.52

Think

Draw the structure of the $C_8H_8^{2-}$ dianion. How many π electrons are present? What must be true of the shape of aromatic compounds?

Solve

(a) The anion is easy to make because the reactant, cyclooctatetraene, is nonaromatic (due to nonplanarity), but becomes aromatic when two more electrons are added. That's because the two additional electrons must go in the π system, giving it a total of 10 π electrons (a Hückel number).

(b) The dianion product should be planar to make it aromatic.

+ K(s) → 2- + 2 K^+

8π electrons
Nonaromatic (nonplanar)

10π electrons
Aromatic

Problem 14.53

Think

Does a lower pK_a indicate a stronger or weaker acid? After the acid donates its proton, what is left on the carbon? These acids are all uncharged, so should acid strength be governed by the stabilities of the acids or of their conjugate bases? Are the conjugate bases aromatic, nonaromatic, or antiaromatic?

Solve

All acids are uncharged, so the difference in acid strength is due primarily to differences in conjugate base stability once the H leaves as a proton. All acidic protons are on an sp^3-hybridized C atom. Different stabilities of the anions are due largely to resonance delocalization of the negative charge. The strongest acid is cyclopentadiene (**D**) because the conjugate base is an aromatic anion. Cyclopropene (**E**) is the weakest acid because the conjugate base is an antiaromatic anion. The stabilities of the other conjugate bases decrease with fewer resonance structures that delocalize the negative charge.

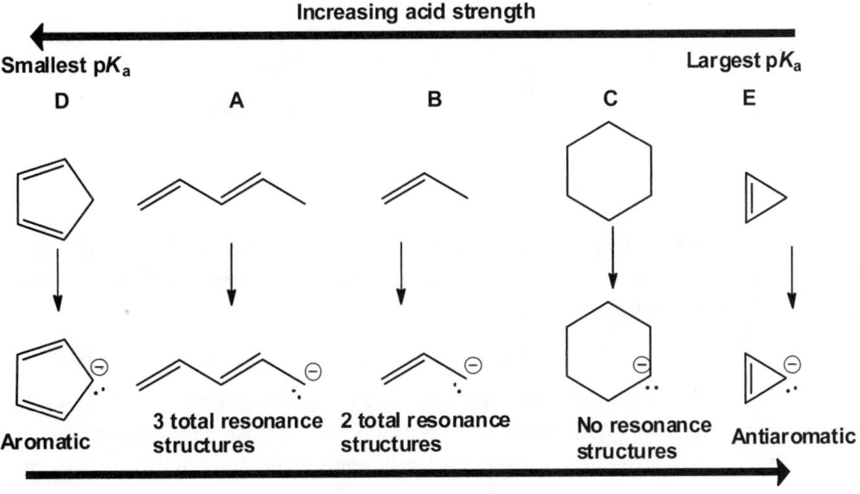

Problem 14.54

Think

Consider an $S_N1/S_N2/E1/E2$ competition. Is there a sufficient leaving group? On what type of carbon is the leaving group located? Is the attacking species a strong or weak nucleophile? A strong or weak base? Is the solvent protic or aprotic? What reactions does heat favor? What is the major mechanism? How many adjacent H atoms are present? How many alkene products could result? Which alkene product is most stable?

Solve

A polar aprotic solvent with a strong nucleophile and strong base will favor the S_N2 or E2 mechanisms, respectively. In a heated solution, CH_3O^- acts as a base to remove a proton. The double bond that forms in the major product is more stable because it is conjugated to the pre-existing double bond in the substrate.

:Cl:

H

CH_3O:⊖

DMSO

E2

Conjugated alkene: major product

Problem 14.55

Think

Is there a significant difference in stability of the two charged acids? Is there a significant difference in stability of the uncharged conjugate bases after the acids donate their protons? How do these stability differences affect acid strength?

Solve

The two charged acids are very similar in stability because, in both cases, the positive charge is located on N, neither charge is delocalized by resonance, and there are similar inductive effects for both. **F** is a stronger acid, however, because its conjugate base is an uncharged, aromatic molecule. The conjugate base of the molecule on the right is uncharged, too, but not aromatic. The acid reaction is shown below.

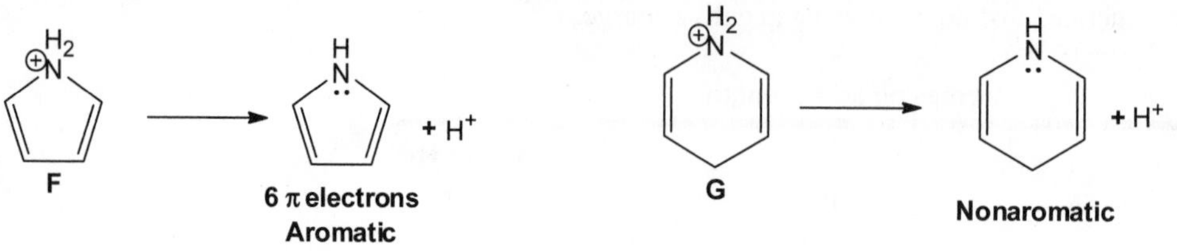

F → + H⁺ (6 π electrons, Aromatic)   G → + H⁺ (Nonaromatic)

Problem 14.56

Think

Is there another resonance structure possible? Does this resonance structure increase the electron density at one O atom over the other? Is that resonance structure aromatic, antiaromatic, or nonaromatic? How does this correspond to the importance of that resonance structure?

Solve

The carbonyl O is the more basic oxygen. As shown below, there is a relatively strong resonance contributor that places a negative charge on the carbonyl oxygen, making it relatively highly basic. That resonance structure has a strong contribution because its ring is aromatic, being completely conjugated and having six π electrons.

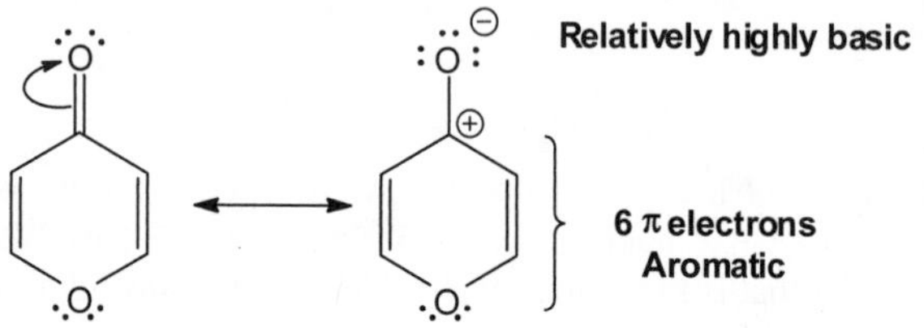

Relatively highly basic

6 π electrons
Aromatic

Problem 14.57

Think

Is there a significant difference in stability of the two uncharged acids? Is there a significant difference in stability of the charged conjugate bases after the acids donate their protons? How do these stability differences affect acid strength?

Solve

(a) Compound **X** should be more acidic because its resulting enolate anion is more stable. The enolate anion of the ketone on the left, **W**, is antiaromatic due to the four π electrons occupying the completely conjugated ring.

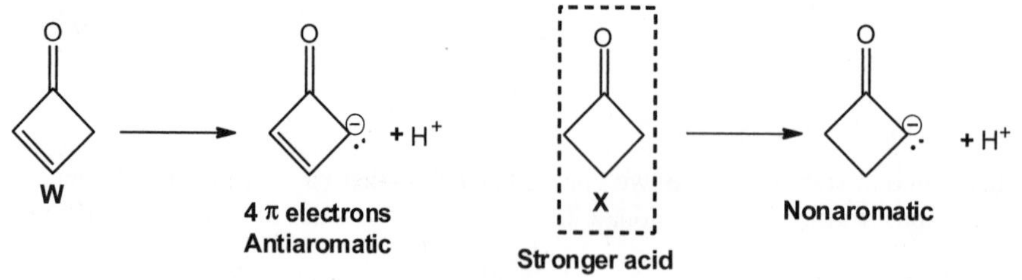

W → + H⁺ (4 π electrons, Antiaromatic)   X (Stronger acid) → + H⁺ (Nonaromatic)

(b) Compound **Y** is more acidic because its resulting enolate anion is aromatic, having six π electrons conjugated around the ring, whereas the enolate anion of the ketone on the right is nonaromatic.

Y
Stronger acid
$+ H^+$
6 π electrons
Aromatic
Z
$+ H^+$
Nonaromatic

Problem 14.58

Think

Is a nucleophile an electron pair donor or acceptor? If the electron pair is part of the aromatic π system, what does that do to the availability of those electrons? What does that do to the nucleophile strength?

Solve

Pyridine is a stronger nucleophile than pyrrole due to the orbital in which each lone pair is located. Pyridine's lone pair is located in an sp^2 orbital and, therefore, is not part of the conjugated π system. The lone pair on pyrrole is located in an unhybridized *p* orbital and is part of the conjugated π system. Each compound has a lone pair on N that can be used to form a bond, thus enabling it to act as a nucleophile. However, the lone pair on pyrrole is required to make the compound aromatic. So that lone pair is more stabilized, making it difficult to be usable to make bonds. The lone pair in pyridine, however, is not part of the aromatic π system, leaving it more available for bonding.

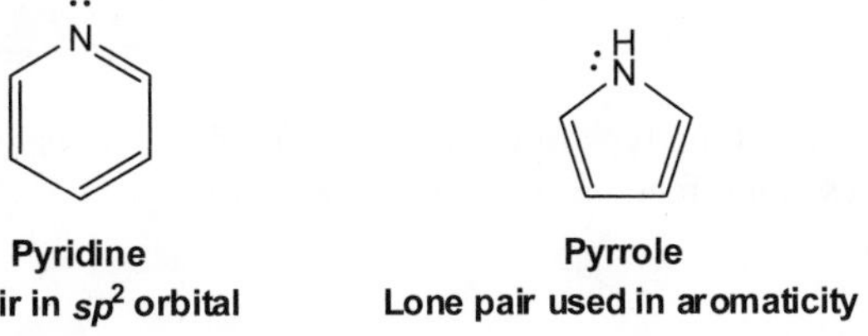

Problem 14.59

Think

What is the rate-limiting step in an S_N1 mechanism? What intermediate forms? Is the cation aromatic, antiaromatic, or nonaromatic? How does the stability of the intermediate affect the rate of the reaction?

Solve

(a) Molecule **R** will undergo S_N1 reactions faster. In an S_N1 reaction, the departure of the leaving group is the rate-determining step. These steps are shown for the two substrates below. **Q** produces an antiaromatic carbocation, which is less stable than the nonaromatic carbocation produced from the substrate **R**.

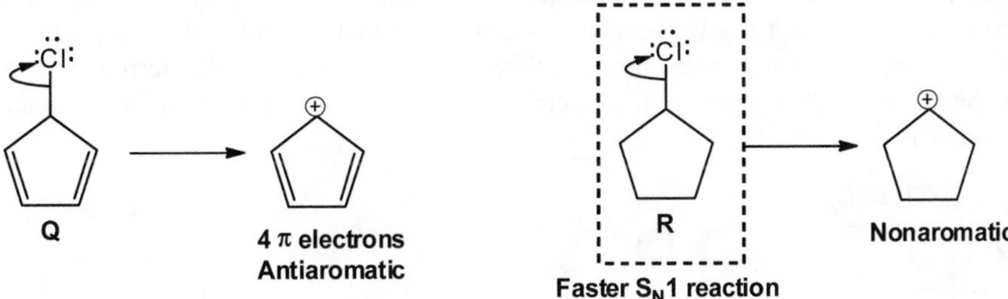

(b) Molecule **S** will undergo S_N1 reactions faster. As shown below, its carbocation is aromatic and, therefore, more stable than the carbocation that is formed from molecule **T**.

S
Faster S_N1 reaction
6 π electrons
Aromatic
T
Nonaromatic

Problem 14.60

Think

For a dehydration reaction to occur on these compounds, what is the leaving group? What is the final product? Is the compound antiaromatic, or nonaromatic?

Solve

Dehydration occurs faster with the molecule **V** because the product is aromatic.

OH
$- H_2O$
U
Nonaromatic
OH
$- H_2O$
V
Faster E1 dehydration reaction
6 π electrons
Aromatic

Problem 14.61

Think

Once the two carbocation intermediates are drawn, label the type of carbocation intermediate. Is either of these intermediates capable of resonance? If so, how does that affect the stability of the cation?

Solve

(a) The carbocation intermediates are shown below.

$H_2C{=}C{=}CH_2$ + HBr → $H_2C{=}\overset{\oplus}{C}{-}CH_3$ (**Vinylic C⁺**) and $H_2C{=}C(H){-}\overset{\oplus}{C}H_2$ (**Allylic C⁺**)

Propadiene

Carbocation intermediates

(b) In the vinylic carbocation, the carbon on the far right is sp^3 hybridized and is an electron-donating group (EDG). This stabilizes the carbocation. At first glance, the allylic carbocation might appear to be more stable due to resonance delocalization of the positive charge. However, even though all three C atoms are sp^2 hybridized, the CH_2 on the far right is not in the same plane as the CH_2 on the left. Therefore, the charge cannot be delocalized. As a result, the vinylic carbocation that is produced upon protonation is the more stable carbocation. Resonance does not stabilize the cation because the terminal *p* AO, being perpendicular to the other p AOs, is not conjugated and does not interact (see Section 14.2a for a discussion regarding conjugation.)

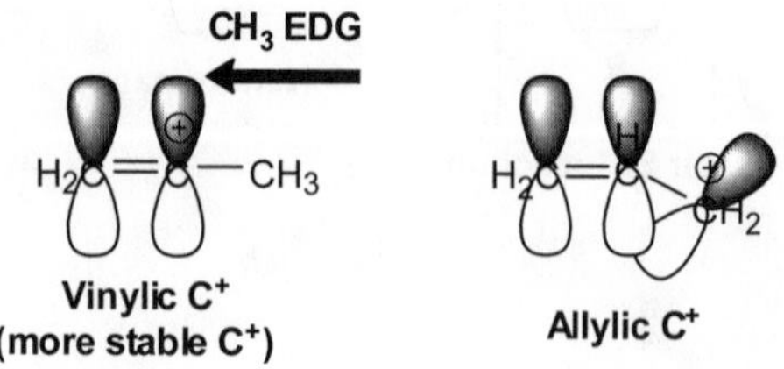

Vinylic C⁺
(more stable C⁺)
Allylic C⁺

Problem 14.62

Think

In the product from the attempted alkyne addition with peroxyacid (RCO_3H), how many electrons from the O atom are part of the conjugated π system? Does an aromatic, nonaromatic, or antiaromatic compound result? Is this stable or unstable?

Solve

Two of the four electrons on the O atom are part of the cyclic π system. This leads to a total of four π electrons and an antiaromatic compound. Due to the unstable nature of an antiaromatic compound, the product does not result.

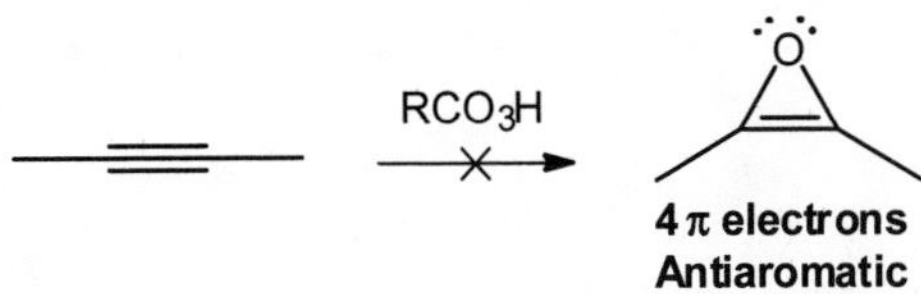

CHAPTER 15 | Structure Determination 1: Ultraviolet–Visible and Infrared Spectroscopies

Your Turn Exercises

Your Turn 15.1

Think

Consult Equation 15-1. With a higher %T, does more light or less light passing through the sample arrive at the detector?

Solve

With a higher %T, more light arrives at the detector, meaning that less light is absorbed by the sample: %T = 20% represents more light being absorbed than %T = 40%. Absorbance, on the other hand, increases with the amount of light absorbed by the sample, so $A = 0.750$ represents more light being absorbed than $A = 0.500$.

Your Turn 15.2

Think

What variable in Equation 15-3 represents molar absorptivity? How is that variable related to absorbance, A?

Solve

Molar absorptivity is represented by ε and is directly proportional to absorbance, A. Therefore, if molar absorptivity increases by a factor of 3, absorbance also increases by a factor of 3.

Your Turn 15.3

Think

In Equation 15-5, which variable represents the wavelength of light? According to the equation, is that variable directly proportional or inversely proportional to the energy of the photon?

Solve

The wavelength of light is represented by λ_{photon}, which is inversely proportional to E_{photon}. So E_{photon} increases as λ_{photon} decreases. Therefore, a 375-nm photon has more energy than a 530-nm photon.

Your Turn 15.4

Think

According to Figure 15-4, which MOs are occupied by electrons? Which MOs are empty? How many electrons can occupy a single MO?

Solve

According to Figure 15-4, the bottom three MOs are occupied with two electrons each. The top three MOs are empty. Up to two electrons can occupy the same MO. Therefore, an electron transition can take place from a lower-energy MO to a higher-energy MO if the lower-energy MO has at least one electron and the higher-energy MO has either zero electrons or one electron. All of the transitions that fit these criteria are shown below.

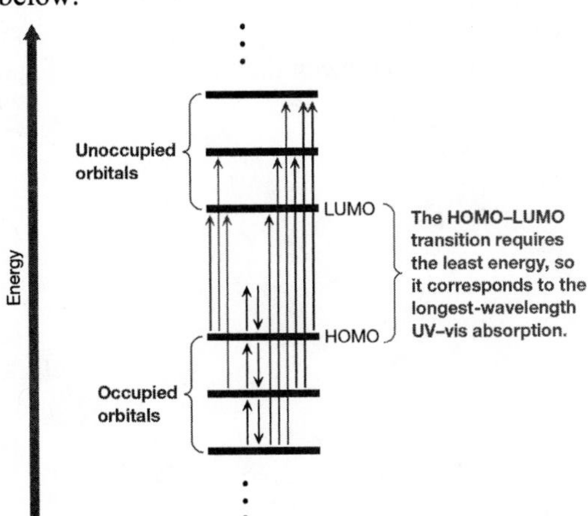

Your Turn 15.5

Think

In the figure, how is the required photon energy represented? How do those representations of photon energies you drew for Your Turn 15.4 compare to that for the HOMO–LUMO transition?

Solve

The required photon energies are represented by the lengths of the arrows in Your Turn 15.4. The longer the arrow, the greater the energy difference between the MOs involved in the electron transition and the greater the photon energy that is required. Of all the arrows drawn to represent these transitions, the one representing the HOMO–LUMO transition is the shortest, meaning that the HOMO–LUMO transition corresponds to the smallest-energy absorbed photon.

Your Turn 15.6

Think

Which orbital in Figure 15.5a lost an electron? Which orbital in Figure 15.5b gained an electron? Relative to the *p* AO energies, are bonding MOs lower energy, higher energy, or the same energy? Antibonding MOs? Nonbonding MOs?

Solve

The electron is circled below, and the transition is indicated by the curved arrow. The lowest two π MOs are bonding because they are below the energy of the isolated *p* orbitals (the horizontal dashed line), whereas the highest two π MOs are antibonding because they are above that line.

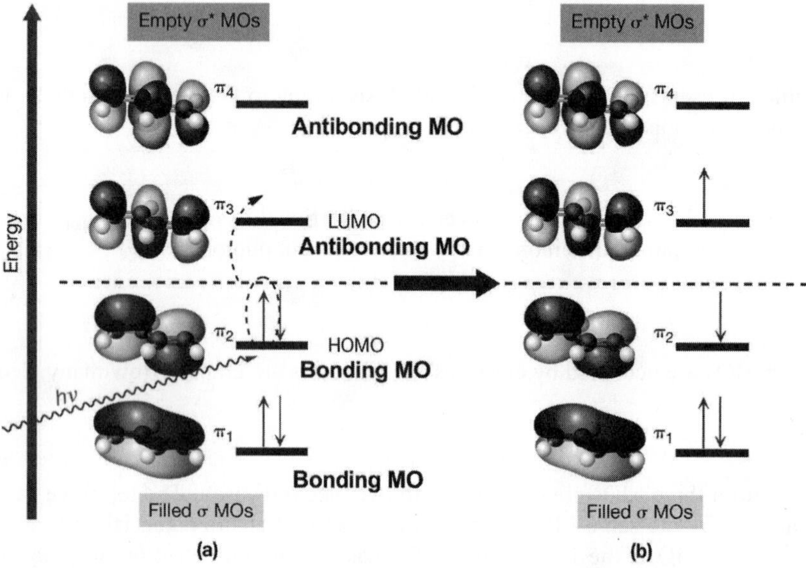

Your Turn 15.7

Think

What is required for π bonds to be conjugated? What trend do you notice between the number of conjugated π bonds and the wavelength?

Solve
See the table below. One C=C π bond (no conjugation) is ~180 nm, two conjugated C=C π bonds are ~225 nm, three conjugated C=C π bonds are ~275 nm, and four conjugated C=C π bonds are ~290 nm. As the number of conjugated π bonds increases, the λ_{max} value increases. The HOMO–LUMO energy gap, therefore, decreases.

Compound	λ_{max} (nm)	Conjugated π Bonds	Compound	λ_{max} (nm)	Conjugated π Bonds
Alkanes and cycloalkanes	<150	0	*cis*-Penta-1,3-diene	223	2
Ethene	161	0	*trans*-Penta-1,3-diene	223.5	2
Hex-1-ene	177	0	2-Methylbuta-1,3-diene (isoprene)	224	2
Penta-1,4-diene	178	0	Cyclopentadiene	239	2
Cyclohexene	182	0	Cyclohexa-1,3-diene	256	2
Hex-1-yne	185	0	Hexa-1,3,5-triene	274	3
Buta-1,3-diene	217	2	Octa-1,3,5,7-tetraene	290	4
β-Carotene				455	11

Your Turn 15.8

Think
Locate another peak in the spectrum. How do you determine the frequency value for the photon that is absorbed? How is the photon frequency of an absorbed photon related to the vibrational mode responsible for absorption of that photon? What are the units?

Solve
The peak at 1320 cm^{-1} is labeled on the next page. However, there are several other peaks that could have just as equally been selected. The frequency value of the absorbed photon is the value of the peak at the *x*-axis and the units are wavenumbers, cm^{-1}. These are the same frequencies for the mode of vibration responsible for photon absorption.

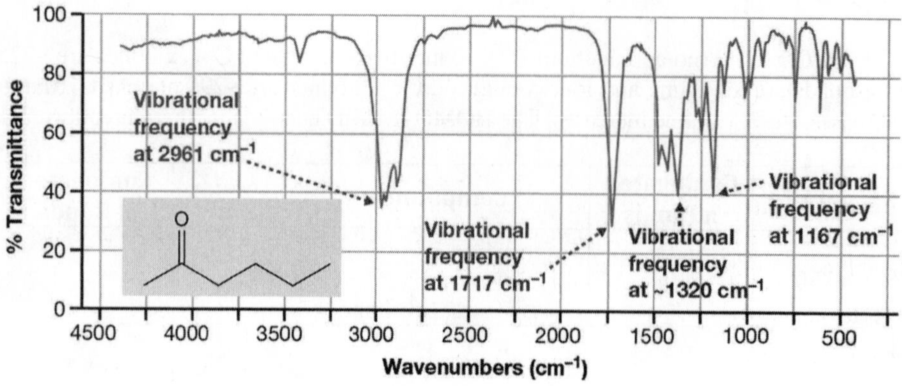

Your Turn 15.9

Think

Around what frequency does the C=O carbonyl peak occur? Is the peak strong or weak? Broad or narrow?

Solve

The C=O peak generally appears at ~1720 cm^{-1} as a strong sharp peak. The benzaldehyde C=O peak specifically occurs at ~1700 cm^{-1}.

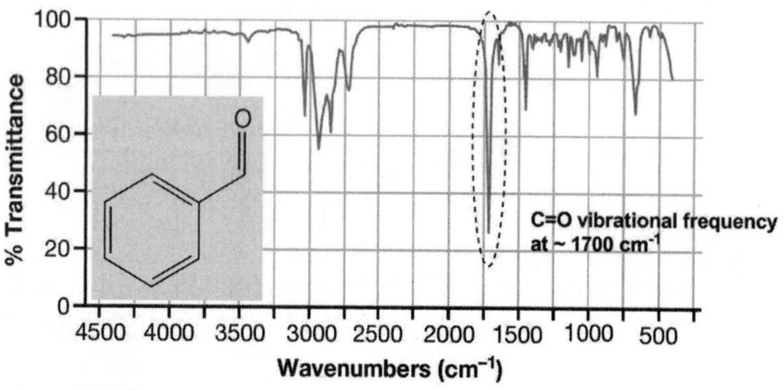

Your Turn 15.10

Think

What wavenumber range does the fingerprint region encompass? Which regions in Figure 15-12 are included in that range?

Solve

The fingerprint region encompasses the frequencies below ~1400 cm^{-1}. In Figure 15-12, the two regions that are in this range are the single-bond stretches and the bending modes.

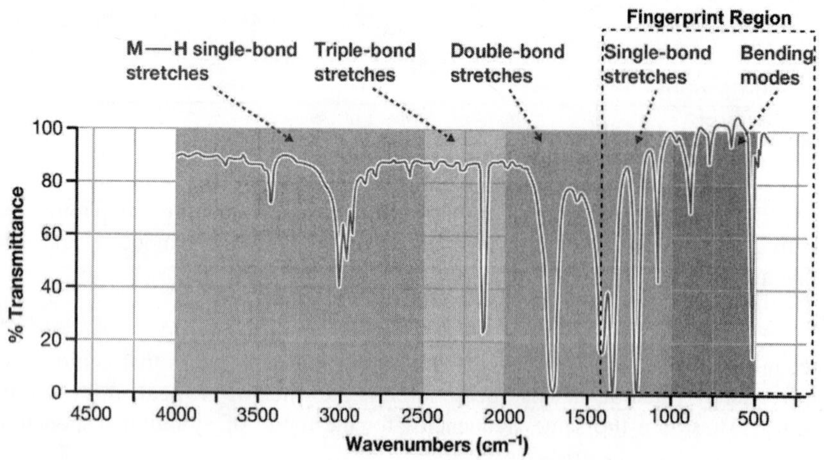

Your Turn 15.11

Think

Consult the IR spectrum in Figure 15-14a for hept-1-ene and Table 15-2 to help you determine where a C=C stretch could appear. Do you see a peak arising from a C=C stretch for hep-3-ene? Why or why not?

Solve

The general region for a C=C stretch is 1620–1680 cm^{-1}. Unlike the IR spectrum for hept-1-ene, hept-3-ene contains no peak in this region. This is due to the larger extent of symmetry about the C=C bond in hept-3-ene. (The annotation at the left of the spectrum corresponds to Your Turn 15.13.)

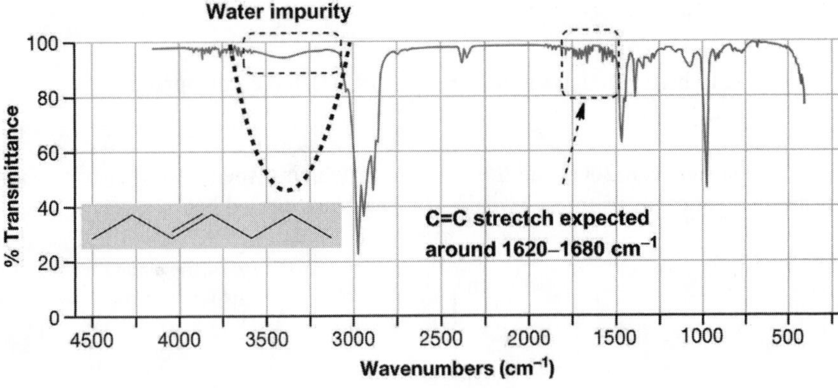

Your Turn 15.12

Think

In what frequency range does the O–H stretch peak occur? Is the peak strong or weak? Broad or narrow?

Solve

The H–O stretch bands are intense, broad, and centered around 3300 cm^{-1} for RO–H (alcohols) and 3000 cm^{-1} for ROO–H (carboxylic acid). The OH stretches are circled below. Notice that in a carboxylic acid, the OH stretch is shifted to a lower frequency and is much broader than in an alcohol; therefore, it overlaps the alkane C–H stretches (2800–3000 cm^{-1}).

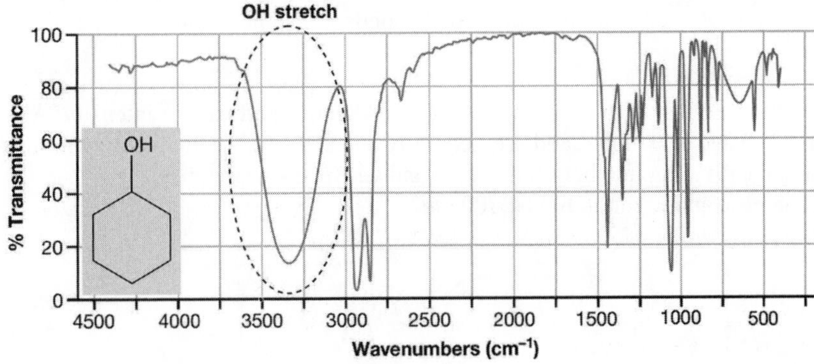

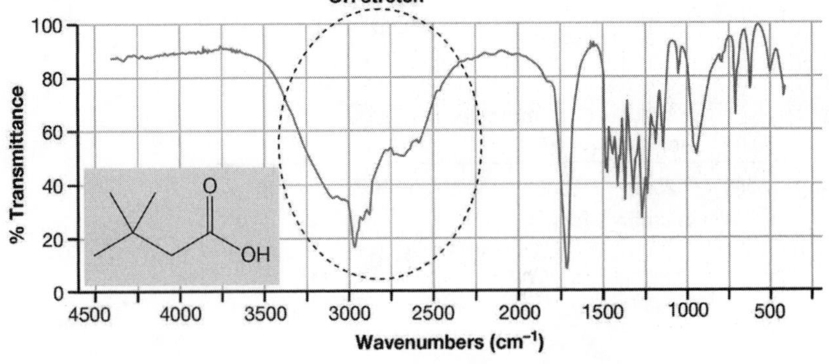

Your Turn 15.13

Think

In what frequency range does the O–H stretch peak occur? How is the strength of the signal affected if H_2O is just an impurity and not part of the compound?

Solve

In Figure 15-14b, the small bump around 3300 cm^{-1} is due to a water impurity in the hept-3-ene sample when the IR spectrum was taken. The OH stretch would appear much more intense if the OH were part of the molecule itself instead of an impurity, as indicated by the bold dashed peak in the solution to Your Turn 15.11.

Your Turn 15.14

Think

Consult Figure 15-12 and Table 15-2 to help you determine where a C≡C stretch could appear.

Solve

Triple-bond stretching modes appear between 2000 and 2500 cm^{-1}. In the first spectrum, that triple bond has moderate intensity, whereas in the second spectrum, it is weak. The peaks are circled below.

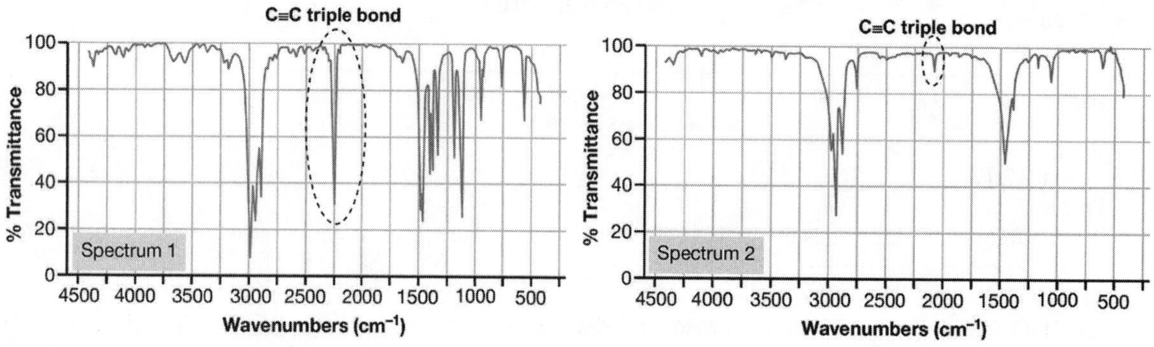

Your Turn 15.15

Think

Consult Table 15-2 to help you determine where alkane sp^3 C–H and alkene sp^2 C–H stretches appear. Do the C–H stretches present in the spectra correspond to the type of C–H bonds in the molecule?

Solve

From Table 15-2, alkane sp^3 C–H stretches occur between 3000 and 2800 cm^{-1} and vary in intensity. Alkene sp^2 C–H stretches occur between 3100 and 3000 cm^{-1} and are generally weak. A dashed line is drawn on each IR spectrum to assist in visualizing the dividing line between alkane and alkene C–H stretches.

(a) Alkane C–H stretches are intense and occur at 3000–2800 cm^{-1}.

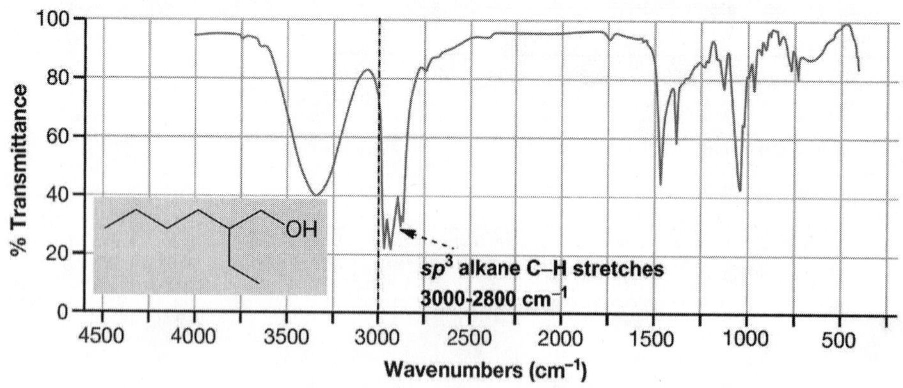

(b) Alkane C–H stretches are moderately intense and occur at 3000–2800 cm^{-1}.

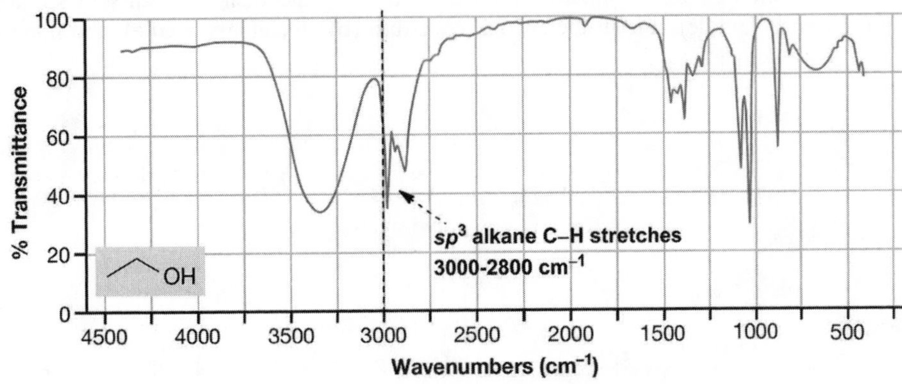

(c) Alkene C–H stretches are weak and occur around 3100 cm^{-1}, no alkane C–H stretches are present.

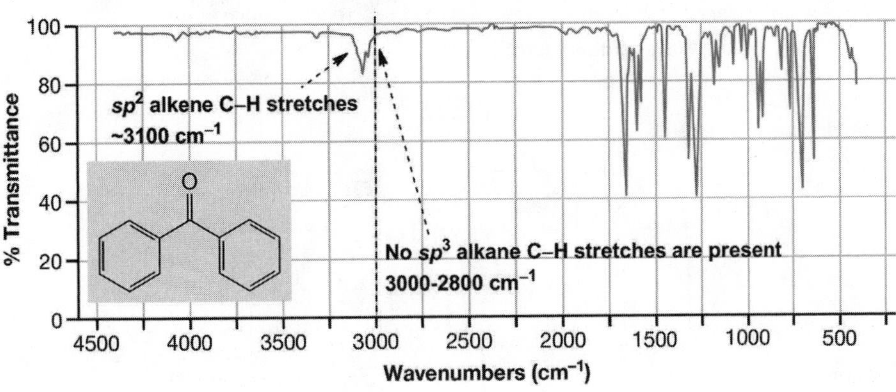

Your Turn 15.16

Think

Consult Table 15-2 to help you determine where an aldehyde C–H stretch should appear.

Solve

The aldehyde C–H stretch appears as two bands—one at ~2820 cm^{-1} and the other ~2720 cm^{-1}. The band at ~2720 cm^{-1} is generally easy to spot on the IR spectrum (see below). However, the band at ~2820 cm^{-1} is often masked by the alkane CH stretches, but can be a shoulder.

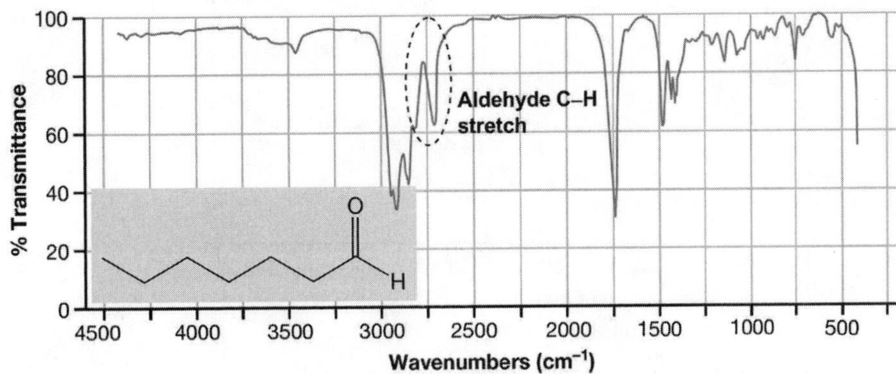

Your Turn 15.17

Think

Consult Table 15-2 to help you determine where an amine/amide N–H stretch should appear. How is the IR absorption band of an N–H stretch different for primary, secondary, and tertiary amines/amides?

Solve

The N–H stretching modes are the moderately intense, moderately broad peaks near 3300 cm^{-1}. There are two in the first spectrum **(a)** (primary amide), one in the second spectrum **(b)** (secondary amide), and none in the third spectrum **(c)**.

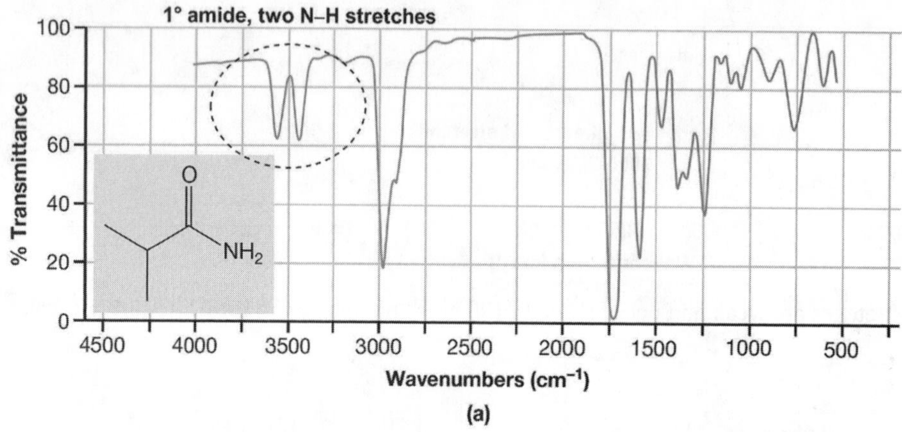

(a)

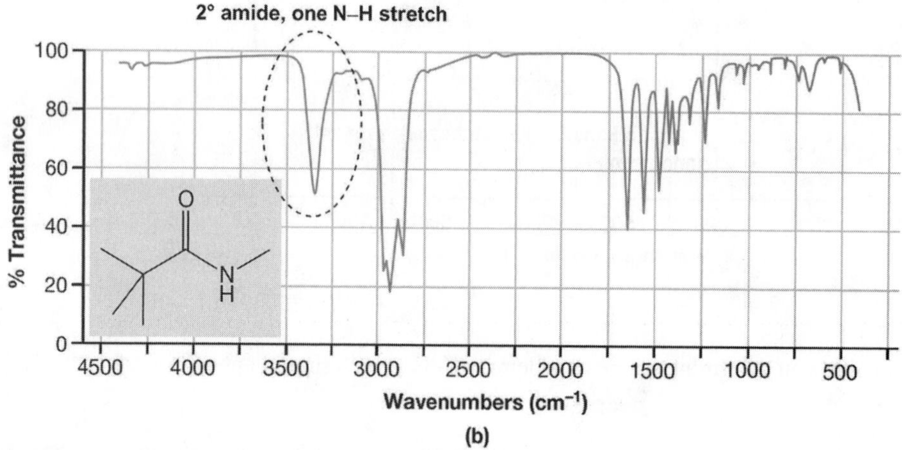

(b)

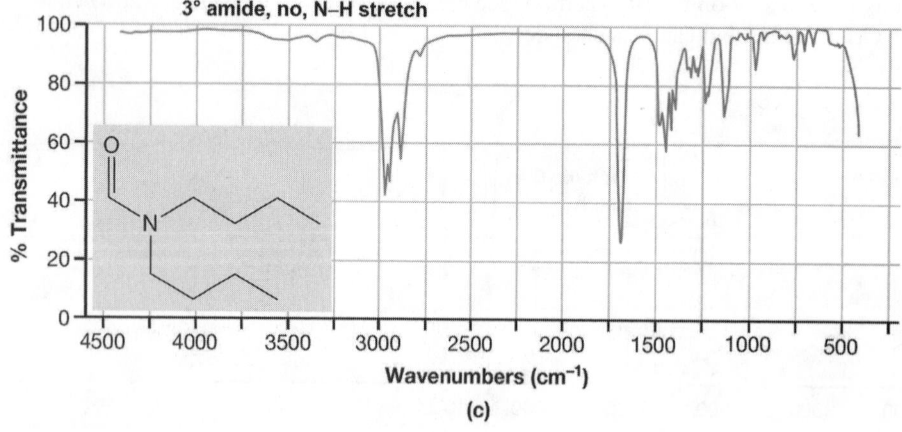

(c)

Your Turn 15.18

Think

Consult Figure 15-9 to help you identify the symmetric N–H and the asymmetric N–H stretch. In a symmetric stretch, should the two bonds stretch at the same time or different times? In an asymmetric stretch?

Solve

Primary amines/amides generally have two N–H stretches in the IR due to the presence of a symmetric and asymmetric stretch. The symmetric and asymmetric stretches are indicated by the arrows below. In the symmetric stretch, the two NH bonds stretch and compress together. In the asymmetric stretch, one NH bond stretches while the other shortens.

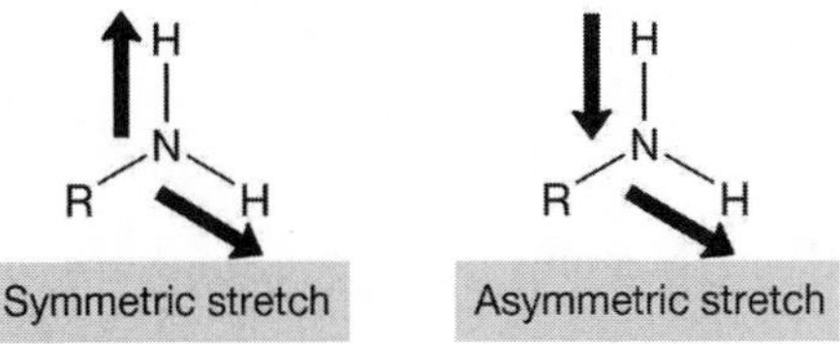

Your Turn 15.19, 15.20, 15.21, 15.22

Think

Consult Tables 15-2 and 15-3 to identify the presence and/or absence of the stretches/bends indicated.

Solve

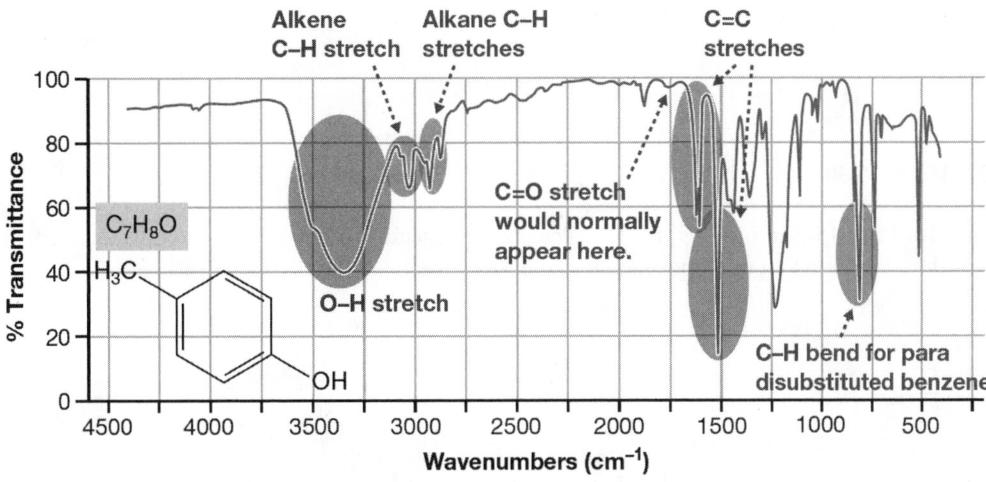

Your Turn 15.23, 15.24, 15.25, 15.26

Think

Consult Tables 15-2 and 15-3 to identify the presence and/or absence of the stretches/bends indicated.

Solve

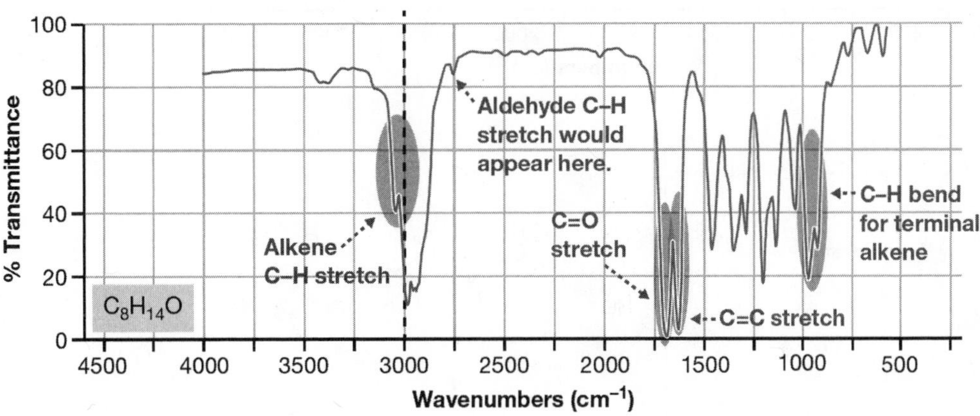

Two possibilities for Unknown 2

Your Turn 15.27

Think

In a saturated compound, how many bonds do each of the C and N atoms have? Are rings possible? Are multiple bonds possible?

Solve

A saturated molecule with six C atoms and one N atom is shown below. Each C atom should have four bonds and no lone pairs, and the N should have two bonds and one lone pair. The saturated molecule has 15 H atoms.

H_3C $\overset{H_2}{C}$ $\underset{H_2}{C}$ $\overset{H_2}{C}$ $\underset{H_2}{C}$ $\overset{H_2}{C}$ NH_2

Your Turn 15.28, 15.29, 15.30

Think

Consult Tables 15-2 and 15-3 to identify the presence and/or absence of the stretches/bends indicated. What is the IHD? How does this help you in proposing a possible structure?

Solve

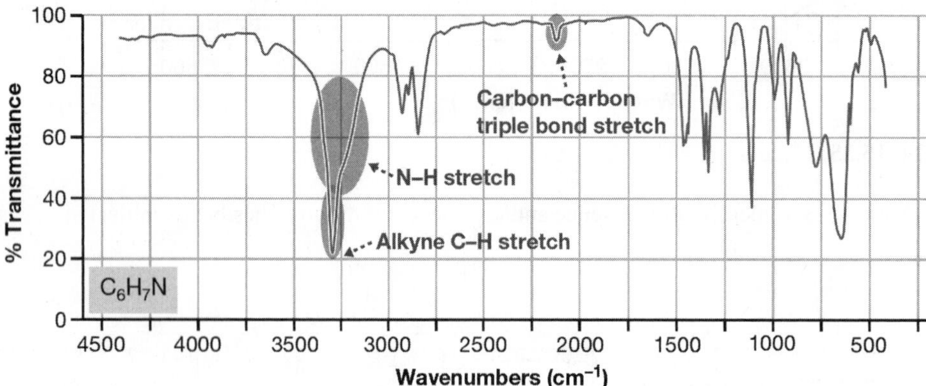

Zoom in of N—H and C≡C—H region

N—H stretch

C≡C—H stretch

$HC{\equiv}C$ $\overset{H_2}{C}$ $\underset{H}{N}$ $\overset{H_2}{C}$ $C{\equiv}CH$

Possible structure for Unknown 3

Chapter Problems

Problem 15.1

Think

What is the intensity of the light source (I_{source})? What is the intensity of the detected light ($I_{detected}$)? How do these values relate to transmittance? See Equation 15-1. What is the relationship between the absorbed light and the percentage of light transmitted? See Equation 15-2.

Solve

Using the formula from Equation 15-1, $\%T = (I_{detected}/I_{source}) \times 100\% = (2.5 \times 10^{-5}/1.0 \times 10^{-4}\ \text{W}) \times 100\% = 25\%$. Then, using the formula from Equation 15-2, $A = 2 - \log(\%T) = 2 - \log(25) = 0.60$.

Problem 15.2

Think

What is Beer–Lambert's law? How do you solve this equation for molar absorptivity (ε)? What units are given for the absorbance, concentration, and path length?

Solve

Beer–Lambert's law is given in Equation 15-3: $A = \varepsilon lc$, where ε is the molar absorptivity. Solve the equation for molar absorptivity.

$$\varepsilon = \frac{A}{lc} = \frac{0.78}{(1.0\ \text{cm})(0.600 \times 10^{-5}\ \text{M})} = 130{,}000\ \text{M}^{-1}\text{cm}^{-1}$$

Problem 15.4

Think

Which is the longest-wavelength absorption? What is its λ_{max}? What energy does a photon with that wavelength possess? How does this energy relate to the energy difference between the HOMO and the LUMO?

Solve

The HOMO–LUMO transition is indicated by the longest-wavelength UV-vis absorption. In the spectrum, that absorption appears at 280 nm, whose photon energy is

$$E = \frac{hc}{\lambda} = \frac{(6.626 \times 10^{-34}\ \text{J}\cdot\text{s})(3.00 \times 10^{-8}\ \frac{\text{m}}{\text{s}})}{280 \times 10^{-9}\ \text{m}} = 7.00 \times 10^{-19}\ \text{J}$$

Problem 15.5

Think

What does the MO energy diagram look like for $CH_2{=}CH_2$ (see Chapter 3)? How many π electrons are present? What happens to the electron in the HOMO when it absorbs a photon of energy?

Solve

The absorption of a UV–vis photon in ethene is shown below. The ground state configuration is shown on the left. The electron in the bonding π_1 HOMO is promoted to the antibonding π_2 orbital. This molecule underwent a $\pi \rightarrow \pi^*$ transition.

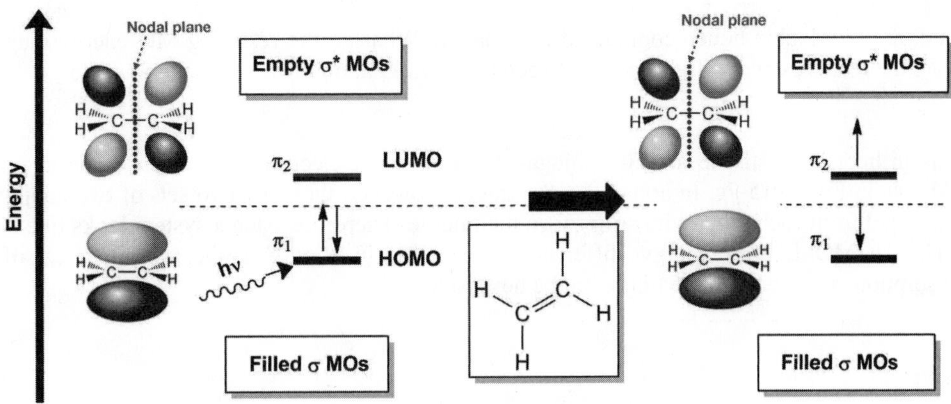

Problem 15.6

Think

How many conjugated π bonds are present? What trend do you observe in λ_{max} with the number of conjugated π bonds? What is the value of λ_{max} in the table that has the closest number of conjugated π bonds compared to deca-1,3,5,7,9-pentaene? Would you expect the λ_{max} value to be higher or lower?

Solve

A trend can be seen:

# Conjugated C=C Bonds	UV λ_{max}	Change in λ_{max}
1	177	–
2	224	47
3	274	50
4	290	16

The compound given, deca-1,3,5,7,9-pentaene, has five conjugated C=C double bonds, so we should expect the wavelength to be longer than 290 nm. But with each additional conjugated double bond, the change is less significant. Since going from three to four conjugated double bonds was about a 16-nm change, going from four to five conjugated double bonds will be about 10-nm change, giving about 300 nm.

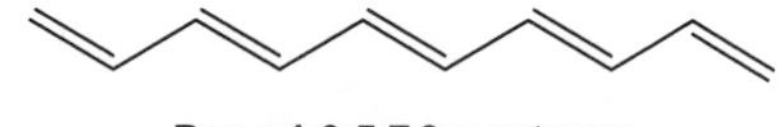

Deca-1,3,5,7,9-pentaene
5 conjugated π bonds
λ_{max} ~ 300 nm

Problem 15.7

Think

How many conjugated π bonds are present? What trend do you observe in λ_{max} with the number of conjugated π bonds in aldehydes? What is the value of λ_{max} of the aldehyde in the table that has the closest number of conjugated π bonds compared to penta-2,4-dienal? Would you expect the λ_{max} value to be higher or lower?

Solve

This compound has two C=C double bonds conjugated to an aldehyde. With no C=C double bonds conjugated, the wavelength is 280 nm. Adding one C=C double bond increases the wavelength by 60 nm, to 340 nm. Adding the second conjugated double bond will increase the wavelength even more, but by <60 nm (e.g., perhaps by 40 nm). So we estimate that the wavelength would be roughly 380 nm.

Penta-2,4-dienal
2 conjugated π bonds to C=O
λ_{max} ~ 380 nm

Problem 15.9

Think

In each compound, are the double bonds conjugated or isolated? What do the resulting MO energy diagrams look like? What are the relative energy differences between the HOMO and LUMO in each molecule?

Solve

The double bonds in hepta-1,3,5-triene are all conjugated, so all six π electrons are in the same π system, resembling that shown in Figure 15-7c. In nona-1,3,6,8-tetraene, however, there are two sets of two conjugated π bonds that are isolated from each other with an sp^3 C in the middle. Therefore, each π system looks like that in Figure 15-7b. The HOMO–LUMO energy difference is smaller in Figure 15-7c, so the λ_{max} of the corresponding absorption is longer. See the figure on the next page.

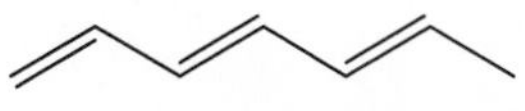

Hepta-1,3,5-triene
3 conjugated π bonds
Smaller HOMO–LUMO gap
Longer λ_{max}

Nona-1,3,6,8-tetraene
2 sets of 2 conjugated π bonds

Problem 15.11

Think

For each compound, does the HOMO–LUMO transition correspond to a $\pi \rightarrow \pi^*$ transition or an $n \rightarrow \pi^*$ transition? In general, what are the relative energies of these types of transitions?

Solve

The HOMO–LUMO transition for $H_2C{=}NH_2^+$ is $\pi \rightarrow \pi^*$, as shown in Figure 15-8a. The transition for $H_2C{=}OH^+$, however, is $n \rightarrow \pi^*$, similar to Figure 15-8b, given the presence of both a lone pair on O and the π bond. Accordingly, the HOMO–LUMO transition for $H_2C{=}OH^+$ should require less energy, and will thus appear at a longer wavelength.

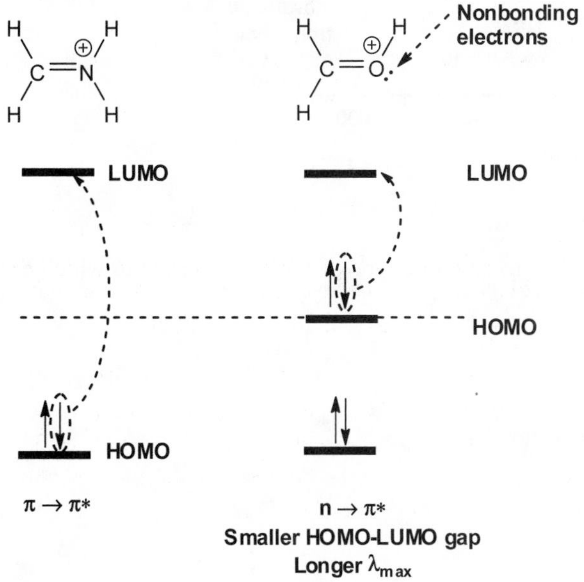

Nonbonding electrons

LUMO

HOMO

LUMO

HOMO

$\pi \rightarrow \pi^*$

$n \rightarrow \pi^*$
Smaller HOMO-LUMO gap
Longer λ_{max}

Problem 15.12

Think

What is the geometry about the central C atom where the vibrational motion takes place? Do bond lengths change? Do bond angles or dihedral angles change? Does the vibrational motion break the plane of the paper?

Solve

This is an in-plane bend. It is not a stretching mode because no bond lengths are changing during the period of the vibration. It is bending because the bond angles are changing—namely, the H–C–O bond angles are changing. It is in-plane bending because all of the atoms that initially lie in the same plane remain in that plane (C atom is trigonal planar) throughout the course of the vibration.

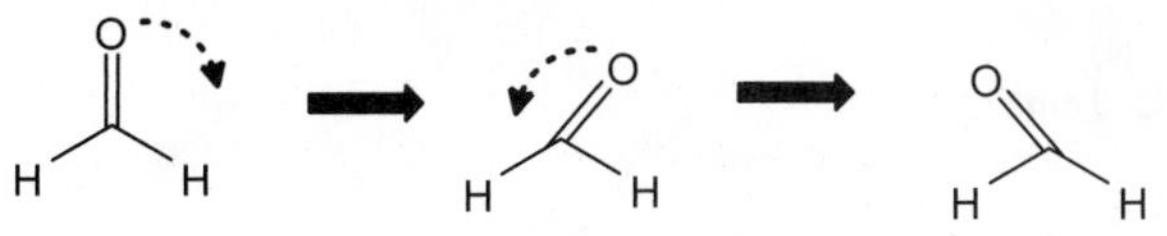

In-plane bend

Problem 15.13

Think

What is the relationship between wavenumbers and frequency? How is frequency related to wavelength?

Solve

According to Equations 15-6 and 15-7, wavenumbers (cm^{-1}) = $\nu(Hz)/100\ c = (c/\lambda)/(100\ c) = 1/(100\ \lambda)$. So $\lambda = 1/(100 \times \text{wavenumbers})$. For the three absorptions in Figure 15-10, we solve

$\lambda = 1/(100 \times 2961) = 3.4 \times 10^{-6}$ m

$\lambda = 1/(100 \times 1717) = 5.8 \times 10^{-6}$ m

$\lambda = 1/(100 \times 1167) = 8.6 \times 10^{-6}$ m.

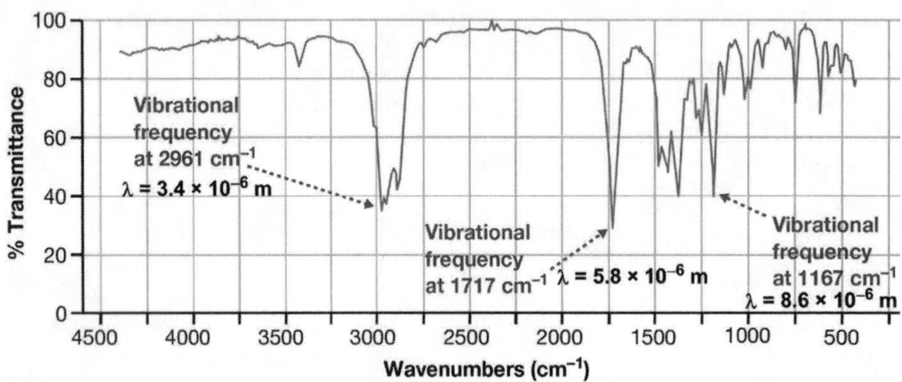

Problem 15.15

Think

What type of bond does each arrow indicate? Is there more than one type of functional group to which each of those bonds could belong?

Solve

(a) Aromatic C=C, 1450–1550 cm^{-1}
(b) Aldehyde C–H, 2820 and 2720 cm^{-1}
(c) Aldehyde C=O, 1720 cm^{-1}
(d) Alkyne C≡C, 2200 cm^{-1}
(e) Alkyne C–H, 3300 cm^{-1}
(f) Secondary amine, N–H, 3400 cm^{-1}
(g) Alcohol C–O, 1100 cm^{-1}
(h) Alcohol O–H, 3400 cm^{-1}
(i) Nitrile C≡N, 2200 cm^{-1}

(a) 1450-1550 cm⁻¹
(b) 2720 & 2820 cm⁻¹
(c) 1720 cm⁻¹
(d) 2200 cm⁻¹
(e) 3300 cm⁻¹
(f) 3400 cm⁻¹
(g) 1100 cm⁻¹
(h) 3400 cm⁻¹
(i) 2200 cm⁻¹

Problem 15.17

Think

Considering the model of masses connected by a spring, are the masses the same? Is the stiffness of each spring the same? How do these factors govern vibrational frequency? How does vibrational frequency affect IR absorption frequencies?

Solve

(a) Because it is stronger than the double bond, the triple bond acts as a stiffer spring and has the higher absorption frequency, and it vibrates faster.

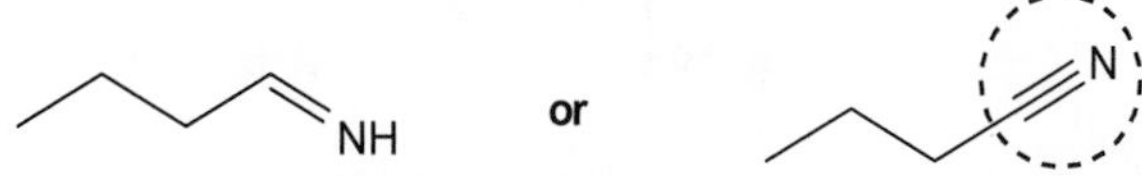

(b) The compound with the C=N double bond has the CN stretch at a higher frequency. Because a C=N double bond is stronger than a C–N single bond, it acts as a stiffer spring and has the faster vibration.

(c) The triple bond with ^{12}C has the higher absorption frequency because ^{12}C has slightly less mass than ^{13}C, allowing the bond to vibrate faster.

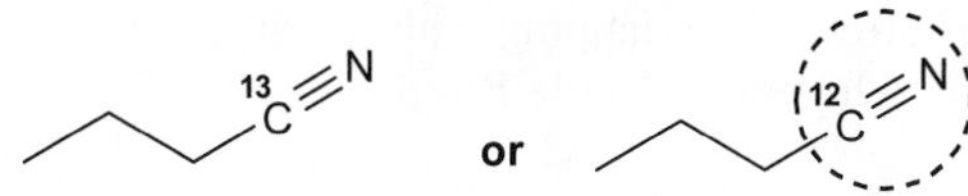

Problem 15.18

Think

Where is the C=C bond located in each spectrum? Which C=C peak has a higher intensity? Which alkene is more symmetrical? Which alkene, when stretched, has a larger change in dipole? What is the relationship between peak intensity and dipole change?

Solve

The spectrum on the left corresponds to 1-methylcyclohexene and the spectrum on the right corresponds to methylenecyclohexane. The small intensity of the C=C stretch around 1650 cm^{-1} in the spectrum on the left is indicative of a C=C stretch for which the dipole moment of the molecule does *not* change much during the period of the stretching vibration. In the spectrum on the right, the intensity of that peak is much greater, signifying a more profound change in dipole moment during the vibration. The C=C stretch of 1-methylcyclohexene should bring about a smaller change in dipole moment during the vibration because the molecule is more symmetric about the double bond than is methylenecyclohexane. 1-Methylcyclohexene is of the form $R_2C{=}CRH$, whereas methylenecyclohexane is of the form $R_2C{=}CH_2$. See the figure on the next page.

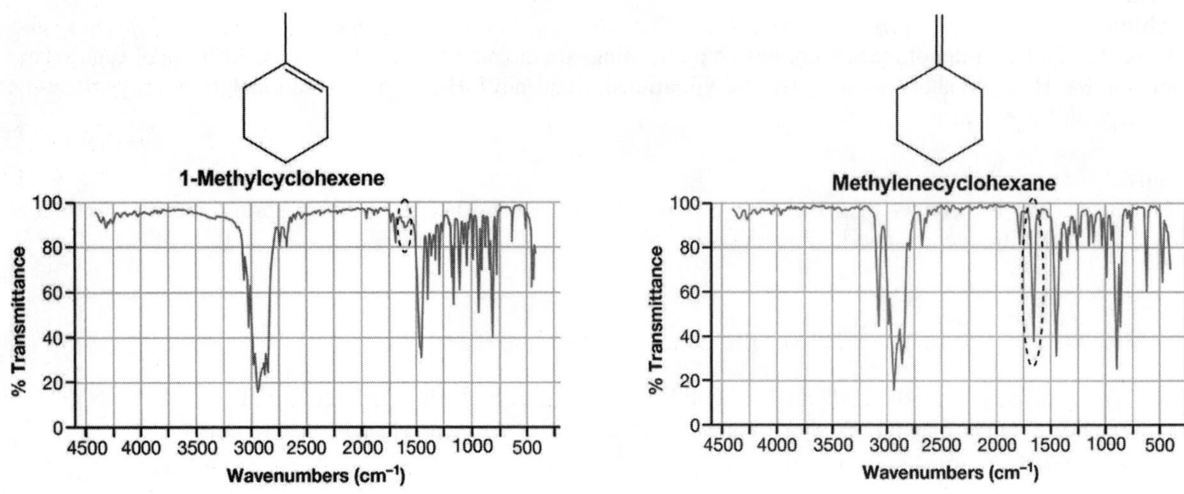

Problem 15.19

Think

What similarities and differences do you see in the structure of each isomer? What functional groups are present? How is the H–O stretch different for an alcohol versus a carboxylic acid?

Solve

The spectrum is consistent with isomer **A** because the OH stretch, overlapping with the alkane C–H stretches, indicates a carboxylic acid. Molecule **A** is a carboxylic acid, whereas molecule **B** is not.

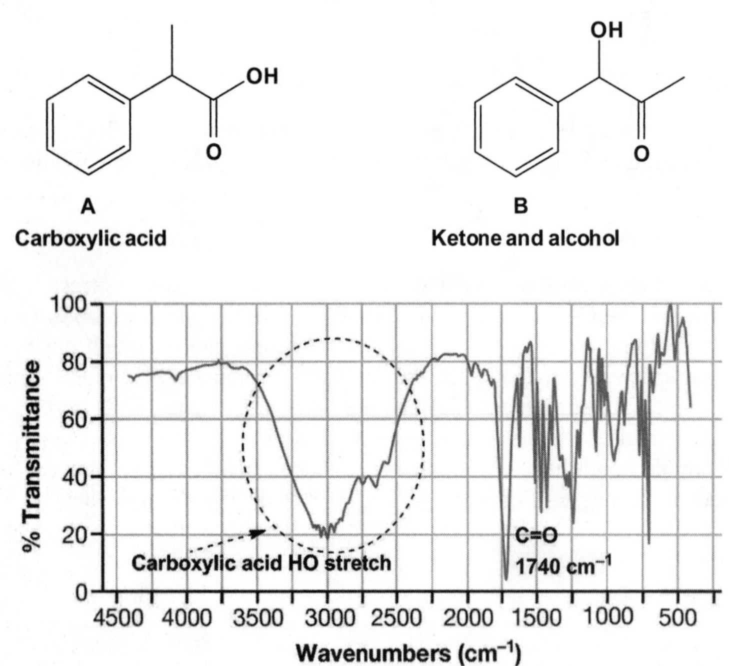

Problem 15.20

Think

What similarities and differences do you see in the structure of each isomer? Where is the C=O peak typically located in an IR spectrum? Which C=O is conjugated to the aromatic ring? How does conjugation affect the frequency of the IR stretch?

Solve

It corresponds to the compound **A**. The absorption at 1686 cm^{-1} corresponds to a C=O stretch that is conjugated to a C=C double bond. An isolated C=O of a ketone normally appears around 1720 cm^{-1}. In the compound on the left, the C=O group is conjugated to the benzene ring, but in the compound on the right, it is not.

A
Conjugated C=O, 1686 cm^{-1}

B
Unconjugated C=O, ~1720 cm^{-1}

Problem 15.21

Think

In which region of the IR spectrum does a triple-bond C≡N stretch occur? Would this stretch be more or less intense than a C≡C? Which bond has a larger change in dipole when stretched?

Solve

The spectrum on the left in Your Turn 15.14 most likely corresponds to a nitrile. This is because the C≡N bond has a relatively large bond dipole, and as it stretches and compresses, the bond dipole oscillates. As mentioned in the chapter, such an oscillation of the bond dipole enhances the probability that a photon will be absorbed. By contrast, the weak absorption in the spectrum on the right is likely a C≡C triple bond, which often has a relatively small bond dipole.

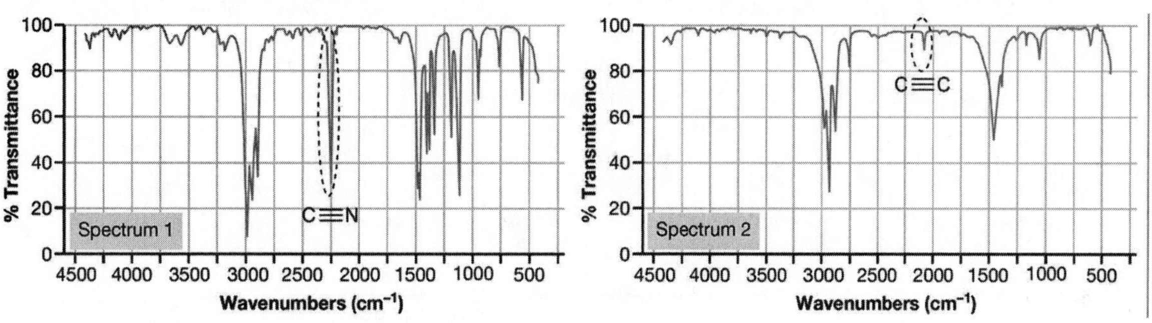

Problem 15.22

Think

What are the characteristic peaks for an aromatic ring? What are the characteristic peaks for the C=C of an alkene?

Solve

Spectrum 3 corresponds to the nonaromatic compound. The C=C stretch appears around 1650 cm^{-1}, indicative of a regular alkene C=C stretch. The C=C stretch in Spectrum 4 appears between 1450 and 1600 cm^{-1}, typical of aromatic rings.

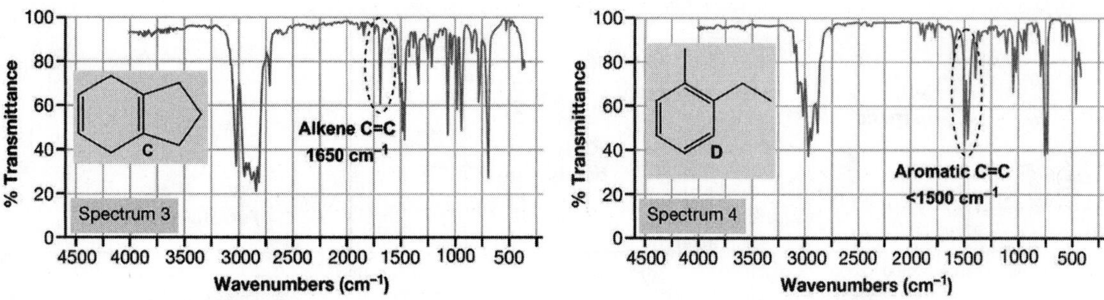

Problem 15.23

Think

How many C–H bonds are in each molecule? Are the C–H stretches from *sp*^{3}-, *sp*^{2}-, or *sp*-hybridized C atoms? How is the intensity of the C–H stretches related to the number of C–H stretches?

Solve

In all cases, there are C=C double bonds and C–O single bonds. The difference is in the ratio of alkane C–H bonds (stretches slightly below 3000 cm^{-1}, marked as a dashed line) to alkene C–H bonds (stretches slightly above 3000 cm^{-1}). In Spectrum 7, there are no alkane C–H stretches, corresponding to no alkane C–H bonds. Therefore, Spectrum 7 corresponds to molecule **E**. In the other two spectra, we see both types of stretches. Spectrum 6 shows a greater proportion of alkane C–H stretches than Spectrum 5, corresponding to a greater proportion of H atoms on *sp*^{3}-hybridized C atoms. In molecule **G**, there are 15 alkane C–H bonds and three alkene C–H bonds. In molecule **F**, that ratio drops to three alkane C–H bonds and three alkene C–H bonds. Therefore, Spectrum 5 corresponds to molecule **F** and Spectrum 6 corresponds to molecule **G**.

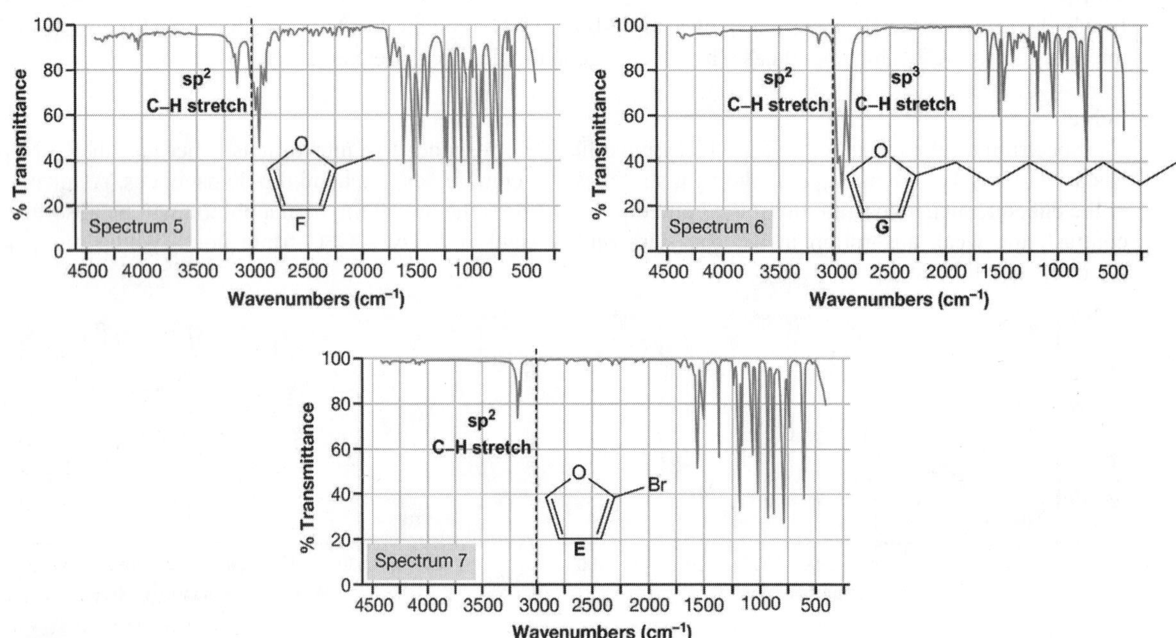

Problem 15.24

Think

Is there a peak suggestive of a C=C stretch? Are there peaks suggestive of *sp*^{2} C–H stretches? Are there strong *sp*^{2} C–H bending stretches?

Solve

Spectrum 8 indicates the presence of a C=C double bond. It has a clear C–H stretch above 3000 cm^{-1}, a C=C stretch at 1650 cm^{-1}, and strong C–H bending bands near 1000 cm^{-1}. The other spectrum, although it has a peak that is consistent with a C=C stretch, does not have a C–H stretch above 3000 cm^{-1}.

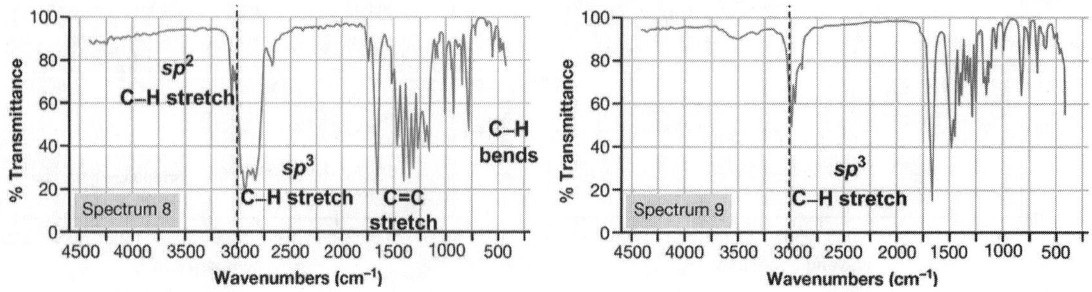

Problem 15.26

Think

In Solved Problem 15.25, what is the basic structural unit that we derived from the IR spectrum? How can the remaining atoms be attached without disturbing that basic structure? What kinds of bonds must connect those atoms?

Solve

The first structure below is the basic structural unit we derived from the spectrum, which accounts for four C atoms, three H atoms, and one O atom.

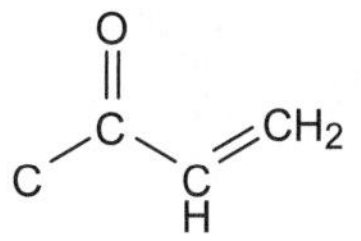

That leaves four C atoms and 11 H atoms yet to add, all of which can be attached only by single bonds—otherwise, we would exceed the compound's IHD of 2. Other possibilities include the following:

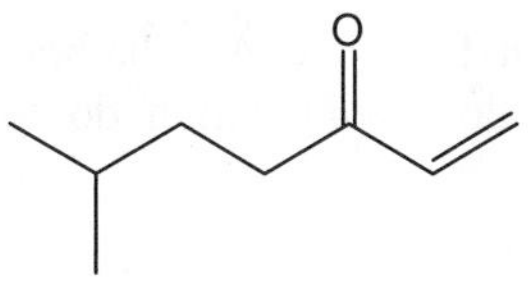

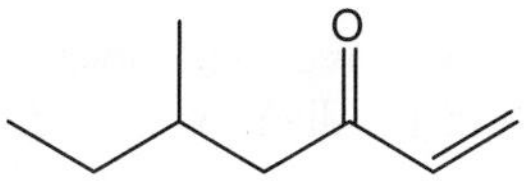

Problem 15.27

Think

What is the basic structural unit that we derived from the IR spectrum? How can the remaining atoms be attached without disturbing that basic structure? What kinds of bonds must connect those atoms?

Solve

From the IR spectrum we derived that the compound must have a terminal alkyne, a secondary NH, and an IHD of 4. From the λ_{max} given, we know that the π bonds should not be conjugated with each other or with the nonbonding electrons on N. The molecule below is possible because it, too, has these features.

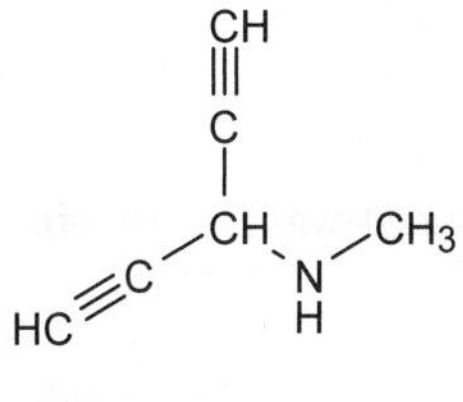

Problem 15.28

Think

Which structure has more conjugated C=C bonds? How is the number of conjugated C=C bonds related to the longest-wavelength of UV–vis absorption?

Solve

The β isomer has the longer-wavelength UV–vis absorption because it has a greater extent of conjugation. In the β isomer, there are 11 conjugated double bonds. In the α isomer, 10 double bonds are conjugated, and 1 is isolated. See the figure on the next page.

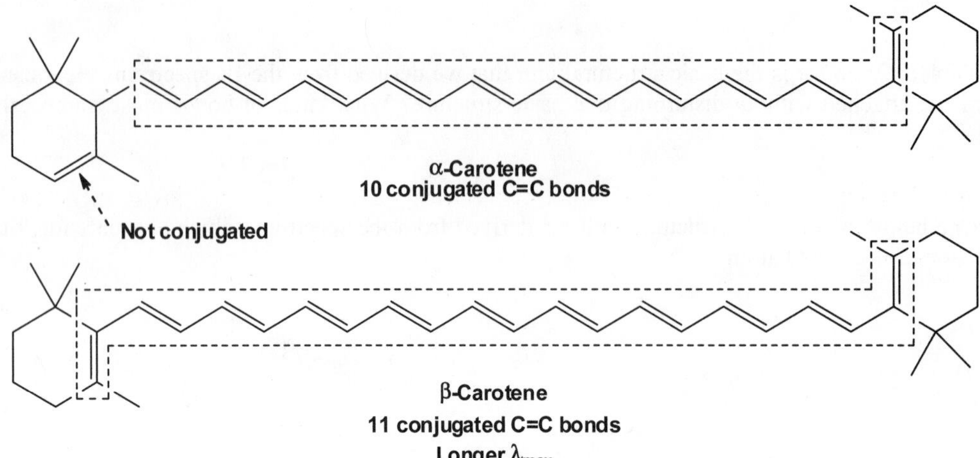

Problem 15.29

Think

Consult Table 15-1 to obtain the value of the longest-wavelength absorption for but-1,3-diene. Are the bonds in but-1,3-diene conjugated? Are the bonds in buta-1,2-diene conjugated? What does conjugation do to the wavelength of maximum absorption?

Solve

Buta-1,3-diene's longest-wavelength absorption is at 217 nm. Buta-1,2-diene's longest wavelength absorption appears at 178 nm. The difference comes from the fact that the double bonds in buta-1,2-diene are not conjugated, whereas those in buta-1,3-diene are. The adjacent double bonds in buta-1,2-diene are formed from perpendicular *p* orbitals, and to be conjugated, they must be parallel.

$H_3C-CH=C=CH_2$

Buta-1,2-diene
λ_{max} = 178 nm
Not conjugated

$H_2C=CH-HC=CH_2$

Buta-1,3-diene
λ_{max} = 217 nm
Conjugated

Problem 15.30

Think

What is the IHD? How many conjugated C=C bonds are likely present with a longest-wavelength absorption at 215 nm?

Solve

The IHD is 3 because the saturated compound with five C atoms has 12 H atoms. Three π bonds can account for this IHD, but only two of them are to be conjugated, because buta-1,3-diene, which has two conjugated C=C bonds, absorbs at 217 nm, which is very close to the wavelength of our unknown compound. Conjugation of two π bonds can be accomplished either by isolating the third π bond (as in a triple bond or adjacent double bonds) or by having the third IHD unit accounted for by a ring. See possible structures below.

$H_2C=CH-C\equiv C-CH_3$ $HC\equiv C-CH=CH-CH_3$ $H_2C=CH-HC=C=CH_2$

Problem 15.31

Think

What is the IHD? How many conjugated C=C bonds are likely present with a longest-wavelength absorption at 191 nm?

Solve

A compound with the formula C_7H_{12} has an IHD of 2. The longest-wavelength UV–vis absorption appears at 191 nm, indicating that there must be a π bond, but it cannot be conjugated. Along with the required ring, that π bond accounts for the IHD.

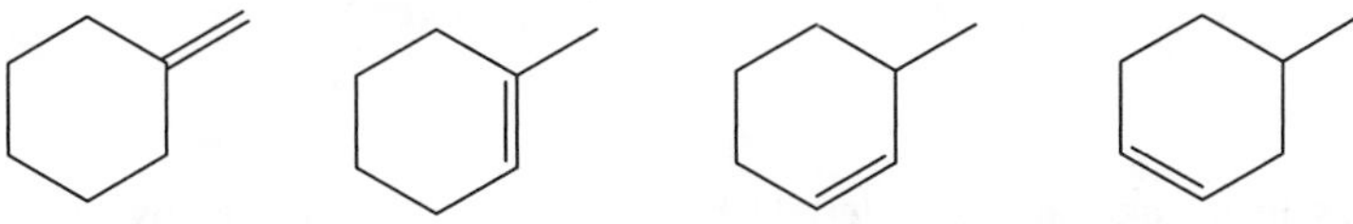

Problem 15.32

Think

How many conjugated C=C are present in each structure? What is likely the longest wavelength of UV–vis absorption? Which of these absorptions is in the visible region? Would this correspond to an observed color by our eyes?

Solve

When the pH is <8 or >14, the π systems on the rings are not conjugated through the central carbon, due to the sp^3 hybridization of that C atom. So the longest-wavelength absorption remains relatively short, lying in the UV region. When absorption takes place in the UV region, our eyes do not register any wavelengths missing (our eyes only detect light in the visible region), so the solution does not appear colored. Between a pH of 9 and 13, however, the central C atom is sp^2 hybridized, so all three π ring systems are conjugated. The much greater extent of conjugation increases the wavelength of the longest-wavelength absorption, placing it in the visible region of the spectrum. When light is absorbed from the visible region, our eyes detect the missing light, which registers as a color.

Phenolphthalein
pH <8
λ_{max} = UV region
Colorless

pH 9–13
λ_{max} = Visible region
Colored

pH >14
λ_{max} = UV region
Colorless

Problem 15.33

Think

What is the difference in structure between propyne and acetonitrile? How do the lone pairs on the N atom in acetonitrile affect the longest-wavelength absorption?

Solve

Acetonitrile has the longer wavelength absorption. Both have isolated triple bonds, but acetonitrile also has a lone pair. So whereas propyne's longest-wavelength absorption corresponds to a π-to-π* transition, that in acetonitrile corresponds to an n-to-π* transition. All else equal, the n-to-π* transition requires lower energy, or longer wavelength light.

$H_3C{-}C{\equiv}CH$
Propyne
π-to-π* transition
Shorter λ_{max}

$H_3C{-}C{\equiv}N{:}$
Acetonitrile
***n*-to-π* transition**
Longer λ_{max}

Problem 15.34

Think

How many conjugated π bonds are present in the starting material? In the product? Are there any lone pairs conjugated to π bonds? How does the number of conjugated π bonds affect the longest-wavelength absorption?

Solve

In reactions **(b)** and **(c)**, the λ_{max} would shift to a shorter wavelength because the product is less conjugated. In reactions **(a)** and **(e)**, λ_{max} would shift to a longer wavelength because the product is more conjugated. In **(d)**, the λ_{max} would shift to a longer wavelength because an n-to-π* is introduced as a result of the lone pair of electrons on the N atom of NH_2.

(a) Base → **Shift to longer λ_{max}**

(b) 1. $LiAlH_4$ 2. H_3O^+ → OH **Shift to shorter λ_{max}**

(c) Li, NH_3/EtOH → **Shift to shorter λ_{max}**

(d) NO_2 → Fe, HCl → NH_2 **Shift to longer λ_{max}**

(e) Cl, $AlCl_3$ → **Shift to longer λ_{max}**

Problem 15.35

Think

What is the structure of propenal? What does the energy diagram of the π MOs look like? What transition in propenal occurs at 340 nm? At 202 nm, is that a shorter or longer wavelength? Higher- or lower-energy photons? How do the lone pairs on the oxygen atom affect the energy required for a transition?

Solve

The molecule and energy diagram of the π system of MOs is shown below.

$H_2C{=}CH{-}CH{=}O$

π_4

π_3

Lone pairs

π_2

π_1

The longest-wavelength absorption at 340 nm corresponds to the n-to-π* transition, in which an electron is promoted from a nonbonding MO to π_3 (lower energy, longer wavelength). The absorption at 202 nm, being lower in wavelength, corresponds to a higher-energy transition, consistent with a regular π-to-π* transition from the conjugated pair of double bonds. This would specifically be from π_2 to π_3. Notice that in 1,3-butadiene, which does not have any lone pairs and, therefore, does not have an n-to-π* transition possible, the longest-wavelength absorption is at 217 nm, consistent with the 202 nm π-to-π* transition in acrolein.

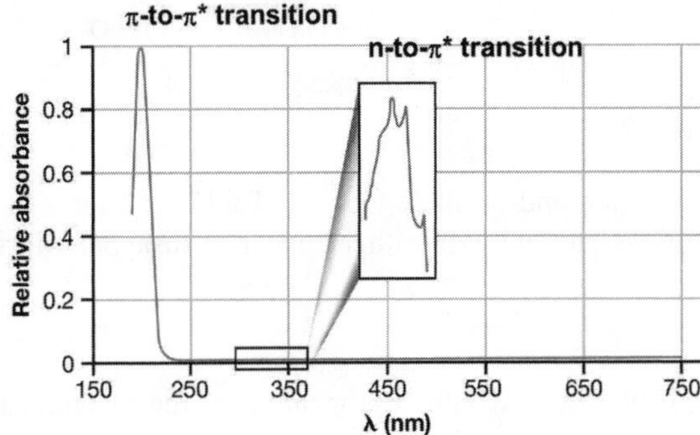

Problem 15.36

Think

How many conjugated π bonds are present in both benzene and hexa-1,3,5-triene? Which compound has a longer-wavelength absorption? Which compound has a lower-energy π-to-π* transition? How does the aromaticity of benzene affect the HOMO–LUMO transition?

Solve

Benzene and hexa-1,3,5-triene both have π systems consisting of three conjugated double bonds. The longest-wavelength absorption of hexa-1,3,5-triene is at 274 nm, which is significantly longer than that for benzene. The shorter-wavelength light necessary for the HOMO–LUMO transition in benzene indicates that the energy difference between benzene's HOMO and LUMO is greater than that for hexa-1,3,5-triene. In turn, this reflects the unusual stability of benzene's π system from its aromaticity. By contrast, hexa-1,3,5-triene is nonaromatic.

Problem 15.37

Think

Identify the functional group to which each indicated bond belongs. Consult Table 15-2 for a listing of characteristic absorption frequencies in IR spectroscopy.

Solve

The stretching frequencies are given below.

(a) Alkene C=C, 1650 cm^{-1}
(b) Ketone C=O, 1720 cm^{-1}
(c) Alcohol C–O, 1100 cm^{-1}
(d) Alcohol O–H, 3400 cm^{-1}
(e) Alkyne C–H, 3300 cm^{-1}
(f) Alkyne C≡C, 2200 cm^{-1}
(g) Amide C=O, 1650 cm^{-1}
(h) Amide N–H, 3300 cm^{-1}
(i) Aromatic C=C, 1500 cm^{-1}
(j) Aldehyde C–H, 2720 and 2820 cm^{-1}
(k) Aromatic sp^2 C–H, 3050 cm^{-1}
(l) Carboxylic acid C=O, 1740 cm^{-1}
(m) Carboxylic acid O–H, 2800 cm^{-1}

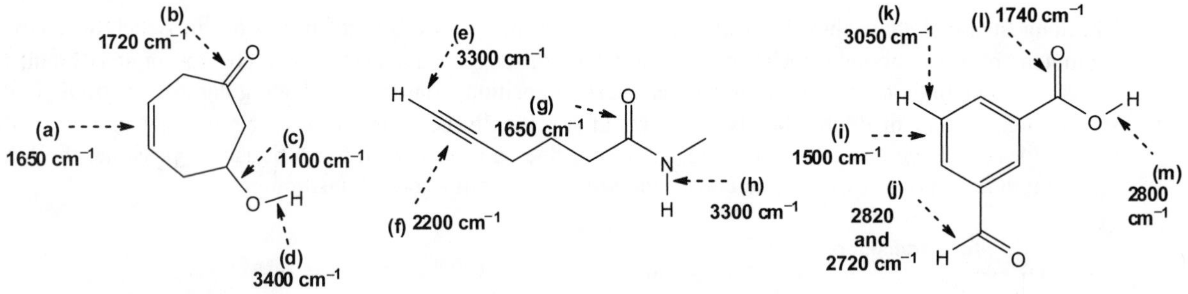

(b) 1720 cm^{-1} (a) 1650 cm^{-1} (c) 1100 cm^{-1} (d) 3400 cm^{-1} (e) 3300 cm^{-1} (g) 1650 cm^{-1} (f) 2200 cm^{-1} (h) 3300 cm^{-1} (k) 3050 cm^{-1} (l) 1740 cm^{-1} (i) 1500 cm^{-1} (j) 2820 and 2720 cm^{-1} (m) 2800 cm^{-1}

Problem 15.38

Think

Identify the different functional groups present in the reactant and product. Consult Table 15-2 for a listing of characteristic absorption frequencies in IR spectroscopy. What peak will either appear or disappear during the course of the reaction?

Solve

(a) An OH stretch (3400 cm^{-1}) and a C–O stretch (1100 cm^{-1}) would disappear from the alcohol, and the characteristic IR bands for an alkene (3050 cm^{-1}, 1650 cm^{-1}) would appear.

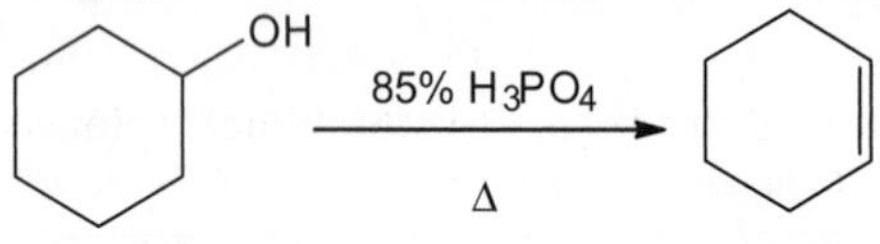

Alcohol —85% H_3PO_4, Δ→ Alkene

(b) The very broad OH stretch (3000–2500 cm^{-1}) from the carboxylic acid would disappear.

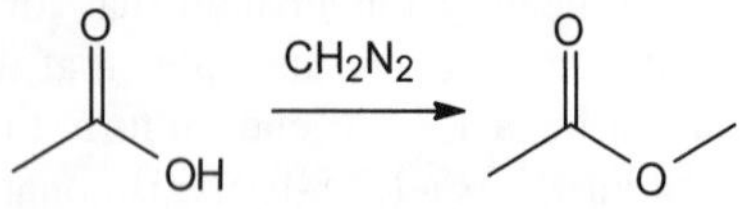

Carboxylic acid —CH_2N_2→ Ester

(c) Two NH stretching bands (3400 cm^{-1}) from the primary amine would become a single band from the secondary amine.

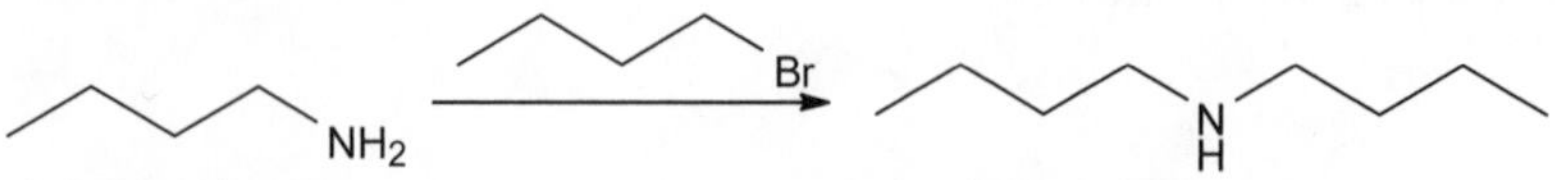

Primary amine —(butyl bromide)→ Secondary amine

(d) A single NH stretching band (3400 cm^{-1}) would disappear from the secondary amine. Tertiary amines have no characteristic IR absorption bands above 3,000 cm^{-1} because they have no N–H bonds.

H N → CH_3Br → N

Secondary amine **Tertiary amine**

(e) An OH stretching band (3400 cm^{-1}) would appear for the alcohol.

O → NaOH / H_2O → OH, OH

Epoxide (ether) **Alcohol**

(f) A C=C stretching band (1600 cm^{-1}) and an alkene C–H stretch (3050 cm^{-1}) from the conjugated ketone would disappear. Also the C=O stretch would move higher by about 30 cm^{-1} because the conjugation in the reactant is not present in the product.

O → H_2 / Pd → O

Conjugated ketone **Ketone**

(g) The C=C stretching bands for an aromatic compound (1450–1550 cm^{-1}) would become a simpler C=C stretching band (1650 cm^{-1}) for the alkene. Also sp^3 C–H stretches would appear, 2800–3000 cm^{-1}.

Li / NH_3/EtOH

Aromatic ring **Alkene**

(h) The strong alkyne *sp* C–H stretch would disappear at (3300 cm^{-1}). Also, the C≡C stretch would become less intense because the two C atoms of the terminal alkyne are more distinct compared to the internal alkyne and would give rise to a less significant bond dipole change.

≡CH → NaOH / Δ

Terminal alkyne **Internal alkyne**

(i) The conjugated ketone C=O stretch (1680 cm^{-1}) would disappear, and an OH stretch (3400 cm^{-1}) and C–O stretch (~1100 cm^{-1}) from the alcohol would appear.

O → 1. $LiAlH_4$ / 2. H_3O^+ → OH

Conjugated ketone **Alcohol + alkene**

Problem 15.39

Think

What functional group is present in each compound? How do those functional groups differ in their characteristic IR stretches?

Solve

The easiest way to distinguish the two compounds is by the appearance of the OH stretching band. In the compound on the left, the OH group is part of an alcohol, which is relatively broad and appears between 3200 and 3600 cm^{-1}. The OH group in the molecule on the right is part of a carboxylic acid, so the OH stretching band is significantly broader, and appears between 2500 and 3000 cm^{-1}.

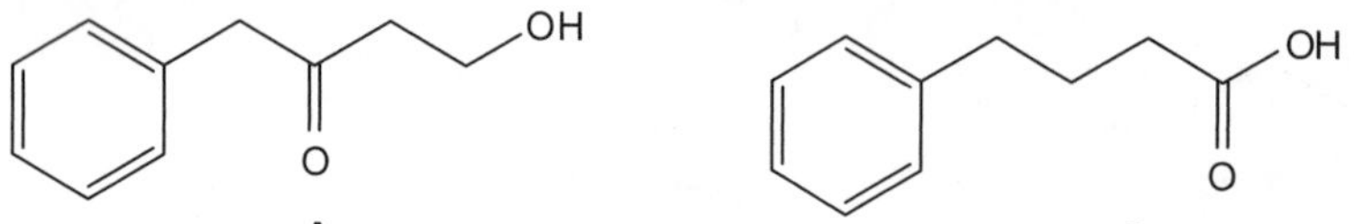

OH
O
A
Aromatic ring
Ketone
Alcohol

OH
O
B
Aromatic ring
Carboxylic acid

Problem 15.40

Think

What similarities do the two isomers possess in terms of functional groups? What differences are present between these two isomers? How many conjugated C=C bonds are present in each isomer? Is this difference evident in the IR spectrum? In the UV–vis spectrum?

Solve

In the IR spectrum, we would expect to see C=C stretching bands for both molecules. Also, neither spectrum would contain an alkene C–H stretch. UV–vis spectroscopy, however, can distinguish between the two compounds because the extent of conjugation is different. In molecule **C**, the three double bonds are conjugated, whereas in molecule **D**, two are conjugated and one is isolated. So the longest-wavelength UV–vis absorption would be longer for **C** than for **D**.

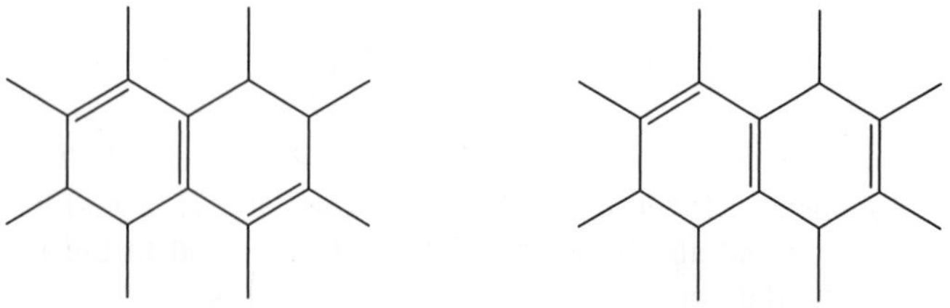

C
C=C
No sp^2 CH
3 conjugated π bonds
Longer λ_{max}

D
C=C
No sp^2 CH
2 conjugated π bonds
Shorter λ_{max}

Problem 15.41

Think

What structural feature is present that indicates there should be another resonance structure? How many curved arrows must be applied to convert to the other resonance structure? In that other resonance structure, what type of bond connects the C and O? Is that bond stronger or weaker than a C=O double bond?

Solve

The higher frequency reflects the stronger C=O bond in a ketone and the lower frequency reflects a weaker C=O bond in an amide. The amide N atom has a lone pair, which can participate in resonance with the C=O bond, as shown below. The amide resonance structure introduces significant C–O single-bond character to the hybrid and lowers the C=O stretching frequency.

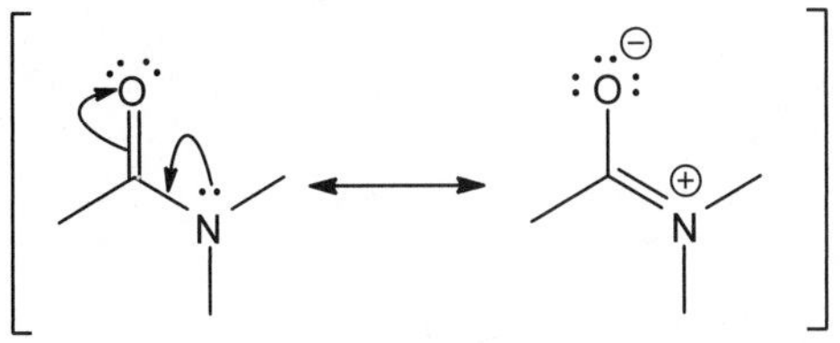

Problem 15.42

Think

Consult the solution to Problem 15.41. How is the resonance structure similar for the amide and the ester? Is the O atom or N atom more electronegative? On which atom is the positive charge more stable? How does this affect the contribution of the resonance structure? How is this reflected in the C=O stretching frequency?

Solve

(a) The resonance structure for methyl acetate is shown below.

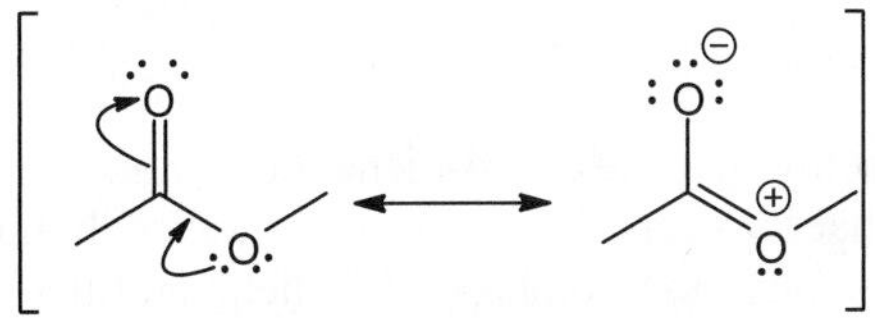

(b) Since the C=O stretching frequency of the ester is higher than that of the amide, it suggests that the resonance structure that introduces C–O character has a *less* significant contribution. Oxygen is more electronegative than nitrogen, and the positive charge is less stable on the O atom compared to the N atom. This results in less contribution of the resonance structure.

Problem 15.43

Think

Consult the solution to Problems 15.41. How is the resonance structure similar for the acid chloride and the amide? From which orbital in Cl do the π electrons in the resonance structure originate? How effective is the overlap in the $C=Cl^+$ bond? Is this more or less effective than in the $C=N^+$ bond?

Solve

The C=O in the acid chloride is stronger than the C=O bond in the amide, evidenced by the higher stretching frequency (1806 cm^{-1} compared to 1650 cm^{-1}). This suggests that the lone pair on the Cl atom is less able to contribute to the resonance hybrid than the N atom lone pairs. The lone pair on Cl does not contribute as much because the electrons on the Cl atom originate in the 3*p* orbital, which does not overlap effectively with the 2*p* orbital on C.

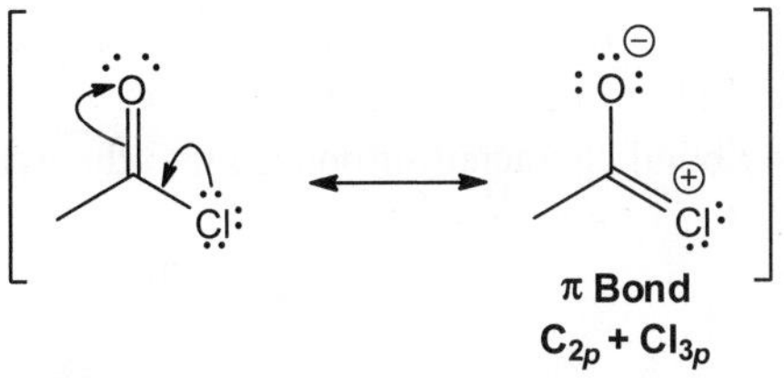

π Bond

$C_{2p} + Cl_{3p}$

Problem 15.44

Think

If the C=O bond is conjugated to a C=C, is a resonance structure possible? What kind of bond connects C and O in that resonance structure? How does this lower the stretching frequency of the C=O?

Solve

The resonance structure is shown below.

The resonance structure introduces some C–O character into the resonance hybrid. This lowers the C=O stretching frequency because a single bond is weaker compared to a double bond.

Problem 15.45

Think

Draw the structures of both cyclohepta-1,3-diene and 3-methylcyclohexa-1,4-diene. Which compound has conjugated C=C bonds? What does conjugation do to the stretching frequency of the C=C bonds in the IR spectrum?

Solve

Cyclohepta-1,3-diene has conjugated C=C bonds and 3-methylcylcohexa-1,4-diene does not. Conjugation lowers the stretching frequency of the C=C in the same way that it lowers the frequency of a C=O bond (see the solution to Problem 15.44). Thus, the stretch at 1618 cm^{-1} belongs to cyclohepta-1,3-diene and the stretch at 1648 cm^{-1} belongs to 3-methylcyclohexa-1,4-diene.

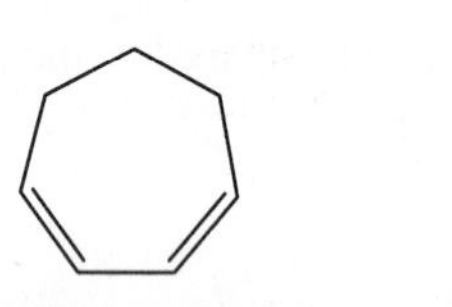

Cyclohepta-1,3-diene
1618 cm^{-1}

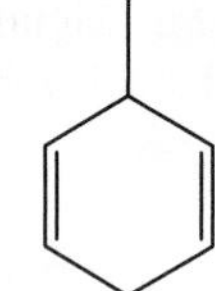

3-Methylcyclohexa-1,4-diene
1648 cm^{-1}

Problem 15.46

Think

Which compound has conjugated C=C bonds? What does conjugation do to the stretching frequency of the C=C bonds in the IR spectrum?

Solve

Aromatic rings have three conjugated C=C bonds and are aromatic. Conjugation lowers the vibrational frequency of a C=C bond in the same way that it does a C=O bond (see the solution to Problem 15.44).

Problem 15.47

Think

Which functional groups can participate in resonance? Is there C=O bond character introduced into the hybrid? If so, how does this affect the C–O stretching frequency?

Solve

The ester and carboxylic acid both have resonance structures that introduce C=O bond character into the C–O. This increases the stretching frequency of the C–O.

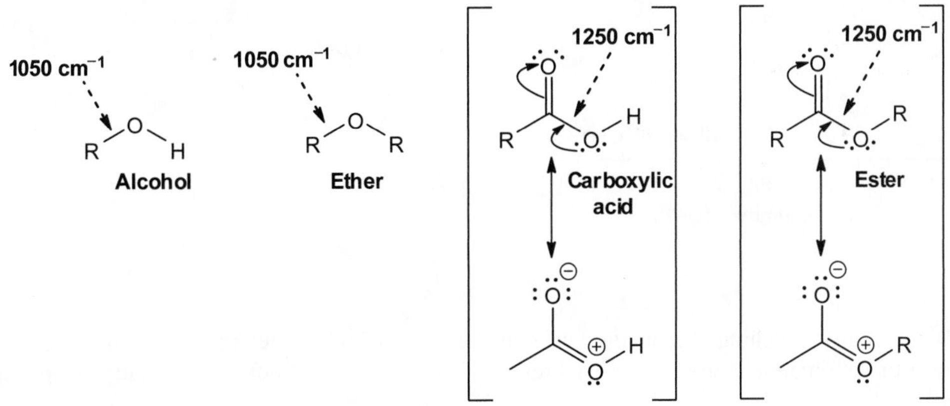

Problem 15.48

Think

Is there a resonance structure involving the C=O bond? In that resonance structure, what type of bond connects the C and O? Does contribution from that resonance contributor increase or decrease the C=O stretch frequency? In which functional group—a ketone or aldehyde—is that resonance structure better stabilized?

Solve

Both the ketone and aldehyde have resonance contributors as shown below. The contribution by that resonance structure is greater for the ketone because of the additional stabilization from the electron-donating alkyl groups attached to C$^+$. With greater contribution from that resonance structure, the hybrid has more single-bond character in the C=O bond, which weakens the bond and lowers the vibrational frequency.

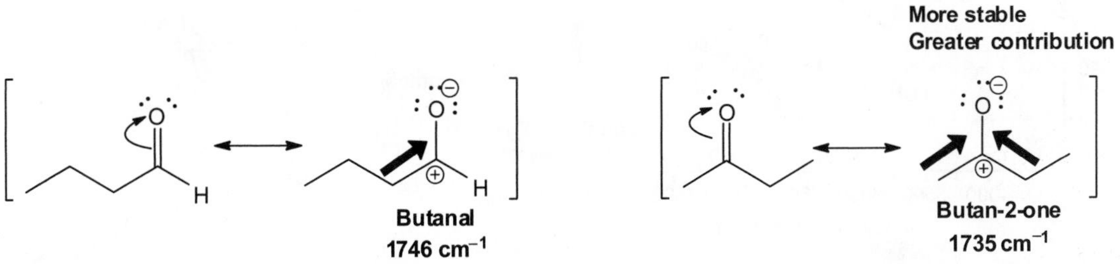

Problem 15.49

Think

What is the IHD? Are any stretching frequencies present that identify characteristic functional groups (see Table 15-2)? Draw those functional groups. What atoms are left in the formula? How can you piece the structure together?

Solve

The IHD of the molecular formula is 4. An IHD of 4 should lead to you consider a benzene ring. A reasonable structure would be aniline, C_6H_5–NH_2. The IR indicates a primary amine NH stretch with two bands (3300 cm^{-1} and 3400 cm^{-1}); furthermore, the C–H stretch (3050 cm^{-1}) indicates an alkene CH and no sp^3 CH bonds exist based on the IR spectrum. The C=C stretch frequencies (1600 and 1500 cm^{-1}) are too low to be normal alkenes but are consistent with aromatic C=C stretches. Finally, the C–H bends (700 and 750 cm^{-1}) indicate a monosubstituted benzene. See the figure on the next page.

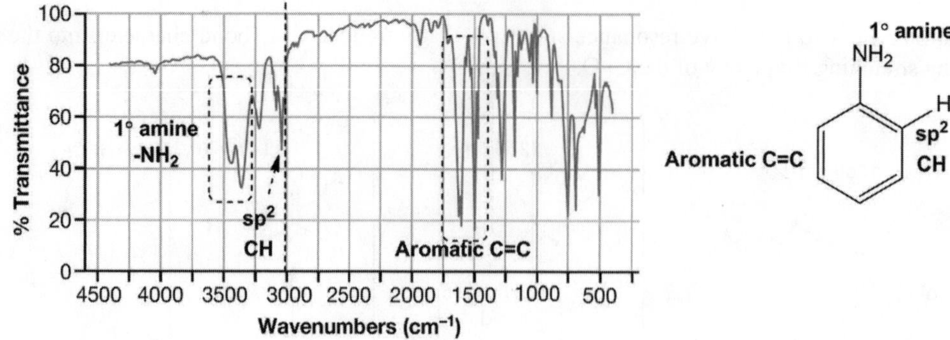

Problem 15.50

Think

What is the IHD? Are any stretching frequencies present that identify characteristic functional groups (see Table 15-2)? Draw those functional groups. What atoms are left in the formula? How can you piece the structure together?

Solve

The formula C_8H_8O has an IHD of 5, which would accommodate an aromatic ring and either another ring or another double bond. A C=O bond (1720 cm^{-1}) is indicated, and the unique C–H stretch of an aldehyde (2820 and 2720 cm^{-1}) is also visible. An aromatic structure is consistent with the C–H stretch (3050 cm^{-1}), C=C stretch (1600 and 1500 cm^{-1}), and C–H bend (700 and 770 cm^{-1}; monosubstituted aromatic ring). A reasonable structure is $C_6H_5CH_2CHO$. Notice also that the C=O stretch at 1720 cm^{-1} indicates that it is *not* conjugated to a double bond.

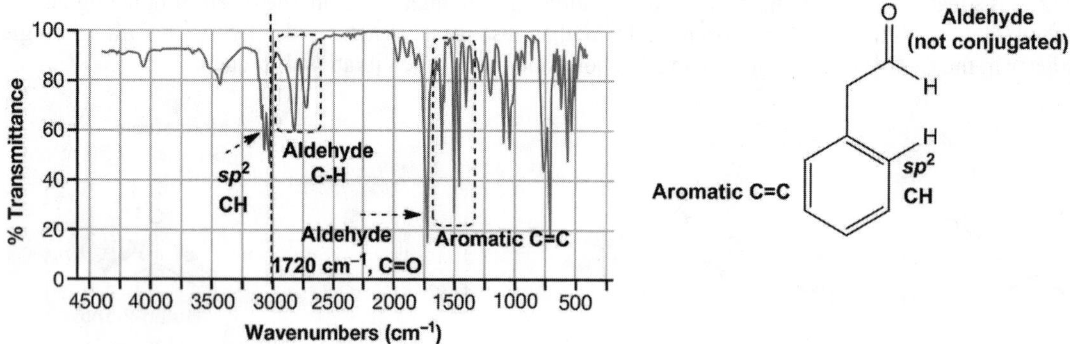

Problem 15.51

Think

What is the IHD? Are any stretching frequencies present that identify characteristic functional groups (see Table 15-2)? Draw those functional groups. What atoms are left in the formula? How can you piece the structure together? Are there any acidic protons present? What would be an estimated pK_a for the functional group?

Solve

The compound's IHD is 3. The compound clearly contains a carboxylic acid, having a C=O stretch at 1700 cm^{-1} and a very broad OH stretch that extends down to ~2500 cm^{-1}. It also has a triple bond, indicated by the absorption around 2100 cm^{-1}. The C≡C triple bond and the C=O bond account for the total IHD. We also see indication of a C≡C triple bond by the alkyne C–H stretch that appears as a bump around 3300 cm^{-1}. The only possibility is the following. The carboxylic acid is conjugated to the alkyne, so an estimated pK_a is ~4. See the figure on the next page.

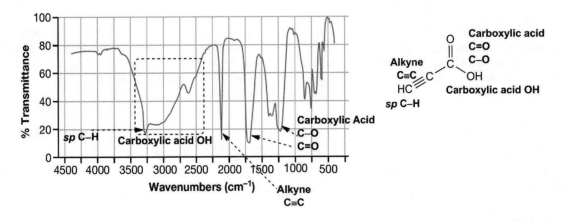

Problem 15.52

Think

What similarities are present in each structure? What differences are present? Which compound is ortho, which is meta, and which is para? Which region of the IR is examined to determine differences in substitution on the benzene ring? Consult Table 15-3.

Solve

The major difference we should expect in the IR spectra for these compounds is indication of the ortho, meta, or para substitution on the benzene ring. Information about these substitutions appears as absorptions by bending vibrational modes in the fingerprint regions. The ortho isomer should have one band at 750 cm^{-1}, which is consistent with Spectrum 2. The meta isomer should have two bands, at 700 and 780 cm^{-1}, consistent with Spectrum 3. And the para isomer should have one band at 830 cm^{-1}, consistent with Spectrum 1.

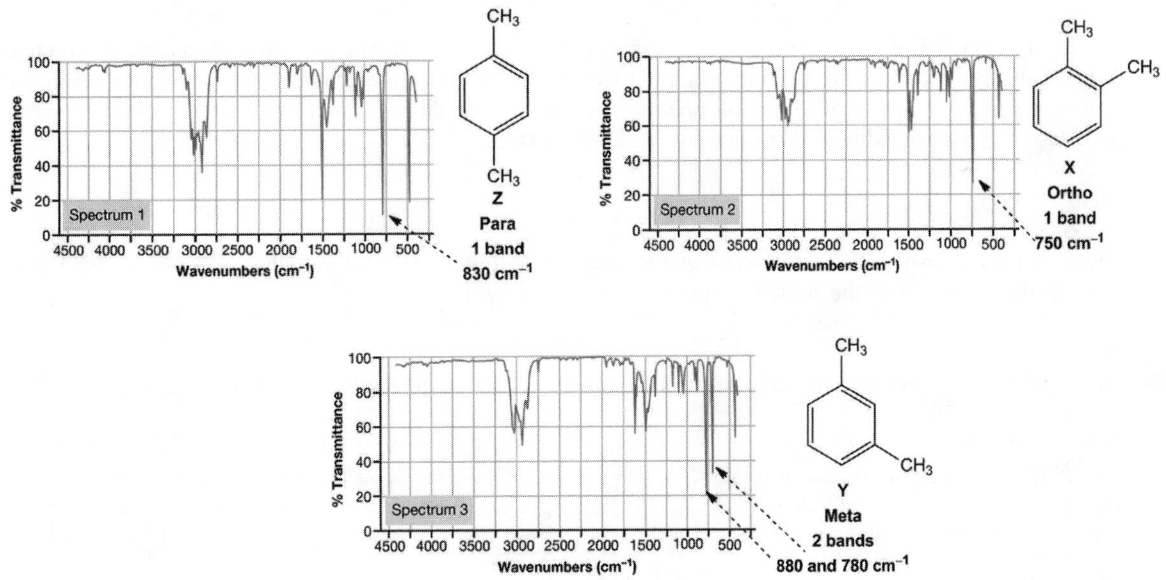

Problem 15.53

Think

In which molecule is resonance more prominent? Are either one of the structures aromatic? If so, how does this affect the resonance structures? Which molecule has more C=O character?

Solve

In both compounds **A** and **B**, the conjugation of the lone pair contributes to strengthening the C–O bond, as suggested by the resonance structure with the C=O bond. In furan (**A**), however, the lone pair is involved in aromaticity, whereas in pyran (**B**) it is not. Aromaticity serves to stabilize the π system more than the regular conjugation, which we associate with a stronger π bond. The stronger π bond gives rise to a higher vibrational frequency. Compound **A** will have a higher-frequency C–O stretch.

Stronger C=O π bond character
Increased IR C–O frequency

Problem 15.54

Think

What affects the intensity of the peak? Which C=C, when stretched, has a larger change in dipole? How does having a C=O near the C=C affect the dipole change of the C=C stretch?

Solve

The C=C stretch in molecule **C** should have a higher intensity. This is because the nearby O atom, which is highly electronegative, places a partial positive charge on the closer of the two alkene carbons, as shown below.

Therefore, as the C=C stretches and compresses, that partial positive charge moves back and forth, creating a significant oscillation in the C=C bond dipole. As we learned in the chapter, a greater oscillation in a dipole moment enhances the probability of photon absorption and thus enhances the absorption intensity. Such oscillation in the bond dipole is not present in molecule **D**.

Problem 15.55

Think

What is the value of the C=O stretching frequency in both compounds? Which one has a higher frequency? How is the frequency of the stretch related to the bond strength?

Solve

The C=O bond in cyclobutanone is stronger because its C=O stretch occurs at a higher frequency—1780 cm^{-1} compared to 1720 cm^{-1}.

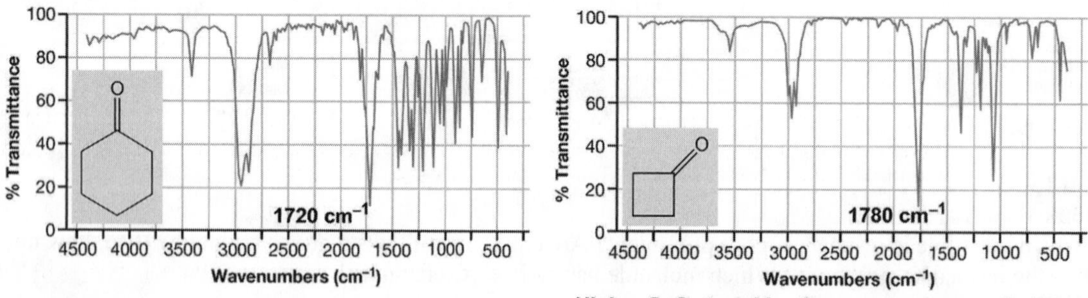

Higher C=O stretching frequency, stronger C=O bond

Problem 15.56

Think

Which orbital is more compact, *s* or *p*? Which gives rise to a shorter bond: more *s* character or less *s* character? Does a shorter bond correspond to a stronger or weaker bond?

Solve

There is more *s* character in the C=O bond of cyclobutanone, higher IR stretch, and stronger bond (Problem 15.55). As we learned in Chapter 3, a greater *s* character gives rise to a shorter, stronger bond.

Problem 15.57

Think

Does the involvement of the N atom's lone pair in resonance give rise to a stronger or weaker C=O bond (consult the solution to Problem 15.41)? Does that result in a higher or lower C=O vibrational frequency? In which molecule does the resonance involving the lone pair on N have a greater impact on a given C=O?

Solve

The C=O stretch will be at a higher frequency for molecule **E**. In an amide, the lone pair on N is involved in resonance with the C=O and, therefore, is responsible for giving the C=O bond more single-bond character, thus weakening the C=O bond and lowering the frequency of the C=O stretch. In molecule **E**, the lone pair on N is involved in resonance with two different C=O groups, and so the effect is diminished.

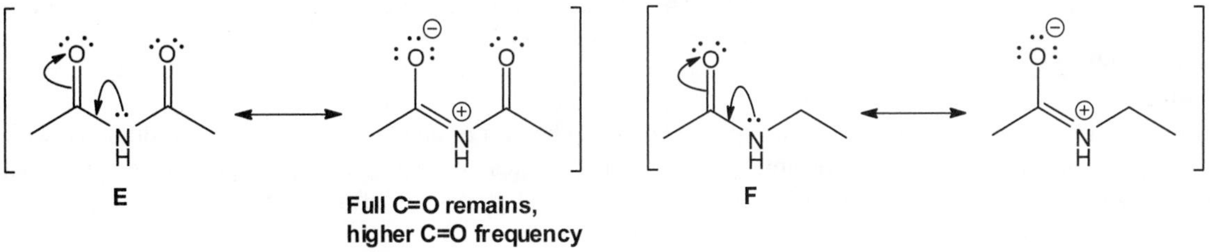

E

Full C=O remains,
higher C=O frequency

F

Problem 15.58

Think

What type of bond exists between the O atom and the Na atom? Is there a resonance structure possible? Is the contribution by that resonance structure greater in sodium acetate or in an ester?

Solve

Sodium acetate has an ionic bond between the O atom and the Na atom. The carboxylate anion has another resonance structure possible with equal contribution. This gives the bond half C–O and half C=O character. Thus, the C=O is significantly weaker than a pure C=O and shows up as a lower frequency in the IR. An ester is not ionic, and the resonance structure with two charges has less contribution to the hybrid.

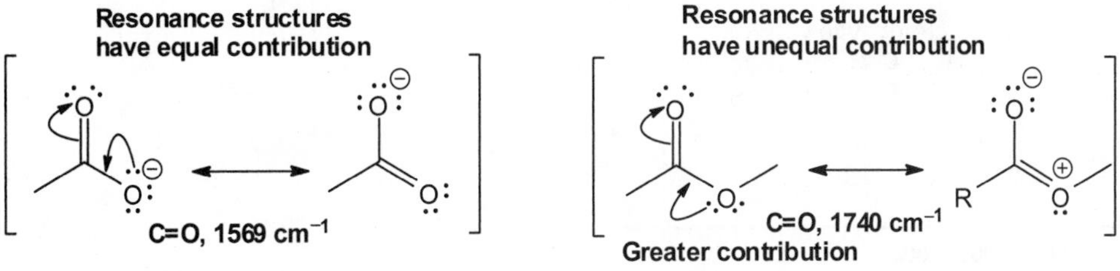

Resonance structures
have equal contribution

C=O, 1569 cm^{-1}

Resonance structures
have unequal contribution

C=O, 1740 cm^{-1}

Greater contribution

Problem 15.59

Think

A strong base in the presence of heat results in what reaction mechanism? Which adjacent H atoms are able to be attacked by the base to form the alkene? Which base is bulkier? Which H atom is more likely to be attacked by the bulky base? Less bulky base? Which product results in a conjugated diene?

Solve

The reaction mechanism is an E2, which will lead to the formation of a diene. Lithium diisopropyl amide (LDA) is a bulky base and irreversibly deprotonates the less substituted C. This leads to a diene that is not conjugated. Sodium hydroxide is a less bulky base and reversibly deprotonates the more substituted C. This leads to a diene that is conjugated. Conjugation increases the wavelength of the longest-wavelength absorption in the UV–vis spectrum, as a result of a smaller HOMO–LUMO gap. Therefore, the reaction with LDA has a λ_{max} = 180 nm and the reaction with NaOH has a λ_{max} = 220 nm. See the mechanism on the next page.

C=C not conjugated
λ_{max} = 180 nm

C=C conjugated
λ_{max} = 220 nm

Problem 15.60

Think

If the pK_a is >40, are there any acidic protons? Can the O be part of an alcohol? A ketone or aldehyde with an α hydrogen? What is the IHD? Are any stretching frequencies present that identify characteristic functional groups (see Table 15-2)? Draw those functional groups. What atoms are left in the formula? How can you piece the structure together?

Solve

The IHD is 1. There are no sp^2 C–H stretches (i.e., nothing >3000 cm^{-1}). There are peaks between 3000 and 2850 cm^{-1}, suggestive of sp^3 C–H stretches. There are also two peaks around 2820 and 2720 cm^{-1}, suggestive of the aldehyde C–H. There is a peak around 1730 cm^{-1}, which comes from the aldehyde C=O. This takes care of one H, C, and O atom each and leaves C_4H_9. Since the pK_a is >40, the molecule is not very acidic, and there is no α C–H. Ketone, aldehyde, and esters have α H–C pK_a values in the range of 20–25. Therefore, the α C must have no H atoms. The structure is $(CH_3)_3CC(O)H$.

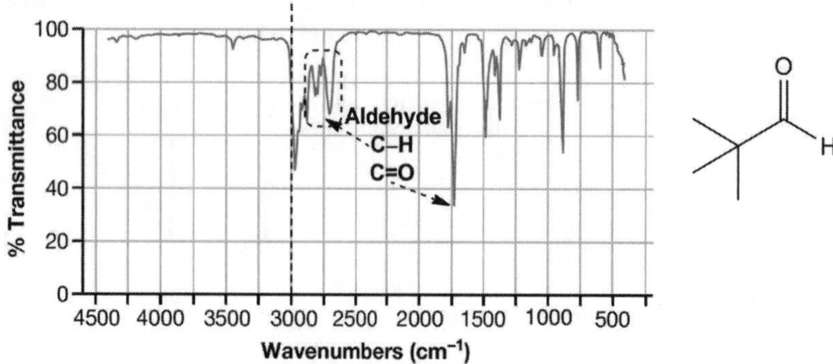

Problem 15.61

Think

A strong base in the presence of heat results in what mechanism? How many leaving groups are present? How many times can the E2 reaction occur? What functional group results? Is this supported by the IR spectrum? Are any stretching frequencies present that identify characteristic functional groups (see Table 15-2)?

Solve

There are two leaving groups present, and the E2 reaction occurs two times to form the terminal alkyne. The formation of the terminal alkyne is supported by the sharp, intense peak at 3300 cm^{-1} (*sp* C–H) and the small peak at 2200 cm^{-1} (C≡C).

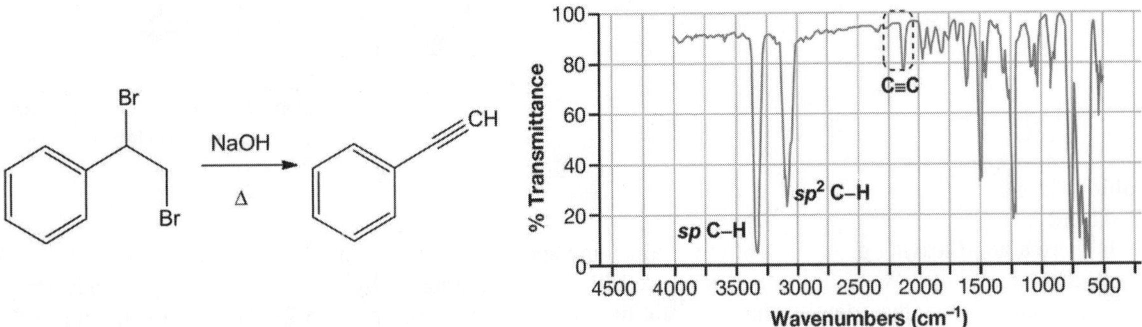

Problem 15.62

Think

Is there an acidic proton to react with NaH? What forms? How does the resulting nucleophilic carbon react with CH_3Br? What bond is formed? What functional group results? Is this supported by the IR spectrum? Are any stretching frequencies present that identify characteristic functional groups (see Table 15-2)?

Solve

The *sp* C≡CH is deprotonated by the NaH to form the alkynide anion, C≡C:$^-$. This is a nucleophilic carbon that reacts with CH_3Br in an S_N2 reaction to form a new C–C bond. There are now sp^3 C–H bonds present. Notice, too, that there is no alkyne C–H stretch near 3300 cm^{-1}.

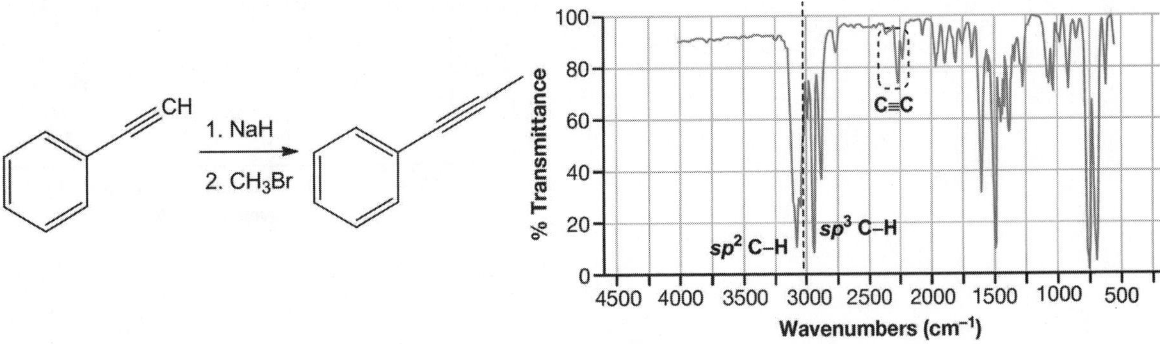

Problem 15.63

Think

Is there a good leaving group? What is the attacking species? What bond is formed? What functional group results? Is this supported by the IR spectrum? Are any stretching frequencies present that identify characteristic functional groups (see Table 15-2)?

Solve

OTs serves as a good leaving group on a primary C atom and NaCN is a strong nucleophile. Thus, an S_N2 reaction mechanism results, and a new C–CN bond is formed. There are only sp^3 C–H stretches (3000–2850 cm^{-1}), and the C≡N stretch is also apparent (2250 cm^{-1}). The nitrile triple-bond stretch is more intense compared to the alkyne triple-bond stretch due to a larger change in dipole moment when stretched. See the figure on the next page.

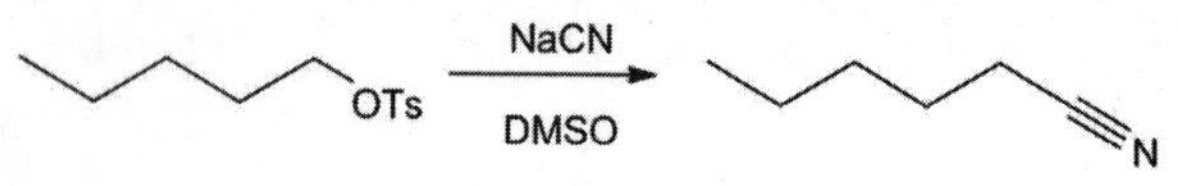

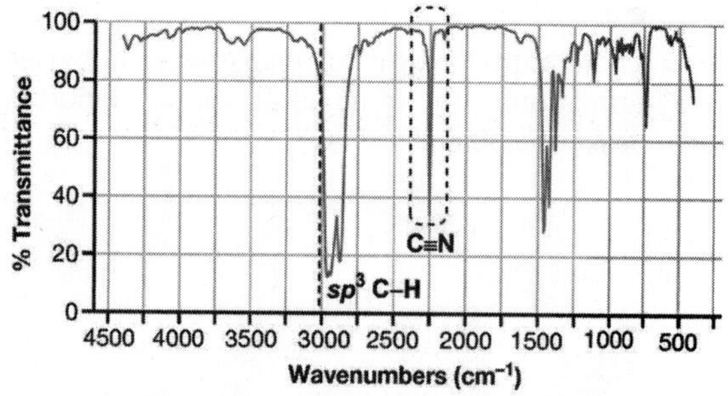

Problem 15.64

Think

Is there a good leaving group? What is the attacking species? With a strong base in the presence of heat, what mechanism results? What functional group results? Is this supported by the IR spectrum? Are any stretching frequencies present that identify characteristic functional groups (see Table 15-2)? How can you tell if the cis or trans isomer results (see Table 15-3)?

Solve

Br serves as a good leaving group on a secondary C atom, and CH_3–ONa is a strong nucleophile and a strong base. When heated, E2 wins over an S_N2 reaction mechanism, and an alkene is formed. A small peak >3000 cm^{-1} indicates an alkene C–H stretch. There is no strong evidence of an alkene C=C stretch, but this should not be surprising, due to the symmetry about the C=C bond. In the fingerprint region, the C–H bend peak at 970 cm^{-1} suggests that the trans isomer forms.

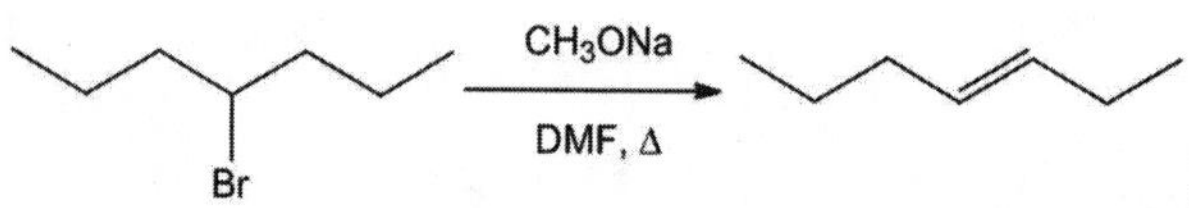

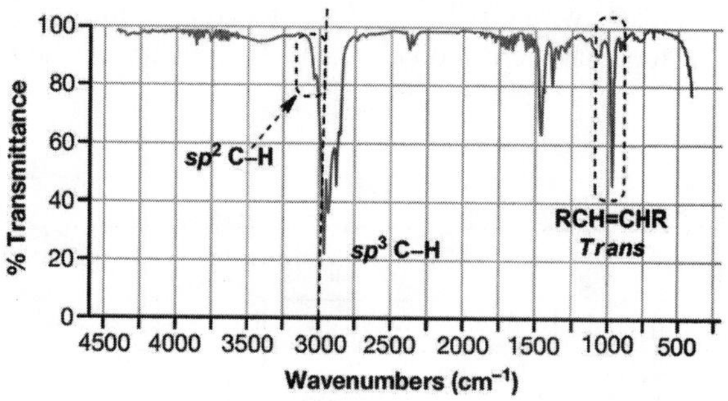

CHAPTER 16 | Structure Determination 2: Nuclear Magnetic Resonance Spectroscopy and Mass Spectrometry

Your Turn Exercises

Your Turn 16.1

Think

Does a larger ΔE_{spin} occur going left or right on the x axis? At the new location on the x axis, where do you draw the new horizontal line to represent the new α spin state? The new β spin state? How does the vertical distance between them compare to the bracket already shown in the figure?

Solve

A stronger magnetic field appears to the right of the states already shown in the figure. As shown below, this creates a larger energy difference between the α and β spin states.

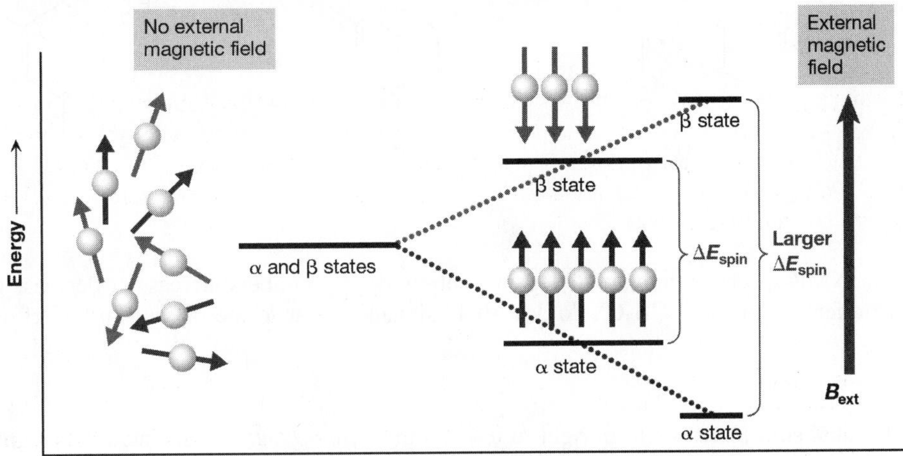

Your Turn 16.2

Think

Does increased shielding increase or decrease B_{eff}? How does the magnitude of B_{eff} correspond to ΔE_{spin}?

Solve

As shown below, the nucleus at the left is shielded to a greater extent. This is reflected by the fact that its spin states are closer together in energy (smaller ΔE_{spin}), meaning that the nucleus experiences less of the external magnetic field.

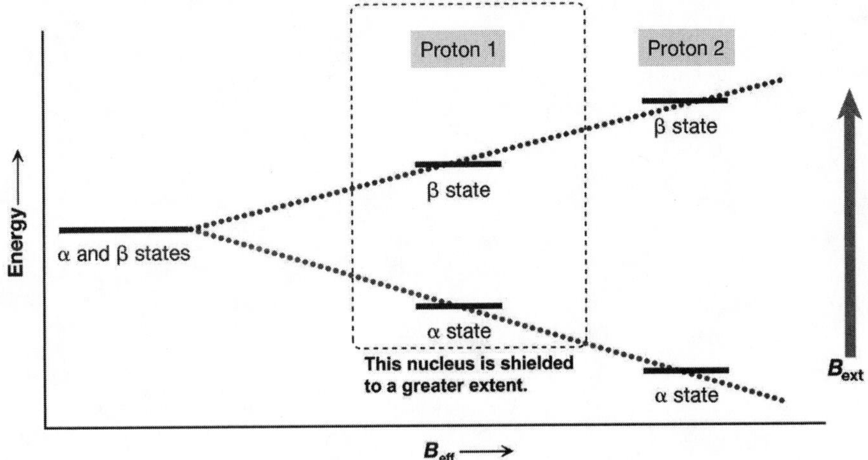

Your Turn 16.3

Think

Can you divide the benzene molecule in two such that one half is the mirror image of the other half? Can you rotate the benzene molecule less than one full rotation so that the H atoms appear to occupy the same locations as they did before rotation?

Solve

One plane of symmetry that benzene has is indicated by the dashed line below. As shown, with respect to that plane of symmetry, H2 and H6 reflect through the plane of symmetry, as do H3 and H5. As indicated below on the right, after benzene is rotated 60° in the plane of the paper, the molecule appears unchanged, but all six H atoms end up occupying a location that a different H atom occupied before rotation. So the answer to question **(b)** is yes.

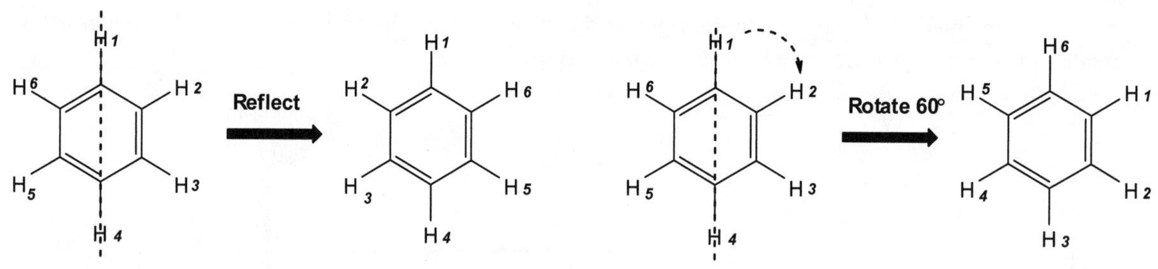

H1 H6 H2 H5 H3 H4 Reflect H1 H2 H6 H3 H5 H4

H1 H6 H2 H5 H3 H4 Rotate 60° H6 H5 H1 H4 H2 H3

Your Turn 16.4

Think

How is the *x* axis on the NMR spectrum numbered? By convention, do the numbers increase or decrease going left to right? From the reference point of CH_3Cl, do downfield signals appear to the right or to the left of the signal at 3.1 ppm?

Solve

By convention, the chemical shift increases from right to left and the term *upfield* is associated with signals to the right and *downfield* is associated with signals to the left. Thus, 0 ppm is on the far right (most upfield) and 12 ppm is on the far left (most downfield). When comparing signals in Table 16-1 to the CH_3Cl proton signal, the chemical shifts with smaller values are upfield relative to CH_3Cl and the chemical shifts with larger values are downfield relative to CH_3Cl. See the table on the next page.

	Type of Proton	Chemical Shift (ppm)		Type of Proton*	Chemical Shift (ppm)
1.	$H_3C-Si(CH_3)_2-CH_2$**H**	0	11.	$R-O-CH_2-$**H**	3.3
2.	$R-CH_2$**H**	0.9	12.	$F-CH_2-$**H**	4.1
3.	R_2CH-**H**	1.3	13.	$R_2C=CH$**H**	4.7
4.	R_3C-**H**	1.4	14.	$R_2C=C(R)$**H**	5.3
5.	$R_2C=CH-H_2C-$**H**	1.7	15.	C_6H_5-**H**	7.3
6.	$R-C(=O)-CH_2-$**H**	2.1	16.	$R-C(=O)-$**H**	9–10
7.	$C_6H_5-CH_2-$**H**	2.3	17.	$R-NH-$**H**	1.5–4
8.	$R-C\equiv C-$**H**	2.4	18.	$R-O-$**H**	2–5
9.	$Br-CH_2-$**H**	2.7	19.	$Ar-O-$**H**	4–7
10.	$Cl-CH_2-$**H**	3.1	20.	$R-C(=O)-O-$**H**	10–12

Protons **upfield** from CH_3Cl (rows 1–9)

Protons **downfield** from CH_3Cl (rows 11–16)

Protons **downfield** from CH_3Cl (rows 19–20)

Your Turn 16.5

Think

How does increasing the electronegativity of an atom attached to CH affect the electron density on that H? How does that relate to the strength of the deshielding (more or less deshielded)? What relationship do you observe between electronegativity and chemical shift?

Solve

The extent of deshielding depends, in part, on the electronegativity of a nearby atom. Increased electronegativity leads to increased deshielding. The electronegative atom is electron withdrawing and removes electron density from the nearby proton. Increased deshielding causes the signal to shift downfield.

Structure	Chemical Shift	Electronegativity of halogen
FCH_3	4.1	3.98
$ClCH_3$	3.1	3.16
$BrCH_3$	2.7	2.96

Your Turn 16.6

Think

In which direction is the external magnetic field pointing? How are the local magnetic field lines represented in the figure? Which local magnetic field line is oriented in the same direction? Where is this local magnetic field line relative to the benzene H atoms?

Solve

The local magnetic field lines are the blue lines in Figure 16-10. The ones that are in the same direction as the external magnetic field are circled below. Notice that these local magnetic field lines appear where benzene's H atoms are located. This leads to additional deshielding of the protons on benzene and the signal appears more downfield than expected.

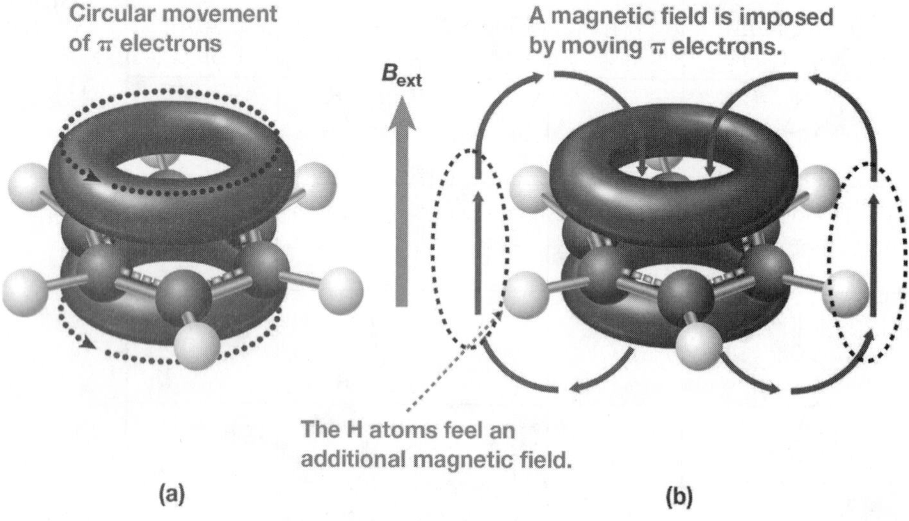

Your Turn 16.7

Think

How are the local magnetic field lines represented in the figure? How is the magnetic field (B_{loc}) aligned where the alkyne H atoms are located? In which direction is the external magnetic field pointing?

Solve

The alkyne protons lie where B_{loc} opposes B_{ext}. This results in a shielding, not deshielding, effect.

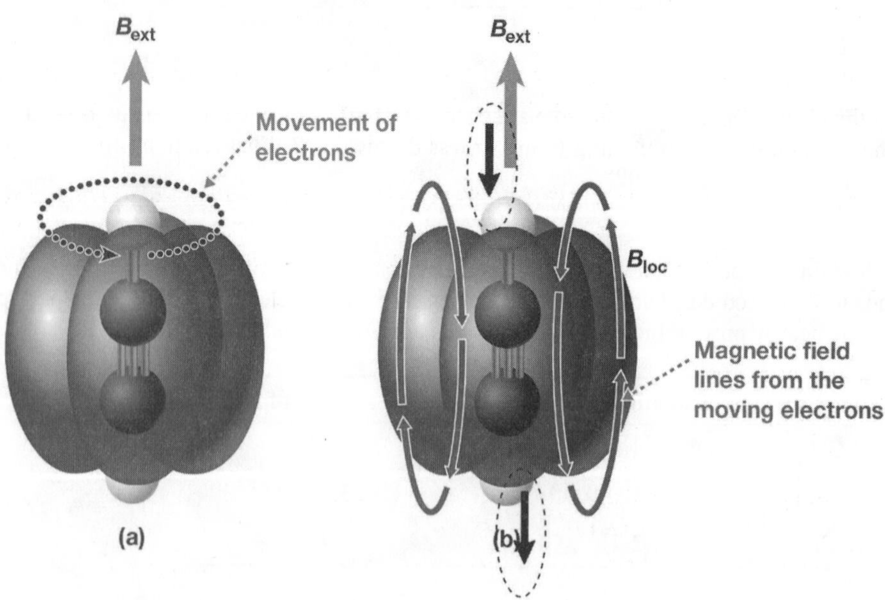

Your Turn 16.8

Think

Does an electronegative atom shield or deshield a nearby CH proton? What is the effect on the chemical shift? What is the effect when the number of these electronegative atoms increases?

Solve

CH_2Cl_2 has two electronegative Cl atoms that deshield the CH proton. This results in a downfield chemical shift (5.3 ppm) compared to CH_3Cl (3.1 ppm [see Table 16-1]). $CHCl_3$ has three electronegative Cl atoms that deshield the proton even more and result in a chemical shift further downfield (7.3 ppm).

Your Turn 16.9

Think

Use a ruler to help measure the height of each signal in mm, and divide each by the smallest height to determine the ratio. How does the ratio of the signal heights inform you about the number of protons to which each corresponds?

Solve

In the figure below, dashed lines are used to illustrate the 1.5:1 ratio. The taller integration is ~19 mm (~0.75 in) and the smaller one is ~13 mm (~0.5 in). Dividing both numbers by 13 gives a 1.46:1 ratio, or about 1.5:1. The molecule has 10 protons total and thus a ratio of 6:4, which correlates to a 1.5:1 ratio.

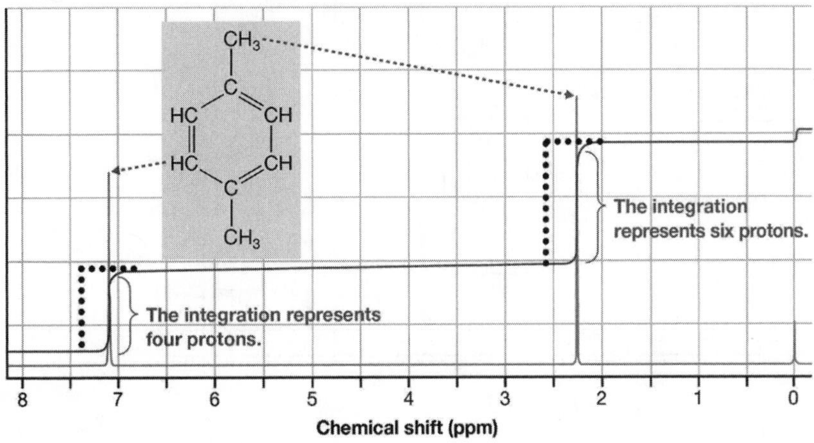

Your Turn 16.10

Think

When a proton is coupled to N protons that are distinct from itself, how many peaks result in the splitting pattern? How many peaks do each of these split signals have?

Solve

When a proton is coupled to N protons that are distinct from itself, the resulting splitting pattern has $N + 1$ peaks. In **(a)**, there are two peaks (a doublet), so $N + 1 = 2$, making $N = 1$ coupled proton. In **(b)**, there are five peaks (quintet), so $N + 1 = 5$, making $N = 4$ coupled protons. In **(c)**, there is one peak (a singlet), so $N + 1 = 1$, making $N = 0$ coupled protons.

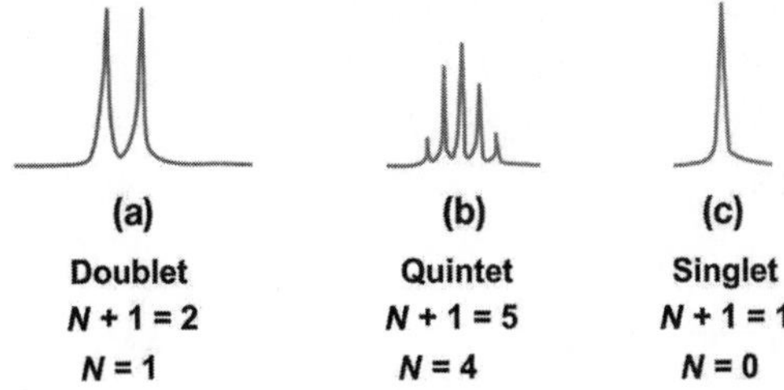

Your Turn 16.11

Think

For protons that are coupled together, should the peaks in their split signals be separated by the same coupling constant or different coupling constants? What are the coupling constants for the signals that correspond to protons **A**, **B**, **C**, and **D**?

Solve

Signals of protons that are coupled together exhibit the same coupling constant. The doublet at 7.36 ppm (protons **A**) and doublet at 7.05 ppm (protons **B**) both have a coupling constant equal to 8.5 Hz and, therefore, are coupled. The quartet at 2.58 ppm (protons **C**) and triplet at 1.20 ppm (protons **D**) both have a coupling constant equal to 7.6 Hz and, therefore, are coupled. The signals for protons **A** and **C** do not have the same coupling constant, so those protons are not coupled together. This is evident from the structure because those protons are separated by five bonds.

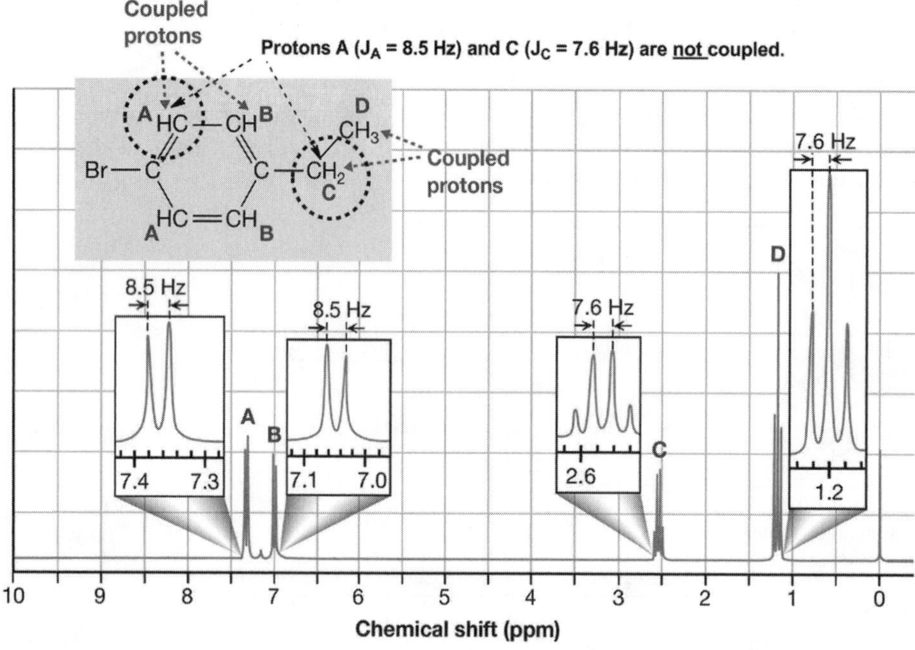

Your Turn 16.12

Think

For protons that are coupled together, should the peaks in their split signals be separated by the same coupling constant or different coupling constants?

Solve

Signals of protons that are coupled together exhibit the same coupling constant. The quartet at 4.15 ppm and triplet at 1.25 ppm both have a coupling constant equal to 7.1 Hz and, therefore, are coupled. The quartet at 2.33 ppm and triplet at 1.15 ppm both have a coupling constant equal to 7.5 Hz and, therefore, are coupled.

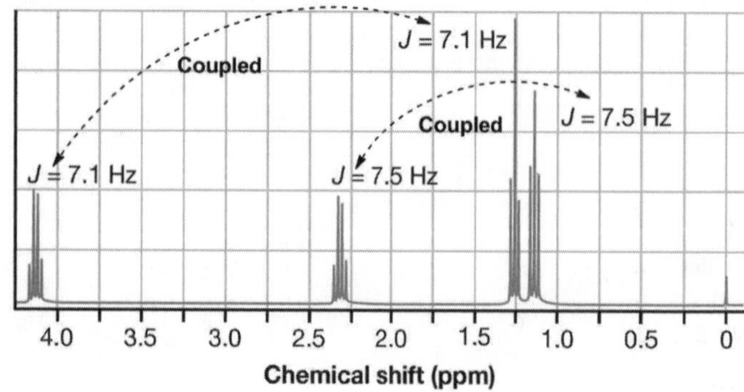

Your Turn 16.13

Think

How many total protons are coupled to H_c? How many signals would the $N + 1$ rule predict?

Solve

There are a total of two H atoms coupled to the H_c proton (both protons H_A and H_B), so, according to the $N + 1$ rule, the H_c signal should be a triplet if H_A and H_B are not chemically distinct. No signals in the spectrum appear to consist of three peaks in a 1:2:1 ratio. So the $N + 1$ rule does not hold for this case because H_A and H_B are chemically distinct. H_A is trans to H_c and H_B is cis to H_c.

Your Turn 16.14

Think

Which protons are coupled to the CH_2? Are those protons equivalent to each other or are they distinct? What is the expected coupling constant for each type of coupling? What does the $N + 1$ rule predict about the splitting of the CH_2 signal by each type of proton? What does the resulting signal look like?

Solve

The CH_2 signal is coupled to CH and CH_3. These protons are not equivalent to each other, so they will split the CH_2 signal differently. The CH would split the CH_2 into a doublet with a coupling constant value ~2 Hz. The CH_3 would split the CH_2 into a quartet with a coupling constant value ~6 Hz. Therefore, the signal is a quartet of doublets.

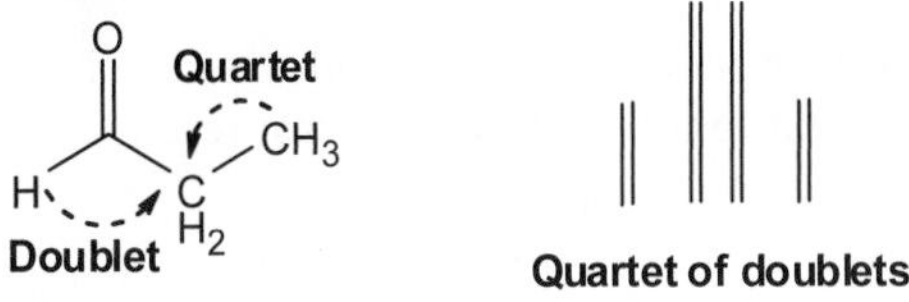

Your Turn 16.15

Think

How many signals do you observe? What is the relationship in a carbon NMR spectrum of the number of signals to the number of chemically distinct carbon atoms? Which signal is from the solvent? Which signal is from TMS?

Solve

There are eight signals in the ^{13}C NMR spectrum, but the one at 77 ppm is from the solvent, and the one at 0 ppm is from TMS. Therefore, the compound gives rise to six signals, so it has six chemically distinct C atoms.

Your Turn 16.16

Think

Consult Table 16-1 and Table 16-4. What similarities do you notice between the two tables? Select two specific examples of molecules whose chemical shifts of C atoms are in the same order as the chemical shifts for the H atoms.

Solve

Several examples are possible. For example, CH_3R has a proton chemical shift of 0.9 ppm and a carbon chemical shift of 10–25 ppm. CH_2R_2 has a proton chemical shift of 1.3 ppm and a carbon chemical shift of 20–45 ppm. Thus, CH_3R and CH_2R_2 have the same order for H and C chemical shifts (different scales). Other examples include $BrCH_3$ and $ClCH_3$, alkenes, and alkynes.

Your Turn 16.17

Think

From the DEPT-135 spectrum (Fig. 16-30c), what signals are positive? Negative? What are the signals that appear in the DEPT-90 spectrum (Fig. 16-30b)? What then can you deduce about the two unlabeled ^{13}C signals?

Solve

In DEPT-135 spectra, CH_3 and CH appear as normal positive signals and CH_2 appear as negative (below the baseline) signals. In DEPT-90 spectra, only CH signals appear. Comparison of both a DEPT-135 spectrum and DEPT-90 spectrum reveals: the signals that correspond to CH, which appear only in the DEPT-90 spectrum and are positive in DEPT-135; the signals that correspond to CH_2, which are negative in the DEPT-135 spectrum; and the signals that correspond to CH_3, which are positive in the DEPT-135 spectrum and missing from the DEPT-90 spectrum.

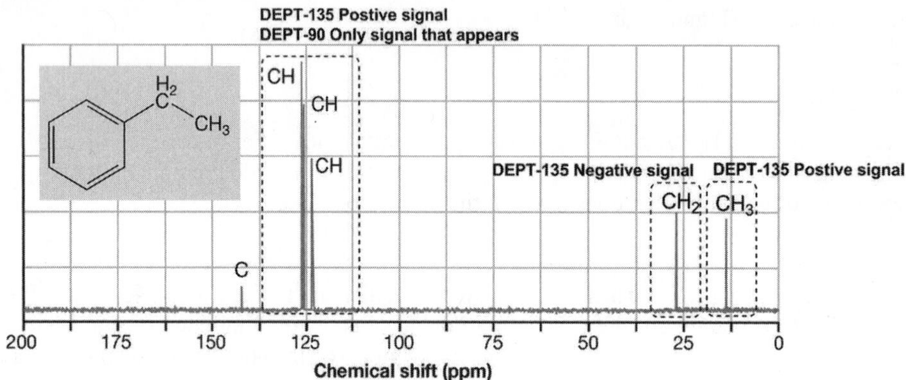

Your Turn 16.18

Think

Which protons are chemically distinct? It is permissible to assign all benzene protons with one label? Which signals are furthest downfield in the structure? Which protons are next to the O atom and what results in the ^{1}H NMR spectrum?

Solve

The benzene protons are labeled H_A and correlate to the signals between 7.5 and 8 ppm. The H_B protons are next to the O atom and the electron-withdrawing group deshields the proton and causes a downfield shift to appear at 4.3 ppm. The remaining H_C and H_D signals correspond to secondary and primary CH signals, respectively.

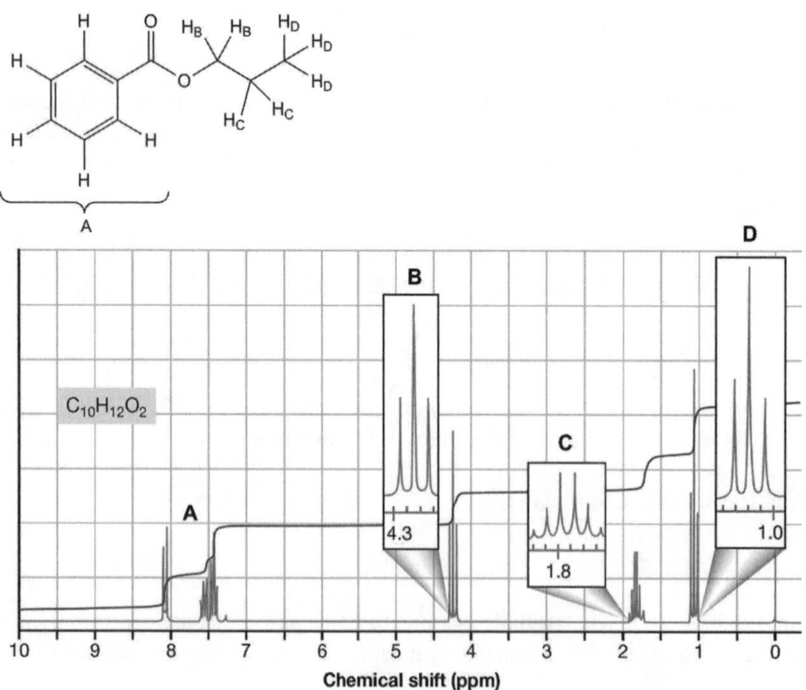

Your Turn 16.19

Think

Where does the H–O stretch signal occur in the IR spectrum of an alcohol? Is it broad or sharp? Weak or strong?

Solve

In general, the H–O signal occurs between 3600 and 3200 cm^{-1} and gives rise to a broad and strong signal. In this example, the H–O signal is centered around 3400 cm^{-1}.

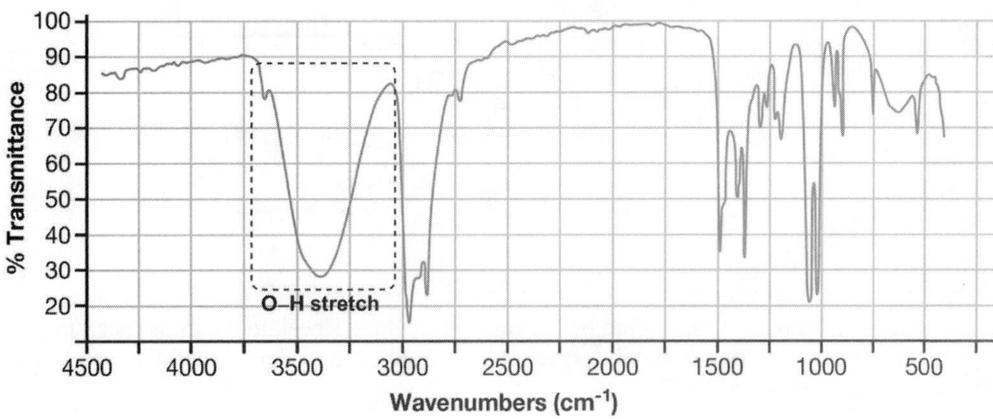

Your Turn 16.20

Think

If an ion is too light, is it deflected too much or not enough? How is this represented in the figure? What about ions that are too massive?

Solve

Ions that are too light are deflected too much (sharper arc) and ions that are too massive are not deflected enough (arc not as sharp). Neither of these types of ions will reach the detector.

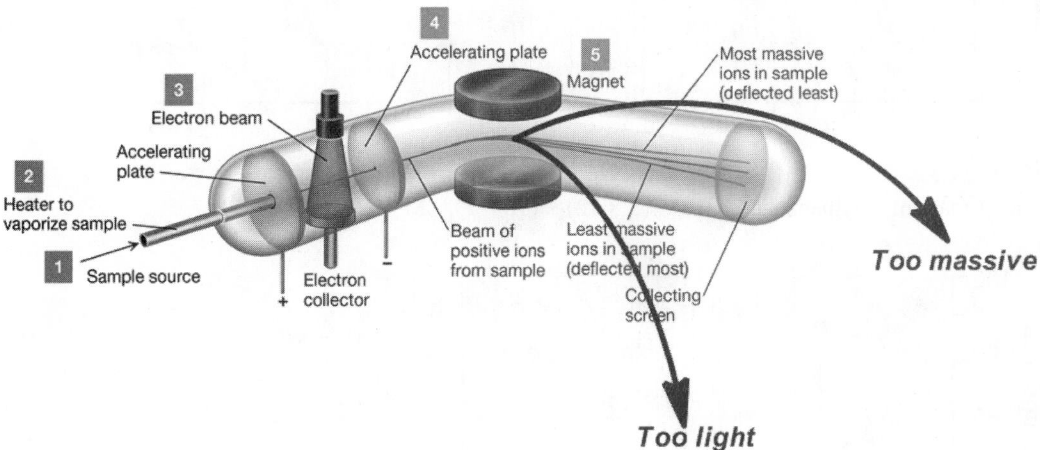

Your Turn 16.21

Think

Where does the fragmentation occur? What part of the original structure is missing that correlates to the *m/z* values indicated in the box?

Solve

The fragmentation pathways are shown below.

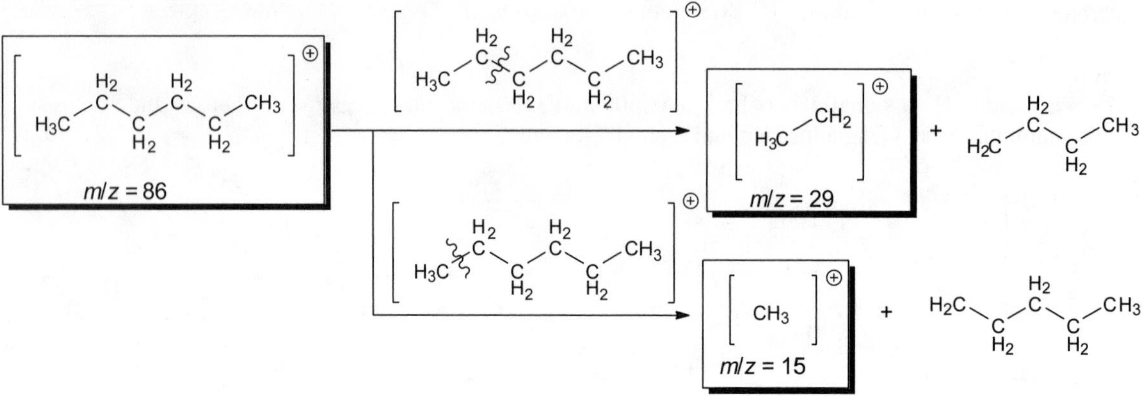

Your Turn 16.22

Think

What is the molar mass of hexane? What would be the value of the M + 1 peak in the mass spectrum for hexane (Fig. 16-35)? What is the molar mass of ethylbenzene? What would be the value of the M + 1 peak in the mass spectrum for ethylbenzene (Problem 16.33)?

Solve

The molar mass of hexane, C_6H_{14} = (6 × 12) + (14 × 1) = M = 86. M + 1 = 87.

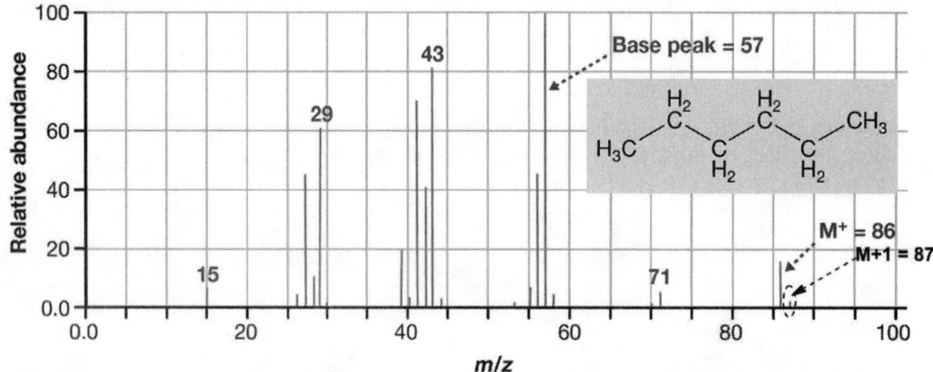

The molar mass of ethylbenzene, C_8H_{10} = (8 × 12) + (10 × 1) = M = 106. M + 1 = 107.

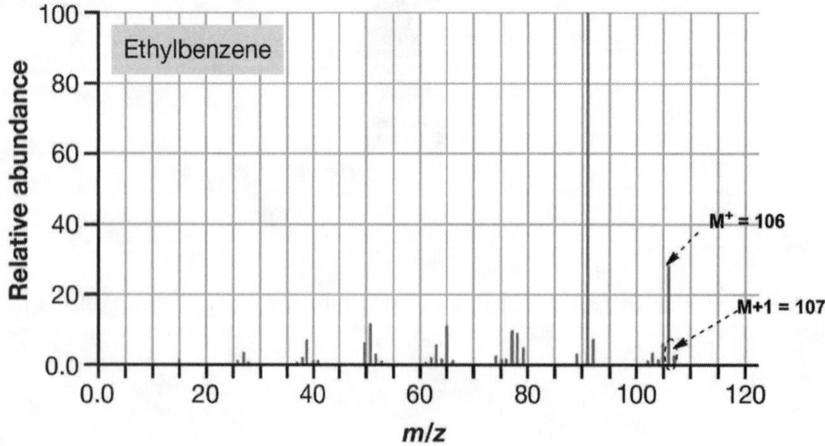

Your Turn 16.23

Think

What is the mass of ^{12}C compared to ^{13}C? How many ^{12}C atoms are present and what is the mass of all ^{12}C atoms? How many ^{13}C atoms are present and what is the mass of all ^{13}C atoms? What is the mass of 14 H atoms?

Solve

The mass of ^{12}C is 12 amu and ^{13}C is 13 amu. There are five ^{12}C atoms (= 60 amu) and one ^{13}C atom (= 13 amu). The mass of the 14 H atoms is 14 amu. The sum of these is 87. This is the M + 1 peak in the mass spectrum of hexane (C_6H_{14}).

Your Turn 16.24

Think

How can you estimate the peak intensity from Figure 16-35 (hexane) and Figure 16-37 (dodecane)? Locate the M and M + 1 peaks by calculating the molar mass of each.

Solve

Hexane: M^+ = 86 (15.5%) and M + 1 = 87 (~1.0%). (1.0/15.5) × (100%/1.1%) = 5.87, which rounds to 6.
Dodecane: M^+ = 160 (5.9%) and M + 1 = 0.8%. (0.8/5.9) × (100%/1.1%) = 12.33, which rounds to 12.

Your Turn 16.25

Think

What is the mass of ^{79}Br compared to ^{81}Br? Calculate the mass of the C_6H_5 portion.

Solve

The mass of ^{79}Br is 79 amu and the mass of ^{81}Br is 81 amu. The mass of the C_6H_5 portion is 77 amu, which is added to either 79 amu or 81 amu, depending on the isotope of Br. The mass of $C_6H_5{}^{79}Br$ is 156 amu and corresponds to the M^+ peak. The mass of $C_6H_5{}^{81}Br$ is 158 amu and corresponds to the M + 2 peak.

Your Turn 16.26

Think

Does the compound have an odd or even molar mass? According to the nitrogen rule, should this correspond to an odd or even number of N atoms?

Solve

CH_3–CO–$N(CH_3)_2$ = C_4H_9NO = 12(4) + 9(1) + 14 + 16 = 87 amu. There is one N atom, which is an odd number. This is consistent with the nitrogen rule.

Chapter Problems

Problem 16.1

Think

How is the energy difference between the α and β spin states, ΔE_{spin}, related to the applied external magnetic field?

Solve

The energy difference will be greater at 5.0 T than 0.5 T. According to Equation 16-1, ΔE_{spin} is directly proportional to B_{ext}.

Problem 16.3

Think

In Equation 16-1, what value should be substituted for γ? For *h*? For B_{ext}?

Solve

As mentioned in the text, γ for a proton is $2.67512 \times 10^8\ T^{-1}s^{-1}$. *h* is 6.626×10^{-34} J·s. In the problem, the value for B_{ext} is given as 2.2 T.

Therefore,
$$\Delta E_{spin} = \gamma h B_{ext}/2\pi$$
$$= (2.67512 \times 10^8\ T^{-1}s^{-1})(6.626 \times 10^{-34}\ J{\cdot}s)(2.2\ T)/2\pi$$
$$= 6.2 \times 10^{-26}\ J$$

The units s and T both cancel, leaving energy in units of J.

Problem 16.4

Think

Do the two signals occur at the same or similar chemical shifts? If not, what does this mean about the chemical equivalence of the two protons?

Solve

No, the protons in benzene are chemically distinct from those in TMS. We know this because the protons from the two compounds have different chemical shifts—7.3 ppm and 0.0 ppm, respectively.

Problem 16.6

Think

How many signals appear in the proton NMR spectrum? Of those signals, which are generated by *trans*-1,2-dichloroethene?

Solve

The compound has just one type of chemically distinct proton. There are two signals total in the spectrum, but the one at 0 ppm belongs to the protons in TMS, which is the reference compound that was added. So the sample gives rise to just one signal, and we know that the number of signals equals the number of chemically distinct types of protons.

Problem 16.8

Think

What molecules are generated by substituting each H atom separately by "X"? Which of those molecules are identical? Enantiomers? Diastereomers? Constitutional isomers? How many signals result?

Solve

(a) Molecules **I** and **II** are enantiomers—they are mirror images. According to Step 3 in the chemical distinction test, these H atoms are *not* chemically distinct. There are *two* signals in the ^{1}H NMR spectrum, one for the CH_a and one for the methyl CH_3. Molecules **I/II** and **III** are constitutional isomers. According to Step 3 in the chemical distinction text, these H atoms are chemically distinct.

Enantiomers

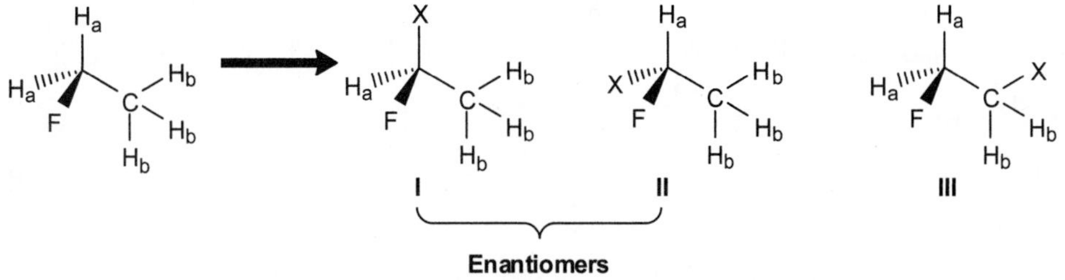

(a) Molecules **I**, **II/III**, **IV/V**, and **VI** are constitutional isomers. According to Step 3 in the chemical distinction test, these H atoms are chemically distinct. Molecules **II** and **III** are enantiomers, as are **IV** and **V**—they are mirror images. According to Step 3 in the chemical distinction test, these H atoms are *not* chemically distinct. This gives rise to *four* signals in the ^{1}H NMR spectrum.

Enantiomers

Enantiomers

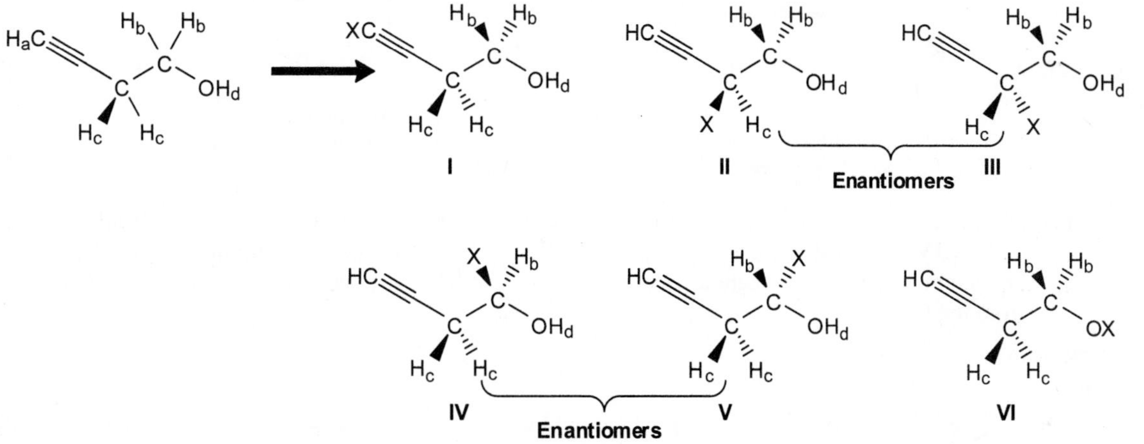

(b) Molecules **I** and **II** are the same molecule. Molecules **III** and **IV** are the same molecule. Molecules **I/II**, **III/IV**, and **V** are constitutional isomers. According to Step 3 in the chemical distinction test, these H atoms are chemically distinct. There are *three* signals in the ^{1}H NMR spectrum.

Same

Same

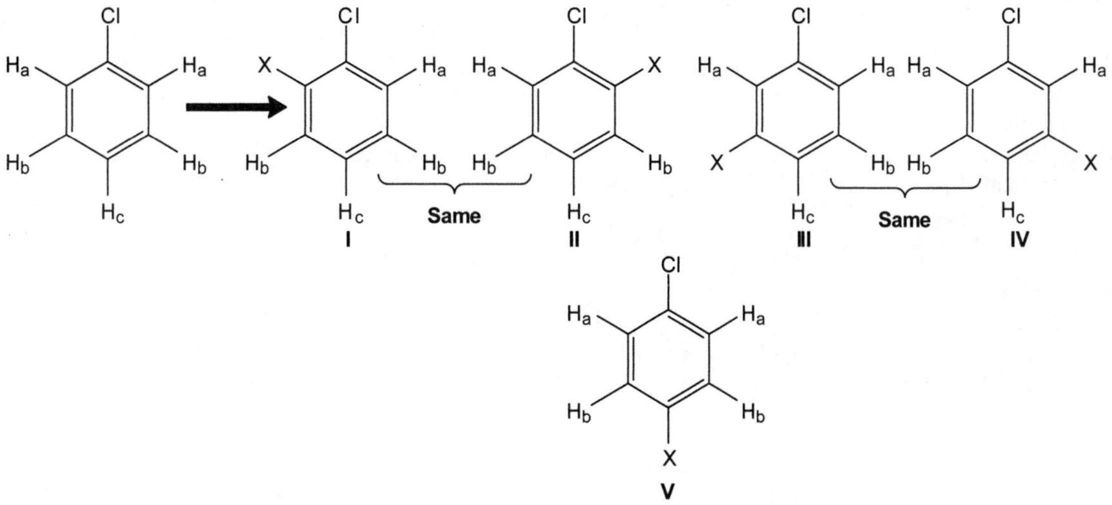

(c) Molecules **II** and **III** are the same molecule. Molecules **I**, **II/III**, and **IV** are constitutional isomers. According to Step 3 in the chemical distinction test, these H atoms are chemically distinct. There are *three* signals in the ^{1}H NMR spectrum.

Same

I II III IV

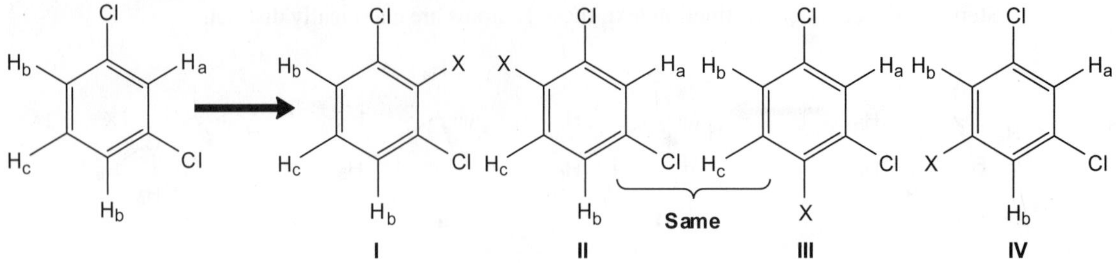

(d) Molecules **I**, **II**, **III**, and **IV** are the same. According to Step 3 in the chemical distinction test, these H atoms are *not* chemically distinct. There is *one* signal in the ^{1}H NMR spectrum.

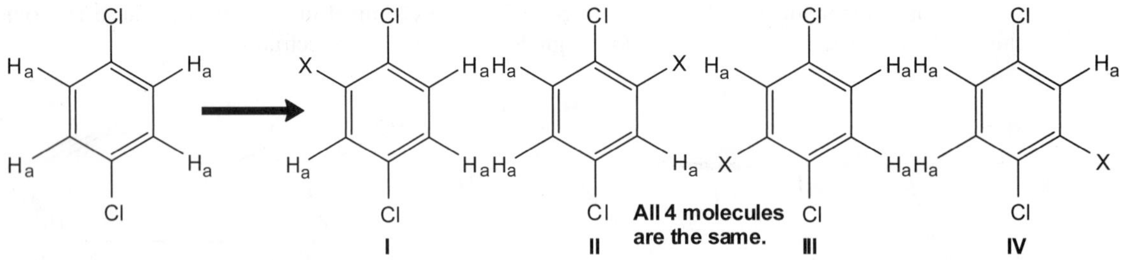

All 4 molecules are the same.

I II III IV

(e) Molecules **I** and **II** are the same molecule. Molecules **III** and **IV** are the same molecule. Molecules **I/II**, and **III/IV** are constitutional isomers. According to Step 3 in the chemical distinction test, these H atoms are chemically distinct. There are *three* signals in the ^{1}H NMR spectrum, two from aromatic C–H and one from the OH.

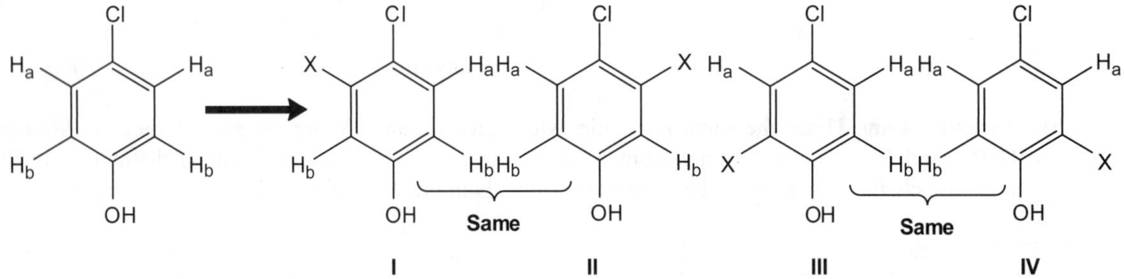

Same Same

I II III IV

Problem 16.9

Think

What molecules are generated by substituting each H atom separately by "X"? Which of those molecules are identical? Enantiomers? Diastereomers? Constitutional isomers? How many signals result?

Solve

Molecule **D** is the only isomer with symmetry that would cause the three protons to give rise to two signals in the ^{1}H NMR spectrum. All other molecules give three signals in the ^{1}H NMR spectrum. This can be seen with the chemical distinction test below.

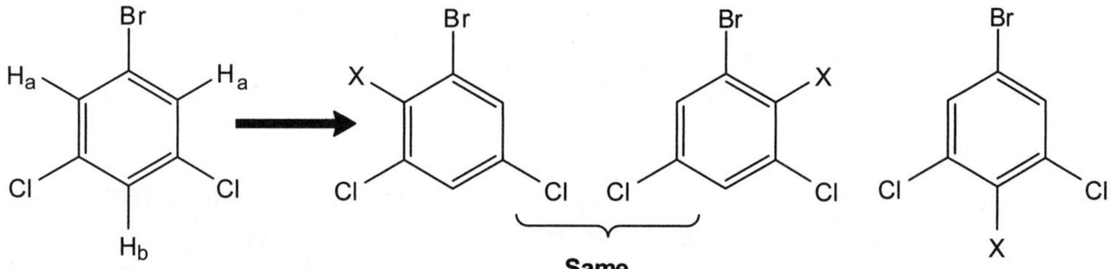

Same

D
2 signals

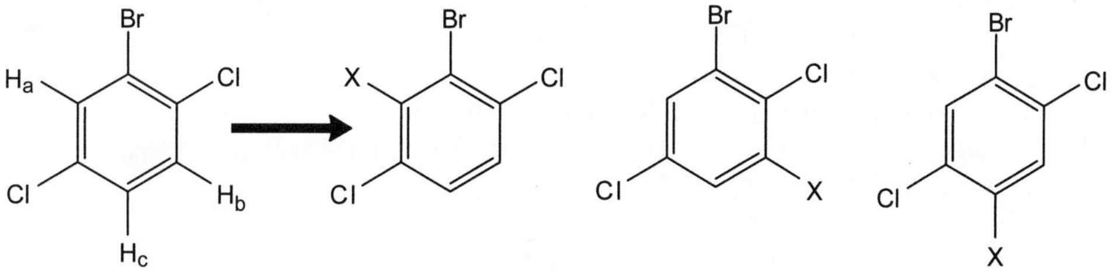

E
3 signals

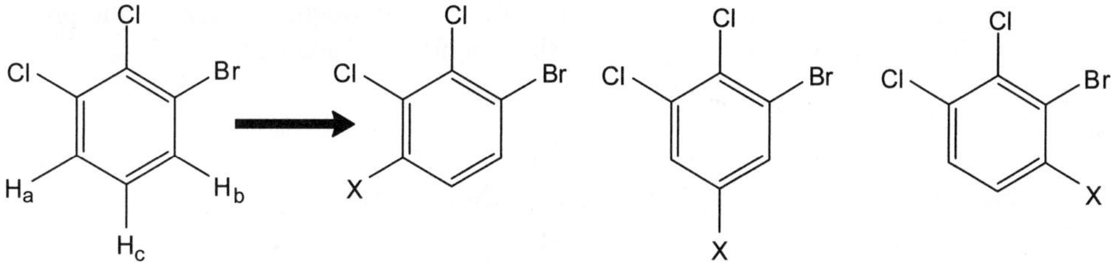

F
3 signals

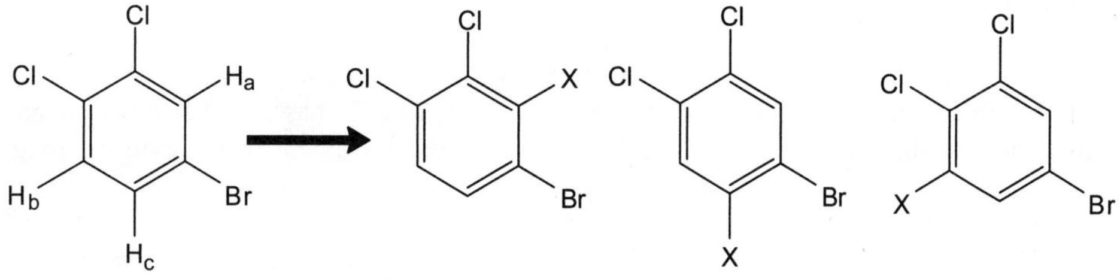

G
3 signals

Problem 16.10

Think

In Equation 16-4, what value should be substituted for γ? For B_{ext}?

Solve

As mentioned in the text, γ for a proton is $2.67512 \times 10^8\ T^{-1}s^{-1}$. In the problem, the value for B_{ext} is given as 11.74 T.

Therefore,
$$\nu_{op} = \gamma B_{ext}/2\pi$$
$$= (2.67512 \times 10^8\ T^{-1}s^{-1})(11.74\ T)/2\pi$$
$$= 500 \times 10^6\ s^{-1} = 500\ MHz$$

The unit T cancels, leaving the frequency units, s^{-1} or Hz.

Problem 16.11

Think

What is operating frequency of the NMR spectrometer using a 7.046-T magnet? In Equation 16-3, what value should be substituted for (ν_{sample} - ν_{TMS}), and ν_{op}? What is the resulting chemical shift in ppm?

Solve

In the problem, the value for B_{ext} is given as 7.064 T.

Therefore,
$$\nu_{op} = \gamma B_{ext}/2\pi$$
$$= (2.67512 \times 10^8\ T^{-1}s^{-1})(7.046\ T)/2\pi$$
$$= 300 \times 10^6\ s^{-1} = 300\ MHz$$

At that magnetic field, the operating frequency is 300 MHz, or 3×10^8 Hz.

Because the signal of the proton of interest is 2200 Hz higher than those in TMS, $(\nu_{sample} - \nu_{TMS}) = 2200$ Hz. Using Equation 16-3, we can solve for the chemical shift in ppm:

$$= [(\nu_{sample} - \nu_{TMS})/\nu_{op}] \times 10^6$$
$$= [(2200\ Hz/(3 \times 10^8\ Hz)] \times 10^6 = 7.33\ ppm$$

Problem 16.12

Think

Using Equation 16-3, how can you rearrange the equation to solve for the frequency (Hz) of the proton? In Equation 16-3, what value should be substituted for chemical shift (ppm), ν_{TMS}, and ν_{op}?

Solve

$$2.4\ ppm = [(\nu_{sample} - \nu_{TMS})/\nu_{op}] \times 10^6 = [(\nu_{sample} - \nu_{TMS})/3 \times 10^8\ Hz] \times 10^6$$

So $= (\nu_{sample} - \nu_{TMS}) = (2.4\ ppm)(3 \times 10^8\ Hz)/(10^6\ Hz) = 720$ Hz higher.

Problem 16.13

Think

Which atom, Br or I, is more electronegative? Which H atoms are more deshielded? Which atoms are, therefore, farther downfield?

Solve

The chemical shift will be greater for the protons in CH_3Br. This is because Br has a higher electronegativity than I, so Br will cause more deshielding, exposing the H atoms to a greater extent of the external magnetic field.

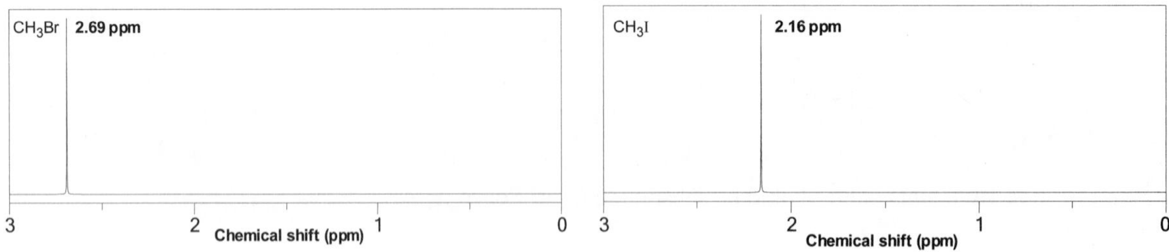

Problem 16.14

Think

Are the effects that cause deshielding additive? If so, which compound has more groups that deshield the H atom?

Solve

Each C=O group deshields the adjacent CH proton via both inductive effects and magnetic anisotropy, which shifts the signal downfield. Protons **C**, therefore, have a higher chemical shift than protons **B**. Proton **A** has an even higher chemical shift, because the CH proton has three adjacent C=O groups.

Lowest chemical shift

Highest chemical shift

B C A

Problem 16.16

Think

What groups present are responsible for deshielding nearby protons? Do those groups deshield protons to the same extent? How severely do those deshielding effects fall off with distance?

Solve

There are two groups present that cause significant deshielding: the C=C group and the CF_3 group. From Table 16-1, we can see that the C=C group is likely to have a much greater effect on deshielding because the chemical shift of an H–CR=CR_2 is 5.3 ppm. Therefore, **L** will have the most downfield peak, attached to C=C and closer to CF_3. **K** would be next. **L** and **K** are so close in chemical shift that the signals overlap. **M** is adjacent to the electron-withdrawing CF_3 group and, therefore, would come next. These deshielding effects fall off very quickly with distance, so proton **J** is next, followed by **I.** The chemical shifts of these protons decrease in the order: **L** > **K** > **M** > **J** > **I.**

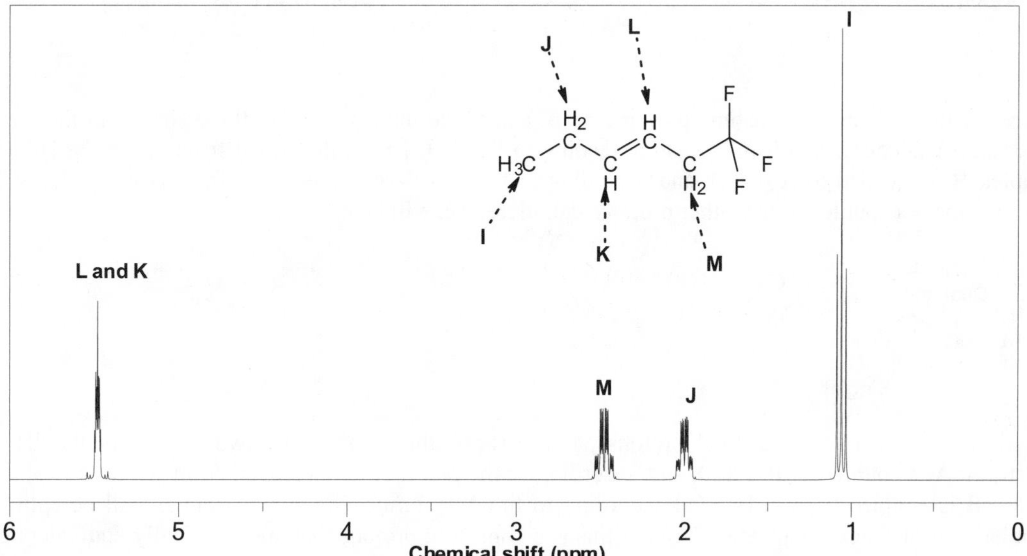

Problem 16.17

Think

How is the area under the absorption peak related to the number of signals that generate that peak? How can you use the integral trace to determine the relative areas under the peaks? Which peak has the largest area under the curve?

Solve

The signal at 1.9 ppm (signal **C**) represents the most H atoms because the stairstep in the integral trace is the largest. The signal at 6.1 ppm (signal **A**) represents the smallest number of H atoms because the height of the stairstep is the smallest. The ratio of the stairstep heights, and thus the corresponding protons, is roughly 3:2:1 for C:B:A.

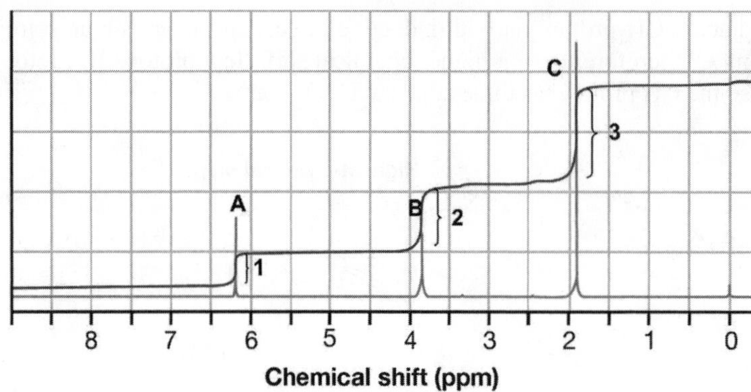

Problem 16.18

Think

What is the ratio of the stairstep heights for the peaks? Do those numbers add up to six H atoms? If not, by what factor can you multiply those numbers so that the values to add up to six H atoms?

Solve

The values in the 3:2:1 ratios already add up to six, so the signals at 6.1, 3.9, and 1.9 are due to one H, two H, and three H atoms, respectively. See Problem 16.17.

Problem 16.20

Think

How many chemically distinct protons are in each compound? How many signals should each chemically distinct proton generate? Which of those protons are coupled to other protons that are chemically distinct from themselves?

Solve

(a) There are three chemically distinct protons, **A**, **B**, and **C**, so there should be three signals in the ^{1}H NMR spectrum. **A** is coupled to **B** (CH), and according to the $N + 1$ rule, the signal from **A** will be split into a doublet. **B** is coupled to **A** (CH_3), and according to the $N + 1$ rule, the signal from **B** will be split into a quartet. **C** is not coupled to any other protons and, therefore, will be a singlet.

(B) Quartet
(A) Doublet
(C) Singlet

(b) There are two chemically distinct protons, **A** and **B**, so there should be two signals in the ^{1}H NMR spectrum. **A** is coupled to **B** (CH_2), and according to the $N + 1$ rule, the signal from **A** will be split into a triplet. **B** is coupled to **A** (CH), and according to the $N + 1$ rule, the signal from **B** will be split into a doublet. **B** is also next to another CH_2, but this is a **B** proton; protons that are chemically equivalent do not split each other.

(A) Triplet
(B) Doublet
(A)
(B)

(c) There are three chemically distinct protons, **A**, **B**, and **C**, so there should be three signals in the ^{1}H NMR spectrum. **A** is coupled to **B** (CH_2), and according to the $N + 1$ rule, the signal from **A** will be split into a triplet. **B** is coupled to **A** (CH_2), and according to the $N + 1$ rule, the signal from **B** will be split into a triplet. **C** is not coupled to any other protons and, therefore, will be a singlet.

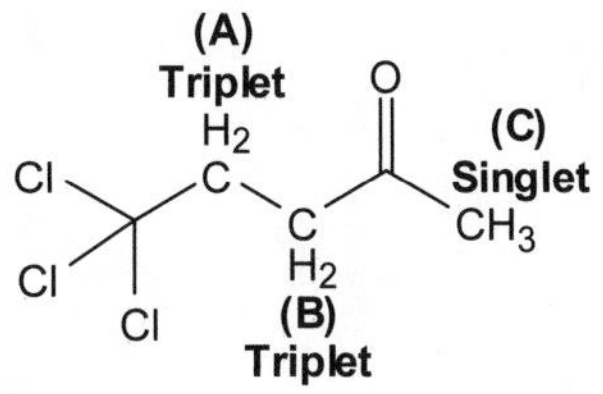

(A)
Triplet
H_2
Cl C O
C (C)
Singlet
CH_3
Cl H_2
Cl (B)
Triplet

(d) There is one chemically distinct proton, **A**, so there should be one signal in the ^{1}H NMR spectrum. **A** is next to another CH_2, but this is an **A** proton. Chemically equivalent protons do not split each other, so there will be one singlet signal.

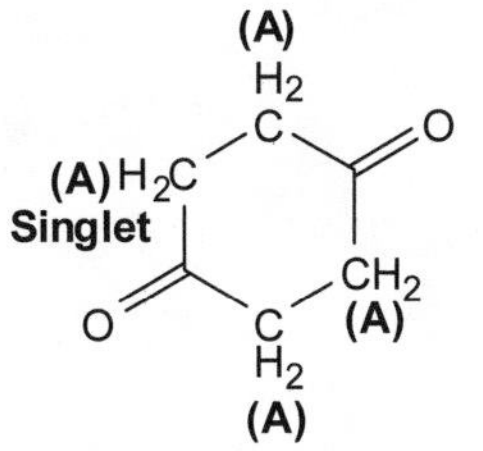

(A)
H_2
C O
(A) H_2C
Singlet
CH_2
O C (A)
H_2
(A)

Problem 16.22

Think

What should the first and last numbers in the set of ratios be? How is each remaining number derived from the line above it in Pascal's triangle?

Solve

The first and last numbers should be 1. In Pascal's triangle, the ratios for a septet would appear just below those for a sextet. So, each remaining number in the septet's ratios is computed by adding each adjacent pair of numbers in the sextet's ratios. That is, 1 + 5 = 6, 5 + 10 = 15, 10 + 10 = 20, 10 + 5 = 15, and 5 + 1 = 6. Therefore, the ratios are 1:6:15:20:15:6:1.

Problem 16.23

Think

How many signals will appear in each ^{1}H NMR spectrum? What does the aromatic H region look like? How many signals are present in the alkene region? How is each signal split? Where does the CH_3 group appear in each? Is the CH_3 coupled to other H atoms? Is there an O–H signal in both? What happens to an O–H signal when D_2O is added to the sample?

Solve

In both compounds, we would expect the aromatic H atoms to have similar chemical shifts and similar splitting patterns because both rings are para-disubstituted, and the substituents are attached by an O and an alkene carbon in both cases. The CH_3 signals would be slightly different in chemical shift and would also be different in splitting, because in **A**, the CH_3 carbon is bonded to an alkene CH, whereas in **B**, it is bonded to an O. The alkene H signals would have different splitting patterns because alkene **B** consists of $CH{=}CH_2$ and alkene **A** consists of $CH{=}CH{-}CH_3$. A major difference is that in the first molecule, the OH signal will be a broad singlet, and no such signal exists for the second molecule. If D_2O were added, that OH signal would diminish in size, and with enough D_2O, would disappear entirely as a result of H/D exchange. See the figure on the next page.

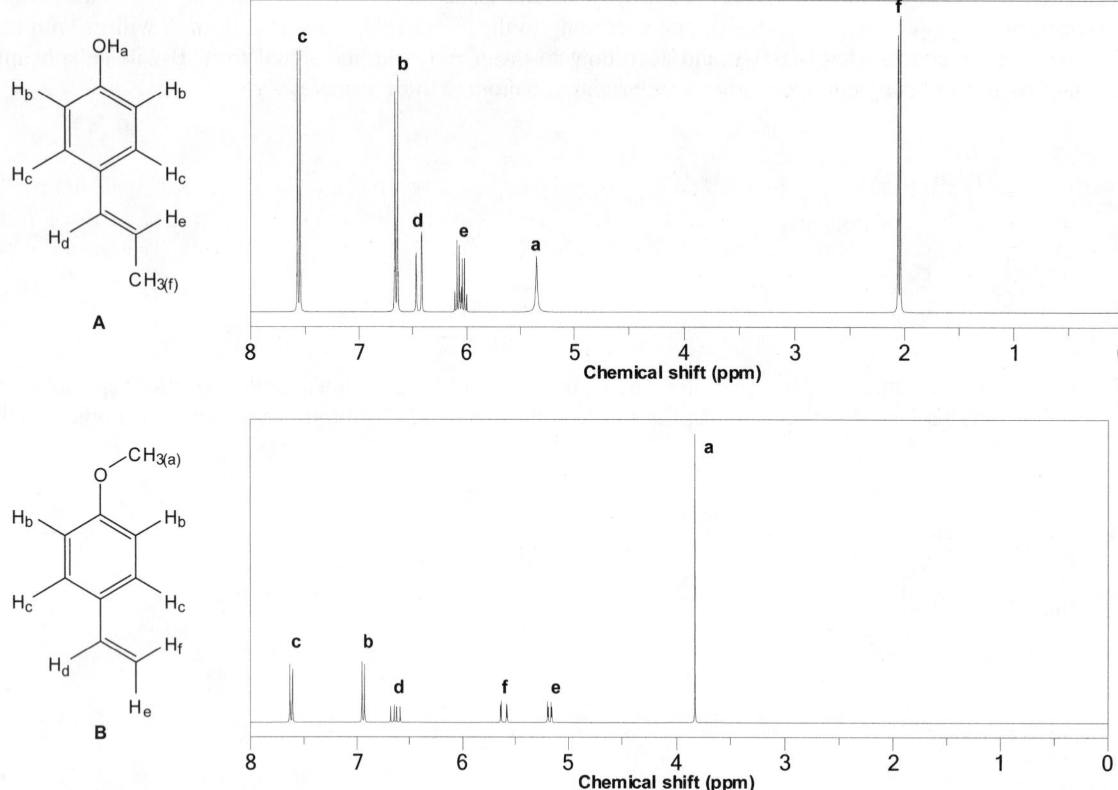

Problem 16.24

Think

If signals are coupled together, what is true about the value of their coupling constants? How can you determine the coupling constant value given the frequency of each absorption in the spectrum?

Solve

The signal at 4.1 ppm exhibits a coupling constant of about 7 Hz, which is computed as the difference in frequency between two adjacent peaks in the signal (e.g., 1647.71 Hz − 1640.63 Hz). The signals at 2.3 ppm, 1.2 ppm, and 1.1 ppm exhibit coupling constants of about 7.5 Hz, 7 Hz, and 7.5 Hz, respectively. So the H atoms giving rise to the signal at 4.1 ppm are coupled to the protons giving rise to the signal at 1.2 ppm.

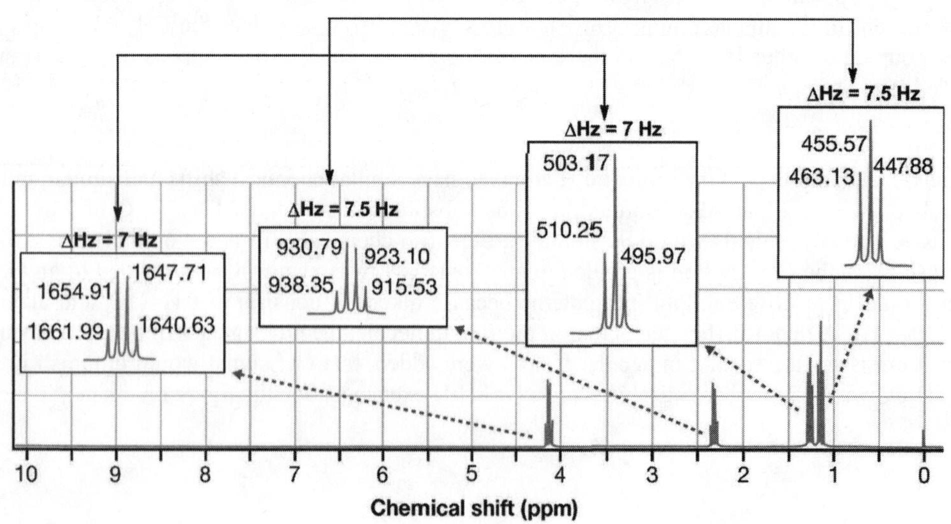

Problem 16.25

Think

To how many unique protons is H_A coupled? To how many unique protons is H_C coupled? What is the coupling constant for each peak? Does it matter in which order you draw the splitting diagram?

Solve

For the signals from both proton H_A and H_C, we should see a doublet of doublets, as shown below. The difference is in the spacing between those four peaks due to the difference in coupling constant values. It does not matter in which order you draw the splitting diagram (see Fig. 16-25); however, it typically is easier to draw the splitting with the *larger* coupling constant first.

Signal A: Split by H_C into a doublet with a coupling constant of 14.9 Hz.
Split by H_B into a doublet with a coupling constant of 1.8 Hz.

Signal C: Split by H_A into a doublet with a coupling constant of 14.9 Hz.
Split by H_B into a doublet with a coupling constant of 7.1 Hz.

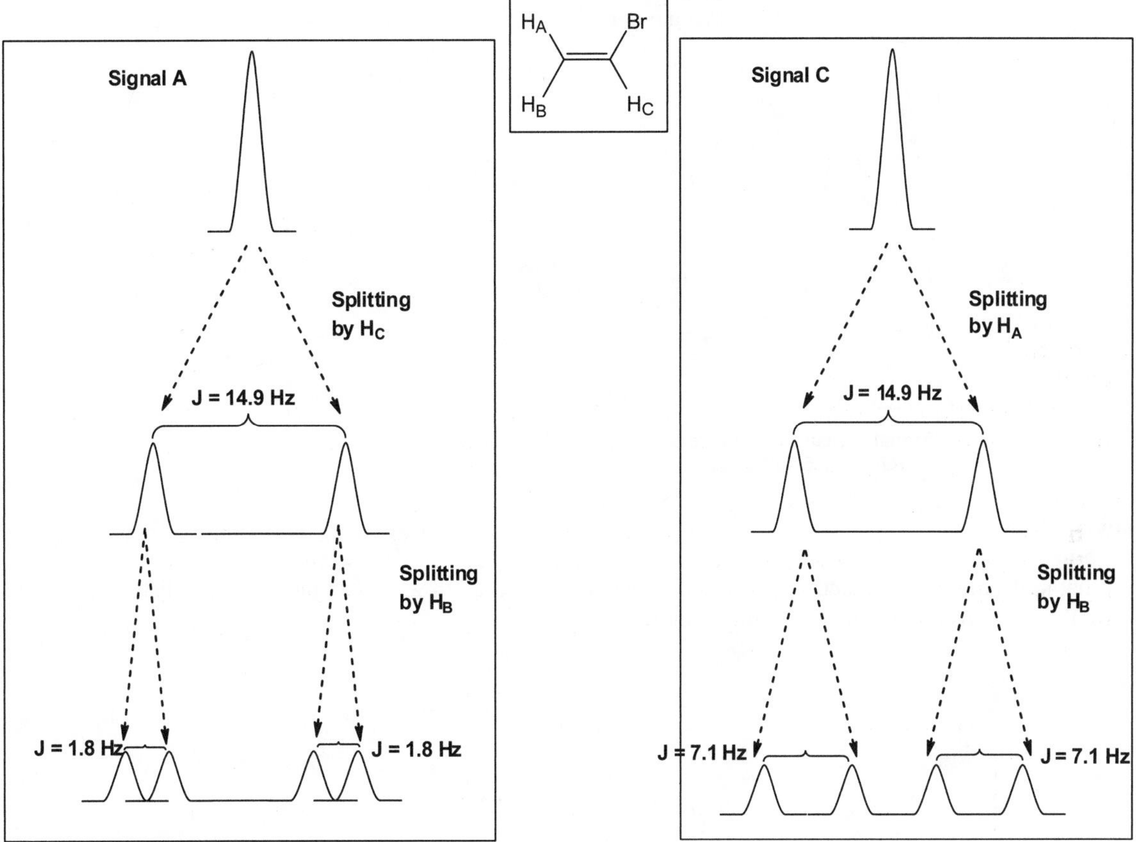

Problem 16.26

Think

If the coupling constants are the same, does overlap occur in the splitting diagram? If so, is a clear quartet of triplets observed? If not, what is the splitting of the resulting signal?

Solve

Because the two J values are the same, the 12-line pattern collapses into a 6-line pattern. This is essentially the same as we would obtain using the $N + 1$ rule with $N = 5$ (a sextet), because protons that give rise to a quartet of triplets are coupled to five protons (three of one type and two of another). Note that the greater area under the peaks toward the center of the "sextet" reflects how the area from the quartet is distributed.

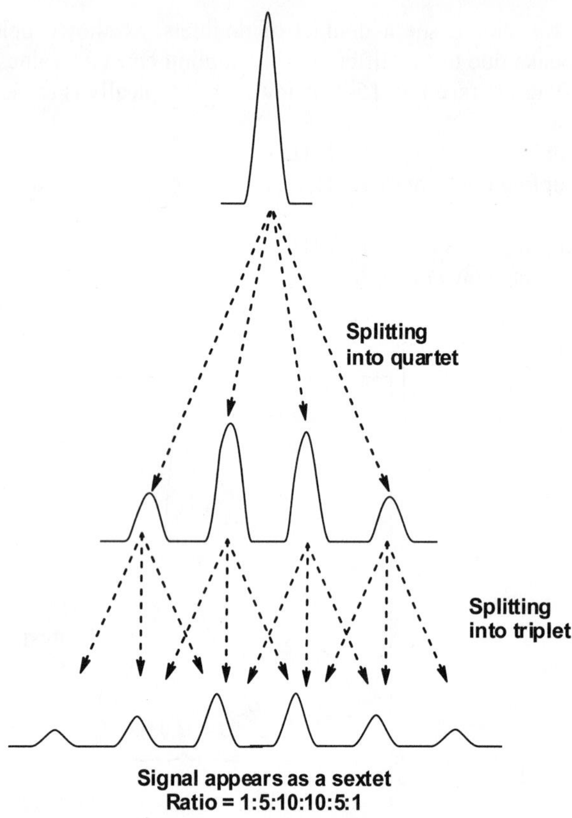

Problem 16.28

Think

What molecules are obtained by separately replacing each C atom with X in the chemical distinction test? What are the relationships among the resulting molecules?

Solve

Replacing each of the C atoms with X, we get compounds **A–J**, as follows:

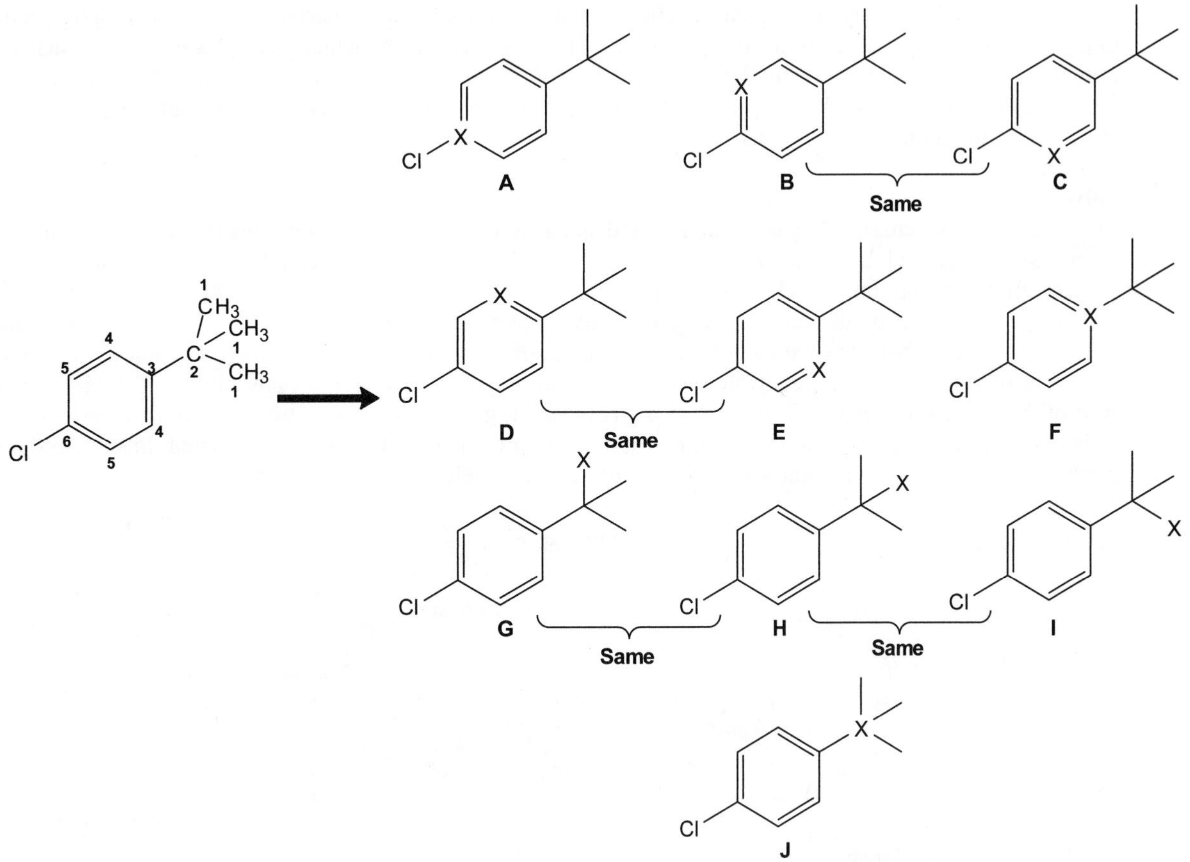

Compounds **B** and **C** are the same. Compounds **D** and **E** are the same. Compounds **G**, **H**, and **I** are the same. Compounds **A**, **B/C**, **D/E**, **F**, **G–I**, and **J** are constitutional isomers, so the corresponding six C atoms are chemically distinct; therefore, six ^{13}C signals are expected.

Problem 16.29

Think

How are the trends in chemical shift similar for ^{1}H and ^{13}C NMR spectroscopy? Which C atoms in the molecule shown experience ring current from the aromatic π electrons? Which C atom in the ring also experiences a deshielding effect due to the Cl atom? Refer to Table 16-4.

Solve

As we can see in Table 16-4, aromatic carbons have chemical shifts around 130 ppm, whereas alkyl halide carbons have chemical shifts of 35 to 55 ppm. Not all the aromatic carbons have the same chemical shift, however. The one closest to the Cl atom will be most deshielded, due to inductive effects, so its signal will be the most downfield shifted, and it will have the highest chemical shift.

Highest chemical shift

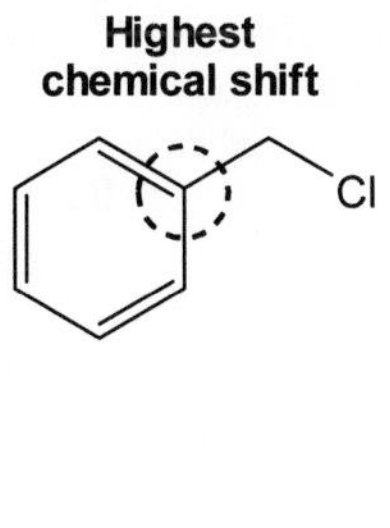

Problem 16.30

Think

What is the IHD? How many signals appear in the ^{1}H and ^{13}C NMR spectra? Is there any symmetry suggested by the total number of signals? From the chemical shifts in each NMR spectrum, what functional groups are likely present? From the splitting patterns in the ^{1}H NMR spectrum, which groups are next to each other? Sometimes it is helpful to draw the NMR spectra from the data given. Start to put a proposed structure together. (*Note*: this can be a trial-and-error process. Propose a structure and then check the NMR data to make sure your structure is consistent with the data.)

Solve

The IHD is 5. A benzene ring is apparent, evidenced by the aromatic proton signals around 7 ppm and the carbon signals around 130 ppm, which accounts for an IHD of 4. The other IHD is from a carboxylic acid, evident from the proton signal at 12.9 ppm and the carbon signal at 170.9 ppm. The benzene ring is trisubstituted, evident by the fact that there are only three aromatic H signals. The ring must be substituted at three successive carbons to give rise to the splitting patterns of the aromatic signals (doublet and triplet). Also, two pairs of ring carbons are equivalent to give rise to only four aromatic signals in the ^{13}C NMR spectrum. And of the three nonaromatic C atoms, two are equivalent, giving rise to only two nonaromatic carbon signals. This suggests that the signal at 2.3 ppm, 6H, is from two equivalent CH_3 groups attached directly to the ring, giving rise to singlets. The unknown has the structure shown below.

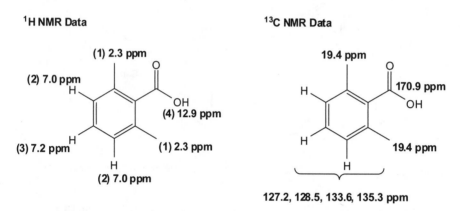

^{1}H NMR Data

(1) 2.3 ppm
(2) 7.0 ppm
(4) 12.9 ppm
(3) 7.2 ppm
(1) 2.3 ppm
(2) 7.0 ppm

^{13}C NMR Data

19.4 ppm
170.9 ppm
19.4 ppm
127.2, 128.5, 133.6, 135.3 ppm

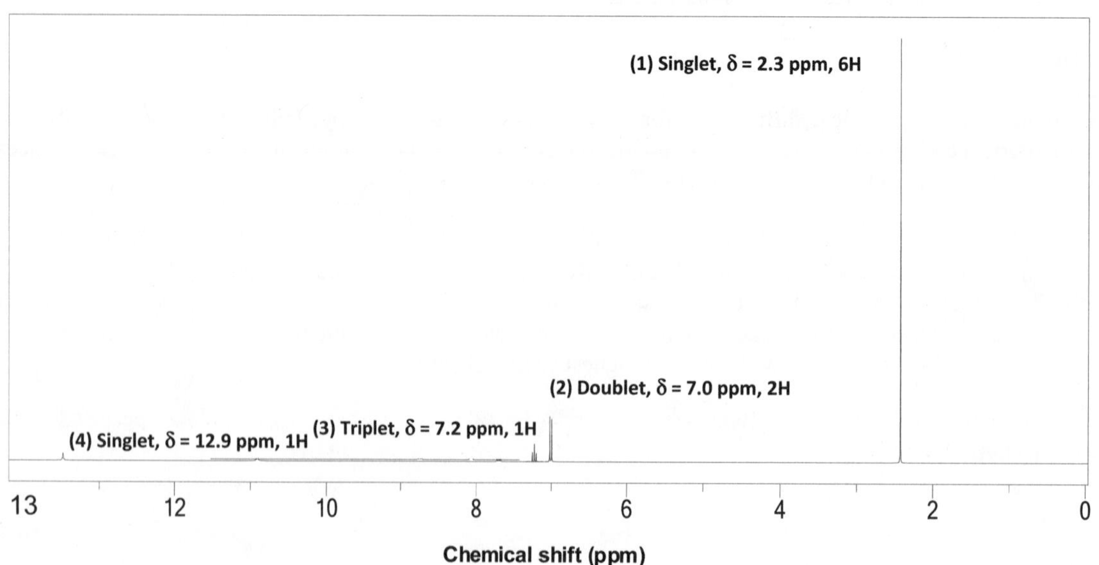

Problem 16.31

Think

What is the IHD? How many signals appear in the ^{1}H and ^{13}C NMR spectra? Is there any symmetry suggested by the number of total signals? From the chemical shifts in each NMR spectrum, what functional groups are likely present? From the splitting patterns in the ^{1}H NMR spectrum, which groups are next to each other? Sometimes it is helpful to draw the NMR spectra from the data given. Start to put a proposed structure together. (*Note*: this can be a trial-and-error process. Propose a structure and then check the NMR data to make sure your structure is consistent with the data.)

Solve

The IHD is 4. A benzene ring accounts for this IHD, consistent with the proton signals around 7.2–7.4 ppm. No other multiple bonds or rings can be present. The benzene ring must be monosubstituted to have five aromatic H atoms. The signal at 2.4 ppm must be due to an OH group for it to disappear when D_2O is added. The signal at 4.8 ppm must come from a CH whose C is attached directly to the ring, because the deshielding of the ring causes such signals to be at least 2.3 ppm, according to Table 16.1. To have a chemical shift of 4.8 ppm, there must be significant additional deshielding, explained by an O atom directly attached to that same carbon. The spectrum, therefore, is consistent with the structure below.

^{1}H NMR Data

(2) 2.4 ppm
OH
(3) 4.8 ppm
(1) 1.4 ppm
(4) 7.2 - 7.4 ppm

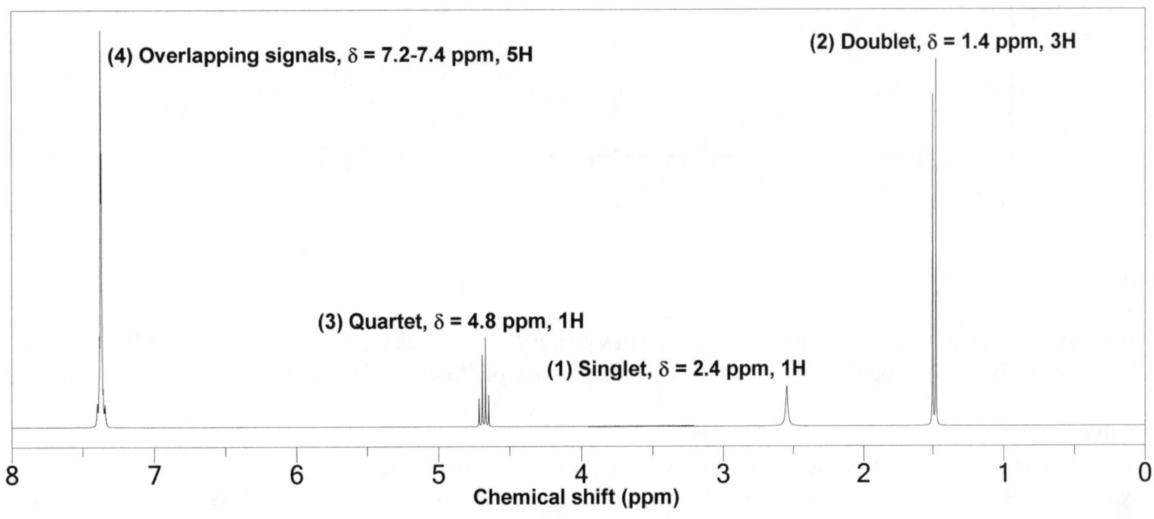

Problem 16.32

Think

What is the molecular mass of ethylbenzene? Is there a peak that matches that mass? Which peak is the most intense? What is the difference in mass between those two peaks?

Solve

The mass of ethylbenzene is 106 amu, so the M^+ peak is the one having the second highest *m/z* value. The base peak is the tallest peak, at 91 amu. The difference in mass between those two peaks is 15 amu. As shown below, that difference can come from fragmentation of a CH_3 group from M^+.

Problem 16.33

Think

What is the intensity of the M + 1 peak? What is the intensity of the M^+ peak? How many carbon atoms are predicted from the formula in Equation 16-7?

Solve

The M + 1 peak appears to be about 8% of the M^+ peak. So,

$$\# \text{ C atoms} \approx [(\text{intensity of M} + 1)/(\text{intensity of M}^+)] \times (100\%/1.1\%)$$
$$= (8\%/100\%) \times (100\%/1.1\%) = 7.3, \text{ or } {\sim}7 \text{ C atoms.}$$

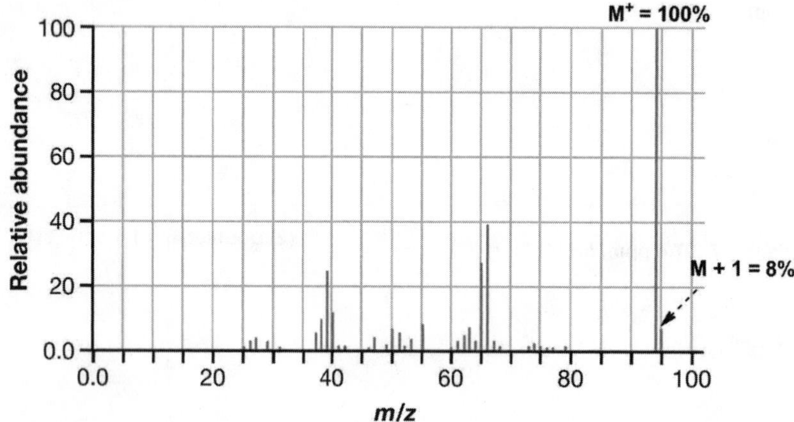

Problem 16.34

Think

What is the molecular mass of the compound? How is the M^+ peak related to the molar mass? What would the M + 2 value be (^{34}S isotope)? What is the relative abundance of ^{32}S and ^{34}S (refer to Table 16-5).

Solve

The M^+ peak should be at a *m/z* value that is the molar mass of the compound, which is 110. The ^{34}S isotope would give an M + 2 peak around 112. Because the ratio of abundance of ^{34}S to ^{32}S is about 4%, we should expect the M + 2 peak to be have an intensity of about 4%.

Problem 16.35

Think

What is the *m/z* value of the M^+ peak in Figure 16-41? What does the odd mass suggest? What formula is consistent with these data? If there are no peaks >3000 cm^{-1} in the IR, are N–H bonds present? What bond stretch gives the strong absorption at 1670 cm^{-1}?

Solve

The formula is given to us as $C_7H_{15}NO$ or $C_7H_{12}NF$. The absorption at 1670 cm^{-1} is indicative of a C=O amide stretch. This confirms the formula as $C_7H_{15}NO$. If we examine Table 16-2, we see that this carbonyl stretch is at a lower frequency than normal ketones or aldehydes, but it is in the middle of the range in which we normally find amide carbonyl stretches. Because we find no IR absorptions above 3000 cm^{-1}, there should be no N–H bonds. Three molecules that are consistent with this are shown below.

Problem 16.36

Think

What is molar mass of this compound? What does the M^+ and M + 2 having a 1:1 ratio suggest? Having four signals in the ^{13}C NMR spectrum calculates to how much of the mass? What does the IR signal at 1720 cm^{-1} suggest? What is the total mass of the atoms now accounted for? What is the formula of the compound?

Solve

The mass spectrum indicates the presence of a Br atom, evidenced by the pairs of peaks that are equal in height but separated by 2 amu. We know that there are at least four carbon atoms in the compound, due to the fact that there are four carbon NMR signals. There is also a C=O bond, evidenced by the IR peak at 1720 cm^{-1}. The total mass of these four C atoms and one O atom is 4(12) + 16 + 79 = 143 amu. Because M^+ is 150 amu, the compound cannot have another heavy atom such as C or O, because that would then exceed the mass of the M^+. The rest of the mass, therefore, must come from H atoms. So the compound's formula is C_4H_7OBr. One possible ketone and aldehyde structure consistent with this information are shown below. Other bromo-substituted ketones and aldehydes with four carbons are also possible.

Possible ketone structure **Possible aldehyde structure**

Problem 16.37

Think

In Equation 16-4, what value should be substituted for γ_H? For B_{ext}? For ν_{op} for ^1H and ^{13}C? If you have two equations and two unknowns, how can you cancel B_{ext}? Solve for γ_C.

Solve

For H and C nuclei, Equation 16-4 gives: $\nu_{op}(H) = \gamma_H B_{ext}/2\pi$ and $\nu_{op}(C) = \gamma_C B_{ext}/2\pi$

If we divide the first equation by the second we get: $\nu_{op}(H)/\nu_{op}(C) = \gamma_H/\gamma_C$, which cancels B_{ext}.

Substituting in for the values we know: 300 MHz/75 MHz = 2.67512×10^8 T^{-1} s^{-1})/γ_C

Solving: $\gamma_C = 6.69 \times 10^7$ T^{-1} s^{-1}

Problem 16.38

Think

In Equation 16-3, what values should be substituted for (ν_{sample} - ν_{TMS}) and ν_{op}?

Solve

Equation 16-3 is: Chemical shift (ppm) = $[(\nu_{sample} - \nu_{TMS})/\nu_{op}] \times 10^6$

The values needed are:

(ν_{sample} - ν_{TMS}) = 11,250 Hz ν_{op} = 75 MHz = 75×10^6 Hz

Chemical shift (ppm) = $[(11{,}250 \text{ Hz})/(75 \times 10^6 \text{ Hz})] \times 10^6 = 150$ ppm

Problem 16.39

Think

Consult Equation 16-1 for the relationship between ΔE_{spin} and B_{ext}. For a photon to be absorbed by a nucleus with spin, how must ΔE_{spin} relate to the energy of the photon? How does the energy of a photon relate to the photon frequency?

Solve

We are given Equation 16-1: $\Delta E_{spin} = \gamma h B_{ext}/2\pi$

For a photon to be absorbed, the energy of the photon must equal the energy difference between two spin states: $\Delta E_{spin} = E_{photon} = h\nu$

Setting the right sides equal to each other: $\gamma h B_{ext}/2\pi = h\nu$

The h cancels out and we get $\gamma B_{ext}/2\pi = \nu$. For a bare proton, ν is defined as ν_{op}, so $\gamma B_{ext}/2\pi = \nu_{op}$, which is Equation 16-4.

Problem 16.40

Think

If you divide the stairstep height of each signal by the smallest value, what values are obtained? To what integer value does each number round? Do these integer values add up to 14 H atoms?

Solve

Dividing each signal by the smallest value (9) gives:

37 mm/9 mm = 4.11, which rounds to 4.

9 mm/9 mm = 1.00.

26 mm/9 mm = 2.89, which rounds to 3.

52 mm/9 mm = 5.78, which rounds to 6.

Sum: 4 H + 1 H + 3 H + 6 H = 14 H.

Problem 16.41

Think

Does a higher signal frequency correspond to a larger or smaller chemical shift? (Consult Equation 16-3.) Which signal is farther downfield? Does a signal farther downfield correspond to a proton that is more shielded or deshielded?

Solve

According to Equation 16-3, the signal that is 450 Hz higher than TMS has a smaller chemical shift, and is therefore less downfield shifted, than the signal 755 Hz higher than TMS. Thus, the signal at 450 Hz represents H atoms that are more shielded and the signal at 755 Hz represents H atoms that are more deshielded. A greater signal frequency represents a greater ΔE_{spin}, which is a result of a larger B_{eff}. A larger B_{eff} is the result of greater exposure of a nucleus to B_{ext}.

Problem 16.42

Think

If the peaks were generated from one chemically distinct H atom, how many signals would be present? What is the ratio of the intensities of the peaks shown? Can that be the result of the signal being split by one chemically distinct type of proton? Does this match the ratio for a simple quartet? Could the signal be the result of complex splitting? What type of splitting gives rise to peaks of essentially equal height?

Solve

If the peaks came from one chemically distinct type of proton, they would represent one signal. That signal cannot be a simple quartet because, according to Pascal's triangle, the height ration should be 1:3:3:1. These four peaks, however, have essentially the same height/area. The signal could be the result of complex splitting, where it is split into a doublet twice—once by each of two distinct types of protons. Each splitting into a doublet would produce two peaks of equal height/area. Thus, the signal could be a doublet of doublets.

Problem 16.43

Think

How many H atoms are present? Does the molecule possess any symmetry? Are there any diastereotopic protons? Are any proton signals in different chemical environments due to rotation restrictions (double bonds or rings)?

Solve

(a) **Two signals**: The molecule has two planes of symmetry and all aromatic H atoms are chemically equivalent, as are the two O–H groups.

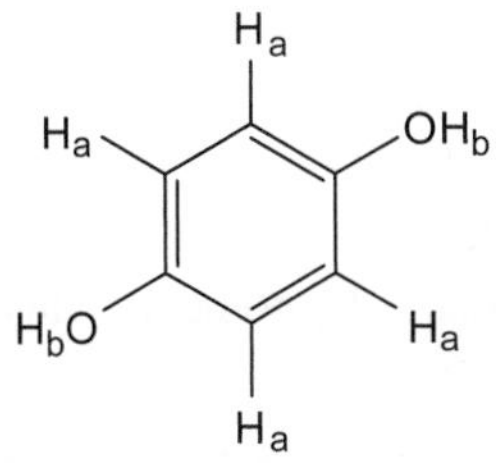

(b) **Five signals**: Protons that are cis to the chlorines are distinct from those that are trans. H atoms that are on the same C in this molecule are diastereotopic; their substitution by X in the chemical distinction test produces diastereomers.

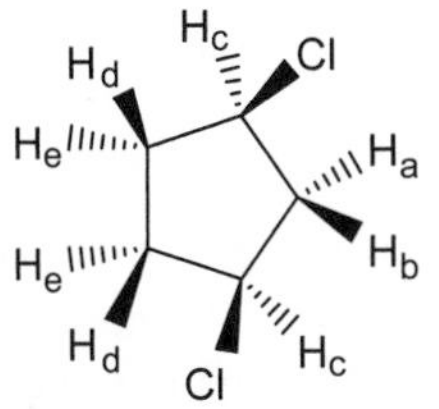

(c) **Eight signals**: All protons are chemically distinct. Protons that are cis to the Cl and Br are distinct from those that are trans. H atoms that are on the same C in this molecule are diastereotopic; their substitution by X in the chemical distinction test produces diastereomers. [Note: **(b)** is a meso compound, but **(c)** is not.]

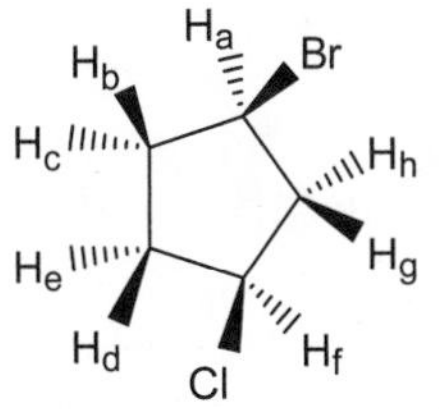

(d) **Four signals**: There are no stereocenter C atoms and rotation is not limited by a ring or a double bond, so H atoms bonded to the same C atom are homotopic.

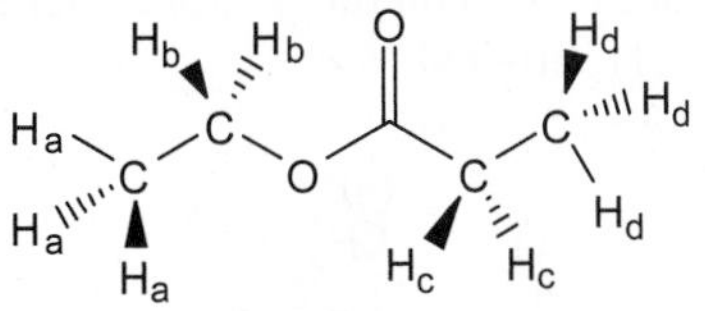

(e) **Two signals**: The molecule has one plane of symmetry perpendicular to the plane of the page along the C=O bond, so the CH_3 groups are in the same chemical environment, as are the CH_2 groups.

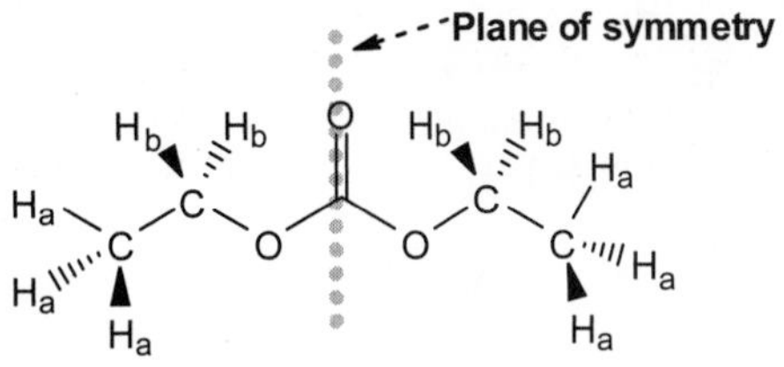

Plane of symmetry

(f) Three signals: The molecule has one plane of symmetry perpendicular to the ring, which contains the Br–C and the C–O bonds, so there are two H signals from the aromatic H atoms, and one from the OH.

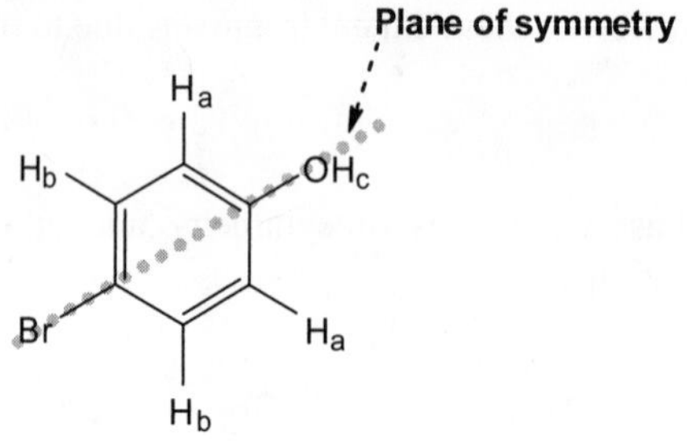

(g) Five signals: All protons are chemically distinct. The molecule has no planes of symmetry other than the plane that contains the ring, so all aromatic H atoms are chemically distinct (four signals). An additional signal arises from the O–H group.

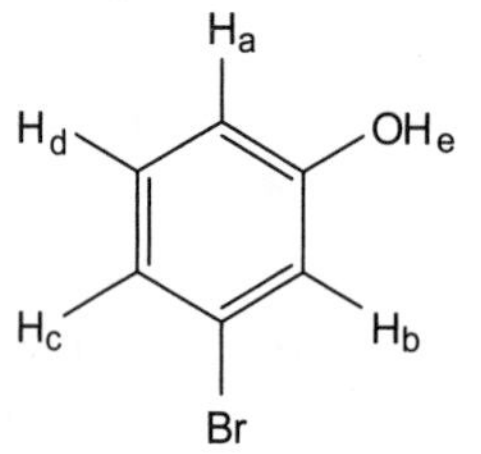

(h) Five signals: All protons are chemically distinct. The molecule has no planes of symmetry other than the one that contains the ring, so all aromatic H atoms are chemically distinct (four signals). An additional signal arises from the O–H group.

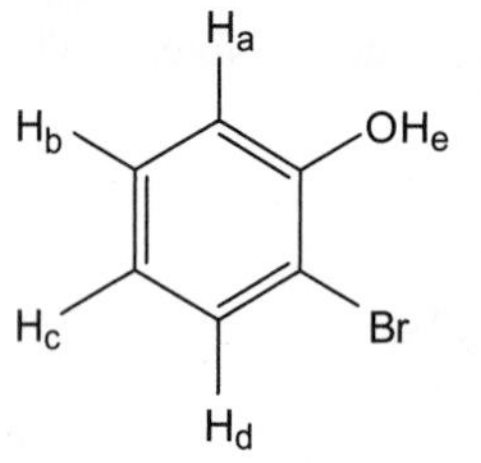

(i) Six signals: The molecule has a plane of symmetry perpendicular to the plane of the ring, which contains the bond that connects the alkyl substituent to the ring. There are three signals from the aromatic H atoms. There are no stereogenic carbon atoms and rotation about the C–C single bond takes place freely, so H atoms bonded to the same carbon atom are homotopic (one signal from the three CH_3 groups) or enantiotopic (two signals from distinct CH_2 groups).

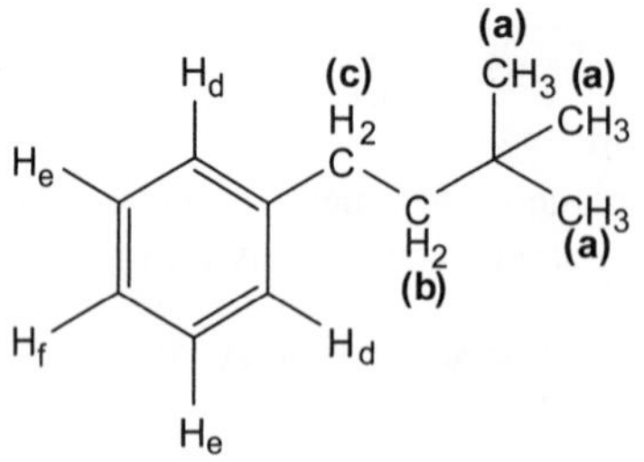

Problem 16.44

Think

How many C atoms are present? Does the molecule possess any symmetry?

Solve

(a) **Two signals:** one for the carbons bearing the OH groups, and one for the other aromatic ring carbons.
(b) **Three signals:** Note the difference from protons in the previous problem: cis/trans relationships don't apply to carbons in the ring.
(c) **Five signals:** All carbons are chemically distinct.
(d) **Five signals:** Note that the carbonyl carbon gives a signal that has no corresponding proton signal.
(e) **Three signals:** Note that the carbonyl carbon gives a signal that has no corresponding proton signal.
(f) **Four signals:** Note that two ring carbons give signals that have no corresponding proton signals.
(g) **Six signals:** All carbons are chemically distinct. Note that some carbons give signals that have no corresponding proton signals.
(h) **Six signals:** All carbons are chemically distinct. Note that some carbons give signals that have no corresponding proton signals.
(i) **Eight signals:** Note that some carbons give signals that have no corresponding proton signals.

Problem 16.45

Think

Which is more electronegative, C or Si? Which atom shields the CH_3 groups to a larger extent? Does shielding or deshielding lower the chemical shift?

Solve

Si is less electronegative than C, so the CH_3 groups take more electron density from the central Si atom in TMS than they do from the central C atom in $(CH_3)_4C$. Thus, there is more electron density around the CH_3 groups in TMS, making those H atoms (and also the C atoms) more shielded, giving rise to a lower chemical shift.

CH_3 groups have greater electron density; more shielded.

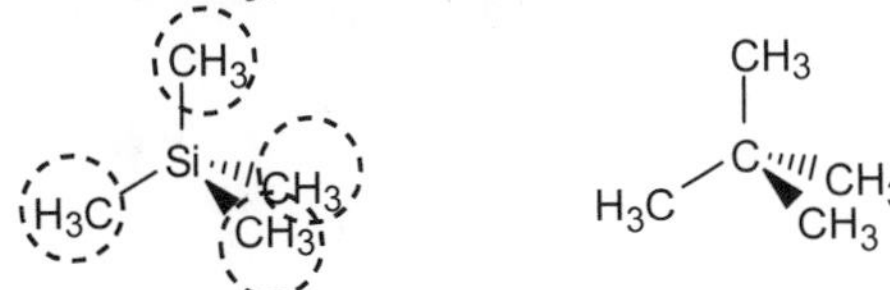

Si less electronegative

Problem 16.46

Think

How many neighboring H atoms are three bonds away or closer? Are those H atoms distinct from the ones highlighted? What type of splitting does the $N + 1$ rule predict?

Solve

(a) **Singlet.** The only H atom within three bonds is another aromatic H, which is equivalent to itself. The O–H bond is more than three bonds away.

(b) Multiplet (overlapped triplet of septets). This proton has eight neighboring protons that are distinct from itself, giving a multiplet. (More specifically, the six methyl protons split the signal into a septet, and the CH_2 protons split each of those peaks into a triplet, for 21 total peaks. But because the CH_2 protons and the methyl protons are have similar coupling constants, the signal will look more like the result of being split by eight protons, resulting in nine peaks—a nonet.)

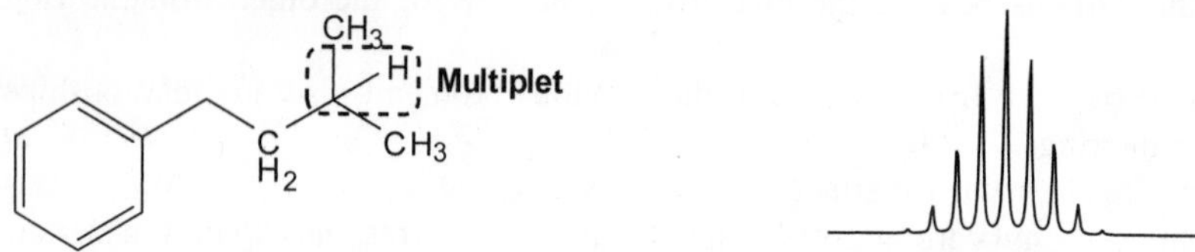

(c) Triplet (left). CH_3 protons (left) are split into a triplet by two adjacent protons.
Quartet (right). The CH_2 protons (right) are split into a quartet by three adjacent protons.

(d) Doublet. The neighboring H on the ring is distinct from the highlighted one.

(e) Triplet (left). The CH_2 protons (left) are split into a triplet by two adjacent protons.
Singlet (right). The CH_3 protons are a singlet (look closely: there are no protons on the **adjacent** carbon).

(f) Triplet of doublet (or doublet of triplet—depends on the J value). There are two H atoms that are in the same chemical environment (H_c), which splits the peak into a triplet. There is also another H atom (attached to the same carbon, H_b) that splits the peak into a doublet. H_a and H_b are diastereotopic protons and, therefore, are distinct. If the coupling constants are the same, $J_{ac} = J_{ab}$ and the splitting would result in a quartet. If $J_{ac} > J_{ab}$, the splitting pattern would look like a triplet of doublets; and if $J_{ab} > J_{ac}$ then it would look like a doublet of triplets. The actual peak (shown below) looks like two adjacent triplets, and looks more like a doublet of triplets.

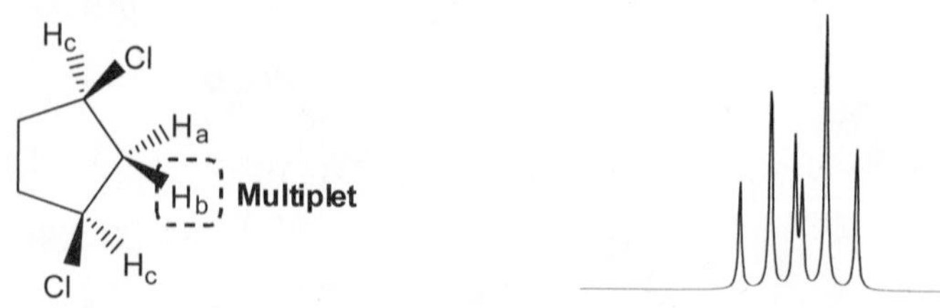

Problem 16.47

Think

What is the structure for octane, 2,5-dimethylhexane, and 4-methylheptane? Are there any planes of symmetry in each compound? How many ^{13}C signals are expected for each compound? How many ^{13}C signals are present in each spectrum shown?

Solve

The easiest way to do this problem is to focus on the number of chemically distinct C atoms that each isomer must have. Octane has four chemically distinct sets of C atoms (C1 and C8, C2 and C7, C3 and C6, C4 and C5). 2,5-Dimethylhexane has three chemically distinct sets of C atoms. 4-Methylheptane has five chemically distinct sets of C atoms. So the first spectrum belongs to octane, the second to 4-methylheptane, and the third to 2,5-dimethylhexane. See below.

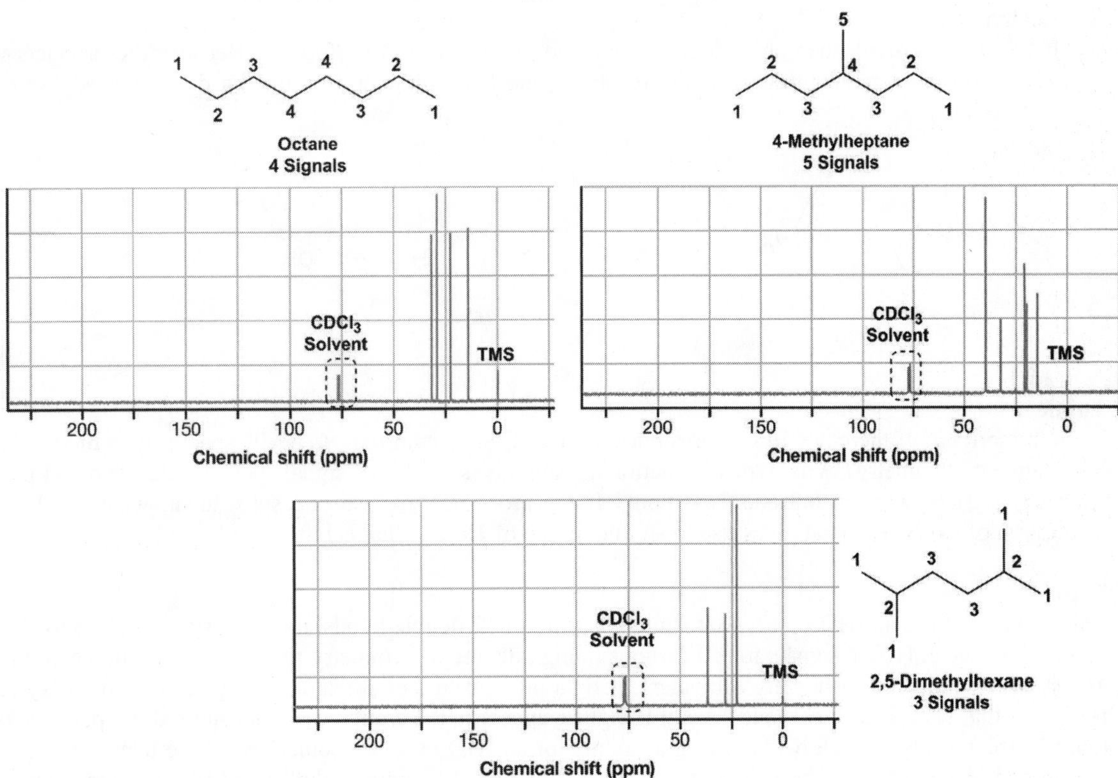

Problem 16.48

Think

What is the mass of the four C atoms? What is the total mass? If there is at least one O atom, what is now the total mass? What is the mass of a N atom? How many N atoms are possible?

Solve

If the compound contains four carbons (48 amu) and at least one O (16 amu), then the mass yet to account for is (87 − 48 − 16) = 23 amu. Because a N atom's mass is 14 amu, the compound can have only a single N atom.

Problem 16.49

Think

Look at the M^+ and M + 2 peak. What is the relative abundance of each? Which isotope matches this relative abundance pattern?

Solve

The compound most likely contains S. The M^+ peak appears to be at $m/z = 146$. The M + 2 peak appears to be about 5% of the M^+ peak, which is consistent with it containing S. If the compound contained Br, the ratio would be roughly 1:1; if it were to contain Cl, it would be in roughly a 3:1 ratio.

Problem 16.50

Think

What are the identities of the organic reactant and product? Would these compounds have different or the same ^1H, ^{13}C, and mass spectra? If the HO^- nucleophile were isotopically labeled with $H^{18}O^-$, how would that change the mass of the product?

Solve

(a) Isotopically labeled atoms would be necessary because the products would otherwise be identical to the reactants.

(b) If $H^{18}O^-$ were used, then an isotopically labeled product would be formed. Because the products would have a different molar mass than the reactants, the two compounds could be distinguished using mass spectrometry.

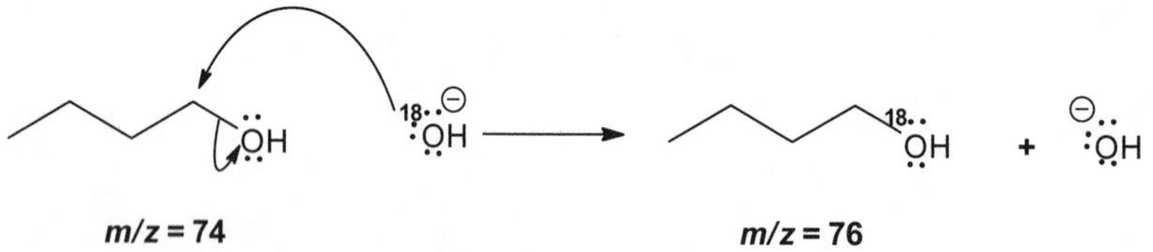

$m/z = 74$ $\qquad$ $m/z = 76$

Problem 16.51

Think

What is the IHD? If there are 19 C atoms but only five C signals in the ^{13}C NMR spectrum, is there a high or low degree of symmetry? What type of functional group gives ^{13}C NMR signals between 126 and 144 ppm and ^1H NMR signals between 6.9 and 7.44 ppm? If the aromatic ring has one substituent, what would be the integration of the peak? What gives rise to an integration of 15 H? What is left in the formula?

Solve

This compound has an IHD of 12, so it can contain up to 12 double bonds, up to 12 rings, and up to six triple bonds. To have only five signals for 19 carbons, a high degree of symmetry must be present in the compound. Many carbons must be chemically equivalent. Insight into the structural feature that gives rise to this symmetry is the fact that several C signals and several H signals appear in the aromatic region, suggesting, perhaps, one or more benzene rings. If each benzene ring accounts for an IHD of 4, we could have three benzene rings. Only one carbon is distinct from these aromatic carbons—the one at 57 ppm in the ^{13}C NMR spectrum. This carbon possesses one chemically distinct proton, which gave the proton NMR signal at 5.5 ppm. Subtracting CH from $C_{19}H_{16}$ gives $C_{18}H_{15}$, which is consistent with three equivalent phenyl groups. The compound is triphenylmethane, $(C_6H_5)_3CH$.

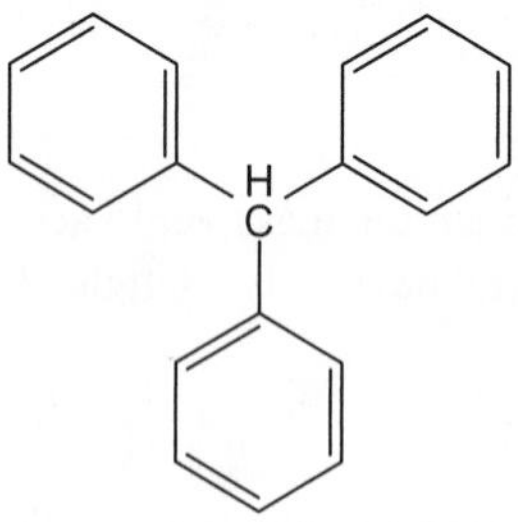

Problem 16.52

Think

The M + 2 peak at one-third intensity of the M^+ peak indicates the presence of what atom? What is the remaining mass? How many C atoms does this suggest? If there is only one signal in the ^{1}H NMR spectrum, is there a high or low degree of symmetry? If the ^{1}H NMR signal is at 1.8 ppm, is the substituent that is causing the deshielding attached directly to the CH or farther away?

Solve

The ratio of the M+ and M + 2 peaks corresponds to the presence of a Cl atom. Taking the mass of Cl (35 amu) away, we are left with an *m/z* of 57 amu. There can be at most four C atoms, because five C atoms would give a mass of 60 amu, which would exceed the 57 amu limit. All the H atoms present must be equivalent because only one signal is present and it is a singlet. Furthermore, these H atoms must be attached to a C atom that is not bonded to any highly electronegative atom, because otherwise the chemical shift would be much higher. This is consistent with a C–Cl that is attached to three equivalent CH_3 groups, as shown below.

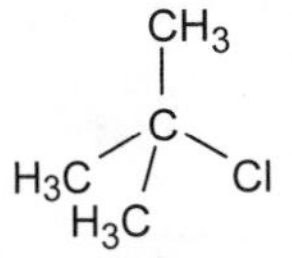

2-Chloro-2-methylpropane

Problem 16.53

Think

What is the structure of heptane? What is the molar mass of heptane? What is the formula of the molecular ion, M^+? What bonds can be severed during fragmentation? What is the molar mass of peaks **A–E**? What structure corresponds to each peak? Which peak is the most intense?

Solve

Heptane has an *m/z* value of 100. The fragmentation patterns occur from a C–C bond breaking, as shown below and on the next page. The M^+ occurs at *m/z* = 100, so the M + 1 peak occurs at *m/z* = 101. The base peak is the peak with the highest intensity. This occurs at *m/z* = 43 ($CH_3CH_2CH_2^+$).

Base peak

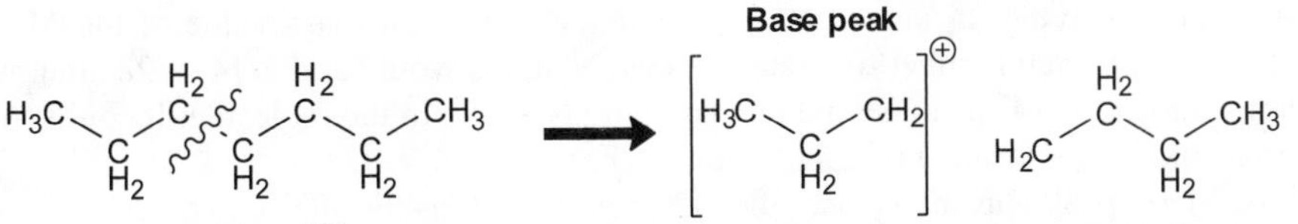

***m/z* = 100** → ***m/z* = 43** **A** + **Uncharged species, not detected**

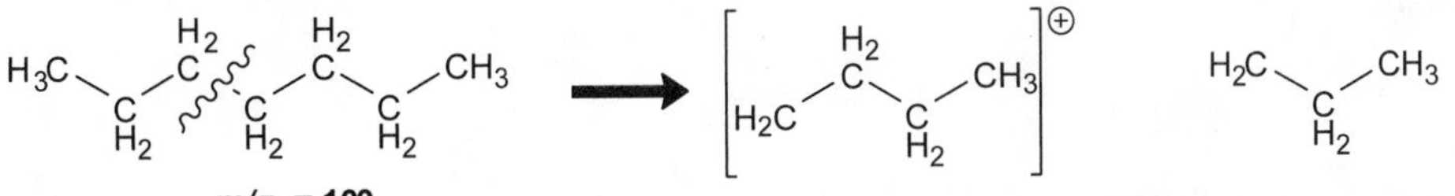

***m/z* = 100** → ***m/z* = 57** **B** + **Uncharged species, not detected**

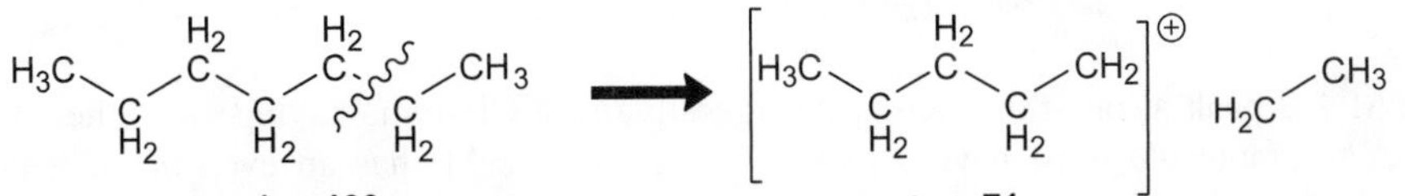

***m/z* = 100** → ***m/z* = 71** **C** + **Uncharged species, not detected**

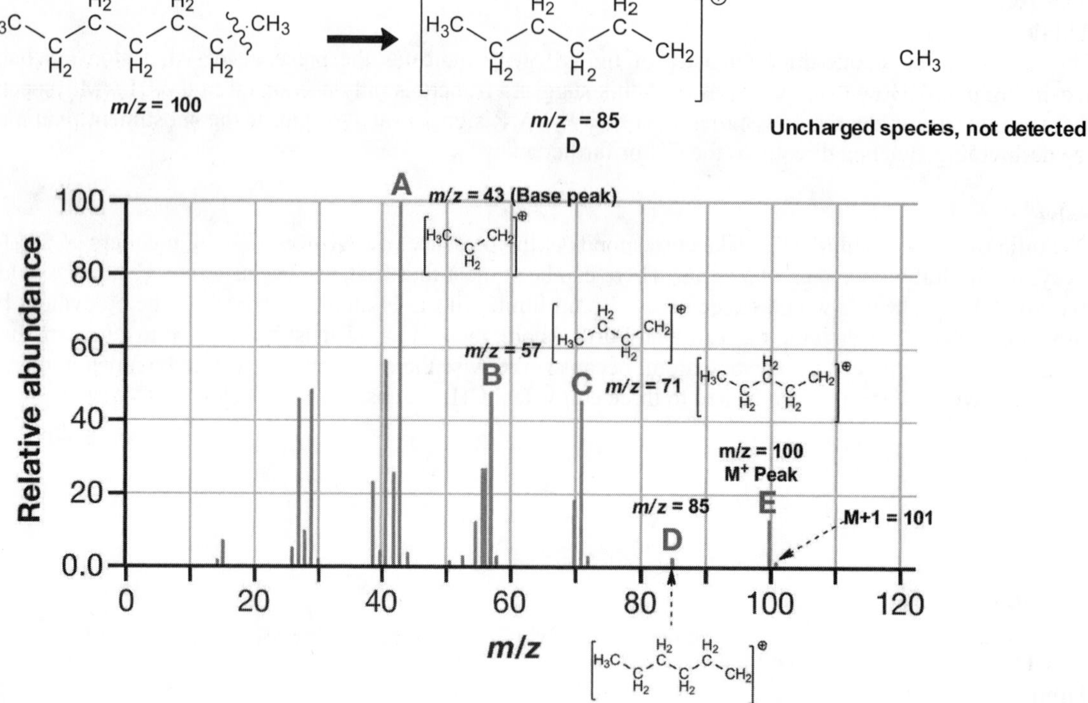

Problem 16.54

Think

What is the M^+ value? What do the M + 1 or M + 2 relative abundance percentages indicate? Are there any easily identifiable atoms from the isotope abundance? Is the mass value odd or even? What is the *m/z* value for the base peak? Is this mass part of an easily identifiable fragmentation pattern?

Solve

(a) The M^+ peak appears at $m/z = 72$ and the M + 1 peak appears at $m/z = 73$. The intensity of the M + 1 peak is used to estimate the number of C atoms: $(1.0/25.0) \times (100\%/1.1\%) = 3.64$, which rounds to 4. That accounts for $4(12) = 48$ amu, leaving 24 amu yet to account for. The even mass value of the M^+ peak suggests either no N atoms or an even number of N atoms. Two N atoms would add $2(14) = 28$ amu, which would exceed the mass value of the M^+ peak, so there must be no N atoms in the molecule. It could contain an O or F atom, however, yielding a formula of C_4H_8O or C_4H_5F.

(b) The M^+ value is 85. The M + 1 peak appears at $m/z = 86$. The relative intensities of the two peaks is used to estimate the number of C atoms: $(2.5/40.5) \times (100\%/1.1\%) = 5.6$, suggesting at least five C atoms. The M^+ has an odd mass and suggests the presence of a nitrogen atom. The mass of five C atoms and one N atom is 74 amu, which is 11 amu less than M^+. To avoid exceeding the mass of M^+, the remaining atoms must be H and the formula is $C_5H_{11}N$.

Problem 16.55

Think

What is the M^+ value? What do the M + 1 or M + 2 relative abundance percentages indicate? Are there any easily identifiable atoms from the isotope abundance? Is the mass value odd or even? What is the *m/z* value for the base peak? Is this mass part of an easily identifiable fragment?

Solve

The M^+ value is 90. The M + 2 peak at one-third height is suggestive of a Cl atom. Taking away the 35 amu from Cl, the remainder of the molecule must have a mass of 55 amu. The M^+ has an even mass, which is consistent with either no N atoms or an even number of N atoms. If the molecule has no N atoms, then it can contain up to four C atoms, and the formula can be C_4H_7Cl. Although it is possible, it is not likely for the molecule to contain an O or F atom, because the formula would have very few H atoms or none at all: C_3H_3OCl or C_3FCl. If the compound contains two N atoms, the formula would be $C_2H_3N_2Cl$, which would be unlikely.

Problem 16.56

Think

How many signals should there be in the ^1H and ^{13}C NMR spectra? How many neighboring H atoms are present for each distinct type of proton? What does the $N + 1$ formula predict? What is the chemical shift of each peak? Are there electron-donating/withdrawing groups nearby? Are there π electrons present? How many H atoms are represented by each peak?

Solve

(a) There are two ^1H NMR signals and three ^{13}C NMR signals. The representative spectra are shown below.

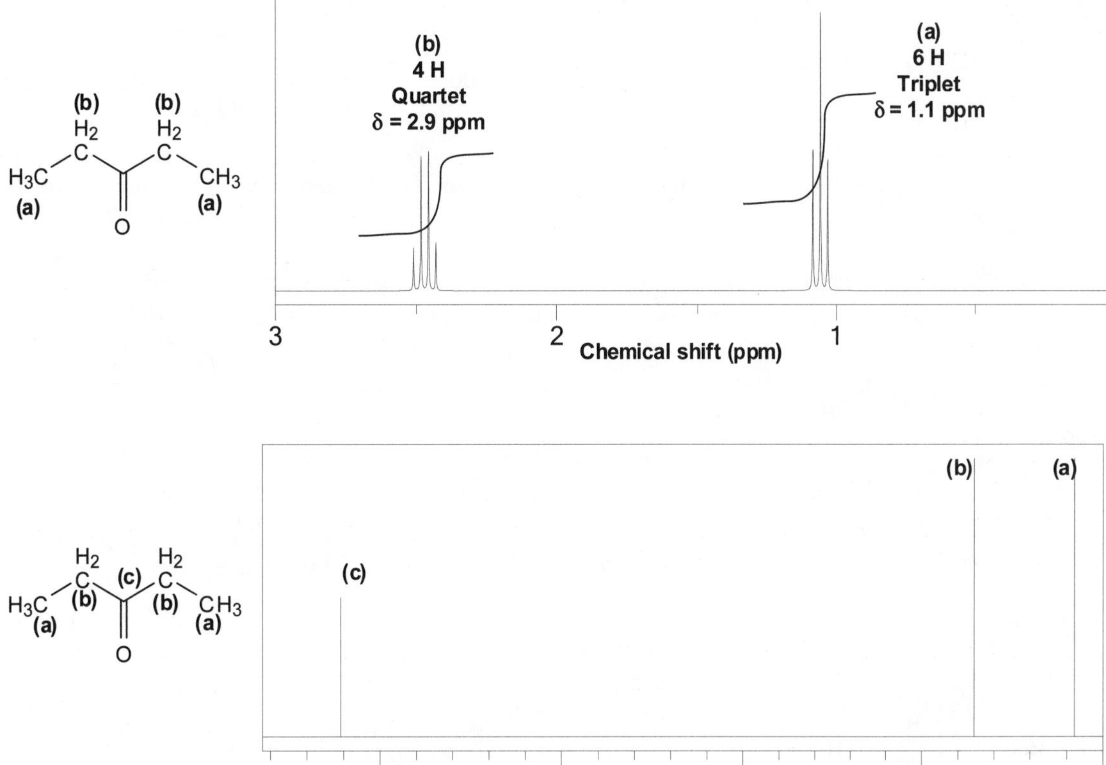

(b) There are three ^1H NMR signals and five ^{13}C NMR signals. The representative spectra are shown below and on the next page.

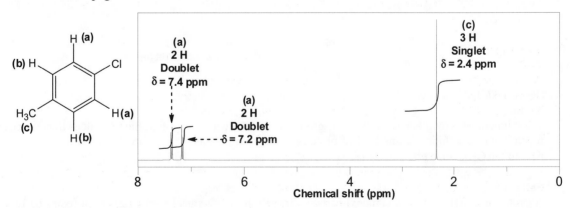

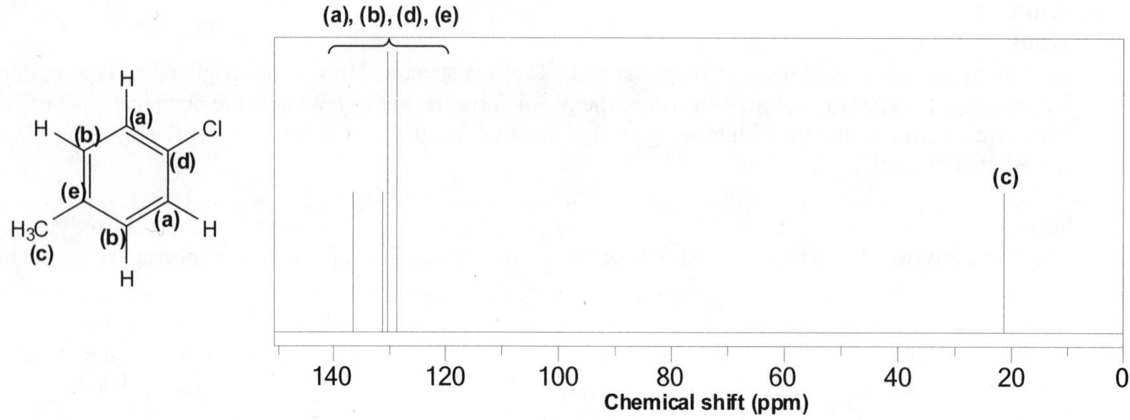

(c) There are five ^{1}H NMR signals and six ^{13}C NMR signals. The representative spectra are shown below.

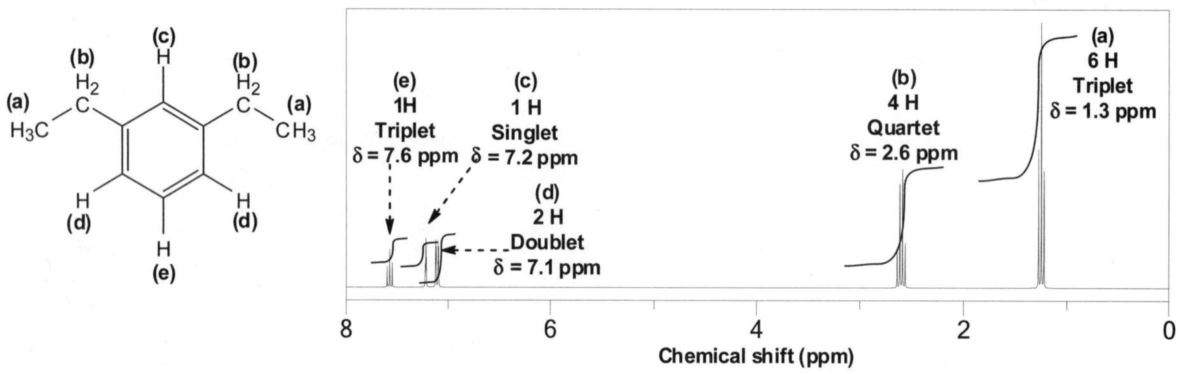

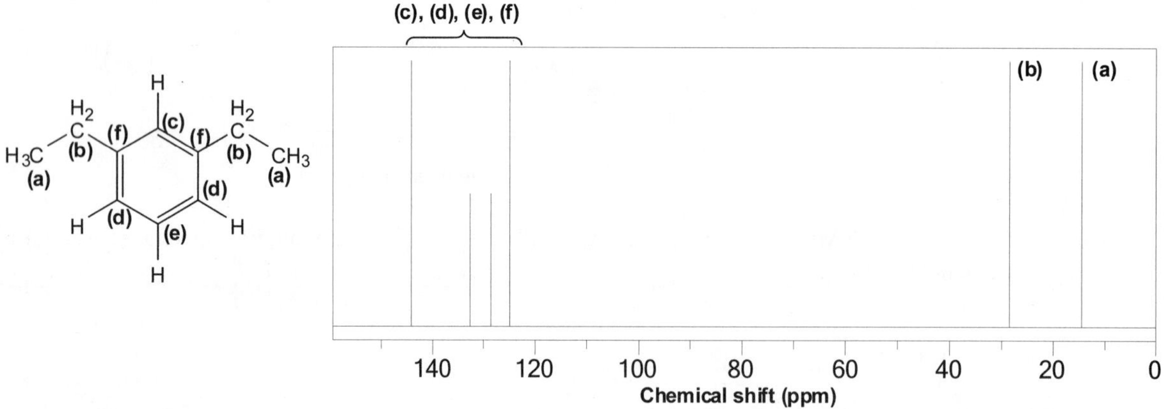

Problem 16.57

Think

What functional group is next to the CH_2 group on C2? To what entry in Table 16-1 does this correlate? How much downfield shifted is this kind of proton relative to an alkane CH_3 proton? If there are two C=O groups next to the CH_2, is the effect additive?

Solve

A proton in a CH_2 next to a carbonyl carbon normally has a chemical shift of 2.1 ppm (entry 6, Table 16-1), which is an increase of 1.2 ppm from the nominal value of 0.9 ppm for a CH_3 group. Two carbonyl groups will shift the signal about twice as much. So, adding 2.4 to 0.9 gives a value of 3.3 ppm.

Problem 16.58

Think

How many signals would be present in the ^{1}H and ^{13}C NMR spectra for compounds **A**–**C**? Would the splitting pattern of each ^{1}H NMR spectrum differ? What would be the integrations for each ^{1}H NMR signal?

Solve

Proton NMR spectroscopy would be better to distinguish among the three compounds. The carbon spectrum of each would exhibit three signals, given that there are three chemically distinct types of carbons. Moreover, the chemical shifts of those carbons in one molecule will be about the same as the chemical shifts in the other two. A similar story exists with the number of signals and chemical shifts in the proton spectra. However, integration of the signals will allow us to clearly distinguish among the compounds. In compound **A**, the aldehyde H signal will have an integration of 2H and the aromatic H signal will have an integration of 4H. In compound **B**, the aldehyde H signal will have an integration of 3H, and the aromatic H signal will have an integration of 3H. In compound **C**, the aldehyde H signal will have an integration of 4H and the aromatic H signal will have an integration of 2H.

Problem 16.59

Think

How many ^{1}H NMR signals are present? How many aromatic H signals are present? What is the splitting pattern of each? Can the number of signals and the splitting pattern distinguish these compounds? What is the chemical shift of each signal?

Solve

The only major difference is in the chemical shift of the CH_3 group. In the first compound, the chemical shift will be around 3.3, given the adjacent O atom. In the second compound, it will be about 2.1 ppm, given the adjacent C=O group. The remaining signals will not be very distinguishable based on chemical shift, splitting, or integration.

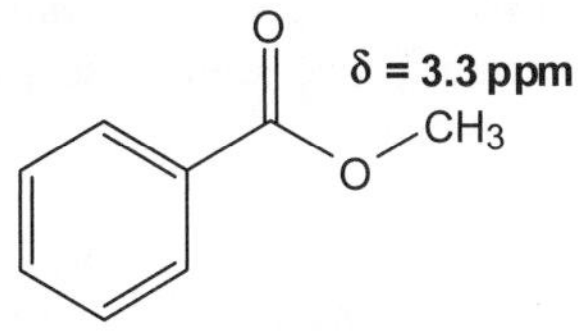

δ = 3.3 ppm
CH_3
D

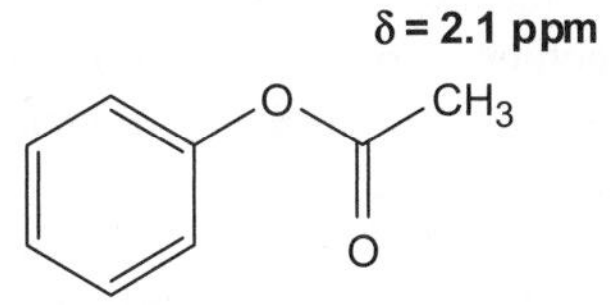

δ = 2.1 ppm
CH_3
E

Problem 16.60

Think

Is oxygen more or less electronegative compared to carbon? Does this have a shielding or deshielding effect? Is oxygen a π resonance donor? If so, what resonance structures are possible? How does this affect the electron density on the para carbon? Will that serve to shield or deshield the para carbon?

Solve

Oxygen is electronegative and will have an inductive deshielding effect on nearby nuclei. However, oxygen also donates electrons via resonance, indicated by the negative charge on the C atoms in the resonance structures below. Thus, resonance serves to shield the C atoms that share the formal −1 charge. At short range, the inductive effect is stronger, explaining the change from 125 ppm to 153 ppm. At long range, the resonance effect predominates, explaining the (smaller) change from 138 to 130 ppm.

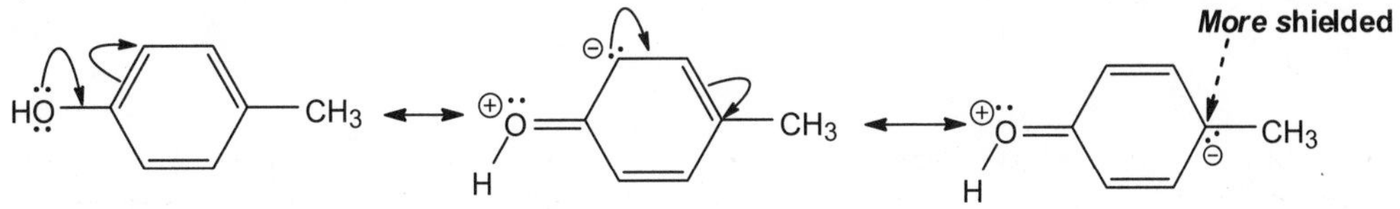

HO — CH_3 ⟷ CH_3 ⟷ **More shielded** CH_3

Problem 16.61

Think

What is the IHD? With only one signal present, is there a high or low degree of symmetry? If the signal is a singlet, are any H atoms within three bonds that are able to split the signal? Are there any deshielding groups nearby?

Solve

The compound has an IHD of 1. The O must not be bonded to a C that has any H atoms; otherwise, the chemical shift would be around 3.3 ppm or higher. With only one signal, all H atoms must be equivalent. They cannot be of the CH_2 variety, because one of the CH_2 groups would have to be attached to O. This is not possible, because the H atoms of that CH_2 group would have a chemical shift around 3.3 ppm. They can't be CH groups, because there would not be enough C atoms to account for 18 H atoms. So the H atoms must all be from CH_3 groups. There must be six such equivalent CH_3 groups to give a total of 18 H atoms, which would account for 6 C atoms and 18 H atoms. We must still account for three C atoms and one O atom, along with an IHD of 1. To make all six CH_3 groups equivalent, we could have two *t*-butyl groups, $(CH_3)_3C$, and each could be bonded to a C=O group to give the following compound:

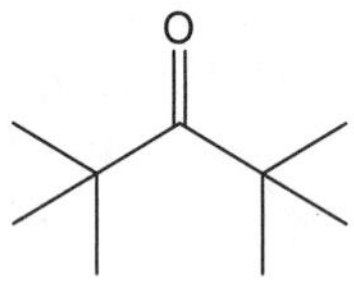

O

Problem 16.62

Think

What is the IHD? How many signals appear in the ^1H and ^{13}C NMR spectra? Is there any symmetry suggested by the number of NMR signals? From the chemical shifts in each NMR spectrum and from the absorption bands in the IR spectrum, what functional groups are likely present? From the splitting patterns in the ^1H NMR spectrum, which groups are next to each other? Sometimes it is helpful to draw the NMR spectra from the data given. Start to put a proposed structure together. (*Note*: This can be a trial-and-error process. Propose a structure and then check the NMR data to make sure your structure is consistent with the data.)

Solve

The IHD is 5. A C=O is evident, with a C=O stretch at around 1700 cm^{-1} in the IR spectrum. There are also C=C stretches around 1450 to 1600 cm^{-1}, which indicate an aromatic ring. Also, the very broad peak that extends from ~2500 cm^{-1} to ~3200 cm^{-1}, which overlaps the C–H stretches, is indicative of a carboxylic acid. In the proton NMR spectrum, the chemical shifts at 7 and 8 ppm indicate the presence of a benzene ring, consistent with the C=C bands in the IR spectrum. We also see further evidence of a carboxylic acid, with a broad OH signal around 12 ppm. The integrations of the signals appear to be in the ratio of 1:2:2:2:3, which totals 10 H atoms. That gives a total of four H atoms on the benzene ring, suggesting a disubstituted benzene. The fact that the aromatic H signals are doublets suggests that the ring is para-disubstituted and the two groups are different. This is consistent with the fact that there are seven carbon signals—four from the para-disubstituted ring, and three from the two attached groups. The quartet and triplet are indicative of a CH_2CH_3 group, and the large chemical shift of the quartet suggests that the CH_2 group is bonded to O. This accounts for all of the atoms, so what remains is to attach the groups to the ring.

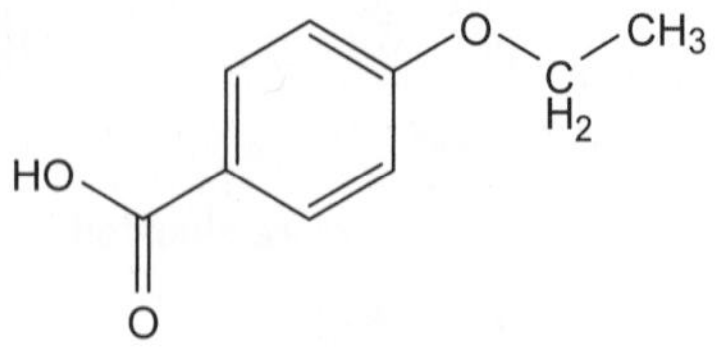

O CH3 C H2

HO

O

Problem 16.63

Think

What is the IHD? How many signals appear in the ^{1}H and ^{13}C NMR spectra? Is there any symmetry suggested by the number of NMR signals? From the chemical shifts in each NMR spectrum and the absorption bands in the IR spectrum, what functional groups are likely present? From the splitting patterns in the ^{1}H NMR spectrum, which groups are next to each other? Sometimes it is helpful to draw the NMR spectra from the data given. Start to put a proposed structure together. (*Note*: This can be a trial-and-error process. Propose a structure and then check the NMR data to make sure your structure is consistent with the data.)

Solve

The IHD is 6. A benzene ring is apparent from the ^{1}H NMR spectrum, with chemical shifts around 7.5. There is also a C=O group present, evidenced by the intense stretching band just below 1750 cm^{-1} in the IR spectrum. Finally, there appears to be an alkene group, indicated by the signal near 6.6 ppm. These groups account for the IHD of 6. The signal at 9.8 ppm is from an aldehyde H, which is consistent with the two peaks at 2700 and 2800 cm^{-1} in the IR spectrum. Because the aldehyde H signal in the ^{1}H NMR spectrum is a doublet, it must be coupled to a single H on an adjacent C, giving rise to a group of atoms as follows: O=CH–CH. To have an alkene group also present, that CH group on the right must be part of a C=C bond, giving an O=CH–CH=C group. That accounts for all of the non-H atoms, so the C on the very right of that group must be bonded to the benzene ring. Notice that for this to be the compound, one of the CH proton signals in the ^{1}H NMR spectrum must be overlapping with the aromatic signals. Notice also that this compound has seven distinct C atoms, consistent with the seven signals in the carbon NMR spectrum. *Note*: Without coupling constant information the cis isomer is also possible.

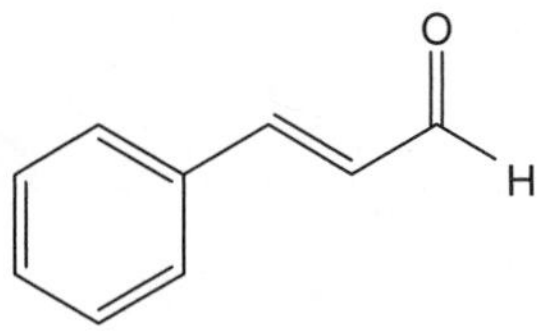

O

H

Problem 16.64

Think

What is the IHD? How many signals appear in the ^{1}H and ^{13}C NMR spectra? Is there any symmetry from the number of NMR signals? From the chemical shifts in each NMR spectrum and IR the absorption bands in the spectrum, what functional groups are likely present? From the splitting patterns in the ^{1}H NMR spectrum, which groups are next to each other? Sometimes it is helpful to draw the NMR spectra from the data given. Start to put a proposed structure together. (*Note*: This can be a trial-and-error process. Propose a structure and then check the NMR data to make sure your structure is consistent with the data.)

Solve

The formula $C_4H_{10}O_2$ has an IHD of 0, so there can be no double bonds, triple bonds, or rings. The IR clearly shows an OH stretch at 3400 cm^{-1}— an alcohol. Because a second oxygen atom is in the formula, there must be an ether oxygen or a second alcohol group. Because the ^{13}C NMR spectrum has only three signals, there are two carbons that must be equivalent. If the largest ^{13}C NMR peak is these equivalent carbons, then both must be attached to oxygen. We consider two equivalent $–CH_2OH$ fragments as very likely. Subtracting $C_2H_6O_2$ from the molecular formula leaves only C_2H_4. This is either two CH_2 groups or a $CH_3–CH$ fragment. Two possible structures are $HOCH_2–CH_2–CH_2–CH_2OH$ or $CH_3–CH(CH_2OH)_2$. The proton NMR spectrum shows four signals integrating to 2H, 4H, 1H, and 3H. This rules out the first structure, which would give only three ^{1}H NMR signals, which integrate to 2H, 4H, and 4H. The signal farthest upfield, integrating to 3H, is split into a doublet, confirming the $CH_3–CH$ fragment. The compound is 2-methyl-1,3-propanediol, $CH_3–CH(CH_2OH)_2$.

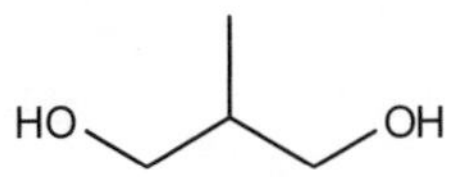

HO OH

Problem 16.65

Think

Which compound has a signal that is farther downfield? Which signal is more deshielded? Do electronegative groups cause nearby protons to have more shielded or deshielded signals? What is the hybridization of each α carbon? What is the ideal bond angle of an atom with this hybridization? What is the actual bond angle of each C–C_{α}–C (think about the ring angle). What atomic orbitals make a 90° angle?

Solve

(a) The α carbon in cyclobutanone has a greater effective electronegativity because the protons (being further downfield) are more deshielded.

(b) The α carbon to which the proton is attached should be sp^3-hybridized in each molecule, giving it 25% *s* character. But to have a 90° C–C–C bond angle, the C–C bonds must use orbitals that have more *p* character than normal. That leaves more *s* character than normal for the orbitals used for the C–H bonds. As we learned in Chapter 3, greater *s* character corresponds to higher effective electronegativity.

Problem 16.66

Think

Consider the resonance structures of *N*,*N*-dimethylformamide. If there is significant C=N character, are the two CH_3 signals chemically equivalent? What happens to the energy available for bond rotation as the temperature is increased?

Solve

The lone pair on the N atom of the amide group participates in resonance with the C=O group, as shown below.

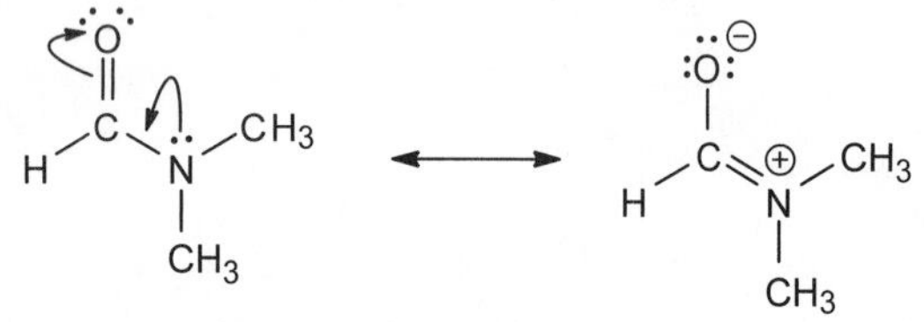

To the extent that the resonance structure on the right contributes to the hybrid, the C–N bond gains more double-bond character. Thus the C–N bond, although formally a single bond, has significantly hindered rotation about it. If that C–N bond rotation is slower than the NMR timescale at room temperature, the two CH_3 sets of H atoms will give rise to different proton signals because they are in different chemical environments. One CH_3 group is closer to the O atom than the other CH_3 group is. But when the sample is heated, the C–N rotation rate increases, and when the rate of rotation surpasses the NMR timescale, the two CH_3 signals blend into one.

Problem 16.67

Think

Draw each C–H bond. How many C–H bonds are *inside* the ring? How many C–H bonds are *outside* the ring? Consider the magnetic field lines from ring current.

Solve

An external magnetic field causes the π electrons in the aromatic ring to circulate (solid curved arrows). This circulation induces a magnetic field (dashed arrows) that is in the same direction as the external magnetic field at the locations outside the ring, and induces a magnetic field that is in the opposite direction inside the ring. Similar to benzene, the 12 equivalent protons outside the ring are deshielded, and appear downfield at 9.3 ppm. The six protons inside the ring are shielded and appear upfield at −2.9 ppm. See the figure on the next page.

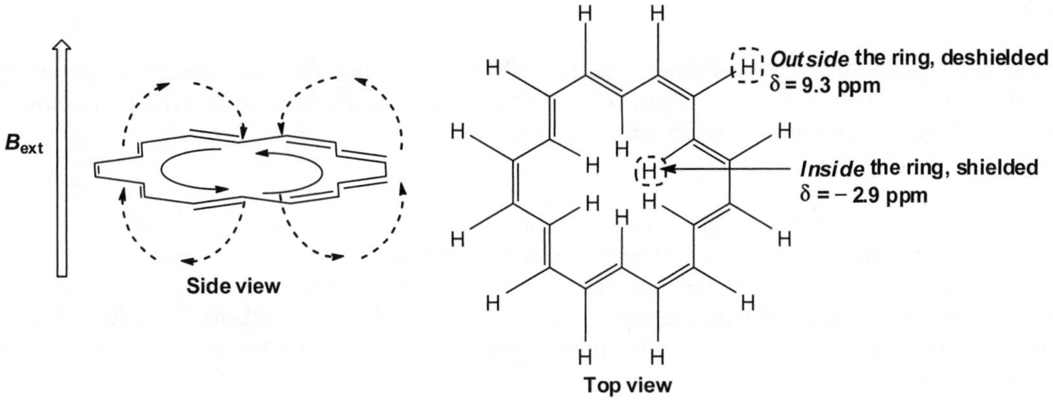

Problem 16.68

Think

What is the structure of cylcohexane-d_{11}? In how many environments can the H atom be located? At low temperature is this exchange fast or slow? How does the rate of exchange affect the number of signals observed?

Solve

The structure of cylcohexane-d_{11} is shown below. The H atom can be located in the equatorial or axial position. These positions are in two distinct chemical environments and give rise to two signals. At lower temperatures, this exchange between the axial and equatorial positions is slow. At higher temperatures, this exchange is fast and the two signals coalesce into one signal.

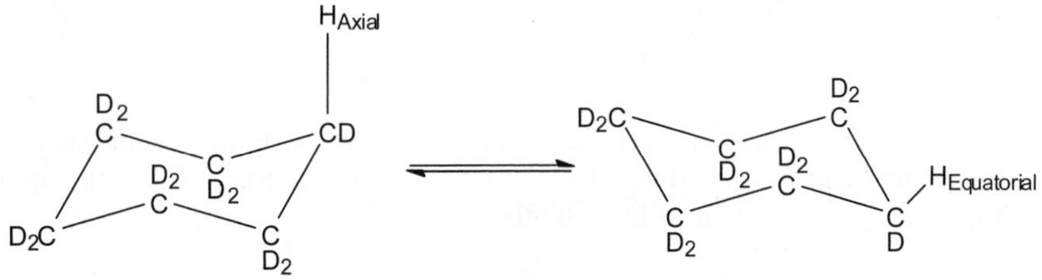

Problem 16.69

Think

What is the chair structure of *cis*-1,2,3,4,5,6-cyclohexanehexacarboxylic acid ? In how many environments can the H atoms bonded to the ring be located? At low temperature, is this exchange fast or slow? How does the rate of exchange affect the number of signals observed?

Solve

The structure of *cis*-1,2,3,4,5,6-cyclohexanehexacarboxylic acid is shown below. The H atoms on the ring can be located in the equatorial or axial position. These positions are in two distinct chemical environments and give rise to two signals if ring flipping is difficult.

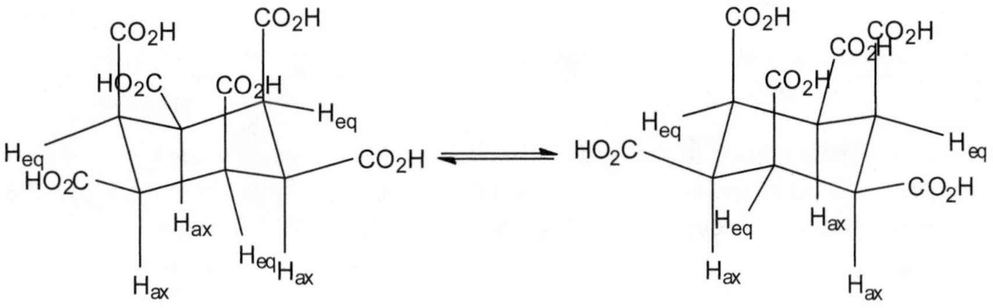

Problem 16.70

Think

What is the IHD? How many signals appear in the ^{1}H NMR spectrum? Is there any symmetry suggested by the number of signals? From the chemical shifts, what functional groups are likely present? From the splitting patterns in the ^{1}H NMR spectrum, which groups are next to each other?

Solve

For $C_6H_{10}O$, the IHD is 2. The peaks between 6 and 7 ppm suggest an alkene that has downfield shifted signals. The neighboring C=O causes the alkene CH peaks to shift from their expected 4.7–5.3 ppm to 6–7 ppm. The quartet at 2.6 ppm is likely coupled to the triplet at 1.1 ppm and appears to be an isolated CH_2CH_3 group. The peak at 1.9 appears to be a doublet, which means that it is directly attached to the alkene C–H. The other IHD is likely from a C=O because there is no O–H bond present and no ^{1}H NMR signals that have a chemical shift ~3–4 suggestive of an O–CH bonding interaction.

Three possible structures are shown below.

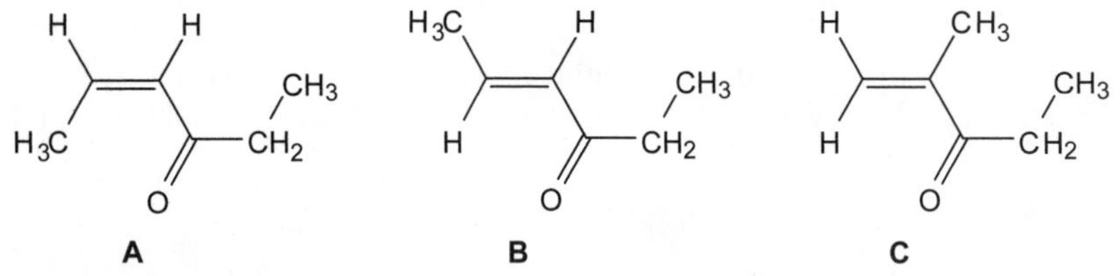

H H CH3 H3C CH2 O **A** H3C H CH3 H CH2 O **B** H CH3 CH3 H CH2 O **C**

C is ruled out because the CH_3 group is four bonds away from the alkene C–H, and would not be split and would show up as a singlet in the ^{1}H NMR spectrum. From Table 16-3, we can see the coupling constants for cis and trans alkene C–H bonds: 7–12 and 13–18 Hz, respectively. Because the coupling constants are not given for the alkene C–H peaks at 6.1 and 6.9 ppm, **A** and **B** are both possible structures.

Problem 16.71

Think

What is the IHD? How many signals appear in the ^{1}H NMR spectrum? Is there any symmetry suggested by the number of signals? From the chemical shifts, what functional groups are likely present? From the splitting patterns in the ^{1}H NMR spectrum, which groups are next to each other?

Solve

The IHD is 1. There are no alkene C–H peaks ~5–6 ppm; therefore, the IHD is likely due to a C=O bond. With two O atoms and no O–H bond, an ester is a reasonable choice for a functional group. The ratio of H atom integrations is 2:2:1:3:6. The peak at 3.75 is a CH_2 next to the O atom and next to a CH as it is a doublet. There is a signature ethyl pattern: splitting of a quartet (2H) with a triplet (3H). The 1:6 ratio with the peak at 0.9 ppm (6H) split into a doublet is due to an isopropyl peak. Therefore, the structure is isobutyl propionate:

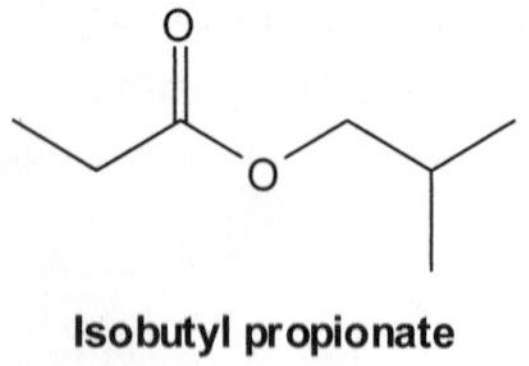

O O

Isobutyl propionate

Problem 16.72

Think

What is the IHD? How many signals appear in the ^{1}H NMR spectrum? Is there any symmetry suggested by the number of signals? From the chemical shifts, what functional groups are likely present? From the splitting patterns in the ^{1}H NMR spectrum, which groups are next to each other?

Solve

The IHD is 1. There are no alkene C–H peaks ~5–6 ppm; therefore, the IHD is likely due to a C=O bond. All three peaks are singlets and the integration is 1:2:6. Because the three signals are singlets, none of the CH groups can be next to each other. The peak at 3.7 is a CH_2 next to the N and next to the C=O. The signal at 4.9 ppm is relatively broad, and can be part of an OH group. The signal at 2.9 ppm has an integration of 6H and can be two CH_3 groups, each attached to N. Therefore, the structure is as shown below.

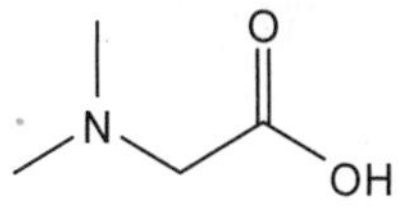

N OH O

Problem 16.73

Think

What is the IHD? How many signals appear in the ^{13}C NMR spectrum? Is there any symmetry suggested by the number of signals? From the chemical shifts, what functional groups are likely present?

Solve

The IHD is 4, which is likely due to an aromatic ring. There are four signals in the ^{13}C NMR spectrum, which means that there is a high degree of symmetry. The peak at 77 ppm is due to $CDCl_3$. Para substitution on the benzene ring with the same two groups leads to two aromatic ^{13}C signals. Because there are no other atoms or functional groups present other than sp^3 C atoms, the two groups are CH_2CH_3. Therefore, the structure is 1,4-diethylbenzene.

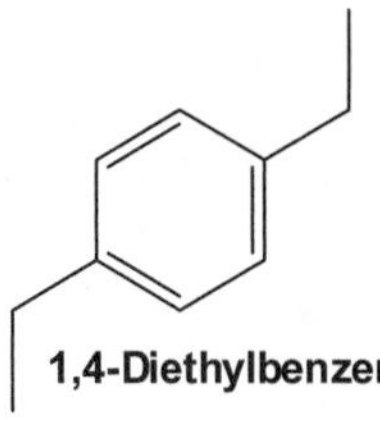

1,4-Diethylbenzene

Problem 16.74

Think

Is there a leaving group initially present? If not, how does the reaction with excess methyl bromide lead to the formation of a good leaving group? When the solution is heated in the presence of a strong base, what mechanism results? Does the more or less substituted alkene form? How is this product supported by the ^1H NMR spectrum?

Solve

This is an example of a Hofmann elimination reaction, yielding $C_6H_5CH_2CH=CH_2$. The five aromatic H signals appear above 7 ppm. The signal at 6 ppm, which integrates to 1H, is the alkene CH, and the one at 5 ppm that integrates to 2H is the alkene CH_2. The doublet is the alkane CH_2, which is split by the alkene CH proton. It is significantly downfield shifted because of the adjacent phenyl ring and alkene group. The complete mechanism is shown below.

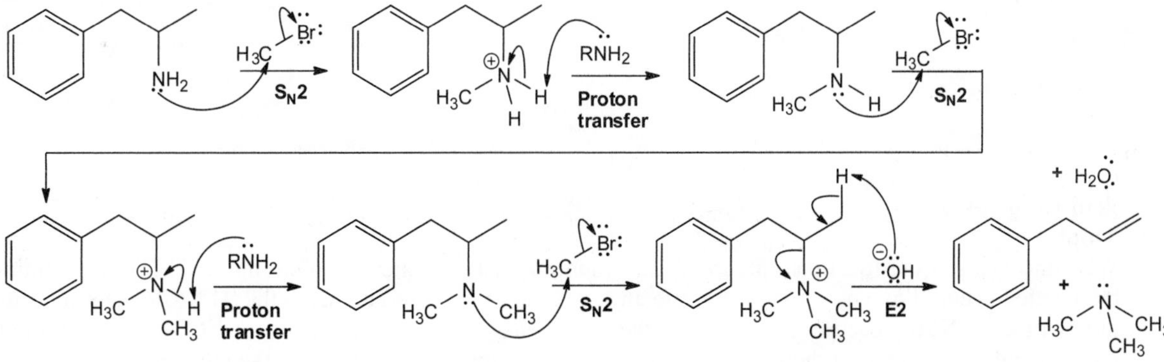

NH_2 H_3C–Br S_N2 → H_3C N^+ H H; RNH_2 Proton transfer → H_3C N H; H_3C–Br S_N2 →

H_3C N^+ H CH_3; RNH_2 Proton transfer → H_3C N CH_3; H_3C–Br S_N2 → H_3C N^+ CH_3 CH_3 H; $^-$:OH E2 → + H_2O + H_3C N CH_3 CH_3

Problem 16.75

Think

Is the reaction an example of addition, elimination, or substitution? Is there a good leaving group? What are the identities of the two products? On which C atom is the leaving group likely located? Can both products be produced from the same carbocation intermediate? If so, is the mechanism most likely S_N1 or S_N2?

Solve

Because the OH group is replaced by Br, the reaction must be an S_N1 or S_N2 reaction. The product alkenes are shown below.

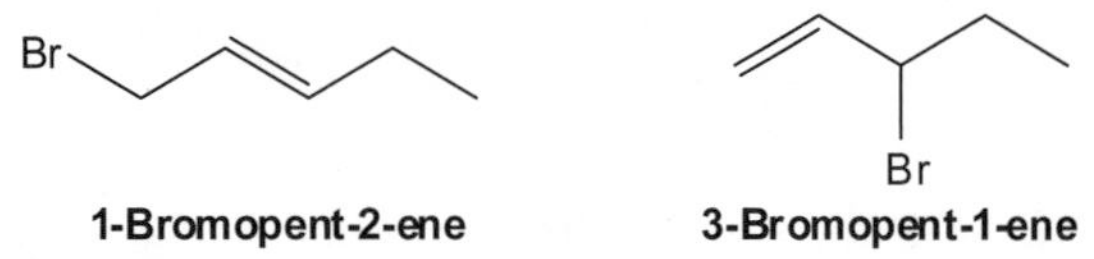

1-Bromopent-2-ene   3-Bromopent-1-ene

These isomers could be produced from the allylic carbocation intermediate shown below, which has two resonance structures.

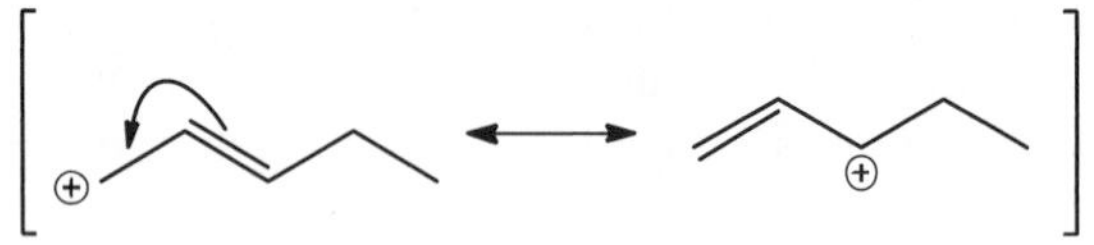

The allylic cation can be produced under S_N1 conditions from the following alcohols.

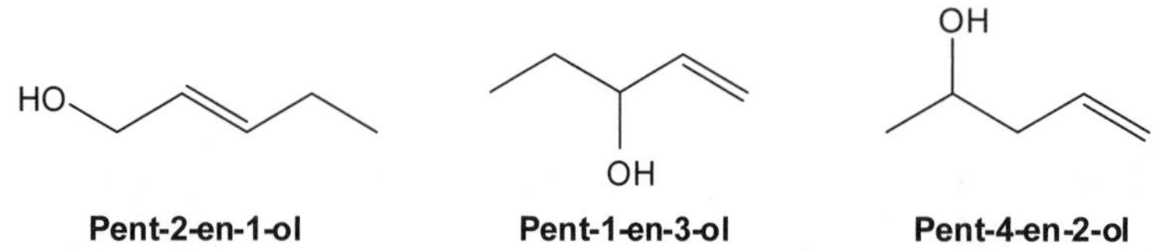

Pent-2-en-1-ol   Pent-1-en-3-ol   Pent-4-en-2-ol

The pent-4-en-2-ol is more likely the starting alcohol, due to the two alkene proton signals (5.8 ppm and 5.1 ppm) integrating to 1H and 2H, respectively. In the compound pent-2-en-1-ol, the two signals would have been 1H and 1H. The peak at 1.2 is also a doublet which is due to the CH_3 that is next to a CH. Pent-1-en-3-ol would have the CH_3 split as a triplet. The complete mechanism is shown below.

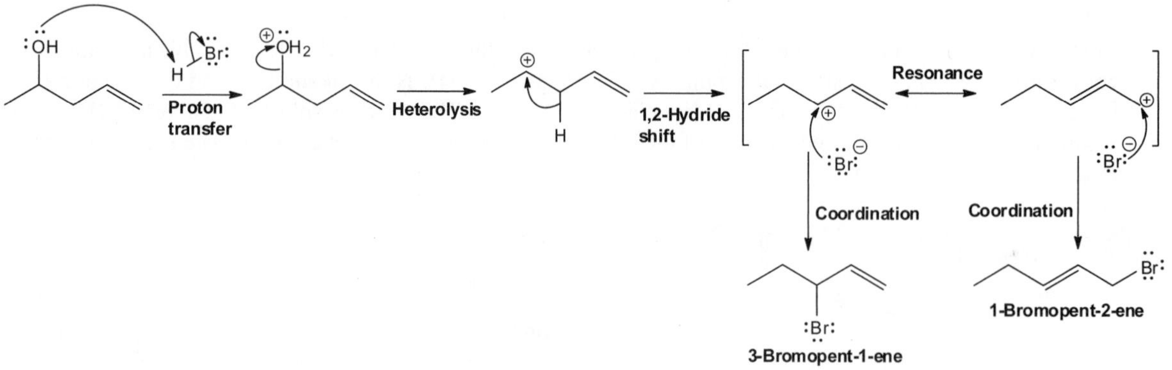

Proton transfer → Heterolysis → 1,2-Hydride shift → [Resonance] → Coordination → 3-Bromopent-1-ene; Coordination → 1-Bromopent-2-ene

Problem 16.76

Think

What mechanism and functional group results when an alcohol is heated in strong acid? Did substitution or elimination occur? From the IR spectrum, are any sp^2 C–H groups present? Does the ^1H NMR spectrum support this? In the ^1H NMR spectrum, what causes the downfield signal at 3.6 ppm? In the ^{13}C NMR spectrum, how many signals are present? Is there a high or low degree of symmetry suggested by the number of signals?

Solve

The ^{13}C–NMR spectrum has only two signals, indicating only two distinct types of carbons in the molecule. One C atom at 68 ppm is clearly attached to an electronegative atom—oxygen—whereas the other is not. The proton NMR spectrum shows only two signals with a relative integral of 1:6 hydrogens. The smaller signal, farther downfield, indicates that proton is on a C atom attached to O. Further, the signal is split into seven peaks, so it is split by six neighboring protons that are equivalent. We consider the fragment $(CH_3)_2CH–O$. If these are the only types of carbons in the molecule, then the rest of the molecule must have the same carbon structure. Recall from Chapter 10 that heating an alcohol with acid will dehydrate it to form either an alkene or an ether. The former is not indicated by the ^{13}C NMR spectrum, so the compound must be diisopropyl ether, $(CH_3)_2CH–O–CH–(CH_3)_2$. The mechanism for its formation is shown below.

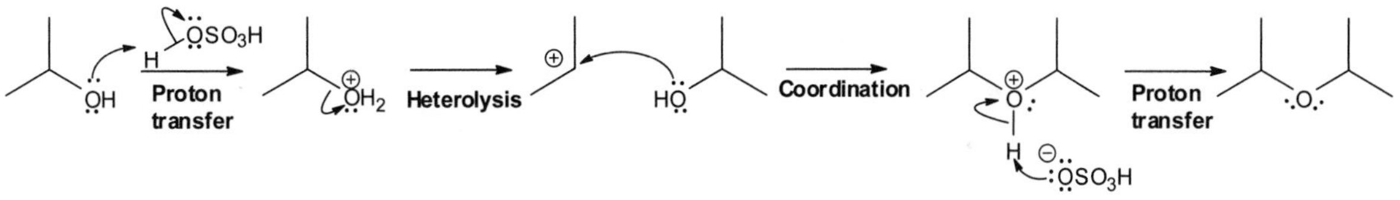

CHAPTER 17 | Nucleophilic Addition to Polar π Bonds: Addition of Strong Nucleophiles

Your Turn Exercises

Your Turn 17.1

Think

Use Equation 17-1 as a guide. What kinds of charges characterize an electron-rich atom? Are there any formal charges present? Any strong partial charges? Which bonds are broken and which bonds are formed? How is a curved arrow used to denote the flow of electrons? Consult Chapters 6 and 7 to review the 10 most common elementary steps. Are any protons or π bonds involved?

Solve

The C in a C=O bond bears a large partial positive charge, δ^+, due to the electronegativity difference of C and O. The electrons flow from the electron-rich Nu:$^-$ to the electron-poor C atom of the C=O bond. A second curved arrow (to illustrate the breaking of the π bond between the C and O) is necessary to avoid exceeding an octet on the C. The pair of electrons in the π bond becomes a lone pair on the O and forms an electron-rich basic product. The second reaction is a proton transfer where the electron-rich negative O deprotonates the H–A acid. Overall, the nucleophile is added to the C and a proton is added to the O of the C=O. The ketone is transformed into an alcohol.

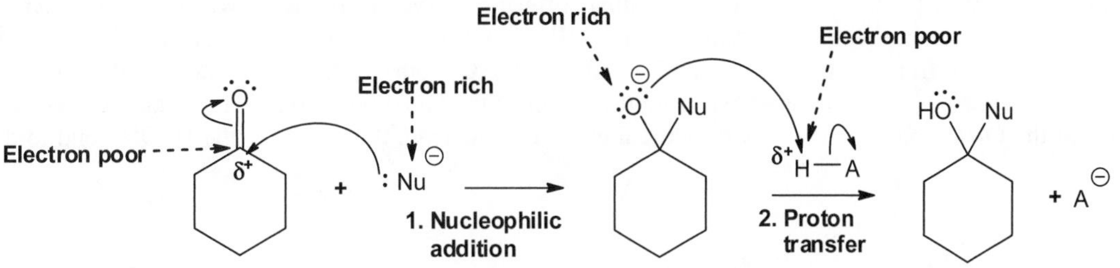

Your Turn 17.2

Think

How many different groups must be bonded to a tetrahedral stereocenter? What is the geometry of the carbonyl C=O carbon? From how many directions can the nucleophile attack?

Solve

The carbon of the C=O forms a new tetrahedral stereocenter from a trigonal planar C atom. The nucleophile can attack the C atom from two different sides: from behind the plane and in front of the plane. See below.

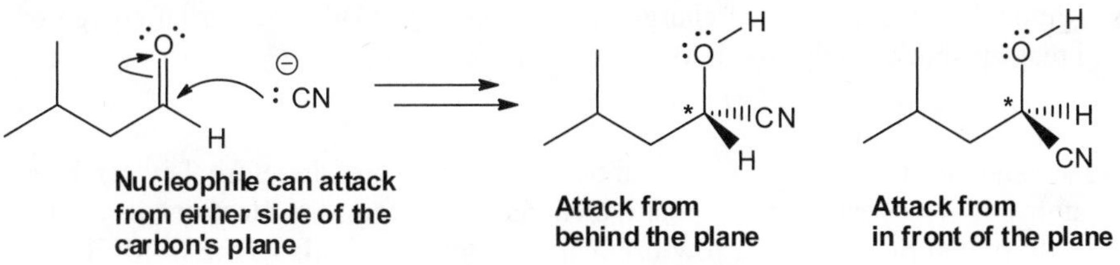

Your Turn 17.3

Think

Are CH_3 groups electron-donating or electron-withdrawing groups? What effect does each CH_3 have on the partial positive charge of the C atom of the C=NH? How does the bulkiness of the CH_3 groups affect the ability of a nucleophile to attack C=NH?

Solve

Alkyl groups are electron donating and decrease the concentration of positive charge at the C atom of the C=NH group. This is noted below by the size of the δ^+ on each C. Alkyl groups also add more steric hindrance surrounding C and decrease the reactivity at the C=NH. Therefore, the most reactive imine is **B**, followed by **A**, followed by **C**.

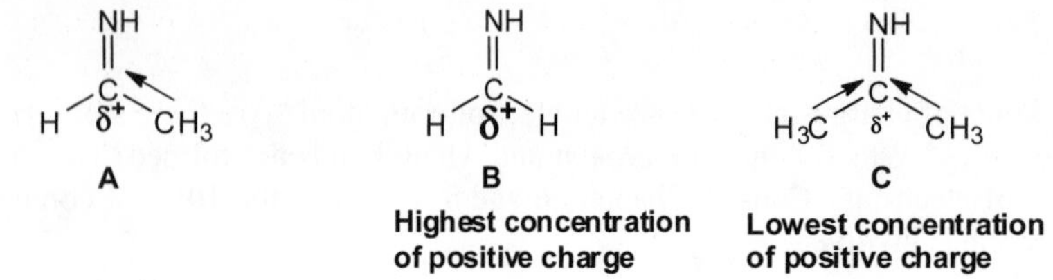

A

B

Highest concentration of positive charge

C

Lowest concentration of positive charge

Your Turn 17.4

Think

Use Equation 17-6 as a guide. Identify the similarities between BH_4^- and AlH_4^-. How do you show a pair of electrons from the C=O bond becoming an O–Al bond? How do you show a pair of electrons from the Al–H bond becoming a C–H bond?

Solve

BH_4^- and AlH_4^- are similar and the mechanism for the reduction of butanone is the same for the two reducing agents. In the first step, to show the O–Al bond being formed, a curved arrow is drawn from the center of the C=O bond to the Al atom. At the same time, the H–AlH_3^- bond is broken and the C–H bond is formed, so a curved arrow is drawn from the center of the Al–H bond to the C atom. In the second step, to show the RO–H bond being formed, a curved arrow is drawn from the center of the O–Al bond to the H atom. At the same time, formation of the O–Al bond is shown by drawing a curved arrow from the center of the H–OH bond to the Al atom.

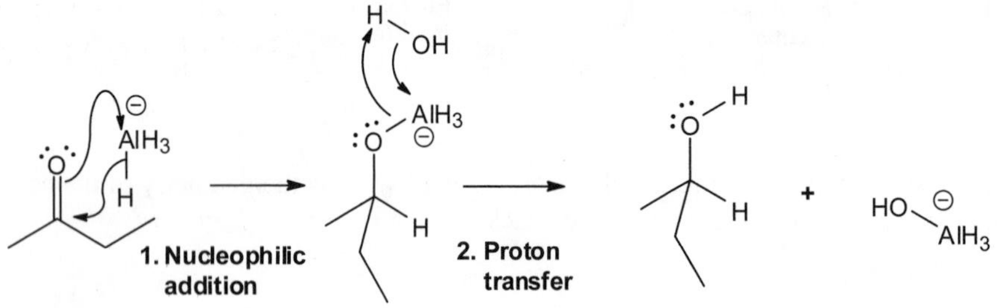

1. Nucleophilic addition

2. Proton transfer

Your Turn 17.5

Think

Use Equation 17-14 as a guide. What is the partial charge on C in the C–MgBr? The partial charge on H of H_2O? How is the alkyl–Li reagent similar to the alkyl–MgBr reagent?

Solve

The alkyl–metal reagents are similar in that C bears a partial negative charge and therefore they can be thought of as *R^- donors*. The H atoms of H_2O bear a partial positive charge. Therefore, the electrons in the C–MgBr bond attack the H in H_2O, illustrated by a curved arrow originating from the middle of the C–MgBr bond and pointing to H. At the same time, the H–O bond breaks, leading to a new C–H bond and HO^- and $MgBr^+$ as products.

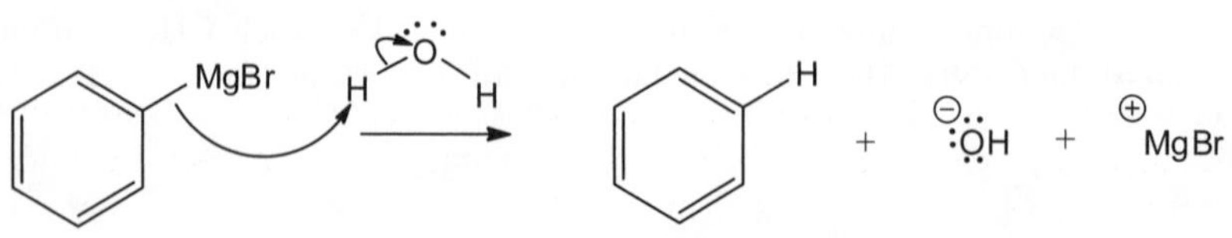

Your Turn 17.6

Think

What kinds of charges characterize an electron-rich atom? Are there any formal charges present? Any strong partial charges?

Solve

In the first step, the C in the C=O bears a large partial positive δ^+ due to the electronegativity difference of C and O. The C of the Wittig reagent is a highly nucleophilic carbon and is, therefore, very electron rich. Thus, a curved arrow is drawn from the electron-rich carbon to the electron-poor C of the C=O. In the second step, the alkoxide O is electron rich and the positive P in $^+PR_4$ is electron poor; a curved arrow is drawn from O to P.

Electron poor Electron rich 1. Nucleophilic addition A betaine Electron rich Electron poor 2. Coordination Strained 4-membered ring An oxaphosphetane

Your Turn 17.7

Think

Consult Table 9-10 to verify this statement. Is Br^- considered to be a good nucleophile? What are the relative nucleophilicities of triphenylphosphine and Br^-?

Solve

Br^- is a very good nucleophile, due to its full negative charge. In Table 9-10, the nucleophilicity of Ph_3P is listed as 10,000,000, and that of Br^- is listed as 620,000. Although Ph_3P is uncharged, it is a good nucleophile because the P atom can accommodate the positive charge in the product rather well, due both to its large size and its modest electronegativity.

Your Turn 17.8

Think

Consult Table 1-6 to review the functional groups. What arrangement of atoms forms an epoxide?

Solve

An epoxide is a three-membered-ring ether, consisting of two C atoms and an O atom.

Epoxide ring 25 °C, 1.5 h 84%

Your Turn 17.9

Think

Identify any polar bonds and consider how bond polarity leads to an electron-deficient site. Draw a possible resonance structure for the molecule on the next page that has a C=C conjugated to a polar C=O. How does this resonance structure illustrate the presence of an electron-deficient site?

Solve

The carbonyl C is made electron deficient by the adjacent electronegative O atom. The β C atom is made electron deficient by the resonance structure that places a positive charge on that C atom, as shown below.

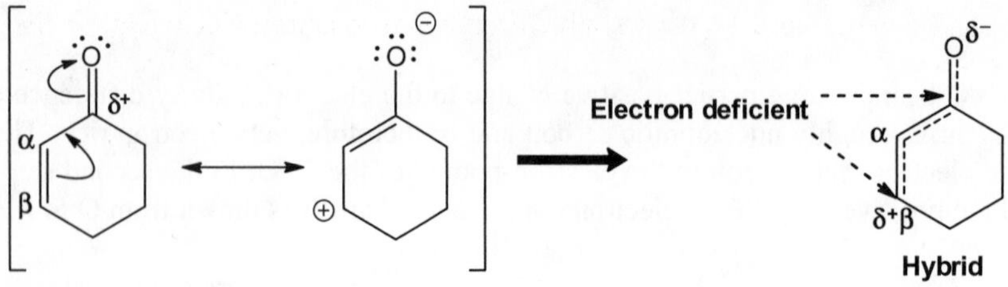

Your Turn 17.10

Think

Consult Section 7.9. In what type of medium—acidic or basic—can the enol convert to the keto form? What type of atom is moved in such a tautomerization? Which elementary steps are involved? Can this take place in a single step?

Solve

The enol can convert to the keto form in either acidic or basic conditions. Both are shown below. A proton must move from the O to the alkene C in two separate proton transfer steps. A single step is not reasonable because it would constitute an intramolecular proton transfer.

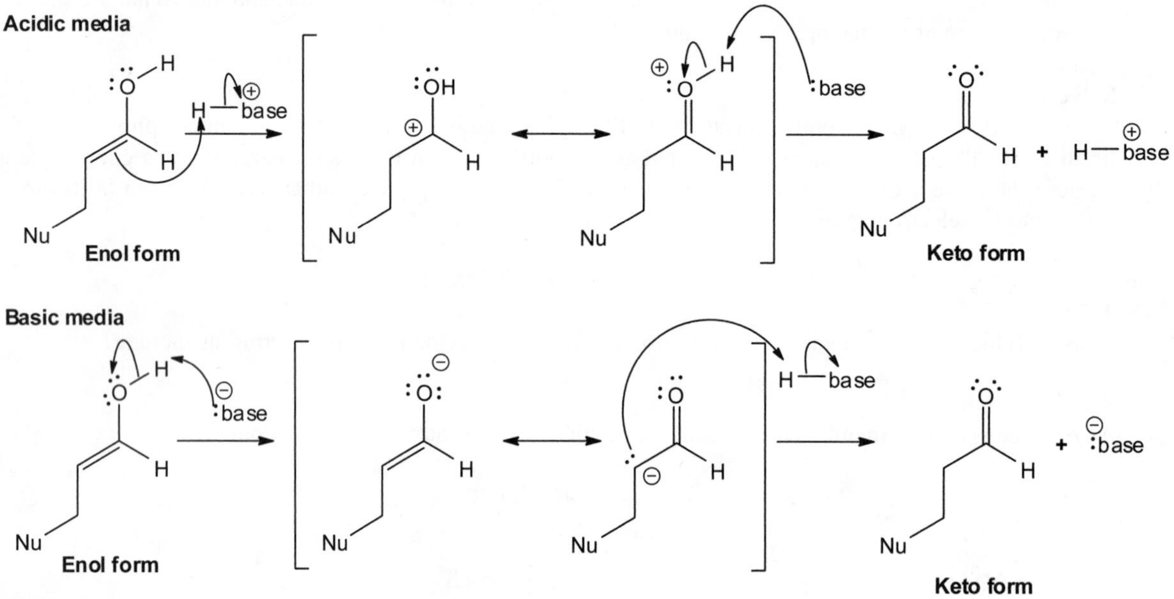

Your Turn 17.11

Think

Which product is produced from the transition state that has the nucleophile attracted to the C atom with a larger partial positive charge? Which activation energy is lower? Which product stabilizes the negative charge better? Consider possible resonance structures.

Solve

The first product is a result of conjugate addition and the second product is a result of direct addition. Direct addition has a lower-energy transition state because the carbonyl C has a larger partial positive charge to which the nucleophile is attracted. This leads to lower activation energy and is, therefore, the kinetic product. The first product stabilizes the negative charge better because the negative charge is resonance delocalized, so the first product is the thermodynamic product.

Conjugate addition
Thermodynamic product

Direct addition
Kinetic product

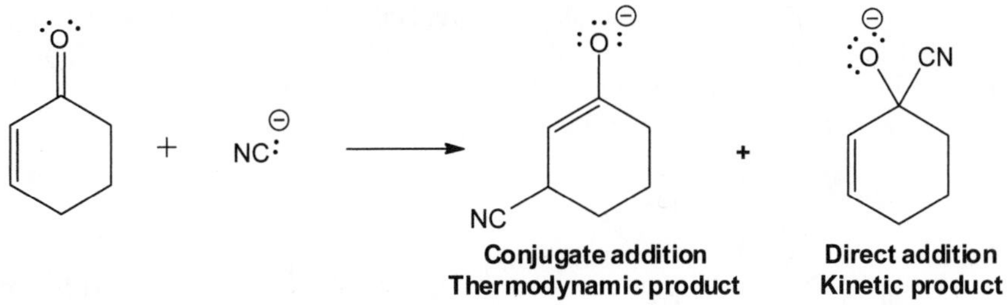

Chapter Problems

Problem 17.2

Think

What nucleophile is generated when $NaSCH_3$ dissolves in ethanol? Which atom will it attack? Which bond is most easily broken in the process? What is the role of the protic solvent?

Solve

$NaSCH_3$ is ionic, consisting of Na^+ and H_3CS^- ions. In ethanol, therefore, $NaSCH_3$ dissolves as Na^+ and H_3CS^-, with H_3CS^- being a strong nucleophile. H_3CS^- will subsequently attack the electron-poor carbonyl C, breaking the π bond of the double bond. The resulting O^- is protonated by the protic ethanol solvent.

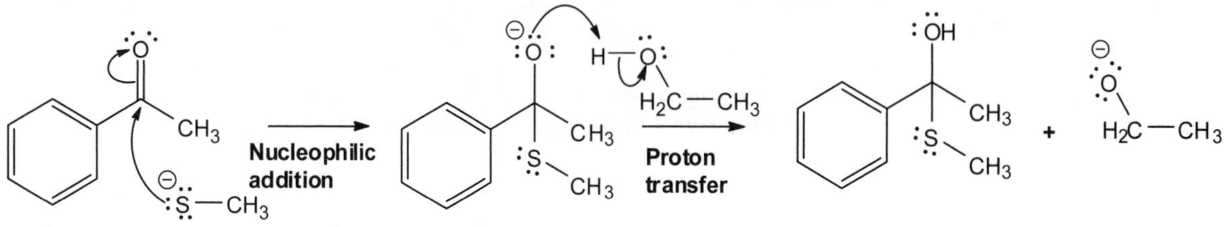

Nucleophilic addition

Proton transfer

Problem 17.3

Think

Is the CH_3 group electron donating or electron withdrawing? Is the CCl_3 group electron donating or withdrawing? Which carbonyl C has a larger concentration of positive charge? Which carbonyl C is more likely to react with the water nucleophile?

Solve

The inductive effect of three electronegative Cl atoms makes the carbonyl C atom extremely electron poor—more electron poor than the carbonyl C in ethanal. So a nucleophile will be attracted to the carbon in chloral more strongly.

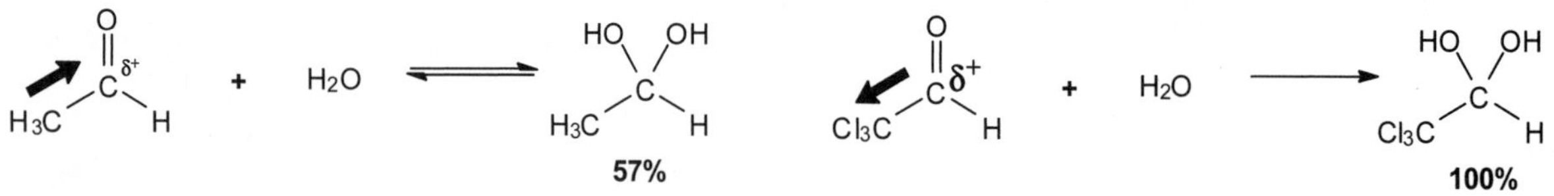

57%

100%

Problem 17.4

Think

Compare the steric repulsion of the hydrate formed in each reaction. Which hydrate is more crowded? How does this affect the extent to which the hydrate is produced? Is there a difference in the inductive effect between the methyl and isopropyl groups?

Solve

Molecule **D** (acetone) will be hydrated to a greater extent. The carbonyl C of molecule **E** (2,4-dimethyl-3-pentanone) will be more sterically hindered, making hydration more difficult. The inductive effect does not differ between the two alkyl groups to a significant extent.

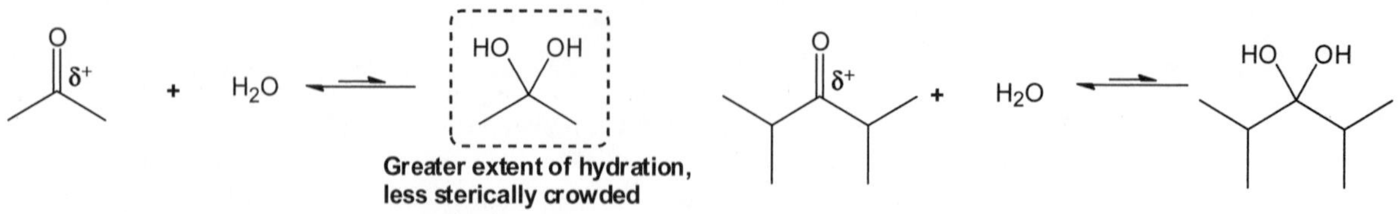

Greater extent of hydration, less sterically crowded

Problem 17.6

Think

Can $LiAlH_4$ and $NaBH_4$ be treated as a simpler nucleophile? Why can the reaction with $NaBH_4$ be conducted in a protic solvent? Why must the reaction with $LiAlH_4$ be complete before H_3O^+ is added?

Solve

In each reaction, H^- adds, followed by protonation. The result is the reduction of the aldehyde or ketone to an alcohol.

(a) $NaBH_4$ can be treated simply as H^-, which attacks the carbonyl C. Ethanol is a protic solvent and will protonate the O^- to form an uncharged final alcohol product.

NaBH4
CH3CH2OH

(b) $LiAlH_4$ can be treated simply as H^-, which attacks the carbonyl C. Protonation in a separate acid workup step is required because $LiAlH_4$ will deprotonate the weakly acidic proton from solvents like water and alcohols.

1. LiAlH4
2. H3O⊕

Problem 17.7

Think

What are the electron-rich and electron-poor species in the starting reagents? How is the C=N similar to the C=O in terms of how it might react with a nucleophile? What elementary step takes place first? What is the role of the H_3O^+? Are proton transfer reactions involved?

Solve

H^- adds to the imine carbon in a nucleophilic addition elementary step. The N^- is protonated by the H_3O^+. The imine, as a result, is reduced to an amine.

Nucleophilic addition

Proton transfer

Problem 17.8

Think

What are the electron-rich and electron-poor species in the starting reagents? How is the C≡N similar to the C=O in terms of how it might react with a nucleophile? What elementary step takes place first? What is the role of the H_3O^+? Are proton transfer reactions involved? How many times does the nucleophile add?

Solve
$LiAlH_4$ is a source of a very strong H^- nucleophile and can add twice to the triple bond of the nitrile. Subsequently, the N atom is protonated twice, producing a primary amine. The complete mechanism is shown below.

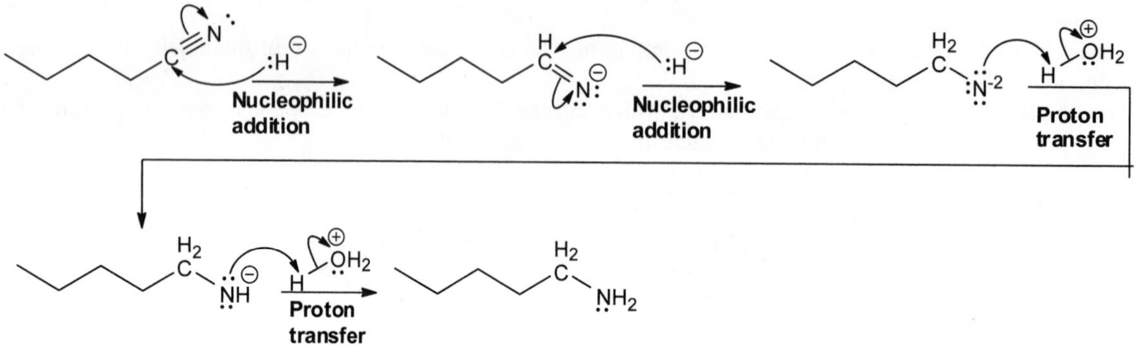

Nucleophilic addition

Nucleophilic addition

Proton transfer

Proton transfer

Problem 17.9

Think
Can $LiAlH_4$ be treated as a simpler nucleophile? How does a nucleophile tend to react with C=N and C≡N groups? How many times does the nucleophilic addition reaction occur? Why must the reaction with $LiAlH_4$ be complete before H_3O^+ is added? What functional group is produced?

Solve
The mechanism for **(a)** is the same as that shown in the solution of Problem 17.8, and a primary amine is produced. The reduction reaction occurs twice and the proton transfer reaction occurs twice.

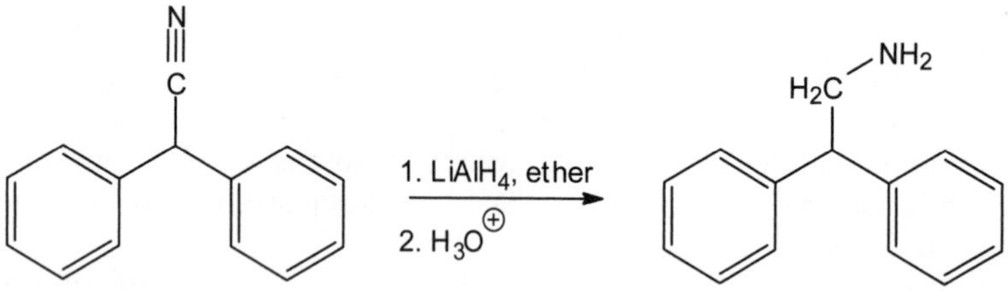

1. $LiAlH_4$, ether
2. H_3O^+

The mechanism for **(b)** is the same as the one shown in the solution for Problem 17.7, and a secondary amine is produced.

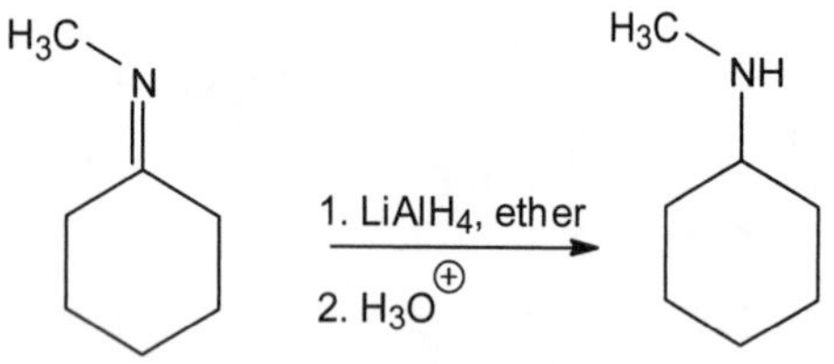

1. $LiAlH_4$, ether
2. H_3O^+

Problem 17.11

Think
Will the hydride anion from NaH act as a base or a nucleophile? Which atom will it attack? How will the resulting species behave in the presence of an alkyl halide?

Solve
NaH is a strong base but a poor nucleophile, so it will deprotonate at the α carbon. As we learned in Section 10.3, the resulting enolate anion is a strong nucleophile and will displace X from RX in an S_N2 reaction, yielding an alpha-alkylated ketone. See the figure on the next page.

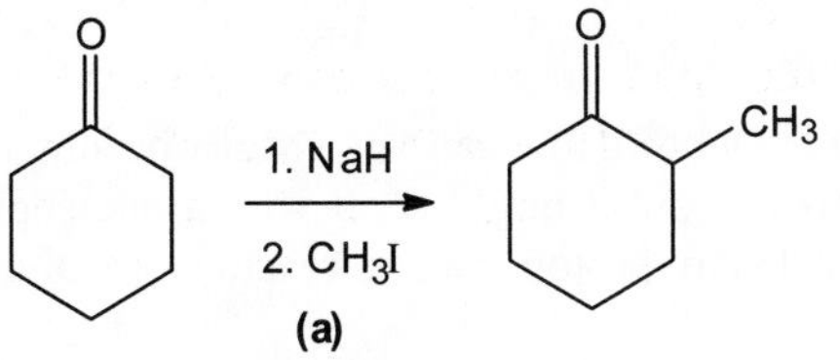

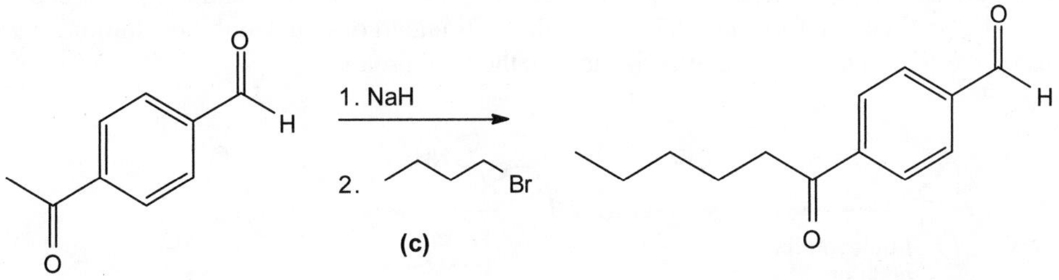

1. NaH
2. CH_3I
(a)

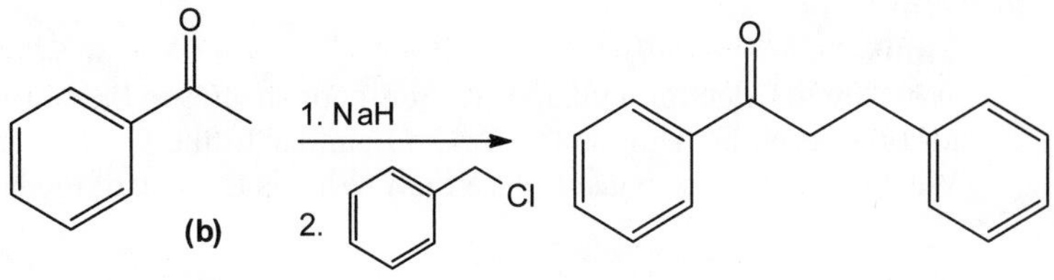

1. NaH
2. (benzyl chloride)
(b)

1. NaH
2. (1-bromobutane)
(c)

Problem 17.13

Think

To what R^- nucleophile can the Grignard or alkyllithium reagent be simplified? Which atom will it attack? What is the role of H_3O^+ or CH_3OH?

Solve

(a) The C_6H_5MgBr reagent can be simplified to $C_6H_5^-$, which will undergo nucleophilic addition at the carbonyl C atom. Once this is complete, H_3O^+ is added in an acid workup to protonate the strongly basic O^- generated in the first step.

1. C_6H_5MgBr
2. $H_3O^{\oplus}$

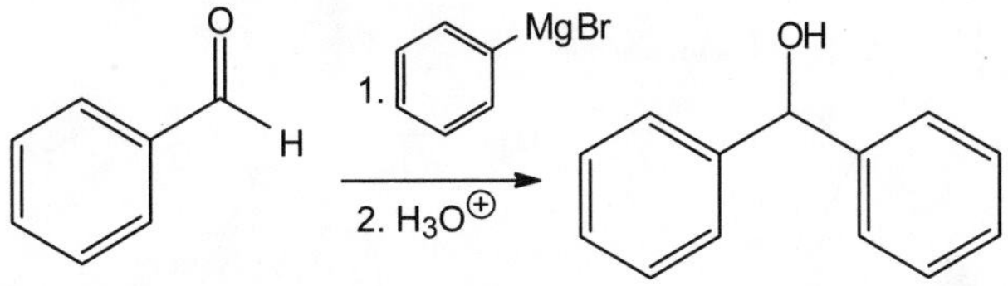

(b) The CH_3Li reagent can be simplified to CH_3^-, which will undergo nucleophilic addition at the carbonyl C. Once this is complete, H_3O^+ is added in an acid workup to protonate the strongly basic O^- generated in the first step.

1. CH_3Li
2. $H_3O^{\oplus}$

+ Enantiomer

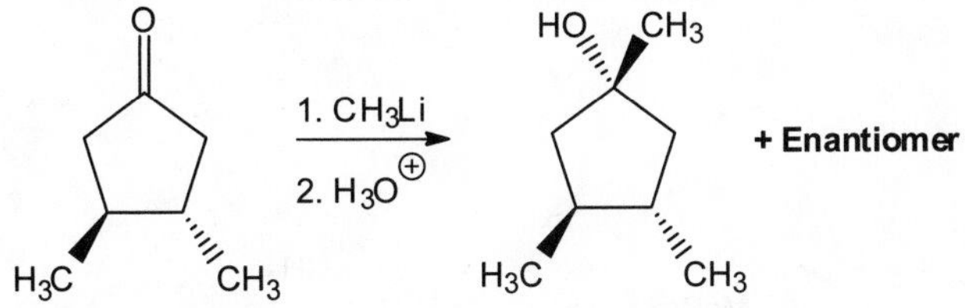

(c) The CH_3Li reagent can be simplified to CH_3^-, which will undergo nucleophilic addition at the nitrile C. Once this is complete, CH_3OH is added in a workup step to protonate the strongly basic N^- generated in the first step. Methanol is used as the proton source instead of H_3O^+ to avoid the formation of the ketone (see Equation 17-16).

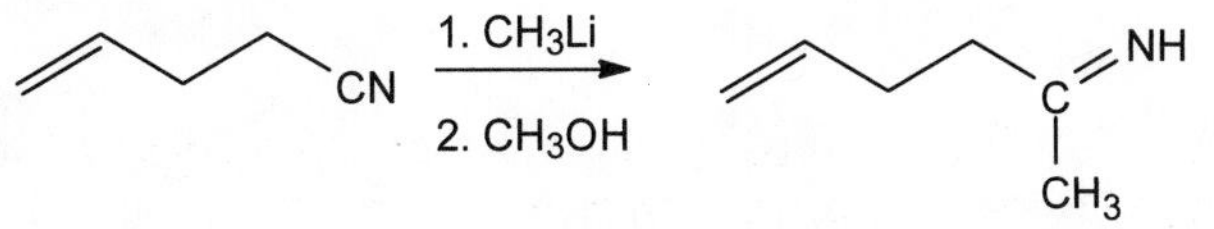

1. CH_3Li
2. CH_3OH

Problem 17.14

Think

What are the electron-rich and electron-poor species in the starting reagents? How can the RMgBr be simplified to a nucleophile? How is the O=C=O similar to the C=O in terms of how it might react with a nucleophile? What elementary step takes place first? What is the role of the NH_4Cl? Are proton transfer reactions involved?

Solve

The RMgBr can be treated as $H_2C{=}CHCH_2^-$. The C of the CO_2 is very electron poor and highly susceptible to nucleophilic attack. In the presence of a strong nucleophile, the CO_2 undergoes nucleophilic addition, followed by proton transfer using the NH_4Cl to form a carboxylic acid as the final product.

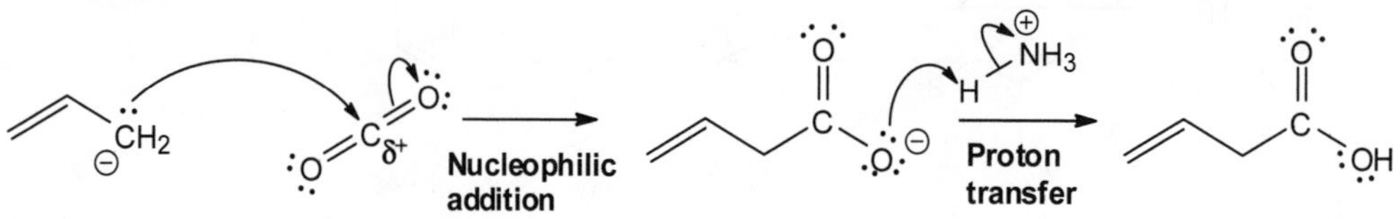

Nucleophilic addition

Proton transfer

Problem 17.15

Think

What is the nucleophile and electrophile in the first step? Which new bonds are formed? Are proton transfer reactions involved?

Solve

The C of CO_2 is the electrophile that is attacked by the nucleophilic C of the Grignard reagent. A new C–C bond is formed. A proton transfer reaction follows to protonate O^- to form an uncharged carboxylic acid product.

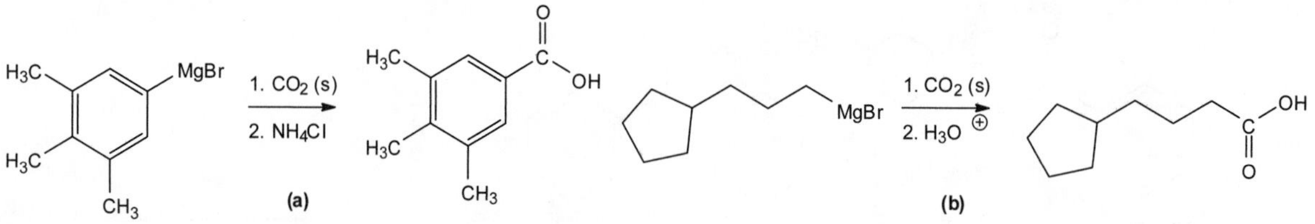

(a) 1. CO_2 (s) 2. NH_4Cl

(b) 1. CO_2 (s) 2. H_3O^+

Problem 17.16

Think

What other groups are susceptible to deprotonation or nucleophilic attack by the electron-rich C atom of the RMgBr? If the reagent reacts with itself, what product forms? Is the Grignard reagent destroyed?

Solve

(a) The OH group is susceptible to deprotonation.

Proton transfer

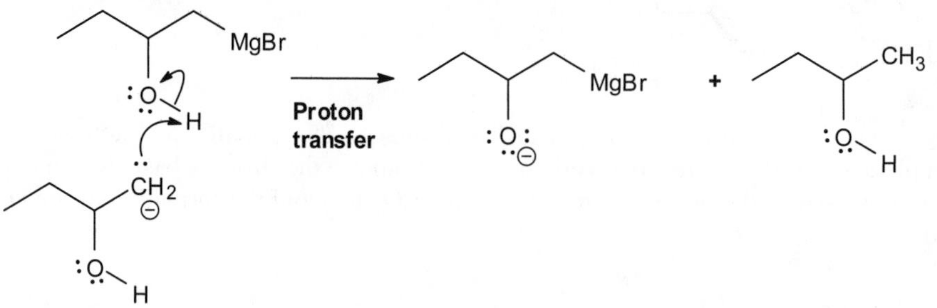

(b) The C=O group is susceptible to nucleophilic attack (nucleophilic addition).

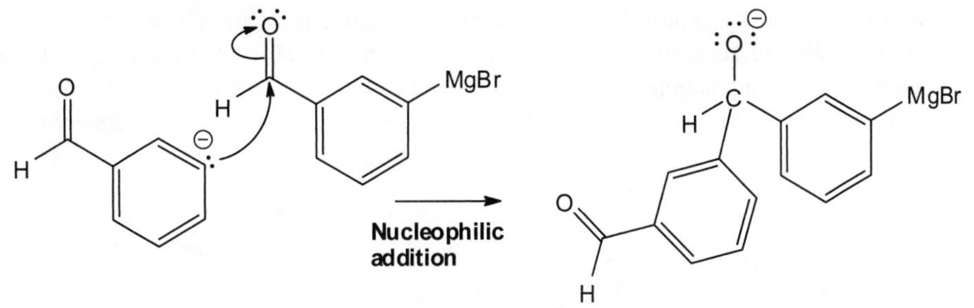

(c) The C–Br carbon is susceptible to elimination (E2).

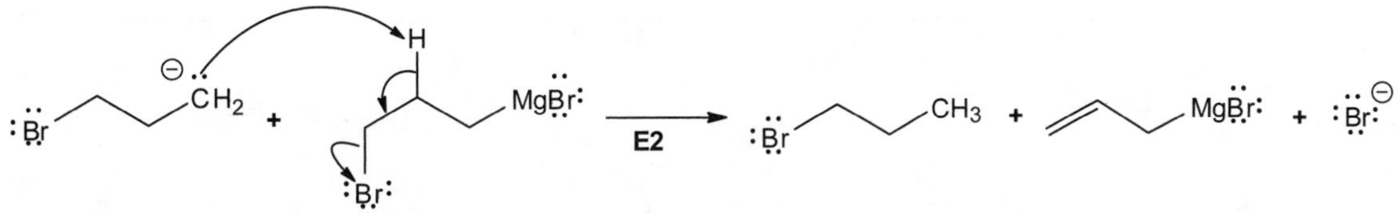

(d) The less crowded C of the epoxide is susceptible to nucleophilic attack (S_N2).

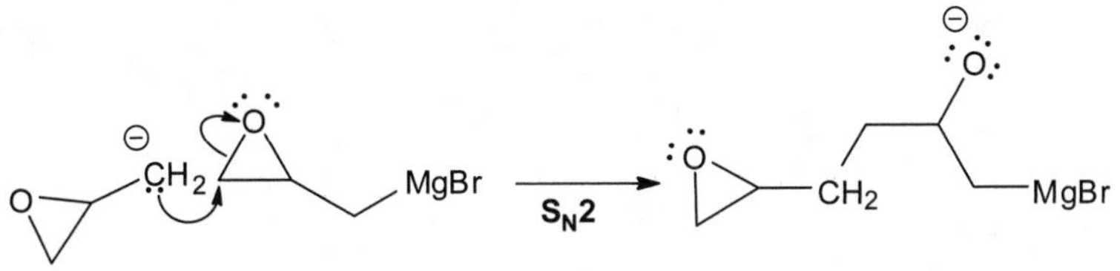

Problem 17.17

Think

What are the electron-rich and electron-poor species in the first step? When the nucleophile attacks, what bond is broken? In how many steps does this mechanism occur? What functional group is produced?

Solve

This is an example of a Wittig reaction. The carbonyl carbon of the aldehyde is electron poor and the Wittig reagent has an electron-rich C atom. This forms a new C–C bond in a nucleophilic addition mechanism, and the π bond breaks to form O^-. The negative O atom coordinates to the positive P atom. An elimination elementary step follows to form the C=C and the oxaphosphetane. Both the cis and the trans isomers are possible.

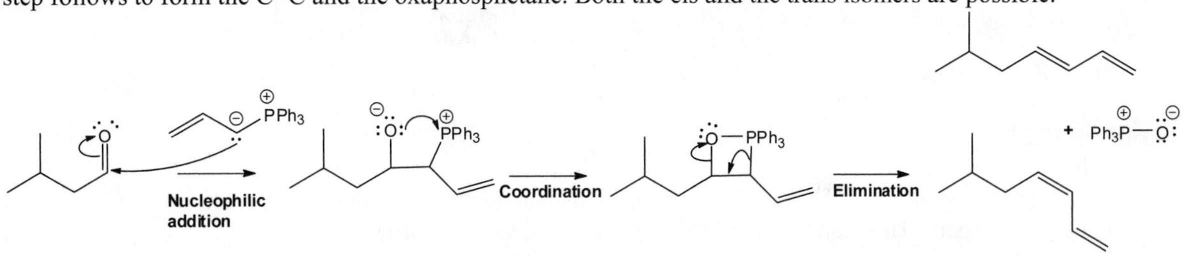

Problem 17.18

Think

What are the electron-rich and electron-poor species in the first reaction? Is triphenylphosphine a good nucleophile? What is the role of the strongly basic Bu–Li? What elementary step takes place between a Wittig reagent and a C=O group? In how many steps does this mechanism occur? What functional group is produced?

Solve

This is an example of a Wittig reagent synthesis followed by a Wittig reaction. The alkyl halide is attacked by PPh_3 in an S_N2 mechanism followed by proton transfer to form the Wittig reagent. The aldehyde C=O carbon atom is electron poor and the Wittig reagent has an electron-rich C atom. This forms a new C–C bond in a nucleophilic addition mechanism. The negative O atom coordinates to the positive P atom. An elimination elementary step follows to form the C=C and the oxaphosphetane. Both the *E* and the *Z* isomers are possible.

Problem 17.19

Think

Which C atom is charged and a strong nucleophile? To which C atom was the halide attached?

Solve

The negatively charged C atom of the Wittig reagent is initially bonded to a halogen atom of an alkyl halide. Benzyl bromide is shown but benzyl chloride or benzyl iodide would also be reasonable choices.

Problem 17.21

Think

What Wittig reagent would you need to carry out this reaction? What alkyl halide could serve as a precursor to that Wittig reagent? How can that alkyl halide be synthesized from benzaldehyde?

Solve

We can begin a retrosynthetic analysis by applying a transform that undoes a Wittig reaction, so we disconnect the C=C bond to give us a carbonyl compound and a Wittig reagent. The Wittig reagent can ultimately be produced from benzaldehyde. See the figure on the next page.

Undo a Wittig reaction.

Undo formation of Wittig reagent.

Undo bromination.

Undo Grignard reaction.

Benzaldehyde

The synthesis in the forward direction appears as follows:

1. CH_3Li
2. H_3O^+

PBr_3

1. PPh_3
2. Bu—Li

Problem 17.22

Think

What are the electron-rich and electron-poor species in the first step? Is there a good leaving group in the product of that step? In how many steps does this mechanism occur? What functional group is produced?

Solve

(a) This mechanism is identical to the mechanism shown in Equation 17-30. The ketone C=O carbon is electron poor and the sulfonium ylide C atom is electron rich. A new C–C bond forms. The negative O atom is a good nucleophile and the $(CH_3)_2S$ group is a good leaving group. The second step is an example of an internal S_N2 reaction. An epoxide functional group results.

Nucleophilic addition

S_N2

(b) This mechanism is identical to the combination of mechanisms shown in Equations 17-28 and 17-30. The sulfonium ylide is synthesized via an S_N2 mechanism followed by a proton transfer. The ketone C=O carbon is electron poor and the sulfonium ylide C atom is electron rich. A new C–C bond forms. The negative O atom is a good nucleophile and the $(CH_3)_2S$ group is a good leaving group. The last step is an example of an intramolecular S_N2 reaction. An epoxide functional group results. Because the sulfonium ylide has methyl and ethyl R groups, the proton transfer can occur at both the CH_3 and the CH_2CH_3. This leads to two different epoxide products. See the second mechanism on the next page.

S_N2

Proton transfer (at the ethyl)

Nucleophilic addition

S_N2

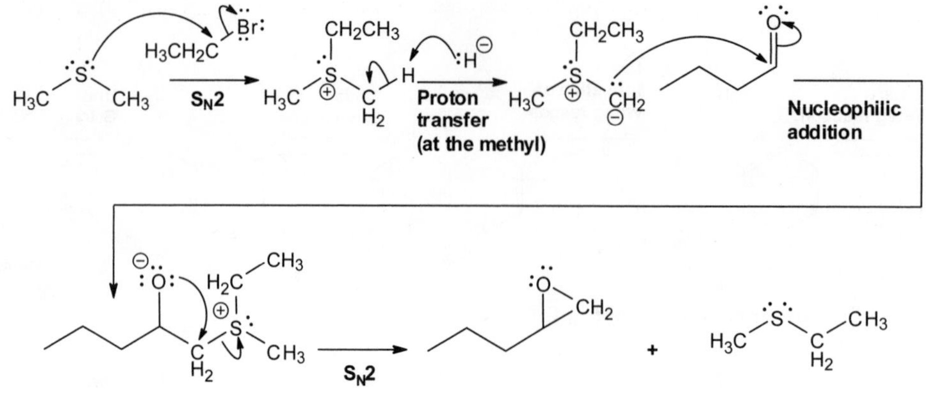

H3C–S–CH3 + H3CH2C–Br →(SN2)→ sulfonium salt →(Proton transfer (at the methyl))→ ylide →(Nucleophilic addition)→ alkoxide →(SN2)→ epoxide + H3C–S–CH2–CH3

Problem 17.23

Think

Which C–C bond formed from the sulfonium ylide reaction? On which C was the C=O and on which C was the sulfonium ylide?

Solve

In the sulfonium ylide reaction, the C–C bond in the epoxide is the new bond that forms. The sulfonium ylide and ketone or aldehyde starting materials are shown below.

(a) Undo sulfonium ylide epoxide.

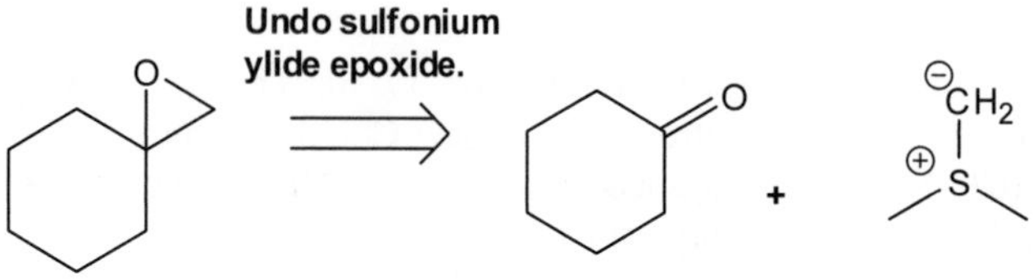

(b) Undo sulfonium ylide epoxide.

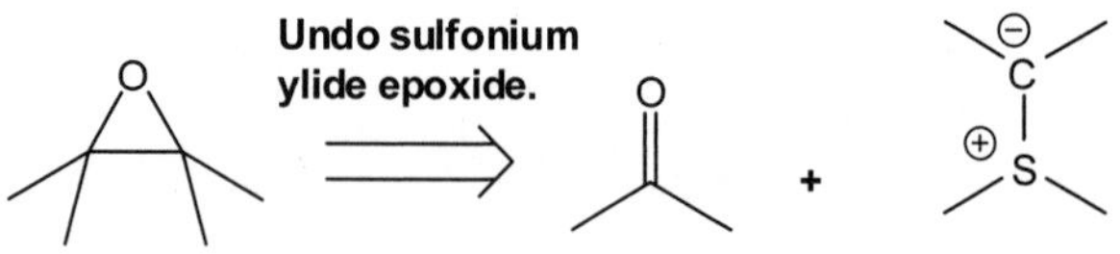

(c) Undo sulfonium ylide epoxide.

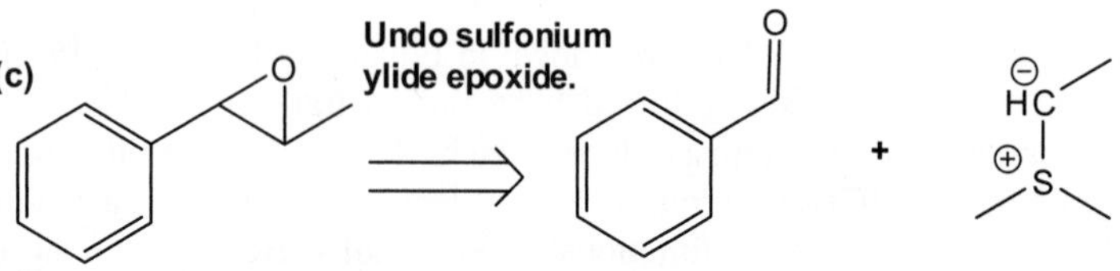

Problem 17.24

Think

What are the two electron-poor sites on the α,β-unsaturated ketone? What C atom does the nucleophile attack in a 1,2-addition mechanism? In a 1,4-addition mechanism? What are the final products in each?

Solve

Addition of the nucleophile at the carbonyl C atom is called 1,2-addition or direct addition, and addition of the nucleophile at the β C atom is called 1,4-addition or conjugate addition. This addition occurs via the nucleophilic addition elementary step. The negative O atom is protonated to form an uncharged alcohol product. In 1,4-addition, the enol tautomerizes to the aldehyde or ketone.

(a) The HO^- acts as the nucleophile at either the 2 or 4 site in 1,2-addition or 1,4-addition.

Direct addition (1,2-addition)

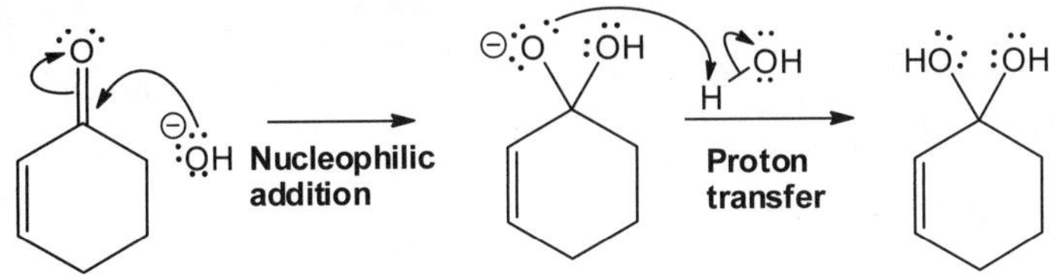

Nucleophilic addition

Proton transfer

Conjugate addition (1,4-addition)

Nucleophilic addition

Proton transfer

Tautomerization

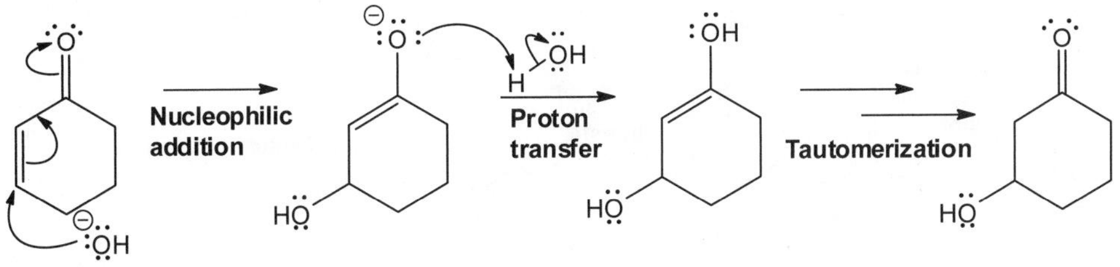

(b) The NC^- acts as the nucleophile at either the 2 or 4 site in 1,2-addition or 1,4-addition.

Direct addition (1,2-addition)

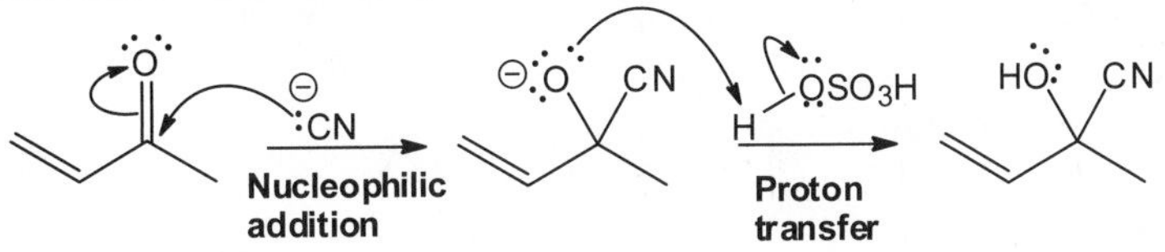

Nucleophilic addition

Proton transfer

Conjugate addition (1,4-addition)

Nucleophilic addition

Proton transfer

Tautomerization

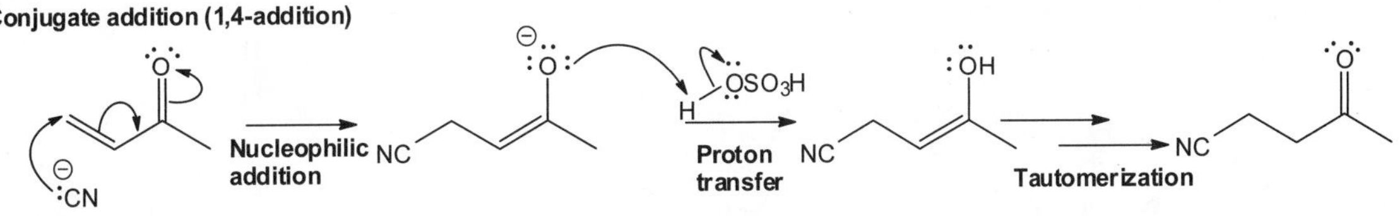

Problem 17.25

Think

Is the nucleophile one that leads to reversible or irreversible nucleophilic addition? Does reversible addition favor 1,2-addition or 1,4-addition? Does irreversible addition favor 1,2-addition or 1,4-addition?

Solve

Only the highly reactive R^- and H^- nucleophiles add irreversibly to a carbonyl C atom: **(b)**, **(d)**, and **(e)**. Nucleophiles that add irreversibly to the carbonyl C yield the direct 1,2-addition product. The remaining nucleophiles, **(a)** and **(c)**, are significantly less reactive, and add reversibly. Nucleophiles that add reversibly to the carbonyl C yield the conjugate 1,4-addition product. The mechanisms for reactions **(a)**–**(e)** are given below and on the next page.

(a) Conjugate addition (1,4-addition)

Nucleophilic addition

Proton transfer

Tautomerization

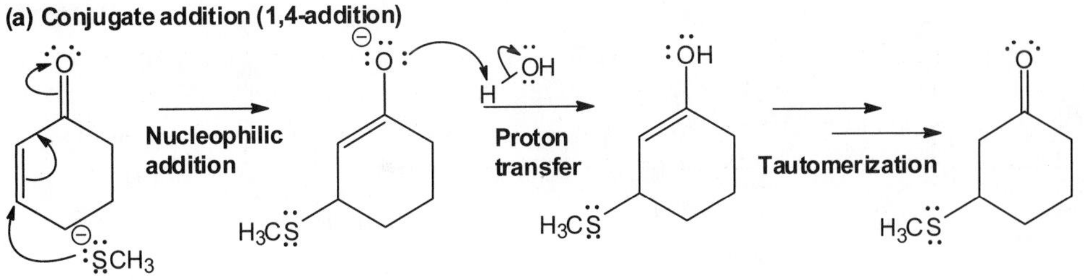

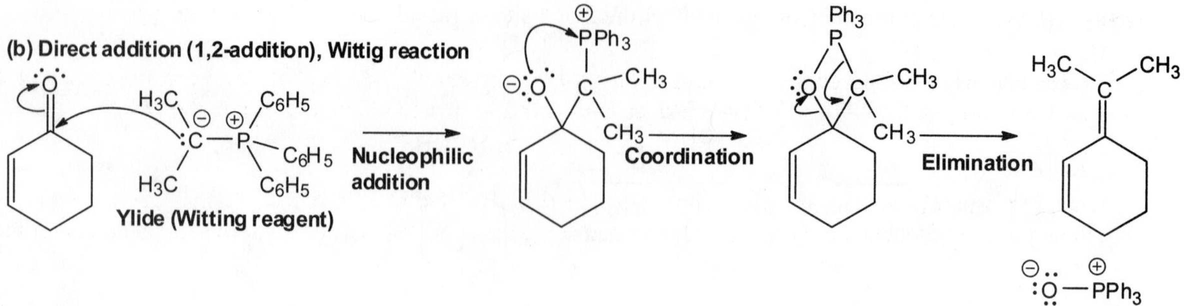

(c) Conjugate addition (1,4-addition)

Nucleophilic addition

Proton transfer

Tautomerization

(d) Direct addition (1,2-addition)

Nucleophilic addition

Proton transfer

(e) Direct addition (1,2-addition)

Nucleophilic addition

Proton transfer

Problem 17.27

Think

Which group has the polar π bond? Will the nucleophile add reversibly or irreversibly? How does that affect whether direct addition or conjugate addition takes place?

Solve

The C≡N group has the polar π bond. The nucleophile is NC^-, which, as we know from Table 17-2, adds reversibly. As a result, the thermodynamic product, which is formed via conjugate addition, is favored. The mechanism is essentially the same as that in Equation 17-32.

Nucleophilic addition

Proton transfer

Tautomerization

Problem 17.29

Think

To what R^- nucleophile can the lithium dialkylcuprate be simplified? Will it add predominantly via direct addition or conjugate addition? What role does H_2O play after the addition of R^- is complete?

Solve

In **(a)**, **(b)**, and **(c)**, the lithium dialkylcuprate is a source of $^-CH_3$, $^-CH_2CH_3$, or $^-CH(CH_3)_2$, respectively, which will add to the α,β-unsaturated carbonyl group via conjugate addition. The resulting enolate anion is protonated by H_2O.

(a)

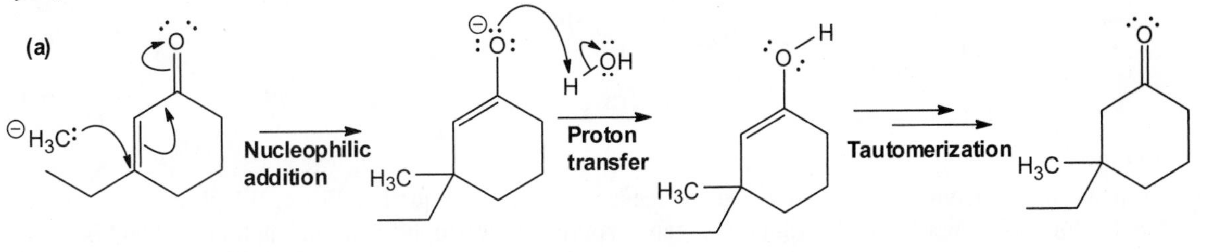

Nucleophilic addition → Proton transfer → Tautomerization

(b)

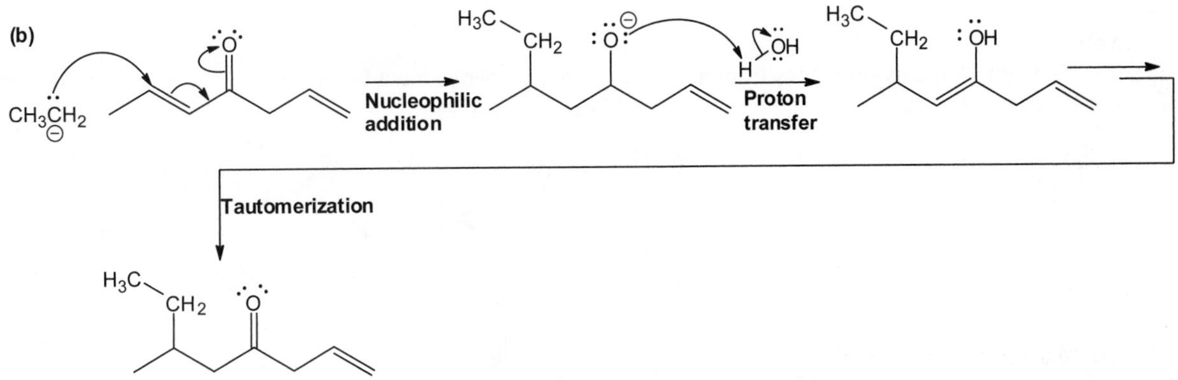

Nucleophilic addition → Proton transfer → Tautomerization

(c)

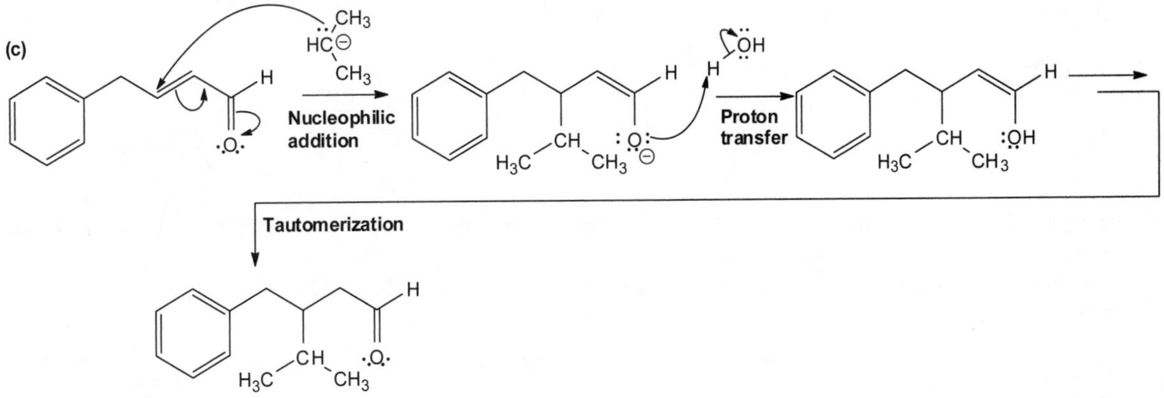

Nucleophilic addition → Proton transfer → Tautomerization

Problem 17.30

Think

To what R^- nucleophile can the Grignard and alkyl lithium reagent in each example be simplified? Will it add predominantly via direct addition or conjugate addition? What role does H_3O^+ or CH_3OH play after the addition of R^- is complete?

Solve

In all three examples, the R^- group adds to the carbonyl C atom in a direct 1,2-addition reaction. The major products are given on the next page. The steps with H_3O^+ and CH_3OH are used to protonate the negatively charged product from nucleophilic addition. In **(c)**, methanol is used instead of water to avoid hydrolyzing the imine that is produced.

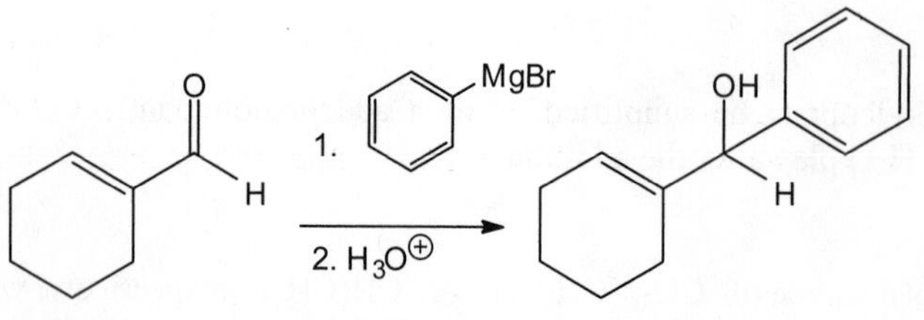

(a) 1. MgBr (phenylmagnesium bromide) 2. $H_3O^{\oplus}$

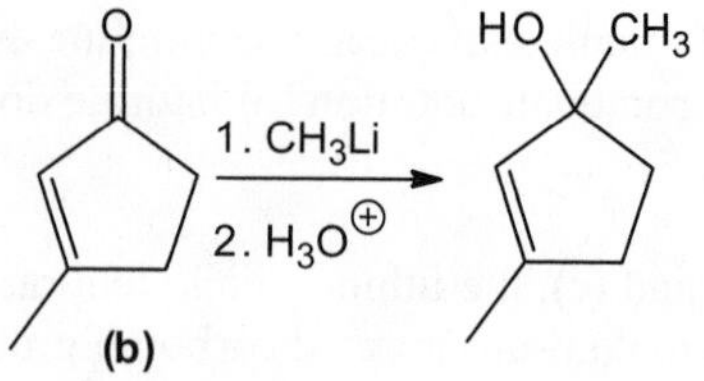

(b) 1. CH_3Li 2. $H_3O^{\oplus}$

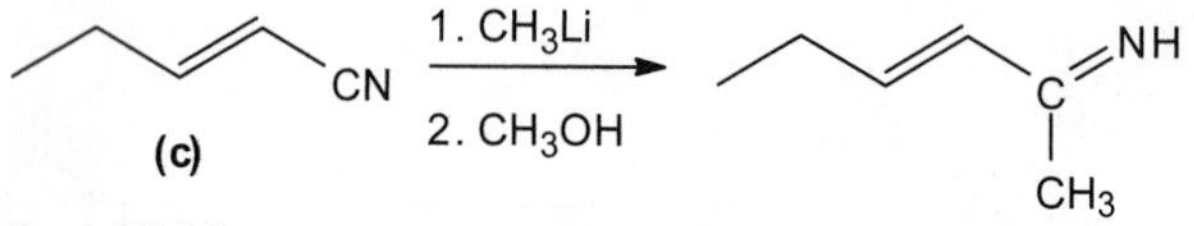

(c) 1. CH_3Li 2. CH_3OH

Problem 17.31

Think

Compare the structure given to the Grignard reagents shown in Equation 17-44. Which bond do you need to "undo" for each reagent? Is the Grignard reagent a source of electrophilic or nucleophilic carbon? Is the product listed an alcohol? If not, how can this functional group be synthesized from an alcohol?

Solve

The final ether product originates from the same tertiary alcohol shown below, which can be produced from the three different Grignard reaction routes indicated in Equation 17-44.

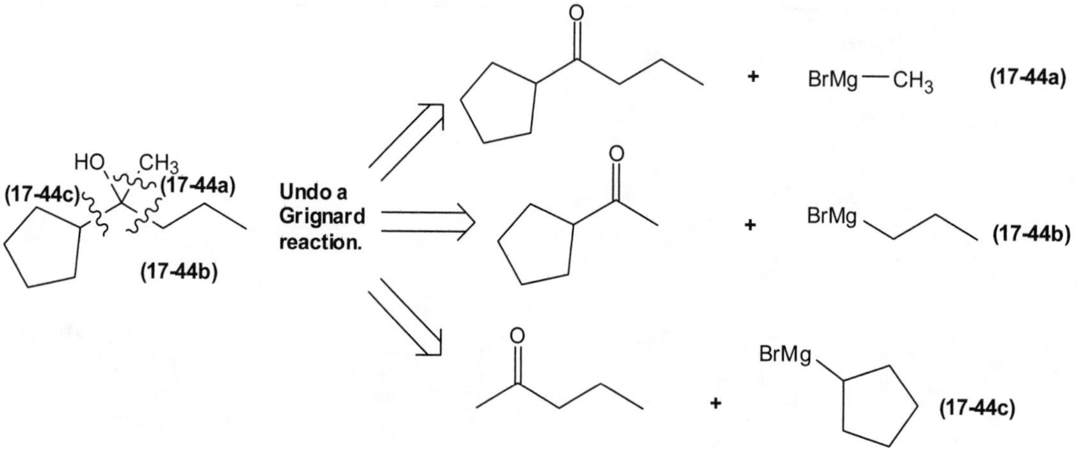

(17-44c) (17-44a) (17-44b) **Undo a Grignard reaction.**

\+ BrMg—CH_3 **(17-44a)**

\+ BrMg **(17-44b)**

\+ BrMg **(17-44c)**

An ether is formed from an alcohol first by deprotonating the alcohol with NaH to form a strong anionic nucleophile followed by an S_N2 reaction.

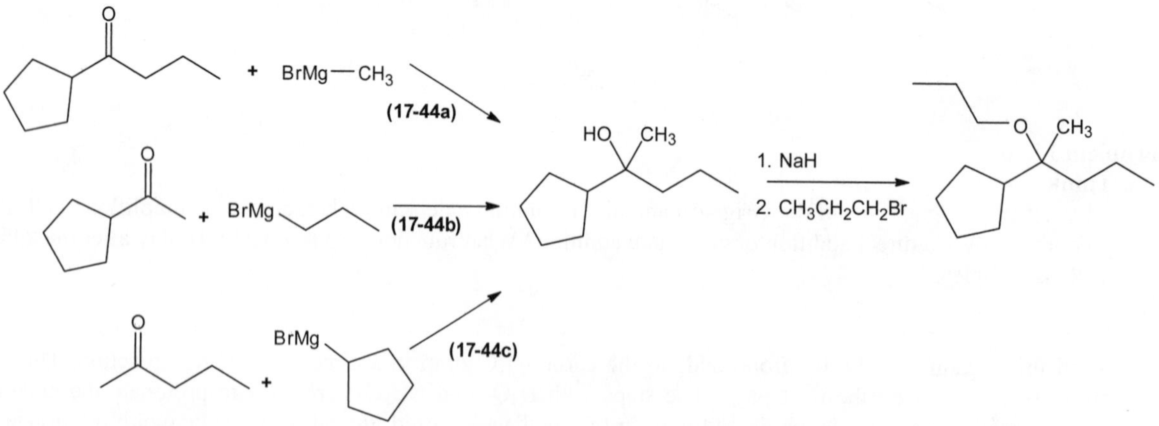

\+ BrMg—CH_3 (17-44a)

\+ BrMg (17-44b)

\+ BrMg (17-44c)

1. NaH
2. $CH_3CH_2CH_2Br$

Problem 17.32

Think

On which C atom in the product must the carbonyl C have been located? What R groups are attached to that C atom? Can that bond be disconnected to arrive at a carbonyl precursor and an alkyllithium precursor? What, then, is the identity of the LiR reagent?

Solve

Disconnecting the C–C bond on either side of the alcohol gives the following:

Undo an alkyllithium addition.

(a) (b) OH Li + O (a)

Undo an alkyllithium addition.

O + Li (b)

In the forward direction, the synthesis would appear as follows:

(a) O 1. Li 2. H_3O^+ OH

(b) O 1. Li 2. H_3O^+ OH

Problem 17.33

Think

What is the structure of hex-4-en-3-one? How many C atoms does it possess? What bonds were formed to synthesize the compounds given from hex-4-en-3-one? What nucleophile was used? Did 1,2-addition or 1,4-addition occur? Which R^- nucleophiles favor 1,2-addition and which ones favor 1,4-addition?

Solve

(a) The new C–C bond that was formed is highlighted below.

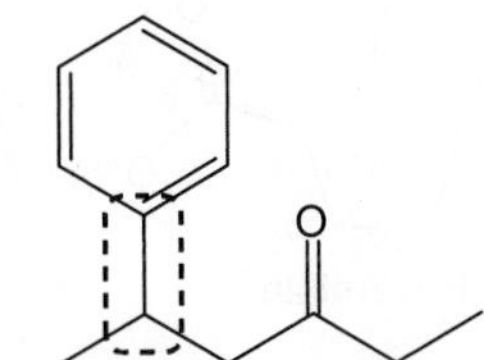

This means that the nucleophile is $H_5C_6^-$ and the addition occurred via 1,4-addition. To have 1,4-addition, the reagent needs to be a dialkylcuprate compound, R_2CuLi. The enolate produced from nucleophilic addition is then protonated to form an enol that tautomerizes back to the ketone, giving the product listed.

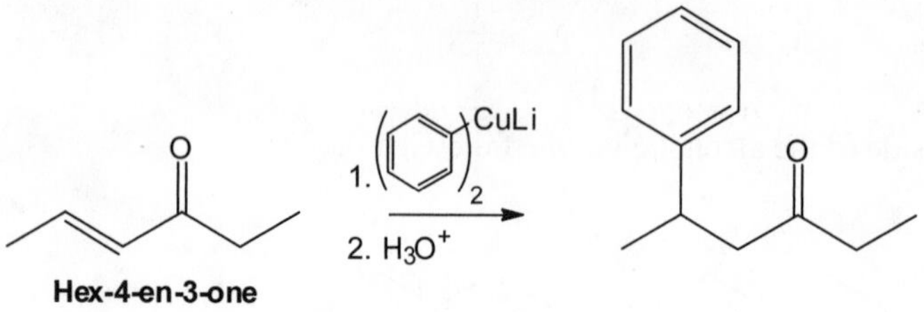

1. $(C_6H_5)_2CuLi$
2. H_3O^+

Hex-4-en-3-one

(b) The new C–C bond that was formed is highlighted below.

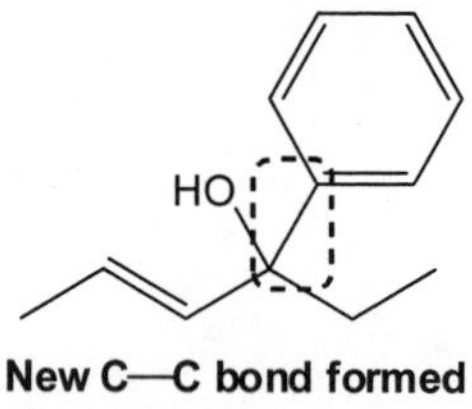

HO

New C—C bond formed

This means that the nucleophile is $H_5C_6^-$ and the addition occurred via 1,2-addition. To have 1,2-addition, the reagent can be an alkyllithium compound, LiR. The O^- atom produced upon nucleophilic addition is then protonated to form the uncharged alcohol product shown below.

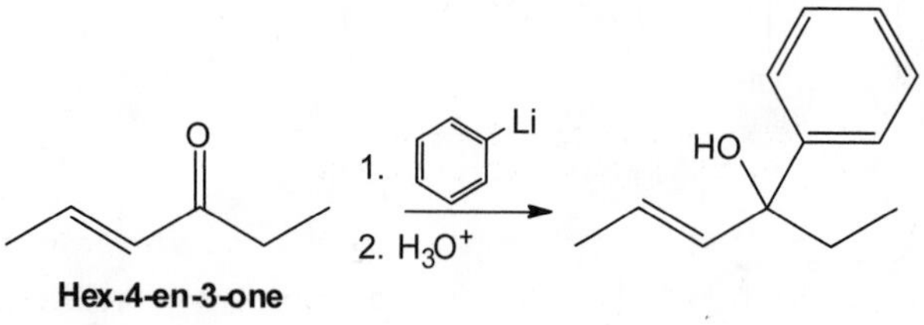

1. C_6H_5Li
2. H_3O^+

HO

Hex-4-en-3-one

Problem 17.34

Think

Which nucleophiles are negatively charged? Which nucleophiles are strong nucleophiles? Which nucleophiles are sources of R^- or H^-? Refer to Table 17-2 for a list of reversible and irreversible nucleophiles.

Solve

Only the highly reactive R^- and H^- nucleophiles add irreversibly to a carbonyl C: **(a)** and **(b)**. The remaining nucleophiles are significantly less reactive and add reversibly. See below and the next page.

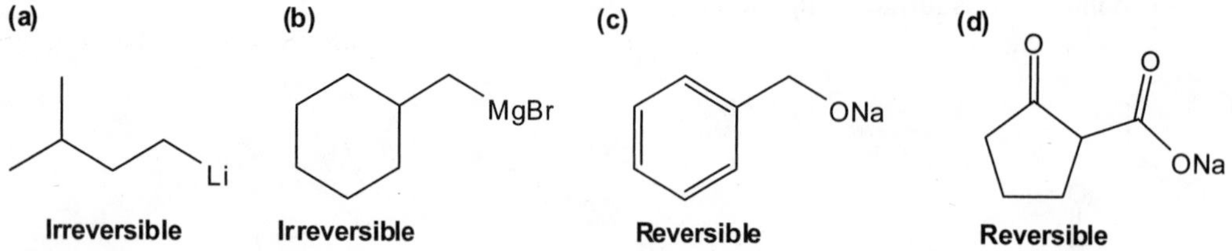

(a) Li — Irreversible

(b) MgBr — Irreversible

(c) ONa — Reversible

(d) ONa — Reversible

(e) $\left(\text{CH}_3\text{CH}_2\text{CH}_2\text{CH}_2\right)_2$CuLi — Reversible

(f) N Li — Reversible

(g) H_2O — Reversible

(h) H N — Reversible

(i) O — Reversible

(j) SNa — Reversible

(k) SH — Reversible

Problem 17.35

Think

Are the groups attached electron donating or electron withdrawing? Which C atom is more electron poor? Which tetrahedral product would be more sterically crowded?

Solve

(a) The second one, because with fewer alkyl groups attached to the carbonyl C, there is less bulkiness and greater concentration of positive charge.

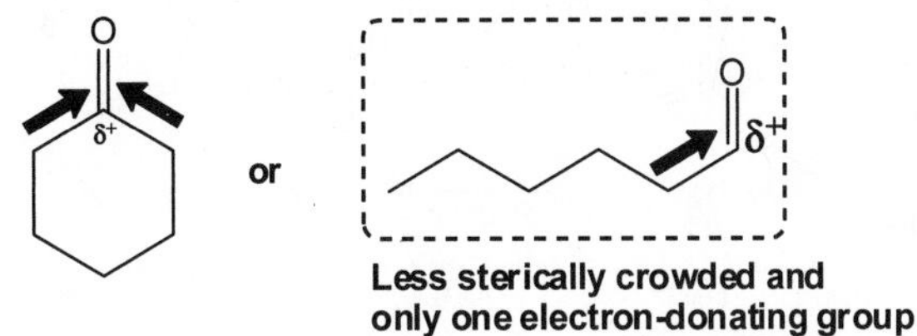

(b) The second one, because the bulkiness at the carbonyl C is about the same for both. However, the CF_3 is electron withdrawing, whereas a CH_3 group is electron donating, so the CF_3 group induces a greater positive charge on the carbonyl C.

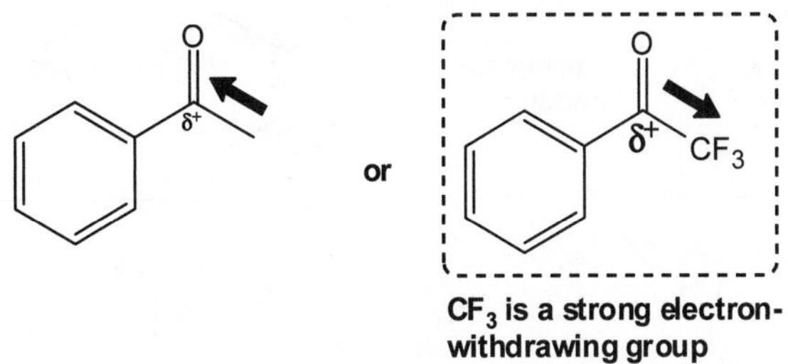

(c) The first one, because with the Cl atom closer to the carbonyl C, the carbonyl C bears a greater concentration of positive charge.

O δ+ Cl — or — Cl O δ+

Electron-withdrawing group, Cl, closer to C=O

(d) The first one, because the nitrile carbon bears a higher concentration of positive charge due to the presence of the electron-withdrawing CCl_3 group. The CH_3 group is electron donating.

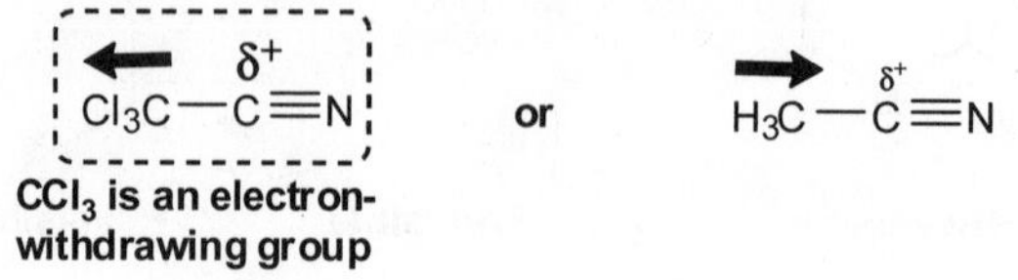

$Cl_3C-\overset{\delta+}{C}\equiv N$ or $H_3C-\overset{\delta+}{C}\equiv N$

CCl_3 is an electron-withdrawing group

Problem 17.36

Think

What is the nucleophile in each reaction? Can the nucleophile be simplified? Which C atom is electrophilic? Are proton transfer reactions involved? How many times does the nucleophilic addition take place?

Solve

The nucleophiles in each reaction are the metal-containing species and can be simplified by treating the metal portions as spectator cations. In each reaction mechanism for **(a)**–**(c)** and **(f)**, the carbonyl C is electrophilic and undergoes nucleophilic addition. The O^- atom is then protonated in a separate proton transfer reaction to form an uncharged alcohol or carboxylic acid. In **(d)**, the nitrile C atom undergoes nucleophilic addition twice, and two proton transfer reactions are involved to form the uncharged amine. In **(e)**, the imine C is electrophilic and undergoes nucleophilic addition. A proton transfer reaction follows to form the uncharged amine. The mechanisms are given below and on the next page.

(a)

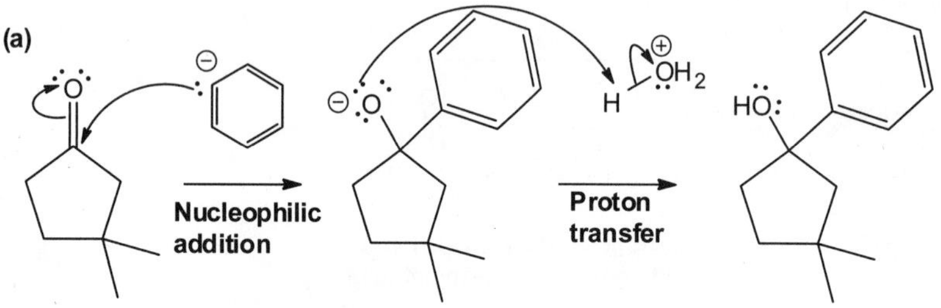

Nucleophilic addition → Proton transfer ($H-OH_2^+$) → HO:

(b)

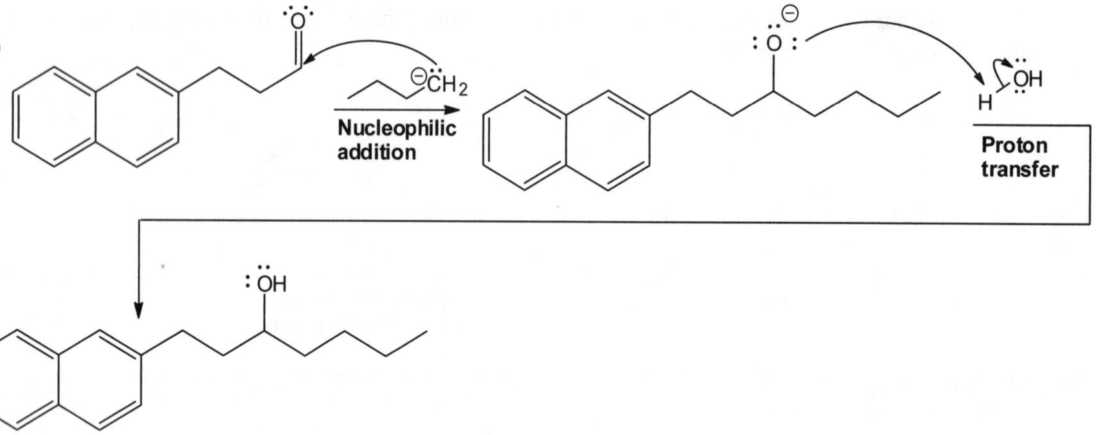

CH_2^- Nucleophilic addition → Proton transfer (H–OH) → :OH

(c)

CH_3^- :O=C=O: → Nucleophilic addition → Proton transfer ($H-OH_2^+$) → OH

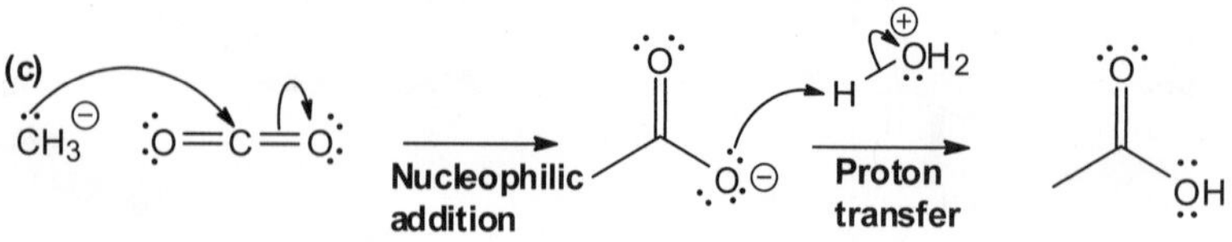

(d)

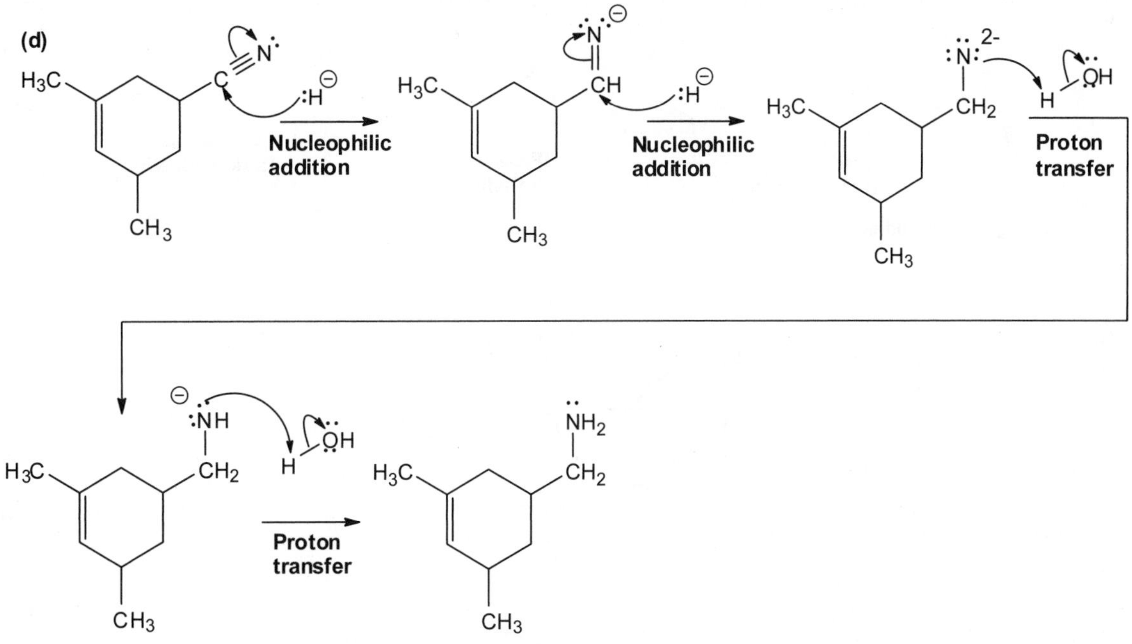

Nucleophilic addition → Nucleophilic addition → Proton transfer → Proton transfer

(e)

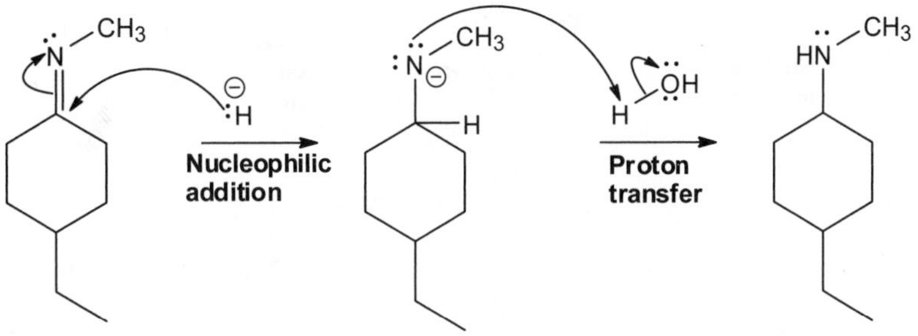

Nucleophilic addition → Proton transfer

(f)

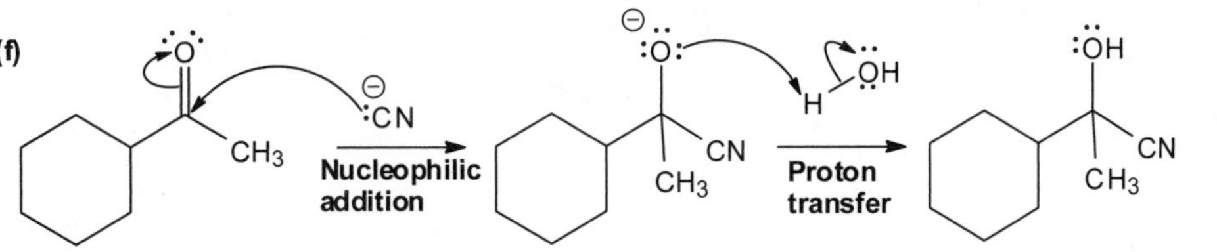

Nucleophilic addition → Proton transfer

Problem 17.37

Think

What is the nucleophile in each reaction? Which C atom is electrophilic? Are proton transfer reactions involved? Does the nucleophilic addition occur via 1,2-addition or 1,4-addition?

Solve

In each reaction mechanism for **(a)**–**(e)**, the carbonyl C is electrophilic and can undergo nucleophilic addition. In **(a)**, **(c)**, **(d)**, and **(e)**, the β carbon is also electrophilic, so you must consider both 1,2- and 1,4-addition. **(a)**, **(d)**, and **(e)** undergo conjugate 1,4-addition because the nucleophile adds reversibly. **(c)** undergoes 1,2-addition because the nucleophile adds irreversibly. In all cases, the O^- atom that forms in the nucleophilic addition step is then protonated in a subsequent proton transfer reaction to form an uncharged alcohol. In the conjugate addition mechanism, the enol tautomerizes back to the ketone or aldehyde. See the mechanisms on the next page.

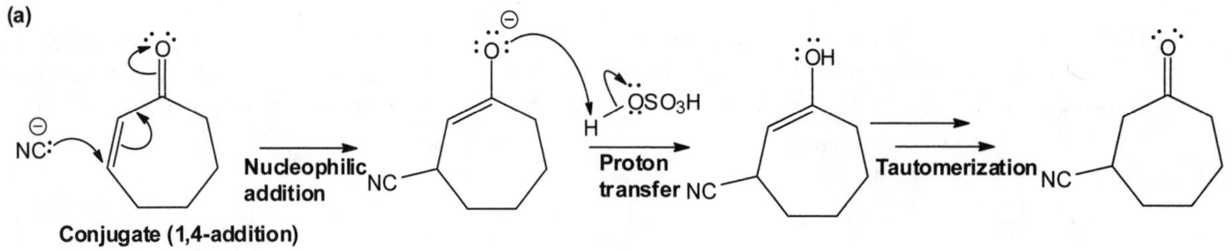

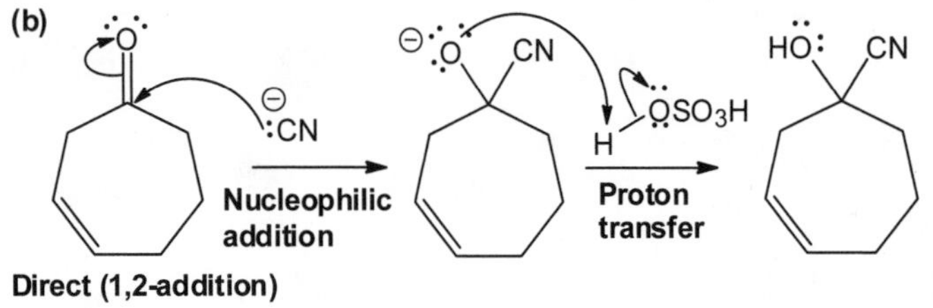

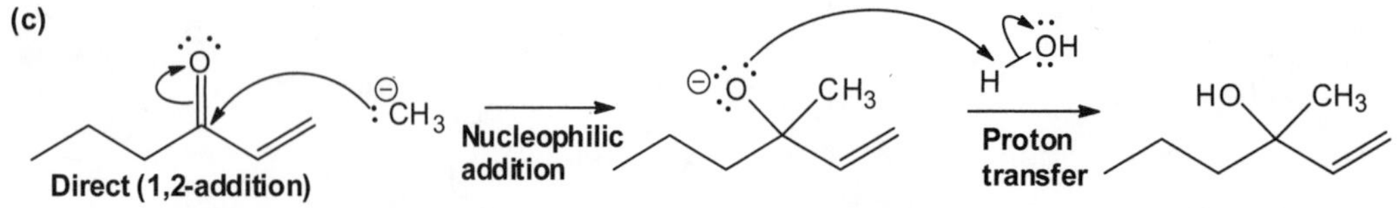

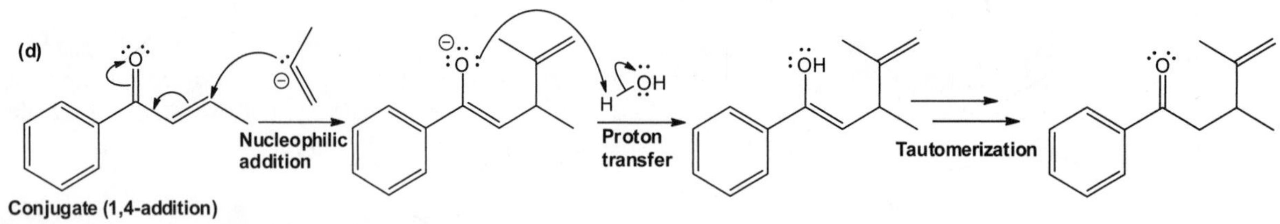

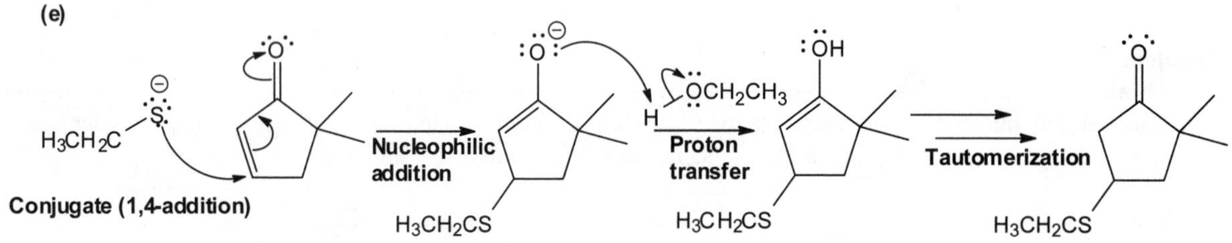

(a)

Nucleophilic addition

Conjugate (1,4-addition)

Proton transfer

Tautomerization

(b)

Nucleophilic addition

Direct (1,2-addition)

Proton transfer

(c)

Nucleophilic addition

Direct (1,2-addition)

Proton transfer

(d)

Nucleophilic addition

Conjugate (1,4-addition)

Proton transfer

Tautomerization

(e)

Nucleophilic addition

Conjugate (1,4-addition)

Proton transfer

Tautomerization

Problem 17.38

Think

What is the nucleophile in each reaction? Which C atom is electrophilic? Are proton transfer reactions involved? Does the nucleophilic addition occur via 1,2-addition or 1,4-addition?

Solve

(a) Hydroxide is not a good leaving group, and PBr_3 turns HO^- into $HOPBr_2$, which is a good leaving group. The Br^- attacks the electrophilic C atom in Step 2 to form the alkybromide. The ylide Wittig reagent is then formed via an S_N2 reaction followed by a proton transfer. The three-step Wittig reaction follows. The mechanism is shown below.

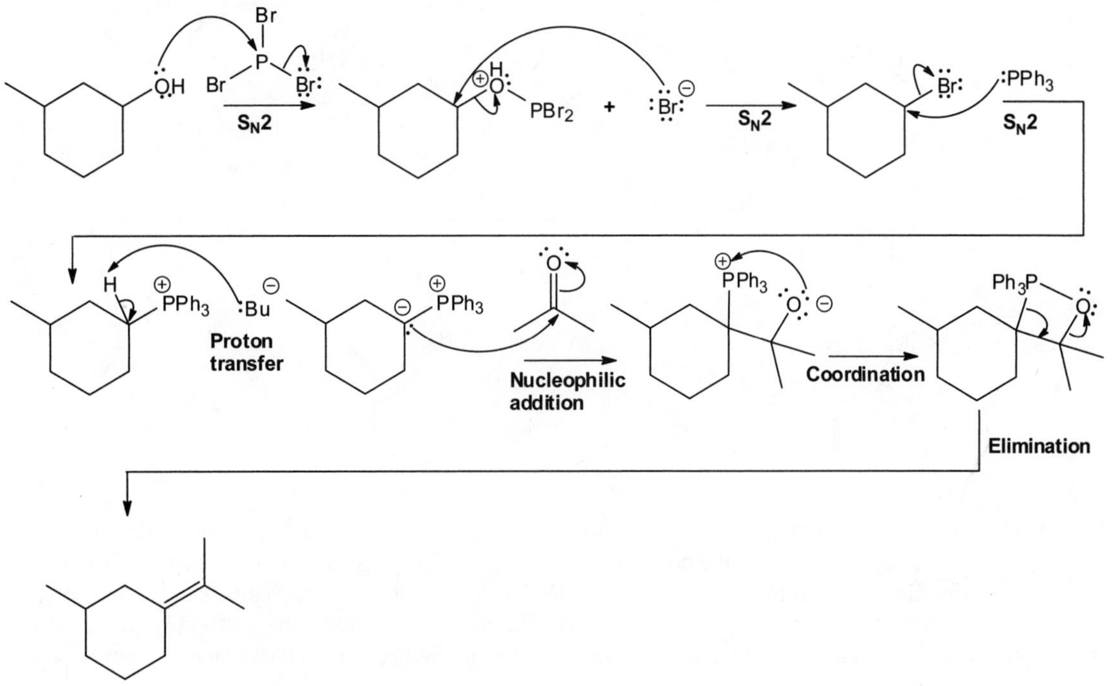

S_N2 S_N2 S_N2

Proton transfer

Nucleophilic addition

Coordination

Elimination

(b) The reagent $(CH_3)_2S^+{-}CH_2^-$ adds to a carbonyl C in a two-step mechanism to form an epoxide. In the presence of a strong anionic nucleophile, the less substituted carbon of the epoxide is attacked via an S_N2 mechanism. A proton transfer reaction follows to form an uncharged alcohol.

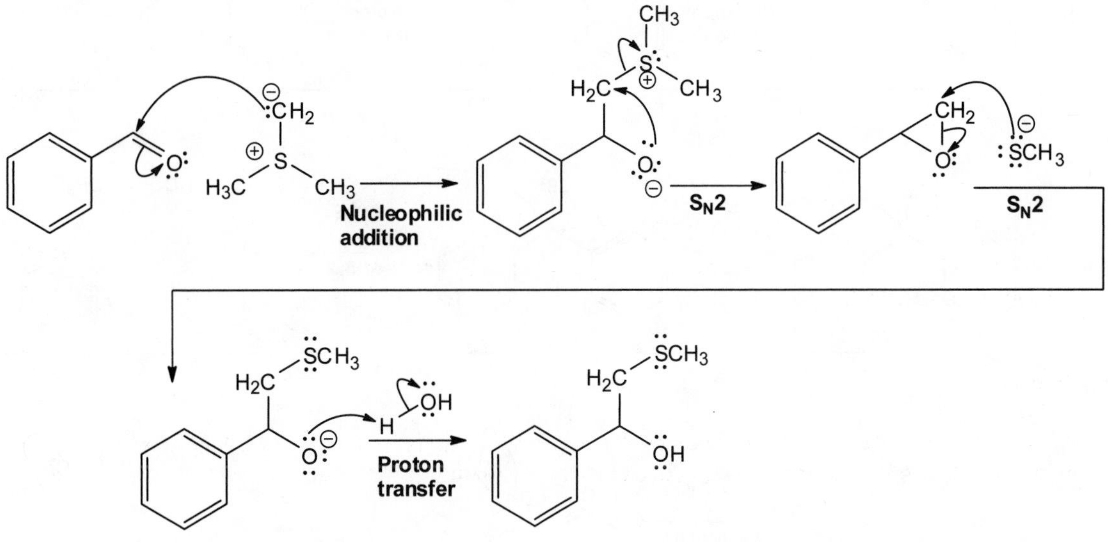

Nucleophilic addition

S_N2 S_N2

Proton transfer

(c) The first two steps are a Grignard reaction, in which the Grignard reagent adds to the C=O, followed by proton transfer to form a tertiary alcohol. The second two steps are a deprotonation by NaH to form a negatively charged alkoxide. The alkoxide is a strong nucleophile and attacks the CH_3CH_2Br via an S_N2 mechanism to form the ether functional group.

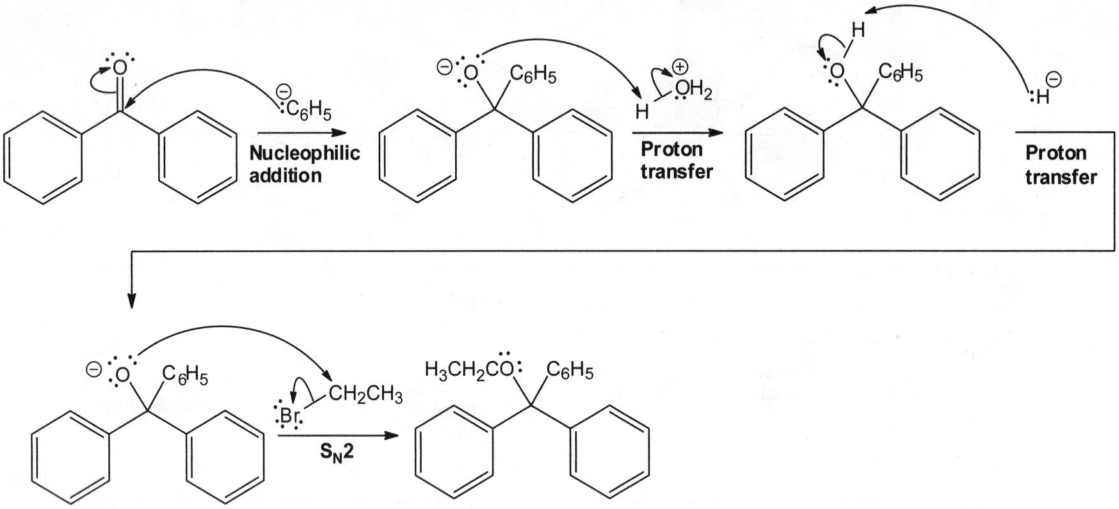

Nucleophilic addition

Proton transfer

Proton transfer

S_N2

(d) The first reaction is reduction of the ketone to the alcohol via $NaBH_4$ (nucleophilic addition) and EtOH (proton transfer). PBr_3 turns the poor HO^- leaving group into a good leaving group, and Br^- then attacks the electrophilic carbon to form the alkyl bromide. The alkyl bromide is then attacked by the sulfur of $(CH_3)_2S$. NaH deprotonates the benzylic C–H to form the sulfonium ylide. The sulfonium ylide then attacks the aldehyde carbon of the C=O, and a subsequent internal S_N2 reaction takes place to form the epoxide as the final product.

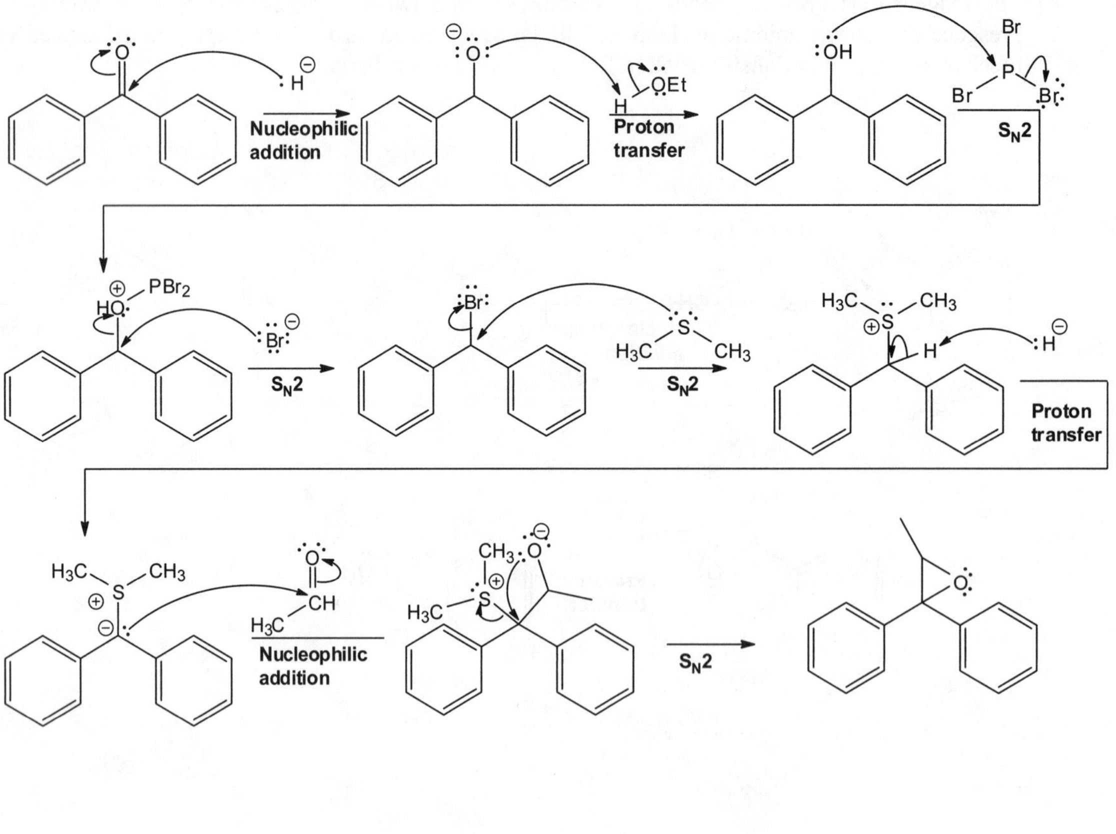

Nucleophilic addition

Proton transfer

S_N2

S_N2

S_N2

Proton transfer

Nucleophilic addition

S_N2

Problem 17.39

Think

Do alkyllithium cuprate compounds undergo 1,2-addition or 1,4-addition? What is the structure of the α,β-unsaturated ketone starting material if you disconnect the bond between the β carbon and CH_3? What would it be if you disconnected the bond to the CH_3CH_2 group instead?

Solve

Lithium dialkylcuprate compounds undergo 1,4-addition. Disconnecting the CH_3CH_2 group yields 1-phenylbut-2-en-1-one as the α,β-unsaturated ketone precursor. Disconnecting the CH_3 group yields 1-phenylpent-2-en-1-one as the α,β-unsaturated ketone precursor. To form the enol that tautomerizes to the ketone, a solvent like water is necessary.

O
1. (⁄)₂CuLi
2. H_2O
O

O
1. $(CH_3)_2CuLi$
2. H_2O
O

Problem 17.40

Think

On which C atom was the C=O located? What is the identity of the three R groups off the alcohol C atom in the target? What is the identity of the three Grignard reagents that can be used as precursors if each of those bonds is disconnected? What other two R groups must attach to the C=O ketone when each of those bonds is disconnected?

Solve

The alcohol C atom was the site of the C=O ketone. The three R groups off the tertiary alcohol are CH_3, C_6H_5, and $CH_3CH{=}CH_2$, and thus each could have come from a RMgBr Grignard reagent precursor. The three different syntheses are given below.

O
1. MgBr
2. H_3O^+
OH
H_3C

O
1. CH_3MgBr
2. H_3O^+
OH
H_3C

O
CH_3
1. BrMg
2. H_3O^+
OH
H_3C

Problem 17.41

Think

In each reaction, is the deuterium a source of D^+ (like a proton), $D:^-$ (like a hydride), or not involved in the reaction mechanism? How are these reaction mechanisms similar to the nucleophilic addition reactions you have already completed? Are proton transfer reactions involved?

Solve

(a) The deuterium in D_2O is a source of D^+ (like a proton) and thus adds to the O^- atom to form an OD alcohol.

Nucleophilic addition

Proton transfer

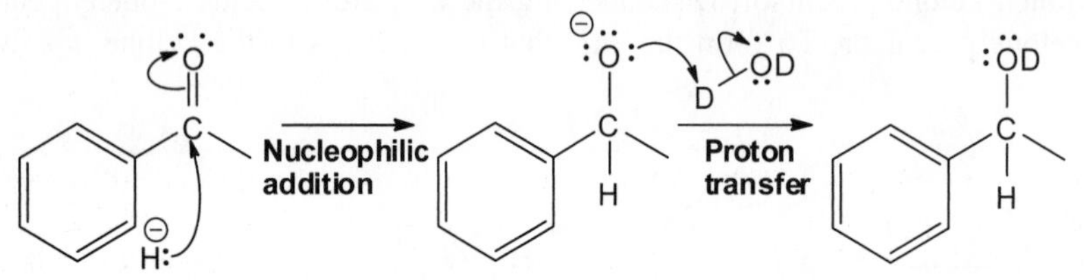

(b) The deuterium in $LiAlD_4$ is a source of $D:^-$ (like a hydride) and thus adds to the carbonyl C atom to form a C–D bond.

Nucleophilic addition

Proton transfer

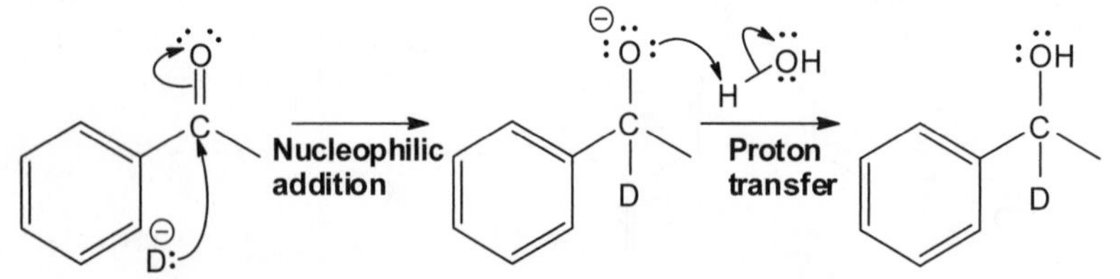

(c) The deuterium in $LiAlD_4$ is a source of $D:^-$ (like a hydride) and thus adds to the carbonyl C atom to form a C–D bond. The deuterium in D_2O is a source of D^+ (like a proton) and thus adds to the O^- atom to form an OD alcohol. The product has two D atoms.

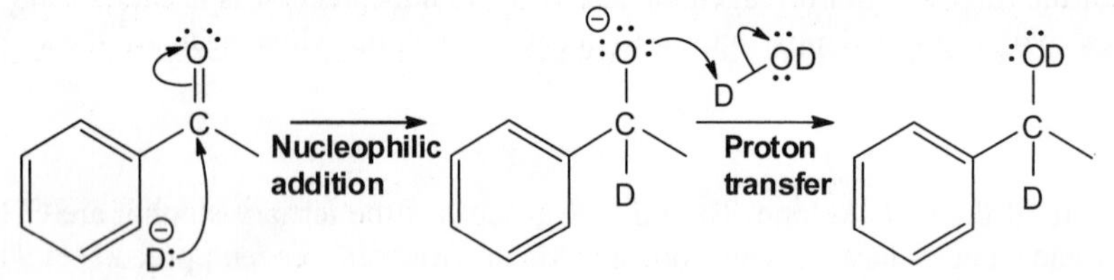

Nucleophilic addition

Proton transfer

(d) The deuterium in D_2O is a source of D^+ (like a proton) and thus adds to the O^- atom to form an OD alcohol.

Nucleophilic addition

Proton transfer

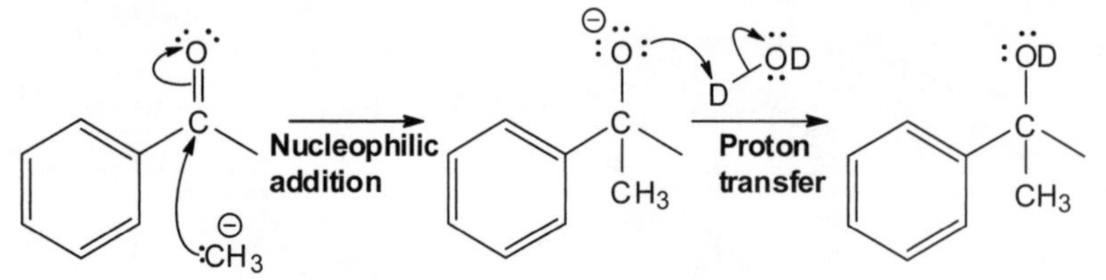

(e) The deuterium in D_3C^- is not directly involved in the reaction mechanism. The C–D bonds are not broken, and the nucleophile is the C atom in D_3C^-.

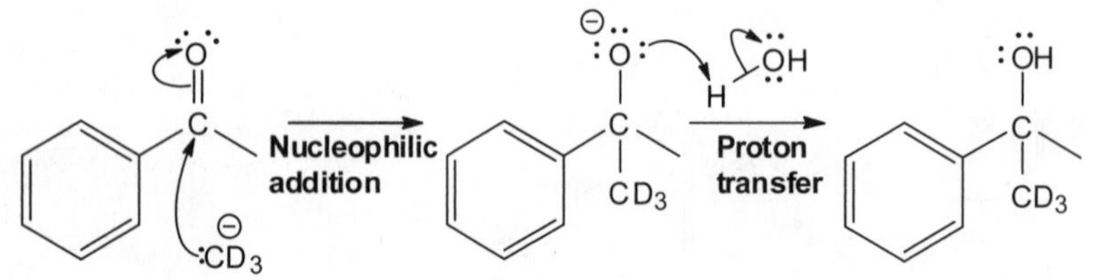

Nucleophilic addition

Proton transfer

(f) The deuterium in D_3C^- is not directly involved in the reaction mechanism. The C–D bonds are not broken, and the nucleophile is the C atom in D_3C^-. The deuterium in D_2O is a source of D^+ (like a proton) and thus adds to the O^- atom to form an OD alcohol.

Nucleophilic addition

Proton transfer

Problem 17.42

Think

In the first step of the reaction, what is electron rich and what is electron poor? What is the substitution product? What is the identity of the sulfonium ylide after deprotonation by the strong base? What percentage of the C atoms in the sulfonium ylide contain the ^{13}C? How likely is it that this ^{13}C atom will be part of the epoxide product?

Solve

The first step in the formation of the sulfonium ylide is shown below. The product is completely symmetrical, with three CH_3 groups. One of them will be deprotonated to make the ylide. The one that is deprotonated will be the one that ends up as part of the epoxide ring. There is a one-third chance of this happening at the $^{13}CH_3$, so one-third of product will contain the labeled carbon.

S_N2

Sulfonium ylide, 33% $^{13}CH_3$

Problem 17.43

Think

What are the electron-rich and electron-poor species in each reaction? Does the reaction mechanism occur via substitution, elimination, or addition? Are proton transfer reactions involved? Be mindful of regio- and stereochemistry where necessary.

Solve

A is formed from the reaction of PBr_3 with the alcohol (two S_N2 reactions), which converts the OH to a Br. **B** is formed from an S_N2 reaction; NC^- is the nucleophile that attacks the electrophilic C–Br. $LiAlH_4$ is a source of hydride, $H:^-$, and undergoes nucleophilic addition to the C of the nitrile. H_2O is a source of protons to form the uncharged amine product, **C**. The final reaction is an example of the Hofmann elimination, in which the NH_2 group is first converted to a $^+N(CH_3)$ leaving group.

PBr_3 → **A**; KCN → **B**; 1. $LiAlH_4$, ether 2. H_2O → **C**; CH_3I (excess), NaOH, Δ

D is formed from a proton transfer reaction to form the strong enolate anion nucleophile that reacts with CH_3CH_2Br in an S_N2 reaction. $NaBH_4$ is a source of hydride that reacts via nucleophilic addition to the aldehyde C. Ethanol is a source of protons to yield **E**, the uncharged alcohol. The same reaction mechanism that formed **D** also forms **F**, but this time the acidic OH is deprotonated to make a strong alkoxide nucleophile.

1. NaH 2. CH_3CH_2Br → **D**; $NaBH_4$, CH_3CH_2OH → **E**; 1. NaH 2. CH_3CH_2Br → **F**

G forms from 1,4-addition of the CH_2=$CHCH_2^-$ followed by proton transfer from NH_4Cl (source of H^+). The enol tautomerizes to the ketone. $NaBH_4$ is a source of hydride that reacts via nucleophilic addition to the ketone C. Ethanol is a source of protons to yield **H**, the uncharged alcohol. **I** is formed from the reaction of PBr_3 with the alcohol (two S_N2 reactions) to convert OH to Br. The last reaction is an E2 (strong bulky base in the presence of heat). There are two adjacent protons, and thus **J** has two isomers possible. The major product is the top one, because the base is very bulky and will attack the least sterically hindered C–H.

1. $(H_2C{=}CH{-}CH_2)_2CuLi$; 2. NH_4Cl → **G**; $NaBH_4$, CH_3CH_2OH → **H**; PBr_3 → **I**; $(CH_3)_3CONa$, Δ → **J**

Problem 17.44

Think

What are the electron-rich and electron-poor species in each reaction? Does the reaction mechanism occur via substitution, elimination, or addition? Are proton transfer reactions involved? Be mindful of regio- and stereochemistry where necessary.

Solve

The imine shown was synthesized from the reaction of a nitrile, **K**, with a Grignard reagent, followed by proton transfer using a weak acid. $LiAlH_4$ is a source of hydride, H:$^-$, and undergoes nucleophilic addition to the carbon of the imine. H_2O is a source of protons to form the uncharged amine product, **L**. Heating **L** with H_3O^+ hydrolyzes the ether to produce a phenol, **M**. NaOH deprotonates the alcohol to form the alkoxide, which then undergoes S_N2 with CH_3CH_2Br to form the ether, **N**.

K → 1. PhMgBr; 2. CH_3OH → (imine, NH) → 1. $LiAlH_4$, ether; 2. H_2O → **L** → $H_3O^{\oplus}$, Δ → **M** (HO) → 1. NaOH; 2. CH_3CH_2Br → **N** (H_3CH_2CO)

$LiAlH_4$ is a source of hydride, H:$^-$, and undergoes nucleophilic addition to the carbon of the ketone. H_2O is a source of protons to form the uncharged alcohol product, **O**. **P** is formed from the reaction of PBr_3 with the alcohol (two S_N2 reactions), converting OH to Br. **Q** is the phosphorous ylide that forms from the S_N2 reaction of the alkybromide and PPh_3. A proton transfer reaction follows to deprotonate the C–H and form a carbanion. The phosphorous ylide reacts with formaldehyde in a Wittig reaction to form the alkene product, **R**.

1. $LiAlH_4$, ether; 2. H_2O → **O** → PBr_3 → **P** → 1. $P(C_6H_5)_3$; 2. Bu-Li → **Q** ($^{\oplus}PPh_3$) → $H_2C{=}O$ → **R** (CH_2)

S is formed from the reaction of PBr_3 with the alcohol (two S_N2 reactions), converting OH to Br. **T** is formed from an S_N2 reaction; NC^- is the nucleophile that attacks the electrophilic C–Br. The CH_3^- (from the Grignard reagent CH_3MgBr) adds via nucleophilic addition to the C of the nitrile. A proton transfer reaction follows to form the uncharged imine product, **U**. $LiAlH_4$ is a source of hydride, $H:^-$, and undergoes nucleophilic addition to the carbon of the imine. H_2O is a source of protons to form the uncharged amine product, **V**. The final reaction is an example of the Hofmann elimination, which first converts the NH_2 group to a $^+N(CH_3)_3$ leaving group. Hofmann elimination produces the less alkyl-substituted alkene.

OH PBr3 Br NaCN CN 1. CH3MgBr, ether 2. CH3OH C NH CH3 1. LiAlH4, ether 2. H2O

S T U

H CH3 C NH2 CH3I (excess) NaOH, Δ

V

W forms from 1,4-addition of the cyclohexyl group ($C_6H_{11}^-$), followed by proton transfer from NH_4Cl (the source of H^+). The enol tautomerizes to the ketone. $(CH_3)_2S^+–^-C(CH_3)_2$ is a sulfonium ylide that yields the epoxide **X** when reacted with ketone **W**. In the last step below, the only plausible reaction is an S_N2 followed by a proton transfer to yield **Y**.

O 1. Li Cu 2. NH4Cl O H3C CH3 C S H3C CH3 O NaOCH2CH3 CH3CH2OH HO OCH2CH3

W X Y

Problem 17.45

Think

What is the structure of acetone? What functional group is present? How does this functional group react with alkyllithium and Grignard reagents?

Solve

Acetone is a ketone and the carbonyl C of acetone can be attacked by an alkyllithium or Grignard reagent, initiating a nucleophilic addition reaction. Students who dry their glassware with acetone before running a Grignard reaction frequently find this out the hard way.

O :R⁻ Nucleophilic addition :O:⁻ R

Acetone

Problem 17.46

Think

What bonds broke and formed in each reaction? What new atoms are present? Did the reaction require an acid workup step?

Solve

The first reaction forms a new C–C bond, and a butyl group is added. Therefore, the reagent **A** is either butylmagnesium bromide ($CH_3CH_2CH_2CH_2MgBr$) or butyllithium ($CH_3CH_2CH_2CH_2Li$). The uncharged alcohol is the final product and, therefore, an acid workup step is necessary. The second reaction requires acid and heat to promote an E1 reaction. The (*E*) isomer is the major product.

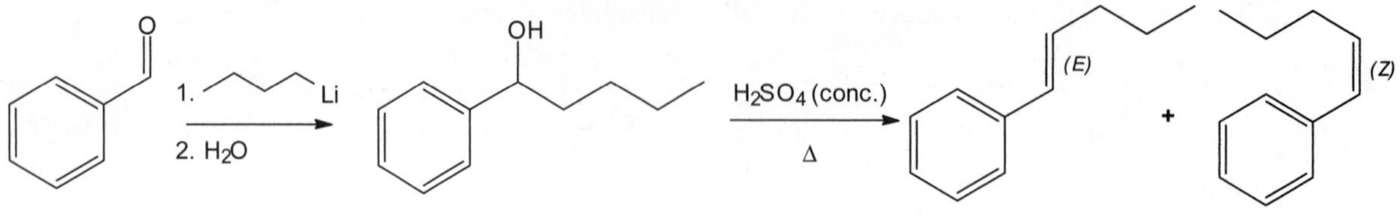

1. Li; 2. H_2O — OH — H_2SO_4 (conc.) Δ — (E) + (Z)

Problem 17.47

Think

Which compound is a stronger reducing agent? Which reducing agent has a larger electronegativity difference? Which reagent, therefore, would be attracted to the carbonyl carbon more strongly?

Solve

$LiAlH_4$ is a stronger reducing agent due to the larger difference in electronegativity of Al and H compared to B and H. The greater concentration of charge on H makes $LiAlH_4$ more attracted to the partial positive charge on the carbonyl C, which has a larger partial positive charge than the β carbon. In the same way that RLi compounds are more reactive and more selective for 1,2-addition, $LiAlH_4$ will also be more selective for 1,2-addition compared to $NaBH_4$.

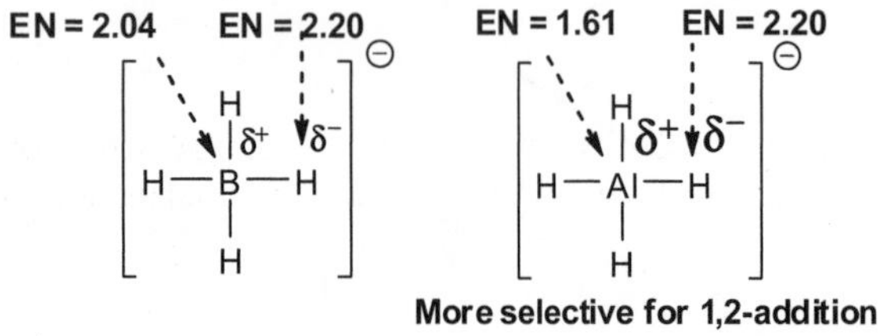

EN = 2.04 EN = 2.20 EN = 1.61 EN = 2.20

δ^+ δ^- H—B—H δ^+ δ^- H—Al—H

More selective for 1,2-addition

Problem 17.48

Think

What is the product of a conjugate addition reaction? How does $NaBH_4$ react with this product? How does a lithium dialkylcuprate react with this product? Why is there a difference?

Solve

In the product of conjugate addition, a polar π bond remains. And $NaBH_4$ adds hydride to such bonds via direct addition. A lithium dialkylcuprate, on the other hand, doesn't react with polar π bonds via direct addition, so no second reaction can take place.

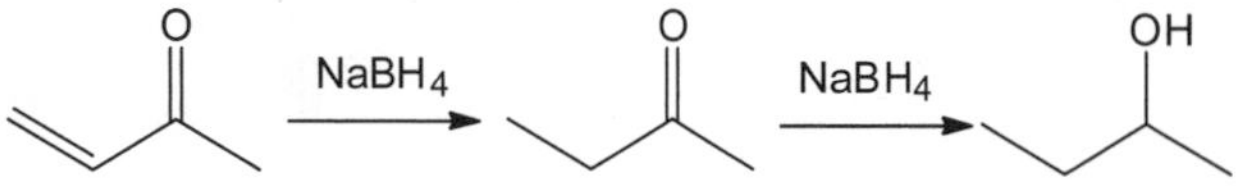

$NaBH_4$ $NaBH_4$ OH

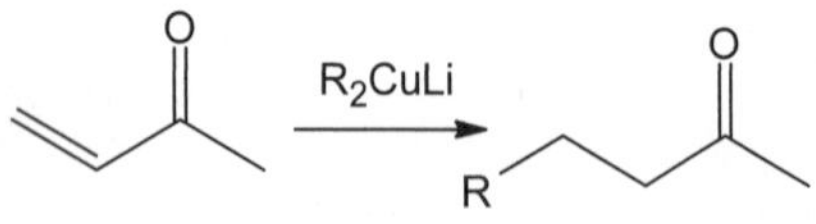

R_2CuLi R

Problem 17.49

Think

What is the product of direct addition? What is the product of conjugate addition? If direct addition takes place first, can the product that is given be produced? If conjugate addition takes place first, can the product that is given be produced?

Solve

$NaBH_4$ is a source of hydride and CH_3CH_2OH is a source of protons. If direct addition were to take place first, an allylic alcohol would be produced and no further reaction would take place, so the given product couldn't be produced. On the other hand, the conjugate addition of $NaBH_4$ to but-3-en-2-one (α,β-unsaturated ketone) results in a ketone, which can be reduced a second time to the alcohol.

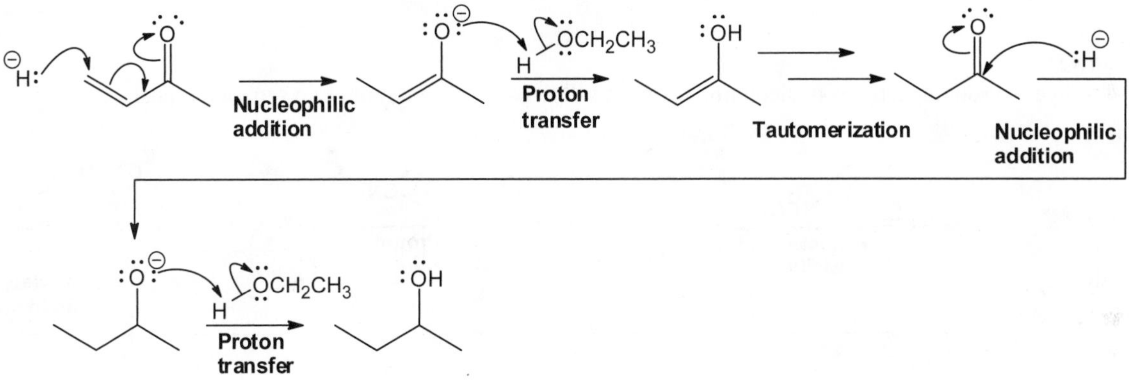

Nucleophilic addition

Proton transfer

Tautomerization

Nucleophilic addition

Proton transfer

Problem 17.50

Think

What is the structure of cyclopropanone? What is the structure of the hydrate of cyclopropanone? Compare the ring strain of the cyclopropanone to the hydrate of cylcopropanone. What is the hybridization and angle of the C atom in the C=O and $C(OH)_2$?

Solve

The hydrate is shown below, alongside cyclopropanone itself. The reason the hydrate is stable is that in the hydrate, there is much less ring strain than in the reactant. The interior angle of the three-membered ring is 60°. The ideal angle at the carbonyl C is 120°, whereas that for the hydrate carbon is 109.5°. The angles, therefore, are better matched in the hydrate.

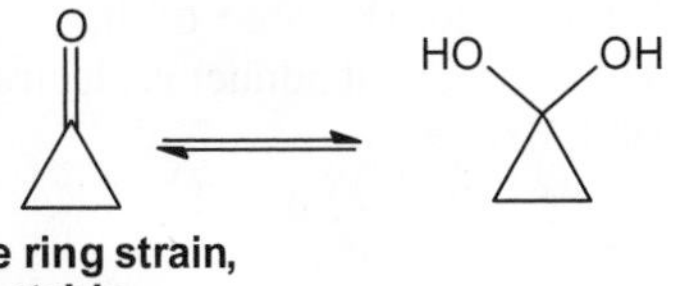

More ring strain, less stable

Problem 17.51

Think

Is the C atom in CS_2 electron rich or electron poor? What reaction occurs between ROH and NaOH? What is the identity of the nucleophile in the nucleophilic addition reaction? How can the xanthate salt react with RBr? What is electron rich and electron poor? What elementary mechanism step occurs?

Solve

HO^- is used to convert the weak ROH nucleophile into a strong RO^- nucleophile, as shown in the first step on the next page. RO^- attacks CS_2 in a nucleophilic addition, as shown in the second step. Once the nucleophilic addition is complete, the xanthate salt can do an S_N2 reaction on an alkyl halide (if present) to make the xanthate ester. See the mechanism on the next page.

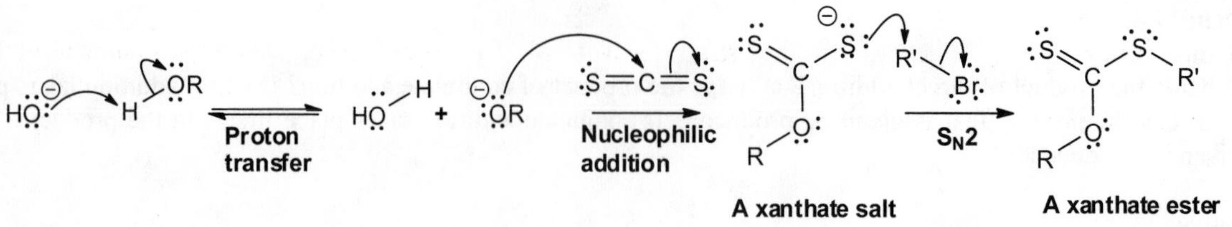

Proton transfer

Nucleophilic addition

S_N2

A xanthate salt

A xanthate ester

Problem 17.52

Think

Is the C atom in the nitrile electron rich or electron poor? What is the identity of the nucleophile in the nucleophilic addition reaction? How many times does the addition reaction occur? Are proton transfer reactions involved?

Solve

The cyano group undergoes nucleophilic addition followed by protonation, two separate times.

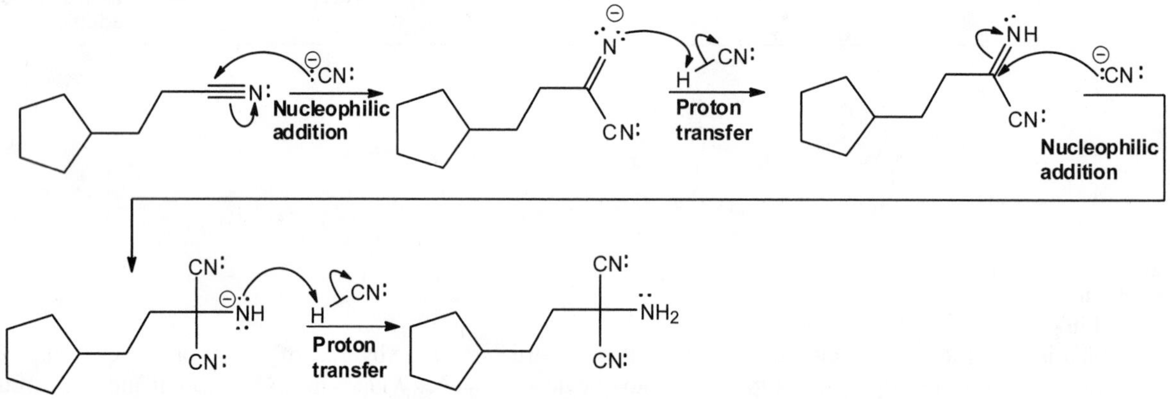

Nucleophilic addition

Proton transfer

Nucleophilic addition

Proton transfer

Problem 17.53

Think

What is the Lewis structure of SO_2? Is the S atom in SO_2 electron rich or electron poor? What is the identity of the nucleophile in the nucleophilic addition reaction? Is the C atom in CH_3Br electron rich or electron poor? What is the identity of the nucleophile in the second reaction?

Solve

Sulfur has a lone pair of electrons. Despite this, sulfur is sufficiently electron poor (because of the neighboring electronegative O atoms) that it can be attacked by a Grignard reagent. The subsequent adduct is electron rich; it can be nucleophilic and reacts with the electrophilic CH_3Br in an S_N2 reaction.

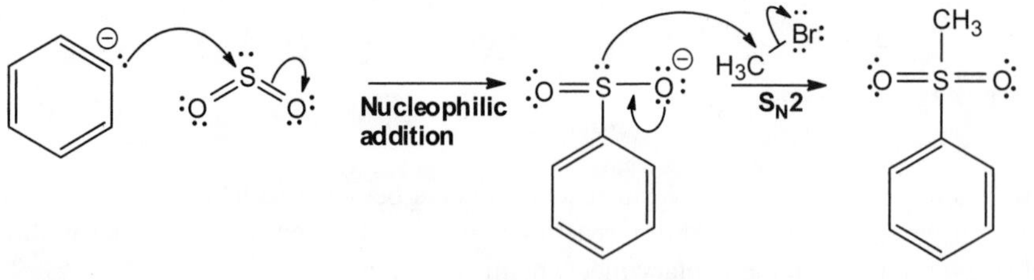

Nucleophilic addition

S_N2

Problem 17.54

Think

Does a nucleophile with a resonance-delocalized negative charge undergo 1,2- or 1,4-addition? Are proton transfer reactions involved? From the intermediate shown, what is electron rich and what is electron poor? Can a Wittig reaction occur in the intermediate to give the final product? At which C does the nucleophilic addition occur?

Solve

To make the intermediate, the resonance structure of the Wittig reagent adds to the α,β-unsaturated carbonyl group in a conjugate addition fashion, followed by some proton transfers. Then an intramolecular Wittig reaction takes place. The mechanism is shown below.

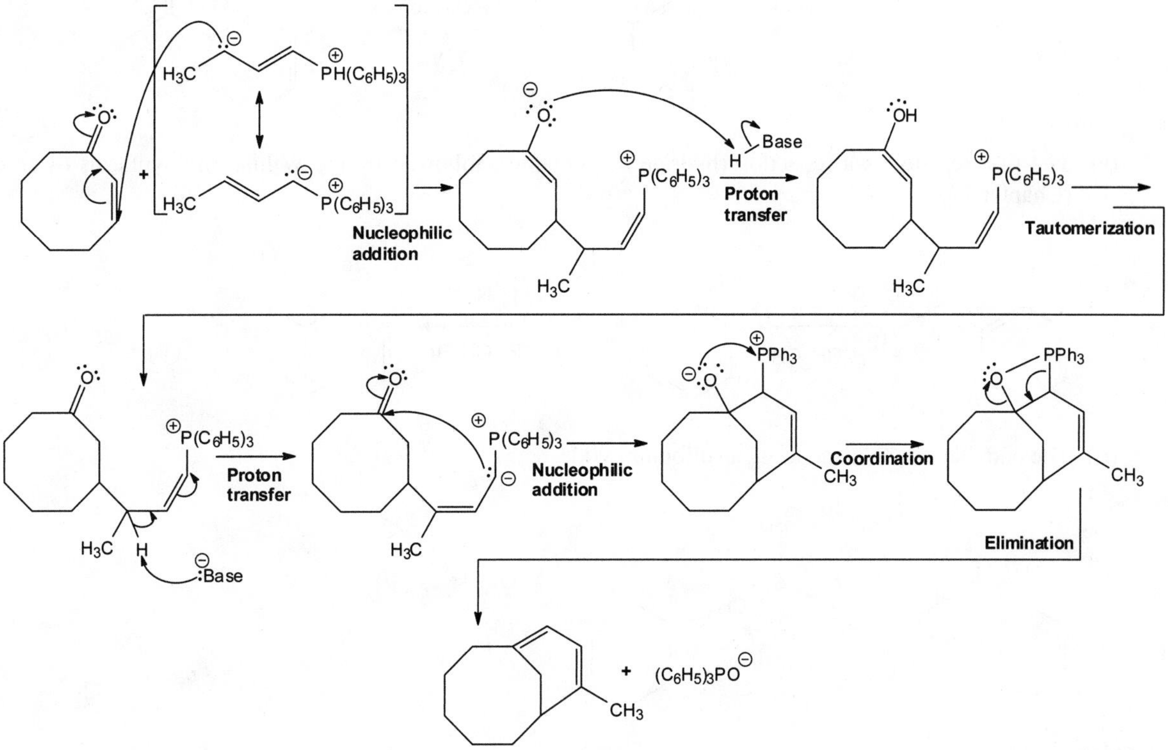

Problem 17.55

Think

What new atoms are present in the products given? What is the identity of the nucleophile that would be necessary to participate in each nucleophilic addition reaction with phenylethanone to produce the products shown in **(a)**–**(d)**? Are proton transfer reactions involved? What other reagents are necessary?

Solve

(a) The first synthesis is a reduction (hydride nucleophile), followed by replacement of O with a halogen (Chapter 10) using PBr_3 (two S_N2 steps).

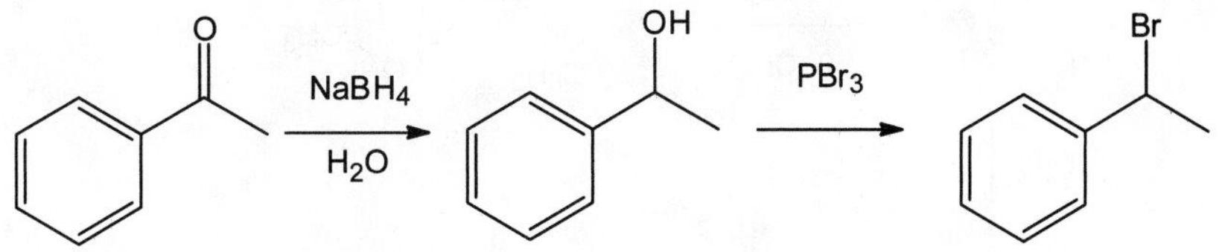

(b) The product can be made by dehydration of an alcohol (Chapter 9), which can be made via a Grignard or alkyllithium reaction (adds on H_3C^- as the nucleophile).

1. CH_3Li
2. H_3O^+

HO

H_3PO_4 (conc.)

Δ

(c) The first reaction is a reduction (hydride nucleophile), followed by the Williamson synthesis of an ether (Chapter 10).

$NaBH_4$

CH_3CH_2OH

OH

1. NaH
2. CH_3Br

(d) The epoxide can be made using a sulfonium ylide.

$^{\ominus}CH_2$ — $^{\oplus}S(CH_3)$ — H_3C

CH_2

Problem 17.56

Think

What new atoms are present in the products given? How can you react a Grignard reagent to produce a carboxyl group? How can you convert a carboxylic acid into an ester?

Solve

A Grignard addition to CO_2 (followed by the usual acid workup) makes a carboxylic acid. Next, a base can generate the moderately strong carboxylate nucleophile, which can attack CH_3Br in an S_N2 step. Alternatively, diazomethane can be used to make the methyl ester (Chapter 10).

C_6H_5—MgBr → (1. CO_2 (s); 2. H_3O^+) → $C_6H_5C(=O)OH$ → (1. NaOH; 2. CH_3Br) → $C_6H_5C(=O)OCH_3$

$C_6H_5C(=O)OH$ → (CH_2N_2) → $C_6H_5C(=O)OCH_3$

Problem 17.57

Think

What new atoms are present in the product given? Which C–C bond needs to be formed? Can the C–OH be produced from a C=O? How can you form a C–C bond at the α carbon of a ketone?

Solve

The five-carbon backbone from the original molecule is shown below. So we need to add a two-carbon piece. Deprotonating the α C atom with a strong base, then performing an alkylation, are sufficient.

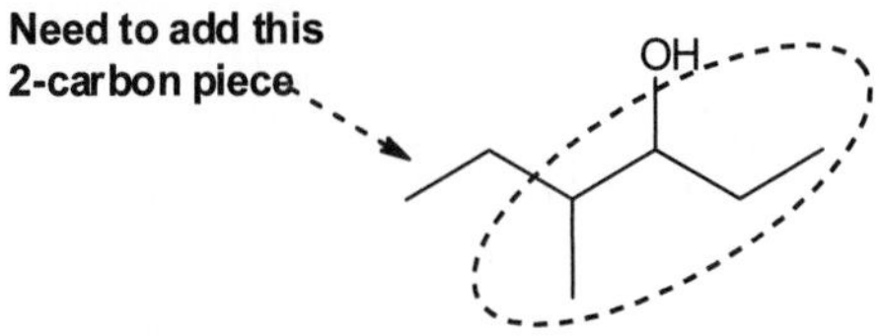

Need to add this 2-carbon piece

OH

We can make the α C nucleophilic by treating the ketone with a strong base like NaH, and attach the two-carbon piece through an S_N2 reaction. The C=O bond can be reduced to the alcohol using $NaBH_4$.

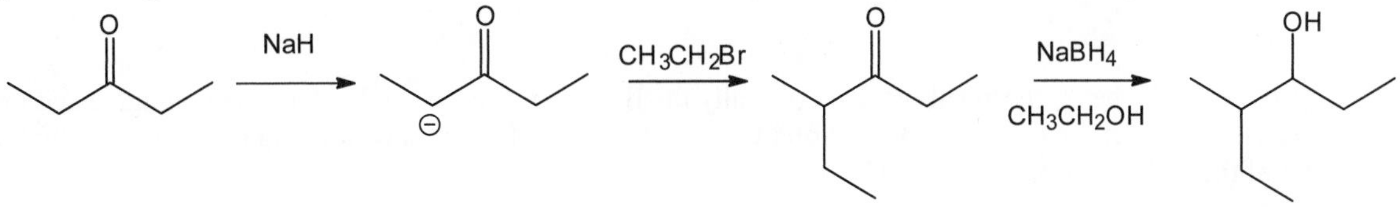

NaH; CH_3CH_2Br; $NaBH_4$, CH_3CH_2OH

Problem 17.58

Think

What new atoms are present in the product given? What C–C bonds need to be formed? Can the C=C be produced from a C=O using a Wittig reaction? How can you add an alkyl group to the β carbon of an α,β-unsaturated ketone? How can you add a methyl group to an α carbon?

Solve

As shown below, this synthesis calls for an α alkylation, a Wittig reaction, and a conjugate addition. The α alkylation can take place first, which would still leave the α,β-unsaturated carbonyl group for subsequent conjugate addition. Then conjugate addition can take place, which still leaves the ketone for a final Wittig reaction.

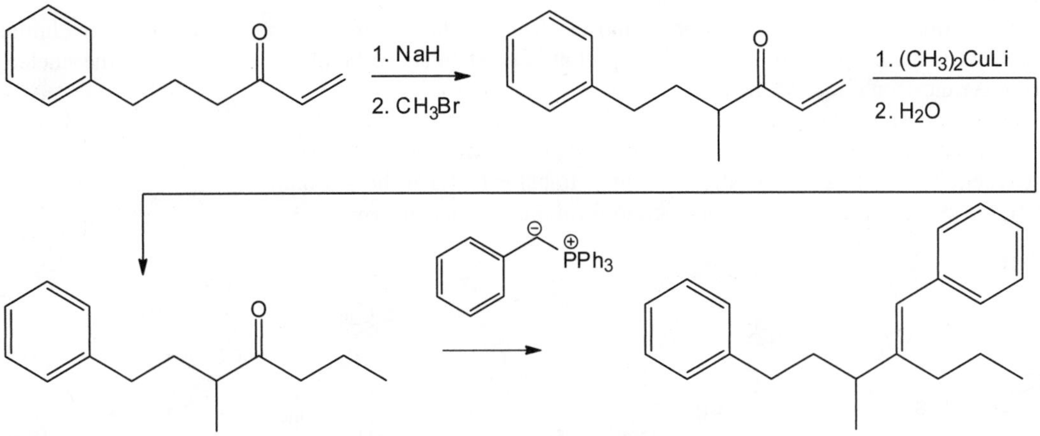

1. NaH 2. CH_3Br; 1. $(CH_3)_2CuLi$ 2. H_2O; PPh_3

Problem 17.59

Think

What C–C bonds must be formed? Does that bond formation require direct addition or conjugate addition? How can you remove an O atom and leave a C=C bond?

Solve

We can spot the original carbon backbone, which is circled below.

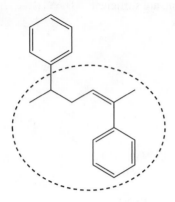

We need to add the benzene ring at what is originally the β C atom, which can be done via conjugate addition—in this case, Ph_2CuLi. That leaves a C=O bond that can be reduced to the alcohol, and subsequent dehydration gets rid of the OH group, leaving a C=C double bond.

1. Ph_2CuLi
2. H_2O

1. $NaBH_4$
2. CH_3CH_2OH

H_3PO_4 (conc.)
Δ

Problem 17.60

Think

In the first reaction, identify the nucleophile and electrophile. Is this an addition, substitution, or elimination reaction? What is the purpose of NaH? Which proton is most likely to be abstracted? What is the nucleophile that adds to the electrophilic C of the imine?

Solve

The first reaction is an S_N2 followed by a proton transfer to form the sulfonium ylide. The sulfonium ylide reacts with the imine in a two-step mechanism to form the product shown.

S_N2 — Proton transfer — Sulfonium ylide — Nucleophilic addition — S_N2

Problem 17.61

Think

In the first reaction, identify the nucleophile and electrophile. Is this an addition, substitution, or elimination reaction? What is the purpose of NaH? Which proton is most likely to be abstracted? Does the nucleophile react with the α,β-unsaturated ketone via 1,2- or 1,4-addition? What is the identity of the nucleophile?

Solve

The first reaction is an S_N2, followed by a proton transfer to form a negative charge on the C atom. The carbanion is the nucleophile that adds to the α,β-unsaturated ketone in a 1,4-addition. The intermediate rearranges to form the cyclopropane ring and regenerate the starting material.

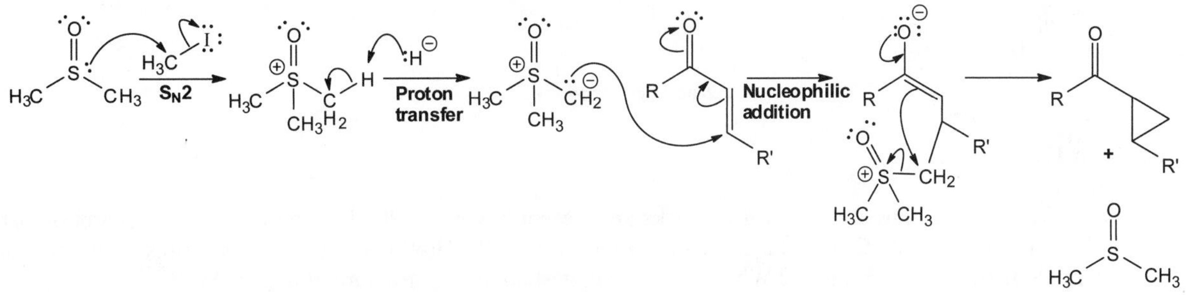

Problem 17.62

Think

What is the mechanism for each reaction? What steps are the same and what step is different? What is the bond strength of the P–O bond? What is the bond strength of the S–O bond?

Solve

The two mechanisms are given below. The first nucleophilic addition step is the same. However, the phosphorous ylide mechanism is a coordination step to form the four-membered ring followed by an elimination step to form the C=C and the P–O bond (bond strength: 537 kJ/mol). The second step in the sulfonium ylide mechanism is an S_N2 to form the $(CH_3)_2S$ and an epoxide. The formation of the double bond doesn't take place because it would require the formation of a S–O bond (362 kJ/mol), which is 175 kJ/mol weaker than a P–O bond. The difference in mechanism is due to the bond strength differences of the P–O and S–O bond.

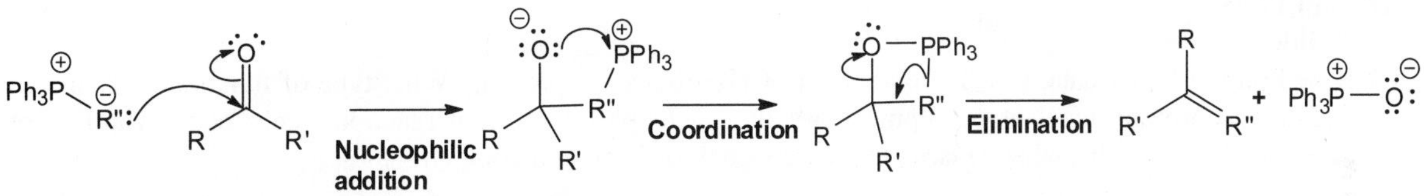

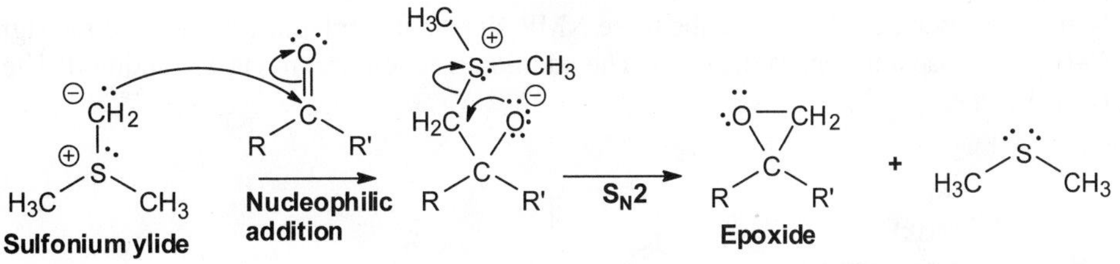

Problem 17.63

Think

What functional group is identified from the broad peak from 3200–3600 cm^{-1}? If the peak at 1650 cm^{-1} is not present in the IR, is the carbonyl group still present in the product? Did the nucleophilic addition occur via 1,2-addition or 1,4-addition?

Solve

(a) The broad peak from 3200–3600 cm^{-1} suggests the presence of an alcohol OH group, and the absence of a peak at 1650 cm^{-1} suggests that the conjugated C=O is no longer present. This leads to the conclusion that the addition reaction occurred via 1,2-addition. 1,4-Addition leads to an enol that tautomerizes back to the aldehyde.

(b) If 1,2-addition occurs, the nucleophile reacts irreversibly.

1. HC≡CNa
2. NH_4Cl

1,2-addition product

Problem 17.64

Think

What functional groups are absent if no peaks are present beyond 3000 cm^{-1} and there is no absorption band at 1700 cm^{-1}? How many C atoms are present in the ^{13}C NMR? How many C atoms are present in the starting material? Is there any symmetry? What does this suggest about the structure of the product?

Solve

The absence of a peak above 3000 cm^{-1} suggests that there are no H–O bonds and no sp^2 or sp C–H atoms. The absence of a peak at 1700 cm^{-1} suggests that there is no carbonyl group. There are five C atoms in the starting material and only three C signals in the ^{13}C NMR. This suggests that there is symmetry in the compound and that is the result of a ring. Therefore, an intramolecular S_N2 reaction occurred after the nucleophilic addition reaction.

Nucleophilic addition

S_N2

Problem 17.65

Think

Based on the NMR data, how many different CH_3 groups are present? What type of functional group can give rise to the proton signal at 6.1 ppm? Does the nucleophilic addition reaction likely occur via 1,2- or 1,4-addition? From the IR, what causes the strong signals at 1650 and 1600 cm^{-1}?

Solve

There are three CH_3 groups present, evidenced by the three NMR signals that integrate to 3H, and the signal at 6.1 ppm is the sp^2 C–H that is also adjacent to the C=O. The addition reaction occurs via 1,4-addition. The enol tautomerizes back to the ketone.

1. $(CH_3)_2CuLi$
2. NH_4Cl

Problem 17.66

Think

What is the mechanism by which a Wittig reagent is formed? Can a similar reaction occur here? What is the identity of the nucleophilic C atom in the Wittig reagent? Where is the electrophilic C atom? Are these atoms on the same compound?

Solve

Once the Wittig reagent is produced, an internal Wittig reaction takes place because the nucleophile and electrophile are on the same compound.

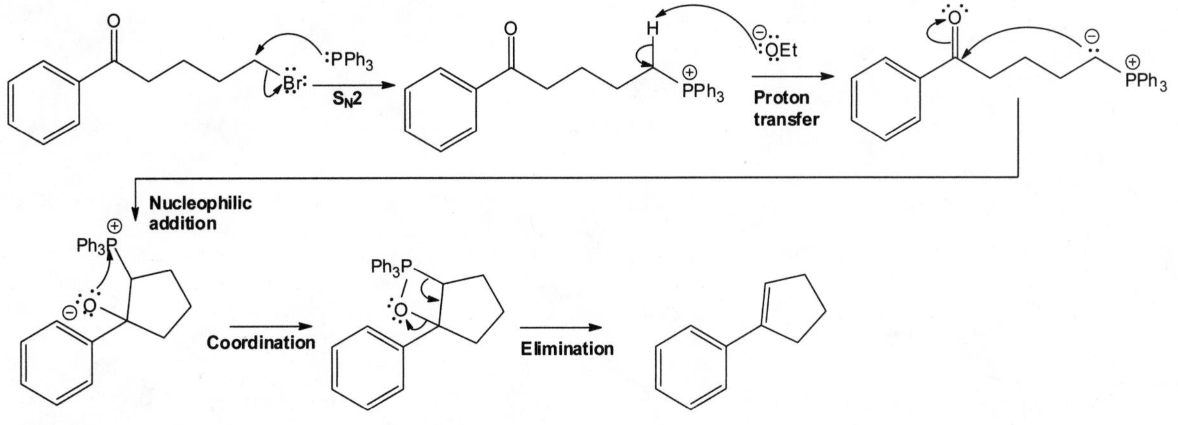

:PPh3
S_N2
Proton transfer
Nucleophilic addition
Coordination
Elimination

Problem 17.67

Think

What is the identity of the ylide? Is the compound aromatic, nonaromatic, or antiaromatic? How does this affect the ability of the anion to behave as a nucleophile?

Solve

The ylide formation is straightforward; however, close examination of the ylide shows a conjugated ring that has six π electrons (a Hückel number) and is aromatic. The large aromatic stability it has, therefore, makes it unreactive—nucleophilic addition would destroy the aromaticity.

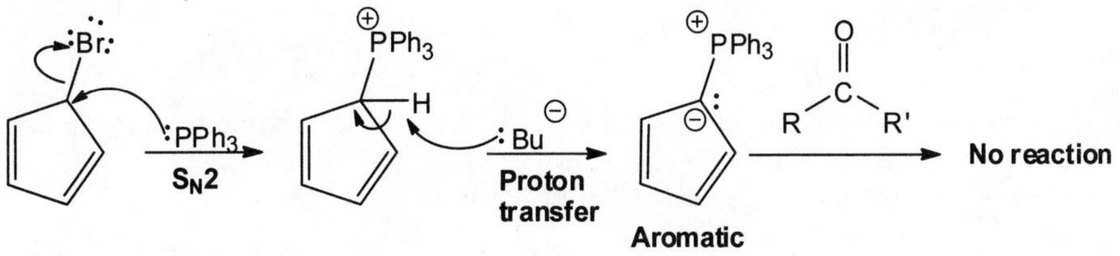

S_N2
Proton transfer
Aromatic
No reaction

CHAPTER 18 | Nucleophilic Addition to Polar π Bonds 2: Addition of Weak Nucleophiles and Acid and Base Catalysis

Your Turn Exercises

Your Turn 18.1

Think

Count the *number* of formal charges present in each step for all reagents involved in the reaction.

Solve

In Equation 18-3, the first stage has no charges and the second stage (after the first reaction step) has two charges. Stage 3 has one charge and stage 4 has no charges. In Equation 18-4, each of the stages has one charge. In Equation 18-5, each of the stages has one charge.

Slow

1. Nucleophilic addition

2. Proton transfer

0 charges

2 charges

3. Proton transfer

1 charge

0 charges

Base:

1 charge

1. Proton transfer

2. Nucleophilic addition

3. Proton transfer

1 charge

1 charge

1 charge

1 charge

1. Proton transfer

2. Nucleophilic addition

3. Proton transfer

1 charge

1 charge

1 charge

Your Turn 18.2

Think

What type of species (electron rich or electron poor) does a nucleophile attack? In which species is that type of charge greater on the C≡N carbon?

Solve

The positively charged species is activated toward nucleophilic attack, because nucleophiles attack electron-poor species. It has a resonance structure that puts a positive charge on the carbon, making the C more susceptible to nucleophilic attack.

$$\left[H_3C{-}CH_2{-}CH_2{-}CH_2{-}C{\equiv}\overset{\oplus}{N}H \longleftrightarrow H_3C{-}CH_2{-}CH_2{-}CH_2{-}\overset{\oplus}{C}{=}\ddot{N}H \right]$$

Your Turn 18.3

Think

Consult Table 6-1. What is the pK_a of HCN? Of H_3O^+? What does a lower pK_a indicate about acid strength and the direction of the reaction?

Solve

HCN has a pK_a = 9.2 and H_3O^+ has a pK_a = −1.7. The stronger acid (lower pK_a) is on the product side, so the reactants are favored.

$$N{\equiv}C{-}H + H_2\ddot{O}: \rightleftharpoons N{\equiv}C:^{\ominus} + H{-}\overset{\oplus}{O}H_2$$

pK_a = 9.2 pK_a = −1.7

Your Turn 18.4

Think

Consult Table 6-1. What is the pK_a of HCN? Of H_2O? What does a lower pK_a indicate about acid strength and the direction of the reaction? How many pK_a units separate the reactant and product acids? How does that translate into a K_{eq}?

Solve

HCN has a pK_a = 9.2 and H_2O has a pK_a = 15.7. The stronger acid (lower pK_a) is on the reactant side, so the products are favored. The difference in pK_a between the reactant and product acids is: 15.7 − 9.2 = 6.5, so the product side is favored by $10^{6.5}$, making $K_{eq} = 3.2 \times 10^6$. This large K_{eq} means that the acid will be deprotonated quantitatively.

$$N{\equiv}C{-}H + :\ddot{O}H^{\ominus} \rightleftharpoons N{\equiv}C:^{\ominus} + H_2O$$

pK_a = 9.2 pK_a = 15.7

Your Turn 18.5

Think

Consult Table 1-6 to identify what defines a hemiacetal and acetal. Locate these two functional groups in the mechanism in Equation 18-13.

Solve

A hemiacetal is characterized by $R_2C(OH)(OR)$ or $RCH(OH)(OR)$ grouping. An acetal is characterized by $R_2C(OR)_2$ or $RCH(OR)_2$ grouping. The hemiacetal forms after Step 3 and the acetal forms after Step 7 (dashed ovals on the next page).

1. Proton transfer
2. Nucleophilic addition
3. Proton transfer

Hemiacetal

Good H_2O leaving group

4. Proton transfer
5. Heterolysis
6. Coordination

Resonance-stabilized carbocation

7. Proton transfer

Acetal

Your Turn 18.6

Think

Consult Section 1.9 to review how to calculate oxidation states. Compare the electronegativity of C versus H and C versus O. Which one is more electronegative? How are covalently bonded electrons assigned when the atoms bonded together have different electronegativities? The same electronegativities?

Solve

In a given covalent bond, all electrons are assigned to the more electronegative atom. If the atoms are identical, the electrons are split up evenly. Hydrogen is *less* electronegative compared to C, and O is *more* electronegative compared to C. In 1-phenyl-1-propanone, the ketone C is assigned one electron from each C–C bond and no electrons from the C=O bond, for a total of two valence electrons. That C has two fewer than an uncharged isolated C, giving it an oxidation state of +2. In propyl benzene, the C is assigned two electrons from each C–H bond and one electron from each C–C bond, giving it a total of six valence electrons. This is two more valence electrons than in an uncharged isolated C, giving that C atom an oxidation state of −2.

Wolff–Kishner reduction

Oxidation state = +2

The carbonyl group is reduced to a CH_2 group.

Oxidation state = −2

$H_2N\text{–}NH_2$, KOH

H_2O/triethylene glycol

Δ

\+ $N_2(g)$ + H_2O

82%

1-Phenyl-1-propanone

Propylbenzene

Your Turn 18.7

Think

Review the functional groups in Table 1-6, if necessary. How many carbons away from the C=O is the α carbon located? The β carbon?

Solve

The aldehyde and alcohol groups are noted below by dashed circles. The α carbon is located one carbon away from the C=O and the β carbon is located two carbons away.

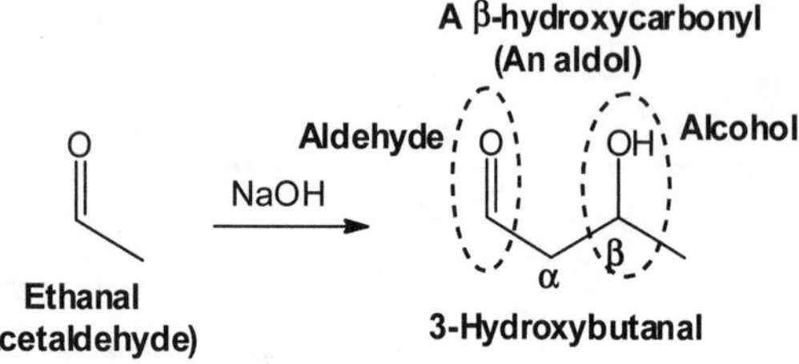

Your Turn 18.8

Think

Consult Table 6-1 to determine the pK_a values and Equation 6-11 to calculate K_{eq}. What does a lower pK_a indicate about acid strength and the side of the reaction that is favored? What does the magnitude of K_{eq} indicate about the extent to which that side of the reaction is favored?

Solve

pK_a ($CH_3CH{=}O$) = 20; pK_a (H_2O) = 15.7; $K_{eq} = 10^{\Delta pK_a} = 10^{15.7-20} = 5 \times 10^{-5}$

K_{eq} <<1, so the reactants are favored heavily.

Your Turn 18.9

Think

Label the α carbons on both butanal and the three compounds listed. Do those C atoms have attached protons? If a compound has no α protons, how is this advantageous when performing a crossed aldol reaction?

Solve

Only compound **A** has no α protons. This is helpful in performing a crossed aldol reaction because the compound without α protons cannot form an enolate ion.

A B C

Therefore, this compound can act as an electrophile only, producing a single crossed aldol product with butanal, as shown below:

HO O

Crossed aldol product with A and butanal

Compounds **B** and **C** have α protons, so they can behave as nucleophiles or electrophiles, and can each produce two different crossed aldol products with butanal.

Crossed aldol products with B and butanal

Crossed aldol products with C and butanal

Your Turn 18.10

Think

Review Section 10.3. Which base (NaOH or LDA) favors deprotonating the more highly substituted α carbon? Is this the kinetic or thermodynamic enolate anion?

Solve

The thermodynamic enolate anion is formed by deprotonating the more highly substituted α carbon using NaOH. The kinetic enolate is formed by deprotonating the less highly substituted α carbon using LDA (a very strong base).

NaOH

LDA

Your Turn 18.11

Think

What functional groups characterize an aldol condensation reaction product? What should the relative locations be for those functional groups? Review Section 18.5.

Solve

The product of an aldol condensation is an α,β-unsaturated carbonyl. This is what we see in the product, where the carbonyl group is conjugated to a C=C.

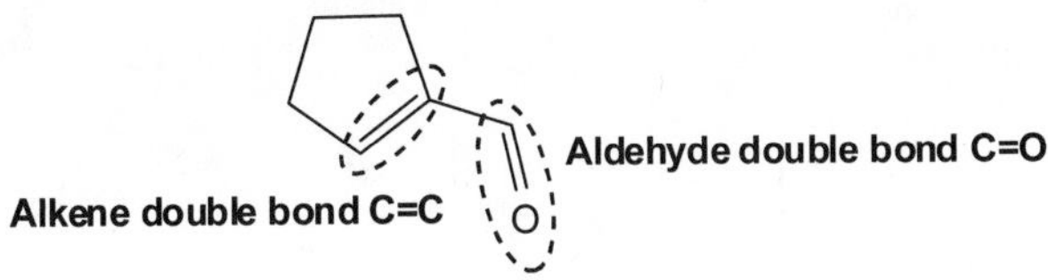

Chapter Problems

Problem 18.2

Think

What nucleophiles are present under acidic conditions? What is the total number of charges before and after the nucleophilic addition step in the mechanism?

Solve

Under acidic conditions, water acts as the nucleophile *after* the ketone O atom is protonated to increase the electrophilic nature of the carbonyl carbon.

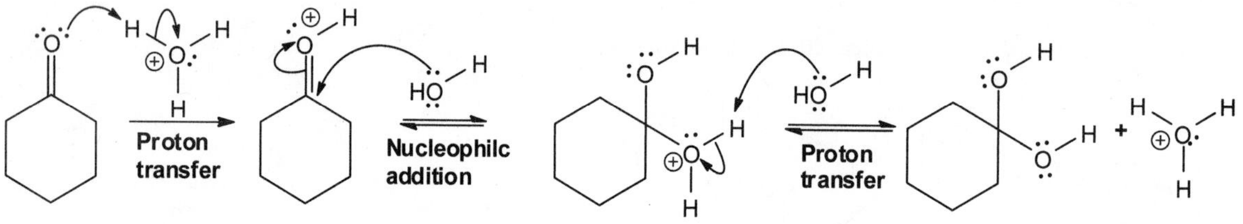

Proton transfer

Nucleophilc addition

Proton transfer

Under neutral conditions, the nucleophilic addition step increases the total number of charges by two, whereas the total number of charges remains the same under acidic conditions; before and after nucleophilic addition there is a single positive charge. Thus, nucleophilic addition is faster under acidic conditions.

Problem 18.3

Think

What nucleophiles are present under basic conditions? Acidic conditions? In what order do the nucleophilic addition and proton transfer reactions need to occur to avoid incompatible species in solution? How can the acid and base catalysts be regenerated?

Solve

(a) In basic conditions, the HO^- deprotonates the phenol O–H to form the phenoxide anion, which acts as the nucleophile. In basic conditions, the nucleophilic addition to the C=O occurs before the proton transfer reaction to the carbonyl O atom. Under basic conditions, no strong acids should appear.

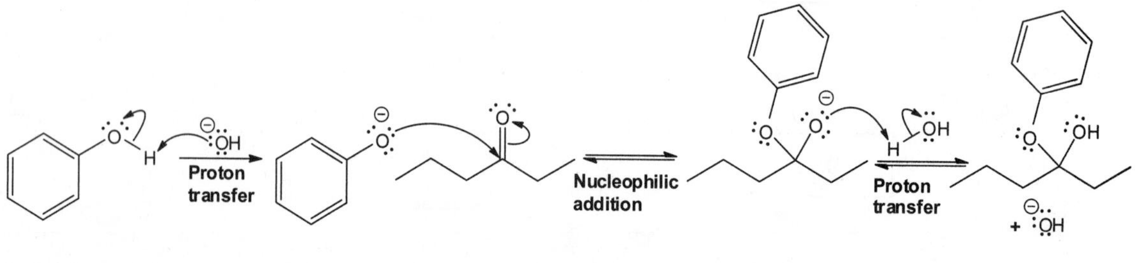

Proton transfer

Nucleophilic addition

Proton transfer

(b) In acidic conditions, phenol acts as the nucleophile *after* the ketone O atom is protonated to increase the electrophilic nature of the carbonyl C. Under acidic conditions, no strong bases should appear.

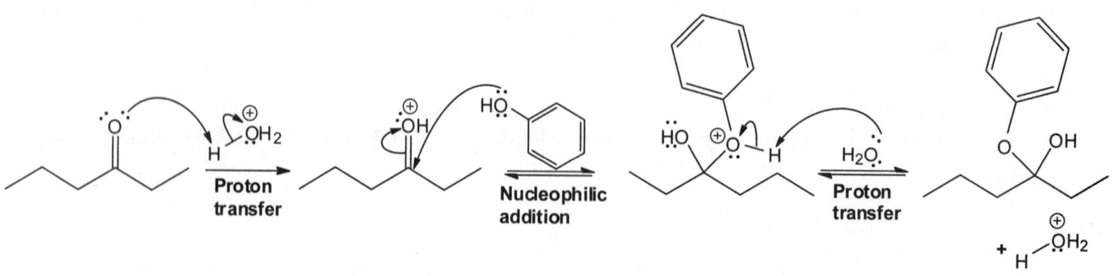

Proton transfer

Nucleophilic addition

Proton transfer

Problem 18.4

Think

What nucleophiles are present under these conditions? Are strong acids or bases present? Does the mechanism follow acidic or basic mechanistic conditions? What product is formed?

Solve

In both reaction conditions, the cyanohydrin product is formed.

(a) In these conditions, CN^- is the strongest nucleophile present. These conditions follow the same mechanism as in Equation 18-7 (nucleophilic addition to the C=O followed by proton transfer).

Nucleophilic addition
Proton transfer
A cyanohydrin

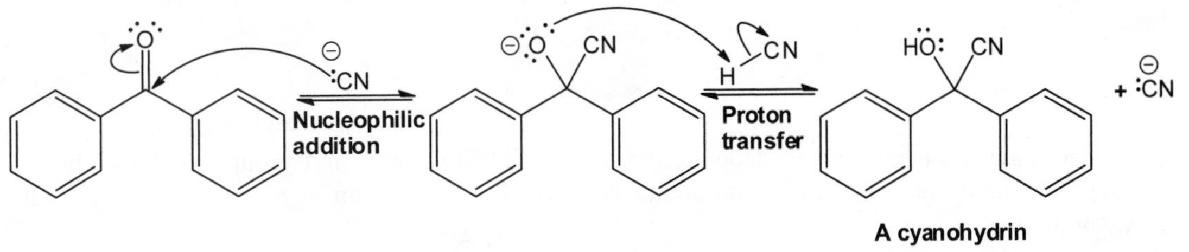

(b) In these conditions, HO^- deprotonates HCN quantitatively to produce NC^- as the major nucleophile. These conditions follow the basic mechanism format (nucleophilic addition to the C=O followed by proton transfer).

Proton transfer
Nucleophilic addition
Proton transfer
A cyanohydrin

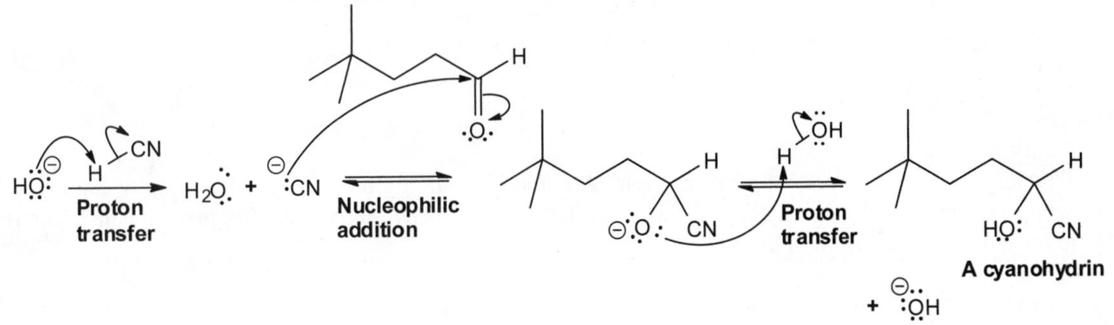

Problem 18.6

Think

Are the nucleophiles charged or uncharged? Does the nucleophile add to the carbonyl group reversibly or irreversibly? Will this result in direct addition or conjugate addition?

Solve

(a) Even though NC^- is a charged nucleophile, it still adds reversibly to a carbonyl group, so it will favor conjugate addition upon attacking an α,β-unsaturated carbonyl. The HO^- below the reaction arrow indicates basic conditions, so no strongly acidic species should appear in the mechanism. Therefore, we do not show the HCN protonating the carbonyl O atom. Nucleophilic addition occurs first.

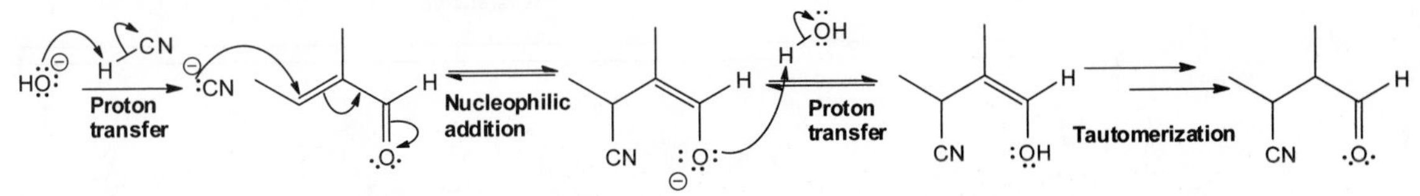

Proton transfer
Nucleophilic addition
Proton transfer
Tautomerization

(b) $PhCH_2SH$, an uncharged nucleophile, adds reversibly to a carbonyl group, so it will favor conjugate addition upon attacking an α,β-unsaturated carbonyl. The reaction is not catalyzed by acidic or basic conditions.

Nucleophilic addition — Proton transfer — Proton transfer — Tautomerization

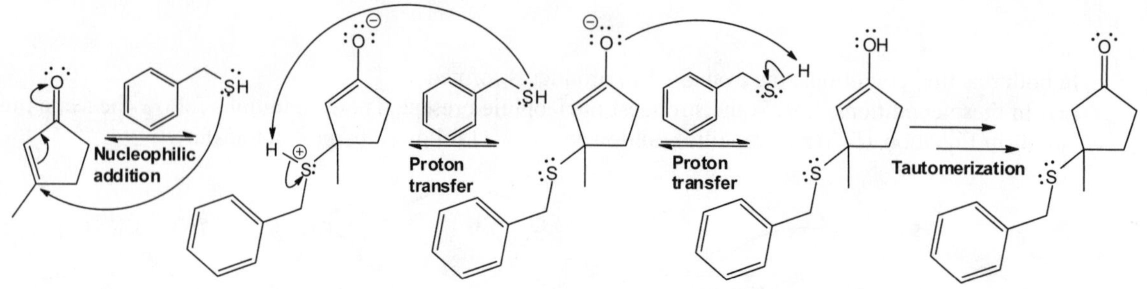

(c) Pyrrolidine, an uncharged nucleophile, adds reversibly to a carbonyl group, so it will favor conjugate addition upon attacking an α,β-unsaturated carbonyl. The reaction is not catalyzed by acidic or basic conditions.

Nucleophilic addition — Proton transfer — Proton transfer — Tautomerization

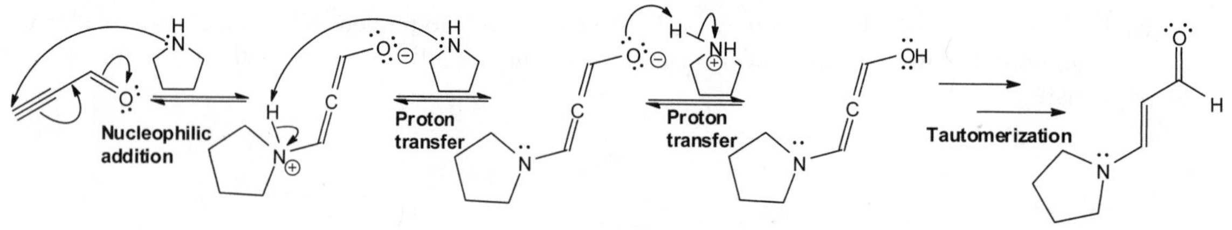

Problem 18.7

Think

This reaction is the reverse of what reaction? Are good leaving groups present? If not, how can a good leaving group be formed? What can act as a nucleophile? In how many steps does this mechanism occur if it is the reverse of the mechanism in Equation 18-13?

Solve

The mechanism for this reaction is the reverse of the mechanism in Equation 18-13. An alkoxide is not a good leaving group. However, in the presence of an acid, H^+, the O is protonated and thus forms ROH, which is a good leaving group. The mechanism is shown below.

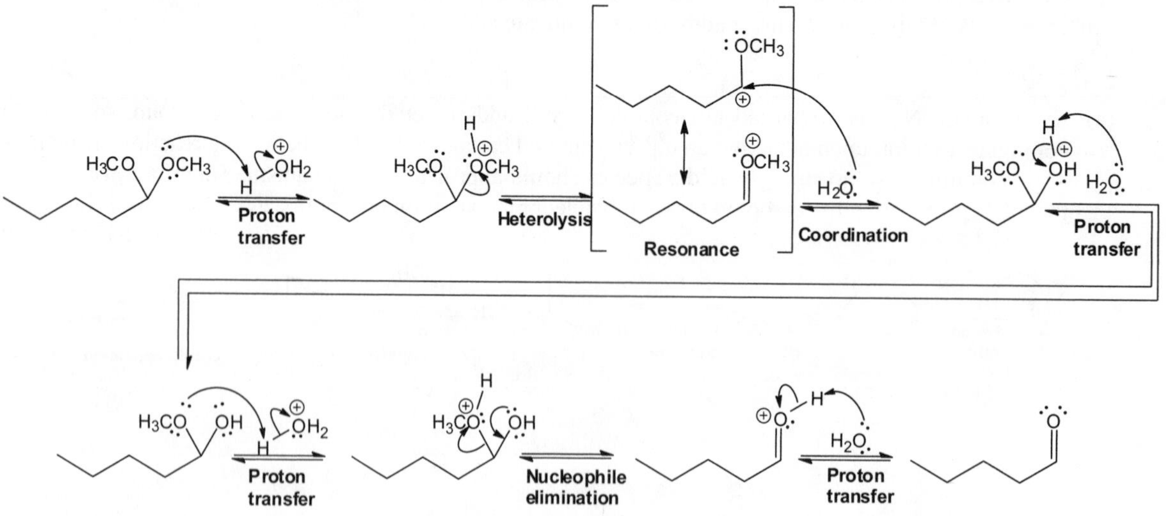

Proton transfer — Heterolysis — Resonance — Coordination — Proton transfer — Proton transfer — Nucleophile elimination — Proton transfer

Problem 18.8

Think

What is the name of the functional group that was produced in this problem? How is this mechanism similar to the mechanism shown in Equation 18-13? How is a good leaving group produced? What acts as the nucleophile?

Solve

An acetal is formed; this mechanism is essentially the same as the mechanism shown in Equation 18-13. The only difference is that the coordination in Step 6 is intramolecular rather than intermolecular.

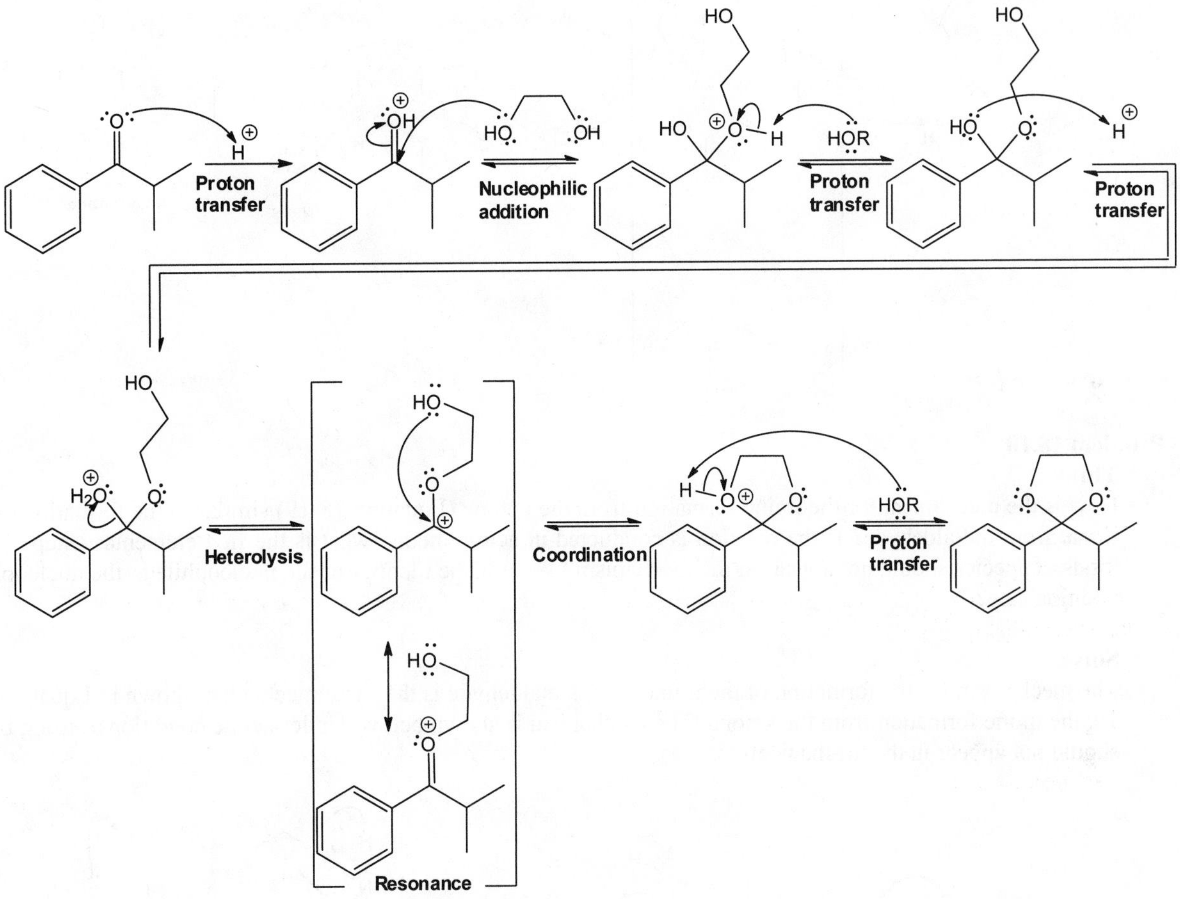

Proton transfer

Nucleophilic addition

Proton transfer

Proton transfer

Heterolysis

Coordination

Proton transfer

Resonance

Problem 18.9

Think

What is the name of the functional group that is produced in this problem? How is this functional group similar to an acetal? How is this mechanism similar to the mechanism shown in Equation 18-13? How is a good leaving group produced under acidic conditions? What is the nucleophile?

Solve

A 1,3-dithiane is formed and is the sulfur analog of an acetal. This mechanism is essentially the same as the mechanism shown in Equation 18-13. See the mechanism on the next page.

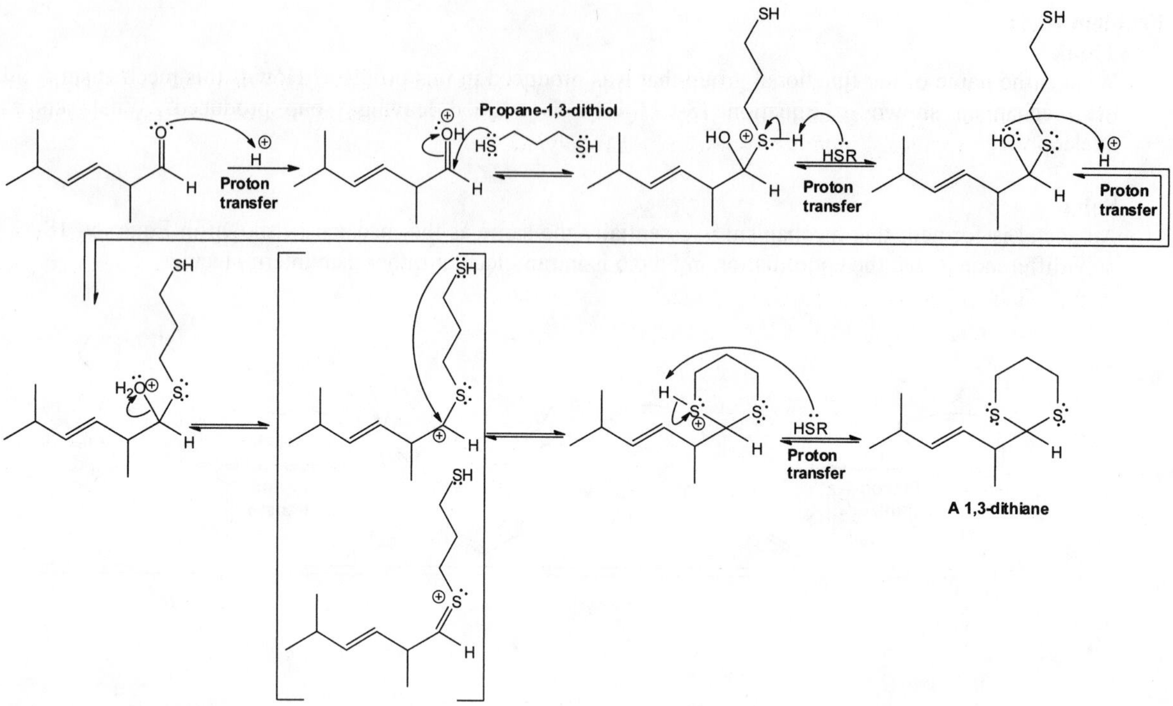

Propane-1,3-dithiol

Proton transfer

HSR

Proton transfer

Proton transfer

HSR

Proton transfer

A 1,3-dithiane

Problem 18.10

Think

How is the mechanism for the imine formation from the ketone (Equation 18-19) similar to the formation of the imine from the aldehyde? If the reaction is conducted in acidic media, what is the first elementary step? What kinds of species should not appear in the mechanism? What is the identity of the nucleophile in the nucleophilic addition step?

Solve

The mechanism for the formation of the imine from the aldehyde is the same mechanism shown in Equation 18-19, the imine formation from the ketone. The mechanism is shown below. Under acidic conditions, strong bases should not appear in the mechanism.

Proton transfer

Nucleophilic addition

Proton transfer

Proton transfer

Heterolysis

Resonance-stabilized carbocation

Proton transfer

(E)

+ (Z) isomer

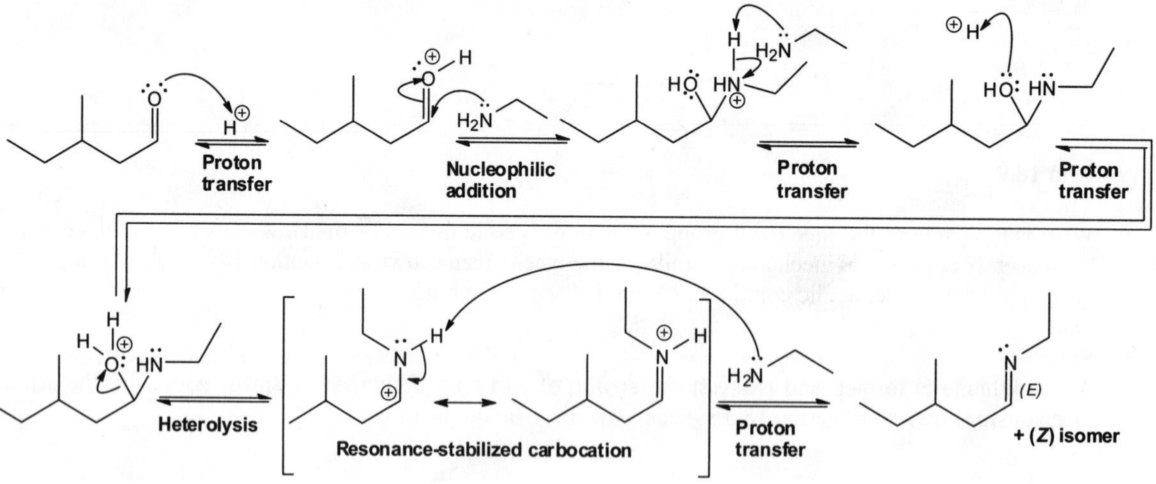

Problem 18.11

Think

This reaction is the reverse of what reaction? In how many steps does this mechanism occur if it is the reverse of the mechanism in Equation 18-19? If the reaction is catalyzed by an acid, what is the first elementary step? What kinds of species should not appear in the mechanism?

Solve

This mechanism is the reverse of the mechanism shown in Equation 18-19. A proton transfer reaction is first, followed by nucleophilic addition of water. The mechanism is shown below. No strong bases should appear in this mechanism, because it takes place under acidic conditions.

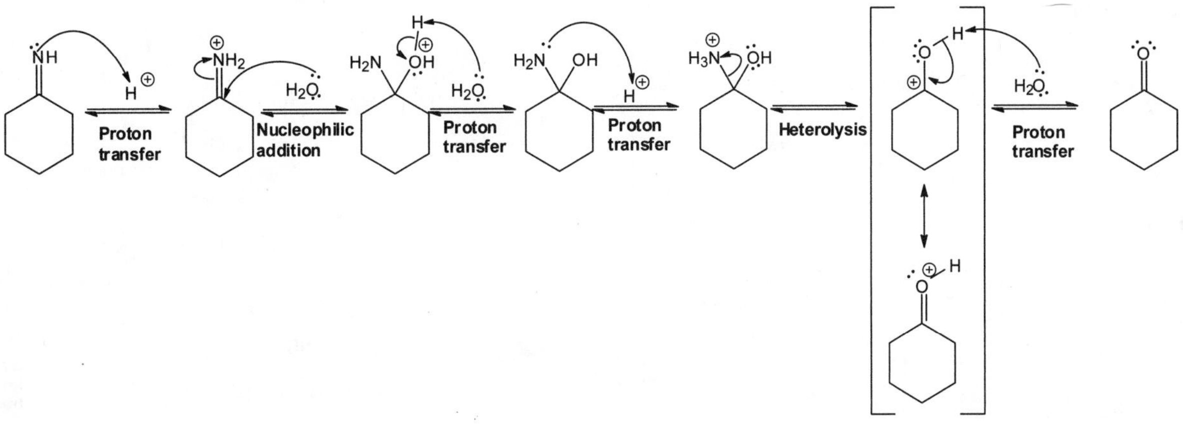

Proton transfer | Nucleophilic addition | Proton transfer | Proton transfer | Heterolysis | Proton transfer

Problem 18.13

Think

Under basic conditions, what types of species should *not* appear in the mechanism? Are all the steps that must happen for imine formation under these conditions reasonable? What species would have to act as the leaving group? Is this reasonable?

Solve

No strong acids can appear in the mechanism under basic conditions. Thus, the N-containing species would have to be eliminated as follows:

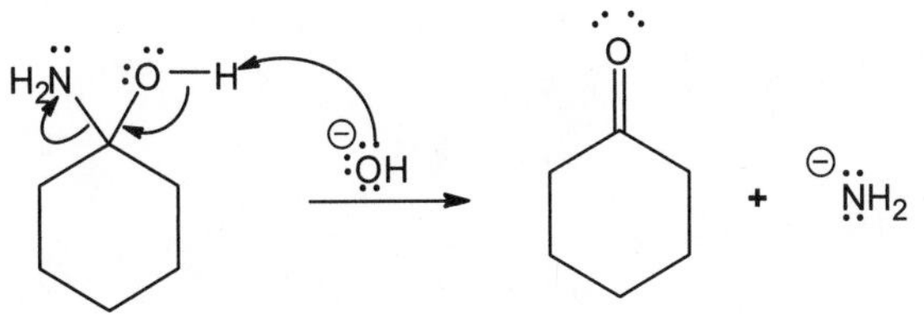

This is unreasonable because H_2N^- is a very poor leaving group.

Problem 18.14

Think

What type of amine is used in each reaction? This mechanism is similar to what other mechanism? At which step do the mechanisms differ? What is the name of the functional group formed? Is the reaction done in acidic or basic conditions?

Solve

In each case, the amine is a secondary amine. The mechanism is essentially the same as the mechanism in Equation 18-19 up through Step 5. After Step 5, the product has no more H atoms on the N atom. Thus, the final proton transfer reaction occurs at the adjacent C–H to form the enamine product.

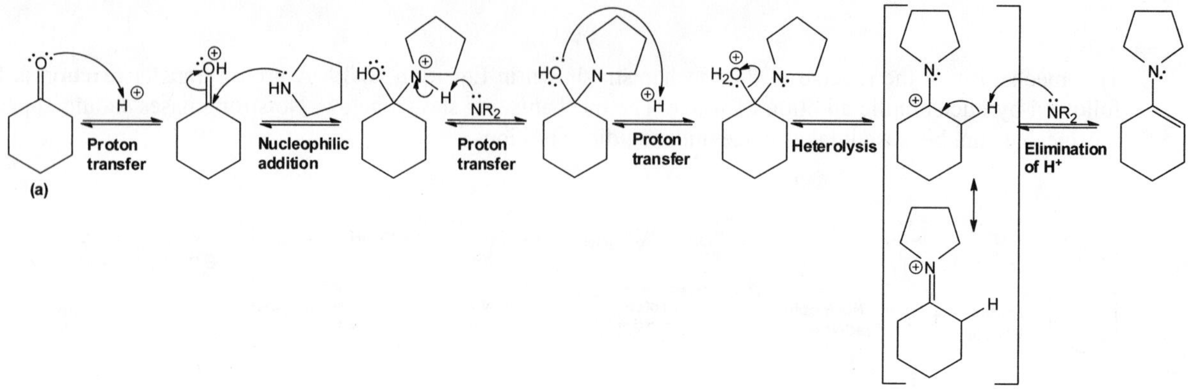

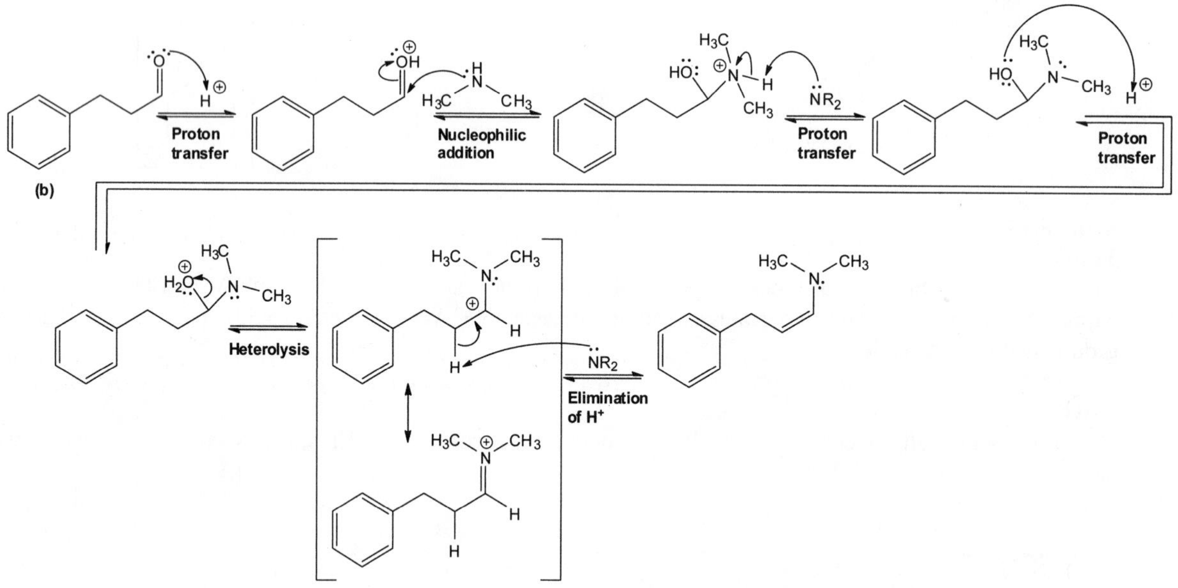

(a) Proton transfer → Nucleophilic addition → Proton transfer → Proton transfer → Heterolysis → Elimination of H⁺

(b) Proton transfer → Nucleophilic addition → Proton transfer → Proton transfer → Heterolysis → Elimination of H⁺

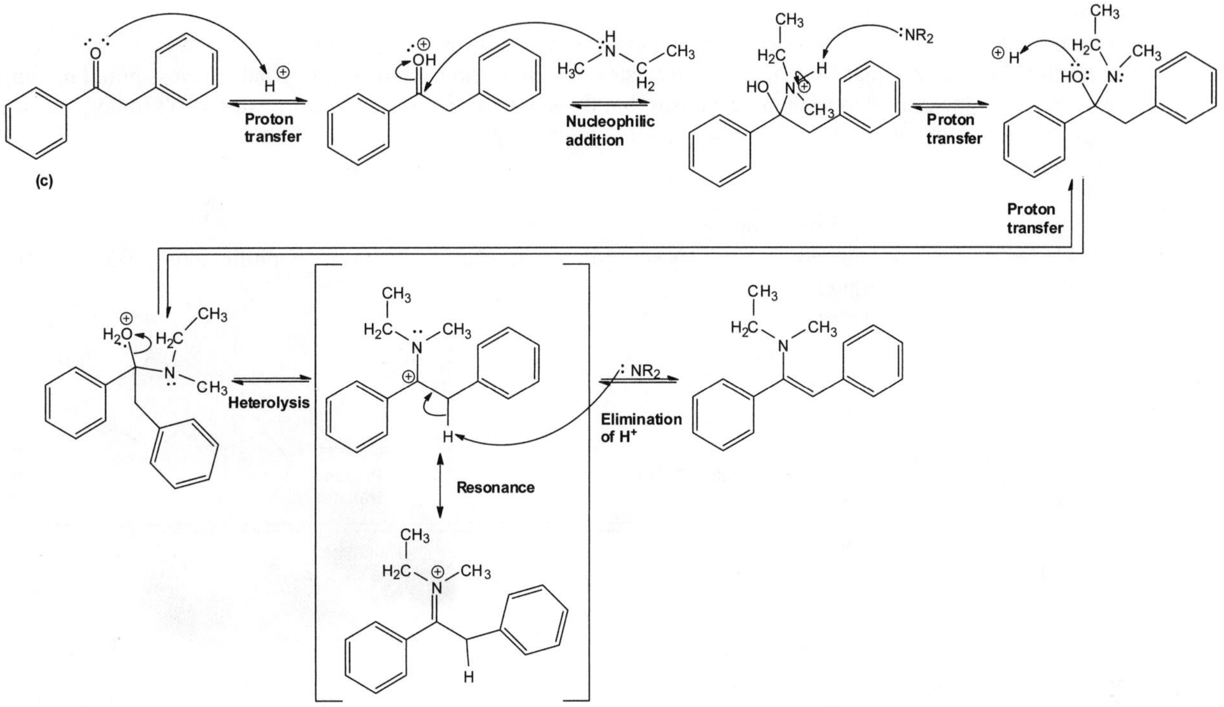

(c)

Proton transfer

Nucleophilic addition

Proton transfer

Proton transfer

Heterolysis

Resonance

Elimination of H^+

Problem 18.16

Think

Which of the steps in Equations 18-19 or 18-22 can the reaction include? Are there any steps that are not possible?

Solve

To form an imine, two H atoms must be attached to N on the initial amine. To form an enamine, a secondary amine must react with a ketone or aldehyde that has at least one α proton. None of these scenarios exist with the reactants given.

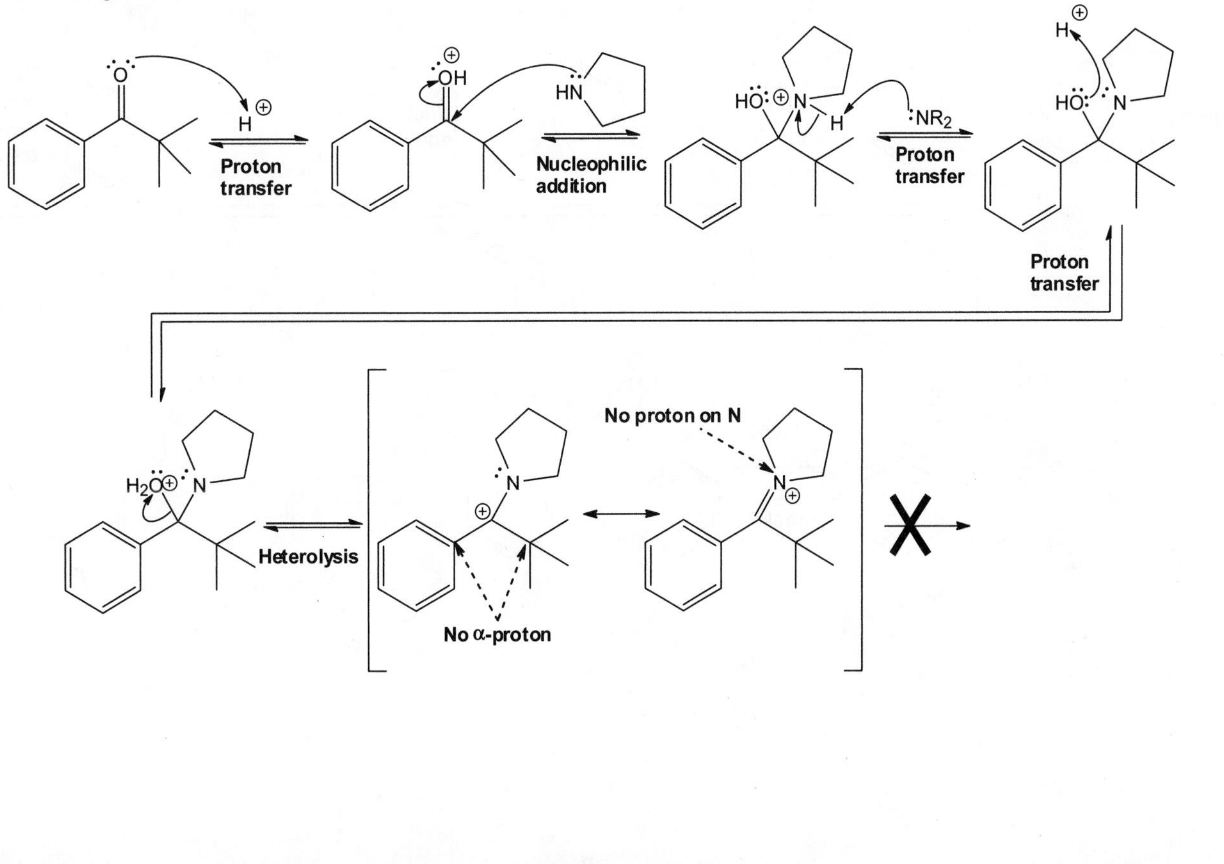

Proton transfer

Nucleophilic addition

Proton transfer

Proton transfer

Heterolysis

No proton on N

No α-proton

Problem 18.17

Think

Does the reaction take place under acidic or basic conditions? In each medium, what species should not appear? Which step is the only step in the mechanism that does not consist of a proton transfer step? In how many steps does the mechanism proceed? What is the functional group produced?

Solve

Both reactions produce a primary amide.

(a) Under basic conditions, no strong acids should appear. The first step (nucleophilic addition) is the only step that is not a proton transfer reaction.

Nucleophilic addition

Proton transfer

Proton transfer

Proton transfer

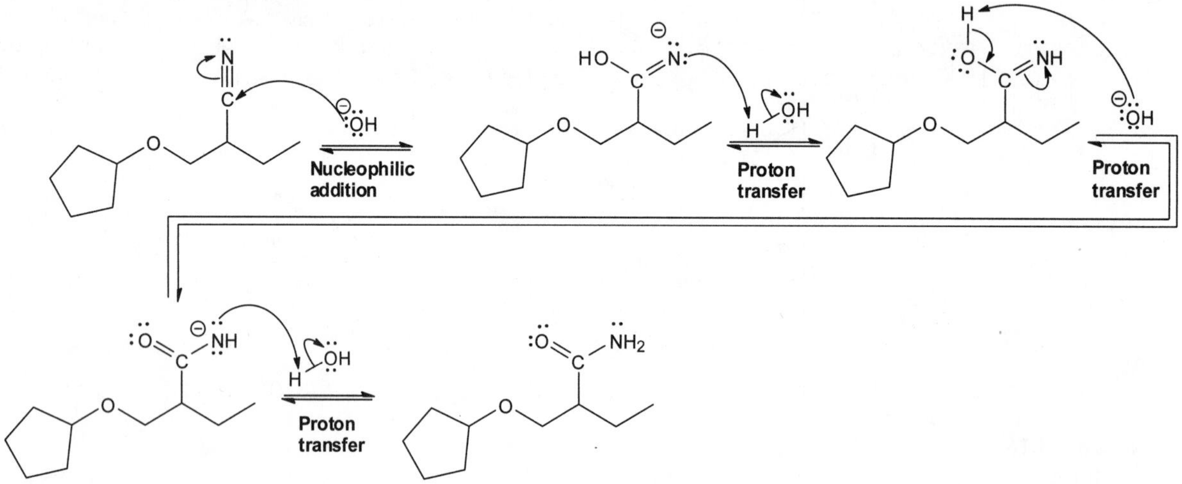

(b) Under acidic conditions, no strong bases should appear. The second step (nucleophilic addition) is the only step that is not a proton transfer reaction. A proton transfer occurs before nucleophilic addition to avoid the formation of a strong base upon nucleophilic addition.

Proton transfer

Nucleophilic addition

Proton transfer

Proton transfer

Proton transfer

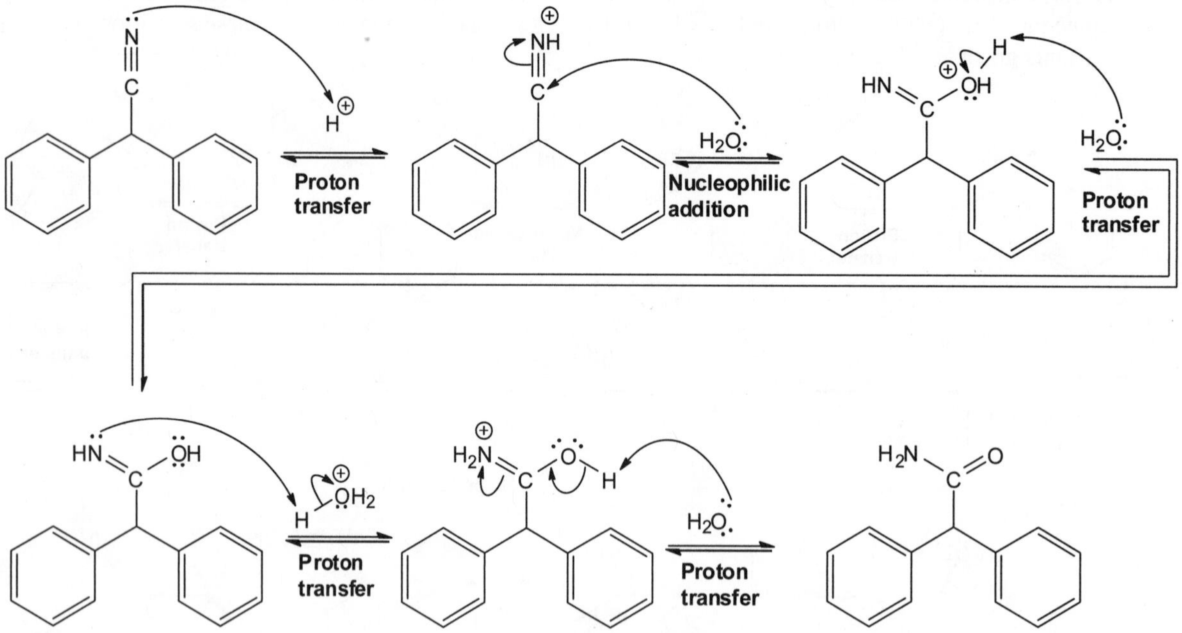

Problem 18.18

Think

Does the reaction take place under acidic or basic conditions? In each medium, what species should not appear? What two mechanisms shown in this chapter combine to show the complete mechanism? What functional group is removed entirely?

Solve

Both of these are Wolff–Kishner reductions, which reduce the C=O bond of a ketone or aldehyde to a methylene (CH_2) group.

(a) First, a nucleophilic addition occurs under acidic addition, followed by elimination of water also under acidic conditions to form the hydrazone. Next, the basic conditions remove two protons from the hydrazone, creating a great N_2 leaving group. The product is methylcyclohexane, which is no longer chiral.

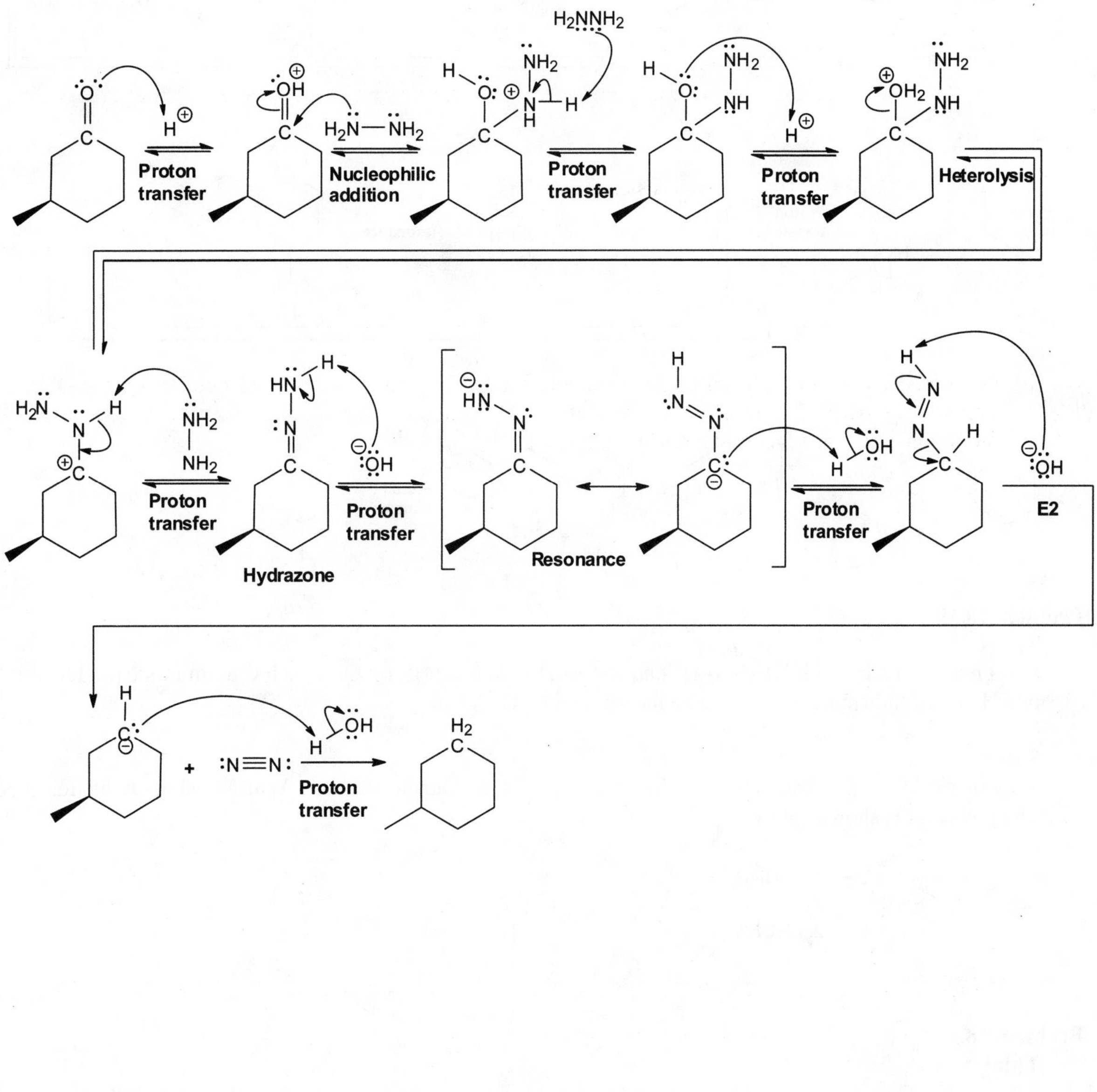

(b) The mechanistic route is the same as **(a)**.

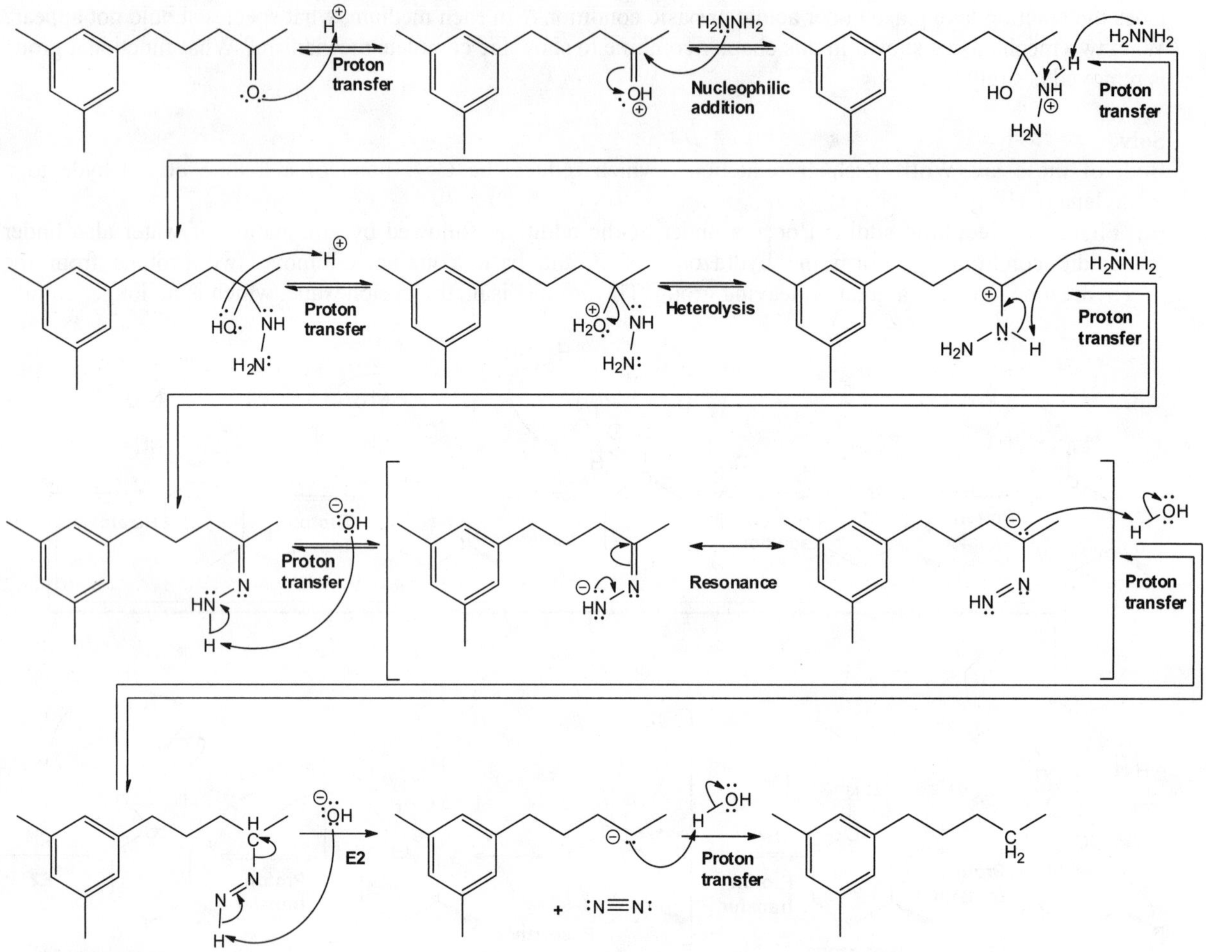

Proton transfer
Nucleophilic addition
Proton transfer
Proton transfer
Heterolysis
Proton transfer
Proton transfer
Resonance
Proton transfer
E2
Proton transfer

Problem 18.19

Think

What reactant reacts with NH_2NH_2/H^+ and NaOH/H_2O to give $-CH_2$? On which C atom in the product are there two H atoms, indicating what could be the original C=O?

Solve

Only one sp^3 C in the target molecule has two H atoms, a characteristic of a Wolff–Kishner reduction product. The precursor is shown below:

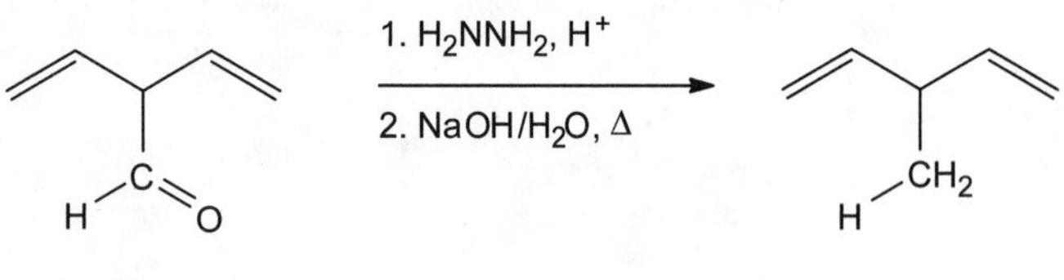

1. H_2NNH_2, H^+
2. NaOH/H_2O, Δ

Problem 18.20

Think

What is the product of every aldol reaction? Which product has the HO group β to the aldehyde?

Solve

The product of every aldol addition is a β-hydroxycarbonyl. As shown below, the only molecule that is a β-hydroxycarbonyl is molecule **(b)**.

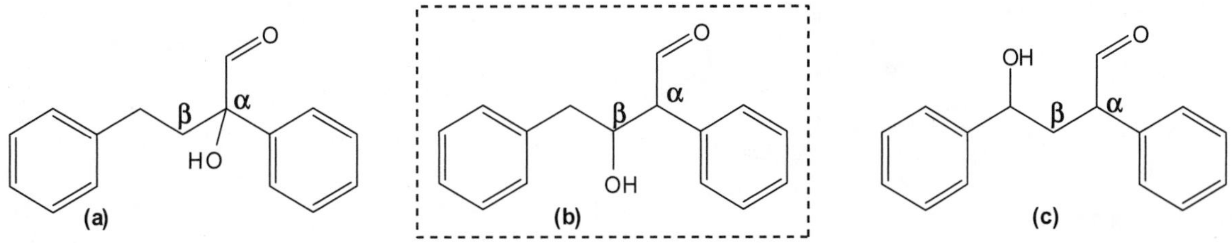

(a) (b) (c)

Problem 18.22

Think

How is this reaction similar to the one in Equation 18-29 (and Equation 18-30)? How is it different? What is the role of NaOH? What can act as a nucleophile to add to the C=O group?

Solve

Just as in Equation 18-29, an aldehyde is treated with hydroxide. The only difference is the carbon backbone; in this case, we have a $PhCH_2CH_2CHO$ aldehyde, whereas it was a two-carbon aldehyde in Equation 18-29. The mechanisms, therefore, are essentially identical.

In the first step, HO^- acts as a base to generate an enolate anion. (Note that HO^- could also act as a nucleophile, but as we saw in Chapter 8, proton transfer reactions are very fast, so the dominant reaction is deprotonation of the α carbon.) In the second step, the enolate anion acts as a nucleophile, attacking a second molecule of aldehyde that still has its proton. This step forms a C–C bond between the two species. Finally, the O^- is protonated to yield the β-hydroxycarbonyl.

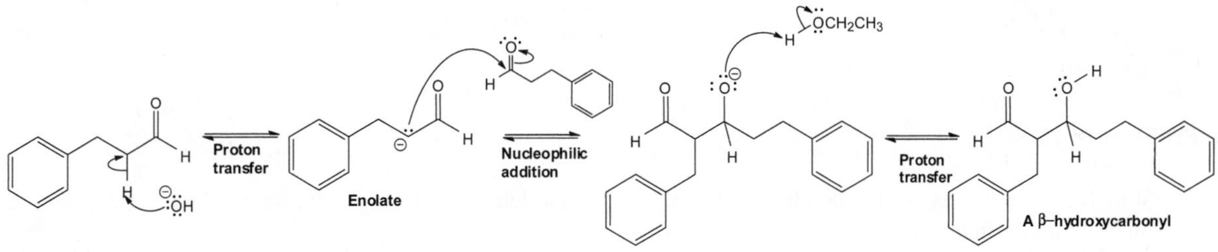

Proton transfer — Enolate — Nucleophilic addition — Proton transfer — A β-hydroxycarbonyl

Problem 18.24

Think

Can you identify the portion of the molecule that characterizes it as a β-hydroxycarbonyl? Which C atoms of the β-hydroxycarbonyl have been joined in the aldol addition? Upon disconnecting that C–C bond in a retrosynthesis, what carbon backbones are required in the precursors?

Solve

To perform a transform that reverses an aldol addition, we can disconnect the carbon–carbon bond between the α and β carbons of the C=O.

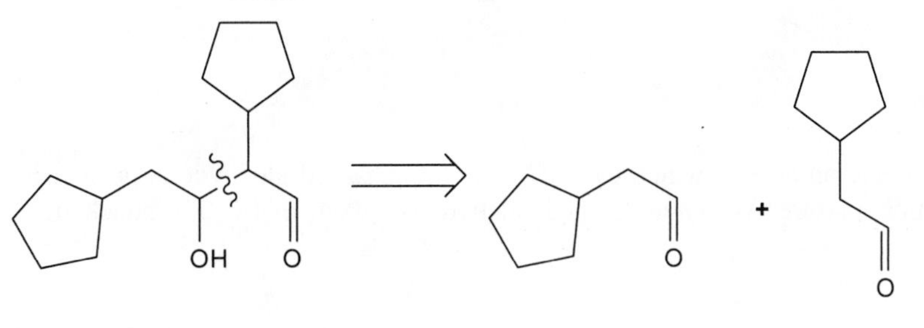

The two aldehyde precursors are the same, so we simply need to treat that aldehyde with base.

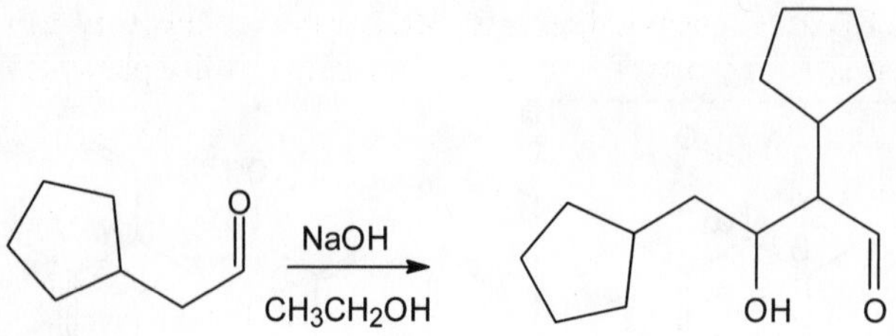

NaOH

CH_3CH_2OH

OH O

Problem 18.25

Think

What is the pK_a of the α hydrogen in the aldehyde? What is the pK_a of the α hydrogen in an alkene? What species is acting as the base? What is the pK_a of the conjugate acid? Which side of the proton transfer reaction does the equilibrium favor?

Solve

Although a mechanism analogous to the reaction in Equation 18-32 can be written, the first step would be very unfavorable. In Equation 18-32 it is feasible because the α hydrogen is reasonably acidic, with a pK_a ~20. But the analogous proton in the compound given in the problem has a pK_a ~40. With the conjugate acid of HO^- (i.e., H_2O) having a pK_a of 15.7, the equilibrium favors the reactants by ~10^{24}, making the reaction unfeasible.

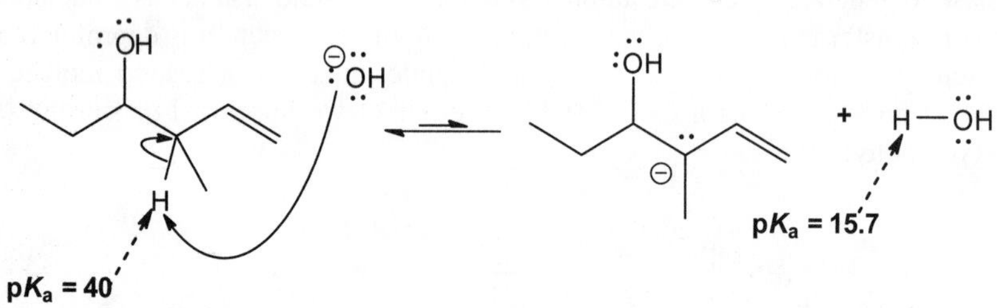

:OH

OH

:OH

H

\+ H—OH

pK_a = 15.7

pK_a = 40

Problem 18.26

Think

Which aldehyde acts as the enolate nucleophile? Which aldehyde is attacked by that enolate nucleophile? Is this an aldol addition or condensation?

Solve

In this example, the ethanal is deprotonated at the α position by HO^- and acts as the enolate that attacks the propanal carbonyl C. This is an example of an aldol addition that forms a β-hydroxycarbonyl. The ethanal is boxed in each step of the reaction.

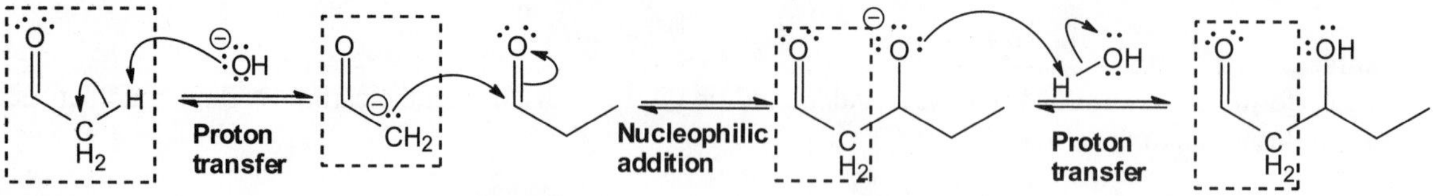

O H C H_2 OH Proton transfer O CH_2 O Nucleophilic addition O O C H_2 H OH Proton transfer O OH C H_2

Problem 18.28

Think

Which aldehyde can form an enolate anion and which cannot? Which uncharged aldehyde is present in only small concentrations in the reaction mixture? Which uncharged aldehyde is the one that will be attacked?

Solve

Dimethyl propanal has no α hydrogens, so it cannot form an enolate anion. Thus, the only enolate nucleophile that is present is derived from phenylethanal. Because phenylethanal is added slowly, there is never a substantial concentration of it in its uncharged form, so the major reaction occurs between the phenylethanal enolate anion and dimethylpropanal.

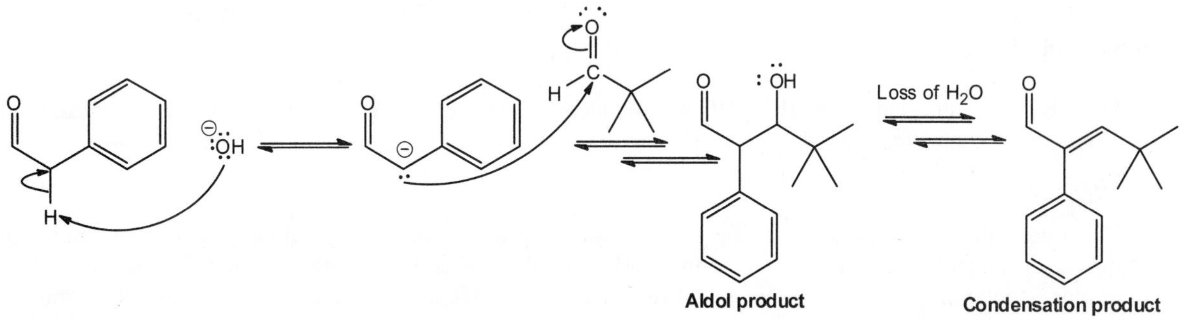

Loss of H_2O

Aldol product

Condensation product

Problem 18.30

Think

Can you identify the portion of the molecule that characterizes it as an α,β-unsaturated carbonyl? From what β-hydroxycarbonyl could it have been generated? Upon applying a transform to that β-hydroxycarbonyl to undo an aldol addition, which C–C bond should be disconnected?

Solve

The α,β-unsaturated carbonyl is produced from a β-hydroxycarbonyl, which, in turn, can be disconnected between the α and β carbons.

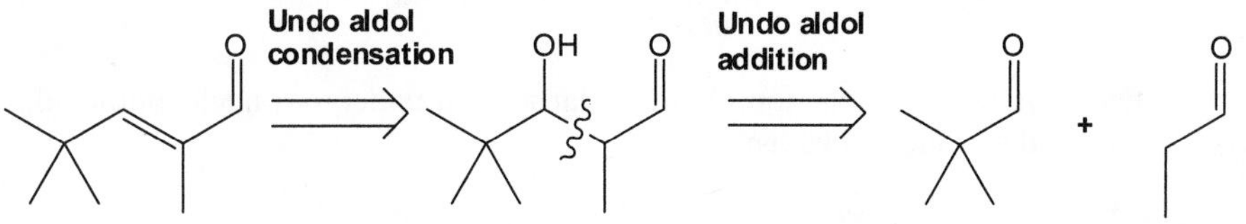

Undo aldol condensation

Undo aldol addition

Because the starting carbonyls are different, we must consider how to execute a crossed aldol reaction. In this case, the first aldehyde has no α hydrogens, so we can treat it with base before adding the second aldehyde.

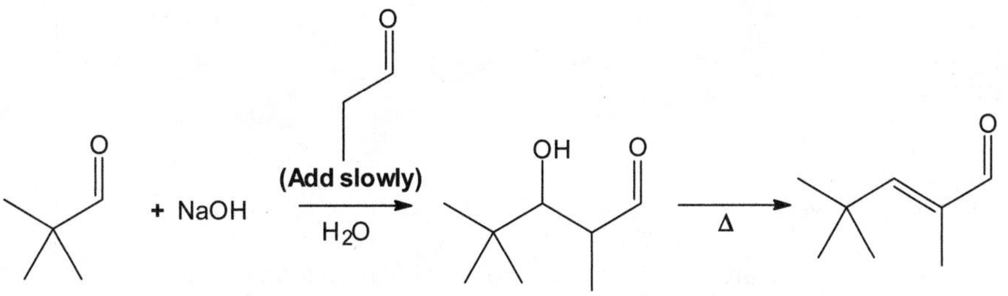

+ NaOH

(Add slowly)

H_2O

Δ

Problem 18.31

Think

What H is attacked by the HO^-? What is the identity of the enolate? What electrophilic atom is attacked by the enolate? How many C atoms are in the ring that forms?

Solve

The mechanism is shown on the next page. The α hydrogen of heptanedial is attacked by the HO^- to form the enolate. The other C of the C=O is attacked by the enolate. The intramolecular aldol reaction is favorable because it produces a six-membered ring.

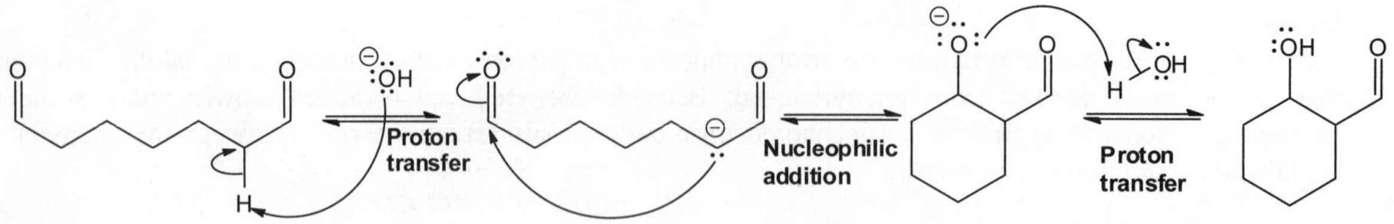

Problem 18.33

Think

What H can be attacked by the HO^-? What is the identity of the enolate? Can the other C=O be attacked to form a five- or six-membered ring?

Solve

There are four α C atoms that can be deprotonated to produce an enolate anion. The two α C atoms in the middle of the molecule won't lead to an intramolecular aldol reaction because they are too close to the carbonyl groups—they will form a three-membered ring if their enolates attack. Either of the terminal C atoms can be deprotonated to form a five-membered ring. The molecule is symmetrical, so it does not matter which end is deprotonated in the first step.

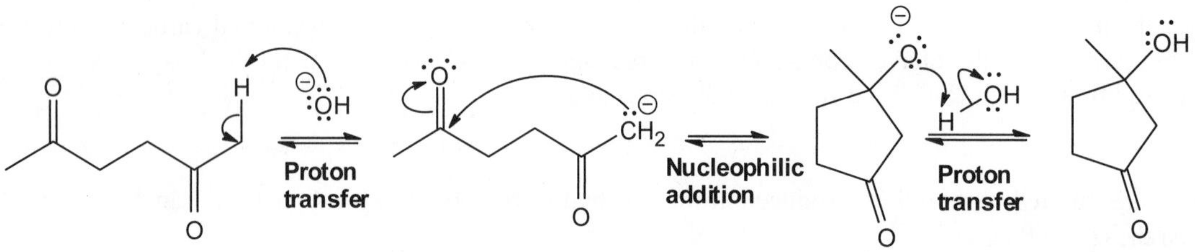

Problem 18.35

Think

Which protons are the most acidic? After deprotonation, which polar π bond undergoes nucleophilic addition? How are proton transfers incorporated into the mechanism?

Solve

The hydrogen that is α to the NO_2 group is the most acidic, so it is removed by the base in the first step. The resulting enolate anion attacks the aldehyde group, and the O^- generated in that step is subsequently protonated.

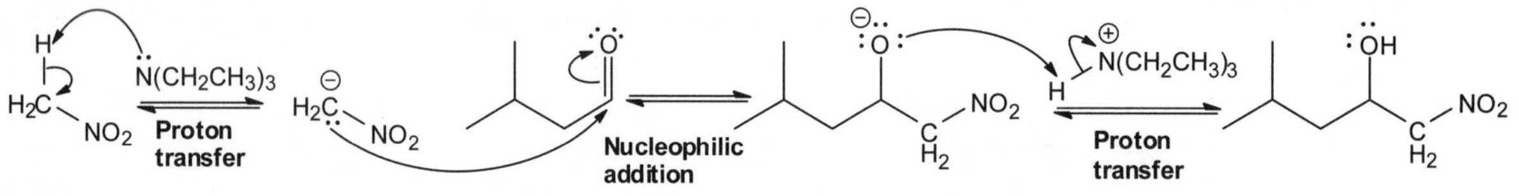

Problem 18.36

Think

Which protons are the most acidic? After deprotonation, which polar π bond undergoes nucleophilic addition? How is a C=C double bond formed in an aldol reaction? Will it be favored in this reaction?

Solve

The H that is α to the CN group is the most acidic, so it is removed by the base in the first step. The resulting enolate anion attacks the aldehyde group, and the O^- generated in that step is subsequently protonated. Due to the conjugation that results, condensation then takes place via an E1cb mechanism.

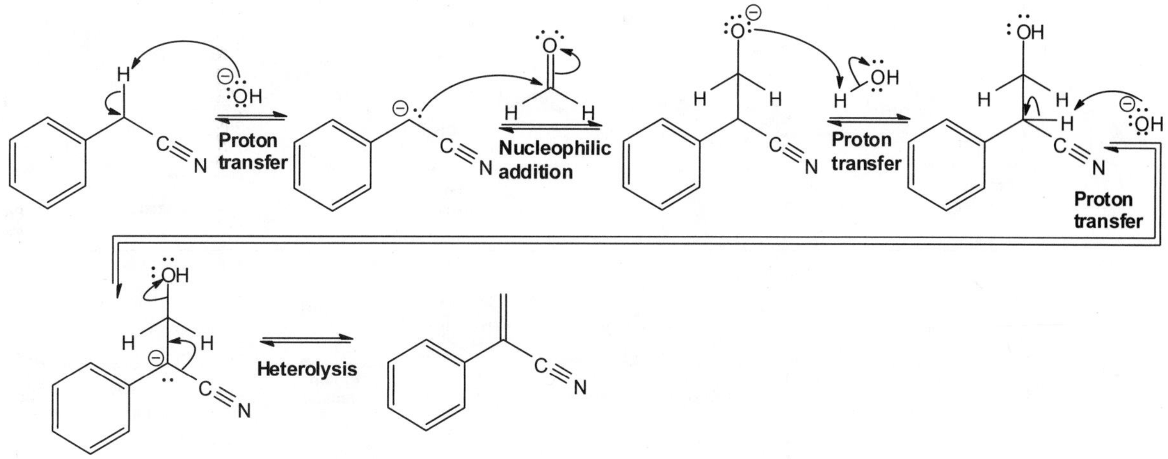

Proton transfer — Nucleophilic addition — Proton transfer — Proton transfer — Heterolysis

Problem 18.37

Think

The Robinson annulation comprises which two mechanisms? What is the nucleophile in the first nucleophilic addition reaction? Does the addition occur via 1,2- or 1,4-addition? Consider the breakdown in Equation 18-55 to guide you through the mechanism.

Solve

The two mechanisms that make up the Robinson annulation are a Michael addition (1,4-addition) followed by an intramolecular aldol addition/dehydration (via E1cb mechanism). The full mechanism is shown below.

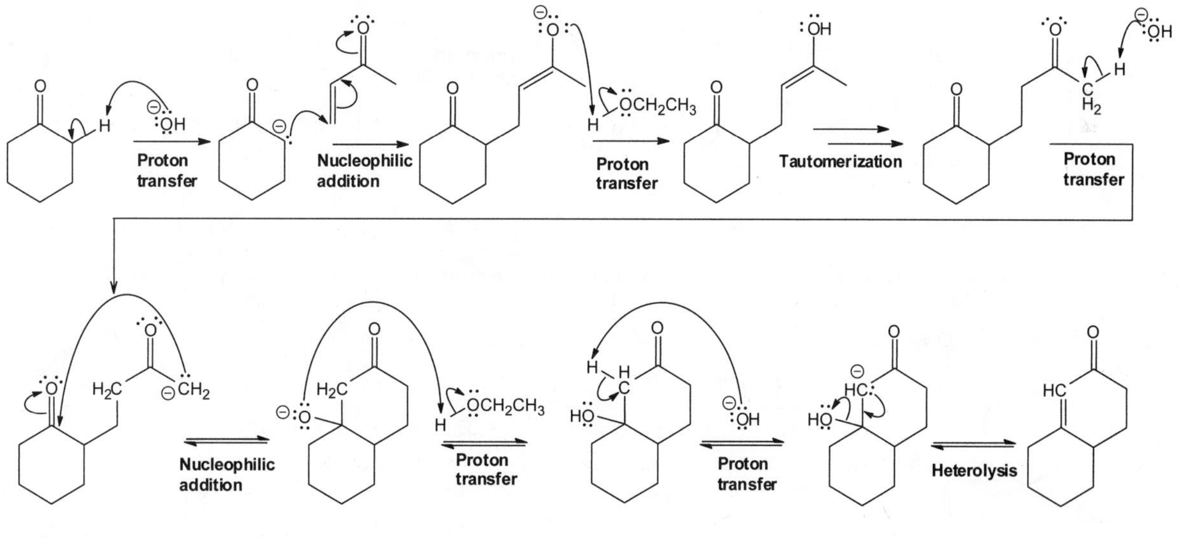

Proton transfer — Nucleophilic addition — Proton transfer — Tautomerization — Proton transfer — Nucleophilic addition — Proton transfer — Proton transfer — Heterolysis

Problem 18.38

Think

What are the requirements for a Robinson annulation? Are any of the reactants an α,β-unsaturated ketone and do they have two acidic α hydrogen atoms? Does another reactant have acidic α hydrogens? If the reagents meet the requirement for Robinson annulation, what two mechanisms make up the Robinson annulation? What is the nucleophile in the first nucleophilic addition reaction? Does the addition occur via 1,2- or 1,4-addition? Consider the breakdown in Equation 18-55 to guide you through the mechanism.

Solve

(a) These reagents will not work because benzaldehyde does not have any α protons.
(b) These reagents will not work because the α,β-unsaturated carbonyl does not have two acidic α protons.
(c) These reagents will not work because neither compound is an α,β-unsaturated carbonyl.
(d) These reagents fit the criteria for a Robinson annulation. The mechanism is shown below.

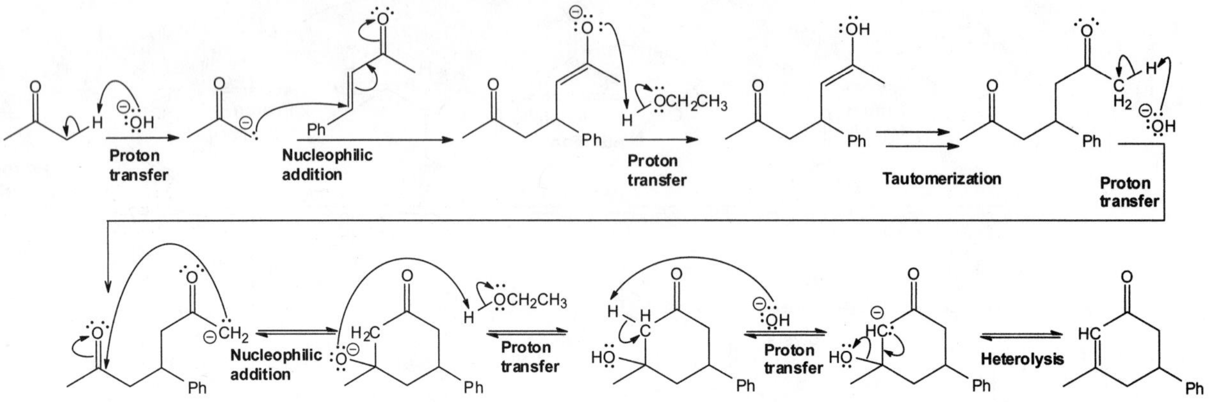

Problem 18.39

Think

Is the product a result of aldol addition or aldol condensation? What are the characteristic features of the products from these reactions? What are the α and β carbon atoms in the product? Which two C atoms formed the new C–C bond? What are the two C=O starting materials?

Solve

(a) This is an α,β-unsaturated carbonyl that results from an aldol condensation and can be produced from a β-hydroxycarbonyl. The β-hydroxycarbonyl is the product of an aldol addition. The retrosynthesis is shown below.

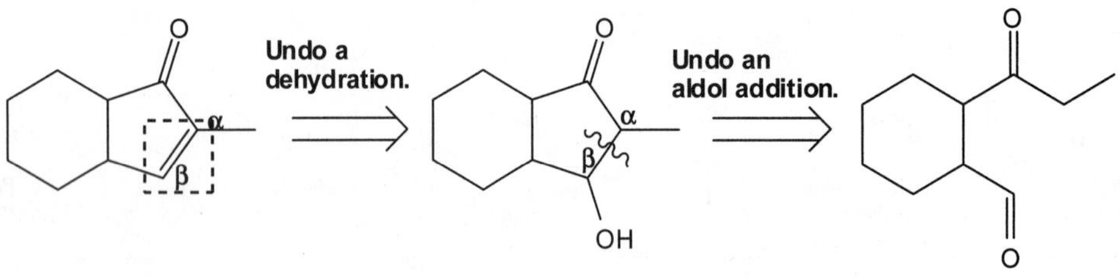

In the forward direction, the synthesis would appear as follows:

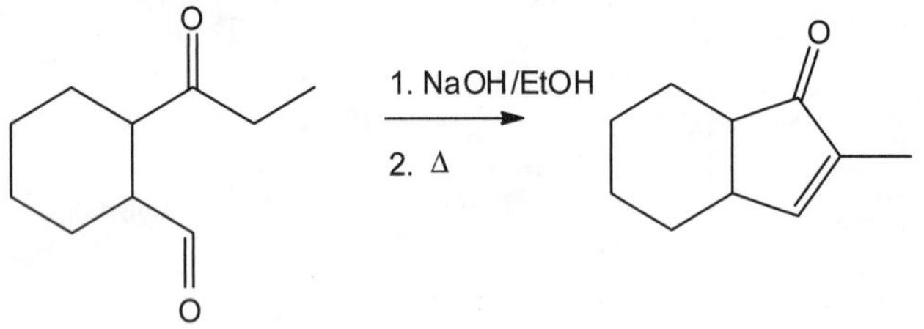

(b) This is a β-hydroxycarbonyl, which can be produced directly via an aldol addition.

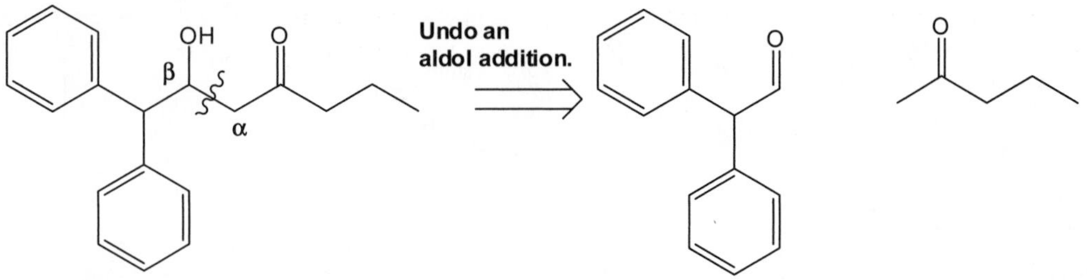

Undo an aldol addition.

This requires a crossed aldol addition, which proceeds through the kinetic enolate anion (the enolate is produced on the less highly substituted α carbon). Thus, we use a strong bulky base like LDA.

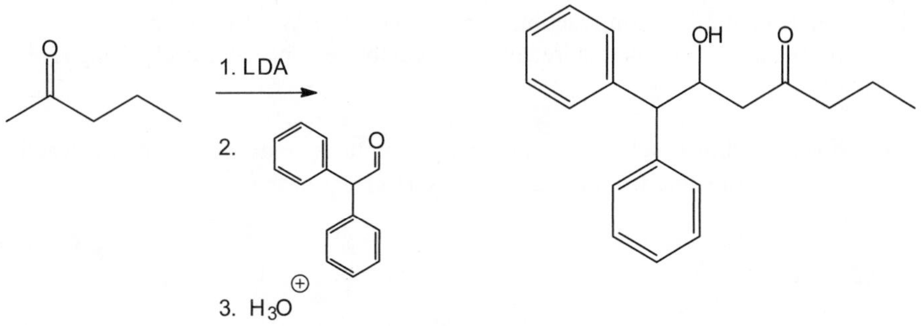

1. LDA
2.
3. H_3O^+

Problem 18.40

Think

On what positions are the Br substituents? Could this product originate from a β-hydroxycarbonyl? If so, which Br was the C=O and which Br was the C–OH? How can you apply a transform to the aldol addition product to undo the aldol addition, yielding the two original reagent(s)?

Solve

The Br substituents are in the same 1,3-positioning as a β-hydroxycarbonyl. The 1,3-dibromo target can be produced by brominating a 1,3-diol, which can be produced by reducing a β-hydroxycarbonyl. The β-hydroxycarbonyl is an aldol addition product. Notice that either of the alcohol groups in the 1,3-diol could have come from the reduction of a C=O, so the target can be produced from either of two different aldol addition reactions, as shown in the retrosynthetic analyses below.

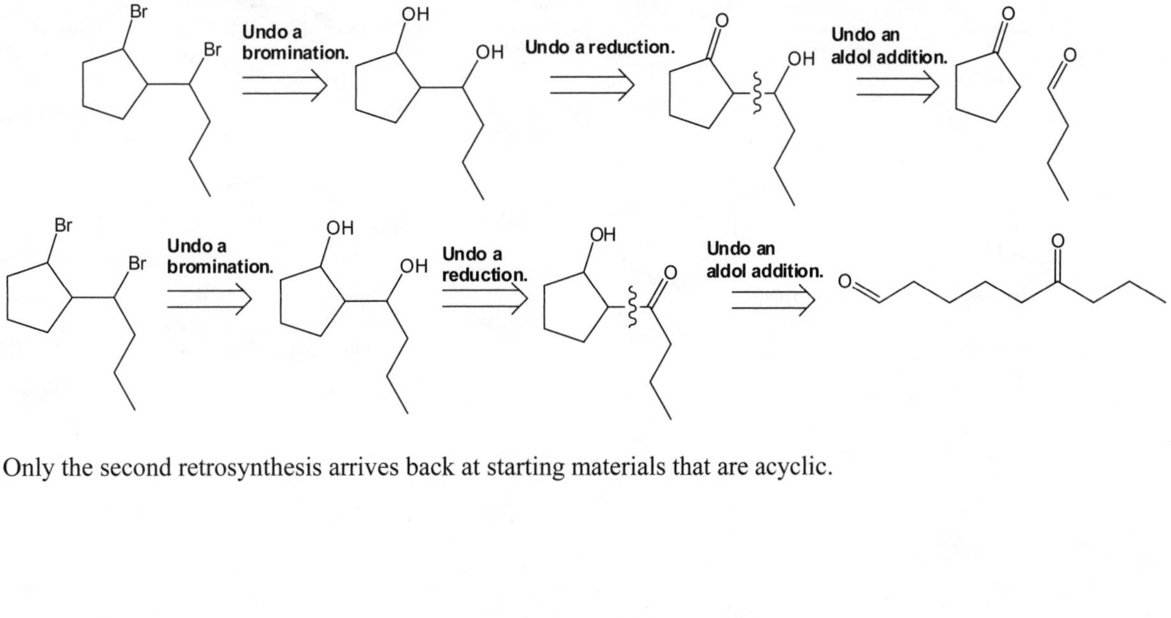

Undo a bromination. Undo a reduction. Undo an aldol addition.

Undo a bromination. Undo a reduction. Undo an aldol addition.

Only the second retrosynthesis arrives back at starting materials that are acyclic.

In the forward direction, the synthesis would appear as follows:

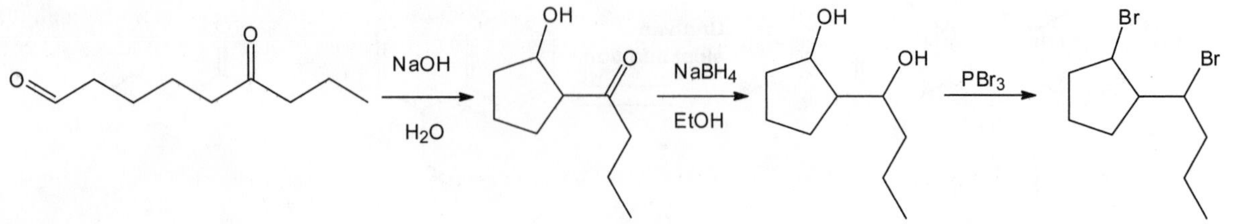

NaOH, H_2O → NaBH4, EtOH → PBr_3 →

Problem 18.41

Think

Consult Equation 18-66 for the structure of D-ribofuranose. What is the Haworth projection (*Hint:* review Section 4-9)? Which C is the anomeric carbon? Which anomer has the CH_2OH group on the same side of the ring as the OH group at the anomeric carbon? Consult Equation 18-67b for the general mechanism. What acts as the nucleophile in the nucleophilic addition reaction to close the ring? What C is electrophilic? Are proton transfer reactions involved? What approach is required of the nucleophile to produce each anomer?

Solve

(a) The Haworth projections are shown below. The anomeric C is the one at the far right in each case. The CH_2OH group is on the opposite side from the anomeric OH group in the α anomer, and they are on the same side in the β anomer.

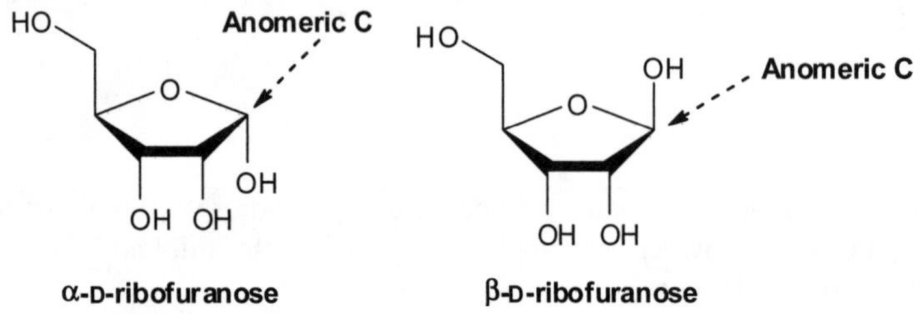

Anomeric C

α-D-ribofuranose

β-D-ribofuranose

(b) The mechanisms showing the production of each anomer are shown below. To form the α anomer, nucleophilic attack must come from the top of the C=O plane, and to form the β anomer, nucleophilic attack must come from the bottom.

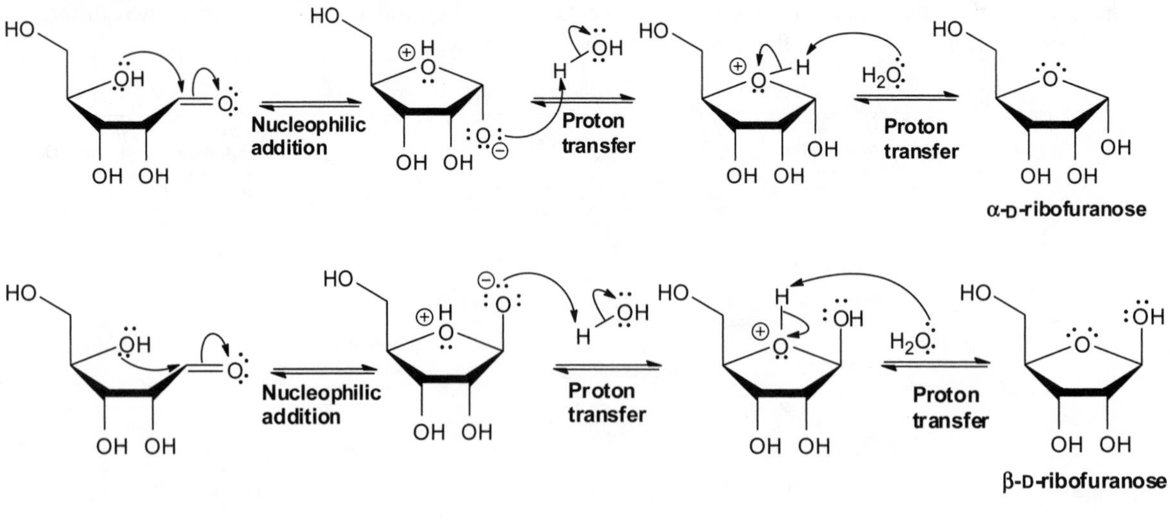

Nucleophilic addition → Proton transfer → Proton transfer → α-D-ribofuranose

Nucleophilic addition → Proton transfer → Proton transfer → β-D-ribofuranose

Problem 18.42

Think

What is the name of the sugar in the acylic form (Fig. 5-34)? To help you name the sugar, can you identify it as a specific epimer of another sugar? Is the anomeric OH and CH_2OH on the same side or opposite sides of the ring?

Solve

(a) α-D-mannopyranose. The CH_2OH group is on the opposite side from the anomeric OH, making it an α anomer. That this sugar is mannose can be seen from the fact that it is a C2 epimer of glucose.

(b) β-D-talopyranose. The CH_2OH group is on the same side as the anomeric OH, making it a β anomer. That this sugar is talose can be seen from the fact that it is a C4 epimer of mannose.

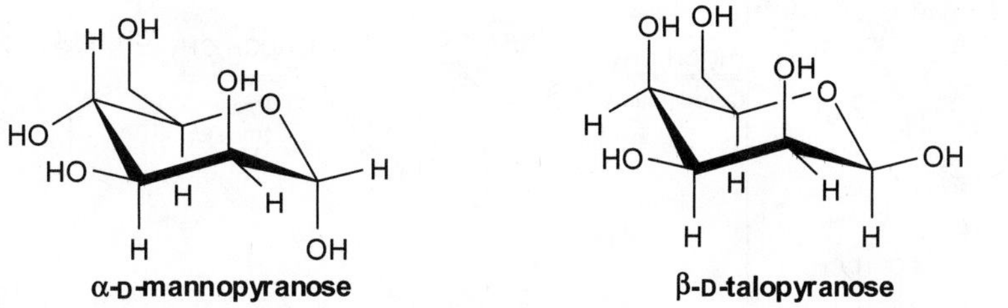

α-D-mannopyranose β-D-talopyranose

Problem 18.43

Think

If the specific rotation is +14.5°, which anomer, α-D-mannopyranose ([α] = +29.3°) or β-D-mannopyranose ([α] = −16.3°) is the dominant species in solution? If you set the percentage of α-D-mannopyranose equal to x in Equation 16-70, then what is the percentage of β-D-mannopyranose?

Solve

Let x be the equilibrium percentage of the α-D-mannopyranose, and $1 - x$ be the equilibrium percentage of β-D-mannopyranose.

$x(+29.3) + (1 - x)(-16.3) = +14.5$

$29.3x - 16.3 + 16.3x = 14.5$

$45.6x = 30.8$

$x = 0.675$, or 67.5% = α-D-mannopyranose

$1 - x = 0.325$, or 32.5% = β-D-mannopyranose

Problem 18.44

Think

What nucleophiles are present under acidic conditions? In what order do the nucleophilic addition and proton transfer reactions occur? How can the acid catalyst be regenerated? What are unreasonable species in each reaction medium? What functional group is produced?

Solve

(a) This is an acetal formation reaction in which the reaction is catalyzed by an acid and the ROH acts as the nucleophile.

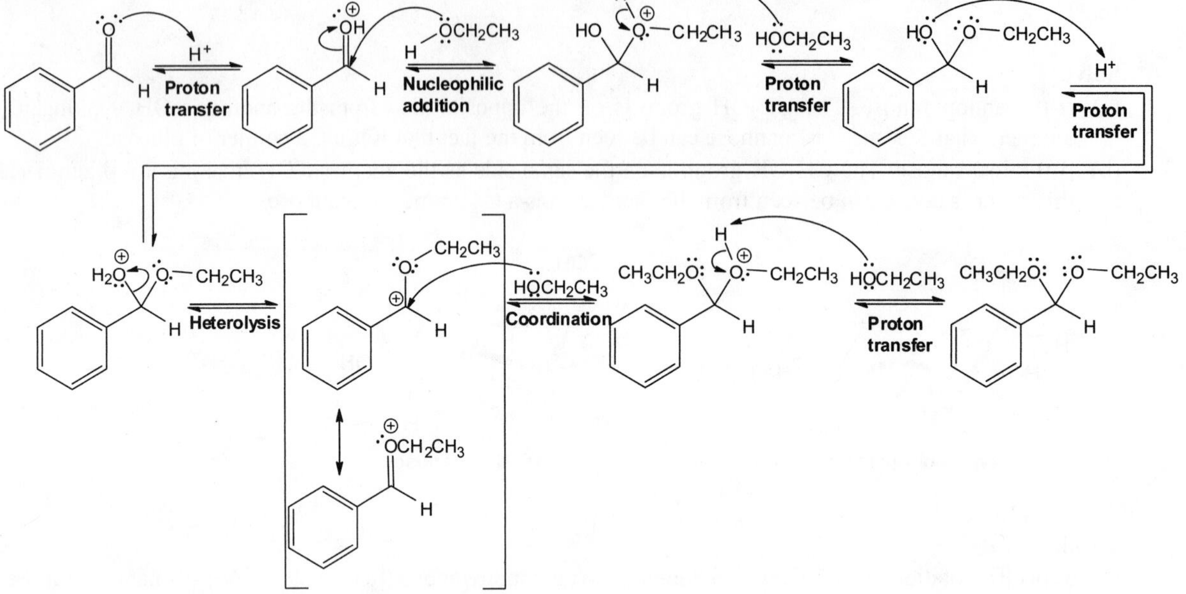

(b) This is an acetal formation reaction in which the reaction is catalyzed by an acid and the diol acts as the nucleophile. The coordination step is intramolecular because it forms a five-membered ring.

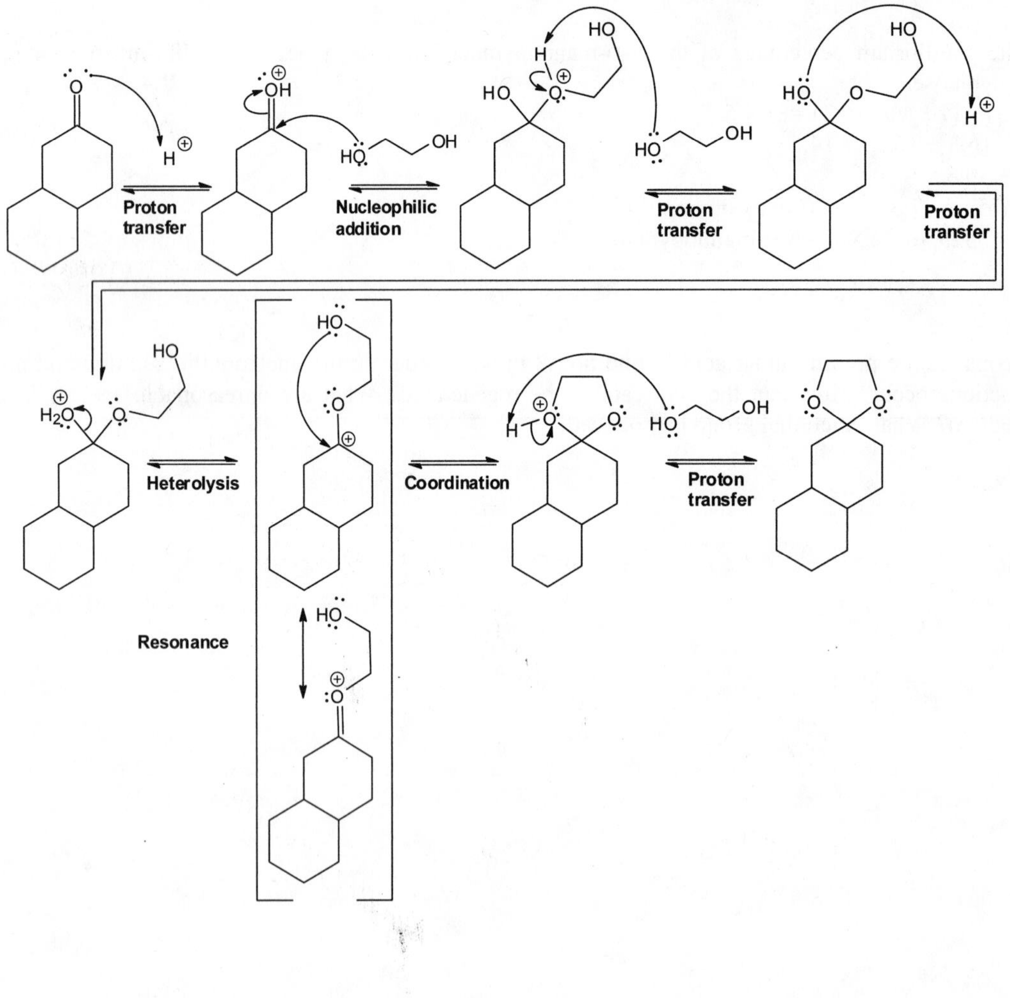

(c) This is an acetal formation reaction in which the reaction is catalyzed by an acid and the ROH acts as the nucleophile.

(d) This mechanism is the same as in **(b)**. The S is more nucleophilic than O because it is larger and can handle the resulting positive charge better, so the S end attacks first. The coordination step is intramolecular because it forms a six-membered ring.

(e) Same mechanism as in **(b)**.

Problem 18.45

Think

What nucleophiles are present under acidic conditions? In what order do the nucleophilic addition and proton transfer steps occur? How can the acid catalyst be regenerated? What are unreasonable species in each reaction medium? What functional group is produced?

Solve

(a) This is an acetal formation reaction in which the reaction is catalyzed by an acid and the diol acts as the nucleophile. The coordination step is intramolecular because it forms a five-membered ring.

Proton transfer
Nucleophilic addition
Proton transfer
Proton transfer
Heterolysis
Resonance
Coordination
Proton transfer

(b) This is an acetal formation reaction in which the reaction is catalyzed by an acid and the diol acts as the nucleophile. The coordination step is intramolecular because it forms a five-membered ring.

Proton transfer
Nucleophilic addition
Proton transfer
Proton transfer
Heterolysis
Coordination
Proton transfer
Resonance

(c) This is an acetal formation reaction in which the reaction is catalyzed by an acid and the diol acts as the nucleophile. The coordination step is intramolecular because it forms a six-membered ring.

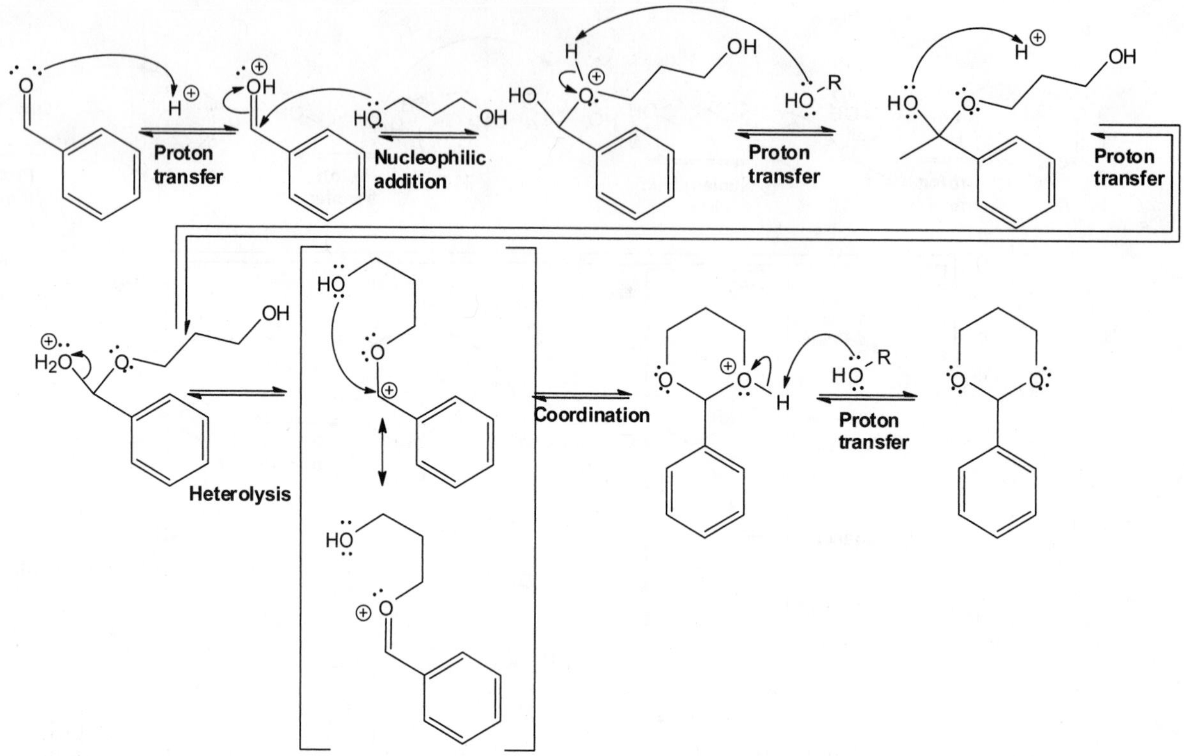

Proton transfer

Nucleophilic addition

Proton transfer

Proton transfer

Heterolysis

Coordination

Proton transfer

Problem 18.46

Think

What nucleophiles are present under acidic conditions? In what order do the nucleophilic addition and proton transfer steps occur? How can the acid catalyst be regenerated? What are unreasonable species in each reaction medium? What functional group is produced when a ketone or aldehyde reacts with NH_3 or a primary amine? When a ketone or aldehyde reacts with a secondary amine? When an imine or enamine reacts with water?

Solve

(a) This is an imine formation. Under acidic conditions, no strong bases should appear. The C=O bond is protonated first, and NH_3 acts as the nucleophile.

H⁺ Proton transfer; NH_3 Nucleophilic addition; NH_3 Proton transfer; H⁺ Proton transfer; Heterolysis; NH_3 Proton transfer

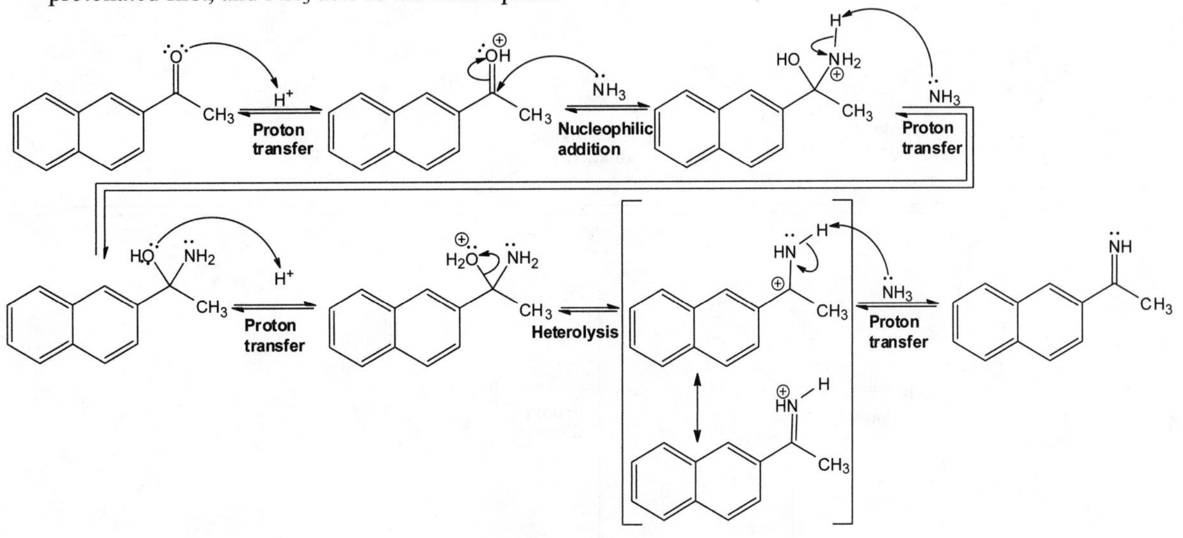

(b) This is an imine formation. No strong bases should appear under acidic conditions. The C=O is protonated first, and the primary amine acts as the nucleophile.

H⁺ Proton transfer; H_2NCH_3 Nucleophilic addition; H_2NCH_3 Proton transfer; H⁺ Proton transfer; Heterolysis; Resonance; H_2NCH_3 Proton transfer

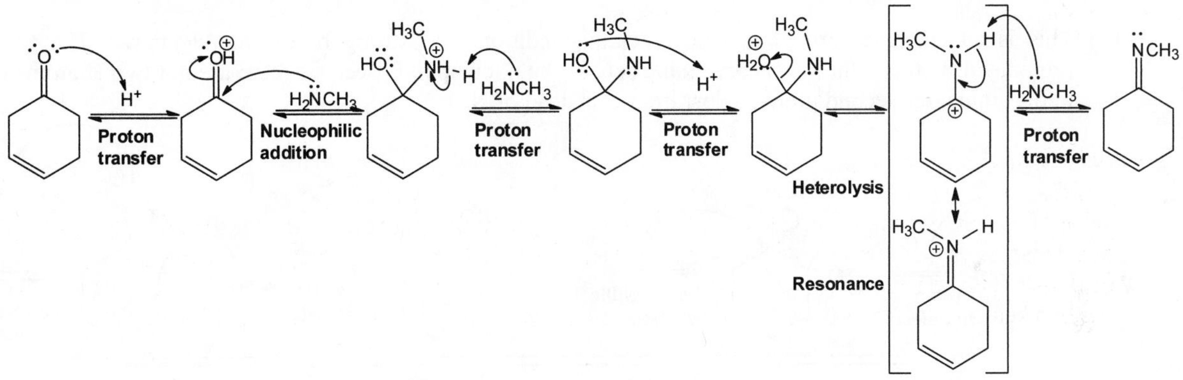

(c) This is an imine hydrolysis to produce a ketone. Under acidic conditions, no strong bases should appear. The C=N is protonated first and water acts as the nucleophile.

H^+ Proton transfer; H_2O: Nucleophilic addition; H_2O: Proton transfer; H^+ Proton transfer; Heterolysis; H_2O: Proton transfer

(d) This is an enamine formation. Under acidic conditions, no strong bases should appear. The C=O is protonated first, and the secondary amine acts as the nucleophile. Because there are not two H atoms on N in the amine, the second proton is lost from C instead of N.

H⊕ Proton transfer; Nucleophilic addition; R_2N Proton transfer; Proton transfer; Heterolysis; R_2N Elimination of H^+

(e) This is an enamine hydrolysis. Under acidic conditions, no strong bases should appear. The C=C is protonated first, and water acts as the nucleophile.

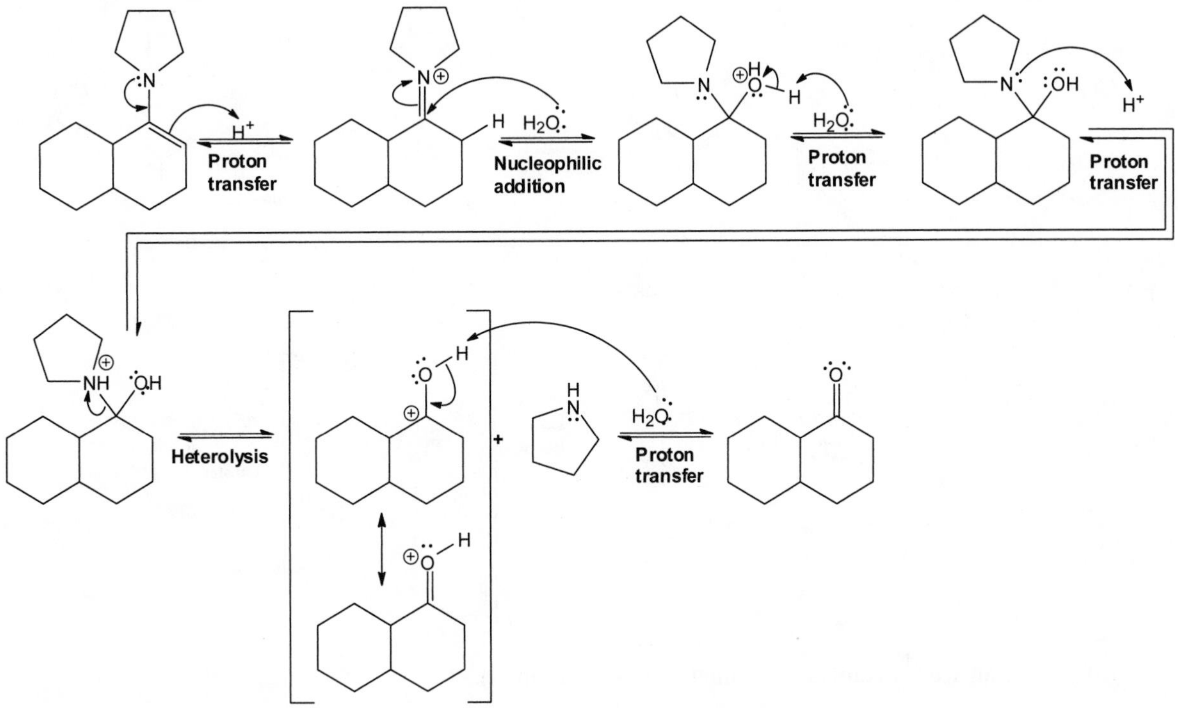

Proton transfer

Nucleophilic addition

Proton transfer

Proton transfer

Heterolysis

Proton transfer

Problem 18.47

Think

What nucleophiles are present under acidic conditions? What are the leaving groups under acidic conditons? In what order do the nucleophilic addition and proton transfer steps occur to avoid unreasonable species in solution? How can the acid catalyst be regenerated? What functional group is produced?

Solve

(a) This is an acetal hydrolysis. Under acidic conditions, no strong bases should appear. The leaving group is an uncharged alcohol, and water acts as the nucleophile.

Proton transfer — Heterolysis — Nucleophilic addition — Proton transfer

Proton transfer — Heterolysis — Proton transfer

(b) This is an acetal hydrolysis, essentially the same as in **(a)**.

Proton transfer — Heterolysis — Coordination — Proton transfer

Proton transfer — Heterolysis — Proton transfer

(c) This is hydrolysis of the nitrile to produce an amide. Under acidic conditions, no strong bases should appear. The C≡N is protonated first, and water acts as the nucleophile.

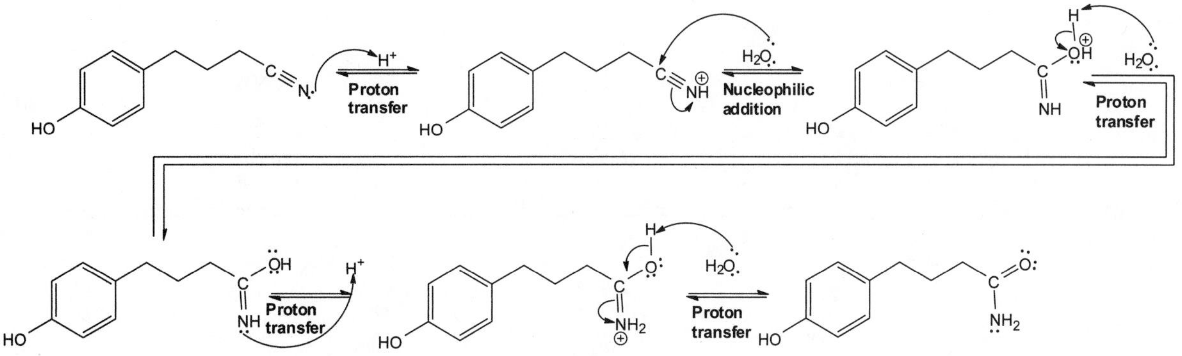

Proton transfer

Nucleophilic addition

Proton transfer

Proton transfer

Proton transfer

(d) This is a hydrolysis of an acetal. Under acidic conditions, no strong bases should appear. The leaving group is an uncharged alcohol, and water is the nucleophile.

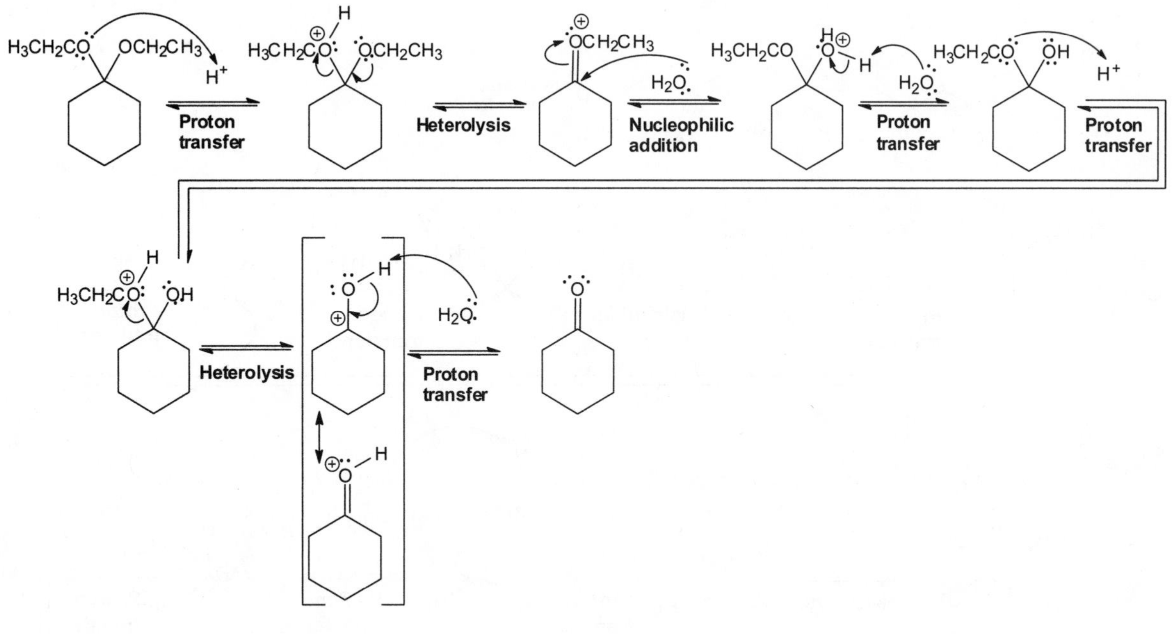

Proton transfer

Heterolysis

Nucleophilic addition

Proton transfer

Proton transfer

Heterolysis

Proton transfer

Problem 18.48

Think

What kind of reaction involves hydrazine and a ketone or aldehyde? What nucleophiles are present under acidic conditions? In what order do the nucleophilic addition and proton transfer steps occur? How can the acid catalyst be regenerated? What are unreasonable species in acidic conditions? Basic conditions? What functional group is produced? How is the base in the second reaction involved?

Solve

Both of these are Wolff–Kishner reductions, which reduce the C=O bond of a ketone or aldehyde to a methylene (CH_2) group. In each case, the first reaction takes place under acidic conditions, so no strong bases should appear. The second reaction takes place under basic conditions, so no strong acids should appear.

(a) Proton transfer; Nucleophilic addition; Proton transfer; Proton transfer; Heterolysis; Proton transfer; add :ÖH⁻ Proton transfer; Proton transfer; E2; Proton transfer

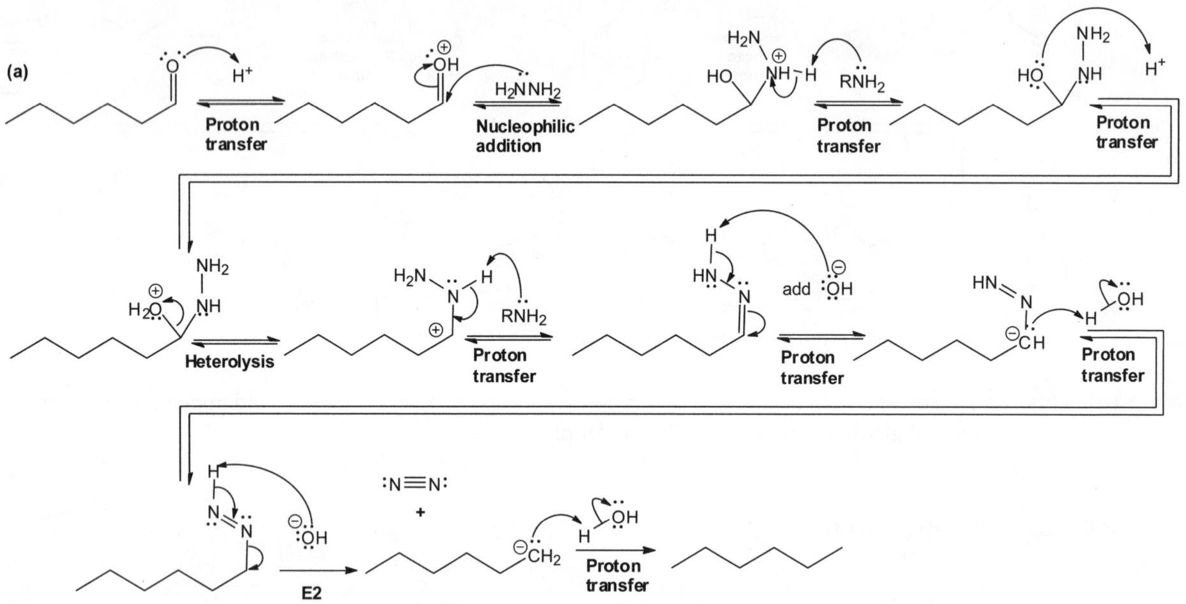

(b) Proton transfer; Nucleophilic addition; Proton transfer; Proton transfer; Heterolysis; Proton transfer; add :ÖH⁻ Proton transfer; Proton transfer; E2; Proton transfer

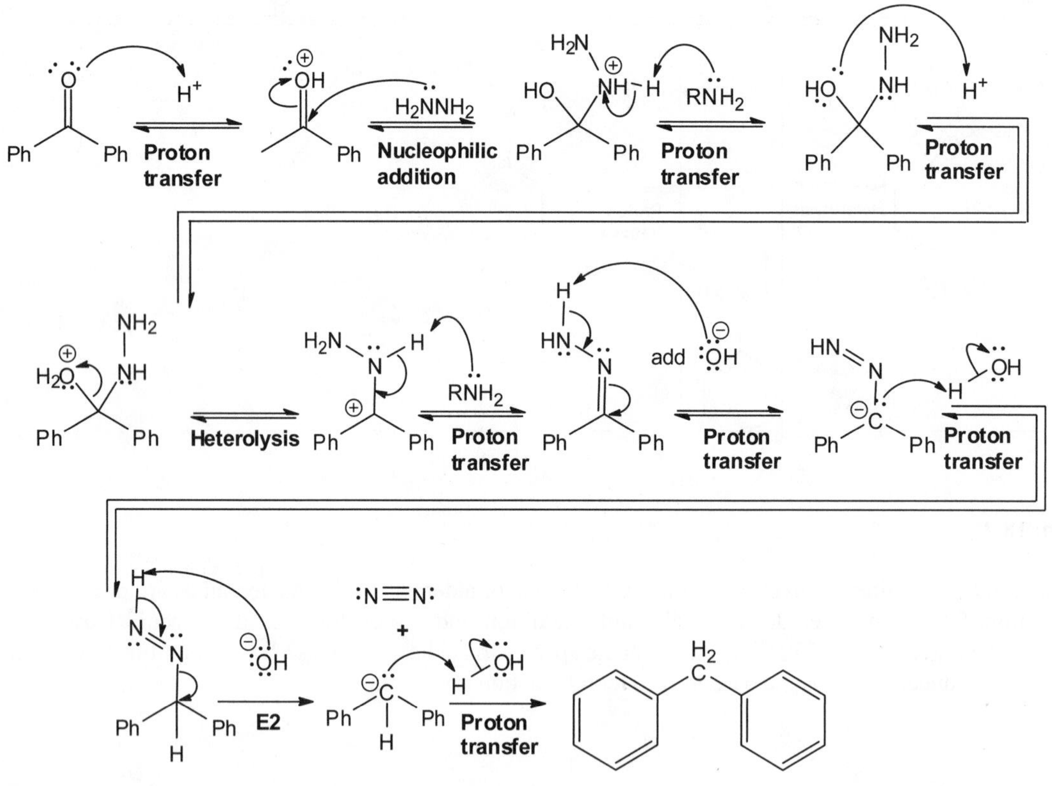

Problem 18.49

Think

What functional group transformation occurs in a Wolff–Kishner reduction? Which C atoms of hexane have at least two H atoms? Can each of those be produced from a C=O using a Wolff–Kishner reduction?

Solve

A Wolff–Kishner reduction reduces a C=O group of a ketone or aldehyde to a CH_2 group. All three distinct C atoms of hexane have at least two H atoms, so any of the three following precursors could be used.

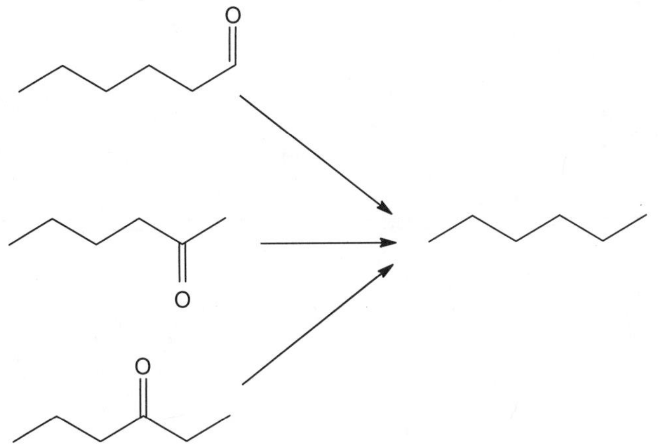

Problem 18.50

Think

What nucleophiles are present under basic conditions? Under acidic conditions? In what order do the nucleophilic addition and proton transfer steps occur to avoid species that are incompatible with the reaction consitions? How can the acid and base catalyst be regenerated?

Solve

In basic solution, $^{18}OH^-$ and $^{18}OH_2$ are present. The reaction is base catalyzed. Notice that the first two steps are nucleophilic addition of $^{18}OH^-$ and a proton transfer. The second two steps are simply the reverse of the first two, just involving a different O atom in each case. This can happen because $^{18}OH^-$ adds reversibly.

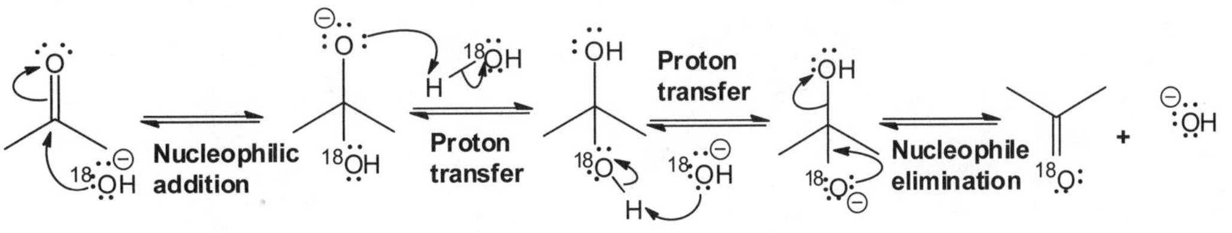

In acidic solution, $H_3{}^{18}O^+$ and $H_2{}^{18}O$ are present. The reaction is acid catalyzed. Again, notice that the last three steps are the reverse of the first three, just involving different O atoms.

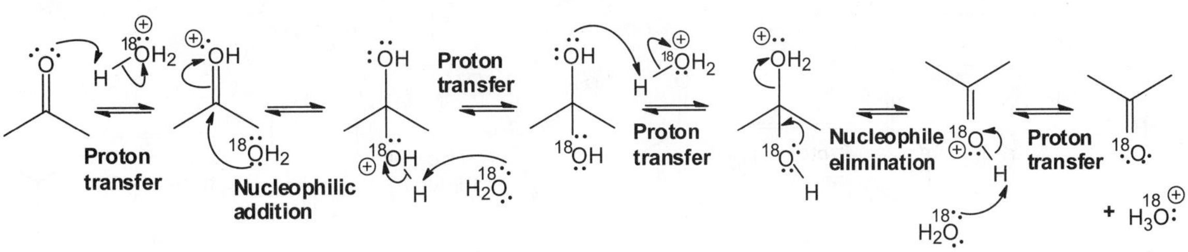

Problem 18.51

Think

What type of reaction takes place among aldehydes under basic conditions? What nucleophiles are present under basic conditions? In what order do the nucleophilic addition and proton transfer steps occur to avoid species that are incompatible with the reaction conditions? How can the base catalyst be regenerated?

Solve

(a) This is an aldol condensation. After the aldol addition to produce a β-hydroxycarbonyl, dehydration occurs via an E1cb mechanism, which is promoted by the added heat. Under basic conditions, no strong acids should appear.

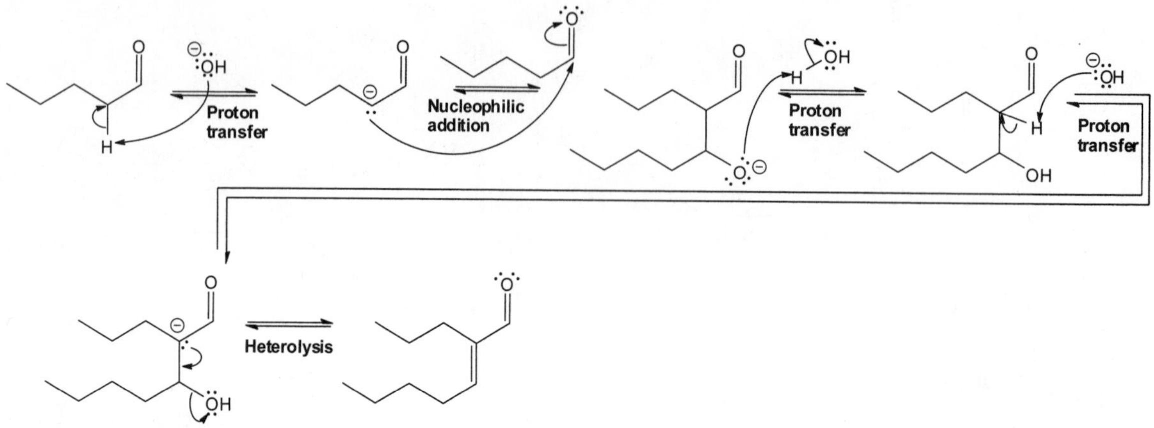

(b) This is an aldol condensation. After the aldol addition to produce a β-hydroxycarbonyl, dehydration occurs via an E1cb mechanism, which is promoted by the added heat. Under basic conditions, no strong acids should appear.

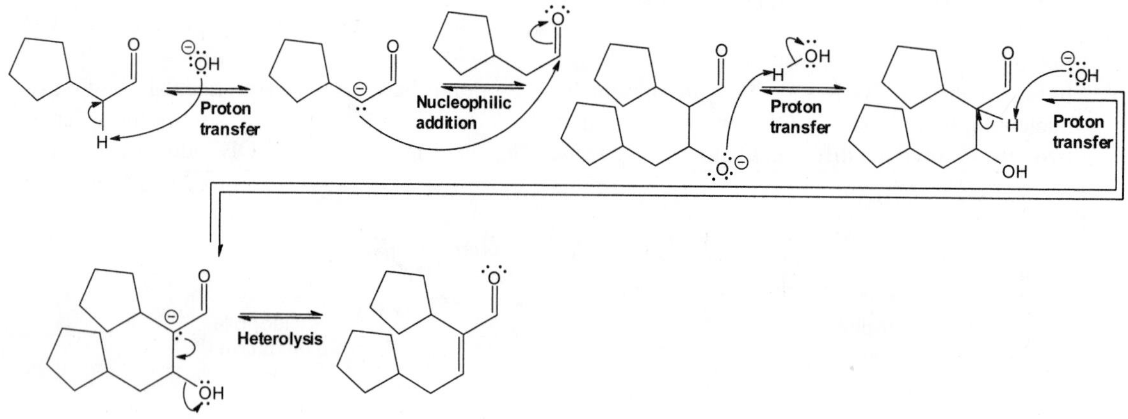

(c) This aldol addition is intramolecular, favored by the formation of the six-membered ring.

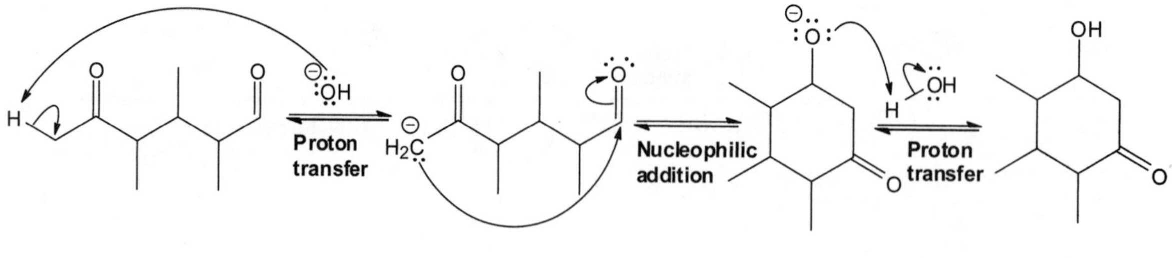

(d) This is an aldol addition. An aldol condensation cannot occur because the α C atom does not have any more hydrogen atoms.

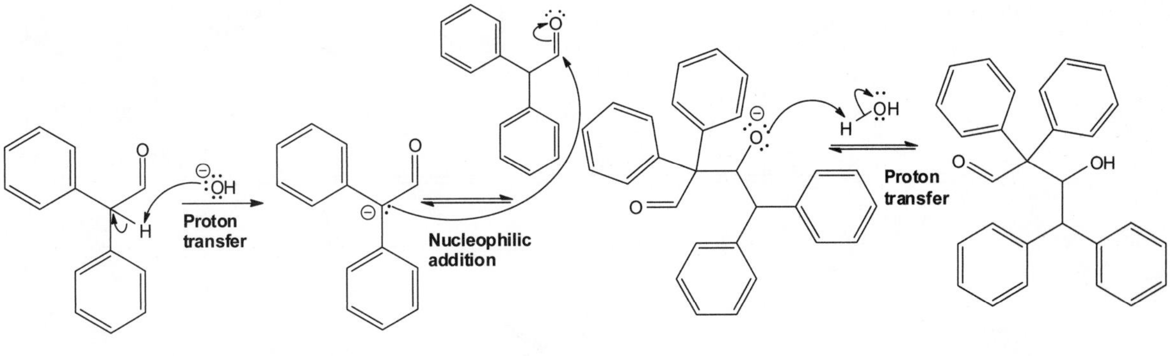

Proton transfer

Nucleophilic addition

Proton transfer

Problem 18.52

Think

What nucleophile is present under acidic conditions? Is it nucleophilic at an α carbon? In what order do the nucleophilic addition and proton transfer steps occur to avoid species that are incompatible with the reaction conditions?

Solve

This aldol condensation reaction takes place under acidic conditions, so no strongly basic species should appear in the mechanism. Therefore, the carbon nucleophile has to be the enol, which is nucleophilic at the α carbon. The enol is produced by back-to-back proton transfer steps.

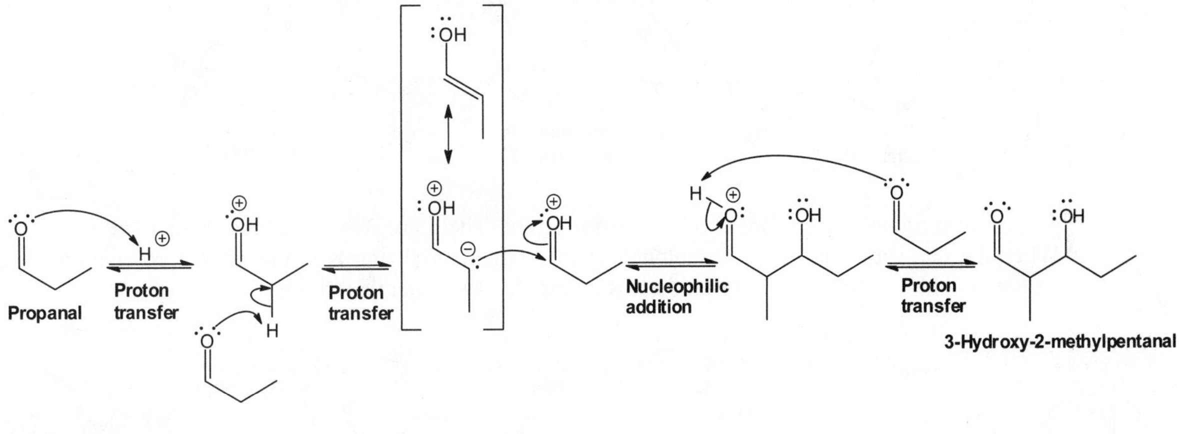

Propanal

Proton transfer

Proton transfer

Nucleophilic addition

Proton transfer

3-Hydroxy-2-methylpentanal

Problem 18.53

Think

Do HO^- and LDA act as bases or nucleophiles in these reactions? What nucleophiles are present under basic conditions? How does the choice of base impact the regiochemistry in the production of an enolate anion? In what order do the nucleophilic addition and proton transfer steps occur to avoid species that are incompatible with the basic conditions? How can the base catalyst be regenerated?

Solve

HO^- and LDA in each of these examples act as bases and deprotonate the α C–H to form the enolate anion.

(a) This is a crossed aldol reaction, where LDA produces the kinetic (less substituted) enolate anion. Condensation occurs, even without heat, due to the conjugation with the aromatic ring in the α,β-unsaturated carbonyl product. See the mechanism on the next page.

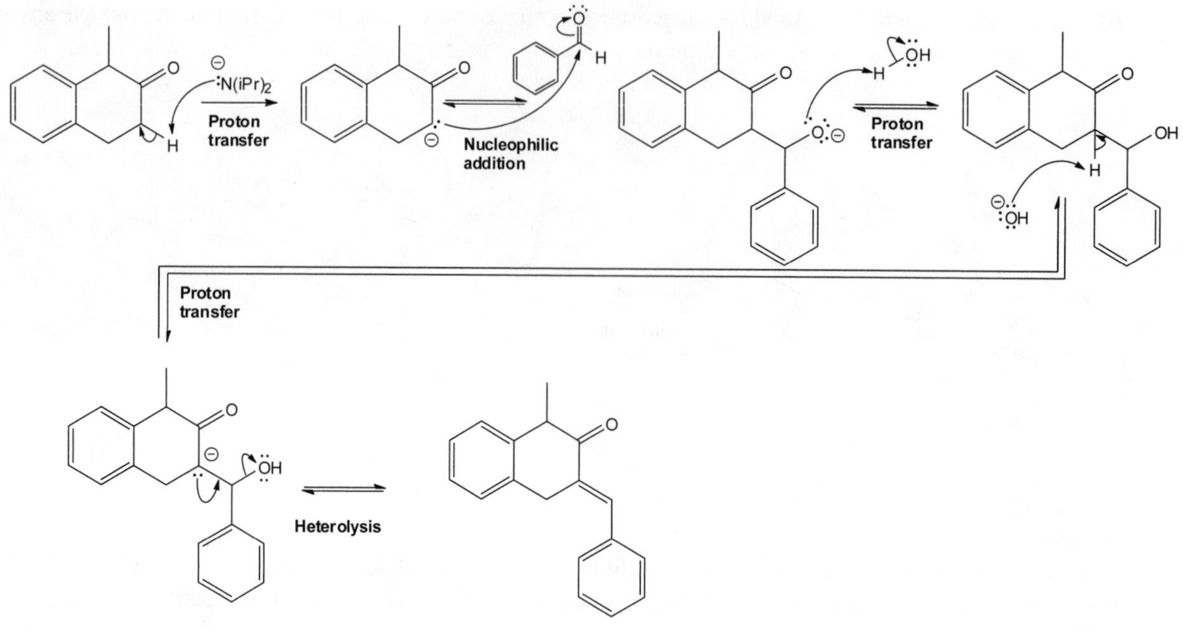

Proton transfer

Nucleophilic addition

Proton transfer

Proton transfer

Heterolysis

(b) This is a crossed aldol reaction, where NaOH produces the thermodynamic (more substituted) enolate anion. An aldol condensation cannot occur because the α C atom does not have any more hydrogen atoms.

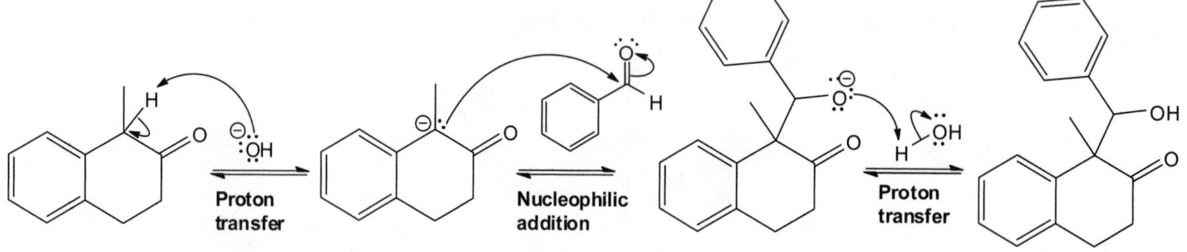

Proton transfer

Nucleophilic addition

Proton transfer

(c) This is an intramolecular aldol condensation reaction. There are two α carbons that can be deprotonated. When the one shown below is deprotonated, a five-membered ring is produced. Deprotonating the other carbon would lead to the formation of a less favorable seven-membered ring.

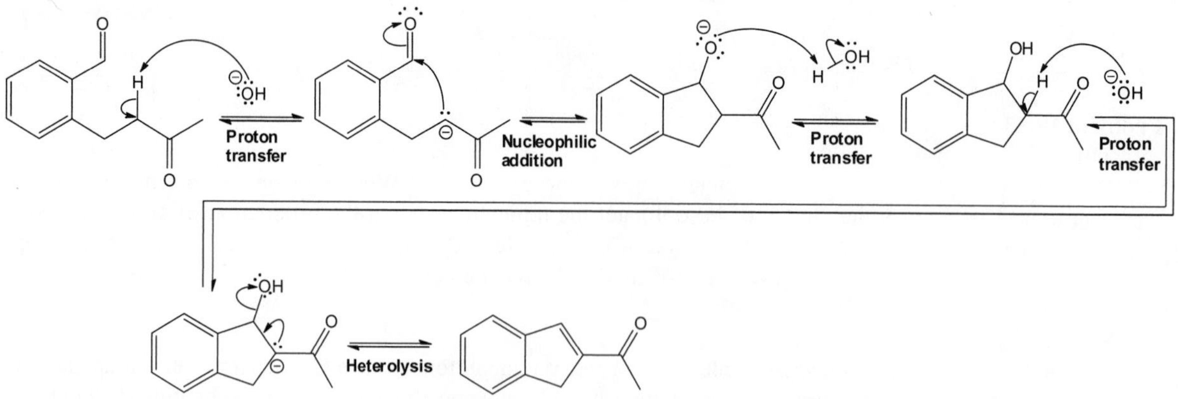

Proton transfer

Nucleophilic addition

Proton transfer

Proton transfer

Heterolysis

(d) LDA ensures that the enolate anion is produced quantitatively in the first reaction. In the second reaction, the enolate anion attacks the aldehyde preferentially over the ketone. Condensation occurs, even without heat, due to the conjugation with the aromatic rings in the α,β-unsaturated carbonyl product.

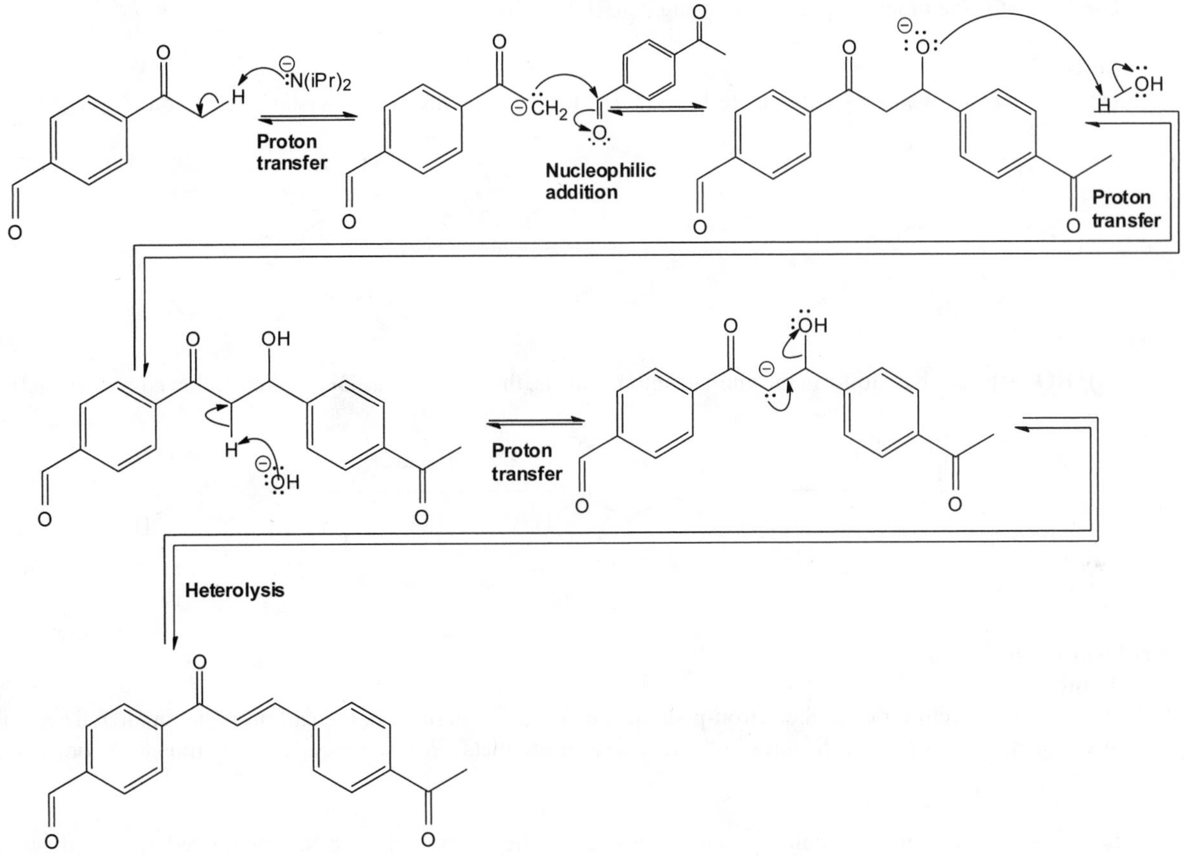

Proton transfer

Nucleophilic addition

Proton transfer

Proton transfer

Heterolysis

Problem 18.54

Think

Compare the sterics of the two products. Is there a significant difference? Which C=O has a more electron-poor C atom? How does this affect the amount of aldol product in equilibrium?

Solve

The difluoroketone **B** will produce more product at equilibrium. F and H atoms are about the same in size (F is small because it is so highly electronegative), so the two molecules have about the same steric bulk surrounding their respective carbonyl carbons. But the F atoms make the carbonyl carbon more electron deficient, which favors attack of a nucleophile.

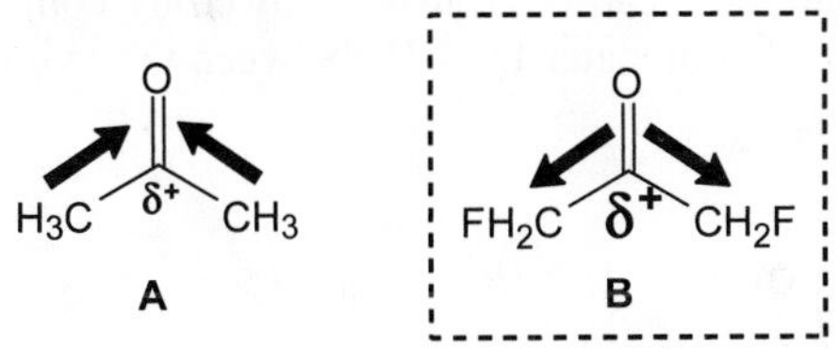

H_3C δ+ CH_3

A

FH_2C δ+ CH_2F

B

Problem 18.55

Think

For **(a)**, does the addition occur via 1,2- or 1,4-addition? What is the product of the Wolff–Kishner reduction? For **(b)**, does the HO^- act as a base or nucleophile? What reaction takes place when an aldehyde is treated with base? What is the product of reduction using $NaBH_4$?

Solve

(a) This reaction involves a conjugate 1,4-addition followed by the Wolff–Kishner reduction.

1. $(CH_3)_2CuLi$

2. H_2NNH_2, HO^-, Δ

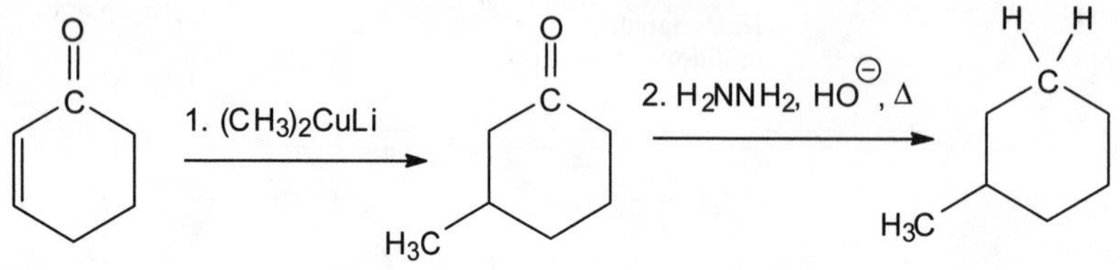

(b) HO^- acts as a base to form the enolate anion. This is, therefore, an aldol addition followed by a reduction.

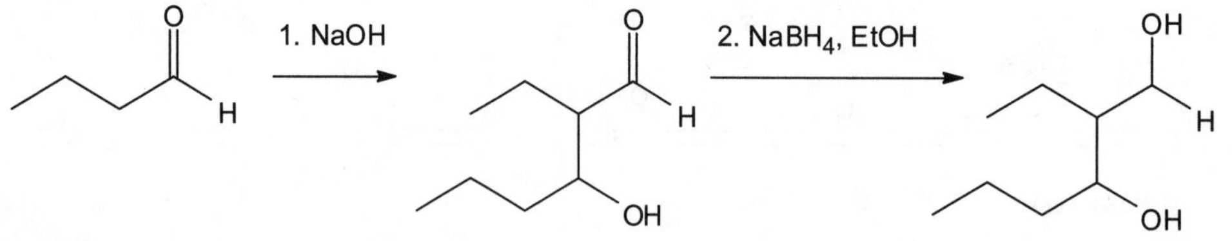

1. NaOH

2. $NaBH_4$, EtOH

Problem 18.56

Think

What are the electron-rich and electron-poor species in each reaction? Is the carbon skeleton altered? Are there any characteristic structural features in the reactants or products? What functional group transformations occur?

Solve

(a) This is a Robinson annulation, characterized by the formation of a six-membered ring with an α,β-unsaturated carbonyl. The carbon skeleton from **A** is boxed below.

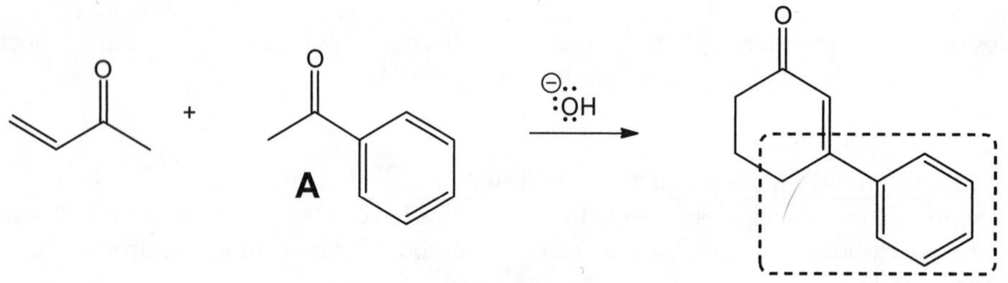

A

$^-$:OH

(b) This is a crossed aldol condensation to produce an α,β-unsaturated carbonyl; it is followed by conjugate addition of an isopropyl group. In the aldol reaction, the HO^- deprotonates the CH_2 between the two C=O groups irreversibly.

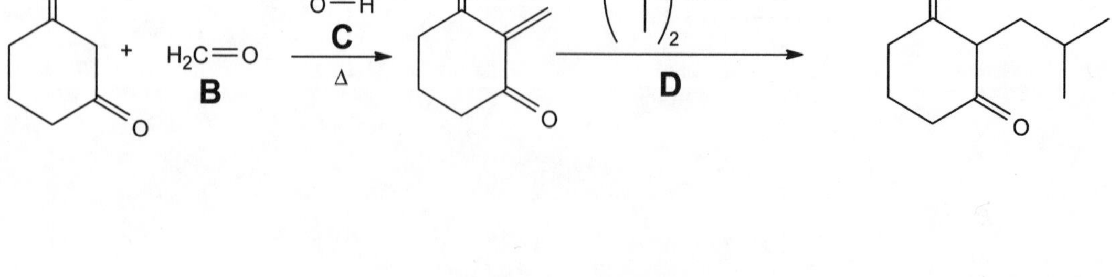

$H_2C{=}O$ **B**

$^-$O–H **C** Δ

D

(c) A Grignard addition to isobutyraldehyde is followed by aqueous workup. If aqueous H_2SO_4 is used in Step 2, then Step 3 follows immediately.

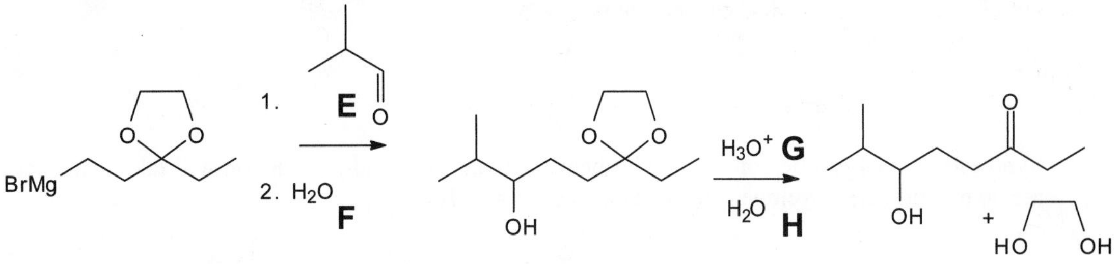

1. **E**

2. H_2O **F**

H_3O^+ **G**

H_2O **H**

(d) This nucleophilic addition of an amine takes place under acidic conditions to form an imine, followed by a reduction to form the amine.

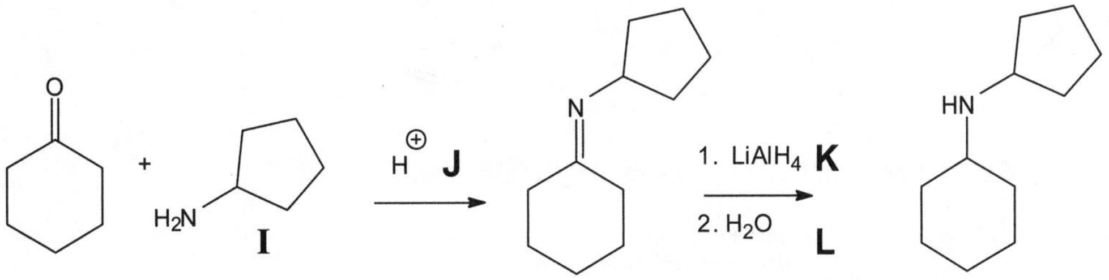

I

$H^{\oplus}$ **J**

1. $LiAlH_4$ **K**

2. H_2O **L**

Problem 18.57

Think

What functional groups are present in each of the five products? Does the relative locations of those functional groups suggest that a particular reaction should be used? What is the structure of cyclopentanone? In addition to cyclopentanone, what other reagents can be used in that reaction to produce the target?

Solve

Each of the products is an α,β-unsaturated ketone or nitrile, which can be produced from an aldol condensation reaction with cyclopentanone. The routes to synthesize **(a)–(e)** are given below.

(a) A simple aldol condensation (with heat to promote dehydration) suffices for the first synthesis. Retrosynthetic analysis: The enolate anion of one cyclopentanone will attack the second molecule.

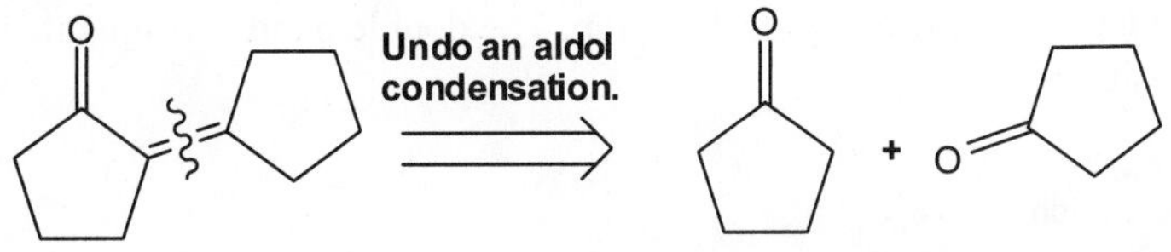

Undo an aldol condensation.

Synthesis:

NaOH, Δ

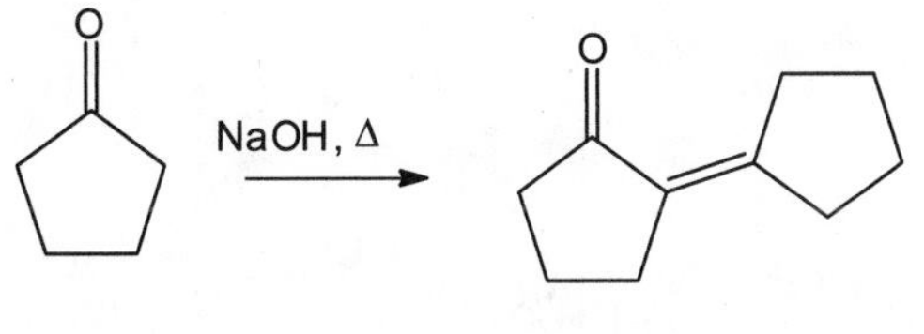

(b) A crossed aldol with dehydration works here. Retrosynthetic analysis:

Undo a crossed aldol condensation.

+

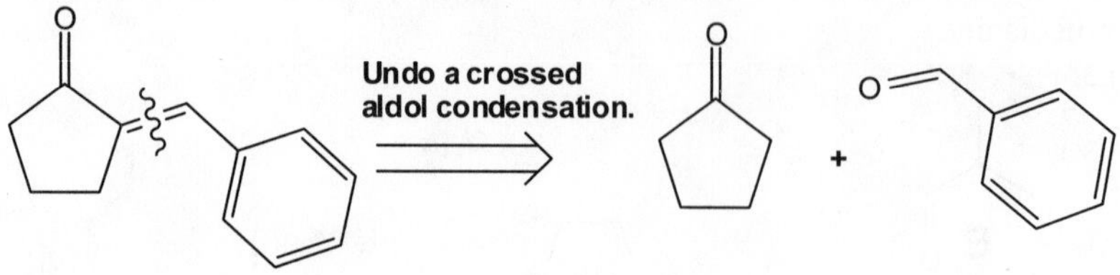

Synthesis: NaOH can be combined with benzaldehyde first, and no reaction will take place because there are no α hydrogens. Cyclopentanone is then added slowly.

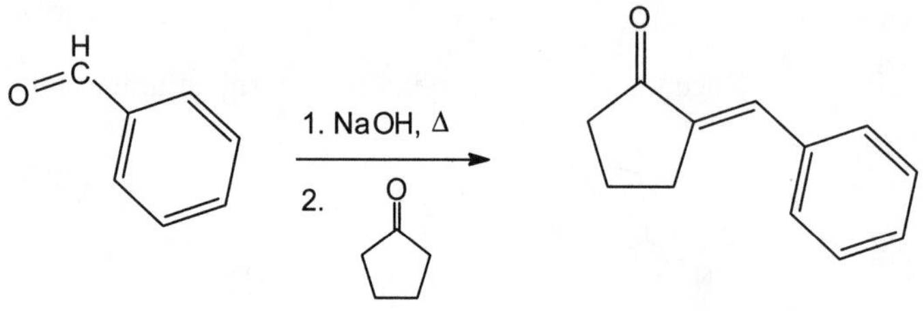

H

1. NaOH, Δ

2.

(c) A crossed aldol reaction, followed by dehydration. Retrosynthetic analysis:

Undo a crossed aldol condensation.

+

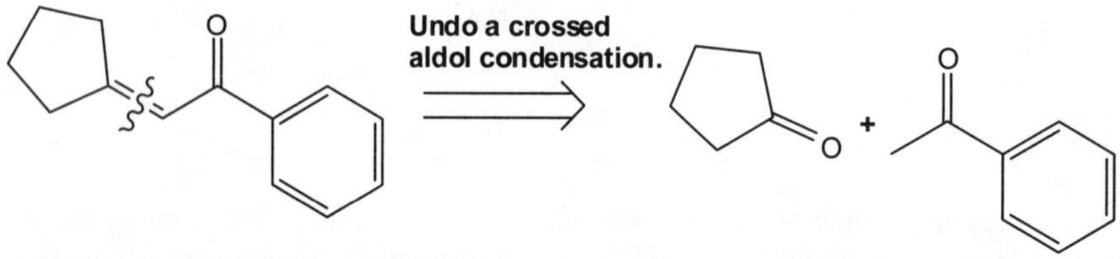

Synthesis: Because both ketones have α carbons, the crossed aldol is achieved by quantitatively deprotonating one ketone first, then adding the second ketone.

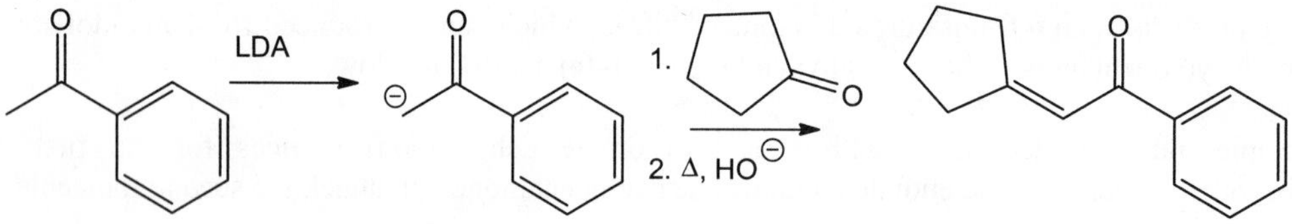

LDA

1.

2. Δ, $HO^{\ominus}$

(d) Aldol reaction involving a nitrile enolate, followed by dehydration. The double bond is conjugated with the nitrile group. Retrosynthetic analysis:

CN

Undo a crossed aldol condensation.

CN

+

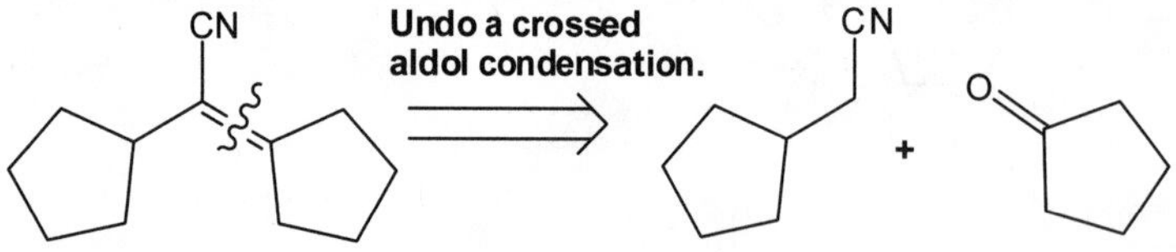

Synthesis:

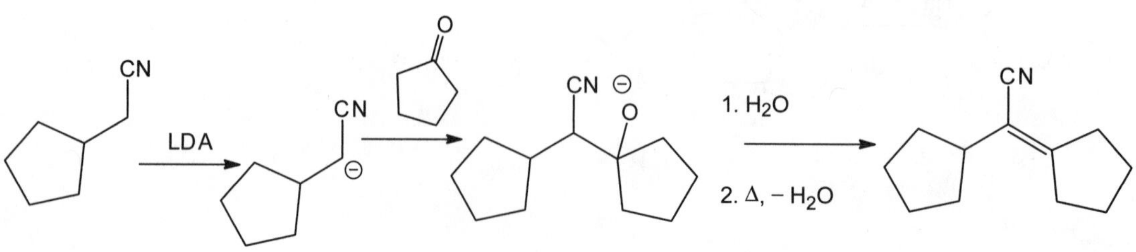

CN

LDA

CN

CN

1. H_2O

2. Δ, $-H_2O$

CN

(e) Retrosynthetic analysis shows that a Robinson annulation is possible. Retrosynthetic analysis:

Undo a Robinson annulation.

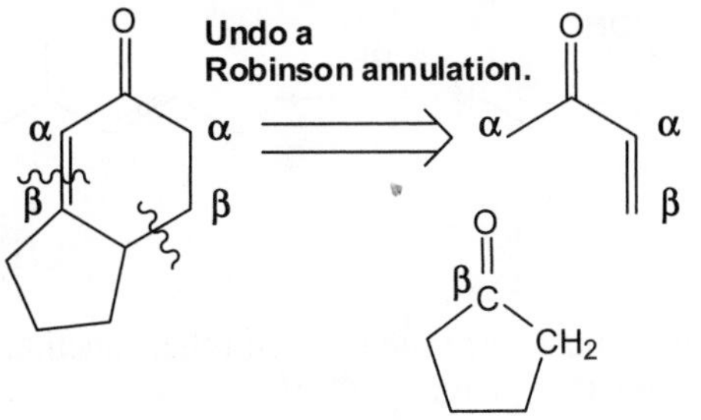

Synthesis: Executing this synthesis involves making the enolate of cyclopentanone and performing a conjugate addition first. To close the ring, an intramolecular aldol reaction is done, followed by dehydration of the adduct.

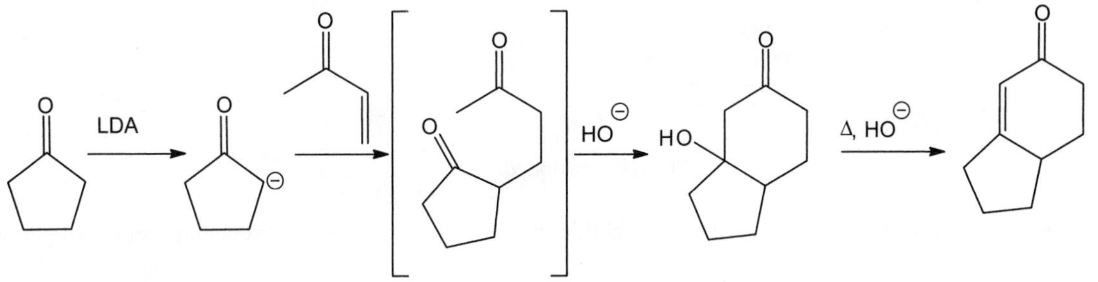

Problem 18.58

Think

How many C atoms are in the product? How many C–C bonds must be formed? What other functional groups are present? What is the relative positioning of those functional groups?

Solve

With only two-carbon precursors allowed, and six C atoms in the product, we must join three molecules of two C atoms each. So in the retrosynthetic analysis, we should disconnect the target C atoms two at a time. We can first try to disconnect the C–C bond below.

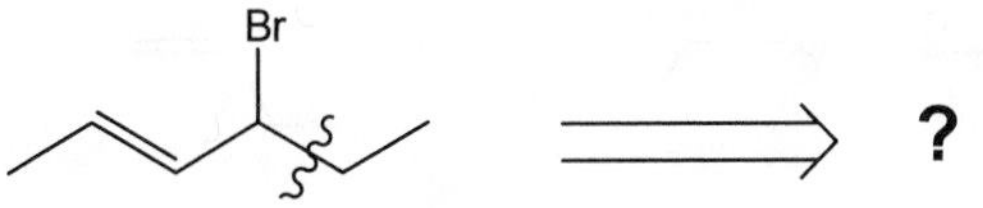

However, we don't know of a reaction that, in the forward direction, generates a C–C bond and leaves us with an alkyl halide on one of the C atoms. On the other hand, a Grignard reaction will make such a bond and leave us with an OH group.

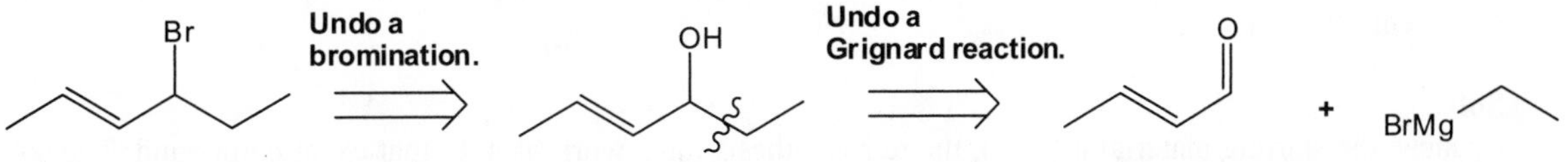

Undo a bromination.

Undo a Grignard reaction.

Synthesis of the unsaturated aldehyde involves an aldol condensation reaction. The retrosynthesis is given below.

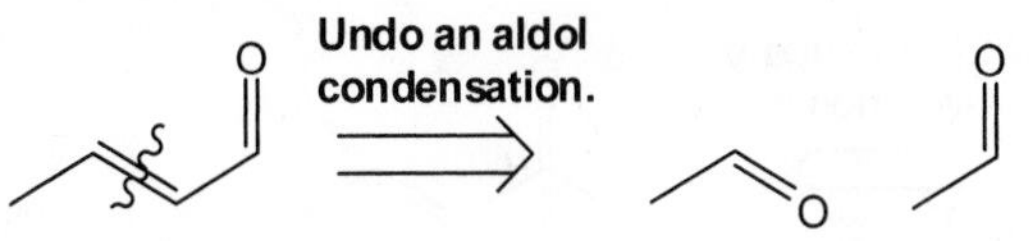

Undo an aldol condensation.

Executing the synthesis involves filling in the necessary reagents for each step.

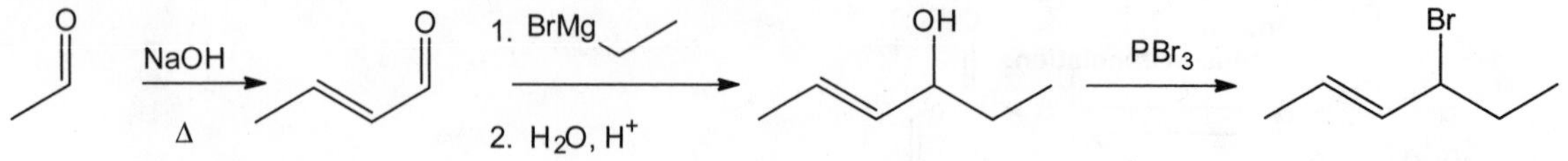

NaOH, Δ; 1. BrMg, 2. H_2O, H^+; PBr_3

Problem 18.59

Think

How many C atoms are in the product? How many C–C bonds must be formed? What other functional groups are present? What reactions do you know that will produce a new C=C bond?

Solve

If we disconnect the C=C double bond, we obtain the same carbon backbone as 2-phenylacetaldehyde, so we should consider undoing a Wittig reaction, giving us 2-phenylacetaldehyde and a Wittig reagent as precursors.

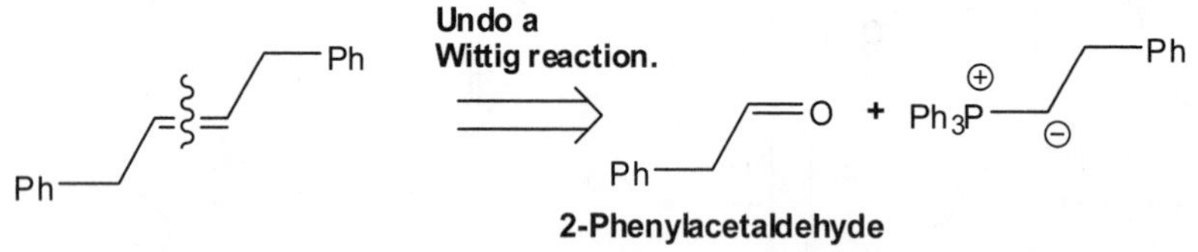

Undo a Wittig reaction.

2-Phenylacetaldehyde

The Wittig reagent is made from an alkyl halide, which can be obtained by reducing benzaldehyde to benzyl alcohol.

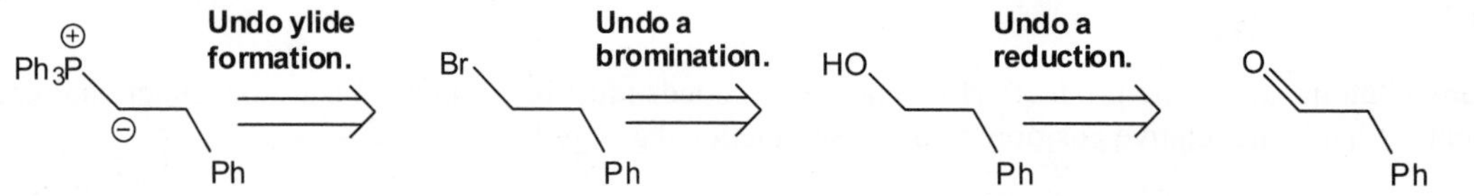

Undo ylide formation. Undo a bromination. Undo a reduction.

Synthesis:

$NaBH_4$; PBr_3; 1. PPh_3, 2. BuLi

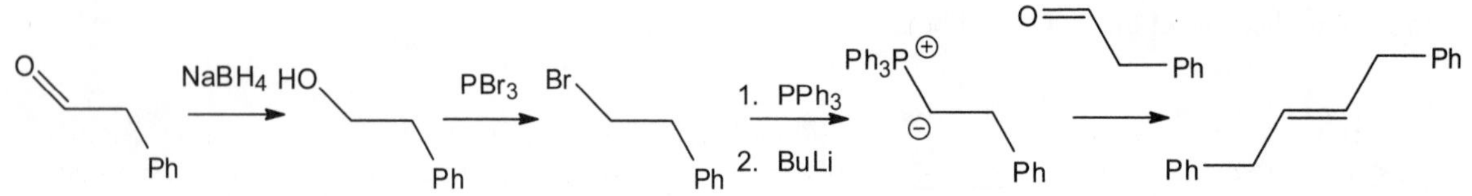

Problem 18.60

Think

How many C atoms are in the product? How many C–C bonds must be formed? What other functional groups are present in the starting material? In the product? What reactions do you know that will remove a functional group entirely?

Solve

Because the starting material is given, the retrosynthesis must work back to that exact compound. The solution becomes evident if you ignore the numbering in the product and realize that you have to add a methyl group to C5 of the ketone. This can be done by a conjugate addition. Then you have to remove all the functional groups. The C=O group can be converted to CH_2 using a Wolff–Kishner reduction.

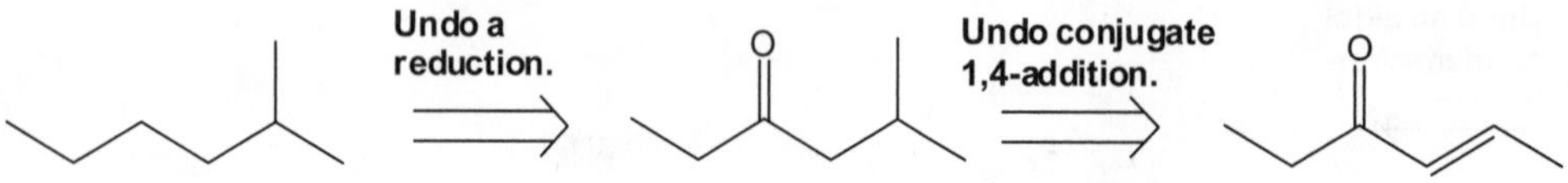

Undo a reduction. Undo conjugate 1,4-addition.

Synthesis:

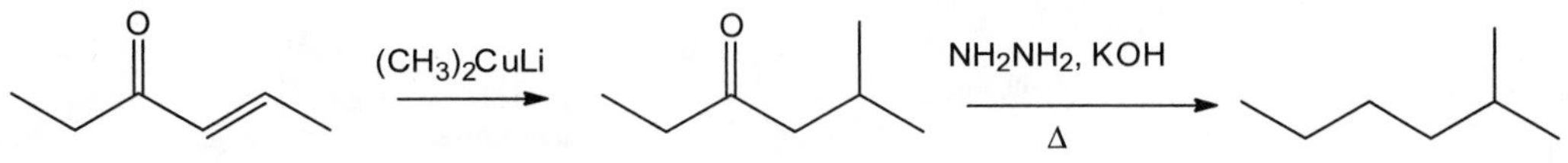

$(CH_3)_2CuLi$; NH_2NH_2, KOH, Δ

Problem 18.61

Think

Which C–C bond was formed? What is the identity of the nucleophile in the nucleophilic addition reaction? Did the addition occur via 1,2- or 1,4-addition? Are proton transfer reactions involved?

Solve

The nucleophilic carbon of the enamine does a conjugate 1,4-addition to the ketone. Water adds to the C=N double bond, starting a hydrolysis that eventually forms the diketone. Under acidic conditions, water is the nucleophile in the hydrolysis reaction, and no strong bases should appear.

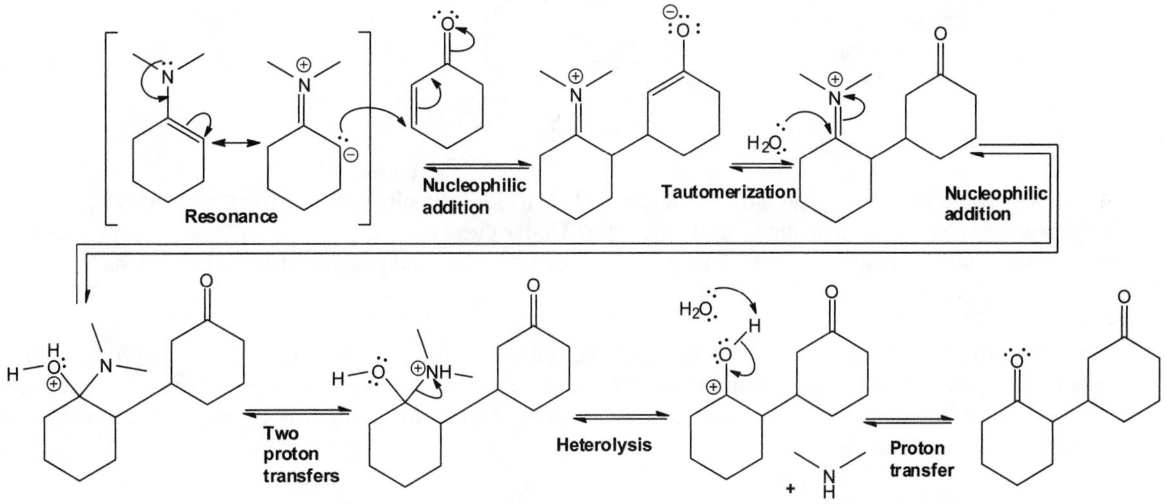

Resonance; Nucleophilic addition; Tautomerization; Nucleophilic addition; Two proton transfers; Heterolysis; Proton transfer

Problem 18.62

Think

What is TsOH? What acts as the nucleophile in the nucleophilic addition reaction? Is oxygen or nitrogen more nucleophilic? Is the reaction conducted in acidic or basic solution? What species are not permitted in the mechanism? Can a new five- or six-membered ring be produced?

Solve

TsOH is an acid catalyst that activates the carbonyl group towards nucleophile addition. Nitrogen is the more nucleophilic atom in aminoethanol, so it adds to the carbonyl C first. The carbonyl O is protonated again to form a good leaving group (water), and an imine forms. Now the oxygen can add to the imino C atom. The final product is a cyclic **hemiaminal**—a functional group that has O and N attached to the same C atom. See the mechanism on the next page.

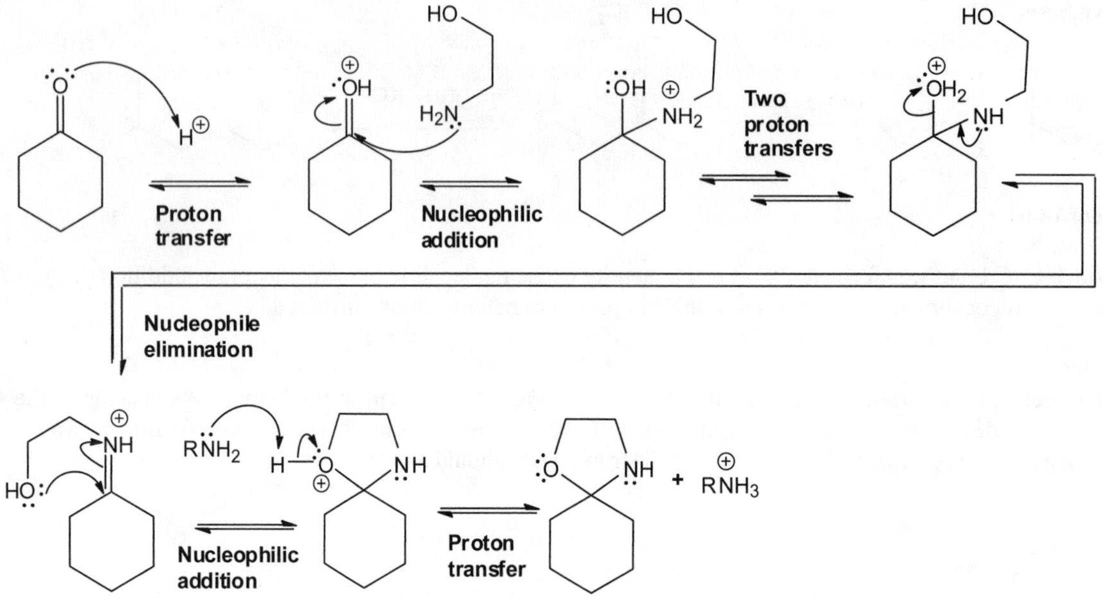

Proton transfer

Nucleophilic addition

Two proton transfers

Nucleophile elimination

Nucleophilic addition

Proton transfer

Problem 18.63

Think

Is the nitro group electron donating or withdrawing? How does the nitro group affect the acidity of the α C–H? Is the reaction conducted in acidic or basic medium? Under these conditions, what acts as the nucleophile in the nucleophilic addition reaction? How can a good leaving group be produced under these conditions?

Solve

Protonation of a nitro group makes it even more electron withdrawing than normal. Loss of a proton from the α carbon forms an *N*,*N*-dihydroxyiminium ion—still quite electron withdrawing. The α carbon is susceptible to nucleophilic addition of water. Two more proton transfers turn N into a good leaving group (the molecule lost is *N*-hydroxyhydroxylamine).

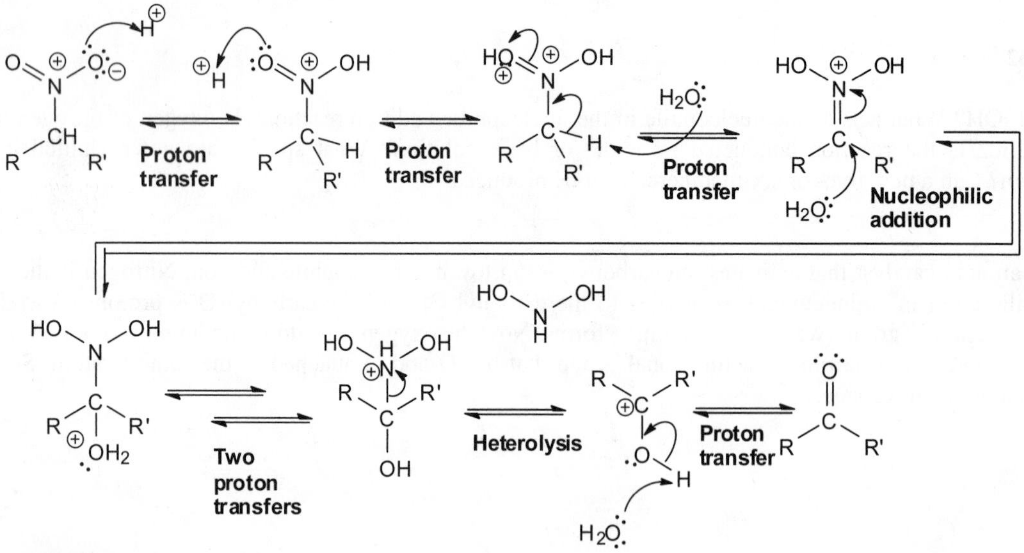

Proton transfer

Proton transfer

Proton transfer

Nucleophilic addition

Two proton transfers

Heterolysis

Proton transfer

Problem 18.64

Think

How is the nitrile similar to a carbonyl? What acts as the nucleophile in the nucleophilic addition reaction? How is the N transformed into a good leaving group? Is the reaction conducted in acidic or basic solution? What species are not permitted in the mechanism?

Solve

Nucleophilic addition of an alcohol to a protonated C≡N triple bond is followed later by nucleophilic addition of water to a protonated C=N double bond. Pay attention to the fact that the reaction takes place under acidic conditions, so no strong bases should appear.

Proton transfer — Nucleophilic addition — Proton transfer — Proton transfer

Nucleophilic addition — Two proton transfers — Heterolysis — Proton transfer

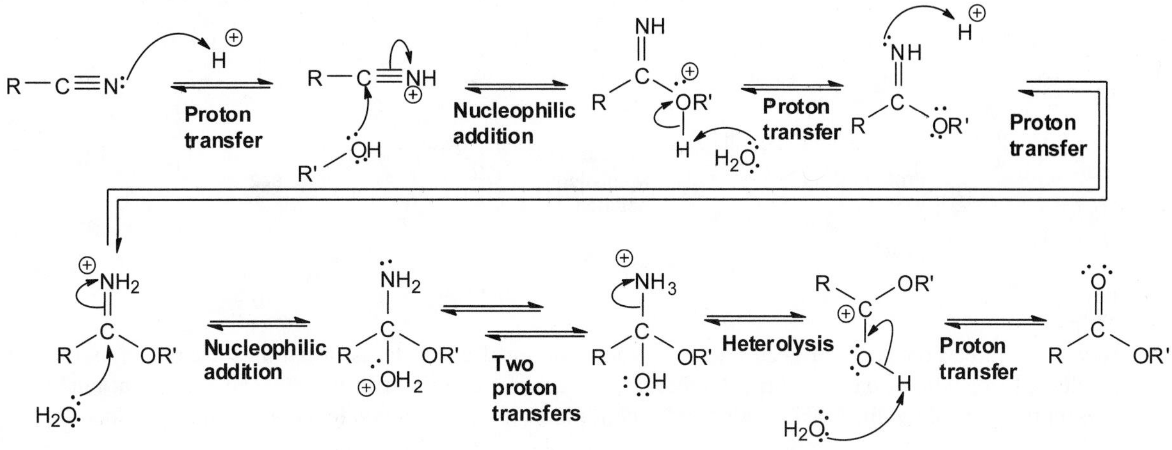

Problem 18.65

Think

Which base is stronger, LDA or HO⁻? Compare the acidity of the α C–H of an imine to that of an aldehyde. What is the mechanism for the aldol addition reaction? What species in this reaction are similar to the aldol addition reaction? Is the reaction conducted in acidic or basic solution? What species are not permitted in the mechanism?

Solve

(a) The α C–H hydrogen of an imine is much less acidic than that of a ketone or aldehyde, because an N atom cannot accommodate a negative charge as well. So hydroxide is simply not a strong enough base to generate the enolate anion.

(b) The mechanism that parallels an aldol addition mechanism is shown in the first three steps below. Under these basic conditions, no strong acids should appear. Then, hydrolysis under acidic conditions takes place to convert the imine to the aldehyde. Under these acidic conditions, no strong bases should appear.

Proton transfer — (Add acid) Proton transfer — Proton transfer — Proton transfer — Nucleophilic addition

Two proton transfers — Heterolysis — Proton transfer

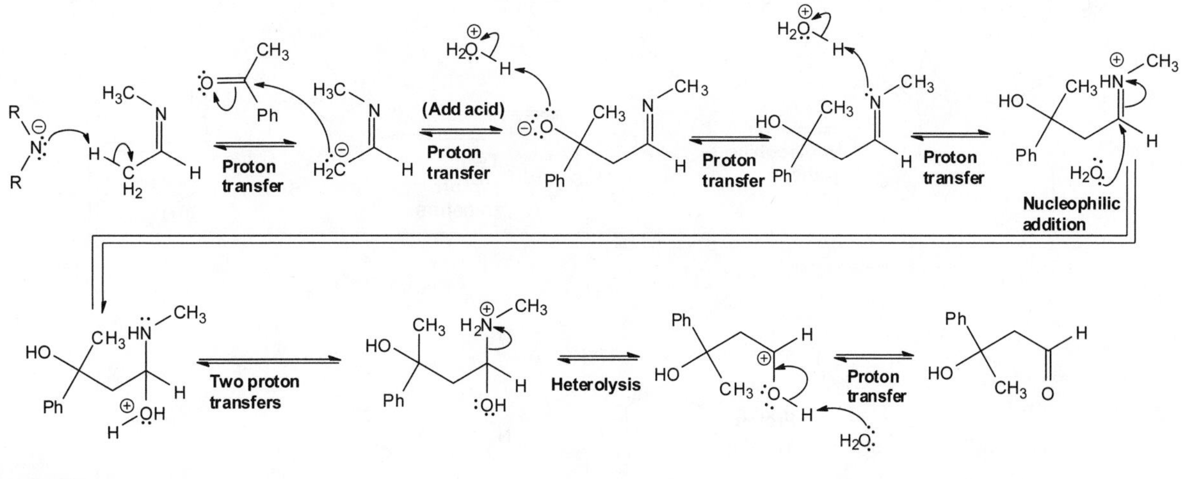

Problem 18.66

Think

How has the carbon skeleton been altered? What functional group transformation took place? Is the reaction conducted in acidic or basic solution? What species are not permitted in the mechanism? What nucleophile can add to the C=O of the aldehyde? In that nucleophilic addition product, are there a good nucleophile and a good leaving group?

Solve

Deprotonation of the chloroester leads to an aldol-type addition. The adduct does an S_N2 reaction on the α carbon, displacing chloride to form the epoxide. The reaction takes place in basic medium; therefore, no strong acids should appear.

Proton transfer

Nucleophilic addition

S_N2

Problem 18.67

Think

How has the carbon skeleton been altered? What functional group transformation took place? Is the reaction conducted in acidic or basic solution? What species are not permitted in the mechanism? What nucleophile is present that can add to the C≡N? Under acidic conditions, how can a good leaving group be produced?

Solve

By counting carbons, we can see that two molecules of the nitrile have joined by an aldol-type reaction. The α carbon of one nitrile is attached to the terminal C1 of the other nitrile molecule. In the mechanism, ethoxide removes an α proton to form an enolate. The enolate attacks another molecule of nitrile. Under these basic conditions, no strong acids should appear. Upon addition of acid, an imine group is formed, which is hydrolyzed to make the ketone. Under these acidic conditions, no strong base should appear.

Proton transfer

Nucleophilic addition

Add acid

Proton transfer

Proton transfer

Nucleophilic addition

Two proton transfers

Heterolysis

Proton transfer

Problem 18.68

Think

How is the isonitrile similar to a carbonyl? What acts as the nucleophile in the nucleophilic addition step? Is the reaction conducted in acidic or basic solution? What species are not permitted in the mechanism?

Solve

Protonation of the isonitrile group forms a group that is reminiscent of a regular nitrile that has been protonated at N. Thus, it is activated toward nucleophilic attack by water. The adduct is the enol form of the final amide product, and tautomerizes in two proton transfer steps.

Resonance
Proton transfer
Nucleophilic addition
Proton transfer
Proton transfer
Proton transfer

Problem 18.69

Think

What happens to an amine at low pH? Is the lone pair still present? Is the amine nucleophilic? How is an alcohol different?

Solve

If the solution pH is too low, then N is protonated to form R–NH_2^+. A protonated N has no lone pair of electrons and, further, carries a +1 formal charge, so it is a terrible nucleophile. This is not a problem for acetal formation, because the nucleophile is an alcohol, which requires the pH to be much more acidic to become protonated.

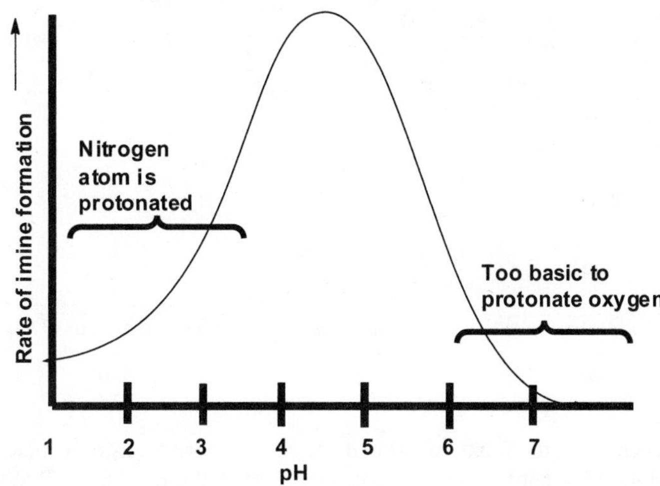

Problem 18.70

Think

What is the Lewis structure of the hydroxylamine? What is the nucleophile in each of the two possible nucleophilic addition reactions? Which atom, N or O, is a stronger nucleophile?

Solve

(a) We will assume a pH of about 5. (See the previous problem for more details as to why.) Acetal formation is shown first.

Proton transfer

Nucleophilic addition

Two proton transfers

Heterolysis

Coordination

Proton transfer

Acetal

Next, formation of the oxime is shown.

Proton transfer

Nucleophilic addition

Two proton transfers

Heterolysis

Proton transfer

Oxime

(b) Formation of the oxime is the favored reaction, because N is more nucleophilic than O.

Problem 18.71

Think

How has the carbon skeleton been altered? What functional group transformation took place? Is the reaction conducted in acidic or basic solution? What species are not permitted in the mechanism? What reaction takes place with NC^- and the aldehyde? In the product of that reaction, what does the presence of CN do to the acidity of the proton attached to the initial aldehyde C?

Solve

All the C atoms are accounted for by combining the two reactants, so cyanide must be acting as a catalyst. In the first two steps of the mechanism, NC^- adds to the C=O group and the resulting O^- is protonated. In that product, the H that was originally the aldehyde proton is much more acidic, due to the electron-withdrawing CN group. Therefore, the third step deprotonates the benzylic C to make an enolate anion.

Nucleophilic addition

Proton transfer

Proton transfer

The resonance stabilized anion attacks another equivalent of benzaldehyde in a nucleophilic addition mechanism. Finally, the catalyst (cyanide) is expelled.

Nucleophilic addition

Proton transfer

Proton transfer

Nucleophile elimination

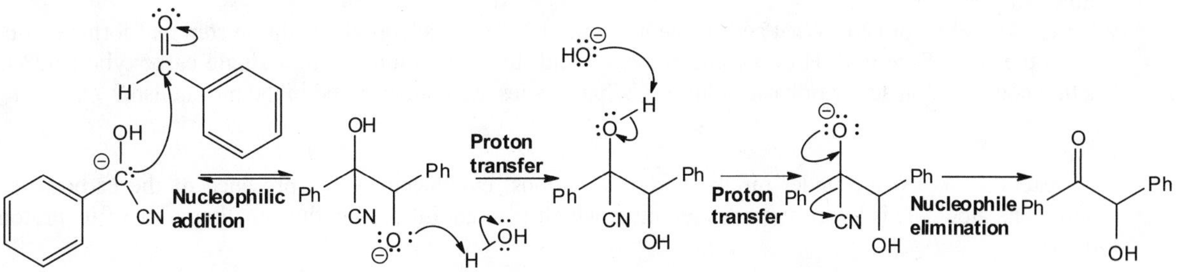

Problem 18.72

Think

How is this reaction similar to the mechanism is Problem 18.71? Which step is different?

Solve

This problem has the same first three steps to form the resonance stabilized enolate anion. The enolate anion then attacks propenenitrile in a conjugate addition. Finally, the catalyst (cyanide) is expelled.

Nucleophilic addition

Proton transfer

Proton transfer

Nucleophile elimination

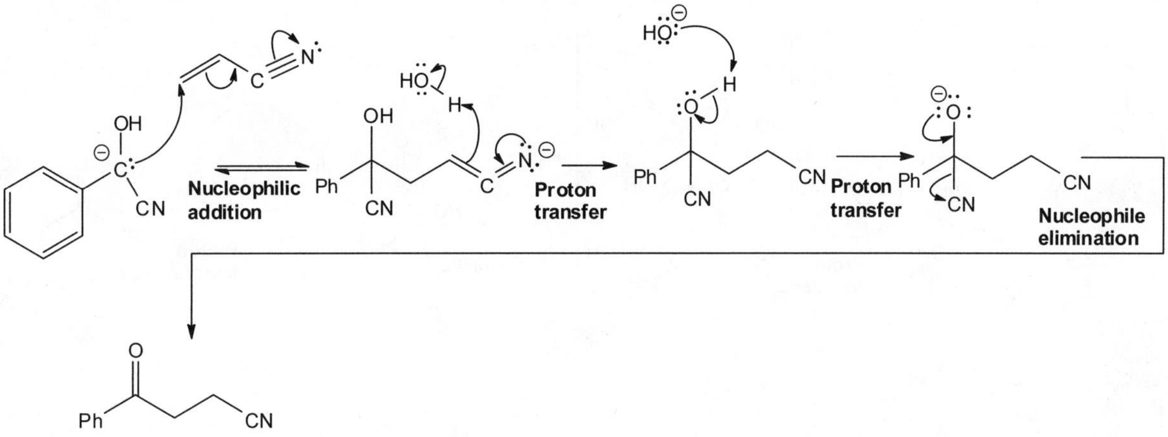

Problem 18.73

Think

What is the mechanism for the crossed aldol reaction? What acts as the nucleophile? Is the reaction conducted in acidic or basic solution? What species are not permitted in the mechanism? How is the pK_a of the α C–H in the β-diester different from an analogous proton in an ester or ketone?

Solve

(a) The mechanism is as follows:

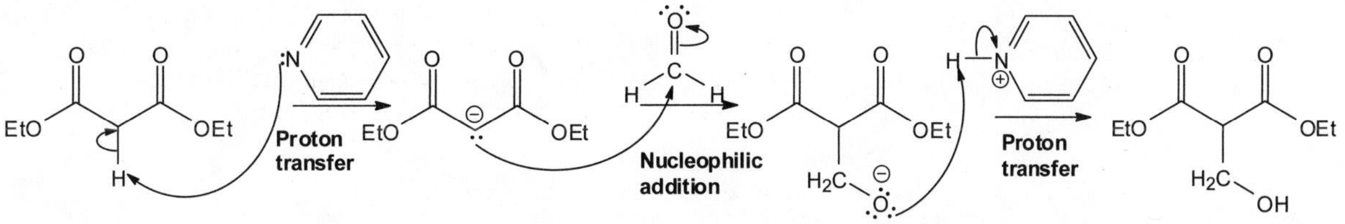

Proton transfer

Nucleophilic addition

Proton transfer

(b) Diethyl malonate is a significantly stronger acid than a typical ketone or aldehyde because the negative charge in its conjugate base is resonance delocalized over two carbonyl O atoms instead of just one. The pK_a of a typical ketone or aldehyde is about 19 to 20. That of diethyl malonate is 13.5.

Problem 18.74

Think

What C–C bond is formed? What acts as the nucleophile in the nucleophilic addition step that forms that bond? How is the pK_a of the α C–H of a β-dicarboxylic acid different compared to a single carboxylic acid? Is the reaction conducted in acidic or basic solution? What species are not permitted in the mechanism?

Solve

This reaction takes place under mildly acidic conditions, evidenced by the presence of the carboxylic acid groups. The aldehyde is hydrolyzed to the imine, which is in equilibrium with a small amount of its protonated form.

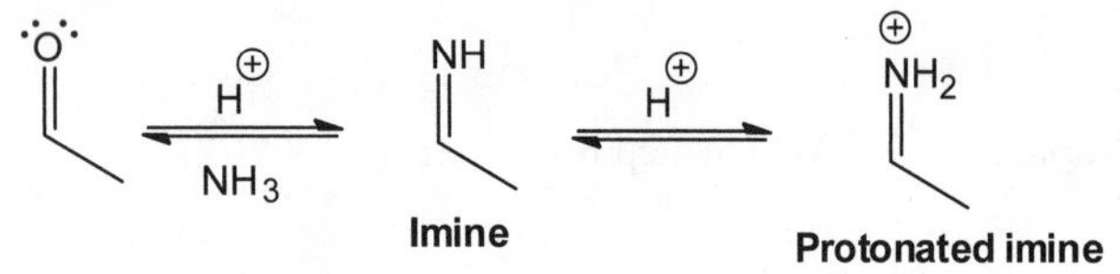

Imine **Protonated imine**

The carbon nucleophile will be the enol. The ammonia deprotonates the α C–H of a β-dicarboxylic acid to produce the enol that is nucleophilic at the C. The mechanism is given below.

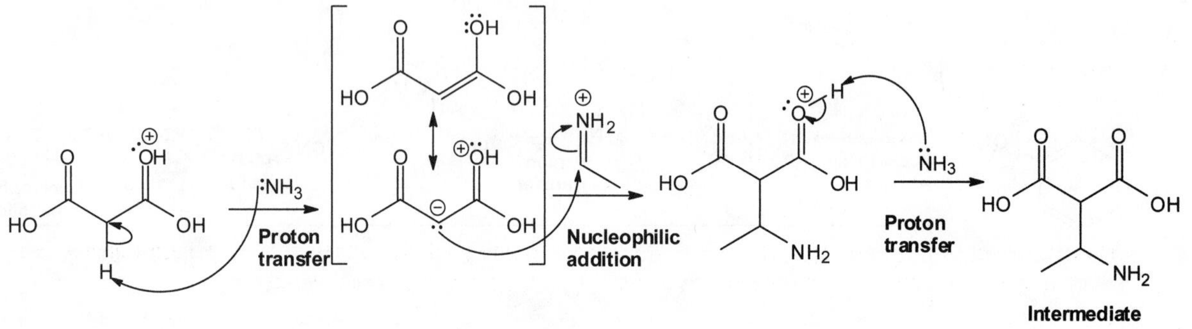

Proton transfer **Nucleophilic addition** **Proton transfer** **Intermediate**

Problem 18.75

Think

How many C atoms are in the 1,5-diketone and the 2,6-dialkyl pyran? Is there any outside source of C atoms? What acts as the nucleophile in the nucleophilic addition step? Is the reaction conducted in acidic or basic solution? Under these conditions, how can a good leaving group be produced? What species are not permitted in the mechanism?

Solve

Because this reaction takes place under acidic conditions, no strong bases should appear. Notice that the last few steps make up an E1 mechanism.

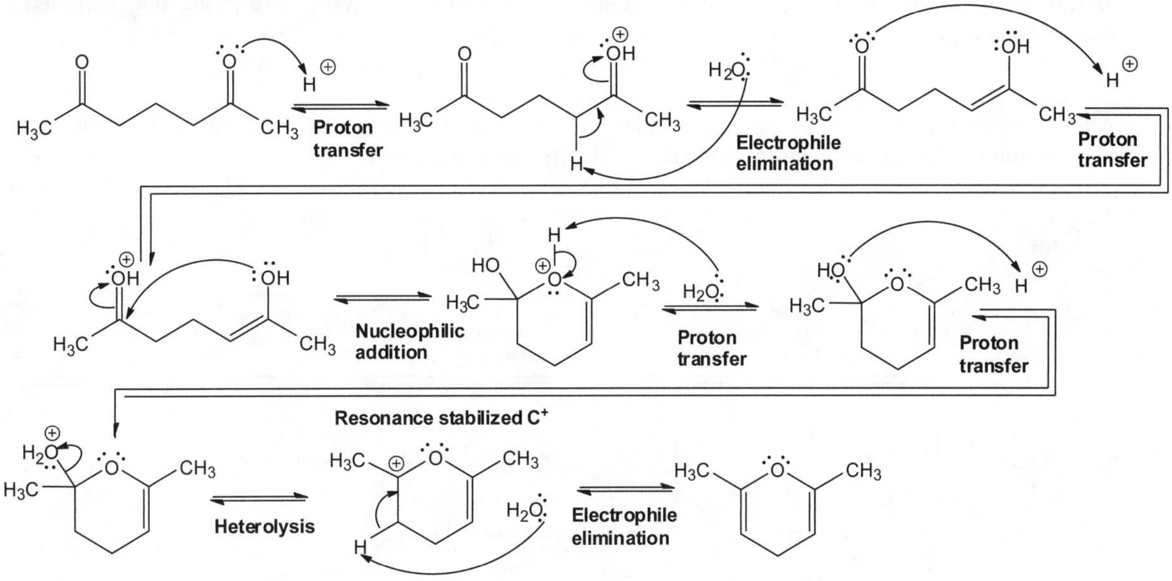

Problem 18.76

Think

What acts as the nucleophile in the nucleophilic addition step? Is the reaction conducted in acidic or basic solution? What species are not permitted in the mechanism?

Solve

The mechanism is a proton transfer followed by a nucleophilic addition, then another proton transfer. The first step has to be protonation to avoid generating any N^- or O^-, which are strongly basic.

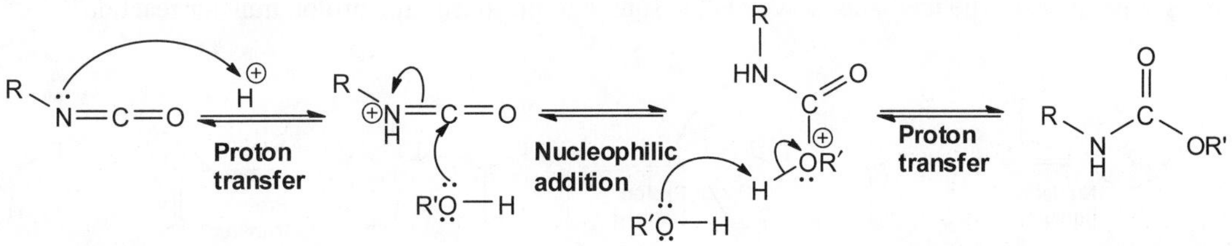

Problem 18.77

Think

What acts as the nucleophile in the nucleophilic addition step? Is the reaction conducted in acidic or basic solution? What species are not permitted in the mechanism? What is the significance of having the reaction free of water?

Solve

The mechanism is a proton transfer followed by a nucleophilic addition, then another proton transfer. The first step has to be proton transfer to avoid generating N^-, which is strongly basic. The reaction is free of water to have the ROH act as the nucleophile and not have water compete for nucleophilic addition.

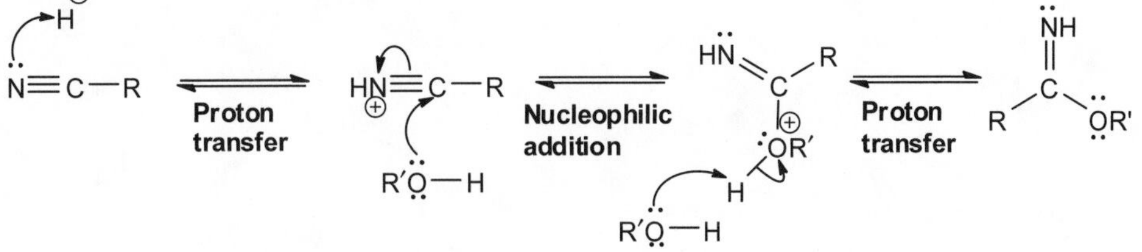

Problem 18.78

Think

What acts as the nucleophile in the nucleophilic addition step? How many times does nucleophilic addition need to take place? Is the reaction conducted in acidic or basic solution? What species are not permitted in the mechanism?

Solve

After being protonated, each cyano group undergoes nucleophilic addition by the same N atom. The nitrile has to be protonated first to avoid strongly basic, negatively charged atoms.

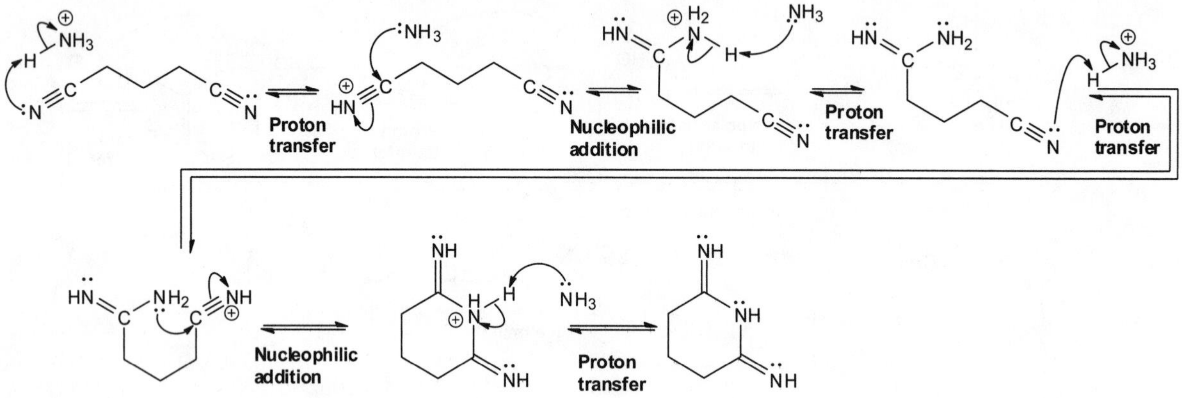

Proton transfer

Nucleophilic addition

Proton transfer

Proton transfer

Nucleophilic addition

Proton transfer

Problem 18.79

Think

What acts as the nucleophile in the nucleophilic addition step? Is the reaction conducted in acidic or basic solution? What species are not permitted in the mechanism?

Solve

The amine acts as the nucleophile. There is no acid or base catalyst, so the mechanism can have a strong acid and a strong base appear. The methanol serves as the source of protons in the proton transfer reaction.

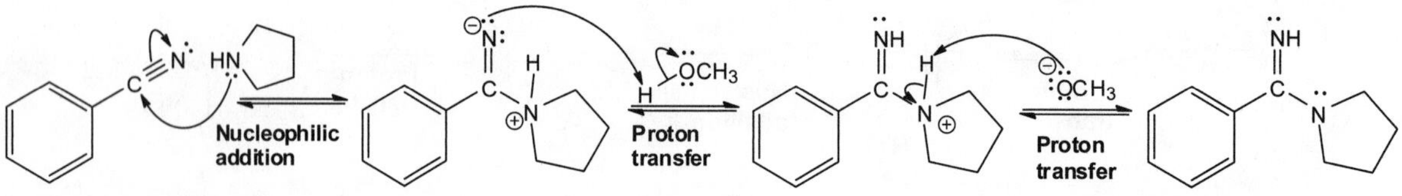

Nucleophilic addition

Proton transfer

Proton transfer

Problem 18.80

Think

Where is the amidine functional group in this problem? What is the structure of benzonitrile? What is the R group in the amine nucleophile?

Solve

Examining the product, we see that the basic structure of benzonitrile appears twice, as shown below.

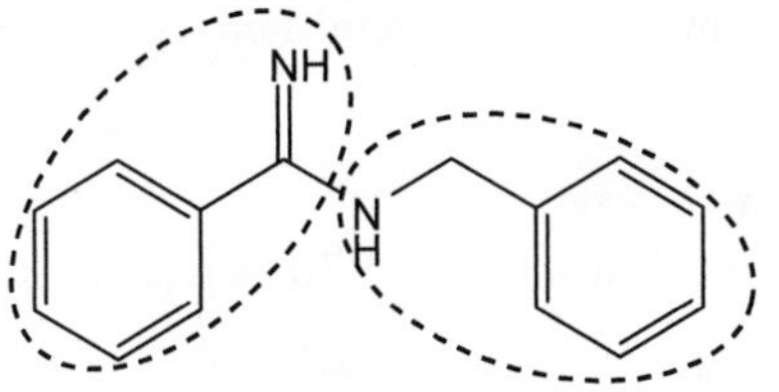

So we can envision the product being the result of a nucleophilic addition of benzylamine to benzonitrile, as shown below.

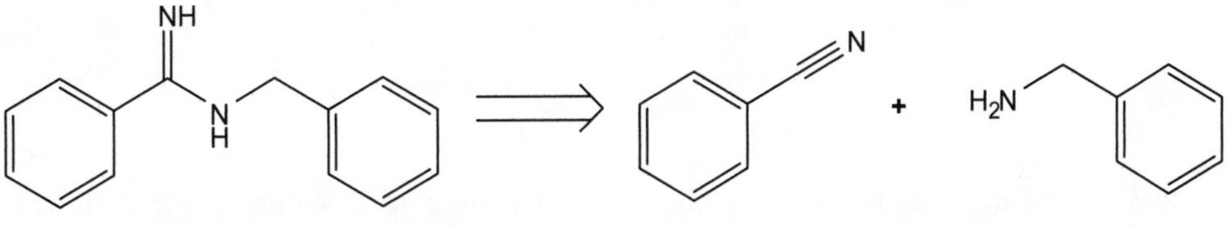

Benzylamine can be made by treating benzonitrile with lithium aluminum hydride. Then, benzylamine can be treated with benzonitrile, according to the reaction in Problem 18.79. We can subsequently protonate using an alcohol (water can convert the C=N to a C=O, as we saw in the text). (Using acidic conditions is not advisable because a protonated amine is not nucleophilic.)

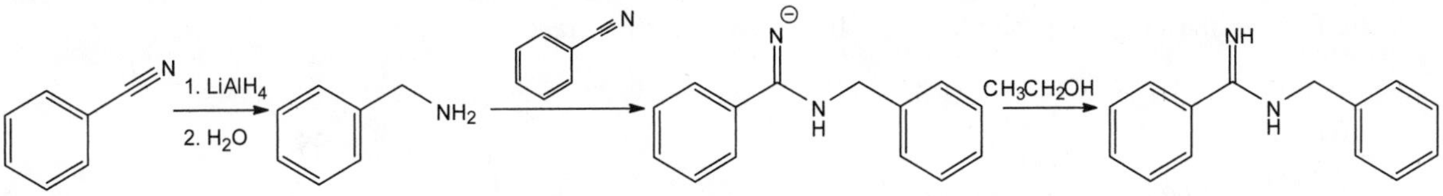

Problem 18.81

Think

What reaction takes place between HO^- and H_2O_2? What nucleophile is produced? Does direct or conjugate nucleophilic addition occur with the α,β-unsaturated ketone? What species are not permitted in the basic solution?

Solve

A nucleophile is required that will add in conjugate fashion to cyclohexenone. If hydroxide reacts with H_2O_2, the hydroperoxide ion, $^-$O–OH, forms. This can add in conjugate fashion. Next, the enolate displaces hydroxide to close the epoxide ring in an internal S_N2 step. Normally, hydroxide is a poor leaving group, but this is feasible under these conditions because the O–O bond is very weak. Notice that hydroxide is *regenerated*, making this a *base-catalyzed* reaction.

Problem 18.82

Think

What acts as the nucleophile in the nucleophilic addition step? Is the reaction conducted in acidic or basic solution? What species are not permitted in the mechanism? Is there a good leaving group in the product of nucleophilic addition?

Solve

The first two steps proceed by the normal aldol addition mechanism. The third step is an S_N2 step instead of a proton transfer.

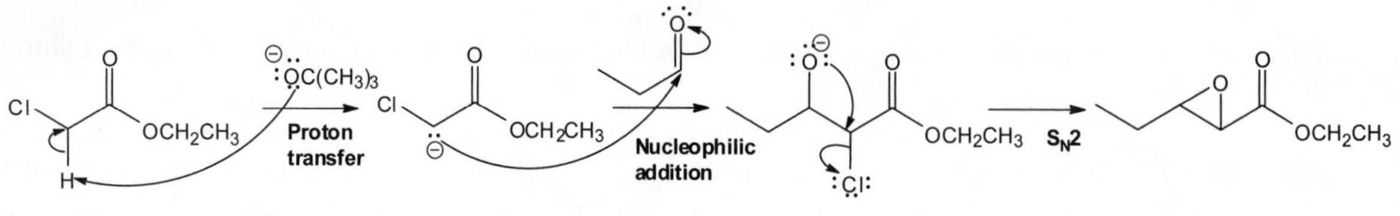

Problem 18.83

Think

What C–H bond does Bu–Li deprotonate? How is this reactant similar to a Wittig reagent? How is this mechanism similar to the Wittig reaction?

Solve

The P-containing starting material resembles a Wittig reagent. Butyllithium is a strong enough base to deprotonate the C, making a carbon nucleophile that can attack the aldehyde in a nucleophilic addition. In the third step, nucleophilic addition to the P=O bond produces the intermediate. From the intermediate, the products are formed in a single step, similar to the final step of a Wittig reaction.

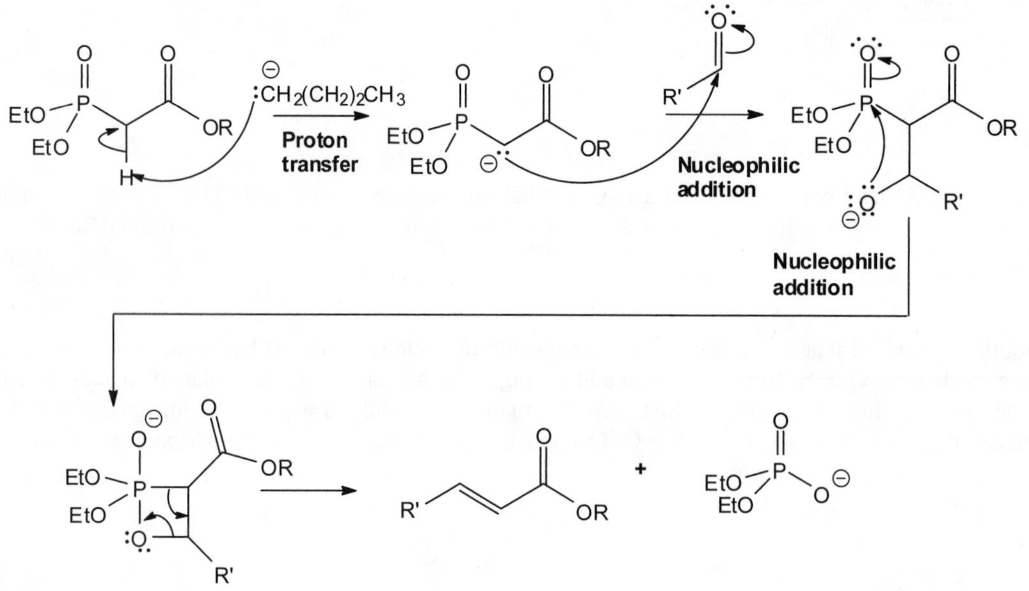

Problem 18.84

Think

What is the IHD? Compare the two formulas. What product is lost when two of the C_5H_8O molecules combine? What functional groups are present in the IR? How many unique ^1H NMR signals are present and at what chemical shift? How many ^{13}C signals are present?

Solve

The IHD of C_5H_8O is 2 and the IHD of $C_{10}H_{14}O$ is 4. Comparing the formulas $C_{10}H_{14}O$ and C_5H_8O, we see that the product results from combination of two reactant molecules, then loss of water occurs. This is a condensation reaction.

Because the compound $C_{10}H_{14}O$ is the product of an aldol condensation reaction, it must be either a β-hydroxycarbonyl or an α,β-unsaturated carbonyl. The first is ruled out because there is no OH band in the IR spectrum and there are not enough O atoms, so it must be an α,β-unsaturated carbonyl. The absence of a proton at 8–10 ppm indicates that there is no aldehyde proton. We conclude that we have an aldol condensation of a ketone:

$C_5H_8O + C_5H_8O \rightarrow C_{10}H_{16}O_2 \rightarrow C_{10}H_{14}O + H_2O$
(a ketone)

It is notable that the proton NMR shows no alkene protons at 5–6 ppm; there are no protons on any C atoms that have a double bond. Notice that no signal in the proton NMR integrates to three protons—all of them integrate to either two protons or four protons. So the product contains no methyl groups, and the same must be true of the ketone reactant. The only ketone of the formula C_5H_8O with no methyl groups is cyclopentanone. A mechanism becomes evident, as does the most likely product.

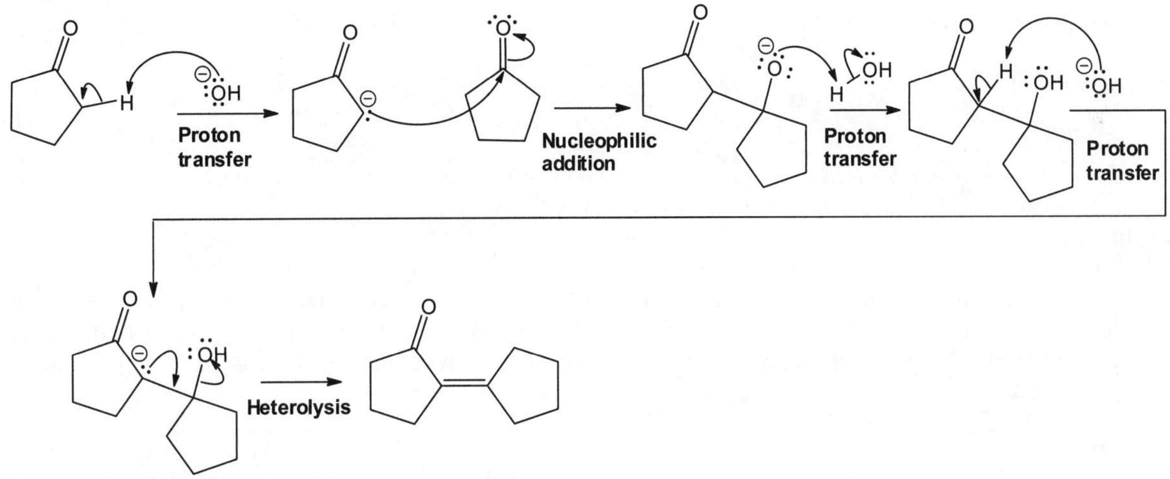

Proton transfer
Nucleophilic addition
Proton transfer
Proton transfer
Heterolysis

Examining the structure of the product, the NMR signals become clear. There are five chemically distinct groups of protons. All the NMR signals in the product are at 1–3 ppm, indicating that the product possesses alkane-type CH_2 protons near a double bond of some type.

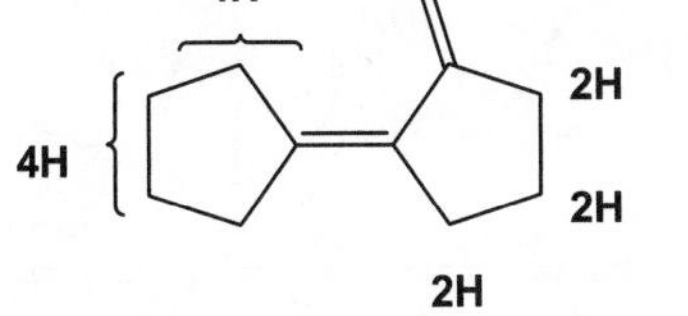

4H
4H
2H
2H
2H

Problem 18.85

Think

What is the IHD? Compare the two formulas. What product is lost when the two reactant molecules combine? What functional groups are indicated by the IR spectrum? How many unique ^{1}H NMR signals are present and at what chemical shift?

Solve

(a) Comparing the formulas of the reactants to the products gives this equation:

$$C_4H_6O + C_9H_8O \rightarrow C_{13}H_{14}O_2 \rightarrow C_{13}H_{12}O + H_2O$$

The proton NMR assignments are 10 ppm, 1H (aldehyde proton), 7–9 ppm, 5H (phenyl protons), and 5–6 ppm, 6H (alkene protons).

(b) Because the carbon structure of the product seems to result from the joining of the two reactants, this reaction is likely an aldol-type reaction. The only acidic C–H atom in the two reactants is a H atom attached to C4 of but-2-enal, due to resonance stabilization of the conjugate base. This C can attack the aldehyde C of 3-phenylpropenal.

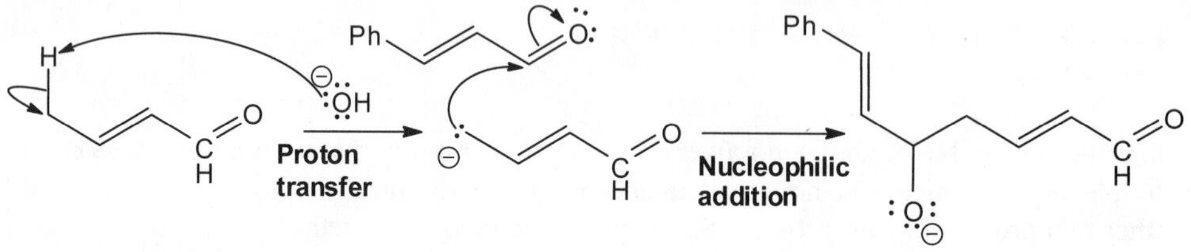

Proton transfer

Nucleophilic addition

The adduct is protonated and loses water to form a product that is completely conjugated. Notice the six alkene protons in the product.

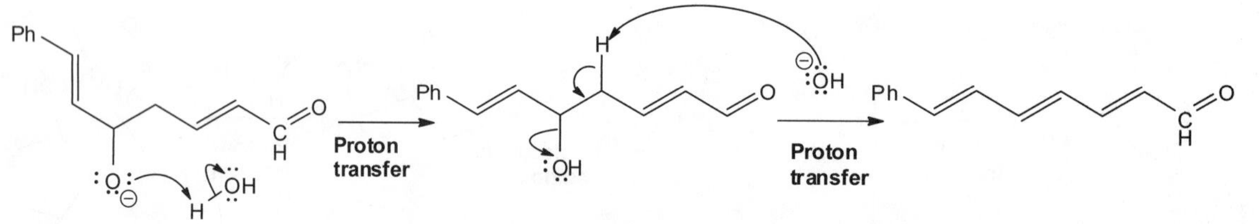

Proton transfer

Proton transfer

Problem 18.86

Think

What is the formula of the starting material? What is the formula of the product? What molecule is lost during the course of the reaction? How many signals are in the ^{1}H NMR? What does this suggest about the symmetry of the molecule? What is the relative ratio of the two signals? What do the chemical shifts tell you about the types of protons?

Solve

The starting material's formula is $C_6H_{10}O_2$, so during the course of the reaction, the molecule must lose one O atom and two H atoms. Also, the proton NMR spectrum shows only two signals, so the product must have significant symmetry. The signals are in a 2:6 ratio, and the 2H signal is indicative of alkene hydrogen atoms. The 6H signal is indicative of two CH_3 groups that are slightly deshielded. The necessary symmetry can be obtained and H_2O can be lost via a cyclization of the intermediate, followed by an E1 reaction.

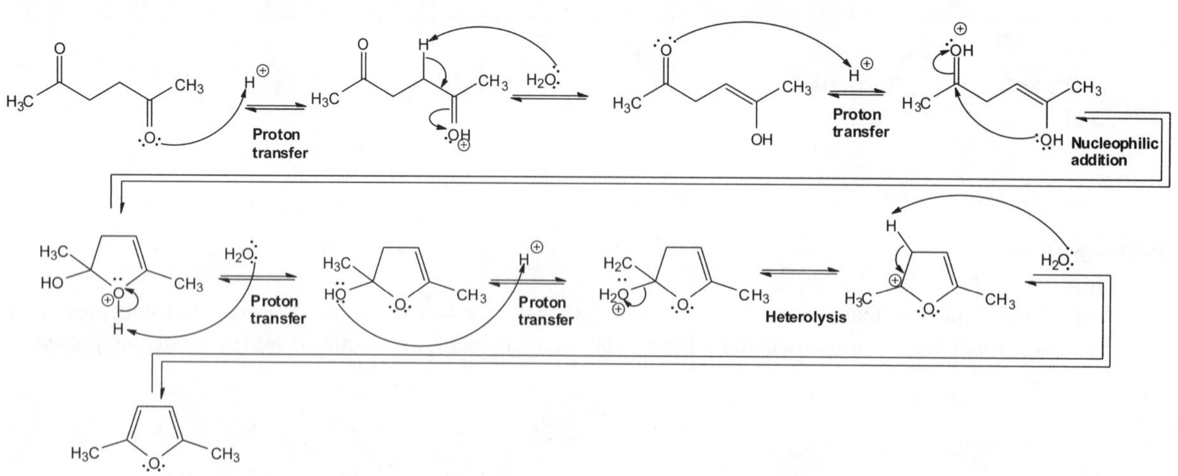

Proton transfer

Proton transfer

Nucleophilic addition

Proton transfer

Proton transfer

Heterolysis

Problem 18.87

Think

What does the 3:9 ratio in the ^{1}H NMR signify? What is the 1H a result of? What functional groups are likely present from the IR signals at 3300–3500 cm^{-1} and 1650 cm^{-1}? Under acidic conditions, what can act as a nucleophile? How can a good leaving group be formed? What types of species should not appear in the mechanism?

Solve

The 9H signal in the proton NMR spectrum signifies the *t*-butyl group. The 3H signal represents the distinct methyl group. The 1H signal is likely from an OH or NH proton, given that it is a broad singlet. The IR spectrum confirms this. The C=O signal at 1650 cm^{-1} is consistent with an amide carbonyl stretch, and the broad absorption between 3300 and 3500 cm^{-1} is consistent with a secondary amide. The key step after the formation of the intermediate is the attack of water on the C attached to N. That forms the C–O bond that will become a C=O bond through some proton transfer steps.

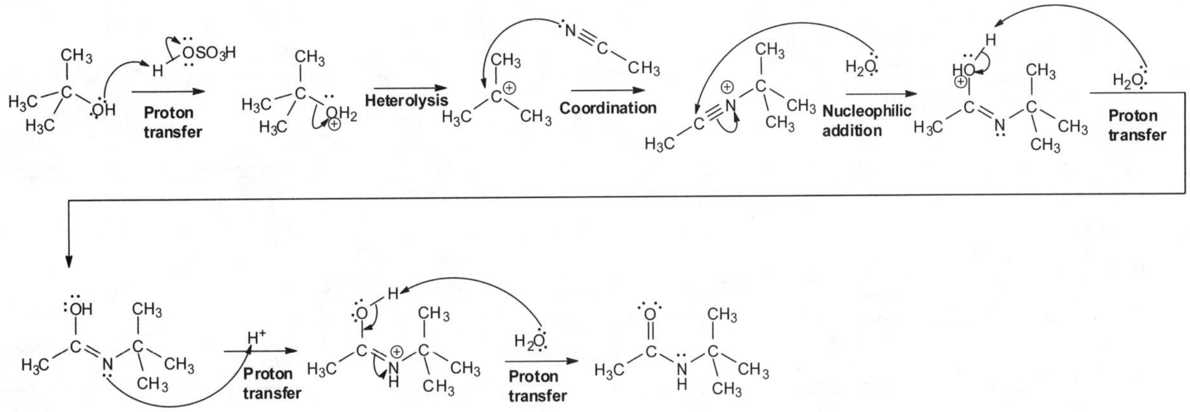

Proton transfer → Heterolysis → Coordination → Nucleophilic addition → Proton transfer → Proton transfer → Proton transfer

Problem 18.88

Think

What are the only type of protons in the product as indicated by the ^{1}H NMR? How many types of C atoms does the ^{13}C NMR spectrum indicate are present? What does the odd molecular mass indicate is present? With those heavy atoms, what molecule must have been eliminated during the course of the reaction?

Solve

The proton NMR spectrum informs us that all protons (via their chemical-shift values) are aromatic. The odd molar mass indicates that N should be present. The nine signals in the ^{13}C NMR spectrum indicates nine distinct carbon atoms. With nine C atoms and one N atom, the mass sums to 9(12) + 1(14) = 122 amu. That leaves seven H atoms, so the formula must be C_9H_7N. Compared to the formula of the reactant, C_9H_9NO, we can see that there is a net loss of H_2O during the course of the reaction. This formula can be produced and all H atoms can be made aromatic if a second ring is formed, giving 10 π electrons total that are part of the same conjugated system.

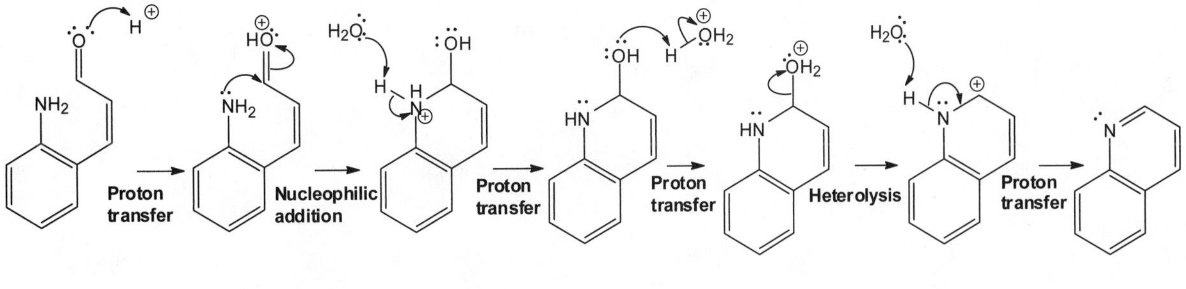

Proton transfer → Nucleophilic addition → Proton transfer → Proton transfer → Heterolysis → Proton transfer

Problem 18.89

Think

What does the relatively few signals in the ^{1}H NMR signify about the symmetry of the product? What are the integrations of the ^{1}H NMR signals? What do the chemical shifts of those signals tell you?

Solve

The relatively low number of carbon and proton NMR signals means that the product has a relatively high degree of symmetry. This symmetry can be achieved by a ring. The proton spectrum shows three signals whose integrations are in a 1:2:6 ratio. The 1H signal is broad, indicative of an NH or OH proton. The 2H signal has a chemical shift consistent with alkene protons. And the 6H signal is consistent with two CH_3 groups that are moderately deshielded. The mechanism below begins with nucleophilic addition of NH_3 to a C=O group. The N subsequently attacks the other C=O group to close the ring, which allows one NH proton to remain. The last several steps are consecutive elimination reactions, which form the alkene groups. Note that for those elimination reactions to occur, the solution needs to be at least slightly acidic.

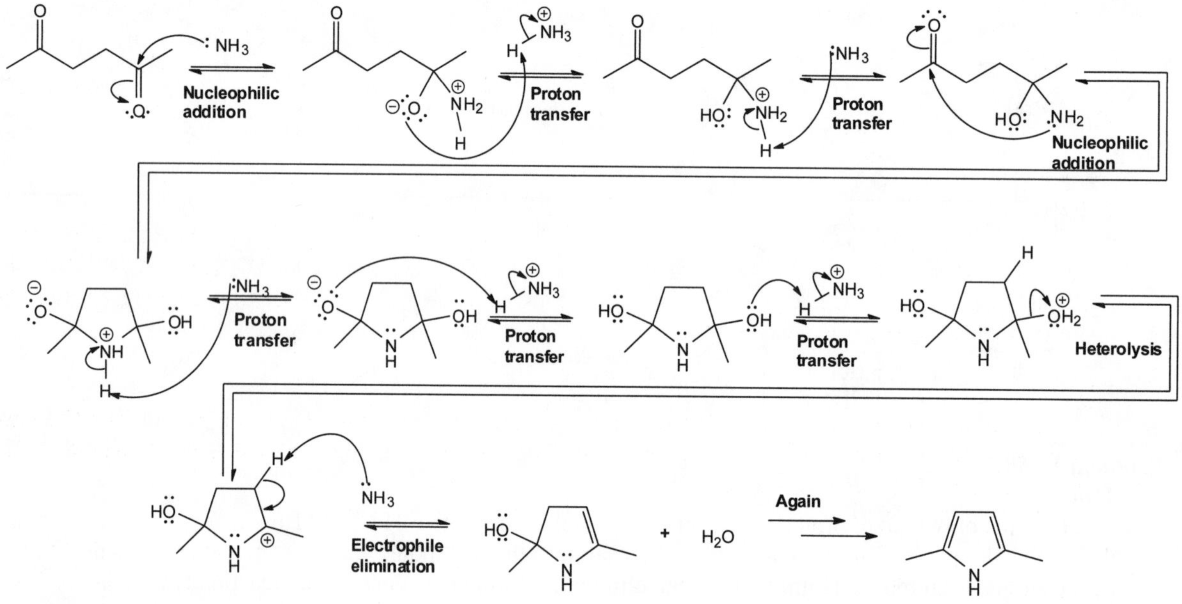

CHAPTER 19 | Organic Synthesis 2: Intermediate Topics of Synthesis Design, and Useful Reduction and Oxidation Reactions

Your Turn Exercises

Your Turn 19.1

Think

Consider the electronegativity of Br compared to that of C, and of Li compared to that of C. Draw a bond dipole and then consider the δ^+ or δ^- charge on C.

Solve

Br is more electronegative compared to C and, therefore, bears the δ^-, leaving a δ^+ on C. Li is less electronegative compared to C and, therefore, bears the δ^+, leaving a δ^- on C.

Br δ^+ —Li(*s*), Ether→ Li δ^-

Your Turn 19.2

Think

What does relative positioning mean? Identify the C atom associated with each functional group in the product. If one of those C atoms is designated "1" and the C atoms in the chain are numbered sequentially, what number does the C atom in the second functional group receive? Review the column labeled "relative positioning" in Table 19-1 to verify your circled C atoms.

Solve

The relative positioning describes the C atoms that are associated with the functional groups. The C atoms are circled below, the carbon chain is numbered from one functional group to the second, and the relative positioning numbers are indicated.

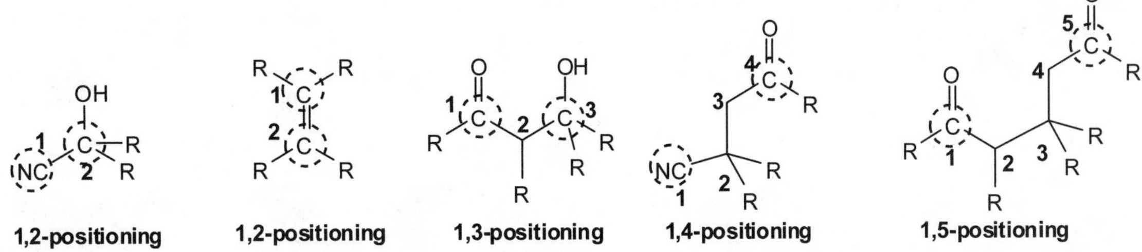

Your Turn 19.3

Think

Review functional group definitions and structures in Section 1.13 and Table 1-6. Locate the acetal group in Entries 2 and 4.

Solve

An acetal is characterized by a $R_2C(OR)_2$ or $RCH(OR)_2$ or $CH_2(OR)_2$ grouping. The acetal group is noted by the dashed ovals below.

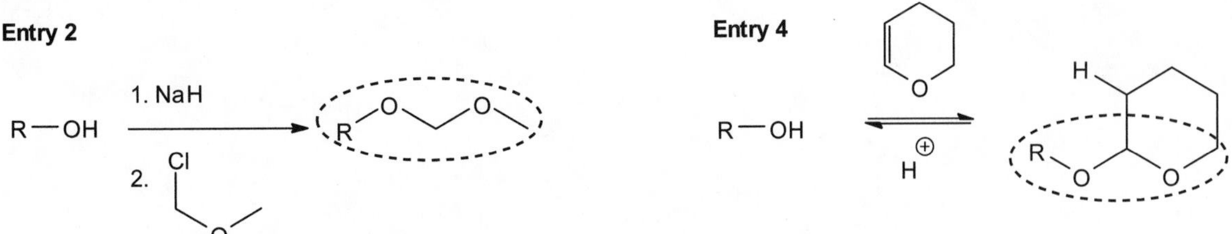

Your Turn 19.4

Think

Consult Section 1.9 to review how to calculate oxidation states. Compare the electronegativity of C versus H and C versus O. Which is more electronegative? How are the covalently bonded electrons assigned when the bonded atoms have different electronegativities? When they have the same electronegativities?

Solve

In a given covalent bond, all electrons are assigned to the more electronegative atom. If the atoms are identical, the electrons are split up evenly. H is *less* electronegative compared to C, and O is *more* electronegative compared to C. In the secondary alcohol, the C atom is assigned two electrons from the C–H bond and one electron from each of the two C–C bonds, for a total of four valence electrons. That is the same number of valence electrons for an isolated uncharged C atom, so that C has an oxidation state of 0. In the ketone, the C is assigned only one electron from each of the two C–C bonds, for a total of two. Because that is two fewer than in an isolated uncharged C atom, that C has an oxidation state of +2. The oxidation state of C increased by two, verifying that C was oxidized.

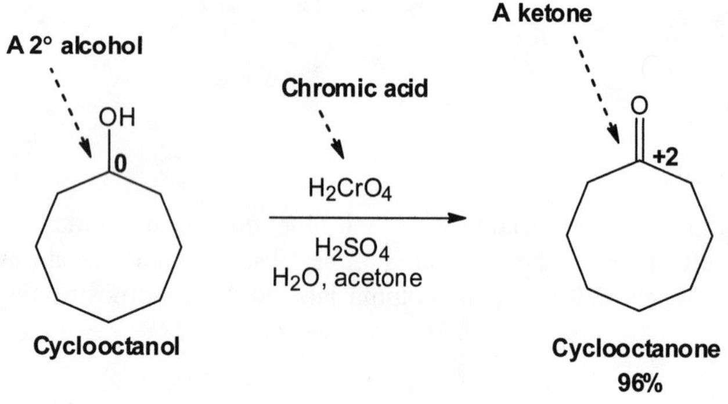

Chapter Problems

Problem 19.1

Think

Is the C atom in the C–I or C–Br bond electron rich or electron poor? What type of reagent forms when the alkyl halide reacts with Mg(*s*) or Li(*s*)? Is the resulting C atom electron rich or poor?

Solve

The C atom in the C–I bond is electron poor, and the same is true of the C–Br bond. When an alkyl halide reacts with Mg(*s*) or Li(*s*), the resulting C atom is electron rich because C is more electronegative than the metals. These products are shown in **(a)**–**(c)**. The electron-rich C atom in **(c)** reacts with the CuI to form a lithium dialkylcuprate reagent.

Mg(*s*), Ether — MgI **(a)**

Mg(*s*), Ether — Br → MgBr **(b)**

I — Li(*s*), THF → Li **(c)**

Product from (c) — Li — CuI, THF → Li Cu **(d)**

Problem 19.2

Think

Which C atom in the product shown was originally bonded to the halide? Did the electronic nature of the C atom change?

Solve

In each case, the C atom bonded to metal was originally bonded to the halide. The C–X bond has an electron-poor C, while the C–M bond (M = metal) has an electron-rich C.

Br — Li → Li **(a)**

Cl — Mg, Ether → MgCl **(b)**

Br — Mg, Ether → MgBr **(c)**

Br — 1. Li, 2. CuI → Li Cu **(d)**

Problem 19.4

Think

Do C–C bonds have to be formed in the synthesis? If so, what should the electron-rich species be? What should the electron-poor species be? What carbon–carbon bond-formation reaction produces an alcohol? Can the synthesis take place in a single reaction? If not, what precursor should we choose?

Solve

Since the source of carbon contains two C atoms and the product contains six C atoms, two C–C bonds must be formed. A C–C bond formation reaction would be difficult to carry out directly between two molecules of ethanal because such a reaction would entail the formation of a bond between two C atoms bearing a partial positive charge. The precursor should bear a C atom with a negative or partial negative charge. The aldehyde can be converted to the alcohol via reduction with $NaBH_4$/ROH. This alcohol can then be converted to an alkyl bromide, which can react to form a Grignard reagent (C atom with a negative charge). Ethanal can then react with an equivalent of CH_3CH_2MgBr, followed by protonation, to form butan-2-ol. This alcohol is dehydrated with H_3PO_4/Δ to yield the alkene, which reacts with Br_2/CCl_4 to give the vicinal dibromide. The vicinal dibromide is converted to the alkyne via a double elimination reaction with excess $NaNH_2$. The alkyne reacts with water and an Hg^{2+} catalyst to produce the ketone. The ketone can then react with another equivalent of CH_3CH_2MgBr, followed by protonation, to form the alcohol product given: 3-methylpentan-3-ol.

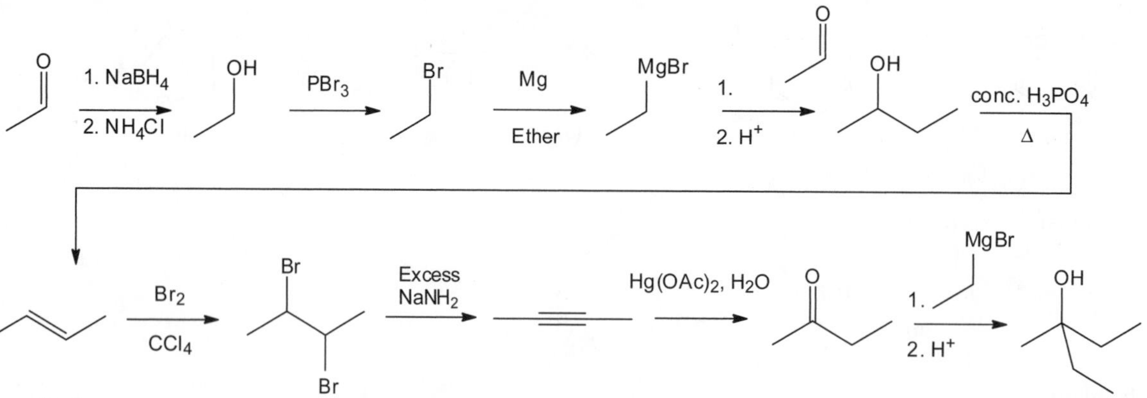

Problem 19.5

Think

Do C–C bonds have to be formed in the synthesis? If so, how many such bonds must form? What should the electron-rich species be? What should the electron-poor species be? Can the synthesis take place in a single reaction? If not, what precursor should we choose?

Solve

Since the other source of carbon other than bromobenzene contains two C atoms and the product contains 10 C atoms, two C–C bonds must be formed. A C–C bond formation reaction would be difficult to carry out directly with bromobenzene and oxirane because such a reaction would entail the formation of a bond between two C atoms bearing a partial positive charge. Bromobenzene can be converted to the Grignard reaction upon reaction with Mg(s). Phenylmagnesium bromide can attack the electron-poor C atom of the epoxide ring and open the compound. The alcohol forms from a proton transfer reaction. The alcohol can be converted to an alkyl bromide via PBr_3, which is not subject to rearrangement. The reaction sequence is then repeated to add the additional two C atoms.

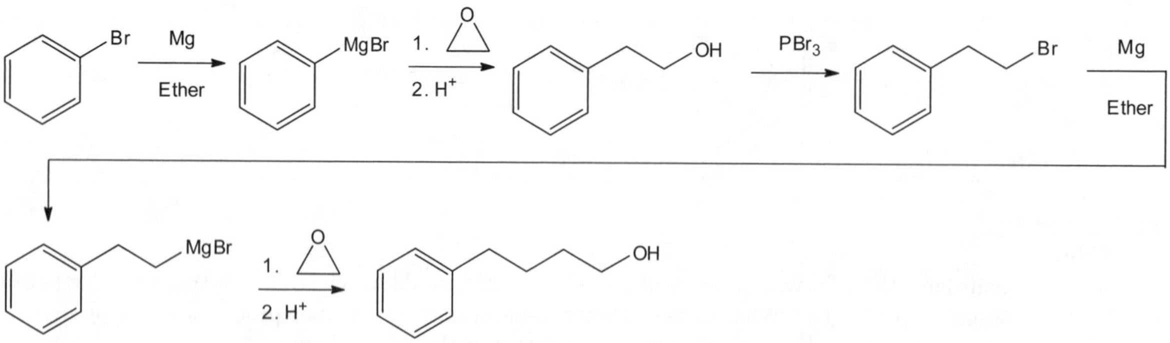

Problem 19.7

Think

Does the target have a 1,2-positioning of the functional groups that could result from the formation of a C–C bond? If so, from what cyanohydrin could this target be synthesized? How could that cyanohydrin be synthesized from the starting compounds?

Solve

The target does indeed have a 1,2-positioning of the CNH_2 and COR functional groups, so we can conceive of obtaining it from a cyanohydrin, as shown in the following retrosynthetic analysis.

Undo a reduction.

Undo an ether synthesis.

Undo a cyanohydrin formation.

+ NaCN

The forward reaction sequence is shown below.

NaCN / HCN

1. NaH 2. $PhCH_2Br$

1. $LiAlH_4$ 2. H_3O^+

Problem 19.9

Think

Will a carbon–carbon bond-formation reaction be necessary? What is the relative positioning of the functional groups in the target? Does the appropriate carbon–carbon bond-formation reaction in Table 19-1 leave us with the correct functional groups, or will we need to carry out an additional functional group conversion?

Solve

We will have to use a carbon–carbon bond-formation reaction because the target's carbon skeleton contains six C atoms bonded together and we are allowed to start with compounds containing only three. The 1,3-positioning of the two hydroxyl groups in the product suggests that we should use an aldol reaction, which yields a β-hydroxycarbonyl. In a retrosynthetic analysis, therefore, our task becomes to apply a transform that takes our target molecule back to a β-hydroxycarbonyl. We can do this by undoing a hydride reduction on C1. From there, we disconnect the appropriate C–C bond to take us back to the aldol reactant, propanal.

Undo a reduction.

Undo an aldol addition.

1,3-positioning

A β-hydroxycarbonyl

The synthesis in the forward direction is as follows:

1. NaOH
2. H^+

$NaBH_4$
EtOH

Problem 19.10

Think

Will a carbon–carbon bond-formation reaction be necessary? Are there any functional groups in the product that are characteristic of a carbon–carbon bond-formation reaction? If not, which C atom was likely reduced to remove the functional group?

Solve

A carbon–carbon bond-formation reaction is necessary. Because the target has no functional groups that result from a C–C bond formation reaction, we can add one back by undoing a Wolff–Kishner reduction. This allows us to use a Grignard reaction to form the C–C bond.

Undo a Wolff–Kishner. **Undo an oxidation.** **Undo a Grignard.**

MgBr +

The synthesis would then appear as follows:

1. MgBr
2. H^+

PCC

1. H_2NNH_2, H^+
2. NaOH, H_2O, Δ

Problem 19.12

Think

What C–C bond could you disconnect in a transform? If that bond is the result of conjugate addition to an α,β-unsaturated ketone or aldehyde, which CH_2 group could we imagine coming from a C=O group?

Solve

One way to carry out the syntheses is to use a conjugate addition of a lithium dialkylcuprate to produce a ketone that can be reduced by a Wolff–Kishner reduction. Four examples are given below and on the next page.

1. Li Cu
2. NH_4Cl

1. H_2NNH_2, H^+
2. NaOH, H_2O, Δ

1. $(\ )_2CuLi$
2. NH_4Cl

1. H_2NNH_2, H^+
2. NaOH, H_2O, Δ

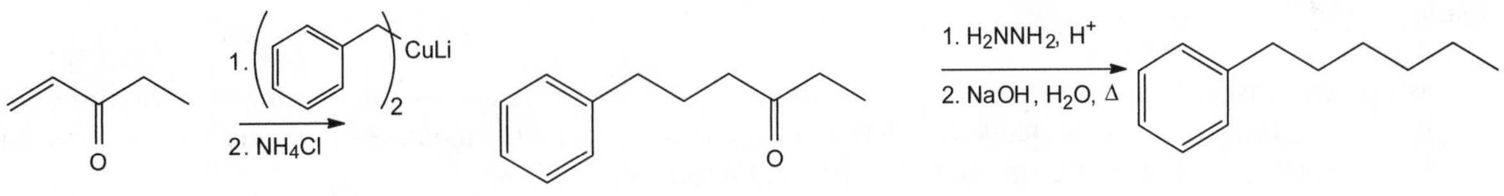

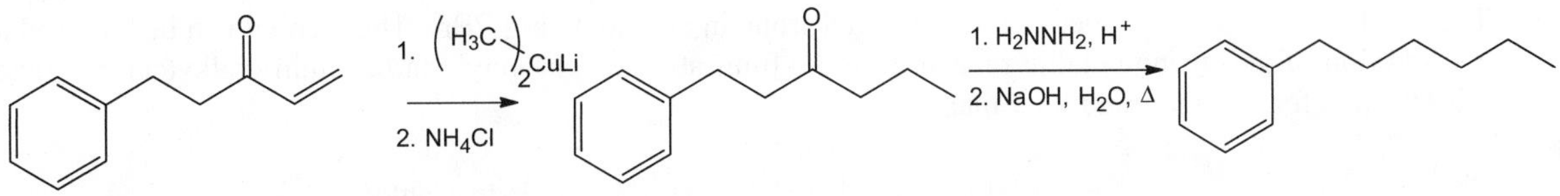

1. $[C_6H_5CH_2CH_2]_2CuLi$ 2. NH_4Cl ; 1. H_2NNH_2, H^+ 2. NaOH, H_2O, Δ

1. $(H_3C)_2CuLi$ 2. NH_4Cl ; 1. H_2NNH_2, H^+ 2. NaOH, H_2O, Δ

Problem 19.14

Think

Are there any functional groups, aside from the carbonyl group, that are susceptible to reaction in the presence of HO^-? HCl? Thiols? $H_2(g)$ and a metal catalyst?

Solve

(a) Only a Wolff–Kishner reduction will work, due to the C=C double bond that is present. Catalytic hydrogenation will reduce the C=C bond, and as we saw in Chapter 11, the acidic, aqueous conditions of the Clemmensen reduction will lead to reaction with the C=C bond.

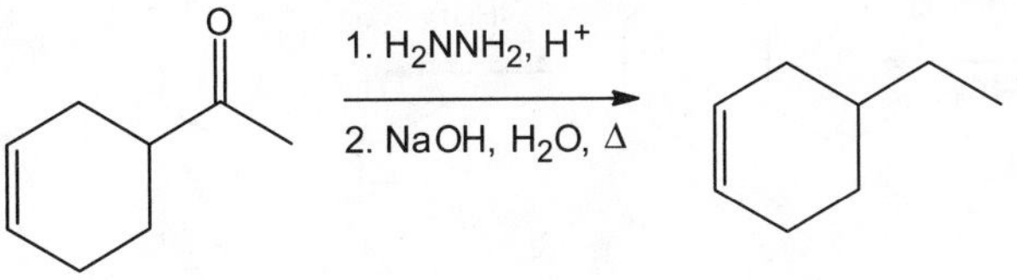

1. H_2NNH_2, H^+ 2. NaOH, H_2O, Δ

(b) Both catalytic hydrogenation and the Wolff–Kishner reduction will work. Hydrolysis of the ether will take place under the acidic, aqueous conditions of the Clemmensen reduction.

1. H_2NNH_2, H^+ 2. NaOH, H_2O, Δ
OR
1. HS–CH₂CH₂CH₂–SH, H^+ 2. Raney Ni (H_2)

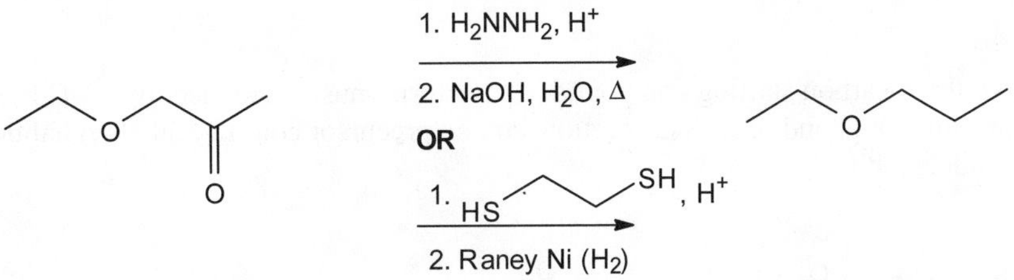

(c) All three reductions will work. The Cl attached to the benzene ring is unreactive toward HO^- in a Wolff–Kishner reduction, because it is attached to an sp^2-hybridized carbon. The aromatic ring is relatively unreactive toward catalytic hydrogenation conditions.

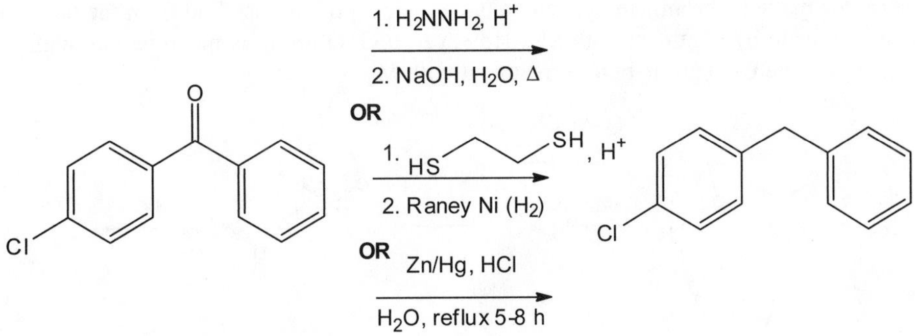

1. H_2NNH_2, H^+ 2. NaOH, H_2O, Δ
OR
1. HS–CH₂CH₂–SH, H^+ 2. Raney Ni (H_2)
OR
Zn/Hg, HCl; H_2O, reflux 5-8 h

Cl Cl

Problem 19.16

Think

What reagents can be used to form a C–C bond to the carbonyl carbon of a ketone? Which ones do so selectively in an α,β-unsaturated ketone? Will such a reaction produce the necessary functional groups that are in the target, or is a functional group transformation necessary?

Solve

The alkyl bromide in the target can be produced from an alcohol using PBr_3. The alcohol can be the product of 1,4-addition of an organometallic reagent to the α,β-unsaturated carbonyl, and lithium dialkylcuprate reagents are selective for 1,4-addition rather than 1,2-addition.

Undo a bromination. Undo selective reduction. Undo a lithium dialkyl cuprate reaction. + LiCu 2

In the forward direction, the synthesis can be written as follows:

1. CuLi 2 2. NH_4Cl $NaBH_4$ EtOH PBr_3

Problem 19.17

Think

How many C atoms are in acetone? How many C atoms are in hexane-2,5-dione? Which C–C bond is formed? What would be the two starting materials? Are all of the functional groups compatible? If not, how can a protecting group be used?

Solve

Because we are limited to a three-carbon starting compound, a transform must disconnect the C–C bond shown below. We can envision forming that bond in an S_N2 reaction, so one precursor could be an alkyl halide, and the other a negatively charged nucleophile.

Undo an S_N2. +

The bromoacetone can be made by brominating acetone under acidic conditions. And from the bromoacetone, a Grignard reagent can be made by treatment with Mg. However, a C=O group is incompatible with a Grignard reagent, so we must protect the C=O group before treatment with Mg.

Br_2 Acetic acid HO OH $H^{\oplus}$ Mg MgBr Br H_3O^+

Problem 19.18

Think

How is the ROH deprotonated in Entries 1 and 2? In the subsequent step, what is the nucleophile and leaving group? How does H^+ react with an alkene in Entries 3 and 4? What is the nucleophile that reacts with the resulting carbocation? For deprotection in Entries 1–4, how do the acidic conditions create a good leaving group? What is the nucleophile?

Solve

The ROH is deprotonated in Entries 1 and 2 using NaH, generating a strong RO^- nucleophile that undergoes a substitution reaction with the alkyl halide. In Entries 3 and 4, H^+ adds to the C=C to produce the more stable carbocation, after which ROH behaves as a nucleophile to add to the C^+. In the deprotection reactions, O is protonated to make a good leaving group and water acts as the nucleophile. The protection and deprotection mechanisms are given below.

Entry 1

Protection step:

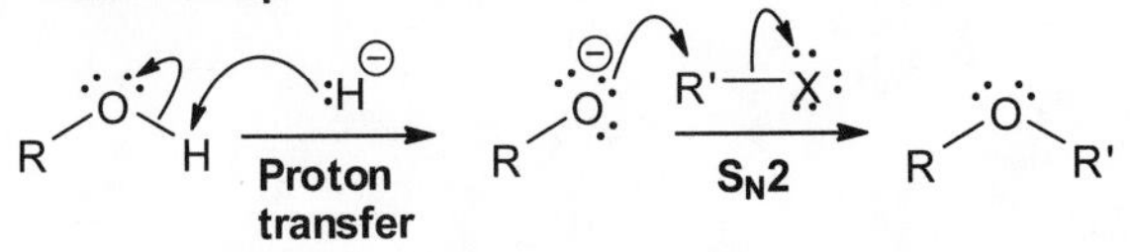

Proton transfer

S_N2

Deprotection step:

Proton transfer

S_N2

Proton transfer

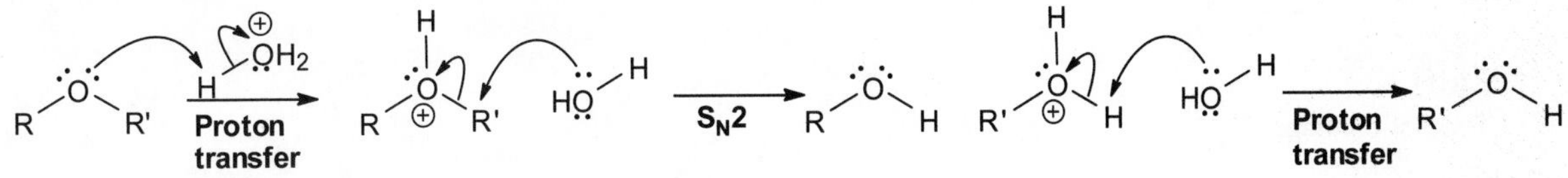

Entry 2

Protection step:

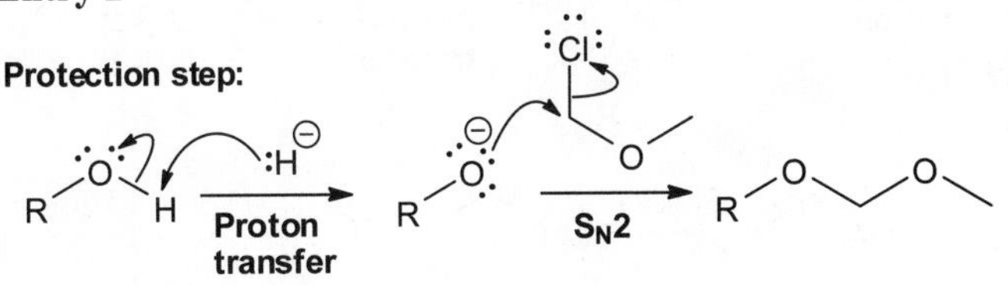

Proton transfer

S_N2

Deprotection step:

Proton transfer

Heterolysis

Nucleophilic addition

Proton transfer

Heterolysis

Proton transfer

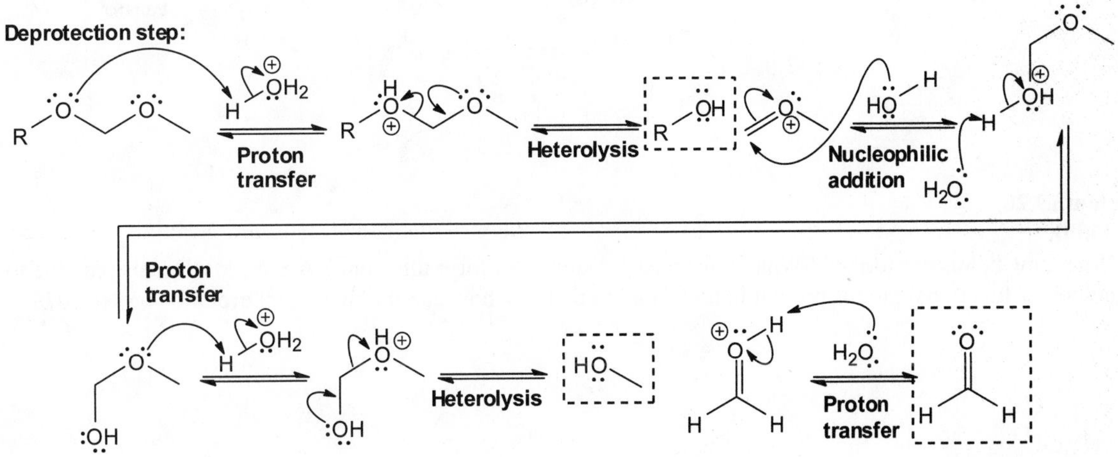

Entry 3

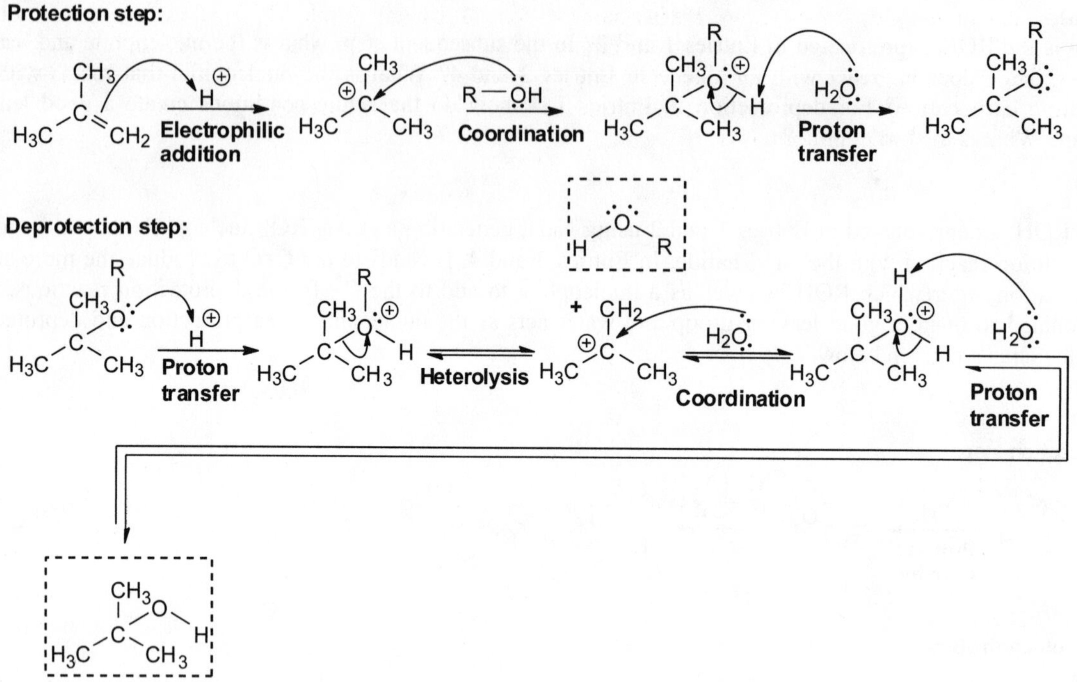

Protection step:

Electrophilic addition → Coordination → Proton transfer

Deprotection step:

Proton transfer → Heterolysis → Coordination → Proton transfer

Entry 4

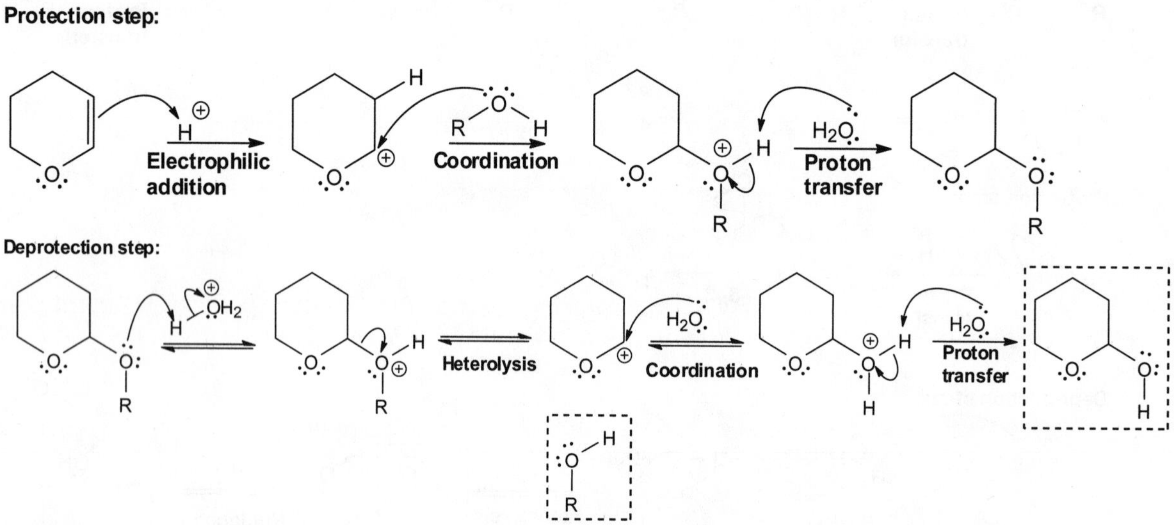

Protection step:

Electrophilic addition → Coordination → Proton transfer

Deprotection step:

Heterolysis → Coordination → Proton transfer

Problem 19.20

Think

What new bonds are formed? What reagents are required to form this bond? Are these reagents reactive toward any other functional groups present in the compound? If so, how can the functional group be protected?

Solve

The C=O is transformed into an ether. This is accomplished via reduction of the C=O followed by a Williamson ether synthesis on the alcohol. The alcohol originally present needs to be protected, because the Williamson ether synthesis can be carried out on either alcohol. The reaction scheme is shown below. Note that TBDMS is used as the protecting group to avoid acidic, aqueous conditions, which would hydrolyze the newly formed ether that is intended to be in the target.

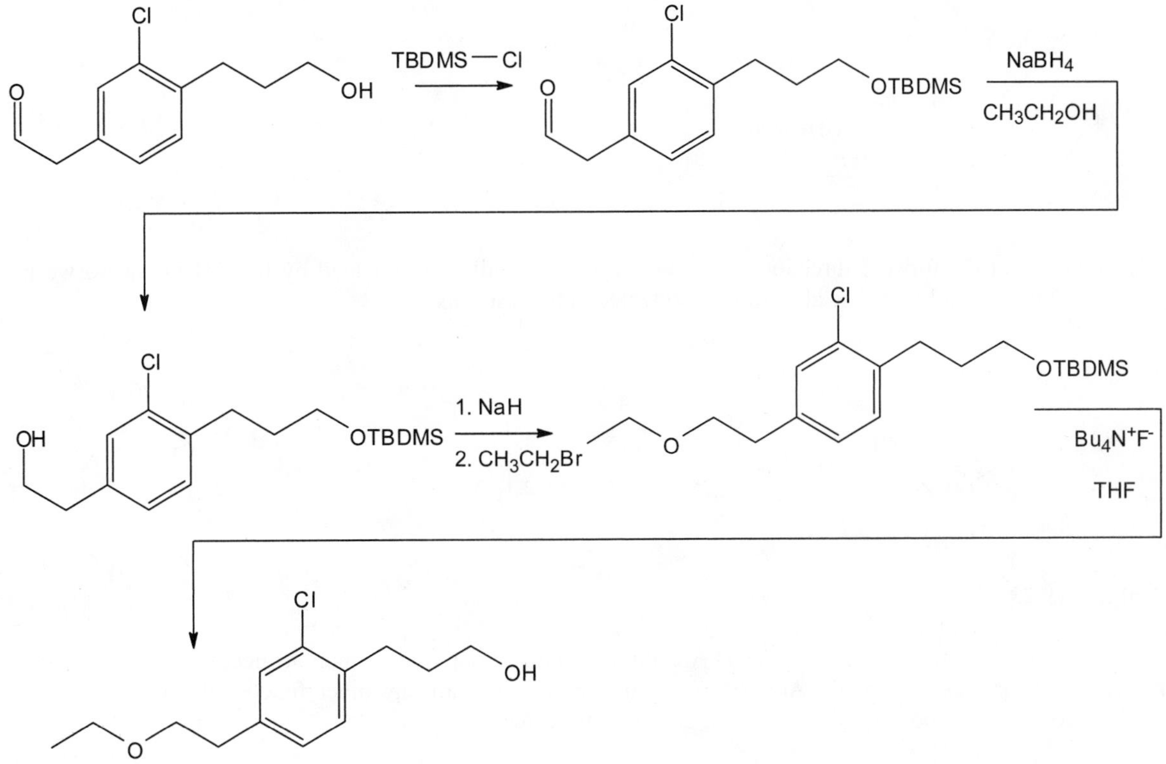

TBDMS—Cl

$NaBH_4$, CH_3CH_2OH

1. NaH
2. CH_3CH_2Br

$Bu_4N^+F^-$, THF

Problem 19.21

Think

What type of reaction occurs in the second reaction sequence? What functional group is formed from an alkylbromide and Mg(s)? Will the alcohol also react?

Solve

(a) The alcohol needs to be protected because in the second reaction sequence the Grignard reaction that is formed will react with the alcohol (see below).

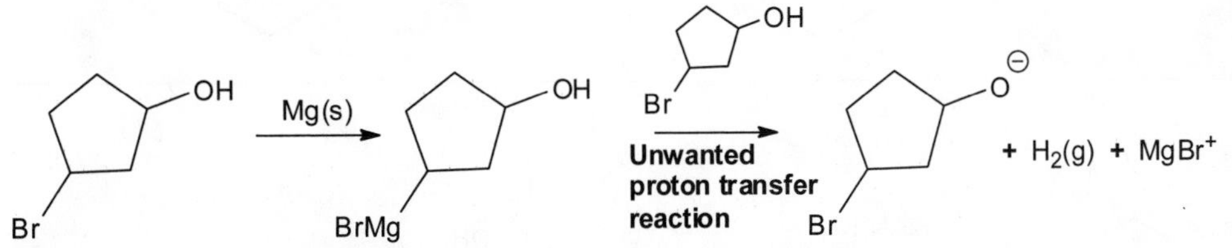

Mg(s)

Unwanted proton transfer reaction

+ $H_2(g)$ + $MgBr^+$

(b) The alcohol could have also been protected prior to performing the Grignard reaction, using dihydropyran (DHP). Once the Grignard reaction is complete, the product can be deprotected. Because the workup step of the Grignard reactionrequires acidic aqueous conditions, the acid workup and deprotection occur together.

Problem 19.22

Think

What new C–C bonds are formed? What reagents are required to form this bond? Are these reagents reactive toward any other functional groups present in the compound? If so, how can the functional group be protected?

Solve

It seems like we should be able to undo a Grignard reaction to arrive at the starting material.

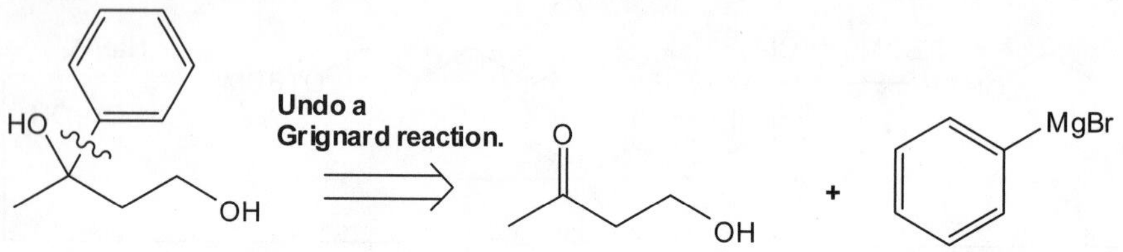

Undo a Grignard reaction.

MgBr

OH

HO

OH

O

However, in the forward direction, the Grignard reagent will be protonated by the OH group, so we need to protect the alcohol first. The alcohol is deprotected in the last step.

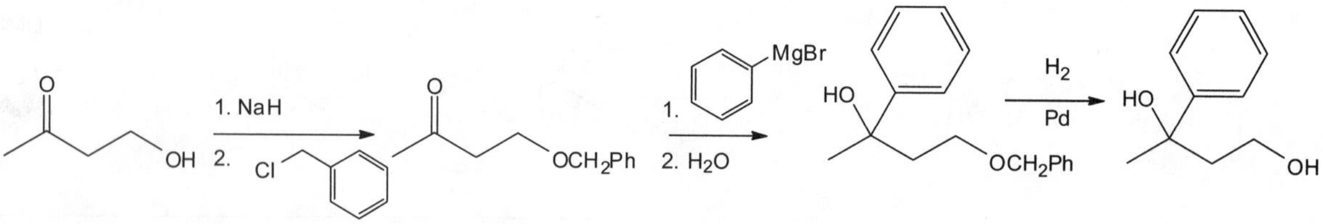

1. NaH

2. Cl

OCH_2Ph

1. MgBr

2. H_2O

HO

OCH_2Ph

H_2

Pd

HO

OH

Problem 19.23

Think

Are there any new C–C bonds formed? Are any functional group transformations necessary? What reagents are required for these reactions? Are these reagents reactive toward any other functional groups present in the compound? If so, how can the functional group be protected?

Solve

The initial ketone can be reduced to an alcohol, which can then be converted to Br. The alcohol on the starting material needs to be protected, however, because of the bromination step. Bromination will convert both alcohol groups to Br if the first one is not protected.

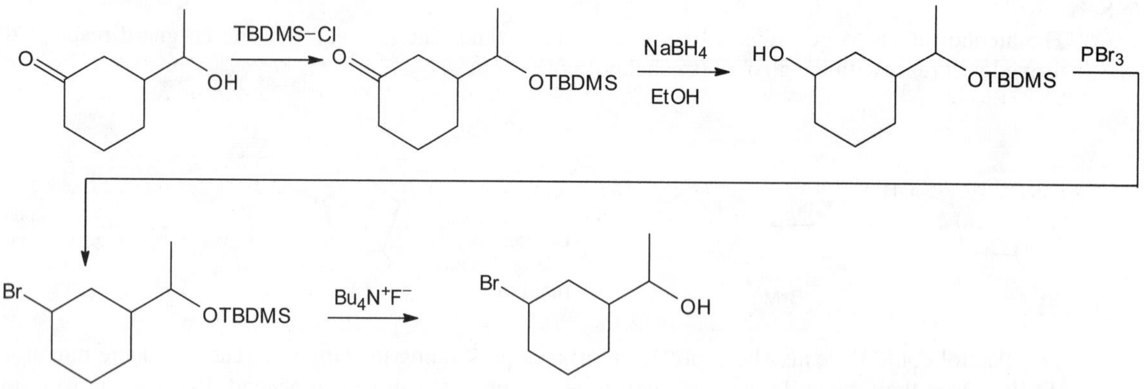

TBDMS–Cl

OH

OTBDMS

$NaBH_4$

EtOH

HO

OTBDMS

PBr_3

Br

OTBDMS

$Bu_4N^+F^-$

Br

OH

Problem 19.24

Think

What is the structure of 1,2,5-pentanetriol? What new C–C bonds are formed? What reagents are required to form this bond? Are these reagents reactive toward any other functional groups present in the compound? If so, how can the functional group be protected? Are any functional group transformations necessary?

Solve

(a) We want to do a simple S_N2 reaction (a Williamson synthesis).

Undo an S_N2

H_3CO OH OH ⟹ CH_3Br + $^{\ominus}O$ OH OH

But to deprotonate only the OH at the left, both at the right have to be protected.

HO OH OH → (H–C(=O)–H, $H^{\oplus}$) HO O–CH_2–O → (1. NaH, 2. CH_3Br) H_3CO O–CH_2–O → (H_3O^+) OH OH OH

(b) For this compound, it looks like we want to carry out a Grignard reaction. However, in the forward direction, the two OH groups at the right of the starting compound are not compatible with the Grignard reagent. So those OH groups should be protected.

HO OH OH → (H–C(=O)–H, $H^{\oplus}$) HO O–CH_2–O → (PBr_3) Br O–CH_2–O → (Mg(s), Ether)

BrMg O–CH_2–O → (1. H–C(=O)–H, 2. H_2O) OH O–CH_2–O → (H_3O^+) OH OH OH

Problem 19.25

Think

Which C–C bond in the final product was previously the C=C? How do you convert a C=C into a C–C? Which C atom in the C=C was previously the ketone C=O? What is the identity of the R group in the ylide?

Solve

Since the Wittig reaction forms an alkene, we must consider an alkene as a critical synthetic intermediate. The ylide can be made from triphenylphosphine and 3-chloropentane.

Undo a hydrogenation. ⟹ Undo a Witting reaction. ⟹ O + $^{\ominus}$ $^{\oplus}PPh_3$

The synthesis in the forward direction would then appear as follows:

⊕PPh3

H2, Pd

Problem 19.26

Think

What is the name of the six C-atom ring that shows three C=C bonds in the starting material? Does this ring have different properties compared to a normal C=C?

Solve

The double bonds in the starting material are part of a benzene ring, whose π system is exceptionally stable. The conditions required to reduce an aromatic ring are more extreme than those required to reduce an alkyne.

H2

Pd / BaSO4

25 °C

H

H

Aromatic ring: inert to hydrogenation under these conditions.

Problem 19.28

Think

Does this synthesis require a carbon–carbon bond-formation reaction? If so, what species possessing an electron-rich carbon could be used? What species containing an electron-poor carbon could be used? Are there any stereochemical issues that must be considered? Can the product be formed in a single reaction, or should we consider making it from a precursor?

Solve

The cis alkene can be the product of reduction using catalytic hydrogenation. The alkyne precursor can be the product of conjugate addition of $CH_3C{\equiv}C^{-}$.

Undo a catalytic hydrogenation.

Undo a conjugate addition.

The synthesis could appear as follows. The reduction involves a poisoned catalyst to stop at the alkene.

1. $H_3C-C{\equiv}CNa$

2. H_2O

H_2

Pd, $BaSO_4$

Problem 19.29

Think

Which side of the bicyclic ring is more sterically crowded? How does steric crowding affect which syn product forms?

Solve

The product **A** is favored. For product **B** to form, the top face of the C=C bond must bind to the catalyst surface, but, as indicated, there is excessive steric hindrance provided by the bridged carbon and its methyl groups. To form product **A**, the bottom face of the C=C binds to the catalyst surface and so do the H atoms.

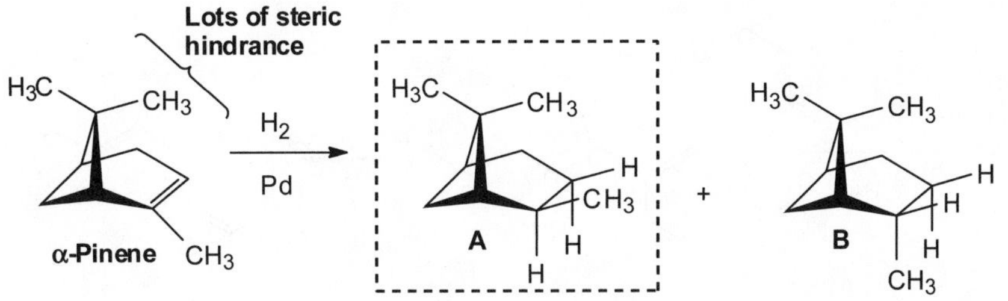

Lots of steric hindrance

α-Pinene $\xrightarrow[\text{Pd}]{H_2}$ **A** + **B**

Problem 19.30

Think

Which C=O is more reactive, an aldehyde or a ketone? Which C=O remains unreacted in the final product? How is it possible to selectively react the aldehyde over the ketone?

Solve

The aldehyde needs to be selectively reduced over the ketone to carry out the subsequent bromination. This selective reduction can take place with catalytic hydrogenation because the aldehyde is more reactive than the ketone. The resulting alcohol can be brominated with PBr_3.

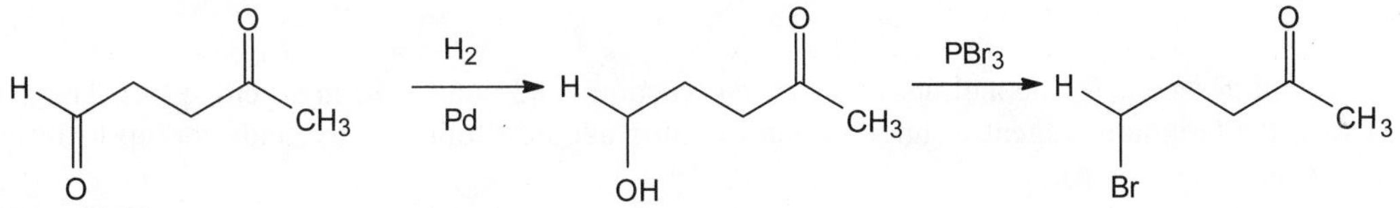

$\xrightarrow[\text{Pd}]{H_2}$ $\xrightarrow{PBr_3}$

Problem 19.32

Think

Which C–C bond can we "disconnect" in a retrosynthetic analysis? By undoing what reaction? Are there any oxidations of alcohols that must be stopped at the aldehyde?

Solve

A Grignard reaction forms a carbon–carbon bond, yielding an alcohol. Our target is a ketone that could be the result of oxidizing an alcohol product, but that alcohol is not the immediate product of a Grignard reaction involving $C_6H_5CH_2MgBr$. In the retrosynthetic analysis below, the third molecule is an alcohol that would be produced from that Grignard reaction.

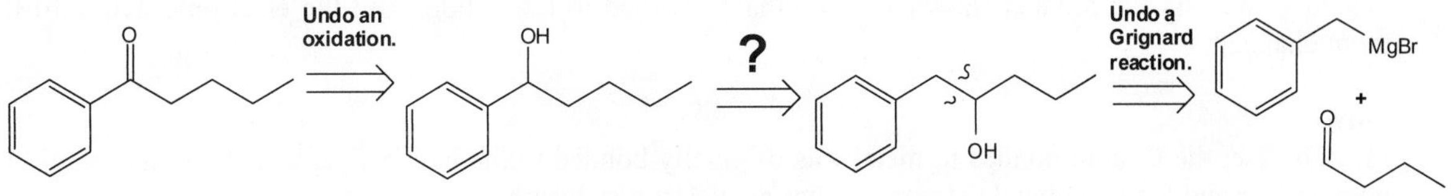

Undo an oxidation. ⟹ ? ⟹ Undo a Grignard reaction. ⟹ MgBr +

The aldehyde can be formed from an oxidation of an alcohol.

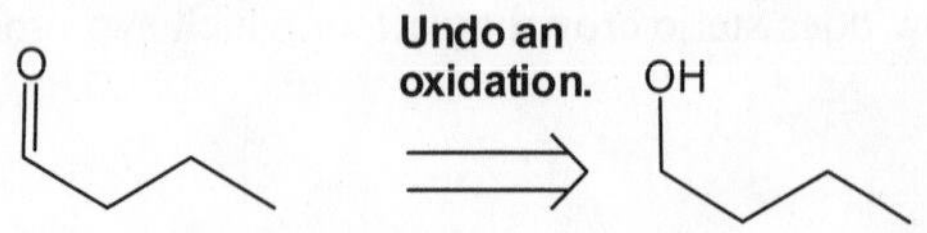

Undo an oxidation.

OH

The reaction in the forward direction is as shown below. Notice that prior to the final oxidation the alcohol group is moved from the 2 position to the 1 position by dehydration to make the alkene, followed by Markovnikov addition of water.

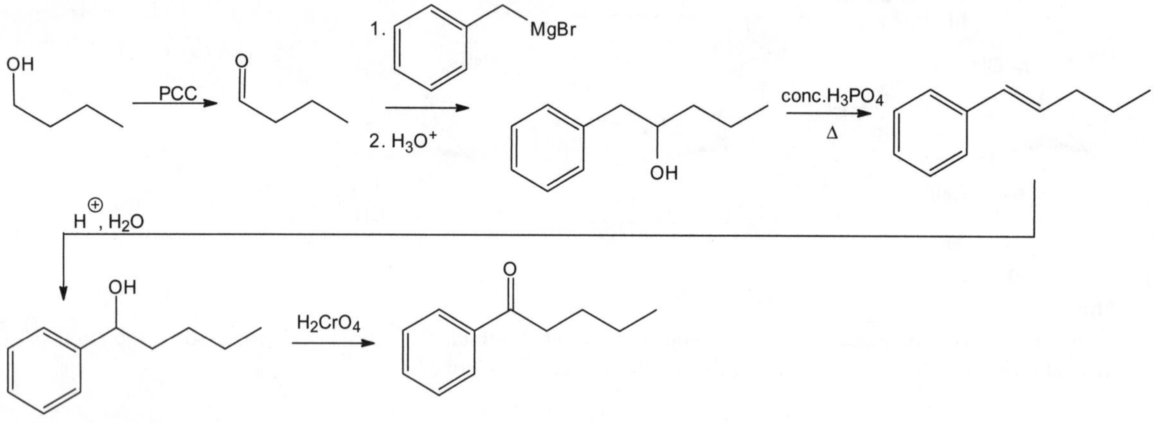

OH PCC 1. MgBr 2. H_3O^+ OH conc.H_3PO_4 Δ

$H^{\oplus}$, H_2O

OH H_2CrO_4

Problem 19.33

Think

What functional group results when a secondary alcohol is oxidized by $KMnO_4$? How can this functional group react with a Grignard reagent? What final uncharged product results?

Solve

$KMnO_4$ reacts with a secondary alcohol under the given reaction conditions to form a ketone (**A**). The ketone then reacts with the Grignard reagent to undergo nucleophilic addition followed by acid workup to form the tertiary alcohol shown below (**B**).

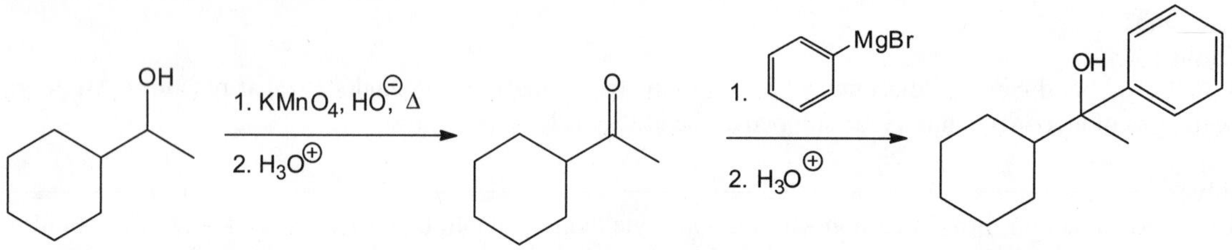

OH 1. $KMnO_4$, $HO^{\ominus}$, Δ 2. $H_3O^{\oplus}$ 1. MgBr 2. $H_3O^{\oplus}$ OH

A B

Problem 19.34

Think

Which C atom in the product shown was originally bonded to the halide? Did the electronic nature of the C atom change?

Solve

In each case, the C atom bonded to metal was originally bonded to the halide. The C–X bond has an electron-poor carbon and the C–M bond (M = metal) has an electron-rich carbon.

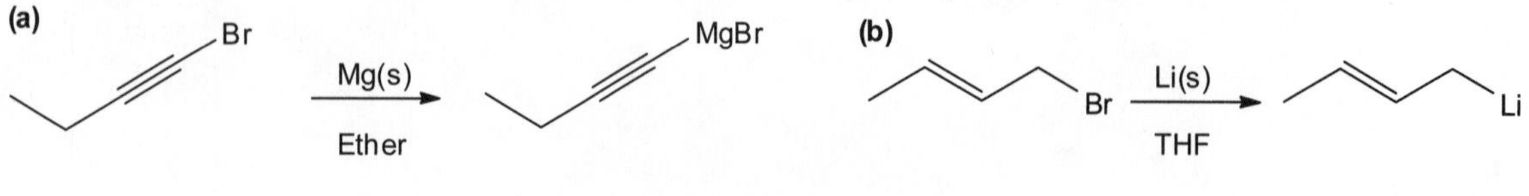

(a) Br Mg(s) Ether MgBr

(b) Br Li(s) THF Li

(c) Mg(s), Ether → MgI

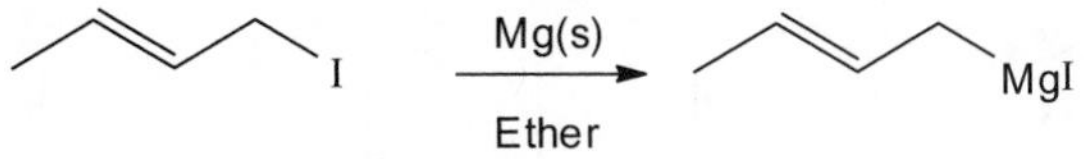

(d) Br; 1. Li(s) 2. CuI → Li Cu; 4; 4; 4

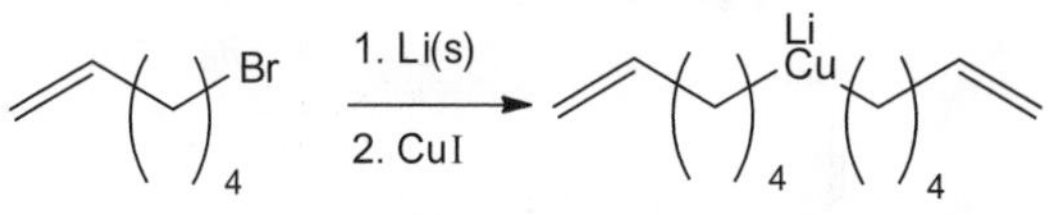

(e) Br; Li(s), THF → Li

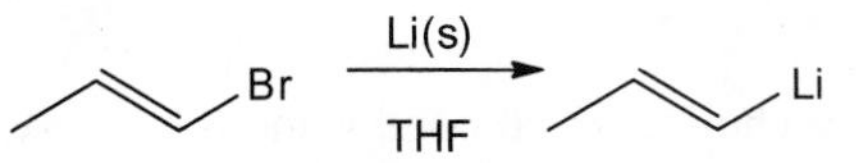

(f) Br; Mg(s), Ether → MgBr

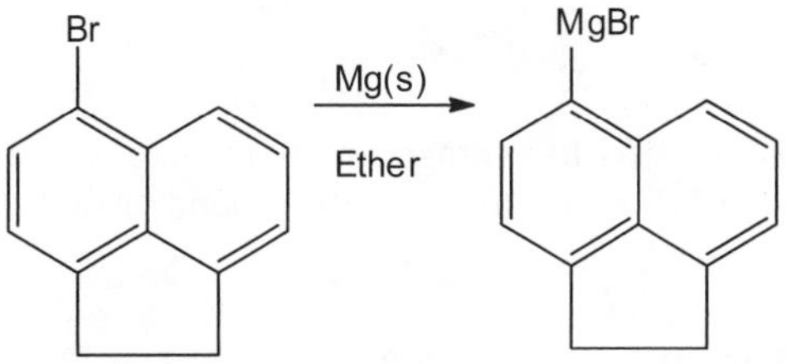

(g) Br; Li(s), THF → Li

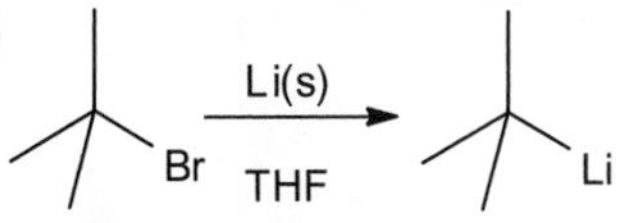

(h) Br; Li(s), THF → Li

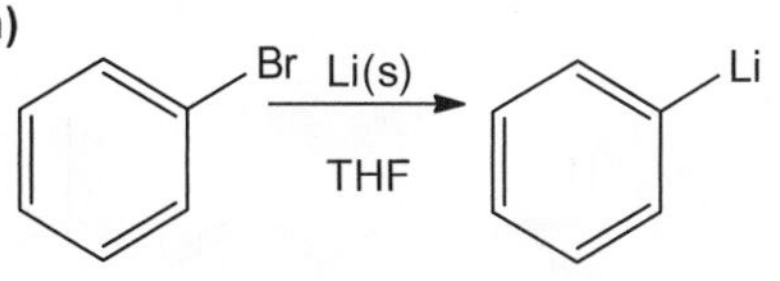

(i) Br; N; Mg(s), Ether → MgBr; N

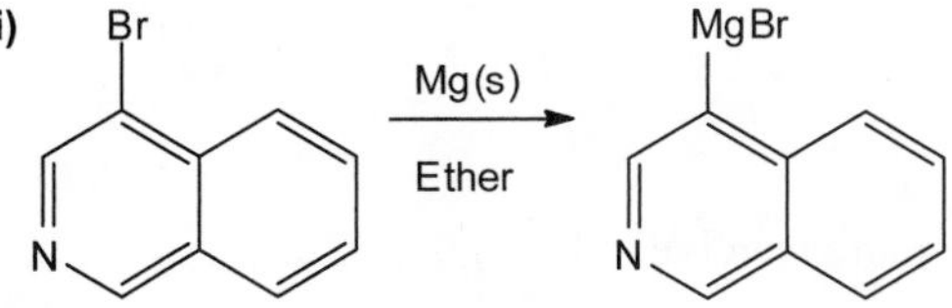

(j) Cl; O; Ph; Mg(s), Ether → ClMg; O; Ph

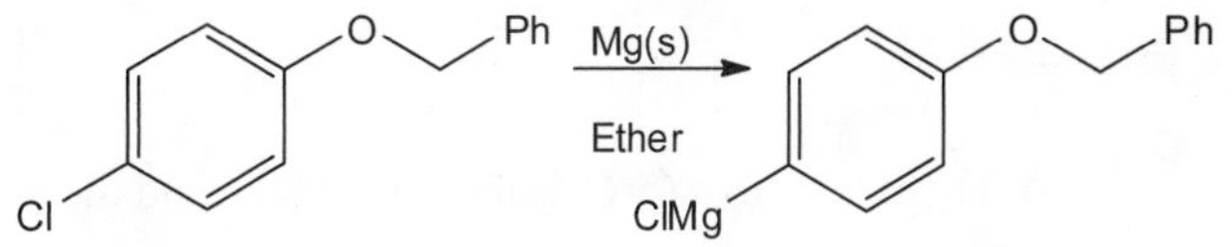

(k) I; Mg(s), Ether → MgI

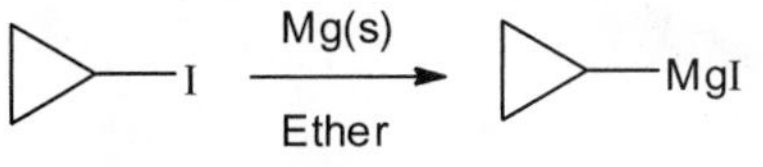

Problem 19.35

Think

Which C atom in the alkyl halide coverts to a carbon–metal bond? Does the electronic nature of the C atom change?

Solve

In each case, the C atom of the alkyl halide group becomes bonded to the metal. The C–X bond has an electron-poor carbon and the C–M bond (M = metal) has an electron-rich carbon.

(a) Br; Li, Ether → Li

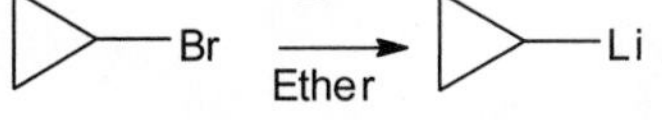

(b) $H_2C{=}C{=}CH$–Br; Mg(s), THF → $H_2C{=}C{=}CH$–MgBr

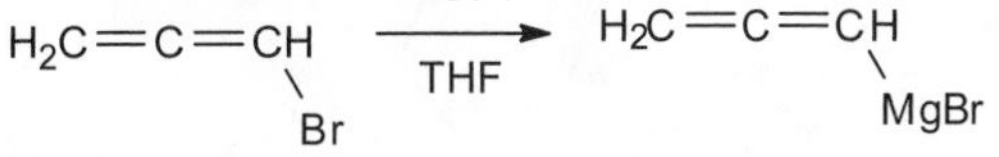

(c) N; Br; Mg(s), THF → N; MgBr

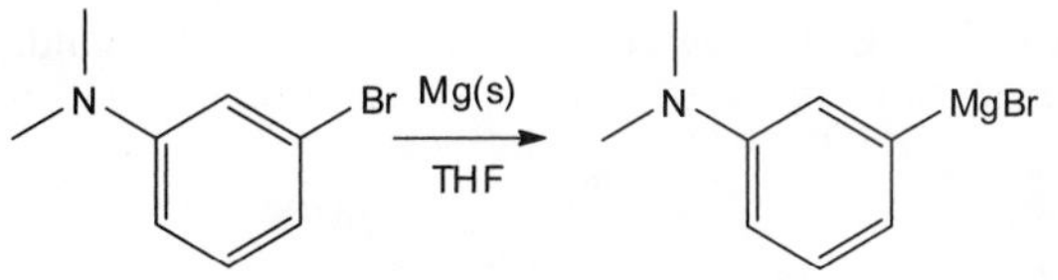

(d) Cl; Mg(s), Ether → MgCl

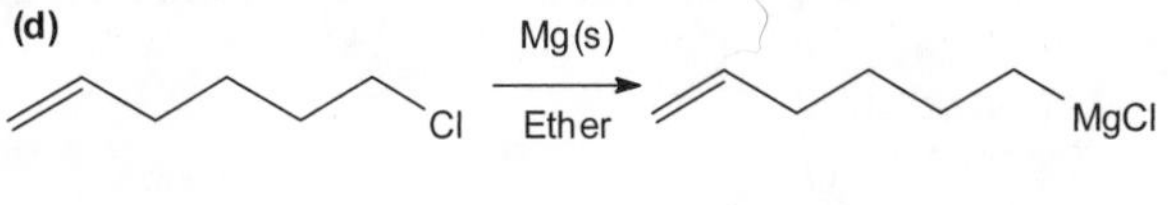

(e) Br; 1. Li(s), THF 2. CuI → Cu Li

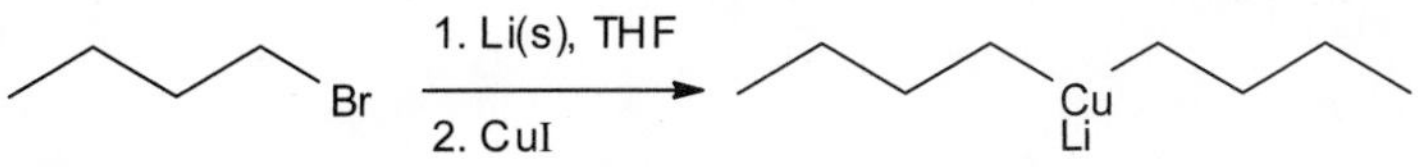

(f) N; Br; Li(s), THF → N; Li

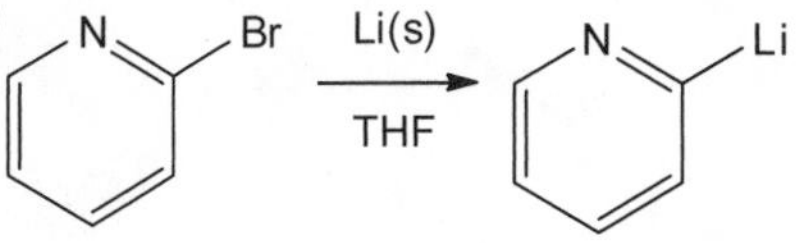

(g)

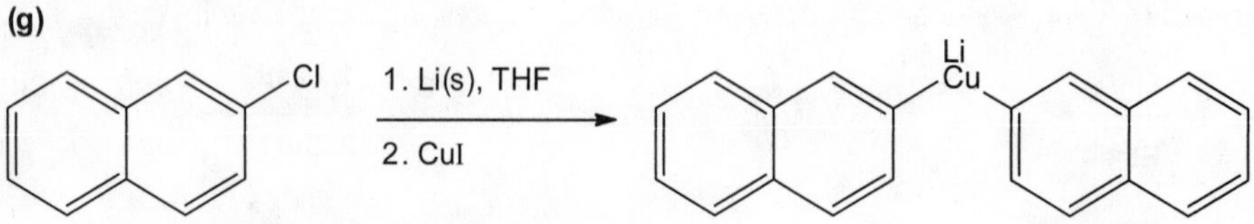

1. Li(s), THF

2. CuI

Problem 19.36

Think

Which bond is subject to hydrogenation? If there is more than one choice, which bond is more selective to the reagents given? What adds to the multiple bond? Are there any stereochemical considerations?

Solve

(a) H_2(g) adds to a C=C to form an alkane and the alkene is more reactive to H_2(g) than the C=O or the aromatic ring.

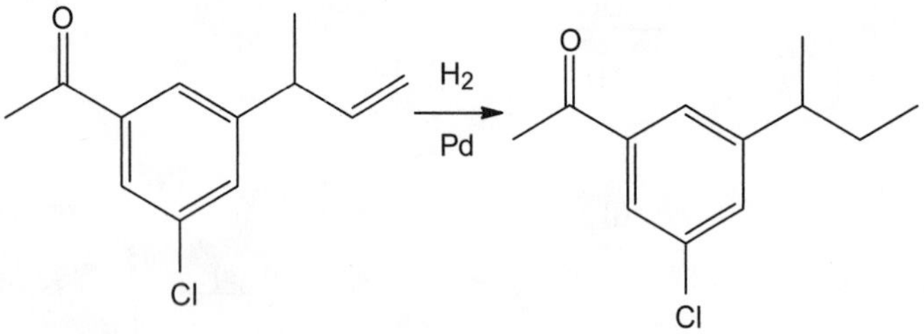

H_2 / Pd

(b) H_2(g) adds to a C=C to form an alkane and the alkene is more reactive to H_2(g) than C≡C.

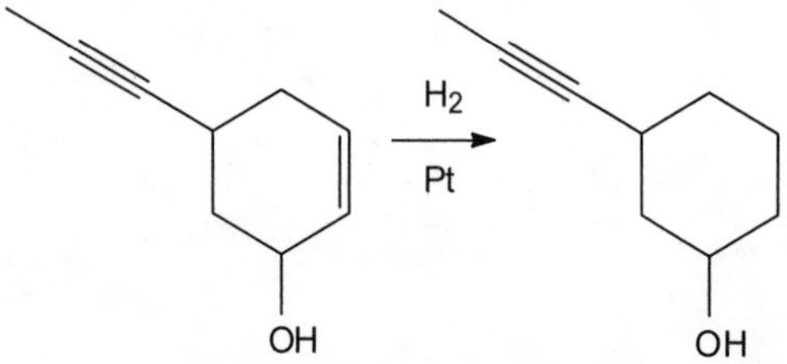

H_2 / Pt

(c) Catalytic hydrogenation is more favored at the *less* sterically hindered C=C bond.

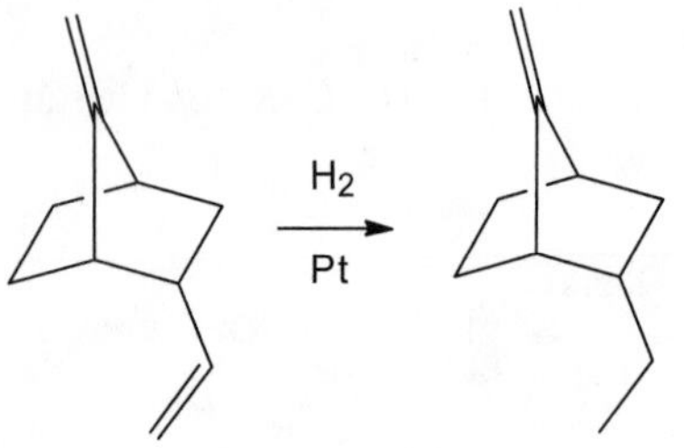

H_2 / Pt

(d) Lindlar's catalyst adds to a C≡C to form a cis alkene. The C≡C is selectively reduced over the amide.

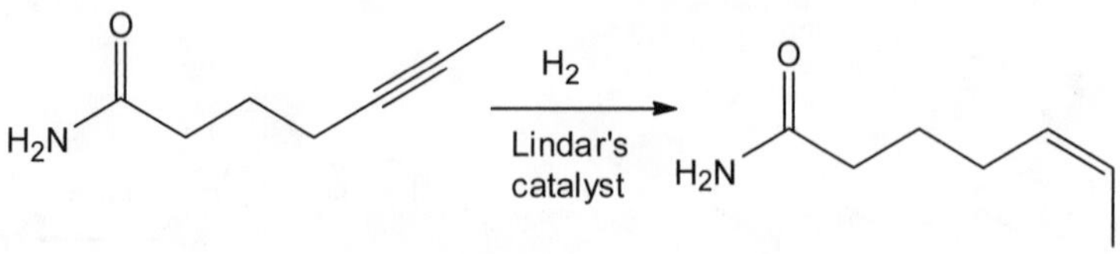

H_2 / Lindar's catalyst

(e) The nitrile is less reactive than the aldehyde. Therefore, H_2 adds preferentially to the aldehyde.

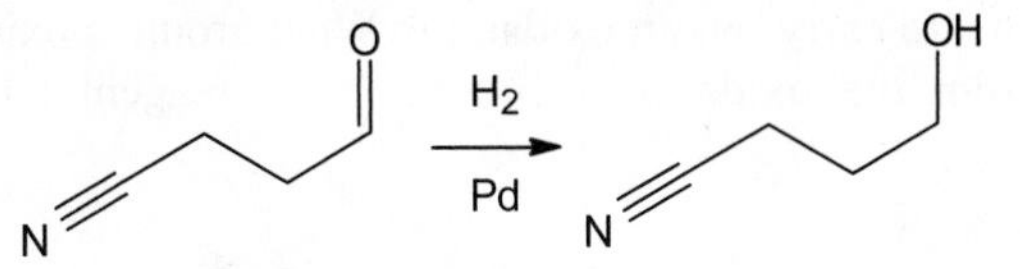

H_2 / Pd

(f) Aldehydes and alkenes have similar reactivity toward catalytic hydrogenation, but the aldehyde is less sterically hindered.

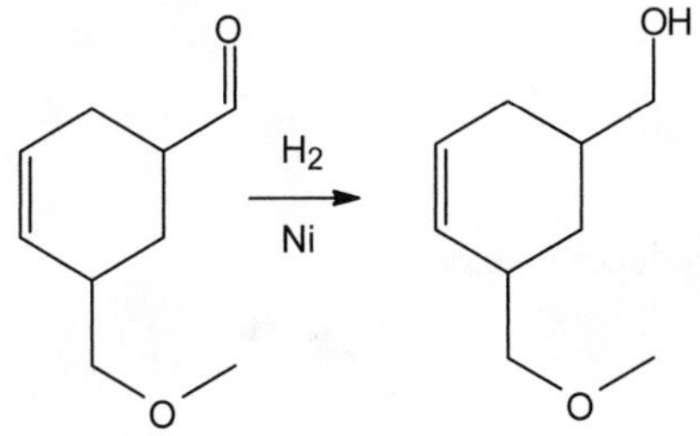

H_2 / Ni

(g) With excess H_2, the alkyne is reduced all the way to the alkane. Without more extreme conditions, the amide does not react.

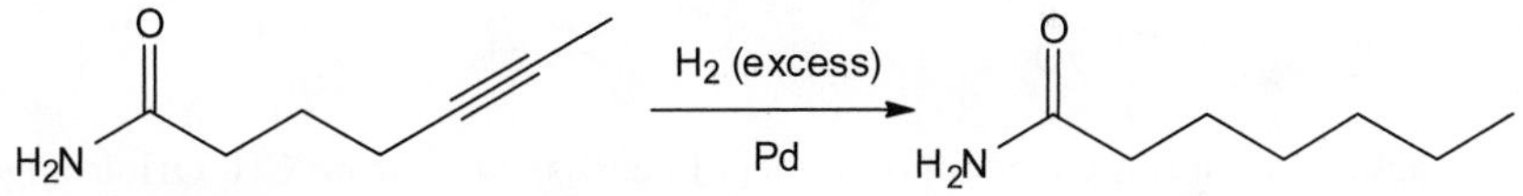

H_2 (excess) / Pd

Problem 19.37

Think

What bond is reduced by the reagents given? What group forms from the reduction reaction?

Solve

(a) The Clemmensen reaction reduces the ketone to a methylene group (CH_2).

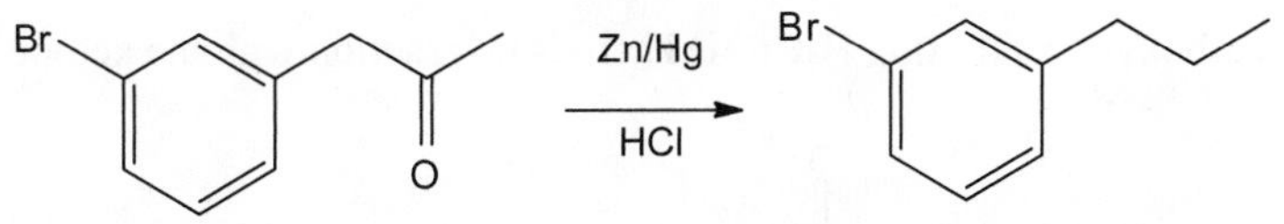

Zn/Hg / HCl

(b) The Wolff–Kishner reaction reduces the ketone to a CH_2 group.

H_2NNH_2 / NaOH, Δ

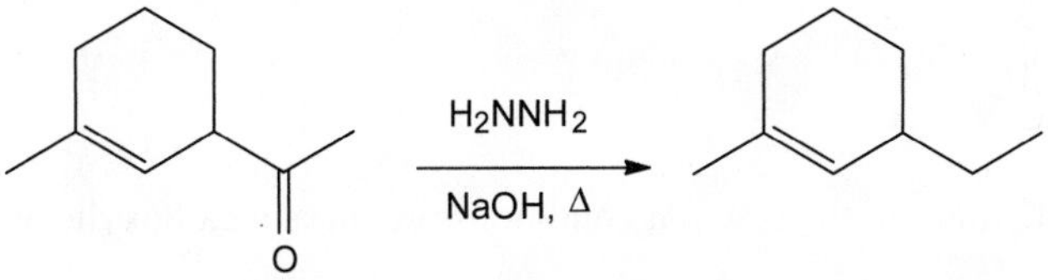

(c) The Raney nickel reaction reduces the ketone to a CH_2 group.

1. HS⁀SH, H^+ / 2. Raney nickel

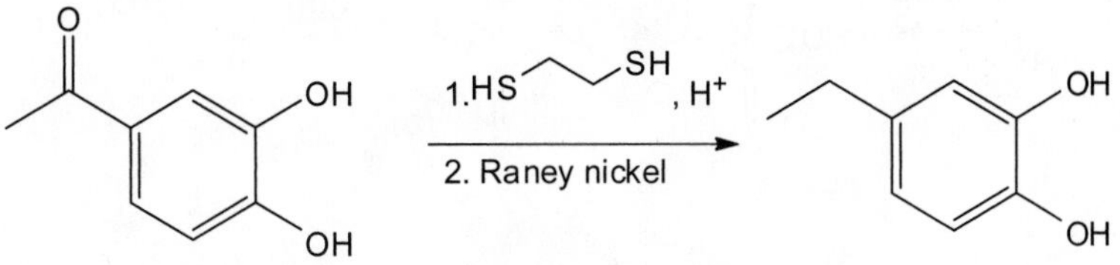

Problem 19.38

Think

What arrangement of atoms is necessary for Cr^{+6} or Mn^{+7} to carry out an oxidation? What group is oxidized by the reagents given? What functional group forms from the oxidation reaction? What reagent allows the oxidation of a primary alcohol to stop at the aldehyde?

Solve

(a) Oxidation of a primary alcohol using $Na_2Cr_2O_7$ in strong acid produces a carboxylic acid.

$Na_2Cr_2O_7$ / H_2SO_4, H_2O

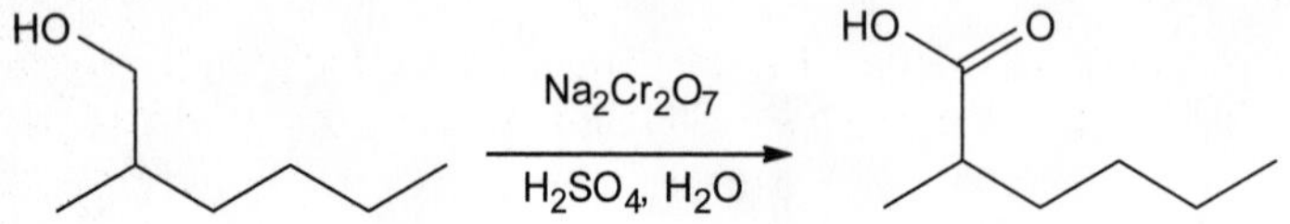

(b) Oxidation of a secondary alcohol by CrO_3 in a strong acid produces a ketone.

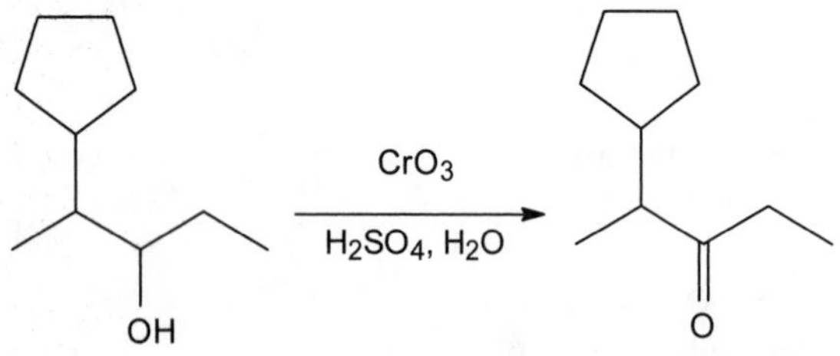

CrO_3 / H_2SO_4, H_2O

(c) There is no oxidation reaction of a tertiary alcohol or a phenol because there is no CH–OH arrangement of atoms.

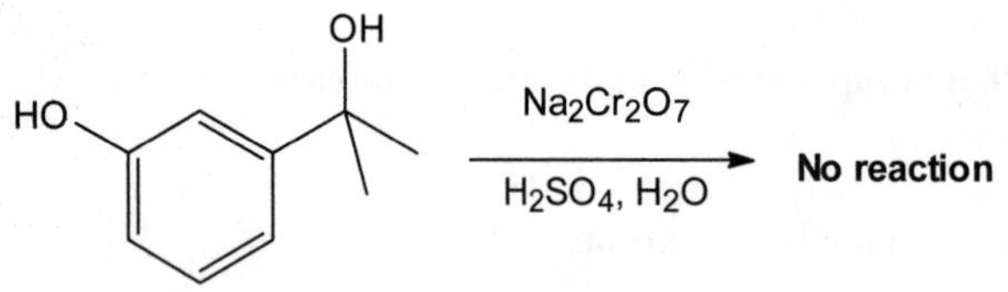

$Na_2Cr_2O_7$ / H_2SO_4, H_2O → **No reaction**

(d) The aldehyde is oxidized to the carboxylic acid using H_2CrO_4. There is no reaction with the ketone.

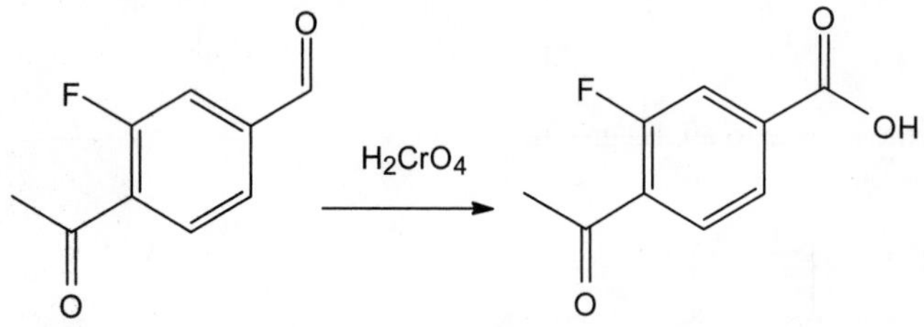

H_2CrO_4

(e) Oxidation of a primary alcohol using $KMnO_4$ in the reaction conditions produces a carboxylic acid. There is no reaction with the ketone.

1. $KMnO_4$, $HO^{\ominus}$ 2. $H^{\oplus}$

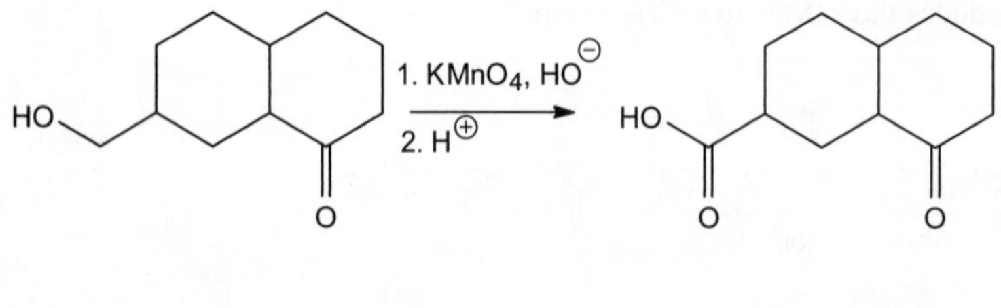

(f) PCC oxidizes the primary alcohol to the aldehyde.

PCC

Problem 19.39

Think

Which C–C bond formed from the reaction of acetone with the other organic precursor? Can the synthesis take place in a single reaction with the alcohol? If not, what precursor should we choose? How can this precursor be formed from the alcohol?

Solve

In each of these reactions, the tertiary alcohol can be formed from a Grignard reaction. The Grignard reagent is formed from the alkylbromide, which can be formed from an alcohol. The syntheses are given below.

(a) Cyclohexanol is the alcohol starting material.

Retrosynthesis

Undo a Grignard reaction.

Reaction in the forward direction

PBr_3 — Mg(s), Ether — 1. acetone; 2. H_3O^+

(b) Pentan-1-ol is the alcohol starting material.

Retrosynthesis

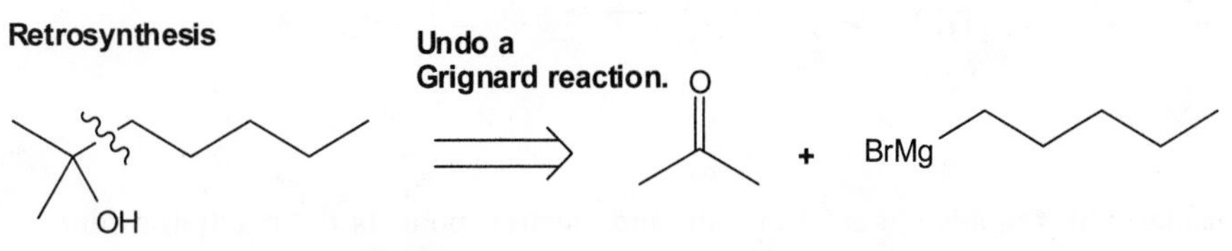

Reaction in the forward direction

1. PBr_3; 2. Mg(s), Ether — 1. acetone; 2. H_3O^+

(c) Prop-2-en-1-ol is the alcohol starting material.

Retrosynthesis **Undo a Grignard reaction.**

HO MgBr + O

Reaction in the forward direction

OH 1. PBr_3 2. Mg(s) Ether MgBr 1. O 2. H_3O^+ HO

(d) The 2,3-dimethylhexan-3-ol target requires a Grigarnd reaction using pentan-2-one, which can be synthesized by the oxidation of penan-2-ol. Acetone is used to form the Grignard reagent.

Retrosynthesis **Undo a Grignard reaction.** **Undo an oxidation.**

OH O + MgBr OH

Reaction in the forward direction

OH H_2CrO_4 O Then EtOH OH

O $NaBH_4$/EtOH OH PBr_3 Br Mg(s) Ether MgBr

Problem 19.40

Think

In each reaction consider which atoms are electron rich and electron poor. Is the reaction a functional group transformation or does the reaction alter the carbon skeleton? Are proton transfer reactions involved? Are oxidations or reductions taking place?

Solve

A is the alcohol product from a reduction reaction of an aldehyde and $LiAlH_4$. The alcohol is then converted to the alkyl bromide, which reacts with Mg(s)/ether to form a Grignard reagent, **B**. A benzyl group is added to alter the carbon skeleton and the identity of **C** is benzaldehyde (source of electrophilic C). Another benzyl group is added to form the benzylic ether in a Williamson ether synthesis, and the identity of **D** is benzylbromide (source of electrophilic C). See the figure on the next page.

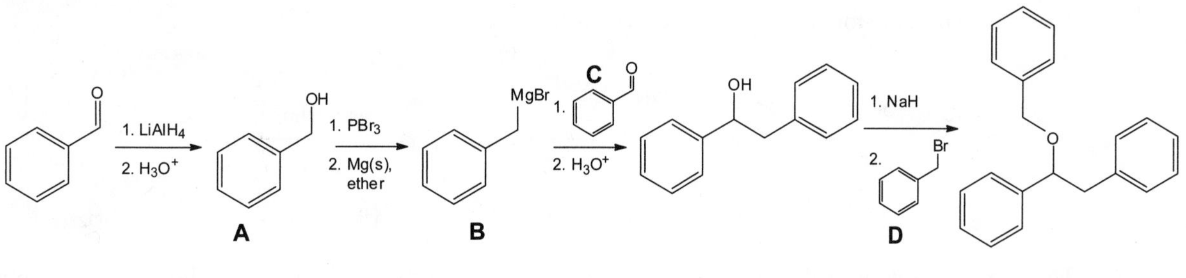

1. $LiAlH_4$ 2. H_3O^+ A; 1. PBr_3 2. Mg(s), ether B; C 1. 2. H_3O^+; 1. NaH 2. D

Problem 19.41

Think

In each reaction consider which atoms are electron rich and electron poor. Is the reaction a functional group transformation or does the reaction alter the carbon skeleton? Are proton transfer reactions involved? Are there any oxidation or reduction reactions?

Solve

An alkylbromide can react with Mg(s)/ether to form a Grignard reagent, **C**. The identity of **A** is PBr_3, which turns the alcohol into alkyl halide, **B**. In the second row, the product of the first reaction sequence is an α,β-unsaturated ketone, where the carbon skeleton was altered by one C atom. This product came from an aldol condensation and, therefore, the identity of **D** is formaldehyde. Notice that LDA produced the kinetic enolate, allowing alkylation to take place at the least hindered α C. The product of the second reaction sequence came from a 1,4-addition of an ethyl (two C atoms) to the α,β-unsaturated ketone. Lithium dialkylcuprate compounds react preferentially via conjugate addition and **E**, therefore, is $(CH_3CH_2)_2CuLi$.

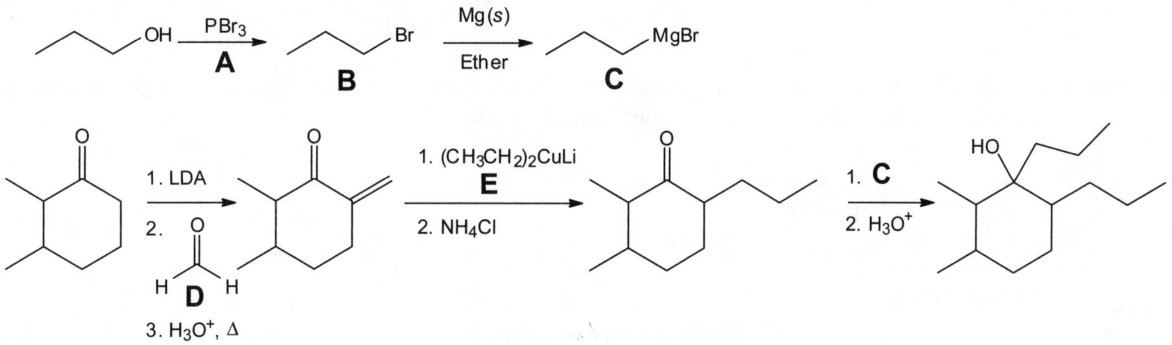

OH PBr_3 A → Br B; Mg(s) Ether → MgBr C

1. LDA 2. H D H 3. H_3O^+, Δ; 1. $(CH_3CH_2)_2CuLi$ E 2. NH_4Cl; 1. C 2. H_3O^+; HO

Problem 19.42

Think

Will a carbon–carbon bond-formation reaction be necessary? What is the relative positioning of the functional groups in the target? Does the appropriate carbon–carbon bond-formation reaction in Table 19-1 leave us with the correct functional groups, or will we need to carry out an additional functional group conversion? Are any oxidation or reduction reactions necessary?

Solve

(a) The relative position of the two functional groups is 1,2, and based on the observation of a N atom, the synthesis can go through a cyanohydrin intermediate.

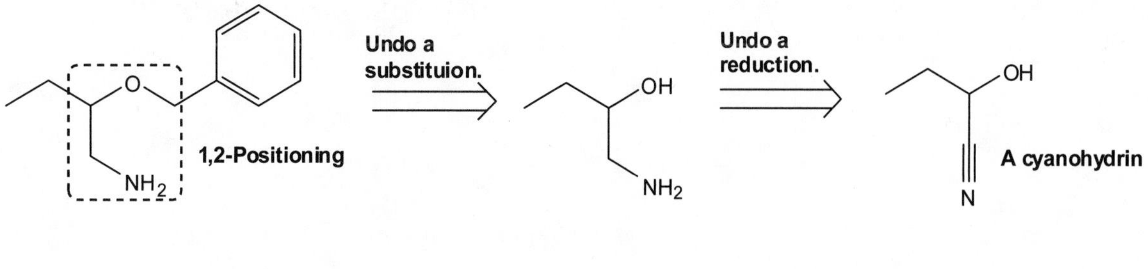

O NH_2 **1,2-Positioning**; **Undo a substituion.** ⟹ OH NH_2; **Undo a reduction.** ⟹ OH N **A cyanohydrin**

The synthesis in the forward direction is given below:

NaCN / HCN → 1. $LiAlH_4$ 2. H_3O^+ → 1. NaH 2. $BrCH_2Ph$

(b) The relative position of the two functional groups is 1,3 and the synthesis likely should include an aldol addition reaction. The aldol addition OH can then be functionalized to the thioether.

1,3-Positioning ⟹ **Undo a substitution.** ⟹ **Undo a bromination.** **A β-hydroxycarbonyl**

The synthesis in the forward direction is given below:

2 → NaOH → PBr_3 → $NaSCH_3$

(c) The relative position of the two functional groups is 1,3 and the synthesis likely should go through an α,β-unsaturated aldehyde, the product of an aldol condensation.

1,3-Positioning ⟹ **Undo an addition.** **An α,β-unsaturated aldehyde**

The synthesis in the forward direction is given below:

$NaBH_4$ / EtOH → NaH

Br_2 / CH_3COOH → NaOH, Δ → NaO / EtOH

(d) The relative position of the two functional groups is 1,5 and the synthesis likely goes through a 1,5-dicarbonyl, which would come from the conjugate addition of an enolate to an α,β-unsaturated ketone.

OH
Undo a reduction.
O
Undo an addition.
O
α,β-Unsaturated ketone
OH
1,5-Dicarbonyl
1,5-Positioning

The synthesis in the forward direction is given below:

O
Br_2
CH_3COOH
O
Br
NaOH, Δ
O

O
NaOH
O
O
$NaBH_4$ (excess)
EtOH
OH
O
OH

Problem 19.43

Think

In each reaction consider which atoms are electron rich and electron poor. Is the reaction a functional group transformation or does the reaction alter the carbon skeleton? Are proton transfer reactions involved? Are there any oxidation or reduction reactions involved?

Solve

The missing reagents and intermediates are filled in below:

O
1. Li
2. H_2O
OH
A
H_2CrO_4
O
B

Br
Li(s)
THF
C
Li
D
1. **B**
2. H_3O^+
HO

Problem 19.44

Think

In each reaction consider which atoms are electron rich and electron poor. Is the reaction a functional group transformation or does the reaction alter the carbon skeleton? Are proton transfer reactions involved? Are there any steps that protect or deprotect functional groups?

Solve

The missing reagents and intermediates are filled in below:

$(CH_3)_2\overset{\oplus}{S}-\overset{\ominus}{C}H_2$ **A**

1. PhMgBr **B**
2. NH_4Cl

TBDMS-Cl **C**

TBDMS

H_3O^+

D

1. MgBr **E**
2. H_3O^+

$Bu_4N^+F^-$

HO

OH

F

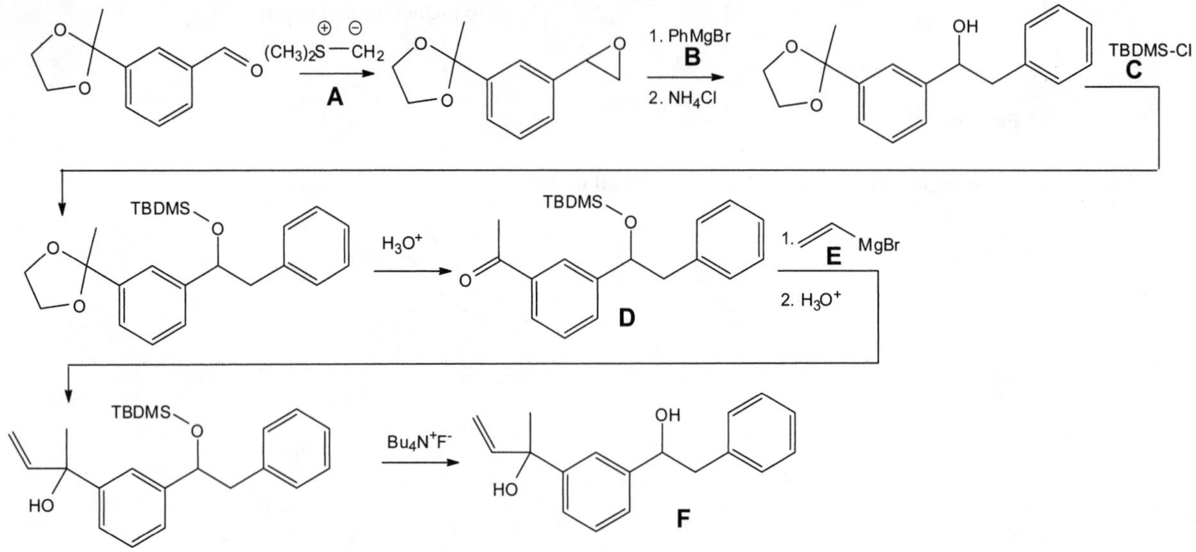

Problem 19.45

Think

In each reaction consider which atoms are electron rich and electron poor. Is the reaction a functional group transformation or does the reaction alter the carbon skeleton? Are proton transfer reactions involved? Are there any oxidation or reduction reactions involved? Is regiochemistry a concern?

Solve

The missing reagents and intermediates are filled in below. Notice that **A** is the result of anti-Markovnikov addition of water, whereas **D** is the result of Markovnikov addition of water. **F** is produced from the kinetic enolate.

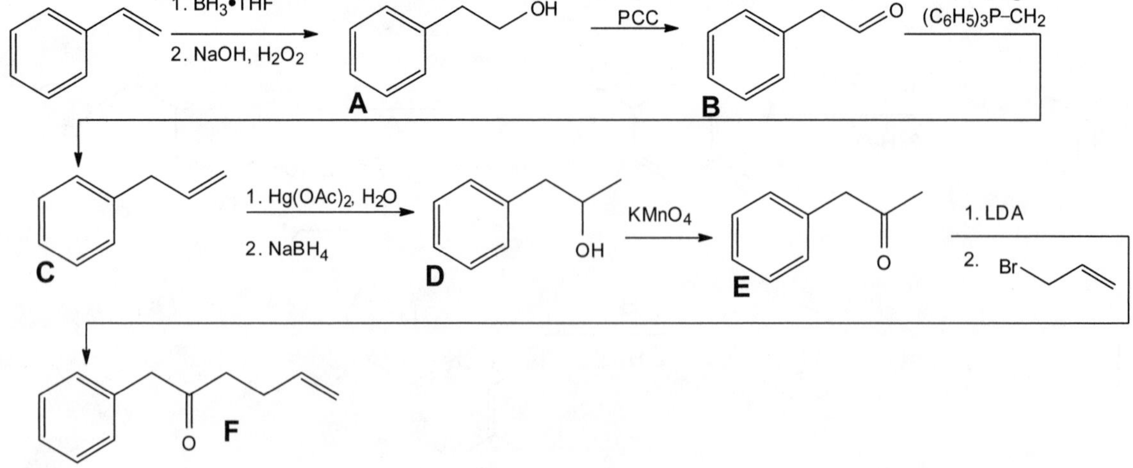

1. $BH_3{\bullet}THF$
2. NaOH, H_2O_2

A OH

PCC

B

$(C_6H_5)_3\overset{\oplus}{P}-\overset{\ominus}{C}H_2$

C

1. $Hg(OAc)_2$, H_2O
2. $NaBH_4$

D OH

$KMnO_4$

E

1. LDA
2. Br

F

Problem 19.46

Think

In each reaction consider which atoms are electron rich and electron poor. Is the reaction a functional group transformation or does the reaction alter the carbon skeleton? Are proton transfer reactions involved? Are there any oxidation or reduction reactions? Is regiochemistry a concern?

Solve

The missing reagents and intermediates are filled in below. There are two products for **F** since both carbon atoms attached to the epoxide are secondary. This leads to two ketones for **G** as well.

Br_2, CCl_4 → **A**; 1. LDA (excess), 2. H_2O → B; 1. NaH, 2. **C**; H_2/Pd, Lindlar's catalyst → **D**; MCPBA → **E** + Enantiomer; 1. CH_3MgBr, 2. H_3O^+ → **F**; H_2CrO_4 → **G**

Problem 19.47

Think

How many carbon–carbon bond-formation reactions are necessary? What is the relative positioning of the functional groups in the target? Does the appropriate carbon–carbon bond-formation reaction in Table 19-1 leave us with the correct functional groups, or will we need to carry out an additional functional group conversion? Are any oxidation or reduction reactions necessary?

Solve

(a) The 1,5-positioning in the target calls for conjugate addition of an enolate anion.

NaOH; $NaBH_4$, EtOH

(b) The 1,4-positioning in the target calls for conjugate addition of cyanide.

KCN, HCN; H_3O^+; CH_2N_2

(c) The 1,2-positioning in the target calls for the formation of a cyanohydrin. The OH of the cyanohydrin can then be protected, and a Grignard reaction can form the necessary C–C bond.

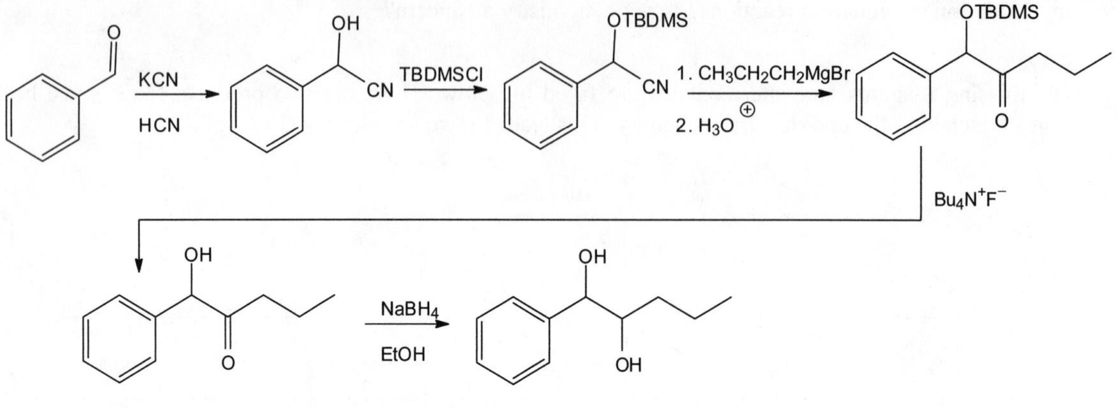

KCN, HCN; TBDMSCl; 1. $CH_3CH_2CH_2MgBr$ 2. $H_3O^{\oplus}$; $Bu_4N^+F^-$; $NaBH_4$, EtOH

(d) The 1,3-positioning in the target calls for an aldol addition.

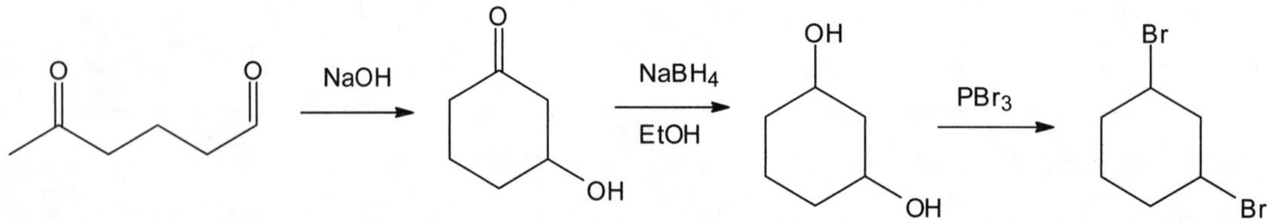

NaOH; $NaBH_4$, EtOH; PBr_3

Problem 19.48

Think

Is the carbon skeleton altered? Does functional group transformation occur? Are protecting groups necessary? Are strong acids or bases needed? What reactions can you use to remove a functional group entirely?

Solve

(a) Requires alkylation via the kinetic enolate, which is why LDA is used as the base. To avoid reaction with the ether, the reduction of C=O to CH_2 should avoid the Clemmensen reduction. A Wolff–Kishner or Raney Ni reduction could be used.

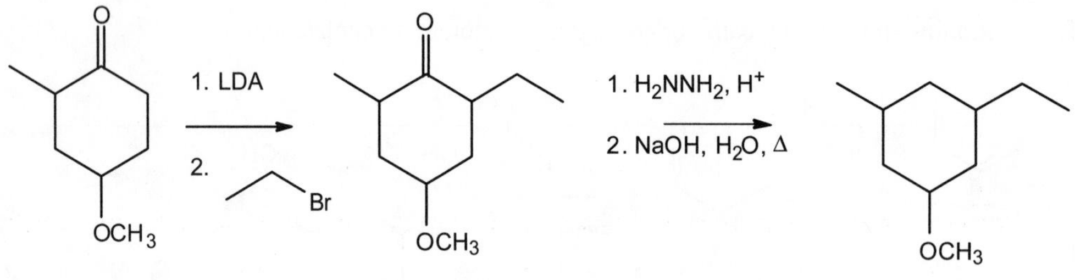

1. LDA 2. ethyl bromide; 1. H_2NNH_2, H^+ 2. NaOH, H_2O, Δ

(b) Requires the thermodynamic enolate, which is why NaOH is used as the base.

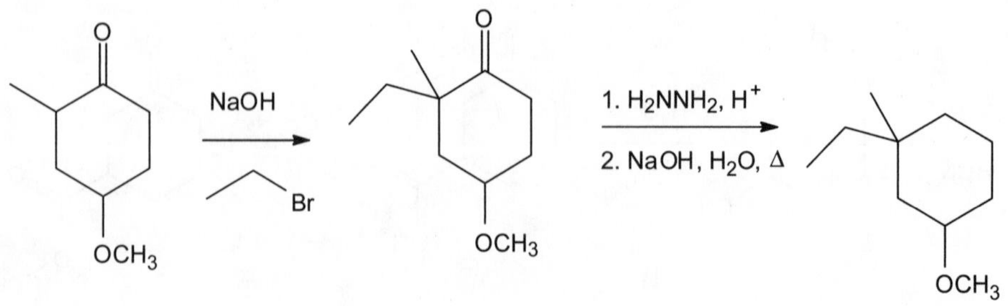

NaOH, ethyl bromide; 1. H_2NNH_2, H^+ 2. NaOH, H_2O, Δ

(c) Conjugate addition is required, so $LiCuR_2$ is used as the alkylating reagent. To keep the alkene unreacted, the Wolff–Kishner reduction conditions are used to avoid the use of strong acids (Clemmensen reduction) or catalytic hydrogenation (Raney Ni).

1. $(\text{cyclopropyl})_2CuLi$ 2. H_3O^+; Br / NaOH; 1. H_2NNH_2, H^+ 2. NaOH, H_2O, Δ

(d) The alcohol is protected to avoid a reaction between it and LDA.

KCN / HCN; H_3O^+/H_2O; TBDMS-Cl; 1. LDA 2. CH_3Br; 1. H_2NNH_2, H^+ 2. NaOH, H_2O, Δ; $Bu_4N^+F^-$

CN, OH, OTBDMS

Problem 19.49

Think

What functional group is formed in this reaction? Which functional group is protected? How do you remove the protecting group? In 1,4-cyclohexanedione, are both ketones reactive toward an ylide? If so, what product would result?

Solve

(a) The acetal is a protected ketone, so we can consider the target as a product of a Wittig reaction.

Undo a Wittig reaction. ⟹ + $\overset{\oplus}{P}Ph_3$ / ⊖

After the Wittig reaction, we can deprotect the ketone. The reaction in the forward direction is as follows:

+ $\overset{\oplus}{P}Ph_3$ ⟶ ; H_3O^+ ⟶

(b) If 1,4-cyclohexanedione were used, the ketone in the product would not remain. Both ketones are susceptible to attack by the ylide to form an alkene. The reaction that would result is given below.

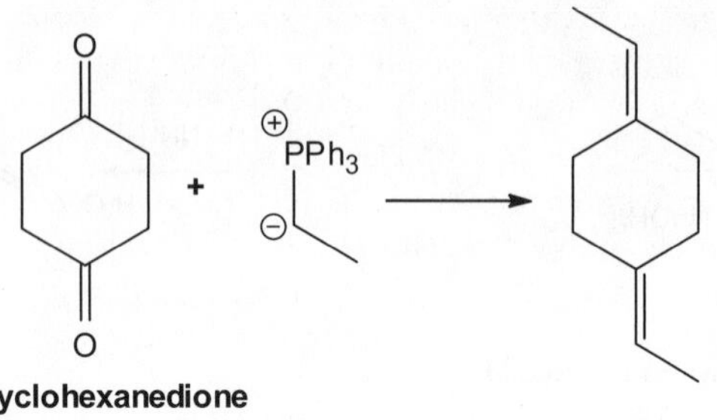

1,4-Cyclohexanedione

Problem 19.50

Think

What type of reaction conditions open an epoxide? If an epoxide reacts under these conditions, is it a good protecting group? Why would an epoxide be more reactive than the five-membered-ring acetal?

Solve

An epoxide has a lot of ring strain and can open easily in an S_N2 reaction with a species like HO^-. So it does a poor job of protecting the C=O group. The five-membered-ring acetal is much more stable and is unreactive in the presence of a strong nucleophile.

Problem 19.51

Think

What type of reagent is LDA? What are the acidic H atoms present in the compound? Is there more than one acidic H that can react with LDA? How can you protect this group?

Solve

LDA is a strong base and selectively produces the kinetic enolate, which leads to alkylation at the less-substituted α C atom. This is what is proposed in the target. However, the alcohol OH proton is acidic and will also be attacked by LDA to form the O^- anion. Both of these anions (C^- and O^-) are subject to reaction with the electrophilic C atom of CH_3I. The actual product from the reaction is given below.

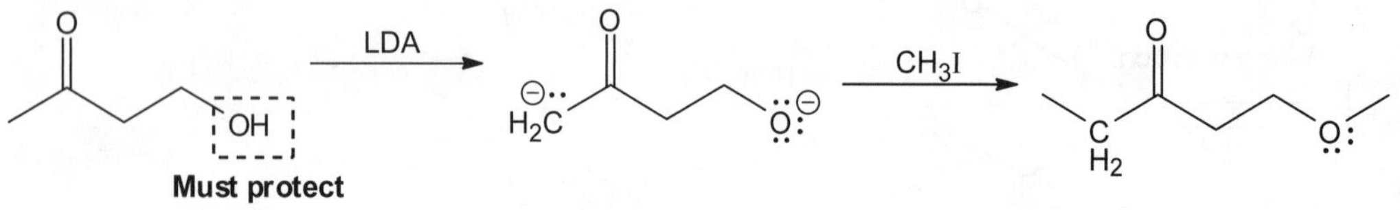

LDA

CH_3I

Must protect

If the OH is to remain unreacted, it must be protected. The corrected reaction conditions to yield the intended product are given below:

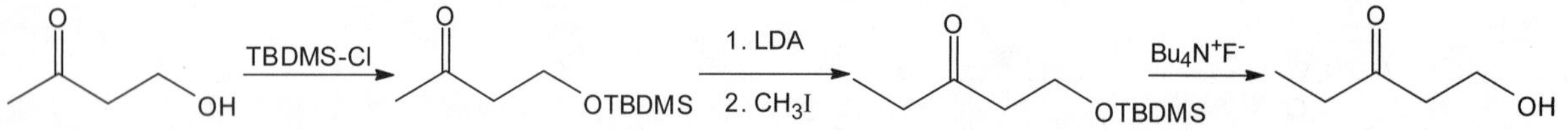

TBDMS-Cl

1. LDA
2. CH_3I

$Bu_4N^+F^-$

Problem 19.52

Think

In addition to the alcohol that is to be protected, what other functional groups are present? What reaction conditions are required for each of the five methods in Table 19-2? Are the other functional groups present reactive under these conditions?

Solve

(a) The TBDMS protection is the only method that will work because the aldehyde will not react with the TBDMS-Cl during protection, or with F^- during deprotection. The ether protection (Entries 1 and 2) will not work because the α C–H is subject to deprotonation by NaH (Section 10-3). Entries 3 and 4 also will likely not work because the aldehyde O atom is subject to protonation in acidic media, which would make the C=O group activated toward nucleophilic attack by weak nucleophiles.

(b) The TBDMS protection will work because the benzene ring and alkyl bromide will not react with the TBDMS-Cl during protection, or with F^- during deprotection. Entries 3 and 4 will likely work because the benzene ring and alkyl halide are unreactive in acidic medium. However, these reactions should probably be avoided because the benzylic bromide could react under S_N1 conditions. The ether protection will not work (Entries 1 and 2) because the RO^- that results after reaction with NaH is reactive with the R–Br.

Problem 19.53

Think

Is the carbon skeleton altered? Does functional group transformation occur? Are protecting groups necessary? Are strong acids or bases needed? What are the stereochemical considerations?

Solve

In both cases, the cis diol must be protected to target just one alcohol group. With the protected diol in **(a)**, we must convert the remaining alcohol into a good leaving group. PBr_3 reverses the stereochemical configuration. KCN promotes another S_N2 reaction that reverses the configuration, and reduction with $LiAlH_4$ yields the primary amine. In **(b)**, the protected intermediate is oxidized to the ketone before deprotection. See the figure on the net page.

(a)

OH, OH, OH; $H_2C{=}O$, H^+ → OH, O, CH_2, O; PBr_3 → Br, O, CH_2, O; KCN → CN, O, CH_2, O; 1. $LiAlH_4$ 2. H_3O^+ → CH_2NH_2, OH, OH

(b)

OH, OH, OH; $H_2C{=}O$, H^+ → OH, O, CH_2, O; H_2CrO_4 → O, O, CH_2, O; H_3O^+ → O, OH, OH

Problem 19.54

Think

Will a carbon–carbon bond-formation reaction be necessary? What is the relative positioning of the functional groups in the target? Does the appropriate carbon–carbon bond-formation reaction in Table 19-1 leave us with the correct functional groups, or will we need to carry out an additional functional group conversion? Are oxidation or reduction reactions needed?

Solve

(a) With a 1,3-positioning of the alcohol groups, we should think of an aldol addition.

OH OH ⇒ Undo an aldol addition. O + O

In the forward direction, we would need to obtain the kinetic enolate, so LDA is a good choice of base. After the aldol, we reduce the carbonyl.

LDA, H–C(=O)–H; NaBH4, EtOH

(b) The target also has a 1,3-positioning of the alcohols, so we should undo an aldol, disconnecting the bond shown.

Undo an aldol addition.

In the forward direction, we want to generate the thermodynamic enolate, so $KOC(CH_3)_3$ is a good choice of base.

1. $KOC(CH_3)_3$ 2. (aldehyde) 3. H_3O^+; NaBH4, EtOH

(c) The target has a 1,5-positioning of the functional groups, so we should consider a conjugate addition with an enolate anion as the nucleophile.

Undo a conjugate addition.

To obtain the kinetic enolate, we use LDA as the base.

1. LDA 2. (cyclohexenone) 3. H_3O^+; NaBH4, EtOH

(d) We can undo a Grignard reaction on the target, as shown below:

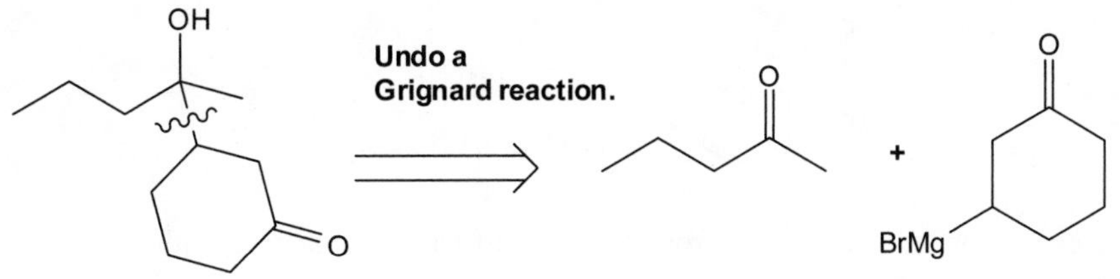

Undo a Grignard reaction.

However, the C=O group is not compatible with a Grignard, so we should use a protecting group before making the Grignard.

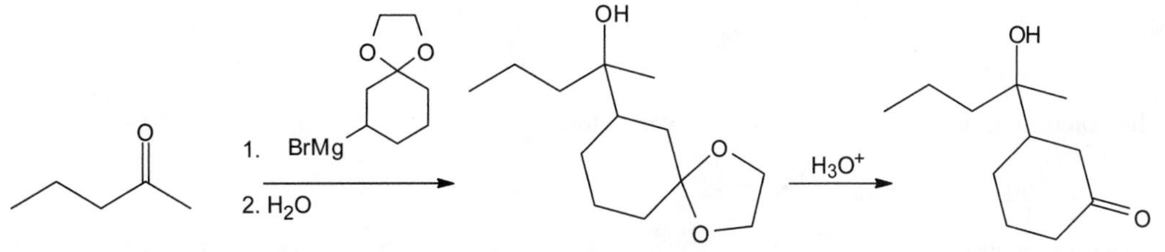

1. BrMg
2. H_2O

H_3O^+

Problem 19.55

Think

Are any C–C bond-formation reactions necessary? Do you need to include oxidation or reduction reactions? Functional group transformations?

Solve

(a) We can undo an oxidation at the benzylic carbon, as shown below:

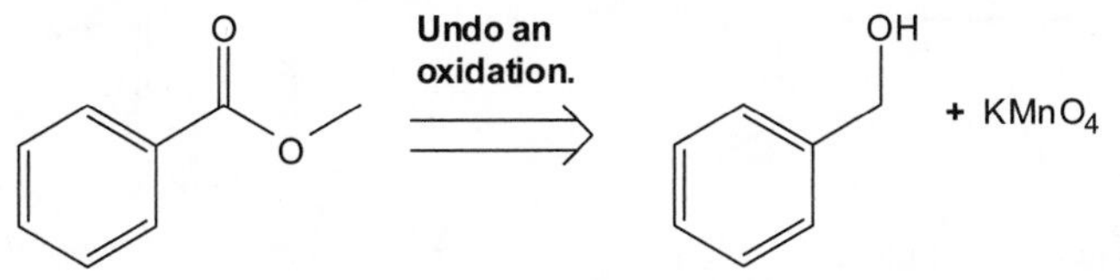

Undo an oxidation.

+ $KMnO_4$

The carboxylic acid is converted to the ether, as shown below:

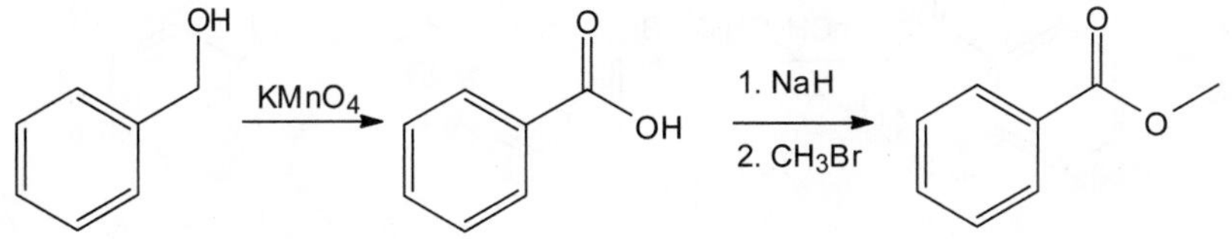

$KMnO_4$

1. NaH
2. CH_3Br

(b) The carbon skeleton is altered by the addition of three C atoms.

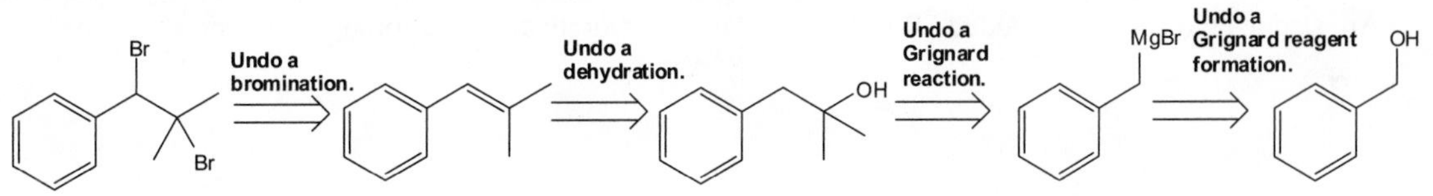

Undo a bromination.

Undo a dehydration.

Undo a Grignard reaction.

Undo a Grignard reagent formation.

The reaction in the forward direction is given below:

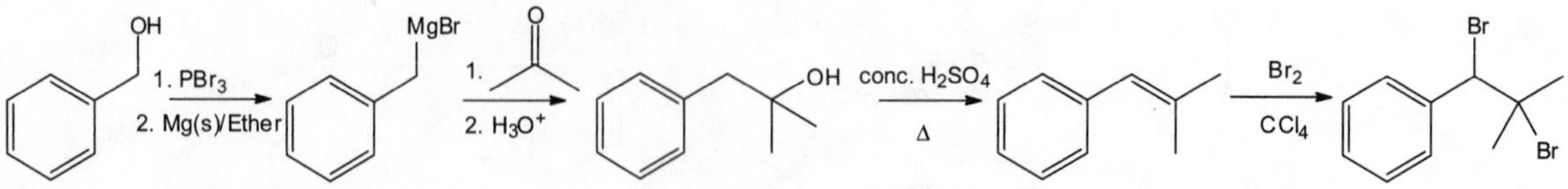

(c) The formation of a cylcopropane ring comes from a source of carbene reacting with an alkene. We can undo the reaction, as shown below:

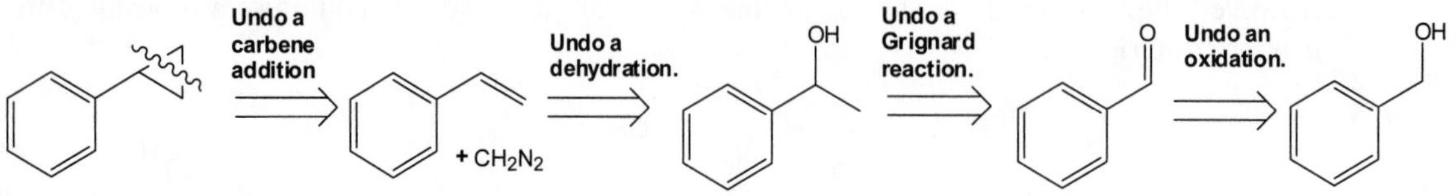

The reaction in the forward direction is given below:

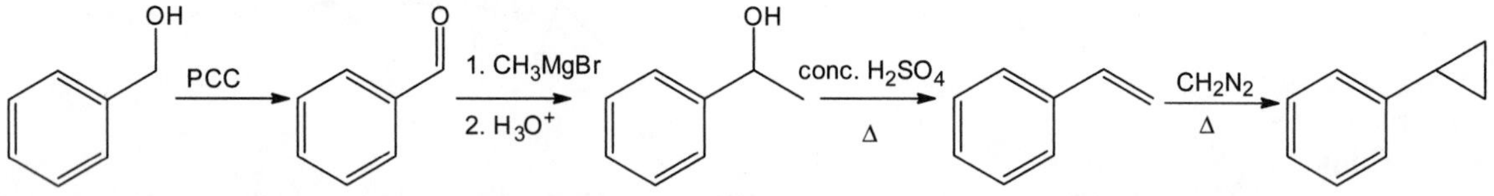

(d) We can undo a Grignard on the target, as shown below:

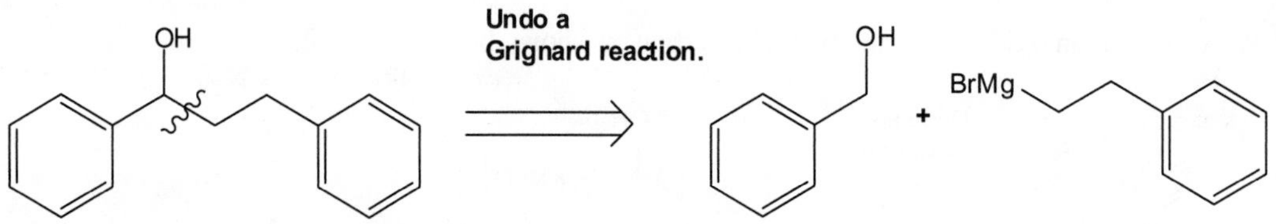

The reaction in the forward direction is shown below:

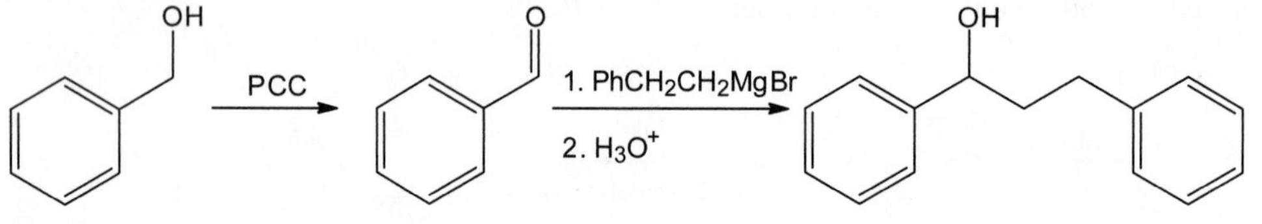

Problem 19.56

Think

Is the carbon skeleton altered? Does functional group transformation occur? Are protecting groups necessary? Are strong acids or bases needed? Will you need to incorporate an oxidation or a reduction reaction?

Solve

Since we are limited to only a starting material with three carbons, we should consider disconnecting the bonds below in a retrosynthetic analysis. The 1,2-diol suggests an epoxidation reaction from an alkene.

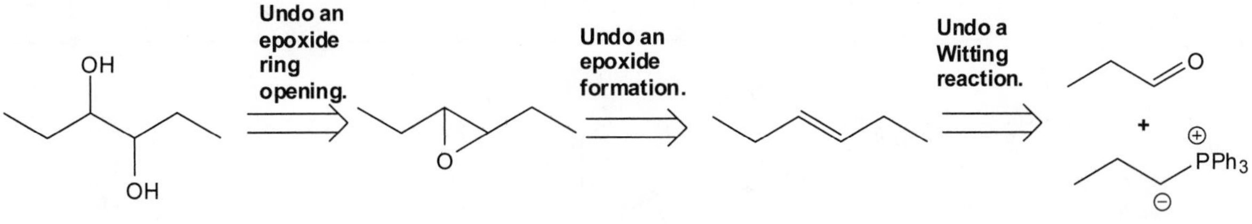
OH
OH
Undo an epoxide ring opening.
O
Undo an epoxide formation.
Undo a Witting reaction.
O
+
PPh3

The reaction in the forward direction is given below:

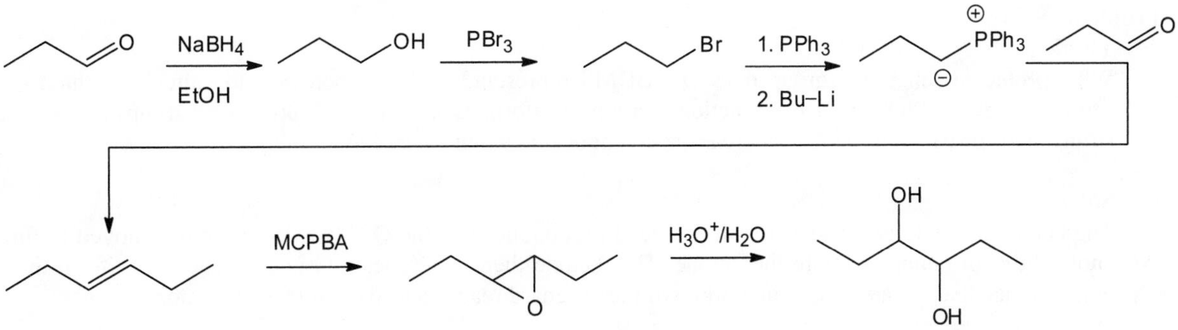
O
NaBH4
EtOH
OH
PBr3
Br
1. PPh3
2. Bu–Li
PPh3
O
MCPBA
O
H3O+/H2O
OH
OH

Problem 19.57

Think

Is the carbon skeleton altered? Does functional group transformation occur? Are protecting groups necessary? Are strong acids or bases needed? Do you need to include any oxidation or reduction reactions?

Solve

(a) The alcohol group must be protected to carry out the Grignard reaction on the C=O to add the ethyl group. The secondary alcohol in the product came from a Grigarnd reaction. The full synthesis is given below:

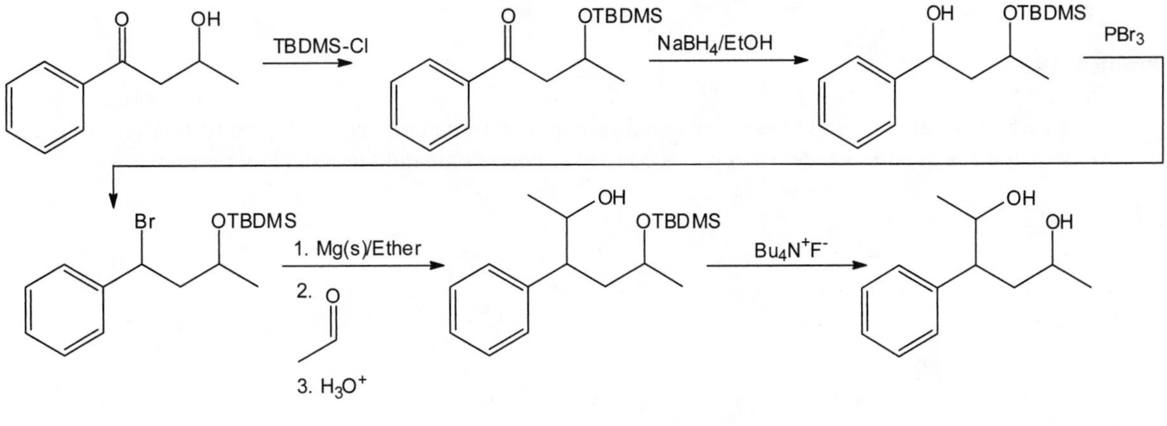
O
OH
TBDMS-Cl
O
OTBDMS
NaBH4/EtOH
OH
OTBDMS
PBr3
Br
OTBDMS
1. Mg(s)/Ether
2. O
3. H3O+
OH
OTBDMS
Bu4N+F-
OH
OH

(b) The alcohol group must be protected to carry out the Grignard reaction on the C=O to add the ethyl group. The alcohol is on the less-substituted carbon and this likely came from the hydroboration–oxidation reaction. The full synthesis is given on the next page.

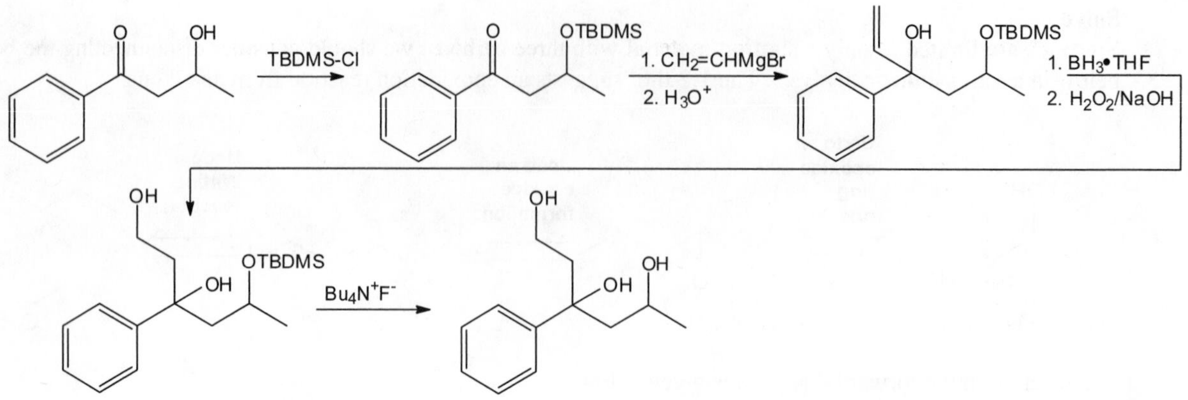

TBDMS-Cl; 1. CH_2=CHMgBr 2. H_3O^+; 1. BH_3•THF 2. H_2O_2/NaOH; $Bu_4N^+F^-$

Problem 19.58

Think

What protected functional group does O–TBDMS represent? Is the carbon skeleton altered? Which C atom forms the new C–Ph bond? Does functional group transformation occur? Are protecting groups necessary? Are strong acids or bases needed? What rearrangement occurs when an enol is formed?

Solve

The ketone needs to be protected with an acetal group, and then the O–TBDMS group is removed to form the enol. The enol tautomerizes to the ketone. This ketone then undergoes a Grignard reaction to form the C–Ph bond. Notice that the aqueous acid workup of the Grignard reaction also deprotects the ketone.

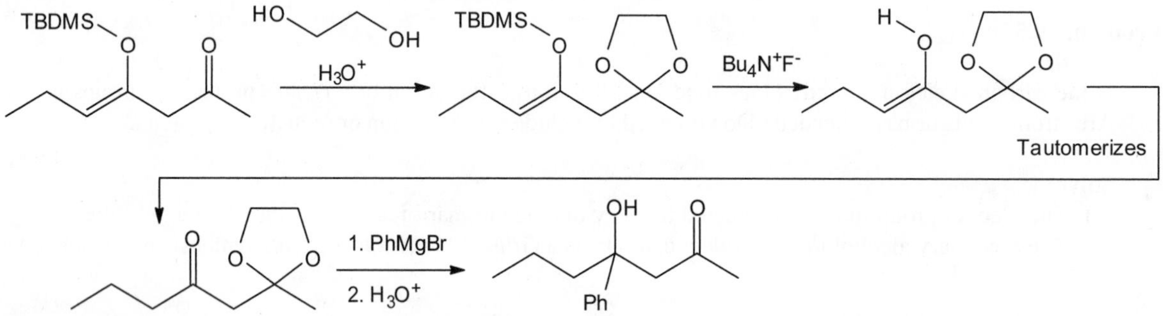

HOCH$_2$CH$_2$OH, H_3O^+; $Bu_4N^+F^-$; Tautomerizes; 1. PhMgBr 2. H_3O^+

Problem 19.59

Think

Is the carbon skeleton altered? Does functional -group transformation occur? Are protecting groups necessary? Are strong acids or bases needed? Do you need to incorporate oxidation or reduction reactions?

Solve

Since we are limited to only a starting material with three carbons, we should consider disconnecting the bonds below in a retrosynthetic analysis.

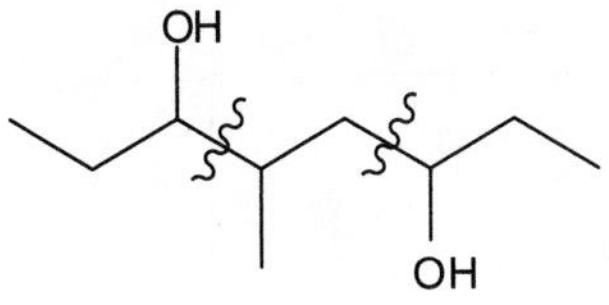

OH

OH

We can disconnect the bond on the left first, undoing a Grignard reaction.

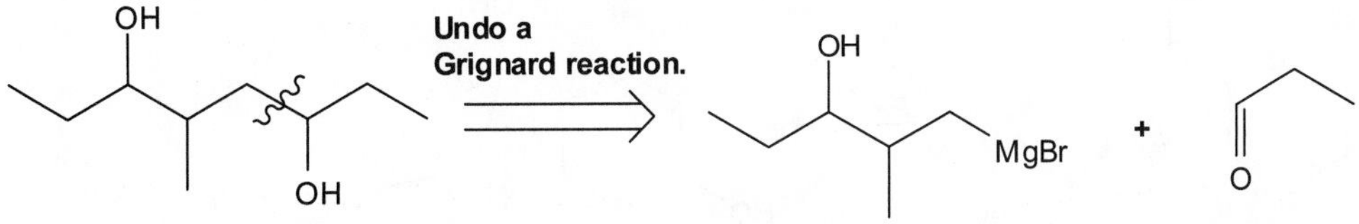

OH **Undo a Grignard reaction.** OH MgBr + O

OH

Not worrying about the incompatibility of the OH group (we'll use a protecting group for it), we see that the Grignard reagent has functional groups in a 1,3-position that could come from an aldol reaction, as shown below:

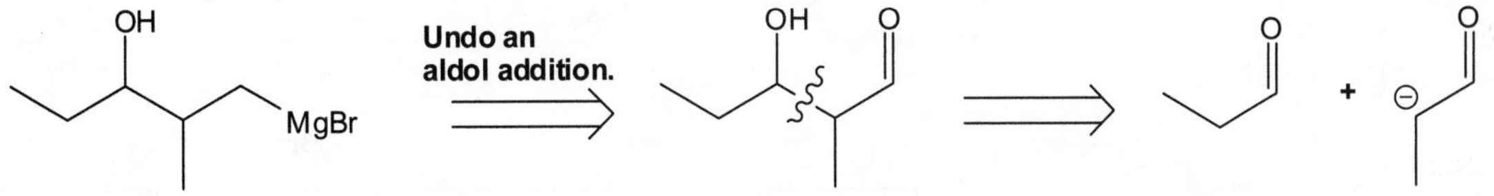

OH **Undo an aldol addition.** OH O O + O

MgBr

In the forward direction, we will need to use protecting groups to prevent incompatibilities.

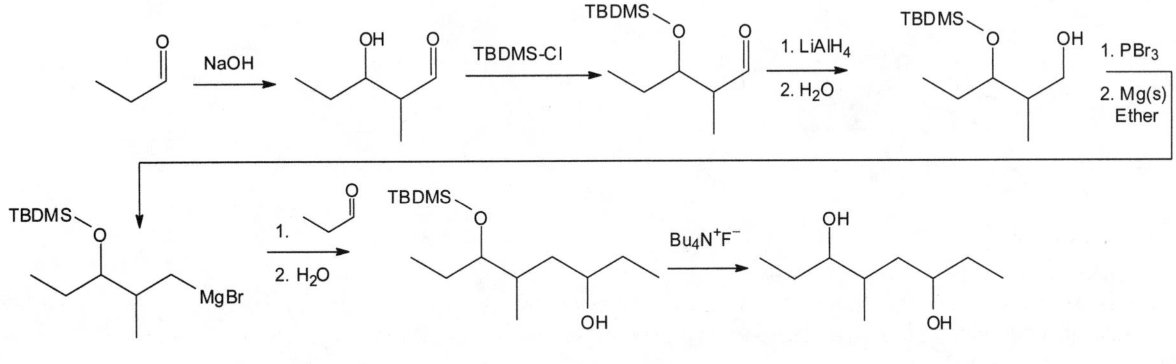

O NaOH OH O TBDMS-Cl TBDMS O O 1. $LiAlH_4$ 2. H_2O TBDMS O OH 1. PBr_3 2. Mg(s) Ether

TBDMS O MgBr 1. O 2. H_2O TBDMS O OH $Bu_4N^+F^-$ OH OH

Problem 19.60

Think

Is the carbon skeleton altered? Does functional group transformation occur? Are protecting groups necessary? Are strong acids or bases needed? Will you need to use oxidation or reduction reactions?

Solve

(a)

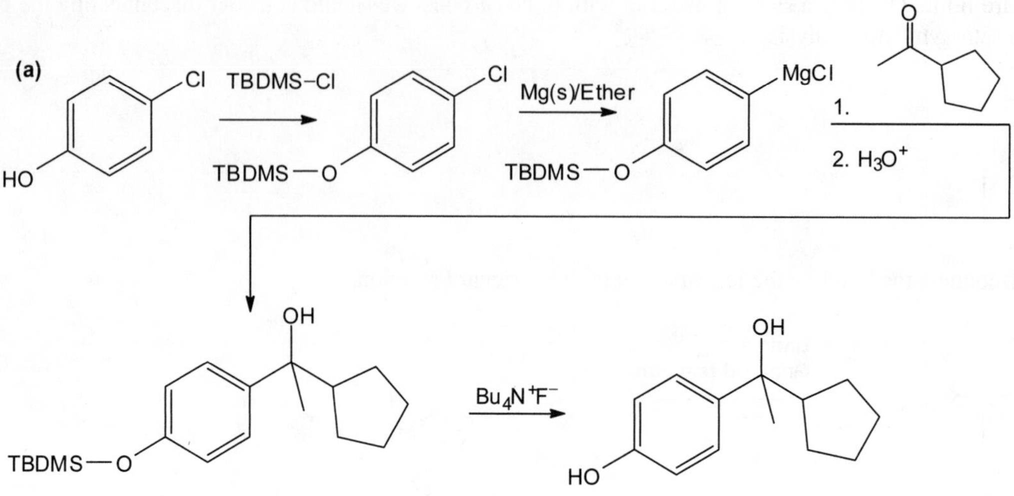

HO–C$_6$H$_4$–Cl → (TBDMS–Cl) → TBDMS—O–C$_6$H$_4$–Cl → (Mg(s)/Ether) → TBDMS—O–C$_6$H$_4$–MgCl → (1. cyclopentyl methyl ketone; 2. H_3O^+) → TBDMS—O–C$_6$H$_4$–C(OH)(CH$_3$)(C$_5$H$_9$) → ($Bu_4N^+F^-$) → HO–C$_6$H$_4$–C(OH)(CH$_3$)(C$_5$H$_9$)

(b)

OH, Br, O → (TBDMS–Cl) → OTBDMS, Br, O → (Mg) → OTBDMS, MgBr, O → (1. $H_2C{=}O$; 2. H_3O^+) → OTBDMS, OH, O → ($Bu_4N^+F^-$) → OH, OH, O

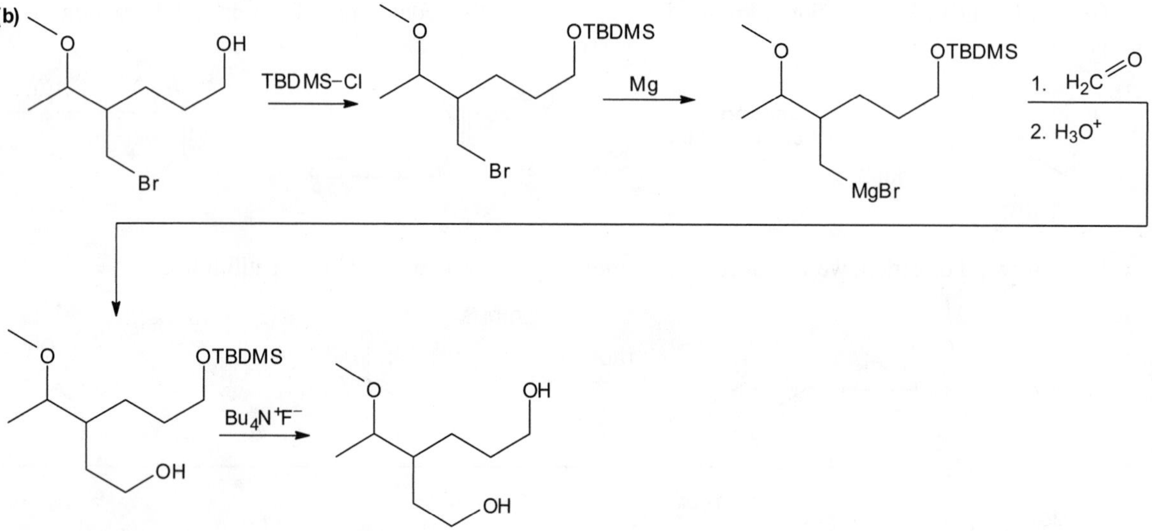

(c)

3-bromocyclohexanone (O, Br) → (HO–CH$_2$CH$_2$–OH, H^+) → ketal (O, O, Br) → (Mg(s), Ether) → ketal (O, O, MgBr) → (1. CO_2; 2. H_3O^+) → 3-oxocyclohexanecarboxylic acid (O, O, OH)

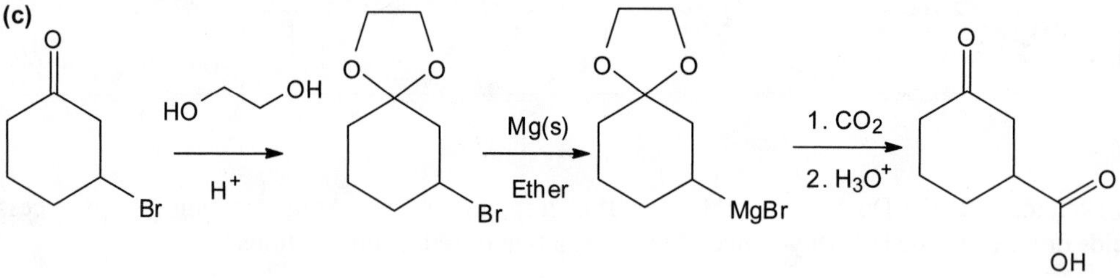

(d)

1. BH_3•THF
2. H_2O_2/NaOH

HO

PBr_3

Br

HO⁀OH

H^+

Br

O O

1. Mg(s)/ether
2. O, H H
3. H_2O

HO

O O

PCC

O

Raney Ni

H_3O^+

$\overset{\oplus}{Ph_3P}—\overset{\ominus}{C}(CH_3)_2$

(e)

Br

O

HO⁀OH

H^+

Br

NaC≡

O

O

H_2/Lindlar's Catalyst

O

O

H_3O^+

O

CHAPTER 20 | Nucleophilic Addition–Elimination Reactions 1: The General Mechanism Involving Strong Nucleophiles

Your Turn Exercises

Your Turn 20.1

Think

In each step, identify the bonds that are broken and the bonds that are formed. How can you use curved arrows to show the electron movement? Review the 10 elementary steps in Chapter 7. Which elementary step involves attack of a nucleophile at a polar π bond? Which elementary step involves regeneration of the polar π bond?

Solve

Step 1 is nucleophilic addition where the nucleophile attacks the electron-poor C of the C=O. This forces the electrons in the C=O π bond onto the O atom. The product of this step is the *tetrahedral* intermediate. The tetrahedral intermediate in Step 2 undergoes nucleophile elimination, in which one of the lone pairs on the O of the tetrahedral intermediate regenerates the C=O and the leaving group, $CH_3CH_2O^-$, departs.

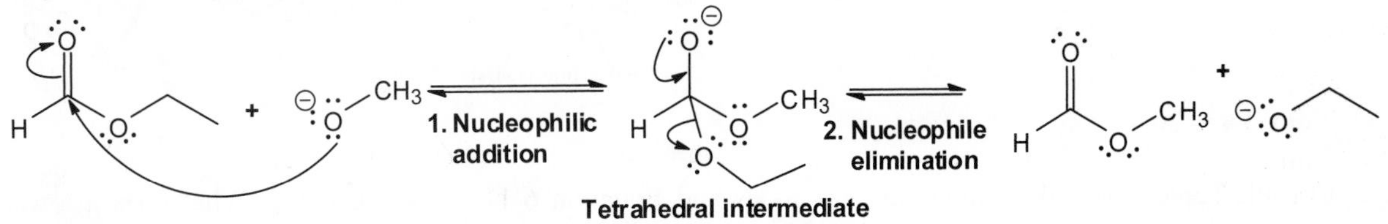

1. Nucleophilic addition

2. Nucleophile elimination

Tetrahedral intermediate

Your Turn 20.2

Think

Use Figure 20-1 as a guide. How do the charges impact the stability of the reactants, products, and intermediates? Resonance? How you do represent the energy barrier in each step?

Solve

In each stage of the reaction, there exists one negative charge that is localized on O, so charge stability is roughly the same for all three. The C=O bond in both the reactants and products is resonance stabilized but the tetrahedral intermediate is not resonance stabilized, so the tetrahedral intermediate is higher in energy. Each energy barrier is identified as the difference in energy between a reactant and the subsequent transition state, and is represented by the double-headed arrows.

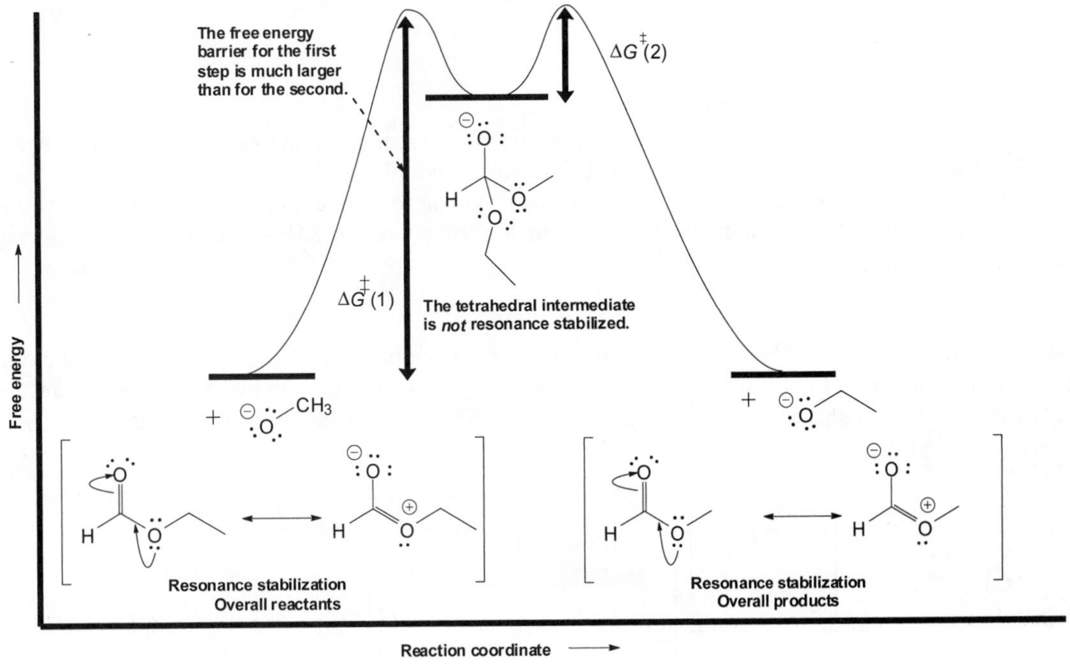

The free energy barrier for the first step is much larger than for the second.

$\Delta G^\ddagger(2)$

$\Delta G^\ddagger(1)$

The tetrahedral intermediate is *not* resonance stabilized.

Free energy

Resonance stabilization Overall reactants

Resonance stabilization Overall products

Reaction coordinate

Your Turn 20.3

Think

Use the mechanism in Equation 20-2 as guide. Write the species/steps appearing in Equation 20-2 in the reverse order, and in each step, identify the bonds that are broken and the bonds that are formed. How can you use curved arrows to show the electron movement?

Solve

Step 1 is nucleophilic addition, in which the nucleophile attacks the electron-poor C of the C=O. This forces the electrons in the C=O π bond onto the O atom. The product of this step is the *tetrahedral* intermediate. The tetrahedral intermediate in Step 2 undergoes nucleophile elimination, in which one of the lone pairs on the O of the tetrahedral intermediate regenerates the C=O and the leaving group, $(CH_3)_3CO^-$, departs.

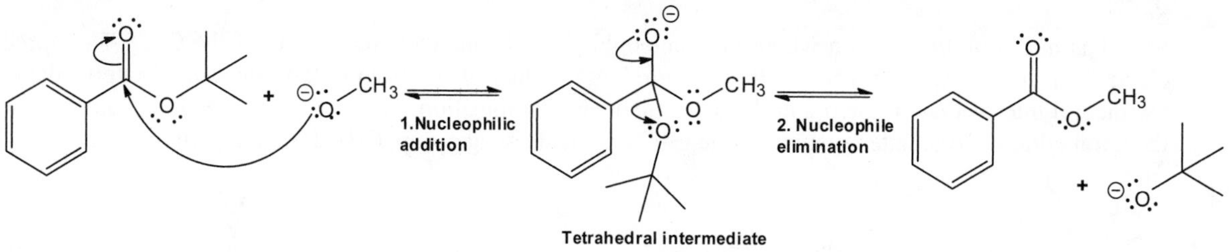

CH3 1.Nucleophilic addition CH3 2. Nucleophile elimination CH3

Tetrahedral intermediate

Your Turn 20.4

Think

Consult Table 6-1 to determine the pK_a values and Equation 6-11 to calculate K_{eq}. What does a lower pK_a indicate about acid strength and the side of the reaction favored? What does the magnitude of K_{eq} indicate about the extent that the reaction is favored?

Solve

CH_3CO_2H has a pK_a = 4.75 and CH_3CH_2OH has a pK_a = 16. The stronger acid (lower pK_a) is on the reactant side, so the products are favored. $K_{eq} = 10^{\Delta pK_a} = 10^{16-4.75} = 1.8 \times 10^{11}$. This shows that the reaction very heavily favors the products, making the reaction irreversible.

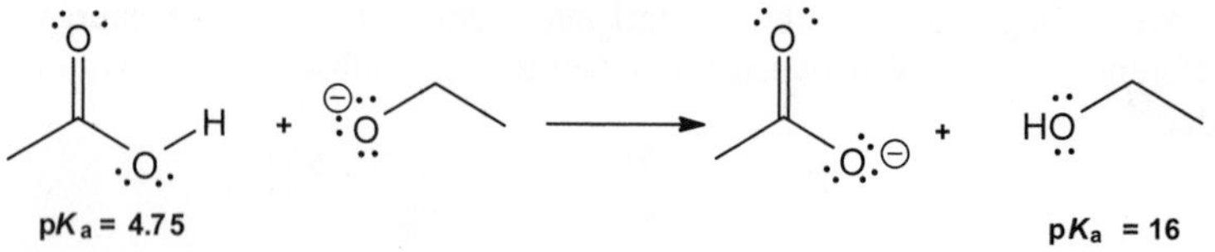

pK_a = 4.75 pK_a = 16

Your Turn 20.5

Think

Review Sections 1.10 and 1.11 to review how to draw resonance structures and resonance hybrids. Are there any lone pairs adjacent to a double or triple bond? Do any of the double bonds convert into lone pairs? How can you show via a curved arrow the electrons in the lone pair moving to form a double bond? How can you show via a curved arrow the electrons in a double bond moving to form a lone pair? How can you draw the average of the two resonance structures?

Solve

The lone pairs on the negative O atom are moved to form a C=O π bond. The π bond electrons in the original C=O are moved onto the O to give the top O a −1 formal charge. The hybrid is an average of the two resonance structures. Each O atom sharing the −1 charge receives a δ^-, and each bonding region sharing the π bond receives a partial π bond.

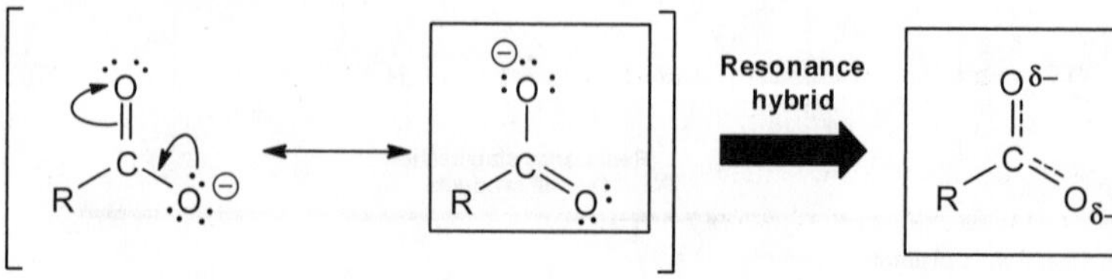

Resonance hybrid

Your Turn 20.6

Think

In each step, identify the bonds that are broken and the bonds that are formed. Which sites are electron rich and which are electron poor? How can you use curved arrows to show the electron movement? Review the 10 elementary steps in Chapter 7. Which elementary step involves attack of a nucleophile at a polar π bond? Which elementary step involves regeneration of the polar π bond?

Solve

Step 1 is nucleophilic addition, in which the $^-NH_2$ nucleophile attacks the electron-poor C of the C=O. This forces the electrons in the C=O π bond onto the O atom. The product of this step is the tetrahedral intermediate. The tetrahedral intermediate in Step 2 undergoes nucleophile elimination, in which one of the lone pairs on the O of the tetrahedral intermediate regenerates the C=O and the leaving group, CH_3O^-, departs.

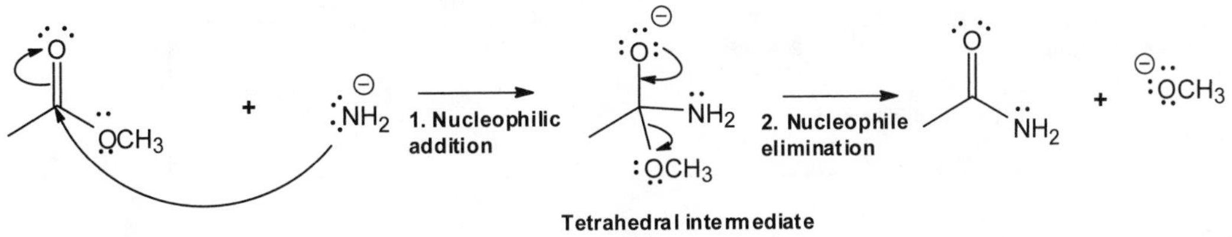

+ :NH2 ⊖ 1. Nucleophilic addition → Tetrahedral intermediate → 2. Nucleophile elimination → NH2 + ⊖:OCH3

Your Turn 20.7

Think

Review the "stability ladder" in Figure 20-7. What is the functional group in the reactant? What is the functional group in the product? Consider the position of the product relative to the reactant. Is going up or down the ladder energetically favorable?

Solve

Whether each conversion is favorable or not is determined by whether the conversion represents going up or down the stability ladder (Fig. 20-7). Going up the ladder is an energetically unfavorable process, whereas going down is favorable.

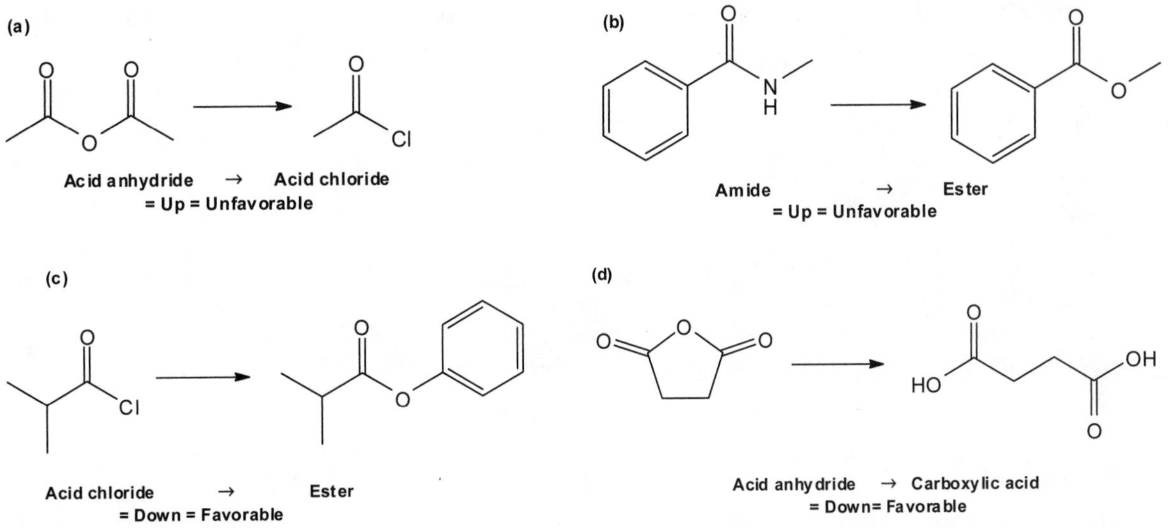

(a) Acid anhydride → Acid chloride = Up = Unfavorable

(b) Amide → Ester = Up = Unfavorable

(c) Acid chloride → Ester = Down = Favorable

(d) Acid anhydride → Carboxylic acid = Down= Favorable

Your Turn 20.8

Think

In each step, identify the bonds that are broken and the bonds that are formed. Which sites are electron rich? Electron poor? How can you use curved arrows to show the electron movement? Review the 10 elementary steps in Chapter 7. Which elementary step involves attack of a nucleophile at a polar π bond? Which elementary step involves regeneration of the polar π bond? When are protons transferred? How does charge stability play a role in reversibility?

Solve

Step 1 is nucleophilic addition where the $^{-}$OH nucleophile attacks the electron-poor C of the C=O. This forces the electrons in the C=O π bond onto the O atom. The product of this step is the tetrahedral intermediate. The tetrahedral intermediate in Step 2 undergoes nucleophile elimination where one of the lone pairs on the O of the tetrahedral intermediate regenerates the C=O and the leaving group, $(CH_3)_2N^-$, departs. In Step 3, the carboxylic acid is quickly deprotonated in the basic conditions. Step 4 is an acid workup to regenerate the carboxylic acid. Steps that significantly increase charge stability tend to be irreversible. This does not happen in the first two steps, so the first two steps are reversible. Charge stability is significantly increased in Step 3, going from a localized N^- to a resonance-delocalized O^-, so that step is irreversible.

1. Nucleophilic addition

Tetrahedral intermediate

2. Nucleophile elimination

3. Proton transfer

Add H^+

4. Proton transfer

Your Turn 20.9

Think

Use Equation 20-15 as a guide. What is the organic product of the haloform reaction? What is the inorganic product when the halogen molecule changes?

Solve

The organic product of the haloform reaction is a carboxylic acid and the inorganic product is a haloform (HCX_3). The identity of the haloform produced depends on the identity of the halogen used as a reagent.

(a)

1. Cl_2, NaOH/H_2O
2. HCl

+ $HCCl_3$

(b)

1. I_2, NaOH/H_2O
2. HCl

+ HCI_3

Your Turn 20.10

Think

What is the requirement for a C=O compound to undergo a haloform reaction? Why is this requirement present?

Solve

Only methyl ketones or aldehydes can undergo a haloform reaction, which includes compounds **A**, **B**, and **D**. This requirement is present because CX_3^- is the suitable leaving group in the mechanism, according to Equation 20-16, Step 8. This leaving group can be produced only when three H atoms can be replaced by X on the same α carbon.

Your Turn 20.11

Think

In each step, identify the bonds that are broken and the bonds that are formed. Which sites are electron rich? Electron poor? How can you use curved arrows to show the electron movement? Review the 10 elementary steps in Chapter 7. Which elementary step involves attack of a nucleophile at a polar π bond? Which elementary step involves regeneration of the polar π bond? When are protons transferred?

Solve

Step 1 is nucleophilic addition, in which the H:$^-$ nucleophile attacks the electron-poor C of the C=O. This forces the electrons in the C=O π bond onto the O atom. The product of this step is the tetrahedral intermediate. The tetrahedral intermediate in Step 2 undergoes nucleophile elimination, in which one of the lone pairs on the O of the tetrahedral intermediate regenerates the C=O and the leaving group Cl$^-$ departs. Step 3 is nucleophilic addition where the H:$^-$ nucleophile attacks the electron-poor C of the C=O. This forces the electrons in the C=O π bond onto the O atom. Step 4 is an acid workup to protonate the alkoxide and form the primary alcohol product.

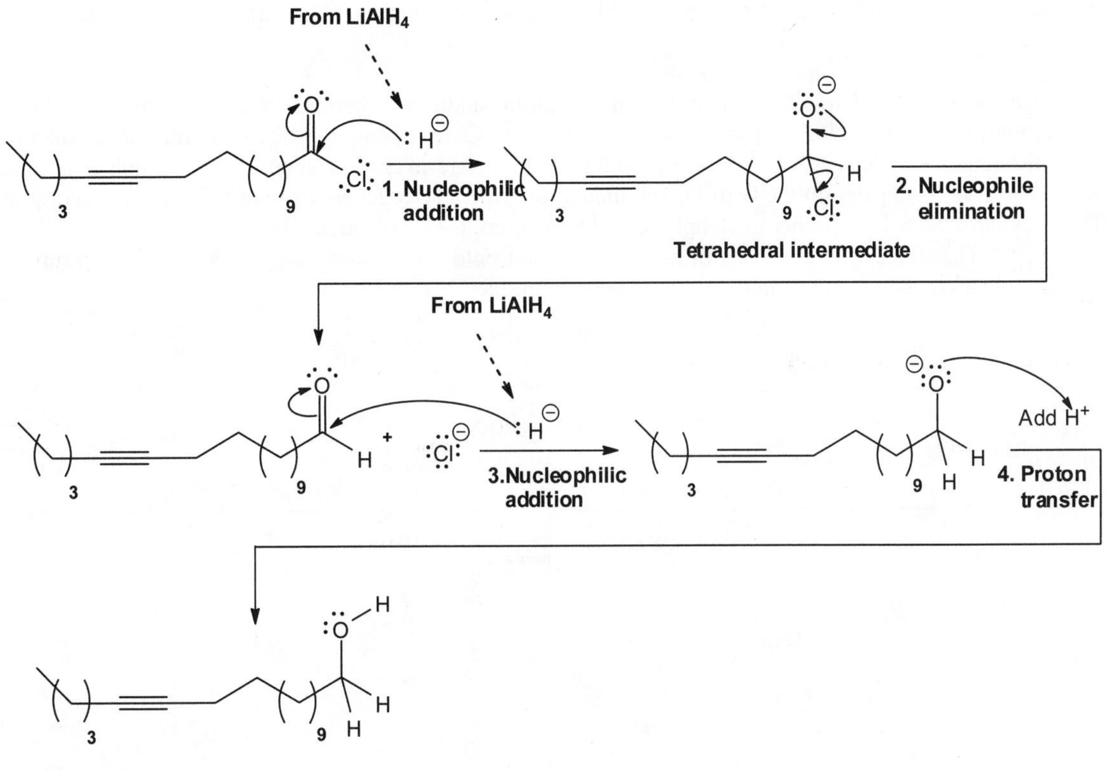

From LiAlH4

1. Nucleophilic addition

Tetrahedral intermediate

2. Nucleophile elimination

From LiAlH4

3.Nucleophilic addition

Add H+

4. Proton transfer

Your Turn 20.12

Think

In each step, identify the bonds that are broken and the bonds that are formed. Which sites are electron rich? Electron poor? How can you use curved arrows to show the electron movement? Review the 10 elementary steps in Chapter 7. Which elementary step involves attack of a nucleophile at a polar π bond? Which elementary step involves regeneration of the polar π bond?

Solve

Step 1 is nucleophilic addition, in which the AlH_4^- (H:$^-$) nucleophile attacks the electron-poor C of the C=O at the same time the O–Al bond forms. The product of this step is the tetrahedral intermediate. The tetrahedral intermediate in Step 2 undergoes nucleophile elimination, in which the lone pair on the N generates the C=N iminium ion as the leaving group (OAl^-H_3) departs. Step 3 is nucleophilic addition where the H:$^-$ nucleophile attacks the electron-poor C of the C=N. This forces the electrons in the C=N π bond onto the N atom. The product of this step is the amine. See the mechanism on the next page.

From LiAlH4

1. Nucleophilic addition

2. Nucleophile elimination

3. Nucleophilic addition

Tetrahedral intermediate

Your Turn 20.13

Think

Use Equation 20-35 as a guide, which is the mechanism for Equation 20-34. How can CH_3Li be simplified? In each step, identify the bonds that are broken and the bonds that are formed. Which sites are electron rich? Electron poor? How can you use curved arrows to show the electron movement? Review the 10 elementary steps in Chapter 7. Which elementary step involves attack of a nucleophile at a polar π bond? Which elementary step involves regeneration of the polar π bond? When are proton transfers involved?

Solve

CH_3Li can be simplified to CH_3^-. Step 1 is nucleophilic addition where the CH_3^- nucleophile attacks the electron-poor C of the C=O. This forces the electrons in the C=O π bond onto the O atom. The product of this step is the tetrahedral intermediate. The tetrahedral intermediate in Step 2 undergoes nucleophile elimination where one of the lone pairs on the O of the tetrahedral intermediate regenerates the C=O and the leaving group, CH_3O^-, departs. Step 3 is another nucleophilic addition where the CH_3^- nucleophile attacks the electron-poor C of the C=O. This forces the electrons in the C=O π bond onto the O atom. Step 4 is an acid workup step to protonate the alkoxide O^- and form the tertiary alcohol product.

Methyl hept-6-enoate

1. Nucleophilic addition

Tetrahedral intermediate

2. Nucleophile elimination

3. Nucleophilic addition

Two equivalents of CH_3^- have added in.

Add H^+

4. Proton transfer

2-Methyloct-7-en-2-ol
99%

Chapter Problems

Problem 20.1

Think

In each step, identify the bonds that are broken and the bonds that are formed. Which sites are electron rich? Electron poor? How can you use curved arrows to show the electron movement? Review the 10 elementary steps in Chapter 7. Which elementary step involves attack of a nucleophile at a polar π bond? Which elementary step involves regeneration of the polar π bond?

Solve

In both of these reactions, the base RO^- attacks the electrophilic carbonyl carbon via nucleophilic addition. The C=O bond is regenerated as the leaving group leaves.

(a) $:\ddot{O}CH_2CH_3$ Nucleophilic addition; CH_2CH_3 Nucleophile elimination; CH_2CH_3 +

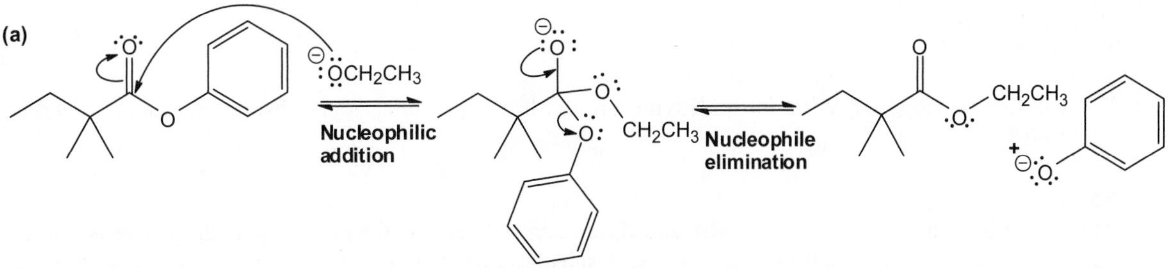

(b) THF; H; +

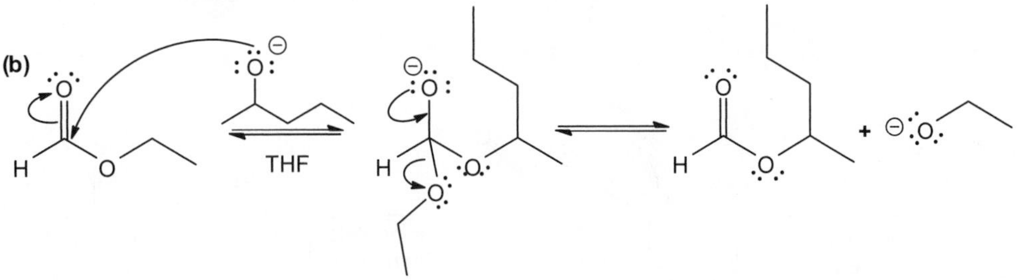

Problem 20.2

Think

What is the identity of the nucleophile in the nucleophilic addition reaction? What must be the leaving group in the nucleophile elimination step for a carboxylic acid to form as the final product? Which step is irreversible? Why is an acid workup necessary?

Solve

These are saponification reactions. After nucleophilic addition–elimination to produce an initial carboxylic acid, the carboxylic acid is deprotonated irreversibly. Acid workup adds the proton back.

(a) $:\ddot{O}H$ Nucleophilic addition; $:\ddot{O}H$ Nucleophile elimination; Proton transfer

Add H^+ Proton transfer; $\ddot{O}H$

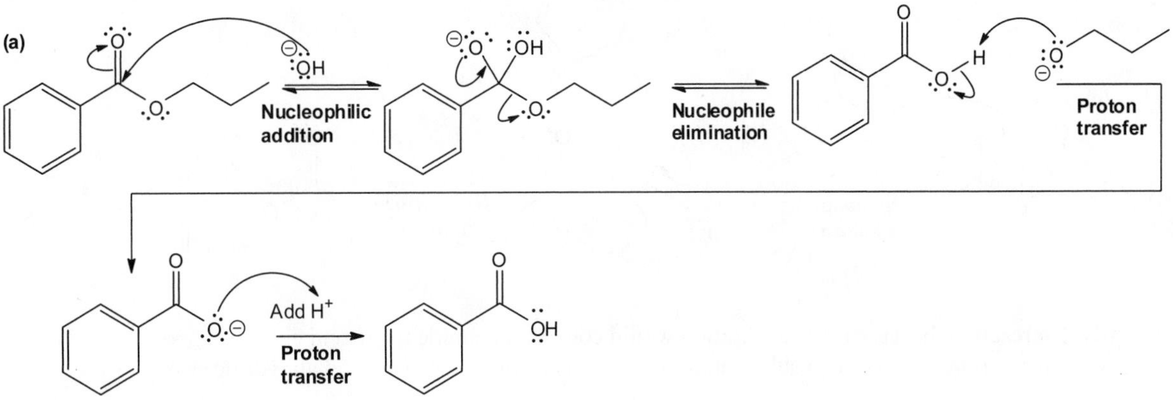

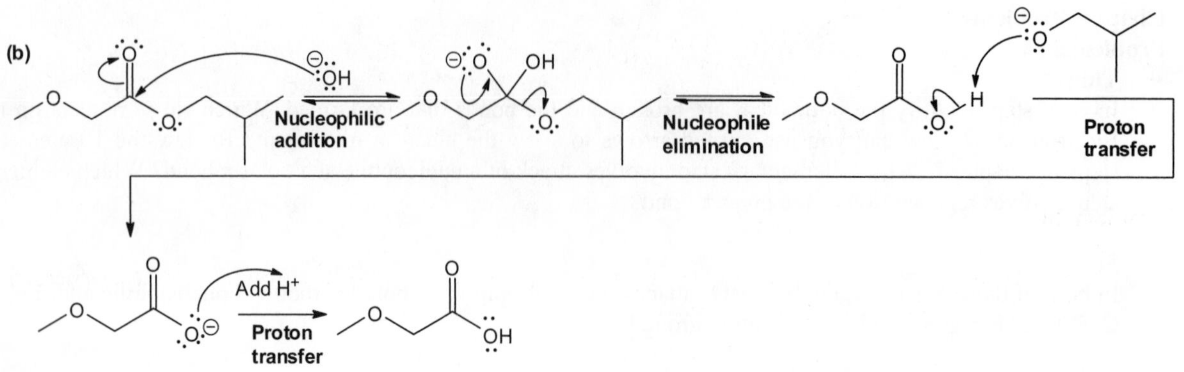

(b) Nucleophilic addition → Nucleophile elimination → Proton transfer → Add H^+ Proton transfer

Problem 20.4

Think

What is the nucleophile? What is the leaving group? Is the potential acyl substitution reaction energetically favorable?

Solve

The acyl substitution reactions in **(a)**, **(b)**, and **(f)** are favored because the product acid derivative is lower on the stability ladder than the starting one. The acyl substitutions in **(d)** and **(e)** are not favorable because they represent going up the stability ladder. Finally, **(c)** converts one carboxylic acid anhydride into another and is reversible.

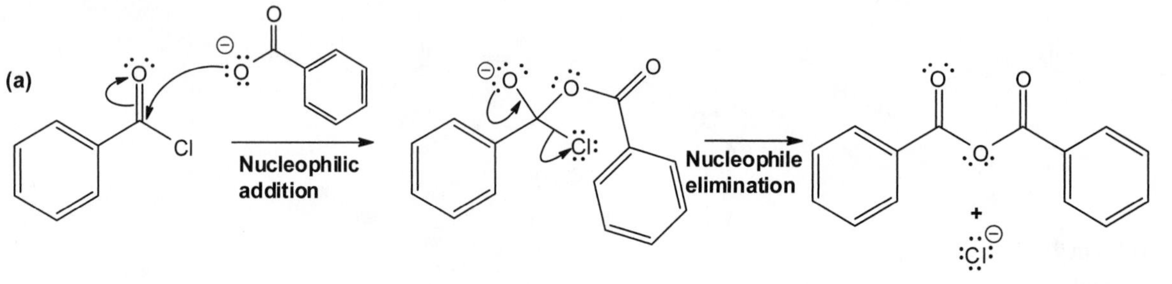

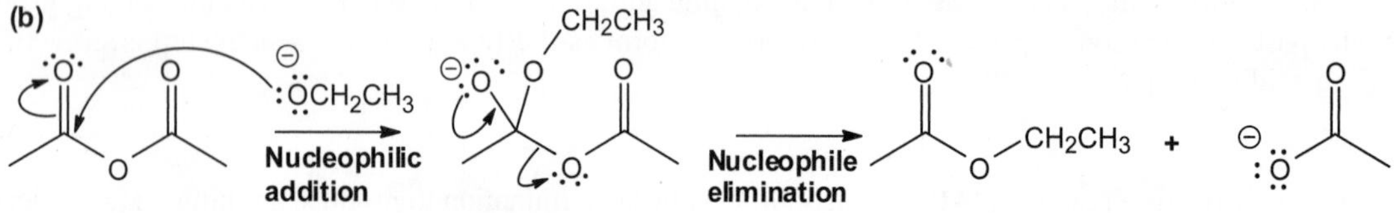

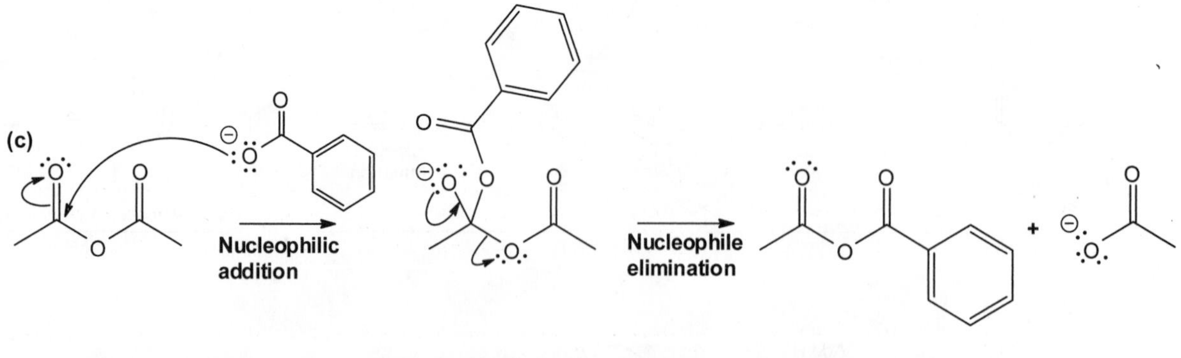

(a) Nucleophilic addition → Nucleophile elimination

(b) Nucleophilic addition → Nucleophile elimination

(c) Nucleophilic addition → Nucleophile elimination

(d) No reaction, because acyl substitution would convert the amide to an acid chloride.
(e) No reaction, because acyl substitution would convert the ester to an acid anhydride.

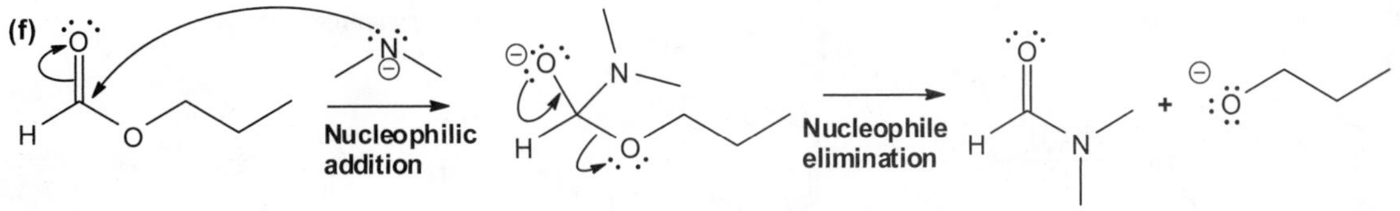

(f) Nucleophilic addition

Nucleophile elimination

Problem 20.5

Think

What is the nucleophile? What is the leaving group? Is the potential acyl substitution reaction energetically favorable? Which step is irreversible? Why is an acid workup necessary?

Solve

The HO^- is the nucleophile and attacks the electron-poor C of the amide. After nucleophilic addition–elimination to produce an initial carboxylic acid, the carboxylic acid is deprotonated irreversibly. Acid workup adds the proton back.

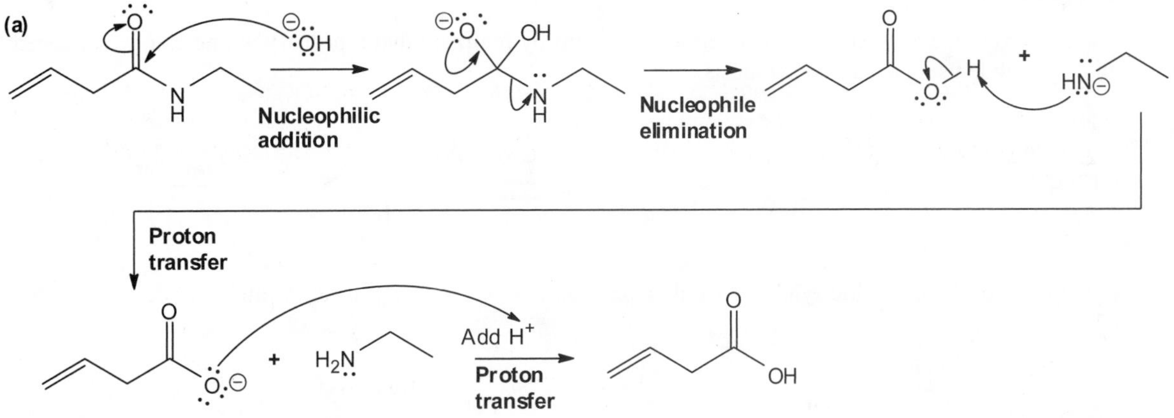

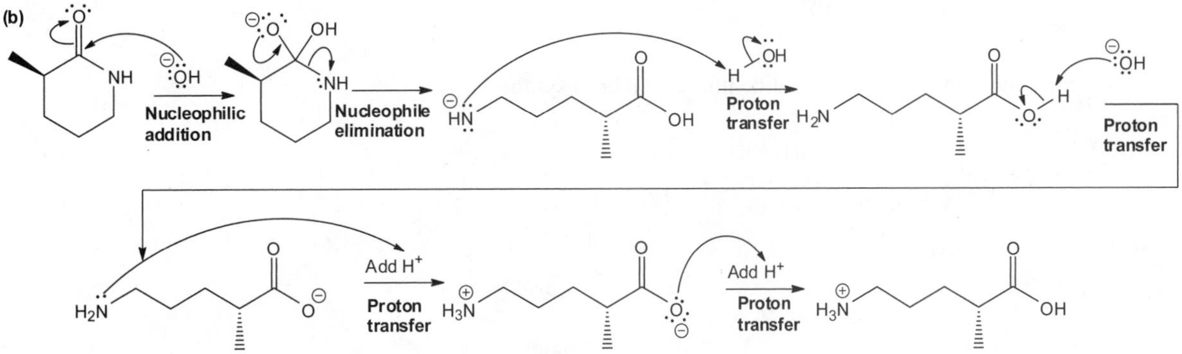

(a) Nucleophilic addition

Nucleophile elimination

Proton transfer

Add H^+

Proton transfer

(b) Nucleophilic addition

Nucleophile elimination

Proton transfer

Proton transfer

Add H^+

Proton transfer

Add H^+

Proton transfer

Problem 20.6

Think

Does the HO^- act as a base or nucleophile in the first reaction? What then acts as the nucleophile and electrophile in the second reaction? If the final reaction is basic, what types of species should not appear in the mechanism? Compare these reagents to the ones given in Equation 20-12. What functional groups form?

Solve

This is a Gabriel synthesis, which yields a primary amine from an alkyl halide and phthalate as a by-product. The aromatic C–Br is inert to these reaction conditions. The sp^3 CH_2–Br is the electrophilic C atom susceptible to nucleophilic attack. The reaction and products are given on the next page.

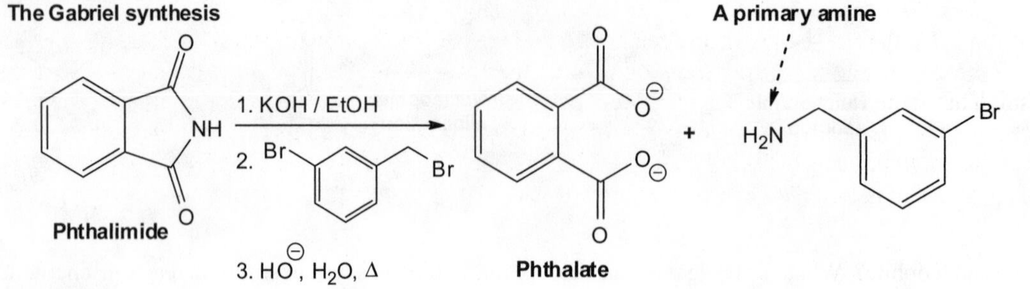

The Gabriel synthesis

Phthalimide; 1. KOH / EtOH; 2. Br–CH₂–C₆H₄–CH₂–Br; 3. $HO^{\ominus}$, H_2O, Δ → Phthalate + A primary amine (H_2N–CH₂–C₆H₄–CH₂–Br)

Problem 20.8

Think

Is the target a primary amine that can be the product of a Gabriel synthesis? What alkyl halide can be used as a precursor in a Gabriel synthesis of the target amine? How can that alkyl halide be produced from the starting material?

Solve

The retrosynthetic analysis is shown below. We begin by realizing that a primary amine can be produced from an alkyl halide using a Gabriel synthesis.

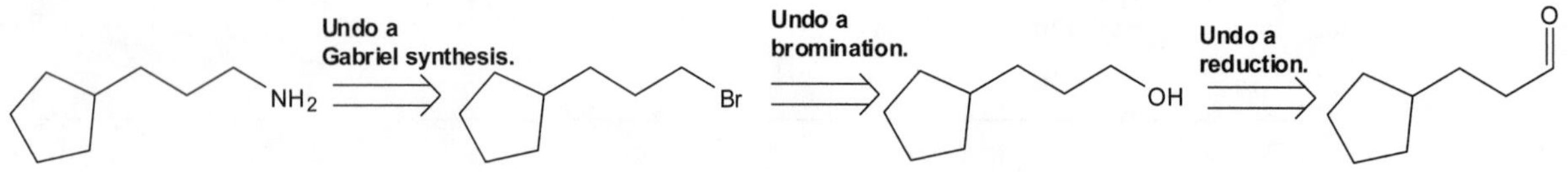

NH_2 ⟹ (Undo a Gabriel synthesis.) Br ⟹ (Undo a bromination.) OH ⟹ (Undo a reduction.) O

In the forward direction, the synthesis would appear as follows. First we make the alkyl halide:

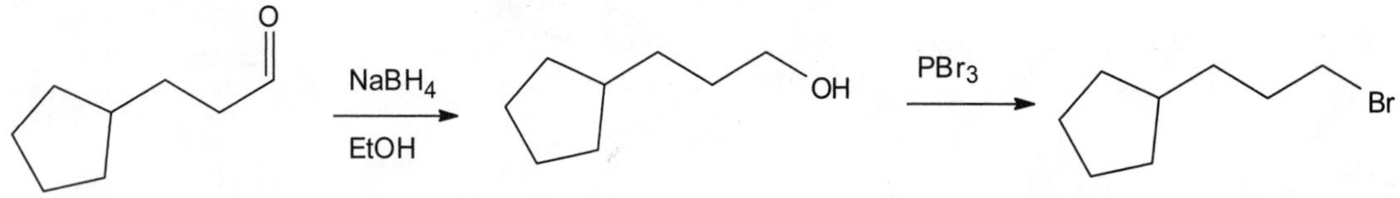

O → ($NaBH_4$, EtOH) → OH → (PBr_3) → Br

Then we incorporate the alkyl halide into the Gabriel synthesis:

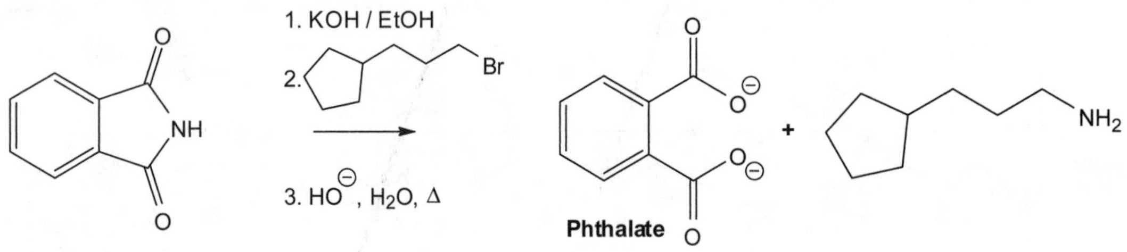

NH (phthalimide) → 1. KOH / EtOH; 2. (cyclopentyl bromide); 3. $HO^{\ominus}$, H_2O, Δ → Phthalate + NH_2

Problem 20.9

Think

Review Section 10.4. How are the α C–H atoms replaced by the halogen? Once the $-CX_3$ group is formed, is it a suitable leaving group? What are the nucleophile and leaving group in the nucleophilic addition–elimination mechanism? What steps are irreversible? What does the acid workup accomplish?

Solve

In both **(a)** and **(b)**, the α C–H bonds are replaced by X (i.e., Br or I) through use of HO^- acting as a base to deprotonate the α C–H and form the carbanion. The C^- then acts as a nucleophile to attack the X–X via an S_N2 mechanism. This same reaction occurs two more times until the CX_3 is formed at the α carbon. The resulting trihalomethylketone then undergoes nucleophilic addition by attack of HO^- to form the tetrahedral intermediate, which subsequently eliminates CX_3^-. The carboxylic acid that is formed is then irreversibly deprotonated, and acid workup adds that proton back. In the mechanism below Br_2 is shown for X_2.

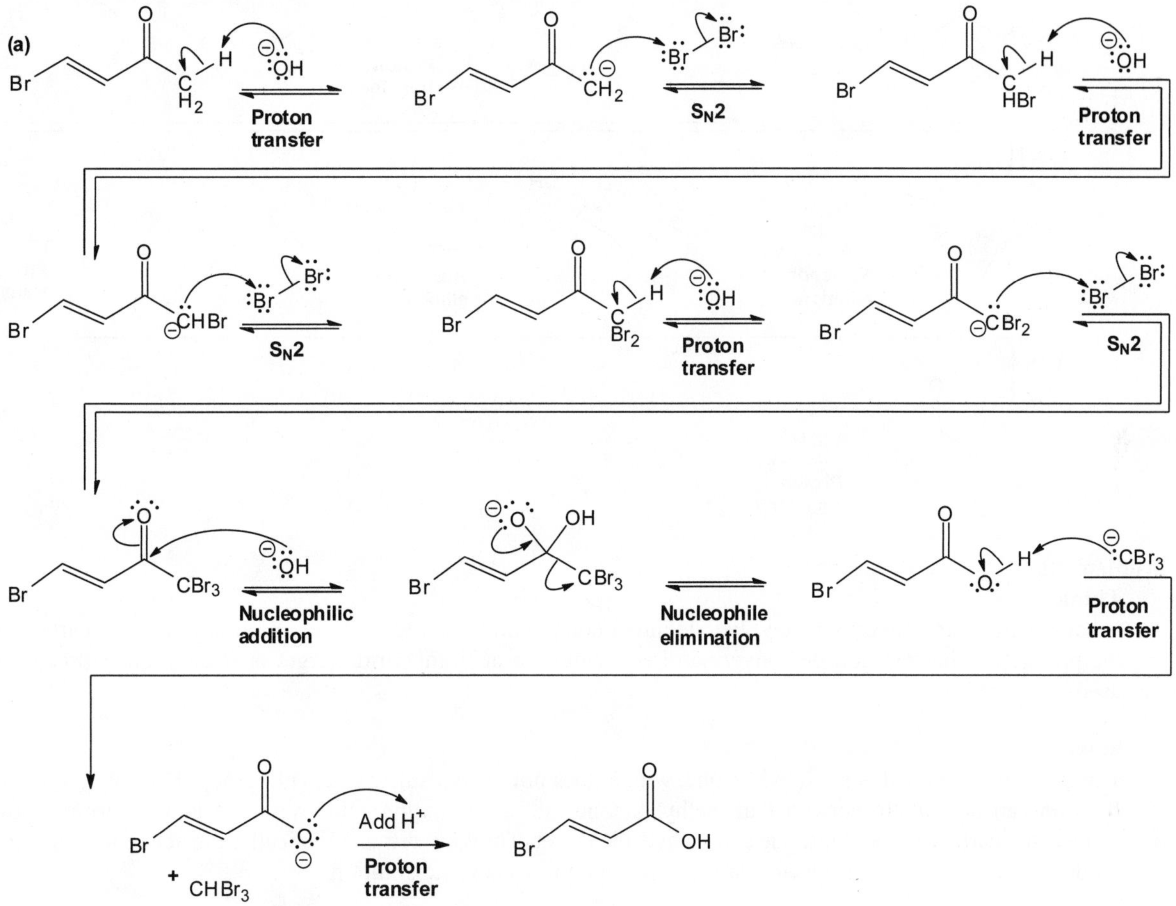

(a)

Proton transfer

S_N2

Proton transfer

S_N2

Proton transfer

S_N2

Nucleophilic addition

Nucleophile elimination

Proton transfer

+ $CHBr_3$

Add H^+

Proton transfer

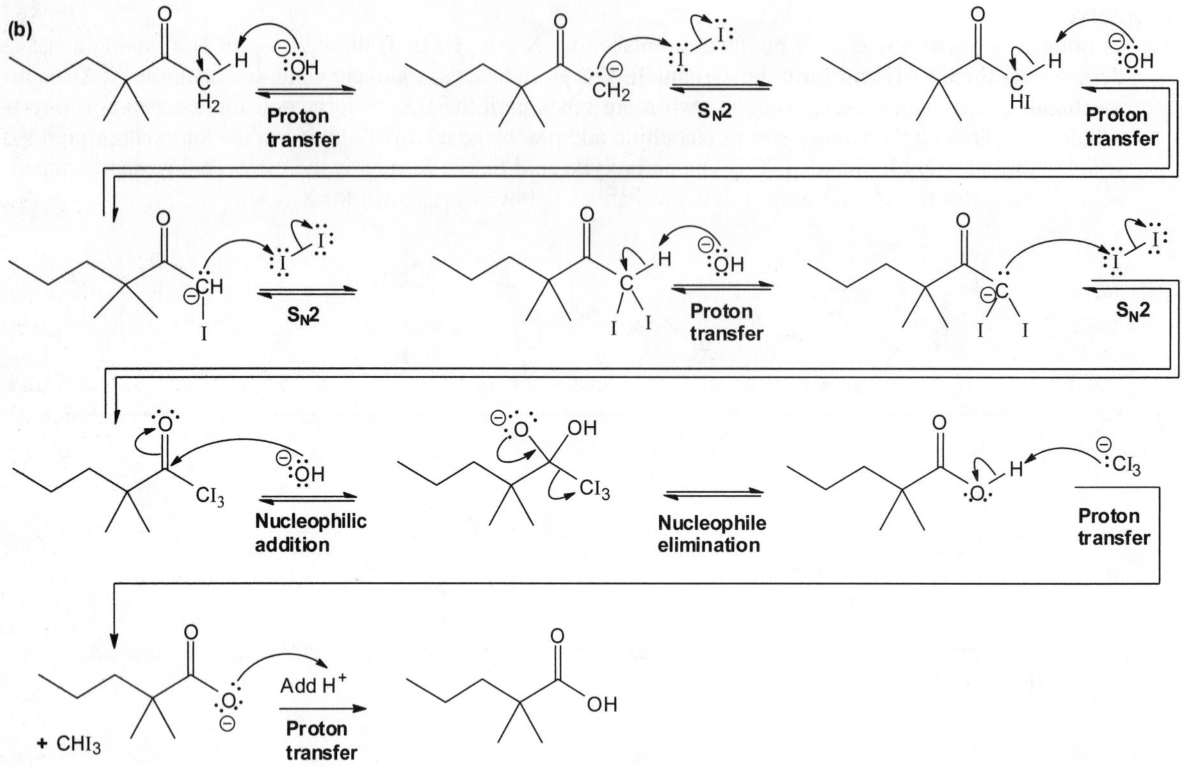

(b)

Proton transfer

S_N2

Proton transfer

S_N2

Proton transfer

S_N2

Nucleophilic addition

Nucleophile elimination

Proton transfer

+ CHI_3

Add H^+

Proton transfer

Problem 20.11

Think

What reaction can convert a methyl ketone into a compound with a leaving group on the carbonyl carbon? Can the product of that reaction be converted directly into the acid anhydride target or must another precursor be used?

Solve

The starting compound is a methyl ketone, which does not have a suitable leaving group. However, a haloform reaction can be used to convert that methyl ketone into a carboxylic acid. We must now determine how to convert the carboxylic acid into the acid anhydride target. Thinking retrosynthetically, the acid anhydride can be produced from an acid chloride and carboxylate anion in an acyl substitution.

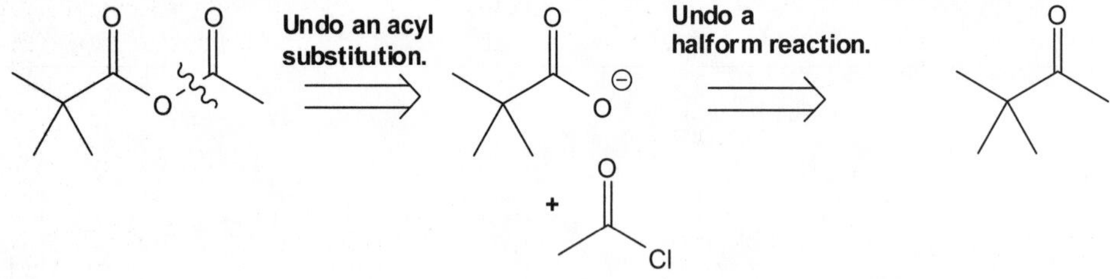

Undo an acyl substitution.

Undo a halform reaction.

The synthesis in the forward direction is shown below:

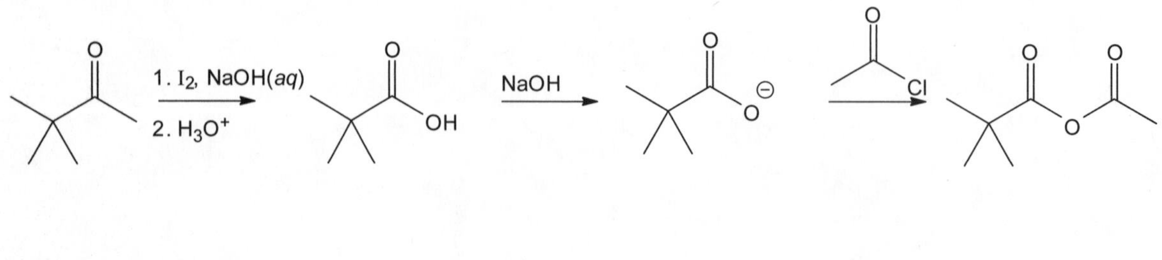

1. I_2, NaOH(*aq*)
2. H_3O^+

NaOH

Problem 20.12

Think

What is the index of hydrogen deficiency (IHD)? Based on this number and the molecular formula, what functional group is likely giving the IR signal at 1708 cm^{-1}? If there are 12 H atoms in the formula and the ^1H NMR gives rise to only two signals, is there symmetry in the structure? What functional group reacts with I_2 to give the bright, yellow solid?

Solve

Notice that IHD = 1, meaning the presence of either a double bond or a ring. The IR peak is indicative of a C=O double bond, which accounts for the IHD.

With a total of 12 H atoms in a 3:1 ratio, one NMR signal must account for nine equivalent H atoms, whereas the other must account for three equivalent H atoms. This is accomplished by a *t*-butyl group, $C(CH_3)_3$, and a methyl group, CH_3. The CH_3 must be adjacent to the C=O bond to give a methyl ketone, which will react in an iodoform reaction to produce a bright, yellow solid precipitate. The methyl ketone also is consistent with the signal at 2.2 ppm, giving the following molecule:

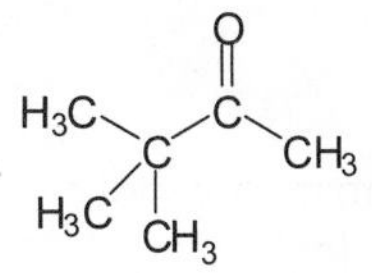

The reaction proceeds as follows. The yellow solid is the HCI_3.

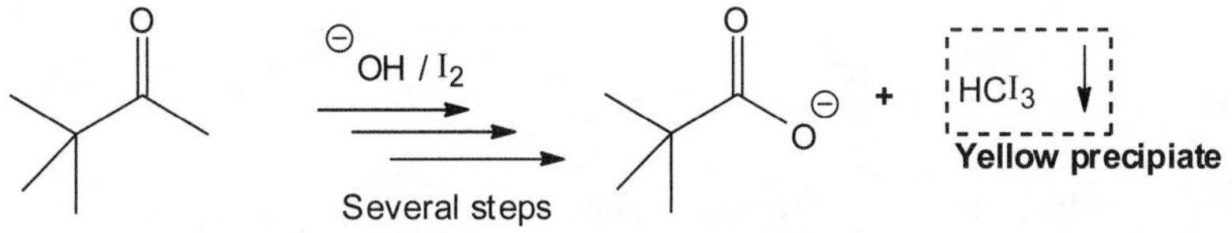

Problem 20.13

Think

Is the reaction a nucleophile addition–elimination mechanism? What acts as the nucleophile in each case? To what functional group does the hydride reduce the acid chloride? Is this functional group susceptible to further reaction?

Solve

Steps 1 and 2 make up the nucleophilic addition–elimination mechanism, which produces an aldehyde intermediate. Under these reaction conditions, the aldehyde reacts rapidly with another equivalent of hydride (Section 17.3) to produce the alkoxide anion. An acid workup step yields the uncharged alcohol. See the mechanisms below **(a)** and on the next page **(b)**.

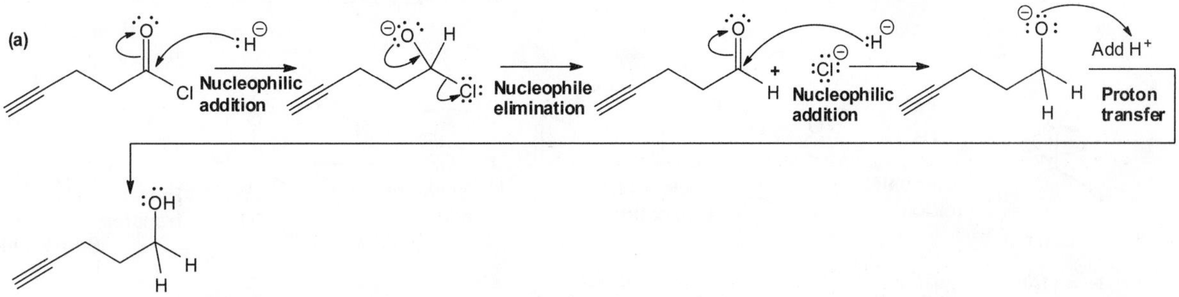

(b) Ph–CH2CH2–C(=O)Cl + :H⁻ —Nucleophilic addition→ tetrahedral alkoxide —Nucleophile elimination→ Ph–CH2CH2–CHO + :Cl:⁻ ; + :H⁻ —Nucleophilic addition→ alkoxide —Add H⁺, Proton transfer→ Ph–CH2CH2–CH2OH

Problem 20.14

Think

Refer to Table 20-1 to see which functional groups are reduced to primary alcohols. Is the reaction a nucleophile addition–elimination mechanism? What acts as the nucleophile in each case? To what functional group does the hydride initially reduce the carboxylic acid derivative given? Is this functional group susceptible to further reaction?

Solve

Acid chlorides, acid anhydrides, and esters react with $NaBH_4$ to form a primary alcohol. Acid chlorides, acid anhydrides, esters, and carboxylic acids react with $LiAlH_4$ to form a primary alcohol.

Steps 1 and 2 make up the nucleophilic addition–elimination mechanism, which produces an aldehyde intermediate. Under these reaction conditions, the aldehyde reacts rapidly with another equivalent of hydride (Section 17.3) to produce the alkoxide anion. An acid workup step yields the uncharged alcohol.

Nucleophilic addition — Nucleophile elimination — Nucleophilic addition — Add H⁺ — Proton transfer

Acid chloride

:H⁻ = $LiAlH_4$ or $NaBH_4$

Primary alcohol

Nucleophilic addition — Nucleophile elimination — Nucleophilic addition — Add H⁺ — Proton transfer

Acid anhydride

:H⁻ = $LiAlH_4$ or $NaBH_4$

Primary alcohol

Nucleophilic addition — Nucleophile elimination — Nucleophilic addition — Add H⁺ — Proton transfer

Ester

:H⁻ = $LiAlH_4$ or $NaBH_4$

Primary alcohol

Carboxylic acids have acidic hydrogen atoms. This functional group reacts with a hydride first via proton transfer. The mechanism is shown below. Carboxylic acids only produce the primary alcohol when reacted with $LiAlH_4$.

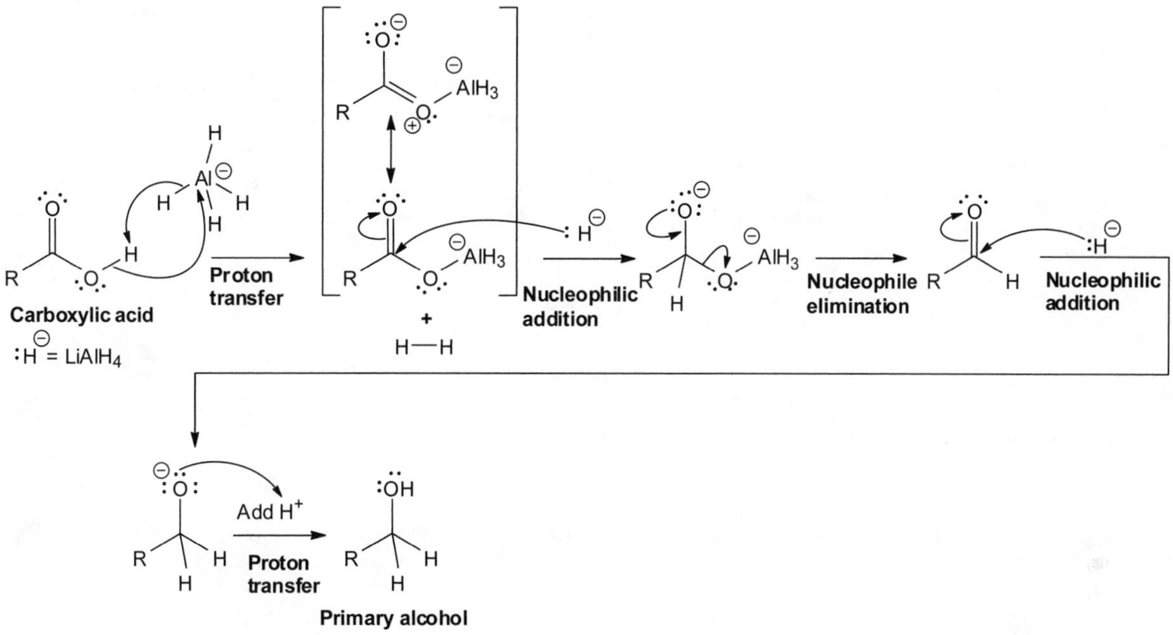

Problem 20.16

Think

Which functional groups are present in each compound? Will they both react with $NaBH_4$? Are the functional groups equally reactive? Which one is more likely to react with $NaBH_4$ in a nucleophilic addition–elimination reaction?

Solve

Sodium borohydride is a selective reducing agent.

(a) $NaBH_4$ will reduce the ketone over the ester. The ketone does not have a suitable leaving group and will undergo only nucleophilic addition followed by proton transfer.

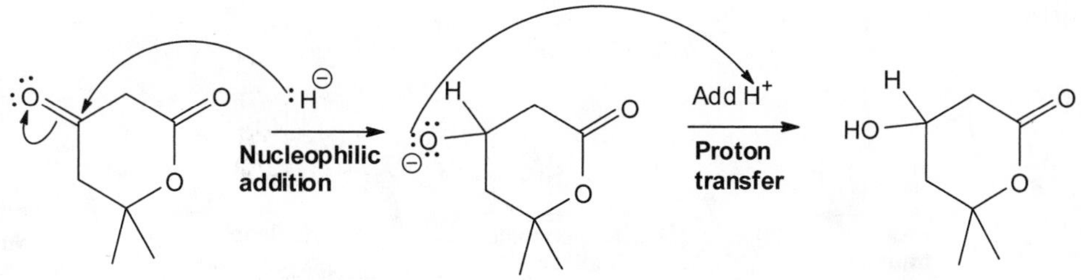

(b) $NaBH_4$ will reduce the aldehyde over the amide. The aldehyde does not have a suitable leaving group and will undergo only nucleophilic addition followed by proton transfer.

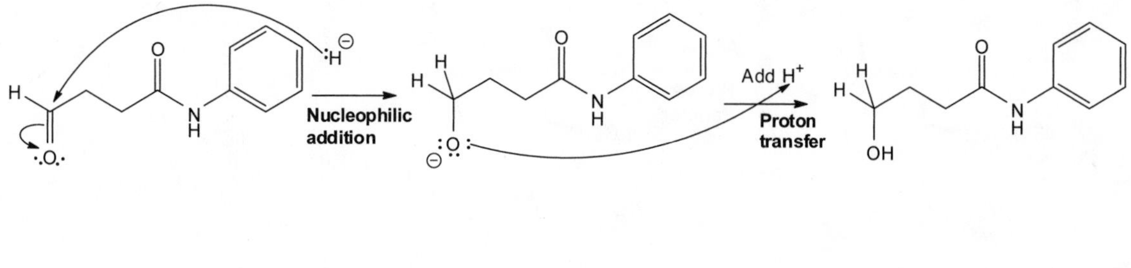

Problem 20.17

Think

Is the carboxylic acid reactive toward $NaBH_4$? If not, can the carboxylic acid be converted to another acid derivative that will be reduced by $NaBH_4$?

Solve

$NaBH_4$ will not reduce a carboxylic acid because deprotonation of the carboxylic acid will first produce a resonance-stabilized carboxylate anion. We know that sodium borohydride can reduce an ester, although slowly.

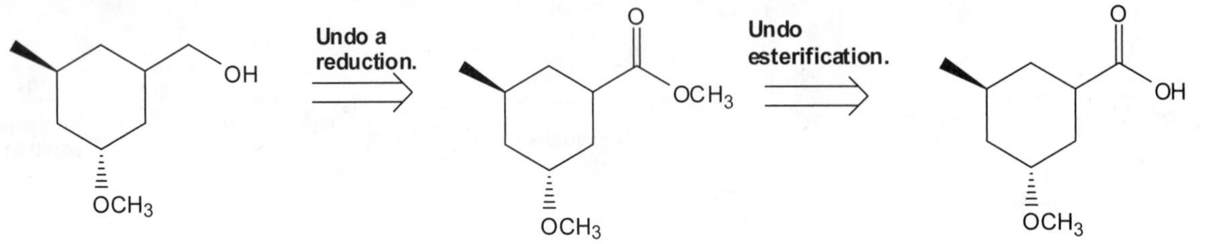

Undo a reduction. Undo esterification.

The synthesis in the forward direction appears as follows:

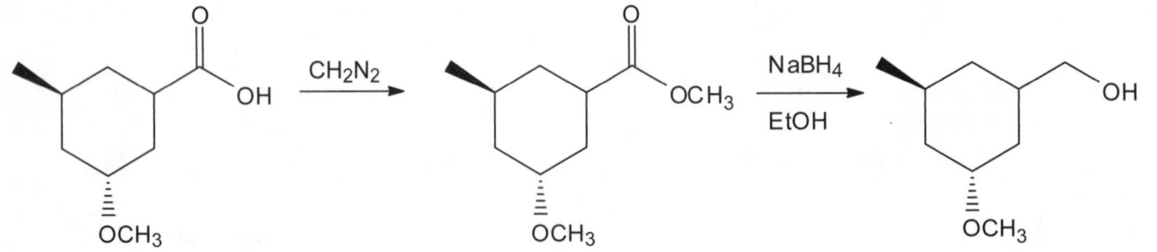

CH_2N_2 $NaBH_4$ EtOH

Problem 20.18

Think

How is the mechanism for the reduction of a carboxylic acid different from that for other acid derivatives in Table 20-1? How does the acidic proton affect the mechanism? How is the nucleophilic addition of the amide different than other acid derivatives where only H^- adds to the substrate? How is the product different?

Solve

(a) Carboxylic acids have acidic hydrogen atoms. This functional group reacts with a hydride first via proton transfer. The mechanism is shown below. Carboxylic acids only produce the primary alcohol when reacted with $LiAlH_4$.

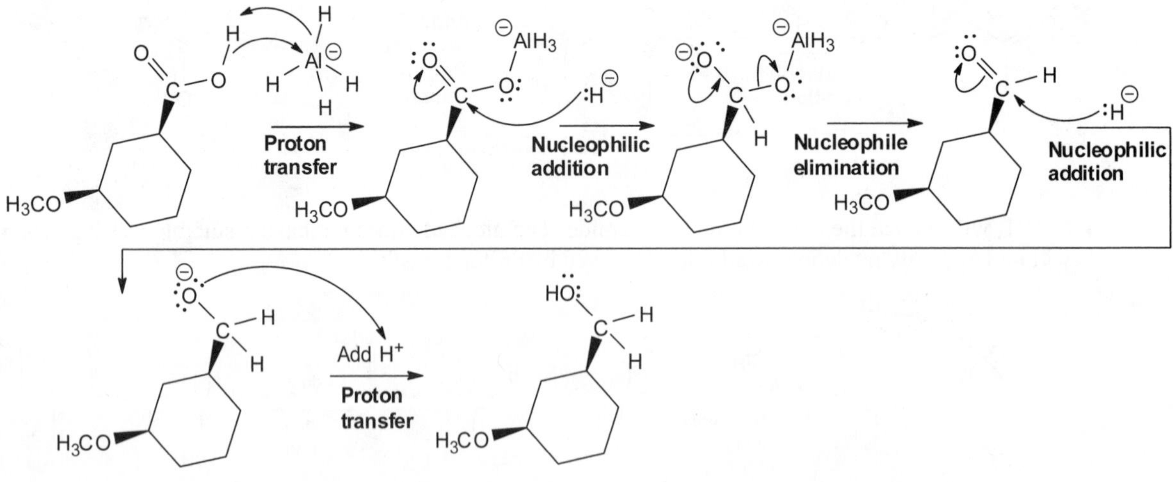

Proton transfer — Nucleophilic addition — Nucleophile elimination — Nucleophilic addition — Add H^+ Proton transfer

(b) Amides react with $LiAlH_4$ to form amines, not alcohols. The $^{-}OAlH_3$ is the leaving group rather than the R_2N^{-}.

Nucleophilic addition

Nucleophile elimination

Nucleophilic addition

Problem 20.20

Think

Which of the groups in the starting compound must be reduced? Which should remain unchanged? Will the desired reduction also reduce the latter functional group? How can a protecting group be used to prevent the undesired reductions?

Solve

The goal is to reduce the amide to an amine, which can be done by using $LiAlH_4$. However, that would also reduce the ketone. We can protect the ketone first, then carry out the reduction as planned, and finally deprotect the ketone.

Protect the ketone.

Reduce the amide.

Deprotect the ketone.

Problem 20.22

Think

What reactions do we know that can convert an acid derivative into an aldehyde? Can this be accomplished directly from an amide?

Solve

An aldehyde can be produced from an ester using diisobutylaluminum hydride (DIBAH). Thus, we must convert the initial amide into a carboxylic acid and then into an ester. We have not yet learned a way to convert a carboxylic acid into an acid chloride, but we have seen that diazomethane can convert the carboxylic acid into a methyl ester.

Undo a reduction.

Undo an esterification.

Undo a hydrolysis.

The synthesis in the forward direction is given below:

Problem 20.23

Think

What are the nucleophile and leaving group in the nucleophilic addition–elimination mechanism? How many times does the nucleophilic addition reaction occur? Are proton transfer reactions involved?

Solve

The nucleophile is the R^- group from the organometallic reagent (RMgBr or RLi) and the leaving group is Cl^- in **(a)** and RO^- in **(b)**. The nucleophilic addition reaction of the R^- occurs twice and proton transfer reactions are involved in an acid workup to yield the uncharged tertiary alcohol product. See the mechanisms below **(a)** and on the next page **(b)**.

(b) Nucleophilic addition → Nucleophile elimination → Nucleophilic addition → Add H^+ / Proton transfer

Problem 20.24

Think

What reactions do you know that can convert a C=O into a tertiary alcohol? What should the nucleophile be? How many times does the nucleophile need to add? Which acid derivatives accomplish this transformation? Can a carboxylic acid accomplish this transformation or would a side reaction occur? How can you convert the carboxylic acid into a functional group that will react with an organometallic reagent?

Solve

Notice that the tertiary alcohol target has two identical alkyl groups attached to the original carbonyl carbon. This is the expected product of a Grignard reaction involving an ester. So we can first convert the carboxylic acid to an ester. The ketone intermediate in brackets does not last long enough to be isolated; it continues to react with more Grignard reagent.

CH_2N_2 → $CH_3CH_2CH_2MgBr$ → 1. $CH_3CH_2CH_2MgBr$ 2. H_3O^+

Problem 20.25

Think

Which acid derivative is reactive enough to react with the lithium dialkylcuprate reagent? What is the nucleophile and leaving group in the nucleophilic addition–elimination mechanism? How many times does the nucleophilic addition reaction occur? Are proton transfer reactions involved?

Solve

Only the acid chloride **(a)** is reactive enough to undergo an acyl substitution reaction with a lithium dialkylcuprate. So no reaction will take place in **(b)** or **(c)**.

The nucleophile is the R^- group from the lithium dialkylcuprate reagent and the leaving group is Cl^-. The nucleophilic addition reaction of the R^- occurs only once and proton transfer reactions are not involved. The product in the reaction is a ketone.

(a) Cl, :CH_3, CH_3, Cl:, CH_3 + :Cl:

Problem 20.27

Think

What reaction converts a C=O into a CH–OH? What precursor can form a ketone from an acid chloride? What reagent must be used to stop at the ketone? Which carbonyl-containing functional group in the starting material must gain an additional carbon–carbon bond? What reagent can be used to selectively react with that functional group?

Solve

The target can be the product of reducing both carbonyls of a diketone. In turn, the diketone can be produced from the starting material using a lithium dialkylcuprate.

OH OH **Undo a reduction.** O O **Undo an acyl substitution.** O O Cl

The synthesis in the forward direction would appear as follows:

O O Cl CuLi O O 1. $LiAlH_4$ 2. H_3O^+ OH OH

Problem 20.28

Think

What is the identity of the nucleophile in the nucleophilic addition reaction? What is the leaving group in the nucleophile elimination? Will nucleophilic addition–elimination produce a derivative lower on the stability ladder? How many times does the nucleophilic addition reaction occur? Are proton transfer reactions involved?

Solve

The mechanisms are given below and on the next page:

(a) Nucleophilic addition → Nucleophile elimination → Proton transfer → Proton transfer

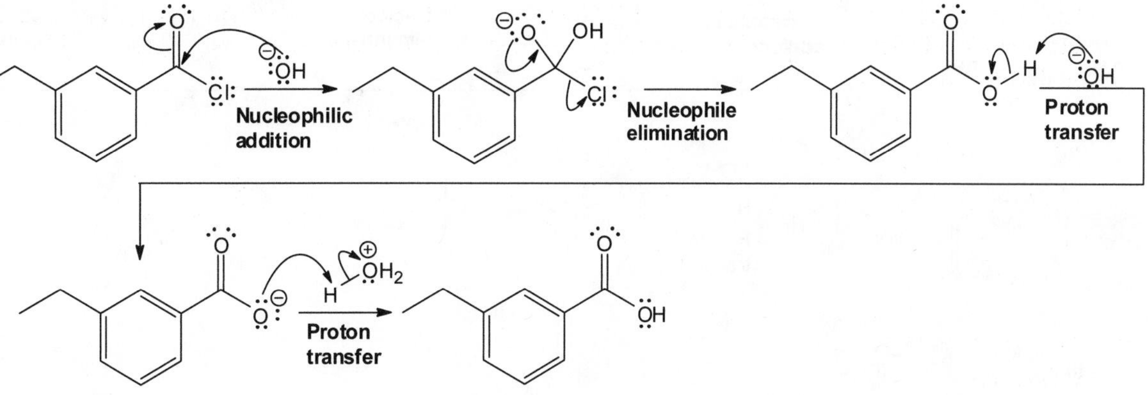

(b) Nucleophilic addition → Nucleophile elimination

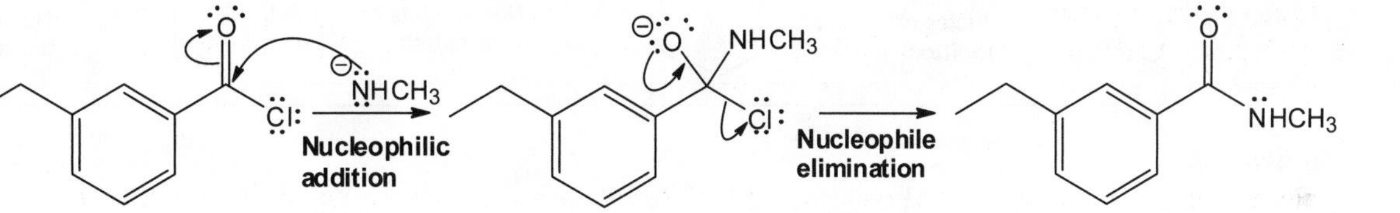

(c) Nucleophilic addition → Nucleophile elimination

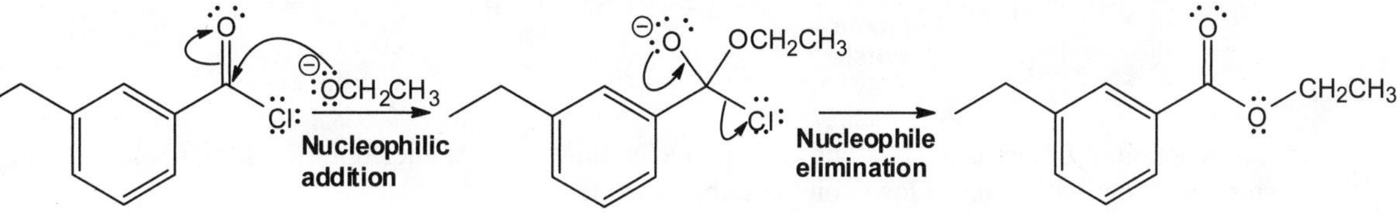

(d) Nucleophilic addition → Nucleophile elimination

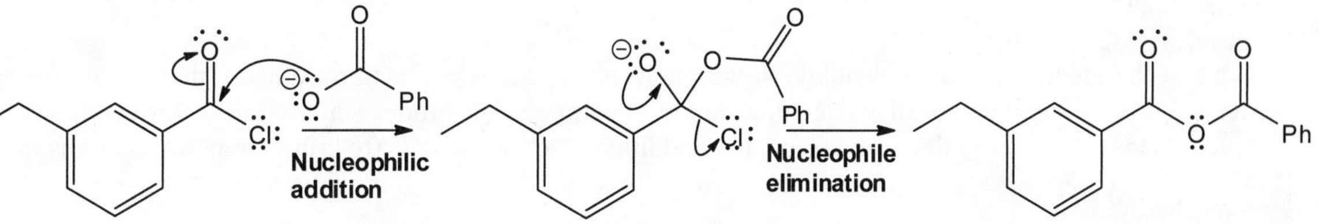

(e) Nucleophilic addition → Nucleophile elimination

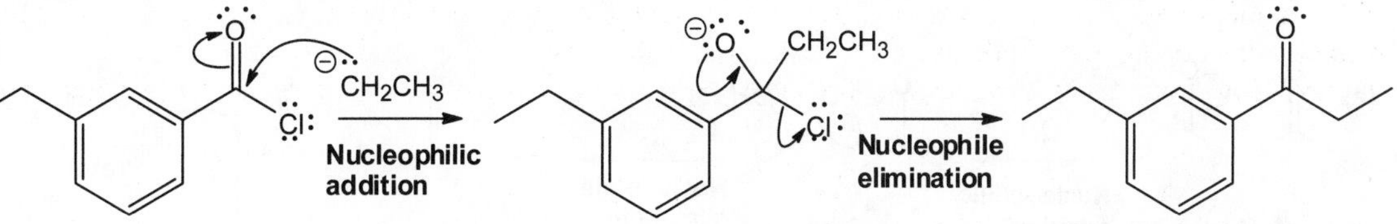

(f) Nucleophilic addition → Nucleophile elimination

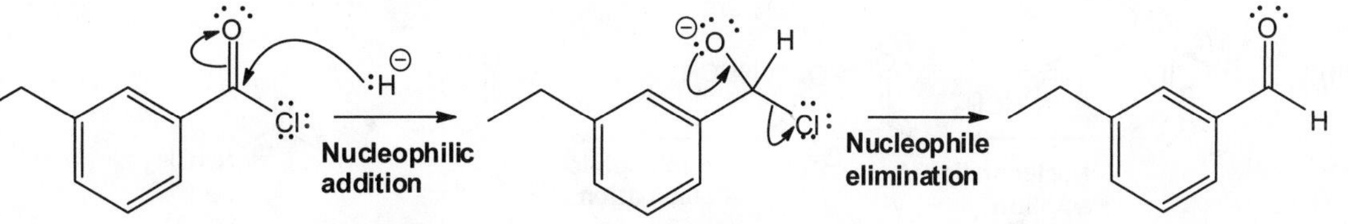

:H⁻ = LiAlH(O-*t*-Bu)$_3$ at −78 °C (stops at aldehyde)

(g) No reaction. CH_3Cl is a source of electrophilic carbon and there is no nucleophile in the reaction.

(h) Nucleophilic addition → Nucleophile elimination → Nucleophilic addition → Proton transfer

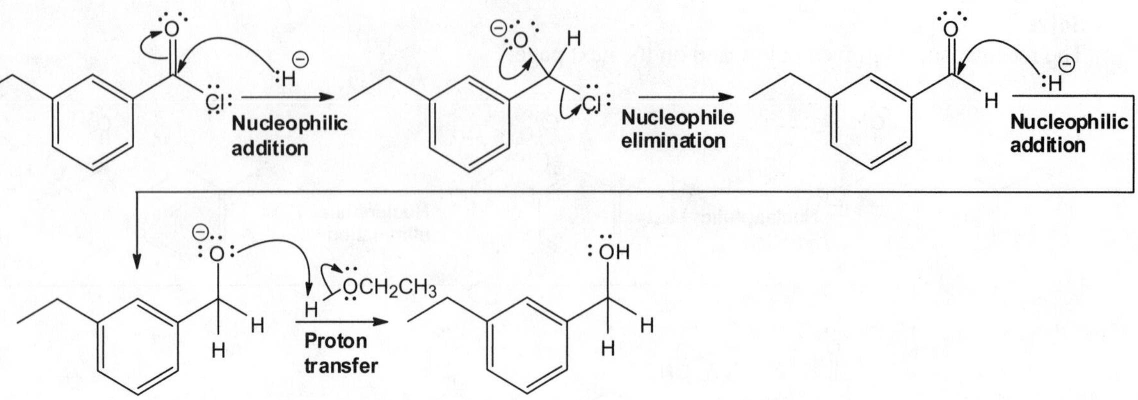

(i) Nucleophilic addition → Nucleophile elimination → Nucleophilic addition → Proton transfer

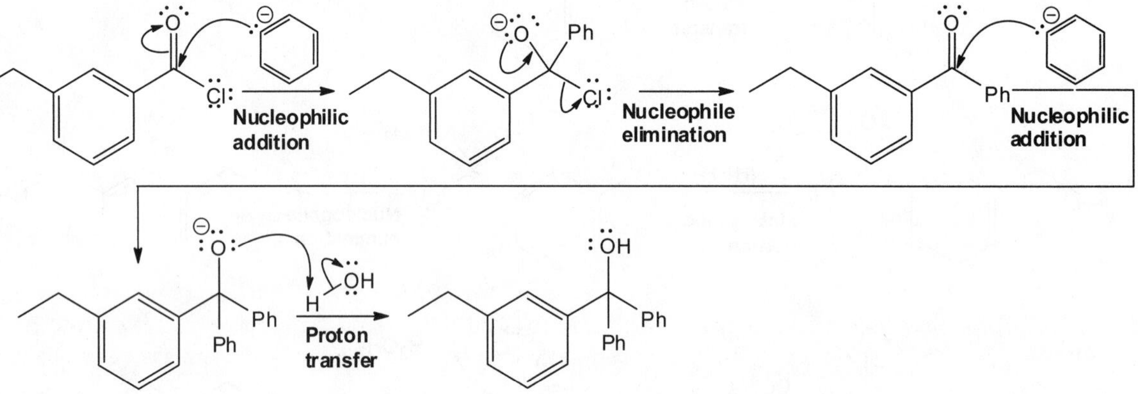

(j) No reaction. An ether is not a strong enough nucleophile. Also, a nucleophilic addition–elimination will not produce a derivative that is lower on the stability ladder.

Problem 20.29

Think

What is the identity of the nucleophile in the nucleophilic addition reaction? What is the leaving group in the nucleophile elimination? Will nucleophilic addition–elimination produce a derivative lower on the stability ladder? How many times does the nucleophilic addition reaction occur? Are proton transfer reactions involved?

Solve

The mechanisms are given below and on the next page:

(a) Nucleophilic addition → Nucleophile elimination → Proton transfer

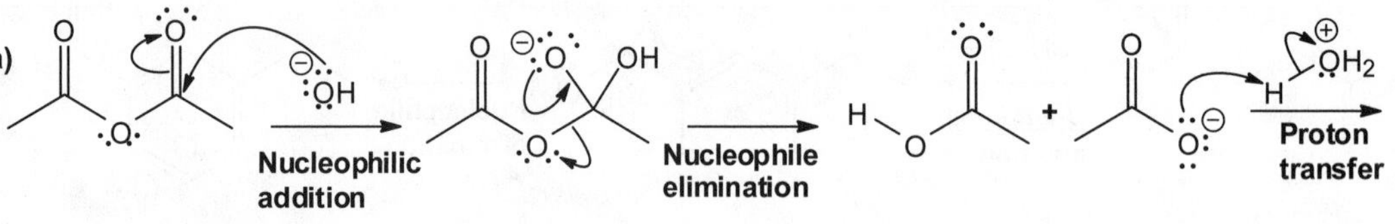

(b) Nucleophilic addition → Nucleophile elimination

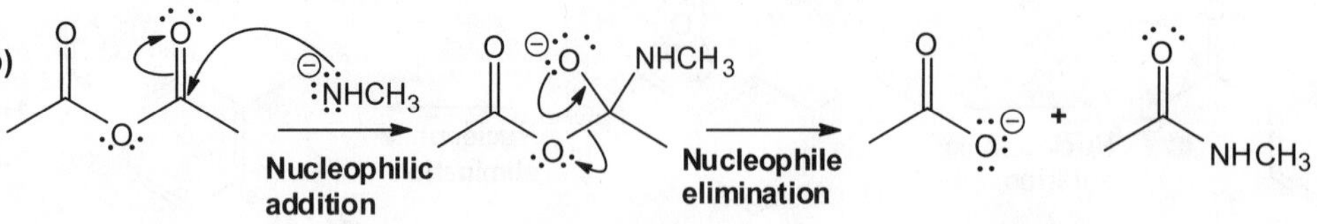

(c) Nucleophilic addition; Nucleophile elimination

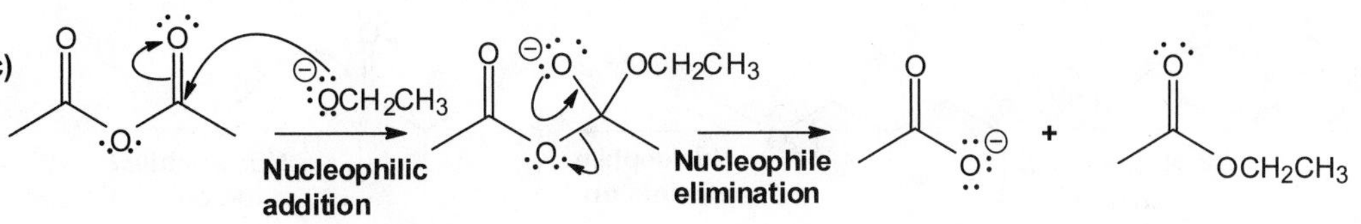

(d) Nucleophilic addition; Nucleophile elimination

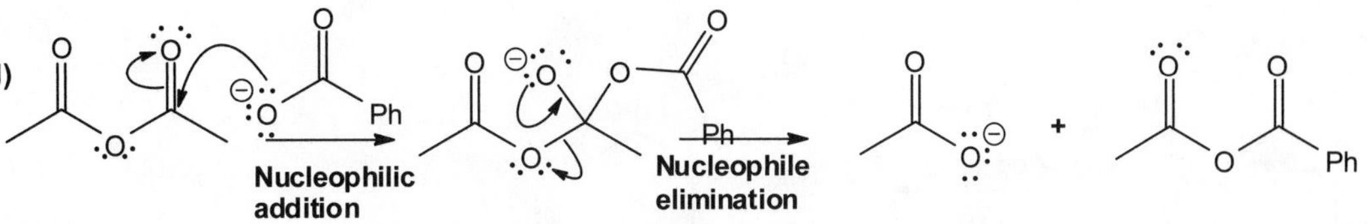

(e) No reaction. This nucleophilic addition–elimination will not produce a derivative that is lower on the stability ladder.

(f) No reaction. An ether is not a strong enough nucleophile. Also, this nucleophilic addition–elimination will not produce a derivative that is lower on the stability ladder.

(g) No reaction. $CH_3CH_2CH(Cl)CH_2CH_3$ is a source of electrophilic carbon and there is no nucleophile in the reaction.

(h) No reaction. An aldehyde is is a source of electrophilic carbon and there is no nucleophile in the reaction.

Problem 20.30

Think

What is the identity of the nucleophile in the nucleophilic addition reaction? What is the leaving group in the nucleophile elimination? Will nucleophilic addition–elimination produce a derivative lower on the stability ladder? How many times does the nucleophilic addition reaction occur? Are proton transfer reactions involved?

Solve

The mechanisms are given below and on the next page:

(a) Nucleophilic addition; Nucleophile elimination; Proton transfer; Add H; Proton transfer

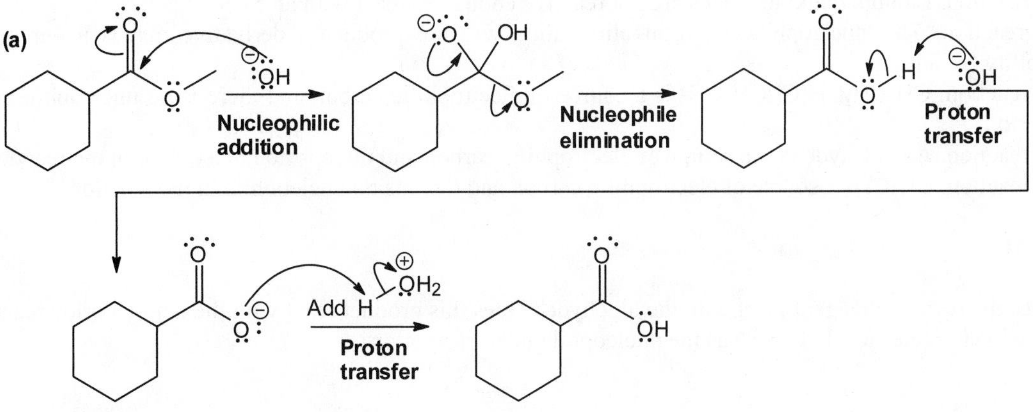

(b) Nucleophilic addition; Nucleophile elimination

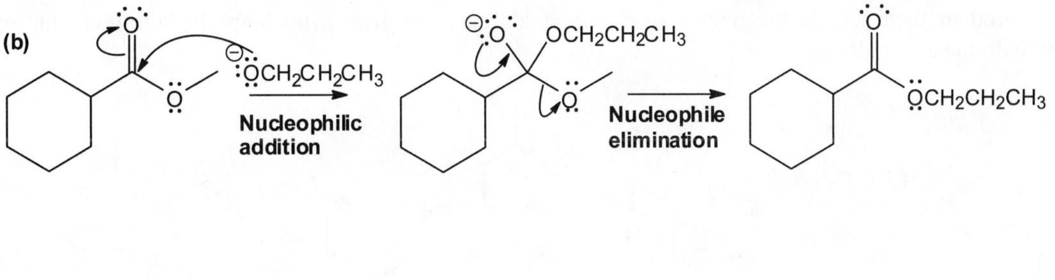

(c)

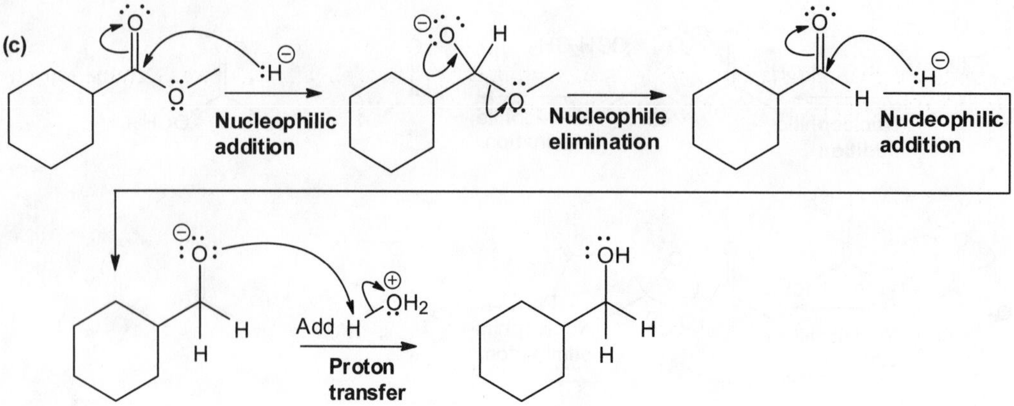

Nucleophilic addition → Nucleophile elimination → Nucleophilic addition → Add H (⊕OH2) Proton transfer

(d)

Nucleophilic addition → Nucleophile elimination → Nucleophilic addition → Add H (⊕OH2) Proton transfer

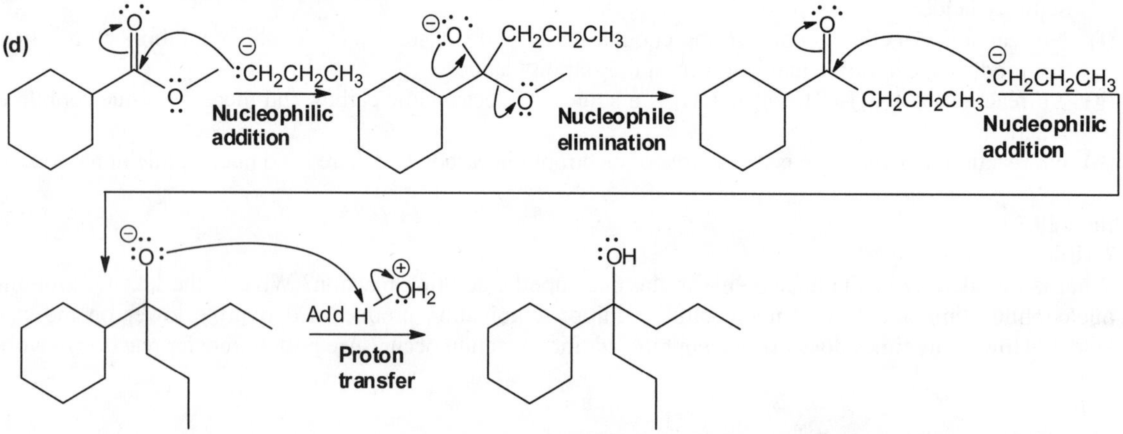

(e) No reaction. Lithium dialkylcuprates are not reactive enough to react with an ester.

(f) No reaction. This nucleophilic addition–elimination will not produce a derivative that is lower on the stability ladder.

(g) No reaction. $CH_3CH_2CH(Cl)CH_2CH_3$ is a source of electrophilic carbon and there is no nucleophile in the reaction.

(h) No reaction. An aldehyde is is a source of electrophilic carbon and there is no nucleophile in the reaction.

(i) No reaction. CH_3Cl is a source of electrophilic carbon and there is no nucleophile in the reaction.

Problem 20.31

Think

How does the reducing reagent react with the aldehyde? Does this product react with the acid chloride reactant? Identify which species would behave as the nucleophile in each reaction.

Solve

$LiAlH_4$ is a reducing agent and hydride, $H:^-$, acts as the nucleophile to attack the C=O of the aldehyde. The alkoxide product then acts as the nucleophile and yields an ester after a nucleophilic addition–elimination reaction with the acid chloride.

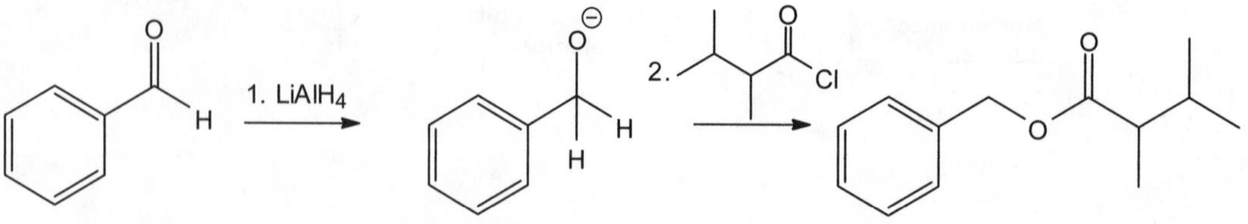

1. $LiAlH_4$ → 2. (acid chloride) →

Problem 20.32

Think

To form the product given, what nucleophile is used to undergo a nucleophilic addition–elimination reaction with the acid chloride? Are proton transfer reactions involved?

Solve

The reactions are given below:

(a) ONa

(b) NaO

(c) $LiAlH_4$ Cl

(d) 1. LiHN 2. H^+ N H $LiAlH_4$ N H

(e) 1. $LiAlH_4$ 2. H_2O OH H_2SO_4 Δ O

(f) $LiAl(Ot\text{-}Bu)_3H$ H

(g) $(CH_3)_2CuLi$

(h) O

Problem 20.33

Think

To form the product given, what nucleophile is used to undergo a nucleophilic addition–elimination reaction with the acid chloride? Are any carbon–carbon bond-formation reactions necessary? Are proton transfer reactions involved?

Solve

The reactions are given below:

(a)

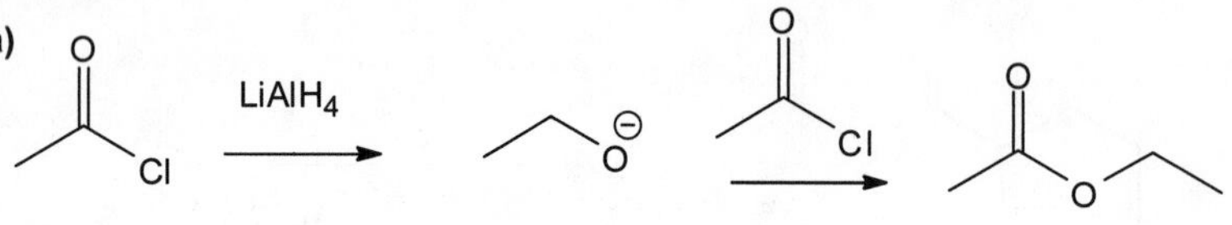

Cl
LiAlH4
Cl

(b)

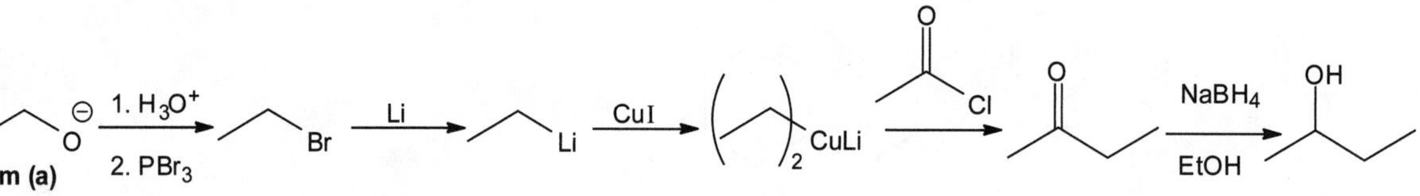

1. H3O+
2. PBr3
Br
Li
Li
CuI
CuLi
2
Cl
NaBH4
EtOH
OH

From (a)

(c) Use the ester from **(a)** and ethyllithium from part **(b)** (or a Grignard reagent) to produce the target.

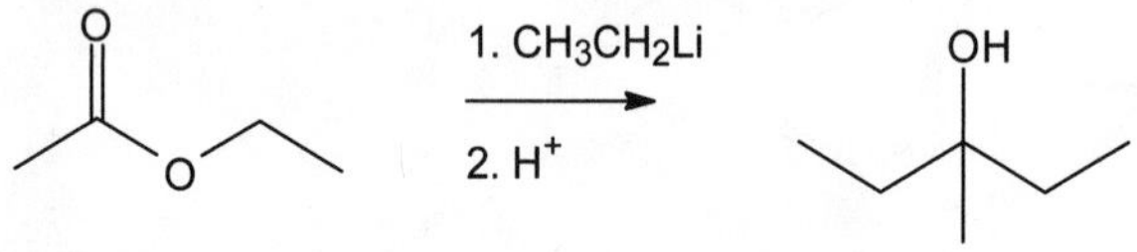

1. CH3CH2Li
2. H+
OH

(d) Butan-2-one is made in **(b)**.

(e)

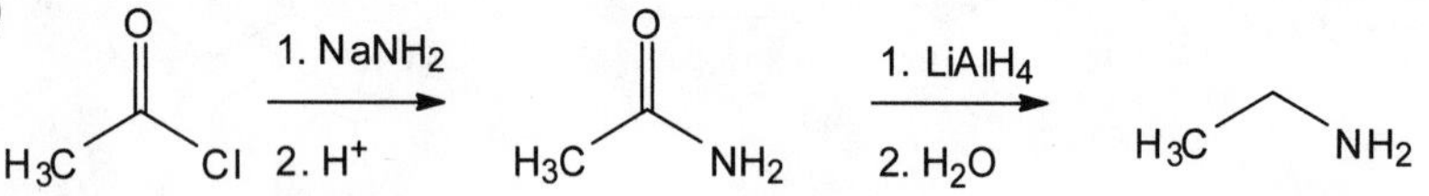

H3C
Cl
1. NaNH2
2. H+
H3C
NH2
1. LiAlH4
2. H2O
H3C
NH2

(f)

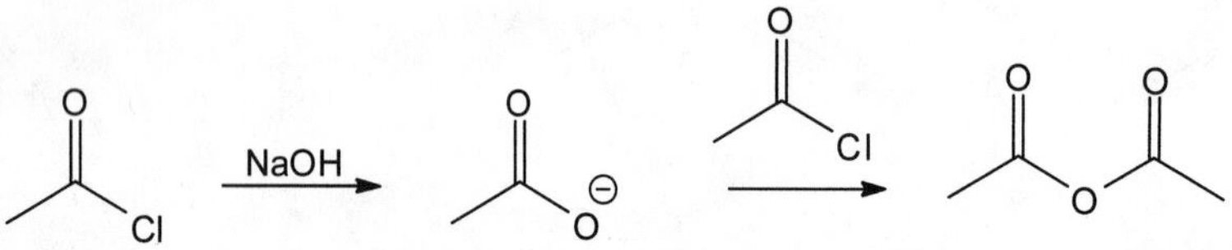

Cl
NaOH
Cl

Problem 20.34

Think

Which group, CH_3 or CF_3, is electron withdrawing? Which carbonyl C atom is more electrophilic? How does the electronic nature of the carbonyl carbon affect the rate of nucleophilic addition?

Solve

The fluorinated ester has a carbonyl carbon that is more electron poor. It will react faster.

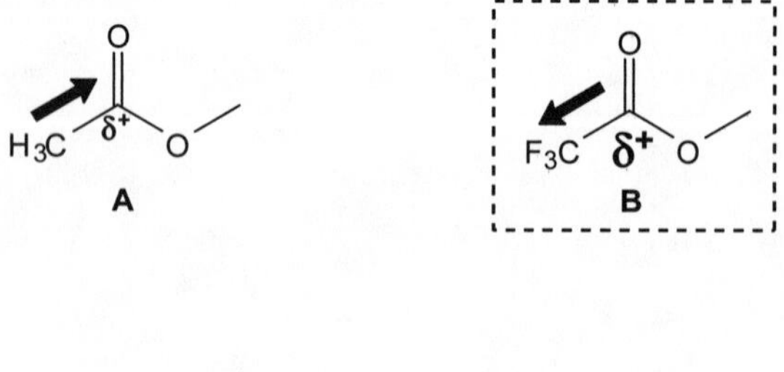

H3C
δ+
A
F3C
δ+
B

Problem 20.35

Think

How many C atoms does the product possess? What functional group does the target possess? Which bonds are likely formed?

Solve

The product is a diester with four C atoms. Each ester group can be made from an acid chloride functional group and an alkoxide functional group. Oxalyl chloride (the acid chloride of oxalic acid, HOOC–COOH) and ethylene glycol dianion will combine in one reaction.

Nucleophilic addition

Nucleophile elimination

Nucleophilic addition

Nucleophile elimination

Problem 20.36

Think

How many C atoms does the product possess? Are carbon–carbon bond-formation reactions necessary? What reactions do you know that will connect two molecular pieces by a C=C bond? What functional groups are required for that reaction?

Solve

The C=C bond can be the last bond formed from a Wittig reaction.

Undo a Wittig reaction.

Ph_3P

The two reagents from the Wittig reaction can be synthesized from a carboxylic acid.

1. $LiAlH_4$ 2. H_3O^+

H_3PO_4, Δ

H_3O^+/H_2O

PCC

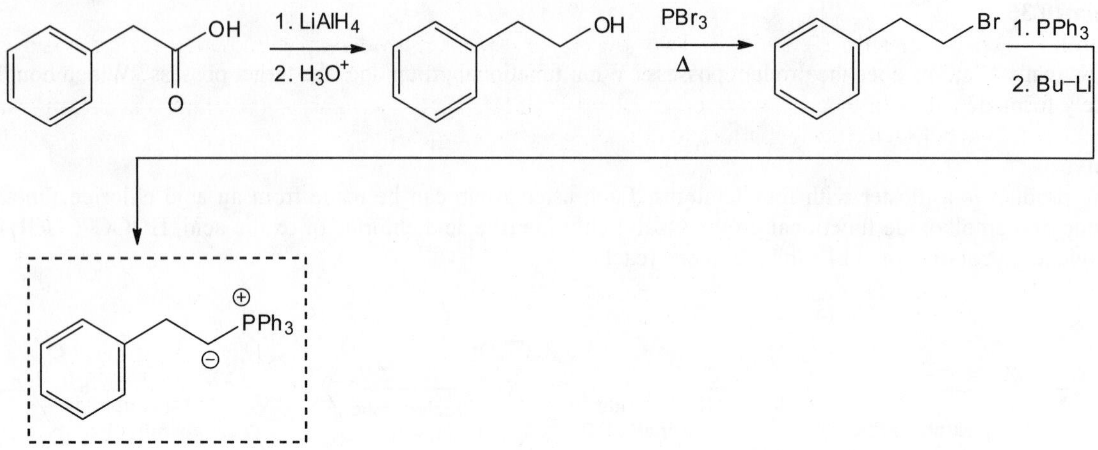

The Wittig reaction in the forward direction is shown below:

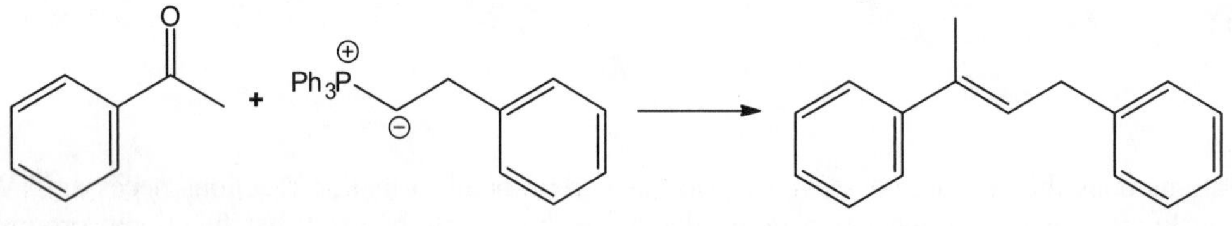

Problem 20.37

Think

What is the charge on the nitrogen in LiN_3? What is the nucleophile in the nucleophilic addition–elimination reaction? After the first nucleophilic addition–elimination reaction, is the resulting product nucleophilic?

Solve

Li_3N can be treated as the nitride ion (N^{3-}), and this nucleophile reacts with three equivalents of acid chloride to undergo three nucleophilic addition–elimination reactions.

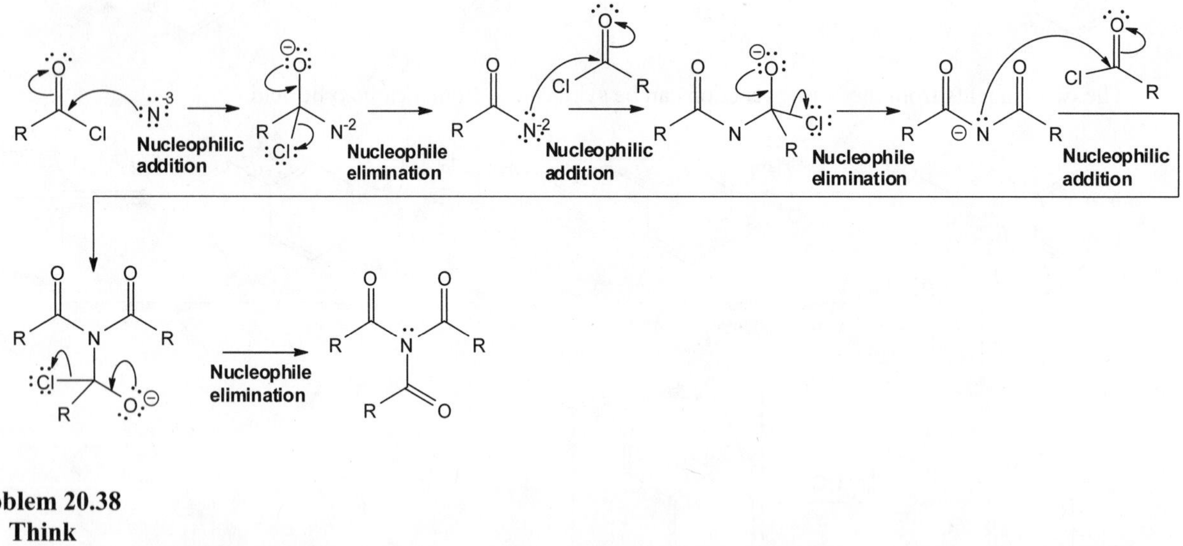

Problem 20.38

Think

What types of species should not appear in the mechanism for a reaction that takes place under basic conditions? What is the nucleophile in the nucleophilic addition–elimination reaction? How many times does the nucleophilic addition–elimination reaction occur? Are proton transfer reactions involved?

Solve

The EtO^- deprotonates the amide N–H. Two successive nucleophilic addition–elimination reactions take place in which the RHN^- is the nucleophile. The second one is intramolecular because it forms a six-membered ring. Note that in basic solution, no strong acids should appear in the mechanism; therefore, the proton transfer reaction occurs before the nucleophilic addition–elimination mechanism does.

Proton transfer
Nucleophilic addition
Nucleophile elimination
Proton transfer
Nucleophilic addition
Nucleophile elimination

Problem 20.39

Think

Consider another resonance structure for the enamine. What is the nucleophile in the nucleophilic addition–elimination reaction? In the hydrolysis reaction, what kinds of species should not appear under acidic conditions? What is the nucleophile in the hydrolysis?

Solve

The enamine has a resonance structure in which a negative charge appears on C, so that carbon is nucleophilic. It attacks the acid chloride in a nucleophilic addition–elimination reaction. When H_3O^+ is added, hydrolysis of the C=N bond takes place in a mechanism that is essentially identical to hydrolysis of imines (see Chapter 18). Because the C=N bond is already activated, that hydrolysis is initiated by attack of the C=N carbon by water.

Resonance
Nucleophilic addition
Nucleophile elimination
Nucleophilic addition
Proton transfer
Proton transfer
Heterolysis
Proton transfer

Problem 20.40

Think

How is the structure of the amidines similar to a C=O? What is the nucleophile in the nucleophilic addition–elimination reaction? How many times does the nucleophilic addition–elimination reaction occur? Under basic conditions, what types of species should not appear in the mechanism? Are proton transfer reactions involved?

Solve

The amidine carbon is electrophilic, just like a carbonyl carbon and, therefore, is susceptible to nucleophilic attack. The mechanism takes place in basic solution, so no strong acids should appear.

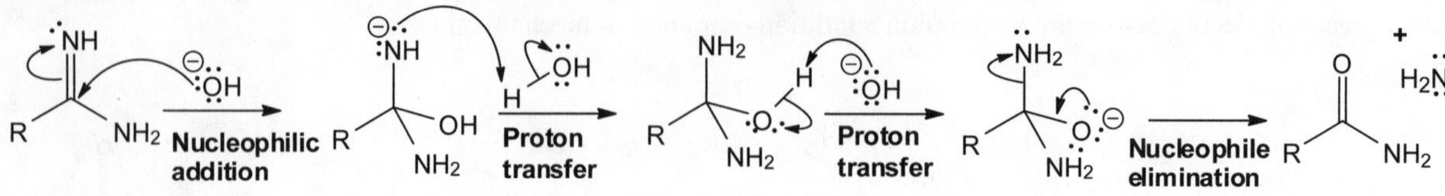

Nucleophilic addition — Proton transfer — Proton transfer — Nucleophile elimination

Problem 20.41

Think

On what type of carbon is the Cl located? Is an S_N2 mechanism likely? What other elementary step involves a nucleophilic attack? How does the nucleophile elimination occur?

Solve

(a) Although an S_N2 mechanism could be written, it is not feasible because the leaving group is on an sp^2-hybridized C. Instead, it is a nucleophilic addition–elimination mechanism, in which conjugate addition takes place instead of direct addition.

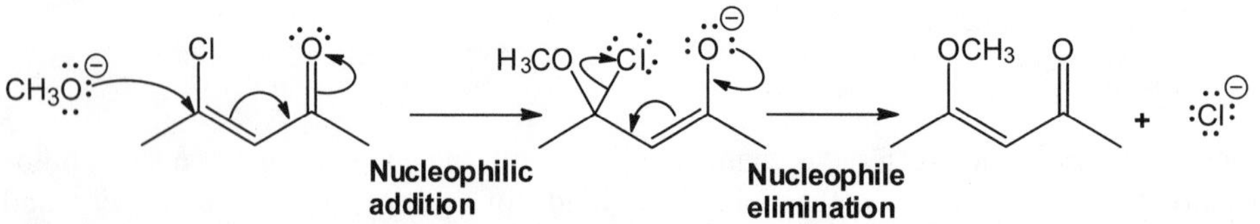

Nucleophilic addition — Nucleophile elimination

(b) The same mechanism is not feasible for the second reaction because it would require a negative charge to develop on C, which is very unstable.

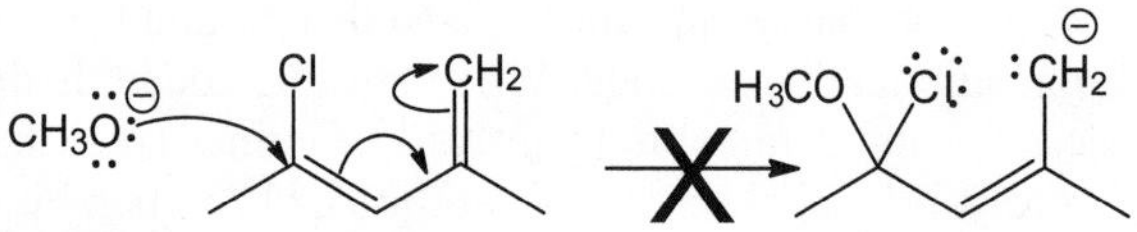

Problem 20.42

Think

Is a new C–C bond formed? How can a carbon nucleophile be produced under these conditions? What carbon atom would it attack? After the epoxide ring is opened, can an intramolecular nucleophilic addition–elimination reaction produce the five-membered ring?

Solve

The central C atom of the diester is mildly acidic, so it is deprotonated to make an enolate anion that is nucleophilic at that C. It attacks the epoxide to ring open the epoxide in an S_N2 mechanism. The alkoxide formed attacks the ester in an intramolecular nucleophilic addition-elimination reaction.

Proton transfer — S_N2 — Nucleophilic addition — Nucleophile elimination

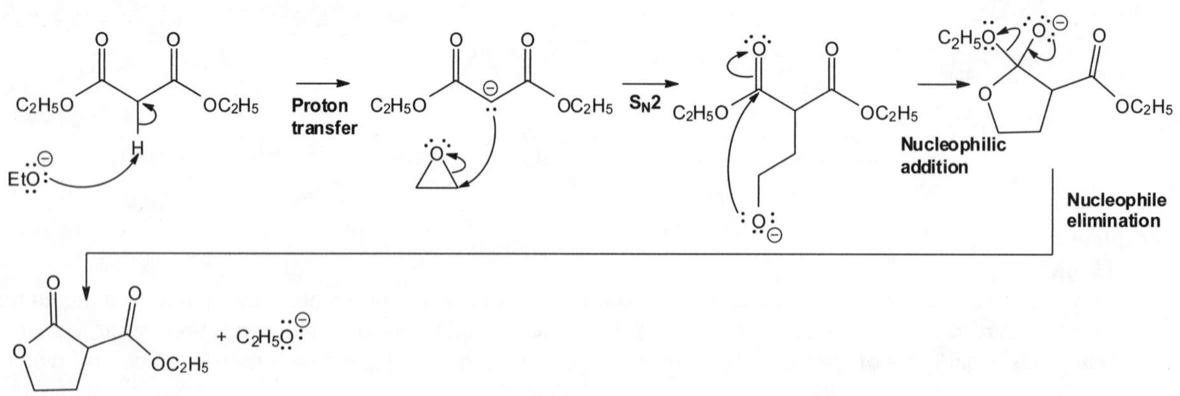

Problem 20.43

Think

Under basic conditions, what types of species should not appear in the mechanism? What is the nucleophile in the nucleophilic addition–elimination reaction? How many times does the nucleophilic addition–elimination reaction occur? Are proton transfer reactions involved?

Solve

LDA converts the NH_2 group into a strong nucleophile that can attack the C=O group. The mechanism takes place in basic solution, so no strong acids should appear.

Proton transfer — Nucleophilic addition — Nucleophile elimination — Proton transfer

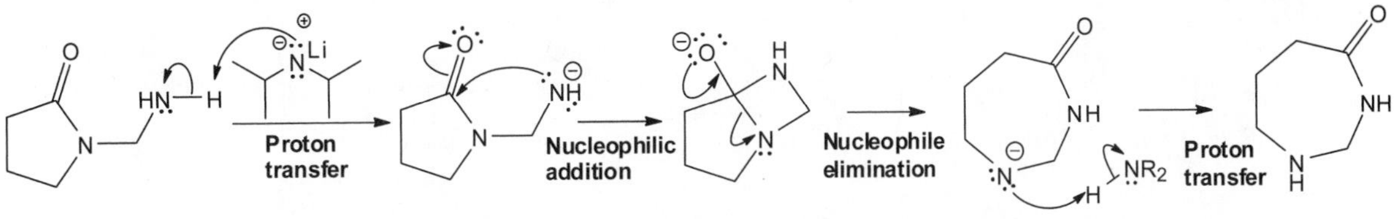

Problem 20.44

Think

Under basic conditions, what types of species should not appear in the mechanism? Is CCl_3 a suitable leaving group? What is the nucleophile in the nucleophilic addition–elimination reaction? How many times does the nucleophilic addition–elimination reaction occur? Are proton transfer reactions involved?

Solve

CH_3NO_2 is moderately acidic, so it can be deprotonated by *t*-butoxide to produce a strong carbon nucleophile, which acts as the nucleophile in the nucleophilic addition–elimination reaction. CCl_3 is a suitable leaving group. The mechanism takes place in basic solution, so no strong acids should appear.

Proton transfer — Nucleophilic addition — Nucleophile elimination

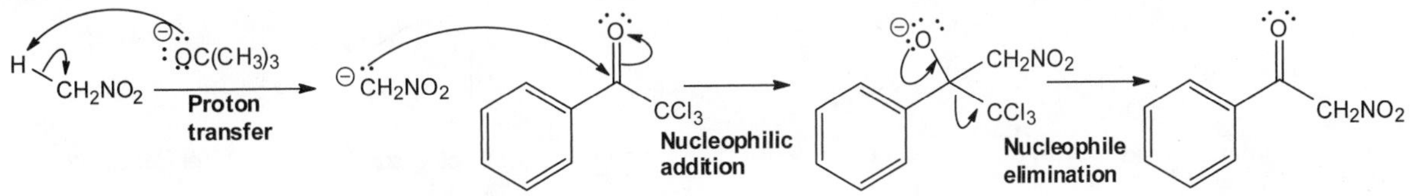

Problem 20.45

Think

Under basic conditions, what types of species should not appear in the mechanism? What is the nucleophile in the nucleophilic addition–elimination reaction? How many times does the nucleophilic addition–elimination reaction occur? Are proton transfer reactions involved?

Solve

The reaction takes place in basic solution, so no strong acids should appear. The $H-^{18}OH$ protonates the carboxylate anion to form the carboxylic acid. The carboxylic acid undergoes nucleophilic addition by the $H^{18}O^-$ anion. The full mechanism is shown on the next page:

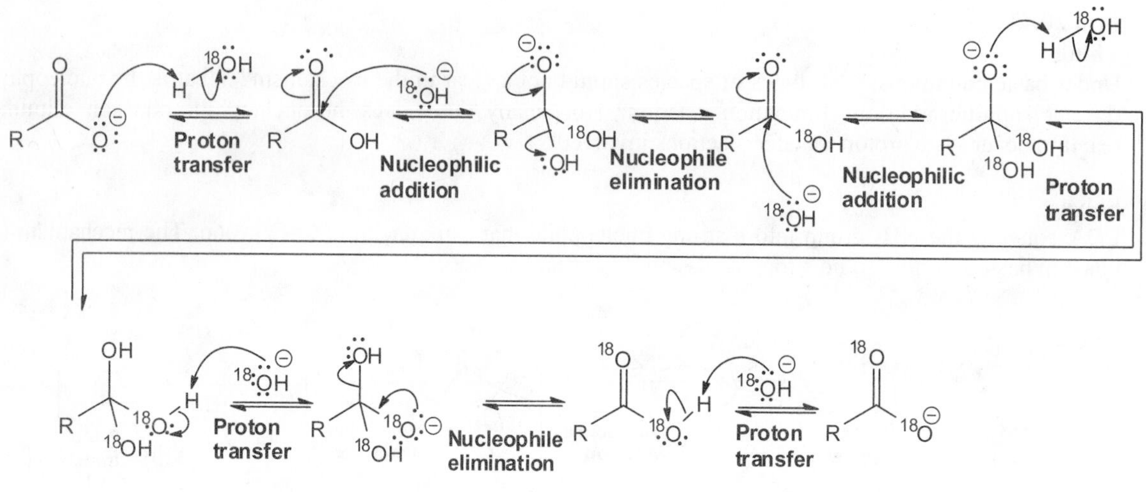

Proton transfer
Nucleophilic addition
Nucleophile elimination
Nucleophilic addition
Proton transfer
Proton transfer
Nucleophile elimination
Proton transfer

Problem 20.46

Think

What is the identity of the yellow precipitate? What type of reaction does it signify? What is the requirement of the α carbon for this reaction to take place?

Solve

The identity of the yellow precipitate is HCI_3. Compound **(b)** is the only one that can form the HCI_3 precipitate. The reaction described is a haloform reaction, which takes place with methyl ketones and methyl aldehydes. The α carbon must be primary.

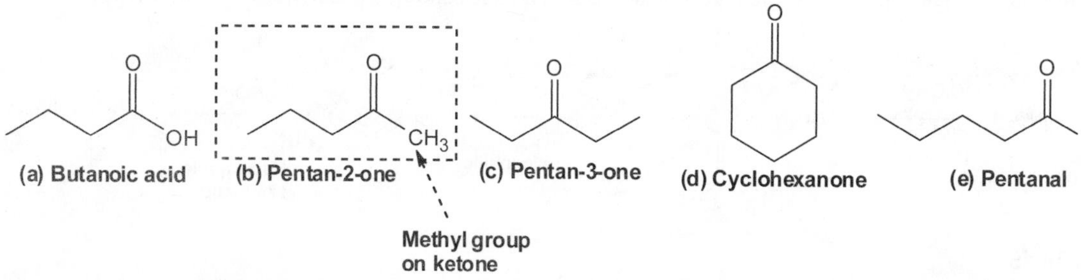

(a) Butanoic acid (b) Pentan-2-one (c) Pentan-3-one (d) Cyclohexanone (e) Pentanal

Methyl group on ketone

Problem 20.47

Think

What functional group is the target in each of these compounds? Is it primary, secondary, or tertiary? What method do you know to form this functional group? What reagents are necessary?

Solve

In a Gabriel synthesis, an alkyl halide is converted to a primary amine. So we must simply choose the appropriate alkyl halide that is analogous to the primary amine target. Notice in **(c)** that the stereochemical configuration is inverted because the second reaction is an S_N2 that takes place at a carbon stereocenter.

(a)

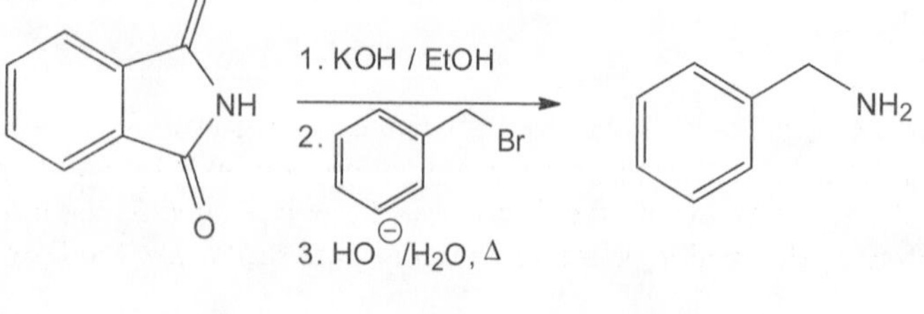

1. KOH / EtOH
2. Br
3. $HO^{\ominus}/H_2O$, Δ

(b)

1. KOH / EtOH
2.
3. $HO^{\ominus}/H_2O$, Δ

(c)

1. KOH / EtOH
2.
3. $HO^{\ominus}/H_2O$, Δ

Problem 20.48

Think

In each reaction, identify the electron-rich and the electron-poor species. Is the carbon skeleton altered? What functional group transformations occur? Is the carbon in each reaction the electrophile or the nucleophile? Are proton transfer reactions involved?

Solve

The methyl ketone starting material undergoes a haloform reaction to produce carboxylic acid **A**, which is then reduced with $LiAlH_4$ to produce the primary alcohol **B**. Bromination with PBr_3 converts the OH to a Br to make alkyl halide **C**. That alkyl halide is then used in a Gabriel synthesis to produce primary amine **D**.

1. I_2, NaOH; 2 H_3O^+ → **A**; 1. $LiAlH_4$, THF; 2. H_3O^+ → **B**; PBr_3 → **C**

1. NaOH; 2. **C**; 3. NaOH, Δ → **D**

Problem 20.49

Think

In each reaction identify the electron-rich and the electron-poor species. Is the carbon skeleton altered? What functional group transformations occur? Is the carbon in each reaction the electrophile or the nucleophile? Are proton transfer reactions involved?

Solve

A is produced by reducing the starting material to the aldehyde. **A** then reacts with ketone **B** in an aldol condensation reaction. Lithium dimethylcuprate then undergoes conjugate addition, followed by reduction to produce **D**. PBr_3 then converts the OH to a Br to produce **E**. In the Gabriel synthesis involving **E**, the primary amine **F** is produced, which undergoes Hofmann elimination in the final step to produce **G**. See the figure on the next page.

1. DIBAH
OCH_3 2 H_3O^+
A
H
B
NaOH, Δ
1. $(CH_3)_2CuLi$
C
2. NH_4Cl
NaBH4
EtOH
OH
D
PBr3
Br
E
NH
1. NaOH
2.
Br
E
3. NaOH, Δ
NH_2
F
CH_3Br (excess)
NaOH, D
G

Problem 20.50

Think

In each reaction identify the electron-rich and the electron-poor species. Is the carbon skeleton altered? What functional group transformations occur? Is the carbon in each reaction the electrophile or the nucleophile? Are proton transfer reactions involved?

Solve

The starting amide undergoes hydrolysis to produce carboxylic acid **A**. Diazomethane converts the carboxylic acid to a methyl ester, which is reduced to the aldehyde **B** by DIBAH. The aldehyde then undergoes a Grignard reaction to make alcohol **C**. The alcohol is then deprotonated to make the alkoxide anion, which is then acylated with acetic anhydride, yielding the target compound **D**.

1. NaOH, Δ
2. H_3O^+
A
OH
CH_2N_2
OCH_3 DIBAH
B
H
1. MgBr
2. H_3O^+
C
OH
1. NaH
2.
D

Problem 20.51

Think

In each reaction identify the electron-rich and the electron-poor species. Is the carbon skeleton altered? What functional group transformations occur? Is the carbon in each reaction the electrophile or the nucleophile? Are proton transfer reactions involved?

Solve

The starting alkene undergoes hydroboration–oxidation to form the primary alcohol **A**. The alcohol is oxidized to the carboxylic acid **B** by $KMnO_4$. The carboxylic acid reacts with CH_2N_2 to form the methyl ester **C**. The ester undergoes a Grignard reaction (addition of two $C_6H_5^-$ nucleophiles) to form the tertiary alcohol **D.** This alcohol undergoes a dehydration reaction to form the trisubstituted alkene. The alkene then undergoes hydroboration–oxidation to form the less substituted alcohol **F**. The alcohol is deprotonated by NaH to form an O^- nucleophile that reacts with the ester **C** to form the new ester **G.**

1. BH_3•THF
2. NaOH, H_2O_2
A
$KMnO_4$
B
CH_2N_2
C
1. C_6H_5MgBr
2. H_3O^+
D
conc. H_3PO_4, Δ
E
1. BH_3•THF
2. NaOH, H_2O_2
F
1. NaH
2. C
G

Problem 20.52

Think

How is this reaction similar to the reaction of an ester with an alkoxide? What acts as the nucleophile in the nucleophilic addition–elimination reaction? Which anion, O^- or S^-, stabilizes the negative charge better?

Solve

(a) The mechanism would appear as follows:

$^-SCH_3$
SCH_3
S
+ $^-:\ddot{O}CH_3$

(b) The reaction would *not* be energetically favorable, because the product anion is less stable than the reactant anion. In the product, a negative charge appears on O, whereas in the reactants, the negative charge appears on S. The S atom is bigger and can accommodate the negative charge better.

Problem 20.53

Think

What functional group is present in this triglyceride? How does this functional group react with HO^-? Are there any irreversible steps? How many times does this reaction occur? Is there a hydrophilic part of the product? Is there a hydrophobic part of product? How can a micelle form?

Solve

The triglyceride is an ester and undergoes base-catalyzed hydrolysis three times to form triglycerol and three fatty-acid carboxylate anions. These fatty-acid carboxylates can serve as soap because there is a very hydrophobic/nonpolar end and a very hydrophilic/polar end that can form a micelle around grease.

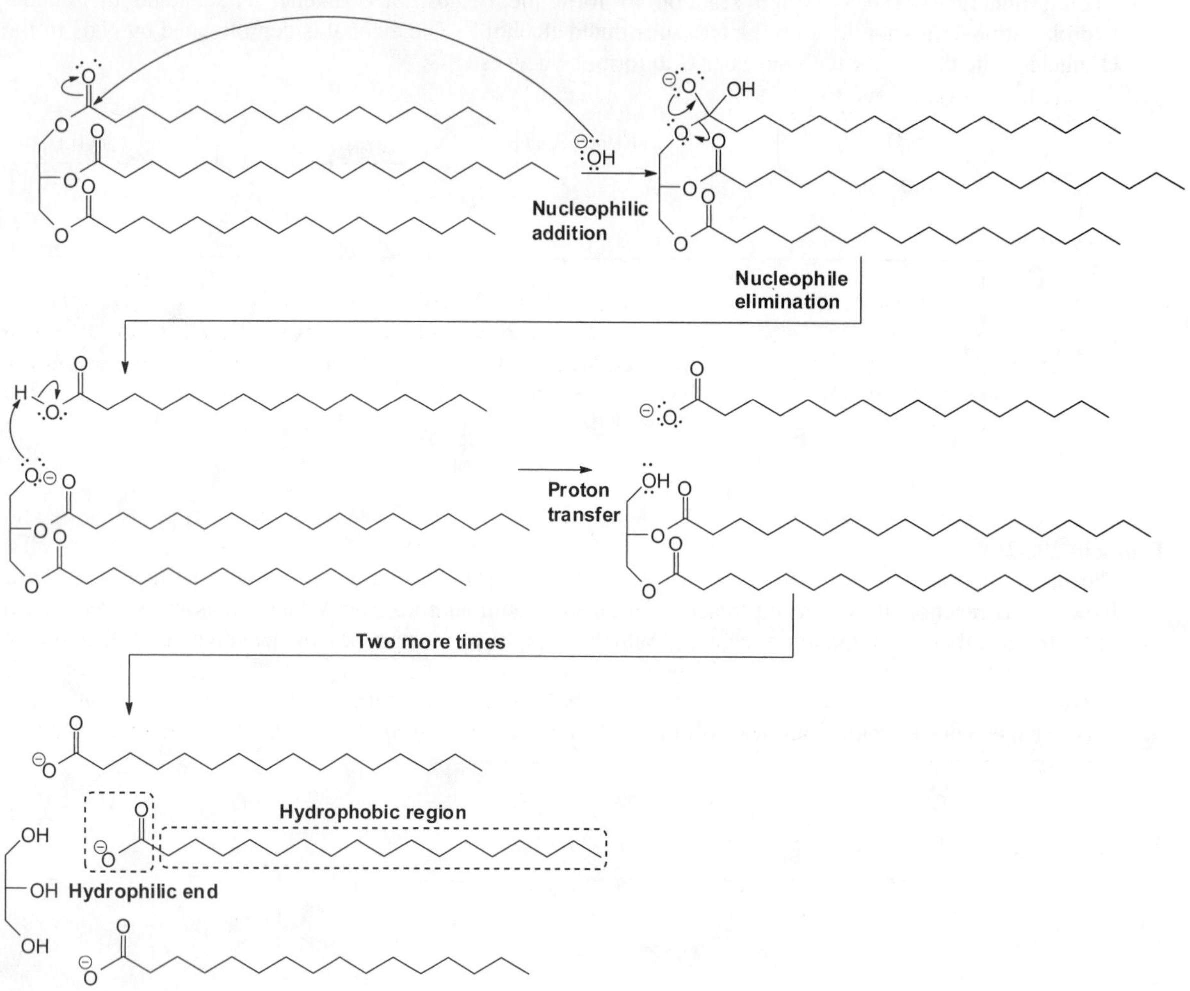

Problem 20.54

Think

In Steps 6–8 of the mechanism in Equation 20-13, what elementary steps need to occur? Is there a competing proton transfer under strongly basic conditions, which would compromise the steps that need to occur?

Solve

In Step 6 of the mechanism, hydroxide needs to act as a nucleophile to add to the C=O group of the amide.

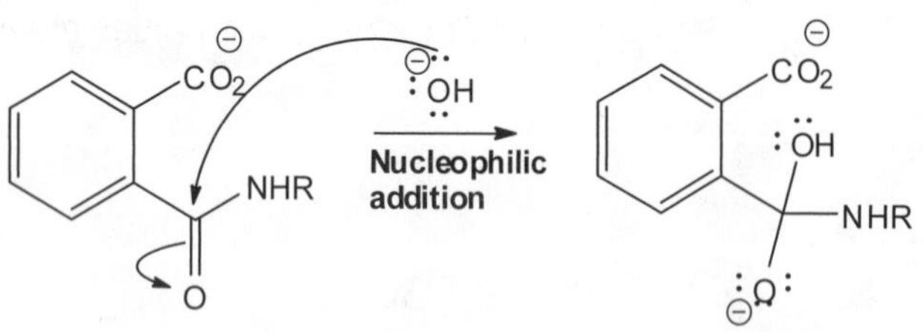

However, the amide proton is somewhat acidic and can be deprotonated under strongly basic conditions, as shown below. When this happens, the resulting anion stabilizes the C=O group via resonance and hinders nucleophilic attack by hydroxide.

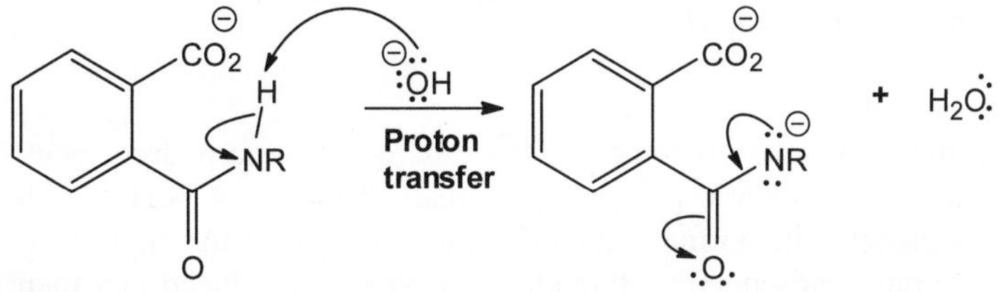

Proton transfer

Problem 20.55

Think

What is the IHD? What functional group do the two broad IR peaks from 3400–3200 cm^{-1} suggest is present? The several IR peaks from 1600–1400 cm^{-1} suggest the presence of what functional group? In the NMR spectrum of the product, what could account for the peaks around 6.5 ppm? The broad signals around 8.4 and 4.3 ppm? What is the structure of the product? What acid derivative yields this product and acetic acid from the hydrolysis reaction in base?

Solve

The IHD = 5. The IR peaks from 3400–3200 cm^{-1} suggest the presence of a primary amine and/or an alcohol. The several peaks from 1600–1400 cm^{-1} are indicative of an aromatic ring, which accounts for an IHD = 4. Therefore, this hydrolysis product contains an amine and is aromatic. This is further indicated by the peaks around 6.5 ppm in the ^{1}H NMR spectrum. The other two peaks are broad. One arises from the protons on the amide group, at 4.3 ppm, which integrates to 2 H. The other, at 8.4 ppm, is consistent with an OH group as part of a phenol, as it integrates to 1 H. The C=O group also contributes one to the IHD to bring it to a total of 5. Acetic acid is a carboxylic acid. Therefore, an amide starting material would yield these two products from the hydrolysis reaction in base followed by an acid workup. The structure of the starting material pain reliever and product are given below:

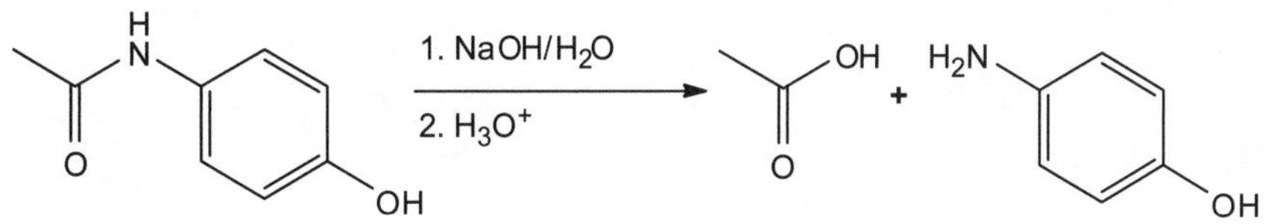

1. NaOH/H_2O
2. H_3O^+

Problem 20.56

Think

What is the IHD? What functional group does the intense absorption at 1740 cm^{-1} suggest? What functional group does the weaker absorption at 1650 cm^{-1} suggest? How many ^{13}C signals are present? How many C atoms are present in the formula? Is there any symmetry? Which C=O in the compound is more reactive? Which one will remain in the product and which one reacts with the Grignard reagent?

Solve

The IHD = 3. The sharp peak at 1740 cm^{-1} is indicative of a C=O due to an ester. The weaker peak at 1650 cm^{-1} is indicative of a C=C. There is no symmetry in the compound because the ^{13}C NMR has seven peaks and there are seven C atoms. The aldehyde is attacked by the $H_2C{=}CH^-$, and the addition–elimination reaction results in the cyclic ester (lactone) shown below.

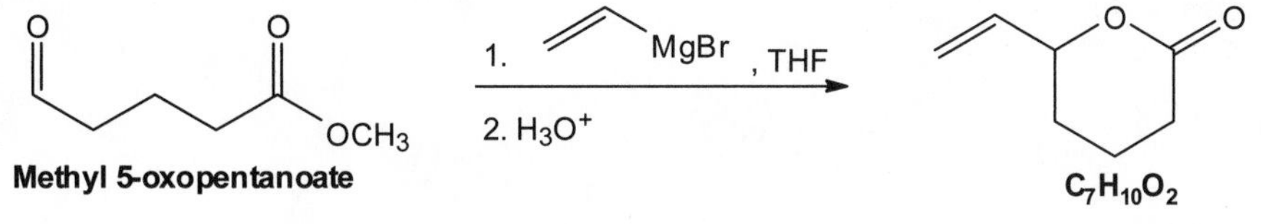

Methyl 5-oxopentanoate

1. MgBr, THF
2. H_3O^+

$C_7H_{10}O_2$

Problem 20.57

Think

What functional group is present in dimethyl phthalate? How does this acid derivative react with $LiAlH_4$? How many signals are in the ^1H NMR? What is the relative ratio of the integration of the peaks? Does the aromatic ring remain in the product? What functional group is formed?

Solve

Notice that the NMR spectrum of the product shows that there are twice as many aromatic hydrogens as there are aliphatic ones. So the product must have two aliphatic hydrogens, instead of the six it begins with. The reaction that produces such a compound begins with the normal reduction of an ester with $LiAlH_4$, in which two equivalents of hydride add to the ester carbonyl carbon. The alkoxide anion that is produced can then be a nucleophile in a nucleophilic addition–elimination mechanism with the other ester group.

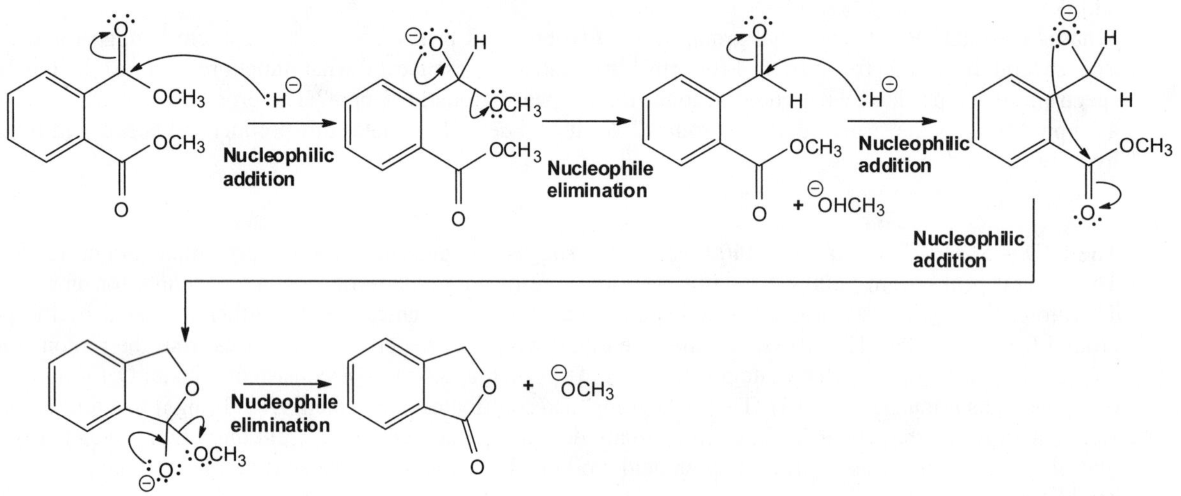

OCH3 :H⁻ OCH3 Nucleophilic addition → OCH3 OCH3 Nucleophile elimination → H :H⁻ OCH3 + ⁻OHCH3 Nucleophilic addition → H H OCH3 Nucleophilic addition → OCH3 Nucleophile elimination → O + ⁻OCH3

CHAPTER 21 | Nucleophilic Addition—Elimination Reactions 2: Weak Nucleophiles

Your Turn Exercises

Your Turn 21.1

Think

Use Equation 21-4 as a guide. What bonds are broken and formed in each step? Which sites are electron rich? Electron poor? How can you use curved arrows to show that electron movement? Review the nucleophilic addition–elimination mechanism from Chapter 20. What similarities do you observe? In what steps are protons involved in the reaction?

Solve

Steps 1 and 2 of the mechanism consist of the usual nucleophilic addition and elimination steps. The alcohol $CH_2{=}CH(CH_2)_3{-}OH$ attacks the carbonyl carbon in Step 1 to produce the tetrahedral intermediate. A lone pair of electrons on the O in the tetrahedral intermediate folds down to re-form the C=O and expel the Cl^- leaving group. Step 3 involves a proton transfer to produce an uncharged product.

1. Nucleophilic addition
2. Nucleophile elimination
3. Proton transfer

Tetrahedral intermediate

Your Turn 21.2

Think

Using Figure 20-3 and Equation 21-4 as guides, what is the intermediate formed in Step 1 and Step 2? Where do intermediates appear on energy diagrams? Which step has the largest free energy of activation, and how does that correspond to its rate constant?

Solve

The species are added below. The tetrahedral intermediate is the product of nucleophilic addition and has a relatively high energy. The first step has the largest activation energy, $\Delta G^{\ddagger}(1)$, which is consistent with the first step (nucleophilic addition) being the slow step.

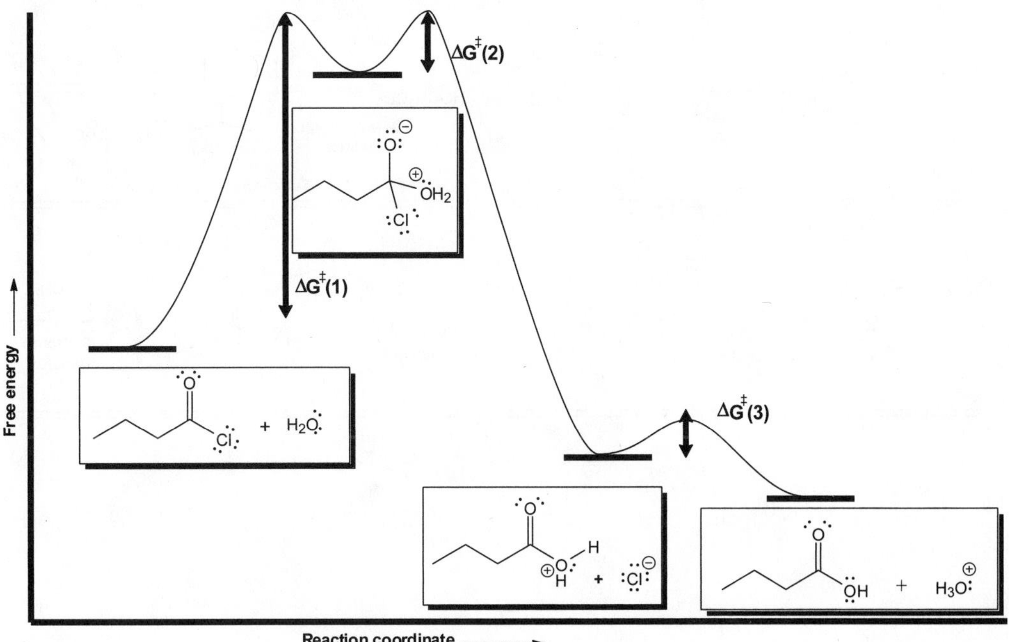

Your Turn 21.3

Think

What bonds are broken and formed in each step? Which sites are electron rich? Electron poor? How can you use curved arrows to show that electron movement? Review the 10 elementary steps from Chapters 6 and 7. In the product of the first step, is there a group that subsequently departs so that it is no longer part of the organic compound? Label this group as the leaving group.

Solve

Step 1 is an S_N2 step in which RCO_2H acts as a nucleophile to attack the electron-poor P in PCl_3 and the P–Cl bond breaks at the same time the O–P bond forms. Steps 2 and 3 make up the nucleophilic addition–elimination mechanism. The leaving group in the product of Step 1 is circled.

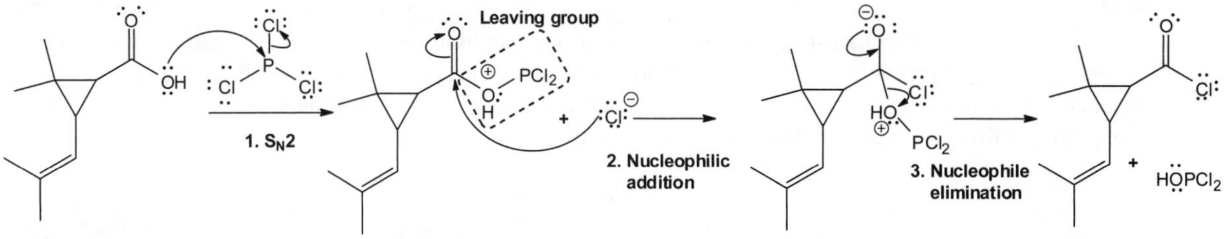

Leaving group

1. S_N2

2. Nucleophilic addition

3. Nucleophile elimination

$HOPCl_2$

Your Turn 21.4

Think

What was the overall reaction? Was the reaction base or acid catalyzed? What bonds are broken and formed in each step? Which sites are electron rich? Electron poor? How can you use curved arrows to show that electron movement? Review the 10 elementary steps from Chapters 6 and 7. Are there steps in which a nucleophile adds to a polar π bond? Are there steps in which a leaving group departs? In which steps are protons involved?

Solve

Step 1 is a proton transfer to the carboxylic acid's carbonyl group to activate the carbonyl. Step 2 is nucleophilic addition of the alcohol to yield the tetrahedral intermediate. Step 3 is a proton transfer to remove the charge from O on the incoming nucleophile. Step 4 is a proton transfer in which the singly bonded O atom of the original carboxylic acid gains a positive charge. This increases the ability of the leaving group to depart. In Step 5, the leaving group departs in a nucleophile elimination step. In Step 6, the carbonyl O is deprotonated, yielding the uncharged product.

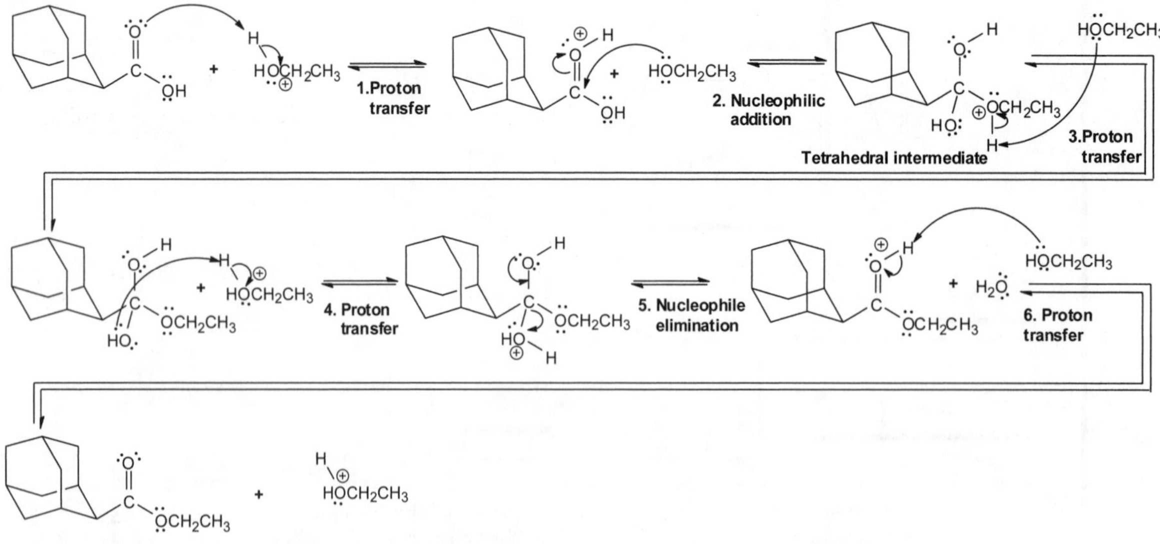

1. Proton transfer

2. Nucleophilic addition

Tetrahedral intermediate

3. Proton transfer

4. Proton transfer

5. Nucleophilic elimination

6. Proton transfer

Your Turn 21.5

Think

Review Sections 20.1 and 20.2 to consider the reactivity of an ester with an alkoxide and a hydroxide. In this example, how can the RO^- and HO^- species act as nucleophiles? As bases? Are there any steps that are irreversible?

Solve

The mechanism for transesterification (Equation 21-39) consists of the usual nucleophilic addition–elimination steps, with an RO^- nucleophile and a separate RO^- leaving group. The mechanism for saponification (Equation 21-40) begins with the same two steps, in which HO^- is the nucleophile and RO^- is the leaving group, followed by an irreversible proton transfer.

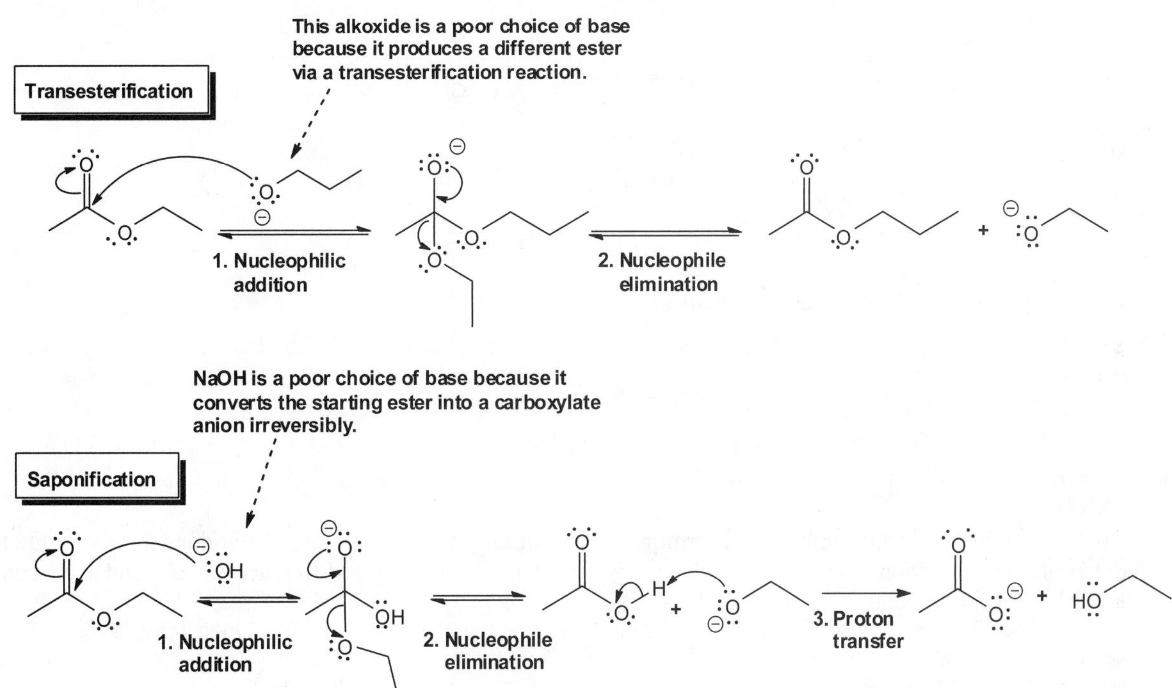

Your Turn 21.6

Think

Review the transesterification mechanism from Section 20.1. How can the $CH_3CH_2O^-$ species act as a nucleophile with a polar π bond? Is there a separate RO^- leaving group? How do these reaction conditions avoid undesired products?

Solve

$CH_3CH_2O^-$ acts as a nucleophile in Step 1, adding to the C atom of the C=O group. In Step 2, a separate $CH_3CH_2O^-$ species is kicked out by the reformation of the C=O to produce the overall product. Transesterification using ethoxide as the base avoids undesired products because the ester product of the reaction is the same as the ester starting material.

Transesterification

1. Nucleophilic addition

2. Nucleophile elimination

Your Turn 21.7

Think

In each proton transfer reaction, what is the base and what is the acid? How do you use curved arrows to show bond formation and breaking? What products form? If the product is nucleophilic, how does that pose a problem to the desired reaction?

Solve

In each case, $CH_3CH_2O^-$ is the base and picks up a proton from each of the uncharged acids. The products are different from the reactants unless ethanol is the acid. When propan-1-ol is used, the propoxide anion is produced, which can react with the ester in a transesterification reaction. When water is used, hydroxide anion is produced, which can react with the ester in a saponification reaction.

Propan-1-ol

Water

Ethanol (notice that the products are identical to the reactants)

Your Turn 21.8

Think

Consult Table 6-1 or Appendix A to determine the pK_a of an ester and a ketone. How does the pK_a value help determine the direction in which the reaction is favored? Calculate the difference in pK_a and then consult Equation 6-11 to calculate K_{eq}.

Solve

The pK_a for an ester is 25 and the pK_a for a ketone is 20, so $K_{eq} = 10^{(20 - 25)} = 10^{-5}$. The p$K_a$ of the product is lower and, therefore, the reactants are favored, so this reaction will not interfere with the intended Claisen condensation reaction.

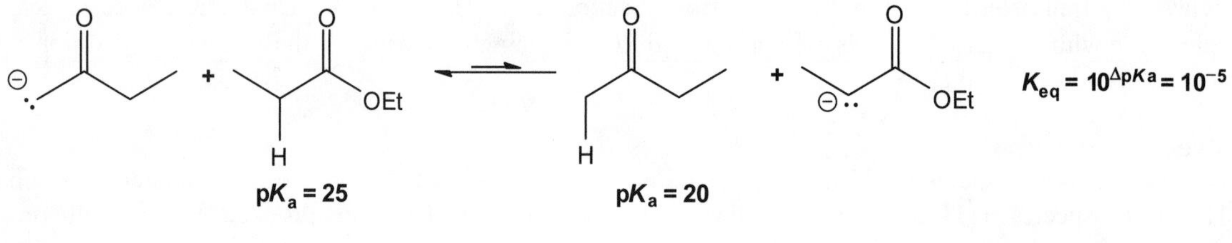

OEt OEt $K_{eq} = 10^{\Delta pKa} = 10^{-5}$

pK_a = 25 pK_a = 20

Your Turn 21.9

Think

Consult Appendix A to determine the pK_a of the diester and an alcohol. How does the value of the pK_a help determine the direction in which the reaction is favored? Calculate the difference in pK_a and then consult Equation 6-11 to calculate K_{eq}.

Solve

Malonic ester has a pK_a of 13.5. The pK_a of ethanol is 16, so $K_{eq} = 10^{(16-13.5)} = 10^{2.5} = 3 \times 10^2$, which favors products significantly.

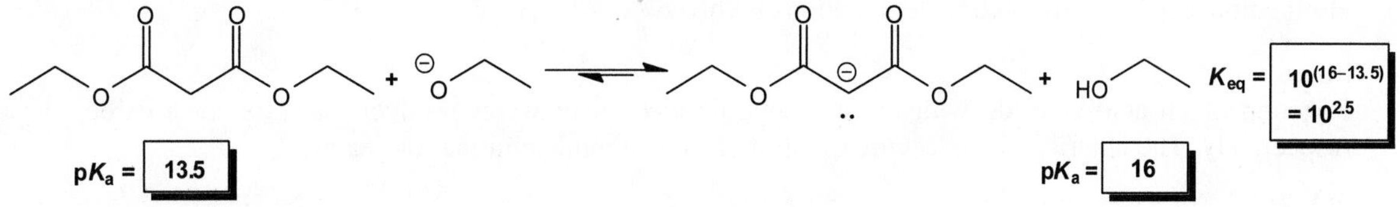

pK_a = 13.5

pK_a = 16

$K_{eq} = 10^{(16-13.5)} = 10^{2.5}$

Your Turn 21.10

Think

Is an alkyl group electron donating or electron withdrawing? What impact does that have on a nearby negative charge?

Solve

The alkyl group is electron donating, so it increases the concentration of negative charge on the nearby carbon, which destabilizes that anion.

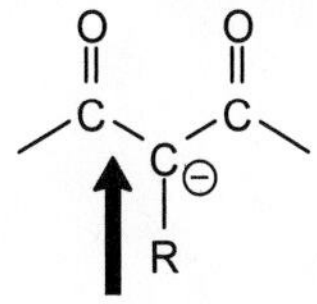

Chapter Problems

Problem 21.1

Think

What acts as the nucleophile in the nucleophilic addition step? What is the leaving group in the nucleophile elimination step? Are proton transfer reactions involved?

Solve

Reaction of an acid chloride with the nucleophilic alcohol or water produces an ester or a carboxylic acid, respectively. The chloride is the leaving group in the nucleophile elimination reaction.

(a) Nucleophilic addition → Nucleophile elimination → Proton transfer

(b) Nucleophilic addition → Nucleophile elimination → Proton transfer

Problem 21.2

Think

Using Figure 20-3 and Equation 21-4 as guides, what is the intermediate formed in Step 1 and Step 2? Where do intermediates appear on energy diagrams? Which step has the largest free energy of activation, and how does that correspond to its rate constant?

Solve

The species are added below. The tetrahedral intermediate is the product of nucleophilic addition and has a relatively high energy. The first step has the largest activation energy, $\Delta G^{\ddagger}(1)$, which is consistent with the first step (nucleophilic addition) being the slow step.

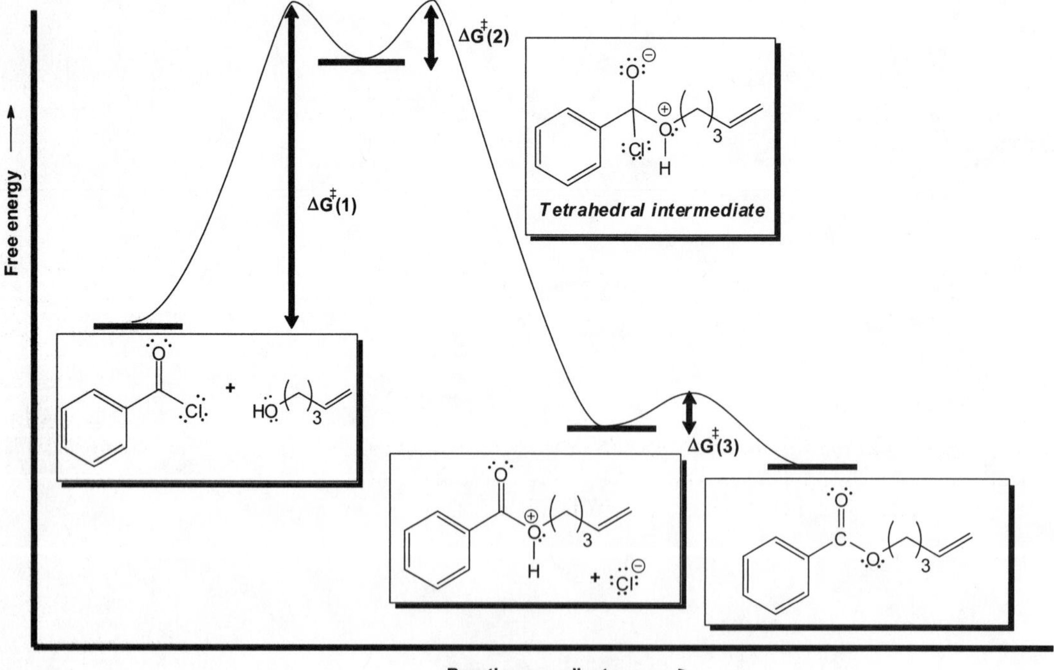

Problem 21.3

Think

What acts as the nucleophile in the nucleophilic addition step? What is the leaving group in the nucleophile elimination step? Are proton transfer reactions involved?

Solve

Reaction of an acid anhydride with a nucleophilic alcohol produces an ester and a carboxylic acid, and reaction of an acid anhydride with the nucleophile water produces two carboxylic acids. In each case, a carboxylate anion is the leaving group in the nucleophile elimination step.

The mechanism for Equation 21-5 is shown below:

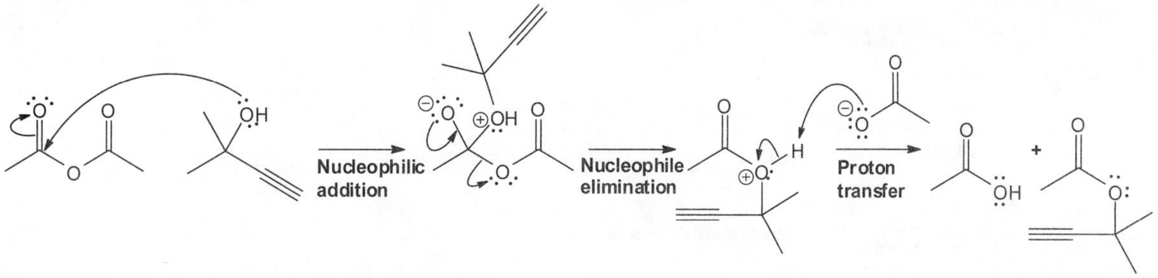

Nucleophilic addition → Nucleophile elimination → Proton transfer

The mechanism for Equation 21-6 is shown below:

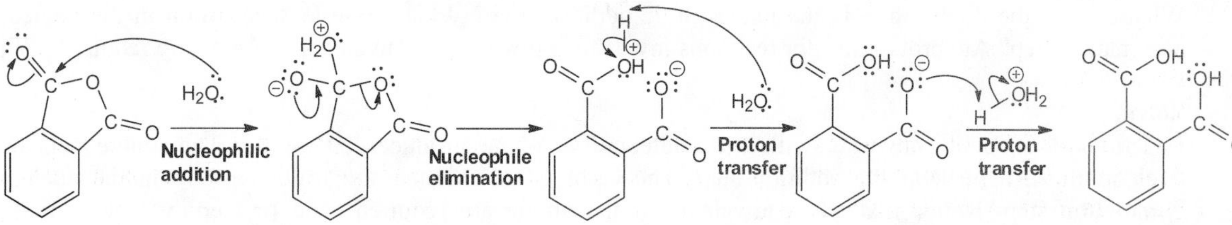

Nucleophilic addition → Nucleophile elimination → Proton transfer → Proton transfer

Problem 21.5

Think

How do resonance and inductive effects play a role in the reactivity of carboxylic acid derivatives? Are the carbonyl groups stabilized differently by resonance? How do inductive effects alter the concentration of partial positive charge on the carbonyl C?

Solve

(a) **A** undergoes hydrolysis more quickly. The difference is whether a CH_3 or CF_3 is attached to the carbonyl carbon. A CF_3 group is inductively electron withdrawing and thus increases the concentration of positive charge on the carbonyl carbon.

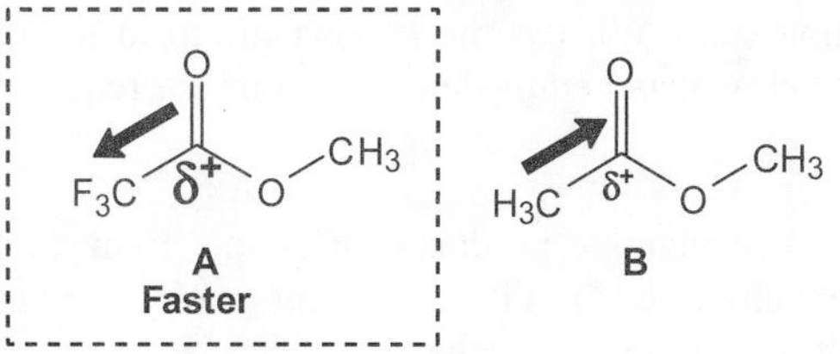

A
Faster

B

(b) **D** undergoes hydrolysis more quickly. The difference is whether a benzene ring or cyclohexane ring is attached to the carbonyl carbon. The benzene ring stabilizes the carbonyl group via resonance, as shown on the next page, which makes the acid derivative less reactive.

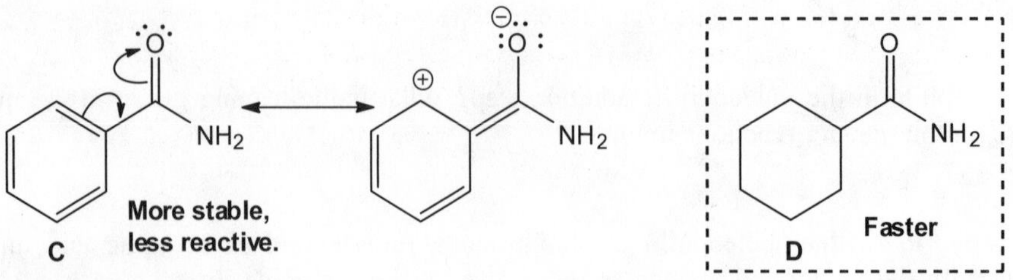

More stable, less reactive.

C

Faster

D

(c) **E** undergoes hydrolysis more quickly. Each RO– group stabilizes the carbonyl group via resonance, decreasing reactivity. Because the derivative on the right has two RO– groups attached to the carbonyl carbon, it is less reactive.

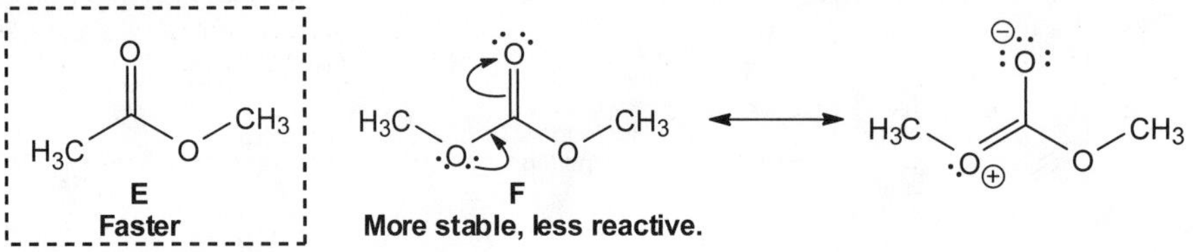

E Faster

F More stable, less reactive.

Problem 21.6

Think

What acts as the nucleophile in the nucleophilic addition step? What is the leaving group in the nucleophile elimination step? Are proton transfer reactions involved? How many equivalents of amine are required? Why?

Solve

Reaction of an acid anhydride with the nucleophilic amine produces an amide. The amine acts as the nucleophile in a nucleophilic addition step. The carboxylate anion is the leaving group in the nucleophile elimination step. Notice that two equivalents of the amine are required. The first equivalent is used as a nucleophile and the second is used as a base.

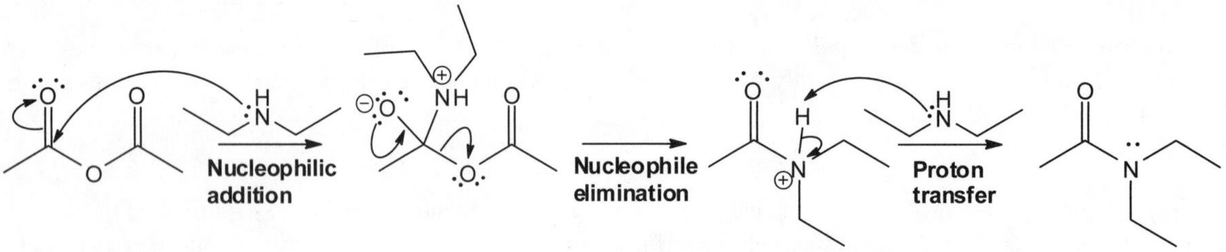

Nucleophilic addition → Nucleophile elimination → Proton transfer

Problem 21.7

Think

What acts as the nucleophile in the nucleophilic addition step? What is the leaving group in the nucleophile elimination step? Are proton transfer reactions involved? How many equivalents of amine are required? Why?

Solve

(a) Reaction of an acid chloride with the nucleophilic benzylamine produces an amide. Benzylamine is the nucleophile in the nucleophilic addition step, and the chloride anion is the leaving group in the nucleophile elimination step. Notice that triethylamine is present, which serves as a base.

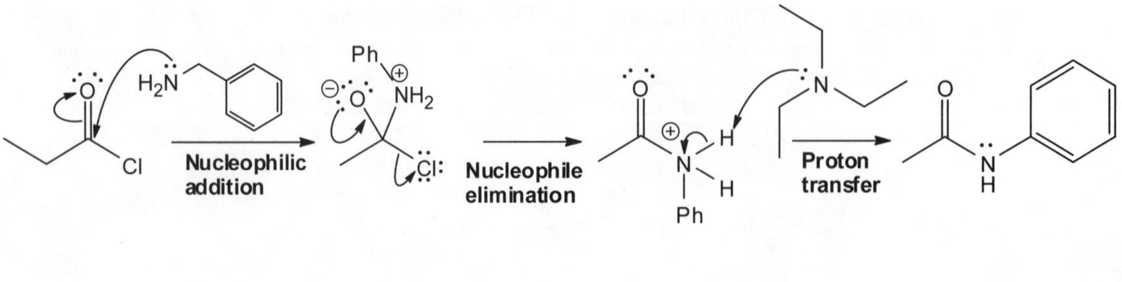

Nucleophilic addition → Nucleophile elimination → Proton transfer

(b) Reaction of an acid anhydride with nucleophilic ammonia produces an amide and a carboxylic acid. Ammonia is the nucleophile in the nucleophilic addition step, and the carboxylate anion is the leaving group in the nucleophile elimination step. Notice that two equivalents of ammonia are required. The first equivalent is used as a nucleophile and the second is used as a base.

Nucleophilic addition

Nucleophile elimination

Proton transfer

Proton transfer

Problem 21.9

Think

What reactions do we know that transform an acid derivative lower on the stability ladder to one higher on the ladder? Can this transformation be done in a single step? Is there an acid chloride that can be used, which is different from the one in Solved Problem 21.8?

Solve

Since $SOCl_2$ can transform any carboxylic acid derivative into another, we can treat acetyl chloride with benzoate to form the target compound. The precursor for benzoate is benzamide.

Undo an acyl substitution.

Undo a hydrolysis.

In the forward direction the reaction would proceed as follows:

NaOH

H_2O, Δ

Problem 21.10

Think

What is the structure of butanoic acid? What new functional group is added to the α carbon? From what intermediate did this functional group originate?

Solve

All reactions require the Hell–Volhard–Zelinsky (HVZ) reaction to produce the α-bromo acid intermediate. From there, different choices of nucleophiles produce the different targets.

(a)

1. Br_2, PCl_3
2. H_2O

NH_3

OH OH OH Br NH_2

(b)

1. Br_2, PCl_3
2. H_2O

CH_3NH_2

OH OH OH Br HN CH_3

(c)

1. Br_2, PCl_3
2. H_2O

1. NaOH
2. H_3O^+

OH OH OH Br OH

Problem 21.12

Think

Is there a good leaving group on the starting organic compound? If not, how can a good leaving group be formed? What functional group is formed in the target compound? Is this the Z or E isomer? How is this product different from the product in Solved Problem 21.11? What is the stereochemistry associated with each of the above reactions?

Solve

The same strategy can be used here as was used in Solved Problem 21.11, in which the OH must first be converted to a good leaving group, followed by an E2 reaction. However, the target here has the opposite configuration about the C=C bond than does the target in the Solved Problem. Thus, the substrate from which the E2 product is produced must have the opposite configuration at the carbon bonded to the leaving group (see Chapter 8). That substrate can be produced by treating the starting material with PBr_3 instead of TsCl, given that PBr_3 leads to inversion of stereochemistry.

OH CH_3 PBr_3 Br H CH_3 $NaOC(CH_3)_3$ (Z) CH_3

Inversion of stereochemistry

Z isomer formed.

Problem 21.13

Think

What functional group is the ester hydrolyzed to? What acts as the nucleophile in the nucleophilic addition step? What is the leaving group in the nucleophile elimination step? Is the reaction conducted in basic or acidic conditions? What types of species should not appear in a mechanism in this medium?

Solve

The ester is hydrolyzed to the carboxylic acid in acidic conditions, thus no strong bases should appear in the mechanism.

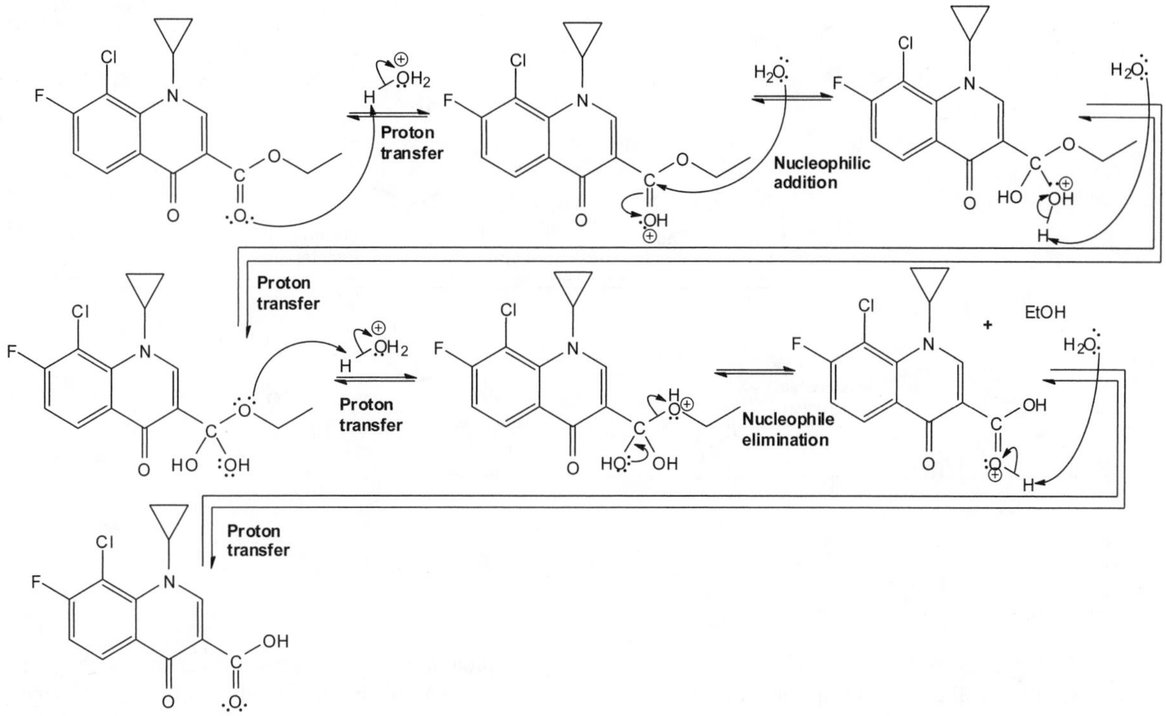

Problem 21.14

Think

What functional group is produced? What acts as the nucleophile in the nucleophilic addition step? What is the leaving group in the nucleophile elimination step? Is the reaction conducted in basic or acidic conditions? What species should not appear in a mechanism in this medium?

Solve

All of these reactions are conducted in acidic conditions, thus no strong bases should appear in the mechanism.

(a) Water acts as the nucleophile in this acid-catalyzed nucleophilic addition-elimination mechanism. The alcohol is the leaving group (formed after proton transfer) in the nucleophile elimination step. The carboxylic acid functional group is formed.

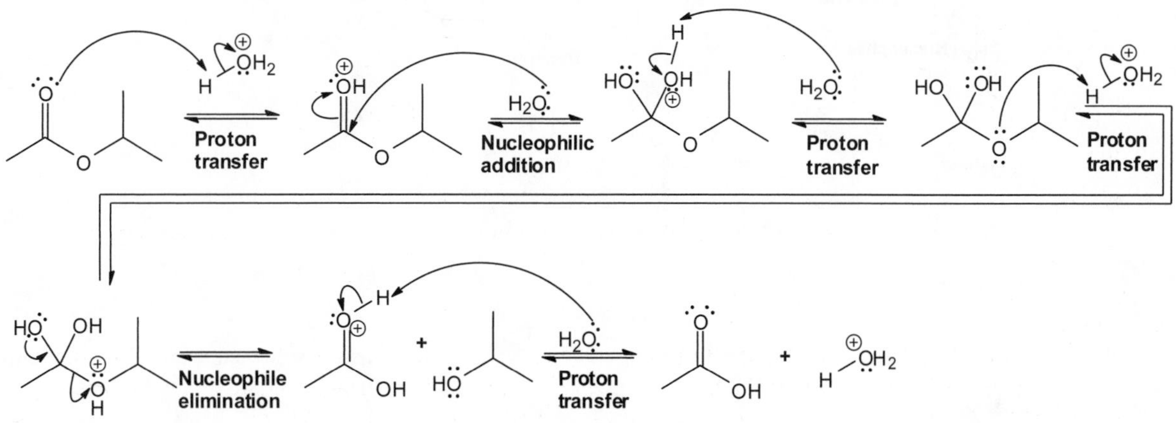

(b) Cyclohexanol acts as the nucleophile in this acid-catalyzed nucleophilic addition–elimination mechanism. The alcohol is the leaving group (formed after proton transfer) in the nucleophile elimination step. A new ester functional group is formed.

Proton transfer

Nucleophilic addition

Proton transfer

Proton transfer

Nucleophile elimination

Proton transfer

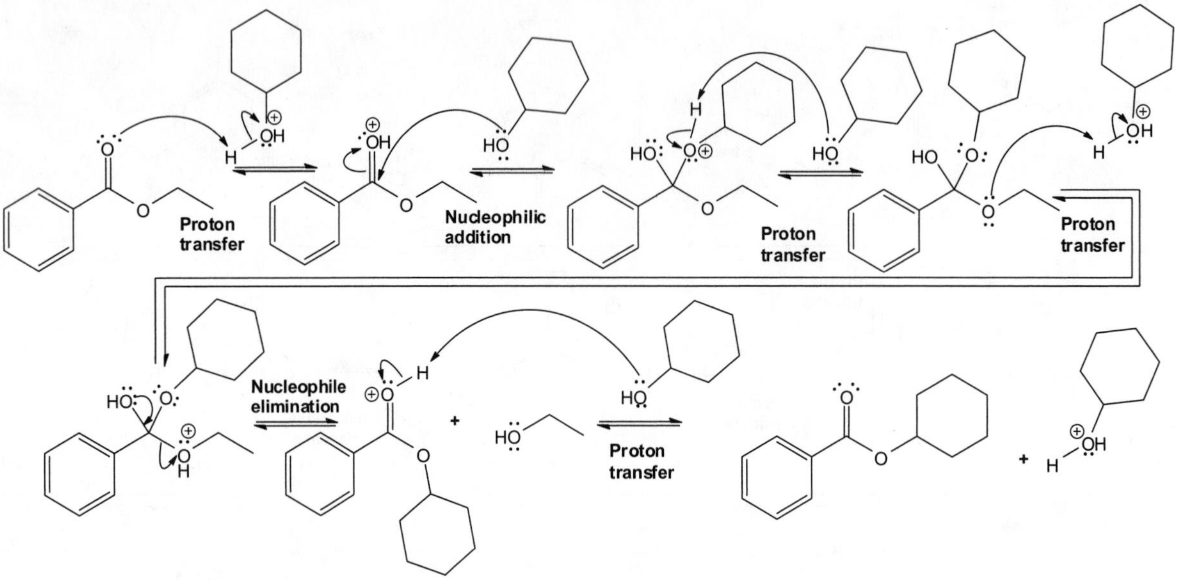

(c) Water acts as the nucleophile in this acid-catalyzed nucleophilic addition–elimination mechanism. The amine is the leaving group (formed after proton transfer) in the nucleophile elimination step. The carboxylic acid functional group is formed.

Proton transfer

Nucleophilic addition

Proton transfer

Proton transfer

Nucleophile elimination

Proton transfer

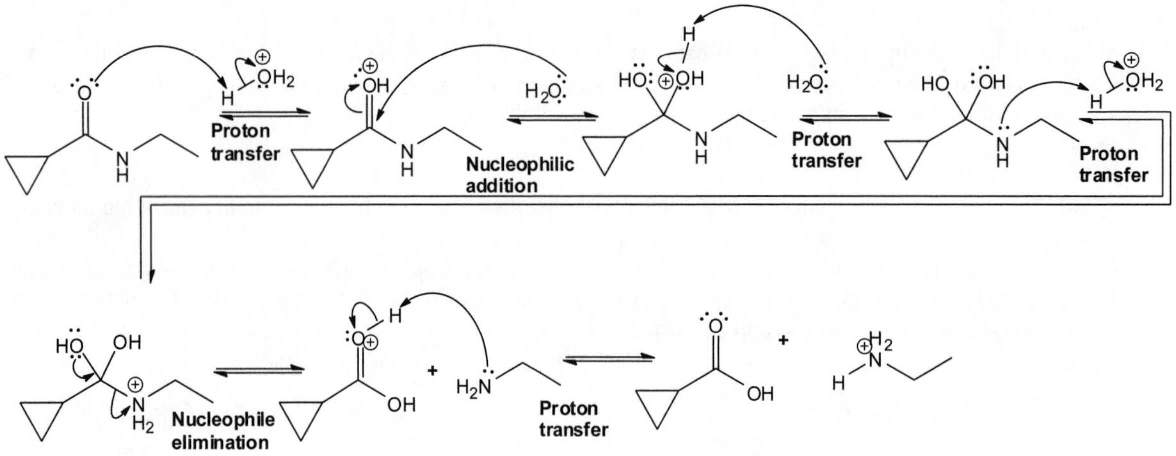

(d) Water acts as the nucleophile in this acid-catalyzed nucleophilic addition–elimination mechanism. The alcohol is the leaving group (formed after proton transfer) in the nucleophile elimination step. The carboxylic acid functional group is formed.

Proton transfer
Nucleophilic addition
Proton transfer
Proton transfer
Nucleophile elimination
Proton transfer

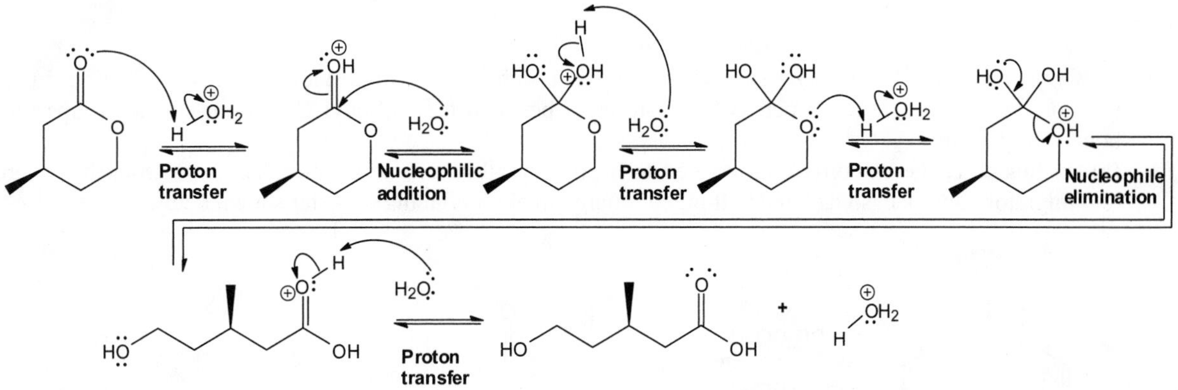

(e) Ethanol acts as the nucleophile in this acid-catalyzed nucleophilic addition–elimination mechanism. Water is the leaving group (formed after proton transfer) in the nucleophile elimination step. The ester functional group is formed.

Proton transfer
Nucleophilic addition
Proton transfer
Proton transfer
Nucleophile elimination
Proton transfer

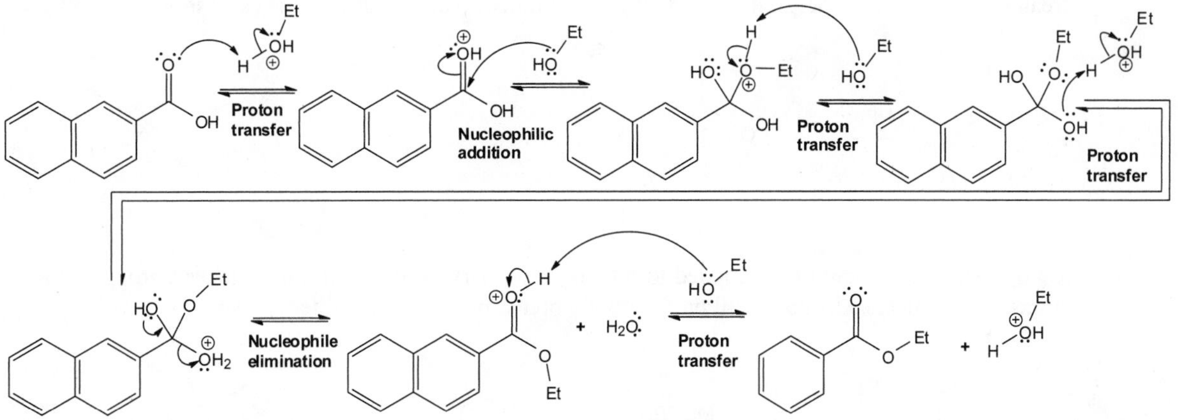

(f) Water acts as the nucleophile in this acid-catalyzed nucleophilic addition–elimination mechanism. The amine is the leaving group (formed after proton transfer) in the nucleophile elimination step. The carboxylic acid functional group is formed.

Proton transfer
Nucleophilic addition
Proton transfer
Proton transfer
Nucleophile elimination
Proton transfer

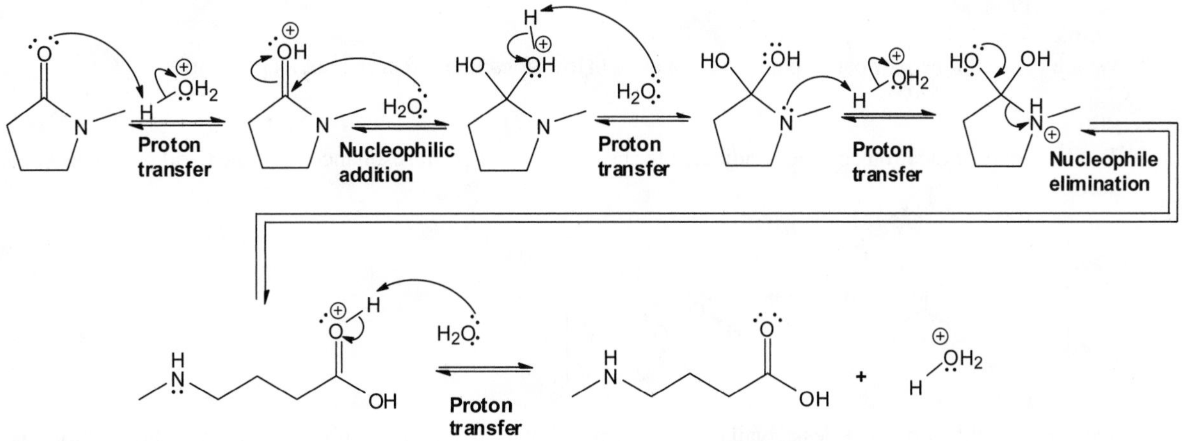

Problem 21.16

Think

What type of reagent is CF_3CO_3H or MCPBA? How does it tend to react with a ketone? Does one side of the ketone favor reaction over the other? If so, which side?

Solve

CF_3CO_3H and MCPBA are peroxyacids and will react with a ketone in a Baeyer–Villiger oxidation. In such a reaction, an O atom from the peroxyacid is inserted between the carbonyl C and one of the groups initially bonded to the carbonyl C.

(a) In this case, the carbonyl C is bonded to a primary alkyl and aryl group. The aryl group has a greater migratory aptitude, so its bond will preferentially break, producing the ester shown below:

Aryl group

CF_3CO_3H

CH_2Cl_2

1° Alkyl group

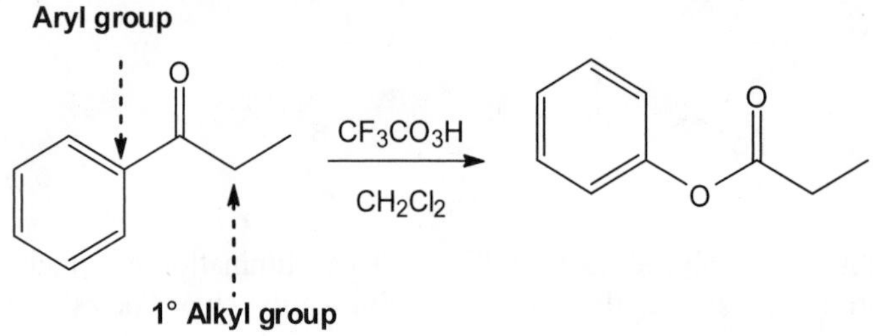

(b) In this case, the carbonyl C is bonded to a primary and tertiary alkyl group. The tertiary alkyl group has a greater migratory aptitude, so its bond will preferentially break, producing the ester shown below:

CF_3CO_3H

CH_2Cl_2

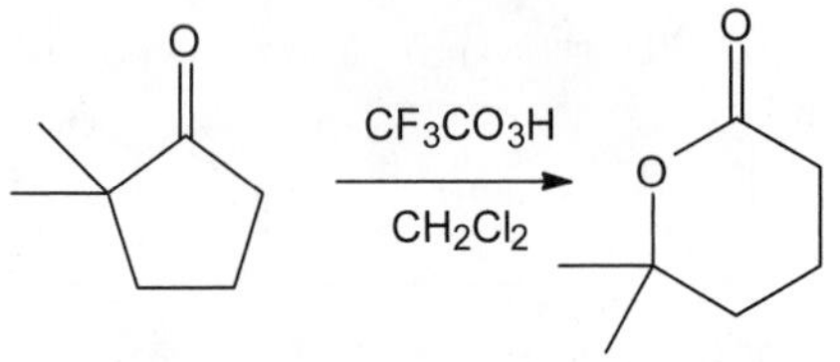

(c) In this case, the carbonyl C is bonded to a benzyl (primary) and aryl group. The aryl group has a greater migratory aptitude, so its bond will preferentially break, producing the ester shown below:

MCPBA

CF_3SO_3H, CH_2Cl_2

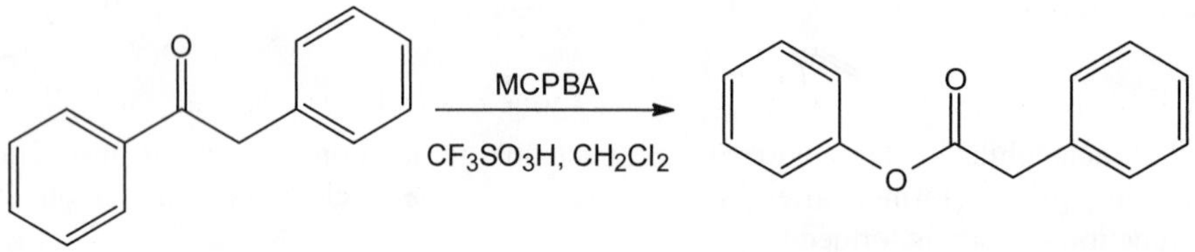

Problem 21.17

Think

Which oxygen lone pair participates in resonance? How does this affect the nucleophile strength?

Solve

The lone pair of electrons on the O adjacent to the C=O is tied up in resonance with the C=O group, as shown here:

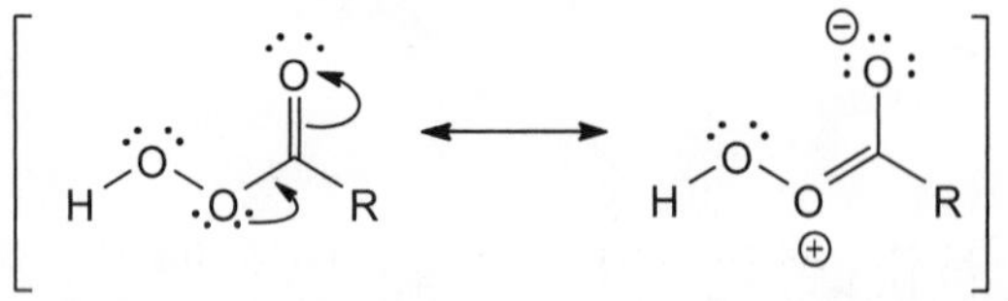

This makes the lone pair less available to form a bond to an electrophile, such that the O atom is less nucleophilic.

Problem 21.18

Think

How does the strong base, CH_3O^-, react with an ester? What product is formed? What acts as the nucleophile in the nucleophilic addition step? What is the leaving group in the nucleophile elimination step? Is the reaction conducted in basic or acidic conditions? What species should not appear in the mechanism for a reaction in this medium? Are any steps irreversible? Why is an acid workup necessary?

Solve

The reaction is conducted under basic conditions, so strong acids should not appear in the mechanism. The base, CH_3O^-, deprotonates the α C–H to form the enolate anion. The enolate anion acts as the nucleophile and attacks another equivalent of ester. The nucleophile elimination step forms the β-ketoester. However, all of these steps are reversible. CH_3O^- then deprotonates the α C–H to form the enolate anion in an irreversible step. The enolate anion is protonated in the acid workup to re-form the β-ketoester.

Proton transfer

Nucleophilic addition

Nucleophile elimination

Proton transfer

Irreversible

Add acid

Proton transfer

Problem 21.20

Think

Which C–C bond in the β-keto ester product would have been formed in a Claisen condensation? Which C atom would have been bonded to the alkyoxy leaving group?

Solve

The new C–C bond is the one between the α and β carbons, as shown in the following structure. Moreover, the leaving group would have been attached to the β C atom.

An alkoxy leaving group would have been attached to this C.

This is the C—C bond that would have formed in a Claisen condensation.

Therefore, we can apply a transform that undoes a Claisen condensation by disconnecting that C–C bond and reattaching the leaving group. Doing so yields two ester precursors that are identical.

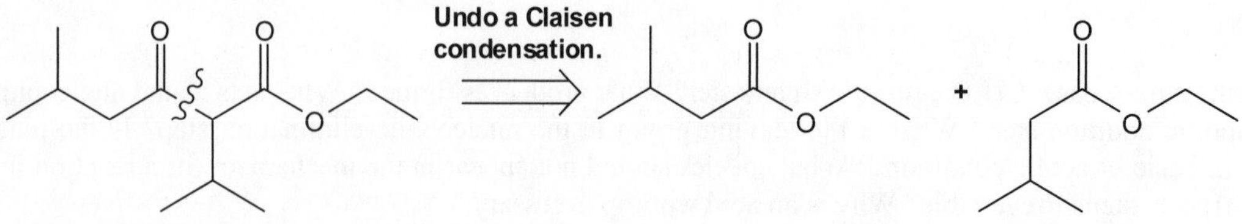

Undo a Claisen condensation.

The reaction in the forward direction would proceed as follows:

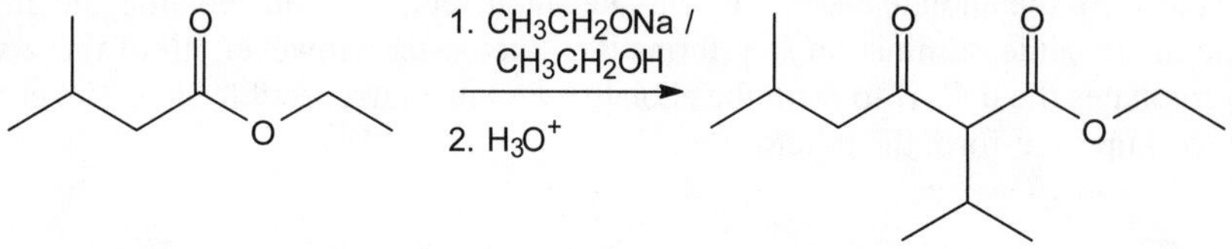

1. CH_3CH_2ONa / CH_3CH_2OH
2. H_3O^+

Problem 21.22

Think

What is the leaving group on the ester? What choice of base would ensure that a nucleophilic addition–elimination would leave us with the same ester? What solvent could be used so that, if it were deprotonated, the base that would form would be the same as the base that is added?

Solve

The appropriate base should be the same as the leaving group of the ester. The appropriate solvent is the conjugate acid of that base. The Claisen condensation product is shown.

(a) CH_3ONa / CH_3OH

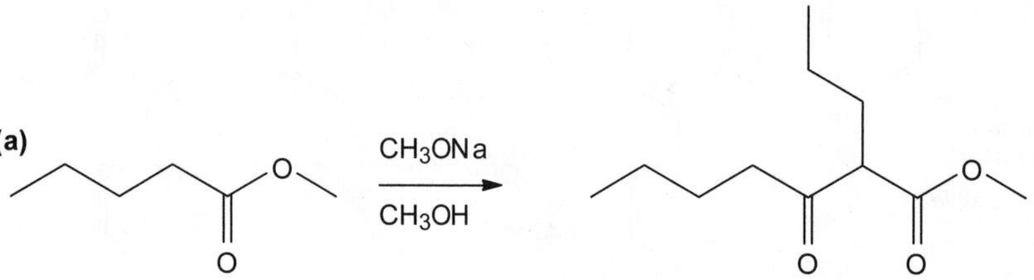

(b) ONa / OH

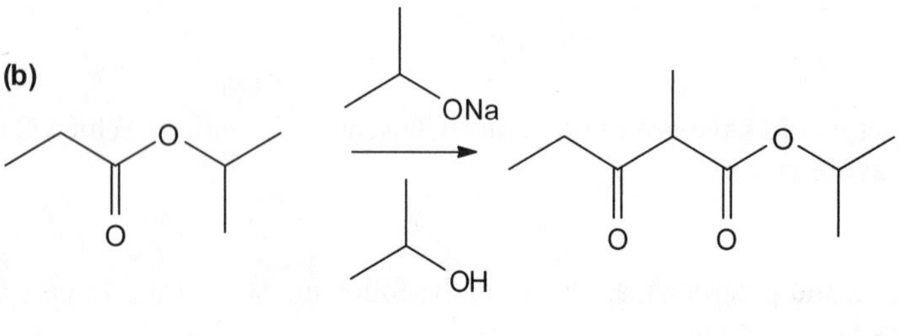

Problem 21.23

Think

Is $(CH_3)_3CO^-$ a good nucleophile? How does the nucleophile strength of the base affect the outcome of the reaction?

Solve

$(CH_3)_3CO^-$ is a relatively poor nucleophile because of the bulkiness provided by the three methyl groups. Therefore, it doesn't attack the ester's carbonyl group to initiate a transesterification.

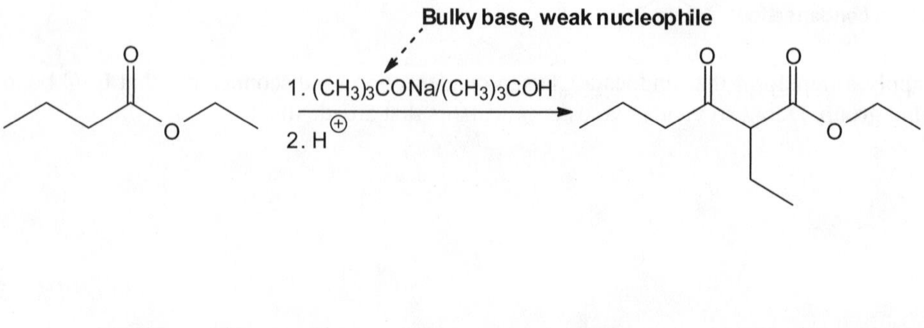

Bulky base, weak nucleophile

1. $(CH_3)_3CONa/(CH_3)_3COH$
2. $H^{\oplus}$

Problem 21.24

How does the strong base, CH_3O^-, react with each ester? How many nucleophile and electrophile choices are possible? What is the leaving group in the nucleophile elimination step?

Solve

The reaction is conducted under basic conditions, so strong acids should not appear in the mechanism. The base, CH_3O^-, deprotonates each ester's α C–H to form the enolate anion. The enolate anion acts as the nucleophile and attacks another equivalent of ester. The nucleophile elimination step forms the β-ketoester. All of these steps are reversible, but CH_3O^- then deprotonates the α C–H to form the enolate anion in an irreversible step. The enolate anion is protonated to re-form the β-ketoester.

There are two possible ester compounds to form two possible enolate anions (nucleophile) and two possible electrophilic C=O compounds. Therefore, four products are possible. The mechanisms are shown below:

Claisen condensation: Methyl acetate (enolate) + methyl acetate

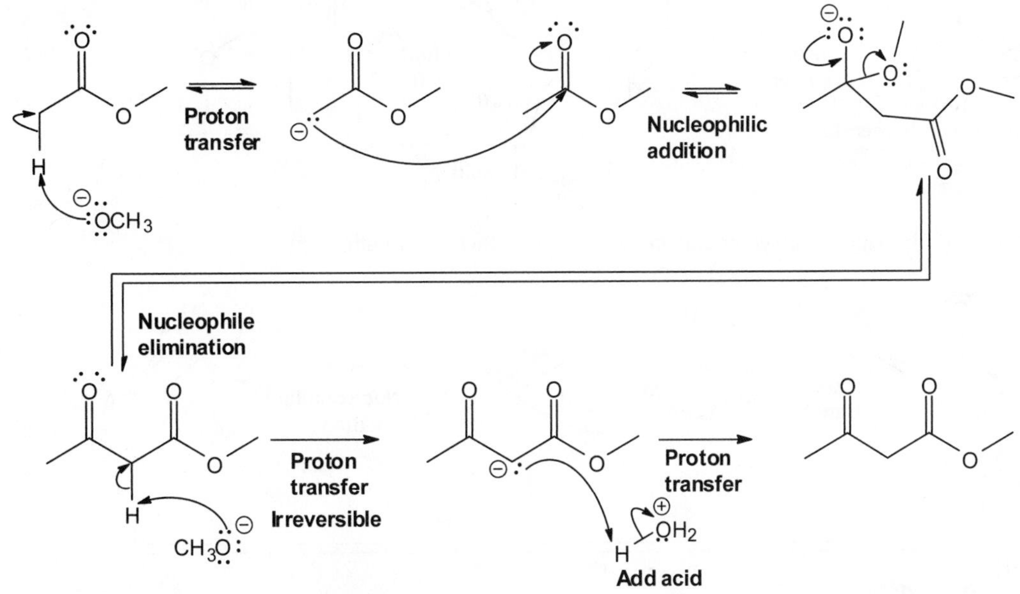

Proton transfer

Nucleophilic addition

Nucleophile elimination

Proton transfer

Irreversible

Proton transfer

Add acid

Claisen condensation: Methyl acetate (enolate) + methyl propionate

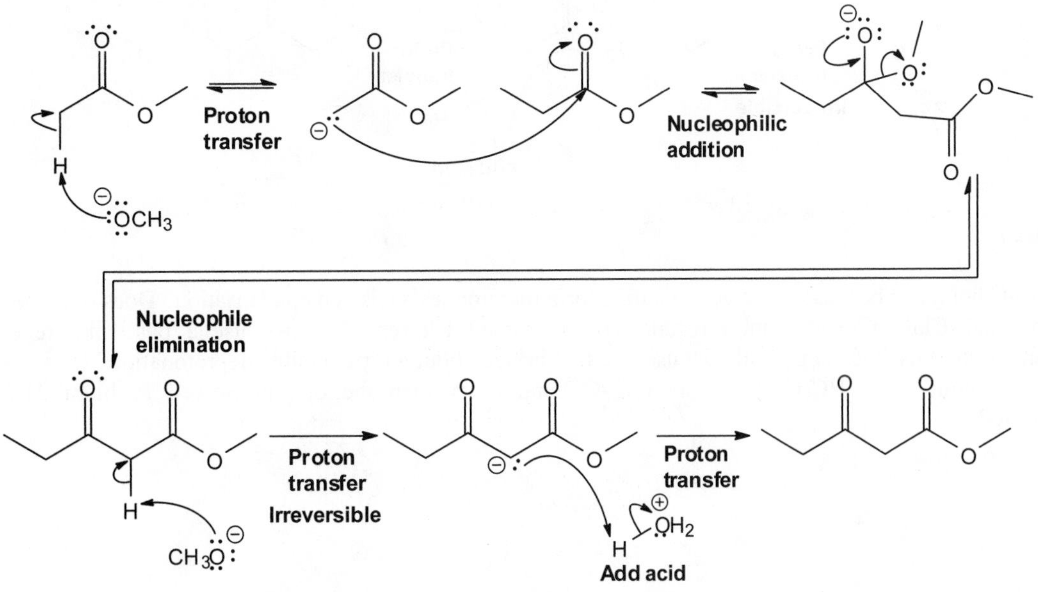

Proton transfer

Nucleophilic addition

Nucleophile elimination

Proton transfer

Irreversible

Proton transfer

Add acid

Claisen condensation: Methyl propionate (enolate) + methyl acetate

Proton transfer

Nucleophilic addition

Nucleophile elimination

Proton transfer

Irreversible

Proton transfer

Add acid

Claisen condensation: Methyl propionate (enolate) + methyl propionate

Proton transfer

Nucleophilic addition

Nucleophile elimination

Proton transfer

Irreversible

Proton transfer

Add acid

Problem 21.26

Think

Which C–C bond can be "disconnected" in a transform that undoes a Claisen condensation? Does this transform suggest a self-Claisen condensation reaction or a crossed Claisen? Can we use a base that reversibly deprotonates an α hydrogen or should we use one that brings about an irreversible deprotonation? How can this be done without using HC(O)OR? Can a C–C bond other than the one in Solved Problem 21.25 be disconnected?

Solve

In applying a transform that reverses a Claisen condensation, one precursor will be an ester and the other will be a ketone. Instead of the ketone and ester precursors in Solved Problem 21.25, we can use the ones below:

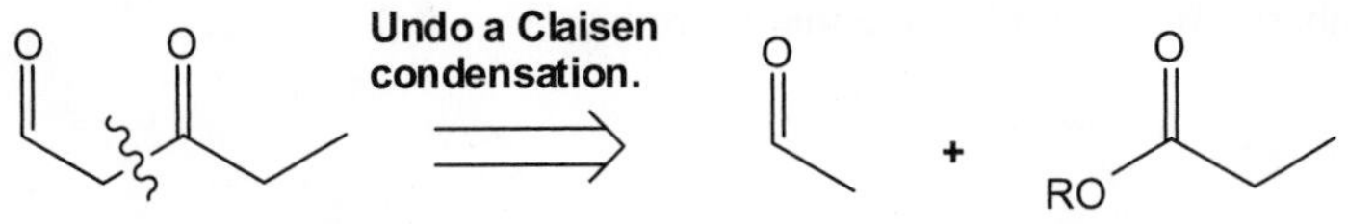

Undo a Claisen condensation.

+

RO

The synthesis in the forward direction would then appear as follows:

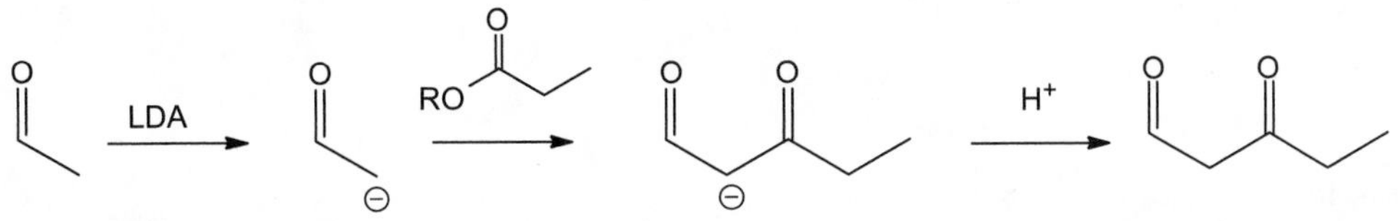

LDA

RO

H^+

Problem 21.27

Think

What are the elementary steps for a Claisen condensation reaction? How is this reaction similar/different? What acts as the nucleophile in the nucleophilic addition step? What is the leaving group in the nucleophile elimination step? Is the reaction conducted in basic or acidic conditions? What species should not appear in the mechanism for a reaction in this medium?

Solve

This is an example of a Dieckmann condensation, which is an intramolecular Claisen condensation. The mechanism is essentially the same as that shown in Equation 21-37.

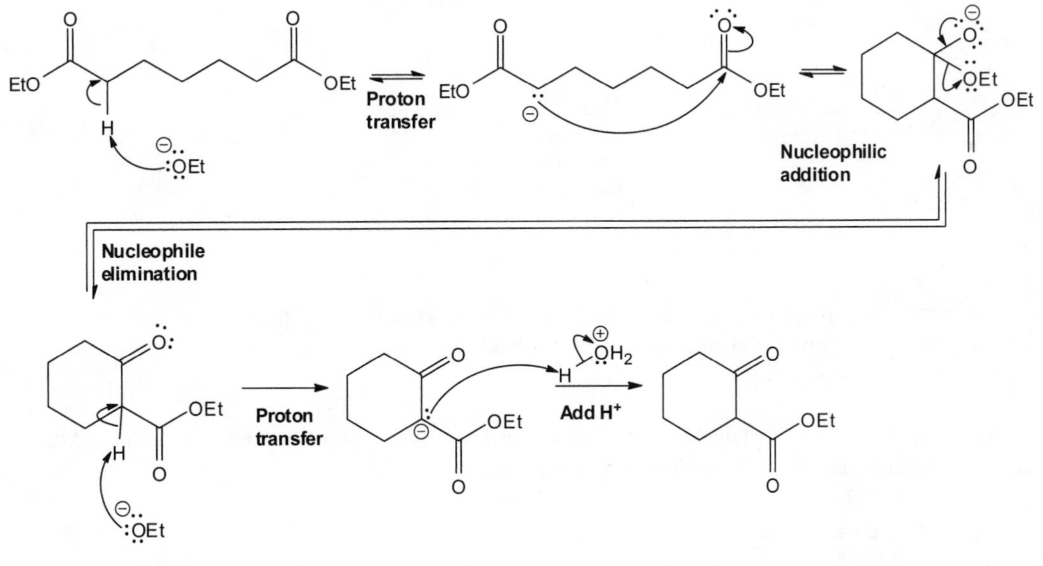

EtO OEt H :OEt **Proton transfer** EtO OEt **Nucleophilic addition** OEt OEt

Nucleophile elimination

OEt H :OEt **Proton transfer** OEt H OH_2 **Add H^+** OEt

Problem 21.28

Think

Identify the two functional groups present. Which α C–H is more acidic? Which carbon forms the enolate? Which carbon is the electrophile?

Solve

The ketone's α C–H is more acidic compared to the ester's. Thus the α C–H of the ketone will be deprotonated and form the enolate anion. Thus an intramolecular nucleophilic addition–elimination mechanism follows. There are two distinct ketone α C–H atoms and thus two possible enolates. Deprotonation of the α C–H on the left leads to the formation of the six-membered ring, which is the major product.

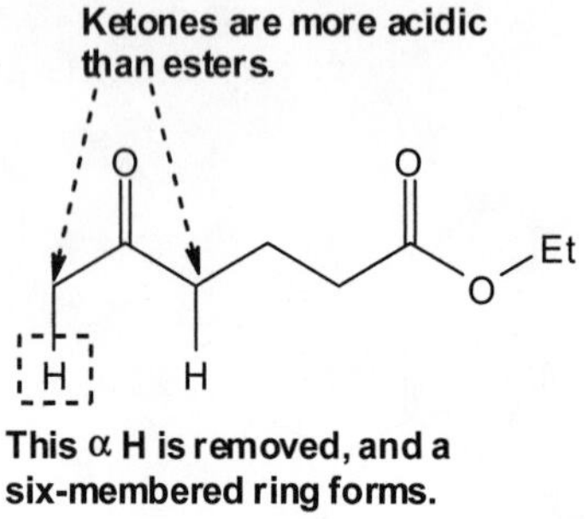

Ketones are more acidic than esters.

This α H is removed, and a six-membered ring forms.

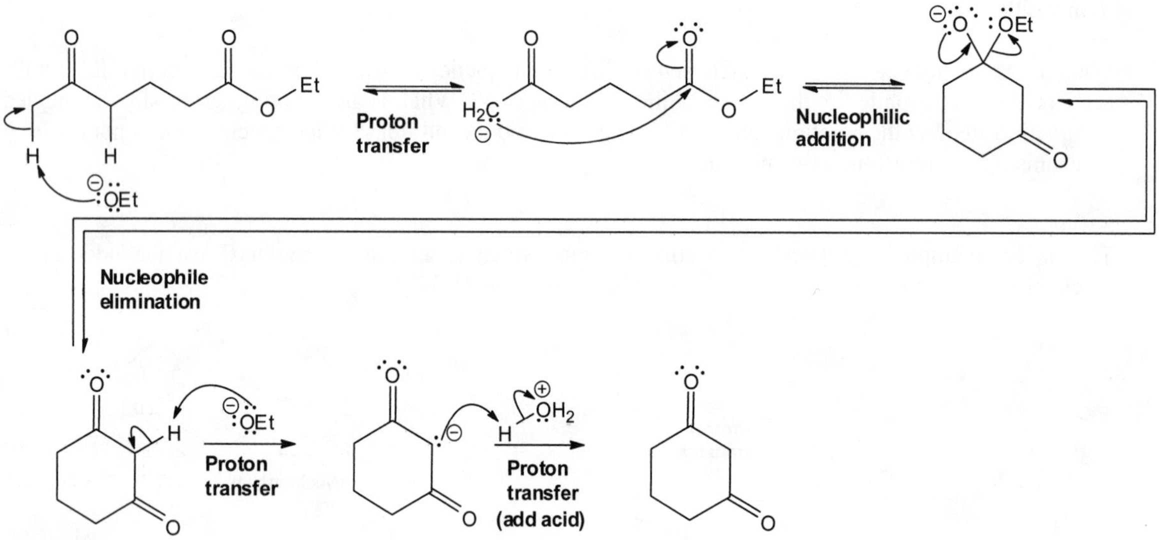

Proton transfer

Nucleophilic addition

Nucleophile elimination

Proton transfer

Proton transfer (add acid)

Problem 21.29

Think

What was the new C–C bond that formed? What functional groups were present in the original acyclic dicarbonyl? Where should those functional groups be put back?

Solve

To apply a transform that undoes a Dieckmann condensation, we must disconnect the C–C bond indicated. The original compound consisted of an aldehyde and an ester.

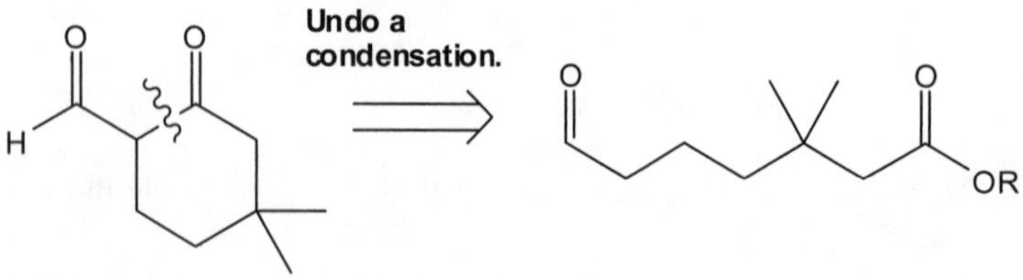

Undo a condensation.

Problem 21.31

Think

What would the precursor look like before decarboxylation? Which precursor, acetoacetic ester or malonic ester, will leave the carboxylic acid functional group in the target? What R group is added to the α carbon?

Solve

A malonic ester synthesis results in an α-substituted carboxylic acid because it has two ester groups. Both esters are hydrolyzed to the carboxylic acid, but only one undergoes decarboxylation. The retrosynthesis is given below:

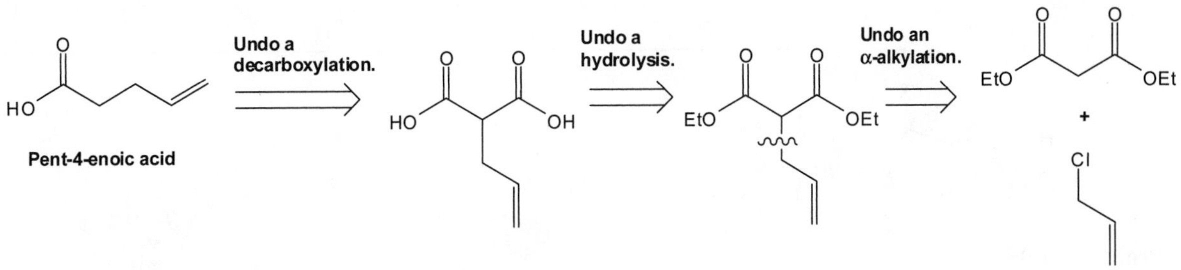

The synthesis in the forward direction would be written as follows:

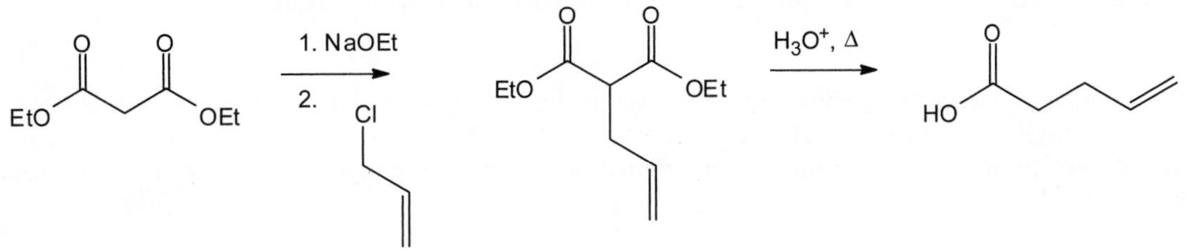

Problem 21.32

Think

What is the structure of 2-ethylhexanoic acid? Does an α-alkylated acetic acid call for a malonic ester synthesis or an acetoacetic ester synthesis? How many α-alkylation reactions occurred? What base is needed for the second alkylation?

Solve

The target is a dialkylated acetic acid, so we can begin with malonic ester. In the second alkylation, we use a stronger alkoxide base than in the first alkylation, because the acidity of the α carbon decreases with the presence of an additional alkyl group.

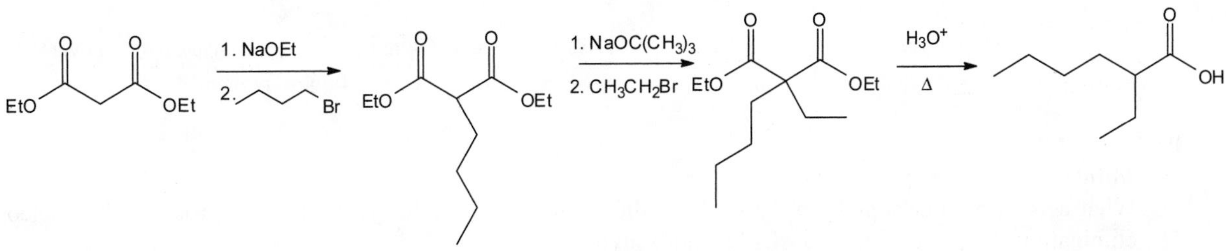

Problem 21.34

Think

Without considering undesired side reactions, how would you normally carry out this synthesis? Under the conditions for that synthesis, are there any functional groups that would undergo an undesired side reaction? If so, how can that functional group be protected?

Solve

The amine and the alcohol are subject to oxidation in the presence of a strong oxidizing agent. The amine therefore has to be protected in order to only oxidize the alcohol to the carboxylic acid. Refer to Table 21-2 Entry 4 for the protection of an amine.

HO, NH_2; 1. Cl (acyl chloride), 2. $H_3O^{\oplus}$; HO, NH; $K_2Cr_2O_7$, H_3O^+; HO, NH; $H_3O^{\oplus}$, Δ; HO, NH_2

Problem 21.35

Think

Are there certain sequences in the peptides that are the same? Can you assume those identical sequences came from the same portion of the original protein? How can you piece together the sequence?

Solve

There are certain sequences in each peptide with overlapping amino acids—specifically, the Asp-Ser sequence in **A** and **C**, the Pro-Asn in **C** and **D**, and the Ile-Val-Met-Pro-Val in **D** and **B**. You can assume that these sequences come from the same portions of the original protein and piece together the overall sequence, shown below:

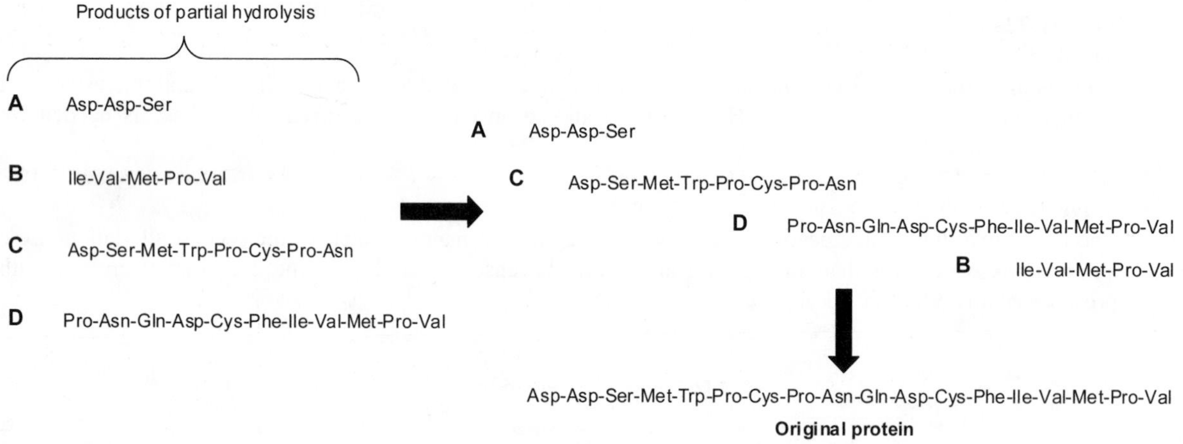

Problem 21.36

Think

What acts as the nucleophile in the nucleophilic addition step? What is the leaving group in the nucleophile elimination step? Are proton transfer reactions involved?

Solve

In each of these reactions, the Cl^- acts as the leaving group in each nucleophile elimination step. The nucleophiles in the nucleophilic addition step differ depending on what functional group is formed. All of these reactions require a proton transfer step to form the uncharged product in the final step. See the mechanisms for **(a)**–**(b)** on the next page.

(a) The carboxylic acid formation requires water as the nucleophile in the nucleophilic addition step and water acts as the base in the proton transfer step.

H_2O
Nucleophilic addition
Nucleophile elimination
H_2O
Proton transfer

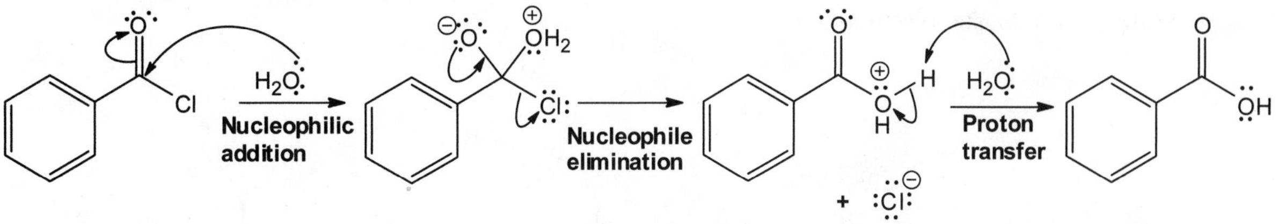

(b) The ester formation requires isopropanol as the nucleophile in the nucleophilic addition step and the alcohol acts as the base in the proton transfer step.

Nucleophilic addition
Nucleophile elimination
Proton transfer

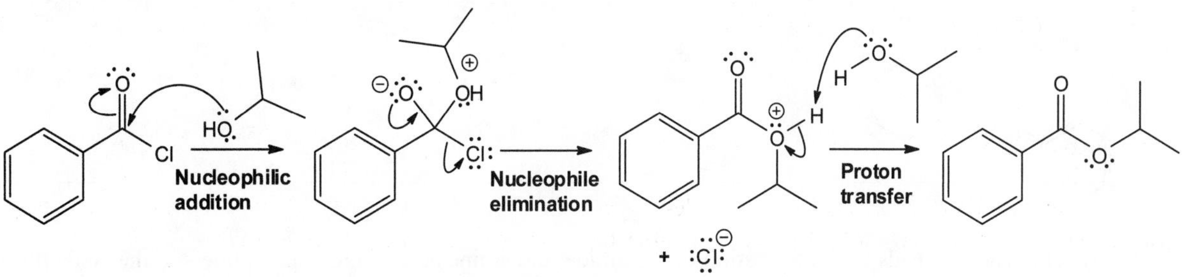

(c) The amide formation requires the amine as the nucleophile in the nucleophilic addition step and pyridine can be added to act as the base in the proton transfer step.

Nucleophilic addition
Nucleophile elimination
Proton transfer

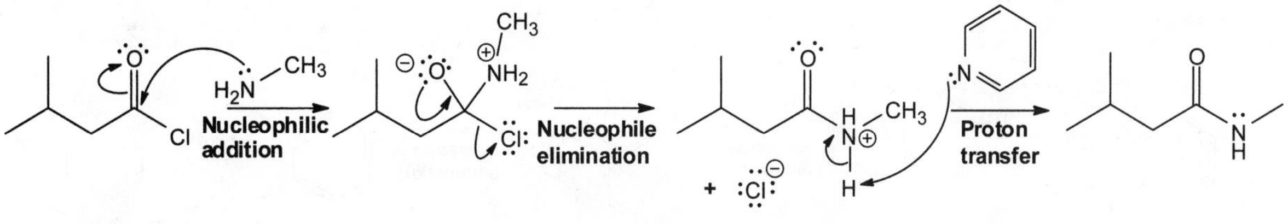

(d) The amide formation requires the amine as the nucleophile in the nucleophilic addition step and pyridine can be added to act as the base in the proton transfer step.

Nucleophilic addition
Nucleophile elimination
Proton transfer

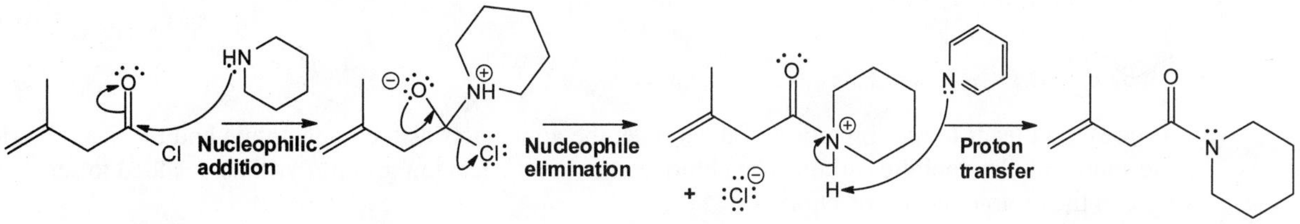

Problem 21.37

Think

What acts as the nucleophile in the nucleophilic addition step? What is the leaving group in the nucleophile elimination step? Are proton transfer reactions involved?

Solve

(a) The reaction leads to the formation of the ester; the alcohol acts as the nucleophile and the anhydride is the source of electrophilic carbon. The carboxylate anion is the leaving group. The alcohol serves as the base in the proton transfer reaction.

Nucleophilic addition

Nucleophile elimination

Proton transfer

ROH

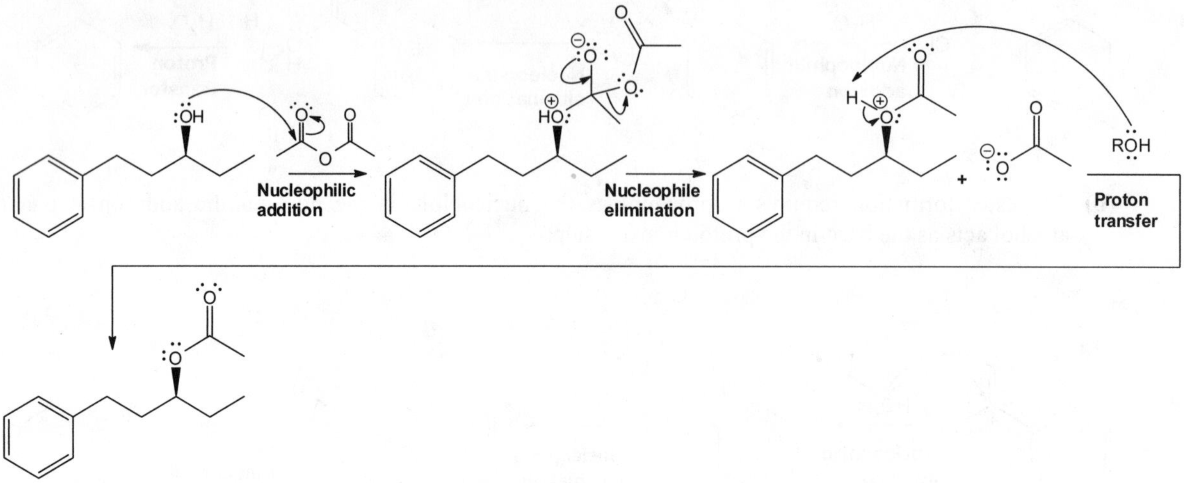

(b) The reaction leads to the formation of the amide; the amine acts as the nucleophile and the acid chloride is the source of electrophilic carbon. The chloride anion is the leaving group. The amine serves as the base in the proton transfer reaction.

Nucleophilic addition

Nucleophile elimination

Proton transfer

RNH_2

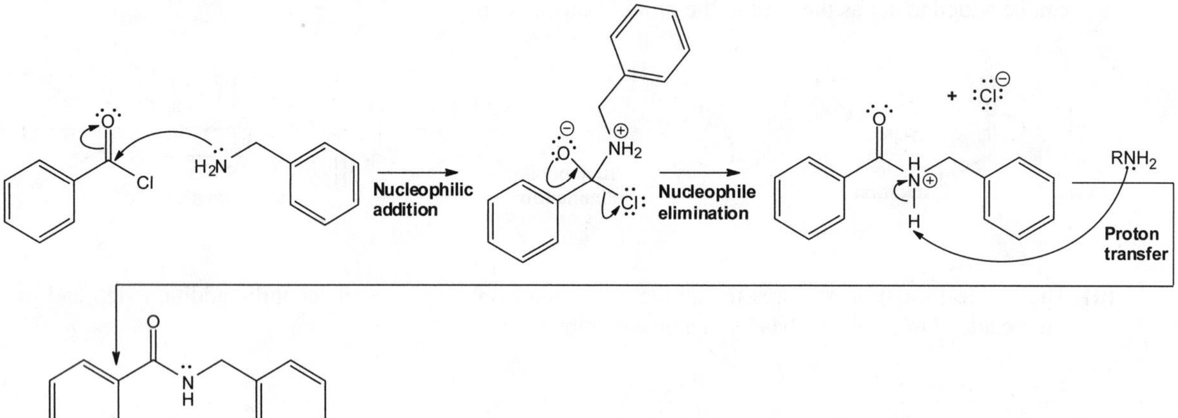

(c) The reaction leads to the formation of the amide; the amine acts as the nucleophile and the acid chloride is the source of electrophilic carbon. The chloride anion is the leaving group. Pyridine is added to serve as the base in the proton transfer reaction.

Nucleophilic addition

Nucleophile elimination

Proton transfer

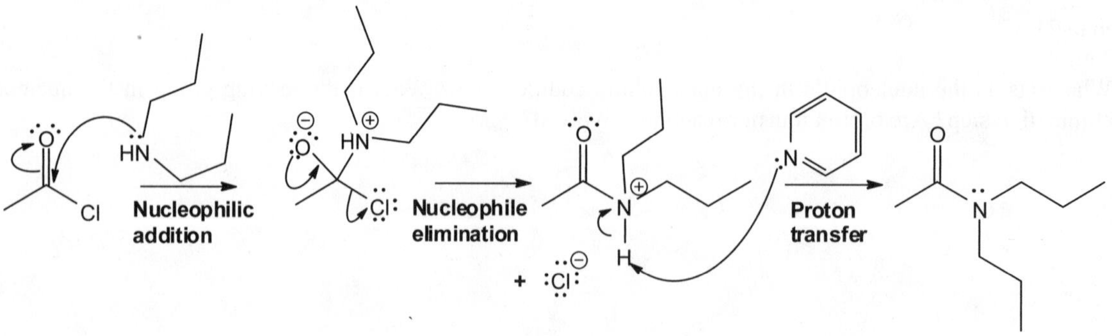

(d) The reaction leads to the formation of the ester; the alcohol acts as the nucleophile and the anhydride is the source of electrophilic carbon. The carboxylate anion is the leaving group. The alcohol serves as the base in the proton transfer reaction.

Nucleophilic addition

Nucleophile elimination

Proton transfer

+ $C_6H_5CO_2^{\ominus}$

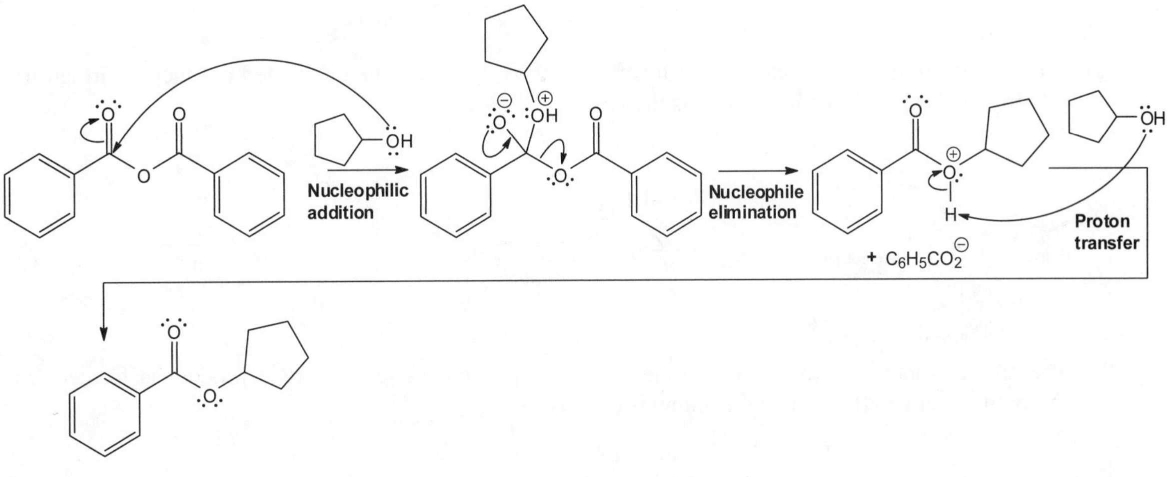

Problem 21.38

Think

What atom is involved in resonance with the C=O group in an ester? In a thioester? Are those atoms in the same row of the periodic table? How does that affect resonance stabilization? How does the stability of the reactant affect the reaction rate?

Solve

The resonance structures for a thioester and an ester are shown below. In a thioester, S is involved in resonance with the C=O group, whereas in an ester, an O atom is involved in resonance with the C=O group. Because S is in the third shell and the other atoms are in the second shell, resonance is less effective in a thioester, making the thioester less stable and more reactive.

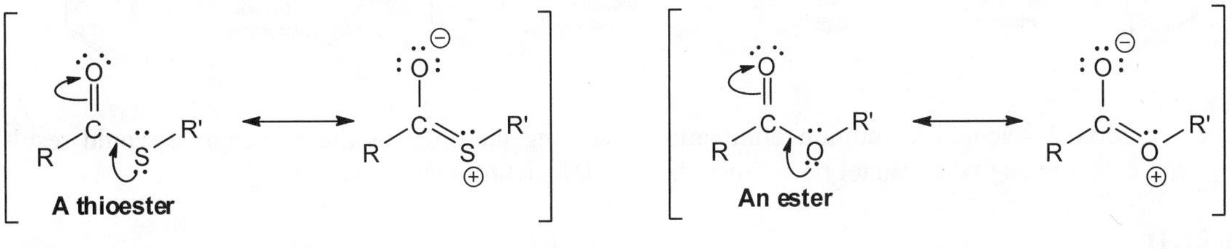

A thioester

An ester

Problem 21.39

Think

What type of reagent is sodium borohydride, $NaBH_4$? In the first reaction, what is the nucleophile? Is there a leaving group? Are proton transfer reactions involved? How does this product react with the acid chloride?

Solve

Sodium borohydride is the source of nucleophilic hydride and reduces an aldehyde to a primary alcohol. In the subsequent treatment with an acid chloride, the alcohol produced in the first reaction will act as the nucleophile in a nucleophilic addition–elimination mechanism, producing the ester.

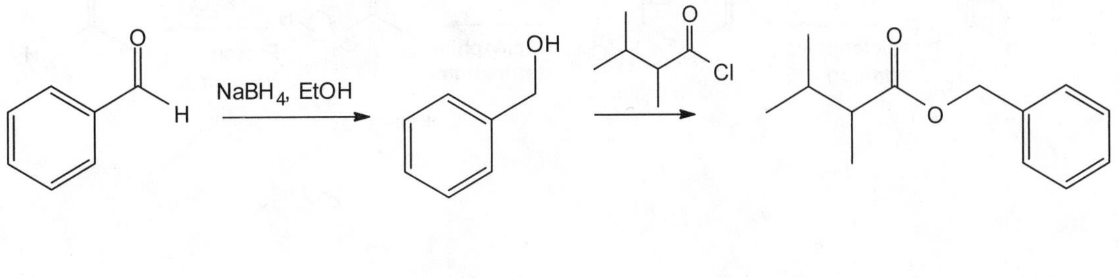

$NaBH_4$, EtOH

Problem 21.40

Think

What acts as the nucleophile in the nucleophilic addition step? What is the leaving group in the nucleophile elimination step? How can an uncharged product form? Is this possible for each nucleophile?

Solve

(a) Water reacts with the acid chloride to form the carboxylic acid. The uncharged product is formed from the proton transfer reaction where water is the base.

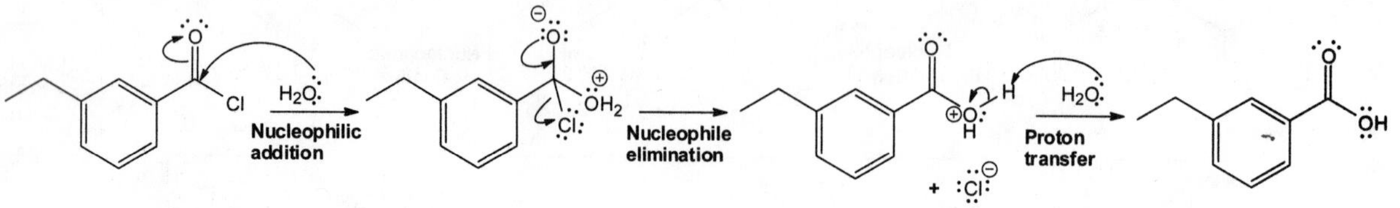

Nucleophilic addition (H_2O) → Nucleophile elimination (+ Cl^-) → Proton transfer (H_2O)

(b) The amine reacts with the acid chloride to form the amide. The uncharged product is formed from the proton transfer reaction where the amine is the base.

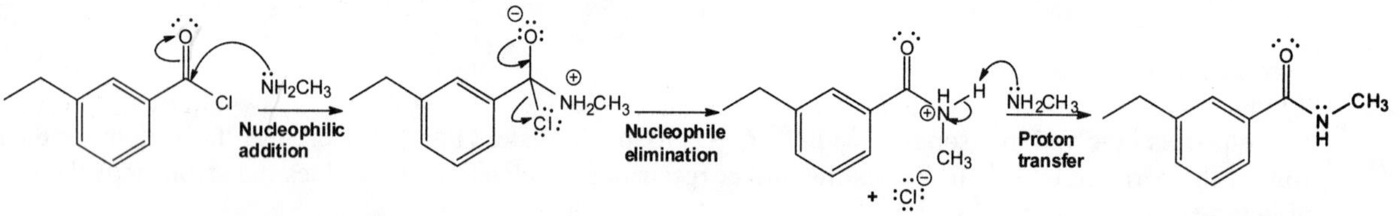

Nucleophilic addition (NH_2CH_3) → Nucleophile elimination (+ Cl^-) → Proton transfer (NH_2CH_3)

(c) The alcohol reacts with the acid chloride to form the ester. The uncharged product is formed from the proton transfer reaction where the alcohol is the base.

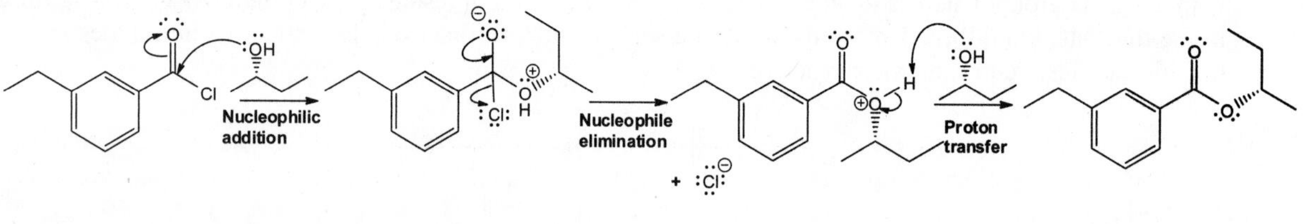

Nucleophilic addition → Nucleophile elimination (+ Cl^-) → Proton transfer

(d) No reaction. Nucleophilic addition–elimination involving the ether as the nucleophile would result in a positive charge on O that cannot be removed by a simple deprotonation.

Problem 21.41

Think

What acts as the nucleophile in the nucleophilic addition step? What is the leaving group in the nucleophile elimination step? How can an uncharged product form? Is this possible for each nucleophile?

Solve

(a) Water reacts with the acid anhydride to form the carboxylic acid. The uncharged product is formed from the proton transfer reaction in which water is the base.

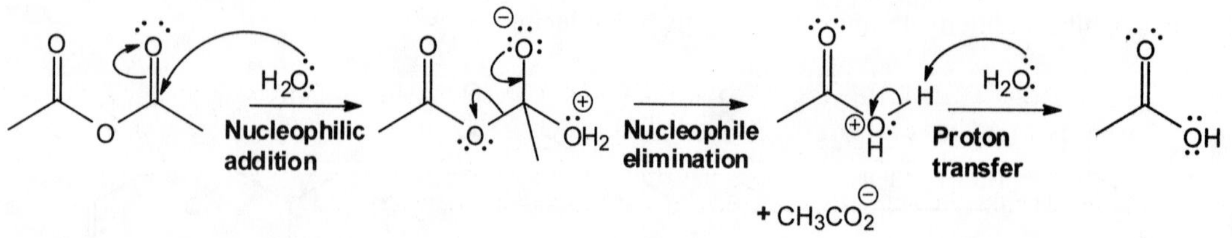

Nucleophilic addition (H_2O) → Nucleophile elimination (+ $CH_3CO_2^-$) → Proton transfer (H_2O)

(b) The amine reacts with the acid anhydride to form the amide. The uncharged product is formed from the proton transfer reaction in which the pyridine is the base.

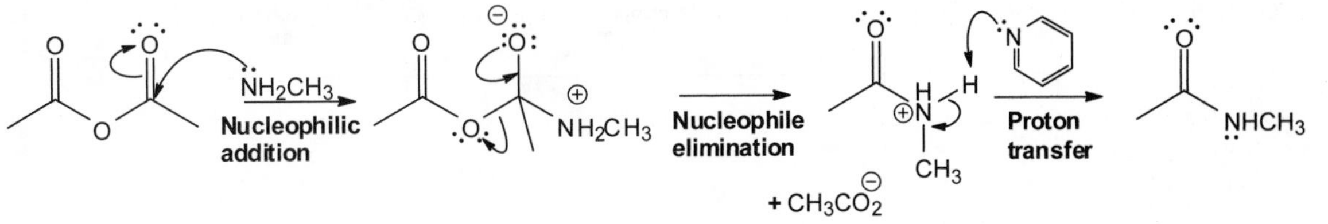

(c) The alcohol reacts with the acid anhydride to form the ester. The uncharged product is formed from the proton transfer reaction in which the alcohol is the base.

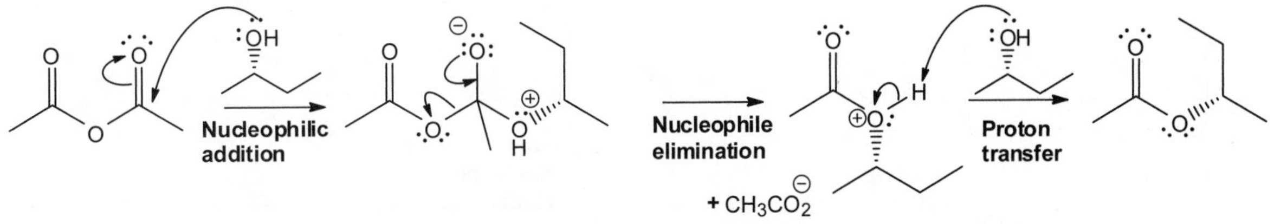

(d) No reaction. Nucleophilic addition–elimination involving the ether as the nucleophile would result in a positive charge on O that cannot be removed by a simple deprotonation.

Problem 21.42

Think

What acts as the nucleophile in the nucleophilic addition step? What is the leaving group in the nucleophile elimination step? How can an uncharged product form? Is this possible for each nucleophile? Is the reaction conducted in basic or acidic solution? What types of species should not appear in the mechanism under those conditions? Are there any irreversible proton transfers?

Solve

(a) This is an acid-catalyzed ester hydrolysis. Therefore, no strong bases should appear in the mechanism. The protonated ester reacts with water to form the carboxylic acid. The uncharged product is formed from the proton transfer reaction in which water is the base.

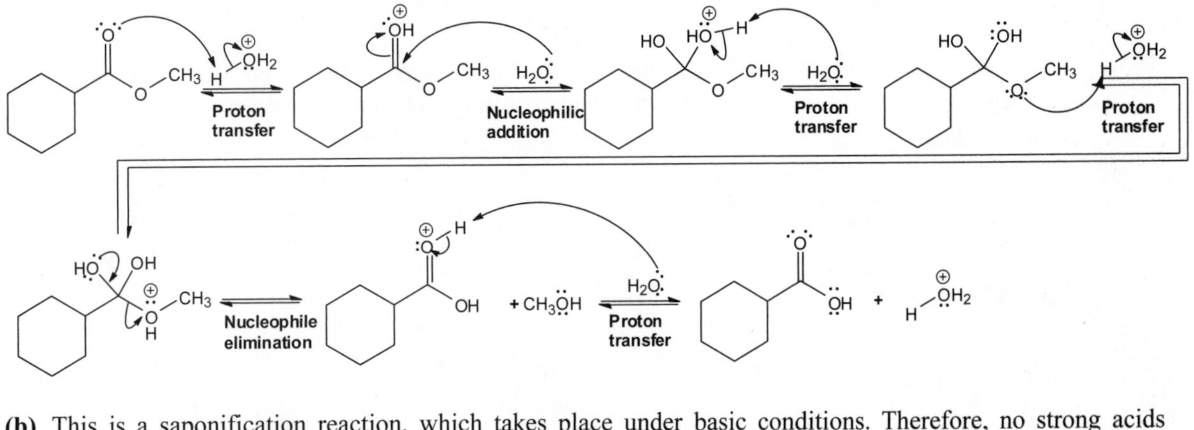

(b) This is a saponification reaction, which takes place under basic conditions. Therefore, no strong acids should appear in the mechanism. The ester reacts with HO^- to form the carboxylic acid. The carboxylic acid is deprotonated irreversibly by HO^- to form the carboxylate anion. Acid is then added to form the uncharged carboxylic acid product. See the mechanism on the next page.

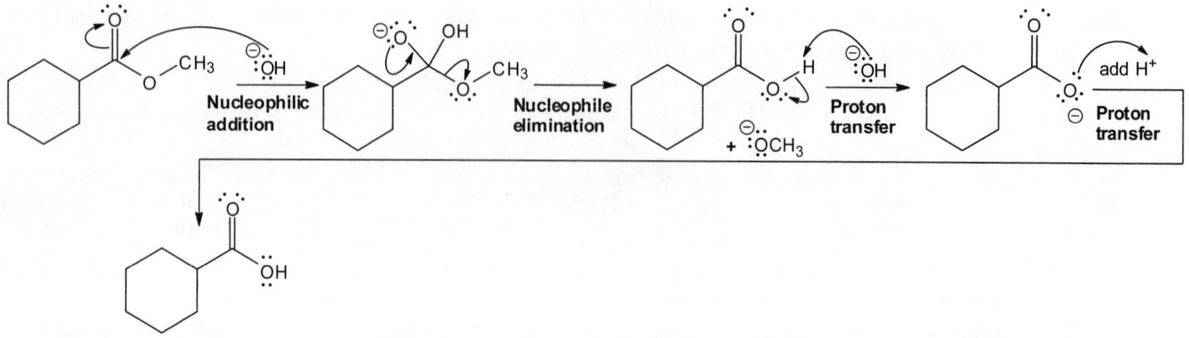

Nucleophilic addition
Nucleophile elimination
Proton transfer
add H^+
Proton transfer

(c) This is a base-catalyzed transesterification. The ester reacts with $CH_3CH_2CH_2O^-$ to form the ester.

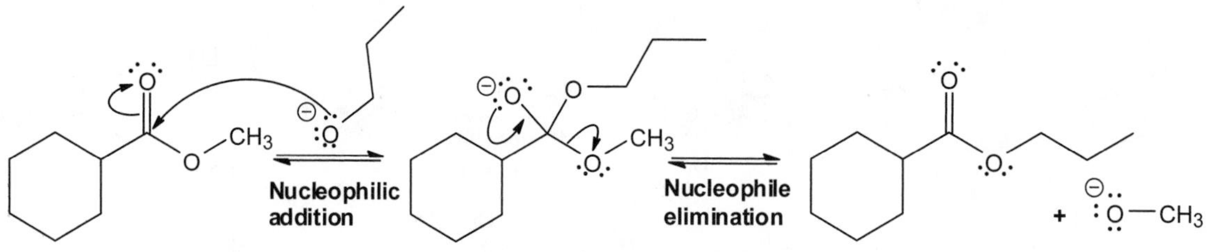

Nucleophilic addition
Nucleophile elimination

(d) This is an acid-catalyzed transesterification. The protonated ester reacts with the alcohol in a nucleophilic addition–elimination reaction. The uncharged product is formed from the proton transfer reaction in which the alcohol is the base.

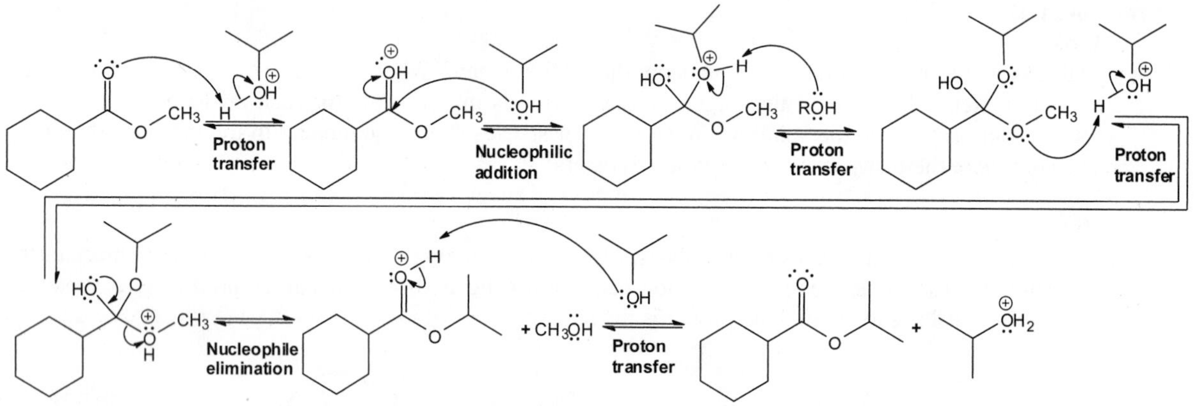

Proton transfer
Nucleophilic addition
Proton transfer
Proton transfer
Nucleophile elimination
Proton transfer

(e) This is an acid-catalyzed aminolysis. The protonated ester reacts with the amine to form the amide. The uncharged product is formed from the proton transfer reaction in which the amine is the base.

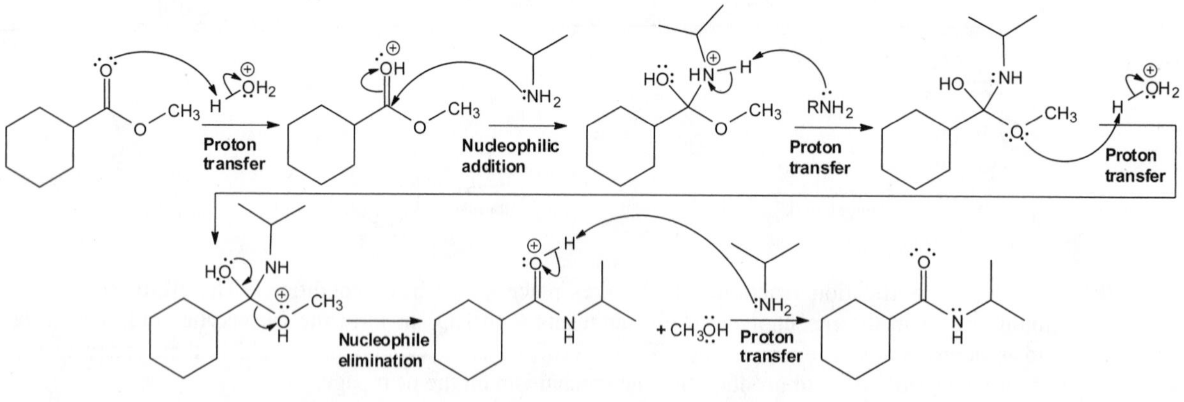

Proton transfer
Nucleophilic addition
Proton transfer
Proton transfer
Nucleophile elimination
Proton transfer

(f) No reaction will take place. Propan-1-ol is a weak nucleophile and the reaction is not catalyzed by acid or base.

Problem 21.43

Think

In a Claisen condensation reaction, are the nucleophilic addition and elimination steps reversible or irreversible? Does that equilibrium favor reactants or products? To be a successful Claisen condensation reaction, what must be true of a proton transfer involving the β-keto ester that is initially produced? Is that type of proton transfer possible in the self-Claisen reaction involving methyl propanoate? Involving methyl 2-methylpropanoate?

Solve

In each case we can envision a self-Claisen condensation reaction. In the second reaction, however, the β-keto ester that is produced does not have an acidic proton on the carbon between the two C=O groups. Therefore, there is no available irreversible proton transfer that can drive the reaction toward products. The nucleophilic addition–elimination equilibrium favors reactants instead.

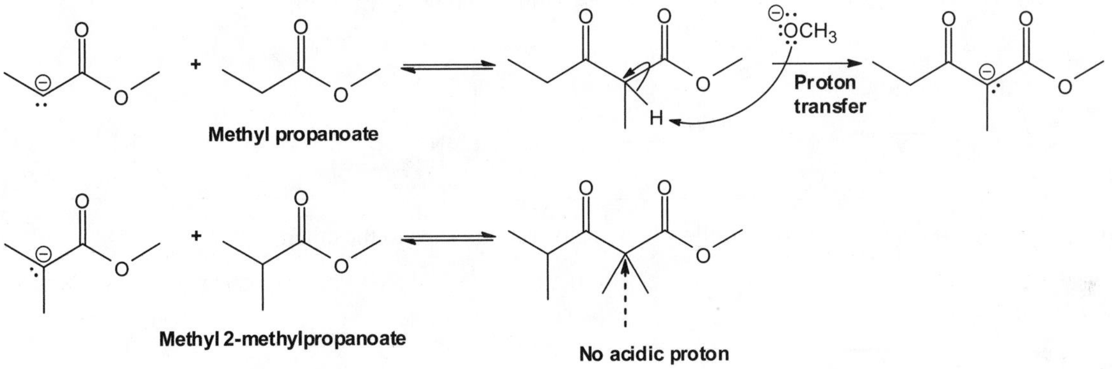

Methyl propanoate

$:OCH_3$ Proton transfer

Methyl 2-methylpropanoate

No acidic proton

Problem 21.44

Think

Which atoms came from acetic acid? What new functional group is formed? What was the identity of the nucleophile in the nucleophilic addition step? Does the acyl substitution require going up or down the stability ladder? What reactions do you know that can accomplish going up the ladder?

Solve

(a) The reaction requires converting a carboxylic acid to an acid anhydride, which represents going up the stability ladder. This is difficult to do directly, but can be carried out by first producing an acid chloride. The acid chloride and the salt of acetic acid will make the desired anhydride.

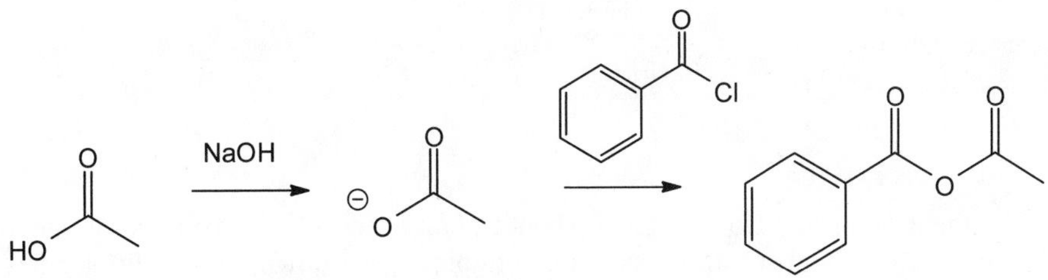

NaOH

(b) Addition of phenol forms the ester under acidic conditions—a Fischer esterification. This is a reversible reaction and can be carried out directly because the reactant and product species are on the same rung of the stability ladder.

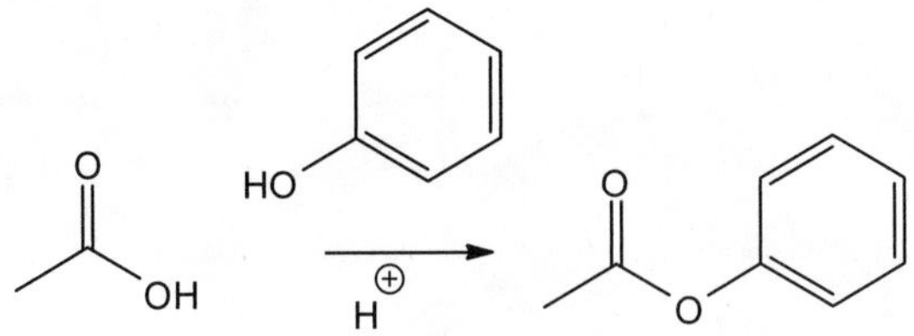

(c) Reaction of the anhydride from **(a)** with ethanol makes the ester. This represents going down the stability ladder and is done relatively easily.

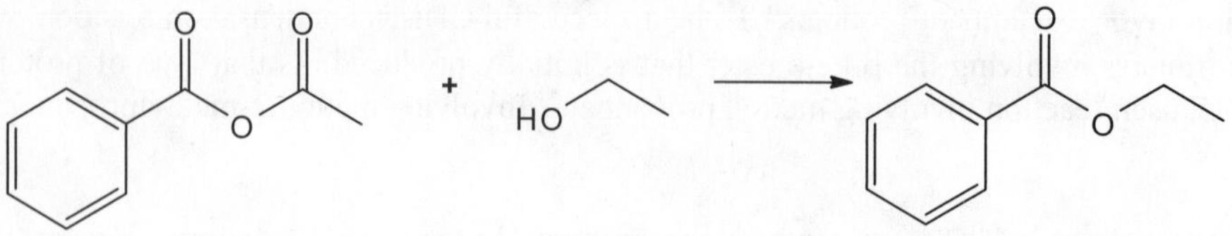

(d) The amine can be produced by reducing the corresponding amide. It is difficult to produce the amide directly from acetic acid because the amine that would be required for the nucleophile would deprotonate the carboxylic acid. The amide can be made relatively easily, however, by first converting acetic acid to acetyl chloride and then treating acetyl chloride with the amine.

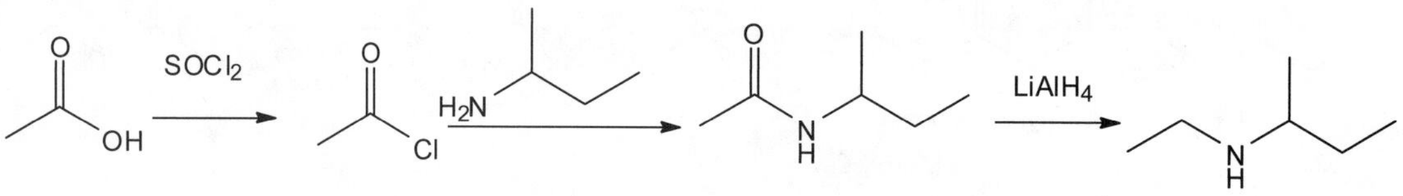

$SOCl_2$ H_2N $LiAlH_4$

(e) Reducing the acid to an alcohol, followed by dehydration, leads to the ether.

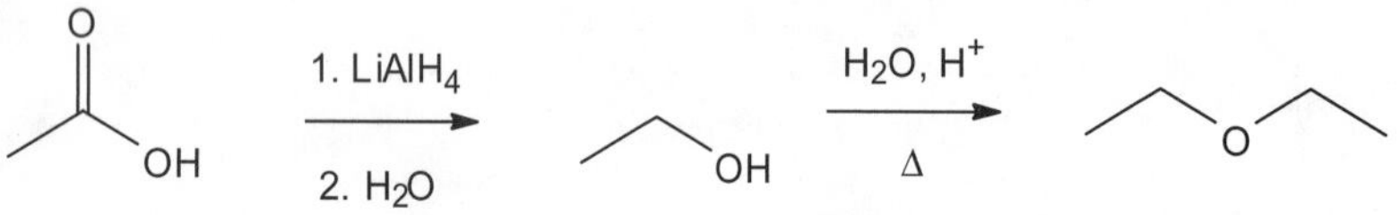

1. $LiAlH_4$ 2. H_2O H_2O, H^+, Δ

(f) The acid chloride plus the specialized reducing agent, LiAlH(O-tBu)$_3$, forms the aldehyde.

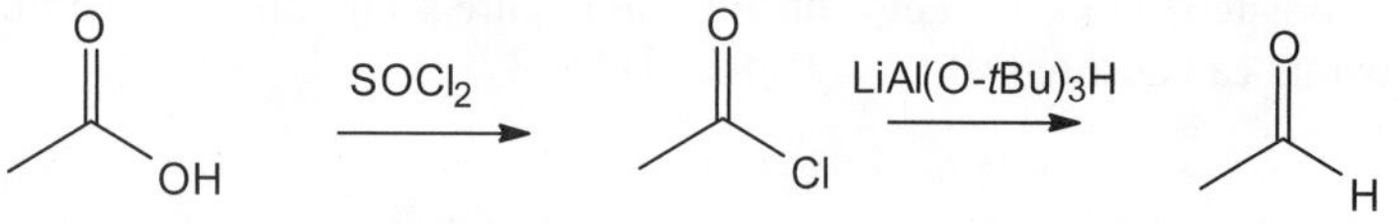

$SOCl_2$ LiAl(O-*t*Bu)$_3$H

(g) A lithium dialkylcuprate forms the ketone from the acid chloride.

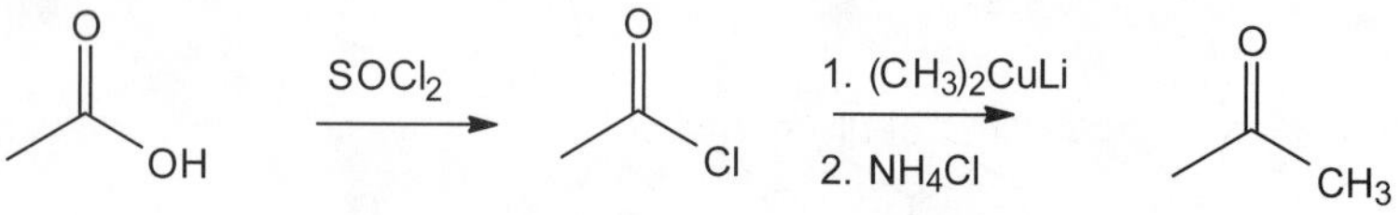

$SOCl_2$ 1. $(CH_3)_2CuLi$ 2. NH_4Cl CH_3

Problem 21.45

Think

Which atoms came from the acetic acid? Do any new carbon–carbon need to be formed? What new functional group is formed? Do the necessary acyl substitutions represent going up or down the stability ladder? Are any oxidations or reductions necessary? What was the identity of the nucleophile in the nucleophilic addition step?

Solve

(a) Reduction produces ethanol, which can react with the acid chloride (formed from the reaction of acetic acid and $SOCl_2$).

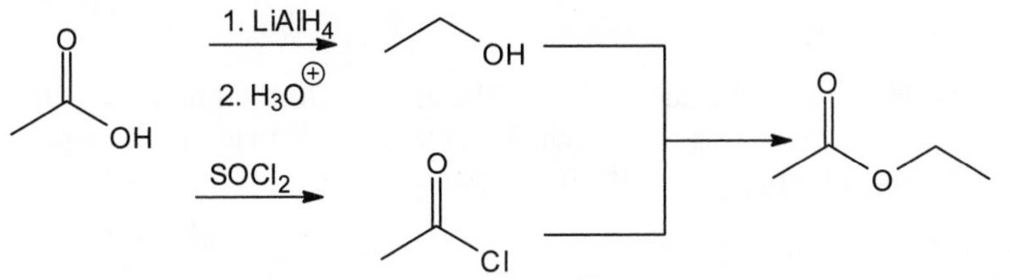

1. $LiAlH_4$
2. $H_3O^{\oplus}$

OH

$SOCl_2$

Cl

O

(b) We can start with ethanol and ethanoyl chloride from **(a)**.

OH $\xrightarrow{PBr_3}$ Br $\xrightarrow{Li}$ Li $\xrightarrow{CuI}$ $(CH_3CH_2)_2CuLi$

Cl $\xrightarrow{(CH_3CH_2)_2CuLi}$ $\xrightarrow[EtOH]{NaBH_4}$ OH

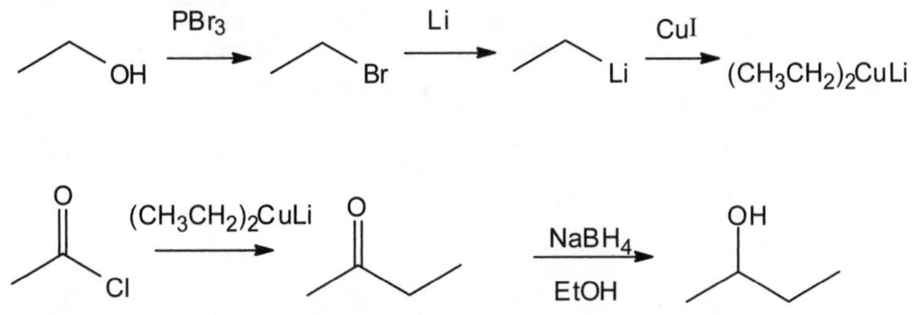

(c) Use of the ester from **(a)** and ethyllithium from **(b)** (or a Grignard reagent) is sufficient.

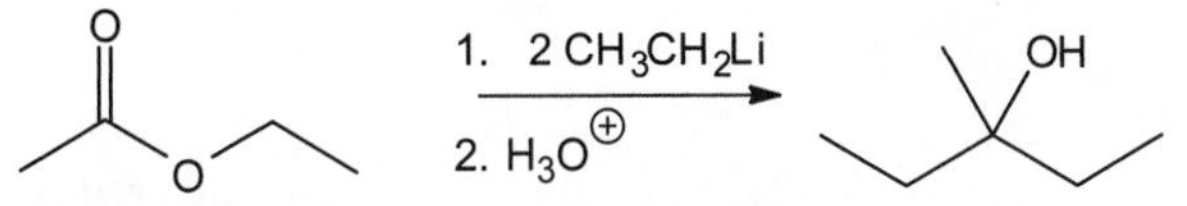

1. $2\ CH_3CH_2Li$
2. $H_3O^{\oplus}$

OH

(d) Butan-2-one is made in **(b)**.

(e) Ethanamine is made using ammonia to form the amide followed by reduction.

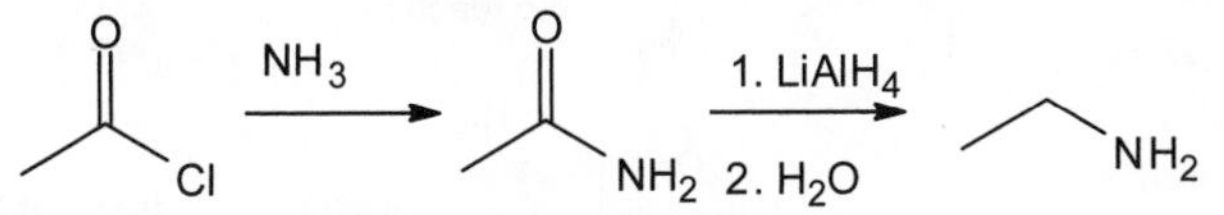

Cl $\xrightarrow{NH_3}$ NH_2 $\xrightarrow[2.\ H_2O]{1.\ LiAlH_4}$ NH_2

(f) We can use ethanamine from **(e)** and the ethanoyl chloride from **(a)**.

Cl + H_2N $\xrightarrow{\text{pyridine}}$ N H

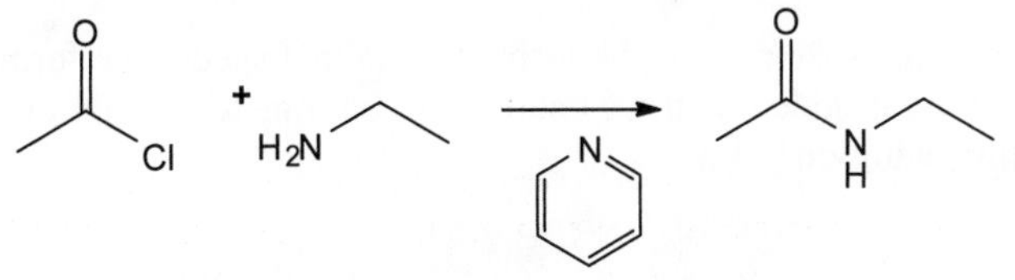

(g) Synthesis of diethylamine requires reduction of the amine produced in **(f)**; then reaction with the acid chloride leads to the desired product.

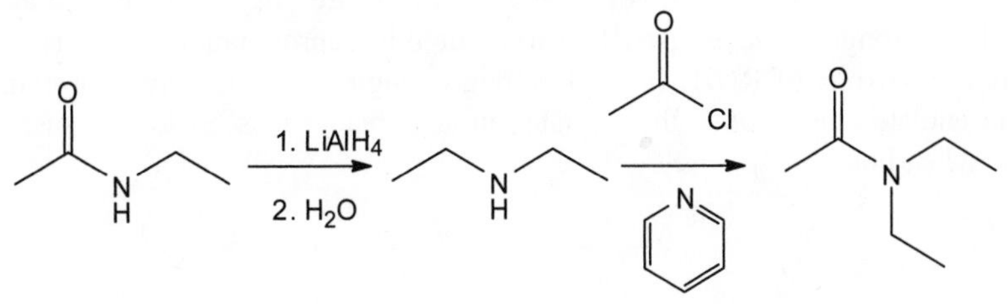

N H $\xrightarrow[2.\ H_2O]{1.\ LiAlH_4}$ N H $\xrightarrow{\text{Cl, pyridine}}$ N

Problem 21.46

Think

What is the structure of acetic anhydride? Which HO group on salicylic acid serves as the nucleophile in the nucleophilic addition step? What is the leaving group in the nucleophile elimination step?

Solve

The alcohol, not the carboxylic acid HO, acts as the nucleophile in the nucleophilic addition step. The alcohol is more nucleophilic because the C=O group of the carboxylic acid is electron withdrawing via both resonance and inductive effects. The carboxylate is the leaving group in the nucleophile elimination step.

Salicylic acid

Nucleophilic addition

Nucleophile elimination

Proton transfer

Acetylsalicylic acid

Problem 21.47

Think

What size ring would the second compound form? Even though the carboxylic acid and alcohol are separated by the same number of carbons in each reaction, what in the product of the second reaction would disfavor the formation of the ring?

Solve

In the second compound, the alcohol and carboxyl groups are too far apart to form a five-membered ring without substantial strain.

Forms 5-membered ring

Too far away to react.

No reaction

Strained ring, does not form

Problem 21.48

Think

To speed up an aminolysis of an ester in base, what species must be formed? Can that species be formed when HO^- or RO^- is added as a base? If a stronger base is added, is there another proton transfer reaction that would take place and prevent the intended nucleophilic addition?

Solve

To speed up a nucleophilic addition–elimination reaction under basic conditions, the job of the base is to convert the weak nucleophile into a strong nucleophile by deprotonation. So under basic conditions, the nucleophile in a hydrolysis reaction, H_2O, would be converted to HO^-. For an aminolysis reaction, the RNH_2 nucleophile would have to be converted to RNH^-. However, HO^- and RO^- are not strong enough as bases to carry out this deprotonation. If a stronger base is used to carry out this deprotonation, there is a separate problem: RNH^- is a very strong base (pK_a of RNH_2 is ~38), strong enough to irreversibly deprotonate the α carbon of an ester to make an enolate anion. Once the enolate anion is made, it is no longer susceptible to nucleophilic attack at the carbonyl carbon.

Problem 21.49

Think

What is the nucleophile in the nucleophilic addition step involving 2-pyridinethiol? Is the reaction promoted by acid or base? What types of species should not appear in the mechanism? In the reaction that produces the lactone, how is the stability of the tetrahedral intermediate affected by the N atom? Does the N atom lead to the formation of a better or worse leaving group?

Solve

(a) The mechanism for the formation of the thioester is shown first. The long carbon chain is abbreviated R.

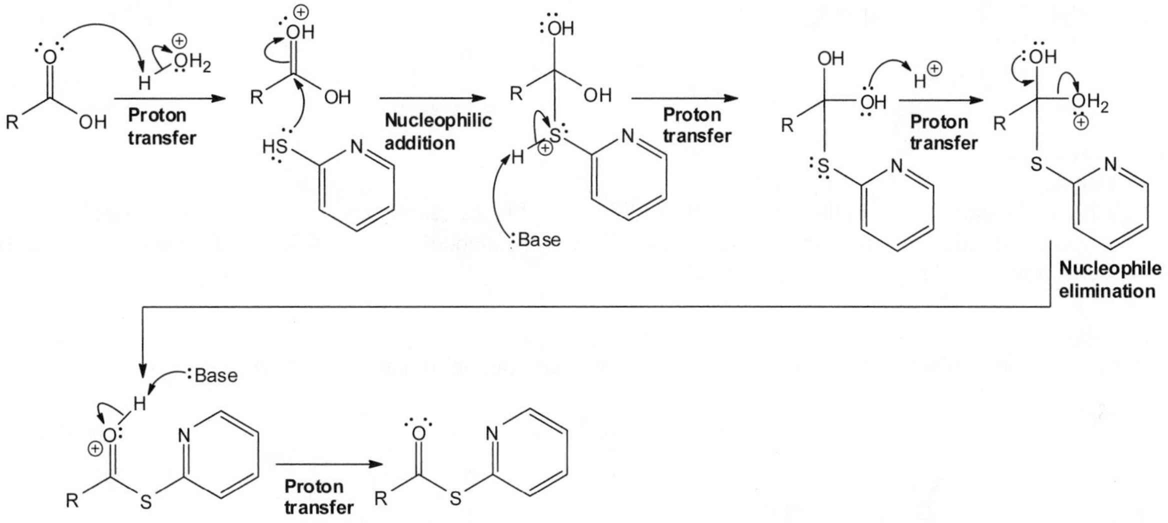

Proton transfer → Nucleophilic addition → Proton transfer → Proton transfer → Nucleophile elimination → Proton transfer

Next comes the formation of the cyclic ester.

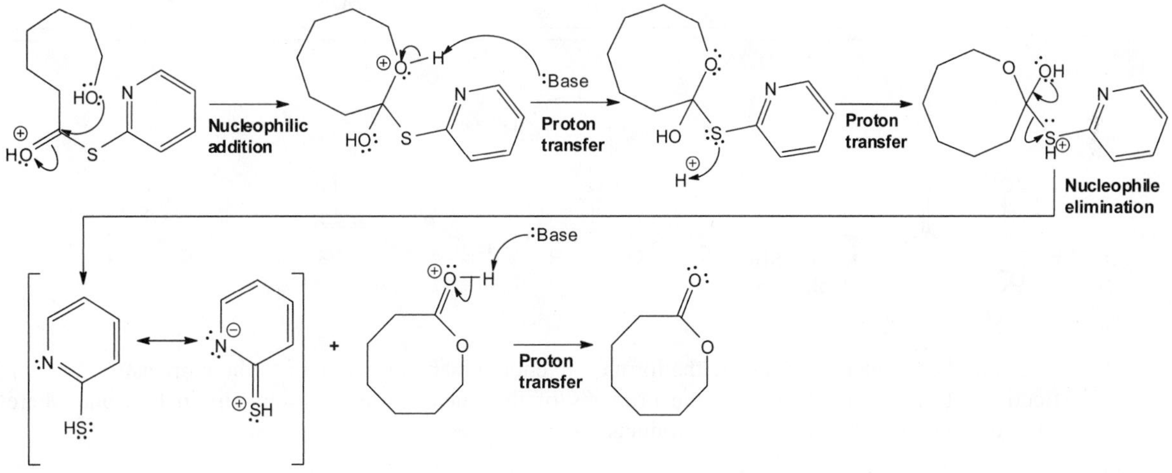

Nucleophilic addition → Proton transfer → Proton transfer → Nucleophile elimination → Proton transfer

(b) There are two reasons that the reaction is less effective. First, the product of nucleophilic addition step, shown below, has an intramolecular hydrogen bond involving the N atom. This stabilization of the tetrahedral intermediate wouldn't be possible if the N atom were a C atom instead. Second, if the N atom were instead a C, the RSH leaving group would not be as good. With the N present, there is greater resonance stabilization in the leaving group because a negative charge is more stable on N than on C (shown on the previous page).

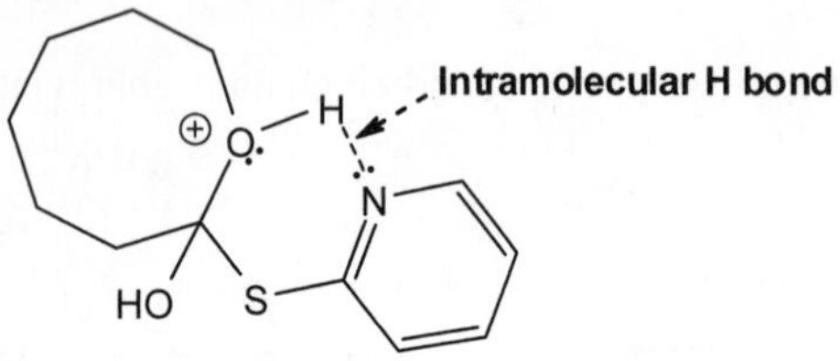

Intramolecular H bond

Problem 21.50

Think

What is the nucleophile in the nucleophilic addition step? Is the reaction promoted by acid or base? What types of species should not appear in the mechanism? How is the alcohol product different if the ester shown in **(b)** is used? Why would this favor the products?

Solve

(a) The mechanism is an acid-catalyzed nucleophilic addition–elimination reaction.

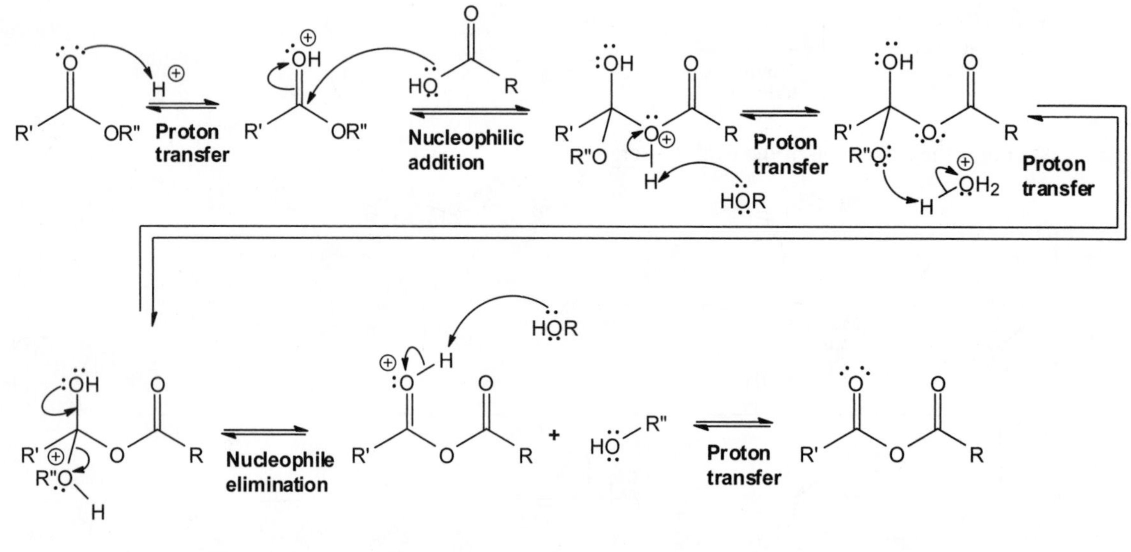

Proton transfer

Nucleophilic addition

Proton transfer

Proton transfer

Nucleophile elimination

Proton transfer

(b) R″–OH in the products would have the form of an enol, which tautomerizes to the more stable ketone. This effectively removes R″–OH from the products of the nucleophilic addition–elimination and, therefore, drives that equilibrium more toward products.

Problem 21.51

Think

By examining the products, which bonds broke and formed in the reaction? Which O atom in each reactant species, therefore, must have been protonated by the $H_3{}^{18}O^+$ acid? How does this change the nature of the mechanism? How does the alkyl group bonded to O favor one mechanism over the other?

Solve

In the first case, $H_3{}^{18}O^+$ attacks the carbonyl carbon in an acid-catalyzed nucleophilic addition–elimination reaction. The initial O–CH_3 bond stays intact and the ^{18}O label ends up in the carboxylic acid after departure of protonated methanol.

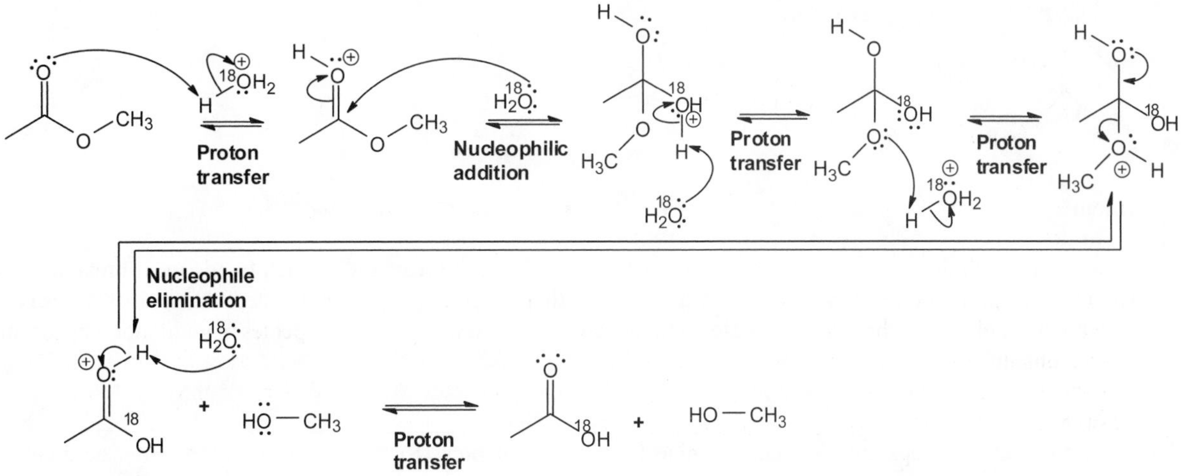

Proton transfer

Nucleophilic addition

Proton transfer

Proton transfer

Nucleophile elimination

Proton transfer

In the second reaction, the *t*-butyl group hinders the nucleophilic attack at the carbonyl carbon, but an S_N1 reaction is favorable due to the generation of a 3° carbocation; $^{18}OH_2$ attacks the carbocation to form the protonated and labeled alcohol.

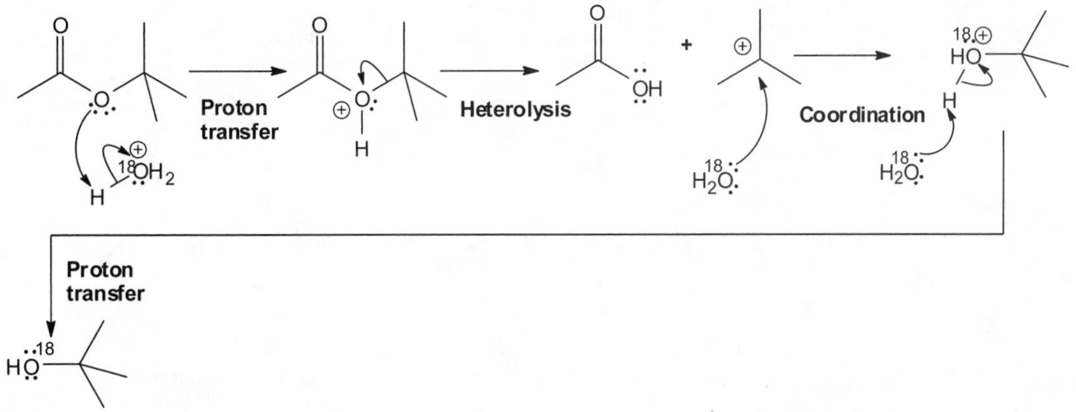

Proton transfer

Heterolysis

Coordination

Proton transfer

Problem 21.52

Think

What mechanism occurs in the first reaction? How does ring strain affect the mechanism in the second reaction? Which bonds in the second reaction must have formed and broken?

Solve

The first reaction is a normal addition–elimination mechanism.

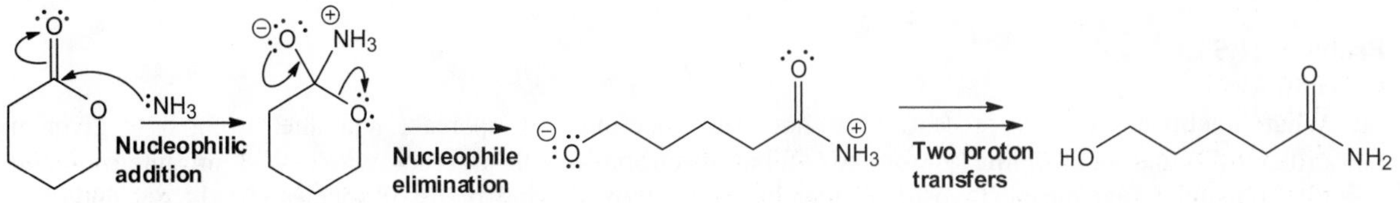

Nucleophilic addition

Nucleophile elimination

Two proton transfers

In the second reaction, the nucleophile must form a bond to the β carbon and the C–O bond must break. Thus, the strained ring opens in an S_N2 reaction, much like an epoxide ring opens. The opening of the ring is facilitated further by the resonance stabilization of the negative charge in the carboxylate leaving group.

S_N2

H_3N

$:NH_3$

Problem 21.53

Think

Can you match the carbon atoms in the product with those in the reactant? Which bonds must break and form? How does an amine react typically with a ketone? What functional group is formed? How does an ester react with an alcohol? Is the reaction base or acid catalyzed? What types of species should not appear in the mechanism?

Solve

The first several steps are identical to imine formation (Chapter 18). Next is an intramolecular acid-catalyzed transesterification reaction.

OEt Proton transfer OEt H_2N–OH Nucleophilic addition OEt H_2N–OH Proton transfer OEt Proton transfer

OEt Heterolysis OEt H_2N–OH Proton transfer OEt Proton transfer OEt Nucleophilic addition

H_2O Proton transfer Proton transfer Nucleophile elimination Proton transfer H_2O

Problem 21.54

Think

What functional group is present? How does this functional group react with the strong base given in the problem? What nucleophile is formed? What electrophile will it react with? Will an intermolecular or intramolecular reaction be favored? Under basic conditions, what types of species should not appear in the mechanism?

Solve

(a) This is an example of a Claisen condensation reaction. Under these basic conditions, no strong acids should appear in the mechanism. The CH_3O^- deprotonates the α C–H to form the enolate, which undergoes nucleophilic addition to the other ester C=O. An intramolecular reaction is favored because of the formation of a five-membered ring. Nucleophile elimination followed by proton transfer forms the enolate anion in an irreversible reaction. An acid workup is necessary to form the uncharged product.

Proton transfer — Nucleophilic addition — Nucleophile elimination — Proton transfer

Add H^+ — Proton transfer

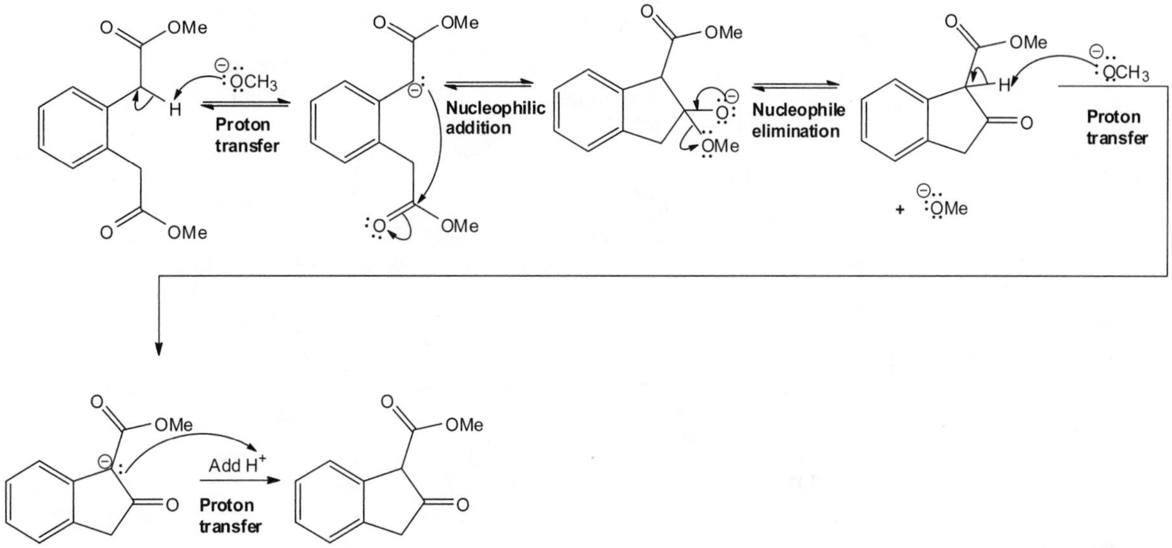

(b) This is a crossed Claisen condensation reaction. LDA deprotonates the α C–H to form the enolate quantitatively, which undergoes nucleophilic addition to the ethyl acetate ester C=O. Nucleophile elimination followed by proton transfer forms the enolate anion in an irreversible reaction. An acid workup is necessary to form the uncharged product.

Proton transfer — Nucleophilic addition — Nucleophile elimination — Proton transfer

Add H^+ — Proton transfer

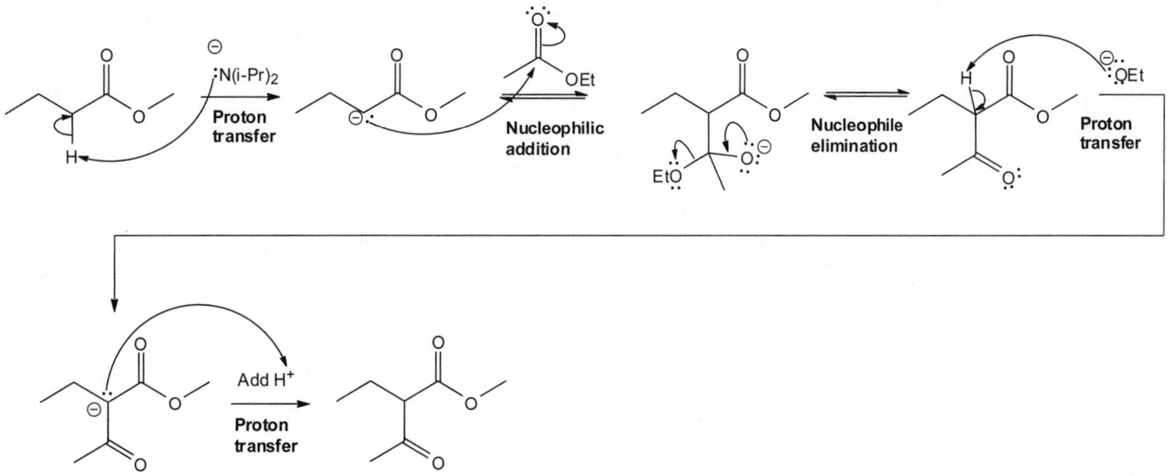

Problem 21.55

Think

What functional group is present? How does this functional group react with the strong base given in the problem? What nucleophile is formed? What electrophile will it react with? Under basic conditions, what types of species should not appear in the mechanism?

Solve

These are all examples of the Claisen condensation reaction. The mechanisms for **(a)–(c)** are given below. Notice that **(a)** is a self-Claisen condensation reaction, whereas **(b)** and **(c)** are crossed Claisen condensations. These crossed Claisen condensations are feasible because one of the esters has no α hydrogen.

(a)

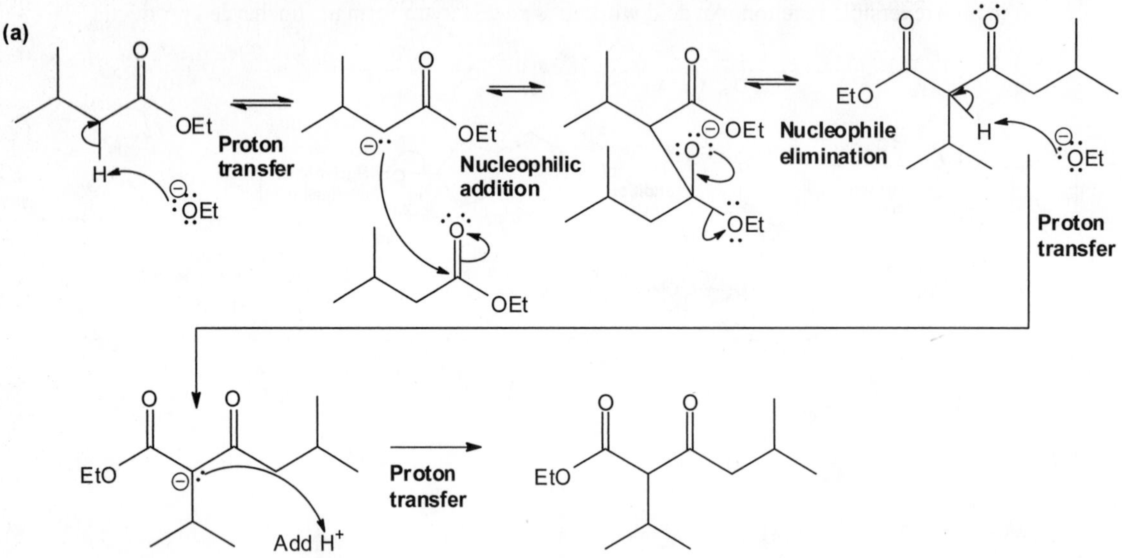

OEt **Proton transfer** OEt **Nucleophilic addition** OEt OEt OEt **Nucleophile elimination** EtO OEt **Proton transfer**

EtO Add H^+ **Proton transfer** EtO

(b)

EtO **Proton transfer** EtO EtO EtO **Nucleophilic addition** OEt OEt

Nucleophile elimination

EtO OEt **Proton transfer** EtO Add H^+ EtO **Proton transfer**

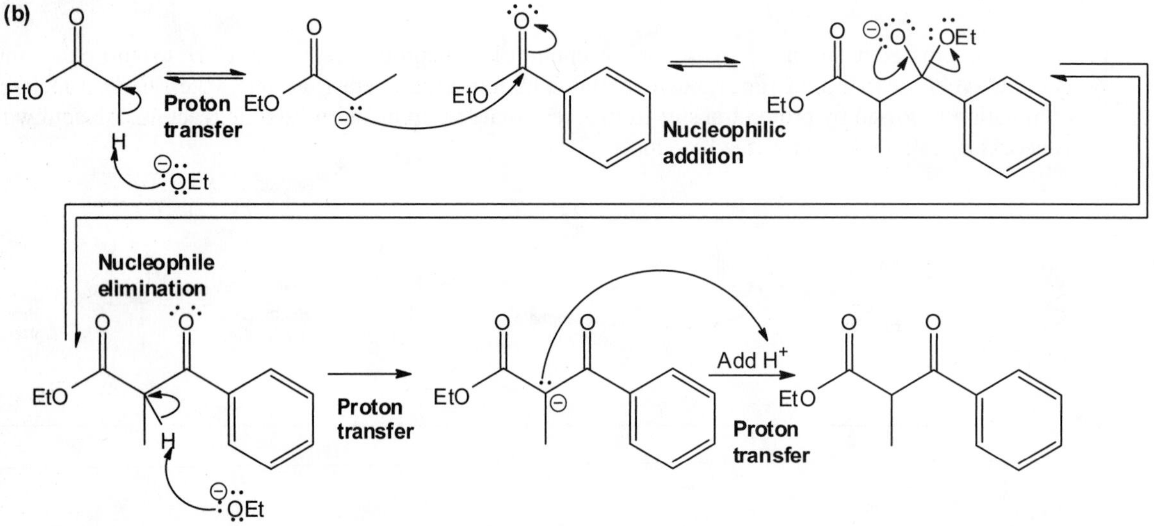

(c)

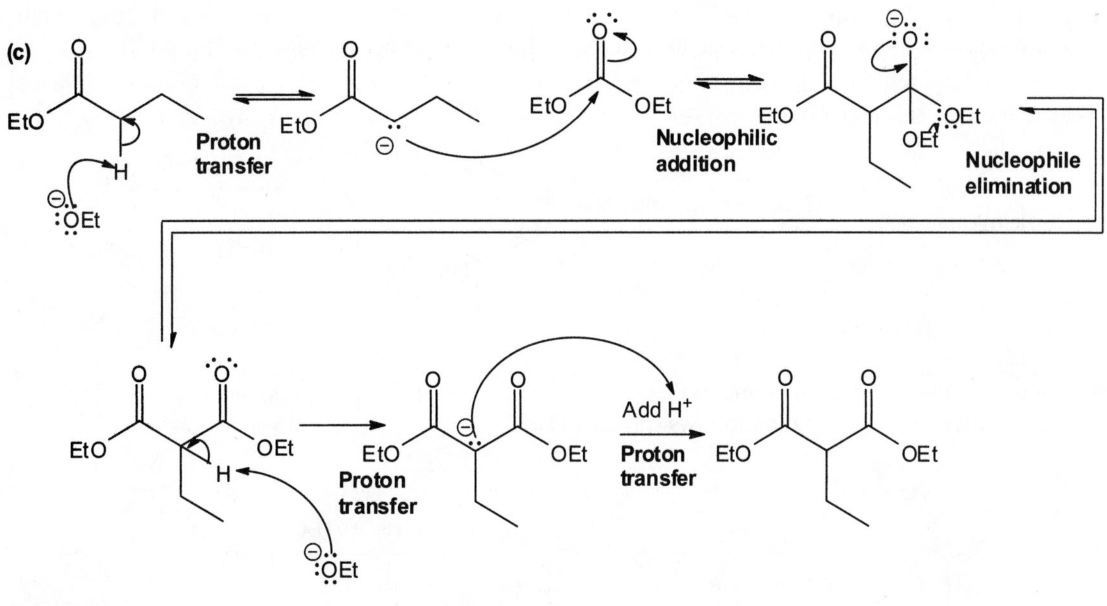

Proton transfer

Nucleophilic addition

Nucleophile elimination

Proton transfer

Add H^+

Proton transfer

Problem 21.56

Think

In each reaction, what are the electron-rich sites and electron-poor sites? Are there any nucleophiles? What electrophiles will they attack? What functional groups are formed in each reaction? What types of reactions occur—substitution, elimination, addition, oxidation, reduction?

Solve

(a) The α C–H is deprotonated quantitatively by EtO^- to form the enolate, which undergoes an S_N2 reaction with 1-chloro-4-methylpentane to alkylate the α C (product **A**). The β-keto ester undergoes hydrolysis, followed by a decarboxylation reaction to form **B**.

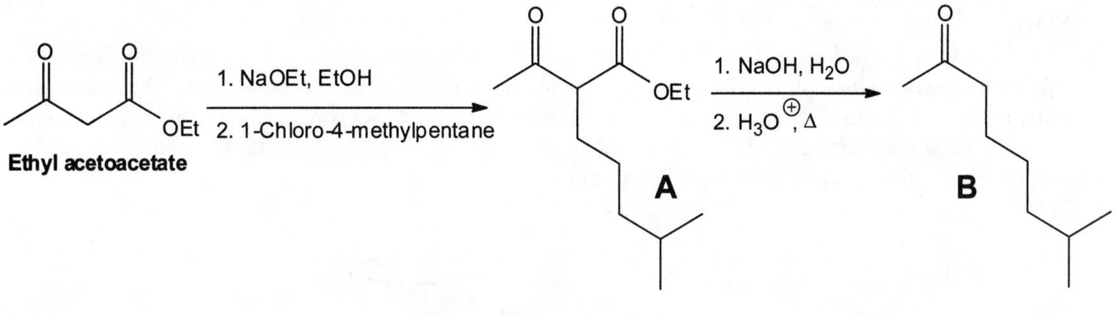

Ethyl acetoacetate → 1. NaOEt, EtOH; 2. 1-Chloro-4-methylpentane → **A** → 1. NaOH, H_2O; 2. H_3O^+, Δ → **B**

(b) The reaction sequence is the same as that in **(a)**: the β-keto ester product **A** is alkylated at the α C (product **C**) and hydrolysis and decarboxylation follow (product **D**).

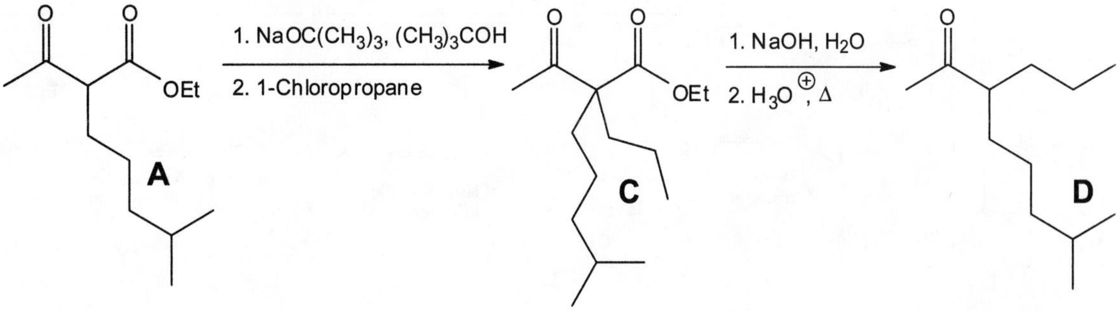

A → 1. NaOC(CH_3)$_3$, (CH_3)$_3$COH; 2. 1-Chloropropane → **C** → 1. NaOH, H_2O; 2. H_3O^+, Δ → **D**

(c) The β-diester diethylmalonate is alkylated at the α C (product **E**). The resulting β-diester undergoes hydrolysis to form two carboxylic acids, and a decarboxylation reaction follows to form **F**.

Diethyl malonate → 1. NaOEt, EtOH; 2. Bromophenylmethane → **E** → 1. NaOH, H_2O; 2. $H_3O^{\oplus}$, Δ → **F**

(d) The reaction sequence is the same as that in **(c)**: the β-diester product **E** is alkylated at the α C (product **G**), and hydrolysis/decarboxylation follows (product **H**).

E → 1. $NaOC(CH_3)_3$, $(CH_3)_3COH$; 2. 1-Iodopentane → **G** → 1. NaOH, H_2O; 2. $H_3O^{\oplus}$, Δ → **H**

Problem 21.57

Think

How many C atoms are in the reactants and products? Were any C atoms lost? Which C–C bonds formed? Is the reaction conducted under basic or acidic conditions? Which α C–H is most acidic? Does the nucleophilic addition occur via 1,2- or 1,4-addition?

Solve

The α C–H on the β-diester diethyl malonate is more acidic than but-3-en-2-one. The enolate is formed from diethyl malonate, which undergoes conjugate nucleophilic addition to but-3-en-2-one. The next step is a proton transfer to the C atom of the enolate anion. Simultaneously, the C=O is re-formed using a lone pair of electrons from the negatively charged O. The next few steps make up an intramolecular Claisen condensation reaction, favored by the formation of a six-membered ring.

Proton transfer → Nucleophilic addition → Proton transfer → Proton transfer → Nucleophilic addition → Nucleophile elimination

Problem 21.58

Think

How does $LiAlH_4$ react with a ketone? In the reaction with TsCl, what is the nucleophile and what is the electrophile? In the reaction with NaCN, is there a good leaving group?

Solve

The carbonyl carbon on the ketone is reduced to the alcohol using $LiAlH_4$ with an acid workup. The alcohol OH is not a good leaving group and is transformed into the OTs, which is a good leaving group. An S_N2 reaction follows with NaCN to form the nitrile.

1. $LiAlH_4$ 2. $H^{\oplus}$; TsCl, Pyridine ; NaCN, DMSO

Problem 21.59

Think

In each reaction, what are the electron-rich sites and electron-poor sites? Are there any nucleophiles? What electrophiles will they attack? Are there any good leaving groups? What functional groups are formed in each reaction? What types of reactions occur—substitution, elimination, addition, oxidation, reduction?

Solve

(a) The ester is hydrolyzed to the carboxylic acid **A**. $SOCl_2$ transforms carboxylic acid **A** into the acid chloride **B**. The acid chloride is transformed into the amide **C**, which is reduced by $LiAlH_4$ into the amine **D**.

$H_3O^{\oplus}$, Δ → **A** ; $SOCl_2$ → **B** ; NH_3 → **C** ; $LiAlH_4$ → **D**

(b) The carboxylic acid **A** is brominated at the α C, via a Hell–Volhard–Zelinski reaction, to form **E**. Br is a good leaving group and an S_N2 reaction with NaCN yields the nitrile **F**. The nitrile is hydrolyzed to the carboxylic acid **G**.

A 1. PBr_3, Br_2 2. $H_3O^{\oplus}$ → **E** ; NaCN → **F** ; $H_3O^{\oplus}$, Δ → **G**

(c) DIBAH reduces the ester to the aldehyde **H**, which undergoes a Grignard reaction to form a secondary alcohol **I**. The alcohol is oxidized to the ketone **J**.

1. DIBAH, −78 °C
2. $H_3O^{\oplus}$
H
1. (phenyl)MgBr
2. $H^{\oplus}$
I
H_2CrO_4
J

(d) $LiAlH_4$ reduces the ester to the alcohol **K**. The HO group is not a good leaving group and is transformed into OTs (product **L**), which can undergo an S_N2 reaction to form the thioether product **M**.

1. $LiAlH_4$
2. $H^{\oplus}$
K
TsCl
Pyridine
L
CH_3CH_2SH
DMSO
SCH_2CH_3
M

Problem 21.60

Think

How many C atoms are in the target compound? How many are in the starting material? Which α C–H is deprotonated? Does that require a reversible or irreversible deprotonation? What is the relative positioning of the functional groups in the target? What reaction can produce a new C–C bond and result in that relative positioning? Will a functional group transformation be necessary after that reaction takes place?

Solve

(a) The starting material has six C atoms and the target has seven C atoms. The target has a 1,3-positioning of the carbonyl groups, suggesting that an aldol addition reaction can be used. The ketone enolate is formed using LDA to deprotonate the *less*-substituted C atom. Reaction with formaldehyde leads to the seven-carbon alcohol intermediate. The intermediate is oxidized, using PCC, to the aldehyde target.

LDA
1. $H_2C{=}O$
2. $H_3O^{\oplus}$
PCC

(b) The starting material has six C atoms and the target has seven C atoms. The ketone enolate is formed using NaOH to deprotonate the *more*-substituted C atom. Reaction with formaldehyde leads to the seven-carbon alcohol intermediate. The intermediate is oxidized, using PCC, to the aldehyde target.

NaOH; 1. $HCHO$; 2. H_3O^+; PCC

Problem 21.61

Think

How many C atoms are in the target compound? How many are permitted in the starting material? How was the carbon skeleton altered? Does the relative positioning of the functional groups in the target suggest which C–C bond-formation reaction might be used? What functional group transformations must take place? Are protecting groups needed?

Solve

The target has six C atoms; therefore, one reaction must be used to alter the carbon skeleton, due to the limitation of a three-carbon atom starting material. The alcohol groups are in the 1,3 position, which suggests that an aldol addition reaction occurred. The synthesis in the forward direction is shown below.

PCC; 1. NaOEt, 2. H_3O^+/H_2O; $NaBH_4$, EtOH

2-Methylpentane-1,3-diol

Problem 21.62

Think

How many C atoms are in the target compound? How many are permitted in the starting material? How was the carbon skeleton altered? Does the relative positioning of the functional groups in the target suggest which C–C bond-formation reaction might be used? What functional group transformations must take place? Are protecting groups needed?

Solve

The target has six C atoms; therefore, one reaction altered the carbon skeleton, due to the limitation of a three-carbon atom starting material. The ketone and aldehyde groups are in the 1,3 position, which suggests that a crossed Claisen condensation reaction could have occurred. The synthesis in the forward direction is shown below.

PCC; $K_2Cr_2O_7$; $SOCl_2$; EtOH

NaOEt

3-Oxo-2-methylpentanal

Problem 21.63

Think

How many C atoms are in the target compound? How many are in the starting material? How was the carbon skeleton altered? What functional group transformations took place? What sequence of reactions can produce an alkyl-substituted acetic acid? Are protecting groups needed?

Solve

The target is an alkyl substituted acetic acid that can be produced from a malonic ester synthesis. Because three C atoms need to be added to the carbon chain, 1-bromopropane can be used as the alkyl halide. The dicarboxylic acid is first transformed into the diester. The diester is alkylated at the α C, which then undergoes decarboxylation to give the desired target compound.

EtOH (excess), H^+ ; 1. NaOEt, EtOH 2. Br⁀ ; 1. NaOH, H_2O 2. H_3O^+, Δ

1,3-Propanedioic acid → Pentanoic acid

Problem 21.64

Think

How many C atoms are in the target compound? How many are in the starting material? How was the carbon skeleton altered? What functional group transformations took place? Are protecting groups needed?

Solve

The target compound has four C atoms and the starting material has three C atoms. The α,β-unsaturated aldehyde undergoes 1,4-addition by NaCN to form the nitrile. The nitrile is hydrolyzed to the carboxylic acid. The aldehyde is reduced to the alcohol, which undergoes an acid-catalyzed intramolecular Fischer esterification reaction.

NaCN ; H_3O^+ ; $NaBH_4$ (excess) ; H_2SO_4, Δ

Problem 21.65

Think

If pyridine is the nucleophile in the nucleophilic addition–elimination reaction, what is the product that forms? Are proton transfer reactions involved? Does this product have a good leaving group? How would that product react with an amine?

Solve

(a) The nucleophilic addition–elimination reaction using pyridine as the nucleophile is shown below. Notice that the pyridine does not have any N–H bonds and, therefore, an uncharged amide product is not possible.

Nucleophilic addition ; Nucleophile elimination ; + $:\ddot{Cl}:^-$

(b) The product formed in **(a)** would react with an amine to form an amide in another nucleophilic addition–elimination reaction, because the positive pyridine is a good leaving group for nucleophile elimination. Therefore, the pyridine does not interfere with the amide formation.

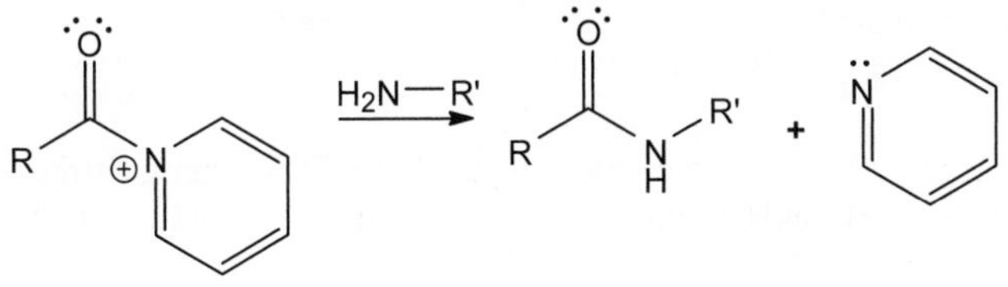

Problem 21.66

Think

In a Claisen condensation reaction, what should the nucleophile be? Can that nucleophile be produced effectively with C_6H_5ONa used as the base?

Solve

C_6H_5ONa is not a strong enough base to remove the α C–H, evidenced by the fact that its conjugate acid is much stronger than an ester. The enolate is not formed and, therefore, the Claisen condensation reaction will not occur.

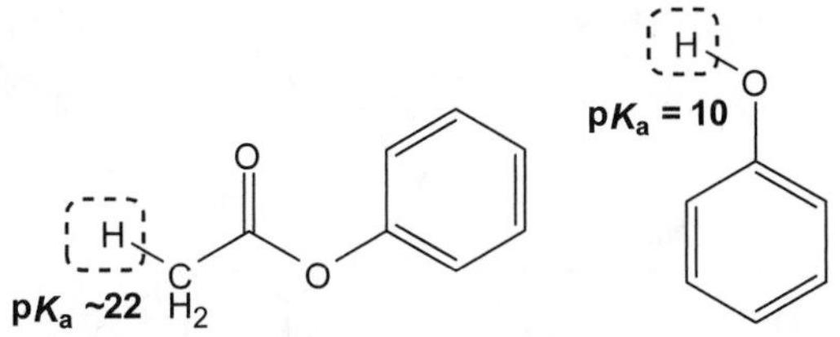

Problem 21.67

Think

How does NH_2NH_2 undergo nucleophilic addition–elimination with the acid chloride? Are proton transfer reactions involved? What is the structure of benzenesulfonyl chloride? How can it participate in a nucleophilic addition–elimination reaction?

Solve

The hydrazine NH_2NH_2 acts in the same manner as an amine in a nucleophilic addition–elimination mechanism to form an amide. The other N of hydrazine acts as the nucleophile in another nucleophilic addition–elimination mechanism with the sulfonyl chloride to yield 1-benzoyl-2-benzenesulfonyl hydrazide.

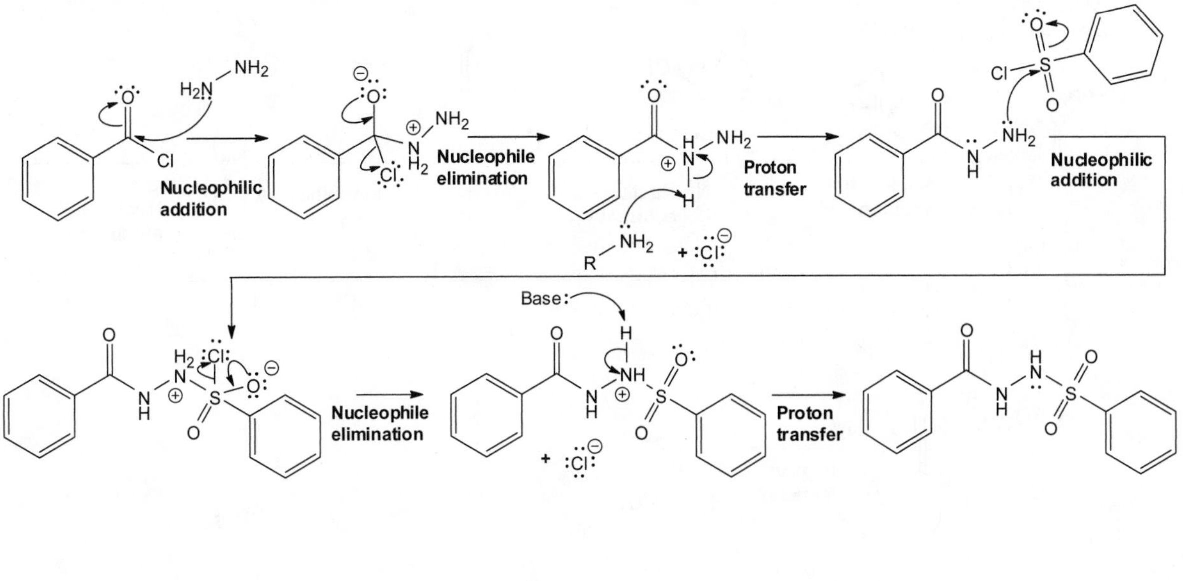

Problem 21.68

Think

What is the structure of 3-methylpentanoic acid? What new functional group is added to the α carbon? From what intermediate did this functional group originate?

Solve

All reactions require the HVZ reaction to produce the α-bromo acid intermediate. From there, different choices of nucleophiles produce the different targets—$(CH_3)_2NH$ yields the tertiary amine **(a)**, CH_3CH_2OH yields the ether **(b)**, and NaSH yields the thiol **(c)**.

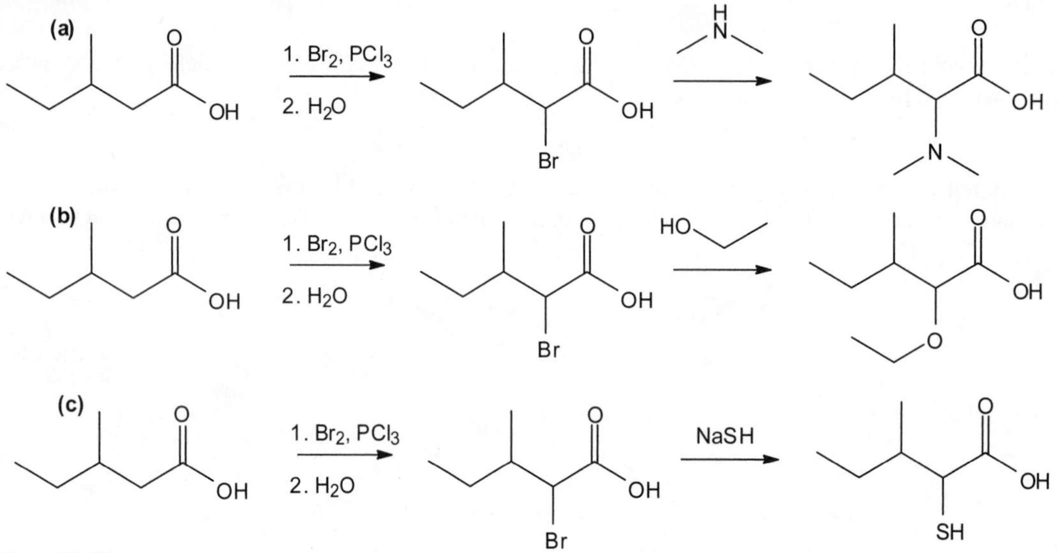

(a) 1. Br_2, PCl_3 2. H_2O → OH, Br; $(CH_3)_2NH$ → OH, N

(b) 1. Br_2, PCl_3 2. H_2O → OH, Br; HO(ethyl) → OH, O

(c) 1. Br_2, PCl_3 2. H_2O → OH, Br; NaSH → OH, SH

Problem 21.69

Think

What is the identity of the nucleophile in each nucleophilic addition? What are the leaving groups? Are proton transfer reactions involved?

Solve

The mechanism consists of a nucleophilic addition–elimination mechanism with the C=O oxygen as the nucleophile and the sulfur of $SOCl_2$ serving as the electrophilic atom. The Cl^- is the nucleophile in the second nucleophilic addition–elimination reaction.

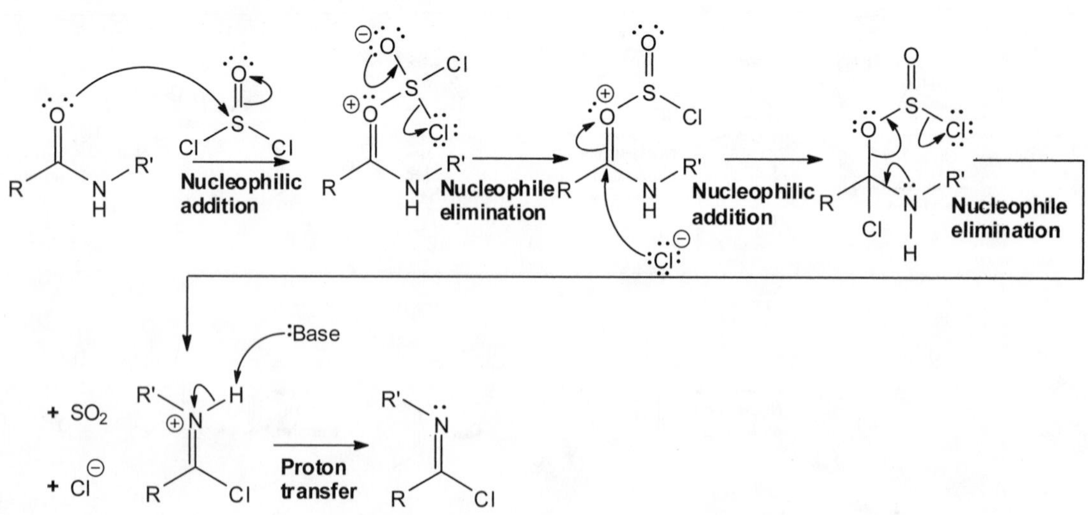

Nucleophilic addition → Nucleophile elimination → Nucleophilic addition → Nucleophile elimination → + SO_2 + Cl^-; :Base; Proton transfer

Problem 21.70

Think

How many C atoms are in the starting material? In the target? Does CH_3O^- behave as a nucleophile or a base? Can a nucleophilic addition–elimination reaction ensue? In the product of that reaction, what is the nucleophile? What is the electrophile? Is there a good leaving group?

Solve

CH_3O^- behaves as a nucleophile and attacks the ester C=O in the nucleophilic addition–elimination mechanism. The resulting enolate anion then acts as the nucleophile in a second nucleophilic addition–elimination mechanism.

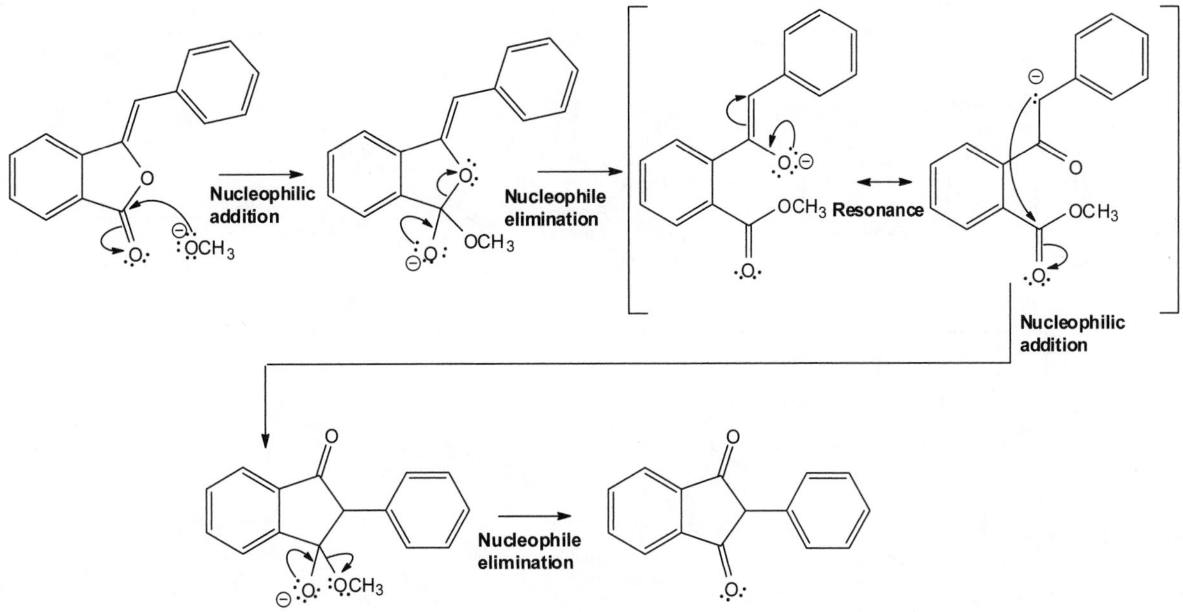

Problem 21.71

Think

What is the purpose of NaOH in the first reaction? What nucleophile is produced? What electrophile will it attack? In the hydrolysis step, how many nucleophilic addition–elimination reactions occur? With hydrolysis taking place under acidic conditions, what species should not appear in the mechanism?

Solve

The initial reaction is a proton transfer to form the N^- nucleophile, which undergoes an S_N2 reaction with the alkyl halide to form the alkyl-substituted imide. In acid, the carbonyl oxygen is protonated and then water is the nucleophile in the nucleophilic addition step. The full mechanism is given below.

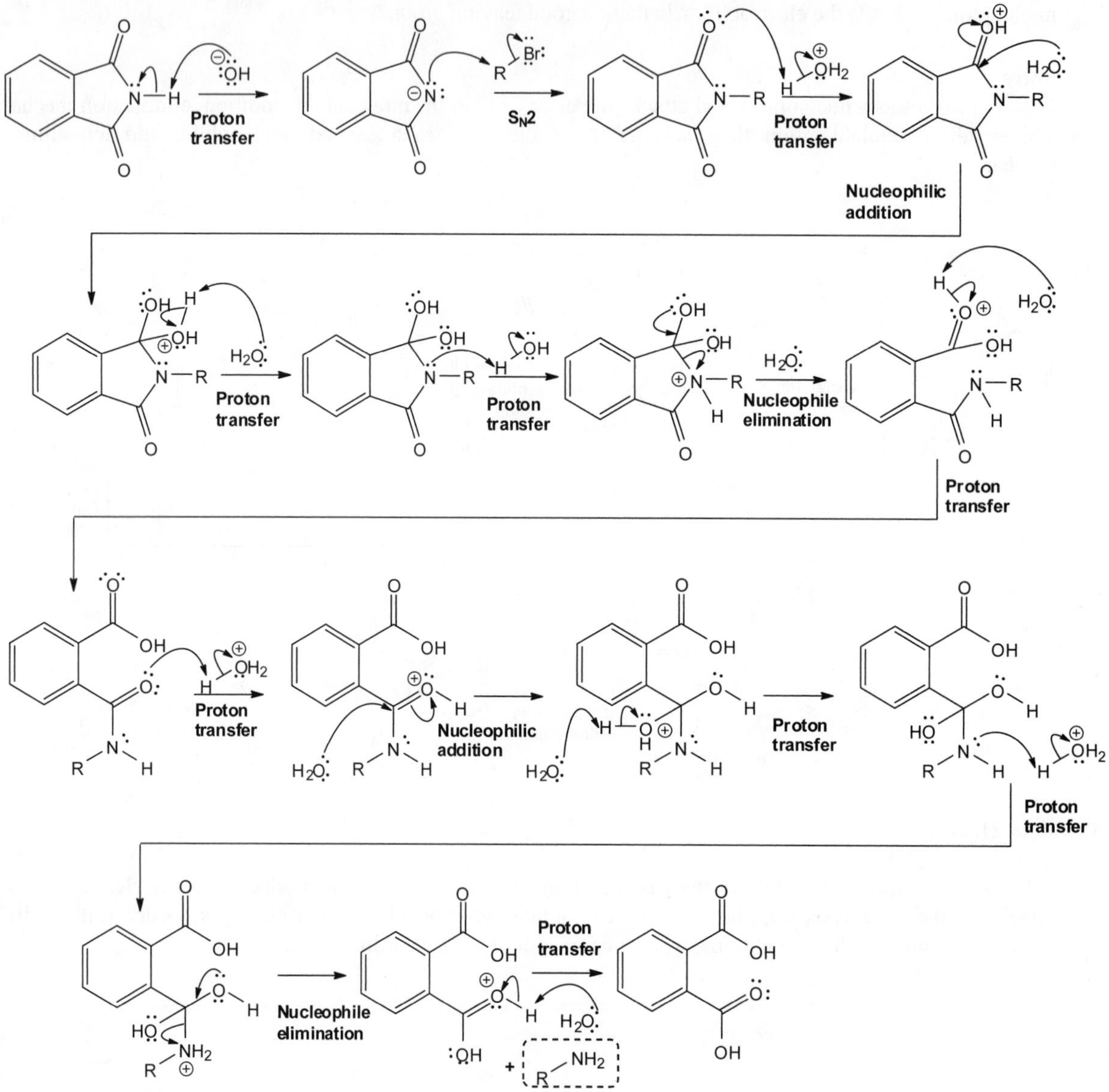

Proton transfer

S_N2

Proton transfer

Nucleophilic addition

Proton transfer

Proton transfer

Nucleophile elimination

Proton transfer

Proton transfer

Nucleophilic addition

Proton transfer

Proton transfer

Nucleophile elimination

Proton transfer

Problem 21.72

Think

What is the acidic proton that reacts with EtO^-? How is the C–R bond formed? What is the identity of intermediate **A**? With the hydrolysis taking place under acidic conditions, what types of species should not appear in the mechanism? How many times do the nucleophilic addition–elimination reactions occur?

Solve

The mechanistic steps that yield **A** are shown. Four acid-catalyzed hydrolysis reactions then take place (two on the imide and one on each of the two esters), forming the four carboxylic acid (RCO_2H) groups and freeing up the amine (NH_2) group. The dicarboxylic acid that is formed is a β-diacid and therefore undergoes decarboxylation with heat.

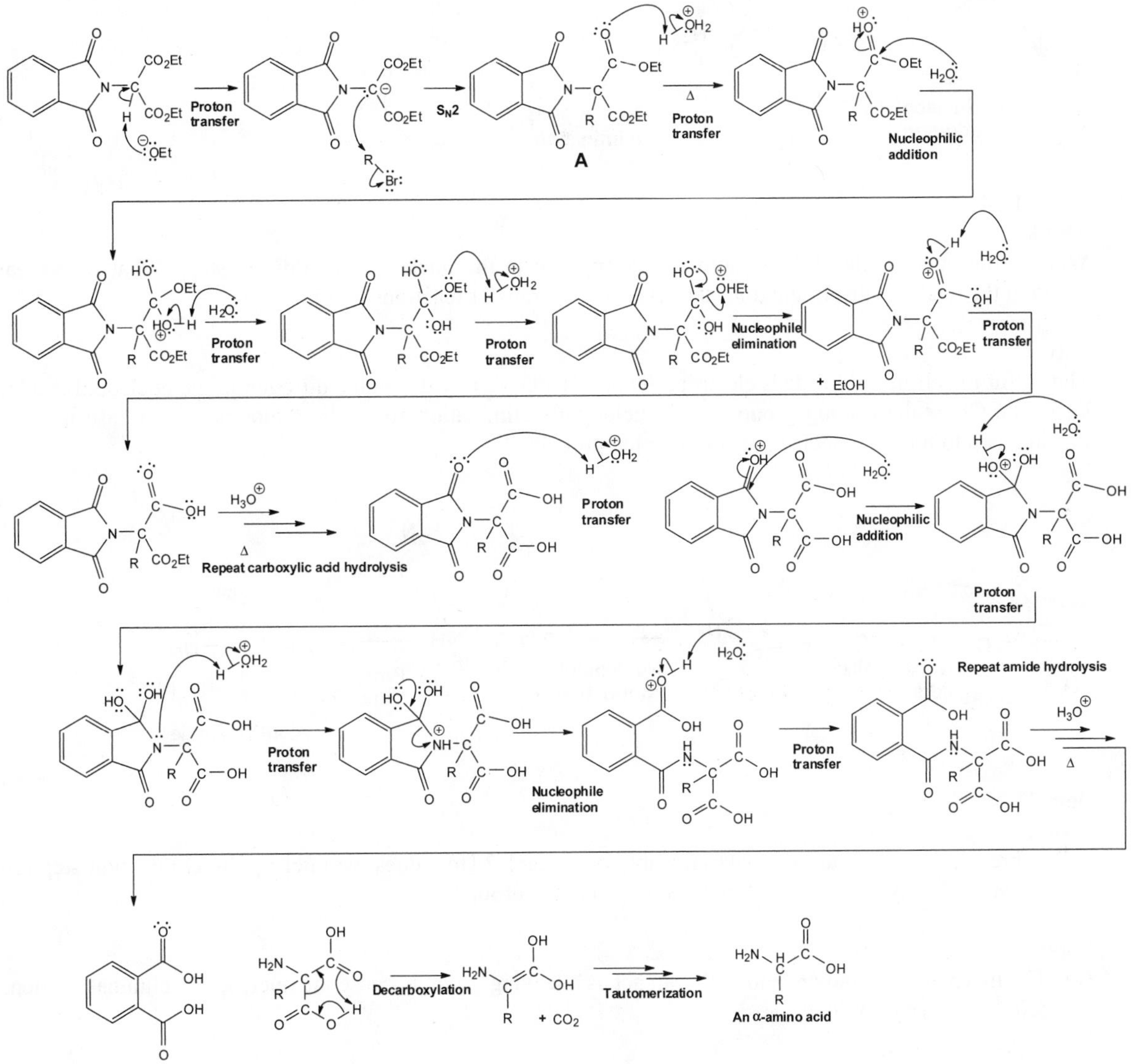

Proton transfer → S_N2 → **A** → Δ Proton transfer → Nucleophilic addition → Proton transfer → Proton transfer → Nucleophile elimination + EtOH → Proton transfer → $H_3O^{\oplus}$, Δ, Repeat carboxylic acid hydrolysis → Proton transfer → Nucleophilic addition → Proton transfer → Proton transfer → Nucleophile elimination → Proton transfer → Repeat amide hydrolysis, $H_3O^{\oplus}$, Δ → Decarboxylation, + CO_2 → Tautomerization → An α-amino acid

Problem 21.73

Think

What is the Lewis structure for CH_2N_2? What acts as the nucleophile in the nucleophilic addition step? What is the leaving group in the nucleophile elimination step? How does the formation of this leaving group drive the reaction forward?

Solve

The mechanism is much like a nucleophilic addition–elimination mechanism, except that in the elimination step, R^- is never truly formed but rather is shifted via an alkyl shift. This resembles a Baeyer–Villiger oxidation and is driven by the formation of a very stable N_2 molecule that leaves irreversibly as a gas.

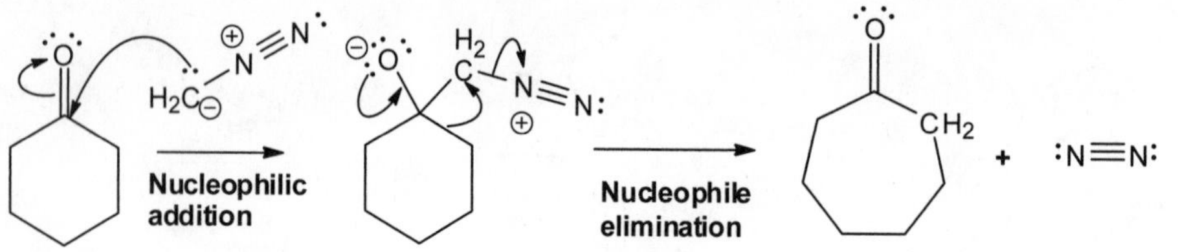

Nucleophilic addition

Nucleophile elimination

Problem 21.74

Think

What is the nucleophile? Which atom is electrophilic in the nucleophilic addition step? What is the leaving group in the nucleophile elimination step? Are proton transfer reactions involved?

Solve

The sulfur in sulfonyl chloride is electrophilic and is attacked by the amine nitrogen in the nucleophilic addition step. The Cl^- is the leaving group in the nucleophile elimination step. The amine acts as a base in a proton transfer step to form the uncharged sulfonamide product.

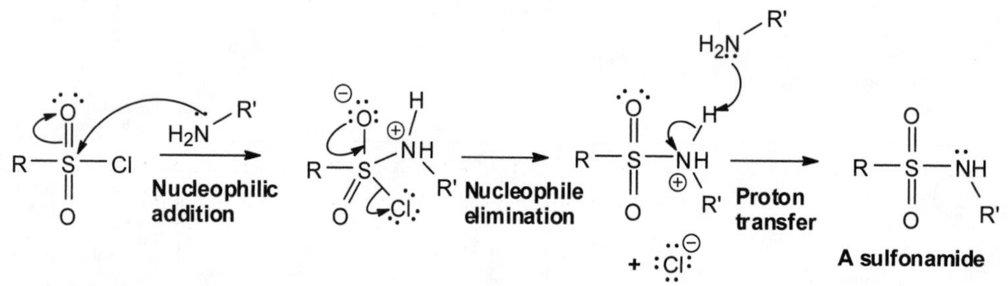

Nucleophilic addition

Nucleophile elimination

Proton transfer

A sulfonamide

Problem 21.75

Think

Is the three-carbon ring starting material stable or unstable? How does the nucleophile elimination step relieve the ring strain? Why would this allow R^- to be a leaving group?

Solve

(a) The mechanism is shown below. R^- can act as a leaving group because the nucleophile elimination step relieves the ring strain.

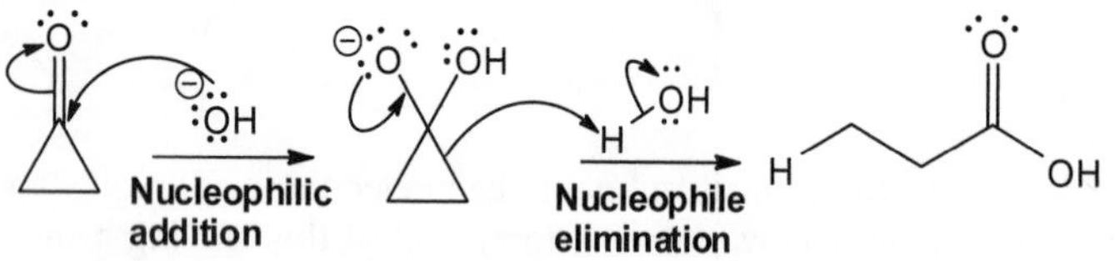

Nucleophilic addition

Nucleophile elimination

(b) To make an ester using this reaction, we can use RO^- as the nucleophile instead of HO^-.

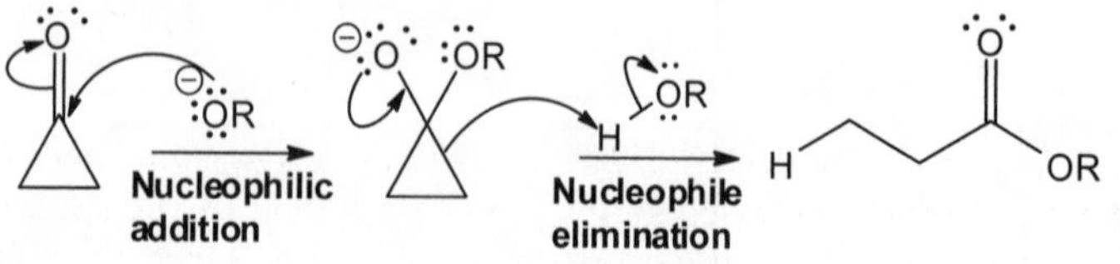

Nucleophilic addition

Nucleophile elimination

Problem 21.76

Think

What functional groups are present on the side chains of these two amino acids? What happens to these functional groups in hot, concentrated HCl? What is the nucleophile under these conditions? What is the leaving group?

Solve

The side chains of asparagine and glutamine contain amide groups. Like any amide group, the amide group in the respective amino acids is hydrolyzed under acidic, aqueous conditions to carboxylic acids.

Amide

H_2N OH NH$_2$ Asparagine $\xrightarrow[\Delta]{H_3O^+}$ HO OH NH$_2$

Amide

H_2N OH NH$_2$ Glutamine $\xrightarrow[\Delta]{H_3O^+}$ HO OH NH$_2$

Problem 21.77

Think

What acts as the nucleophile and the leaving group in the nucleophilic addition–elimination mechanism? How many times does this mechanism occur? What product forms? How many unique H atoms are in the product? Does the chemical environment of these H atoms support the appearance of a signal at 3.8 ppm?

Solve

There are two chloride leaving groups that can be replaced by methanol. The product is dimethylcarbonate, which has six equivalent H atoms, giving rise to a singlet in the NMR spectrum at 3.8 ppm. The moderate chemical shift reflects the deshielding that occurs from the O atom attached to CH_3.

Cl Cl CH_3OH → Nucleophilic addition → Nucleophile elimination → $+ :Cl:^-$ → CH_3OH Proton transfer → CH_3OH → Nucleophilic addition → Nucleophile elimination → $+ :Cl:^-$ CH_3OH → Proton transfer → H_3C O O CH_3

Problem 21.78

Think

How does diethyl malonate react with a base and an alkyl halide? How many times does this kind of reaction occur? Is an intramolecular reaction favored? What functional groups are evident from the IR spectrum? How many ^{13}C signals are present? Does the number of ^{13}C signals suggest there should be symmetry? What gives rise to the signal at 180 ppm?

Solve

This reaction sequence is a malonic ester synthesis. The second alkylation is intramolecular, producing **B**. Then, hydrolysis and decarboxylation result in the final product, **C**. For product **C**, the IR spectrum is consistent with a carboxylic acid (i.e., a broad OH stretch at 3100 cm^{-1} and a C=O stretch at ~1740 cm^{-1}). There are five C signals in the NMR spectrum, consistent with the five distinct C atoms in the product. The ^{13}C signal at 180 ppm is due to the carbonyl carbon. The reaction sequence is shown below:

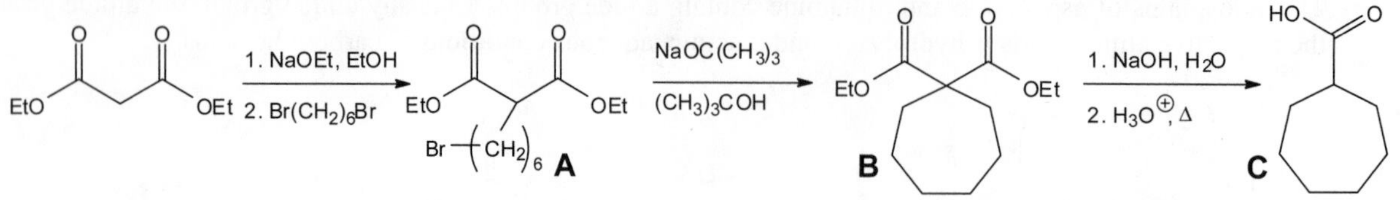

EtO–C(=O)CH$_2$C(=O)–OEt → 1. NaOEt, EtOH; 2. $Br(CH_2)_6Br$ → **A** → $NaOC(CH_3)_3$, $(CH_3)_3COH$ → **B** → 1. NaOH, H_2O; 2. $H_3O^{\oplus}$, Δ → **C**

Problem 21.79

Think

What is the structure of methyl 2-methylpropanoate? What is the product of the condensation reaction of an ester? How many signals are in each NMR spectrum? What structural information is evident from the NMR spectra? What functional groups are evident from the IR? Why did the reaction not go as planned?

Solve

The student expected this reaction, which is a Claisen condensation:

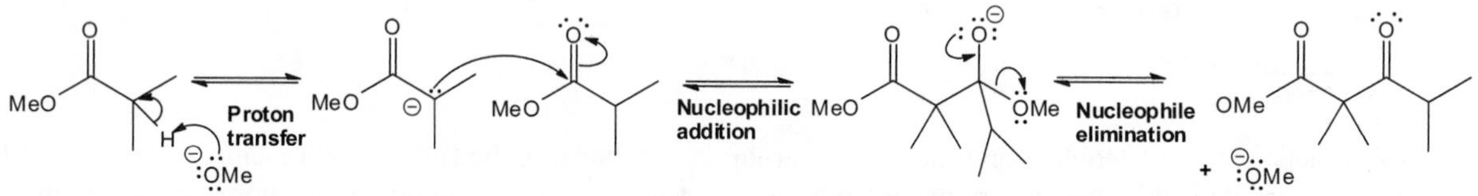

Proton transfer ⇌ Nucleophilic addition ⇌ Nucleophile elimination + $^{\ominus}$:OMe

However, Claisen condensation reactions are reversible and favor the reactants unless the product β-keto ester has an acidic proton in between the two carbonyl groups. This one doesn't. Therefore, the student simply isolated the starting material. This is consistent with the M^+ peak at 102 in the mass spectrum, and three signals of 1H, 3H, and 6H in the ^1H NMR spectrum.

Problem 21.80

Think

What functional group is insoluble in water but soluble in basic solution? What functional groups are evident from the IR? How many aromatic H signals are there? What are the relative ratios? What causes the signal at 4.3 ppm to have a downfield shift? What does the quartet:triplet in a 2:3 ratio signify?

Solve

An insoluble compound that is soluble under basic conditions suggests a carboxylic acid. That carboxylic acid could be made by hydrolysis of either an ester or an amide, which would also be insoluble in water. Indeed, the absorbance around 1700 cm^{-1} in the IR spectrum suggests a C=O bond. The signals at ~7–8 ppm in the ^1H NMR spectrum suggest a benzene ring. In the ^{13}C NMR spectrum, there are three aromatic signals, so we probably have a monosubstituted benzene ring. In the ^1H NMR spectrum, the triplet and quartet are consistent with a CH_3CH_2 group attached to an electronegative atom like N or O. The integration of those signals suggests that the number of ethyl hydrogens is the same as the aromatic hydrogens: five. So there is only one ethyl group, which rules out an *N,N*-diethyl amide. And since the IR spectrum does not indicate any NH bonds, we likely have an ester with the following structure:

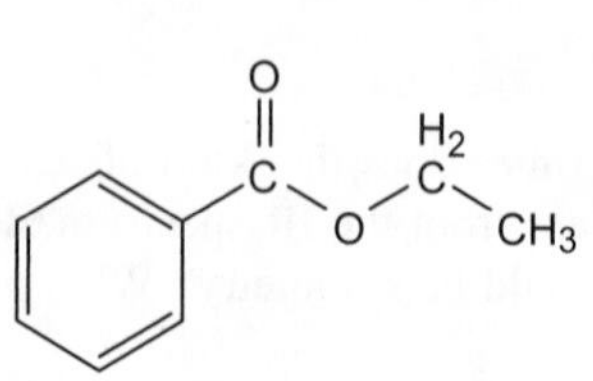

C_6H_5–C(=O)–O–CH_2–CH_3

Problem 21.81

Think

What functional groups are evident from the IR spectrum? What is the IHD? Based on the given molecular formula and the reactant ethanol, how many C atoms are in the reactant? How many ^1H NMR signals are present? What are the splitting patterns, integration ratios, and chemical shifts?

Solve

We are faced with the following question:

? → (ethanol, OH) $C_6H_{10}O_4$

In the IR spectrum of the product, there are no OH groups or NH stretches, but there is a C=O stretch near 1700 cm^{-1}. So we could have a ketone, ester, or *N,N*-dialkyl amide. The C=O stretch appears to be around 1740 cm^{-1}, so it is likely an ester. Because there are four O atoms, we could have a diester. The IHD is 2, which is accounted for by the two ester groups, so there should be no rings or other double or triple bonds. The NMR spectrum has a quartet and a triplet, this suggests the presence of an ethyl group. The downfield signal is 4 H, whereas the upfield one is 6 H. The 6 H is due to two symmetrical CH_3 groups coupled to a CH_2, and the 4 H signal is due to two symmetrical CH_2 groups coupled to a CH_3. Also, the CH_2 methyl groups are downfield, so they should be attached to the oxygen atom of the ester. The structure could be as follows:

$H_3C-CH_2-O-C(=O)-C(=O)-O-CH_2-CH_3$

And this compound could be formed from reaction of acetyl chloride with ethanol, as follows:

$Cl-C(=O)-C(=O)-Cl$ —(H_3C-CH_2-OH, Excess)→ $H_3C-CH_2-O-C(=O)-C(=O)-O-CH_2-CH_3$

CHAPTER 22 | Electrophilic Aromatic Substitution 1: Substitution on Benzene; Useful Accompanying Reactions

Your Turn Exercises

Your Turn 22.1

Think

What would the product be for the HCl addition to a C=C of benzene? What would the product of Cl_2 addition be? Review the requirements for aromaticity in Section 14.4. Is there still a cyclic, fully conjugated π system containing a Hückel number of electrons? Are these products still aromatic?

Solve

The products of HCl and Cl_2 addition are shown below. In both cases, aromaticity has been lost because the π system is no longer fully conjugated around the ring and no longer contains an odd number of pairs of electrons.

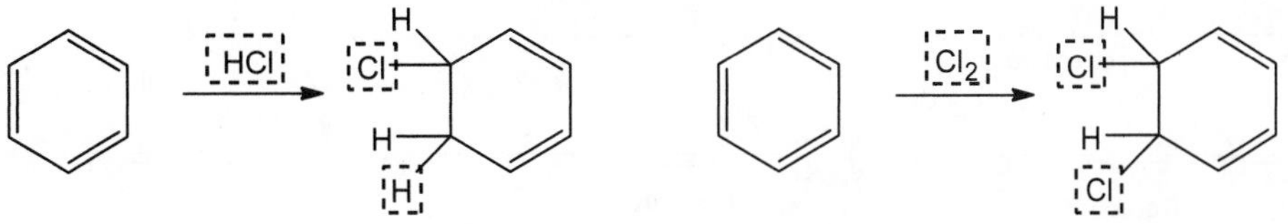

Your Turn 22.2

Think

Review atomic hybridization in Section 3.4. How many electron groups does each C atom have? How does that number of electron groups correspond to electron geometry and to hybridization?

Solve

Each doubly bonded C has three electron groups, and so does the positively charged C. These C atoms have a trigonal planar electron geometry, which corresponds to an sp^2 hybridization. The C atom attached to E has four electron groups, which corresponds to a tetrahedral geometry and an sp^3 hybridization. The arenium ion intermediate has five sp^2-hybridized C atoms and one sp^3-hybridized C atom.

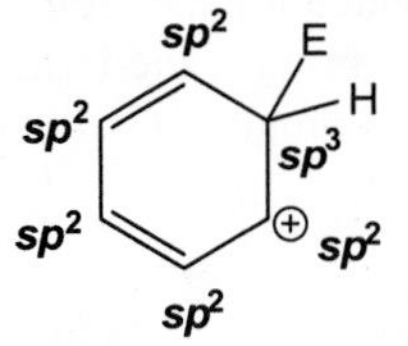

Your Turn 22.3

Think

Are there any incomplete octets adjacent to a double or triple bond? Does the sp^3-hybridized C participate in resonance?

Solve

The carbocation C^+ is sp^2 hybridized and has an incomplete octet. Resonance is possible among all five sp^2-hybridized C atoms. The sp^3-hybridized C is a dead-end for resonance.

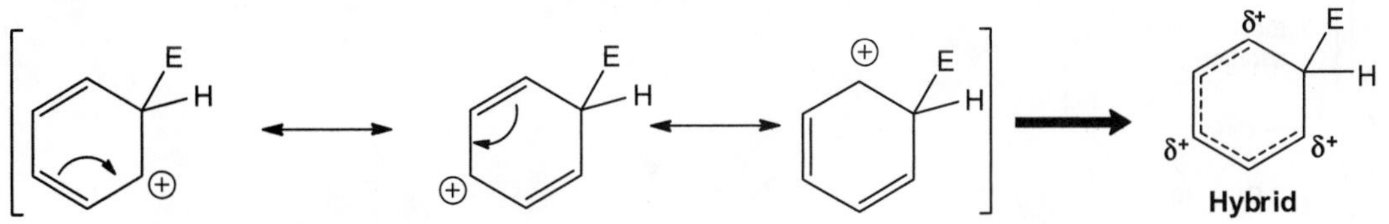

Hybrid

Your Turn 22.4

Think

Which bonds were broken and formed in each step? Which sites are electron rich and which are electron poor? How can you use curved arrows to show the flow of electrons in each step? Review the 10 elementary steps from Chapters 6 and 7. Which steps involve the formation of the electrophile? What is the electrophile? Which steps temporarily break and reform aromaticity in the ring, resulting in substitution of H^+ by an electrophile?

Solve

Steps 1 and 2 involve the formation of the Cl^+ electrophile. In Step 1, $FeCl_3$ acts as a Lewis acid (electron-pair acceptor) and complexes with Cl_2 in a coordination step. In Step 2, heterolysis of Cl–Cl takes places slowly to produce Cl^+ and $FeCl_3^-$. Steps 3 and 4 formally make up the electrophilic aromatic substitution mechanism with Cl^+ as the electrophile. Step 3 is electrophilic addition and Step 4 is electrophile elimination. $FeCl_3$ is regenerated at the end of the reaction and is, therefore, a catalyst. It should be noted that electrophilic addition and heterolysis steps probably occur simultaneously instead of separately but the electrophile Cl^+ is shown explicitly for emphasis.

Formation of the Cl⁺ electrophile

1. Coordination

2. Heterolysis

3. Electrophilic addition

4. Electrophile elimination

(+ $FeCl_3$ + HCl)

Your Turn 22.5

Think

Which bonds were broken and formed in each step? Which sites are electron rich and which are electron poor? How can you use curved arrows to show the flow of electrons in each step? Which steps involve the formation of the electrophile? What is the electrophile? Which steps temporarily break and re-form aromaticity in the ring, resulting in substitution of H^+ by an electrophile? Review the mechanism in Equation 22-4.

Solve

Steps 1 and 2 involve the formation of the cyclohexyl C^+ electrophile. Steps 3 and 4 formally make up the electrophilic aromatic substitution mechanism with the cyclohexyl C^+ as the electrophile.

Formation of R⁺ electrophile

1. Coordination

2. Heterolysis

3. Electrophilic addition

4. Electrophile elimination

(+ HCl + $AlCl_3$)

Your Turn 22.6

Think

Consult Figure 3-33 for the C–H bond strengths of an sp^3 C–H bond compared to an sp^2 C–H bond.

Solve

The C–H bond energy for an sp^3-hybridized C is 410 kJ/mol, and that for an sp^2-hybridized C is 431 kJ/mol.

Your Turn 22.7

Think

Which bonds were broken and formed in each step? Which sites are electron rich and which are electron poor? How can you use curved arrows to show the flow of electrons in each step? Which steps involve the formation of the electrophile? What is the electrophile? Which steps temporarily break and reform aromaticity in the ring, resulting in substitution of H^+ by an electrophile? Review the mechanism in Equation 22-4.

Solve

Steps 1 and 2 involve the formation of the acylium ion electrophile. Steps 3 and 4 formally make up the electrophilic aromatic substitution mechanism with the acylium ion as the electrophile.

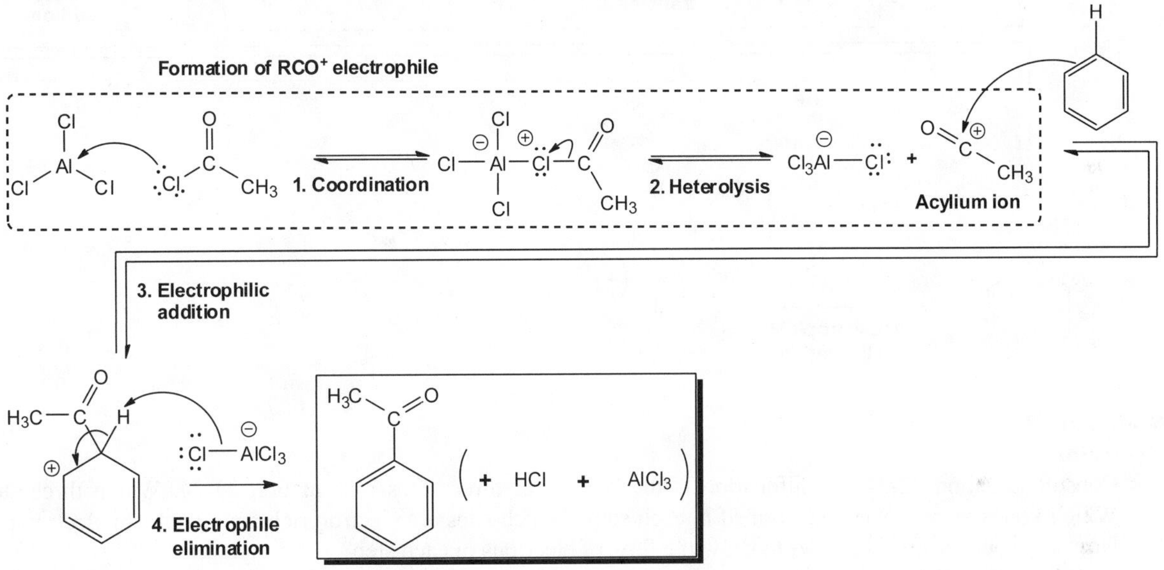

Formation of RCO^+ electrophile

1. Coordination
2. Heterolysis

Acylium ion

3. Electrophilic addition
4. Electrophile elimination

(+ HCl + $AlCl_3$)

Your Turn 22.8

Think

Review Section 1.10 to draw a resonance hybrid. How do you represent a bond intermediate between a double and a triple bond? Which atoms share the positive charge?

Solve

A resonance hybrid is an average of all resonance structures. The carbon–oxygen bond is between a double and triple bond, represented as two solid bonds and a dashed bond. The positive charge is dispersed on both C and O, so each receives a δ^+ in the hybrid.

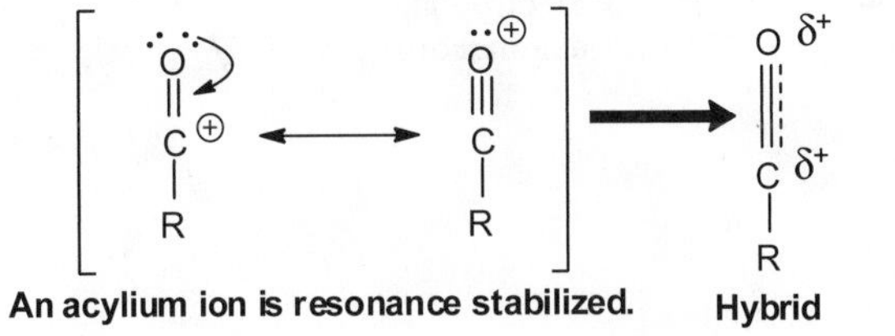

An acylium ion is resonance stabilized. **Hybrid**

Your Turn 22.9

Think

Consult Equation 22-22 in writing a complete mechanism for the nitration of benzene in the presence of sulfuric acid, H_2SO_4. How is the mechanism similar/different compared to Equation 22-22? What is the purpose of H_2SO_4?

Solve

Sulfuric acid is a stronger acid compared to HNO_3 and serves to protonate nitric acid in Step 1. This generates a substantially higher concentration of the nitronium electrophile, NO_2^+. Other than that, the mechanism is identical to the mechanism is Equation 22-22.

Formation of the nitronium ion NO_2^+

1. Proton transfer

2. Heterolysis

+ H_2O

3. Electrophilic addition

H_2O:

4. Electrophile elimination

NO_2

(+ H_3O^+ + HSO_4^-)

Your Turn 22.10

Think

Consult Equation 22-24. Desulfonation is the reverse of three steps in Equation 22-24. Which three steps? Which bonds were broken and formed in each step? Which sites are electron rich and which are electron poor? How can you use curved arrows to show the flow of electrons in each step?

Solve

Desulfonation is the reverse of Steps 4–6 in Equation 22-24.

H_2O:

1. Proton transfer

+ H–OH_2^+

2. Electrophilic addition

H_2O:

3. Electrophile elimination

Chapter Problems

Problem 22.1

Think

What is the rate law for the reaction? What is the order of each reactant? How does the reaction rate change upon doubling (electrophile) or tripling (benzene)?

Solve

The rate law for the reaction is: **Rate = k[E$^+$][Benzene]**

The reaction would be six times faster. Because the reaction is first order with respect to benzene, tripling the concentration of benzene triples the reaction rate. Also, the reaction is first order with respect to the electrophile, so doubling the concentration of the electrophile doubles the reaction rate.

$$\textbf{Rate} = k[2 \times E^+][3 \times \text{Benzene}] \rightarrow 6\times \text{ the rate}$$

Problem 22.3

Think

What reaction can be used to form the necessary C–C bond? Does the product of that reaction contain the appropriate functional groups? What precursors are necessary to carry out the C–C bond formation reaction? How must halogenation be incorporated to produce the appropriate precursor?

Solve

The product can be made from the alcohol produced in Solved Problem 22.2, but the OH group is not a good leaving group for a substitution reaction. PBr_3 can be used to convert it into a good leaving group, and then a substitution reaction can be carried out, using $NaSCH_3$ as the nucleophile.

SCH_3 Undo an S_N2 ⟹ Br Undo a bromination. ⟹ OH

Alcohol product from Solved Problem 22.2

In the forward direction, the synthesis would be written as follows:

Br_2, $FeBr_3$ → Br; Mg(s) → MgBr; 1. (propanal) 2. H_3O^+ → OH; PBr_3 → Br; $NaSCH_3$ → SCH_3

Problem 22.4

Think

What is the electrophile? What two elementary steps make up the electrophilic aromatic substitution reaction? What acts as the base to remove the H$^+$ and regenerate the aromatic ring?

Solve

The I^+ cation is the electrophile; once formed, it can be attacked by the aryl ring. I^- acts as the base and can remove a proton from the arenium ion intermediate.

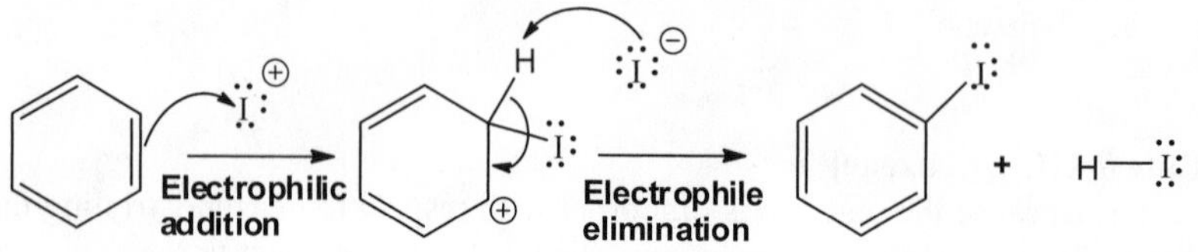

Electrophilic addition

Electrophile elimination

Problem 22.5

Think

Does the $(CH_3)_3COH$ have a good leaving group? How does a reaction with H_3PO_4 generate a good leaving group? What is the strong electrophile that is produced? What two elementary steps make up the electrophilic aromatic substitution reaction? What acts as the base to remove the H^+ and regenerate the aromatic ring?

Solve

$(CH_3)_3COH$ does not have a good leaving group. Reaction in an acid (H_3PO_4) protonates the alcohol and turns it into $(CH_3)_3CO^+H_2$. Water now acts as the leaving group and thus generates the carbocation via heterolysis. Once the strongly electrophilic carbocation is formed, the electrophilic aromatic substitution mechanism is identical to the corresponding steps in Equation 22-12. The ROH acts as the base to remove the proton and regenerate the aromaticity.

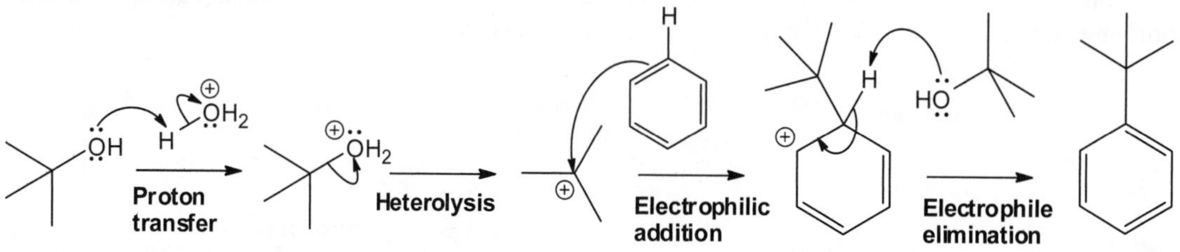

Proton transfer

Heterolysis

Electrophilic addition

Electrophile elimination

Problem 22.6

Think

What reaction occurs between a strong acid and an alkene? What is the strong electrophile that results? What two elementary steps make up the electrophilic aromatic substitution reaction? What acts as the base to remove the H^+ and regenerate the aromatic ring?

Solve

Attack of cyclohexene on a proton forms the carbocation electrophile (Chapter 11). Then the electrophilic aromatic substitution mechanism occurs via the usual electrophilic addition–elimination route.

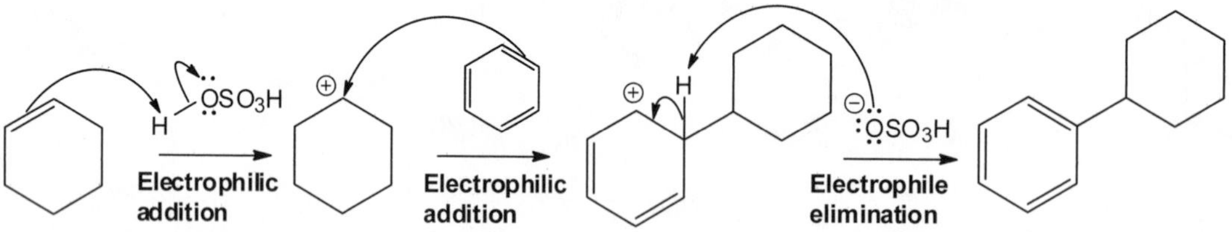

Electrophilic addition

Electrophilic addition

Electrophile elimination

Problem 22.8

Think

What is the electrophile that must substitute for H^+? How can that electrophile be produced from an alkyl halide, an alkene, or an alcohol?

Solve

The electrophile that must substitute for H^+ as a carbocation is as follows:

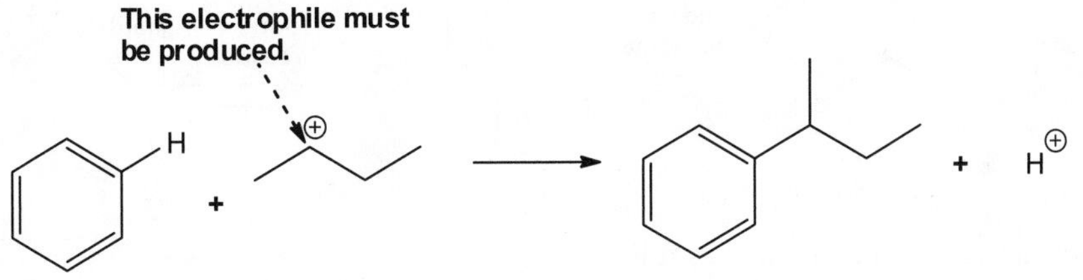

This electrophile must be produced.

The carbocation can be produced from an alkyl halide using $AlCl_3$, a strong Lewis acid catalyst, or it can be produced from an alkene or alcohol under acidic conditions.

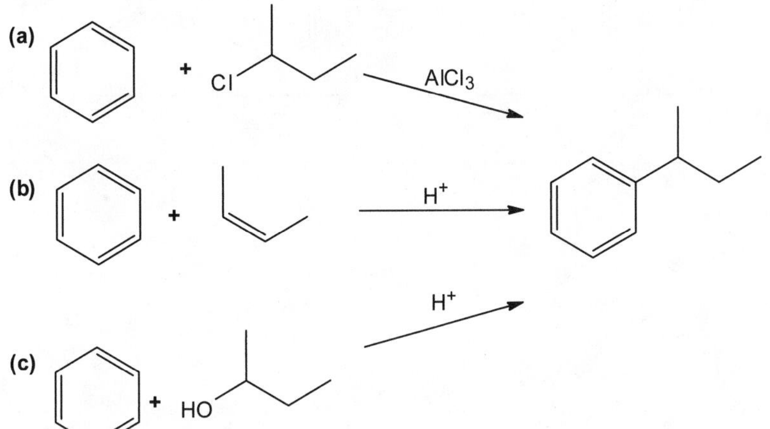

(a) $AlCl_3$

(b) H^+

(c) H^+

Problem 22.10

Think

Can the target be produced directly from a simple Friedel–Crafts alkylation? What carbocation intermediate would be produced? Would that carbocation intermediate rearrange? Is there another reaction that can be used to form the appropriate C–C bond?

Solve

The product forms the C–C bond from the aromatic ring and a 1° carbon. This cannot be formed from a Friedel–Crafts alkylation due to a carbocation rearrangement that would take place prior to electrophilic addition.

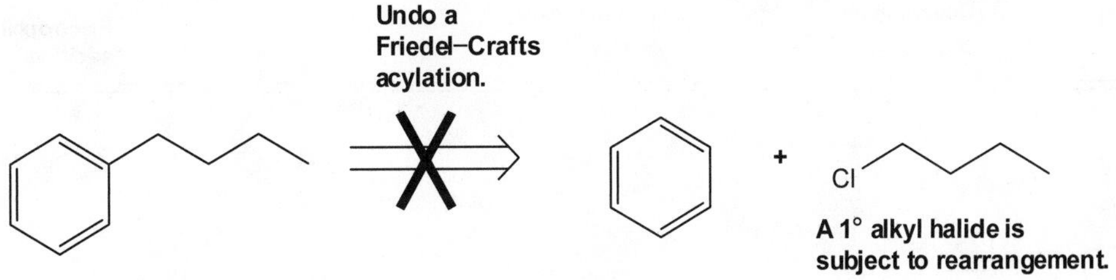

Undo a Friedel–Crafts acylation.

A 1° alkyl halide is subject to rearrangement.

We can use a Grignard addition to form the C–C. This is followed by dehydration of the resulting alcohol, then hydrogenation, to remove the functional groups on the chain. See the figure on the next page.

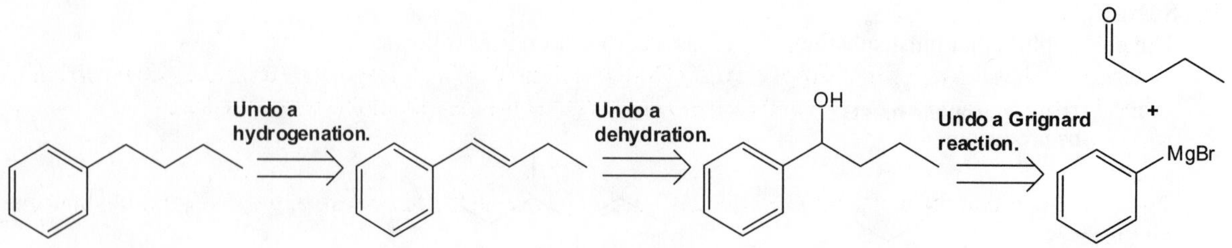

In the forward direction, the synthesis would be written as follows:

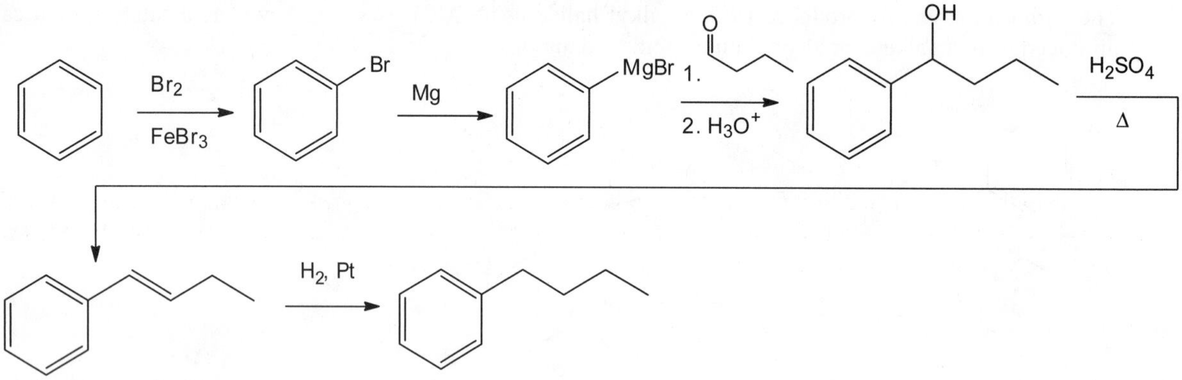

Problem 22.11

Think

Which atom in the anhydride forms a complex to the strong Lewis acid? What can act as the leaving group in that complex? What is the strong electrophile that results? What two elementary steps make up the electrophilic aromatic substitution reaction? What acts as the base to remove the H^+ and regenerate the aromatic ring?

Solve

The mechanism is identical to that with the acid chloride (see Equation 22-19), with the $[CH_3CO_2–AlCl_3]^-$ as the leaving group instead of $[Cl–AlCl_3]^-$.

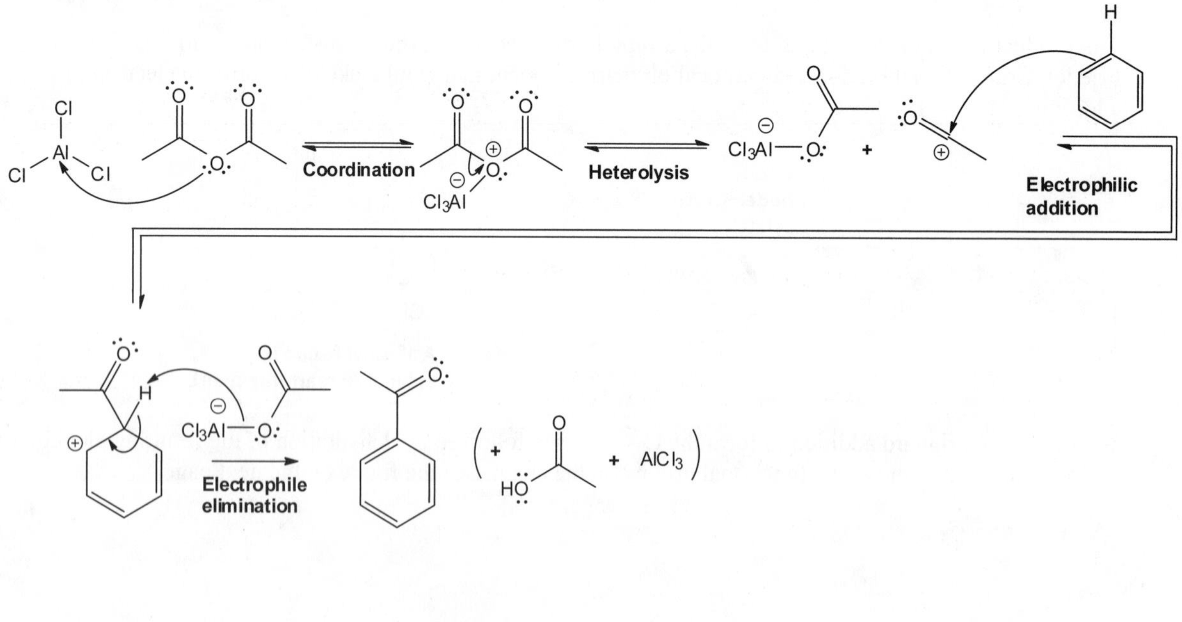

Problem 22.12

Think

Which C–C bond was formed from Friedel–Crafts acylation? Which ring came from the benzene ring? What other functional group is present on the other ring? What is the identity of the acid anhydride?

Solve

The C–C bond that formed from the Friedel–Crafts acylation is the one between the C=O and the benzene ring on the right.

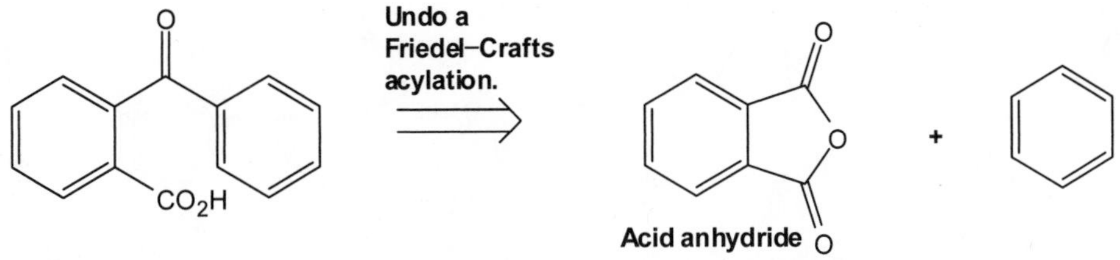

Undo a Friedel–Crafts acylation.

CO_2H

Acid anhydride

The anhydride is shown above. If this anhydride enters into the mechanism shown above for Problem 22-8, we arrive at the product shown. This is because when the anhydride dissociates in the second step, the two carbon-containing portions are attached to the same ring.

Problem 22.14

Think

The product is an aromatic ketone, so you should consider a Friedel–Crafts acylation. What precursors are necessary to produce the target from a Friedel–Crafts acylation reaction? How can those precursors be produced from the starting material?

Solve

In a Friedel–Crafts acylation, an aromatic carbon forms a bond to an acyl carbon. We begin our retrosynthetic analysis by disconnecting that bond to arrive at the appropriate precursors—an aromatic ring and an acyl chloride. The acyl chloride can be produced by chlorinating the corresponding carboxylic acid, which can be produced, in turn, by oxidizing a primary alcohol.

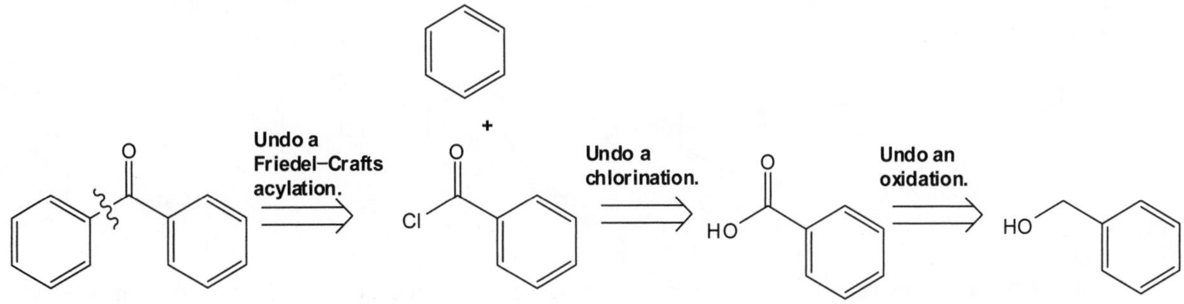

Undo a Friedel–Crafts acylation.

Undo a chlorination.

Undo an oxidation.

In the forward direction, the synthesis would be written as follows:

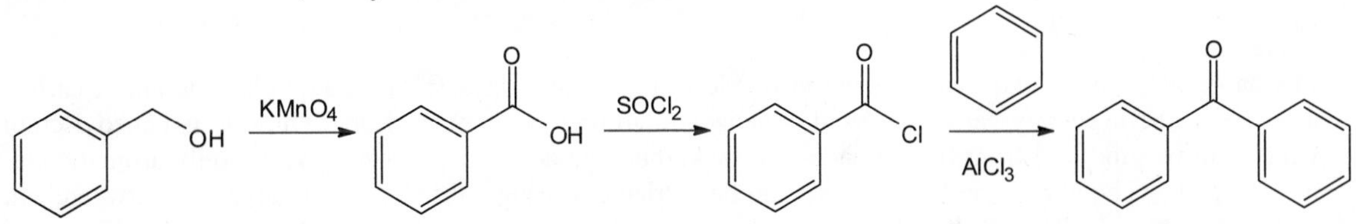

$KMnO_4$

$SOCl_2$

$AlCl_3$

Problem 22.16

Think

The product is an alkylbenzene. Can a Friedel–Crafts alkylation be used to form the C–C bond to benzene? If not, why not? How can a Friedel–Crafts acylation reaction be incorporated into this synthesis? Which step in Solved Problem 22.15 uses the Clemmensen reduction? Are there other methods to reduce a ketone?

Solve

The reduction could be carried out using a Wolff–Kishner or Raney-nickel reduction. Alternatively, it could be carried out with a hydride reduction, dehydration, and then a hydrogenation, as shown below.

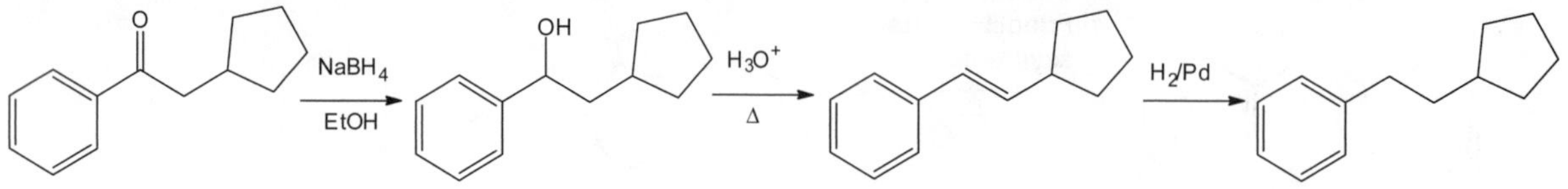

Problem 22.17

Think

Did Friedel–Crafts alkylation or acylation occur to form the new C–C bonds? How many Friedel–Crafts reactions had to occur? What was the identity of the Friedel–Crafts starting material? What other reaction occurred to form the target? Are oxidation or reduction reactions necessary?

Solve

Friedel–Crafts alkylation cannot be done because the carbocation formed will be a primary carbocation and will rearrange to a secondary carbocation. The synthesis, however, can be carried out with a Friedel–Crafts acylation. Friedel–Crafts acylation occurs twice with the following reagent as the starting material:

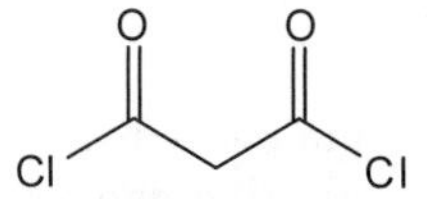

The Friedel–Crafts acylation followed by the Clemmensen (or Wolff–Kishner) reduction is sufficient.

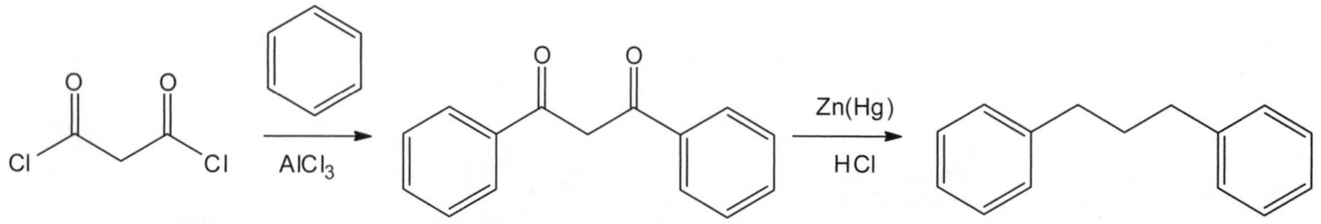

Problem 22.19

Think

From what precursors can an acid anhydride be produced? To produce those precursors from the available starting materials, are carbon–carbon bond-formation reactions necessary? Can a Friedel–Crafts alkylation or acylation be used to form those bonds? Are oxidation or reduction reactions necessary?

Solve

In Chapter 21, we learned that an acid anhydride can be synthesized from an acid chloride and a carboxylic acid. Here, the necessary carboxylic acid is benzoic acid and the necessary acid chloride is benzoyl chloride, which can be produced from benzoic acid. To make benzoic acid using benzene as the only aromatic starting material, a C–C bond must be formed. This can be carried out using a Friedel–Crafts acylation between benzene and acetyl chloride. The resulting ketone can be oxidized to benzoic acid using $KMnO_4$. See the figure on the next page.

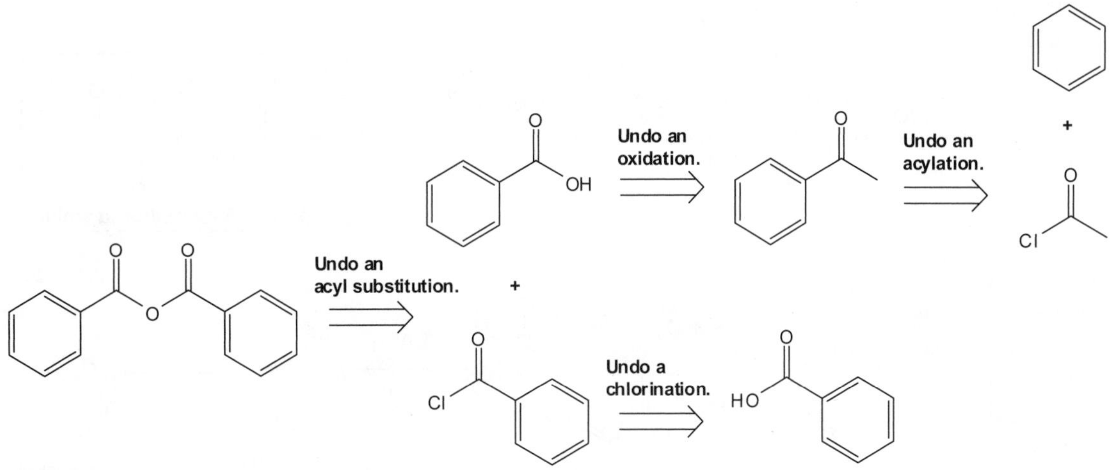

In the forward direction, the synthesis would proceed as follows:

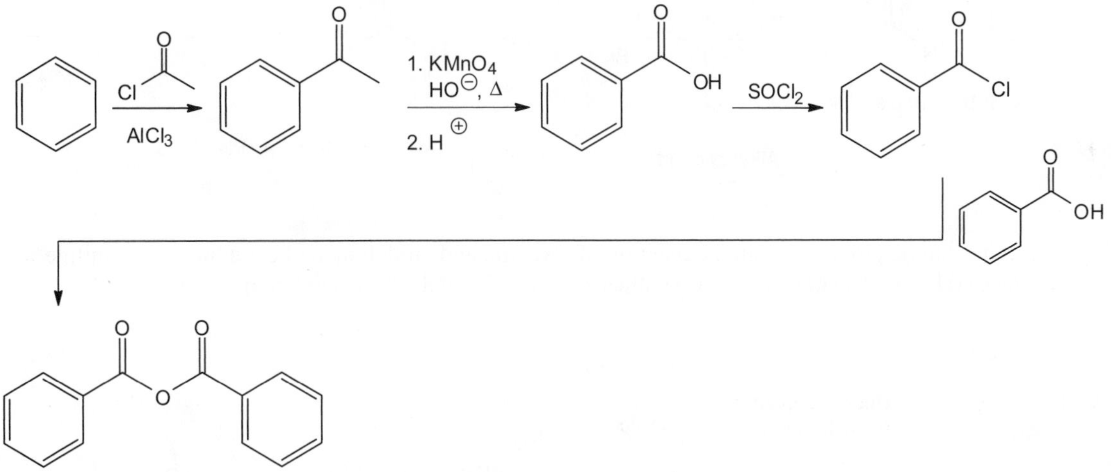

Problem 22.21

Think

What acid derivative and amine precursors will produce the amide target? Does each of those precursors have the same carbon backbone as the benzene ring, or will you need to produce a new carbon–carbon bond? What reactions add a carbon substituent to an aromatic ring?

Solve

(a) An *N,N*-disubstituted amide can be produced from an *N*-substituted amide and an alkyl halide. The *N*-substituted amide can be produced from an amine and an acid chloride—in this case, aniline and ethanoyl chloride. Aniline can be produced by the nitration of benzene, followed by reduction of the nitrobenzene product. The alkyl halide can be produced from Friedel–Crafts alkylation followed by oxidation to form benzoic acid. Benzoic acid is then reduced to form the alcohol, which is then brominated.

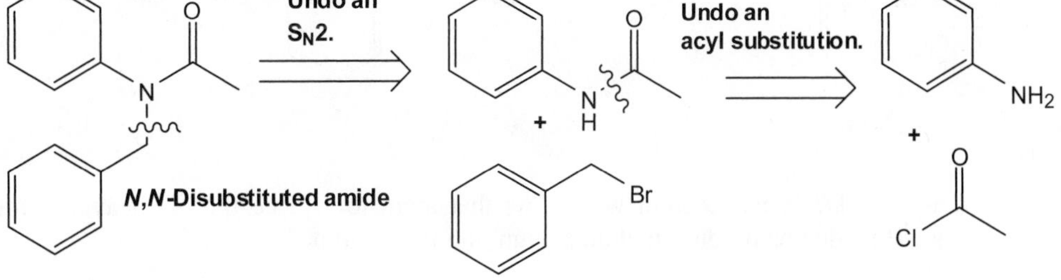

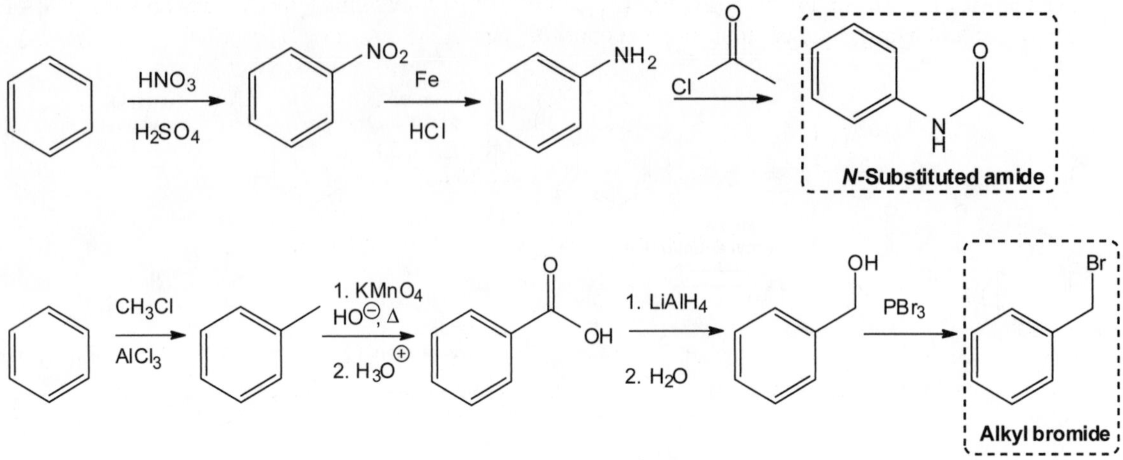

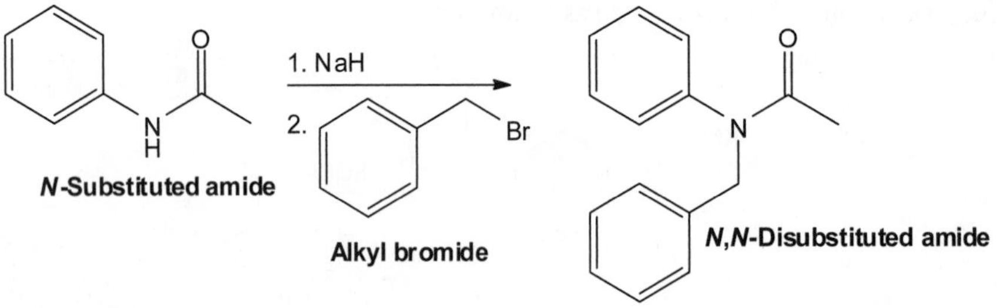

(b) The imine can be produced from the reaction of a ketone and aniline in acidic medium. The aniline was synthesized in **(a)**. The ketone can be synthesized from Friedel–Crafts acylation.

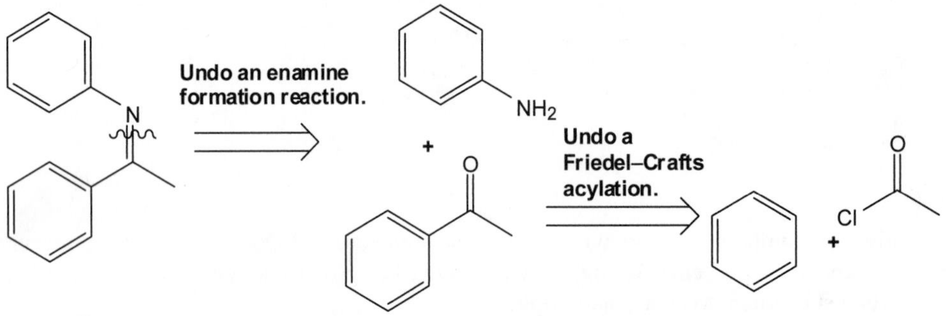

The forward synthesis would proceed as follows:

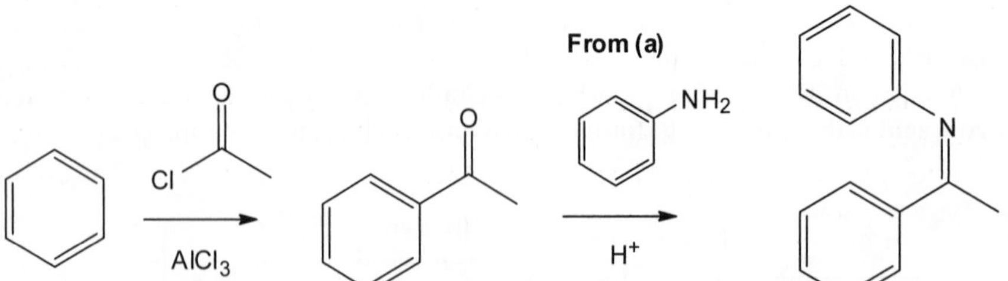

Problem 22.22

Think

What atom has the isotope D? What reaction will allow that atom to replace the N_2^+ leaving group in an benzenediazonium ion? How do you produce that diazonium ion from aniline?

Solve

Deuterium, D, is an isotope of hydrogen. The aniline is first transformed into the diazonium ion, which can be converted to the target compound using deuterated hypophosphorous acid, D_3PO_2.

Aniline — $NaNO_2$, HBr, <5 °C → The benzene diazonium ion — D_3PO_2 → Deuterobenzene

Problem 22.24

Think

Can an alcohol group replace H on a benzene ring directly? If not, what precursor is required? Can that precursor be synthesized from benzene using electrophilic aromatic substitution?

Solve

We have not encountered a reaction in which an alcohol group directly replaces an H on benzene. However, phenol can be made from a benzenediazonium ion precursor, $C_6H_5\text{–}N_2^+$. The precursor to the benzenediazonium ion is an aromatic amine, which can be made, in turn, by reducing nitrobenzene. Finally, nitrobenzene can be made directly from benzene via nitration.

OH ⟹ (Undo phenol formation.) $N_2^{\oplus}$ ⟹ (Undo a diazotization.) NH_2 ⟹ (Undo a reduction.) NO_2 ⟹ (Undo a nitration.) benzene

The synthesis in the forward direction would proceed as follows:

benzene — HNO_3, H_2SO_4 → NO_2 — HCl, Fe → NH_2 — HONO, H_2SO_4 → $N{\equiv}N^{\oplus}$ — H_2O, Cu_2O, $Cu(NO_3)_2$ → OH

Problem 22.25

Think

In each step, consider whether electrophilic aromatic substitution occurred or some other reaction on the side chain (i.e., oxidation, reduction, substitution, elimination, addition). What acts as the nucleophile and electrophile in each step? Was the carbon skeleton altered? Did functional group transformation occur?

Solve

Compound **A** is formed from an electrophilic aromatic substitution bromination reaction. **A** (bromobenzene) is then transformed into the Grignard reagent **B**, which can open an epoxide ring to form the primary alcohol **C** after acid workup. In the presence of the mild oxidizing agent PCC, the primary alcohol **C** is oxidized to the aldehyde **D**.

benzene — Br_2, $FeBr_3$ → Br (**A**) — Mg(s), Ether → MgBr (**B**) — 1. epoxide, 2. H_3O^+ → OH (**C**) — PCC → O (**D**)

Using the primary alcohol **C** as the starting material, **C** reacts with PBr_3 to form the primary alkyl halide. **E** reacts with PPh_3 and Bu–Li to form the ylide Wittig reagent **F**. The ylide reacts with the aldehyde **D** to form a new C=C shown in product **G.**

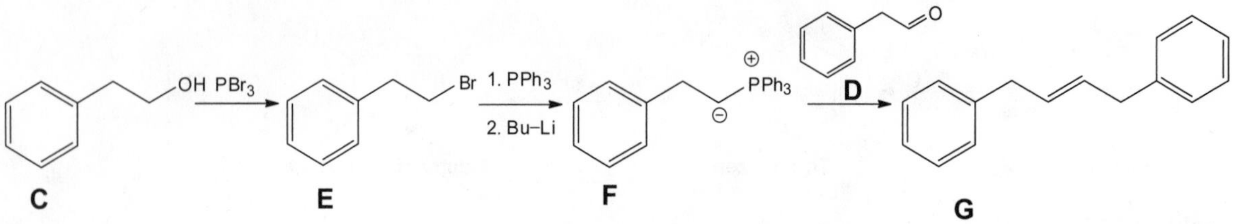

OH PBr3 → Br 1. PPh3 2. Bu–Li → PPh3 D →

C E F G

Problem 22.26

Think

In each step, consider whether electrophilic aromatic substitution occurred or some other reaction on the side chain (i.e., oxidation, reduction, substitution, elimination, addition). What acts as the nucleophile and electrophile in each step? Was the carbon skeleton altered? Did functional group transformation occur?

Solve

Compound **A** is formed from an electrophilic aromatic substitution Friedel–Crafts acylation reaction. The next reaction is a haloform reaction. The primary methyl ketone is chlorinated three times at the α carbon to form CCl_3. This intermediate can undergo acyl substitution with HO^- followed by an acid workup to form the carboxylic acid **B**, which is then transformed by $SOCl_2$ to the acyl chloride **C**.

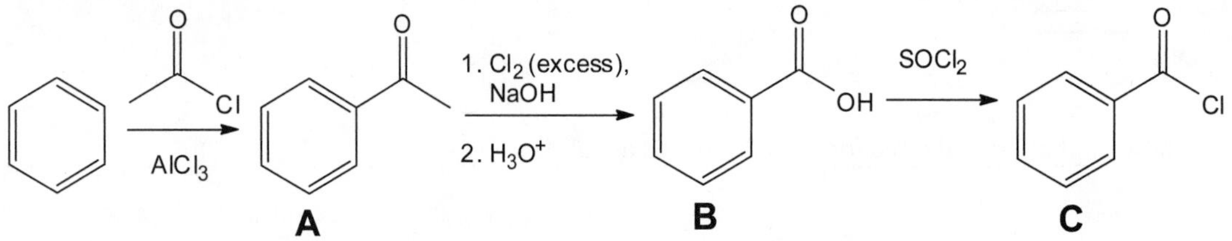

O Cl AlCl3 → O 1. Cl2 (excess), NaOH 2. H3O+ → O OH SOCl2 → O Cl

A B C

Nitrobenzene **D** is produced from the electrophilic aromatic substitution reaction of benzene. **D** is reduced to form aniline **E**. The aniline reacts with the acid chloride **C** to form the amide **F.** The amide is reduced to the secondary amine.

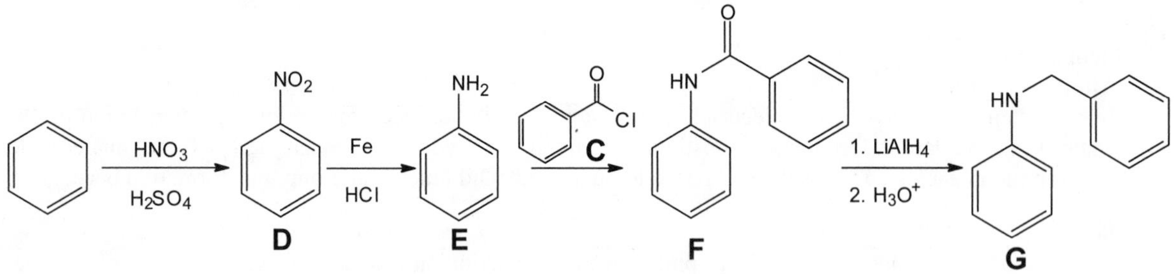

HNO3 H2SO4 → NO2 Fe HCl → NH2 O Cl C → HN O 1. LiAlH4 2. H3O+ → HN

D E F G

Problem 22.27

Think

In each step, consider whether electrophilic aromatic substitution occurred or some other reaction on the side chain (i.e., oxidation, reduction, substitution, elimination, addition). What acts as the nucleophile and electrophile in each step? Was the carbon skeleton altered? Did functional group transformation occur?

Solve

Nitrobenzene **A** is produced from the electrophilic aromatic substitution reaction of benzene. **A** is reduced to form aniline **B**. The aniline reacts with the $NaNO_2/HCl$ to form **C**, the benzenediazonium ion intermediate, $C_6H_5–N_2^+$. The benzenediazonium ion undergoes a Sandmeyer reaction with CuCN to form the benzonitrile **D**.

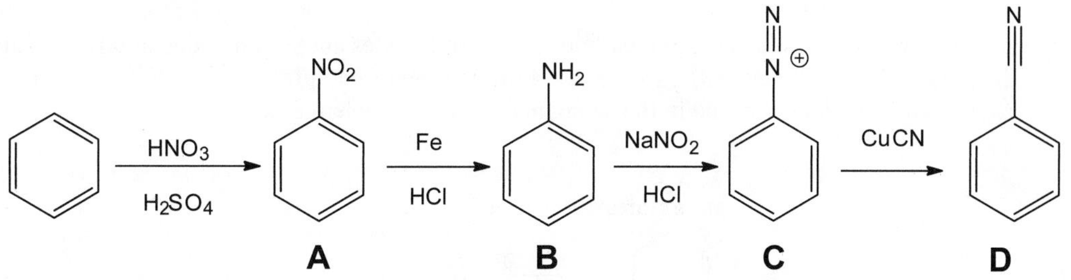

Compound **E** is formed from an electrophilic aromatic substitution bromination reaction. **E** (bromobenzene) is then transformed into the Grignard reagent **F**, which reacts with benzonitrile **D** to form an imine that is hydrolyzed to the ketone **G**. **G** is reduced by $NaBH_4$/EtOH to form the alcohol **H**. The alcohol O–H is deprotonated to form the alkoxide anion, which acts as a nucleophile in an S_N2 reaction (a Williamson ether synthesis) to form the ether **I**.

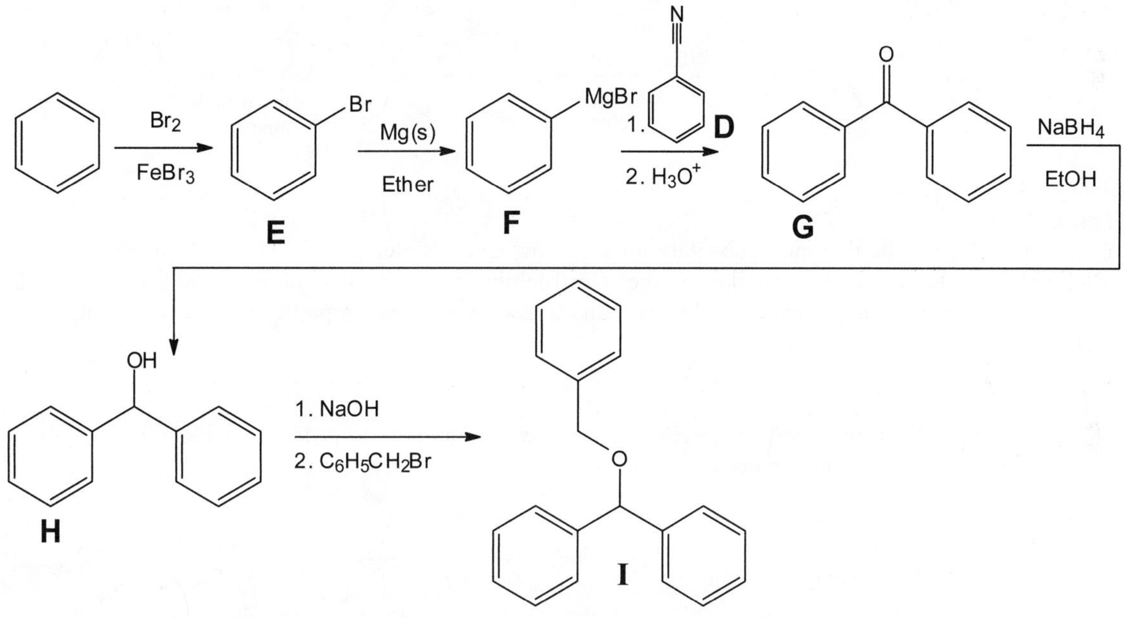

Problem 22.28

Think

Is this reaction an example of electrophilic aromatic substitution? What acts as the electrophile in the electrophilic addition step? What acts as the base to regenerate the aromatic ring?

Solve

Electrophilic addition of H^+ occurs first. In the second step, the first step is simply reversed, but D^+ is eliminated instead of H^+.

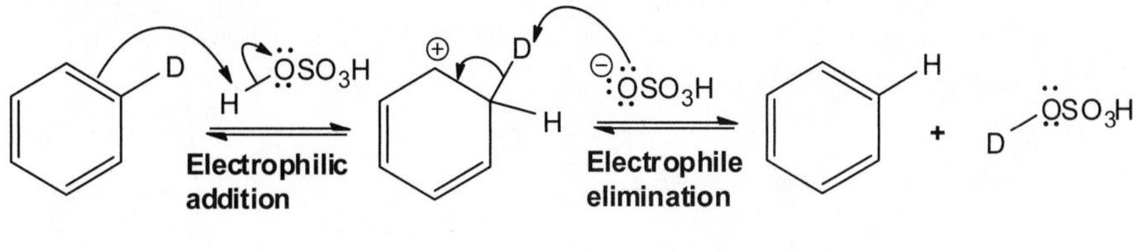

Problem 22.29

Think

What is the electrophile that must substitute for H^+? How can that electrophile be produced from the alkene in acid? How many carbocation intermediates are possible? Which one is more stable?

Solve

When a proton adds to 1-methylcyclohexene, two carbocation intermediates are possible, depending on which alkene C gains the proton. The more stable 3° carbocation, however, is formed preferentially. This carbocation is a strong electrophile and alkylates benzene in the same manner as in Problem 22.6.

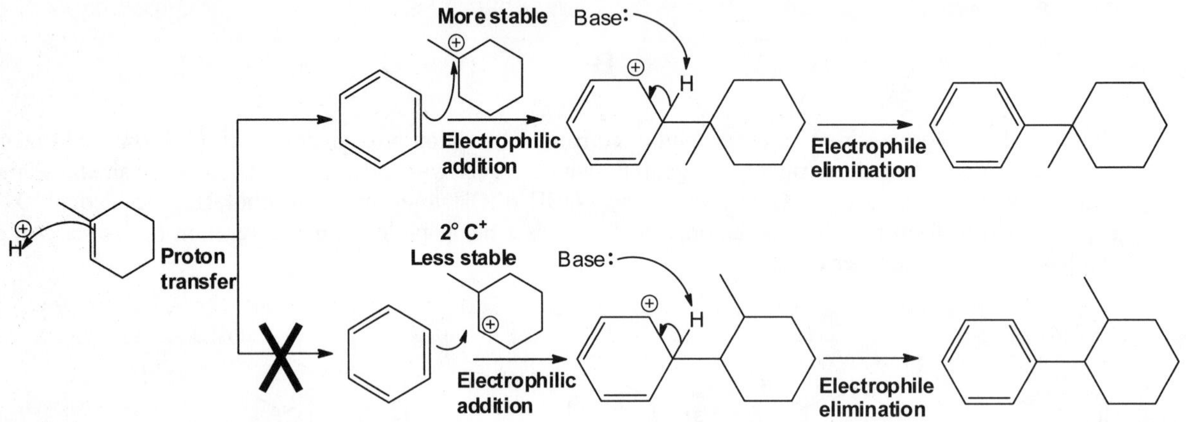

Problem 22.30

Think

What is the electrophile that must substitute for H^+? How can that electrophile be produced from the SCl_2 and $AlCl_3$? How is this mechanism similar to other electrophilic aromatic substitution mechanisms? What acts as the base to regenerate the aromatic ring? How many times does the electrophilic aromatic substitution reaction occur?

Solve

The electrophile S^+ is generated in precisely the same way as the R^+ is in Friedel–Crafts alkylation. The electrophilic aromatic substitution occurs twice.

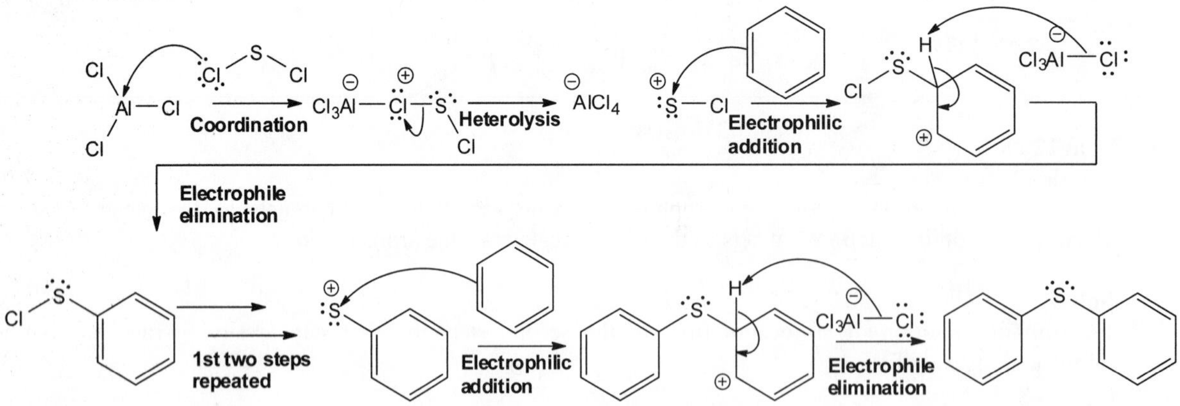

Problem 22.31

Think

How is HSO_3Cl similar in structure to H_2SO_4? How is this mechanism similar to the sulfonation electrophilic aromatic substitution mechanism in Equation 22-23?

Solve

HSO_3Cl is similar in structure to H_2SO_4 in that it can undergo a proton transfer reaction followed by a heterolysis reaction with itself to form the SO_2Cl^+ electrophile cation. The mechanism is a normal electrophilic addition–elimination mechanism for an electrophilic aromatic substitution.

Proton transfer

Heterolysis

Electrophilic addition

Electrophile elimination

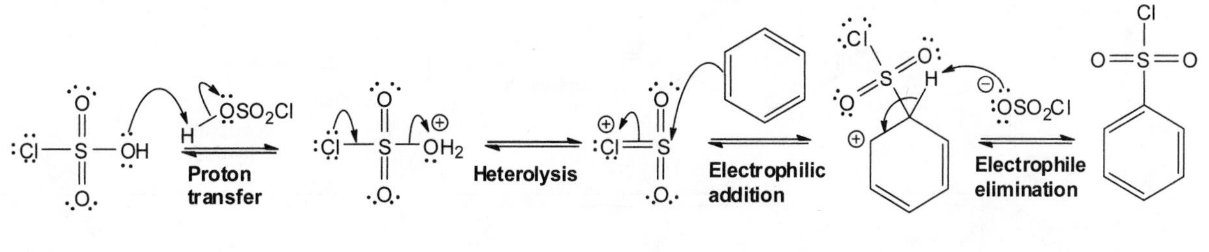

Problem 22.32

Think

What is the electrophile that must substitute for H^+? How does the epoxide interact with $AlCl_3$? How can that electrophile be produced from the Lewis acid–base complex? How is this mechanism similar to other electrophilic aromatic substitution mechanisms? What acts as the base to regenerate the aromatic ring? Are proton transfer reactions involved?

Solve

The $AlCl_3$ makes the epoxide very electrophilic, so the ring can be opened by even weak nucleophiles like benzene. Note that $AlCl_3$ is regenerated in the final step, so it is a catalyst. The purpose of the water is to serve as the base to regenerate the aromatic ring and to facilitate a proton transfer reaction to form the alcohol product.

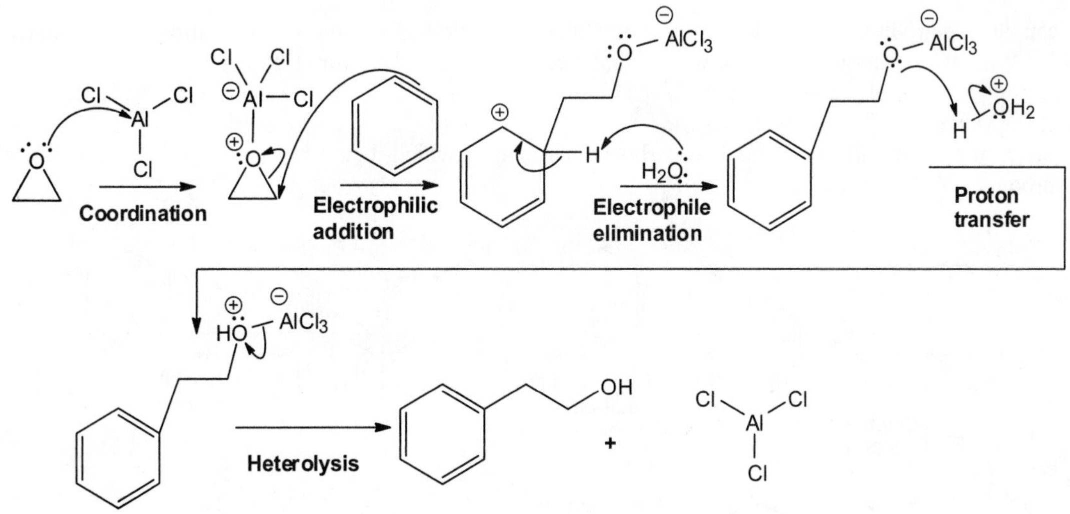

Coordination

Electrophilic addition

Electrophile elimination

Proton transfer

Heterolysis

Problem 22.33

Think

How does an aldehyde interact with a strong acid? What is the electrophile that must substitute for H^+? How is this mechanism similar to other electrophilic aromatic substitution mechanisms?

Solve

First, the aldehyde oxygen is protonated, which activates it toward nucleophilic attack. Then electrophilic aromatic substitution occurs—electrophilic addition followed by electrophile elimination. The oxygen is protonated again to make a good water leaving group, and water leaves in the first of the two steps for an E1 mechanism, generating a relatively stable benzylic carbocation. A final deprotonation yields a larger aromatic system.

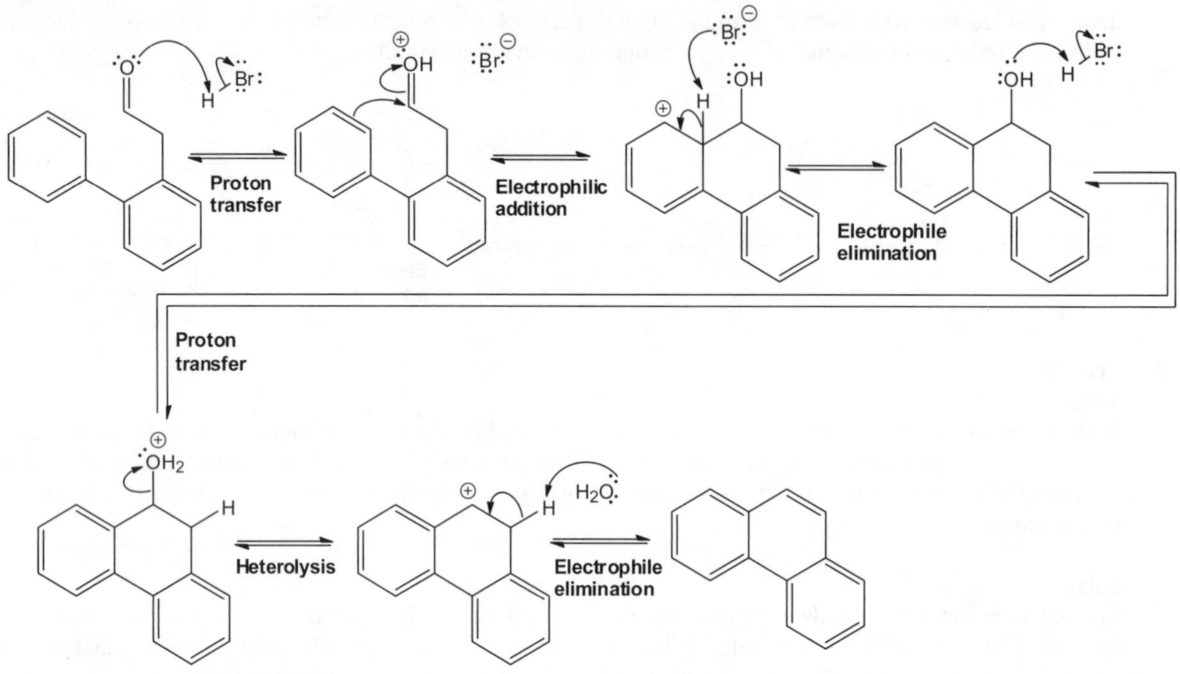

Proton transfer

Electrophilic addition

Electrophile elimination

Proton transfer

Heterolysis

Electrophile elimination

Problem 22.34

Think

How can an electrophile be produced from the diene in acid? How many carbocation intermediates are possible? Which one is more stable? How does that electrophile substitute for H^+?

Solve

Protonation of cyclohexadiene forms two carbocations, one of which, being resonance stabilized, leads to the major product.

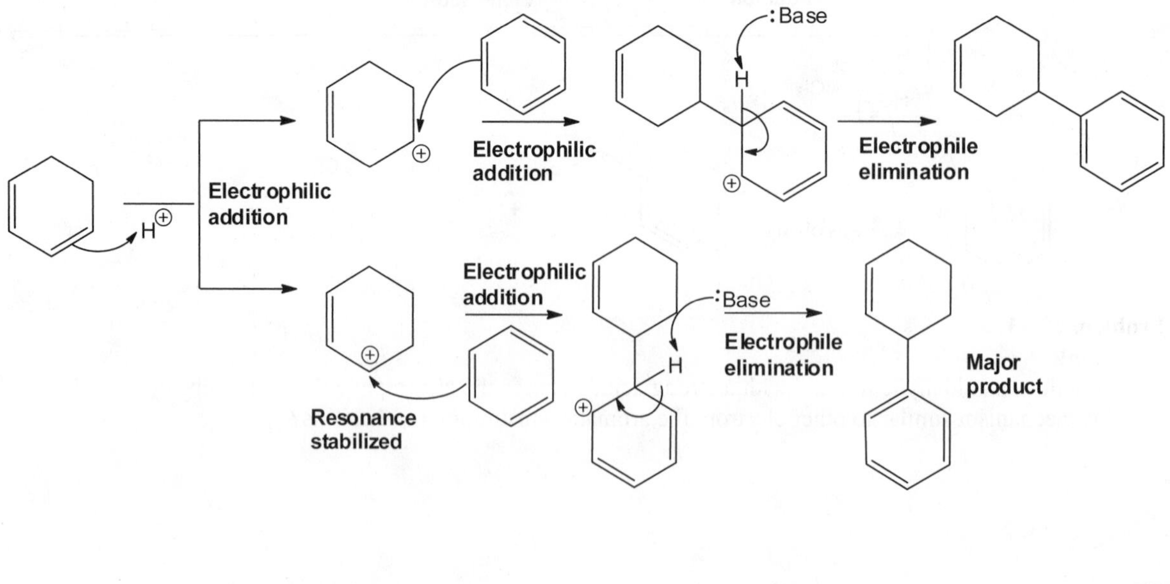

Electrophilic addition

Electrophilic addition

Electrophile elimination

Electrophilic addition

Electrophile elimination

Resonance stabilized

Major product

Problem 22.35

Think

How does I–Cl interact with Fe(s)? What is the electrophile that must substitute for H^+? Which atom stabilizes a positive charge better? Is the reaction under kinetic or thermodynamic control?

Solve

These conditions generate a small amount of an iron trihalide catalyst, FeX_3, so the mechanism for each reaction is essentially identical to the bromination reaction we saw earlier in the chapter. The electrophilic aromatic substitution reaction runs under kinetic control, and the reaction rate is governed by the first electrophilic aromatic substitution step. The rate of that first step is governed by the concentration of X^+—either Cl^+ or I^+. Formation of X^+ is reversible and the extent of X^+ that is formed depends on the stability of the ion. I^+ is larger and, therefore, more stable than Cl^+, so there should be more of it available to react with benzene. Thus, the iodobenzene product is favored.

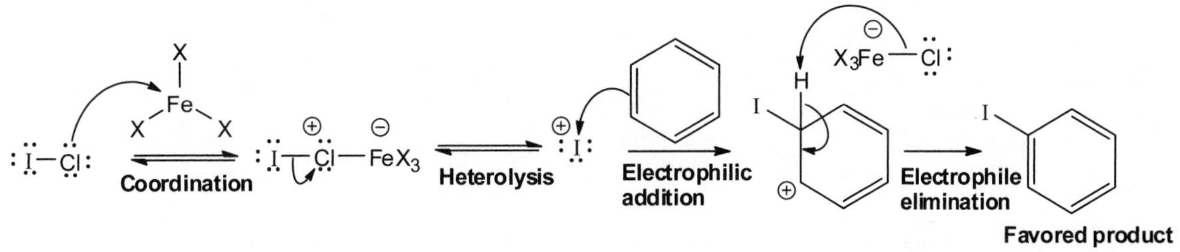

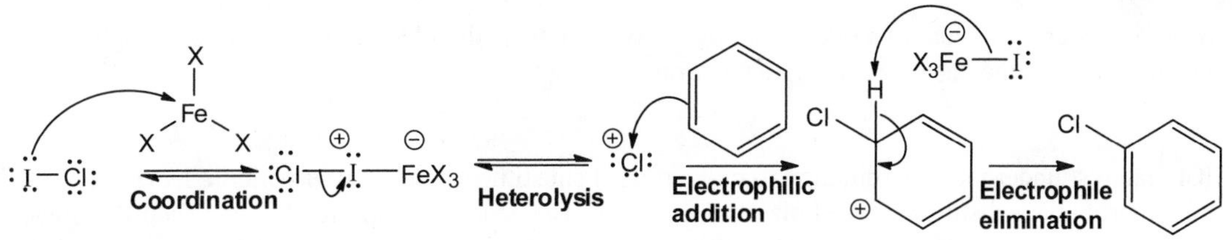

Problem 22.36

Think

What intermediate forms when an alkyl bromide reacts with $AlCl_3$? Is the carbocation chiral? What is the electrophile that must substitute for H^+? How is this mechanism similar to other electrophilic aromatic substitution mechanisms?

Solve

This is a Friedel–Crafts alkylation. There is a stereocenter in the product, but because the halide goes through an achiral carbocation intermediate, the product is formed in a racemic mixture, and will not be optically active.

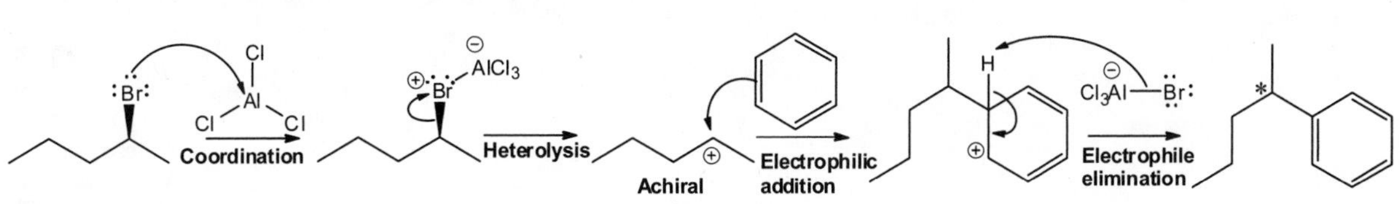

Problem 22.37

Think

In the production of the formyl chloride intermediate, is the carbon of C≡O electron rich or poor? Which atom in CO is protonated by HCl? Which atom in CO is attacked by the Cl^-? In the electrophilic aromatic substitution part of the mechanism, what is the electrophile that must substitute for H^+? How is this mechanism similar to other electrophilic aromatic substitution mechanisms?

Solve

The carbon in C≡O is electron rich (with three bonds and a lone pair, it is negatively charged) and is protonated by H^+. The carbon in $HC{\equiv}O^+$ then undergoes nucleophilic addition by the Cl^- to yield formyl chloride. Friedel–Crafts acylation then follows.

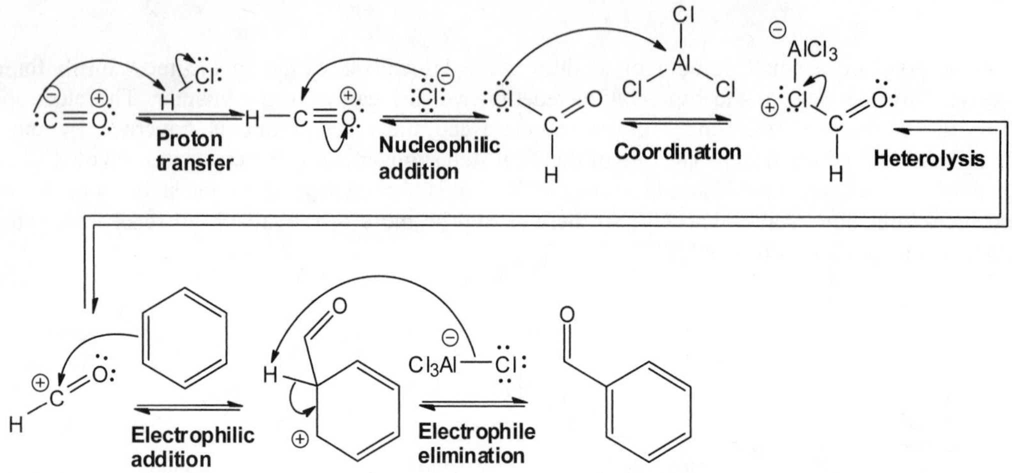

Proton transfer
Nucleophilic addition
Coordination
Heterolysis
Electrophilic addition
Electrophile elimination

Problem 22.38

Think

What reaction occurs between a halogen and $AlCl_3$? What acts as the electrophile in the electrophilic addition reaction? What acts as the base to regenerate the aromatic ring?

Solve

The $AlCl_3$ catalyst undergoes coordination to convert the I substituent into a better cationic leaving group. Then electrophilic aromatic substitution takes place in two steps. Benzene first captures H^+, and then I^- displaces the benzene ring from the I^+. This is analogous to the second step of a typical electrophilic aromatic substitution mechanism, in which a base removes an H^+ from benzene.

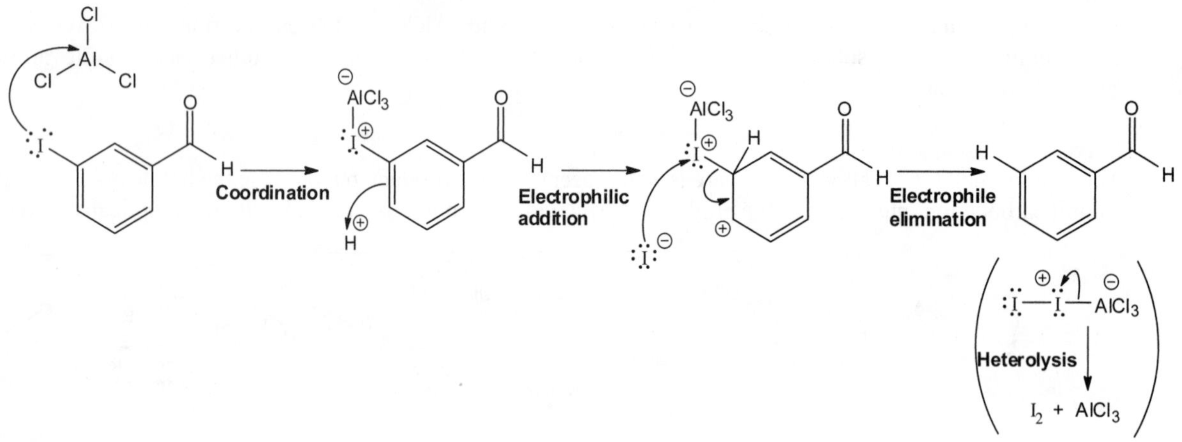

Coordination
Electrophilic addition
Electrophile elimination
Heterolysis
$I_2 + AlCl_3$

Problem 22.39

Think

In Problem 22.38, what is the leaving group? Which halogen best stabilizes the positive charge?

Solve

We can see in the mechanism in Problem 22.38 that the leaving group leaves in the form in which the halogen has a positive charge. Stability decreases in the order $I^+ > Br^+ > Cl^+ > F^+$.

Problem 22.40

Think

Is the product aromatic? If Br_2 were to add to just benzene, would the product be aromatic? How does this affect whether the reaction will go forward or not?

Solve

Addition to the middle ring allows the two outside rings to remain (independently) aromatic. A similar reaction with just benzene isn't feasible, because addition of Br_2 destroys the aromaticity.

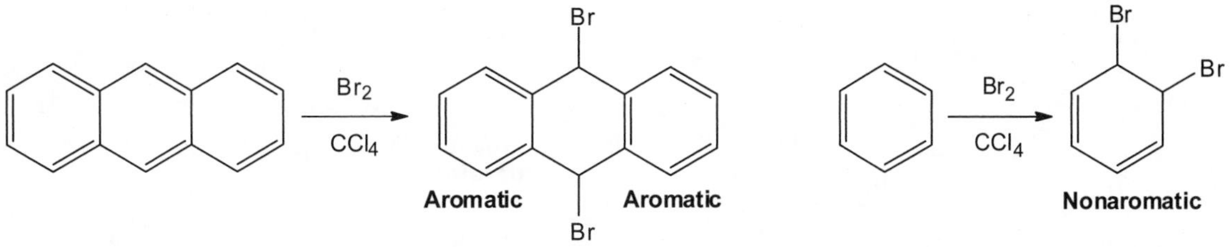

Br
Br2
CCl4
Aromatic
Aromatic
Br
Br
Br
Br2
CCl4
Nonaromatic

Problem 22.41

Think

What intermediate forms when the alkene reacts with a strong acid? What acts as the electrophile in the electrophilic addition reaction? What acts as the base to regenerate the aromatic ring?

Solve

This is a Friedel–Crafts alkylation. Protonation generates a benzylic carbocation that benzene attacks. Note that under these acidic conditions, the amine N is protonated, but that protonated amine is unreactive.

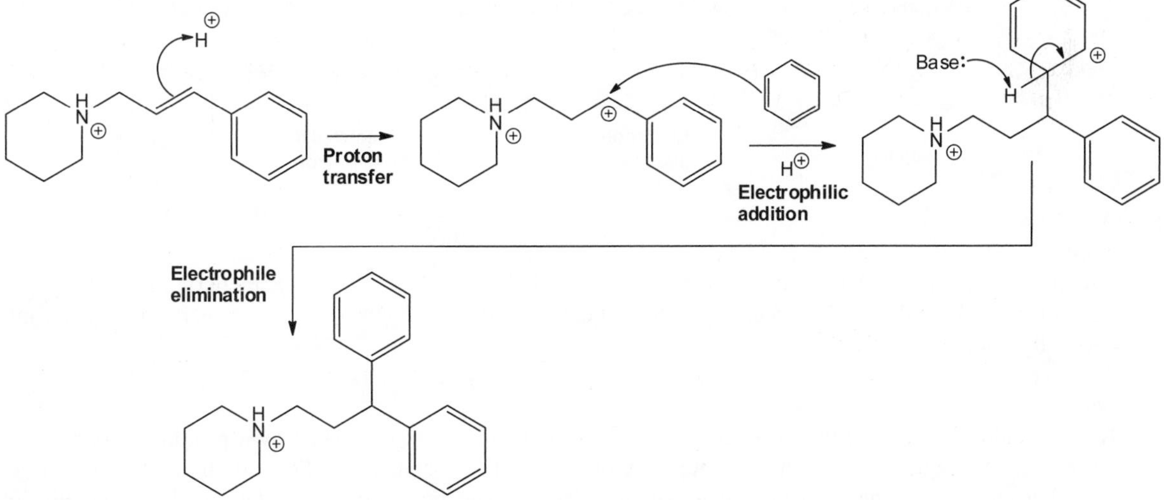

Proton transfer
Electrophilic addition
Base:
Electrophile elimination

Problem 22.42

Think

What functional group is present? How does this functional group react with an acid? What acts as the electrophile in the electrophilic addition step? What acts as the base to regenerate the aromatic ring? How does ring strain help drive the reaction?

Solve

This reaction combines an electrophilic aromatic substitution mechanism with an acid-catalyzed nucleophilic addition–elimination mechanism. The carbonyl group becomes activated by protonation of O. An electrophilic aromatic substitution occurs in two steps, and the ring opens upon elimination of the N-containing leaving group. See the mechanism on the next page.

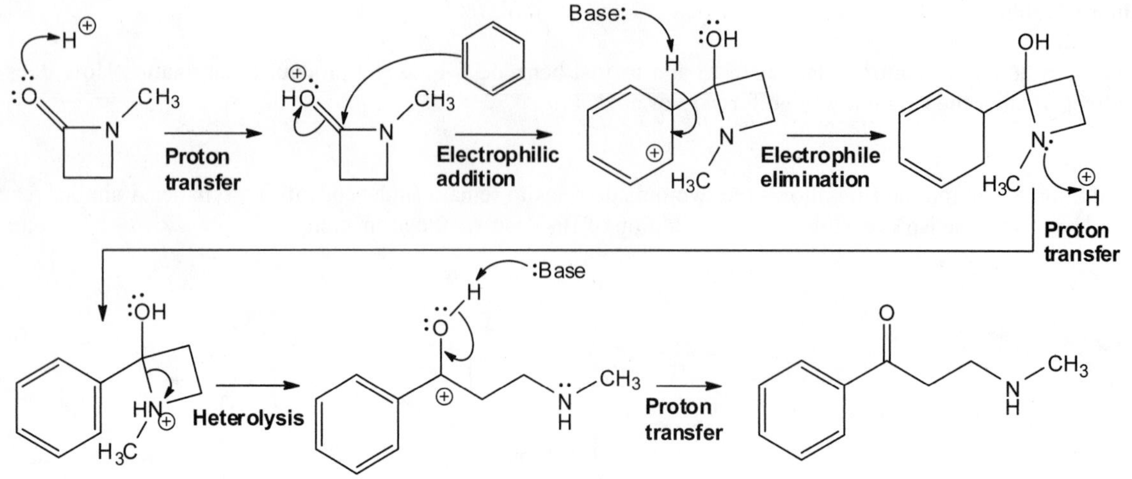

Problem 22.43

Think

What is the product of the proton transfer reaction between H_2O_2 and TfOH? What is the electrophile that must substitute for H^+? How is this mechanism similar to other electrophilic aromatic substitution mechanisms? What acts as the base to regenerate the aromatic ring?

Solve

H_2O_2 and TfOH react in a proton transfer reaction to form $HOOH_2^+$, which acts as the electrophile to add HO to the benzene ring. TfO^- acts as the base to regenerate the aromatic ring.

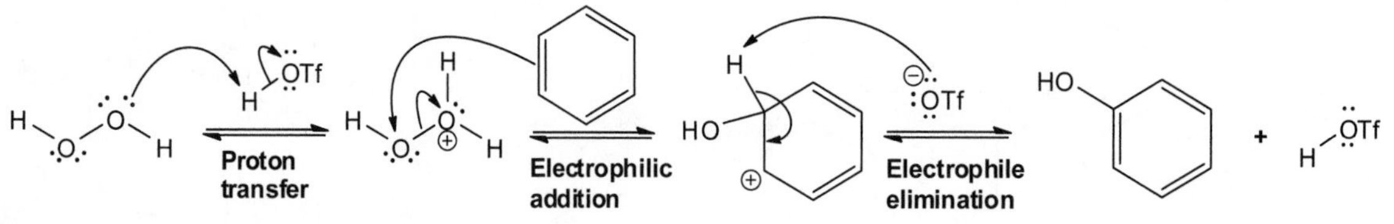

Problem 22.44

Think

Recall from Chapter 20 how to form a tertiary alcohol from a carboxylic acid derivative. How can you transform an alcohol into the necessary carboxylic acid derivative? How can you transform benzene into the necessary nucleophilic carbon?

Solve

An acid chloride reacts with two equivalents of a Grignard reagent (followed by a proton transfer) to form a tertiary alcohol where two of the R groups are the same. The precursor to the acid chloride is the carboxylic acid that came from the oxidation of a primary alcohol. The precursor to the Grignard reagent PhMgBr is bromobenzene, which can be formed from benzene.

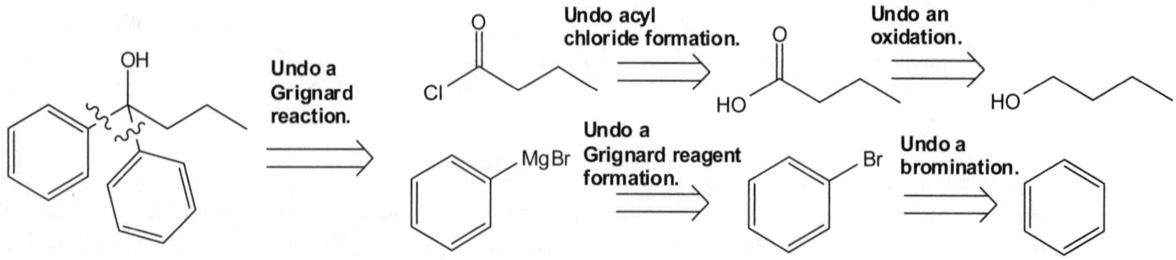

In the forward direction, the synthesis would proceed as follows:

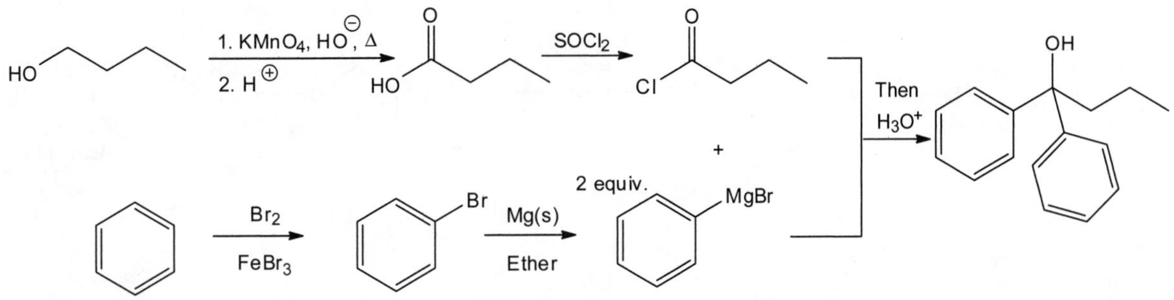

1. $KMnO_4$, HO^-, Δ
2. H^+

$SOCl_2$

Then H_3O^+

2 equiv.

Br_2 / $FeBr_3$

Mg(s) / Ether

Problem 22.45

Think

Recall from Chapter 20 how to form an amide from a carboxylic acid derivative. How can you transform an alkyl halide into the necessary carboxylic acid derivative? What are the two functional groups formed on benzene?

Solve

The *N*-substituted amide is formed from the reaction of an acid chloride and an amine (aniline). The acid chloride's precursor can be the carboxylic acid that forms from the oxidation of $PhCH_2OH$. Benzyl alcohol is formed from a substitution reaction of $PhCH_2Br$ and HO^-. Aniline's precursor is nitrobenzene, which forms from the nitration reaction between HNO_3 and benzene.

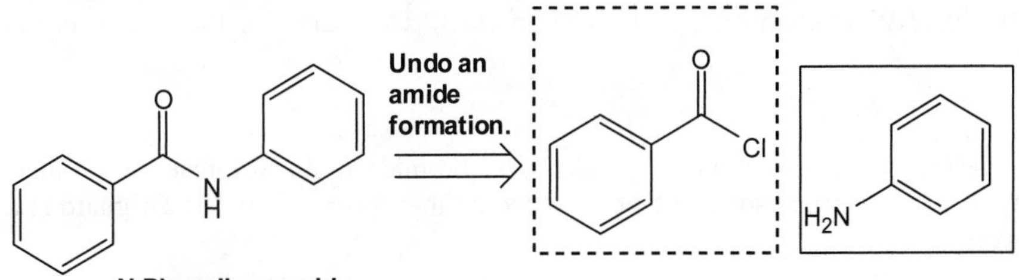

Undo an amide formation.

N-Phenylbenzamide (Benzanilide)

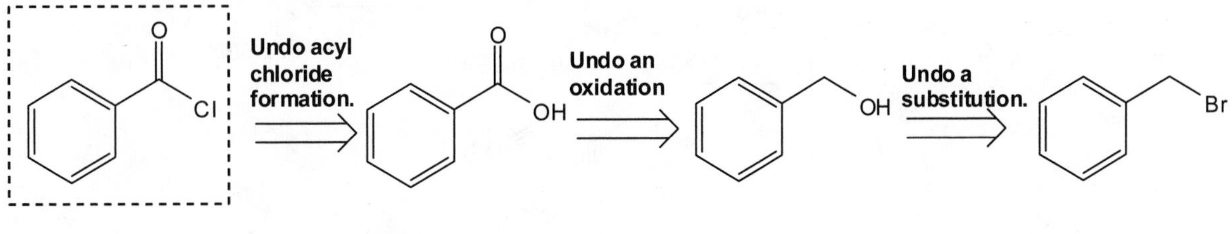

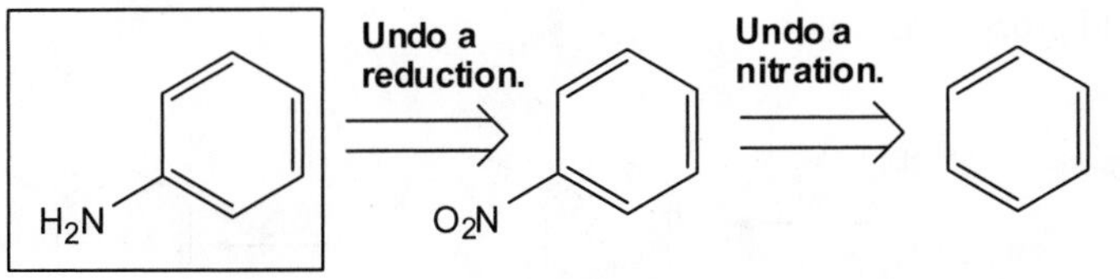

Undo acyl chloride formation.

Undo an oxidation

Undo a substitution.

Undo a reduction.

Undo a nitration.

In the forward direction, the synthesis would proceed as follows:

Acid chloride

Aniline

Aniline + Acid chloride → *N*-Phenylbenzamide (Benzanilide)

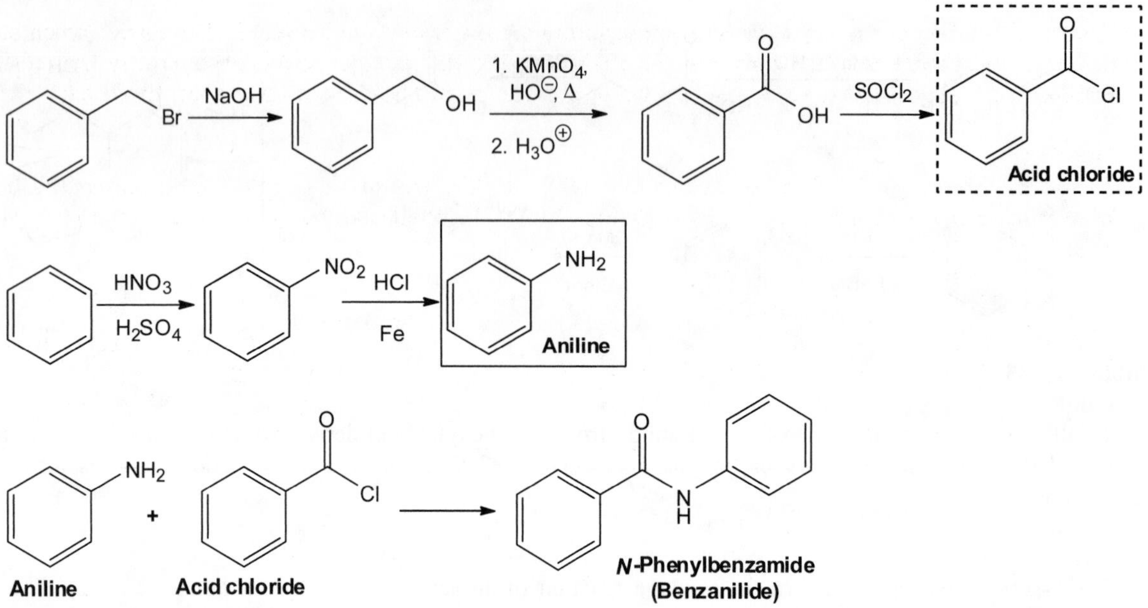

Problem 22.46

Think

How many carbon atoms in the target compound did not originate from benzene? How many carbon atoms are, therefore, in the C=O compound? What precursors will make an ether? How can you form those precursors from benzene and that C=O?

Solve

Dibenzyl ether can be synthesized from benzyl alcohol and benzyl bromide in an S_N2 reaction (Williamson ether synthesis). The benzyl bromide's precursor is the benzyl alcohol that formed from the Grignard reaction of PhMgBr and formaldehyde.

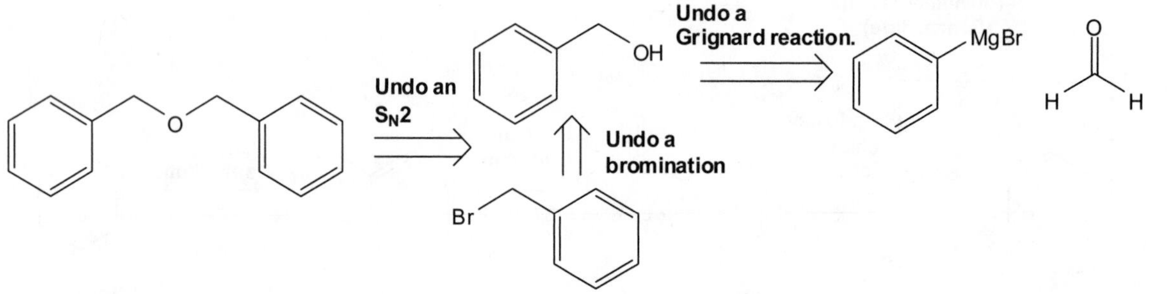

Undo an S_N2

Undo a bromination

Undo a Grignard reaction.

The synthesis in the forward direction would proceed as follows:

Benzyl alcohol

Benzyl bromide

Benzyl alcohol

Benzyl bromide

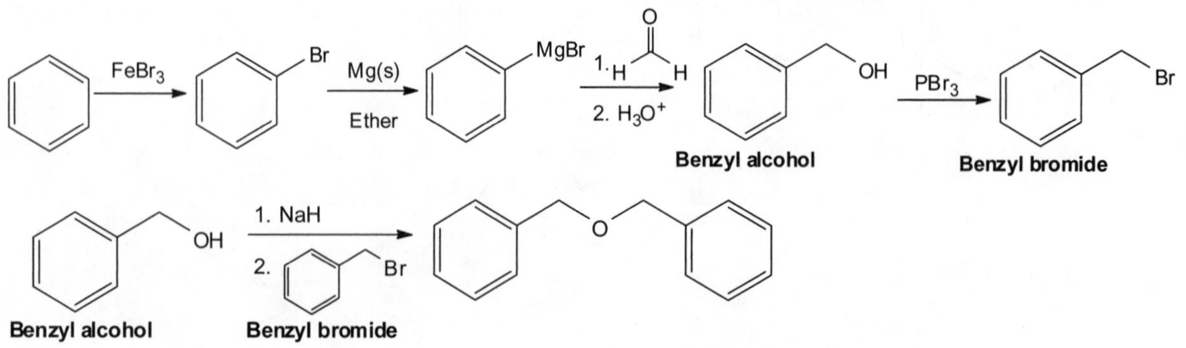

Problem 22.47

Think

How many carbon atoms in the target compound did not originate from benzene? How many carbon atoms, therefore, must come from the carboxylic acid compound? Can the target be produced directly from a simple Friedel–Crafts alkylation? Is there another reaction that can be used to form the appropriate C=C bond?

Solve

The alkene can be formed from the dehydration of an alcohol. This alcohol could have been reduced from a ketone that came from a Friedel–Crafts acylation. The acid chloride can be synthesized from 2-phenylacetic acid.

Undo a dehydration

OH

Undo a reduction

O

Undo a Friedel–Crafts acylation.

Cl

O

Undo an acyl chloride formation.

OH

O

2-Phenylacetic acid

The synthesis in the forward direction would proceed as follows:

OH

O

2-Phenylacetic acid

$SOCl_2$

Cl

O

$AlCl_3$

O

$NaBH_4$/EtOH

OH

H_3PO_4

Δ

Problem 22.48

Think

Which C atoms on benzene are available to undergo electrophilic addition? Are those sites chemically distinct?

Solve

Examine the three isomers of trimethylbenzene. **A** has two chemically distinct H atoms, so it can give rise to two distinct bromination products. **B** has three distinct H atoms, so it gives rise to three possible bromination products. **C** has just one distinct H, giving rise to one bromination product.

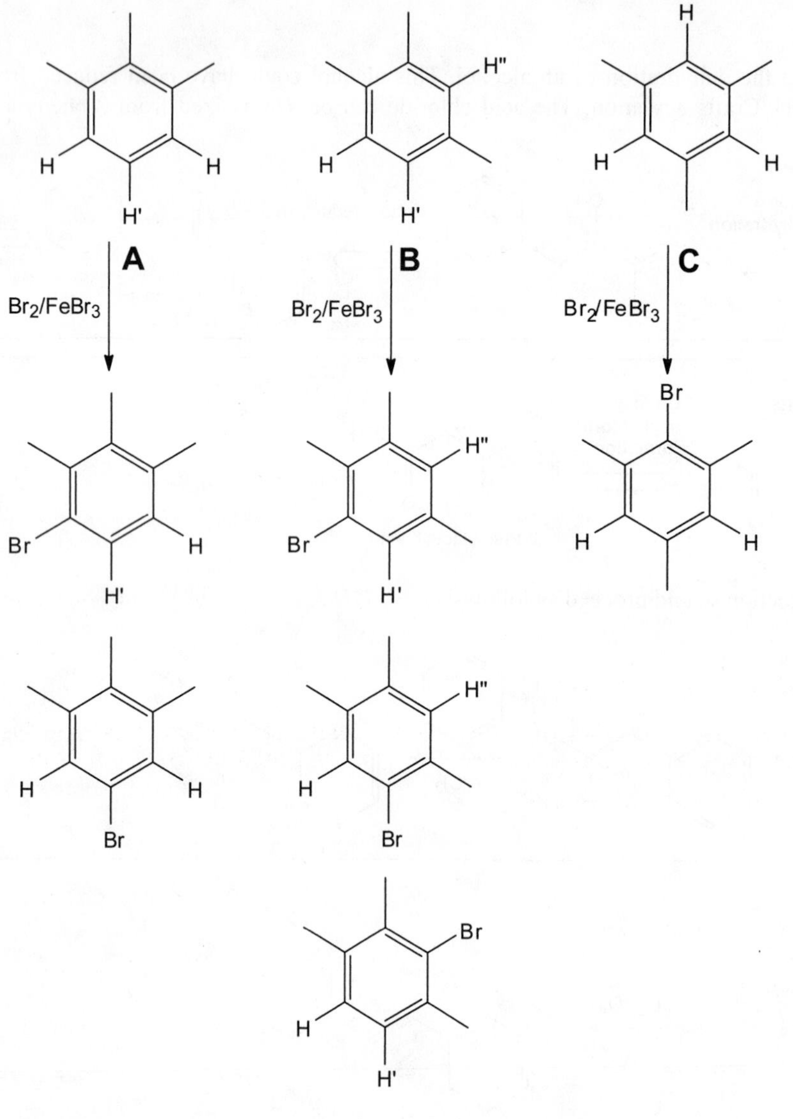

A: $Br_2/FeBr_3$

B: $Br_2/FeBr_3$

C: $Br_2/FeBr_3$

Problem 22.49

Think

What are the possible isomers of tetramethylbenzene? Which one has only one distinct H atom?

Solve
Examine the three isomers of tetramethylbenzene. All three isomers, **A**, **B**, and **C**, have just one distinct H, giving rise to only one nitration product for each, so it can be any one of these.

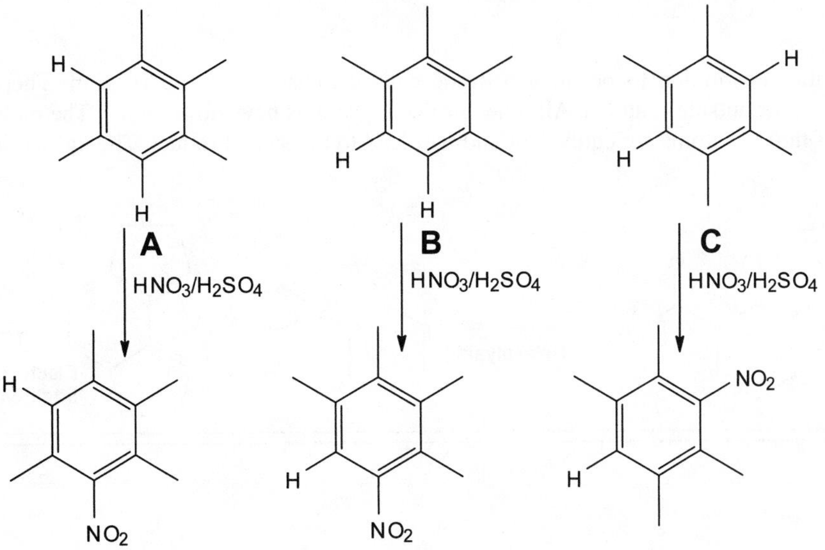

H H H H H H

A **B** **C**

HNO_3/H_2SO_4 HNO_3/H_2SO_4 HNO_3/H_2SO_4

H NO_2 H H NO_2 NO_2

Problem 22.50

Think
What are the structures of the three isomers of dimethylbenzene? Which one has only one distinct H atom? Which one has two distinct H atoms? Which one has three distinct H atoms?

Solve
Examine the three isomers of dimethylbenzene. If **A** produces only one nitration product, it must have only one distinct H atom and therefore must be the para isomer. If **B** produces only two products, it must only have two distinct H atoms and must be the ortho isomer. If **C** produces three products, it must have three distinct H atoms and must be the meta isomer.

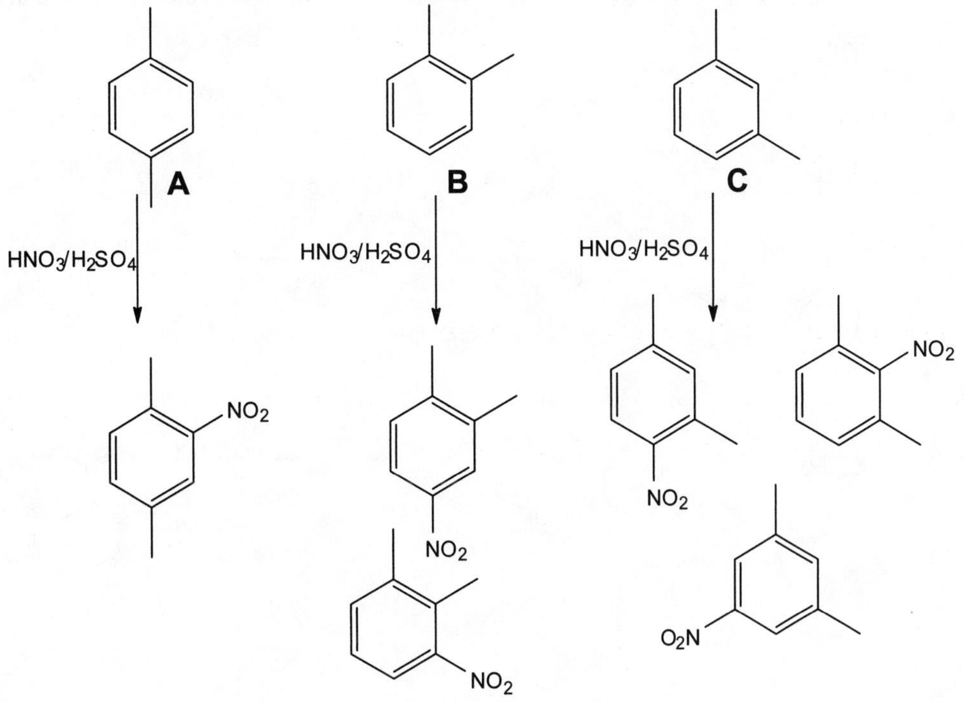

A **B** **C**

HNO_3/H_2SO_4 HNO_3/H_2SO_4 HNO_3/H_2SO_4

NO_2 NO_2 NO_2 NO_2 NO_2 O_2N

Problem 22.51

Think

For the given electrophilic aromatic substitution to occur, what must be present on the benzene carbon? Which carbon atoms are available to participate in that substitution? Is there symmetry?

Solve

For this electrophilic aromatic substitution to occur, a benzene carbon must possess an H atom. There are aromatic H atoms on only the four outside C atoms. All other aromatic C atoms have no H atoms. The molecule is symmetrical, so all four of those C atoms are equivalent and will lead to the same product. The mechanism is shown below:

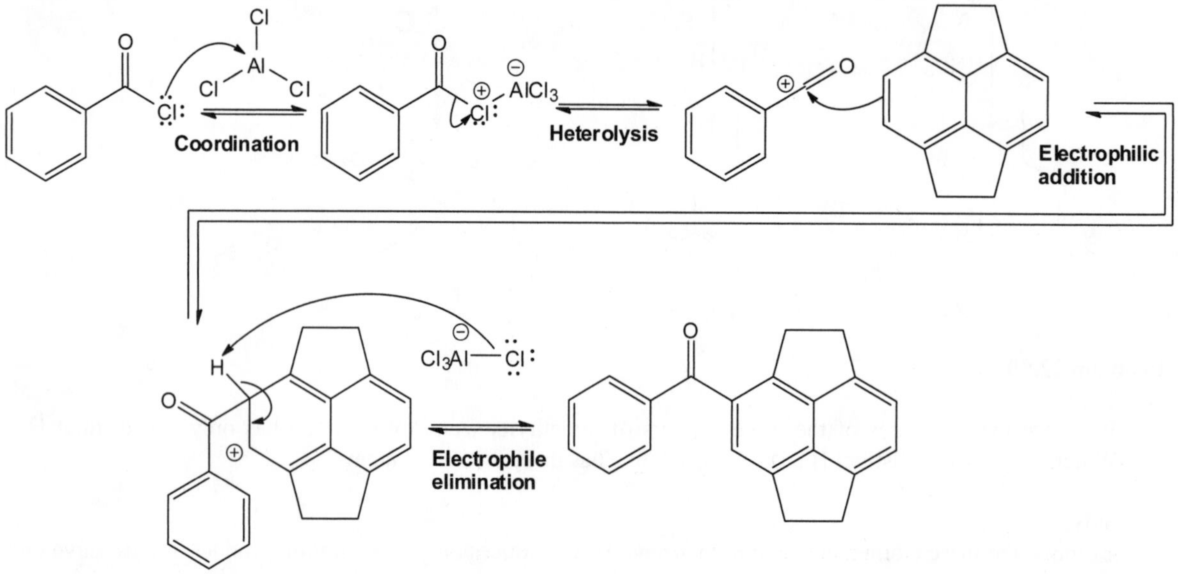

Coordination

Heterolysis

Electrophilic addition

Electrophile elimination

Problem 22.52

Think

What site on the aromatic ring is subject to electrophilic aromatic substitution? What is the electrophile in each reaction?

Solve

The products are given below:

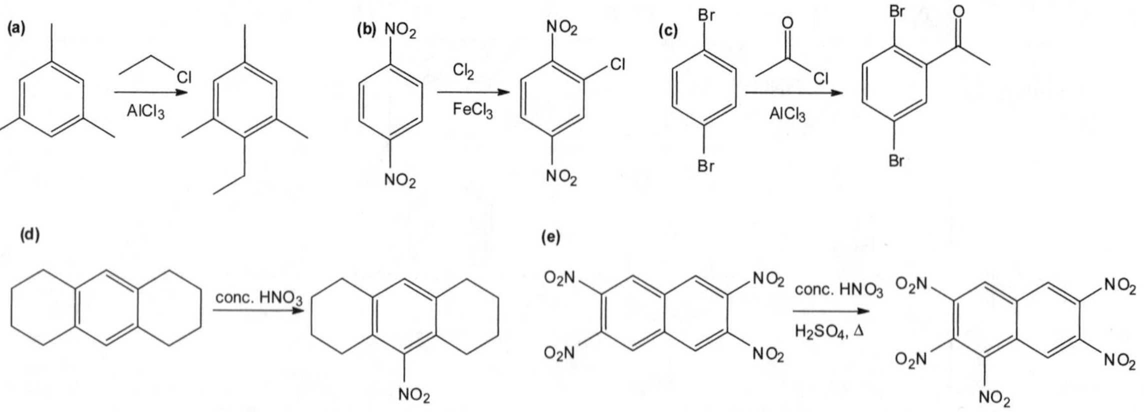

(a) $AlCl_3$

(b) Cl_2, $FeCl_3$

(c) $AlCl_3$

(d) conc. HNO_3

(e) conc. HNO_3, H_2SO_4, Δ

Problem 22.53

Think

Did electrophilic aromatic substitution occur? If so what was the electrophile? What reagents are necessary to produce that electrophile? Did any other reactions occur?

Solve

(a) This is a Friedel–Crafts acylation.

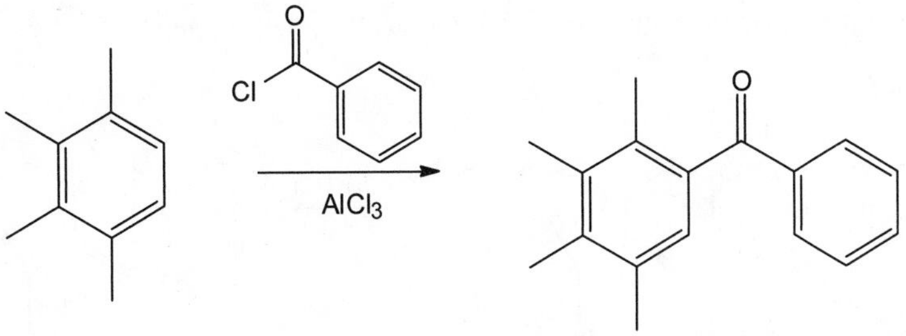

Cl
AlCl3
O
O

(b) This is a bromination using electrophilic aromatic substitution (Br_2/$FeBr_3$).

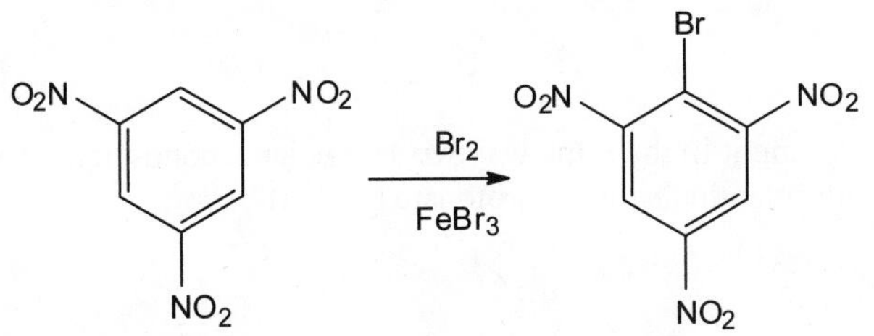

O2N
NO2
NO2
Br2
FeBr3
Br
O2N
NO2
NO2

(c) This is a sulfonation using electrophilic aromatic substitution.

SO3
H2SO4
SO3H

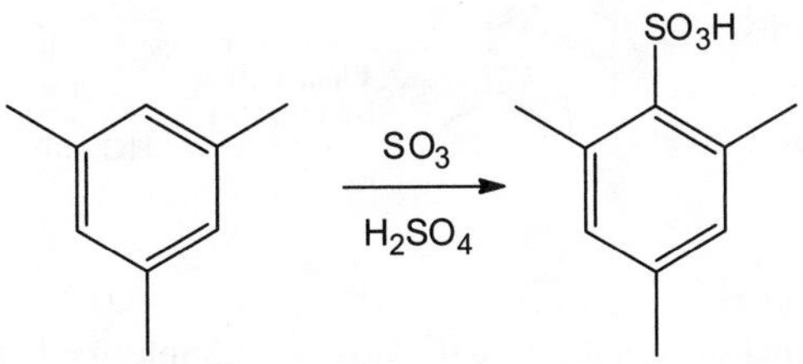

(d) This is a nitration using electrophilic aromatic substitution.

Br
Cl
Cl
Br
HNO3
H2SO4, Δ
NO2
Br
Cl
Cl
Br

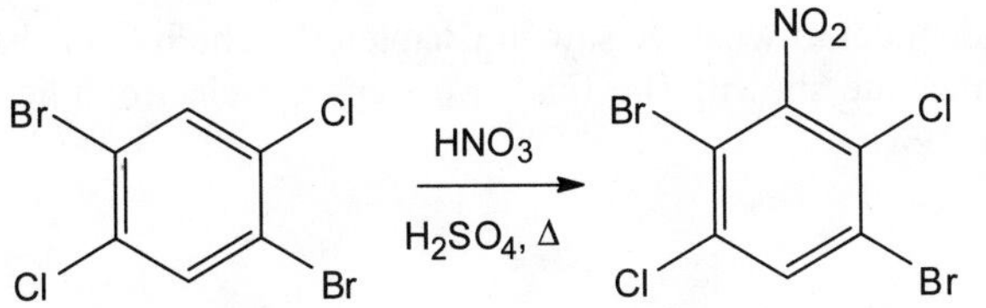

(e) The aniline formed from an electrophilic aromatic substitution nitration followed by a reduction.

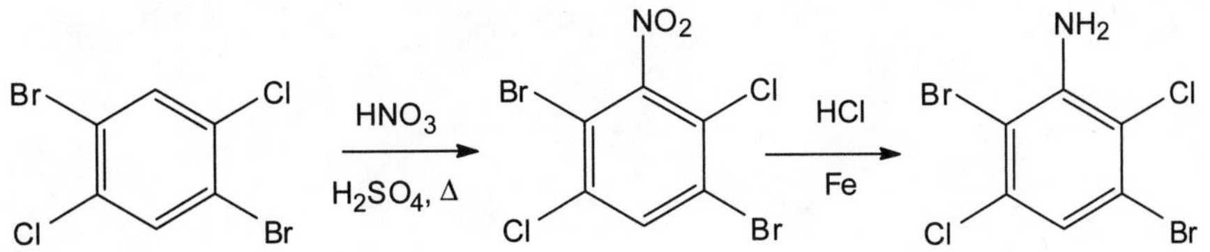

Br
Cl
Cl
Br
HNO3
H2SO4, Δ
NO2
Br
Cl
Cl
Br
HCl
Fe
NH2
Br
Cl
Cl
Br

(f) The carboxylic acid formed from a Friedel–Crafts alkylation followed by an oxidation.

CH_3Cl, $AlCl_3$; 1. $KMnO_4$, HO^-, Δ; 2. H_3O^+

(g) The primary alkyl group formed from Friedel–Crafts acylation followed by a reduction.

$AlCl_3$; Zn(Hg), HCl

Problem 22.54

Think

What happens to a nitrile in the presence of a strong acid? How does this compound behave as an electrophile in the electrophilic addition step?

Solve

Protonation makes the nitrile electron poor, activating it in the same way we saw polar π bonds activated in Chapter 18. The $ZnCl_2$ may serve to complex chloride; this helps to protonate the nitrile.

Proton transfer; Electrophilic addition; Electrophile elimination

Problem 22.55

Think

How does an amine react with an aldehyde in the presence of a strong acid? How is the intermediate shown formed? What is the electrophile in the electrophilic addition step?

Solve

The first several steps follow imine formation, identical to what we saw in Chapter 18. The intermediate given in the problem is the resonance-stabilized intermediate shown. The final two steps are electrophilic aromatic substitution. See the mechanism on the next page.

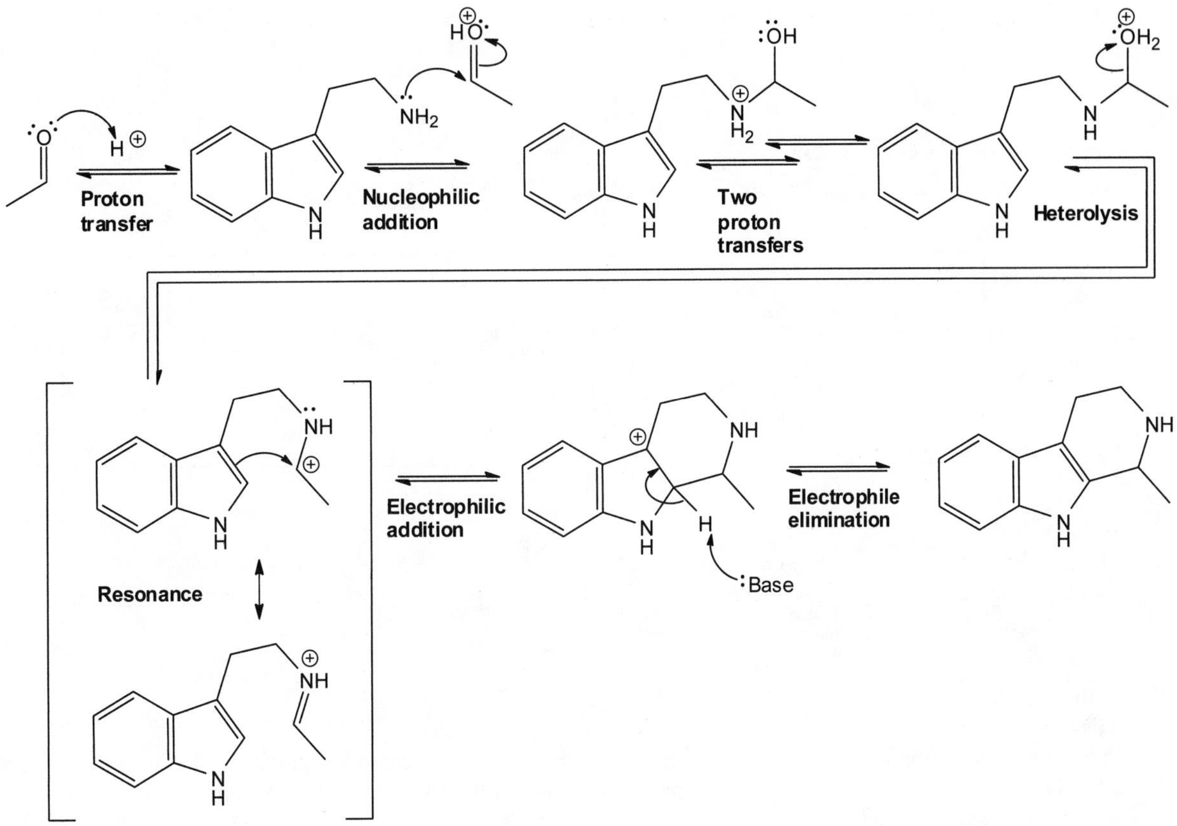

Proton transfer

Nucleophilic addition

Two proton transfers

Heterolysis

Resonance

Electrophilic addition

Electrophile elimination

:Base

Problem 22.56

Think

What reactive species is produced when an aromatic amine reacts with HONO? Consider the mechanism shown in Equation 22-31. How does this intermediate react with CuCN?

Solve

The intermediate is an arenediazonium ion. In the presence of CuCN, the N_2^+ group is replaced by CN.

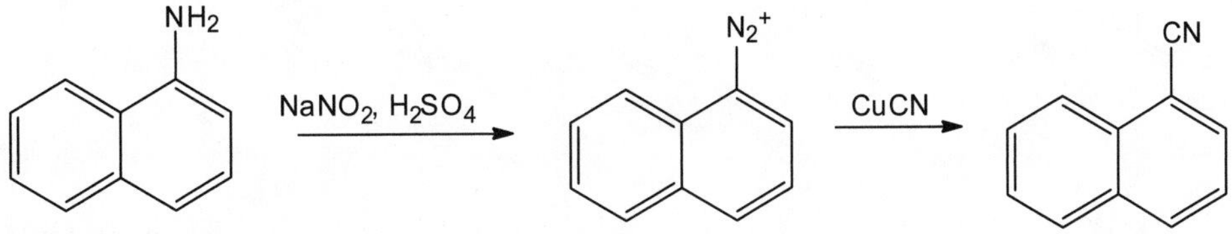

NH_2 — $NaNO_2$, H_2SO_4 → N_2^+ — CuCN → CN

The mechanism is shown on the next page. First, the *N*-nitrosamine is generated. Then, three proton transfers generate a good water leaving group that leaves in the final step to yield the aryldiazonium ion.

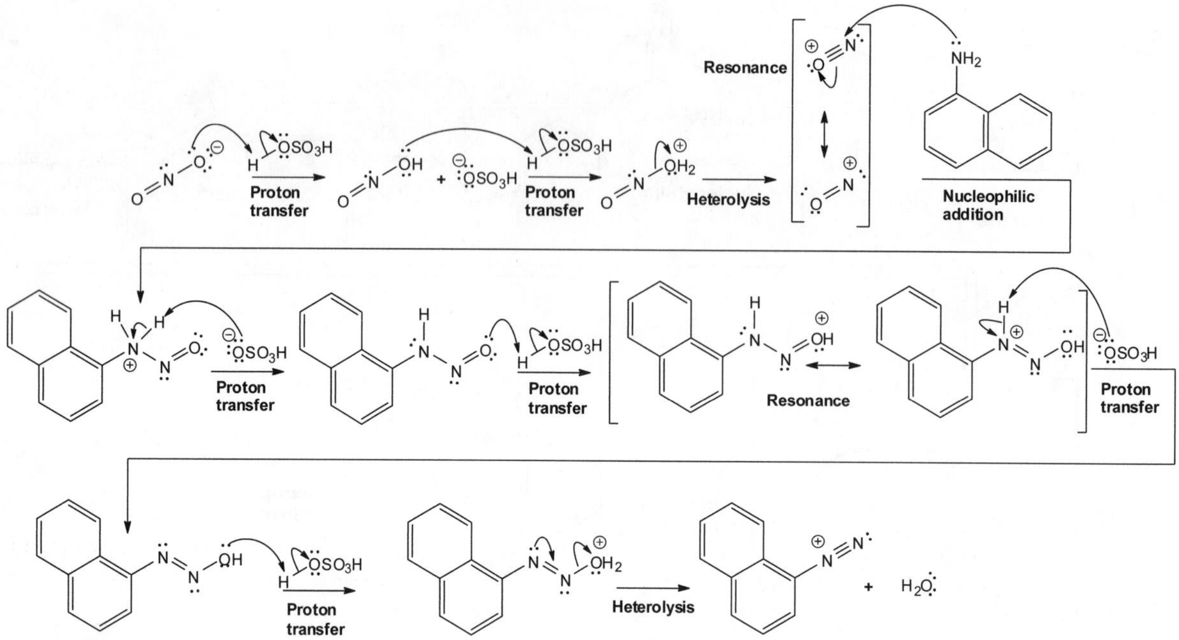

Problem 22.57

Think

What are the reaction conditions in Problem 22.30? What sites are available on diphenyl ether for electrophilic addition? To give rise to only six carbon signals, is symmetry present? How can electrophilic aromatic substitution take place without giving rise to more than six signals?

Solve

If a substitution occurs on only one of the two benzene rings in diphenyl ether, the product would have at least seven different carbon signals. The only way for the two separate rings to give rise to fewer signals is to have symmetry. This can be accomplished if substitution takes place on both rings at the position ortho to the oxygen. The following product is consistent with the spectrum:

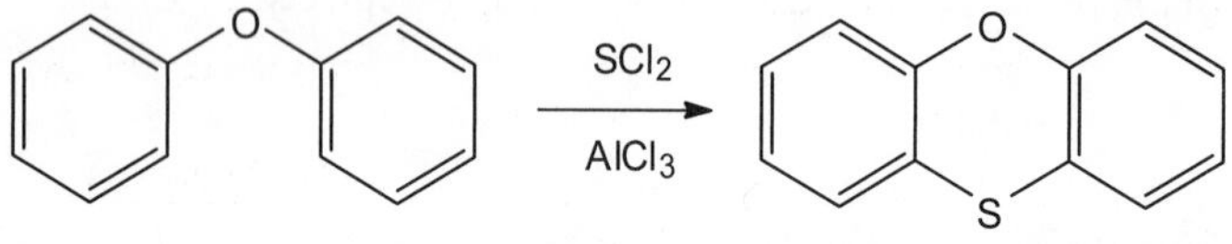

Problem 22.58

Think

What reaction occurs between an aromatic ring and an acid chloride in the presence of $AlCl_3$? Is an intramolecular reaction favored? What type of reagent is HCl, Zn(Hg)? From the ^1H NMR spectrum, how many aliphatic signals are present? What is the relative integration of the three signals? Is symmetry present? From the IR spectrum, is the C=O still present in the product?

Solve

The first reaction is a Friedel–Crafts acylation. Because it can form a six-membered ring, it takes place intramolecularly. The second reaction is a Clemmensen reduction, which removes the C=O bond. This is consistent with the spectral data. The NMR spectrum shows three signals (one aromatic) whose integrations are equal. There are four aromatic protons, four benzylic protons, and eight aliphatic protons. The IR spectrum shows no C=O bond (normal for a Clemmensen reduction), and clear aromatic absorption bands at 3050, 1600, and 1500 cm^{-1}, as well as a strong bending band at 750 cm^{-1}, indicating an ortho-substituted aromatic ring. See the mechanism on the next page.

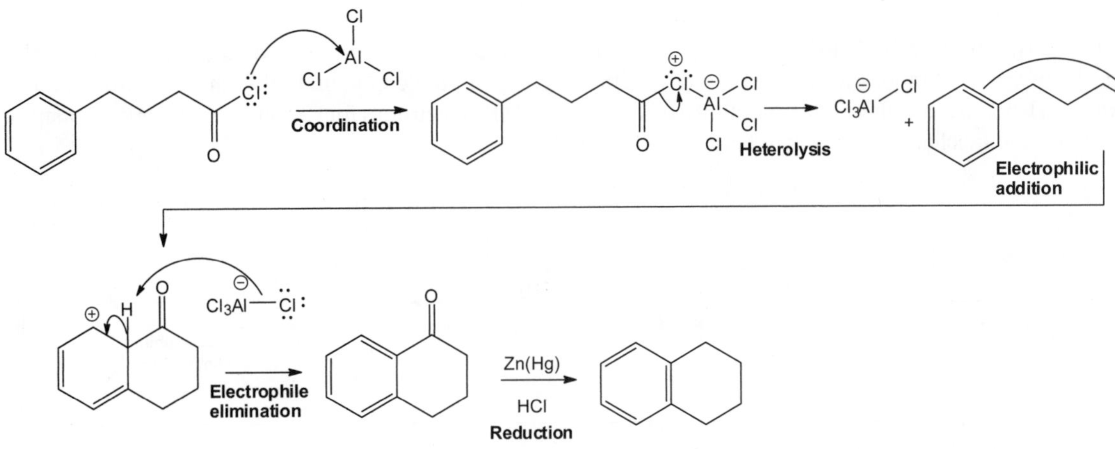

Coordination
Heterolysis
Electrophilic addition
Electrophile elimination
Zn(Hg)
HCl
Reduction

Problem 22.59

Think

How does an alcohol react with $SOCl_2$? How does this product react with $AlCl_3$? What is the electrophile in the electrophilic addition step? Is an intramolecular reaction favored? Predict the product of the reaction. Does the product have eight distinct C atoms that would give rise to eight ^{13}C NMR signals?

Solve

The first reaction converts the OH group to a Cl. The second reaction is a Friedel–Crafts alkylation. The product has five chemically distinct H atoms. (Note that although a primary carbocation is shown being generated, to be consistent with the mechanism shown in the chapter, it probably does not form because it would undergo carbocation rearrangement to a more stable secondary carbocation. Rather, the leaving of $AlCl_4^-$ and the electrophilic addition that follows it probably happen in a single, concerted step.)

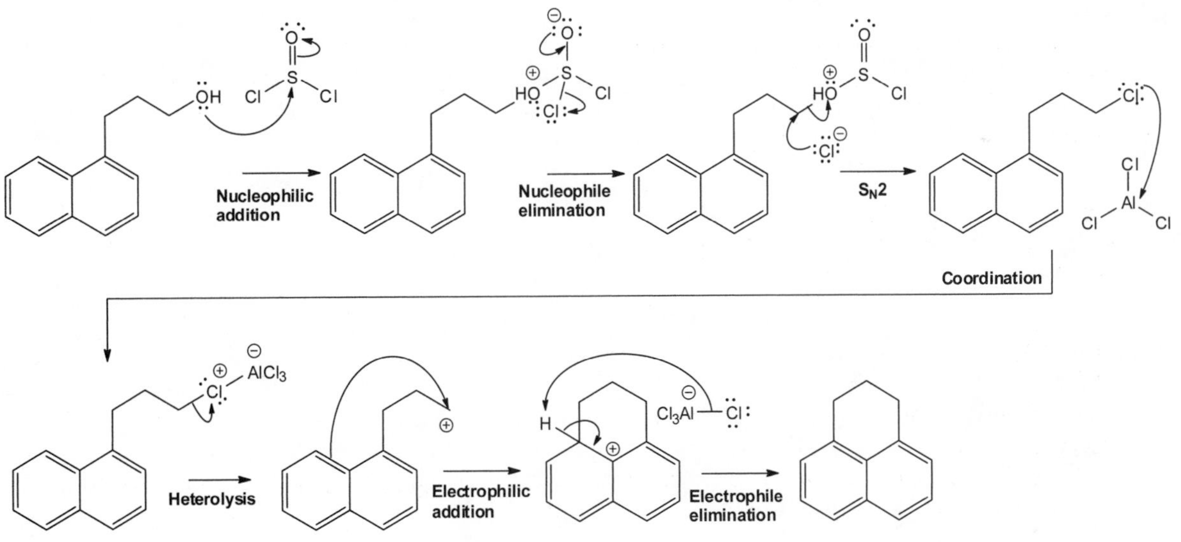

Nucleophilic addition
Nucleophile elimination
S_N2
Coordination
Heterolysis
Electrophilic addition
Electrophile elimination

Problem 22.60

Think

What is the product of the first Friedel–Crafts alkylation? Can that product undergo another Friedel–Crafts alkylation? What are the relative integrations of the aromatic and aliphatic signals? How many signals are in the ^{13}C NMR spectrum? What functional groups are evident from the IR spectrum?

Solve

This is two back-to-back Friedel–Crafts alkylation reactions. The integrations of the aromatic and aliphatic peaks give a 10:2 ratio. There are no other functional groups in the compound other than the aromatic ring and sp^3 C–H bonds, and there are five signals in the ^{13}C NMR spectrum (four aromatic and one aliphatic). The product and mechanism are shown below:

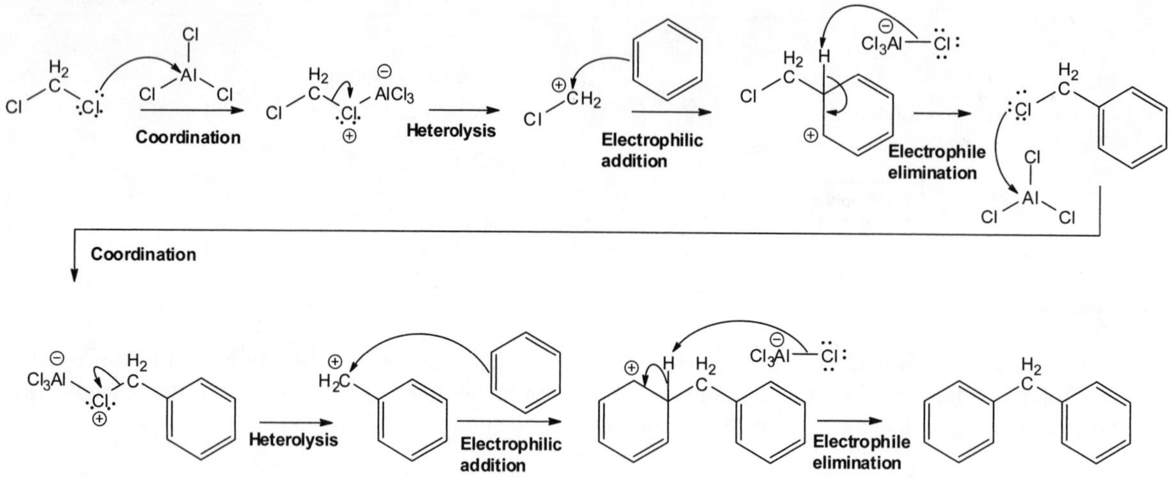

Coordination

Heterolysis

Electrophilic addition

Electrophile elimination

Coordination

Heterolysis

Electrophilic addition

Electrophile elimination

CHAPTER 23 | Electrophilic Aromatic Substitution 2: Substitution Involving Mono- and Disubstituted Benzene and Other Aromatic Rings

Your Turn Exercises

Your Turn 23.1

Think

If the reaction mixture only contains ortho, meta, and para products, what percentage of the product mixture is left for the para product? In comparing the relative percentages, which percentage group is the highest: ortho/para or meta?

Solve

The ortho/para percentages are combined, showing that –I is an ortho/para director and –CHO is a meta director.

Sub	O	M	P	O+P	Type of director
–I	45	1	54	99	Ortho/para
–CHO	19	72	9	28	Meta

Your Turn 23.2

Think

How many bonds and lone pairs does a positively charged C have? Does that add up to an octet?

Solve

Each positively charged C (circled) does not possess an octet. Each has three bonds and no lone pairs, and thus only has a share of six electrons. Notice that the ortho intermediate has a resonance structure in which all atoms have octets, but the meta intermediate does not.

Ortho intermediate

All atoms have octets, making this resonance structure especially stable.

(23-4)

Meta intermediate

The lone pair on O cannot participate in resonance to stabilize the positive charge.

(23-5)

Your Turn 23.3

Think

Are alkyl groups electron-donating or electron-withdrawing groups? In which resonance structure does C^+ have the smallest concentration of positive charge?

Solve

The second resonance structure is especially stable due to the electron-donating CH_3 group attached directly to the C^+. This is because the CH_3 group decreases the concentration of positive charge there.

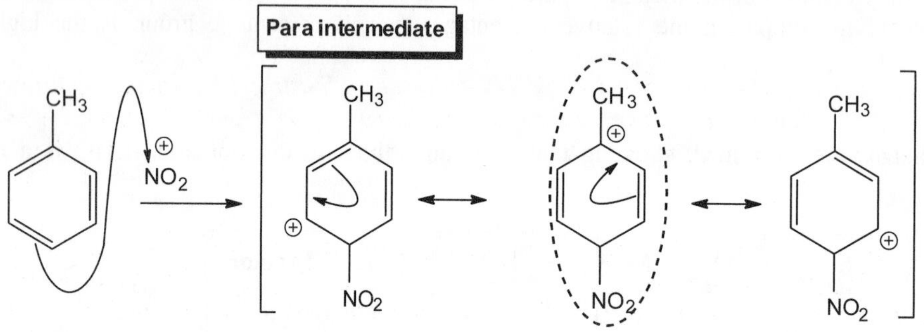

Para intermediate

Your Turn 23.4

Think

In comparing the stability of the three resonance structures, what do you notice about the proximity of the positive charge on the C and the positive charge on the N? How do you think this affects the stability of the arenium ion?

Solve

When the positive charge on the ring is adjacent to the positive charge on the N, the arenium ion is destabilized (circled). The resonance structure, as a result, is much higher in energy than the other two and does not contribute as significantly to the resonance hybrid.

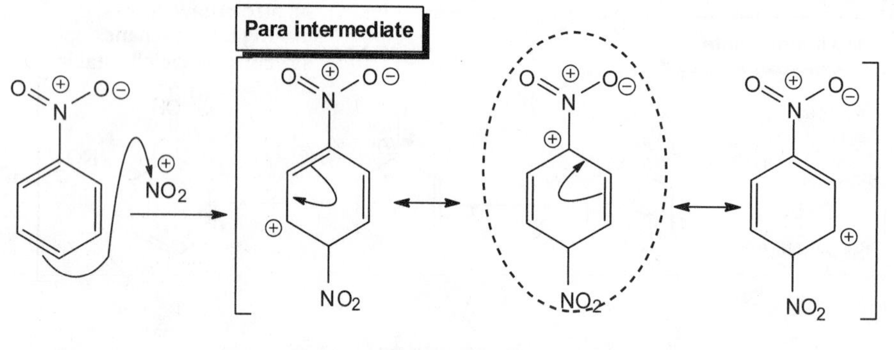

Para intermediate

Your Turn 23.5

Think

Are there nonbonding electrons adjacent to a carbon lacking an octet? In the second resonance structure, what new bond is formed? How many electrons surround each C and the Cl atom? Do any lack an octet? Is Cl more or less electronegative compared to C? How does this group affect the positive charge on the ring?

Solve

The lone pairs on the Cl fold down to the C^+ lacking an octet to form the Cl=C bond. In that resonance structure, all atoms have an octet. Chlorine is an inductive electron-withdrawing group and pulls electron density *away* from the ring. This increases the positive charge on the ring. These two reasons are why Cl is an ortho/para director (leads to a resonance structure with a full octet via resonance donation of electrons) but is a deactivating group (the Cl is an inductive electron-withdrawing group). See the figure on the next page.

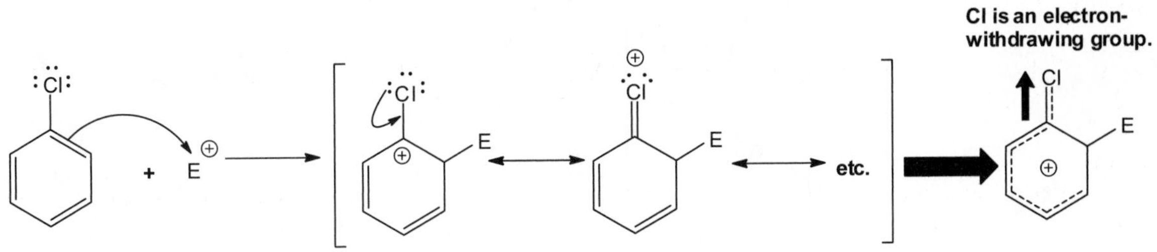

Cl is an electron-withdrawing group.

Your Turn 23.6

Think

Which color (red or blue) indicates a larger amount of electron density? In electrophilic aromatic substitution, should the ring be more electron rich or electron poor to be considered activated? Think about the charge needed on the ring and the charge needed on an electrophile.

Solve

Red indicates greater electron density (electron rich). In electrophilic aromatic substitution, the ring should be sufficiently electron rich compared to the incoming electron-poor electrophile. The ring of the given substituted benzene is deactivated, therefore, because the ring is displaying less negative charge (i.e., less red).

Your Turn 23.7

Think

Are NO_2 groups electron donating or withdrawing? Are CH_3 groups electron donating or withdrawing? How do the electron-donating or electron-withdrawing properties of the substituent affect the stability of an attached C^+? Does a resonance structure contribute more to the resonance hybrid when the resonance structure is more stable or less stable?

Solve

NO_2 groups are electron withdrawing and the structure where the C^+ is adjacent to the positive charge on the N is the least stable resonance structure. CH_3 groups are electron donating and the resonance structure that has C^+ adjacent to the CH_3 is the most stable. Overall, this intermediate is more stable, due to the stabilization provided by the third structure; the contribution by the least stable resonance structure is minimized.

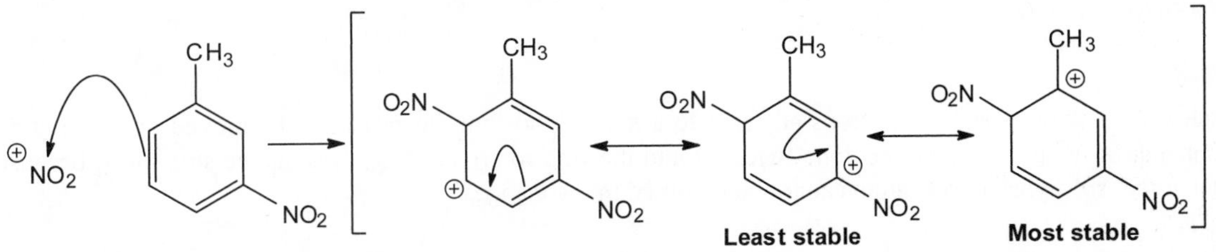

Least stable

Most stable

Your Turn 23.8

Think

Consult Equation 23-31a and 23-31b as a guide for electrophilic aromatic substitution of a five-membered aromatic ring with a heteroatom. What steps make up the usual electrophilic aromatic substitution mechanism? Can substitution take place at C2? At C3?

Solve

The usual electrophilic aromatic substitution mechanism consists of two steps: electrophilic addition followed by elimination of H^+. Substitution can occur at both the 2 and 3 positions on the ring. See the mechanism on the next page.

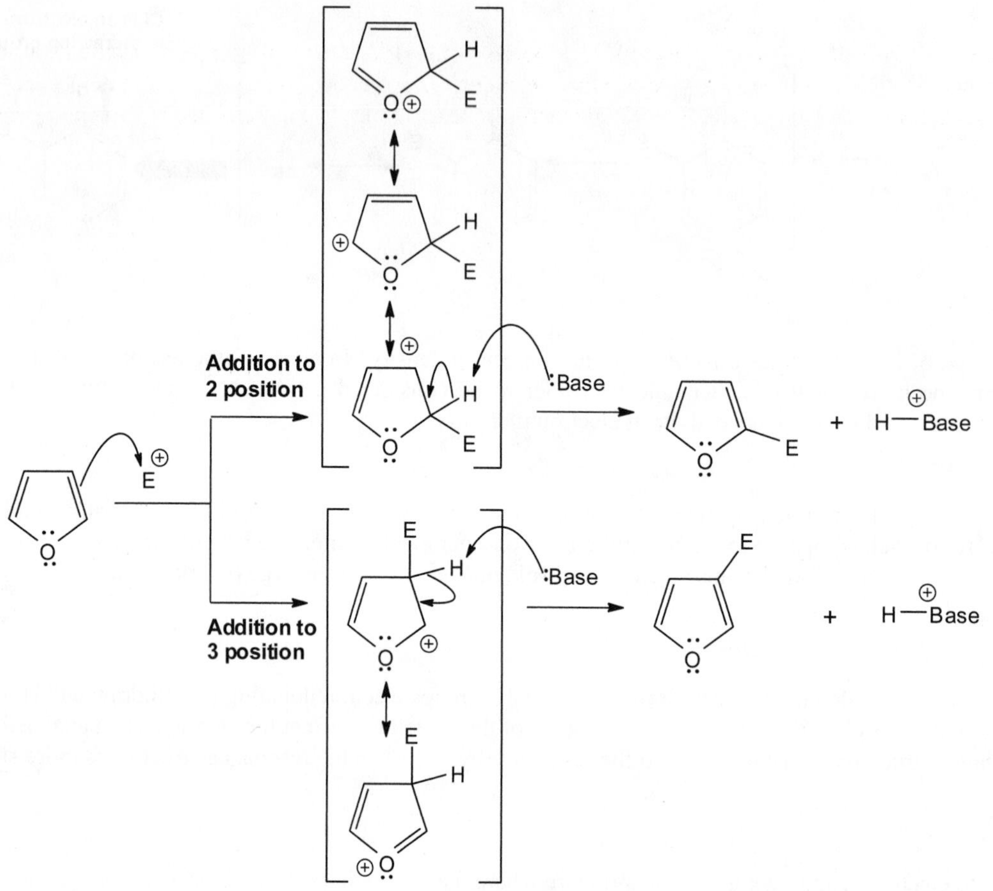

Your Turn 23.9

Think

Are there any lone pair electrons adjacent to π bonds? Is a resonance structure possible involving an sp^3-hybridized C atom?

Solve

Each C⁻ has a lone pair of electrons adjacent to a π bond, so four electrons can be moved to arrive at a new resonance structure. This can be done twice around the ring to arrive at new resonance structures, because the uncharged, sp^3-hybridized C atom cannot be involved in resonance.

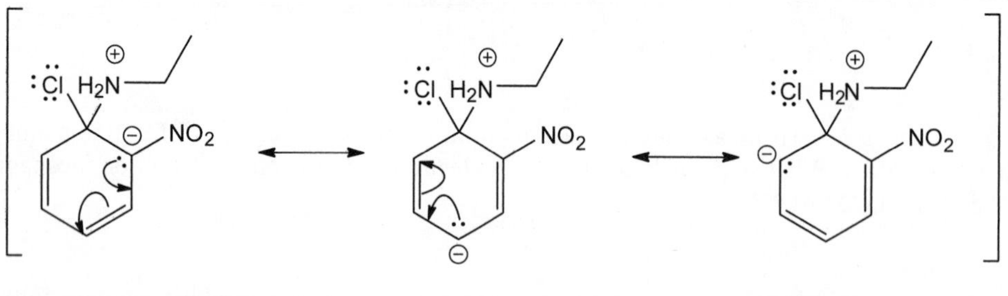

Your Turn 23.10

Think

Are there any lone pair electrons adjacent to π bonds? Is a resonance structure possible at an sp^3-hybridized C atom?

Solve

Each C^- has a lone pair of electrons adjacent to a π bond, so four electrons can be moved to arrive at a new resonance structure. This can be done twice around the ring to arrive at new resonance structures because the uncharged, sp^3-hybridized C atom cannot be involved in resonance. In the second resonance structure, the C^- is attached to NO_2.

Your Turn 23.11

Think

What bonds are broken and formed in each step? How can you use curved arrows to show the electron movement? Review the 10 elementary steps from Chapters 6 and 7. In which steps are protons involved? Which steps involve the formation of a π bond? Which steps involve breaking a π bond?

Solve

Step 1 is a proton transfer reaction between the basic NH_2^- anion and the proton ortho to the Cl leaving group. Step 2 is a nucleophile elimination in which the lone pair on the ring folds down to form the C≡C of the benzyne intermediate and the Cl leaves. Step 3 is nucleophilic addition in which the NH_2^- anion acts as a nucleophile to add to the C≡C and folds the electrons back to the C to break the triple bond. Step 4 is a proton transfer step to protonate the C^-.

1. Proton transfer

2. Nucleophile elimination

3. Nucleophilic addition

4. Proton transfer

Your Turn 23.12

Think

Is the CO_2H group meta or ortho/para directing? How would the product change if you oxidized the CH_3 into CO_2H prior to the nitration step? Is *m*-nitrobenzoic acid a possible product if nitration is the first step? Is more than one route possible?

Solve

The CO_2H group is meta directing and oxidation of the alkyl group prior to nitration will direct the NO_2 group meta. If nitration were performed first, the ring would be deactivated enough that Friedel–Crafts alkylation would not occur as shown on the next page. Thus, the first route shown below is the best route to *m*-nitrobenzoic acid.

R–Cl, $AlCl_3$; 1. $KMnO_4$, HO^-, Δ; 2. H_3O^+; HNO_3, H_2SO_4

HNO_3 / H_2SO_4 → NO_2 **Deactivated** → R–Cl / $AlCl_3$ → R, NO_2 (crossed out)

Chapter Problems

Problem 23.1

Think

Does the monosubstituted benzene in each example have an ortho/para or meta director (consult Table 23-1)? How is each electrophile formed? What are the two steps that make up the general electrophilic aromatic substitution mechanism (Chapter 22)?

Solve

(a) The HO is an ortho/para director. This is an example of Friedel–Crafts alkylation. The electrophile in the reaction is $CH_3CH_2^+$. It should be noted that the electrophilic addition and heterolysis steps probably occur simultaneously instead of separately but the electrophile R^+ is shown explicitly for emphasis. The mechanism is given below:

Formation of the electrophile

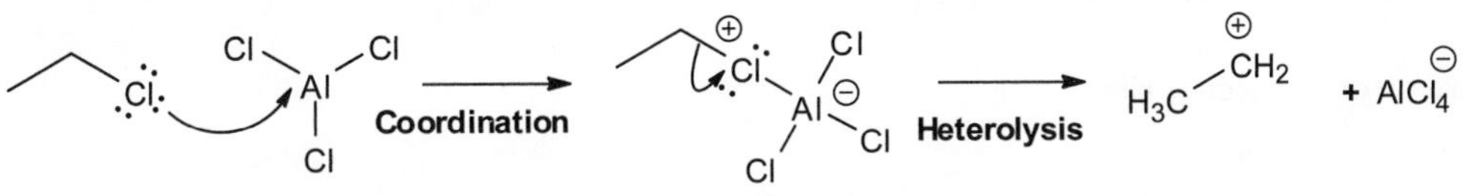

Coordination

Heterolysis

$+ AlCl_4^-$

Ortho substitution

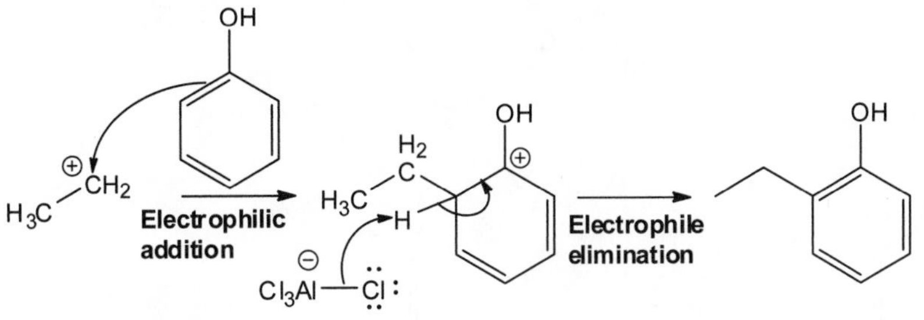

Electrophilic addition

Electrophile elimination

Para substitution

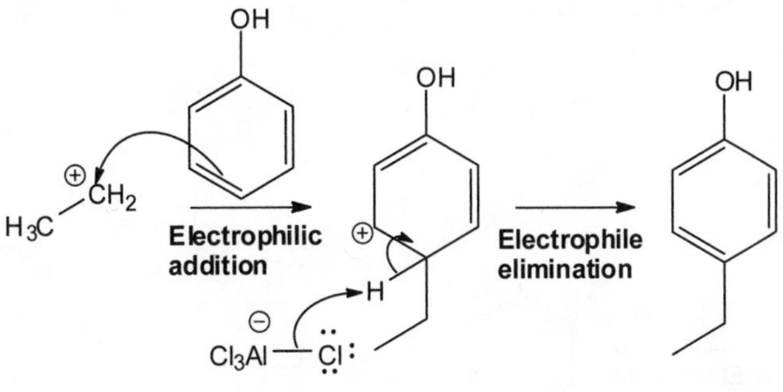

Electrophilic addition

Electrophile elimination

(b) The NO_2 is a meta director. This is a chlorination reaction and the electrophile in the reaction is Cl^+. It should be noted that the electrophilic addition and heterolysis steps probably occur simultaneously instead of separately but the electrophile Cl^+ is shown explicitly for emphasis. The mechanism is given below and on the next page:

Formation of the electrophile

Coordination

Heterolysis

$+ FeCl_4^-$

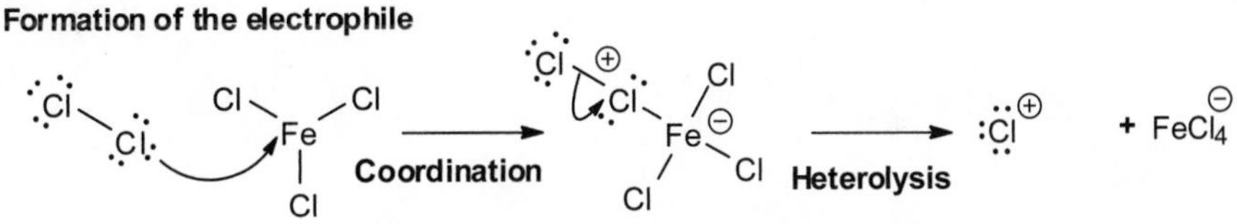

Meta substitution

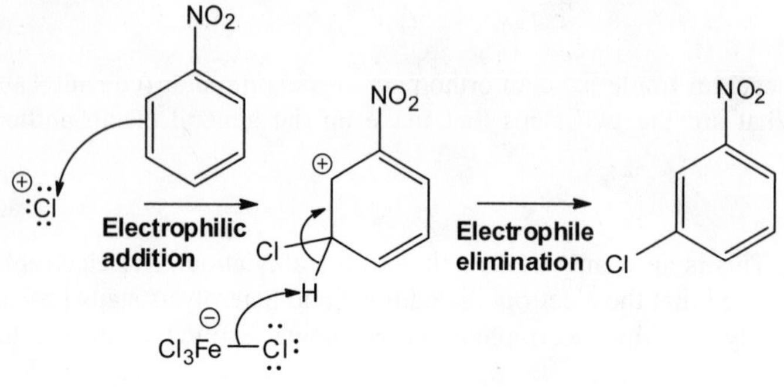

NO_2 NO_2 NO_2

Cl   Electrophilic addition   Cl   H Cl_3Fe — Cl   Electrophile elimination   Cl

(c) The CH_3 is an ortho/para director. This is an example of Friedel–Crafts acylation. The electrophile in the reaction is CH_3CO^+. The mechanism is given below:

Formation of the electrophile

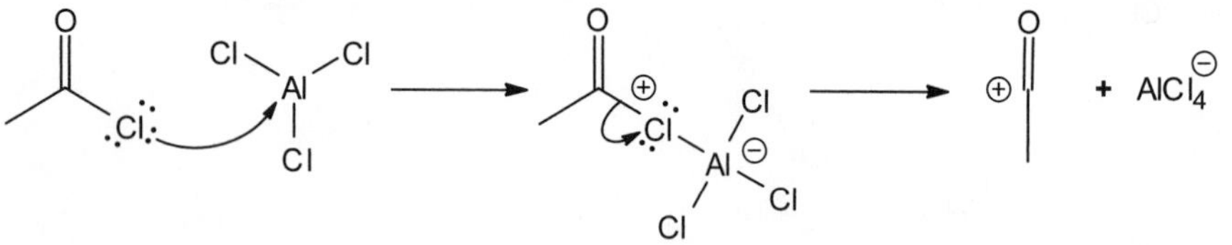

O   Cl   Cl   Al   Cl   Cl   →   O   Cl   Cl   Al   Cl   Cl   →   O   + $AlCl_4^-$

Ortho substitution

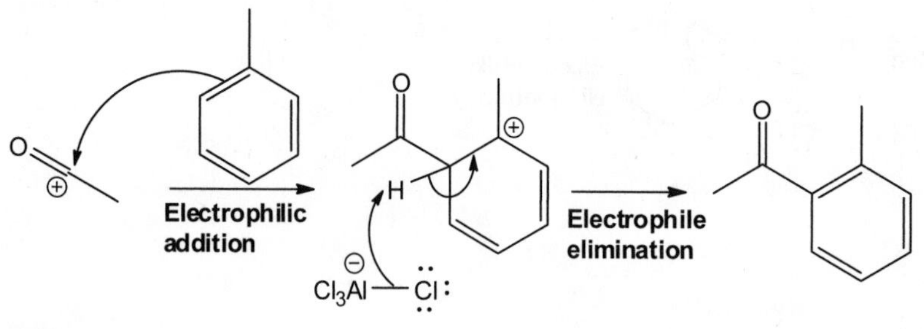

O   Electrophilic addition   O   H Cl_3Al — Cl   Electrophile elimination   O

Para substitution

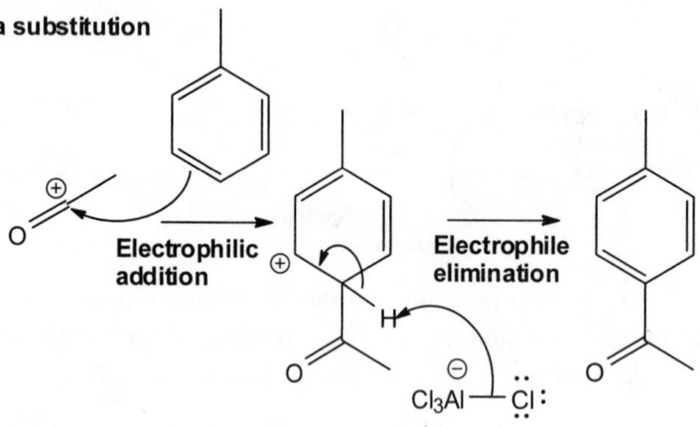

O   Electrophilic addition   H   O Cl_3Al — Cl   Electrophile elimination   O

(d) The $CH_3C(O)$ is a meta director. This is a bromination reaction and the electrophile in the reaction is Br^+. It should be noted that the electrophilic addition and heterolysis steps probably occur simultaneously instead of separately but the electrophile Br^+ is shown explicitly for emphasis. The mechanism is given below:

Formation of the electrophile

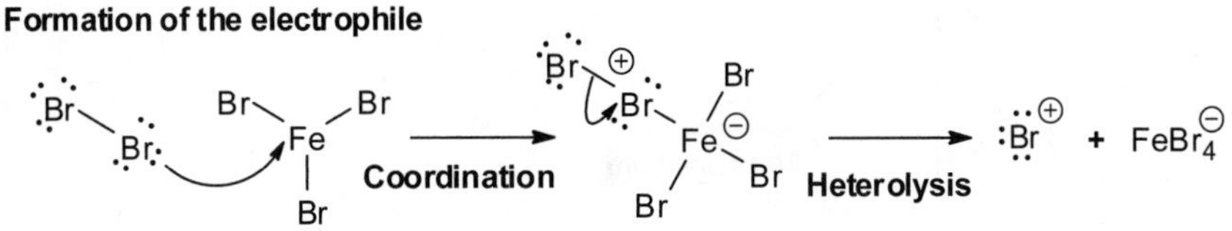

Coordination

Heterolysis

$:Br^{\oplus}$ + $FeBr_4^{\ominus}$

Meta substitution

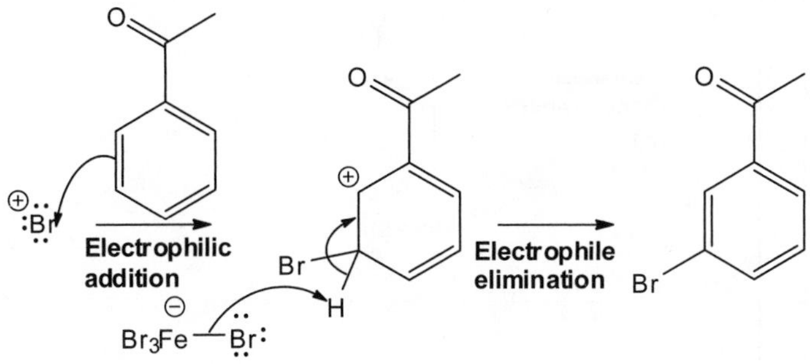

Electrophilic addition

Electrophile elimination

Problem 23.2

Think

Consult Table 23-1. Which product is produced in the greatest abundance: ortho, meta, or para? The least abundance? How does the relative amount of each product correspond to the stability of the arenium ion intermediate from which it was produced?

Solve

According to Table 23-1, the para product is the most abundant, so the para arenium ion is produced the fastest. The meta product is the least abundant, so the meta arenium ion is produced the slowest.

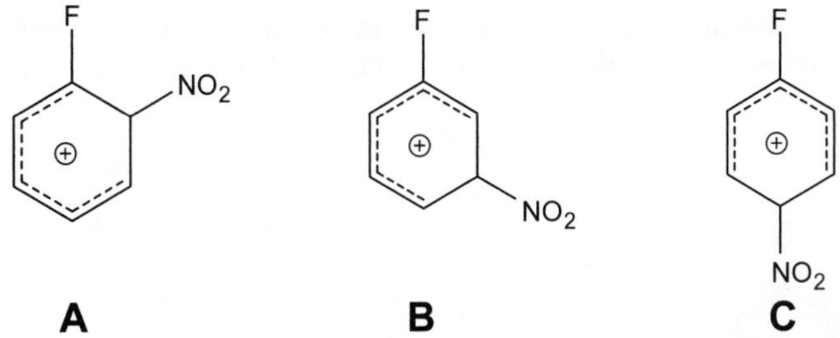

A

B

Meta product:
Least abundant product
Least stable intermediate

C

Para product:
Most abundant product
Most stable intermediate

Problem 23.4

Think

Which of the isomeric products—ortho or meta—is formed faster? What does that say about the stability of the respective intermediates? What role is played by resonance delocalization of charge?

Solve

According to Table 23-1, the ortho arenium ion is formed faster than the meta arenium ion, so the ortho intermediate must be more stable. This gives rise to the following energy diagram:

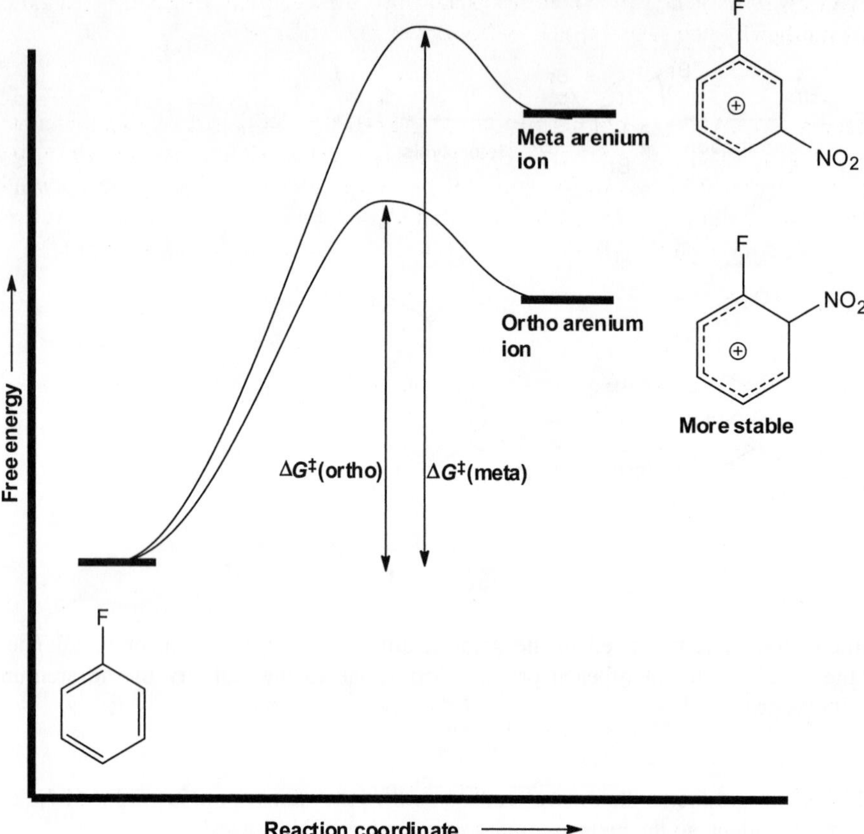

Whereas the meta intermediate only has three resonance structures, the ortho intermediate has *four*, similar to the para intermediate. Furthermore, the one that is boxed is the most important of the four because all of the atoms have complete octets.

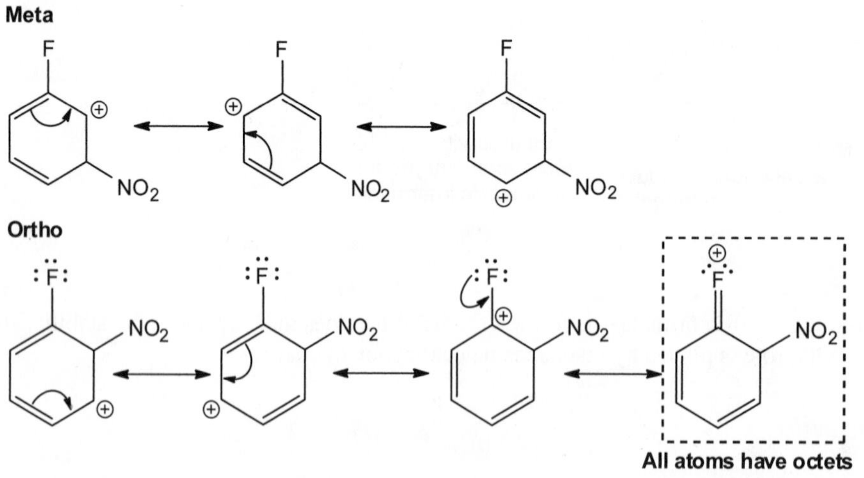

Meta

Ortho

All atoms have octets

Problem 23.5

Think

What is the identity of the electrophile in the electrophilic addition–elimination reaction? From which C=C bond should a curved arrow originate to show the π electrons attacking the electrophile in ortho, meta, and para electrophilic aromatic substitution? Is the substituent an ortho/para or meta director?

Solve

(a) This is a chlorination reaction and the electrophile is the Cl^+ cation. The electrophilic aromatic substitution mechanism is shown for ortho, meta, and para substitution products. OCH_3 is an ortho/para director because the O atom that attaches it to the ring has a lone pair of electrons, so the major products will be the ortho and para isomers. It should be noted that the electrophilic addition and heterolysis steps probably occur simultaneously instead of separately but the electrophile Cl^+ is shown explicitly for emphasis.

Formation of the electrophile

Coordination

Heterolysis

$:Cl^{\oplus}$ + $FeCl_4^{\ominus}$

Ortho product

$:OCH_3$

Electrophilic addition

$Cl_3Fe^{\ominus}$—Cl:

Electrophile elimination

Major

Meta product

$:OCH_3$

Electrophilic addition

$Cl_3Fe^{\ominus}$—Cl:

Electrophile elimination

Para product

$:OCH_3$

Electrophilic addition

$Cl_3Fe^{\ominus}$—Cl:

Electrophile elimination

Major

(b) The $CH_3CH_2^+$ cation is the electrophile in this Friedel–Crafts alkylation reaction. The electrophilic aromatic substitution mechanism is shown for ortho, meta, and para substitution products. SC(O)CH_3 is an ortho/para director because the S atom that attaches it to the ring has a lone pair of electrons, so the major products will be the ortho and para isomers. It should be noted that the electrophilic addition and heterolysis steps probably occur simultaneously instead of separately but the electrophile R^+ is shown explicitly for emphasis.

Formation of the electrophile

Coordination

Heterolysis

$H_3C-CH_2^{\oplus}$ + $AlCl_4^{\ominus}$

Ortho product

Electrophilic addition

Electrophile elimination

Cl_3Al-Cl

Major

Meta product

Electrophilic addition

Electrophile elimination

Cl_3Al-Cl

Para product

Electrophilic addition

Electrophile elimination

Cl_3Al-Cl

Major

Problem 23.6

Think

What is the identity of the electrophile in the electrophilic addition–elimination reaction? From which C=C bond should a curved arrow originate to show the π electrons attacking the electrophile in ortho, meta, and para electrophilic aromatic substitution? Is CH_3CH_2 an ortho/para or meta director?

Solve

The NO_2^+ cation is the electrophile in this nitration reaction. CH_3CH_2 is an ortho/para director because it is electron donating, so the major products will be the ortho and para isomers.

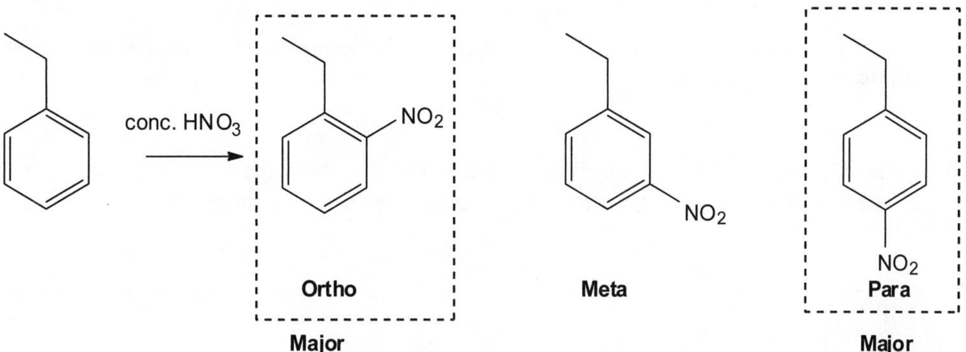

Problem 23.8

Think

Draw the arenium ion intermediate produced when E^+ adds to the para position and consider all of the resonance structures of that arenium ion intermediate. Do any of the arenium ion resonance structures exhibit the positive charge on the ring directly adjacent to the CF_3 substituent? What effect does the CF_3 have on an adjacent positive charge? Is it stabilizing or destabilizing?

Solve

The resonance structures of the para intermediate are shown below. It is less stable than the meta intermediate because the second resonance structure has the positive charge on the C directly attached to CF_3. The CF_3 group is electron withdrawing, so it destabilizes that positive charge.

Positive charge destabilized

CF_3 CF_3 CF_3

E E E

Problem 23.9

Think

In which group number is Br located in the periodic table? Are there other such substituents in Table 23-2? Relative to those substituents, where should Br fall on the list?

Solve

Because Br is between Cl and I in the periodic table, and in the same group, we should expect its influence to be intermediate between the two. So we should expect the relative rate of nitration of bromobenzene to be between 0.18 and 0.033, or ~0.1. This would make it weakly deactivating relative to H.

Problem 23.11

Think

Is CF_3 an inductively electron-donating or electron-withdrawing group? Which type of group activates the benzene ring toward electrophilic aromatic substitution?

Solve

The CF_3 group is inductively electron withdrawing, so it will destabilize the arenium ion intermediate, causing the reaction to be slower. Thus, the ring in trifluoromethylbenzene is deactivated.

Problem 23.12

Think

Is the monosubstituted ring electron rich or electron poor? Activated or deactivated? Which group, CH_3 or CF_3, is an inductive electron-withdrawing group?

Solve

Trifluoromethylbenzene. The ring is less red, signifying less electron density in the π system, consistent with a deactivated ring. CF_3 is a deactivating group, whereas CH_3 in toluene is an activating group.

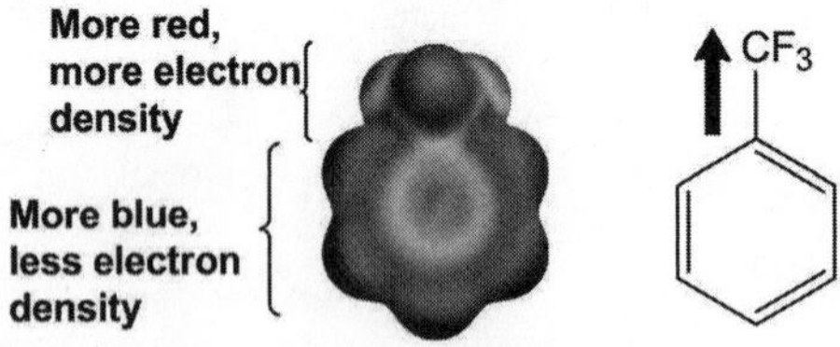

Problem 23.13

Think

Is benzenesulfonic acid activated or deactivated toward electrophilic aromatic substitution sulfonation? How does this affect the rate of sulfonation? What does the addition of SO_3 do to the rate of sulfonation?

Solve

Benzenesulfonic acid is strongly deactivated toward electrophilic aromatic substitution. The addition of SO_3 increases the concentration of the electrophile that adds to the aromatic ring, which serves to speed up the reaction. The addition of SO_3, therefore, counteracts the deactivation from the sulfonyl group already attached.

SO_3H — conc. H_2SO_4, SO_3, 4 h, <90 °C ⇌ HO_3S … SO_3H

Strongly deactivated

1,3-Benzenedisulfonic acid
90%

Problem 23.14

Think

What type of elementary step yields the phenoxide anion? What is the two-step mechanism by which bromination takes places on the phenoxide anion? How many times does this reaction mechanism occur? How is the uncharged product produced?

Solve

A proton transfer reaction yields the phenoxide anion, which then undergoes three electrophilic addition–elimination mechanisms to give the tribromo phenoxide product. The uncharged phenol product is produced by a proton transfer reaction.

Proton transfer

Electrophilic addition

Electrophile elimination

Electrophilic addition

Electrophile elimination

Electrophilic addition

Electrophile elimination

Proton transfer

Problem 23.16

Think

Identify each substituent as a weak, moderate, or strong activator or deactivator. How many activating and deactivating groups does each ring have? Are they all the same strength?

Solve

According to Table 23-3, the NO_2 group is strongly deactivating, the $CH_3C{=}O$ group is moderately deactivating, and the Cl and Br groups are weakly deactivating. **C** will react the slowest because it is the only one with three deactivating groups. **A**, **E**, and **F** have two deactivating groups, and the rate is based on the strength of the deactivating group, **A** < **F** < **E**. Structure **D** only has one deactivating group, so it is next. Benzene has no deactivating groups and reacts the fastest. Therefore, the order of increasing reaction rate is **C** < **A** < **F** < **E**< **D** < **B**.

A: NO_2 Strongly deactivating; Moderately deactivating

B

C: NO_2 Strongly deactivating; Cl Weakly deactivating; Moderately deactivating

D: Cl Weakly deactivating

E: Br Weakly deactivating; Cl Weakly deactivating

F: NO_2 Strongly deactivating; Cl Weakly deactivating

C < A < F < E < D < B

Slowest → Fastest

Problem 23.17

Think

What is the leaving group? Is this mechanism nucleophilic addition–elimination or electrophilic addition–elimination? Is there a strong electron-withdrawing group ortho or para to the leaving group?

Solve

Cl is the leaving group and there is a strong electron-withdrawing group ortho (and meta) to the leaving group. This mechanism is a nucleophilic addition–elimination mechanism because the amine serves as a nucleophile to attack an electrophilic carbon. This reaction is *slower* than the one shown in Equation 23-35 because the strongly electron-withdrawing groups are ortho and meta and not ortho and para.

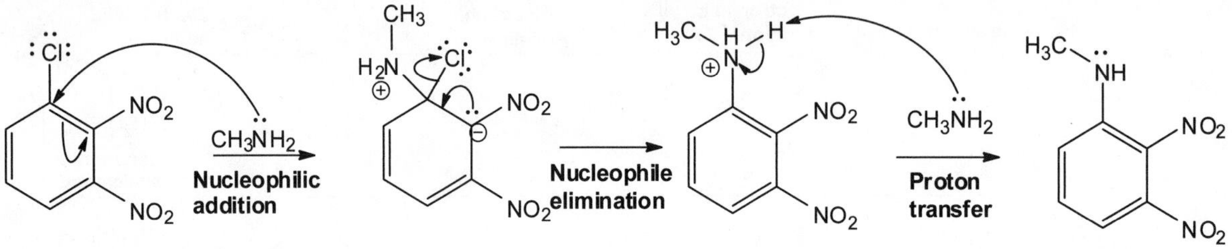

CH_3NH_2 Nucleophilic addition → Nucleophile elimination → CH_3NH_2 Proton transfer

Problem 23.18

Think

What is the leaving group? Is this mechanism nucleophilic addition–elimination or electrophilic addition–elimination? Is there a strong electron-withdrawing group ortho or para to the leaving group?

Solve

(a) Cl is the leaving group and the amine is the nucleophile in the nucleophilic addition–elimination mechanism. Notice that the leaving group is ortho to one electron-withdrawing group and para to another.

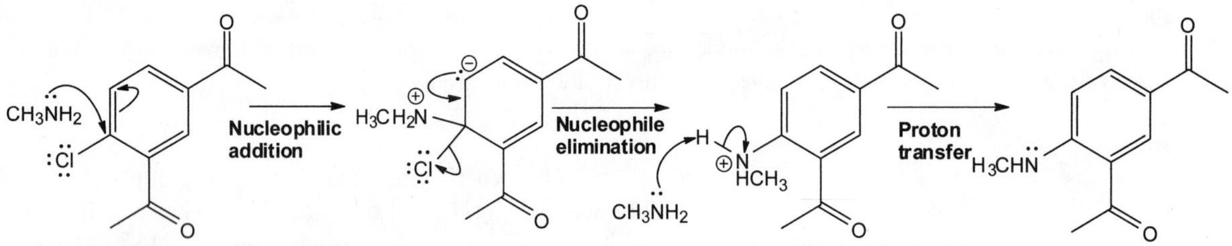

Nucleophilic addition → Nucleophile elimination → Proton transfer

(b) F is the leaving group and the alkoxide is the nucleophile in the nucleophilic addition–elimination mechanism. Notice that the leaving group is indeed ortho to a strongly electron-withdrawing group.

Nucleophilic addition → Nucleophile elimination

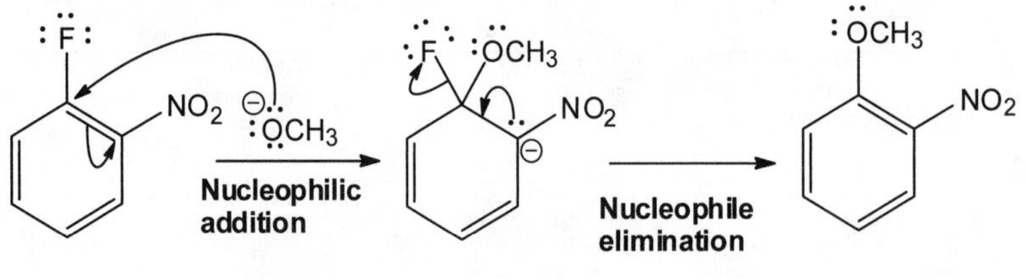

(c) There are two F atoms that could serve as leaving groups. One F is meta to the NO_2 and the other F is ortho to the NO_2. The Meisenheimer complex is stabilized when the electron-withdrawing group is ortho (or para) to the leaving group.

Nucleophilic addition → Nucleophile elimination

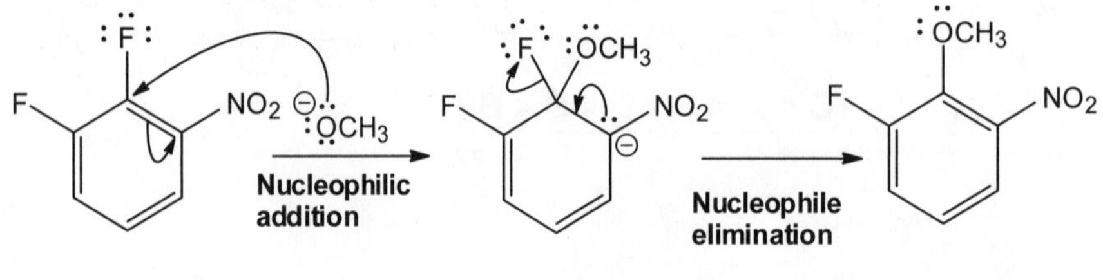

Problem 23.19

Think

How does the bromobenzene interact with a strong base? How does the Br leaving group leave? What intermediate is formed? How does a nucleophile react with that intermediate? In the deuterated compound, which ortho carbon—C–H, C–D, or both—can be deprotonated by the NaOH base? How does this affect the distribution of the D isotope in the product?

Solve

Hydroxide acts as a base to remove a proton from the aromatic ring to produce a carbanion. In Step 2, the leaving group departs in a nucleophile elimination step. The product of Step 2 is a benzyne intermediate, which can undergo nucleophilic addition by hydroxide to produce a deprotonated form of phenol. A final proton transfer step occurs to yield the uncharged product.

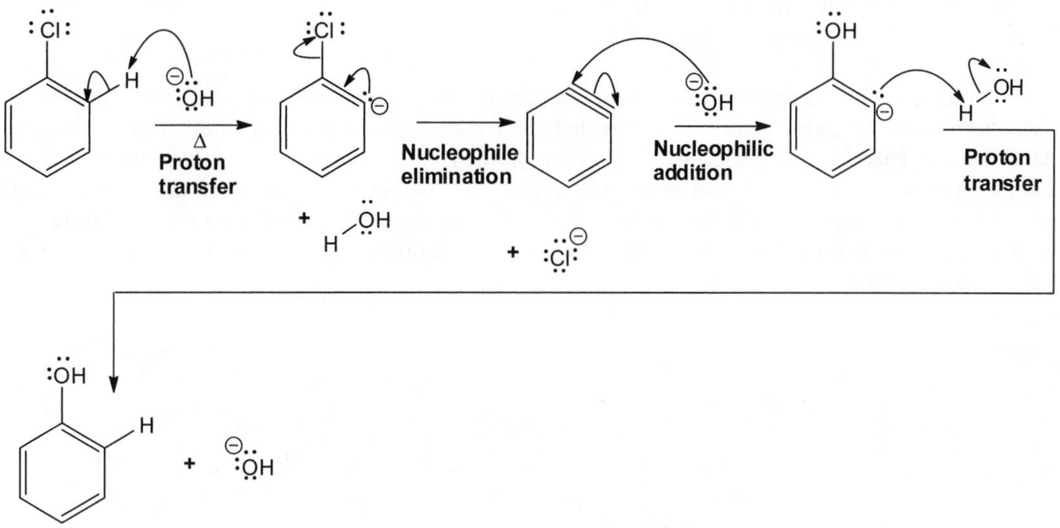

Proton transfer (Δ) · Nucleophile elimination · Nucleophilic addition · Proton transfer

H and D can both be eliminated with essentially equal likelihood; therefore, 50% of the product is expected to contain D.

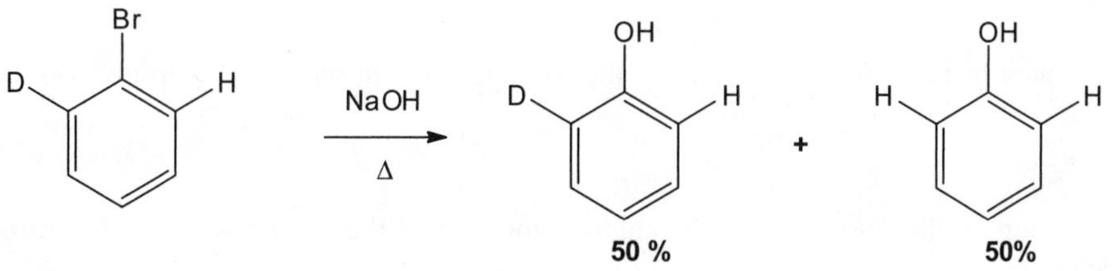

NaOH, Δ → 50 % + 50%

Problem 23.21

Think

Is the Cl substituent an ortho/para director or a meta director? Is the alkanoyl group an ortho/para director or a meta director? Which group should be on the ring prior to the second substitution?

Solve

The Cl substituent is an ortho/para director and the alkanoyl group is a meta director. Because our target is a meta isomer, the alkanoyl group should be on the ring before the second substitution is carried out. Therefore, a Friedel–Crafts acylation should take place first, followed by a chlorination as shown on the next page.

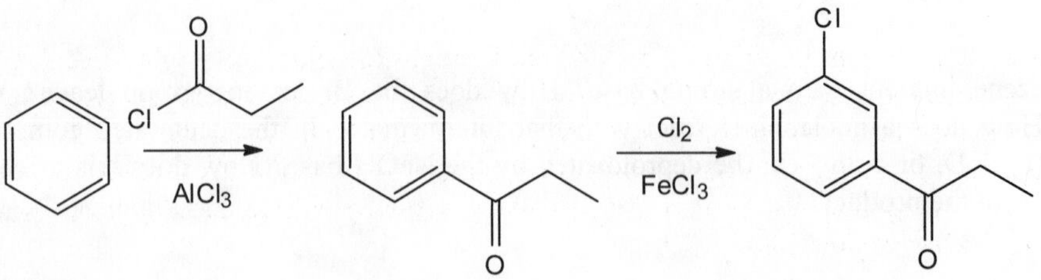

Problem 23.22

Think

Are the CO_2H groups ortho/para or meta directors? What is the route for adding on the CO_2H group? Does it go through a synthetic intermediate that has an ortho/para director? How can you transform ethanol into the necessary reagent for Friedel–Crafts alkylation?

Solve

A carboxyl group is a meta director, but the target is a para isomer. Adding a carboxyl group to a benzene ring, however, can proceed through a synthetic intermediate in which the substituent is an alkyl group, which is an ortho/para director. Therefore, with the first alkyl group on the ring, the second alkyl group can be added to the para position prior to oxidizing both groups to carboxylic acids. To begin the synthesis, first transform ethanol into the alkyl bromide reagent to conduct a Friedel–Crafts alkylation. The Friedel–Crafts alkylation needs to occur two times to form *p*-diethylbenzene. Then use $KMnO_4$ followed by acid workup to oxidize the alkyl groups on benzene to the carboxylic acids.

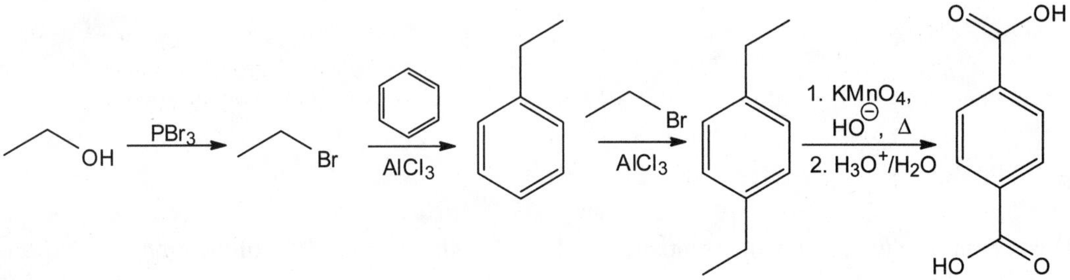

Problem 23.23

Think

Are the substitutents ortho/para or meta directors? What is the route for adding on the NH_2 group? Does it go through a meta director?

Solve

We can envision an NO_2 group on the ring to direct the chlorination toward the *meta* position. Afterward, the NO_2 group can be reduced to NH_2.

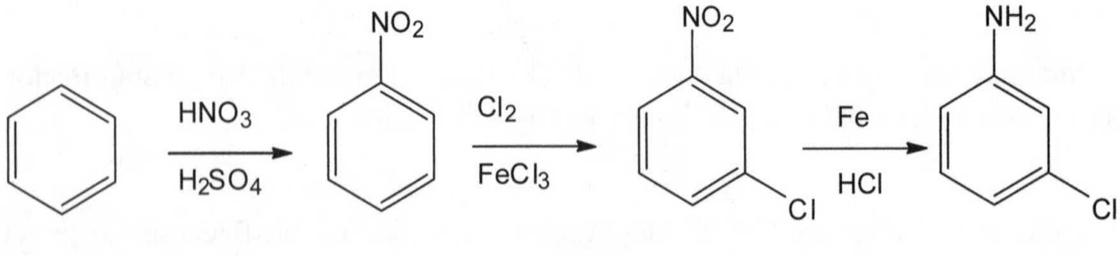

Problem 23.24

Think

Which group is the most activating? Where will it direct incoming electrophiles? What reaction occurs with aniline and a strong acid? Will this interfere with the desired reactions in your synthesis? How can this be avoided by protecting the aniline? What type of director is the protected aniline compound?

Solve

We can envision having an NH_2 group on the ring to direct the alkyl groups ortho and para in a Friedel–Crafts alkylation. The NH_2 group can come from nitration of benzene, followed by reduction of the NO_2. However, we must protect the NH_2 group before the introduction of the strong Lewis acid. Otherwise, the Lewis acid will complex to the NH_2 group, turning it into a strong deactivating group. This would prevent the necessary Friedel–Crafts alkylations from happening. We protect the NH_2 using acetic anhydride and deprotect it in the final step using hydrolysis under basic conditions.

NO_2 NH_2 O O HN O HN O NH_2 HNO_3 H_2SO_4 Fe HCl O Cl $AlCl_3$ NaOH H_2O, Δ

Problem 23.25

Think

How can you block the para position? Are alkyl groups meta or ortho/para directors? Do you add the second alkyl group before or after reduction of the ketone?

Solve

We can envision a Friedel–Crafts acylation taking place with a sulfonyl group temporarily blocking the para position. The sulfonyl group can be placed at the para position after the acetyl group is reduced to an ethyl group to make it an ortho/para director. A final reduction step is necessary to form the propyl group.

O HCl Zn(Hg) SO_3 H_2SO_4 O Cl $AlCl_3$ O H_3O^+ O HCl Zn(Hg) SO_3H SO_3H

Problem 23.26

Think

What is the electrophile in each case? How is the electrophile formed? Is the group on the ring ortho/para or meta directing? Is ortho or para favored? What are the elementary steps for the electrophilic addition–elimination mechanism?

Solve

(a) The electrophile is the NO_2^+ cation. HO_3S is a meta director. The mechanism is shown below:

SO_3H OSO_3H H HO N O O Proton transfer H_2O N O O Heterolysis O N O Electrophilic addition SO_3H O N O H H_2O Electrophile elimination SO_3H O N O

(b) The electrophile is the Cl^+ cation. CH_3CH_2 is an ortho/para director. It should be noted that the electrophilic addition and heterolysis steps probably occur simultaneously instead of separately but the electrophile Cl^+ is shown explicitly for emphasis. The mechanism is shown below:

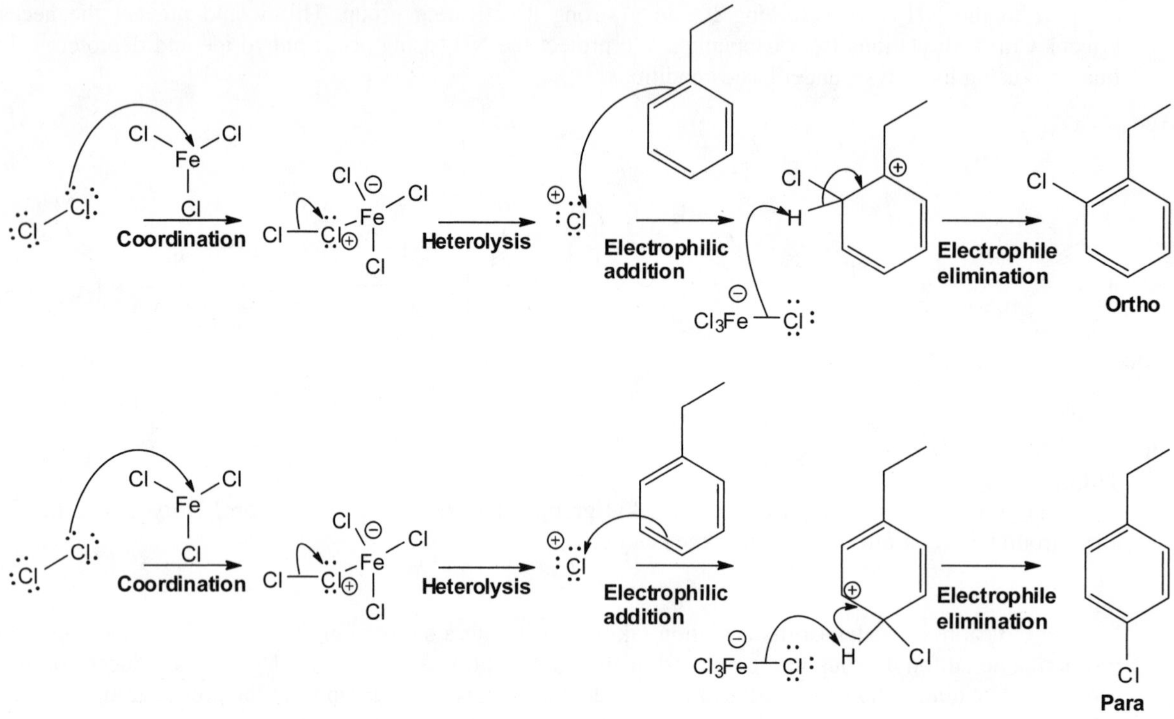

(c) The electrophile is the $PhCO^+$ cation. CH_3O is an ortho/para director. The mechanism is shown below:

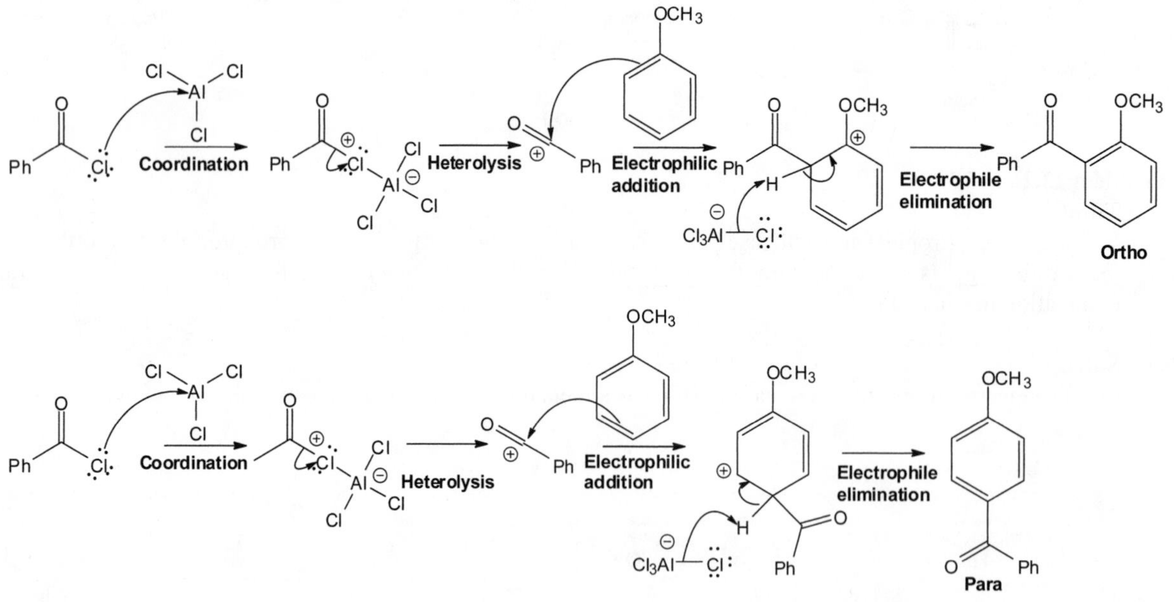

(d) The electrophile is the Br^+ cation. RCO is a meta director. It should be noted that the electrophilic addition and heterolysis steps probably occur simultaneously instead of separately but the electrophile Br^+ is shown explicitly for emphasis. The mechanism is shown below:

Coordination → Heterolysis → Electrophilic addition → Electrophile elimination

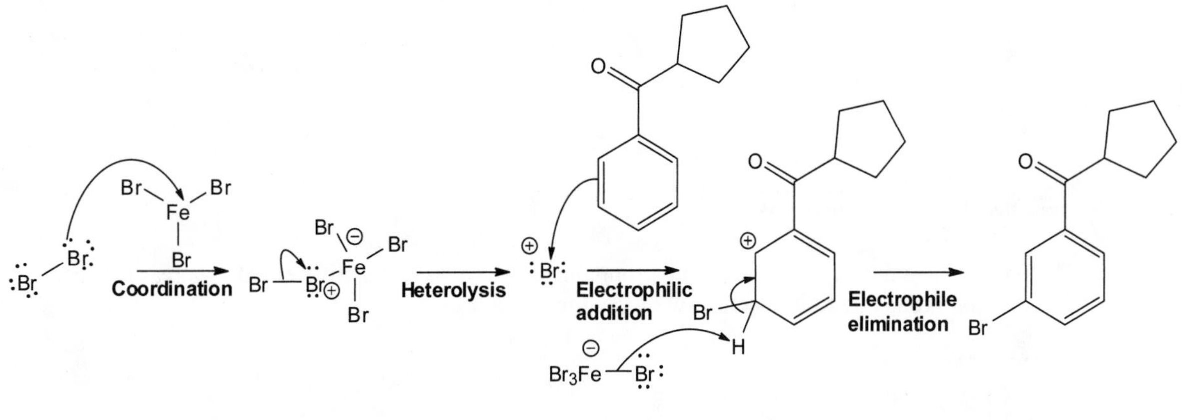

(e) The electrophile is the NO_2^+ cation. $(CH_3)_2CH$ is an ortho/para director. The mechanism is shown below:

Proton transfer → Heterolysis → Electrophilic addition → Electrophile elimination

Ortho

Proton transfer → Heterolysis → Electrophilic addition → Electrophile elimination

Para

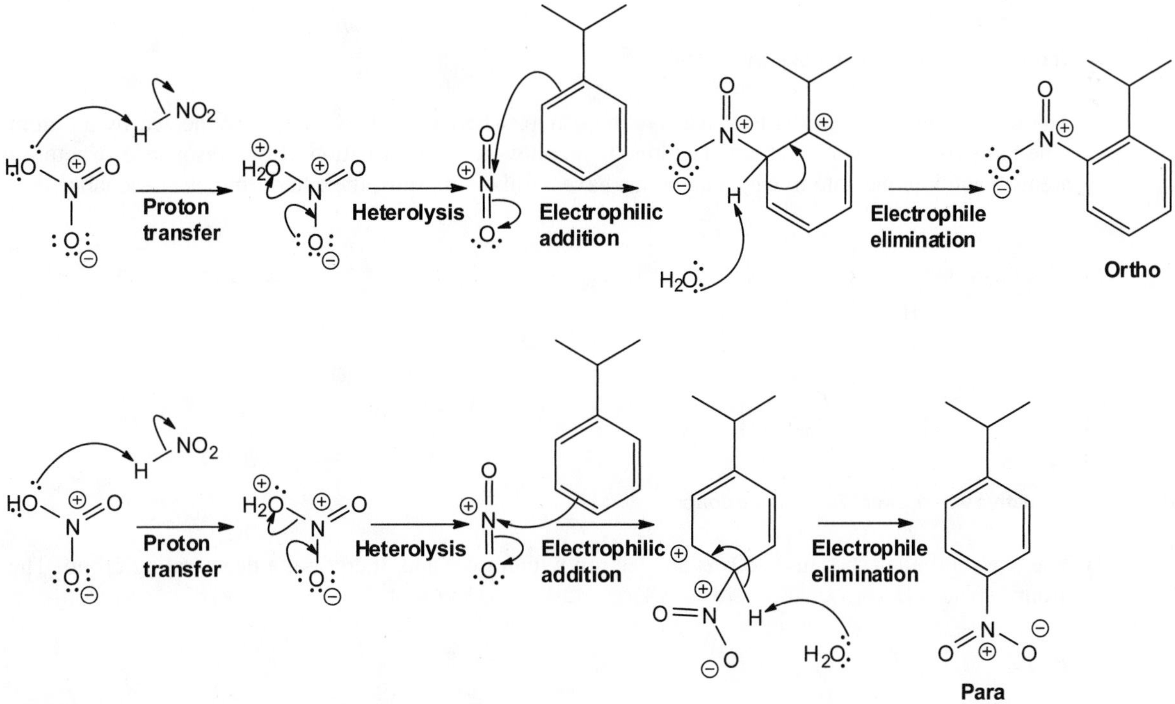

Problem 23.27

Think

Consult Table 23-3 to identify each substituent as a weak, moderate, or strong activator or deactivator. Which ring is most activated? Which ring is most deactivated? How does the activated/deactivated nature of the ring correspond to the rate of electrophilic aromatic substitution?

Solve

(a) The one on the right; OH is a stronger activator than CH_3. This is because OH donates via resonance, whereas CH_3 donates via the weaker inductive electron-donating effects.

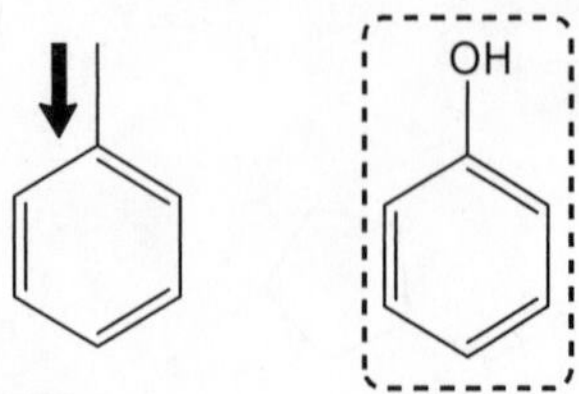

(b) The one on the left; CH_3 is an activating group due to its electron-donating effect, whereas CF_3 is a deactivating group due to its electron-withdrawing effect.

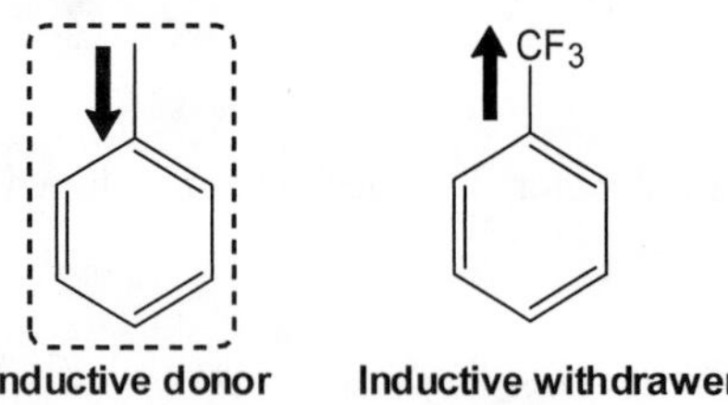

(c) The one on the right; CO_2H is a deactivating group, whereas OCH=O, being connected by an atom with lone pairs, is an activating group. The lone pairs allow for the formation of one resonance structure of the arenium ion intermediate in which all atoms have a full octet. Thus, the ether group is a resonance donor.

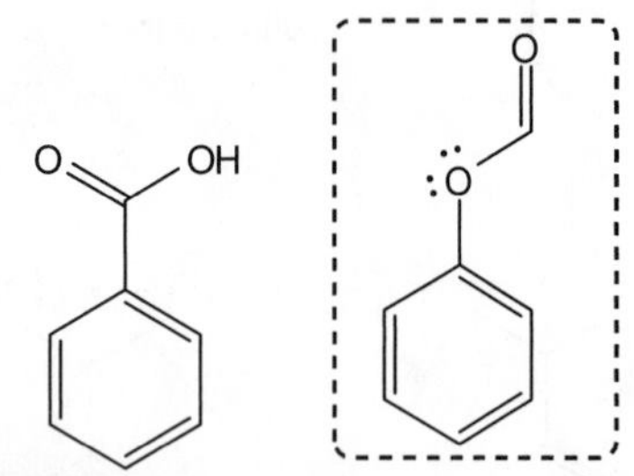

(d) The one on the left, because NO_2 is an electron withdrawer and, therefore, a deactivating group. The NO_2 group replaces H.

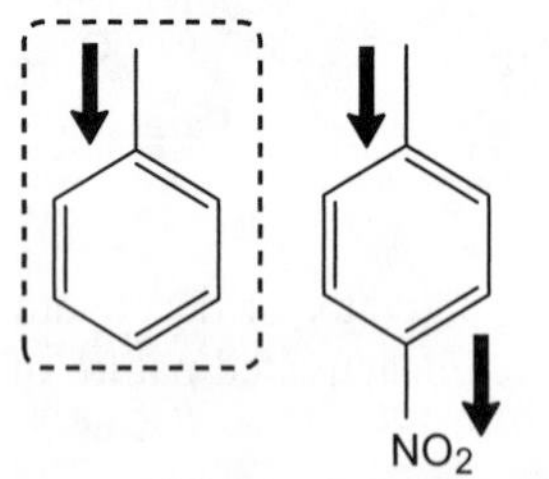

(e) The one on the left, because Cl is a deactivating group and two deactivating groups slow the reaction more than one.

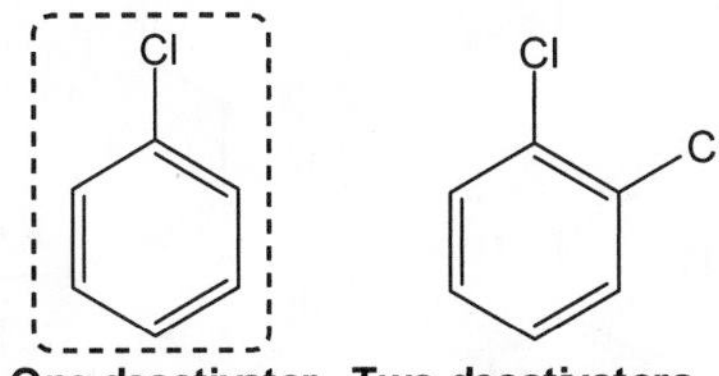

(f) The one on the right. The S atom in the structure on the right has lone pairs that make it an activating group, whereas SO_3H is a deactivating group.

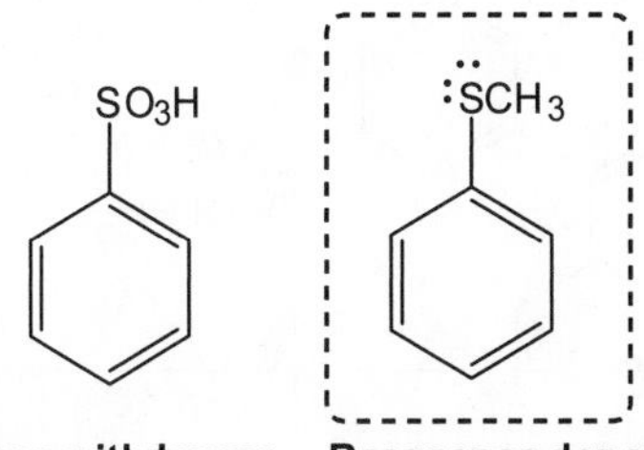

(g) The one on the left, because OH is a stronger activating group than CH_3.

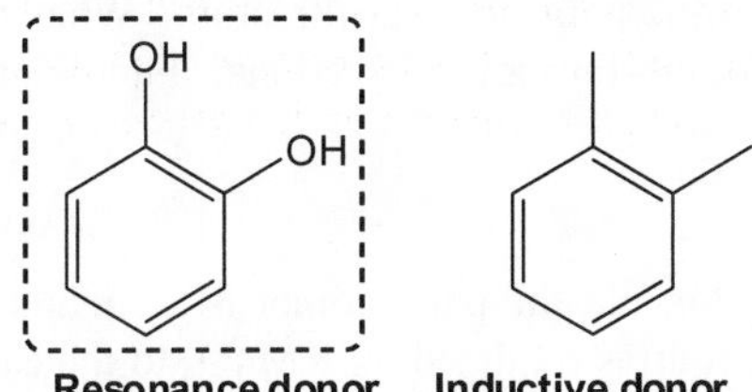

(h) The one on the right, because it has two activating groups, whereas the one on the left has just one.

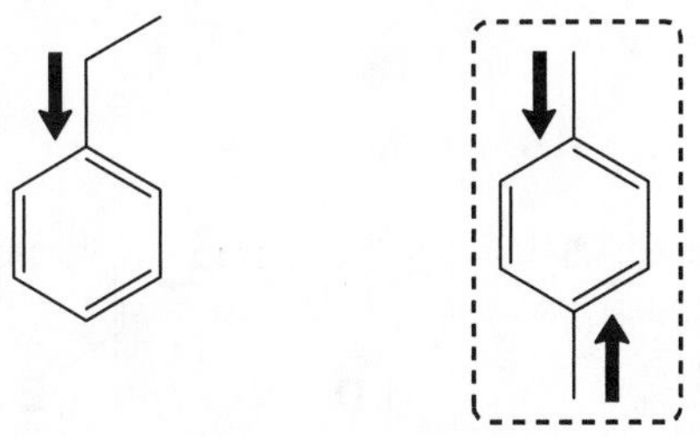

Problem 23.28

Think

What groups are added to the benzene ring? Are these groups ortho/para or meta directors? What reactions occur on functional groups off the ring? Are there any C–C bond-formation reactions? Functional group transformations? Oxidations or reductions?

Solve

Friedel–Crafts acylation forms the ketone product **A**. An acyl group is a meta director and $Cl_2/FeCl_3$ chlorinates the ring at the meta position to form **B**. A Wittig reaction occurs with the ketone and $CH_3CH^{-}{-}^{+}PPh_3$ to form the alkene **C**. The alkene reacts with the peroxyacid MCPBA to form the epoxide **D**. The epoxide is attacked at the *less* crowded carbon by the Grignard reagent C_6H_5MgBr to form the alcohol **E**. The alcohol undergoes a dehydration reaction to form the tetrasubstituted alkene **F**. See the figure on the next page.

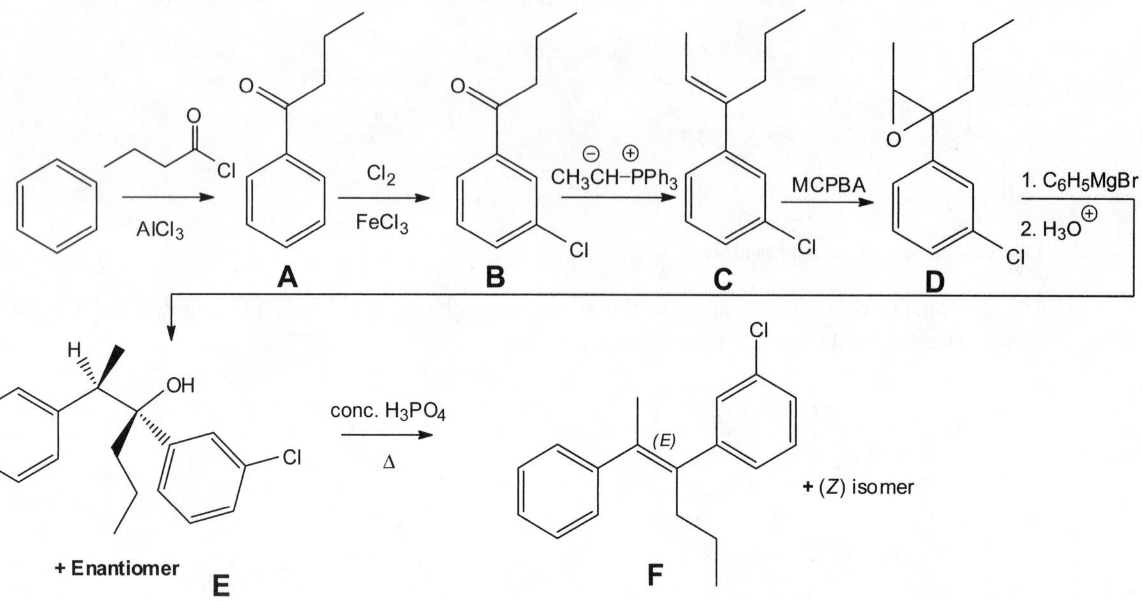

Problem 23.29

Think

What groups are added to the benzene ring? Are these groups ortho/para or meta directors? What reactions occur on functional groups off the ring? Are there any C–C bond-formation reactions? Functional group transformations? Oxidations or reductions?

Solve

Benzene is brominated to form **A**. Bromine is an ortho/para director, but the para isomer is the major product because the CH_3 has some steric bulk. This leads to **B**. The CH_3 group is oxidized by $KMnO_4$ to the carboxylic acid **C** after acid workup. The carboxylic acid is reduced by $LiAlH_4$ to the alcohol **D**, which is oxidized using PCC to the aldehyde **E**. The aldehyde reacts with acetone in a crossed aldol addition to produce the α,β-unsaturated carbonyl **F**, which undergoes conjugate nucleophilic addition to form **G**. The cyano group is hydrolyzed to the carboxylic acid **H**.

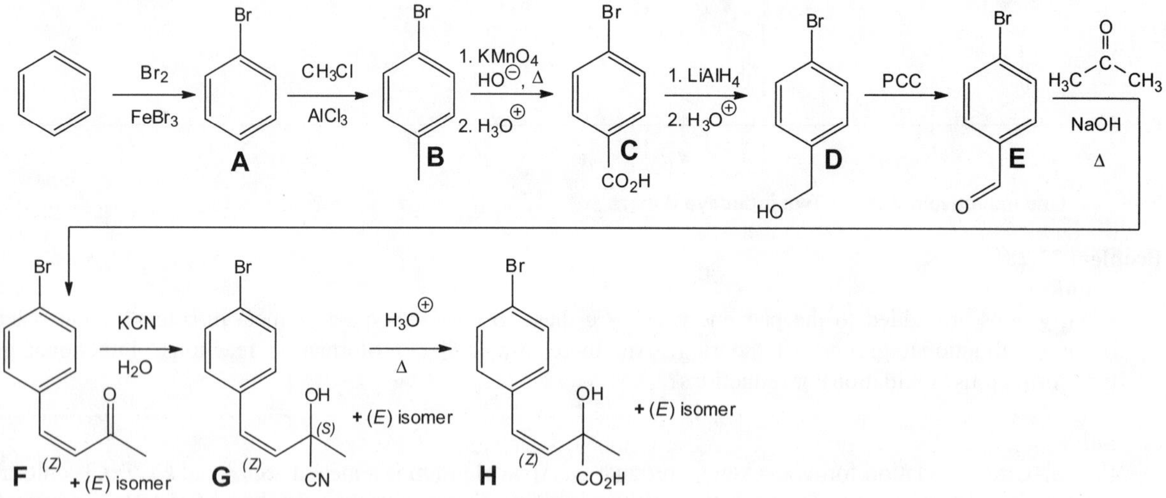

Problem 23.30

Think

What groups are added to the benzene ring? Are these groups ortho/para or meta directors? What reactions occur on functional groups off the ring? Are there any C–C bond-formation reactions? Functional group transformations? Oxidations or reductions?

Solve

Friedel–Crafts acylation occurs to form the ketone **A**. The ketone is reduced to the isopentylbenzene **B**. The alkyl group is an ortho/para director and, due to sterics, nitration occurs primarily at the para position to form **C**. The nitro group is reduced to the amine **D**, which is then transformed into the diazonium ion **E**. The diazonium ion reacts with CuCN to form the nitrile **F**. The nitrile undergoes a Grignard reaction to form the ketone **G**, which is reduced to the secondary alcohol **H**.

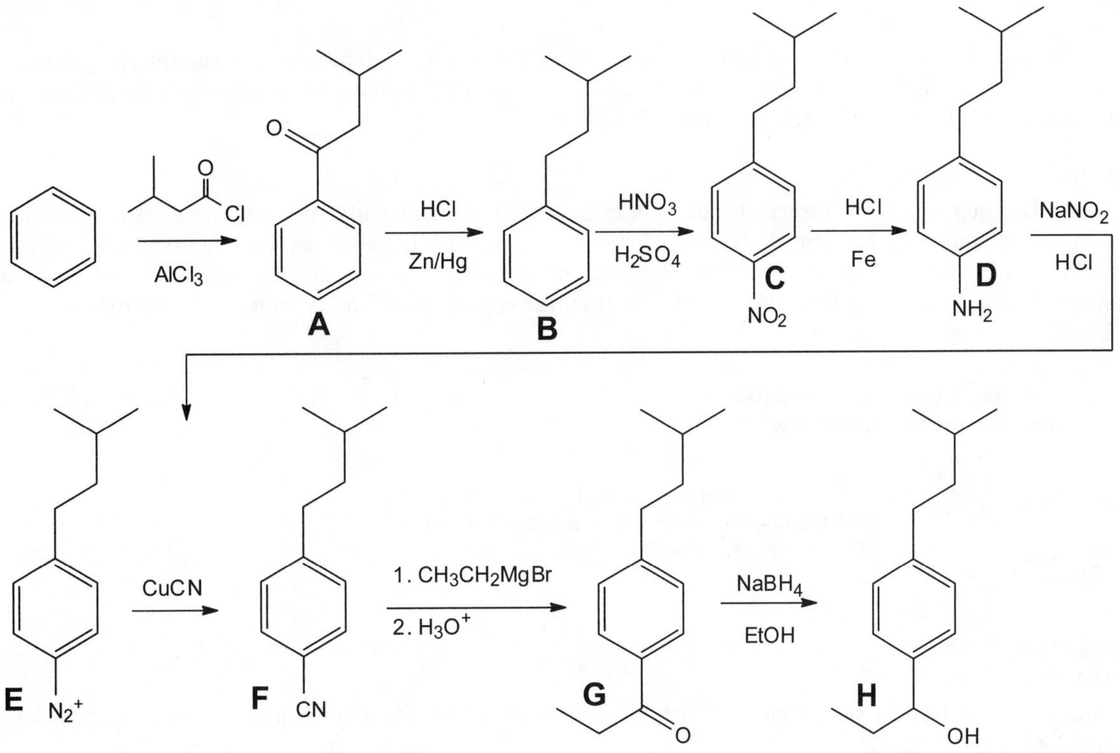

O Cl AlCl$_3$ A HCl Zn/Hg B HNO_3 H_2SO_4 C NO_2 HCl Fe D NH_2 $NaNO_2$ HCl

E N_2^+ CuCN F CN 1. CH_3CH_2MgBr 2. H_3O^+ G O $NaBH_4$ EtOH H OH

Problem 23.31

Think

What is the structure of the –NHCOR substituent? Are there lone pairs on the N atom? Are additional arenium ion resonance structures possible as a result? Are the additional resonance structures possible when an electrophile adds to the ortho and para positions or the meta position? Is the C=O electron donating or withdrawing? How does the nearby C=O affect the strength of the –NHCOR group's activating/deactivating nature?

Solve

(a) It is an ortho/para directing group because the lone pair on N participates in resonance in the arenium ion intermediate only when E^+ attacks an ortho or para position. This leads to a resonance structure where all atoms have a full octet.

(b) It is an activating group because the participation of the lone pair on N in resonance in the arenium ion intermediate provides more stability to the intermediate than H does.

(c) It is less activating than NH_2 because the C=O group in NHCOR is an inductive electron withdrawing group, which does not allow the lone pair to participate in resonance to the extent that the lone pair in $-NH_2$ does. Because the C=O group is electron withdrawing, it contributes to destabilizing the arenium ion intermediate.

See the figure on the next page.

Inductive effects from carbonyl group destabilize intermediate.

O
R NH

Lone pair can participate in resonance to stabilize the arenium ion intermediate.

Problem 23.32

Think

In phenylmethanol, are there lone pairs on the O atom? What type of atom is in between the HO group and the ring? Are additional arenium ion resonance structures possible? Is the HO inductively electron donating or withdrawing? How does this affect the ring's activation?

Solve

If an OH group is attached *directly* to the ring, as in phenol, the lone pairs on O can participate in resonance to stabilize an arenium ion intermediate. In phenylmethanol, the O is not attached directly to the ring and an *sp*^{3} C atom is in between. Therefore, the lone pairs cannot participate in resonance to provide stabilization of the arenium ion. Instead, the O atom is highly electronegative and, therefore, is inductively electron withdrawing and destabilizes an arenium ion.

Inductive effects from O destabilize the arenium ion intermediate.

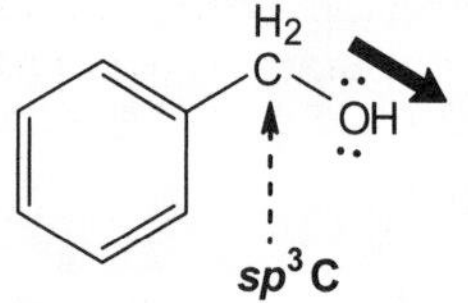

Lone pairs cannot participate in resonance with the ring and, therefore, can't stabilize the arenium ion intermediate.

Problem 23.33

Think

Does red mean that the ring is electron rich or electron poor? Which ring is more electron poor? Which ring, therefore, is deactivated?

Solve

The pyridine ring is *deactivated* compared to the benzene ring because the nitrogen is an electronegative atom and pulls electron density away from the aromatic ring. In the electrostatic potential maps, red signifies a buildup of negative charge and blue signifies a buildup of positive charge. Notice the greater intensity of red in the benzene ring than in the pyridine ring.

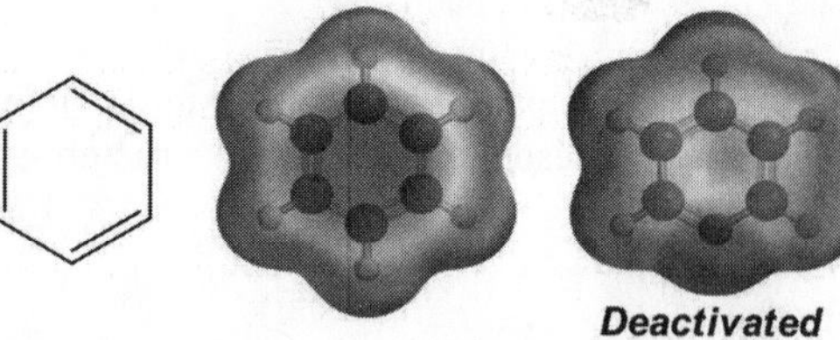

N

Electron rich (red)

Less electron rich (more blue)

Problem 23.34

Think

Does red mean that the ring is electron rich or electron poor? In electrophilic aromatic substitution, is the ring activated when it is electron rich or electron poor? Which ring, therefore, is deactivated?

Solve

There appears to be more red in pyrrole's ring than in benzene's ring, signifying additional concentration of negative charge. Therefore, this is consistent with pyrrole's ring being activated relative to benzene's because the nitrogen electron pair is part of the aromatic system and donates electron density via resonance.

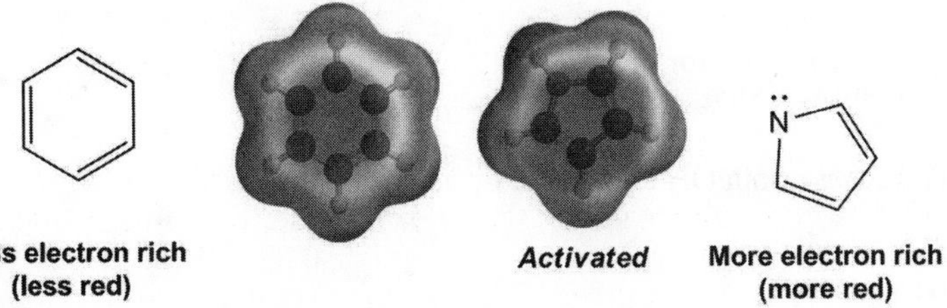

Problem 23.35

Think

How many para and ortho sites are present? Statistically, which is the more likely site of reaction? Where is the ortho site located relative to the R group? How does steric crowding affect the product distribution?

Solve

There are two ortho sites and only one para site; statistically, therefore, it would be expected that the ortho site would have twice the product formation. However, the R group and electrophile E are next to each other in the ortho-substituted product and are on opposite sides of the ring in the para-substituted product. The electrophile's ability to undergo electrophilic addition is hindered as the sterics of the R group increases. Therefore, as the steric bulk of the R group increases, the amount of ortho product decreases and the amount of para product increases.

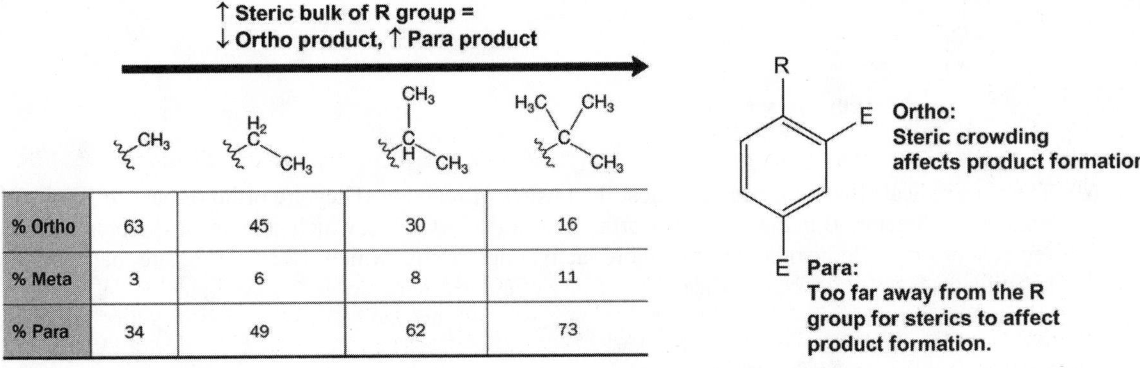

	CH_3	CH_2CH_3	$CH(CH_3)_2$	$C(CH_3)_3$
% Ortho	63	45	30	16
% Meta	3	6	8	11
% Para	34	49	62	73

Problem 23.36

Think

Identify each substituent as a weak, moderate, or strong activator or deactivator. To what positions (ortho, meta, or para) does each substituent direct? Are the two substituents in competition? If so, which one wins? How does steric hindrance come into play?

Solve

(a) The sites indicated with an * are the most likely sites of reaction. OCH_3 is an ortho/para director. The sites indicated with an * are ortho or para to each OCH_3 group; so is the site in between the OCH_3 groups, but steric hindrance prevents attack there.

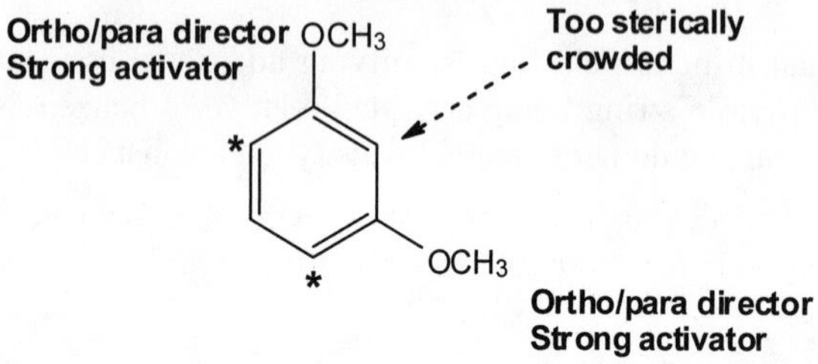

Ortho/para director OCH_3
Strong activator

Too sterically crowded

OCH_3
Ortho/para director
Strong activator

(b) Precisely the same reason as in **(a)**, because both OH and Cl are ortho/para directors.

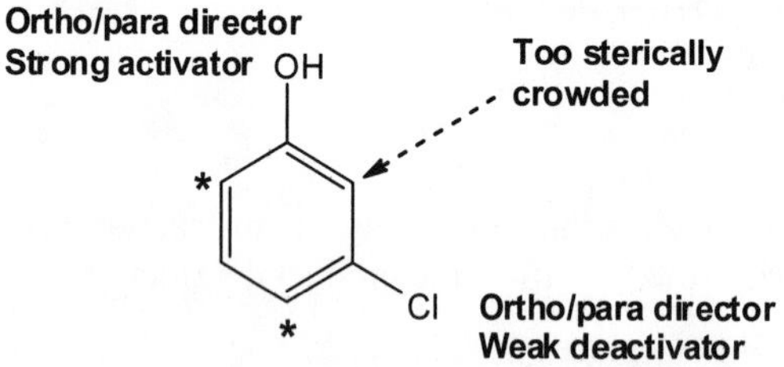

Ortho/para director
Strong activator OH

Too sterically crowded

Cl Ortho/para director
Weak deactivator

(c) The most likely sites of reaction are indicated with an *. Br is an ortho/para director, whereas $COCH_3$ is a meta director. The sites indicated are ortho and para to Br and are meta to $COCH_3$, so those are the most likely sites of reaction.

Meta director
Moderate deactivator

Br Ortho/para director
Weak deactivator

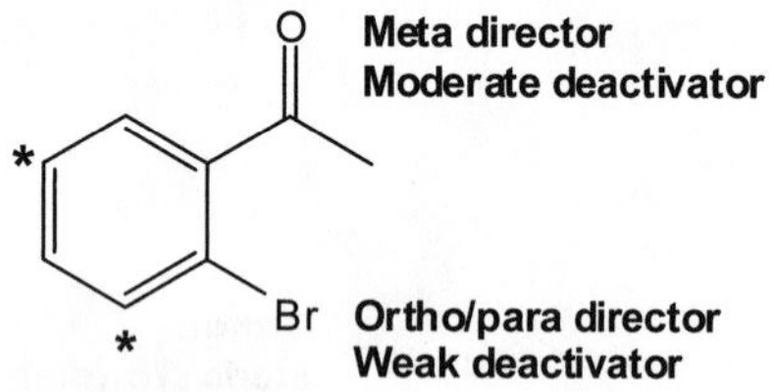

(d) The sites indicated with an * are the most likely sites of reaction. They are ortho or para to Br, which is an ortho/para director. But they are also ortho or para to $COCH_3$, which is a meta director. In this case, regiochemistry is governed by the more activating group, which is Br. The site between the two substituents is not attacked significantly, because of steric hindrance.

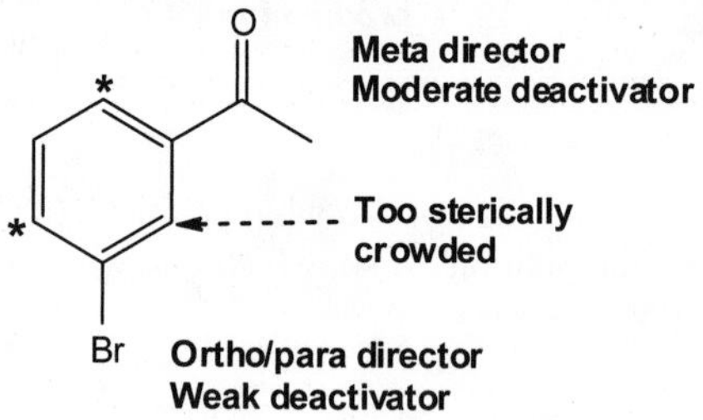

Meta director
Moderate deactivator

Too sterically crowded

Br Ortho/para director
Weak deactivator

(e) The most likely sites of reaction are indicated with an *. These sites are ortho or para to the CH_3 group (an ortho/para director) and meta to the NO_2 group (a meta director).

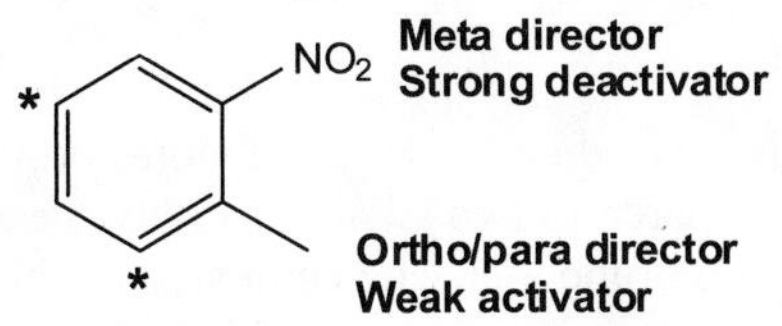

NO_2 Meta director
Strong deactivator

Ortho/para director
Weak activator

(f) The most likely sites of reaction are indicated with an *. All four possible sites for attack are activated by an alkyl group. The ones adjacent to the *t*-butyl group, however, are sterically hindered.

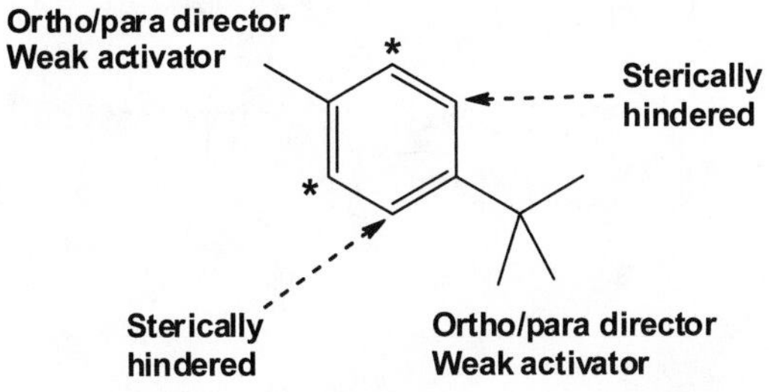

Ortho/para director
Weak activator

Sterically hindered

Sterically hindered

Ortho/para director
Weak activator

(g) The most likely site of reaction is indicated by an *. The NH_2 group is the most activating group on the benzene ring, and so attack should be either ortho or para to the NH_2. The para position is occupied by an alkyl group. The position that is ortho to the NH_2 group that is not marked is not favored because of steric hindrance.

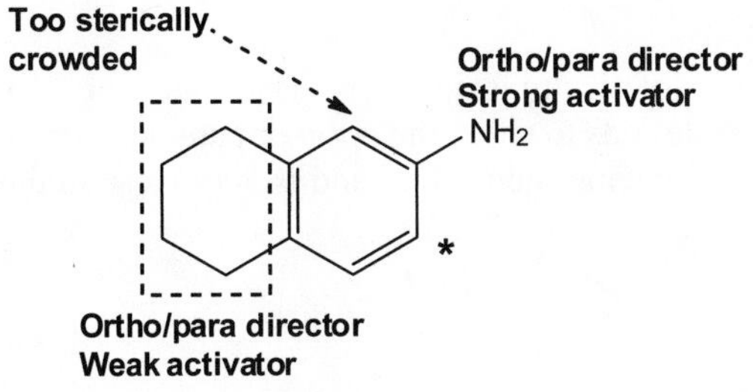

Too sterically crowded

Ortho/para director
Strong activator
NH_2

Ortho/para director
Weak activator

Problem 23.37

Think

Which atom is larger, Br or Cl? How does the size of the atom affect the ortho/para product distribution?

Solve

Br is a larger atom compared to Cl. Br atoms ortho to each other exhibit more steric hindrance during electrophilic addition compared to when Cl and Br ortho to each other. This leads to an increased amount of para product for the bromination of bromobenzene compared to chlorination of bromobenzene.

Steric strain

Br —(Br_2 / $FeBr_3$)→ Br, Br (13%) + Br, Br (85%)

Less steric strain

Br —(Cl_2 / $FeCl_3$)→ Br, Cl (42%) + Br, Cl (53%)

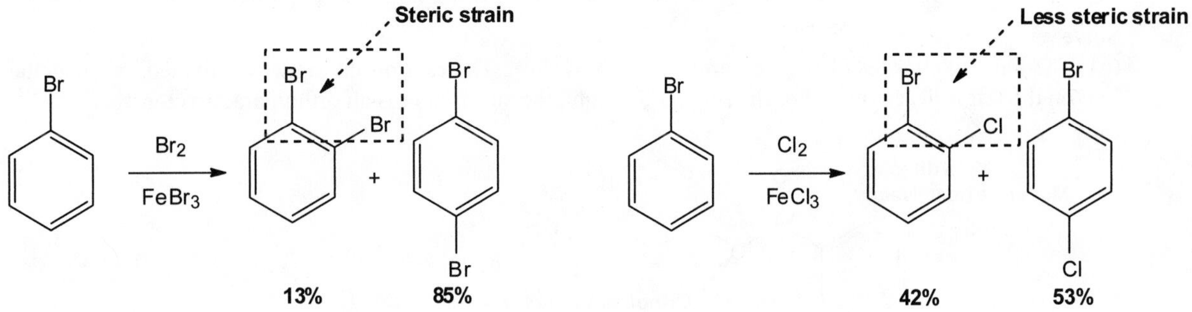

Problem 23.38

Think

Compare the sizes of I, Br, and Cl. How does the size of the atom affect the ortho/para product distribution?

Solve

A would have more para product compared to **B**. I and Br are both larger than Cl. Br and I atoms ortho to each other exhibit more steric hindrance during electrophilic addition compared to two Cl atoms ortho to each other. This leads to an increased amount of para product for the bromination of iodobenzene compared to chlorination of chlorobenzene.

Problem 23.39

Think

Are there lone pairs on the N atom? How does that lone pair affect the stability of the arenium ion intermediate when an electrophile adds ortho/para or meta? Is the –N=O group an electron-withdrawing or electron-donating group? How does that affect whether the group is activating or deactivating?

Solve

The –N=O is ortho/para directing because of the lone pair of electrons on the N. That lone pair stabilizes the arenium ion intermediate via resonance when an electrophile adds to either the ortho or para position. The group is deactivating because the O atom is highly electron withdrawing inductively and is very close to the ring. Its effect is similar to those from a C=O group.

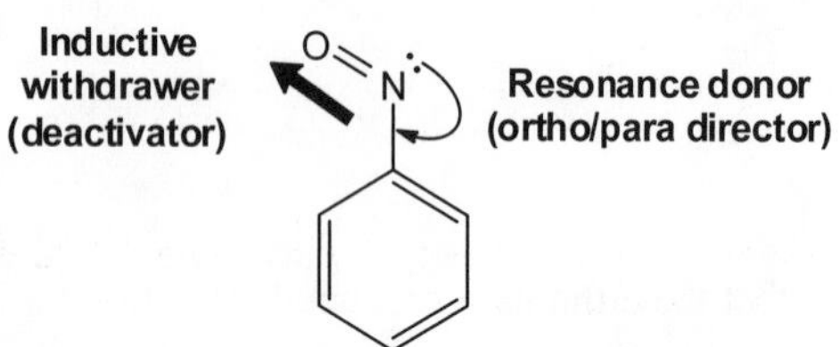

Problem 23.40

Think

Identify each substituent as a weak, moderate, or strong activator or deactivator. Which ring is most activated? Which ring is most deactivated? Are the substituents on that ring ortho/para or meta directors?

Solve

(a) The most likely sites of reaction are indicated by an *. The ring on the right is activated, whereas the ring on the left is deactivated. For the ring on the right, the substituent is an ortho/para director.

(b) The most likely sites of reaction are indicated by an *. All three rings are activated, but the ring in the middle is activated by two different substituents. Both substituents are ortho/para directors.

Ortho/para director
Strong activator

(c) The most likely sites of reaction are indicated by an *. The ring on the left is activated by the CH_3 group and is deactivated by the C=O group. The ring on the right is deactivated only. The sites indicated are ortho or para to the methyl group (an ortho/para director) and meta to the C=O group (a meta director).

Ortho/para director
Weak activator

Meta director
Moderate deactivator

Meta director
Moderate deactivator

(d) The most likely sites of attack are indicated by an *. Both rings are activated, but the lone pair on N activates the ring on the left more than the lone pair on O activates the ring on the right. The ring on the left has an ortho/para directing substituent.

Ortho/para director
Moderate activator
(better resonance donor)

Ortho/para director
Moderate activator

Problem 23.41

Think

Are there any lone pair electrons on B? Is the group an inductive electron donator or withdrawer? Is the ring activated or deactivated?

Solve

The substituent is a meta director and the most likely sites of attack are indicated by an *. The B has no lone pairs (which would otherwise make it an ortho/para director). The electronegative O atoms make it such that the substituent would destabilize the arenium ion intermediate if attack occurs at an ortho or para position.

*No lone pairs on B
*OH is an inductive electron-withdrawing group

Meta director

Problem 23.42

Think

Which sites on the benzene ring are activated from the CH_3 groups? Are any of the sites activated by more than one CH_3? How does steric hindrance come into play?

Solve

Each CH_3 group is an ortho/para director. The symbols on the ring in each molecule below correspond to the CH_3 group that favors that site of attack. In each case, there are two sites (indicated by boxes) that are favored by two separate CH_3 groups. In 1,2,4-trimethylbenzene, one of those sites is sterically hindered, which slows the reaction.

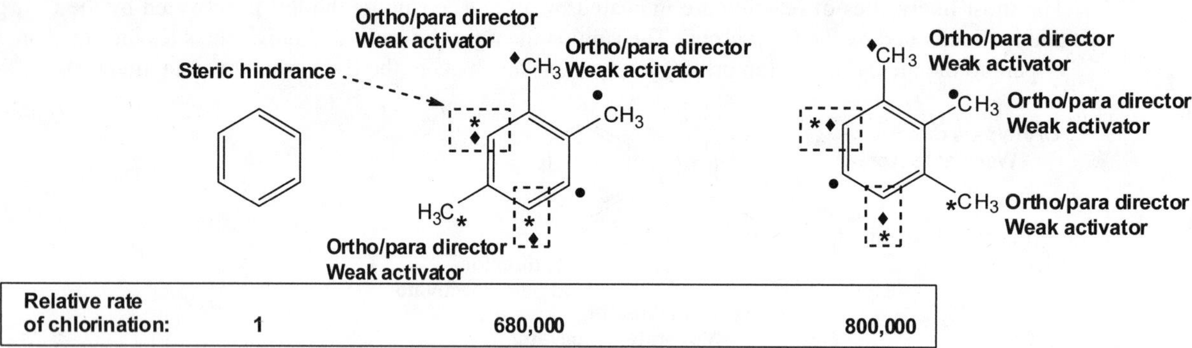

Problem 23.43

Think

Which sites on the benzene ring are activated from the CH_3 groups? Are any of the sites activated by more than one CH_3? How does steric hindrance come into play?

Solve

(a) The methyl groups in *o*-dimethylbenzene activate different carbons on the ring, whereas the methyl groups in *m*-dimethylbenzene activate the same carbons in the ring. So *m*-dimethylbenzene will undergo chlorination faster.

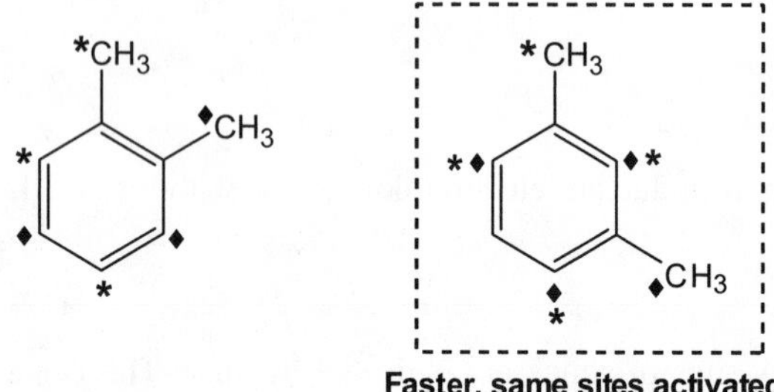

(b) The methyl groups in *p*-dimethylbenzene activate different carbons on the ring, whereas the methyl groups in *m*-dimethylbenzene activate the same carbons on the ring. So *m*-dimethylbenzene will undergo chlorination faster.

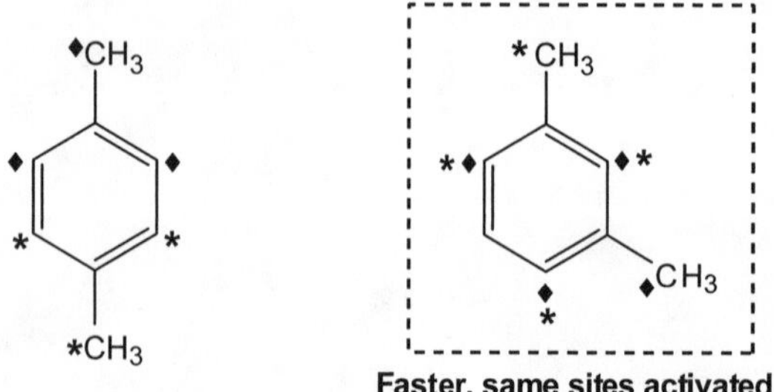

(c) 1,2,3,5-Tetramethylbenzene will undergo chlorination faster because each position is activated by three methyl groups. In 1,2,3,4-tetramethylbenzene, each position is activated by only two methyl groups.

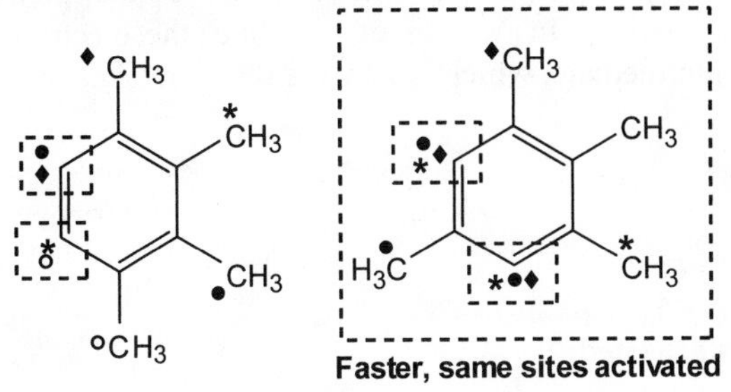

Faster, same sites activated

Problem 23.44

Think

How can the NH_2 group be transformed into the Br group? Where are the two Br groups on the ring relative to the NO_2 and NH_2 groups? Does this match their directing nature?

Solve

The retrosynthetic analysis would appear as follows:

Undo bromination. Undo substitution. Undo diazotization.

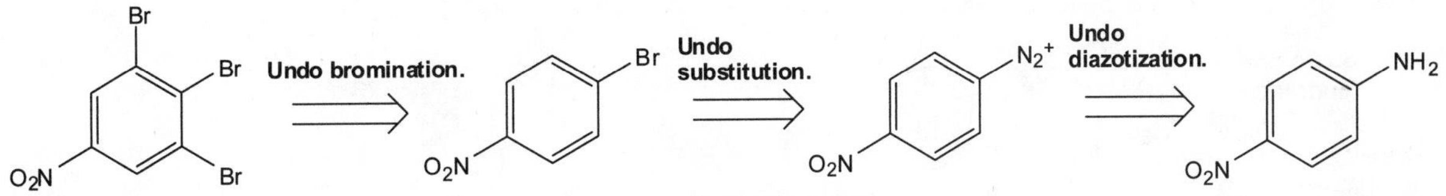

The NH_2 group can be replaced by Br by first converting it to a diazonium salt. Bromination can then occur at the positions ortho to the Br group and meta to the NO_2 group. The reaction in the forward direction would appear as follows:

$NaNO_2$, HCl / H_2O → CuBr → Br_2 / $FeBr_3$

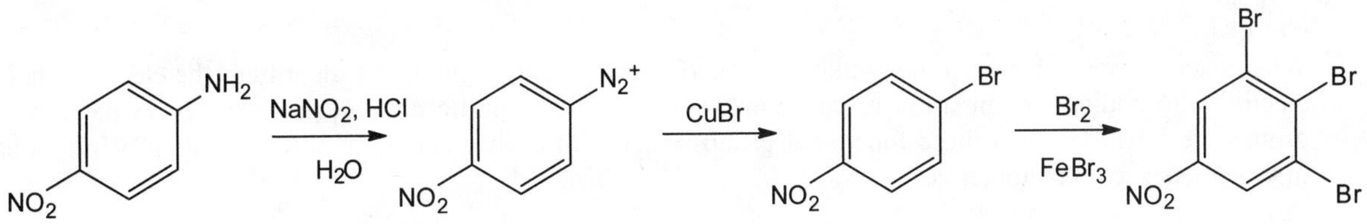

Problem 23.45

Think

How is the H atom different than the CH_3 group in terms of the electron donating effect? How should that effect impact the rate of electrophilic aromatic substitution when NH_2 is replaced by $N(CH_3)_2$? Are there lone pairs on the N atom? For the lone pair to stabilize the arenium ion intermediate, how must the plane of the N be oriented with respect to the plane of the ring? Can that orientation be achieved when there are two CH_3 groups on the ring in the ortho positions?

Solve

The NH_2 group activates the ring toward electrophilic aromatic substitution because its lone pair can participate in resonance to stabilize the arenium ion intermediate if the electrophile attacks either the ortho or para position. A key resonance structure upon para attack is shown below. In *N,N*-dimethylaniline, the electron-donating capability of the methyl groups further stabilizes this intermediate, which is why the reaction takes place faster.

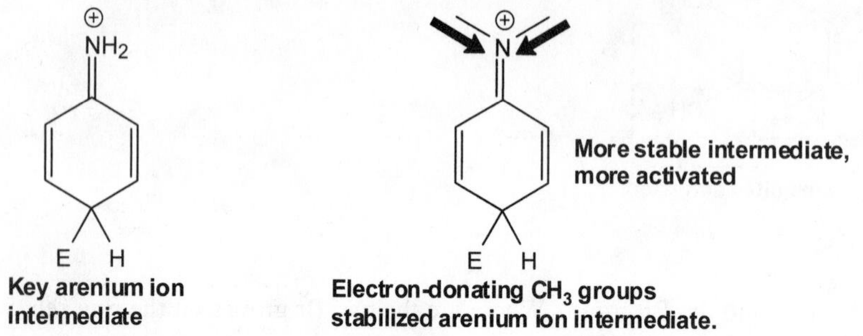

Key arenium ion intermediate

More stable intermediate, more activated

Electron-donating CH_3 groups stabilized arenium ion intermediate.

Notice that this resonance structure can contribute only if the NH_2 or the $N(CH_3)_2$ group is in the same plane as the ring. If, however, two methyl groups are also on the ring at the ortho positions, steric repulsion forces the $N(CH_3)_2$ to be perpendicular to the ring, effectively destroying the resonance stabilization.

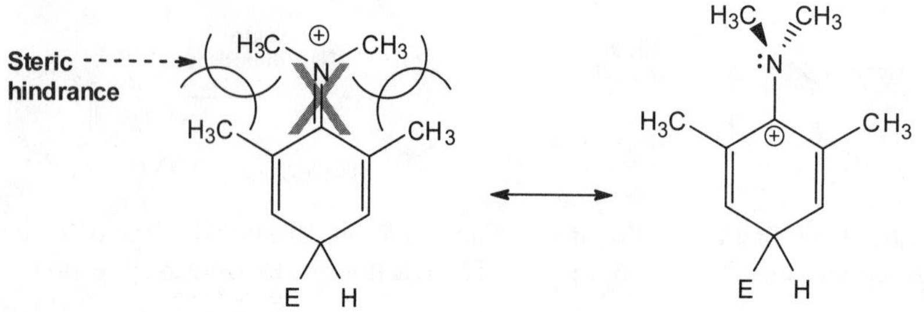

Steric hindrance

Problem 23.46

Think

What reaction occurs between the β-diketoester and a strong acid? What is the identity of the electrophile in the electrophilic addition–elimination mechanism? After the electrophilic aromatic substitution, what functional groups are left? How can these functional groups react with each other in an acidic solution? How does the intermediate shown tautomerize?

Solve

A protonated carbonyl from the ketone acts as the electrophile, which is attacked by the activated benzene ring of phenol to initiate electrophilic aromatic substitution. After the electrophilic aromatic substitution reaction, an acid-catalyzed transesterification takes place to produce a new cyclic ester, which is favored by the formation of the six-membered ring. The final three steps make up an acid-catalyzed dehydration reaction to produce the new C=C. See the mechanism on the next page.

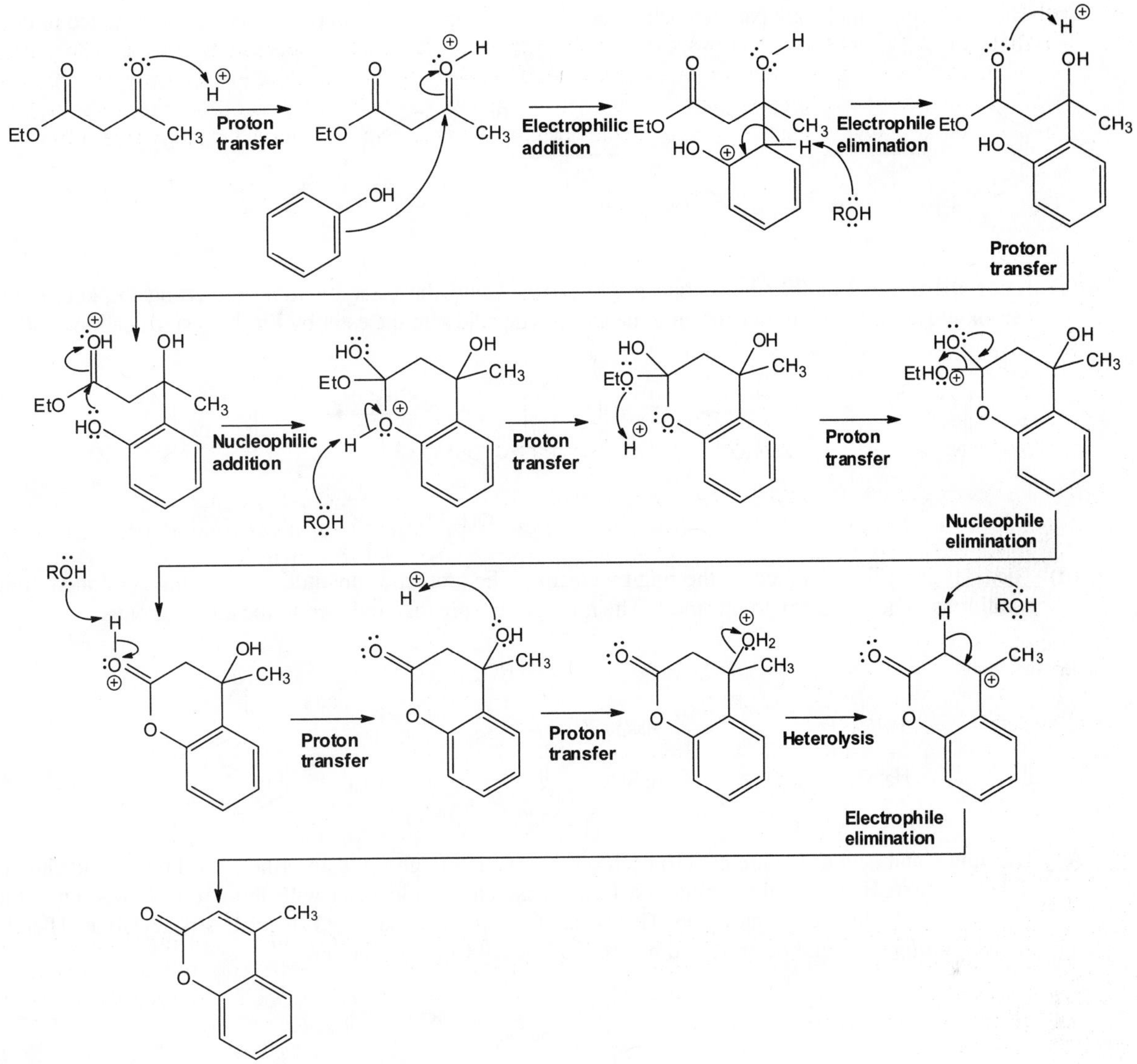

Proton transfer

Electrophilic addition

Electrophile elimination

Proton transfer

Nucleophilic addition

Proton transfer

Proton transfer

Nucleophile elimination

Proton transfer

Proton transfer

Heterolysis

Electrophile elimination

Problem 23.47

Think

How are the groups positioned relative to one another? Does the positioning match their directing ability? What reactions are a result of electrophilic aromatic substitution? What functional group transformations occur? Are any C–C bond-formation reactions necessary? Oxidation or reduction reactions?

Solve

(a) The acyl group and Br are meta to each other. The acyl group is a meta director and is added to the ring first.

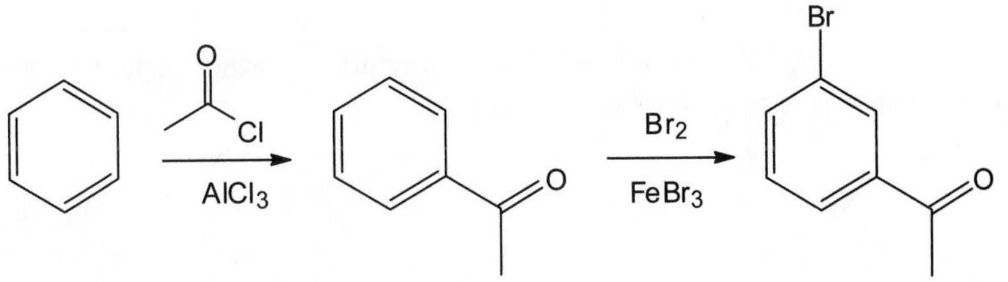

$AlCl_3$

Br_2

$FeBr_3$

(b) The acyl group and Br are para to each other. The Br group is an ortho/para director and is added to the ring first.

Br_2, $FeBr_3$; then acetyl chloride, $AlCl_3$

(c) Using the product in **(b)**, the ketone is transformed into the ester by first converting the ketone to the carboxylic acid, and then transforming the carboxylic acid into the ester by Fischer esterification.

1. $KMnO_4$, HO^-, Δ 2. H_3O^+; then CH_3OH, H^+

(d) The NH_2 group is not added on the ring by electrophilic aromatic substitution. The ring is nitrated in harsh conditions to get the nitro groups meta. The nitro groups are then reduced to the amine groups.

HNO_3, H_2SO_4; HNO_3, H_2SO_4; Fe, HCl

(e) The amine and acyl group are para to each other. The amine group is an ortho/para director and should be added on first. However, the amine is a Lewis base and would react with the strong Lewis acid, $AlCl_3$, resulting in a highly deactivated ring. This would prevent the necessary Friedel-Crafts acylation. Therefore, the NH_2 group must first be protected before the Friedel–Crafts acylation.

HNO_3, H_2SO_4; Fe, HCl; **Need to protect amine**: acetic anhydride, Pyridine; propanoyl chloride, $AlCl_3$; H_3O^+, Δ

Problem 23.48

Think

Where are the ortho and para sites on the ring? Which arenium ion intermediate—ortho, meta, or para—allows for the charge delocalization to occur on both rings? Why is this more stable?

Solve

(a) In this bromination reaction, H^+ is replaced by Br^+. Because the phenyl ring is an ortho/para director, this occurs at the ortho and para positions to yield:

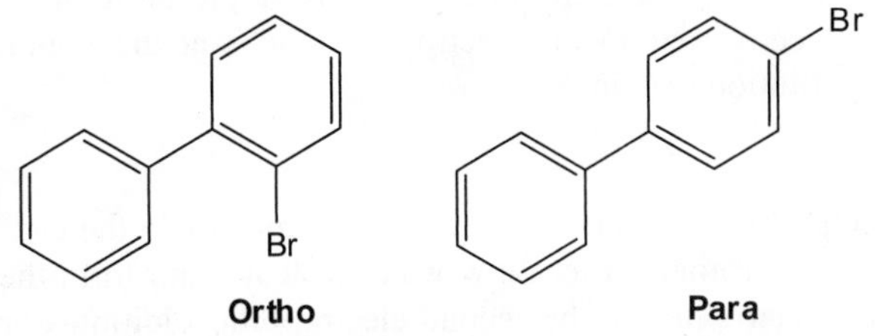

Br Br
Ortho **Para**

(b) We can justify why the phenyl ring is an ortho/para director by drawing the three isomeric intermediates, as shown below. When the Br^+ attacks either the ortho or para position, the resulting positive charge is delocalized onto both rings, whereas when the Br^+ attacks at the meta position, it is delocalized only over one ring.

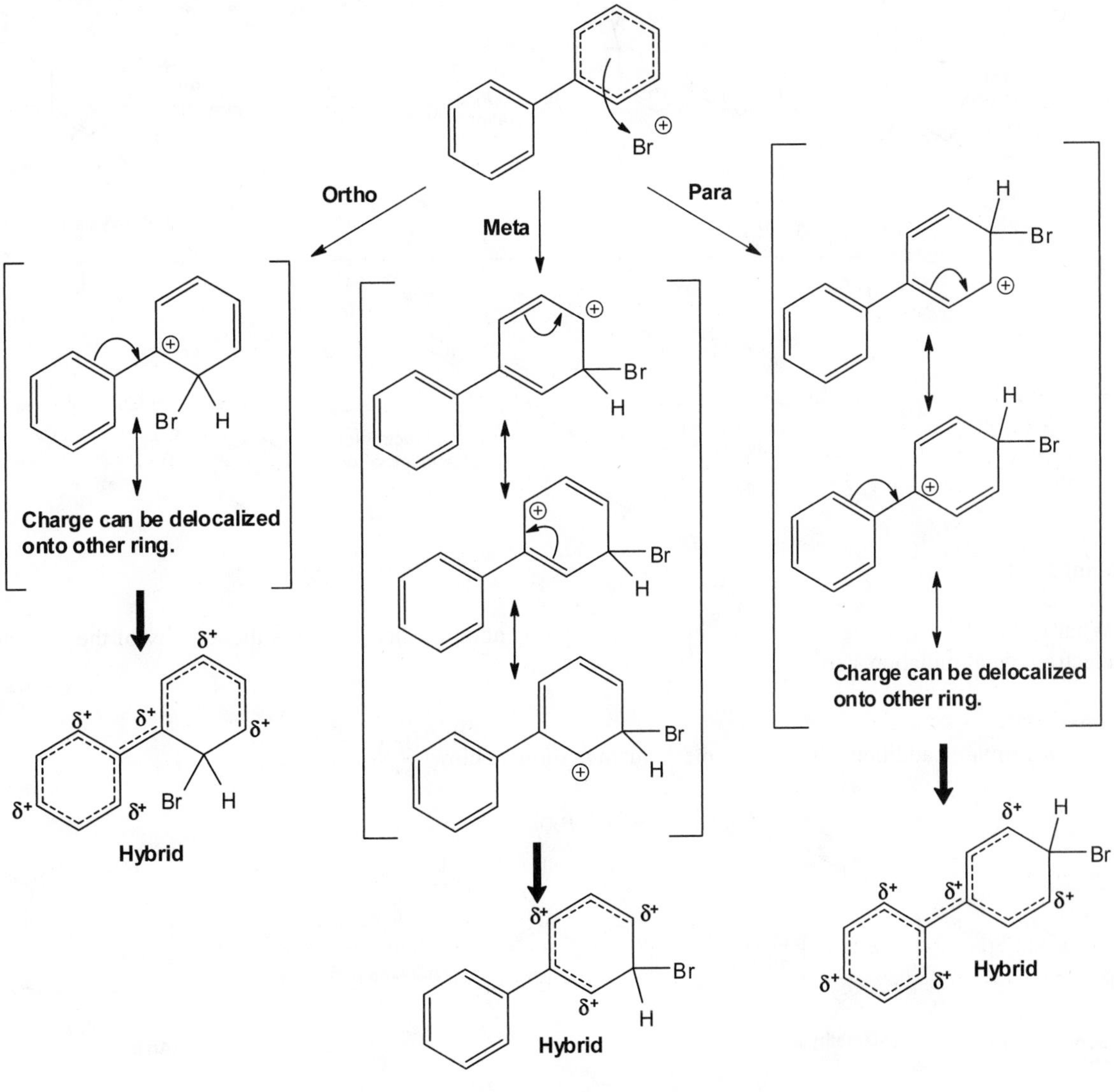

Br
Ortho **Meta** **Para**

Br H

Charge can be delocalized onto other ring.

δ+ δ+ δ+ δ+ δ+ Br H

Hybrid

Br H

Br H

Br H

δ+ δ+ Br δ+ H

Hybrid

H Br

H Br

Charge can be delocalized onto other ring.

δ+ H Br δ+ δ+ δ+ δ+ δ+

Hybrid

Problem 23.49

Think

How many times does the electrophilic aromatic substitution mechanism need to occur? In the first electrophilic aromatic substitution reaction, what is the identity of the electrophile? How is it produced under acidic conditions? How is the product of that reaction converted to an electrophile under acidic conditions, to participate in the second electrophilic aromatic substitution reaction?

Solve

The ketone O atom is protonated to form the electrophile that the para C of phenol attacks in the electrophilic addition–elimination mechanism. The alcohol is then protonated to allow water to leave and form the tertiary benzylic carbocation. This carbocation acts as the electrophile in the second electrophilic addition–elimination mechanism.

Proton transfer → Electrophilic addition → Electrophile elimination → Proton transfer → Heterolysis → Electrophilic addition → Electrophile elimination

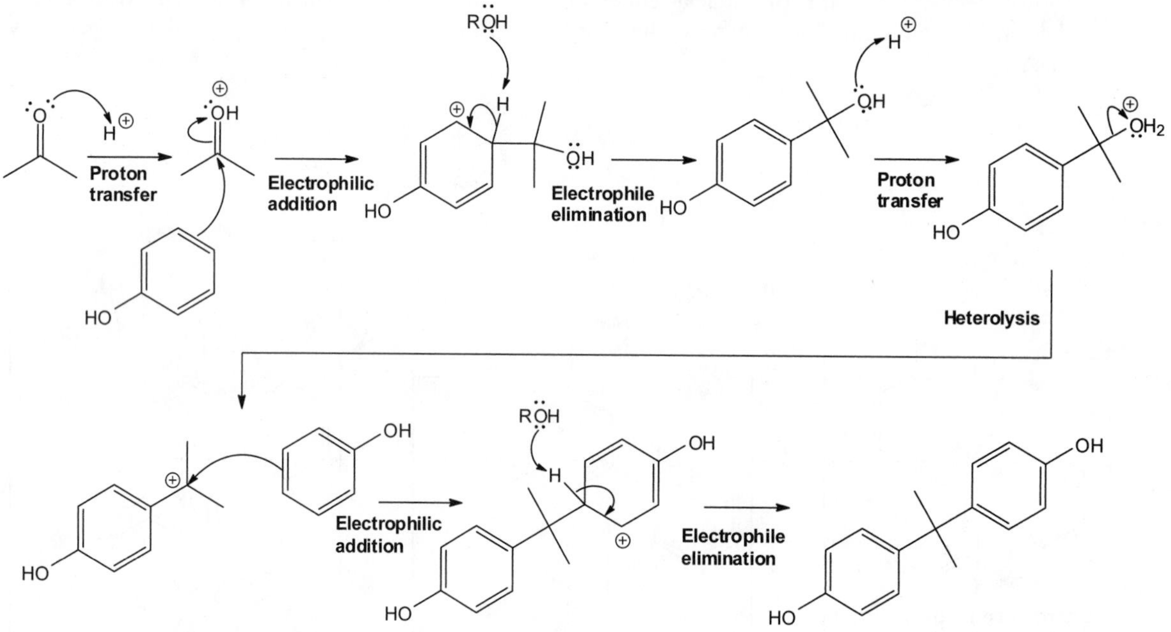

Problem 23.50

Think

What is the identity of the electrophile in the electrophilic addition step? What is the identity of the base in the electrophile elimination step?

Solve

The electrophilic addition–elimination mechanism is shown below:

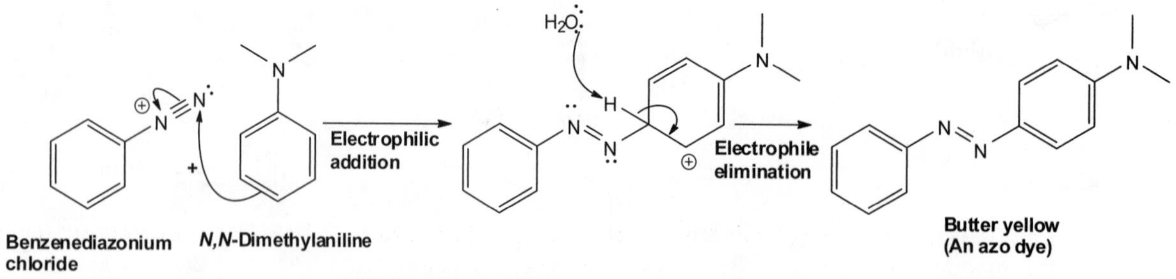

Benzenediazonium chloride + *N,N*-Dimethylaniline → Electrophilic addition → Electrophile elimination → Butter yellow (An azo dye)

The $N(CH_3)_2$ group is an ortho/para director, but the para product should be favored significantly over the ortho due to steric hindrance by the CH_3 groups at the ortho positions.

Problem 23.51

Think

What is the position of the two groups on the ring? Does that match the ortho/para or meta directing capabilities of the substituents? What is the order of electrophilic aromatic substitution for each group? How can a primary alkyl group get added to the ring? Are carbocation rearrangements a concern? From what functional group did the ester originate? Is an oxidation or reduction reaction necessary? Are protecting groups necessary?

Solve

To attach an alkyl group, the alcohol is oxidized to the carboxylic acid, which is transformed into the acid chloride. Friedel–Crafts acylation followed by reduction yields propylbenzene. The propylbenzene is brominated. In that reaction, the propyl group is an ortho/para director and, due to steric hindrance, the para isomer is the major product. The PhBr is transformed into the Grignard reagent and reaction with CO_2 followed by an acid workup yields the CO_2H group. A Fischer esterification of the carboxylic acid with propanol yields the target compound.

OH
1. H_2CrO_4
2. $SOCl_2$
O
Cl
O
Zn/Hg
HCl
Br_2
$FeBr_3$
Br
Mg(s)
Ether
MgBr
1. CO_2
2. $H_3O^{\oplus}$
O
OH
OH
$H_3O^{\oplus}$
O
O

Problem 23.52

Think

From what type of reaction did the C=C result? How can you prepare those necessary reagents from butan-1-ol? What is the positioning of the Cl and the group containing the C=C? Which one was added first?

Solve

The C=C in both reactions could have come from a Wittig reaction between an ylide and a ketone. The ylide is the $CH_3CH_2CH_2CH_2^{-}–^{+}PPh_3$.

Ylide prep:

OH
PBr_3
Br
1. PPh_3
2. BuLi
$\ominus$
$\oplus$
PPh_3

The aromatic ketone with which this ylide would react should be produced by reacting an aromatic ring with butanoyl chloride, which can be prepared as shown on the next page.

Acid chloride prep:

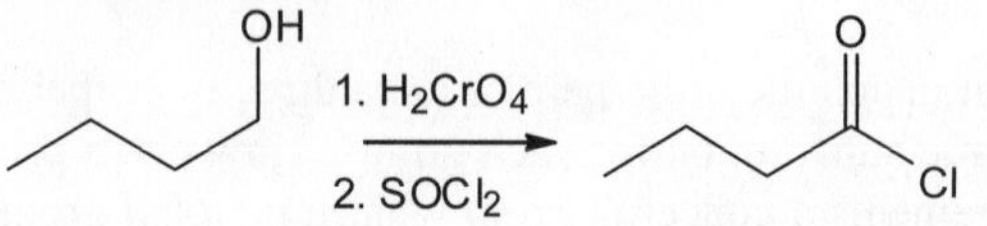

In **(a)**, the Cl and the group containing the C=C group are para to each other, which suggests that chlorination occurred prior to Friedel–Crafts acylation.

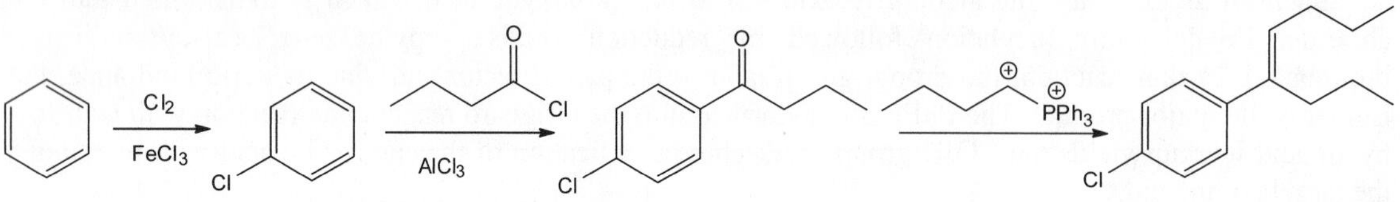

In **(b)**, the Cl and C=C groups are meta to each other, which suggests that chlorination occurred after Friedel–Crafts acylation.

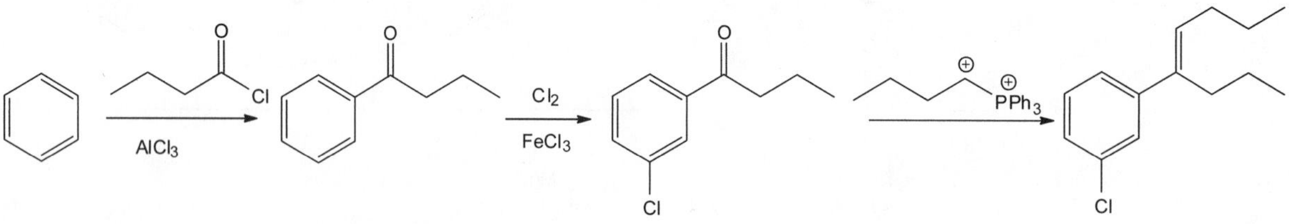

Problem 23.53

Think

How is phenol synthesized from benzene? What type of director is HO? How can the para position be blocked? How can a carboxylic acid be added to a benzene ring? How can an alcohol be transformed into an ester?

Solve

Phenol is formed from the diazonium salt. The carboxylic is added ortho from the Friedel–Crafts alkylation to add CH_3, which is then oxidized. Notice that a reversible sulfonation is used prior to the alkylation step to block the para position, and the sulfonyl group is removed after the alkylation is complete. The Ph–OH is transformed into the ester by reaction with acetic anhydride.

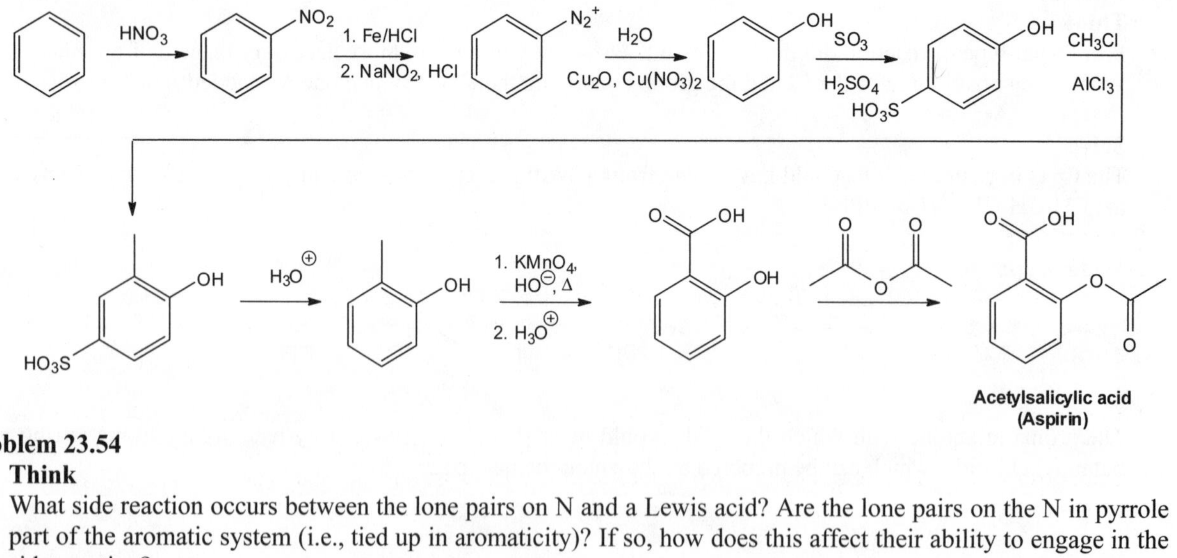

Acetylsalicylic acid (Aspirin)

Problem 23.54

Think

What side reaction occurs between the lone pairs on N and a Lewis acid? Are the lone pairs on the N in pyrrole part of the aromatic system (i.e., tied up in aromaticity)? If so, how does this affect their ability to engage in the side reaction?

Solve

The coordination of N to a Lewis acid catalyst is a problem if the lone pair on N is readily available to form a bond. In aniline, the lone pair on N remains relatively available because it is only weakly involved in resonance with the benzene ring—any resonance structures involving the lone pair will not exhibit aromaticity. In pyrrole, the lone pair is not readily available, because it is part of the six π electrons that establish aromaticity in the ring.

Lone pairs not part of aromatic π system

NH_2 + acetyl chloride → (Lewis acid catalyst) → Lewis acid–$\overset{\oplus}{N}H_2$

No Friedel–Crafts acylation reaction

Lone pairs are part of aromatic π system

pyrrole (N–H) + acetyl chloride → (Lewis acid catalyst) → 2-acetylpyrrole

Friedel–Crafts acylation reaction

Problem 23.55

Think

What fraction of the ortho product came from reaction involving the C–D bond and what fraction came from reaction involving the C–H bond? How does this affect the product distribution?

Solve

Half of the ortho product came from reaction involving the C–D bond and half came from reaction involving the C–H bond. Half of 50% is 25%. Therefore, the product distribution is 50% para, 25% ortho from C–D, and 25% ortho from C–H.

OH, D (2-deuteriophenol) → (HNO_3, H_2SO_4) →

Para NO_2 **50%** | O_2N **25%** | D, NO_2 **25%** — **Ortho**

Problem 23.56

Think

What is the structure of the arenium ion intermediate? Is the naphthalene ring more or less activated compared to benzene? Why might this affect the need for an acid catalyst?

Solve

The naphthalene ring is more activated compared to benzene. The arenium ion intermediate that forms in the reaction of naphthalene with Br_2 results in an ion that is still aromatic. Since the aromaticity of one of the rings is not destroyed, milder conditions are permitted. See the mechanism on the next page.

Br H

Br

120 °C, 1 h

No catalyst

Arenium ion (aromatic)

87 %

Problem 23.57

Think

Which product has less steric crowding? How does this affect the product formation?

Solve

The β-substituted product has less steric crowding and, therefore is the more stable product and the one formed as the major product.

Steric strain

conc. H_2SO_4

SO_3

SO_3H

β-substituted

α-substituted

Problem 23.58

Think

Draw all resonance structures for each arenium ion intermediate. Do they both have the same number of resonance structures? How many resonance structures of each intermediate preserve the aromaticity?

Solve

The relevant resonance structures for α and β substitution by Friedel–Crafts acylation of naphthalene are shown below. Both arenium ion intermediates have five total resonance structures. The α-substituted product, however, is the major product because two of the resonance structures maintain aromaticity.

$AlCl_3$

α Substitution (major)

Aromatic

Aromatic

Three other resonance structures destroy aromaticity.

$AlCl_3$

β Substitution (minor)

Aromatic

Four other resonance structures destroy aromaticity.

Problem 23.59

Think

When the substituent on naphthalene is an activator like CH_3, which ring is more activated? Which ring is more susceptible to electrophilic aromatic substitution? When the substituent is a deactivator like NO_2, which ring is more activated? Which ring is more susceptible to electrophilic aromatic substitution?

Solve

When the substituent on naphthalene is an activator like CH_3, electrophilic aromatic substitution takes place on the substituted ring because it is the more activated ring. When the substituent is a deactivator like NO_2, electrophilic aromatic substitution takes place on the unsubstituted ring because it is the more activated ring. In electrophilic aromatic substitution, the ring acts as a nucleophile, so when the ring has an electron donating group, it is more electron density, but when a ring has an electron withdrawing group, the other ring is more electron rich.

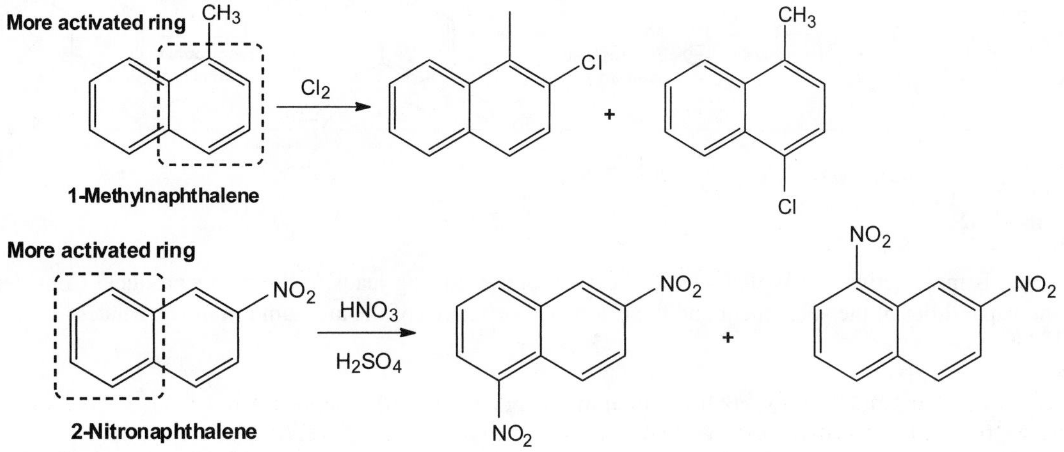

Problem 23.60

Think

Draw all resonance structures for each arenium ion intermediate. Do they both have the same number of resonance structures? How many resonance structures of each intermediate preserve the aromaticity?

Solve

The arenium ion from which each product is formed has the same number of resonance structures. When an electrophile attacks at the 1 position, however, there are more resonance structures that allow the aromaticity to be preserved in the benzene ring.

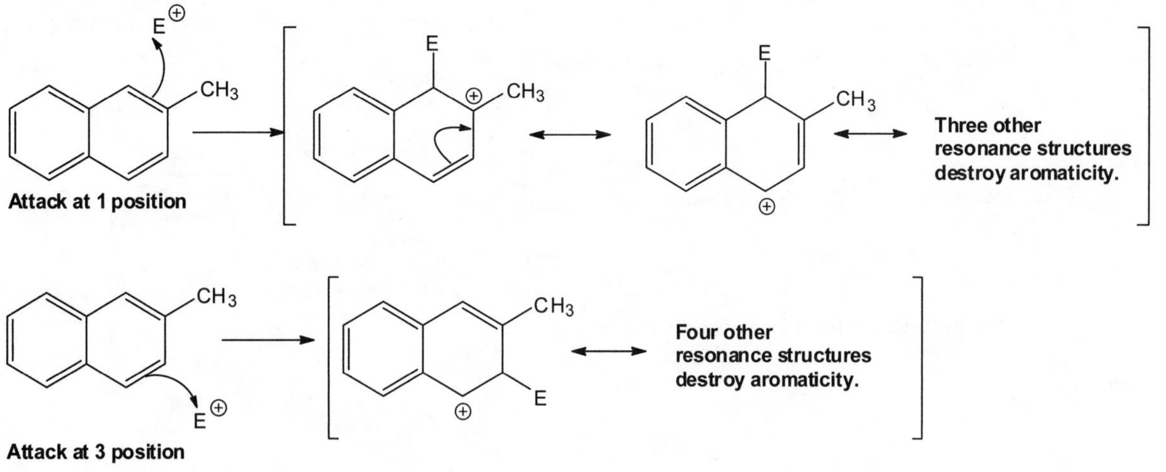

Problem 23.61

Think

Which ring is more activated? On which carbons in that ring can the electrophile be added? Which group, CH_3 or HO, is a stronger activator?

Solve

Electrophilic aromatic substitution will take place in the ring on the right, because it is activated compared to the ring on the left. The HO group is a stronger activator compared to CH_3 because of the lone pair electrons on the O atom. Therefore, the incoming electrophile will be directed primarily to the positions that are ortho and para to the OH group. Only the position ortho to the OH group has a proton, however.
The mechanism is given below.

Proton transfer → Heterolysis → Electrophilic addition → Electrophile elimination

Problem 23.62

Think

Which ring is more activated? Which site on the more activated ring leads to the major product? Consider the directing capabilities of the substituent and the resonance structures of the arenium ion intermediates.

Solve

(a) The ring with the HO is more activated and nitration occurs at the 1 position.

Proton transfer → Heterolysis → Electrophilic addition → Electrophile elimination

(b) The ring with the CH_3 is more activated and Friedel–Crafts acylation occurs at the 2 position, which is ortho to the CH_3.

Coordination → Heterolysis → Electrophilic addition → Electrophile elimination

(c) The unsubstituted ring is more activated and bromination occurs at the 2 position.

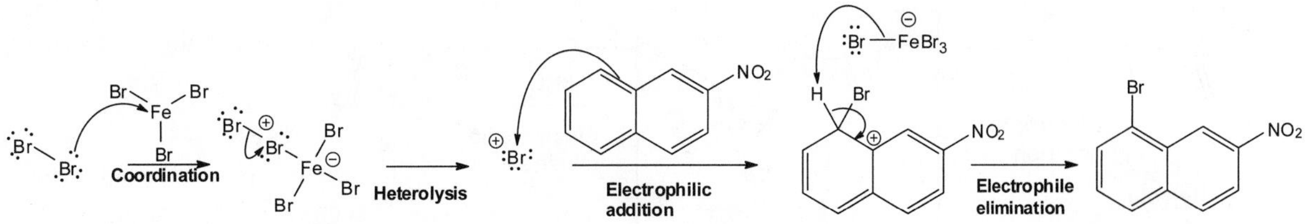

Coordination Heterolysis Electrophilic addition Electrophile elimination

(d) The ring with the amide is more activated and chlorination occurs at the 4 position. Not only is the 4 position para to the ortho/para directing group, but also its arenium ion intermediate has more resonance structures that preserve aromaticity than the arenium ion produced from reaction at the 2 position. It should be noted that the electrophilic addition and heterolysis steps probably occur simultaneously instead of separately but the electrophile Cl^+ is shown explicitly for emphasis.

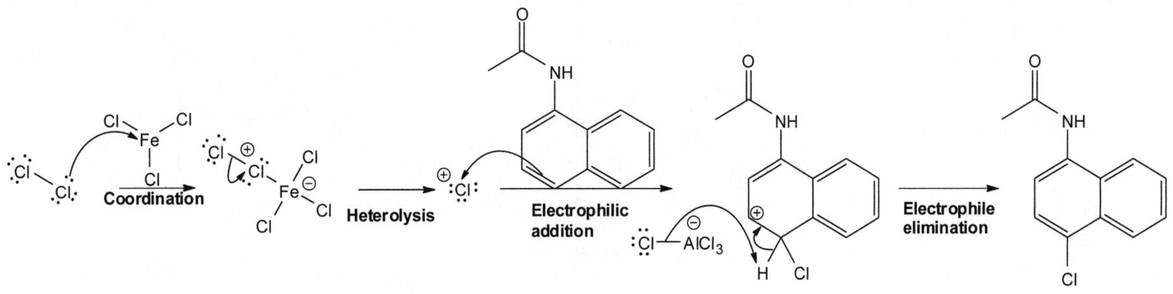

Coordination Heterolysis Electrophilic addition Electrophile elimination

Problem 23.63

Think

Draw all resonance structures for the 2- and the 3-substituted arenium ion intermediates. Do the arenium ion intermediates have the same number of resonance structures? Which substitution forms a more stable intermediate?

Solve

The 2-substituted pyrrole forms an arenium ion intermediate with three resonance structures, while the 3-substituted arenium ion intermediate forms two. Therefore, the 2-substituted arenium ion intermediate is more stable and will lead to the major product. See the figure on the next page.

2 Substitution

Pyrrole — Electrophilic addition → [resonance structures] — H_2O, Electrophile elimination → 2-nitropyrrole (NO_2)

3 Substitution

Pyrrole — Electrophilic addition → [resonance structures] — H_2O, Electrophile elimination → 3-nitropyrrole (NO_2)

Problem 23.64

Think

Draw all resonance structures for the 2-, 3-, and 4-substituted arenium ion intermediates. Do the arenium ion intermediates have the same number of resonance structures? Which substitution forms a more stable intermediate?

Solve

The major resonance contributors of the 2-, 3-, and 4-substituted arenium ion intermediates are shown on the next page. In none of the cases is a resonance structure shown in which the N lacks an octet because N is more electronegative than C and cannot handle the positive charge as well. Notice that the 2-substituted pyridine forms an arenium ion intermediate with two resonance structures, 3-substituted pyridine forms an arenium ion intermediate with three resonance structures, and 4-substituted pyridine forms an arenium ion intermediate with no resonance structures. Therefore, the 3-substituted arenium ion intermediate is the most stable and will lead to the major product.

2 Substitution

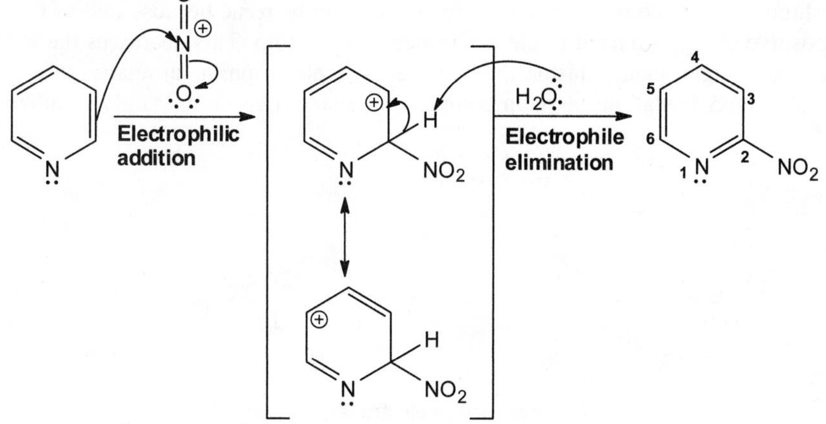

3 Substitution

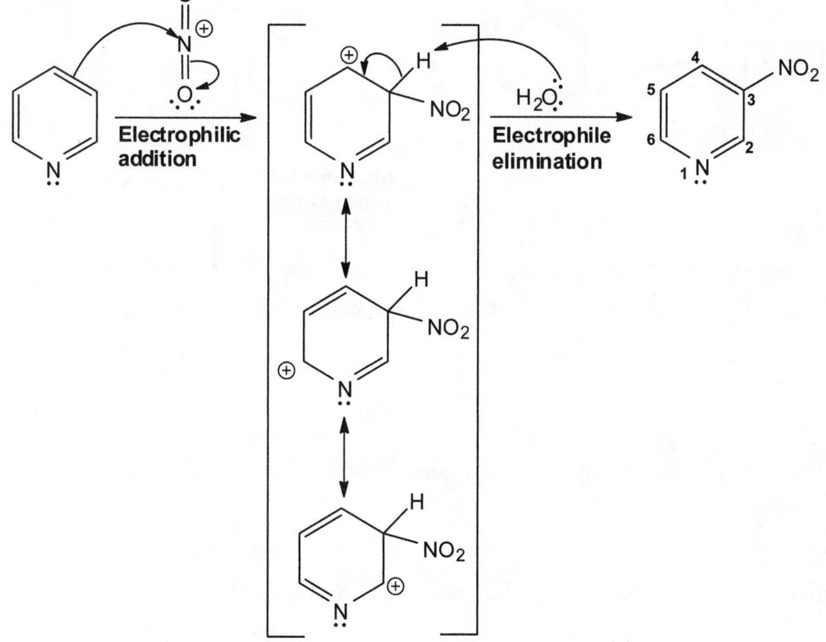

4 Substitution

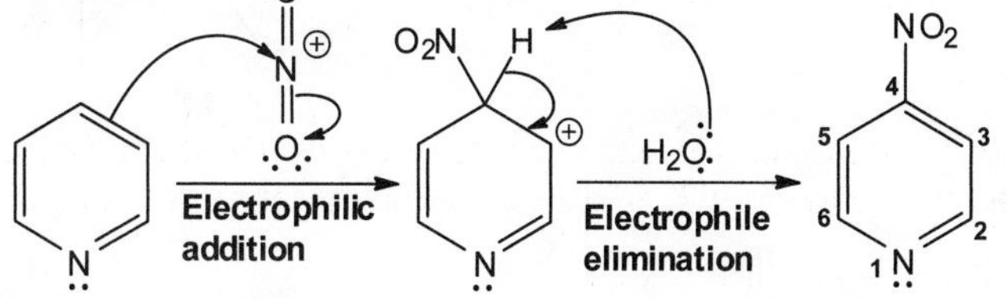

Problem 23.65

Think

Draw all resonance structures for each compound's most stable arenium ion intermediate. In pyridine's intermediate, can you identify resonance structures that are destabilized relative to those of benzene's intermediate? In pyrrole's intermediate, can you identify any resonance structures that are particularly stable?

Solve

In electrophilic aromatic substitution, pyridine reacts much more slowly than benzene because two of the three resonance structures have a positive charge adjacent to the electronegative N atom. This decreases the stability of the intermediate. Pyrrole reacts significantly faster than benzene in electrophilic aromatic substitution because one of the resonance structures has all atoms with complete valences. This significantly stabilizes the intermediate.

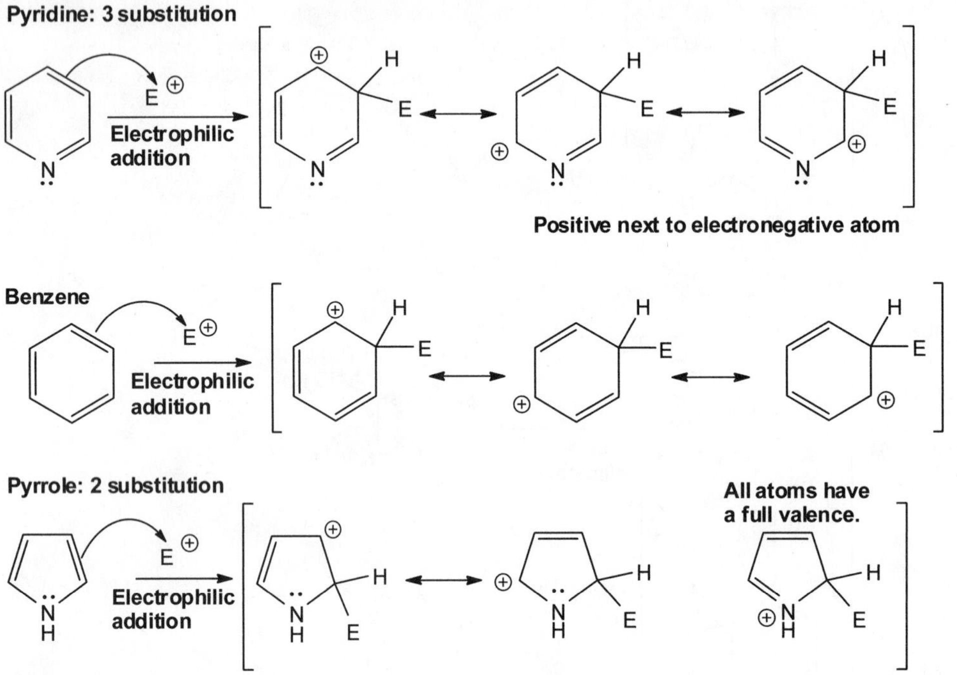

Pyridine: 3 substitution

Electrophilic addition

Positive next to electronegative atom

Benzene

Electrophilic addition

Pyrrole: 2 substitution

Electrophilic addition

All atoms have a full valence.

Problem 23.66

Think

In comparing furan and thiophene to pyrrole, at what position does nitration take place? Which atom, O or S, better stabilizes a positive charge?

Solve

Upon attack of NO_2^+, the arenium ion intermediates are as follows, where X is either O or S:

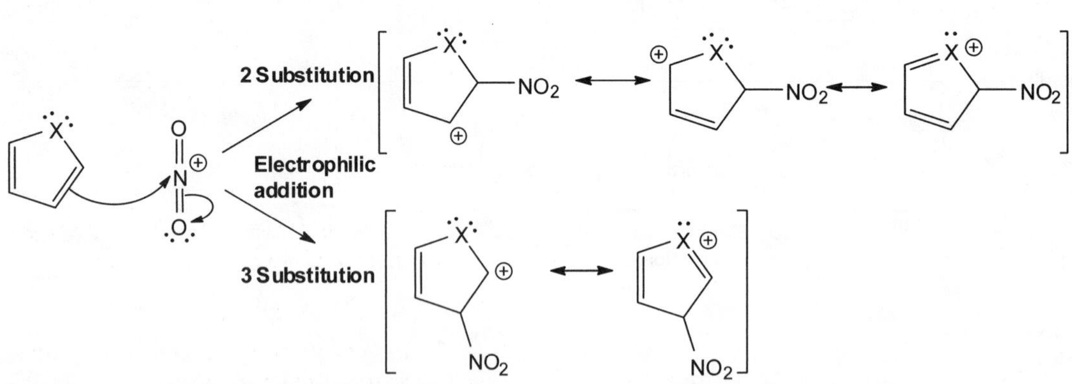

2 Substitution

Electrophilic addition

3 Substitution

Attack at the C adjacent to X allows for more resonance delocalization of the positive charge, making the 2-substituted nitro-isomer the major product in both cases. Also, because the most important resonance contributor is that with the positive charge on X (because all octets are fulfilled), substitution is faster on thiophene than on furan—the positive charge is better stabilized on S than on O.

Problem 23.67

Think

In comparing thiophene to pyrrole, at what position does nitration take place? What acts as the electrophile in the electrophilic addition reaction?

Solve

Just an in pyrrole, the arenium ion intermediate is more stable if attack of the electrophile occurs adjacent to the heteroatom (S) than at two carbons away—that is, attack is directed toward the 2 and 5 positions. Because a methyl group already exists at the 2 position, attack is directed toward the 5 position.

Electrophilic addition

Electrophile elimination

Problem 23.68

Think

Which ring is more activated? On the more activated ring, which carbon site most likely reacts with the electrophile? What is the identity of the electrophile in each electrophilic addition reaction?

Solve

(a) The pyrrole ring is more activated than the benzene ring, so substitution will take place on the pyrrole ring, replacing an H^+ for Br^+. The electrophilic aromatic substitution takes place at the 2 position for a pyrrole.

Electrophilic addition

Electrophile elimination

(b) The pyridine ring is deactivated relative to the benzene ring, so substitution will take place on the ring on the left. In an irreversible reaction such as this one, H atoms at the α positions are replaced preferentially over those at the β positions. It should be noted that the electrophilic addition and heterolysis steps probably occur simultaneously instead of separately but the electrophile Br^+ is shown explicitly for emphasis.

Coordination

Heterolysis

Electrophilic addition

Electrophile elimination

(c) The C atoms adjacent to S are activated toward substitution. In this nitration reaction, NO_2^+ is the electrophile.

Proton transfer

Heterolysis

Electrophilic addition

Electrophile elimination

(d) Similar to the pyridine ring, the C atoms adjacent to N are deactivated, directing substitution toward the position two C atoms away. NO_2^+ is the electrophile.

Problem 23.69

Think

Is there a site (or sites) of electrophilic addition on these compounds where the arenium ion intermediate can remain aromatic? If so, why would this be the site that undergoes electrophilic aromatic substitution?

Solve

If the arenium ion intermediate remains aromatic, this leads to a more stable ion, which is lower in energy and, therefore, the more likely position of substitution.

(a) Substitution at these positions, which allows the two outer rings to remain aromatic in the arenium ion intermediate:

(b) Substitution at these positions, which allows aromaticity in the seven-membered ring to be preserved in the arenium ion intermediate.

Problem 23.70

Think

What is the mechanism for nucleophilic addition–elimination? Which halogen is most electronegative? Which halogen is the best leaving group?

Solve

Iodine is the best leaving group. If Step 2 were the rate-limiting step, Ar–I would have the fastest rate. This suggests that the first step (nucleophilic addition) is the rate-limiting step. This is in agreement with the relative rates of reaction because fluorine is the most electronegative atom and makes the carbon the most electrophilic. This draws in the incoming nucleophile more strongly. Moreover, the electron-withdrawing nature of F helps stabilize the negatively charged intermediate.

Problem 23.71

Think

Is the mechanism electrophilic or nucleophilic addition–elimination? Which halogen is more electronegative? Which site is more likely attacked by the nucleophile?

Solve

(a) Cl and Br are both possible leaving groups because they are both ortho to the electron-withdrawing NO_2 group. But in nucleophilic aromatic substitution, reaction at the Cl carbon is faster, due to Cl being more electron withdrawing.

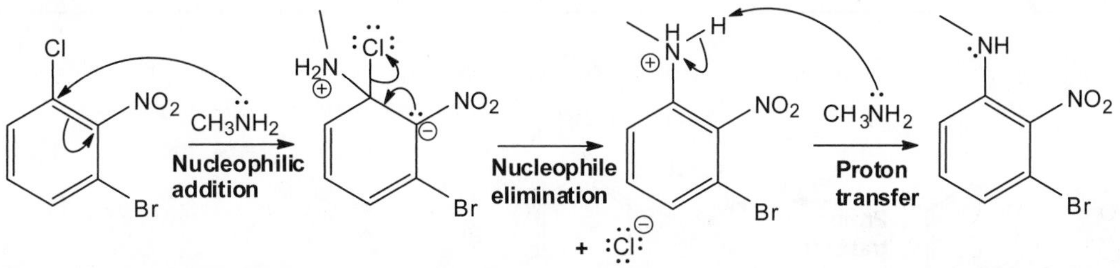

(b) F and I are both possible leaving groups, and attack at each of their C atoms would lead to an intermediate anion that is resonance stabilized by both NO_2 groups. But because F is more electron withdrawing, nucleophilic attack will take place more favorably at that C.

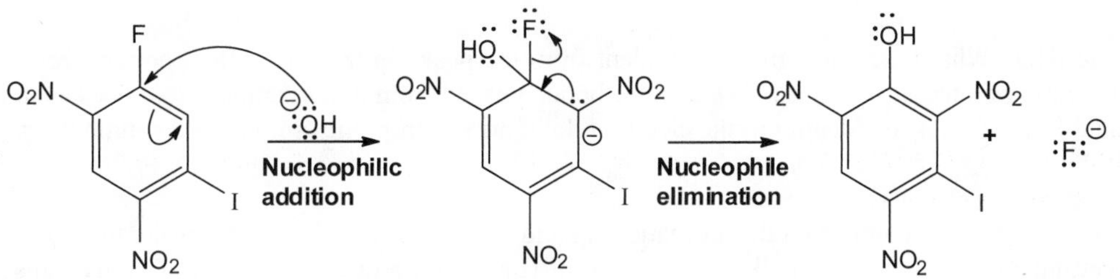

Problem 23.72

Think

What is the index of hydrogen deficiency (IHD)? How did the carbon skeleton change? What functional group gives rise to the broad peak around 11.7 ppm in the ^{1}H NMR spectrum? How many unique aromatic H atoms are present? How does NaOH react with the phenol O–H?

Solve

The IHD is 5, 4 of which can come from the aromatic ring and 1 likely from a C=O. Note that the proton signals between about 6 ppm and 8 ppm are the aromatic protons. The broad peak around 11.7 ppm in the ^{1}H NMR spectrum is due to the overlap of the carboxylic acid OH and the two phenol OH groups. This is indicated by the integration that appears to correspond to the 3H. This peak is significantly broadened due to H bonding. There are seven carbon signals, indicating that all seven carbons are distinct. The carbonyl carbon signal from the carboxylic acid is around 172 ppm. Sodium hydroxide is a strong enough base to deprotonate a phenolic proton. The O^- is a very activating group, allowing the ring to react with the weak electrophile, CO_2, in an electrophilic aromatic substitution reaction analogous to nitration. See the mechanism on the next page.

Problem 23.73

Think

What is the IHD? What functional group is evident from the peaks in the ^{1}H NMR spectrum from 5–6 ppm? What is the ratio of integration values? Which C–C bonds formed from dimerization? How does an alkene react with an acid HA? What is the identity of the electrophile in the electrophilic aromatic substitution?

Solve

The IHD is 5, 4 of which come from the aromatic ring and 1 likely from a C=C. The peaks from 5–6 in the ^{1}H NMR spectrum are suggestive of an alkene. The ratio of H atoms is 5 (aromatic):1:1:3. The 3H signal is due to a CH_3. Therefore, the structure is likely prop-1-en-2-ylbenzene.

The alkene reacts with H–A in an electrophilic addition reaction to form the tertiary benzylic cation. This electrophile then reacts with another equivalent of prop-1-en-ylbenzene to form another tertiary benzylic cation. This cation serves as the electrophile in the electrophilic addition–elimination mechanism that takes place intramolecularly.

Problem 23.74

Think

What is the IHD? If the signals are in a ratio of 1:3 and there are 12 H atoms, what is the actual ratio? If there are 14 signals between 120 and 140 in the ^{13}C NMR spectrum, how many C atoms are left in the structure that are not in this range? Is there any symmetry? How does a ketone react with an H–A acid? What is the identity of the electrophile in the electrophilic addition–elimination mechanism?

Solve

The IHD is 10, 8 of which can be from the two initial benzene rings. The actual H ratio is 3:9. The signal representing the three H atoms is likely from a CH_3 group, because all of the C atoms are aromatic except one. The ketone reacts with the H–A acid via a proton transfer reaction. This protonated ketone is the electrophile in the electrophilic addition–elimination mechanism. The final three steps make up an acid-catalyzed dehydration reaction. The formation of the new ring and the C=C double bond make up the additional IHD of 2.

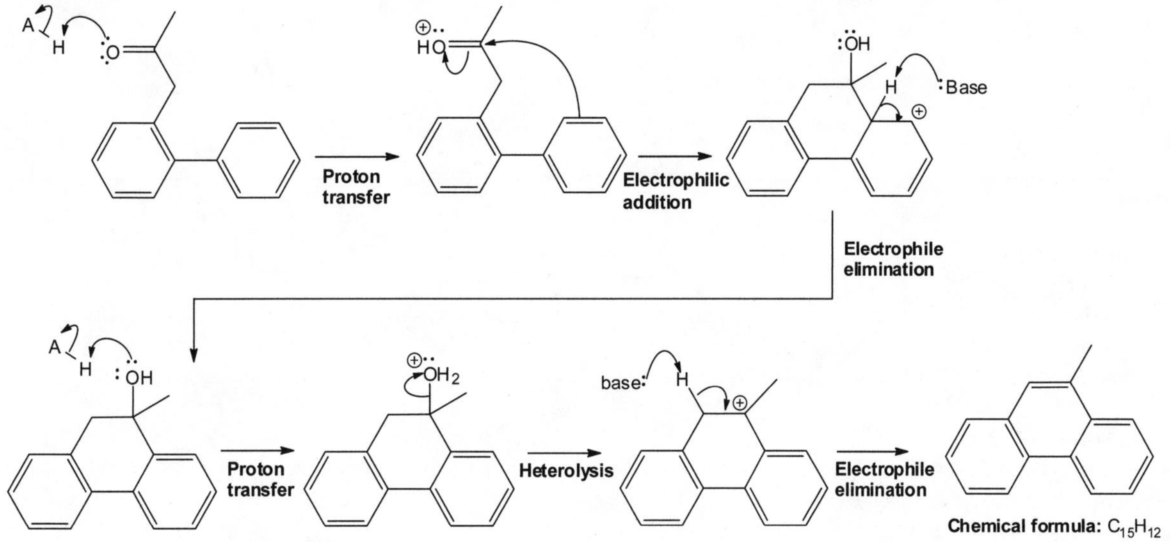

Proton transfer

Electrophilic addition

Electrophile elimination

Proton transfer

Heterolysis

Electrophile elimination

Chemical formula: $C_{15}H_{12}$

CHAPTER 24 | The Diels–Alder Reaction and Other Pericyclic Reactions

Your Turn Exercises

Your Turn 24.1

Think

In how many steps does the Diels–Alder reaction take place? Which reactant is the conjugated diene? The dienophile? Which has more electrons in a single π system, the diene or the dienophile?

Solve

The Diels–Alder is a one-step concerted reaction. The diene has two conjugated C=C bonds and is relatively electron rich due to the presence of four electrons in a single π system. The dienophile is the C≡C and is relatively electron poor due to the presence of only two electrons in a single π system. The π bonds of the triple bond belong to two different π systems.

The **diene** is relatively electron **rich**.

The **dienophile** is relatively electron **poor**.

Your Turn 24.2

Think

How many π electrons are involved in the cycloaddition reaction? Does this correspond to a Hückel number or an anti-Hückel number?

Solve

Antiaromatic, because there are four electrons involved in the cyclic transition state, and four is an anti-Hückel number (it is $4n$ with $n = 1$, and is an even number of pairs).

Your Turn 24.3

Think

What does the "*s*" stand for? What groups are being described as cis or trans to each other?

Solve

The "*s*" stands for single, as in single bond. Conformation **B** is *s*-cis, illustrated by the fact that the double bonds are on the same side of the single bond connecting them. Conformation **A** is *s*-trans.

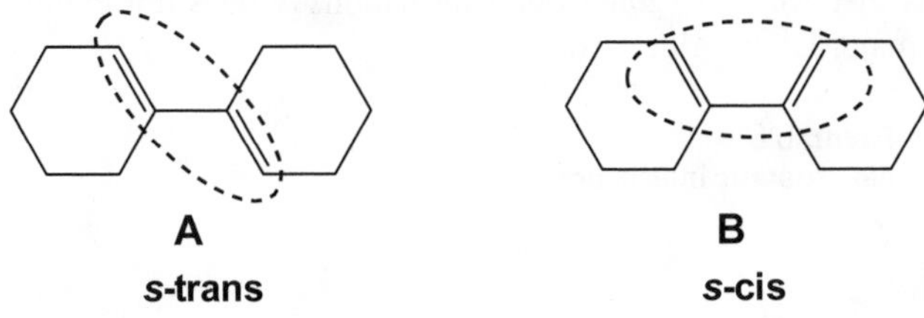

Your Turn 24.4

Think

If a C atom is sp^3 hybridized, can it participate in π bonding? Count the electron groups belonging to each C atom. How many electron groups does an sp^2 atom have? An sp^3 atom?

Solve

If a C atom is *sp*^{3} hybridized, it cannot participate in π bonding; it has four electron groups. A C that is *sp*^{2} hybridized has three electron groups. The two terminal C atoms of the diene and both C atoms of the dienophile are transformed from *sp*^{2} to *sp*^{3} hybridization, as indicated below.

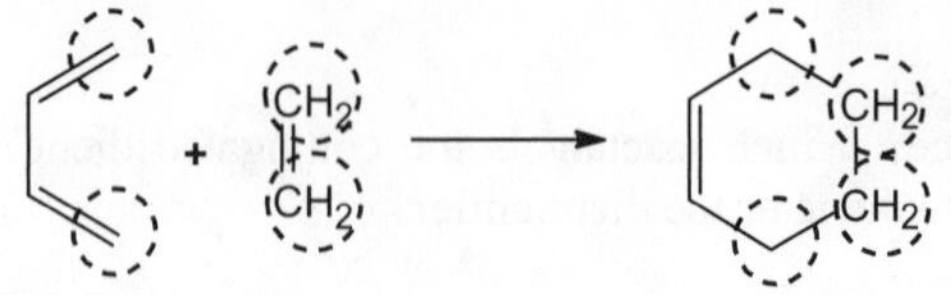

Circled C atoms are sp^2 Circled C atoms are sp^3

Your Turn 24.5

Think

Draw in the missing C–H bonds to have a bond drawn for W, X, Y, and Z. Tilt the diene into the paper to have the same orientation as the diene in Equation 24-20. Which groups are wedge and which groups are dash? How do the terminal C atoms in the diene rotate to form the product?

Solve

The H atoms in Equation 24-18 match up with W and Y in Equation 24-20, whereas the CH_3 groups match up with X and Z. Because X and Z end up on opposite sides of the ring in Equation 24-20, so, too, do the CH_3 groups in Equation 24-18, resulting in the trans product.

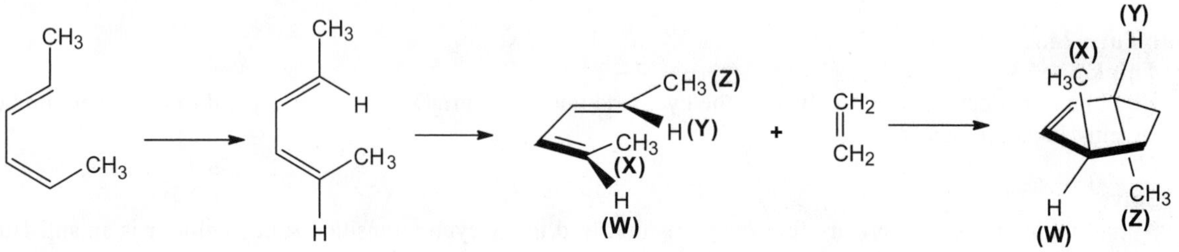

Draw in C–H bonds. Tilt diene into the paper.

Your Turn 24.6

Think

Use Figure 24-4 as a guide. What constitutes a favorable electrostatic interaction? Which orientation, **A** or **B**, shows a favorable electrostatic interaction among atoms undergoing bond formation?

Solve

Attraction between opposite charges constitutes a favorable electrostatic interaction. **B** has a favorable interaction, exhibited by the δ^+ and δ^- on atoms that are undergoing bond formation. **A** does not exhibit this kind of interaction. Therefore, **B** will lead to the major Diels–Alder product.

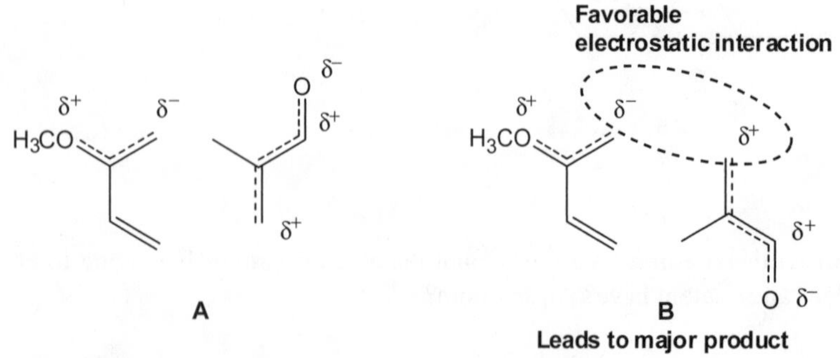

Favorable electrostatic interaction

A

B

Leads to major product

Your Turn 24.7

Think

Refer to Figure 24-4. How do you evaluate favorable versus nonfavorable interactions? Draw the reactant molecules approaching with the nonfavorable interaction. With that orientation of the diene and dienophile, how do you arrive at the product?

Solve

The orientation with no favorable interaction is the 180° flip of the orientation of one reactant in Solved Problem 24.15, as shown below. In that orientation, none of the pairs of atoms undergoing bond formation exhibits a favorable interaction between opposite partial charges. The product is obtained by adding the three curved arrows and moving the electron pairs accordingly.

No favorable electrostatic interaction

Minor product

Your Turn 24.8

Think

Fill in the bond energies for C=C and C–C single bonds. Be mindful of the number of each type of bond and the negative sign.

Solve

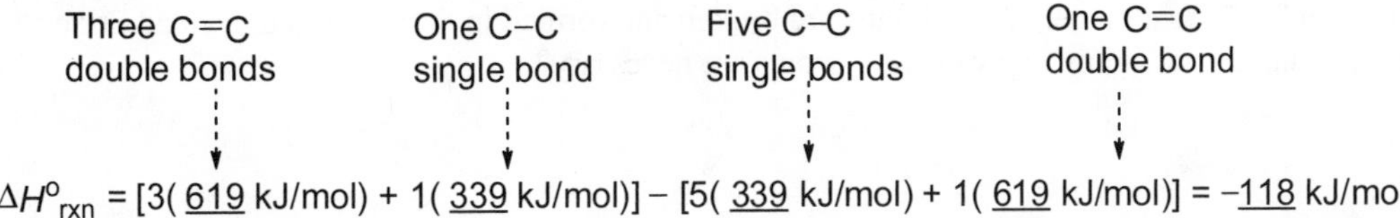

$$\Delta H^{o}_{rxn} = [3(\underline{619}\text{ kJ/mol}) + 1(\underline{339}\text{ kJ/mol})] - [5(\underline{339}\text{ kJ/mol}) + 1(\underline{619}\text{ kJ/mol})] = \underline{-118}\text{ kJ/mol}$$

This value is significantly negative, in agreement with the experimentally measured value of −168 kJ/mol, but is 50 kcal/mol more positive.

Your Turn 24.9

Think

Can you identify one molecule of cylcopentadiene as the diene and the other as the dienophile? Can you identify the six-membered ring that has formed in the product? How can you show the movement of electrons to show the concerted Diels–Alder reaction?

Solve

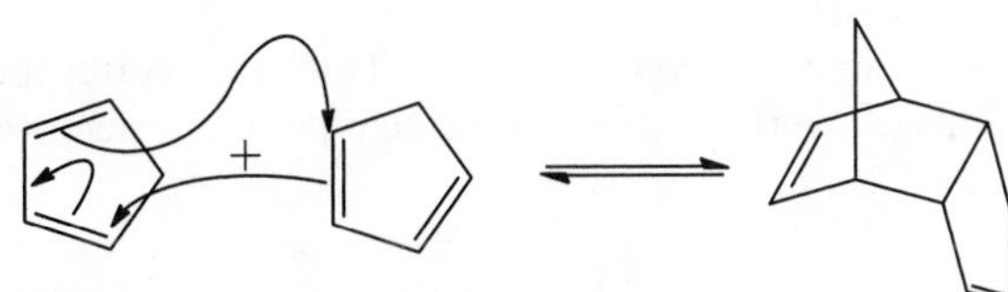

Diene **Dienophile**

(*Note:* Arrows are drawn clockwise, but counterclockwise arrows would be equally correct.)

Your Turn 24.10

Think

Compare this Diels–Alder reaction to the one in Equation 24-28, which is reversible. Are the reaction temperatures the same? How does the ΔH°_{rxn} for this reaction compare to the one in Equation 24-28? How does that impact the ΔG°_{rxn}? How does the ΔG°_{rxn} relate to reversibility?

Solve

This Diels–Alder reaction takes place at the same temperature as the one in Equation 24-28, but the ΔH°_{rxn} is much more negative (−121 kJ/mol vs. −77 kJ/mol). With the ΔH°_{rxn} much more negative, the ΔG°_{rxn} is much more negative. A significantly negative ΔG°_{rxn} is associated with an irreversible reaction, so whereas the reaction in Equation 24-28 is reversible, this one is irreversible.

Your Turn 24.11

Think

Refer to Equation 24-29 as an example of a retro Diels–Alder mechanism. Which bonds are broken and formed to produce the diene and the dienophile?

Solve

The C=C π bond electrons move to form one new C=C bond of the diene product. The adjacent C–C σ bond breaks to form the dienophile C=C. The next C–C σ bond breaks to form the second C=C of the diene product.

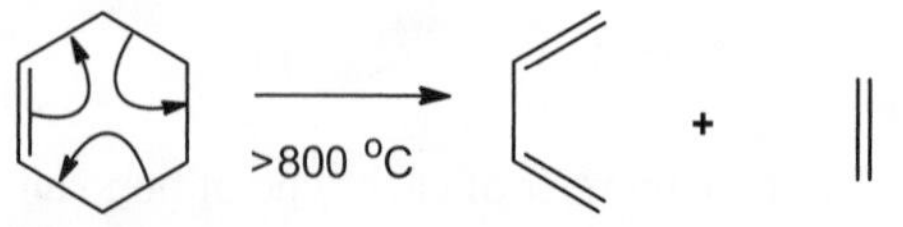

>800 °C

Your Turn 24.12

Think

Refer to Equation 24-31 as a guide. Which bonds are broken and formed in the concerted cycloaddition reaction to form the manganate ester? How many curved arrows are necessary?

Solve

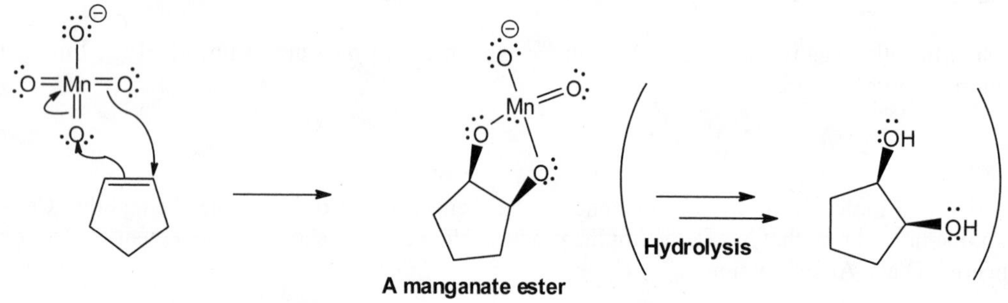

A manganate ester

Hydrolysis

Your Turn 24.13

Think

In the regions of orbital overlap, do the orbitals have the same or opposite phases? Does that correspond to constructive or destructive interference? Do the two regions of overlap exhibit the same type of interference?

Solve

In both regions of overlap, the orbitals have the same phase, which leads to constructive interference. Because both regions exhibit the same kind of interference, the orbitals have the appropriate symmetries to interact.

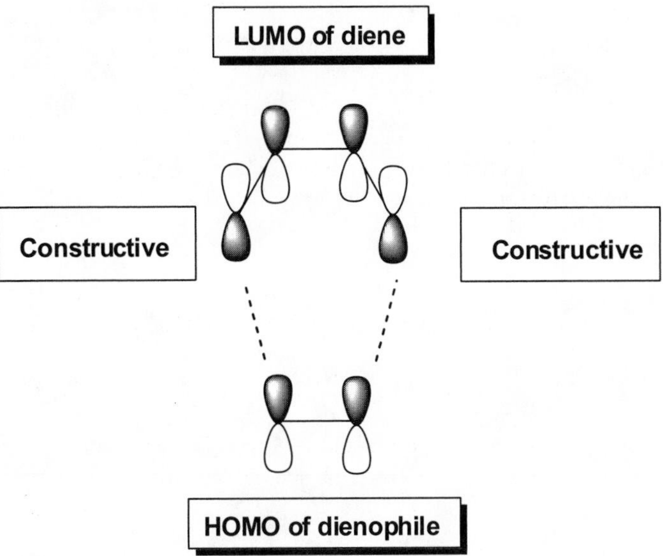

Your Turn 24.14

Think

In the regions of orbital overlap, do the orbitals have the same or opposite phases? Does that correspond to constructive or destructive interference? Do the two regions of overlap exhibit the same type of interference?

Solve

On the left, the orbitals have opposite phases, resulting in destructive interference. On the right, the orbitals have the same phase, resulting in constructive interference. Therefore, there is no net overlap and the orbitals do not have the appropriate symmetries to interact.

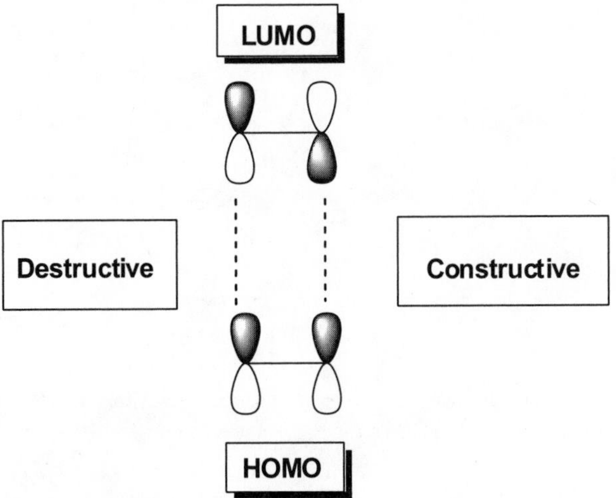

Your Turn 24.15

Think

Referring to Figure 24-11a, which C atoms of the diene and dienophile are involved in secondary orbital overlap? Which atoms do those correspond to in the figure given?

Solve

The *p* orbitals on C2 and C3 of the dienophile are involved in primary orbital overlap in both Figure 24-11a and 24-11b. In Figure 24-11a, the *p* orbital on C1 of the dienophile is involved in secondary orbital overlap with the leftmost *p* orbital of the diene. The same two orbitals in the figure below (circled) are farther away than they are in Figure 24-11a.

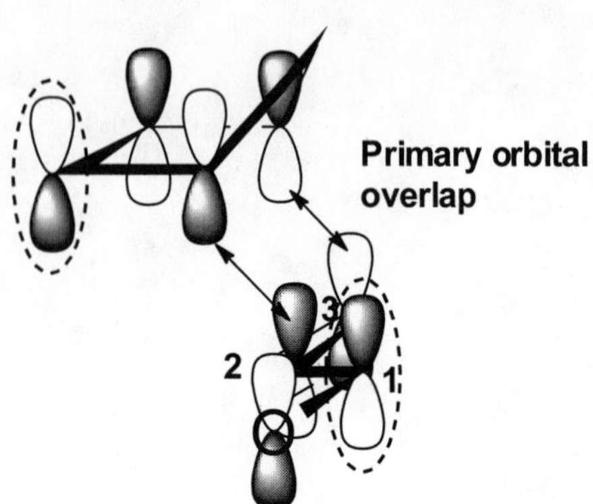

Chapter Problems

Problem 24.2

Think

In the transition state, are there electrons delocalized over an entire ring? If so, how many electrons? How can you use the number of curved arrows to tell? Is that a Hückel number or an anti-Hückel number? How does that correspond to whether the reaction is allowed or forbidden?

Solve

Eight electrons are delocalized over an entire ring in the transition state (two electrons are represented by each of the four curved arrows). Since eight is an anti-Hückel number of electrons, the transition state is antiaromatic, so the reaction does not take place readily; it is thermally forbidden.

Antiaromatic

Reaction does NOT readily occur.

Problem 24.4

Think

Are the dienes in an *s*-cis conformation? If not, can they attain an *s*-cis conformation? Which is more stable in the *s*-cis conformation?

Solve

Neither diene is in the *s*-cis conformation; both are in the *s*-trans conformation as written. The single bond in the molecule that connects the two double bonds can rotate to attain the *s*-cis conformation, as shown below. **D** will react faster. The methyl groups help destabilize the *s*-trans conformation of **D**, due to steric strain, thus helping to favor the *s*-cis conformation.

s-Trans C — Rotation about the single bond — *s*-Cis

s-Trans D — Too much steric strain — Rotation about the single bond — *s*-Cis

Problem 24.5

Think

If benzene reacted in ethene, what is the [4+2] cycloaddition product? Evaluate the aromaticity of the reactants and products. Does the product exhibit ring strain? Why might this prohibit the [4+2] cycloaddition reaction?

Solve

First, the aromaticity of benzene is destroyed if it reacts, because the product is nonaromatic. Second, there is significant ring strain in the product; the alkene carbons, being sp^2 hybridized, want to have bond angles of 120°. The bicyclic system does not allow this.

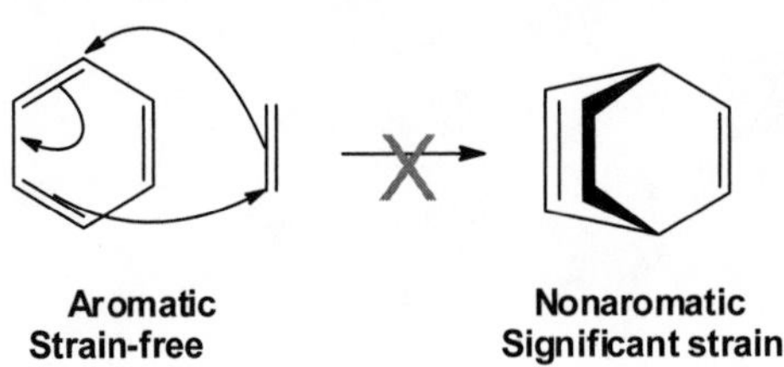

Problem 24.6

Think

What is the mechanism of a [4+2] cycloaddition? What size ring contains the diene? Do you think the diene in this reaction will be locked in the *s*-cis conformation in the same way as it is in Equation 24-10?

Solve

The mechanism for the [4+2] cycloaddition is shown below. The diene in this reaction is part of an eight-membered ring, which is significantly more flexible than the five-membered ring in Equation 24-10. Therefore, the diene will not be locked in the *s*-cis conformation as rigidly as the dienophile in Equation 24-10, making the reaction slower.

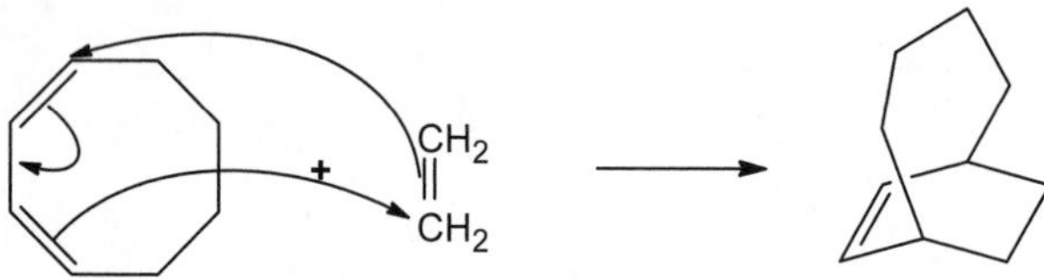

Problem 24.8

Think

Is the dienophile relatively electron rich or electron poor in a Diels–Alder reaction? Will electron-donating groups (EDG) or electron-withdrawing groups (EWG) on the dienophile speed up the reaction? What is the relative electron-donating/electron-withdrawing capability of each substituent?

Solve

The dienophile is relatively electron poor in a Diels–Alder reaction, so the more electron withdrawing a substituent on the dienophile is, the faster the reaction. In dienophile **C**, the aldehyde substituents are electron withdrawing. In **D**, the alkoxy substituents are electron donating via resonance. Therefore, **C** would react faster.

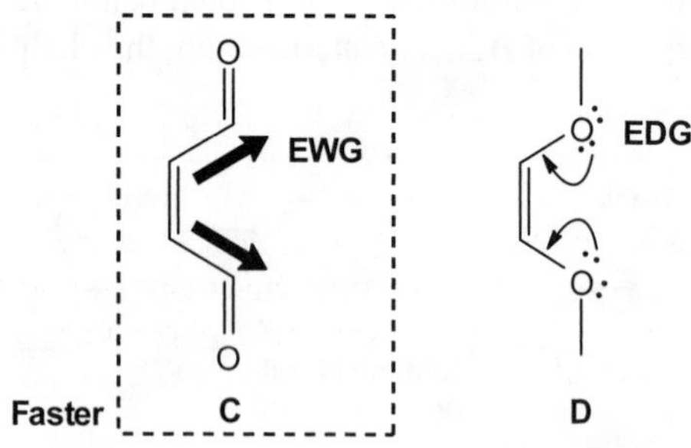

Problem 24.9

Think

What is the configuration of the NO_2 groups in the alkene dienophile? Is that stereochemistry conserved in a Diels–Alder reaction?

Solve

The cis alkene dienophile becomes a cis product because the [4+2] cycloaddition reaction is a concerted reaction and conserves the stereochemistry.

Problem 24.10

Think

Which carbons came from the diene? Which came from the dienophile? Do the carbon atoms that came from the dienophile have substituents that are cis or trans to each other in the ring? How does that translate to the stereochemistry of the dienophile?

Solve

(a) The carbon atoms that came from the dienophile are attached to aldehyde and methyl groups in the product. Those aldehyde and methyl groups are cis in the cyclohexene Diels–Alder reaction product and, therefore, must have been cis in the original alkene dienophile reactant.

Undo a Diels–Alder reaction.

CH_3

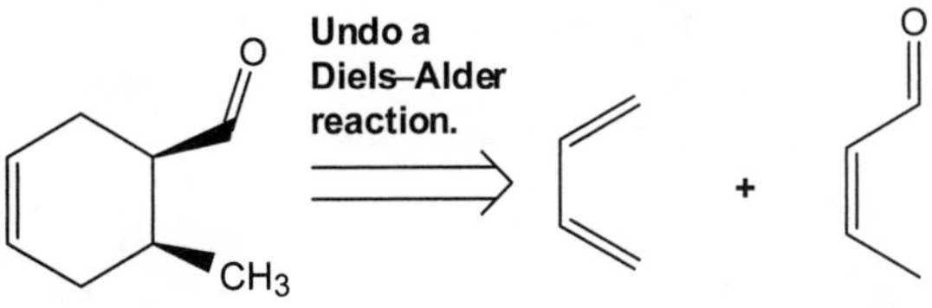

Cis

(b) The aldehyde and methyl group are trans in the cyclohexene Diels–Alder reaction product and, therefore, must have been trans in the original alkene dienophile reactant.

Undo a Diels-Alder reaction.

CH_3

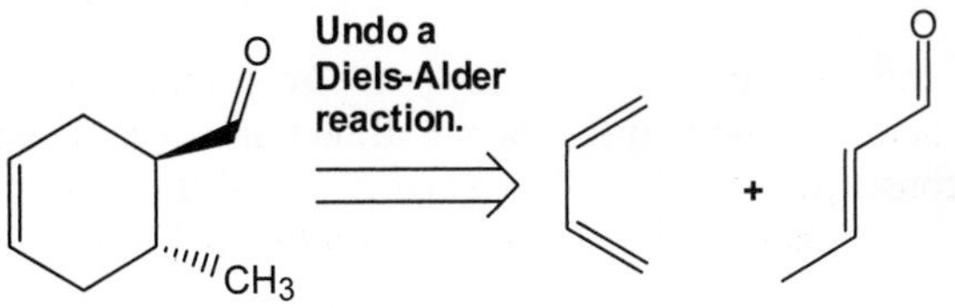

Trans

Problem 24.12

Think

What must the conformation of the diene be to react in a Diels–Alder reaction? Do the carbons at the ends of the diene become tetrahedral stereocenters in the products? What are the configurations about each double bond in the diene? Do the alkene carbons of the dienophile become tetrahedral stereocenters? What is the configuration about the C=C double bond in the dienophile?

Solve

The dienes must achieve an *s*-cis conformation prior to reaction. This conformational change is shown for each reaction.

(a) If we draw the product without worrying about stereochemistry, notice that each end carbon in the diene becomes a tetrahedral stereocenter, and so does each alkene carbon in the dieneophile (noted with an *).

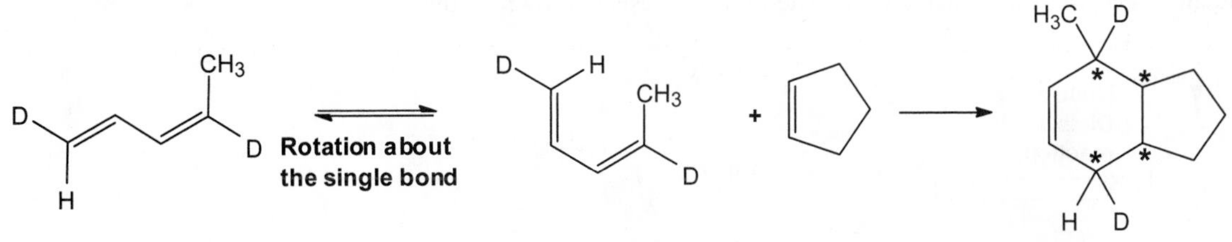

Rotation about the single bond

The H–C=C–C configuration is cis along the diene skeleton, the two D–C=C–C configurations are trans, and the CH_3–C=C–C is cis. Therefore, the two D atoms will be cis to each other in the product. Overall there are four results to consider:

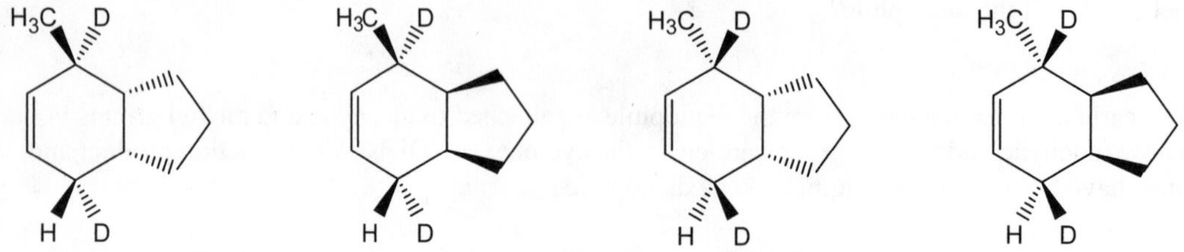

(b) If we draw the product without worrying about stereochemistry, notice that each end carbon in the diene becomes a tetrahedral stereocenter but the two C atoms of the dienophile do not.

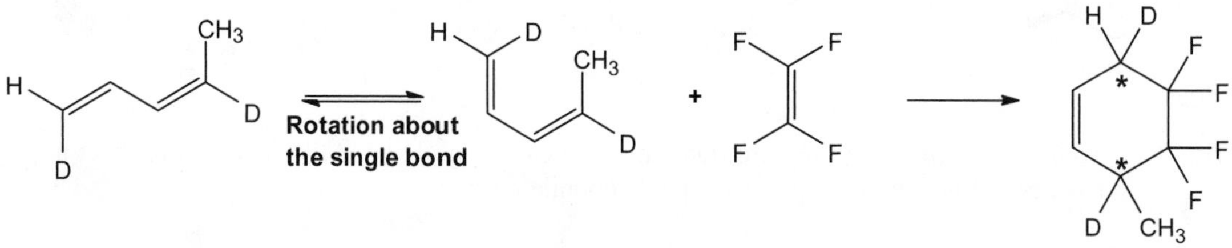

The H–C=C–C configuration is trans along the diene skeleton, the one D–C=C–C configuration is trans and the other is cis, and the CH_3–C=C–C is cis. Therefore, the two D atoms will be trans to each other in the product. Overall then, there are two results to consider:

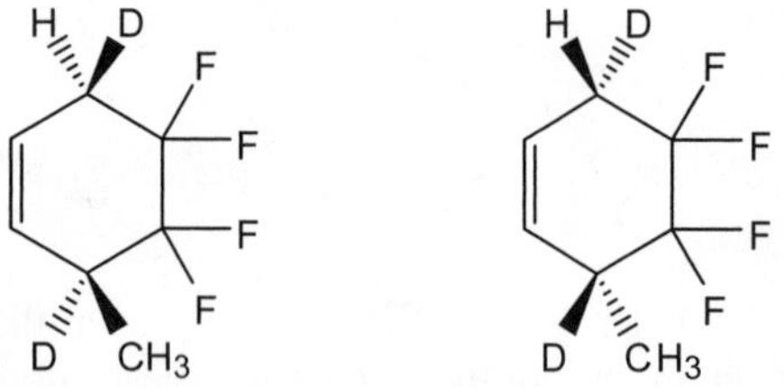

Problem 24.13

Think

Which carbon atoms in the product came from the diene? From the dienophile? Are the methoxy OCH_3 groups cis or trans to each other with respect to the plane of the ring? What does this mean about the configuration of the two double bonds in the diene?

Solve

The methoxy OCH_3 groups in the product are trans to each other. This means that one diene double bond from the diene is cis and the other is trans. The retrosynthesis is shown below:

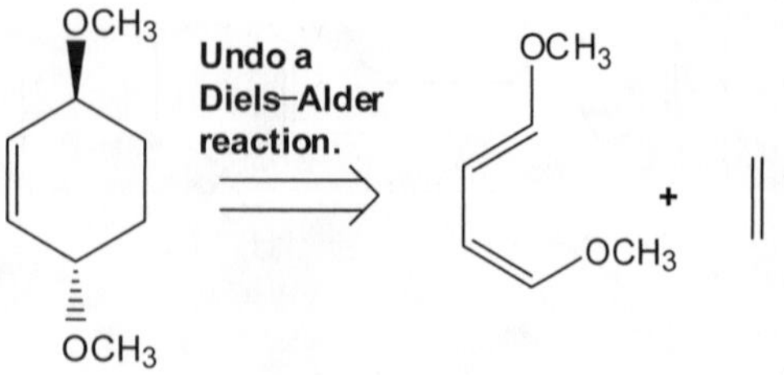

Problem 24.14

Think

Is the major product the endo or the exo? Which groups are endo? How many C atoms are in the bicyclic bridge?

Solve

The major product is the endo product. There are two C atoms in the bicyclic bridge.

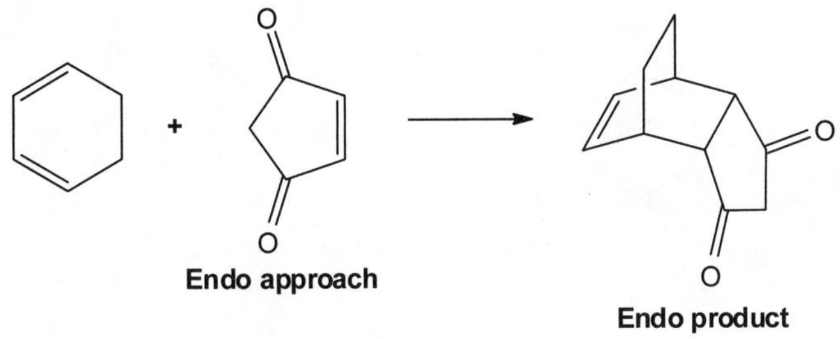

Endo approach

Endo product

Problem 24.16

Think

Where are the regions of excess negative charge and excess positive charge in these molecules? How can you use resonance structures to determine this? Which approach leads to favorable electrostatic interactions among the atoms undergoing bond formation?

Solve

The partial charges from resonance hybrids of the diene and dienophile are shown below. The major product comes from the most favorable electrostatic interaction among the atoms undergoing bond formation.

(a)

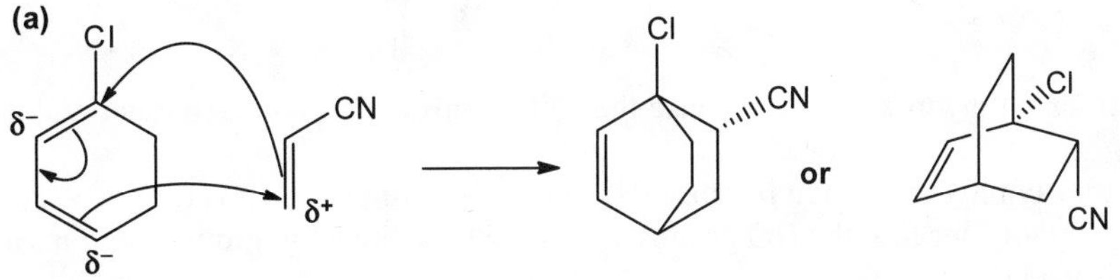

or

(b)

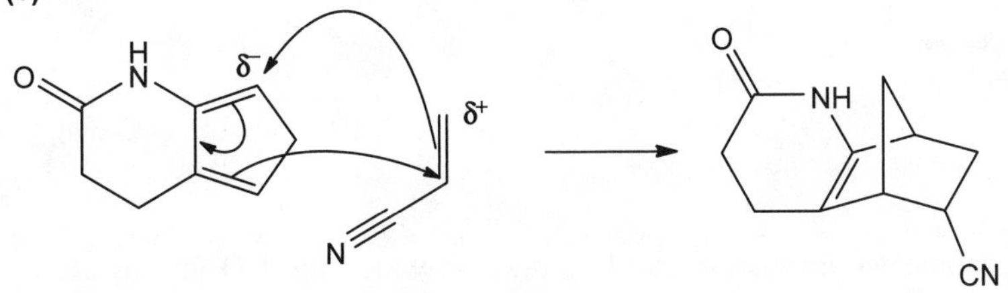

Problem 24.17

Think

What are the values for $\Delta H°$ and $\Delta S°$ for Equation 24-27 and how can you obtain the value for $\Delta G°_{rxn}$? What equation relates temperature, $\Delta G°_{rxn}$, and K_{eq}? When you solve for temperature, what value do you obtain?

Solve

The values for $\Delta H°$ and $\Delta S°$ are −168 kJ/mol and −0.184 kJ/mol (from $T\Delta S° = 55$ kJ/mol at 298 K). Therefore, the value for temperature can be solved using the following two $\Delta G°$ relationships:

$$\Delta G°_{rxn} = \Delta H°_{rxn} - T\Delta S°_{rxn} = -RT(\ln K_{eq})$$

Solving for T, we get:

$$T = \frac{\Delta H°}{\Delta S° - R[lnK_{eq}]} = \frac{-168\frac{\text{kJ}}{\text{mol}}}{-0.184\frac{\text{kJ}}{\text{mol}} - 0.008314\frac{\text{kJ}}{\text{mol}\cdot\text{K}} * [\ln(120)]} = 750.7\text{ K} = 478\text{ °C}$$

Problem 24.18

Think

Can the reactant be formed from a Diels–Alder reaction? If so, will heating the reactant facilitate a retro Diels–Alder reaction? If so, what bonds are broken to form the diene and dienophile?

Solve

This is an example of a retro Diels–Alder reaction. The mechanism and products are shown below.

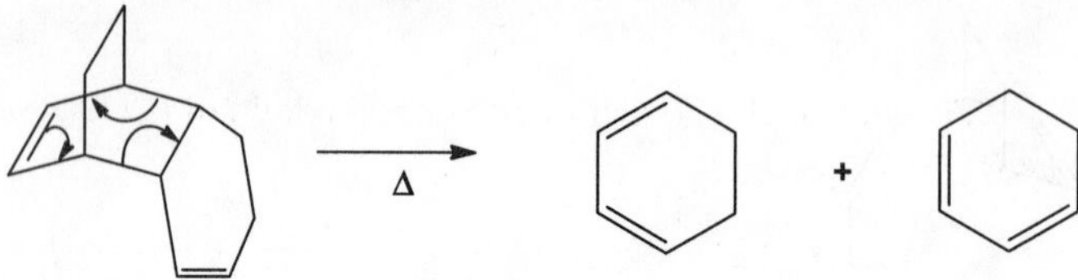

Problem 24.19

Think

Which diols appear to be *cis*-1,2 diols in the conformation shown? What about in other configurations about the C–C bond. In the conformation in which the OH groups are syn to each other, what are the relative orientations of the other groups attached to the C–C bond?

Solve

(a) This is an example of a *cis*-1,2 diol. The alkene is cyclohexene.

OH
OH
Undo syn dihydroxylation.

(b) This product did not form from syn hydroxylation because the HO groups are anti to each other and a ring does not have free rotation.

(c) This product did not form from syn hydroxylation because the HO groups are in a 1,3 position.

(d) This is an example of a *cis*-1,2 diol. Because the HO groups are already cis, the ethyl groups are trans about the alkene from which it was made.

OH
OH
Undo syn dihydroxylation.

(e) The HO groups are trans, but another conformation can be drawn in which the HO groups are *cis*-1,2. Therefore, the ethyl groups are cis in the alkene.

OH
OH
Rotation about C–C bond
OH
OH
Undo syn dihydroxylation.

Problem 24.21

Think

What was the possible original structure given in Solved Problem 24-20? What is the configuration about the C=C bond? Would changing that configuration affect the oxidative cleavage products? How can you connect the C atoms of the ketone and aldehyde groups differently to arrive at the original structure?

Solve

The original structure given in Solved Problem 24-20 is the (*E*) isomer shown below:

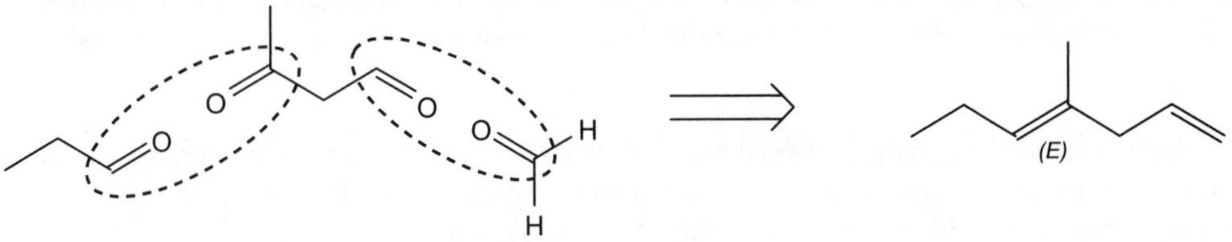

The starting material also could have been the (*Z*) isomer, as shown below. The configuration of the alkene does not impact the identities of the oxidative cleavage products.

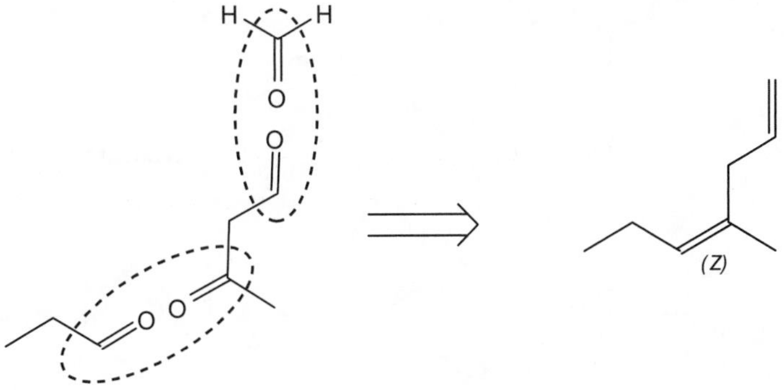

Alternatively, the C=O carbons could have been connected together differently, as shown below.

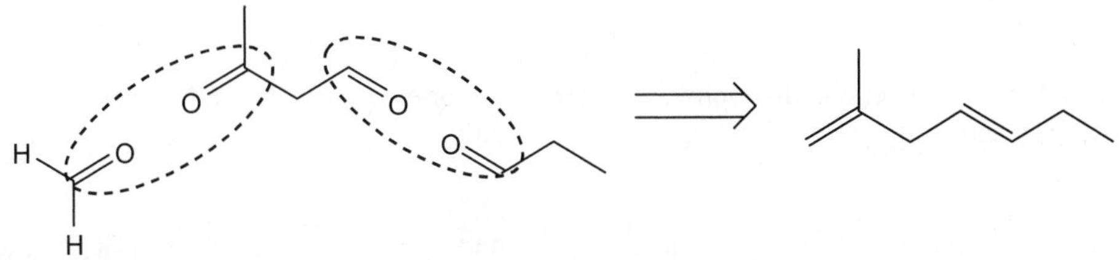

Problem 24.22

Think

What position does the *t*-butyl group occupy on the cyclohexane ring? What positions do the OH groups occupy in each structure? With the OH groups in those positions, can the cyclic periodate ester form upon treatment with IO_4^-?

Solve

In both structures, the most stable conformation of the cyclohexane ring has the *t*-butyl group in the equatorial position. In **A**, the OH groups are also equatorial, and, as shown below, are close enough to each other to form the cyclic periodate ester upon treatment with IO_4^-. In **B**, the OH groups occupy axial positions and are too far apart to form the cyclic periodate ester.

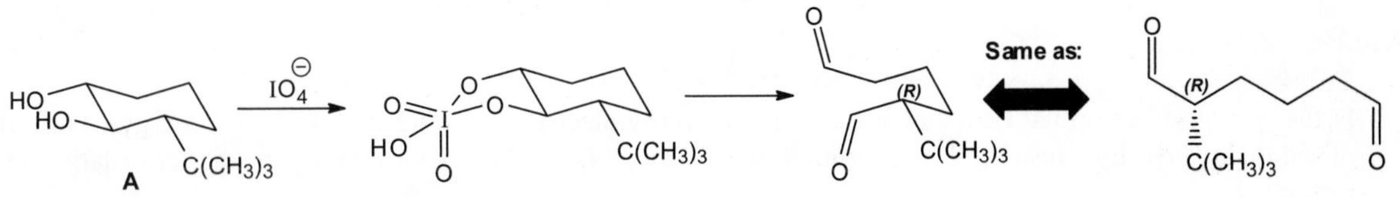

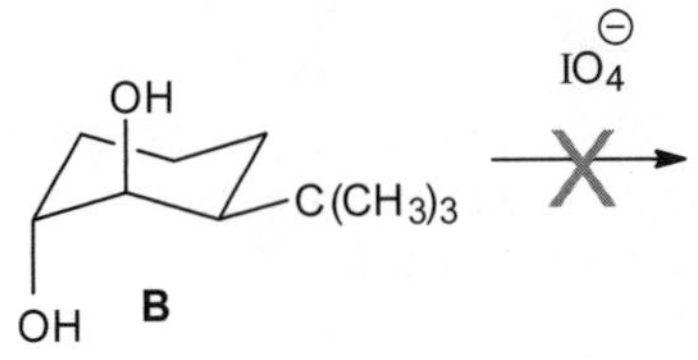

Problem 24.23

Think

Refer to the mechanism in Equation 24-45 to review the structure of the molozonide and the ozonide. What bond is cleaved in the ozonolysis reaction? Is the workup done under oxidative or reducing conditions?

Solve

The structure of the molozonide, ozonide, and final products from the ozonolysis reactions are shown below. In **(a)**, the workup is done under oxidative conditions, so $H_2C{=}O$ is oxidized to CO_2. In **(b)**, the workup is done under reducing conditions, so the aldehyde that is initially formed is not oxidized further.

(a)

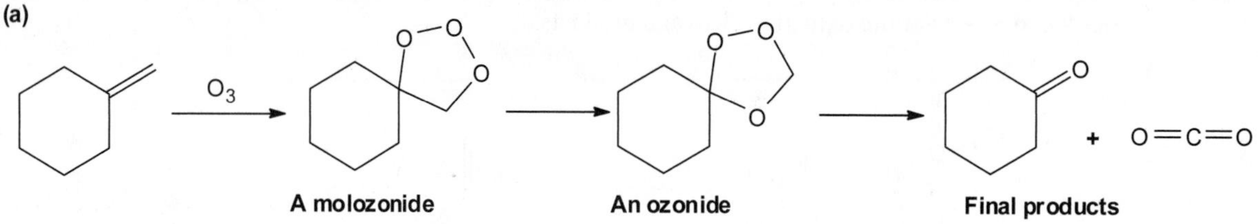

O_3

A molozonide

An ozonide

Final products

+ O=C=O

(b)

O_3

A molozonide

An ozonide

Final products

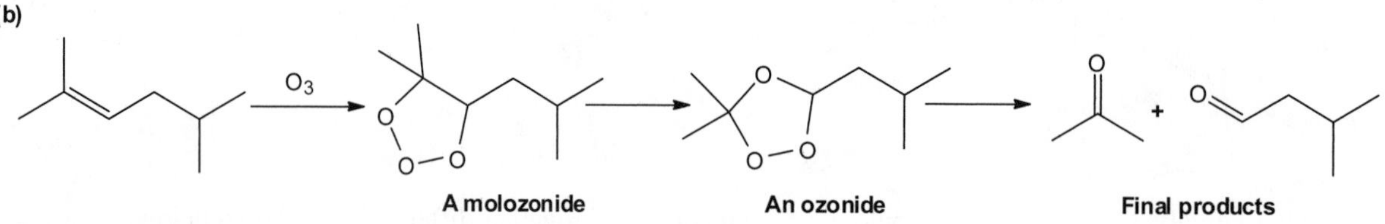

Problem 24.24

Think

Is the Diels–Alder reaction favored when the dienophile is electron rich or electron poor? What types of groups are CH_3 and OCH_3? Are they activating or deactivating? Are they equally so?

Solve

The Diels–Alder reaction is favored when the dienophile is electron poor—namely, when electron-withdrawing groups are on the dienophile. The OCH_3 and CH_3 are both electron-donating groups (EDGs), however, but CH_3 is a weaker EDG compared to the OCH_3. Given the choice, the diene will react with the dienophile that has the weaker EDG.

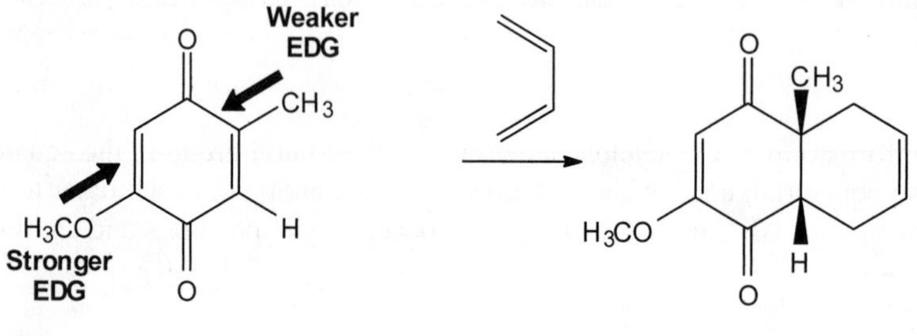

Weaker EDG

CH_3

H_3CO

Stronger EDG

Problem 24.25

Think

Is the energy of an orbital raised or lowered by a nearby electron-donating group? Is the energy of an orbital raised or lowered by a nearby electron-withdrawing group? Is the HOMO–LUMO gap smaller or larger in this example?

Solve

The energy of an orbital is usually *raised* by a nearby electron-donating group. The energy of an orbital is usually *lowered* by a nearby electron-withdrawing group. Therefore, the diene's MOs are lowered and the dienophile's MOs are raised. This increases the HOMO–LUMO energy gap between the HOMO of the diene and the LUMO of the dienophile, as shown below. Notice, however, that it lowers the HOMO–LUMO energy gap between the HOMO of the dienophile and the LUMO of the diene.

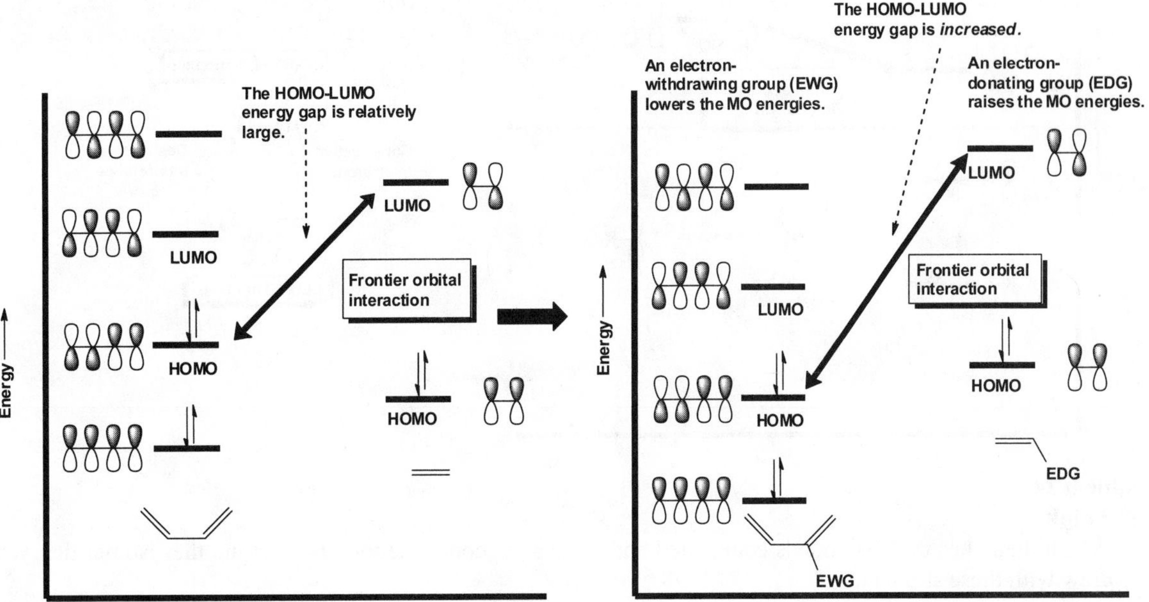

Problem 24.26

Think

If the diene is in the excited state, what is the new HOMO and LUMO? What are the HOMO and LUMO for ethene in its ground state? Does this new HOMO–LUMO interaction result in constructive or destructive interference? Is this forbidden or allowed? How does this all change if the diene is in its ground state and the dienophile is in its excited state?

Solve

If the diene is in the excited state, the LUMO is the fourth MO level and the HOMO is the third. The MO diagram for the dienophile does not change. As shown below, the overlap between these HOMO and LUMO orbitals leads to no net interaction and shows why the [4+2] photochemical cycloaddition reaction is forbidden.

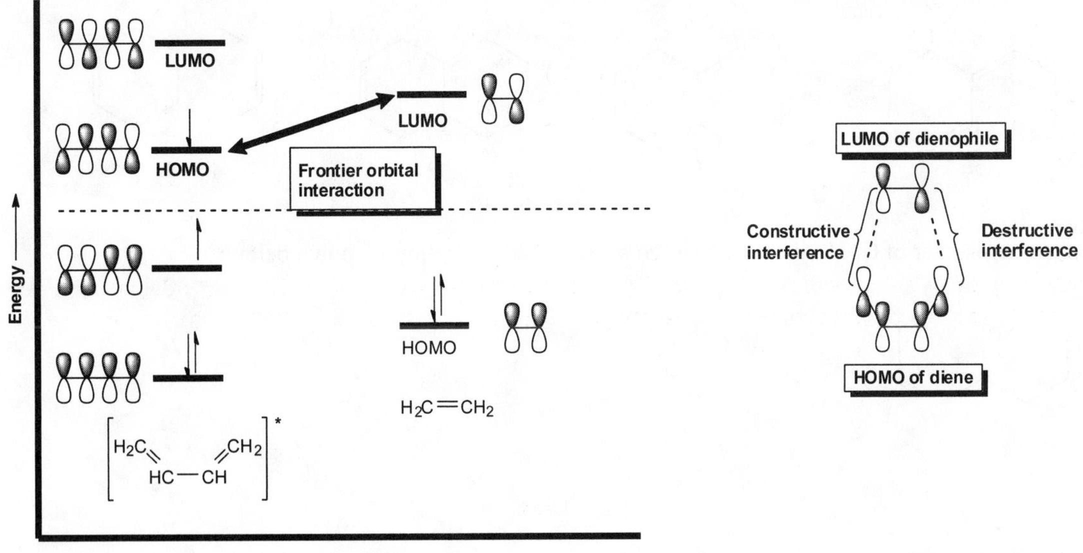

The figure below shows how the energy diagrams change when the dienophile is in its excited state and the diene is in its ground state. Notice, however, that the HOMO–LUMO interaction involves the same two orbitals as above, resulting in no net interaction and a forbidden reaction.

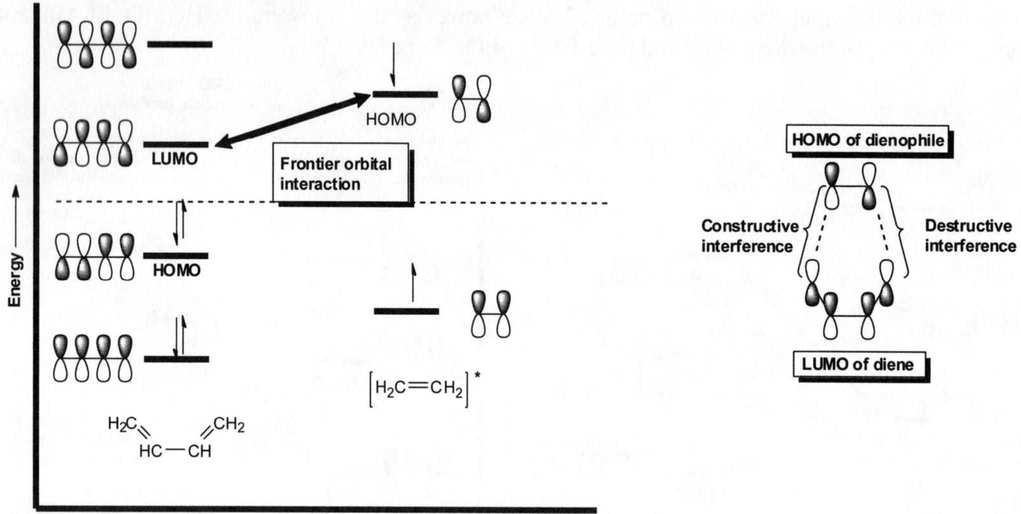

Problem 24.27

Think

Which diene has the C=C bonds conjugated and in an *s*-cis conformation? Is there another isomer that you can draw with these stipulations?

Solve

(a) Only the three circled isomers will react. The others have double bonds that are not conjugated dienes, or cannot assume the *s*-cis conformation needed to react.

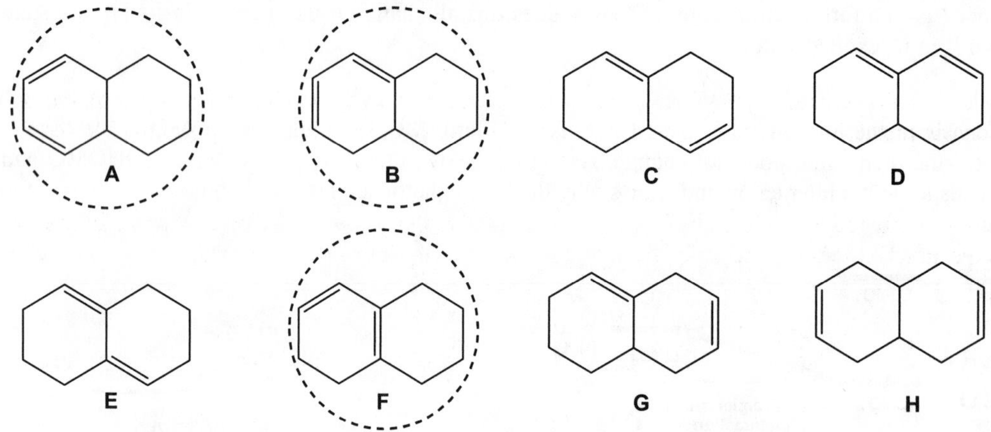

(b) Another isomer of $C_{10}H_{14}$ that can undergo a Diels–Alder reaction is shown below:

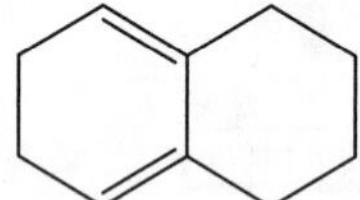

Problem 24.28

Think

What is the necessary conformation of the diene in the Diels–Alder reaction? In what conformation is each diene shown? Can rotation about a single bond convert the dienes to the necessary conformation? Which diene is most stable in this conformation? Most unstable?

Solve

The diene must be in the *s*-cis conformation to react in a Diels–Alder reaction. Compounds **I**, **J**, and **K** are listed below in increasing stability in the *s*-cis conformation (due to decreasing steric strain) and thus also listed in increasing rate of Diels–Alder reaction. The order is **K** < **J** < **I.**

Rotate to *s*-cis conformation.

Increasing rate of Diels–Alder reaction.

Problem 24.29

Think

In a Diels–Alder reaction, is the dienophile relatively electron rich or electron poor? Are the substituents attached to the dienophile electron-donating groups (EDG) or electron-withdrawing groups (EWG)?

Solve

In a Diels–Alder reaction, the dienophile is relatively electron poor and thus the reaction is favored with EWG substituents attached to the dienophile. The order of reactivity is **P** < **O** < **L** < **N** < **M**.

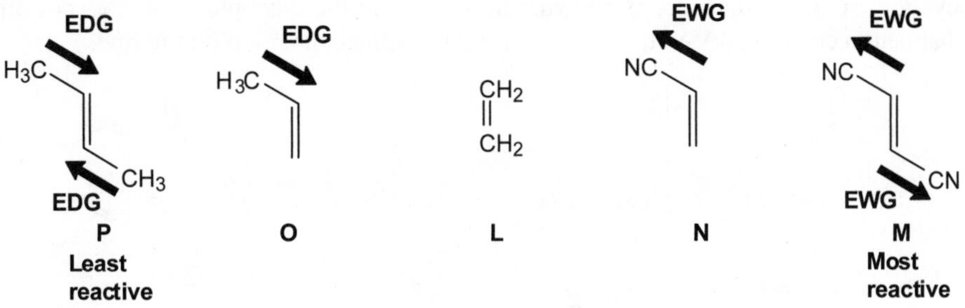

Problem 24.30

Think

In a Diels–Alder reaction, is the diene relatively electron rich or electron poor? Are the substituents attached to the diene electron-donating groups or electron-withdrawing groups?

Solve

In a Diels–Alder reaction, the diene is relatively electron rich and thus the reaction is favored with electron-donating group attached to the diene. In compound **Q**, the O atom can donate electrons via resonance, making the diene more electron rich. In compound **R**, only the inductive effect of the O atom is felt since an sp^3 C atom separates the CH_3O from the diene. Therefore, the order is **R** < **S** < **Q**.

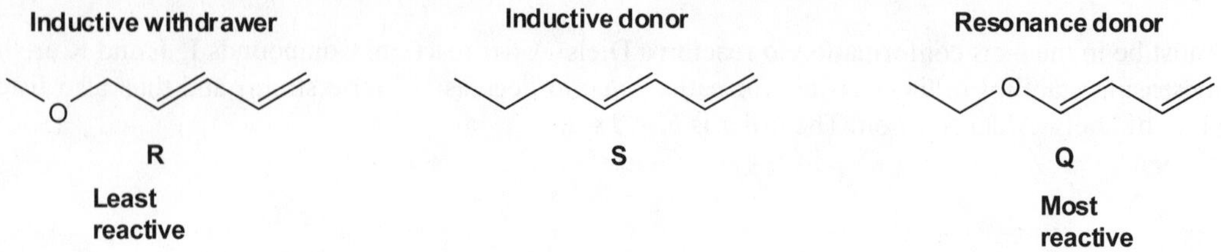

Inductive withdrawer — R — Least reactive; Inductive donor — S; Resonance donor — Q — Most reactive

Problem 24.31

Think

What is the necessary conformation of the diene in the Diels–Alder reaction? In what conformation is each diene shown? Can rotation about a single bond convert the dienes to the necessary conformation? Which diene is most stable in this conformation? Most unstable?

Solve

The diene must be in the *s*-cis conformation. Compound **U** will react faster as a diene because it has less steric hindrance in, and can more easily adopt an *s*-cis conformation, as shown below:

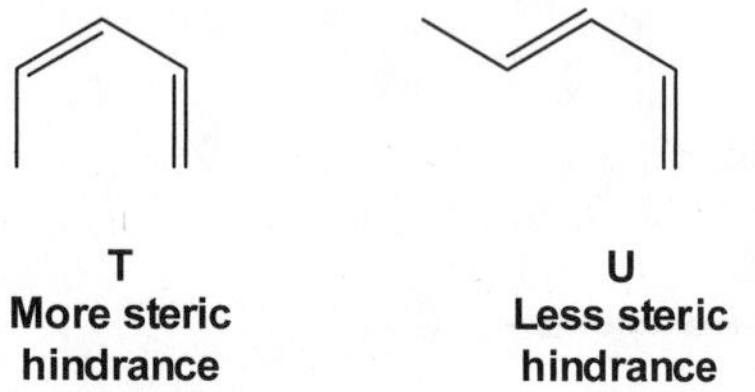

T — More steric hindrance; U — Less steric hindrance

Problem 24.32

Think

In a Diels–Alder reaction, is the dienophile relatively electron rich or electron poor? Are the substituents attached to the dienophile electron-donating groups (EDG) or electron-withdrawing groups (EWG)? How is this property of the attached group affected when the NO_2 group on the ring occupies the meta versus para position?

Solve

In the Diels–Alder reaction, the dienophile is relatively electron poor and thus the reaction is favored with electron-withdrawing group substituents. Compound **W** will react faster. The nitro group can withdraw electrons from the exocyclic C=C via resonance, as shown below, making the dienophile more electron poor. Such resonance cannot happen in compound **V** with the nitro group in the meta position on the ring.

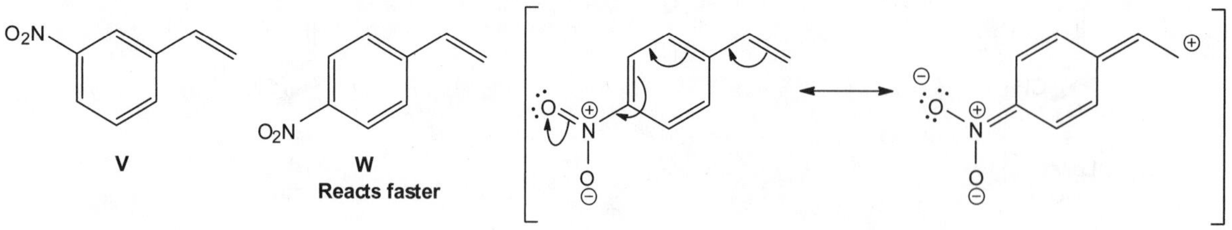

O_2N — V; O_2N — W — Reacts faster

Problem 24.33

Think

In a Diels–Alder reaction, is the dienophile relatively electron rich or electron poor? Are the substituents attached to the dienophile electron-donating groups (EDG) or electron-withdrawing groups (EWG)? How is this property of the attached group affected when the OH group on the ring occupies the meta versus para position?

Solve

In the Diels–Alder reaction, the dienophile is relatively electron poor and thus the reaction is favored with electron-withdrawing group substituents. Compound **X** will react faster. The OH group can donate electrons to the exocyclic C=C via resonance when the C=C and the OH are para to each other, thus making the dienophile less electron-poor. This resonance cannot happen in compound **X**, in which the OH group and the C=C are meta to each other.

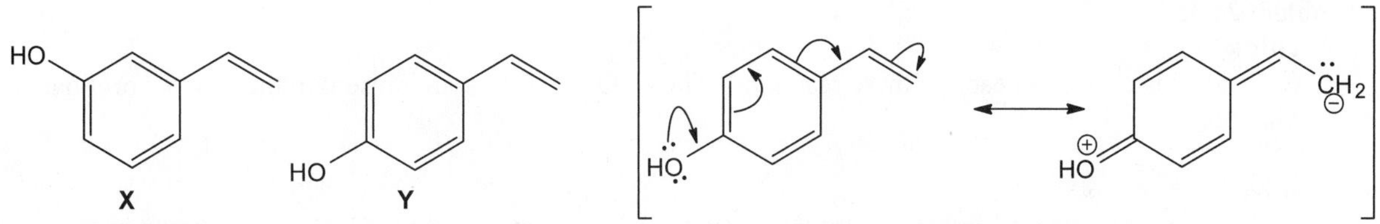

HO X Reacts faster

HO Y

HO ⟷ HO CH2

Problem 24.34

Think

What bond is cleaved in each of these reactions? Are these oxidation or reduction reactions? What product is formed with $KMnO_4/HO^-/\Delta$? With OsO_4? With O_3?

Solve

In each of these reactions, the C=C or C≡C bond is cleaved, indicated by a squiggly line. The ketone or carboxylic acid is formed with $KMnO_4/HO^-/\Delta$. Aldehydes are formed with OsO_4 followed by periodate. With O_3, the ketone or aldehyde is formed when the workup is carried out under reducing conditions, whereas a ketone or carboxylic acid (or CO_2) is formed when the workup is carried out under oxidizing conditions. The products for reactions **(a)**–**(f)** are given below. In the case of the C≡C, reaction with $KMnO_4/HO^-/\Delta$ produces the carboxylic acid and CO_2. Notice in **(f)** that the initial secondary alcohol is oxidized to a ketone under these conditions.

(a) 1. $KMnO_4$, NaOH, Δ; 2. H_3O^+ → O, OH, O, OH

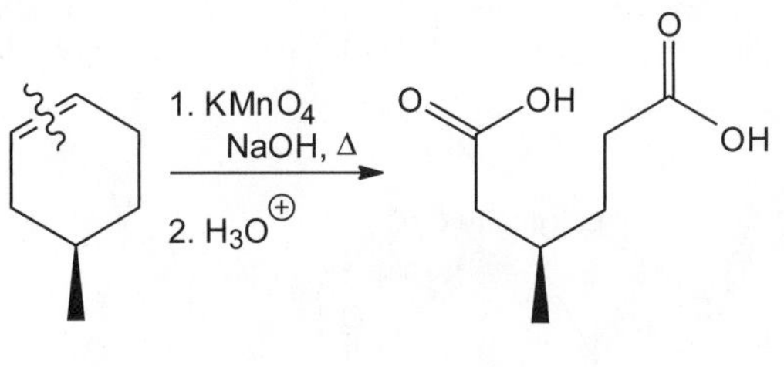

(b) O; 1. $KMnO_4$, NaOH, Δ; 2. H_3O^+ → O, O, OH + $CO_2(g)$

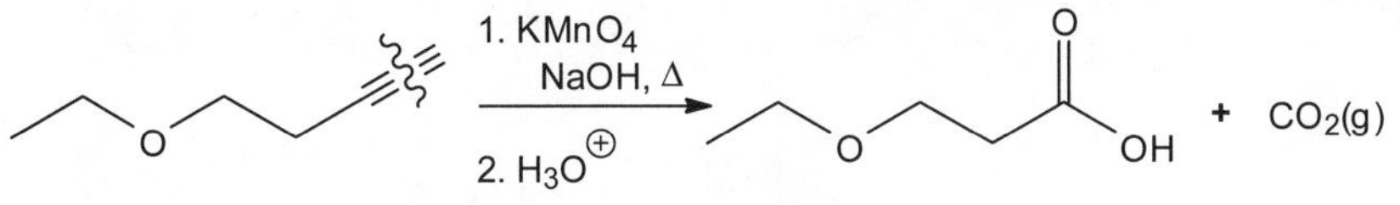

(c) 1. O_3, −78 °C; 2. Zn, HOAc → O + O

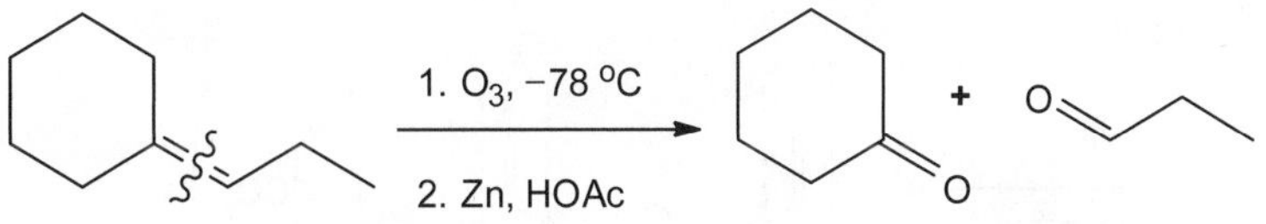

(d) 1. O_3, −78 °C; 2. H_2O_2 → O + O, OH

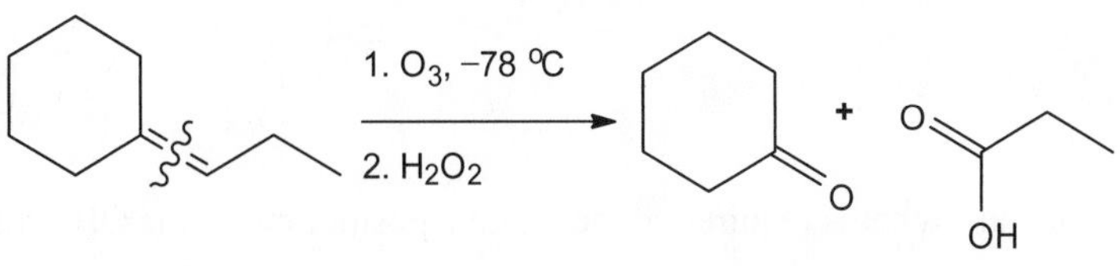

(e) 1. OsO_4, H_2O_2 2. HIO_4 +

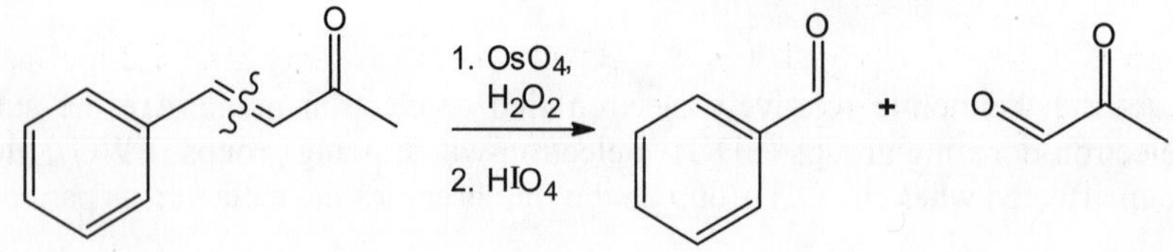

(f) HO 1. $KMnO_4$ NaOH, Δ 2. $H_3O^{\oplus}$ +

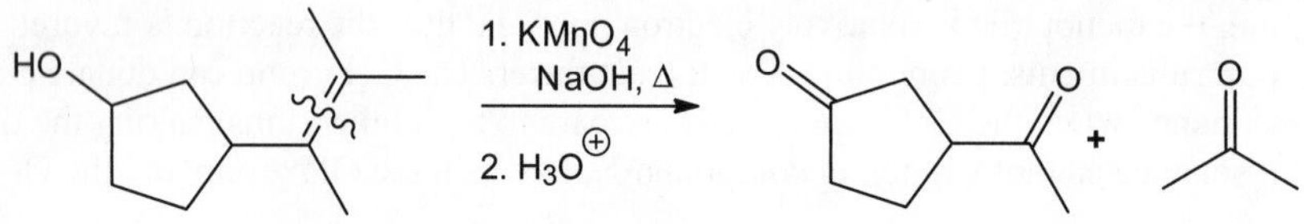

Problem 24.35

Think

What bond is cleaved in each of these reactions? The C=O bonds that are present in the product originated from what bond?

Solve

The C=C bond is cleaved in each of these reactions and the C atoms of the C=O bonds shown in the products were originally the C=C bond in the reactant. The starting materials for reactions **(a)–(d)** are shown below:

(a) C_9H_{16} 1. $KMnO_4$ NaOH, Δ 2. $H_3O^{\oplus}$ Originally C=C OH

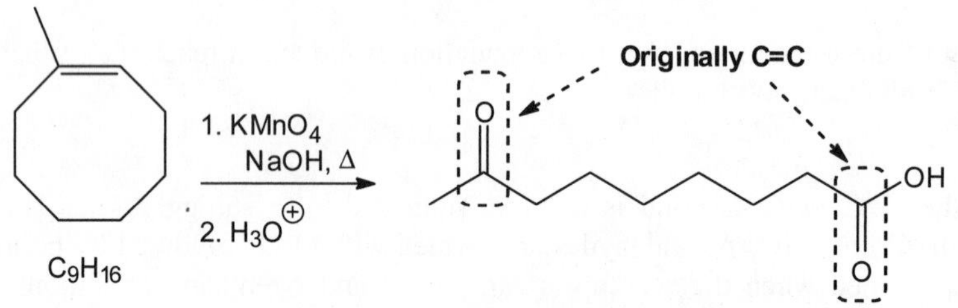

(b) C_6H_{10} $KMnO_4$ NaOH Δ Originally C=C + CO_2

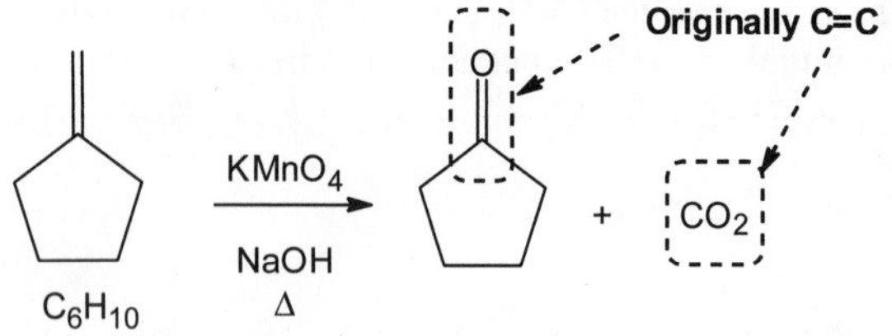

(c) $C_{10}H_{18}$ 1. O_3, −78 °C 2. $(CH_3)_2S$ Originally C=C +

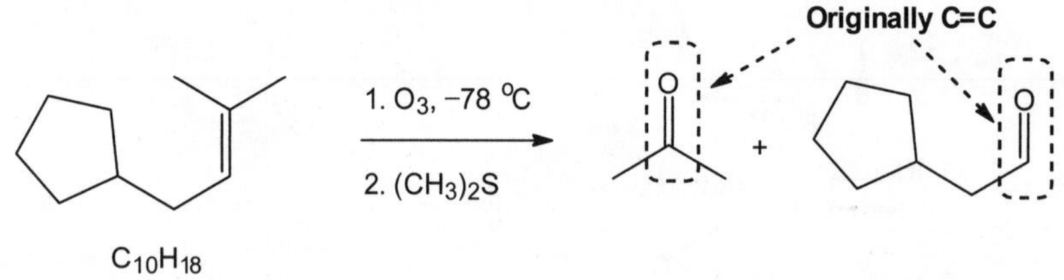

(d) C_7H_{12} 1. O_3, −78 °C 2. H_2O_2 Originally C=C HO OH + CO_2

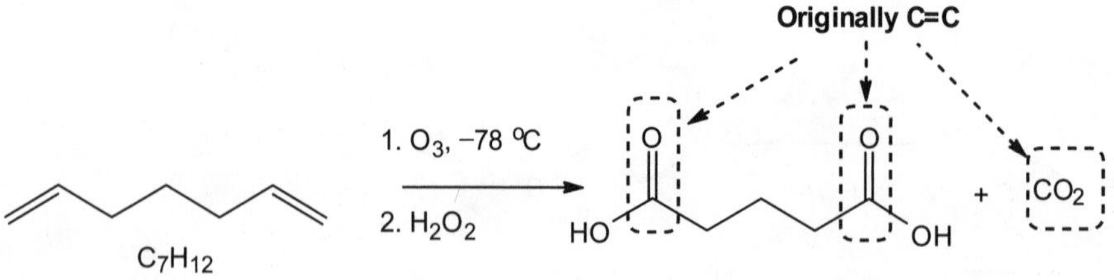

Problem 24.36

Think

How does HIO_4 react with a diol compound? What is required of the relative positioning of the OH groups?

Solve

The 1,2-diol OH groups react and the HO group that is at carbon #4 does not react. The 1,2-diol reacts with HIO_4 to cleave the C–C bond and form the dialdehyde.

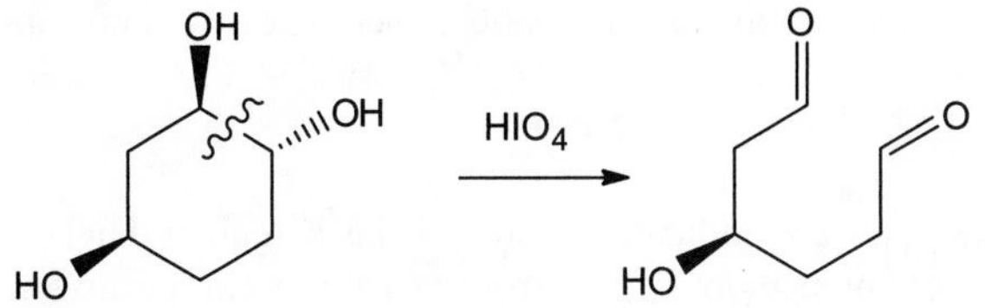

$$\xrightarrow{HIO_4}$$

Problem 24.37

Think

How does ozone react with an alkene? What product forms? Is the structure still a polymer?

Solve

Ozone, O_3, reacts with an alkene to cleave the C=C bond and form two C=O bonds. When rubber's alkene groups react with ozone, the polymer breaks down as shown below:

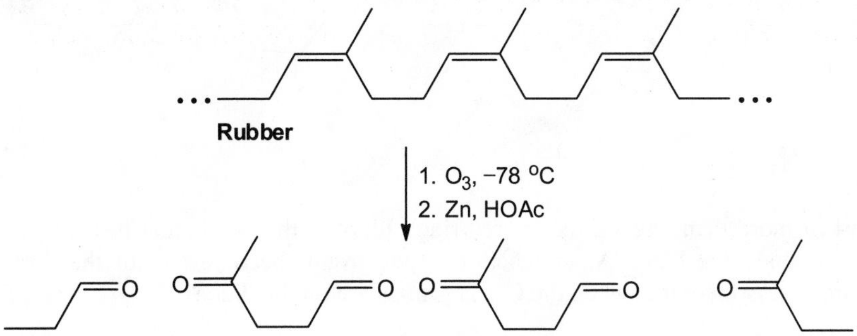

Rubber

1. O_3, −78 °C
2. Zn, HOAc

Problem 24.38

Think

Which diene in anthracene will react with the dienophile and still have an aromatic compound? Are the dienes in the correct conformation?

Solve

Anthracene can react in one ring as a dienophile, and the aromaticity of the two other rings can be preserved. Either the middle ring or an outer ring can react.

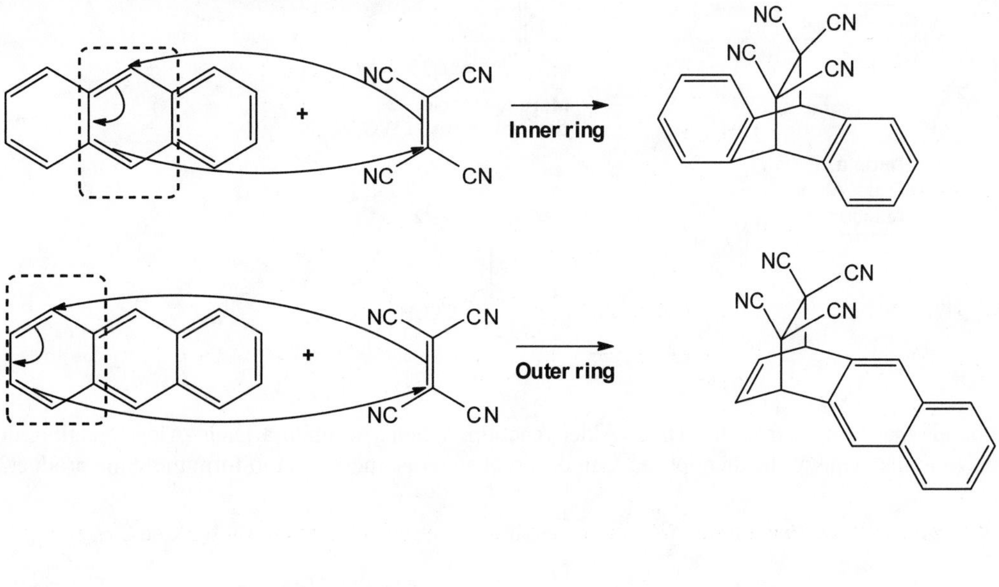

Inner ring

Outer ring

Problem 24.39

Think

Which pair of alkynes is analogous to a conjugated diene in a regular Diels–Alder reaction? Which alkyne is analogous to an alkene in a regular Diels–Alder reaction? How do you add the curved arrows? Which bonds form and which ones break? Does the product of that reaction appear to be a benzyne? Is there another resonance contributor?

Solve

The alkynes that are analogous to a diene and dienophile are indicated below. Like a regular Diels–Alder reaction, this reaction involves the cyclic movement of three pairs of electrons, so three curved arrows are necessary. The product of this curved arrow notation doesn't immediately appear to be a benzyne intermediate, but another resonance contributor of the ring does.

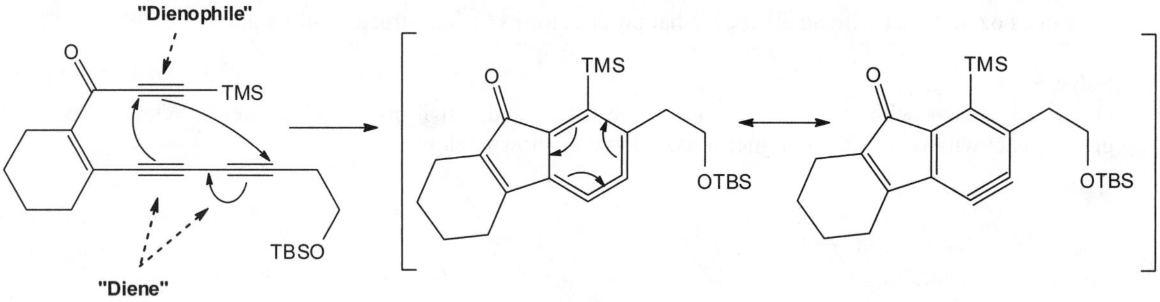

"Dienophile"

TMS

TBSO

"Diene"

TMS

OTBS

TMS

OTBS

Problem 24.40

Think

Is the C=C in the product part of more than one six-membered ring? Identify the bonds that formed from the Diels–Alder reaction. When you undo the Diels–Alder reaction, what groups become part of the diene and dienophile? Is the reaction more or less favorable when the CO_2H groups are on the diene?

Solve

The C=C in the product is part of two different six-membered rings, so we can undo a Diels–Alder reaction two different ways, as shown below. The carboxyl groups can be on either the dienophile or the diene in the precursors. The first reaction will work better because the carboxyl groups make the dienophile electron poor, leaving the diene electron rich.

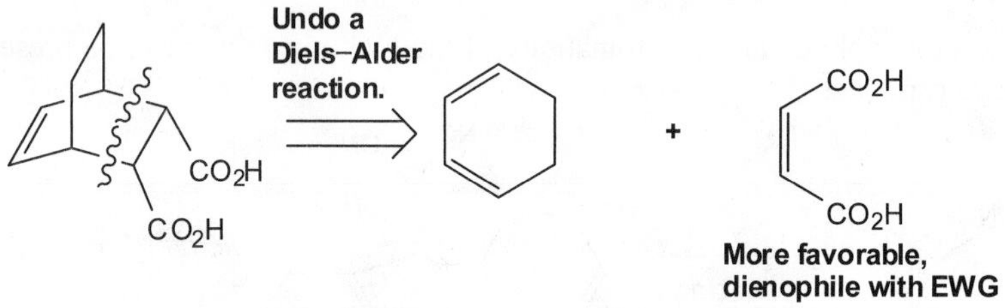

CO_2H CO_2H **Undo a Diels–Alder reaction.** + CO_2H CO_2H

More favorable, dienophile with EWG

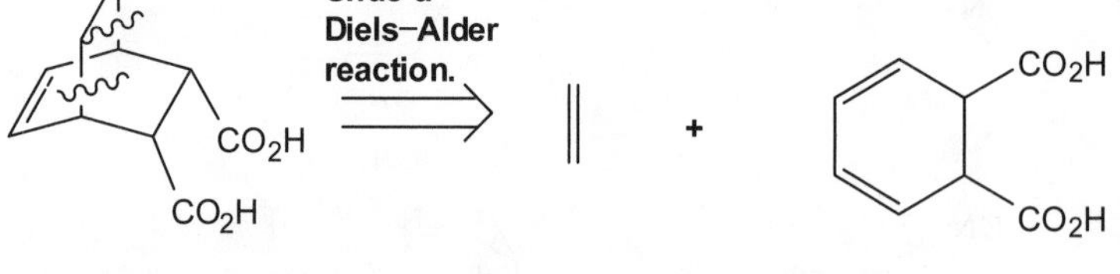

CO_2H CO_2H **Undo a Diels–Alder reaction.** + CO_2H CO_2H

Problem 24.41

Think

Identify the bonds that formed from the Diels–Alder reaction. When you undo a Diels–Alder reaction, do the OCH_3 groups go to the diene or the dienophile? Can different stereoisomers react to form the same product?

Solve

The bonds that should be disconnected when undoing a Diels–Alder reaction are shown below. The ether belongs to the diene in the precursors. To achieve the cis configuration in the ring product, both the C=C bonds in the diene must be trans or both must be cis. Both possibilities are shown below. The first is better, because in the second diene, the ether groups induce steric strain that would force the π orbitals to be not completely coplanar and would thus decrease the effective concentration of the diene in the *s*-cis conformation.

OCH3 Undo a Diels–Alder reaction. OCH3 + OCH3

More favorable, less steric strain

OCH3 Undo a Diels–Alder reaction. OCH3 OCH3 + OCH3

Problem 24.42

Think

Are Diels–Alder reactions intrinsically reversible? How does the added heat affect reversibility of this reaction? What are the products of the Diels–Alder and retro Diels–Alder reactions? Is there symmetry in the Diels–Alder product? How does this account for the isotope scrambling?

Solve

The Diels–Alder reaction is made reversible with the added heat. The mechanism is a Diels–Alder reaction followed by a retro-Diels–Alder reaction. Notice that the Diels–Alder product is symmetrical except for the isotopic labeling.

Diels–Alder — Retro Diels–Alder — H–C≡C–H + (13, 13)

Problem 24.43

Think

Is this a [4+2] cycloaddition? How many curved arrows should be drawn for this type of reaction? What bonds are broken and what bonds are formed? In how many steps does the mechanism occur?

Solve

Three curved arrows are necessary in the Diels–Alder reaction. The mechanism is shown below:

Hydroquinone, benzene, 25 °C, 90 min — 86%

Problem 24.44

Think

Identify the diene and dienophile for reach reaction. What orientation of the diene and dienophile in **(a)** leads to a favorable electrostatic interaction? What is the configuration of the alkene dienophile in **(b)**? How do you show that the stereochemistry of the dienophile is retained?

Solve

(a) The partial charges from the resonance hybrids are shown for the diene and dienophile, and the two orient in a way to show favorable electrostatic interaction among the atoms forming bonds. The endo product is also the major product.

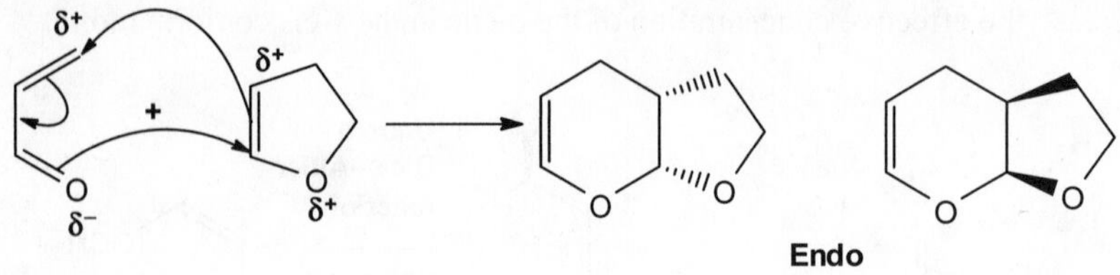

δ+ δ+ + δ− δ+ Endo

(b) The electron-withdrawing ester groups on the dienophile will be trans in the product, as they were in the reactant.

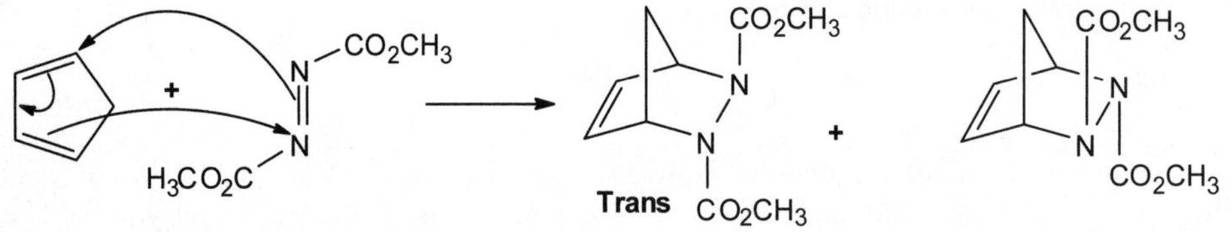

CO_2CH_3 + H_3CO_2C CO_2CH_3 + CO_2CH_3 Trans CO_2CH_3 CO_2CH_3

Problem 24.45

Think

How many ways can the dienophile approach the diene? Is the phenyl ring electron rich or electron poor? What atoms have partial charges in the resonance hybrid of the diene and dienophile? Which orientation leads to favorable electrostatic interactions among the atoms undergoing bond formation?

Solve

The two possible constitutional isomers are shown below, differing by the approach of the dienophile to the diene.

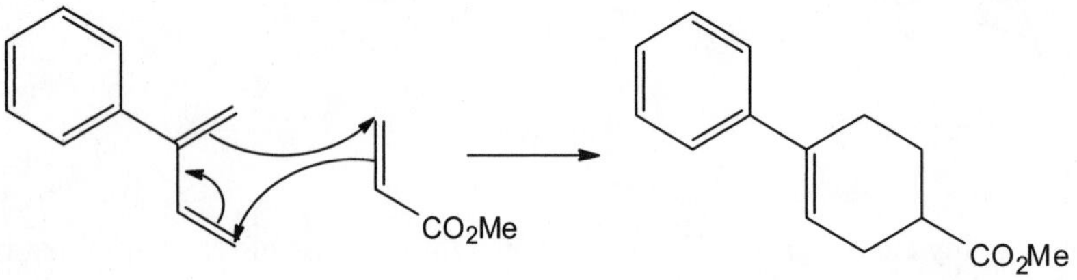

CO_2Me CO_2Me

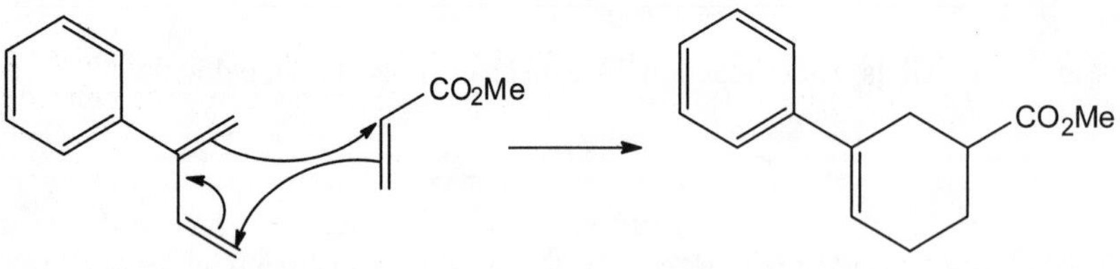

CO_2Me CO_2Me

The first one is favored because the electron-rich benzene ring induces a partial negative charge on the top-right C atom of the diene via resonance, and the CO_2Me group induces a partial positive charge on the terminal carbon of the dienophile. That leads to significant electrostatic attraction among the atoms undergoing bond formation, as shown on the next page.

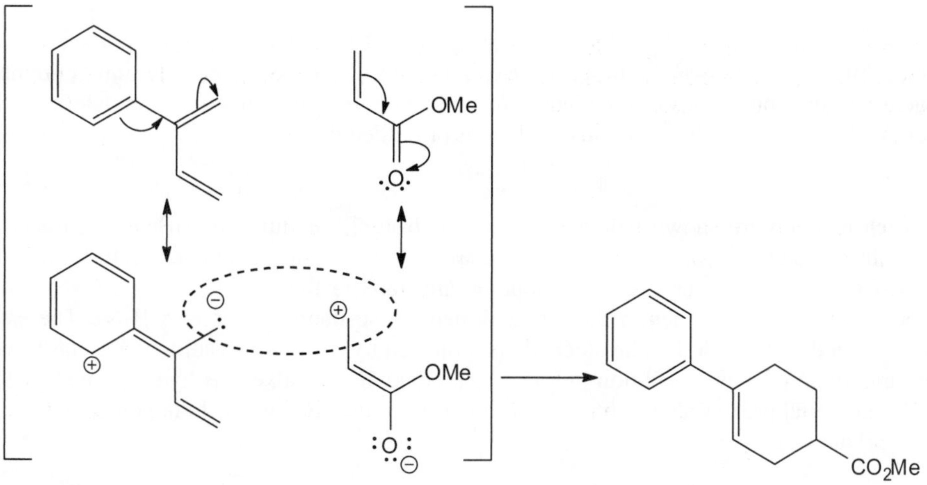

OMe
OMe
CO_2Me

Problem 24.46

Think

What is the structure of *cis*-1,3,9-decatriene? How can this compound react with itself? What is the index of hydrogen deficiency (IHD) of the product of the Br_2 reaction? What does this mean for the IHD for the intermediate product? What functional group is likely present in the intermediate product if it gains two Br atoms in the reaction with Br_2?

Solve

The structure of *cis*-1,3,9-decatriene is shown below. When heated, it can undergo an intramolecular Diels–Alder reaction to form a bicyclic alkene. This alkene then reacts with Br_2 to form a product with the formula $C_{10}H_{16}Br_2$.

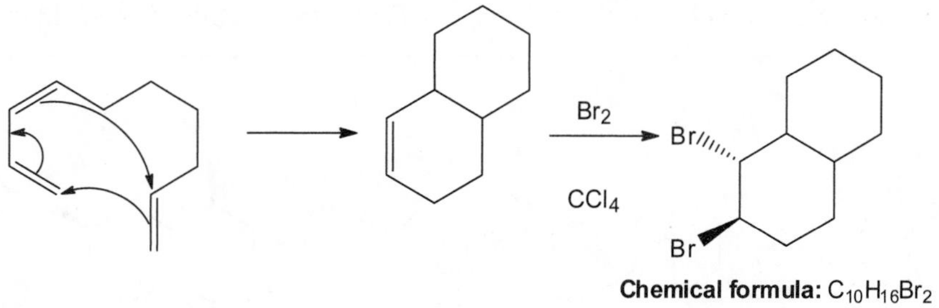

Br_2
CCl_4
Br
Br
Chemical formula: $C_{10}H_{16}Br_2$

Problem 24.47

Think

What is the structure of the benzyne intermediate? How is it formed in this reaction? How might benzyne react with cyclopentadiene to form a product with the formula $C_{11}H_{10}$? Will the benzyne intermediate behave as a diene or dienophile?

Solve

The benzyne intermediate is formed from a proton transfer reaction followed by a nucleophile elimination. The benzyne intermediate will behave as a dienophile in a [4+2] cycloaddition reaction (a Diels–Alder reaction) to form the bicyclic compound shown below:

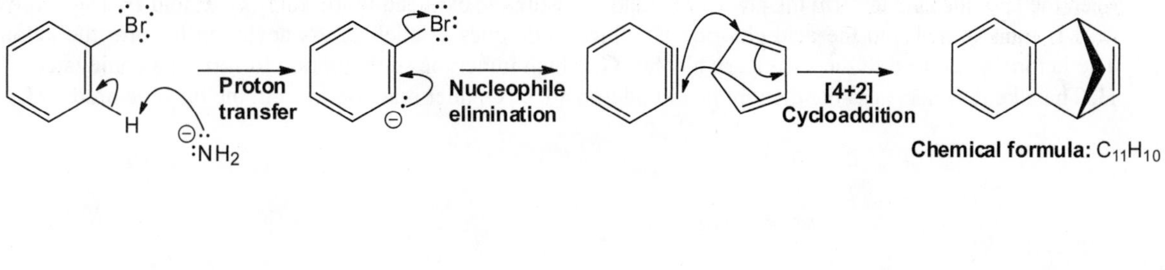

Br
H
$:NH_2$
Proton transfer
Br:
Nucleophile elimination
[4+2] Cycloaddition
Chemical formula: $C_{11}H_{10}$

Problem 24.48

Think

What are the nucleophile and electrophile in each reaction? Are these reactions addition, elimination, substitution, cycloaddition, or proton transfer reactions? Are the reactions functional group transformations or do they form or break C–C bonds? Are any reactions oxidations or reductions?

Solve

The products from each reaction are shown below. **A** is formed from 1,4-addition of HBr to the diene. **B** is formed from an S_N2 substitution reaction, which then undergoes a [4+2] cycloaddition reaction to form **C**. Note that the CH_3 and CH_2CN are trans in the alkene dienophile and trans in the product **C**. The C≡N nitrile is hydrolyzed to the carboxylic acid **D**, which is then transformed to the methyl ester by CH_2N_2. The ester is reduced by DIBAH to form the aldehyde **E**. The aldehyde is protected by ethylene glycol to the acetal **F**, which can now react with furan in a [4+2] cycloaddition to form the product **G**. The alkene is hydrogenated by H_2/Pd to form the alkane **H**. The acetal protecting group is removed to reform the aldehyde **I**. In the process, the cyclic ether is hydrolyzed to produce the diol.

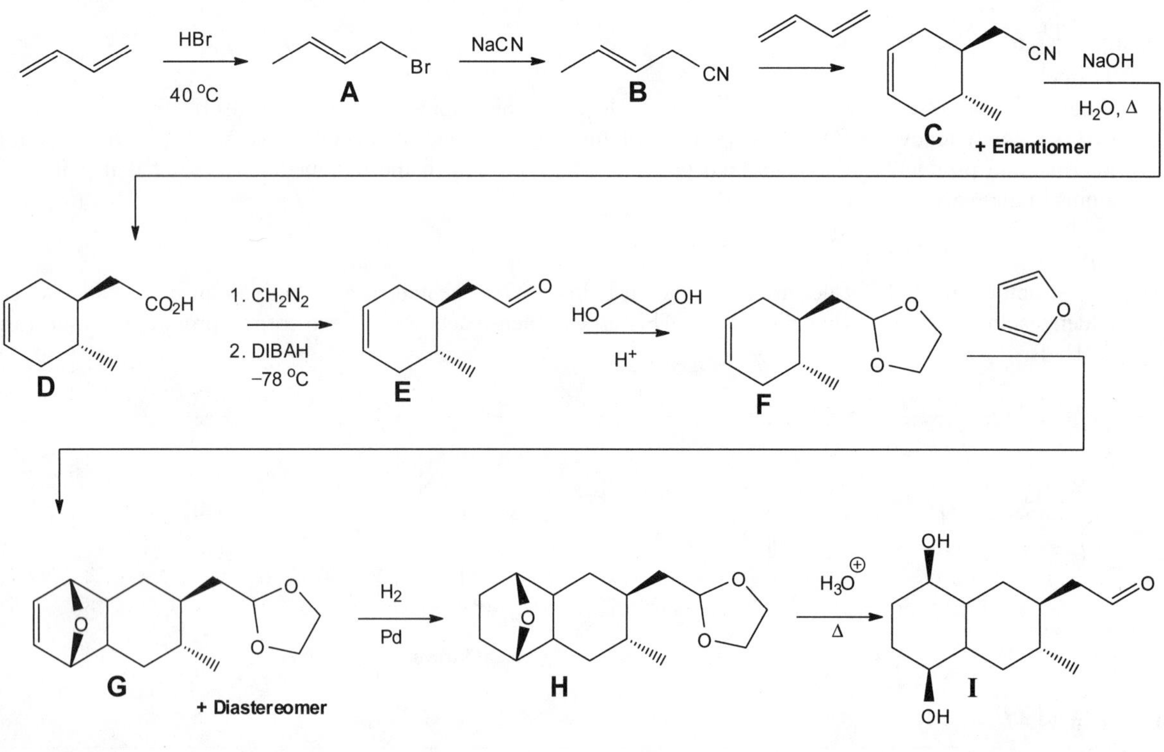

Problem 24.49

Think

What are the nucleophile and electrophile in each reaction? Are these reactions addition, elimination, substitution, cycloaddition, or proton transfer reactions? Are the reactions functional group transformations or do they form or break C–C bonds? Are any reactions oxidations or reductions?

Solve

The products from each reaction are shown on the next page. **A** is formed from an electrophilic aromatic substitution bromination reaction. **A** then reacts with Mg(s) to form the Grignard reagent **B**, which can then open the epoxide ring to form the primary alcohol **C**, which is oxidized to the carboxylic acid **D**. The carboxylic acid is transformed into the acid chloride **E**, which undergoes Friedel–Crafts acylation to form the ketone **F**. The ketone is reduced to the secondary alcohol **G**, which undergoes dehydration to form the conjugated alkene **H**. The alkene **H** serves as the dienophile to undergo a Diels–Alder reaction to form the bicyclic product **I**.

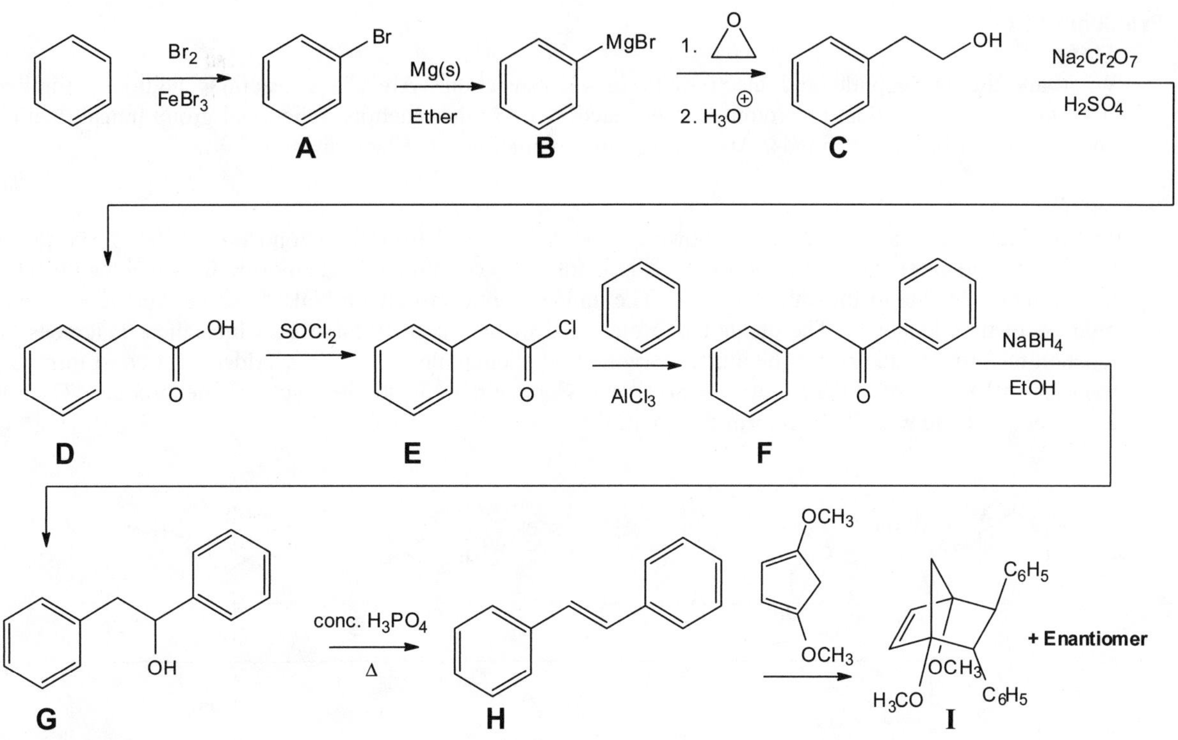

Br$_2$, FeBr$_3$ → A; Mg(s), Ether → B; 1. (epoxide) 2. H_3O^+ → C; $Na_2Cr_2O_7$, H_2SO_4 → D; $SOCl_2$ → E; benzene, $AlCl_3$ → F; $NaBH_4$, EtOH → G; conc. H_3PO_4, Δ → H; (1,4-dimethoxycyclopentadiene) → I + Enantiomer

Problem 24.50

Think

What are the nucleophile and electrophile in each reaction? Are these reactions addition, elimination, substitution, cycloaddition, or proton transfer reactions? Are the reactions functional group transformations or do they form or break C–C bonds? Are any reactions oxidations or reductions?

Solve

The products from each reaction are shown below. **A** is formed in an E2 reaction to form the alkene, which undergoes oxymercuration–reduction (reagents **B**) to form the secondary alcohol. The alcohol is oxidized by $KMnO_4$ to the ketone **C**, which reacts with the Grignard reagent to form the tertiary alcohol **D**. The alcohol undergoes a dehydration reaction to form alkene **E**. The alkene reacts with OsO_4 to form the syn diol **F**.

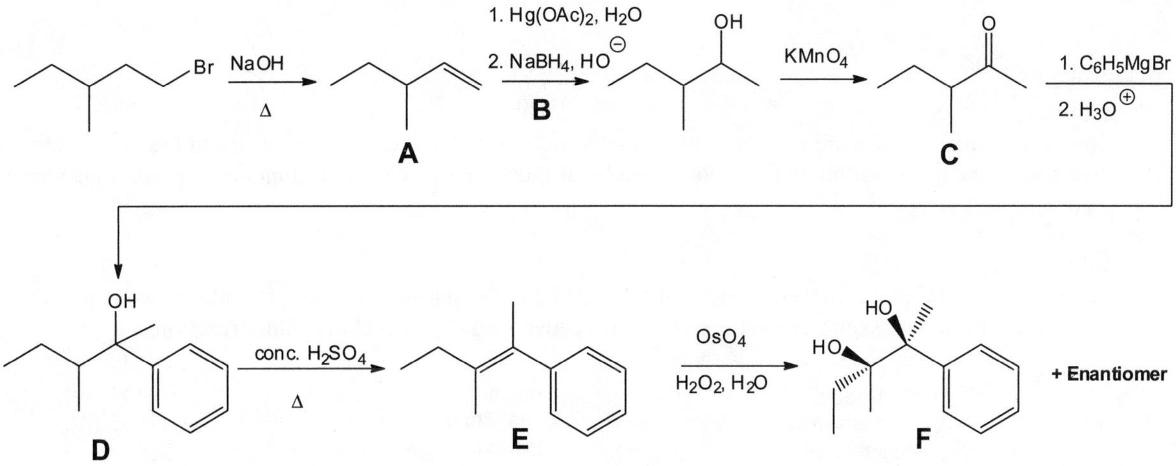

NaOH, Δ → A; 1. $Hg(OAc)_2$, H_2O 2. $NaBH_4$, HO^- (B); $KMnO_4$ → C; 1. C_6H_5MgBr 2. H_3O^+ → D; conc. H_2SO_4, Δ → E; OsO_4, H_2O_2, H_2O → F + Enantiomer

Problem 24.51

Think

What are the nucleophile and electrophile in each reaction? Are these reactions addition, elimination, substitution, cycloaddition, or proton transfer reactions? Are the reactions functional group transformations or do they form or break C–C bonds? Are any reactions oxidations or reductions?

Solve

The products from each reaction are shown below. **A** is formed from the ozonolysis of the alkene to form a carboxylic acid and ketone. The carboxylic acid is transformed into the acid chloride **B**, which then reacts with the amine $(CH_3)_2NH$ to form the amide **C**. The amide is unreactive with $NaBH_4$, so the ketone is selectively reduced to the alcohol **D**. The alcohol is brominated to the alkyl bromide **E**, which then undergoes an E2 reaction to form the alkene **F**. The alkene serves as the dienophile in the Diels–Alder reaction to form **G**. The trans configuration of the substituents in alkene **F** remain trans in the cyclohexene product. The alkene undergoes reaction with OsO_4 to form the syn diol **H**.

1. O_3, −78 °C
2. H_2O_2
A
$SOCl_2$
B
$(CH_3)_2NH$
Pyridine
C
$NaBH_4$
D
PBr_3
E
NaOH
Δ
F
Δ
G
+ Enantiomer
OsO_4
H_2O_2, H_2O
H
+ Diastereomer

Problem 24.52

Think

How many carbon atoms are in your starting materials? How many carbon atoms are in the target compound? How many reactions that altered the carbon skeleton must take place? What functional group transformations need to occur?

Solve

(a) There are 10 C atoms in the product and four or two in the starting material. Therefore, two reactions that altered the carbon skeleton must occur—namely, two back-to-back Diels–Alder reactions.

Undo a Diels–Alder reaction.

Undo a Diels–Alder reaction.

The synthesis in the forward direction would proceed as follows:

+

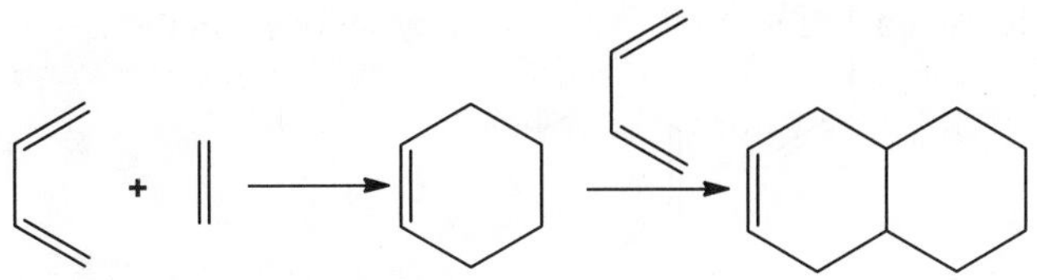

(b) There are six C atoms in the product and four or two in the starting material. Therefore, one reaction that alters the carbon skeleton occurs—namely, one Diels–Alder reaction. The cyclohexene product alkene is cleaved by $KMnO_4$ to form the dicarboxylic acid, which is reduced to the primary alcohol.

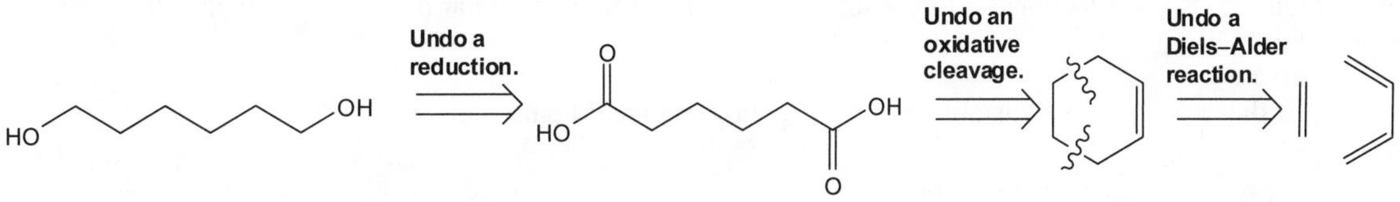

HO OH
Undo a reduction.
HO OH
O
O
Undo an oxidative cleavage.
Undo a Diels–Alder reaction.

The synthesis in the forward would proceed as follows:

1. $KMnO_4$, $HO^{\ominus}$, Δ
2. $H_3O^{\oplus}$
HO OH
O
O
1. $LiAlH_4$
2. H_3O^+
HO OH

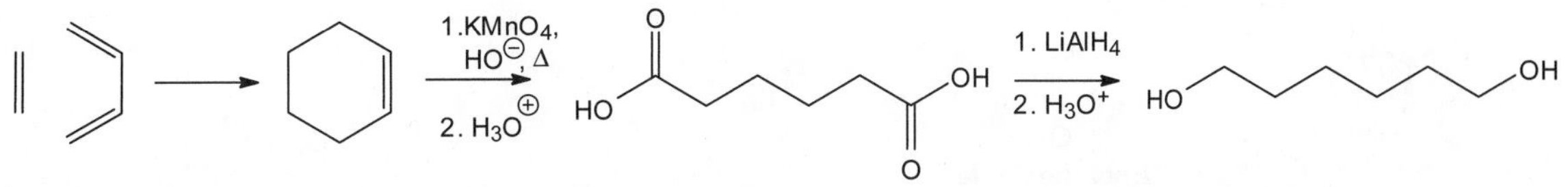

(c) There are six C atoms in the product and four or two in the starting material. The target is a β-hydroxyaldehyde, which results from an aldol addition reaction. The dialdehyde is the product of an ozonolysis reaction, and the cyclohexene precursor can be produced from a Diels–Alder reaction.

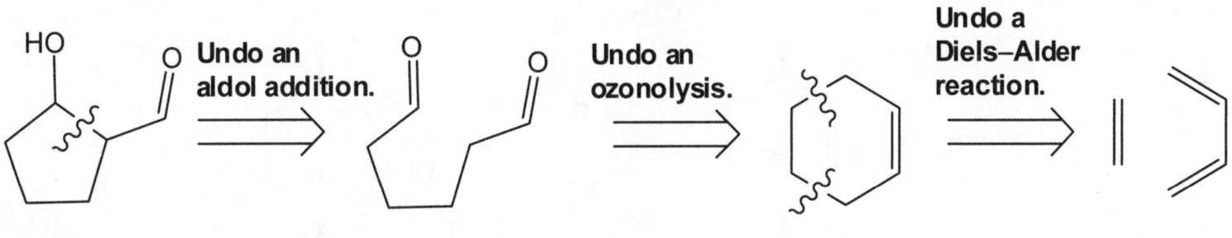

HO
O
Undo an aldol addition.
O
O
Undo an ozonolysis.
Undo a Diels–Alder reaction.

The synthesis in the forward would proceed as follows:

O_3, then Zn, HOAc
(ozonolysis)
O
O
NaOH
(aldol)
EtOH
HO
O

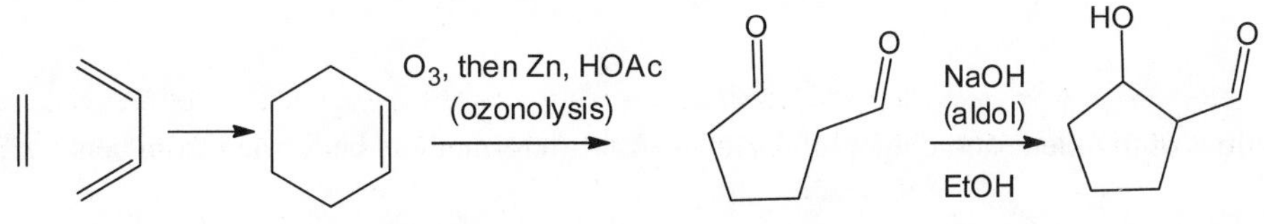

(d) There are 10 C atoms in the product and four or two in the starting material; therefore, two reactions that altered the carbon skeleton occurred. The dialdehyde from **(c)** undergoes an aldol condensation reaction to form the α,β-unsaturated carbonyl, which reacts with buta-1,3-diene in a Diels–Alder reaction.

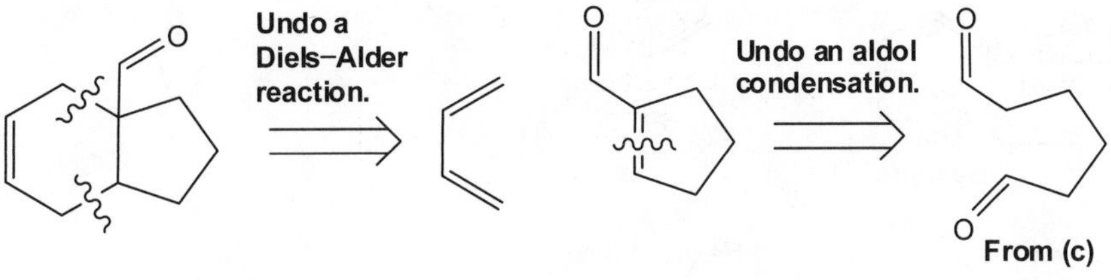

O
Undo a Diels–Alder reaction.
O
Undo an aldol condensation.
O
O
From (c)

The synthesis direction would proceed as follows:

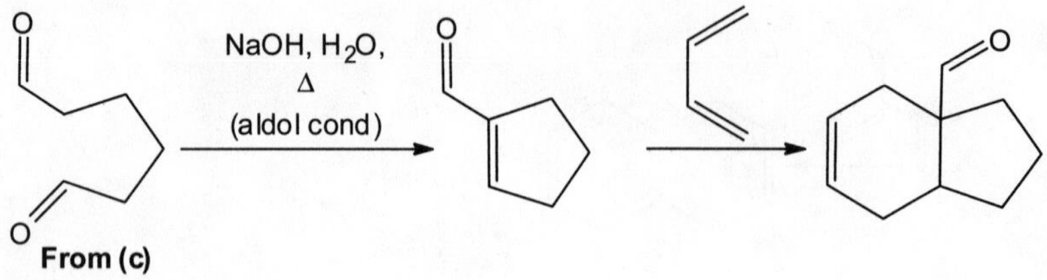

NaOH, H_2O, Δ (aldol cond)

From (c)

Problem 24.53

Think

What bonds need to be disconnected to undo a Diels–Alder reaction? Examine the reactants that would be required. Is this functional group stable? How can you form a similar diene that is stable?

Solve

The diene would have unstable enol C=C bonds and would tautomerize.

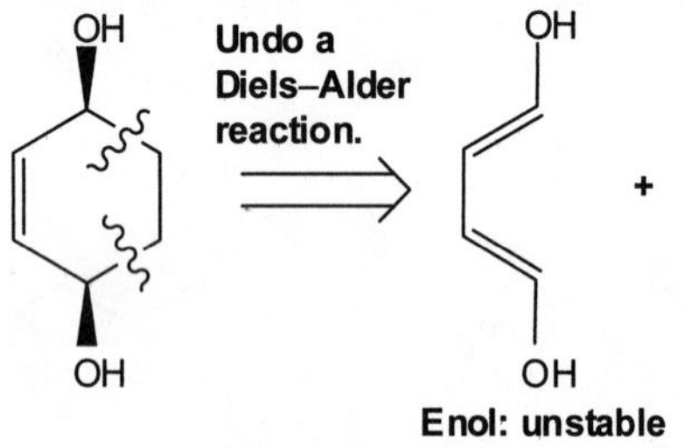

OH OH

Undo a Diels–Alder reaction.

OH OH

Enol: unstable

However, if a diether were to be synthesized with alkyl groups that could be removed after the Diels–Alder reaction was over, the synthesis would be feasible. A bis[benzyl ether] is a possibility.

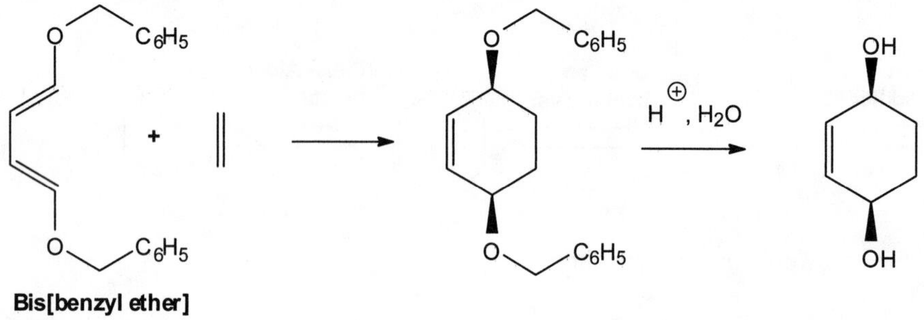

C_6H_5 C_6H_5 $H^{\oplus}$, H_2O OH OH

Bis[benzyl ether]

Problem 24.54

Think

What are the identities of the diene and dienophile? How can this intermediate be formed from benzene?

Solve

If the Diels–Alder reaction is undone, it shows that the diene is (2*E*, 4*E*)-hexa-2,4-diene and the dienophile is benzyne.

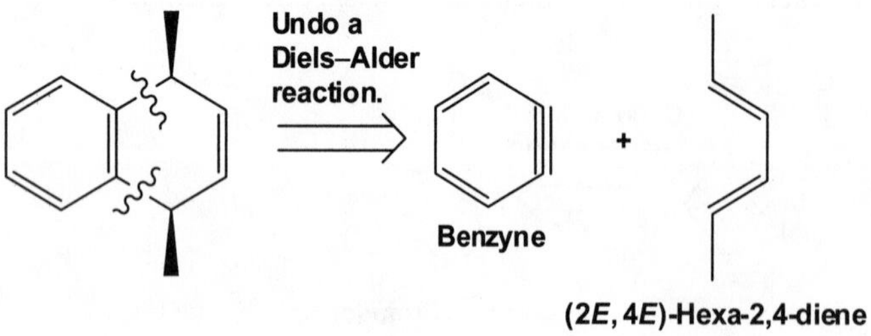

Undo a Diels–Alder reaction.

Benzyne

(2*E*, 4*E*)-Hexa-2,4-diene

The synthesis in the forward would proceed as follows:

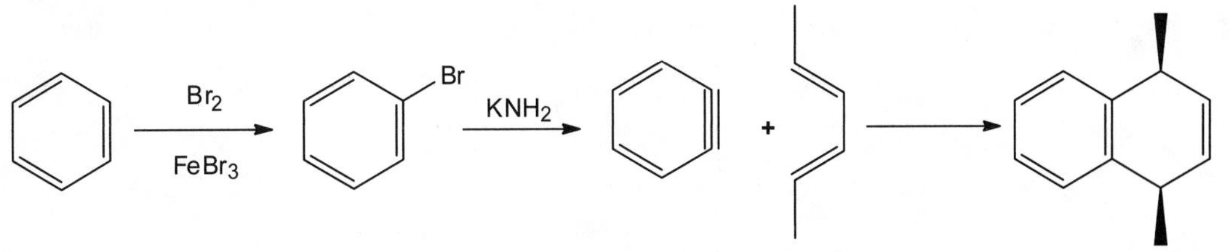

Br_2 / $FeBr_3$; Br; KNH_2; +

Problem 24.55

Think

Can you identify the bonds that were broken and formed? How many π electrons move? In the transition state, are there electrons delocalized over an entire ring? If so, how many electrons? Is that a Hückel number or an anti-Hückel number? How does that correspond to whether the reaction is allowed or forbidden?

Solve

(i) With 10 π electrons moving, the transition state is aromatic (Hückel number = 10). The reaction is allowed.

‡ Aromatic O

O +

O

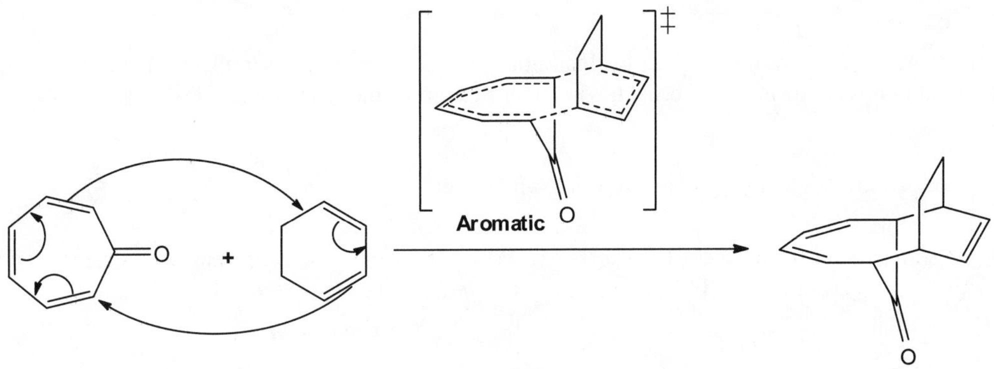

(ii) For this reaction to occur, an eight-electron cyclic transition state would be required, which is an antiaromatic transition state. This is not allowed under thermal conditions.

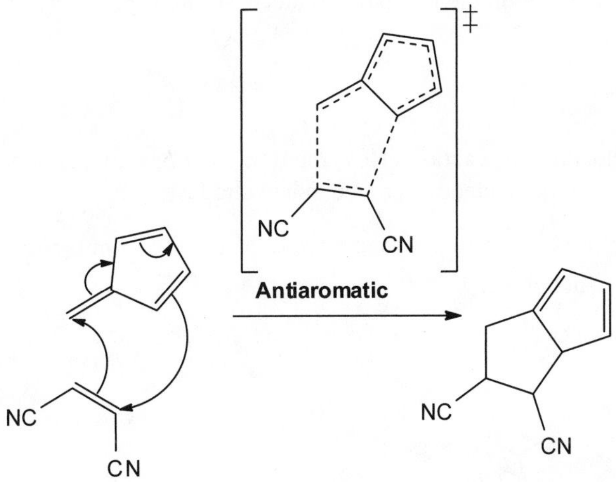

‡ NC CN Antiaromatic

NC CN

NC CN

(iii) This reaction proceeds through a 10-electron cyclic transition state, which is an aromatic transition state and will be allowed under thermal conditions.

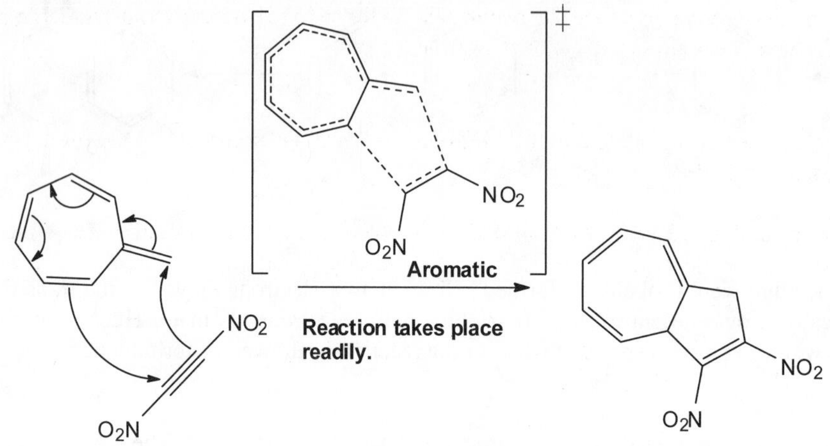

NO_2

O_2N

Aromatic

NO_2

Reaction takes place readily.

O_2N

NO_2

O_2N

Problem 24.56

Think

How many π electrons move in an [8+2] cycloaddition reaction? How many π electrons are supplied by each reactant? What bonds are formed and broken when these electrons move in a cyclic fashion?

Solve

The [8+2] cycloaddition reaction mechanism is shown below:

[8+2] Cycloaddition

+ Enantiomer

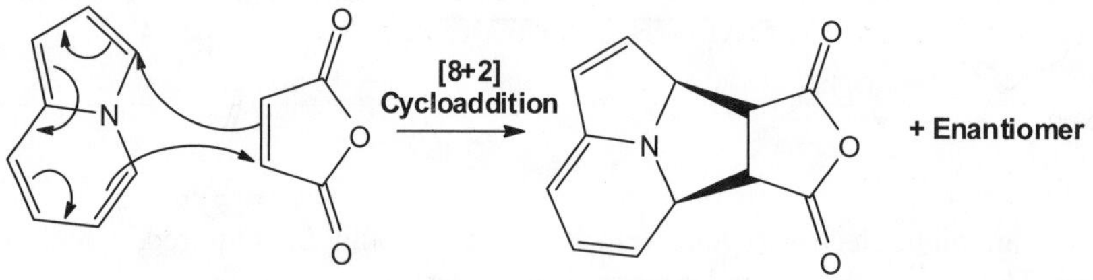

Notice that eight electrons are supplied by the reactant on the left and two electrons are supplied by the reactant on the right.

Problem 24.57

Think

How many π electrons move in an [6+4] cycloaddition reaction? How many π electrons are supplied by each reactant? What bonds are formed and broken when these electrons move in a cyclic fashion?

Solve

The [6+4] cycloaddition reaction mechanism is shown below:

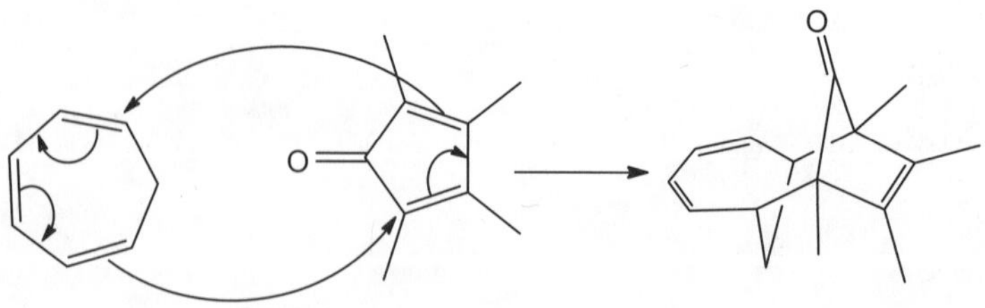

Notice that six electrons are supplied by the reactant on the left and four electrons are supplied by the reactant on the right.

Problem 24.58

Think

What does the HOMO of the larger molecule look like? What does the LUMO of the smaller molecule look like? Does constructive or destructive interference occur among the orbitals of the atoms undergoing bond formation? Do the orbitals have the appropriate symmetries to interact?

Solve

In all cases, the HOMO of the larger molecule and the LUMO of the smaller molecule will be used. (Doing the reverse still leads to the same phase relationship.)

(i) The HOMO of the triene and the LUMO of the alkene are in phase at the top but out of phase at the bottom, so the orbitals will have no net interaction. The reaction is thermally forbidden.

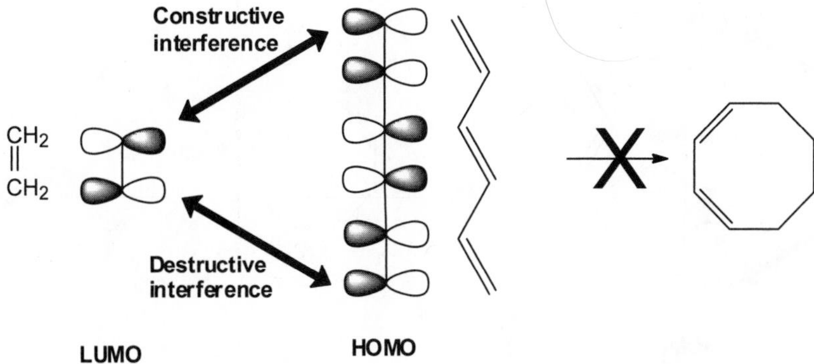

(ii) The HOMO of one diene and the LUMO of the other are in phase at the bottom but out of phase at the top, so the orbitals will have no net interaction. The reaction is thermally forbidden.

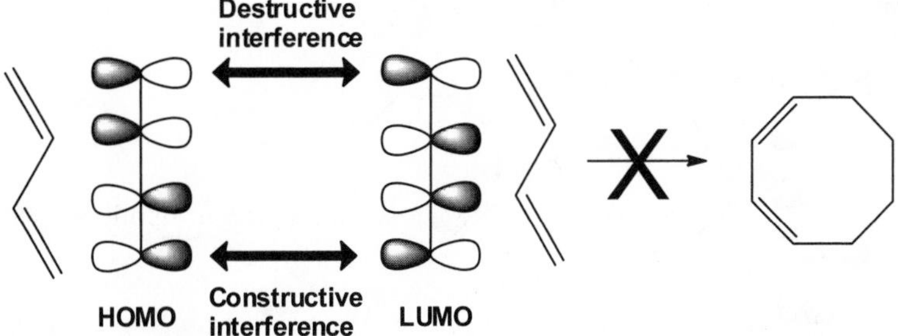

(iii) The HOMO of the triene and the LUMO of the diene are in phase at both the top and the bottom, so the orbitals will interact. This reaction is thermally allowed.

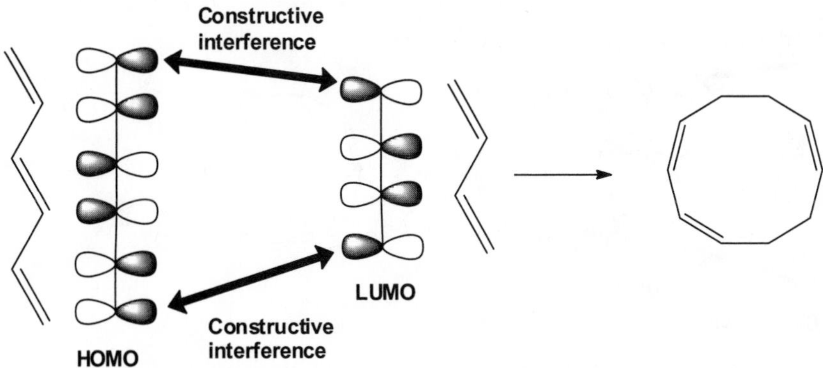

Problem 24.59

Think

Assume that the larger molecule absorbs the photon. How does the HOMO change when the molecule is in the excited state? Does the LUMO of the smaller molecule change if the other molecule didn't absorb a photon? Does constructive or destructive interference occur among the orbitals of the atoms undergoing bond formation? Do the orbitals have the appropriate symmetries to interact?

Solve

For each HOMO, exciting with a photon will promote an electron to the next highest-energy MO, which has one additional node.

(i) The HOMO of the triene and the LUMO of the alkene are in phase, so the orbitals will interact. This reaction is photochemically allowed.

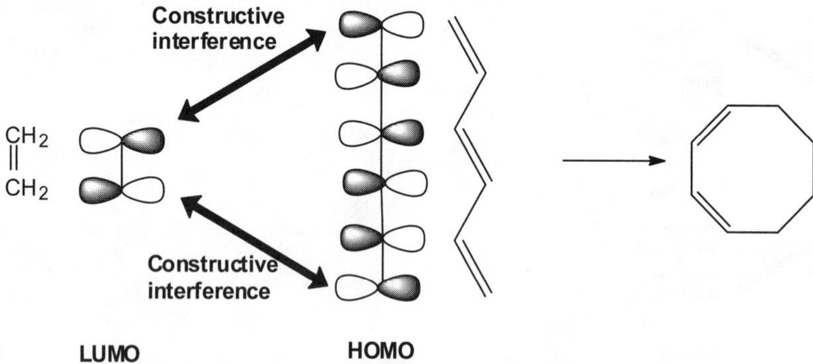

(ii) The HOMO of one diene and the LUMO of the other are in phase, so the orbitals will interact. This reaction is photochemically allowed.

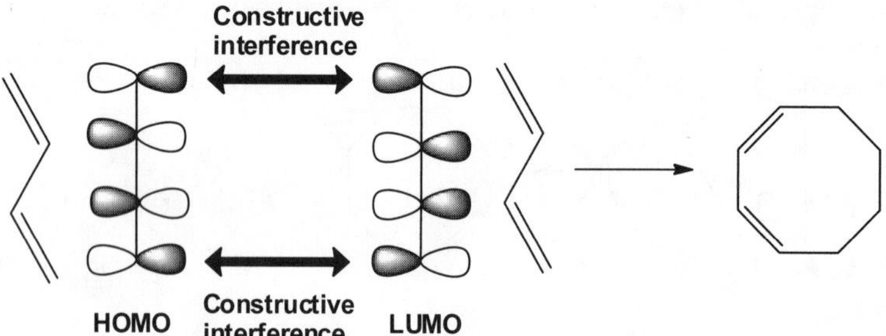

(iii) The HOMO of the triene and the LUMO of the diene are in phase at the top but out of phase at the bottom, so the orbitals will not interact. This reaction is photochemically forbidden.

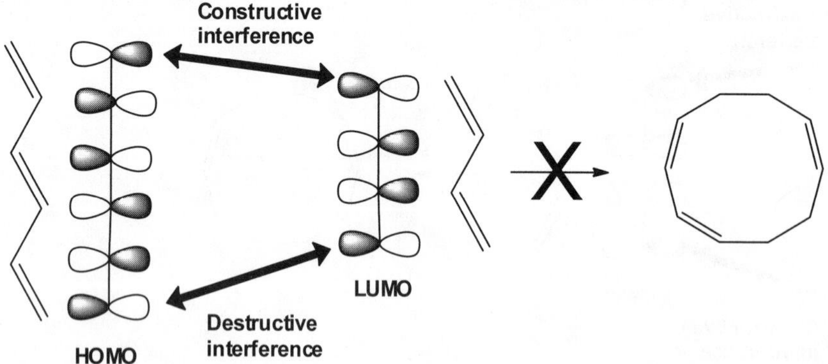

Problem 24.60

Think

What functional group is present in the target compound? How many carbon atoms are in the starting material and target compound? Is the carbon skeleton altered? What reactions do you know that will break a C=C bond? What functional group transformations must occur?

Solve

The target compound is an α,β-unsaturated ketone. This product is a result of an aldehyde–ketone condensation reaction. The pentene ring undergoes oxidative cleavage to break the C=C bond, which opens the ring and forms the ketone and carboxylic acid. The full synthesis is shown below:

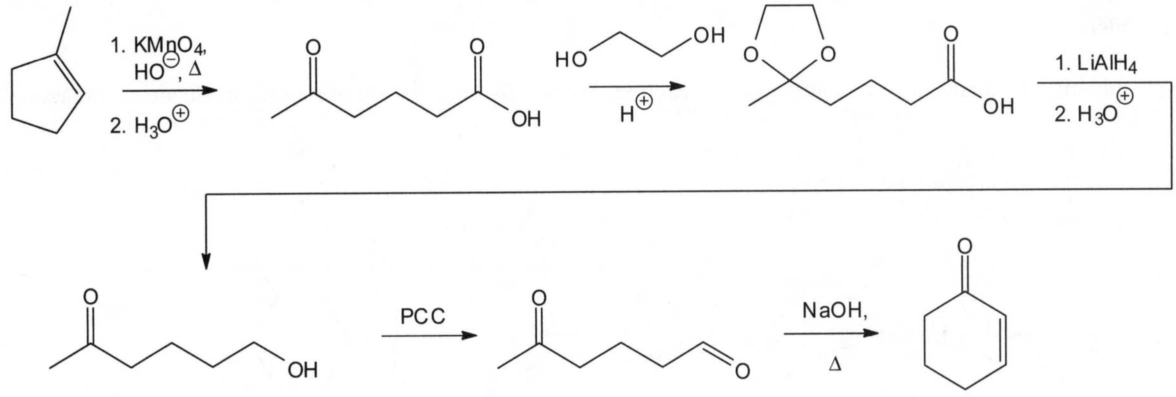

1. $KMnO_4$, $HO^{\ominus}$, Δ
2. $H_3O^{\oplus}$

HO–CH2CH2–OH, $H^{\oplus}$

1. $LiAlH_4$
2. $H_3O^{\oplus}$

PCC

NaOH, Δ

Problem 24.61

Think

Is the Diels–Alder reaction favored when the dienophile is electron rich or electron poor? Which alkene is more electron poor? Identify the diene and dieneophile in each reaction. What is the mechanism of a [4+2] cycloaddition reaction?

Solve

(a) The Diels–Alder reaction is favored when the dienophile is electron poor—namely, electron-withdrawing groups are on the dienophile. Therefore, the alkene with the OCH_3 group is more electron rich and will be less likely to react as the dienophile. The mechanism is given below:

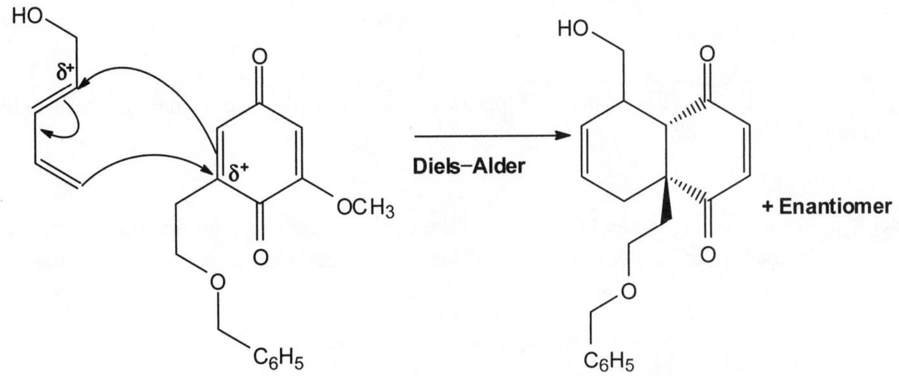

Diels–Alder

+ Enantiomer

(b) The diene is located in the middle of the structure and is already in the *s*-cis conformation as written. The dienophile is the Cl-substituted alkene, which has the substituents trans.

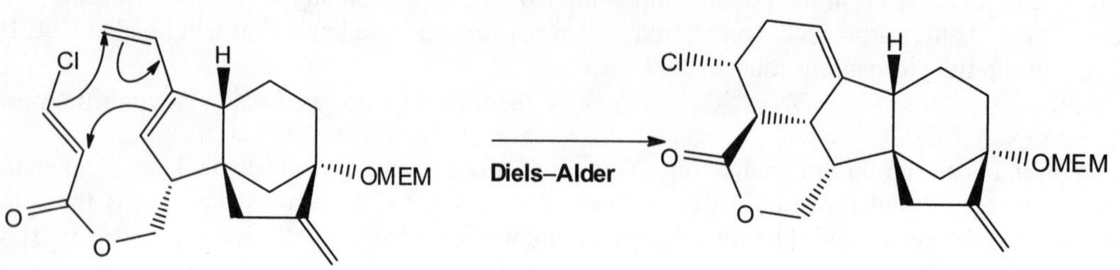

Diels–Alder

Problem 24.62

Think

Identify the diene and dienophile in each reaction. What is the mechanism of a [4+2] cycloaddition reaction?

Solve

The mechanisms are given below:

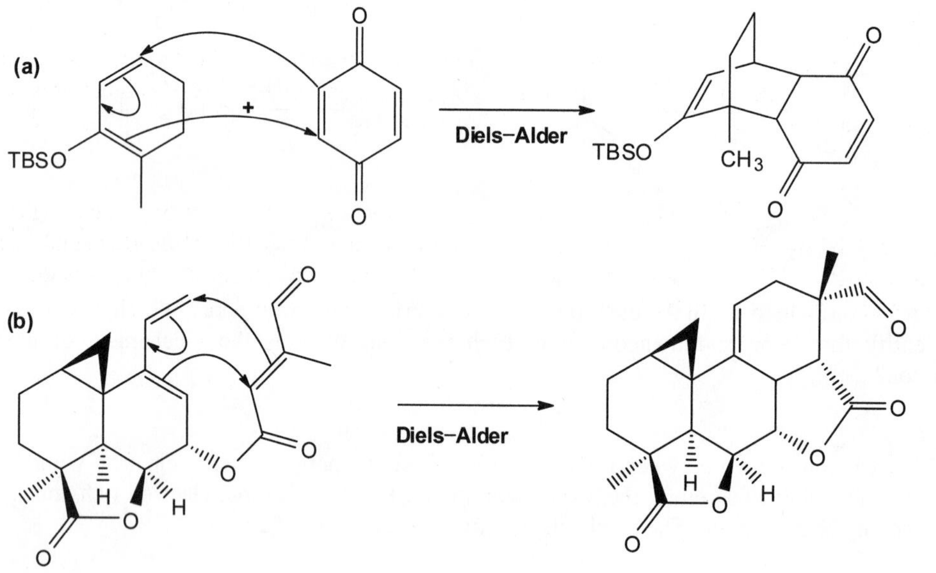

(a)

Diels–Alder

(b)

Diels–Alder

Problem 24.63

Think

What functional group has ^{1}H NMR signals around 7.3 ppm? If there is only one signal, is there a lot of symmetry? Is the starting material strained?

Solve

(a) and **(b)**. The product is benzene. To break the two bonds of the cyclobutane ring, the reaction must proceed through a transition state in which four π electrons are cyclically delocalized, making it an antiaromatic transition state.

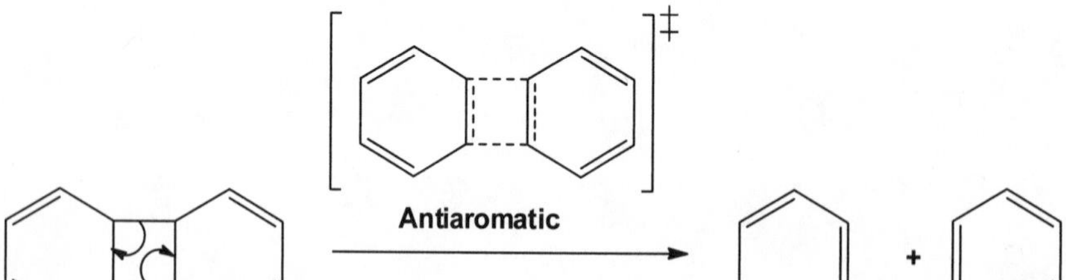

Antiaromatic

(c) The reaction relieves substantial ring strain and also produces two aromatic rings. This compensates for the high energy in the antiaromatic transition state.

Problem 24.64

Think

How many carbon atoms are in the reactant? Product? What functional group has ^{1}H NMR signals between 5–7 ppm and ^{13}C NMR signals between 120–140 ppm? Is there symmetry in the product?

Solve

The product only has four C atoms; therefore, a retro Diels–Alder reaction occurred. An alkene has ^{1}H NMR signals between 5–7 ppm and ^{13}C NMR signals between 120–140 ppm. The mechanism and product are given below:

700 °C

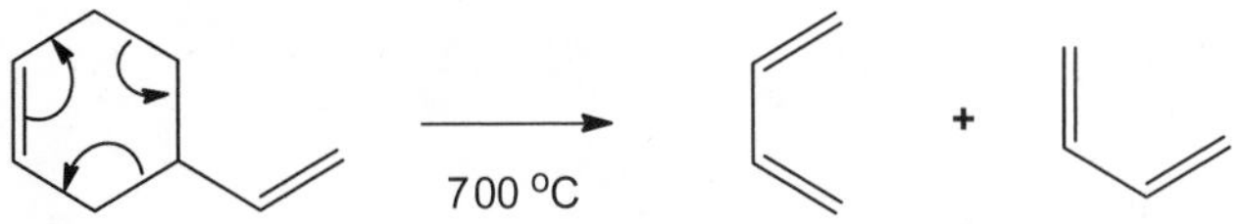

Problem 24.65

Think

How many carbon atoms are in the reactant? Product? What functional group has ^{1}H NMR signals between 7–9 ppm and ^{13}C NMR signals between 120–140 ppm? What functional group has a ^{13}C NMR signal around 190 ppm? Is there symmetry in the product?

Solve

The product only has 14 C atoms and the starting material has 16. It lost four C atoms from two equivalents of ethylene. Thus, a retro Diels–Alder reaction occurred on both outer rings. Aromatic ^{1}H NMR signals appear between 7–9 ppm and their ^{13}C NMR signals appear between 120–140 ppm. The signal at 190 ppm is due to the C=O. The mechanism and product are given below.

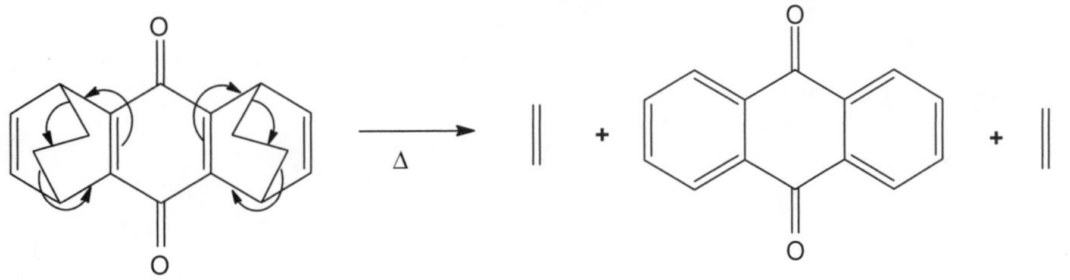

Problem 24.66

Think

What is the index of hydrogen deficiency (IHD)? What functional group reacts both with Br_2 and $KMnO_4/HO^-$? What functional group causes a broad absorption peak in the infrared spectrum (IR) from 2500–3300 cm^{-1} and a sharp peak at 1700 cm^{-1}? What information about functional group and symmetry is given in the NMR spectra?

Solve

The IHD is 2. The functional group that reacts with Br_2 and $KMnO_4/HO^-$ is an alkene. The IHD of 2 is likely due to a C=C and a ring. The IR suggests a carboxylic acid. The structure of the reactants and products are given below:

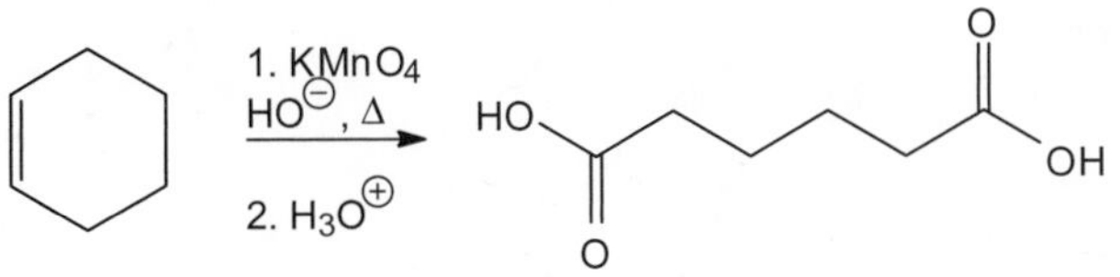

1. $KMnO_4$
$HO^{\ominus}$, Δ
2. $H_3O^{\oplus}$

Problem 24.67

Think

How does the formula of the product compare to the formula of the reactant? What molecule must have been lost? What is a reasonable [4+2] cycloaddition involving the reactant? What is a reasonable [4+2] cycloelimination involving the product of that reaction, which can result in the loss of N_2? What types of protons have NMR signals between 7–9 ppm? What types of protons could give rise to the 2H singlet at 4.56 ppm?

Solve

The signals between 7–9 ppm are aromatic protons and there are 14 of them. The 2H singlet at 5.46 ppm must be from the CH_2 group, which is substantially deshielded. This product has two fewer N atoms than the starting compound ($C_{23}H_{16}N_4O$), so the reaction resulted in the loss of N_2. This can be achieved in the sequence of reactions below, in which the C≡N bond serves as the dienophile in the [4+2] cycloaddition, and N_2 is ejected in the [4+2] cycloelimination.

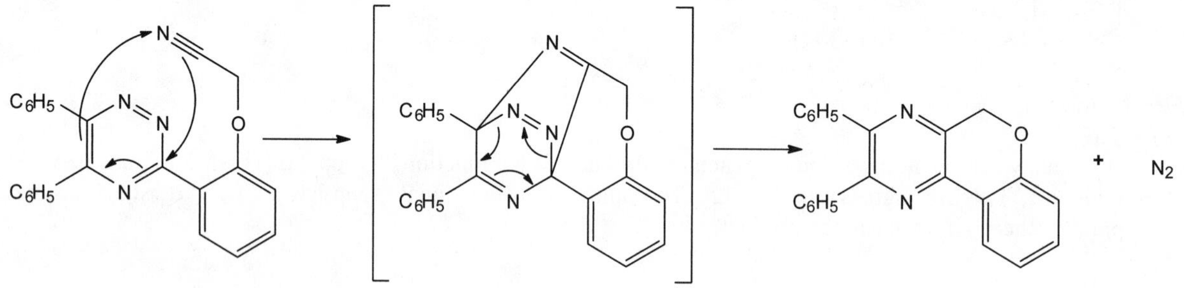

CHAPTER 25 | Reactions Involving Free Radicals

Your Turn Exercises

Your Turn 25.1

Think

In homolysis, what type of curved arrow is used (single or double barbed) to show bond breaking? How many electrons are in a bond? How many electrons go to each atom connected to the bond? How many curved arrows are required?

Solve

Homolysis uses the single-barbed curved arrow. Two electrons are in each bond and one electron goes to each atom in the bond, so two arrows are required.

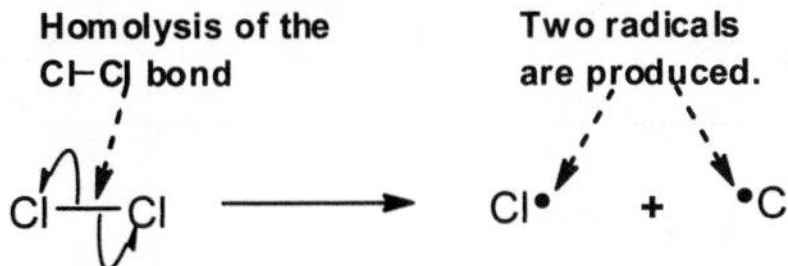

Your Turn 25.2

Think

Refer to Tables 25-1 and 25-3 to determine the bond energies of the bonds indicated. What does bond energy indicate about the strength of the bond?

Solve

Cl–Cl = 243 kJ/mol; Br–Br = 192 kJ/mol; I–I = 151 kJ/mol; HO–OH = 213 kJ/mol; H_3C–H = 435 kJ/mol. The larger the bond energy, the stronger the bond. Thus, the C–H bond is much stronger than any other bond in this list.

Your Turn 25.3

Think

Refer to Table 25-1 to determine the H–Br and H–Cl bond energies. Which bond is more difficult to break? Does that correspond to products that are higher in energy or lower?

Solve

The H–Br and H–Cl bond energies are 368 and 431 kJ/mol, respectively. The H–Cl bond is stronger and requires more energy to cleave homolytically. The product chlorine radical (•Cl), therefore, is higher in energy than the bromine radical (•Br).

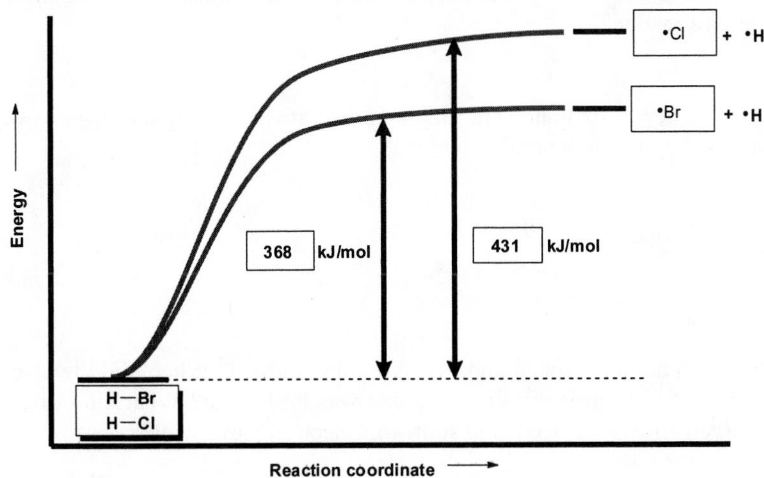

Your Turn 25.4

Think

What is the hybridization of each of the C atoms in the benzyl cation? Are there any atoms with an incomplete valence adjacent to π bonds? How do you use curved arrow notation to show the movement of a pair of electrons? How many times can that electron movement be carried out?

Solve

All C atoms in this structure are sp^2 hybridized. The carbocation C^+ has an incomplete valence and is adjacent to a π bond. One double-barbed arrow is used to show the movement of a pair of electrons to arrive at the next resonance structure. This can be repeated two more times to arrive at a total of five resonance structures.

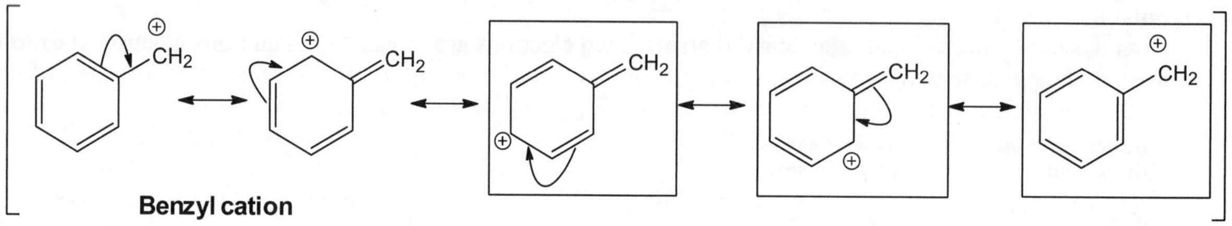

Benzyl cation

Your Turn 25.5

Think

Where does the unpaired electron end up in the final resonance structure? Do you need to use single-barbed curved arrows or double-barbed curved arrows to show that? Where do the electrons from the adjacent π bond end up? How many curved arrows are necessary to show that electron movement? Do those arrows point in the same direction?

Solve

The unpaired electron ends up as part of a π bond involving the same C atom, and one single-barbed arrow is needed to show that movement. The two electrons from the adjacent π bond move in opposite directions, requiring two single-barbed arrows. One of those electrons joins the unpaired electron, and the other moves to the CH_2 carbon to become unpaired.

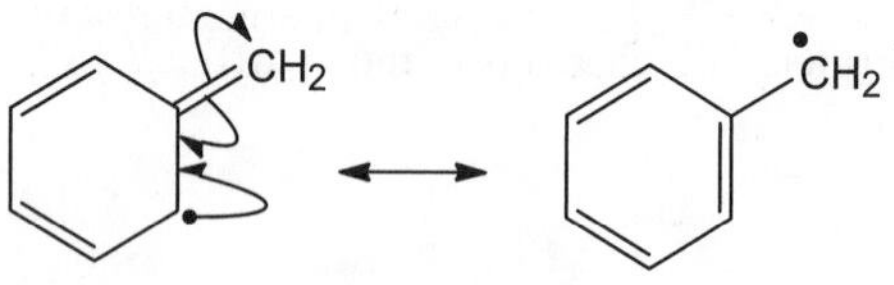

Your Turn 25.6

Think

Where do the unpaired electrons end up in the product? What kind of curved arrows must be used to show that movement? How many curved arrows are needed?

Solve

The two unpaired electrons join to form a new C–C bond. The single-barbed arrow is used and two arrows are needed. Each originates from a radical and points to the region where the new bond forms.

$$H_3C\cdot + \cdot CH_3 \longrightarrow H_3C-CH_3$$

Your Turn 25.7

Think

Refer to Equation 25-12 as a guide. Which bond is broken and which bond is formed? Where did the electrons from the new bond originate? Where did the electrons in the bond that was broken go? What kind of curved arrows should be used? How many curved arrows are needed to show this electron movement?

Solve

The C–H bond is broken and the H–Cl bond is formed. The electrons in the C–H bond are cleaved homolytically. One electron from the C–H bond goes to the C to form the radical and the other electron pairs with the Cl radical to form the H–Cl bond. Three single-barbed arrows are needed to show this electron movement.

$H_3C{-}CH_2{-}H + \bullet Cl \longrightarrow H_3C{-}\dot{C}H_2 + H{-}Cl$

Your Turn 25.8

Think

Refer to Equation 25-14 as a guide. Which bond is broken and which bond is formed? Where did the electrons from the new bond originate? Where did the electrons in the bond that was broken go? What kind of curved arrows should be used? How many curved arrows are needed to show this electron movement?

Solve

The C=C π bond is broken and the C–Cl bond is formed. The electrons in the C=C π bond are homolytically cleaved. One electron from the C=C π bond goes to the C to form the radical and the other electron joins with the Cl radical to form the C–Cl bond. Three single-barbed arrows are needed to show this electron movement.

Your Turn 25.9

Think

What does it mean to be an initiation step? What is the weakest bond present? What supplies the energy necessary to carry out the homolysis of that bond? Refer to Equation 25-18 as a guide.

Solve

An initiation step produces radicals from a closed-shell species. The UV light ($h\nu$) supplies the energy to break the weakest bond homolytically—in this case, the Br–Br bond. The products of the initiation step are two equivalents of Br•.

Initiation $Br{-}Br \xrightarrow{h\nu} Br\bullet + Br\bullet$

Your Turn 25.10

Think

What is the main difference between the reaction in Equation 25-19 and the one in Your Turn 25.9? What is meant by a propagation step? Are radicals present in the products, reactants, or both? How many propagation steps are necessary? Which bonds are broken and which bonds are formed?

Solve

The difference between this reaction and the one in Equation 25-19 is the presence of Br in place of Cl. A propagation step has radicals present in the reactants and products. The breaking of the C–H bond and formation of the C–Br bond requires two propagation steps (as shown on the next page). In the first step, a hydrogen atom is abstracted to form the carbon radical (•C). In the second step, the •C forms the C–Br bond, while the Br–Br bond breaks to form a new bromine radical (•Br).

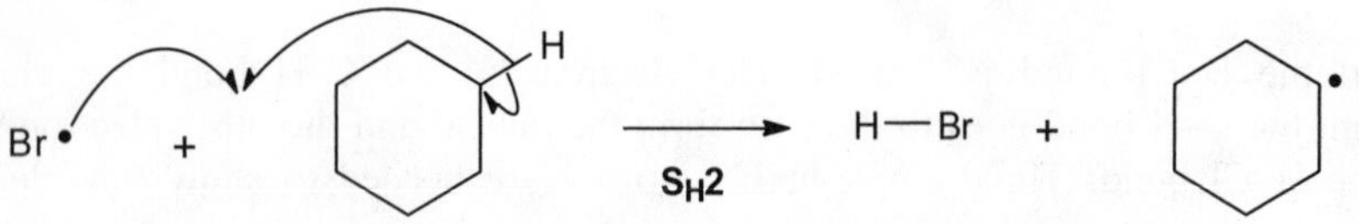

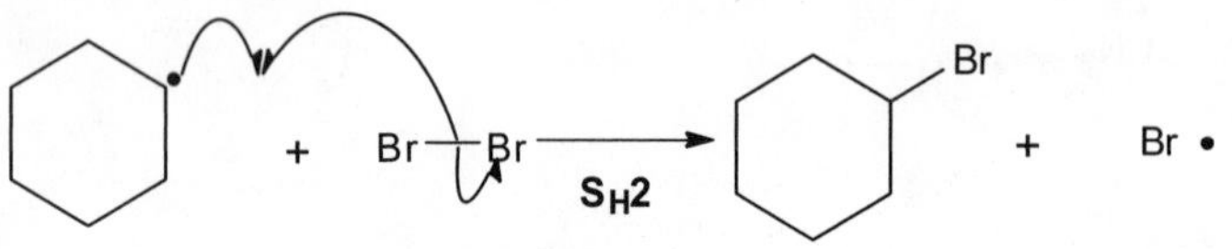

Your Turn 25.11

Think

What is meant by a termination step? Are radicals present in the products, reactants, or both? Use Equation 25-20 as a guide. Which bonds are formed?

Solve

A termination step decreases the number of free radicals. Two free radicals come together in the reactants and no radicals are present on the product side. The Br–Br bond or the C–Br bond is formed from the two radicals present in solution: •C and •Br.

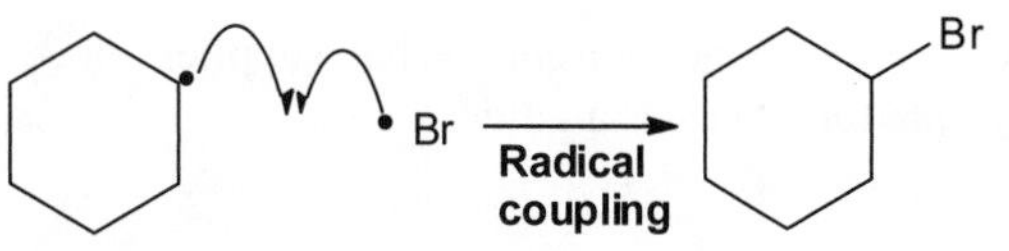

Your Turn 25.12

Think

What are the overall products and reactants for the free radical bromination of methane, CH_4? Which bond in the reactants is the weakest and can undergo homolysis via an initiation step, and what is the product of initiation? What are the products of the first propagation step and of the second propagation step (see the solution to Your Turn 25.10)?

Solve

The overall reaction is:

$$CH_4 + Br_2 \xrightarrow{h\nu} H_3C{-}Br + H{-}Br$$

The Br–Br bond is the weakest bond and undergoes homolysis in an initiation step to produce two •Br radicals. The •Br reacts with CH_3–H in the first propagation step to homolytically cleave the C–H bond and form the •C. The •C reacts with Br–Br in the second propagation step to produce CH_3–Br and another •Br. The CH_4/Br• and •CH_3/Br_2 species continue to react through this propagation cycle. See the figure on the next page.

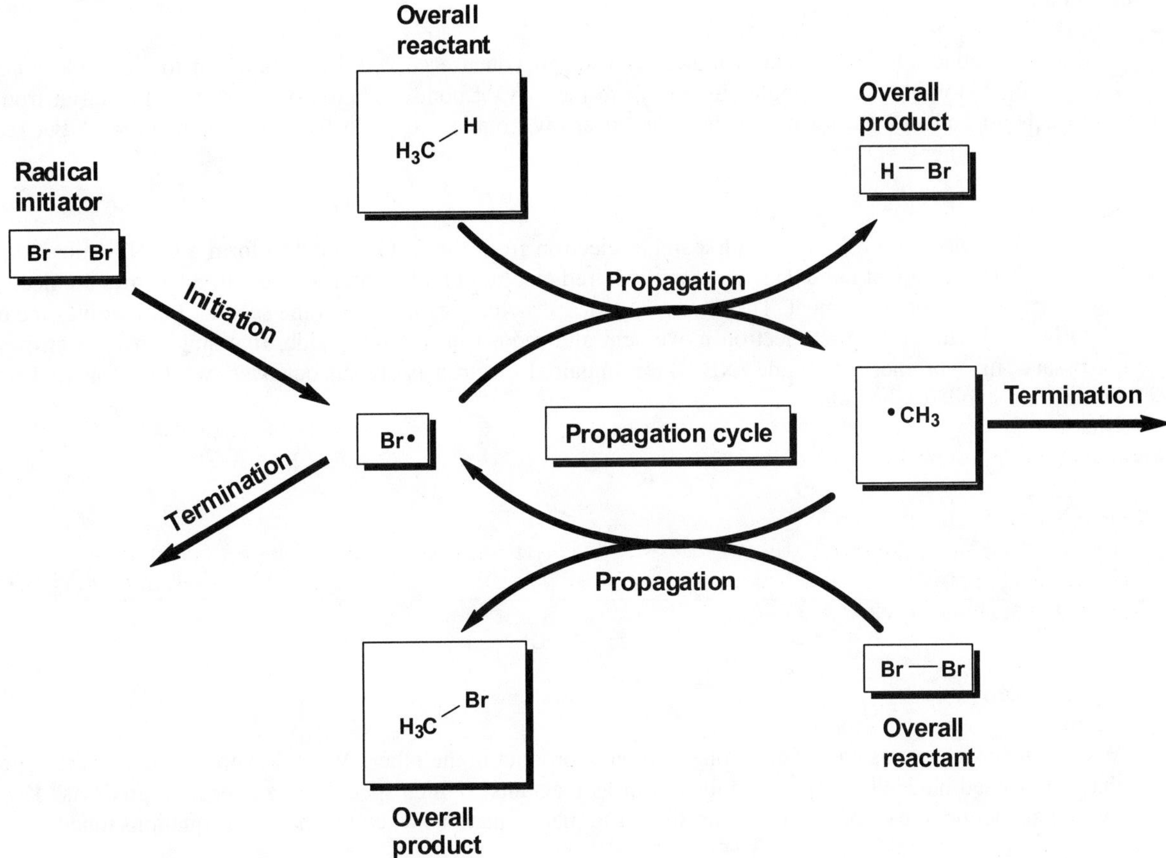

Your Turn 25.13

Think

Which radical, primary or primary allylic, is more stable? Why? How does the stability of the radical formed contribute to the rate of the reaction?

Solve

The allylic radical that is produced in (**b**) is more stable due to resonance delocalization of the unpaired electron, as shown below. The rate of free radical halogenation increases with increasing stability of the radical that is formed.

Resonance of allylic radical

Your Turn 25.14

Think

Refer to Table 25-1 to look up bond energies. Does a greater bond energy correspond to a stronger or weaker bond?

Solve

H–Br = 368 kJ/mol and H–Cl = 431 kJ/mol. With the lower bond energy, the HBr bond is weaker than the HCl bond.

Your Turn 25.15

Think

Is there a π bond adjacent to the N radical? How can you use curved arrow notation to show the unpaired electron on N joining with a single electron from the C=O π bond? Where does the second electron from the C=O π bond end up? How can you use curved arrow notation to undo this electron movement but redo it involving the other C=O π bond?

Solve

The unpaired electron on N joins with a single electron from the C=O π bond to form a C=N, while the other electron from the C=O π bond ends up as an unpaired electron on O•. Three single-barbed arrows are necessary to show this electron movement. The third resonance structure is identical to the second, but involves the other C=O bond. To undo the first electron movement and redo it on the other side, five single-barbed arrows are necessary. In *N*-bromosuccinimide (NBS), the unpaired electron is spread out over two O atoms and one N atom. This stabilizes the radical.

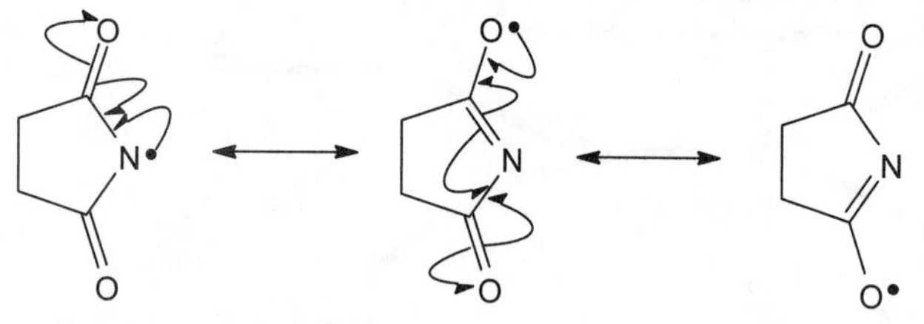

Your Turn 25.16

Think

Which species appear as a reactant in one step but a product in the other? What do you notice about the species that get crossed out? Which species only appear as reactants? Which species only appear as products? Keep in mind that chemical reaction equations are similar to math equations when you add two equations together.

Solve

The two species that are crossed out are both radicals. The species that remain as overall reactants are $H_2C{=}CHCH_3$ and HBr, and the one that remains as an overall product is $BrCH_2CH_2CH_3$. H–Br adds across the C=C to yield an alkyl bromide as a final product. The radicals are intermediates and do not appear in the overall chemical equation.

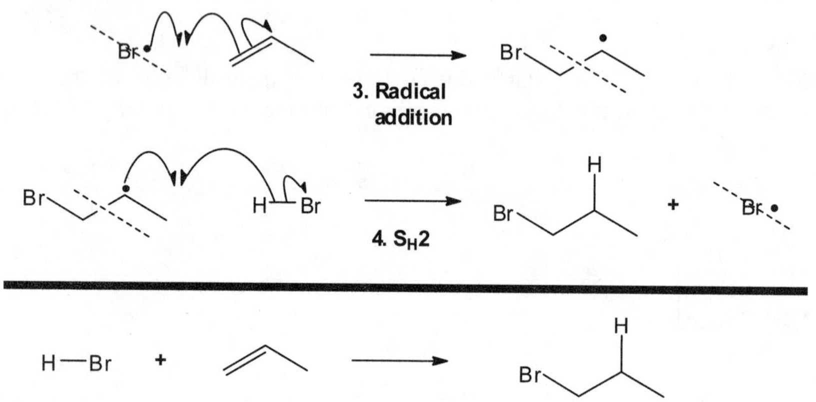

3. Radical addition

4. S_H2

Your Turn 25.17

Think

What radicals are involved in the propagation steps of the mechanism for Equation 25-32? What is meant by a termination step? For a termination step, are radicals present in the products, reactants, or both?

Solve

A termination reaction destroys free radicals that participate in the propagation cycle. Equation 25-33 shows that the radicals involved in the propagation cycle include Br• and the C• produced after radical addition. In a termination step, two of these free radicals come together as reactants and no radicals are present in the product. The Br–Br bond, C–C bond, or C–Br bond is formed from radical coupling of any of those radicals.

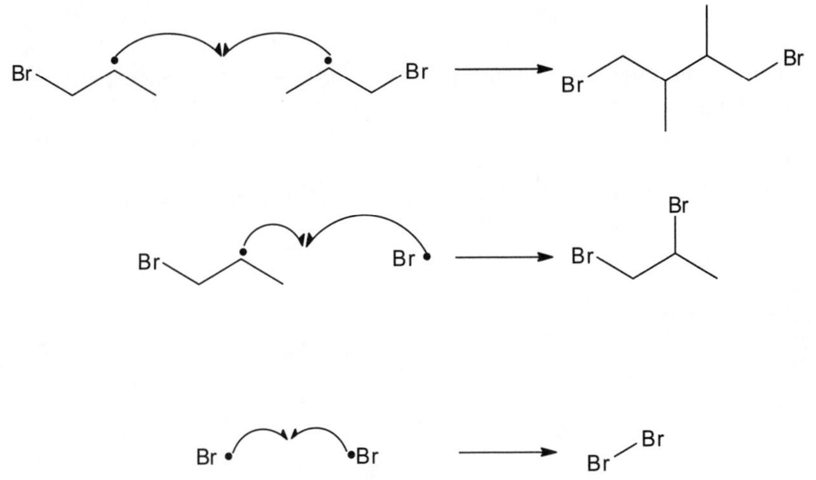

Br Br → Br Br

Br Br• → Br Br

Br• •Br → Br Br

Your Turn 25.18

Think

Refer to Section 19.5 to review catalytic hydrogenation. In catalytic hydrogenation of an alkyne, where do the H atoms add? What functional group is produced as a result? Which configuration, cis or trans, is formed? Why?

Solve

Catalytic hydrogenation involves $H_2(g)$ on a metal surface, and the $H_2(g)$ adds syn to the π bond. This converts an alkyne to an alkene, and the cis configuration of the alkene is formed exclusively. The poisoned catalyst helps ensure that the reduction stops at the alkene stage.

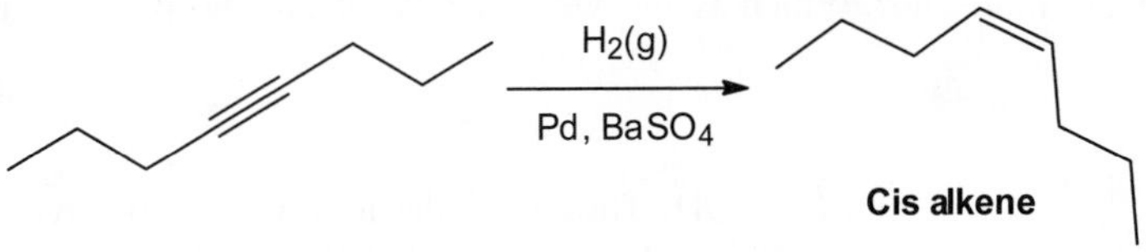

$H_2(g)$ / Pd, $BaSO_4$ →

Cis alkene

Chapter Problems

Problem 25.1

Think

In homolysis, what type of curved arrow is used (single or double barbed) to show bond breaking? How many electrons are in a bond? How many electrons go to each atom connected to the bond? How many curved arrows are required?

Solve

Homolysis uses the single-barbed arrow. Two electrons are in each bond, and one electron goes to each atom in the bond, so two curved arrows are required.

a) $H_3C-CH_3 \longrightarrow H_3C\cdot + \cdot CH_3$

b) $H_3C-CH_2-H \longrightarrow H_3C-\dot{C}H_2 + \dot{H}$

c) $H_3C-CH(Br)-CH_3 \longrightarrow H_3C-\dot{C}H-CH_3 + \cdot Br$

d) $C_6H_5-H \longrightarrow C_6H_5\cdot + \dot{H}$

Problem 25.3

Think

What are the distinct types of bonds in the molecule? Which is the weakest? What are the products upon homolysis of the bond?

Solve

There are three distinct bonds: H–OCH_3, H–CH_2OH, and H_3C–OH. They have the following respective bond dissociation energies: 427 kJ/mol (102 kcal/mol), 385 kJ/mol (92 kcal/mol), and 383 kJ/mol (92 kcal/mol). The weakest bond is the C–O bond, so it is most likely to break in a homoloysis reaction. (Because of their similar strengths, however, the C–H and C–O bonds will have similar likelihoods of undergoing homolysis.)

385 kJ/mol (C–H), 427 kJ/mol (O–H), 383 kJ/mol (C–O)

$H_2C(H)-O-H \longrightarrow H_3C\cdot + \cdot OH$

Problem 25.4

Think

What is the structure of 1,3,5-trimethylcyclohexane? How many distinct H–C bonds are present? What is the type of carbon (methyl, 1°, 2°, 3°)? Which type of C atom stabilizes an unpaired electron the most? The least? Why?

Solve

There are three chemically distinct H atoms that can be removed. Radical stability decreases in the order 3° > 2° > 1°. This is because a C• is electron poor and each additional alkyl group is electron donating.

Primary Tertiary Secondary

Primary
Least stable radical

Secondary

Tertiary
Most stable radical

Problem 25.5

Think

Use Figure 25-3 and Your Turn 25.3 as a guide. What is the identity of the ethyl radical and the allyl radical? Which bond dissociation energy is lower? Why?

Solve

Bond dissociation energies are from Table 25-1: ethyl radical = 410 kJ/mol (104 kcal/mol) and allyl radical = 356 kJ/mol (85 kcal/mol). Because the allyl radical requires less energy to be formed, it is more stable than the primary ethyl radical. The allyl radical is more stable due to the presence of resonance structures that delocalize the unpaired electron over two C atoms.

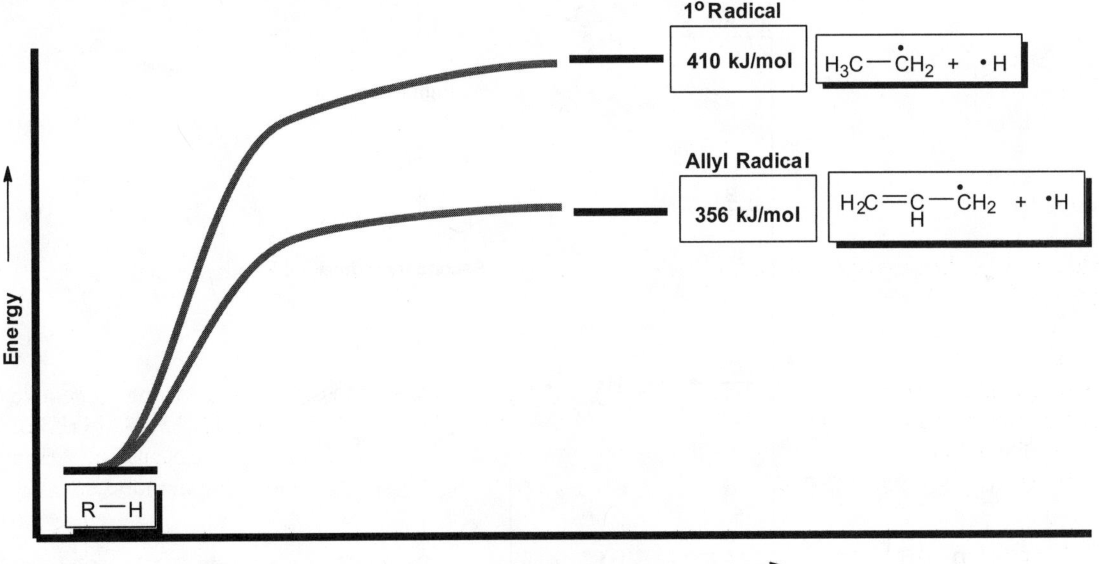

Problem 25.6

Think

Which carbon in Figure 25-5 has the most electron spin density (deepest blue)? Examine the resonance structures for the benzyl radical. What is special about the resonance structure in which the unpaired electron is on that carbon?

Solve

The primary (CH_2) carbon possesses the most spin density (it has the deepest blue color in Fig. 25-5). Examination of the resonance contributors (Equation 25-10) shows that the first radical retains the aromatic character of the benzene ring, while the others do not. This contributor is more stable and contributes more to the resonance hybrid.

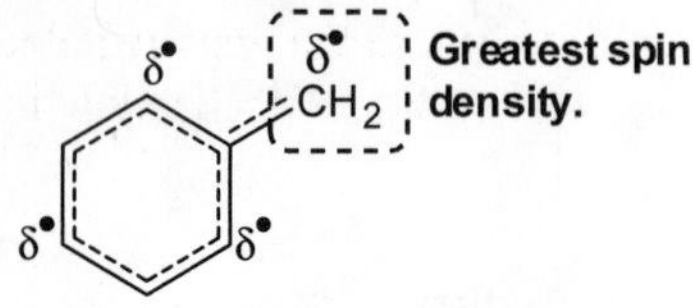

$\delta^\bullet$ CH_2 Greatest spin density.

Problem 25.8

Think

What are the distinct types of C–H bonds? What are the products upon homolysis of each of those bonds? Can the stabilities of the product radicals be differentiated based on resonance and/or alkyl substitution?

Solve

There are six distinct C–H single bonds, labeled **A–F** below. The corresponding homolysis products are shown on the right. Only one C–H bond can be broken to produce a resonance-stabilized alkyl radical. Thus, homolysis of bond C–H **D** gives the most stable product organic radical, making bond C–H **D** the weakest of the C–H bonds.

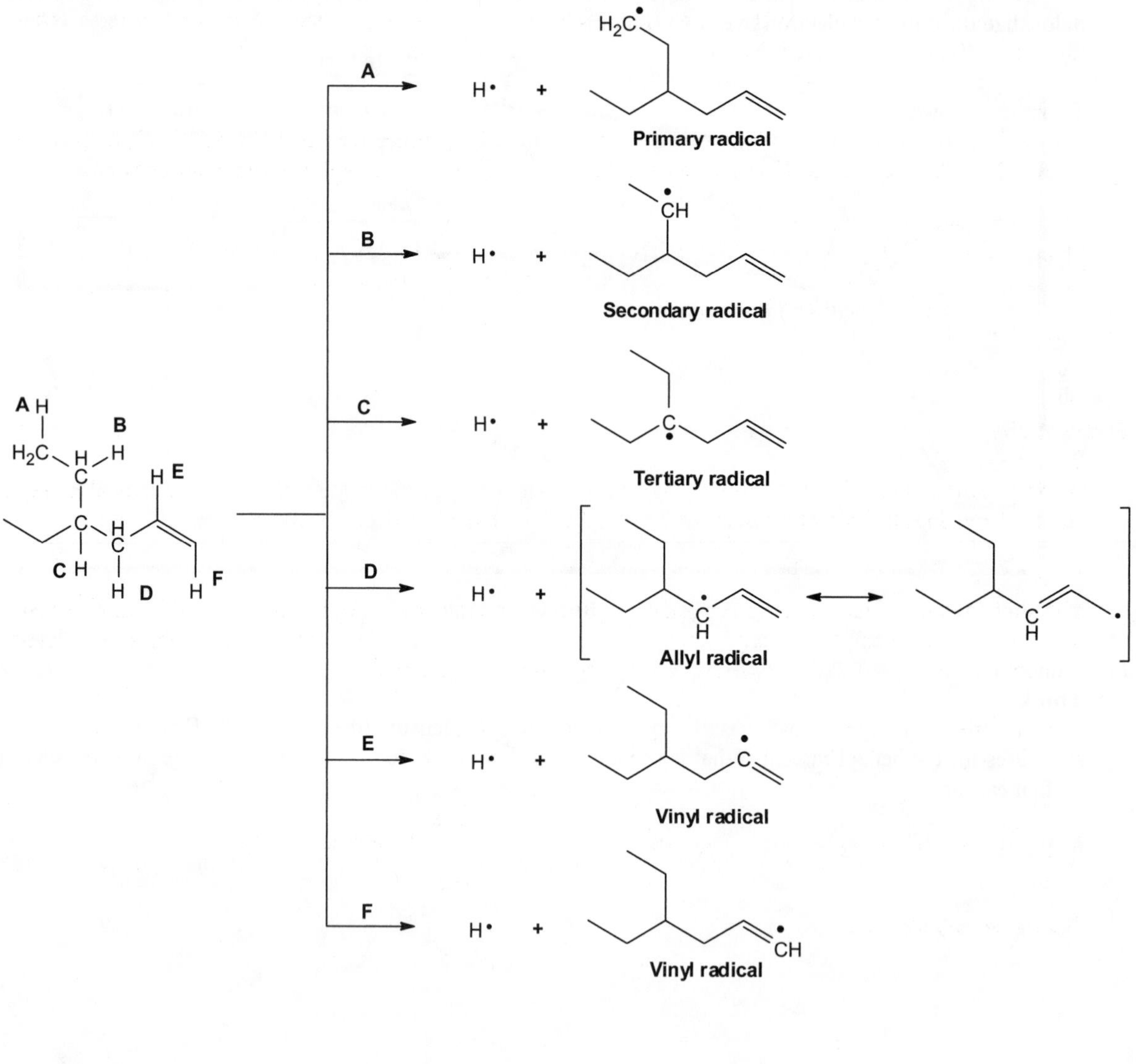

A → H• + Primary radical

B → H• + Secondary radical

C → H• + Tertiary radical

D → H• + Allyl radical

E → H• + Vinyl radical

F → H• + Vinyl radical

Problem 25.9

Think

What are the distinct types of C–H bonds? What are the products upon homolysis of each of those bonds? Can the stabilities of the product radicals be differentiated based on resonance and/or alkyl substitution? Which allyl radical has more resonance structures?

Solve

The allylic hydrogens circled below are the weakest in their respective molecules because the unpaired electron of the resulting radical can be delocalized by resonance. There is greater resonance delocalization in the resulting radical of the second molecule, so its C–H bond energy is lower.

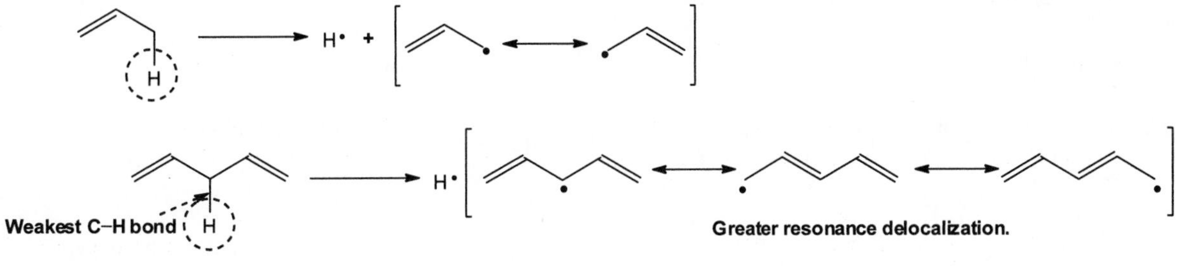

H• +

H•

Weakest C–H bond H

Greater resonance delocalization.

Problem 25.10

Think

What bond in cyclohexene is the weakest? In HBr? In H_2O_2? What is the product for each homolysis reaction? Which bond is the weakest of the three molecules? How does bond strength correspond to the likelihood of undergoing a homolysis reaction?

Solve

The weakest bond in each molecule is indicated below. The respective bond dissociation energies are from Tables 25-1, 25-3, and 25-3. H_2O_2 has the weakest of those bonds, so it will dissociate most easily.

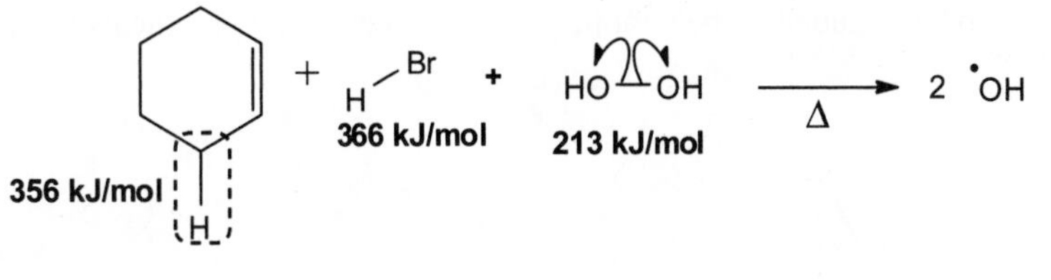

356 kJ/mol + H–Br 366 kJ/mol + HO–OH 213 kJ/mol →(Δ) 2 •OH

Problem 25.11

Think

Is the unpaired electron able to be delocalized over the two Ph rings? Does this make the radical more or less stable? How does this affect the extent to which the closed-shell product is favored?

Solve

The diphenylamino radical is exceptionally stable for a radical species because of the extensive resonance delocalization over the two phenyl rings. This can be seen in the hybrid drawn below. Thus, the closed-shell product is less favored in this reaction, making the reaction reversible.

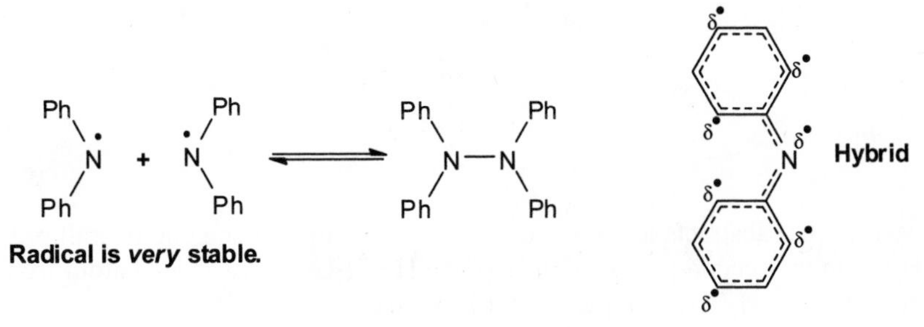

$Ph_2N\cdot + \cdot NPh_2 \rightleftharpoons Ph_2N{-}NPh_2$

Radical is *very* stable.

Hybrid

Problem 25.12

Think

Which atom(s) can be attacked by Cl•? Which bond(s) can be broken?

Solve

We can think of the Cl• attacking either an H atom or a C atom. If it attacks an H atom, then an H atom is abstracted and the ethyl radical is displaced. If it attacks a C atom, either H is displaced or CH_3 is displaced.

$H_3C-CH_2(H)-H$ + •Cl ⟶ $H_3C-\dot{C}H_2$ + H—Cl

H_3C-CH_3 + •Cl ⟶ •CH_3 + H_3C-Cl

H_3C-CH_2-H + •Cl ⟶ H_3C-CH_2Cl + •H

Problem 25.13

Think

What radicals are possible in Equation 25-17? What radicals couple in the termination steps shown in Equations 25-20a and 25-20b? Which radical coupling is left?

Solve

There are two radicals that are part of the propagation cycle: the Cl• and the cyclohexyl radical. In Equation 25-20a, the two Cl• radicals couple, and in Equation 25-20b, the Cl• radical combines with the cyclohexyl radical. The radical coupling reaction that has yet to be included is the coupling of the two cyclohexyl radicals.

Radical coupling

Problem 25.15

Think

For the initiation step, what is the weakest bond in the reactants that can undergo homolysis? For the propagation steps, what S_H2 step can the initial radical undergo to produce the HCl product? What S_H2 step can the resulting radical undergo to produce the CH_3CH_2Cl product and regenerate the initial radical? For the termination steps, what possible radical coupling steps can take place?

Solve

The weakest bond present is the Cl–Cl bond, which undergoes homolysis in an initiation step to produce two Cl• radicals.

Initiation:

Cl—Cl $\xrightarrow{h\nu}$ Cl• + •Cl

In the first of two propagation steps, a Cl• abstracts a hydrogen from CH_3CH_3 to produce one overall product, HCl, and an ethyl radical, H_3CH_2C•. In the second propagation step, the H_3CH_2C• abstracts a Cl atom from Cl_2 to produce the second overall product, CH_3CH_2Cl, and regenerate the initial Cl•.

Propagation:

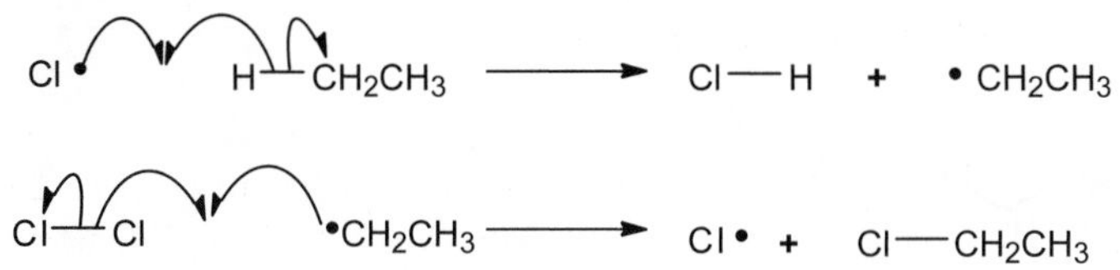

$$Cl\bullet + H-CH_2CH_3 \longrightarrow Cl-H + \bullet CH_2CH_3$$

$$Cl-Cl + \bullet CH_2CH_3 \longrightarrow Cl\bullet + Cl-CH_2CH_3$$

Termination steps can be a radical coupling between any two radicals that appear, such as the following two:

Termination:

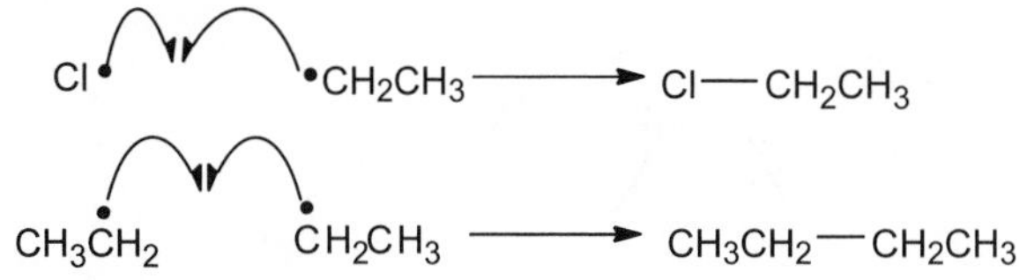

$$Cl\bullet + \bullet CH_2CH_3 \longrightarrow Cl-CH_2CH_3$$

$$CH_3\dot{C}H_2 + \dot{C}H_2CH_3 \longrightarrow CH_3CH_2-CH_2CH_3$$

Problem 25.17

Think

What is the initial radical that is formed? By what process? How will that radical interact with the uncharged molecules present? Is regiochemistry a concern? Which hydrogen can be abstracted to produce the most stable alkyl radical?

Solve

Br_2 will undergo homolysis, as suggested by the presence of light, so the initial radical is Br•.

Initiation:

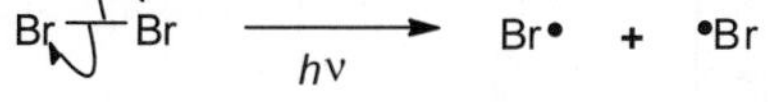

$$Br-Br \xrightarrow{h\nu} Br\bullet + \bullet Br$$

With the benzyl compound present, radical halogenation will occur via two propagation steps. The Br• abstracts a H atom to yield a benzyl radical, R•. That benzyl radical then abstracts a Br atom from Br_2 to yield $PhCH(Br)CH_2CH_3$ and Br•.

Propagation:

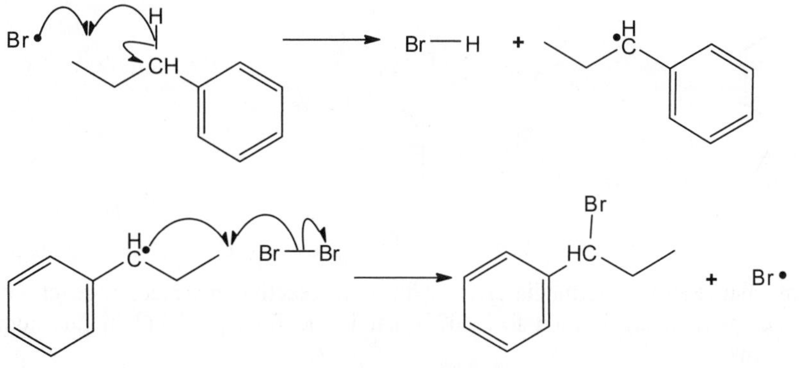

$$Br\bullet + PhCH(H)CH_2CH_3 \longrightarrow Br-H + Ph\dot{C}HCH_2CH_3$$

$$Ph\dot{C}HCH_2CH_3 + Br-Br \longrightarrow PhCH(Br)CH_2CH_3 + Br\bullet$$

Regiochemistry is a concern because there are multiple, distinct H atoms that can be abstracted by Br•. Recall, however, that Br• is highly selective in hydrogen abstractions, favoring the production of the most stable alkyl radical. In this case, the most stable alkyl radical is produced from abstracting a hydrogen atom attached to the benzylic carbon, as shown above.

Termination steps can be a radical coupling between any two radicals that appear, such as the following two:

Termination:

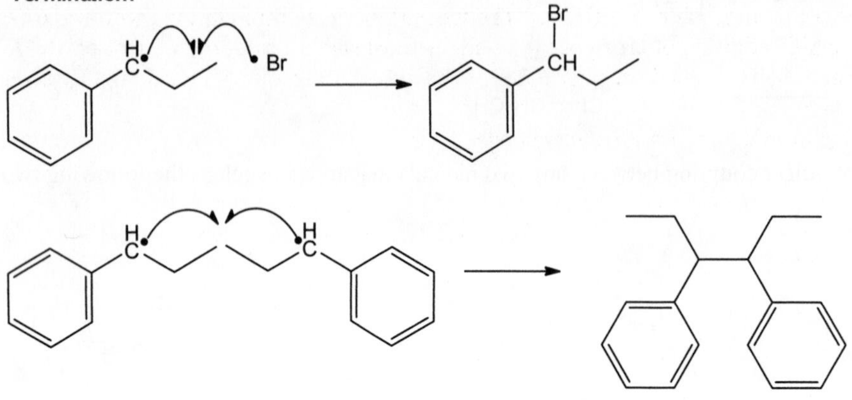

Br
CH
Br

Problem 25.18

Think

What type of reagent is NBS? Which C–H bond is most likely to undergo homolysis? Why? What intermediate is formed? Is resonance possible?

Solve

NBS is a source of a small, steady concentration of Br_2, which produces some Br• upon irradiation with light. The tertiary allylic C–H bond will undergo hydrogen abstration to form the most stable allyl radical. The allyl radical has another resonance structure, and each can abstract a Br atom from Br_2 to form the final product, as shown below.

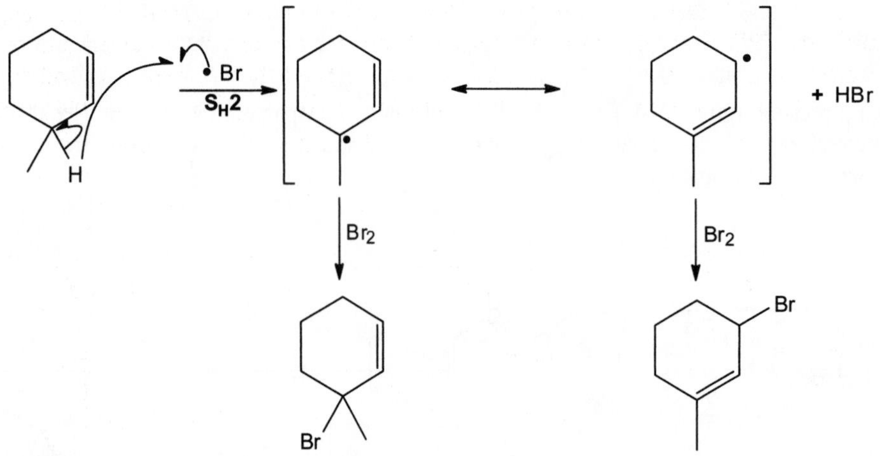

Br
S_H2
+ HBr
H
Br_2
Br_2
Br
Br

Problem 25.19

Think

HBr and peroxides react to form what reactive intermediate? How does this reactive intermediate react with an alkene C=C bond? What organic intermediate is then formed? What is the final product? In this addition reaction, is regiochemistry a concern?

Solve

HBr and peroxides are a source of Br• radical, which reacts with an alkene via radical addition to form the most stable C• intermediate. The C• intermediate reacts with an equivalent of H–Br to form a new C–Br bond. This leads to anti-Markovnikov addition of HBr to the C=C. The mechanism goes through the most stable radical intermediate, just as electrophilic addition of HBr goes through the most stable carbocation intermediate. In this reaction, the Br• adds first and the H• second. In electrophilic addition of an alkene with H–Br, the H^+ adds first and the Br^- adds second.

(a) The major product came from the tertiary C• intermediate.

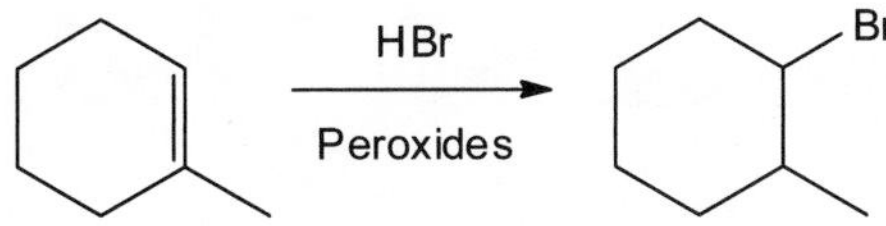

HBr
Peroxides
Br

(b) The major product came from the benzylic C• intermediate.

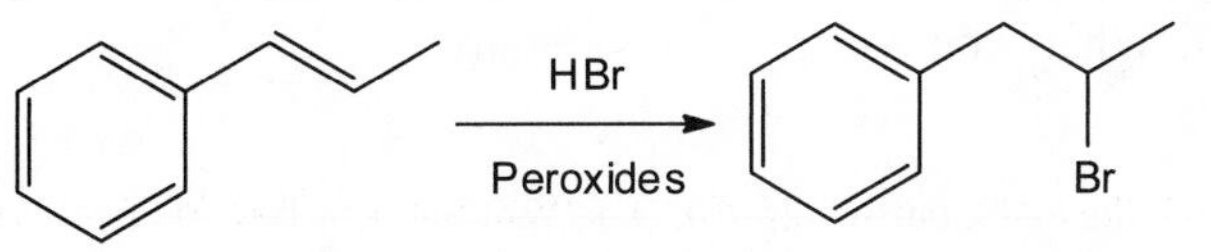

HBr
Peroxides
Br

Problem 25.20

Think

Are there any tetrahedral stereocenters in the organic reactant? How many tetrahedral stereocenters are in the organic product? Are those stereocenters formed in the same elementary step or in different steps? When each one is formed, is one configuration produced or is a mixture of configurations produced?

Solve

(a) An organic radical intermediate is formed from radical addition of Br• to the C=C bond. In that step, the carbon that gains the new C–Br bond becomes a tetrahedral stereocenter and both configurations are produced. The C• radical then reacts with H–Br to form a new C–H bond. In that step, a second carbon becomes a tetrahedral stereocenter, and both of its configurations are produced. This leads to four possible stereoisomer products.

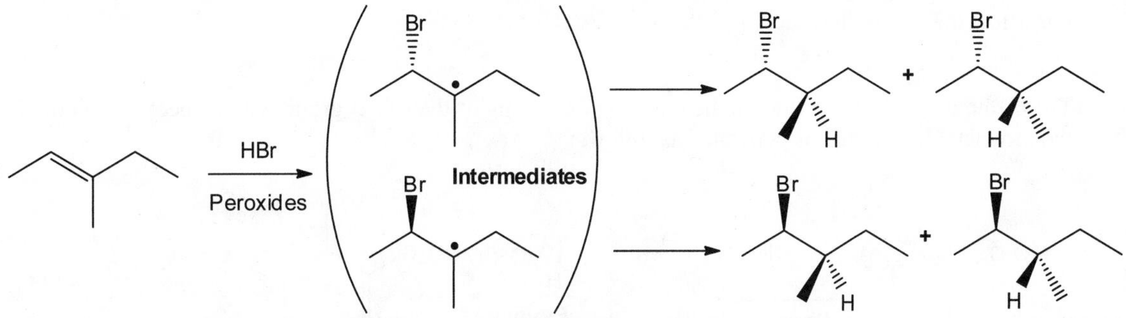

HBr
Peroxides
Br
Intermediates
Br
Br
H
+
Br
H
Br
H
+
Br
H

(b) An organic radical intermediate is formed when Br• abstracts a hydrogen to form the secondary allylic C•. This intermediate has two resonance structures, but due to symmetry, they both lead to the same enantiomeric products. When the C• gains a Br from Br_2, that C becomes a tetrahedral stereocenter and both configurations are produced.

H
Br_2
$h\nu$
Br
+
Br

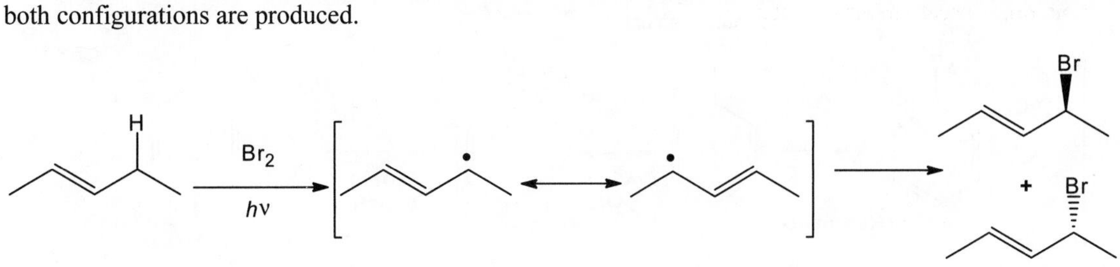

Problem 25.21

Think

Is the cis or trans alkene produced from the reaction of a dissolving metal reduction with an alkyne? Which alkene, therefore, was the C≡C?

Solve

The trans alkene is the one that is produced from dissolving metal reduction of an alkyne, so that was the C≡C in the starting compound.

Li(*s*)

$NH_3(l)$, –78 °C

Problem 25.22

Think

How is this mechanism similar to the one in Equation 25-40? How does the radical addition reaction occur? How many H atoms are added across the C=O? What is the source of the H atoms?

Solve

This mechanism is essentially identical to the one in Equation 25-40. The mechanism begins with a solvated electron adding via radical addition to the C=O. This converts the C=O to a C• and O^-. A proton transfer reaction follows to form the O–H bond. The C• couples with another solvated electron to form the C:$^-$ carbanion, which is a strong base. Another proton transfer reaction results.

H_3C H_3C C=O: e⊖ Radical addition H_3C H_3C C•–O:⊖ H–OR Proton transfer H_3C H_3C C•–OH e⊖ Radical coupling H_3C H_3C ⊖C–OH H–OR Proton transfer H_3C H_3C CH–OH

Problem 25.23

Think

How is this mechanism similar to the one in Equation 25-45? In cyclohexa-1,4-diene, where can the CO_2H group end up relative to the two C=C bonds in the product? Is CO_2H an electron-donating or electron-withdrawing group? How does this impact the stability of the anionic intermediate that is produced? How does this affect the rate of the reaction?

Solve

The different products depend on the relative positioning of the CO_2H group with respect to the double bonds in the product. One product is generated as follows:

HO_2C e⊖ Radical addition HO_2C H–OR Proton transfer HO_2C H e⊖ Radical coupling

Most stable resonance structure

[HO_2C H ⊖ ⟷ HO_2C H ⊖ ⟷ HO_2C H ⊖] Resonance H–OR Proton transfer HO_2C H H

The other product is generated as follows:

The second product should be the major product because the negative charge produced on C in the first step is stabilized by the attached electron-withdrawing CO_2H group. For this reason, the presence of a CO_2H group should also *speed up* the reaction compared to unsubstituted benzene.

Problem 25.24

Think

Which acid, ROH or NH_3, is stronger? In each reduction mechanism, in which step is ROH or NH_3 deprotonated? In that step, which mechanism has the stronger base?

Solve

An alcohol (ROH) with a pK_a of ~16 is a much stronger acid than ammonia with a pK_a of ~36 (see Chapter 6 for pK_a values). When an alkyne is reduced, the species that deprotonates the proton source is a carbanion (R:$^-$) in which the negative charge is localized on an sp^2-hybridized C. That carbanion is a strong enough base to deprotonate NH_3. In a Birch reduction, the carbanion has a resonance-stabilized negative charge, so it is more stable and thus a weaker base. Therefore, it requires a stronger acid, like ROH.

Problem 25.25

Think

Which C–H bond is most likely to undergo radical abstraction? Which C• would be most stable? Where does the C–Br bond form? How many times does this substitution occur? What functional group is present in the product that is not present in the reactant? How is that functional group produced from the synthetic intermediate in the presence of a base?

Solve

(a) Bromination occurs at two benzylic positions. The following are possible isomers that can form:

(b) Both isomers undergo E2 or E1 reactions twice to form two new C=C bonds. The amine serves as the base for such reactions. The resulting product has a new naphthalene ring.

Problem 25.27

Think

What bonds are formed and broken in this S_H2 mechanism? What kinds of curved arrows should be used? How many curved arrows are needed? How is this similar to an S_N2 mechanism?

Solve

The Cl• radical forms a new C–Cl bond with the less substituted C atom in the most electron-rich (i.e., most alkyl substituents) C–C bond in the ring. The C• then results on the tertiary carbon. The dynamics of an S_H2 reaction are similar to an S_N2 but the reaction arrows differ.

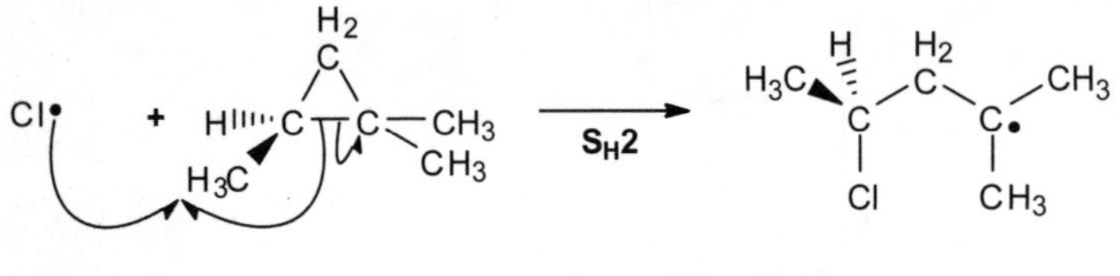

Problem 25.28

Think

Identify which bonds are broken and formed in this rearrangement reaction. On which C atom does the radical end up? How is the new C=C bond formed? What kinds of curved arrows should be used? How many of them are needed?

Solve

The mechanism for the McLafferty rearrangement is given below. The mechanism goes through a six-atom ring conformation where the radical from the O• forms a new O–H bond, thereby homolytically cleaving the H–C bond and the adjacent C–C bond.

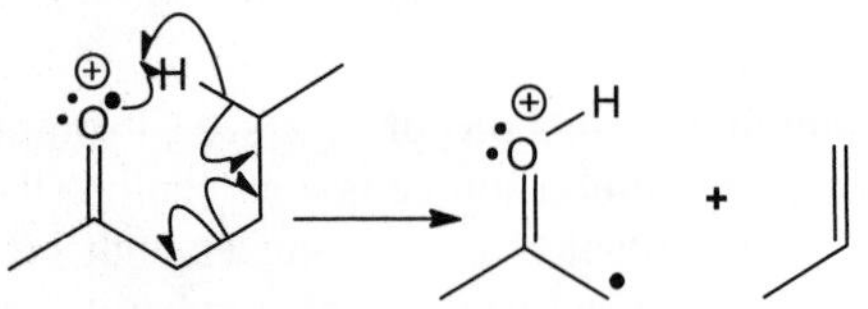

Problem 25.29

Think

Is there a π bond adjacent to the C• radical? If so, to which atom is the unpaired electron moved? Where does the π bond move? How can the π bond and the unpaired electron be shown to move to the new locations? What kinds of curved arrows are needed to show this electron movement?

Solve

There is a π bond adjacent to the C• radical. The resonance structure therefore has a C=C bond and the radical then ends up on the O atom. Three single-barbed curved arrows are necessary to show this electron movement.

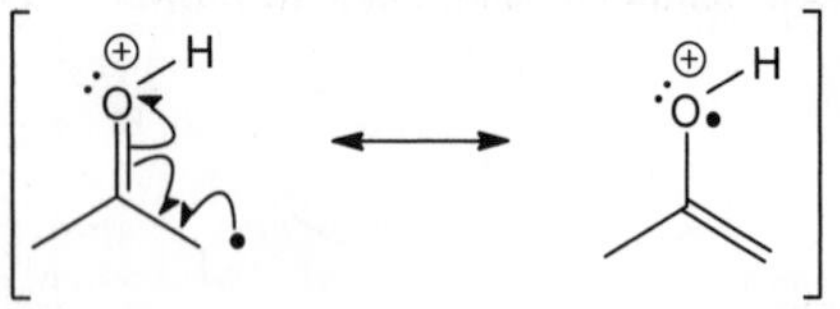

Problem 25.30

Think

What are the products of the various H atom abstractions? What types of carbon atoms are present (methyl, 1°, 2°, 3°, allylic, benzylic)? Which C• would be most stable? How does that affect the likelihood of the H atom being abstracted?

Solve

The most reactive H in each molecule, shown below, is the one with the weakest bond and leads to the formation of the most stable C•. In **(a)** and **(b)**, the H is on a 3° carbon, which forms a 3° carbon radical. A 3° radical is more stable than a 1° or 2° radical. In **(c)**, the H is on a 2° benzylic C. A benzylic radical has the unpaired electron delocalized around the ring, making it more stable than a localized radical. Notice that another benzylic radical could form by abstracting a hydrogen from the benzylic CH_3, but the resulting radical would be less stable because the radical would be 1°. And in **(d)**, the H is on an allylic C, so the resulting radical would be resonance stabilized.

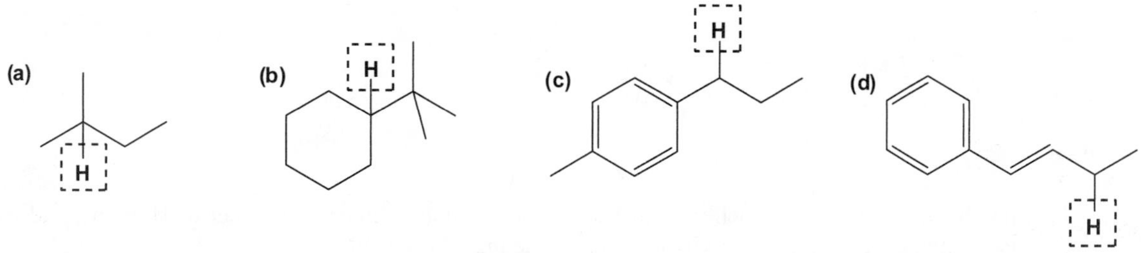

(a) H (b) H (c) H (d) H

Problem 25.31

Think

What is the index of hydrogen deficiency (IHD) of the compound? How many distinct types of reactive C–H bonds are present if only one monochloride isomer is formed? Is there symmetry in the molecule?

Solve

The IHD is 4, so a benzene ring is possible, which would account for six C atoms. Aromatic H atoms are typically not reactive under these conditions, but benzylic H atoms are. All H atoms that are sufficiently reactive must be equivalent, so the remaining three C atoms can be benzylic CH_3 groups, and they must be attached at the 1, 3, and 5 positions.

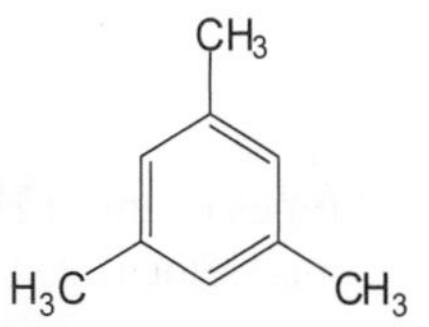

CH_3 H_3C CH_3

Problem 25.32

Think

Is a resonance structure possible for the allyl radical ($H_2C{=}CH{-}CH_2{\bullet}$)? What is the structure of the hybrid? Is the single bond 100% single? What does this mean for the rotation barrier?

Solve

The $CH_3CH_2{-}CH_3$ bond in propane has free rotation because all C atoms are sp^3 hybridized and all bonds are single bonds. The $H_2C{=}CH{-}CH_2{\bullet}$ bond in the allyl radical is not 100% a single bond, due to the presence of resonance. The hybrid shows partial double-bond character and thus has limited rotation.

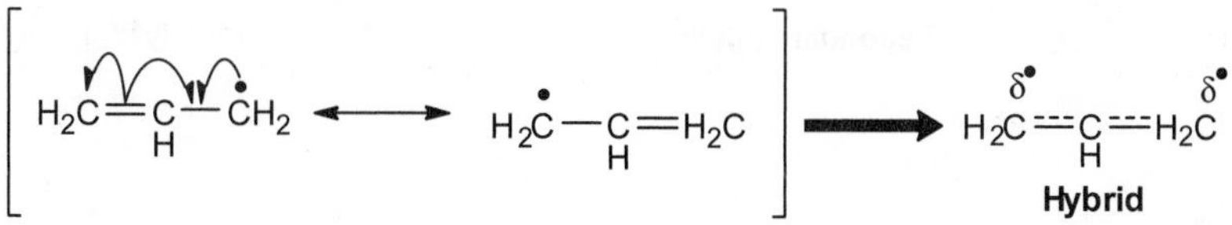

$$\left[H_2C{=}\underset{H}{C}{-}\dot{C}H_2 \longleftrightarrow H_2\dot{C}{-}\underset{H}{C}{=}H_2C \right] \longrightarrow \overset{\delta\bullet}{H_2C}{\cdots}\underset{H}{C}{\cdots}\overset{\delta\bullet}{H_2C}$$

Hybrid

Problem 25.33

Think

What types of carbon atoms are present (methyl, 1°, 2°, 3°, benzylic)? How many of each type of C–H bond are present? How does the number of C–H bonds affect the actual product distribution? How does the 5:1 selectivity affect the actual product distribution?

Solve

(a) and **(b)** The radical and curved arrow notation are given below for the two isomeric propyl radicals. One radical comes from the 1° C atom and the other comes from the 2° C atom.

1° C–H Bond

2° C–H Bond

(c) The likelihood of abstracting each type of H increases with an increasing number of H atoms, and it also increases with an increasing selectivity toward abstracting that type of H.

Relative probability of abstracting a 1° H = six 1° H atoms × reactivity of 1 = 6
Relative probability of abstracting a 2° H = two 2° H atoms × reactivity of 5 = 10

Total = 16

Percentage of abstraction of 1° H: (6/16) × 100% = 37.5%
Percentage of abstraction of 2° H: (10/16) × 100% = 62.5%

Problem 25.34

Think

What types of carbon atoms are present (methyl, 1°, 2°, 3°, allylic)? How many of each type of C–H bond are present? How does the number of C–H bonds affect the actual product distribution?

Solve

An H atom at the allylic position will be most easily abstracted. There are two distinct types of these H atoms, at C1 and C4. Abstraction at C4 leads to a radical in which the unpaired electron is delocalized on two secondary C atoms. Abstraction at C1 leads to a radical in which the unpaired electron is delocalized on one primary and one secondary C. So abstraction at C4 leads to the more stable radical.

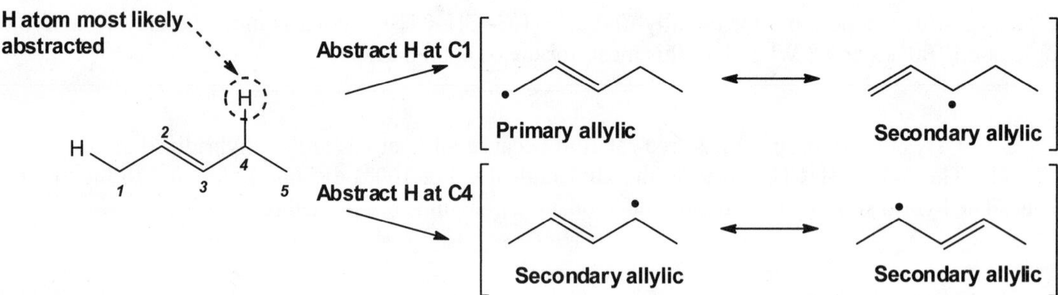

Problem 25.35

Think

What type of reactive organic intermediate appears in the mechanism? How does this intermediate react with the RC≡C–H group differently than it reacts with a RC≡C⁻ group? Why does this prevent the alkene from forming?

Solve

The reason is that a very strongly basic R^- species appears in the mechanism, which will irreversibly deprotonate a terminal alkyne. Reduction of such a deprotonated terminal alkyne would not be reasonable because it would proceed through an intermediate with adjacent negatively charged C atoms.

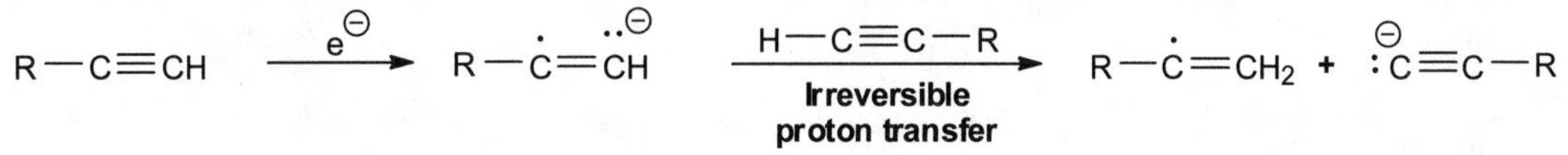

Irreversible proton transfer

Problem 25.36

Think

Are radicals formed? If so, what type of radical reaction occurs: free radical bromination or radical addition? Is resonance possible? What stereochemical considerations must be made? What regiochemical considerations must be made?

Solve

For **(a)** and **(b)**, initiation steps (as shown in Solved Problem 25-14 or Equations 25-18 or 25-28) will form Br• radicals from Br_2. These are all examples of free radical bromination. From there, propagation steps (only) are shown:

(a) H abstraction takes place most readily at the 3° C. The product is achiral.

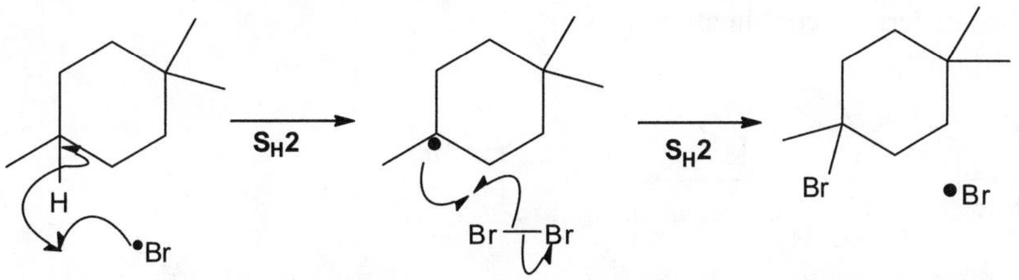

(b) H abstraction takes place most easily at the allylic H, and the resulting radical has two resonance contributors. Br can thus add at two different C atoms, giving rise to two constitutional isomers. Each constitutional isomer produced gains a new tetrahedral stereocenter, and a mixture of both stereochemical configurations will be produced.

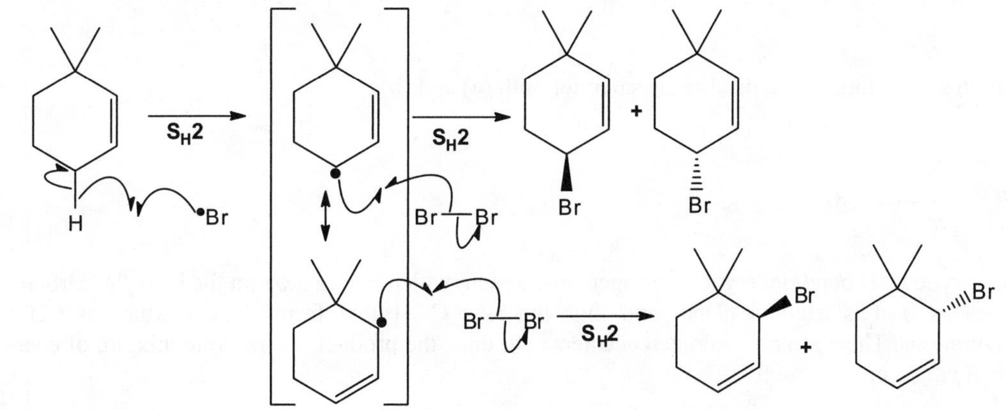

(c) No radicals are produced, so this mechanism involves only closed-shell species. This is an electrophilic addition of Br_2 across the C=C bond, going through a bromonium ion intermediate (Chapter 12).

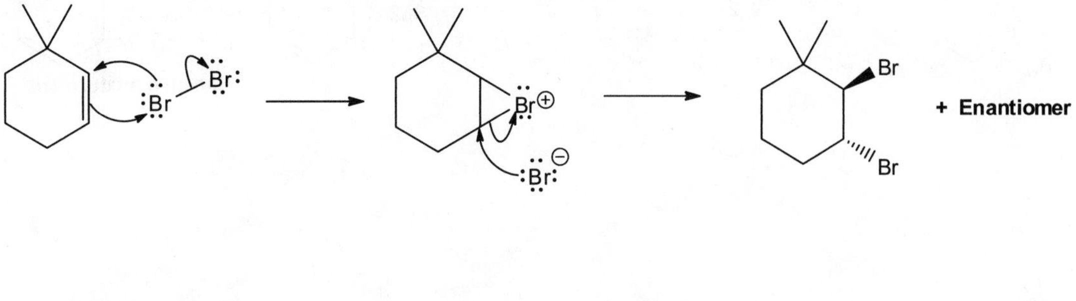

+ Enantiomer

(d) This is an example of a radical addition mechanism where the Br• radical reacts with the C=C to form a new C–Br bond so that the C• radical forms on the 3° carbon. The new tetrahedral stereocenter is produced as a mixture of both configurations. The C• then abstracts a H atom from HBr to form a new C–H bond. That step also forms a new stereocenter in which both configurations are produced and, thus, four products are possible.

Br

Radical addition

Br

H–Br

S_H2

+

Br

Br

OR

H–Br

S_H2

+

Br

Br

Br

(e) This not a radical mechanism, since no initiator (heat, light, or peroxides) is present. This mechanism is electrophilic addition (Chapter 11). A proton adds in the first step to produce the more stable carbocation intermediate, which then undergoes coordination with Br^-.

H–Br:

Electrophilic addition

H

:Br:⊖

Coordination

Br

Problem 25.37

Think

What types of carbon atoms are present (methyl, 1°, 2°, 3°, benzylic)? Which C–H bond undergoes hydrogen abstraction most readily? Is a new stereocenter produced? What is the relationship between the isomers produced?

Solve

The initiation step to form Br• radical is the same for both **(a)** and **(b)**.

Initiation:

Br–Br $\xrightarrow{h\nu}$ Br• + •Br

(a) The benzylic C–H bond undergoes hydrogen abstraction, which leaves a C• on the benzylic carbon. The C• behaves as if it is trigonal planar and thus the new C–Br that forms is a mixture of (R) and (S) configurations. There are no other stereocenters and, thus, the product is a racemic mixture of enantiomers and will not be optically active.

H

•Br

S_H2

Br–Br

S_H2

Br

(R)

+

Br

(S)

Racemic mixture

(b) The tertiary C–H bond undergoes hydrogen abstraction, which leaves a C• on the 3° carbon. The C• behaves as if it is trigonal planar and thus the new C–Br that forms is a mixture of (R) and (S) configurations. The other stereocenter remains unchanged during the course of the reaction. Thus, the products are diastereomers, which would form an optically active mixture.

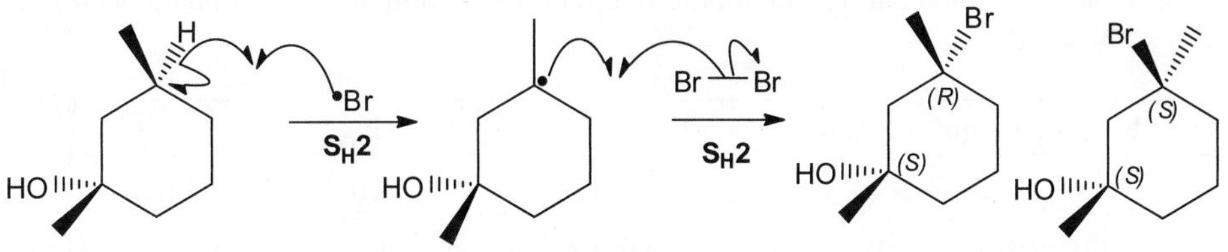

Diastereomers

Problem 25.38

Think

Which bond is the weakest and undergoes homolysis? What radical is formed? How can this radical react with H–H? What radicals are present in the reaction mixture that can couple?

Solve

The weakest bond present is the X–X bond, which undergoes homolysis in an initiation step to produce two X• radicals.

Initiation

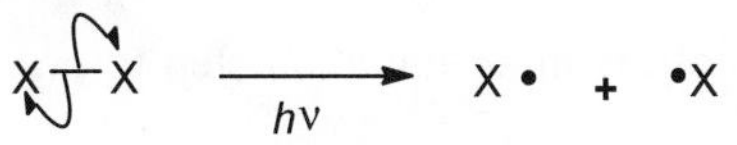

In the first of two propagation steps, an X• abstracts a hydrogen from H–H to produce the overall product, HX, and an H• radical. In the second propagation step, an H• radical abstracts an X atom from X_2 to produce the second equivalent of HX and regenerate the initial X•.

Propagation

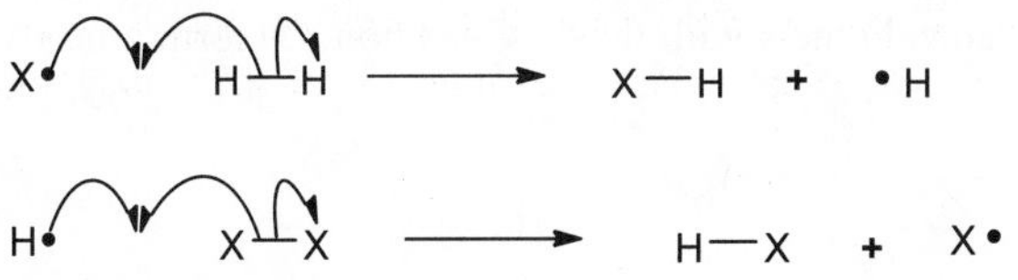

Termination steps can be a radical coupling between any two radicals that appear in the propagation steps, such as the following two:

Termination

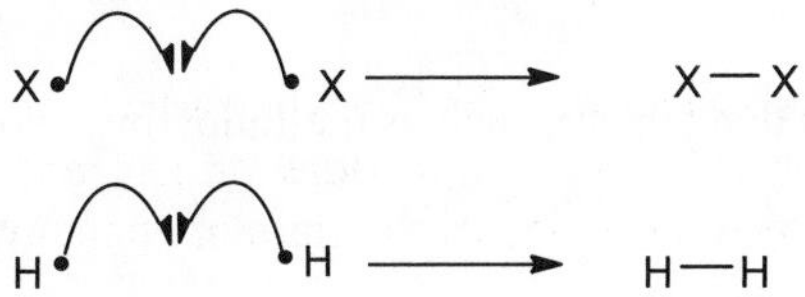

Problem 25.39

Think

What are the heats of reaction for the first propagation step of each reaction (see the solution to Problem 25.38 for the propagation steps)? How does the heat of reaction for that step govern the rate of the overall reaction?

Solve

Just as with the radical halogenation of alkanes, the main contribution to the energy barrier for the radical halogenation of H_2 comes from the first propagation step. The $\Delta H°_{rxn}$ of this step for each halogenation is shown below and is calculated by subtracting the energy of the X-H bond from that of the H-H bond. Because this step is most exothermic with fluorination and most endothermic with iodination, the rate of halogenation increases in the order: $I_2 < Br_2 < Cl_2 < F_2$.

First propagation step	$\Delta H°_{rxn}$
F• + H–H → F–H + H•	436 kJ/mol - 569 kJ/mol = -133 kJ/mol
Cl• + H–H → Cl–H + H•	436 kJ/mol - 431 kJ/mol = +5 kJ/mol
Br• + H–H → Br–H + H•	436 kJ/mol - 368 kJ/mol = +68 kJ/mol
I• + H–H → I–H + H•	436 kJ/mol - 297 kJ/mol = +139 kJ/mol

Problem 25.40

Think

Which bond is the weakest? Which bond undergoes homolysis? How does this radical react with the other organic reagent? What propagation and termination steps are possible?

Solve

The weakest bond present is the Cl_3C–Cl bond, which undergoes homolysis in an initiation step to produce a Cl• radical and Cl_3C• radical.

Initiation

Cl–CCl3 —hν→ •Cl + •CCl3

In the first of two propagation steps, a Cl_3C• adds to the C=C via radical addition to form a new C–CCl_3 bond and a C• radical. In the second propagation step, a C• radical reacts with the Cl_3C–Cl bond to form a new C–Cl bond and a Cl_3C• radical.

Propagation

H2C=CH2 + •CCl3 ⟶ H2C•–CH2–CCl3

Cl3C–Cl + H2C•–CH2–CCl3 ⟶ •CCl3 + Cl–CH2–CH2–CCl3

Termination steps can be a radical coupling between any two radicals that appear, such as the following two:

Termination

Cl• + •Cl ⟶ Cl–Cl

•CCl3 + •Cl ⟶ Cl3C–Cl

Problem 25.41

Think

Oxygen is a diradical. Which C–H bond undergoes H atom abstraction? How does the resulting radical react with another oxygen molecule? Do the propagation steps make a cycle? What termination steps are possible that can reduce the number of radicals?

Solve

Oxygen is a diradical, so it can abstract an H atom to generate the initial radical that enters the propagation cycle. The propagation cycle is defined by the second and third steps below. Notice that the initial ether radical is regenerated.

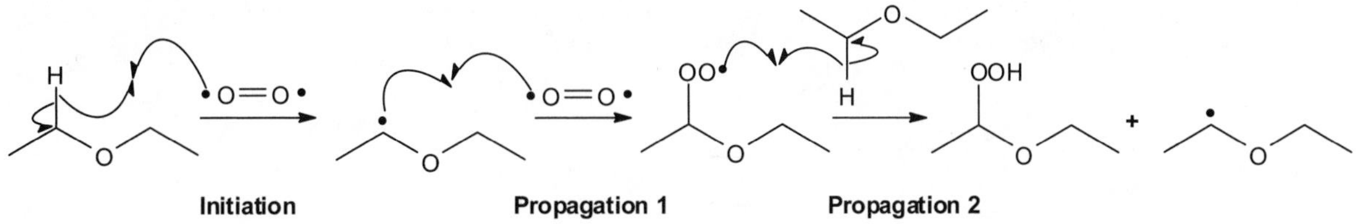

Initiation Propagation 1 Propagation 2

Two possible termination steps include:

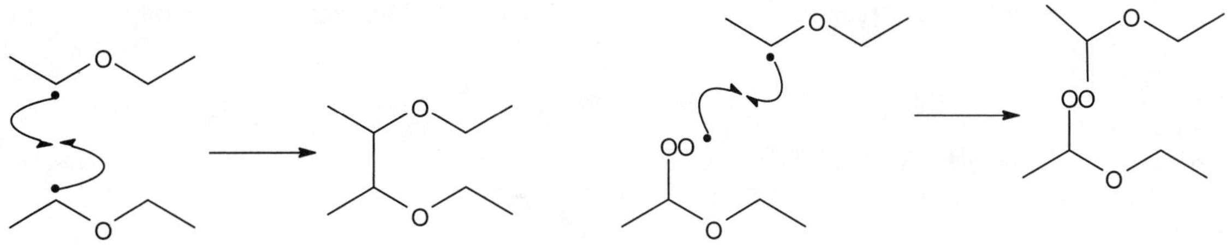

Problem 25.42

Think

What is the identity of the two possible radicals that can form from reaction of •O_2 and diethyl ether? Are there any inductive and/or resonance factors that stabilize the radical from the C atom that is α to the O?

Solve

The main reason is that the resulting radical is resonance stabilized, as shown below, involving the lone pair of electrons on O. Abstraction of a H atom from the terminal C results in a localized radical.

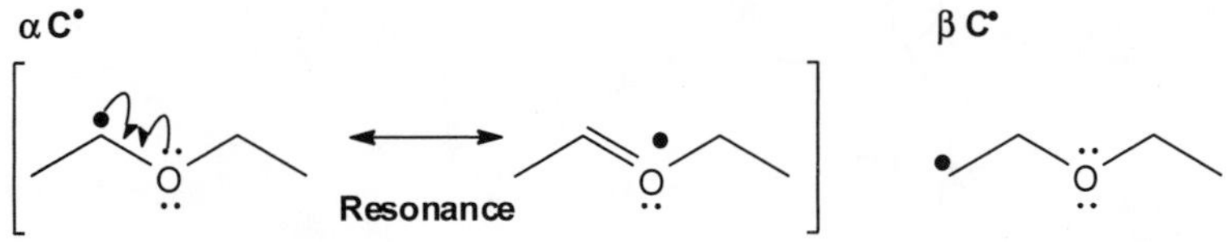

α C• Resonance β C•

Problem 25.43

Think

Which step has radicals only in the product? Which step has radicals in the product and reactant? Which step has radicals only in the reactant? What two radicals could come together in a termination step?

Solve

(a) The first step is the initiation; the remaining steps are propagation steps. Since each step forms a new radical or radical ion, there are no termination steps shown.

Initiation	$ArI + e^{\ominus} \longrightarrow ArI^{\bullet\ominus}$
Propagation	$ArI^{\bullet\ominus} \longrightarrow Ar^{\bullet} + I^{\ominus}$
Propagation	$Ar^{\bullet} + H_2N^{\ominus} \longrightarrow ArNH_2^{\bullet\ominus}$
Propagation	$ArNH_2^{\bullet\ominus} + ArI \longrightarrow ArNH_2 + ArI^{\bullet\ominus}$

(b) To arrive at the net reaction, we sum the propagation steps only, crossing out redundant species on either side of the reactions.

$$ArI^{\bullet -} \longrightarrow Ar^{\bullet} + I^{-}$$
$$Ar^{\bullet} + H_2N^{-} \longrightarrow ArNH_2^{\bullet -}$$
$$ArNH_2^{\bullet -} + ArI \longrightarrow ArNH_2 + ArI^{\bullet -}$$

Net reaction $H_2N^{-} + ArI \longrightarrow ArNH_2 + I^{-}$

(c) The only conceivable termination is where two uncharged radicals combine or where an uncharged radical combines with a charged radical, as shown below. Combination of two radical anions (like charge) is not probable.

Ar• + Ar• → Ar–Ar
Ar• + •$ArNH_2^-$ → Ar–$ArNH_2^-$

Problem 25.44

Think

Review the mechanism and arrow usage for initiation, propagation, and termination. Which step has radicals only in the product? Which step has radicals in the product and reactant? Which step has radicals only in the reactant? What two radicals could come together in a termination step?

Solve

(a) and **(b)** The arrows and labels are below. The initiation step generates new radicals. The propagation steps neither increase nor decrease the number of radicals. The termination steps decrease the number of radicals. Which types of steps should be summed to arrive at the net reaction?

Propagation #1: Cl• + cyclopropane → Cl–CH₂CH₂CH₂•

Initiation: Cl–Cl —hν→ 2 Cl•

Termination: Cl–CH₂CH₂CH₂• + •Cl → Cl–CH₂CH₂CH₂–Cl

Termination: Cl–CH₂CH₂CH₂• + •CH₂CH₂CH₂–Cl → Cl–(CH₂)₆–Cl

Propagation #2: Cl–CH₂CH₂CH₂• + Cl–Cl → Cl–CH₂CH₂CH₂–Cl + Cl•

(c) The net reaction is obtained by summing the propagation steps only.

Propagation #1: Cl• + cyclopropane → Cl–CH₂CH₂CH₂•

Propagation #2: Cl–CH₂CH₂CH₂• + Cl–Cl → Cl–CH₂CH₂CH₂–Cl + Cl•

Net reaction cyclopropane + Cl–Cl → Cl–CH₂CH₂CH₂–Cl

Problem 25.45

Think

Review the mechanism and arrow usage for initiation, propagation, and termination. Which step has radicals only in the product? Which step has radicals in the product and reactant? Which step has radicals only in the reactant? What two radicals could come together in a termination step? What steps are summed to arrive at the net reaction?

Solve

(a) and **(b)** All steps are propagation steps, evidenced by the fact that all the radicals present are redundant and can be crossed out. [The Cu(I) ion is actually a radical, possessing an unpaired electron during the reaction.]

R'—C(=O)—O—Cu⊕ + R• ⟶ R'C(=O)OR + Cu⊕•

R'C(=O)O—OC(CH3)3 + Cu⊕• ⟶ R'C(=O)OCu⊕ + •OC(CH3)3

R—H + •OC(CH3)3 ⟶ R• + HOC(CH3)3

(c) The net reaction is obtained by summing all three of these propagation steps.

Net reaction

R'C(=O)O—OC(CH3)3 + R—H ⟶ R'C(=O)OR + HOC(CH3)3

(d) Two termination steps:

R• + •OC(CH3)3 ⟶ R—O—C(CH3)3

R• + Cu⁺• ⟶ R—Cu⁺

Problem 25.46

Think

Review the mechanism and arrow usage for initiation, propagation, and termination. Which step has radicals only in the product? Which step has radicals in the product and reactant? Which step has radicals only in the reactant? What two radicals could come together in a termination step? Which steps make up the propagation cycle? Which steps are summed to obtain the net reaction?

Solve

(a) and **(b)** Steps 1–3 make up the propagation cycle, evidenced by the fact that all the radicals present in those steps are redundant and can be crossed out when summed. The other two steps are initiation, as they are responsible for producing the first radical that enters the propagation cycle. See the figure on the next page.

Propagation $Bu_3Sn\bullet$ + Br–(hex-5-enyl) ⟶ $Bu_3Sn—Br$ + •(hex-5-enyl)

Propagation (cyclopentylmethyl radical) + H—$SnBu_3$ ⟶ methylcyclopentane + $\bullet SnBu_3$

Propagation (hex-5-enyl radical) ⟶ (cyclopentylmethyl radical)

Initiation AIBN ⟶ 2 $(CH_3)_2\dot{C}CN$ + N≡N

Initiation $(CH_3)_2\dot{C}CN$ + H—$SnBu_3$ ⟶ $(CH_3)_2CHCN$ + $\bullet SnBu_3$

(c) The net reaction is obtained by summing all the propagation steps.

Net reaction

H—$SnBu_3$ + Br–(hex-5-enyl) ⟶ $Bu_3Sn—Br$ + methylcyclopentane

(d) Two termination steps:

2 (cyclopentylmethyl radical) ⟶ 1,2-dicyclopentylethane

2 $(CH_3)_2\dot{C}CN$ ⟶ $NC—C(CH_3)_2—C(CH_3)_2—CN$

Problem 25.47

Think

Recall that radical chlorination is less selective than radical bromination. Is that because the steps in chlorination are more endothermic or exothermic than in bromination? Does a more stable reactant species lead to a more endothermic or exothermic reaction?

Solve

Chlorination is less selective than bromination because the chlorination steps are more exothermic. The complexation described in this problem is indicated to stabilize chlorine, which makes its reaction less exothermic than without the complexation. Thus, the complexation makes the reaction *more* selective.

Problem 25.48

Think

Review the mechanism and arrow usage for initiation, propagation, and termination. Which step has radicals only in the product? Which step has radicals in the product and reactant? Which step has radicals only in the reactant? Which steps make up the propagation cycle? Which steps are summed to obtain the net reaction?

Solve

(a) The chain-reaction mechanism for the elimination of HI from 2-iodopropane to form propene is shown below.

(b)

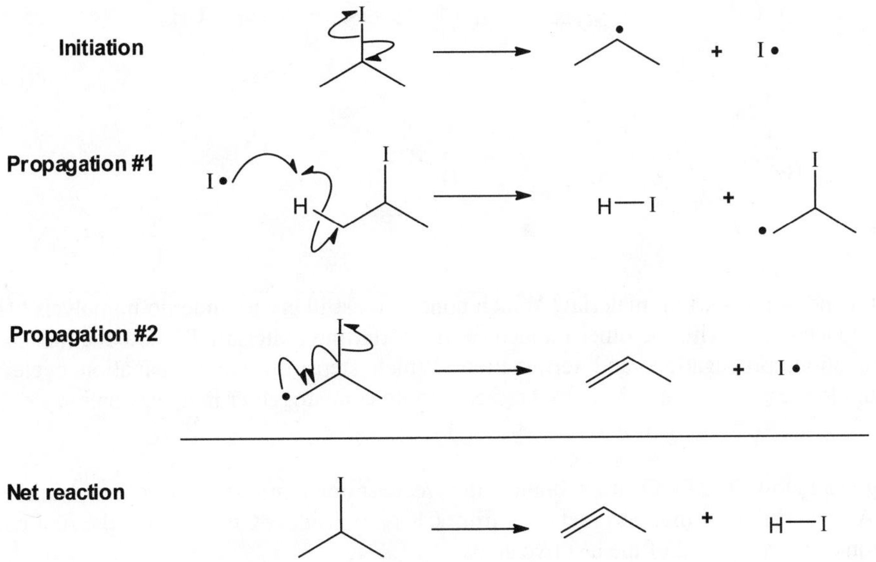

Initiation

Propagation #1

Propagation #2

Net reaction

(c) The $\Delta H°_{rxn}$ of each propagation step and the net reaction are given below.

Reaction	$\Delta H°_{rxn}$
Propagation #1 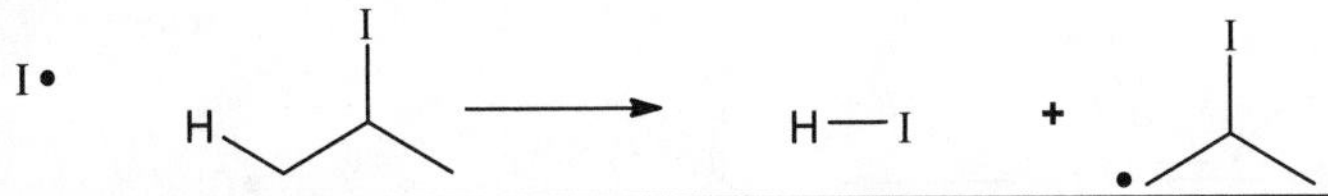	435 kJ/mol − 297 kJ/mol = <u>+138 kJ/mol</u>
Propagation #2 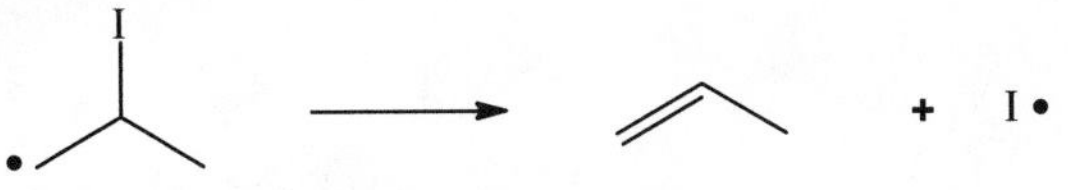	234 kJ/mol − 464 kJ/mol = <u>−230 kJ/mol</u>
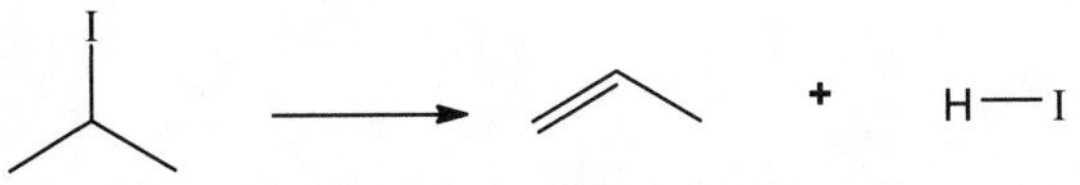	$\Delta H°_{rxn}$ **= −92 kJ/mol**

Problem 25.49

Think

What is the weakest bond in the starting material? Which bond is most likely to undergo homolysis? How does this radical react with the starting material? What are the propagation steps? Do they make a cycle? Do the propagation steps sum to the net reaction?

Solve

The C–O bond is weaker than a CH bond, so the C–O bond is the one broken in the initiation step. In the first propagation step, a •CH_3 abstracts a H atom from the ether to produce CH_4 as an overall product and a new radical. In the second step, the new radical undergoes homolysis to produce a molecule of $H_2C{=}O$ as an overall product and replenishes the •CH_3 radical for further reaction.

Initiation $H_3C{-}O{-}CH_3 \xrightarrow{h\nu} H_3C{-}O\bullet + \bullet CH_3$

Propagation 1 $H_3C{-}O{-}CH_2{-}H + \bullet CH_3 \longrightarrow H_3C{-}O{-}\dot{C}H_2 + CH_4$

Propagation 2 $H_3C{-}O{-}CH_2\bullet \longrightarrow H_2C{=}O + \bullet CH_3$

Problem 25.50

Think

What is the weakest bond in the starting material? Which bond is most likely to undergo homolysis? How does the radical that is produced react with the other molecules in the starting material? Review the mechanism and arrow usage for initiation, propagation, and termination. Which steps form a propagation cycle? Do the propagation steps sum to the net reaction? What two radicals could come together in a termination step?

Solve

The mechanism is given below. The O–O single bond is the weakest bond and undergoes homolysis to produce the initial radicals. An initial radical then abstracts CN from ClCN to product Cl•, which is the first radical that enters the propagation cycle composed of the last two steps.

Initiation $C_6H_5C(O)O{-}OC(O)C_6H_5 \xrightarrow{h\nu} 2\ C_6H_5C(O)O\bullet$

Initiation $C_6H_5C(O)O\bullet + Cl{-}C{\equiv}N \longrightarrow C_6H_5C(O)O{-}CN + \bullet Cl$

Propagation $CH_3CH_2O{-}CH{=}CH_2 + \bullet Cl \longrightarrow CH_3CH_2O{-}\dot{C}H{-}CH_2Cl$

Propagation $CH_3CH_2O{-}\dot{C}H{-}CH_2Cl + Cl{-}C{\equiv}N \longrightarrow CH_3CH_2O{-}CH(CN){-}CH_2Cl + \bullet Cl$

Notice that to obtain the observed regiochemistry, Cl• must add to the C=C first. Specifically, Cl• adds to the terminal C to produce the more stable (resonance stabilized) carbon radical. Then, CN is abstracted from ClCN to complete the overall product and regenerate Cl• for further reaction in the propagation cycle.

Two possible termination steps:

Cl• • Cl → Cl—Cl

Problem 25.51

Think

It may help to build molecular models of each of these compounds. What is the ideal geometry about the C• radical intermediate? Is this possible in bicyclo[2.2.1]heptane?

Solve

The smaller bicyclic system does not allow the tertiary carbon to become sufficiently planar to allow radical formation. Thus, a tertiary radical that would be formed would be highly strained, making the secondary radical more stable than the tertiary in this case.

Bicyclo[3.3.1]nonane —$BrCCl_3$, hν→ [3° C• Intermediate] → Br (100%)

Bicyclo[2.2.1]heptane —$BrCCl_3$, hν→ X [3° C• Intermediate Too strained to form] → Br (0%)

Problem 25.52

Think

What is the weakest bond in the starting material? Which bond is most likely to undergo homolysis? How does the radical that is produced react with the other molecules of the starting material? Review the mechanism and arrow usage for initiation, propagation, and termination. Do the propagation steps make a cycle? What two radicals could come together in a termination step?

Solve

The initiation and propagation steps are given on the next page. The Br–C bond is the weakest, so it undergoes homolysis for the initiation step. Then, in the first propagation step, the •CCl_3 radical abstracts a H atom from the tertiary C–H bond to produce a carbon radical. Then, in the second propagation step, the C• abstracts a Br atom from $BrCCl_3$, producing another •CCl_3 that can react further in the propagation cycle.

Initiation Br—CCl_3 —hν→ Br• + •CCl_3

Propagation + $CHCl_3$

•CCl_3

Propagation + Cl_3C•

Cl_3C—Br

Br

Two possible termination steps:

Br

Br•

Cl_3C• •CCl_3 ⟶ Cl_3C—CCl_3

Problem 25.53

Think

What is the weakest bond? Which bond undergoes homolysis in an initiation step? How can this radical react with itself to form CO_2 and R• radical?

Solve

The key first step is the breaking of the O–O bond, which is the weakest in the molecule. Generation of the alkyl radical in the second step is facilitated by the production of CO_2, which is very stable and is a gas, so it irreversibly leaves the reaction mixture.

hν

2 ⟶ 2 R• + 2 O═C═O

Problem 25.54

Think

What is the initiation step? Which C–H bond is abstracted? What is the identity of the C• radical intermediate? How can the nearby Br atom interact with the C• radical to prevent the production of both configurations when the new tetrahedral stereocenter is formed? (*Hint:* Consider the intermediate in bromination of an alkene that is responsible for the reaction's stereoselectivity [Chapter 12].)

Solve

As the H atom is abstracted, the terminal bromine can form a cyclic bromo radical (analogous to the cyclic bromonium ion in Chapter 12). Subsequent attack of Br_2 can only occur *from the side opposite* the attached bromine, which reverses the stereochemical configuration, just as in an S_N2 reaction.

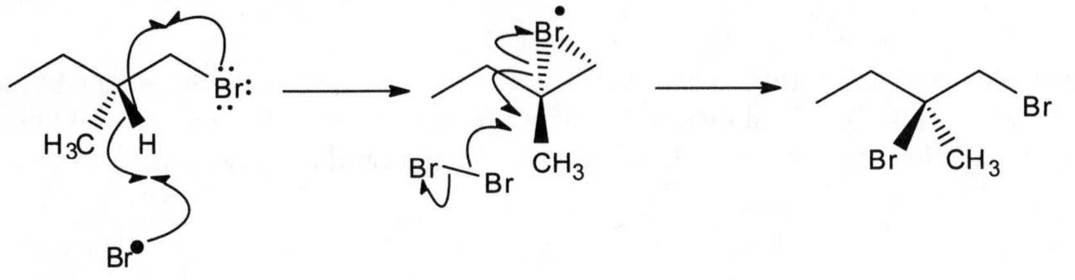

Problem 25.55

Think

Which bond homolytically cleaves to form the initial radical? What radical reacts with the C=C? On which C atom does the C• radical form? Why?

Solve

The initiation steps are shown below and will form Br• radicals. If Br• adds to C3 as shown, then the radical on C2 can be stabilized by resonance with the pre-existing Br atom.

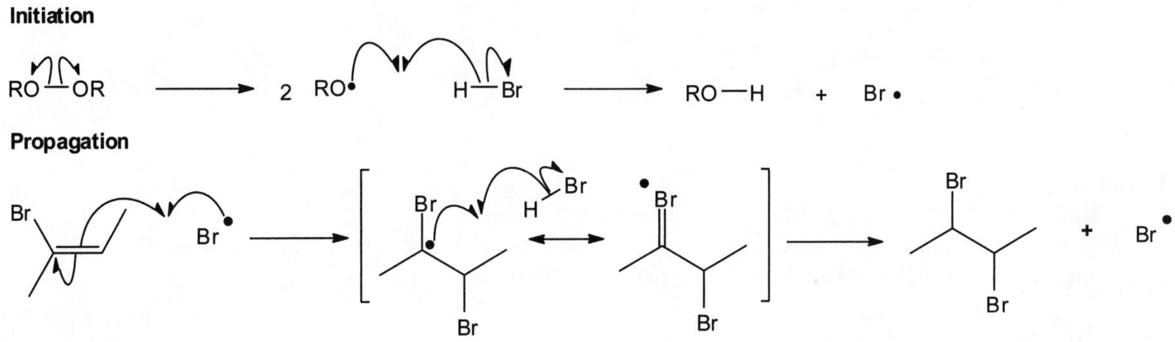

Initiation

RO—OR ⟶ 2 RO• H—Br ⟶ RO—H + Br•

Propagation

Problem 25.56

Think

Which bond homolytically cleaves to form the initial radical? What radical reacts with the C=C? On which C atom does the C• radical form? How does the trans alkene form from that intermediate?

Solve

The I–I bond is the weakest present and is cleaved to form 2 I• radicals. After the iodine radical forms, it adds to the double bond. This converts the double bond to a single bond temporarily, which can freely rotate. Because of the bulkiness of the rings, the single bond rotates so they are trans to each other. Subsequently, the iodine radical is eliminated to reform the C=C double bond.

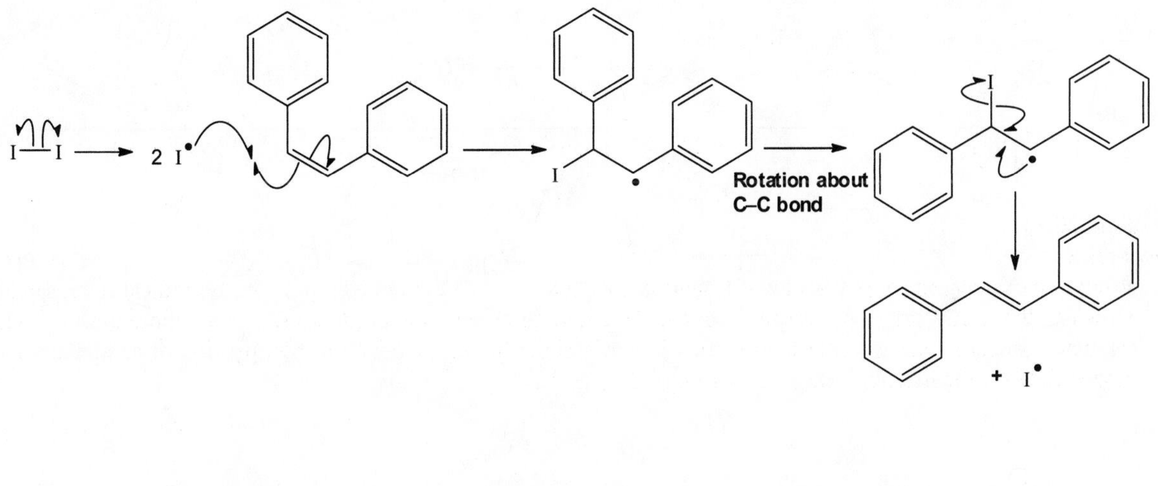

I—I ⟶ 2 I•

Rotation about C–C bond

+ I•

Problem 25.57

Think

Which bond homolytically cleaves to form the initial radical? What radical reacts with the C=C? On which C atom does the C• radical form? How does the five-carbon ring form from that intermediate?

Solve

The initiation steps are shown below and will form Br• radicals. The Br• radical reacts with the alkene to form a new C–Br bond and a secondary C•. This C• radical then reacts with the other C=C to form a new C–C bond in a cyclopentane ring. Another propagation step yields the target compound and produces another Br• to complete the propagation cycle.

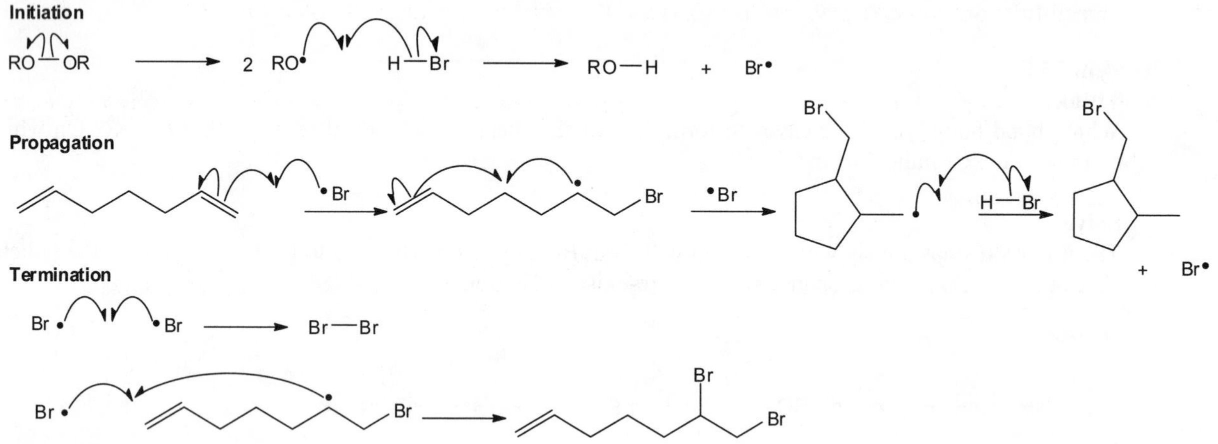

Initiation

RO—OR ⟶ 2 RO• H—Br ⟶ RO—H + Br•

Propagation

Br Br Br •Br H—Br + Br•

Termination

Br• •Br ⟶ Br—Br

Br• Br ⟶ Br Br

Problem 25.58

Think

Which bond homolytically cleaves to form the initial radical? What radical reacts with the C=C? On which C atom does the C• radical form? How do the two five-carbon rings form?

Solve

The initiation steps are shown above in Problem 25.57 and will form Br• radicals. From there, two radical cyclizations occur; numbering the carbons indicates that C2 joins C6 and C7 joins C11.

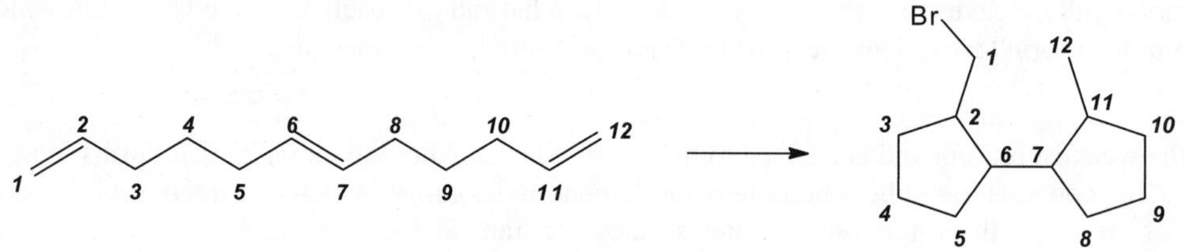

1 2 3 4 5 6 7 8 9 10 11 12 ⟶ Br 1 12 2 11 3 10 6 7 4 9 5 8

The mechanism involves a 1° radical in the next to last step.

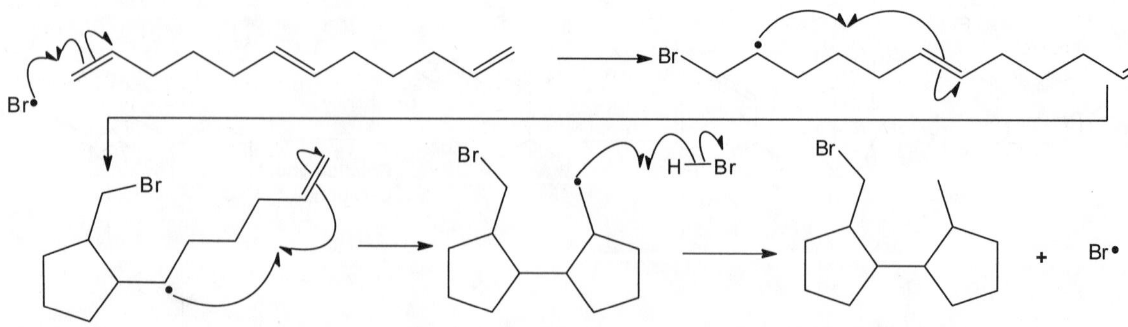

Br• ⟶ Br Br Br H—Br Br + Br•

Two possible termination steps are shown below.

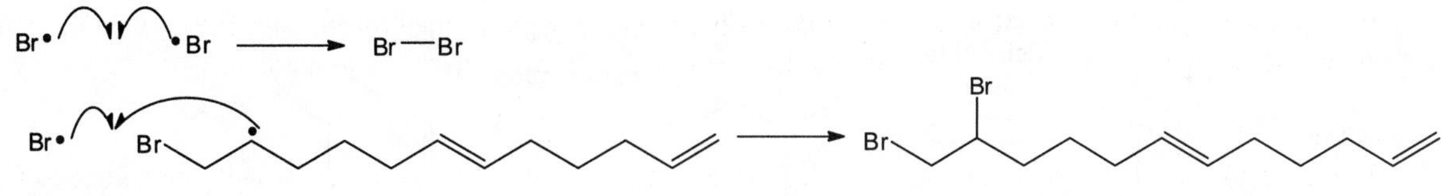

Br• •Br ⟶ Br—Br

Br• Br~ ⟶ Br~

Problem 25.59

Think

Which C atom in propylbenzene is most likely to undergo free radical bromination? What functional group transformations can take place on the product? How must the carbon skeleton be altered?

Solve

Retrosynthesis and synthesis: We can functionalize the benzyl radical by bromination, so this is a logical step in each of the syntheses.

(a) Retrosynthesis

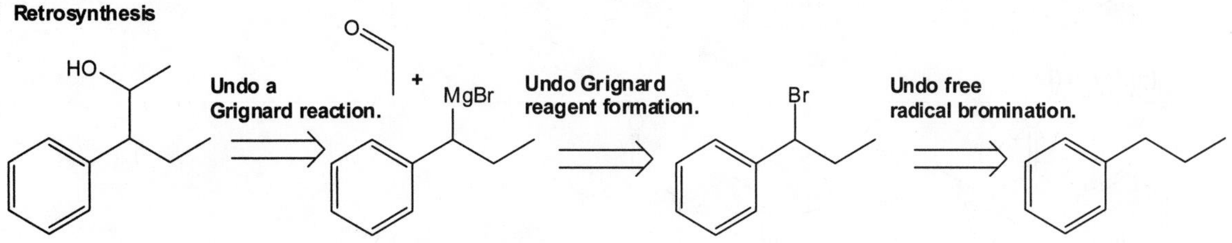

HO — Undo a Grignard reaction. ⟹ O= + MgBr — Undo Grignard reagent formation. ⟹ Br — Undo free radical bromination. ⟹

Synthesis in the forward direction

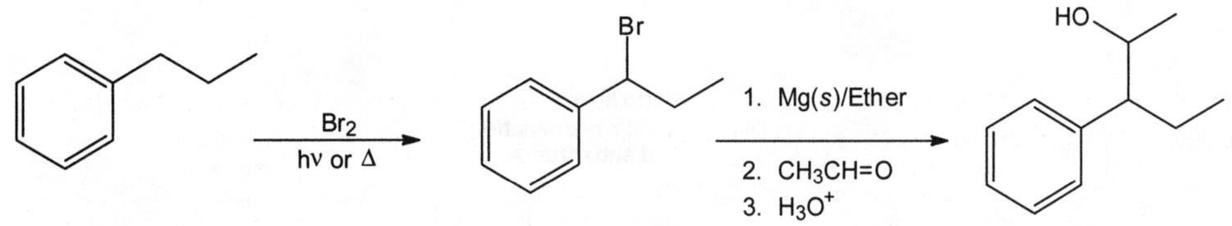

$\xrightarrow[h\nu \text{ or } \Delta]{Br_2}$ Br $\xrightarrow{}$ 1. Mg(*s*)/Ether 2. $CH_3CH{=}O$ 3. H_3O^+ → HO

(b) Retrosynthesis

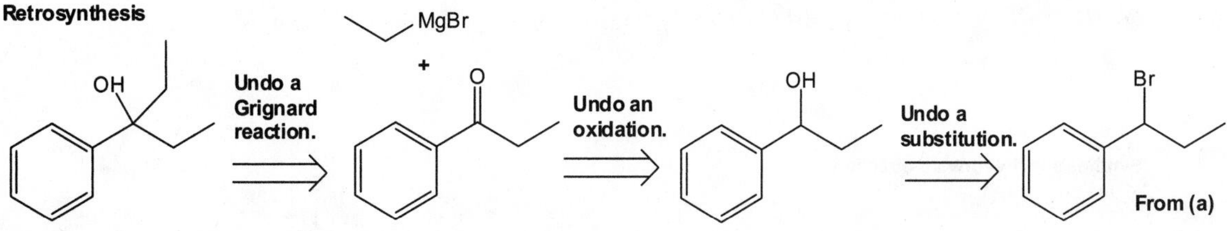

OH — Undo a Grignard reaction. ⟹ MgBr + O — Undo an oxidation. ⟹ OH — Undo a substitution. ⟹ Br — From (a)

Synthesis in the forward direction

From (a) Br $\xrightarrow[(S_N1)]{\text{dil NaOH}}$ OH $\xrightarrow{CrO_3}$ O $\xrightarrow{}$ 1. CH_3CH_2MgBr 2. H_3O^+ → HO

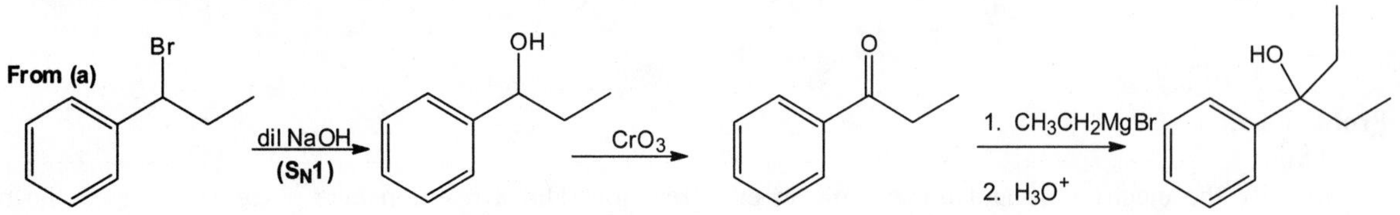

(c) Retrosynthesis

Undo a Diels–Alder reaction.

+

Undo a dehydration.

OH

From (b)

Synthesis in the forward direction

OH

H_3PO_4

Δ

(d) From (b)

O

Cl_2, $FeCl_3$

O

Cl

(e) Retrosynthesis

O

O_2N

Undo an oxidation.

OH

O_2N

Undo a free radical bromination and substitution.

O_2N

Undo a nitration.

Synthesis in the forward direction

HNO_3, H_2SO_4

O_2N

1. NBS
2. dil NaOH

OH

O_2N

CrO_3

O

O_2N

Problem 25.60

Think

Identify the electrophile and nucleophile in each reaction. Did a reaction take place that altered the carbon skeleton or functional group? Is the reaction an example of substitution, elimination, addition, or redox? Are radicals involved?

Solve

A is formed from free radical bromination. The alkybromide undergoes an E2 reaction to form cyclohexene, **B**. The cyclohexene undergoes free radical bromination at the allylic C–H, which yields **C**. **C** then undergoes a substitution reaction to give the alcohol, **D**. **D** is oxidized to the α,β-unsaturated ketone, **E**, which then undergoes 1,4-addition.

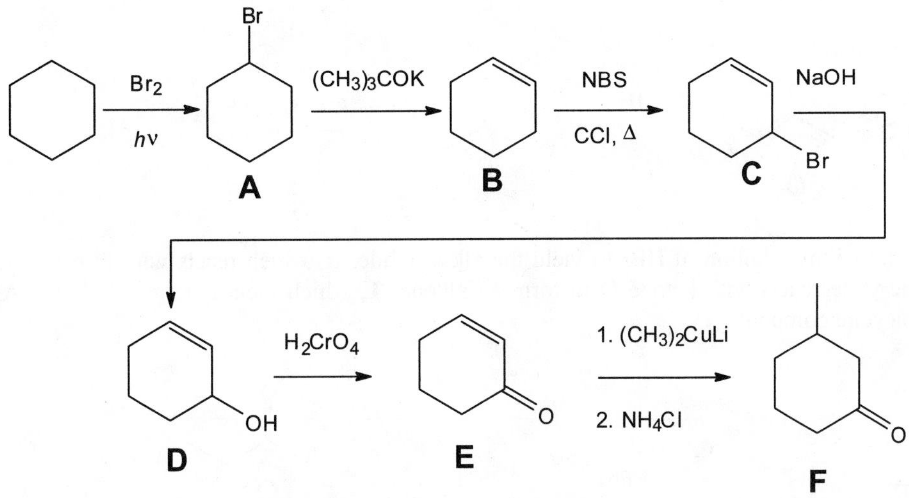

Br

Br_2 / $h\nu$ → **A** → $(CH_3)_3COK$ → **B** → NBS / CCl, Δ → **C** (Br) → NaOH → **D** (OH) → H_2CrO_4 → **E** (O) → 1. $(CH_3)_2CuLi$ 2. NH_4Cl → **F** (O)

Alkyl bromide **C** reacts with Mg(s) to form a Grignard reagent, **G**, which then reacts with **F** to form the tertiary alcohol, **H**. Br_2 adds to the C=C in an electrophilic addition reaction to form the dibromo compound **I**.

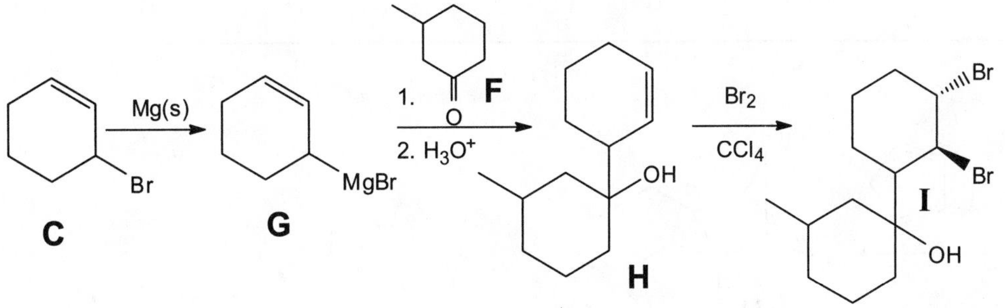

C (Br) → Mg(s) → **G** (MgBr) → 1. **F** (O) 2. H_3O^+ → **H** (OH) → Br_2 / CCl_4 → **I** (Br, Br, OH)

Problem 25.61

Think

Identify the electrophile and nucleophile in each reaction. Did a reaction take place that altered the carbon skeleton or functional group? Is the reaction an example of substitution, elimination, addition, or redox? Are radicals involved?

Solve

J is formed from free radical bromination of the secondary C–H. The alkyl bromide undergoes E2 to form propene **K**, which reacts with Br_2 in an electrophilic addition reaction to form a dibromo compound, **L**. **L** then undergoes a double E2 followed by acid workup to form the alkyne **M**. The alkyne is deprotonated and then a new C–C bond is formed by an S_N2 reaction to form **N**. The alkyne reacts with $Na(s)/NH_3(l)$ to form the trans alkene, **O**. The trans alkene undergoes an electrophilic addition reaction with H_3O^+ to form an alcohol, **P**, which is then oxidized to ketone **Q**. See the figure on the next page.

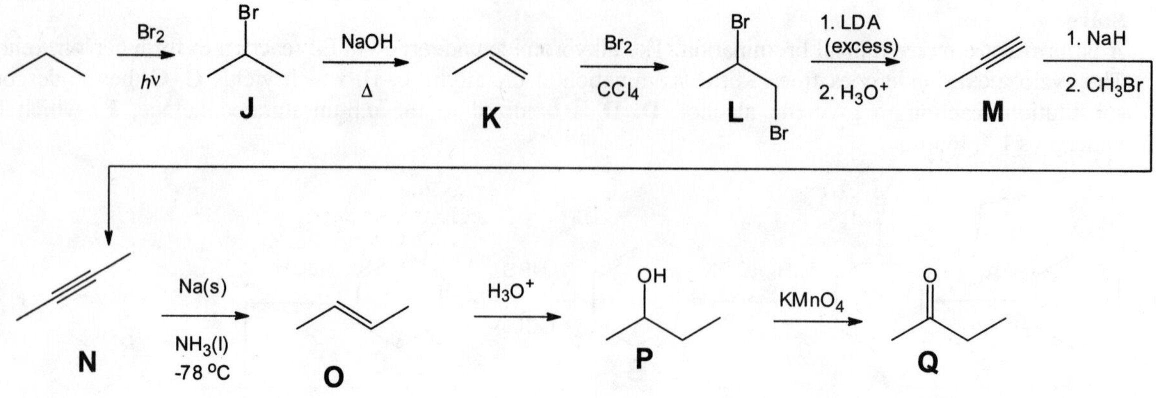

K undergoes anti-Markovnikov addition of HBr to yield the alkyl halide, **R**, which reacts with PPh_3/Bu–Li to form the ylide, **S**. The ylide reacts with ketone **Q** to form an alkene, **T**, which then undergoes a Diels–Alder reaction to form the bicyclic compound **U**.

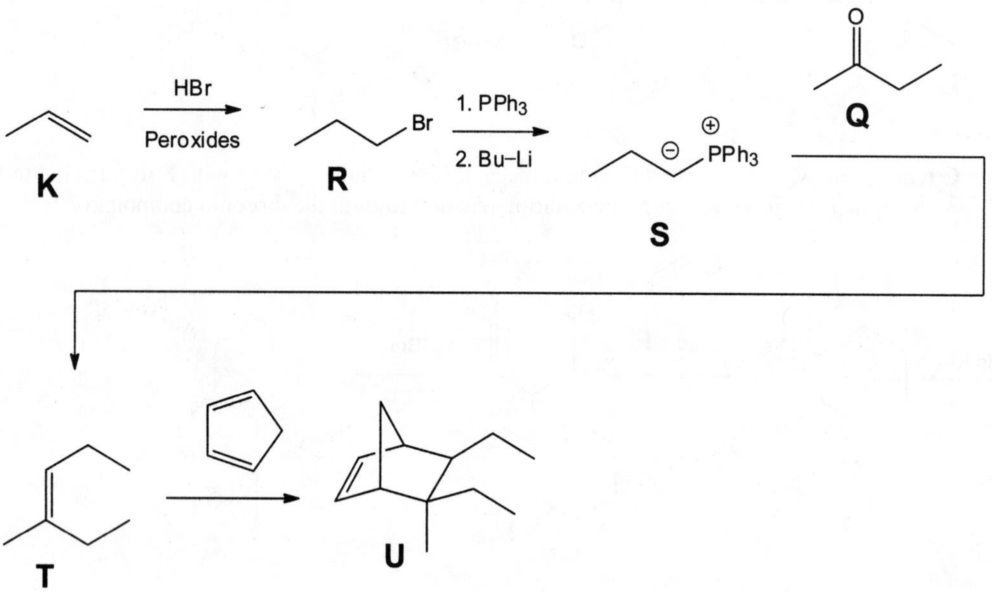

Problem 25.62

Think

Which C atom on 2-methylpropane is most likely to undergo free radical bromination? How can that product react to carry out functional group conversions? How can the carbon skeleton be altered? Are any oxidations or reductions necessary?

Solve

Retrosynthesis and synthesis: We can functionalize the tertiary carbon of 2-methylpropane by bromination, so this is a logical step in each of the syntheses.

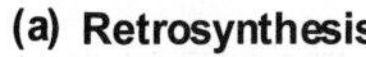

(a) Retrosynthesis

Undo an S_N1. ⟹ OH + Br ⟹ Undo free radical bromination.

Synthesis in the forward direction

Br_2, hν or Δ → Br; OH, S_N1 → O

(b) Retrosynthesis

S ⟹ Undo a substitution. ⟹ H_3C–SNa + Br ⟹ Undo an anti-Markovnikov free radical addition. ⟹ Undo an E_2. ⟹ Br

From (a)

Synthesis in the forward direction

Br (From (a)) → KOH, Δ → HBr, Peroxides → Br → NaS–CH_3 → S

(c) Retrosynthesis

Undo a Wittig reaction. ⟹ O (benzaldehyde) + ⊖ ⊕PPh_3 ⟹ Undo an ylide synthesis. ⟹ Br

From (b)

Synthesis in the forward direction

Br (From (b)) → 1. PPh_3, 2. BuLi → ⊖ ⊕PPh_3 → O (benzaldehyde) →

(d) Retrosynthesis

Undo Diels–Alder

From (b)

Synthesis in the forward direction

From (b)

(e)

$\overset{\oplus}{P}Ph_3$

From (c)

(f)

D_2, Pt

D D

From (b)

Problem 25.63

Think

How is the carbon skeleton altered? What functional group transformations took place? How is the alkyne functional group transformed? Does the alkyne group need to be hydrogenated? If so, does the cis or trans alkene need to be produced?

Solve

Retrosynthesis and synthesis: We can functionalize the alkyne by transformation into a cis or trans alkene.

(a) Retrosynthesis

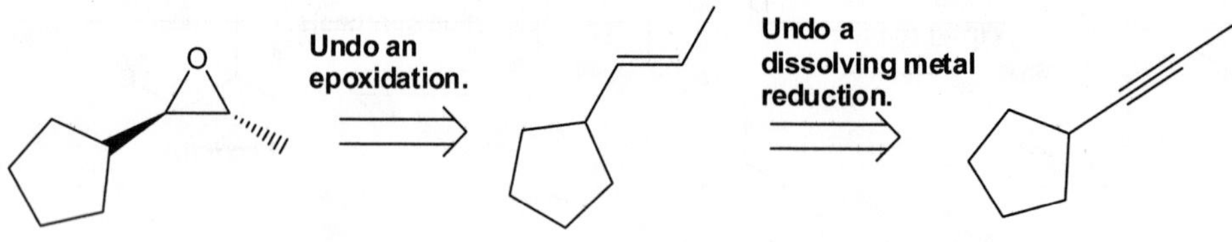
O
Undo an epoxidation.
Undo a dissolving metal reduction.

Synthesis in the forward direction

Na, NH_3

MCPBA

(b) Retrosynthesis

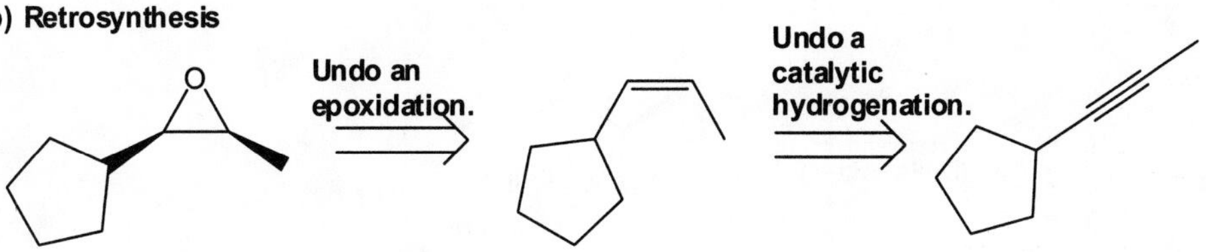

Undo an epoxidation. ⟹ **Undo a catalytic hydrogenation.** ⟹

Synthesis in the forward direction

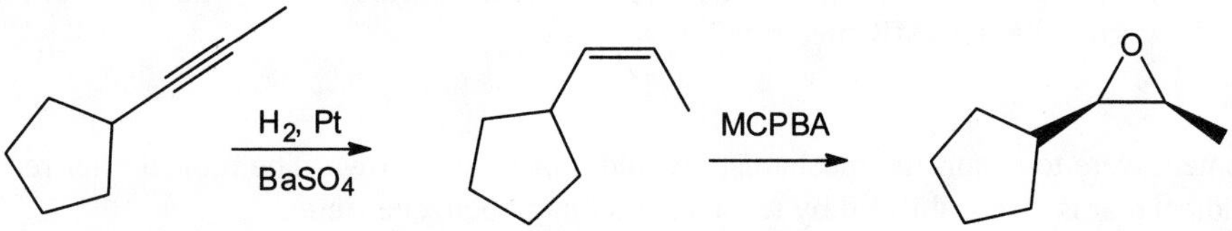

H_2, Pt / $BaSO_4$ → MCPBA →

(c) Retrosynthesis

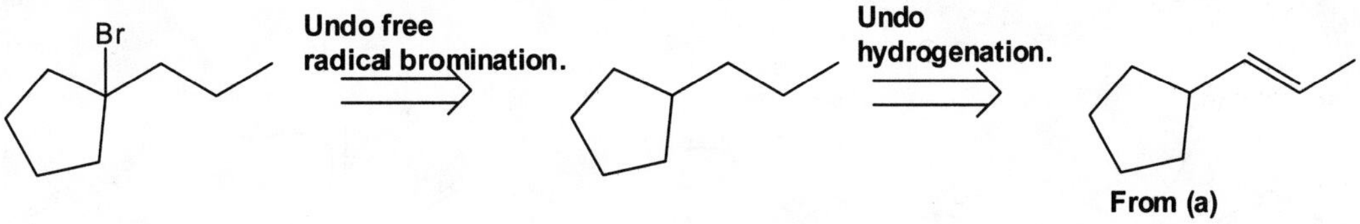

Br

Undo free radical bromination. ⟹ **Undo hydrogenation.** ⟹

From (a)

Synthesis in the forward direction

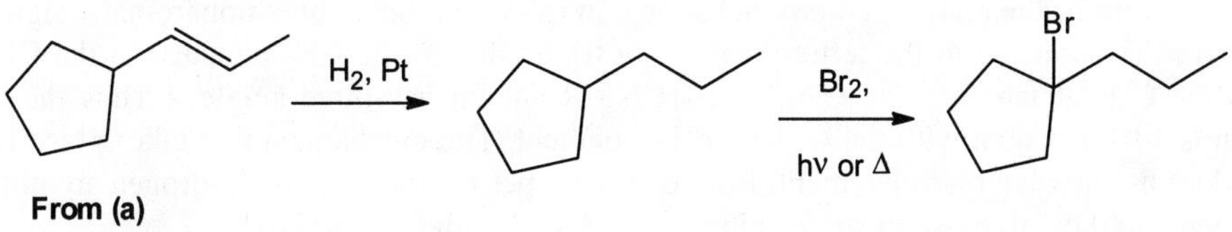

H_2, Pt → Br_2, hν or Δ → Br

From (a)

(d) Br_2, (Anti addition) → Br Br

From (a)

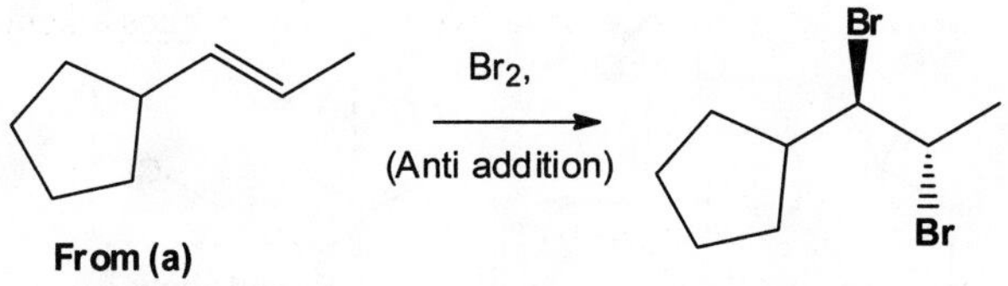

(e) Δ (Diels–Alder) →

From (b)

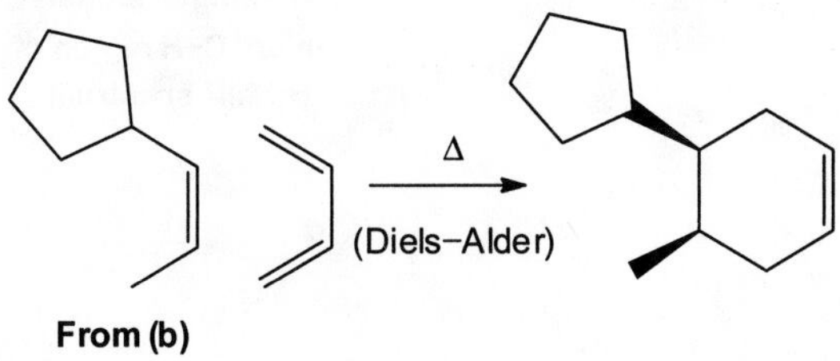

(f)

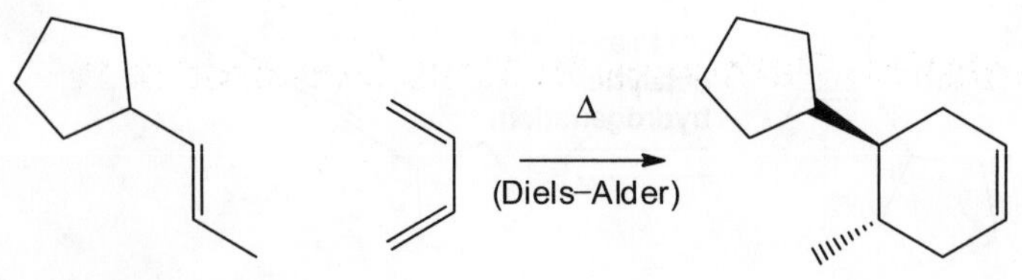

(Diels–Alder)

From (a)

Problem 25.64

Think

What types of carbons are present? What C–H bond undergoes homolysis? What type of radical forms? Theoretically, is there a more stable radical possible if rearrangement occurs? If so, what product forms? What conclusion do you draw from the ^{1}H NMR spectrum?

Solve

(a) If a rearrangement were to occur, the mechanism would appear as follows. The radical after rearrangement is a tertiary radical that is also stabilized by resonance with the benzene ring.

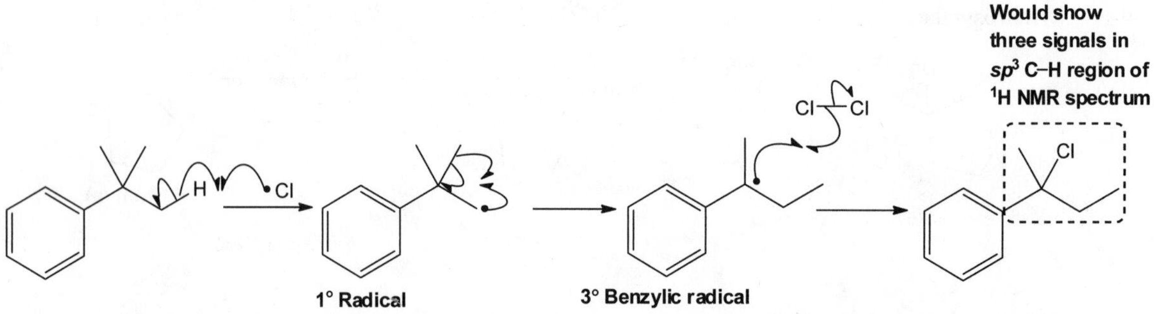

Would show three signals in sp^3 C–H region of ^{1}H NMR spectrum

1° Radical

3° Benzylic radical

(b) The ^{1}H NMR spectrum for the rearranged product shown in **(a)** would have three nonaromatic signals—one singlet due to the CH_3 attached to the tertiary carbon, a triplet due to the CH_3 attached to the CH_2, and a quartet due to the CH_2 attached to the CH_3. The ^{1}H NMR shown has three singlets. Thus the ^{1}H NMR spectrum given is not consistent with the radical rearrangement. The mechanism that takes place is the one below, which does not involve rearrangement. Both distinct types of nonaromatic hydrogen are not coupled to other hydrogens, so they give rise to singlets in the NMR spectrum.

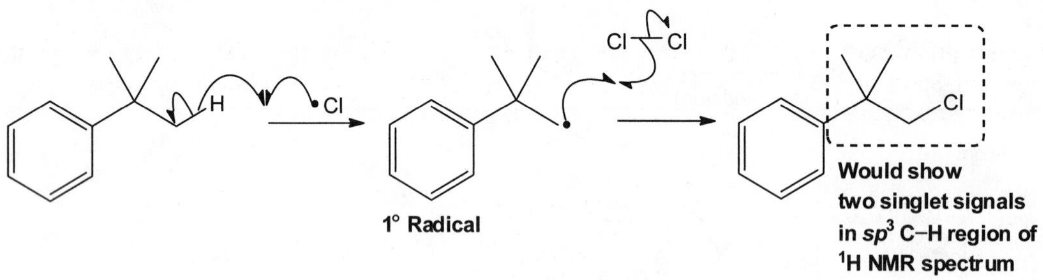

1° Radical

Would show two singlet signals in sp^3 C–H region of ^{1}H NMR spectrum

INTERCHAPTER 2 | Fragmentation Pathways in Mass Spectrometry

Chapter Problems
Problem IC2.1

Think

What is the identity of the ion that corresponds to the peaks at 29 u and 71 u? Which ion is more stable (Section 6.6c)? Does cation or radical stability win out?

Solve

Figure IC2-3 is the mass spectrum of 2-methylpentane. The peak at 29 u is from an ethyl cation $CH_3CH_2^+$ and the peak at 71 u is from the fragmentation of the methyl group to give $CH_3CH^+CH_2CH_2CH_3$. The fragmentation pathways are shown below:

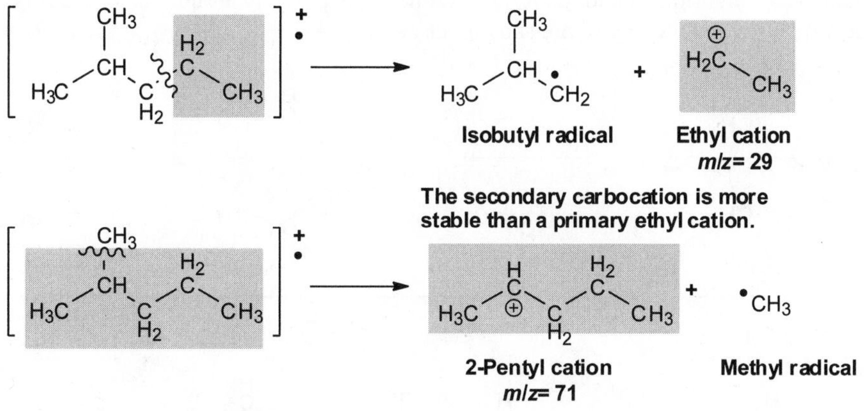

Isobutyl radical

Ethyl cation
m/z= 29

The secondary carbocation is more stable than a primary ethyl cation.

2-Pentyl cation
m/z= 71

Methyl radical

The secondary 2-pentyl cation is more stable than the primary ethyl cation; therefore, the peak is more intense.

Problem IC2.2

Think

What peaks are unique to each spectrum? What is the structure of butylcyclopentane and *tert*-butylcyclopentane? What are likely fragmentation ions for each? What are the corresponding masses of each cation fragment?

Solve

The major differences in the two spectra are the peaks at 126, 97, and 83 u for Spectrum **A** and the peak at 111 u for Spectrum **B**. The structures for butylcyclopentane and *tert*-butylcyclopentane and possible fragmentation ions are shown below (solution continues on the next page):

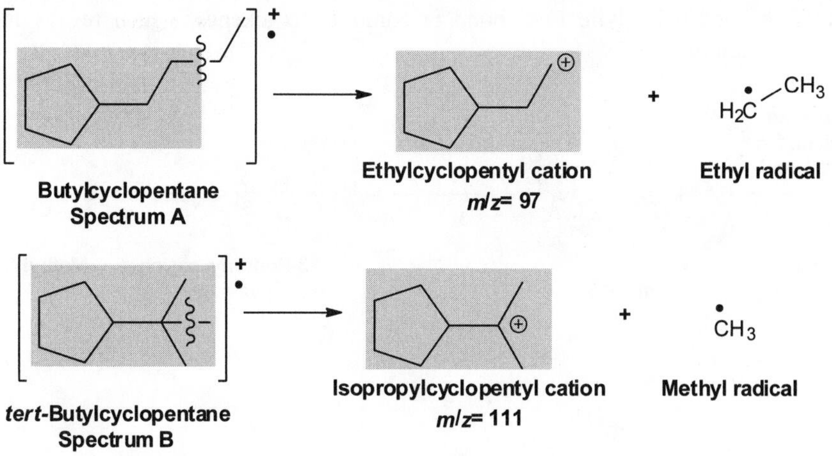

Butylcyclopentane
Spectrum A

Ethylcyclopentyl cation
m/z= 97

Ethyl radical

tert-Butylcyclopentane
Spectrum B

Isopropylcyclopentyl cation
m/z= 111

Methyl radical

The peak in Spectrum **A** at 97 u is from the loss of an ethyl radical fragment and this is only possible for butylcyclopentane. The peak in Spectrum **B** at 111 u is from the loss of a methyl, which is more likely to occur in *tert*-butylcyclopentane due to the stability of the tertiary cation. Therefore, Spectrum **A** is from butylcyclopentane and Spectrum **B** is from *tert*-butylcyclopentane. Notice that the M^+ peak for **B** is absent because the 3° carbocation that is produced upon fragmentation is a relatively stable cation.

Problem IC2.3

Think

Is there a multiple bond adjacent to an atom lacking an octet? What moves upon going from one resonance structure to another? What does not move? What is the hybridization of each carbon atom around the ring? How many resonance structures are possible?

Solve

Only nonbonding electrons and π electrons move in resonance structures. Atoms and σ bonds do not move. All of the atoms in the tropylium are sp^2 hybridized and, therefore, resonance is possible around the entire ring. The seven equivalent resonance structures and the hybrid are shown below:

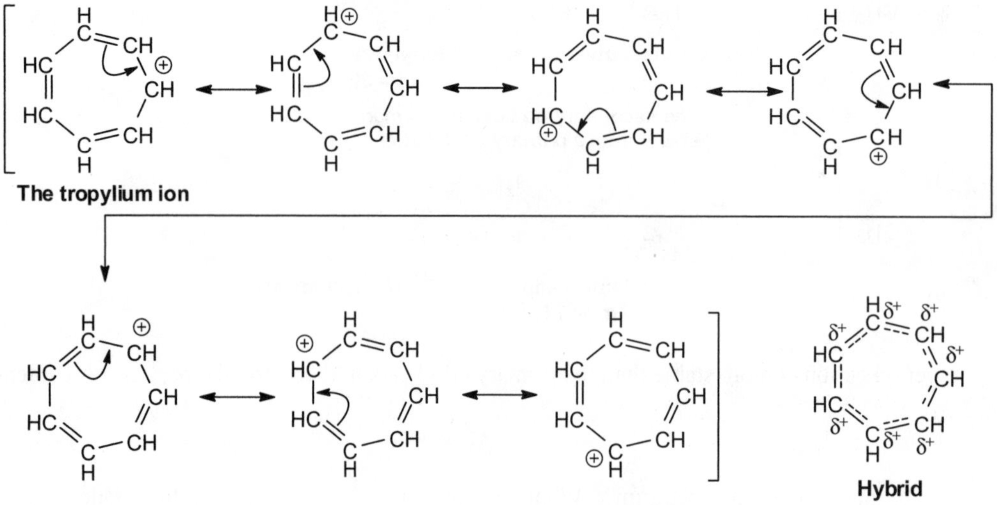

The tropylium ion

Hybrid

Problem IC2.4

Think

What is the mass of hex-3-ene? What mass is lost to give the peak at 69 u? What common fragmentation pathways are available to the molecular ion of an alkene?

Solve

The mass of hex-3-ene is 84 u, so the peak at 69 u is from the loss of a fragment that is 15 u, which is a methyl radical. This fragmentation of the original allylic C–C bond is common to alkenes, which results in the formation of the allylic 3-pentenyl cation.

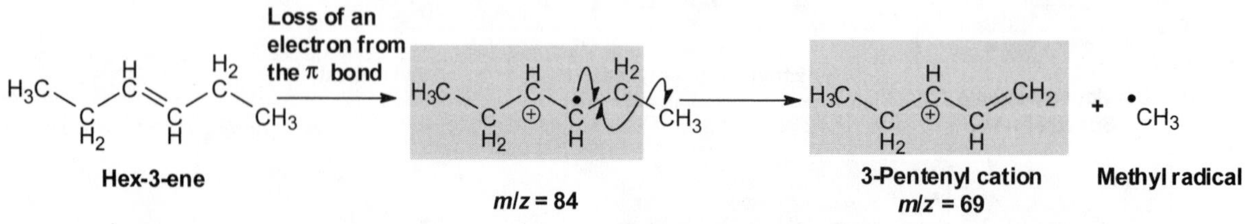

Loss of an electron from the π bond

Hex-3-ene

$m/z = 84$

3-Pentenyl cation
$m/z = 69$

Methyl radical

Problem IC2.5

Think

What is the mass of 1,4-diethylbenzene and 1-methyl-4-propylbenzene? What radical is lost and gives rise to the peaks at 105 u for Spectrum **A** and 119 u for Spectrum **B**? Which compound is capable of losing that radical?

Solve

Both 1,4-diethylbenzene and 1-methyl-4-propylbenzene have a mass of 134 u. The base peak in Spectrum **A** at 105 u is due to loss of an ethyl fragment and the base peak in Spectrum **B** at 119 u is due to a loss of a methyl fragment. Therefore 1,4-diethylbenzene is Spectrum **B** and 1-methyl-4-propylbenzene is Spectrum **A**. The M^+ of **A** can lose a CH_3 radical via the cleavage of an original benzylic C–C bond, and the M^+ of **B** can lose an ethyl radical via the cleavage of an original benzylic C–C bond.

Loss of an electron from the π bond

1,4-Diethylbenzene
Spectrum B

***m/z* = 134**

Benzylic cation
***m/z* = 119**
Spectrum B

Methyl radical

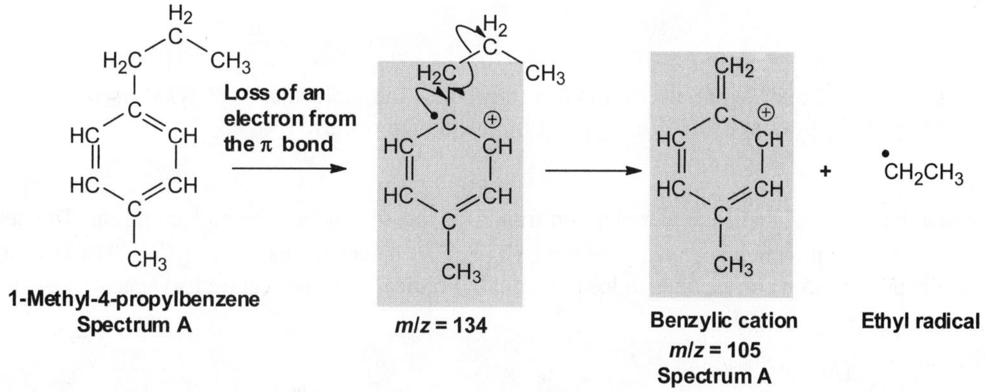

Loss of an electron from the π bond

1-Methyl-4-propylbenzene
Spectrum A

***m/z* = 134**

Benzylic cation
***m/z* = 105**
Spectrum A

Ethyl radical

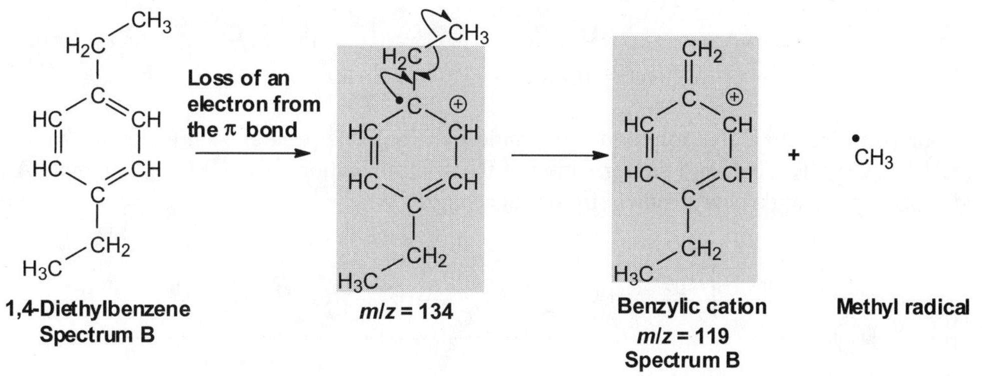

Problem IC2.6

Think

What is the mass of pentan-1-ol? What electron is lost to produce the molecular ion? What mass is lost to give the peak at 31 u? What type of fragmentation is common to the molecular ions of alcohols?

Solve

Pentan-1-ol has a mass of 88 u. The parent ion is produced from the loss of a lone pair electron on O. The peak at 31 u is due to the loss of 57 u from the molecular ion, which is a butyl radical, $CH_3CH_2CH_2CH_2$•. This leaves the cation $^+CH_2$–OH at 31 u. Therefore, the peak at 31 u corresponds to loss of a butyl radical from M^+ via α-cleavage.

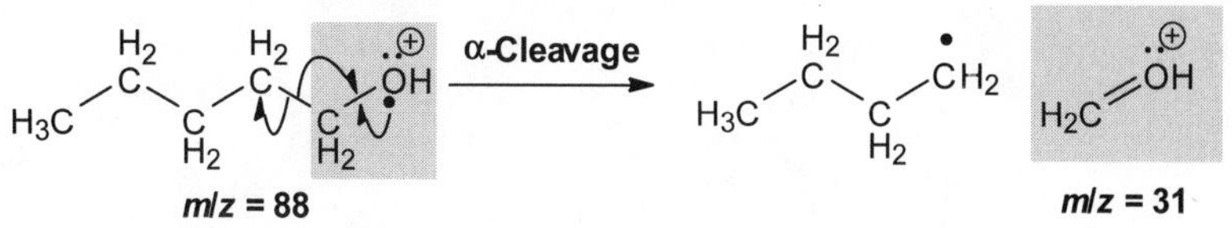

α-Cleavage

***m/z* = 88**

***m/z* = 31**

Problem IC2.7

Think

What is the mass of compounds **A**–**C**? What electron is lost from an amine to produce the molecular ion? What mass is lost from the molecular ion to give the base peak at m/z = 30 u? Which compound is capable of losing the radical of the appropriate mass? Can that fragmentation occur via α-cleavage?

Solve

The mass of each compound is 101 u.

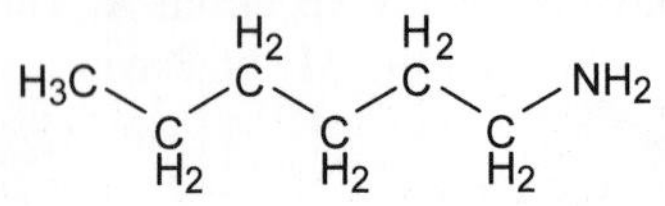

A 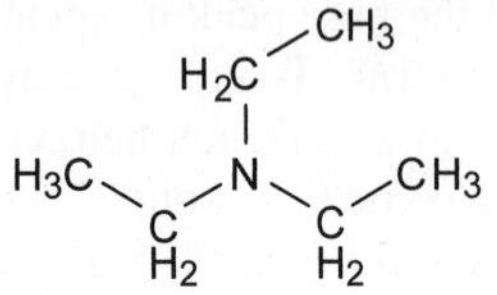 B 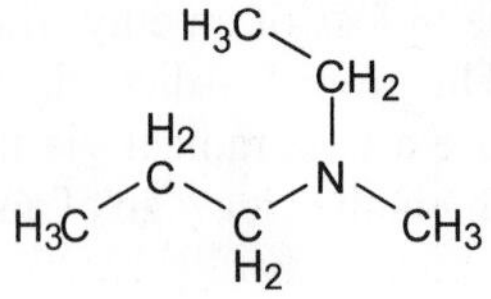 C

m/z = 101 u

In each case, a lone pair electron is lost to produce the molecular ion. The peak at 30 u is due to the loss of a pentyl radical, $CH_3CH_2CH_2CH_2CH_2\bullet$, and the formation of the iminium cation, $H_2C{=}^{+}NH_2$. Compound **A** is the only compound capable of losing a pentyl radical from α-cleavage.

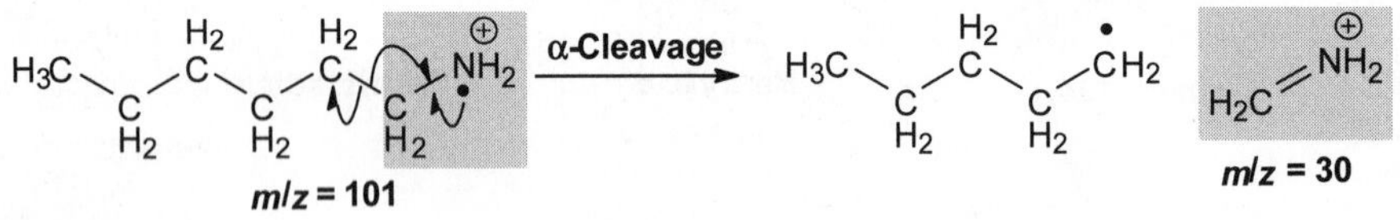

α-Cleavage

m/z = 101 m/z = 30

Problem IC2.8

Think

What is the mass of hexan-2-one? What electron is lost to produce the molecular ion? What mass is lost from the molecular ion to give the peak at 85 u? What type of fragmentation occurred?

Solve

Hexan-2-one has a mass of 100 u. A lone pair electron from O is lost to produce the molecular ion. The peak at 85 u is due to the loss of 15 u, which is a methyl radical, $CH_3\bullet$. This leaves the cation $CH_3CH_2CH_2CH_2C{\equiv}O^+$ at 85 u. Therefore, the peak at 85 u corresponds to loss of a methyl radical from M^+ via α-cleavage.

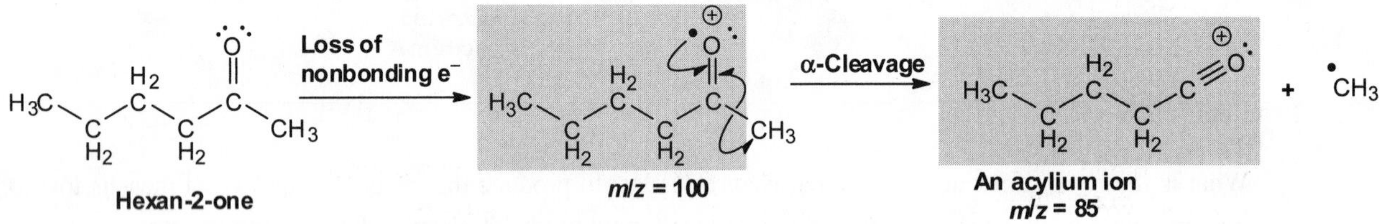

Hexan-2-one — Loss of nonbonding e⁻ → m/z = 100 — α-Cleavage → An acylium ion m/z = 85 + $\bullet CH_3$

Problem IC2.9

Think

What are the structures of heptan-2-one, heptan-3-one, and heptan-4-one? What is the mass of these three compounds? Which electron is lost to produce the molecular ion? What mass is lost to give the peak at 57 u? Which compound is capable of losing the radical of the appropriate mass through a common fragmentation pathway of ketones?

Solve

The mass of each compound is 114 u.

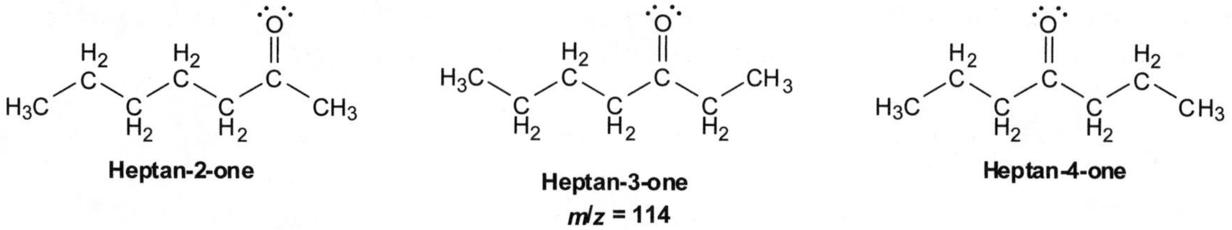

Heptan-2-one **Heptan-3-one** $m/z = 114$ **Heptan-4-one**

In each case, the molecular ion is produced from the loss of a lone pair electron on O. The peak at 57 u is due to the loss of a butyl radical, $CH_3CH_2CH_2CH_2$•, and the formation of the acylium cation, $CH_3CH_2C{\equiv}O^+$. Hepan-3-one is the only compound capable of losing a butyl radical from α-cleavage.

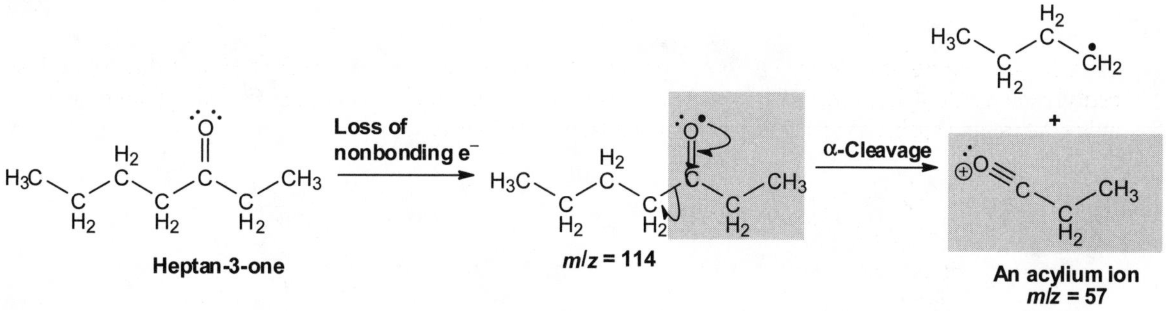

Heptan-3-one — Loss of nonbonding e⁻ → $m/z = 114$ — α-Cleavage → **An acylium ion** $m/z = 57$ + $CH_3CH_2CH_2\dot{C}H_2$

Problem IC2.10

Think

What is the mass of hexanamide? Which electron is lost to produce the molecular ion? What mass is lost to give the peak at 59 u? What type of fragmentation occurred?

Solve

The mass of hexanamide is 115. The peak at 59 u is due to the loss of a but-1-ene, $CH_3CH_2CH{=}CH_2$, and the formation of the enol radical cation $CH_2{=}C(^+OH)NH_2$. This is a McLafferty rearrangement of an amide.

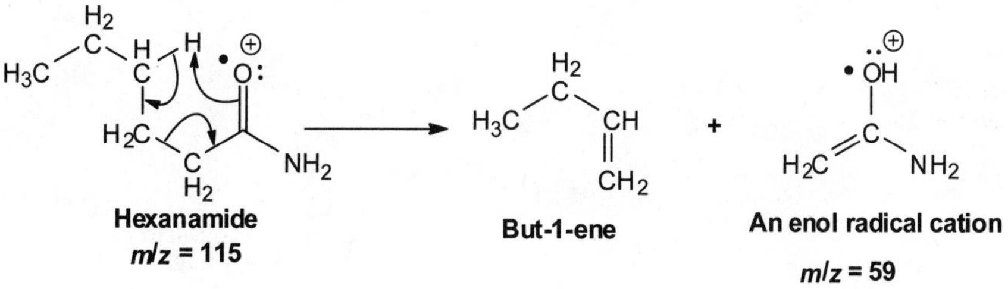

Hexanamide $m/z = 115$ → **But-1-ene** + **An enol radical cation** $m/z = 59$

CHAPTER 26 | Polymers

Your Turn Exercises

Your Turn 26.1

Think

Which carbon atoms make up the main chain? Which carbon groups are not part of this main chain?

Solve

The carbon atoms that make up the main chain or backbone are the carbon atoms that are connected to each other, as shown below (boxed). The side groups are the groups "hanging" off the main chain. In this example, the side groups are CH_3, or methyl groups.

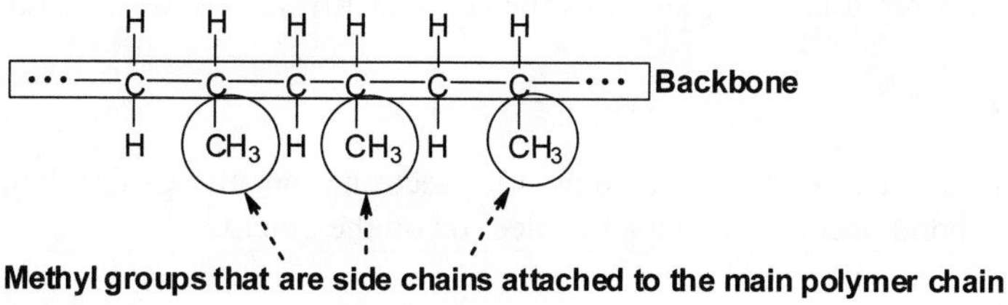

Backbone

Methyl groups that are side chains attached to the main polymer chain

Your Turn 26.2

Think

What atoms and bonds make up the main chain? Are the groups attached to the second carbon the same as the ones attached to the first? What about the third carbon? The fourth? What pattern is established, and how many times is it shown to repeat?

Solve

The main chain consists of C–C single bonds only. Two H atoms are attached to the first C shown, whereas the second C has an attached H and CH_3. That pattern is repeated for each of the next pairs of C atoms along the main chain, so the repeating unit involves just two atoms of the main chain. In the structure given, the repeating unit appears three times, so $n = 3$.

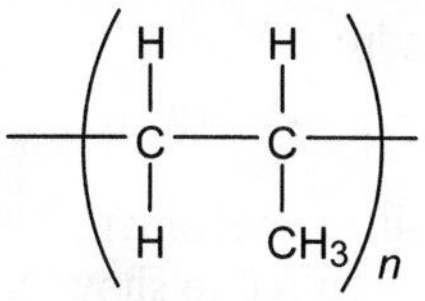

Your Turn 26.3

Think

What type of curved arrows is used in forming a radical via homolysis? How many electrons are represented by each curved arrow? Which bonds are broken and formed? To draw a resonance structure of the resulting radical, can you identify a π bond adjacent to the unpaired electron? Where do the unpaired electron and the electrons from the π bond end up, and how many curved arrows must be used to show that movement?

Solve

To show electron movement involving radicals, use single-barbed curved arrows; the movement of one electron is represented by each curved arrow. Thus, to move two electrons, two arrows are needed, and to move three electrons, three arrows are needed. In the first equation on the next page, the O–O bond is homolytically cleaved to form the oxygen radical, •O. In that radical, an unpaired electron on O is attached to a C=O bond. To arrive at the other resonance structure, the unpaired electron joins with one electron from the C=O π bond, and the other electron from the π bond ends up on the second O. Three single-barbed arrows are used to show this electron movement.

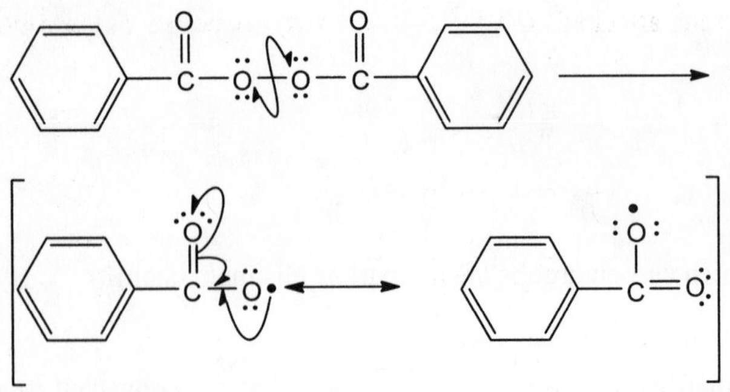

Your Turn 26.4

Think

Which atoms form the new bond? How is this represented by single-barbed curved arrows? On which atom will an unpaired electron appear?

Solve

A new C–C bond is formed when the unpaired electron on •C joins one electron from the C=C π bond on styrene. The other electron from the C=C π bond becomes an unpaired electron on the other C.

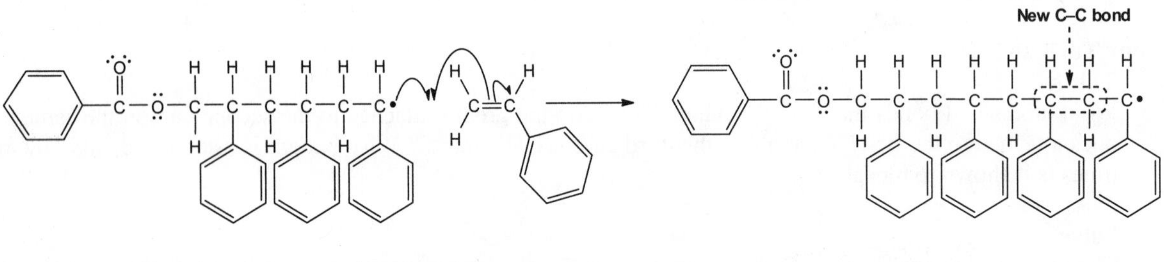

Your Turn 26.5

Think

On the polymer and monomer, which part is the head and which part is the tail? Making sure that the two species are aligned head to tail, draw the curved arrows to show the next radical addition step. How many curved arrows are necessary? In the product, is there an unpaired electron adjacent to a π bond?

Solve

In the growing polymer chain, the head is the C atom with the unpaired electron. In a monomer of ethyl acrylate, the tail is the C atom attached to two H atoms. Three single-barbed arrows are required to show radical addition, as shown below for the head-to-tail bond formation.

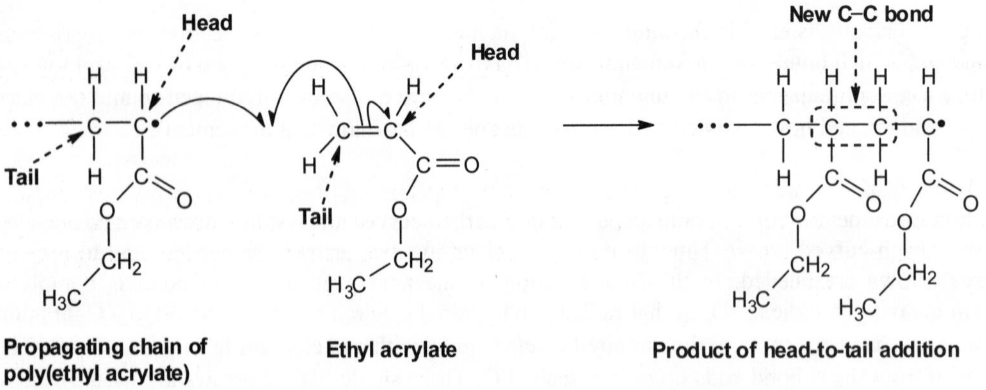

Propagating chain of poly(ethyl acrylate)

Ethyl acrylate

Product of head-to-tail addition

In the product, the unpaired electron is on an atom attached to the C=O π bond, so it has the additional resonance structure as shown below:

Product of head-to-tail addition

In head-to-head addition shown below, there is no resonance stabilization of the product. In addition, the steric interactions between the acrylate groups (boxed) will hinder head-to-head addition.

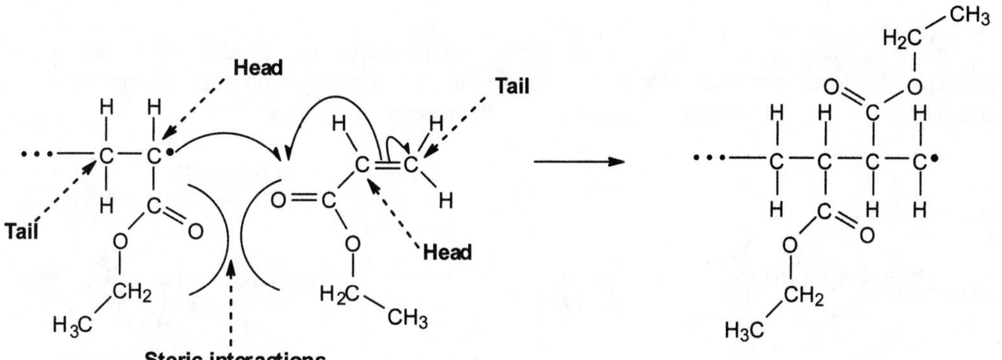

Product of head-to-head addition

Your Turn 26.6

Think

How many atoms of each kind appear on the reactant side? On the product side? What atoms seem to be missing from the products? What molecule can be made out of that collection of atoms?

Solve

The atoms that appear on the reactant side are 12 C, 16 H, and 6 O. The atoms that appear on the product side are 10 C, 8 H, and 4 O. Therefore, on the product side, there appear to be missing 2 C, 8 H, and 2 O. These come from two molecules of CH_3OH.

Your Turn 26.7

Think

What functional group is present in the repeating unit of nylon? Does that functional group appear more than once? Is this functional group part of the main chain?

Solve

Nylon is a polyamide. The amide group appears twice in the repeating unit as shown on the next page. One complete amide appears in the center of the repeating unit. The second amide is split between two repeating units—the N at the right of one repeating unit is connected to a C=O group that is part of the next repeating unit (represented by the C=O at the far left).

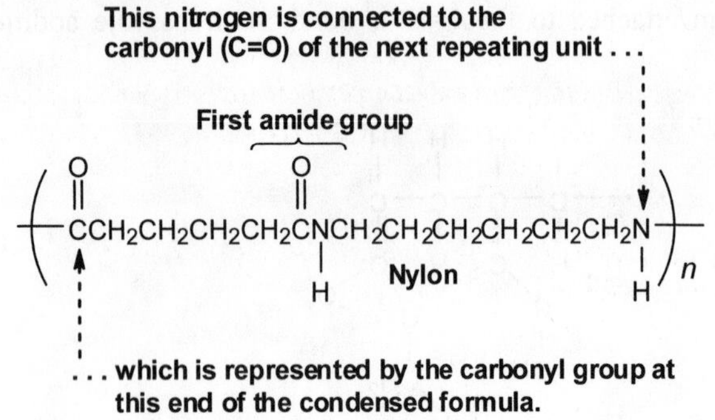

This nitrogen is connected to the carbonyl (C=O) of the next repeating unit . . .

First amide group

$$\left(\overset{O}{\overset{\|}{C}}CH_2CH_2CH_2CH_2\overset{O}{\overset{\|}{C}}\underset{H}{N}CH_2CH_2CH_2CH_2CH_2CH_2\underset{H}{N} \right)_n$$

Nylon

. . . which is represented by the carbonyl group at this end of the condensed formula.

Your Turn 26.8

Think

What are stabilizing factors for a negative charge? Are the negative charges on the same atom? Are the negative charges equally delocalized via resonance? Are inductive effects present?

Solve

The negative charge appears on C in both the reactants and products. The product in Equation 26-16 is resonance stabilized due to the presence of π bonds next to the carbanion electron pair. Resonance is possible around the entire benzene ring. No such resonance exists in the reactants.

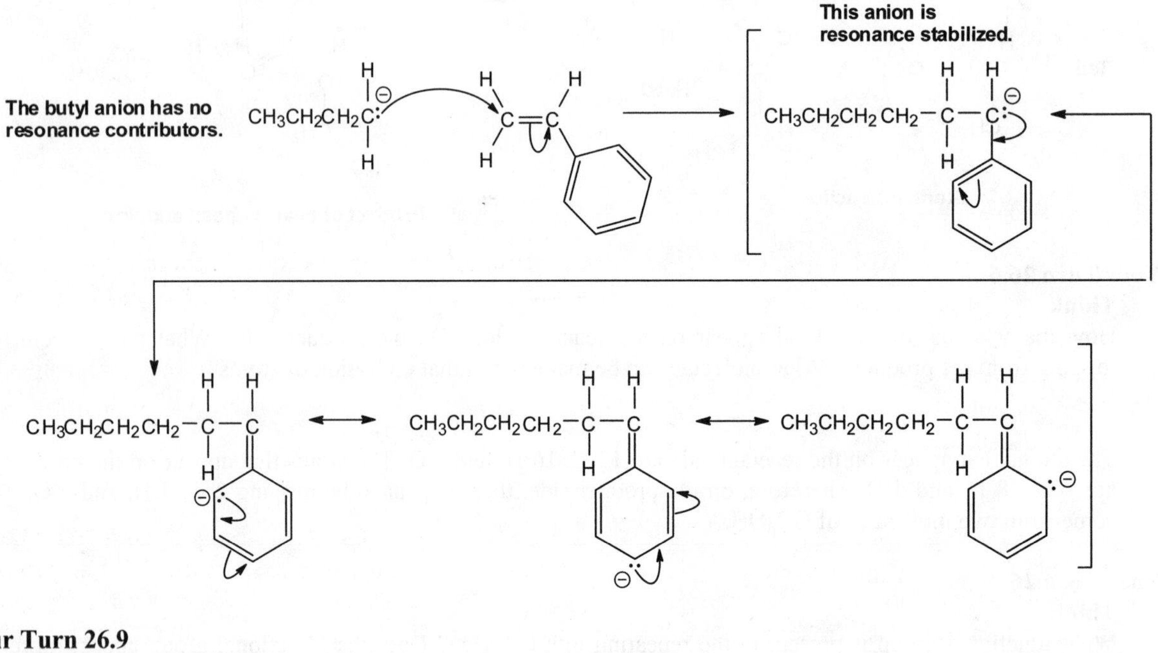

The butyl anion has no resonance contributors.

This anion is resonance stabilized.

Your Turn 26.9

Think

In a free radical mechanism, what types of species react in a combination or disproportionation step? What types of species would be analogous to these radicals in an anionic mechanism? What factor would prevent these species from coming together?

Solve

In a radical mechanism, combination and disproportionation each require two radicals reacting. The analogous species in an anionic mechanism are anions. Neither of these steps is likely because the like charges of the two anions would repel each other.

Your Turn 26.10

Think

What type of compound is added to induce termination? What are the reactive parts of the product in Equation 26-21? What elementary step will take place as a result?

Solve

An acid is added to induce termination. Each negatively charged O is strongly basic and, therefore, will be protonated. Because there are two O^- sites, two proton transfer steps will take place.

$$^{-}:\ddot{O}{+}CH_2CH_2O{+}_n CH_2CH_2\ddot{O}:^{-} \xrightarrow{H-B} :B^- + \ ^{-}:\ddot{O}{+}CH_2CH_2O{+}_n CH_2CH_2\ddot{O}H \xrightarrow{H-B} :B^- + H\ddot{O}{+}CH_2CH_2O{+}_n CH_2CH_2\ddot{O}H$$

Your Turn 26.11

Think

Do the keto and enol tautomers have different charge stability? Different total bond strengths? Which bond is particularly strong?

Solve

Vinyl alcohol is an enol and will tautomerize to an aldehyde, as shown below. Both molecules are uncharged, so they do not exhibit differences in charge stability. The C=O bond in the keto form, however, is particularly strong, which helps give the keto form a greater total bond strength. Thus, the keto form is significantly more stable than the enol form.

$$H_2C{=}CH{-}OH \rightleftharpoons H_3C{-}CH{=}O$$

Vinyl alcohol
Enol form

Acetaldehyde or ethanal
Keto form

Your Turn 26.12

Think

What functional groups are present? What type of bond breaks during the course of the reaction? What type of species should CH_3O^- behave like to break that bond? Can that bond break in a single step, or will it require multiple steps?

Solve

An ester group is part of the repeating unit of the polymer, and during the course of the reaction, the bond between O and C=O must break. This can occur when CH_3O^- acts as a nucleophile and RO^- (as part of the polymer) acts as a leaving group from the C=O bond as shown on the next page. Because the leaving group is attached to a C=O, this reaction takes place in a nucleophilic addition–elimination mechanism. The second product will be methyl acetate.

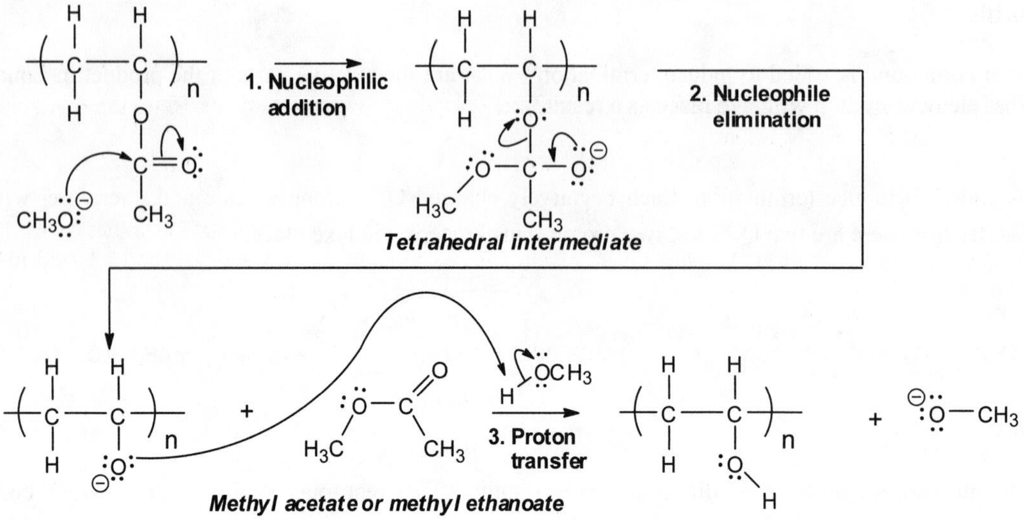

1. Nucleophilic addition

2. Nucleophile elimination

Tetrahedral intermediate

3. Proton transfer

Methyl acetate or methyl ethanoate

Your Turn 26.13

Think

If radicals are present, what kinds of curved arrows are used to show electron movement? What bonds are formed and broken in the first step? In the second step? What are the names of each elementary step, and how many curved arrows are needed to show the proper electron movement?

Solve

Because radicals are present, single-barbed curved arrows are used to show electron movement. The first step is a hydrogen atom abstraction, whereby one C–H bond is formed and a second one is broken simultaneously. The second step is a radical addition, whereby a C–C single bond is formed and a C=C π bond is broken. Both steps require three single-barbed curved arrows.

(a)

(b)

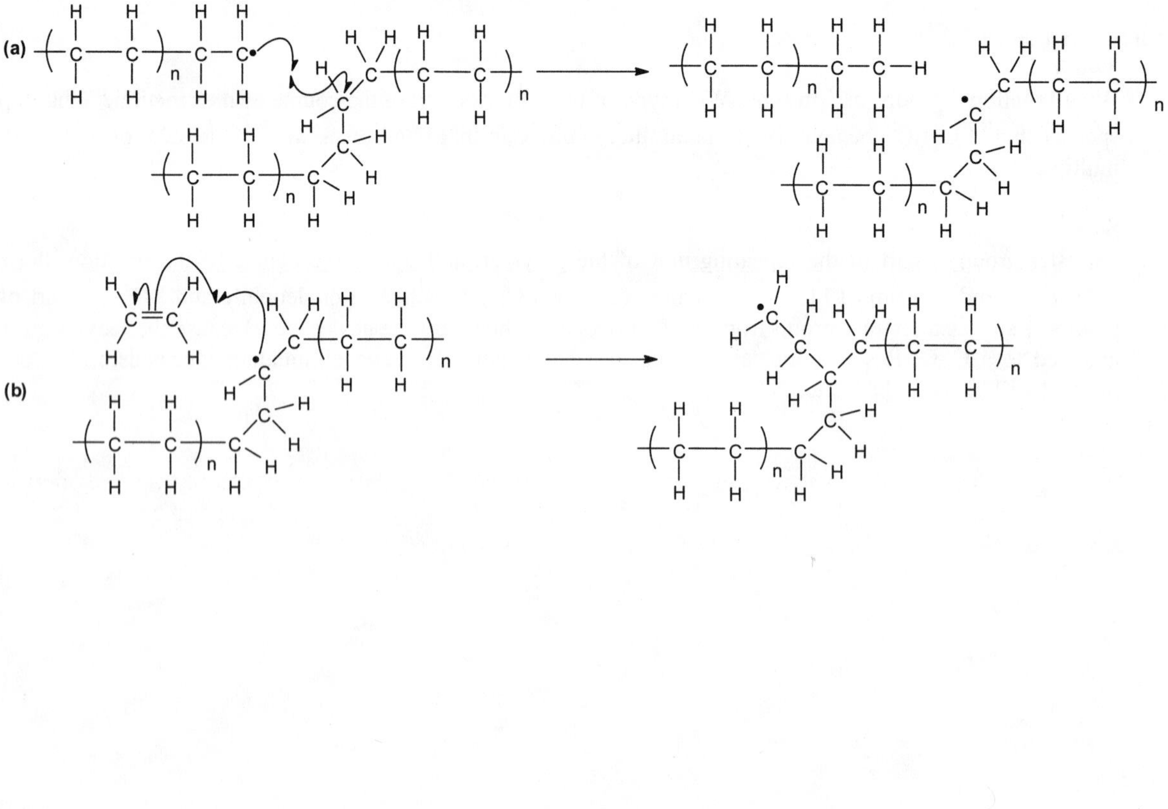

Chapter Problems

Problem 26.1

Think

What is the repeating unit? What bonds are formed to yield a polymer? Which atoms make up the main chain? Which atoms are part of the substituent?

Solve

The repeating unit is shown in parentheses. In both cases, a new C–C bond is formed on either side of the repeating unit. In **(a)**, there are four F substituents attached to the main chain of the repeating unit, and in **(b)**, there are two Cl substituents. The structures below show the repeating units repeated three times.

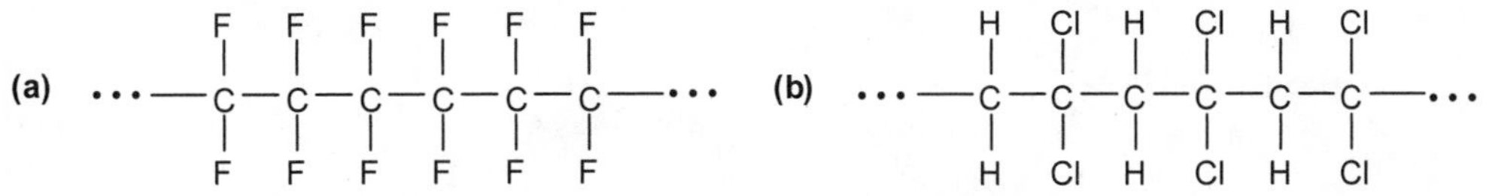

Problem 26.3

Think

What is the repeating unit of poly(methyl methacrylate)? How is the repeating unit of a vinyl polymer related to the monomer? What bonds exist in the monomer that do not exist in the repeating unit, and vice versa?

Solve

The C atoms in the structure make up the main chain of the molecule. The CH_3 and CO_2CH_3 groups are substituents, or side groups, analogous to the phenyl rings in polystyrene. The repeating unit is as follows:

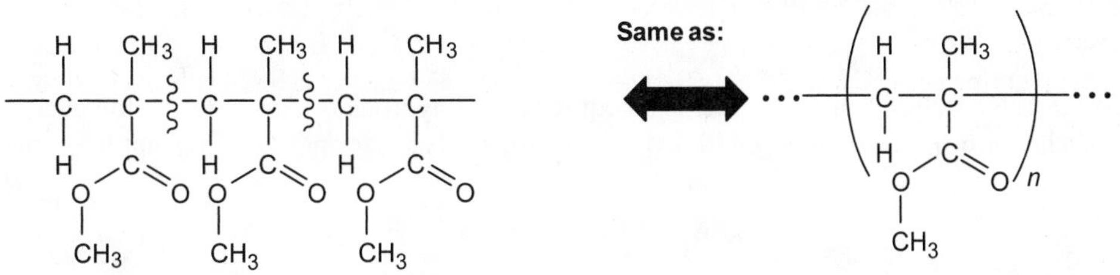

Same as:

To determine the monomer from the repeating unit, remove the single bonds on the outside the repeating unit and change the C–C single bond to a C=C double bond. The monomer is methyl methacrylate.

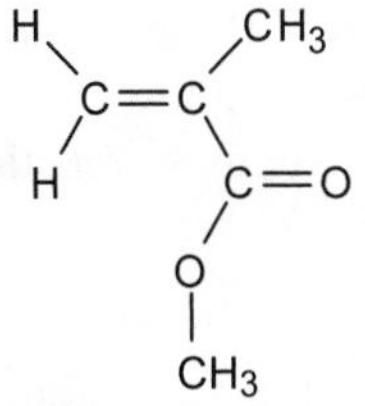

Problem 26.4

Think

What is the difference in structure between the two monomers? Is head-to-tail or tail-to-head polymerization favored? What are the head and tail in each monomer? How does the extra methyl group in crotonic acid affect the ability of the monomer and the growing polymer chain to come together?

Solve

Both monomers will undergo head-to-tail addition. The extra methyl group in crotonic acid creates steric hindrance and slows the rate of polymerization.

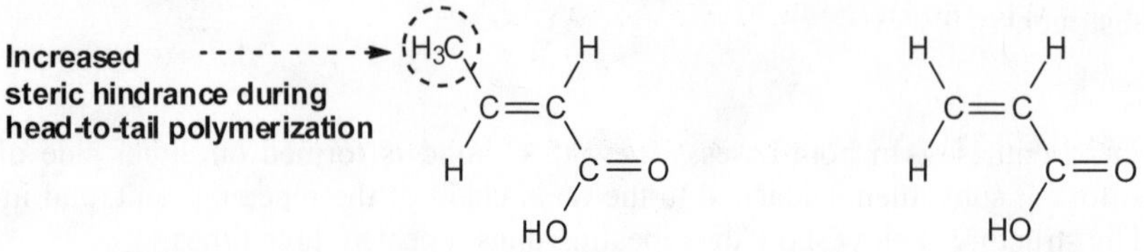

Increased steric hindrance during head-to-tail polymerization

Crotonic acid

Acrylic acid

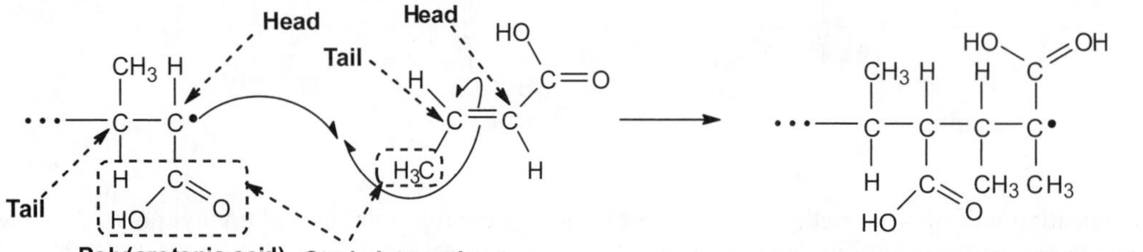

Head

Head

Tail

Tail

Poly(crotonic acid)

Steric interactions

Product of head-to-head addition

Problem 26.5

Think

What functional group results from the reaction of an acid chloride and an amine? What new bond is formed? What leaves from the acid chloride and the amine?

Solve

One product will be the nylon polymer and the other is an equivalent of hydrochloric acid, HCl. However, a base in the system (such as 1,6-hexanediamine) will deprotonate the HCl as it forms, so HCl is not the actual product.

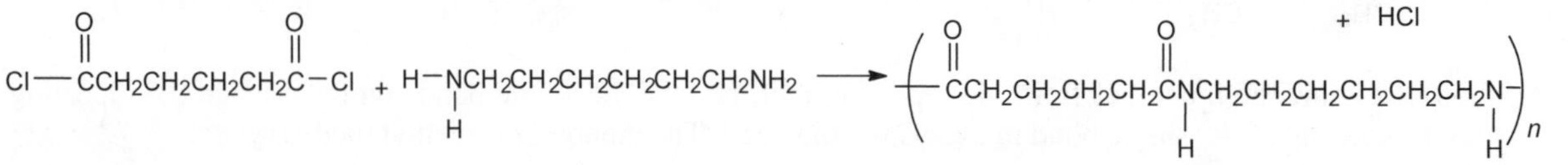

$$Cl{-}\overset{O}{\overset{\|}{C}}CH_2CH_2CH_2CH_2\overset{O}{\overset{\|}{C}}{-}Cl + H{-}\underset{H}{\underset{|}{N}}CH_2CH_2CH_2CH_2CH_2CH_2NH_2 \longrightarrow \left(\overset{O}{\overset{\|}{C}}CH_2CH_2CH_2CH_2\overset{O}{\overset{\|}{C}}\underset{H}{\underset{|}{N}}CH_2CH_2CH_2CH_2CH_2CH_2\underset{H}{\underset{|}{N}} \right)_n + HCl$$

Nylon

Problem 26.7

Think

What functional group makes up the backbone? What other functional groups are present? Are these functional groups part of the polymer's repeating unit?

Solve

Amide and alcohol functional groups are both present in the polymer's repeating unit; thus, polyamide or polyalcohol could be used to describe polyserine.

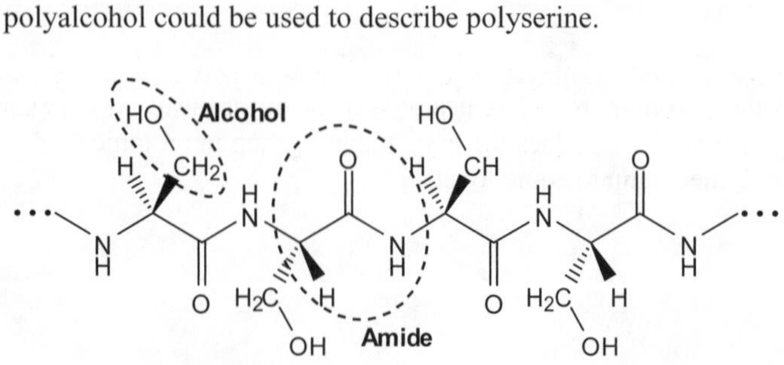

Alcohol

Amide

Polyserine

Problem 26.8

Think

How is the structure of a monomer related to the structure of the repeating unit? How many words are in the name of the monomer? Are parentheses necessary in the name of the polymer? What are the initial letters of the different portions of the monomer? What bonds are formed in the polymer?

Solve

The condensed formula is given below. The C–C bonds form from the C=C carbon atoms.

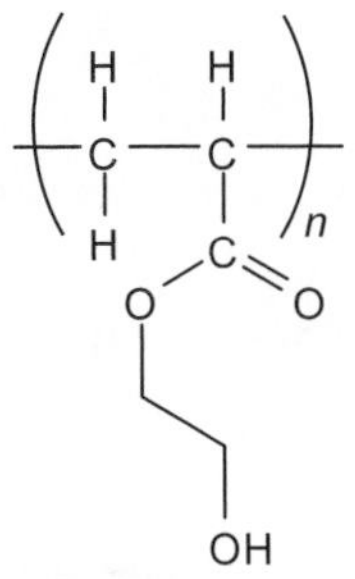

The polymer comes from 2-hydroxyethyl acrylate, which is two words and thus requires parentheses. The polymer is, therefore, poly(2-hydroxyethyl acrylate), or PHEA.

Problem 26.9

Think

What is the structure of poly(ethylene terephthalate) (PET)? What functional group forms from an acid chloride and an alcohol? What is the significance of having a diacid chloride and a diol? How are the starting materials similar/different compared to the ones shown in Equations 26-24 and 26-25?

Solve

One starting material is the same: ethylene glycol. The other material is similar to the one shown in Equations 26-24 and 26-25, with the exception of the carboxylic acid derivative functional group. The new starting material is a diacid chloride rather than a dicarboxylic acid or diester. The reaction is given below:

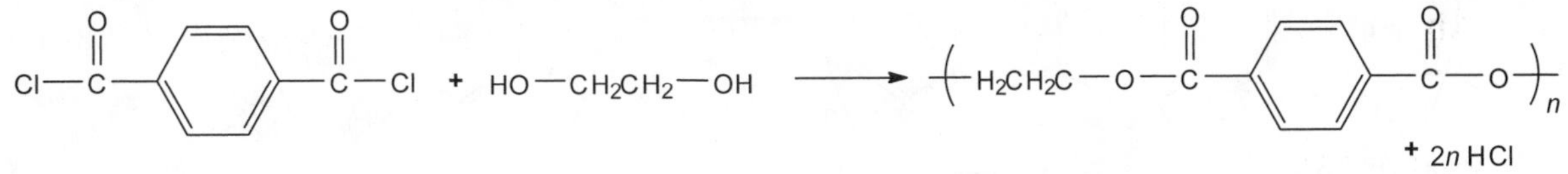

$$\text{Cl–C(=O)–C}_6\text{H}_4\text{–C(=O)–Cl} + \text{HO–CH}_2\text{CH}_2\text{–OH} \longrightarrow \left[\text{H}_2\text{CH}_2\text{C–O–C(=O)–C}_6\text{H}_4\text{–C(=O)–O}\right]_n + 2n\,\text{HCl}$$

Problem 26.10

Think

Given that the reaction occurs via one nucleophilic addition to a polar double bond followed by two proton transfers, identify the nucleophile and electrophile in the first step. What species facilitates the proton transfer reactions?

Solve

As shown on the next page, the electron-rich oxygen in ethylene glycol donates a pair of electrons to the electron-poor carbon in the isocyanate group, and a shared pair of electrons becomes a lone pair. Then the positively charged O is deprotonated and the negatively charged N is protonated:

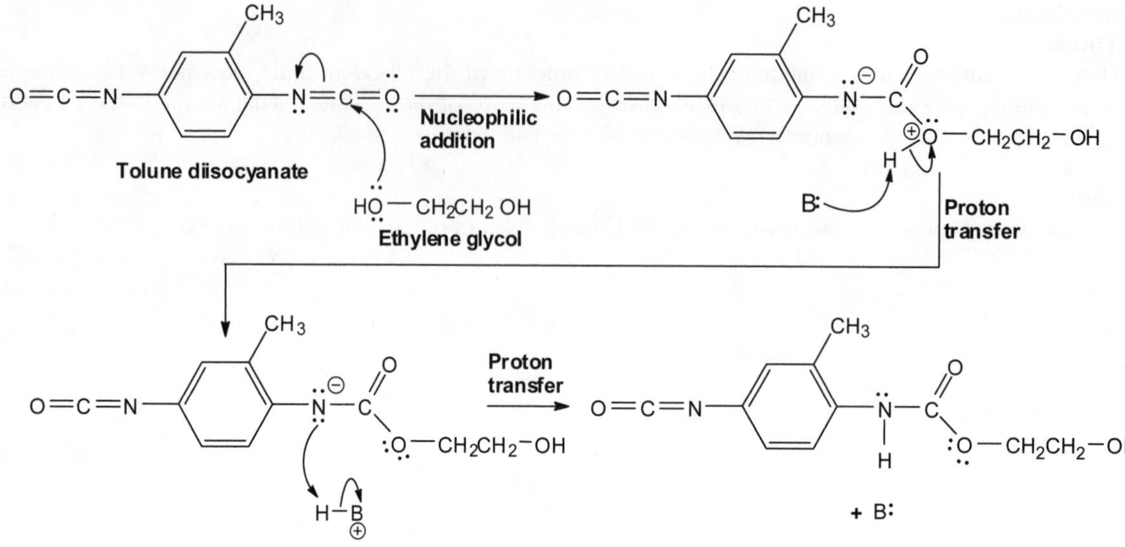

Tolune diisocyanate

Nucleophilic addition

Ethylene glycol

Proton transfer

Proton transfer

Problem 26.11

Think

What is the hybridization of the neighboring C atoms? Is resonance possible between the O atom and adjacent benzene rings? If so, what is the hybridization of the O atom in the resonance structure? What is the ideal C–O–C bond angle that corresponds to that hybridization?

Solve

When a heteroatom with a lone pair is adjacent to a benzene ring, the lone pair of electrons is involved in resonance with the ring. For example, a lone pair on the phenolic O atom can be delocalized into the aromatic ring:

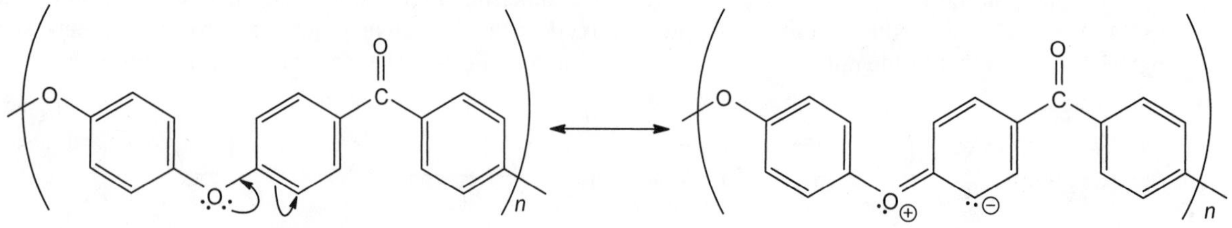

Poly(ether ether ketone) or PEEK

In this resonance structure, the O atom has three electron groups, consistent with an sp^2 hybridization and a ~120° bond angle. Experimental evidence that supports this view includes the bond angles in the following two ethers:

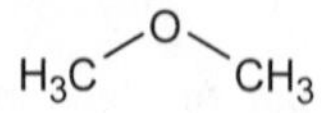

Dimethyl ether

C–O–C bond angle: 111°

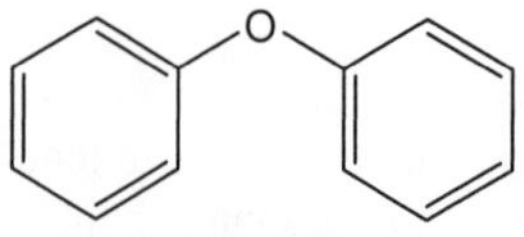

Diphenyl ether

C–O–C bond angle: 119–129°

Dimethyl ether contains an O that is sp^3 hybridized, and the angle is close to the 109.5° that we expect. Diphenyl ether contains an O that is sp^2 hybridized, so a lone pair on O can be included in the conjugated system and extend the conjugation of the system throughout the whole molecule.

Problem 26.12

Think

What is the mechanism for electrophilic aromatic substitution? What electrophile is lost from the aromatic ring in PEEK with a *tert*-butyl substituent? What is the stability of this cationic electrophile? Is a linear carbocation as stable?

Solve

Recall the steps for electrophilic aromatic substitution: electrophilic addition followed by electrophile elimination. The mechanism for the trans *tert*-butylation, therefore, is two back-to-back electrophilic aromatic substitution reactions. The electrophile lost in the first electrophile elimination step is the electrophile that reacts with toluene in the second electrophilic aromatic substitution reaction.

CF_3SO_3H · PEEK with a *tert*-butyl substituent · Electrophilic addition · Electrophile elimination · Electrophilic addition · Electrophile elimination · CH_3 · B: · BH

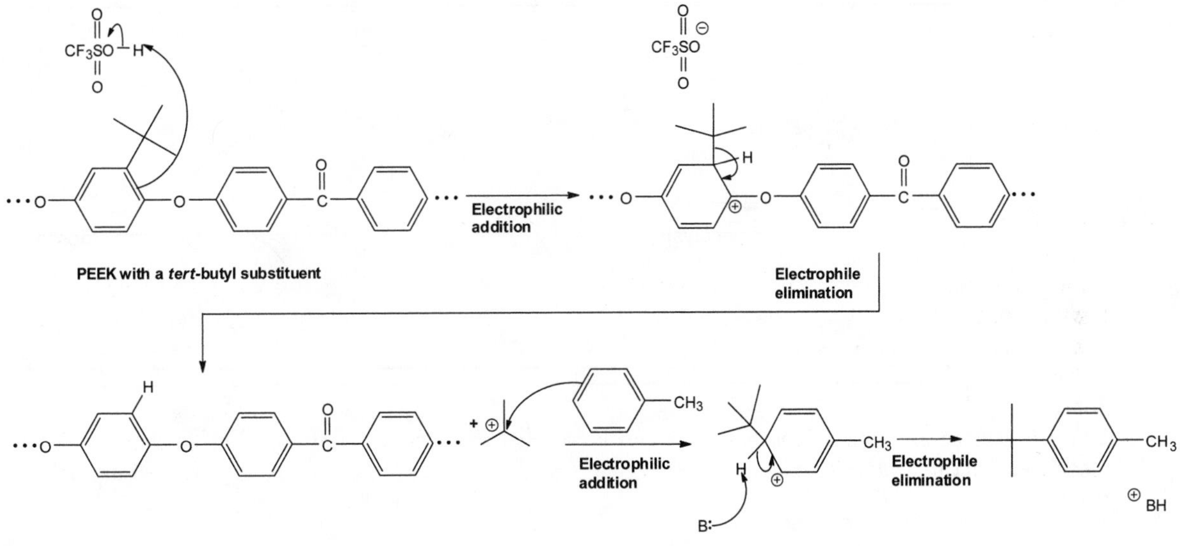

Problem 26.13

Think

What functional group is formed in the product? What is the purpose of the acid catalyst? Is the aldehyde or alcohol the electrophile? Which is the nucleophile? Are proton transfer reactions involved? What are reasonable leaving groups in the mechanism?

Solve

The functional group formed is an acetal (Chapter 18), which forms via a nucleophilic addition reaction of the alcohol to the C=O of the aldehyde. The reaction is acid catalyzed, so in the first step the C=O group is protonated to make the carbon of the C=O more electrophilic. Also, under these acidic conditions, a good leaving group, H_2O, is eventually produced. The full mechanism is given below and on the next page:

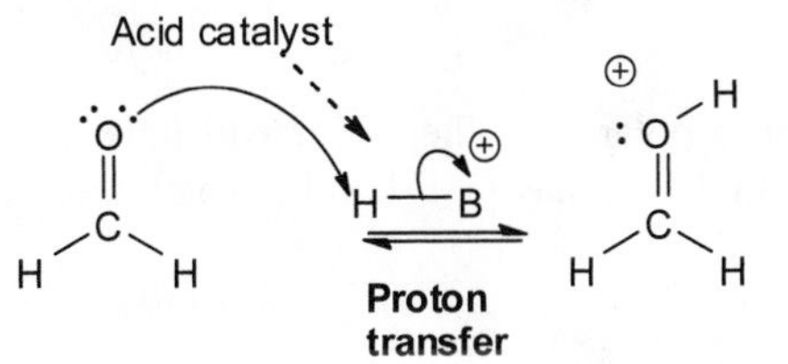

Acid catalyst · Proton transfer

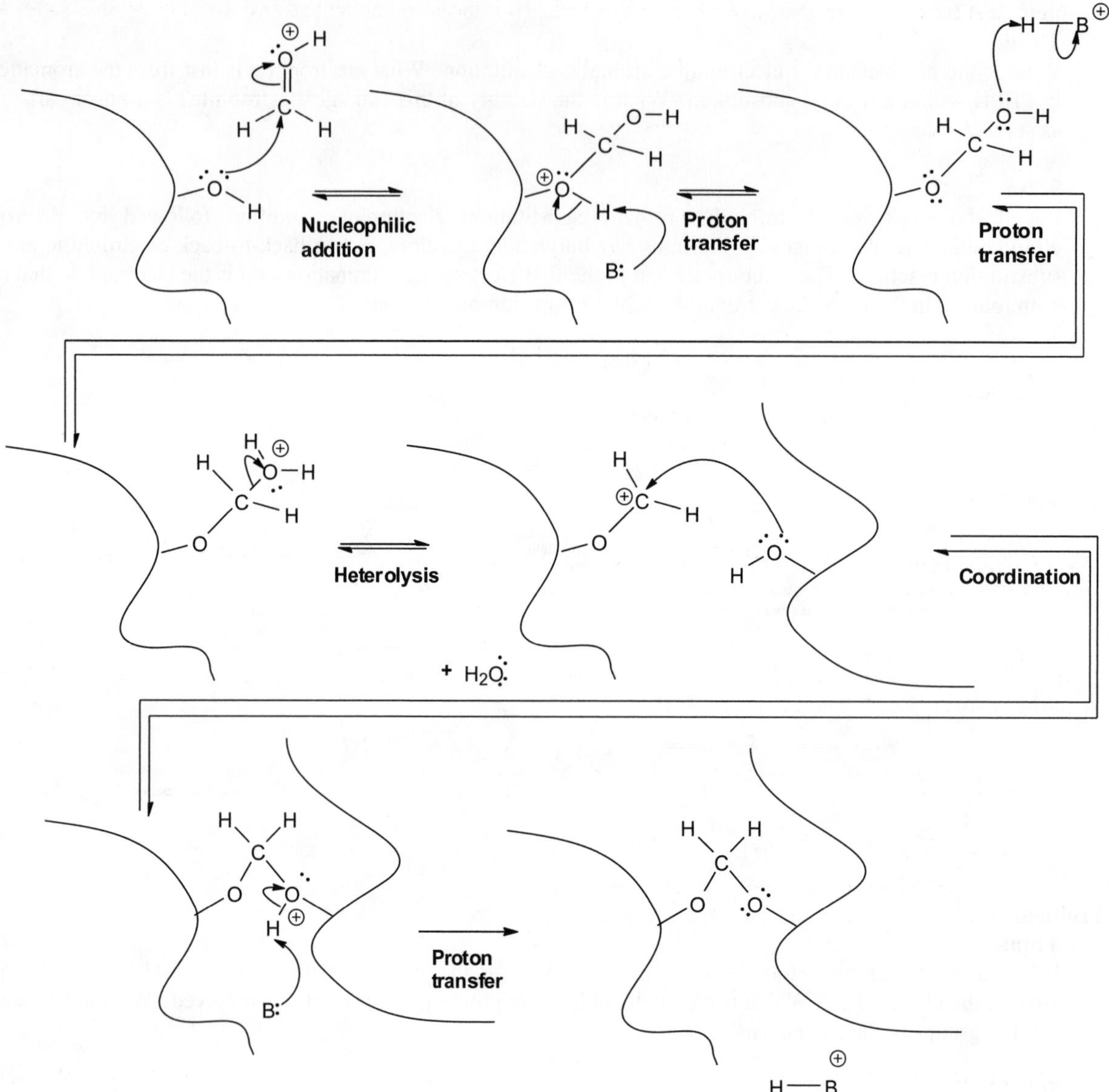

Problem 26.14

Think

Review the seven factors that affect the heat required to reach the T_g. What are the structural differences in each monomer? What intermolecular forces will differ?

Solve

(a) The second polymer will have a higher T_g because it is more polar and, therefore, will have stronger dipole–dipole interactions (factor #2). Even though the first molecule has C–F bonds, which have large bond dipoles, those bond dipoles effectively cancel.

$$\left(\mathrm{CF_2{-}CF_2}\right)_n \qquad \left(\mathrm{CH_2{-}CH_2{-}O}\right)_n$$

Higher T_g
More polar

(b) The second polymer will have a higher T_g because it is has ions and will have ionic interactions (factor #7).

Higher T_g
ion–ion interactions

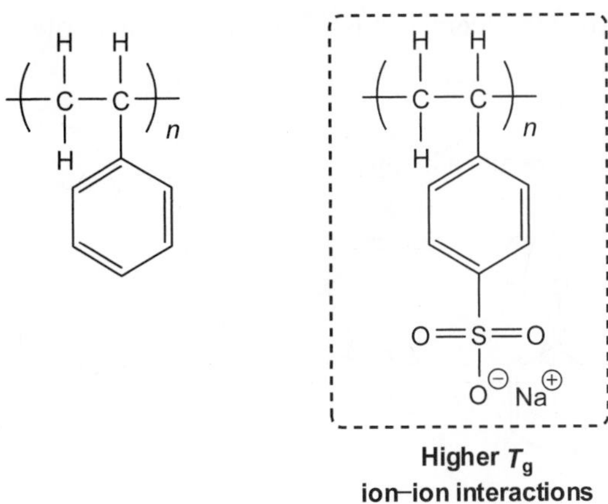

Problem 26.15

Think

Review the seven factors that affect the heat required to reach the T_g. What are the structural differences in each monomer? What intermolecular forces will differ?

Solve

(a) The second polymer will have a higher T_g because it is more polar and, therefore, will have stronger dipole–dipole interactions (factor #2).

Higher T_g
More polar

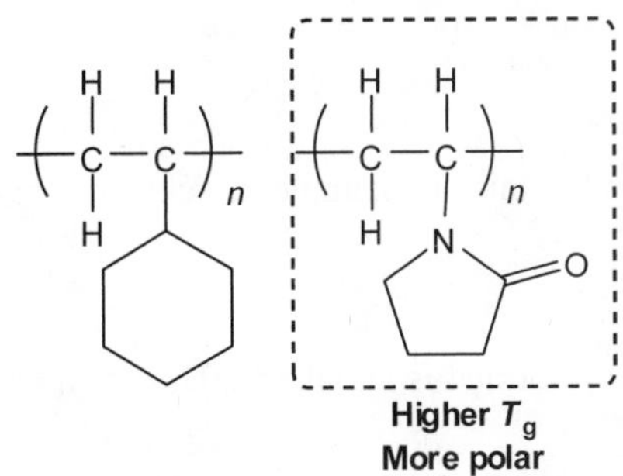

(b) The first polymer will have a higher T_g because it has more branching (factor #4). In the second polymer, however, the additional rigidity contributes to increasing its *T*g. As a result, the *T*g values of the two polymers are not very different: −70° for the first polymer and −73° for the second.

Similar T_g
More branching **More rigid**

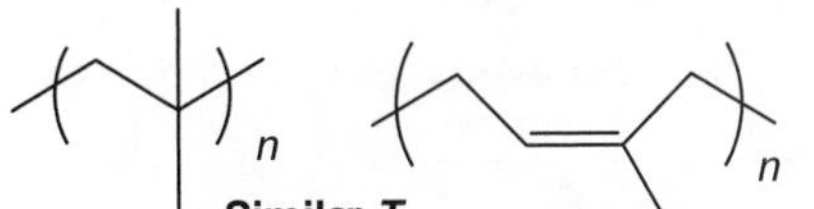

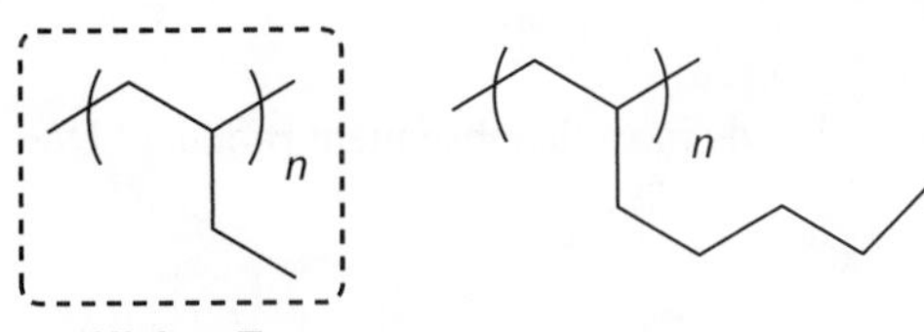

(c) The first polymer will have a higher T_g because the longer side groups of the second polymer push the polymer chains apart and decrease the strength of the dispersion forces (factor #3).

Higher T_g
Increased dispersion forces

Problem 26.16

Think

Are polar or nonpolar compounds water soluble? Are ionic compounds likely to be water soluble? Which monomers have polar or ionic repeating units?

Solve

The more polar and ionic polymers are water soluble and are given below.

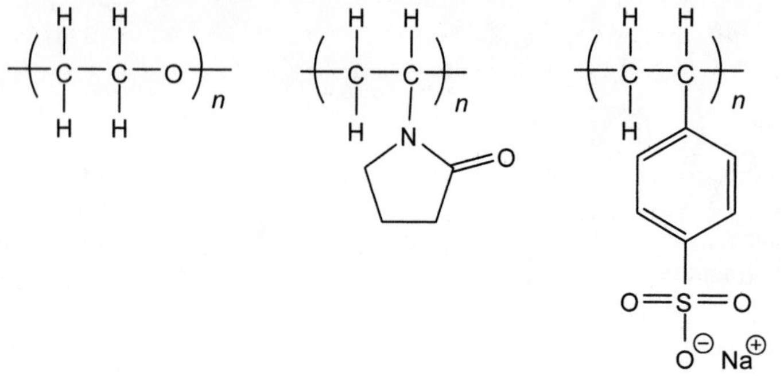

Problem 26.17

Think

Which polymer is rigid? Which polymer is flexible? Which type of polymer is better for each application?

Solve

HDPE is not sufficiently rigid for a CD case. PS is too rigid and would crack if dropped, making it inappropriate for a milk container.

Problem 26.18

Think

What is the leaving group? What new functional group is formed? Is this an example of substitution, addition, or elimination?

Solve

The chloride Cl^- is the leaving group and the new functional group formed is an alkene. This is an example of E1 elimination. The mechanism is given below:

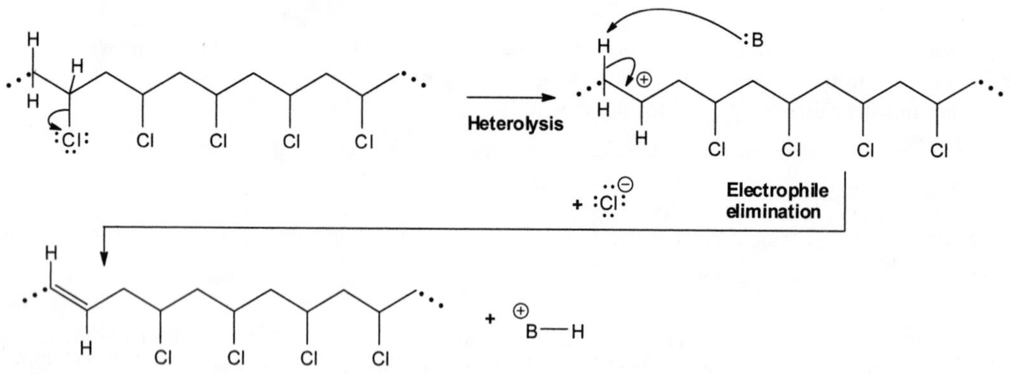

Heterolysis

Electrophile elimination

Problem 26.19

Think

What do you notice about the arrangement of the double bonds in each subsequent product? Does this lead to a more or less stable product?

Solve

The second double bond is conjugated with the first double bond; subsequent double bonds increase the length of the conjugated system. Conjugated double bonds are more stable than isolated double bonds and will form more readily.

Problem 26.20

Think

How many possibilities are there for the identity of the first amino acid? The second amino acid? The third? How many combinations, therefore, are possible for a dipeptide? By what factor does that number increase for a tripeptide? How can you extend this trend for a protein that is 300 amino acids long?

Solve

There are 20 possibilities for the first amino acid, 20 for the second, 20 for the third, and so on. There are $20 \times 20 = 400$ possibilities, therefore, for a dipeptide. That number is multiplied by 20 for a tripeptide, which is $20^3 = 8000$. In general, there are 20^Y combinations for a protein that is Y amino acids long. Therefore, if the protein has 300 amino acids, the number of primary sequences possible is: $20^{300} = 2.0 \times 10^{390}$. This is much larger than the number of atoms in the entire universe, suggested by some estimates to be $\sim 10^{80}$.

Problem 26.21

Think

What atoms make up a disulfide bond? Where is this atom in the monomer? What does the dimer look like when those atoms are connected together?

Solve

A disulfide bond is an S–S bond. The dimer is below:

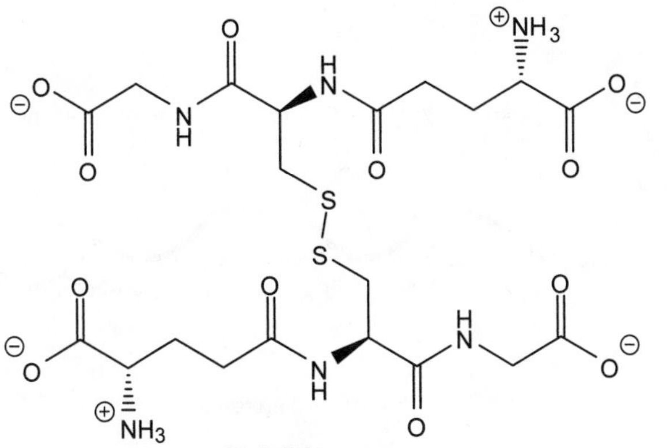

Glutathione

Problem 26.22

Think

What are the structures of alanine and isoleucine? Which one has a bulkier side chain? Is the formation of an α-helix favored or disfavored with increased sterics?

Solve

The structures of alanine and isoleucine are shown on the next page. Alanine has only a methyl group as the R group and isoleucine has a *sec*-butyl group. The *sec*-butyl group is bulkier than the methyl group. An α-helix tends to be disrupted when adjacent or nearby amino acids in a protein's sequence are bulky. Therefore, alanine should be found more commonly in an α-helix.

Alanine

$H_2N-CH(CH_3)-C(=O)-OH$

Less bulky, better for α-helix

Isoleucine

$H_2N-CH(CH(CH_3)CH_2CH_3)-C(=O)-OH$

Problem 26.23

Think

What holds the secondary and tertiary structures in a protein together? How does sodium dodecyl sulfate (SDS) disrupt these interactions?

Solve

The protein's secondary and tertiary structures are held together by intermolecular forces, such as dispersion forces (among the nonpolar regions), hydrogen bonding, and ion–ion interactions. SDS has a long hydrophobic region and an ionic hydrophilic head. The hydrophobic tail can interact favorably with the nonpolar portions of a protein, which tend to be on the interior of the protein when it is properly folded. The ionic head group interacts favorably with the polar regions of a protein. Therefore, SDS disrupts the intermolecular forces that a protein normally exhibits between its various portions when it is properly folded, making it much less favorable for the protein to remain folded.

Sodium dodecyl sulfate

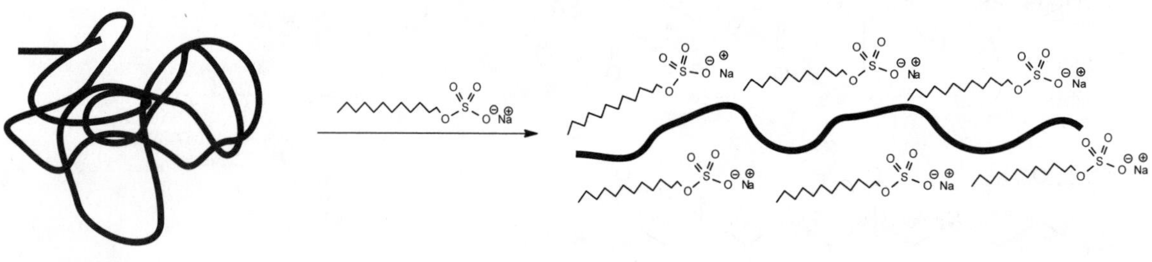

Folded protein

Unfolded protein

Problem 26.24

Think

Hemoglobin is a biological protein in the blood, and biological environments are generally aqueous. Therefore, would you expect hydrophilic or hydrophobic groups to be on the outside? How does this affect the solubility of hemoglobin in blood?

Solve

Hemoglobin is more soluble in blood if hydrophilic or polar amino acids are on the surface. Blood is a water-based liquid, and having a greater number of polar groups on the exterior of the protein makes the protein more soluble in the polar medium.

Problem 26.25

Think

What is the structure of chitin? Can you draw a portion of its structure by substituting an acetamide group for the OH group at each 2 position of cellulose? What intermolecular force leads to more rigid structures? How does the addition of the acetamide group contribute to the rigidity?

Solve

The partial structure of chitin is shown below. Notice the acetamide groups at the 2 position of each ring. The acetamide group contributes to the rigidity by increasing the possible hydrogen bonding interactions. Whereas the OH group at the 2 position has one hydrogen-bond donor and one hydrogen-bond acceptor, the acetamide group has one hydrogen-bond donor (the N–H bond) and two hydrogen-bond acceptors (the N and O atoms). Not only can there be more extensive hydrogen bonding among monomers within a single polymer chain, but this also enhances the hydrogen bonding between separate polymer chains.

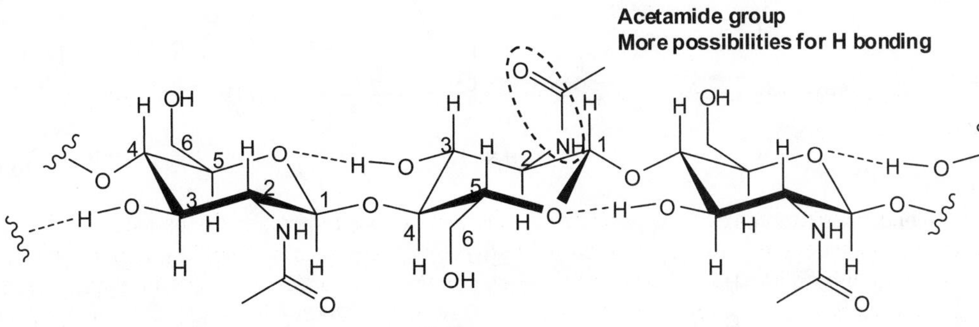

Acetamide group
More possibilities for H bonding

Chitin

Problem 26.26

Think

What is the structure of benzoyl peroxide and what is the identity of the initial radical that is produced? What bond in the monomer will react with that initial radical? How will that resulting radical react with another molecule of the monomer? After two monomers have been added to the growing polymer chain, can you identify the structural pattern that is being repeated?

Solve

The initiation step is the same for each reaction.

Intitiation:

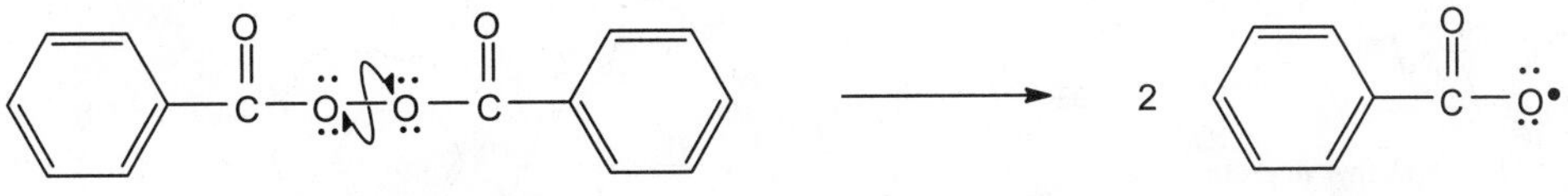

Benzoyl peroxide → 2 Radical initiator

(a) Answers for the vinylidene chloride monomer are given below:

(i) The mechanism for the vinylidene chloride monomer's two propagation steps in polymerization is given below. In the first step, the initial radical adds to the C=C bond of the monomer, resulting in a new radical. The new radical reacts with another molecule of the monomer in the same type of radical addition to the C=C.

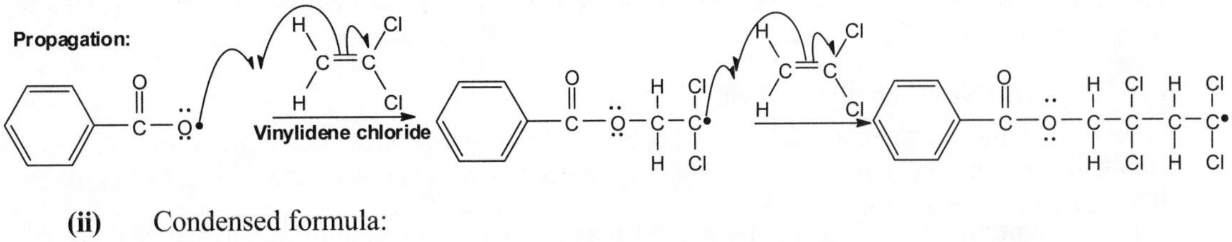

Propagation:

Vinylidene chloride

(ii) Condensed formula:

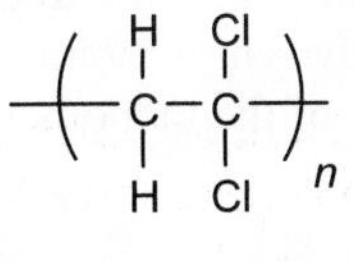

$$\left(CH_2 - CCl_2 \right)_n$$

(iii) Name: Poly(vinylidene chloride)

(b) Answers for the acrylamide monomer are given below:

(i) The mechanism for the acrylamide monomer's two propagation steps in polymerization is given below. In the first step, the initial radical adds to the C=C bond of the monomer, resulting in a new radical. The new radical reacts with another molecule of the monomer in the same type of radical addition to the C=C.

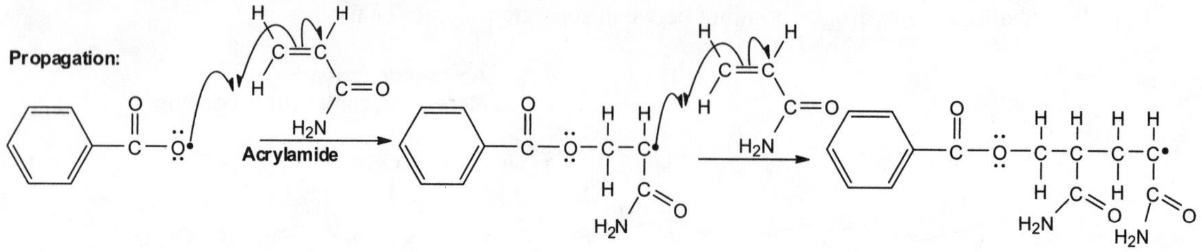

Propagation:

Acrylamide

(ii) Condensed formula:

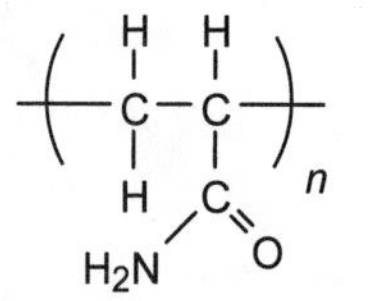

(iii) Name: Polyacrylamide

(c) Answers for the methyl vinyl ether monomer are given below:

(i) The mechanism for the methyl vinyl ether monomer's two propagation steps in polymerization is given below. In the first step, the initial radical adds to the C=C bond of the monomer, resulting in a new radical. The new radical reacts with another molecule of the monomer in the same type of radical addition to the C=C.

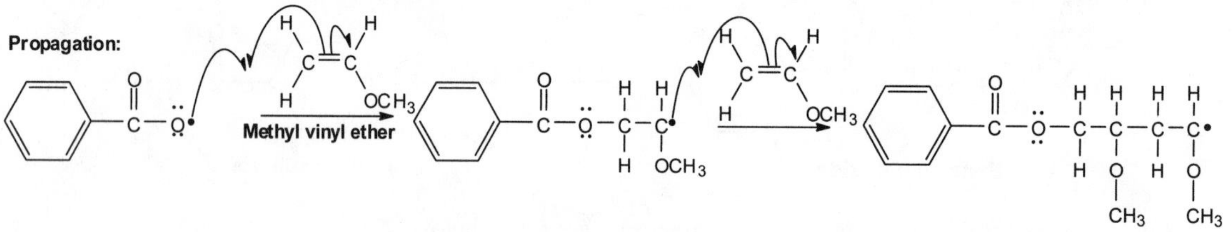

Propagation:

Methyl vinyl ether

(ii) Condensed formula:

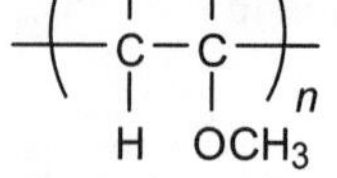

(iii) Name: Poly(methyl vinyl ether)

Problem 26.27

Think

What is the repeating unit of each polymer? What substituents are on the main polymer chain of the polymer? How many C atoms along the main chain must be traversed before the attached substituents repeat? How is a repeating unit of a vinyl polymer related the structure of the vinyl monomer? What bonds are the same, and what bonds are different? How many words are in the name of the monomer? Are parentheses necessary? What are the initial letters of the different portions of the monomer?

Solve

(a) Polymer:

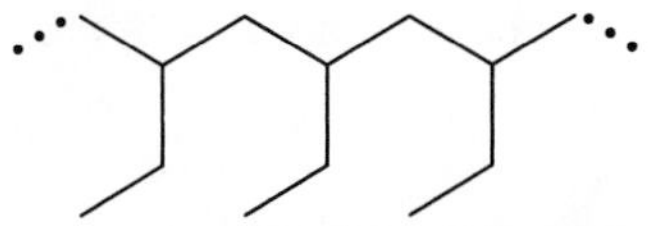

(i) Condensed formula:

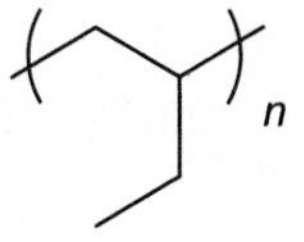

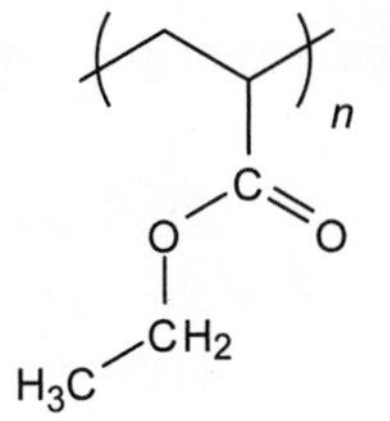

or

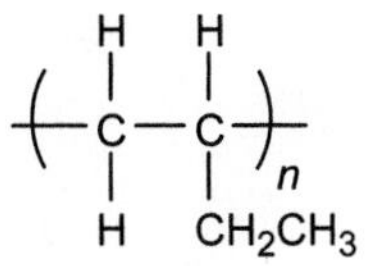

$$-\!\!\left(\begin{array}{cc} H & H \\ | & | \\ C - & C \\ | & | \\ H & CH_2CH_3 \end{array}\right)\!\!-_n$$

(ii) Monomer structure:

(iii) Monomer name: but-1-ene or butylene
Polymer name: polybut-1-ene or polybutylene

(b) Polymer:

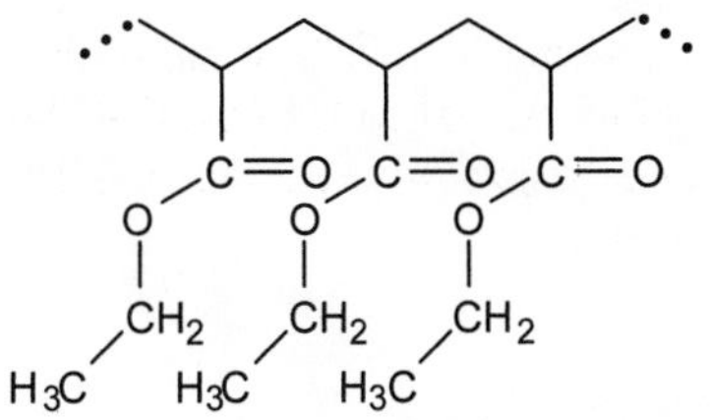

C=O C=O C=O

O O O

CH_2 CH_2 CH_2

H_3C H_3C H_3C

(i) Condensed formula:

n

C=O

O

CH_2

H_3C

(ii) Monomer structure:

C=O

O

CH_2

H_3C

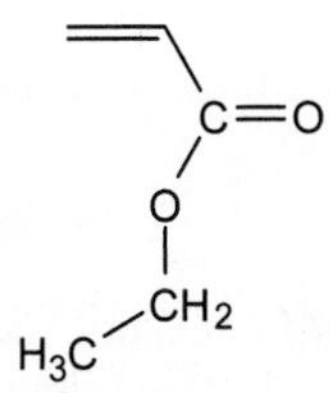

(iii) Monomer name: ethyl acrylate
Polymer name: poly(ethyl acrylate)

(c) Polymer:

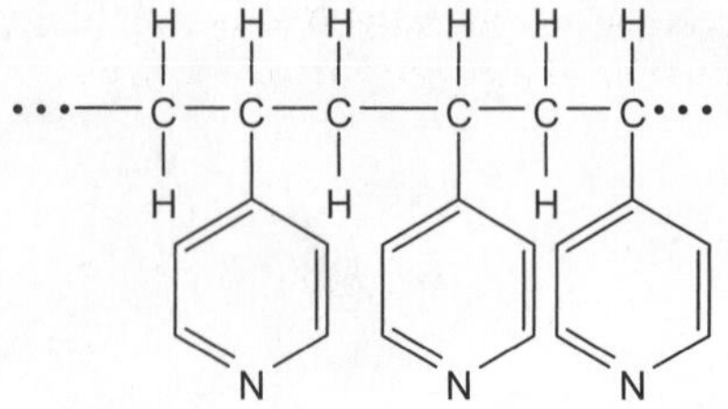

(i) Condensed formula:

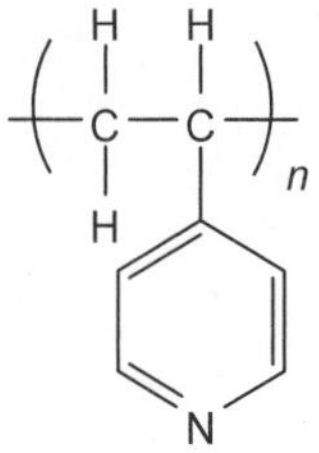

(ii) Monomer structure:

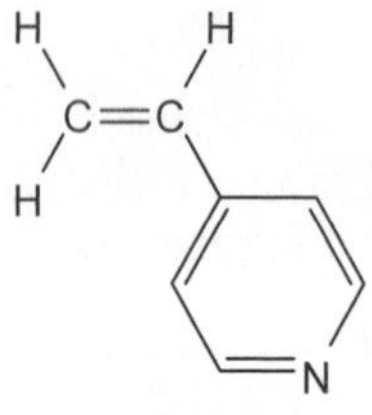

(iii) Monomer name: 4-vinylpyridine
Polymer name: poly(4-vinylpyridine)

Problem 26.28

Think

Which polymers are capable of hydrogen bonding? How does hydrogen bonding assist in a polymer's solubility in ethanol?

Solve

The polymer of ethyl acrylate, poly(ethyl acrylate), **(b)**, will be most soluble in ethanol. The oxygen atoms in the pendant groups are capable of hydrogen bonding with ethanol molecules. The oxygen atoms are hydrogen-bond acceptors.

Problem 26.29

Think

Which polymer is nonpolar? Is hexane polar or nonpolar? Which polymer has similar intermolecular forces to hexane?

Solve

Polybut-1-ene, or **(a)**, will be most soluble in hexane. It is a hydrocarbon, as is hexane, and most similar to hexane in polarity. The polar side chains of **(b)** and **(c)** will make them unlikely to dissolve in nonpolar hexane because, upon doing so, significant intermolecular interactions between polymer molecules would be lost.

Problem 26.30

Think

Identify the main polymer chain. What are the substituents? How many atoms must be traversed along the main polymer chain before the substituents begin to repeat? Which bond links each repeating unit? How is the condensed formula written?

Solve

(a) Poly(lactic acid) or polylactide

The main chain is composed of repeating C–C–O atoms, and the substituents attached to those atoms (CH_3 and =O) also repeat every three atoms.

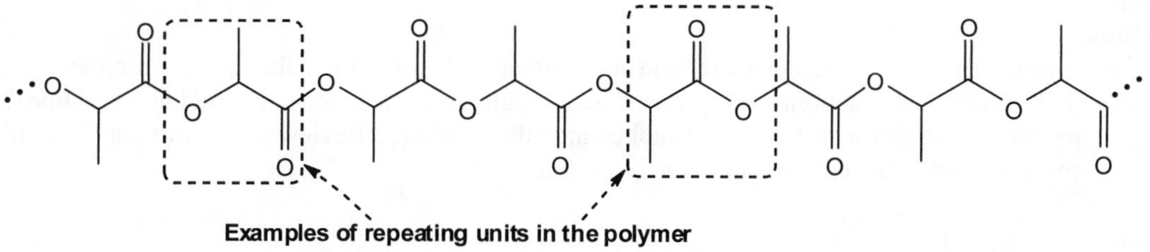

Examples of repeating units in the polymer

Condensed formula: or

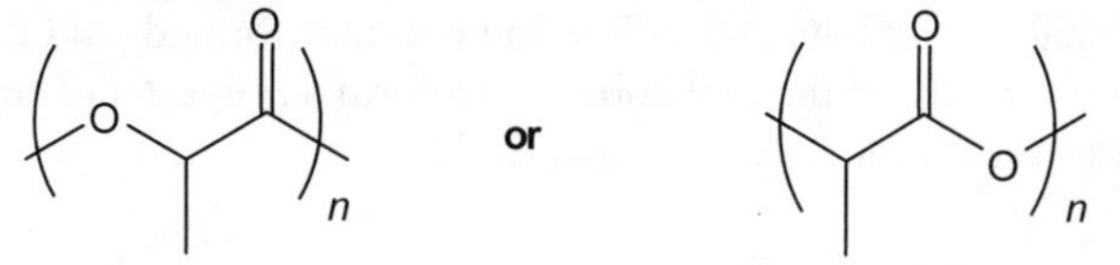

(b) Poly(propylene oxide)

The main chain is composed of repeating C–C–O atoms, and the substituent attached to those atoms (CH_3) also repeats every three atoms. When the polymer is made, a new bond forms between oxygen and carbon, not between carbon and carbon. Consequently, we show only condensed formulas with oxygen at one end of the formula.

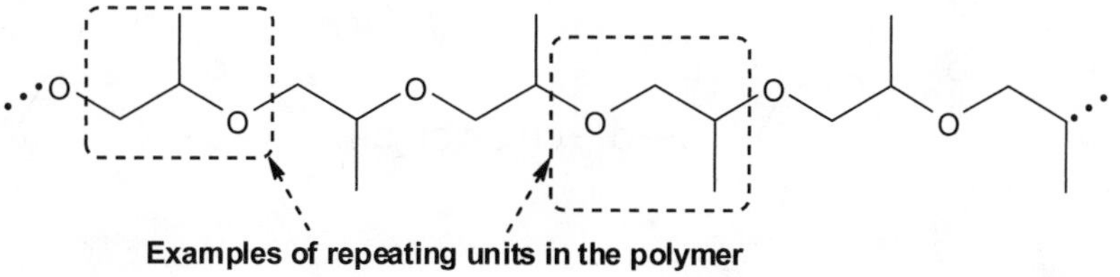

Examples of repeating units in the polymer

Condensed formula: or but usually not

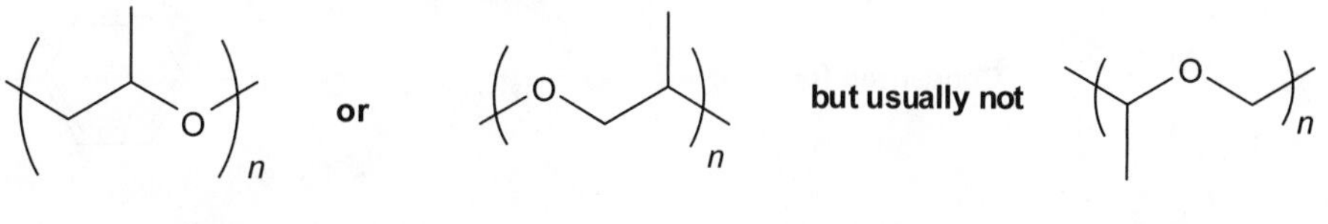

(c) Polyglycine

The main chain is composed of repeating C–C–N atoms, and the substituent attached to those atoms (=O) also repeats every three atoms.

Examples of repeating units in the polymer

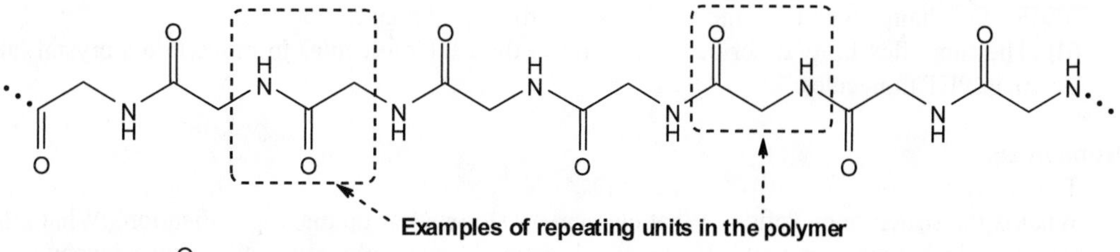

Condensed formula:

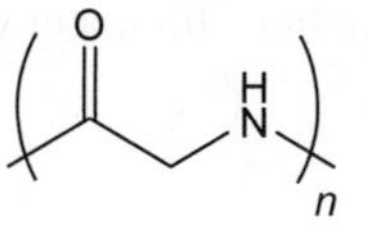

Problem 26.31

Think

What functional groups are present in each condensed formula? How are the functional groups related to the class of polymer?

Solve

(a) Polylactide has ester functional groups and, therefore, is a polyester.
(b) Poly(propylene oxide) has ether functional groups and, therefore, is a polyether.
(c) Polyglycine has amide functional groups and, therefore, is a polyamide.

Problem 26.32

Think

What are the structures of terephthalic acid and butane-1,4-diol? What functional group is present in each compound? What new functional group forms as a result of the reaction under acidic conditions? How does replacing ethane-1,2-diol with butane-1,4-diol change the structure? Review the factors for T_m. Which polymer has a longer portion of its repeating unit that is flexible?

Solve

(a) Structures for terephthalic acid, butane-1,4-diol, the polymer formed, and the condensed formula are given below. The dicarboxylic acid and the diol combine to form a polyester. The reaction produces each ester is essentially a Fischer esterification reaction (Chapter 21).

HO—C(=O)—C₆H₄—C(=O)—OH + HO—CH₂CH₂CH₂CH₂—OH

Terephthalic acid

Butane-1,4-diol

Acid catalyst | Δ

$\cdots CH_2CH_2CH_2CH_2—O—C(=O)—C_6H_4—C(=O)—O—CH_2CH_2CH_2CH_2—O—C(=O)—C_6H_4—C(=O)—O\cdots + H_2O$

Condensed formula: $\left(CH_2CH_2CH_2CH_2—O—C(=O)—C_6H_4—C(=O)—O \right)_n$

(b) PBT stands for poly(butylene terephthalate). As 1,2-ethanediol, or ethylene glycol, is replaced by 1,4-butanediol, or butylene glycol, the "ethylene" in poly(ethylene terephthalate) is replaced by butylene.
(c) PBT will have a lower melting point. Increasing the length of the flexible alkyl chains in the polymer will allow easier mobility of the polymer molecules. A lower temperature will supply sufficient kinetic energy for the chains to go from the solid phase to the liquid phase.
(d) The same flexibility described in **(c)** allows the PBT molecules to move into a crystal lattice more quickly than PET molecules.

Problem 26.33

Think

What is the structure of Teflon? What elementary step makes up disproportionation? What atom is abstracted in this step? Is that atom present? What is combination? Can combination still occur between two radicals?

Solve

Two portions of growing Teflon polymers are shown below. Disproportionation generally involves H atom abstraction from the atom adjacent to the C• on one of the growing polymers. This does not occur because there are no hydrogen atoms for a radical to abstract. Even without such H atoms, the two carbon radicals can undergo combination to form a new C–C in a termination step.

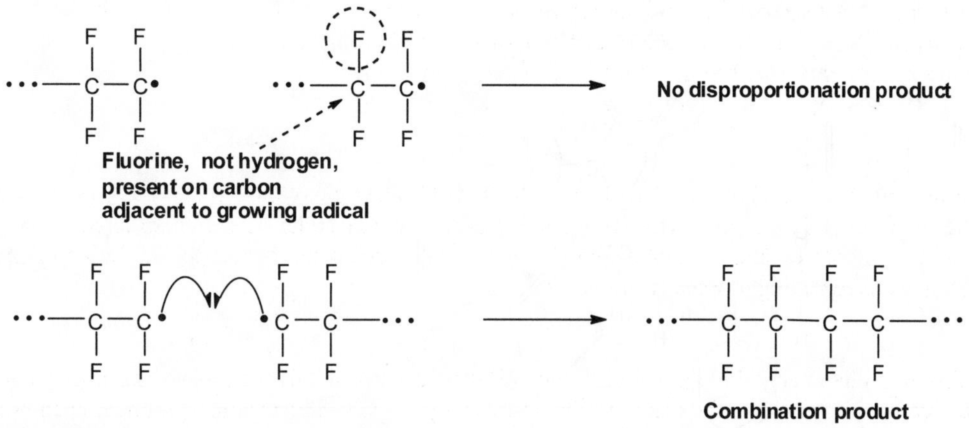

No disproportionation product

Fluorine, not hydrogen, present on carbon adjacent to growing radical

Combination product

Problem 26.34

Think

What type of elementary step is necessary for C• to appear in the middle of the polymer's main chain? For that step to occur, do the polymers need to approach each other closely? Which polymer exhibits more steric hindrance? How does steric hindrance affect the above elementary step?

Solve

Hydrogen abstraction is what causes C• to appear in the middle of a polymer's main chain, and this occurs readily during chain growth of polyethylene:

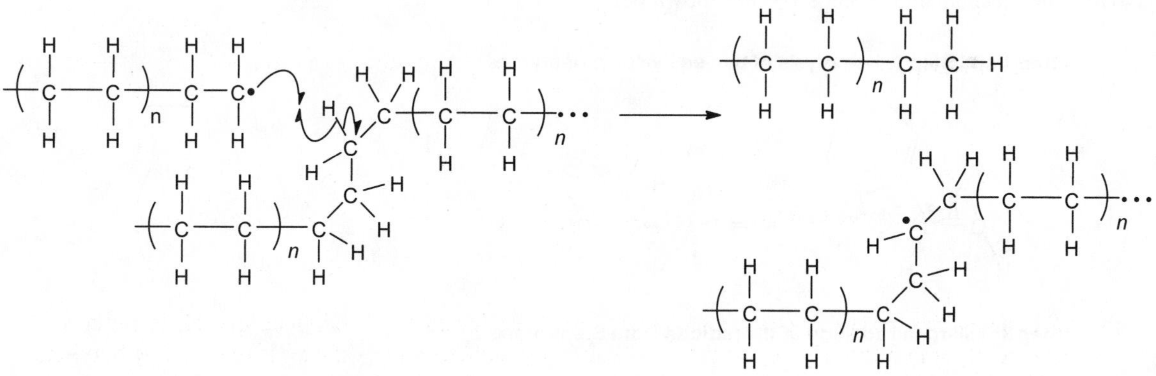

The phenyl groups in polystyrene, on the other hand, sterically hinder radicals that may abstract H atoms. Therefore, the bulky phenyl rings prevent branching:

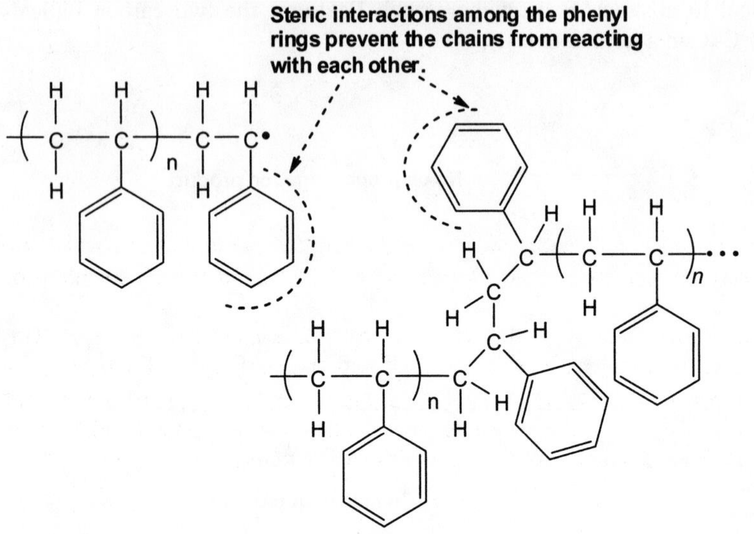

Steric interactions among the phenyl rings prevent the chains from reacting with each other.

Problem 26.35

Think

What is the vinyl π bond? How many single-barbed curved arrows are needed to show a homolysis step? Which part of the radical monomer is the tail and which part is the head? How can you show tail-to-tail addition? What are the steps in a free-radical polymerization? How does Step 4 interfere with polymer growth? Does initiation with benzoyl peroxide suffer from the same problem?

Solve

(a) The mechanism for Steps 1–4 are shown below:

Step 1 and Step 2: Homolysis of two equivalents of styrene

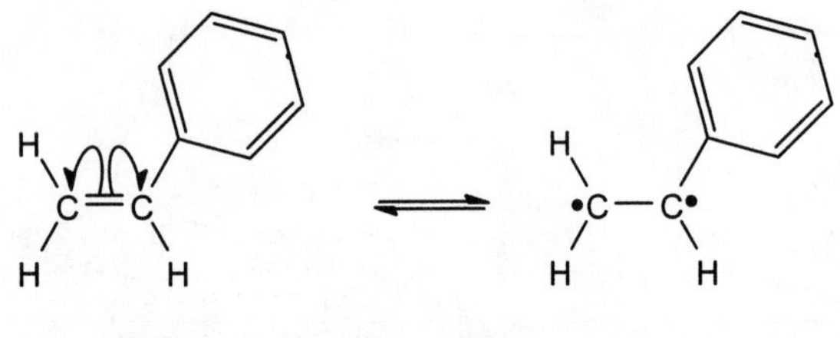

Step 3: Tail-to-tail addition of the radicals from Steps 1 and 2

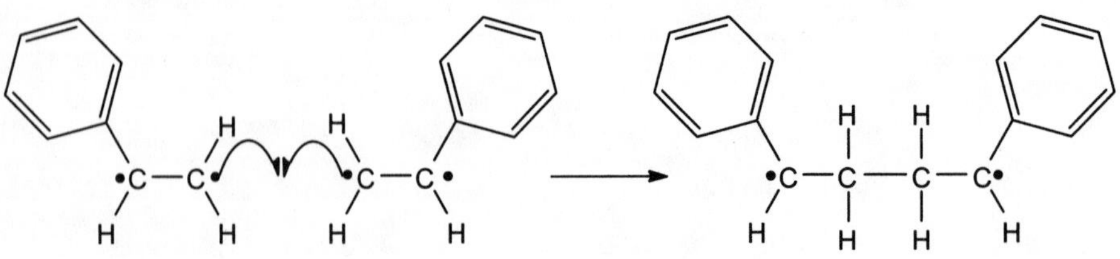

Step 4: Radical coupling of the diradical formed in Step 3

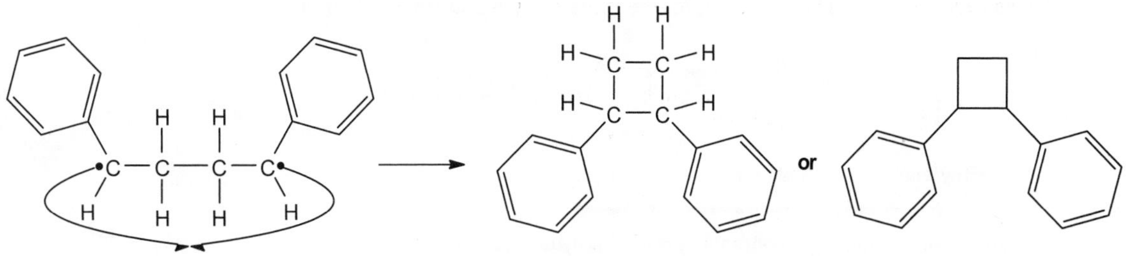

or

(b) No, this mechanism is not consistent with a free-radical polymerization. The steps in which the free radicals are involved do not make up a propagation cycle. Furthermore, the mechanism does not produce a polymer.

(c) In the autoinitiation reaction described here, the reaction can terminate too quickly. The 1,2-phenylcyclobutane forms because two styrene molecules react and the product, a diradical, undergoes termination when the two radicals in the molecule combine. This prevents growth of a long chain of molecules. With benzoyl peroxide, the radical that forms after initiation is a monoradical, and is more likely to continue propagation because it is not likely to react with itself in a termination step.

Problem 26.36

Think

Does oxetane have a strained ring? When the O^{2-} initiator is added, what reaction will take place to open the ring of the oxetane? How will the resulting product of that reaction react with another molecule of the oxetane? Compare the repeating units of poly(ethylene oxide) and polyoxetane. Which one has a greater number of O atoms? How does that affect water solubility? How do the O atoms affect the flexibility of the polymer chain, and what is the effect on melting point?

Solve

(a) The mechanism for the polymerization of polyoxetane from oxetane is shown below.
Step 1: Oxide anion acts as a nucleophile and initiates the reaction and forms an alkoxide anion.
Step 2: The resulting alkoxide anion reacts with another oxetane molecule.
Step 3: The propagation step (Step 2) repeats.

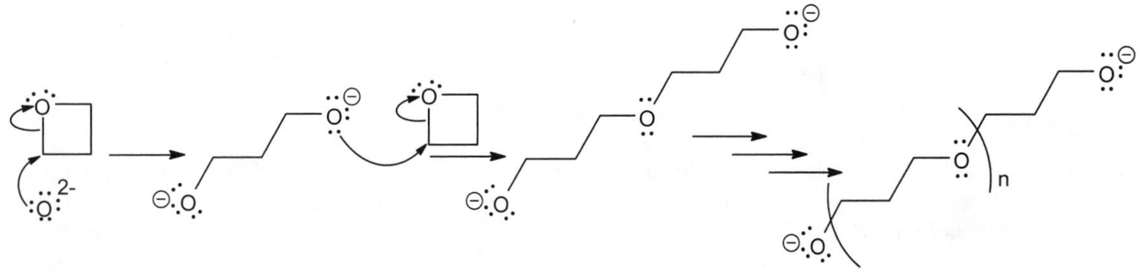

(b) Condensed formula:

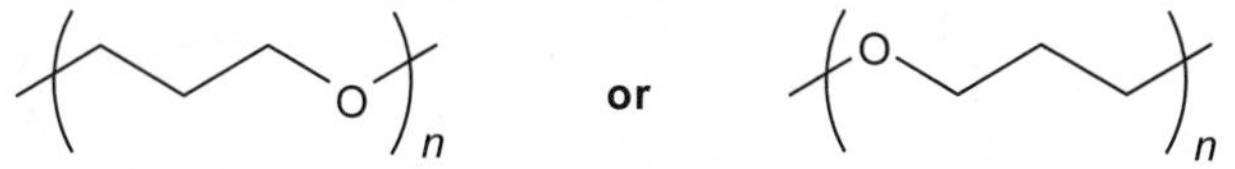

or

(c) Polyoxetane (PO) will be less soluble in water than poly(ethylene oxide) (PEO). The oxygens occur less frequently in the polymer chain in PO, so there will be less hydrogen bonding along a chain of PO than along a chain of PEO of the same length.

(d) PO will have a higher melting point. The $-CH_2-$ segments are not as flexible as the –O– segments, and the higher barrier of rotation will require a higher temperature to give the PO molecules sufficient kinetic energy to move freely.

(e) Both ethylene oxide and oxetane have ring strain because they are, respectively, three-membered and four-membered rings. THF does not have significant ring strain and is much less reactive.

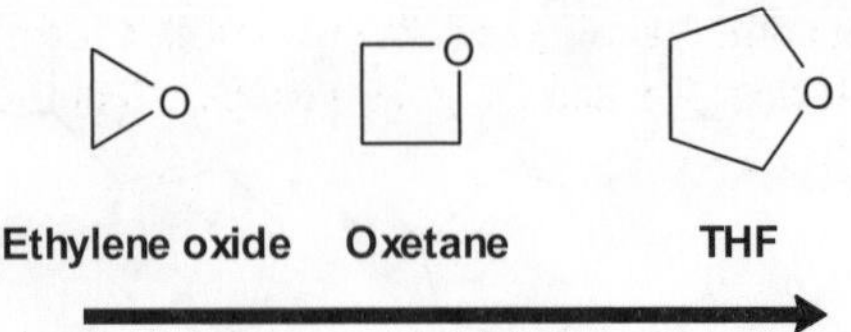

Decreasing ring strain, decreasing rate of polymerization

Problem 26.37

Think

How does an alkene π bond react with an acid, HA (Chapter 11)? How does the electrophile formed in the initiation step react with another equivalent of isobutylene? What is the repeating unit? Do the methyl groups stabilize or destabilize positive charges? Negative charges? How does that impact whether isobutylene would be a good candidate for anionic polymerization?

Solve

(a) and **(b)** The mechanism for the polymerization of polyisobutylene from isobutylene is shown below.

Step 1: The alkene reacts with H–A in an electrophilic addition step to form a tertiary carbocation.

Step 2: The tertiary carbocation reacts with another equivalent of isobutylene in a second electrophilic addition step.

Step 3: Propagation step (Step 2) repeats the electrophilic addition reaction.

Isobutylene — Electrophilic addition — Electrophilic addition

(c) Condensed formula for polyisobutylene:

(d) Isobutylene forms a tertiary carbocation as the intermediate. Ethylene and propylene form a primary and a secondary carbocation intermediate, respectively. The tertiary carbocation is more stable and thus leads to a faster polymerization.

$H_2C{=}CH_2 \xrightarrow{H-A} H_3C{-}\overset{\oplus}{C}H_2$ **Primary C⁺**

$H_2C{=}CH{-}CH_3 \xrightarrow{H-A} H_3C{-}\overset{\oplus}{C}H{-}CH_3$ **Secondary C⁺**

$H_2C{=}C(CH_3){-}CH_3 \xrightarrow{H-A} H_3C{-}\overset{\oplus}{C}(CH_3){-}CH_3$ **Tertiary C⁺**

(e) Isobutylene is not likely to polymerize through an anionic polymerization mechanism because the attached CH_3 groups are electron donating and destabilize nearby negative charges—the type of charge that develops in anionic polymerization.

Problem 26.38

Think

Which carbon of the C=C is most electrophilic in methyl 2-cyanoacrylate? How does HO^- react with that C, and what anionic intermediate forms? Can the product of that reaction react in the same way with another equivalent of the monomer? What is the repeating unit? How do the attached CN and CO_2CH_3 groups affect the stability of the charge that develops on the growing polymer chain?

Solve

(a) and **(b)** The mechanism for the polymerization of poly(methyl 2-cyanoacrylate) from methyl 2-cyanoacrylate is shown below.

Step 1: The alkene reacts with HO^- in a nucleophilic addition step to form a carbanion.

Step 2: The carbanion reacts with another equivalent of methyl 2-cyanoacrylate in a second nucleophilic addition step.

Step 3: Propagation step (Step 2) repeats the nucleophilic addition reaction.

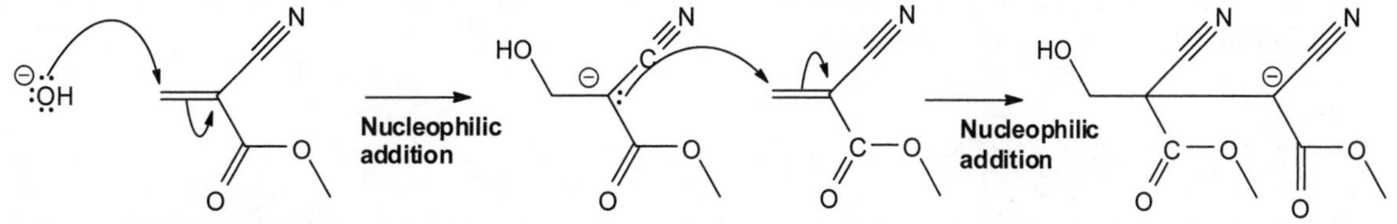

Nucleophilic addition

Methyl 2-cyanoacrylate

Nucleophilic addition

(c) Condensed formula for poly(methyl 2-cyanoacrylate):

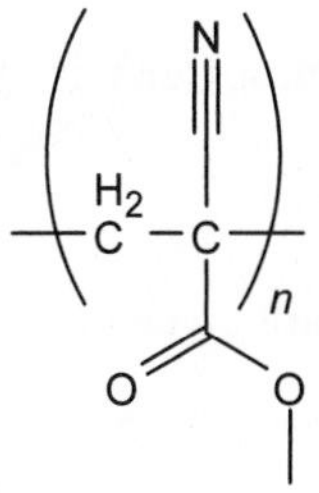

(d) Methyl 2-cyanoacrylate forms a carbanion intermediate that is stabilized by the C≡N group via resonance and inductive effects. Methyl acrylate has less resonance stabilization because the C≡N is not adjacent to the lone pair. Thus, methyl 2-cyanoacrylate polymerizes faster.

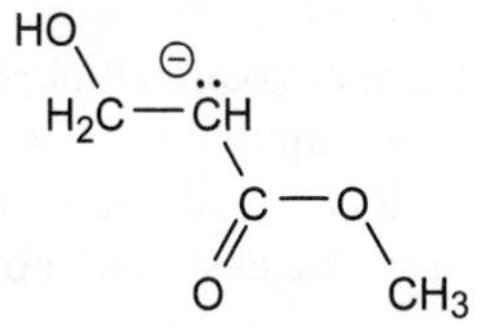

Methyl acrylate

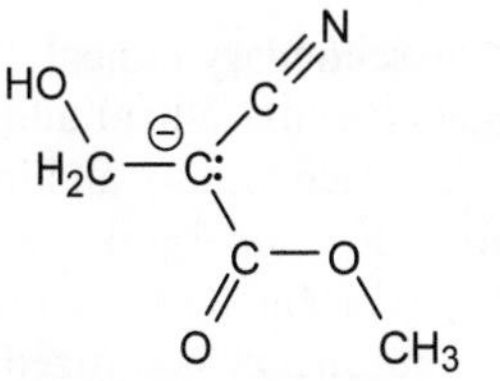

Methyl 2-cyanoacrylate
More stable anion

(e) Methyl 2-cyanoacrylate is not likely to polymerize through a cation polymerization mechanism because the cation would be unstable due to the electron-withdrawing nitrile and ester groups.

Problem 26.39

Think

What would be the structure of each anion produced upon nucleophilic addition to the C=C? What would be the structure of each cation produced upon electrophilic addition to the C=C? Are the neighboring groups stabilizing or destabilizing groups for an anion? For a cation?

Solve

(a) The anions produced after nucleophilic addition to the C=C for ethyl vinyl ether, but-1-ene, and acrolein are shown below. The aldehyde of acrolein is both an inductive and resonance-stabilizing group for the anion, so acrolein is most likely to undergo anionic polymerization.

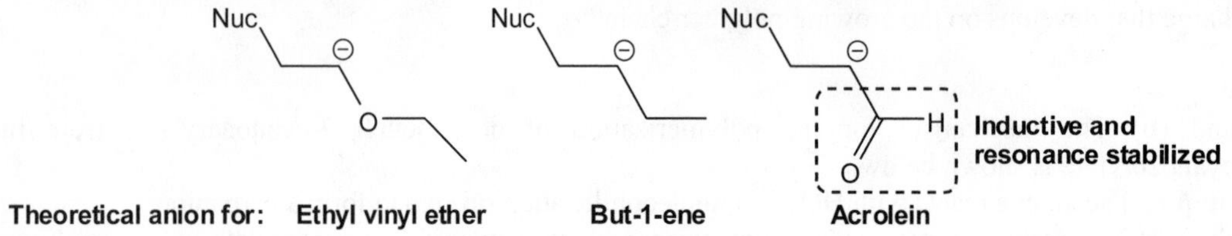

Nuc Nuc Nuc

Inductive and resonance stabilized

Theoretical anion for: Ethyl vinyl ether But-1-ene Acrolein

(b) The cations produced after electrophilic addition to the C=C for ethyl vinyl ether, but-1-ene, and acrolein are shown below. The ether O of ethyl vinyl ether is a resonance-stabilizing group for the C^+, so ethyl vinyl ether is most likely to undergo cationic polymerization.

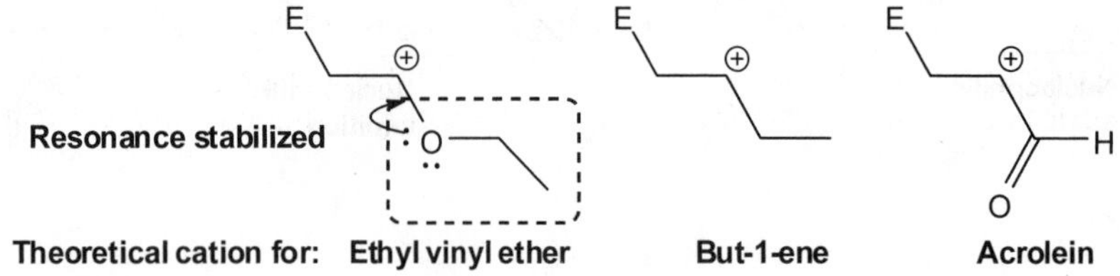

E E E

Resonance stabilized

Theoretical cation for: Ethyl vinyl ether But-1-ene Acrolein

Problem 26.40

Think

What propagation step is possible for R• and propylene? What atom in propylene is most likely to be abstracted? Which radical is most stable?

Solve

(a) The propagation and H atom abstraction steps involving R• and propylene are given below:

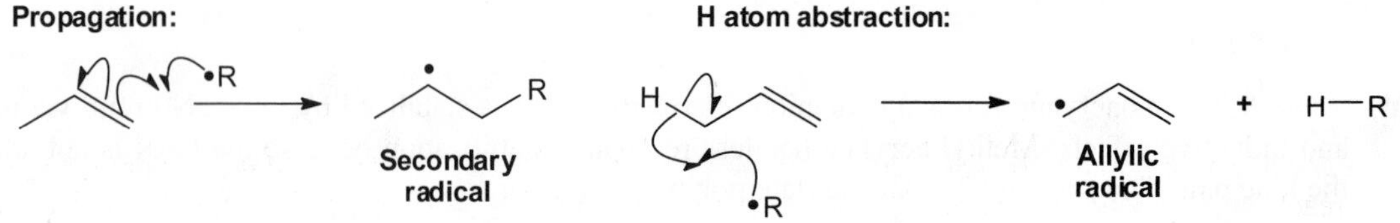

Propagation: H atom abstraction:

•R → Secondary radical; H, •R → Allylic radical + H—R

(b) An allylic radical is more stable than the secondary radical due to resonance.

(c) The step that is necessary for propagation (i.e., radical addition) produces the less stable alkyl radical. H atom abstraction, therefore, tends to take place rapidly and interferes with the formation of the polymer.

(d) But-1-ene will be a poor candidate for free-radical polymerization because it, too, has allylic C–H bonds. Ethyl vinyl ether will also suffer from the problem of H atom abstraction because the abstraction of H from the C that is α to the ether O produces a resonance-stabilized radical.

Problem 26.41

Think

Does the presence of more OH groups usually lead to greater or less water solubility? If there are more OH atoms adjacent to each other, is intramolecular hydrogen bonding likely? How does this affect the availability of those OH groups to form hydrogen bonds with water? How does that affect the water solubility of the polymer?

Solve

Usually, the presence of more HO groups increases the water solubility of a molecule (including polymers), due to increased intermolecular hydrogen bonding capability. However, in the case of poly(vinyl alcohol), the presence of more HO groups leads to more *intra*molecular hydrogen bonding and decreased *inter*molecular hydrogen bonding with water. The decreased hydrogen bonding with water decreases the polymer's water solubility.

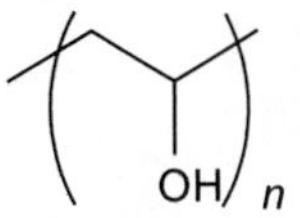

OH n

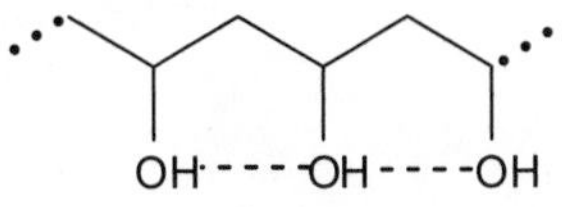

OH----OH----OH

Intramolecular hydrogen bonding

Problem 26.42

Think

What reaction occurs between formaldehyde and an NH_2 group from melamine? Can the resulting functional group react with an NH_2 group from another melamine molecule? How many amines are there in melamine that can react? In how many directions does polymerization occur with respect to a single monomer? Are network polymers reactive?

Solve

(a) The NH_2 groups on melamine are nucleophilic and can undergo nucleophilic addition to formaldehyde. As shown below, this can occur three times to produce trimethylol melamine.

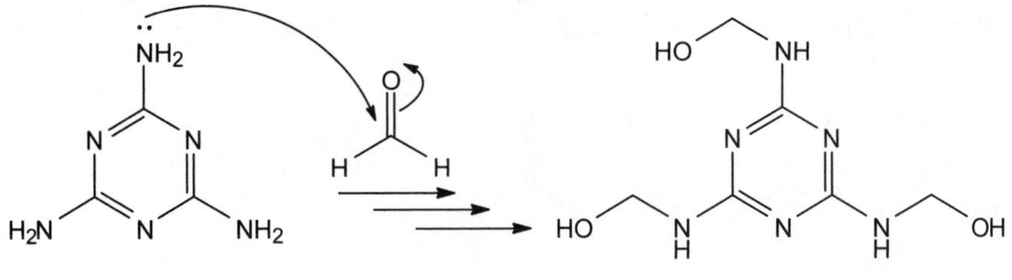

Trimethylol melamine

Under acidic conditions, the OH groups of trimethylol melamine become good leaving groups and can react in a substitution reaction with the NH_2 group of another melamine molecule. This creates a NH–CH_2–NH bridge between two aromatic rings, as shown below.

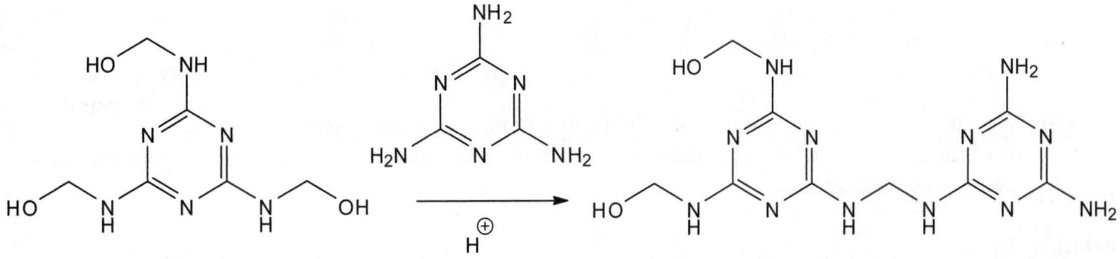

Further reaction results in bridges at three locations on each ring, as shown below.

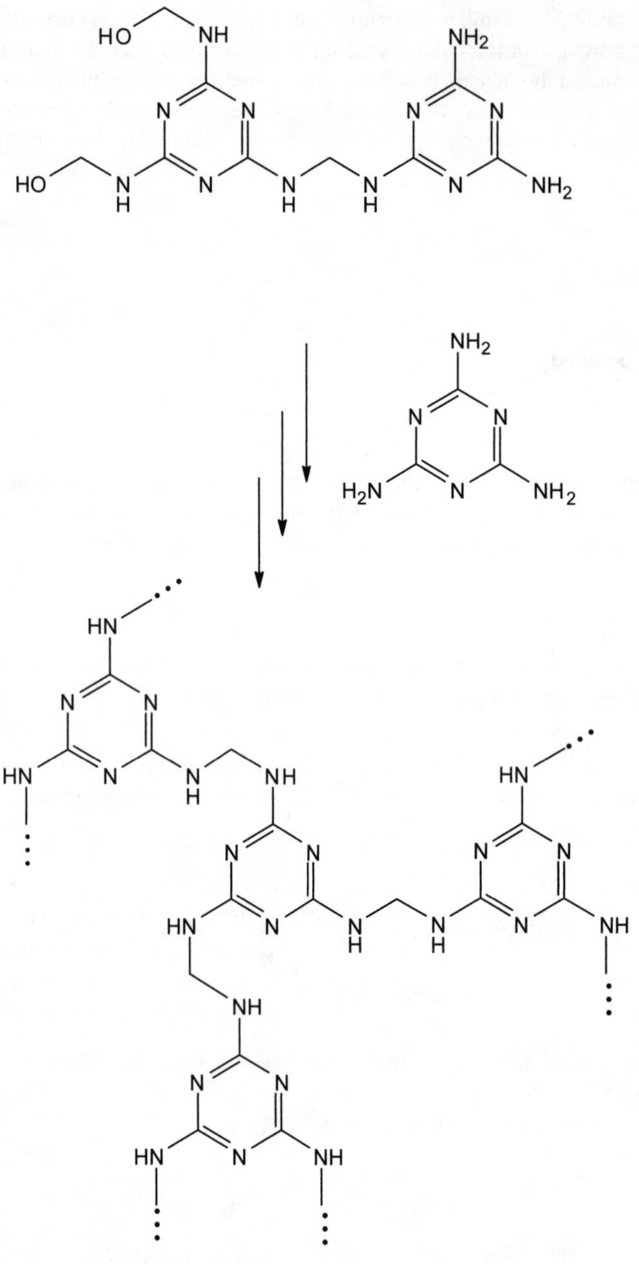

HO NH NH2 N N N N HO N N N N N NH2 H H H

NH2 N N H2N N NH2

HN N N HN N N NH HN N N N N HN N N N N NH H H NH N N HN N NH

(b) Network polymers are generally insoluble and, therefore, are unaffected by solvents. The polymer will be a good surface to write on because the solvent in the markers will not penetrate the surface, nor will solvents used to clean the surface.

Problem 26.43

Think

Review the seven factors that affect T_g. How does the introduction of CH_3 groups on the ring in **B** affect the ease in which the polymer chain can rotate? When one of those methyl groups is replaced by a phenyl ring, as in **C**, how does that impact the space between polymer chains? What about when the substituent is even longer, as in **D**?

Solve

In **A**, the absence of large groups on the benzene ring allows the polymer chains to rotate fairly easily, giving it the lowest T_g. The introduction of the two methyl groups makes rotation of the polymer chains and movement of one past another more difficult, and raises T_g. However, when a phenyl ring replaces a methyl group, the phenyl ring pushes neighboring chains farther away, and creates space between the polymer molecules and weakens the intermolecular interactions. This allows for easier movement of polymer molecules and lowers T_g. Lengthening the spacer (a benzyl group instead of a phenyl group in **D**) pushes the molecules even farther apart and lowers T_g.

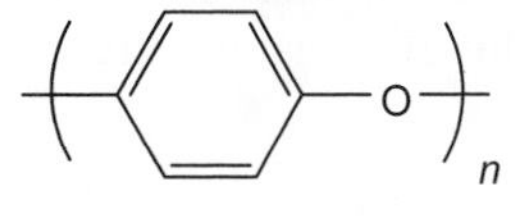

A T_g = 82 °C
Lowest T_g

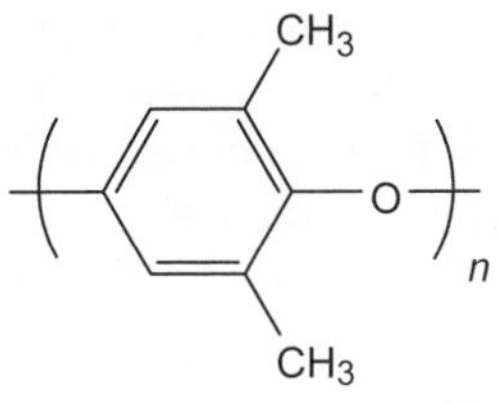

B T_g = 211 °C
Highest T_g
Two CH_3 groups, limited rotation

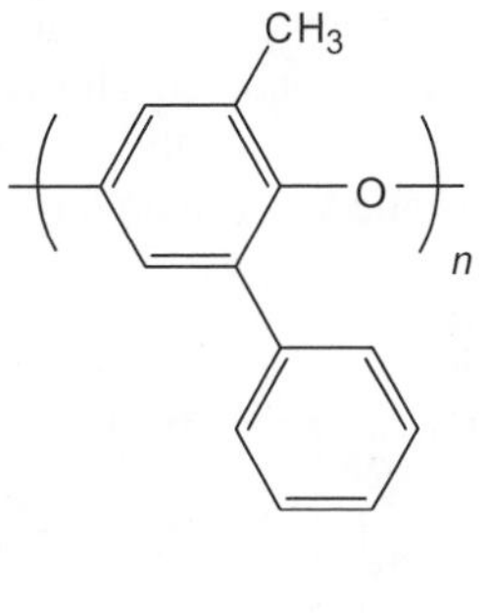

C T_g = 169 °C

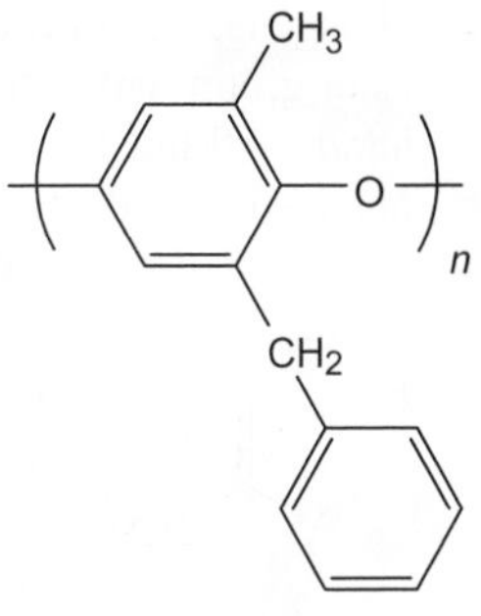

D T_g = 99 °C

Problem 26.44

Think

Review the seven factors that affect T_g. How does the introduction of CH_3 groups along the chain affect the ease in which the polymer chain can rotate? How is that rotation affected when the bulkier phenyl groups are attached?

Solve

Compound **F** will have the lowest T_g, as the hydrogen atoms attached to the sp^3-hybridized carbon give the lowest barrier to rotation around the bonds to that carbon. When the hydrogens are replaced with methyl groups, as in compound **E**, the T_g will increase, because the bulkier groups will increase the amount of energy required for rotation, which makes the chain more rigid. The two phenyl groups in compound **G** will give the highest barrier to rotation, and that compound will have the highest T_g.

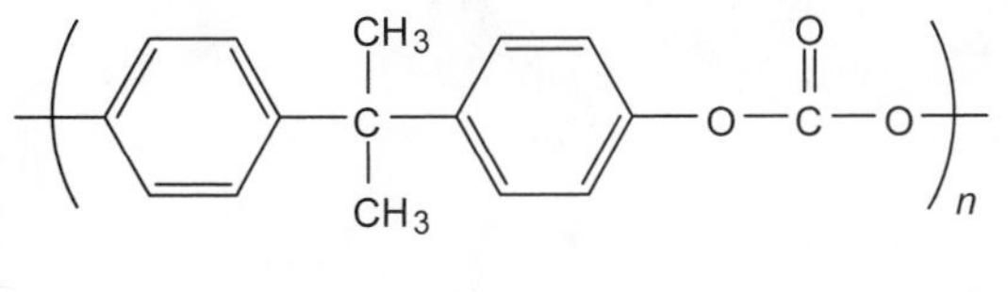

E

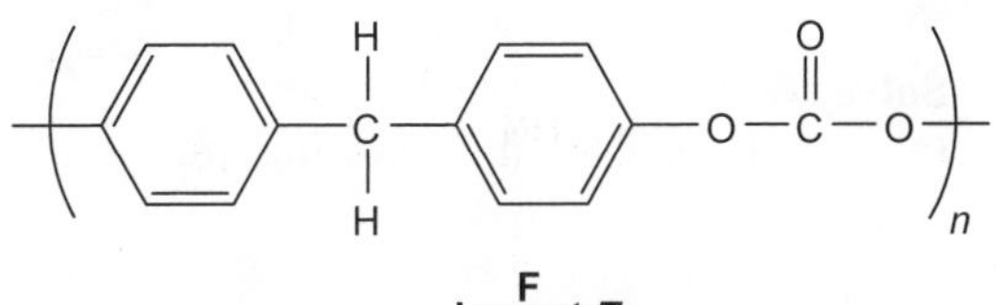

F
Lowest T_g
Lowest barrier of rotation

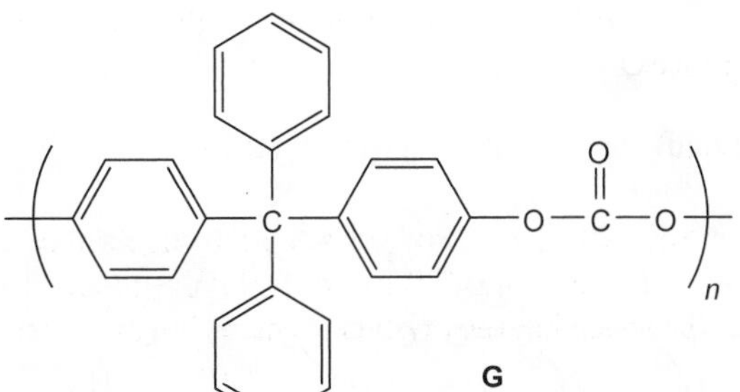

G
Highest T_g
Highest barrier of rotation

Problem 26.45

Think

How does an alcohol react with an acid chloride? What functional group does it produce? For a polymer to form, should phosgene be able to react with two separate monomers? Which monomer will prevent reaction with phosgene in such a way as to prevent phosgene from being able to react with a second monomer of the diol?

Solve

An alcohol will react with an acid chloride to form an ester. This can happen twice for each diol and twice for phosgene. To form a polymer, however, phosgene should form one ester with one diol monomer and the other ester group with a separate diol monomer. This can happen for Compounds **Y** and **Z**. Compound **X**, on the other hand, will not form a polymer because the second –OH group is close enough to the first that both react with phosgene, producing a molecule that can no longer react with the monomers.

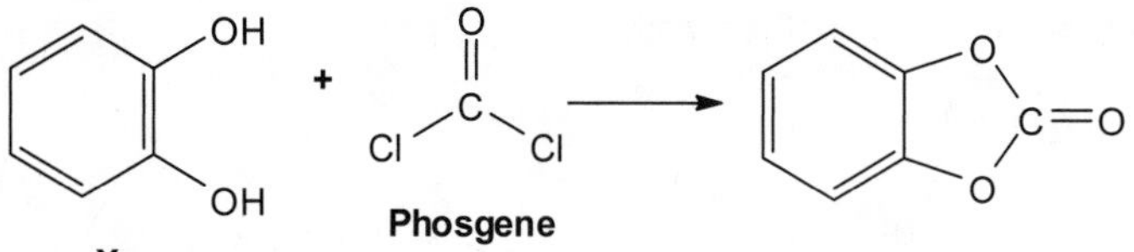

OH OH + O C Cl Cl → O O C=O

X Phosgene

Problem 26.46

Think

Is polystyrene polar or nonpolar? Which solvent, benzene or methanol, is nonpolar?

Solve

Benzene is a good solvent for polystyrene (PS), because it is an aromatic hydrocarbon similar to the phenyl groups that are substituents (or side groups) on the PS chain. Both benzene and PS are nonpolar. Methanol, however, is a polar solvent, and the addition of methanol to the solution of benzene and PS decreases the solubility of PS until the PS precipitates as a solid.

Problem 26.47

Think

What is the structure of acrylic acid? What is the structure of poly(acrylic acid)? In the copolymerization with divinyl glycol, how does one C=C bond of divinyl glycol react with the growing polymer chain of poly(acrylic acid)? After a molecule of divinyl glycol is incorporated, how can the other C=C on that monomer react with another growing polymer chain? What does the structure of the crosslinked polymer look like? Is the crosslinked polymer more or less water soluble than the linear polymer?

Solve

(a) Acrylic acid has the following structure:

H H C=C H C=O HO

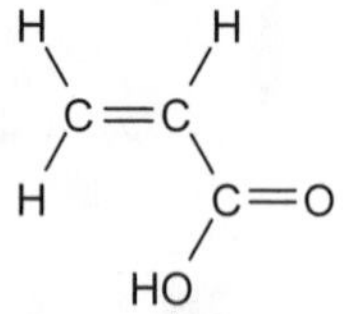

Linear poly(acrylic acid) has the following structure:

C=O C=O C=O C=O C=O C=O C=O C=O C=O
HO HO HO HO HO HO HO HO HO

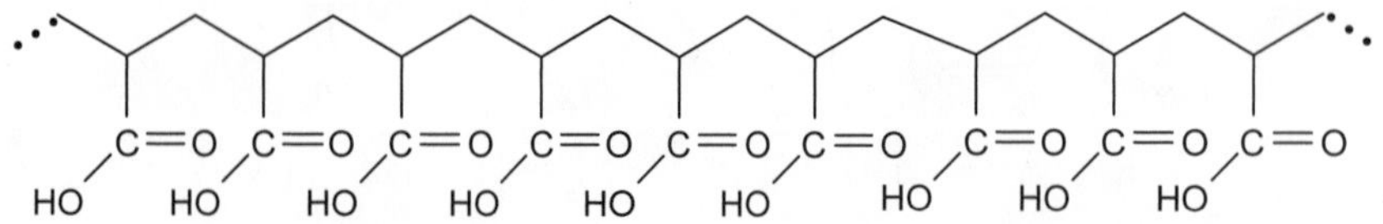

Incorporation of divinyl glycol into poly(acrylic acid) will leave pendant vinyl groups that can react with other growing polymer chains:

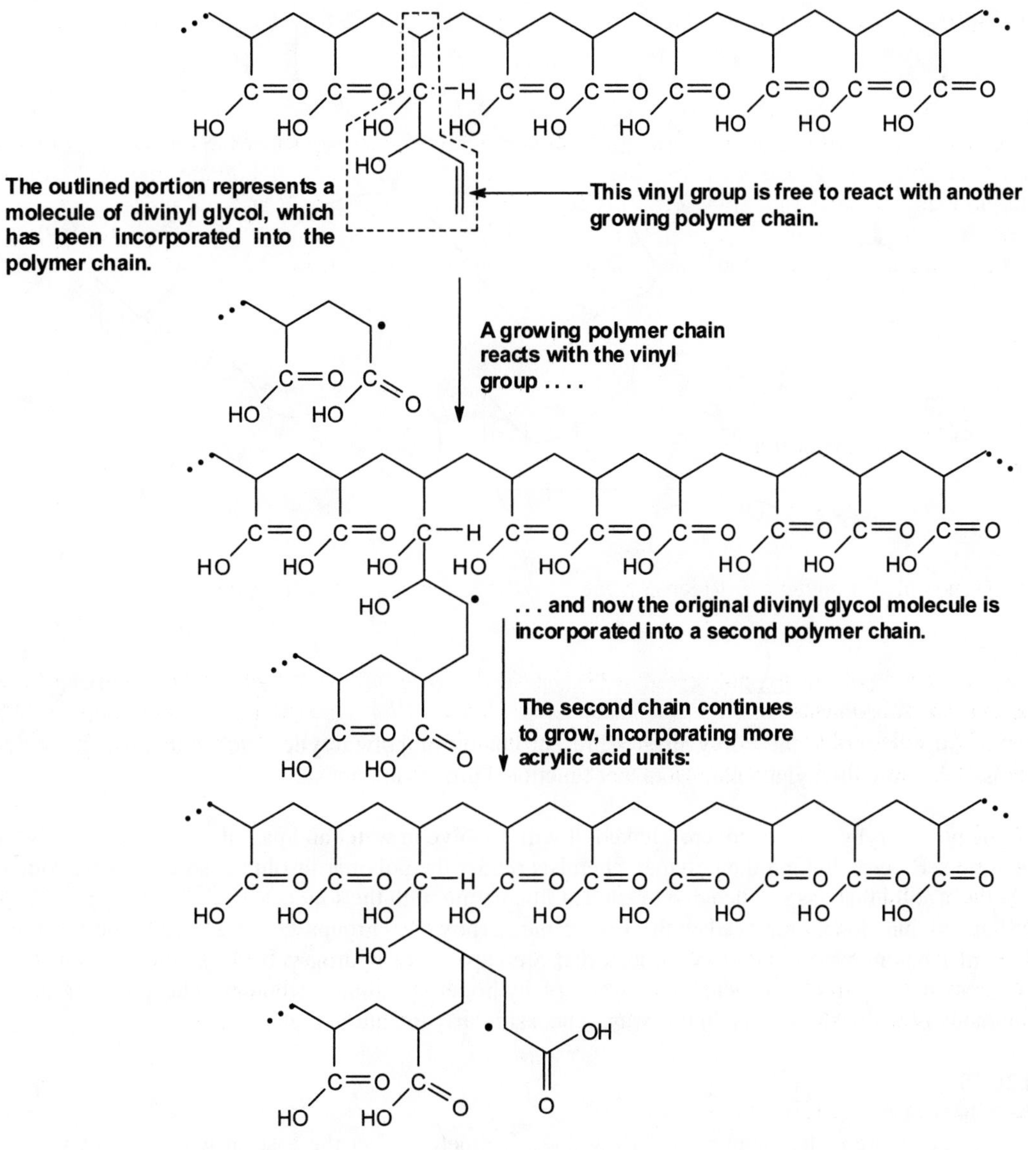

The outlined portion represents a molecule of divinyl glycol, which has been incorporated into the polymer chain.

This vinyl group is free to react with another growing polymer chain.

A growing polymer chain reacts with the vinyl group . . . .

. . . and now the original divinyl glycol molecule is incorporated into a second polymer chain.

The second chain continues to grow, incorporating more acrylic acid units:

The vinyl glycol acts as a crosslinking agent. Conceptually, the polymer will look like this:

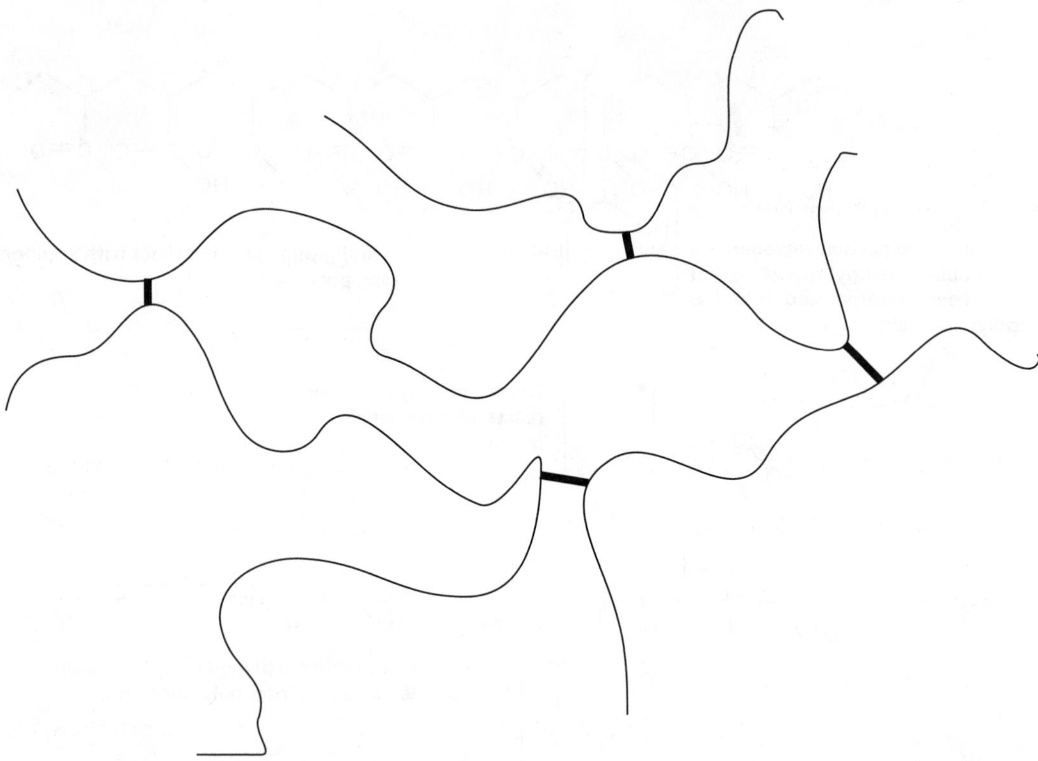

In the figure above, the curvy lines represent chains of poly(acrylic acid) and the bold represents the crosslinks from divinyl alcohol. Note that functional groups are not shown.

(b) If the poly(acrylic acid) is not crosslinked, it will dissolve in water and pass through the digestive system in solution. By crosslinking the polymer, chemists render the polymer insoluble, so it passes through as gel—that is, a solid that has swollen due to solvent interacting with the solid.

(c) After calcium ion is replaced with H^+ and the carboxylate groups become carboxylic acid groups, the polymer has numerous hydroxyl groups that are capable of hydrogen bonding. In addition, the carbonyl groups in the carboxylic acid groups act as hydrogen bonding acceptors. The polymer absorbs large amounts of water and swells. In this way, it acts similarly to natural dietary fiber.

Problem 26.48

Think

Is the polymer more or less symmetric? How does symmetry affect the ease in which a polymer can form a crystal lattice? Will this favor the polymer existing in a dissolved state or in a crystalline state?

Solve

The addition of a second *tert*-butyl group increases the symmetry of the polymer. This makes it easier for the polymer to form a repeating pattern in its crystal lattice and therefore favors the solid form. Just as this increases the T_m of the polymer (factor 6 on p. 1289 of the textbook), it decreases the polymer's solubility.

Problem 26.49

Think

What functional group repeats in polyurethane? How does that functional group compare to the structure of urea? What atoms are different? What does that suggest needs to be changed in the monomer to produce a polyurea?

Solve

(a) To synthesize a polyurea, you replace the diol in the polyurethane synthesis (shown below) with a diamine.

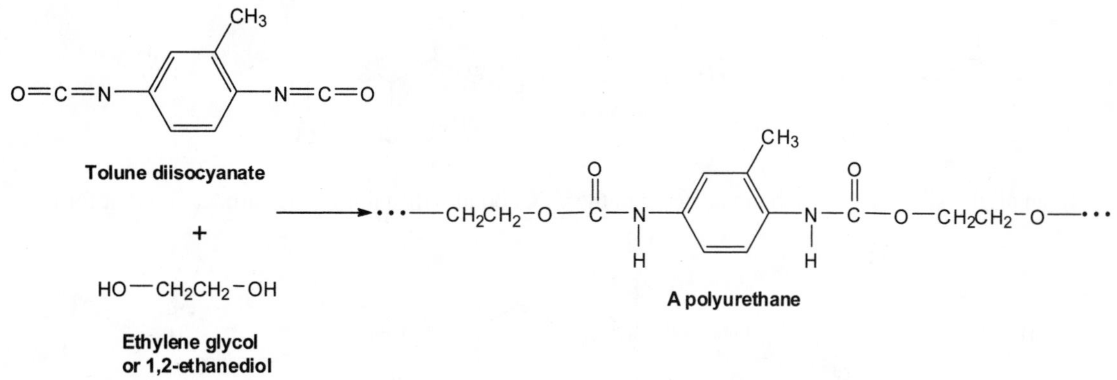
CH3
O=C=N
N=C=O
Tolune diisocyanate
+
HO—CH2CH2—OH
Ethylene glycol
or 1,2-ethanediol
CH3
O
O
•••—CH2CH2—O—C—N
N—C—O—CH2CH2—O—•••
H
H
A polyurethane

The resulting polymer will have a carbonyl with a nitrogen on either side, yielding a polyurea:

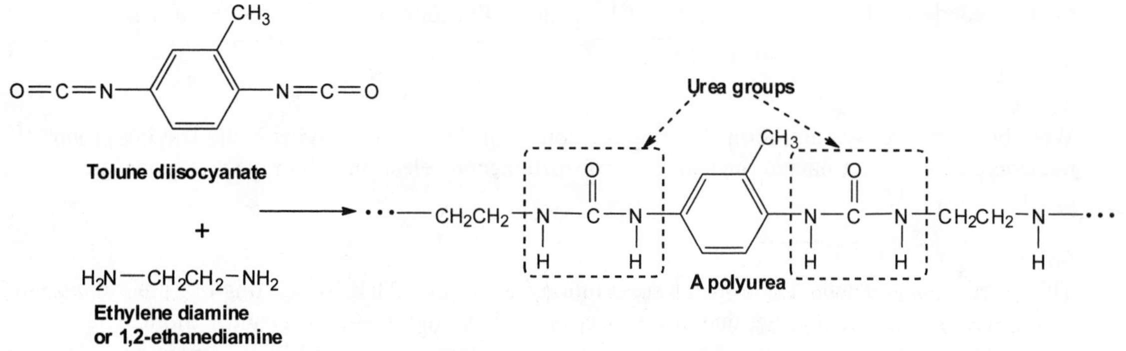
CH3
O=C=N
N=C=O
Tolune diisocyanate
+
H2N—CH2CH2—NH2
Ethylene diamine
or 1,2-ethanediamine
Urea groups
CH3
O
O
•••—CH2CH2—N—C—N
N—C—N—CH2CH2—N—•••
H
H
H
H
H
A polyurea

(b) Condensed formula:

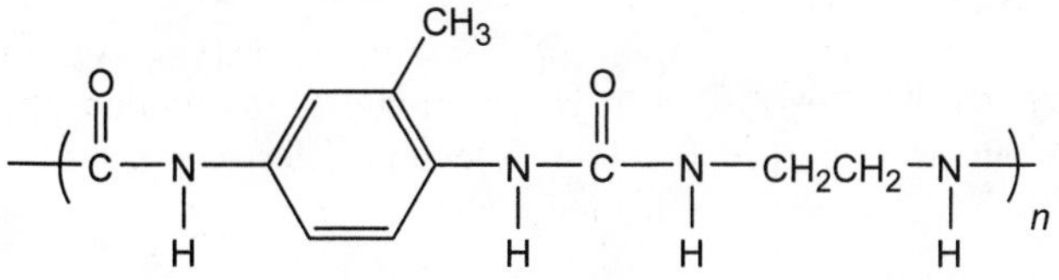
CH3
O
O
C—N
N—C—N—CH2CH2—N
H
H
H
H
n

Problem 26.50

Think

Review the mechanism in the solution to Problem 26.18 to review how PVC unzips. What is the difference in stability of the intermediate of PVC unzipping compared to PVC with tertiary or allylic chlorides? How will that affect the rate of the reaction?

Solve

The mechanism by which PVC "unzips" is shown below and in the solution to Problem 26.18:

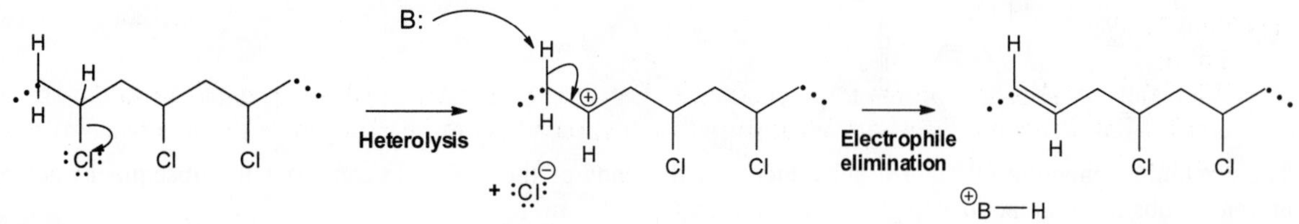
B:
H
H
H
Cl
Cl
Cl
Heterolysis
H
H
H
Cl
Cl
+ :Cl:
Electrophile
elimination
H
H
Cl
Cl
B—H

Note that the carbocation formed in the first step is secondary. If the PVC contains sites with tertiary chlorides, the resulting carbocation is tertiary, and more stable than a secondary carbocation:

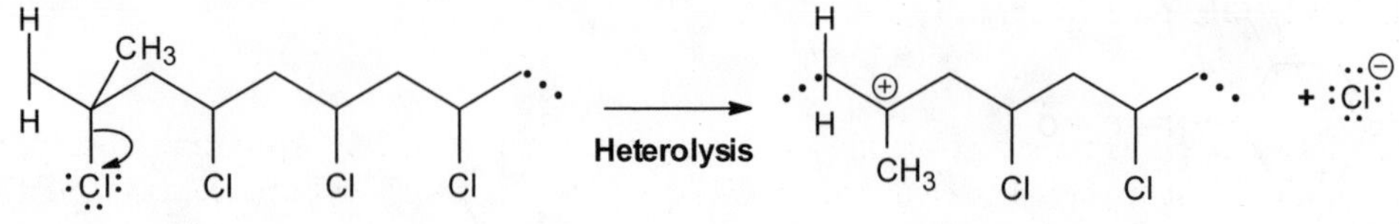

In an allylic site, the double bond adjacent to the C–Cl bond will stabilize the carbocation that forms:

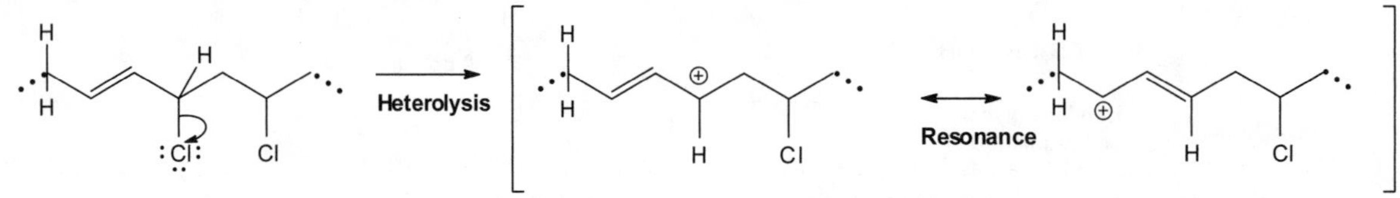

The Cl^- will leave more readily from either a tertiary or allylic chloride, and the resulting cation is more stable than a secondary cation. Consequently, an E1 reaction will occur more quickly at these sites.

Problem 26.51

Think

What bonds are broken and formed during the course of the reaction? What is the leaving group? What is the nucleophile? Are the aromatic rings in difluorobenzophenone electron rich or electron poor? Is the reaction acid or base catalyzed?

Solve

This reaction is a nucleophilic aromatic substitution (see Section 23.10). Each ring of difluorobenzophenone has an electron-poor aromatic ring, due to the electron-withdrawing C=O group and F atom. The rings, therefore, are susceptible to nucleophilic attack and have F^- leaving groups. The first step is an acid–base reaction between the phenolic hydrogen and the carbonate anion, a base:

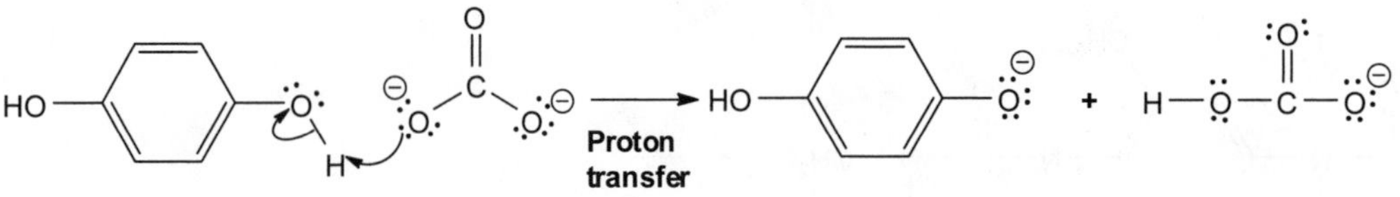

The phenoxide anion then acts as a nucleophile in the subsequent nucleophilic addition–elimination:

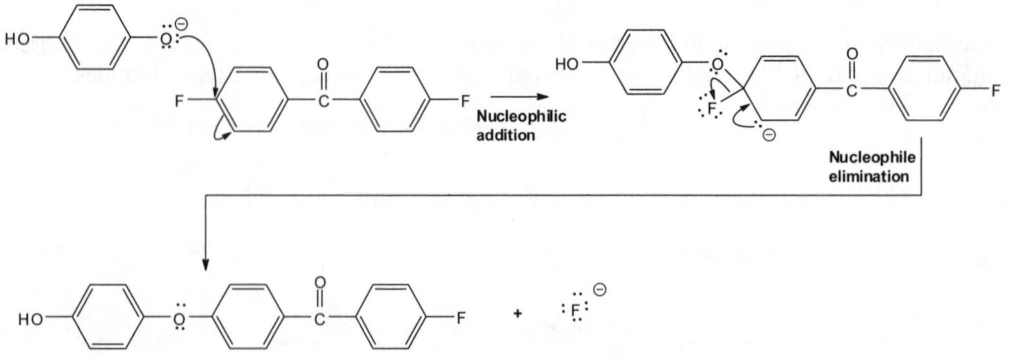

The remaining phenolic OH and F substituents at the ends of the molecule can react in subsequent nucleophilic aromatic substitution reactions to grow the polymer.

Problem 26.52

Think

What are the structures of the two polymers? What functional groups are present in each? What differences do these functional groups have in their IR spectra?

Solve

Poly(vinyl acetate) has ester groups attached to the main polymer chain, so the IR spectrum of poly(vinyl acetate) will have a peak for the carbonyl stretch around 1740 cm^{-1}. As the poly(vinyl acetate) is converted to poly(vinyl alcohol), an absorption stretch of the hydroxyl group will appear around 3400 cm^{-1}. As the conversion proceeds, the carbonyl stretch will decrease in intensity and the hydroxyl stretch will increase in intensity. The absence of the carbonyl group will indicate full conversion of poly(vinyl acetate) to poly(vinyl alcohol).

Problem 26.53

Think

What contributes to the radical stability of **B** and **D**? Which radical is more stable? Does an increase in product stability contribute to a faster or slower reaction of the reactant?

Solve

D has an additional CH_3 group attached to the C•, which stabilizes the radical due to the electron-donating ability of alkyl groups. This greater stability makes **D** less reactive than **B**. Moreover, because **D** is at a lower energy than **B**, **D** is produced more readily from **C** than **B** is produced from **A**. This makes **C** more reactive than **A**.

R• + $H_2C=CH-C(=O)-O-CH_3$ (Methyl acrylate, **A**) → $R-CH_2-\dot{C}H-C(=O)-O-CH_3$ (**B**)

R• + $H_2C=C(CH_3)-C(=O)-O-CH_3$ (More reactive; Methyl methacrylate, **C**) → $R-CH_2-\dot{C}(CH_3)-C(=O)-O-CH_3$ (**D**, More stable C• radical product)

CREDITS

p. 588: Fig. YTA15.8: © Sigma-Aldrich Co. LLC. Used with permission.
p. 588: Fig. YTA15.9: © Sigma-Aldrich Co. LLC. Used with permission.
p. 588: Fig. YTA15.10: © Sigma-Aldrich Co. LLC. Used with permission.
p. 589: Fig. YTA15.11: © Sigma-Aldrich Co. LLC. Used with permission.
p. 589: Fig. YTA15.12: © Sigma-Aldrich Co. LLC. Used with permission.
p. 590: Fig. YTA15.14: © Sigma-Aldrich Co. LLC. Used with permission.
p. 590–91: Fig. YTA15.15: © Sigma-Aldrich Co. LLC. Used with permission.
p. 591: Fig. YTA15.16: © Sigma-Aldrich Co. LLC. Used with permission.
p. 592: Fig. YTA15.17: © Sigma-Aldrich Co. LLC. Used with permission.
p. 593: Fig. YTA15.19–22: © Sigma-Aldrich Co. LLC. Used with permission.
p. 593: Fig. YTA15.23–26: © Sigma-Aldrich Co. LLC. Used with permission.
p. 594: Fig. YTA15.28–30: © Sigma-Aldrich Co. LLC. Used with permission.
p. 598: Fig. PA15.13: © Sigma-Aldrich Co. LLC. Used with permission.
p. 600: Fig. PA15.18: © Sigma-Aldrich Co. LLC. Used with permission.
p. 600: Fig. PA15.19: © Sigma-Aldrich Co. LLC. Used with permission.
p. 601: Fig. PA15.21: © Sigma-Aldrich Co. LLC. Used with permission.
p. 601: Fig. PA15.22: © Sigma-Aldrich Co. LLC. Used with permission.
p. 602: Fig. PA15.23: © Sigma-Aldrich Co. LLC. Used with permission.
p. 602: Fig. PA15.24: © Sigma-Aldrich Co. LLC. Used with permission.
p. 614: Fig. PA15.49: © Sigma-Aldrich Co. LLC. Used with permission.
p. 614: Fig. PA15.50: © Sigma-Aldrich Co. LLC. Used with permission.
p. 615: Fig. PA15.51: © Sigma-Aldrich Co. LLC. Used with permission.
p. 615: Fig. PA15.52: © Sigma-Aldrich Co. LLC. Used with permission.
p. 616: Fig. PA15.55: © Sigma-Aldrich Co. LLC. Used with permission.
p. 618: Fig. PA15.60: © Sigma-Aldrich Co. LLC. Used with permission.
p. 619: Fig. PA15.61: © Sigma-Aldrich Co. LLC. Used with permission.
p. 619: Fig. PA15.62: © Sigma-Aldrich Co. LLC. Used with permission.
p. 620: Fig. PA15.63: © Sigma-Aldrich Co. LLC. Used with permission.
p. 620: Fig. PA15.64: © Sigma-Aldrich Co. LLC. Used with permission.